D0379020

BUILDING DESIGN AND CONSTRUCTION HANDBOOK

Other McGraw-Hill Books Edited by Frederick S. Merritt

Merritt • STANDARD HANDBOOK FOR CIVIL ENGINEERS
Brockenbrough & Merritt • STRUCTURAL STEEL DESIGNER'S HANDBOOK

Other McGraw-Hill Books of Interest

Beall • MASONRY DESIGN AND DETAILING
Breyer • DESIGN OF WOOD STRUCTURES
Brown • FOUNDATION BEHAVIOR AND REPAIR
Faherty & Williamson • WOOD ENGINEERING AND CONSTRUCTION HANDBOOK
Gaylord & Gaylord • STRUCTURAL ENGINEERING HANDBOOK
Harris • NOISE CONTROL IN BUILDINGS
Kubal • WATERPROOFING THE BUILDING ENVELOPE
Newman • STANDARD HANDBOOK OF STRUCTURAL DETAILS FOR BUILDING CONSTRUCTION
Sharp • BEHAVIOR AND DESIGN OF ALUMINUM STRUCTURES
Waddell & Dobrowolski • CONCRETE CONSTRUCTION HANDBOOK

BUILDING DESIGN AND CONSTRUCTION HANDBOOK

Frederick S. Merritt **Editor**
Consulting Engineer, West Palm Beach, Florida

Jonathan T. Ricketts **Editor**
Consulting Engineer, Palm Springs, Florida

Fifth Edition

McGRAW-HILL, INC.
New York San Francisco Washington, D.C. Auckland Bogotá Caracas Lisbon London Madrid Mexico City Milan Montreal New Delhi San Juan Singapore Sydney Tokyo Toronto

Library of Congress Cataloging-in-Publication Data

Building design and construction handbook / Frederick S. Merritt, editor, Jonathan T. Ricketts, editor. — 5th ed.
p. cm.
Includes bibliographical references and index.
ISBN 0-07-041596-X
1. Building—Handbooks, manuals, etc. I. Merritt, Frederick S. II. Ricketts, Jonathan T.
TH151.B825 1994
690—dc20 94-777
CIP

1 2 3 4 5 6 7 8 9 0 DOC/DOC 9 0 9 8 7 6 5 4

ISBN 0-07-041596-X

The sponsoring editor for this book was Larry Hager, the editing supervisor was Peggy Lamb, and the production supervisor was Suzanne W. Babeuf. It was set in Times Roman by Techna Type, Inc.

Printed and bound by R. R. Donnelley & Sons Company.

This book is printed on acid-free paper.

CONTENTS

PREENGINEERED STEEL BUILDINGS

OPEN-WEB STEEL JOISTS

CONCRETE AND ITS INGREDIENTS

QUALITY CONTROL

FORMWORK

REINFORCEMENT

CONCRETE PLACEMENT

Builders' Hardware

Acoustics

Roof Materials

Section 13. Heating, Ventilation, and Air Conditioning
Frank C. Yanocha **13.1**

Section 14. Water-Supply, Sprinkler, and Wastewater Systems
Gregory P. Gladfelter **14.1**

Section 18. Surveying for Building Construction *John H. Baker* 18.1

Section 19. Construction Cost Estimating *Coleman J. Mullin* and *Robert H. Kantor* 19.1

CONTRIBUTORS

Akers, David J. *Civil Engineer, San Diego, California* (SEC. 4)

Baker, John H., P.E. *Adjunct Professor of Civil Engineering, Baker Consulting Engineers, San Jose, California* (SEC. 18)

Bannon, James Michael, P.E. *Electrical Engineer, STV Group, Pottstown, Pennsylvania* (SEC. 15)

Borg, Kiri J. *Fishman Construction Corporation of New York, New York, New York* (SEC. 17)

Borg, Robert F. *President, Kreisler Borg Horman General Construction Company, Scarsdale, New York* (SEC. 17)

Edgett, Stephen D. *Edgett Williams Consulting Group, Inc., Mill Valley, California* (SEC. 16)

Gladfelter, Gregory P. *Howard Needles Tammen & Bergendoff, Kansas City, Missouri and Miami, Florida* (SEC. 14)

Glidden, Bruce *President, Glidden & Co. LTD., Upper St. Clair, Pennsylvania* (SEC. 7)

Gupton, Charles P. *Principal, Dames & Moore, Inc., Boca Raton, Florida* (SEC. 6)

Gustafson, David P. *Technical Director, Concrete Reinforcing Steel Institute, Schaumburg, Illinois* (SEC. 9)

Hinklin, Alan D. *Director, Skidmore, Owings & Merrill, Chicago, Illinois* (SEC. 2)

Hoffman, Edward S. *President, Edward S. Hoffman, Ltd., Structural Engineers, Chicago, Illinois* (SEC. 9)

Merritt, Frederick S. *Consulting Engineer, West Palm Beach, Florida* (SECS. 1, 3, 5, & 11)

Mullin, Colman J. *Project Estimator, Bechtel Corporation, San Francisco, California* (SEC. 19)

Smith, Thomas Lee *Director of Technology and Research, National Roofing Contractors Association, Rosemont, Illinois* (SEC. 12)

Williamson, Thomas G., P.E. *American Plywood Association, Tacoma, Washington* (SEC. 10)

Wolford, Don S. *Consulting Engineer, Middletown, Ohio* (SEC. 8)

Yanocha, Frank C., P.E. *Vice President, STV Group, Pottstown, Pennsylvania* (SEC. 13)

ABOUT THE EDITORS

Frederick S. Merritt is a consulting engineer with many years of experience in building and bridge design, structural analysis, and construction management. He is a Fellow of the American Society of Civil Engineers, and a Senior Member of ASTM. Formerly senior editor of *Engineering News-Record,* he is also the author/editor of many books, including McGraw-Hill's *Standard Handbook for Civil Engineers* and *Structural Steel Designer's Handbook*.

Jonathan T. Ricketts is a consulting engineer with broad experience in general civil engineering, environmental design and construction management. A registered engineer in several states, he is an active member in the American Society of Civil Engineers, the National Society of Professional Engineers, and the Solid Waste Association of North America.

PREFACE

The fifth edition of the *Building Design and Construction Handbook* adopts the original objectives of its editors and contributors, which gained broad readership for preceding editions. It provides a comprehensive overview of the building process. It places special emphasis on information on the latest recommended practices, including proven innovations in design and construction, with the aim of meeting:

1. Demands brought about by global competition
2. The need to reduce the environmental impact of new buildings and construction operations
3. The necessity of achieving greater energy efficiency
4. The mandate not only to protect property against hazards but also to ensure the health, welfare, and safety of building occupants, including the disabled, in both ordinary and emergency situations

Accordingly, the book presents in a single volume the best of current building design and construction practices, information that would be of the greatest usefulness to those who have to make decisions affecting selection of building materials and construction methods. Emphasis is placed on fundamental principles and practical applications of them, with special attention to simplified procedures. Frequent reference is made to other sources where additional authoritative information may be obtained. The book provides an extensive index as well as a detailed table of contents to assist the reader in locating topics.

The new edition may well be considered a new book. New contributors and the addition of a coeditor bring a fresh viewpoint and new ideas. The book contains several new or completely rewritten sections, or chapters, and much additional subject matter in the sections that have been retained, to cover new developments.

There have been many far-reaching developments in the building field since publication of the preceding edition. These include load-factor-and-resistance design (LRFD) of structural materials, improvements in HVAC and elevator conveyance, greater use of computers, refinements in management of design and construction, and more productive surveying equipment. The authors have incorporated the most significant of these and other important developments in their sections. To present the necessary information in a single volume, obsolete and less-important information in the earlier editions has been deleted. Also, for data that can be readily obtained from other sources, such as architectural and engineering societies and manufacturers associations, the reader is referred to the appropriate publications.

In preparing this reference work, the contributors drew heavily on numerous sources. Many of these are credited or given as reference throughout the book, but space limitations preclude mentioning them all. The editors and contributors wish to acknowledge their indebtedness to these sources and to express their gratitude.

The editors are especially grateful to the contributors, not only because they appreciate the great value of the contributions but also because they are keenly aware of the considerable sacrifices involved in taking time to prepare the sections.

Frederick S. Merritt
Jonathan T. Ricketts

SECTION ONE
BUILDING SYSTEMS

Frederick S. Merritt
Consulting Engineer
West Palm Beach, Florida

Sociological changes, new developments in industry and commerce, new building codes, other new laws and regulations, inflationary economies of nations, and advances in building technology place an ever-increasing burden on building designers and constructors. They need more and more knowledge and skill to cope with the demands placed on them.

The public continually demands more complex buildings than in the past. They must serve more purposes, last longer, and require less maintenance and repair. As in the past, they must look pretty. Yet, both building construction and operating costs must be kept within acceptable limits or new construction will cease.

To meet this challenge successfully, continual improvements in building design and construction must be made. Building designers and constructors should be alert to these advances and learn how to apply them skillfully.

One advance of note is the adaptation to building design of operations research, or systems design, developed around the middle of the twentieth century and originally applied with noteworthy results to design of machines and electronic equipment. In the past, design of a new building was mainly an imitation of the design of an existing building. Innovations were often developed fortuitously and by intuition and were rare occurrences. In contrast, systems design encourages innovation. It is a precise procedure that guides creativity toward the best decisions. As a result, it can play a significant role in meeting the challenges posed by increasing building complexity and costs. The basic principles of systems design are presented in this section.

1.1 PRINCIPLES OF ARCHITECTURE

A building is an assemblage that is firmly attached to the ground and that provides total or nearly total shelter for machines, processing equipment, performance of human activities, storage of human possessions, or any combination of these.

Building design is the process of providing all information necessary for construction of a building that will meet its owner's requirements and also satisfy public health, welfare, and safety requirements. **Architecture** is the art and science of building design. **Building construction** is the process of assembling materials to form a building.

Building design may be legally executed only by persons deemed competent to do so by the state in which the building is to be constructed. Competency is determined on the basis of education, experience, and ability to pass a written test of design skills.

Architects are persons legally permitted to practice architecture. **Engineers** are experts in specific scientific disciplines and are legally permitted to design parts of buildings; in some cases, complete buildings. In some states, persons licensed as **building designers** are permitted to design certain types of buildings.

Building construction is generally performed by laborers and craftspeople engaged for the purpose by an individual or organization, called a **contractor.** The contractor signs an agreement, or contract, with the building owner under which the contractor agrees to construct a specific building on a specified site and the owner agrees to pay for the materials and services provided. Often, building construction is restricted only to contractors granted a license for the purpose by the community in which the building is to be constructed.

In design of a building, architects should be guided by the following principles:

1. The building should be constructed to serve purposes specified by the client.
2. The design should be constructable by known techniques and with available labor and equipment, within an acceptable time.
3. The building should be capable of withstanding the elements and normal usage for a reasonably long period of time.
4. Both inside and outside, the building should be visually pleasing.
5. No part of the building should pose a hazard to the safety or health of its occupants under normal usage, and the building should provide for safe evacuation or refuge in emergencies.
6. The building should provide the degree of shelter from the elements and of control of the interior environment—air, temperature, humidity, light, and acoustics—specified by the client and not less than the minimums required for safety and health of the occupants.
7. The building should be constructed to minimize adverse environmental impact on its neighbors.
8. Operation of the building should consume a minimum of energy while permitting the structure to serve its purposes.
9. The sum of costs of construction, operation, maintenance, repair, and anticipated future alterations should be kept within the limit specified by the client.

To provide all the information necessary for the construction of a building, the ultimate objective of design is production of **drawings,** or **plans,** showing what is to be constructed, **specifications** stating what materials and equipment are to be incorporated in the building, and a **construction contract** between the client and a contractor. Designers also should observe construction of the building while it is in process. This should be done not only to assist the client in ensuring that the building is being constructed in accordance with plans and specifications but also to obtain information that will be useful in design of future buildings.

1.2 SYSTEMS DESIGN AND ANALYSIS

Systems design comprises a rational, orderly series of steps that leads to the best decision for a given set of conditions. The procedure requires:

Analysis of a building as a system.

Synthesis, or selection of components, to form a system that meets specific objectives while subject to constraints, or variables controllable by designers.

Appraisal of system performance, including comparisons with alternative systems.

Feedback to analysis and synthesis of information obtained in system evaluation, to improve the design.

The prime advantage of the procedure is that, through comparisons of alternatives and data feedback to the design process, systems design converges on an optimum, or best, system for the given conditions. Another advantage is that the procedure enables designers to clarify the requirements for the building being designed. Still another advantage is that the procedure provides a common basis of understanding and promotes cooperation between the specialists in various aspects of building design.

For a building to be treated as a system, as required in systems design, it is necessary to know what a system is and what its basic characteristics are.

A system is an assemblage formed to satisfy specific objectives and subject to constraints and restrictions and consisting of two or more components that are interrelated and compatible, each component being essential to the required performance of the system.

Because the components are required to be interrelated, operation, or even the mere existence, of one component affects in some way the performance of other components. Also, the required performance of the system as a whole, as well as the constraints on the system, imposes restrictions on each component.

A building meets the preceding requirements. By definition, it is an assemblage (Art. 1.1). It is constructed to serve specific purposes. It is subject to constraints while doing so, inasmuch as designers can control properties of the system by selection of components (Art. 1.9). Building components, such as walls, floors, roofs, windows, and doors, are interrelated and compatible with each other. The existence of any of these components affects to some extent the performance of the others. And the required performance of the building as a whole imposes restrictions on the components. Consequently, a building has the basic characteristics of a system, and systems-design procedures should be applicable to it.

Systems Analysis. A group of components of a system may also be a system. Such a group is called a **subsystem.** It, too, may be designed as a system, but its goal must be to assist the system of which it is a component to meet its objectives. Similarly, a group of components of a subsystem may also be a system. That group is called a **subsubsystem.**

For brevity, the major subsystems of a building are referred to as systems in this book.

In a complex system, such as a building, subsystems and other components may be combined in a variety of ways to form different systems. For the purposes of building design, the major systems are usually defined in accordance with the construction trades that will assemble them, for example, structural framing, plumbing, electrical systems, and heating, ventilation, and air conditioning.

In systems analysis, a system is resolved into its basic components. Subsystems are determined. Then, the system is investigated to determine the nature, interaction, and performance of the system as a whole. The investigation should answer such questions as:

What does each component (or subsystem) do?

What does the component do it to?

How does the component serve its function?

What else does the component do?

Why does the component do the things it does?

What must the component really do?

Can it be eliminated because it is not essential or because another component can assume its tasks?

See also Art. 1.8.

1.3 TRADITIONAL DESIGN PROCEDURES

Systems design of buildings requires a different approach to design and construction than that used in traditional design (Art. 1.9). Because traditional design and construction procedures are still widely used, however, it is desirable to incorporate as much of those procedures in systems design as is feasible without destroying its effectiveness. This will make the transition from traditional design to systems design easier. Also, those trained in systems design of buildings will then be capable of practicing in traditional ways, if necessary.

There are several variations of traditional design and construction. These are described throughout this book. For the purpose of illustrating how they may be modified for systems design, however, one widely used variation, which will be called basic traditional design and construction, is described in the following and in Art. 1.4.

In the basic traditional design procedure, design usually starts when a client recognizes the need for and economic feasibility of a building and engages an architect, a professional with a broad background in building design. The architect, in turn, engages consulting engineers and other consultants.

For most buildings, structural, mechanical, and electrical consulting engineers are required. A structural engineer is a specialist trained in the application of scientific principles to design of load-bearing walls, floors, roofs, foundations, and skeleton framing needed for the support of buildings and building components. A mechanical engineer is a specialist trained in the application of scientific principles to design of plumbing, elevators, escalators, horizontal walkways, dumbwaiters, conveyors, installed machinery, and heating, ventilation, and air conditioning. An electrical engineer is a specialist trained in the application of scientific principles to design of electric circuits, electric controls and safety devices, electric motors and generators, electric lighting, and other electric equipment.

For buildings on a large site, the architect may engage a landscape architect as a consultant. For a concert hall, an acoustics consultant may be engaged; for a hospital, a hospital specialist; for a school, a school specialist.

The architect does the overall planning of the building and incorporates the output of the consultants into the contract documents. The architect determines what internal and external spaces the client needs, the sizes of these spaces, their relative locations, and their interconnections. The results of this planning are shown in floor plans, which also diagram the internal flow, or circulation, of people and supplies. Major responsibilities of the architect are enhancement of the appearance inside and outside of the building and keeping adverse environmental impact of the structure to a minimum. The exterior of the building is shown in drawings, called elevations. The location and orientation of the building is shown in a site plan. The architect also prepares the specifications for the building. These describe in detail the materials and equipment to be installed in the structure. In addition, the architect, usually with the aid of an attorney engaged by the client, prepares the construction contract.

The basic traditional design procedure is executed in several stages. In the first stage, the architect develops a **program,** or list of the client's requirements. In the second stage, or **conceptual phase,** the architect translates requirements into spaces, relates the spaces and makes sketches, called schematics, to illustrate the concepts. When sufficient information is obtained on the size and general construction of the building, a rough estimate is made of construction cost. If this cost does not exceed the cost budgeted by the client for construction, the next stage, **design development,** proceeds. In this stage, the architect and consultants work out more details and show the results in preliminary construction drawings and outline specifications. A preliminary cost estimate utilizing the greater amount of information on the building now available is then prepared. If this cost does not exceed the client's budget, the final stage, the **contract documents phase,** starts. It culminates in production of working, or construction, drawings and specifications, which are incorporated in the contract between the client and a builder and therefore become legal documents. Before the documents are completed, however, a final cost estimate is prepared. If the cost exceeds the client's budget, the design is revised to achieve the necessary cost reduction.

After the estimated cost is brought within the budget and the client has approved the contract documents, the architect helps the owner in obtaining bids from contractors or in negotiating a construction price with a qualified contractor. For private work, construction not performed for a governmental agency, the owner generally awards the construction contract to a contractor, called a **general contractor.** Assigned the responsibility for construction of the building, this contractor may perform some, all, or none of the work. Usually, much of the work is let out to specialists, called **subcontractors.** For public work, there may be a legal requirement that bids be taken and the contract awarded to the lowest responsible bidder. Sometimes also, separate contracts have to be awarded for the major specialists, such as mechanical and electrical trades, and to a general contractor, who is assigned responsibility for coordinating the work of the trades and performance of the work. (See also Art. 1.4.)

Building design should provide for both normal and emergency conditions. The latter includes fire, explosion, power cutoffs, hurricanes, and earthquakes. The design should include facilities for disabled persons.

1.4 TRADITIONAL CONSTRUCTION PROCEDURES

As mentioned in Art. 1.8, construction is performed by contractors. While they would like to satisfy the owner and the building designers, contractors have the main objective of making a profit. Hence, their initial task is to prepare a bid price based on an accurate estimate of construction costs. This requires development of a concept for performance of the work and a construction time schedule. After a contract has been awarded, contractors must furnish and pay for all materials, equipment, power, labor, and supervision required for construction. The owner compensates the contractors for construction costs and services.

A **general contractor** assumes overall responsibility for construction of a building. The contractor engages **subcontractors** who take responsibility for the work of the various trades required for construction. For example, a plumbing contractor installs the plumbing, an electrical contractor installs the electrical system, a steel erector erects structural steel, and an elevator contractor installs elevators. Their contracts are with the general contractor, and they are paid by the general contractor.

Sometimes, in addition to a general contractor, the owner contracts separately with specialty contractors, such as electrical and mechanical contractors, who perform a substantial amount of the work required for a building. Such contractors are called **prime contractors.** Their work is scheduled and coordinated by the general contractor, but they are paid directly by the owner.

Sometimes also, the owner awards a contract to an organization for both the design and construction of a building. Such organizations are called **design-build contractors.** One variation of this type of contract is employed by developers of groups of one-family homes or low-rise apartment buildings. The **homebuilder** designs and constructs the dwellings, but the design is substantially completed before owners purchase the homes.

Administration of the construction procedure often is difficult. Consequently, some owners seek assistance from an expert, called a **professional construction manager,** with extensive construction experience, who receives a fee. The construction manager negotiates with general contractors and helps select one to construct the building. Managers usually also supervise selection of subcontractors. During construction, they help control costs, expedite equipment and material deliveries, and keep the work on schedule (see Art. 17.9). In some cases, instead, the owner may prefer to engage a **construction program manager,** to assist in administering both design and construction.

Construction contractors employ labor that may or may not be unionized. Unionized craftspeople are members of unions that are organized by construction trades, such as carpenter, plumber, and electrician unions. Union members will perform only the work assigned to their trade. On the job, groups of workers are supervised by **crew supervisors,** all of whom report to a **superintendent.**

During construction, all work should be inspected. For the purpose, the owner, often through the architect and consultants, engages inspectors. The field inspectors may be placed under the control of an owner's representative, who may be titled *clerk of the works, architect's superintendent, engineer's superintendent,* or *resident engineer.* The inspectors have the responsibility of ensuring that construction meets the requirements of the contract documents and is performed under safe conditions. Such inspections may be made at frequent intervals.

In addition, inspections also are made by representatives of one or more governmental agencies. They have the responsibility of ensuring that construction meets legal requirements and have little or no concern with detailed conformance with the contract documents. Such legal inspections are made periodically or at the end of certain stages of construction. One agency that will make frequent inspections is the local or state building department, whichever has jurisdiction.

The purpose of these inspections is to ensure conformance with the local or state building code.

During construction, standards, regulations, and procedures of the Occupational Safety and Health Administration should be observed. These are given in detail in "Construction Industry. OSHA Safety and Health Standards (29CFR1926/1910)," Government Printing Office, Washington, DC 20402.

Following is a description of the basic traditional construction procedure for a multistory building:

After the award of a construction contract to a general contractor, the owner may ask the contractor to start work before signing of the contract by giving the contractor a *letter of intent* or after signing of the contract by issuing a *written notice to proceed*. The contractor then obtains construction permits, as required, from governmental agencies, such as the local building, water, sewer, and highway departments.

The general contractor plans and schedules construction operations in detail and mobilizes equipment and personnel for the project. Subcontractors are notified of the contract award and issued letters of intent or awarded subcontracts, then are given, at appropriate times, notices to proceed.

Before construction starts, the general contractor orders a survey to be made of adjacent structures and terrain, both for the record and to become knowledgeable of local conditions. A survey is then made to lay out construction.

Field offices for the contractor are erected on or near the site. If desirable for safety reasons to protect passersby, the contractor erects a fence around the site and an overhead protective cover, called a bridge. Structures required to be removed from the site are demolished and the debris is carted away.

Next, the site is prepared to receive the building. This work may involve grading the top surface to bring it to the proper elevations, excavating to required depths for basement and foundations, and shifting of utility piping. For deep excavations, earth sides are braced and the bottom is drained.

Major construction starts with the placement of foundations, on which the building rests. This is followed by the erection of load-bearing walls and structural framing. Depending on the height of the building, ladders, stairs, or elevators may be installed to enable construction personnel to travel from floor to floor and eventually to the roof. Also, hoists may be installed to lift materials to upper levels. If needed, temporary flooring may be placed for use of personnel.

As the building rises, pipes, ducts, and electric conduit and wiring are installed. Then, permanent floors, exterior walls, and windows are constructed. At the appropriate time, permanent elevators are installed. If required, fireproofing is placed for steel framing. Next, fixed partitions are built and the roof and its covering, or roofing, are put in place.

Finishing operations follow. These include installation of the following: ceilings; tile; wallboard; wall paneling; plumbing fixtures; heating furnaces; air-conditioning equipment; heating and cooling devices for rooms; escalators; floor coverings; window glass; movable partitions; doors; finishing hardware; electrical equipment and apparatus, including lighting fixtures, switches, outlets, transformers, and controls; and other items called for in the drawings and specifications. Field offices, fences, bridges, and other temporary construction must be removed from the site. Utilities, such as gas, electricity, and water, are hooked up to the building. The site is landscaped and paved. Finally, the building interior is painted and cleaned.

The owner's representatives then give the building a final inspection. If they find that the structure conforms with the contract documents, the owner accepts the project and gives the general contractor final payment on issuance by the building department of a certificate of occupancy, which indicates that the completed building meets building-code requirements.

1.5 ROLE OF THE CLIENT IN DESIGN AND CONSTRUCTION

Article 1.4 points out that administration of building construction is difficult, as a result of which some clients, or owners, engage a construction manager or construction program manager to act as the owner's authorizing agent and project overseer. The reasons for the complexity of construction administration can be seen from an examination of the owner's role before and during construction.

After the owner recognizes the need for a new building, the owner establishes project goals and determines the economic feasibility of the project. If it appears to be feasible, the owner develops a building program (list of requirements), budget, and time schedule for construction. Next, preliminary arrangements are made to finance construction. Then, the owner selects a construction program manager or an architect for design of the building. Later, a construction manager may be chosen, if desired.

The architect may seek from the owner approval of the various consultants that will be needed for design. If a site for the building has not been obtained at this stage, the architect can assist in site selection. When a suitable site has been found, the owner purchases it and arranges for surveys and subsurface explorations to provide information for locating the building, access, foundation design and construction, and landscaping. It is advisable at this stage for the owner to start developing harmonious relations with the community in which the building will be erected.

During design, the owner assists with critical design decisions; approves schematic drawings, rough cost estimates, preliminary drawings, outline specifications, preliminary cost estimates, contract documents, and final cost estimate; pays designers' fees in installments as design progresses; and obtains a construction loan. Then, the owner awards the general contract for construction and orders construction to start. Also, the owner takes out liability, property, and other desirable insurance.

At the start of construction, the owner arranges for construction permits. As construction proceeds, the owner's representatives inspect the work to ensure compliance with the contract documents. Also, the owner pays contractors in accordance with the terms of the contract. Finally, the owner approves and accepts the completed project.

One variation of the preceding procedure is useful when time available for construction is short. It is called **phased, or fast-track, construction.** In this variation, the owner engages a construction manager and a general contractor before design has been completed, to get an early start on construction. Work then proceeds on some parts of the building while other parts are still being designed. For example, excavation and foundation construction are carried out while design of the structural framing is being finished. The structural framing is erected, while heating, ventilation, and air-conditioning, electrical, plumbing, wall, and finishing details are being developed. For tall buildings, the lower portion can be constructed while the upper part is still being designed. For large, low-rise buildings, one section can be built while another is under design.

1.6 BUILDING COSTS

Construction cost of a building usually is a dominant design concern. One reason is that if construction cost exceeds the owner's budget, the owner may cancel the project. Another reason is that costs, such as property taxes and insurance, that occur after completion of the building often are proportional to the initial cost. Hence, owners usually try to keep that cost low. Designing a building to minimize construction cost, however, may not be in the owner's best interests. There are many other costs that the owner incurs during the anticipated life of the building that should be taken into account.

Before construction of a building starts, the owner generally has to make a sizable investment in the project. The major portion of this expenditure usually goes for purchase of the site and building design. Remaining preconstruction costs include those for feasibility studies, site selection and evaluation, surveys, and program definition.

The major portion of the construction cost is the sum of the payments to the general contractor and prime contractors. Remaining construction costs usually consist of interest on the construction loan, permit fees, and costs of materials, equipment, and labor not covered by the construction contracts.

The **initial cost** to the owner is the sum of preconstruction, construction, and occupancy costs. The latter covers costs of moving possessions into the building and start-up of utility services, such as water, gas, electricity, and telephone.

After the building is occupied, the owner incurs costs for operation and maintenance of the buildings. Such costs are a consequence of decisions made during building design.

Often, postconstruction costs are permitted to be high so that initial costs can be kept low. For example, operating the building may be expensive because the design makes artificial lighting necessary when daylight could have been made available or because extra heating and air conditioning are necessary because of inadequate insulation of walls and roof. As another example, maintenance may be expensive because of the difficulty of changing electric lamps or because cleaning the building is time-consuming and laborious. In addition, frequent repairs may be needed because of poor choice of materials during design. Hence, operation and maintenance costs over a specific period of time, say 10 or 20 years, should be taken into account in optimizing the design of a building.

Life-cycle cost is the sum of initial, operating, and maintenance costs. Generally, it is life-cycle cost that should be minimized in building design rather than construction cost. This would enable the owner to receive the greatest return on the investment in the building. ASTM has promulgated a standard method for calculating life-cycle costs of buildings, E917, Practice for Measuring Life-Cycle Costs of Buildings and Building Systems, as well as a computer program and user's guide to improve accuracy and speed of calculation.

Nevertheless, a client usually establishes a construction budget independent of life-cycle cost. This often is necessary because the client does not have adequate capital for an optimum building and places too low a limit on construction cost. The client hopes to have sufficient capital later to pay for the higher operating and maintenance costs or for replacement of undesirable building materials and installed equipment. Sometimes, the client establishes a low construction budget because the client's goal is a quick profit on early sale of the building, in which case the client has little or no concern with future high operating and maintenance costs for the building. For these reasons, construction cost frequently is a dominant concern in design.

1.7 MAJOR BUILDING SYSTEMS

The simplest building system consists of only two components. One component is a floor, a flat, horizontal surface on which human activities can take place. The other component is an enclosure that extends over the floor and generally also around it to provide shelter from the weather for human activities.

The ground may serve as the floor in primitive buildings. In better buildings, however, the floor may be a structural deck laid on the ground or supported above ground on structural members, such as the joist and walls in Fig. 1.1. Use of a deck and structural members adds at least two different types of components, or two subsystems, to the simplest building system. Also, often, the enclosure over the floor requires supports, such as the rafter and walls in Fig. 1.1, and the walls, in turn, are seated on foundations in the ground. Additionally, **footings** are required at the base of the foundations to spread the load over a large area of the ground, to prevent the building from sinking (Fig. 1.2*a*). Consequently, even slight improvements in a primitive building introduce numerous additional components, or subsystems, into a building.

More advanced buildings consist of numerous subsystems, which are referred to as systems in this book when they are major components. Major subsystems generally include structural framing and foundations, enclosure systems, plumbing, lighting, acoustics, safety systems, vertical-circulation elements, electric power and signal systems, and heating, ventilation, and air conditioning (HVAC).

Structural System. The portion of a building that extends above the ground level outside it is called the **superstructure.** The portion below the outside ground level is called the **substructure.** The parts of the substructure that distribute building loads to the ground are known as **foundations.**

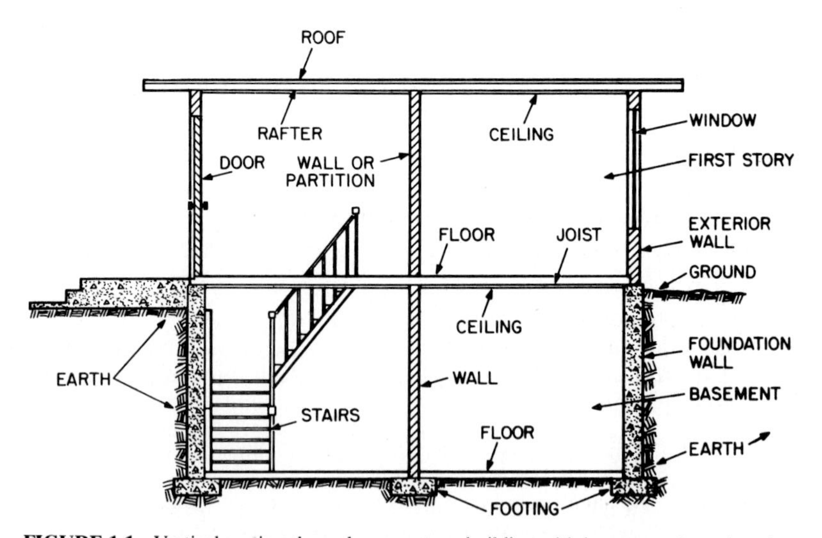

FIGURE 1.1 Vertical section through a one-story building with basement shows location of some major components. *(Reprinted with permission from F. S. Merritt and J. Ambrose, "Building Engineering and Systems Design," 2d ed., Van Nostrand Reinhold, New York.)*

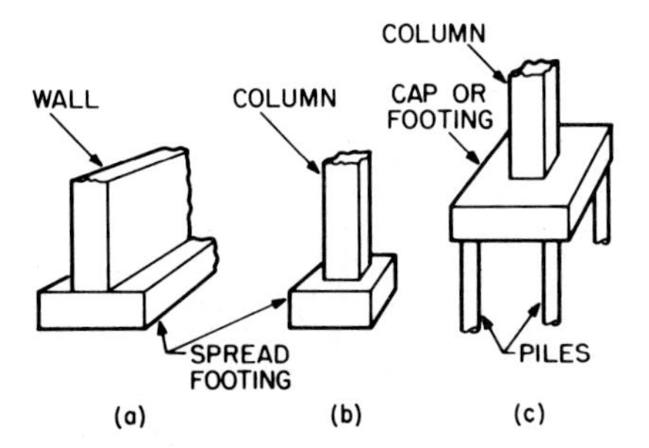

FIGURE 1.2 Commonly used foundations: (*a*) foundation wall on continuous footing; (*b*) individual spread footing for a column; (*c*) pile footing for a column.

Foundations may take the form of walls. When the ground under the building is excavated for a cellar, or basement, the foundation walls have the additional task of retaining the earth along the outside of the building (Fig. 1.1). The superstructure in such cases is erected atop the foundation walls.

The footing under a wall (Fig. 1.2*a*) is called a **continuous spread footing.** A slender structural member, such as a column (Fig. 1.2*b*), usually is seated on an **individual spread footing.** When the soil is so weak, however, that the spread footings for columns become very large, it often is economical to combine the footings into a single footing under the whole building. Such a footing is called a **raft,** or **mat, footing** or a **floating foundation.** For very weak soils, it generally is necessary to support the foundations on **piles** (Fig. 1.2*c*). These are slender structural members that are hammered or otherwise driven through the weak soil, often until the tips seat on rock or a strong layer of soil.

The foundation system must be designed to transmit the loads from the superstructure structural system directly to the ground in such a manner that settlement of the completed building as the soil deflects will be within acceptable limits. The superstructure structural system, in turn, should be designed to transmit its loads to the foundation system in the manner anticipated in the design of the foundations. (See also Sec. 6.)

In most buildings, the superstructure structural system consists of floor and roof decks, horizontal members that support them, and vertical members that support the other components.

The horizontal members are generally known as **beams,** but they also are called by different names in specific applications. For example:

Joists are closely spaced to carry light loads.

Stringers support stairs.

Headers support structural members around openings in floors, roofs, and walls.

Purlins are placed horizontally to carry level roof decks.

Rafters are placed on an incline to carry sloping roof decks.

Girts are light horizontal members that span between columns to support walls.

Lintels are light horizontal beams that support walls at floor levels in multistory buildings or that carry the part of walls above openings for doors and windows.

Girders may be heavily loaded beams or horizontal members that support other beams (Fig. 1.3).

Spandrels carry exterior walls and support edges of floors and roofs in multistory buildings.

Trusses serve the same purposes as girders but consist of slender horizontal, vertical, and inclined components with large open spaces between them. The spaces are triangular in shape. Light beams similarly formed are called **open-web joists** (Fig. 1.6*d*).

FIGURE 1.3 Structural-steel skeleton framing for a multistory building. *(Courtesy of the American Institute of Steel Construction.)*

Floor and roof decks or the beams that support them are usually seated on load-bearing walls or carried by columns, which carry the load downward. (The horizontal members also may be suspended on hangers, which transmit the load to other horizontal members at a higher level.) The system comprising decks, beams, and bearing walls is known as load-bearing construction (Fig. 1.1). The system composed of decks, beams, and columns is known as skeleton framing (Fig. 1.3).

Both types of systems must be designed to transmit to the foundations vertical (gravity) loads, vertical components of inclined loads, horizontal (lateral) loads, and horizontal components of inclined loads. Vertical walls and columns have the appropriate alignment for carrying vertical loads downward. But acting alone, these structural members are inadequate for resisting lateral forces.

One way to provide lateral stability is to incorporate in the system diagonal members, called **bracing** (Fig. 1.3). Bracing, columns, and beams then work together to carry the lateral loads downward. Another way is to rigidly connect beams to columns to prevent a change in the angle between the beams and columns, thus making them work together as a **rigid frame** to resist lateral movement. Still another way is to provide long walls, known as **shear walls,** in two perpendicular directions. Lateral forces on the building can be resolved into forces in each of these directions. The walls then act like vertical beams (**cantilevers**) in transmitting the forces to the foundations. (See also Arts. 3.5 and 3.6.)

Because of the importance of the structural system, the structural members should be protected against damage, especially from fire. For fire protection, bracing may be encased in fire-resistant floors, roofs, or walls. Similarly, columns may be encased in walls, and beams may be encased in floors. Or a fire-resistant material, such as concrete, mineral fiber, or plaster, may be used to box in the structural members (Fig. 1.6*c*).

See also Secs. 7 to 11.

Systems for Enclosing Buildings. Buildings are enclosed for privacy, to exclude wind, rain, and snow from the interior, and to control interior temperature and humidity. A single-enclosure type of system is one that extends continuously from the ground to enclose the floor. Simple examples are cone-like tepees and dome igloos. A multiple-enclosure type of system consists of a horizontal or inclined top covering, called a **roof** (Fig. 1.1), and vertical or inclined side enclosures called **walls.**

Roofs may have any of a wide variety of shapes. A specific shape may be selected because of appearance, need for attic space under the roof, requirements for height between roof and floor below, desire for minimum enclosed volume, structural economy, or requirements for drainage of rainwater and shedding of snow. While roofs are sometimes given curved surfaces, more often roofs are composed of one or more plane surfaces. Some commonly used types are shown in Fig. 1.4.

The flat roof shown in Fig. 1.4*a* is nearly horizontal but has a slight pitch for drainage purposes. A more sloped roof is called a shed roof (Fig. 1.4*b*). A pitched roof (Fig. 1.4*c*) is formed by a combination of two inclined planes. Four inclined planes may be combined to form either a hipped roof (Fig. 1.4*d*) or a gambrel roof (Fig. 1.4*e*). A mansard roof (Fig. 1.4*f*) is similar to a hipped roof but, composed of additional planes, encloses a larger volume underneath. Any of the preceding roofs may have glazed openings, called **skylights** (Fig. 1.4*b*), for daylighting the building interior. The roofs shown in Fig. 1.4*c* to *f* are often used to enclose attic space. Windows may be set in **dormers** that project from a sloped roof (Fig. 1.4*c*). Other alternatives, often used to provide large areas free of walls or columns, include flat-plate and arched or dome roofs.

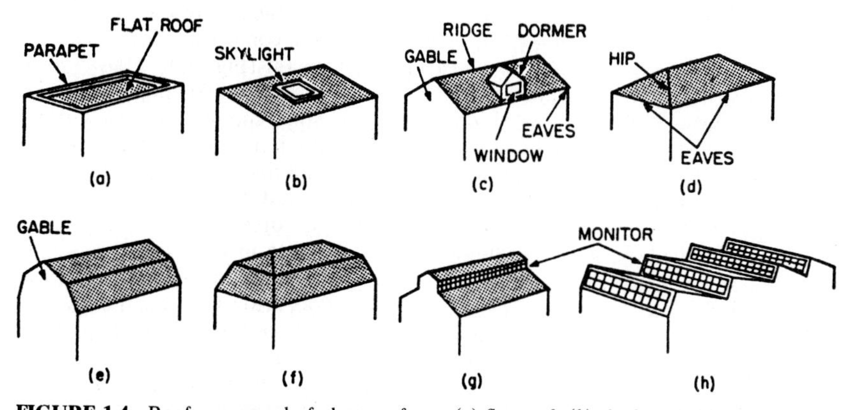

FIGURE 1.4 Roofs composed of plane surfaces: (*a*) flat roof; (*b*) shed roof; (*c*) pitched roof; (*d*) hipped roof; (*e*) gambrel roof; (*f*) mansard roof; (*g*) monitored roof; (*h*) sawtooth roof. *(Reprinted with permission from F. S. Merritt and J. Ambrose, "Building Engineering and Systems Design," 2d ed., Van Nostrand Reinhold, New York.)*

Monitored roofs are sometimes used for daylighting and ventilating the interior. A **monitor** is a row of windows installed vertically, or nearly so, above a roof (Fig. 1.4*g*). Figure 1.4*h* illustrates a variation of a monitored roof that is called a sawtooth roof.

The basic element in a roof is a thin, waterproof covering, called **roofing** (Sec. 12). Because it is thin, it is usually supported on **sheathing,** a thin layer, or **roof deck,** a thick layer, which in turn, is carried on structural members, such as beams or trusses. The roof or space below should contain thermal insulation (Fig. 1.6*c* and *d*).

Exterior walls enclose a building below the roof. The basic element in the walls is a strong, durable, water-resistant facing. For added strength or lateral stability, this facing may be supplemented on the inner side by a backing or sheathing (Fig. 1.5*b*). For esthetic purposes, an interior facing usually is placed on the inner side of the backing. A layer of insulation should be incorporated in walls to resist passage of heat.

Generally, walls may be built of unit masonry, panels, framing, or a combination of these materials.

Unit masonry consists of small units, such as clay brick, concrete block, glass block, or clay tile, held together by a cement such as mortar. Figure 1.5*a* shows a wall built of concrete blocks.

Panel walls consist of units much larger than unit masonry. Made of metal, concrete, glass, plastics, or preassembled bricks, a panel may extend from foundation to roof in single-story buildings, or from floor to floor or from window header in one story to window sill of floor above in multistory buildings. Large panels may incorporate one or more windows. Figure 1.5*c* shows a concrete panel with a window.

Framed walls consist of slender, vertical, closely spaced structural members, tied together with horizontal members at top and bottom, and interior and exterior facings. Thermal insulation may be placed between the components. Figure 1.5*b* shows a wood-framed exterior wall.

Combination walls are constructed of several different materials. Metal, brick, concrete, or clay tile may be used as the exterior facing because of strength, durability, and water and fire resistance. These materials, however, are relatively expensive. Consequently, the exterior facing is made thin and backed up with a less expensive material. For example, brick may be used as an exterior facing with wood framing or concrete block as the backup.

Exterior walls may be classified as curtain walls or bearing walls. **Curtain walls** serve primarily as an enclosure. Supported by the structural system, such walls need to be strong enough to carry only their own weight and wind pressure on the exterior face. **Bearing walls,** in contrast, serve not only as an enclosure but also to transmit to the foundations loads from other building components, such as beams, floors, roofs, and other walls (Fig. 1.5*a* and *b*). (See also Sec. 11.)

Openings are provided in exterior walls for a variety of purposes, but mainly for windows and doors. Where openings occur, structural support must be provided over them to carry the weight of the wall above and any other loads on that portion of the wall. Usually, a beam called a lintel is placed over openings in masonry walls (Fig. 1.5*a*) and a beam called a top header is set over openings in wood-framed walls.

A **window** usually consists of transparent glass or plastics (**glazing**) held in place by light framing, called **sash.** The window is fitted into a frame secured to the walls (Fig. 1.5*a*). For sliding windows, the frame carries guides in which the sash slides.

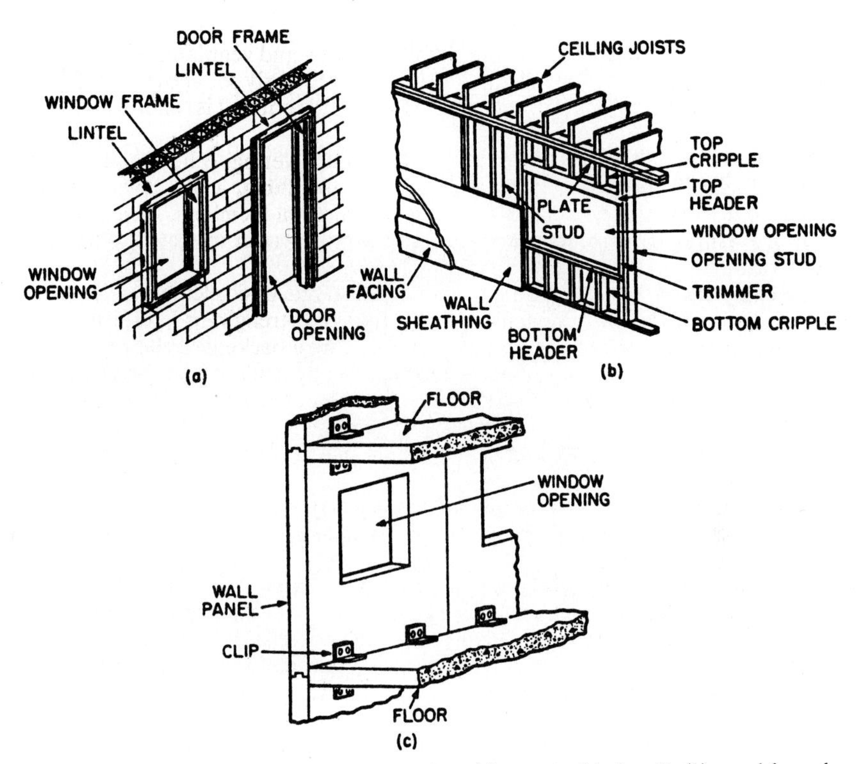

FIGURE 1.5 Types of exterior wall construction: (*a*) concrete-block wall; (*b*) wood-framed wall; (*c*) precast-concrete curtain wall.

For swinging windows, stops against which the window closes are built into the frame.

Hardware is provided to enable the window to function as required. For movable windows, the hardware includes grips for moving them, locks, hinges for swinging windows, and sash balances and pulleys for vertically sliding windows.

The main purposes of windows are to illuminate the building interior with daylight, to ventilate the interior, and to give occupants a view of the outside. For retail stores, windows may have the major purpose of giving passersby a view of items displayed inside. (See also Sec. 11.)

Doors are installed in exterior walls to give access to or from the interior or to prevent such access. For similar reasons, doors are also provided in interior walls and partitions. Thus, a door may be part of a system for enclosing a building or a component of a system for enclosing interior spaces.

Systems for Enclosing Interior Spaces. The interior of a building usually is compartmented into spaces or rooms by horizontal dividers (floor-ceiling or roof-ceiling systems) and vertical dividers (interior walls and partitions). (The term partitions is generally applied to non-load-bearing walls.)

Floor-Ceiling Systems. The basic element of a floor is a load-carrying deck. For protection against wear, esthetic reasons, foot comfort, or noise control, a

floor covering often is placed over the deck, which then may be referred to as a **subfloor**. Figure 1.6*a* shows a concrete subfloor with a flexible-tile floor covering. A hollow-cold-formed steel deck is incorporated in the subfloor to house electric wiring.

In some cases, a subfloor may be strong and stiff enough to span, unaided, long distances between supports provided for it. In other cases, the subfloor is closely

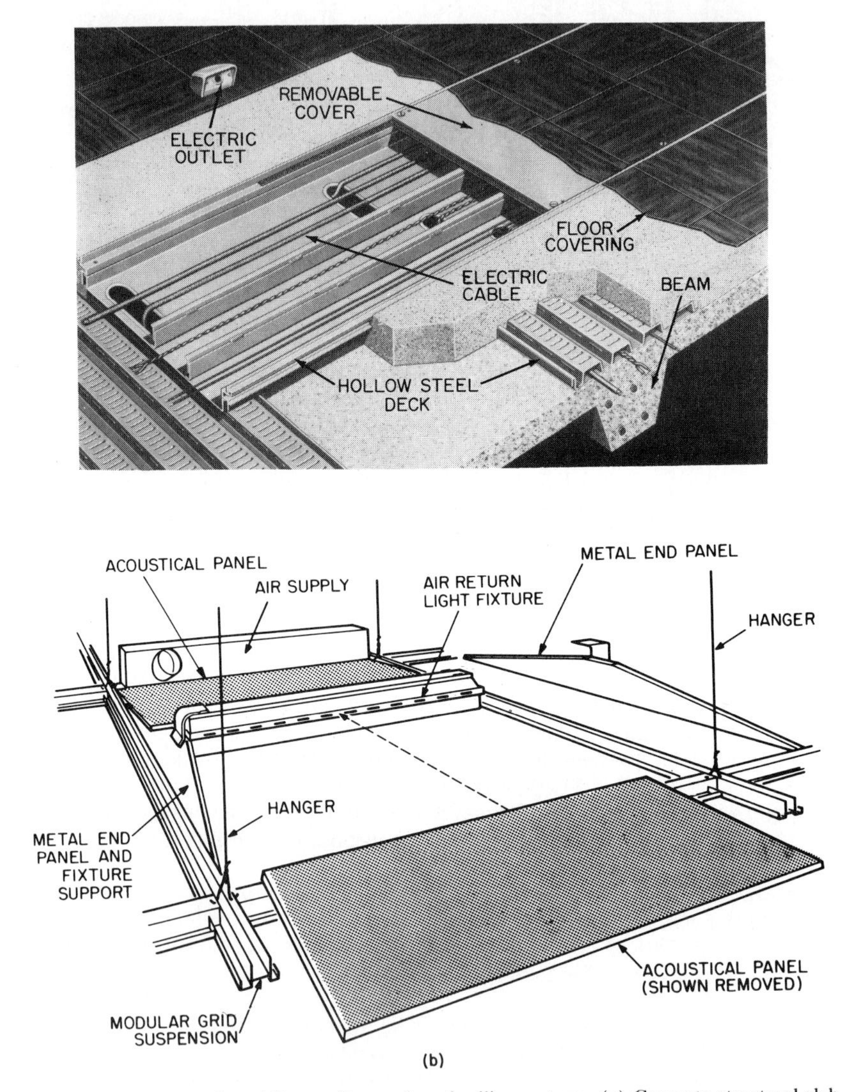

FIGURE 1.6 Examples of floor-ceiling and roof-ceiling systems. (*a*) Concrete structural slab carries hollow-steel deck, concrete fill, and flexible tile flooring. (*b*) Acoustical-tile ceiling incorporating a lighting fixture with provisions for air distribution is suspended below a floor. (*c*) Insulated roof and steel beams are sprayed with mineral fiber for fire protection. (*d*) Insulated roof and open-web joists are protected by a fire-rated suspended ceiling.

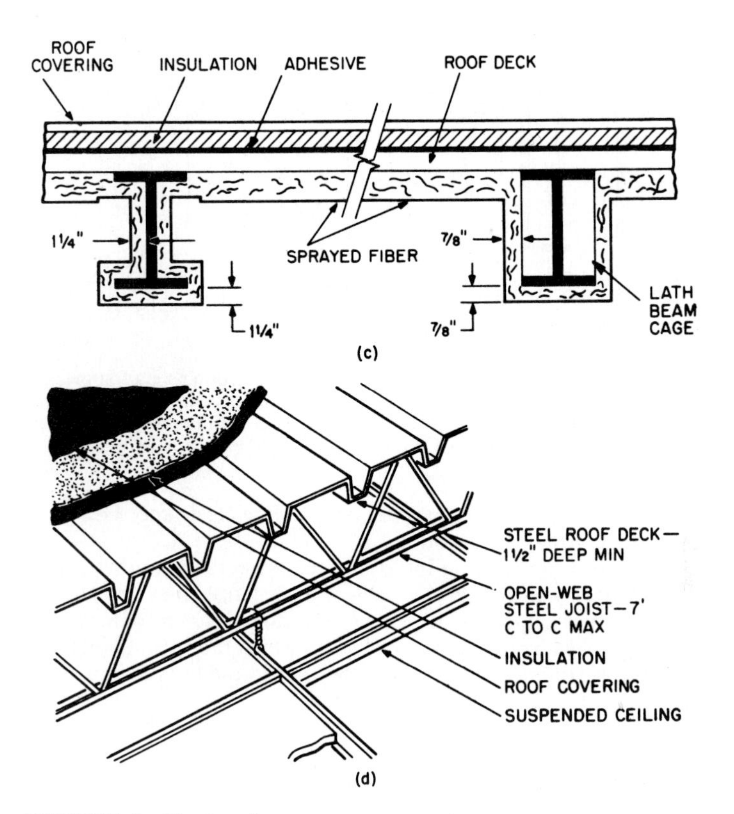

FIGURE 1.6 *(Continued)*

supported on beams. The subfloor in Fig. 1.6*a*, for example, is shown constructed integrally with concrete beams, which carry the loads from the subfloor to bearing walls or columns.

The underside of a floor or roof and of beams supporting it, including decorative treatment when applied to that side, is called a **ceiling.** Often, however, a separate ceiling is suspended below a floor or roof for esthetic or other reasons. Figure 1.6*b* shows such a ceiling. It is formed with acoustical panels and incorporates a lighting fixture and air-conditioning inlets and outlets.

Metal and wood subfloors and beams require fire protection. Figure 1.6*c* shows a roof and its steel beams protected on the underside by a sprayed-on mineral fiber. Figure 1.6*d* shows a roof and open-web steel joists protected on the underside by a continuous, suspended, fire-resistant ceiling. As an alternative to encasement in or shielding by a fire-resistant material, wood may be made fire-resistant by treatment with a fire-retardent chemical.

Fire Ratings. Tests have been made, usually in conformance with E119, "Standard Methods of Tests of Building Construction and Materials," developed by ASTM, to determine the length of time specific assemblies of materials can withstand a standard fire, specified in E119. On the basis of test results, each construction is assigned a fire rating, which gives the time in hours that the assembly can withstand the fire. Fire ratings for various types of construction may be obtained

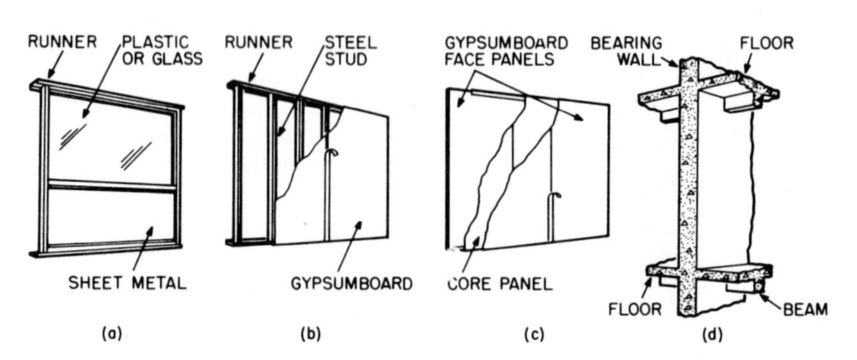

FIGURE 1.7 Types of partitions: (*a*) non-load-bearing; (*b*) gypsumboard on metal studs; (*c*) gypsumboard face panels laminated to a gypsum core panel; (*d*) concrete bearing wall, floors, and beams. *(Reprinted with permission from F. S. Merritt and J. Ambrose, "Building Engineering and Systems Design," 2d ed., Van Nostrand Reinhold, New York.)*

from local, state, or model building codes or the "Fire Resistance Design Manual," published by the Gypsum Association.

Interior Walls and Partitions. Interior space dividers do not have to withstand such severe conditions as do exterior walls. For instance, they are not exposed to rain, snow, and solar radiation. Bearing walls, however, must be strong enough to transmit to supports below them the loads to which they are subjected. Usually, such interior walls extend vertically from the roof to the foundations of a building and carry floors and roof. The basic element of a bearing wall may be a solid core, as shown in Fig. 1.7*d*, or closely spaced vertical framing (**studs**), as shown in Fig. 1.7*b*.

Non-load-bearing partitions do not support floors or roof. Hence, partitions may be made of such thin materials as sheet metal (Fig. 1.7*a*), brittle materials as glass (Fig. 1.7*a*), or weak materials as gypsum (Fig. 1.7*c*). Light framing may be used to hold these materials in place. Because they are non-load-bearing, partitions may be built and installed to be easily shifted or to be foldable, like a horizontally sliding door. (See also Sec. 11.)

Wall Finishes. Walls are usually given a facing that meets specific architectural requirements for the spaces enclosed. Such requirements include durability under indoor conditions, ease of maintenance, attractive appearance, fire resistance, water resistance, and acoustic properties appropriate to the occupancy of the space enclosed. The finish may be the treated surface of the exposed wall material, such as the smooth, painted face of a sheet-metal panel, or a separate material, such as plaster, gypsumboard, plywood, or wallpaper. (See also Sec. 11.)

Doors. Openings are provided in interior walls and partitions to permit passage of people and equipment from one space to another. Doors are installed in the openings to control passage and also to provide privacy.

Usually, a door frame is set around the perimeter of the opening to hold the door in place (Fig. 1.8). Depending on the purpose of the door, size, and other factors, the door may be hinged to the frame at top, bottom, or either side. Or the door may be constructed to slide vertically or horizontally or to rotate about a vertical axis in the center of the opening (revolving door). (See also Sec. 11.)

Hardware is provided to enable the door to function as required. For example, hinges are provided for swinging doors, and guides are installed for sliding doors. Locks or latches are placed in or on doors to prevent them from being opened. Knobs or pulls are attached to doors for hand control.

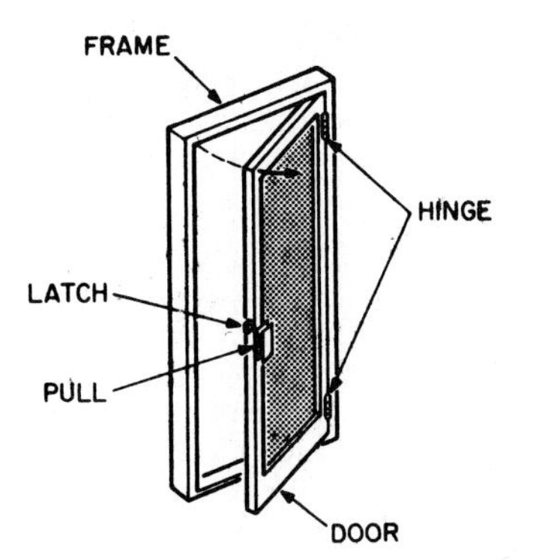

FIGURE 1.8 Example of door and frame.

Builders' Hardware. This is a general term applied to fastenings and devices, such as nails, screws, locks, hinges, and pulleys. These items generally are classified as either finishing hardware or rough hardware (Sec. 11).

Plumbing. The major systems for conveyance of liquids and gases in pipes within a building are classified as plumbing. Plumbing pipes usually are connected to others that extend outside the building to a supply source, such as a public water main or utility gas main, or to a disposal means, such as a sewer. For health, safety, and other reasons, pipes of different types of plumbing systems must not be interconnected, and care must be taken to prevent flow from one system to another.

The major purposes of plumbing are: (1) to convey water and heating gas, if desired, from sources outside a building to points inside where the fluid or gas is needed, and (2) to collect wastewater and storm water in the building, on the roof, or elsewhere on the site and convey the liquid to sewers outside the building.

For these purposes, plumbing requires fixtures for collecting discharged water and wastes; pipes for supply and disposal; valves for controlling flow; drains, and other accessories. For more details, see Sec. 14.

Heating, Ventilation, and Air Conditioning (HVAC). Part of the environmental control systems within buildings, along with lighting and sound control, HVAC is often necessary for the health and comfort of building occupants. Sometimes, however, HVAC may be needed for manufacturing processes, product storage, or operation of equipment, such as computers. HVAC usually is used to control temperature, humidity, air movement, and air quality in the interior of buildings.

Ventilation is required to supply clean air for breathing, to furnish air for operation of combustion equipment, and to remove contaminated air. Ventilation, however, also can be used for temperature control by bringing outside air into a building when there is a desirable temperature difference between that air and the interior air.

The simplest way to ventilate is to open windows. When this is not practicable, mechanical ventilation is necessary. This method employs fans to draw outside air into the building and distribute the air, often through ducts, to interior spaces. The method, however, can usually be used only in mild weather. To maintain comfort conditions in the interior, the fresh air may have to be heated in cold weather and cooled in hot weather.

Heating and cooling of a building interior may be accomplished in any of a multitude of ways. Various methods are described in Sec. 13.

Lighting. For health, safety, and comfort of occupants, a building interior should be provided with an adequate quantity of light, good quality of illumination, and proper color of light. The required illumination may be supplied by natural or artificial means.

Daylight is the source of natural illumination. It enters a building through fenestration, such as windows in the exterior walls or monitors or skylights on the roof.

Artificial illumination can be accomplished in primitive fashion by burning of candles or of oil or gas in lamps. Generally, however, artificial illumination is more conveniently obtained through consumption of electrical energy in incandescent, fluorescent, electroluminescent, or other electric lamps. In such cases, the light source is housed in a **luminaire,** or **lighting fixture.** More details are given in Sec. 15.

Acoustics. The science of sound, its production, transmission, and effects are applied in building design for sound and vibration control.

A major objective of acoustics is provision of an environment that enhances communication in the building interior, whether the sound is created by speech or music. This is accomplished by installation of enclosures with appropriate acoustic properties around sound sources and receivers. Another important objective is reduction or elimination of noise—unwanted sound—from building interiors. This may be accomplished by elimination of the noise at the source, by installation of sound barriers, or by placing sound-absorbing materials on the surfaces of enclosures.

Still another objective is reduction or elimination of vibrations that can annoy occupants, produce noise by rattling loose objects, or crack or break parts or contents of a building. The most effective means of preventing undesirable vibrations is correction of the source. Otherwise, the source should be isolated from the building structure and potential transmission paths should be interrupted with carefully designed discontinuities.

Electric Power and Communication Systems. Electric power is generally bought from a nearby utility and often supplemented for emergency purposes by power from batteries or a generating plant on the site. Purchased power is brought from the power lines connected to the generating source to an entrance control point and a meter in the building. From there, conductors distribute the electricity throughout the building to outlets where the power can be tapped for lighting, heating, and operating electric devices.

Two interrelated types of electrical systems are usually provided within a building. One type is used for communications, including telephone, television, background music, paging, signal and alarm systems. The second type serves the other electrical needs of the building and its occupants.

In addition to conductors and outlets, an electrical system also incorporates devices and apparatus for controlling electric voltage and current. Because electricity can be hazardous, the system must be designed and installed to prevent injury to occupants and damage to building components.

For more details, see Sec. 15.

Vertical-Circulation Elements. In multistory buildings, provision must be made for movement of people, supplies, and equipment between the various levels. This may be accomplished with ramps, stairs, escalators, elevators, dumbwaiters, vertical conveyors, pneumatic tubes, mail chutes, or belt conveyors. Some of the mechanical equipment, however, may not be used for conveyance of people.

A **ramp,** or sloping floor, is often used for movement of people and vehicles in such buildings as stadiums and garages. In most buildings, however, stairs are installed because they can be placed on a steeper slope and therefore occupy less space than ramps. Nevertheless, access to at least one entrance to a building should be provided by a ramp for use by handicapped persons.

A **stairway** consists of a series of steps and landings. Each **step** consists of a horizontal platform, or **tread,** and a vertical separation or enclosure, called a **riser** (Fig. 1.9*a*). Railings are placed along the sides of the stairway and floor openings for safety reasons. Also, structural members may be provided to support the stairs and the floor edges. Often, in addition, the stairway must be enclosed for fire protection.

Escalators, or powered stairs, are installed in such buildings as department stores and transportation terminals, or in the lower stories of office buildings and hotels, where there is heavy pedestrian traffic between floors. Such powered stairs consist basically of a conveyor belt with steps attached; an electric motor for moving the belt, and steps, controls, and structural supports.

Elevators are installed to provide speedier vertical transportation, especially in tall buildings. Transportation is provided in an enclosed car that moves along guides, usually within a fire-resistant vertical shaft but sometimes unenclosed along the exterior of a building. The shaft, or the exterior wall, has openings, protected by doors, at each floor to provide access to the elevator car. The car may be suspended on and moved by cables (Fig. 1.9*b*) or set atop a piston moved by hydraulic pressure (Fig. 1.9*c*).

More information on vertical-circulation elements is given in Sec. 16.

Intelligent Buildings. In addition to incorporating the major systems previously described, intelligent buildings, through the use of computers and communication equipment, have the ability to control the total building environment. The equipment and operating personnel usually are stationed in a so-called **control center.** Various sensors and communication devices, feeding information to and from the control center, are located in key areas throughout the building for the purposes of analyzing and adjusting the environment, delivering messages during emergencies, and dispatching repair personnel and security guards, as needed.

To conserve energy, lighting may be operated by sensors that detect people movement. HVAC may be adjusted in accordance with temperature changes. El-

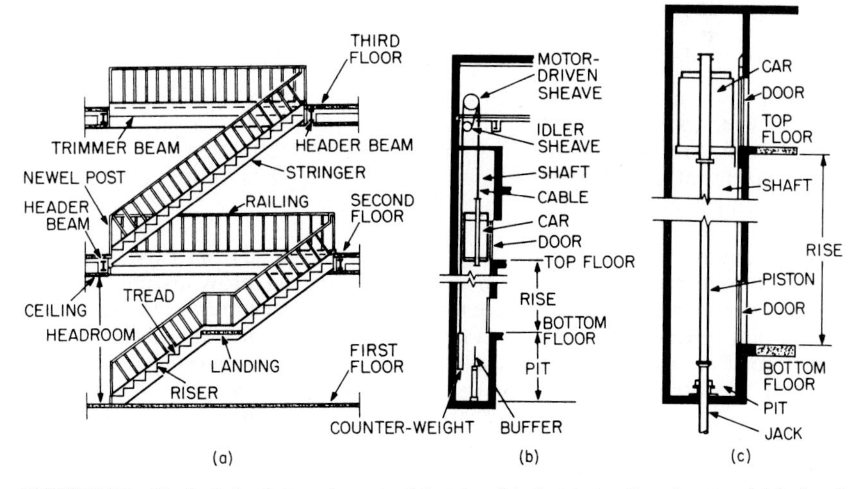

FIGURE 1.9 Vertical-circulation elements: (*a*) stairs; (*b*) electric traction elevator; (*c*) hydraulic elevator.

evators may be programmed for efficient handling of variations in traffic patterns and may be equipped with voice synthesizers to announce floor stops and give advice in emergencies. In addition, intelligent buildings are designed for ease and flexibility in providing for changes in space use, piping, electrical conductors, and installed equipment. See also Arts. 3.5.12 and 3.7.2.

(F. S. Merritt and J. Ambrose, "Building Engineering and Systems Design," 2nd Ed., Van Nostrand Reinhold, New York.)

1.8 VALUE ENGINEERING

As indicated in Art. 1.3, the client in the initial design phase develops a program, or list of requirements. The goal of the designers is to select a system that meets these requirements. Before the designers do this, however, it is advisable for them to question whether the requirements represent the client's actual needs. Can the criteria and standards affecting the design be made less stringent? After the program has been revised to answer these questions, the designers select a system. Next, it is advisable for the designers to question whether the system provides the best value at the lowest cost. Value engineering is a useful procedure for answering this question and selecting a better alternative if the answer indicates this is desirable.

Value engineering is the application of the scientific method to the study of values of systems. The major objective of value engineering in building design and construction is reduction of initial and life-cycle costs (Art. 1.6). Thus, value engineering has one of the objectives of systems design, in which the overall goal is production of an optimum building, and should be incorporated in the systems-design procedure.

The **scientific method,** which is incorporated in the definitions of value engineering and systems design, consists of the following steps:

1. Collection of data and observations of natural phenomena
2. Formulation of a hypothesis capable of predicting future observations
3. Testing of the hypothesis to verify the accuracy of its predictions and abandonment or improvement of the hypothesis if it is inaccurate

Those who conduct or administer value studies are often called **value engineers,** or **value analysts.** They generally are organized into an interdisciplinary team for value studies for a specific project. Sometimes, however, an individual, such as an experienced contractor, performs value engineering services for the client for a fee or a percentage of savings achieved by the services.

Value Analysis. Value is a measure of benefits anticipated from a system or from the contribution of a component to system performance. This measure must be capable of serving as a guide in a choice between alternatives in evaluations of system performance. Because generally in comparisons of systems only relative values need be considered, value takes into account both advantages and disadvantages, the former being considered positive and the latter negative. It is therefore possible in comparisons of systems that the value of a component of a system may be negative and subtracts from the overall performance of the system.

System evaluations would be relatively easy if a monetary value could always be placed on performance. Then, benefits and costs could be compared directly. Value, however, often must be based on a subjective decision of the client. For example, how much extra is an owner willing to pay for beauty, prestige, or better community relations? Will the owner accept gloom, glare, draftiness, or noise for a savings in cost? Consequently, other values than monetary must be considered in value analysis. Such considerations require determination of the relative importance of the client's requirements and weighting of values accordingly.

Value analysis is the part of the value-engineering procedure devoted to investigation of the relation between costs and values of components and systems and alternatives to these. The objective is to provide a rational guide for selection of the lowest-cost system that meets the client's actual needs.

Measurement Scales. For the purposes of value analysis, it is essential that characteristics of a component or system on which a value is to be placed be distinguishable. An analyst should be able to assign different numbers, not necessarily monetary, to values that are different. These numbers may be ordinates of any one of the following four measurement scales: ratio, interval, ordinal, nominal.

Ratio Scale. This scale has the property that, if any characteristic of a system is assigned a value number k, any characteristic that is n times as large must be assigned a value number nk. Absence of the characteristic is assigned the value zero. This type of scale is commonly used in engineering, especially in cost comparisons. For example, if a value of $10,000 is assigned to system A and of $5000 to system B, then A is said to cost twice as much as B.

Interval Scale. This scale has the property that equal intervals between assigned values represent equal differences in the characteristic being measured. The scale zero is assigned arbitrarily. The Celsius scale of temperature measurement is a good example of an interval scale. Zero is arbitrarily established as the temperature at which water freezes; the zero value does not indicate absence of heat. The boiling point of water is arbitrarily assigned the value of 100. The scale between 0 and 100 is then divided into 100 equal intervals called degrees (°C). Despite the arbitrariness of the selection of the zero point, the scale is useful in heat measurement. For example, changing the temperature of an object from 40°C to 60°C, an increase of 20°C, requires twice as much heat as changing the temperature from 45°C to 55°C, an increase of 10°C.

Ordinal Scale. This scale has the property that the magnitude of a value number assigned to a characteristic indicates whether a system has more, or less, of the characteristic than another system has or is the same with respect to that characteristic. For example, in a comparison of the privacy afforded by different types of partitions, each may be assigned a number that ranks it in accordance with the degree of privacy that it provides. Partitions with better privacy are given larger numbers. Ordinal scales are commonly used when values must be based on subjective judgments of nonquantifiable differences between systems.

Nominal Scale. This scale has the property that the value numbers assigned to a characteristic of systems being compared merely indicate whether the systems differ in this characteristic. But no value can be assigned to the difference. This type of scale is often used to indicate the presence or absence of a characteristic or component. For example, the absence of a means of access to equipment for maintenance may be represented by zero or a blank space, whereas the presence of such access may be denoted by 1 or X.

Weighting. In practice, construction cost usually is only one factor, perhaps the only one with a monetary value, of several factors that must be evaluated in a comparison of systems. In some cases, some of the other characteristics of the system may be more important to the owner than cost. Under such circumstances, the comparison may be made by use of an ordinal scale for ranking each characteristic and then weighting the rankings in accordance with the importance of the characteristic to the owner.

As an example of the use of this procedure, calculations for comparison of two partitions are shown in Table 1.1. Alternative 1 is an all-metal partition and alternative 2 is made of glass and metal.

In Table 1.1, characteristics of concern in the comparison are listed in the first column. The numbers in the second column indicate the relative importance of each characteristic to the owner: 1 denotes lowest priority and 10 highest priority. These are the weights. In addition, each of the partitions is ranked on an ordinal scale, with 10 as the highest value, in accordance with the degree to which it possesses each characteristic. These rankings are listed as relative values in Table 1.1. For construction cost, for instance, the metal partition is assigned a relative value of 10 and the glass-metal partition a value of 8, because the metal partition costs a little less than the other one. In contrast, the glass-metal partition is given a relative value of 8 for visibility, because the upper portion is transparent, whereas the metal partition has a value of zero, because it is opaque.

To complete the comparison, the weight of each characteristic is multiplied by the relative value of the characteristic for each partition and entered in Table 1.1 as a weighted value. For construction cost, for example, the weighted values are $8 \times 10 = 80$ for the metal partition and $8 \times 8 = 64$ for the glass-metal partition. The weighted values for each partition are then added, yielding 360 for alternative

TABLE 1.1 Comparison of Alternative Partitions*

		Alternatives			
		1 All metal		2 Glass and metal	
Characteristics	Relative importance	Relative value	Weighted value	Relative value	Weighted value
Construction cost	8	10	80	8	64
Appearance	9	7	63	9	81
Sound transmission	5	5	25	4	20
Privacy	3	10	30	2	6
Visibility	10	0	0	8	80
Movability	2	8	16	8	16
Power outlets	4	0	0	0	0
Durability	10	9	90	9	90
Low maintenance	8	7	56	5	40
Total weighted values			360		397
Cost			$12,000		$15,000
Ratio of values to cost			0.0300		0.0265

*Reprinted with permission from F. S. Merritt, "Building Engineering and Systems Design," Van Nostrand Reinhold Company, New York.

1 and 397 for alternative 2. While this indicates that the glass-metal partition is better, it may not be the best for the money. To determine whether it is, the weighted value for each partition is divided by its cost, yielding 0.0300 for the metal partition and 0.0265 for the other. Thus, the metal partition appears to offer more value for the money and would be recommended.

Economic Comparisons. In a choice between alternative systems, only the differences between system values are significant and need to be compared.

Suppose, for example, the economic effect of adding 1 in of thermal insulation to a building is to be investigated. In a comparison, it is not necessary to compute the total cost of the building with and without the insulation. Generally, the value analyst need only subtract the added cost of 1 in of insulation from the decrease in HVAC cost to obtain the net saving or cost increase resulting from addition of insulation. A net saving would encourage addition of insulation. Thus, a decision can be reached without the complex computation of total building cost.

In evaluating systems, value engineers must take into account not only initial and life-cycle costs but also the return the client wishes to make on the investment in the building. Generally, a client would like not only to maximize profit, the difference between revenue from use of the building and total costs, but also to ensure that the rate of return, the ratio of profit to investment, is larger than all of the following:

Rate of return expected from the type of business

Interest rate for borrowed money

Rate for government bonds or notes

Rate for highly rated corporate bonds

The client is concerned with interest rates because all costs represent money that must be borrowed or that could otherwise be invested at a current interest rate. The client also has to be concerned with time, measured from the date at which an investment is made, because interest cost increases with time. Therefore, in economic comparisons of systems, interest rates and time must be taken into account. (Effects of monetary inflation can be taken into account in much the same way as interest.)

An economic comparison usually requires evaluation of initial capital investments, salvage values after several years, annual disbursements and annual revenues. Because each element in such a comparison may have associated with it an expected useful life different from that of the other elements, the different types of costs and revenues must be made commensurable by reduction to a common basis. This is commonly done by either:

1. Converting all costs and revenues to equivalent uniform annual costs and income
2. Converting all costs and revenues to present worth of all costs and revenues at time zero

Present worth is the money that, invested at time zero, would yield at later times required costs and revenues at a specified interest rate. In economic comparisons, the conversions should be based on a rate of return on investment that is attractive to the client. It should not be less than the interest rate the client would have to pay if the amount of the investment had to be borrowed. For this reason, the desired rate of return is called interest rate in conversions. Calculations also should be based on actual or reasonable estimates of time periods. Salvage values, for instance, should be taken as the expected return on sale or trade-in of an item after a specific number of years that it has been in service. Interest may be considered compounded annually.

Future Value. Based on the preceding assumptions, a sum invested at time zero increases in time to

$$S = P(1 + i)^n \tag{1.1}$$

where S = future amount of money, equivalent to P, at the end of n periods of time with interest i

i = interest rate

n = number of interest periods, years

P = sum of money invested at time zero = present worth of S

Present Worth. Solution of Eq. (1.1) for P yields the present worth of a sum of money S at a future date:

$$P = S(1 + i)^{-n} \tag{1.2}$$

The present worth of payments R made annually for n years is

$$P = R \frac{1 - (1 + i)^{-n}}{i} \tag{1.3}$$

The present worth of the payments R continued indefinitely can be obtained from Eq. (1.3) by making n infinitely large:

$$P = \frac{R}{i} \tag{1.4}$$

Capital Recovery. A capital investment P at time zero can be recovered in n years by making annual payments of

$$R = P \frac{i}{1 - (1 + i)^{-n}} = P \left[\frac{i}{(1 + i)^n - 1} + i \right] \tag{1.5}$$

When an item has salvage value V after n years, capital recovery R can be computed from Eq. (1.5) by subtraction of the present worth of the salvage value from the capital investment P.

$$R = [P - V(1 + i)^{-n}] \left[\frac{i}{(1 + i)^n - 1} + i \right] \tag{1.6}$$

Example. To illustrate the use of these formulas, an economic comparison is made in the following for two air-conditioning units being considered for an office building. Costs are estimated as follows:

	Unit 1	Unit 2
Initial cost	\$300,000	\$500,000
Life, years	10	20
Salvage value	\$50,000	\$100,000
Annual costs	\$30,000	\$20,000

Cost of operation, maintenance, repairs, property taxes, and insurance are included in the annual costs. The present-worth method is used for the comparison, with interest rate $i = 8\%$.

Conversion of all costs and revenues to present worth must be based on a common service life, although the two units have different service lives, 10 and 20 years, respectively. For the purpose of the conversion, it may be assumed that replacement assets will repeat the investment and annual costs predicted for the initial asset. (Future values, however, should be corrected for monetary inflation.) In some cases, it is convenient to select for the common service life the least common multiple of the lives of the units being compared. In other cases, it may be more convenient to assume that the investment and annual costs continue indefinitely. The present worth of such annual costs is called **capitalized cost.**

For this example, a common service life of 20 years, the least common multiple of 10 and 20, is selected. Hence, it is assumed that unit 1 will be replaced at the end of the tenth period at a cost of $300,000 less the salvage value. Similarly, the replacement unit will be assumed to have the same salvage value after 20 years.

The calculations in Table 1.2 indicate that the present worth of the net cost of unit 2 is less than that for unit 1. If cost were the sole consideration, purchase of unit 2 would be recommended.

ASTM has developed several standard procedures for making economic studies of buildings and building systems, in addition to ASTM E917 for measuring life-cycle costs, mentioned previously. For example, ASTM E964 is titled Practice for Measuring Benefit-to-Cost and Savings-to-Investment Ratios for Buildings and Building Systems. Other standards available present methods for measuring internal rate of return, net benefits, and payback. ASTM also has developed computer programs for these calculations.

Value Analysis Procedure. In building design, value analysis generally starts with a building system or subsystem proposed by the architect and consultants. The client or the client's representative appoints an interdisciplinary team to study the system or subsystem and either recommend its use or propose a more economical alternative. The team coordinator sets goals and priorities for the study and may

TABLE 1.2 Example Comparison of Two Air-Conditioning Units

	Unit 1	Unit 2
Initial investment	$300,000	$500,000
Present worth of replacement cost in 10 years $P - V$ at 8% interest [Eq. (1.2)]	115,800	
Present worth of annual costs for 20 years at 8% interest [Eq. (1.3)]	294,540	196,360
Present worth of all costs	710,340	696,360
Revenue:		
Present worth of salvage value after 20 years at 8% interest [Eq. (1.2)]	10,730	21,450
Net cost:		
Present worth of net cost in 20 years at 8% interest	$699,610	$674,910

appoint task groups to study parts of the building in accordance with the priorities. The value analysts should follow a systematic, scientific procedure for accomplishing all the necessary tasks that comprise a value analysis. The procedure should provide an expedient format for recording the study as it progresses, assure that consideration has been given to all information, some of which may have been overlooked in development of the proposed system, and logically resolve the analysis into components that can be planned, scheduled, budgeted, and appraised.

The greatest cost reduction can be achieved by analysis of every component of a building. This, however, is not practical, because of the short time usually available for the study and because the cost of the study increases with time. Hence, it is advisable that the study concentrate on those building systems (or subsystems) whose cost is a relatively large percentage of the total building (or system) cost, because those components have possibilities for substantial cost reduction.

During the initial phase of value analysis, the analysts should obtain a complete understanding of the building and its major systems by rigorously reviewing the program, proposed design and all other pertinent information. They should also define the functions, or purposes, of each building component to be studied and estimate the cost of accomplishing the functions. Thus, the analysts should perform a systems analysis, as indicated in Art. 1.2, answer the questions listed in Art. 1.2 for the items to be studied, and estimate the initial and life-cycle costs of the items.

In the second phase of value analysis, the analysts should question the cost-effectiveness of each component to be studied. Also, by use of imagination and creative techniques, they should generate several alternative means for accomplishing the required functions of the component. Then, in addition to answers to the questions in Art. 1.2, the analysts should obtain answers to the following questions:

Do the original design and each alternative meet performance requirements?

What does each cost installed and over the life cycle?

Will it be available when needed? Will skilled labor be available?

Can any components be eliminated?

What other components will be affected by adoption of an alternative? What will the resulting changes in the other components cost? Will there be a net saving in cost?

In investigating the possibility of elimination of a component, the analysts also should see if any part of it can be eliminated, if two parts or more can be combined into one, and if the number of different sizes and types of an element can be reduced. If costs might be increased by use of a nonstandard or unavailable item, the analysts should consider substitution of a more appropriate alternative. In addition, consideration should be given to simplification of construction or installation of components and to ease of maintenance and repair.

In the following phase of value analysis, the analysts should critically evaluate the original design and alternatives. The ultimate goal should be recommendation of the original design or an alternative, whichever offers the greatest value and cost-savings potential. The analysts also should submit estimated costs for the original design and the alternative.

In the final phase, the analysts should prepare and submit to the client or the client's representative who appointed them a written report on the study and resulting recommendations. Also, they should submit a workbook containing detailed backup information.

Value engineering should start during the conceptual phase of design. Then, it has the greatest impact on cost control and no cost is involved in making design changes. During later design phases, design changes involve some cost, especially when substitution of major subsystems is involved, but the cost is nowhere near as great as when changes are made during construction. Such changes should be avoided if possible. Value engineering, however, should be applied to the project specifications and construction contract. Correction of unnecessary and overconservative specifications and contract provisions offers considerable potential for cost reduction.

(E. D. Heller, "Value Management: Value Engineering and Cost Reduction," Addison-Wesley, Reading, Mass.; L. D. Miles, "Techniques of Value Analysis and Engineering," McGraw-Hill Publishing Co., New York; A. Mudge, "Value Engineering," McGraw-Hill Publishing Company, New York; M. C. Macedo, P. V. Dobrow, and J. J. O'Rourke, "Value Management for Construction," John Wiley & Sons, Inc., New York.)

1.9 EXECUTION OF SYSTEMS DESIGN

The basic traditional design procedure (Art. 1.3), which has been widely used for many years, and commonly used variations of it have resulted in many excellent buildings. It needs improvement, however, because clients cannot be certain that its use gives the best value for the money or that the required performance could not have been attained at lower cost. The uncertainty arises because historically:

1. Actual construction costs often exceed low bids or negotiated prices, because of design changes during construction; unanticipated delays during construction, which increase costs; and unforeseen conditions, such as unexpectedly poor subsurface conditions that make excavation and foundation construction more expensive.
2. Construction, operation, or maintenance costs are higher than estimated, because of design mistakes or omissions.
3. Separation of design and construction into different specialties leads to underestimated or overestimated construction costs and antagonistic relations between designers and builders.
4. Construction costs are kept within the client's budget at the expense of later higher operating, maintenance, and repair costs.
5. Coordination of the output of architects and consultants is not sufficiently close for production of an optimum building for the client's actual needs.

One objective of systems design is to correct these defects. This can be done while retaining the desirable features of traditional procedures, such as development of building design in stages, with progressively more accurate cost estimates and frequent client review. Systems design therefore should at least do the following:

1. Question the cost effectiveness of proposed building components and stimulate generation of lower-cost alternatives that achieve the required performance. This can be done by incorporating value engineering in systems design.
2. More closely coordinate the work of various design specialists and engage building construction and operation experts to assist in design.

3. Take into account both initial and life-cycle costs.
4. Employ techniques that will reduce the number of design mistakes and omissions that are not discovered until after construction starts.

Systems Design Procedure. Article 1.2 defines systems and explains that systems design comprises a rational, orderly series of steps that leads to the best decision for a given set of conditions. Article 1.2 also lists the basic components of the procedure as analysis, synthesis, appraisal, and feedback. Following is a more formal definition:

Systems design is the application of the scientific method to selection and assembly of components or subsystems to form the optimum system to attain specified goals and objectives while subject to given constraints and restrictions.

The scientific method is defined in Art. 1.8. Goals, objectives, and constraints are discussed later.

Systems design of buildings, in addition to correcting defects in traditional design, must provide answers to the following questions:

1. What does the client actually want the building to accomplish (goals, objectives, and associated criteria)?
2. What conditions exist, or will exist after construction, that are beyond the designers' control?
3. What requirements for the building or conditions affecting system performance does design control (constraints and associated standards)?
4. What performance requirements and time and cost criteria can the client and designers use to appraise system performance?

Collection of information necessary for design of the building starts at the inception of design and may continue through the contract documents phase. Data collection is an essential part of systems design but because it is continuous throughout design it is not listed as one of the basic steps.

For illustrative purposes, the systems design procedure is shown resolved into nine basic steps in Fig. 1.10. Because value analysis is applied in step 5, steps 4 through 8 covering synthesis, analysis, and appraisal may be repeated several times. Each iteration should bring the design closer to the optimum.

In preparation for step 1, the designers should secure a building program and information on existing conditions that will affect building design. In step 1, the designers use the available information to define goals to be met by the system.

Goals. These state what the building is to accomplish, how it will affect the environment and other systems, and how other systems and the environment will affect the building. Goals should be generalized but brief statements, encompassing all the design objectives. They should be sufficiently specific, however, to guide generation of initial and alternative designs and control selection of the best alternative.

A simple example of a goal is: Design a branch post-office building with 100 employees to be constructed on a site owned by the client. The building should harmonize with neighboring structures. Design must be completed within 90 days and construction within 1 year. Construction cost is not to exceed $500,000.

When systems design is applied to a subsystem, goals serve the same purpose as for a system. They indicate the required function of the subsystem and how it affects and is affected by other systems.

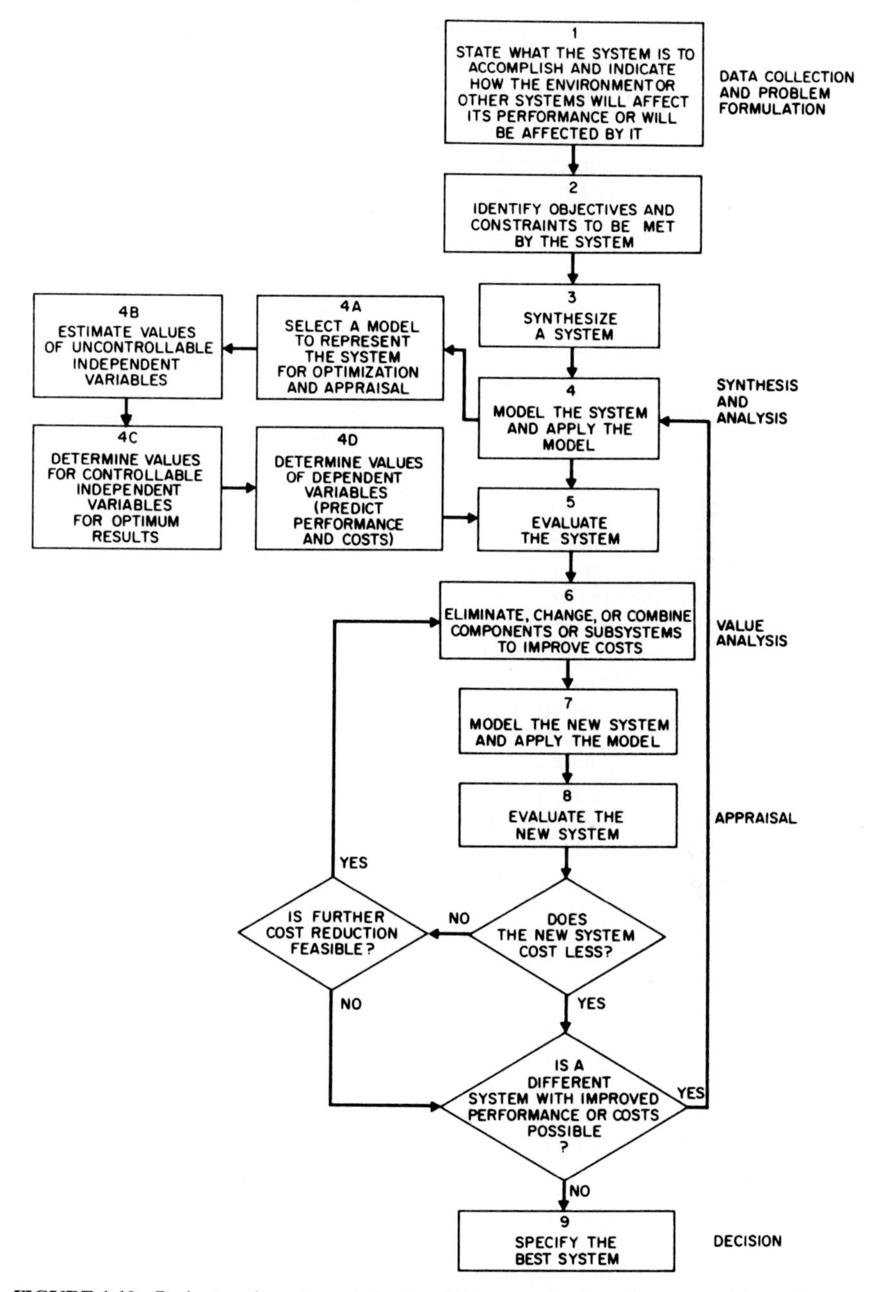

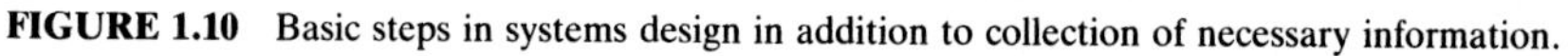
FIGURE 1.10 Basic steps in systems design in addition to collection of necessary information.

Objectives. With the goals known, the designers can advance to step 2 and define the system objectives. These are similar to goals but supply in detail the requirements that the system must satisfy to attain the goals.

In listing objectives, the designers may start with broad generalizations that they later develop at more detailed levels to guide design of the system. Some objectives, such as minimization of initial costs, life-cycle costs and construction time, should be listed. Other objectives that apply to the design of almost every building, such as the health, safety, and welfare objectives of the building, zoning, and Occupational Safety and Health Administration regulations, are too numerous to list and may be adopted by reference. Objectives should be sufficiently specific to guide the planning of building interior spaces and selection of specific characteristics for the building and its components: appearance, strength, durability, stiffness, operational efficiency, maintenance, and fire resistance. Also, objectives should specify the degree of control needed for operation of systems provided to meet the other objectives.

At least one criterion must be associated with each objective. The criterion is a range of values within which the performance of the system must lie for the objective to be met. The criterion should be capable of serving as a guide in evaluations of alternative systems. For example, for fire resistance of a wall, the criterion might be 2-hr fire rating.

In addition to establishing criteria, the designers should weight the objectives in accordance with the relative importance of the objectives to the client (Art. 1.8). These weights should also serve as guides in comparisons of alternatives.

System Constraints. In step 2 of systems design, the designers should also define constraints on the system. Constraints are restrictions on the values of design variables that represent properties of the system and are controllable by the designers. Designers are seldom completely free to choose any values desired for controllable variables because of various restrictions, which may be legal ones such as building or zoning code requirements, or may be economic, physical, chemical, temporal, psychological, sociological, or esthetic requirements. Such restrictions may fix the values of the controllable variables or establish a range in which they must lie.

At least one standard must be associated with each constraint. A standard is a value or range of values governing a property of the system. The standard specifying a fixed value may be a minimum or maximum value.

For example, a designer may be seeking to determine the thickness of a load-bearing brick wall. The local building code may state that such a wall may not be less than 8 in thick. This requirement is a minimum standard. The designer may then select a wall thickness of 8 in or more. The requirements of other systems, however, may indicate that the wall thickness may not exceed 16 in. This is a maximum standard. Furthermore, bricks may be available only in nominal widths of 4 in. Hence, the constraints limit the values of the controllable variable, in this case wall thickness, to 8, 12, or 16 in.

Synthesis. In step 3, the designers must conceive at least one system that satisfies the objectives and constraints. For this, they rely on their past experience, knowledge, imagination, and creative skills and on advice from consultants, including value engineers, construction experts, and experienced operators of the type of facilities to be designed.

In addition, the designers should select systems that are cost-effective and can be erected speedily. To save design time in selection of a system, the designers should investigate alternative systems in a logical sequence for potential for achieving optimum results. The following is a possible sequence:

1. Selection of an available industrialized building, a system that is preassembled in a factory. Such a system is likely to be low cost, because of the use of mass-production techniques and factory wages, which usually are lower than those for field personnel. Also, the quality of materials and construction may be better than for custom-built structures, because of assembly under controlled conditions and close supervision.
2. Design of an industrialized building (if the client needs several of the same type of structure).
3. Assembling a building with prefabricated components or systems. This type of construction is similar to that used for industrialized buildings except that the components preassembled are much smaller parts of the building system.
4. Specification of as many prefabricated and standard components as feasible. Standard components are off-the-shelf items, readily available from building supply companies.
5. Repetition of the same component as many times as possible. This may permit mass production of some nonstandard components. Also, repetition may speed construction, because field personnel will work faster as they become familiar with the components.
6. Design of components for erection so that building trades will be employed on the site continuously. Work that compels one trade to wait for completion of work by another trade delays construction and is costly.

Models. In step 4, the designers should represent the system by a model that will enable them to analyze the system and evaluate its performance. The model should be simple, consistent with the role for which it is selected, for practical reasons. The cost of formulating and using the model should be negligible compared with the cost of assembling and testing the actual system.

For every input to a system, there must be a known, corresponding input to the model such that the responses (output) of the model to that input are determinable and correspond to the response of the system to its input. The correlation may be approximate but nevertheless close enough to serve the purposes for which the model is to be used. For example, for cost estimates during the conceptual phase of design, use may be made of a cost model that yields only reasonable guesses of construction costs. The cost model used in the contract documents phase, however, should be accurate.

Models may be classified as iconic, symbolic, or analog. The **iconic type** may be the actual system or a part of it or merely bear a physical resemblance to the actual system. This type is often used for physical tests of performance, such as load or wind-tunnel tests or adjustments of controls. **Symbolic models** represent by symbols the input and output of a system and are usually amenable to mathematical analysis of a system. They enable relationships to be generally, yet compactly, expressed, are less costly to develop and use than other types of models, and are easy to manipulate. **Analog models** are real systems but with physical properties different from those of the actual system. Examples include dial watches for measuring time, thermometers for measuring heat changes, slide rules for multiplying numbers, flow of electric current for measuring heat flow through a metal plate, and soap membranes for measuring torsion in an elastic shaft.

Variables representing input and properties of a system may be considered independent variables. These are of two types:

1. Variables that the designers can control or constraints: $x_1, x_2, x_3, \ldots$
2. Variables that are uncontrollable: $y_1, y_2, y_3, \ldots$

Variables representing system output or performance may be considered dependent variables: $z_1, z_2, z_3. \ldots$

The dependent variables are functions of the independent variables. These functions also contain parameters, which can be adjusted in value to calibrate the model to the behavior of the actual system.

Step 4 of systems design then may be resolved into four steps, as indicated in Fig. 1.10:

1. Select and calibrate a model to represent the system for optimization and appraisal.
2. Estimate values for the uncontrollable, independent variables.
3. Determine values for the controllable variables.
4. Determine the output or performance of the system from the relationship of dependent and independent variables by use of the model.

Cost Models. As an example of the use of models in systems design, consider the following cost models:

$$C = Ap \tag{1.7}$$

where C = construction cost of building
A = floor area, ft^2, in the building
p = unit construction cost, dollars per square foot

This is a symbolic model applicable only in the early stages of design when systems and subsystems are specified only in general form. Both A and p are estimated, usually on the basis of past experience with similar types of buildings.

$$C = \Sigma A_i p_i \tag{1.8}$$

where A_i = convenient unit of measurement for the ith system
p_i = cost per unit for the ith system

This symbolic cost model is suitable for estimating building construction cost in preliminary design stages after types of major systems have been selected. Equation (1.8) gives the cost as the sum of the cost of the major systems, to which should be added the estimated costs of other systems and contractor's overhead and profit. A_i may be taken as floor or wall area, square feet, pounds of steel, cubic yards of concrete, or any other applicable parameter for which the unit cost may be reasonably accurately estimated.

$$C = \Sigma A_j p_j \tag{1.9}$$

where A_j = convenient unit of measurement for the jth subsystem
p_j = cost per unit for the jth subsystem

This symbolic model may be used in the design development phase and later after components of the major systems have been selected and greater accuracy of the cost estimate is feasible. Equation (1.9) gives the construction cost as the sum of the costs of all the subsystems, to which should be added contractor's overhead and profit.

For more information on cost estimating, see Sec. 19.

Optimization. The objective of systems design is to select the single best system for a given set of conditions, a process known as optimization. When more than one property of the system is to be optimized or when there is a single characteristic to be optimized but it is nonquantifiable, an optimum solution may or may not exist. If it does exist, it may have to be found by trial and error with a model or by methods such as those described in Art. 1.8.

When one characteristic, such as construction cost, of a system is to be optimized, the criterion may be expressed as

$$\text{Optimize } z_r = f_r(x_1, x_2, x_3, \ldots y_1, y_2, y_3, \ldots) \tag{1.10}$$

where z_r = dependent variable to be maximized or minimized
x = controllable variable, identified by a subscript
y = uncontrollable variable, identified by a subscript
f_r = objective function

Generally, however, there are restrictions on values of the independent variables. These restrictions may be expressed as

$$\begin{aligned} &f_1(x_1, x_2, x_3, \ldots y_1, y_2, y_3, \ldots) \geq 0 \\ &f_2(x_1, x_2, x_3, \ldots y_1, y_2, y_3, \ldots) \geq 0 \\ &\ldots\ldots\ldots\ldots\ldots\ldots\ldots\ldots \\ &f_n(x_1, x_2, x_3, \ldots y_1, y_2, y_3, \ldots) \geq 0 \end{aligned} \tag{1.11}$$

Simultaneous solution of Eqs. (1.10) and (1.11) yields the optimum values of the variables. The solution may be obtained by use of such techniques as calculus, linear programming, or dynamic programming depending on the nature of the variables and the characteristics of the equations.

Direct application of Eqs. (1.10) and (1.11) to a whole building, its systems, and its larger subsystems usually is impractical, because of the large number of variables and the complexity of their relationships. Hence optimization generally has to be attained in a different way, generally by such methods as suboptimization or simulation.

Systems with large numbers of variables may sometimes be optimized by a process called **simulation,** which involves trial and error with the actual system or a model. In simulation, the properties of the system or model are adjusted with a specific input or range of inputs to the system, and outputs or performance are measured until an optimum result is obtained. When the variables are quantifiable and models are used, the solution usually can be expedited by use of computers. The actual system may be used when it is available and accessible and changes in it will have little or no effect on construction costs. For example, after installation of air ducts, an air-conditioning system may be operated for a variety of conditions to determine the optimum damper position for control of airflow for each condition.

Suboptimization is a trial-and-error process in which designers try to optimize a system by first optimizing its subsystems. It is suitable when components influence each other in series. For example, consider a structural system consisting only of roof, columns, and footings. The roof has a known load (input), exclusive of its own weight. Design of the roof affects the columns and footings, because its output equals the load on the columns. Design of the columns loads only the footings. Design of the footings, however, has no effect on any of the other structural components. Therefore, the structural components are in series and they may be designed by suboptimization to obtain the minimum construction cost or least weight of the system.

Suboptimization of the system may be achieved by first optimizing the footings, for example, designing the lowest-cost footings. Next, the design of both the columns and the footings should be optimized. (Optimization of the columns alone will not yield an optimum structural system, because of the effect of the column weight on the footings.) Finally, roof, columns, and footings together should be optimized. (Optimization of the roof alone will not yield an optimum structural system, because of the effect of its weight on columns and footings. A low-cost roof may be very heavy, requiring costly columns and footings, whereas the cost of a lightweight roof may be so high as to offset any savings from less-expensive columns and footings. An alternative roof may provide optimum results.)

Appraisal. In step 5 of systems design, the designers should evaluate the results obtained in step 4, modeling the system and applying the model. The designers should verify that construction and life-cycle costs will be acceptable to the client and that the proposed system satisfies all objectives and constraints.

During the preceding steps, value analysis may have been applied to parts of the building. In step 6, however, value analysis should be applied to the whole building system. This process may result in changes only to parts of the system, producing a new system, or several alternatives to the original design may be proposed. In steps 7 and 8, therefore, the new systems, or at least those with good prospects, should be modeled and evaluated. During and after this process, completely different alternatives may be conceived. As a result, steps 4 through 8 should be repeated for the new concepts. Finally, in step 9, the best of the systems studied should be selected.

(R. J. Aguilar, "Systems Analysis and Design in Engineering, Architecture, Construction and Planning," Prentice-Hall, Inc., Englewood Cliffs, N.J.; R. L. Ackoff and M. W. Saseini, "Fundamentals of Operations Research," John Wiley & Sons, Inc., New York; K. I. Majid, "Optimum Design of Structures," Halsted Press/Wiley, New York; E. J. McCormick, "Human Factors in Engineering," McGraw-Hill Publishing Company, New York; F. S. Merritt and J. A. Ambrose, "Building Engineering and Systems Design," 2nd Ed., Van Nostrand Reinhold, New York; R. DeNeufville and J. H. Stafford, "Systems Analysis for Engineers and Managers," McGraw-Hill Publishing Company, New York; L. Spunt, "Optimum Structural Design," Prentice-Hall, Englewood Cliffs, N.J.)

1.10 BUILDING CODES

Many of the restrictions encountered in building design are imposed by legal regulations. While all must be met, those in building codes are the most significant because they affect almost every part of a building.

Building codes are established under the police powers of a state to protect the health, welfare, and safety of communities. A code is administered by a building official of the municipality or state that adopts it by legislation. Development of a local code may be guided by a model code, such as those promulgated by the

International Conference of Building Officials, Inc., Building Officials and Code Administrators International, Inc., and Southern Building Code Congress International, Inc.

In general, building-code requirements are the minimum needed for public protection. Design of a building must satisfy these requirements. Often, however, architects and engineers must design more conservatively, to meet the client's needs, produce a more efficient building system, or take into account conditions not covered fully by code provisions.

Construction drawings for a building should be submitted to the building-code administrator before construction starts. If the building will meet code requirements, the administrator issues a building permit, on receipt of which the contractor may commence building. During construction, the administrator sends inspectors periodically to inspect the work. If they discover a violation, they may issue an order to remove it or they may halt construction, depending on the seriousness of the violation. On completion of construction, if the work conforms to code requirements, the administrator issues to the owner a certificate of occupancy.

Forms of Codes. Codes often are classified as specifications type or performance type. A specification-type code names specific materials for specific uses and specifies minimum or maximum dimensions, for example, "a brick wall may not be less than 6 in thick." A performance-type code, in contrast, specifies required performance of a construction but leaves materials, methods, and dimensions for the designers to choose. Performance-type codes are generally preferred, because they give designers greater design freedom in meeting clients' needs, while satisfying the intent of the code. Most codes, however, are neither strictly specifications nor performance type but rather a mixture of the two. The reason for this is that insufficient information is currently available for preparation of an entire enforceable performance code.

The organization of building codes varies with locality. Generally, however, they consist of two parts, one dealing with administration and enforcement and the other specifying requirements for design and construction in detail.

Part 1 usually covers licenses, permits, fees, certificates of occupancy, safety, projections beyond street lines, alterations, maintenance, applications, approval of drawings, stop-work orders, and posting of buildings to indicate permissible live loads and occupant loads.

Part 2 gives requirements for structural components, lighting, HVAC, plumbing, gas piping and fixtures, elevators and escalators, electrical distribution, stairs, corridors, walls, doors, and windows. This part also defines and sets limits on occupancy and construction-type classifications. In addition, the second part contains provisions for safety of public and property during construction operations and for fire protection and means of egress after the building is occupied.

Many of the preceding requirements are *adopted by reference* in the code from nationally recognized standards or codes of practice. These may be promulgated by agencies of the federal government or by such organizations as the American National Standards Institute, ASTM, American Institute of Steel Construction, American Concrete Institute, and American Institute of Timber Construction.

Code Classifications of Buildings. Building codes usually classify a building in accordance with the fire zone in which it is located, the type of occupancy, and the type of construction, which is an indication of the fire protection offered.

The **fire zone** in which a building is located may be determined from the community's fire-district zoning map. The building code specifies the types of construction and occupancy groups permitted or prohibited in each fire zone.

The **occupancy group** to which a building official assigns a building depends on the use to which the building is put. Typical classifications include one- and two-story dwellings; apartment buildings, hotels, dormitories; industrial buildings with noncombustible, combustible, or hazardous contents; schools; hospitals and nursing homes; and places of assembly, such as theaters, concert halls, auditoriums, and stadiums.

Type of construction of a building is determined, in general, by the fire ratings assigned to its components. A code usually establishes two major categories: combustible and noncombustible construction. The combustible type may be subdivided in accordance with the fire protection afforded major structural components and the rate at which they will burn; for example, heavy timber construction is considered slow-burning. The noncombustible type may be subdivided in accordance with the fire-resistive characteristics of components.

Building codes may set allowable floor areas for fire-protection purposes. The limitations depend on occupancy group and type of construction. The purpose is to delay or prevent spread of fire over large portions of the building. For the same reason, building codes also may restrict building height and number of stories. In addition, to permit rapid and orderly egress in emergencies, such as fire, codes limit the occupant load, or number of persons allowed in a building or room. In accordance with permitted occupant loads, codes indicate the number of exits of adequate capacity and fire protection that must be provided.

1.11 ZONING CODES

Like building codes, zoning codes are established under the police powers of the state, to protect the health, welfare, and safety of the public. Zoning, however, primarily regulates land use by controlling types of occupancy of buildings, building height, and density and activity of population in specific parts of a jurisdiction.

Zoning codes are usually developed by a planning commission and administered by the commission or a building department. Land-use controls adopted by the local planning commission for current application are indicated on a zoning map. It divides the jurisdiction into districts, shows the type of occupancy, such as commercial, industrial, or residential, permitted in each district, and notes limitations on building height and bulk and on population density in each district.

The planning commission usually also prepares a master plan as a guide to the growth of the jurisdiction. A future land-use plan is an important part of the master plan. The commission's objective is to steer changes in the zoning map in the direction of the future land-use plan. The commission, however, is not required to adhere rigidly to the plans for the future. As conditions warrant, the commission may grant variances from any of the regulations.

In addition, the planning commission may establish land subdivision regulations, to control development of large parcels of land. While the local zoning map specifies minimum lot area for a building and minimum frontage a lot may have along a street, subdivision regulations, in contrast, specify the level of improvements to be installed in new land-development projects. These regulations contain criteria for location, grade, width, and type of pavement of streets, length of blocks, open spaces to be provided, and right of way for utilities.

A jurisdiction may also be divided into fire zones in accordance with population density and probable degree of danger from fire. The fire-zone map indicates the limitations on types of construction that the zoning map would otherwise permit.

In the vicinity of airports, zoning may be applied to maintain obstruction-free approach zones for aircraft and to provide noise-attenuating distances around the airports. Airport zoning limits building heights in accordance with distance from the airport.

Control of Building Height. Zoning places limitations on building dimensions to limit population density and to protect the rights of occupants of existing buildings to light, air, and esthetic surroundings. Various zoning ordinances achieve these objectives in a variety of ways, including establishment of a specific maximum height or number of stories, limitation of height in accordance with street width, setting minimums for distances of buildings from lot lines, or relating total floor area in a building to the lot area or to the area of the lot occupied by a building. Applications of some of these limitations are illustrated in Fig. 1.11.

Figure 1.11*a* shows a case where zoning prohibits buildings from exceeding 12 stories or 150 ft in height. Figure 1.11*b* illustrates a case where zoning relates building height to street width. In this case, for the specific street width, zoning permits a building to be erected along the lot boundary to a height of six stories or 85 ft. Greater heights are permitted, however, so long as the building does not penetrate *sky-exposure planes*. For the case shown in Fig. 1.11*b*, these planes start at the lot line at the 85-ft height and incline inward at a slope of 3:1. Some zoning codes will permit the upper part of the building to penetrate the planes if the floor area of the tower at any level does not exceed 40% of the lot area and the ratio of floor area to lot area (**floor-area ratio**) of the whole building does not exceed 15. To maximize the floor area in the building and maintain verticality of exterior walls, designers usually set back the upper parts of a building in a series of steps (Fig. 1.11*b*).

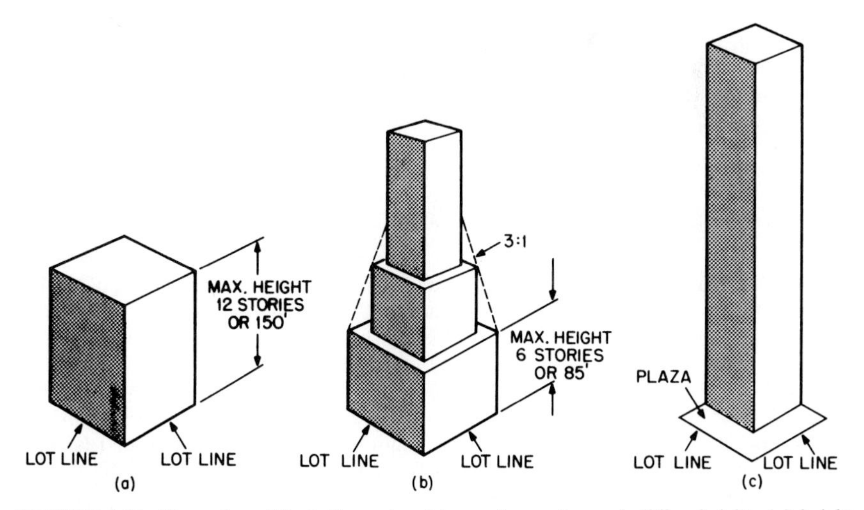

FIGURE 1.11 Examples of limitations placed by zoning codes on building height: (*a*) height limitations for buildings constructed along lot boundaries; (*b*) setbacks required by a 3:1 sky exposure plane; (*c*) height of a sheer tower occupying only part of a lot is limited by the total floor area permitted. *(Reprinted with permission from F. S. Merritt and J. Ambrose, "Building Engineering and Systems Design," 2d ed., Van Nostrand Reinhold, New York.)*

Some zoning ordinances, however, permit an alternative that many designers prefer. If the building is set back from the lot lines at the base to provide a street-level plaza, which is a convenience to the public and reduces building bulk, zoning permits the building to be erected as a sheer tower (Fig. 1.11*c*). The code may set a maximum floor-area ratio of 15 or 18, depending on whether the floor area at any level of the tower does not exceed 50 or 40%, respectively, of the lot area.

1.12 OTHER REGULATIONS

In addition to building and zoning codes, building design and construction must comply with many other regulations. These include those of the local or state health, labor, and fire departments; local utility companies; and local departments of highways, streets, sewers, and water. These agencies may require that drawings for the building be submitted for review and that a permit be granted before construction starts.

Also, building construction and conditions in buildings after completion must comply with regulations of the U.S. Occupational Safety and Health Administration (OSHA) based on the Occupational Safety and Health Act originally passed by Congress in 1970. There is, however, no provision in this law for reviewing building plans before construction starts. OSHA usually inspects buildings only after an accident occurs or a complaint has been received. Therefore, building owners, designers, and contractors should be familiar with OSHA requirements and enforce compliance with them.

Other government agencies also issue regulations affecting buildings. For example, materials used in military construction must conform with federal specifications. Another example: Buildings must provide access and facilities for disabled persons, in accordance with requirements of the Americans with Disabilities Act (ADA).

["Construction Industry: OSHA Safety and Health Standards (29CFR 1926/1910)," Superintendent of Documents, Government Printing Office, Washington, D.C. 20402; "ADA Compliance Guidebook," Building Owners and Managers Association International," 1201 New York Ave., N.W., Washington, D.C. 20005.]

1.13 SYSTEMS DESIGN BY TEAM

For efficient and successful execution of systems design of buildings, a design organization superior to that used for traditional design (Art. 1.3) is highly desirable. For systems design, the various specialists required should form a building team, to contribute their skills in concert.

One reason why the specialists should work closely together is that in systems design account must be taken of the effects of each component on the performance of the building and of the interaction of building components. Another reason is that for cost effectiveness, unnecessary components should be eliminated and, where possible, two or more components should be combined. When the components are the responsibility of different specialists, these tasks can be accomplished with facility only when the specialists are in direct and immediate communication.

In addition to the design consultants required for traditional design, the building team should be staffed with value engineers, cost estimators, construction experts, and building operators and users experienced in operation of the type of building to be constructed. Because of the diversity of skills present on such a team, it is highly probable that all ramifications of a decision will be considered and chances for mistakes and omissions will be reduced. See also Sec. 2.

(W. W. Caudill, "Architecture by Team," and F. S. Merritt and J. Ambrose, "Building Engineering and Systems Design," 2nd Ed., Van Nostrand Reinhold, New York.)

1.14 PROJECT PEER REVIEW

The building team should make it standard practice to have the output of the various disciplines checked at the end of each design step and especially before incorporation in the contract documents. Checking of the work of each discipline should be performed by a competent practitioner of that discipline other than the original designer and reviewed by principals and other senior professionals. Checkers should seek to ensure that calculations, drawings, and specifications are free of errors, omissions, and conflicts between building components.

For projects that are complicated, unique, or likely to have serious effects if failure should occur, the client or the building team may find it advisable to request a peer review of critical elements of the project or of the whole project. In such cases, the review should be conducted by professionals with expertise equal to or greater than that of the original designers, that is, by peers; and they should be independent of the building team, whether part of the same firm or an outside organization. The review should be paid for by the organization that requests it. The scope may include investigation of site conditions, applicable codes and governmental regulations, environmental impact, design assumptions, calculations, drawings, specifications, alternative designs, constructibility, and conformance with the building program. The peers should not be considered competitors or replacements of the original designers, and there should be a high level of respect and communication between both groups. A report of the results of the review should be submitted to the authorizing agency and the leader of the building team.

("The Peer Review Manual," American Consulting Engineers Council, 1015 15th St., NW, Washington, D.C. 20005, and "Peer Review, a Program Guide for Members of the Association of Soil and Foundation Engineers," ASFE, Silver Spring, Md.)

1.15 APPLICATION OF SYSTEMS DESIGN

Systems design may be used profitably in all phases of building design. Systems design, however, is most advantageous in the early design stages. One system may be substituted for another, and components may be eliminated or combined in those stages with little or no cost.

Systems design should be preferably applied in the contract documents stage only to the details being worked out then. Major changes are likely to be costly.

Value analysis, though, should be applied to the specifications and construction contract, because such studies may achieve significant cost savings.

Systems design should be applied in the construction stage only when design is required because of changes necessary in plans and specifications at that time. Time available at that stage, however, may not be sufficient for thorough studies. Nevertheless, value analysis should be applied to the extent feasible.

(F. S. Merritt and J. Ambrose, "Building Engineering and Systems Design," 2nd Ed., Van Nostrand Reinhold, New York.)

SECTION TWO
THE BUILDING TEAM—MANAGING THE BUILDING PROCESS

Alan D. Hinklin
Director
Skidmore, Owings & Merrill
Chicago, Illinois

Since the beginning of time, mankind has been in the business of **building.** Technology and construction methods continually evolve: from the Egyptian post and lintel system, the Greek pediment, the Roman arch and dome, the Byzantine basilica, and the new Renaissance perspective to the School of the Bauhaus and the International Style leading us into modern times. Over time, societies change, clients change, construction methods change, and the architect's tools change; however, the excitement and energy inherent in the building process does not change, because of one factor only—the process itself. To begin this process, two elements are necessary: an idea and a client. Creative minds then carry the process forward. With the idea comes the development of a building concept. A sketch or drawing, created through personal interaction with the client, develops the vocabulary for the physical construction of the concept. A builder and labor force turn the concept into reality.

Many processes have been used to manage this interaction. Continual evolution of the management process has turned it into an independent discipline which, coupled with the computer, is a major focus of the building industry today. From the beginning, individuals generating the concepts, preparing drawings, and building the project were considered part of what we now call the "service industry." This section outlines the various complex components and professionals involved in the building process with respect primarily to the architectural profession. Despite the changes that have occurred, the basics of the building team and the building process remain unchanged.

2.1 PROFESSIONAL AND BUSINESS REQUIREMENTS OF ARCHITECTS AND ENGINEERS

Management of the building process is best performed by the individuals educated and trained in the profession, that is, architects. While the laws of various states and foreign countries differ, they are consistent relative to the registration requirements for practicing architecture. No individual may legally indicate to the public that he or she is entitled to practice as an architect without a professional certificate of registration as an architect registered in the locale in which the project is to be constructed. This individual is the **registered architect.** In addition to the requirements for individual practice of architecture, most states and countries require a certificate of registration for a single practitioner and a certificate of authorization for an entity such as a corporation or partnership to conduct business in that locale.

An architect is a person who is qualified by education, training, experience, and examination and who is registered under the laws of the locale to practice architecture there. The practice of architecture within the meaning and intent of the law includes:

Offering or furnishing of professional services such as environmental analysis, feasibility studies, programming, planning, and aesthetic and structural design

Preparation of construction documents, consisting of drawings and specifications, and other documents required in the construction process

Administration of construction contracts and project representation in connection with the construction of building projects or addition to, alteration of, or restoration of buildings or parts of buildings

All documents intended for use in construction are required to be prepared and administered in accordance with the standards of reasonable skill and diligence of the profession. Care must be taken to reflect the requirements of country and state statutes and county and municipal building ordinances. Inasmuch as architects are licensed for the protection of the public health, safety, and welfare, documents prepared by architects must be of such quality and scope and be so administered as to conform to professional standards.

Nothing contained in the law is intended to prevent drafters, students, project representatives, and other employees of those lawfully practicing as registered architects from acting under the instruction, control, or supervision of their employers, or to prevent employment of project representatives from acting under the immediate personal supervision of the registered architect who prepared the construction documents.

2.2 CLIENT OBJECTIVES FOR BUILDINGS

Building types, time schedules, building attitudes, and legal and economic conditions affect relations with the four major client types for whom an architect may provide services. These are known as the traditional, developer, turnkey, and design/build client base.

Traditional client is usually an individual or company building a one-time project with no in-house building expertise. The client, however, possesses the innate excitement for the process of witnessing the transformation of plans into the built environment and seeks an architect to assert total control of the process. In most cases, this includes the architect's definition of the client's space needs, program and physical plant requirements. A more sophisticated traditional client might be a large corporation, university or other institutional entity that may or may not have an architect on staff, but still looks to a selected architect to guide the development process. In this case, the client may have more input into the client's program definition based on the in-house capabilities. In both cases, the architect plays the lead role in the management process and normally provides programming, design, construction documents, bidding, and construction administration in the role of the traditional architect.

Developer client offers building process management that reduces some of the architect's management role in managing the overall project and provides alternative methods for approaching design and construction. Development processes such as scope documentation, fast track, and bid packages are construction methodologies resulting from the developer client's need to accelerate the total process due to fluctuating interest rates and the need to be first in providing office space in the market place. Through this client base the acceptance of a construction consultant as a necessary part of the design team evolved. The construction consultant enables accelerated schedules to be met, provides for the compression of time, and allows a contractor to be selected by the client to build while the architect is still designing.

Turnkey client is interchangeable with the design/build client in concept. Both are based on a complete project being turned over to the owner by a single entity that is responsible for designing and constructing the project. The owner has little input in the process until it is turned over. The turnkey developer or contractor employs the services of an architect, or has an on-staff registered architect, who designs the project in accordance with the owner's program requirements. Bids are usually taken on turnkey developer designs and cost proposals to meet these requirements. Once a turnkey developer is selected, the owner may sell the property to the developer or authorize its purchase from a third party under option. From this point forward the owner has little or no participation in the project; the developer is the turnkey client of an externally employed architect. The architect is then working on the developer team and is not an independent voice for the real owner. All decisions are then made by the turnkey developer relative to the architect's services.

Design/build client also has the architect on the developer team and not performing services for the owner. Designers/builders offer to design and construct a facility for a fixed lump-sum price. They bid competitively to provide this service or provide free design services prior to commitment to the project and as a basis for negotiation. Their design work is not primarily aimed at cost-performance tradeoffs, but at reduced cost for acceptable quality.

The design/build approach to facilities is best employed when the owner requires a relatively straightforward building and does not want to participate in detailed decision making regarding the various building systems and materials. This does not mean that the owner has no control over these items. On the contrary, the owner is often permitted a wide range of selection. But the range of choices is affected by the fixed-cost restraints imposed by the designer/builder and accepted by the owner. When the facilities required are within the range of relatively standard industry-wide prototypes, this restriction may have little significance.

A common misconception regarding design/build is that poor-quality work inevitably results. While there is a general benefit to the builder for reductions in material and labor costs, the more reputable designer/builder may be relied on to deliver a building within acceptable industry standards. Facilities where higher-quality systems, more sensitive design needs, or atypical technical requirements occur deserve the services of an independent design professional.

2.3 PROGRAM DEFINITION

Usually when the term "program definition" is used relative to an architect, it is understood to mean the client's program for physical space requirements in a building. With the decline in the office market in the late 1980s came the loss of, or minimum use of, the traditional developer and construction management/construction consultant roles. As an outgrowth of the developer client era, certain developers and construction consultants turned their emphasis to "program management." In this process, a firm is engaged by the client to manage the total development process, acting as the client's agent throughout the total process. The program management approach expanded the meaning of the word "program" beyond that normally associated with only the physical space program requirements. The term "program" in this new context defines the process of organizing and executing a project from inception to completion. This process takes into account legal, financial, funding, insurance, architecture, engineering, specialist consulting, design administration, land acquisition, construction administration, and facilities operation and/or management. The client, instead of managing portions of the process as in the traditional client and developer client scenarios, looks to one firm for managing the total process.

2.4 ORGANIZATION OF THE BUILDING TEAM

Architecture is a process involving multidisciplinary input by many professionals. Comprehensive design services in the professional disciplines of planning, architecture, landscape architecture, interior design, and civil, structural, mechanical, electrical, plumbing, and fire protection engineering are offered within one organization by some large architect-engineer (A/E) and engineer-architect (E/A) firms. Smaller architectural firms retain these services by contract with consultants. Single-source design responsibility, coordinated via a common, integrated management structure, is a requirement in either case for successful development of a project.

In the performance of professional A/E services on any project, a design team charged with successful completion of the project in a dedicated professional manner is essential. This team provides continuous service to the project from start to finish, establishing and maintaining the quality and integrity of each design. A project leader should be selected to coordinate and manage all the professional disciplines and consultants involved in the project and to act as liaison with the client. This leader should work closely with the client to provide policy direction and set goals and objectives for the professional team. Day-to-day management and direction of the project's technical development should be provided by an individual, usually identified as the architect's project manager, who performs the key administrative duties, establishes and maintains design services budgets and schedules, and coordinates the entire A/E effort. A senior designer supervises daily organization and progress of design development and directs the design efforts of the project team. As a project's specific needs or schedule require, additional architects, planners, engineers, interior architects, and consultants are involved in the project to augment the team or to provide specialized consultation.

2.4.1 Architects and Engineering Consultants

The major distinctions between architects and engineers run along generalist and specialist lines. The generalists are ultimately responsible for the overall planning.

It is for this reason that an architect is generally employed as the prime professional by a client. On some special projects, such as dams, power plants, wastewater treatment, and research or industrial installations, where one of the engineering specialties becomes the predominant feature, a client may select an engineering professional or an A/E firm to assume responsibility for design and construction and take on the lead role. On certain projects, it is the unique and imaginative contribution of the engineer that may make the most significant total impact on the architectural design. The overall strength of a dynamic, exposed structure, the sophistication of complex lighting systems, or the quiet efficiency of a well-designed mechanical system may prove to be the major source of the client's pride in a facility. In any circumstance, the responsibilities of the professional engineer for competence and contribution are just as great and important to the project as those of the architect.

Engineers, for example, play a major role in **intelligent building system design,** which involves mechanical-electrical systems. However, a building's intelligence is also measured by the way it responds to people, both on the inside and outside. The systems of the building must meet the functional needs of the occupants as well as respect the human response to temperature, humidity, airflow, noise, light, and air quality. To achieve the multifaceted goals, an intelligent building requires an intelligent design process with respect to design and system formulation as well as efficient and coordinated execution of design and technical documentation within the management structure.

An intelligent building begins with intelligent architecture—the shape, the building enclosure, and the way the building appears and functions. Optimal design solutions can be achieved through a design process that explores and compares varying architectural and engineering options in concert. Sophisticated visualization and analytical tools using three-dimensional computer modeling techniques permit architects and engineers to rapidly evaluate numerous alternatives. Options can be carefully studied both visually and from a performance standpoint, identifying energy and life-cycle cost impact. This enables visualization and technical evaluation of multiple schemes early in the design phase, setting the basis for an intelligent building.

In all cases, the architect's or engineer's legal responsibilities to the client remain firm. The prime professional is fully responsible for the services delivered. The consultants, in turn, are responsible to the architect or engineer with whom they contract. Following this principle, the architect or engineer is responsible to clients for performance of each consultant. Consequently, it is wise for architects and engineers to evaluate their expertise in supervising others before retaining consultants in other areas of responsibility.

2.4.2 Other Consultants

A building team may require the assistance of specialists. These specialty consultants provide skills and expertise not normally found in an architectural or engineering firm. The prime professional should define the consultants required and assist the client in selecting those consultants. The architect or engineer should define and manage their services even if the specialty consultant contracts directly with the client for liability purposes, with the understanding that the client has the ultimate say in decision making.

While several consultants may be required, depending on the complexity of the project, the cost for each may be minimal since their services are provided over short periods of time during the development process, and all consultants are usually not servicing the project at any one time. The following consultant services, most of which are not normally provided by architects and engineers, are provided by various firms:

- Acoustical
- Audiovisual
- Communications
- Exterior wall maintenance
- Fire and life safety
- Food service
- Geotechnical engineering and subsurface exploration
- Graphics
- Space-usage operations
- Independent research and testing
- Landscaping
- Marketing and leasing
- Materials handling
- Parking
- Preconstruction survey
- Schedule
- Security
- Site surveyor
- Special foundation systems
- Special structures
- Specialty lighting
- Traffic
- Vertical transportation
- Water feature
- Wind tunnel testing
- Computers

2.5 CLIENT-A/E AGREEMENT

Although verbal contracts can be considered legal, a formal written document is the preferred way to contract for professional services to be provided by an architect. Purchase orders are not an acceptable means, since they are not applicable to a service arrangement but rather only provide a financial accounting system for purchasing a product, which is normally required internally by a client. A purchase order should not be used as a client-A/E agreement.

Most professionals use the *AIA Standard Form of Agreement for Architect and Owner* (client). Some larger firms, however, have their own form of agreement which augments or further defines that of the AIA. The basic elements of the agreement establish the definition and identification of project phases and define the specific scope and compensation for the architect's basic services. Flexibility is built into this agreement to accommodate supplementary services that may be considered. In addition, the agreement should define the understandings of the two parties as well as of any third parties that may be involved in the process and stipulate how the third parties are to be managed and compensated.

Furthermore, the client-A/E agreement should define items considered as direct costs that may be reimbursed under the agreement. Other items also to be addressed include project terminology, project terms and definitions, and the architect's status as it relates to the profession such that the standard of care is clearly understood. The definition of additional services, changes, and compensation for such services, as well as the method and timing of payment, the responsibility for client-furnished information, ownership of documents, confidentiality provisions, insurance requirements, termination provisions by either party, and dispute resolution may also be addressed. A/E agreements may also define the documents to be delivered at the conclusion of each development phase and, in certain cases, the time estimated for completion of each phase of service.

Compensation for Professional Services. A major concern of an architect is to arrive at an accurate assessment of the scope of services to be performed. The nature of the project, the degree of professional involvement, and the skills required should be considered in arriving at an equitable fee arrangement. Types of fees that may be used are

- Percentage of the construction cost of the project
- Fee plus reimbursement of expenses
- Multiple of direct personnel expense
- Multiple of technical personnel hourly rates
- Stipulated or lump sum
- Billing rates for personnel classification

For a project requiring what could be described as standard services, the percentage-of-construction-cost fee is a safe standard. Years of experience with the relationship between the scope of architectural services required for various sizes of standard construction contracts provide a basis for such rule-of-thumb fee agreements.

For projects where atypical services are required, other arrangements are more suitable. For example, for projects where the scope of service is indefinite, a cost-plus fee is often best. It permits services to proceed on an as-authorized basis, without undue gambling for either party to the agreement. Under such an arrangement, the architect is reimbursed for costs and also receives an agreed-on fee for each unit of effort the architect expended on the project. Special studies, consultations, investigations, and unusual design services are often performed under such an arrangement.

For projects where the scope can be clearly defined, a lump-sum fee is often appropriate. In such cases, however, architects should know their own costs and be able to accurately project the scope of service required to accomplish fixed tasks. Architects should take care, for the protection of their own, their staff's, and the client's interests, that fees cover the costs adequately. Otherwise, the client's interests will suffer, and the architect's own financial stability may be undermined.

Fee agreements should be accompanied by a well-defined understanding in the form of a written agreement for services between architect and client.

2.6 A/E LIABILITY AND INSURANCE

Architecture and engineering firms normally maintain professional liability insurance. This requires payment of annual premiums based on the coverage provided. (Architects and engineers should maintain coverage in connection with their foreign operations as well as with their domestic operations.) Various types of insurance usually carried by architects and engineers are listed in Table 2.1.

2.6.1 "Service" vs. "Work"

The building industry generally recognizes that the professional architect, engineer, or design consultant provides **service,** whereas the contractor, subcontractor, or material supplier provides **work.** In providing work, the contractor delivers a product and then warrants or guarantees the work. These distinctions are important to understand with respect to insurance. In the architect's case, professional liability insurance provides coverage for the judgment the professional provides while using reasonable care and therefore does not normally have liquidated damages provisions. Professional liability insurance does not cover the work itself or items undertaken by the contractor in pursuit of the work but does cover negligent errors and omissions of the architect or engineer. This insurance is a means of managing the risk associated with the architect's judgment; it is not product-related. Most claims against professionals in the building industry are made by clients. Fewer claims are made by contractors and workers.

2.6.2 Risk Management

So that the architect's or engineer's business goals can be accomplished, professional liability insurance is offered through various underwriters and managed by profes-

TABLE 2.1 Types of Architect and Engineer Insurance

Type of insurance	Coverage
Commercial general liability	According to occurrence and aggregate
Commercial automobile liability	Bodily injury and property damage
Workers' compensation	Statutory limits
Employer's liability	Medical care and time lost as a result of injuries incurred during the performance of the services
Professional liability	Errors and omissions
Valuable papers	Loss of drawings, models, computer-produced data, etc.
Umbrella liability	Provides coverage in excess of professional liability coverage

sionals. Such professionals should not dictate or limit architectural practice, but rather should support it; neither should they tell architects to turn away from risk, but instead they should help manage it.

Insurance allows the architect or engineer to transfer the risk of financial uncertainty to an insurance company for a known premium. The professional should calculate how much risk to assume. The risk the individual retains is the deductible. The risk the insurance company accrues is the limit of liability over and above the deductible. By choosing a higher deductible, the professional retains more risk but pays a lower premium.

Professional liability protection for the architectural and engineering profession has been designed with the help of the American Institute of Architects (AIA) and the National Society of Professional Engineers (NSPE)/Professional Engineers in Private Practice (PEPP). In addition to errors and omissions coverage, the protection incorporates liability coverage for on-time performance, cost estimating, interior design, asbestos, and pollution.

Liability programs vary widely from company to company. In general, the insurance industry recommends that architects and engineers:

- Select a program with flexible limits of liability and deductible options
- Carefully review the insurance coverage
- Compare competitive costs
- Consider the insurance company's experience
- Examine the insurance company's criteria for accepting risk
- Compare loss prevention services
- Assure that the company shares its loss information

The AIA and NSPE/PEPP can also provide architects and engineers with valuable information on what to look for in a professional liability insurance program.

2.6.3 Project Insurance

Project insurance permits the architect to be responsive to the client who has particular insurance demands. Suppose, for example, that the client wants 3 times the coverage the architect carries. Project insurance can respond to this requirement. Project insurance costs are often reimbursable costs and considered a common element of the construction cost, similar to the cost of the contractor's insurance coverage and performance bonds. Project insurance can sometimes reduce the architect's policy costs because project billings are not included in the architect's billings when the architect's practice policy premium is calculated. Project insurance may provide long-term coverage guarantees to the day of substantial or final completion and up to 5 years thereafter with no annual renewals. Project insurance permits clients to take control in the design of an insurance package to protect their investment and provides clients with stability, security, and risk management.

2.7 DEFINITION OF PROJECT PHASES

The definition of the various phases of development for a particular project from initial studies through postconstruction should be understood by the client and outlined thoroughly in the client-A/E agreement. The most-often-used phases of development include the following:

Feasibility Studies. To assist the client in determining the scope of the project and the extent of services to be performed by various parties, the architect may enter into an interim agreement for services relating to feasibility studies, environmental impact studies or reports, master planning, site selection, site analysis, code and zoning review, programming, and other predesign services.

Environmental Impact Studies. Determination of environmental studies and reports required for a project and preparation of such reports, special drawings, or other documents that may be required for governmental approvals are normally performed under separate agreements. Attention should be given to zoning, soils, and the potential of hazardous materials in any form. If any impermissible hazardous materials are encountered, clients should be advised so that they can obtain the services of a specialty consultant to determine what course of action to take.

Programming. If the architect is required to prepare the program of space requirements for a project, the program should be developed in consultation with the client to help the client recognize particular needs. Space requirements, interrelationships of spaces and project components, organization subdivision of usage, special provision and systems, flexibility, constraints, future expansion, phasing, site requirements, budgetary and scheduling limitations, and other pertinent data should all be addressed.

Conceptual Design. During this phase of development, the architect evaluates the client's program requirements and develops alternatives for design of the project and overall site development. A master plan may also be developed during this phase. The plan serves as the guide and philosophy for the remainder of the development of the project or for phasing, should the project be constructed in various phases or of different components.

Schematic Design. During this phase the project team, including all specialty consultants, prepares schematic design documents based on the conceptual design alternative selected by the client. Included are schematic drawings, a written description of the project, and other documents that can establish the general extent and scope of the project and the interrelationships of the various project components, sufficient for a preliminary estimate of probable construction costs to be prepared. Renderings and finished scale models may also be prepared at this time for promotional and marketing purposes.

Design Development. After client approval of the schematic design, the architect and the specialty consultants prepare design development documents to define further the size and character of the project. Included are applicable architectural, civil, structural, mechanical, and electrical systems, materials, specialty systems, interior development, and other such project components that can be used as a basis for working drawing development.

Construction Documents. After approval of the design development documents, the architectural-engineering team, together with the applicable specialty consultants, prepares construction documents, consisting of working drawings and technical specifications for the project components. These include architectural, structural, mechanical, electrical, hydraulic, and civil work, together with general and supplementary conditions of the construction contract for use in preparing a final detailed estimate of construction costs and for bidding purposes.

Construction Administration. Diligent construction administration is essential to translate design into a finished project. The A/E team continues with the development process by issuing clarifications of the bid documents and assisting in contractor selection. Also, during the construction period, the team reviews shop drawings, contractor payment requests, and change-order requests, and visits to construction sites to observe the overall progress and quality of the work. Architect and engineer personnel involved in the design of the project should be available during construction to provide continuity in the design thought process until project completion and occupancy.

Postconstruction Services. Follow-up with the client after construction completion is essential to good client relations. Periodic visits to the project by the architect through the contractor's warranty period is considered good business.

2.8 SCHEDULING AND PERSONNEL ASSIGNMENTS

The effective coordination of any project relies on management's ability to organize the project into a series of discreet efforts, with deadlines and milestones identified in advance. The interdependence of these milestones should be clearly understood by the client and the project team so that the project can be structured yet still be flexible to respond to changes and unforeseen delays without suffering in overall coordination and completion.

Experience is the basis on which architects and engineers establish major project milestones that form the framework for project development. The critical path method (CPM) of scheduling can be used to confirm intermediate milestones corresponding to necessary review and approvals, program and budget reconciliation, and interdisciplinary coordination. CPM consultants can also assist contractors in establishing overall shop drawings and fabrication and installation schedules for efficient phasing and coordination of construction. Schedules can be maintained in a project management computer database. They should be updated on a regular basis for the duration of the project, since critical path items change from time to time depending on actual progress of construction. See also Art. 2.9.

2.9 ACCELERATED DESIGN AND CONSTRUCTION

The traditional process of design and construction and the roles and responsibilities of the various parties need not be changed when fast track, an accelerated design and construction process, is required. However, this process can affect scheduling and personnel assignments.

In the traditional process, the entire facility moves phase by phase through the entire development process, that is, programming, design, design development, construction documents, bid and award of contracts, construction and acceptance of completed project (Art. 2.7). With any form of accelerated design and construction, the final phases remain substantially the same, but the various building systems or subsystems move through the development process at different times and result in the release of multiple construction contracts at various times throughout the process.

For any project, basic building siting is determined early in the design process. Therefore, at an early stage in design, a construction contract can be awarded for demolition and excavation work. Similarly, basic structural decisions can be made before all details of the building are established. This permits early award of foundation and structural work contracts. Under such circumstances, construction can be initiated early in the design process, rather than at the conclusion of a lengthy design and contract preparation period. Months and even years can be taken out of the traditional project schedule, depending on the scale and complexity of the project. Purchase of preengineered, commercially available building systems can be integrated into the accelerated design and construction process when standard system techniques are employed, reducing time even more.

The major requirements for a project in which design and construction occur simultaneously are

- Accurate cost management to maintain project budgets.
- Full understanding of the construction process by the client, contractor, and design professionals so that design decisions and contract documents for each building system or subsystem can be completed in a professional manner that addresses the requirements of the ongoing construction process.
- Organized and efficient management of the construction process with feedback into the design process to maintain a clear definition of the required contract packages and schedule.
- Overall project cost control and project construction responsibilities, including interface management of independent prime contracts, should also be established.

Often the major purpose of accelerated design and construction is to reduce the effect of rapidly increasing construction costs and inflation over the extended project design and construction period. For projects extending over several years, for example, contractors and subcontractors have to quote costs for providing material and labor that may be installed several years later. In most cases, the costs associated with such work are uncertain. Bid prices for such work, especially when it is of large magnitude, therefore, must be conservative. Accelerated design and construction, however, brings all the financial benefits of a shortened project duration and early occupancy and reduces the impact of cost escalation. Also, bid prices can be closer to the actual costs, thus reducing bidding risk to the contractor. The combination of phased bidding, shortened contract duration, reduced escalation, smaller bid packages, and a greater number of bidders can produce substantial savings in overall construction costs.

A major objection to accelerated design and construction is that project construction is initiated before bids are obtained for the total project and assurance is secured that the total project budget can be maintained. In this regard, the reliability of early cost estimating becomes even more critical. It is the experience of most clients and architects involved with multiple contracts, however, that such contracts, bid one at a time, can be readily compared with a total budget line item or trade breakdown and thus provide safeguards against budget overruns. The ability to design, bid, and negotiate each contract as a separate entity provides optimum cost control.

For accelerated design and construction programs to work effectively, services of a professional construction manager are normally required. This cost, however, can be offset by the overall saving in the total project cost due to the reduction in construction time.

Normally, the client is responsible for the various construction contracts when multiple contracts are used. The construction manager acts as the client's agent in administration of the contracts. If the architect is to administer the contracts, additional compensation will be required beyond that associated with one general contractor who holds all subcontracts, as is the case in the traditional client-contractor relationship.

2.10 DESIGN MANAGEMENT

Architects manage all aspects of project design simultaneously, their own internal resources, relations with the specialty consultants, the processes that deliver service to the client, and through that service, the programs of client needs through the development process to the creation of a built environment. The requirement that architects be capable businesspersons is, therefore, far-reaching. The need for good business sense and a thorough knowledge of the architect's own cost is reinforced by the need to manage these costs throughout the duration of the project. Allocation, commitment, and monitoring of the expenditure of resources are of critical importance to the financial success of every project. Only when these are properly managed can quality services, proper advice, appropriate design, and best contract documents be delivered to clients.

As a businessperson, an architect is faced with acquiring personnel, advancing those who are outstanding, and removing those who are unacceptable. The firm should keep records of business expenses, file tax returns, provide employee benefits, distribute and account for profits, and keep accurate cost records for project planning and to satisfy government requirements. The architect must meet legal requirements for practice as an individual, partnership, or corporation. In many of these areas, the architect will be assisted by experts. It is impossible for an architect to practice effectively or successfully without a thorough understanding and complete concern for the business of architecture.

Once the resources required to deliver services are assured, the architect should provide management skills to see that these services are kept timely, well-coordinated, accurate, and closely related to the client's needs. This is especially important for work on large projects, in large design offices, or when dealing with the architect's employees and consultants. The best talent must be secured, appropriately organized, directed, and coordinated to see that the project receives well-integrated and well-directed professional service.

The objective is to produce the facility the client needs, within budget, and on schedule. While the contractor has the front-line responsibility for budgeted construction cost and schedule, the architect's resources and the services provided should be helpful in managing the construction process for the benefit of the client. The architect's management of materials and technology and relationship with the client and contractors will account in good measure for the success of the project.

2.11 INTERNAL RECORD KEEPING

Part of good office management is document control and record keeping. Much information is received, disseminated, and collated in an architect's office. Included are project directories, contractual correspondence, client correspondence, consultant correspondence, minutes of meetings, insurance certifications, in-progress drawings, drawing release for owner review, and building permit and construction issues. Also dealt with are computer tapes, calculations, shop drawings, specifications, material samples, renderings, photography, slides, field reports, specifications addenda, contract modifications, invoices, financial statements, audit records, and time cards. In addition, there are contractor payment requests, change orders, personnel records, client references and more. Certain clients may have particular formats or record-keeping controls they impose on a project in addition to the architect's standard procedures.

A multitude of data is transferred among many parties during the progress of the architect's services. The data should be maintained in an organized manner for future reference and archival purposes. The architect should establish an office procedure for document control, record keeping, and document storage beyond the life of the project to ensure easy retrieval. There are many computerized systems that can aid the architect in catalog filing and information retrieval. Record keeping can typically be subdivided into the following categories: contractual, financial, personnel, marketing and publicity, legal, correspondence, project documentation, drawings, shop drawings, warehousing, and archival records. These should not only be supervised but also controlled, inasmuch as some files require limited access for reasons of confidentiality and legalities.

2.12 CODES AND REGULATIONS

Various statutory codes, regulations, statutes, laws, and guidelines affect design and construction of projects. In most jurisdictions, the architect and engineer are required by law to design to applicable building codes and regulations, which vary from one jurisdiction to another. Some jurisdictions that do not have sophisticated codes usually follow recognized national or international codes, which should be agreed on at the onset of a project so that the client and architect understand the rules for design and construction. All codes are intended for the health, welfare, and safety of the public and occupants of buildings.

Affirmative-Action Program. The objective of equal employment opportunity and affirmative-action programs should be to ensure that individuals are recruited, hired, and promoted for all job classifications without regard to race, color, religion, national origin, sex, age, handicap, or veteran status. Employment decisions should be based solely on an individual's qualifications for the position for which the individual is considered.

Affirmative action means more than equal employment opportunity. It means making a concentrated effort to inform the community of the architect's desire to foster equal employment opportunity. It also means making a special effort to attract individuals to the profession and to engage them in a program of professional development. Furthermore, architects should be committed to a meaningful minority business enterprise (MBE) and women businesss enterprise (WBE) participation program. Initial contact with local MBE/WBE firms should be pursued for each applicable project to respond to this important requirement. Architects should be prepared to review this requirement with clients to achieve participation targets consistent with client goals and objectives.

2.13 PERMITS

Most jurisdictions require a building permit for construction or remodeling. The building permit, for which a fee is paid by the contractor or client, is an indication that drawings showing the work to be done have been prepared by a registered professional and submitted to the authority having jurisdiction over design and construction of the project. Furthermore, it is an indication that this authority stipulates that the documents meet the intent of the applicable building codes and regulations. Issuance of a permit, however, does not relieve the governing agency of the right to inspect the project during and after construction and to require minor modifications. In addition, while most locales do not provide for a written permit by the fire department, this agency is involved in the review process relative to life-safety provisions. It also has the right to inspect the project when constructed and to require modifications if they are considered appropriate to meet the intent of the code or the department's specific requirements. Major items reviewed by both the permit-issuing agencies relate to occupancy classifications, building population, fire separations, exiting, travel paths for exiting, areas of refuse, and other general life safety and public health issues.

Occupancy Permits. Many jurisdictions require that a permit be obtained by the client or tenant of a multitenant building indicating that the building or tenant space has been reviewed by the applicable agency and fire department. This permit indicates that the building meets the requirements of the building codes and is appropriate for occupancy for the intended use and classification for which the building or space was designed and constructed.

In addition, elevator usage certificates are issued by certain building authorities. These certificates indicate that the elevators have been inspected and found to be acceptable for use based on the size, loading, and number of occupants posted on the certificate.

Furthermore, certain spaces within a project may have a maximum-occupancy limitation for which a notice is posted in those spaces by the applicable building authority. Examples of this type of usage include restaurants, ballrooms, convention centers, and indoor sports facilities where a large number of occupants might be gathered for the intended use.

2.14 ENERGY CONSERVATION

In response to the national need for energy conservation and in recognition of the high consumption of energy in buildings, the U.S. Department of Energy gave a grant to the American Society of Heating, Refrigeration, and Air-Conditioning Engineers (ASHRAE) for development of a national energy conservation standard for new buildings. The resulting standard, ASHRAE 90-75, establishes thermal design requirements for exterior walls and roofs. It is incorporated in some building codes.

Seeking greater energy-use reduction, Congress passed the Energy Conservation Standards for New Buildings Act of 1976, mandating development of energy performance standards for new buildings (BEPS). Accordingly, the Department of

Energy develops such standards, for adoption by federal agencies and state and local building codes. BEPS consists of three fundamental elements:

1. Energy budget levels for different classifications of buildings in different climates, expressed as rate of energy consumption, Btu/ft^2-yr.
2. A method for applying these energy budget levels to a specific building design to obtain a specific annual rate of energy consumption, or design energy budget, for the proposed building.
3. A method for calculating the estimated annual rate of energy consumption, or design energy consumption, of the proposed building.

The design energy consumption may not exceed the design energy budget of a new building. Even without these regulations, energy conservation for buildings makes good sense, for a reduction in energy usage also reduces building operating costs. It is worthwhile, therefore, to spend more on a building initially to save energy over its service life, at least to the point where the amortized annual value of the increased investment equals the annual savings in energy costs. As a consequence, life-cycle cost, considered the sum of initial, operating, and maintenance costs, may be given preference over initial cost in establishment of a cost budget for a proposed building.

Energy use and conservation are key elements in an architect's approach to design. Aided by computer simulation, engineers can develop system concepts and evaluate system performance, deriving optimal operation schedules and procedures. During the initial design phase, the computer can be used in feasibility studies involving energy programs, preliminary load calculations for the selection of heating, ventilating, and air-conditioning (HVAC) systems and equipment, technical and economic evaluation of conservation alternatives. Using solar heating and cooling systems for new and existing facilities, modeling energy consumption levels, forecasting probable operating costs, and developing energy recovery systems can be investigated during the early design phases of a project.

2.15 THE INTERIOR ENVIRONMENT

Architects have long been leaders in building design that is sensitive to environmental issues. Several areas of general concern for all buildings are described in the following paragraphs; they support the basic philosophy that the environment within buildings is as critical a concern as esthetics.

Indoor Air Quality. Many factors, such as temperature, air velocity, fresh-air ventilation rates, relative humidity, and noise, affect indoor air quality. The fresh-air ventilation rate has the greatest influence on indoor air quality in many buildings. Fresh-air ventilation rates in a building is the flow of outside air brought into the building for the well-being of the occupants and the dilution of odors and other internally generated air pollutants. The outside air may vary in its "freshness" depending on the location of the building, its surrounding conditions, and the location of the fresh-air intakes for the building. Therefore, careful studies should be made by the architect to ensure the optimum internal air quality.

Ventilation is required to combat not only occupant-generated odors, as has been traditionally the case, but also to provide ventilation for materials used and stored in buildings. ASHRAE Standard 62-1989, American Society of Heating,

Refrigeration, and Air-Conditioning Engineers, recommends a rate of 20 cfm per person as a minimum ventilation rate for office buildings. Air-handling systems for numerous buildings provide not only this minimum recommended level but also often increased fan capacity (available when outdoor temperatures and humidity levels are favorable) through an air-side economizer control.

Environmental Pollution. In response to current concern for the effect of chlorofluorocarbons (CFCs, fully halogenated refrigerants) on the earth's ozone layer, the refrigerant for mechanical systems should have the lowest ozone depletion potential compatible with commercial building cooling systems.

Noise Control. The acoustical environment within a building is a result of the noise entering the space from outdoors, or from adjacent interior areas, or most importantly, from the mechanical, electrical, and elevator systems of the building. This is in addition to the noise generated within the space by people and equipment. Mechanical systems should be designed to limit equipment noise and to maintain the transmission of noise via mechanical systems to occupied spaces within a range necessary for efficient and enjoyable use of the building. Occupied space noise should normally be limited to NC-35 or less if desired, through the use of state-of-the-art air-distribution equipment and appropriate use of materials within the finished spaces.

Safe Building Materials. The technical specifications provided by the architect should be continually updated to eliminate any materials that are potential health hazards to occupants or construction workers, such as materials that give off gas within the occupied spaces. In addition, requirements in local, national, and international building codes to reduce fire and smoke hazards should be met.

Occupational Health and Safety Issues. As discussed in the preceding, architects should exercise professional care in design and specification of all architectural and building systems to create a state-of-the-art building offering a safe, healthy environment for all occupants, visitors, and users.

2.16 COST ESTIMATING AND VALUE ENGINEERING

During development of a project the client normally looks to the architect for construction cost estimates. It is advisable to provide a probable cost of construction at completion of the schematic design, design development, and construction document phases. A design contingency is usually carried in cost estimates. It can be reduced as the documents are further developed. At completion of the construction documents, the architect prepares, or has a consultant prepare, a final and most accurate estimate of construction cost, which can be used for comparison with the bids submitted to perform the work.

Value engineering may be performed by consultants and construction managers during the development of the construction documents. (This is a misnomer for cost-reduction engineering, since value engineering often occurs after a design has been established.) To be effective, value engineering should be undertaken prior to design of the building systems.

Value engineering should address operating and maintenance costs as well as first costs, to provide true life-cycle cost estimates for comparative analysis. This can be accomplished as early as the conceptual design phase of the project and should use the expertise of cost consultants, if such service is not offered directly by the architect or engineer.

Cost analysis should be performed concurrently with technical evaluation of the systems proposed by the architects or engineers, to provide the client with proper information to make an informed decision. The architects and engineers should address cost without compromising the building program, building safety, or desired performance of the facility and respond to the client in a professional manner regarding cost estimating and value engineering.

2.17 TECHNICAL SPECIFICATIONS

Specifications for a building project are written descriptions, and the drawings are a diagrammatic presentation of the construction work required for that project. The drawings and specifications are complementary.

Specifications are addressed to the prime contractor. Presenting a written description of the project in an orderly and logical manner, they are organized into divisions and sections representing, in the opinion of the specification writer, the trades that will be involved in construction. Proper organization of the specifications facilitates cost estimating and aids in preparation of bids.

2.17.1 Content of Specifications

It is not practical for an architect or engineer to include sufficient notes on the drawings to describe in complete detail all of the products and methods required of a construction project. Detailed descriptions should be incorporated in specifications. For example, workmanship required should be stated in the specifications.

Contractors study specifications to determine details or materials required, sequence of work, quality of workmanship, and appearance of the end product. From this information, contractors can estimate costs of the various skills and labor required. If workmanship is not determined properly, unrealistic costs will result and quality will suffer. Good specifications expand or clarify drawing notes, define quality of materials and workmanship, establish the scope of the work, and describe the responsibilities of the contractor.

The terms of the contract documents should obligate each contractor to guarantee to the client and the architect or engineer that all labor and materials furnished and the work performed are in accordance with the requirements of the contract documents. In addition, a guarantee should also provide that if any defects develop from use of inferior materials, equipment, or workmanship during the guarantee period (1 year or more from the date of final completion of the contract or final occupancy of the building by the client, whichever is earlier), the contractor must, as required by the contract, restore all unsatisfactory work to a satisfactory condition or replace it with acceptable materials. Also, the contractor should repair or replace any damage resulting from the inferior work and should restore any work or equipment or contents disturbed in fulfilling the guarantee.

Difficult and time-consuming to prepare, technical specifications supply a written description of the project, lacking only a portrayal of its physical shape and its dimensions. The specifications describe in detail the material, whether concealed or exposed, in the project and fixed equipment needed for the normal functioning of the project. If they are properly prepared, well-organized, comprehensive, and indexed, the applicable requirements for any type of work, kind of material, or piece of equipment in a project can be easily located.

The technical specifications cover the major types of work—architectural, civil, structural, mechanical, and electrical. Each of these types is further divided and subdivided in the technical specifications and given a general title that describes work performed by specific building trades or technicians, such as plasterers, tile setters, plumbers, carpenters, masons, and sheet-metal workers.

The prime contractor has the responsibility to perform all work, to furnish all materials, and to complete the project within a schedule. The contractor, therefore, has the right to select subcontractors or perform the work with the contractor's own forces. In recognition of this, each specification should contain a statement either in the General Conditions or in the Special Conditions, that, regardless of the subdivision of the technical specifications, the contractor shall be responsible for allocation of the work to avoid delays due to conflict with local customs, rules, and union jurisdictional regulations and decisions.

Standard forms for technical specifications can be obtained from the Construction Specifications Institute (CSI). The CSI publishes a Master List of Section Titles and Numbers, which is the generally accepted industry standard. In it, technical specifications are organized into 16 divisions, each with titles that identify a major class of work. Each division contains basic units of work, called sections, related to the work described by the division title. Following is the division format developed by CSI:

1. General requirements
2. Site work
3. Concrete
4. Masonry
5. Metals
6. Woods and plastics
7. Thermal and moisture protection
8. Doors and windows
9. Finishes
10. Specialties
11. Equipment
12. Furnishings
13. Special construction
14. Conveying systems
15. Mechanical
16. Electrical

Language should be clear and concise. Good specifications contain as few words as necessary to describe the materials and the work. The architect or engineer should use the term "shall" when specifying the contractor's duties and responsibilities under the contract and use the term "will" to specify the client's or architect's responsibilities.

Phrases "as directed by the architect," ". . . to the satisfaction of the architect," or ". . . approved by the architect" should be avoided. The specification should be comprehensive and adequate in scope to eliminate the necessity of using these phrases. "Approved by the architect" may be used, however, if it is accompanied by a specification that indicates what the architect would consider in a professional evaluation. The term "by others" is not clear or definite and, when used, can result in extra costs to the client. The word "any" should not be used when "all" is meant.

2.17.2 Types of Specifications

Technical requirements may be specified in different ways, depending on what best meets the client's requirements. One or more of the following types of technical specifications may be used for a building project.

Descriptive Specifications. These describe the components of a product and how they are assembled. The specification writer specifies the physical and chemical properties of the materials, size of each member, size and spacing of fastening devices, exact relationship of moving parts, sequence of assembly, and many other requirements. The contractor has the responsibility of constructing the work in accordance with this description. The architect or engineer assumes total responsibility for the function and performance of the end product. Usually, architects and engineers do not have the resources, laboratory, or technical staff capable of conducting research on the specified materials or products. Therefore, unless the specification writer is very sure the assembled product will function properly, descriptive specifications should not be used.

Reference Specifications. These employ standards of recognized authorities to specify quality. Among these authorities are ASTM, American National Standards Institute, National Institute of Standards and Technology, Underwriters Laboratories, Inc., American Institute of Steel Construction, American Concrete Institute, and American Institute of Timber Construction.

An example of a reference specification is: *Cement shall be portland cement conforming to ASTM C150, "Specification for Portland Cement," using Type I or Type II for general concrete construction.*

Reputable companies state in their literature that their products conform to specific recognized standards and furnish independent laboratory reports supporting their claims. The buyer is assured that the products conform to minimum requirements and that the buyer will be able to use them consistently and expect the same end result. Reference specifications generally are used in conjunction with one or more of the other types of specifications.

Proprietary Specifications. These specify materials, equipment, and other products by trade name, model number, and manufacturer. This type of specification simplifies the specification writer's task, because commercially available products set the standard of quality acceptable to the architect or engineer.

Sometimes proprietary specifications can cause complications because manufacturers reserve the right to change their products without notice, and the product incorporated in the project may not be what the specifier believed would be installed. Another disadvantage of proprietary specifications is that they may permit use of alternative products that are not equal in every respect. Therefore, the specifier should be familiar with the products and their past performance under similar use and should know whether they have had a history of satisfactory service.

The specifier should also take into consideration the reputation of the manufacturers or subcontractors for giving service and their attitude toward repair or replacement of defective or inferior work.

Under a proprietary specification, the architect or engineer is responsible to the client for the performance of the material or product specified and for checking the installation to see that it conforms with the specification. The manufacturer of the product specified by the model number has the responsibility of providing the performance promised in its literature.

In general, the specification writer has the responsibility of maintaining competition between manufacturers and subcontractors to help keep costs in line. Naming only one supplier may result in a high price. Two or more names are normally supplied for each product to enhance competition.

Use of "or equal" should be avoided. It is not fully satisfactory in controlling quality of materials and equipment, though it saves time in preparing the specification. Only one or two products need to be investigated and research time needed to review other products is postponed.

Base-Bid Specifications. These establish acceptable materials and equipment by naming one or more (often three) manufacturers and fabricators. The bidder is required to prepare a proposal with prices submitted from these suppliers. Usually, base-bid specifications permit the bidder to submit substitutions or alternatives for the specified products. When this is done, the bidder should state in the proposal the price to be added to, or deducted from, the base bid and include the name, type, manufacturer, and descriptive data for the substitutions. Final selection rests with the client. Base-bid specifications often provide the greatest control of quality of materials and equipment, but there are many pros and cons for the various types of specifications, and there are many variations of them.

2.17.3 Automated Specifications

For building projects, specification writers normally maintain a library of *master* documents that are used as a basis for creating project specifications with a computer. Typically, they employ the industry-standard Construction Specifications Institute format (Art. 2.17.1). Computers are used to facilitate and speed production of specifications and other technical documents.

Although computer systems can be complex, requiring an experienced person for setup and maintenance, they are cost-effective, saving time and effort. For example, one program used for preparing specifications has a point-and-click graphics user interface with directories and files represented by icons and manipulated by a mouse. Multiple files are viewed and edited on the screen simultaneously, and each file is seen as a full-page display exactly as it will be printed. The graphics and document layout capabilities of the program are suitable for producing technical manuals and for publishing periodicals. Documents displayed on the computer permit the architect to eliminate the editing of drafts on paper or markups. Instead, editing is performed directly on the computer screen, thus reducing the amount of paper filing and printing that would otherwise be required.

2.18 UPFRONT DOCUMENTS

The contract documents prepared by the architect, engineer, or client's legal counsel include the contract between the client and contractor; the bidding requirements, which contain the invitation to bid, instruction to bidders, general information, bid forms, and bid bond; the contract forms, which may include the agreement (contract) format between the client and contractor, performance bond, and payment bond and certificates; the contract conditions identified as the general and supplementary conditions; the list of technical specifications; drawings; addenda; and contract modifications. The bidding requirements, contract forms, and contract conditions are sometimes referred to as the upfront documents.

Bidding Requirements. These explain the procedures bidders are to follow in preparing and submitting their bid. They assist all bidders in following established guidelines so that bids can be submitted for comparative purposes and not be disqualified because of technicalities. The bidding requirements address all prospective bidders, whereas the final contract documents address only the successful bidder, who, after signing the client-contractor agreement, becomes the **contractor.**

Contract Forms. The agreement (contract) is the written document, signed by the client and contractor, which is the legal instrument binding the two parties. This contract defines the relationships and obligations that exist between the client and contractor. It incorporates other contract documents by reference.

The contract may require a construction performance bond for financial protection of the client in the event the contractor is unable to complete the work in accordance with the contract. Not all clients require performance bonds, but the architect should review its necessity with the client and prepare the bidding documents in accordance with the client's decision.

The contract usually requires a contractor payment bond from the contractor to ensure that a surety will pay the labor force and material suppliers should the contractor fail to pay them. The use of this bond precludes the need for the labor force or suppliers to seek payment directly from the client, through liens or otherwise, because of nonpayment by the contractor.

Certificates include those project forms that may be required for insurance, certificate of compliance, guarantees or warranties, or compliance with applicable laws and regulations. Contract forms vary, depending on the type and usage of the project.

Contract Conditions. These define the rights, responsibilities, and relationships of the various parties involved in the construction process. Two types of contract conditions exist, General Conditions and Supplementary Conditions.

The General Conditions have general clauses that establish how the project is to be administered. They normally contain provisions that are common practice. Definitions of project terms, temporary provisions, site security, management processes required, and warranties and guarantees are among those items addressed in the General Conditions.

The Supplementary Conditions modify or supplement the general conditions to provide for requirements unique to a specific project and not normally found in standard General Conditions.

2.19 QUALITY CONTROL FOR ARCHITECTS AND ENGINEERS

To maintain a consistently high level of quality in design and construction documentation, a rigorous internal review of the documents prepared by the architect or engineer, which draws on the full depth and experience of resources available, should be undertaken during the contract document phase. Quality control can begin in the earliest stages of design, when criteria are established and developed as design guidelines for use throughout the project. At each stage of development, a coordination checklist, based on previous experience, can be utilized for the project through an independent internal or external technical checking program.

Computer file management may be used to enable the various technical disciplines to share graphic data and check for interference conditions, thereby enhancing technical coordination of the documents. Quality control should also continue throughout the construction phase with architect and engineer review of shop drawings and on-site observation of the work.

Quality Management Program. To have a truly meaningful quality management program, all personnel must be committed to it. To help the professional staff understand the quality program, quality systems should be developed, updated, maintained, and administered to assist the architect and professional staff in providing quality service to clients. An individual in each office may be assigned to assist in the quality management program. This person should undertake to instill in all personnel the importance of such a program in every aspect of the daily conduct of business.

The quality management program should set quality goals; develop professional interaction for meeting these goals among peers and peer groups; review building systems, specifications, and drawings to ensure quality; and see that these objectives are known to the public. Such a program will result in a client base that will communicate the quality level of the architect to others in the community, profession, and international marketplace. The architect's image is of extreme importance in acquiring and maintaining clients, and the best quality management program focuses on client service and dedication to the profession.

2.20 BIDDING AND CONTRACT AWARD

Competitive bidding is one method of determining the least cost for performing work defined by the construction documents. The bid states the price that the bidder will charge to perform the work based on the work shown and described in the bidding documents. Bids are prepared in confidence by each bidder. They are usually sealed when submitted to the client (or, in the case of subcontractors, to the bidding contractors). At a specified time and date, all bids are opened, competitively examined, and compared. Unless there are compelling reasons to do otherwise, the client (contractor in the case of subcontractors) usually enters into an agreement to have the work performed by the bidder submitting the lowest price.

Before bids may be received, prospective bidders need to be identified and made aware of the project. Sufficient data should be furnished to potential bidders to allow preparation of their bids. The client may or may not wish to prequalify bidders. In those cases where prequalification is required, the architect can have meaningful input in the process based on past experience with potential bidders.

The terms *bid* and *proposal* are synonymous. Although proposal may imply an opportunity for more consideration and discussion with the client, architect, or engineer, *bid*, *bidder*, and *bid form* are preferable, to prevent misunderstanding by the bidders.

After client approval of the construction documents and selection of a construction bidding method, the architect may assist in the selection of contractors to bid the work; preparation of bid forms; issuance of bidding documents for competitive bidding; answering inquiries from bidders; and preparing and issuing any necessary addenda to the bidding documents. Furthermore, the architect may assist in analyzing bid proposals and making recommendations to the client as to the award of the construction contract. The architect can also assist in preparation of the construction contract.

Bidders may elect to change their bid on the basis of certain conditions, such as errors in the bid, changes in product cost, changes in labor rates, or nonavailability of labor because of other work or strikes. Each bidder is responsible for providing for any eventuality during the period the bid is open for acceptance. Unless provided for otherwise, bidders may withdraw their bid before acceptance by the client, unless the client consents to a later withdrawal. If all conditions of the instructions to bidders have been met, then after the bids have been opened, the bids should be evaluated. The low bid especially should be analyzed to ensure that it reflects accurately the cost of the work required by the contract documents. The bids may be compared with the architect's construction cost estimate that was prepared on completion of the contract documents. The client can accept a bid and award the contract to the selected bidder, who then becomes the *contractor* for the work.

2.21 CONSTRUCTION SCHEDULING

Normally, a client asks the architect for an estimate of the construction time for the project. The client can then incorporate this estimate in the overall development schedule.

The contractor should prepare a detailed construction schedule for use in administering the work of subcontractors and the contractor's own forces. The contractor should be requested to submit the schedule to the architect and the client within 30 days of contract award. The schedule will also form the basis for the contractor's development of a shop drawing schedule.

A construction schedule can consist simply of a bar chart for each item of work or a breakdown for the major trades on the project. Alternatively, the schedule can be highly detailed; for example, a critical-path-method (CPM) schedule. This is recommended for large projects for monitoring the critical-path item at any point in time, since the critical path can change, depending on actual construction conditions. The contractor should monitor and update the schedule monthly during the construction phase so that the anticipated completion and move-in date can be verified or adjusted. If the completion date cannot be adjusted and the schedule appears to be of concern, more work time (overtime) may be required to maintain the nonadjusted schedule. This could have an impact on cost, depending on how the client-contract agreement was structured.

The construction schedule is an extremely meaningful tool in monitoring the construction process. It can assist the architect's ongoing role in quality control during the construction phase, when the management of the building process is transferred to, and becomes the responsibility of, the contractor. The schedule also is a meaningful tool for use by all trades involved in the building process. The schedule affects trades in different ways, depending on the size of the labor force, availability of material and personnel hoisting equipment, access to the work, coordination of subcontractors' work with material suppliers, material testing agencies involved, preparation of mock-ups, shop-drawing submittals, and general overall construction coordination issues.

2.22 SHOP DRAWING REVIEW

After the construction contract is awarded, the contractor should submit a proposed schedule for submission of shop drawings to meet the construction schedule. This permits the architect to anticipate submissions and plan manpower requirements accordingly, based on the number and complexity of each submission.

As an ongoing part of quality control, the architect should review the shop drawings, product literature, and samples and observe material and mock-up testing. This is considered part of the shop drawing submittal process. The architect should be an independent agent and side neither with the client nor the contractor in acceptance or rejection of a submittal. Rather, based on professional judgment, the architect should render a decision as to whether the submittal is in general accordance with the construction documents and design intent. All submittals should be properly identified and recorded when received by the architect, as part of document control. The architect should review the submittal expeditiously and return it to the contractor with the appropriate action.

The architect's action shown on the submittal usually records that the contractor can proceed, proceed as noted, or not proceed. A copy of the proceed and proceed-as-noted submittal should be maintained in the architect's and contractor's site office for reference. The client should also be provided with the transmittal associated with submittals. This helps keep the client informed regarding the progress of the work relative to the schedule for submission of shop drawings.

2.23 ROLE OF ARCHITECT OR ENGINEER DURING CONSTRUCTION

After award of the construction contract, the architect or engineer generally continues to assist the client in relations with the contractor.

2.23.1 Site Observation

As part of their ongoing services during construction, and depending on the scale and complexity of the project, architects and engineers may make periodic site visits or maintain full-time representation on site during a portion or all of the construction period. The professional's role is to expedite day-to-day communication and decision making by having on-site personnel available to respond to required drawing and specification clarifications.

Site-observation requirements for the project should be discussed with the client at the onset of the project and be outlined in the architect-client agreement. Many clients prefer periodic or regularly scheduled site visits by the design professional. A provision for additional or full-time on-site representation, however, can be addressed in the agreement, and compensation for this additional service can be outlined in the agreement for discussion with the client later in the development process or during the construction phase. The client and the architect and engineer should agree on the appropriate amount of site visitation provided in the architect's basic services to allow adequate site-observation services based on specific project conditions.

If periodic site observations are made, the architect should report such observations to the client in written form. This should call attention to items observed that do not meet the intent of the construction documents. It is normally left to the client to reject or replace work unless such defective work involves life safety, health, or welfare of the building occupants or is a defect involving structural integrity. If the architect provides full-time site observation services, daily or weekly reports should be issued to the client outlining items observed that are not in accordance with the construction documents or design intent.

2.23.2 Site Record Keeping

Depending on contractual requirements for service during the construction phase, the architect may establish a field office. In this event, dual record keeping is suggested between the site and architect's office so that records required for daily administration of construction are readily accessible on site. Contractor correspondence, field reports, testing and balancing reports, shop drawings, record documents, contractor payment requests, change orders, bulletin issues, field meeting minutes, and schedules are used continually during construction. Computer systems and electronic mail make the communication process somewhat easy to control.

2.23.3 Inspection and Testing

Technical specifications require testing and inspection of various material and building systems during construction to verify that the intent of the design and construction documents is being fulfilled under field conditions. Testing is required where visual observations cannot verify actual conditions. Subsurface conditions, concrete and steel testing, welding, air infiltration, and air and water balancing of mechanical systems are such building elements that require inspection and testing services. Normally, these services are performed by an independent testing agency employed directly by the client so that third-party evaluation can be obtained.

Although the architect does not become involved in the conduct of work or determine the means or methods of construction, the architect has the general responsibility to the client to see that the work is installed in general accordance with the contract documents.

Other areas of inspection and testing involve establishing and checking benchmarks for horizontal and vertical alignment, examining soils and backfill material, compaction testing, examining subsurface retention systems, inspecting connections to public utilities, verifying subsoil drainage, verifying structural column centerlines and base-plate locations (if applicable), checking alignment and bracing of concrete formwork, verifying concrete strength and quality, and other similar items.

2.23.4 Payment Requests

The contractor normally submits a consolidated payment request monthly to the architect and client for review and certification. The payment request should be subdivided by trade and compared with the schedule of values for each trade that would have been submitted with the subcontractor bid if required by the instructions to bidders and bid form. The architect should review the payment request with respect to the percentage of completion of the pertinent work item or trade.

Some clients or lending institutions require that a partial waiver of lien be submitted for each work item or trade with each payment request. This partial waiver of lien can either be for the prior monthly request, which will indicate that the prior month's payment has been received, or in certain cases for the current monthly request. If the latter procedure is followed, the waiver may require revision, depending on the architect's review, if a work-item or trade-payment request is modified. The architect is not expected to audit the payment request or check the mathematical calculations for accuracy.

2.23.5 Change Orders

Contractor's change-order requests require the input of the architect, engineer, and client and are usually acted on as part of the payment request procedure. A change order is the instrument for amending the original contract amount and schedule, as submitted with the bid and agreed on in the client-contractor contract. Change orders can result from departures from the contract documents ordered during construction, by the architect, engineer, or client; errors or omissions; field conditions; unforeseen subsoil; or other similar conditions.

A change order outlines the nature of the change and the effect, if any, on the contract amount and construction schedule. Change orders can occur with both a zero cost and zero schedule change. Nevertheless, they should be documented in writing and approved by the contractor, architect, and client to acknowledge that the changes were made, with no impact. Change orders are also used to permit a material substitution when a material or system not included in the contract documents is found acceptable by the client and architect. For material substitutions proposed by the contractor, schedule revisions are not normally recognized as a valid change.

The sum of the change-order amounts is added or deducted from the original contract amount. Then, the revised contract amount is carried forward on the contractor's consolidated application for payment after the change orders have been signed by all parties. The normal contractor payment request procedure is then followed, on the basis of the new contract amount. If the schedule is changed because of a change order, the subsequent issue of the construction schedule should indicate the revised completion or move-in date, or both, that result from the approved change.

2.23.6 Project Closeout

Project closeout involves all parties, including subcontractors and material suppliers. It should be addressed early in the construction phase so that the closeout can be expedited and documented in an organized and meaningful manner. At this point in the construction process, the attention of the contractor and architect is focused on accomplishing the necessary paperwork and administrative functions required for final acceptance of the work and issuance of the contractor's final consolidated application for payment and final waiver of lien.

The normal project closeout proceeds as follows:

1. The contractor formally notifies the architect and the client that the contracted work is substantially complete.

2. From on-site observations and representations made by the contractor, the architect documents substantial completion with the client and the contractor. In some cases, this may trigger the start of certain guarantees or warranties, depending on the provisions of the general and supplementary conditions of the contract.

3. For some projects that are phased, some but not all the building systems may be recognized by the architect and the client as being substantially complete. This should be well-documented, since start dates for warranty and guarantee periods for various building systems or equipment may vary.

4. On-site visits are made by the architect and representatives of the client, sometimes called a walk-through, and a final *punchlist* is developed by the architect to document items requiring remedial work or replacement to meet the requirement of the construction documents.

5. A complete keying schedule, with master, submaster, room, and specialty keys, is documented by the contractor and delivered to the client.

6. The contractor submits all record drawings, **as-builts,** testing and balancing reports, and other administrative paperwork required by the contract documents.

7. The contractor should submit all required guarantees, warranties, certificates, and bonds required by the general and supplementary conditions of the contract or technical specifications for each work item or trade outlined in the breakdown of the contractor's consolidated final payment request.

8. The contractor corrects all work noted on the punchlist. A final observation of the corrected work may then be made by the architect and client.

9. If the client accepts the work, the architect sends a certificate of completion to the contractor with a copy to the client. The certificate documents that final completion of the work has occurred. All required operating manuals and maintenance instructions are given to the architect for document control and forwarding to the client.

10. The contractor submits final waivers of lien from each subcontractor or material supplier. Also provided is an affidavit stating that all invoices have been paid, with the exception of those amounts shown on the final waiver of lien. With these documents, the contractor submits the final consolidated payment request, including all change orders.

11. The architect sends a final certificate of payment to the client, with a copy to the contractor.

12. The contractor provides any required certificate of occupancy, indicating that the building authorities having jurisdiction over the project approve occupancy of the space for the intended use.

13. The client makes final payment to the contractor and notifies the architect of this.

This process is important inasmuch as it can trigger the transfer of risk from the contractor's insurance program during construction to the client's insurance program for the completed project.

2.24 TESTING AND BALANCING OF BUILDING SYSTEMS

It is normal for projects to go through what is known as a *shakedown period* after final acceptance and occupancy by the client or building tenant. The warranty and guarantee period (normally 1 year) is the contractor's representation and recognition that certain building elements and systems may need adjustment or slight modification, depending on actual occupancy conditions or normal maintenance and usage of such systems. The heating, ventilating, air conditioning, and systems unique to a project require testing and balancing and potential minor modifications and adjustments during this warranty and guarantee period, even though they were tested and balanced by the contractor's testing agency prior to project closeout. An independent testing and balancing contractor who was employed prior to final project closeout normally returns on an as-needed, on-call basis to adjust, test, and balance systems during the first year. In addition, the building engineer will become familiar with the systems during this first year of operation and may also adjust and balance systems.

2.25 POSTCONSTRUCTION OPERATION AND MAINTENANCE

The technical specifications for a building project normally require that some time be devoted prior to project closeout for instruction and training of the client's building operating personnel and building engineer, who will be responsible for operating and maintaining the various building systems. Manufacturers' operating procedures, manuals, and inventory of spare parts and attic stock should be reviewed with the client, building engineer, and the contractor installing the work. The building engineer should thus gain a general understanding of the individual systems and their interaction in the operation of the building. During the warranty and guarantee period, the contractor or applicable subcontractor may be requested to assist the building engineer further in operation and maintenance of a system, including testing, balancing, and minor adjustment. After the shakedown period and when the engineer thoroughly understands system operation, the client's personnel assume full responsibility and deal directly with the manufacturers of various building components for maintenance. Or the client may subcontract maintenance, a normal procedure for such systems as elevators and escalators where specialty expertise in maintenance is required.

2.26 RECORD DRAWINGS

The normal procedure for submission of record drawings rests primarily with the contractor. These are edited drawings and specifications submitted by the contractor that describe actual installed conditions based on the contractor's field coordination of the work.

In some instances, the client may request that the architect revise the original construction documents or prepare new drawings to reflect the as-built conditions. This is normally an additional service in the architect-client agreement. It should be made clear to the client that the architect, if brought into this process, is acting only in a drafting role, inasmuch as the as-built documentation, including dimensions and details, is furnished by, and is the responsibility of, the contractor.

As-built and record drawings are helpful to the client in remodeling, maintenance, building-system modification, or making future additions to the project. The client should retain the drawings with maintenance manuals and operations procedures.

2.27 FOLLOW-UP INTERVIEWS

It is advisable that the architect or engineer have follow-up interviews with the client and occupants of the building or tenant spaces to help ascertain the success of the project and learn where certain materials, details, equipment, or systems may be improved for future use in other projects. Good client relations demand this type of exchange. It is also helpful for the architect or engineer to disseminate the interview results throughout the office and professional community, to improve problem solving, design, and construction.

2.28 MANAGEMENT OF DISPUTES

Even in the best of relationships, disputes can arise between the client and architect, client and contractor, or architect and contractor, even though the architect and contractor do not normally have a written agreement with each other. Disputes should be quickly addressed and resolved for the well-being of the project and to minimize disruption of the design and building process. If the dispute cannot be resolved by the parties, various methods of resolution are offered that include settlement, mediation, arbitration, and litigation. To maintain insurance coverage and protect appropriate interests, proper notification to insurers or involvement of legal counsel is required.

Settlement of Disputes. Disputes between two parties should be addressed quickly and, if at all possible, a settlement should be rendered and recorded. Settlement can be in the form of monetary adjustments or payments, free services on behalf of the architect to remedy or correct an error, or such other agreement between the two parties. It is recommended that this method of dispute resolution be used whenever possible to avoid time, cost, and anguish, which can occur as a result of mediation, arbitration, and litigation.

Mediation. In mediation, the parties in dispute agree on a third independent party to act as a mediator and hear each side's position in the dispute in an attempt to mediate a resolution. Mediation is not binding on either party but helps resolve certain disputes due to a third party's focus on, and question of, the issues.

Arbitration. This is a method of handling disputes in which an arbitrator or arbitration panel, often consisting of three members, is selected to hear the positions of the parties in the dispute and decide on a potential resolution. The resolution is binding on the parties. Cost and time for arbitration is usually, but not always, less than that required for litigation. The arbitrators usually consist of professionals (architects and engineers), lawyers, contractors, or other parties involved in the building industry.

Litigation. In the event settlement or mediation cannot resolve a dispute and the parties do not wish to arbitrate, the only remaining course of action is to litigate the dispute. This requires that much time and money be expended for depositions, document and other discovery, and preparation for trial. The final results are rendered by a group of individuals (the jury) or judge not involved in the building industry. Therefore, a possession of a thorough knowledge and understanding of issues affecting the architectural and engineering profession and construction industry become the responsibility of each party's legal counsel to establish a true and accurate picture of each party's position and the facts in the case. See also Art. 17.14.

2.29 PROFESSIONAL ETHICS

The American Institute of Architects has formulated the following basic principles for guidance of architects:

> Advice and counsel constitute the service of the profession. Given in verbal, written, or graphic form, they are normally rendered in order that buildings with their equipment and the areas about them, in addition to being well suited to their purposes, well planned for health, safety, and efficient operation and economical maintenance, and soundly constructed of materials and by methods most appropriate and economical for their particular uses, shall have a beauty and distinction that lift them above the commonplace. It is the purpose of the profession of architecture to render such services from the beginning to the completion of a project.

The fulfillment of that purpose is advanced every time architects render the highest quality of service they are capable of giving. In particular, the architect's drawings, specifications, and other documents should be complete, definite, and clear concerning the architect's intentions, the scope of the contractor's work, the materials to be employed, and the conditions under which the construction is to be completed and the work paid for. The relation of architects to their clients depends on good faith. Architects should explain the exact nature and extent of their services and the conditional character of construction cost estimates made before final drawings and specifications are complete.

The contractor depends on the architect to guard the contractor's interests as well as those of the client. The architect should reject workmanship and materials that are determined not to be in conformity with the contract documents, but it is also the architect's duty to give reasonable aid toward a complete understanding of those documents so that errors may be avoided. An exchange of information between architects and those who supply and handle building materials should be encouraged.

Architects, in their investments and business relations outside the profession, should avoid financial or personal activities that tend to weaken or discredit their standing as an unprejudiced and honest adviser, free to act in the client's best interests. Permitting use of free architectural or engineering services to be offered by manufacturers; suppliers of building materials, appliances, and equipment; or contractors may imply an obligation that can become detrimental to the best interest of the client.

Architects may offer their services to anyone for commission, salary, or fee as architect, consultant, adviser, or assistant, provided the architect rigidly maintains professional integrity, disinterestedness, and freedom to act.

Architects should work together through their professional organizations to promote the welfare of the physical environment. They should share in the interchange of technical information and experience.

Architects should seek opportunities to be of service in civic affairs. To the best of their ability, they should endeavor to advance the safety, health, and well-being of the community in which they reside by promoting appreciation of good design, good construction, proper placement of facilities, and harmonious development of the areas surrounding the facility.

Architects should take action to advance the interests of their personnel, providing suitable working conditions for them, requiring them to render competent and efficient services, and paying them adequate and just compensation. Architects should also encourage and sponsor those who are entering the profession, assisting them to a full understanding of the functions, duties, and responsibilities of the architectural profession.

Every architect should contribute toward justice, courtesy, and sincerity in the profession. In the conduct of their practice, architects should maintain a totally professional attitude toward those served, toward those who assist in the practice, toward fellow architects, and toward the members of other professions. Daily performance should command respect to the extent that the profession will benefit from the example architects set to other professionals and to the public in general.

SECTION THREE
PROTECTION AGAINST HAZARDS

Frederick S. Merritt
Consulting Engineer,
West Palm Beach, Florida

A hazard poses the threat that an unwanted event, possibly a catastrophe, may occur. Risk is the probability that the event will occur. Inasmuch as all buildings are subject to hazards such as hurricanes, earthquakes, flood, fire, and lightning strikes, both during and after construction, building designers and contractors have the responsibility of estimating the risks of these hazards and the magnitudes of the consequences should the events be realized.

3.1 RISK MANAGEMENT

After the risk of a hazard has been assessed, the building designers and contractors, guided by building-code, zoning-code, and health-agency specifications and exercising their best judgment, should decide on an acceptable level for the risk. With this done, they should then select a cost-effective way of avoiding the hazard, if possible, or protecting against it so as to reduce the risk of the hazard's occurring to within the acceptable level.

Studies of building failures provide information that building designers should use to prevent similar catastrophes. Many of the lessons learned from failures have led to establishment of safety rules in building codes. These rules, however, generally are minimum requirements and apply to ordinary structures. Building designers, therefore, should use judgment in applying code requirements and should adopt more stringent design criteria where conditions dictate.

Such conditions are especially likely to exist for buildings in extreme climates or in areas exposed to natural hazards, such as high winds, earthquakes, floods, landslides, and lightning. Stricter criteria should also be used for buildings that are tall and narrow, are low but very large, have irregular or unusual shapes, house hazardous materials, or are of novel construction. Furthermore, building codes may not contain provisions for some hazards against which building designers nevertheless should provide protection. Examples of such hazards are vandalism, trespass, and burglary. In addition, designers should anticipate conditions that may exist in buildings in emergencies and provide refuge for occupants or safe evacuation routes.

Building designers also should use judgment in determining the degree of protection to be provided against specific hazards. Costs of protection should be commensurate with probable losses from an incident. In many cases, for example, it is uneconomical to construct a building that will be immune to extreme earthquakes, high winds of tornadoes, arson, bombs, burst dams, or professional burglars. Full protection, however, should always be provided against hazards with a high probability of occurrence accompanied by personal injuries or high property losses. Such hazards include hurricanes and gales, fire, and vandals.

3.1.1 Design Life of Buildings

For natural phenomena, design criteria may be based on the probability of occurrence of extreme conditions, as determined from statistical studies of events in specific localities. These probabilities are often expressed as mean recurrence intervals.

A **mean recurrence interval** of an extreme condition is the average time, in years, between occurrences of a condition equal to or worse than the specified extreme condition. For example, the mean recurrence interval of a wind of 60 mi/hr or more may be recorded for Los Angeles as 50 years. Thus, after a building has been erected in Los Angeles, chances are that in the next 50 years it will be subjected only once to a wind of 60 mi/hr or more. Consequently, if the building was assumed to have a 50-year life, designers might logically design it basically for a 60-mi/hr wind, with a safety factor included in the design to protect against low-probability faster winds. Mean recurrence intervals are the basis for minimum design loads for high winds, snowfall, and earthquake in many building codes.

3.1.2 Safety Factors

Design of buildings for both normal and emergency conditions should always incorporate a safety factor against failure. The magnitude of the safety factor should be selected in accordance with the importance of a building, the extent of personal injury or property loss that may result if a failure occurs, and the degree of uncertainty as to the magnitude or nature of loads and the properties and behavior of building components.

As usually incorporated in building codes, a safety factor for quantifiable system variables is a number greater than unity. The factor may be applied in either of two ways.

One way is to relate the maximum permissible load, or demand, on a system under service conditions to design capacity. This system property is calculated by dividing by the safety factor the ultimate capacity, or capacity at failure, for sustaining that type of load. For example, suppose a structural member assigned a safety factor of 2 can carry 1000 lb before failure occurs. The service load then is $1000/2 = 500$ lb.

The second way in which codes apply safety factors is to relate the ultimate capacity of a system to a design load. This load is calculated by multiplying the maximum load under service conditions by a safety factor, often referred to as a **load factor.** For example, suppose a structural member assigned a load factor of 2 is required to carry a service load of 500 lb. Then, the member should be designed to have a capacity for sustaining a design load of $500 \times 2 = 1000$ lb, without failing.

While both methods achieve the objective of providing reserve capacity against unexpected conditions, use of load factors offers the advantage of greater flexibility in design of a system for a combination of different loadings, because a different load factor can be assigned to each type of loading in accordance with probability of occurrence and effects of other uncertainties.

Safety factors for various building systems are discussed in following sections of the book. This section presents general design principles for protection of buildings and occupants against high winds, earthquakes, water, fire, lightning, and intruders.

3.2 WIND PROTECTION

For practical design, wind and earthquakes may be treated as horizontal, or lateral, loads. Although wind and seismic loads may have vertical components, these generally are small and readily resisted by columns and bearing walls.

The variation with height of the magnitude of a wind load for a multistory building differs from that of a seismic load. Nevertheless, provisions for resisting either type of load are similar.

In areas where the probability of either a strong earthquake or a high wind is small, it is nevertheless advisable to provide in buildings considerable resistance to both types of load. In many cases, such resistance can be incorporated with little or no increase in costs over designs that ignore either high wind or seismic resistance.

3.2.1 Wind Characteristics

Because wind loads are considered horizontal forces, wind pressure, for design purposes, should be assumed to be applied to the gross area of the vertical projection of that portion of the building above the average level of the adjoining ground. Although the loads are assumed to be horizontal, they may nevertheless apply either inward pressures or suctions to inclined and horizontal surfaces. In any case, wind loads should be considered to act normal to the exposed building surfaces. Furthermore, wind should be considered to be likely to come from any direction, unless it is known for a specific locality that extreme winds may come only from one direction. As a consequence of this assumption, each wall of a rectangular building should be considered in design to be subject to the maximum wind load.

Winds generally strike a building in gusts. Consequently, the building is subjected to dynamic loading. Nevertheless, except for unusually tall or narrow buildings, it is common practice to treat wind as a static loading, even though wind pressures are not constant.

Estimation of design wind pressures is complicated by several factors. One factor is the effect of natural and man-made obstructions along the ground. Another factor is the variation of wind velocity with height above ground. Still another factor complicating wind-pressure calculation is the effect of building or building-component shape or geometry (relationship of height or width to length) on pressures. For important buildings, it is advisable to base design wind pressures on the results of wind tunnel tests of a model of a building, neighboring buildings, and nearby terrain.

3.2.2 Wind Pressures and Suctions

Pressures are considered positive when they tend to push a building component toward the building interior. They are treated as negative for suctions or uplifts, which tend to pull components outward.

Figure 3.1*a* illustrates wind flow over the sloping roof of a low building. For roofs with inclines up to about 30°, the wind may create an uplift over the entire roof (Fig. 3.1*b*). Also, as shown in Fig. 3.1*b* and *c*, the pressure on the external face of the windward wall is positive and on the leeward wall, negative (suction). If there are openings in the walls, the wind will impose internal pressures on the walls, floors, and roof. The net pressure on any building component, therefore, is the vector sum of the pressures acting on opposite faces of the component.

Because of the wind characteristics described in Art. 3.2.1 and the dependence of wind pressures on building geometry, considerable uncertainty exists as to the magnitude, direction, and duration of the maximum wind loads that may be imposed on any portion of a specific building. Consequently, numerous assumptions, based to some extent on statistical evidence, generally are made to determine design wind loads for buildings. Minimum requirements for wind loads are presented in local and model building codes.

Codes usually permit design wind loads to be determined either by mathematical calculations in accordance with an analytical procedure specified in the code or by wind-tunnel tests. Such tests are advisable for structures with unusual shapes,

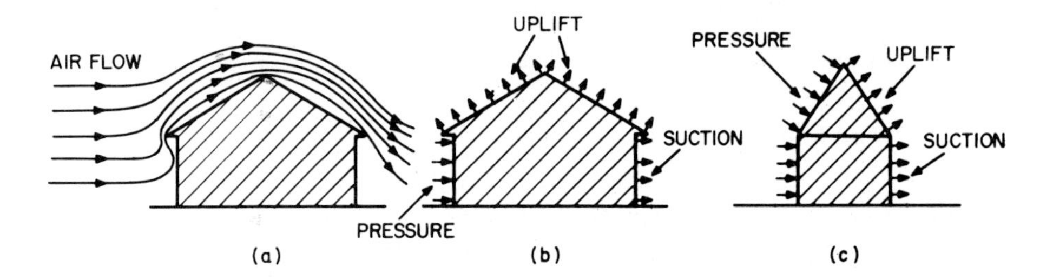

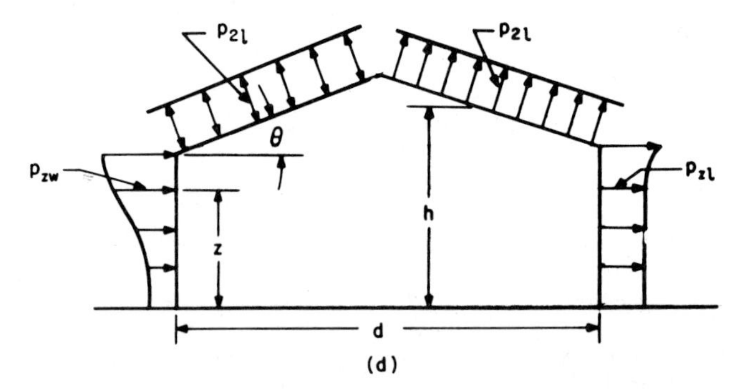

FIGURE 3.1 Effects of wind on a low building with pitched roof. (*a*) Airflow at the building. (*b*) Wind applies inward pressure against the windward wall, suction on the leeward wall, and uplift over all of a roof with slight slopes. (*c*) With a steep roof, inward pressure acts on the windward side of the roof and uplift only on the leeward side. (*d*) Pressure distribution along walls and roof assumed for design of wind bracing of a building.

unusual response to lateral loading, or location where channeling effects or buffeting in the wake of upwind obstructions are likely to occur. Tests also are desirable where wind records are not available or when more accurate information is needed. Codes often require that the following conditions be met in execution of wind-tunnel tests:

1. Air motion should be modeled to account for variation of wind speed with elevation and the intensity of the longitudinal component of turbulence.
2. The geometric scale of the model should not be greater than 3 times that of the longitudinal component of turbulence.
3. Instruments used should have response characteristics consistent with the required accuracy of measurements to be recorded.
4. Account should be taken of the dependence of forces and pressures on the Reynolds number of the air motion.
5. Tests for determining the dynamic response of a structure should be conducted on a model scaled with respect to dimensions, mass distribution, stiffness, and damping of the proposed structure.

In the analytical methods specified by building codes, maximum wind speeds observed in a region are converted to velocity pressures. These are then multiplied by various factors, to take into account building, site, and wind characteristics, to obtain design static wind loads. Bear in mind, however, that, in general, code requirements are applicable to pressures considerably smaller than those created by tornadoes, which may have wind speeds up to 600 mi/hr. For more information on wind loads, see Art. 5.1.2.

3.2.3 Failure Modes

Consideration of the ways in which winds may damage or destroy buildings suggests provisions that should be made to prevent failures. Past experience with building damage by winds indicates buildings are likely to fail by overturning; sliding; separation of components; excessive sway, or drift; or structural collapse.

Subjected to lateral forces W, a building may act as a rigid body and overturn. It would tend to rotate about the edge of its base on the leeward side (Fig. 3.2a).

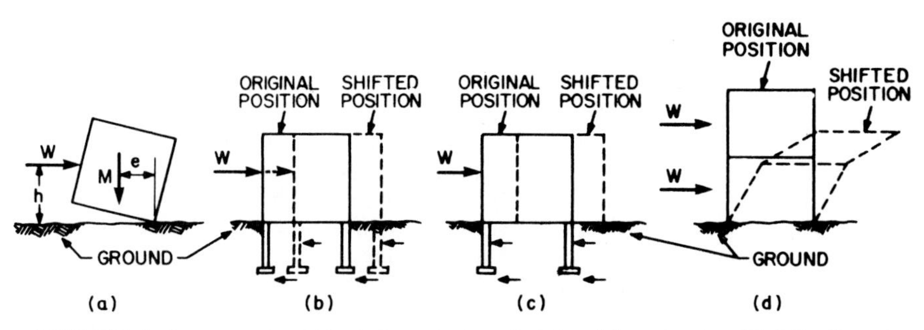

FIGURE 3.2 Some ways in which wind may destroy a building: (a) overturning; (b) sliding through the ground; (c) sliding off the foundations; (c) excessive drift (sidesway).

Overturning is resisted by the weight of the building M with a lever arm e measured from the axis of rotation. Building codes usually require that

$$Me \geq 1.5Wh \tag{3.1}$$

where Wh is the overturning moment.

Resistance to overturning may be increased by securely anchoring buildings to foundations. When this is done, the weight of earth atop the footings and pressing against foundation walls may be included with the weight of the building.

In addition to the danger of overturning, there is the risk of a building being pushed laterally by high winds. Sliding is resisted by friction at the base of the footings and earth pressure against foundation walls (Fig. 3.2*b*). (Consideration should be given to the possibility that soil that is highly resistant to building movement when dry may become weak when wet.) Another danger is that a building may be pushed by wind off the foundations (Fig. 3.2*c*). Consequently, to prevent this, a building should be firmly anchored to its foundations.

Buildings also may be damaged by separation of other components from each other. Therefore, it is essential that all connections between structural members and between other components and their supports be capable of resisting design wind loads. The possibility of separation of components by uplift or suction should not be overlooked. Such pressures can slide a roof laterally or lift it from its supports, tear roof coverings, rip off architectural projections, and suck out windows.

Another hazard is drift (sway) or collapse without overturning or sliding. Excessive drift when the wind rocks a building can cause occupant discomfort, induce failure of structural components by fatigue, or lead to complete collapse of the structure. The main resistance to drift usually is provided by structural components, such as beams, columns, and walls that are also assigned the task of supporting gravity loads. Some means must be provided to transmit wind or seismic loads from these members to the foundations and thence to the ground. Otherwise, the building may topple like a house of cards (Fig. 3.2*d*).

3.2.4 Limitation of Drift

There are no generally accepted criteria for maximum permissible lateral deflections of buildings. Some building codes limit drift of any story of a building to a maximum of 0.25% of the story height for wind and 0.50% of the story height for earthquake loads. Drift of buildings of unreinforced masonry may be restricted to half of the preceding values. The severer limitation of drift caused by wind loads is applied principally because they are likely to occur more frequently than earthquakes and will produce motions that will last much longer.

Three basic methods are commonly used, separately or in combination with each other, to prevent collapse of buildings under lateral loads, limit drift and transmit the loads to the foundations. These methods are illustrated in Fig. 3.3. One method is to incorporate shear walls in a building. A shear wall is a vertical cantilever with high resistance to horizontal loads parallel to its length (Fig. 3.3*a*). A pair of perpendicular walls can resist wind from any direction, because any wind load can be resolved into components in the planes of the walls (Fig. 3.3*b*).

A second method of providing resistance to lateral loads is to incorporate diagonal structural members to carry lateral forces to the ground (Fig. 3.3*c*). (The diagonals in Fig. 3.3*c* are called X bracing. Other types of bracing are illustrated in Fig. 3.6.) Under lateral loads, the braced bays of a building act like cantilever

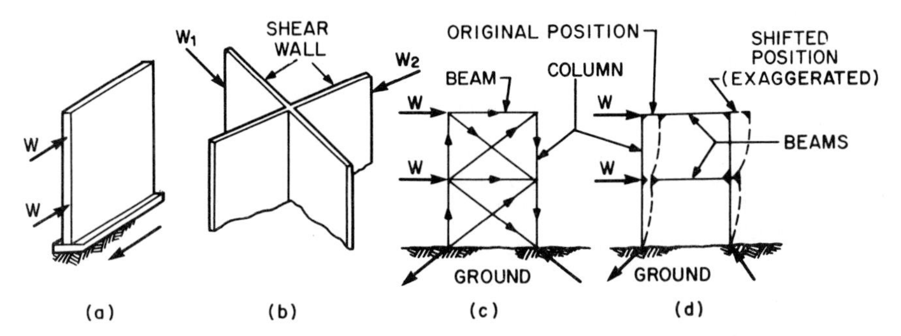

FIGURE 3.3 Some ways of restricting drift of a building; (*a*) shear wall; (*b*) pair of perpendicular shear walls; (*c*) diagonal bracing; (*d*) rigid frames.

vertical trusses. The arrows in Fig. 3.3*c* show the paths taken by wind forces from points of application to the ground. Note that the lateral loads impose downward axial forces on the leeward columns, causing compression, and uplift on the windward columns, causing tension.

A third method of providing resistance to lateral loads is to integrate the beams, or girders, and columns into rigid frames (Fig. 3.3*d*). In a rigid frame, connections between horizontal and vertical components prevent any change of angle between the members under loads. (Drift can occur only if beams and columns bend.) Such joints are often referred to as rigid, moment, or wind connections. They prevent the frame from collapsing in the manner shown in Fig. 3.2*d* until the loads are so large that the strength of the members and connections is exhausted. Note that in a rigid frame, leeward columns are subjected to bending and axial compression and windward columns are subjected to bending and axial tension.

In addition to using one or more of the preceding methods, designers can reduce drift by proper shaping of buildings, arrangements of structural components, and selection of members with adequate dimensions and geometry to withstand changes in dimensions. Shape is important because low, squat buildings have less sidesway than tall, narrow buildings, and buildings with circular or square floor plans have less sidesway than narrow, rectangular buildings with the same floor area per story.

Low Buildings. Figure 3.4*a* illustrates the application of diagonal bracing to a low, industrial-type building. Bracing in the plane of the roof acts with the rafters, ridge beam, and an edge roof beam as an inclined truss, which resists wind pressures on the roof. Each truss transmits the wind load to the ends of the building. Diagonals in the end walls transmit the load to the foundations. Wind pressure on the end walls is resisted by diagonal bracing in the end panels of the longitudinal walls. Wind pressure on the longitudinal walls, like wind on the roof, is transmitted to the end walls.

For large buildings, rigid frames are both structurally efficient and economic.

Alternatively, for multistory buildings, shear walls may be used. Figure 3.4*b* shows shear walls arranged in the shape of a T in plan, to resist wind from any direction. Figure 3.4*c* illustrates the use of walls enclosing stairwells and elevator shafts as shear walls. In apartment buildings, closet enclosures also can serve as shear walls if designed for the purpose. Figure 3.4*d* shows shear walls placed at the ends of a building to resist wind on its longitudinal walls. Wind on the shear walls, in turn, is resisted by girders and columns in the longitudinal direction acting as rigid frames. (See also Art. 5.12.)

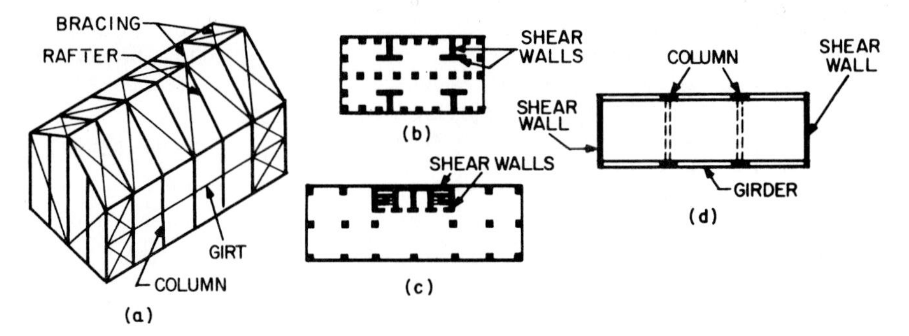

FIGURE 3.4 Bracing of low buildings: (*a*) diagonal bracing in roofs and walls; (*b*) isolated pairs of shear walls in a T pattern; (*c*) service-core enclosure used as shear walls; (*d*) shear walls at ends of building and rigid frames in the perpendicular direction.

Tall Buildings. For low buildings, structural members sized for gravity loads may require little or no enlargement to also carry stresses due to lateral loads. For tall buildings, however, structural members often have to be larger than sizes necessary only for gravity loads. With increase in height, structural material requirements increase rapidly. Therefore, for tall buildings, designers should select wind-bracing systems with high structural efficiency to keep material requirements to a minimum.

While shear walls, diagonal bracing, and rigid frames can be used even for very tall buildings, simple framing arrangements, such as planar systems, are not so efficient in high structures as more sophisticated framing. For example, shear walls or rigid frames in planes parallel to the lateral forces (Fig. 3.5*a*) may sway considerably at the top if the building is tall (more than 30 stories) and slender. Resistance to drift may be improved, however, if the shear walls are arranged in the form of

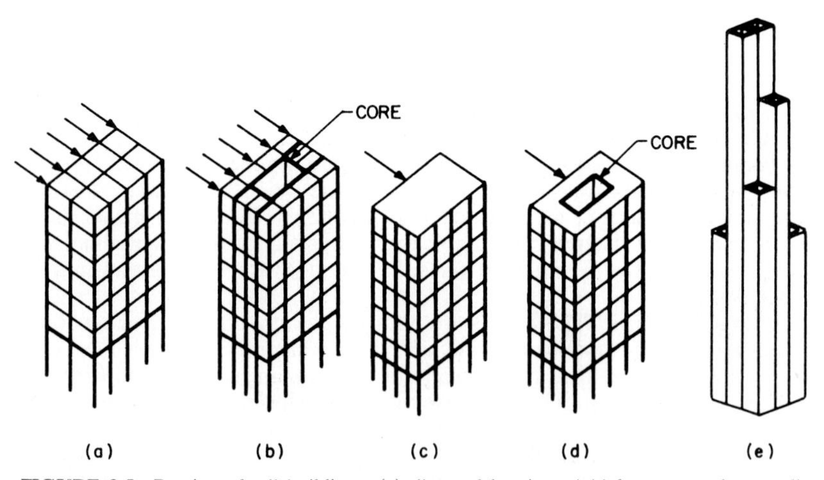

FIGURE 3.5 Bracing of tall buildings: (*a*) diagonal bracing, rigid frames, or shear walls placed in planes (bents) parallel to the lateral forces; (*b*) interior tube enclosing service core; (*c*) perforated tube enclosing the building; (*d*) tube within a tube; (*e*) bundled tubes.

a tube within the building (Fig. 3.5*b*). (The space within the tube can be utilized for stairs, elevators, and other services. This space is often referred to as the **service core.**) The cantilevered tube is much more efficient in resisting lateral forces than a series of planar, parallel shear walls containing the same amount of material. Similarly, rigid frames and diagonal bracing may be arranged in the form of an internal tube to improve resistance to lateral forces.

The larger the size of the cantilevered tube for a given height, the greater will be its resistance to drift. For maximum efficiency of a simple tube, it can be arranged to enclose the entire building (Fig. 3.5*c*). For the purpose, bracing or a rigid frame may be erected behind or in the exterior wall, or the exterior wall itself may be designed to act as a perforated tube. Floors act as horizontal diaphragms to brace the tube and distribute the lateral forces to it.

For very tall buildings, when greater strength and drift resistance are needed than can be provided by a simple tube, the tube around the exterior may be augmented by an internal tube (Fig. 3.5*d*) or by other arrangements of interior bracing, such as shear walls attached and perpendicular to the exterior tube. As an alternative, a very tall building may be composed of several interconnected small tubes, which act together in resisting lateral forces (Fig. 3.5*e*). Known as bundled tubes, this type of framing offers greater flexibility in floor-area reduction at various levels than a tube-within-tube type, because the tubes in a bundle can differ in height.

Diagonal bracing is more efficient in resisting drift than the other methods, because the structural members carry the lateral loads to the foundations as axial forces, as shown in Fig. 3.3*c*, rather than as a combination of bending, shear, and axial forces. Generally, the bracing is arranged to form trusses composed of triangular configurations, because of the stability of such arrangements. The joints between members comprising a triangle cannot move relative to each other unless the length of the members changes. Figure 3.6*a* illustrates the use of X bracing in the center bay of a multistory building to form a vertical cantilever truss to resist lateral forces.

Other forms of bracing, however, may be used as an alternative to reduce material requirements. Figure 3.6*b* shows how a single diagonal can be used in the

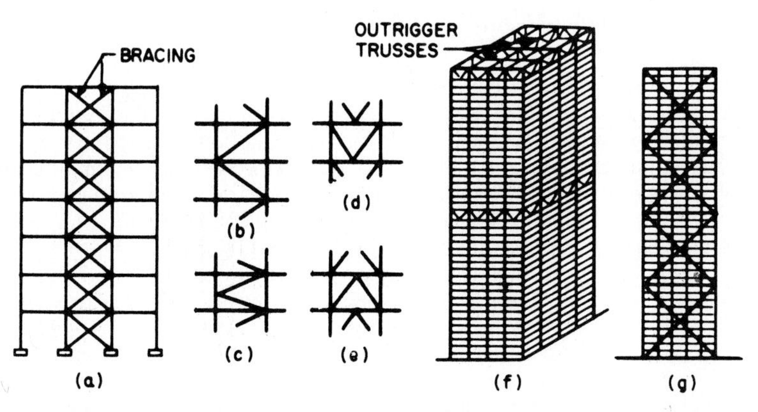

FIGURE 3.6 Some types of diagonal bracing: (*a*) X bracing in an interior bent; (*b*) single diagonal; (*c*) K bracing; (*d*) V bracing; (*e*) inverted V bracing; (*f*) horizontal trusses at the roof and intermediate levels to restrict drift; (*g*) X bracing on the exterior of a building.

center bay to form a vertical truss. In large bays, however, the length of the diagonal may become too long for structural efficiency. Hence, two or more diagonals may be inserted in the bay instead, as shown in Fig. 3.6*c* to *e*. The type of bracing in Fig. 3.6*c* is known as K bracing; that in Fig. 3.6*d*, as V bracing; and that in Fig. 3.6*e*, as inverted V bracing. The V type, however, has the disadvantage of restricting deflection of the beams to which the diagonals are attached and thus compelling the diagonals to carry gravity loads applied to the beams.

The bracing shown in Fig. 3.6*a* to *e* has the disadvantage of obstructing the bay and interfering with placement of walls, doors, passageways, and, for bracing along the building exterior, placement of windows. Accordingly, the inverted V type often is converted to knee bracing, short diagonals placed near beam-to-column joints. When knee bracing also is architecturally objectionable because of interference with room arrangements, an alternative form of wind bracing, such as rigid frames or shear walls, has to be adopted.

Trusses also can be placed horizontally to stiffen buildings for less drift. For example, Fig. 3.6*f* shows a building with wind bracing provided basically by an internal vertical cantilever tube. A set of horizontal trusses at the roof and a similar set at an intermediate level tie the tube to the exterior columns. The trusses reduce the drift at the top of the building by utilizing bending resistance of the columns. A belt of horizontal trusses around the building exterior at the roof and the intermediate level also helps resist drift of the building by utilizing bending resistance of the exterior columns.

When not considered architecturally objectionable, diagonal bracing may be placed on the building exterior to form a braced tube. The bracing may also serve as columns to transmit floor and roof loads to the ground. Figure 3.6*g* shows how multistory X bracing has been used to create a braced tube for a skyscraper.

See also Arts. 3.3.5 and 5.18.12, and Secs. 7 through 10.

(Council on Tall Buildings and Urban Habitat, "Planning and Design of Tall Buildings," Vols. SC, SB, and CB, American Society of Civil Engineers, New York; E. Simiu and R. H. Scanlon, "Wind Effects on Structures," John Wiley & Sons, Inc., New York.)

3.3 PROTECTION AGAINST EARTHQUAKES

Buildings should be designed to withstand minor earthquakes without damage, because they may occur almost everywhere. For major earthquakes, it may not be economical to prevent all damage but collapse should be precluded.

Because an earthquake and a high wind are not likely to occur simultaneously, building codes usually do not require that buildings be designed for a combination of large seismic and wind loads. Thus, designers may assume that the full strength of wind bracing is also available to resist drift caused by earthquakes.

The methods of protecting against high winds described in Art. 3.2.4 may also be used for protecting against earthquakes. Shaking of buildings produced by temblors, however, is likely to be much severer than that caused by winds. Consequently, additional precautions must be taken to protect against earthquakes. Because such protective measures will also be useful in resisting unexpectedly high winds, such as those from tornadoes, however, it is advisable to apply aseismic design principles to all buildings.

These principles require that collapse be avoided, oscillations of buildings damped, and damage to both structural and nonstructural components minimized.

Nonstructural components are especially liable to damage from large drift. For example, walls are likely to be stiffer than structural framing and therefore subject to greater seismic forces. The walls, as a result, may crack or collapse. Also, they may interfere with planned actions of structural components and cause additional damage. Consequently, aseismic design of buildings should make allowance for large drift, for example, by providing gaps between adjoining buildings and between adjoining building components not required to be rigidly connected together and by permitting sliding of such components. Thus, partitions and windows should be free to move in their frames so that no damage will occur when an earthquake wracks the frames.

3.3.1 Earthquake Characteristics

Earthquakes are produced by sudden release of tremendous amounts of energy within the earth by a sudden movement at a point called the **hypocenter.** (The point on the surface of the earth directly above the hypocenter is called the **epicenter.**) The resulting shock sends out longitudinal and transverse vibrations in all directions, both through the earth's crust and along the surface, and at different velocities. Consequently, the shock waves arrive at distant points at different times.

As a result, the first sign of the advent of an earthquake at a distant point is likely to be faint surface vibration of short duration as the first longitudinal waves arrive at the point. Then, severe shocks of longer duration occur there, as other waves arrive.

Movement at any point of the earth's surface during a temblor may be recorded with seismographs and plotted as seismograms, which show the variation with time of displacements. Seismograms of past earthquakes indicate that seismic wave forms are very complex.

Ground accelerations are also very important, because they are related to the inertial forces that act on building components during an earthquake. Accelerations are recorded in accelerograms, which are a plot of the variation with time of components of the ground accelerations. Newton's law relates acceleration to force:

$$F = Ma = \frac{W}{g} a \tag{3.2}$$

where F = force, lb
M = mass accelerated
a = acceleration of the mass, ft/s^2
W = weight of building component accelerated, lb
g = acceleration due to gravity = 32.2 ft/s^2

3.3.2 Seismic Scales

For study of the behavior of buildings in past earthquakes and application of the information collected to contemporary aseismic design, it is useful to have some quantitative means for comparing earthquake severity. Two scales, the Modified Mercalli and the Richter, are commonly used in the United States.

The Modified Mercalli scale compares earthquake intensity by assigning values to human perceptions of the severity of oscillations and extent of damage to buildings. The scale has 12 divisions. The severer the reported oscillations and damage, the higher is the number assigned to the earthquake intensity (Table 3.1).

TABLE 3.1 Modified Mercalli Intensity Scale (Abridged)

Intensity	Definition
I	Detected only by sensitive instruments.
II	Felt by a few persons at rest, especially on upper floors. Delicate suspended objects may swing.
III	Felt noticeably indoors; not always recognized as an earthquake. Standing automobiles rock slightly. Vibration similar to that caused by a passing truck.
IV	Felt indoors by many, outdoors by few; at night some awaken. Windows, dishes, doors rattle. Standing automobiles rock noticeably.
V	Felt by nearly everyone. Some breakage of plaster, windows, and dishes. Tall objects disturbed.
VI	Felt by all; many frightened and run outdoors. Falling plaster and damaged chimneys.
VII	Everyone runs outdoors. Damage of buildings negligible to slight, depending on quality of construction. Noticeable to drivers of automobiles.
VIII	Damage slight to considerable in substantial buildings, great in poorly constructed structures. Walls thrown out of frames; walls, chimneys, monuments fall; sand and mud ejected.
IX	Considerable damage to well-designed structures; structures shifted off foundations; buildings thrown out of plumb; underground pipes damaged. Ground cracked conspicuously.
X	Many masonry and frame structures destroyed; rails bent; water splashed over banks; landslides; ground cracked.
XI	Bridges destroyed; rails bent greatly; most masonry structures destroyed; underground service pipes out of commission; landslides; broad fissures in ground.
XII	Total damage. Waves seen in surface level; lines of sight and level distorted; objects thrown into air.

The Richter scale assigns numbers M to earthquake intensity in accordance with the amount of energy released, as measured by the maximum amplitude of ground motion:

$$M = \log A - 1.73 \log \frac{100}{D} \tag{3.3}$$

where M = earthquake magnitude 100 km from epicenter
A = maximum amplitude of ground motion, micrometers
D = distance, km, from epicenter to point where A is measured

The larger the ground displacement at a given location, the higher the value of the number assigned on the Richter scale. A Richter magnitude of 8 corresponds approximately to a Modified Mercalli intensity of XI, and for smaller intensities, Richter scale digits are about one unit less than corresponding Mercalli Roman numerals.

3.3.3 Effects of Ground Conditions

The amplitude of ground motion at a specific location during an earthquake depends not only on distance from the epicenter but also on the types of soil at the location. (Some soils suffer a loss of strength in an earthquake and allow large, uneven foundation settlements, which cause severe property damage.) Ground motion usually is much larger in alluvial soils (sands or clays deposited by flowing water) than in rocky areas or diluvial soils (material deposited by glaciers). Reclaimed land or earth fills generally undergo even greater displacements than alluvial soils. Consequently, in selection of sites for buildings in zones where severe earthquakes are highly probable during the life of the buildings, preference should be given to sites with hard ground or rock to considerable depth, with sand and gravel as a less desirable alternative and clay as a poor choice.

3.3.4 Seismic Forces

During an earthquake, the ground may move horizontally in any direction and up and down, shifting the building foundations correspondingly. Inertial forces, or seismic loads, on the building resist the displacements. Major damage usually is caused by the horizontal components of these loads, inasmuch as vertical structural members generally have adequate strength to resist the vertical components. Hence, as for wind loads, buildings should be designed to resist the maximum probable horizontal component applied in any direction.

Seismic forces vary rapidly with time. Therefore, they impose a dynamic loading on buildings. Calculation of the building responses to such loading is complex (Art. 5.18.6) and is usually carried out only for important buildings that are very tall and slender. For other types of buildings, building codes generally permit use of an alternative static loading for which structural analysis is much simpler (Art. 5.19).

3.3.5 Aseismic Design

The basic methods for providing wind resistance—shear walls, diagonal bracing, and rigid frames (Art. 3.24) are also suitable for resisting seismic loads. Ductile rigid frames, however, are preferred because of large energy-absorbing capacity. Building codes encourage their use by permitting them to be designed for smaller seismic loads than those required for shear walls and diagonal bracing. (Ductility is a property that enables a structural member to undergo considerable deformation without failing. The more a member deforms, the more energy it can absorb and therefore the greater is the resistance it can offer to dynamic loads.)

For tall, slender buildings, use of the basic methods alone in limiting drift to an acceptable level may not be cost-effective. In such cases, improved response to the dynamic loads may be improved by installation of heavy masses near the roof, with their movements restricted by damping devices. Another alternative is installation of energy-absorbing devices at key points in the structural framing, such as at the bearings of bottom columns or the intersections of cross bracing.

Designers usually utilize floors and roofs, acting as horizontal diaphragms, to transmit lateral forces to the resisting structural members. Horizontal bracing, however, may be used instead. Where openings occur in floors and roofs, for example for floors and elevators, structural framing should be provided around the openings to bypass the lateral forces.

As for wind loads, the weight of the building and of earth adjoining foundations are the only forces available to prevent the horizontal loads from overturning the building. [See Eq. (3.1) in Art. 3.2.3.] Also, as for wind loads, the roof should be firmly anchored to the superstructure framing, which, in turn, should be securely attached to the foundations. Furthermore, individual footings, especially pile and caisson footings, should be tied to each other to prevent relative movement.

Building codes often limit the drift per story under the equivalent static seismic load (see Art. 5.18.12). Connections and intersections of curtain walls and partitions with each other or with the structural framing should allow for a relative movement of at least twice the calculated drift in each story. Such allowances for displacement may be larger than those normally required for dimensional changes caused by temperature variations.

See also Art. 5.19.

(N. M. Newmark and E. Rosenblueth, "Fundamentals of Earthquake Engineering," and J. S. Stratta, "Manual of Seismic Design," Prentice-Hall, Englewood Cliffs, N.J.; "Standard Building Code," Southern Building Code Congress International, Inc., 900 Montclair Road, Birmingham, AL 35213-1206; "Uniform Building Code," International Conference of Building Officials, Inc., 5360 South Workman Mill Road, Whittier, CA 90601.)

3.4 PROTECTION AGAINST WATER

Whether thrust against and into a building by a flood, driven into the interior by a heavy rain, leaking from plumbing, or seeping through the exterior enclosure, water can cause costly damage to a building. Consequently, designers should protect buildings and their contents against water damage.

Protective measures may be divided into two classes: floodproofing and waterproofing. Floodproofing provides protection against flowing surface water, commonly caused by a river overflowing its banks. Waterproofing provides protection against penetration through the exterior enclosure of buildings of groundwater, rainwater, and melting snow.

3.4.1 Floodproofing

A flood occurs when a river rises above an elevation, called flood stage, and is not prevented by enclosures from causing damage beyond its banks. Buildings constructed in a flood plain, an area that can be inundated by a flood, should be protected against a flood with a mean recurrence interval of 100 years. Maps showing flood-hazard areas in the United States can be obtained from the Federal Insurance Administrator, Department of Housing and Urban Development, who administers the National Flood Insurance Program. Minimum criteria for floodproofing are given in National Flood Insurance Rules and Regulations (*Federal Register,* vol. 41, no. 207, Oct. 26, 1976).

Major objectives of floodproofing are to protect fully buildings and contents from damage from a 100-year flood, reduce losses from more devastating floods, and lower flood insurance premiums. Floodproofing, however, would be unnecessary if buildings were not constructed in flood plains. Building in a flood plain should be avoided unless the risk to life is acceptable and construction there can be economically and socially justified.

Some sites in flood plains possess some ground high enough to avoid flood damage. If such sites must be used, buildings should be clustered on the high areas. Where such areas are not available, it may be feasible to build up an earth fill, with embankments protected against erosion by water, to raise structures above flood levels. Preferably, such structures should not have basements, because they would require costly protection against water pressure.

An alternative to elevating a building on fill is raising it on stilts (columns in an unenclosed space). In that case, utilities and other services should be protected against damage from flood flows. The space at ground level between the stilts may be used for parking automobiles, if the risk of water damage to them is acceptable or if they will be removed before flood waters reach the site.

Buildings that cannot be elevated above flood stage should be furnished with an impervious exterior. Windows should be above flood stage, and doors should seal tightly against their frames. Doors and other openings may also be protected with a flood shield, such as a wall. Openings in the wall for access to the building may be protected with a movable flood shield, which for normal conditions can be stored out of sight and then positioned in the wall opening when a flood is imminent.

To prevent water damage to essential services for buildings in flood plains, important mechanical and electrical equipment should be located above flood level. Also, auxiliary electric generators to provide some emergency power are desirable. In addition, pumps should be installed to eject water that leaks into the building. Furthermore, unless a building is to be evacuated in case of flood, an emergency water supply should be stored in a tank above flood level, and sewerage should be provided with cutoff valves to prevent backflow.

3.4.2 Waterproofing*

In addition to protecting buildings against floods, designers also should adopt measures that prevent groundwater, rainwater, snow, or melted snow from penetrating into the interior through the exterior enclosure. Water may leak through cracks, expansion joints or other openings in walls and roofs, or through cracks around windows and doors. Also, water may seep through solid but porous exterior materials, such as masonry. Leakage generally may be prevented by use of weatherstripping around windows and doors, impervious waterstops in joints, or calking of cracks and other openings. Methods of preventing seepage, however, depend on the types of materials used in the exterior enclosure.

Definitions of Terms Related to Water Resistance

Permeability. Quality or state of permitting passage of water and water vapor into, through, and from pores and interstices, without causing rupture or displacement.

Terms used in this section to describe the permeability of materials, coatings, structural elements, and structures follow in decreasing order of permeability:

Pervious or Leaky. Cracks, crevices, leaks, or holes larger than capillary pores, which permit a flow or leakage of water, are present. The material may or may not contain capillary pores.

*Excerpted with minor revisions from Sec. 12 of the third edition of this handbook, authored by Cyrus C. Fishburn, formerly with the Division of Building Technology, National Bureau of Standards.

Water-resistant. Capillary pores exist that permit passage of water and water vapor, but there are few or no openings larger than capillaries that permit leakage of significant amounts of water.

Water-repellent. Not "wetted" by water; hence, not capable of transmitting water by capillary forces alone. However, the material may allow transmission of water under pressure and may be permeable to water vapor.

Waterproof. No openings are present that permit leakage or passage of water and water vapor; the material is impervious to water and water vapor, whether under pressure or not.

These terms also describe the permeability of a surface coating or a treatment against water penetration, and they refer to the permeability of materials, structural members, and structures whether or not they have been coated or treated.

Permeability of Concrete and Masonry. Concrete contains many interconnected voids and openings of various sizes and shapes, most of which are of capillary dimensions. If the larger voids and openings are few in number and not directly connected with each other, there will be little or no water penetration by leakage and the concrete may be said to be water-resistant.

Concrete in contact with water not under pressure ordinarily will absorb it. The water is drawn into the concrete by the surface tension of the liquid in the wetted capillaries.

Water-resistant concrete for buildings should be a properly cured, dense, rich concrete containing durable, well-graded aggregate. The water content of the concrete should be as low as is compatible with workability and ease of placing and handling. Resistance of concrete to penetration of water may be improved, however, by incorporation of a water-repellent admixture in the mix during manufacture. (See also Art. 9.9.)

Water-repellent concrete is permeable to water vapor. If a vapor-pressure gradient is present, moisture may penetrate from the exposed face to an inner face. The concrete is not made waterproof (in the full meaning of the term) by the use of an integral water repellent. Note also that water repellents may not make concrete impermeable to penetration of water under pressure. They may, however, reduce absorption of water by the concrete.

Most masonry units also will absorb water. Some are highly pervious under pressure. The mortar commonly used in masonry will absorb water too but usually contains few openings permitting leakage.

Masonry walls may leak at the joints between the mortar and the units, however. Except in single-leaf walls of highly pervious units, leakage at the joints results from failure to fill them with mortar and poor bond between the masonry unit and mortar. As with concrete, rate of capillary penetration through masonry walls is small compared with the possible rate of leakage.

Capillary penetration of moisture through above-grade walls that resist leakage of wind-driven rain is usually of minor importance. Such penetration of moisture into well-ventilated subgrade structures may also be of minor importance if the moisture is readily evaporated. However, long-continued capillary penetration into some deep, confined subgrade interiors frequently results in an increase in relative humidity, a decrease in evaporation rate, and objectionable dampness.

3.4.3 Roof Drainage

Many roof failures have been caused by excessive water accumulation. In most cases, the overload that caused failure was not anticipated in design of those roofs, because the designers expected rainwater to run off the roof. But because of inadequate drainage, the water ponded instead.

On flat roofs, ponding of rainwater causes structural members to deflect. The resulting bowing of the roof surface permits more rainwater to accumulate, and the additional weight of this water causes additional bowing and collection of even more water. This process can lead to roof collapse. Similar conditions also can occur in the valleys of sloping roofs.

To avoid water accumulation, roofs should be sloped toward drains and pipes that have adequate capacity to conduct water away from the roofs, in accordance with local plumbing codes. Minimum roof slope for drainage should be at least ¼ in/ft, but larger slopes are advisable.

The primary drainage system should be supplemented by a secondary drainage system at a higher level to prevent ponding on the roof above that level. The overflow drains should be at least as large as the primary drains and should be connected to drain pipes independent of the primary system. The roof and its structural members should be capable of sustaining the weight of all rainwater that could accumulate on the roof if part or all of the primary drainage system should become blocked.

3.4.4 Drainage for Subgrade Structures

Subgrade structures located above groundwater level in drained soil may be in contact with water and wet soil for periods of indefinite duration after long-continued rains and spring thaws. Drainage of surface and subsurface water, however, may greatly reduce the time during which the walls and floor of a structure are subjected to water, may prevent leakage through openings resulting from poor workmanship and reduce the capillary penetration of water into the structure. If subsurface water cannot be removed by drainage, the structure must be made waterproof or highly water-resistant.

Surface water may be diverted by grading the ground surface away from the walls and by carrying the runoff from roofs away from the building. The slope of the ground surface should be at least ¼ in/ft for a minimum distance of 10 ft from the walls. Runoff from high ground adjacent to the structure should also be diverted.

Proper subsurface drainage of ground water away from basement walls and floors requires a drain of adequate size, sloped continuously, and, where necessary, carried around corners of the building without breaking continuity. The drain should lead to a storm sewer or to a lower elevation that will not be flooded and permit water to back up in the drain.

Drain tile should have a minimum diameter of 6 in and should be laid in gravel or other kind of porous bed at least 6 in below the basement floor. The open joints between the tile should be covered with a wire screen or building paper to prevent clogging of the drain with fine material. Gravel should be laid above the tile, filling the excavation to an elevation well above the top of the footing. Where considerable water may be expected in heavy soil, the gravel fill should be carried up nearly to the ground surface and should extend from the wall a distance of at least 12 in (Fig. 3.7).

3.4.5 Concrete Floors at Grade

Floors on ground should preferably not be constructed in low-lying areas that are wet from ground water or periodically flooded with surface water. The ground

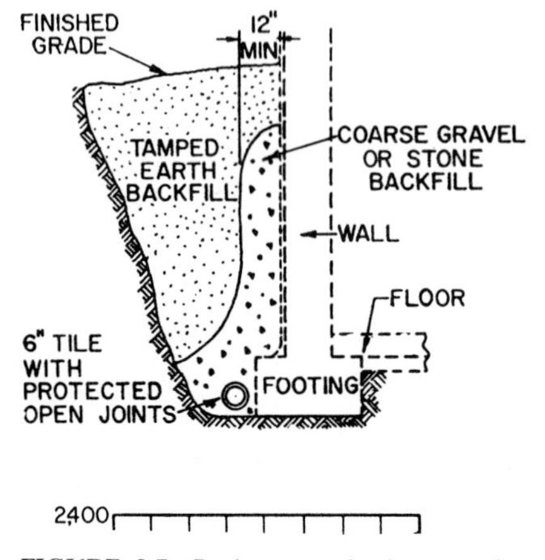

FIGURE 3.7 Drainage at the bottom of a foundation wall.

should slope away from the floor. The level of the finished floor should be at least 6 in above grade. Further protection against ground moisture and possible flooding of the slab from heavy surface runoffs may be obtained with subsurface drains located at the elevation of the wall footings.

All organic material and topsoil of poor bearing value should be removed in preparation of the subgrade, which should have a uniform bearing value to prevent unequal settlement of the floor slab. Backfill should be tamped and compacted in layers not exceeding 6 in in depth.

Where the subgrade is well-drained, as where subsurface drains are used or are unnecessary, floor slabs of residences should be insulated either by placing a granular fill over the subgrade or by use of a lightweight-aggregate concrete slab covered with a wearing surface of gravel or stone concrete. The granular fill, if used, should have a minimum thickness of 5 in and may consist of coarse slag, gravel, or crushed stone, preferably of 1-in minimum size. A layer of 3-, 4-, or 6-in-thick hollow masonry building units is preferred to gravel fill for insulation and provides a smooth, level bearing surface.

Moisture from the ground may be absorbed by the floor slab. Floor coverings, such as oil-base paints, linoleum, and asphalt tile, acting as a vapor barrier over the slab, may be damaged as a result. If such floor coverings are used and where a complete barrier against the rise of moisture from the ground is desired, a two-ply bituminous membrane or other waterproofing material should be placed beneath the slab and over the insulating concrete or granular fill (Fig. 3.8). The top of the lightweight-aggregate concrete, if used, should be troweled or brushed to a smooth level surface for the membrane. The top of the granular fill should be

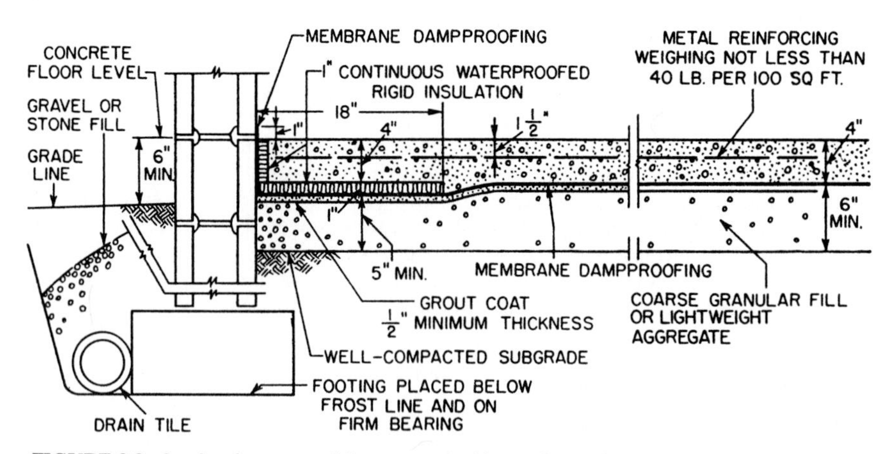

FIGURE 3.8 Insulated concrete slab on ground with membrane dampproofing.

covered with a grout coating, similarly finished. (The grout coat, ½ to 1 in thick, may consist of a 1:3 or a 1:4 mix by volume of portland cement and sand. Some ⅜- or ½-in maximum-sized coarse aggregate may be added to the grout if desired.) After the top surface of the insulating concrete or grout coating has hardened and dried, it should be mopped with hot asphalt or coal-tar pitch and covered before cooling with a lapped layer of 15-lb bituminous saturated felt. The first ply of felt then should be mopped with hot bitumen and a second ply of felt laid and mopped on its top surface. Care should be exercised not to puncture the membrane, which should preferably be covered with a coating of mortar, immediately after its completion. If properly laid and protected from damage, the membrane may be considered to be a waterproof barrier.

Where there is no possible danger of water reaching the underside of the floor, a single layer of 55-lb smooth-surface asphalt roll roofing or an equivalent waterproofing membrane may be used under the floor. Joints between the sheets should be lapped and sealed with bituminous mastic. Great care should be taken to prevent puncturing of the roofing layer during concreting operations. When so installed, asphalt roll roofing provides a low-cost and adequate barrier against the movement of excessive amounts of moisture by capillarity and in the form of vapor.

("A Guide to the Use of Waterproofing, Dampproofing, Protective and Decorative Barrier Systems for Concrete," ACI 515.1R, American Concrete Institute.)

3.4.6 Basement Floors

Where a basement is to be used in drained soils as living quarters or for the storage of things that may be damaged by moisture, the floor should be insulated and should preferably contain the membrane waterproofing described in Art. 3.4.5. In general, the design and construction of such basement floors are similar to those of floors on ground.

If passage of moisture from the ground into the basement is unimportant or can be satisfactorily controlled by air conditioning or ventilation, the waterproof membrane need not be used. The concrete slab should have a minimum thickness of 4 in and need not be reinforced, but should be laid on a granular fill or other insulation placed on a carefully prepared subgrade. The concrete in the slab should have a minimum compressive strength of 2000 psi and may contain an integral water repellent.

A basement floor below the water table will be subjected to hydrostatic upward pressures. The floor should be made heavy enough to counteract the uplift.

An appropriate sealant in the joint between the basement walls and a floor over drained soil will prevent leakage into the basement of any water that may occasionally accumulate under the slab. Space for the joint may be provided by use of beveled siding strips, which are removed after the concrete has hardened. After the slab is properly cured, it and the wall surface should be in as dry a condition as is practicable before the joint is filled to ensure a good bond of the filler and to reduce the effects of slab shrinkage on the permeability of the joint.

("Guide to Joint Sealants for Concrete Structures," ACI 504R, American Concrete Institute.)

3.4.7 Monolithic Concrete Basement Walls

These should have a minimum thickness of 6 in. Where insulation is desirable, as where the basement is used for living quarters, lightweight aggregate, such as those prepared by calcining or sintering blast-furnace slag, clay, or shale that meet the requirements of ASTM Standard C330 may be used in the concrete. The concrete should have a minimum compressive strength of 2000 psi.

For the forms in which concrete for basement walls is cast, form ties of an internal-disconnecting type are preferable to twisted-wire ties. Entrance holes for the form ties should be sealed with mortar after the forms are removed. If twisted-wire ties are used, they should be cut a minimum distance of 1½ in inside the face of the wall and the holes filled with mortar.

The resistance of the wall to capillary penetration of water in temporary contact with the wall face may be increased by the use of a water-repellent admixture. The water repellent may also be used in the concrete at and just above grade to reduce the capillary rise of moisture from the ground into the superstructure walls.

Where it is desirable to make the wall resistant to passage of water vapor from the outside and to increase its resistance to capillary penetration of water, the exterior wall face may be treated with an impervious coating. The continuity and the resultant effectiveness in resisting moisture penetration of such a coating is dependent on the smoothness and regularity of the concrete surface and on the skill and technique used in applying the coating to the dry concrete surface. Some bituminous coatings that may be used are listed below in increasing order of their resistance to moisture penetration:

Spray- or brush-applied asphalt emulsions

Spray- or brush-applied bituminous cutbacks

Trowel coatings of bitumen with organic solvent, applied cold

Hot-applied asphalt or coal-tar pitch, preceded by application of a suitable primer

Cementitious brush-applied paints and grouts and trowel coatings of a mortar increase moisture resistance of monolithic concrete, especially if such coatings contain a water repellent. However, in properly drained soil, such coatings may not be justified unless needed to prevent leakage of water through openings in the concrete resulting from segregation of the aggregate and bad workmanship in casting the walls. The trowel coatings may also be used to level irregular wall surfaces in preparation for the application of a bituminous coating. For information on other waterproofing materials, see "A Guide to the Use of Waterproofing, Dampproofing, Protective and Decorative Barrier Systems for Concrete," ACI 515.1R, American Concrete Institute.

3.4.8 Unit-Masonry Basement Walls

Water-resistant basement walls of masonry units should be carefully constructed of durable materials to prevent leakage and damage due to frost and other weathering exposure. Frost action is most severe at the grade line and may result in structural damage and leakage of water. Where wetting followed by sudden severe freezing may occur, the masonry units should meet the requirements of the following specifications:

Building brick (solid masonry units made from clay or shale), ASTM Standard C62, Grade SW

Facing brick (solid masonry units made from clay or shale), ASTM Standard C216, Grade SW

Structural clay load-bearing wall tile, ASTM Standard C34, Grade LBX

Hollow load-bearing concrete masonry units, ASTM Standard C90, Grade N

For such exposure conditions, the mortar should be a Type S mortar (Table 12.1) having a minimum compressive strength of 1800 psi when tested in accordance with the requirements of ASTM Standard C270. For milder freezing exposures and where the walls may be subjected to some lateral pressure from the earth, the mortar should have a minimum compressive strength of 1000 psi.

Leakage through an expansion joint in a concrete or masonry foundation wall may be prevented by insertion of a waterstop in the joint. Waterstops should be of the bellows type, made of 16-oz copper sheet, which should extend a minimum distance of 6 in on either side of the joint. The sheet should be embedded between wythes of masonry units or faced with a 2-in-thick cover of mortar reinforced with welded-wire fabric. The outside face of the expansion joint should be filled flush with the wall face with a joint sealant, as recommended in ACI 504R.

Rise of moisture, by capillarity, from the ground into the superstructure walls may be greatly retarded by use of an integral water-repellent admixture in the mortar. The water-repellent mortar may be used in several courses of masonry located at and just above grade.

The use of shotcrete or trowel-applied mortar coatings, ¾ in or more in thickness, to the outside faces of both monolithic concrete and unit-masonry walls greatly increases their resistance to penetration of moisture. Such plaster coatings cover and seal construction joints and other vulnerable joints in the walls against leakage. When applied in a thickness of 2 in or more, they may be reinforced with welded-wire fabric to reduce the incidence of large shrinkage cracks in the coating. However, the cementitious coatings do not protect the walls against leakage if the walls, and subsequently the coatings, are badly cracked as a result of unequal foundation settlement, excessive drying shrinkage, and thermal changes. ("Guide to Shotcrete," ACI 506, American Concrete Institute.)

Two trowel coats of a mortar containing 1 part portland cement to 3 parts sand by volume should be applied to the outside faces of basement walls built of hollow masonry units. One trowel coat may suffice on the outside of all-brick and of brick-faced walls.

The wall surface and the top of the wall footing should be cleansed of dirt and soil, and the masonry should be thoroughly wetted with water. While still damp, the surface should be covered with a thin scrubbed-on coating of portland cement tempered to the consistency of thick cream. Before this prepared surface has dried, a ⅜-in-thick trowel-applied coating of mortar should be placed on the wall and over the top of the footing; a fillet of mortar may be placed at the juncture of the wall and footing.

Where a second coat of mortar is to be applied, as on hollow masonry units, the first coat should be scratched to provide a rough bonding surface. The second coat should be applied at least 1 day after the first, and the coatings should be cured and kept damp by wetting for at least 3 days. A water-repellent admixture in the mortar used for the second or finish coat will reduce the rate of capillary penetration of water through the walls. If a bituminous coating is not to be used, the mortar coating should be kept damp until the backfill is placed.

Thin, impervious coatings may be applied to the plaster if resistance to penetration of water vapor is desired. (See ACI 515.1R.) The plaster should be dry and clean before the impervious coating is applied over the surfaces of the wall and the top of the footing.

3.4.9 Impervious Membranes

These are waterproof barriers providing protection against penetration of water under hydrostatic pressure and water vapor. To resist hydrostatic pressure, a membrane should be made continuous in the walls and floor of a basement. It also should be protected from damage during building operations and should be laid by experienced workers under competent supervision. It usually consists of three or more alternate layers of hot, mopped-on asphalt or coal-tar pitch and plies of treated glass fabric, or bituminous saturated cotton or woven burlap fabric. The number of moppings exceeds the number of plies by one.

Alternatives are cold-applied bituminous systems, liquid-applied membranes, and sheet-applied membranes, similar to those used for roofing. In installation, manufacturers' recommendations should be carefully followed. See also ACI 515.1R and "The NRCA Waterproofing Manual," National Roofing Manufacturers Association.

Bituminous saturated cotton fabric is stronger and is more extensible than bituminous saturated felt but is more expensive and more difficult to lay. At least one or two of the plies in a membrane should be of saturated cotton fabric to provide strength, ductility, and extensibility to the membrane. Where vibration, temperature changes, and other conditions conducive to displacement and volume changes in the basement are to be expected, the relative number of fabric plies may be increased.

The minimum weight of bituminous saturated felt used in a membrane should be 13 lb per 100 ft^2. The minimum weight of bituminous saturated woven cotton fabric should be 10 oz/yd^2.

Although a membrane is held rigidly in place, it is advisable to apply a suitable primer over the surfaces receiving the membrane and to aid in the application of the first mopped-on coat of hot asphalt or coal-tar pitch.

Materials used in the hot-applied system should meet the requirements of the following current ASTM standards:

Creosote primer for coal-tar pitch—D43

Primer for asphalt—D41

Coal-tar pitch—D450, Type II

Asphalt—D449, Type A

Cotton fabric, bituminous saturated—D173

Woven burlap fabric, bituminous saturated—D1327

Treated glass fabric—D1668

Coal-tar saturated felt—D227

Asphalt saturated organic felt—D226

The number of plies of saturated felt or fabric should be increased with increase in the hydrostatic head to which the membrane is to be subjected. Five plies is the maximum commonly used in building construction, but 10 or more plies have been recommended for pressure heads of 35 ft or greater. The thickness of the membrane crossing the wall footings at the base of the wall should be no greater than necessary, to keep very small the possible settlement of the wall due to plastic flow in the membrane materials.

The amount of primer to be used may be about 1 gal per 100 ft^2. The amount of bitumen per mopping should be at least 4½ gal per 100 ft^2. The thickness of the first and last moppings is usually slightly greater than the thickness of the moppings between the plies.

The surfaces to which the membrane is to be applied should be smooth, dry, and at a temperature above freezing. Air temperature should be not less than 50°F. The temperature of coal-tar pitch should not exceed 300°F and asphalt, 350°F.

If the concrete and masonry surfaces are not sufficiently dry, they will not readily absorb the priming coat, and the first mopping of bitumen will be accompanied by bubbling and escape of steam. Should this occur, application of the membrane should be stopped and the bitumen already applied to damp surfaces should be removed.

The membrane should be built up ply by ply, the strips of fabric or felt being laid immediately after each bed has been hot-mopped. The lap of succeeding plies or strips over each other depends on the width of the roll and the number of plies. In any membrane there should be a lap of the top or final ply over the first, initial ply of at least 2 in. End laps should be staggered at least 24 in, and the laps between succeeding rolls should be at least 12 in.

For floors, the membrane should be placed over a concrete base or subfloor whose top surface is troweled smooth and which is level with the tops of the wall footings. The membrane should be started at the outside face of one wall and extend over the wall footing, which may be keyed. It should cover the floor and tops of other footings to the outside faces of the other walls, forming a continuous horizontal waterproof barrier. The plies should project from the edges of the floor membrane and lap into the wall membrane.

The loose ends of felt and fabric must be protected; one method is to fasten them to a temporary vertical wood form about 2 ft high, placed just outside the wall face. Immediately after the floor membrane has been laid, its surface should be protected and covered with a layer of portland-cement concrete, at least 2 in thick.

For walls, the installed membrane should be protected against damage and held in position by protection board or a facing of brick, tile, or concrete block. A brick facing should have a minimum thickness of 2½ in. Facings of asphalt plank, asphalt block, or mortar require considerable support from the membrane itself and give protection against abrasion of the membrane from lateral forces only. Protection against downward forces such as may be produced by settlement of the backfill is given only by the self-supporting masonry walls.

The kind of protective facing may have some bearing on the method of constructing the membrane. The membrane may be applied to the exterior face of the wall after its construction, or it may be applied to the back of the protective facing before the main wall is built. The first of these methods is known as the outside application; the second is known as the inside application.

For the inside application, a protective facing of considerable stiffness against lateral forces must be built, especially if the wall and its membrane are to be used as a form for the casting of a main wall of monolithic concrete. The inner face of the protecting wall must be smooth or else leveled with mortar to provide a suitable base for the membrane. The completed membrane should be covered with a ⅜-in-thick layer of mortar to protect it from damage during construction of the main wall.

Application of wall membranes should be started at the bottom of one end of the wall and the strips of fabric or felt laid vertically. Preparation of the surfaces and laying of the membrane proceed much as they do with floor membranes. The surfaces to which the membrane is attached must be dry and smooth, which may require that the faces of masonry walls be leveled with a thin coat of grout or mortar. The plies of the wall membrane should be lapped into those of the floor membrane.

If the outside method of application is used and the membrane is faced with masonry, the narrow space between the units and the membrane should be filled with mortar as the units are laid. The membrane may be terminated at the grade line by a return into the superstructure wall facing.

Waterstops in joints in walls and floors containing a bituminous membrane should be the metal-bellows type. The membrane should be placed on the exposed face of the joint and it may project into the joint, following the general outline of the bellows.

The protective facing for the membrane should be broken at the expansion joint and the space between the membrane and the line of the facing filled with a joint sealant, as recommended in ACI 504R.

Details at pipe sleeves running through the membrane must be carefully prepared. The membrane should be reinforced with additional plies and may be calked at the sleeve. Steam and hot-water lines should be insulated to prevent damage to the membrane.

3.4.10 Above-Grade Walls

The rate of moisture penetration through capillaries in above-grade walls is low and usually of minor importance. However, such walls should not permit leakage of wind-driven rain through openings larger than those of capillary dimension.

Precast-concrete or metal panels are usually made of dense, highly water-resistant materials. However, walls made of these panels are vulnerable to leakage at the joints. In such construction, edges of the panels may be recessed and the interior of vertical joints filled with grout or other sealant after the panels are aligned.

Calking compound is commonly used as a facing for the joints. Experience has shown that calking compounds often weather badly; their use as a joint facing creates a maintenance problem and does not prevent leakage of wind-driven rain after a few years' exposure.

The amount of movement to be expected in the vertical joints between panels is a function of the panel dimensions and the seasonal fluctuation in temperature and, for concrete, the moisture content of the concrete. For panel construction, it may be more feasible to use an interlocking water-resistant joint. For concrete, the joint may be faced on the weather side with mortar and backed with either a compressible premolded strip or calking. See ACI 504R.

Brick walls 4 in or more in thickness can be made highly water-resistant. The measures that need to be taken to ensure there will be no leakage of wind-driven rain through brick facings are not extensive and do not require the use of materials other than those commonly used in masonry walls. The main factors that need to be controlled are the rate of suction of the brick at the time of laying and filling of all joints with mortar (Art. 12.5).

In general, the greater the number of brick leaves, or wythes, in a wall, the more water-resistant the wall.

Walls of hollow masonry units are usually highly permeable, and brick-faced walls backed with hollow masonry units are greatly dependent upon the water resistance of the brick facing to prevent leakage of wind-driven rain. For exterior concrete masonry walls without facings of brick, protection against leakage may be obtained by facing the walls with a cementitious coating of paint, stucco, or shotcrete.

For walls of rough-textured units, a portland cement–sand grout provides a highly water-resistant coating. The cement may be either white or gray.

Factory-made portland-cement paints containing a minimum of 65%, and preferably 80%, portland cement may also be used as a base coat on concrete masonry. Application of the paint should conform with the requirements of ACI 515.1R. The paints, stuccos, and shotcrete should be applied to dampened surfaces. Shotcrete should conform with the requirements of ACI 506R.

Cavity walls, particularly brickfaced cavity walls, may be made highly resistant to leakage through the wall facing. However, as usually constructed, facings are highly permeable, and the leakage is trapped in the cavity and diverted to the outside of the wall through conveniently located weep holes. This requires that the inner tier of the cavity be protected against the leakage by adequate flashings, and weep holes should be placed at the bottom of the cavities and over all wall openings. The weep holes may be formed by the use of sash-cord head joints or ⅜-in-diameter rubber tubing, withdrawn after the wall is completed.

Flashings should preferably be hot-rolled copper sheet of 10-oz minimum weight. They should be lapped at the ends and sealed either by solder or with bituminous plastic cement. Mortar should not be permitted to drop into the flashings and prevent the weep holes from functioning.

Prevention of Cracking. Shrinkage of concrete masonry because of drying and a drop in temperature may result in cracking of a wall and its cementitious facing. Such cracks readily permit leakage of wind-driven rain. The chief factor reducing incidence of shrinkage cracking is the use of dry block. When laid in the wall, the block should have a low moisture content, preferably one that is in equilibrium with the driest condition to which the wall will be exposed.

The block should also have a low potential shrinkage. See moisture-content requirements in ASTM C90 and method of test for drying shrinkage of concrete block in ASTM C426.

Formation of large shrinkage cracks may be controlled by use of steel reinforcement in the horizontal joints of the masonry and above and below wall openings. Where there may be a considerable seasonal fluctuation in temperature and moisture content of the wall, high-yield-strength, deformed-wire joint reinforcement should be placed in at least 50% of all bed joints in the wall.

Use of control joints faced with calking compound has also been recommended to control shrinkage cracking; however, this practice is marked by frequent failures to keep the joints sealed against leakage of rain. Steel joint reinforcement strengthens a concrete masonry wall, whereas control joints weaken it, and the calking in the joints requires considerable maintenance.

Water-Resistant Surface Treatments for Above-Grade Walls. Experience has shown that leakage of wind-driven rain through masonry walls, particularly those of brick, ordinarily cannot be stopped by use of an inexpensive surface treatment or coating that will not alter the appearance of the wall. Such protective devices either have a low service life or fail to stop all leakage.

Both organic and cementitious pigmented coating materials, properly applied as a continuous coating over the exposed face of the wall, do stop leakage. Many of the organic pigmented coatings are vapor barriers and are therefore unsuitable for use on the outside, "cold" face of most buildings. If vapor barriers are used on the cold face of the wall, it is advisable to use a better vapor barrier on the warm face to reduce condensation in the wall and behind the exterior coating.

Coatings for masonry may be divided into four groups, as follows: (1) colorless coating materials; (2) cementitious coatings; (3) pigmented organic coatings; and (4) bituminous coatings.

Colorless Coating Materials. The colorless "waterproofings" are often claimed to stop leakage of wind-driven rain through permeable masonry walls. Solutions of oils, paraffin wax, sodium silicate, chlorinated rubber, silicone resins, and salts of fatty acids have been applied to highly permeable test walls and have been tested at the National Institute of Standards and Technology under exposure conditions simulating a wind-driven rain. Most of these solutions contained not more than 10% of solid matter. These treatments reduced the rate of leakage but did not stop all leakage through the walls. The test data show that colorless coating materials applied to permeable walls of brick or concrete masonry may not provide adequate protection against leakage of wind-driven rain.

Solutions containing oils and waxes tended to seal the pores exposed in the faces of the mortar joints and masonry units, thereby acting more or less as vapor barriers, but did not seal the larger openings, particularly those in the joints.

Silicone water-repellent solutions greatly reduced leakage through the walls as long as the treated wall faces remained water-repellent. After an exposure period of 2 or 3 hr, the rate of leakage gradually increased as the water repellency of the wall face diminished.

Coatings of the water-repellent, breather type, such as silicone and "soap" solutions, may be of value in reducing absorption of moisture into the wall surface. They may be of special benefit in reducing the soiling and disfiguration of stucco facings and light-colored masonry surfaces. They may be applied to precast-concrete panels to reduce volume changes that may otherwise result from changes in moisture content of the concretes. However, it should be noted that a water-repellent treatment applied to the surface may cause water, trapped in the masonry, to evaporate beneath the surface instead of at the surface. If the masonry is not water-resistant and contains a considerable amount of soluble salts, as evidenced by efflorescence, application of a water repellent may cause salts to be deposited beneath the surface, thereby causing spalling of the masonry. The water repellents therefore should be applied only to walls having water-resistant joints. Furthermore, application of a colorless material makes the treated face of the masonry water-repellent and may prevent the proper bonding of a cementitious coating that could otherwise be used to stop leakage.

Cementitious Coatings. Coatings of portland-cement paints, grouts, and stuccos and of pneumatically applied mortars are highly water-resistant. They are preferred above all other types of surface coatings for use as water-resistant base coatings on above-grade concrete masonry. They may also be applied to the exposed faces of brick masonry walls that have not been built to be water-resistant.

The cementitious coatings absorb moisture and are of the breather type, permitting passage of water vapor. Addition of water repellents to these coatings does not greatly affect their water resistance but does reduce the soiling of the surface from the absorption of dirt-laden water. If more than one coating is applied, as in a two-coat paint or stucco facing job, the repellent is preferably added only to the finish coat, thus avoiding the difficulty of bonding a cementitious coating to a water-repellent surface.

The technique used in applying the cementitious coatings is highly important. The backing should be thoroughly dampened. Paints and grouts should be scrubbed into place with stiff fiber brushes and the coatings should be properly cured by wetting. Properly applied, the grouts are highly durable; some grout coatings applied to concrete masonry test walls were found to be as water-resistant after 10 years out-of-doors exposure as when first applied to the walls.

Pigmented Organic Coatings. These include textured coatings, mastic coatings, conventional paints, and aqueous dispersions. The thick-textured and mastic coatings are usually spray-applied but may be applied by trowel. Conventional paints and aqueous dispersions are usually applied by brush or spray. Most of these coatings are vapor barriers but some textured coatings, conventional paints, and aqueous dispersions are breathers. Except for the aqueous dispersions, all the coatings are recommended for use with a primer.

Applied as a continuous coating, without pinholes, the pigmented organic coatings are highly water-resistant. They are most effective when applied over a smooth backing. When they are applied with paintbrush or spray by conventional methods to rough-textured walls, it is difficult to level the surface and to obtain a continuous water-resistant coating free from holes. A scrubbed-on cementitious grout used as a base coat on such walls will prevent leakage through the masonry without the use of a pigmented organic coating.

The pigmented organic coatings are highly decorative but may not be so water-resistant, economical, or durable as the cementitious coatings.

Bituminous Coatings. Bituminous cutbacks, emulsions, and plastic cements are usually vapor barriers and are sometimes applied as "dampproofers" on the inside faces of masonry walls. Plaster is often applied directly over these coatings, the bond of the plaster to the masonry being only of a mechanical nature. Tests show that bituminous coatings applied to the inside faces of highly permeable masonry walls, not plastered, will readily blister and permit leakage of water through the coating. It is advisable not to depend on such coatings to prevent the leakage of wind-driven rain unless they are incorporated in the masonry or held in place with a rigid self-sustaining backing.

Even though the walls are resistant to wind-driven rain, but are treated on their inner faces with a bituminous coating, water may be condensed on the warm side of the coating and damage to the plaster may result, whether the walls are furred or not. However, the bituminous coating may be of benefit as a vapor barrier in furred walls, if no condensation occurs on the warm side.

See also Secs. 9 and 11.

("Admixtures for Concrete," ACI 212.1R; "Guide for Use of Admixtures for Concrete," ACI 212.2R; "Guide to Joint Sealants for Concrete Structures," ACI 504R; "Specification for Materials, Proportioning and Application of Shotcrete," ACI 506.2; "A Guide to the Use of Waterproofing, Dampproofing, Protective and Decorative Barrier Systems for Concrete," ACI 515.1R; "Specification for Concrete Masonry Construction," ACI 531.1; "Polymers in Concrete," ACI 548R; "Guide for the Use of Polymers in Concrete," ACI 548.1R, American Concrete Institute, P.O. Box 19150, Redford Station, Detroit, MI 48219.)

3.5 PROTECTION AGAINST FIRE

There are two distinct aspects of fire protection: life safety and property protection. Although providing for one aspect generally results in some protection for the other, the two goals are not mutually inclusive. A program that provides for prompt notification and evacuation of occupants meets the objectives for life safety, but provides no protection for property. Conversely, it is possible that adequate property protection might not be sufficient for protection of life.

Absolute safety from fire is not attainable. It is not possible to eliminate all combustible materials or all potential ignition sources. Thus, in most cases, an adequate fire protection plan must assume that unwanted fires will occur despite the best efforts to prevent them. Means must be provided to minimize the losses caused by the fires that do occur.

The first obligation of designers is to meet legal requirements while providing the facilities required by the client. In particular, the requirements of the applicable building code must be met. The building code will contain fire safety requirements, or it will specify some recognized standard by reference. Many owners will also require that their own insurance carrier be consulted—to obtain the most favorable insurance rate, if for no other reason.

3.5.1 Fire-Protection Standards

The standards most widely adopted are those published by the National Fire Protection Association (NFPA), Batterymarch Park, Quincy, MA 02269. The NFPA "National Fire Codes" comprise several volumes containing numerous standards, updated annually. (These are also available separately.) The standards are supplemented by the NFPA "Fire Protection Handbook," which contains comprehensive and detailed discussion of fire problems and much valuable statistical and engineering data.

Underwriters Laboratories, Inc. (UL), 333 Pfingsten Road, Northbrook, IL 60062, publishes testing laboratory approvals of devices and systems in its "Fire Protection Equipment List," updated annually and by bimonthly supplements. The publication outlines the tests that devices and systems must pass to be listed. The UL "Building Materials List" describes and lists building materials, ceiling-floor assemblies, wall and partition assemblies, beam and column protection, interior finish materials, and other pertinent data. UL also publishes lists of "Accident Equipment," "Electrical Equipment," "Electrical Construction Materials," "Hazardous Location Equipment," "Gas and Oil Equipment," and others.

Separate standards for application to properties insured by the Factory Mutual System are published by the Factory Mutual Engineering Corporation (FM), Norwood, MA 02062. FM also publishes a list of devices and systems it has tested and approved.

The General Services Administration, acting for the federal government, has developed many requirements that must be considered, if applicable. Also, the federal government encourages cities to adopt some uniform code. In addition, buildings must comply with provisions of the Americans with Disability Act (ADA). (See Department of Justice final rules, *Federal Register,* 28 CFR Part 36, July 26, 1991; American National Standards Institute "Accessibility Standard," ANSI A117.1; "ADA Compliance Guidebook," Building Owners and Managers Association International, 1201 New York Ave., Washington, D.C. 20005.)

The Federal Occupational Safety and Health Act (OSHA) sets standards for protecting the health and safety of nearly all employees. It is not necessary that a business be engaged in interstate commerce for the law to apply. OSHA defines **employer** as "a person engaged in a business affecting commerce who has employees, but does not include the United States or any State or political subdivision of a State."

An employer is required to "furnish to each of his employees employment and a place of employment which are free from recognized hazards that are causing or are likely to cause death or serious physical harm to his employees." Employers are also required to "comply with occupational safety and health standards promulgated under the Act."

Building codes consist of a set of rules aimed at providing reasonable safety to the community, to occupants of buildings, and to the buildings themselves. The codes may adopt the standards mentioned previously and other standards concerned with fire protection by reference or adapt them to the specific requirements of the community. In the absence of a municipal or state building code, designers may apply the provisions of the Uniform Building Code, promulgated by the International Conference of Building Officials, or other national model code.

Many states have codes for safety to life in commercial and industrial buildings, administered by the Department of Labor, the State Fire Marshal's Office, the State Education Department, or the Health Department. Some of these requirements are drastic and must always be considered.

Obtaining optimum protection for life and property can require consultation with the owner's insurance carrier, municipal officials, and the fire department. If the situation is complicated enough, it can require consultation with a specialist in all phases of fire protection and prevention. In theory, municipal building codes are designed for life safety and for protection of the public, whereas insurance-oriented codes (except for NFPA 101, "Life Safety Code") are designed to minimize property fire loss. Since about 70% of any building code is concerned with fire protection, there are many circumstances that can best be resolved by a fire protection consultant.

3.5.2 Fire-Protection Concepts

Although fires in buildings can be avoided, they nevertheless occur. Some of the reasons for this are human error, arson, faulty electrical equipment, poor maintenance of heating equipment, and natural causes, such as lightning. Consequently, buildings should be designed to minimize the probability of a fire and to protect life and limit property damage if a fire should occur. The minimum steps that should be taken for the purpose are as follows:

1. Limit potential fire loads, with respect to both combustibility and ability to generate smoke and toxic gases.
2. Provide means for prompt detection of fires, with warnings to occupants who may be affected and notification of the presence of fire to fire fighters.
3. Communication of instructions to occupants as to procedures to adopt for safety, such as to staying in place, proceeding to a designated refuge area, or evacuating the building.
4. Provide means for early extinguishment of any fire that may occur, primarily by automatic sprinklers but also by trained fire fighters.
5. Make available also for fire fighting an adequate water supply, appropriate chemicals, adequate-size piping, conveniently located valves on the piping, hoses, pumps, and other equipment necessary.

6. Prevent spread of fire from building to building, either through adequate separation or by enclosure of the building with incombustible materials.
7. Partition the interior of the building with fire barriers, or divisions, to confine a fire to a limited space.
8. Enclose with protective materials structural components that may be damaged by fire (fireproofing).
9. Provide refuge areas for occupants and safe evacuation routes to outdoors.
10. Provide means for removal of heat and smoke from the building as rapidly as possible without exposing occupants to these hazards, with the air-conditioning system, if one is present, assisting the removal by venting the building and by pressurizing smokeproof towers, elevator shafts, and other exits.
11. For large buildings, install standby equipment for operation in emergencies of electrical systems and elevators.

These steps are discussed in the following articles.

3.5.3 Fire Loads and Resistance Ratings

The nature and potential magnitude of fire in a building are directly related to the amount and physical arrangement of combustibles present, as contents of the building or as materials used in its construction. Because of this, all codes classify buildings by **occupancy** and **construction,** because these features are related to the amount of combustibles.

The total amount of combustibles is called the **fire load** of the building. Fire load is expressed in pounds per square foot (psf) of floor area, with an assumed calorific value of 7000 to 8000 Btu/lb. (This Btu content applies to organic materials similar to wood and paper. Where other materials are present in large proportion, the weights must be adjusted accordingly. For example, for petroleum products, fats, waxes, alcohol, and similar materials, the weights are taken at twice their actual weights, because of the Btu content.)

National Institute of Standards and Technology burnout tests presented in Report BMS92 indicate a relation between fire load and fire severity as shown in Table 3.2.

The temperatures used in standard fire tests of building components are indicated by the internationally recognized time-temperature curve shown in Fig. 3.9. Fire resistance of construction materials, determined by standard fire tests, is expressed in hours. The Underwriters Laboratories "Building Materials List" tabulates fire ratings for materials and assemblies it has tested.

Every building code specifies required fire-resistance ratings for structural members, exterior walls, fire divisions, fire separations, ceiling-floor assemblies, and any other constructions for which a fire rating is necessary. (Fire protection for structural steel is discussed in Arts. 7.45 to 7.49. Design for fire resistance of steel deck in Arts. 8.21.5 and 8.22.4. Design for fire safety with wood construction is covered in Art. 10.28.)

Building codes also specify the ratings required for interior finish of walls, ceilings, and floors. These are classified as to flame spread, fuel contributed, and smoke developed, determined in standard tests performed according to ASTM E84 or ASTM E119.

TABLE 3.2 Relation between Weight of Combustibles and Fire Severity*

Average weight of combustibles, psf	Equivalent fire severity, hr
5	½
7½	¾
10	1
15	1½
20	2
30	3
40	4½
50	6
60	7½

*Based on National Institute of Standards and Technology Report BMS92, "Classifications of Building Constructions," Government Printing Office, Washington, D.C. 20402.

3.5.4 Fire and Smoke Barriers

A major consideration in building design is safety of the community. Hence, buildings should be designed to control fires and smoke so that they will not spread from building to building.

One way that building codes try to achieve this objective is to establish fire zones or fire limits that restrict types of construction or occupancy that can be used. Additional zoning regulations establish minimum distances between buildings. Another way to achieve the objective is to specify the types of construction that can be used for enclosing the exterior of buildings. The distance between adjoining buildings, fire rating, and stability when exposed to fire of exterior walls, windows, and doors, and percent of window area are some of the factors taken into account in building codes for determination of the construction classification of a building.

To prevent spread of fire from roof to roof, building codes also often require that exterior walls extend as a parapet at least 3 ft above the roof level. Parapets also are useful in shielding fire fighters who may be hosing a fire from roofs of buildings adjoining the one on fire. In addition, buildings should be topped with roof coverings that are fire-resistant.

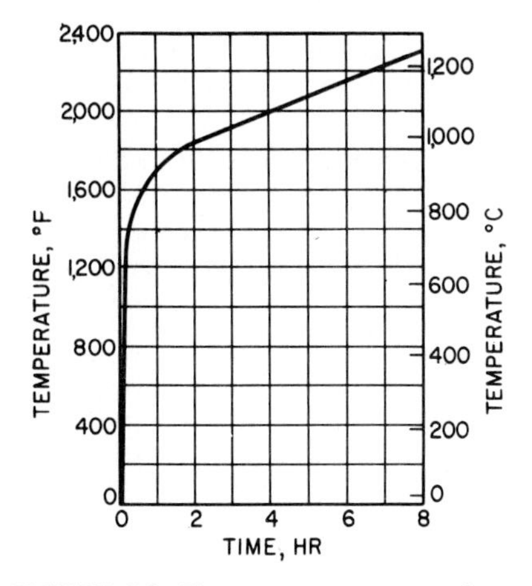

FIGURE 3.9 Time-temperature curve for a standard fire test.

Fire Divisions. To prevent spread of fire vertically in building interiors, building codes generally require that floor-ceiling and roof-ceiling assemblies be fire-resistant. The fire rating of such assemblies is one of the factors considered in determination of the construction classification of a building. Also, openings in floors and roofs should be fire-protected, although building codes do not usually require this for one-story or two-story dwellings. For the purpose, an opening, such as that for a stairway, may be protected with a fire-resistant enclosure and fire doors. In particular, stairways and escalator and elevator shafts should be enclosed, not only to prevent spread of fire and smoke but also to provide a protected means of egress from the building for occupants and of approach to the fire source by fire fighters.

To prevent spread of fire and smoke horizontally in building interiors, it is desirable to partition interiors with fire divisions. A **fire division** is any construction with the fire-resistance rating and structural stability under fire conditions required for the type of occupancy and construction of the building to bar the spread of fire between adjoining buildings or between parts of the same building on opposite sides of the division. A fire division may be an exterior wall, fire window, fire door, fire wall, ceiling, or firestop.

A fire wall should be built of incombustible material, have a fire rating of at least 4 hr, and extend continuously from foundations to roof. Also, the wall should have sufficient structural stability in a fire to allow collapse of construction on either side without the wall collapsing. Building codes restrict the size of openings that may be provided in a fire wall and require the openings to be fire-protected (Art. 14.3).

To prevent spread of fire through hollow spaces, such spaces should be firestopped. A **firestop** is a solid or compact, tight closure set in a hollow, concealed space in a building to retard spread of flames, smoke, or hot gases. All partitions and walls should be firestopped at every floor level, at the top-story ceiling level, and at the level of support for roofs. Also, very large unoccupied attics should be subdivided by firestops into areas of 3000 ft^2 or less. Similarly, any large concealed space between a ceiling and floor or roof should be subdivided. For the purpose, firestops extending the full depth of the space should be placed along the line of supports of structural members and elsewhere, if necessary, to enclose areas not exceeding 1000 ft^2 when situated between a floor and ceiling or 3000 ft^2 when located between a ceiling and roof.

Openings between floors for pipes, ducts, wiring, and other services should be sealed with the equal of positive firestops. Partitions between each floor and a suspended ceiling above are not generally required to be extended to the slab above unless this is necessary for required compartmentation. But smoke stops should be provided at reasonable intervals to prevent passage of smoke to noninvolved areas.

3.5.5 Height and Area Restrictions

Limitations on heights and floor areas included between fire walls in any story of a building are given in every building code and are directly related to occupancy and construction. From the standpoint of fire protection, these provisions are chiefly concerned with safety to life. They endeavor to ensure this through requirements determining minimum number of exits, proper location of exits, and maximum travel distance (hence escape time) necessary to reach a place of refuge. The limitations are also aimed at limiting the size of fires.

Unlimited height and area are permitted for the most highly fire-resistant type of construction. Permissible heights and areas are decreased with decrease in fire resistance of construction. Area permitted between fire walls in any story reduces to 6000 ft^2 for a one-story, wood-frame building.

Installation of automatic sprinklers increases permissible heights and areas in all classes, except those allowed unlimited heights and areas.

Permissible unlimited heights and areas in fire-resistive buildings considered generally satisfactory in the past may actually not be safe. A series of fires involving loss of life and considerable property damage opened the fire safety of such construction to question. As a result, some cities have made more stringent the building-code regulations applicable to high-rise buildings.

Many building codes prohibit floor areas of unlimited size unless the building is sprinklered. Without automatic sprinklers, floor areas must be subdivided into fire-wall-protected areas of from 7500 to 15,000 ft^2 and the enclosing fire walls must have 1- or 2-hr fire ratings, depending on occupancy and construction.

("Life Safety Handbook" and "Fire Protection Handbook," National Fire Protection Association, Quincy, Mass.)

3.5.6 Fire-Resistance Classification of Buildings

Although building codes classify buildings by occupancy and construction, there is no universal standard for number of classes of either occupancy or construction. Table 3.3 lists some typical occupancy classifications and associates approximate

TABLE 3.3 Approximate Fire Loads for Various Occupancies*

Occupancy class	Typical average fire load including floors and trim, psf
Assembly	10.0
Business	12.6
Educational	7.6
High hazard	†
Industrial	25.0
Institutional	5.7
Mercantile	15–20
Residential	8.8
Storage	30.0

*From National Institute of Standards and Technology Report BMS92, "Classifications of Building Constructions," Government Printing Office, Washington, D.C. 20402.

†Special provisions are made for this class, and hazards are treated for the specific conditions encountered, which might not necessarily be in proportion to the actual fire load.

fire loads with them. This table should be used only as a guide. For a specific project refer to the applicable local code. Note, however, that codes do not relate life-safety hazards to the actual fire load, but deal with them through requirements for exit arrangements, interior finishes, and ventilation.

Types of construction may be classified by a local building code as follows but may have further subdivisions, depending on fire-resistance requirements:

1. Fire-resistive construction
2. Protected noncombustible construction
3. Unprotected noncombustible construction
4. Heavy-timber construction
5. Ordinary construction
6. Wood-frame construction

The required fire resistance varies from 4 hr for exterior bearing walls and interior columns in the highest fire resistive class to 1 hr for walls and none for columns in the wood-frame construction class.

Type of construction affects fire-protection-system design through requirements that structural members as well as contents of buildings be protected.

3.5.7 Extinguishment of Fires

Design of all buildings should include provisions for prompt extinguishment of fires. Apparatus installed for the purpose should take into account the nature and amount of combustible and smoke-producing materials that may be involved in a fire. Such apparatus may range from small, hand-held extinguishers for small fires to hoses attached to a large, pressurized water supply and automatic fire sprinklers. Also desirable are fire and smoke detectors and a protective signaling system that sounds an alarm to alert building occupants and calls fire fighters.

Classes of Fires. For convenience in defining effectiveness of extinguishing media, Underwriters Laboratories, Inc., has developed a classification that separates combustible materials into four types:

1. *Class A fires* involve ordinary combustibles and are readily extinguishable by water or cooling, or by coating with a suitable chemical powder.
2. *Class B fires* involve flammable liquids where smothering is effective and where a cooling agent must be applied with care.
3. *Class C fires* are those in live electrical equipment where the extinguishing agent must be nonconductive. Since a continuing electrical malfunction will keep the fire source active, circuit protection must operate to cut off current flow, after which an electrically conductive agent can be used with safety.
4. *Class D fires* involve metals that burn, such as magnesium, sodium, and powdered aluminum. Special powders are necessary for such fires, as well as special training for operators. These fires should never be attacked by untrained personnel.

Automatic Sprinklers. The most widely used apparatus for fire protection in buildings is the automatic sprinkler system. In one or more forms, automatic sprinklers are effective protection in all occupancy classes. Special treatment and use of additional extinguishing agents, though, may be required in many high-hazard, industrial, and storage occupancies.

Basically, a sprinkler system consists of a network of piping installed at the ceiling or roof and supplied with water from a suitable source. On the piping at systematic intervals are placed heat-sensitive heads, which discharge water when a predetermined temperature is reached at any head. A gate valve is installed in the main supply, and drains are provided. An alarm can be connected to the system so that local and remote signals can be given when the water flows.

Sprinkler systems are suitable for extinguishing all Class A fires and, in many cases, also Class B and C fires. For Class B fires, a sealed (fusible) head system may be used if the flammable liquid is in containers or is not present in large quantity. Sprinklers have a good record for extinguishing fires in garages, for example. An oil-spill fire can be extinguished or contained when the water is applied in the form of spray, as from a sprinkler head. When an oil spill or process-pipe rupture can release flammable liquid under pressure, an open-head (deluge) system may be required to apply a large volume of water quickly and to keep surrounding equipment cool.

For Class C fires, water can be applied to live electrical equipment if it is done in the form of a nonconducting foglike spray. This is usually the most economical way to protect outdoor oil-filled transformers and oil circuit breakers.

Fire protection should be based on complete coverage of the building by the sprinkler system. Partial coverage is rarely advisable, because extinguishing capacity is based on detecting and extinguishing fires in their incipiency, and the system must be available at all times in all places. Systems are not designed to cope with fires that have gained headway after starting in unsprinklered areas.

See also Arts. 14.27 to 14.30.

Standpipes. Hoses supplied with water from standpipes are the usual means of manual application of water to interior building fires. Standpipes are usually designed for this use by the fire department, but they can be used by building fire fighters also.

Standpipes are necessary in buildings higher than those that ground-based fire department equipment can handle effectively. The Standard Building Code requires standpipes in buildings higher than 50 ft. The Uniform Building Code requirement starts at four stories or occupancies over 5000 ft^2 in area and depends on whether automatic sprinklers are installed.

See also Art. 14.31.

Chemical Extinguishment. Fires involving some materials may not be readily extinguished with water alone. When such materials may be present in a building, provision should be made for application of appropriate chemicals.

Foamed chemicals, mostly masses of air- or gas-filled bubbles, formed by chemical or mechanical means, may be used to control fires in flammable liquids. Foam is most useful in controlling fires in flammable liquids with low flash points and low specific gravity, such as gasoline. The mass of bubbles forms a cohesive blanket that extinguishes fire by excluding air and cooling the surface.

Foam clings to horizontal surfaces and can also be used on vertical surfaces of process vessels to insulate and cool. It is useful on fuel-spill fires, to extinguish and confine the vapors.

For fire involving water-soluble liquids, such as alcohol, a special foam concentrate must be used. Foam is not suitable for use on fires involving compressed gases, such as propane, nor is it practical on live electrical equipment. Because of the water content, foam cannot be used on fires involving burning metals, such as sodium, which reacts with water. It is not effective on oxygen-containing materials.

Three distinct types of foam are suitable for fire control: chemical foam, air foam (mechanical foam), and high-expansion foam.

Chemical foam was the first foam developed for fire fighting. It is formed by the reaction of water with two chemical powders, usually sodium bicarbonate and aluminum sulfate. The reaction forms carbon dioxide, which is the content of the bubbles. This foam is the most viscous and tenacious of the foams. It forms a relatively tough blanket, resistant to mechanical or heat disruption. The volume of expansion may be as much as 10 times that of the water used in the solution.

Chemical foam is sensitive to the temperature at which it is formed, and the chemicals tend to deteriorate during long storage periods. It is not capable of being transported through long pipe lines. For these reasons, it is not used as much as other foams. National Fire Protection Association standard NEPA 11 covers chemical foam.

Air foam (mechanical foam) is made by mechanical mixing of water and a protein-based chemical concentrate. There are several methods of combining the components, but essentially the foam concentrate is induced into a flowing stream of water through a metering orifice and a suitable device, such as a venturi. The volume of foam generated is from 16 to 33 times the volume of water used. Several kinds of mixing apparatus are available, choice depending on volume required, availability of water, type of hazard, and characteristics of the protected area or equipment.

Air foam can be conducted through pipes and discharged through a fixed chamber mounted in a bulk fuel storage tank, or it can be conducted through hoses and discharged manually through special nozzles. This foam can also be distributed through a sprinkler system of special design to cover small equipment, such as process vessels, or in multisystem applications, over an entire airplane hangar. The standard for use and installation of air foam is NFPA 11, and for foam-water sprinkler systems, NFPA 16.

High-expansion foam was developed for use in coal mines, where its extremely high expansion rate allowed it to be generated quickly in sufficient volume to fill mine galleries and reach inaccessible fires. This foam can be generated in volumes of from 100 to 1000 times the volume of water used, with the latter expansion in most general use. The foam is formed by passage of air through a screen constantly wetted by a solution of chemical concentrate, usually with a detergent base. The foam can be conducted to a fire area by ducts, either fixed or portable, and can be applied manually by small portable generators. Standard for equipment and use of high-expansion foam is NFPA 11A.

High-expansion foam is useful for extinguishing fires by totally flooding indoor confined spaces, as well as for local application to specific areas. It extinguishes by displacing air from the fire and by the heat-absorbing effect of converting the foam water content into steam. The foam forms an insulating barrier for exposed equipment or building components.

High-expansion foam is more fragile than chemical or air foam. Also, it is not generally reliable when used outdoors where it is subject to wind currents. High-expansion foam is not toxic, but it has the effect of disorienting people who may be trapped in it.

Carbon dioxide is useful as an extinguishing agent, particularly on surface fires, such as those involving flammable liquids in confined spaces. It is nonconductive and is effective on live electrical equipment. Because carbon dioxide requires no clean-up, it is desirable on equipment such as gasoline or diesel engines. The gas can be used on Class A fires. But when a fire is deep-seated, an extended discharge period is required to avoid rekindling.

Carbon dioxide provides its own pressure for discharge and distribution and is nonreactive with most common industrial materials. Because its density is 1½ times that of air, carbon dioxide tends to drop and to build up from the base of a fire. Extinguishment of a fire is effected by reduction of the oxygen concentration surrounding a fire.

Carbon dioxide may be applied to concentrated areas or machines by hand-held equipment, either carried or wheeled. Or the gas may be used to flood totally a room containing a hazard. The minimum concentrations for total flooding for fires involving some commercial liquids are listed in "Standard on Carbon-Dioxide Extinguishing Systems," NFPA 12.

Carbon dioxide is not effective on fires involving burning metals, such as magnesium, nor is it effective on oxygen-containing materials, such as nitrocellulose. Hazard to personnel is involved to the extent that a concentration of 9% will cause suffocation in a few minutes, and concentrations of 20% can be fatal. When used in areas where personnel are present, a time delay before discharge is necessary to permit evacuation.

For use in total flooding systems, carbon dioxide is available in either high-pressure or low-pressure equipment. Generally, it is more economical to use low-pressure equipment for large volumes, although there is no division point applicable in all cases.

Halon 1301 is one of a series of halogenated hydrocarbons, bromotrifluoromethane ($CBrF_3$), used with varying degrees of effectiveness as fire-extinguishing agents. Its use, in general, is the same as that of carbon dioxide. It also needs no clean-up. Its action is quicker than carbon dioxide, however, because Halon 1301 does not have to displace the air from around the fire.

At room temperatures, Halon 1301 is a colorless, odorless gas with a density 5 times that of air. The gas is stored in pressure vessels, similar to those for carbon dioxide. These vessels can be manifolded for large systems. Because the vapor pressure varies widely with temperature of Halon 1301, the containers are superpressurized with dry nitrogen to 600 psig or 360 psig.

Action of the gas as an extinguisher is considered to be that of a chemical inhibitor, interfering with the chain reaction necessary to maintain combustion. The action is almost instantaneous. Required concentrations are substantially less than those for carbon dioxide.

Because of the low concentration provided, and the density of Halon 1301, the discharge system must provide a thorough mixing of the agent with the air surrounding the fire. Flame extinguishment is very rapid. This characteristic is aided by quick detection of fires, indicating the need for an automatically actuated system.

If rapid egress is possible, there should be no prohibition on use of Halon 1301 in normally occupied areas in concentrations less than 7%. It is considered by Underwriters Laboratories to be about half as toxic as carbon tetrachloride, which was used for many years as a fire-extinguishing agent. The decomposition products of Halon 1301, however, are irritating, and prolonged exposure of the gas to temperatures above 900°F will cause the formation of toxic fluorides and bromides.

Dry chemical extinguishing agents were used originally to extinguish Class B fires. One type consisted of a sodium bicarbonate base with additives to prevent caking and to improve fluid flow characteristics. Later, multipurpose dry chemicals effective on Class A, B, and C fires were developed. These chemicals are distinctly different from the dry powder extinguishing agents used on combustible metals described below.

Dry chemicals are effective on surface fires, especially on flammable liquids. When used on Class A fires, they do not penetrate into the burning material. So when a fire involves porous or loosely packed material, water is used as a backup. The major effect of dry chemicals is due almost entirely to ability to break the chain reaction of combustion. A minor effect of smothering is obtained on Class A fires.

Fires that are likely to rekindle are not effectively controlled by dry chemicals. When these chemicals are applied to machinery or equipment at high temperatures, caking can cause some difficulty in cleaning up after the fire.

Dry chemicals can be discharged in local applications by hand-held extinguishers, wheeled portable equipment, or nozzles on hose lines. These chemicals can also be used for extinguishing fires by total flooding, when they are distributed through a piped system with special discharge nozzles. The expellant gas is usually dry nitrogen.

Dry powder extinguishing agents are powders effective in putting out combustible-metal fires. There is no universal extinguisher that can be used on all fires involving combustible metals. Such fires should never be fought by untrained personnel.

There are several proprietary agents effective on several metals, but none should be used without proper attention to the manufacturer's instructions and the specific metal involved. For requirements affecting handling and processing of combustible metals, reference should be made to National Fire Protection Association standards NFPA 48 and 652 for magnesium, NFPA 481 for titanium, NFPA 482M for zirconium, and NFPA 65 and 651 for aluminum.

("The SFPE Handbook of Fire Protection Engineering," and "Automatic Sprinkler Systems Handbook," National Fire Protection Association, Quincy, Mass.)

3.5.8 Fire Detection

Every fire-extinguishing activity must start with detection. To assist in this, many types of automatic detectors are available, with a wide range of sensitivity. Also, a variety of operations can be performed by the detection system. It can initiate an alarm, local or remote, visual or audible; notify a central station; actuate an extinguishing system; start or stop fans or processes, or perform any other operation capable of automatic control.

There are five general types of detectors, each employing a different physical means of operation. The types are designated *fixed-temperature, rate-of-rise, photoelectric, combustion-products,* and *ultraviolet* or *infrared* detectors.

A wide variety of detectors has been tested and reported on by Underwriters Laboratories, Inc. See Art. 3.5.1.

Fixed-Temperature Detectors. In its approval of any detection device, UL specifies the maximum distance between detectors to be used for area coverage. This spacing should not be used without competent judgment. In arriving at the permitted spacing for any device, UL judges the response time in comparison with that of automatic sprinkler heads spaced at 10-ft intervals. Thus, if a device is more sensitive than a sprinkler head, the permitted spacing is increased until the response times are nearly equal. If greater sensitivity is desired, the spacing must be reduced.

With fixed-temperature devices, there is a thermal lag between the time the ambient temperature reaches rated temperature and the device itself reaches that temperature. For thermostats having a rating of 135°F, the ambient temperature can reach 206°F.

Disk thermostats are the cheapest and most widely used detectors. The most common type employs the principle of unequal thermal expansion in a bimetallic assembly to operate a snap-action disk at a preset temperature, to close electrical contacts. These thermostats are compact. The disk, ½ in in diameter, is mounted on a plastic base 1¾ in in diameter. The thermostats are self-resetting, the contacts being disconnected when normal temperature is restored.

Thermostatic cable consists of two sheathed wires separated by a heat-sensitive coating which melts at high temperature, allowing the wires to contact each other. The assembly is covered by a protective sheath. When any section has functioned, it must be replaced.

Continuous detector tubing is a more versatile assembly. This detector consists of a small-diameter Inconel tube, of almost any length, containing a central wire, separated from the tube by a thermistor element. At elevated temperatures, the resistance of the thermistor drops to a point where a current passes between the wire and the tube. The current can be monitored, and in this way temperature changes over a wide range, up to 1000°F, can be detected. The detector can be assembled to locate temperature changes of different magnitudes over the same length of detector. It is self-restoring when normal temperature is restored. This detector is useful for industrial applications, as well as for fire detection.

Fusible links are the same devices used in sprinkler heads and are made to operate in the same temperature range. Melting or breaking at a specific temperature, they are used to restrain operation of a fire door, electrical switch, or similar mechanical function, such as operation of dampers. Their sensitivity is substantially reduced when installed at a distance below a ceiling or other heat-collecting obstruction.

Rate-of-Rise Detectors. Detectors and detector systems are said to operate on the rate-of-rise principle when they function on a rapid increase in temperature, whether the initial temperature is high or low. The devices are designed to operate when temperature rises at a specified number of degrees, usually 10 or 15°F, per minute. They are not affected by normal temperature increases and are not subject to thermal lag, as are fixed-temperature devices.

Photoelectric Detectors. These indicate a fire condition by detecting the smoke. Sensitivity can be adjusted to operate when obscuration is as low as 0.4% per ft. In these devices, a light source is directed so that it does not impinge on a photoelectric cell. When sufficient smoke particles are concentrated in the chamber, their reflected light reaches the cell, changing its resistance and initiating a signal.

These detectors are particularly useful when a potential fire is likely to generate a substantial amount of smoke before appreciable heat and flame erupt. A fixed-temperature, snap-action disk is usually included in the assembly.

Combustion-Products Detectors. Two physically different means, designated ionization type and resistance-bridge type, are used to operate combustion-products detectors.

The **ionization type,** most generally used, employs ionization of gases by alpha particles emitted by a small quantity of radium or americum. The detector contains two ionization chambers, one sealed and the other open to the atmosphere, in electrical balance with a cold-cathode tube or transistorized amplifier. When sufficient combustion products enter the open chamber, the electrical balance is upset, and the resulting current operates a relay.

The **resistance-bridge type** of detector operates when combustion products change the impedance of an electric bridge grid circuit deposited on a glass plate.

Combustion-products detectors are designed for extreme early warning, and are most useful when it is desirable to have warning of impending combustion when combustion products are still invisible. These devices are sensitive in some degree to air currents, temperature, and humidity, and should not be used without consultation with competent designers.

Flame Detectors. These discriminate between visible light and the light produced by combustion reactions. Ultraviolet detectors are responsive to flame having wavelengths up to 2850 Å. The effective distance between flame and detectors is about 10 ft for a 5-in-diam pan of gasoline, but a 12-in-square pan fire can be detected at 30 ft.

Infrared detectors are also designed to detect flame. These are not designated by range of wavelength because of the many similar sources at and above the infrared range. To identify the radiation as a fire, infrared detectors usually employ the characteristic flame flicker, and have a built-in time delay to eliminate accidental similar phenomena.

("The SFPE Handbook of Fire Detection Engineering," National Fire Protection Association, Quincy, Mass.)

3.5.9 Smoke and Heat Venting

In extinguishment of any building fire, the heat-absorption capacity of water is the principal medium of reducing the heat release from the fire. When, however, a fire is well-developed, the smoke and heat must be released from confinement to make the fire approachable for final manual action. If smoke and heat venting is not provided in the building design, holes must be opened in the roof or building sides by the fire department. In many cases, it has been impossible to do this, with total property losses resulting.

Large-area, one-story buildings can be provided with venting by use of monitors, or a distribution of smaller vents. Multistory buildings present many problems, particularly since life safety is the principal consideration in these buildings.

Ventilation facilities should be provided in addition to the protection afforded by automatic sprinklers and hose stations.

Large One-Story Buildings. For manufacturing purposes, low buildings are frequently required to be many hundreds of feet in each horizontal dimension. Lack of automatic sprinklers in such buildings has proven to be disastrous where adequate smoke and heat venting has not been provided. Owners generally will not permit fire division walls, because they interfere with movement and processing of materials. With the whole content of a building subject to the same fire, fire protection and venting are essential to prevent large losses in windowless buildings, underground structures, and buildings housing hazardous operations.

There is no accepted formula for determining the exact requirements for smoke and heat venting. Establishment of guide lines is the nearest approach that has been made to venting design, and these must be adapted to the case at hand. Consideration must be given to quantity, shape, size, and combustibility of contents.

Venting Ratios. The ratio of effective vent opening to floor area should be at least that given in Table 3.4.

TABLE 3.4 Minimum Ratios of Effective Vent Area to Floor Area

Low-heat-release contents	1:150
Moderate-heat-release contents	1:100
High-heat-release contents	1:30–1:50

TABLE 3.5 Maximum Distance between Vents, Ft

Low-heat-release contents	150
Moderate-heat-release contents	120
High-heat-release contents	75–100

Venting can be accomplished by use of monitors, continuous vents, unit-type vents, or sawtooth skylights. In moderate-sized buildings exterior-wall windows may be used if they are near the eaves.

Monitors must be provided with operable panels or other effective means of providing openings at the required time.

Continuous gravity vents are continuous narrow slots provided with a weather hood above. Movable shutters can be provided and should be equipped to open automatically in a fire condition.

Vent Spacing. Unit-type vents are readily adapted to flat roofs, and can be installed in any required number, size, and spacing. They are made in sizes from 4 × 4 ft to 10 × 10 ft, with a variety of frame types and means of automatic opening. In arriving at the number and size of vents, preference should be given to a large number of small vents, rather than a few large vents. Because it is desirable to have a vent as near as possible to any location where a fire can start, a limit should be placed on the distance between units. Table 3.5 lists the generally accepted maximum distance between vents.

Releasing Methods. Roof vents should be automatically operated by means that do not require electric power. They also should be capable of being manually operated. Roof vents approved by Underwriters Laboratories, Inc., are available from a number of manufacturers.

Refer to National Fire Protection Association standard NFPA 204 in designing vents for large, one-story buildings. Tests conducted prior to publication of NFPA 231C indicated that a sprinkler system designed for adequate density of water application will eliminate the need for roof vents, but the designers would be well advised to consider the probable speed of fire and smoke development in making a final decision. NFPA 231C covers the rack storage of materials as high as 20 ft.

High-Rise Buildings. Building codes vary in their definition of high-rise buildings, but the intent is to define buildings in which fires cannot be fought successfully by ground-based equipment and personnel. Thus, ordinarily, high-rise means buildings 100 ft or more high. In design for smoke and heat venting, however, any multistory building presents the same problems.

Because smoke inhalation has been the cause of nearly all fatalities in high-rise buildings, some building codes require that a smoke venting system be installed and made to function independently of the air-conditioning system. Also, smoke detectors must be provided to actuate exhaust fans and at the same time warn the fire department and the building's control center. The control center must have two-way voice communication, selectively, with all floors and be capable of issuing instructions for occupant movement to a place of safety.

Because the top story is the only one that can be vented through the roof, all other stories must have the smoke conducted through upper stories to discharge safely above the roof. A separate smoke shaft extending through all upper stories will provide this means. It should be provided with an exhaust fan and should be connected to return-air ducts with suitable damper control of smoke movement, so that smoke from any story can be directed into the shaft. The fan and dampers should be actuated by smoke detectors installed in suitable locations at each inlet to return-air ducts. Operation of smoke detectors also should start the smoke-vent-shaft fan and stop supply-air flow. Central-station supervision (Art. 3.5.12) should be provided for monitoring smoke-detector operation. Manual override controls should be installed in a location accessible under all conditions.

Windows with fixed sash should be provided with means for emergency opening by the fire department.

Pressurizing stair towers to prevent the entrance of smoke is highly desirable but difficult to accomplish. Most standpipe connections are usually located in stair towers, and it is necessary to open the door to the fire floor to advance the hose stream toward the fire. A more desirable arrangement would be to locate the riser in the stair tower, if required by code, and place the hose valve adjacent to the door to the tower. Some codes permit this, and it is adaptable to existing buildings.

("The SFPE Handbook of Fire Protection Engineering," National Fire Protection Association, Quincy, Mass.)

3.5.10 Emergency Egress

In addition to providing means for early detection of fire, preventing its spread, and extinguishing it speedily, building designers should also provide the appropriate number, sizes, and arrangements of exits to permit quick evacuation of occupants if fire or other conditions dangerous to life occur. Buildings should be designed to preclude development of panic in emergencies, especially in confined areas where large numbers of persons may assemble. Hence, the arrangement of exit facilities should permit occupants to move freely toward exits that they can see clearly and that can be reached by safe, unobstructed, uncongested paths. Redundancy is highly desirable; there should be more than one path to safety, so that loss of a single path will not prevent escape of occupants from a danger area. The paths should be accessible to and usable by handicapped persons, including those in wheelchairs, if they may be occupants.

Building codes generally contain requirements for safe, emergency egress from buildings. Such requirements also are concisely presented in the "Life Safety Code" of the National Fire Protection Association.

Egress Components. Many building codes define an exit as a safe means of egress from the interior of a building to an open exterior space beyond the reach of a building fire or give an equivalent definition. Other codes consider an exterior door or a stairway leading to access to such a door to be an exit. To prevent misunderstandings, the "Life Safety Code" defines a means of egress composed of three parts.

Accordingly, a **means of egress** is a continuous, unobstructed path for evacuees from any point in a building to a public way. Its three parts are:

Exit access—that portion that leads to an entrance to an exit

Exit—the portion that is separated from all other building spaces by construction or equipment required to provide a protected path to the exit discharge

Exit discharge—the portion that connects the termination of an exit to a public way

Means of egress may be provided by exterior and interior doors and enclosed horizontal and vertical passageways, including stairs and escalators. (Elevators and exterior fire escapes are not generally recognized as reliable means of egress in a fire.) Exit access includes the space from which evacuation starts and passageways and doors that must be traversed to reach an exit.

Types of Exits. Building codes generally recognize the following as acceptable exits when they meet the codes' safety requirements:

Corridors—enclosed horizontal or slightly inclined public passageways, which lead from interior spaces toward an exit discharge. Minimum floor-to-ceiling height permitted is generally 80 in. Minimum width depends on type of occupancy and passageway (Table 3.7 and Art. 3.5.11). Codes may require subdivision of corridors into lengths not exceeding 300 ft for educational buildings and 150 ft for institutional buildings. Subdivision should be accomplished with noncombustible partitions incorporating smokestop doors. In addition, codes may require the corridor enclosures to have a fire rating of 1 or 2 hr.

Exit passageways—horizontal extensions of vertical passageways. Minimum floor-to-ceiling height is the same as for corridors. Width should be at least that of the vertical passageways. Codes may require passageway enclosures to have a 2-hr fire rating. A street-floor lobby may serve as an exit passageway if it is sufficiently wide to accommodate the probable number of evacuees from all contributing spaces at the lobby level.

Exit doors—doors providing access to streets or to stairs or exit passageways. Those at stairs or passageways should have a fire rating of at least ¾ hr.

Horizontal exit—passageway to a refuge area. The exit may be a fire door through a wall with a 2-hr fire rating, a balcony providing a path around a fire barrier, or a bridge or tunnel between two buildings. Doors in fire barriers with 3- or 4-hr fire ratings should have a 1½-hr rated door on each face of the fire division. Walls permitted to have a lower fire rating may incorporate a single door with a rating of at least 1½ hr. Balconies, bridges, and tunnels should be at least as wide as the doors providing access to them, and enclosures or sides of these passageways should have a fire rating of 2 hr or more. Exterior-wall openings below or within 30 ft of an open bridge or balcony should have at least ¾-hr fire protection.

Interior stairs—stairs that are inside a building and that serve as an exit. Except in one-story or two-story low-hazard buildings, such stairs should be built of noncombustible materials. Stairway enclosures generally should have a 2-hr fire rating. Building codes, however, may exempt low dwellings from this requirement.

Exterior stairs—stairs that are open to the outdoors and that serve as an exit to ground level. Height of such stairs is often limited to 75 ft or six stories. The stairs should be protected by a fire-resistant roof and should be built of noncombustible materials. Wall openings within 10 ft of the stairs should have ¾-hr fire protection.

Smokeproof tower—a continuous fire-resistant enclosure protecting a stairway from fire or smoke in a building. At every floor, a passageway should be provided by vestibules or balconies directly open to the outdoors and at least 40 in wide. Tower enclosures should have a 2-hr fire rating. Access to the vestibules or balconies and entrances to the tower should be provided by doorways at least 40 in wide, protected by self-closing fire doors.

Escalators—moving stairs. Building codes may permit their use as exits if they meet the safety requirements of interior stairs and if they move in the direction of exit travel or stop gradually when an automatic fire-detection system signals a fire.

Moving walks—horizontal or inclined conveyor belts for passengers. Building codes may permit their use as exits if they meet the safety requirements for exit passageways and if they move in the direction of exit travel or stop gradually when an automatic fire-detection system signals a fire.

Refuge Areas. A refuge area is a space protected against fire and smoke. When located within a building, the refuge should be at about the same level as the areas served and separated from them by construction with at least a 2-hr fire rating. Access to the refuge areas should be protected by fire doors with a fire rating of 1½ hr or more.

A refuge area should be large enough to shelter comfortably its own occupants plus those from other spaces served. The minimum floor area required may be calculated by allowing 3 ft^2 of unobstructed space for each ambulatory person and 30 ft^2 per person for hospital or nursing-home patients. Each refuge area should be provided with at least one horizontal or vertical exit, such as a stairway, and in locations more than 11 stories above grade, with at least one elevator.

Location of Exits. Building codes usually require a building to have at least two means of egress from every floor. Exits should be remote from each other, to reduce the chance that both will be blocked in an emergency.

All exit access facilities and exits should be located so as to be clearly visible to building occupants or signs should be installed to indicate the direction of travel to the exits. Signs marking the locations of exits should be illuminated with at least 5 ft-c of light. Floors of means of egress should be illuminated with at least 1 ft-c of artificial light whenever the building is occupied.

If an open floor area does not have direct access to an exit, a protected, continuous passageway should be provided directly to an exit. The passageway should be kept open at all times. Occupants using the passageway should not have to pass any high-hazard areas not fully shielded.

To ensure that occupants will have sufficient escape time in emergencies, building codes limit the travel distance from the most remote point in any room or space to a door that opens to an outdoor space, stairway, or exit passageway. The maximum travel distance permitted depends on the type of occupancy and whether the space is sprinklered. For example, for corridors not protected by sprinklers, maximum permitted length may range from 100 ft for storage and institutional buildings to 150 ft for residential, mercantile, and industrial occupancies. With sprinkler protection, permitted length may range from 150 ft for high-hazard and storage buildings to 300 ft for commercial buildings, with 200 ft usually permitted for other types of occupancies.

Building codes also may prohibit or limit the lengths of passageways or courts that lead to a dead end. For example, a corridor that does not terminate at an exit is prohibited in high-hazard buildings. For assembly, educational, and institutional buildings, the maximum corridor length to a dead end may not exceed 30 ft, whereas the maximum such length is 40 ft for residential buildings and 50 ft for all other occupancies, except high-hazard.

3.5.11 Required Exit Capacity

Minimum width of a passageway for normal use is 36 in. This is large enough to accommodate one-way travel for persons on crutches or in wheelchairs. For two-way travel, a 60-in width is necessary. (A corridor, however, need not be 60 in wide for its full length, if 60 × 60-in passing spaces, alcoves, or corridor intersections are provided at short intervals.) Building codes, however, may require greater widths to permit rapid passage of the anticipated number of evacuees in emergencies. This number depends on a factor called the occupant load, but the minimum width should be ample for safe, easy passage of handicapped persons. Running slope should not exceed 1:20, and cross slope, 1:50.

Occupant load of a building space is the maximum number of persons that may be in the space at any time. Building codes may specify the minimum permitted capacity of exits in terms of occupant load, given as net floor area, square feet, per person, for various types of occupancy (Table 3.6). The number of occupants permitted in a space served by the exits then can be calculated by dividing the floor area, square feet, by the specified occupant load.

The occupant load of any space should include the occupant load of other spaces if the occupants have to pass through that space to reach an exit.

With the occupant load known, the required width for an exit or an exit door can be determined by dividing the occupant load on the exit by the capacity of the exit.

Capacities of exits and access facilities generally are measured in units of width of 22 in, and the number of persons per unit of width is determined by the type of occupancy. Thus, the number of units of exit width for a doorway is found by dividing by 22 the clear width of the doorway when the door is in the open position. (Projections of stops and hinge stiles may be disregarded.) Fractions of a unit of width less than 12 in should not be credited to door capacity. If, however, 12 in or more is added to a multiple of 22 in, one-half unit of width can be credited. Building codes indicate the capacities in persons per unit of width that may be assumed for various means of egress. Recommendations of the "Life Safety Code" of the National Fire Protection Association, Batterymarch Park, Quincy, MA 02269, are summarized in Table 3.7.

3.5.12 Building Operation in Emergencies

For buildings that will be occupied by large numbers of persons, provision should be made for continuation of services essential to safe, rapid evacuation of occupants in event of fire or other emergencies and for assisting safe movement of fire fighters, medical personnel, or other aides.

Standby electric power, for example, should be available in all buildings to replace the basic power source if it should fail. The standby system should be equipped with a generator that will start automatically when normal power is cut off. The emergency power supply should be capable of operating all emergency electric equipment at full power within 1 min of failure of normal service. Such equipment includes lights for exits, elevators for fire fighters' use, escalators and moving walks designated as exits, exhaust fans and pressurizing blowers, communication systems, fire detectors, and controls needed for fire fighting and life safety during evacuation of occupants.

In high-rise buildings, at least one elevator should be available for control by fire fighters and to give them access to any floor from the street-floor lobby. Also, elevator controls should be designed to preclude elevators from stopping automatically at floors affected by fire.

Supervision of emergency operations can be efficiently provided by personnel at a control center placed in a protected area. This center may include a computer, supplemented by personnel performing scheduled maintenance, and should be

TABLE 3.6 Typical Occupant Load Requirements for Types of Occupancy

Occupancy	Net floor area per occupant, ft^2
Auditoriums	7
Billiard rooms	50
Bowling alleys	50
Classrooms	20
Dance floors	7
Dining spaces (nonresidential)	12
Exhibition spaces	10
Garages and open parking structures	250
Gymnasiums	15
Habitable rooms	200
Industrial shops	200
In schools	50
Institutional sleeping rooms	120
Kindergartens	35
Kitchens (nonresidential)	200
Laboratories	50
Preparation rooms	100
Libraries	25
Locker rooms	12
Offices	100
Passenger terminals or platforms	1.5C*
Sales areas (retail)	
First floor or basement	30
Other floors	60
Seating areas (audience) in places of assembly	
Fixed seats	D†
Movable seats	10
Skating rinks	15
Stages	S‡
Storage rooms	300

*C = capacity of all passenger vehicles that can be unloaded simultaneously.

†D = number of seats or occupants for which space is to be used.

‡S = 75 persons per unit of width of exit openings serving a stage directly, or one person per 15 ft^2 of performing area plus one person per 50 ft^2 of remaining area plus number of seats that may be placed for an audience on stage.

TABLE 3.7 Capacities, Persons per Unit of Width, for Means of Egress

Level egress components, including doors	100
Stairway	60
Ramps 44 in or more wide, slope not more than 10%	100
Narrower or steeper ramps	
Up	60
Down	100

capable of continuously monitoring alarms, gate valves on automatic fire sprinklers, temperatures, air and water pressures, and perform other pertinent functions. Also, the center should be capable in emergencies of holding two-way conversations with occupants and notifying police and fire departments of the nature of the emergencies. In addition, provision should be made for the control center to dispatch investigators to sources of potential trouble or send maintenance personnel to make emergency repairs when necessary. Standards for such installations are NFPA 72A, "Local Protective Signaling Systems," NFPA 72B, "Auxiliary Protective Signaling Systems," NFPA 72C, "Remote Station Protective Signaling Systems," and NFPA 72D, "Proprietary Protective Signaling Systems." See also Art. 3.7.2.

For economical building operation, the emergency control center may be made part of a control center used for normal building operation and maintenance. Thus, the control center may normally control HVAC to conserve energy, turn lights on and off, and schedule building maintenance and repair. When an emergency occurs, emergency control should be activated in accordance with prepared plans for handling each type of emergency.

The control center need not be located within the building to be supervised nor operated by in-house personnel. Instead, an external central station may provide the necessary supervision. Such services are available in most cities and are arranged by contract, usually with an installation charge and an annual maintenance charge. Requirements for such systems are in National Fire Protection Association standard NFPA 71.

3.5.13 Safety during Construction

Most building codes provide specific measures that must be taken for fire protection during construction of buildings. But when they do not, fundamental fire-safety precautions must be taken. Even those structures that will, when completed, be noncombustible contain quantities of forming and packing materials that present a serious fire hazard.

Multistory buildings should be provided with access stairways and, if applicable, an elevator for fire department use. Stairs and elevator should follow as closely as possible the upward progress of the structure and be available within one floor of actual building height. In buildings requiring standpipes, the risers should be placed in service as soon as possible, and as close to the construction floor as practicable. Where there is danger of freezing, the water supply can consist of a Siamese connection for fire department use.

In large-area buildings, required fire walls should be constructed as soon as possible. Competent watchman service also should be provided.

The greatest source of fires during construction is portable heaters. Only the safest kind should be used, and these safeguarded in every practical way. Fuel supplies should be isolated and kept to a minimum.

Welding operations also are a source of fires. They should be regulated in accordance with building-code requirements.

Control of tobacco smoking is difficult during building construction, so control of combustible materials is necessary. Good housekeeping should be provided, and all combustible materials not necessary for the work should be removed as soon as possible.

Construction offices and shanties should be equipped with adequate portable extinguishers. So should each floor in a multistory building.

3.6 LIGHTNING PROTECTION

Lightning, a high-voltage, high-current electrical discharge between clouds and the ground, may strike and destroy life and property anywhere thunderstorms have occurred in the past. Buildings and their occupants, however, can be protected against this hazard by installation of a special electrical system. Because an incomplete or poor installation can cause worse damage or injuries than no protection at all, a lightning-protection system should be designed and installed by experts.

As an addition to other electrical systems required for a building, a lightning-protection system increases the construction cost of a building. A building owner therefore has to decide whether potential losses justify the added expenditure. In doing so, the owner should take into account the importance of the building, danger to occupants, value and nature of building contents, type of construction, proximity of other structures or trees, type of terrain, height of building, number of days per year during which thunderstorms may occur, costs of disruption of business or other activities and the effects of loss of essential services, such as electrical and communication systems. (Buildings housing flammable or explosive materials generally should have lightning protection.) Also, the owner should compare the cost of insurance to cover losses with the cost of the protection system.

3.6.1 Characteristics of Lightning

Lightning strikes are associated with thunderstorms. In such storms, the base of the clouds generally develops a negative electrical charge, which induces a positive charge in the earth directly below. As the clouds move, the positive charges, being attracted by the negative charges, follow along the surface of the earth and climb up buildings, antennas, trees, power transmission towers, and other conducting or semiconducting objects along the path. The potential between clouds and earth may build up to 10^6 to 10^9 V. When the voltage becomes great enough to overcome the electrical resistance of the air between the clouds and the ground or an object on it, current flows in the form of a lightning flash. Thus, the probability of a building being struck by lightning depends not only on the frequency of occurrence of thunderstorms but also on building height relative to nearby objects and the intensity of cloud charges.

Destruction at the earth's surface may result not only at points hit by lightning directly but also by electrostatic induction at points several feet away. Also, lightning striking a tall object may flash to a nearby object that offers a suitable path to the ground.

Lightning often shatters nonconductors or sets them on fire if they are combustible. Conductors struck may melt. Living things may be burned or electrocuted. Also, lightning may induce overvoltages in electrical power lines, sending electrical charges along the lines in both directions from the stricken point to ground. Direct-stroke overvoltages may range up to several million volts and several hundred thousand amperes. Induced strokes, which occur more frequently, may be on the order of several hundred thousand volts with currents up to 2000 A. Such overvoltages may damage not only electric equipment connected to the power lines but also buildings served by them. Consequently, lightning protection is necessary for outdoor conductors as well as for buildings.

3.6.2 Methods for Protecting against Lightning

Objectives of lightning protection are life safety, prevention of property damage, and maintenance of essential services, such as electrical and communication systems. Lightning protection usually requires installation of electrical conductors that extend from points above the roof of a building to the ground, for the purpose of conducting to the ground lightning that would otherwise strike the building. Such an installation, however, possesses the potential hazard that, if not done properly, lightning may flash from the lightning conductors to other building components. Hence, the system must ensure that the lightning discharge is diverted away from the building and its contents. Lightning protection systems should conform to the standards of the American National Standards Institute, National Fire Protection Association (NFPA 78, "Lightning Protection Code") and Underwriters Laboratories (UL 96A, "Master Labeled Lightning-Protection Systems").

The key element in diverting lightning away from a building is an air terminal or lightning rod, a conductor that projects into the air at least 12 in above the roof. Air terminals should be spaced at intervals not exceeding 25 ft. Alternatively, a continuous wire conductor or a grid of such conductors may be placed along the highest points of a roof. If the tallest object on a roof is a metal mast, it can act as an air terminal. A metal roof also can serve as an air terminal, but only if all joints are made electrically continuous by soldering, welding, or interlocking. Arranged to provide a cone of protection over the entire building, all the air terminals should be connected by conductors to each other and, by the same or other conductors, to the ground along at least two separated paths.

For roof and down conductors, copper, copper-clad steel, galvanized steel or a metal alloy that is as resistant to corrosion as copper may be used. (A solid copper conductor should be at least ¼ in in diameter.) Direct connections between dissimilar metals should be avoided to prevent corrosion. Metal objects and non-current-carrying components of electrical systems should be kept at least 6 ft away from the lightning conductors or should be bonded to the nearest lightning conductor. Sharp bends in the conductors are not desirable. If a 90° bend must be used, the conductor should be firmly anchored, because the high current in a lightning stroke will tend to straighten the bend. If the conductor has a U bend, the high current may induce an electric arc to leap across the loop while also exerting forces to straighten out the bend.

In steel-frame buildings, the steel frame can be used as a down conductor. In such cases, the top of the frame should be electrically connected to air terminals and the base should be electrically connected to grounding electrodes. Similarly, the reinforcing steel of a reinforced concrete building can be used as down conductors if the reinforcing steel is bonded together from foundations to roof.

Damage to the electrical systems of buildings can be limited or prevented by insertion of lightning arresters, safety valves that curtail overvoltages and bypass the current surge to a ground system, at the service entrance. Further protection can be afforded electrical equipment, especially sensitive electronic devices, by installing surge protectors, or spark gaps, near the equipment.

The final and equally important elements of a lightning-protection system are grounding electrodes and the earth itself. The type and dimensions of the grounds, or grounding electrodes, depends on the electrical resistance, or resistivity, of the earth, which can be measured by technicians equipped with suitable instruments. The objective of the grounding installation, which should be electrically bonded to the down conductors, should be an earth-system resistance of 10 Ω or less. Underground water pipes can serve as grounds if they are available. If not, long metal rods can be driven into the ground to serve as electrodes. Where earth resistivity is poor, an extensive system of buried wires may be required.

(J. L. Marshall, "Lightning Protection," John Wiley & Sons, Inc., New York.)

3.7 PROTECTION AGAINST INTRUDERS

Prevention of illegal entry into buildings by professional criminals determined to break in is not practical. Hence, the prime objective of security measures is to make illegal entry difficult. If this is done, it will take an intruder longer to gain entry or will compel the intruder to make noise, thus increasing the chances of detection and apprehension. Other objectives of security measures are detection of break-in attempts and intruders, alarming intruders so that they leave the premises before they cause a loss or injury, and alerting building occupants and the police of the break-in attempt. Also, an objective is to safeguard valuable assets by placing them in a guarded, locked, secure enclosure with access limited only to approved personnel.

Some communities have established ordinances setting minimum requirements for security and incorporated them in the building code. (Communities that have done this include Los Angeles, Oakland, and Concord in California; Indianapolis, Ind.; Trenton, N.J.; Arlington Heights, Ill.; Arlington County, Va.; and Prince George's County, Md.) Provisions of these codes cover security measures for doors and windows and associated hardware, accessible transoms, roof openings, safes, lighting of parking lots, and intrusion-detection devices. For buildings requiring unusual security measures, owners and designers should obtain the advice of a security expert.

3.7.1 Security Measures

Basic security for a building is provided by commonly used walls and roofs with openings protected by doors with key-operated locks or windows with latches. The degree of protection required for a building and its occupants beyond basic security and privacy needs depends on the costs of insurance and security measures relative to potential losses from burglary and vandalism.

For a small building not housing small items of great value (these can be placed in a safety deposit box in a bank), devices for detecting break-in attempts are generally the most practical means for augmenting basic security. Bells, buzzers, or sirens should be installed to sound an alarm and notify the police when an intruder tries to enter the locked building or a security area.

For a large building or a building requiring tight security, defense should be provided in depth. Depending on the value of assets to be protected, protection should start at the boundary of the property, with fences, gates, controlled access, guard patrols, exterior illumination, alarms, or remote surveillance by closed-circuit television. This defense should be backed up by similar measures at the perimeter of the building and by security locks and latches on doors and windows. Openings other than doorways or windows should be barred or made too small for human entry and screened. Within the building, valuables should be housed in locked rooms or a thick, steel safe, with controlled access to those areas.

For most types of occupancy, control at the entrance often may be provided by a receptionist who records names of visitors and persons visited, notifies the latter and can advise the police of disturbances. When necessary, the receptionist can be augmented by a guard at the control point or in a security center and, in very large or high-rise buildings, by a roving guard available for emergencies. If a large security force is needed, facilities should be provided in the building for an office for the security administrator and staff, photographic identification, and squad room and lockers—all in or adjoining a security center.

3.7.2 Security Center

The security center may be equipped with or connected to electronic devices that do the following:

1. Detect a break-in attempt and sound an alarm.
2. Identify the point of intrusion.
3. Turn on lights.
4. Display the intruder on closed-circuit television and record observations on videotape.
5. Notify the police.
6. Limit entry to specific spaces only to approved personnel and only at permitted times.
7. Change locks automatically.

In addition, the center may be provided with emergency reporting systems, security guard tour reporting systems, fire detection and protection systems, including supervision of automatic fire sprinklers, HVAC controls, and supervision of other life safety measures. See also Art. 3.5.12.

(P. S. Hopf, "Handbook of Building Security Planning and Design," McGraw-Hill Publishing Company, New York.)

SECTION FOUR
BUILDING MATERIALS

Frederick S. Merritt
Consulting Engineer, West Palm Beach, Florida

and

David J. Akers*
Civil Engineer, San Diego, California

This section describes the basic materials used in building construction and discusses their common applications. For discussion purposes, materials used in similar applications are grouped and discussed in sequence, for example, masonry materials, wood, metals, plastics, etc.

CEMENTITIOUS MATERIALS

Cementitious materials include the many products that are mixed with either water or some other liquid or both to form a cementing paste that may be formed or molded while plastic but will set into a rigid shape. When sand is added to the paste, mortar is formed. A combination of coarse and fine aggregate (sand) added to the paste forms concrete.

4.1 TYPES OF CEMENTITIOUS MATERIALS

There are many varieties of cements and numerous ways of classification. One of the simplest classifications is by the chemical constituent that is responsible for the setting or hardening of the cement. On this basis, the silicate and aluminate cements, wherein the setting agents are calcium silicates and aluminates, constitute the most important group of modern cements. Included in this group are the portland, aluminous, and natural cements.

*Revised and updated Sec. 4, Building Materials, by Albert G. H. Dietz, in the 4th Edition.

Limes, wherein the hardening is due to the conversion of hydroxides to carbonates, were formerly widely used as the sole cementitious material, but their slow setting and hardening are not compatible with modern requirements. Hence, their principal function today is to plasticize the otherwise harsh cements and add resilience to mortars and stuccoes. Use of limes is beneficial in that their slow setting promotes healing, the recementing of hairline cracks.

Another class of cements is composed of calcined gypsum and its related products. The gypsum cements are widely used in interior plaster and for fabrication of boards and blocks; but the solubility of gypsum prevents its use in construction exposed to any but extremely dry climates.

Oxychloride cements constitute a class of specialty cements of unusual properties. Their cost prohibits their general use in competition with the cheaper cements; but for special uses, such as the production of sparkproof floors, they cannot be equaled.

Masonry cements or mortor cements are widely used because of their convenience. While they are, in general, mixtures of one or more of the above-mentioned cements with some admixtures, they deserve special consideration because of their economies.

Other cementitious materials, such as polymers, fly ash, and silica fume, may be used as a cement replacement in concrete. Polymers are plastics with long-chain molecules. Concretes made with them have many qualities much superior to those of ordinary concrete.

Silica fume, also known as microsilica, is a waste product of electric-arc furnaces. The silica reacts with lime in concrete to form a cementitious material. A fume particle has a diameter only 1% of that of a cement particle.

4.2 PORTLAND CEMENTS

Portland cement, the most common of the modern cements, is made by carefully blending selected raw materials to produce a finished material meeting the requirements of ASTM C150 for one of eight specific cement types. Four major compounds [lime (CaO), iron (Fe_2O_3), silica (SiO_2), and alumina (Al_2O_3)] and two minor compounds [gypsum ($CaSO_4 \cdot 2H_2O$) and magnesia (MgO)] constitute the raw materials. The calcareous (CaO) materials typically come from limestone, calcite, marl, or shale. The argillaceous (SiO_2 and Al_2O_3) materials are derived from clay, shale, and sand. The materials used for the manufacture of any specific cement are dependent on the manufacturing plant's location and availability of raw materials. Portland cement can be made of a wide variety of industrial by-products.

In the manufacture of cement, the raw materials are first mined and then ground to a powder before blending in predetermined proportions. The blend is fed into the upper end of a rotary kiln heated to 2600 to 3000°F by burning oil, gas, or powdered coal. Because cement production is an energy-intensive process, reheaters and the use of alternative fuel sources, such as old tires, are used to reduce the fuel cost. (Burning tires provide heat to produce the clinker and the steel belts provide the iron constituent.) Exposure to the elevated temperature chemically fuses the raw materials together into hard nodules called cement clinker. After cooling, the clinker is passed through a ball mill and ground to a fineness where essentially all of it will pass a No. 200 sieve (75 μm). During the grinding, gypsum is added in small amounts to control the temperature and regulate the cement setting time. The material that exits the ball mill is portland cement. It is normally sold in bags containing 94 lb of cement.

Concrete, the most common use for portland cement, is a complex material consisting of portland cement, aggregates, water, and possibly chemical and mineral admixtures. Only rarely is portland cement used alone, such as for a cement slurry for filling well holes or for a fine grout. Therefore, it is important to examine the relationship between the various portland cement properties and their potential effect upon the finished concrete. Portland cement concrete is generally selected for structural use because of its strength and durability. Strength is easily measured and can be used as a general directly proportional indicator of overall durability. Specific durability cannot be easily measured but can be specified by controlling the cement chemistry and aggregate properties.

4.2.1 Specifications for Portland Cements

ASTM C150 defines requirements for eight types of portland cement. The pertinent chemical and physical properties are shown in Table 4.1. The chemical composition of portland cement is expressed in a cement-chemistry shorthand based on four phase compounds: tricalcium silicate (C_2S), dicalcium silicate (C_2S), tricalcium aluminate (C_3A), and tetracalcium aluminum ferrite (C_4AF). C_2S and C_3S are termed the calcium silicate hydrates (CSH).

Most cements will exceed the requirements shown in Table 4.1 by a comfortable margin. Note that the required compressive strengths are minimums. Almost without exception, every portland cement will readily exceed these minimum values. However, a caution must be attached to compressive strengths that significantly exceed the minimum values. While there is not a one-to-one correlation between a cement cube strength and the strength of concrete made with that cement (5000-psi cement does not equate to 5000-psi concrete), variations in cube strengths will be reflected in the tested concrete strength. It is imperative that, as the designed concrete strength reaches 5000 psi and greater, the cement cube strength be rigorously monitored. Any lowering of the running average will have a negative effect on the strength of concrete if the concrete mix design is not altered.

The basic types of portland cement covered by ASTM C150 are as follows:

Type I, general-purpose cement, is the one commonly used for many structural purposes. Chemical requirements for this type of cement are limited to magnesia and sulfur-trioxide contents and loss on ignition, since the cement is adequately defined by its physical characteristics.

Type II is a modified cement for use in general concrete where a moderate exposure to sulfate attack may be anticipated or where a moderate heat of hydration is required. These characteristics are attained by placing limitations on the C_3S and C_3A content of the cement. Type II cement gains strength a little more slowly than Type I but ultimately will achieve equal strength. It is generally available in most sections of the country and is preferred by some engineers over Type I for general construction. Type II cement may also be specified as a low-alkali cement for use where alkali reactive aggregates are present. To do so requires that optional chemical requirements (Table 4.2) be included in the purchase order. Type II low-alkali cement is commonly specified in California.

Type III cement attains high early strength. In 7 days, strength of concrete made with it is practically equal to that made with Type I or Type II cement at 28 days. This high early strength is attained by finer grinding (although no minimum is placed on the fineness by specification) and by increasing the C_3S and C_3A content of the cement. Type III cement, however, has high heat evolution and therefore should not be used in large masses. Because of the higher C_3A content, Type III

TABLE 4.1 Chemical and Physical Requirements for Portland Cement*

Type: Name:	I and IA General-purpose	II and IIA Modified	III and IIIA High early	IV Low-heat	V Sulfate-resisting
C_3S, max %				35	
C_2S, min %				40	
C_3A, max %		8	15	7	5
SiO_2, min %		20			
Al_2O_3, max %		6			
Fe_2O_3, max %		6		6.5	
MgO, max %	6	6	6	6	6
SO_3, max %:					
When $C_3A \leq 8\%$	3	3	3.5	2.3	2.3
When $C_3A > 8\%$	3.5		4.5		
$C_4AF + 2(C_3A)$, max %					25
Fineness, specific surface, m^2/kg					
Average min, by turbidimeter	160	160		160	160
Average min, by air permeability test	280	280		280	280
Compressive strength, psi, mortar cubes of 1 part cement and 2.75 parts graded standard sand after:					
1 day min					
Standard			1800		
Air-entraining			1450		
3 days min					
Standard	1800	1500	3500		1200
Air-entraining	1450	1200	2800		
7 days min					
Standard	2800	2500		1000	2200
Air-entraining	2250	2000			
28 days min					
Standard				2500	3000

*Based on requirements in "Standard Specification for Portland Cement," ASTM C150. See current edition of C150 for exceptions, alternatives, and changes in requirements.

cement also has poor sulfate resistance. Type III cement is not always available from building materials dealers' stocks but may be obtained by them from the cement manufacturer on short notice. Ready-mix concrete suppliers generally do not stock Type III cement because its shorter set time makes it more volatile to transport and discharge, especially in hot weather.

Type IV is a low-heat cement that has been developed for mass concrete construction. Normal Type I cement, if used in large masses that cannot lose heat by radiation, will liberate enough heat during the hydration of the cement to raise the temperature of the concrete as much as 50 or 60°F. This results in a relatively large

TABLE 4.2 Optional Chemical Requirements for Portland Cement*

Cement type	I and IA	II and IIA	III and IIIA	IV	V
Tricalcium aluminate (C_3A), max %					
For moderate sulfate resistance			8		
For high sulfate resistance			5		
Sum of tricalcium silicate and tricalcium aluminate, max %†		58			
Alkalies ($Na_2O + 0.658K_2O$), max %‡	0.60	0.60	0.60	0.60	0.60

*These optional requirements apply only if specifically requested. Availability should be verified.
†For use when moderate heat of hydration is required.
‡Low-alkali cement. This limit may be specified when cement is to be used in concrete with aggregates that may be deleteriously reactive. See "Standard Specification for Concrete Aggregates," ASTM C33.

increase in dimensions while the concrete is still soft and plastic. Later, as the concrete cools after hardening, shrinkage causes cracks to develop, weakening the concrete and affording points of attack for aggressive solutions. The potential-phase compounds that make the largest contribution to the heat of hydration are C_3S and C_3A; so the amounts of these that are permitted to be present are limited. Since these compounds also produce the early strength of cement, the limitation results in a cement that gains strength relatively slowly. This is of little importance, however, in the mass concrete for which this type of cement is designed.

Type V is a portland cement intended for use when high sulfate resistance is required. Its resistance to sulfate attack is attained through the limitation on the C_3A content. It is particularly suitable for structures subject to attack by liquors containing sulfates, such as liquids in wastewater treatment plants, seawaters, and some other natural waters.

Both Type IV and Type V cements are specialty cements. They are not normally available from dealer's stock but are usually obtainable for use on a large job if arrangements are made with the cement manufacturer in advance.

4.2.2 Air-Entraining Portland Cements

For use in the manufacture of air-entraining concrete, agents may be added to the cement by the manufacturer, thereby producing air-entraining portland cements ("Air-Entraining Additions for Use in the Manufacture of Air-Entraining Portland Cement," ASTM C226). These cements are available as Types IA, IIA, and IIIA.

4.3 ALUMINOUS CEMENTS

These are prepared by fusing a mixture of aluminous and calcareous materials (usually bauxite and limestone) and grinding the resultant product to a fine powder. These cements are characterized by their rapid-hardening properties and the high strength developed at early ages. Table 4.3 shows the relative strengths of 4-in

TABLE 4.3 Relative Strengths of Concrete Made from Portland and Aluminous Cements*

	Compressive strength, psi		
Days	Normal portland	High-early portland	Aluminous
1	460	790	5710
3	1640	2260	7330
7	2680	3300	7670
28	4150	4920	8520
56	4570	5410	8950

*Adapted from F. M. Lea, "Chemistry of Cement and Concrete," St. Martin's Press, Inc., New York.

cubes of 1:2:4 concrete made with normal portland, high-early-strength portland, and aluminous cements.

Since a large amount of heat is liberated with rapidity by aluminous cement during hydration, care must be taken not to use the cement in places where this heat cannot be dissipated. It is usually not desirable to place aluminous-cement concretes in lifts of over 12 in; otherwise the temperature rise may cause serious weakening of the concrete.

Aluminous cements are much more resistant to the action of sulfate waters than are portland cements. They also appear to be much more resistant to attack by water containing aggressive carbon dioxide or weak mineral acids than the silicate cements. Their principal use is in concretes where advantage may be taken of their very high early strength or of their sulfate resistance, and where the extra cost of the cement is not an important factor.

Another use of aluminous cements is in combination with firebrick to make refractory concrete. As temperatures are increased, dehydration of the hydration products occurs. Ultimately, these compounds create a ceramic bond with the aggregates.

4.4 NATURAL CEMENTS

Natural cements are formed by calcining a naturally occurring mixture of calcareous and argillaceous substances at a temperature below that at which sintering takes place. The "Specification for Natural Cement," ASTM C10, requires that the temperature be no higher than necessary to drive off the carbonic acid gas. Since natural cements are derived from naturally occurring materials and no particular effort is made to adjust the composition, both the composition and properties vary rather widely. Some natural cements may be almost the equivalent of portland cement in properties; others are much weaker. Natural cements are principally used in masonry mortars and as an admixture in portland-cement concretes.

4.5 LIMES

These are made principally of calcium oxide (CaO), occurring naturally in limestone, marble, chalk, coral, and shell. For building purposes, they are used chiefly in mortars.

4.5.1 Hydraulic Limes

These are made by calcining a limestone containing silica and alumina to a temperature short of incipient fusion so as to form sufficient free lime to permit hydration and at the same time leave unhydrated sufficient calcium silicates to give the dry powder its hydraulic properties (see "Specification for Hydraulic Hydrated Lime for Structural Purposes," ASTM C141).

Because of the low silicate and high lime contents, hydraulic limes are relatively weak. They find their principal use in masonry mortars. A hydraulic lime with more than 10% silica will set under water.

4.5.2 Quicklimes

When limestone is heated to a temperature in excess of 1700°F, the carbon dioxide content is driven off and the remaining solid product is quicklime. It consists essentially of calcium and magnesium oxides plus impurities such as silica, iron, and aluminum oxides. The impurities are usually limited to less than 5%. If they exceed 10%, the product may be a hydraulic lime.

Two classes of quicklime are recognized, high-calcium and dolomitic. A high-calcium quicklime usually contains less than 5% magnesium oxide. A dolomitic quicklime usually contains from 35 to 40% magnesium oxide. A few quicklimes are found that contain 5 to 35% magnesium oxide and are called magnesian limes.

The outstanding characteristic of quicklime is its ability to slake with water. When quicklime is mixed with from two to three times its weight of water, a chemical reaction takes place. The calcium oxide combines with water to form calcium hydroxide, and sufficient heat is evolved to bring the entire mass to a boil. The resulting product is a suspension of finely divided calcium hydroxide (and magnesium hydroxide or oxide if dolomitic lime is used) in water. On cooling, the semifluid mass stiffens to a putty of such consistency that it may be shoveled or carried in a hod. This slaked quicklime putty, when cooled and preferably screened, is the material used in construction. Quicklime should always be thoroughly slaked.

The yield of putty will vary, depending on the type of quicklime, its degree of burning, and slaking conditions, and will usually be from 70 to 100 ft^3 of putty per ton of quicklime. The principal use of the putty is in masonry mortars, where it is particularly valuable because of the high degree of plasticity or workability it imparts to the mortar. It is used at times as an admixture in concrete to improve workability. It also is used in some localities as finish-coat plaster where full advantage may be taken of its high plasticity.

4.5.3 Mason's Hydrated Lime

Hydrated limes are prepared from quicklimes by addition of a limited amount of water. After hydration ceases to evolve heat, the resulting product is a fine, dry powder. It is then classified by air-classification methods to remove undesirable oversize particles and packaged in 50-lb sacks. It is always a factory-made product, whereas quicklime putty is almost always a job-slaked product.

Mason's hydrated limes are those hydrates suitable for use in mortars, base-coat plasters, and concrete. They necessarily follow the composition of the quicklime. High-calcium hydrates are composed primarily of calcium hydroxide. Normal dolomitic hydrates are composed of calcium hydroxide plus magnesium oxide.

Plasticity of mortars made from normal mason's hydrated limes (Type N) is fair. It is better than that attained with most cements, but not nearly so high as that of mortars made with an equivalent amount of slaked putty.

The normal process of hydration of a dolomitic quicklime at atmospheric pressure results in the hydration of the calcium fraction only, leaving the magnesium-oxide portion substantially unchanged chemically. When dolomitic quicklime is hydrated under pressure, the magnesium oxide is converted to magnesium hydroxide. This results in the so-called special hydrates (Type S), which not only have their magnesia contents substantially completely hydrated but also have a high degree of plasticity immediately on wetting with water. Mortars made from Type S hydrates are more workable than those made from Type N hydrates. In fact, Type S hydrates are nearly as workable as those made from slaked quicklime putties. The user of this type of hydrate may therefore have the convenience of a bagged product and a high degree of workability without having the trouble and possible hazard of slaking quicklime.

4.5.4 Finishing Hydrated Limes

Finishing hydrated limes are particularly suitable for use in the finishing coat of plaster. They are characterized by a high degree of whiteness and plasticity. Practically all finishing hydrated limes are produced in the Toledo district of Ohio from dolomitic limestone. The normal hydrate is composed of calcium hydroxide and magnesium oxide. When first wetted, it is no more plastic than Type N mason's hydrates. It differs from the latter, however, in that, on soaking overnight, the finishing hydrated lime develops a very high degree of plasticity, whereas the mason's hydrate shows relatively little improvement in plasticity on soaking.

4.6 LOW-TEMPERATURE GYPSUM DERIVATIVES

When gypsum rock ($CaSO_4 \cdot 2H_2O$) is heated to a relatively low temperature, about 130°C, three-fourths of the water of crystallization is driven off. The resulting product is known by various names such as hemihydrate, calcined gypsum, and first-settle stucco. Its common name, however, is **plaster of paris.** It is a fine powder, usually white. While it will set under water, it does not gain strength and ultimately, on continued water exposure, will disintegrate.

Plaster of paris, with set retarded or unretarded, is used as a molding plaster or as a gaging plaster. The molding plaster is used for preparing ornamental plaster objects. The gaging plaster is used for finishing hydrated lime to form the smooth white-coat finish on plaster walls. The unretarded plaster of paris is used by manufacturers to make gypsum block, tile, and gypsumboard (wallboard, lath, backerboard, coreboard, etc.).

When plaster of paris is retarded and mixed with fiber such as sisal, it is marketed under the name of hardwall plaster or cement plaster. (The latter name is misleading, since it does not contain any portland cement.) Hardwall plaster, mixed with water and with from two to three parts of sand by weight, is widely used for base-coat plastering. In some cases wood fiber is used in place of sand, making a "wood-fibered" plaster.

Special effects are obtained by combining hardwall plaster with the correct type of aggregate. With perlite or vermiculite aggregate, a lightweight plaster is obtained.

Gypsum plasters, in general, have a strong set, gain their full strength when dry, do not have abnormal volume changes, and have excellent fire-resistance characteristics. They are not well adapted, however, for use under continued damp conditions or intermittent wet conditions. See also Arts. 4.26 to 4.30.

4.7 OXYCHLORIDE CEMENTS

Lightly calcined magnesium oxide mixed with a solution of magnesium chloride forms a cement known as magnesium oxychloride cement, or **Sorel cement.** It is particularly useful in making flooring compositions in which it is mixed with colored aggregates. Floors made of oxychloride cement are sparkproof and are more resilient than floors of concrete.

Oxychloride cement has very strong bonding power and, because of its higher bonding power, may be used with greater quantities of aggregate than are possible with portland cement. Oxychloride cement also bonds well with wood and is used in making partition block or tile with wood shavings or sawdust as aggregate. It is moderately resistant to water but should not be used under continually wet conditions.

4.8 MASONRY CEMENTS

Masonry cements, or—as they are sometimes called—mortar cements, are intended to be mixed with sand and used for setting unit masonry, such as brick, tile, and stone. They may be any one of the hydraulic cements already discussed or mixtures of them in any proportion.

Many commercial masonry cements are mixtures of portland cement and pulverized limestone, often containing as much as 50 or 60% limestone. They are sold in bags containing from 70 to 80 lb, each bag nominally containing a cubic foot. Price per bag is commonly less than that of portland cement, but because of the use of the lighter bag, cost per ton is higher than that of portland cement.

Since there are no limits on chemical content and physical requirements, masonry cement specifications are quite liberal. Some manufacturers vary the composition widely, depending on competition, weather conditions, or availability of materials. Resulting mortars may vary widely in properties.

4.9 FLY ASHES

Fly ash meeting the requirements of ASTM C618, "Specification for Fly Ash and Raw or Calcined Natural Pozzolan for Use as a Mineral Admixture in Portland Cement Concrete," is generally used as a cementitious material as well as an admixture.

Natural pozzolans are derived from some diatomaceous earths, opaline cherts and shales, and other materials. While part of a common ASTM designation with fly ash, they are not as readily available as fly ashes and thus do not generate the same level of interest or research.

Fly ashes are produced by coal combustion, generally in an electrical generating station. The ash that would normally be released through the chimney is captured by various means, such as electrostatic precipitators. The fly ash may be sized prior to shipment to concrete suppliers.

All fly ashes possess pozzolanic properties, the ability to react with calcium hydroxide at ordinary temperatures to form compounds with cementitious properties. When cement is mixed with water, a chemical reaction (hydration) occurs. The product of this reaction is calcium silicate hydrate (CSH) and calcium hydroxide [$Ca(OH)_2$]. Fly ashes have high percentages of silicon dioxide (SiO_2). In the presence of moisture, the $Ca(OH)_2$ will react with the SiO_2 to form another CSH.

Type F ashes are the result of burning anthracite or bituminous coals and possess pozzolanic properties. They have been shown by research and practice to provide usually increased sulfate resistance and to reduce alkali-aggregate expansions. Type C fly ashes result from burning lignite or subbituminous coals. Because of the chemical properties of the coal, the Type C fly ashes have some cementitious properties in addition to their pozzolanic properties. Type C fly ashes may reduce the durability of concretes into which they are incorporated.

4.10 SILICA FUME (MICROSILICA)

Silica fume, or microsilica, is a condensed gas, the by-product of metallic silicon or ferrosilicon alloys produced by electric arc furnaces. (While both terms are correct, microsilica (MS) is a less confusing name.) The Canadian standard CAN/CSA-A23.5-M86, "Supplementary Cementing Materials," limits amorphous SiO_2 to a maximum of 85% and oversize to 10%. Many MS contain more than 90% SiO_2.

MS has an average diameter of 0.1 to 0.2 μm, a particle size of about 1% that of portland cement. Because of this small size, it is not possible to utilize MS in its raw form. Manufacturers supply it either densified, in a slurry (with or without water-reducing admixtures), or pelletized. Either the densified or slurried MS can be utilized in concrete. The pelletized material is densified to the point that it will not break down during mixing.

Because of its extremely small size, MS imparts several useful properties to concrete. It greatly increases long-term strength. It very efficiently reacts with the $Ca(OH)_2$ and creates a beneficial material in place of a waste product. MS is generally used in concrete with a design strength in excess of 12,000 psi. It provides increased sulfate resistance to concrete, and it significantly reduces the permeability of concrete. Also, its small size allows MS to physically plug microcracks and tiny openings.

AGGREGATES

Aggregate is a broad term encompassing boulders, cobbles, crushed stone, gravel, air-cooled blast furnace slag, native and manufactured sands, and manufactured and natural lightweight aggregates. Aggregates may be further described by their respective sizes.

4.11 NORMAL-WEIGHT AGGREGATES

These typically have specific gravities between 2.0 and 3.0. They are usually distinguished by size as follows:

Boulders	Larger than 6 in
Cobbles	6 to 3 in
Coarse aggregate	3 in to No. 4 sieve
Fine aggregate	No. 4 sieve to No. 200 sieve
Mineral filler	Material passing No. 200 sieve

Used in most concrete construction, normal-weight aggregates are obtained by draining riverbeds or mining and crunching formational material. Concrete made with normal-weight fine and coarse aggregates generally weighs about 144 lb/ft^3.

Boulders and cobbles are generally not used in their as-mined size but are crushed to make various sizes of coarse aggregate and manufactured sand and mineral filler. Gravels and naturally occurring sand are produced by the action of water and weathering on glacial and river deposits. These materials have round, smooth surfaces and particle-size distributions that require minimal processing. These materials can be supplied in either coarse or fine-aggregate sizes.

Fine aggregates have 100% of their material passing the ⅜-in sieve. Coarse aggregates have the bulk of the material retained on the No. 4 sieve.

Aggregates comprise the greatest volume percentage in portland-cement concrete, mortar, or asphaltic concrete. In a portland-cement concrete mix, the coarse and fine aggregates occupy about 60 to 75% of the total mix volume. For asphaltic concrete, the aggregates represent 75 to 85% of the mix volume. Consequentially, the aggregates are not inert filler materials. The individual aggregate properties have demonstrable effects on the service life and durability of the material system in which the aggregate is used, such as portland-cement concrete, asphaltic concrete, mortar, or aggregate base.

The acceptability of a coarse or fine aggregate for use in concrete or mortar is judged by many properties including gradation, amount of fine material passing the No. 200 sieve, hardness, soundness, particle shape, volume stability, potential alkali reactivity, resistance to freezing and thawing, and organic impurities. For aggregates used in general building construction, property limits are provided in ASTM C33, "Specification for Concrete Aggregates," C637, "Specification for Aggregates for Radiation-Shielding Concrete," and C330, "Specification for Lightweight Aggregates for Structural Concrete." For other types of construction, such as highways and airports, standards written by various trade or governmental organizations are available.

4.11.1 Gradation of Aggregates

The distribution of aggregate sizes in a concrete mix is important because it directly influences the amount of cement required for a given strength, workability of the mix (and amount of effort to place the mix in the forms), in-place durability, and overall economy. ASTM C33 provides ranges of fine- and coarse-aggregate grading limits. The latter are listed from Size 1 (3½ to 1½ in) to Size 8 (⅜ to No. 8). The National Stone Association specifies a gradation for manufactured sand that differs from that for fine aggregate in C33 principally for the No. 100 and 200 sieves. The

NSA gradation is noticeably finer (greater percentages passing each sieve). The fine materials, composed of angular particles, are rock fines, as opposed to silts and clays in natural sand, and contribute to concrete workability.

The various gradations provide standard sizes for aggregate production and quality-control testing. They are conducive to production of concrete with acceptable properties. Caution should be exercised, however, when standard individual grading limits are used. If the number of aggregate sizes are limited or there is not sufficient overlap between aggregate sizes, an acceptable or economical concrete may not be attainable with acceptably graded aggregates. The reason for this is that the combined gradation is gap graded. The ideal situation is a dense or well-graded size distribution that optimizes the void content of the combined aggregates (Art. 4.17). It is possible, however, to produce acceptable concrete with individual aggregates that do not comply with the standard limits but that can be combined to produce a dense gradation.

4.11.2 Amount of Fine Material Passing the No. 200 Sieve

The material passing the No. 200 sieve is clay, silt, or a combination of the two. It increases the water demand of the aggregate. Large amounts of materials smaller than No. 200 may also indicate the presence of clay coatings on the coarse aggregate that would decrease bond of the aggregate to the cement matrix. A test method is given in ASTM C117, "Materials Finer than 75 μm Sieve in Mineral Aggregates by Washing."

4.11.3 Hardness

Coarse-aggregate hardness is measured by the Los Angeles Abrasion Test, ASTM C131 or C595. These tests break the aggregate down by impacting it with steel balls in a steel tumbler. The resulting breakdown is not directly related to the abrasion an aggregate receives in service, but the results can be empirically related to concretes exhibiting acceptable service lives.

4.11.4 Soundness

Aggregate soundness is measured by ASTM C88, "Test Method for Soundness of Aggregates by Use of Sodium Sulfate or Magnesium Sulfate." This test measures the amount of aggregate degradation when exposed to alternating cycles of wetting and drying in a sulfate solution.

4.11.5 Particle Shape

Natural sand and gravel have a round, smooth particle shape. Crushed aggregate (coarse and fine) may have shapes that are flat and elongated, angular, cubical, disk, or rodlike. These shapes result from the crushing equipment employed and the aggregate mineralogy. Extreme angularity and elongation increase the amount of cement required to give strength, difficulty in finishing, and effort required to pump the concrete. Flat and elongated particles also increase the amount of required mixing water.

The bond between angular particles is greater than that between smooth particles. Properly graded angular particles can take advantage of this property and offset the increase in water required to produce concrete with cement content and strength equal to that of a smooth-stone mix.

4.11.6 Potential Alkali Reactivity

Aggregates that contain certain forms of silicas or carbonates may react with the alkalies present in portland cement (sodium oxide and potassium oxide). The reaction product cracks the concrete or may create pop-outs at the concrete surface. The reaction is more pronounced when the concrete is in a warm, damp environment.

The potential reactivity of an aggregate with alkalies can be determined either by a chemical test (ASTM C289) or by a mortar-bar method (ASTM C227). The mortar-bar method is the more rigorous test and provides more reliable results but it requires a much longer time to perform.

4.11.7 Resistance to Freezing and Thawing

The pore structure, absorption, porosity, and permeability of aggregates are especially important if they are used to make concrete exposed to repeated cycles of freezing and thawing. Aggregates that become critically saturated and then freeze cannot accommodate the expansion of the frozen water. Empirical data show that freeze-thaw deterioration is caused by the coarse aggregates and not the fine. A method prescribed in "Test Method for Resistance of Concrete to Rapid Freezing and Thawing," ASTM C666, measures concrete performance by weight changes, a reduction in the dynamic modulus of elasticity, and increases in sample length.

4.11.8 Impurities in Aggregates

Erratic setting times and rates of hardening may be caused by organic impurities in the aggregates, primarily the sand. The presence of these impurities can be investigated by a method given in "Test Method for Organic Impurities in Fine Aggregates for Concrete," ASTM C40.

Pop-outs and reduced durability can be caused by soft particles, chert, clay lumps and other friable particles, coal, lignite, or other lightweight materials in the aggregates. Coal and lignite may also cause staining of exposed concrete surfaces.

4.11.9 Volume Stability

Volume stability refers to susceptibility of aggregate to expansion when heated or to cyclic expansions and contractions when saturated and dried. Aggregates that are susceptible to volume change due to moisture should be avoided.

4.12 HEAVYWEIGHT AND LIGHTWEIGHT AGGREGATES

Heavyweight aggregates include magnetite, with a specific gravity δ of 4.3; barite, $\delta = 4.2$; limonite, $\delta = 3.8$; ferrophosphorus, $\delta = 6.3$; and steel shot or punchings,

$\delta = 7.6$. Such heavyweight aggregates may be used instead of gravel or crushed stone to produce a dense concrete; for example, for shielding of nuclear reactors as specified in ASTM C637.

Lightweight Aggregates. These can be divided into two categories: structural and nonstructural. The structural lightweight aggregates are defined by ASTM C330 and C331. They are either manufactured (expanded clay, shale, or slate, or blast-furnace slag) or natural (scoria and pumice). These aggregates produce concretes generally in the strength range of 3000 to 4000 psi; higher strengths are attainable and are discussed in Art. 4.17. The unit weight of lightweight concrete ranges from 100 to 115 lb/ft^3.

The common nonstructural lightweight aggregates (ASTM C332) are vermiculite and perlite, although scoria and pumice can also be used. These materials are used in insulating concretes for soundproofing and nonstructural floor toppings.

Lightweight aggregates produce concrete with low thermal conductivities, which equate to good fire protection. When concrete is exposed to extreme heat, the moisture present within the concrete rapidly changes from a liquid to steam having a volume of up to 15 times larger. The large number and large size of pores within lightweight aggregates create pressure-relief regions.

ADMIXTURES FOR CONCRETE

Admixtures are anything other than portland cement, water, and aggregates that are added to a concrete mix to modify its properties. Included in this definition are chemical admixtures (ASTM C494 and C260), mineral admixtures such as fly ash (C618) and silica fume, corrosion inhibitors, colors, fibers, and miscellaneous (pumping aids, dampproofing, gas-forming, permeability-reducing agents).

4.13 CHEMICAL AND MINERAL ADMIXTURES

Chemical admixtures used in concrete generally serve as water reducers, accelerators, set retarders, or a combination. ASTM C494, "Standard Specification for Chemical Admixtures for Concrete," contains the following classification:

Type	Property
A	Water reducer
B	Set retarder
C	Set accelerator
D	Water reducer and set retarder
E	Water reducer and set accelerator
F	High-range water reducer
G	High-range water reducer and set retarder

High-range admixtures reduce the amount of water needed to produce a concrete of a specific consistency by 12% or more.

4.13.1 Water-Reducing Admixtures

These decrease water requirements for a concrete mix by chemically reacting with early hydration products to form a monomolecular layer of admixture at the cement-water interface. This layer isolates individual particles of cement and reduces the energy required to cause the mix to flow. Thus, the mix is "lubricated" and exposes more cement particles for hydration.

The Type A admixture allows the amount of mixing water to be reduced while maintaining the same mix slump. Or at a constant water-cement ratio, this admixture allows the cement content to be decreased without loss of strength. If the amount of water is not reduced, slump of the mix will be increased and also strength will be increased because more of the cement surface area will be exposed for hydration. Similar effects occur for Type D and E admixtures. Typically, a reduction in mixing water of 5 to 10% can be expected.

Type F and G admixtures are used where there is a need for high-workability concrete. A concrete without an admixture typically has a slump of 2 to 3 in. After the admixture is added, the slump may be in the range of 8 to 10 in without segregation of mix components. These admixtures are especially useful for mixes with a low water-cement ratio. Their 12 to 30% reduction in water allows a corresponding reduction in cementitious material.

The water-reducing admixtures are commonly manufactured from lignosulfonic acids and their salts, hydroxylated carboxylic acids and their salts, or polymers of derivatives of melamines or naphthalenes or sulfonated hydrocarbons. The combination of admixtures used in a concrete mix should be carefully evaluated and tested to ensure that the desired properties are achieved. For example, depending on the dosage of admixture and chemistry of the cement, it is possible that a retarding admixture will accelerate the set. Note also that all normal-set admixtures will retard the set if the dosage is excessive. Furthermore, because of differences in percentage of solids between products from different companies, there is not always a direct correspondence in dosage between admixtures of the same class. Therefore, it is important to consider the chemical composition carefully when evaluating competing admixtures.

Superplasticizers are high-range water-reducing admixtures that meet the requirements of ASTM C494 Type F or G. They are often used to achieve high-strength concrete by use of a low water-cement ratio with good workability and low segregation. They also may be used to produce concrete of specified strengths with less cement at constant water-cement ratio. And they may be used to produce self-compacting, self-leveling flowing concretes, for such applications as long-distance pumping of concrete from mixer to formwork or placing concrete in forms congested with reinforcing steel. For these concretes, the cement content or water-cement ratio is not reduced, but the slump is increased substantially without causing segregation. For example, an initial slump of 3 to 4 in for an ordinary concrete mix may be increased to 7 to 8 in without addition of water and decrease in strength.

Superplasticizers may be classified as sulfonated melamine-formaldehyde condensates, sulfonated naphthaline-formaldehyde condensates, modified lignosulfonates, or synthetic polymers.

4.13.2 Air-Entraining Admixtures

These create numerous microscopic air spaces within concrete to protect it from degradation due to repeated freezing and thawing or exposure to aggressive chemicals. For concrete exposed to repeated cycles of freezing and thawing, the air gaps provide room for expansion of external and internal water, which otherwise would damage the concrete.

Since air-entrained concrete bleeds to a lesser extent than non-air-entrained, there are fewer capillaries extending from the concrete matrix to the surface. Therefore, there are fewer avenues available for ingress of aggressive chemicals into the concrete.

The "Standard Specification for Air-Entraining Admixtures for Concrete," ASTM C260, covers materials for use of air-entraining admixtures to be added to concrete in the field. Air entrainment may also be achieved by use of Types IIA and IIIA portland cements (Art. 4.2.2).

4.13.3 Set-Accelerating Admixtures

These are used to decrease the time from the start of addition of water to cement to initial set and to increase the rate of strength gain of concrete. The most commonly used set-accelerating admixture is calcium chloride. Its use, however, is controversial in cases where reinforcing or prestressing steel is present. The reason is that there is a possibility that the accelerator will introduce free chloride ions into the concrete, thus contributing to corrosion of the steel. An alternative is use of one of many admixtures not containing chloride that are available.

4.13.4 Retarding Admixtures

To some extent, all normal water-reducing admixtures retard the initial set of concrete. A Type B or D admixture will allow transport of concrete for a longer time before initial set occurs. Final set also is delayed. Hence, precautions should be taken if retarded concrete is to be used in walls.

Depending on the dosage and type of base chemicals in the admixture, initial set can be retarded for several hours to several days. A beneficial side effect of retardation of initial and final sets is an increase in the compressive strength of the concrete. A commonly used Type D admixture provides higher 7- and 28-day strengths than a Type A when used in the same mix design.

4.13.5 Mineral Admixtures

Fly ashes, pozzolans, and microsilicates are included in the mineral admixture classification (Arts. 4.9 and 4.10). Natural cement (Art. 4.4) is sometimes used as an admixture.

4.13.6 Corrosion Inhibitors

Reinforcing steel in concrete usually is protected against corrosion by the high alkalinity of the concrete, which creates a passivating layer at the steel surface. This layer is composed of ferric oxide, a stable compound. Within and at the surface of the ferric oxide, however, are ferrous-oxide compounds, which are more reactive. When the ferrous-oxide compounds come into contact with aggressive substances, such as chloride ions, they react with oxygen to form solid, iron-oxide corrosion products. These produce a fourfold increase in volume and create an expansion force greater than the concrete tensile strength. The result is deterioration of the concrete.

For corrosion to occur, chloride in the range of 1.0 to 1.5 lb/yd^3 must be present. If there is a possibility that chlorides may be introduced from outside the concrete matrix, for example, by deicing salts, the concrete can be protected by lowering the water-cement ratio, or increasing the amount of cover over the reinforcing steel, or entraining air in the concrete, or adding a calcium-nitrate admixture, or adding an internal-barrier admixture, or cathodic protection, or a combination of these methods.

To inhibit corrosion, calcium-nitrite admixtures are added to the concrete at the time of batching. They do not create a physical barrier to chloride ion ingress. Rather, they modify the concrete chemistry near the steel surface. The nitrite ions oxidize ferrous oxide present, converting it to ferric oxide. The nitrite is also absorbed at the steel surface and fortifies the ferric-oxide passivating layer. For a calcium-nitrite admixture to be effective, the dosage should be adjusted in accordance with the exposure condition of the concrete to corrosive agents. The greater the exposure, the larger should be the dosage. The correct dosage can only be determined on a project-by-project basis with data for the specific admixture proposed.

Internal-barrier admixtures come in two groups. One comprises waterproofing and dampproofing compounds (Art. 4.15). The second consists of agents that create an organic film around the reinforcing steel, supplementing the passivating layer. This type of admixture is promoted for addition at a fixed rate regardless of expected chloride exposure.

4.13.7 Coloring Admixtures

Colors are added to concrete for architectural reasons. They may be mineral oxides or manufactured pigments. Raw carbon black, a commonly used material for black color, greatly reduces the amount of entrained air in a mix. Therefore, if black concrete is desired for concrete requiring air-entrainment (for freeze-thaw or aggressive chemical exposure), either the carbon black should be modified to entrain air or an additional air-entraining agent may be incorporated in the mix. The mix design should be tested under field conditions prior to its use in construction. Use of color requires careful control of materials, batching, and water addition in order to maintain a consistent color at the jobsite.

4.14 FIBERS FOR CONCRETE MIXES

As used in concrete, fibers are discontinuous, discrete units. They may be described by their aspect ratio, the ratio of length to equivalent diameter. Fibers find their greatest use in crack control of concrete flatwork, especially slabs on grade.

The most commonly used types of fibers in concrete are synthetics, which include polypropylene, nylon, polyester, and polyethylene materials. Specialty synthetics include aramid, carbon, and acrylic fibers. Glass-fiber-reinforced concrete is made using E-glass and alkali-resistant (AR) glass fibers. Steel fibers are chopped high-tensile or stainless steel.

Fibers should be dispersed uniformly throughout a mix. Orientation of the fibers in concrete generally is random. Conventional reinforcement, in contrast, typically is oriented in one or two directions, generally in planes parallel to the surface. Further, welded-wire fabric or reinforcing steel bars must be held in position as concrete is placed. Regardless of the type, fibers are effective in crack control because they provide omnidirectional reinforcement to the concrete matrix. With steel fibers, impact strength and toughness of concrete may be greatly improved and flexural and fatigue strengths enhanced.

Synthetic fibers are typically used to replace welded-wire fabric as secondary reinforcing for crack control in concrete flatwork. Depending on the fiber length, the fiber can limit the size and spread of plastic shrinkage cracks or both plastic and drying shrinkage cracks. Although synthetic fibers are not designed to provide structural properties, slabs tested in accordance with ASTM E72, "Standard Methods of Conducting Strength Tests of Panels for Building Construction," showed that test slabs reinforced with synthetic fibers carried greater uniform loads than slabs containing welded wire fabric. While much of the research for synthetic fibers has used reinforcement ratios greater than 2%, the common field practice is to use 0.1% (1.5 lb/yd^3). This dosage provides more cross-sectional area than 10-gage welded-wire fabric. The empirical results indicate that cracking is significantly reduced and is controlled. A further benefit of fibers is that after the initial cracking, the fibers tend to hold the concrete together.

Aramid, carbon, and acrylic fibers have been studied for structural applications, such as wrapping concrete columns to provide additional strength. Other possible uses are for corrosion-resistance structures. The higher costs of the specialty synthetics limit their use in general construction.

Glass-fiber-reinforced concrete (GFRC) is used to construct many types of building elements, including architectural wall panels, roofing tiles, and water tanks. The full potential of GFRC has not been attained because the E-glass fibers are alkali reactive and the AR-glass fibers are subject to embrittlement, possibly from infiltration of calcium-hydroxide particles.

Steel fibers can be used as a structural material and replace conventional reinforcing steel. The volume of steel fiber in a mix ranges from 0.5 to 2%. Much work has been done to develop rapid repair methods using thin panels of densely packed steel fibers and a cement paste squeegeed into the steel matrix. American Concrete Institute Committee 544 states in "Guide for Specifying, Mixing, Placing, and Finishing Steel Fiber Reinforced Concrete," ACI 544.3R, that, in structural members such as beams, columns, and floors not on grade, reinforcing steel should be provided to support the total tensile load. In other cases, fibers can be used to reduce section thickness or improve performance. See also ACI 344.1R and 344.2R.

4.15 MISCELLANEOUS ADMIXTURES

There are many miscellaneous concrete additives for use as pumping aids and as dampproofing, permeability-reducing, gas-forming agents.

Pumping aids are used to decrease the viscosity of harsh or marginally pumpable mixes. Organic and synthetic polymers, fly ash, bentonite, or hydrated lime may be used for this purpose. Results depend on concrete mix, including the effects of increased water demand and the potential for lower strength resulting from the increased water-cement ratio. If sand makes the mix marginally pumpable, fly ash is the preferred pumping additive. It generally will not increase the water demand and it will react with the calcium hydroxide in cement to provide some strength increase.

Dampproofing admixtures include soaps, stearates, and other petroleum products. They are intended to reduce passage of water and water vapor through concrete. Caution should be exercised when using these materials inasmuch as they may increase water demand for the mix, thus increasing the permeability of the concrete. If dense, low-permeable concrete is desired, the water-cement ratio should be kept to a maximum of 0.50 and the concrete should be well vibrated and damp cured.

Permeability of concrete can be decreased by the use of fly ash and silica fume as admixtures. Also, use of a high-range water-reducing admixture and a water-cement ratio less than 0.50 will greatly reduce permeability.

Gas-forming admixtures are used to form lightweight concrete. They are also used in masonry grout where it is desirable for the grout to expand and bond to the concrete masonry unit. They are typically an aluminum powder.

MORTARS AND CONCRETES

4.16 MORTARS

Mortars are composed of a cementitious material, fine aggregate, sand, and water. They are used for bedding unit masonry, for plasters and stuccoes, and with the addition of coarse aggregate, for concrete. Here consideration is given primarily to those mortars used for unit masonry and plasters.

Properties of mortars vary greatly, being dependent on the properties of the cementitious material used, ratio of cementitious material to sand, characteristics and grading of the sand, and ratio of water to solids.

4.16.1 Packaging and Proportioning of Mortar

Mortars are usually proportioned by volume. A common specification is that not more than 3 ft^3 of sand be used with 1 ft^3 of cementitious material. Difficulty is sometimes encountered, however, in determining just how much material constitutes a cubic foot: a bag of cement (94 lb) by agreement is called a cubic foot in proportioning mortars or concretes, but an actual cubic foot of lime putty may be used in proportioning mortars. Since hydrated limes are sold in 50-lb bags (Art. 4.5.3), each of which makes somewhat more than a cubic foot of putty, weights of 40, 42, and 45 lb of hydrated lime have been used as a cubic foot in laboratory studies; but on the job, a bag is frequently used as a cubic foot. Masonry cements are sold in bags containing 70 to 80 lb (Art. 4.8), and a bag is considered a cubic foot.

4.16.2 Properties of Mortars

Table 4.4 lists types of mortars as a guide in selection for unit masonry.

Workability is an important property of mortars, particularly of those used in conjunction with unit masonry of high absorption. Workability is controlled by the character of the cement and amount of sand. For example, a mortar made from 3 parts sand and 1 part slaked lime putty will be more workable than one made from

TABLE 4.4 Types of Mortar

Mortar type	Parts by volume: Portland cement	Masonry cement	Hydrated lime or lime putty	Aggregate measured in damp, loose condition	Min avg compressive strength of three 2-in cubes at 28 days, psi
M	1	1			2500
	1		¼		
S	½	1			1800
	1		Over ¼ to ½	Not less than 2¼	
N		1		and not more than	750
	1		Over ½ to 1¼	3 times the sum of	
O		1		the volumes of the	350
	1		Over 1¼ to 2½	cements and limes	
K	1		Over 2½ to 4	used	75
PL	1		¼ to ½		2500
PM	1	1			2500

2 parts sand and 1 part portland cement. But the 3:1 mortar has lower strength. By proper selection or mixing of cementitious materials, a satisfactory compromise may usually be obtained, producing a mortar of adequate strength and workability.

Water retention—the ratio of flow after 1-min standard suction to the flow before suction—is used as an index of the workability of mortars. A high value of water retention is considered desirable for most purposes. There is, however, a wide variation in water retention of mortars made with varying proportions of cement and lime and with varying limes. The "Standard Specification for Mortar for Unit Masonry," ASTM C270, requires mortar mixed to an initial flow of 100 to 115, as determined by the test method of ASTM C109, to have a flow after suction of at least 75%.

Strength of mortar is frequently used as a specification requirement, even though it has little relation to the strength of masonry. (See, for example, ASTM C270, C780, and C476). The strength of mortar is affected primarily by the amount of cement in the matrix. Other factors of importance are the ratio of sand to cementing material, curing conditions, and age when tested.

Volume change of mortars constitutes another important property. Normal volume change (as distinguished from unsoundness) may be considered as the shrinkage during early hardening, shrinkage on drying, expansion on wetting, and changes due to temperature.

After drying, mortars expand again when wetted. Alternate wetting and drying produces alternate expansion and contraction, which apparently continues indefinitely with portland-cement mortars.

Coefficients of thermal expansion of several mortars, reported in "Volume Changes in Brick Masonry Materials," *Journal of Research of the National Bureau of Standards,* Vol. 6, p. 1003, ranged from 0.38×10^{-5} to 0.60×10^{-5} for masonry-cement mortars; from 0.41×10^{-5} to 0.53×10^{-5} for lime mortars, and from 0.42×10^{-5} to 0.61×10^{-5} for cement mortars. Composition of the cementitious material apparently has little effect on the coefficient of thermal expansion of a mortar.

4.16.3 High-Bond Mortars

When polymeric materials, such as styrene-butadiene and polyvinylidene chloride, are added to mortar, greatly increased bonding, compressive, and shear strengths result. To obtain high strength, the other materials, including sand, water, Type I or III portland cement, and a workability additive, such as pulverized ground limestone or marble dust, must be of quality equal to that of the ingredients of standard mortar. The high strength of the mortar enables masonry to withstand appreciable bending and tensile stresses. This makes possible thinner walls and prelaying of single-wythe panels that can be hoisted into place.

4.17 PORTLAND-CEMENT CONCRETE

Portland-cement concrete is a mixture of portland cement, water, coarse and fine aggregates, and admixtures proportioned to form a plastic mass capable of being cast, placed, or molded into forms that will harden to a solid mass. The desirable properties of plastic concrete are that it be workable, placeable and nonsegregating, and that it set in the desired time. The hardened concrete should provide the desired service properties:

1. Strength (compressive and flexural)
2. Durability (lack of cracks, resistance to freezing and thawing and to chemical attacks, abrasion resistance, and air content)
3. Appearance (color, lack of surface imperfections)

Each of these properties affects the final cost of the mix design and the cost of the in-place concrete. These properties are available from normal-weight, lightweight, and heavyweight concretes.

4.17.1 Normal-Weight Concrete

The nominal weight of normal concrete is 144 lb/ft^3 for non-air-entrained concrete, but is less for air-entrained concrete. (The weight of concrete plus steel reinforcement is often assumed as 150 lb/ft^3.)

Strength for normal-weight concrete ranges from 2000 to 20,000 psi. It is generally measured using a standard test cylinder 6 in in diameter by 12 in high. The strength of a concrete is defined as the average strength of two cylinders taken from the same load and tested at the same age. Flexural beams 6 × 6 × 20 in may be used for concrete paving mixes. The strength gains of air-entrained and non-air-entrained concretes are graphically shown in Fig. 9.2.

As illustrated in Fig. 9.2, the strength of a given mix is determined by the water-cement ratio (W/C), and whether or not air entraining is used. Other factors are the maximum-size aggregate and the desired fluidity (slump) of the concrete at the point of placement. When no historical record is available for the aggregates and cements to be used, the water-cement ratios in Table 9.2 can provide guidance for the initial designs.

Each combination of coarse and fine aggregates has a specific water demand for a given mix fluidity, or slump. Two general guidelines are:

1. For a constant slump, the water demand increases with increase in maximum-size aggregate.
2. For a constant maximum-size aggregate, as the slump increases, the water demand increases.

There are many different methods for designing a normal-weight concrete mix. A standard method is given in ACI 211, "Standard Practice for Selecting Proportions for Normal, Heavyweight, and Mass Concrete." See also Art. 9.10.

Workability of a concrete is the property most important to contractors who must place the concrete into forms and finish it. Workability includes the properties of cohesiveness, plasticity, and nonsegregation. It is greatly influenced by aggregate shape and gradation. Mixes that are hard to pump, place, and finish include those deficient in fines, those with flat and elongated aggregates, and those with an excessive amount of fines (sand and cement). If the sand is deficient in fines, workability can be increased by addition of 30 to 50 lb/yd^3 of fly ash. The most effective method of producing workable concrete is to employ a well graded, combined aggregate gradation.

Modulus of elasticity of normal-weight concrete is between 2,000,000 and 6,000,000 psi. An estimate of the modulus of elasticity for normal-weight concrete with compressive strengths f'_c between 3000 and 5000 psi can be obtained by multiplying the square root of f'_c by 57,000. Above 5000 psi, the modulus should be determined using the procedure of ASTM C469. [See also Eq. (4.1) in Art. 4.17.2.]

Volume changes occur as either drying shrinkage, creep, or expansion due to external thermal sources. Drying shrinkage causes the most problems, because it produces cracks in the concrete surface. The primary cause of drying shrinkage cracks is an excessive amount of water in the mix. The water has two effects. First, it increases the water-cement ratio (W/C), weakening the concrete. Second, additional water beyond that needed for hydration of the cement creates an excessive number of bleed channels to exposed surfaces. When the cement paste undergoes its normal drying shrinkage, these channels cannot provide any resistance to penetration of water or aggressive chemicals.

Creep is a time-dependent deformation of concrete that occurs after an external load is applied to the concrete. It is an important consideration in design of prestressed concrete.

Concrete expands when heated and contracts when cooled. Coefficients of thermal expansion range from 3.2 to 7.0 millionths per °F. The most notable result of the response of concrete to thermal changes is the movement of external walls, which may bow because of temperature differentials.

Normal-weight concrete that is not designed for fire exposure expands on being heated. A side effect is some strength loss and a reduction in the modulus of elasticity.

Resistance to freezing and thawing can be accomplished by proper air entrainment in the concrete, use of a mix with a minimum water content, and proper curing of the concrete. Table 9.3 provides guidelines for the amount of air to use based upon exposure and maximum aggregate size.

Chemical attack may be internal (alkali-aggregate reaction) or external (sulfate attack or an aggressive service environment). In either case, the basic concerns are the characteristics of the available materials and the environment in which the concrete will be used. Alkali-reactive aggregates should be avoided, but if they must be used, a low-alkali cement complying with ASTM C150 Type II Modified should be selected. If sulfate attack is a concern, a low W/C (0.45 maximum) and air entrainment should be used with either a C150 Type V cement or a C150 Type II cement with C618 Type F fly ash. For protection from attack by other chemicals, a low W/C (0.45 maximum), more concrete cover over the reinforcing steel, a corrosion-protection additive, or a latex-modified concrete should be used. The American Concrete Institute "Building Code Requirements for Reinforced Concrete," ACI 318, contains requirements for special exposure conditions.

Abrasion resistance is a concern with pavements and hydraulic structures. Both require use of sound, durable, hard-rock aggregates, low W/C, and well-cured concrete.

Acceptable appearance depends on good workmanship and a supply of consistent materials. The formwork should be watertight and properly oiled before concrete placement. Forms should not be made of wood that will release sugars into the concrete and create a retarded surface finish. During concrete placement, the concrete should have consistent workability. The forms should be uniformly and consistently vibrated to consolidate the concrete.

("Standard Practice for Selecting Proportions for Normal Heavyweight, and Mass Concrete," ACI 211.1, and "Guide for Use of Normal Weight Aggregates in Concrete," ACI 221.)

4.17.2 Lightweight Concrete

Concrete weighing considerably less than the 144 lb/ft^3 of normal-weight concrete may be produced by use of lightweight aggregates or by expanding or foaming the concrete. Lightweight concrete is used principally to reduce the dead load of a structure and lower the cost of foundations. The light weight of the aggregates used for this type of concrete derives from the cellular structure of the particles. Hence, lightweight-aggregate concrete as well as foamed and expanded concretes have excellent fire-protection capabilities because of the internal voids in the aggregates or the concrete itself. When lightweight aggregates are used, they may be both fine and coarse, or lightweight coarse and normal-weight fine (sand), or normal-weight coarse and lightweight fine. The last combination is the least often used. Unit weights range from 90 lb/ft^3 (all aggregates lightweight) to 115 lb/ft^3 (sand lightweight). Typically, compressive strengths range from 2500 to 4000 psi. High-strength lightweight concretes, however, have been produced with maximum unit weights of 125 lb/ft^3 and strengths from 6000 to 9000 psi. Structural lightweight concretes are defined by the ACI as concretes with a 28-day compressive strength more than 2500 psi and air-dry unit weight of 115 lb/ft^3 or less.

The variable amount of water absorbed in the voids of lightweight aggregates makes use of W/C difficult in design of a lightweight-aggregate mix (Table 4.5). Air entrainment of 4 to 6% is desirable to prevent segregation. Maximum size of the coarse aggregate should not exceed half the depth of cover over the reinforcing steel.

Lightweight-aggregate concrete exposed to sulfates should have a compressive strength ranging from 3750 to 4750 psi (see ACI 318). For marine structures, the W/C should not exceed 0.40 and at least seven bags of cement should be used per cubic yard of concrete.

TABLE 4.5 Approximate Relationship between Cement Content and Compressive Strength

Compressive strength f'_c, psi	Aggregates all lightweight, lb/yd³	Sand aggregate lightweight, lb/yd³
2500	400–510	400–510
3000	440–560	420–560
4000	530–660	490–660

The modulus of elasticity E_c of lightweight concrete generally ranges from 1,500,000 to 3,000,000 psi. It may be estimated from

$$E_c = w^{1.5}\sqrt{f'_c} \tag{4.1}$$

where w = unit weight of concrete, lb/ft³
f'_c = 28-day compressive strength of concrete, psi

Volume changes occur in lightweight concrete as in normal-weight concrete, but lightweight concrete is stabler when exposed to heat. Drying shrinkage causes the most undesirable volume changes, because it produces cracks in the surfaces of the concrete. The primary cause of drying-shrinkage cracks is excessive water in the mix. The water has two effects. First, it increases the W/C and weakens the concrete. Second, the additional water beyond that needed for hydration of the cement creates an excessive number of bleed channels to the exposed surfaces. When the cement paste undergoes normal drying shrinkage, these channels cannot provide any resistance to ingress of aggressive chemicals.

Creep is an important concern for lightweight concrete, as it is for normal-weight concrete, especially for prestressed concrete.

("Standard Practice for Selecting Proportions for Structural Lightweight Concrete," ACI 211.2, and "Guide for Structural Lightweight Aggregate Concrete," ACI 213.)

4.17.3 Heavyweight Concrete

Concretes made with heavyweight aggregates are used for shielding and structural purposes in construction of nuclear reactors and other structures exposed to high-intensity radiation (see Art. 4.12). Heavyweight aggregates are used where heavyweight is needed, such as ship's ballast and encasement of underwater pipes, and for making shielding concretes because absorption of such radiation is proportional to density, and, consequently, these aggregates have greater capacity for absorption than those ordinarily used for normal concrete. With such aggregates, concrete weighing up to about 385 lb/ft³ can be produced.

Concrete made with limonite or magnetite can develop densities of 210 to 224 lb/ft³ and compressive strengths of 3200 to 5700 psi. With barite, concrete may weigh 230 lb/ft³ and have a strength of 6000 psi. With steel punchings and sheared bars as coarse aggregate and steel shot as fine aggregate, densities of 250 to 288 lb/ft³ and strengths of about 5600 psi can be attained. Generally, grading of aggregates and mix proportions are similar to those used for normal concrete.

The properties of heavyweight concrete are similar to those of normal-weight concrete. Mixing and placing operations, however, are more difficult than those for normal-weight concrete, because of segregation. Good grading, high cement content, low W/C, and air entrainment should be employed to prevent segregation. Sometimes, heavyweight aggregates are grouted in place to avoid segregation. Heavyweight concretes usually do not have good resistance to weathering or abrasion.

("Recommended Practice for Selecting Proportions for Normal, Heavyweight, and Mass Concrete," ACI 211.1.)

4.17.4 High-Performance Concretes

These concretes either have a high design strength (more than 6000 psi for normal-weight concrete and 5000 psi for lightweight concrete) or will be subjected to severe service environments. The differences between high-performance concretes and normal-weight concretes is that the former have lower W/C and smaller maximum aggregate sizes. ACI 318 specifies the W/C and compressive strengths for concrete in severe exposures and the maximum chloride-ion content of concrete. (See also Art. 4.17.1)

High-strength, portland-cement concretes generally incorporate in the mix fly ash, silica fume, or superplasticizers, or a combination of these admixtures. A retarder is often beneficial in controlling early hydration. The W/C may be as small as 0.25. The maximum size of aggregate should generally be limited to ½ in.

With superplasticizers, relatively high strengths can be achieved at early ages, such as 7-day strengths of normal concrete in 3 days and 28-day strengths in 7 days. Compressive strengths exceeding 10,000 psi can be achieved in 90 days.

Aside from reduction in W/C, the use of superplasticizers in production of high-strength concretes does not require significant changes in mix proportioning. An increase in the range of sand content of about 5%, however, may help avoid a harsh mix. Curing is very important, because strength gain halts when water is no longer available for hydration. Also, it is important that proper quantities of air-entraining admixtures be determined by trial. Some air loss may result when melamine- or naphthalene-based superplasticizers are used, whereas lignosulfonate-based water reducers may actually increase air content. Larger amounts of air-entraining agent may be needed for high-strength concretes, especially for low-slump mixes with high cement content and mixes with large amounts of some types of fly ash. Furthermore, some types of superplasticizers and air-entraining admixtures may not be compatible with each other.

("State-of-the-Art Report on High-Strength Concrete," ACI 363.)

4.17.5 Nonstructural or Foamed Cellular Concretes

These are formed by the use of admixtures that generate or liberate gas bubbles in concrete in the plastic stage. Aluminum powder, which reacts with the alkalies in cement to release hydrogen, is generally used for this purpose, although hydrogen peroxide, which generates oxygen, or activated carbon, which liberates absorbed air, can be used. These foaming agents create stable, uniformly dispersed air spaces within the concrete when it sets. Perlite and vermiculite are most frequently used as aggregates. The resulting concrete may weigh 50 lb/ft^3 or less and have a compressive strength up to 2500 psi. Applications of such lightweight concretes include topping and soundproofing barriers over structural concrete slabs.

The effectiveness of the admixture is controlled by the duration of mixing, handling, and placing of the mix relative to the gas-generation rate. The amount of unpolished aluminum powder to be added to a mix may range from 0.005 to 0.02% by weight of cement under normal conditions. Larger quantities, however, may be used to produce lower-strength concretes. More aluminum may be needed at low temperatures to achieve the same amount of concrete expansion, for example, twice as much at 40°F as at 70°F. Furthermore, at low temperatures, to speed up gas generation, it may be necessary to add to the mix alkalies such as sodium hydroxide, hydrated lime, or trisodium phosphate. Also, to prevent the powder from floating on the surface of mixing water, the aluminum may be premixed with sand or combined with other admixtures.

Curing is very important. If good curing practices and jointing are not followed, extensive drying shrinkage may result.

4.18 POLYMER CONCRETES

Plastics with long-chain molecules, called polymers, are used in several ways to enhance concrete properties: replacement of portland cement, incorporation in a mix as an admixture, and impregnating hardened concrete.

Polymer concretes, such as methyl methacrylate and unsaturated polyester, in which a polymer replaces portland cement may have more than double the strength and modulus of elasticity of portland-cement concrete. Creep is less and resistance to freezing and thawing cycles is higher with the polymer concretes. After curing for a very short time, for example, overnight at room temperature, polymer concretes are ready for use, whereas ordinary concrete may have to cure for about a week before exposure to service loads.

Monomers and polymers may be used as admixtures for restoring and resurfacing deteriorated concrete surfaces. Latexes of methyl methacrylate, polyester, styrene, epoxy-styrene, furans, styrene-butadiene, and vinylidene chloride have been employed for these purposes. The resulting concrete hardens more rapidly than normal concrete. A polymer admixture may also be used to improve the bonding properties of portland cement. Inserted in a mix as an emulsion for this purpose, the admixture supplies a significant amount of water to the mix, which becomes available for hydration of the cement. The release of the water also sets the emulsion. Hence, moist curing is not desirable, inasmuch as the emulsion needs to dry to develop the desired strength. A grout or mortar containing the bonding admixture develops a higher bond strength when applied as a thin layer than as a thick one and the bond may be stronger than materials being joined.

Impregnation of concrete with polymers is sometimes used to harden surfaces exposed to heavy traffic. Strength and other properties of the impregnated concrete are similar to those of concrete in which polymers replace portland cement. Impregnation is achieved by first drying the concrete surface with heat and then soaking the surface with a monomer, such as methyl methacrylate, styrene, acrylonitrile, or tert-butyl styrene. It is subsequently cured with heat.

Slab Toppings. At least partly because of excellent adhesion, epoxies are formulated with sand and other fillers to provide surfacing materials for concrete. Unlike standard concrete topping, epoxy-based surfacing materials can be thin. They are especially useful for smoothing uneven, irregular surfaces. The epoxy cures quickly, allowing use of the surface in a short time.

Grout. Cracked concrete can be repaired with an epoxy grout. The grout is forced into cracks under pressure for deep penetration. Because of its good bonding strength, the epoxy grout can largely restore strength, while, at the same time, sealing the crack against penetration by liquids.

("Polymers in Concrete," ACI 548; "Guide for the Use of Polymers in Concrete," ACI 548.1; and "Polymer Modified Concrete," SP-99, American Concrete Institute.)

4.19 CONCRETE MASONRY UNITS

A wide variety of manufactured products are produced from concrete and used in building construction. These include such items as concrete brick, concrete block or tile, concrete floor and roof slabs, precast wall panels, precast beams, and cast stone. These items are made both from normal dense concrete mixes and from mixes with lightweight aggregates. Concrete blocks are made with holes through them to reduce their weight and to enable masons to grip them.

Nominal size (actual dimensions plus width of mortar joint) of hollow concrete block usually is 8 × 8 × 16 in. Solid blocks often are available with nominal size of 4 × 8 × 16 in or 4 × 2½ × 8 in. For a list of modular sizes, see "Standard Sizes of Clay and Concrete Modular Units," ANSI A62.3.

Properties of the units vary tremendously—from strong, dense load-bearing units used under exposed conditions to light, relatively weak, insulating units used for roof and fire-resistant construction.

Many types of concrete units have not been covered by adequate standard specifications. For these units, reliance must be placed upon the manufacturer's specifications. Requirements for strength and absorption of concrete brick and block established by ASTM for Type I, Grades N-I and S-I (moisture-controlled), and Type II, Grades N-II and S-II (non-moisture-controlled), units are summarized in Table 4.6.

Manufactured concrete units have the advantage (or sometimes disadvantage) that curing is under the control of the manufacturer. Many methods of curing are used, from simply stacking the units in a more or less exposed location to curing under high-pressure steam. The latter method appears to have considerable merit in reducing ultimate shrinkage of the block. Shrinkage may be as small as ¼ to ⅜ in per 100 ft for concrete units cured with high-pressure steam. These values are about one-half as great as those obtained with normal atmospheric curing. Tests for moisture movement in blocks cured with high-pressure and high-temperature steam indicate expansions of from ¼ to ½ in per 100 ft after saturation of previously dried specimens.

BURNED-CLAY UNITS

Use of burned-clay structural units dates from prehistoric times. Hence durability of well-burned units has been adequately established through centuries of exposure in all types of climate.

Modern burned-clay units are made in a wide variety of sizes, shapes, colors, and textures to suit the requirements of modern architecture. They include such widely diverse units as common and face brick; hollow clay tile in numerous shapes,

TABLE 4.6 Summary of ASTM Specification Requirements for Concrete Masonry Units

	Compressive strength, min, psi		Moisture content for Type I units, max, % of total absorption (average of 5 units)			Moisture absorption, max, lb/ft³ (average of 5 units)		
			Avg annual relative humidity, %			Oven-dry weight of concrete, lb/ft³		
	Avg of 5 units	Individual min	Over 75	75 to 50	Under 50	125 or more	105 to 125	Under 105
Concrete building brick, ASTM C55:								
N-I, N-II (high strength severe exposures)	3500	3000				10	13	15
S-I, S-II (general use, moderate exposures)	2500	2000				13	15	18
Linear shrinkage, %:								
0.03 or less			45	40	35			
0.03 to 0.45			40	35	30			
Over 0.045			35	30	25			
Solid, load-bearing units, ASTM C145:								
N-I, N-II (unprotected exterior walls below grade or above grade exposed to frost)	1800	1500				13	15	18
S-I, S-II (protected exterior walls below grade or above grade exposed to frost)	1200	1000						20*
Linear shrinkage, %: (Same as for brick)								
Hollow, load-bearing units, ASTM C90:								
N-I, N-II (general use)	1000	800				13	15	18
S-I, S-II (above grade, weather protected)	700	600						20*
Linear shrinkage, %: (Same as for brick)								
Hollow, non-load-bearing units, ASTM C129	600	500						
Linear shrinkage, %: (Same as for brick)								

*For units weighing less than 85 lb/ft³.

sizes, and designs for special purposes; ceramic tile for decorative and sanitary finishes, and architectural terra cotta for ornamentation.

Properties of burned-clay units vary with the type of clay or shale used as raw material, method of fabrication of the units, and temperature of burning. As a consequence, some units, such as salmon brick, are underburned, highly porous, and of poor strength. But others are almost glass hard, have been pressed and burned to almost eliminate porosity, and are very strong. Between these extremes lie most of the units used for construction.

4.20 BRICK—CLAY OR SHALE

Brick have been made in a wide range of sizes and shapes, from the old Greek brick, which was practically a 23-in cube of 12,650 in^3 volume, to the small Belgian brick, about 1¾ × 3⅜ × 4½ in with a total volume of only 27 in.^{3} The present common nominal sizes in the United States are 4 or 6 in thick by 2⅔ or 4 in high by 8 or 12 in long. For a list of modular sizes, see "Standard Sizes of Clay and Concrete Modular Masonry Units," ANSI A62.3. Actual dimensions are smaller, usually by the amount of the width of the mortar joint. Current specification requirements for strength and absorption of building brick are given in Table 4.7 (see ASTM C652, C62, and C216). Strength and absorption of brick from different producers vary widely.

Thermal expansion of brick may range from 0.0000017 per °F for fire-clay brick to 0.0000069 per °F for surface-clay brick. Wetting tests of brick indicated expansions varying from 0.0005 to 0.025%.

The thermal conductivity of dry brick as measured by several investigators ranges from 1.29 to 3.79 Btu/(hr)(ft^3)(°F)(in). The values are increased by wetting.

4.21 STRUCTURAL CLAY TILE

Structural clay tiles are hollow burned-clay masonry units with parallel cells. Such units have a multitude of uses: as a facing tile for interior and exterior unplastered walls, partitions, or columns; as load-bearing tile in masonry constructions designed

TABLE 4.7 Physical Requirements for Clay or Shale Solid Brick

Grade	Compressive strength, flat, min, psi		Water absorption, 5-hr boil, max—%		Saturation* coefficient, max—%	
	Avg of 5	Individual	Avg of 5	Individual	Avg of 5	Individual
SW—Severe weathering	3000	2500	17.0	20.0	0.78	0.80
MW—Moderate weathering	2500	2200	22.0	25.0	0.88	0.90
NW—No exposure	1500	1250	No limit	No limit	No limit	No limit

*Ratio of 24-hr cold absorption to 5-hr boil absorption.

TABLE 4.8 Physical Requirement Specification for Structural Clay Tile

Type and grade	Absorption, % (1 hr boiling)		Compressive strength, psi (based on gross area)			
			End-construction tile		Side-construction tile	
	Avg of 5 tests	Individual max	Min avg of 5 tests	Individual min	Min avg of 5 tests	Individual min
Load-bearing (ASTM C34):						
LBX	16	19	1400	1000	700	500
LB	25	28	1000	700	700	500
Non-load-bearing (ASTM C56):						
NB		28				
Floor tile (ASTM C57):						
FT1		25	3200	2250	1600	1100
FT2		25	2000	1400	1200	850
Facing tile (ASTM C212):						
FTX	9 (max)	11				
FTS	16 (max)	19				
Standard			1400	1000	700	500
Special duty			2500	2000	1200	1000
Glazed units (ASTM C126)			3000	2500	2000	1500

LBX. Tile suitable for general use in masonry construction and adapted for use in masonry exposed to weathering. They may also be considered suitable for direct application of stucco.

LB. Tile suitable for general use in masonry where not exposed to frost action, or in exposed masonry where protected with a facing of 3 in or more of stone, brick, terra cotta, or other masonry.

NB. Non-load-bearing tile made from surface clay, shale, or fired clay.

FT 1 and FT 2. Tile suitable for use in flat or segmental panels or in combination tile and concrete ribbed-slab construction.

FTX. Smooth-face tile suitable for general use in exposed exterior and interior masonry walls and partitions, and adapted for use where tiles low in absorption, easily cleaned, and resistant to staining are required and where a high degree of mechanical perfection, narrow color range, and minimum variation in face dimensions are required.

FTS. Smooth or rough-texture face tile suitable for general use in exposed exterior and interior masonry walls and partitions and adapted for use where tile of moderate absorption, moderate variation in face dimensions, and medium color range may be used, and where minor defects in surface finish, including small handling chips, are not objectionable.

Standard. Tile suitable for general use in exterior or interior masonry walls and partitions.

Special duty. Tile suitable for general use in exterior or interior masonry walls and partitions and designed to have superior resistance to impact and moisture transmission, and to support greater lateral and compressive loads than standard tile construction.

Glazed units. Ceramic-glazed structural clay tile with a glossy or satin-mat finish of either an opaque or clear glaze, produced by the application of a coating prior to firing and subsequently made vitreous by firing.

to carry superimposed loads; as partition tile for interior partitions carrying no superimposed load; as fireproofing tile for protection of structural members against fire; as furring tile for lining the inside of exterior walls; as floor tile in floor and roof construction; and as header tiles, which are designed to provide recesses for header units in brick or stone-faced walls. Units are available with the following ranges in nominal dimensions: 8 to 16 in in length, 4 in for facing tile to 12 in for load-bearing tile in height, and 2 in for facing tile to 12 in for load-bearing tile in thickness.

Two general types of tile are available—side-construction tile, designed to receive its principal stress at right angles to the axis of the cells, and end-construction tile designed to receive its principal stress parallel to the axis of the cells.

Tiles are also available in a number of surface finishes, such as opaque glazed tile, clear ceramic-glazed tile, nonlustrous glazed tile, and scored, combed, or roughened finishes designed to receive mortar, plaster, or stucco.

Requirements of the appropriate ASTM specifications for absorption and strength of several types of tile are given in Table 4.8 (see ASTM C34, C56, C57, C212, and C126 for details pertaining to size, color, texture, defects, etc.). Strength and absorption of tile made from similar clays but from different sources and manufacturers vary widely. The modulus of elasticity of tile may range from 1,620,000 to 6,059,000 psi.

4.22 CERAMIC TILES

Ceramic tile is a burned-clay product used primarily for decorative and sanitary effects. It is composed of a clay body on which is superimposed a decorative glaze.

The tiles are usually flat but vary in size from about ½ in square to more than 6 in. Their shape is also widely variable—squares, rectangles, and hexagons are the predominating forms, to which must be added coved moldings and other decorative forms. These tiles are not dependent on the color of the clay for their final color, since they are usually glazed. Hence, they are available in a complete color gradation from pure whites through pastels of varying hue to deep solid colors and jet blacks.

Properties of the base vary somewhat. In particular, absorption ranges from almost zero to about 15%. The glaze is required to be impervious to liquids and should not stain, crack, or craze.

Ceramic tiles are applied on a solid backing by means of a mortar or adhesive. They are usually applied with the thinnest possible mortar joint; consequently accuracy of dimensions is of greatest importance. Since color, size, and shape of tile are important, selection of tile should be based on the current literature of the manufacturer.

4.23 ARCHITECTURAL TERRA COTTA

The term "terra cotta" has been applied for centuries to decorative molded-clay objects whose properties are similar to brick. The molded shapes are fired in a manner similar to brick.

Terra cotta is frequently glazed to produce a desired color or finish. This introduces the problem of cracking or crazing of the glaze, particularly over large areas.

Structural properties of terra cotta are similar to those of clay or shale brick.

TABLE 4.9 Strength Characteristics of Commercial Building Stones

Stone	Compressive strength, psi, range	Modulus of rupture, psi, range	Shear strength, psi, range	Tensile strength, psi, range	Elastic modulus, psi, range	Toughness		Wear resistance	
						Range	Avg	Range	Avg
Granite	7,700–60,000	1,430–5,190	2,000–4,800	600–1,000	5,700,000–8,200,000	8–27	13	43.9–87.9	60.8
Marble	8,000–50,000	600–4,900	1,300–6,500	150–2,300	7,200,000–14,500,000	2–23	6	6.7–41.7	18.9
Limestone	2,600–28,000	500–2,000	800–4,580	280–890	1,500,000–12,400,000	5–20	7	1.3–24.1	8.4
Sandstone	5,000–20,000	700–2,300	300–3,000	280–500	1,900,000–7,700,000	2–35	10	1.6–29.0	13.3
Quartzite	16,000–45,000					5–30	15		
Serpentine	11,000–28,000	1,300–11,000		800–1,600	4,800,000–9,600,000			13.3–111.4	46.9
Basalt	28,000–67,000					5–40	20		
Diorite	16,000–35,000					6–38	23		
Syenite	14,000–28,000								
Slate		6,000–15,000	2,000–3,600	3,000–4,300	9,800,000–18,000,000	10–56		5.6–11.7	7.7
Diabase						6–50	19		
Building limestone						3–8	4.4		

BUILDING STONES

Principal building stones generally used in the United States are limestones, marbles, granites, and sandstones. Other stones such as serpentine and quartzite are used locally but to a much lesser extent. Stone, in general, makes an excellent building material, if properly selected on the basis of experience; but the cost may be relatively high.

Properties of stone depend on what nature has provided. Therefore, the designer does not have the choice of properties and color available in some of the manufactured building units. The most the stone producer can do for purchasers is to avoid quarrying certain stone beds that have been proved by experience to have poor strength or poor durability.

4.24 PROPERTIES OF BUILDING STONES

Data on the strength of building stones are presented in Table 4.9, summarized from *U.S. National Bureau of Standards Technical Papers,* No. 123, B. S. Vol. 12; No. 305, Vol. 20, p. 191; No. 349, Vol. 21, p. 497; *Journal of Research of the National Bureau of Standards,* Vol. 11, p. 635; Vol. 25, p. 161). The data in Table 4.9 pertain to dried specimens. Strength of saturated specimens may be either greater or less than that of completely dry specimens.

The modulus of rupture of dry slate is given in Table 4.9 as ranging from 6000 to 15,000 psi. Similar slates, tested wet, gave moduli ranging from 4700 to 12,300 psi. The ratio of wet modulus to dry modulus varied from 0.42 to 1.12 and averaged 0.73.

Data on the true specific gravity, bulk specific gravity, unit weights, porosity, and absorption of various stones are given in Table 4.10.

TABLE 4.10 Specific Gravity and Porosity of Commercial Building Stones

Stone	Specific gravity		Unit weight, lb per cu ft	Porosity, %	Absorption, %	
	True	Bulk			By weight	By volume
Granite	2.599–3.080	2.60–3.04	157–187	0.4–3.8	0.02–0.58	0.4–1.8
Marble	2.718–2.879	2.64–2.86	165–179	0.4–2.1	0.01–0.45	0.04–1.2
Limestone	2.700–2.860	1.87–2.69	117–175	1.1–31.0		6–15
Slate	2.771–2.898	2.74–2.89	168–180	0.1–1.7	0.00–1.63	0.3–2.0
Basalt		2.9–3.2				
Soapstone		2.8–3.0				
Gneiss		2.7–3.0				
Serpentine		2.5–2.8	158–183			
Sandstone		2.2–2.7	119–168	1.9–27.3		6–18
Quartzite			165–170	1.5–2.9		

Permeability of stones varies with types of stone, thickness, and driving pressure that forces water through the stone. Table 4.11 presents data for the more common stones at three different pressures, as reported in "Permeability of Stone," *U.S. National Bureau of Standards Technical Papers,* No. 305, Vol. 20, p. 191. The units of measurement of permeability are cubic inches of water that will flow through a square foot of a specimen ½ in thick in 1 hr.

TABLE 4.11 Permeability of Commercial Building Stones

[*$in^3/(ft^2)(hr)$ for ½-in thickness*]

Pressure, psi	1.2	50	100
Granite	0.06–0.08	0.11	0.28
Slate	0.006–0.008	0.08–0.11	0.11
Marble	0.06–0.35	1.3–16.8	0.9–28.0
Limestone	0.36–2.24	4.2–44.8	9.0–109
Sandstone	4.2–174.0	51.2	221

Data on thermal expansion of building stones as given in Table 4.12 show that limestones have a wide range of expansion as compared with granites and slates.

Marble loses strength after repeated heating and cooling. A marble that had an original strength of 9174 psi had a strength after 50 heatings to 150°C of 8998 psi—a loss of 1.9%. After 100 heatings to 150°C, the strength was only 8507 psi, or a loss of 7.3%. The latter loss in strength was identical with that obtained on freezing and thawing the same marble for 30 cycles. Also, marble retains a permanent expansion after repeated heating.

TABLE 4.12 Coefficient of Thermal Expansion of Commercial Building Stones

Stone	Range of coefficient
Limestone	$(4.2–22) \times 10^{-6}$
Marble	$(3.6–16) \times 10^{-6}$
Sandstone	$(5.0–12) \times 10^{-6}$
Slate	$(9.4–12) \times 10^{-6}$
Granite	$(6.3–9) \times 10^{-6}$

4.25 FREEZING AND THAWING OF STONE

In freezing and thawing tests of 89 different marbles ("Physical and Chemical Tests of Commercial Marbles of U.S.," *U.S. National Bureau of Standards Technical Papers,* No. 123, Vol. 12), after 30 cycles, 66 marbles showed loss of strength ranging from 1.2 to 62.1% and averaging 12.3% loss. The other 23 marbles showed increases in strength ranging from 0.5 to 43.9% and averaging 11.2% increase.

Weight change was also determined in this investigation to afford another index of durability. Of 86 possible comparisons after 30 cycles of freezing and thawing, 16 showed no change in weight, 64 showed decreases in weight ranging from 0.01 to 0.28% and averaging 0.04% loss, while 6 showed increases in weight ranging from 0.01 to 0.08% and averaging 0.04%.

GYPSUM PRODUCTS

Gypsum is a cementitious material composed of at least 70% of $CaSO_4 \cdot 2H_2O$ by weight (Art. 4.6). It is a main ingredient of many building products.

4.26 GYPSUMBOARD

This product consists of a core of set gypsum surfaced with specially manufactured paper firmly bonded to the core. It is designed to be used without addition of plaster for walls, ceilings, or partitions and provides a surface suitable to receive either paint or paper (see also Sec. 11). Gypsumboard is extensively used in "dry-wall" construction, where plaster is eliminated. It is also available with one surface covered with aluminum or other heat-reflecting type of foil, or with imitation wood-grain or other patterns on the exposed surface so that no additional decoration is required.

The types of gypsumboard generally available include wallboard, backing board, coreboard, fire-resistant gypsumboard, water-resistant gypsumboard, gypsum sheathing, and gypsum formboard.

Gypsum Wallboard. This type is used for the surface layer on interior walls and ceilings. Regular gypsum wallboard comes with gray liner paper on the back and a special paper covering, usually cream-colored, on facing side and edges. This covering provides a smooth surface suitable for decoration. Foil-backed gypsum wallboard has aluminum foil bonded to the liner paper to serve as a vapor barrier and, when contiguous to an airspace, as thermal insulation. Predecorated gypsum wallboard does not require decorative treatment after installation because it comes with a finished surface, often a decorative vinyl or paper sheet. Wallboard should conform with ASTM C36.

Wallboard usually is available 4 ft wide in the following thicknesses and lengths:

¼ in—for covering and rehabilitating old walls and ceilings, 4 to 12 ft long

5⁄16 in—where thickness greater than ¼ in is desired, 4 to 14 ft long

⅜ in—mainly for the outer face in two-layer wall systems, 4 to 16 ft long

½ in—for single-layer new construction with supports 16 to 24 in c to c, 4 to 16 ft long

⅝ in—for better fire resistance and sound control than ½ in provides, 4 to 16 ft long

Standard edges are rounded, beveled, tapered, or square.

Backing Board. This type is used as a base layer in multi-ply construction, where several layers of gypsumboard are desired for high fire resistance, sound control, and strength in walls. It has gray liner paper on front and back faces. Also available is backing board with aluminum foil bonded to the back face. Gypsum backing board should conform with ASTM C442. The boards come 16 to 48 in wide, 4 to 16 ft long, and ¼ to 1 in thick.

Gypsum Coreboard. To save space, this type is used as a base in multi-ply construction of self-supporting (studless) gypsum walls. Coreboard may be supplied as 1-in-thick, solid backing board or as two factory-laminated, ½-in-thick layers of backing board. Coreboard too should conform with C442.

Type X Gypsumboard. For use in fire-rated assemblies, Type X may be gypsum wallboard, backing board, or coreboard with core made more fire resistant by addition of glass fiber or other reinforcing materials.

Water-Resistant Gypsum Backing Board. This type comes with a water-resistant gypsum core and water-repellant face paper. It may be used as a base for wall tile in baths, showers, and other areas subject to wetting. The board should conform with ASTM C630.

Gypsum Sheathing. This type is used as fire protection and bracing of exterior frame walls. It must be protected from the weather by an exterior facing. Sheathing should conform with ASTM C79. It comes 24 to 48 in wide, 6 to 12 ft long, and ⅜, 4⁄10, ½, and ⅝ in thick.

Gypsum Formboard. This type is used as a permanent form in the casting of gypsum-concrete roof decks.

("Architect Data Book—Construction Products and Systems," Gold Bond Building Products, a National Gypsum Division, 2001 Rexford Road, Charlotte, NC 28211; "Gypsum Products Design Data," Gypsum Association, 1603 Orrington Ave., Evanston, IL 60201; "Gypsum Construction Handbook," United States Gypsum, 101 South Wacker Drive, Chicago, IL 60606.)

4.27 GYPSUM LATH

Gypsum lath is similar to gypsumboard in that it consists of a core of set gypsum surfaced with paper. The paper for gypsumboard, however, is produced so that it is ready to receive paint or paper, while that for gypsum lath is specially designed or treated so that plaster will bond tightly to the paper. In addition, some lath provides perforations or other mechanical keying to assist in holding the plaster firmly on the lath. It is also available with reflective foil backing (see also Art. 11.25.5).

Gypsum lath should conform with ASTM C37. It comes in 16-, 16½-, 24-, and 32-in widths, lengths of 32, 36, and 48 in, and ⅜- and ½-in widths.

Veneer plasters, special proprietary compositions for thin plaster surfaces, are best applied over veneer plaster base, similar to gypsum lath, but produced to accommodate the veneer plaster compositions. Both gypsum lath and veneer base are made as regular, X-rated (fire-retardant), and insulating (foil-backed) types. These bases should conform with ASTM G588. They come 48 in wide, 6 to 16 ft long, and ⅜, ½, and ⅝ in thick.

4.28 GYPSUM SHEATHING BOARD

Gypsum sheathing boards are similar in construction to gypsumboard (Art. 4.26), except that they are provided with a water-repellent paper surface. They are commonly made ¾ to ⅝ in thick, 6 to 12 ft long, and with a nominal width of 24 or 48 in in conformance with ASTM C79. They are made with either square edges or with V tongue-and-groove edges. Sheathing boards also are available with a water-repellent core or fire-resistant Type X.

4.29 GYPSUM PARTITION TILE OR BLOCK

Gypsum tiles or blocks are used for non-load-bearing partition walls and for protection of columns, elevator shafts, etc., against fire. They have been essentially replaced by dry-wall systems.

4.30 GYPSUM PLANK

A precast gypsum product used particularly for roof construction is composed of a core of gypsum cast in the form of a plank, with wire-fabric reinforcement and usually with tongue-and-groove metal edges and ends. The planks are available in two thicknesses—a 2-in plank, which is 15 in wide and 10 ft long, and a 3-in plank, which is 12 in wide and 30 in long. (See ASTM C377.)

GLASS AND GLASS BLOCK

Glass is so widely used for decorative and utilitarian purposes in modern construction that it would require an encyclopedia to list all the varieties available. Clear glass for windows and doors is made in varying thicknesses or strengths, also in double layers to obtain additional thermal insulation. Safety glass, laminated from sheets of glass and plastic, or made with embedded wire reinforcement, is available for locations where breakage might be hazardous. For ornamental work, glass is available in a wide range of textures, colors, finishes, and shapes.

4.31 WINDOW GLASS

Various types and grades of glass are used for glazing:

Clear Window Glass. This is the most extensively used type for windows in all classes of buildings. A range of grades, as established by Federal Government Standard DD-G-451c, classifies quality according to defects. The more commonly used grades are A and B. A is used for the better class of buildings where appearance is important, and B is used for industrial buildings, some low-cost residences, basements, etc.

With respect to thickness, clear window glass is classified as "single-strength" about 3⁄32 in thick; "double-strength," about 1⁄8 in thick; and "heavy-sheet," up to 7⁄32 in thick. Maximum sizes are as follows: single-strength, 40 × 50 in; double-strength, 60 × 80 in; and heavy sheet, 76 × 120 in. Because of flexibility, single strength and double strength should never be used in areas exceeding 12 ft^2, and for appearance's sake areas should not exceed 7 ft^2.

Plate and Float Glass. These have, in general, the same performance characteristics. They are of superior quality, more expensive, and have better appearance, with no distortion of vision at any angle. Showcase windows, picture windows, and exposed windows in offices and commercial buildings are usually glazed with polished plate or float glass. Thicknesses range from 1⁄8 to 7⁄8 in. There are two standard qualities, *silvering* and *glazing,* the latter being employed for quality glazing.

Processed Glass and Rolled Figured Sheet. These are general classifications of obscure glass. There are many patterns and varying characteristics. Some provide true obscurity with a uniform diffusion and pleasing appearance, while others may give a maximum transmission of light or a smoother surface for greater cleanliness. The more popular types include a clear, polished surface on one side with a pattern for obscurity on the other side.

Obscure Wired Glass. This usually is specified for its fire-retarding properties, although it is also used in doors or windows where breakage is a problem. It should not be used in pieces over 720 in^2 in area (check local building code).

Polished Wired Glass. More expensive than obscure wired glass, polished wired glass is used where clear vision is desired, such as in school or institutional doors.
There are also many special glasses for specific purposes:

Heat-Absorbing Glass. This reduces heat, glare, and a large percentage of ultraviolet rays, which bleach colored fabrics. It often is used for comfort and reduction of air-conditioning loads where large areas of glass have a severe sun exposure. Because of differential temperature stresses and expansion induced by heat absorption under severe sun exposure, special attention should be given to edge conditions. Glass having clean-cut edges is particularly desirable, because these affect the edge strength, which, in turn must resist the central-area expansion. A resilient glazing material should be used.

Corrugated Glass, Wired Glass, and Plastic Panels. These are used for decorative treatments, diffusing light, or as translucent structural panels with color.

Laminated Glass. This consists of two or more layers of glass laminated together by one or more coatings of a transparent plastic. This construction adds strength. Some types of laminated glass also provide a degree of security, sound isolation, heat absorption, and glare reduction. Where color and privacy are desired, fade-proof opaque colors can be included. When fractured, a laminated glass tends to adhere to the inner layer of plastic and, therefore, shatters into small splinters, thus minimizing the hazard of flying glass.

Bullet-Resisting Glass. This is made of three or more layers of plate glass laminated under heat and pressure. Thicknesses of this glass vary from 3⁄4 to 3 in. The more common thicknesses are 1 3⁄16 in, to resist medium-powered small arms; 1 1⁄2 in, to resist high-powered small arms; and 2 in, to resist rifles and submachine guns. (Underwriters Laboratories lists materials having the required properties for various degrees of protection.) Greater thicknesses are used for protection against armor-piercing projectiles. Uses of bullet-resisting glass include cashier windows, bank teller cages, toll-bridge booths, peepholes, and many industrial and military applications. Transparent plastics also are used as bullet-resistant materials, and some of these materials have been tested by the Underwriters Laboratories. Thicknesses of 1¼ in or more have met UL standards for resisting medium-powered small arms.

Tempered Glass. This is produced by a process of reheating and sudden cooling that greatly increases strength. All cutting and fabricating must be done before tempering. Doors of ½- and ¾-in-thick tempered glass are commonly used for commercial building. Other uses, with thicknesses from ⅛ to ⅞ in, include backboards for basketball, showcases, balustrades, sterilizing ovens, and windows, doors, and mirrors in institutions. Although tempered glass is 4½ to 5 times as strong as annealed glass of the same thickness, it is breakable, and when broken, disrupts into innumerable small fragments of more or less cubical shape.

Tinted and Coated Glasses. These are available in several types and for varied uses. As well as decor, these uses can provide for light and heat reflection, lower light transmission, greater safety, sound reduction, reduced glare, and increased privacy.

Transparent Mirror Glass. This appears as a mirror when viewed from a brightly lighted side, and is transparent to a viewer on the darker opposite side. This one-way-vision glass is available as a laminate, plate or float, tinted, and in tempered quality.

Plastic Window Glazing. Made of such plastics as acrylic or polycarbonate, plastic glazing is used for urban school buildings and in areas where high vandalism might be anticipated. These plastics have substantially higher impact strength than glass or tempered glass. Allowance should be made in the framing and installation for expansion and contraction of plastics, which may be about 8 times as much as that of glass. Note also that the modulus of elasticity (stiffness) of plastics is about one-twentieth that of glass. Standard sash, however, usually will accommodate the additional thickness of plastic and have sufficient rabbet depth.

Suspended Glazing. This utilizes metal clamps bonded to tempered plate glass at the top edge, with vertical glass supports at right angles for resistance to wind pressure (Fig. 4.1). These vertical supports, called stabilizers, have their exposed edges polished. The joints between the large plates and the stabilizers are sealed with a bonding cement. The bottom edge or sill is held in position by a metal channel, and sealed with resilient waterproofing. Suspended glazing offers much greater latitude in use of glass and virtually eliminates visual barriers.

Safety Glazing. A governmental specification Z-97, adopted by many states, requires entrance-way doors and appurtenances glazed with tempered, laminated, or plastic material.

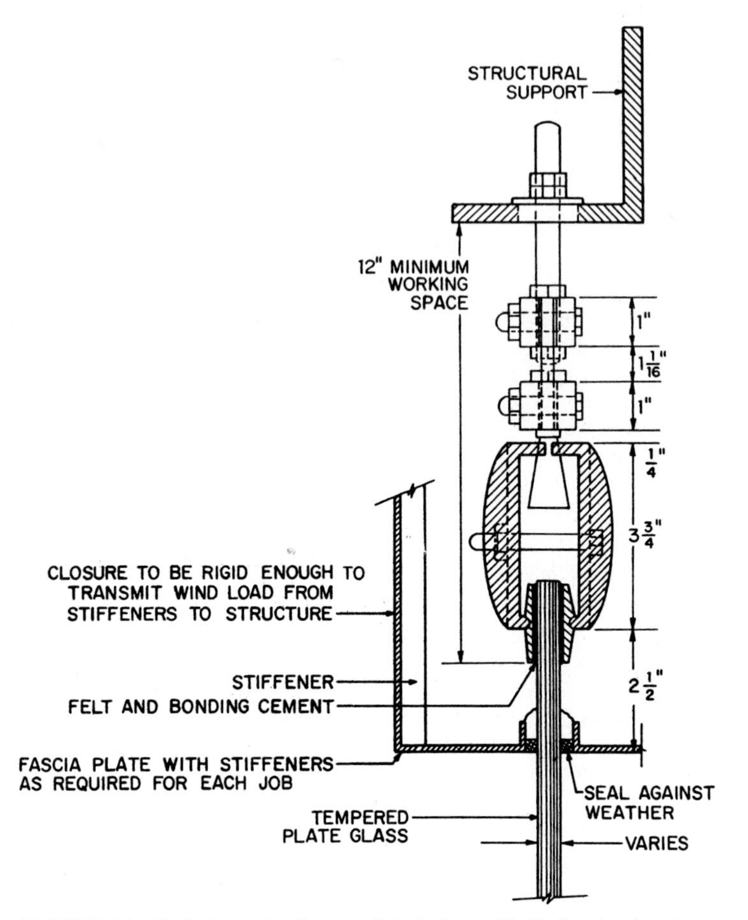

FIGURE 4.1 Typical details of suspended glazing. *(F. H. Sparks, Co., Inc., New York.)*

4.32 GLASS BLOCK

Glass blocks are made by first pressing or shaping half blocks to the desired form, then fusing the half blocks to form a complete block. A block is usually 3⅞ in thick and 5¾, 7¾, or 11¾ in square. The center of the block is hollow and is under a partial vacuum, which adds to the insulating properties of the block. Corner and radial blocks are also available to produce desired architectural effects.

Glass block is commonly laid up in a cement or a cement-lime mortar. Since there is no absorption by the block to facilitate bond of mortar, various devices are employed to obtain a mechanical bond. One such device is to coat the sides of the block with a plastic and embed therein particles of sand. The difficulty in obtaining permanent and complete bond sometimes leads to the opening up of mortar joints. A wall of glass block, exposed to the weather, may leak badly in a rainstorm unless unusual precautions have been taken during the setting of the block to obtain full and complete bond.

Glass blocks have a coefficient of thermal expansion that is from 1½ to 2 times that of other masonry. For this reason, large areas of block may expand against solid masonry and develop sufficient stress so that the block will crack. Manufacturers usually recommend an expansion joint every 10 ft or so, to prevent building up of pressure sufficient to crack the block. With adequate protection against expansion and with good workmanship, or with walls built in protected locations, glass-block walls are ornamental, sanitary, excellent light transmitters, and have rather low thermal conductivity.

WOOD

Wood is a building material made from trees. It is a natural polymer composed of cells in the shape of long, thin tubes with tapered ends. The cell wall consists of cellulose crystals, which are bonded together by a complex amorphous lignin composed of carbohydrates. Most of the cells in a tree trunk are oriented vertically. Consequently, properties of wood in the direction of cell axes, usually referred to as longitudinal, or parallel to grain, differ from those in the other (radial or circumferential) directions, or across the grain.

4.33 MECHANICAL PROPERTIES OF WOOD

Because of its structure, wood has different strength properties parallel and perpendicular to the grain. Tensile, bending, and compressive strengths are greatest parallel to the grain and least across the grain, whereas shear strength is least parallel to the grain and greatest across the grain. Except in plywood, the shearing strength of wood is usually governed by the parallel-to-grain direction.

The compressive strength of wood at an angle other than parallel or perpendicular to the grain is given by the following formula:

$$C_\theta = \frac{C_2 C_2}{C_1 \sin^2 \theta + C_2 \cos^2 \theta} \tag{4.2}$$

in which C_θ is the strength at the desired angle θ with the grain, C_1 is the compressive strength parallel to grain, and C_2 is the compressive strength perpendicular to the grain.

Increasing moisture content reduces all strength properties except impact bending, in which green wood is stronger than dry wood. The differences are brought out in Table 4.13. In practice, no differentiation is made between the strength of green and dry wood in engineering timbers, because of seasoning defects that may occur in timbers as they dry and because large timbers normally are put into service without having been dried. This is not true of laminated timber, in which dry wood must be employed to obtain good glued joints. For laminated timber, higher stresses can be employed than for ordinary lumber. In general, compression and bending parallel to the grain are affected most severely by moisture, whereas modulus of elasticity, shear, and tensile strength are affected less. In practice, tensile strength parallel to the grain is taken equal to the bending strength of wood.

**TABLE 4.13 Strength of Some Commercially Important Woods Grown in the United States*
(Results of Tests on Small, Clear Specimens†)**

Commercial name of species	Specific gravity	Modulus of		Compression parallel to grain, maximum crushing strength, psi	Compression perpendicular to grain, fiber stress at proportional limit, psi	Shear parallel to grain, maximum shearing strength, psi	Side hardness, load perpendicular to grain, lb
		Rupture, ksi	Elasticity, ksi				
Ash, white	0.55	9.6	1440	3,990	670	1380	960
	0.60	15.4	1740	7,410	1160	1950	1320
Beech, American	0.56	8.6	1380	3,550	540	1290	850
	0.64	14.9	1720	7,300	1010	2010	1300
Birch, yellow	0.55	8.3	1500	3,380	430	1110	780
	0.62	16.6	2010	8,170	970	1880	1260
Cedar, western red	0.31	5.2	940	2,770	240	770	260
	0.32	7.5	1110	4,560	460	990	350
Chestnut, American	0.40	5.6	930	2,470	310	800	420
	0.43	8.6	1230	5,230	620	1080	540
Cypress, bald	0.42	6.6	1180	3,580	400	810	390
	0.46	10.6	1440	6,360	730	1000	510
Douglas fir, coast	0.45	7.7	1560	3,780	380	900	500
	0.48	12.4	1950	7,240	800	1130	710
Douglas fir, interior, west	0.46	7.7	1510	3,870	420	940	510
	0.50	12.6	1820	7,440	760	1290	660
Elm, American	0.46	7.2	1110	2,910	360	1000	620
	0.50	11.8	1340	5,520	690	1510	830

Hemlock, eastern	0.38	6.4	1070	3,080	360	850	400
	0.40	8.9	1200	5,410	650	1060	500
Hemlock, western	0.42	6.6	1310	3,360	280	860	410
	0.45	11.3	1640	7,110	550	1250	540
Hickory, pecan	0.60	9.8	1370	3,990	780	1480	1310
	0.66	13.7	1730	7,850	1720	2080	1820
Locust, black	0.66	13.8	1850	6,800	1160	1760	1570
	0.69	19.4	2050	10,180	1830	2480	1700
Larch, western	0.48	4.9	960	3,760	400	870	510
	0.52	13.1	1870	7,640	930	1360	830
Maple, sugar	0.56	9.4	1550	4,020	640	1460	970
	0.63	15.8	1830	7,830	1470	2330	1450
Oak, northern red	0.56	8.3	1350	3,440	610	1210	1000
	0.63	14.3	1820	6,760	1010	1780	1290
Oak, white	0.60	8.3	1250	3,560	670	1250	1060
	0.68	15.2	1780	7,440	1070	2000	1360
Pine shortleaf	0.47	7.4	1390	3,530	350	910	440
	0.51	13.1	1750	7,270	820	1390	690
Pine, longleaf	0.54	8.5	1590	4,320	480	1040	590
	0.59	14.5	1980	8,470	960	1510	870
Pine, sugar	0.34	4.9	1030	2,460	210	720	270
	0.36	8.2	1190	4,460	500	1130	380
Pine, western white	0.35	4.7	1190	2,430	190	680	260
	0.38	9.7	1460	5,040	470	1040	420
Yellow poplar	0.40	6.0	1220	2,660	270	790	440
	0.42	10.1	1580	5,540	500	1190	540
Redwood, old growth	0.38	7.5	1180	4,200	420	800	410
	0.40	10.0	1340	6,150	700	940	480
Spruce, white	0.37	5.6	1070	2,570	240	690	320
	0.40	9.8	1340	5,470	460	1080	480
Tupelo, black	0.46	7.0	1030	3,040	480	1100	640
	0.50	9.6	1200	5,520	930	1340	810

*From U.S. Forest Products Laboratory.
†Values in first line are for green material. Values in second line are adjusted to 12% moisture content.

In Table 4.13 are summarized also the principal mechanical properties of the most important American commercial species.

Values given in the table are average ultimate strengths. To obtain working stresses from these, the following must be considered: (1) Individual pieces may vary 25% above and below the average. (2) Values given are for standard tests that are completed in a few minutes. Over a period of years, however, wood may fail under a continuous load about 9/16 that sustained in a standard test. (3) The modulus of rupture of a standard 2-in-deep flexural-test specimen is greater than that of a deep beam. In deriving working stresses, therefore, variability, probable duration of load, and size are considered, and reduction factors are applied to the average ultimate strengths to provide basic stresses, or working stresses, for blemishless lumber. These stresses are still further reduced to account for such blemishes as knots, wane, slope of grain, shakes, and checks, to provide working stresses for classes of commercial engineering timbers. (See Sec. 10 for engineering design in timber.)

4.34 EFFECTS OF HYGROSCOPIC PROPERTIES OF WOOD

Because of its nature, wood tends to absorb moisture from the air when the relative humidity is high, and to lose it when the relative humidity is low. Moisture imbibed into the cell walls causes the wood to shrink and swell as the moisture content changes with the relative humidity of the surrounding air. The maximum amount of imbibed moisture the cell walls can hold is known as the fiber-saturation point, and for most species is in the vicinity of 25 to 30% of the oven-dry weight of the wood. Free water held in the cell cavities above the fiber-saturation point has no effect upon shrinkage or other properties of the wood. Changes in moisture content below the fiber-saturation point cause negligible shrinkage or swelling along the grain, and such shrinkage and swelling are normally ignored; but across the grain, considerable shrinkage and swelling occur in both the radial and tangential direction. Tangential shrinkage (as in flat-cut material) is normally approximately 50% greater than radial shrinkage (as in edge-grain material). See also Art. 10.1.

Separation of grain, or checking, is the result of rapid lowering of surface moisture content combined with a difference in moisture content between inner and outer portions of the piece. As wood loses moisture to the surrounding atmosphere, the outer cells of the member lose at a more rapid rate than the inner cells. As the outer cells try to shrink, they are restrained by the inner portion of the member. The more rapid the drying, the greater will be the differential in shrinkage between outer and inner fibers, and the greater the shrinkage stresses. As a result, checks may develop into splits.

Checks are radial cracks caused by nonuniform drying of wood. A **split** is a crack that results from complete separation of the wood fibers across the thickness of a member and extends parallel to the grain. (**Shakes** are another type of defect. Usually parallel to an annular ring, they develop in standing trees, whereas checks and splits are seasoning defects.) Lumber grading rules limit these types of defects.

Checks affect the horizontal shear strength of timber. A large reduction factor is applied to test values in establishing design values, in recognition of stress concentrations at the ends of checks. Design values for horizontal shear are adjusted for permissible checking in the various stress grades at the time of the grading. Since strength properties of wood increase with dryness, checks may enlarge with increasing dryness after shipment, without appreciably reducing shear strength.

Cross-grain checks and splits that tend to run out the side of a piece, or excessive checks and splits that tend to enter connection areas, may be serious and may require servicing. Provisions for controlling the effects of checking in connection areas may be incorporated in design details.

To avoid excessive splitting between rows of bolts caused by shrinkage during seasoning of solid-sawn timbers, rows should not be spaced more than 5 in apart, or a saw kerf, terminating in a bored hole, should be provided between lines of bolts. Whenever possible, maximum end distances for connections should be specified to minimize the effect of checks running into the joint area. Some designers require stitch bolts in members, with multiple connections loaded at an angle to the grain. Stitch bolts, kept tight, will reinforce pieces where checking is excessive.

One of the principal advantages of glued-laminated timber construction is relative freedom from checking. Seasoning checks may, however, occur in laminated members for the same reasons that they exist in solid-sawn members. When laminated members are glued within the typical range of moisture contents of 7 to 16% for the laminating lumber at the time of gluing, they will approximate the moisture content in normal-use conditions, thereby minimizing checking. Moisture content of the lumber at the time of gluing is thus of great importance to the control of checking in service. However, rapid changes in moisture content of large wood sections after gluing will result in shrinkage or swelling of the wood, and during shrinking, checking may develop in both glued joints and wood.

Differentials in shrinkage rates of individual laminations tend to concentrate shrinkage stresses at or near the glue line. For this reason, when checking occurs, it is usually at or near glue lines. The presence of wood-fiber separation indicates adequate glue bonds, and not delamination.

In general, checks have very little effect on the strength of glued-laminated members. Laminations in such members are thin enough to season readily in kiln drying without developing checks. Since checks lie in a radial plane, and the majority of laminations are essentially flat grain, checks are so positioned in horizontally laminated members that they will not materially affect shear strength. When members are designed with laminations vertical (with wide face parallel to the direction of load application), and when checks may affect the shear strength, the effect of checks may be evaluated in the same manner as for checks in solid-sawn members.

Seasoning checks in bending members affect only the horizontal shear strength (Art. 10.5.13). They are usually not of structural importance unless the checks are significant in depth and occur in the midheight of the member near the support, and then only if shear governs the design of the members. The reduction in shear strength is nearly directly proportional to the ratio of depth of check to width of beam. Checks in columns are not of structural importance unless the check develops into a split, thereby increasing the slenderness ratio of the columns.

Minor checking may be disregarded, since there is ample safety factor in allowable design values. The final decision as to whether shrinkage checks are detrimental to the strength requirements of any particular design or structural member should be made by a competent engineer experienced in timber construction.

4.35 COMMERCIAL GRADES OF WOOD

Lumber is graded by the various associations of lumber manufacturers having jurisdiction over various species. Two principal sets of grading rules are employed: (1) for softwoods, and (2) for hardwoods.

Softwoods. Softwood lumber is classified as dry, moisture content 19% or less; and green, moisture content above 19%.

According to the American Softwood Lumber Standard, softwoods are classified according to use as:

Yard Lumber. Lumber of grades, sizes, and patterns generally intended for ordinary construction and general building purposes.

Structural Lumber. Lumber 2 in or more nominal thickness and width for use where working stresses are required.

Factory and Shop Lumber. Lumber produced or selected primarily for manufacturing purposes.

Softwoods are classified according to extent of manufacture as:

Rough Lumber. Lumber that has not been dressed (surfaced) but has been sawed, edged, and trimmed.

Dressed (Surfaced) Lumber. Lumber that has been dressed by a planing machine (for the purpose of attaining smoothness of surface and uniformity of size) on one side (S1S), two sides (S2S), one edge (S1E), two edges (S2E), or a combination of sides and edges (S1S1E, S1S2, S2S1E, S4S).

Worked Lumber. Lumber that, in addition to being dressed, has been matched, shiplapped or patterned:

Matched Lumber. Lumber that has been worked with a tongue on one edge of each piece and a groove on the opposite edge.

Shiplapped Lumber. Lumber that has been worked or rabbeted on both edges, to permit formation of a close-lapped joint.

Patterned Lumber. Lumber that is shaped to a pattern or to a molded form.

Softwoods are also classified according to nominal size:

Boards. Lumber less than 2 in in nominal thickness and 2 in or more in nominal width. Boards less than 6 in in nominal width may be classified as strips.

Dimension. Lumber from 2 in to, but not including, 5 in in nominal thickness, and 2 in or more in nominal width. Dimension may be classified as framing, joists, planks, rafters, studs, small timbers, etc.

Timbers. Lumber 5 in or more nominally in least dimension. Timber may be classified as beams, stringers, posts, caps, sills, girders, purlins, etc.

Actual sizes of lumber are less than the nominal sizes, because of shrinkage and dressing. In general, dimensions of dry boards, dimension lumber, and timber less than 2 in wide or thick are ¼ in less than nominal; from 2 to 7 in wide or thick, ½ in less, and above 6 in wide or thick, ¾ in less. Green-lumber less than 2 in wide or thick is 1⁄32 in more than dry; from 2 to 4 in wide or thick, 1⁄16 in more, 5 and 6 in wide or thick, ⅛ in more, and 8 in or above in width and thickness, ¼ in more than dry lumber. There are exceptions, however.

Yard lumber is classified on the basis of quality as:

Appearance. Lumber of good appearance and finishing qualities, often called *select.*

- Suitable for natural finishes
  - Practically clear
  - Generally clear and of high quality
- Suitable for paint finishes
  - Adapted to high-quality paint finishes
  - Intermediate between high-finishing grades and common grades, and partaking somewhat of the nature of both

Common. Lumber suitable for general construction and utility purposes, often given various commercial designations.

For standard construction use
Suitable for better-type construction purposes
Well adapted for good standard construction
Designed for low-cost temporary construction

For less exacting purposes
Low quality, but usable

Structural lumber is assigned modulus of elasticity values and working stresses in bending, compression parallel to grain, compression perpendicular to grain, and horizontal shear in accordance with ASTM procedures. These values take into account such factors as sizes and locations of knots, slope of grain, wane, and shakes or checks, as well as such other pertinent features as rate of growth and proportions of summerwood.

Factory and shop lumber is graded with reference to its use for doors and sash, or on the basis of characteristics affecting its use for general cut-up purposes, or on the basis of size of cutting. The grade of factory and shop lumber is determined by the percentage of the area of each board or plank available in cuttings of specified or of given minimum sizes and qualities. The grade of factory and shop lumber is determined from the poor face, although the quality of both sides of each cutting must be considered.

Hardwoods. Because of the great diversity of applications for hardwood both in and outside the construction industry, hardwood grading rules are based on the proportion of a given piece that can be cut into smaller pieces of material clear on one or both sides and not less than a specified size. Grade classifications are therefore based on the amount of clear usable lumber in a piece.

Special grading rules of interest in the construction industry cover hardwood interior trim and moldings, in which one face must be practically free of imperfections and in which Grade A may further limit the amount of sapwood as well as stain. Hardwood dimension rules, in addition, cover clears, which must be clear both faces; clear one face; paint quality, which can be covered with paint; core, which must be sound on both faces and suitable for cores of glued-up panels; and sound, which is a general-utility grade.

Hardwood flooring is graded under two separate sets of rules: (1) for maple, birch, and beech; and (2) for red and white oak and pecan. In both sets of rules, color and quality classifications range from top-quality to the lower utility grades. Oak may be further subclassified as quarter-sawed and plain-sawed. In all grades, top-quality material must be uniform in color, whereas other grades place no limitation on color.

Shingles are graded under special rules, usually into three classes: Numbers 1, 2, and 3. Number 1 must be all edge grain and strictly clear, containing no sapwood. Numbers 2 and 3 must be clear to a distance far enough away from the butt to be well covered by the next course of shingles.

4.36 DESTROYERS AND PRESERVATIVES

The principal destroyers of wood are decay, caused by fungus, and attack by a number of animal organisms of which termites, carpenter ants, grubs of a wide variety of beetles, teredo, and limnoria are the principal offenders. In addition, fire annually causes widespread destruction of wood structures.

Decay will not occur if wood is kept well ventilated and air-dry or, conversely, if it is kept continuously submerged so that air is excluded.

Most termites in the United States are subterranean and require contact with the soil. The drywood and dampwood termites found along the southern fringes of the country and along the west coast, however, do not require direct soil contact and are more difficult to control.

Teredo, limnoria, and other water-borne wood destroyers are found only in salt or brackish waters.

Various wood species vary in natural durability and resistance to decay and insect attack. The sapwood of all species is relatively vulnerable; only the heartwood can be considered to be resistant. Table 4.14 lists the common species in accordance with heartwood resistance. Such a list is only approximate, and individual pieces deviate considerably.

Preservatives employed to combat the various destructive agencies may be subdivided into oily, water-soluble salts, and solvent-soluble organic materials. The principal oily preservatives are coal-tar creosote and creosote mixed with petroleum. The most commonly employed water-soluble salts are acid copper chromate, chromated copper arsenate and arsenite, fluor chrome arsenate phenol, chromated zinc chloride, and other materials that are often sold under various proprietary names. The principal solvent-soluble organic materials are chlorinated phenols, such as pentachlorphenol, and copper naphthenate.

Preservatives may be applied in a variety of ways, including brushing and dipping, but for maximum treatment, pressure is required to provide deep side-grain penetration. Butts of poles and other parts are sometimes placed in a hot boiling creosote or salt solution, and after the water in the wood has been converted to steam, they are quickly transferred to a cold vat of the same preservative. As the steam condenses, it produces a partial vacuum, which draws the preservative fairly deeply into the surface.

Pressure treatments may be classified as full-cell and empty-cell. In the full-cell treatment, a partial vacuum is first drawn in the pressure-treating tank to withdraw most of the air in the cells of the wood. The preservative is then let in without breaking the vacuum, after which pressure is applied to the hot solution. After treatment is completed, the individual cells are presumably filled with preservative. In the empty-cell method, no initial vacuum is drawn, but the preservative is pumped in under pressure against the back pressure of the compressed air in the wood. When the pressure is released, the air in the wood expands and forces out excess preservative, leaving only a coating of preservative on the cell walls.

Retentions of preservative depend on the application. For teredo-infestation, full-cell creosote treatment to refusal may be specified, ranging from 16 to 20 lb per cubic foot of wood. For ordinary decay conditions and resistance to termites and other destroyers of a similar nature, the empty-cell method may be employed with retentions in the vicinity of 6 to 8 lb of creosote per cubic foot of wood. Salt retentions generally range in the vicinity of 1½ to 3 lb of dry salt retained per cubic food of wood.

Solvent-soluble organic materials, such as pentachlorphenol, are commonly employed for the treatment of sash and door parts to impart greater resistance to decay. This is commonly done by simply dipping the parts in the solution and then allowing them to dry. As the organic solvent evaporates, it leaves the water-insoluble preservative behind in the wood.

These organic materials are also employed for general preservative treatment, including fence posts and structural lumber. The water-soluble salts and solvent-soluble organic materials leave the wood clean and paintable. Creosote in general

TABLE 4.14 Resistance to Decay of Heartwood of Domestic Woods

Resistant or very resistant	Moderately resistant	Slightly or nonresistant
Baldcypress (old growth)*	Baldcypress (young growth)*	Alder
Catalpa	Douglas fir	Ashes
Cedars	Honeylocust	Aspens
Cherry, black	Larch, western	Basswood
Chestnut	Oak, swamp chestnut	Beech
Cypress, Arizona	Pine, eastern white*	Birches
Junipers	Southern pine:	Buckeye
Locust, black†	Longleaf*	Butternut
Mesquite	Slash*	Cottonwood
Mulberry, red†	Tamarack	Elms
Oak:		Hackberry
Bur		Hemlocks
Chestnut		Hickories
Gambel		Magnolia
Oregon white		Maples
Post		Oak (red and black species)
White		Pines (other than longleaf, slash, and eastern white)
Osage orange†		Poplars
Redwood		Spruces
Sassafras		Sweetgum
Walnut, black		True firs (western and eastern)
Yew, Pacific†		Willows
		Yellow poplar

*The southern and eastern pines and baldcypress are now largely second growth with a large proportion of sapwood. Consequently, substantial quantities of heartwood lumber of these species are not available.

†These woods have exceptionally high decay resistance.

From U.S. Forest Products Laboratory.

cannot be painted over, although partial success can be achieved with top-quality aluminum-flake pigment paints.

Treatment against fire consists generally of applying salts containing ammonium and phosphates, of which monoammonium phosphate and diammonium phosphate are widely employed. At retentions of 3 to 5 lb of dry salt per cubic foot, the wood does not support its own combustion, and the afterglow when fire is removed is short. A variety of surface treatments is also available, most of which depend on the formation of a blanket of inert-gas bubbles over the surface of the wood in the presence of flame or other sources of heat. The blanket of bubbles insulates the wood beneath and retards combustion.

See also Art. 10.6.

4.37 GLUES AND ADHESIVES FOR WOOD

A variety of adhesives is now available for use with wood, depending on the final application. The older adhesives include animal glue, casein glue, and a variety of vegetable glues, of which soybean is today the most important. Animal glues provide strong, tough, easily made joints, which, however, are not moisture-resistant. Casein mixed with cold water, when properly formulated, provides highly moisture-resistant glue joints, although they cannot be called waterproof. The vegetable glues have good dry strength but are not moisture-resistant.

The principal high-strength glues today are synthetic resins, of which phenol formaldehyde, urea formaldehyde, resorcinol formaldehyde, melamine formaldehyde, and epoxy are the most important. Phenol, resorcinol, and melamine provide glue joints that are completely waterproof and will not separate when properly made even on boiling. Urea formaldehyde provides a glue joint of high moisture resistance, although not quite so good as the other three. Phenol and melamine require application of heat, as well as pressure, to cure the adhesive. Urea and resorcinol, however, can be formulated to be mixed with water at ordinary temperatures and hardened without application of heat above room temperature. Waterproof plywood is commonly made in hot-plate presses with phenolic or melamine adhesive. Resorcinol is employed where heat cannot be applied, as in a variety of assembly operations and the manufacture of laminated parts like ships' keels, which must have the maximum in waterproof qualities. Epoxide resins provide strong joints. Adhesives containing an elastomeric material, such as natural or synthetic rubber, may be classified as contact or mastic. The former, applied to both mating surfaces and allowed to partly dry, permit adhesion on contact. Mastics are very viscous and applied with a trowel or putty knife. They may be used to set wood-block flooring.

An emulsion of polyvinyl acetate serves as a general-purpose adhesive, for general assembly operations where maximum strength and heat or moisture resistance are not required. This emulsion is merely applied to the surfaces to be bonded, after which they are pressed together and the adhesive is allowed to harden.

4.38 PLYWOOD AND OTHER FABRICATED WOOD BOARDS

As ordinarily made, plywood consists of thin sheets, or veneers, of wood glued together. The grain is oriented at right angles in adjacent plies. To obtain plywood with balance—that is, which will not warp, shrink, or twist unduly—the plies must be carefully selected and arranged to be mirror images of each other with respect to the central plane. The outside plies or faces are parallel to each other and are of species that have the same shrinkage characteristics. The same holds true of the cross bands. As a consequence, plywood has an odd number of plies, the minimum being three.

Principal advantages of plywood over lumber are its more nearly equal strength properties in length and width, greater resistance to checking, greatly reduced shrinkage and swelling, and resistance to splitting.

The approach to equalization of strength of plywood in the various directions is obtained at the expense of strength in the parallel-to-grain direction; i.e., plywood is not so strong in the direction parallel to its face plies as lumber is parallel to the grain. But plywood is considerably stronger in the direction perpendicular to its face plies than wood is perpendicular to the grain. Furthermore, the shearing strength of plywood in a plane perpendicular to the plane of the plywood is very much greater than that of ordinary wood parallel to the grain. In a direction parallel to the plane of the plywood, however, the shearing strength of plywood is less than that of ordinary wood parallel to the grain, because in this direction rolling shear occurs in the plywood; i.e., the fibers in one ply tend to roll rather than to slide.

Depending on whether plywood is to be used for general utility or for decorative purposes, the veneers employed may be cut by peeling from the log, by slicing, or today very rarely, by sawing. Sawing and slicing give the greatest freedom and versatility in the selection of grain. Peeling provides the greatest volume and the most rapid production, because logs are merely rotated against a flat knife and the veneer is peeled off in a long continuous sheet.

Plywood is classified as interior or exterior, depending on the type of adhesive employed. Interior-grade plywood must have a reasonable degree of moisture resistance but is not considered to be waterproof. Exterior-grade plywood must be completely waterproof and capable of withstanding immersion in water or prolonged exposure to outdoor conditions.

In addition to these classifications, plywood is further subclassified in a variety of ways depending on the quality of the surface ply. Top quality is clear on one or both faces, except for occasional patches. Lower qualities permit sound defects, such as knots and similar blemishes, which do not detract from the general utility of the plywood but detract from its finished appearance.

Particle Board. Wood chips, sawdust, and flakes are pressed with a binder (urea-formaldehyde or phenol-formaldehyde) to form boards (sheathing, underlayment, corestock), having uniform strength and low shrinkage in the plane of the board.

Hardboard. Wood chips (exploded by high-pressure steam into wood fibers) and lignin are pressed to form boards of various densities. Additives may add weather resistance and other properties.

4.39 WOOD BIBLIOGRAPHY

Forest Products Laboratory, Forest Service, U.S. Department of Agriculture: "Wood Handbook," Government Printing Office, Washington, D.C.

National Hardwood Lumber Association, Chicago, Ill.: "Rules for the Measurement and Inspection of Hardwood Lumber, Cypress, Veneer, and Thin Lumber."

National Forest Products Association, Washington, D.C.: "National Design Specifications for Stress-grade Lumber and Its Fastenings."

U.S. Department of Commerce, National Bureau of Standards, Washington, D.C.: American Softwood Lumber Standard, Voluntary Practice Standard PS20; Douglas Fir Plywood, Commercial Standard CS 45; Hardwood Plywood, Commercial Standard CS 35.

Western Wood Products Association, Portland, Ore.: "Western Woods Use Book."
K. F. Faherty and T. G. Williamson, "Wood Engineering and Construction Handbook," McGraw-Hill Publishing Company, New York.

STEEL AND STEEL ALLOYS

Iron and its alloys are generally referred to as **ferrous metals.** Even small amounts of alloy change the properties of ferrous metals significantly. Also, the properties can be changed considerably by changing the atomic structure of these metals by heating and cooling.

4.40 TYPES OF IRONS AND STEELS

Steel is a solution of carbon in iron. Various types of steel are produced by varying the percentage of carbon added to molten iron and controlling the cooling, which affects the atomic structure of the product, and hence its properties. Some of the structural changes can be explained with the aid of an iron-carbon equilibrium diagram (Fig. 4.2).

4.40.1 Iron-Carbon Equilibrium Diagram

The iron-carbon equilibrium diagram in Fig. 4.2 shows that, under equilibrium conditions (slow cooling) if not more than 2.0% carbon is present, a solid solution of carbon in gamma iron exists at elevated temperatures. This is called austenite.

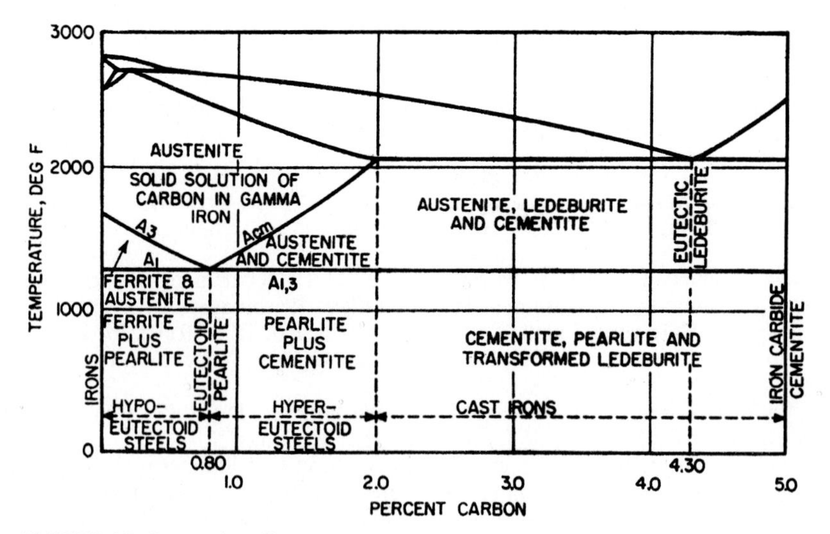

FIGURE 4.2 Iron-carbon diagram.

If the carbon content is less than 0.8%, cooling below the A_3 temperature line causes transformation of some of the austenite to ferrite, which is substantially pure alpha iron (containing less than 0.01% carbon in solution). Still further cooling to below the A_1 line causes the remaining austenite to transform to pearlite—the eutectoid mixture of fine plates, or lamellas, of ferrite and cementite (iron carbide) whose iridescent appearance under the microscope gives it its name.

If the carbon content is 0.8%, no transformation on cooling the austenite occurs until the A_1 temperature is reached. At that point, all the austenite transforms to pearlite, with its typical "thumbprint" microstructure.

At carbon contents between 0.80 and 2.0%, cooling below the A_{cm} temperature line causes iron carbide, or cementite, to form in the temperature range between A_{cm} and $A_{1,3}$. Below $A_{1,3}$, the remaining austenite transforms to pearlite.

4.40.2 Types of Irons

Metals containing substantially no carbon (several hundredths of 1%) are called irons, of which wrought iron, electrolytic iron, and "ingot" iron are examples.

Wrought iron, whether made by the traditional puddling method or by mixing very low carbon iron and slag, contains a substantial amount of slag. Because it contains very little carbon, it is soft, ductile, and tough and, like low-carbon ferrous metals generally, is relatively resistant to corrosion. It is easily worked. When broken, it shows a fibrous fracture because of the slag inclusions. "Ingot" iron is a very low carbon iron containing no slag, which is also soft, ductile, and tough.

Above 2.0% carbon content is the region of the cast irons. Above the $A_{1,3}$ temperature, austenite, the eutectic ledeburite and cementite occur; below the $A_{1,3}$ temperature, the austenite transforms to pearlite, and a similar transformation of the ledeburite occurs.

When the silicon content is kept low, and the metal is cooled rapidly, **white cast iron** results. It is hard and brittle because of the high cementite content. White cast iron as such has little use; but when it is reheated and held a long time in the vicinity of the transformation temperature, then cooled slowly, the cementite decomposes to ferrite and nodular or temper carbon. The result is black-heart **malleable iron.** If the carbon is removed during malleabilization, white-heart malleable iron results.

If the silicon content is raised, and the metal is cooled relatively slowly, **gray cast iron** results. It contains cementite, pearlite, ferrite, and some free carbon, which gives it its gray color. Gray iron is considerably softer and tougher than white cast iron and is generally used for castings of all kinds. Often, it is alloyed with elements like nickel, chromium, copper, and molybdenum.

At 5.0% carbon, the end product is hard, brittle iron carbide or cementite.

4.40.3 Types of Carbon Steels

Most of the steel used for construction is low- to medium-carbon, relatively mild, tough, and strong, fairly easy to work by cutting, punching, riveting, and welding. Table 4.15 summarizes the most important carbon steels and low-alloy steels used in construction as specified by ASTM.

The plain iron-carbon metals with less than 0.8% carbon content consist of ferrite and pearlite and provide the low-carbon (0.06 to 0.30%), medium-carbon (0.30 to 0.50%), and high-carbon (0.50 to 0.80%) steels called hypoeutectoid steels. The higher-carbon or hypereutectoid tool steels contain 0.8 to 2.0% carbon and

TABLE 4.15 ASTM Requirements for Structural, Reinforcing, and Fastening Steels*

	ASTM specification	Tensile strength, min, ksi†	Yield point, min, ksi†	Elongation in 8 in, min, %	Elongation in 2 in, min, %‡	Bend test, ratio of bend diameter, in, to specimen thickness, in§				
						0–¾	¾–1	1–1½	1½–2	Over 2
Structural steel	A36	58–80	36	20	23–21	½	1	1½	2½	3
Welded or seamless pipe	A53	45–60	25–35							
High-strength, low-alloy, structural steel	A242	63–70	42–50	18	21	1	1½	2	2½	3
High-strength, low-alloy columbium-vanadium steels	A572	60–80	42–65	20–15	24–17	Depends on grade*				
High-strength, low-alloy structural steel	A588	63–70	42–50	18	21	1	1½	2	2½	3
High-yield-strength, quenched and tempered alloy steel	A514	110–130	90–100		17–18	2	2	3	4	4
Structural steel	A529	60–85	42	19		1				
High-strength, quenched and tempered alloy steel	A852	90–110	70	19						
Normalized high-strength low-alloy steel	A633	63–100	42–60	18	23	2	2	2½	2½	3
Quenched and tempered steel plate	A678	70–110	50–75		22–18	1–2	2–3	2–3	2½–3	2–2½
Cold-formed, welded and seamless tubing	A500	45–62	33–46		25–14					
Hot-formed, welded and seamless tubing	A501	58	36	20	23					

High-strength steel bolts	A325	105	81		14	
High-strength, alloy steel bolts	A490	150–170	115–130		14	
Bolts and nuts, machine	A307	60–100			18	
Sheetpiling	A328	70	39	17		2
Cast steel, 65–35, annealed	A27	60–70	30–40		22–24	
High-strength cast steel, 80–50	A148	80–260	40–210		18–3	
						180° bend test; ratio of pin diameter to specimen diameter
Reinforcing steel for concrete:						
Billet-steel bars	A615					
Grade 40		70	40	7–11		Under No. 6: 4; Nos. 6, 7, 8, 9, 10, 11: 5
Grade 60		90	60	7–9		Under No. 6: 4; No. 6: 5; Nos. 7, 8: 6; Nos. 9, 10, 11: 8
Rail-steel bars	A616					
Grade 50		80	50	5–6		Under No. 8: 6; Nos. 9, 10, 11: 8¶
Grade 60		90	60	4.5–6		Under No. 8: 6; Nos. 9, 10, 11: 8¶

*The following are appropriate values for all the steels:
Modulus of elasticity—29,000 ksi
Shear modulus—11,000 ksi
Poisson's ratio—0.30
Yield stress in shear—$0.57F_t$, where F_t = tensile stress
Ultimate strength in shear—$0.67F_t$ to $0.75F_t$
Coefficient of thermal expansion—0.0000065 in/in·°F for temperatures between −60 and 150°F
Density—490 lb/ft^3

†Where two values are given, the first is the minimum and the second is the maximum. See the relevant specification for the values for each grade and applicable thicknesses.

‡The minimum elongations are modified for some thicknesses in accordance with the specification for the steel.

§Optional. See ASTM A6, "General Requirements for Rolled Steel Plates, Shapes, Sheet Piling, and Bars for Structural Use."

¶90° bend for No. 11 bars.

TABLE 4.16 Standard Steels for Hot-Rolled Bars (Basic open-hearth and acid bessemer carbon steels)

SAE and AISI No.	Chemical composition limits, %			
	Carbon	Manganese	Max phosphorus	Max sulfur
1008	0.10 max	0.30/0.50	0.040	0.050
1010	0.08/0.13	0.30/0.60	0.040	0.050
1015	0.13/0.18	0.30/0.60	0.040	0.050
1020	0.18/0.23	0.30/0.60	0.040	0.050
1025	0.22/0.28	0.30/0.60	0.040	0.050
1030	0.28/0.34	0.60/0.90	0.040	0.050
1040	0.37/0.44	0.60/0.90	0.040	0.050
1050	0.48/0.55	0.60/0.90	0.040	0.050
1070	0.65/0.75	0.60/0.90	0.040	0.050
1084	0.80/0.93	0.60/0.90	0.040	0.050
1095	0.90/1.03	0.30/0.50	0.040	0.050

consist of pearlite and cementite. The eutectoid steels occurring in the vicinity of 0.8% carbon are essentially all pearlite.

The American Iron and Steel Institute and the Society of Automotive Engineers have designated standard compositions for various steels including plain carbon steels and alloy steels. AISI and SAE numbers and compositions for several representative hot-rolled carbon-steel bars are given in Table 4.16.

Prestressed concrete imposes special requirements for reinforcing steel. It must be of high strength with a high yield point and minimum creep in the working range. Tables 4.15 and 4.17 give ASTM specification requirements for bars, wires, and strands.

4.40.4 Types of Structural Steels

Structural steels are low- to medium-carbon steels used in elements ¼ in thick or more to form structural framing. The American Institute of Steel Construction (AISC) "Code of Standard Practice for Steel Buildings and Bridges" lists the elements that are included in the scope of the work in contract documents for structural steel. The list includes flexural members, columns, trusses, bearings and bearing plates, bracing, hangers, bolts and nuts, shear connectors, wedges, and shims. The AISC "Specification for Structural Steel Buildings" (ASD and LRFD) tabulates the types of structural steel that are approved for use in buildings. These steels are given in Table 4.15.

In accordance with present practice, the steels described in this section and in Sec. 7 are given the names of the corresponding ASTM specifications for the steels. For example, all steels conforming with ASTM A588, "Specification for High-Strength Low-Alloy Structural Steel," are called A588 steel. Further identification may be given by a grade, which usually indicates the steel yield strength.

Structural steels may be classified as carbon steels; high-strength, low-alloy steels; heat-treated, high-strength carbon steels; or heat-treated, constructional alloy steels.

TABLE 4.17 **ASTM Requirements for Prestressing Bars and Wires**

Material	ASTM designation	Tensile strength, ksi	Minimum yield strength
Seven-wire steel strand	A416		
Grade 250		250	85% of breaking strength, at 1% extension
Grade 270		170	
Uncoated steel wire	A421		
Type BA		235–240	85% of breaking strength, at 1% extension
Type WA		235–250	
High-strength bar	A722		
Type I		150	85% of tensile strength
Type II		150	80% of tensile strength

Carbon steels satisfy all of the following requirements:

1. The maximum content specified for alloying elements does not exceed the following: manganese, 1.65%; silicon, 0.60%; copper, 0.60%.
2. The specified minimum for copper does not exceed 0.40%.
3. No minimum content is specified for other elements added to obtain a desired alloying effect.

A36 and A529 steels are included in this category.

High-strength, low-alloy steels have specified minimum yield strengths larger than 40 ksi, which are attained without heat treatment. A242, A572, and A588 steels are included in this category. A242 and A572 steel are often referred to as weathering steels, because they have higher resistance to corrosion than carbon steels. On exposure to ordinary atmospheric conditions, they develop a protective oxide surface.

Heat-treated, high-strength carbon steels are heat-treated to achieve specified high strength and toughness. A633, A678, and A852 steels are included in this category.

Heat-treated, constructional alloy steels contain alloying elements in excess of the limits for carbon steels and are heat-treated to obtain a combination of high strength and toughness. These are the strongest steels in general structural use. The various grades of A514 steel, with yield strengths up to 100 ksi, are in this category.

4.41 PROPERTIES OF STRUCTURAL STEELS

Figure 4.3 shows a typical stress-strain curve for each classification of structural steels defined in Art. 4.40.4. The diagram illustrates the higher-strength levels achieved with heat treatment and addition of alloys.

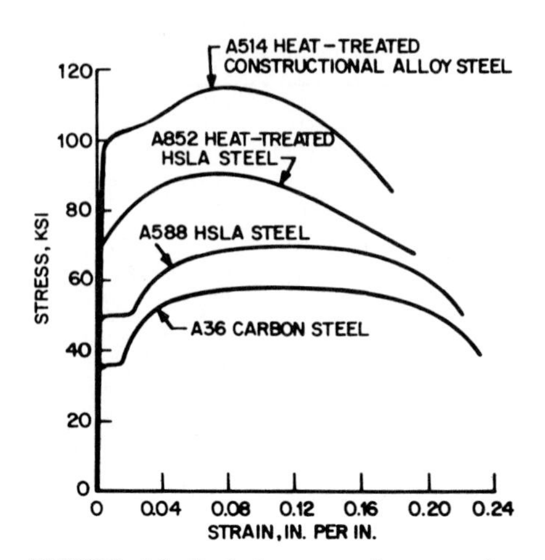

FIGURE 4.3 Typical stress-strain curves for structural steels.

4.41.1 Tensile Properties of Structural Steels

The curves in Fig. 4.3 were derived from tensile tests. The yield points, strengths, and modulus of elasticity obtained from compression tests would be about the same.

The initial portion of the curves in Fig. 4.3 is shown to a magnified scale in Fig. 4.4. It indicates that there is an initial elastic range for the structural steels in which there is no permanent deformation on removal of the load. The modulus of elasticity

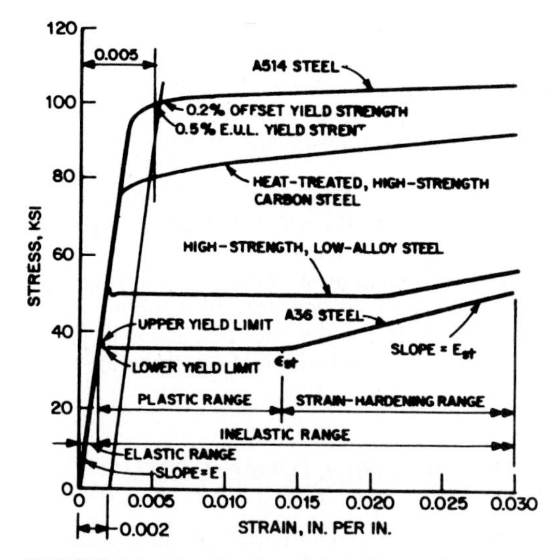

FIGURE 4.4 Magnification of the initial portions of the stress-strain curves for structural steels.

E, which is given by the slope of the curves, is nearly a constant 29,000 ksi for all the steels. For carbon and high-strength, low-alloy steels, the inelastic range, where strains exceed those in the elastic range, consists of two parts: Initially, a plastic range occurs in which the steels yield; that is, strain increases with no increase in stress. Then follows a strain-hardening range in which increase in strain is accompanied by a significant increase in stress.

The curves in Fig. 4.4 also show an upper and lower yield point for the carbon and high-strength, low-alloy steels. The upper yield point is the one specified in standard specifications for the steels. In contrast, the curves do not indicate a yield point for the heat-treated steels. For these steels, ASTM 370, "Mechanical Testing of Steel Products," recognizes two ways of indicating the stress at which there is a significant deviation from the proportionality of stress to strain. One way, applicable to steels with a specified yield point of 80 ksi or less, is to define the yield point as the stress at which a test specimen reaches a 0.5% extension under load (0.5% *EUL*). The second way is to define the yield strength as the stress at which a test specimen reaches a strain (offset) 0.2% greater than that for elastic behavior. Yield point and yield strength are often referred to as yield stress.

Ductility is measured in tension tests by percent elongation over a given gage length—usually 2 or 8 in—or percent reduction of cross-sectional area. Ductility is an important property because it permits redistribution of stresses in continuous members and at points of high local stresses.

Poisson's ratio, the ratio of transverse to axial strain, also is measured in tension tests. It may be taken as 0.30 in the elastic range and 0.50 in the plastic range for structural steels.

Cold working of structural steels, that is, forming plates or structural shapes into other shapes at room temperature, changes several properties of the steels. The resulting strains are in the strain-hardening range. Yield strength increases but ductility decreases. (Some steels are cold rolled to obtain higher strengths.) If a steel element is strained into the strain-hardening range, then unloaded and allowed to age at room or moderately elevated temperatures (a process called **strain aging**), yield and tensile strengths are increased, whereas ductility is decreased. Heat treatment can be used to modify the effects of cold working and strain aging.

Residual stresses remain in structural elements after they are rolled or fabricated. They also result from uneven cooling after rolling. In a welded member, tensile residual stresses develop near the weld and compressive stresses elsewhere. Plates with rolled edges have compressive residual stresses at the edges, whereas flame-cut edges have tensile residual stresses. When loads are applied to such members, some yielding may take place where the residual stresses occur. Because of the ductility of steel, however, the effect on tensile strength is not significant but the buckling strength of columns may be lowered.

Strain rate also changes the tensile properties of structural steels. In the ordinary tensile test, load is applied slowly. The resulting data are appropriate for design of structures for static loads. For design for rapid application of loads, such as impact loads, data from rapid tension tests are needed. Such tests indicate that yield and tensile strengths increase but ductility and the ratio of tensile strength to yield strength decrease.

High temperatures too affect properties of structural steels. As temperatures increase, the stress-strain curve typically becomes more rounded and tensile and yield strengths, under the action of strain aging, decrease. Poisson's ratio is not significantly affected but the modulus of elasticity decreases. Ductility is lowered until a minimum value is reached. Then, it rises with increase in temperature and becomes larger than the ductility at room temperature.

Low temperatures in combination with tensile stress and especially with geometric discontinuities, such as notches, bolt holes, and welds, may cause a brittle failure. This is a failure that occurs by cleavage, with little indication of plastic deformation. A ductile failure, in contrast, occurs mainly by shear, usually preceded by large plastic deformation. One of the most commonly used tests for rating steels on their resistance to brittle fracture is the Charpy V-notch test. It evaluates notch toughness at specific temperatures.

Toughness is defined as the capacity of a steel to absorb energy; the greater the capacity, the greater the toughness. Determined by the area under the stress-strain curve, toughness depends on both strength and ductility of the metal. Notch toughness is the toughness in the region of notches or other stress concentrations. A quantitative measure of notch toughness is fracture toughness, which is determined by fracture mechanics from relationships between stress and flaw size.

4.41.2 Shear Properties of Structural Steels

The shear modulus of elasticity G is the ratio of shear stress to shear strain during initial elastic behavior. It can be computed from Eq. (5.25) from values of modulus of elasticity and Poisson's ratio developed in tension stress-strain tests. Thus G for structural steels is generally taken as 11,000 ksi.

The shear strength, or shear stress at failure in pure shear, ranges from $0.67F_t$ to $0.75F_t$ for structural steels, where F_t is the tensile strength. The yield strength in shear is about $0.57F_t$.

4.41.3 Creep and Relaxation

Creep, a gradual change in strain under constant stress, is usually not significant for structural steel framing in buildings, except in fires. Creep usually occurs under high temperatures or relatively high stresses, or both.

Relaxation, a gradual decrease in load or stress under a constant strain, is a significant concern in the application of steel tendons to prestressing (Art. 9.104). With steel wire or strand, relaxation can occur at room temperature. To reduce relaxation substantially, stabilized, or low-relaxation, strand may be used. This is produced by pretensioning strand at a temperature of about 600°F. A permanent elongation of about 1% remains and yield strength increases to about 5% over stress-relieved (heat-treated but not tensioned) strand.

4.41.4 Hardness of Structural Steels

Hardness is used in production of steels to estimate tensile strength and to check the uniformity of tensile strength in various products. Hardness is determined as a number related to resistance to indentation. Any of several tests may be used, the resulting hardness numbers being dependent on the type of penetrator and load. These should be indicated when a hardness number is given. Commonly used hardness tests are the Brinell, Rockwell, Knoop, and Vickers. ASTM A370, "Mechanical Testing of Steel Products," contains tables that relate hardness numbers from the different tests to each other and to the corresponding approximate tensile strength.

4.41.5 Fatigue of Structural Steels

Under cyclic loading, especially when stress reversal occurs, a structural member may eventually fail because cracks form and propagate. Known as a fatigue failure, this can take place at stress levels well below the yield stress. Fatigue resistance may be determined by a rotating-beam test, flexure test, or axial-load test. In these tests, specimens are subjected to stresses that vary, usually in a constant stress range between maximum and minimum stresses until failure occurs. Results of the tests are plotted on an *S-N* diagram, where *S* is the maximum stress (fatigue strength) and *N* is the number of cycles to failure (fatigue life). Such diagrams indicate that the fatigue strength of a structural steel decreases with increase in the number of cycles until a minimum value is reached, the **fatigue limit.** Presumably, if the maximum stress does not exceed the fatigue limit, an unlimited number of cycles of that ratio of maximum to minimum stress can be applied without failure. With tension considered positive and compression, negative, tests also show that as the ratio of maximum to minimum stress is decreased, fatigue strength is lowered significantly.

Since the tests are made on polished specimens and steel received from mills has a rough surface, fatigue data for design should be obtained from tests made on as-received material.

Tests further indicate that steels with about the same tensile strength have about the same fatigue strength. Hence the *S-N* diagram obtained for one steel may be used for other steels with about the same tensile strength.

4.42 HEAT TREATMENT AND HARDENING OF STEELS

Heat-treated and hardened steels are sometimes required in building operations. The most familiar heat treatment is annealing, a reheating operation in which the metal is usually heated to the austenitic range (Fig. 4.2) and cooled slowly to obtain the softest, most ductile state. Cold working is often preceded by annealing. Annealing may be only partial, just sufficient to relieve internal stresses that might cause deformation or cracking, but not enough to reduce markedly the increased strength and yield point brought about by the cold working, for example.

Another type of heat treatment that may be used is normalizing. It requires heating steel to 100 to 150°F above the A_3 temperature line in Fig. 4.2. Then, the steel is allowed to cool in still air. (The rate of cooling is much more rapid than that used in annealing.) Normalizing may be used to refine steel grain size, which depends on the finishing temperature during hot rolling, or to obtain greater notch toughness.

Thick plates have a coarser grain structure than thin plates and thus can benefit more from normalizing. This grain structure results from the fewer rolling passes required for production of thick plates, consequent higher finishing temperature, and slower cooling.

Sometimes, a hard surface is required on a soft, tough core. Two principal casehardening methods are employed. For **case carburizing,** a low- to medium-carbon steel is packed in carbonaceous materials and heated to the austenite range. Carbon diffuses into the surface, providing a hard, high-carbon case when the part is cooled. For **nitriding,** the part is exposed to ammonia gas or a cyanide at moderately elevated temperatures. Extremely hard nitrides are formed in the case and provide a hard surface.

4.43 EFFECTS OF GRAIN SIZE

When a low-carbon steel is heated above the A_3 temperature line (Fig. 4.2), for example, to hot rolling and forging temperatures, the steel may grow coarse grains. For some applications, this structure may be desirable; for example, it permits relatively deep hardening, and if the steel is to be used in elevated-temperature service, it will have higher load-carrying capacity and higher creep strength than if the steel had fine grains.

Fine grains, however, enhance many steel properties: notch toughness, bendability, and ductility. In quenched and tempered steels, higher yield strengths are obtained. Furthermore, fine-grain, heat-treated steels have less distortion, less quench cracking, and smaller internal stresses.

During the production of a steel, grain growth may be inhibited by an appropriate dispersion of nonmetallic inclusions or by carbides that dissolve slowly or remain undissolved during cooling. The usual method of making fine-grain steel employs aluminum deoxidation. In such steels, the inhibiting agent may be a submicroscopic dispersion of aluminum nitride or aluminum oxide. Fine grains also may be produced by hot working rolled or forged products, which otherwise would have a coarse-grain structure. The temperature at the final stage of hot working determines the final grain size. If the finishing temperature is relatively high and the grains after air cooling are coarse, the size may be reduced by normalizing (Art. 4.42). Fine- or coarse-grain steels may be heat treated to be coarse- or fine-grain.

4.44 STEEL ALLOYS

Plain carbon steels can be given a great range of properties by heat treatment and by working; but addition of alloying elements greatly extends those properties or makes the heat-treating operations easier and simpler. For example, combined high tensile strength and toughness, corrosion resistance, high-speed cutting, and many other specialized purposes require alloy steels. However, the most important effect of alloying is the influence on hardenability.

4.44.1 Effects of Alloying Elements

Important alloying elements from the standpoint of building, and their principal effects, are summarized below:

Aluminum restricts grain growth during heat treatment and promotes surface hardening by nitriding.

Chromium is a hardener, promotes corrosion resistance (see Art. 4.44.2), and promotes wear resistance.

Copper promotes resistance to atmospheric corrosion and is sometimes combined with molybdenum for this purpose in low-carbon steels and irons. It strengthens steel and increases the yield point without unduly changing elongation or reduction of area.

Manganese in low concentrations promotes hardenability and nondeforming, nonshrinking characteristics for tool steels. In high concentrations, the steel is austenitic under ordinary conditions, is extremely tough, and work-hardens readily. It is therefore used for teeth of power-shovel dippers, railroad frogs, rock crushers, and similar applications.

Molybdenum is usually associated with other elements, especially chromium and nickel. It increases corrosion resistance, raises tensile strength and elastic limit without reducing ductility, promotes casehardening, and improves impact resistance.

Nickel boosts tensile strength and yield point without reducing ductility; increases low-temperature toughness, whereas ordinary carbon steels become brittle; promotes casehardening; and in high concentrations improves corrosion resistance under severe conditions. It is often used with chromium (see Art. 4.44.2). **Invar** contains 36% nickel.

Silicon strengthens low-alloy steels; improves oxidation resistance; with low carbon yields transformer steel, because of low hysteresis loss and high permeability; in high concentrations provides hard, brittle castings, resistant to corrosive chemicals, useful in plumbing lines for chemical laboratories.

Sulfur promotes free machining, especially in mild steels.

Titanium prevents intergranular corrosion of stainless steels by preventing grain-boundary depletion of chromium during such operations as welding and heat treatment.

Tungsten, vanadium, and **cobalt** are all used in high-speed tool steels, because they promote hardness and abrasion resistance. Tungsten and cobalt also increase high-temperature hardness.

The principal effects of alloying elements are summarized in Table 4.18.

4.44.2 Stainless Steels

Stainless steels of primary interest in building are the wrought stainless steels of the austenitic type. The austenitic stainless steels contain both chromium and nickel. Total content of alloy metals is not less than 23%, with chromium not less than 16% and nickel not less than 7%. Commonly used stainless steels have a tensile strength of 75 ksi and yield point of 30 ksi when annealed. Cold-finished steels may have a tensile strength as high as 125 ksi with a yield point of 100 ksi.

Austenitic stainless steels are tough, strong, and shock-resistant, but work-harden readily; so some difficulty on this score may be experienced with cold working and machining. These steels can be welded readily but may have to be stabilized (e.g., AISI Types 321 and 347) against carbide precipitation and intergranular corrosion due to welding unless special precautions are taken. These steels have the best high-temperature strength and resistance to scaling of all the stainless steels.

Types 303 and 304 are the familiar 18-8 stainless steels widely used for building applications. These and Types 302 and 316 are the most commonly employed stainless steels. Where maximum resistance to corrosion is required, such as resistance to pitting by seawater and chemicals, the molybdenum-containing Types 316 and 317 are best.

For resistance to ordinary atmospheric corrosion, some of the martensitic and ferritic stainless steels, containing 15 to 20% chromium and no nickel, are employed. The martensitic steels, in general, range from about 12 to 18% chromium and from 0.08 to 1.10% carbon. Their response to heat treatment is similar to that of the plain carbon steels. When chromium content ranges from 15 to 30% and carbon content is below 0.35%, the steels are ferritic and nonhardenable. The high-chromium steels are resistant to oxidizing corrosion and are useful in chemical plants.

TABLE 4.18 Effects of Alloying Elements in Steel*

Element	Solid solubility		Influence on ferrite	Influence on austenite (hardenability)	Influence exerted through carbide		Principal functions
	In gamma iron	In alpha iron			Carbide-forming tendency	Action during tempering	
Aluminum (Al)	1.1% (increased by C)	36%	Hardens considerably by solid solution	Increases hardenability mildly, if dissolved in austenite	Negative (graphitizes)		1. Deoxides efficiently 2. Restricts grain growth (by forming dispersed oxides or nitrides) 3. Alloying element in nitriding steel
Chromium (Cr)	12.8% (20% with 0.5% C)	Unlimited	Hardens slightly; increases corrosion resistance	Increases hardenability moderately	Greater than Mn; less than W	Mildly resists softening	1. Increases resistance to corrosion and oxidation 2. Increases hardenability 3. Adds some strength at high temperatures 4. Resists abrasion and wear (with high carbon)
Cobalt (Co)	Unlimited	75%	Hardens considerably by solid solution	Decreases hardenability as dissolved	Similar to Fe	Sustains hardness by solid solution	1. Contributes to red hardness by hardening ferrite
Manganese (Mn)	Unlimited	3%	Hardens markedly; reduces plasticity somewhat	Increases hardenability moderately	Greater than Fe; less than Cr	Very little, in usual percentages	1. Counteracts brittleness from the sulfur 2. Increases hardenability inexpensively

Molybdenum (Mo)	3% ± (8% with 0.3% C)	37.5% (less with lowered temp)	Provides age-hardening system in high Mo-Fe alloys	Increases hardenability strongly (Mo > Cr)	Strong; greater than Cr	Opposes softening by secondary hardening	1. Raises grain-coarsening temperature of austenite 2. Deepens hardening 3. Counteracts tendency toward temper brittleness 4. Raises hot and creep strength, red hardness 5. Enhances corrosion resistance in stainless steel 6. Forms abrasion-resisting particles
Nickel (Ni)	Unlimited	10% (irrespective of carbon content)	Strengthens and toughens by solid solution	Increases hardenability mildly, but tends to retain austenite with higher carbon	Negative (graphitizes)	Very little in small percentages	1. Strengthens unquenched or annealed steels 2. Toughens pearlitic-ferritc steels (especially at low temperature) 3. Renders high-chromium iron alloys austenitic
Phosphorus (P)	0.5%	2.8% (irrespective of carbon content)	Hardens strongly by solid solution	Increases hardenability	Nil		1. Strengthens low-carbon steel 2. Increases resistance to corrosion 3. Improves machinability in free-cutting steels.
Silicon (Si)	2% ± (9% with 0.35% C)	18.5% (not much changed by carbon)	Hardens with loss in plasticity (Mn < Si < P)	Increases hardenability moderately	Negative (graphitizes)	Sustains hardness by solid solution	1. Used as general-purpose deoxidizer 2. Alloying element for electrical and magnetic sheet 3. Improves oxidation resistance 4. Increases hardenability of steel carrying nongraphitizing elements 5. Strengthens low-alloy steels

TABLE 4.18 Effects of Alloying Elements in Steel* (Continued)

Element	Solid solubility		Influence on ferrite	Influence on austenite (hardenability)	Influence exerted through carbide		Principal functions
	In gamma iron	In alpha iron			Carbide-forming tendency	Action during tempering	
Titanium (Ti)	0.75% (1% ± with 0.20% C)	6% ± (less with lowered temp)	Provides age-hardening system in high Ti-Fe alloys	Probably increases hardenability very strongly as dissolved. The carbide effects reduce hardenability	Greatest known (2% Ti renders 0.50% carbon steel unhardenable)	Persistent carbides probably unaffected. Some secondary hardening	1. Fixes carbon in inert particles a. Reduces martensitic hardness and hardenability in medium-chromium steels b. Prevents formation of austenite in high-chromium steels c. Prevents localized depletion of chromium in stainless steel during long heating.
Tungsten (W)	6% (11% with 0.25% C)	33% (less with lowered temp)	Provides age-hardening system in high W-Fe alloys	Increases hardenability strongly in small amounts	Strong	Opposes softening by secondary hardening	1. Forms hard, abrasion-resistant particles in tool steels 2. Promotes hardness and strength at elevated temperature
Vanadium (V)	1% (4% with 0.20% C)	Unlimited	Hardens moderately by solid solution	Increases hardenability very strongly, as dissolved	Very strong (V < Ti or Cb)	Max for secondary hardening	1. Elevates coarsening temperature of austenite (promotes fine grain) 2. Increases hardenability (when dissolved) 3. Resists tempering and causes marked secondary hardening

*"Metals Handbook," American Society for Metals.

4.45 WELDING FERROUS METALS

General welding characteristics of the various types of ferrous metals are as follows:

Wrought iron is ideally forged but may be welded by other methods if the base metal is thoroughly fused. Slag melts first and may confuse unwary operators.

Low-carbon iron and steels (0.30%C or less) are readily welded and require no preheating or subsequent annealing unless residual stresses are to be removed.

Medium-carbon steels (0.30 to 0.50%C) can be welded by the various fusion processes. In some cases, especially in steel with more than 0.40% carbon, preheating and subsequent heat treatment may be necessary.

High-carbon steels (0.50 to 0.90%C) are more difficult to weld and, especially in arc welding, may have to be preheated to at least 500°F and subsequently heated between 1200 and 1450°F. For gas welding, a carburizing flame is often used. Care must be taken not to destroy the heat treatment to which high-carbon steels may have been subjected.

Tool steels (0.80 to 1.50%C) are difficult to weld. Preheating, postannealing, heat treatment, special welding rods, and great care are necessary for successful welding.

Welding of structural steels is governed by the American Welding Society "Structural Welding Code," AWS D1.1, the American Institute of Steel Construction Specification for the Design, Fabrication and Erection of Structural Steel for Buildings, or a local building code. AWS D1.1 specifies tests to be used in qualifying welders and types of welds. The AISC Specification and many building codes require, in general, that only qualified welds be used and that they be made only by qualified welders.

Structural steels may be welded by shielded metal arc, submerged arc, gas metal arc, flux-cored arc, electroslag, electrogas, or stud-welding processes.

Shielded-metal-arc welding fuses parts to be joined by the heat of an electric arc stuck between a coated metal electrode and the material being joined, or base metal. The electrode supplies filler material for making the weld, gas for shielding the molten metal from the air, and flux for refining this metal.

Submerged-arc welding fuses the parts to be joined by the heat of an electric arc struck between a bare metal electrode and base metal. The weld is shielded from the air by flux. The electrode or a supplementary welding rod supplies metal filler for making the weld.

Gas-metal-arc welding produces fusion by the heat of an electric arc struck between a filler-metal electrode and base metal, while the molten metal is shielded by a gas or mixture of gas and flux. For structural steels, the gas may be argon, argon with oxygen, or carbon dioxide.

Electroslag welding uses a molten slag to melt filler metal and surfaces of the base metal and thus make a weld. The slag, electrically conductive, is maintained molten by its resistance to an electric current that flows between an electrode and the base metal. The process is suitable only for welding in the vertical position. Moving, water-cooled shoes are used to contain and shape the weld surface. The slag shields the molten metal.

Electrogas welding is similar to the electroslag process. The electrogas process, however, maintains an electric arc continuously, uses an inert gas for shielding, and the electrode provides flux.

Stud welding is used to fuse metal studs or similar parts to other steel parts by the heat of an electric arc. A welding gun is usually used to establish and control the arc, and to apply pressure to the parts to be joined. At the end to be welded, the stud is equipped with a ceramic ferrule, which contains flux and which also partly shields the weld when molten.

Preheating before welding reduces the risk of brittle failure. Initially, its main effect is to lower the temperature gradient between the weld and adjoining base metal. This makes cracking during cooling less likely and gives entrapped hydrogen, a possible source of embrittlement, a chance to escape. A later effect of preheating is improved ductility and notch toughness of base and weld metals and lower transition temperature of weld. When, however, welding processes that deposit weld metal low in hydrogen are used and suitable moisture control is maintained, the need for preheat can be eliminated. Such processes include use of low-hydrogen electrodes and inert-arc and submerged-arc welding.

Rapid cooling of a weld can have an adverse effect. One reason that arc strikes that do not deposit weld metal are dangerous is that the heated metal cools very fast. This causes severe embrittlement. Such arc strikes should be completely removed. The material should be preheated, to prevent local hardening, and weld metal should be deposited to fill the depression.

Pronounced segregation in base metal may cause welds to crack under certain fabricating conditions. These include use of high-heat-input electrodes, such as the ¼-in E6020, and deposition of large beads at slow speeds, as in automatic welding. Cracking due to segregation, however, is rare with the degree of segregation normally occurring in hot-rolled carbon-steel plates.

Welds sometimes are peened to prevent cracking or distortion, though there are better ways of achieving these objectives. Specifications often prohibit peening of the first and last weld passes. Peening of the first pass may crack or punch through the weld. Peening of the last pass makes inspection for cracks difficult. But peening is undesirable because it considerably reduces toughness and impact properties of the weld metal. (The adverse effects, however, are eliminated by a covering weld layer.) The effectiveness of peening in preventing cracking is open to question. And for preventing distortion, special welding sequences and procedures are simpler and easier.

Failures in service rarely, if ever, occur in properly made welds of adequate design. If a fracture occurs, it is initiated at a notchlike defect. Notches occur for various reasons. The toe of a weld may form a natural notch. The weld may contain flaws that act as notches. A welding-arc strike in the base metal may have an embrittling effect, especially if weld metal is not deposited. A crack started at such notches will propagate along a path determined by local stresses and notch toughness of adjacent material.

Weldability of structural steels is influenced by their chemical content. Carbon, manganese, silicon, nickel, chromium, and copper, for example, tend to have an adverse effect, whereas molybdenum and vanadium may be beneficial. To relate the influence of chemical content on structural steel properties to weldability, the use of a carbon equivalent has been proposed. One formula suggested is

$$C_{eq} = C + \frac{Mn}{4} + \frac{Si}{4} \tag{4.3}$$

where C = carbon content, %
Mn = manganese content, %
Si = silicon content, %

Another proposed formula includes more elements:

$$C_{eq} = C + \frac{Mn}{6} + \frac{Ni}{20} + \frac{Cr}{10} - \frac{Mo}{50} - \frac{V}{10} + \frac{Cu}{40} \tag{4.4}$$

where Ni = nickel content, %
Cr = chromium content, %
Mo = molybdenum content, %
V = vanadium content, %
Cu = copper content, %

Carbon equivalent appears to be related to the maximum rate at which a weld and adjacent base metal may be cooled after welding without underbead cracking occurring. The higher the carbon equivalent, the lower will be the allowable cooling rate. Also, the higher the carbon equivalent, the more important use of low-hydrogen electrodes and preheating becomes.

4.46 EFFECTS OF STEEL PRODUCTION METHODS

The processing of steels after conversion of pig iron to steel in a furnace has an important influence on the characteristics of the final products. The general procedure is as follows: The molten steel at about 2900°F is fed into a steel ladle, a refractory-lined open-top vessel. Alloying materials and deoxidizers may be added during the tapping of the heat or to the ladle. From the ladle, the liquid steel is poured into molds, where it solidifies. These castings, called ingots, then are placed in special furnaces, called soaking pits. There, the ingots are held at the desired temperature for rolling until the temperature is uniform throughout each casting.

Ideally, an ingot should be homogeneous, with a fine, equiaxial crystal structure. It should not contain nonmetallic inclusions or cavities and should be free of chemical segregation. In practice, however, because of uneven cooling and release of gases in the mold, an ingot may develop any of a number of internal and external defects. Some of these may be eliminated or minimized during the rolling operation. Prevention or elimination of the others often adds to the cost of steels.

Steel cools unevenly in a mold, because the liquid at the mold walls solidifies first and cools more rapidly than metal in the interior of the ingot. Gases, chiefly oxygen, dissolved in the liquid, are released as the liquid cools. Four types of ingot may result—killed, semikilled, capped, and rimmed—depending on the amount of gases dissolved in the liquid, the carbon content of the steel, and the amount of deoxidizers added to the steel.

A fully killed ingot develops no gas; the molten steel lies dead in the mold. The top surface solidifies relatively fast. Pipe, an intermittently bridged shrinkage cavity, forms below the top. Fully killed steels usually are poured in big-end-up molds with "hot tops" to confine the pipe to the hot top, which is later discarded. A semikilled ingot develops a slight amount of gas. The gas, trapped when the metal solidifies, forms blowholes in the upper portion of the ingot. A **capped ingot** develops rimming action, a boiling caused by evolution of gas, forcing the steel to rise. The action is stopped by a metal cap secured to the mold. Strong upward currents along the sides of the mold sweep away bubbles that otherwise would form blowholes in the upper portion of the ingot. Blowholes do form, however, in the lower portion, separated by a thick solid skin from the mold walls. A **rimmed ingot** develops a violent rimming action, confining blowholes to only the bottom quarter of the ingot.

Rimmed or capped steels cannot be produced if too much carbon is present (0.30% or more), because insufficient oxygen will be dissolved in the steels to cause the rimming action. Killed and semikilled steels require additional costs for deoxidizers if carbon content is low, and the deoxidation products form nonmetallic inclusions in the ingot. Hence, it often is advantageous for steel producers to make low-carbon steels by rimmed or capped practice, and high-carbon steels by killed or semikilled practice.

Pipe, or shrinkage cavities, generally is small enough in most steels to be eliminated by rolling. **Blowholes** in the interior of an ingot, small voids formed by entrapped gases, also usually are eliminated during rolling. If they extend to the surface, they may be oxidized and form seams when the ingot is rolled, because the oxidized metal cannot be welded together. Properly made ingots have a thick enough skin over blowholes to prevent oxidation.

Segregation in ingots depends on the chemical composition and on turbulence from gas evolution and convection currents in the molten metal. Killed steels have less segregation than semikilled steels, and these types of steels have less segregation than capped or rimmed steels. In rimmed steels, the effects of segregation are so marked that interior and outer regions differ enough in chemical composition to appear to be different steels. The boundary between these regions is sharp.

Rimmed steels are made without additions of deoxidizers to the furnace and with only small additions to the ladle, to ensure sufficient evolution of gas. When properly made, rimmed ingots have little pipe and a good surface. Such steels are preferred where surface finish is important and the effects of segregation will not be harmful.

Capped steels are made much like rimmed steels but with less rimming action. Capped steels have less segregation. They are used to make sheet, strip, skelp, tinplate, wire, and bars.

Semikilled steel is deoxidized less than killed steel. Most deoxidation is accomplished with additions of a deoxidizer to the ladle. Semikilled steels are used in structural shapes and plates.

Killed steels usually are deoxidized by additions to both furnace and ladle. Generally, silicon compounds are added to the furnace to lower the oxygen content of the liquid metal and stop oxidation of carbon (block the heat). This also permits addition of alloying elements that are susceptible to oxidation. Silicon or other deoxidizers, such as aluminum, vanadium, and titanium, may be added to the ladle to complete deoxidation. Aluminum, vanadium, and titanium have the additional beneficial effect of inhibiting grain growth when the steel is normalized. (In the hot-rolled conditions, such steels have about the same ferrite grain size as semikilled steels.) Killed steels deoxidized with aluminum and silicon (made to fine-grain practice) often are specified for construction applications because of better notch toughness and lower transition temperatures than semikilled steels of the same composition.

4.47 EFFECTS OF HOT ROLLING

While plates and shapes for construction applications can be obtained from processes other than casting and rolling of ingots, such as continuous casting, most plates and shapes are made by hot-rolling ingots (Art. 4.46). But usually, the final products are not rolled directly from ingots. First, the ingots are generally reduced in cross section by rolling into billets, slabs, and blooms. These forms permit correction of defects before finish rolling, shearing into convenient lengths for final rolling, reheating for further rolling, and transfer to other mills. if desired, for that processing.

Plates produced from slabs or directly from ingots, are distinguished from sheet, strip, and flat bars by size limitations in ASTM A6. Generally, plates are heavier, per linear foot, than these other products. Sheared plates, or sheared mill plates, are made with straight horizontal rolls and later trimmed on all edges. Universal plates, or universal mill plates, are formed between vertical and horizontal rolls and are trimmed on the ends only.

Some of the plates may be heat-treated, depending on grade of steel and intended use. For carbon steel, the treatment may be annealing, normalizing, or stress relieving. Plates of high-strength, low-alloy constructional steels may be quenched and tempered. See Art. 4.42.

Shapes are rolled from blooms that first are reheated to 2250°F. Rolls gradually reduce the plastic blooms to the desired shapes and sizes. The shapes then are cut to length for convenient handling with a hot saw.

ASTM A6 requires that material for delivery "shall be free from injurious defects and shall have a workmanlike finish." The specification permits manufacturers to condition plates and shapes "for the removal of injurious surface imperfections or surface depressions by grinding, or chipping and grinding. . . ."

Internal structure and many properties of plates and shapes are determined largely by the chemistry of the steel, rolling practice, cooling conditions after rolling, and heat treatment, where used. The interior of ingots consists of large crystals, called dendrites, characterized by a branching structure. Growth of individual dendrites occurs principally along their longitudinal axes perpendicular to the ingot surfaces. Heating for rolling tends to eliminate dendritic segregation, so that the rolled products are more homogeneous than ingots. Furthermore, during rolling, the dendritic structure is broken up. Also, recrystallization occurs. The final austenitic grain size is determined by the temperature of the steel during the last passes through the rolls (Art. 4.43). In addition, dendrites and inclusions are reoriented in the direction of rolling. As a result, ductility and bendability are much better in the longitudinal direction than in the transverse, and these properties are poorest in the thickness direction. The cooling rate after rolling determines the distribution of ferrite and the grain size of the ferrite.

In addition to the preceding effects, rolling also may induce residual stresses in plates and shapes (Art. 4.41.1). Still other effects are a consequence of the final thickness of the hot-rolled material.

Thicker material requires less rolling, the finish rolling temperature is higher, and the cooling rate is slower than for thin material. As a consequence, thin material has a superior microstructure. Furthermore, thicker material can have a more unfavorable state of stress because of stress raisers, such as tiny cracks and inclusions, and residual stresses. Consequently, thin material develops higher tensile and yield strengths than thick material of the same steel. ASTM specifications for structural steels recognize this usually by setting lower yield points for thicker material. A36 steel, however, has the same yield point for all thicknesses. To achieve this, the chemistry is varied for plates and shapes and for thin and thick plates. Thicker plates contain more carbon and manganese to raise the yield point. This cannot be done for high-strength steels because of the adverse effect on notch toughness, ductility, and weldability.

Thin material has greater ductility than thick material of the same steel. Since normalizing refines the grain structure, thick material improves relatively more with normalizing than does thin material. The improvement is even greater with silicon-aluminum-killed steels.

4.48 EFFECTS OF PUNCHING AND SHEARING

Punching holes and shearing during fabrication are cold-working operations that can cause brittle failure. Bolt holes, for example, may be formed by drilling, punching, or punching followed by reaming. Drilling is preferable to punching, because punching drastically cold-works the material at the edge of a hole. This makes the steel less ductile and raises the transition temperature. The degree of embrittlement depends on type of steel and plate thickness. Furthermore, there is a possibility that punching can produce short cracks extending radially from the hole. Consequently, brittle failure can be initiated at the hole when the member is stressed.

Should the material around the hole become heated, an additional risk of failure is introduced. Heat, for example, may be supplied by an adjacent welding operation. If the temperature should rise to the 400 to 850°F range, strain aging will occur in material susceptible to it. The result will be a loss in ductility.

Reaming a hole after punching can eliminate the short radial cracks and the risks of embrittlement. For the purpose, the hole diameter should be increased by $^1/_{16}$ to $^1/_4$ in by reaming, depending on material thickness and hole diameter.

Shearing has about the same effects as punching. If sheared edges are to be left exposed, $^1/_{16}$ in or more material, depending on thickness, should be trimmed by gas cutting. Note also that rough machining, for example, with edge planers making a deep cut, can produce the same effects as shearing or punching.

4.49 CORROSION OF IRON AND STEEL

Corrosion of ferrous metals is caused by the tendency of iron (anode) to go into solution in water as ferrous hydroxide and displace hydrogen, which in turn combines with dissolved oxygen to form more water. At the same time, the dissolved ferrous hydroxide is converted by more oxygen to the insoluble ferric hydroxide, thereby allowing more iron to go into solution. Corrosion, therefore, requires liquid water (as in damp air) and oxygen (which is normally present dissolved in the water).

Alloying elements can increase the resistance of steel considerably. For example, addition of copper to structural steels A36 and A529 can about double their corrosion resistance. Other steels, such as A242 and A588, are called weathering steels, because they have three to four times the resistance of A36 steel (Art. 4.40.4).

Protection against corrosion takes a variety of forms:

Deaeration. If oxygen is removed from water, corrosion stops. In hot-water heating systems, therefore, no fresh water should be added. Boiler feedwater is sometimes deaerated to retard corrosion.

Coatings

1. **Paints.** Most paints are based on oxidizing oil and a variety of pigments, of which oxides of iron, zinc sulfate, graphite, aluminum, and various hydrocarbons are a few. No one paint is best for all applications. Other paints are coatings of asphalt and tar. The AISC "Specification for Structural Steel Buildings" (ASD and LRFD) states that, in general, steelwork to be concealed within a building need not be painted and that steel to be encased in concrete should not be painted. Inspections of old buildings have revealed that concealed steelwork withstands corrosion virtually to the same degree whether or not it is painted.
2. **Metallic.** Zinc is applied by hot dipping (**galvanizing**) or powder (**sherardizing**), hot tin dip, hot aluminum dip, and electrolytic plates of tin, copper, nickel, chromium, cadmium, and zinc. A mixture of lead and tin is called **terneplate.** Zinc is anodic to iron and protects, even after the coating is broken, by sacrificial protection. Tin and copper are cathodic and protect as long as the coating is unbroken but may hasten corrosion by pitting and other localized action once the coating is pierced.
3. **Chemical.** Insoluble phosphates, such as iron or zinc phosphate, are formed on the surface of the metal by treatment with phosphate solutions. These have some protective action and also form good bases for paints. Black oxide coatings are formed by treating the surface with various strong salt solutions. These coatings are good for indoor use but have limited life outdoors. They provide a good base for rust-inhibiting oils.

Cathodic Protection. As corrosion proceeds, electric currents are produced as the metal at the anode goes into solution. If a sufficient countercurrent is produced, the metal at the anode will not dissolve. This is accomplished in various ways, such as connecting the iron to a more active metal like magnesium (rods suspended in domestic water heaters) or connecting the part to be protected to buried scrap iron and providing an external current source such as a battery or rectified current from a power line (protection of buried pipe lines).

4.50 STEEL AND STEEL ALLOY BIBLIOGRAPHY

American Iron and Steel Institute, 1000 16th St., N.W., Washington, DC 20036: "Carbon Steels, Chemical Composition Limits," "Constructional Alloys, Chemical Composition Limits"; "Steel Products Manuals."

American Society for Testing and Materials, Philadelphia, Pa.: "Standards."

American Society for Metals, Cleveland, Ohio: "Metals Handbook."

M. E. Shank, "Control of Steel Construction to Avoid Brittle Failure," Welding Research Council, New York.

R. L. Brockenbrough and F. S. Merritt, "Structural Steel Designers Handbook," 2nd ed., McGraw-Hill, Inc., New York.

ALUMINUM AND ALUMINUM-BASED ALLOYS

Pure aluminum and aluminum alloys are used in buildings in various forms. High-purity aluminum (at least 99% pure) is soft and ductile but weak. It has excellent corrosion resistance and is used in buildings for such applications as bright foil for heat insulation, roofing, flashing, gutters and downspouts, exterior and interior architectural trim, and as pigment in aluminum-based paints. Its high heat conductivity recommends it for cooking utensils. The electrical conductivity of the electrical grade is 61% of that of pure copper on an equal-volume basis and 201% on an equal-weight basis.

Aluminum alloys are generally harder and stronger than the pure metal. Furthermore, pure aluminum is difficult to cast satisfactorily, whereas many of the alloys are readily cast.

Pure aluminum is generally more corrosion resistant than its alloys. Furthermore, its various forms—pure and alloy—have different solution potentials; that is, they are anodic or cathodic to each other, depending on their relative solution potentials. A number of alloys are therefore made with centers or "cores" of aluminum alloys, overlaid with layers of metal, either pure aluminum or alloys, which are anodic to the core. If galvanic corrosion conditions are encountered, the cladding metal protects the core sacrificially.

4.51 ALUMINUM-ALLOY DESIGNATIONS

The alloys may be classified: (1) as cast and wrought, and (2) as heat-treatable and non-heat-treatable. Wrought alloys can be worked mechanically by such processes as rolling, extruding, drawing, or forging. Alloys are heat-treatable if the dissolved constituents are less soluble in the solid state at ordinary temperatures than at elevated temperatures, thereby making age-hardening possible. When heat-treated to obtain complete solution, the product may be unstable and tend to age spontaneously. It may also be treated to produce stable tempers of varying degree. Cold working or strain hardening is also possible, and combinations of tempering and strain hardening can also be obtained.

Because of these various possible combinations, a system of letter and number designations has been worked out by the producers of aluminum and aluminum alloys to indicate the compositions and the tempers of the various metals. Wrought alloys are designated by a four-digit index system. 1xxx is for 99.00% aluminum minimum. The last two digits indicate the minimum aluminum percentage. The second digit represents impurity limits. (EC is a special designation for electrical conductors.) 2xxx to 8xxx represent alloy groups in which the first number indicates the principal alloying constituent, and the last two digits are identifying numbers in the group. The second digit indicates modification of the basic alloy. The alloy groups are listed in Table 4.19.

TABLE 4.19 Aluminum Association Designations for Wrought Aluminum Alloys

Copper	2xxx
Manganese	3xxx
Silicon	4xxx
Magnesium	5xxx
Magnesium and silicon	6xxx
Zinc	7xxx
Other elements	8xxx
Unused series	9xxx

TABLE 4.20 Basic Temper Designations for Wrought Aluminum Alloys*

-F	**As fabricated.** This designation applies to the products of shaping processes in which no special control over thermal conditions or strain hardening is employed. For wrought products, there are no mechanical property limits.
-O	**Annealed.** This designation applies to wrought products annealed to obtain the lowest-strength temper, and to cast products annealed to improve ductility and dimensional stability.
-H†	**Strain hardened (wrought products only).** This designation applies to products that have their strength increased by strain hardening, with or without supplementary thermal treatments to produce some reduction in strengths. The H is always followed by two or more digits.
-W	**Solution heat treated.** An unstable temper applicable only to alloys that spontaneously age at room temperature after solution heat treatment. This designation is specific only when the period of natural aging is indicated: for example, W ½ hr.
-T‡	**Thermally treated to produce stable tempers other than F, O, or H.** This designation applies to products that are thermally treated, with or without supplementary strain hardening, to produce stable tempers. The T is always followed by one or more digits.

*Recommended by the Aluminum Association.

†A digit after H represents a specific combination of basic operations, such as H1—strain hardened only, H2—strain hardened and partly annealed, and H3—strain hardened and stabilized. A second digit indicates the degree of strain hardening, which ranges from 0 for annealing to 9 in the order of increasing tensile strength.

‡A digit after T indicates a type of heat treatment, which may include cooling, cold working, and aging.

For cast alloys, a similar designation system is used. The first two digits identify the alloy or its purity. The last digit, preceded by a decimal point, indicates the form of the material; for example, casting or ingot. Casting alloys may be sand or permanent-mold alloys.

Among the wrought alloys, the letters F, O, H, W, and T indicate various basic temper designations. These letters in turn may be followed by numerals to indicate various degrees of treatment. Temper designations are summarized in Table 4.20.

The structural alloys generally employed in building fall in the 2xxx, 5xxx, and 6xxx categories. Architectural alloys often used include 3xxx, 5xxx, and 6xxx groups.

4.52 FINISHES FOR ALUMINUM

Almost all finishes used on aluminum may be divided into three major categories in the system recommended by the The Aluminum Association: mechanical finishes, chemical finishes, and coatings. The last may be subdivided into anodic coatings, resinous and other organic coatings, vitreous coatings, electroplated and other metallic coatings, and laminated coatings.

In The Aluminum Association system, mechanical and chemical finishes are designated by M and C, respectively, and each of the five classes of coating is also designated by a letter. The various finishes in each category are designated by two-digit numbers after a letter. The principal finishes are summarized in Table 4.21.

TABLE 4.21 Finishes for Aluminum and Aluminum Alloys

Type of finish	Designation*
Mechanical finishes:	
As fabricated	M1*Y*
Buffed	M2*Y*
Directional textured	M3*Y*
Nondirectional textured	M4*Y*
Chemical finishes:	
Nonetched cleaned	C1*Y*
Etched	C2*Y*
Brightened	C3*Y*
Chemical conversion coatings	C4*Y*
Coatings:	
Anodic	
General	A1*Y*
Protective and decorative (less than 0.4 mil thick)	A2*Y*
Architectural Class II (0.4–0.7 mil thick)	A3*Y*
Architectural Class I (0.7 mil or more thick)	A4*Y*
Resinous and other organic coatings	R1*Y*
Vitreous coatings	V1*Y*
Electroplated and other metallic coatings	E1*Y*
Laminated coatings	L1*Y*

**Y* represents digits (0, 1, 2, . . . 9) or *X* (to be specified) that describe the surface, such as specular, satin, matte, degreased, clear anodizing or type of coating.

4.53 STRUCTURAL ALUMINUM

Structural aluminum shapes are produced by extrusion. Angles, I beams, and channels are available in standard sizes and in lengths up to 85 ft. Plates up to 6 in thick and 200 in wide also may be obtained.

There are economic advantages in selecting structural aluminum shapes more efficient for specific purposes than the customary ones. For example, sections such as hollow tubes, shapes with stiffening lips on outstanding flanges, and stiffened panels can be formed by extrusion.

Aluminum alloys generally weigh about 170 lb/ft^3, about one-third that of structural steel. The modulus of elasticity in tension is about 10,000 ksi, compared with 29,000 ksi for structural steel. Poisson's ratio may be taken as 0.50. The coefficient of thermal expansion in the 68 to 212°F range is about 0.000013 in/in·°F, about double that of structural steel.

Alloy 6061-T6 is often used for structural shapes and plates. ASTM B308 specifies a minimum tensile strength of 38 ksi, minimum tensile yield strength of 35 ksi, and minimum elongation in 2 in of 10%, but 8% when the thickness is less than ¼ in.

The preceding data indicate that, because of the low modulus of elasticity, aluminum members have good energy absorption. Where stiffness is important, however, the effect of the low modulus should be taken into account. Specific data for an application should be obtained from the producers.

4.54 WELDING AND BRAZING OF ALUMINUM

Weldability and brazing properties of aluminum alloys depend heavily on their composition and heat treatment. Most of the wrought alloys can be brazed and welded, but sometimes only by special processes. The strength of some alloys depends on heat treatment after welding. Alloys heat treated and artificially aged are susceptible to loss of strength at the weld, because weld is essentially cast. For this reason, high-strength structural alloys are commonly fabricated by riveting or bolting, rather than by welding.

Brazing is done by furnace, torch, or dip methods. Successful brazing is done with special fluxes.

Inert-gas shielded-arc welding is usually used for welding aluminum alloys. The inert gas, argon or helium, inhibits oxide formation during welding. The electrode used may be consumable metal or tungsten. The gas metal arc is generally preferred for structural welding, because of the higher speeds that can be used. The gas tungsten arc is preferred for thicknesses less than ½ in.

Butt-welded joints of annealed aluminum alloys and non-heat-treatable alloys have nearly the same strength as the parent metal. This is not true for strain-hardened or heat-tempered alloys. In these conditions, the heat of welding weakens the metal in the vicinity of the weld. The tensile strength of a butt weld of alloy 6061-T6 may be reduced to 24 ksi, about two-thirds that of the parent metal. Tensile yield strength of such butt welds may be only 15 to 20 ksi, depending on metal thickness and type of filler wire used in welding.

Fillet welds similarly weaken heat-treated alloys. The shear strength of alloy 6061-T6 decreases from about 27 ksi to 17 ksi or less for a fillet weld.

Welds should be made to meet the requirements of the American Welding Society, "Structural Welding Code—Aluminum," AWS D1.2.

4.55 BOLTED AND RIVETED ALUMINUM CONNECTIONS

Aluminum connections also may be bolted or riveted. Bolted connections are bearing type. Slip-critical connections, which depend on the frictional resistance of joined parts created by bolt tension, are not usually employed because of the relatively low friction and the potential relaxation of the bolt tension over time.

Bolts may be aluminum or steel. Bolts made of aluminum alloy 7075-T73 have a minimum expected shear strength of 40 ksi. Cost per bolt, however, is higher than that of 2024-T4 or 6061-T6, with tensile strengths of 37 and 27 ksi, respectively. Steel bolts may be used if the bolt material is selected to prevent galvanic corrosion or the steel is insulated from the aluminum. One option is use of stainless steel. Another alternative is to galvanize, aluminize, or cadmium plate the steel bolts.

Rivets typically are made of aluminum alloys. They are usually driven cold by squeeze-type riveters. Alloy 6053-T61, with a shear strength of 20 ksi, is preferred for joining relatively soft alloys, such as 6063-T5, Alloy 6061-T6, with a shear strength of 26 ksi, is usually used for joining 6061-T6 and other relatively hard alloys.

4.56 PREVENTION OF CORROSION OF ALUMINUM

Although aluminum ranks high in the electromotive series of the metals, it is highly corrosion resistant because of the tough, transparent, tenacious film of aluminum oxide that rapidly forms on any exposed surface. It is this corrosion resistance that recommends aluminum for building applications. For most exposures, including industrial and seacoast atmospheres, the alloys normally recommended are adequate, particularly if used in usual thicknesses and if mild pitting is not objectionable.

Pure aluminum is the most corrosion resistant of all and is used alone or as cladding on strong-alloy cores where maximum resistance is wanted. Of the alloys, those containing magnesium, manganese, chromium, or magnesium and silicon in the form of $MgSi_2$ are highly resistant to corrosion. The alloys containing substantial proportions of copper are more susceptible to corrosion, depending markedly on the heat treatment.

Certain precautions should be taken in building. Aluminum is subject to attack by alkalies, and it should therefore be protected from contact with wet concrete, mortar, and plaster. Clear methacrylate lacquers or strippable plastic coatings are recommended for interiors and methacrylate lacquer for exterior protection during construction. Strong alkaline and acid cleaners should be avoided and muriatic acid should not be used on masonry surfaces adjacent to aluminum. If aluminum must be contiguous to concrete and mortar outdoors, or where it will be wet, it should be insulated from direct contact by asphalts, bitumens, felts, or other means. As is true of other metals, atmospheric-deposited dirt must be removed to maintain good appearance.

Electrolytic action between aluminum and less active metals should be avoided, because the aluminum then becomes anodic. If aluminum must be in touch with other metals, the faying surfaces should be insulated by painting with asphaltic or similar paints, or by gasketing. Steel rivets and bolts, for example, should be insulated. Drainage from copper-alloy surfaces onto aluminum must be avoided. Frequently, steel surfaces can be galvanized or cadmium-coated where contact is expected with aluminum. The zinc or cadmium coating is anodic to the aluminum and helps to protect it.

4.57 ALUMINUM BIBLIOGRAPHY

"Aluminum Standards and Data," "Engineering Data for Aluminum Structures," "Designation Systems for Aluminum Finishes," and "Specifications for Aluminum Structures," The Aluminum Association, Washington, D.C.

E. H. Gaylord, Jr., and C. N. Gaylord, "Structural Engineering Handbook," 3rd ed., McGraw-Hill Publishing Company, New York.

COPPER AND COPPER-BASED ALLOYS

Copper and its alloys are widely used in the building industry for a large variety of purposes, particularly applications requiring corrosion resistance, high electrical conductivity, strength, ductility, impact resistance, fatigue resistance, or other special characteristics possessed by copper or its alloys. Some of the special characteristics of importance to building are ability to be formed into complex shapes, appearance, and high thermal conductivity, although many of the alloys have low thermal conductivity and low electrical conductivity as compared with the pure metal.

4.58 COPPER

The excellent corrosion resistance of copper makes it suitable for such applications as roofing, flashing, cornices, gutters, downspouts, leaders, fly screens, and similar applications. For roofing and flashing, soft-annealed copper is employed, because it is ductile and can easily be bent into various shapes. For gutters, leaders, downspouts, and similar applications, cold-rolled hard copper is employed, because its greater hardness and stiffness permit it to stand without large numbers of intermediate supports.

Copper and copper-based alloys, particularly the brasses, are employed for water pipe in buildings, because of their corrosion resistance. Electrolytic tough-pitch copper is usually employed for electrical conductors, but for maximum electrical conductivity and weldability, oxygen-free high-conductivity copper is used.

When arsenic is added to copper, it appears to form a tenacious adherent film, which is particularly resistant to pitting corrosion. Phosphorus is a powerful deoxidizer and is particularly useful for copper to be used for refrigerator tubing and other applications where flaring, flanging, and spinning are required. Arsenic and phosphorus both reduce the electrical conductivity of the copper.

For flashing, copper is frequently coated with lead to avoid the green patina formed on copper that is sometimes objectionable when it is washed down over adjacent surfaces, such as ornamental stone. The patina is formed particularly in industrial atmospheres. In rural atmospheres, where industrial gases are absent, the copper normally turns to a deep brown color.

Principal types of copper and typical uses are:

Electrolytic tough pitch (99.90% copper) is used for electrical conductors—bus bars, commutators, etc.; building products—roofing, gutters, etc.; process equipment—kettles, vats, distillery equipment; forgings. General properties are high electrical conductivity, high thermal conductivity, and excellent working ability.

Deoxidized (99.90% copper and 0.025% phosphorus) is used, in tube form, for water and refrigeration service, oil burners, etc.; in sheet and plate form, for welded construction. General properties include higher forming and bending qualities than electrolytic copper. They are preferred for coppersmithing and welding (because of resistance to embrittlement at high temperatures).

4.59 BRASS

A considerable range of brasses is obtainable for a large variety of end uses. The high ductility and malleability of the copper-zinc alloys, or brasses, make them suitable for operations like deep drawing, bending, and swaging. They have a wide range of colors. They are generally less expensive than the high-copper alloys.

Grain size of the metal has a marked effect upon its mechanical properties. For deep drawing and other heavy working operations, a large grain size is required, but for highly finished polished surfaces, the grain size must be small.

Like copper, brass is hardened by cold working. Hardnesses are sometimes expressed as quarter hard, half hard, hard, extra hard, spring, and extra spring, corresponding to reductions in cross section during cold working ranging from approximately 11 to 69%. Hardness is strongly influenced by alloy composition, original grain size, and form (strip, rod, tube, wire).

4.59.1 Plain Brass

Brass compositions range from higher copper content to zinc contents as high as 40% or more. Brasses with less than 36% zinc are plain alpha solid solutions; but Muntz metal, with 40% zinc, contains both alpha and beta phases.

The principal plain brasses of interest in building, and their properties are:

Commercial bronze, 90% (90.0% copper, 10.0% zinc). Typical uses are forgings, screws, weatherstripping, and stamped hardware. General properties include excellent cold working and high ductility.

Red brass, 85% (85.0% copper, 15.0% zinc). Typical uses are dials, hardware, etched parts, automobile radiators, and tube and pipe for plumbing. General properties are higher strength and ductility than copper, and excellent corrosion resistance.

Cartridge brass, 70% (70.0% copper, 30.0% zinc). Typical uses are deep drawing, stamping, spinning, etching, rolling—for practically all fabricating processes—cartridge cases, pins, rivets, eyelets, heating units, lamp bodies and reflectors, electrical sockets, drawn shapes, etc. General properties are best combination of ductility and strength of any brass, and excellent cold-working properties.

Muntz metal (60.0% copper, 40.0% zinc). Typical uses are sheet form, perforated metal, architectural work, condenser tubes, valve stems, and brazing rods. General properties are high strength combined with low ductility.

4.59.2 Leaded Brass

Lead is added to brass to improve its machinability, particularly in such applications as automatic screw machines where a freely chipping metal is required. Leaded brasses cannot easily be cold-worked by such operations as flaring, upsetting, or cold heading. Several leaded brasses of importance in the building field are the following:

High-leaded brass (64.0% copper, 34.0% zinc, 2.0% lead). Typical uses are engraving plates, machined parts, instruments (professional and scientific), nameplates, keys, lock parts, and tumblers. General properties are free machining and good blanking.

Forging brass (60.0% copper, 38.0% zinc, 2.0% lead). Typical uses are hot forging, hardware, and plumbing goods. General properties are extreme plasticity when hot and a combination of good corrosion resistance with excellent mechanical properties.

Architectural bronze (56.5% copper, 41.25% zinc, 2.25% lead). Typical uses are handrails, decorative moldings, grilles, revolving door parts, miscellaneous architectural trim, industrial extruded shapes (hinges, lock bodies, automotive parts). General properties are excellent forging and free-machining properties.

4.59.3 Tin Brass

Tin is added to a variety of basic brasses to obtain hardness, strength, and other properties that would otherwise not be available. Two important alloys are:

Admiralty (71.0% copper, 28.0% zinc, 1.0% tin, 0.05% arsenic). Typical uses are condenser and heat-exchanger plates and tubes, steam-power-plant equipment, chemical and process equipment, and marine uses. General properties are excellent corrosion resistance, combined with strength and ductility.

Manganese bronze (58.5% copper, 39.0% zinc, 1.4% iron, 1.0% tin, 0.1% manganese). Typical uses are forgings, condenser plates, valve stems, and coal screens. General properties are high strength combined with excellent wear resistance.

4.60 NICKEL SILVERS

These are alloys of copper, nickel, and zinc. Depending on the composition, they range in color from a definite to slight pink cast through yellow, green, whitish green, whitish blue, to blue. A wide range of nickel silvers is made, of which only one typical composition will be described. Those that fall in the combined alpha-beta phase of metals are readily hot-worked and therefore are fabricated without difficulty into such intricate shapes as plumbing fixtures, stair rails, architectural shapes, and escalator parts. Lead may be added to improve machining.

Nickel silver, 18% (A) (65.0% copper, 17.0% zinc, 18.0% nickel). Typical uses are hardware, architectural panels, lighting, electrical and plumbing fixtures. General properties are high resistance to corrosion and tarnish, malleable, and ductile. Color: silver-blue-white.

4.61 CUPRONICKEL

Copper and nickel are alloyed in a variety of compositions of which the high-copper alloys are called the cupronickels. Typical commercial types of cupronickel contain 10 or 30% nickel (Table 4.15):

Cupronickel, 10% (88.5% copper, 10% nickel, 1.5% iron). Recommended for applications requiring corrosion resistance, especially to salt water, as in tubing for condensers, heat exchangers, and formed sheets.

Cupronickel, 30% (70.0% copper, 30.0% nickel). Typical uses are condenser tubes and plates, tanks, vats, vessels, process equipment, automotive parts, meters, refrigerator pump valves. General properties are high strength and ductility and resistance to corrosion and erosion. Color: white-silver.

4.62 BRONZE

Originally, the bronzes were all alloys of copper and tin. Today, the term bronze is generally applied to engineering metals having high mechanical properties and the term brass to other metals. The commercial wrought bronzes do not usually contain more than 10% tin because the metal becomes extremely hard and brittle. When phosphorus is added as a deoxidizer, to obtain sound, dense castings, the alloys are known as phosphor bronzes. The two most commonly used tin bronzes contain 5 or 8% tin. Both have excellent cold-working properties.

4.62.1 Silicon Bronze

These are high-copper alloys containing percentages of silicon ranging from about 1% to slightly more than 3%. In addition, they generally contain one or more of the four elements, tin, manganese, zinc, and iron. A typical one is high-silicon bronze, type A.

High-silicon bronze, A (96.0% copper, 3.0% silicon, 1.0% manganese). Typical users are tanks—pressure vessels, vats; weatherstrips, forgings. General properties are corrosion resistance of copper and mechanical properties of mild steel.

4.62.2 Aluminum Bronze

Like aluminum, these bronzes form an aluminum oxide skin on the surface, which materially improves resistance to corrosion, particularly under acid conditions. Since the color of the 5% aluminum bronze is similar to that of 18-carat gold, it is used for costume jewelry and other decorative purposes. Aluminum-silicon bronzes are used in applications requiring high tensile properties in combination with good corrosion resistance in such parts as valves, stems, air pumps, condenser bolts, and similar applications. Their wear-resisting properties are good; consequently, they are used in slide liners and bushings.

4.63 COPPER BIBLIOGRAPHY

"Alloy Data," Copper Development Association, New York, N.Y.

G. S. Brady and H. R. Clauser, "Materials Handbook," 13th ed., and J. H. Callender, "Time-Saver Standards for Architectural Design Data," 6th ed., McGraw-Hill Publishing Company, New York.

LEAD AND LEAD-BASED ALLOYS

Lead is used primarily for its corrosion resistance. Lead roofs 2000 years old are still intact.

4.64 APPLICATIONS OF LEAD

Exposure tests indicate corrosion penetrations of sheet lead ranging from less than 0.0001 in to less than 0.0003 in in 10 years in atmospheres ranging from mild rural to severe industrial and seacoast locations. Sheet lead is therefore used for roofing, flashing, spandrels, gutters, and downspouts.

Because the green patina found on copper may wash away sufficiently to stain the surrounding structure, lead-coated copper is frequently employed. ASTM B101-78 covers two classes, defined by the weight of coating.

Lead pipe should not be used for the transport of drinking water. Distilled and very soft waters slowly dissolve lead and may cause cumulative lead poisoning. Hard waters apparently deposit a protective coating on the wall of the pipe and little or no lead is subsequently dissolved in the water.

Principal alloying elements used with building leads are antimony (for hardness and strength) and tin. But copper, arsenic, bismuth, nickel, zinc, silver, iron, and manganese are also added in varying proportions.

Soft solders consist of varying percentages of lead and tin. For greater hardness, antimony is added, and for higher-temperature solders, silver is added in small amounts. ASTM Standard B32 specifies properties of soft solders.

Low-melting alloys and many bearing metals are alloys of lead, bismuth, tin, cadmium, and other metals including silver, zinc, indium, and antimony. The fusible links used in sprinkler heads and fire-door closures, made of such alloys, have a low melting point, usually lower than the boiling point of water. Yield (softening) temperatures range from 73 to 160°F and melting points from about 80 to 480°F, depending on the composition.

4.65 LEAD BIBLIOGRAPHY

American Society for Metals, Cleveland, Ohio: "Metals Handbook."

NICKEL AND NICKEL-BASED ALLOYS

Nickel is used mostly as an alloying element with other metals, but it finds use in its own right, largely as electroplate or as cladding metal. Among the principal high-nickel alloys are Monel and Inconel. The nominal compositions of these metals are given in Table 4.22.

4.66 PROPERTIES OF NICKEL AND ITS ALLOYS

Nickel is resistant to alkaline corrosion under nonoxidizing conditions but is corroded by oxidizing acids and oxidizing salts. It is resistant to fatty acids, other

TABLE 4.22 Composition of Nickel Alloys

	Nickel alloy, low-carbon NO2201	Nickel alloy NO2200	Monel NO4400	Inconel NO6600	70–30 cupro-nickel C71500	90–10 cupro-nickel C70600
Content	ASTM B160	ASTM B160	ASTM B127	ASTM B168	ASTM B171	ASTM B171
Carbon	0.02	0.15	0.3	0.15 max		
Manganese	0.35	0.35	2.00 max	1.0 max	1.0 max	1.0 max
Sulfur	0.01	0.01	0.024 max	0.015 max		
Silicon	0.35	0.35	0.5	0.5 max		
Chromium				14–17		
Nickel	99 min	99 min	63–70	72 min	29–33	9–11
Copper	0.25	0.25	Remainder	0.5 max	65 min	86.5 min
Iron	0.40 max	0.40 max	2.5 max	6–10	0.40–1.0	1.0–1.8
Lead					0.05 max	0.05 max
Zinc					1.0	1.0

mildly acid conditions, such as food processing and beverages, and resists oxidation at temperatures as high as 1600°F.

Monel is widely used in kitchen equipment. It is better than nickel in reducing conditions like warm unaerated acids, and better than copper under oxidizing conditions, such as aerated acids, alkalies, and salt solutions. It is widely used for handling chlorides of many kinds.

Inconel is almost completely resistant to corrosion by food products, pharmaceuticals, biologicals, and dilute organic acids. It is superior to nickel and Monel in resisting oxidizing acid salts like chromates and nitrates but is not resistant to ferric, cupric, or mercuric chlorides. It resists scaling and oxidation in air and furnace atmospheres at temperatures up to 2000°F.

4.67 NICKEL BIBLIOGRAPHY

International Nickel Co., New York: "Nickel and Nickel Alloys."

Albert Hoerson, Jr.: "Nonferrous-clad Plate Steels," Chap. 13 in A. G. H. Dietz, "Composite Engineering Laminates," M.I.T. Press, Cambridge, Mass.

PLASTICS

The synonymous terms plastics and synthetic resins denote synthetic organic high polymers, all of which are plastic at some stage in their manufacture. Plastics fall into two large categories—thermoplastic and thermosetting materials.

4.68 GENERAL PROPERTIES OF PLASTICS

Thermoplastics may be softened by heating and hardened by cooling any number of times. Thermosetting materials are either originally soft or liquid, or they soften once upon heating; but upon further heating, they harden permanently. Some thermosetting materials harden by an interlinking mechanism in which water or other by-product is given off, by a process called condensation; but others, like the unsaturated polyesters, harden by a direct interlinking of the basic molecules without release of a by-product.

Most plastics are modified with plasticizers, fillers, or other ingredients. Consequently, each base material forms the nucleus for a large number of products having a wide variety of properties. This section can only indicate generally the range of properties to be expected.

Because plastics are quite different in their composition and structure from other materials, such as metals, their behavior under stress and under other conditions is likely to be different from other materials. Just as steel and lead are markedly different and are used for different applications, so the various plastics materials—some hard and brittle, others soft and extensible—must be designed on different bases and used in different ways. Some plastics show no yield point, because they fail before a yield point can be reached. Others have a moderately

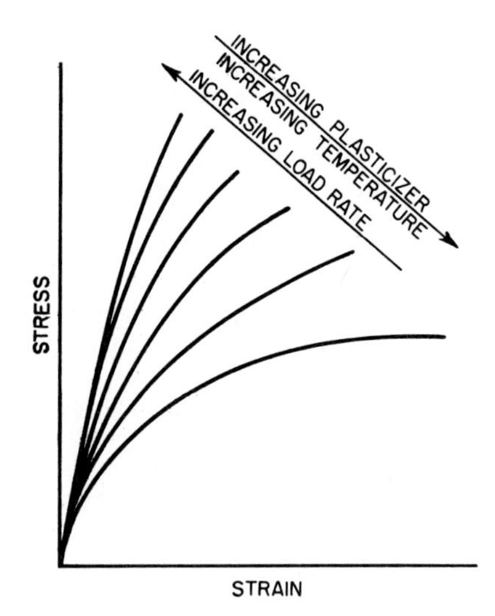

FIGURE 4.5 Stress-strain diagram shows the influence of temperature, plasticizer, and rate of loading on behavior of plastics.

high elastic range, followed by a highly plastic range. Still others are highly extensible and are employed at stresses far beyond the yield point.

More than many other materials, plastics are sensitive to temperature and to the rate and time of application of load. How these parameters influence the properties is indicated in a general way in Fig. 4.5, which shows that for many plastics an increase in temperature, increase in plasticizer content, and decrease in rate of load application mean an increase in strain to fracture, accompanied by a decrease in maximum stress. This viscoelastic behavior, combining elastic and viscous or plastic reaction to stress, is unlike the behavior of materials which are traditionally considered to behave only elastically.

4.69 FILLERS AND PLASTICIZERS

Fillers are commonly added, particularly to the thermosetting plastics, to alter their basic characteristics. For example, wood flour converts a hard, brittle resin, difficult to handle, into a cheaper, more easily molded material for general purposes. Asbestos fibers provide better heat resistance; mica gives better electrical properties; and a variety of fibrous materials, such as chopped fibers, chopped fabric, and chopped tire cords, increase the strength and impact properties.

Plasticizers are added to many thermoplastics, primarily to transform hard and rigid materials into a variety of forms having varying degrees of softness, flexibility, and strength. In addition, dyes or pigments, stabilizers, and other products may be added.

4.70 MOLDING AND FABRICATING METHODS FOR PLASTICS

Both thermosetting and thermoplastic molding materials are formed into final shape by a variety of molding and fabricating methods.

Thermosetting materials are commonly formed by placing molding powder or molded preform in heated dies and compressing under heat and pressure into the final infusible shape. Or they are formed by forcing heat-softened material into a heated die for final forming into the hard infusible shape.

Thermoplastics are commonly formed by injection molding, that is, by forcing soft, hot plastic into a cold die, where it hardens by cooling. Continuous profiles of thermoplastic materials are made by extrusion. Thermoplastic sheets, especially transparent acrylics, are frequently formed into final shape by heating and then blowing to final form under compressed air or by drawing a partial vacuum against the softened sheet.

Foamed plastics are employed for thermal insulation in refrigerators, buildings, and many other applications. In buildings, plastics are either prefoamed into slabs, blocks, or other appropriate shapes, or they are foamed in place.

Prefoamed materials, such as polystyrene, are made by adding a blowing agent and extruding the mixture under pressure and at elevated temperatures. As the material emerges from the extruder, it expands into a large "log" that can be cut into desired shapes. The cells are "closed"; that is, they are not interconnecting and are quite impermeable.

Foamed-in-place plastics are made with pellets or liquids. The pellets, made, for example, of polystyrene, are poured into the space to be occupied, such as a mold, and heated, whereupon they expand and occupy the space. The resulting mass may be permeable between pellets. Liquid-based foams, exemplified by polyurethane, are made by mixing liquid ingredients and immediately casting the mixture into the space to be occupied. A quick reaction results in a foam that rises and hardens by a thermosetting reaction. When blown with fluorocarbon gases, such foams have exceptionally low thermal conductivities.

All the plastics can be machined, if proper allowance is made for the properties of the materials.

Plastics are often combined with sheet or mat stocks, such as paper, cotton muslin, glass fabric, glass filament mats, nylon fabric, and other fabrics, to provide laminated materials in which the properties of the combined plastic and sheet stock are quite different from the properties of either constituent by itself. Two principal varieties of laminates are commonly made: (1) High-pressure laminates employing condensation-type thermosetting materials, which are formed at elevated temperatures and pressures. (2) Reinforced plastics employing unsaturated polyesters and epoxides, from which no by-products are given off, and consequently, either low pressures or none at all may be required to form combinations of these materials with a variety of reinforcing agents, like glass fabric or mat.

4.71 THERMOSETTING PLASTICS

General properties of thermosetting plastics are described in Art. 4.68. Following are properties of several thermosetting plastics used in buildings:

Phenol Formaldehyde. These materials provide the greatest variety of thermosetting molded plastic articles. They are used for chemical, decorative, electrical, mechanical, and thermal applications of all kinds. Hard and rigid, they change slightly, if at all, on aging indoors but, on outdoor exposure, lose their bright surface gloss. However, the outdoor-exposure characteristics of the more durable formulations are otherwise generally good. Phenol formaldehydes have good electrical properties, do not burn readily, and do not support combustion. They are strong, light in weight, and generally pleasant to the eye and touch, although light colors by and large are not obtainable because of the fairly dark-brown basic color of the resin. They have low water absorption and good resistance to attack by most commonly found chemicals.

Epoxy and Polyester Casting Resins. These are used for a large variety of purposes. For example, electronic parts with delicate components are sometimes cast completely in these materials to give them complete and continuous support, and resistance to thermal and mechanical shock. Some varieties must be cured at elevated temperatures; others can be formulated to be cured at room temperatures. One of the outstanding attributes of the epoxies is their excellent adhesion to a variety of materials, including such metals as copper, brass, steel, and aluminum.

Polyester Molding Materials. When compounded with fibers, particularly glass fibers, or with various mineral fillers, including clay, the polyesters can be formulated into putties or premixes that are easily compression- or transfer-molded into parts having high impact resistance. Polyesters are often used in geotextiles (Art. 6.12).

Melamine Formaldehyde. These materials are unaffected by common organic solvents, greases, and oils, as well as most weak acids and alkalies. Their water absorption is low. They are insensitive to heat and are highly flame-resistant, depending on the filler. Electrical properties are particularly good, especially resistance to arcing. Unfilled materials are highly translucent and have unlimited color possibilities. Principal fillers are alpha cellulose for general-purpose compounding; minerals to improve electrical properties, particularly at elevated temperatures; chopped fabric to afford high shock resistance and flexural strength; and cellulose, mainly for electrical purposes.

Cellulose Acetate Butyrate. The butyrate copolymer is inherently softer and more flexible than cellulose acetate and consequently requires less plasticizer to achieve a given degree of softness and flexibility. It is made in the form of clear transparent sheet and film, or in the form of molding powders, which can be molded by standard injection-molding procedures into a wide variety of applications. Like the other cellulosics, this material is inherently tough and has good impact resistance. It has infinite colorability, like the other cellulosics. Cellulose acetate butyrate tubing is used for such applications as irrigation and gas lines.

Cellulose Nitrate. One of the toughest of the plastics, cellulose nitrate is widely used for tool handles and similar applications requiring high impact strength. The high flammability requires great caution, particularly in the form of film. Most commercial photographic film is cellulose nitrate as opposed to safety film.

Polyurethane. This plastic is used in several ways in building. As thermal insulation, it is used in the form of foam, either prefoamed or foamed in place. The latter is particularly useful in irregular spaces. When blown with fluorocarbons, the foam has an exceptionally low *K*-factor and is, therefore, widely used in thin-walled refrigerators. Other uses include field-applied or baked-on clear or colored coatings and finishes for floors, walls, furniture, and casework generally. The rubbery form is employed for sprayed or troweled-on roofing, and for gaskets and calking compounds.

Urea Formaldehyde. Like the melamines, these offer unlimited translucent to opaque color possibilities, light-fastness, good mechanical and electrical properties, and resistance to organic solvents as well as mild acids and alkalies. Although there is no swelling or change in appearance, the water absorption of urea formaldehyde is relatively high, and it is therefore not recommended for applications involving long exposure to water. Occasional exposure to water is without deleterious effect. Strength properties are good, although special shock-resistant grades are not made.

Silicones. Unlike other plastics, silicones are based on silicon rather than carbon. As a consequence, their inertness and durability under a wide variety of conditions are outstanding. As compared with the phenolics, their mechanical properties are poor, and consequently glass fibers are added. Molding is more difficult than with other thermosetting materials. Unlike most other resins, they may be used in continuous operations at 400°F; they have very low water absorption; their dielectric properties are excellent over an extremely wide variety of chemical attack; and under outdoor conditions their durability is particularly outstanding. In liquid solutions, silicones are used to impart moisture resistance to masonry walls and to fabrics. They also form the basis for a variety of paints and other coatings capable of maintaining flexibility and inertness to attack at high temperatures in the presence of ultraviolet sunlight and ozone. Silicone rubbers maintain their flexibility at much lower temperatures than other rubbers.

4.72 THERMOPLASTIC RESINS

Materials under this heading in general can be softened by heating and hardened by cooling.

Acrylics. In the form of large transparent sheets, these are used in aircraft enclosures and building construction. Although not so hard as glass, they have perfect clarity and transparency. Among the most resistant of the transparent plastics to sunlight and outdoor weathering, they possess an optimum combination of flexibility and sufficient rigidity with resistance to shattering. A wide variety of transparent, translucent, and opaque colors can be produced. The sheets are readily formed to complex shapes. They are used for such applications as transparent windows, outdoor and indoor signs, parts of lighting equipment, decorative and functional automotive parts, reflectors, household-appliance parts, and similar applications. They can be used as large sheets, molded from molding powders, or cast from the liquid monomer.

Acrylonitrile-Butadiene-Styrene (ABS). This three-way copolymer provides a family of tough, hard, chemically resistant resins with many grades and varieties, depending on variations in constituents. The greatest use is for pipes and fittings, especially drain-waste-vent (DWV). Other uses include buried sewer and water lines, mine pipe, well casings, conduit, and appliance housings.

Polyethylene. In its unmodified form, this is a flexible, waxy, translucent plastic. It maintains flexibility at very low temperatures, in contrast with many other thermoplastic materials.

Polyethylene may be provided as low-density, or standard, or as high-density, or linear material. High-density polyethylene has greater strength and stiffness, withstands somewhat higher temperatures, and has a more sharply defined softening temperature range. The heat-distortion point of the low-density polyethylenes is low; these plastics are not recommended for uses above 150°F. Unlike most plastics, polyethylene is partly crystalline. It is highly inert to solvents and corrosive chemicals of all kinds at ordinary temperatures. Usually low moisture permeability and absorption are combined with excellent electrical properties. Its density is lower than that of any other commercially available nonporous plastic. It is widely used as a primary insulating material on wire and cable and has been used as a replacement for the lead jacket in communication cables and other cables. It is widely used also in geogrids, geonets, and geomembranes (Art. 6.12) and as corrosionproof lining for tanks and other chemical equipment.

Polypropylene. This polyolefin is similar in many ways to its counterpart, polyethylene, but is generally harder, stronger, and more temperature-resistant. It finds a great many uses, among them piping, geotextiles, and geogrids (Art. 6.12), and complete water cisterns for water closets in plumbing systems.

Polycarbonate. Excellent transparency, high impact resistance, and good resistance to weathering combine to recommend this plastic for safety glazing and for general illumination and shatter-resistant fixtures. It is available in large, clear, tinted, and opaque sheets that can be formed into shells, domes, globes, and other forms. It can be processed by standard molding methods.

Polytetrafluorethylene. This is a highly crystalline linear-type polymer, unique among organic compounds in its chemical inertness and resistance to change at high and low temperatures. Its electrical properties are excellent. Its outstanding property is extreme resistance to attack by corrosive agents and solvents of all kinds. Waxy and self-lubricating, polytetrafluoroethylene is used in buildings where resistance to extreme conditions or low friction is desired. In steam lines, for example, supporting pads of this plastic permit the lines to slide easily over the pads. The temperatures involved have little or no effect. Other low-friction applications include, for example, bearings for girders and trusses. Mechanical properties are only moderately high, and reinforcement may be necessary to prevent creep and squeezeout under heavy loads. These fluorocarbons are difficult to wet; consequently, they are often used as parting agents, or where sticky materials must be handled.

Polyvinylfluoride. This has much of the superior inertness to chemical and weathering attack typical of the fluorocarbons. Among other uses, it is used as thin-film overlays for building boards to be exposed outdoors.

Polyvinyl Formal and Polyvinyl Butyral. Polyvinyl formal resins are principally used as a base for tough, water-resistant insulating enamel for electric wire. Polyvinyl butyral is the tough interlayer in safety glass. In its cross-linked and plasticized form, polyvinyl butyral is extensively used in coating fabrics for raincoats, upholstery, and other heavy-duty moisture-resistant applications.

Vinyl Chloride Polymers and Copolymers. Polyvinyl chloride is naturally hard and rigid but can be plasticized to any required degree of flexibility as in raincoats and shower curtains. Copolymers, including vinyl chloride plus vinyl acetate, are naturally flexible without plasticizers. Nonrigid vinyl plastics are widely used as insulation and jacketing for electric wire and cable because of their electrical properties and their resistance to oil and water. Thin films are used in geomembranes (Art. 6.12). Vinyl chlorides also are used for floor coverings in the form of tile and sheet because of their abrasion resistance and relatively low water absorption. The rigid materials are used for tubing, pipe, and many other applications where their resistance to corrosion and action of many chemicals, especially acids and alkalies, recommends them. They are attacked by a variety of organic solvents, however. Like all thermoplastics, they soften at elevated temperatures.

Vinylidene Chloride. This material is highly resistant to most inorganic chemicals and to organic solvents generally. It is impervious to water on prolonged immersion, and its films are highly resistant to moisture-vapor transmission. It can be sterilized, if not under load, in boiling water. It is used as pipe for transporting chemicals and geomembranes (Art. 6.12).

Nylon. Molded nylon is used in increasing quantities for impact and high resistance to abrasion. It is employed in small gears, cams, and other machine parts, because even when unlubricated they are highly resistant to wear. Its chemical resistance, except to phenols and mineral acids, is excellent. Extruded nylon is coated onto electric wire, cable, and rope for abrasion resistance. Applications like hammerheads indicate its impact resistance.

Polystyrene. This is one of the lightest of the presently available commercial plastics. It is relatively inexpensive, easily molded, has good dimensional stability, and good stability at low temperatures; it is brilliantly clear when transparent and has an infinite range of colors. Water absorption is negligible even after long immersion. Electrical characteristics are excellent. It is resistant to most corrosive chemicals, such as acids, and to a variety of organic solvents, although it is attacked by others. Polystyrenes as a class are considerably more brittle and less extensible than many other thermoplastic materials, but these properties are markedly improved in copolymers. Under some conditions, they have a tendency to develop fine cracks, known as craze marks, on exposure, particularly outdoors. This is true of many other thermoplastics, especially when highly stressed. It is widely used in synthetic rubbers.

4.73 ELASTOMERS, OR SYNTHETIC RUBBERS

Rubber for construction purposes is both natural and synthetic. Natural rubber, often called crude rubber in its unvulcanized form, is composed of large complex molecules of isoprene. Synthetic rubbers, also known as elastomers, are generally rubber-like only in their high elasticity. The principal synthetic rubbers are the following:

GR-S is the one most nearly like crude rubber and is the product of styrene and butadiene copolymerization. It is the most widely used of the synthetic rubbers. It is not oil-resistant but is widely used for tires and similar applications.

Nitril is a copolymer of acrylonitrile and butadiene. Its excellent resistance to oils and solvents makes it useful for fuel and solvent hoses, hydraulic-equipment parts, and similar applications.

Butyl is made by the copolymerization of isobutylene with a small proportion of isoprene or butadiene. It has the lowest gas permeability of all the rubbers and consequently is widely used for making inner tubes for tires and other applications in which gases must be held with a minimum of diffusion. It is used for gaskets in buildings.

Neoprene is made by the polymerization of chloroprene. It has very good mechanical properties and is particularly resistant to sunlight, heat, aging, and oil; it is therefore used for making machine belts, gaskets, oil hose, insulation on wire cable, and other applications to be used for outdoor exposure, such as roofing, and gaskets for building and glazing.

Sulfide rubbers—the polysulfides of high molecular weight—have rubbery properties, and articles made from them, such as hose and tank linings and glazing compounds, exhibit good resistance to solvents, oils, ozone, low temperature, and outdoor exposure.

Silicone rubber, which also is discussed in Art. 4.71, when made in rubbery consistency forms a material exhibiting exceptional inertness and temperature resistance. It is therefore used in making gaskets, electrical insulation, and similar products that maintain their properties at both high and low temperatures.

Additional elastomers include polyethylene, cyclized rubber, plasticized polyvinyl chloride, and polybutene. A great variety of materials enters into various rubber compounds and therefore provide a wide range of properties. In addition, many elastomeric products are laminated structures of rubber-like compounds combined with materials like fabric and metals (Art. 4.76).

COMBINATIONS OF PLASTICS AND OTHER MATERIALS

Plastics often are used as part of a composite construction with other materials. The composites may be in the form of laminates, matrix systems, sandwich structures, or combinations of these.

4.74 HIGH-PRESSURE LAMINATES

Laminated thermosetting products consist of fibrous sheet materials combined with a thermosetting resin, usually phenol formaldehyde or melamine formaldehyde. The commonly used sheet materials are paper, cotton fabric, asbestos paper or fabric, nylon fabric, and glass fabric. The usual form is flat sheet, but a variety of rolled tubes and rods is made.

Decorative Laminates. These high-pressure laminates consist of a base of phenolic resin-impregnated kraft paper over which a decorative overlay, such as printed paper, is applied. Over all this is laid a thin sheet of melamine resin. When the entire assemblage is pressed in a hot-plate press at elevated temperatures and pressures, the various layers are fused together and the melamine provides a completely transparent finish, resistant to alcohol, water, and common solvents. This material is widely used for tabletops, counter fronts, wainscots, and similar building applications. It is customarily bonded to a core of plywood to develop the necessary thickness and strength. In this case, a backup sheet consisting of phenolic resin and paper alone, without the decorative surface, is employed to provide balance to the entire sandwich.

4.75 REINFORCED PLASTICS

These are commonly made with phenolic, polyester, and epoxide resins combined with various types of reinforcing agents, of which glass fibers in the form of mats or fabrics are the most common. Because little or no pressure is required to form large complex parts, rather simple molds can be employed for the manufacture of such things as boat hulls and similar large parts. In buildings, reinforced plastics have been rather widely used in the form of corrugated sheet for skylights and side lighting of buildings, and as molded shells, concrete forms, sandwiches, and similar applications.

These materials may be formulated to cure at ordinary temperatures, or they may require moderate temperatures to cure the resins. Customarily, parts are made by laying up successive layers of the glass fabric or the glass mat and applying the liquid resin to them. The entire combination is allowed to harden at ordinary temperatures, or it is placed in a heated chamber for final hardening. It may be placed inside a rubber bag and a vacuum drawn to apply moderate pressure, or it may be placed between a pair of matching molds and cured under moderate pressure in the molds.

The high impact resistance of these materials combined with good strength properties and good durability recommends them for building applications. When the quantity of reinforcing agent is kept relatively low, a high degree of translucence may be achieved, although it is less than that of the acrylics and the other transparent thermoplastic materials.

Fabrics for Air-Supported Roofs. Principal requirements for fabrics and coatings for air-supported structures are high strip tensile strength in both warp and fill directions, high tear resistance, good coating adhesion, maximum weathering resistance, maximum joint strength, good flexing resistance, and good flame resistance. Translucency may or may not be important, depending on the application. The most commonly used fabrics are nylon, polyester, and glass. Neoprene and Hypalon have commonly been employed for military and other applications where opacity is desired. For translucent application, vinyl chloride and fluorocarbon polymers are more common. Careful analysis of loads and stresses, especially dynamic wind loads, and means of joining sections and attaching to anchorage is required.

4.76 LAMINATED RUBBER

Rubber is often combined with various textiles, fabrics, filaments, and metal wire to obtain strength, stability, abrasion resistance, and flexibility. Among the laminated materials are the following:

V Belts. These consist of a combination of fabric and rubber, frequently combined with reinforcing grommets of cotton, rayon, steel, or other high-strength material extending around the central portion.

Flat Rubber Belting. This laminate is a combination of several plies of cotton fabric or cord, all bonded together by a soft-rubber compound.

Conveyor Belts. These, in effect, are moving highways used for transporting such material as crushed rock, dirt, sand, gravel, slag, and similar materials. When the belt operates at a steep angle, it is equipped with buckets or similar devices and becomes an elevator belt. A typical conveyor belt consists of cotton duct plies alternated with thin rubber plies; the assembly is wrapped in a rubber cover, and all elements are united into a single structure by vulcanization. A conveyor belt to withstand extreme conditions is made with some textile or metal cords instead of the woven fabric. Some conveyor belts are especially arranged to assume a trough form and made to stretch less than similar all-fabric belts.

Rubber-Lined Pipes, Tanks, and Similar Equipment. The lining materials include all the natural and synthetic rubbers in various degrees of hardness, depending on the application. Frequently, latex rubber is deposited directly from the latex solution onto the metal surface to be covered. The deposited layer is subsequently vulcanized. Rubber linings can be bonded to ordinary steel, stainless steel, brass, aluminum, concrete, and wood. Adhesion to aluminum is inferior to adhesion to steel. Covering for brass must be compounded according to the composition of the metal.

Rubber Hose. Nearly all rubber hose is laminated and composed of layers of rubber combined with reinforcing materials like cotton duck, textile cords, and metal wire. Typical hose consists of an inner rubber lining, a number of intermediate layers consisting of braided cord or cotton duck impregnated with rubber, and outside that, several more layers of fabric, spirally wound cord, spirally wound metal, or in some cases, spirally wound flat steel ribbon. Outside of all this is another layer of rubber to provide resistance to abrasion. Hose for transporting oil, water, wet concrete under pressure, and for dredging purposes is made of heavy-duty laminated rubber.

Vibration Insulators. These usually consist of a layer of soft rubber bonded between two layers of metal. Another type of insulator consists of a rubber tube or cylinder vulcanized to two concentric metal tubes, the rubber being deflected in shear. A variant of this consists of a cylinder of soft rubber vulcanized to a tubular or solid steel core and a steel outer shell, the entire combination being placed in torsion to act as a spring. Heavy-duty mounts of this type are employed on trucks, buses, and other applications calling for rugged construction.

4.77 PLASTICS BIBLIOGRAPHY

American Concrete Institute, "Polymer Modified Concrete," SP-99; "Polymers in Concrete," ACI 548; and "Guide for the Use of Polymers in Concrete," ACI 548.1.

American Society of Civil Engineers, "Structural Plastics Design Manual," and "Structural Plastics Selection Manual."

"Modern Plastics Encyclopedia," Plastics Catalog Corp., New York.

A. G. H. Dietz, "Plastics for Architects and Engineers," M.I.T. Press, Cambridge, Mass.

C. A. Harper, "Handbook of Plastics and Elastomers," McGraw-Hill Publishing Company, New York.

R. M. Koerner, "Designing with Geosynthetics," 2nd ed., Prentice-Hall, Englewood Cliffs, N.J.

I. Skeist, "Plastics in Building," Van Nostrand Reinhold, New York.

PORCELAIN-ENAMELED PRODUCTS

Porcelain enamel, also known as vitreous enamel, is an aluminum-silicate glass, which is fused to metal under high heat. Porcelain-enameled metal is used for indoor and outdoor applications because of its hardness, durability, washability, and color possibilities. For building purposes, porcelain enamel is applied to sheet metal and cast iron, the former for a variety of purposes including trim, plumbing, and kitchen fixtures, and the latter almost entirely for plumbing fixtures. Most sheet metal used for porcelain enameling is steel—low in carbon, manganese, and other elements. Aluminum is also used for vitreous enamel.

4.78 PORCELAIN ENAMEL ON METAL

Low-temperature softening glasses must be employed, especially with sheet metal, to avoid the warping and distortion that would occur at high temperatures. To obtain lower softening temperatures than would be attainable with high-silica glasses, boron is commonly added. Fluorine may replace some of the oxygen, and lead may also be added to produce easy-flowing brilliant enamels; but lead presents an occupational health hazard.

Composition of the enamel is carefully controlled to provide a coefficient of thermal expansion as near that of the base metal as possible. If the coefficient of the enamel is greater than that of the metal, cracking and crazing are likely to occur, but if the coefficient of the enamel is slightly less, it is lightly compressed upon cooling, a desirable condition because glass is strong in compression.

To obtain good adhesion between enamel and metal, one of the so-called transition elements used in glass formulation must be employed. Cobalt is favored. Apparently, the transition elements promote growth of iron crystals from base metal into the enamel, encourage formation of an adherent oxide coating on the iron, which fuses to the enamel, or develop polar chemical bonds between metal and glass.

Usually, white or colored opaque enamels are desired. Opacity is promoted by mixing in, but not dissolving, finely divided materials possessing refractive indexes widely different from the glass. Tin oxide, formerly widely used, has been largely displaced by less expensive and more effective titanium and zirconium compounds. Clay adds to opacity. Various oxides are included to impart color.

Most enameling consists of a ground coat and one or two cover coats fired on at slightly lower temperatures; but one-coat enameling of somewhat inferior quality can be accomplished by first treating the iron surface with soluble nickel salts.

The usual high-soda glasses used to obtain low-temperature softening enamels are not highly acid-resistant and therefore stain readily and deeply when iron-containing water drips on them. Enamels highly resistant to severe staining conditions must be considerably harder; i.e., have higher softening temperatures and therefore require special techniques to avoid warping and distorting of the metal base.

Interiors of refrigerators are often made of porcelain-enameled steel sheets for resistance to staining by spilled foods, whereas the exteriors are commonly baked-on synthetic-resin finishes.

4.79 PORCELAIN BIBLIOGRAPHY

F. H. Norton, "Elements of Ceramics," Addison-Wesley Publishing Company, Cambridge, Mass.

W. D. Kingery, H. K. Bowen, and D. R. Uhlmann, "Introduction to Ceramics," John Wiley & Sons, Inc., New York.

G. S. Brady and H. R. Clauser, "Materials Handbook," 13th ed., and J. H. Callender, "Time-Saver Standards for Architectural Design Data," McGraw-Hill Publishing Company, New York.

ASPHALT AND BITUMINOUS PRODUCTS

Asphalt, because of its water-resistant qualities and good durability, is used for many building applications to exclude water, provide a cushion against vibration and expansion, and serve as pavement.

4.80 ASPHALTS FOR DAMPPROOFING AND WATERPROOFING

Dampproofing is generally only a mopped-on coating, whereas waterproofing usually is a built-up coating of one or more plies. Bituminous systems used for dampproofing and waterproofing may be hot applied or cold applied.

ASTM D449, "Asphalt Used in Dampproofing and Waterproofing," specifies three types of asphalt. Type I, a soft, adhesive, easy-flowing, self-healing bitumen, is intended for use for underground construction, such as foundations, or where similar moderate temperature conditions exist. The softening point of Type I may range from 115 to 140°F. Type II may be used above ground; for example, on retaining walls or where temperatures will not exceed 122°F. The softening point of Type II may range from 145 to 170°F.

D449 asphalts are suitable for use with an asphalt primer meeting the requirements of ASTM D41. In construction of membrane waterproofing systems with these asphalts, felts should conform to ASTM D226 or D250, fabrics to D173, D1327, or D1668, and asphalt-impregnated glass mats to D2178.

For cold-applied systems, asphalt emulsions or cut-back asphalt mastic reinforced with glass fabric may be used. ASTM D1187 specifies asphalt-based emulsions for protective coatings for metal. D491 contains requirements for asphalt mastic for use in waterproofing building floors but not intended as pavement. The mastic is a mixture of asphalt cement, mineral filler, and mineral aggregate. D1668 covers glass fabric for roofing and waterproofing membranes.

4.81 BITUMINOUS ROOFING

Hot asphalt or coal tar are used for conventional built-up roofing. The bitumens are heated to a high enough temperature to fuse with saturant bitumen in roofing felts, thus welding the plies together. The optimum temperature at the point of application for achieving complete fusion, optimum mopping properties, and the desirable interply mopping weight is called the equiviscous temperature (EVT). Information on EVT should be obtained from the manufacturer.

4.81.1 Built-Up Roofing

For constructing built-up roofing, four grades of asphalt are recognized (ASTM D312): Type I, for inclines up to ½ in/ft; Type II, for inclines up to 1½ in/ft; Type III, for inclines up to 3 in/ft; and Type IV, suited for inclines up to 6 in/ft, generally in areas with relatively high year-round temperatures. Types I through IV may be either smooth or surfaced with slag or gravel. Softening ranges are 135 to 150°F, 158 to 176°F, 180 to 200°F and 210 to 225°F, respectively. Heating of the asphalts should not exceed the flash point, the finished blowing temperature, or 475°F for Type I, 500°F for Type II, 525°F for Types III and IV.

Coal-tar pitches for roofing, dampproofing, and waterproofing are of three types (ASTM D450): Type I, for built-up roofing systems; Type II, for dampproofing and membrane waterproofing systems; Type III, for built-up roofing, but containing less volatiles than Type I. Softening ranges are 126 to 140°F, 106 to 126°F, and 133 to 147°F, respectively.

4.81.2 Roofing Felts

For built-up waterproofing and roofing, types of membranes employed include felt (ASTM D226, D227) and cotton fabrics (ASTM D173). Felts are felted sheets of inorganic or organic fibers saturated with asphalt or coal tar conforming to ASTM D312 and D450.

Standard asphalt felts weigh 15, 20, or 30 lb per square (100 ft^2), and standard coal-tar felts weigh 13 lb per square.

Cotton fabrics are open-weave materials weighing at least 3½ oz/yd^2 before saturation, with thread counts of 24 to 32 per inch. The saturants are either asphalts or coal tars. The saturated fabric must weigh at least 10 oz/yd^2.

4.81.3 Roll Roofing

Asphalt roll roofing, shingles, and siding consist basically of roofing felt, first uniformly impregnated with hot asphaltic saturant and then coated on each side with at least one layer of a hot asphaltic coating and compounded with a water-insoluble mineral filler. The bottom or reverse side, in each instance, is covered with some suitable material, like powdered mica, to prevent sticking in the package or roll.

Granule-surfaced roll roofing (ASTM D249) is covered uniformly on the weather side with crushed mineral granules, such as slate. Minimum weight of the finished roofing should be 81 to 83 lb per square (100 ft^2), and the granular coating should weigh at least 18.5 lb per square.

Roll roofing (ASTM 224), surfaced with powdered talc or mica, is made in two grades, 39.8 and 54.6 lb per square, of which at least 18 lb must be the surfacing material.

4.82 ASPHALT SHINGLES

There are three standard types: Type I, uniform or nonuniform thickness; Type II, thick butt; and Type III, uniform or nonuniform thickness (ASTM D225). Average weights must be 95 lb per square (100 ft^2). For Types I and III, the weather-side coating must weigh 23.0 lb per square; for Type II, 30.0 lb per square. The material in these shingles is similar to that in granule-surfaced roll roofing.

4.83 ASPHALT MASTICS AND GROUTS

Asphalt mastics used for waterproofing floors and similar structures, but not intended for pavement, consist of mixtures of asphalt cement, mineral filler, and mineral aggregate, which can be heated at about 400°F to a sufficiently soft condition to be poured and troweled into place. The raw ingredients may be mixed on the job or may be premixed, formed into cakes, and merely heated on the job (ASTM D491).

Bituminous grouts are suitable for waterproofing above or below ground level as protective coatings. They also can be used for membrane waterproofing or for bedding and filling the joints of brickwork. Either asphaltic or coal-tar pitch materials of dampproofing and waterproofing grade are used, together with mineral aggregates as coarse as sand.

4.84 BITUMINOUS PAVEMENTS

Asphalts for pavement (ASTM D946) contain petroleum asphalt cement, derived by the distillation of asphaltic petroleum. Various grades are designated as 40-50, 60-70, 85-100, 120-150, and 200-300, depending upon the depth of penetration of a standard needle in a standard test (ASTM D5).

Emulsions range from low to high viscosity and quick- to slow-setting (ASTM D977).

4.85 ASPHALT BIBLIOGRAPHY

"The NRCA Roofing and Waterproofing Manual," National Roofing Contractors Association, Rosemont, IL 60018-5607.

JOINT SEALS

Calking compounds, sealants, and gaskets are employed to seal the points of contact between similar and dissimilar building materials that cannot otherwise be made completely tight. Such points include glazing, the joints between windows and walls, the many joints occurring in the increasing use of panelized construction, the copings of parapets, and similar spots.

The requirements of a good joint seal are: (1) good adhesion to or tight contact with the surrounding materials, (2) good cohesive strength, (3) elasticity to allow for compression and extension as surrounding materials retract or approach each other because of changes in moisture content or temperature, (4) good durability or the ability to maintain their properties over a long period of time without marked deterioration, and (5) no staining of surrounding materials such as stone.

4.86 CALKING COMPOUNDS

These sealers are used mostly with traditional materials such as masonry, with relatively small windows, and at other points where motion of building components is relatively small. They are typically composed of elastomeric polymers or bodied linseed or soy oil, or both, combined with calcium carbonate (ground marble or limestone), tinting pigments, a gelling agent, drier, and mineral spirits (thinners).

Two types are commonly employed, gun grade and knife grade. Gun grades are viscous semiliquids suitable for application by hand or air-operated calking guns. Knife grades are stiffer and are applied by knife, spatula, or mason's pointing tools.

Because calking compounds are based on drying oils that eventually harden in contact with the air, the best joints are generally thick and deep, with a relatively small portion exposed to the air. The exposed surface is expected to form a tough protective skin for the soft mass underneath, which in turn provides the cohesiveness, adhesiveness, and elasticity required. Thin shallow beads cannot be expected to have the durability of thick joints with small exposed surface areas.

4.87 SEALANTS

For joints and other points where large movements of building components are expected, elastomeric materials may be used as sealants. Whereas traditional calking compounds should not be used where movements of more than 5% or at most 10% are expected, larger movements, typically 10 to 25%, can be accommodated by the rubbery sealants.

Some elastomeric sealants consist of two components, mixed just before application. Polymerization occurs, leading to conversion of the viscous material to a rubbery consistency. The working time or pot life before this occurs varies, depending upon formulation and temperature, from a fraction of an hour to several hours or a day. Other formulations are single-component and require no mixing. They harden upon exposure to moisture in the air.

Various curing agents, accelerators, plasticizers, fillers, thickeners, and other agents may be added, depending on the basic material and the end-use requirements.

Among the polymeric materials employed are:

Acrylics: solvent-release type, water-release type, latex

Butyls: skinning and nonskinning

Polysulfide: two-part and one-part

Silicone: one-part

Polyurethane: two-part and one-part

Chlorosulfonated polyethylene: one-part

Polyurethane-polyepoxide: two-part

Characteristics of the preceding formulations vary. Hence, the proper choice of materials depends upon the application. A sealant with the appropriate hardness, extensibility, useful temperature ranges, expected life, dirt pickup, staining, colorability, rate of cure to tack-free condition, toxicity, resistance to ultraviolet light, and other attributes should be chosen for the specific end use.

In many joints, such as those between building panels, it is necessary to provide backup; that is, a foundation against which the compound can be applied. This serves to limit the thickness of the joint, to provide the proper ratio of thickness to width, and to force the compound into intimate contact with the substrate, thereby promoting adhesion. For the purpose, any of various compressible materials, such as polyethylene or polyurethane rope, or oakum, may be employed.

To promote adhesion to the substrate, various primers may be needed. (To prevent adhesion of the compound to parts of the substrate where adhesion is not wanted, any of various liquid and tape bond-breakers may be employed.) Generally, good adhesion requires dry, clean surfaces free of grease and other deleterious materials.

4.88 GASKETS

Joint seals described in Arts. 4.86 and 4.87 are formed in place; that is, soft masses are put into the joints and conform to their geometry. A gasket, on the other hand, is preformed and placed into a joint whose geometry must conform with the gasket in such a way as to seal the joint by compression of the gasket. Gaskets, however, are cured under shop-controlled conditions, whereas sealants cure under variable and not always favorable field conditions.

Rubbery materials most commonly employed for gaskets are cellular or noncellular (dense) neoprene, EPDM (ethylene-propylene polymers and terpolymers), and polyvinylchloride polymers.

Gaskets are generally compression types or lock-strip (*zipper*) types. The former are forced into the joint and remain tight by being kept under compression. With lock-strip gaskets, a groove in the gasket permits a lip to be opened and admit glass or other panel, after which a strip is forced into the groove, tightening the gasket in place. If the strip is separable from the gasket, its composition is often harder than the gasket itself.

For setting large sheets of glass and similar units, setting or supporting spacer blocks of rubber are often combined with gaskets of materials such as vulcanized synthetic rubber and are finally sealed with the elastomeric rubber-based sealants or glazing compounds.

4.89 JOINT SEALS BIBLIOGRAPHY

"Building Seals and Sealants," STP 606, ASTM, Philadelphia, Pa.

J. P. Cook, "Construction Sealants and Adhesives," John Wiley & Sons, Inc., New York.

A. Damusis, "Sealants," Van Nostrand Reinhold Company, New York.

PAINTS AND OTHER COATINGS

Protective and decorative coatings generally employed in building are the following:

Oil Paint. Drying-oil vehicles or binders plus opaque and extender pigments.

Water Paint. Pigments plus vehicles based on water, casein, protein, oil emulsions, and rubber or resin latexes, separately or in combination.

Calcimine. Water and glue, with or without casein, plus powdered calcium carbonate and any desired colored pigments.

Varnish. Transparent combination of drying oil and natural or synthetic resins.

Enamel. Varnish vehicle plus pigments.

Lacquer. Synthetic-resin film former, usually nitrocellulose, plus plasticizers, volatile solvents, and other resins.

Shellac. Exudations of the lac insect, dissolved in alcohol.

Japan. Solutions of metallic salts in drying oils, or varnishes containing asphalt and opaque pigments.

Aluminum Paint. Fine metallic aluminum flakes suspended in drying oil plus resin, or in nitrocellulose.

4.90 VEHICLES OR BINDERS

Following are descriptions of the most commonly used vehicles and binders for paint:

Natural Drying Oils. Drying oils harden by absorbing oxygen. The most important natural oils are linseed from flax seed (for many years the standard paint vehicle), tung oil (faster drying, good compatibility with varnish), oiticica oil (similar to tung), safflower (best nonyellowing oil), soybean (flexible films), dehydrated caster (good adhesion, fast drying), and fish oil (considered inferior but cheap).

Alkyds. These, the most widely used paint vehicles, are synthetic resins that are modified with various vegetable oils to produce clear resins that are harder than natural oils. Properties of the film depend on relative proportions of oil and resin. The film is both air drying and heat hardening.

Latexes. Latex paints are based on emulsions of various polymers including acrylics, polyvinyl acetate, styrene-butadiene, polyvinyl chloride, and rubber. They are easy to apply, dry quickly, have no solvent odor, and application tools are easily cleaned with soap and water. The films adhere well to various surfaces, have good color retention, and have varying degrees of flexibility.

Epoxy and Epoxy-Polyester. Catalyzed two-part, all-epoxy coatings are formed by addition of a catalyst to the liquid epoxy just before application (pot life a few minutes to a day). Films are as hard as many baked-on coatings and are resistant to solvents and traffic. Oil-modified epoxy esters, in contrast, harden on oxidation without a catalyst. They are less hard and chemically resistant than catalyzed epoxies, but dry fast and are easily applied. Epoxy-polyesters mixed just before use produce smooth finishes suitable for many interior surfaces and are chemically resistant.

Polyurethanes. These produce especially abrasion-resistant, fast-hardening coatings. Two-component formulations, of variable pot life, are mixed just before use. One-component formulations cure by evaporation and reaction with moisture in air (30 to 90% relative humidity). Oils and alkyds may be added.

Vinyl Solutions. Solutions of polyvinyl chloride and vinyl esters dry rapidly and are built up by successive, sprayed thin coatings. They characteristically have low gloss, high flexibility, and inertness to water but are sensitive to some solvents. Adhesion may be a problem. Weather resistance is excellent.

Dryers. These are catalysts that hasten the hardening of drying oils. Most dryers are salts of heavy metals, especially cobalt, manganese, and lead, to which salts of zinc and calcium may be added. Iron salts, usable only in dark coatings, accelerate hardening at high temperatures. Dryers are normally added to paints to hasten hardening, but they must not be used too liberally or they cause rapid deterioration of the oil by overoxidation.

Thinners. These are volatile constituents added to coatings to promote their spreading qualities by reducing viscosity. They should not react with the other constituents and should evaporate completely. Commonly used thinners are turpentine and mineral spirits, i.e., derivatives of petroleum and coal tar.

4.91 PIGMENTS FOR PAINTS

Pigments may be classified as white and colored, or as opaque and extender pigments. The hiding power of pigments depends on the difference in index of refraction of the pigment and the surrounding medium—usually the vehicle of a protective coating. In opaque pigments, these indexes are markedly different from those of the vehicles (oil or other); in extender pigments, they are nearly the same. The comparative hiding efficiencies of various pigments must be evaluated on the basis of hiding power per pound and cost per pound.

Principal white pigments, in descending order of relative hiding power per pound, are approximately as follows: rutile titanium dioxide, anatase titanium dioxide, zinc sulfide, titanium-calcium, titanium-barium, zinc sulfide-barium, titanated lithopone, lithopone, antimony oxide, zinc oxide.

Zinc oxide is widely used by itself or in combination with other pigments. Its color is unaffected by many industrial and chemical atmospheres. It imparts gloss and reduces chalking but tends to crack and alligator instead.

Zinc sulfide is a highly opaque pigment widely used in combination with other pigments.

Titanium dioxide and extended titanium pigments have high opacity and generally excellent properties. Various forms of the pigments have different properties. For example, anatase titanium dioxide promotes chalking, whereas rutile inhibits it.

Colored pigments for building use are largely inorganic materials, especially for outdoor use, where the brilliant but fugitive organic pigments soon fade. The principal inorganic colored pigments are:

Metallic. Aluminum flake or ground particle, copper bronze, gold leaf, zinc dust

Black. Carbon black, lampblack, graphite, vegetable black, and animal blacks

Earth colors. Yellow ocher, raw and burnt umber, raw and burnt sienna; reds and maroons

Blue. Ultramarine, iron ferrocyanide (Prussian, Chinese, Milori)

Brown. Mixed ferrous and ferric oxide

Green. Chromium oxide, hydrated chromium oxide, chrome greens

Orange. Molybdated chrome orange

Red. Iron oxide, cadmium red, vermilion

Yellow. Zinc chromate, cadmium yellows, hydrated iron oxide

Extender pigments are added to extend the opaque pigments, increase durability, provide better spreading characteristics, and reduce cost. The principal extender pigments are silica, china clay, talc, mica, barium sulfate, calcium sulfate, calcium carbonate, and such materials as magnesium oxide, magnesium carbonate, barium carbonate, and others used for specific purposes.

4.92 RESINS FOR PAINTS

Natural and synthetic resins are used in a large variety of air-drying and baked finishes. The natural resins include both fossil resins, which are harder and usually superior in quality, and recent resins tapped from a variety of resin-exuding trees. The most important fossil resins are amber (semiprecious jewelry), Kauri, Congo, Boea Manila, and Pontianak. Recent resins include Damar, East India, Batu, Manila, and rosin. Shellac, the product of the lac insect, may be considered to be in this class of resins.

The synthetic resins, in addition to the ones discussed in Art. 4.90, are used for applications requiring maximum durability. Among them are phenol formaldehyde, melamine formaldehyde, urea formaldehyde, silicones, fluorocarbons, and cellulose acetate-butyrate.

Phenolics in varnishes are used for outdoor and other severe applications on wood and metals. They are especially durable when baked.

Melamine and urea find their way into a large variety of industrial finishes, such as automobile and refrigerator finishes.

Silicones are used when higher temperatures are encountered than can be borne by the other finishes.

Fluorocarbons are costly but provide high-performance coatings, industrial siding, and curtain walls with excellent gloss retention, stain resistance, and weather resistance.

Cellulose acetate-butyrate provides shop-applied, high-gloss finishes.

4.93 COATINGS BIBLIOGRAPHY

A. Banov, "Paints and Coatings Handbook," Structures Publishing Company, Farmington, Mich.

R. M. Burns and W. Bradley, "Protective Coatings for Metals," Van Nostrand Reinhold Company, New York.

C. R. Martens, "The Technology of Paints, Varnishes and Lacquers," Van Nostrand Reinhold Company, New York.

W. C. Golton, "Analysis of Paints and Related Materials: Current Techniques for Solving Coatings Problems," STP 1119, ASTM, Philadelphia, Pa.

SECTION FIVE
STRUCTURAL THEORY

Frederick S. Merritt
Consulting Engineer
West Palm Beach, Florida

Structural theory establishes symbolic (mathematical) models for describing the behavior of structures. These models, when verified by laboratory and field tests and observations of structures under service conditions, form the basis for analysis and design methods for structures.

Structural design is the application of structural theory to ensure that buildings and other structures are built to support all loads and resist all constraining forces that may be reasonably expected to be imposed on them during their expected service life, without hazard to occupants or users and preferably without dangerous deformations, excessive sidesway (drift), or annoying vibrations (see Sec. 3). In addition, good design requires that this objective be achieved economically.

Provision should be made in application of structural theory to design for abnormal as well as normal service conditions. Abnormal conditions may arise as a result of accidents, fire, explosions, tornadoes, severer-than-anticipated earthquakes, floods, and inadvertent or even deliberate overloading of building components. Under such conditions, parts of a building may be damaged. The structural system, however, should be so designed that the damage will be limited in extent and undamaged portions of the building will remain stable. For the purpose, structural elements should be proportioned and arranged to form a stable system under normal service conditions. In addition, the system should have sufficient continuity and ductility, or energy-absorption capacity, so that if any small portion of it should sustain damage, other parts will transfer loads (at least until repairs can be made) to remaining structural components capable of transmitting the loads to the ground.

(D. M. Schultz, F. F. P. Burnett and M. Fintel, "A Design Approach to General Structural Integrity," in "Design and Construction of Large-Panel Concrete Structures," U.S. Department of Housing and Urban Development, 1977; E. V. Leyendecker and B. R. Ellingwood, "Design Methods for Reducing the Risk of Progressive Collapse in Buildings," NBS Buildings Science Series 98, National Institute of Standards and Technology, 1977.)

5.1 DESIGN LOADS

Loads are the external forces acting on a structure. Stresses are the internal forces that resist them. Depending on the manner in which the loads are applied, they tend to deform the structure and its components—tensile forces tend to stretch, compressive forces to squeeze together, torsional forces to twist, and shearing forces to slide parts of the structure past each other.

5.1.1 Types of Loads

External loads on a structure may be classified in several different ways. In one classification, they may be considered as static or dynamic.

Static loads are forces that are applied slowly and then remain nearly constant. One example is the weight, or dead load, of a floor or roof system.

Dynamic loads vary with time. They include repeated and impact loads.

Repeated loads are forces that are applied a number of times, causing a variation in the magnitude, and sometimes also in the sense, of the internal forces. A good example is an off-balance motor.

Impact loads are forces that require a structure or its components to absorb energy in a short interval of time. An example is the dropping of a heavy weight on a floor slab, or the shock wave from an explosion striking the walls and roof of a building.

External forces may also be classified as distributed and concentrated.

Uniformly distributed loads are forces that are, or for practical purposes may be considered, constant over a surface area of the supporting member. Dead weight of a rolled-steel I beam is a good example.

Concentrated loads are forces that have such a small contact area as to be negligible compared with the entire surface area of the supporting member. A beam supported on a girder, for example, may be considered, for all practical purposes, a concentrated load on the girder.

Another common classification for external forces labels them axial, eccentric, and torsional.

An **axial load** is a force whose resultant passes through the centroid of a section under consideration and is perpendicular to the plane of the section.

An **eccentric load** is a force perpendicular to the plane of the section under consideration but not passing through the centroid of the section, thus bending the supporting member (see Arts. 5.4.2, 5.5.17, and 5.5.19).

Torsional loads are forces that are offset from the shear center of the section under consideration and are inclined to or in the plane of the section, thus twisting the supporting member (see Arts. 5.42 and 5.5.19).

Also, building codes classify loads in accordance with the nature of the source. For example:

Dead loads include materials, equipment, constructions, or other elements of weight supported in, on, or by a building, including its own weight, that are intended to remain permanently in place.

Live loads include all occupants, materials, equipment, constructions, or other elements of weight supported in, on, or by a building and that will or are likely to be moved or relocated during the expected life of the building.

Impact loads are a fraction of the live loads used to account for additional stresses and deflections resulting from movement of the live loads.

Wind loads are maximum forces that may be applied to a building by wind in a mean recurrence interval, or a set of forces that will produce equivalent stresses.

Snow loads are maximum forces that may be applied by snow accumulation in a mean recurrence interval.

Seismic loads are forces that produce maximum stresses or deformations in a building during an earthquake.

5.1.2 Service Loads

In designing structural members, designers should use whichever is larger of the following:

1. Loadings specified in the local or state building code.
2. Probable maximum loads, based not only on current site conditions and original usage of proposed building spaces but also on possible future events. Loads that are of uncertain magnitude and that may be treated as statistical variables should be selected in accordance with a specific probability that the chosen magnitudes will not be exceeded during the life of the building or in accordance with the corresponding mean recurrence interval. The mean recurrence interval generally used for ordinary permanent buildings is 50 years. The interval, however, may be set at 25 years for structures with no occupants or offering negligible risk to life, or at 100 years for permanent buildings with a high degree of sensitivity to the loads and an unusually high degree of hazard to life and property in case of failure.

In the absence of a local or state building code, designers can be guided by loads specified in a national model building code or by the following data:

Loads applied to structural members may consist of the following, alone or in combination: dead, live, impact, earth pressure, hydrostatic pressure, snow, ice, rain, wind, or earthquake loads; constraining forces, such as those resulting from restriction of thermal, shrinkage, or moisture-change movements; or forces caused by displacements or deformations of members, such as those caused by creep, plastic flow, differential settlement, or sidesway (drift).

Dead Loads. Actual weights or masses of materials and installed equipment should be used. See Tables 5.1 and 5.2*c*.

Live Loads. These may be concentrated or distributed loads and should be considered placed on the building to produce maximum effects on the structural member being designed. Minimum live loads to be used in building design are listed in Table 5.2. These include an allowance for impact, except as noted in the footnote of Table 5.2*b*.

Partitions generally are considered to be live loads, because they may be installed at any time, almost anywhere, to subdivide interior spaces, or may be shifted from original places to other places in the future. Consequently, unless a floor is designed for a large live load, for example, 80 lb/ft^2, the weight of partitions should be added to other live loads, whether or not partitions are shown on the working drawings for building construction.

Because of the low probability that a large floor area contributing load to a specific structural member will be completely loaded with maximum design live loads, building codes generally permit these loads to be reduced for certain types of occupancy. Usually, however, codes do not permit any reduction for places of public assembly, dwellings, garages for trucks and buses, or one-way slabs. For areas with a minimum required live load exceeding 100 lb/ft^2 and for passenger-car garages, live loads on columns supporting more than one floor may be decreased 20%. Except for the preceding cases, a reduced live load L, lb/ft^2, may be computed from

$$L = \left(0.25 + \frac{15}{\sqrt{A_I}}\right) L_o \tag{5.1}$$

TABLE 5.1 Minimum Design Dead Loads

Walls	lb/ft²
Clay brick	
High-absorption, per 4-in wythe	34
Medium-absorption, per 4-in wythe	39
Low-absorption, per 4-in wythe	46
Sand-lime brick, per 4-in wythe	38
Concrete brick	
4-in, with heavy aggregate	46
4-in, with light aggregate	33
Concrete block, hollow	
8-in, with heavy aggregate	55
8-in, with light aggregate	35
12-in, with heavy aggregate	85
12-in, with light aggregate	55
Clay tile, loadbearing	
4-in	24
8-in	42
12-in	58
Clay tile, nonloadbearing	
2-in	11
4-in	18
8-in	34
Furring tile	
1½-in	8
2-in	10
Glass block, 4-in	18
Gypsum block, hollow	
2-in	9.5
4-in	12.5
6-in	18.5

Floor Finishes	lb/ft²
Asphalt block, 2-in	24
Cement, 1-in	12
Ceramic or quarry tile, 1-in	12
Hardwood flooring, ⅞-in	4
Plywood subflooring, ½-in	1.5
Resilient flooring, such as asphalt tile and linoleum	2
Slate, 1-in	15
Softwood subflooring, per in of thickness	3
Terrazzo, 1-in	13
Wood block, 3-in	4

Wood joists, double wood floor, joist size	lb/ft²	
	12-in spacing	16-in spacing
2 × 6	6	5
2 × 8	6	6
2 × 10	7	6
2 × 12	8	7
3 × 6	7	6
3 × 8	8	7
3 × 10	9	8
3 × 12	11	9
3 × 14	12	10

Concrete Slabs	lb/ft²
Stone aggregate, reinforced, per in of thickness	12.5
Slag, reinforced, per in of thickness	11.5
Lightweight aggregate, reinforced, per in of thickness	6 to 10

Material	Weight
Masonry	lb/ft³
Cast-stone masonry	144
Concrete, stone aggregate, reinforced	150
Ashlar:	
Granite	165
Limestone, crystalline	165
Limestone, oölitic	135
Marble	173
Sandstone	144
Roof and Wall Coverings	lb/ft²
Clay tile shingles	9 to 14
Asphalt shingles	2
Composition:	
3-ply ready roofing	1
4-ply felt and gravel	5.5
5-ply felt and gravel	6
Copper or tin	1
Corrugated steel	2
Sheathing (gypsum), ½-in	2
Sheathing (wood), per in thickness	3
Slate, ¼-in	10
Wood shingles	2
Waterproofing	lb/ft²
Five-ply membrane	5
Ceilings	lb/ft²
Plaster (on tile or concrete)	5
Suspended metal lath and gypsum plaster	10
Suspended metal lath and cement plaster	15
Suspended steel channel supports	2
Gypsumboard per ¼-in thickness	1.1
Floor Fill	lb/ft²
Cinders, no cement, per in of thickness	5
Cinders, with cement, per in of thickness	9
Sand, per in of thickness	8
Partitions	lb/ft²
Plaster on masonry	
Gypsum, with sand, per in of thickness	8.5
Gypsum, with lightweight aggregate, per in	4
Cement, with sand, per in of thickness	10
Cement, with lightweight aggregate, per in	5
Plaster, 2-in solid	20
Metal studs	
Plastered two sides	18
Gypsumboard each side	6
Wood studs, 2 × 4-in	
Unplastered	3
Plastered one side	11
Plastered two sides	19
Gypsumboard each side	7
Glass	lb/ft²
Single-strength	1.2
Double-strength	1.6
Plate, ⅛-in	1.6
Insulation	lb/ft²
Cork, per in of thickness	1.0
Foamed glass, per in of thickness	0.8
Glass-fiber bats, per in of thickness	0.06
Polystyrene, per in of thickness	0.2
Urethane	0.17
Vermiculite, loose fill, per in of thickness	0.5

TABLE 5.2 Minimum Design Live Loads

a. Uniformly distributed live loads, lb/ft², impact included[a]

Occupancy or use	Load
Assembly spaces:	
Auditoriums[b] with fixed seats	60
Auditoriums[b] with movable seats	100
Ballrooms and dance halls	100
Bowling alleys, poolrooms, similar recreational areas	75
Conference and card rooms	50
Dining rooms, restaurants	100
Drill rooms	150
Grandstand and reviewing-stand seating areas	100
Gymnasiums	100
Lobbies, first-floor	100
Roof gardens, terraces	100
Skating rinks	100
Bakeries	150
Balconies (exterior)	100
Up to 100 ft² on one- and two-family houses	60
Bowling alleys, alleys only	40
Broadcasting studios	100
Catwalks	30
Corridors:	
Areas of public assembly, first-floor lobbies	100
Other floors same as occupancy served, except as indicated elsewhere in this table	
Fire escapes:	
Multifamily housing	40
Others	100
Garages:	
Passenger cars	50
Trucks and buses	[c]
Hospitals:	
Operating rooms, laboratories, service areas	60
Patients' rooms, wards, personnel areas	40
Kitchens other than domestic	150
Laboratories, scientific	100
Libraries:	
Corridors above first floor	80
Reading rooms	60
Stack rooms, books and shelving at 65 lb/ft³, but at least	150
Manufacturing and repair areas:	
Heavy	250
Light	125
Marquees	75
Morgue	125
Office buildings:	
Corridors above first floor	80
Files	125
Offices	50
Penal institutions:	
Cell blocks	40
Corridors	100
Residential:	
Dormitories	
Nonpartitioned	60
Partitioned	40
Dwellings, multifamily:	
Apartments	40
Corridors	80
Hotels:	
Guest rooms, private corridors	40
Public corridors	80
Housing, one- and two-family:	
First floor	40
Storage attics	80
Uninhabitable attics	20
Upper floors, habitable attics	30
Schools:	
Classrooms	40
Corridors	80
Shops with light equipment	60
Stairs and exitways	100
Handrails, vertical and horizontal thrust, lb/lin ft	50
Storage warehouse:	
Heavy	250
Light	125
Stores:	
Retail:	
Basement and first floor	100
Upper floors	75
Wholesale	100
Telephone equipment rooms	80
Theaters:	
Aisles, corridors, lobbies	100
Dressing rooms	40
Projection rooms	100
Stage floors	150
Toilet areas	40

[a]See Eqs. (5.1) and (5.2).
[b]Including churches, schools, theaters, courthouses, and lecture halls.
[c]Use American Association of State Highway and Transportation Officials highway lane loadings.

TABLE 5.2 Minimum Design Live Loads (Continued)

b. Concentrated live loads[d]

Location	Load, lb
Elevator machine room grating (on 4-in^2 area)	300
Finish, light floor-plate construction (on 1-in^2 area)	200
Garages:	
Passenger cars:	
Manual parking (on 20-in^2 area)	2,000
Mechanical parking (no slab), per wheel	1,500
Trucks, buses (on 20-in^2 area), per wheel	16,000
Office floors (on area 2.5 ft square)	2,000
Roof-truss panel point over garage, manufacturing, or storage floors	2,000
Scuttles, skylight ribs, and accessible ceilings (on area 2.5 ft square)	200
Sidewalks (on area 2.5 ft square)	8,000
Stair treads (on 4-in^2 area at center of tread)	300

[d]Use instead of uniformly distributed live load, except for roof trusses, if concentrated loads produce greater stresses or deflections. Add impact factor for machinery and moving loads: 100% for elevators, 20% for light machines, 50% for reciprocating machines, 33% for floor or balcony hangers. For craneways, add a vertical force equal to 25% of maximum wheel load; a lateral force equal to 10% of the weight of trolley and lifted load, at the top of each rail; and a longitudinal force equal to 10% of maximum wheel loads, acting at top of rail.

c. Minimum design loads for materials

Material	Load, lb/ft^3	Material	Load, lb/ft^3
Aluminum, cast	165	Gravel, dry	104
Bituminous products:		Gypsum, loose	70
Asphalt	81	Ice	57.2
Petroleum, gasoline	42	Iron, cast	450
Pitch	69	Lead	710
Tar	75	Lime, hydrated, loose	32
Brass, cast	534	Lime, hydrated, compacted	45
Bronze, 8 to 14% tin	509	Magnesium alloys	112
Cement, portland, loose	90	Mortar, hardened:	
Cement, portland, set	183	Cement	130
Cinders, dry, in bulk	45	Lime	110
Coal, anthracite, piled	52	Riprap (not submerged):	
Coal, bituminous or lignite, piled	47	Limestone	83
Coal, peat, dry, piled	23	Sandstone	90
Charcoal	12	Sand, clean and dry	90
Copper	556	Sand, river, dry	106
Earth (not submerged):		Silver	656
Clay, dry	63	Steel	490
Clay, damp	110	Stone, ashlar:	
Clay and gravel, dry	100	Basalt, granite, gneiss	165
Silt, moist, loose	78	Limestone, marble, quartz	160
Silt, moist, packed	96	Sandstone	140
Sand and gravel, dry, loose	100	Shale, slate	155
Sand and gravel, dry, packed	110	Tin, cast	459
Sand and gravel, wet	120	Water, fresh	62.4
Gold, solid	1205	Water, sea	64

where L_o = unreduced live load, lb/ft^2 (see Table 5.1*a*)
A_I = influence area, or floor area over which the influence surface for structural effects is significantly different from zero
= area of four surrounding bays for an interior column, plus similar areas from supported floors above, if any
= area of two adjoining bays for an interior girder or for an edge column, plus similar areas from supported floors above, if any
= area of one bay for an edge girder or for a corner column, plus similar areas from supported floors above, if any

The reduced live load L, however, should not be less than $0.5L_o$ for members supporting one floor or $0.4L_o$ for members supporting two or more floors.

Roofs used for promenades should be designed for a minimum live load of 60 lb/ft^2, and those used for gardens or assembly, for 100 lb/ft^2. Ordinary roofs should be designed for a minimum live load L, lb/ft^2, computed from

$$L = 20R_1R_2 \geq 12 \tag{5.2}$$

where R_1 = $1.2 - 0.001A_t$ but not less than 0.6 or more than 1.0
A_t = tributary area, ft^2, for structural member being designed
R_2 = $1.2 - 0.05r$ but not less than 0.6 or more than 1.0
r = rise of roof in 12 in for a pitched roof or 32 times the ratio of rise to span for an arch or dome

This minimum live load need not be combined with snow load for design of a roof but should be designed for the larger of the two.

Subgrade Pressures. Walls below grade should be designed for lateral soil pressures and the hydrostatic pressure of subgrade water, plus the load from surcharges at ground level. Design pressures should take into account the reduced weight of soil because of buoyancy when water is present. In design of floors at or below grade, uplift due to hydrostatic pressures on the underside should be considered.

Wind Loads. Horizontal pressures produced by wind are assumed to act normal to the faces of buildings for design purposes and may be directed toward the interior of the buildings or outward (Arts. 3.2.1 and 3.2.2). These forces are called velocity pressures because they are primarily a function of the velocity of the wind striking the buildings. Building codes usually permit wind pressures to be either calculated or determined by tests on models of buildings and terrain if the tests meet specified requirements (see Art. 3.2.2). Codes also specify procedures for calculating wind loads, such as the following:

Velocity pressures due to wind to be used in building design vary with type of terrain, distance above ground level, importance of building, likelihood of hurricanes, and basic wind speed recorded near the building site. The wind pressures are assumed to act horizontally on the building area projected on a vertical plane normal to the wind direction.

The basic wind speed used in design is the fastest-mile wind speed recorded at a height of 10 m (32.8 ft) above open, level terrain with a 50-year mean recurrence interval.

Unusual wind conditions often occur over rough terrain and around ocean promontories. Basic wind speeds applicable to such regions should be selected with the aid of meteorologists and the application of extreme-value statistical analysis to anemometer readings taken at or near the site of the proposed building. Generally, however, minimum basic wind velocities are specified in local building codes and in national model building codes but should be used with discretion, because actual velocities at a specific site and on a specific building may be significantly larger. In the absence of code specifications and reliable data, basic wind speed at a height of 10 m above grade may be approximated for preliminary design from the following:

Coastal areas, northwestern and southeastern United States and mountainous areas	110 mph
Northern and central United States	90 mph
Other parts of the contiguous states	80 mph

For design purposes, wind pressures should be determined in accordance with the degree to which terrain surrounding the proposed building exposes it to the wind. Exposures may be classified as follows:

Exposure A applies to centers of large cities, where for at least one-half mile upwind from the building the majority of structures are over 70 ft high and lower buildings extend at least one more mile upwind.

Exposure B applies to wooded or suburban terrain or to urban areas with closely spaced buildings mostly less than 70 ft high, where such conditions prevail upwind for a distance from the building of at least 1500 ft or 10 times the building height.

Exposure C exists for flat, open country or exposed terrain with obstructions less than 30 ft high.

Exposure D applies to flat unobstructed areas exposed to wind blowing over a large expanse of water with a shoreline at a distance from the building of not more than 1500 ft or 10 times the building height.

For design purposes also, the following formulas may be used to determine, for heights z (in feet) greater than 15 ft above ground, a pressure coefficient K for converting wind speeds to pressures.

For Exposure A, for heights up to 1500 ft above ground level,

$$K = 0.000517 \left(\frac{z}{32.8}\right)^{2/3} \tag{5.3}$$

For z less than 15 ft, $K = 0.00031$.

For Exposure B, for heights up to 1200 ft above ground level,

$$K = 0.00133 \left(\frac{z}{32.8}\right)^{4/9} \tag{5.4}$$

For z less than 15 ft, $K = 0.00095$.

For Exposure C, for heights up to 900 ft above ground level,

$$K = 0.00256 \left(\frac{z}{32.8}\right)^{2/7} \tag{5.5}$$

For z less than 15 ft, $K = 0.0020$.

For Exposure D, for heights up to 700 ft above ground level,

$$K = 0.00357 \left(\frac{z}{32.8}\right)^{1/5} \tag{5.6}$$

For z less than 15 ft, $K = 0.0031$.

For ordinary buildings not subject to hurricanes, the velocity pressure q_z, psf, at height z may be calculated from

$$q_z = KV^2 \tag{5.7}$$

where V = basic wind speed, mi/hr, but not less than 70 mi/hr.

For important buildings, such as hospitals and communication buildings, for buildings sensitive to wind, such as slender skyscrapers, and for buildings presenting a high degree of hazard to life and property, such as auditoriums, q_z computed from Eq. (5.7) should be increased 15%.

To allow for hurricanes, q_z should be increased 10% for ordinary buildings and 20% for important, wind-sensitive or high-risk buildings along coastlines. These increases may be assumed to reduce uniformly with distance from the shore to zero for ordinary buildings and 15% for the more important or sensitive buildings at points 100 mi inland.

Wind pressures on low buildings are different at a specific elevation from those on tall buildings. Hence, building codes may give different formulas for pressures for the two types of construction. In any case, however, design wind pressure should be a minimum of 10 psf.

Multistory Buildings. For design of the main wind-force resisting system of ordinary, rectangular, multistory buildings, the design pressure at any height z, ft, above ground may be computed from

$$p_{zw} = G_o C_{pw} q_z \tag{5.8}$$

where p_{zw} = design wind pressure, psf, on windward wall
G_o = gust response factor
C_{pw} = external pressure coefficient
q_z = velocity pressure computed from Eq. (5.7) and modified for hurricanes and building importance, risks, and wind sensitivity

For windward walls, C_{pw} may be taken as 0.8. For side walls, C_{pw} may be assumed as -0.7 (suction). For roofs and leeward walls, the design pressure at elevation z is

$$p_{zl} = G_o C_p q_h \tag{5.9}$$

where p_{zl} = design pressure, psf, on roof or leeward wall
C_p = external pressure coefficient for roof or leeward wall
q_h = velocity pressure at mean roof height h (see Fig. 3.1d)

In these equations, the gust response factor may be taken approximately as

$$G_o = 0.65 + \frac{8.58D}{(h/30)^n} \geq 1 \tag{5.10}$$

where D = 0.16 for Exposure A, 0.10 for Exposure B, 0.07 for Exposure C, and 0.05 for Exposure D
n = $\frac{1}{3}$ for Exposure A, $\frac{2}{9}$ for Exposure B, $\frac{1}{7}$ for Exposure C, and 0.1 for Exposure D
h = mean roof height, ft

TABLE 5.3 External Pressure Coefficients C_p for Roofs*

Flat roofs	−0.7
Wind parallel to ridge of sloping roof	
h/b or $h/d \leq 2.5$	−0.7
h/b or $h/d > 2.5$	−0.8
Wind perpendicular to ridge of sloping roof, at angle θ with horizontal	
Leeward side	−0.7
Windward side	

	Slope of roof θ, deg					
h/d	10	20	30	40	50	60 or more
0.3 or less	0.2	0.2	0.3	0.4	0.5	
0.5	−0.9	−0.75	−0.2	0.3	0.5	0.01θ
1.0	−0.9	−0.75	−0.2	0.3	0.5	
1.5 or more	−0.9	−0.9	−0.9	0.35	0.21	

*h = height of building, ft; d = depth, ft, of building in direction of wind; b = width, ft, of building transverse to wind.
Based on data in ANSI A58.1-1981.

For leeward walls, subjected to suction, C_p depends on the ratio of the depth d to width b of the building and may be assumed as follows:

d/b =	1 or less	2	4 or more
C_p =	−0.5	−0.3	−0.2

The negative sign indicates suction. Table 5.3 lists values of C_p for pressures on roofs.

Flexible Buildings. These are structures with a fundamental natural frequency less than 1 Hz or with a ratio of height to least horizontal dimension (measured at mid-height for buildings with tapers or setbacks) exceeding 5. For such buildings, the main wind-force resisting system should be designed for a pressure on windward walls at any height z, ft, above ground computed from

$$p_{zw} = G_f C_{pw} q_z \tag{5.11}$$

where G_f = gust response factor determined by analysis of the system taking into account its dynamic properties. For leeward walls of flexible buildings,

$$p_{zl} = G_f C_p q_h \tag{5.12}$$

(In ASCE A7-88, G_f is represented by $\overline{G}$. Requiring a knowledge of the fundamental frequency, structural damping characteristics, and type of exposure of the building, the formula for G_f is complicated, but computations may be simplified somewhat by use of tables and charts in the standard.)

One-Story Buildings. For design of the main wind-force resisting system of rectangular, one-story buildings, the design pressure at any height z, ft, above ground may be computed for windward walls from

$$p_{zw} = (G_o C_p + C_{p1}) q_z \tag{5.13}$$

where C_{p1} = 0.75 if the percentage of openings in one wall exceeds that of other walls by 10% or more
= 0.25 for all other cases

For roofs and leeward walls, the design pressure at elevation z is

$$p_{zl} = G_o C_p q_h - C_{p2} q_z \tag{5.14}$$

where C_{p2} = $+0.75$ or -0.25 if the percentage of openings in one wall exceeds that of other walls by 10% or more
= ± 0.25 for all other cases

(Positive signs indicate pressures acting toward a wall; negative signs indicate pressures acting away from the wall.)

Snow, Ice, and Rain Loads. These, in effect, are nonuniformly distributed, vertical, live loads that are imposed by nature and hence are generally uncertain in magnitude and duration. They may occur alone or in combination. Design snow loads preferably should be determined for the site of the proposed building with the advice of meteorologists and application of extreme-value statistical analysis to rain and snow records for the locality.

Rain loads depend on drainage and may become large enough to cause roof failure when drainage is blocked (see Art. 3.4.3).

Ice loads are created when snow melts, then freezes, or when rain follows a snow storm and freezes. These loads should be considered in determining the design snow load. Snow loads may consist of pure snow or a mixture of snow, ice, and water.

Design snow loads on roofs may be assumed to be proportional to the maximum ground snow load p_g, lb/ft^2, measured in the vicinity of the building, with a 50-year mean recurrence interval. Determination of the constant of proportionality should take into account:

1. Appropriate mean recurrence interval.
2. Roof exposure. Wind may blow snow off the roof or onto the roof from nearby higher roofs or create nonuniform distribution of snow.
3. Roof thermal conditions. Heat escaping through the roof melts the snow. If the water can drain off, the snow load decreases. Also, for sloped roofs, if they are warm, there is a tendency for snow to slide off. Insulated roofs, however, restrict heat loss from the interior and therefore are subjected to larger snow loads.
4. Type of occupancy and uses of building. More conservative loading should be used for public-assembly buildings, because of the risk of great loss of life and injury to occupants if overloads should cause the roof to collapse.
5. Roof slope. The steeper a roof, the greater is the likelihood of good drainage and that snow will slide off.

In addition, roof design should take into account not only the design snow load uniformly distributed over the whole roof area but also possible unbalanced loading. Snow may be blown off part of the roof, and snow drifts may pile up over a portion of the roof.

For **flat roofs,** in the absence of building-code requirements, the basic snow load when the ground snow load p_g is 20 lb/ft^2 or less may be taken as

$$p_{\min} = p_g \tag{5.15}$$

When p_g is between 20 and 25 lb/ft^2, the minimum allowable design load is $p_{\min} = 20$ lb/ft^2, and when p_g exceeds 25 lb/ft^2, the basic snow load may be taken as

$$p_f = 0.8 p_g \tag{5.16}$$

where p_f = design snow load, lb/ft^2, for a flat roof that may have unheated space underneath and that may be located where the wind cannot be relied on to blow snow off, because of nearby higher structures or trees

p_g = ground snow load, lb/ft^2

For roofs sheltered from the wind, increase p_f computed from Eq. (5.16) by 20%, and for windy sites, reduce p_f 10%. For a poorly insulated roof with heated space underneath, decrease p_f by 30%.

Increase p_f 10% for large office buildings and public-assembly buildings, such as auditoriums, schools, factories. Increase p_f 20% for essential buildings, such as hospitals, communication buildings, police and fire stations, power plants, and for structures housing expensive objects or equipment. Decrease p_f 20% for structures with low human occupancy, such as farm buildings.

The **ground snow load** p_g should be determined from an analysis of snow depths recorded at or near the site of the proposed building. For a rough estimate in the absence of building-code requirements, p_g may be taken as follows for the United States, except for mountainous regions:

0–5 lb/ft^2—southern states from about latitude N32° southward
10–15 lb/ft^2—Pacific coast between latitudes N32° and N40° and other states between latitudes N32° and N37°
20–30 lb/ft^2—Pacific coast from latitude N40° northward and other states between latitudes N37° and N40°
40–50 lb/ft^2—north Atlantic and central states between latitudes N40° and N43°
60–80 lb/ft^2—northern New England between latitudes N43° and N45° and central states from N43° northward
80–120 lb/ft^2—Maine above latitude N45°

For **sloping roofs,** the snow load depends on whether the roof will be warm or cold. In either case, the load may be assumed to be zero for roofs making an angle θ of 70° or more with the horizontal. Also, for any slope, the load need not be taken greater than p_f given by Eq. (5.16). For slopes θ, deg, between 0° and 70°, the snow load, lb/ft^2, acting vertically on the projection of the roof on a horizontal plane, may be computed for warm roofs from

$$p_s = \left(\frac{70 - \theta}{40}\right) p_f \leq p_f \tag{5.17}$$

and for cold roofs from

$$p_s = \left(\frac{70 - \theta}{25}\right) p_f \leq p_f \tag{5.18}$$

Hip and gable roofs should be designed for the condition of the whole roof loaded with p_s, and also with the windward wide unloaded and the leeward side carrying $1.5p_s$.

For **curved roofs,** the snow load on the portion that is steeper than 70° may be taken as zero. For the less-steep portion, the load p_s may be computed as for a sloped roof, with θ taken as the angle with the horizontal of a line from the crown to points on the roof where the slope starts to exceed 70°. Curved roofs should be designed with the whole area fully loaded with p_s. They also should be designed for the case of snow only on the leeward side, with the load varying uniformly from $0.5p_s$ at the crown to $2p_s$ at points where the roof slope starts to exceed 30° and then decreasing to zero at points where the slope starts to exceed 70°.

Multiple folded-plate, sawtooth, and barrel-vault roofs similarly should be designed for unbalanced loads increasing from $0.5p_s$ at ridges to $3p_s$ in valleys.

Snow drifts may form on a roof near a higher roof that is less than 20 ft horizontally away. The reason for this is that wind may blow snow from the higher roof onto the lower roof. Drifts also may accumulate at projections above roofs, such as at parapets, solar collectors, and penthouse walls. Drift loads accordingly should be taken into account when:

1. The ground snow load p_g exceeds 10 lb/ft^2.
2. A higher roof exists (or may be built in the future) within 20 ft of the building, if the height differential, ft, exceeds $1.2p_f/\gamma$, where p_f is computed from Eq. (5.16) and γ is the snow density, lb/ft^3.
3. A projection extends a distance, ft, exceeding $1.2p_f/\gamma$ above the roof and is more than 15 ft long.

In computation of drift loads, the snow density γ, lb/ft^3, may be taken as follows:

$p_g =$	11–30	31–60	60 or more
$\gamma =$	15	20	25

The drift may be assumed to be a triangular prism with maximum height, located adjacent to a higher roof or along a projection, taken as $h_d = 2p_g/\gamma$, modified by factors for risk and exposure, described for flat roofs. Width of the prism should be at least 10 ft and may be taken as $3h_d$ for projections up to 50 ft long and as $4h_d$ for projections more than 50 ft long. Accordingly, the load varies uniformly with distance from a projection, from $h_d\gamma$ at the projection to zero. For drifts due to snow load from a higher roof at a horizontal distance S, ft, away horizontally ($S \leq 20$ ft), the maximum drift intensity may be taken as $h_d\gamma(20 - S)/20$.

Rain-Snow Load Combination. In roof design, account should be taken of the combination of the design snow load with a temporary water load from an intense rainstorm, including the effects of roof deflection on ponding. The added water load depends on the drainage characteristics of the roof, which, in turn, depend on the roof slope. For a flat roof, the rain surcharge may be taken as 8 lb/ft^2 for slopes less ¼ in/ft and as 5 lb/ft^2 for steeper slopes, except where the minimum allowable design snow load p_{min} exceeds p_f computed from Eq. (5.16). In such cases, these water surcharges may be reduced by $p_{min} - p_f$.

(W. Tobiasson and R. Redfield, "Snow Loads for the United States," Part II, and S. C. Colbeck, "Snow Loads Resulting from Rain on Snow," U.S. Army Cold Regions Research and Engineering Laboratory, Hanover, N.H.)

Seismic Loads. These are the result of horizontal and vertical movements imposed on a building by earth vibrations during an earthquake. Changing accelerations of the building mass during the temblor create changing inertial forces. These are assumed in building design to act as seismic loads at the various floor and roof levels in proportion to the portion of the building mass at those levels. Because analysis of building response to such dynamic loading generally is very complex, building codes permit, for design of ordinary buildings, substitution of equivalent static loading for the dynamic loading (see Art. 5.18.7).

("Minimum Design Loads for Buildings and Other Structures," ASCE 7-93, American Society of Civil Engineers, 345 E. 47th St., New York, NY 10164-0619; "Uniform Building Code," International Conference of Building Officials, 5360 South Workman Mill Road, Whittier, CA 90601; "Standard Building Code," Southern Building Code Congress International, Inc., 900 Montclair Road, Birmingham, AL 35213-1206.)

5.1.3 Factored Loads

Structural members must be designed with sufficient capacity to sustain without excessive deformation or failure those combinations of service loads that will produce the most unfavorable effects. Also, the effects of such conditions as ponding of water on roofs, saturation of soils, settlement, and dimensional changes must be included. In determination of the structural capacity of a member or structure, a safety margin must be provided and the possibility of variations of material properties from assumed design values and of inexactness of capacity calculations must be taken into account.

Building codes may permit either of two methods, allowable-stress design or load–and–resistance factor design (also known as ultimate-strength design), to be used for a structural material. In both methods, design loads, which determine the required structural capacity, are calculated by multiplying combinations of service loads by factors. Different factors are applied to the various possible load combinations in accordance with the probability of occurrence of the loads.

In allowable-stress design, required capacity is usually determined by the load combination that causes severe cracking or excessive deformation. For the purpose, dead, live, wind, seismic, snow, and other loads that may be imposed simultaneously are added together, then multiplied by a factor equal to or less than 1. Load combinations usually considered in allowable-stress design are

$$(1)\ D \qquad (2)\ D + G \qquad (3)\ D + (W \text{ or } E) \qquad (4)\ D + G + (W \text{ or } E)$$

where D = dead load
$G = L + L_r$ or $L + S$ or $L + R$
L = live loads due to intended use of occupancy, including partitions
L_r = roof live loads
S = snow loads
R = rain loads
W = wind loads
E = seismic loads

Building codes usually permit a smaller factor when the probability is small that combinations of extreme loads, such as dead load plus maximum live load plus maximum wind or seismic forces, will occur. Generally, for example, a factor of 0.75 is applied to load-combination sums (3) and (4) and 0.66 when dimensional-change effects are added to (4). Such factors are equivalent to permitting higher allowable unit stresses for the applicable loading conditions than for load combinations (1) and (2). The allowable stress is obtained by dividing the unit stress causing excessive deformation or failure by a factor greater than 1.

In load-and-resistance factor design, the various types of loads are each multiplied by a load factor, the value of which is selected in accordance with the probability of occurrence of each type of load. The factored loads are then added to obtain the total load a member or system must sustain. A structural member is selected to provide a load-carrying capacity exceeding that sum. This capacity is determined by multiplying the ultimate-load capacity by a resistance factor, the value of which reflects the reliability of the estimate of capacity. Load criteria generally used are as follows:

1. $1.4D$
2. $1.2D + 1.6L + 0.5(L_r \text{ or } S \text{ or } R)$
3. $1.2D + 1.6(L_r \text{ or } S \text{ or } R) + (0.5L \text{ or } 0.8W)$
4. $1.2D + 1.3W + 0.5G$
5. $1.2D + 1.5E + (0.5L \text{ or } 0.2S)$
6. $0.9D - (1.3W \text{ or } 1.5E)$

For garages, places of public assembly, and areas for which live loads exceed 100 lb/ft^2, the load factor usually is taken as unity for L in combinations 3, 4, and 5. The load factor should be taken as 1.3 for liquid loads, 1.6 for loads from soils, and 1.2 for ponding loads, forces due to differential settlement, and restraining forces due to prevention of dimensional changes. The recommended load factors recognize the greater certainty of the magnitude of dead loads but provide a larger safety factor against overloads due to dead loads alone.

5.2 STRESS AND STRAIN

Structural capacity, or ultimate strength, is that property of a structural member that serves as a measure of its ability to support all potential loads without severe cracking or excessive deformations. To indicate when the limit on load-carrying usefulness has been reached, design specifications for the various structural materials establish allowable unit stresses or design strengths that may not be exceeded under maximum loading. Structural theory provides methods for calculating unit stresses and for estimating deformations. Many of these methods are presented in the rest of this section.

5.2.1 Static Equilibrium

If a structure and its components are so supported that, after a very small deformation occurs, no further motion is possible, they are said to be in equilibrium. Under such circumstances, internal forces, or stresses, exactly counteract the loads.

Several useful conclusions may be drawn from the state of static equilibrium: Since there is no translatory motion, the sum of the external forces must be zero; and since there is no rotation, the sum of the moments of the external forces about any point must be zero.

For the same reason, if we consider any portion of the structure and the loads on it, the sum of the external and internal forces on the boundaries of that section must be zero. Also, the sum of the moments of these forces must be zero.

In Fig. 5.1, for example, the sum of the forces R_L and R_R needed to support the roof truss is equal to the 20-kip load on the truss (1 kip = 1 kilopound = 1000 lb = 0.5 ton). Also, the sum of moments of the external forces is zero about any point. About the right end, for instance, it is $40 \times 15 - 30 \times 20 = 600 - 600$.

In Fig. 5.2 is shown the portion of the truss to the left of section AA. The internal forces at the cut members balance the external load and hold this piece of the truss in equilibrium.

Generally, it is convenient to decompose the forces acting on a structure into components parallel to a set of perpendicular axes that will simplify computations. For example, for forces in a single plane—a condition commonly encountered in building design—the most useful technique is to resolve all forces into horizontal and vertical components. Then, for a structure in equilibrium, if H represents the horizontal components, V the vertical components, and M the moments of the components about any point in the plane,

$$\Sigma H = 0 \qquad \Sigma V = 0 \quad \text{and} \qquad \Sigma M = 0 \tag{5.19}$$

These three equations may be used to evaluate three unknowns in any nonconcurrent coplanar force system, such as the roof truss in Figs. 5.1 and 5.2. They may determine the magnitude of three forces for which the direction and point of application already are known, or the magnitude, direction, and point of application of a single force.

Suppose, for the truss in Fig. 5.1, the reactions at the supports are to be computed. Taking moments about the right end and equating to zero yields $40\,R_L - 30 \times 20 = 0$, from which left reaction $R_L = 600/40 = 15$ kips. Equating the sum of the vertical forces to zero gives $20 - 15 - R_R = 0$, from which the right reaction $R_R = 5$ kips.

5.2.2 Unit Stress and Strain

To ascertain whether a structural member has adequate load-carrying capacity, the designer generally has to compute the maximum unit stress produced by design loads in the member for each type of internal force—tensile, compressive, or shearing—and compare it with the corresponding allowable unit stress.

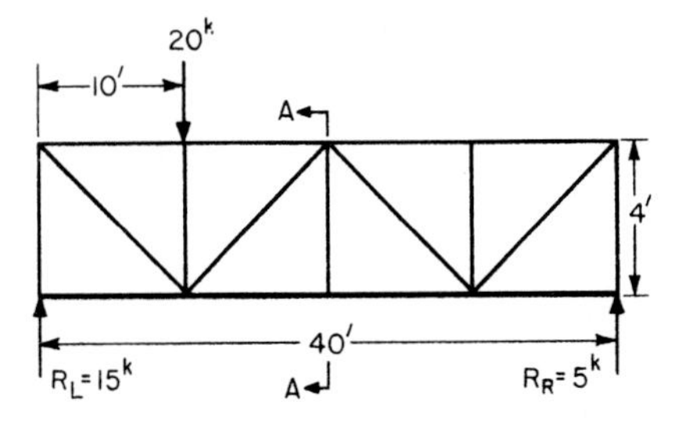

FIGURE 5.1 Truss in equilibrium under load. Upward-acting forces equal those acting downward.

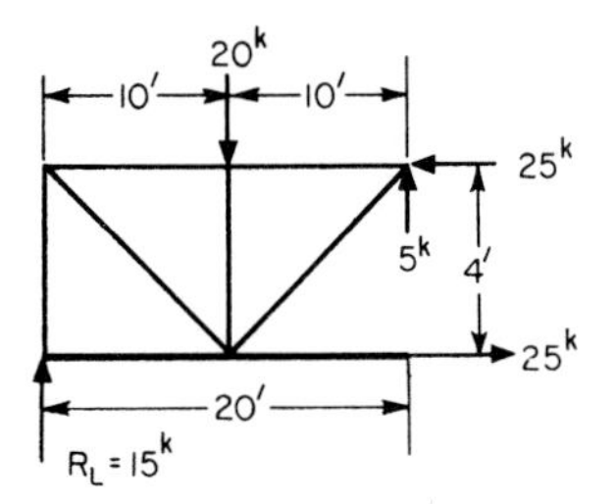

FIGURE 5.2 Portion of a truss is held in equilibrium by stresses in its components.

When the loading is such that the unit stress is constant over a section under consideration, the stress may be obtained by dividing the force by the area of the section. But in general, the unit stress varies from point to point. In that case, the unit stress at any point in the section is the limiting value of the ratio of the internal force on any small area to that area, as the area is taken smaller and smaller.

Sometimes in the design of a structure, unit stress may not be the prime consideration. The designer may be more interested in limiting the deformation or strain.

Deformation in any direction is the total change in the dimension of a member in that direction.

Unit strain in any direction is the deformation per unit of length in that direction.

When the loading is such that the unit strain is constant over a portion of a member, it may be obtained by dividing the deformation by the original length of that portion. In general, however, the unit strain varies from point to point in a member. Like a varying unit stress, it represents the limiting value of a ratio.

5.2.3 Hooke's Law

For many materials, unit strain is proportional to unit stress, until a certain stress, the **proportional limit,** is exceeded. Known as Hooke's law, this relationship may be written as

$$f = E\epsilon \qquad \text{or} \qquad \epsilon = \frac{f}{E} \tag{5.20}$$

where f = unit stress
ϵ = unit strain
E = modulus of elasticity

Hence, when the unit stress and modulus of elasticity of a material are known, the unit strain can be computed. Conversely, when the unit strain has been found, the unit stress can be calculated.

When a member is loaded and the unit stress does not exceed the proportional limit, the member will return to its original dimensions when the load is removed. The **elastic limit** is the largest unit stress that can be developed without a permanent deformation remaining after removal of the load.

Some materials possess one or two **yield points.** These are unit stresses in the region of which there appears to be an increase in strain with no increase or a small decrease in stress. Thus, the materials exhibit plastic deformation. For materials that do not have a well-defined yield point, the offset yield strength is used as a measure of the beginning of plastic deformation.

The **offset yield strength,** or **proof stress** as it is sometimes referred to, is defined as the unit stress corresponding to a permanent deformation, usually 0.01% (0.0001 in/in) or 0.20% (0.002 in/in).

5.2.4 Constant Unit Stress

The simplest cases of stress and strain are those in which the unit stress and strain are constant. Stresses due to an axial tension or compression load or a centrally applied shearing force are examples; also an evenly applied bearing load. These loading conditions are illustrated in Figs. 5.3 to 5.6.

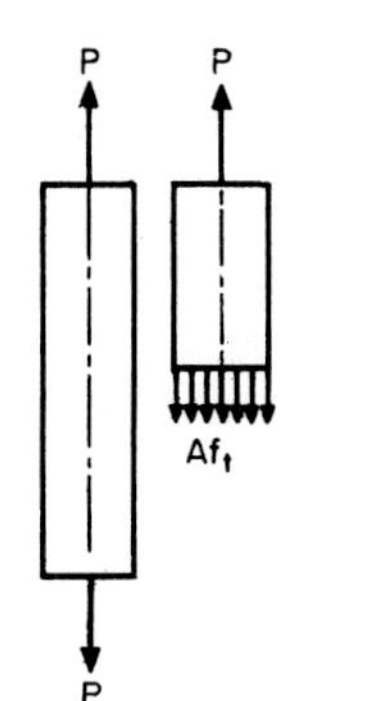

FIGURE 5.3 Tension member.

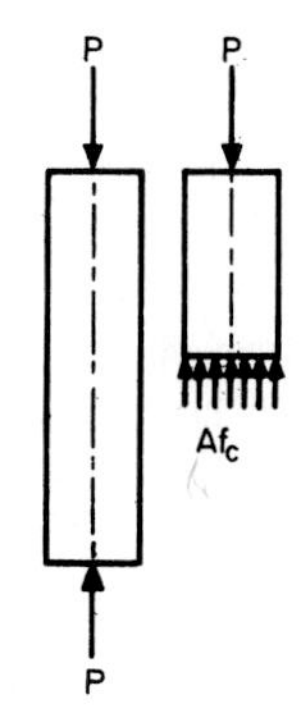

FIGURE 5.4 Compression member.

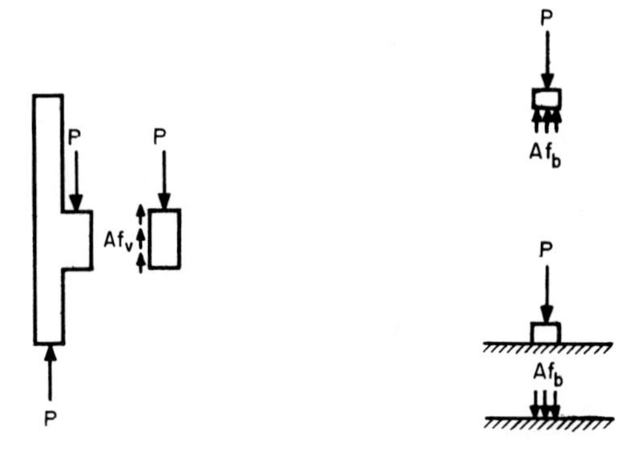

FIGURE 5.5 Bracket in shear. **FIGURE 5.6** Bearing load and pressure.

For the axial tension and compression loadings, we take a section normal to the centroidal axis (and to the applied forces). For the shearing load, the section is taken along a plane of sliding. And for the bearing load, it is chosen through the plane of contact between the two members.

Since for these loading conditions, the unit stress is constant across the section, the equation of equilibrium may be written

$$P = Af \tag{5.21}$$

where P = load
f = a tensile, compressive, shearing, or bearing unit stress
A = cross-sectional area for tensile or compressive forces, or area on which sliding may occur for shearing forces, or contact area for bearing loads

For torsional stresses, see Art. 5.4.2.

The unit strain for the axial tensile and compressive loads is given by the equation

$$\epsilon = \frac{e}{L} \tag{5.22}$$

where ϵ = unit strain
e = total lengthening or shortening of the member
L = original length of the member

Applying Hooke's law and Eq. (5.22) to Eq. (5.21) yields a convenient formula for the deformation:

$$e = \frac{PL}{AE} \tag{5.23}$$

where P = load on the member
A = its cross-sectional area
E = modulus of elasticity of the material

[Since long compression members tend to buckle, Eqs. (5.21) to (5.23) are applicable only to short members.]

While tension and compression strains represent a simple stretching or shortening of a member, shearing strain represents a distortion due to a small rotation. The load on the small rectangular portion of the member in Fig. 5.5 tends to distort it into a parallelogram. The unit shearing strain is the change in the right angle, measured in radians.

Modulus of rigidity, or shearing modulus of elasticity, is defined by

$$G = \frac{v}{\gamma} \tag{5.24}$$

where G = modulus of rigidity
v = unit shearing stress
γ = unit shearing strain

It is related to the modulus of elasticity in tension and compression E by the equation

$$G = \frac{E}{2\,(1 + \mu)} \tag{5.25}$$

where μ is a constant known as Poisson's ratio.

5.2.5 Poisson's Ratio

Within the elastic limit, when a material is subjected to axial loads, it deforms not only longitudinally but also laterally. Under tension, the cross section of a member decreases, and under compression, it increases. The ratio of the unit lateral strain to the unit longitudinal strain is called Poisson's ratio.

For many materials, this ratio can be taken equal to 0.25. For structural steel, it is usually assumed to be 0.3.

Assume, for example, that a steel hanger with an area of 2 in^2 carries a 40-kip (40,000-lb) load. The unit stress is 40,000/2, or 20,000 psi. The unit tensile strain, taking the modulus of elasticity of the steel as 30,000,000 psi, is 20,000/30,000,000, or 0.00067 in/in. With Poisson's ratio as 0.3, the unit lateral strain is -0.3×0.00067, or a shortening of 0.00020 in/in.

5.2.6 Thermal Stresses

When the temperature of a body changes, its dimensions also change. Forces are required to prevent such dimensional changes, and stresses are set up in the body by these forces.

If α is the coefficient of expansion of the material and T the change in temperature, the unit strain in a bar restrained by external forces from expanding or contracting is

$$\epsilon = \alpha T \tag{5.26}$$

According to Hooke's law, the stress f in the bar is

$$f = E\alpha T \tag{5.27}$$

where E = modulus of elasticity.

5.2.7 Strain Energy

Whan a bar is stressed, energy is stored in it. If a bar supporting a load P undergoes a deformation e the energy stored in it is

$$U = \tfrac{1}{2}Pe \tag{5.28}$$

This equation assumes the load was applied gradually and the bar is not stressed beyond the proportional limit. It represents the area under the load-deformation curve up to the load P. Applying Eqs. (5.20) and (5.21) to Eq. (5.28) gives another useful equation for energy:

$$U = \frac{f^2}{2E} AL \tag{5.29}$$

where f = unit stress
E = modulus of elasticity of the material
A = cross-sectional area
L = length of the bar

Since AL is the volume of the bar, the term $f^2/2E$ indicates the energy stored per unit of volume. It represents the area under the stress-strain curve up to the stress f. Its value when the bar is stressed to the proportional limit is called the **modulus of resilience.** This modulus is a measure of the capacity of the material to absorb energy without danger of being permanently deformed and is of importance in designing members to resist energy loads.

Equation (5.28) is a general equation that holds true when the principle of superposition applies (the total deformation produced by a system of forces is equal to the sum of the elongations produced by each force). In the general sense, P in Eq. (5.28) represents any group of statically interdependent forces that can be completely defined by one symbol, and e is the corresponding deformation.

The strain-energy equation can be written as a function of either the load or the deformation.

For axial tension or compression:

$$U = \frac{P^2L}{2AE} \qquad U = \frac{AEe^2}{2L} \tag{5.30}$$

where P = axial load
e = total elongation or shortening
L = length of the member
A = cross-sectional area
E = modulus of elasticity

For pure shear:

$$U = \frac{V^2L}{2AG} \qquad U = \frac{AGe^2}{2L} \tag{5.31}$$

where V = shearing load
e = shearing deformation
L = length over which deformation takes place
A = shearing area
G = shearing modulus

For torsion:

$$U = \frac{T^2L}{2JG} \qquad U = \frac{JG\phi^2}{2L} \tag{5.32}$$

where T = torque
ϕ = angle of twist
L = length of shaft
J = polar moment of inertia of the cross section
G = shearing modulus

For pure bending (constant moment):

$$U = \frac{M^2L}{2EI} \qquad U = \frac{EI\theta^2}{2L} \tag{5.33}$$

where M = bending moment
θ = angle of rotation of one end of the beam with respect to the other
L = length of beam
I = moment of inertia of the cross section
E = modulus of elasticity

For beams carrying transverse loads, the strain energy is the sum of the energy for bending and that for shear.

See also Art. 5.10.4.

5.3 STRESSES AT A POINT

Tensile and compressive stresses are sometimes referred to also as **normal stresses,** because they act normal to the cross section. Under this concept, tensile stresses are considered as positive normal stresses and compressive stresses as negative.

5.3.1 Stress Notation

Suppose a member of a structure is acted upon by forces in all directions. For convenience, let us establish a reference set of perpendicular coordinate x, y, and z axes. Now let us take at some point in the member a small cube with sides parallel to the coordinate axes. The notations commonly used for the components of stress acting on the sides of this element and the directions assumed as positive are shown in Fig. 5.7.

For example, for the sides of the element perpendicular to the z axis, the normal component of stress is denoted by f_z. The shearing stress v is resolved into two components and requires two subscript letters for a complete description. The first letter indicates the direction of the normal to the plane under consideration. The second letter indicates the direction of the component of the stress. For the sides perpendicular to the z axis, the shear component in the x direction is labeled v_{zx} and that in the y direction v_{zy}.

5.3.2 Stress and Strain Components

If, for the small cube in Fig. 5.7, moments of the forces acting on it are taken about the x axis, considering the cube's dimensions as dx, dy, and dz, the equation of equilibrium requires that

$$v_{zy}\,dx\,dy\,dz = v_{yz}\,dx\,dy\,dz$$

(Forces are taken equal to the product of the area of the face and the stress at the center.) Two similar equations can be written for moments taken about the y axis and the z axis. These equations show that

$$v_{xy} = v_{yx} \qquad v_{zx} = v_{xz} \qquad \text{and} \qquad v_{zy} = v_{yz} \tag{5.34}$$

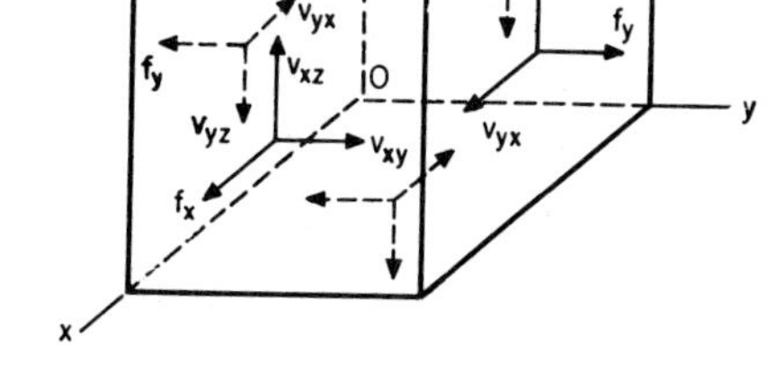

FIGURE 5.7 Normal and shear stresses in an orthogonal coordinate system.

In words, the components of shearing stress on two perpendicular faces and acting normal to the intersection of the faces are equal.

Consequently, to describe the stresses acting on the coordinate planes through a point, only six quantities need be known. These stress components are f_x, f_y, f_z $v_{xy} = v_{yx}$, $v_{yz} = v_{zy}$, and $v_{zx} = v_{xz}$.

If the cube in Fig. 5.7 is acted on only by normal stresses f_x, f_y, and f_z, from Hooke's law and the application of Poisson's ratio, the unit strains in the x, y, and z directions, in accordance with Arts. 5.2.3 and 5.2.4, are, respectively,

$$\begin{aligned} \epsilon_x &= \frac{1}{E}[f_x - \mu(f_y + f_z)] \\ \epsilon_y &= \frac{1}{E}[f_y - \mu(f_x + f_z)] \\ \epsilon_z &= \frac{1}{E}[f_z - \mu(f_x + f_y)] \end{aligned} \tag{5.35}$$

where μ = Poisson's ratio. If only shearing stresses act on the cube in Fig. 5.7, the distortion of the angle between edges parallel to any two coordinate axes depends only on shearing-stress components parallel to those axes. Thus, the unit shearing strains are (see Art. 5.2.4)

$$\gamma_{xy} = \frac{1}{G}v_{xy} \qquad \gamma_{yz} = \frac{1}{G}v_{yz} \qquad \text{and} \qquad \gamma_{zx} = \frac{1}{G}v_{zx} \tag{5.36}$$

5.3.3 Two-Dimensional Stress

When the six components of stress necessary to describe the stresses at a point are known (Art. 5.3.2), the stress on any inclined plane through the same point can be determined. For the case of two-dimensional stress, only three stress components need be known.

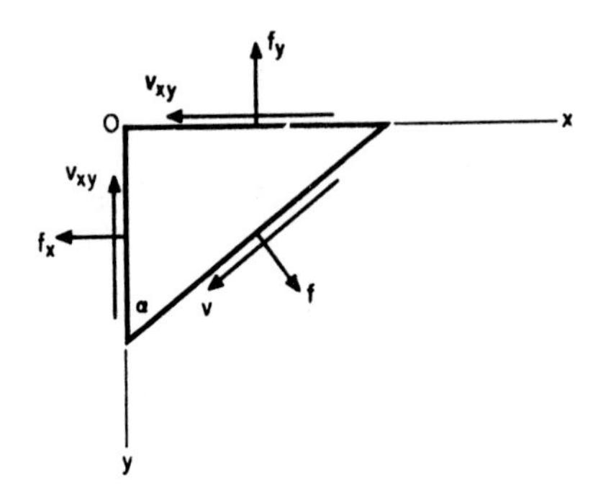

FIGURE 5.8 Normal and shear stresses at a point on a plane inclined to the axes.

Assume, for example, that at a point O in a stressed plate, the components f_x, f_y, and v_{xy} are known (Fig. 5.8). To find the stresses for any plane through the z axis, take a plane parallel to it close to O. This plane and the coordinate planes form a triangular prism. Then, if α is the angle the normal to the plane makes with the x axis, the normal and shearing stresses on the inclined plane, obtained by application of the equations of equilibrium, are

$$f = f_x \cos^2 \alpha + f_y \sin^2 \alpha + 2v_{xy} \sin \alpha \cos \alpha \tag{5.37}$$

$$v = v_{xy}(\cos^2 \alpha - \sin^2 \alpha) + (f_y - f_x) \sin \alpha \cos \alpha \tag{5.38}$$

Note. All structural members are three-dimensional. While two-dimensional-stress calculations may be sufficiently accurate for most practical purposes, this is not always the case. For example, although loads may create normal stresses on two perpendicular planes, a third normal stress also exists, as computed with Poisson's ratio. [See Eq. (5.35).]

5.3.4 Principal Stresses

A plane through a point on which stresses act may be assigned a direction for which the normal stress is a maximum or a minimum. There are two such positions, perpendicular to each other. And on those planes, there are no shearing stresses.

The direction in which the normal stresses become maximum or minimum are called principal directions and the corresponding normal stresses principal stresses.

To find the principal directions, set the value of v given by Eq. (5.38) equal to zero. The resulting equation is

$$\tan 2\alpha = \frac{2v_{xy}}{f_x - f_y} \tag{5.39}$$

If the x and y axes are taken in the principal directions, v_{xy} is zero. Consequently, Eqs. (5.37) and (5.38) may be simplified to

$$f = f_x \cos^2 \alpha + f_y \sin^2 \alpha \tag{5.40}$$

$$v = \tfrac{1}{2} \sin 2\alpha(f_y - f_x) \tag{5.41}$$

where f and v are, respectively, the normal and shearing stress on a plane at an angle α with the principal planes and f_x and f_y are the principal stresses.

Pure Shear. If on any two perpendicular planes only shearing stresses act, the state of stress at the point is called pure shear or simple shear. Under such conditions, the principal directions bisect the angles between the planes on which these shearing stresses occur. The principal stresses are equal in magnitude to the unit shearing stresses.

5.3.5 Maximum Shearing Stress

The maximum unit shearing stress occurs on each of two planes that bisect the angles between the planes on which the principal stresses act. The maximum shear is equal to one-half the algebraic difference of the principal stresses:

$$\max v = \frac{f_1 - f_2}{2} \tag{5.42}$$

where f_1 is the maximum principal stress and f_2 the minimum.

5.3.6 Mohr's Circle

The relationship between stresses at a point may be represented conveniently on Mohr's circle (Fig. 5.9). In this diagram, normal stress f and shear stress v are taken as coordinates. Then, for each plane through the point, there will correspond a point on the circle, whose coordinates are the values of f and v for the plane.

To construct the circle given the principal stresses, mark off the principal stresses f_1 and f_2 on the f axis (points A and B in Fig. 5.9). Tensile stresses are measured to the right of the v axis and compressive stresses to the left. Construct a circle

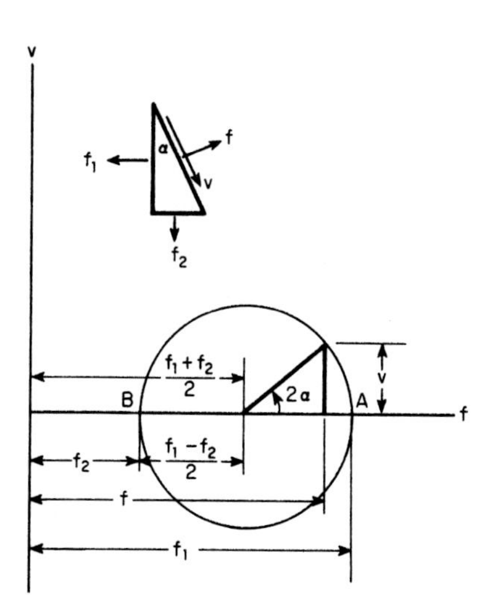

FIGURE 5.9 Mohr's circle for stresses at a point—constructed from known principal stresses.

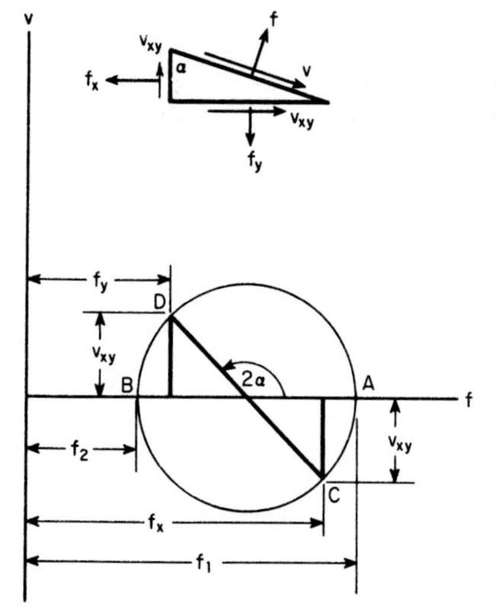

FIGURE 5.10 Stress circle constructed from two known positive stresses f_x and f_y and a shear stress v_{xy}.

with its center on the f axis and passing through the two points representing the principal stresses. This is the Mohr's circle for the given stresses at the point under consideration.

Suppose now, we wish to find the stresses on a plane at an angle α to the plane of f_1. If a radius is drawn making an angle 2α with the f axis, the coordinates of its intersection with the circle represent the normal and shearing stresses acting on the plane.

Mohr's circle can also be plotted when the principal stresses are not known but the stresses f_x, f_y, and v_{xy}, on any two perpendicular planes, are. The procedure is to plot the two points representing these known stresses with respect to the f and v axes (points C and D in Fig. 5.10). The line joining these points is a diameter of Mohr's circle. Constructing the circle on this diameter, we find the principal stresses at the intersection with the f axis (points A and B in Fig. 5.10).

For more details on the relationship of stresses and strains at a point, see Timoshenko and Goodier, "Theory of Elasticity," McGraw-Hill Publishing Company, New York.

5.4 TORSION

Forces that cause a member to twist about a longitudinal axis are called torsional loads. Simple torsion is produced only by a couple, or moment, in a plane perpendicular to the axis.

If a couple lies in a nonperpendicular plane, it can be resolved into a torsional moment, in a plane perpendicular to the axis, and bending moments, in planes through the axis.

5.4.1 Shear Center

The point in each normal section of a member through which the axis passes and about which the section twists is called the shear center. The location of the shear center depends on the shape and dimensions of the cross section. If the loads on a beam do not pass through the shear center, they cause the beam to twist. See also Art. 5.5.19.

If a beam has an axis of symmetry, the shear center lies on it. In doubly symmetrical beams, the shear center lies at the intersection of the two axes of symmetry and hence coincides with the centroid.

For any section composed of two narrow rectangles, such as a T beam or an angle, the shear center may be taken as the intersection of the longitudinal center lines of the rectangles.

For a channel section with one axis of symmetry, the shear center is outside the section at a distance from the centroid equal to $e(1 + h^2A/4I)$, where e is the distance from the centroid to the center of the web, h is the depth of the channel, A the cross-sectional area, and I the moment of inertia about the axis of symmetry. (The web lies between the shear center and the centroid.)

Locations of shear centers for several other sections are given in Friedrich Bleich, "Buckling Strength of Metal Structures," Chap. III, McGraw-Hill Publishing Company, New York.

5.4.2 Stresses Due to Torsion

Simple torsion is resisted by internal shearing stresses. These can be resolved into radial and tangential shearing stresses, which being normal to each other also are equal (see Art. 5.3.2). Furthermore, on planes that bisect the angles between the planes on which the shearing stresses act, there also occur compressive and tensile stresses. The magnitude of these normal stresses is equal to that of the shear. Therefore, when torsional loading is combined with other types of loading, the maximum stresses occur on inclined planes and can be computed by the methods of Arts. 5.3.3 and 5.3.6.

Circular Sections. If a circular shaft (hollow or solid) is twisted, a section that is plane before twisting remains plane after twisting. Within the proportional limit, the shearing unit stress at any point in a transverse section varies with the distance from the center of the section. The maximum shear, psi, occurs at the circumference and is given by

$$v = \frac{Tr}{J} \tag{5.43}$$

where T = torsional moment, in-lb
r = radius of section, in
J = polar moment of inertia, in^4

Polar moment of inertia of a cross section is defined by

$$J = \int \rho^2 \, dA \tag{5.44}$$

where ρ = radius from shear center to any point in the section
dA = differential area at the point

In general, J equals the sum of the moments of inertia above any two perpendicular axes through the shear center. For a solid circular section, $J = \pi r^4/2$. For a hollow circular section with diameters D and d, $J = \pi(D^4 - d^4)/32$.

Within the proportional limits, the angular twist between two points L inches apart along the axis of a circular bar is, in radians (1 rad = 57.3°):

$$\theta = \frac{TL}{GJ} \tag{5.45}$$

where G is the shearing modulus of elasticity (see Art. 5.2.4).

Noncircular Sections. If a shaft is not circular, a plane transverse section before twisting does not remain plane after twisting. The resulting warping increases the shearing stresses in some parts of the section and decreases them in others, compared with the shearing stresses that would occur if the section remained plane. Consequently, shearing stresses in a noncircular section are not proportional to distances from the shear center. In elliptical and rectangular sections, for example, maximum shear occurs on the circumference at a point nearest the shear center.

For a solid rectangular section, this maximum may be expressed in the following form:

$$v = \frac{T}{kb^2d} \tag{5.46}$$

where b = short side of rectangle, in
d = long side, in
k = constant depending on ratio of these sides;

d/b =	1.0	1.5	2.0	2.5	3	4	5	10	∞
k =	0.208	0.231	0.246	0.258	0.267	0.282	0.291	0.312	0.333

(S. Timoshenko and J. N. Goodier, "Theory of Elasticity," McGraw-Hill Publishing Company, New York.)

Hollow Tubes. If a thin-shell hollow tube is twisted, the shearing force per unit of length on a cross section (**shear flow**) is given approximately by

$$H = \frac{T}{2A} \tag{5.47}$$

where A is the area enclosed by the mean perimeter of the tube, in^2, and the unit shearing stress is given approximately by

$$v = \frac{H}{t} = \frac{T}{2At} \tag{5.48}$$

where t is the thickness of the tube, in. For a rectangular tube with sides of unequal thickness, the total shear flow can be computed from Eq. (5.47) and the shearing stress along each side from Eq. (5.48), except at the corners, where there may be appreciable stress concentration.

Channels and I Beams. For a narrow rectangular section, the maximum shear is very nearly equal to

$$v = \frac{t}{\frac{1}{3}b^2d} \tag{5.49}$$

This formula also can be used to find the maximum shearing stress due to torsion in members, such as I beams and channels, made up of thin rectangular components. Let $J = \frac{1}{3}\Sigma b^3d$, where b is the thickness of each rectangular component and d the corresponding length. Then, the maximum shear is given approximately by

$$v = \frac{Tb'}{J} \tag{5.50}$$

where b' is the thickness of the web or the flange of the member. Maximum shear will occur at the center of one of the long sides of the rectangular part that has the greatest thickness. (A. P. Boresi, O. Sidebottom, F. B. Seely, and J. O. Smith, "Advanced Mechanics of Materials," 3d ed., John Wiley & Sons, Inc., New York.)

5.5 STRAIGHT BEAMS

Beams are the horizontal members used to support vertically applied loads across an opening. In a more general sense, they are structural members that external loads tend to bend, or curve. Usually, the term *beam* is applied to members with top continuously connected to bottom throughout their length, and those with top and bottom connected at intervals are called *trusses.* See also Structural System, Art. 1.7.

5.5.1 Types of Beams

There are many ways in which beams may be supported. Some of the more common methods are shown in Figs. 5.11 to 5.16.

The beam in Fig. 5.11 is called a simply supported, or **simple beam.** It has supports near its ends, which restrain it only against vertical movement. The ends of the beam are free to rotate. When the loads have a horizontal component, or when change in length of the beam due to temperature may be important, the supports may also have to prevent horizontal motion. In that case, horizontal restraint at one support is generally sufficient.

The distance between the supports is called the **span.** The load carried by each support is called a **reaction.**

The beam in Fig. 5.12 is a **cantilever.** It has only one support, which restrains it from rotating or moving horizontally or vertically at that end. Such a support is called a **fixed end.**

If a simple support is placed under the free end of the cantilever, the **propped beam** in Fig. 5.13 results. It has one end fixed, one end simply supported.

The beam in Fig. 5.14 has both ends fixed. No rotation or vertical movement can occur at either end. In actual practice, a fully fixed end can seldom be obtained. Some rotation of the beam ends generally is permitted. Most support conditions are intermediate between those for a simple beam and those for a fixed-end beam.

In Fig. 5.15 is shown a beam that overhangs both its simple supports. The overhangs have a free end, like a cantilever, but the supports permit rotation.

When a beam extends over several supports, it is called a **continuous beam** (Fig. 5.16).

Reactions for the beams in Figs. 5.11, 5.12, and 5.15 may be found from the equations of equilibrium. They are classified as **statically determinate beams** for that reason.

The equations of equilibrium, however, are not sufficient to determine the reactions of the beams in Figs. 5.13, 5.14, and 5.16. For those beams, there are

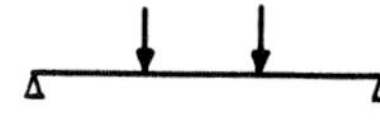

FIGURE 5.11 Simple beam.

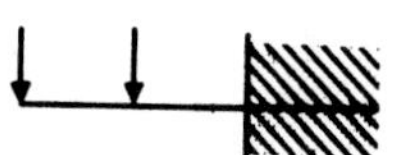

FIGURE 5.12 Cantilever beam.

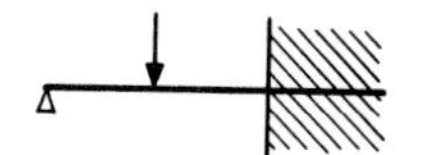

FIGURE 5.13 Beam with one end fixed.

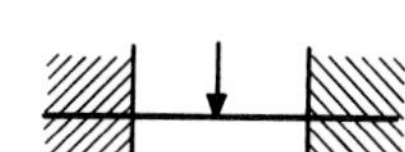

FIGURE 5.14 Fixed-end beam.

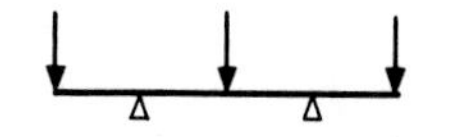

FIGURE 5.15 Beam with overhangs.

FIGURE 5.16 Continuous beam.

more unknowns than equations. Additional equations must be obtained on the basis of deformations permitted; on the knowledge, for example, that a fixed end permits no rotation. Such beams are classified as **statically indeterminate.** Methods for finding the stresses in that type of beam are given in Arts. 5.10.4, 5.10.5, 5.11, and 5.13.

5.5.2 Reactions

As an example of the application of the equations of equilibrium (Art. 5.2.1) to the determination of the reactions of a statically determinate beam, we shall compute the reactions of the 60-ft-long beam with overhangs in Fig. 5.17. This beam carries a uniform load of 200 lb/lin ft over its entire length and several concentrated loads. The supports are 36 ft apart.

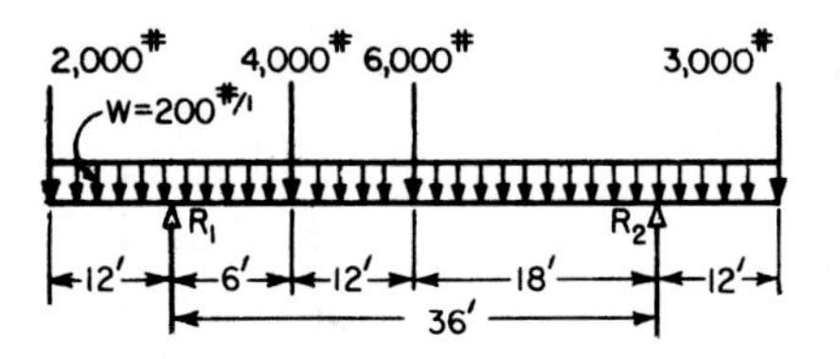

FIGURE 5.17 Beam with overhangs loaded with both uniform and concentrated loads.

To find reaction R_1, we take moments about R_2 and equate the sum of the moments to zero (clockwise rotation is considered positive, counterclockwise, negative):

$$-2000 \times 48 + 36R_1 - 4000 \times 30 - 6000 \times 18 + 3000 \times 12 - 200 \times 60 \times 18 = 0$$

$$R_1 = 14{,}000 \text{ lb}$$

In this calculation, the moment of the uniform load was found by taking the moment of its resultant, which acts at the center of the beam.

To find R_2, we can either take moments about R_1 or use the equation $\Sigma V = 0$. It is generally preferable to apply the moment equation and use the other equation as a check.

$$3000 \times 48 - 36R_2 + 6000 \times 18 + 4000 \times 6 - 2000 \times 12 + 200 \times 60 \times 18 = 0$$

$$R_2 = 13{,}000 \text{ lb}$$

As a check, we note that the sum of the reactions must equal the total applied load:

$$14{,}000 + 13{,}000 = 2000 + 4000 + 6000 + 3000 + 12{,}000$$

$$27{,}000 = 27{,}000$$

5.5.3 Internal Forces

Since a beam is in equilibrium under the forces applied to it, it is evident that at every section internal forces are acting to prevent motion. For example, suppose we cut the beam in Fig. 5.17 vertically just to the right of its center. If we total the external forces, including the reaction, to the left of this cut (see Fig. 5.18*a*), we find there is an unbalanced downward load of 4000 lb. Evidently, at the cut

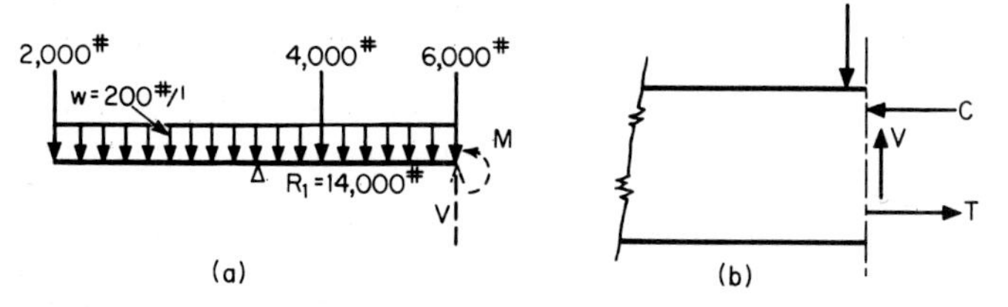

FIGURE 5.18 Portions of a beam are held in equilibrium by internal stresses.

section, an upward-acting internal force of 4000 lb must be present to maintain equilibrium. Again, if we take moments of the external forces about the section, we find an unbalanced moment of 54,000 ft-lb. So there must be an internal moment of 54,000 ft-lb acting to maintain equilibrium.

This internal, or resisting, moment is produced by a couple consisting of a force C acting on the top part of the beam and an equal but opposite force T acting on the bottom part (Fig. 5.18*b*). The top force is the resultant of compressive stresses acting over the upper portion of the beam, and the bottom force is the resultant of tensile stresses acting over the bottom part. The surface at which the stresses change from compression to tension—where the stress is zero—is called the **neutral surface.**

5.5.4 Shear Diagrams

The unbalanced external vertical force at a section is called the shear. It is equal to the algebraic sum of the forces that lie on either side of the section. Upward-acting forces on the left of the section are considered positive, downward forces negative; signs are reversed for forces on the right.

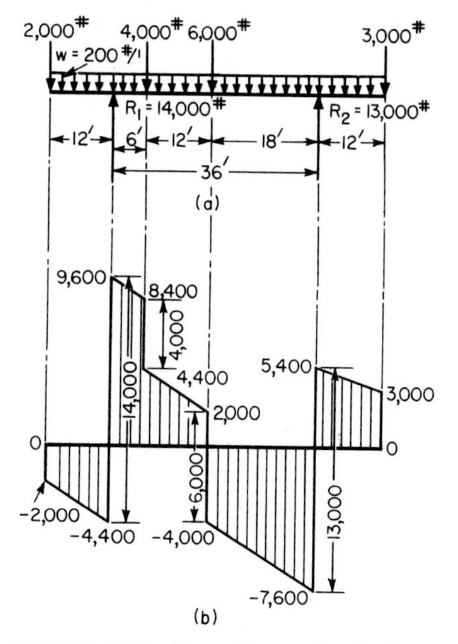

FIGURE 5.19 Shear diagram for the beam with loads shown in Fig. 5.17.

A diagram in which the shear at every point along the length of a beam is plotted as an ordinate is called a shear diagram. The shear diagram for the beam in Fig. 5.17 is shown in Fig. 5.19*b*.

The diagram was plotted starting from the left end. The 2000-lb load was plotted downward to a convenient scale. Then, the shear at the next concentrated load—the left support—was determined. This equals $-2000 - 200 \times 12$, or -4400 lb. In passing from just to the left of the support to a point just to the right, however, the shear changes by the magnitude of the reaction. Hence, on the right-hand side of the left support the shear is $-4400 + 14{,}000$, or 9600 lb. At the next concentrated load, the shear is $9600 - 200 \times 6$, or 8400 lb. In passing the 4000-lb load, however, the shear changes to $8400 - 4000$, or 4400 lb. Proceeding in this manner to the right end of the beam, we terminate with a shear of 3000 lb, equal to the load on the free end there.

It should be noted that the shear diagram for a uniform load is a straight line sloping downward to the right (see Fig. 5.21). Therefore, the shear diagram was completed by connecting the plotted points with straight lines.

Shear diagrams for commonly encountered loading conditions are given in Figs. 5.30 to 5.41.

5.5.5 Bending-Moment Diagrams

The unbalanced moment of the external forces about a vertical section through a beam is called the bending moment. It is equal to the algebraic sum of the moments about the section of the external forces that lie on one side of the section. Clockwise moments are considered positive, counterclockwise moments negative, when the forces considered lie on the left of the section. Thus, when the bending moment is positive, the bottom of the beam is in tension.

A diagram in which the bending moment at every point along the length of a beam is plotted as an ordinate is called a bending-moment diagram.

Figure 5.20*c* is the bending-moment diagram for the beam loaded with concentrated loads only in Fig. 5.20*a*. The bending moment at the supports for this simply

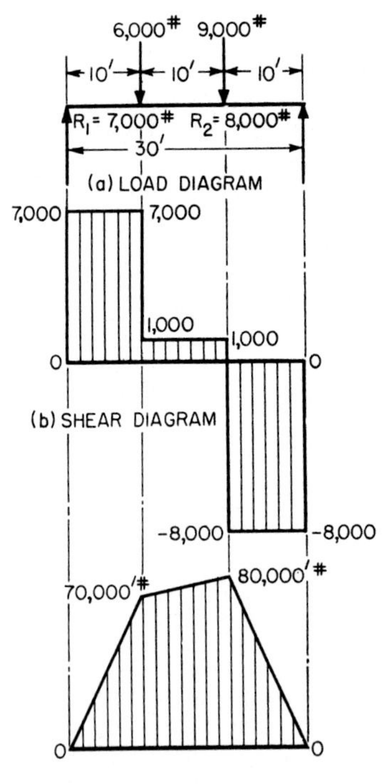

FIGURE 5.20 Shear and moment diagrams for a simply supported beam with concentrated loads.

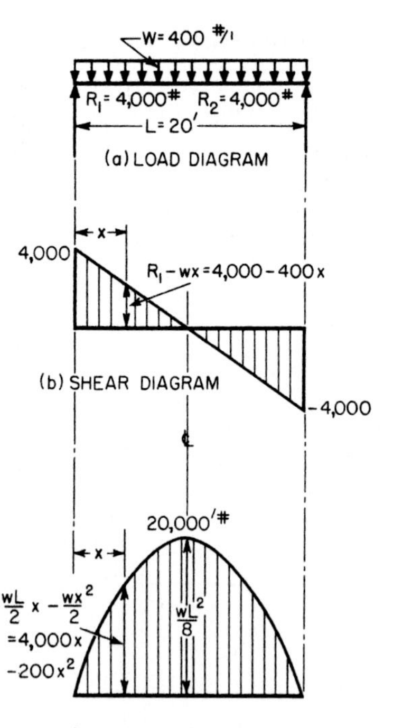

FIGURE 5.21 Shear and moment diagrams for a simply supported, uniformly loaded beam.

supported beam obviously is zero. Between the supports and the first load, the bending moment is proportional to the distance from the support, since it is equal to the reaction times the distance from the support. Hence the bending-moment diagram for this portion of the beam is a sloping straight line.

The bending moment under the 6000-lb load in Fig. 5.20*a* considering only the force to the left is 7000×10, or 70,000 ft-lb. The bending-moment diagram, then, between the left support and the first concentrated load is a straight line rising from zero at the left end of the beam to 70,000 ft-lb, plotted to a convenient scale, under the 6000-lb load.

The bending moment under the 9000-lb load, considering the forces on the left of it, is $7000 \times 20 - 6000 \times 10$, or 80,000 ft-lb. (It could have been more easily obtained by considering only the force on the right, reversing the sign convention: $8000 \times 10 = 80{,}000$ ft-lb.) Since there are no loads between the two concentrated loads, the bending-moment diagram between the two sections is a sloping straight line.

If the bending moment and shear are known at any section of a beam, the bending moment at any other section may be computed, providing there are no unknown forces between the two sections. The rule is:

The bending moment at any section of a beam is equal to the bending moment at any section to the left, plus the shear at that section times the distance between sections, minus the moments of intervening loads. If the section with known moment and shear is on the right, the sign convention must be reversed.

For example, the bending moment under the 9000-lb load in Fig. 5.20*a* could also have been obtained from the moment under the 6000-lb load and the shear to the right of the 6000-lb load given in the shear diagram (Fig. 5.20*b*). Thus, $80{,}000 = 70{,}000 + 1000 \times 10$. If there had been any other loads between the two concentrated loads, the moment of these loads about the section under the 9000-lb load would have been subtracted.

Bending-moment diagrams for commonly encountered loading conditions are given in Figs. 5.30 to 5.41. These may be combined to obtain bending moments for other loads.

5.5.6 Moments in Uniformly Loaded Beams

When a beam carries a uniform load, the bending-moment diagram does not consist of straight lines. Consider, for example, the beam in Fig. 5.21*a*, which carries a uniform load over its entire length. As shown in Fig. 5.21*c*, the bending-moment diagram for this beam is a parabola.

The reactions at both ends of a simply supported, uniformly loaded beam are both equal to $wL/2 = W/2$, where w is the uniform load in pounds per linear foot, $W = wL$ is the total load on the beam, and L is the span.

The shear at any distance x from the left support is $R_1 - wx = wL/2 - wx$ (see Fig. 5.21*b*). Equating this expression to zero, we find that there is no shear at the center of the beam.

The bending moment at any distance x from the left support is

$$M = R_1 x - wx\left(\frac{x}{2}\right) = \frac{wLx}{2} - \frac{wx^2}{2} = \frac{w}{2}x(L - x) \tag{5.51}$$

Hence:

The bending moment at any section of a simply supported, uniformly loaded beam is equal to one-half the product of the load per linear foot and the distances to the section from both supports.

The maximum value of the bending moment occurs at the center of the beam. It is equal to $wL^2/8 = WL/8$.

5.5.7 Shear-Moment Relationship

The slope of the bending-moment curve for any point on a beam is equal to the shear at that point; i.e.,

$$V = \frac{dM}{dx} \tag{5.52}$$

Since maximum bending moment occurs when the slope changes sign, or passes through zero, maximum moment (positive or negative) occurs at the point of zero shear.

After integration, Eq. (5.52) may also be written

$$M_1 - M_2 = \int_{x_2}^{x_1} V\,dx \tag{5.53}$$

This equation indicates that the change in bending moment between any two sections of a beam is equal to the area of the shear diagram between ordinates at the two sections.

5.5.8 Moving Loads and Influence Lines

One of the most helpful devices for solving problems involving variable or moving loads is an influence line. Whereas shear and moment diagrams evaluate the effect of loads at all sections of a structure, an influence line indicates the effect at a given section of a unit load placed at any point on the structure.

For example, to plot the influence line for bending moment at some point *A* on a beam, a unit load is applied at some point *B*. The bending moment at *A* due to the unit load at *B* is plotted as an ordinate to a convenient scale at *B*. The same procedure is followed at every point along the beam and a curve is drawn through the points thus obtained.

Actually, the unit load need not be placed at every point. The equation of the influence line can be determined by placing the load at an arbitrary point and computing the bending moment in general terms. (See also Art. 5.10.5.)

Suppose we wish to draw the influence line for reaction at *A* for a simple beam *AB* (Fig. 5.22*a*). We place a unit load at an arbitrary distance of xL from *B*. The reaction at *A* due to this load is $1\,xL/L = x$. Then, $R_A = x$ is the equation of the influence line. It represents a straight line sloping upward from zero at *B* to unity at *A* (Fig. 5.22*a*). In other words, as the unit load moves across the beam, the reaction at *A* increases from zero to unity in proportion to the distance of the load from *B*.

Figure 5.22*b* shows the influence line for bending moment at the center of a beam. It resembles in appearance the bending-moment diagram for a load at the center of the beam, but its significance is entirely different. Each ordinate gives the moment at midspan for a load at the corresponding location. It indicates that, if a unit load is placed at a distance xL from one end, it produces a bending moment of ½ xL at the center of the span.

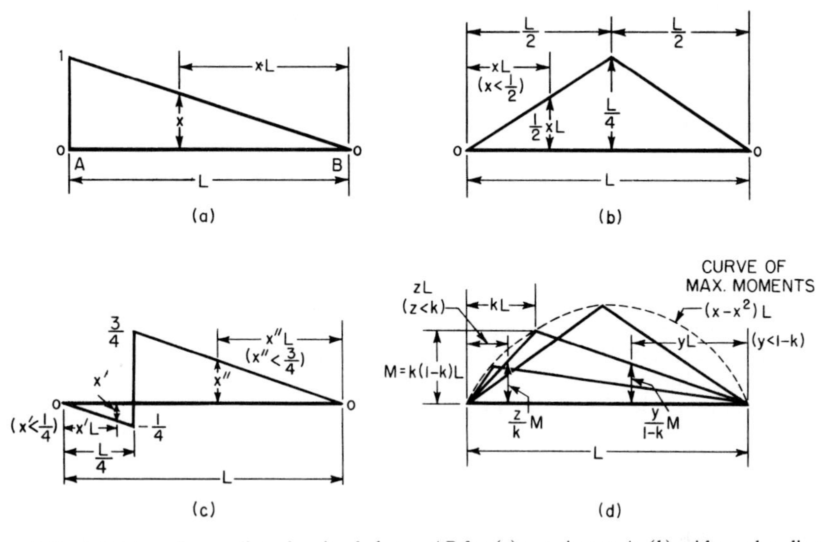

FIGURE 5.22 Influence lines for simple beam *AB* for (*a*) reaction at *A*; (*b*) midspan bending moment; (*c*) quarter-point shear; and (*d*) bending moments for unit load at several points on the beam.

Figure 5.22*c* shows the influence line for shear at the quarter point of a beam. When the load is to the right of the quarter point, the shear is positive and equal to the left reaction. When the load is to the left, the shear is negative and equal to the right reaction.

The diagram indicates that, to produce maximum shear at the quarter point, loads should be placed only to the right of the quarter point, with the largest load at the quarter point, if possible. For a uniform load, maximum shear results when the load extends from the right end of the beam to the quarter point.

Suppose, for example, that the beam is a crane girder with a span of 60 ft. The wheel loads are 20 and 10 kips, respectively, and are spaced 5 ft apart. For maximum shear at the quarter point, the wheels should be placed with the 20-kip wheel at that point and the 10-kip wheel to the right of it. The corresponding ordinates of the influence line (Fig. 5.22*c*) are $3/4$ and $40/45 \times 3/4$. Hence, the maximum shear is $20 \times 3/4 + 10 \times 40/45 \times 3/4 = 21.7$ kips.

Figure 5.22*d* shows influence lines for bending moment at several points on a beam. It is noteworthy that the apexes of the diagrams fall on a parabola, as shown by the dashed line. This indicates that the maximum moment produced at any given section by a single concentrated load moving across a beam occurs when the load is at that section. The magnitude of the maximum moment increases when the section is moved toward midspan, in accordance with the equation shown in Fig. 5.22*d* for the parabola.

5.5.9 Maximum Bending Moment

When there is more than one load on the span, the influence line is useful in developing a criterion for determining the position of the loads for which the bending moment is a maximum at a given section.

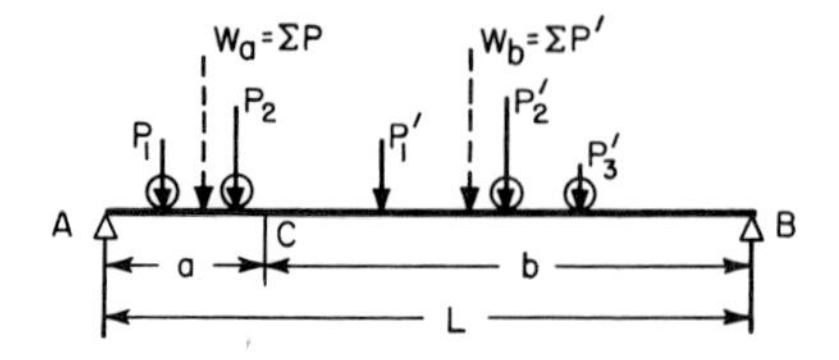

FIGURE 5.23 Moving loads on simple beam *AB* are placed for maximum bending moment at point *C* on the beam.

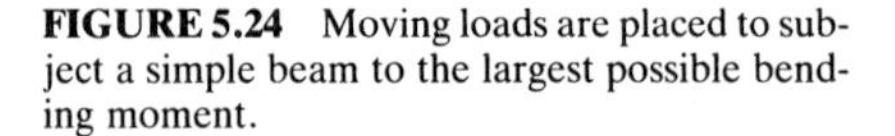

FIGURE 5.24 Moving loads are placed to subject a simple beam to the largest possible bending moment.

Maximum bending moment will occur at a section *C* of a simple beam as loads move across it when one of the loads is at *C*. The proper load to place at *C* is the one for which the expression $W_a/a - W_b/b$ (Fig. 5.23) changes sign as that load passes from one side of *C* to the other.

When several loads move across a simple beam, the maximum bending moment produced in the beam may be near but not necessarily at midspan. To find the maximum moment, first determine the position of the loads for maximum moment at midspan. Then shift the loads until the load P_2 that was at the center of the beam is as far from midspan as the resultant of all the loads on the span is on the other side of midspan (Fig. 5.24). Maximum moment will occur under P_2.

When other loads move on or off the span during the shift of P_2 away from midspan, it may be necessary to investigate the moment under one of the other loads when it and the resultant are equidistant from midspan.

5.5.10 Bending Stresses in a Beam

To derive the commonly used flexure formula for computing the bending stresses in a beam, we have to make the following assumptions:

1. The unit stress at a point in any plane parallel to the neutral surface of a beam is proportional to the unit strain in the plane at the point.
2. The modulus of elasticity in tension is the same as that in compression.
3. The total and unit axial strain in any plane parallel to the neutral surface are both proportional to the distance of that plane from the neutral surface. (Cross sections that are plane before bending remain plane after bending. This requires that all planes have the same length before bending; thus, that the beam be straight.)
4. The loads act in a plane containing the centroidal axis of the beam and are perpendicular to that axis. Furthermore, the neutral surface is perpendicular to the plane of the loads. Thus, the plane of the loads must contain an axis of symmetry of each cross section of the beam. (The flexure formula does not apply to a beam loaded unsymmetrically. See Arts. 5.5.18 and 5.5.19.)
5. The beam is proportioned to preclude prior failure or serious deformation by torsion, local buckling, shear, or any cause other than bending.

Equating the bending moment to the resisting moment due to the internal stresses at any section of a beam yields

$$M = \frac{fI}{c} \tag{5.54}$$

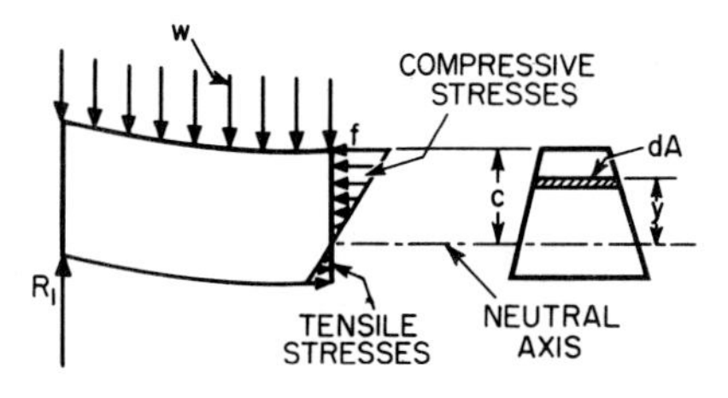

FIGURE 5.25 Unit stresses on a beam cross section caused by bending of the beam.

M is the bending moment at the section, f is the normal unit stress in a plane at a distance c from the neutral axis (Fig. 5.25), and I is the moment of inertia of the cross section with respect to the neutral axis. If f is given in pounds per square inch (psi), I in in^4, and c in inches, then M will be in inch-pounds. For maximum unit stress, c is the distance to the outermost fiber. See also Arts. 5.5.11 and 5.5.12.

5.5.11 Moment of Inertia

The neutral axis in a symmetrical beam is coincidental with the centroidal axis; i.e., at any section the neutral axis is so located that

$$\int y \, dA = 0 \tag{5.55}$$

where dA is a differential area parallel to the axis (Fig. 5.25), y is its distance from the axis, and the summation is taken over the entire cross section.

Moment of inertia with respect to the neutral axis is given by

$$I = \int y^2 \, dA \tag{5.56}$$

Values of I for several common types of cross section are given in Fig. 5.26. Values for structural-steel sections are presented in manuals of the American Institute of Steel Construction, Chicago, Ill. When the moments of inertia of other types of sections are needed, they can be computed directly by application of Eq. (5.56) or by breaking the section up into components for which the moment of inertia is known.

If I is the moment of inertia about the neutral axis, A the cross-sectional area, and d the distance between that axis and a parallel axis in the plane of the cross section, then the moment of inertia about the parallel axis is

$$I' = I + Ad^2 \tag{5.57}$$

With this equation, the known moment of inertia of a component of a section about the neutral axis of the component can be transferred to the neutral axis of the complete section. Then, summing up the transferred moments of inertia for all the components yields the moment of inertia of the complete section.

When the moments of inertia of an area with respect to any two perpendicular axes are known, the moment of inertia with respect to any other axis passing through the point of intersection of the two axes may be obtained through the use of Mohr's circle, as for stresses (Fig. 5.10). In this analog, I_x corresponds with f_x, I_y with f_y, and the **product of inertia** I_{xy} with v_{xy} (Art. 5.3.6).

$$I_{xy} = \int xy \, dA \tag{5.58}$$

The two perpendicular axes through a point about which the moments of inertia are a maximum and a minimum are called the principal axes. The products of inertia are zero for the principal axes.

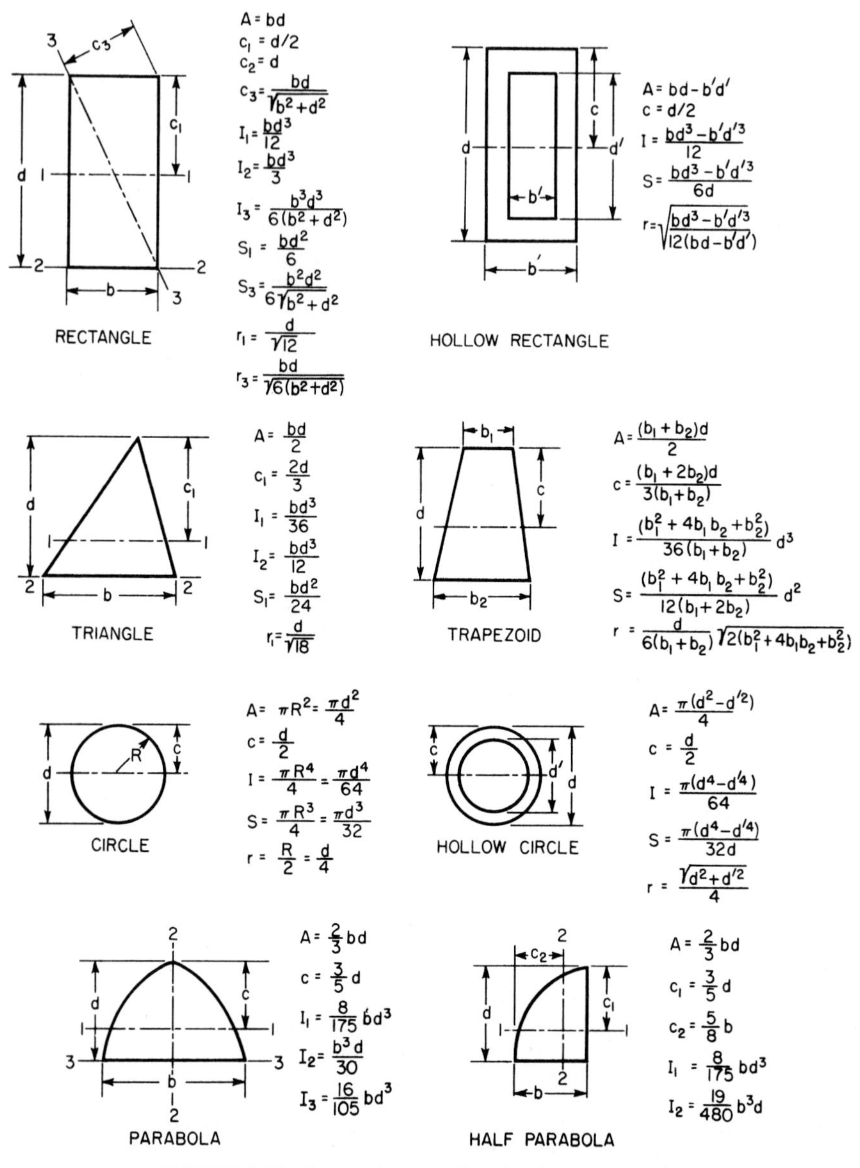

FIGURE 5.26 Geometric properties of various cross sections.

5.5.12 Section Modulus

The ratio $S = I/c$ in Eq. (5.54) is called the section modulus. I is the moment of inertia of the cross section about the neutral axis and c the distance from the neutral axis to the outermost fiber. Values of S for common types of sections are given in Fig. 5.26.

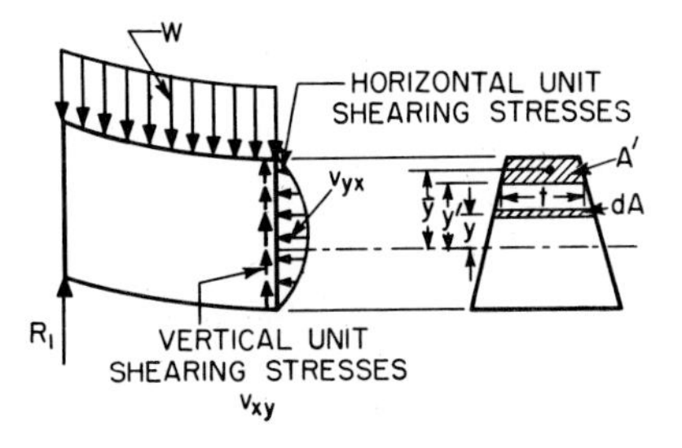

FIGURE 5.27 Unit shearing stresses on a beam cross section.

5.5.13 Shearing Stresses in a Beam

The vertical shear at any section of a beam is resisted by nonuniformly distributed, vertical unit stresses (Fig. 5.27). At every point in the section, there is also a horizontal unit stress, which is equal in magnitude to the vertical unit shearing stress there [see Eq. (5.34)].

At any distance y' from the neutral axis, both the horizontal and vertical shearing unit stresses are equal to

$$v = \frac{V}{It} A'\bar{y} \tag{5.59}$$

where V = vertical shear at the cross section
t = thickness of beam at distance y' from neutral axis
I = moment of inertia about neutral axis
A' = area between the outermost fiber and the fiber for which the shearing stress is being computed
$\bar{y}$ = distance of center of gravity of this area from the neutral axis (Fig. 5.27)

For a rectangular beam with width b and depth d, the maximum shearing stress occurs at middepth. Its magnitude is

$$v = \frac{12V}{bd^3b} \frac{bd^2}{8} = \frac{3}{2} \frac{V}{bd}$$

That is, the maximum shear stress is 50% greater than the average shear stress on the section. Similarly, for a circular beam, the maximum is one-third greater than the average. For an I beam, however, the maximum shearing stress in the web is not appreciably greater than the average for the web section alone, if it is assumed that the flanges take no shear.

5.5.14 Combined Shear and Bending Stress

For deep beams on short spans and beams made of low-strength materials, it is sometimes necessary to determine the maximum stress f' on an inclined plane caused by a combination of shear and bending stress—v and f, respectively. This stress f', which may be either tension or compression, is greater than the normal stress. Its value may be obtained by application of Mohr's circle (Art. 5.3.6), as indicated in Fig. 5.10, but with $f_y = 0$, and is

$$f' = \frac{f}{2} + \sqrt{v^2 + \left(\frac{f}{2}\right)^2} \tag{5.60}$$

5.5.15 Beam Deflections

When a beam is loaded, it deflects. The new position of its longitudinal centroidal axis is called the **elastic curve.**

At any point of the elastic curve, the radius of curvature is given by

$$R = \frac{EI}{M} \tag{5.61}$$

where M = bending moment at the point
E = modulus of elasticity
I = moment of inertia of the cross section about the neutral axis

Since the slope dy/dx of the curve is small, its square may be neglected, so that, for all practical purposes, $1/R$ may be taken equal to d^2y/dx^2, where y is the deflection of a point on the curve at a distance x from the origin of coordinates. Hence, Eq. (5.61) may be rewritten

$$M = EI\frac{d^2y}{dx^2} \tag{5.62}$$

To obtain the slope and deflection of a beam, this equation may be integrated, with M expressed as a function of x. Constants introduced during the integration must be evaluated in terms of known points and slopes of the elastic curve.

Equation (5.62), in turn, may be rewritten after one integration as

$$\theta_B - \theta_A = \int_A^B \frac{M}{EI}\,dx \tag{5.63}$$

in which θ_A and θ_B are the slopes of the elastic curve at any two points A and B. If the slope is zero at one of the points, the integral in Eq. (5.63) gives the slope of the elastic curve at the other. It should be noted that the integral represents the area of the bending-moment diagram between A and B with each ordinate divided by EI.

The **tangential deviation** t of a point on the elastic curve is the distance of this point, measured in a direction perpendicular to the original position of the beam, from a tangent drawn at some other point on the elastic curve.

$$t_B - t_A = \int_A^B \frac{Mx}{EI}\,dx \tag{5.64}$$

Equation (5.64) indicates that the tangential deviation of any point with respect to a second point on the elastic curve equals the moment about the first point of the M/EI diagram between the two points. The **moment-area method** for determining the deflection of beams is a technique in which Eqs. (5.63) and (5.64) are utilized.

Suppose, for example, the deflection at midspan is to be computed for a beam of uniform cross section with a concentrated load at the center (Fig. 5.28).

Since the deflection at midspan for this loading is the maximum for the span, the slope of the elastic curve at the center of the beam is zero; i.e., the tangent is parallel to the undeflected position of the beam. Hence, the deviation of either support from the midspan tangent is equal to the deflection at the center of the beam. Then, by the moment-area theorem [Eq. (5.64)], the deflection y_c is given by the moment about either support of the area of the M/EI diagram included between an ordinate at the center of the beam and that support.

$$y_c = \frac{1}{2}\frac{PL}{4EI}\frac{L}{2}\frac{2}{3}\frac{L}{2} = \frac{PL^3}{48EI}$$

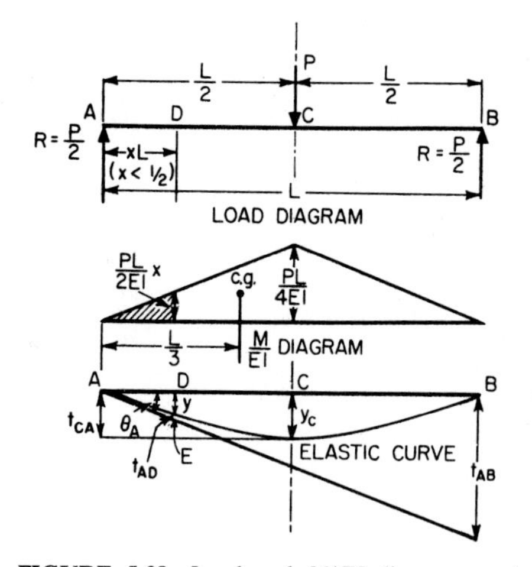

FIGURE 5.28 Load and M/EI diagrams and elastic curve for a simple beam with midspan load.

Suppose, now, the deflection y at any point D at a distance xL from the left support (Fig. 5.28) is to be determined. Referring to the sketch, we note that the distance DE from the undeflected point of D to the tangent to the elastic curve at support A is given by

$$y + t_{AD} = xt_{AB}$$

where t_{AD} is the tangential deviation of D from the tangent at A and t_{AB} is the tangential deviation of B from that tangent. This equation, which is perfectly general for the deflection of any point of a simple beam, no matter how loaded, may be rewritten to give the deflection directly:

$$y = xt_{AB} - t_{AD} \tag{5.65}$$

But t_{AB} is the moment of the area of the M/EI diagram for the whole beam about support B. And t_{AD} is the moment about D of the area of the M/EI diagram included between ordinates at A and D. Hence

$$y = x\frac{1}{2}\frac{PL}{4EI}\frac{L}{2}\left(\frac{2}{3} + \frac{1}{3}\right)L - \frac{1}{2}\frac{PLx}{2EI}xL\frac{xL}{3} = \frac{PL^3}{48EI}x(3 - 4x^2)$$

It is also noteworthy that, since the tangential deviations are very small distances, the slope of the elastic curve at A is given by

$$\theta_A = \frac{t_{AB}}{L} \tag{5.66}$$

This holds, in general, for all simple beams regardless of the type of loading.

The procedure followed in applying Eq. (5.65) to the deflection of the loaded beam in Fig. 5.28 is equivalent to finding the bending moment at D with the

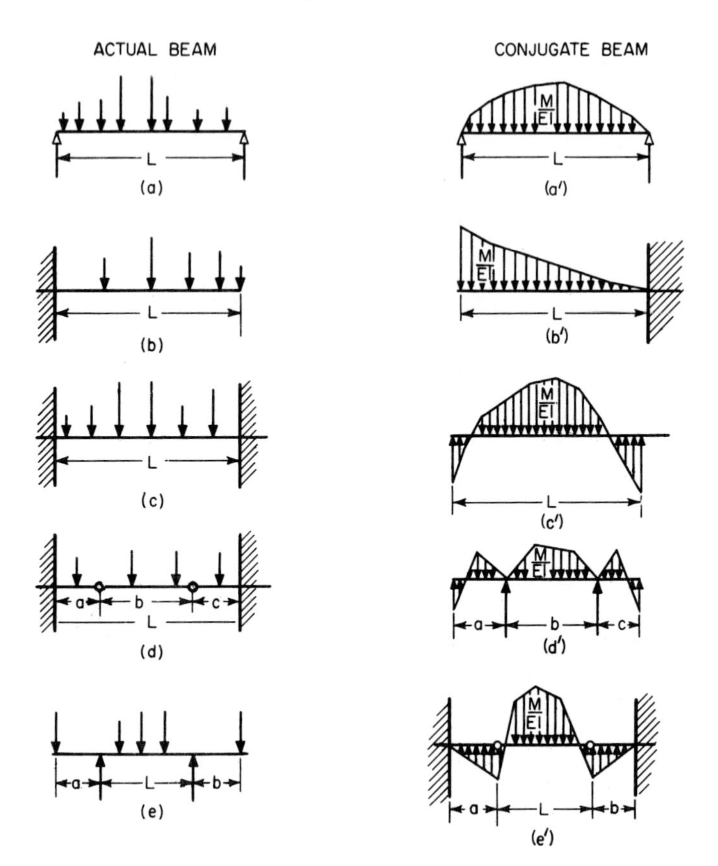

FIGURE 5.29 Various types of beams and corresponding conjugate beams.

M/*EI* diagram serving as the load diagram. The technique of applying the *M*/*EI* diagram as a load and determining the deflection as a bending moment is known as the **conjugate-beam method.**

The conjugate beam must have the same length as the given beam; it must be in equilibrium with the *M*/*EI* load and the reactions produced by the load; and the bending moment at any section must be equal to the deflection of the given beam at the corresponding section. The last requirement is equivalent to requiring that the shear at any section of the conjugate beam with the *M*/*EI* load be equal to the slope of the elastic curve at the corresponding section of the given beam. Figure 5.29 shows the conjugates for various types of beams.

Deflections for several types of loading on simple beams are given in Figs. 5.30 to 5.35 and for overhanging beams and cantilevers in Figs. 5.36 to 5.41.

When a beam carries a number of loads of different types, the most convenient method of computing its deflection generally is to find the deflections separately for the uniform and concentrated loads and add them up.

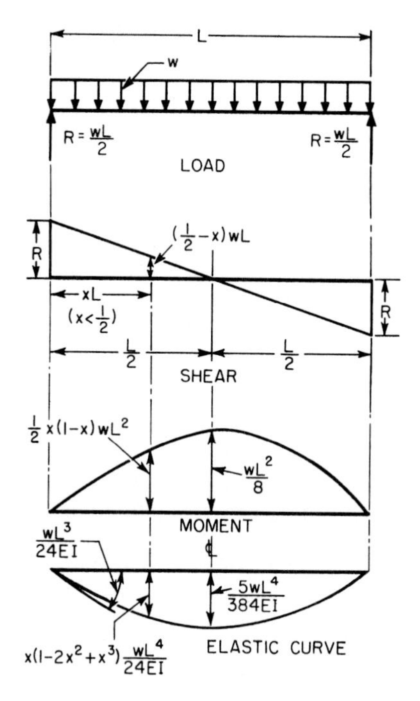

FIGURE 5.30 Uniform load over the whole span of a simple beam.

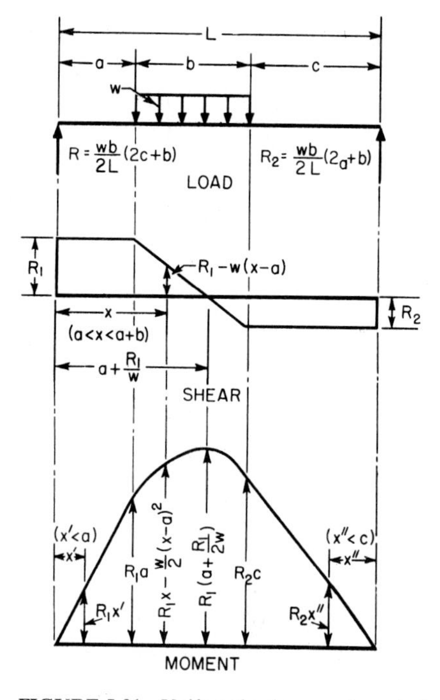

FIGURE 5.31 Uniform load over only part of a simple beam.

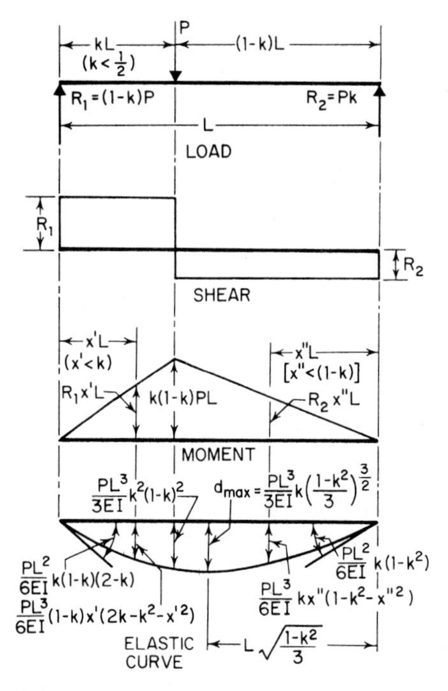

FIGURE 5.32 Concentrated load at any point of a simple beam.

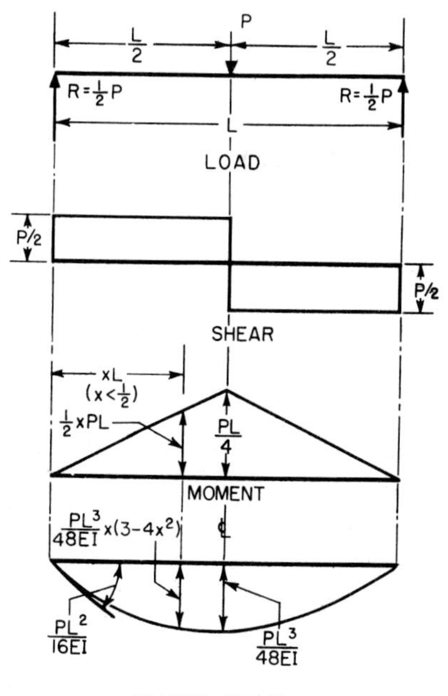

FIGURE 5.33 Concentrated load at midspan of a simple beam.

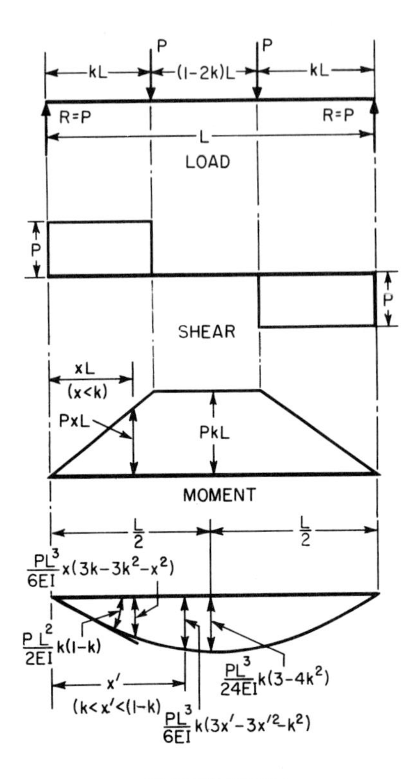

FIGURE 5.34 Two equal concentrated loads on a simple beam.

For several concentrated loads, the easiest solution is to apply the reciprocal theorem (Art. 5.10.5). According to this theorem, if a concentrated load is applied to a beam at a point A, the deflection it produces at point B is equal to the deflection at A for the same load applied at $B(d_{AB} = d_{BA})$.

Suppose, for example, the midspan deflection is to be computed. Then, assume each load in turn applied at the center of the beam and compute the deflection at the point where it originally was applied from the equation of the elastic curve given in Fig. 5.33. The sum of these deflections is the total midspan deflection.

Another method for computing deflections of beams is presented in Art. 5.10.4. This method may also be applied to determining the deflection of a beam due to shear.

5.5.16 Combined Axial and Bending Loads

For stiff beams, subjected to both transverse and axial loading, the stresses are given by the principle of superposition if the deflection due to bending may be neglected without serious error. That is, the total stress is given with sufficient accuracy at any section by the sum of the axial stress and the bending stresses. The maximum stress equals

$$f = \frac{P}{A} + \frac{Mc}{I} \tag{5.67}$$

where P = axial load
A = cross-sectional area
M = maximum bending moment
c = distance from neutral axis to outermost surface at the section where maximum moment occurs
I = moment of inertia of cross section about neutral axis at that section

When the deflection due to bending is large and the axial load produces bending stresses that cannot be neglected, the maximum stress is given by

$$f = \frac{P}{A} + (M + Pd)\frac{c}{I} \tag{5.68}$$

where d is the deflection of the beam. For axial compression, the moment Pd should be given the same sign as M, and for tension, the opposite sign, but the minimum value of $M + Pd$ is zero. The deflection d for axial compression and

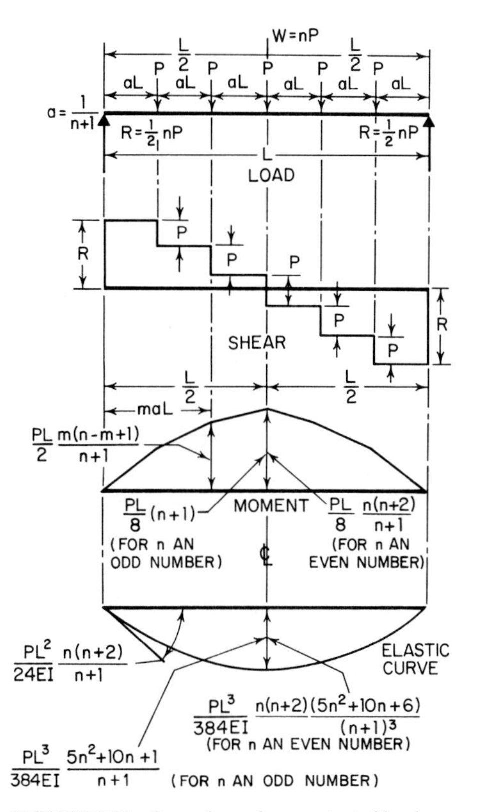

FIGURE 5.35 Several equal concentrated loads on a simple beam.

bending can be obtained by applying Eq. (5.62). (S. Timoshenko and J. M. Gere, "Theory of Elastic Stability," McGraw-Hill Publishing Company, New York; Friedrich Bleich, "Buckling Strength of Metal Structures," McGraw-Hill Publishing Company, New York.) However, it may be closely approximated by

$$d = \frac{d_o}{1 - (P/P_c)} \tag{5.69}$$

where d_o = deflection for the transverse loading alone
P_c = the critical buckling load $\pi^2 EI/L^2$ (see Art. 5.7.2)

5.5.17 Eccentric Loading

An eccentric longitudinal load in the plane of symmetry produces a bending moment Pe where e is the distance of the load from the centroidal axis. The total unit stress is the sum of the stress due to this moment and the stress due to P applied as an axial load:

$$f = \frac{P}{A} \pm \frac{Pec}{I} = \frac{P}{A}\left(1 \pm \frac{ec}{r^2}\right) \tag{5.70}$$

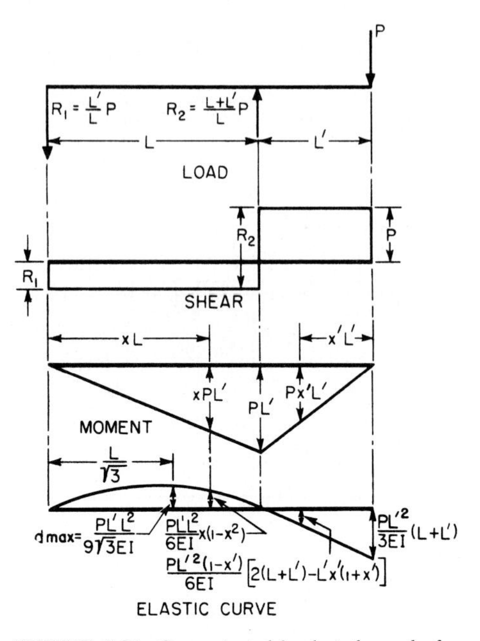

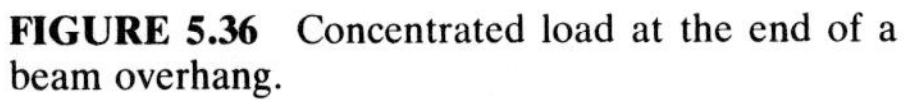

FIGURE 5.36 Concentrated load at the end of a beam overhang.

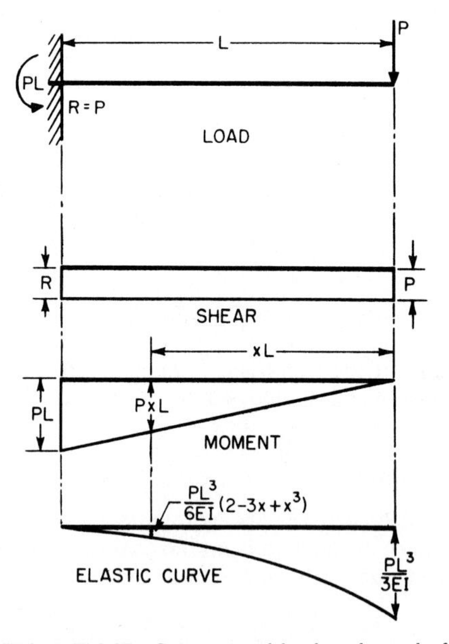

FIGURE 5.37 Concentrated load at the end of a cantilever.

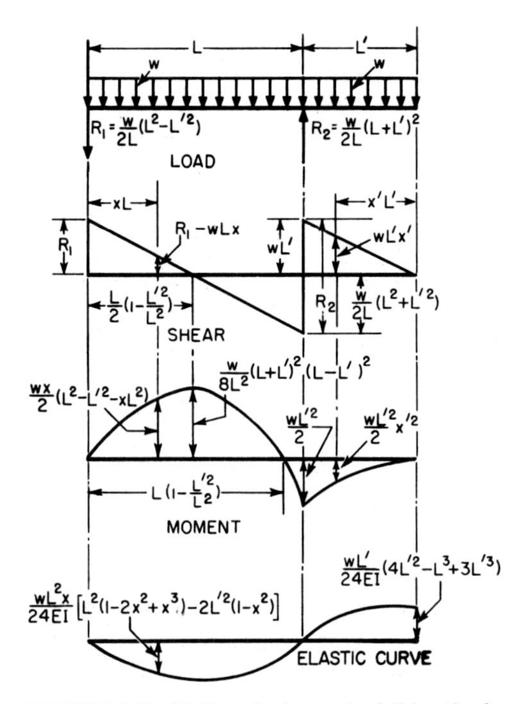

FIGURE 5.38 Uniform load over the full length of a beam with overhang.

where A = cross-sectional area
c = distance from neutral axis to outermost fiber
I = moment of inertia of cross section about neutral axis
r = **radius of gyration,** which is equal to $\sqrt{I/A}$

Figure 5.26 gives values of the radius of gyration for some commonly used cross sections.

For an axial compression load, if there is to be no tension on the cross section, e should not exceed r^2/c. For a rectangular section with width b and depth d, the eccentricity, therefore, should be less than $b/6$ and $d/6$; i.e., the load should not be applied outside the middle third. For a circular cross section with diameter D, the eccentricity should not exceed $D/8$.

When the eccentric longitudinal load produces a deflection too large to be neglected in computing the bending stress, account must be taken of the additional bending moment Pd, where d is the deflection. This deflection may be computed by employing Eq. (5.62) or closely approximated by

$$d = \frac{4eP/P_c}{\pi(1 - P/P_c)} \tag{5.71}$$

P_c is the critical buckling load $\pi^2 EI/L^2$ (see Art. 5.7.2).

If the load P does not lie in a plane containing an axis of symmetry, it produces bending about the two principal axes through the centroid of the cross section.

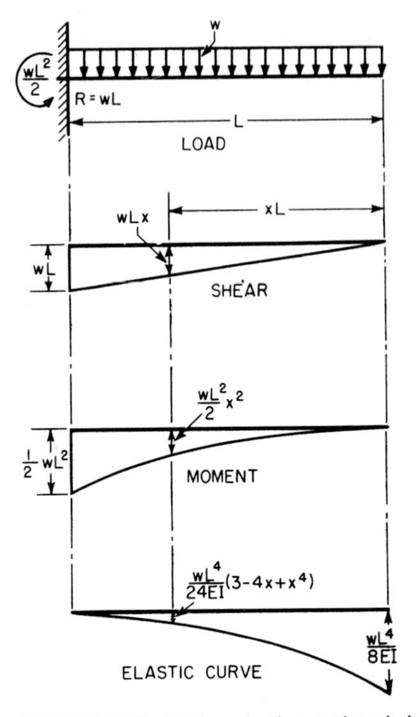

FIGURE 5.39 Uniform load over the whole length of a cantilever.

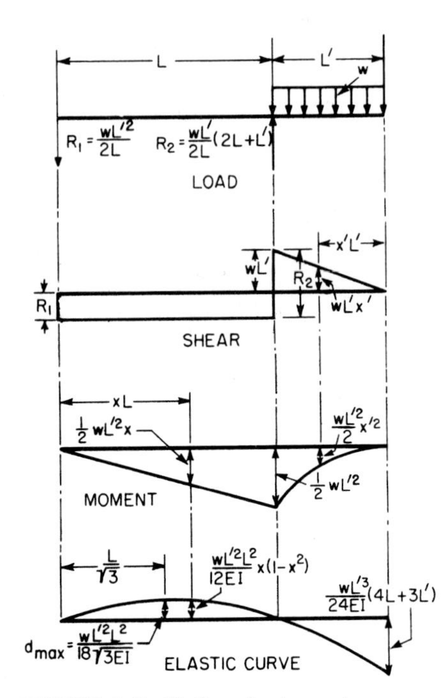

FIGURE 5.40 Uniform load on a beam overhang.

The stresses are given by

$$f = \frac{P}{A} \pm \frac{Pe_xc_x}{I_y} \pm \frac{Pe_yc_y}{I_x} \tag{5.72}$$

where A = cross-sectional area
e_x = eccentricity with respect to principal axis YY
e_y = eccentricity with respect to principal axis XX
c_x = distance from YY to outermost fiber
c_y = distance from XX to outermost fiber
I_x = moment of inertia about XX
I_y = moment of inertia about YY

The principal axes are the two perpendicular axes through the centroid for which the moments of inertia are a maximum or a minimum and for which the products of inertia are zero.

5.5.18 Unsymmetrical Bending

Bending caused by loads that do not lie in a plane containing a principal axis of each cross section of a beam is called unsymmetrical bending. If the bending axis of the beam lies in the plane of the loads, to preclude torsion (see Art. 5.4.1), and if the loads are perpendicular to the bending axis, to preclude axial components, the stress at any point in a cross section is given by

$$f = \frac{M_x y}{I_x} \pm \frac{M_y x}{I_y} \tag{5.73}$$

where M_x = bending moment about principal axis XX
M_y = bending moment about principal axis YY
x = distance from point for which stress is to be computed to YY axis
y = distance from point to XX axis
I_x = moment of inertia of the cross section about XX
I_y = moment of inertia about YY

If the plane of the loads makes an angle θ with a principal plane, the neutral surface will form an angle α with the other principal plane such that

$$\tan \alpha = \frac{I_x}{I_y} \tan \theta \tag{5.74}$$

5.5.19 Beams with Unsymmetrical Sections

In the derivation of the flexure formula $f = Mc/I$ [Eq. (5.54)], the assumption is made that the beam bends, without twisting, in the plane of the loads and that the neutral surface is perpendicular to the plane of the loads. These assumptions are correct for beams with cross sections symmetrical about two axes when the plane of the loads contains one of these axes. They are not necessarily true for beams that are not doubly symmetrical. The reason is that in beams that are doubly symmetrical the bending axis coincides with the centroidal axis, whereas in unsymmetrical sections the two axes may be separate. In the latter case, if the plane of the loads contains the centroidal axis but not the bending axis, the beam will be subjected to both bending and torsion.

The **bending axis** may be defined as the longitudinal line in a beam through which transverse loads must pass to preclude the beam's twisting as it bends. The point in each section through which the bending axis passes is called the **shear center,** or center of twist. The shear center is also the center of rotation of the section in pure torsion (Art. 5.4.1).

Computation of stresses and strains in members subjected to both bending and torsion is complicated, because

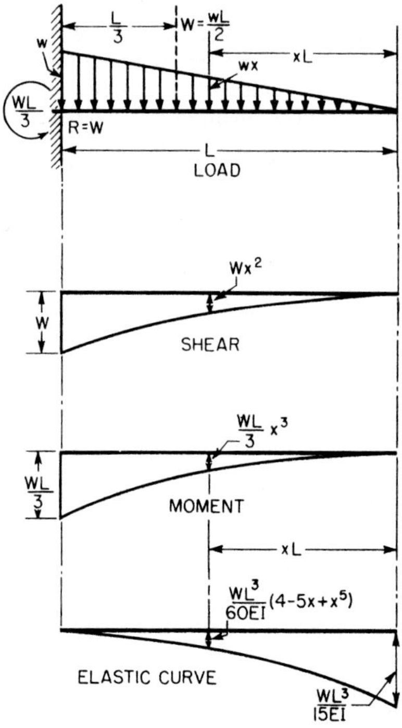

FIGURE 5.41 Triangular loading on a cantilever.

warping of the cross section and buckling effects should be taken into account. Preferably, twisting should be prevented by use of bracing or avoided by selecting appropriate shapes for the members and by locating and directing loads to pass through the bending axis.

(F. Bleich, "Blucking Strength of Metal Structures," McGraw-Hill Publishing Company, New York.)

5.6 CURVED BEAMS

Structural members, such as arches, crane hooks, chain links, and frames of some machines, that have considerable initial curvature in the plane of loading are called curved beams. The flexure formula of Art. 5.5.10, $f = Mc/I$, cannot be applied to them with any reasonable degree of accuracy unless the depth of the beam is small compared with the radius of curvature.

Unlike the condition in straight beams, unit strains in curved beams are not proportional to the distance from the neutral surface, and the centroidal axis does not coincide with the neutral axis. Hence the stress distribution on a section is not linear but more like the distribution shown in Fig. 5.42*c*.

5.6.1 Stresses in Curved Beams

Just as for straight beams, the assumption that plane sections before bending remain plane after bending generally holds for curved beams. So the total strains are proportional to the distance from the neutral axis. But since the fibers are initially of unequal length, the unit strains are a more complex function of this distance. In Fig. 5.42*a*, for example, the bending couples have rotated section AB of the curved beam into section $A'B'$ through an angle $\Delta d\theta$. If ϵ_o is the unit strain at the centroidal axis and ω is the angular unit strain $\Delta d\theta / d\theta$, then the unit strain at a

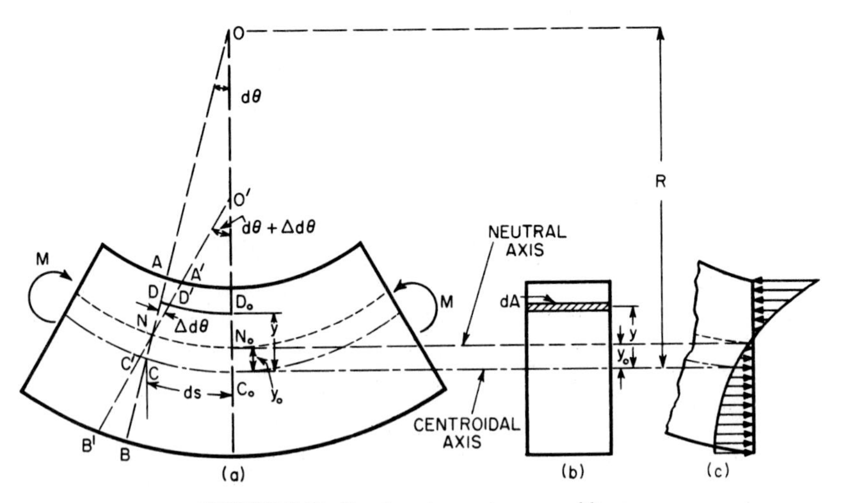

FIGURE 5.42 Bending stresses in a curved beam.

distance y from the centroidal axis (measured positive in the direction of the center of curvature) is

$$\epsilon = \frac{DD'}{DD_o} = \frac{\epsilon_o R\, d\theta - y \Delta d\theta}{(R - y)\, d\theta} = \epsilon_o - (\omega - \epsilon_o)\frac{y}{R - y} \tag{5.75}$$

where R = radius of curvature of centroidal axis.

Equation (5.75) can be expressed in terms of the bending moment if we take advantage of the fact that the sum of the tensile and compressive forces on the section must be zero and the moment of these forces must be equal to the bending moment M. These two equations yield

$$\epsilon_o = \frac{M}{ARE} \qquad \text{and} \qquad \omega = \frac{M}{ARE}\left(1 + \frac{AR^2}{I'}\right) \tag{5.76}$$

where A is the cross-sectional area, E the modulus of elasticity, and

$$I' = \int \frac{y^2\, dA}{1 - y/R} = \int y^2 \left(1 + \frac{y}{R} + \frac{y^2}{R^2} + \cdots\right) dA \tag{5.77}$$

It should be noted that I' is very nearly equal to the moment of inertia I about the centroidal axis when the depth of the section is small compared with R, so that the maximum ratio of y to R is small compared with unity. M is positive when it decreases the radius of curvature.

Since the stress $f = E\epsilon$, we obtain the stresses in the curved beam from Eq. (5.75) by multiplying it by E and substituting ϵ_o and ω from Eq. (5.76):

$$f = \frac{M}{AR} - \frac{My}{I'}\frac{1}{1 - y/R} \tag{5.78}$$

The distance y_o of the neutral axis from the centroidal axis (Fig. 5.42) may be obtained from Eq. (5.78) by setting $f = 0$:

$$y_o = \frac{I'R}{I' + AR^2} \tag{5.79}$$

Since y_o is positive, the neutral axis shifts toward the center of curvature.

5.6.2 Curved Beams with Various Cross Sections

Equation (5.78) for bending stresses in curved beams subjected to end moments in the plane of curvature can be expressed for the inside and outside beam faces in the form:

$$f = K\frac{Mc}{I} \tag{5.80}$$

where c = distance from the centroidal axis to the inner or outer surface. Table 5.4 gives values of K calculated from Eq. (5.78) for circular, elliptical, and rectangular cross sections.

If Eq. (5.78) is applied to I or T beams or tubular members, it may indicate circumferential flange stresses that are much lower than will actually occur. The

TABLE 5.4 Values of K for Curved Beams

Section	$\frac{R}{c}$	K Inside face	K Outside face	y_o
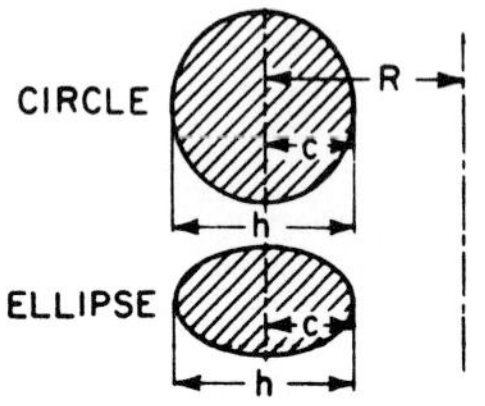 CIRCLE; ELLIPSE	1.2	3.41	0.54	$0.224R$
	1.4	2.40	0.60	$0.151R$
	1.6	1.96	0.65	$0.108R$
	1.8	1.75	0.68	$0.084R$
	2.0	1.62	0.71	$0.069R$
	3.0	1.33	0.79	$0.030R$
	4.0	1.23	0.84	$0.016R$
	6.0	1.14	0.89	$0.0070R$
	8.0	1.10	0.91	$0.0039R$
	10.0	1.08	0.93	$0.0025R$
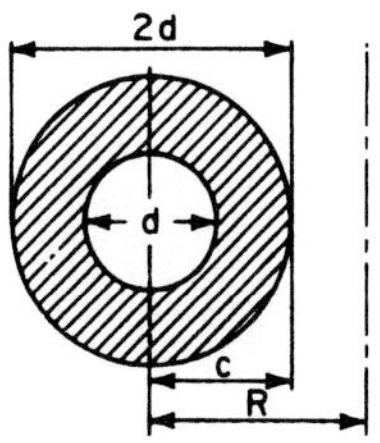	1.2	3.28	0.58	$0.269R$
	1.4	2.31	0.64	$0.182R$
	1.6	1.89	0.68	$0.134R$
	1.8	1.70	0.71	$0.104R$
	2.0	1.57	0.73	$0.083R$
	3.0	1.31	0.81	$0.038R$
	4.0	1.21	0.85	$0.020R$
	6.0	1.13	0.90	$0.0087R$
	8.0	1.10	0.92	$0.0049R$
	10.0	1.07	0.93	$0.0031R$
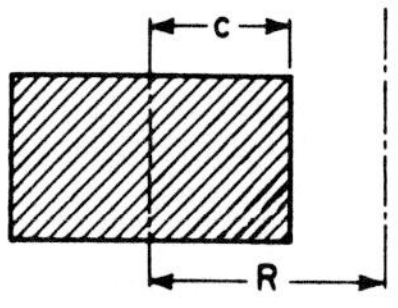	1.2	2.89	0.57	$0.305R$
	1.4	2.13	0.63	$0.204R$
	1.6	1.79	0.67	$0.149R$
	1.8	1.63	0.70	$0.112R$
	2.0	1.52	0.73	$0.090R$
	3.0	1.30	0.81	$0.041R$
	4.0	1.20	0.85	$0.021R$
	6.0	1.12	0.90	$0.0093R$
	8.0	1.09	0.92	$0.0052R$
	10.0	1.07	0.94	$0.0033R$

error is due to the fact that the outer edges of the flanges deflect radially. The effect is equivalent to having only part of the flanges active in resisting bending stresses. Also, accompanying the flange deflections, there are transverse bending stresses in the flanges. At the junction with the web, these reach a maximum, which may be greater than the maximum circumferential stress. Furthermore, there are radial stresses (normal stresses acting in the direction of the radius of curvature) in the web that also may have maximum values greater than the maximum circumferential stress.

A good approximation to the stresses in I or T beams is as follows: for circumferential stresses, Eq. (5.78) may be used with a modified cross section, which is obtained by using a reduced flange width. The reduction is calculated from $b' =$

αb, where b is the length of the portion of the flange projecting on either side from the web, b' is the corrected length, and α is a correction factor determined from equations developed by H. Bleich. α is a function of b^2/rt, where t is the flange thickness and r the radius of the center of the flange:

$b^2/rt =$	0.5	0.7	1.0	1.5	2	3	4	5
$\alpha =$	0.9	0.6	0.7	0.6	0.5	0.4	0.37	0.33

When the parameter b^2/rt is greater than 1.0, the maximum transverse bending stress is approximately equal to 1.7 times the stress obtained at the center of the flange from Eq. (5.78) applied to the modified section. When the parameter equals 0.7, that stress should be multiplied by 1.5, and when it equals 0.4, the factor is 1.0. In Eq. (5.78), I' for I beams may be taken for this calculation approximately equal to

$$I' = I\left(1 + \frac{c^2}{R^2}\right) \tag{5.81}$$

where I = moment of inertia of modified section about its centroidal axis
R = radius of curvature of centroidal axis
c = distance from centroidal axis to center of the more sharply curved flange

Because of the high stress factor, it is advisable to stiffen or brace curved I-beam flanges.

The maximum radial stress will occur at the junction of web and flange of I beams. If the moment is negative, that is, if the loads tend to flatten out the beam, the radial stress is tensile, and there is a tendency for the more sharply curved flange to pull away from the web. An approximate value of this maximum stress is

$$f_r = -\frac{A_f}{A}\frac{M}{t_w c_g r'} \tag{5.82}$$

where f_r = radial stress at junction of flange and web of a symmetrical I beam
A_f = area of one flange
A = total cross-sectional area
M = bending moment
t_w = thickness of web
c_g = distance from centroidal axis to center of flange
r' = radius of curvature of inner face of more sharply curved flange

(A. P. Boresi, O. Sidebottom, F. B. Seely, and J. O. Smith, "Advanced Mechanics of Materials," John Wiley & Sons, Inc., New York.)

5.6.3 Axial and Bending Loads on Curved Beams

If a curved beam carries an axial load P as well as bending loads, the maximum unit stress is

$$f = \frac{P}{A} \pm \frac{Mc}{I}K \tag{5.83}$$

where K is a correction factor for the curvature [see Eq. (5.80)]. The sign of M is taken positive in this equation when it increases the curvature, and P is positive when it is a tensile force, negative when compressive.

5.6.4 Slope and Deflection of Curved Beams

If we consider two sections of a curved beam separated by a differential distance ds (Fig. 5.42), the change in angle $\Delta d\theta$ between the sections caused by a bending moment M and an axial load P may be obtained from Eq. (5.76), noting that $d\theta = ds/R$.

$$\Delta d\theta = \frac{M\,ds}{EI'}\left(1 + \frac{I'}{AR^2}\right) + \frac{P\,ds}{ARE} \tag{5.84}$$

where E is the modulus of elasticity, A the cross-sectional area, R the radius of curvature of the centroidal axis, and I' is defined by Eq. (5.77).

If P is a tensile force, the length of the centroidal axis increases by

$$\Delta ds = \frac{P\,ds}{AE} + \frac{M\,ds}{ARE} \tag{5.85}$$

The effect of curvature on shearing deformations for most practical applications is negligible.

For shallow sections (depth of section less than about one-tenth the span), the effect of axial forces on deformations may be neglected. Also, unless the radius of curvature is very small compared with the depth, the effect of curvature may be ignored. Hence, for most practical applications, Eq. (5.84) may be used in the simplified form:

$$\Delta d\theta = \frac{M\,ds}{EI} \tag{5.86}$$

For deeper beams, the action of axial forces, as well as bending moments, should be taken into account; but unless the curvature is sharp, its effect on deformations may be neglected. So only Eq. (5.86) and the first term in Eq. (5.85) need be used. (S. Timoshenko and D. H. Young, "Theory of Structures," McGraw-Hill Publishing Company, New York.) See also Arts. 5.14.1 to 5.14.3.

5.7 BUCKLING OF COLUMNS

Columns are compression members whose cross-sectional dimensions are relatively small compared with their length in the direction of the compressive force. Failure of such members occurs because of instability when a certain axial load P_c (called critical or **Euler load**) is equaled or exceeded. The member may bend, or buckle, suddenly and collapse.

Hence the strength P of a column is not determined by the unit stress in Eq. (5.21) ($P = Af$) but by the maximum load it can carry without becoming unstable. The condition of instability is characterized by disproportionately large increases in lateral deformation with slight increase in axial load. Instability may occur in slender columns before the unit stress reaches the elastic limit.

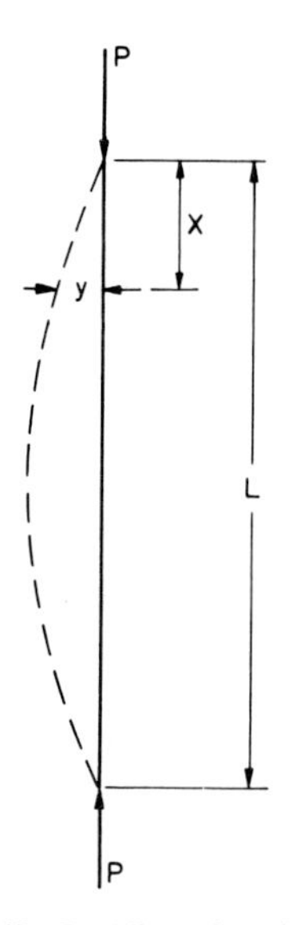

FIGURE 5.43 Buckling of a pin-ended long column.

5.7.1 Stable Equilibrium

Consider, for example, an axially loaded column with ends unrestrained against rotation, shown in Fig. 5.43. If the member is initially perfectly straight, it will remain straight as long as the load P is less than the critical load P_c. If a small transverse force is applied, the column will deflect, but it will return to the straight position when this force is removed. Thus, when P is less than P_c, internal and external forces are in stable equilibrium.

5.7.2 Unstable Equilibrium

If $P = P_c$ and a small transverse force is applied, the column again will deflect, but this time, when the force is removed, the column will remain in the bent position (dashed line in Fig. 5.43). The equation of this elastic curve can be obtained from Eq. (5.62):

$$EI\frac{d^2y}{dx^2} = -P_c y \tag{5.87}$$

in which E = modulus of elasticity
I = least moment of inertia
y = deflection of the bent member from the straight position at a distance x from one end

This assumes, of course, that the stresses are within the elastic limit. Solution of Eq. (5.87) gives the smallest value of the Euler load as

$$P_c = \frac{\pi^2 EI}{L^2} \tag{5.88}$$

Equation (5.88) indicates that there is a definite finite magnitude of an axial load that will hold a column in equilibrium in the bent position when the stresses are below the elastic limit. Repeated application and removal of small transverse forces or small increases in axial load above this critical load will cause the member to fail by buckling. Internal and external forces are in a state of unstable equilibrium.

It is noteworthy that the Euler load, which determines the load-carrying capacity of a column, depends on the stiffness of the member, as expressed by the modulus of elasticity, rather than on the strength of the material of which it is made.

By dividing both sides of Eq. (5.88) by the cross-sectional area A and substituting r^2 for I/A (r is the radius of gyration of the section), we can write the solution of Eq. (5.87) in terms of the average unit stress on the cross section:

$$\frac{P_c}{A} = \frac{\pi^2 E}{(L/r)^2} \tag{5.89}$$

This holds only for the elastic range of buckling; i.e., for values of the slenderness ratio L/r above a certain limiting value that depends on the properties of the material. For inelastic buckling, see Art. 5.7.4.

5.7.3 Effect of End Conditions

Equation (5.89) was derived on the assumption that the ends of the column are free to rotate. It can be generalized, however, to take into account the effect of end conditions:

$$\frac{P_c}{A} = \frac{\pi^2 E}{(kL/r)^2} \tag{5.90}$$

where k is the factor that depends on the end conditions. For a pin-ended column, $k = 1$; for a column with both ends fixed, $k = \frac{1}{2}$; for a column with one end fixed and one end pinned, k is about 0.7; and for a column with one end fixed and one end free from all restraint, $k = 2$.

5.7.4 Inelastic Buckling

Equations (5.88) and (5.90) are derived from Eq. (5.87), the differential equation for the elastic curve. They are based on the assumption that the critical average stress is below the elastic limit when the state of unstable equilibrium is reached. In members with slenderness ratio L/r below a certain limiting value, however, the elastic limit is exceeded before the column buckles. As the axial load approaches the critical load, the modulus of elasticity varies with the stress. Hence Eqs. (5.88) and (5.90), based on the assumption that E is a constant, do not hold for these short columns.

After extensive testing and analysis, prevalent engineering opinion favors the Engesser equation for metals in the inelastic range:

$$\frac{P_t}{A} = \frac{\pi^2 E_t}{(kL/r)^2} \tag{5.91}$$

This differs from Eqs. (5.88) to (5.90) only in that the tangent modulus E_t (the actual slope of the stress-strain curve for the stress P_t/A) replaced the modulus of elasticity E in the elastic range. P_t is the smallest axial load for which two equilibrium positions are possible, the straight position and a deflected position.

5.7.5 Column Curves

Curves obtained by plotting the critical stress for various values of the slenderness ratio are called column curves. For axially loaded, initially straight columns, the column curve consists of two parts: (1) the Euler critical values, and (2) the Engesser, or tangent-modulus critical values.

The latter are greatly affected by the shape of the stress-strain curve for the material of which the column is made, as shown in Fig. 5.44. The stress-strain curve for a material, such as an aluminum alloy or high-strength steel, which does not have a sharply defined yield point, is shown in Fig. 5.44*a*. The corresponding

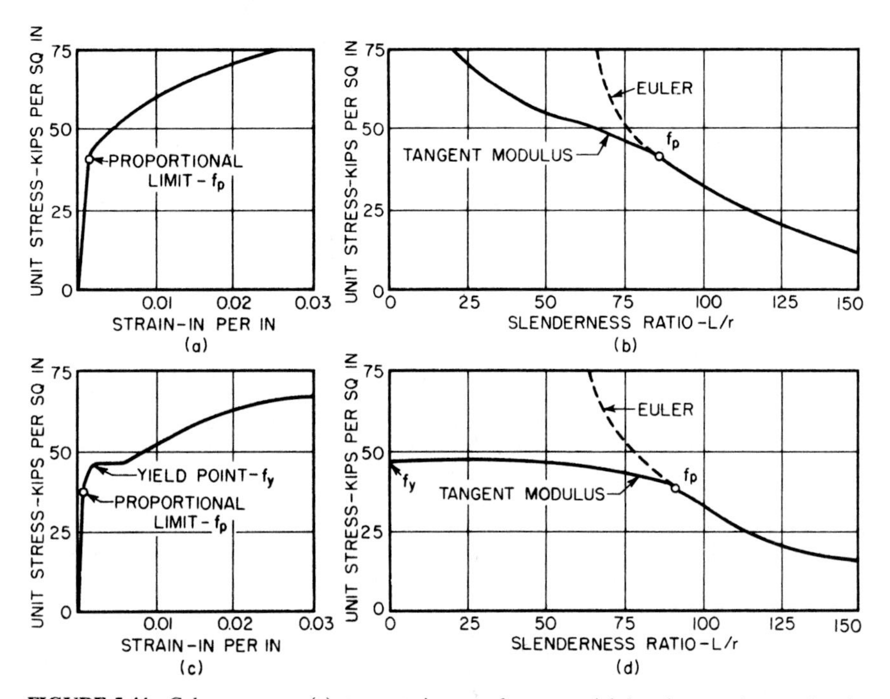

FIGURE 5.44 Column curves: (*a*) stress-strain curve for a material that does not have a sharply defined yield point; (*b*) column curve for this material; (*c*) stress-strain curve for a material with a sharply defined yield point; (*d*) column curve for that material.

column curve is drawn in Fig. 5.44*b*. In contrast, Fig. 5.44*c* presents the stress-strain curve for structural steel, with a sharply defined point, and Fig. 5.44*d* the related column curve. This curve becomes horizontal as the critical stress approaches the yield strength of the material and the tangent modulus becomes zero, whereas the column curve in Fig. 5.44*b* continues to rise with decreasing values of the slenderness ratio.

Examination of Fig. 5.44*d* also indicates that slender columns, which fall in the elastic range, where the column curve has a large slope, are very sensitive to variations in the factor k, which represents the effect of end conditions. On the other hand, in the inelastic range, where the column curve is relatively flat, the critical stress is relatively insensitive to changes in k. Hence the effect of end conditions on the stability of a column is of much greater significance for long columns than for short columns.

5.7.6 Local Buckling

A column may not only fail by buckling of the member as a whole but as an alternative, by buckling of one of its components. Hence, when members like I beams, channels, and angles are used as columns or when sections are built up of plates, the possibility of the critical load on a component (leg, half flange, web, lattice bar) being less than the critical load on the column as a whole should be investigated.

Similarly, the possibility of buckling of the compression flange or the web of a beam should be looked into.

Local buckling, however, does not always result in a reduction in the load-carrying capacity of a column. Sometimes, it results in a redistribution of the stresses enabling the member to carry additional load.

5.7.7 Behavior of Actual Columns

For many reasons, columns in structures behave differently from the ideal column assumed in deriving Eqs. (5.88) and (5.91). A major consideration is the effect of accidental imperfections, such as nonhomogeneity of materials, initial crookedness, and unintentional eccentricities of the axial load, since neither field nor shopwork can be perfect. These and the effects of residual stresses usually are taken into account by a proper choice of safety factor.

There are other significant conditions, however, that must be considered in any design rule: continuity in frame structures and eccentricity of the axial load. Continuity affects column action in two ways. The restraint at column ends determines the value of k, and bending moments are transmitted to the column by adjoining structural members.

Because of the deviation of the behavior of actual columns from the ideal, columns generally are designed by empirical formulas. Separate equations usually are given for short columns, intermediate columns, and long columns. For specific materials—steel, concrete, timber—these formulas are given in Secs. 7 to 10.

For more details on column action, see F. Bleich, "Buckling Strength of Metal Structures," McGraw-Hill Publishing Company, New York, 1952; S. Timoshenko and J. M. Gere, "Theory of Elastic Stability," McGraw-Hill Publishing Company, New York, 1961; and T. V. Galambos, "Guide to Stability Design Criteria for Metal Structures," 4th ed., John Wiley & Sons, Inc., Somerset, N.J., 1988.

5.8 GRAPHIC-STATICS FUNDAMENTALS

A force may be represented by a straight line of fixed length. The length of line to a given scale represents the magnitude of the force. The position of the line parallels the line of action of the force. And an arrowhead on the line indicates the direction in which the force acts.

Forces are concurrent when their lines of action meet. If they lie in the same plane, they are coplanar.

5.8.1 Parallelogram of Forces

The resultant of several forces is a single force that would produce the same effect on a rigid body. The resultant of two concurrent forces is determined by the parallelogram law:

If a parallelogram is constructed with two forces as sides, the diagonal represents the resultant of the forces (Fig. 5.45*a*).

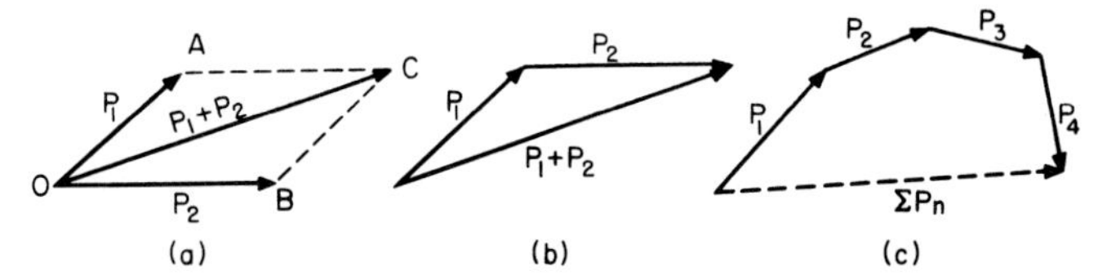

FIGURE 5.45 Addition of forces by (*a*) parallelogram law; (*b*) triangle construction; (*c*) polygon construction.

The **resultant** is said to be equal to the sum of the forces, sum here meaning, of course, addition by the parallelogram law. Subtraction is carried out in the same manner as addition, but the direction of the force to be subtracted is reversed.

If the direction of the resultant is reversed, it becomes the **equilibrant,** a single force that will hold the two given forces in equilibrium.

5.8.2 Resolution of Forces

To resolve a force into two components, a parallelogram is drawn with the force as a diagonal. The sides of the parallelogram represent the components. The procedure is: (1) Draw the given force. (2) From both ends of the force draw lines parallel to the directions in which the components act. (3) Draw the components along the parallels through the origin of the given force to the intersections with the parallels through the other end. Thus, in Fig. 5.45*a,* P_1 and P_2 are the components in directions OA and OB of the force represented by OC.

5.8.3 Force Polygons

Examination of Fig. 5.45*a* indicates that a step can be saved in adding the two forces. The same resultant could be obtained by drawing only the upper half of the parallelogram. Hence, to add two forces, draw the first force; then draw the second force beginning at the end of the first one. The resultant is the force drawn from the origin of the first force to the end of the second force, as shown in Fig. 5.45*b*. Again, the equilibrant is the resultant with direction reversed.

From this diagram, an important conclusion can be drawn: **If three forces meeting at a point are in equilibrium, they will form a closed force triangle.**

The conclusions reached for addition of two forces can be generalized for several concurrent forces: To add several forces, P_1, P_2, P_3, . . . , P_n, draw P_2 from the end of P_1, P_3 from the end of P_2, etc. The force required to close the force polygon is the resultant (Fig. 5.45*c*).

If a group of concurrent forces are in equilibrium, they will form a closed force polygon.

5.9 *ROOF TRUSSES*

A truss is a coplanar system of structural members joined together at their ends to form a stable framework. If small changes in the lengths of the members due to loads are neglected, the relative positions of the joints cannot change.

5.9.1 Characteristics of Trusses

Three bars pinned together to form a triangle represent the simplest type of truss. Some of the more common types of roof trusses are shown in Fig. 5.46.

The top members are called the upper chord; the bottom members, the lower chord; and the verticals and diagonals, web members.

The purpose of roof trusses is to act like big beams, to support the roof covering over long spans. They not only have to carry their own weight and the weight of the roofing and roof beams, or purlins, but cranes, wind loads, snow loads, suspended ceilings, and equipment, and a live load to take care of construction, maintenance, and repair loading. These loads are applied at the intersection of the members, or panel points, so that the members will be subjected principally to axial stresses—tension or compression.

Methods of computing stresses in trusses are presented in Arts. 5.9.3 and 5.9.4. A method of computing truss deflections is described in Art. 5.10.4.

5.9.2 Bow's Notation

For simple designation of loads and stresses, capital letters are placed in the spaces between truss members and between forces. Each member and load is then designated by the letters on opposite sides of it. For example, in Fig. 5.47*a,* the upper

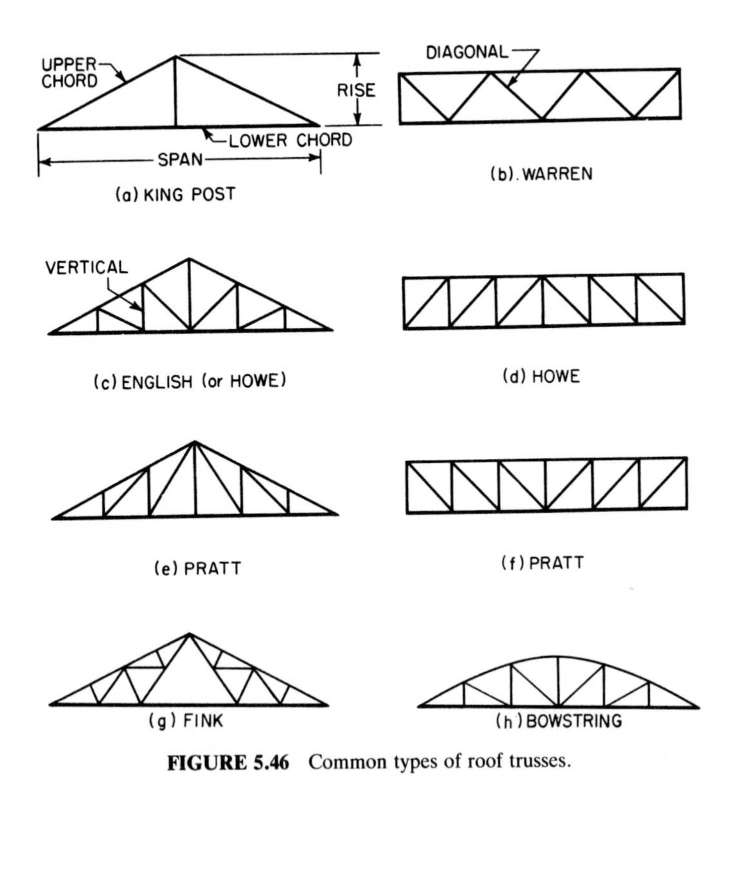

FIGURE 5.46 Common types of roof trusses.

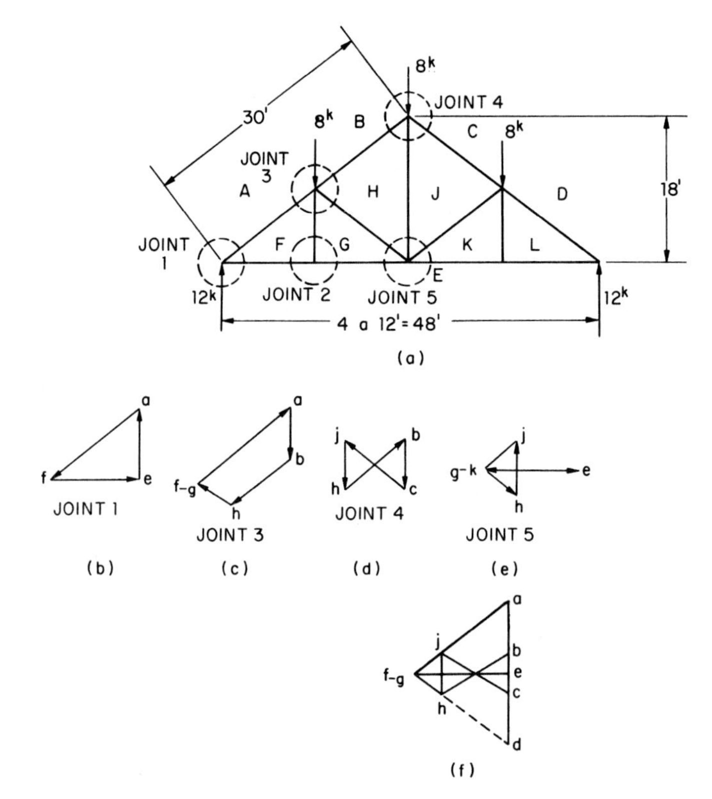

FIGURE 5.47 Method of joints applied to the roof truss shown in (*a*). Stresses in members at each joint are determined graphically in succession (*b*) to (*e*).

chord members are *AF, BH, CJ,* and *DL*. The loads are *AB, BC,* and *CD,* and the reactions are *EA* and *DE*. Stresses in the members generally are designated by the same letters but in lowercase.

5.9.3 Method of Joints

A useful method for determining the stresses in truss members is to select sections that isolate the joints one at a time and then apply the laws of equilibrium to each. Considering the stresses in the cut members as external forces, the sum of the horizontal components of the forces acting at a joint must be zero, and so must be the sum of the vertical components. Since the lines of action of all the forces are known, we can therefore compute two unknown magnitudes at each joint by this method. The procedure is to start at a joint that has only two unknowns (generally at the support) and then, as stresses in members are determined, analyze successive joints.

Let us, for illustration, apply the method to joint 1 of the truss in Fig. 5.47*a*. Equating the sum of the vertical components to zero, we find that the vertical component of the top chord must be equal and opposite to the reaction, 12 kips

(12,000 lb). The stress in the top chord at this joint, then, must be a compression equal to $12 \times 30/18 = 20$ kips. From the fact that the sum of the horizontal components must be zero, we find that the stress in the bottom chord at the joint must be equal and opposite to the horizontal component of the top chord. Hence the stress in the bottom chord must be a tension equal to $20 \times 24/30 = 16$ kips.

Moving to joint 2, we note that, with no vertical loads at the joint, the stress in the vertical is zero. Also, the stress is the same in both bottom chord members at the joint, since the sum of the horizontal components must be zero.

Joint 3 now contains only two unknown stresses. Denoting the truss members and the loads by the letters placed on opposite sides of them, as indicated in Fig. 5.47*a*, the unknown stresses are S_{BH} and S_{HG}. The laws of equilibrium enable us to write the following two equations, one for the vertical components and the second for the horizontal components:

$$\Sigma V = 0.6S_{FA} - 8 - 0.6S_{BH} + 0.6S_{HG} = 0$$

$$\Sigma H = 0.8S_{FA} - 0.8S_{BH} - 0.8S_{HG} = 0$$

Both unknown stresses are assumed to be compressive; i.e., acting toward the joint. The stress in the vertical does not appear in these equations, because it was already determined to be zero. The stress in *FA*, S_{FA}, was found from analysis of joint 1 to be 20 kips. Simultaneous solution of the two equations yields $S_{HG} = 6.7$ kips and $S_{BH} = 13.3$ kips. (If these stresses had come out with a negative sign, it would have indicated that the original assumption of their directions was incorrect; they would, in that case, be tensile forces instead of compressive forces.) See also Art. 5.9.4.

All the force polygons in Fig. 5.47 can be conveniently combined into a single stress diagram. The combination (Fig. 5.47*f*) is called a **Maxwell diagram.**

5.9.4 Method of Sections

An alternative method to that described in Art. 5.9.3 for determining the stresses in truss members is to isolate a portion of the truss by a section so chosen as to cut only as many members with unknown stresses as can be evaluated by the laws of equilibrium applied to that portion of the truss. The stresses in the cut members are treated as external forces. Compressive forces act toward the panel point and tensile forces away from the joint.

Suppose, for example, we wish to find the stress in chord *AB* of the truss in Fig. 5.48*a*. We can take a vertical section *XX* close to panel point *A*. This cuts not only *AB* but *AD* and *ED* as well. The external 10-kip (10,000-lb) loading and 25-kip reaction at the left are held in equilibrium by the compressive force *C* in *AB*, tensile force *T* in *ED*, and tensile force *S* in *AD* (Fig. 5.48*b*). The simplest way to find *C* is to take moments about *D*, the point of intersection of *S* and *T*, eliminating these unknowns from the calculation.

$$-9C + 36 \times 25 - 24 \times 10 - 12 \times 10 = 0$$

from which *C* is found to be 60 kips.

Similarly, to find the stress in *ED*, the simplest way is to take moments about *A*, the point of intersection of *S* and *C*:

$$-9T + 24 \times 25 - 12 \times 10 = 0$$

from which *T* is found to be 53.3 kips.

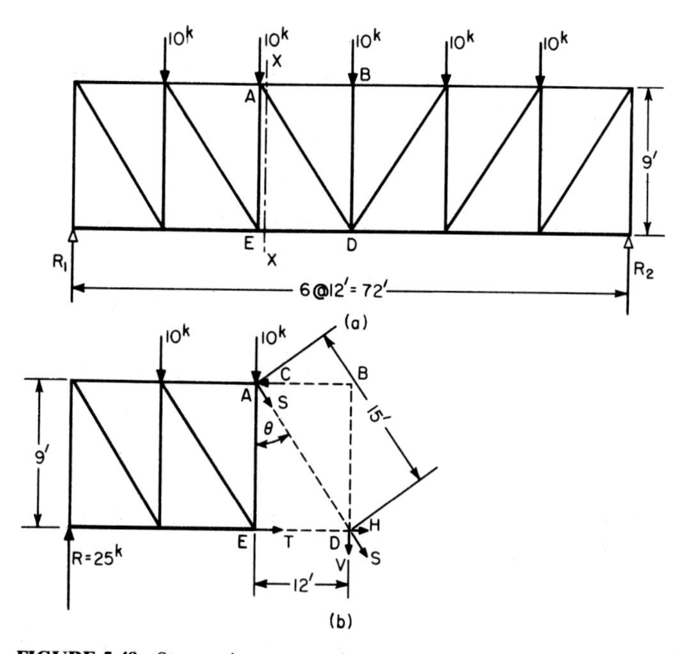

FIGURE 5.48 Stresses in truss members cut by section *XX*, shown in (*a*), are determined by method of sections (*b*).

On the other hand, the stress in *AD* can be easily determined by two methods. One takes advantage of the fact that *AB* and *ED* are horizontal members, requiring *AD* to carry the full vertical shear at section *XX*. Hence we know that the vertical component V of $S = 25 - 10 - 10 = 5$ kips. Multiplying V by sec θ (Fig. 5.48*b*), which is equal to the ratio of the length of *AD* to the rise of the truss ($^{15}/_{9}$), S is found to be 8.3 kips. The second method—presented because it is useful when the chords are not horizontal—is to resolve S into horizontal and vertical components at D and take moments about E. Since both T and the horizontal component of S pass through E, they do not appear in the computations, and C already has been computed. Equating the sum of the moments to zero gives $V = 5$, as before.

Some trusses are complex and require special methods of analysis. (Norris et al., "Elementary Structural Analysis," 4th ed., McGraw-Hill Book Company, New York).

5.10 GENERAL TOOLS FOR STRUCTURAL ANALYSIS

For some types of structures, the equilibrium equations are not sufficient to determine the reactions or the internal stresses. These structures are called **statically indeterminate.**

For the analysis of such structures, additional equations must be written on the basis of a knowledge of the elastic deformations. Hence methods of analysis that enable deformations to be evaluated in terms of unknown forces or stresses are important for the solution of problems involving statically indeterminate structures. Some of these methods, like the method of virtual work, are also useful in solving complicated problems involving statically determinate systems.

5.10.1 Virtual Work

A virtual displacement is an imaginary small displacement of a particle consistent with the constraints upon it. Thus, at one support of a simply supported beam, the virtual displacement could be an infinitesimal rotation $d\theta$ of that end but not a vertical movement. However, if the support is replaced by a force, then a vertical virtual displacement may be applied to the beam at that end.

Virtual work is the product of the distance a particle moves during a virtual displacement by the component in the direction of the displacement of a force acting on the particle. If the displacement and the force are in opposite directions, the virtual work is negative. When the displacement is normal to the force, no work is done.

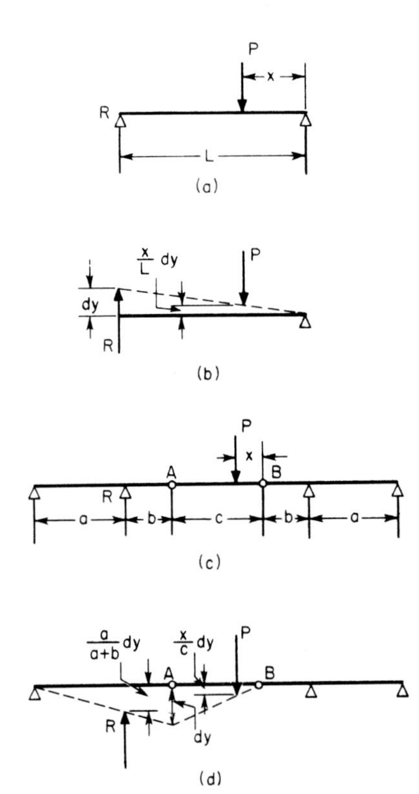

FIGURE 5.49 Principle of virtual work applied to determination of a simple-beam reaction (*a*) and (*b*) and to the reaction of a beam with a suspended span (*c*) and (*d*).

Suppose a rigid body is acted upon by a system of forces with a resultant R. Given a virtual displacement ds at an angle α with R, the body will have virtual work done on it equal to $R \cos \alpha\, ds$. (No work is done by internal forces. They act in pairs of equal magnitude but opposite direction, and the virtual work done by one force of a pair is equal but opposite in sign to the work done by the other force.) If the body is in equilibrium under the action of the forces, then $R = 0$ and the virtual work also is zero.

Thus, the principle of virtual work may be stated: **If a rigid body in equilibrium is given a virtual displacement, the sum of the virtual work of the forces acting on it must be zero.**

As an example of how the principle may be used to find a reaction of a statically determinate beam, consider the simple beam in Fig. 5.49*a*, for which the reaction R is to be determined. First, replace the support by an unknown force R. Next, move that end of the beam upward a small amount dy as in Fig. 5.49*b*. The displacement under the load P will be $x\, dy/L$, upward. Then, by the principle of virtual work, $R\, dy - Px\, dy/L = 0$, from which $R = Px/L$.

The principle may also be used to find the reaction R of the more complex beam in Fig. 5.49*c*. The first step again is to replace the support by an unknown force R. Next, apply a virtual downward displacement dy at hinge A (Fig. 5.49d). Displacement under load P is $x\, dy/c$, and at the reaction R, $a\, dy/(a + b)$. According to the principle of virtual work, $-Ra\, dy/(a + b) + Px\, dy/c = 0$, from which reaction $R = Px(a + b)/ac$. In this type of problem, the method has the advantage that only one reaction need be considered at a time and internal forces are not involved.

5.10.2 Strain Energy

When an elastic body is deformed, the virtual work done by the internal forces is equal to the corresponding increment of the strain energy dU, in accordance with the principle of virtual work.

Assume a constrained elastic body acted upon by forces $P_1, P_2, \ldots$, for which the corresponding deformations are $e_1, e_2 \ldots$. Then, $\Sigma P_n\, de_n = dU$. The increment of the strain energy due to the increments of the deformations is given by

$$dU = \frac{\partial U}{\partial e_1} de_1 + \frac{\partial U}{\partial e_2} de_2 + \cdots$$

In solving a specific problem, a virtual displacement that is most convenient in simplifying the solution should be chosen. Suppose, for example, a virtual displacement is selected that affects only the deformation e_n corresponding to the load P_n, other deformations being unchanged. Then, the principle of virtual work requires that

$$P_n\, de_n = \frac{\partial U}{\partial e_n} de_n$$

This is equivalent to

$$\frac{\partial U}{\partial e_n} = P_n \tag{5.92}$$

which states that the partial derivative of the strain energy with respect to any specific deformation gives the corresponding force.

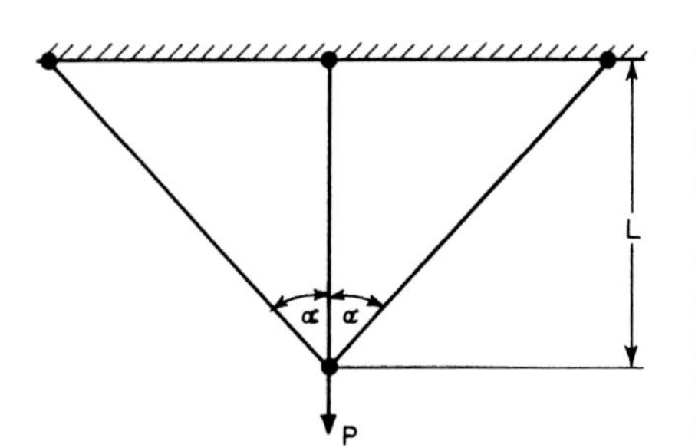

FIGURE 5.50 Statically indeterminate truss.

Suppose, for example, the stress in the vertical bar in Fig. 5.50 is to be determined. All bars are made of the same material and have the same cross section. If the vertical bar stretches an amount e under the load P, the inclined bars will each stretch an amount $e \cos \alpha$. The strain energy in the system is [from Eq. (5.30)]

$$U = \frac{AE}{2L}(e^2 + 2e^2 \cos^3 \alpha)$$

and the partial derivative of this with respect to e must be equal to P; that is

$$P = \frac{AE}{2L}(2e + 4e\cos^3\alpha)$$

$$= \frac{AEe}{L}(1 + 2\cos^3\alpha)$$

Noting that the force in the vertical bar equals AEe/L, we find from the above equation that the required stress equals $P/(1 + 2\cos^3\alpha)$.

Castigliano's Theorems. It can also be shown that, if the strain energy is expressed as a function of statically independent forces, the partial derivative of the strain energy with respect to one of the forces gives the deformation corresponding to that force. (See Timoshenko and Young, "Theory of Structures," McGraw-Hill Publishing Company, New York.)

$$\frac{\partial U}{\partial P_n} = e_n \tag{5.93}$$

This is known as Castigliano's first theorem. (His second theorem is the principle of least work.)

5.10.3 Method of Least Work

If displacement of a structure is prevented, as at a support, the partial derivative of the strain energy with respect to that supporting force must be zero, according to Castigliano's first theorem. This establishes his second theorem:

The strain energy in a statically indeterminate structure is the minimum consistent with equilibrium.

As an example of the use of the method of least work, we shall solve again for the stress in the vertical bar in Fig. 5.50. Calling this stress X, we note that the stress in each of the inclined bars must be $(P - X)/2\cos\alpha$. With the aid of Eq. (5.30), we can express the strain energy in the system in terms of X as

$$U = \frac{X^2L}{2AE} + \frac{(P - X)^2L}{4AE\cos^3\alpha}$$

Hence, the internal work in the system will be a minimum when

$$\frac{\partial U}{\partial X} = \frac{XL}{AE} - \frac{(P - X)L}{2AE\cos^3\alpha} = 0$$

Solving for X gives the stress in the vertical bar as $P/(1 + 2\cos^3\alpha)$, as before (Art. 5.10.1).

5.10.4 Dummy Unit-Load Method

In Art. 5.2.7, the strain energy for pure bending was given as $U = M^2L/2EI$ in Eq. (5.33). To find the strain energy due to bending stress in a beam, we can apply this equation to a differential length dx of the beam and integrate over the entire span. Thus,

$$U = \int_0^L \frac{M^2\,dx}{2EI} \tag{5.94}$$

If M represents the bending moment due to a generalized force P, the partial derivative of the strain energy with respect to P is the deformation d corresponding to P. Differentiating Eq. (5.94) under the integral sign gives

$$d = \int_0^L \frac{M}{EI}\frac{\partial M}{\partial P}\,dx \tag{5.95}$$

The partial derivative in this equation is the rate of change of bending moment with the load P. It is equal to the bending moment m produced by a unit generalized load applied at the point where the deformation is to be measured and in the direction of the deformation. Hence, Eq. (5.95) can also be written

$$d = \int_0^L \frac{Mm}{EI}\,dx \tag{5.96}$$

To find the vertical deflection of a beam, we apply a vertical dummy unit load at the point where the deflection is to be measured and substitute the bending moments due to this load and the actual loading in Eq. (5.96). Similarly, to compute a rotation, we apply a dummy unit moment.

Beam Deflections. As a simple example, let us apply the dummy unit-load method to the determination of the deflection at the center of a simply supported, uniformly loaded beam of constant moment of inertia (Fig. 5.51*a*). As indicated in Fig. 5.51*b*, the bending moment at a distance x from one end is $(wL/2)x - (w/2)x^2$. If we apply a dummy unit load vertically at the center of the beam (Fig. 5.51*c*), where the vertical deflection is to be determined, the moment at x is $x/2$, as indicated in Fig. 5.51*d*. Substituting in Eq. (5.96) and taking advantage of the symmetry of the loading gives

$$d = 2\int_0^{L/2}\left(\frac{wL}{2}x - \frac{w}{2}x^2\right)\frac{x}{2}\frac{dx}{EI} = \frac{5wL^4}{384EI}$$

Beam End Rotations. As another example, let us apply the method to finding the end rotation at one end of a simply supported, prismatic beam produced by a moment applied at the other end. In other words, the problem is to find the end rotation at B, θ_B, in Fig. 5.52*a*, due to M_A. As indicated in Fig. 5.52*b*, the bending moment at a distance x from B caused by M_A is M_Ax/L. If we applied a dummy unit moment at B (Fig. 5.52*c*), it would produce a moment at x of $(L - x)/L$ (Fig. 5.52*d*). Substituting in Eq. (5.96) gives

$$\theta_B = \int_0^L M_A \frac{x}{L}\frac{L - x}{L}\frac{dx}{EI} = \frac{M_AL}{6EI}$$

Shear Deflections. To determine the deflection of a beam caused by shear, Castigliano's theorems can be applied to the strain energy in shear

$$V = \int\int \frac{v^2}{2G}\,dA\,dx$$

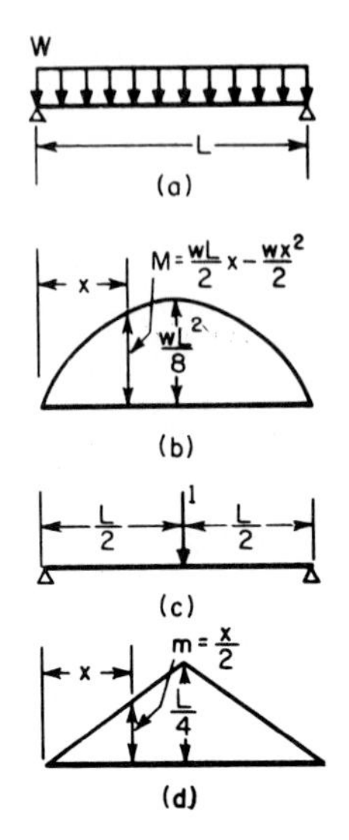

FIGURE 5.51 Dummy unit-load method applied to a uniformly loaded, simple beam (*a*) to find midspan deflection; (*b*) moment diagram for the uniform load; (*c*) unit load at midspan; (*d*) moment diagram for the unit load.

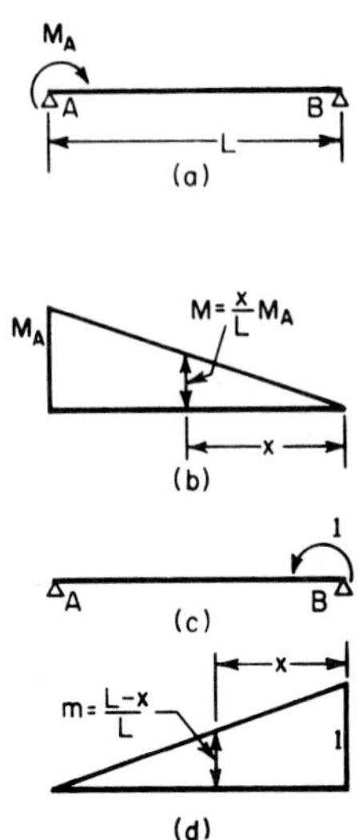

FIGURE 5.52 End rotation of a simple beam due to an end moment: (*a*) by dummy unit-load method; (*b*) moment diagram for the end moment; (*c*) unit moment applied at beam end; (*d*) moment diagram for the unit moment.

where v = shearing unit stress
G = modulus of rigidity
A = cross-sectional area

Truss Deflections. The dummy unit-load method may also be adapted for the determination of the deformation of trusses. As indicated by Eq. (5.30), the strain energy in a truss is given by

$$U = \sum \frac{S^2L}{2AE} \tag{5.97}$$

which represents the sum of the strain energy for all the members of the truss. S is the stress in each member caused by the loads. Applying Castigliano's first theorem and differentiating inside the summation sign yield the deformation:

$$d = \sum \frac{SL}{AE}\frac{\partial S}{\partial P} \tag{5.98}$$

The partial derivative in this equation is the rate of change of axial stress with the load P. It is equal to the axial stress u in each bar of the truss produced by a unit load applied at the point where the deformation is to be measured and in the direction of the deformation. Consequently, Eq. (5.98) can also be written

$$d = \sum \frac{Sul}{AE} \tag{5.99}$$

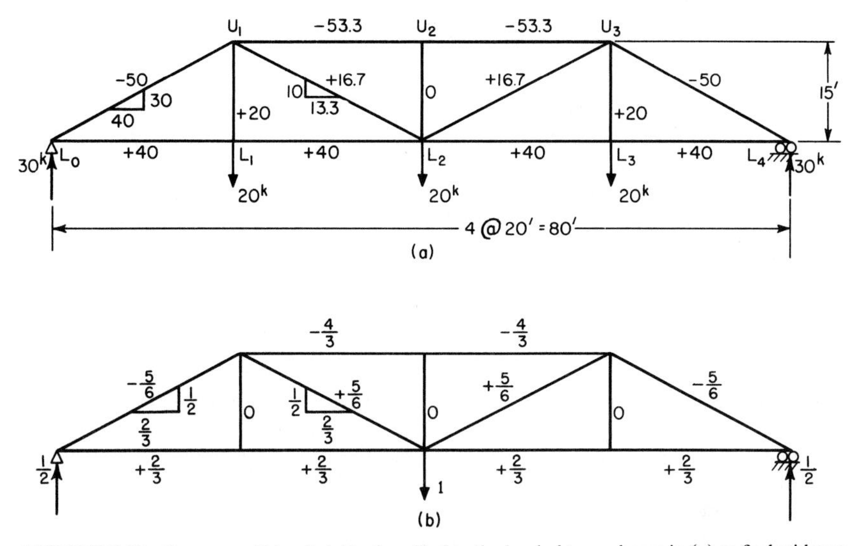

FIGURE 5.53 Dummy unit-load method applied to the loaded truss shown in (*a*) to find midspan deflection; (*b*) unit load applied at midspan.

To find the deflection of a truss, apply a vertical dummy unit load at the panel point where the deflection is to be measured and substitute in Eq. (5.99) the stresses in each member of the truss due to this load and the actual loading. Similarly, to find the rotation of any joint, apply a dummy unit moment at the joint, compute the stresses in each member of the truss, and substitute in Eq. (5.99). When it is necessary to determine the relative movement of two panel points, apply dummy unit loads in opposite directions at those points.

It is worth noting that members that are not stressed by the actual loads or the dummy loads do not enter into the calculation of a deformation.

As an example of the application of Eq. (5.99), let us compute the deflection of the truss in Fig. 5.53. The stresses due to the 20-kip load at each panel point are shown in Fig. 5.53*a,* and the ratio of length of members in inches to their cross-sectional area in square inches is given in Table 5.5. We apply a vertical dummy unit load at L_2, where the deflection is required. Stresses u due to this load are shown in Fig. 5.53*b* and Table 5.5.

The computations for the deflection are given in Table 5.5. Members not stressed by the 20-kip loads or the dummy unit load are not included. Taking advantage of

TABLE 5.5 Deflection of a Truss

Member	L/A	S	u	SuL/A
L_0L_2	160	+40	+⅔	4,267
L_0U_1	75	−50	−⅚	3,125
U_1U_2	60	−53.3	−4/3	4,267
U_1L_2	150	+16.7	+⅚	2,083
				13,742

the symmetry of the truss, we tabulate the values for only half the truss and double the sum.

$$d = \frac{SuL}{AE} = \frac{2 \times 13{,}742{,}000}{30{,}000{,}000} = 0.916 \text{ in}$$

Also, to reduce the amount of calculation, we do not include the modulus of elasticity E, which is equal to 30,000,000, until the very last step, since it is the same for all members.

5.10.5 Reciprocal Theorem and Influence Lines

Consider a structure loaded by a group of independent forces A, and suppose that a second group of forces B are added. The work done by the forces A acting over the displacements due to B will be W_{AB}.

Now, suppose the forces B had been on the structure first, and then load A had been applied. The work done by the forces B acting over the displacements due to A will be W_{BA}.

The reciprocal theorem states that $W_{AB} = W_{BA}$.

Some very useful conclusions can be drawn from this equation. For example, there is the reciprocal deflection relationship: **The deflection at a point *A* due to a load at *B* is equal to the deflection at *B* due to the same load applied at *A*. Also, the rotation at *A* due to a load (or moment) at *B* is equal to the rotation at *B* due to the same load (or moment) applied at *A*.**

Another consequence is that deflection curves may also be influence lines to some scale for reactions, shears, moments, or deflections **(Muller-Breslau principle).** (Influence lines are defined in Art. 5.5.8.) For example, suppose the influence line for a reaction is to be found; that is, we wish to plot the reaction R as a unit load moves over the structure, which may be statically indeterminate. For the loading condition A, we analyze the structure with a unit load on it at a distance x from some reference point. For loading condition B, we apply a dummy unit vertical load upward at the place where the reaction is to be determined, deflecting the structure off the support. At a distance x from the reference point, the displacement in d_{xR} and over the support the displacement is d_{RR}. Hence $W_{AB} = -1(d_{xR}) + Rd_{RR}$. On the other hand, W_{BA} is zero, since loading condition A provides no displacement for the dummy unit load at the support in condition B. Consequently, from the reciprocal theorem,

$$R = \frac{d_{xR}}{d_{RR}}$$

Since d_{RR} is a constant, R is proportional to d_{xR}. Hence the influence line for a reaction can be obtained from the deflection curve resulting from a displacement of the support (Fig. 5.54). The magnitude of the reaction is obtained by dividing each ordinate of the deflection curve by the displacement of the support.

Similarly, the influence line for shear can be obtained from the deflection curve produced by cutting the structure and shifting the cut ends vertically at the point for which the influence line is desired (Fig. 5.55).

The influence line for bending moment can be obtained from the deflection curve produced by cutting the structure and rotating the cut ends at the point for which the influence line is desired (Fig. 5.56).

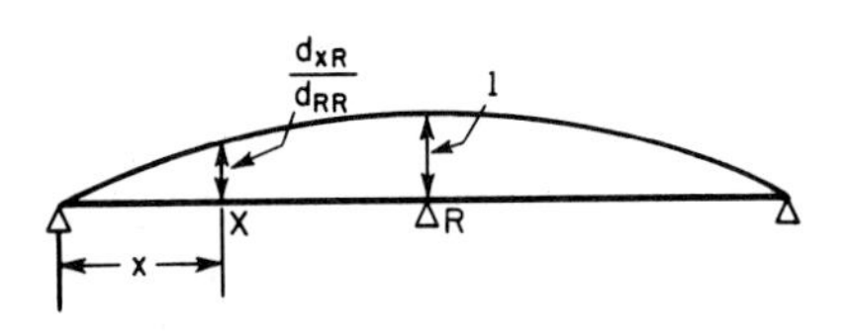

FIGURE 5.54 Reaction-influence line for a continuous beam.

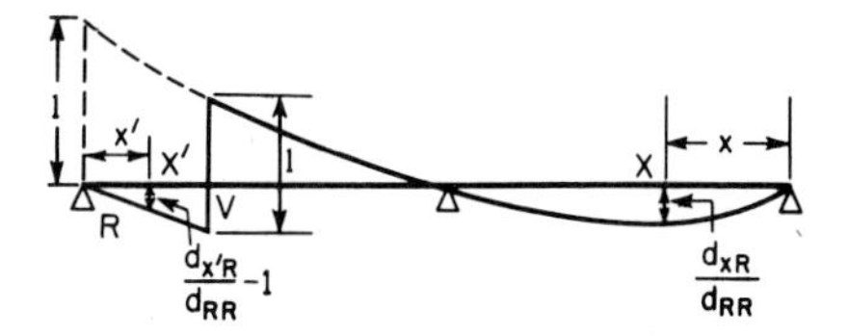

FIGURE 5.55 Shear-influence line for a continuous beam.

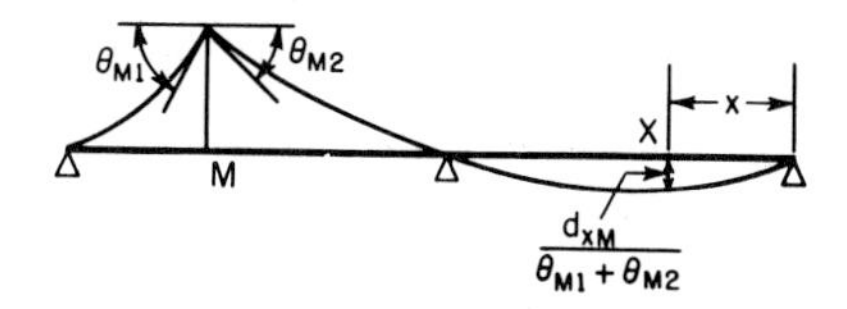

FIGURE 5.56 Moment-influence line for a continuous beam.

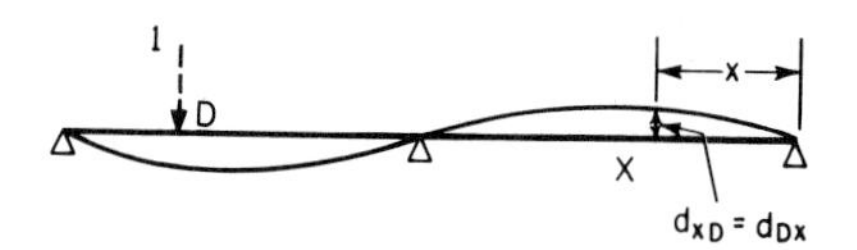

FIGURE 5.57 Deflection-influence line for a continuous beam.

And finally, it may be noted that the deflection curve for a load of unity at some point of a structure is also the influence line for deflection at that point (Fig. 5.57).

5.10.6 Superposition Methods

The principle of superposition applies when the displacement (deflection or rotation) of every point of a structure is directly proportional to the applied loads. The principle states that the displacement at each point caused by several loads equals the sum of the displacements at the point when the loads are applied to the structure individually in any sequence. Also, the bending moment (or shear) at every point induced by applied loads equals the sum of the bending moments (or shears) induced at the point by the loads applied individually in any sequence.

The principle holds for linearly elastic structures, for which unit stresses are proportional to unit strains, when displacements are very small and calculations can be based on the undeformed configuration of the structure without significant error.

As a simple example, consider a bar with length L and cross-sectional area A loaded with n axial loads $P_1, P_2 \ldots P_n$. Let F equal the sum of the loads. From Eq. (5.23), F causes an elongation $\delta = FL/AE$, where E is the modulus of elasticity of the bar. According to the principle of superposition, if e_1 is the elongation caused by P_1 alone, e_2 by P_2 alone, . . . and e_n by P_n alone, then regardless of the sequence in which the loads are applied, when all the loads are on the bar,

$$\delta = e_1 + e_2 + \cdots + e_n$$

This simple case can be easily verified by substituting $e_1 = P_1L/AE$, $e_2 = P_2L/AE$, . . ., and $e_n = P_nL/AE$ in this equation and noting that $F = P_1 + P_2 + \cdots + P_n$:

$$\delta = \frac{P_1L}{AE} + \frac{P_2L}{AE} + \cdots + \frac{P_nL}{AE} = (P_1 + P_2 + \cdots + P_n)\frac{L}{AE} = \frac{FL}{AE}$$

In the preceding equations, L/AE represents the elongation induced by a unit load and is called the **flexibility** of the bar.

The reciprocal, AE/L, represents the force that causes a unit elongation and is called the **stiffness** of the bar.

Analogous properties of beams, columns, and other structural members and the principle of superposition are useful in analysis of many types of structures. Calculation of stresses and displacements of statically indeterminate structures, for example, often can be simplified by resolution of bending moments, shears, and displacments into components chosen to supply sufficient equations for the solution from requirements for equilibrium of forces and compatibility of displacements.

Consider the continuous beam *ALRBC* shown in Fig. 5.58*a*. Under the loads shown, member *LR* is subjected to end moments M_L and M_R (Fig. 5.58*b*) that are initially unknown. The bending-moment diagram for *LR* for these end moments is shown at the left in Fig. 5.58*c*. If these end moments were known, *LR* would

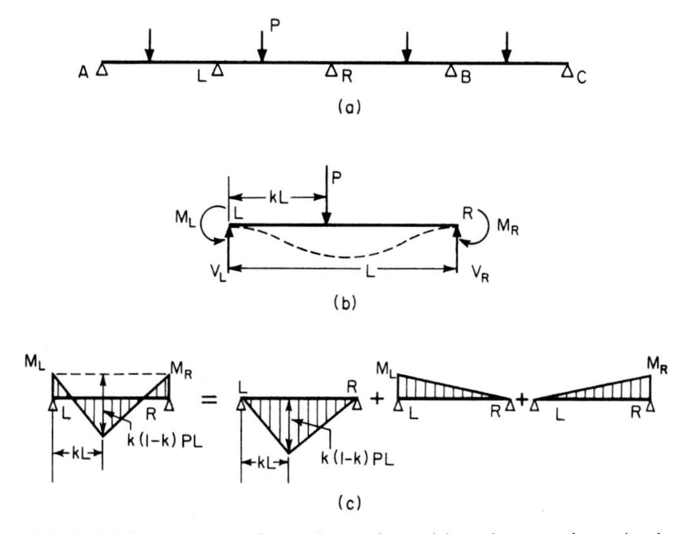

FIGURE 5.58 Any span of a continuous beam (*a*) can be treated as a simple beam, as shown in (*b*) and (*c*). In (*c*), the moment diagram is resolved into basic components.

be statically determinate; that is, *LR* could be treated as a simply supported beam subjected to known end moments M_L and M_R. The analysis can be further simplified by resolution of the bending-moment diagram into the three components shown to the right of the equal sign in Fig. 5.58*c*. This example leads to the following conclusion:

The bending moment at any section of a span *LR* of a continuous beam or frame equals the simple-beam moment due to the applied loads, plus the simple-beam moment due to the end moment at *L*, plus the simple-beam moment due to the end moment at *R*.

When the moment diagrams for all the spans of *ALRBC* in Fig. 5.58 have been resolved into components so that the spans may be treated as simple beams, all the end moments (moments at supports) can be determined from two basic requirements:

1. The sum of the moments at every support equals zero.
2. The end rotation (angular change at the support) of each member rigidly connected at the support is the same.

5.10.7 Influence-Coefficient Matrices

A matrix is a rectangular array of numbers in rows and columns that obeys certain mathematical rules known generally as matrix algebra and matrix calculus. A matrix consisting of only a single column is called a **vector.** In this book, matrices and vectors are represented by boldfaced letters and their elements by lightface symbols, with appropriate subscripts. It often is convenient to use numbers for the subscripts to indicate the position of an element in the matrix. Generally, the first digit indicates the row and the second digit the column. Thus, in matrix **A,** A_{23} represents the element in the second row and third column:

$$\mathbf{A} = \begin{bmatrix} A_{11} & A_{12} & A_{13} \\ A_{21} & A_{22} & A_{23} \\ A_{31} & A_{32} & A_{33} \end{bmatrix} \tag{5.100}$$

Methods based on matrix representations often are advantageous for structural analysis and design of complex structures. One reason is that matrices provide a compact means of representing and manipulating large quantities of numbers. Another reason is that computers can perform matrix operations automatically and speedily. Computer programs are widely available for this purpose.

Matrix Equations. Matrix notation is especially convenient in representing the solution of simultaneous linear equations, which arise frequently in structural analysis. For example, suppose a set of equations is represented in matrix notation by $\mathbf{AX} = \mathbf{B}$, where $\mathbf{X}$ is the vector of variables $X_1, X_2, \ldots, X_n$, $\mathbf{B}$ is the vector of the constants on the right-hand side of the equations, and $\mathbf{A}$ is a matrix of the coefficients of the variables. Multiplication of both sides of the equation by $\mathbf{A}^{-1}$, the inverse of $\mathbf{A}$, yields $\mathbf{A}^{-1}\mathbf{AX} = \mathbf{A}^{-1}\mathbf{B}$. Since $\mathbf{A}^{-1}\mathbf{A} = \mathbf{I}$, the identity matrix, and $\mathbf{IX} = \mathbf{X}$, the solution of the equations is represented by $\mathbf{X} = \mathbf{A}^{-1}\mathbf{B}$. The matrix inversion $\mathbf{A}^{-1}$ can be readily performed by computers. For large matrices, however, it often is more practical to solve the equations, for example, by the Gaussian procedure of eliminating one unknown at a time.

In the application of matrices to structural analysis, loads and displacements are considered applied at the intersection of members (joints, or nodes). The loads may be resolved into moments, torques, and horizontal and vertical components. These may be assembled for each node into a vector and then all the node vectors may be combined into a force vector $\mathbf{P}$ for the whole structure.

$$\mathbf{P} = \begin{bmatrix} P_1 \\ P_2 \\ \vdots \\ P_n \end{bmatrix} \tag{5.101}$$

Similarly, displacement corresponding to those forces may be resolved into rotations, twists, and horizontal and vertical components and assembled for the whole structure into a vector $\mathbf{\Delta}$.

$$\mathbf{\Delta} = \begin{bmatrix} \Delta_1 \\ \Delta_2 \\ \vdots \\ \Delta_n \end{bmatrix} \tag{5.102}$$

If the structure meets requirements for application of the principle of superposition (Art. 5.10.6) and forces and displacements are arranged in the proper sequence, the vectors of forces and displacements are related by

$$\mathbf{P} = \mathbf{K\Delta} \tag{5.103a}$$

$$\mathbf{\Delta} = \mathbf{FP} \tag{5.103b}$$

where $\mathbf{K}$ = stiffness matrix of the whole structure
$\mathbf{F}$ = flexibility matrix of the whole structure = $\mathbf{K}^{-1}$

The stiffness matrix $\mathbf{K}$ transforms displacements into loads. The flexibility matrix $\mathbf{F}$ transforms loads into displacements. The elements of $\mathbf{K}$ and $\mathbf{F}$ are functions of material properties, such as the modulus of elasticity; geometry of the structure; and sectional properties of members of the structure, such as area and moment of inertia. $\mathbf{K}$ and $\mathbf{F}$ are square matrices; that is, the number of rows in each equals the number of columns. In addition, both matrices are symmetrical; that is, in each matrix, the columns and rows may be interchanged without changing the matrix. Thus, $K_{ij} = K_{ji}$, and $F_{ij} = F_{ji}$, where i indicates the row in which an element is located and j the column.

Influence Coefficients. Elements of the stiffness and flexibility matrices are influence coefficients. Each element is derived by computing the displacements (or forces) occurring at nodes when a unit displacement (or force) is imposed at one node, while all other displacements (or forces) are taken as zero.

Let Δ_i be the ith element of matrix $\mathbf{\Delta}$. Then, a typical element F_{ij} of $\mathbf{F}$ gives the displacement of a node i in the direction of Δ_i when a unit force acts at a node j in the direction of force P_j and no other forces are acting on the structure. The jth column of $\mathbf{F}$, therefore, contains all the nodal displacements induced by a unit force acting at node j in the direction of P_j.

Similarly, let P_i be the ith element of matrix $\mathbf{P}$. Then, a typical element K_{ij} of $\mathbf{K}$ gives the force at a node i in the direction of P_i when a node j is given a unit displacement in the direction of displacement Δ_j and no other displacements are permitted. The jth column of $\mathbf{K}$, therefore, contains all the nodal forces caused by a unit displacement of node j in the direction of Δ_j.

Application to a Beam. A general method for determining the forces and moments in a continuous beam is as follows: Remove as many redundant supports or members as necessary to make the structure statically determinant. Compute for the actual loads the deflections or rotations of the statically determinate structure in the direction of the unknown forces and couples exerted by the removed supports and members. Then, in terms of these forces and couples, treated as variables, compute the corresponding deflections or rotations the forces and couples produce in the statically determinate structure (see Arts. 5.5.16 and 5.10.4). Finally, for each redundant support or member write equations that give the known rotations or deflections of the original structure in terms of the deformations of the statically determinate structure.

For example, one method of finding the reactions of the continuous beam AC in Fig. 5.59*a* is to remove supports 1, 2, and 3 temporarily. The beam is now simply supported between A and C, and the reactions and moments can be computed from the laws of equilibrium. Beam AC deflects at points 1, 2, and 3, whereas we know that the continuous beam is prevented from deflecting at these points by the supports there. This information enables us to write three equations in terms of the three unknown reactions that were eliminated to make the beam statically determinate.

To determine the equations, assume that nodes exist at the location of the supports 1, 2, and 3. Then, for the actual loads, compute the vertical deflections d_1, d_2, and d_3 of simple beam AC at nodes 1, 2, and 3, respectively (Fig. 5.59*b*). Next, form two vectors, $\mathbf{d}$ with elements d_1, d_2, d_3 and $\mathbf{R}$ with the unknown reactions R_1 at node 1, R_2 at node 2, and R_3 at node 3 as elements. Since the beam may be assumed to be linearly elastic, set $\mathbf{d} = \mathbf{FR}$, where $\mathbf{F}$ is the flexibility matrix for simple beam AC. The elements y_{ij} of $\mathbf{F}$ are influence coefficients. To determine them, calculate column 1 of $\mathbf{F}$ as the deflections y_{11}, y_{21}, and y_{31} at nodes 1, 2, and

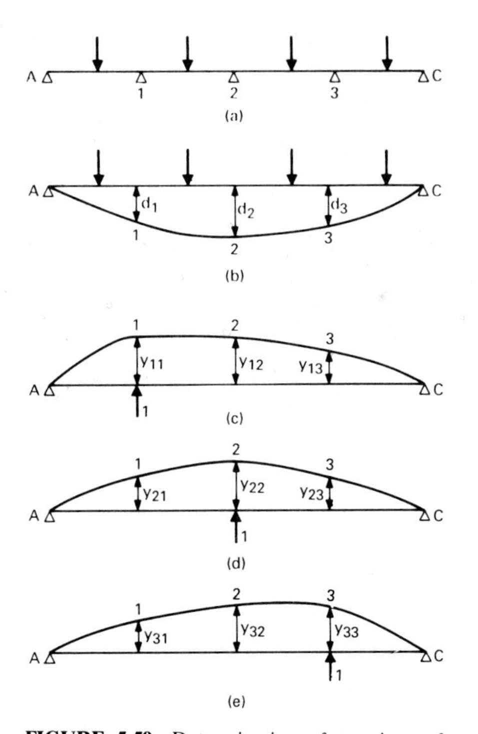

FIGURE 5.59 Determination of reactions of continuous beam AC: (*a*) Loaded beam with supports at points 1, 2, and 3. (*b*) Deflection of beam when supports are removed. (*c*) to (*e*) Deflections when a unit load is applied successively at points 1, 2, and 3.

3, respectively, when a unit force is applied at node 1 (Fig. 5.59*c*). Similarly, compute column 2 of **F** for a unit force at node 2 (Fig. 5.59*d*) and column 3 for a unit force at node 3 (Fig. 5.59*e*). The three equations then are given by

$$\begin{bmatrix} y_{11} & y_{12} & y_{13} \\ y_{21} & y_{22} & y_{23} \\ y_{31} & y_{32} & y_{33} \end{bmatrix} \begin{bmatrix} R_1 \\ R_2 \\ R_3 \end{bmatrix} = \begin{bmatrix} d_1 \\ d_2 \\ d_3 \end{bmatrix} \tag{5.104}$$

The solution may be represented by $\mathbf{R} = \mathbf{F}^{-1}\mathbf{d}$ and obtained by matrix or algebraic methods. See also Art. 5.13.

5.11 CONTINUOUS BEAMS AND FRAMES

Fixed-end beams, continuous beams, continuous trusses, and rigid frames are statically indeterminate. The equations of equilibrium are not sufficient for the determination of all the unknown forces and moments. Additional equations based on a knowledge of the deformation of the member are required.

Hence, while the bending moments in a simply supported beam are determined only by the loads and the span, bending moments in a statically indeterminate member are also a function of the geometry, cross-sectional dimensions, and modulus of elasticity.

5.11.1 Sign Convention

For computation of end moments in continuous beams and frames, the following sign convention is most convenient: A moment acting at an end of a member or at a joint is positive if it tends to rotate the joint clockwise, negative if it tends to rotate the joint counterclockwise.

Similarly, the angular rotation at the end of a member is positive if in a clockwise direction, negative if counterclockwise. Thus, a positive end moment produces a positive end rotation in a simple beam.

For ease in visualizing the shape of the elastic curve under the action of loads and end moments, bending-moment diagrams should be plotted on the tension side of each member. Hence, if an end moment is represented by a curved arrow, the arrow will point in the direction in which the moment is to be plotted.

5.11.2 Carry-Over Moments

When a member of a continuous beam or frame is loaded, bending moments are induced at the ends of the member as well as between the ends. The magnitude of the end moments depends on the magnitude and location of the loads, the geometry of the member, and the amount of restraint offered to end rotation of the member by other members connected to it. Because of the restraint, end moments are induced in the connecting members, in addition to end moments that may be induced by loads on those spans.

If the far end of a connecting member is restrained by support conditions against rotation, a resisting moment is induced at that end. That moment is called a carry-over moment. The ratio of the carry-over moment to the other end moment is

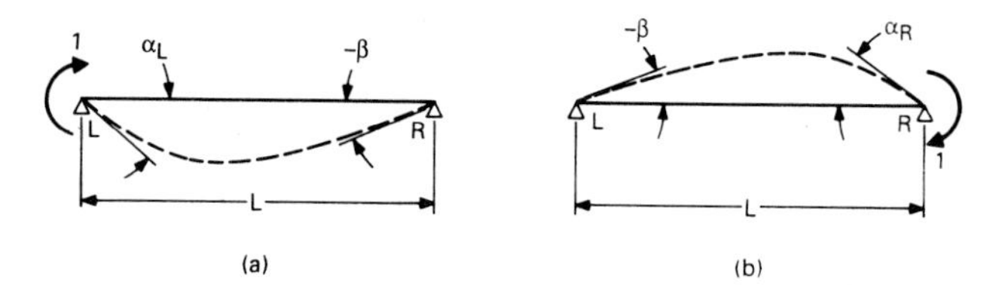

FIGURE 5.60 End rotations of a simple beam *LR* when a unit moment is applied (*a*) at end *L* and (*b*) at end *R*.

called carry-over factor. It is a constant for the member, independent of the magnitude and direction of the moments to be carried over. Every beam has two carry-over factors, one directed toward each end.

As pointed out in Art. 5.10.6, analysis of a continuous span can be simplified by treating it as a simple beam subjected to applied end moments. Thus, it is convenient to express the equations for carry-over factors in terms of the end rotations of simple beams: Convert a continuous member *LR* to a simple beam with the same span *L*. Apply a unit moment to one end (Fig. 5.60). The end rotation at the support where the moment is applied is α, and at the far end, the rotation is β. By the dummy-load method (Art. 5.10.4), if x is measured from the β end,

$$\alpha = \frac{1}{L^2} \int_0^L \frac{x^2}{EI_x}\, dx \tag{5.105}$$

$$\beta = \frac{1}{L^2} \int_0^L \frac{x(L - x)}{EI_x}\, dx \tag{5.106}$$

in which I_x = moment of inertia at a section a distance of x from the β end
E = modulus of elasticity

In accordance with the reciprocal theorem (Art. 5.10.5), β has the same value regardless of the beam end to which the unit moment is applied (Fig. 5.60). For prismatic beams (I_x = constant),

$$\alpha_L = \alpha_R = \frac{L}{3EI} \tag{5.107}$$

$$\beta = \frac{L}{6EI} \tag{5.108}$$

Carry-Over Factors. The preceding equations can be used to determine carry-over factors for any magnitude of end restraint. The carry-over factors toward fixed ends, however, are of special importance.

The bending-moment diagram for a continuous span *LR* that is not loaded except for a moment *M* applied at end *L* is shown in Fig. 5.61*a*. For determination of the carry-over factor C_R toward *R*, that end is assumed fixed (no rotation can occur there). The carry-over moment to *R* then is $C_R M$. The moment diagram in Fig. 5.61*a* can be resolved into two components: a simple beam with *M* applied at *L* (Fig. 5.61*b*) and a simple beam with $C_R M$ applied at *R* (Fig. 5.61*c*). As indicated in Fig. 5.61*d*, *M* causes an angle change at *R* of $-\beta$. As shown in Fig. 5.61*e*, $C_R M$

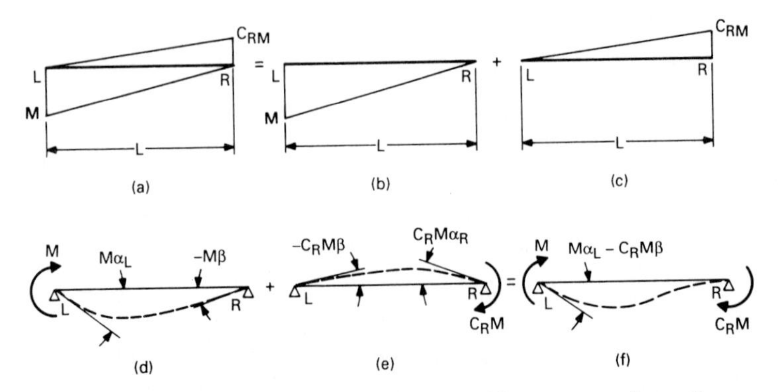

FIGURE 5.61 Effect of applying an end moment M to any span of a continuous beam: (*a*) An end moment $C_R M$ is induced at the opposite end. (*b*) and (*c*) The moment diagram in (*a*) is resolved into moment diagrams for a simple beam. (*d*) and (*e*) Addition of the end rotations corresponding to conditions (*b*) and (*c*) yields (*f*), the end rotations induced by M in the beam shown in (*a*).

induces an angle change at R of $C_R M \alpha_R$. Since the net angle change at R is zero (Fig. 5.61*f*), $C_R M \alpha_R - M\beta = 0$, from which

$$C_R = \frac{\beta}{\alpha_R} \tag{5.109}$$

Similarly, the carry-over factor toward support L is given by

$$C_L = \frac{\beta}{\alpha_L} \tag{5.110}$$

Since the carry-over factors are positive, the moment carried over has the same sign as the applied moment. For prismatic beams, $\beta = L/6EI$ and $\alpha = L/3EI$. Hence,

$$C_L = C_R = \frac{L}{6EI}\frac{3EI}{L} = \frac{1}{2} \tag{5.111}$$

For beams with variable moment of inertia, β and α can be determined from Eqs. (5.105) and (5.106) and the carry-over factors from Eqs. (5.109) and (5.110).

If an end of a beam is free to rotate, the carry-over factor toward that end is zero.

5.11.3 Fixed-End Stiffness

The fixed-end stiffness of a beam is defined as the moment that is required to induce a unit rotation at the support where it is applied while the other end of the beam is fixed against rotation. Stiffness is important because, in the moment-distribution method, it determines the proportion of the total moment applied at a joint, or intersection of members, that is distributed to each member of the joint.

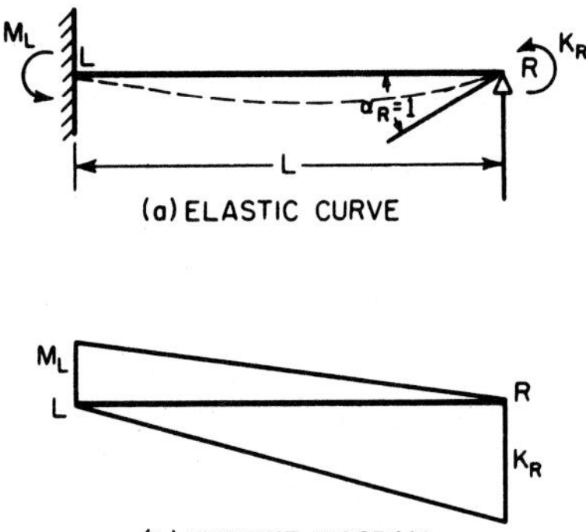

FIGURE 5.62 Determination of fixed-end stiffness: (*a*) elastic curve for moment K_R causing a unit end rotation; (*b*) the moment diagram for condition (*a*).

In Fig. 5.62*a,* the fixed-end stiffness of beam *LR* at end *R* is represented by K_R. When K_R is applied to beam *LR* at *R,* a moment $M_L = C_L K_R$ is carried over to end *L,* where C_L is the carry-over factor toward *L* (see Art. 5.11.2). K_R induces an angle change α_R at *R,* where α_R is given by Eq. (5.105). The carry-over moment induces at *R* an angle change $-C_L k_R \beta$, where β is given by Eq. (5.106). Since, by the definition of stiffness, the total angle change at *R* is unity, $K_R \alpha_R - C_L K_R \beta = 1$, from which

$$K_R = \frac{1/\alpha_R}{1 - C_R C_L} \tag{5.112}$$

when C_R is substituted for β/α_R [see Eq. (5.109)].

In a similar manner, the stiffness at *L* is found to be

$$K_L = \frac{1/\alpha_L}{1 - C_R C_L} \tag{5.113}$$

With the use of Eqs. (5.107) and (5.111), the stiffness of a beam with constant moment of inertia is given by

$$K_L = K_R = \frac{3EI/L}{1 - 1/2 \times 1/2} = \frac{4EI}{L} \tag{5.114}$$

where L = span of the beam
E = modulus of elasticity
I = moment of inertia of beam cross section

Beam with Hinge. The stiffness of one end of a beam when the other end is free to rotate can be obtained from Eqs. (5.112) or (5.113) by setting the carry-over factor toward the hinged end equal to zero. Thus, for a prismatic beam with one end hinged, the stiffness of the beam at the other end is given by

$$K = \frac{3EI}{L} \tag{5.115}$$

This equation indicates that a prismatic beam hinged at only one end has three-fourths the stiffness, or resistance to end rotation, of a beam fixed at both ends.

5.11.4 Fixed-End Moments

A beam so restrained at its ends that no rotation is produced there by the loads is called a fixed-end beam, and the end moments are called fixed-end moments. Fixed-end moments may be expressed as the product of a coefficient and *WL,* where *W* is the total load on the span *L.* The coefficient is independent of the properties of

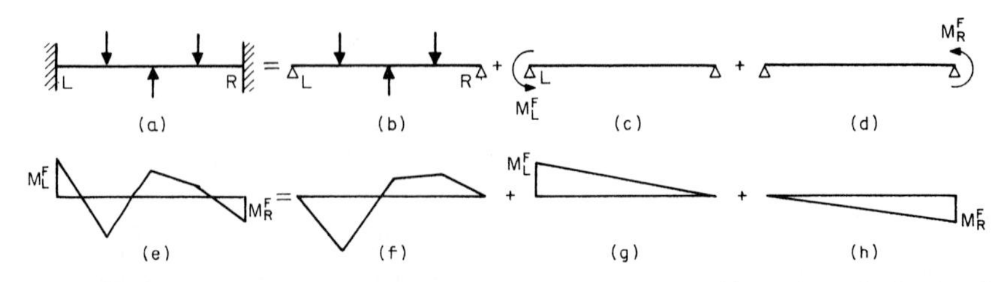

FIGURE 5.63 Determination of fixed-end moments in beam LR: (*a*) Loads on the fixed-end beam are resolved (*b*) to (*d*) into the sum of loads on a simple beam. (*e*) to (*h*) Bending-moment diagrams for conditions (*a*) to (*d*), respectively.

other members of the structure. Thus, any member can be isolated from the rest of the structure and its fixed-end moments computed.

Assume, for example, that the fixed-end moments for the loaded beam in Fig. 5.63*a* are to be determined. Let M_L^F be the moment at the left end L and M_R^F the moment at the right end R of the beam. Based on the condition that no rotation is permitted at either end and that the reactions at the supports are in equilibrium with the applied loads, two equations can be written for the end moments in terms of the simple-beam end rotations, θ_L at L and θ_R, at R for the specific loading.

Let K_L be the fixed-end stiffness at L and K_R the fixed-end stiffness at R, as given by Eqs. (5.112) and (5.113). Then, by resolution of the moment diagram into simple-beam components, as indicated in Fig. 5.63*f* to *h*, and application of the superposition principle (Art. 5.10.6), the fixed-end moments are found to be

$$M_L^F = -K_L(\theta_L + C_R\theta_R) \tag{5.116}$$

$$M_R^F = -K_R(\theta_R + C_L\theta_L) \tag{5.117}$$

where C_L and C_R are the carry-over factors to L and R, respectively [Eqs. (5.109) and (5.110)]. The end rotations θ_L and θ_R can be computed by a method described in Art. 5.5.15 or 5.10.4.

Prismatic Beams. The fixed-end moments for beams with constant moment of inertia can be derived from the equations given above with the use of Eqs. (5.111) and (5.114):

$$M_L^F = -\frac{4EI}{L}\left(\theta_L + \frac{1}{2}\theta_R\right) \tag{5.118}$$

$$M_R^F = -\frac{4EI}{L}\left(\theta_R + \frac{1}{2}\theta_L\right) \tag{5.119}$$

where L = span of the beam
E = modulus of elasticity
I = moment of inertia

For horizontal beams with gravity loads only, θ_R is negative. As a result, M_L^F is negative and M_R^F positive.

For propped beams (one end fixed, one end hinged) with variable moment of inertia, the fixed-end moments are given by

$$M_L^F = \frac{-\theta_L}{\alpha_L} \quad \text{or} \quad M_R^F = \frac{-\theta_R}{\alpha_R} \tag{5.120}$$

where α_L and α_R are given by Eq. (5.105). For prismatic propped beams, the fixed-end moments are

$$M_L^F = \frac{-3EI\theta_L}{L} \quad \text{or} \quad M_R^F = \frac{-3EI\theta_R}{L} \tag{5.121}$$

Deflection of Supports. Fixed-end moments for loaded beams when one support is displaced vertically with respect to the other support may be computed with the use of Eqs. (5.116) to (5.121) and the principle of superposition: Compute the fixed-end moments induced by the deflection of the beam when not loaded and add them to the fixed-end moments for the loaded condition with immovable supports.

The fixed-end moments for the unloaded condition can be determined directly from Eqs. (5.116) and (5.117). Consider beam LR in Fig. 5.64, with span L and support R deflected a distance d vertically below its original position. If the beam were simply supported, the angle change caused by the displacement of R would be very nearly d/L. Hence, to obtain the fixed-end moments for the deflected condition, set $\theta_L = \theta_R = d/L$ and substitute these simple-beam end rotations in Eqs. (5.116) and (5.117):

$$M_L^F = -K_L(1 + C_R)d/L \tag{5.122}$$

$$M_R^F = -K_R(1 + C_L)d/L \tag{5.123}$$

If end L is displaced downward with respect to R, d/L would be negative and the fixed-end moments positive.

For beams with constant moment of inertia, the fixed-end moments are given by

$$M_L^F = M_R^F = -\frac{6EI}{L}\frac{d}{L} \tag{5.124}$$

The fixed-end moments for a propped beam, such as beam LR shown in Fig. 5.65, can be obtained similarly from Eq. (5.120). For variable moment of inertia,

$$M^F = \frac{d}{L}\frac{1}{\alpha_L} \tag{5.125}$$

For a prismatic propped beam,

$$M^F = -\frac{3EI}{L}\frac{d}{L} \tag{5.126}$$

Reverse signs for downward displacement of end L.

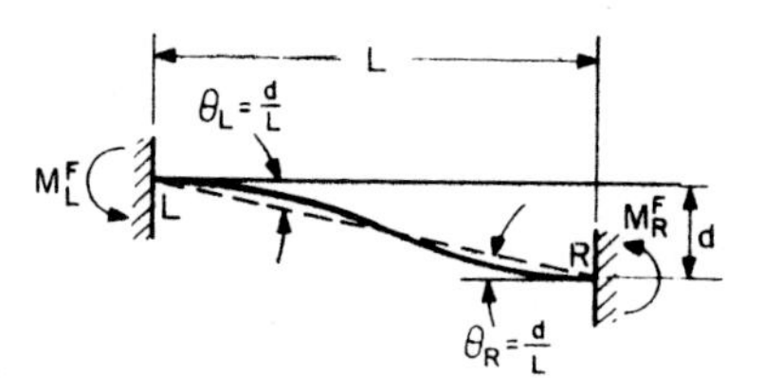

FIGURE 5.64 End moments caused by displacement d of one end of a fixed-end beam.

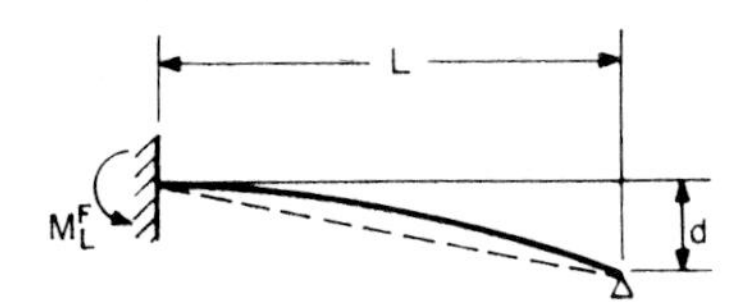

FIGURE 5.65 End moment caused by displacement d of one end of a propped beam.

Computation Aids for Prismatic Beams. Fixed-end moments for several common types of loading on beams of constant moment of inertia (prismatic beams) are given in Figs. 5.66 to 5.69. Also, the curves in Fig. 5.71 enable fixed-end moments

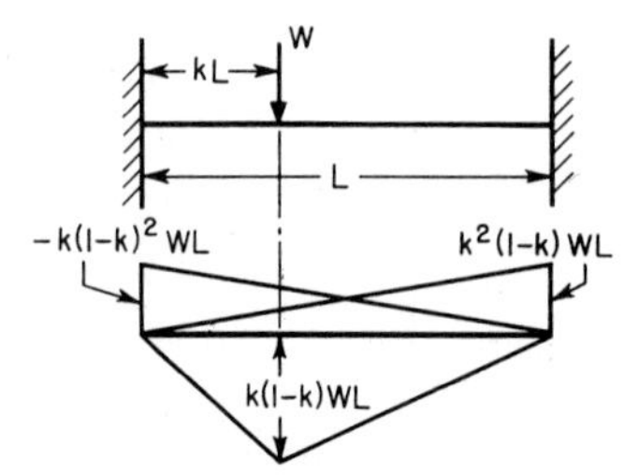

FIGURE 5.66 Moments for concentrated load on a prismatic fixed-end beam.

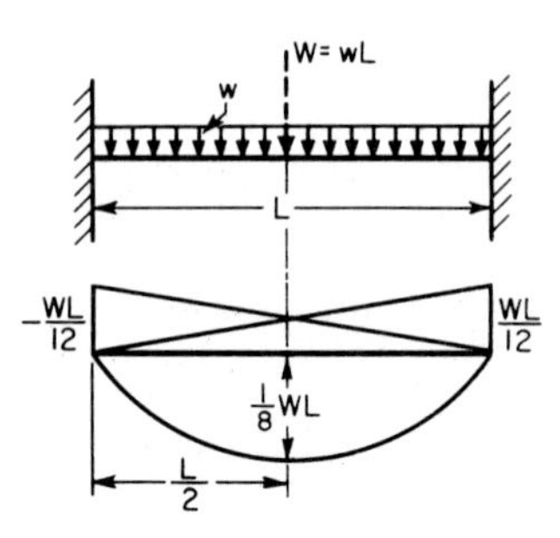

FIGURE 5.67 Moments for a uniform load on a prismatic fixed-end beam.

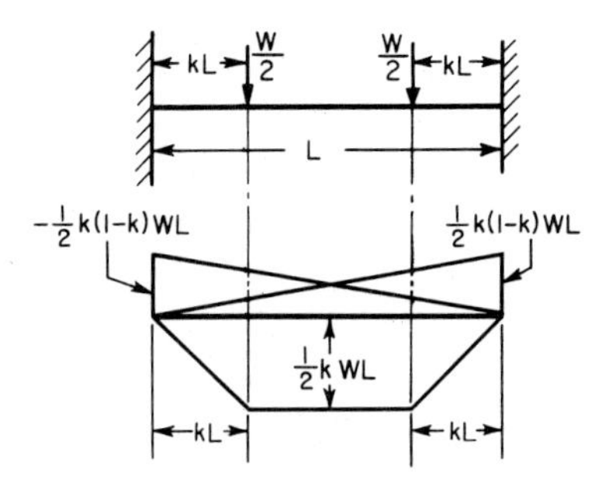

FIGURE 5.68 Moments for two equal loads on a prismatic fixed-end beam.

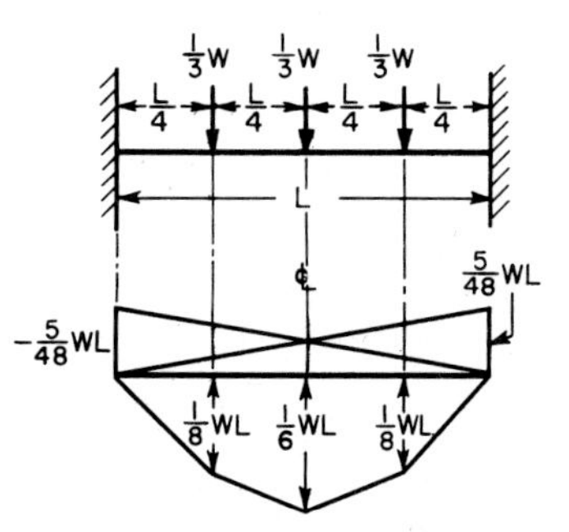

FIGURE 5.69 Moments for several equal loads on a prismatic fixed-end beam.

to be computed easily for any type of loading on a prismatic beam. Before the curves can be entered, however, certain characteristics of the loading must be calculated. These include $\bar{x}L$, the location of the center of gravity of the loading with respect to one of the loads; $G^2 = \Sigma b_n^2 P_n / W$, where $b_n L$ is the distance from each load P_n to the center of gravity of the loading (taken positive to the right); and $S^3 = \Sigma b_n^3 P_n / W$. (See Case 9, Fig. 5.70.) These values are given in Fig. 5.70 for some common types of loading.

The curves in Fig. 5.71 are entered with the location a of the center of gravity with respect to the left end of the span. At the intersection with the proper G curve, proceed horizontally to the left to the intersection with the proper S line, then vertically to the horizontal scale indicating the coefficient m by which to multiply WL to obtain the fixed-end moment. The curves solve the equations:

$$m_L = \frac{M_L^F}{WL} = G^2[1 - 3(1 - a)] + a(1 - a)^2 + S^3 \tag{5.127}$$

$$m_R = \frac{M_R^F}{WL} = G^2(1 - 3a) + a^2(1 - a) - S^3 \tag{5.128}$$

where M_L^F is the fixed-end moment at the left support and M_R^F at the right support.

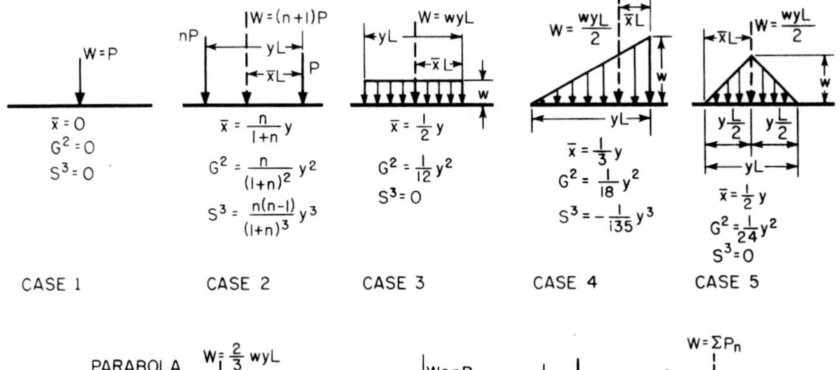

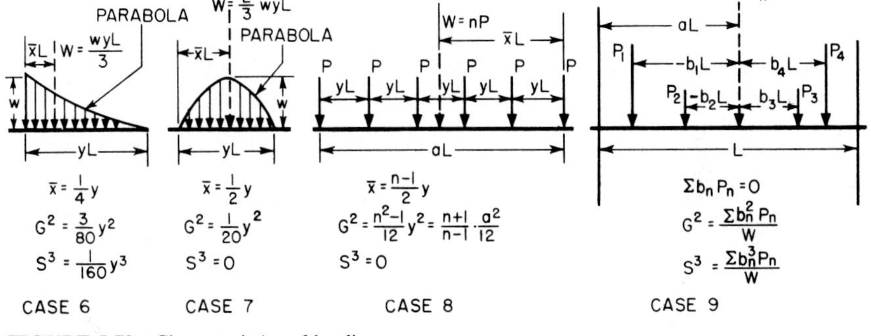

FIGURE 5.70 Characteristics of loadings.

As an example of the use of the curves, find the fixed-end moments in a prismatic beam of 20-ft span carrying a triangular loading of 100 kips, similar to the loading shown in Case 4, Fig. 5.70, distributed over the entire span, with the maximum intensity at the right support.

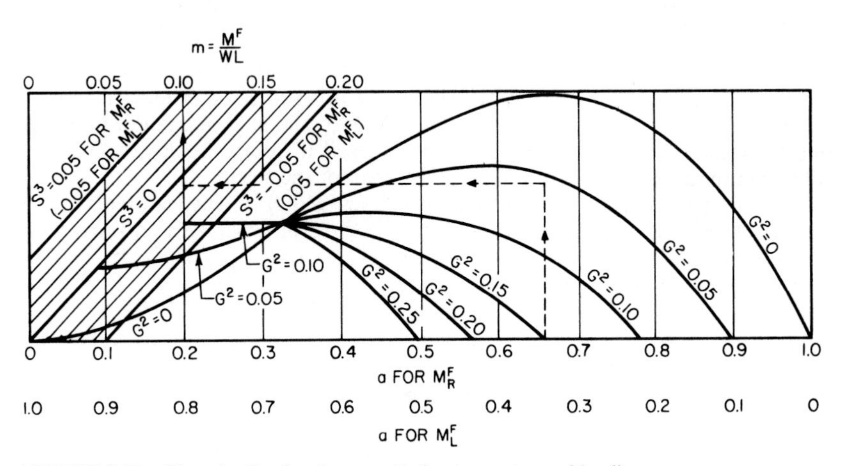

FIGURE 5.71 Chart for fixed-end moments due to any type of loading.

Case 4 gives the characteristics of the loading: $y = 1$; the center of gravity is $0.33L$ from the right support, so $a = 0.667$; $G^2 = 1/18 = 0.056$; and $S^3 = -1/135 = -0.007$. To find M_R^F, enter Fig. 5.71 with $a = 0.67$ on the upper scale at the bottom of the diagram, and proceed vertically to the estimated location of the intersection of the coordinate with the $G^2 = 0.06$ curve. Then, move horizontally to the intersection with the line for $S^3 = -0.007$, as indicated by the dash line in Fig. 5.71. Referring to the scale at the top of the diagram, find the coefficient m_R to be 0.10. Similarly, with $a = 0.67$ on the lowest scale, find the coefficient m_L to be 0.07. Hence, the fixed-end moment at the right support is $0.10 \times 100 \times 20 = 200$ ft-kips, and at the left support $-0.07 \times 100 \times 20 = -140$ ft-kips.

5.11.5 Slope-Deflection Equations

In Arts. 5.11.2 and 5.11.4, moments and displacements in a member of a continuous beam or frame are obtained by addition of their simple-beam components. Similarly, moments and displacements can be determined by superposition of fixed-end-beam components. This method, for example, can be used to derive relationships between end moments and end rotations of a beam known as slope-deflection equations. These equations can be used to compute end moments in continuous beams.

Consider a member LR of a continuous beam or frame (Fig. 5.72). LR may have a moment of inertia that varies along its length. The support R is displaced

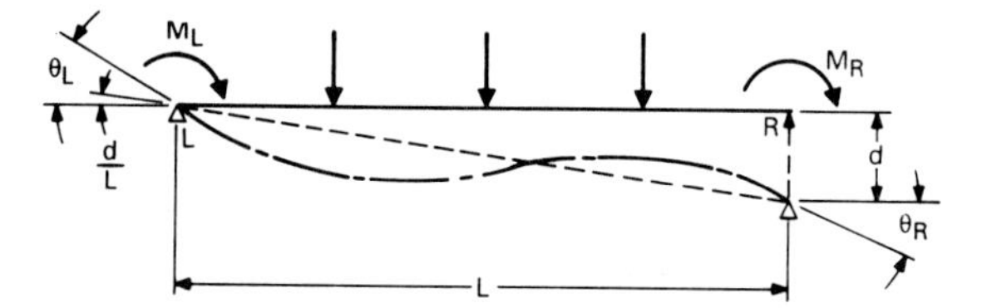

FIGURE 5.72 Elastic curve for a span LR of a continuous beam subjected to end moments and displacement of one end.

vertically downward a distance d from its original position. Because of this and the loads on the member and adjacent members, LR is subjected to end moments M_L at L and M_R at R. The total end rotation at L is θ_L and at R, θ_R. All displacements are so small that the member can be considered to rotate clockwise through an angle nearly equal to d/L, where L is the span of the beam.

Assume that rotation is prevented at ends L and R by end moments m_L at L and m_R at R. Then, by application of the principle of superposition (Art. 5.10.6) and Eqs. (5.122) and (5.123),

$$m_L = M_L^F - K_L(1 + C_R)\frac{d}{L} \tag{5.129}$$

$$m_R = M_R^F - K_R(1 + C_L)\frac{d}{L} \tag{5.130}$$

where M_L^F = fixed-end moment at L due to the load on LR

M_R^F = fixed-end moment at R due to the load on LR

K_L = fixed-end stiffness at end L
K_R = fixed-end stiffness at end R
C_L = carry-over factor toward end L
C_R = carry-over factor toward end R

Since ends L and R are not fixed but actually undergo angle changes θ_L and θ_R at L and R, respectively, the joints must now be permitted to rotate while an end moment M'_L is applied at L and an end moment M'_R at R to produce those angle changes (Fig. 5.73). With the use of the definitions of carry-over factor (Art. 5.11.2) and fixed-end stiffness (Art. 5.11.3), these moments are found to be

$$M'_L = K_L(\theta_L + C_R\theta_R) \tag{5.131}$$

$$M'_R = K_R(\theta_R + C_L\theta_L) \tag{5.132}$$

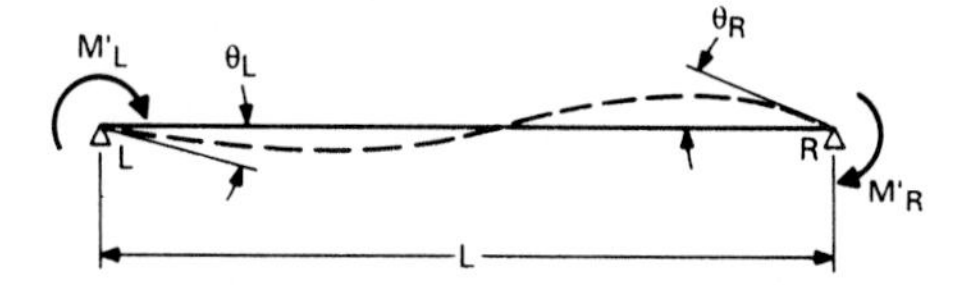

FIGURE 5.73 Elastic curve for a simple beam LR subjected to end moments.

The slope-deflection equations for LR then result from addition of M'_L to m_L, which yields M_L, and of M'_R to m_R, which yields M_R:

$$M_L = K_L(\theta_L + C_R\theta_R) + M_L^F - K_L(1 + C_R)\frac{d}{L} \tag{5.133}$$

$$M_R = K_R(\theta_R + C_L\theta_L) + M_R^F - K_R(1 + C_L)\frac{d}{L} \tag{5.134}$$

For beams with constant moment of inertia, the slope-deflection equations become

$$M_L = \frac{4EI}{L}\left(\theta_L + \frac{1}{2}\theta_R\right) + M_L^F - \frac{6EI}{L}\frac{d}{L} \tag{5.135}$$

$$M_R = \frac{4EI}{L}\left(\theta_R + \frac{1}{2}\theta_L\right) + M_R^F - \frac{6EI}{L}\frac{d}{L} \tag{5.136}$$

where E = modulus of elasticity
I = moment of inertia of the cross section

Note that if end L moves downward with respect to R, the sign for d in the preceding equations is changed.

If the end moments M_L and M_R are known and the end rotations are to be determined, Eqs. (5.131) to (5.134) can be solved for θ_L and θ_R or derived by superposition of simple-beam components, as is done in Art. 5.11.4. For beams with moment of inertia varying along the span:

$$\theta_L = (M_L - M_L^F)\,\alpha_L - (M_R - M_R^F)\,\beta + \frac{d}{L} \tag{5.137}$$

$$\theta_R = (M_R - M_R^F)\,\alpha_R - (M_L - M_L^F)\,\beta + \frac{d}{L} \tag{5.138}$$

where α is given by Eq. (5.105) and β by Eq. (5.106). For beams with constant moment of inertia:

$$\theta_L = \frac{L}{3EI}(M_L - M_L^F) - \frac{L}{6EI}(M_R - M_R^F) + \frac{d}{L} \tag{5.139}$$

$$\theta_R = \frac{L}{3EI}(M_R - M_R^F) - \frac{L}{6EI}(M_L - M_L^F) + \frac{d}{L} \tag{5.140}$$

The slope-deflection equations can be used to determine end moments and rotations of the spans of continuous beams by writing compatibility and equilibrium equations for the conditions at each support. For example, the sum of the moments at each support must be zero. Also, because of continuity, the member must rotate through the same angle on both sides of every support. Hence, M_L for one span, given by Eq. (5.133) or (5.135), must be equal to $-M_R$ for the adjoining span, given by Eq. (5.134) or (5.136), and the end rotation θ at that support must be the same on both sides of the equation. One such equation with the end rotations at the supports as the unknowns can be written for each support. With the end rotations determined by simultaneous solution of the equations, the end moments can be computed from the slope-deflection equations and the continuous beam can now be treated as statically determinate.

See also Arts. 5.11.9 and 5.13.2.

(C. H. Norris et al., "Elementary Structural Analysis," 4th ed., McGraw-Hill Book Company, New York.)

5.11.6 Moment Distribution

The frame in Fig. 5.74 consists of four prismatic members rigidly connected together at O and fixed at ends A, B, C, and D. If an external moment U is applied at O, the sum of the end moments in each member at O must be equal to U. Furthermore, all members must rotate at O through the same angle θ, since they are assumed to be rigidly connected there. Hence, by the definition of fixed-end stiffness, the proportion of U induced in the end of each member at O is equal to the ratio of the stiffness of that member to the sum of the stiffnesses of all the members at the joint (Art. 5.11.3).

Suppose a moment of 100 ft-kips is applied at O, as indicated in Fig. 5.74*b*. The relative stiffness (or I/L) is assumed as shown in the circle on each member. The distribution factors for the moment at O are computed from the stiffnesses and shown in the boxes. For example, the distribution factor for OA equals its stiffness divided by the sum of the stiffnesses of all the members at the joint: $3/(3 + 2 + 4 + 1) = 0.3$. Hence, the moment induced in OA at O is $0.3 \times 100 = 30$ ft-kips. Similarly, OB gets 10 ft-kips, OC 40 ft-kips, and OD 20 ft-kips.

Because the far ends of these members are fixed, one-half of these moments are carried over to them (Art. 5.11.2). Thus $M_{AO} = 0.5 \times 30 = 15$; $M_{BO} = 0.5 \times 10 = 5$; $M_{CO} = 0.5 \times 40 = 20$; and $M_{DO} = 0.5 \times 20 = 10$.

Most structures consist of frames similar to the one in Fig. 5.74, or even simpler, joined together. Though the ends of the members are not fixed, the technique employed for the frame in Fig. 5.74*b* can be applied to find end moments in such continuous structures.

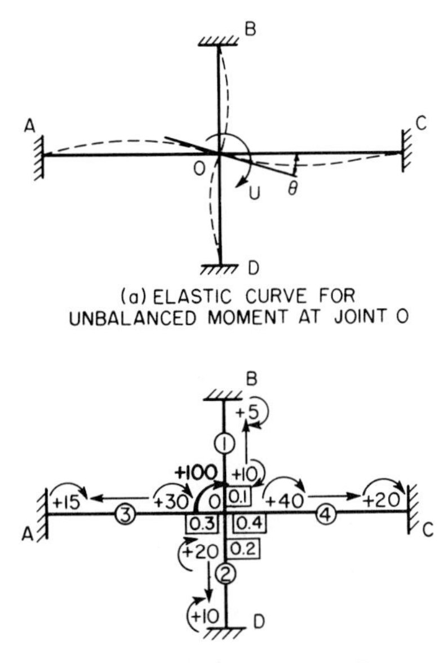

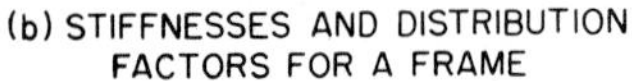
(b) STIFFNESSES AND DISTRIBUTION FACTORS FOR A FRAME

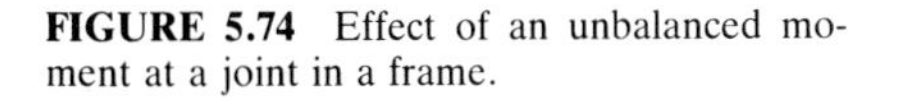
FIGURE 5.74 Effect of an unbalanced moment at a joint in a frame.

Before the general method is presented, one short cut is worth noting. Advantage can be taken when a member has a hinged end to reduce the work of distributing moments. This is done by using the true stiffness of a member instead of the fixed-end stiffness. (For a prismatic beam with one end hinged, the stiffness is three-fourths the fixed-end stiffness; for a beam with variable I, it is equal to the fixed-end stiffness times $1 - C_L C_R$, where C_L and C_R are the carry-over factors for the beam.) Naturally, the carry-over factor toward the hinge is zero.

When a joint is neither fixed nor pinned but is restrained by elastic members connected there, moments can be distributed by a series of converging approximations. All joints are locked against rotation. As a result, the loads will create fixed-end moments at the ends of every member. At each joint, a moment equal to the algebraic sum of the fixed-end moments there is required to hold it fixed. Then, one joint is unlocked at a time by applying a moment equal but opposite in sign to the moment that was needed to prevent rotation. The unlocking moment must be distributed to the members at the joint in proportion to their fixed-end stiffnesses and the distributed moments carried over to the far ends.

After all joints have been released at least once, it generally will be necessary to repeat the process—sometimes several times—before the corrections to the fixed-end moments become negligible. To reduce the number of cycles, the unlocking of joints should start with those having the greatest unbalanced moments.

Suppose the end moments are to be found for the prismatic continuous beam $ABCD$ in Fig. 5.75. The I/L values for all spans are equal; therefore, the relative

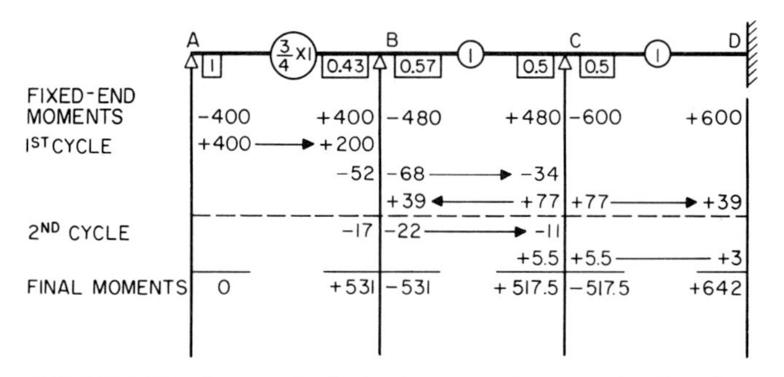

FIGURE 5.75 Moment distribution by converging approximations for a continuous beam.

fixed-end stiffness for all members is unity. However, since *A* is a hinged end, the computation can be shortened by using the actual relative stiffness, which is ¾. Relative stiffnesses for all members are shown in the circle on each member. The distribution factors are shown in boxes at each joint.

The computation starts with determination of fixed-end moments for each member (Art. 5.11.4). These are assumed to have been found and are given on the first line in Fig. 5.75. The greatest unbalanced moment is found from inspection to be at hinged end *A;* so this joint is unlocked first. Since there are no other members at the joint, the full unlocking moment of +400 is distributed to *AB* at *A* and one-half of this is carried over to *B*. The unbalance at *B* now is +400 − 480 plus the carry-over of +200 from *A*, or a total of +120. Hence, a moment of −120 must be applied and distributed to the members at *B* by multiplying by the distribution factors in the corresponding boxes.

The net moment at *B* could be found now by adding the entries for each member at the joint. However, it generally is more convenient to delay the summation until the last cycle of distribution has been completed.

The moment distributed to *BA* need not be carried over to *A,* because the carry-over factor toward the hinged end is zero. However, half the moment distributed to *BC* is carried over to *C.*

Similarly, joint *C* is unlocked and half the distributed moments carried over to *B* and *D,* respectively. Joint *D* should not be unlocked, since it actually is a fixed end. Thus, the first cycle of moment distribution has been completed.

The second cycle is carried out in the same manner. Joint *B* is released, and the distributed moment in *BC* is carried over to *C.* Finally, *C* is unlocked, to complete the cycle. Adding the entries for the end of each member yields the final moments.

5.11.7 Maximum Moments in Continuous Frames

In design of continuous frames, one objective is to find the maximum end moments and interior moments produced by the worst combination of loading. For maximum moment at the end of a beam, live load should be placed on that beam and on the beam adjoining the end for which the moment is to be computed. Spans adjoining these two should be assumed to be carrying only dead load.

For maximum midspan moments, the beam under consideration should be fully loaded, but adjoining spans should be assumed to be carrying only dead load.

The work involved in distributing moments due to dead and live loads in continuous frames in buildings can be greatly simplified by isolating each floor. The tops of the upper columns and the bottoms of the lower columns can be assumed fixed. Furthermore, the computations can be condensed considerably by following the procedure recommended in "Continuity in Concrete Building Frames," EB033D, Portland Cement Association, Skokie, IL 60077, and indicated in Fig. 5.74.

Figure 5.74 presents the complete calculation for maximum end and midspan moments in four floor beams *AB, BC, CD,* and *DE.* Building columns are assumed to be fixed at the story above and below. None of the beam or column sections is known to begin with; so as a start, all members will be assumed to have a fixed-end stiffness of unity, as indicated on the first line of the calculation.

On the second line, the distribution factors for each end of the beams are shown, calculated from the stiffnesses (Arts. 5.11.3 and 5.11.4). Column stiffnesses are not shown, because column moments will not be computed until moment distribution to the beams has been completed. Then the sum of the column moments at each joint may be easily computed, since they are the moments needed to make the sum of the end moments at the joint equal to zero. The sum of the column moments at each joint can then be distributed to each column there in proportion to its stiffness. In this example, each column will get one-half the sum of the column moments.

Fixed-end moments at each beam end for dead load are shown on the third line, just above the heavy line, and fixed-end moments for live plus dead load on the fourth line. Corresponding midspan moments for the fixed-end condition also are shown on the fourth line and, like the end moments, will be corrected to yield actual midspan moments.

For maximum end moment at *A*, beam *AB* must be fully loaded, but *BC* should carry dead load only. Holding *A* fixed, we first unlock joint *B*, which has a total-load fixed-end moment of +172 in *BA* and a dead-load fixed-end moment of −37 in *BC*. The releasing moment required, therefore, is −(172 − 37), or − 135. When *B* is released, a moment of −135 × ¼ is distributed to *BA*. One-half of this is carried over to *A*, or −135 × ¼ × ½ = −17. This value is entered as the carry-over at *A* on the fifth line in Fig. 5.76. Joint *B* is then relocked.

At *A*, for which we are computing the maximum moment, we have a total-load fixed-end moment of −172 and a carry-over of −17, making the total −189, shown on the sixth line. To release *A*, a moment of +189 must be applied to the joint. Of this, 189 × ⅓, or 63, is distributed to *AB*, as indicated on the seventh line of the calculation. Finally, the maximum moment at *A* is found by adding lines 6 and 7: −189 + 63 = −126.

For maximum moment at *B*, both *AB* and *BC* must be fully loaded, but *CD* should carry only dead load. We begin the determination of the moment at *B* by first releasing joints *A* and *C*, for which the corresponding carry-over moments at *BA* and *BC* are +29 and −(+78 −70) × ¼ × ½ = −1, shown on the fifth line in Fig. 5.76. These bring the total fixed-end moments in *BA* and *BC* to +201 and −79, respectively. The releasing moment required is −(201 − 79) = −122. Multiplying this by the distribution factors for *BA* and *BC* when joint *B* is released, we find the distributed moments, −30, entered on line 7. The maximum end moments finally are obtained by adding lines 6 and 7: +171 at *BA* and −109 at *BC*. Maximum moments at *C*, *D*, and *E* are computed and entered in Fig. 5.76 in a similar manner. This procedure is equivalent to two cycles of moment distribution.

The computation of maximum midspan moments in Fig. 5.76 is based on the assumption that in each beam the midspan moment is the sum of the simple-beam

	A		B			C			D			E
1. RELATIVE STIFFNESS		$K^F=1$			$K^F=1$			$K^F=1$			$K^F=1$	
2. DISTRIBUTION FACTOR	⅓		¼	¼		¼	¼		¼	¼		⅓
3. F.E.M. DEAD LOAD	—		+91	−37		+37	−70		+70	−59		—
4. F.E.M. TOTAL LOAD	−172	+99	+172	−78	+73	+78	−147	+85	+147	−126	+63	+126
5. CARRY-OVER	−17 →	+11	+29	−1 →	+1	−2	−11 →	+7	+14	−21 →	+13	+7
6. ADDITION	−189	+18	+201	−79	−1	+76	−158	+9	+161	−147	+5	+133
7. DISTRIBUTION	+63		−30	−30		+21	+21		−4	−4		−44
8. MAX. MOMENTS	−126	+128	+171	−109	+73	+97	−137	+101	+157	−151	+81	+89

FIGURE 5.76 Bending moments in a continuous frame obtained by moment distribution.

midspan moment and one-half the algebraic difference of the final end moments (the span carries full load but adjacent spans only dead load). Instead of starting with the simple-beam moment, however, we begin with the midspan moment for the fixed-end condition and apply two corrections. In each span, these corrections are equal to the carry-over moments entered on line 5 for the two ends of the beam multiplied by a factor.

For beams with variable moment of inertia, the factor is $\pm\frac{1}{2}[(1/C^F) + D - 1]$ where C^F is the fixed-end carry-over factor toward the end for which the correction factor is being computed and D is the distribution factor for that end. The plus sign is used for correcting the carry-over at the right end of a beam, and the minus sign for the carry-over at the left end. For prismatic beams, the correction factor becomes $\pm\frac{1}{2}(1 + D)$.

For example, to find the corrections to the midspan moment in AB, we first multiply the carry-over at A on line 5, -17, by $-\frac{1}{2}(1 + \frac{1}{3})$. The correction, $+11$, is also entered on the fifth line. Then, we multiply the carry-over at B, $+29$, by $+\frac{1}{2}(1 + \frac{1}{4})$ and enter the correction, $+18$, on line 6. The final midspan moment is the sum of lines 4, 5, and 6: $+99 + 11 + 18 = +128$. Other midpsan moments in Fig. 5.74 are obtained in a similar manner.

See also Arts. 5.11.9 and 5.11.10.

5.11.8 Moment-Influence Factors

In certain types of framing, particularly those in which different types of loading conditions must be investigated, it may be convenient to find maximum end moments from a table of moment-influence factors. This table is made up by listing for the end of each member in the structure the moment induced in that end when a moment (for convenience, $+1000$) is applied to every joint successively. Once this table has been prepared, no additional moment distribution is necessary for computing the end moments due to any loading condition.

For a specific loading pattern, the moment at any beam end M_{AB} may be obtained from the moment-influence table by multiplying the entries under AB for the various joints by the actual unbalanced moments at those joints divided by 1000, and summing (see also Art. 5.11.9 and Table 5.6).

5.11.9 Procedure for Sidesway

Computation of moments due to sidesway, or drift, in rigid frames is conveniently executed by the following method:

1. Apply forces to the structure to prevent sidesway while the fixed-end moments due to loads are distributed.
2. Compute the moments due to these forces.
3. Combine the moments obtained in Steps 1 and 2 to eliminate the effect of the forces that prevented sidesway.

Suppose the rigid frame in Fig. 5.77 is subjected to a 2000-lb horizontal load acting to the right at the level of beam BC. The first step is to compute the moment-influence factors (Table 5.6) by applying moments of $+1000$ at joints B and C, assuming sidesway prevented.

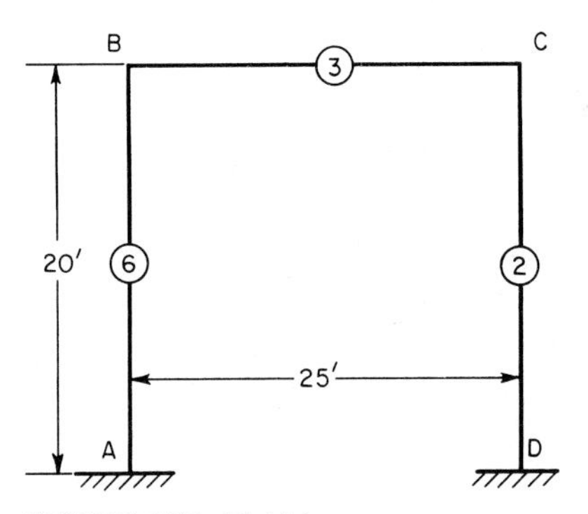

FIGURE 5.77 Rigid frame.

Since there are no intermediate loads on the beams and columns, the only fixed-end moments that need be considered are those in the columns resulting from lateral deflection of the frame caused by the horizontal load. This deflection, however, is not known initially. So assume an arbitrary deflection, which produces a fixed-end moment of $-1000M$ at the top of column CD. M is an unknown constant to be determined from the fact that the sum of the shears in the deflected columns must be equal to the 2000-lb load. The same deflection also produces a moment of $-1000M$ at the bottom of CD [see Eq. (5.126)].

From the geometry of the structure, furthermore, note that the deflection of B relative to A is equal to the deflection of C relative to D. Then, according to Eq. (5.126) the fixed-end moments in the columns are proportional to the stiffnesses of the columns and hence are equal in AB to $-1000M \times 6/2 = -3000M$. The column fixed-end moments are entered in the first line of Table 5.7, which is called a moment-collection table.

In the deflected position of the frame, joints B and C are unlocked. First, apply a releasing moment of $+3000M$ at B and distribute it by multiplying by 3 the entries in the column marked "+1000 at B" in Table 5.6. Similarly, a releasing moment of $+1000M$ is applied at C and distributed with the aid of Table 5.6. The distributed

TABLE 5.6 Moment-Influence Factors for Fig. 5.77

Member	+1000 at B	+1000 at C
AB	351	−105
BA	702	−210
BC	298	210
CB	70	579
CD	−70	421
DC	−35	210

moments are entered in the second and third lines of Table 5.7. The final moments are the sum of the fixed-end moments and the distributed moments and are given in the fifth line.

Isolating each column and taking moments about one end, we find that the overturning moment due to the shear is equal to the sum of the end moments. There is one such equation for each column. Addition of these equations, noting that the sum of the shears equals 2000 lb, yields

$$-M(2052 + 1104 + 789 + 895) = -2000 \times 20$$

from which $M = 8.26$. This value is substituted in the sidesway totals in Table 5.7 to yield the end moments for the 2000-lb horizontal load.

TABLE 5.7 Moment-Collection Table for Fig. 5.77

Remarks	AB +	AB −	BA +	BA −	BC +	BC −	CB +	CB −	CD +	CD −	DC +	DC −
Sidesway, *FEM*		3,000*M*		3,000*M*						1,000*M*		1,000*M*
B moments	1,053*M*		2,106*M*		894*M*		210*M*			210*M*		105*M*
C moments		105*M*		210*M*	210*M*		579*M*		421*M*		210*M*	
Partial sum	1,053*M*	3,105*M*	2,106*M*	3,210*M*	1,104*M*		789*M*		421*M*	1,210*M*	210*M*	1,105*M*
Totals		2,052*M*		1,104*M*	1,104*M*		789*M*			789*M*		895*M*
For 2000-lb load		17,000		9,100	9,100		6,500			6,500		7,400
4000-lb load, *FEM*						12,800	3,200					
B moments	4,490		8,980		3,820		897			897		448
C moments	336		672			672		1,853		1,347		672
Partial sum	4,826		9,652		3,820	13,472	4,097	1,853		2,244		1,120
No-sidesway sum	4,826		9,652			9,652	2,244			2,244		1,120
Sidesway *M*		4,710		2,540	2,540		1,810			1,810		2,060
Totals	120		7,110			7,110	4,050			4,050		3,180

Suppose now a vertical load of 4000 lb is applied to *BC* of the rigid frame in Fig. 5.77, 5 ft from *B*. Tables 5.6 and 5.7 can again be used to determine the end moments with a minimum of labor:

The fixed-end moment at *B*, with sidesway prevented, is −12,800, and at *C* +3200. With the joints locked, the frame is permitted to move laterally an arbitrary amount, so that in addition to the fixed-end moments due to the 4000-lb load, column fixed-end moments of $-3000M$ at *B* and $-1000M$ at *C* are induced. Table 5.7 already indicates the effect of relieving these column moments by unlocking joints *B* and *C*. We now have to superimpose the effect of releasing joints *B* and *C* to relieve the fixed-end moments for the vertical load. This we can do with the aid of Table 5.6. The distribution is shown in the lower part of Table 5.7. The sums of the fixed-end moments and distributed moments for the 4000-lb load are shown on the line "No-sidesway sum."

The unknown *M* can be evaluated from the fact that the sum of the horizontal forces acting on the columns must be zero. This is equivalent to requiring that the sum of the column end moments equals zero:

$$-M(2052 + 1104 + 789 + 895) + 4826 + 9652 - 2244 - 1120 = 0$$

from which $M = 2.30$. This value is substituted in the sidesway total in Table 5.7 to yield the sidesway moments for the 4000-lb load. The addition of these moments to the totals for no sidesway yields the final moments.

This procedure enables one-story bents with straight beams to be analyzed with the necessity of solving only one equation with one unknown regardless of the number of bays. If the frame is several stories high, the procedure can be applied to each story. Since an arbitrary horizontal deflection is introduced at each floor or roof level, there are as many unknowns and equations as there are stories.

The procedure is more difficult to apply to bents with curved or polygonal members between the columns. The effect of the change in the horizontal projection of the curved or polygonal portion of the bent must be included in the calculations. In many cases, it may be easier to analyze the bent as a curved beam (arch).

(A. Kleinlogel, "Rigid Frame Formulas," Frederick Ungar Publishing Co., New York.)

5.11.10 Rapid Approximate Analysis of Multistory Frames

Exact analysis of multistory rigid frames subjected to lateral forces, such as those from wind or earthquakes, involves lengthy calculations, and they are time-consuming and expensive, even when performed with computers. Hence, approximate methods of analysis are an alternative, at least for preliminary designs and, for some structures, for final designs.

It is noteworthy that for some buildings even the "exact" methods, such as those described in Arts. 5.11.8 and 5.11.9, are not exact. Usually, static horizontal loads are assumed for design purposes, but actually the forces exerted by wind and earthquakes are dynamic. In addition, these forces generally are uncertain in intensity, direction, and duration. Earthquake forces, usually assumed as a percentage of the mass of the building above each level, act at the base of the structure, not at each floor level as is assumed in design, and accelerations at each level vary nearly linearly with distance above the base. Also, at the beginning of a design, the sizes of the members are not known. Consequently, the exact resistance to lateral deformation cannot be calculated. Furthermore, floors, walls, and partitions help resist the lateral forces in a very uncertain way. See Art. 5.12 for a method of calculating the distribution of loads to rigid-frame bents.

Portal Method. Since an exact analysis is impossible, most designers prefer a wind-analysis method based on reasonable assumptions and requiring a minimum of calculations. One such method is the so-called "portal method."

It is based on the assumptions that points of inflection (zero bending moment) occur at the midpoints of all members and that exterior columns take half as much shear as do interior columns. These assumptions enable all moments and shears throughout the building frame to be computed by the laws of equilibrium.

Consider, for example, the roof level (Fig. 5.78*a*) of a tall building. A wind load of 600 lb is assumed to act along the top line of girders. To apply the portal method, we cut the building along a section through the inflection points of the top-story columns, which are assumed to be at the column midpoints, 6 ft down from the top of the building. We need now consider only the portion of the structure above this section.

Since the exterior columns take only half as much shear as do the interior columns, they each receive 100 lb, and the two interior columns, 200 lb. The moments at the tops of the columns equal these shears times the distance to the inflection point. The wall end of the end girder carries a moment equal to the moment in the column. (At the floor level below, as indicated in Fig. 5.78*b*, that end of the end girder carries a moment equal to the sum of the column moments.) Since the inflection point is at the midpoint of the girder, the moment at the inner end of the girder must be the same as at the outer end. The moment in the adjoining girder can be found by subtracting this moment from the column moment, because

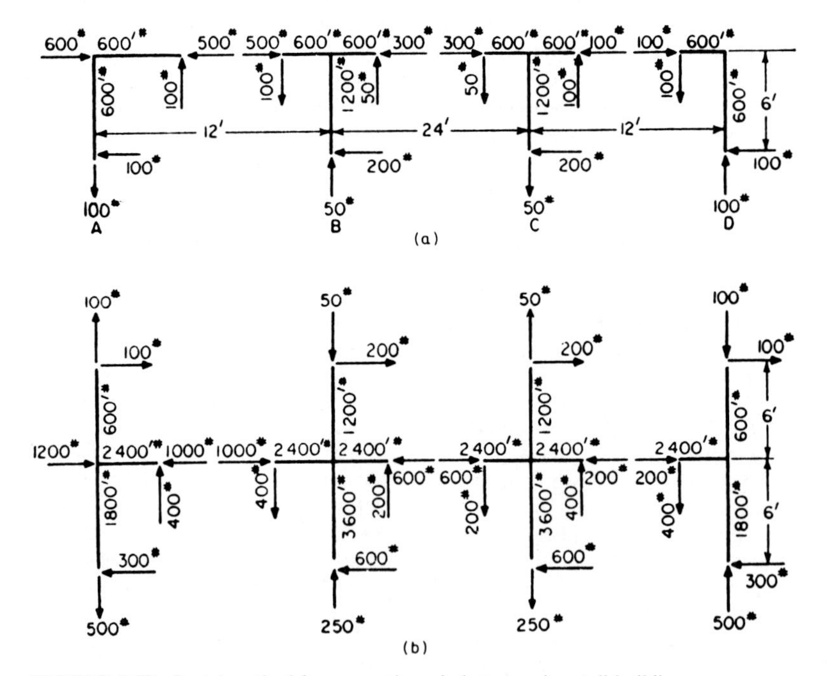

FIGURE 5.78 Portal method for computing wind stresses in a tall building.

the sum of the moments at the joint must be zero. (At the floor level below, as shown in Fig. 5.78*b,* the moment in the interior girder is found by subtracting the moment in the exterior girder from the sum of the column moments.)

Girder shears then can be computed by dividing girder moments by the half span. When these shears have been found, column loads can be easily computed from the fact that the sum of the vertical loads must be zero, by taking a section around each joint through column and girder inflection points. As a check, it should be noted that the column loads produce a moment that must be equal to the moments of the wind loads above the section for which the column loads were computed. For the roof level (Fig. 5.78*a*), for example, $-50 \times 24 + 100 \times 48 = 600 \times 6$.

Cantilever Method. Another wind-analysis procedure that is sometimes employed is the cantilever method. Basic assumptions here are that inflection points are at the midpoints of all members and that direct stresses in the columns vary as the distances of the columns from the center of gravity of the bent. The assumptions are sufficient to enable shears and moments in the frame to be determined from the laws of equilibrium.

For multistory buildings with height-to-width ratio of 4 or more, the Spurr modification is recommended ("Welded Tier Buildings," U.S. Steel Corp.). In this method, the moments of inertia of the girders at each level are made proportional to the girder shears.

The results obtained from the cantilever method generally will be different from those obtained by the portal method. In general, neither solution is correct, but the answers provide a reasonable estimate of the resistance to be provided against lateral deformation. (See also *Transactions of the ASCE,* Vol. 105, pp. 1713–1739, 1940.)

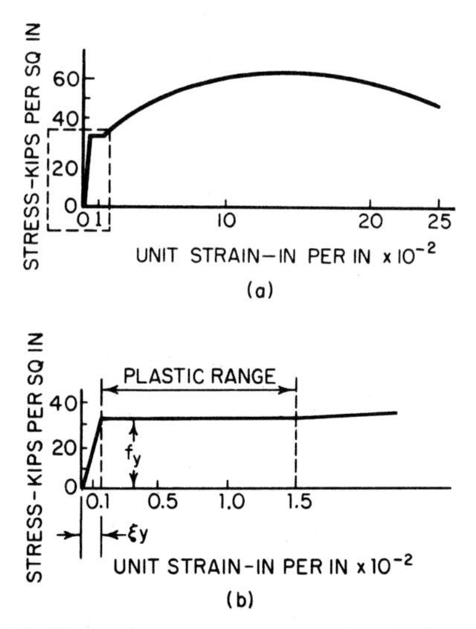

FIGURE 5.79 Stress-strain relationship for a ductile material generally is similar to the curve shown in (*a*). To simplify plastic analysis, the portion of (*a*) enclosed by the dash lines is approximated by the curve in (*b*), which extends to the range where strain hardening begins.

5.11.11 Beams Stressed into the Plastic Range

When an elastic material, such as structural steel, is loaded in tension with a gradually increasing load, stresses are proportional to strains up to the proportional limit (near the yield point). If the material, like steel, also is ductile, then it continues to carry load beyond the yield point, though strains increase rapidly with little increase in load (Fig. 5.79*a*).

Similarly, a beam made of a ductile material continues to carry more load after the stresses in the outer surfaces reach the yield point. However, the stresses will no longer vary with distance from the neutral axis, so the flexure formula [Eq. (5.54)] no longer holds. However, if simplifying assumptions are made, approximating the stress-strain relationship beyond the elastic limit, the load-carrying capacity of the beam can be computed with satisfactory accuracy.

Modulus of rupture is defined as the stress computed from the flexure formula for the maximum bending moment a beam sustains at failure. This is not a true stress but it is sometimes used to compare the strength of beams.

For a ductile material, the idealized stress-strain relationship in Fig. 5.79*b* may be assumed. Stress is proportional to strain until the yield-point stress f_y is reached, after which strain increases at a constant stress.

For a beam of this material, the following assumptions will also be made:

1. Plane sections remain plane, strains thus being proportional to distance from the neutral axis.
2. Properties of the material in tension are the same as those in compression.
3. Its fibers behave the same in flexure as in tension.
4. Deformations remain small.

Strain distribution across the cross section of a rectangular beam, based on these assumptions, is shown in Fig. 5.80*a*. At the yield point, the unit strain is ϵ_y and

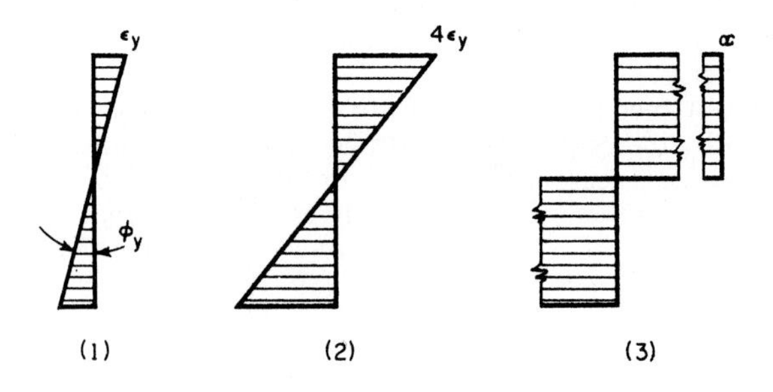

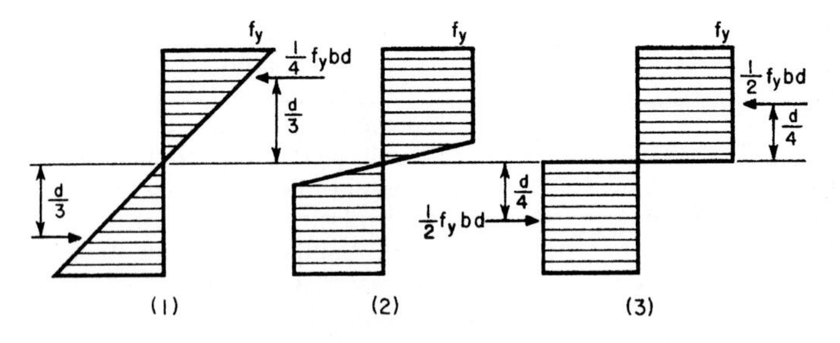

FIGURE 5.80 Strain distribution is shown in (*a*) and stress distribution in (*b*) for a cross section of a beam as it is loaded beyond the yield point, for the idealized stress-strain relationship in Fig. 5.79*b*: stage (1) shows the condition at the yield point of the outer surface; (2) after yielding starts; (3) at ultimate load.

the curvature ϕ_y, as indicated in (1). In (2), the strain has increased several times, but the section still remains plane. Finally, at failure, (3), the strains are very large and nearly constant across upper and lower halves of the section.

Corresponding stress distributions are shown in Fig. 5.80*b*. At the yield point, (1), stresses vary linearly and the maximum is f_y. With increase in load, more and more fibers reach the yield point, and the stress distribution becomes nearly constant, as indicated in (2). Finally, at failure, (3), the stresses are constant across the top and bottom parts of the section and equal to the yield-point stress.

The resisting moment at failure for a rectangular beam can be computed from the stress diagram for stage 3. If b is the width of the member and d its depth, then the ultimate moment for a rectangular beam is

$$M_P = \frac{bd^2}{4} f_y \tag{5.141}$$

Since the resisting moment at stage 1 is $M_y = f_y bd^2/6$, the beam carries 50% more moment before failure than when the yield-point stress is first reached at the outer surfaces.

A circular section has an M_P/M_y ratio of about 1.7, while a diamond section has a ratio of 2. The average wide-flange rolled-steel beam has a ratio of about 1.14.

Plastic Hinges. The relationship between moment and curvature in a beam can be assumed to be similar to the stress-strain relationship in Fig. 5.80*b*. Curvature ϕ varies linearly with moment until $M_y = M_P$ is reached, after which ϕ increases indefinitely at constant moment. That is, a plastic hinge forms.

Moment Redistribution. This ability of a ductile beam to form plastic hinges enables a fixed-end or continuous beam to carry more load after M_P occurs at a section, because a redistribution of moments takes place. Consider, for example, a uniformly loaded, fixed-end, prismatic beam. In the elastic range, the end moments are $M_L = M_R = WL/12$, while the midspan moment M_C is $WL/24$. The load when the yield point is reached at the outer surfaces at the beam ends is $W_y = 12M_y/L$. Under this load the moment capacity of the ends of the beam is nearly exhausted; plastic hinges form there when the moment equals M_P. As load is increased, the ends then rotate under constant moment and the beam deflects like a simply supported beam. The moment at midspan increases until the moment capacity at that section is exhausted and a plastic hinge forms. The load causing that condition is the ultimate load W_u since, with three hinges in the span, a link mechanism is formed and the member continues to deform at constant load. At the time the third hinge is formed, the moments at ends and center are all equal to M_P. Therefore, for equilibrium, $2M_P = W_u L/8$, from which $W_u = 16M_P/L$. Since for the idealized moment-curvature relationship, M_P was assumed equal to M_y, the carrying capacity due to redistribution of moments is 33% greater than W_y.

5.12 LOAD DISTRIBUTION TO BENTS AND SHEAR WALLS

Buildings must be designed to resist horizontal forces as well as vertical loads. In tall buildings, the lateral forces must be given particular attention, because if they are not properly provided for, they can collapse the structure (Art. 3.2.3). The usual procedure for preventing such disasters is to provide structural framing capable of transmitting the horizontal forces from points of application to the building foundations.

Because the horizontal loads may come from any direction, they generally are resolved into perpendicular components, and correspondingly the lateral-force-resisting framing is also placed in perpendicular directions. The maximum magnitude of load is assumed to act in each of those directions. Bents or shear walls, which act as vertical cantilevers and generally are often also used to support some of the building's gravity loads, usually are spaced at appropriate intervals for transmitting the loads to the foundations.

A bent consists of vertical trusses or continuous rigid frames located in a plane. The trusses usually are an assemblage of columns, horizontal girders, and diagonal bracing (Art. 3.2.4). The rigid frames are composed of girders and columns, with so-called **wind connections** between them to establish continuity. Shear walls are thin cantilevers braced by floors and roofs (Art. 3.2.4).

5.12.1 Diaphragms

Horizontal distribution of lateral forces to bents and shear walls is achieved by the floor and roof systems acting as diaphragms (Fig. 5.81).

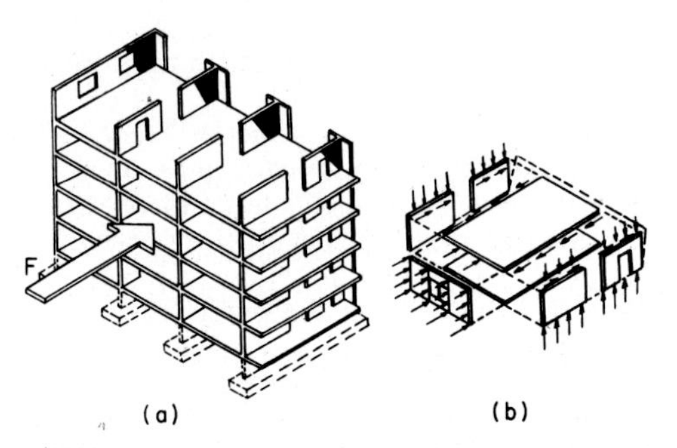

FIGURE 5.81 Floors of building distribute horizontal loads to shear walls (diaphragm action).

To qualify as a diaphragm, a floor or roof system must be able to transmit the lateral forces to bents and shear walls without exceeding a horizontal deflection that would cause distress to any vertical element. The successful action of a diaphragm also requires that it be properly tied into the supporting framing. Designers should ensure this action by appropriate detailing at the juncture between horizontal and vertical structural elements of the building.

Diaphragms may be considered analogous to horizontal (or inclined, in the case of some roofs) plate girders. The roof or floor slab constitutes the web; the joists, beams, and girders function as stiffeners; and the bents and shear walls act as flanges.

Diaphragms may be constructed of structural materials, such as concrete, wood, or metal in various forms. Combinations of such materials are also possible. Where a diaphragm is made up of units, such as plywood, precast-concrete planks, or steel decking, its characteristics are, to a large degree, dependent on the attachments of one unit to another and to the supporting members. Such attachments must resist shearing stresses due to internal translational and rotational actions.

The stiffness of a horizontal diaphragm affects the distribution of the lateral forces to the bents and shear walls. For the purpose of analysis, diaphragms may be classified into three groups—rigid, semirigid or semiflexible, and flexible—although no diaphragm is actually infinitely rigid or infinitely flexible.

A **rigid diaphragm** is assumed to distribute horizontal forces to the vertical resisting elements in proportion to the relative rigidities of these elements (Fig. 5.82).

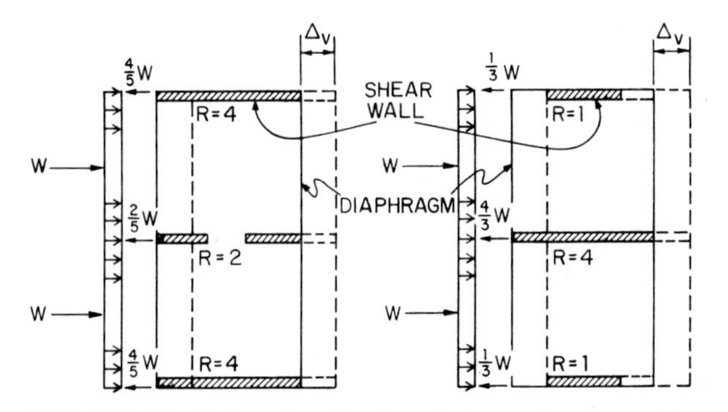

FIGURE 5.82 Horizontal section through shear walls connected by a rigid diaphragm. R = relative rigidity and Δ_v = shear-wall deflection.

Semirigid or semiflexible diaphragms are diaphragms that deflect significantly under load, but have sufficient stiffness to distribute a portion of the load to the vertical elements in proportion to the rigidities of these elements. The action is analogous to a continuous beam of appreciable stiffness on yielding supports (Fig. 5.83). Diaphragm reactions are dependent on the relative stiffnesses of diaphragm and vertical resisting elements.

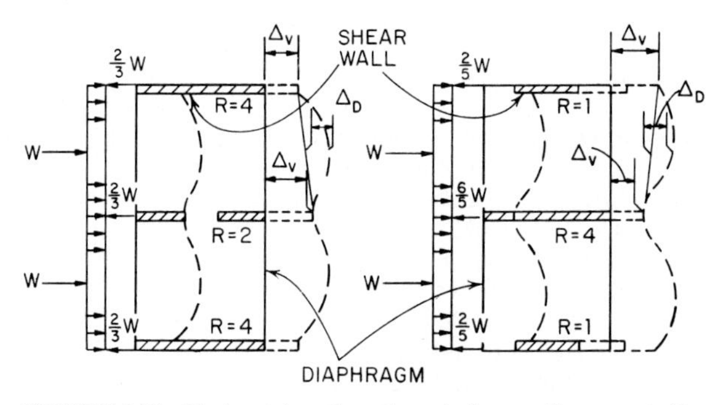

FIGURE 5.83 Horizontal sections through shear walls connected by a semirigid diaphragm. Δ_D = diaphragm horizontal deflection.

A **flexible diaphragm** is analogous to a continuous beam or series of simple beams spanning between nondeflecting supports. Thus, a flexible diaphragm is considered to distribute the lateral forces to the vertical resisting elements in proportion to the exterior-wall tributary areas (Fig. 5.84).

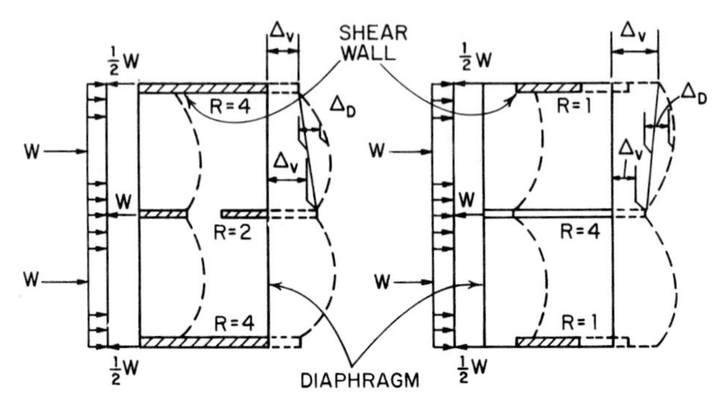

FIGURE 5.84 Horizontal section through shear walls connected by a flexible diaphragm.

A rigorous analysis of lateral-load distribution to shear walls or bents is sometimes very time-consuming, and frequently unjustified by the results. Therefore, in many cases, a design based on reasonable limits may be used. For example, the load may be distributed by first considering the diaphragm rigid, and then by considering it flexible. If the difference in results is not great, the shear walls can then be safely designed for the maximum applied load. (See also Art. 5.12.2.)

5.12.2 Torque Distribution to Shear Walls

When the line of action of the resultant of lateral forces acting on a building does not pass through the center of rigidity of a vertical, lateral-force-resisting system, distribution of the rotational forces must be considered as well as distribution of the translational forces. If rigid or semirigid diaphragms are used, the designer may assume that torsional forces are distributed to the shear walls in proportion to their relative rigidities and their distances from the center of rigidity. A flexible diaphragm should not be considered capable of distributing torsional forces.

See also Art. 5.12.5.

Example of Torque Distribution to Shear Walls. To illustrate load-distribution calculations for shear walls with rigid or semirigid diaphragms, Fig. 5.85 shows a horizontal section through three shear walls *A*, *B*, and *C* taken above a rigid floor. Wall *B* is 16 ft from wall *A*, and 24 ft from wall *C*. Rigidity of *A* is 0.33, of *B* 0.22, and of *C* 0.45 (Art. 5.12.5). A 20-kip horizontal force acts at floor level parallel to the shear walls and midway between *A* and *C*.

The center of rigidity of the shear walls is located, relative to wall *A*, by taking moments about *A* of the wall rigidities and dividing the sum of these moments by the sum of the wall rigidities, in this case 1.00.

$$x = 0.22 \times 16 + 0.45 \times 40 = 21.52 \text{ ft}$$

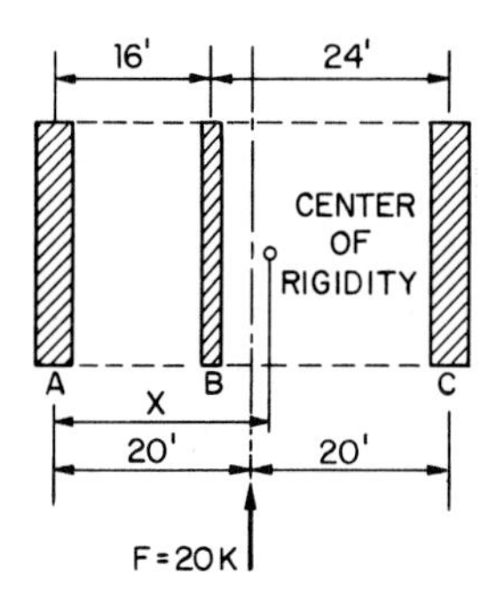

FIGURE 5.85 Rigid diaphragm distributes 20-kip horizontal force to shear walls *A*, *B*, and *C*.

Thus, the 20-kip lateral force has an eccentricity of 21.52 − 20 = 1.52 ft. The eccentric force may be resolved into a 20-kip force acting through the center of rigidity and not producing torque, and a couple producing a torque of 20 × 1.52 = 30.4 ft-kips.

The nonrotational force is distributed to the shear walls in proportion to their rigidities:

Wall *A*: 0.33 × 20 = 6.6 kips

Wall *B*: 0.22 × 20 = 4.4 kips

Wall *C*: 0.45 × 20 = 9.0 kips

For distribution of the torque to the shear walls, the equivalent of moment of inertia must first be computed:

$$I = 0.33(21.52)^2 + 0.22(5.52)^2 + 0.45(18.48)^2 = 313$$

Then, the torque is distributed in direct proportion to shear-wall rigidity and distance from center of rigidity and in inverse proportion to *I*.

Wall *A*: 30.4 × 0.33 × 21.52/313 = 0.690 kips

Wall *B*: 30.4 × 0.22 × 5.52/313 = 0.118 kips

Wall *C*: 30.4 × 0.45 × 18.48/313 = 0.808 kips

The torsional forces should be added to the nonrotational forces acting on walls *A* and *B*, whereas the torsional force on wall *C* acts in the opposite direction to the nonrotational force. For a conservative design, the torsional force on wall *C* should not be subtracted. Hence, the walls should be designed for the following forces:

Wall *A*: 6.6 + 0.7 = 7.3 kips

Wall *B*: 4.4 + 0.1 = 4.5 kips

Wall *C*: 9 kips

5.12.3 Deflections of Bents and Shear Walls

When parallel bents or shear walls are connected by rigid diaphragms (Art. 5.12.1) and horizontal loads are distributed to the vertical resisting elements in proportion to their relative rigidities, the relative rigidity of the framing depends on the combined horizontal deflections due to shear and flexure. For the dimensions of lateral-force-resisting framing used in many high-rise buildings, however, deflections due to flexure greatly exceed those due to shear. In such cases, only flexural rigidity need be considered in determination of relative rigidity of the bents and shear walls (Art. 5.12.5).

Horizontal deflections can be determined by treating the bents and shear walls as cantilevers. Deflections of braced bents can be calculated by the dummy-unit-load method (Art. 5.10.4) or a matrix method (Art. 5.13.3). Deflections of rigid frames can be obtained by summing the drifts of the stories, as determined by moment distribution (Art. 5.11.9) or a matrix method. And deflections of shear walls can be computed from formulas given in Art. 5.5.15, the dummy-unit-load method, or a matrix method.

For a shear wall with a solid, rectangular cross section, the flexural deflection at the top under uniform loading is given by the formula for a cantilever in Fig. 5.39:

$$\delta_c = \frac{wH^4}{8EI} \tag{5.142}$$

where w = uniform lateral load
H = height of the wall
E = modulus of elasticity of the wall material
I = moment of inertia of wall cross section = $tL^3/12$
t = wall thickness
L = length of wall

The cantilever shear deflection under uniform loading may be computed from

$$\delta_v = \frac{0.6wH^2}{E_v A} \tag{5.143}$$

where E_v = modulus of rigidity of wall cross section
= $E/2(1 + \mu)$
μ = Poisson's ratio for the wall material (0.25 for concrete and masonry)
A = cross-sectional area of the wall = tL

The total deflection then is

$$\delta_c + \delta_v = \frac{1.5wH}{Et}\left[\left(\frac{H}{L}\right)^3 + \frac{H}{L}\right] \tag{5.144}$$

For a cantilever wall subjected to a concentrated load P at the top, the flexural deflection at the top is

$$\delta_c = \frac{PH^3}{3EI} \tag{5.145}$$

The shear deflection at the top of the wall is

$$\delta_v = \frac{1.2PH}{E_v A} \tag{5.146}$$

Hence, the total deflection of the cantilever is

$$\delta = \frac{4P}{Et}\left[\left(\frac{H}{L}\right)^3 + 0.75\,\frac{H}{L}\right] \tag{5.147}$$

For a wall fixed against rotation at the top and subjected to a concentrated load P at the top, the flexural deflection at the top is

$$\delta_c = \frac{PH^3}{12EI} \tag{5.148}$$

The shear deflection for the fixed-end wall is given by Eq. (5.145). Hence, the total deflection for the wall is

$$\delta = \frac{P}{Et}\left[\left(\frac{H}{L}\right)^3 + 3\frac{H}{L}\right] \tag{5.149}$$

5.12.4 Diaphragm-Deflection Limitations

As indicated in Art. 5.12.1, horizontal deflection of diaphragms plays an important role in determining lateral-load distribution to bents and shear walls. Another design consideration is the necessity of limiting diaphram deflection to prevent excessive stresses in walls perpendicular to shear walls. Equation (5.150) was suggested by the Structural Engineers Association of Southern California for allowable story deflection Δ, in, of masonry or concrete building walls.

$$\Delta = \frac{h^2 f}{0.01Et} \tag{5.150}$$

where h = height of wall between adjacent horizontal supports, ft
t = thickness of wall, in
f = allowable flexural compressive stress of wall material, psi
E = modulus of elasticity of wall material, psi

This limit on deflection must be applied with engineering judgment. For example, continuity of wall at floor level is assumed, and in many cases is not present because of through-wall flashing. In this situation, the deflection may be based on the allowable compressive stress in the masonry, if a reduced cross section of wall is assumed. The effect of reinforcement, which may be present in a reinforced

TABLE 5.8 Maximum Span-Width or Span-Depth Ratios for Diaphragms—Roofs or Floors*

Diaphragm construction	Masonry and concrete walls	Wood and light steel walls
Concrete	Limited by deflection	
Steel deck (continuous sheet in a single plane)	4:1	5:1
Steel deck (without continuous sheet)	2:1	2½:1
Cast-in-place reinforced gypsum roofs	3:1	4:1
Plywood (nailed all edges)	3:1	4:1
Plywood (nailed to supports only—blocking may be omitted between joists)	2½:1†	3½:1
Diagonal sheathing (special)	3:1†	3½:1
Diagonal sheathing (conventional construction)	2:1†	2½:1

*From California Administrative Code, Title 21, Public Works.

†Use of diagonal sheathed or unblocked plywood diaphragms for buildings having masonry or reinforced concrete walls shall be limited to one-story buildings or to the roof of a top story.

brick masonry wall or as a tie to the floor system in a nonreinforced or partly reinforced masonry wall, was not considered in development of Eq. (5.150). Note also that the limit on wall deflection is actually a limit on differential deflection between two successive floor, or diaphragm, levels.

Maximum span-width or span-depth ratios for diaphragms are usually used to control horizontal diaphragm deflection indirectly. Normally, if the diaphragm is designed with the proper ratio, the diaphragm deflection will not be critical. Table 5.8 may be used as a guide for proportioning diaphragms.

5.12.5 Shear-Wall Rigidity

Where shear walls are connected by rigid diaphragms so that they must deflect equally under horizontal loads, the proportion of total horizontal load at any level carried by a shear wall parallel to the load depends on the relative rigidity, or stiffness, of the wall in the direction of the load (Art. 5.12.1). Rigidity of a shear wall is inversely proportional to its deflection under unit horizontal load. This deflection equals the sum of the shear and flexural deflections under the load (Art. 5.12.3).

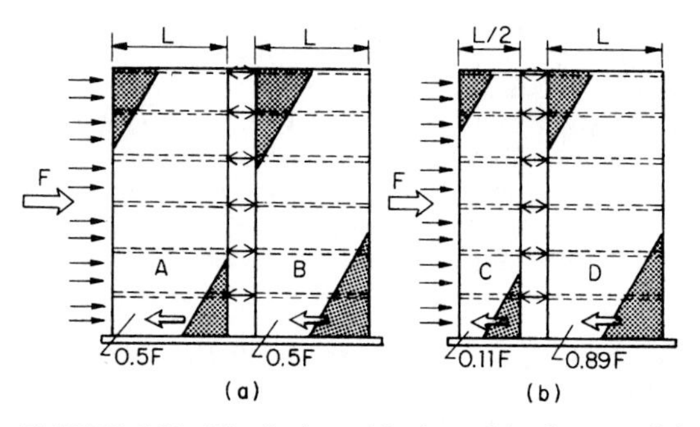

FIGURE 5.86 Distribution of horizontal load to parallel shear walls: (*a*) walls with the same length and rigidity share the load equally; (*b*) wall half the length of another carries less than one-eighth of the load.

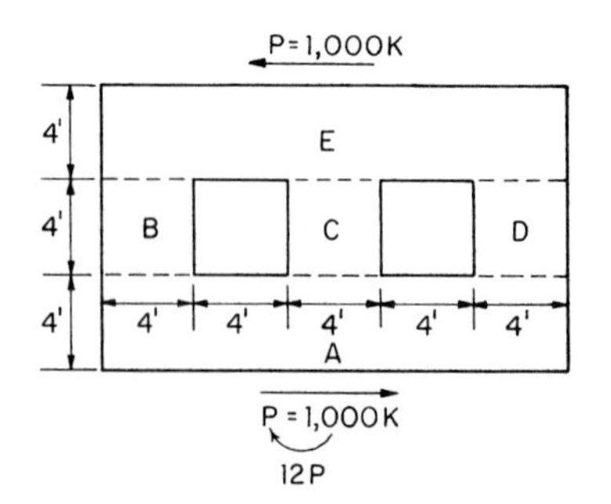

FIGURE 5.87 Shear wall, 8 in thick, with openings.

Where a shear wall contains no openings, computations for deflection and rigidity are simple. In Fig. 5.86*a*, each of the shear walls has the same length and rigidity. So each takes half the total load. In Fig. 5.86*b*, length of wall *C* is half that of wall *D*. By Eq. (5.142), *C* therefore receives less than one-eighth the total load.

Walls with Openings. Where shear walls contain openings, such as doors and windows, computations for deflection and rigidity are more complex. But approximate methods may be used.

For example, the wall in Fig. 5.87, subjected to a 1000-kip load at the top, may be treated in parts. The wall is 8 in thick, and its modulus of elasticity $E = 2400$ ksi. Its height-length ratio H/L is $12/20 = 0.6$. The wall is perforated by two, symmetrically located, 4-ft-square openings.

Deflection of this wall can be estimated by subtracting from the deflection it would have if it were solid the deflection of a solid, 4-ft-deep, horizontal midstrip, and then adding the deflection of the three coupled piers B, C, and D.

Deflection of the 12-ft-high solid wall can be obtained from Eq. (5.147):

$$\delta = \frac{4 \times 10^3}{2.4 \times 10^3 \times 8} [(0.6)^3 + 0.75 \times 0.6] = 0.138 \text{ in}$$

Rigidity of the solid wall then is

$$R = \frac{1}{0.138} = 7.22$$

Similarly, the deflection of the 4-ft-deep solid midstrip can be computed from Eq. (5.147), with $H/L = 4/20 = 0.20$.

$$\delta = \frac{4 \times 10^3}{2.4 \times 10^3 \times 8} [(0.20)^3 + 0.75 \times 0.20] = 0.033 \text{ in}$$

Deflection of the piers, which may be considered fixed top and bottom, can be obtained from Eq. (5.149), with $H/L = 4/4 = 1$. For any one of the piers, the deflection is

$$\delta'v = \frac{10^3}{2.4 \times 10^3 \times 8} (1 + 3) = 0.208 \text{ in}$$

The rigidity of a single pier is $1/0.208 = 4.81$, and of the three piers, $3 \times 4.81 = 14.43$. Therefore, the deflection of the three piers when coupled is

$$\delta = \frac{1}{14.43} = 0.069 \text{ in}$$

The deflection of the whole wall, with openings, then is approximately

$$\delta = 0.138 - 0.033 + 0.069 = 0.174 \text{ in}$$

And its rigidity is

$$R = \frac{1}{0.174} = 5.74$$

5.12.6 Effects of Shear-Wall Arrangements

To increase the stiffness of shear walls and thus their resistance to bending, intersecting walls or flanges may be used. Often in the design of buildings, Z-, T-, U-, L-, and I-shaped walls in plan develop as natural parts of the design. Shear walls with these shapes have better flexural resistance than a single, straight wall.

In calculation of flexural stresses in masonry shear walls for symmetrical T or I sections, the effective flange width may not exceed one-sixth the total wall height

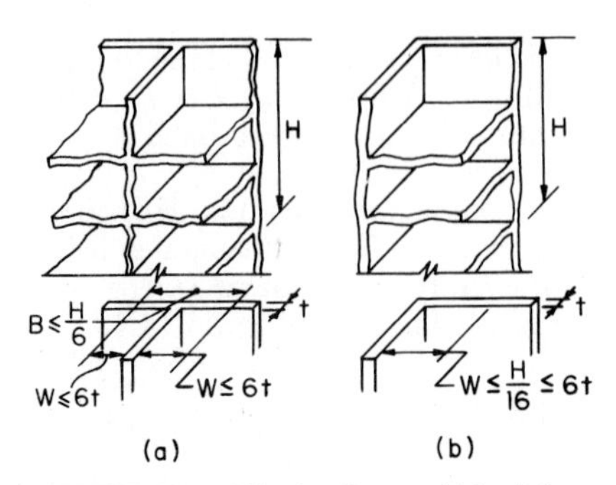

FIGURE 5.88 Effective flange width of shear walls may be less than the actual width: (*a*) limits for flanges of I and T shapes; (*b*) limits for C and L shapes.

above the level being analyzed. For unsymmetrical L or C sections, the width considered effective may not exceed one-sixteenth the total wall height above the level being analyzed. In either case, the overhang for any section may not exceed six times the flange thickness (Fig. 5.88).

The shear stress at the intersection of the walls should not exceed the permissible shear stress.

5.12.7 Coupled Shear Walls

Another method than that described in Art. 5.12.6 for increasing the stiffness of a bearing-wall structure and reducing the possibility of tension developing in masonry shear walls under lateral loads is coupling of coplanar shear walls.

Figures 5.89 and 5.90 indicate the effect of coupling on stress distribution in a pair of walls under horizontal forces parallel to the walls. A flexible connection

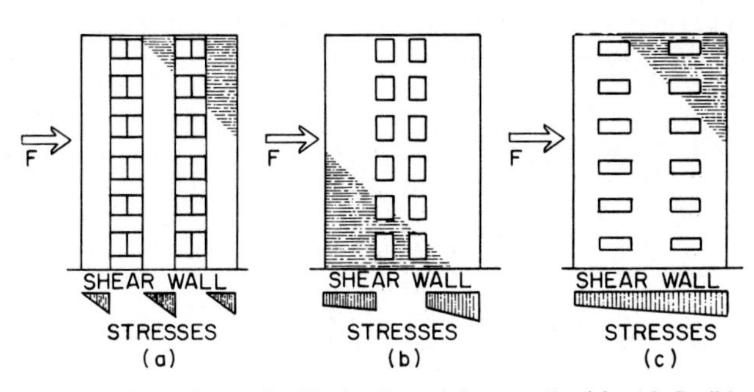

FIGURE 5.89 Stress distribution in end shear walls: (*a*) with flexible coupling; (*b*) with rigid-frame-type action; (*c*) with plate-type action.

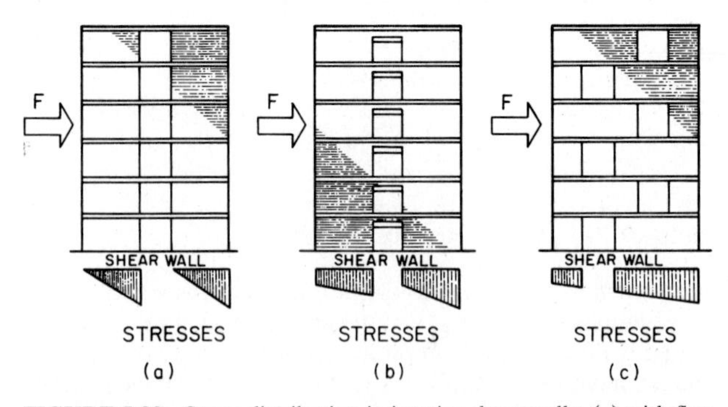

FIGURE 5.90 Stress distribution in interior shear walls: (*a*) with flexible coupling; (*b*) with rigid-frame-type action; (*c*) with plate-type action.

between the walls is assumed in Figs. 5.89*a* and 5.90*a*, so that the walls act as independent vertical cantilevers in resisting lateral loads. In Figs. 5.89*b* and 5.90*b*, the walls are assumed to be connected with a more rigid member, which is capable of shear and moment transfer. A rigid-frame-type action results. This can be accomplished with a steel-reinforced concrete, or reinforced brick masonry coupling. A plate-type action is indicated in Figs. 5.89*c* and 5.90*c*. This assumes an extremely rigid connection between walls, such as full story-height walls or deep rigid spandrels.

5.13 FINITE-ELEMENT METHODS

From the basic principles given in preceding articles, systematic procedures have been developed for determining the behavior of a structure from a knowledge of the behavior under load of its components. In these methods, called finite-element methods, a structural system is considered an assembly of a finite number of finite-size components, or elements. These are assumed to be connected to each other only at discrete points, called nodes. From the characteristics of the elements, such as their stiffness or flexibility, the characteristics of the whole system can be derived. With these known, internal stresses and strains throughout can be computed.

Choice of elements to be used depends on the type of structure. For example, for a truss with joints considered hinged, a natural choice of element would be a bar, subjected only to axial forces. For a rigid frame, the elements might be beams subjected to bending and axial forces, or to bending, axial forces, and torsion. For a thin plate or shell, elements might be triangles or rectangles, connected at vertices. For three-dimensional structures, elements might be beams, bars, tetrahedrons, cubes, or rings.

For many structures, because of the number of finite elements and nodes, analysis by a finite-element method requires mathematical treatment of large amounts of data and solution of numerous simultaneous equations. For this purpose, the use of computers is advisable. The mathematics of such analyses is usually simpler and more compact when the data are handled in matrix form. (See also Art. 5.10.7.)

5.13.1 Force and Displacement Methods

The methods used for analyzing structures generally may be classified as force (flexibility) or displacement (stiffness) methods.

In analysis of statically indeterminate structures by force methods, forces are chosen as redundants, or unknowns. The choice is made in such a way that equilibrium is satisfied. These forces are then determined from the solution of equations that ensure compatibility of all displacements of elements at each node. After the redundants have been computed, stresses and strains throughout the structure can be found from equilibrium equations and stress-strain relations.

In displacement methods, displacements are chosen as unknowns. The choice is made in such a way that geometric compatibility is satisfied. These displacements are then determined from the solution of equations that ensure that forces acting at each node are in equilibrium. After the unknowns have been computed, stresses and strains throughout the structure can be found from equilibrium equations and stress-strain relations.

In choosing a method, the following should be kept in mind: In force methods, the number of unknowns equals the degree of indeterminancy. In displacement methods, the number of unknowns equals the degrees of freedom of displacement at nodes. The fewer the unknowns, the fewer the calculations required.

Both methods are based on the force-displacement relations and utilize the stiffness and flexibility matrices described in Art. 5.10.7. In these methods, displacements and external forces are resolved into components—usually horizontal, vertical, and rotational—at nodes, or points of connection of the finite elements. In accordance with Eq. (5.103*a*), the stiffness matrix transforms displacements into forces. Similarly, in accordance with Eq. (5.103*b*), the flexibility matrix transforms forces into displacements. To accomplish the transformation, the nodal forces and displacements must be assembled into correspondingly positioned elements of force and displacement vectors. Depending on whether the displacement or the force method is chosen, stiffness or flexibility matrices are then established for each of the finite elements and these matrices are assembled to form a square matrix, from which the stiffness or flexibility matrix for the structure as a whole is derived. With that matrix known and substituted into equilibrium and compatibility equations for the structure, all nodal forces and displacements of the finite elements can be determined from the solution of the equations. Internal stresses and strains in the elements can be computed from the now known nodal forces and displacements.

5.13.2 Element Flexibility and Stiffness Matrices

The relationship between *independent* forces and displacements at nodes of finite elements comprising a structure is determined by flexibility matrices **f** or stiffness matrices **k** of the elements. In some cases, the components of these matrices can be developed from the defining equations:

The *j*th column of a flexibility matrix of a finite element contains all the nodal displacements of the element when one force $\mathbf{S}_j$ is set equal to unity and all other independent forces are set equal to zero.

The *j*th column of a stiffness matrix of a finite element consists of the forces acting at the nodes of the element to produce a unit displacement of the node at which displacement δ_j occurs and in the direction of δ_j but no other nodal displacements of the element.

Bars with Axial Stress Only. As an example of the use of the definitions of flexibility and stiffness, consider the simple case of an elastic bar under tension applied by axial forces P_i and P_j at nodes *i* and *j*, respectively (Fig. 5.91). The bar might be the finite element of a truss, such as a diagonal or a hanger. Connections to other members are made at nodes *i* and *j*, which can transmit only forces in the directions *i* to *j* or *j* to *i*.

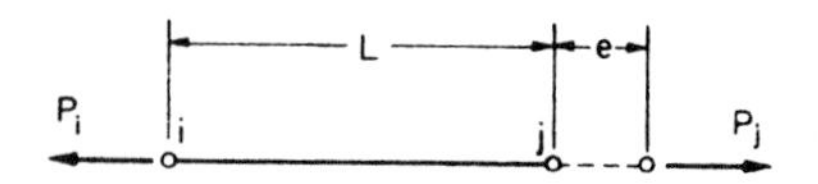

FIGURE 5.91 Elastic bar in tension.

For equilibrium, $P_i = P_j = P$. Displacement of node *j* relative to node *i* is *e*. From Eq. (5.23), $e = PL/AE$, where L is the initial length of the bar, A the bar cross-sectional area, and E the modulus of elasticity. Setting $P = 1$ yields the flexibility of the bar,

$$f = \frac{L}{AE} \tag{5.151}$$

Setting $e = 1$ gives the stiffness of the bar,

$$k = \frac{AE}{L} \tag{5.152}$$

Beams with Bending Only. As another example of the use of the definition to determine element flexibility and stiffness matrices, consider the simple case of an elastic prismatic beam in bending applied by moments M_i and M_j at nodes i and j, respectively (Fig. 5.92*a*). The beam might be a finite element of a rigid frame. Connections to other members are made at nodes i and j, which can transmit moments and forces normal to the beam.

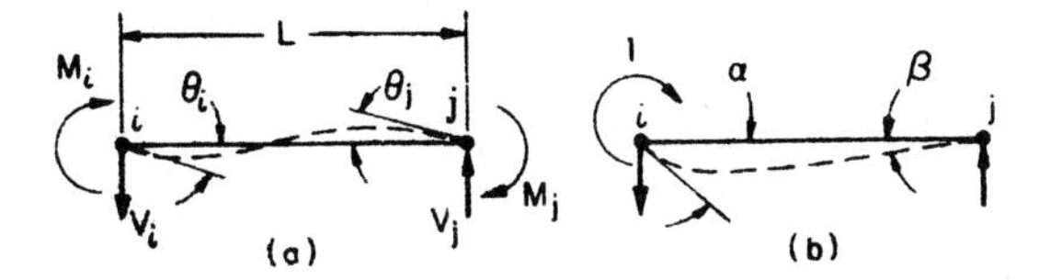

FIGURE 5.92 Beam subjected to end moments and shears.

Nodal displacements of the element can be sufficiently described by rotations θ_i and θ_j relative to the straight line between nodes i and j. For equilibrium, forces $V_j = -V_i$ normal to the beam are required at nodes j and i, respectively, and $V_j = (M_i + M_j)/L$, where L is the span of the beam. Thus, M_i and M_j are the only independent forces acting. Hence, the force-displacement relationship can be written for this element as

$$\theta = \begin{bmatrix} \theta_i \\ \theta_j \end{bmatrix} = \mathbf{f} \begin{bmatrix} M_i \\ M_j \end{bmatrix} = \mathbf{fM} \tag{5.153}$$

$$\mathbf{M} = \begin{bmatrix} M_i \\ M_j \end{bmatrix} = \mathbf{k} \begin{bmatrix} \theta_i \\ \theta_j \end{bmatrix} = \mathbf{k}\theta \tag{5.154}$$

The flexibility matrix **f** then will be a 2 × 2 matrix. The first column can be obtained by setting $M_i = 1$ and $M_j = 0$ (Fig. 5.92*b*). The resulting angular rotations are given by Eqs. (5.107) and (5.108): For a beam with constant moment of inertia I and modulus of elasticity E, the rotations are $\alpha = L/3EI$ and $\beta = -L/6EI$. Similarly, the second column can be developed by setting $M_i = 0$ and $M_j = 1$.

The flexibility matrix for a beam in bending then is

$$\mathbf{f} = \begin{bmatrix} \dfrac{L}{3EI} & -\dfrac{L}{6EI} \\ -\dfrac{L}{6EI} & \dfrac{L}{3EI} \end{bmatrix} = \frac{L}{6EI} \begin{bmatrix} 2 & -1 \\ -1 & 2 \end{bmatrix} \tag{5.155}$$

The stiffness matrix, obtained in a similar manner or by inversion of **f**, is

$$\mathbf{k} = \begin{bmatrix} \dfrac{4EI}{L} & \dfrac{2EI}{L} \\ \dfrac{2EI}{L} & \dfrac{4EI}{L} \end{bmatrix} = \frac{2EI}{L} \begin{bmatrix} 2 & 1 \\ 1 & 2 \end{bmatrix} \tag{5.156}$$

Beams Subjected to Bending and Axial Forces. For a beam subjected to nodal moments M_i and M_j and axial forces P, flexibility and stiffness are represented by 3×3 matrices. The load-displacement relations for a beam of span L, constant moment of inertia I, modulus of elasticity E, and cross-sectional area A are given by

$$\begin{bmatrix} \theta_i \\ \theta_j \\ e \end{bmatrix} = \mathbf{f} \begin{bmatrix} M_i \\ M_j \\ P \end{bmatrix} \qquad \begin{bmatrix} M_i \\ M_j \\ P \end{bmatrix} = \mathbf{k} \begin{bmatrix} \theta_i \\ \theta_j \\ e \end{bmatrix} \tag{5.157}$$

In this case, the flexibility matrix is

$$\mathbf{f} = \frac{L}{6EI} \begin{bmatrix} 2 & -1 & 0 \\ -1 & 2 & 0 \\ 0 & 0 & \eta \end{bmatrix} \tag{5.158}$$

where $\eta = 6I/A$, and the stiffness matrix is

$$\mathbf{k} = \frac{EI}{L} \begin{bmatrix} 4 & 2 & 0 \\ 2 & 4 & 0 \\ 0 & 0 & \psi \end{bmatrix} \tag{5.159}$$

where $\psi = A/I$.

5.13.3 Displacement (Stiffness) Method

With the stiffness or flexibility matrix of each finite element of a structure known, the stiffness or flexibility matrix for the whole structure can be determined, and with that matrix, forces and displacements throughout the structure can be computed (Art. 5.13.2). To illustrate the procedure, the steps in the displacement, or stiffness, method are described in the following. The steps in the flexibility method are similar. For the stiffness method:

Step 1. Divide the structure into interconnected elements and assign a number, for identification purposes, to every node (intersection and terminal of elements). It may also be useful to assign an identifying number to each element.

Step 2. Assume a right-handed cartesian coordinate system, with axes x, y, z. Assume also at each node of a structure to be analyzed a system of base unit vectors, $\mathbf{e}_1$ in the direction of the x axis, $\mathbf{e}_2$ in the direction of the y axis, and $\mathbf{e}_3$ in the direction of the z axis. Forces and moments acting at a node are resolved into components in the directions of the base vectors. Then, the forces and moments at the node may be represented by the vector $P_i\mathbf{e}_i$, where P_i is the magnitude of the force or moment acting in the direction of $\mathbf{e}_i$. This vector, in turn, may be conveniently represented by a column matrix $\mathbf{P}$. Similarly, the displacements—translations and rotation—of the node may be represented by the vector $\Delta_i\mathbf{e}_i$, where Δ_i is the magnitude of the displacement acting in the direction of $\mathbf{e}_i$. This vector, in turn, may be represented by a column matrix Δ.

For compactness, and because, in structural analysis, similar operations are performed on all nodal forces, all the loads, including moments, acting on all the nodes may be combined into a single column matrix $\mathbf{P}$. Similarly, all the nodal displacements may be represented by a single column matrix Δ.

When loads act along a beam, they should be replaced by equivalent forces at the nodes—simple-beam reactions and fixed-end moments, both with signs reversed from those induced by the loads. The final element forces are then determined by adding these moments and reactions to those obtained from the solution with only the nodal forces.

Step 3. Develop a stiffness matrix $\mathbf{k}_i$ for each element i of the structure (see Art. 5.13.2). By definition of stiffness matrix, nodal displacements and forces for the ith element are related by

$$\mathbf{S}_i = \mathbf{k}_i \boldsymbol{\delta}_i \qquad i = 1, 2, \ldots, n \tag{5.160}$$

where $\mathbf{S}_i$ = matrix of forces, including moments and torques acting at the nodes of the ith element
$\boldsymbol{\delta}_i$ = matrix of displacements of the nodes of the ith element

Step 4. For compactness, combine this relationship between nodal displacements and forces for each element into a single matrix equation applicable to all the elements:

$$\mathbf{S} = \mathbf{k}\boldsymbol{\delta} \tag{5.161}$$

where $\mathbf{S}$ = matrix of all forces acting at the nodes of all elements
$\boldsymbol{\delta}$ = matrix of all nodal displacements for all elements

$$\mathbf{k} = \begin{bmatrix} \mathbf{k}_1 & 0 & \cdots & 0 \\ 0 & \mathbf{k}_2 & \cdots & 0 \\ \cdot\cdot & \cdot\cdot & \cdots & \cdot\cdot \\ 0 & 0 & \cdots & \mathbf{k}_n \end{bmatrix} \tag{5.162}$$

Step 5. Develop a matrix b_0 that will transform the displacements Δ of the nodes of the structure into the displacement vector $\boldsymbol{\delta}$ while maintaining geometric compatibility:

$$\boldsymbol{\delta} = \mathbf{b}_0 \boldsymbol{\Delta} \tag{5.163}$$

$\mathbf{b}_0$ is a matrix of influence coefficients. The jth column of $\mathbf{b}_0$ contains the element nodal displacements when the node where Δ_j occurs is given a unit displacement in the direction of Δ_j, and no other nodes are displaced.

Step 6. Compute the stiffness matrix $\mathbf{K}$ for the whole structure from

$$\mathbf{K} = \mathbf{b}_0^T \mathbf{k} \mathbf{b}_0 \tag{5.164}$$

where $\mathbf{b}_0^T$ = transpose of $\mathbf{b}_0$ = matrix $\mathbf{b}_0$ with rows and columns interchanged

This equation may be derived as follows: From energy relationships, $\mathbf{P} = \mathbf{b}_0^T \mathbf{S}$. Substitution of $\mathbf{k}\boldsymbol{\delta}$ for $\mathbf{S}$ [Eq. (5.161)] and then substitution of $\mathbf{b}_0\boldsymbol{\Delta}$ for $\boldsymbol{\delta}$ [Eq. (5.163)] yields $\mathbf{P} = \mathbf{b}_0^T \mathbf{k} \mathbf{b}_0 \boldsymbol{\Delta}$. Comparison of this with Eq. (5.103*a*), $\mathbf{P} = \mathbf{k}\boldsymbol{\Delta}$ leads to Eq. (5.164).

Step 7. With the stiffness matrix $\mathbf{K}$ now known, solve the simultaneous equations

$$\boldsymbol{\Delta} = \mathbf{K}^{-1}\mathbf{P} \tag{5.165}$$

for the nodal displacements $\mathbf{\Delta}$. With these determined, calculate the member forces from

$$\mathbf{S} = \mathbf{k}\mathbf{b}_0\mathbf{\Delta} \tag{5.166}$$

(N. M. Baran, "Finite Element Analysis on Microcomputers," and H. Kardesluncer and D. H. Norris, "Finite Element Handbook," McGraw-Hill Publishing Company, New York; K. Bathe, "Finite Element Procedures in Engineering Analysis," T. R. Hughes, "The Finite Element Method," W. Weaver, Jr., and P. R. Johnston, "Structural Dynamics by Finite Elements," and H. T. Y. Yang, "Finite Element Structural Analysis," Prentice-Hall, Englewood Cliffs, N.J.)

5.14 STRESSES IN ARCHES

An arch is a curved beam, the radius of curvature of which is very large relative to the depth of the section. It differs from a straight beam in that: (1) loads induce both bending and direct compressive stresses in an arch; (2) arch reactions have horizontal components even though loads are all vertical; and (3) deflections have horizontal as well as vertical components (see also Arts. 5.6.1 to 5.6.4). Names of arch parts are given in Fig. 5.93.

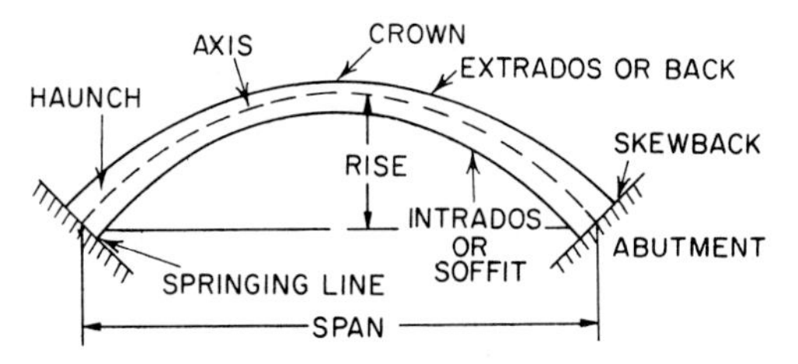

FIGURE 5.93 Components of an arch.

The necessity of resisting the horizontal components of the reactions is an important consideration in arch design. Sometimes these forces are taken by tie rods between the supports, sometimes by heavy abutments or buttresses.

Arches may be built with fixed ends, as can straight beams, or with hinges at the supports. They may also be built with a hinge at the crown.

5.14.1 Three-Hinged Arches

An arch with a hinge at the crown as well as at both supports (Fig. 5.94) is statically determinate. There are four unknowns—two horizontal and two vertical components of the reactions—but four equations based on the laws of equilibrium are available: (1) The sum of the horizontal forces must be zero. (2) The sum of the moments about the left support must be zero. (3) The sum of the moments about the right support must be zero. (4) The bending moment at the crown hinge must be zero (not to be confused with the sum of the moments about the crown, which also must be equal to zero but which would not lead to an independent equation for the solution of the reactions).

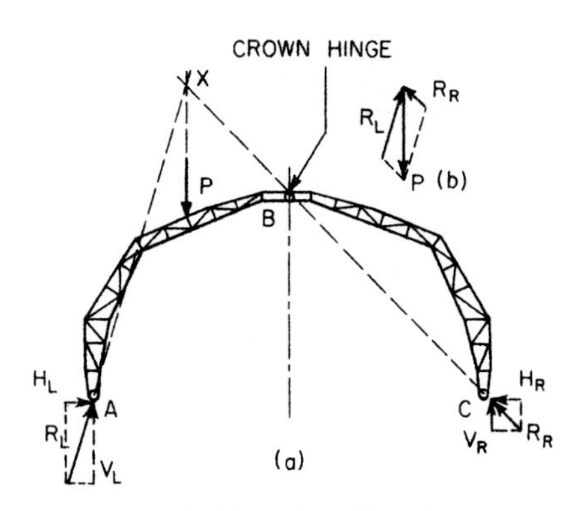

FIGURE 5.94 Three-hinged arch.

Stresses and reactions in three-hinged arches can be determined graphically by taking advantage of the fact that the bending moment at the crown hinge is zero. For example, in Fig. 5.94*a*, a concentrated load P is applied to segment AB of the arch. Then, since the bending moment at B must be zero, the line of action of the reaction at C must pass through the crown hinge. It intersects the line of action of P at X. The line of action of the reaction at A must also pass through X. Since P is equal to the sum of the reactions, and since the directions of the reactions have thus been determined, the magnitude of the reactions can be measured from a parallelogram of forces (Fig. 5.94*b*). When the reactions have been found, the stresses can be computed from the laws of statics (see Art. 5.14.3) or, in the case of a trussed arch, determined graphically.

5.14.2 Two-Hinged Arches

When an arch has hinges at the supports only (Fig. 5.95), it is statically indeterminate, and some knowledge of its deformations is required to determine the reactions. One procedure is to assume that one of the supports is on rollers. This

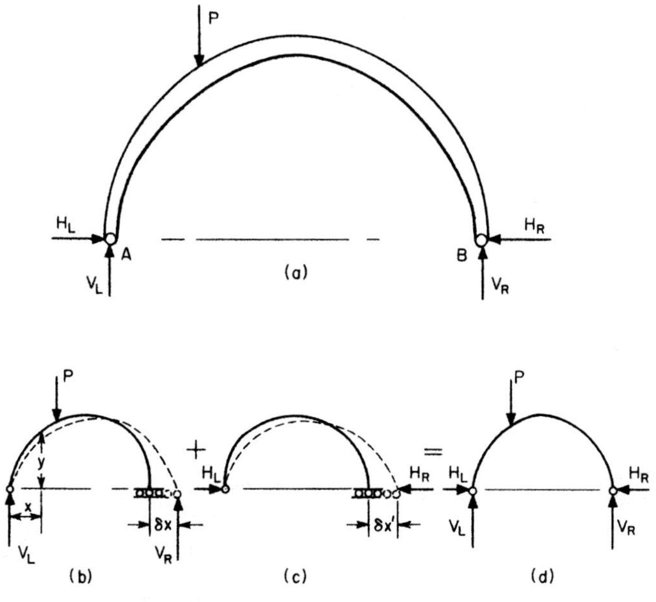

FIGURE 5.95 Two-hinged arch.

makes the arch statically determinate. The reactions and the horizontal movement of the support are computed for this condition (Fig. 5.95*b*). Then, the magnitude of the horizontal force required to return the movable support to its original position is calculated (Fig. 5.95*c*). The reactions for the two-hinged arch are finally found by superimposing the first set of reactions on the second (Fig. 5.95*d*).

For example, if δx is the horizontal movement of the support due to the loads, and if $\delta x'$ is the horizontal movement of the support due to a unit horizontal force applied to the support, then

$$\delta x + H\delta x' = 0 \tag{5.167}$$

$$H = -\frac{\delta x}{\delta x'} \tag{5.168}$$

where H is the unknown horizontal reaction. (When a tie rod is used to take the thrust, the right-hand side of Eq. (5.167) is not zero, but the elongation of the rod, HL/AE.) The dummy unit-load method [Eq. (5.96)] can be used to compute δx and $\delta x'$:

$$\delta x = \int_A^B \frac{My}{EI}\,ds - \int_A^B \frac{N\,dx}{AE} \tag{5.169}$$

where M = moment at any section resulting from loads
N = normal thrust on cross section
A = cross-sectional area of arch
y = ordinate of section measured from A as origin, when B is on rollers
I = moment of inertia of section
E = modulus of elasticity
ds = differential length along axis of arch
dx = differential length along horizontal

$$\delta x' = -\int_A^B \frac{y^2}{EI}\,ds - \int_A^B \frac{\cos^2 \alpha\,dx}{AE} \tag{5.170}$$

where α = the angle the tangent to the axis at the section makes with the horizontal. Unless the thrust is very large and would be responsible for large strains in the direction of the arch axis, the second term on the right-hand side of Eq. (5.169) can usually be ignored.

In most cases, integration is impracticable. The integrals generally must be evaluated by approximate methods. The arch axis is divided into a convenient number of sections and the functions under the integral sign evaluated for each section. The sum is approximately equal to the integral. Thus, for the usual two-hinged arch,

$$H = \frac{\sum_A^B (My\,\Delta s/EI)}{\sum_A^B (y^2\,\Delta s/EI) + \sum_A^B (\cos^2 \alpha\,\Delta x/AE)} \tag{5.171}$$

(S. Timoshenko and D. H. Young, "Theory of Structures," McGraw-Hill Book Company, New York; S. F. Borg and J. J. Gennaro, "Modern Structural Analysis," Van Nostrand Reinhold Company, Inc., New York.)

5.14.3 Stresses in Arch Ribs

When the reactions have been found for an arch (Arts. 5.14.1 to 5.14.2), the principal forces acting on any cross section can be found by applying the equations of equilibrium. For example, consider the portion of an arch in Fig. 5.96, where the forces acting at an interior section X are to be found. The load P, H_L (or H_R), and V_L (or V_R) may be resolved into components parallel to the axial thrust N and

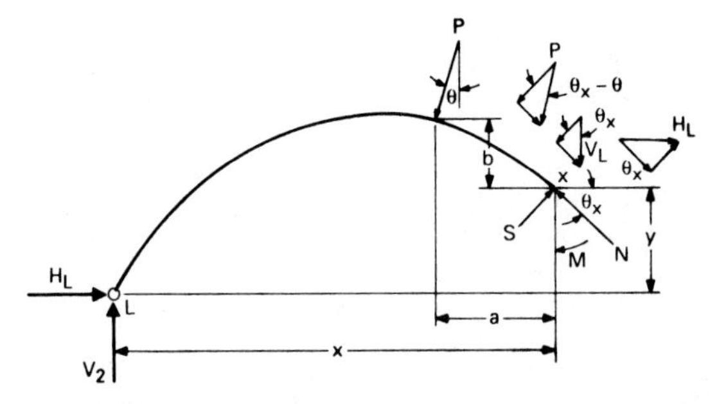

FIGURE 5.96 Interior stresses at X hold portion LX of an arch rib in equilibrium.

the shear S at X, as indicated in Fig. 5.96. Then, by equating the sum of the forces in each direction to zero, we get

$$N = V_L \sin \theta_x + H_L \cos \theta_x + P \sin (\theta_x - \theta) \tag{5.172}$$

$$S = V_L \cos \theta_x - H_L \sin \theta_x + P \cos (\theta_x - \theta) \tag{5.173}$$

And the bending moment at X is

$$M = V_L x - H_L y - Pa \cos \theta - Pb \sin \theta \tag{5.174}$$

The shearing unit stress on the arch cross section at X can be determined from S with the aid of Eq. (5.59). The normal unit stresses can be calculated from N and M with the aid of Eq. (5.67).

In designing an arch, it may be necessary to compute certain secondary stresses, in addition to those caused by live, dead, wind, and snow loads. Among the secondary stresses to be considered are those due to temperature changes, rib shortening due to thrust or shrinkage, deformation of tie rods, and unequal settlement of footings. The procedure is the same as for loads on the arch, with the deformations producing the secondary stresses substituted for or treated the same as the deformations due to loads.

5.15 THIN-SHELL STRUCTURES

A structural membrane or shell is a curved surface structure. Usually, it is capable of transmitting loads in more than two directions to supports. It is highly efficient structurally when it is so shaped, proportioned, and supported that it transmits the loads without bending or twisting.

A membrane or a shell is defined by its middle surface, halfway between its extrados, or outer surface, and intrados, or inner surface. Thus, depending on the geometry of the middle surface, it might be a type of dome, barrel arch, cone, or hyperbolic paraboloid. Its thickness is the distance, normal to the middle surface, between extrados and intrados.

5.15.1 Thin-Shell Analysis

A thin shell is a shell with a thickness relatively small compared with its other dimensions. But it should not be so thin that deformations would be large compared with the thickness.

The shell should also satisfy the following conditions: Shearing stresses normal to the middle surface are negligible. Points on a normal to the middle surface before it is deformed lie on a straight line after deformation. And this line is normal to the deformed middle surface.

Calculation of the stresses in a thin shell generally is carried out in two major steps, both usually involving the solution of differential equations. In the first, bending and torsion are neglected (membrane theory, Art. 5.15.2). In the second step, corrections are made to the previous solution by superimposing the bending and shear stresses that are necessary to satisfy boundary conditions (bending theory, Art. 5.15.3).

Ribbed Shells. For long-span construction, thin shells often are stiffened at intervals by ribs. Usually, the construction is such that the shells transmit some of the load imposed on them to the ribs, which then perform structurally as more than just stiffeners. Stress and strain distributions in shells and ribs consequently are complicated by the interaction between shells and ribs. The shells restrain the ribs, and the ribs restrain the shells. Hence, ribbed shells usually are analyzed by approximate methods based on reasonable assumptions.

For example, for a cylindrical shell with circumferential ribs, the ribs act like arches. For an approximate analysis, the ribbed shell therefore may be assumed to be composed of a set of arched ribs with the thin shell between the ribs acting in the circumferential direction as flanges of the arches. In the longitudinal direction, it may be assumed that the shell transfers load to the ribs in flexure. Designers may adjust the results of a computation based on such assumptions to correct for a variety of conditions, such as the effects of free edges of the shell, long distances between ribs, relative flexibility of ribs and shell, and characteristics of the structural materials.

5.15.2 Membrane Theory for Thin Shells

Thin shells usually are designed so that normal shears, bending moments, and torsion are very small, except in relatively small portions of the shells. In the membrane theory, these stresses are ignored.

Despite the neglected stresses, the remaining stresses are in equilibrium, except possibly at boundaries, supports, and discontinuities. At any interior point, the number of equilibrium conditions equals the number of unknowns. Thus, in the membrane theory, a thin shell is statically determinate.

The membrane theory does not hold for concentrated loads normal to the middle surface, except possibly at a peak or valley. The theory does not apply where boundary conditions are incompatible with equilibrium. And it is inexact where there is geometric incompatibility at the boundaries. The last is a common condition, but the error is very small if the shell is not very flat. Usually, disturbances of membrane equilibrium due to incompatibility with deformations at boundaries, supports, or discontinuities are appreciable only in a narrow region about each source of disturbance. Much larger disturbances result from incompatibility with equilibrium conditions.

To secure the high structural efficiency of a thin shell, select a shape, proportions, and supports for the specific design conditions that come as close as possible to satisfying the membrane theory. Keep the thickness constant; if it must change, use a gradual taper. Avoid concentrated and abruptly changing loads. Change curvature gradually. Keep discontinuities to a minimum. Provide reactions that are tangent to the middle surface. At boundaries, ensure, to the extent possible, compatibility of shell deformations with deformations of adjoining members, or at least keep restraints to a minimum. Make certain that reactions along boundaries are equal in magnitude and direction to the shell forces there.

Means usually adopted to satisfy these requirements at boundaries and supports are illustrated in Fig. 5.97. In Fig. 5.97*a*, the slope of the support and provision for movement normal to the middle surface ensure a reaction tangent to the middle surface. In Fig. 5.97*b*, a stiff rib, or ring girder, resists unbalanced shears and transmits normal forces to columns below. The enlarged view of the ring girder in Fig. 5.97*c* shows gradual thickening of the shell to reduce the abruptness of the change in section. The stiffening ring at the lantern in Fig. 5.97*d*, extending around the opening at the crown, projects above the middle surface, for compatibility of strains, and connects through a transition curve with the shell; often, the rim need merely be thickened when the edge is upturned, and the ring can be omitted. In

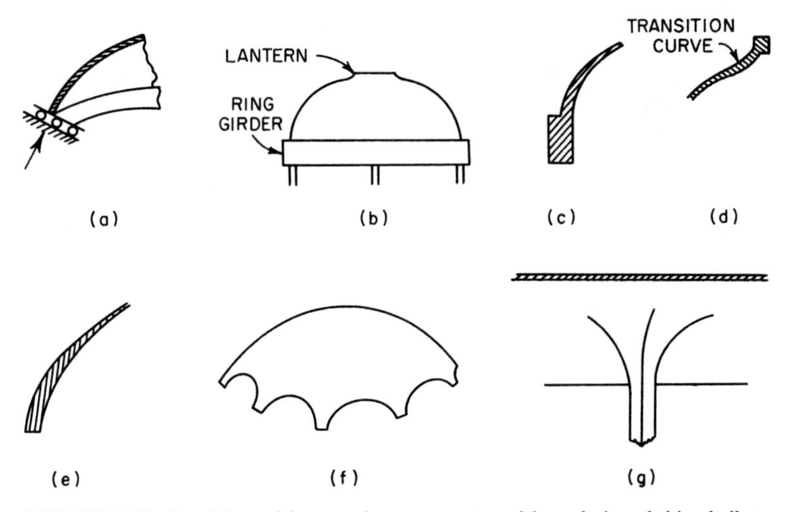

FIGURE 5.97 Special provisions made at supports and boundaries of thin shells to meet requirements of the membrane theory include: (*a*) a device to ensure a reaction tangent to the middle surface; (*b*) stiffened edges, such as the ring girder at the base of a dome; (*c*) gradually increased shell thicknesses at a stiffening member; (*d*) a transition curve at changes in section; (*e*) a stiffening edge obtained by thickening the shell; (*f*) scalloped edges; (*g*) a flared support.

Fig. 5.97*e*, the boundary of the shell is a stiffened edge. In Fig. 5.97*f*, a scalloped shell provides gradual tapering for transmitting the loads to the supports, at the same time providing access to the shell enclosure. And in Fig. 5.97*g*, a column is flared widely at the top to support a thin shell at an interior point.

Even when the conditions for geometric compatibility are not satisfactory, the membrane theory is a useful approximation. Furthermore, it yields a particular solution to the differential equations of the bending theory.

(D. P. Billington, "Thin Shell Concrete Structures," 2d ed., and S. Timoshenko and S. Woinowsky-Krieger, "Theory of Plates and Shells," McGraw-Hill Book Company, New York: V. S. Kelkar and R. T. Sewell, "Fundamentals of the Analysis and Design of Shell Structures," Prentice-Hall, Englewood Cliffs, N.J.)

5.15.3 Bending Theory for Thin Shells

When equilibrium conditions are not satisfied or incompatible deformations exist at boundaries, bending and torsion stresses arise in the shell. Sometimes, the design of the shell and its supports can be modified to reduce or eliminate these stresses (Art. 5.15.2). When the design cannot eliminate them, provision must be made for the shell to resist them.

But even for the simplest types of shells and loading, the stresses are difficult to compute. In bending theory, a thin shell is statically indeterminate; deformation conditions must supplement equilibrium conditions in setting up differential equations for determining the unknown forces and moments. Solution of the resulting equations may be tedious and time-consuming, if indeed solution is possible.

In practice, therefore, shell design relies heavily on the designer's experience and judgment. The designer should consider the type of shell, material of which it is made, and support and boundary conditions, and then decide whether to apply a bending theory in full, use an approximate bending theory, or make a rough estimate of the effects of bending and torsion. (Note that where the effects of a disturbance are large, these change the normal forces and shears computed by the membrane theory.) For concrete domes, for example, the usual procedure is to use as a support a deep, thick girder or a heavily reinforced or prestressed tension ring, and the shell is gradually thickened in the vicinity of this support (Fig. 5.97*c*).

Circular barrel arches, with ratio of radius to distance between supporting arch ribs less than 0.25, may be designed as beams with curved cross section. Secondary stresses, however, must be taken into account. These include stresses due to volume change of rib and shell, rib shortening, unequal settlement of footings, and temperature differentials between surfaces.

Bending theory for cylinders and domes is given in W. Flügge, "Stresses in Shells," Springer-Verlag, New York; D. P. Billington, "Thin Shell Concrete Structures," 2d ed., and S. Timoshenko and S. Woinowsky-Krieger, "Theory of Plates and Shells," McGraw-Hill Book Company, New York; "Design of Cylindrical Concrete Shell Roofs," Manual of Practice No. 31, American Society of Civil Engineers.

5.15.4 Stresses in Thin Shells

The results of the membrane and bending theories are expressed in terms of unit forces and unit moments, acting per unit of length over the thickness of the shell. To compute the unit stresses from these forces and moments, usual practice is to assume normal forces and shears to be uniformly distributed over the shell thickness and bending stresses to be linearly distributed.

Then, normal stresses can be computed from equations of the form

$$f_x = \frac{N_x}{t} + \frac{M_x}{t^3/12} z \tag{5.175}$$

where z = distance from middle surface
t = shell thickness
M_x = unit bending moment about axis parallel to direction of unit normal force N_x

Similarly, shearing stresses produced by central shears and twisting moments may be calculated from equations of the form

$$v_{xy} = \frac{T}{t} \pm \frac{D}{t^3/12} z \tag{5.176}$$

where D = twisting moment and T = unit shear force along the middle surface. Normal shearing stresses may be computed on the assumption of a parabolic stress distribution over the shell thickness:

$$v_{xz} = \frac{V}{t^3/6}\left(\frac{t^2}{4} - z^2\right) \tag{5.177}$$

where V = unit shear force normal to middle surface.

5.15.5 Folded Plates

A folded-plate structure consists of a series of thin planar elements, or flat plates, connected to one another along their edges. Usually used on long spans, especially for roofs, folded plates derive their economy from the girder action of the plates and the mutual support they give one another.

Longitudinally, the plates may be continuous over their supports. Transversely, there may be several plates in each bay (Fig. 5.98). At the edges, or folds, they may be capable of transmitting both moment and shear or only shear.

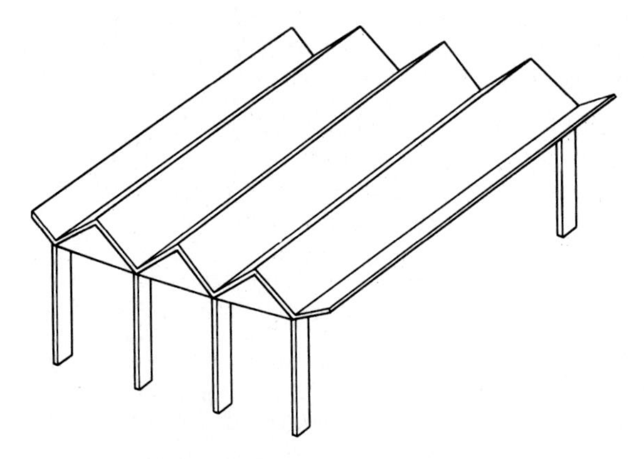

FIGURE 5.98 Folded-plate structure.

A folded-plate structure has a two-way action in transmitting loads to its supports. Transversely, the elements act as slabs spanning between plates on either side. The plates then act as girders in carrying the load from the slabs longitudinally to supports, which must be capable of resisting both horizontal and vertical forces.

If the plates are hinged along their edges, the design of the structure is relatively simple. Some simplification also is possible if the plates, though having integral edges, are steeply sloped or if the span is sufficiently long with respect to other dimensions that beam theory applies. But there are no criteria for determining when such simplification is possible with acceptable accuracy. In general, a reasonably accurate analysis of folded-plate stresses is advisable.

Several good methods are available (D. Yitzhaki, "The Design of Prismatic and Cylindrical Shell Roofs," North Holland Publishing Company, Amsterdam; "Phase I Report on Folded-plate Construction," Proceedings Paper 3741, *Journal of the Structural Division, American Society of Civil Engineers,* December 1963; and A. L. Parme and J. A. Sbarounis, "Direct Solution of Folded Plate Concrete Roofs," EB021D, Portland Cement Association, Skokie, Ill.). They all take into account the effects of plate deflections on the slabs and usually make the following assumptions:

The material is elastic, isotropic, and homogeneous. The longitudinal distribution of all loads on all plates is the same. The plates carry loads transversely only by bending normal to their planes and longitudinally only by bending within their planes. Longitudinal stresses vary linearly over the depth of each plate. Supporting members, such as diaphragms, frames, and beams, are infinitely stiff in their own planes and completely flexible normal to their own planes. Plates have no torsional stiffness normal to their own planes. Displacements due to forces other than bending moments are negligible.

Regardless of the method selected, the computations are rather involved; so it is wise to carry out the work by computer or, when done manually, in a well-organized table. The Yitzhaki method offers some advantages over others in that the calculations can be tabulated, it is relatively simple, it requires the solution of no more simultaneous equations than one for each edge for simply supported plates, it is flexible, and it can easily be generalized to cover a variety of conditions.

Yitzhaki Method. Based on the assumptions and general procedure given above, the Yitzhaki method deals with the slab and plate systems that comprise a folded-plate structure in two ways. In the first, a unit width of slab is considered continuous over supports immovable in the direction of the load (Fig. 5.99*b*). The strip usually is taken where the longitudinal plate stresses are a maximum. Second, the slab reactions are taken as loads on the plates, which now are assumed to be hinged along the edges (Fig. 5.99*c*). Thus, the slab reactions cause angle changes in the plates at each fold. Continuity is restored by applying to the plates an unknown moment at each edge. The moments can be determined from the fact that at each edge the sum of the angle changes due to the loads and to the unknown moments must equal zero.

The angle changes due to the unknown moments have two components. One is the angle change at each slab end, now hinged to an adjoining slab, in the transverse strip of unit width. The second is the angle change due to deflection of the plates. The method assumes that the angle change at each fold varies in the same way longitudinally as the angle changes along the other folds.

For example, for the folded-plate structure in Fig. 5.99*a*, the steps in analysis are as follows:

Step 1. Compute the loads on a 12-in-wide transverse strip at midspan.

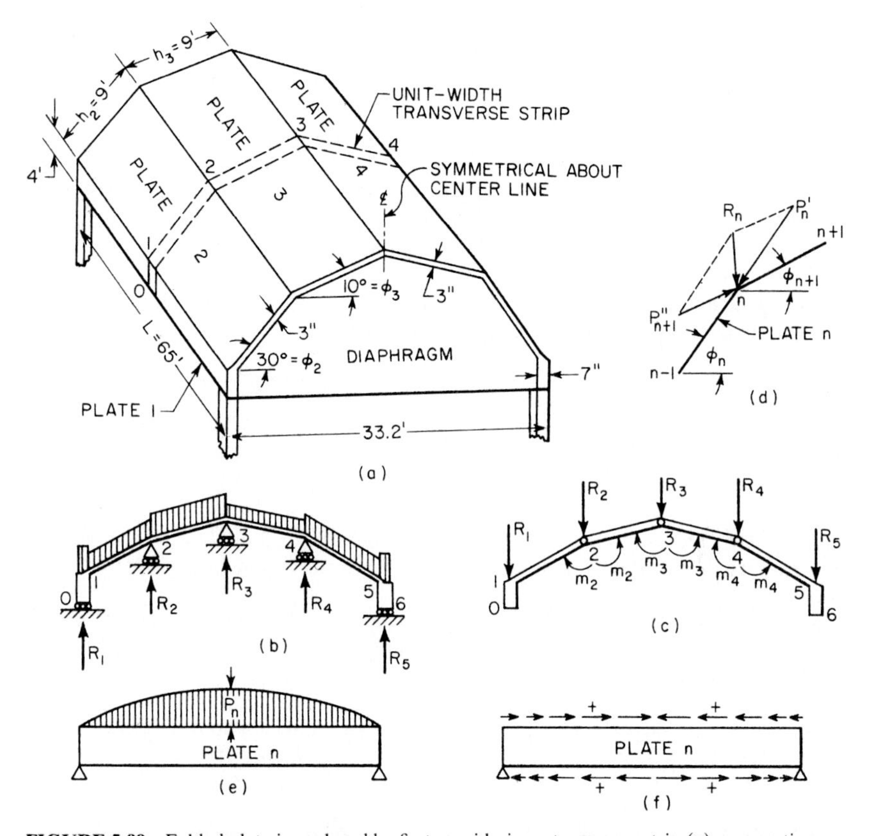

FIGURE 5.99 Folded plate is analyzed by first considering a transverse strip (*a*) as a continuous slab on supports that do not settle (*b*). Then, (*c*) the slabs are assumed hinged and acted upon by the reactions computed in the first step and by unknown moments to correct for this assumption. (*d*) Slab reactions are resolved into plate forces, parallel to the planes of the plates. (*e*) In the longitudinal direction, the plates act as deep girders with shears along the edges. (*f*) Arrows indicate the positive directions for the girder shears.

Step 2. Consider the strip as a continuous slab supported at the folds (Fig. 5.99*b*), and compute the bending moments by moment distribution.

Step 3. From the end moments M found in Step 2, compute slab reactions and plate loads. Reactions (positive upward) at the nth edge are

$$R_n = V_n + V_{n+1} + \frac{M_{n-1} + M_n}{a_n} - \frac{M_n + M_{n+1}}{a_{n+1}} \tag{5.178}$$

where V_n, V_{n+1} = shears at both sides of edge n
M_n = moment at edge n
M_{n-1} = moment at edge $(n - 1)$
M_{n+1} = moment at edge $(n + 1)$
a = horizontal projection of depth h

Let $k = \tan \phi_n - \tan \phi_{n+1}$, where ϕ is positive as shown in Fig. 99*a*. Then, the load (positive downward) on the *n*th plate is

$$P_n = \frac{R_n}{k_n \cos \phi_n} - \frac{R_{n-1}}{k_{n-1} \cos \phi_n} \tag{5.179}$$

(Figure 5.99*d* shows the resolution of forces at edge n; $n - 1$ is similar.) Equation (5.179) does not apply for the case of a vertical reaction on a vertical plate, for R/k is the horizontal component of the reaction.

Step 4. Calculate the midspan (maximum) bending moment in each plate. In this example, each plate is a simple beam and $M = PL^2/8$, where L is the span in feet.

Step 5. Determine the free-edge longitudinal stresses at midspan. In each plate, these can be computed from

$$f_{n-1} = \frac{72M}{Ah} \qquad f_n = -\frac{72M}{Ah} \tag{5.180}$$

where f is the stress in psi, M the moment in ft-lb from Step 4, A = plate cross-sectional area, and tension is taken as positive, compression as negative.

Step 6. Apply a shear to adjoining edges to equalize the stresses there. Compute the adjusted stresses by converging approximations, similar to moment distribution. To do this, distribute the unbalanced stress at each edge in proportion to the reciprocals of the areas of the plates, and use a carry-over factor of −½ to distribute the stress to a far edge. Edge 0, being a free edge, requires no distribution of the stress there. Edge 3, because of symmetry, may be treated the same, and distribution need be carried out only in the left half of the structure.

Step 7. Compute the midspan edge deflections. In general, the vertical component δ can be computed from

$$\frac{E}{L^2}\,\delta_n = \frac{15}{k_n}\left(\frac{f_{n-1} - f_n}{a_n} - \frac{f_n - f_{n+1}}{a_{n+1}}\right) \tag{5.181}$$

where E = modulus of elasticity, psi
$k = \tan \phi_n - \tan \phi_{n+1}$, as in Step 3

The factor E/L^2 is retained for convenience; it is eliminated by dividing the simultaneous angle equations by it. For a vertical plate, the vertical deflection is given by

$$\frac{E}{L^2}\,\delta_n = \frac{15(f_{n-1} - f_n)}{h_n} \tag{5.182}$$

Step 8. Compute the midspan angle change θ_P at each edge. This can be determined from

$$\frac{E}{L^2}\,\theta_P = -\frac{\delta_{n-1} - \delta_n}{a_n} + \frac{\delta_n - \delta_{n+1}}{a_{n+1}} \tag{5.183}$$

Step 9. To correct the edge rotations with a symmetrical loading, apply an unknown moment of $+1000m_n \sin(\pi x/L)$, in-lb (positive when clockwise) to plate n at edge n and $-1000m_n \sin(\pi x/L)$ to its counterpart, plate n' at edge n'. Also, apply $-1000m_n \sin(\pi x/L)$ to plate $(n + 1)$ at edge n and $+1000m_n \sin(\pi x/L)$ to its counterpart; x is the distance along an edge from the end of a plate. (The sine function is assumed to make the loading vary longitudinally in approximately the same manner as the deflections.) At midspan, the absolute value of these moments is $1000m_n$.

The 12-in-wide transverse strip at midspan, hinged at the supports, will then be subjected at the supports to moments of $1000m_n$. Compute the rotations thus caused in the slabs from

$$\frac{E}{L^2}\theta''_{n-1} = \frac{166.7h_n m_n}{L^2 t_n^3}$$

$$\frac{E}{L^2}\theta''_n = \frac{333.3m_n}{L^2}\left(\frac{h_n}{t_n^3} + \frac{h_{n+1}}{t_{n+1}^3}\right) \qquad (5.184)$$

$$\frac{E}{L^2}\theta''_{n+1} = \frac{166.7h_{n+1}m_n}{L^2 t_{n+1}^3}$$

Step 10. Compute the slab reactions and plate loads due to the unknown moments. The reactions are

$$R_{n-1} = \frac{1000m_n}{a_n} \qquad R_n = 1000m_n\left(\frac{1}{a_n} + \frac{1}{a_{n+1}}\right) \qquad R_{n+1} = -\frac{1000m_n}{a_{n+1}} \qquad (5.185)$$

The plate loads are

$$P_n = \frac{1}{\cos\phi_n}\left(\frac{R_n}{k_n} - \frac{R_{n-1}}{k_{n-1}}\right) \qquad (5.186)$$

Step 11. Assume that the loading on each plate is $P_n \sin(\pi x/L)$ (Fig. 5.99*e*), and calculate the midspan (maximum) bending moment. For a simple beam,

$$M = \frac{PL^2}{\pi^2}$$

Step 12. Using Eq. (5.180), compute the free-edge longitudinal stresses at midspan. Then, as in Step 6, apply a shear at each edge to equalize the stresses. Determine the adjusted stresses by converging approximations.

Step 13. Compute the vertical component of the edge deflections at midspan from

$$\frac{E}{L^2}\delta_n = \frac{144}{\pi^2 k_n}\left(\frac{f_{n-1} - f_n}{a_n} - \frac{f_n - f_{n+1}}{a_{n+1}}\right) \qquad (5.187)$$

or for a vertical plate from

$$\frac{E}{L^2}\delta_n = \frac{144(f_{n-1} - f_n)}{\pi^2 h_n} \qquad (5.188)$$

Step 14. Using Eq. (5.183), determine the midspan angle change θ' at each edge.

Step 15. At each edge, set up an equation by putting the sum of the angle changes equal to zero. Thus, after division by E/L^2: $\theta_P + \theta'' + \Sigma\theta' = 0$. Solve these simultaneous equations for the unknown moments.

Step 16. Determine the actual reactions, loads, stresses, and deflections by substituting for the moments the values just found.

Step 17. Compute the shear stresses. The shear stresses at edge *n* (Fig. 5.99*f*) is

$$T_n = T_{n-1} - \frac{f_{n-1} + f_n}{2} A_n \tag{5.189}$$

In the example, $T_o = 0$, so the shears at the edges can be obtained successively, since the stresses *f* are known.

For a uniformly loaded folded plate, the shear stress *S*, psi, at any point on an edge *n* is approximately

$$S = \frac{2T_{max}}{3Lt}\left(\frac{1}{2} - \frac{x}{L}\right) \tag{5.190}$$

with a maximum at plate ends of

$$S_{max} = \frac{T_{max}}{3Lt} \tag{5.191}$$

The shear stress, psi, at middepth (not always a maximum) is

$$v_n = \left(\frac{3P_nL}{2A_n} + \frac{S_{n-1} + S_n}{2}\right)\left(\frac{1}{2} - \frac{x}{L}\right) \tag{5.192}$$

and has its largest value at $x = 0$:

$$v_{max} = \frac{0.75P_nL}{A_n} + \frac{S_{n-1} + S_n}{4} \tag{5.193}$$

For more details, see D. Yitzhaki and Max Reiss, "Analysis of Folded Plates," Proceedings Paper 3303, *Journal of the Structural Division, American Society of Civil Engineers,* October 1962.

5.16 CABLE-SUPPORTED STRUCTURES*

A cable is a linear structural member, like a bar of a truss. The cross-sectional dimensions of a cable relative to its length, however, are so small that it cannot withstand bending or compression. Consequently, under loads at an angle to its longitudinal axis, a cable sags and assumes a shape that enables it to develop tensile stresses that resist the loads.

Structural efficiency results from two cable characteristics: (1) uniformity of tensile stresses over the cable cross section, and (2) usually, small variation of

*Reprinted with permission from F. S. Merritt, "Structural Steel Designers' Handbook," McGraw-Hill Book Company, New York.

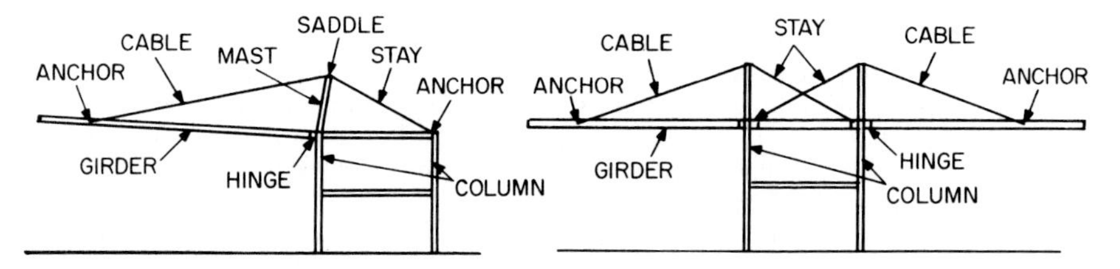

FIGURE 5.100 Two types of cable-stayed girder construction for roofs.

tension along the longitudinal axis. Hence, it is economical to use materials with very high tensile strength for cables.

Cables sometimes are used in building construction as an alternative to such tension members as hangers, ties, or tension chords of trusses. For example, cables are used in a form of long-span cantilever-truss construction in which a horizontal roof girder is supported at one end by a column and near the other end by a cable that extends diagonally upward to the top of a vertical mast above the column support (cable-stayed-girder construction, Fig. 5.100). Cable stress can be computed for this case from the laws of equilibrium.

Cables also may be used in building construction instead of girders, trusses, or membranes to support roofs. For the purpose, cables may be arranged in numerous ways. It is consequently impractical to treat in detail in this book any but the simplest types of such applications of cables. Instead, general procedures for analyzing cable-supported structures are presented in the following.

5.16.1 Simple Cables

An ideal cable has no resistance to bending. Thus, in analysis of a cable in equilibrium, not only is the sum of the moments about any point equal to zero but so is the bending moment at any point. Consequently, the equilibrium shape of the cable corresponds to the funicular, or bending-moment, diagram for the loading (Fig. 5.101*a*). As a result, the tensile force at any point of the cable is tangent there to the cable curve.

The point of maximum sag of a cable coincides with the point of zero shear. (Sag in this case should be measured parallel to the direction of the shear forces.)

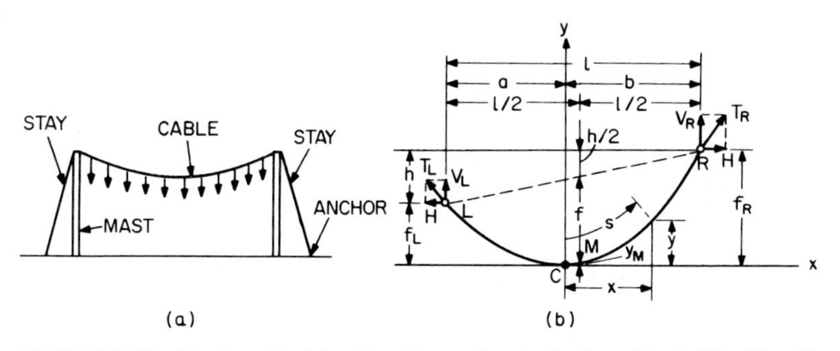

FIGURE 5.101 Simple cable: (*a*) cable with a uniformly distributed load; (*b*) cable with supports at different levels.

Stresses in a cable are a function of the deformed shape. Equations needed for analysis, therefore, usually are nonlinear. Also, in general, stresses and deformations cannot be obtained accurately by superimposition of loads. A common procedure in analysis is to obtain a solution in steps by using linear equations to approximate the nonlinear ones and by starting with the initial geometry to obtain better estimates of the final geometry.

For convenience in analysis, the cable tension, directed along the cable curve, usually is resolved into two components. Often, it is advantageous to resolve the tension T into a horizontal component H and a vertical component V (Fig. 5.100*b*). Under vertical loading then, the horizontal component is constant along the cable. Maximum tension occurs at the support. V is zero at the point of maximum sag.

For a general, distributed vertical load q, the cable must satisfy the second-order linear differential equation

$$Hy'' = q \tag{5.194}$$

where y = rise of cable at distance x from low point (Fig. 5.100*b*)
$y'' = d^2y/dx^2$

Catenary. Weight of a cable of constant cross section represents a vertical loading that is uniformly distributed along the length of cable. Under such a loading, a cable takes the shape of a catenary.

Take the origin of coordinates at the low point C and measure distance s along the cable from C (Fig. 5.100*b*). If q_o is the load per unit length of cable, Eq. (5.194) becomes

$$Hy'' = \frac{q_o\, ds}{dx} = q_o \sqrt{1 + y'^2} \tag{5.195}$$

where $y' = dy/dx$. Solving for y' gives the slope at any point of the cable

$$y' = \sinh \frac{q_o x}{H} = \frac{q_o x}{H} + \frac{1}{3!}\left(\frac{q_o x}{H}\right)^3 + \cdots \tag{5.196}$$

A second integration then yields the equation for the cable shape, which is called a catenary.

$$y = \frac{H}{q_o}\left(\cosh \frac{q_o x}{H} - 1\right) = \frac{q_o}{H}\frac{x^2}{2!} + \left(\frac{q_o}{H}\right)^3 \frac{x^4}{4!} + \cdots \tag{5.197}$$

If only the first term of the series expansion is used, the cable equation represents a parabola. Because the parabolic equation usually is easier to handle, a catenary often is approximated by a parabola.

For a catenary, length of arc measured from the low point is

$$s = \frac{H}{q_o} \sinh \frac{q_o x}{H} = x + \frac{1}{3!}\left(\frac{q_o}{H}\right)^2 x^3 + \cdots \tag{5.198}$$

Tension at any point is

$$T = \sqrt{H^2 + q_o^2 s^2} = H + q_o y \tag{5.199}$$

The distance from the low point C to the left support L is

$$a = \frac{H}{q_o} \cosh^{-1}\left(\frac{q_o}{H} f_L + 1\right) \tag{5.200}$$

where f_L = vertical distance from C to L. The distance from C to the right support R is

$$b = \frac{H}{q_o} \cosh^{-1}\left(\frac{q_o}{H} f_R + 1\right) \tag{5.201}$$

where f_R = vertical distance from C to R.

Given the sags of a catenary f_L and f_R under a distributed vertical load q_o, the horizontal component of cable tension H may be computed from

$$\frac{q_o l}{H} = \cosh^{-1}\left(\frac{q_o f_L}{H} + 1\right) + \cosh^{-1}\left(\frac{q_o f_R}{H} + 1\right) \tag{5.202}$$

where l = span, or horizontal distance between supports L and $R = a + b$. This equation usually is solved by trial. A first estimate of H for substitution in the right-hand side of the equation may be obtained by approximating the catenary by a parabola. Vertical components of the reactions at the supports can be computed from

$$R_L = H \sinh \frac{q_o a}{H} \qquad R_R = H \sinh \frac{q_o b}{H} \tag{5.203}$$

Parabola. Uniform vertical live loads and uniform vertical dead loads other than cable weight generally may be treated as distributed uniformly over the horizontal projection of the cable. Under such loadings, a cable takes the shape of a parabola.

Take the origin of coordinates at the low point C (Fig. 5.100*b*). If w_o is the load per foot horizontally, Eq. (5.194) becomes

$$Hy'' = w_o \tag{5.204}$$

Integration gives the slope at any point of the cable

$$y' = \frac{w_o x}{H} \tag{5.205}$$

A second integration yields the parabolic equation for the cable shape

$$y = \frac{w_o x^2}{2H} \tag{5.206}$$

The distance from the low point C to the left support L is

$$a = \frac{l}{2} - \frac{Hh}{w_o l} \tag{5.207}$$

where l = span, or horizontal distance between supports L and $R = a + b$
h = vertical distance between supports

The distance from the low point C to the right support R is

$$b = \frac{l}{2} + \frac{Hh}{w_o l} \tag{5.208}$$

When supports are not at the same level, the horizontal component of cable tension H may be computed from

$$H = \frac{w_o l^2}{h^2}\left(f_R - \frac{h}{2} \pm \sqrt{f_L f_R}\right) = \frac{w_o l^2}{8f} \tag{5.209}$$

where f_L = vertical distance from C to L
f_R = vertical distance from C to R
f = sag of cable measured vertically from chord LR midway between supports (at $x = Hh/w_o l$)

As indicated in Fig. 5.100*b*,

$$f = f_L + \frac{h}{2} - y_M \tag{5.210}$$

where $y_M = Hh^2/2w_o l^2$. The minus sign should be used in Eq. (5.209) when low point C is between supports. If the vertex of the parabola is not between L and R, the plus sign should be used.

The vertical components of the reactions at the supports can be computed from

$$V_L = w_o a = \frac{w_o l}{2} - \frac{Hh}{l} \qquad V_R = w_o b = \frac{w_o l}{2} + \frac{Hh}{l} \tag{5.211}$$

Tension at any point is

$$T = \sqrt{H^2 + w_o^2 x^2} \tag{5.212}$$

Length of parabolic arc RC is

$$L_{RC} = \frac{b}{2}\sqrt{1 + \left(\frac{w_o b}{H}\right)^2} + \frac{H}{2w_o}\sinh\frac{w_o b}{H} = b + \frac{1}{6}\left(\frac{w_o}{H}\right)^2 b^3 + \cdots \tag{5.213}$$

Length of parabolic arc LC is

$$L_{LC} = \frac{a}{2}\sqrt{1 + \left(\frac{w_o a}{H}\right)^2} + \frac{H}{2w_o}\sinh\frac{w_o a}{H} = a + \frac{1}{6}\left(\frac{w_o}{H}\right)^2 a^3 + \cdots \tag{5.214}$$

When supports are at the same level, $f_L = f_R = f$, $h = 0$, and $a = b = l/2$. The horizontal component of cable tension H may be computed from

$$H = \frac{w_o l^2}{8f} \tag{5.215}$$

The vertical components of the reactions at the supports are

$$V_L = V_R = \frac{w_o l}{2} \tag{5.216}$$

Maximum tension occurs at the supports and equals

$$T_L = T_R = \frac{w_o l}{2}\sqrt{1 + \frac{l^2}{16f^2}} \tag{5.217}$$

Length of cable between supports is

$$\begin{aligned} L &= \frac{1}{2}\sqrt{1 + \left(\frac{w_o l}{2H}\right)^2} + \frac{H}{w_o}\sinh\frac{w_o l}{2H} \\ &= l\left(1 + \frac{8}{3}\frac{f^2}{l^2} - \frac{32}{5}\frac{f^4}{l^4} + \frac{256}{7}\frac{f^6}{l^6} + \cdots\right) \end{aligned} \tag{5.218}$$

If additional uniformly distributed load is applied to a parabolic cable, the change in sag is approximately

$$\Delta f = \frac{15}{16}\frac{l}{f}\frac{\Delta L}{5 - 24f^2/l^2} \tag{5.219}$$

For a rise in temperature t, the change in sag is about

$$\Delta f = \frac{15}{16}\frac{l^2 ct}{f(5 - 24f^2/l^2)}\left(1 + \frac{8}{3}\frac{f^2}{l^2}\right) \tag{5.220}$$

where c = coefficient of thermal expansion.

Elastic elongation of a parabolic cable is approximately

$$\Delta L = \frac{Hl}{AE}\left(1 + \frac{16}{3}\frac{f^2}{l^2}\right) \tag{5.221}$$

where A = cross-sectional area of cable
E = modulus of elasticity of cable steel
H = horizontal component of tension in cable

If the corresponding change in sag is small, so that the effect on H is negligible, this change may be computed from

$$\Delta f = \frac{15}{16}\frac{Hl^2}{AEf}\frac{1 + 16f^2/3l^2}{5 - 24f^2/l^2} \tag{5.222}$$

For the general case of vertical dead load on a cable, the initial shape of the cable is given by

$$y_D = \frac{M_D}{H_D} \tag{5.223}$$

where M_D = dead-load bending moment that would be produced by the load in a simple beam
H_D = horizontal component of tension due to dead load

For the general case of vertical live load on the cable, the final shape of the cable is given by

$$y_D + \delta = \frac{M_D + M_L}{H_D + H_L} \tag{5.224}$$

where δ = vertical deflection of cable due to live load
M_L = live-load bending moment that would be produced by the live load in a simple beam
H_L = increment in horizontal component of tension due to live load

Subtraction of Eq. (5.223) from Eq. (5.224) yields

$$\delta = \frac{M_L - H_L y_D}{H_D + H_L} \tag{5.225}$$

If the cable is assumed to take a parabolic shape, a close approximation to H_L may be obtained from

$$\frac{H_L}{AE} K = \frac{w_D}{H_D} \int_0^l \delta \, dx - \frac{1}{2} \int_0^l \delta'' \delta \, dx \tag{5.226}$$

$$K = l \left[\frac{1}{4} \left(\frac{5}{2} + \frac{16f^2}{l^2} \right) \sqrt{1 + \frac{16f^2}{l^2}} + \frac{3l}{32f} \log_e \left(\frac{4f}{l} + \sqrt{1 + \frac{16f^2}{l^2}} \right) \right] \tag{5.227}$$

where $\delta'' = d^2\delta/dx^2$.

If elastic elongation and δ'' can be ignored, Eq. (5.226) simplifies to

$$H_L = \frac{\int_0^l M_L \, dx}{\int_0^l y_D \, dx} = \frac{3}{2fl} \int_0^l M_L \, dx \tag{5.228}$$

Thus, for a load uniformly distributed horizontally w_L,

$$\int_0^l M_L \, dx = \frac{w_L l^3}{12} \tag{5.229}$$

and the increase in the horizontal component of tension due to live load is

$$H_L = \frac{3}{2fl} \frac{w_L l^3}{12} = \frac{w_L l^2}{8f} = \frac{w_L l^2}{8} \frac{8H_D}{w_D l^2} = \frac{w_L}{w_D} H_D \tag{5.230}$$

When a more accurate solution is desired, the value of H_L obtained from Eq. (5.230) can be used for an initial trial in solving Eqs. (5.225) and (5.226).

(S. P. Timoshenko and D. H. Young, "Theory of Structures," McGraw-Hill Book Company, New York: W. T. O'Brien and A. J. Francis, "Cable Movements under Two-dimensional Loads," *Journal of the Structural Division, ASCE,* Vol. 90, No. ST3, *Proceedings Paper* 3929, June 1964, pp. 89–123; W. T. O'Brien, "General Solution of Suspended Cable Problems," *Journal of the Structural Division, ASCE,* Vol. 93, No. ST1, *Proceedings Paper* 5085, February, 1967, pp. 1–26; W. T. O'Brien, "Behavior of Loaded Cable Systems," *Journal of the Structural*

Division, ASCE, Vol. 94, No. ST10, *Proceedings Paper* 6162, October 1968, pp. 2281–2302; G. R. Buchanan, "Two-dimensional Cable Analysis," *Journal of the Structural Division, ASCE,* Vol. 96, No. ST7, *Proceedings Paper* 7436, July 1970, pp. 1581–1587).

5.16.2 Cable Systems

Analysis of simple cables is described in Art. 5.16.1. Cables, however, may be assembled into many types of systems. One important reason for such systems is that roofs to be supported are two- or three-dimensional. Consequently, three-dimensional cable arrangements often are advantageous. Another important reason is that cable systems can be designed to offer much higher resistance to vibrations than simple cables do.

Like simple cables, cable systems behave nonlinearly. Thus, accurate analysis is difficult, tedious, and time-consuming. As a result, many designers use approximate methods that appear to have successfully withstood the test of time. Because of the numerous types of systems and the complexity of analysis, only general procedures will be outlined in this article.

Cable systems may be stiffened or unstiffened. Stiffened systems, usually used for suspension bridges, are rarely used in buildings. This article will deal only with unstiffened systems, that is, systems where loads are carried to supports only by cables.

Often, unstiffened systems may be classified as a network or as a cable truss, or double-layered plane system.

Networks consist of two or three sets of cables intersecting at an angle (Fig. 5.102). The cables are fastened together at their intersections.

Cable trusses consist of pairs of cables, generally in a vertical plane. One cable of each pair is concave downward, the other concave upward (Fig. 5.103).

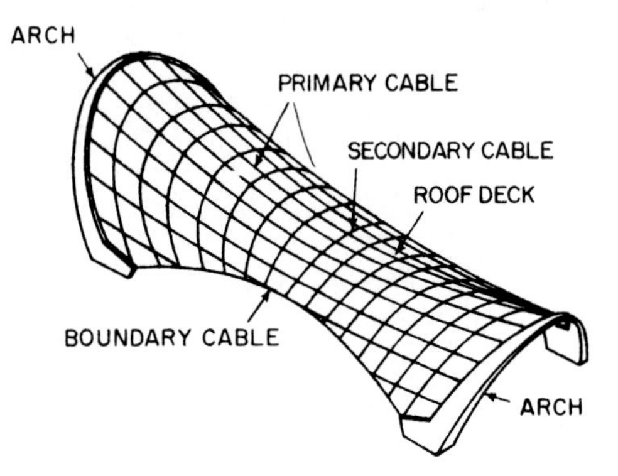

FIGURE 5.102 Cable network.

Cable Trusses. Both cables of a cable truss are initially tensioned, or prestressed, to a predetermined shape, usually parabolic. The prestress is made large enough that any compression that may be induced in a cable by loads only reduces the tension in the cable; thus, compressive stresses cannot occur. The relative vertical position of the cables is maintained by verticals, or spreaders, or by diagonals.

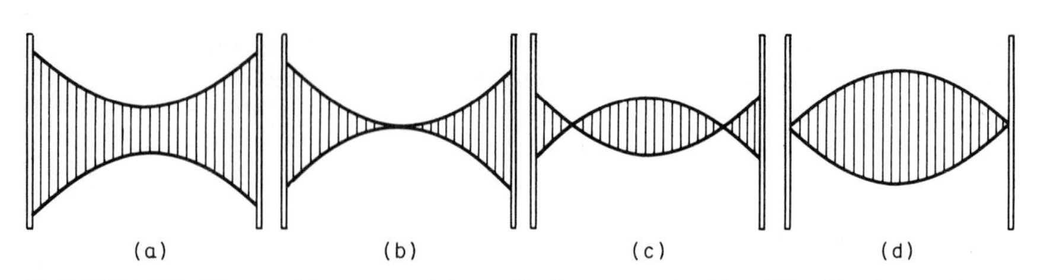

FIGURE 5.103 Planar cable systems: (*a*) completely separated cables; (*b*) cables intersecting at midspan; (*c*) crossing cables; (*d*) cables meeting at supports.

Diagonals in the truss plane do not appear to increase significantly the stiffness of a cable truss.

Figure 5.103 shows four different arrangements of the cables, with spreaders, in a cable truss. The intersecting types (Fig. 5.103*b* and *c*) usually are stiffer than the others, for given size cables and given sag and rise.

For supporting roofs, cable trusses often are placed radially at regular intervals (Fig. 5.104). Around the perimeter of the roof, the horizontal component of the tension usually is resisted by a circular or elliptical compression ring. To avoid a joint with a jumble of cables at the center, the cables usually are also connected to a tension ring circumscribing the center.

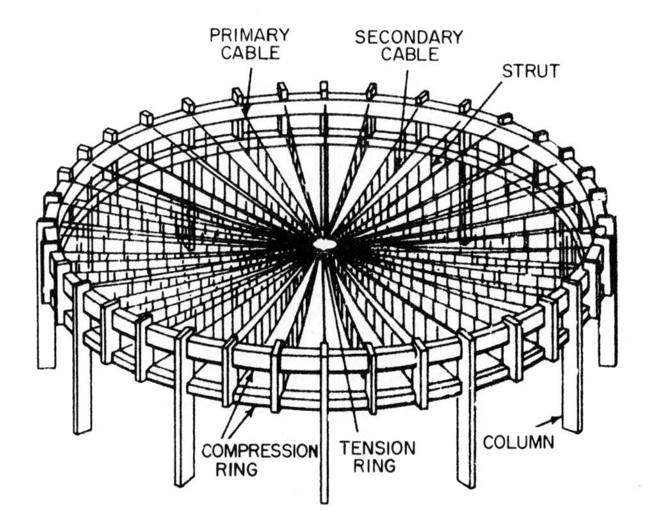

FIGURE 5.104 Cable trusses placed radially to support a round roof.

Properly prestressed, such double-layer cable systems offer high resistance to vibrations. Wind or other dynamic forces difficult or impossible to anticipate may cause resonance to occur in a single cable, unless damping is provided. The probability of resonance occurring may be reduced by increasing the dead load on a single cable. But this is not economical, because the size of cable and supports usually must be increased as well. Besides, the tactic may not succeed, because future loads may be outside the design range. Damping, however, may be achieved economically with interconnected cables under different tensions, for example, with cable trusses or networks.

The cable that is concave downward (Fig. 5.103) usually is considered the load-carrying cable. If the prestress in that cable exceeds that in the other cable, the natural frequencies of vibration of both cables will always differ for any value of live load. To avoid resonance, the difference between the frequencies of the cables should increase with increase in load. Thus, the two cables will tend to assume different shapes under specific dynamic loads. As a consequence, the resulting flow of energy from one cable to the other will dampen the vibrations of both cables.

Natural frequency, cycles per second, of each cable may be estimated from

$$\omega_n = \frac{n\pi}{l}\sqrt{\frac{Tg}{w}} \tag{5.231}$$

where n = integer, 1 for the fundamental mode of vibration, 2 for the second mode, . . .
l = span of cable, ft
w = load on cable, kips per ft
g = acceleration due to gravity = 32.2 ft/s^2
T = cable tension, kips

The spreaders of a cable truss impose the condition that under a given load the change in sag of the cables must be equal. But the changes in tension of the two cables may not be equal. If the ratio of sag to span f/l is small (less than about 0.1), Eq. (5.222) indicates that, for a parabolic cable, the change in tension is given approximately by

$$\Delta H = \frac{16}{3}\frac{AEf}{l^2}\Delta f \tag{5.232}$$

where Δf = change in sag
A = cross-sectional area of cable
E = modulus of elasticity of cable steel

Double cables interconnected with struts may be analyzed as discrete or continuous systems. For a discrete system, the spreaders are treated as individual members. For a continuous system, the spreaders are replaced by a continuous diaphragm that ensures that the changes in sag and rise of cables remain equal under changes in load. Similarly, for analysis of a cable network, the cables, when treated as a continuous system, may be replaced by a continuous membrane.

(C. H. Mollman, "Analysis of Plane Prestressed Cable Structures," *Journal of the Structural Division, ASCE,* Vol. 96, No. ST10, *Proceedings Paper* 7598, October 1970, pp. 2059–2082; D. P. Greenberg, "Inelastic Analysis of Suspension Roof Structures," *Journal of the Structural Division, ASCE,* Vol. 96, No. ST5, *Proceedings Paper* 7284, May 1970, pp. 905–930; H. Tottenham and P. G. Williams, "Cable Net: Continuous System Analysis," *Journal of the Engineering Mechanics Division, ASCE,* Vol. 96, No. EM3, *Proceedings Paper* 7347, June 1970, pp. 277–293; A. Siev, "A General Analysis of Prestressed Nets," *Publications, International Association for Bridge and Structural Engineering,* Vol. 23, pp. 283–292, Zurich, Switzerland, 1963; A. Siev, "Stress Analysis of Prestressed Suspended Roofs,"*Journal of the Structural Division, ASCE,* Vol. 90, No. ST4, *Proceedings*

Paper 4008, August 1964, pp. 103–121; C. H. Thornton and C. Birnstiel, "Three-dimensional Suspension Structures," *Journal of the Structural Division, ASCE,* Vol. 93, No. ST2, *Proceedings Paper* 5196, April 1967, pp. 247–270.)

5.17 AIR-STABILIZED STRUCTURES

A true membrane is able to withstand tension but is completely unable to resist bending. Although it is highly efficient structurally, like a shell, a membrane must be much thinner than a shell and therefore can be made of a very lightweight material, such as fabric, with considerable reduction in dead load compared with other types of construction. Such a thin material, however, would buckle if subjected to compression. Consequently, a true membrane, when loaded, deflects and assumes a shape that enables it to develop tensile stresses that resist the loads.

Membranes may be used for the roof of a building or as a complete exterior enclosure. One way to utilize a membrane for these purposes is to hang it with initial tension between appropriate supports. For example, a tent may be formed by supporting fabric atop one or more tall posts and anchoring the outer edges of the stretched fabric to the ground. As another example, a dish-shaped roof may be constructed by stretching a membrane and anchoring it to the inner surface of a ring girder. In both examples, loads induce only tensile stresses in the membrane. The stresses may be computed from the laws of equilibrium, because a membrane is statically determinate.

Another way to utilize a membrane as an enclosure or roof is to pretension the membrane to enable it to carry compressive loads. For the purpose, forces may be applied, and retained as long as needed, around the edges or over the surface of the membrane to induce tensile stresses that are larger than the largest compressive stresses that loads will impose. As a result, compression due to loads will only reduce the prestress, and the membrane will always be subjected only to tensile stresses.

5.17.1 Pneumatic Construction

A common method of pretensioning a membrane enclosure is to pressurize the interior with air. Sufficient pressure is applied to counteract dead loads, so that the membrane actually floats in space. Slight additional pressurization is also used to offset wind and other anticipated loads. Made of lightweight materials, a membrane thus can span large distances economically. This type of construction, however, has the disadvantage that energy is continuously required for operation of air compressors to maintain interior air at a higher pressure than that outdoors.

Pressure differentials used in practice are not large. They often range between 0.02 and 0.04 psi (3 and 5 psf). Air must be continually supplied, because of leakage. While there may be some leakage of air through the membrane, more important sources of air loss are the entrances and exits to the structure. Air locks and revolving doors, however, can reduce these losses.

An air-stabilized enclosure, in effect, is a membrane bag held in place by small pressure differentials applied by environmental energy. Such a structure is analogous to a soap film. The shape of a bubble is determined by surface-tension forces. The membrane is stressed equally in all directions at every point. Consequently, the film forms shapes with minimum surface area, frequently spherical. Because

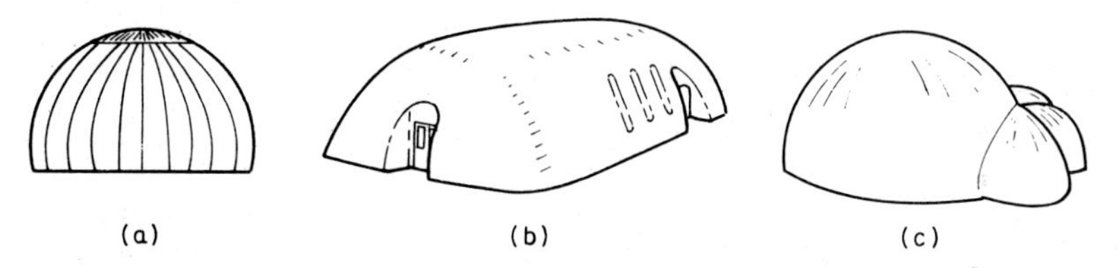

FIGURE 5.105 Some shapes for air-supported structures. *(Reprinted with permission from F. S. Merritt, "Building Engineering and Systems Design," Van Nostrand Reinhold Company, New York.)*

of the stress distribution, any shape that can be obtained with soap films is feasible for an air-stabilized enclosure. Figure 5.105*c* shows a configuration formed by a conglomeration of bubbles as an illustration of a shape that can be adopted for an air-stabilized structure.

In practice, shapes of air-stabilized structures often resemble those used for thin-shell enclosures. For example, spherical domes (Fig. 5.105*a*) are frequently constructed with a membrane. Also, membranes are sometimes shaped as semicircular cylinders with quarter-sphere ends (Fig. 5.105*b*).

Air-stabilized enclosures may be classified as air-inflated, air-supported, or hybrid structures, depending on the type of support.

Air-inflated enclosures are completely supported by pressurized air entrapped within membranes. There are two main types, inflated-rib structures and inflated dual-wall structures.

In inflated-rib construction, the membrane enclosure is supported by a framework of air-pressurized tubes, which serve much like arch ribs in thin-shell construction (Art. 5.15.1). The principle of their action is demonstrated by a water hose. A flexible hose, when empty, collapses under its own weight on short spans or under loads normal to its length; but it stiffens when filled with water. The water pressure tensions the hose walls and enables them to withstand compressive stresses.

In inflated dual-walled construction, pressurized air is trapped between two concentric membranes (Fig. 5.106). The shape of the inner membrane is maintained by suspending it from the outer one. Because of the large volume of air compressed between the membranes, this type of construction can span longer distances than can inflated-rib structures.

Because of the variation of air pressure with changes in temperature, provision must be made for adjustment of the pressure of the compressed air in air-inflated structures. Air must be vented to relieve excessive pressures, to prevent overtensioning of the membranes. Also, air must be added to compensate for pressure drops, to prevent collapse.

Air-supported enclosures consist of a single membrane supported by the difference between internal air pressure and external atmospheric pressure (Fig. 5.107). The pressure differential deflects the membrane outward, inducing tensile

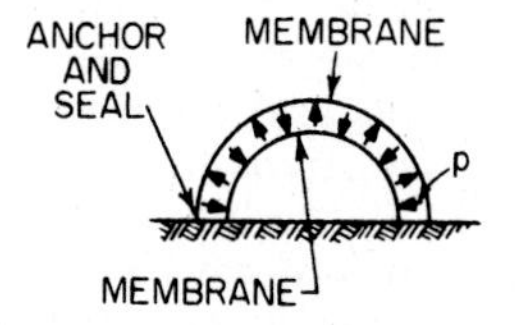

FIGURE 5.106 Inflated dual-wall structure.

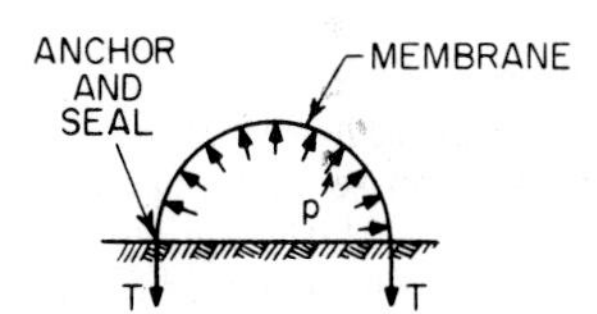

FIGURE 5.107 Air-supported structure.

stresses in it, thus enabling it to withstand compressive forces. To resist the uplift, the construction must be securely anchored to the ground. Also, the membrane must be completely sealed around its perimeter to prevent air leakage.

Hybrid structures consist of one of the preceding types of pneumatic construction augmented by light metal framing, such as cables. The framing may be merely a safety measure to support the membrane if pressure should be lost or a means of shaping the membrane when it is stretched. Under normal conditions, air pressure against the membrane reduces the load on the framing from heavy wind and snow loads.

5.17.2 Membrane Stresses

Air-supported structures are generally spherical or cylindrical because of the supporting uniform pressure.

When a spherical membrane with radius R, in, is subjected to a uniform radial internal pressure, p, psi, the internal unit tensile force, lb/in, in any direction, is given by

$$T = \frac{pR}{2} \tag{5.233}$$

In a cylindrical membrane, the internal unit tensile force, lb/in, in the circumferential direction is given by

$$T = pR \tag{5.234}$$

where R = radius, in, of the cylinder. The longitudinal membrane stress depends on the conditions at the cylinder ends. For example, with immovable end enclosures, the longitudinal stress would be small. If, however, the end enclosure is flexible, a tension about half that given by Eq. (5.234) would be imposed on the membrane in the longitudinal direction.

Unit stress in the membrane can be computed by dividing the unit force by the thickness, in, of the membrane.

(R. N. Dent, "Principles of Pneumatic Architecture," John Wiley & Sons, Inc., New York; J. W. Leonard, "Tension Structures," McGraw-Hill Publishing Company, New York.)

5.18 STRUCTURAL DYNAMICS

Article 5.1.1 notes that loads can be classified as static or dynamic and that the distinguishing characteristic is the rate of application of load. If a load is applied slowly, it may be considered static. Since dynamic loads may produce stresses and deformations considerably larger than those caused by static loads of the same magnitude, it is important to know reasonably accurately what is meant by slowly.

A useful definition can be given in terms of the natural period of vibration of the structure or member to which the load is applied. If the time in which a load rises from zero to its maximum value is more than double the natural period, the load may be treated as static. Loads applied more rapidly may be dynamic. Structural analysis and design for such loads are considerably different from and more complex than those for static loads.

In general, exact dynamic analysis is possible only for relatively simple structures, and only when both the variation of load and resistance with time are a convenient mathematical function. Therefore, in practice, adoption of approximate methods that permit rapid analysis and design is advisable. And usually, because of uncertainties in loads and structural resistance, computations need not be carried out with more than a few significant figures, to be consistent with known conditions.

5.18.1 Properties of Materials under Dynamic Loading

In general, mechanical properties of structural materials improve with increasing rate of load application. For low-carbon steel, for example, yield strength, ultimate strength, and ductility rise with increasing rate of strain. Modulus of elasticity in the elastic range, however, is unchanged. For concrete, the dynamic ultimate strength in compression may be much greater than the static strength.

Since the improvement depends on the material and the rate of strain, values to use in dynamic analysis and design should be determined by tests approximating the loading conditions anticipated.

Under many repetitions of loading, though, a member or connection between members may fail because of "fatigue" at a stress smaller than the yield point of the material. In general, there is little apparent deformation at the start of a fatigue failure. A crack forms at a point of high stress concentration. As the stress is repeated, the crack slowly spreads, until the member ruptures without measurable yielding. Though the material may be ductile, the fracture looks brittle.

Some materials (generally those with a well-defined yield point) have what is known as an **endurance limit.** This is the maximum unit stress that can be repeated, through a definite range, an indefinite number of times without causing structural damage. Generally, when no range is specified, the endurance limit is intended for a cycle in which the stress is varied between tension and compression stresses of equal value.

A range of stress may be resolved into two components—a steady, or mean, stress and an alternating stress. The endurance limit sometimes is defined as the maximum value of the alternating stress that can be superimposed on the steady stress an indefinitely large number of times without causing fracture.

Design of members to resist repeated loading cannot be executed with the certainty with which members can be designed to resist static loading. Stress concentrations may be present for a wide variety of reasons, and it is not practicable to calculate their intensities. But sometimes it is possible to improve the fatigue strength of a material or to reduce the magnitude of a stress concentration below the minimum value that will cause fatigue failure.

In general, avoid design details that cause severe stress concentrations or poor stress distribution. Provide gradual changes in section. Eliminate sharp corners and notches. Do not use details that create high localized constraint. Locate unavoidable stress raisers at points where fatigue conditions are the least severe. Place connections at points where stress is low and fatigue conditions are not severe. Provide structures with multiple load paths or redundant members, so that a fatigue crack in any one of the several primary members is not likely to cause collapse of the entire structure.

Fatigue strength of a material may be improved by cold-working the material in the region of stress concentration, by thermal processes, or by prestressing it in such a way as to introduce favorable internal stresses. Where fatigue stresses are unusually severe, special materials may have to be selected with high energy absorption and notch toughness.

(J. H. Faupel, "Engineering Design," John Wiley & Sons, Inc., New York; C. H. Norris et al., "Structural Design for Dynamic Loads," McGraw-Hill Book Company, New York; W. H. Munse, "Fatigue of Welded Steel Structures," Welding Research Council, 345 East 47th Street, New York, NY 10017.)

5.18.2 Natural Period of Vibration

A preliminary step in dynamic analysis and design is determination of this period. It can be computed in many ways, including by application of the laws of conservation of energy and momentum or Newton's second law, $F = M(dv/dt)$, where F is force, M mass, v velocity, and t time. But in general, an exact solution is possible only for simple structures. Therefore, it is general practice to seek an approximate—but not necessarily inexact—solution by analyzing an idealized representation of the actual member or structure. Setting up this model and interpreting the solution require judgment of a high order.

Natural period of vibration is the time required for a structure to go through one cycle of free vibration, that is, vibration after the disturbance causing the motion has ceased.

To compute the natural period, the actual structure may be conveniently represented by a system of masses and massless springs, with additional resistances provided to account for energy losses due to friction, hysteresis, and other forms of damping. In simple cases, the masses may be set equal to the actual masses; otherwise, equivalent masses may have to be computed (Art. 5.18.6). The spring constants are the ratios of forces to deflections.

For example, a single mass on a spring (Fig. 5.108*b*) may represent a simply supported beam with mass that may be considered negligible compared with the

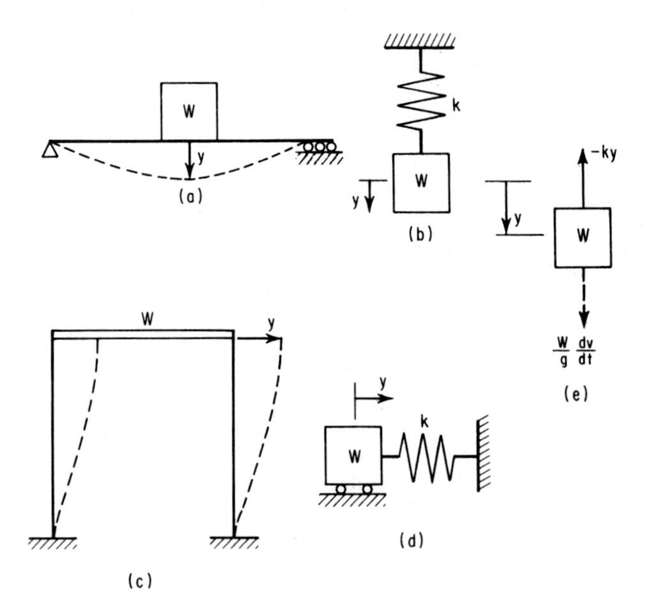

FIGURE 5.108 Mass on a weightless spring (*b*) or (*d*) may represent the motion of (*a*) a beam or (*c*) a rigid frame in free vibration.

load W at midspan (Fig. 5.108*a*). The spring constant k should be set equal to the load that produces a unit deflection at midspan; thus, $k = 48EI/L^3$, where E is the modulus of elasticity, psi; I the moment of inertia, in^4; and L the span, in, of the beam. The idealized mass equals W/g, where W is the weight of the load, lb, and g is the acceleration due to gravity, 386 in/s^2.

Also, a single mass on a spring (Fig. 5.108*d*) may represent the rigid frame in Fig. 5.108*c*. In that case, $k = 2 \times 12EI/h^3$, where I is the moment of inertia, in^4, of each column and h the column height, in. The idealized mass equals the sum of the masses on the girder and the girder mass. (Weight of columns and walls is assumed negligible.)

The spring and mass in Fig. 5.108*b* and *d* form a one-degree system. The **degree of a system** is determined by the least number of coordinates needed to define the positions of its components. In Fig. 5.108, only the coordinate y is needed to locate the mass and determine the state of the spring. In a two-degree system, such as one comprising two masses connected to each other and to the ground by springs and capable of movement in only one direction, two coordinates are required to locate the masses.

If the mass with weight W, lb, in Fig. 5.108 is isolated, as shown in Fig. 5.108*e* it will be in dynamic equilibrium under the action of the spring force $-ky$ and the inertia force $(d^2y/dt^2)(W/g)$. Hence, the equation of motion is

$$\frac{W}{g}\frac{d^2y}{dt^2} + ky = 0 \tag{5.235}$$

where y = displacement of mass, in, measured from rest position. Equation (5.235) may be written in the more convenient form

$$\frac{d^2y}{dt^2} + \frac{kg}{W}y = \frac{d^2y}{dt^2} + \omega^2 y = 0 \tag{5.236}$$

The solution is

$$y = A \sin \omega t + B \cos \omega t \tag{5.237}$$

where A and B are constants to be determined from initial conditions of the system, and

$$\omega = \sqrt{\frac{kg}{W}} \tag{5.238}$$

is the natural circular frequency, rad/s.

The motion defined by Eq. (5.237) is harmonic. Its natural period, s, is

$$T = \frac{2\pi}{\omega} = 2\pi\sqrt{\frac{W}{kg}} \tag{5.239}$$

Its natural frequency, Hz, is

$$f = \frac{1}{T} = \frac{1}{2\pi}\sqrt{\frac{kg}{W}} \tag{5.240}$$

If, at time $t = 0$, the mass has an initial displacement y_0 and velocity v_0, substitution in Eq. (5.237) yields $A = v_0/\omega$ and $B = y_0$. Hence, at any time t, the mass is completely located by

$$y = \frac{v_0}{\omega} \sin \omega t + y_0 \cos \omega t \tag{5.241}$$

The stress in the spring can be computed from the displacement y.

Vibrations of Lumped Masses. In multiple-degree systems, an independent differential equation of motion can be written for each degree of freedom. Thus, in an N-degree system with N masses, weighing $W_1, W_2, \ldots, W_N$ lb, and N^2 springs with constants k_{rj} ($r = 1, 2, \ldots, N$; $j = 1, 2, \ldots, N$), there are N equations of the form

$$\frac{W_r}{g}\frac{d^2y_r}{dt^2} + \sum_{j=1}^{N} k_{rj}y_j = 0 \qquad r = 1, 2, \ldots, N \tag{5.242}$$

Simultaneous solution of these equations reveals that the motion of each mass can be resolved into N harmonic components. They are called the fundamental, second, third, etc., harmonics. Each set of harmonics for all the masses is called a **normal mode** of vibration.

There are as many normal modes in a system as degrees of freedom. Under certain circumstances, the system could vibrate freely in any one of these modes. During any such vibration, the ratio of displacement of any two of the masses remains constant. Hence, the solutions of Eqs. (5.242) take the form

$$y_r = \sum_{n=1}^{N} a_{rn} \sin \omega_n(t + \tau_n) \tag{5.243}$$

where a_{rn} and τ_n are constants to be determined from the initial conditions of the system and ω_n is the natural circular frequency for each normal mode.

To determine ω_n, set $y_1 = A_1 \sin \omega t$; $y_2 = A_2 \sin \omega t \ldots$. Then, substitute these values of y_r and their second derivatives in Eqs. (5.242). After dividing each equation by $\sin \omega t$, the following N equations result:

$$\begin{aligned}
\left(k_{11} - \frac{W_1}{g}\omega^2\right) A_1 + k_{12}A_2 + \cdots + k_{1N}A_N &= 0 \\
k_{21}A_1 + \left(k_{22} - \frac{W_2}{g}\omega^2\right) A_2 + \cdots + k_{2N}A_N &= 0 \\
\cdots\cdots\cdots\cdots\cdots\cdots\cdots\cdots\cdots\cdots\cdots& \\
k_{N1}A_1 + k_{N2}A_2 + \cdots + \left(k_{NN} - \frac{W_N}{g}\omega^2\right) A_N &= 0
\end{aligned} \tag{5.244}$$

If there are to be nontrivial solutions for the amplitudes $A_1, A_2, \ldots, A_N$, the determinant of their coefficients must be zero. Thus,

$$\begin{vmatrix}
k_{11} - \frac{W_1}{g}\omega^2 & k_{12} & \cdots & k_{1N} \\
k_{21} & k_{22} - \frac{W_2}{g}\omega^2 & \cdots & k_{2N} \\
\cdots & \cdots & \cdots & \cdots \\
k_{N1} & k_{N2} & \cdots & k_{NN} - \frac{W_N}{g}\omega^2
\end{vmatrix} = 0 \tag{5.245}$$

Solution of this equation for ω yields one real root for each normal mode. And the natural period for each normal mode can be obtained from Eq. (5.239).

If ω for a normal mode now is substituted in Eqs. (5.244), the amplitudes A_1, $A_2, \ldots, A_N$ for that mode can be computed in terms of an arbitrary value, usually unity, assigned to one of them. The resulting set of modal amplitudes defines the **characteristic shape** for that mode.

The normal modes are mutually orthogonal; that is,

$$\sum_{r=1}^{N} W_r A_{rn} A_{rm} = 0 \tag{5.246}$$

where W_r is the rth mass out of a total of N, A represents the characteristic amplitude of a normal mode, and n and m identify any two normal modes. Also, for a total of S springs

$$\sum_{s=1}^{S} k_s y_{sn} y_{sm} = 0 \tag{5.247}$$

where k_s is the constant for the sth spring and y represents the spring distortion.

When there are many degrees of freedom, this procedure for analyzing free vibration becomes very lengthy. In such cases, it may be preferable to solve Eqs. (5.244) by numerical, trial-and-error procedures, such as the Stodola-Vianello method. In that method, the solution converges first on the highest or lowest mode. Then, the other modes are determined by the same procedure after elimination of one of the equations by use of Eq. (5.246). The procedure requires assumption of a characteristic shape, a set of amplitudes A_{r1}. These are substituted in one of Eqs. (5.244) to obtain a first approximation of ω^2. With this value and with $A_{N1} = 1$, the remaining $N - 1$ equations are solved to obtain a new set of A_{r1}. Then, the procedure is repeated until assumed and final characteristic amplitudes agree.

Because even this procedure is very lengthy for many degrees of freedom, the Rayleigh approximate method may be used to compute the fundamental mode. The frequency obtained by this method, however, may be a little on the high side.

The Rayleigh method also starts with an assumed set of characteristic amplitudes A_{r1} and depends for its success on the small error in natural frequency produced by a relatively large error in the shape assumption. Next, relative inertia forces acting at each mass are computed: $F_r = W_r A_{r1}/A_{N1}$, where A_{N1} is the assumed displacement at one of the masses. These forces are applied to the system as a static load and displacements B_{r1} due to them calculated. Then, the natural frequency can be obtained from

$$\omega^2 = \frac{g \sum_{r=1}^{N} F_r B_{r1}}{\sum_{r=1}^{N} W_r B_{r1}^2} \tag{5.248}$$

where g is the acceleration due to gravity, 386 in/s^2. For greater accuracy, the computation can be repeated with B_{r1} as the assumed characteristic amplitudes.

When the Rayleigh method is applied to beams, the characteristic shape assumed initially may be chosen conveniently as the deflection curve for static loading.

The Rayleigh method may be extended to determination of higher modes by the Schmidt orthogonalization procedure, which adjusts assumed deflection curves to satisfy Eq. (5.246). The procedure is to assume a shape, remove components associated with lower modes, then use the Rayleigh method for the residual deflection curve. The computation will converge on the next higher mode. The method is shorter than the Stodola-Vianello procedure when only a few modes are needed.

For example, suppose the characteristic amplitudes A_{r1} for the fundamental mode have been obtained and the natural frequency for the second mode is to be computed. Assume a value for the relative deflection of the rth mass A_{r2}. Then, the shape with the fundamental mode removed will be defined by the displacements

$$a_{r2} = A_{r2} - c_1 A_{r1} \tag{5.249}$$

where c_1 is the participation factor for the first mode.

$$c_1 = \frac{\sum_{r=1}^{N} W_r A_{r2} A_{r1}}{\sum_{r=1}^{N} W_r A_{r1}^2} \tag{5.250}$$

Substitute a_{r2} for B_{r1} in Eq. (5.248) to find the second-mode frequency and, from deflections produced by $F_r = W_r a_{r2}$, an improved shape. (For more rapid convergence, A_{r2} should be selected to make c_1 small.) The procedure should be repeated, starting with the new shape.

For the third mode, assume deflections A_{r3} and remove the first two modes:

$$a_{r3} = A_{r3} - c_1 A_{r1} - c_2 A_{r2} \tag{5.251}$$

The participation factors are determined from

$$c_1 = \frac{\sum_{r=1}^{N} W_r A_{r3} A_{r1}}{\sum_{r=1}^{N} W_r A_{r1}^2} \qquad c_2 = \frac{\sum_{r=1}^{N} W_r A_{r3} A_{r2}}{\sum_{r=1}^{N} W_r A_{r2}^2} \tag{5.252}$$

Use a_{r3} to find an improved shape and the third-mode frequency.

Vibrations of Distributed Masses. For some structures with mass distributed throughout, it sometimes is easier to solve the dynamic equations based on distributed mass than the equations based on equivalent lumped masses. A distributed mass has an infinite number of degrees of freedom and normal modes. Every particle in it can be considered a lumped mass on springs connected to other particles. Usually, however, only the fundamental mode is significant, though sometimes the second and third modes must be taken into account.

For example, suppose a beam weighs w lb/lin ft and has a modulus of elasticity E, psi, and moment of inertia I, in^4. Let y be the deflection at a distance x from one end. Then, the equation of motion is

$$EI \frac{\partial^4 y}{\partial x^4} + \frac{w}{g} \frac{\partial^2 y}{\partial t^2} = 0 \tag{5.253}$$

(This equation ignores the effects of shear and rotational inertia.) The deflection y_n for each mode, to satisfy the equation, must be the product of a harmonic function of time $f_n(t)$ and of the characteristic shape $Y_n(x)$, a function of x with undetermined amplitude. The solution is

$$f_n(t) = c_1 \sin \omega_n t + c_2 \cos \omega_n t \tag{5.254}$$

where ω_n is the natural circular frequency and n indicates the mode, and

$$Y_n(x) = A_n \sin \beta_n x + B_n \cos \beta_n x + C_n \sinh \beta_n x + D_n \cosh \beta_n x \tag{5.255}$$

where

$$\beta_n = \sqrt[4]{\frac{w\omega_n^2}{EIg}} \tag{5.256}$$

For a simple beam, the boundary (support) conditions for all values of time t are $y = 0$ and bending moment $M = EI\, \partial^2 y/\partial x^2 = 0$. Hence, at $x = 0$ and $x = L$, the span length, $Y_n(x) = 0$ and $d_2Y_n/dx^2 = 0$. These conditions require that

$$B_n = C_n = D_n = 0 \qquad \beta_n = \frac{n\pi}{L}$$

to satisfy Eq. (5.255). Hence, according to Eq. (5.256), the natural circular frequency for a simply supported beam is

$$\omega_n = \frac{n^2\pi^2}{L^2}\sqrt{\frac{EIg}{w}} \tag{5.257}$$

The characteristic shape is defined by

$$Y_n(x) = \sin \frac{n\pi x}{L} \tag{5.258}$$

The constants c_1 and c_2 in Eq. (5.254) are determined by the initial conditions of the disturbance. Thus, the total deflection, by superposition of modes, is

$$y = \sum_{n=1}^{\infty} A_n(t) \sin \frac{n\pi x}{L} \tag{5.259}$$

where $A_n(t)$ is determined by the load (see Art. 5.18.4).

Equations (5.254) to (5.256) apply to spans with any type of end restraints. Figure 5.109 shows the characteristic shape and gives constants for determination of natural circular frequency ω and natural period T for the first four modes of cantilever, simply supported, fixed-end, and fixed-hinged beams. To obtain ω, select the appropriate constant from Fig. 5.109 and multiply it by $\sqrt{EI/wL^4}$, where L = span of beam, ft. To get T, divide the appropriate constant by $\sqrt{EI/wL^4}$.

To determine the characteristic shapes and natural periods for beams with variable cross section and mass, use the Rayleigh method. Convert the beam into a lumped-mass system by dividing the span into elements and assuming the mass of each element to be concentrated at its center. Also, compute all quantities, such as deflection and bending moment, at the center of each element. Start with an assumed characteristic shape and apply Eq. (5.255).

Methods are available for dynamic analysis of continuous beams. (R. Clough and J. Penzien, "Dynamics of Structures," McGraw-Hill Book Company, New

TYPE OF SUPPORT	FUNDAMENTAL MODE	SECOND MODE	THIRD MODE	FOURTH MODE
CANTILEVER	L	0.774L	0.5L 0.132L	0.356L 0.094L 0.644L
$\omega\sqrt{wL^4/EI}$ =	0.480	3.031	8.421	16.504
$T\sqrt{EI/wL^4}$ =	13.090	2.073	0.746	0.381
SIMPLE	L	0.5L	L/3 L/3	L/4 L/4 L/2
$\omega\sqrt{wL^4/EI}$ =	1.347	5.389	12.125	21.556
$T\sqrt{EI/wL^4}$ =	4.665	1.166	0.518	0.292
FIXED	L	L/2	0.359L 0.359L	0.278L L/2 0.278L
$\omega\sqrt{wL^4/EI}$ =	3.031	8.421	16.504	27.283
$T\sqrt{EI/wL^4}$ =	2.073	0.746	0.381	0.230
FIXED-HINGED	L	0.56L	0.384L 0.308L	0.294L 0.235L 0.529L
$\omega\sqrt{wL^4/EI}$ =	2.105	6.821	14.231	24.336
$T\sqrt{EI/wL^4}$ =	2.985	0.921	0.442	0.258

FIGURE 5.109 Coefficients for computing natural circular frequencies ω and natural periods of vibration T, s, of prismatic beams. w = weight of beam, lb/lin ft; L = span, ft; E = modulus of elasticity of the beam material, psi; I = moment of inertia of the beam cross section, in^4.

York; D. G. Fertis and E. C. Zobel, "Transverse Vibration Theory," The Ronald Press Company, New York.) But even for beams with constant cross section, these procedures are very lengthy. Generally, approximate solutions are preferable.

(J. M. Biggs, "Introduction to Structural Dynamics," McGraw-Hill Book Company, New York; N. M. Newmark and E. Rosenblueth, "Fundamentals of Earthquake Engineering," Prentice-Hall, Englewood Cliffs, N.J.)

5.18.3 Impact and Sudden Loads

Under impact, there is an abrupt exchange or absorption of energy and drastic change in velocity. Stresses caused in the colliding members may be several times larger than stresses produced by the same weights applied statically.

An approximation of impact stresses in the elastic range can be made by neglecting the inertia of the body struck and the effect of wave propagation and assuming that the kinetic energy is converted completely into strain energy in that body. Consider a prismatic bar subjected to an axial impact load in tension. The energy absorbed per unit of volume when the bar is stressed to the proportional limit is called the **modulus of resilience.** It is given by $f_y^2/2E$, where f_y is the yield stress and E the modulus of elasticity, both in psi.

Below the proportional limit, the unit stress, psi, due to an axial load U, in-lb, is

$$f = \sqrt{\frac{2UE}{AL}} \tag{5.260}$$

where A is the cross-sectional area, in^2, and L the length of bar, in. This equation indicates that, for a given unit stress, energy absorption of a member may be improved by increasing its length or area. Sharp changes in cross section should be avoided, however, because of associated high stress concentrations. Also, uneven distribution of stress in a member due to changes in section should be avoided. For example, if part of a member is given twice the diameter of another part, the stress under axial load in the larger portion is one-fourth that in the smaller. Since the energy absorbed is proportional to the square of the stress, the energy taken per unit of volume by the larger portion is therefore only one-sixteenth that absorbed by the smaller. So despite the increase in volume caused by doubling of the diameter, the larger portion absorbs much less energy than the smaller. Thus, energy absorption would be larger with a uniform stress distribution throughout the length of the member.

Impact on Short Members. If a static axial load W would produce a tensile stress f' in the bar and an elongation e', in, then the axial stress produced in a short member when W falls a distance h, in, is

$$f = f' + f' \sqrt{1 + \frac{2h}{e'}} \tag{5.261}$$

if f is within the proportional limit. The elongation due to this impact load is

$$e = e' + e' \sqrt{1 + \frac{2h}{e'}} \tag{5.262}$$

These equations indicate that the stress and deformation due to an energy load may be considerably larger than those produced by the same weight applied gradually.

The same equations hold for a beam with constant cross section struck by a weight at midspan, except that f and f' represent stresses at midspan and e and e', midspan deflections.

According to Eqs. (5.261) and (5.262), a sudden load ($h = 0$) causes twice the stress and twice the deflection as the same load applied gradually.

Impact on Long Members. For very long members, the effect of wave propagation should be taken into account. Impact is not transmitted instantly to all parts of the struck body. At first, remote parts remain undisturbed, while particles struck accelerate rapidly to the velocity of the colliding body. The deformations produced move through the struck body in the form of elastic waves. The waves travel with a constant velocity, ft/s,

$$c = 68.1 \sqrt{\frac{E}{\rho}} \tag{5.263}$$

where E = modulus of elasticity, psi
ρ = density of the struck body, lb/ft^3

If an impact imparts a velocity v, ft/s, to the particles at one end of a prismatic bar, the stress, psi, at that end is

$$f = E\frac{v}{c} = 0.0147v\sqrt{E\rho} = 0.000216\rho cv \tag{5.264}$$

if f is in the elastic range. In a compression wave, the velocity of the particles is in the direction of the wave. In a tension wave, the velocity of the particles is in the direction opposite the wave.

In the plastic range, Eqs. (5.263) and (5.264) hold, but with E as the tangent modulus of elasticity. Hence, c is not a constant and the shape of the stress wave changes as it moves. The elastic portion of the stress wave moves faster than the wave in the plastic range. Where they overlap, the stress and irrecoverable strain are constant.

(The impact theory is based on an assumption difficult to realize in practice—that contact takes place simultaneously over the entire end of the bar.)

At the free end of a bar, a compressive stress wave is reflected as an equal tension wave, and a tension wave as an equal compression wave. The velocity of the particles there equals $2v$.

At a fixed end of a bar, a stress wave is reflected unchanged. The velocity of the particles there is zero, but the stress is doubled, because of the superposition of the two equal stresses on reflection.

For a bar with a fixed end struck at the other end by a moving mass weighing W_m lb, the initial compressive stress, psi, is

$$f_o = 0.0147v_o\sqrt{E\rho} \tag{5.265}$$

where v_o is the initial velocity of the particles, ft/s, at the impacted end of the bar and E and ρ, the modulus of elasticity, psi, and density, lb/ft^3, of the bar. As the velocity of W_m decreases, so does the pressure on the bar. Hence, decreasing compressive stresses follow the wave front. At any time $t < 2L/c$, where L is the length of the bar, in, the stress at the struck end is

$$f = f_o e^{-2\alpha t/\tau} \tag{5.266}$$

where e = 2.71828, α is the ratio of W_b, the weight of the bar, to W_m, and $\tau = 2L/c$.

When $t = \tau$, the wave front with stress f_o arrives back at the struck end, assumed still to be in contact with the mass. Since the velocity of the mass cannot change suddenly, the wave will be reflected as from a fixed end. During the second interval, $\tau < t < 2\tau$, the compressive stress is the sum of two waves moving away from the struck end and one moving toward this end.

Maximum stress from impact occurs at the fixed end. For α greater than 0.2, this stress is

$$f = 2f_o(1 + e^{-2\alpha}) \tag{5.267}$$

For smaller values of α, it is given approximately by

$$f = f_o\left(1 + \sqrt{\frac{1}{\alpha}}\right) \tag{5.268}$$

Duration of impact, time it takes for the impact stress at the struck end to drop to zero, is approximately

$$T = \frac{\pi L}{c\sqrt{\alpha}} \tag{5.269}$$

for small values of α.

When W_m is the weight of a falling body, velocity at impact is $\sqrt{2gh}$, when it falls a distance h, in. Substitution in Eq. (5.265) yields $f_o = \sqrt{2EhW_b/AL}$, since $W_b = \rho AL$ is the weight of the bar. Putting $W_b = \alpha W_m$; $W_m/A = f'$, the stress produced by W_m when applied gradually, and $E = f'L/e'$, where e' is the elongation for the static load, gives $f_o = f' \sqrt{2h\alpha/e'}$. Then, for values of α smaller than 0.2, the maximum stress, from Eq. (5.268), is

$$f = f'\left(\sqrt{\frac{2h\alpha}{e'}} + \sqrt{\frac{2h}{e'}}\right) \tag{5.270}$$

For larger values of α, the stress wave due to gravity acting on W_m during impact should be added to Eq. (5.267). Thus, for α larger than 0.2,

$$f = 2f'(1 - e^{-2\alpha}) + 2f'\sqrt{\frac{2h\alpha}{e'}}(1 + e^{-2\alpha}) \tag{5.271}$$

Equations (5.270) and (5.271) correspond to Eq. (5.261), which was developed without wave effects being taken into account. For a sudden load, $h = 0$, Eq. (5.271) gives for the maximum stress $2f'(1 - e^{-2\alpha})$, not quite double the static stress, the result indicated by Eq. (5.261). (See also Art. 5.18.4.)

(S. Timoshenko and J. N. Goodier, "Theory of Elasticity," McGraw-Hill Book Company, New York; S. Timoshenko and D. H. Young, "Engineering Mechanics," McGraw-Hill Book Company, New York; J. H. Faupel, "Engineering Design," John Wiley & Sons, Inc., New York.)

5.18.4 Dynamic Analysis of Simple Structures

Articles 5.18.1 to 5.18.3 present a theoretic basis for analysis of structures under dynamic loads. As noted in Art. 5.18.2, an approximate solution based on an idealized representation of an actual member or structure is advisable for dynamic analysis and design. Generally, the actual structure may be conveniently represented by a system of masses and massless springs, with additional resistances to account for damping. In simple cases, the masses may be set equal to the actual masses; otherwise, equivalent masses may be substituted for the actual masses (Art. 5.18.6). The spring constants are the ratios of forces to deflections (see Art. 5.18.2).

Usually, for structural purposes, the data sought are the maximum stresses in the springs and their maximum displacements and the time of occurrence of the maximums. This time is generally computed in terms of the natural period of vibration of the member or structure, or in terms of the duration of the load. Maximum displacement may be calculated in terms of the deflection that would result if the load were applied gradually.

The term D by which the static deflection e', spring forces, and stresses are multiplied to obtain the dynamic effects is called the **dynamic load factor.** Thus, the dynamic displacement is

$$y = De' \tag{5.272}$$

And the maximum displacement y_m is determined by the maximum dynamic load factor D_m, which occurs at time t_m.

One-Degree Systems. Consider the one-degree-of-freedom system in Fig. 5.110*a*. It may represent a weightless beam with a mass weighing W lb applied at midspan and subjected to a varying force $F_o f(t)$, or a rigid frame with a mass weighing W lb at girder level and subjected to this force. The force is represented by an arbitrarily chosen constant force F_o times $F(t)$, a function of time.

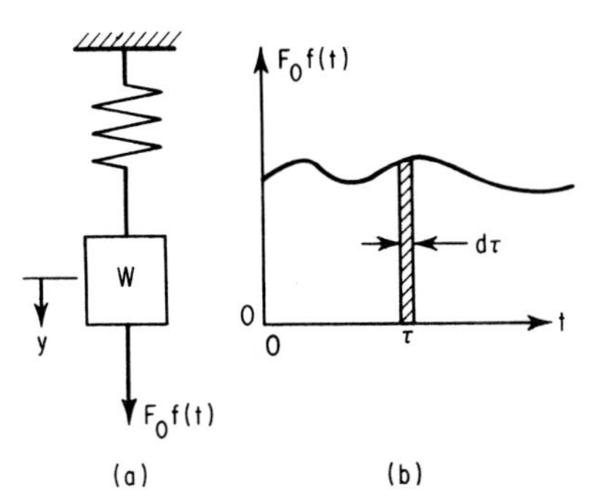

FIGURE 5.110 One-degree system acted on by a force varying with time.

If the system is not damped, the equation of motion in the elastic range is

$$\frac{W}{g}\frac{d^2y}{dt^2} + ky = F_o f(t) \tag{5.273}$$

where k is the spring constant and g the acceleration due to gravity, 386 in/s². The solution consists of two parts. The first, called the complementary solution, is obtained by setting $f(t) = 0$. This solution is given by Eq. (5.237). To it must be added the second part, the particular solution, which satisfies Eq. (5.273).

The general solution of Eq. (5.273), arrived at by treating an element of the force-time curve (Fig. 5.111*b*) as an impulse, is

$$y = y_o \cos \omega t + \frac{v_o}{\omega} \sin \omega t + e'\omega \int_0^t f(\tau) \sin \omega(t - \tau)\, d\tau \tag{5.274}$$

where y = displacement of mass from equilibrium position, in
y_o = initial displacement of mass ($t = 0$), in
$\omega = \sqrt{kg/W}$ = natural circular frequency of free vibration
k = spring constant = force producing unit deflection, lb/in
v_o = initial velocity of mass, in/s
$e' = F_o/k$ = displacement under static load, in

A closed solution is possible if the integral can be evaluated.

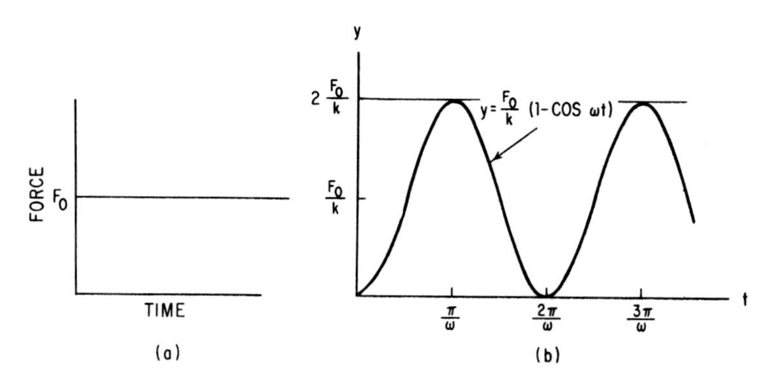

FIGURE 5.111 Harmonic motion. (*a*) Constant force applied to an undamped one-degree system, such as the one in Fig. 5.110*a*. (*b*) Displacements vary with time like a cosine curve.

Assume, for example, the mass is subjected to a suddenly applied force F_o that remains constant (Fig. 5.111*a*). If y_o and v_o are initially zero, the displacement y of the mass at any time t can be obtained from the integral in Eq. (5.274) by setting $f(\tau) = 1$:

$$y = e'\omega \int_0^t \sin \omega(t - \tau)\, d\tau = e'(1 - \cos \omega t) \tag{5.275}$$

This equation indicates that the dynamic load factor $D = 1 - \cos \omega t$. It has a maximum value $D_m = 2$ when $t = \pi/\omega$. Figure 5.111*b* shows the variation of displacement with time.

Multidegree Systems. A multidegree lumped-mass system may be analyzed by the modal method after the natural frequencies of the normal modes have been determined (Art. 5.18.2). This method is restricted to linearly elastic systems in which the forces applied to the masses have the same variation with time. For other cases, numerical analysis must be used.

In the modal method, each normal mode is treated as an independent one-degree system. For each degree of the system, there is one normal mode. A natural frequency and a characteristic shape are associated with each mode. In each mode, the ratio of the displacements of any two masses is constant with time. These ratios define the characteristic shape. The modal equation of motion for each mode is

$$\frac{d^2A_n}{dt^2} + \omega_n^2 A_n = \frac{gf(t) \sum_{r=1}^{j} F_r \phi_{rn}}{\sum_{r=1}^{j} W_r \phi_{rn}^2} \tag{5.276}$$

where A_n = displacement in the nth mode of an arbitrarily selected mass
ω_n = natural frequency of the nth mode
$F_r f(t)$ = varying force applied to the rth mass
W_r = weight of the rth mass
j = number of masses in the system
ϕ_{rn} = ratio of the displacement in the nth mode of the rth mass to A_n
g = acceleration due to gravity

We define the modal static deflection as

$$A'_n = \frac{g \sum_{r=1}^{j} F_r \phi_{rn}}{\omega_n^2 \sum_{r=1}^{j} W_r \phi_{rn}^2} \tag{5.277}$$

Then, the response for each mode is given by

$$A_n = D_n A'_n \tag{5.278}$$

where D_n = dynamic load factor.

Since D_n depends only on ω_n and the variation of force with time $f(t)$, solutions for D_n obtained for one-degree systems also apply to multidegree systems. The total deflection at any point is the sum of the displacements for each mode, $\Sigma A_n \phi_{rn}$, at that point.

Beams. The response of beams to dynamic forces can be determined in a similar way. The modal static deflection is defined by

$$A'_n = \frac{\int_0^L p(x)\phi_n(x)\,dx}{\omega_n^2 \dfrac{w}{g} \int_0^L \phi_n^2(x)\,dx} \tag{5.279}$$

where $p(x)$ = load distribution on the span [$p(x)f(x)$ is the varying force]
$\phi_n(x)$ = characteristic shape of the nth mode (see Art. 5.18.2)
L = span length
w = uniformly distributed weight on the span

The response of the beam then is given by Eq. (5.278), and the dynamic deflection is the sum of the modal components, $\Sigma A_n \phi_n(x)$.

Nonlinear Responses. When the structure does not react linearly to loads, the equations of motion can be solved by numerical analysis if resistance is a unique function of displacement. Sometimes, the behavior of the structure can be represented by an idealized resistance-displacement diagram that makes possible a solution in closed form. Figure 5.112*a* shows such a diagram.

Elastic-Plastic Responses. Resistance is assumed linear ($R = ky$) in Fig. 5.112*a* until a maximum R_m is reached. After that, R remains equal to R_m for increases in y substantially larger than the displacement y_e at the elastic limit. Thus, some portions of the structure deform into the plastic range. Figure 5.112*a*, therefore, may be used for ductile structures only rarely subjected to severe dynamic loads. When this diagram can be used for designing such structures, more economical designs can be produced than for structures limited to the elastic range, because of the high energy-absorption capacity of structures in the plastic range.

For a one-degree system, Eq. (5.273) can be used as the equation of motion for the initial sloping part of the diagram (elastic range). For the second stage, $y_e < y < y_m$, where y_m is the maximum displacement, the equation is

$$\frac{W}{g}\frac{d^2y}{dt^2} + R_m = F_o f(t) \tag{5.280}$$

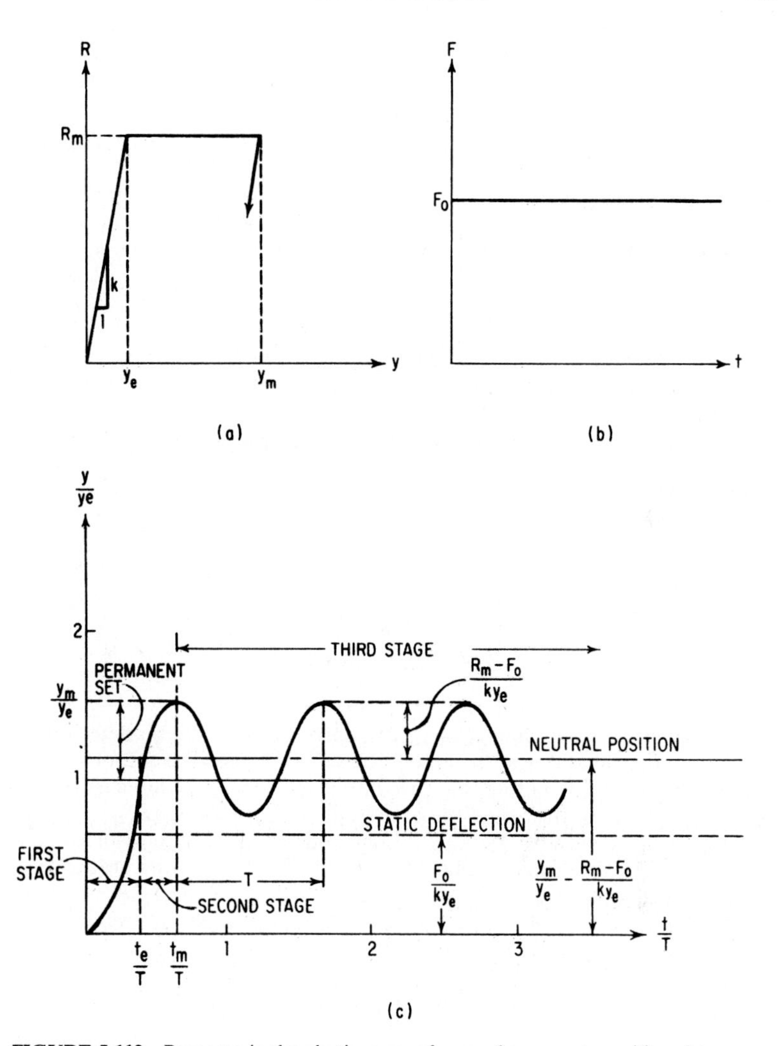

FIGURE 5.112 Response in the plastic range of a one-degree system with resistance characteristics indicated in (*a*) and subjected to a constant force (*b*) is shown in (*c*).

For the unloading stage, $y < y_m$, the equation is

$$\frac{W}{g}\frac{d^2y}{dt^2} + R_m - k(y_m - y) = F_o f(t) \tag{5.281}$$

Suppose, for example, the one-degree undamped system in Fig. 5.109*a* behaves in accordance with the bilinear resistance function of Fig. 5.112*a* and is subjected to a suddenly applied constant load (Fig. 5.112*b*). With zero initial displacement and velocity, the response in the first stage ($y < y_e$), according to Eq. (5.281), is

$$y = e'(1 - \cos \omega t_1) \tag{5.282}$$

$$\frac{dy}{dt} = e'\omega \sin \omega t_1 \tag{5.283}$$

Equation (5.275) also indicates that displacement y_e will be reached at a time t_e such that $\cos \omega t_e = 1 - y_e/e'$.

For convenience, let $t_2 = t - t_e$ be the time in the second stage; thus, $t_2 = 0$ at the start of that stage. Since the condition of the system at that time is the same as at the end of the first stage, the initial displacement is y_e and the initial velocity $e'\omega \sin \omega t_e$.

The equation of motion of the second stage is

$$\frac{W}{g}\frac{d^2y}{dt^2} + R_m = F_o \tag{5.284}$$

The solution, taking into account initial conditions for $y_e < y < y_m$ is

$$y = \frac{g}{2W}(F_o - R_m)t_2^2 + e'\omega t_2 \sin \omega t_e + y_e \tag{5.285}$$

Maximum displacement occurs at the time

$$t_m = \frac{W\omega e'}{g(R_m - F_o)} \sin \omega t_e \tag{5.286}$$

and can be obtained by substituting t_m in Eq. (5.285).

The third stage, unloading after y_m has been reached, can be determined from Eq. (5.281) and conditions at the end of the second stage. The response, however, is more easily found by noting that the third stage consists of an elastic, harmonic residual vibration. In this stage the amplitude of vibration is $(R_m - F_o)/k$, since this is the distance between the neutral position and maximum displacement, and in the neutral position the spring force equals F_o. Hence, the response can be obtained directly from Eq. (5.275) by substituting $y_m - (R_m - F_o)/k$ for e', because the neutral position, located at $y = y_m - (R_m - F_o)/k$, occurs when $\omega t_3 = \pi/2$, where $t_3 = t - t_e - t_m$. The solution is

$$y = y_m - \frac{R_m - F_o}{k} + \frac{R_m - F_o}{k} \cos \omega t_3 \tag{5.287}$$

Response in the three stages is shown in Fig. 5.112*c*. In that diagram, however, to represent a typical case, the coordinates have been made nondimensional by expressing y in terms of y_e and the time in terms of T, the natural period of vibration.

(J. M. Biggs, "Introduction to Structural Dynamics," and R. Clough and J. Penzien, "Dynamics of Structures," McGraw-Hill Book Company, New York; D. G. Fertis and E. C. Zobel, "Transverse Vibration Theory," The Ronald Press Company, New York; N. M. Newmark and E. Rosenblueth, "Fundamentals of Earthquake Engineering," Prentice-Hall, Englewood Cliffs, N.J.)

5.18.5 Resonance and Damping

Damping in structures, resulting from friction and other causes, resists motion imposed by dynamic loads. Generally, the effect is to decrease the amplitude and lengthen the period of vibrations. If damping is large enough, vibration may be eliminated.

When maximum stress and displacement are the prime concern, damping may not be of great significance for short-time loads. These maximums usually occur under such loads at the first peak of response, and damping, unless unusually large, has little effect in a short period of time. But under conditions close to resonance, damping has considerable effect.

Resonance is the condition of a vibrating system under a varying load such that the amplitude of successive vibrations increases. Unless limited by damping or changes in the condition of the system, amplitudes may become very large.

Two forms of damping generally are assumed in structural analysis, viscous or constant (Coulomb). For viscous damping, the damping force is taken proportional to the velocity but opposite in direction. For Coulomb damping, the damping force is assumed constant and opposed in direction to the velocity.

Viscous Damping. For a one-degree system (Arts. 5.18.2 to 5.18.4), the equation of motion for a mass weighing W lb and subjected to a force F varying with time but opposed by viscous damping is

$$\frac{W}{g}\frac{d^2y}{dt^2} + ky = F - c\frac{dy}{dt} \tag{5.288}$$

where y = displacement of the mass from equilibrium position, in
k = spring constant, lb/in
t = time, s
c = coefficient of viscous damping
g = acceleration due to gravity = 386 in/s^2

Let us set $\beta = cg/2W$ and consider those cases in which $\beta < \omega$, the natural circular frequency [Eq. (5.238)], to eliminate unusually high damping (overdamping). Then, for initial displacement y_o and velocity v_o, the solution of Eq. (5.288) with $F = 0$ is

$$y = e^{-\beta t}\left(\frac{v_o + \beta y_o}{\omega_d}\sin\omega_d t + y_o\cos\omega_d t\right) \tag{5.289}$$

where $\omega_d = \sqrt{\omega^2 - \beta^2}$ and $e = 2.71828$. Equation (5.289) represents a decaying harmonic motion with β controlling the rate of decay and ω_d the natural frequency of the damped system.

When $\beta = \omega$

$$y = e^{-\omega t}[v_o t + (1 + \omega t)y_o] \tag{5.290}$$

which indicates that the motion is not vibratory. Damping producing this condition is called critical, and, from the definition of β, the critical coefficient is

$$c_d = \frac{2W\beta}{g} = \frac{2W\omega}{g} = 2\sqrt{\frac{kW}{g}} \tag{5.291}$$

Damping sometimes is expressed as a percent of critical (β as a percent of ω).

For small amounts of viscous damping, the damped natural frequency is approximately equal to the undamped natural frequency minus $\frac{1}{2}\beta^2/\omega$. For example, for 10% critical damping ($\beta = 0.1\omega$), $\omega_d = \omega[1 - \frac{1}{2}(0.1)^2] = 0.995\omega$. Hence, the decrease in natural frequency due to small amounts of damping generally can be ignored.

Damping sometimes is measured by **logarithmic decrement,** the logarithm of the ratio of two consecutive peak amplitudes during free vibration.

$$\text{Logarithmic decrement} = \frac{2\pi\beta}{\omega} \tag{5.292}$$

For example, for 10% critical damping, the logarithmic decrement equals 0.2π. Hence, the ratio of a peak to the following peak amplitude is $e^{0.2\pi} = 1.87$.

The complete solution of Eq. (5.288) with initial displacement y_o and velocity v_o is

$$y = e^{-\beta t}\left(\frac{v_o + \beta y_o}{\omega_d}\sin\omega_d t + y_o\cos\omega_d t\right) + e'\frac{\omega^2}{\omega_d}\int_0^t f(\tau)e^{-\beta(t-\tau)}\sin\omega_d(t-\tau)\,d\tau \tag{5.293}$$

where e' is the deflection that the applied force would produce under static loading. Equation (5.293) is identical to Eq. (5.274) when $\beta = 0$.

Unbalanced rotating parts of machines produce pulsating forces that may be represented by functions of the form $F_o \sin\alpha t$. If such a force is applied to an undamped one-degree system, Eq. (5.274) indicates that if the system starts at rest the response will be

$$y = \frac{F_o g}{W}\left(\frac{1/\omega^2}{1 - \alpha^2/\omega^2}\right)\left(\sin\alpha t - \frac{\alpha}{\omega}\sin\omega t\right) \tag{5.294}$$

And since the static deflection would be $F_o/k = F_o g/W\omega^2$, the dynamic load factor is

$$D = \frac{1}{1 - \alpha^2/\omega^2}\left(\sin\alpha t - \frac{\alpha}{\omega}\sin\omega t\right) \tag{5.295}$$

If α is small relative to ω, maximum D is nearly unity; thus, the system is practically statically loaded. If α is very large compared with ω, D is very small; thus, the mass cannot follow the rapid fluctuations in load and remains practically stationary. Therefore, when α differs appreciably from ω, the effects of unbalanced rotating parts are not too serious. But if $\alpha = \omega$, resonance occurs; D increases with time. Hence, to prevent structural damage, measures must be taken to correct the unbalanced parts to change α, or to change the natural frequency of the vibrating mass, or damping must be provided.

The response as given by Eq. (5.294) consists of two parts, the free vibration and the forced part. When damping is present, the free vibration is of the form of Eq. (5.289) and is rapidly damped out. Hence, the free part is called the **transient response,** and the forced part, the **steady-state response.** The maximum value of the dynamic load factor for the steady-state response D_m is called the **dynamic magnification factor.** It is given by

$$D_m = \frac{1}{\sqrt{(1 - \alpha^2/\omega^2)^2 + (2\beta\alpha/\omega^2)^2}} \tag{5.296}$$

With damping, then, the peak values of D_m occur when $\alpha = \omega\sqrt{1 - \beta^2/\omega^2}$ and are approximately equal to $\omega/2\beta$. For example, for 10% critical damping,

$$D_m = \frac{\omega}{0.2\omega} = 5$$

So even small amounts of damping significantly limit the response at resonance.

Coulomb Damping. For a one-degree system with Coulomb damping, the equation of motion for free vibration is

$$\frac{W}{g}\frac{d^2y}{dt^2} + ky = \pm F_f \tag{5.297}$$

where F_f is the constant friction force and the positive sign applies when the velocity is negative. If initial displacement is y_o and initial velocity is zero, the response in the first half cycle, with negative velocity, is

$$y = \left(y_o - \frac{F_f}{k}\right)\cos\omega t + \frac{F_f}{k} \tag{5.298}$$

equivalent to a system with a suddenly applied constant force. For the second half cycle, with positive velocity, the response is

$$y = \left(-y_o + 3\frac{F_f}{k}\right)\cos\omega\left(t - \frac{\pi}{\omega}\right) - \frac{F_f}{k} \tag{5.299}$$

If the solution is continued with the sign of F_f changing in each half cycle, the results will indicate that the amplitude of positive peaks is given by $y_o - 4nF_f/k$, where n is the number of complete cycles, and the response will be completely damped out when $t = ky_oT/4F_f$, where T is the natural period of vibration, or $2\pi/\omega$.

Analysis of the steady-state response with Coulomb damping is complicated by the possibility of frequent cessation of motion.

(S. Timoshenko, D. H. Young, and W. Weaver, "Vibration Problems in Engineering," 4th ed., John Wiley & Sons, Inc., New York; D. D. Barkan, "Dynamics of Bases and Foundations," McGraw-Hill Book Company; W. C. Hurty and M. F. Rubinstein, "Dynamics of Structures," Prentice-Hall, Englewood Cliffs, N.J.)

5.18.6 Approximate Design for Dynamic Loading

Complex analysis and design methods seldom are justified for structures subject to dynamic loading because of lack of sufficient information on loading, damping, resistance to deformation, and other factors. In general, it is advisable to represent the actual structure and loading by idealized systems that permit a solution in closed form (see Arts. 5.18.1 to 5.18.5).

Whenever possible, represent the actual structure by a one-degree system consisting of an equivalent mass with massless spring. For structures with distributed mass, simplify the analysis in the elastic range by computing the response only for one or a few of the normal modes. In the plastic range, treat each stage—elastic,

elastic-plastic, and plastic—as completely independent; for example, a fixed-end beam may be treated, when in the elastic-plastic stage, as a simply supported beam.

Choose the parameters of the equivalent system to make the deflection at a critical point, such as the location of the concentrated mass, the same as it would be in the actual structure. Stresses in the actual structure should be computed from the deflections in the equivalent system.

Compute an assumed shape factor ϕ for the system from the shape taken by the actual structure under static application of the loads. For example, for a simple beam in the elastic range with concentrated load at midspan, ϕ may be chosen, for $x < L/2$, as $(Cx/L^3)(3L^2 - 4x^2)$, the shape under static loading, and C may be set equal to 1 to make ϕ equal to 1 when $x = L/2$. For plastic conditions (hinge at midspan), ϕ may be taken as Cx/L, and C set equal to 2, to make $\phi = 1$ when $x = L/2$.

For a structure with concentrated forces, let W_r be the weight of the rth mass, ϕ_r the value of ϕ for a specific mode at the location of that mass, and F_r the dynamic force acting on W_r. Then, the equivalent weight of the idealized system is

$$W_e = \sum_{r=1}^{j} W_r \phi_r^2 \tag{5.300}$$

where j is the number of masses. The equivalent force is

$$F_e = \sum_{r=1}^{j} F_r \phi_r \tag{5.301}$$

For a structure with continuous mass, the equivalent weight is

$$W_e = \int w\phi^2 \, dx \tag{5.302}$$

where w is the weight in lb/lin ft. The equivalent force is

$$F_e = \int q\phi \, dx \tag{5.303}$$

for a distributed load q, lb/lin ft.

The resistance of a member or structure is the internal force tending to restore it to its unloaded static position. For most structures, a bilinear resistance function, with slope k up to the elastic limit and zero slope in the plastic range (Fig. 5.112*a*), may be assumed. For a given distribution of dynamic load, maximum resistance of the idealized system may be taken as the total load with that distribution that the structure can support statically. Similarly, stiffness is numerically equal to the total load with the given distribution that would cause a unit deflection at the point where the deflections in the actual structure and idealized system are equal. Hence, the equivalent resistance and stiffness are in the same ratio to the actual as the equivalent forces to the actual forces.

Let k be the actual spring constant, g acceleration due to gravity, 386 in/s^2, and

$$W' = \frac{W_e}{F_e} \Sigma F \tag{5.304}$$

where ΣF represents the actual total load. Then, the equation of motion of an equivalent one-degree system is

$$\frac{d^2y}{dt^2} + \omega^2 y = g\frac{\Sigma F}{W'} \tag{5.305}$$

and the natural circular frequency is

$$\omega = \sqrt{\frac{kg}{W'}} \tag{5.306}$$

The natural period of vibration equals $2\pi/\omega$. Equations (5.305) and (5.306) have the same form as Eqs. (5.236), (5.238), and (5.273). Consequently, the response can be computed as indicated in Arts. 5.18.2 to 5.18.4.

Whenever possible, select a load-time function for ΣF to permit use of a known solution.

For preliminary design of a one-degree system loaded into the plastic range by a suddenly applied force that remains substantially constant up to the time of maximum response, the following approximation may be used for that response:

$$y_m = \frac{y_e}{2(1 - F_o/R_m)} \tag{5.307}$$

where y_e is the displacement at the elastic limit, F_o the average value of the force, and R_m the maximum resistance of the system. This equation indicates that for purely elastic response, R_m must be twice F_o; whereas, if y_m is permitted to be large, R_m may be made nearly equal to F_o, with greater economy of material.

For preliminary design of a one-degree system subjected to a sudden load with duration t_d less than 20% of the natural period of the system, the following approximation can be used for the maximum response:

$$y_m = \frac{1}{2}y_e\left[\left(\frac{F_o}{R_m}\omega t_d\right)^2 + 1\right] \tag{5.308}$$

where F_o is the maximum value of the load and ω the natural frequency. This equation also indicates that the larger y_m is permitted to be, the smaller R_m need be.

For a beam, the spring force of the equivalent system is not the actual force, or reaction, at the supports. The real reactions should be determined from the dynamic equilibrium of the complete beam. This calculation should include the inertia force, with distribution identical with the assumed deflected shape of the beam. For example, for a simply supported beam with uniform load, the dynamic reaction in the elastic range is $0.39R + 0.11F$, where R is the resistance, which varies with time, and $F = qL$ is the load. For a concentrated load F at midspan, the dynamic reaction is $0.78R - 0.28F$. And for concentrated loads $F/2$ at each third point, it is $0.62R - 0.12F$. (Note that the sum of the coefficients equals 0.50, since the dynamic-reaction equations must hold for static loading, when $R = F$.) These expressions also can be used for fixed-end beams without significant error. If high accuracy is not required, they also can be used for the plastic range.

5.19 APPROXIMATE EARTHQUAKE ANALYSIS

Article 3.3 describes the characteristics of earthquakes, scales for measuring the magnitude of temblors, and means for protecting buildings against seismic damage.

To provide a building with sufficient resistance to seismic damage, designers must first make a conservative estimate of the maximum probable seismic loads, guided by a knowledge of the geology of the region in which the building is to be constructed and past experience with earthquakes in that region. Since the loads are dynamic in nature, generally uncertain in direction, magnitude, and vibration period, calculation of their effects on structures is complicated and time-consuming, even by the approximate method described in Art. 5.18.6, and the accuracy of the results may not justify the dynamic analysis. Consequently, as permitted by most building codes, ordinary structures often are designed for equivalent static loads to resist the dynamic forces of earthquakes.

The design must anticipate that seismic loads may act in any horizontal direction. For ordinary structures, however, the design forces may be assumed to act during different seismic events along different principal axes. For structures with irregular framing, such as those with vertical, lateral-load-resisting elements that are not parallel to or symmetric about the major orthogonal axes of that framing, provision should be made in design for the possibility that the seismic loads may not be directed along the principal axes. One way to do this is to design for the full seismic design loads in one direction plus 30% of the seismic design loads in the perpendicular direction. An alternative is to combine the effects of full seismic loads in two directions vectorially, but in no case should the resulting stresses be less than those that would be caused by the loads acting in only one direction.

5.19.1 Equivalent Static Forces

Building codes generally require that computation of static, horizontal forces for aseismic design be derived from a prescribed base shear, or total horizontal force developed at grade. For example, the 1988 Uniform Building Code, promulgated by the International Conference of Building Officials, Whittier, Calif., specifies that the base shear V, kips, be determined from

$$V = \frac{ZICW}{R_w} \tag{5.309}$$

Z provides for variation in design forces with changes in the probability of seismic intensity with geographic zones, as indicated in Fig. 5.113. Since zone 0 indicates negligible probability of significant earthquake damage, Z may be taken as zero for that zone; but it would be prudent to adopt at least a low value, for example, 0.05. For the other zones,

$$\begin{aligned} Z &= 0.075 \text{ for zone 1} \\ &= 0.15 \text{ for zone 2} \\ &= 0.20 \text{ for zone 3} \\ &= 0.30 \text{ for zone 4} \\ &= 0.40 \text{ for zone 5} \end{aligned}$$

W, in general, is the total dead load, kips. For storage and warehouse occupancies, however, W should be taken as the dead load plus 25% of the live load. There are other cases, however, where designers would find it prudent and realistic to include a portion of the live load in W.

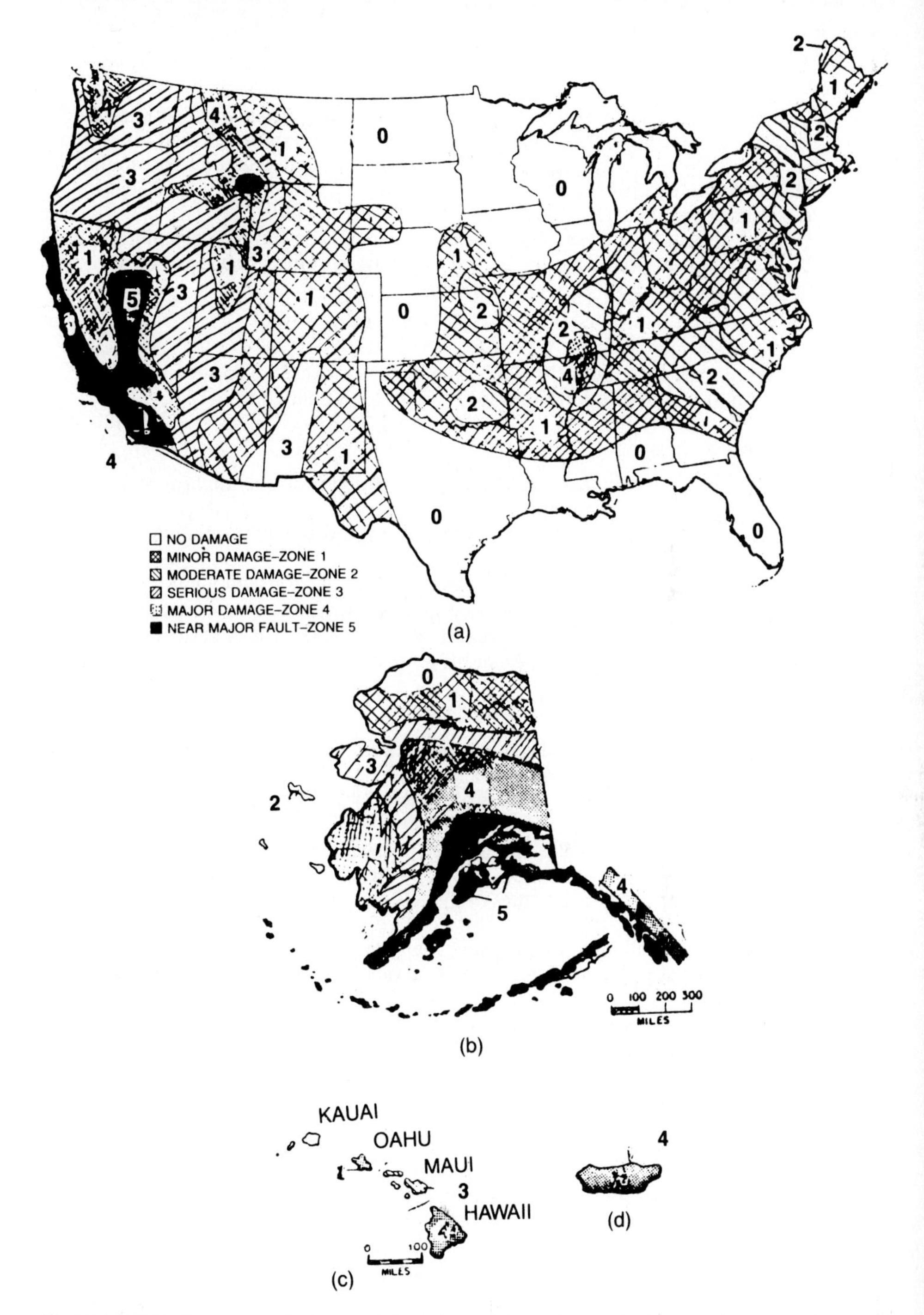

FIGURE 5.113 Zones of probable seismic intensity in (*a*) mainland United States; (*b*) Alaska; (*c*) Hawaii; (*d*) Puerto Rico.

I is a factor that depends on the type of building occupancy. For hazardous facilities, for example, those housing toxic or explosive substances, and for essential facilities, including hospitals, fire and police stations, and structures and equipment required for response to emergencies, the 1988 Uniform Building Code assigns $I = 1.25$. For other occupancies, I may be taken as unity.

The value of the numerical coefficient C depends on the type of soil underlying the building site and the natural period of vibration of the structure in the direction of the lateral forces. C is determined from

$$C = 1.25S/T^{2/3} \tag{5.310}$$

but should not be less than $0.075R_w$ and need not exceed 2.75.

T, the natural period of vibration, s, may be estimated from

$$T = C_t H^{3/4} \tag{5.311}$$

where H = height, ft, from the base of the structure to the uppermost level
C_t = 0.035 for moment-resisting steel frames
= 0.030 for eccentric-braced frames and moment-resisting reinforced-concrete frames
= 0.020 for other types of framing, but may be computed for structures with concrete or masonry shear walls from $0.1/\sqrt{A_c}$, where A_c is the total effective shear area, ft², of shear walls in the first story and parallel to the lateral forces and is given by

$$A_c = \sum A_e[0.2 + (D_e H)^2] \tag{5.312}$$

A_e = minimum cross-sectional shear area, ft², in any horizontal plane in the first story of a shear wall
D_e = length, ft, of a shear wall in the first story and parallel to the lateral forces but not more than $0.9H$

More appropriate values of T may be used to calculate C if they can be substantiated by technical data. For example, since $T = 2\pi/\omega$, it can be determined from ω as given by Eq. (5.248). Such values of T should not exceed 1.40 times T determined from Eq. (5.311).

S is a coefficient equal to or greater than unity that accounts for forces resulting from soil-structure interaction. Some building codes may permit S to be chosen as follows:

S = 1 for rock or similar material characterized by a shear wave velocity exceeding 2500 ft/s or for stiff soil, such as stable deposits of sand, gravel, or stiff clay, overlying rock at a depth of less than 200 ft
= 1.2 for stable deposits of sand, gravel, or stiff clay overlying rock at a depth greater than 200 ft
= 1.5 for sands or weak to medium-stiff clays 30 ft or more deep (the clay strata may incorporate layers of sand or gravel)
= 2.0 for a soil profile with more than a 40-ft depth of soft clay

For sites where soil characteristics are not known in detail or do not fit any of the preceding descriptions, S may be taken as 1.5.

Moment-resisting frames, as defined in the Uniform Building Code, are space frames that support gravity loads and resist lateral loads primarily by flexure of frame components.

R_w takes into account the potential for inelastic energy absorption in moment-resisting frames. It also recognizes the redundancy of framing, or second line of defense, present in most complete frames, whether or not designed to resist lateral loads. Buildings that do not possess at least a complete moment-resisting space frame are penalized by assignment of a low R_w. Table 5.9 lists suggested values for R_w. See the local building code or the Uniform Building Code for values of R_w for other types of construction.

The Uniform Building Code permits use of different types of lateral-force-resisting systems in orthogonal directions in structures less than 160 ft high. For higher structures, there are limitations on types of framing that may be used in perpendicular directions in seismic zones 4 and 5 (Fig. 5.113), but generally combinations with R_w of 8 or more are permitted. Also, in those zones, if a structure has a bearing-wall system in only one direction, R_w for the system in the perpendicular direction should not exceed R_w for the bearing-wall system. The code also establishes restrictions on vertical combinations of lateral-force-resisting systems. For example, R_w for any story may not exceed R_w in the same direction for the story above. (This need not be complied with if the dead weight for the upper stories is less than 10% of the total dead weight of the structure.)

Parts of Structures. Elements of a structure and nonstructural components supported by the structure, including connections, anchorages, and bracing, should be designed to resist a total horizontal seismic force F_p computed from

$$F_p = ZIC_pW_p \tag{5.313}$$

Z and I are the values also used for Eq. (5.309), except that $I = 1.5$ for anchorage of equipment used for life-safety systems. [Equation (5.313), however, has no bearing on determination of lateral forces to be applied to the building as a whole.] W_p is the weight of the component being designed.

For the purpose of establishing values for C_p, the Uniform Building Code distinguishes between rigid and nonrigid, or flexibly supported, elements. Rigid ele-

TABLE 5.9 R_w **for Aseismic Design of Buildings**

Type	Framing system	R_w
1	Moment-resisting space frames that carry gravity loads, resist lateral loads by bending of members and detailed to ensure ductile behavior	12
2	Combinations of Type 1 and concrete shear walls	12
3	Combinations of Type 2 and eccentric-braced steel frames	12
4	Combinations of Type 1 and concentric-braced steel frames	10
5	Eccentric-braced steel frames	10
6	Combinations of Type 1 and concentric-braced concrete frames	9
7	Combinations of Type 1 and masonry shear walls	8
8	Non-load-bearing shear walls	8
9	Concentric-braced frames of steel or heavy timber	8
10	Load-bearing shear walls	6
11	Braced steel frames with bracing carrying gravity loads	6
12	Light-framed, load-bearing walls with shear panels	6
13	Heavy-timber-braced frames with bracing carrying gravity loads or light steel-framed bearing walls with bracing only in tension	4

ments are defined as those components having a fixed-base vibration period not exceeding 0.06 s. Those for which the element or the element plus its supporting structure have longer periods are considered nonrigid.

For most rigid components, including exterior walls above the ground floor; interior walls and partitions; storage racks plus contents; tanks plus contents, support systems, and anchorages; and mechanical, electrical, and plumbing installations, C_p should be taken as 0.75. For such items as cantilevered parapets; ornamentation and appendages; signs and billboards; and chimneys, stacks, trussed towers, and tanks on legs, if they cantilever above the roof for a distance of more than half their total height, C_p should be taken as 2.0. Unless justified by dynamic analysis, C_p for a nonrigid or flexibly supported component should be taken as twice the above values, but not to exceed 2.0. For components supported at or below ground level, C_p may be assumed at two-thirds that for a corresponding rigid element. In no case, however, should the design seismic load for a component be less than the load required when the component is considered an independent structure and computed from Eq. (5.309).

The Uniform Building Code also requires that floor and roof diaphragms and horizontal bracing systems for seismic loads be designed to resist horizontal forces computed from

$$F_{px} = w_{px} \sum_{i=x}^{n} F_i \bigg/ \sum_{i=x}^{n} w_i \leq 0.75ZIw_{px} \qquad \text{but not less than } 0.35ZIw_{px} \tag{5.314}$$

where F_{px} = seismic force on the diaphragm or horizontal bracing system at floor or roof level x

F_i = lateral force applied to level i, as computed from Eq. (5.315)

n = number of levels in structure

w_i = portion of dead load (plus 25% of floor live load in storage and warehouse occupancies) assigned to level i

w_{px} = weight of diaphragm or bracing at level x, plus live load as for w_i, and weight of tributary elements

Z and I are the values also used for Eq. (5.309).

If lateral forces are to be transferred from vertical structural components, such as columns and walls, above a diaphragm or bracing system to lower vertical structural members resisting lateral forces, the transferred forces should be added to those determined from Eq. (5.314). Such transfers may be necessitated by offsets in placement of vertical members in successive stories or by changes in the stiffness of those members.

5.19.2 Distribution of Seismic Loads

Seismic forces are assumed to act at each floor level on vertical planar frames, or bents, or on shear walls extending in the direction of the loads. The seismic loads at each level should be distributed over the floor or roof area in accordance with the distribution of mass on that level.

The Uniform Building Code recommends that the seismic force F_x, to be assigned to any level at a height h_x, ft, above the ground be calculated from

$$F_x = (V - F_t) \frac{w_x h_x}{\sum_{i=1}^{n} w_i h_i} \tag{5.315}$$

where w_x = portion of W located at or assigned to level x
h_x = height, ft, of level x above ground level
w_i = portion of W located at or assigned to level i
h_i = height, ft, of level i above ground level
n = number of levels in structure

V is the base shear computed from Eq. (5.309). F_t is an additional seismic force assigned to the top level of the structure and is calculated from

$$F_t = 0.07TV \tag{5.316}$$

where T = fundamental elastic period of vibration of the structure in the direction of the lateral force, s. F_t need not be more than $0.25V$ and may be taken as zero when $T \leq 0.7$ s. Equation (5.316) recognizes the influence of higher modes of vibration as well as deviations from straight-line deflection patterns, particularly in tall buildings with relatively small dimensions in plan. Consequently, the design seismic shear at any level i is given by

$$V_i = F_t + \sum_{x=i}^{n} F_x \tag{5.317}$$

This shear should be distributed to the bents or shear walls of the lateral-force-resisting system in proportion to their rigidities. The distribution should, however, take into account the rigidities of horizontal bracing and diaphragms (floors and roofs). In lightly loaded structures, for example, diaphragms may be sufficiently flexible to permit independent action of the lateral-force-resisting bents. (See Art. 5.12.) A strong temblor could cause severe distress in frames and diaphragms if relative rigidities were not properly evaluated.

The design seismic force computed from Eq. (5.313) for an element of a structure or a nonstructural component supported by the structure should be distributed in proportion to the distribution of mass of the element or component.

Seismic force distribution for buildings or structural frames with irregular shapes should be determined by dynamic analysis.

5.19.3 Vertical Seismic Forces

Provision should be made in aseismic design for the possibility of uplift due to seismic loads. When design of a structure is based on allowable unit stresses, only 85% of the dead load and no live loads should be considered available to counteract the uplift. Furthermore, the Uniform Building Code requires for structures in seismic zones 4 and 5 (Fig. 5.113) that horizontal cantilever components be designed for a net uplift force of $0.2W_p$, where W_p is the weight of the components.

5.19.4 Horizontal Torsion

For calculating the effects of torsion due to seismic loads on a structure, the rigidity of diaphragms that distribute the seismic loads laterally to lateral-force-resisting framing should be considered. For the purpose, an inflexible diaphragm is defined as one for which the in-plane deflection of its midpoint due to the force F_{px}, computed from Eq. (5.314), is less than twice the average story drift of the stories above and below the diaphragm under the action of seismic forces V_i, calculated from Eq. (5.317). When diaphragms are inflexible, the Uniform Building Code requires that shears at any level i due to horizontal torsion be added to the direct horizontal shears. These are the shears at level i that result from the distribution of V_i, computed from Eq. (5.317), to the components of the vertical, lateral-force-resisting framing in proportion to their rigidities.

The design seismic torsion at any level i consists of two components: (1) The horizontal moment at level i due eccentricities between the design seismic forces at upper levels and the vertical resisting components at level i. (2) An accidental torsion. This torsion is intended to account for uncertainties in location of seismic loads. For the purpose of computing eccentricities, the mass at each level is assumed to be displaced from the calculated center of mass a distance equal to 5% of the building dimension at that level and in the direction in which that dimension is measured. The displacements should be assumed to occur normal to the seismic load under consideration.

When a structure with inflexible diaphragms is torsionally irregular, the Uniform Building Code requires that the accidental torsion at each level i be multiplied by an amplification factor A_i. To determine whether a structure is torsionally irregular, locate the vertical, lateral-force-resisting bents (or shear walls) parallel to the design seismic loads and at or near the sides of the structure. Compute the maximum story drift due to the seismic shears, including accidental torsional shears, for each of those bents. (Story drift is the displacement of a level relative to the level above or below.) Let d_m be the larger of those drifts and d_a the average of the two. Then, if d_m exceeds $1.2d_a$, the structure is torsionally irregular. If it is, multiply the accidental torsion by A_i computed from

$$A_i = \left(\frac{d_m}{1.2d_a}\right)^2 < 3 \tag{5.318}$$

5.19.5 Limitation on Story Drift

To prevent damage to building components that could affect life safety, many building codes place limits on the amount of story drift permissible. For example, for buildings less than 65 ft high, the Uniform Building Code restricts story drift to a maximum of $0.005h$, where h is the story height, or $0.04h/R_w$, where R_w is the value used for Eq. (5.309). For taller buildings, the story drift may not exceed $0.004h$ or $0.03h/R_w$.

Story drift as well as vibrations caused by lateral forces may be restricted by selection of appropriate types and arrangements of structural framing, as described in Arts. 3.2.4 and 3.3.5. Sometimes, it may be necessary or more economical to assist the framing with damping devices. These may be active or passive types.

Active dampers use mechanical means to reduce building motion. One example of the active type is motion control by use of computers fed data from accelerometers on the building. The computers, as data are received, activate equipment, such as jacks, to oppose the movements.

Passive dampers instead resist motion without the need for an external power source. Though not as effective over all potential frequencies as active dampers, the passive type is likely to be less costly initially, as well as in operation, and to require less maintenance. One example of this type is the tuned-mass damper. It requires installation of a spring in parallel with a damper that acts as a shock absorber between a heavy mass, usually concrete or iron, or both, and a structural component. Another example is the viscoelastic damper. Incorporating a layer of viscoelastic material and installed between a structural member and another component with a substantially smaller frequency of vibration, the damper restricts motion by transforming energy of motion to heat.

5.19.6 Overturning

The equivalent static lateral forces applied to a building at various levels induce overturning moments. At any level, the overturning moment equals the sum of the products of each force and its height above that level. The overturning moments acting on the base of the structure and in each story are resisted by axial forces in vertical elements and footings.

At any level, the increment in the design overturning moment should be distributed to the resisting elements in the same proportion as the distribution of shears to those elements. Where a vertical resisting element is discontinued, the overturning moment at that level should be carried down as loads to the foundation.

(J. M. Biggs, "Introduction to Structural Dynamics," and R. Clough and J. Penzien, "Dynamics of Structures," McGraw-Hill Publishing Company, New York; E. Rosenblueth, "Design of Earthquake-Resistant Structures," Halsted/Wiley, Somerset, N.J.; N. M. Newmark and E. Rosenblueth, "Fundamentals of Earthquake Engineering," Prentice-Hall, Englewood Cliffs, N.J.; S. Okamoto, "Introduction to Earthquake Engineering," John Wiley & Sons, Inc., New York.)

5.20 REDUCTION OF VIBRATIONS

In general, vibrations in a building are objectionable from the viewpoint of physical comfort, noise, protection of delicate apparatus, and danger of fatigue failures (Figs. 5.114 and 5.115). The simplest ways to reduce or eliminate them are to isolate vibrating machinery, install supports that transmit little energy, install shock absorbers, or insert mufflers and screens to absorb noise (see Art. 3.18.12 and Sec. 11).

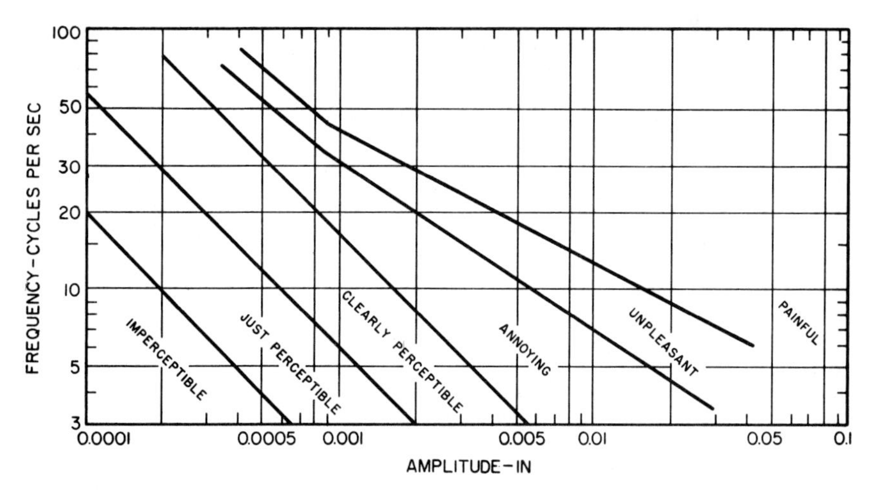

FIGURE 5.114 Chart indicates human sensitivity to vibrations. *(British Research Station Digest No. 78.)*

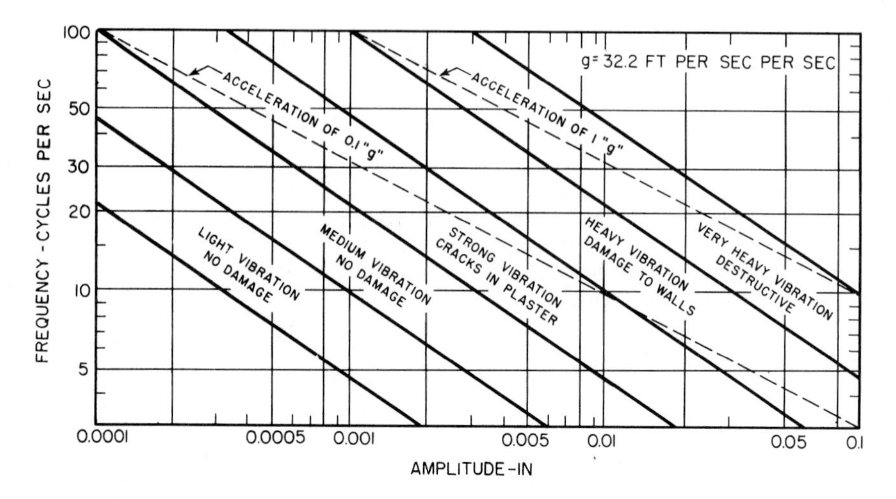

FIGURE 5.115 Chart shows possibility of damage to buildings from vibrations. *(British Research Station Digest No. 78.)*

Sometimes, machines with rotating parts can be made to vibrate less by balancing the rotating parts to reduce the exciting force or the speed can be adjusted to prevent resonance. Also, the masses of the supports can be adjusted to change the natural frequency so that resonance is avoided.

The surfaces of a foundation subjected to vibration should be shaped to preclude the possibility of the vibrations being reflected from them back into the interior in a direction in which a stress buildup can occur. Such a foundation should be made of a homogeneous material, since objectionable stresses may occur in the region of an inhomogeneity. The foundation should be massive—weight comparable with that of the machine it supports—and should be isolated by an air gap or insulating material from the rest of the building.

("Vibration in Buildings," Building Research Station Digest No. 78, Her Majesty's Stationery Office, London, June 1955; D. D. Barkan, "Dynamics of Bases and Foundations," McGraw-Hill Book Company, New York.)

SECTION SIX
SOIL MECHANICS AND FOUNDATIONS

Charles P. Gupton
Principal, Dames & Moore, Inc.
Boca Raton, Florida

Foundations constitute the building subsystems that are used to transmit building loads directly to the earth below. Foundations must be built to distribute the loads over contiguous soils so that they and the underlying materials will have sufficient strength and rigidity to carry the loads without excessive deformations. Because of the interactions of soils and foundations, the characteristics of the soils underlying a building strongly influence selection of types and sizes of foundations. The foundations, in turn, may significantly affect design of the superstructure, the project construction time, and, consequently, building construction costs. For production of a safe, cost-effective building, a knowledge of soil mechanics and foundation design and construction is essential.

The study of soils, their properties, and their behavior for engineering purposes is the province of soil mechanics. This section deals with the applications of soil mechanics to design and construction of foundations for buildings. Structural design of concrete foundations is treated in Arts. 9.72 to 9.80.

6.1 CLASSIFICATIONS OF SOILS

The geologist defines a soil as an altered rock. The engineer defines a soil as the material that supports or loads a structure at its base.

Natural soil materials may be divided into four classifications: sands and gravels, silts, clays, and organics. Sands and gravels are granular, nonplastic materials. Clays composed of much smaller particles, exhibit plasticity and are cohesive. Silts are intermediate in grain size, typically behave as granular materials, but may be slightly plastic. Organic materials are principally vegetable refuse.

The origin and mode of deposition of soils shed much light on their nature and variability in the field. There are two origins of soils: residual and sedimentary. Residual soils form in situ by the chemical weathering of rocks and, having never been physically disturbed, retain the minor geologic features of the parent material. In this geologic setting, changes from soil to rock in the field are usually gradational. Sedimentary soils are those transported and deposited by the action of rivers, seas, glaciers, or wind. The mode of deposition often controls grain size, its variation, and the stratigraphy and uniformity of the soil units.

The engineer needs to know, for complete identification of a soil: the size of grains, gradation, particle shapes, grain orientation, chemical makeup, and colloidal and dust fractions. Even then, the physical properties can be made to vary over a large range with slight chemical additives or electrochemical controls.

Where surface properties are important, grain shape becomes at least equal in influence to size gradation. Normally, an important characteristic is the relative positioning of the grains within the soil body, since this controls the resistance to internal displacement and is at least a qualitative measure of shear and compression strengths.

There have been many attempts to codify all soils into classes of similar and recognizable properties. As more information is collected on soil properties, the systems of classification become more elaborate and complicated. One difficulty is the attempt to use similar classifications for different uses; for instance, a system applicable to highway design has less value when the problem is basically that of building foundations. See also Art. 6.4.

(R. F. Legget and A. W. Hatheway, "Geology and Engineering," 3d ed., McGraw-Hill Publishing Company, New York.)

6.2 STATES OF MATTER AFFECTING SOIL BEHAVIOR

Soils can exhibit solid, viscous, plastic, or liquid action; if the true state can be predicted, the structural design of foundations can be prepared accordingly.

By contrast, solids are materials having constant density, elasticity, and internal resistance, little affected by normal temperature changes, moisture variations, and vibration below seismic values. Deformation by shearing takes place along two sets of parallel planes, with the angle between the sets constant for any material and independent of the nature or intensity of the external forces inducing shearing strain. The harder and more brittle the material, the more this angle differs from a right angle.

If tension is taken as negative compression, Hartmann's law states that the acute angle formed by the shear planes is bisected by the axis of maximum compression, and the obtuse angle by the axis of minimum compression (usually tension). Shearing planes do not originate simultaneously and are not uniformly distributed.

The angle of shear is related to the ultimate compression and tension stresses by Mohr's formula:

$$\cos\theta = \frac{f_1 - f_2}{f_1 + f_2} \tag{6.1}$$

i.e., the cosine of the angle of shear equals the ratio of the difference of the normal stresses to the sum. When lateral pressures are applied externally simultaneously with loads, the angle of shear increases; i.e., the material becomes less brittle.

These basic properties of solids can be used in designs affecting soils only if the soils remain solid. When changes in conditions modify soil structures so that they are not solid, these properties are nonexistent, and a new set of rules governs the actions. Most soils will act like solids, but only up to a certain loading. This limit loading depends on many external factors, such as flow of moisture, temperature, vibration, age, and sometimes the rate of loading.

No marked subdivision exists between liquid, plastic, and viscous action. These three states have the common property that volume changes are difficult to accomplish, but shape changes occur continuously. They differ in the amount of force necessary to start motion. There is a minimum value necessary for the plastic or viscous states, while a negligible amount will start motion in the liquid state. Upon cessation of the force, a plastic material will stop motion, but liquid and viscous materials will continue indefinitely, or until counteracting forces come into play. Usually, division between solid and plastic states is determined by the percentage of moisture in the soil. This percentage, however, is not a constant but decreases with increase in pressure on the material. Furthermore, the entire relationship between moisture percentage and change of state can be deranged by adding chemicals to the water.

Liquids retained in a vessel are almost incompressible. In water-bearing soils, therefore, if movement or loss of water can be presented, volume change and settlement will be avoided. Water loss can be prevented by both physical and chemical changes in the nature of the water.

6.2.1 Soil Moisture

Water is usually present in a soil as a film on its particles or as a liquid between particles, or pore water. If the water content of soil is chiefly film or adsorbed moisture, the mass will not act as a liquid. All solids tend to adsorb or condense upon their surfaces any liquids (and gases) with which they come into contact.

The kind of ion or metal element in the chemical makeup of a solid has a great influence on how much water can be adsorbed. Ion-exchange procedures for soil stabilization and percolation control are therefore an important part of soil mechanics.

Water films are tougher than pore water. (Terzaghi, in 1920, stated that water films less than two-millionths of an inch thick are semisolid in action; they will not boil or freeze at normal temperatures.) Consequently, saturated soil freezes much more easily than damp soil, and ice crystals will grow by sucking free moisture out of the pores. A sudden thaw will then release large concentrations of water, often with drastic results. When liquids evaporate, they first change to films, requiring a large thermal increment for the change from film to vapor. Thus, the effect of temperature on soil action may be explained by the reduction of film thickness with rise of temperature.

The presence of moisture in a soil is essential for controlled compaction. Compaction of soils is best carried out within a fairly narrow range of moisture contents, since redistribution of soil grains to a closely packed volume cannot be accomplished without sufficient moisture to coat each particle. The film acts as a lubricant to permit easier relative movements and by its capillary tension holds the grains in position. Obviously, finer grains require more water for best stabilization than coarser materials.

6.2.2 Resistance of Soils to Pressure

Before 1640, Galileo distinguished between solids, semifluids, and fluids. He stated that semifluids, unlike fluids, maintain their condition when heaped up, and if a hollow or cavity is made, agitation causes the filling up of the hollow, whereas in solids, the hollow does not fill up. This is an early description of the property known as the natural slope, or angle of repose, of granular materials—an easily observed condition for clean, dry sands, but a variable slope for soils containing clays with varying moisture percentages. The angle of natural repose should not be confused with the angle of internal friction, although many writers have followed Woltmann, who, in translating Coulomb's papers, made that mistake.

The fundamental laws of friction were applied to soils by Coulomb. He recognized that the resistance along a surface of failure within a soil is a function of both the load per unit area and the surface of contact.

Resistance of soils to deformation is controlled mainly by their shear resistance. Shear resistance in soils is derived from the sum of two components, friction and cohesion. Frictional resistance develops from the irregularities of grain-to-grain contacts and is proportional to the normal force between the particles. Cohesion, the soil's maximum tensile strength, is a result of the attractive forces operating between grains in close contact and is independent of the normal pressure. Such true cohesion is relatively rare. Some frictional behavior is normal for most soils.

(H. Y. Fang, "Foundation Engineering Handbook," 2d ed., Van Nostrand Reinhold, New York.)

6.3 SOIL PROPERTIES IMPORTANT IN ENGINEERING

The normally important soil properties are:

Density. The amount of solid material in a unit volume is called the dry density. For granular soils and fibrous organics, the dry density is the most important factor controlling their engineering properties. One such property is the state, or degree, of compaction which usually is expressed in terms of relative density. This is defined as the ratio (as percentage) of the difference between dry density of the natural soil and its minimum dry density to the difference between its maximum and minimum dry densities. During construction of engineering fills, however, the degree of compaction is usually specified and measured as the ratio of the actual in-place dry density to the maximum dry density obtained in a laboratory moisture-density relationship test (ASTM D1557 orD698).

Internal Friction. Coulomb pure friction corresponds to simple shear in the theory of elasticity. Internal friction is usually expressed geometrically as the angle of internal friction ϕ, where $\tan \phi = f$, the coefficient of friction. The frictional component of shear T_{max} in a soil mass thus is equal to $N \tan \phi$, where N is the normal force acting on the mass.

Values of ϕ range from about 28° for loose sands and nonplastic silts to about 48° for dense sands and gravels. The value increases with increasing density, angularity, and gradation; decreases if mica is present; is relatively unaffected by rate of loading and grain size; and may be either increased or decreased under repeated or cyclic loading. Many designers use T_{max} as the total shear resistance—an assumption also made in most earth-pressure formulas.

Cohesion. This is the maximum tensile strength of the soil. It results from a complicated interaction of many factors, such as colloidal adhesion of the grain covering, capillary tension of the moisture films, electrostatic attraction of charged surfaces, drainage conditions, and stress history. True cohesion exists only in clays with edge-to-face contact between particles. Apparent cohesion can occur in non-plastic fine-grained soils under partial saturation conditions. The value of cohesion to be used in design depends directly on the drainage conditions under the imposed loading as well as on the method of testing for its determination and must be evaluated carefully.

Compressibility. This property defines the stress-strain characteristics of the soil. Application of added stresses to a soil mass results in volume changes and associated displacements, which can be estimated if the compressibility modulus and stress change are known. Such displacements at the foundation level lead to foundation settlements. Limitation of settlements to allowable values often controls foundation design, particularly for granular soils.

For granular soils, compressibility is expressed in terms of Young's modulus E, taken usually as the secant modulus of the stress-strain curve from a standard triaxial test. The modulus decreases with increasing axial stress and increases with increasing confining pressure and with repeated loading.

Compressibility of saturated clays is expressed as the compression index C_c, together with a determination of the maximum past pressure to which they have been subjected. Both values are determined from standard laboratory one-dimensional consolidation tests (ASTM D2435). C_c is the change in void ratio per logarithmic cycle of stress and is a function of past stress history. For practical applications, its value in the stress range of interest must be known.

Permeability. This is the ability of the soil mass to conduct fluid flow under a unit of hydraulic gradient. In foundation design, only permeability under saturated conditions usually is needed. Permeabilities of most soil formations are highly variable and strongly sensitive to relatively small variations in the soil mass.

Controlled by the size and continuity of the pore space and consequently by the grain size, permeability typically is an anisotropic property, usually being greater in the horizontal direction than in the vertical. Permeability values for the various soil classifications vary by a factor of more than 10 million, as determined directly by laboratory or field permeability tests or indirectly by consolidation tests and grain-size analyses. Field pumping tests with observation wells provide the most reliable values.

Other Properties. There are other minor soil properties that are sometimes of importance. The vegetable-free refuse content of a soil, for example, may affect the fixity of any properties induced by treatment. Soils very heavy in rotted vegetation, containing tannic acids, are not suitable for cement stabilization, and soils with large limestone-dust content can be weakened by water flow through the mass or disintegrated by percolating wastewater and other waste liquors. Sulfate content of soil and groundwater pH are of concern because of the potential for corrosion of buried structures.

(H. Y. Fang, "Foundation Engineering Handbook," 2d ed., Van Nostrand Reinhold, New York.)

IDENTIFICATION, SAMPLING, AND TESTING OF SOILS

To permit the application of previous experience to new soil conditions, a standard system of soil identification is necessary. For the purpose, classification of a soil usually is based on physical properties determined in accordance with standardized testing procedures. Testing for soil properties or reactions of soils to applied loads involves both laboratory and field procedures.

6.4 SOIL IDENTIFICATION

Field investigations to identify soils can be made by surface surveying, aerial surveying, or geophysical or subsurface exploratory analysis. Complete knowledge of the geological structure of an area permits definite identification from a surface reconnaissance. Tied together with a mineralogical classification of the surface layers, the reconnaissance can at least recognize structures of some soils. The surveys, however, cannot alone determine soil behavior unless identical conditions have previously been encountered.

Geological and agronomic soil maps and detailed reports useful for this purpose are available for most populated areas in the United States. They are issued by the U.S. Department of Agriculture, U.S. Geological Survey, and corresponding state offices. Old surveys are of great value in locating original shore lines and stream courses, as well as existence of surface-grade changes.

A complete site inspection is a necessary adjunct to the data collected from maps and surveys and will often clarify the question of uniformity. Furthermore, inspection of neighboring structures will point up some possible difficulties.

Aerial surveying permits rapid determination of soil types at low cost over very large areas. Stereoptic photographic data, correlated with standard charts, identify soil types from color, texture, drainage characteristics, and ground cover.

Soil Classification. The most widely accepted soil classification system is the Unified Soil Classification (Table 6.1). It sets definite criteria for naming soils and lists them in fixed divisions based on grain size and laboratory tests of physical characteristics.

6.5 SUBSURFACE EXPLORATION

This is the field phase of soils analysis and substructure design, and its importance cannot be overemphasized. Inadequate, inaccurate or misleading information at this stage is the most common cause of uneconomical foundation and earthworks designs, and of their failures. The key word is exploration. The purpose of this work is to determine—by exploratory techniques—the nature of the site subsurface conditions and their impact on design. The work must therefore be planned and executed so as to reveal the nature of the soils and not simply conducted in a perfunctory manner. The type and extent of exploration techniques, in-situ testing, and sampling methods thus should be chosen in accordance with the unknowns associated with the site, geologic hazards that may be reasonably expected, the loading intensity imposed by the structure, and the amount of settlement of the structure that can be tolerated.

The intensity and methodology of the exploration effort will vary widely. No standard can be strictly applied universally, nor should one be. In heavily developed areas where the subsurface conditions and engineering characteristics are well-known through prior work for other structures, an appropriate investigation might

TABLE 6.1 Unified Soil Classification Including Identification and Description[a]

Major division	Group symbol	Typical name	Field identification procedures[b]	Laboratory classification criteria[c]	
A. Coarse-grained soils (more than half of material larger than No. 200 sieve)[d]					
1. Gravels (more than half of coarse fraction larger than No. 4 sieve)[e]					
Clean gravels (little or no fines)	GW	Well-graded gravels, gravel-sand mixtures, little or no fines	Wide range in grain sizes and substantial amounts of all intermediate particle sizes	$D_{60}/D_{10} > 4$ $1 < D_{30}^2/D_{10}D_{60} < 3$ D_{10}, D_{30}, D_{60} = sizes corresponding to 10, 30, and 60% on grain-size curve	
	GP	Poorly graded gravels or gravel-sand mixtures, little or no fines	Predominantly one size, or a range of sizes with some intermediate sizes missing	Not meeting all gradation requirements for GW	
Gravels with fines (appreciable amount of fines)	GM	Silty gravels, gravel-sand-silt mixtures	Nonplastic fines or fines with low plasticity (see ML soils)	Atterberg limits below A line or PI < 4	Soils above A line with 4 < PI < 7 are borderline cases, require use of dual symbols
	GC	Clayey gravels, gravel-sand-clay mixtures	Plastic fines (see CL soils)	Atterberg limits above A line with PI > 7	
2. Sands (more than half of coarse fraction smaller than No. 4 sieve)[e]					
Clean sands (little or no fines)	SW	Well-graded sands, gravelly sands, little or no fines	Wide range in grain sizes and substantial amounts of all intermediate particle sizes	$D_{60}/D_{10} > 6$ $1 < D_{30}^2/D_{10}D_{60} < 3$	
	SP	Poorly graded sands or gravelly sands, little or no fines	Predominantly one size, or a range of sizes with some intermediate sizes missing	Not meeting all gradation requirements for SW	
Sands with fines (appreciable amount of fines)	SM	Silty sands, sand-silt mixtures	Nonplastic fines or fines with low plasticity (see ML soils)	Atterberg limits above A line or PI < 4	Soils with Atterberg limits above A line while 4 < PI < 7 are borderline cases, require use of dual symbols
	SC	Clayey sands, sand-clay mixtures	Plastic fines (see CL soils)	Atterberg limits above A line with PI > 7	

TABLE 6.1 Unified Soil Classification Including Identification and Description[a] (*Continued*)

Information required for describing coarse-grained soils:

For undisturbed soils, add information on stratification, degree of compactness, cementation, moisture conditions, and drainage characteristics. Give typical name; indicate approximate percentage of sand and gravel; maximum size, angularity, surface condition, and hardness of the coarse grains; local or geological name and other pertinent descriptive information; and symbol in parentheses. Example: *Silty sand,* gravelly; about 20% hard, angular gravel particles, ½ in maximum size; rounded and subangular sand grains, coarse to fine; about 15% nonplastic fines with low dry strength; well compacted and moist in place; alluvial sand; (SM).

Major division	Group symbol	Typical names	Identification procedure[f]			Laboratory classification criteria[c]
			Dry strength (crushing characteristics)	Dilatancy (reaction to shaking)	Toughness (consistency near PL)	
B. Fine-grained soils (more than half of material smaller than No. 200 sieve)[d]						
Silts and clays with liquid limit less than 50	ML	Inorganic silts and very fine sands, rock flour, silty or clayey fine sands, or clayey silts with slight plasticity	None to slight	Quick to slow	None	Plasticity chart (PLASTICITY INDEX 0–80 vs. LIQUID LIMIT 0–100; regions: CL-ML, CL, ML AND OL, CH, MH AND OH, A LINE)
	CL	Inorganic clays of low to medium plasticity, gravelly clays, sandy clays, silty clays, lean clays	Medium to high	None to very slow	Medium	
	OL	Organic silts and organic silty clays of low plasticity	Slight to medium	Slow	Slight	
Silts and clays with liquid limit more than 50	MH	Inorganic silts, micaceous or diatomaceous fine sandy or silty soils, elastic silts	Slight to medium	Slow to none	Slight to medium	
	CH	Inorganic clays of high plasticity, fat clays	None to very high	None	High	Plasticity chart for laboratory classification of fine-grained soils compares them at equal liquid limit. Toughness and dry strength increase with increasing plasticity index (PI)
	OH	Organic clays of medium to high plasticity	Medium to high	None to very slow	Slight to medium	

C. Highly organic soils				
	Pt	Peat and other highly organic soils	Readily identified by color, odor, spongy feel, and often by fibrous texture	

Field identification procedures for fine-grained soils or fractions[g]:

Dilantancy (reaction to shaking)

After removing particles larger than No. 40 sieve, prepare a pat of moist soil with a volume of about ½ cu in. Add enough water if necessary to make the soil soft but not sticky.

Place the pat in the open palm of one hand and shake horizontally, striking vigorously against the other hand several times. A positive reaction consists of the appearance of water on the surface of the pat, which changes to a livery consistency and becomes glossy. When the sample is squeezed between the fingers, the water and gloss disappear from the surface, the pat stiffens, and finally it cracks or crumbles. The rapidity of appearance of water during shaking and of its disappearance during squeezing assist in identifying the character of the fines in a soil.

Very fine clean sands give the quickest and most distinct reaction, whereas a plastic clay has no reaction. Inorganic silts, such as a typical rock flour, show a moderately quick reaction

Dry strength (crushing characteristics)

After removing particles larger than No. 40 sieve, mold a pat of soil to the consistency of putty, adding water if necessary. Allow the pat to dry completely by oven, sun, or air drying, then test its strength by breaking and crumbling between the fingers. This strength is a measure of character and quantity of the colloidal fraction contained in the soil. The dry strength increases with increasing plasticity.

High dry strength is characteristic of clays of the CH group. A typical inorganic silt possesses only very slight dry strength. Silty fine sands and silts have about the same slight dry strength but can be distinguished by the feel when powdering the dried specimen. Fine sand feels gritty, whereas a typical silt has the smooth feel of flour.

Toughness (consistency near PL)

After particles larger than the No. 40 sieve are removed, a specimen of soil about ½ in^3 in size is molded to the consistency of putty. If it is too dry, water must be added. If it is too sticky, the specimen should be spread out in a thin layer and allowed to lose some moisture by evaporation. Then, the specimen is rolled out by hand on a smooth surface or between the palms into a thread about ⅛ in in diameter. The thread is then folded and rerolled repeatedly. During this manipulation, the moisture content is gradually reduced and the specimen stiffens, finally loses its plasticity, and crumbles when the plastic limit (PL) is reached.

After the thread crumbles, the pieces should be lumped together and a slight kneading action continued until the lump crumbles.

The tougher the thread near the PL and the stiffer the lump when it finally crumbles, the more potent is the colloidal clay fraction in the soil. Weakness of the thread at the PL and quick loss of coherence of the lump below the PL indicate either organic clay of low plasticity or materials such as kaolin-type clays and organic clays that occur below the *A* line.

Highly organic clays have a very weak and spongy feel at PL.

TABLE 6.1 Unified Soil Classification Including Identification and Description[a] (*Continued*)

Information required for describing fine-grained soils:
For undisturbed soils, add information on structure, stratification, consistency in undisturbed and remolded states, moisture, and drainage conditions. Give typical name; indicate degree and character of plasticity; amount and maximum size of coarse grains; color in wet condition; odor, if any; local or geological name and other pertinent descriptive information; and symbol in parentheses. Example: *Clayey silt,* brown; slightly plastic; small percentage of fine sand; numerous vertical root holes; firm and dry in place; loess; (ML).

[a]Adapted from recommendations of Corps of Engineers and U.S. Bureau of Reclamation. All sieve sizes United States standard.

[b]Excluding particles larger than 3 in. and basing fractions on estimated weights.

[c]Use grain-size curve in identifying the fractions as given under field identification.
For coarse-grained soils, determine percentage of gravel and sand from grain-size curve. Depending on percentage of fines (fractions smaller than No. 200 sieve), coarse-grained soils are classified as follows:

Less than 5% fines . . . . GW, GP, SW, SP
More than 12% fines . . . . . . GM, GC, SM, SC
5% to 12% fines . . . . Borderline cases requiring use of dual symbols

Soils processing characteristics of two groups are designated by combinations of group symbols; for example, GW-GC indicates a well-graded, gravel-sand mixture with clay binder.

[d]The No. 200 sieve size is about the smallest particle visible to the naked eye.

[e]For visual classification, the ¼-in. size may be used as equivalent to the No. 4 sieve size.

[f]Applicable to fractions smaller than No. 40 sieve.

[g]These procedures are to be performed on the minus 40-sieve-size particles (about 1/64 in.).
For field classification purposes, screening is not intended. Simply remove by hand the coarse particles that interfere with the tests.

well consist of a limited, confirming-type program, even though the structure under design may be a major facility. In contrast, a lightly loaded structure in a remote area with poor subsurface conditions may require an extensive investigation. Some building codes specify exploration programs, *but these are minimum critera.* Blind adherence to their requirements does not constitute prudent engineering practice. The best guideline is: the geotechnical engineer should be reasonably certain that no *major unknowns* exist at the site and should know the controlling characteristics of the subsurface materials.

In developing the subsurface exploration program, the geotechnical engineer should consider the building code standards and other elements of local experience, the size and complexity of the planned facility and the anticipated variability of subsurface conditions. For low-rise structures of ordinary complexity on sites of anticipated ordinary conditions, borings spaced at about 75 to 125 ft on centers would usually suffice, but should typically be viewed as the minimal acceptable program. Other exploratory techniques, such as test pits or soundings, should be considered as supplementary to the borings (and not as substitutes) in the initial exploration phase.

The exploration scope includes not only the number and spacing of borings and other exploratory techniques, but also their depth and the nature of the data obtained. The guiding philosophy should be that the subsurface conditions be explored to such a depth or to such materials that the stresses imposed by the planned facility at that depth on those materials will have no significant impact on the facility. If this is done, it follows that all materials below that depth can be safely ignored in the foundation engineering analyses. Imposed stresses usually reduce to about 10% of the value applied by the foundation at a depth equal to twice the foundation's least dimension B (for shallow foundations). That stress level (10%) and depth ratio $2B$ for shallow foundations and the equivalent for deep foundations is a satisfactory guideline at most sites for the depth of explorations. If dense, very stiff materials occur at shallower depths and there are no geologic indications of underlying softer, compressible materials, the borings can usually be safely terminated after penetration of the dense materials to depths of 5 to 10 ft below the tip elevation of any deep foundations or to the 10% imposed stress level (from the potential deep foundation).

In accordance with this general philosophy, it is often advisable to conduct subsurface investigations in two or more phases, where successive phases provide increasing detail. Start with a few borings, or other widely spaced explorations. From these, ascertain the stratigraphy and general soil properties. Then, plan the second phase to fill in blind spots, to confirm uniformity or predictability, or to delineate and define anomalies. A third-level detailed effort is usually required only to further define anomalies or to conduct special testing imposed by the requirements of the particular facility under design.

6.5.1 Remote-Sensing Techniques

These methods provide indirect evidence of the subsurface materials. They do not indicate directly the engineering properties but can give depths to strata contacts in many instances and may permit qualitative evaluations of the materials.

Aerial surveys are appropriate where large areas are to be explored. Analyses of conventional aerial stereoscopic photographs; thermal and false-color, infrared imagery; multispectral satellite imagery; or side-looking aerial radar can disclose the surface topography and drainage, linear features that reflect geologic structure, type of surface soil and often the type of underlying rock. These techniques are particularly useful in locating filled-in sinkholes in karst regions, which are often characterized by closely spaced, slight surface depressions.

Geophysical exploration provides information quickly and is economical for supplementing data from borings. The geophysical techniques include seismic reflection and refraction, uphole and crosshole seismic tests, electrical resistivity, microgravimetric measurements, acoustical subbottom profiling, and subsurface radar. Seismic exploration of subsoils is a carry-over from practices standardized in oil-field surveys to determine discontinuities in soil structure. The principles involved are the well-known characteristics of sound-wave transmission, reflection, and refraction in passing through materials of different densities. The method charts arrival time at various points of sound waves induced by explosion or hammer blows.

A similar technique uses the variation in electric conductivity of various soil densities and at discontinuities in layer contacts. The microgravimetric method employs very small variations in the gravity field to locate subsurface voids or changes in rock type; radar systems measure small changes in radar reflectivity for the same purpose. Subbottom profiling is used offshore to provide a continuous record, from seismic reflection, of the subbottom stratigraphic contacts.

In many cases, geophysical methods can give a good picture of certain soil properties, depths of layers, and depths to rock. They cannot be expected to give more than an average or statistical picture of conditions or to give definite identification of the kinds of soils. Information from these methods, however, is an excellent guide for programming a complete exploratory investigation and may be invaluable in providing coverage between widely spaced boreholes.

6.5.2 In-Situ Exploration and Testing Techniques

A wide variety of techniques are available for in-situ investigations of subsurface conditions. Several of these methods have been standardized, and many are utilized worldwide. The most common technique for site exploration worldwide is the mechanically driven boring with Standard Penetration Tests (STPs) performed at depth intervals of usually 5 ft. The STP N value of penetration resistance is an indicator of granular soils' density or the strength of clays, and the sampling spoon recovers a disturbed sample for visual observation and index testing.

Borings. The traditional method of exploring the subsurface is to drill and examine holes and the material removed from them. Many foundation construction difficulties, often of a very expensive nature, result from either unreliable methods of sampling borings or from too much reliance on extrapolation of the results. Consequently, expert engineering judgment should be employed in planning and executing soil explorations with borings.

A boring can be expected to depict truly only the conditions at the boring location. The information obtained there may or may not be indicative of the conditions between borings. Hence, choosing the number and location of borings calls for good judgment and understanding of the local geology.

Many municipal and state building codes specify a minimum number of borings for a specific size of site, but the information obtained from these borings may not be adequate for foundation design and construction. The usual plan for making supplementary borings is to spot an additional boring midway between a pair of borings that shows inconsistent conditions. Then, if the new boring is not a reasonable average of the two, another boring may be placed between the pair with the greatest variation. More borings should be added, as indicated by the results of the preceding explorations. In this way, the location of changes in condition can be closely fixed and overconservative estimates eliminated.

Borings are advanced with continuous flight augers or by rotating drill bits with circulating fluid to remove the cuttings. To avoid contamination of soil samples taken from the borings, it is usually advantageous to use the rotary-drilling method with bentonitic drilling mud as the circulating fluid. Regardless of the method used, great care should be taken that the fluid level inside the borehole never falls below the water table, even during the interval that the drilling tools are being removed from the hole.

During the drilling and sampling of borings, a log should be kept by an engineer, geologist, or trained technician. The log should show the hole depth where samples are taken or tests made, depth to strata changes and to the water table, and the results of any in-situ tests, such as the standard penetration test (ASTM D1586). Additionally, the log should include a complete description of all materials encountered, as judged from the cuttings and samples. Where the borehole encounters rock, the log should note the core recovery, in length and in percentage of the distance drilled into the rock; any weathered seams or solution features; and any joints and their inclination. For many rock formations, the log should report the rock quality designation (RQD), which is a good indication of rock soundness. A typical boring log for soil is shown in Fig. 6.1.

In the Standard Penetration Test (ASTM D1586), a split-barrel sampler, or spoon, with 2-in outside diameter is driven into the borehole bottom by a 140-lb hammer falling 30 in. The number of blows required to drive the spoon 12 in, the *N* value in blows per foot, is recorded. *N* is an indication of the density of granular soils and may reflect the consistency of nonsensitive clays. The SPT provides a moderately disturbed soil sample, but it is suitable for observation, classification, and index testing. The blow count, however, is subject to many influences during drilling and sampling and in the materials penetrated. Therefore, *N* should be carefully evaluated. (See V. F. B. de Mello, "The Standard Penetration Test," Proceedings of the Fourth Panamerican Conference on Soil Mechanics and Foundation Engineering, Vol. 1, 1971, San Juan, Puerto Rico, and J. Schmerttmann, "Measurement of In-situ Shear Strength," Conference on In-situ Measurement of Soil Properties, Vol. II, 1975, American Society of Civil Engineers.)

Other borehole tests include the vane shear, for measuring the in-situ shear strength of clays and organic soils, and the pressuremeter, in which an expanding vertical membrane inside a borehole measures the stiffness and strength of the surrounding soil.

Soil and rock sampling devices should be selected to maximize recovery and limit disturbances of the sample to the degree needed. Sound, unweathered rock should be sampled with core barrels fitted with diamond bits. Several sizes are available, from AX bits to 4.5-in-O.D. bits. The NX bit is most often used, but the larger sizes normally give better recovery in weathered, closely jointed, or porous rock. Recovery can also usually be improved by use of multiple-tube core barrels. A typical rock boring log is shown in Fig. 6.2

The SPT barrel is a relatively good sampler for most soils, although the sample is moderately disturbed. In firm-to-stiff clays, relatively undisturbed samples can be obtained with hydraulically pushed, thin-wall Shelby tubes. In very soft clays and loose sands, a piston sampler gives better results. Dense soils and severely weathered rocks may be sampled with Dennison or Pitcher samplers.

LABORATORY TESTING					IN-SITU TESTING AND SAMPLING					
PASSING NO. 100 SIEVE, %	PASSING NO. 200 SIEVE, %	NATURAL MOISTURE, %	LIQUID LIMIT	PLASTICITY INDEX	SPT DRIVING RECORD, BLOWS/15 CM: 15	15	15	15	RECOVERY, CM	SPT BLOW COUNT N
					2	2	3	-	22	5
					4	5	10		22	15
					12	20	27		19	47
					9	12	15		32	27
					9	10	11		33	21
					7	8	9		30	17
92/32	75/21				6	11	13		25	24
96/26	88/20	25		NP	7	5	10		39	15
90/24	24/16	29		NP	6	22	44		34	66
90/74	52/50				15	16	27		33	43
55	26									
44/92	16/80				11	12	10		32	22
33	19				8	13	13		14	26
24	14				8	10	16		26	26
16	6				12	14	20		33	34
18	4				18	24	27		29	51
19	13				13	15	16		38	31
37/100	17/94	29		NP	8	7	14		36	21
22	6				8	10	10		34	20
12	6				8	12	11		33	23
22	4				10	10	16		27	26
24	6				9	9	9		30	18
97	89	26		NP	6	10	11		28	21
33/19	13/13				9	9	13		22	22
22	14				9	12	11		21	23
95/17	83/9	35		NP	3	8	12		34	20
99	96	40	30	11	4	2	5	11	30	7
22	8				5	11	15	-	24	26
94	88	33		NP	2	2	5	7	51	7
91	79	36		NP	3	3	4	10	49	7

DEPTH, M (0–20) | SAMPLE LOCATION | UNIFIED SOIL CLASSIFICATION

BORING T 12
ZONE 2A
N: 1895 E:1145
ELEV. + 9.50m BPWD

SOIL DESCRIPTION

UNIFIED SOIL CLASSIFICATION	SOIL DESCRIPTION
SP-SM	BROWN FINE SAND WITH SILT AND TRACES OF MICA (DREDGED FILL)
	WITH THIN SEAMS OF BROWN MICACEOUS SILT
SM-ML	GRAY MICACEOUS SILT LAYERED WITH GRAY SILTY FINE SAND AND WITH TRACES OF MICA (UPPER SILTS)
	GRADING WITH BROWN SILT AND SAND
SM	BROWN SILTY FINE SAND WITH TRACES OF MICA (UPPER SILTY SANDS)
ML	GRADING GRAY IN COLOR AND WITH A THIN SEAM OF GRAY SILT
SM	WITH OCCASIONAL BLACK LAMINATIONS OF BIOTITE
ML	GRAY SILT
SM	
ML, SM, ML, SM, ML	GRAY MICACEOUS SILT LAYERED WITH GRAY FINE SAND AND WITH TRACES OF MICA (INTERMEDIATE SANDS AND SILTS)
	WITH BLACK LAMINATIONS OF BIOTITE
	WITH PARTINGS OF GRAY FINE SAND

NOTES:
PERCUSSION BORING COMPLETED TO 20.2M ON FEB. 4, 1994.
SPT = STANDARD PENETRATION TEST.
WATER LEVEL AT APPROXIMATELY 4.2 – METER DEPTH
NP = NOT PLASTIC.

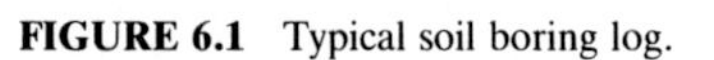
FIGURE 6.1 Typical soil boring log.

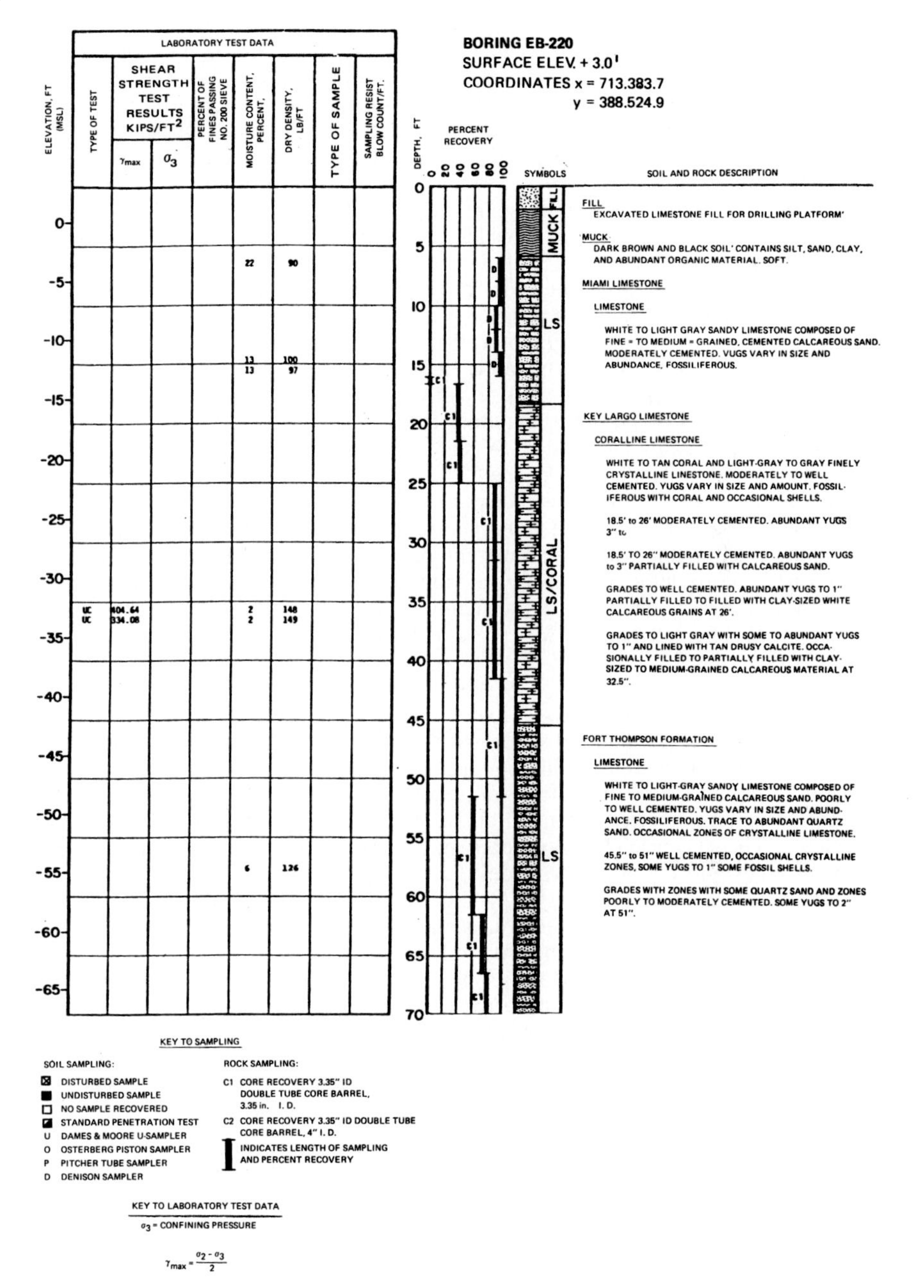

FIGURE 6.2 Typical rock boring log.

Soundings. A sounding with a cone penetrometer is an effective supplementary technique for use in conjunction with samples from borings. The hydraulically or mechanically advanced Dutch cone utilizes a conical point to measure penetration resistance and a follower sleeve to measure frictional resistance. These two values are good indicators of strength and stiffness and relatively reliable indicators of soil type.

Cone Penetration Test (CPT) soundings provide a nearly continuous record of tip resistance and sleeve friction, which can be used to estimate soil type and the stratigraphic profile. Modern CPT devices are often electronically instrumented with computerized output; the more advanced incorporate a porous zone in the tip for instantaneous measurement of pore pressure which further improves the accuracy of the soil-type designation. Tip resistance can be correlated to SPT N values and to relative density of sands and cohesion of soft-to-firm clays, and can provide a basis for estimating the stress-strain (compressibility) modulus of sands.

The Marchetti dilatometer sounding tool is useful where the compressibility modulus at low strains is needed. A plate is driven by a falling hammer like the SPT or pushed hydraulically like the CPT. The dilatometer includes a thin wedge-shaped plate from whose face a circular diaphragm is pushed laterally into the adjacent soil a distance of 1.1 mm. The required force is measured and used to compute the full range of soil parameters for compressibility and strength analyses.

Cone penetrometers may also be advanced dynamically by driving with a hammer, as in the SPT. Obtained in this manner, the test results are analogous to SPT blow counts and can be used for the same purposes when appropriate correction factors are applied.

Test Pits and Trenches. These provide direct visual observation and hand sampling of soils but are limited to practical depths of 10 ft or less. Trenches are advantageous for locating surface contacts between steeply dipping strata. Test pits are an economical and rapid means of obtaining subsurface information but should not be used alone unless the materials lying below the pit are known or are of no concern. The soils in pits and trenches may be sampled by manually pushing thin-walled tubes into the bottom or sides.

Probes. These are drilled or driven soundings, without samples, used to locate top of rock or to obtain cuttings for soil identification. Driven probes give good indication of soil properties. Rock probes can also indicate relative rock soundness by measuring the rate of drill penetration under constant down pressure.

(J. E. Bowles, "Foundation Analysis and Design," 4th ed.; J. A. Franklin and M. Dusseault, "Rock Engineering," McGraw-Hill Publishing Company, New York; H. Y. Fang, "Foundation Engineering Handbook," 2d ed., Van Nostrand Reinhold, New York.)

6.6 LABORATORY TESTS

Basically laboratory tests give more precise values of engineering properties than do interpretations of simple field tests—provided the samples are truly representative of the subsurface conditions. Laboratory testing of soils has developed into a complicated maze of interrelated tests, with many forms of apparatus. A summary of these tests is published periodically by ASTM (see STP 479, "Special Procedures for Testing Soil and Rock for Engineering Purposes"). Many of the test techniques are applicable only to certain restricted soil groups and give misleading data if otherwise used.

If settlement-sensitive structures, major earthworks, dams, or steep slopes are to be founded on soft or uncertain soils, laboratory tests should be performed on representative samples. Laboratory tests, however, are expensive and time-consuming. Therefore, except on large projects, usually few such tests are made. Consequently, the samples for testing should be carefully selected.

Index tests, such as Atterberg limits, density, and grain-size distribution, are used to classify and characterize soils, indicate general engineering characteristics of soils, determine suitability of soils as fill material, and estimate susceptibility of soils to ground-improvement techniques. It is common to make at least several of these tests on each stratum of interest.

Compressibility tests, such as consolidation and triaxial compression tests, provide values for estimating settlements under load. Consolidation tests are performed on plastic soils, and triaxial compression tests are performed on granular materials. Consolidation tests also give time rate of settlements, change in permeability with consolidation, and the maximum past pressure to which the soil has been subjected. Triaxial compression tests are also used to evaluate the stiffness of sands, which increases with increasing confining pressure, and the strength gain of clays under consolidation.

Laboratory strength tests measure soil strength for determination of bearing capacity, resistance to lateral earth pressure, and slope stability. Strengths of sands are measured by triaxial and direct shear tests. Laboratory vane-shear, unconfined-compression, and triaxial-compression tests are applicable to cohesive soils.

In strength testing, it is necessary that drainage conditions during the tests simulate as closely as possible those that will prevail in the soil stratum at the time, and for the structure under consideration. All the tests *except triaxial* are essentially undrained and approximate rapid field loading, such as that for most building foundations and excavations. Triaxial tests permit testing under undrained, consolidated undrained, and consolidated drained conditions. Tests under drained conditions would be used for such applications as checking stability under an embankment after excess pore pressures have dissipated.

If loose, saturated sands may be subjected to earthquake loads, the strength of the sands under cyclic loading and their liquefaction potential may be estimated in the laboratory by cyclic triaxial tests.

(J. E. Bowles, "Foundation Analysis and Design," 4th ed., McGraw-Hill Publishing Company, New York; H. Y. Fang, "Foundation Engineering Handbook," 2d ed., Van Nostrand Reinhold, New York.)

SOIL IMPROVEMENT

Soil as an engineering material differs from stone, timber, and other natural products in that soil can be altered to conform to desired characteristics. Soil improvement is an age-old practice for building on marginal ground and is often used in modern geotechnical-engineering practice. Improvement may be achieved by in-situ methods or by construction of man-made fills. In either case, the objectives are increased bearing capacity or decreased settlements, or both. Many techniques, including densification, surcharging, grading, and fill construction, have been developed and become widely accepted. They are largely responsible for the increased and economical usage of marginal sites.

6.7 MAN-MADE FILLS

The term *fill* is used in this section to mean those earth materials that are placed primarily to level or to raise the ground surface, in contrast with retaining structures, such as earth dams and dikes. Most of the general principles presented in the following, however, apply to both types of earthwork.

Nearly all building sites require some man-made fills, even if for nothing more than the subbase for floor slabs or pavements. Yet, many undesirable conditions, including improper compaction, volume changes, and unanticipated settlements due to weights of fills, often are associated with fills. To prevent such conditions from occurring, fills should be viewed as structural elements of a project and designed accordingly, not haphazardly. Materials and their gradation, placement, degree of compaction, and occasionally their thickness should be selected so as to properly support the loads anticipated.

There are two basic types of fills: those placed dry by conventional earthmoving equipment and techniques, and those placed wet by hydraulic dredges. The latter type is used mainly for filling behind bulkheads or for large-scale areal fills.

A wide range of materials and grain sizes are suitable for fills for most purposes, but all organic matter and refuse should be prohibited. Economics usually requires that the source of fill materials be as close to the site as possible, yet may eliminate some materials. For example, economics may not allow the drying of saturated fine-grained soils. For most fills, maximum particle size in the 18 in below foundations, slabs, or the surface should be limited to 3 in.

The most common test for determining the gross suitability of a soil as fill and for providing a standard for compaction is the moisture-density relationship test (ASTM D698 and D1557), often called the Proctor test. Figure 6.3 shows typical Proctor test results for sands, silts, and clays, as well as the 100% saturation or zero-air-voids line. Several such laboratory tests should be performed on the borrow material, and the standard moisture-density relationship should be established. The peak of the curve indicates the maximum density obtainable in the laboratory by the test method, as well as the optimum moisture content.

The two ASTM tests represent different levels of compactive effort. In the field, however, much higher compactive efforts may be employed than those used in the laboratory. Thus, a different moisture-density relationship may exist on the site. The Proctor test results should, therefore, not be regarded as an inherent property of the material. Nevertheless, the curves indicate the sensitivity to moisture content and the degree of field control that may be required to obtain the specified density.

Fill Compaction. The required degree of compaction for a fill is expressed normally as a minimum percentage of the maximum dry density obtained in the laboratory test, to be accomplished within a stipulated moisture range. Typically, densities representing 90 to 100% of the maximum density, at moisture contents within 2 to 4% of the optimum moisture content, are specified. ASTM D1557 is used as the standard when high bearing capacity and low compressibility are needed; ASTM D698 should be used where the requirements are less, such as beneath parking lots. Minimum densities of 90 to 95% of the maximum density are appropriate for most fills; 100% compaction is often required beneath roadways, footings, or other highly loaded areas.

Note that field densities can be greater than 100% of the maximum laboratory value. Furthermore, with greater compactive effort, such densities can be achieved at moisture contents off the laboratory curve. Care should be taken, however, that

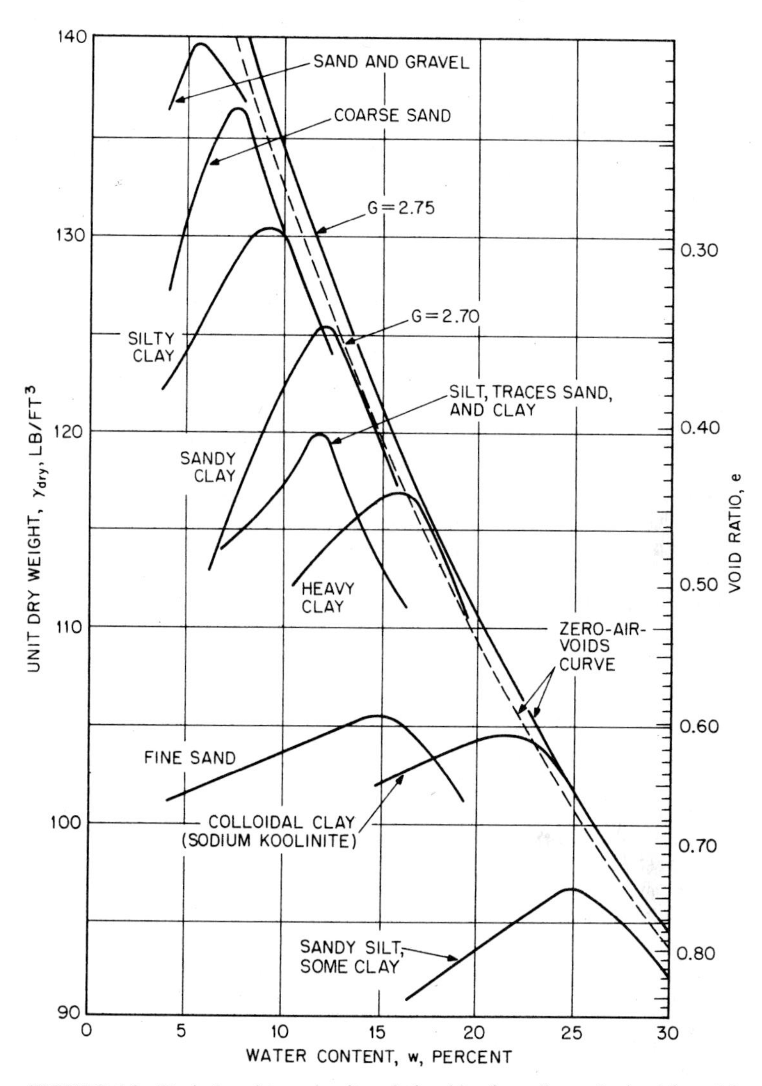

FIGURE 6.3 Typical moisture-density relationships for soils as obtained from laboratory tests made in accordance with ASTM D1557-64T. Method C of this standard was used to determine the curve for sand and gravel; method A was used for all other curves.

fine-grained materials not be overcompacted on the dry side of optimum, because they may swell and soften significantly on saturation.

Conventional fills are placed in lifts of 8 to 24 in. Each lift should be compacted before placement of the next lift. Actual compaction achieved may be determined by field density tests performed on each lift. For the purpose, wet density and moisture content should be measured and the dry density computed. Field densities may be obtained by the sand-cone (ASTM D1556) or balloon volume-meter (ASTM

D2167) method, from an undisturbed sample or with the nuclear moisture-density meter. For most applications, one field density test for each 4000 to 10,000 ft^2 of lift surface is adequate.

On large projects where heavy compaction equipment is used, lift thicknesses of 18 to 24 in are acceptable. For most projects, however, lift thicknesses should be restricted to 8 to 12 in.

Hydraulically placed fills composed of dredged soils usually need not be compacted as they are being placed; many never are compacted. Segregation of the silt and clay fractions of the dredged material is common and generally not harmful; but accumulation of these fines in pockets adjacent to bulkheads or beneath structures should be avoided. Proper use of internal dikes, weirs, and decanting techniques can prevent such accumulations.

Bear in mind that fills are very heavy dead loads and may severely stress even deeply buried strata. A 1-ft depth of compacted fill is equal in load to about 1.5 stories of an ordinary office building. Consequently, the following undesirable conditions may occur: If a building straddles a cut-fill boundary line, there is a potential for sharply delineated differential settlement. Deep hydraulic fills may cause surface subsidence of several feet. Pile-supported structures with slabs on grade over deep fills may, as the fill settles, be damaged because of slab, utility, and entranceway movement relative to the structure. Bulkhead tieback rods that pass through a hydraulic fill may be loaded by subsequent fill settlement unless the rods are sleeved through the fill.

6.8 DENSIFICATION

Any of several different techniques, usually involving some form of vibration, may be used for densification, the in-situ compaction of primarily granular soils to increase their density. Applicability of these methods is sensitive to grain size, as shown in Fig. 6.4. Consequently, grain-size distribution must be carefully considered in selection of the densification method.

Clean sands can be readily densified to depths of about 6 ft by rolling the surface with a heavy, vibratory, steel-drum roller. The vibration frequency is to some extent adjustable but is generally most effective in the range of 25 to 30 Hz. Little densification will be achieved below 6 ft, and the top foot may be actually loosened.

Vibroflotation and Terra-Probe increase densities of sands by the multiple insertions of vibrating probes. The cylindrical voids created are filled with imported sand. Insertions are generally made in clusters, with typical spacing of about 4.5 ft, where building columns will be erected. Relative densities of 85% or higher are obtainable throughout the depth of insertion, which may exceed 40 ft. These techniques will not work if fines content exceeds about 15% or if organic matter is present in colloidal form in amounts greater than about 5% by weight.

Compaction piles are an alternative used to densify sands and permit the subsequent use of shallow foundations. Such piles may be of any material but commonly are wood or a sand replacement (sand pile) in which a shell is driven and the resultant hole filled with sand. The volume displaced by the piles and the vibration accompanying pile driving lead to densification of the surrounding soils. The foundation structural element does not typically bear directly on the compaction pile, but rather on the densified mass. Compaction piles would normally be used under the same structural requirements and subsurface conditions as Vibroflotation or the Terra-Probe technique.

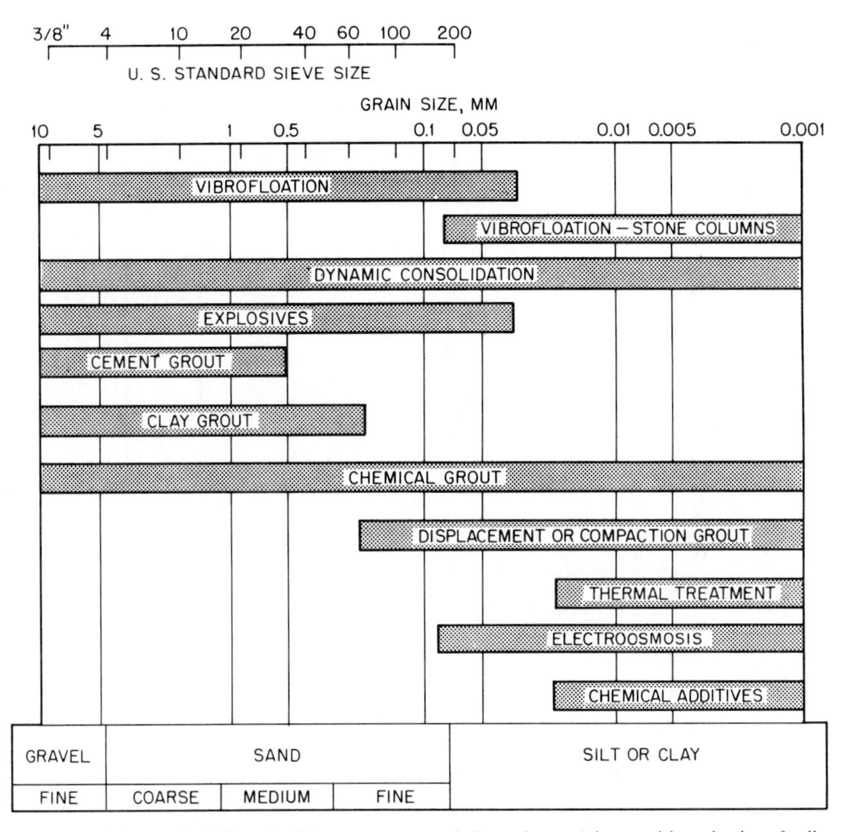

FIGURE 6.4 Applicability of soil-improvement techniques in accordance with grain size of soils.

Another technique for large-scale areal densification is dynamic compaction, as developed by Techniques Louis Menard. It requires that heavy weights be dropped from great heights on the ground surface. Weights may range from 10 to 40 tons. Dropping heights may be as large as 100 ft, and impact spacings up to 60 ft on center may be used. Multiple drops are made at each location, and several such passes across the area are required. This technique is capable of densifying the widest range of grain sizes and materials, as shown in Fig. 6.4.

With 15-ton weights falling 80 ft, loose sands can be compacted to dense or very dense states to depths of 25 to 35 ft. With 40-ton weights, the effective depths can be extended to 40 to 50 ft, with further increase in density. With sands and nonplastic silts, enforced ground settlements on the order of 2 ft can be achieved.

The technique accomplishes densification as if the site had been subjected to numerous miniearthquakes, with the compaction resulting from partial liquefaction (where the soil is saturated) and the passage of the wave train. Elevated pore pressures are produced in saturated masses, and successive passes of pounding must await the dissipation of these pressures if the following pass is to be effective in producing further compaction. Figure 6.5 presents the results of a large-scale dynamic compaction effort in Bangladesh, shown as a plot of mean Standard Penetration Test values, before and after dynamic compaction, as a function of depth.

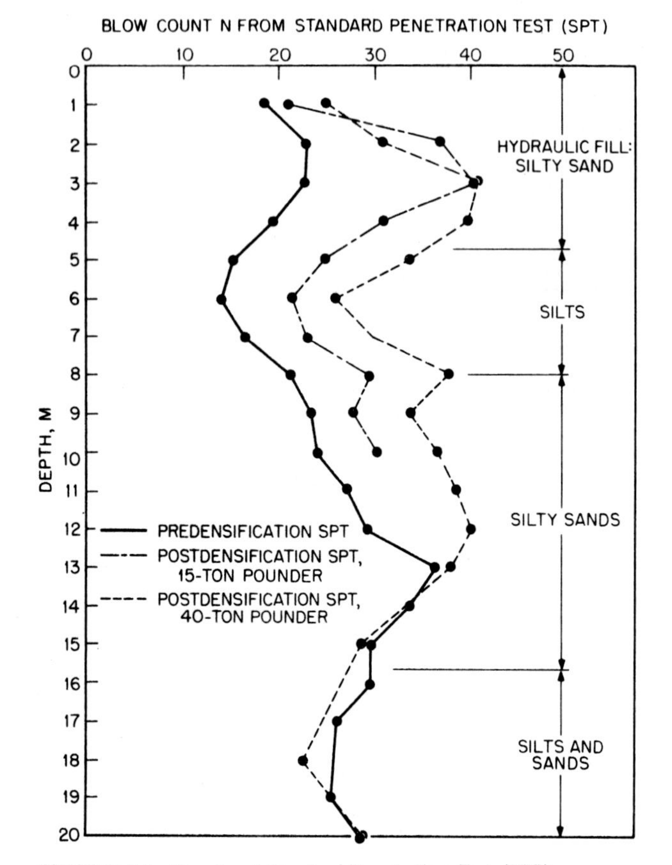

FIGURE 6.5 Results of Standard Penetration Test (SPT) measurements of densification achieved with dynamic compaction.

6.9 SURCHARGES

Sometimes, suitable foundation materials are underlain by soft, compressible clays that would permit unacceptable settlements. In such cases, the site often can be made usable by surcharging the surface. The intent of surcharging is to preload the ground and preconsolidate the clays, thus canceling the settlements that would otherwise have occurred under the structures. A concurrent objective may be to increase the strength of the clay.

In practice, the process is straightforward if the clay is overlain by sufficiently strong soils that bearing capacity is not a concern. The area to be improved is loaded with loose, dumped earth until the weight is equivalent to the load that will later be imposed by the permanent structure. If highly plastic clays or thick layers with little internal drainage are present, sand drains may have to be inserted to achieve consolidation within a reasonable time. Settlements of the original ground surface and of the clay layer should be closely monitored during and after placement of the surcharge. The fill should remain until settlement has about stopped. Then the surcharge may be removed and the structures erected. If surcharging has been done properly, the structures should experience no further settlement due to primary consolidation. Potential settlements due to secondary compression, however, should be evaluated, particularly if the soft soils are highly organic.

The technique is effective and most beneficial for large areas. The need for low-cost temporary fill and the long times occasionally required for settlement are the principal constraints.

6.10 GROUTING

The injection of cementitious materials or chemical agents into soil and rock masses improves them by either filling voids or causing compaction of the masses. Such grouting often is used to fill openings in rock masses, typically carbonate rocks, to restrict seepage or prevent structural collapse. Special materials and methods have been developed for specific purposes. For example, injection of quick-setting gels is effective in stabilizing loose sands, and high-pressure cement grouting can produce localized compaction of loose sands.

6.11 DRAINAGE

Permanent lowering of a near-surface water table in fine sands and silts often improves the near-surface soils markedly, particularly for such applications as roads, parking areas, and low-rise residential construction. Drainage is effective because the strength of soils decreases, in general, with an increase in amount and pressure of pore water.

In granular soils, the water table can be lowered with pumps and vertical wells, and the removed water conducted to off-site disposal basins or sewers. For silts, which are difficult to drain by such means, electroosmosis, or electrical drainage, may be used. This method employs the principle that water flows to the cathode when a direct current passes through saturated soils. The water may be pumped out at the cathode.

Measures should be taken to maintain the lowered water table. Surface and subsurface flows should be intercepted and conducted away from the site. For the purpose, the ground surface should be graded to control the flow of surface runoff and intercepting drains should be installed in the ground. Such drains, laid approximately along contours, are especially useful in stabilizing earth slopes.

6.12 GEOSYNTHETICS

These are fabrics made of plastics, primarily polymers, but sometimes rubber, glass fibers, or other materials, that are incorporated in soils to improve certain geotechnical characteristics. The roles served by geosynthetics may be grouped into five main categories: separation of materials, reinforcement of soil, filtration, drainage within soil masses and barrier to moisture movement. There are several types of geosynthetics:

Geotextiles are flexible, porous fabrics made of synthetic fibers by standard weaving machines or by matting or knitting (nonwoven). They offer the advantages for geotechnical purposes of resistance to biodegration and porosity, permitting flow across and within the fabric.

Geogrids consist of rods or ribs made of plastics and formed into a net or grid. They are used mainly for reinforcement of and anchorage in soils. Aperture sizes for geogrids range from about 1 to 6 in in longitudinal and transverse directions, depending on the manufacturer.

Geonets are netlike fabrics similar to geogrids but with apertures of only about 0.25 in. The ribs generally are extruded polyethylene. Geonets are used as drainage media.

Geomembranes are relatively impervious, polymeric fabrics that are usually fabricated into continuous, flexible sheets. They are used primarily as a liquid or vapor barrier. They can serve as liners for landfills and covers for storage facilities. Some geomembranes are made by impregnating geotextiles with asphalt or elastomerics.

Geocomposites consist of a combination of other types of geosynthetics, formulated to fulfill specific functions.

Design of geosynthetic filters, or earth reinforcement, or an impervious membrane landfill liner requires a clear statement of the geotechnical characteristics to be achieved with geosynthetic application, a thorough understanding of geosynthetic properties, and a knowledge of materials currently available and their properties.

(R. M. Koerner, "Designing with Geosynthetics," 2d ed., McGraw-Hill Publishing Company, New York.)

FOUNDATION DESIGN

The selection of foundation design criteria—type of foundation and its depths and allowable load or bearing pressure—is often an iterative process. To provide proper support, all foundations must satisfy two requirements simultaneously: adequate bearing capacity and structurally tolerable settlements. Although related, the two requirements usually cannot be automatically met simultaneously. A foundation that fails in bearing also settles excessively, but one with normally adequate bearing capacity may also settle excessively. The two factors, bearing or load capacity and settlement, then must be examined and the foundation design based on the limiting condition.

6.13 STEPS IN FOUNDATION DESIGN

In practice, the general procedure for foundation design is:

1. Determine the ultimate bearing capacity of the one or more types of foundation that are feasible, given the subsurface conditions and structural requirements.

2. Reduce the ultimate bearing capacities computed by a factor of safety of 2 to 3. Use the higher factor of safety where there is less certainty about the subsurface conditions.

3. Compute the settlements that may occur for a foundation with the reduced allowable bearing capacity and the given structural loadings.

4. If settlements are structurally acceptable, compute relative costs of the satisfactory foundation types, on a comparative basis such as cost per ton of column load or cost per square foot of building area. These costs should include those for all structural elements of the foundation system, such as caps for piles and any ground improvement works required, and any unusual costs such as dewatering. Consider the time required for construction. All other things being equal, choose the least-cost system.

5. If settlements are unacceptable for all types considered, then examine other alternatives, such as ground improvement, building relocation, reduced bearing pressures or loads, different bearing depths, and revision of superstructure. Repeat steps 3 and 4 until a safe design offering the best economy is selected.

6.14 BEARING OR LOAD CAPACITY OF FOUNDATIONS

Bearing or load capacity is a characteristic of a foundation-soil system, not an intrinsic quality of the soil only. Soils of different type differ in bearing capacity, but also for a specific soil, bearing capacity may vary with the type, shape, size, and depth of the foundation element applying pressure.

There are two basic types of foundations, shallow and deep. Also, there are several varieties of each type.

Shallow foundations include spread, continuous and strapped footings, mats, and compensated rafts. Deep foundations include drilled caissons, or piers, and many varieties of driven and cast-in-place concrete piles.

Note that it is not sufficient to base foundation design on bearing capacity of soil and foundation alone. The design should also limit settlement (see Art. 6.15).

6.14.1 Shallow Foundations

The ultimate bearing capacity q_{ult}, psf, of a soil under an infinitely long strip foundation is given by

$$q_{ult} = cN_c + qN_q + \frac{\gamma B}{2} N_\gamma \tag{6.2}$$

where c = cohesion of the soil (see Art. 6.3)
q = overburden pressure, psf, at level of foundation base
γ = unit weight of soil, lb/ft^3
B = foundation width, ft
N_c, N_q, N_γ = bearing-capacity factors

Variation of the bearing-capacity factors with ϕ, angle of internal friction of the soil, is shown in Fig. 6.6.

The basic equation presumes an infinitely long strip foundation of width B and should be corrected for other shapes. Correction factors by which the bearing-

TABLE 6.2 Shape Corrections for Bearing-Capacity Factors of Shallow Foundations*

Shape of foundation	Correction factor N_c	N_q	N_γ
Rectangle†	$1 + (B/L)(N_q/N_c)$	$1 + (B/L)\tan\phi$	$1 - 0.4(B/L)$
Circle and square	$1 + (N_q/N_c)$	$1 + \tan\phi$	0.60

*After E. E. De Beer, as modified by A. S. Vesić. See H. Y. Fang, "Foundation Engineering Handbook," Van Nostrand Reinhold, 2d ed., New York.

†No correction factor is needed for long strip foundations.

capacity factors should be multiplied are given in Table 6.2, in which L = footing length.

The derivation of Eq. (6.2) presumes the soils to be homogeneous throughout the stressed zone, which is seldom the case. Consequently, adjustments may be required for departures from homogeneity. For stratified clay soils, for example, N_c may be adjusted as indicated in Fig. 6.7. In sands, if there is a moderate variation in strength, it is safe to use Eq. (6.2), but with bearing-capacity factors representing a weighted average strength.

For strongly varied soil profiles or interlayered sands and clays, the bearing capacity of each layer should be determined. This should be done by assuming the foundation bears on each layer successively but at the contact pressure for the depth below the bottom of the foundation of the top of the layer. Examples of

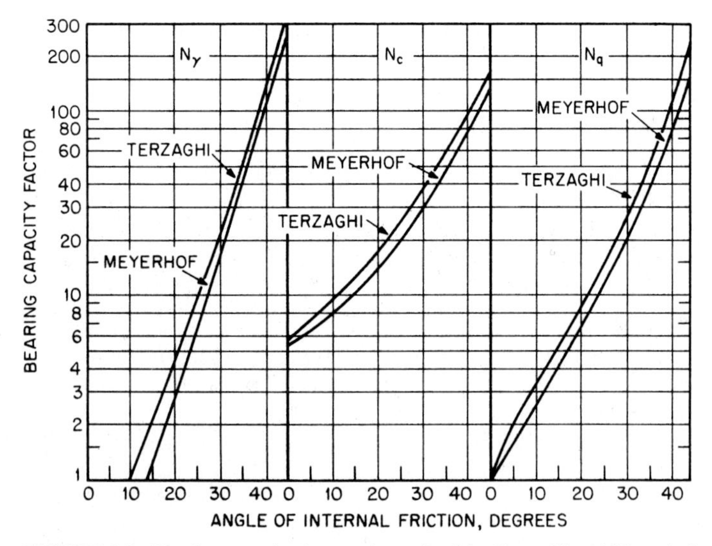

FIGURE 6.6 Bearing-capacity factors determined by Terzaghi and Meyerhof for use in Eq. (6.2).

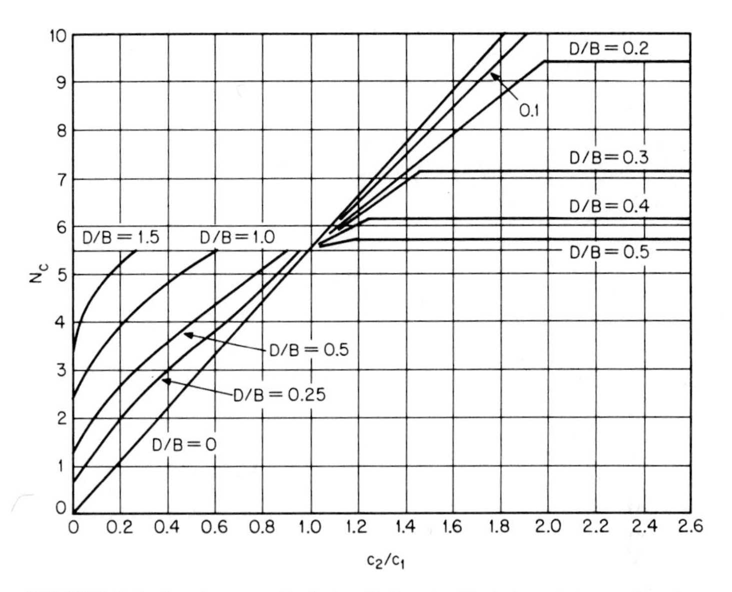

FIGURE 6.7 Bearing-capacity factor N_c for stratified clay plots as a function of the ratio of the undrained shear strength c_2 of the lower layer to that of the upper layer c_1 and of the ratio of the depth D of the top of the lower layer below the foundation to the width B of the foundation.

pressure distributions below the foundation that may be assumed are shown in Figs. 6.8 and 6.9.

6.14.2 Deep Foundations

The load capacity of deep-footing foundations can be evaluated analytically in basically the same manner as described for shallow foundations. In the case of driven piles, there are additional methods for determining the load capacity that are based on pile resistance to driving.

Piles. These are described in Art. 6.17. A relatively large number of static equations have been proposed for computing the capacity of piles. Terzaghi and Peck proposed the formula

$$P_u = Q_{ip} + 2\pi r f_s L_f \tag{6.3}$$

where P_u = ultimate load on pile, kips

Q_{ip} = point bearing capacity, kips, of pile, computed with the same equations as for shallow foundations, except that deep-foundation bearing-capacity factors are used (see J. E. Bowles, "Foundation Analysis and Design," 4th ed., McGraw-Hill Publishing Company, New York)

r = radius, ft, of pile

L_f = embedded length, ft, of the pile

f_s = friction, psf, along the pile (ranges from 0.5 to 2 ksf for cohesionless soils and may be taken as the cohesion for clays)

(See K. Terzaghi and R. B. Peck, "Soil Mechanics in Engineering Practice," John Wiley & Sons, Inc., New York. A more rigorous formula, preferred for tapered piles, is given in R. J. Nordlund, "Bearing Capacity of Piles in Cohesionless Soils," Journal of the Soil Mechanics and Foundations Division, May 1963, American Society of Civil Engineers.)

The load capacity of driven piles is often estimated from their resistance to driving. The most commonly used driving formula is the *Modified Engineering News Formula*

$$P = \frac{2.5E}{S + 0.1} \frac{W_r + n^2 W_p}{W_r + W_p} \qquad (6.4)$$

where P = pile design load, kips
E = manufacturer's pilehammer rated energy, ft-kips
S = average set or tip penetration, inches per blow for the last five blows for gravity hammers and the last 20 blows for other types of hammers
W_r = weight of ram, kips
W_p = weight of pile, kips
n = coefficient of restitution of cap blocks or cushions or butt of pile

The formula gives results such that the actual load capacity and safety factor may vary widely and unpredictably. Pile-driving formulas should never be used blindly, without references to pile-driving experience in the area of the site or to load tests.

An alternative dynamic formula for estimating the load capacity of piles is derived from the wave equation. It relates the velocity of the piledriver hammer at the instant of impact on the pile to displacements associated with compression waves in the pile that are caused by the impact. Stresses can be computed from the displacements. In its simplest derivation, the wave equation ignores the effects of shears along the sides of the pile. More complex forms of the equation take the shears into account. Computations may be expedited with computers.

For projects where piles do not bear directly on rock, the wave equation may be used to predict pile behavior during driving and establish driving criteria that will ensure achievement of design capacity without exceeding allowable pile stresses. For large or complex projects where driving criteria are often established in the field (after wave-equation analyses are performed in the office), use of techniques, such as the pile driving analyzer, for measuring pile acceleration and the energy applied by the hammer to the pile, are often warranted. They are employed during pile installation to establish driving criteria for achieving static-load capacity. They also are used for direct measurement of the dynamic load resistance actually developed. The energy applied to a pile is usually determined with the aid of strain gages, and the resistance (load capacity) is established from measurements of the pile acceleration. As with other techniques that rely on the driving resistance to estimate the static-load capacity of piles, the computerized dynamic-measurement methods do not address the effects of long-term settlements that may occur in sites with compressible soils nor do these methods incorporate pile-group effects.

Piles generally are driven in groups, rather than singly. Consequently, the group efficiency should be checked against the performance of a single pile. For piles supported by friction with the soil, the perimeter friction of the group normally would be less than the sum of the friction forces on the individual piles. In sands, densification due to driving a pile cluster increases the supporting friction forces. For piles supported at the tip, the group efficiency becomes even more important, with respect to the capacity of pile foundations to resist uplift loads.

See also Arts. 6-18 to 6-20.

Drilled Shafts. The bearing capacity of caissons or drilled shafts is computed as if they were deep footings, except that account may be taken of the friction between shaft and soil. End bearing should be computed as for footings, but with deep-foundation bearing-capacity factors (see J. E. Bowles, "Foundation Analysis and Design," 4th ed., McGraw-Hill Publishing Company, New York.) Adhesion or friction developed along the caisson shaft adds to the total bearing capacity and may be significant where the construction technique promotes good contact of the caisson concrete with relatively undisturbed soil or rock around the periphery of the shaft. In clays, the adhesion of soil to the shaft may approach 75% of the soil cohesion. (A value of $0.45c$, with a limit of 2000 psf, is recommended by A. W. Skempton in "Cast-in-situ Bored Piles in London Clay," *Geotechnique,* Vol. 9, No. 4.)

A shaft to be drilled in sands may be designed for total load capacity developed through both end bearing and side shear if strain compatibility is maintained. End-bearing capacity may be computed as for a deep pile foundation. Computation of side shears may be based on the average undrained shear strength of the soil along the length of the shaft, as estimated from field loading tests. Development of end-bearing capacity requires strain (settlement) of the drilled-shaft base. Most authorities estimate that base movement of about 5 to 10% of the shaft diameter is required to develop full bearing capacity. In contrast, the amount of side movement (relative to the adjacent soil surface) required to mobilize side shear fully is estimated at typically less than 1 in. Further strains lead to post-peak residual strengths that are generally less than the maximum peak values. Thus, the developed total load capacity of a shaft equals the side-shear capacity mobilized through the post-peak side movement and the end-bearing capacity mobilized at a given amount of tip displacement. This total can best be determined by plotting composite load capacity as a function of butt deflection (including elastic shortening), selecting the load capacity developed at the tolerable butt deflection, and applying an appropriate safety factor.

This design approach applies where a shaft is significantly stiffer vertically and stronger than the surrounding soil. This condition may not obtain when the shaft is constructed in rock. For drilled shafts embedded in rock, the relative stiffnesses of the rock and concrete, the aspect ratio, and the mode of failure are important considerations in determining the contribution of end bearing to the total load capacity. For embedment lengths in rock more than about twice the shaft diameter, the end-bearing capacity is in the range of 10 to 20% of total capacity. If the shaft penetrates very soft soil and bears immediately on sound rock with little embedment, the load capacity may be considered as essentially due only to end bearing. See also Art. 6.22.

(J. E. Bowles, "Foundation Analysis and Design," 4th ed., McGraw-Hill Publishing Company, New York; R. G. Ahlvin and V. A. Smoots, "Construction Guide for Soils and Foundations," 2d ed.; C. L. Crowther, "Load Testing of Deep Foundations," John Wiley & Sons, Inc., New York; M. J. Tomlinson, "Foundation Design and Construction," 5th ed., John Wiley & Sons, Inc., New York; H. Y.

Fang, "Foundation Engineering Handbook," 2d ed., Van Nostrand Reinhold, New York; R. G. Horvath and T. C. Kenney, "Shaft Resistance of Rock-Socketed Drilled Piers," Symposium on Deep Foundations, 1979, American Society of Civil Engineers, 345 E. 47th St., New York, NY 10017.)

6.15 SETTLEMENT OF FOUNDATIONS

All foundations settle. All of the materials used or soils encountered in ordinary construction are, to some degree, compressible under the range of stresses induced by most structures. The amount of settlement that can be accommodated safely by a structure is a function of the uniformity of the movements, the settlement rate, and the time of occurrence of settlement relative to the construction sequence, as well as the sensitivity of the structure as governed by column spacing, type of construction, and spacing of expansion joints.

In general, nuisance- or damage-causing settlements are those that create differential movements between building components and that occur after construction has been completed. Everything else being equal, differential movements become increasingly difficult to tolerate as column spacing decreases.

Cast-in-place concrete structures are typically far more sensitive to such movements than are steel buildings. For most steel structures, a deflected beam slope of 1:300 is about the steepest acceptable, whereas, for rigid concrete structures the deflected beam slope should not exceed about 1:800 to 1:1000.

Estimation of foundation settlements requires a knowledge of the stress distribution within the strata being analyzed and their compressibilities, for both shallow and deep foundations. Also required for both types of foundations is knowledge of the depth of the foundation base below grade, depth to and thicknesses of the soil layers surrounding and underlying the base, loads imposed by each foundation unit, and distribution of loads between static (dead) and dynamic (live) states.

For structures that may be subjected to unpredictable dynamic loads, settlements under rarely occurring, short-duration maximum live loads, such as wind or earthquakes, are usually not estimated separately from the dead loads imposed by the structure itself. Instead, the foundations are sized for the sum of the design dead and live loads, and an increase in allowable stresses, usually one-third, is permitted for the unlikely simultaneous occurrence of maximum dead loads and rarely occurring, design live loads. For major tall buildings in areas potentially subject to sustained extreme wind loading, such as that imposed during hurricanes, the magnitude and duration of the design wind loading imposed on the structures should be applied to the foundations. The resulting foundation settlements should be estimated, giving consideration to the time required for compression or consolidation, or both, to occur in accordance with the anticipated duration of the storm. Consideration should be given to the possibility that storm-induced settlements might occur during construction as well as after, and in the latter case, they may cause expensive damage to finishes and utility connections.

The procedure for estimating foundation settlements is nearly the same for shallow as for deep foundations. The principal difference is in the stress distribution in the soil beneath the foundation. Shallow foundations bear directly on the soil, and stresses in the underlying soil may be estimated from charts or by computerized stress analyses that enforce strain compatibility. For deep foundations, stresses in the soils or rock depend on the manner in which the loads are shed from the foundations to those materials. If a deep foundation penetrates soft and compressible soils and is embedded in rock or other very stiff material, it may be assumed that all of the load is transmitted to the stiff material in the depth of embedment. In the more common case in which a deep foundation develops its load capacity in both side shear (skin friction) and end bearing, a more accurate approximation of the distribution of load to the surrounding and underlying soil layers requires analyses of the relative stiffnesses of the foundation and the various loaded soil or rock layers. Strain compatibility must be assumed in the analyses and an estimated load transfer curve developed. Computations may be facilitated by use of a computer.

6.15.1 Stress Distribution under Footings

For most applications, stresses may be computed by the pressure-bulb concept with the methods of either Boussinesq or Westergaard. For thick deposits, use the Boussinesq distribution shown in Fig. 6.8; for thinly stratified soils, use the Wes-

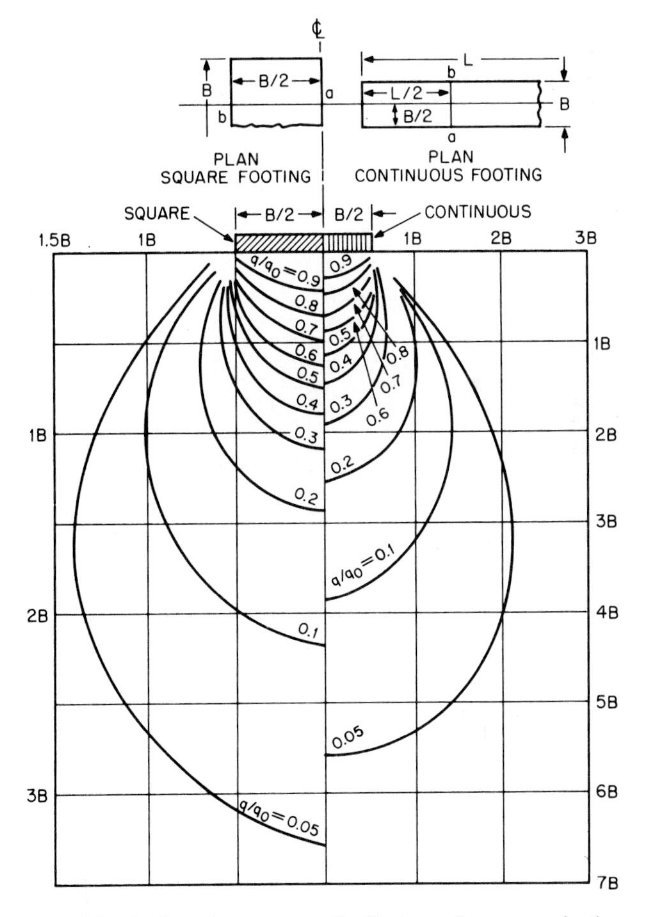

FIGURE 6.8 Boussinesque stress distribution under a square footing with side *B* and under a continuous footing with width *B*.

tergaard approach shown in Fig. 6.9. These charts indicate the stresses q beneath a single foundation unit that applies a pressure at its base of q_o.

Most facilities, however, involve not only multiple foundation units of different sizes, but also floor slabs, perhaps fills, and other elements that contribute to the induced stresses. The stresses used for settlement calculation should include the overlapping and contributory stresses that may arise from these multiple loads.

6.15.2 Settlement Due to Cohesive Soils

Foundation settlement due to clays arises from the phenomenon called consolidation. Evaluation of the settlement of clays requires knowledge of the compressibility as well as the prior stress history of the deposit. Analysis of these parameters can be complex. (See, for example, T. W. Lambe and R. J. Whitman, "Soil Mechanics," John Wiley & Sons, Inc., New York.)

Settlement due to clay may be estimated from

$$S_o = \frac{C_c}{1 + e_o} H \log \frac{P_b + \Delta P}{P_o} \tag{6.5}$$

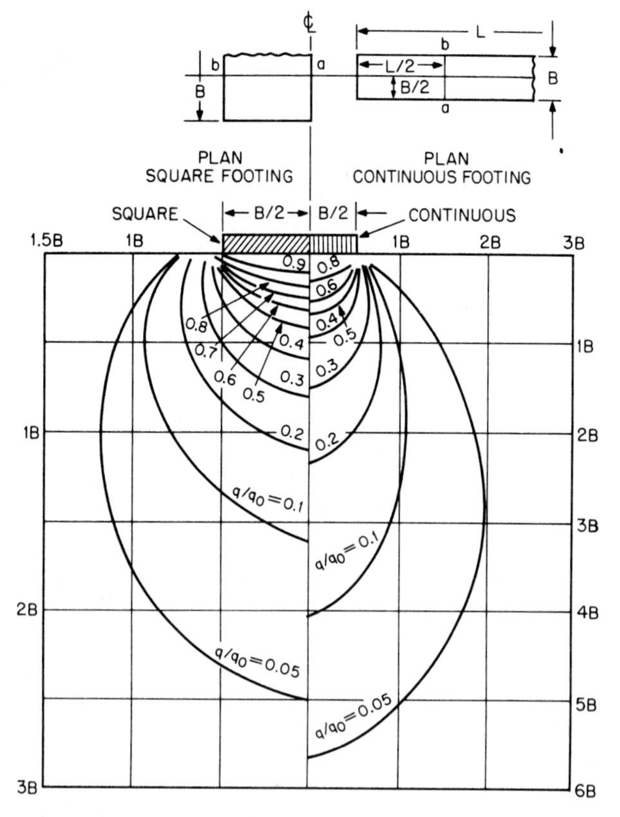

FIGURE 6.9 Westergaard stress distribution under a square footing with side B and under a continuous footing with width B.

where S_o = total settlement
C_c = compression index, taken as the slope of the virgin compression curve
e_o = initial void ratio, or ratio of void volume to volume of solids in a soil mass
H = layer thickness, or one-half the layer thickness if it is drained top and bottom
P_o = preconsolidation pressure
ΔP = induced additional pressure

Values of C_c and P_o are obtained from laboratory consolidation, or odometer, test results plotted for void ratio vs. log P or for unit strain vs. log P, in which case the slope of the virgin curve C_r is called the compression ratio

$$C_r = \frac{C_c}{1 + e_o} \tag{6.6}$$

The portion of the curve obtained with continually increasing P is called the **virgin compression curve.** A different curve is obtained for unloading and for reloading.

Several techniques have been proposed for determination of the maximum past pressure P_o. The most commonly used are those developed by Casagrande and Burmeister, as shown in Fig. 6.10. With the Casagrande construction (Fig. 6.10*a*), to determine P_o, draw a tangent at the point of sharpest curvature (minimum radius of curvature) of the consolidation curve, a horizontal line at the point of tangency, and a line bisecting the angle between the two. Extend the steeply sloped, nearly straight portion of the virgin curve backward, and its intersection with the bisecting line is taken as the maximum past pressure.

The Burmeister technique requires a small unload-reload cycle just beyond the point at which the consolidation curve begins to bend downward sharply. A characteristic triangle is formed between the rebound and reload curves, as shown in Fig. 6.10*b*. Shift the rebound curve upward, parallel to itself, to a point where a geometrically similar triangle of the same vertical height is found. The pressure at that point represents the maximum past pressure. For other techniques, see H. Y. Fang, "Foundation Engineering Handbook," 2d ed., Van Nostrand Reinhold, New York.

The rate of settlement is usually an important consideration where clays are involved. The time, months (or years), to end of primary consolidation is given by

$$t = \frac{H^2}{C_v} \tag{6.7}$$

where C_v = coefficient of consolidation, ft^2 per month (or year)
H = length, ft, of longest drainage path to an outlet, for example, the ground surface or a drained pervious stratum

C_v may be determined from the time-compression curve of a single load increment during the laboratory test. Two graphical methods are commonly used, the square-root-of-time and the log-time fitting methods. (See preceding reference.)

The coefficient of consolidation as determined by laboratory techniques is very sensitive to sample disturbance, testing techniques, and a number of other factors. Hence, there may be substantial error in the estimate of the time required to complete the primary consolidation of clays.

Some very plastic clays and most organic soils are subject to further compression beyond the end of primary consolidation. This creep-like characteristic is termed secondary compression and may represent a significant portion of the total settlement. For evaluation techniques and typical values, see T. W. Lambe and R. J. Whitman, "Soil Mechanics," John Wiley & Sons, Inc., New York.

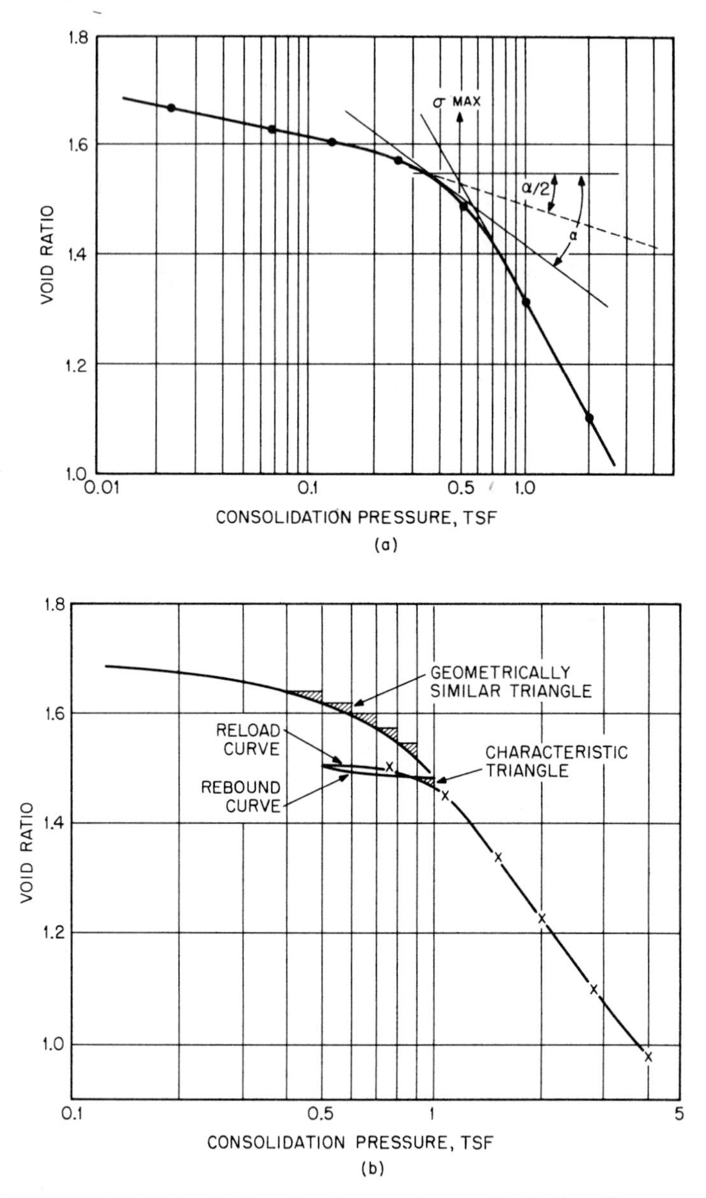

FIGURE 6.10 Determination of maximum past pressure in clay from the curve relating consolidation pressure and void ratio: (*a*) method devised by Casagrande; (*b*) Burmeister's method.

6.15.3 Settlement Due to Granular Soils

The bearing pressure of foundations on granular soils is usually governed by allowable settlements, rather than by bearing capacity. Settlements of foundations on these soils are estimated by empirical techniques utilizing field test data and by methods based on elastic theory.

One method is based on the standard penetration test. In this method, the allowable soil pressure q_{all}, ksf, settlement of 1 in is computed from

$$q_{all} = 0.72(N - 3)\left(\frac{B + 1}{2B}\right)^2 K_d \tag{6.8}$$

where N = representative standard penetration resistance, blows per foot, at depth B below foundation = $N'[8/(p + 2)]$ when corrected for overburden pressure
N' = actual blows per foot
p = effective overburden pressure, ksf
B = footing width, ft
K_d = depth factor = $1 + 0.2D/B \geq 1.2$
D = depth of foundation base, ft

Cone penetration soundings are frequently used to explore sedimentary soils (Art. 6.5.2). The Cone Penetration Test values can be used to predict settlement of footings. The J. Schmertmann technique, for example, which uses CPT values to predict settlement of footings on sand, has proved to be accurate. The technique requires that the cone resistance diagram be plotted from the bottom of the footing to a depth below the footing equal to $2B$ for a square footing or $4B$ for a long footing ($L/B > 10$) or to an assumed boundary layer, if this comes first, where B is the least width of the loaded area and L is the length of footing. Then, the soil within this depth is divided into layers in each of which the cone resistance q_c is nearly constant. The settlement in each layer is calculated using the Young's modulus and vertical strain influence factor, from Fig. 6.11, appropriate to each layer. The footing settlement is then the sum of the settlements in each layer, corrected for depth below the footing and creep:

$$\rho = C_1 C_2 \Delta_p \sum_{1}^{n} (I_z/xq_c)\,\Delta Z \tag{6.9}$$

where ρ = settlement of footing in units of ΔZ
ΔZ = thickness of each of n layers
q_c = cone resistance of a layer
n = number of layers
Δp = net increase in load, in q_c units, at the bottom of the footing due to applied loading, after subtraction of p'_o
p'_o = previous effective overburden pressure at the elevation of the bottom of the footing
I_z = vertical strain influence factor at the center height of each layer (Fig. 6.11)
x = factor by which to multiply q_c to obtain equivalent Young's modulus for sand
= 2.5 for square footings
= 3.5 for long footings
C_1 = correction factor for depth of embedment
= $1 - 0.5(p'_o/\Delta p)$
C_2 = correction factor for secondary creep settlement
= $1 + 0.2 \log_{10}(t_{yr}/0.1)$
t_{yr} = time, years, from application of $p'_o + \Delta p$ on the footing

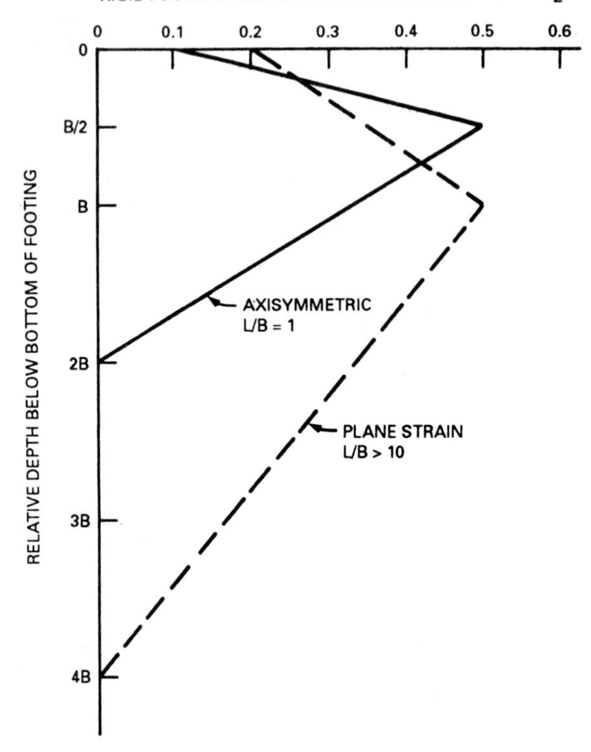

FIGURE 6.11 Modified strain influence diagrams for use in Schmertmann method and Eq. (6.9), for estimating settlement over sand.

6.15.4 Immediate Settlement of Compressible Layers

Immediate settlement, ft, of a foundation may be computed, on the basis of the theory of elasticity, from

$$S = \mu_0 \mu_1 q B \left(\frac{1 - \mu^2)}{E_s} \right) \tag{6.10}$$

where μ_0 and μ_1 = depth and shape factors obtainable from Fig. 6.12
μ = Poisson's ratio for the soil
q = pressure at base of foundation, ksf
E_s = elastic modulus, ksf

S should be computed for each compressible soil layer under the foundation, and the settlements should be added to yield the total immediate settlement. For soils subject to consolidation, total settlement equals the sum of immediate settlement and consolidation settlement.

While E_s may be estimated from the undrained shear strength S_u, determined from a triaxial compression test, it may range from $250S_u$ to $500S_u$. Because of this

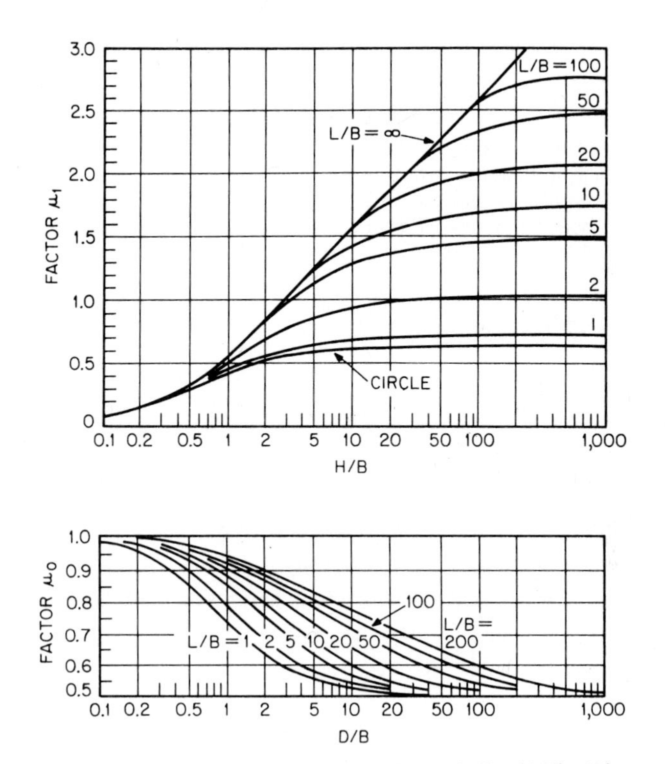

FIGURE 6.12 Shape and depth factors for use in Eq. (6.10). *(After N. Janbu.)*

wide variation, especially in view of the likelihood of sample disturbance, it is advisable to determine E_s from field tests, such as the Standard Penetration Test, Cone Penetration Test, or plate loading tests (see Art. 6.5.2). For example, E_s, ksf, may be approximated from SPT data by the following, for sand

$$E_s = 10(N + 15) \tag{6.11a}$$

and for clayey sands,

$$E_s = 6(N + 5) \tag{6.11b}$$

where N = resistance, blows per foot. From CPT data, E_s may be approximated by $3q_c$ for sand and by $2q_c$ for clay, where q_c is the cone penetration resistance, ksf.

See also Art. 6.16.

6.16 SPREAD FOOTINGS

The most economical foundation for a load is one that transmits the load directly to the soil. In consequence, a spread footing is the most frequently used type of foundation for an isolated column.

6.16.1 Allowable Bearing Pressures

Approximate allowable soil bearing pressures, without tests, for various soils and rocks are given in Table 6.3 for normal conditions. These basic bearing pressures may be increased when the base of the footing is embedded beyond normal depth. Rock values may be increased by 10% for each foot of embedment beyond 4 ft in fully confined conditions, but the values may not exceed twice the basic values.

In any case, bearing pressures should be limited to values such that the proposed construction will be safe against failure of the soil under 100% overload.

6.16.2 Cause of Settlement of Footings

Inasmuch as there are on record many cases of foundation failures but very few footing failures, it must be concluded that the ground on which the footings rest is the usual cause of the trouble, and not the structural footing.

The displacement of the load carried can be one or several of the following movements:

1. Compression of the footing material elastically.
2. Deflection of the footing unit elastically.
3. Compression of the soil directly below the footing elastically.

TABLE 6.3 Allowable Bearing Pressures

Soil material	tons/ft^2	Notes
Unweathered sound rock	60	No adverse seam structure
Medium rock	40	
Intermediate rock	20	
Weathered, seamy, or porous rock	2 to 8	
Hardpan	12	Well cemented
Hardpan	8	Poorly cemented
Gravel soils	10	Compact, well graded
Gravel soils	8	Compact with more than 10% gravel
Gravel soils	6	Loose, poorly graded
Gravel soils	4	Loose, mostly sand
Sand soils	3 to 6	Dense
Fine sand	2 to 4	Dense
Clay soils	5	Hard
Clay soils	2	Medium stiff
Silt soils	3	Dense
Silt soils	1½	Medium dense
Compacted fills		Compacted to 90 to 95% of maximum density (ASTM D1557)
Fills and soft soils	2 to 4	By field or laboratory test only

4. Vertical yield of the soil directly below the footing because of horizontal transfer of stress, with consequent horizontal strain—elastic and plastic—possibly not equally in all directions.

5. Vertical yield of the soil because of change in character of the soil, such as change of water content under pressure, change in volume at contact planes of different materials, change in chemical constituents of soil or rock.

It should be kept in mind that items 3, 4, and 5 may be caused not only by the direct loading but also by loadings on adjacent footings. Also, the loading at the bottom of a footing is not uniform over its entire area.

The action of loads on soil may be conceived as a series of interrelated phases, chiefly:

1. The load compresses the soil directly below it.

2. The compressed soil bulges laterally.

3. The bulge compresses the side soil areas.

4. The side soil areas resist and distort, sometimes showing definite surface bulges.

6.16.3 Amounts of Settlement

From data that have been accumulated, it can be concluded that:

1. In soft rock, hardpan, and dense glacial deposits, settlements of from ¼ to ⅜ in must be expected and will occur almost immediately under loads up to 10 $tons/ft^2$.

2. In dry sands, settlements of ½ to ¾ in must be expected under loads from 3 to 4 $tons/ft^2$ and will occur almost immediately.

3. In wet sands, settlement of ¼ to ½ in will occur immediately under loads from 2 to 3 $tons/ft^2$ and will increase to ¾ in in a period of months, if the sand is drained.

4. In stiff clays mixed with gravel or sand, settlements under loads of 3 to 4 $tons/ft^2$ will be ½ in immediately and may increase to about 2 in in about 2 years.

5. In soft clays, settlements under even small loads may be substantial and will continue with time. Settlements of 6 to 9 in are not unusual.

These settlements are for small footings; large footings will settle more. The settlement of large areas in granular soils will not be uniform, chiefly because the base pressures are not uniform. Furthermore, edge shear on a footing provides some resistance to settlement. The settlement of large areas in plastic soils will not be uniform, because the base pressures are not uniform and the effect of plastic flow is much less in the interior regions.

6.16.4 Footing Loadings

The load to be used for footing design must be at least the dead load, including the weight of all fixed partitions, furniture, finishes, etc. However, it is doubtful whether the live load has any effect on settlements, except in such structures as grandstands, where the live-load increment may be over 150 psf for a few minutes sporadically.

Where building codes are enforced, the loads used in foundation design usually are fixed by a code. Two requirements are common:

1. Footings should be designed to sustain full dead load and a reduced summation of live loads on all floors, the percentage of reduction increasing with the number of stories, but seldom exceeding 50%. Pressures under the footings are limited by listed maximum unit pressures for various classes of soils.

2. Footings should be designed as in 1, but the areas of the footings should be proportioned for equal soil pressure under dead load only.

In method 2, unit base pressures under dead plus live load are greater for interior footings than for wall footings. Since wall footings are usually rectangular—and if at street lines, on soil less affected by adjacent loads—the opposite condition would be more conducive to equal settlements. In method 1, since the computed live loads seldom, if ever, exist, the actual base pressures on the interior footings are less than those on the exterior footings.

6.16.5 Effect of Pressure Distribution on Footing Design

The rigid solution for the stresses in a footing for a pressure distribution on the base that is a function of the settlement is many times indeterminate. As compared with a uniform distribution, the effect of the variable distribution on the computed bending moments in a single loaded footing (whether square, round, or strip shape) may be large, but the effects on computed shear, as well as unit shear and bond stresses in the concrete, may be disregarded from the points of both safety and economy. Where bending moment controls the design, as well as in combination footings of either the mat, cantilever, or combined types, the departure from uniform distribution may affect both safety and economy.

Why, then, have there not been many structural failures of footings constructed from designs based on the assumption of uniform distribution? The usual explanations point to the surplus strength in the materials (factors of safety) as well as the nonexisting live loads included in the design. However, since unequal settlements probably occur, even though the total base pressures may be less than computed, another explanation is necessary. It is found in the theory of "limit design": in a closed system of forces and restraints, local failure cannot occur, because the continuity of the mass causes a transfer of load to the spots which can best take the load.

6.16.6 Helpful Hints for Footing Design

Several practical considerations may be used as a guide in design and construction:

1. The more rigid the footing, the more uniform the pressure on the base.

2. Nonuniform distribution may be disregarded in footings resting on soils having a bearing value over 3 tons/ft^2.

3. On similar soils, with the same loads per unit area, square footings settle more than rectangular footings.

4. On the same soils, for the same shaped footings under the same average load intensity, the larger the footing, the greater the settlement.

5. Footings on the exterior of a building, not adjacent to other loads, will settle less than those affected by adjacent loads.

6. Footings will probably settle again when adjacent soil areas are loaded.

7. Plate loading tests are reliable only when site conditions are duplicated exactly.

8. Flexible footings may be designed for smaller bending moments than are given by the assumption of uniform distribution.

9. Footing failures are seldom structural failures.

10. Uniform settlements of buildings can be obtained by taking into consideration: actual loadings to be expected; greater soil bearing value of exterior locations; provision for slip joints for settlement due to adjacent future loadings; effect of difference in levels of footings in close proximity; changes in portions of the soil area due to water, weathering, and vibration; and differences in bearing value of differently shaped footings.

11. Maximum soil pressure under centers of footings is no criterion of expected settlements, because of the distribution of load due to rigidity.

12. Bearing pressures of soil and expected settlements deduced from load tests in the field or laboratory tests on undisturbed samples are of little value if the conditions in the construction work are not considered.

13. Exposure of soils during excavation may modify the soil structure and change all the characteristics.

14. All structures will settle, and provision must be made for the expected settlement by keeping them free of adjacent structures that have moved into a stable position.

15. Lateral resistance in soils must not be disturbed or reduced below that necessary to balance the lateral stresses resulting from the added loadings.

(J. E. Bowles, "Foundation Analysis and Design," 4th ed., McGraw-Hill Publishing Company, New York; M. J. Tomlinson, "Foundation Design and Construction," 5th ed., John Wiley & Sons, Inc., New York.)

6.17 COMBINED FOUNDATIONS AND MATS

In many cases, it is not possible or economical to provide a footing centrally located under each column and wall load. There are two corrective procedures normally used for such cases: combination of loads for one footing or imposition of balancing loads to neutralize the eccentric relative positions of load and footing. Two or more loads may be supported on a single footing, and in some locations an entire building may be constructed on a mat, or raft. Balancing of loads can be accomplished by means of structural beams acting as levers (often called "pan handles" or "pump handles") between column loads, or by imposition of counterbalancing moment resistance in the form of rigid frames.

Mats are continuous shallow foundations that support multiple columns and walls, often the entire building. Mats are beneficial where soft or highly variable surface soils would lead to excessive differential settlements with isolated footings.

They are used occasionally where loads are heavy and strongly variable so as to lead, for that reason, to excessive differential deflections. (See "Mat Footings" in the following.)

Design of Combined Footings. For isolated footings the form of soil-resistance distribution may not be a determining factor in the economy of structural design, since the bending moment of the pressures about the centrally located column usually does not control the design of either the concrete or the reinforcing. The opposite condition holds, however, when the superimposed loads are not single and centrally located. For unbalanced loads, the usual design procedure is to assume uniform distribution, with sufficient soil resistance to counterbalance eccentricity of loading. The better approach is to use numerical computer solutions that model the variable loading and deflection.

Once a design of this type is prepared, analysis of the deflections resulting from a uniform distribution can be used to revise the shape of the soil-reaction curve. The resulting moments will usually be found less than those computed for the uniform assumption.

For this computation, the subgrade reaction modulus, or relationship between gross deformation of the soil and bearing value, must be known. This can be taken from the results of laboratory or field tests, but the effect of edge shears should be considered.

When more than one load is supported on a footing, the center of gravity of the loads should coincide with the geometric center of the footing. In addition, axes of symmetry through the column loadings should coincide with axes dividing the footing into equal areas. Since column loads are seldom of constant value, and since the variation in uncertain loadings (live load, impact, wind, etc.) is different even for adjacent columns, these requirements can be met only approximately.

If the soil is very weak and highly sensitive to load changes, combined footings should be avoided, if possible, since irreversible tipping may cause structural damage. A study must be made of different possible load combinations, and the shape of the footing should be adjusted so that there will be pressure over the entire base area at all times.

Mat Footings. Design of mat foundations, which almost always are thick, reinforced-concrete slabs, requires consideration of soil-structure interaction. Mat stresses are directly related to mat deflections, which are affected by soil stiffness and stresses. The preferred design technique thus utilizes an iterative computation involving input from both the geotechnical and structural engineer. The geotechnical engineer estimates the bulk modulus of compressibility or subgrade reaction for the average stress or the anticipated range in stress imposed by the loaded mat and computes stresses and strains in the soil. The computation takes stiffness of the mat into account, inasmuch as this will influence the stress distribution and magnitude of settlements. The results of this computation are used by the structural engineer in the first design of the mat, and the computed stresses and strains are input to the soil reaction model. The procedure is repeated until convergence of results is achieved.

For many structures, especially high-rise buildings, utility services have to pass through the concrete mat. Therefore, it is necessary to estimate closely the total settlements and their time of occurrence relative to the connection of utility services at the mat exterior. For very tall buildings, it is advisable to make such connections only after the building has been completed.

Strapped Footings. Used to support two columns, strapped footings consist of a pad under each column with a strap between them. Where soil resistance is assumed only under the footings, the strap connections (levers) are more economical than combined footings. The usual assumptions are that base pressure is uniform under each pad, or bearing area, and that a rigid beam connects the two column loads, so that the center of gravity of the bearing areas coincides with the center of gravity of the column loads.

See also Arts. 6.15 and 6.16.

6.18 PILE FOUNDATIONS

A pile is defined as a structural unit introduced into the ground to trasnsmit loads to deep soil strata or to alter the physical properties of the ground. Usually driven into the ground in groups, piles are of such shape, size, and length that the supporting material cannot be manually inspected. However, where manual inspection of the bottom is possible, as in shallow pits or sufficiently wide, deep caissons, these types of construction are not classified as piles.

Piles support loads by end bearing on rock or other stiff material or by side shear (skin friction), or both. The strength of a pile, however, is only one consideration in design of a pile foundation. Settlement of the foundation also should be evaluated. (See Arts. 6.14.2 and 6.15.)

6.18.1 Pile-Soil Interaction

Piles transfer load to lower strata by frictional resistance along the surface in contact with the soil and by direct bearing at or near the bottom of the pile. The portion of the load carried by friction and by end bearing is known between limits.

The maximum frictional resistance between soil and pile surface is a definite limit to the amount of load that can be transferred by friction. The confined character of the soil considerably increases its bearing value and a compressed volume of soil at the pile tip acts as an integral part of the base, of unknown total area.

Soil Limitations on Piles. Structural piles can be relied upon within the limitations normally set on the materials of which they are made. Maximum pile loading is more dependent on the nature of the soils into which the pile is driven than on the materials composing the pile.

Structures on piles subject to lateral movements should be braced by inclined, or batter, piles or designed to take the load in bending in the piles.

Piles installed in soils subject to future subsidence and consolidation must be investigated for possible future loads when the consolidation occurs and the weight of these upper layers hangs on the piles. Bear in mind that the maximum negative skin friction is equal to the surface frictional resistance and consider this load as part of the total design load.

6.18.2 Development of Pile Load-Carrying Capacity

Piles require a definite embedment in resistant soil. It is wise to require at least 10 ft of embedment in natural soil of sufficient strength to develop theoretically the design load-carrying capacity. The embedment should be below grade at such elevation as will not be affected by future excavations.

Pile Caps. Usually, structural members, such as columns and bearing walls, have to be supported on a group of closely spaced piles. In such cases, the load is transmitted from the structural members to the piles through a thick concrete footing, or cap, in which the pile tops are embedded.

Pile caps are designed as structural members with no allowance for any load support on the soil directly below. It is usual to assume, however, that passive pressure on the vertical faces acts to resist lateral forces.

Pile Bracing. Each pile should be braced against lateral displacement by rigid connection to at least two other piles in radial directions not less than 60° apart. Where fewer than three piles are in a group, or where extra lateral stability is made necessary by the nature of loads or softness of soil layers, concrete ties or braces should be provided between caps. Such bracing should be considered as struts, and all reinforcing bars should be anchored in the caps to develop full tension value. A continuous reinforced-concrete mat 6 in or more in thickness, supported by and anchored to the pile caps, or in which piles are embedded at least 3 in, may be used as the bracing tie, if the slab does not depend on the soil for the direct support of its own weight and loads.

Pile Spacing. Inasmuch as the purpose of piles is to transfer loads into soil layers, pile spacing should be a function of the load to be carried and the capacity of the soil to take the load.

Piles bearing on rock or penetrating into rock should have a minimum spacing, center to center, of twice the average diameter, or, for a square pile, 1.75 times the diagonal of the pile, but not less than 24 in. This spacing will not overload the rock, even if the bearing piles are fully loaded and no load relief occurs from skin friction.

Piles that transfer a major part of the load to the soil by skin friction should have a minimum spacing center to center of twice the average diameter, or 1.75 times the diagonal of the pile, but not less than 30 in. For a pile group consisting of more than four piles, such minimum spacing should be increased by 10% for each interior pile, to a maximum of 40%.

Where lot-line conditions, space restrictions, or obstructions prevent proper spacing of piles, the carrying capacity should be reduced for those piles that are too close to each other or less than half the required spacing from a lot line. The percentage reduction of load-carrying capacity of each pile should be taken as one-half of the percentage reduction in required spacing. If, for example, a pile is 24 in from a lot line instead of 30 in, the percentage reduction of spacing is 20%; so pile capacity is reduced 20%/2, or 10%. Such reduction in values is sometimes economical where closer spacing permits the placing of a pile group below the columns and eliminates eccentricity straps or combined footings.

Minimum Sizes of Piles. Usual restrictions on pile dimensions are as follows: The pile point should be at least 6 in in diameter. Piles of uniform section should have a minimum diameter of 8 in or, if not circular, a minimum thickness of 7½ in. For tapered piles, the butt at cutoff should be at least 8 in in diameter.

Actually, if piles can satisfy stress restrictions as short columns and withstand stresses during driving, there is no justification for dimensional limitations. Generally piles of minimum size are suitable only for end-bearing piles, because they have too little surface to develop sufficient skin friction.

6.18.3 Timber Piles

Kept continuously wet, timber piles may be considered permanent. But a narrow band of alternately wet and dry condition, sometimes caused by seasonal water-level fluctuations, will soon be apparent in pile failures due to rot. For this reason, the pile should be pressure-treated to a final retention of not less than 12 lb of preservative per cubic foot of wood. The tops of all piles as cut off should be below ground level and the cutoff section should be treated with three coats of preservative. Piles above ground level require some replenishment of preservative with age, as well as protection against fire damage.

Almost every kind of timber can be used for a pile. Some species take the shock of driving better than others, and the species have different static compression values. Prime-grade sticks cannot be expected to be sold for piles; yet it is reasonable to insist on timber cut above the ground swell and free from decay, from unsound or grouped knots, from wind shakes, and from short reversed bends. The maximum diameter of any sound knot should be one-third the diameter of the pile section where the knot occurs, but not more than 4 in in the lower half of the pile length or more than 5 in elsewhere. All knots should be trimmed flush with the body of the pile, and ends should be squared with the axis.

Piles should have reasonably uniform taper throughout the length. They should be so straight that a line joining the centers of point and butt should not depart from the body of the pile. All dimensions should be measured inside the bark, which may be left on the pile if untreated but must be removed before treatment with any preservative. The diameter at any section is the average of the maximum and minimum dimensions; piles not exactly circular in section should not be rejected.

No timber should have a point less than 6 in in diameter, except for temporary use. Where the point is reinforced with a steel shoe, the minimum dimension is at the upper end of such shoe. No untreated timber piles should be used unless the cutoff or top level of the pile is below permanent water-table level. This level must not be assumed higher than the invert level of any sewer, drain, or subsurface structure, existing or planned, in the immediate vicinity, or higher than the water level at the site resulting from the lowest drawdown of wells or sumps.

Of the various timber species, cedar, western hemlock, Norway pine, spruce, or similar kinds are restricted to an average unit compression at any section of 600 psi; cypress, Douglas fir, hickory, oak, southern pine, or similar kinds are allowed a value of 800 psi. A maximum load of 20 tons is commonly allowed on timber piles with 6-in points, and 25 tons where the point is 8 in or more, subject, of course, to compliance with indicated or tested driving resistance and maximum stresses.

Timber piles are seldom used as end bearing piles, but there is no real objection to such use, if the piles are properly installed.

6.18.4 Steel H Piles

Where heavy resistances exist in layers that overlie the bearing depths, steel H piles will often take the punishment of the heavy driving much better than any other type. After insertion in the ground, the soil gripped between the web and inside faces of the flanges becomes an integral part of the pile, so that frictional resistance is measured along the surface of the enclosing rectangle and not along the metal surface of the section.

To prevent local crippling during driving, the section should have flange projections not exceeding 14 times the minimum thickness of metal in either web or flange and with total flange width not less than 85% of the depth of the section. All metal thicknesses should be at least 0.40 in.

Other structural-steel sections, or combinations of sections, having flange widths and depths of at least 10 in and metal thickness of at least ½ in may also be used—usually in combination with steel sheeting. Most rolling mills issue special sections for use as piles.

The load-carrying value of a steel pile is usually restricted to a maximum unit stress of 12,600 psi at any cross section but not more than 35% of the yield stress.

6.18.5 Concrete Piles

Precast-concrete piles are most suitable for foundations in which the pile length can be forecast accurately. Otherwise, the added costs of cutoffs and splices tend to make precast piles more expensive than the various cast-in-place types.

Drivability of precast piles is a serious design concern. Difficult driving should be avoided, by predrilling or punching, if necessary. Also, fatigue failure due to excessively hard driving should be avoided.

Precast piles should be cured to full strength before use. They should be reinforced to withstand local stresses during lifting and driving. Instead of ordinary steel reinforcement, however, prestressing may be preferred for precast piles, because it is more economical, inasmuch as the cost of ordinary reinforcing is at least half the total cost of a precast pile before driving.

Cast-in-place concrete piles come in many variations, in which concrete is placed into premolded cavities in the soil. The cavity for a pile can be formed by removal of the soil or by forcing the volume displaced into the adjacent soil, as is done by timber or precast-concrete piles. Soil removal can be done by auger or casing carried to a desired depth. Volume displacement is accomplished by driving a thin, metal, closed-end shell stiffened with a retractable mandrel or a heavier cylindrical metal shell, which is retracted as the concrete filling is installed.

Retracted shells are usually heavy steel pipe, sometimes with reinforced cutting edges and with driving rings welded at the top to provide a grip for retraction.

In some types, the concrete is compressed by a vibrating or drop hammer and tends to squeeze out laterally below the casing to form annular expansions. These greatly increase the bearing value of the pile.

One advantage of cast-in-place pipe or thin-shell piles is the inspector's ability to examine the cavity before placement of the concrete, to check depth of penetration and variation from plumb or desired batter position.

For all cast-in-place piles, care must be exercised to clean the cavity of all foreign matter and to fill the entire volume with concrete. If water enters the shell, tremie tubes may be necessary for casting the concrete.

Structural concrete piles are designed as short concrete columns. For recommended allowable design stresses and loads, see "Recommendations for Design, Manufacture, and Installation of Concrete Piles," ACI 543R-74, American Contrete Institute.

Concrete Displacement Piles. The pressure-injected extrusion pile requires a steel shell set into a pit, the bottom filled with gravel or dry concrete to a depth of two or three pile diameters. The fill is pushed into the soil by dropping a heavy ram inside the shell. The concrete packs and grips the shell sufficiently to pull it along through almost any type of soil. When the shell reaches the desired depth, rapid impact with the ram breaks up the plug and extrudes it into the soil in the shape of an inverted mushroom (Fig. 6.13). Concrete is added before the plug material entirely clears the shell. As the additional concrete is extruded, the shell is withdrawn.

The result is a very rough-surfaced concrete cylinder with a number of annular fins projecting into the soil, with a mushroom-shaped expansion at the bottom. Under load tests, these piles have carried much more load than is indicated by the area of the mushroom multiplied by the normally presumptive bearing capacity of the soil. These piles are suitable for granular soils and develop load capacities of 100 to 150 tons.

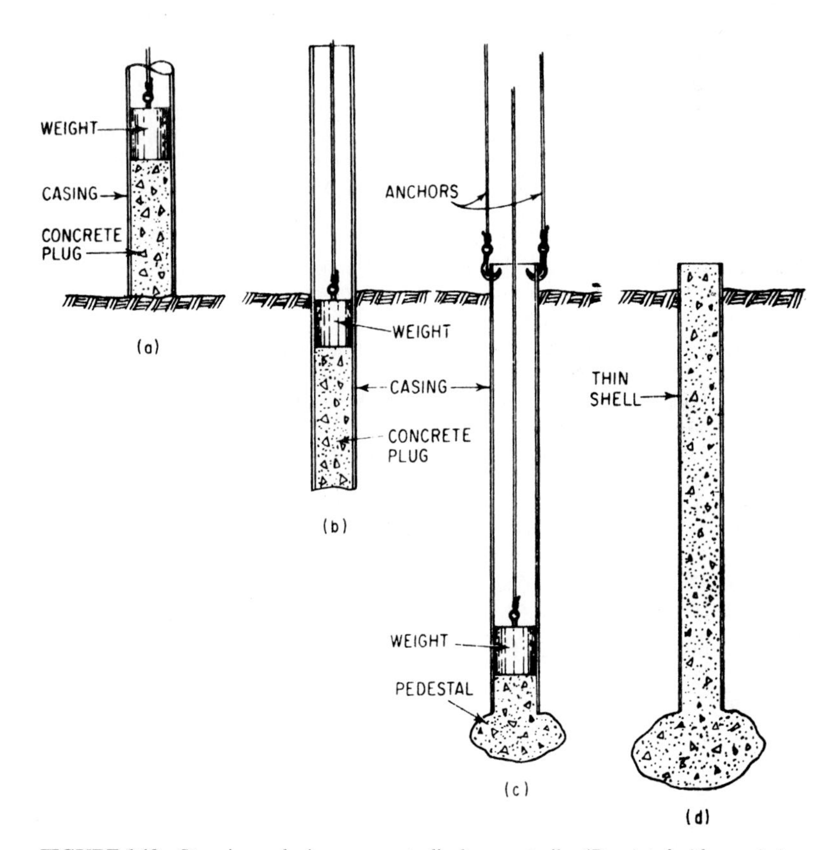

FIGURE 6.13 Steps in producing a concrete displacement pile. *(Reprinted with permission from F. S. Merritt, "Standard Handbook for Civil Engineers," 3d ed., McGraw-Hill Publishing Company, New York.)*

Drilled Injection Piles. These are small-diameter concrete shafts. This type of pile is constructed by first drilling into the soil with an auger, then following with simultaneous extraction of the auger and filling of the cavity with a cement grout that is injected through the hollow shaft of the auger. Piles of this type have been used successfully as friction piles. The method also has been used for building continuous bulkhead cutoffs, similar to tight sheeting.

6.18.6 Reused Piles

Reuse of piles, whether where structures have been demolished or for new structures, should be restricted to places where complete data are available for length and driving conditions of each pile. Where a group of piles is to be reused for new loadings, their supporting capacity should be restricted to about 75% of rated capacity, unless they are first load-tested. Additional piles for the same structure should be of similar type and about the same length as those being reused.

6.18.7 Pile Splices

Piles are often made up of special sections of structural elements. Adjoining sections, if desired, may be composed of different materials. The load capacity of such composite piles is usually set equal to that of the weakest section. Splices should be designed to hold the sections together during driving as well as to resist the later effects from lateral loads or the driving of adjacent piles.

Timber-pile sections are usually spliced with sleeves—short, tight-fitting steel pipes into which the ends of the piles are driven. Sometimes drift pins or through bolts are used to hold the assembly together. Steel rolled-section or pipe piles are often spliced with a full-strength weld. A precast-concrete pile may be extended with prefabricated steel splices or by placing longitudinal reinforcing steel in holes drilled into the concrete and then epoxy grouting the reinforcing steel and mated concrete surfaces.

(J. E. Bowles, "Foundation Analysis and Design," 4th ed., McGraw-Hill Publishing Company, New York; M. J. Tomlinson, "Foundation Design and Construction," 5th ed., and R. G. Ahlvin and V. A. Smoots, "Construction Guide for Soils and Foundations," 2d ed., John Wiley & Sons, Inc., New York; H. Y. Fang, "Foundation Engineering Handbook," Van Nostrand Reinhold, New York.)

6.19 LOAD CAPACITY OF PILES

All types of piles must be able to support safely the maximum combination of the following loads:

1. All dead loads and the weight of the pile cap and superimposed loads.

2. All live loads, reduced as permitted by building code because of floor area or number of stories in the building.

3. Lateral force and moment reactions, including the effect of eccentricity, if any, between the column load and the center of gravity of the pile group.

4. That amount of the vertical, lateral, and moment reactions resulting from wind and seismic loads in excess of one-third of the respective reactions totaled from 1, 2, and 3.

The maximum load permitted on a vertical pile is the allowable pile load, as determined by formula, test, or computation, applied concentrically in the direction of its axis. (See Art. 6.14.2.)

Lateral-load capacity, deflection, and pile bending stresses should be determined by rational analysis. (See H. Y. Fang, "Foundation Engineering Handbook," 2d ed., Van Nostrand Reinhold, New York.) Lateral loads that cause allowable deflections are usually on the order of 3 to 5 tons per pile. Very stiff soils and stiff piles, however, permit substantially greater loads. Where piles are subjected to both axial and lateral loads, the compressive and bending stresses are additive and should be checked to see that the allowable pile stress is not exceeded.

If batter piles are used to take lateral forces, the resultant loads along the axis of the pile should not exceed the allowable pile value. Also, allowable unit stresses in any pile section must not be exceeded.

The total pile load is considered as existing in all end-bearing piles without relief. For a friction pile, the entire load is assumed to exist at the section located at two-thirds the embedded length of the pile, measured from the top, unless the rate of load transfer is estimated or determined otherwise.

6.19.1 Pile Load Tests

The load capacity of piles is often determined directly by full-scale load tests, because most building codes require verification by load tests even where capacity has been estimated by static or dynamic techniques.

When the allowable pile load is to be determined by load tests, subsurface investigations must first prove that the conditions in the test area are fairly uniform. The test piles should be driven close to points where borings were made, to establish a correlation with borehole observations. A complete pile-driving log record should be kept. Driving should be carried to resistances indicated by dynamic formulas for the desired load capacity or to tip elevations indicated by static analyses.

There are four basic types of pile load tests. The test most often used, and required by most building codes, is the "Standard Test." In this method, the pile is loaded in eight increments, usually to twice the design load. Each increment is held until the settlement rate is 0.01 in/h or for 2 h, whichever is less. The final load is held for 24 to 48 h.

The quick maintained-load method uses a constant time interval per load increment and thus minimizes the time-dependent phenomena, which may produce misleading results in the Standard Test. In the quick test, load increments are held for a constant interval of 5 to 15 min and about 25 load increments are applied, which usually results in a better definition of the load-deflection curve. One quick test method is given in ASTM D1143.

Another quick method is the constant-rate-of-penetration test, in which the pile is forced to settle at a constant rate. Tests indicate that this test may provide superior definition of the load-deflection curve.

Occasionally, test piles are subjected to cyclic loading by introduction of unloading-reloading cycles. Tests indicate, however, that this method sheds little light on tip movement (supposedly the principal reason for its use), and, in general, is inferior to the quick tests.

The interpretation of pile load tests to determine failure and design loads is a matter deserving close attention. Building codes often stipulate the criteria by which the failure load should be determined but usually also provide alternative methods or permit some flexibility. Some codes, for example, set the maximum allowable load as one-half that load causing a net settlement of 0.005 in/ton, a gross settlement of 1 in, or a disproportionate increase in settlement.

Failure-load determination techniques that incorporate time-dependent phenomena or that may be altered by the simple changing of scales on a graphic plot may yield results that are misleading. The preferred methods for determining failure load employ graphical techniques. For the purpose, in one method, the failure load is defined as the load corresponding to the butt movement that exceeds the elastic deformation by 0.15 in plus a factor equal to the pile diameter, in, divided by 120. This technique is intended for application to results from quick tests.

In another method, the failure load is obtained from the load-settlement curve. In this case, the failure load is defined as the load corresponding to the intersection of the elastic line of the pile and a line with a slope of 0.05 in/ton drawn tangent to the load settlement curve.

Tip movement may be assumed to occur under the load corresponding to the point where the load-settlement curve becomes tangent to the elastic line of the pile.

In the subsequent driving of piles in the area in which piles were tested, all piles of the same type, driven with the same equipment and showing the same or greater resistance to penetration of the soil as was shown by the piles tested, should be allowed the average capacity obtained from the load tests. Any piles that stop at higher levels and are one-third or more shorter than the average length of the test piles should be reevaluated.

6.19.2 Bearing Piles

Piles driven to end bearing in rock or in extremely tough and dense layers directly overlying bedrock are called bearing piles. They may be of any type of material but are usually steel or precast concrete, to withstand the heavy driving. (If timber piles are used, extreme care should be exercised that the piles are not overdriven.) Steel piles may be either rolled H sections or concrete-filled pipes.

The load capacities of bearing piles may be estimated from driving formulas or static equations, or determined by test loads. If a pile is driven to essential refusal, the allowable load is governed by the structural capacity of the pile. Typical load capacities of rolled section, heavy pipes, and precast concrete range from 100 to 200 tons. (See also Art. 6.14.2.)

Special pile forms may be used to obtain a very high allowable load. One frequently used is known by its proprietary name of **drilled-in caisson.** It is formed with a heavy steel pipe with a reinforced cutting edge, which is driven into bedrock and then cleaned out. A socket is then drilled or chopped into the rock at the bottom of the pipe. Next, the pipe is filled with concrete. For higher load capacity, the pile may be reinforced as a concrete column or with a structural-steel shape that extends into the socket (Fig. 6.14).

The load-carrying capacity is the structural capacity of the pile as a column, but may not exceed the load transfer from the socket to the rock. This load is the sum of the allowable bearing value on the bottom of the socket plus the bond of the concrete to the rock along the surface area of contact in the depth of the socket. Not economical for small loads, such piles have been used to sustain loads of 3000 tons, requiring reinforcing with the heaviest rolled-steel column sections with plates welded to them.

The high cost of the steel pipe shell warrants retraction during concreting. Often, when the shell is to be retracted, a thin shell liner is inserted to act as a form for the concrete.

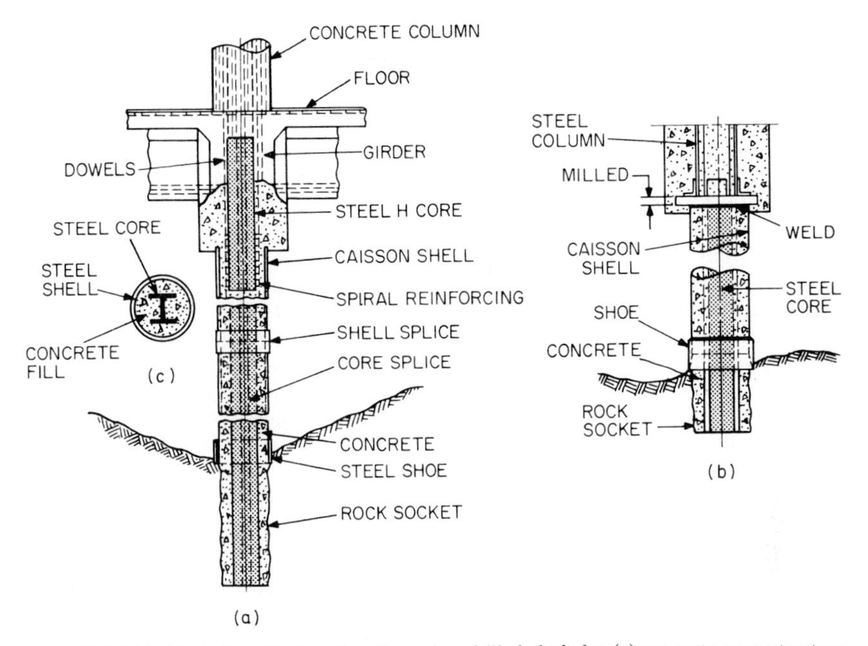

FIGURE 6.14 Typical vertical section through a drilled shaft for (*a*) concrete superstructure and (*b*) steel superstructure. (*c*) Typical horizontal section. *(Reprinted with permission from F. S. Merritt, "Standard Handbook for Civil Engineers," 3d ed., McGraw-Hill Publishing Company, New York.)*

6.19.3 Pile Deterioration

Any analysis of load transfer and determination of pile load-carrying value is based on the premise that the pile structure will not change. That danger comes from many directions; the pile may rust away or rot, and exterior conditions, such as soil acidity, fluctuation in water levels, and loss of lateral support in the soil encasing the piles must be carefully considered and precautionary measures taken.

The rusting of steel surfaces in the ground (the average corrosion loss is about 0.003 in/yr) cannot harm the load value of a pile, unless the mill scale becomes loose and surface friction is seriously affected. The life of steel H and sheet piling is far beyond the normal life of a building.

See also Arts. 6.17 and 6.19.

6.20 INSTALLATION OF PILES

Equipment and methods for installing piles should be such as to produce satisfactory piles in proper position and alignment. This involves the proper choice of equipment for the size and weight of pile, expected load value, and type of soil to be penetrated.

Hammers. Piles usually are driven into place with hammers powered by gravity (very slow), compressed air, steam, or diesel fuel. Generally, heavier piles require heavier hammers. Use of wave equation analyses facilitates proper matching of hammer to pile. Piles also may be driven with vibratory hammers.

Experience with wood piles indicates that a lighter hammer than required by some building codes should be used to reduce breakage of piles, certainly in areas where soil layers change suddenly in resistance to penetration. Although the usual argument that a big spike should not be driven with a tack hammer is true, so is the opposite statement that a long slender finishing nail should not be driven with a sledge.

A pile-driving rig may comprise a rig specially built for pile driving or a specially equipped crane. A specialized rig may consist of a framework and platform for supporting the equipment needed in pile driving, which may include a hammer, boilers, winches, and engine. If a crane is used, it should be equipped with a boom for handling pile leads and the hammer, and with a drum or winch head with a line to the head of the boom for positioning the piles. For underwater driving, an enclosed steam hammer may be used, with a steam-exhaust hose extending above the water surface. Water may be kept out of the casing by maintenance of air pressure at about 0.5 psi in the lower part of the hammer housing. As an alternative, the hammer may be kept above water by insertion of a follower, or pile extension, at the top of the pile.

Jetting Piles into Place. Other methods of pile insertion include water-jet loosening of the soil. This allows the pile to drop by gravity or under the weight of a hammer resting on the butt. Jetting, however, is not necessary or advisable in soils other than fairly coarse, dense sands. Even there, it must be remembered that jetting is excavation of soil; it has been so decided by the New York State Supreme Court, and all legal restrictions concerning excavations along a property line are available protections for the adjacent owner.

Jetting should be stopped at least 3 ft above the final pile-tip elevation, and the piles should then be driven at least 3 ft to the required resistance. The condition of adjacent completed piles should be checked, and if loss of resistance is found, all piles should be redriven.

Jacking. Piles also may be installed with jacks reacting against static loads (in the form of an existing structural mass or a temporary setup of weights). This method is usually limited to addition of piles during underpinning operations or within existing buildings with inadequate headroom for a pile-driving rig.

Tolerance for Piles. In design of a cap, each pile is considered a point load concentrated at the center of the cutoff cross section. However, only in ideal soil conditions and with perfect equipment will the piles all be inserted vertically and exactly in the planned position so that this assumption is true. Some practical rules of acceptance of piles not perfectly positioned must control every pile job:

A tolerance of 3 in in any direction from the designated location should be permitted without reduction in load capacity—provided that no pile in the group will then carry more than 110% of its normal permitted value, even though the average loadings of the piles are much less. The distribution of column loads into the individual piles should be determined either analytically or graphically from a plotted graph of the surveyed locations of the cutoff centers.

If any pile is overloaded, either additional balancing piles should be added to the group, or structural levers should be added for load redistribution or connection with adjacent groups.

Computation of actual pile loads can be simplified by plotting the pile and column locations to scale, then locating the center of gravity of the piles graphically and using the line between this point and the center of the column loads as an axis of symmetry. Multiplying pile values by the offsets of all piles from the normal axis drawn through the center of gravity of the piles gives a total moment equal to the eccentricity moment. Assume a linear variation for reduction or increase of pile value with distance from the axis. If the maximum pile value is just beyond the 110% limit, a second approximation should be tried, relocating the center of gravity of the piles on a weighted basis, using the values from the first solution. This will slightly reduce the eccentricity and may be sufficient to permit approval of the pile group.

If piles are installed out of plumb, a tolerance of 2% of the pile length, measured by plumb bob inside a casing pile or by extending the exposed section of solid piles, is permissible. A greater variation requires special provision to balance the resultant horizontal and vertical forces, using ties to adjacent column caps where necessary. For hollow-pipe piles, which are driven out of plumb or where obstructions have caused deviations from the intended line, the exact shape of the pile can be determined by the use of a magnetic shape-measuring instrument, or an electronic plumb bob. After the deflection curve of a pile has been plotted, the reduction in allowable load, if any, can be determined.

See also Arts. 6.18 and 6.19

6.21 SLURRY TRENCH WALLS

Use of a bentonite-slurry-filled trench dug to or into rock, and later displacement of the slurry by tremie concrete, is often an economical method for foundation-wall construction (Fig. 6.15). Walls, usually a minimum of 24 in thick, can be constructed by this method in any soil condition. Reinforcing cages may be inserted into the slurry before concreting. Good bond between the steel and the concrete has resulted.

Reuse of the slurry, a specially homogenized machine mix, necessitates dividing the length of the wall into sections. On some projects, leakage problems have caused trouble. In the slurry-wall constructions for the Budapest subway, the entire inside face was covered with ⅜-in steel plates welded to form a continuous sheet to prevent leakage. On other jobs, grouting of the joints was found necessary.

Application of the slurry method to enclosure of large, deep cellars, such as those of the World Trade Center in New York, has shown that the method can compete economically with normal cofferdam construction. For such large areas, interior bracing frames are too expensive. Anchors drilled into the exterior rock are usually provided to replace the temporary bracing.

(J. E. Bowles, "Foundation Analysis and Design," 4th ed., McGraw-Hill Publishing Company, New York.)

6.22 DRILLED SHAFTS

A drilled shaft is a large-diameter, cast-in-place concrete foundation with the shape of a round column. Development of this type of foundation, which is also known as a caisson or drilled pier, has progressed rapidly since about 1960 and it now finds wide acceptance.

Several types of bits and drilling rigs are available. The drilling rigs are heavy equipment utilizing square Kelly bars, often telescoping, and capable of high

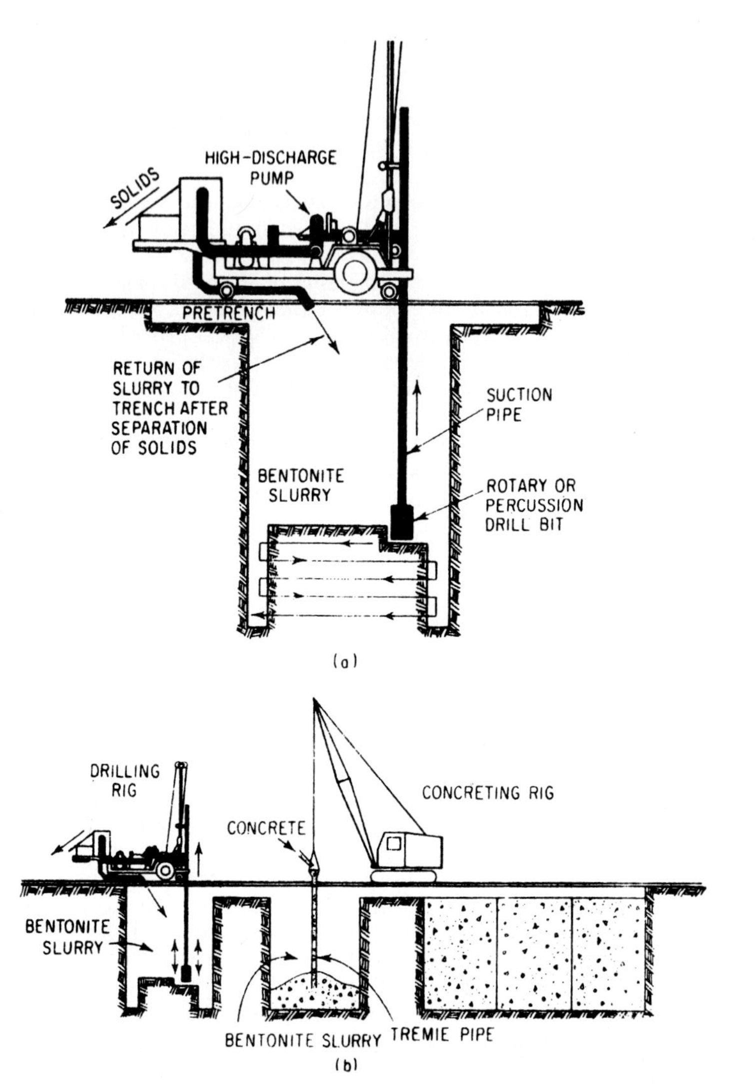

FIGURE 6.15 Slurry-trench method for constructing a continuous concrete wall: (*a*) excavating one section; (*b*) concreting one section while another is excavated. *(Reprinted with permission from F. S. Merritt, "Standard Handbook for Civil Engineers," 3d ed., McGraw-Hill Publishing Company, New York.)*

torques and considerable downpressure. The most commonly used bits are the flight auger, bucket auger, and underreaming bit. The flight and bucket augers advance the shaft, remove loosened material to the surface, and produce a cylindrical shaft of constant diameter. The underream bit is used to enlarge the base by undercutting the shaft and forming what is known as the *bell.* The flight augers, equipped with specially hardened teeth, are capable of cutting many of the sedimentary rocks as well as weathered igneous and metamorphic rocks.

Shaft diameters range from 2 to about 10 ft, with bell diameters ranging up to about 15 ft. The heavy equipment now available permits shaft construction to depths approaching 150 ft.

Drilled shafts are very efficient foundations where high-strength bearing materials can be reached. Satisfactory performance, however, requires not only an adequate bearing material but also that the hole be cleaned before concreting and that there be no mud inclusions or voids in the concrete. Because drilling in unstable ground can encounter severe difficulties, design of a caisson foundation should always be preceded by a detailed evaluation of constructability.

See also Art. 6.14.2.

In foundation construction, three types of earth pressure are commonly encountered:

Active earth pressure. This exerts horizontal and vertical load components against any structure that impedes the tendency of the earth to fall, slide, or creep into its natural state of equilibrium but yields slightly under the pressure.

Passive earth pressure. This pressure is mobilized when a structure tends to compress the earth.

At-rest pressure. This is a horizontal pressure exerted by soil against a nonyielding vertical surface.

Each of these pressures is dependent on many of the physical properties of the soil, as well as the relative rigidity of the soil and structure. The most important properties of the soil appear to be density, angle of internal friction in sands, and cohesion and overconsolidation ratio in clays. Some typical values for various soil types are given in Table 6.4.

TABLE 6.4 Typical Soil Densities and Internal Resistance

Soil type	Average weight, w, lb per ft^3	Coefficient of internal resistance, tan ϕ
Soft flowing mud	105–120	0.18
Wet fine sand	110–120	0.27–0.58
Dry sand	90–110	0.47–0.70
Gravel	120–135	0.58–0.84
Compact loam	90–110	0.58–1.00
Loose loam	75–90	0.27–0.58
Clay	95–120	0.18–1.00
Cinders	35–45	0.47–1.00
Coke	40–50	0.58–1.00
Anthracite coal	52	0.58
Bituminous coal	50	0.70
Ashes	40	0.84
Wheat	50	0.53

6.23 ACTIVE EARTH PRESSURE

The horizontal component of active earth pressure for any material acting against an ordinary wall is closely given by the general wedge theory for the case of a vertical wall and fill with a substantially horizontal top surface. This wall is assumed to be backfilled by usual construction methods and capable of a small rotational movement (of the magnitude of 0.001 its height), to mobilize the internal friction of the backfill. This rotation is equivalent to an outward movement of ¼ in at the top of a 20-ft wall under earth pressure.

The horizontal component of lateral pressure, P_a, psf, at depth h, also known as Coulomb pressure, is given for vertical walls and horizontal fills by

$$P_a = whK_a - 2c\sqrt{K_a} \tag{6.12}$$

where w = average unit weight, lb per ft^3, of the soil
h = depth, ft
c = cohesion of the soil, psf
$K_a = \tan^2(45 - \phi/2)$
ϕ = angle of internal friction, deg

Table 6.5 gives values of K_a for various values of ϕ.

TABLE 6.5 Active and Passive Pressure Coefficients

Values of $K_a = \tan^2\left(45° - \dfrac{\phi}{2}\right) = \dfrac{1 - \sin\phi}{1 + \sin\phi}$

and of $K_p = \tan^2\left(45° + \dfrac{\phi}{2}\right) = \dfrac{1 + \sin\phi}{1 - \sin\phi}$

ϕ	$\tan\phi$	K_a	K_p
0	0	1.00	1.00
10	0.176	0.70	1.42
20	0.364	0.49	2.04
25	0.466	0.41	2.47
30	0.577	0.33	3.00
35	0.700	0.27	3.69
40	0.839	0.22	4.40
45	1.000	0.17	5.83
50	1.192	0.13	7.55
60	1.732	0.07	13.90
70	2.748	0.03	32.40
80	5.671	0.01	132.20
90	∞	0	∞

The general wedge-theory formulas (taking into account friction along the surface of a wall) may also be used for evaluation of the horizontal component for all other conditions of wall batters and sloping fills. However, comparison with experimental results indicates that the results are somewhat too small for negative surcharges and somewhat too large for positive surcharges, but the differences are no greater than 10%. It is quite accurate enough, taking into account the uncertainties of conditions as to actual slope, to use a table of ratios, referred to as the simplified case of vertical wall and horizontal fill (see Table 6.6).

The vertical component of lateral pressure, in all cases, is such that the resultant pressure forms an angle with the normal to the back of the wall equal to the angle

TABLE 6.6 Lateral Pressure Ratios for General Conditions

For horizontal fill	
Wall slope*	%
+1:6	107
+1:12	103
Vertical	100
−1:12	95
−1:6	90

For vertical walls			
$\phi=$	40°	30°	20°
Fill slope†		%	
+10°	123	110	109
0°	100	100	100
−10°	74	83	88

NOTE: Percentages apply to values given by Eq. (6.1) for the case of vertical wall and horizontal fill and for a specific value of ϕ. In computing the above values, the angle of friction between the wall and fill is assumed equal to ϕ, but not greater than 30°.

*+ represents a slope away from the fill; − represents a slope toward the fill.

†+ represents a slope uphill away from the wall; − represents a downhill slope.

of wall friction. However, under no condition can this angle exceed the angle of internal friction of the fill.

The pressure of soils that, because of lack of drainage and because of their nature, may become fluid at any time—whether such fluid material is widespread or only a narrow layer against the wall—is the same as hydrostatic pressure of a liquid having the same density as that soil.

The pressure of submerged soils is given by Eq. (6.12) with the weight of material reduced by buoyancy (for the solid fraction of the soil only) and the coefficient of internal friction evaluated for the submerged condition; in addition, full hydrostatic pressure of water must be included. For granular materials, submergence affects the coefficients of internal and wall friction very little.

The pressure of fills during saturation and prior to complete submergence and the pressure during drainage periods are affected by the rapidity of water movement. Drainage produces a slight temporary decrease in pressure from normal. Submergence produces an expansion of the fill with consequent increase in pressure. Such variations will not occur if adequate provision is made for drainage; if such provision is not made, the soil may become submerged and the pressure should be computed as for a submerged soil.

Surface loading increases the lateral pressure on the wall. For a uniform surcharge, the lateral pressure on the wall is increased by the surcharge pressure multiplied by the appropriate earth-pressure coefficient.

The pressure in pits and bins is not given by the wedge theory unless a correction for sidewall friction is made, in which case actual field observations of pressures are closely checked.

(H. Y. Fang, "Foundation Engineering Handbook," 2d ed., Van Nostrand Reinhold, New York.)

6.24 PASSIVE EARTH PRESSURE

The horizontal component P_a, psf, of passive pressure at depth h may be computed from

$$P_p = whK_p + 2c\sqrt{K_p} \tag{6.13}$$

where w = average unit weight, lb/ft^3, of the soil
h = depth, ft
$K_p = \tan^2(45 - \phi/2)$
c = cohesion of the soil, psf
ϕ = angle of internal friction, deg

Table 6.5 lists values of K_p for various values of ϕ.

See also Art. 6.22.

6.25 AT-REST PRESSURE

Soil at rest develops earth pressures on vertical walls that are nonyielding. Such pressures represent an intermediate case between the limiting cases of active and passive pressures (Arts. 6.22 and 6.23.) Pressures at rest are calculated as for active or passive conditions from Eq. (6.12) or (6.13), except that the coefficient K_o is employed.

$$K_o = 1 - \sin\phi \tag{6.14}$$

for cohesionless soils and

$$K_o = 0.95 - \sin\phi \tag{6.15}$$

for normally consolidated clays.

For clays, K_o will increase with overconsolidation ratio (OCR), the ratio of preconsolidation pressure to overburden pressure, and may approach a value of 3 for very high values of OCR.

6.26 PRESSURES ON UNDERGROUND STRUCTURES

The vertical load reaction due to soil above a buried structure has been determined chiefly on culverts, although there are considerable data available from tunnel-construction experience. The relative rigidity or flexibility of surrounding earth and the intruding culvert, tunnel, or foundation has considerable influence on whether the load to be carried is greater than, equal to, or less than the weight of the soil directly above. (A summary of general relationships is given in several reports of the Iowa State College Engineering Experiment Station, Ames, Iowa, where a succession of fine work has been done on this subject from 1910 to the present.) Rigidity in the backfill has the same effect.

SHEETING AND BRACING OF EXCAVATIONS

Sheeting and bracing should be carefully designed to hold excavation banks in safe position during construction. Nevertheless, trouble due to bank movement or collapse develops more often than necessary, because of inadequate design. Legal responsibility for preventing damage to adjacent property, public streets, or buried utility services during and after foundation construction is usually placed on the builder. Because sheeting and bracing are temporary, however, the tendency of builders is to minimize materials and labor and "take a chance." The results usually are expensive reconstruction, loss of valuable time, and lengthy litigation. Some cities require filing of a carefully prepared design for temporary sheeting. The Occupational Safety and Health Administration promulgates requirements for protection of employees in excavations, including design and installation of shoring, sloping and benching systems. ["Construction Industry: OSHA Safety and Health Standards," (29 CFR 1926/1910) U.S. Government Printing Office, Washington, DC 20402.]

6.27 SHEETING DESIGN PRESSURES

Sheeting without special anchorage or buttresses can be used in several forms or combinations of shapes. Any formula used for stresses in sheeting should be based on assumptions of the shape that the sheeting takes under load.

Measurement of strut loads and sheeting pressures, both in the field and in the laboratory, led to the pressure diagrams in Fig. 6.16. As can be seen, these diagrams result in pressures greater than the Coulomb pressure [Eq. (6.12)] on the upper part of the wall, because of the nature of the sheeting deflection during construction.

(J. E. Bowles, "Foundation Analysis and Design," 4th ed., McGraw-Hill Publishing Company, New York; M. J. Tomlinson, "Foundation Design and Construction," 5th ed., John Wiley & Sons, Inc., New York; H. Y. Fang, "Handbook of Foundation Engineering," 2d ed., Van Nostrand Reinhold, New York.)

6.28 BRACING OF SHEETPILES

Sheetpiling is sheeting formed by driving piles closely together or interlocking them so that they present a continuous wall or sheet at the perimeter of an excavation. The purpose generally is to restrict substantially flow of water through the perimeter into the excavation or to exclude water and soil from an excavation that requires vertical cuts too deep to stand without support. Sheetpiles may be made of wood, concrete, or structural or cold-formed steel.

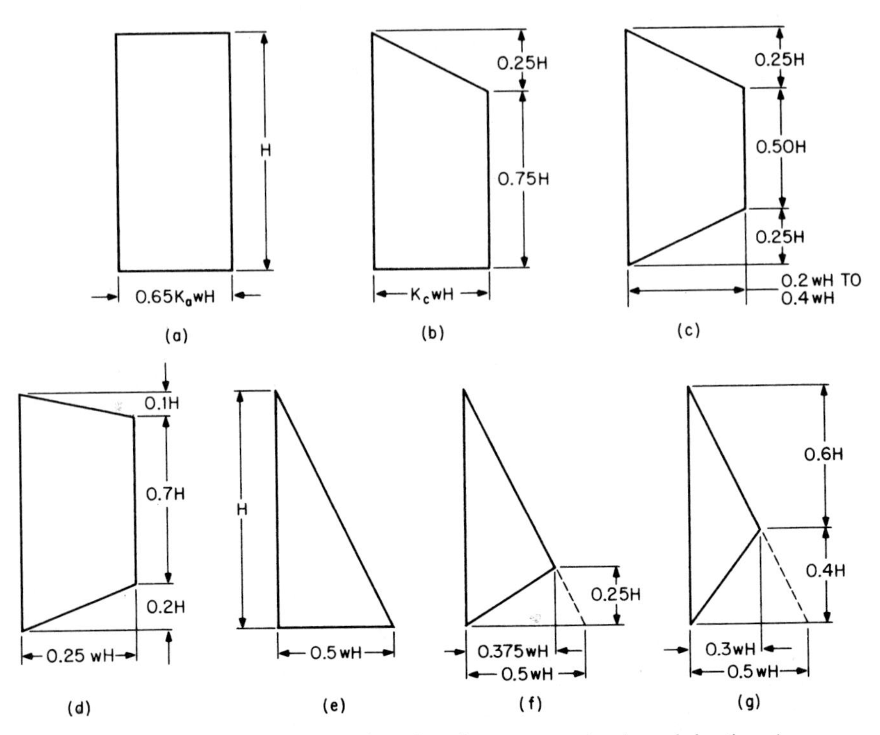

FIGURE 6.16 Diagrams representing lateral earth pressure against braced sheeting. As recommended by R. B. Peck and K. Terzaghi (*a*) for sand, with $K_a = (1 - \sin \phi)/(1 + \sin \phi)$; (*b*) for soft to medium clay, with $K_c = 1 - 4mc$, where c is the cohesion for the clay and $m = 1$ for most clays but can be as low as 0.4 for unstable clays; and (*c*) for stiff, fissured clays. As recommended by G. Tschebotarioff for (*d*) sand, (*e*) soft clays, (*f*) medium clay, and (*g*) stiff clay.

Sheetpiling can be provided with additional supports by either bracing installed inside the cut or anchoring the top to buried **dead men** located outside (Fig. 6.17). Where space is available, the latter procedure is more economical. A dead man may be a battered pile, or a buried mass of concrete, or a short, continuous sheetpile wall to which are attached adjustable lengths of cables or rods. (See also Art. 6.29).

The reactions to be provided at each level of support may be determined from the pressure diagrams in Fig. 6.16 (Art. 6.27). The actual distribution of pressure on the sheeting depends on the flexibility of the sheeting, which, in turn, is affected by the relative stiffness of its supports. A liberal allowance for support at each level is therefore necessary, and the total of all the resistances should be considerably above the total pressure.

Normally, water pressure need not be considered, because the excavation must be kept dry, and any hydrostatic pressures on the back of the sheeting can be easily relieved. If water can accumulate in back of the sheeting, however, such pressure must be considered. Bear in mind also that saturation will reduce considerably the soil-anchorage reaction at the bottom of the sheeting and increase the upper reactions. Saturation of the bottom will also require deeper penetration of the sheet-

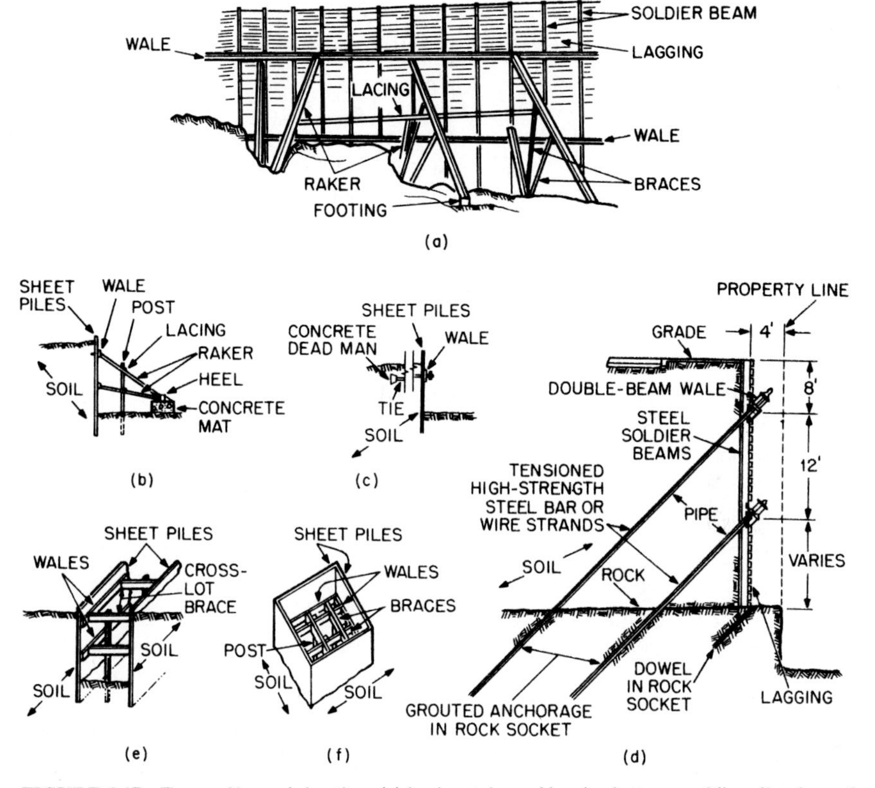

FIGURE 6.17 Types of braced sheeting: (*a*) horizontal wood lagging between soldier piles, braced with wales and rakers; (*b*) continuous sheetpile wall braced with wales and rakers; (*c*) sheetpiles tied back to a concrete dead man; (*d*) soldier piles tied back with tensioned steel bars or wire strands; (*e*) parallel sheetpile walls with cross-lot bracing in perpendicular directions. *(Reprinted with permission from F. S. Merritt and J. Ambrose, "Building Engineering and Systems Design," 2d ed., Van Nostrand Reinhold, New York.)*

ing, to avoid blowouts or boils along the sheeting, either of which will prove diastrous. Drainage is a cheap investment for safety.

Bracing against previously completed footings (Fig. 6.17*a*) is permissible only if the footings are properly braced against undisturbed soil or have enough lateral strength. In evaluation of lateral strength, the frictional resistance on the base of footings (disregarding any cohesive forces) may be added to the passive resistance of the soil against vertical faces.

6.29 ANCHORAGE OF SHEETING

Where exterior buried anchorages are used for sheeting, the pullout resistance of an anchor depends on the type of soil and varies with size of the anchor and its location relative to the bracing. For long, continuous dead men, the allowable anchor pull P, kips, may be computed from

$$P = (P_p - P_a)\frac{L}{\text{SF}} \tag{6.16}$$

where P_p = passive earth force, kips/lin ft, on anchor
P_a = active earth force, kips/lin ft, on anchor
L = length, ft, of anchor
SF = factor of safety (minimum of 2 is recommended)

For short dead men, W. C. Teng has proposed the following. For granular soils,

$$A = L(P_p - P_a) + \tfrac{1}{3}K_o w(K_p + K_a)d_2^3 \tan\phi \tag{6.17}$$

where d_2 = depth to bottom of dead man, ft
A = ultimate anchor capacity, kips
K_a, K_p = pressure coefficients given in Table 6.5
K_o = pressure coefficient given by Eq. (6.14)

For cohesive soils,

$$A = L\,[(P_p - P_a) + q_u d_2] \tag{6.18}$$

where q_u = unconfined compressive strength, psf.

(W. C. Teng, "Foundation Design," Prentice-Hall, Englewood Cliffs, N.J.)

Where possible, anchors should be located so that the failure wedge of the anchor does not intercept the failure wedge of the bulkhead. If the anchor lies entirely within the bulkhead failure zone, it will be ineffective; if the two failure zones do not intercept, the anchor will not load the wall and will be completely effective. For the intermediate case, the allowable anchor pull should be reduced to

$$N_p = [(P_p - P_a) - (P'_p - P'_a)]\frac{L}{\text{SF}} \tag{6.19}$$

where P'_p, P'_a = the forces acting upon the vertical surface extending from the point of intersection of the two failure zones, kips/lin ft.

Tendon Anchors. With the techniques developed for posttensioned concrete, sheeting may be anchored by means of tendons inserted in holes diagonally drilled from the sheeted excavation outward, tensioned against the wales to desired reactions, and grouted (Fig. 6.17*d*). This type of anchorage may be used instead of interior bracing frames. The grouted end of the tendon must be located beyond any possible slip plane in the soil body. Usually, the anchorage is in rock; some successful operations, however, have anchored the tendons or high-strength rods in soil. Tendons are allowed the same stress values as in posttensioned concrete. They should be pretested to 10 to 15% above the desired reaction before the holding anchorage is locked on the wales. Special sloping reaction blocks are needed on the wales, and are usually made of double beams, or else the wales are set on a slope to provide a right-angle bearing surface for the tendon blocks.

Legal clearance from adjacent property owners and public lands is necessary to permit the subsurface encroachment by such anchorage.

Where excavations are necessarily kept open for long periods, say more than a year, some provision may be necessary to avoid possible corrosion loss of the metal in the tendons. The usual wet environment, often with stray electric currents, may cause some serious loss in original strength of the anchorages. Fully grouting the drilled holes in rock or the cased holes in earth after the tendons have been tensioned can reduce the hazard of corrosion.

(J. E. Bowles, "Foundation Analysis and Design," 4th ed., McGraw-Hill Publishing Company, New York.)

6.30 SHEETING REMOVAL

Sheeting should not be removed until all backfill is properly consolidated. Some additional fill and compaction are necessary as sheeting is removed to avoid internal slip and possible damage to adjacent pavements.

CAUSES OF FOUNDATION FAILURES

Structures that show unwanted vertical or lateral movements as a result of foundation settlements, heaves, or displacements are examples where the foundation design is a failure. Rarely does a structural failure occur in the footing; the usual trouble is in the soil and a consequence of the assumption that movements either will not occur or will be uniform.

6.31 FAILURES DUE TO HEAVING

That many structures are affected by the heave of fine-grained soils is now common knowledge. Certain bentonitic clay soils undergo large volume changes with seasonal variation in moisture content—especially when lightly loaded. The phenomenon is further complicated by the shading effect, both under and toward the northerly side of the buildings, causing an inequality in the drying out of the soil.

Tidal intrusion of water along shore lines has caused upward movements of light frame structures.

Evaporation of moisture taken out by trees adjacent to foundations has caused sufficient volume change to tip and crack walls.

6.32 SETTLEMENT FAILURES

Compact, hard, dry clays do not remain in that condition if water is present. Consolidation of soil layers from drainage induced by sewer and other subsurface trenches, as well as the normal slower consolidation from normal pressure of overburden, can cause nonuniform settlement of foundations. Even pile foundations have been known to settle and move laterally where unconsolidated layers—usually not uniform in thickness—start to shrink after the trigger action of pile driving or change in loading conditions.

A very common failure is the dragging down of an existing wall by new construction built in intimate contact; an open gap or sliding plane must be provided between any previously settled-in-place structure and new work. A similiar trouble is the sympathetic settlement of existing structures on a stable foundation underlaid by soft strata, which are compressed and bent by new loading, even if the two structures do not touch.

Settlements can be easily observed, and the trouble can often be recognized from the shape of the masonry cracks. Every design should be examined to see where and how foundation movements can occur, and design modifications should be made to eliminate the effects on other structures and to equalize movements, in amount and direction, in the new structure.

Settlements are possible for pile foundations or even footings on rock, and any assumption made that no settlement will occur is usually inappropriate. Pile foundations will settle as much as the supporting soils under the distributed loads.

Certain shale and siltstone will disintegrate under continuous loading if they become saturated. Unless bottoms of footings are sealed, percolations of ground or bleed water may soften such rocks significantly.

Settlements produced by dissolving of limestone by percolating groundwater can be especially damaging. Such movements are usually catastrophic and must be avoided. Before a structure is built over limestone, there should be adequate indications that such activity is highly unlikely during the life of the facility.

Backfill for basement walls with soils that shrink on dehydration—often caused by suction of the soil water into the concrete—show voids in the fill immediately adjacent to the wall. Such conditions have been known to be the cause of complete collapse of a wall immediately after a heavy rainfall, when enough water entered the voids to provide full hydrostatic pressure, for which the wall was not designed.

Localized footing failures within buildings are often caused by the pipe trenches dug after the footings are completed, without regard to relative depths of excavations. Similiar troubles often occur when new trunk sewers, or even transit tunnels, are built in public streets without proper underpinning of all footings above the surface of influence from the new excavation. Such construction will also affect the normal groundwater table, and together with pumping for industrial and air-conditioning requirements, has in the past lowered the water level sufficiently to cause rotting of untreated wood piles. This difficulty has been so prevalent that many cities now restrict the use of untreated piles for any area where groundwater level may not remain stable. Proximity of natural water areas is no proof of such stability; the groundwater level has been shown to be 35 ft below sea level at only short distances from bulkhead lines where industrial pumping was carried to excess.

A serious foundation failure of a heavy factory bearing on compact fine sand was caused by the sinking of some wells within the building and pumping process water from the subsoil. The underlying soil layers were reduced in thickness, and a dish-shaped floor resulted.

Local overload on ground-floor slabs, transmitted to some footings, explains the odd-shaped roofs on many single-story warehouses.

All these troubles can be controlled, but only by recognition in design that foundations will settle if soil support is altered and that every new foundation will settle as load is applied.

(M. Ordacky, "Lessons from Structural Failures," American Society of Civil Engineers, 345 E. 47th St., New York, NY 10017.)

SECTION SEVEN
STRUCTURAL STEEL CONSTRUCTION

Bruce Glidden*
President, Glidden & Co., Ltd.
Bridgeville, Pennsylvania

Structural steel is an economical construction material for building applications. It offers high ratios of strength to weight and strength to volume. Thus, structural steel has the advantage of permitting long clear spans for horizontal members and requiring less floor space for columns than other common construction materials. It also can be used in combination with reinforced concrete to provide cost-effective building components. For large industrial buildings, where the structural frame can be exposed, it is often the material of choice.

The design of a structural building frame involves the following principal steps:

1. Select the general configuration and type of structure (Sec. 1).
2. Determine the service loads as required by the applicable building code (Art. 5.1.2).
3. Compute the internal forces and moments for the individual members (Sec. 5).
4. Proportion the members and connections.
5. Check performance characteristics, such as deflection, under service conditions.
6. Make a general overall review for economy of function.

Designers, in addition to performing these steps, should also have an appreciation of the complete construction cycle to assure a practical and economical design. This includes understanding the needs of other disciplines and trades, types and availability of the materials used in steel of construction, applicable codes and specifications, the role and responsibilities of the fabricator and the erector, and a designer's own responsibilities in the area of quality assurance.

The other principal parties involved in structural steel construction are fabricators and erectors. Fabrication involves the operations, primarily in a shop, of cutting steel shapes and plates to final size, punching and drilling, and assembling

*Updated and revised Sec. 8, "Structural Steel Construction," by Henry J. Stetina, in "Building Design and Construction Handbook," 4th ed.

the components into finished members ready for shipment. Because each fabrication shop has a unique mix of equipment, fabricators design connection details to satisfy criteria established by the designer and subect to the designer's approval. Shop painting is also a common fabrication responsibility. Erectors, who may be subcontractors to fabricators, employ their equipment together with skilled laborers, called ironworkers, to position and connect the steel into its final location at the project site.

Structural steel consists of hot-rolled steel shapes, steel plates of thickness of ⅛ in or greater, and such fittings as bolts, welds, bracing rods, and turnbuckles. The owner and the engineer should understand fully what will be furnished by the fabricator under a contract to furnish "structural steel." To promote uniformity in bidding practices, the American Institute of Steel Construction (AISC) has adopted a "Code of Standard Practice" (American Institute of Steel Construction, One East Wacker Drive, Chicago, IL 60601.)

7.1 CODES AND SPECIFICATIONS

Codes, specifications, and standards provide steel designers with sound design procedures and guidelines. These documents cover selection of service and design loads, criteria for proportioning members and their connections, procedures for fabrication and erection, requirements for inspections, and standards for protection against corrosion and fire. Use of these documents generally ensures safety, economical designs, and sound operational techniques.

The applicable building code defines the minimum legal requirements for a design. Most building authorities incorporate in their building code one of the model building codes (Art. 1.10), but some write their own code requirements. Usually, the basis for the requirements for steel design and construction in building codes are the American Institute of Steel Construction specifications for structural steel buildings (Table 7.1). Note that two AISC specifications are available, one applicable to allowable stress design and plastic design (ASD) and the second to load and resistance factor design (LRFD).

Table 7.1 also lists codes and specifications most frequently used by steel designers. Requirements for special-function buildings, needs of governmental agencies, and other unique requirements has led to promulgation of many other codes and specifications. Some of the organizations that publish these standards are the General Services Administration, U.S. Department of Commerce, Corps of Engineers, and U.S. Navy Bureau of Yards and Docks.

7.2 MILL MATERIALS

The steel shapes, plates, and bars that make up most of the materials used for structural steel are produced by mills as hot-rolled products. These products are made in a batch process; each grade of steel comes from a "heat." The specific grade of steel in all mill products is identified by reference to the heat number.

Through universal acceptance of ASTM specifications (Table 7.1), mill materials have uniform physical and quality characteristics. There is no significant metallurgical or physical difference between products ordered to a specific ASTM specification and rolled by any U.S. structural mill.

TABLE 7.1 Basic Steel Construction Codes and Specifications

Organization	Document	Scope
American Institute of Steel Construction (AISC) One East Wacker Drive Chicago, IL 60601	Code of Standard Practice for Steel Buildings and Bridges	Defines structural steel Plans and specifications Fabrication Erection Quality control
	Specification for Structural Steel Buildings—Allowable Stress Design and Plastic Design (ASD)	Materials Loads Design criteria Serviceability Fabrication Erection Quality control
	Specification for Structural Steel Buildings—Load and Resistance Factor Design (LRFD)	
American Iron and Steel Institute (AISI) 1000 16th St., NW Washington, DC 20036	Specification for the Design of Cold-Formed Steel Structural Members	Materials Design criteria
ASTM 1916 Race St. Philadelphia, PA 19103	ASTM A6	Delivery-shapes/plates
	Various ASTM material specifications	Physical and chemical requirements
American Welding Society (AWS) 550 N.W. LeJeune Road Miami, FL 33135	Structural Welding Code—Steel (AWS D1.1)	Joint design Workmanship Procedures Inspection
Research Council on Structural Connections Engineering Foundation 345 E. 47th St. New York, NY 10017	Specification for Structural Joints Using ASTM A325 or A490 Bolts	Materials Connection design Installation Inspection
Steel Joist Institute (SJI) 1205 48th Ave., North Myrtle Beach, SC 29577	Standard Specifications and Load Tables, Open-Web Steel Joists	Materials Design
Steel Structures Painting Council (SSPC) 4400 Fifth Ave. Pittsburgh, PA 15213	Steel Structures Painting Manual, Vols. 1 and 2	Good practice Systems Specifications

7.2.1 Grades of Steel

Structural steel grades are referred to by their corresponding ASTM designation. For example, the most commonly used grade of structural steel is A36, which is produced to meet the requirements of the ASTM A36 specification. This grade offers a good mix of strength, weldability, and cost. In many designs, this specification alone will satisfy designers' needs. Other specifications, such as A53 for pipe, provide an equivalent grade of steel for that type of product. However, as loads on the structural elements become larger, other grades of steel may become more economical because of dimensional limitations or simpler fabrication. These grades provide greater strength levels at somewhat higher costs per unit weight.

AISC recommends certain grades of steel, all of which have desirable characteristics, such as weldability and cost-effectiveness, for use where higher strength levels are required. The specifications covering these grades are listed in Table 7.2. Several steels have more than one level of tensile strength and yield stress, the levels being dependent on thickness of material. The listed thicknesses are precise for plates and nearly correct for shapes. To obtain the precise value for shapes, refer to an AISC "Manual of Steel Construction" (ASD or LRFD) or to mill catalogs.

Weathering Steels. The A242 and A588 grades of steel offer enhanced corrosion resistance relative to A36 material. These steels, called weathering steels, form a thin oxidation film on the surfaces that inhibits further corrosion in ordinary atmospheric conditions. However, special treatment of construction details is required. Because of such constraints, and because these grades are more expensive, utilization of weathering steels in building construction is limited. These grades are more commonly used in bridge construction.

Steel Grade Identification. Because of the several grades of steel in use, ASTM specifications require that each piece of hot-rolled steel be properly identified with vital information, including the heat number. The AISC specifications for structural steel buildings require fabricators to be prepared to demonstrate, by written procedure and by actual practice, the visible identification of all main stress-carrying elements at least through shop assembly. Steel identification includes ASTM designation, heat number (if required), and mill test reports when specially ordered.

Availability. Because structural steel is produced in a batch process, the less commonly used shapes and the higher-strength grades are produced less frequently than commonly used A36 shapes. Furthermore, steel service centers stock the smaller A36 shapes. As a result, availability of steels can affect construction schedules. Consequently, steel designers should be aware of the impact of specifying less commonly used materials and shapes if the project has a tight schedule. Fabricator representatives can provide needed information.

7.2.2 Structural Shapes

Steel mills have a standard classification for the many products they make, one of which is *structural shapes (heavy)*. By definition this classification takes in all shapes having at least one cross-sectional dimension of 3 in or more. Shapes of lesser size are classified as *structural shapes (light)* or, more specifically, bars.

Shapes are identified by their cross-sectional characteristics—angles, channels, beams, columns, tees, pipe, tubing, and piles. For convenience, structural shapes

TABLE 7.2 Characteristics of Structural Steels

ASTM specification	Thickness, in	Minimum tensile strength, ksi	Minimum yield stress,* ksi
	Carbon steels		
A36	To 8 in incl.	58–80†	36
A529	To ½ in incl.	60–85†	42
	High-strength, low-alloy steels		
A441	To ¾ incl.	70	50
	Over ¾ to 1½	67	46
	Over 1½ to 4 incl.	63	42
	Over 4 to 8 incl.	60	40
A572	Gr 42: to 4 incl.	60	42
	Gr 45: to 1½ incl.	60	45
	Gr 50: to 1½ incl.	65	50
	Gr 55: to 1½ incl.	70	55
	Gr 60: to 1 incl.	75	60
	Gr 65: to ½ incl.	80	65
A242	To ¾ incl.	70	50
	Over ¾ to 1½	67	46
	Over 1½ to 4 incl.	63	42
A588	To 4 incl.	70	50
	Over 4 to 5	67	46
	Over 5 to 8 incl.	63	42
	Heat-treated low-alloy steels		
A514	To ¾ incl.	115–135	100
	Over ¾ to 2½	115–135	100
	Over 2½ to 4 incl.	105–135	90

*Yield stress or yield strength, whichever shows in the stress-strain curve.

†Minimum tensile strength may not exceed the higher value.

are simply identified by letter symbols as indicated in Table 7.3. The industry recommended standard (adopted 1970) for indicating a specific size of beam or column-type shape on designs, purchase orders, shop drawings, etc., specifies listing of symbol, depth, and weight, in that order. For example, W14 × 30 identifies a wide-flange shape with nominal depth of 14 in and weight of 30 lb/lin ft. The ×, read as "by," is merely a separation.

Each shape has its particular functional use, but the workhorse of building construction is the wide-flange W section. For all practical purposes, W shapes have parallel flange surfaces. The profile of a W shape of a given nominal depth

TABLE 7.3 Symbols for Structural Shapes

Section	Symbol
Wide-flange shapes	W
Standard I shapes	S
Bearing-pile shapes	HP
Similar shapes that cannot be grouped in W, S, or HP	M
Structural tees cut from W, S, or M shapes	WT, ST, MT
American standard channels	C
All other channel shapes	MC
Angles	L

and weight available from different producers is essentially the same, except for the size of fillets between web and flanges.

7.2.3 Tolerances for Structural Shapes and Plates

Mills are granted a tolerance because of variations peculiar to working of hot steel and wear of equipment. Limitations for such variations are established by ASTM specification A6.

Wide-flange beams or columns, for example, may vary in depth by as much as ½ in, i.e., ¼ in over and under the nominal depth. The designer should always keep this in mind. Fillers, shims, and extra weld metal installed during erection may not be desirable, but often they are the only practical solution to dimensional variations from nominal.

Cocked flanges on column members are particularly troublesome to the erector, for it is not until the steel is erected in the field that the full extent of mill variations becomes evident. This is particularly true for a long series of spans or bays, where the accumulating effect of dimensional variation of many columns may require major adjustment. Fortunately, the average variation usually is negligible and nominal erection clearance allowed for by the fabricator will suffice.

Mill tolerances also apply to beams ordered from the mills cut to length. Where close tolerance may be desired, as sometimes required for welded connections, it may be necessary to order the beams long and then finish the ends in the fabricating shop to precise dimensions. This is primarily the concern of structural detailers.

7.2.4 Cambered Beams

Frequently, designers want long-span beams slightly arched (cambered) to offset deflection under load and to prevent a flat or saggy appearance. Such beams may be procured from the mills, the required camber being applied to cold steel. The AISC Manuals give the maximum cambers that mills can obtain and their prediction of the minimum cambers likely to remain permanent. Smaller cambers than these minimums may be specified, but their permanency cannot be guaranteed. Nearly all beams will have some camber as permitted by the tolerance for straightness, and advantage may be taken of such camber in shop fabrication.

A method of cambering, not dependent on mill facilities, is to employ heat. In welded construction, it is commonplace to flame-straighten members that have become distorted. By the same procedure, it is possible to distort or camber a beam to desired dimensions.

7.2.5 Steel Plates

Used by fabricators to manufacture built-up structural members, such as columns and girders, and for detail connection material, plates are identified by the symbol PL. Cross-sectional dimensions are given in inches (or millimeters). A plate ½ in thick and 2 ft wide is billed as PL ½ × 24. Plates may also be specified by weight, although this is unusual in building construction work.

Mill tolerances for plate products for structural applications are also defined by ASTM specification A6. There are provisions for thickness, crown, camber, and length. Consideration of these characteristics are primarily the responsibility of fabricators. However, steel designers should be aware of how these tolerances affect the fabricator's work and permit the design to accommodate these characteristics.

7.2.6 Pipe and Tubular Sections

Pipe meeting the requirements of ASTM specification A53, Types E and S, Grade B, is comparable to A36 steel, with yield strength $F_y = 36$ ksi. It comes in three weight classification: standard, extra strong, and double extra strong, and in diameters ranging up to 26 in.

Several mills produce square and rectangular tubing in sizes from 3 × 2 and 2 × 2 to 12 × 8 and 10 × 10 in, with wall thickness up to ⅝ in. These flat-sided shapes afford easier connections than pipes, not only for connecting beams but also for such items as window and door frames.

The main strength properties of several grades of steel used for pipe and tubular sections are summarized in Table 7.4.

7.3 FASTENERS

Two basic types of fasteners are typically used in construction, bolts and welds. Both are used in the fabricating shop and on the job site in connections joining individual members. Welds are also used to fasten together components of built-up members. Bolts, however, are more commonly used for field connections, and welds, for shop work. Rivets, which were once widely used for main connections, both shop and field, are essentially obsolete.

Many variables affect selection of fasteners. Included among these are economy of fabrication and erection, availability of equipment, inspection criteria, labor supply, and such design considerations as fatigue, size and type of connections, continuity of framing, reuse, and maintenance. It is not uncommon for steel framing to be connected with such combinations as shop welds and field bolts or to be all-welded. It is usual to use field welds for column splices with bolted connections elsewhere. The variables affecting decisions on use of fasteners should be explored with engineers representing the fabricator and the erector.

TABLE 7.4 Characteristics of Pipe and Tubular Steels

ASTM spec.	Grade	Product	Min tensile strength, ksi	Min yield stress, ksi
A53	B	Pipe	60.0	35.0*
A500	A	Round	45.0	33.0
	A	Shaped	45.0	39.0
	B	Round	58.0	42.0
	B	Shaped	58.0	46.0
A501	. . .	All tubing	58.0	36.0
A618	I	All tubing	70.0	50.0
	II	All tubing	70.0	50.0
	III	All tubing	65.0	50.0

*Use 36.0 for purpose of design.

7.3.1 High-Strength Bolts

Development of high-strength bolts is vested in the Research Council on Riveted and Bolted Structural Joints of the Engineering Foundation. Its "Specification for Structural Steel Joints Using A325 or A490 Bolts" (Table 7.1) was adopted by the American Institute for Steel Construction. Bolts conforming to ASTM A449 are acceptable, but their usage is restricted to bearing-type connections (Fig. 7.1)

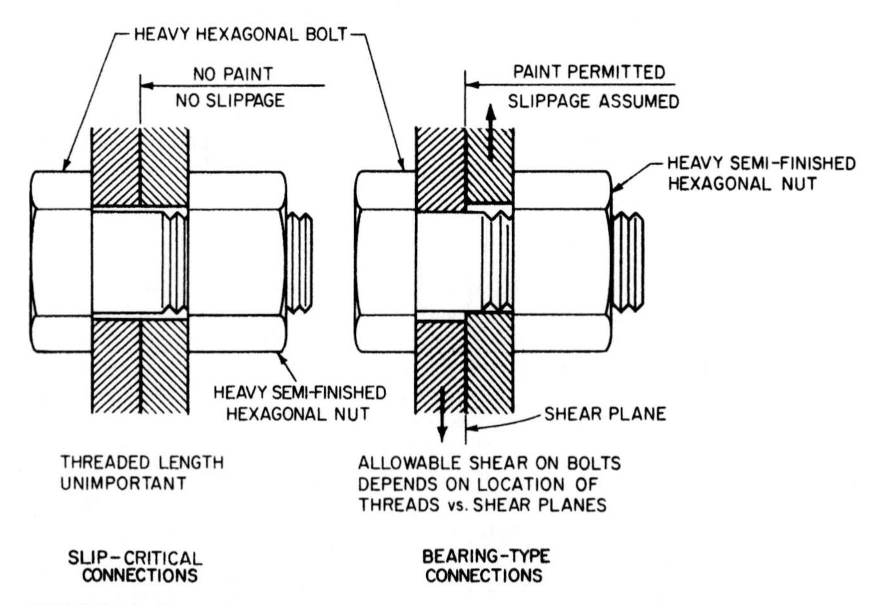

FIGURE 7.1 Two main types of construction with high-strength bolts. Although, in general, no paint is permitted on faying surfaces in slip-critical connections, the following are allowed: scored galvanized coatings, inorganic zinc-rich paint, and metallized zinc or aluminum coatings.

requiring bolt diameters greater than 1½ in. Furthermore, when they are required to be tightened to more than 50% of their specified minimum tensile strength, hardened steel washers should be installed under the heads.

When high-strength bolts are used in a connection, they are highly tensioned by tightening of the nuts and thus tightly clamp together the parts of the connection.

For convenient computation of load capacity, the clamping force and resulting friction are resolved as shear. Bearing between bolt body and connected material is not a factor until loads become large enough to cause slippage between the parts of the connection. The bolts are assumed to function in shear following joint slippage into full bearing.

The clamping and bearing actions lead to the dual concept: slip-critical connections and bearing-type connections. For the latter, the allowable shear depends on the cross-sectional bolt area at the shear plane. Hence, two shear values are assigned, one for the full body area and one for the reduced area at the threads.

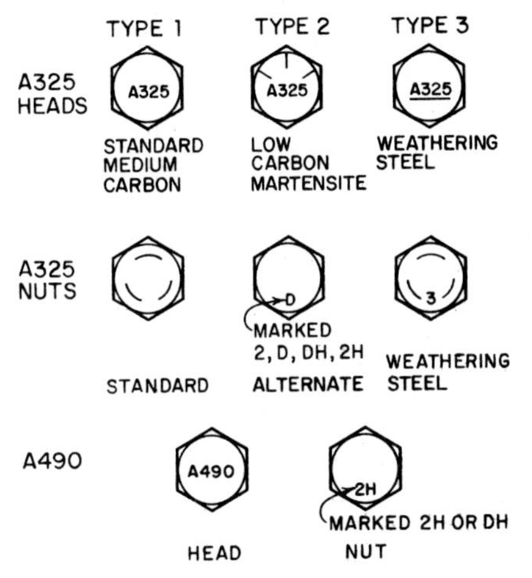

FIGURE 7.2 Identification markings on heads and nuts of high-strength bolts.

Identification. There is no difference in appearance of high-strength bolts intended for either slip-critical or bearing-type connections. To aid installers and inspectors in identifying the several available grades of steel, bolts and nuts are manufactured with permanent markings (Fig. 7.2).

7.3.2 High-Strength Bolt Installation

Washer requirements for high-strength bolted assemblies depend on the method of installation and type of bolt holes in the connected elements. These requirements are summarized in Table 7.5.

TABLE 7.5 Washer Requirements for High-Strength Bolts

Method of tensioning	A325 bolts	A490 bolts: Base material $F_y < 40.0$*	A490 bolts: Base material $F_y > 40.0$*
Calibrated wrench	One washer under turned element	Two washers	One washer under turned element
Turn-of-the-nut	None	Two washers	One washer under turned element
Both methods, slotted and oversized holes	Two washers	Two washers	Two washers

*F_y = specified minimum yield stress, ksi.

Bolt Tightening. Specifications require that all high-strength bolts be tightened to 70% of their specified minimum tensile strength, which is nearly equal to the proof load (specified lower bound to the proportional limit) for A325 bolts, and within 10% of the proof load for A490 bolts. Tightening above these minimum tensile values does not damage the bolts, but it is prudent to avoid excessive uncontrolled tightening. The required minimum tension, kips, for A325 and A490 bolts is given in Table 7.6.

TABLE 7.6 Minimum Tightening Tension, kips, for High-Strength Bolts

Dia, in	A325	A490
5⁄8	19	24
3⁄4	28	35
7⁄8	39	49
1	51	64
1 1⁄8	56	80
1 1⁄4	71	102
1 3⁄8	85	121
1 1⁄2	103	148

There are three methods for tightening bolts to assure the prescribed tensioning:

Turn-of-Nut. By means of a manual or powered wrench, the head or nut is turned from an initial snug-tight position. The amount of rotation, varying from one-third to a full turn, depends on the ratio of bolt length (underside of head to end of point) to bolt diameter and on the disposition of the outer surfaces of bolted parts (normal or sloped not more than 1:20 with respect to the bolt axis). Required rotations are tabulated in the "Specification for Structural Steel Joints Using A325 or A490 Bolts."

Calibrated Wrench. By means of a powered wrench with automatic cutoff and calibration on the job. Control and test are accomplished with a hydraulic device equipped with a gage that registers the tensile stress developed.

Direct Tension Indicator. Special indicators are permitted on satisfactory demonstration of performance. One example is a hardened steel washer with protrusions on one face. The flattening that occurs on bolt tightening is measured and correlated with the induced tension.

7.3.3 Unfinished Bolts

Known in construction circles by several names—ordinary, common, machine, or rough—unfinished bolts are characterized chiefly by the rough appearance of the shank. They are covered by ASTM A307. They fit into holes 1⁄16 in larger in diameter than the nominal bolt diameter.

Unfinished bolts have relatively low load-carrying capacity. This results from the possibility that threads might lie in shear planes. Thus, it is unnecessary to extend the bolt body by use of washers.

One advantage of unfinished bolts is the ease of making a connection; only a wrench is required. On large jobs, however, erectors find they can tighten bolts more economically with a pneumatic-powered impact wrench. Power tightening generally yields greater uniformity of tension in the bolts and makes for a better-balanced connection.

While some old building codes restrict unfinished bolts to minor applications, such as small, secondary (or intermediate) beams in floor panels and in certain parts of one-story, shed-type buildings, the AISC specifications for structural steel buildings, with a basis of many years of experience, permit A307 bolts for main connections on structures of substantial size. For example, these bolts may be used for beam and girder connections to columns in buildings up to 125 ft in height.

There is an economic relation between the strength of a fastener and that of the base material. So while A307 may be economical for connecting steel with a 36-ksi yield point, this type of bolt may not be economical with 50-ksi yield-point steel. The number of fasteners to develop the latter becomes excessive and perhaps impractical due to size of detail material.

A307 bolts should always be considered for use, even in an otherwise all-welded building, for minimum-type connections, such as for purlins, girts, and struts.

Locking Devices for Bolts. Unfinished bolts (ASTM A307) and interference-body-type bolts (Art 7.3.4) usually come with American Standard threads and nuts. Properly tightened, connections with these bolts give satisfactory service under static loads. But when the connections are subjected to vibration or heavy dynamic loads, a locking device is desirable to prevent the nut from loosening.

Locking devices may be classified according to the method employed: special threads, special nuts, special washers, and what may be described as field methods. Instead of conventional threads, bolt may be supplied with a patented self-locking thread called Dardelet. Sometimes, locking features are built into the nuts. Patented devices, the Automatic-Nut, Union-Nut, and Pal-Nut, are among the common ones. Washers may be split rings or specially toothed. Field methods generally used include checking, or distorting, the threads by jamming them with a chisel or locking by tack welding the nuts.

7.3.4 Other Bolt-Type Fasteners

Interference body or bearing-type bolts are characterized by a ribbed or interrupted-ribbed shank and a button-shaped head; otherwise, including strength, they are similar to the regular A325 high-strength bolts. The extreme diameter of the shank is slightly larger than the diameter of the bolt hole. Consequently, the tips of the ribs or knurlings will groove the side of the hole, assuring a tight fit. One useful application has been in high television towers, where minimum-slippage joints are desired with no more installation effort than manual tightening with a spud wrench. Nuts may be secured with lock washers, self-locking nuts, or Dardelet self-locking threads. The main disadvantage of interference body bolts is the need for accurate matching of truly concentric holes in the members being joined; reaming sometimes is necessary.

Huckbolts are grooved (not threaded) and have an extension on the end of the shank. When the bolt is in the hole, a hydraulic machine, similar to a bolting or riveting gun, engages the extension. The machine pulls on the bolt to develop a high clamping force, then swages a collar into the grooved shank and snaps off the extension, all in one quick operation.

7.3.5 Welds

Welding is used to fasten together components of a built-up member, such as a plate girder, and to make connections between members. This technique, which uses fusion in a controlled atmosphere, requires more highly skilled labor than does bolting. However, because of cost advantages, welding is widely used in steel construction, especially in fabricating shops where conditions are more favorable to closely controlled procedures. When field welding is specified, the availability of skilled welders and inspection technicians and the use of more stringent quality-control criteria should be considered.

Any of several welding processes may be used: manual shielded metal arc, submerged arc, flux cored arc, gas metal arc, electrogas, and electroslag. They are not all interchangeable, however; each has its advantageous applications.

Many building codes accept the recommendations of the American Welding Society "Structural Welding Code" (AWS D1.1). The AISC specification incorporates many of this code's salient requirements.

Weld Types. Practically all welds used for connecting structural steel are of either of two types: fillet or groove.

Figure 7.3*a* and *b* illustrates a typical **fillet weld.** As stated in Art. 7.27, all stresses on fillet welds are resolved as shear on the effective throat. The normal throat dimension, as indicated in Fig. 7.3*a* and *b*, is the effective throat for all welding processes, except the submerged-arc method. The deep penetration characteristic of the latter process is recognized by increasing the effective throat dimension, as shown in Fig. 7.3*c*.

Groove welds (Fig. 7.3*d*, *e*, and *f*) are classified in accordance with depth of solid weld metal as either complete or partial penetration. Most groove welds, such as those in Fig. 7.3*d* and *e*, are made complete-penetration welds by the workmanship requirements: use backup strips or remove slag inclusions and imperfections (step called back-gouging) on the unshielded side of the root weld. The partial-penetration groove weld shown in Fig. 7.3*f* is typical of the type of weld used for box-type members and column splices. Effective throat depends on the welding process, welding position, and the chamfer angle α. The indicated effective throat (Fig. 7.3*f*) is proper for the shielded-metal-arc processes and for all welding positions. (See also Art. 7.27.)

Welding Electrodes. Specifications for all welding electrodes, promulgated by the American Welding Society (AWS), are identified as A5.1, A5.5, A5.17, etc., depending on the welding process. Electrodes for manual arc welding, often called stick electrodes, are designated by the letter E followed by four or five digits. The first two or three digits designate the strength level; thus, E70XX means electrodes having a minimum tensile strength of 70.0 ksi. Allowable shear stress on the depositied weld metal is taken as 0.30 times the electrode strength classification; thus, 0.30 times 70 to an E70 results in an allowable stress of 21.0 ksi. The remaining digits provide information on the intended usage, such as the particular welding positions and types of electrode coating.

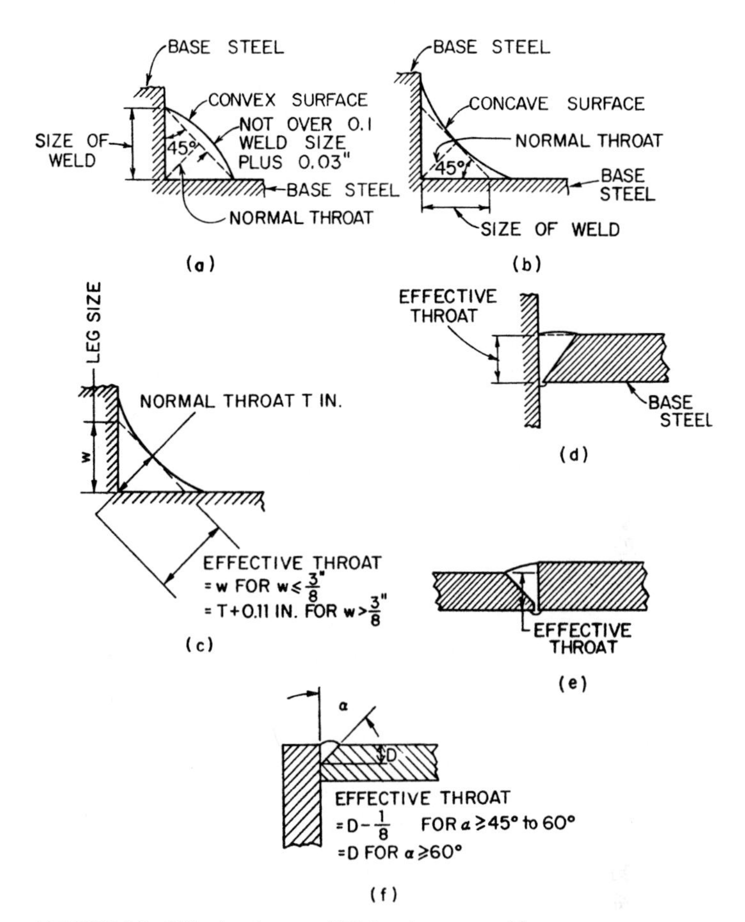

FIGURE 7.3 Effective throats of fillet and groove welds.

Welding Procedures. The variables that affect the quality of a weld are controlled by welding procedures that must be approved by the structural engineer. Specification AWS D1.1 contains several prequalified welding procedures, the use of which permits fabricators and erectors to avoid the need for obtaining approvals for specific routine work. Where unusual conditions exist, the specification requires that formal documentation be submitted for review and approval.

Base-Metal Temperatures. An important requirement in production of quality welds is the temperature of base metal. Minimum preheat and interpass temperature as specified by the AWS and AISC standards must be obtained within 3 in of the welded joint before welding starts and then maintained until completion. Table 7.7 gives the temperature requirements based on thickness (thickest part of joint) and welding process for several structural steels. When base metal temperature is below 32°F, it must be preheated to at least 70° and maintained at that temperature during welding. No welding is permitted when ambient temperature

TABLE 7.7 Minimum Preheat and Interpass Temperatures for Base Metal to Be Welded

	Shielded-metal-arc welding with other than low-hydrogen electrodes		Shielded-metal-arc welding with low-hydrogen electrodes, gas-metal-arc, and flux-cored arc welding	
Steel*	Thickness, in	Temp, °F	Thickness, in	Temp, °F
A36	To ¾ in incl. Over ¾ to 1½ Over 1½ to 2½ Over 2½	32 150 225 300	To ¾ in incl. Over ¾ to 1½ Over 1½ to 2½ Over 2½	32 50 150 225
A242 A441 A588 A572 to $F_y = 50$	Not permitted		To ¾ in incl. Over ¾ to 1½ Over 1½ to 2½ Over 2½	32 70 150 225

*For temperatures for other steels, see AWS D1.1, "Structural Welding Code," American Welding Society.

is below 0°F. Additional information, including temperature requirements for other structural steels, is given in AWS D1.1 and the AISC specifications for structural steel buildings (Table 7.1).

Another quality-oriented requirement applicable to fillet welds is minimum leg size, depending on thickness of steel (Table 7.8). The thicker part connected governs, except that the weld size need not exceed the thickness of the thinner part. This rule is intended to minimize the effects of restraint resulting from rapid cooling due to disproportionate mass relationships.

TABLE 7.8 Minimum Sizes* of Fillet and Partial-Penetration Welds

Base-metal thickness, in	Weld size, in
To ¼ incl.	⅛
Over ¼ to ½	3/16
Over ½ to ¾	¼
Over ¾ to 1½	5/16
Over 1½ to 2¼	⅜
Over 2¼ to 6	½
Over 6	⅝

*Leg dimension for fillet welds; minimum effective throat for partial-penetration groove welds.

7.3.6 Fastener Symbols

Fasteners are indicated on design, shop, and field erection drawings by notes and symbols. A simple note may suffice for bolts; for example: "⅞-in A325 bolts, except as noted." Welds require more explicit information, since their location is not so obvious as that of holes for bolts.

Symbols are standard throughout the industry. Figure 7.4 shows the symbols for bolts, Fig. 7.5 the symbols for welds. The welding symbols (Fig. 7.5*a*) together

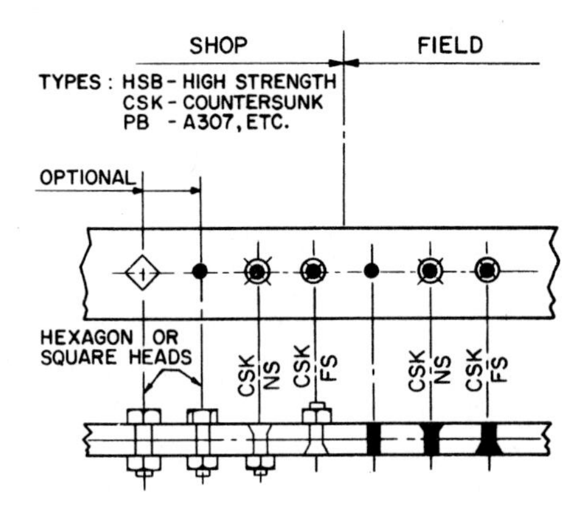

FIGURE 7.4 Symbols for shop and field bolts.

with the information key (Fig. 7.5*b*) are from the American Welding Society "Symbols for Welding and Nondestructive Testing, AWS A2.4.

7.3.7 Erection Clearance for Fasteners

All types of fasteners require clearances for proper installation in both shop and field. Shop connections seldom are a problem, since each member can be easily manipulated for access. Field connections, however, require careful planning, because connections can be made only after all members to be connected are aligned in final position. This is the responsibility of the fabricator's engineering staff and is discharged during the making of shop drawings. However, the basic design configuration must permit the necessary clearances to be developed.

Clearances are required for two reasons: to permit entry, as in the case of bolts entering holes, and to provide access to the connected elements either to allow the tightening of bolts with field tools or to permit the movement of manual electrodes or semiautomatic welding tools in depositing weld metal.

("Structural Steel Detailing," American Institute of Steel Construction.)

7.4 QUALITY ASSURANCE

Concepts for improving and maintaining quality in the constructed project stress the participation of the design professional in the project team consisting of the owner, design professional, and general contractor. While the structural engineer plays a varying role in the major phases of a project—that is, conceptual, preliminary, and final design; bidding; and construction—his or her participation is vital to achieving the appropriate level of quality.

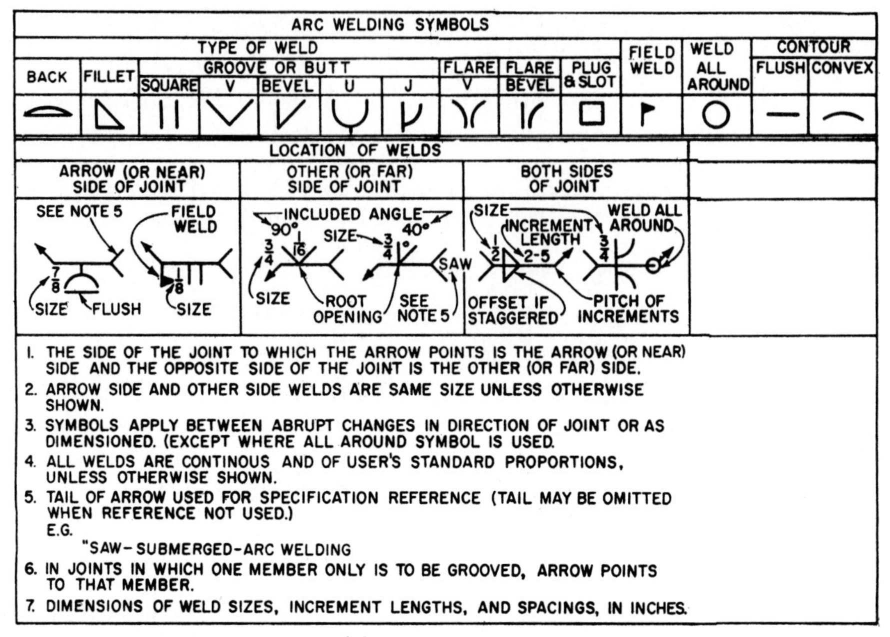

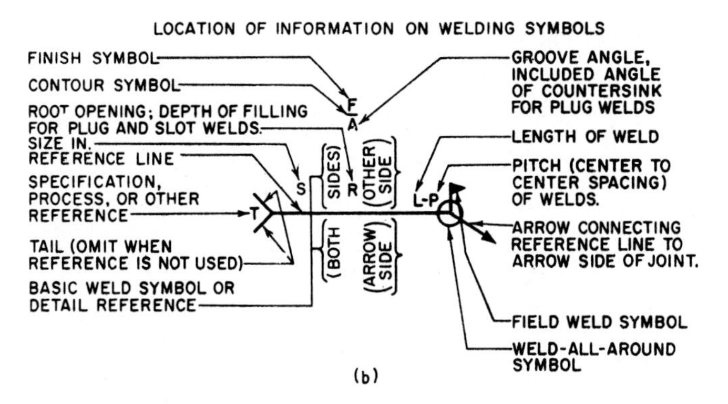

FIGURE 7.5 Symbols for shop and field welds.

Those activities of the structural engineer that have the greatest impact on quality are materials selection, determination of workmanship quality levels, quality control (QC) requirements, preparation of clear and complete contract documents, and review of the contractor's work. One aspect of the last item that is particularly important in steel construction is the review and approval of the fabricator's shop drawings. Because the fabricator's engineers design connections to meet the criteria provided by the design professional, the review and approval process must assure that connection designs and details are compatible with the intent and requirements of the basic design.

("Quality in the Constructed Project," American Society of Civil Engineers.)

STRUCTURAL STEEL SYSTEMS

Steel construction may be classified into three broad categories: wall-bearing, skeleton, and long-span framing. Depending on the needs of the building, one or more of these categories may be incorporated.

In addition to the main building elements—floors, roofs, walls—the structural system must include bracing members that provide lateral support for main members as well as for other bracing members, resistance to lateral loads on the building, redundant load paths, and stiffness to the structure to limit deflections. An economical and safe design properly integrates these systems into a completed structure.

7.5 WALL-BEARING FRAMING

Probably the oldest and commonest type of framing, wall-bearing framing (not to be confused with bearing-wall construction), occurs whenever a wall of a building, interior or exterior, is used to support ends of main structural elements carrying roof or floor loads. The walls must be strong enough to carry the reaction from the supported members and thick enough to ensure stability against any horizontal forces that may be imposed. Such construction often is limited to relatively low structures, because load-bearing walls become massive in tall structures. Nevertheless, a wall-bearing system may be advantageous for tall buildings when designed with reinforcing steel.

A common application of wall-bearing construction may be found in many one-family homes. A steel beam, usually 8 or 10 in deep, is used to carry the interior walls and floor loads across the basement with no intermediate supports, the ends of the beam being supported on the foundation walls. The relatively shallow beam depth affords maximum headroom for the span. In some cases, the spans may be so large that an intermediate support becomes necessary to minimize deflection. Usually a steel pipe column serves this purpose.

Another example of wall-bearing framing is the member used to support masonry over windows, doors, and other openings in a wall. Such members, called **lintels,** may be a steel angle section (commonly used for brick walls in residences) or, on longer spans and for heavier walls, a fabricated assembly. A variety of frequently used types is shown in Fig. 7.6. In types *b*, *c*, and *e*, a continuous plate is used to close the bottom, or soffit, of the lintel, and to join the load-carrying beams and channels into a single shipping unit. The gap between the toes of the channel flanges in type *d* may be covered by a door frame or window trim, to be installed later. Pipe and bolt separators are used to hold the two channels together to form a single member for handling.

Bearing Plates. Because of low allowable pressures on masonry, bearing plates (sometimes called masonry plates) are usually required under the ends of all beams

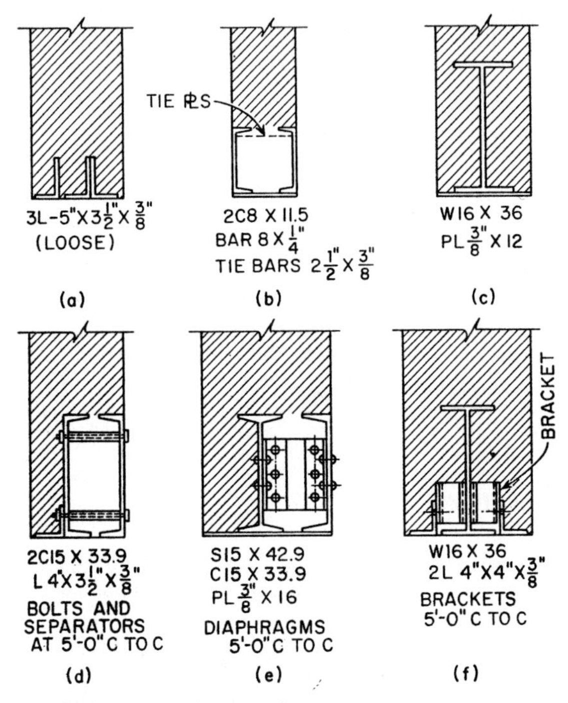

FIGURE 7.6 Lintels supporting masonry.

that rest on masonry walls, as illustrated in Fig. 7.7. Even when the pressure on the wall under a member is such that an area no greater than the contact portion of the member itself is required, wall plates are sometimes prescribed, if the member is of such weight that it must be set by the steel erector. The plates, shipped loose and in advance of steel erection, are then set by the mason to provide a satisfactory seat at the proper elevation.

Anchors. The beams are usually anchored to the masonry. **Government anchors,** as illustrated in Fig. 7.7, are generally preferred.

Nonresidential Uses. Another common application for the wall-bearing system is in one-story commercial and light industrial-type construction. The masonry side walls support the roof system, which may be rolled beams, open-web joists, or

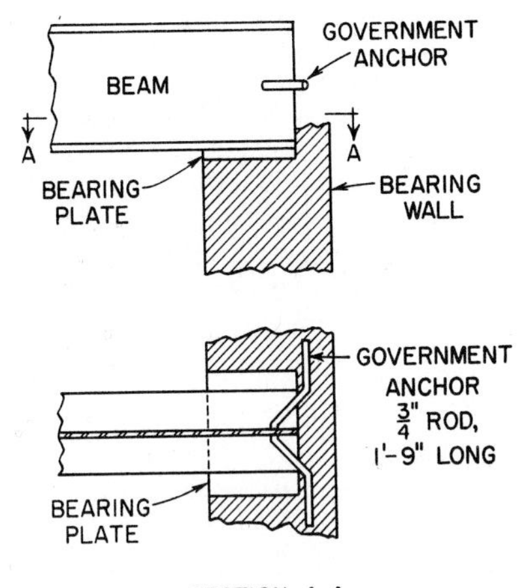

FIGURE 7.7 Wall-bearing beam.

light trusses. Clear spans of moderate size are usually economical, but for longer spans (probably over 40 ft), wall thickness and size of buttresses (pilasters) must be built to certain specified minimum proportions commensurate with the span—a requirement of building codes to assure stability. Therefore, the economical aspect should be carefully investigated. It may cost less to introduce steel columns and keep wall size to the minimum permissable. On the other hand, it may be feasible to reduce the span by introducing intermediate columns and still retain the wall-bearing system for the outer end reactions.

Planning for Erection. One disadvantage of wall-bearing construction needs emphasizing: Before steel can be set by the ironworkers, the masonry must be built up to the proper elevation to receive it. When these elevations vary, as is the case at the end of a pitched or arched roof, then it may be necessary to proceed in alternate stages, progress of erection being interrupted by the work that must be performed by the masons, and vice versa. The necessary timing to avoid delays is seldom obtained. A few columns or an additional rigid frame at the end of a building may cost less than using trades to fit an intermittent and expensive schedule. Remember, too, that labor-union regulations may prevent the trades from handling any material other than that belonging to their own craft. An economical rule may well be: Lay out the work so that the erector and ironworkers can place and connect all the steelwork in one continuous operation.

(F. S. Merritt and R. Brockenbrough, "Structural Steel Designers Handbook," 2d ed., McGraw-Hill Publishing Company, New York.)

7.6 SKELETON FRAMING

In skeleton framing all the gravity loadings of the structure, including the walls, are supported by the steel framework. Such walls are termed **nonbearing** or **curtain**

walls. This system made the skyscraper possible. Steel, being so much stronger than all forms of masonry, is capable of sustaining far greater load in a given space, thus obstructing less of the floor area in performing its function.

With columns properly spaced to provide support for the beams spanning between them, there is no limit to the floor and roof area that can be constructed with this type of framing, merely by duplicating the details for a single bay. Erected tier upon tier, this type of framing can be built to any desired height. Fabricators refer to this type of construction as "beam and column." A typical arrangement is illustrated in Fig. 7.8.

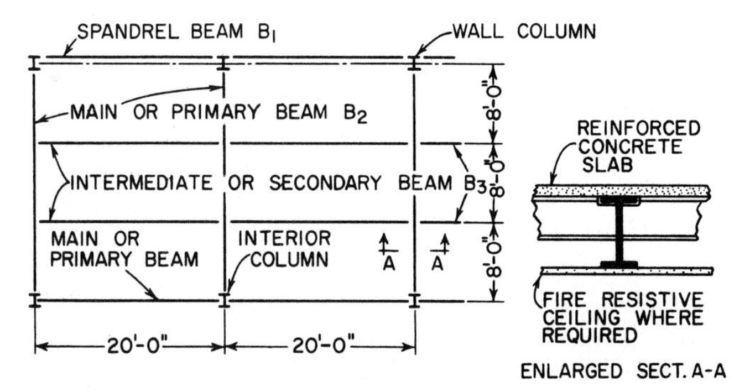

FIGURE 7.8 Typical beam-and-column steel framing, shown in plan.

The spandrel beams, marked B1 in Fig. 7.8, are located in or under the wall so as to reduce eccentricity caused by wall loads. Figure 7.9 shows two methods for connecting to the spandrel beam the shelf angle that supports the outer course of masonry over window openings 6 ft or more in width. In order that the masonry contractor may proceed expeditiously with the work, these shelf angles must be in alignment with the face of the building and at the proper elevation to match a masonry joint. The connection of the angles to the spandrel beams is made by bolting; shims are provided to make the adjustments for line and elevation.

Figure 7.9*a* illustrates a typical connection arrangement when the outstanding leg of the shelf angle is about 3 in or less below the bottom flange of the spandrel beam; Fig. 7.9*b* illustrates the corresponding arrangement when the outstanding leg of the shelf angle is more than about 3 in below the bottom flange of the spandrel beam. In the cases represented by Fig. 7.9*b*, the shelf angles are usually shipped attached to the spandrel beam. If the distance from the bottom flange to the horizontal leg of the shelf angle is greater than 10 in, a hanger may be required.

In some cases, as over door openings, the accurate adjustment features provided by Fig. 7.9*a* and *b* may not be needed. It may then be more economical to simplify the detail, as shown in Fig. 7.9*c*. The elevation and alignment will then conform to the permissible tolerances associated with the steel framework.

(E. H. Gaylord, Jr., et al., "Design of Steel Structures," 3d ed.; R. L. Brockenbrough and F. S. Merritt, "Structural Steel Designers Handbook," 2d ed., McGraw-Hill Publishing Company, New York.)

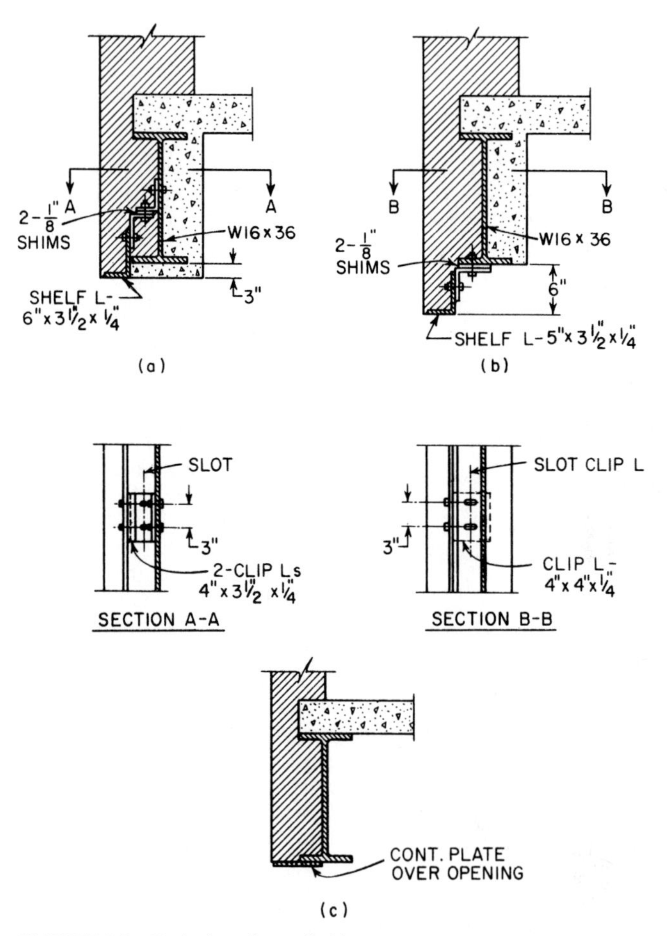

FIGURE 7.9 Typical steel spandrel beams.

7.7 LONG-SPAN FRAMING

Large industrial buildings, auditoriums, gymnasiums, theaters, hangars, and exposition buildings require much greater clear distance between supports than can be supplied by beam and column framing. When the clear distance is greater than can be spanned with rolled beams, several alternatives are available. These may be classified as *girders, simple trusses, arches, rigid frames, cantilever-suspension spans,* and various types of space frames, such as *folded plates, curvilinear grids, thin-shell domes, two-way trusses, and cable networks.*

Girders are the usual choice where depths are limited, as over large unobstructed areas in the lower floors of tall buildings, where column loads from floors above must be carried across the clear area. Sometimes, when greater strength is required

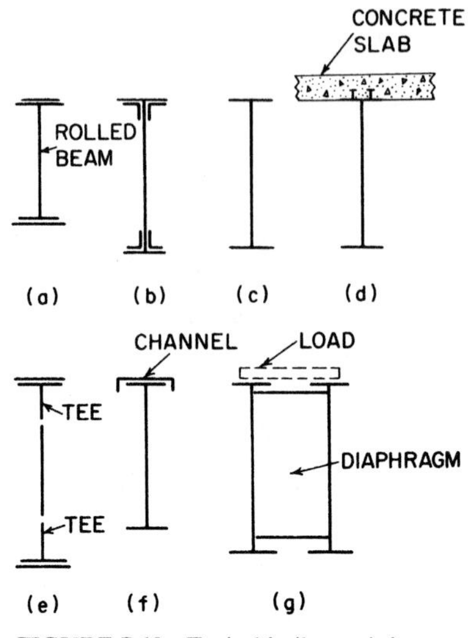

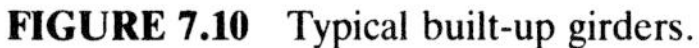
FIGURE 7.10 Typical built-up girders.

than is available in rolled beams, cover plates are added to the flanges (Fig. 7.10*a*) to provide the additional strength.

When depths exceed the limit for rolled beams, i.e., for spans exceeding about 67 ft (based on the assumption of a depth-span ratio of 1:22 with 36-in-deep Ws), the girder must be built up from plates and shapes. Welded girders are used instead of the old-type conventional riveted girders (Fig. 7.10*b*), composed of web plate, angles, and cover plates.

Welded girders generally are composed of three plates (Fig. 7.10*c*). This type offers the most opportunity for simple fabrication, efficient use of material, and least weight. Top and bottom flange plates may be of different size (Fig. 7.10*d*), an arrangement advantageous in composite construction, which integrates a concrete floor slab with the girder flange, to function together.

Heavy girders may use cover-plated tee sections (Fig. 7.10*e*). Where lateral loads are a factor, as in the case of girders supporting cranes, a channel may be fastened to the top flange (Fig. 7.10*f*). In exceptionally heavy construction, it is not unusual to use a pair of girders diaphragmed together to share the load (Fig. 7.10*g*).

The availability of high-strength, weldable steels resulted in development of **hybrid girders.** For example, a high-strength steel, say A572 Grade 50, whose yield stress is 50 ksi, may be used in a girder for the most highly stressed flanges, and the lower-priced A36 steel, whose yield stress is 36 ksi, may be used for lightly stressed flanges and web plate and detail material. The AISC specification for allowable-stress design requires that the top and bottom flanges at any cross section have the same cross-sectional area, and that the steel in these flanges be of the same grade. The allowable bending stress may be slightly less than that for conventional homogeneous girders of the high-strength steel, to compensate for possible overstress in the web at the junction with the flanges. Hybrid girders are efficient and economical for heavy loading and long spans and, consequently, are frequently employed in bridgework.

Trusses. When depth limits permit, a more economical way of spanning long distances is with trusses, for both floor and roof construction. Because of their greater depth, trusses usually provide greater stiffness against deflection when compared pound for pound with the corresponding rolled beam or plate girder that otherwise would be required. Six general types of trusses frequently used in building frames are shown in Fig. 7.11 together with modifications that can be made to suit particular conditions.

Trusses in Fig. 7.11*a* to *d* and *k* may be used as the principal supporting members in floor and roof framing. Types *e* to *j* serve a similar function in the framing of symmetrical roofs having a pronounced pitch. As shown, types *a* to *d* have a top chord that is not quite parallel to the bottom chord. Such an arrangement is used

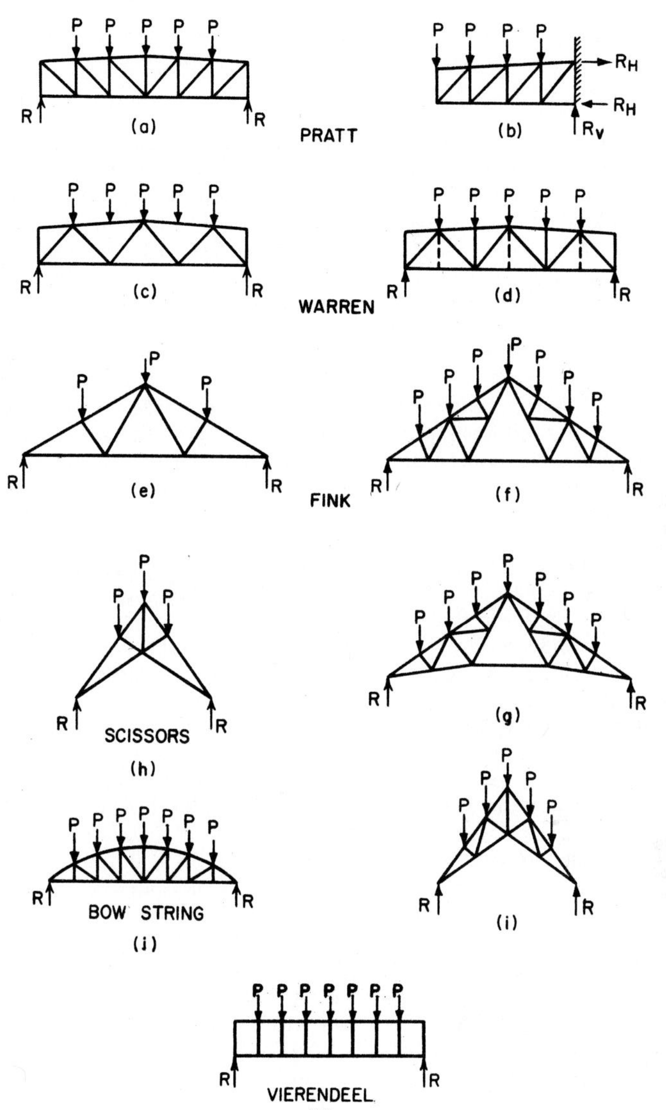

FIGURE 7.11 Types of steel trusses.

to provide for drainage of flat roofs. Most of the connections of the roof beams **(purlins),** which these trusses support, can be identical, which would not be the case if the top chord were dead level and the elevation of the purlins varied. When used in floors, truss types *a* to *d* have parallel chords.

Properly proportioned, bow string trusses (Fig. 7.11*j*) have the unique characteristic that the stress in their web members is relatively small. The top chord, which usually is formed in the arc of a circle, is stressed in compression, and the bottom chord is stressed in tension. In spite of the relatively expensive operation of forming the top chord, this type of truss has proved very popular in roof framing on spans of moderate lengths up to about 100 ft.

The Vierendeel truss (Fig. 7.11*k*) generally is shop welded to the extent possible to develop full rigidity of connections between the verticals and chords. It is useful where absence of diagonals is desirable to permit passage between the verticals.

Trusses also may be used for long spans, as three-dimensional trusses (space frames) or as grids. In two-way grids, one set of parallel lines of trusses is intersected at 90° by another set of trusses so that the verticals are common to both sets. Because of the rigid connections at the intersections, loads are distributed nearly equally to all trusses. Reduced truss depth and weight savings are among the apparent advantages of such grids.

Long-span joists are light trusses closely spaced to support floors and flat roofs. They conform to standard specifications (Table 7.1, p. 7.3) and to standard loading. Both Pratt and Warren types are used, the shape of chords and webs varying with the fabricator. Yet, all joists with the same designation have the same guaranteed load-supporting capacity. The standard loading tables list allowable loads for joists up to 72 in deep and with clear span up to 144 ft. The joists may have parallel or sloping chords or other configuration.

Truss Applications. Cross sections through a number of buildings having roof trusses of the general type just discussed are shown diagrammatically in Fig. 7.12. Cross section *a* might be that of a storage building or a light industrial building. A Fink truss provides a substantial roof slope. Roofs of this type are often designed to carry little loading, if any, except that produced by wind and snow, since the contents of the building are supported on the ground floor. For light construction, the roof and exterior wall covering may consist of thin, cold-formed metal panels. Lighting and ventilation, in addition to that provided by windows in the vertical side walls, frequently are furnished by means of sash installed in the vertical side of a continuous monitor, framing for which is indicated by the dotted lines in the sketch.

Cross section *b* shows a scissors truss supporting the high roof over the nave of a church. This type of truss is used only when the roof pitch is steep, as in ecclesiastical architecture.

A modified Warren truss, shown in cross section *c,* might be one of the main supporting roof members over an auditorium, gymnasium, theater, or other assembly-type building where large, unobstructed floor space is required. Similar trusses, including modified Pratt, are used in the roofs of large garages, terminal buildings, and airplane hangars, for spans ranging from about 80 up to 500 ft.

The Pratt truss (Fig. 7.12*d*) is frequently used in industrial buildings, while *e* depicts a type of framing often used where overhead traveling cranes handle heavy loads from one point on the ground to another.

Arches. When very large clear spans are needed, the bent framing required to support walls and roof may take the form of solid or open-web arches, of the kind shown in Fig. 7.13. A notable feature of bents *a* and *b* is the heavy steel pins at points *A, B,* and *C,* connecting the two halves of the arch together at the crown and supporting them at the foundation. These pins are designed to carry all the reaction from each half arch, and to function in shear and bearing much as a single bolt is assumed to perform when loaded in double shear.

Use of hinge pins offers two advantages in long-span frames of the type shown in Fig. 7.13. In the first place, they simplify design calculations. Second, they

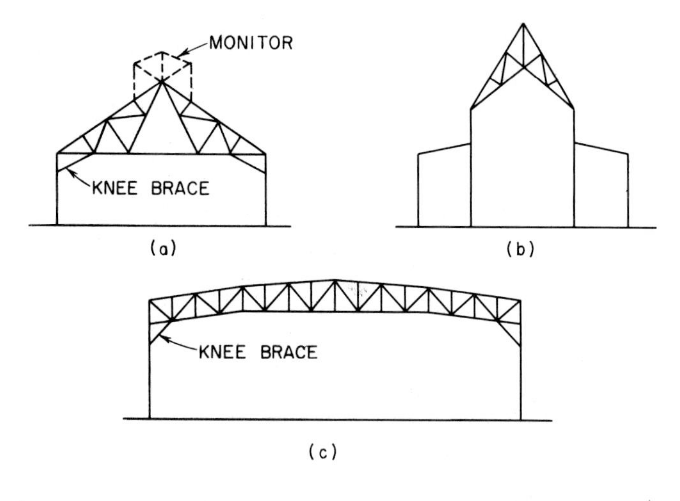

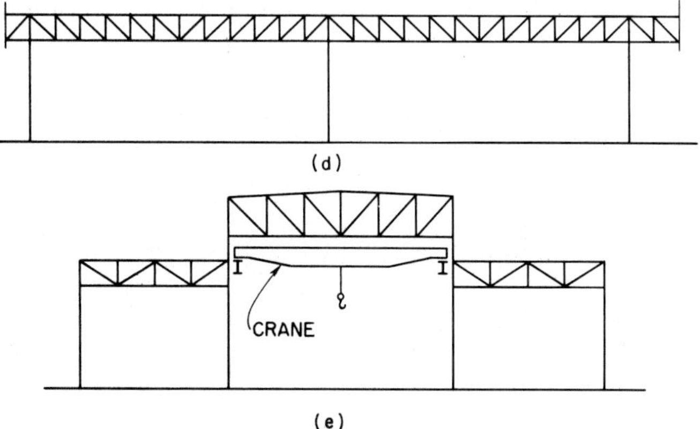

FIGURE 7.12 Some examples of structures with truss roofs.

simplify erection. All the careful fitting can be done and strong connections required to develop the needed strength at the ends of the arch can be made in the shop, instead of high above ground in the field. When these heavy members have been raised in the field about in their final position, the upper end of each arch is adjusted, upward or downward, by means of jacks near the free end of the arch. When the holes in the pin plates line up exactly, the crown pin is slipped in place and secured against falling out by the attachment of keeper plates. The arch is then ready to carry its loading. Bents of the type shown in Fig. 7.13*a* and *b* are referred to as **three-hinged arches.**

When ground conditions are favorable and foundations are properly designed, and if the loads to be carried are relatively light, as, for example, for a large gymnasium, a **hingeless arch** similar to the one shown diagrammatically in Fig. 7.13*c* may offer advantage in overall economy.

In many cases, the arches shown in Fig. 7.13*a* and *b* are designed without the pins at *B* **(two-hinged arch).** Then, the section at *B* must be capable of carrying

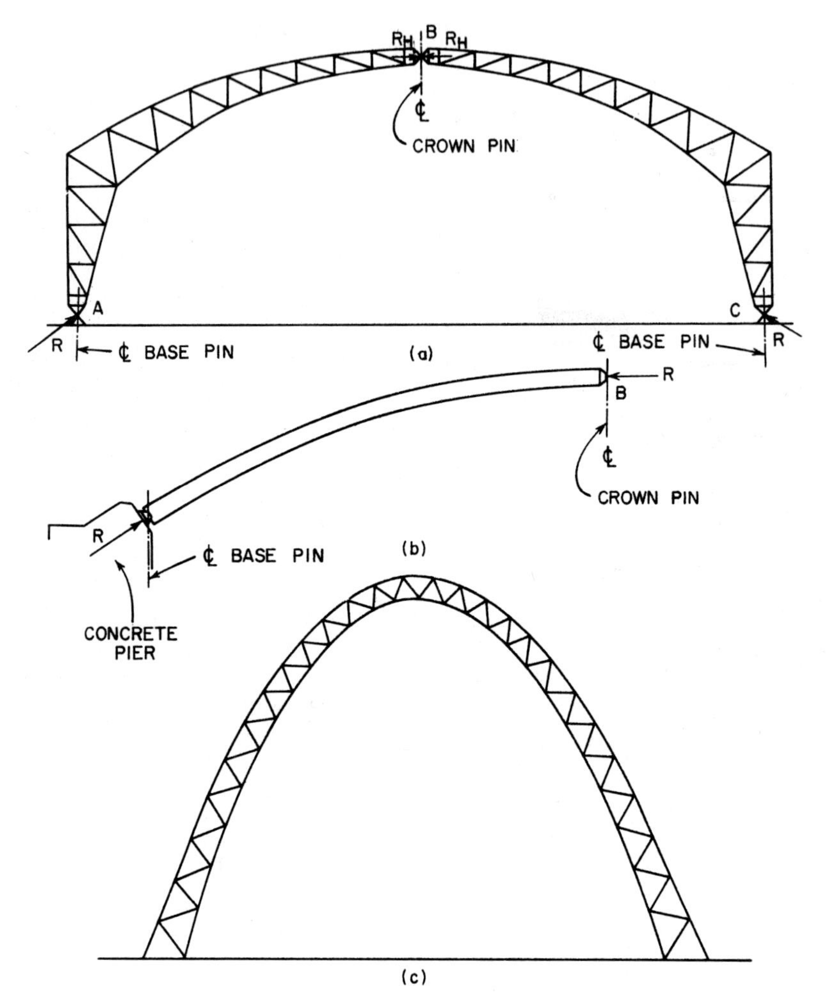

FIGURE 7.13 Steel arches: (*a*) and (*b*) three-hinged; (*c*) fixed.

the moment and shear present. Therefore, the section at *B* may be heavier than for the three-hinged arch, and erection will be more exacting for correct closure.

Rigid Frames. These are another type of long-span bent. In design, the stiffness afforded by beam-to-column connections is carefully evaluated and counted on in the design to relieve some of the bending moment that otherwise would be assumed as occurring with maximum intensity at midspan. Typical examples of rigid frame bents are shown in Fig. 7.14. When completely assembled in place in the field, the frames are fully continuous throughout their entire length and height. A distinguishing characteristic of rigid frames is the absence of pins or hinges at the crown, or midspan.

In principle, single-span rigid-frame bents are either two-hinged or hingeless arches. For hingeless arches, the column bases are fully restrained by large rigid

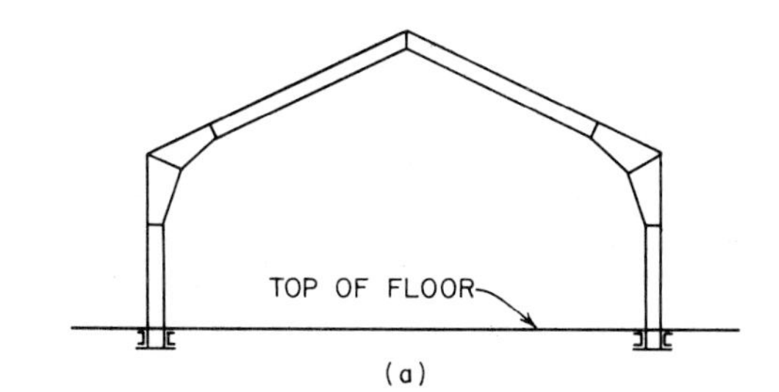

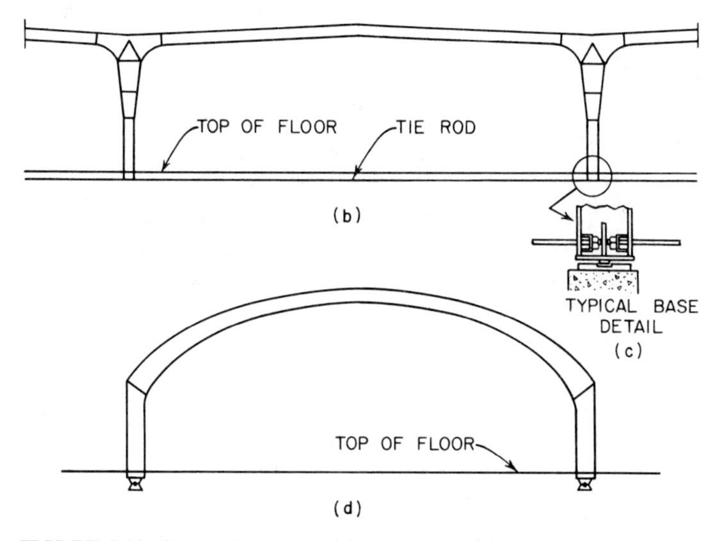

FIGURE 7.14 Steel rigid frames: (*a*) single bent; (*b*) continuous frame with underfloor tie; (*c*) connection of tie to a column; (*d*) two-hinged.

foundations, to which they are attached by a connection capable of transmitting moment as well as shear. Since such foundations may not be economical or even possible when soil conditions are not favorable, the usual practice is to consider the bents hinged at each reaction. However, this does not imply the necessity of expensive pin details; in most cases, sufficient rotation of the column base can be obtained with the ordinary flat-ended base detail and a single line of anchor bolts placed perpendicular to the span on the column center line. Many designers prefer to obtain a hinge effect by concentrating the column load on a narrow bar, as shown in Fig. 7.14*c*; this refinement is worthwhile in larger spans.

Regardless of how the frame is hinged, there is a problem in resisting the horizontal shear that the rigid frame imparts to the foundation. For small spans and light thrusts, it may be feasible to depend on the foundation to resist lateral displacement. However, more positive performance and also reduction in costs are usually obtained by connecting opposite columns of a frame with tie rods, as illustrated in Fig. 7.14*b*, thus eliminating these horizontal forces from the foundation.

For ties on small spans, it may be possible to utilize the reinforcing bars in the floor slab or floor beams, by simply connecting them to the column bases. On larger spans, it is advisable to use tie rods and turnbuckles, the latter affording the opportunity to prestress the ties and thus compensate for elastic elongation of the rods when stressed. Prestressing the rod during erection to 50% of its value has been recommended for some major installations; but the foundations should be checked for resisting some portion of the thrust.

Single-story, welded rigid frames often are chosen where exposed steelwork is desired for such structures as churches, gymnasiums, auditoriums, bowling alleys, and shopping centers, because of attractive appearance and economy. Columns may be tapered, girders may vary in depth linearly or parabolically, haunches (knees) may be curved, field joints may be made inconspicuous, and stiffeners may simply be plates.

Field Splices. One problem associated with long-span construction is that of locating field splices compatible with the maximum sizes of members that can be shipped and erected. Field splices in frames are generally located at or near the point of counterflexure, thus reducing the splicing material to a minimum. In general, the maximum height for shipping by truck is 8 ft, by rail 10 ft. Greater overall depths are possible, but these should always be checked with the carrier; they vary with clearances under bridges and through tunnels.

Individual shipping pieces must be stiff enough to be handled without buckling or other injury, light enough to be lifted by the raising equipment, and capable of erection without interference from other parts of the framework. This suggests a study of the entire frame to ensure orderly erection, and to make provisions for temporary bracing of the members, to prevent jackknifing, and for temporary guying of the frame, to obtain proper alignment.

Hung-Span Beams. In some large one-story buildings, an arrangement of cantilever-suspension (hung) spans (Fig. 7.15) has proved economical and highly efficient. This layout was made so as to obtain equal maximum moments, both negative and positive, for the condition of uniform load on all spans. A minimum of three spans is required; that is, a combination of two end spans (A) and one intermediate span (C). The connection at the end of the cantilever (point D) must be designed as a shear connection only. If the connection is capable of transmitting moment as well as shear, it will change the design to one of continuity and the dimensions in Fig. 7.15 will not apply. This scheme of cantilever and suspended spans is not necessarily limited to one-story buildings.

As a rule, interior columns are separate elements in each story. Therefore, horizontal forces on the building must be taken solely by the exterior columns.

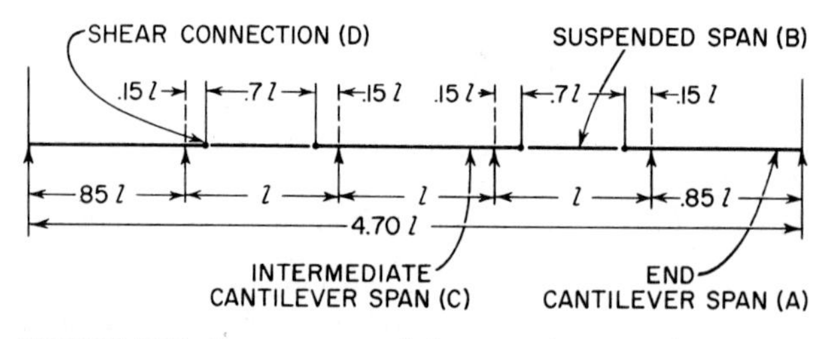

FIGURE 7.15 Hung- or suspended-span steel construction.

(E. H. Gaylord, Jr., et al., "Design of Steel Structures," 3d ed.; and F. S. Merritt and R. L. Brockenbrough, "Structural Steel Designer's Handbook," 2d ed., McGraw-Hill Publishing Company, New York.)

7.8 STEEL AND CONCRETE FRAMING

In another type of framing system, different from those described in Arts. 7.6 and 7.7, a partial use of structural steel has an important role, namely, **composite framing** of reinforced concrete and structural steel.

Composite construction actually occurs whenever concrete is made to assist steel framing in carrying loads. The term composite, however, often is used for the specific cases in which concrete slabs act together with flexural members.

Reinforced-concrete columns of conventional materials when employed in tall buildings and for large spans become excessively large. One method of avoiding this objectionable condition is to use high-strength concrete and high-strength reinforcing bars. Another is to use a structural-steel column core. In principle, the column load is carried by both the steel column and the concrete that surrounds the steel shape. Building codes usually contain an appropriate formula for this condition.

A number of systems employ a combination of concrete and steel in various ways. One method features steel columns supporting a concrete floor system by means of a steel shearhead connected to the columns at each floor level. The shallow grillage is embedded in the floor slab, thus obtaining a smooth ceiling without drops or capitals.

Another combination system is the **lift-slab** method. In this system, the floor slabs are cast one on top of another at ground level. Jacks, placed on the permanent steel columns, raise the slabs, one by one, to their final elevation, where they are made secure to the columns. When fireproofing is required, the columns may be boxed in with any one of many noncombustible materials available for that purpose. The merit of this system is the elimination of formwork and shoring that are essential in conventional reinforced-concrete construction.

For high-rise buildings, structural-steel framing often is used around a central, load-bearing, concrete core, which contains elevators, stairways, and services. The thick walls of the core, whose tubular configuration may be circular, square, or rectangular, are designed as shear walls to resist all the wind forces as well as gravity loads. Sometimes, the surrounding steel framing is cantilevered from the core, or the perimeter members are hung from trusses or girders atop the core and possibly also, in very tall buildings, at midheight of the core.

FLOOR AND ROOF SYSTEMS

7.9 FLOOR-FRAMING DESIGN CONSIDERATIONS

Selection of a suitable and economical floor system for a steel-frame building involves many considerations: load-carrying capacity, durability, fire resistance, dead weight, overall depth, facility for installing power, light, and telephones, facility for installing air conditioning, sound transmission, appearance, maintenance, and construction time.

Building codes specify minimum design live loads for buildings. In the absence of a code regulation, one may use "Minimum Design Loads in Buildings and Other Structures," ASCE 7-88, American Society of Civil Engineers. See also Art. 5.1.2. Floors should be designed to support the actual loading or these minimum loads, whichever is larger. Most floors can be designed to carry any given load. However, in some instances, a building code may place a maximum load limit on particular floor systems without regard to calculated capacity.

Resistance to lateral forces should not be disregarded, especially in areas of seismic disturbances or for perimeter windbents. In designs for such conditions, floors may be employed as horizontal diaphragms to distribute lateral forces to walls or the framing may be designed to transmit them to the ground.

Durability becomes a major consideration when a floor is subject to loads other than static or moderately kinetic types of forces. For example, a light joist system may be just the floor for an apartment or an office building but may be questionable for a manufacturing establishment where a floor must resist heavy impact and severe vibrations. Shallow floor systems deflect more than deep floors; the system selected should not permit excessive or objectionable deflections.

Fire resistance and fire rating are very important factors, because building codes, in the interest of public safety, specify the degree of resistance that must be provided. Many floor systems are rated by the codes or by fire underwriters for purposes of satisfying code requirements or basing insurance rates.

The dead weight of the floor system, including the framing, is an important factor affecting economy of construction. For one thing, substantial saving in the weight and cost of a steel frame may result with lightweight floor systems. In addition, low dead weight may also reduce foundation costs.

Joist systems, either steel or concrete, require no intermediate support, since they are obtainable in lengths to meet normal bay dimensions in tier building construction. On the other hand, concrete arch and cellular-steel floors are usually designed with one or two intermediate beams within the panel. The elimination of secondary beams does not necessarily mean overall economy just because the structural-steel contract is less. These beams are simple to fabricate and erect and allow much duplication. An analysis of contract prices shows that the cost per ton of secondary beams will average 20% under the cost per ton for the whole steel structure; or viewed another way, the omission of secondary beams increases the price per ton on the balance of the steelwork by 3½% on the average. This fact should be taken into account when making a cost analysis of several systems.

Sometimes, the depth of a floor system is important. For example, the height of a building may be limited for a particular type of fire-resistant construction or by zoning laws. The thickness of the floor may well be the determining factor limiting the number of stories that can be built. Also, the economy of a deep floor is partly offset by the increase in height of walls, columns, pipes, etc.

Another important consideration, particularly for office buildings and similar-type occupancies, is the need for furnishing an economical and flexible electrical wiring system. With the accent on movable partitions and ever-changing office arrangements, the readiness and ease with which telephones, desk lights, computers, and other electric-powered business machines can be relocated are of major importance. Therefore, the floor system that by its makeup provides large void spaces or cells for concealing wiring possesses a distinct advantage over competitive types of solid construction. Likewise, accommodation of recessed lighting in ceilings may disclose an advantage for one system over another. Furthermore, for economical air conditioning and ventilation, location of ducts and method of support warrant study of several floor systems.

Sound transmission and acoustical treatments are other factors that need to be evaluated. A wealth of data are available in reports of the National Institute of Standards and Technology. In general, floor systems of sandwich type with air spaces between layers afford better resistance to sound transmission than solid systems, which do not interrupt sound waves. Although the ideal soundproof floor is impractical, because of cost, several reasonably satisfactory systems are available. Much depends on type of occupancy, floor coverings, and ceiling finish—acoustical plaster or tile.

Appearance and maintenance also should be weighed by the designer and the owner. A smooth, neat ceiling is usually a prerequisite for residential occcupany; a less expensive finish may be deemed satisfactory for an institutional building.

Speed of construction is essential. Contractors prefer systems that enable the following-up trades to work immediately behind the erector and with unimpeded efficiency.

In general, the following constructions are commonly used in conjunction with steel framing: concrete joists (removable pan), steel joists, cellular steel flooring, and composite concrete-steel beams (Arts. 7.9.1 to 7.9.4). In addition, there are numerous adaptations of these and proprietary systems.

7.9.1 Concrete-Pan Floors

Concrete floors cast on removable metal forms or pans, which form the joists, are frequently used with steel girders. Since the joists span the distance between columns, intermediate steel beams are not needed (Fig. 7.16). This floor generally weighs less than the arch system (reinforced concrete slabs on widely spaced beams), but still considerably more than the lightest types.

There are a number of variations of the concrete-joist system, such as the "grid" or "waffle" system, where the floor is cast on small, square, removable pans, or domes, so that the finished product becomes a two-way joist system. Other systems employ permanent filler blocks—usually a lightweight tile. Some of these variations fall in the heaviest floor classification: also the majority require substantial forms and shoring.

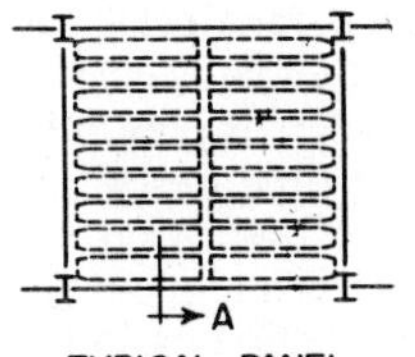

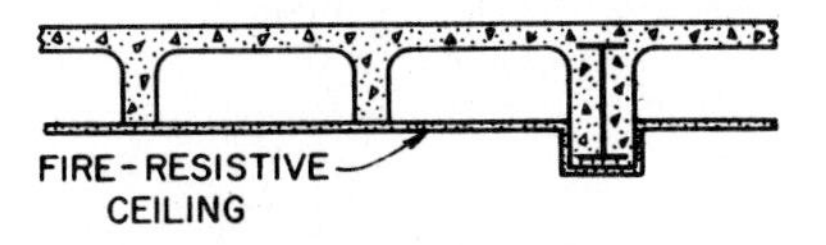

FIGURE 7.16 Concrete joist floor.

7.9.2 Steel Joist Floors

The lightest floor system in common use is the open-web steel joist construction shown in Fig. 7.17. It is popular for all types of light occupancies, principally because of initial low cost.

Many types of open-web joists are available. Some employ bars in their makeup, while others are entirely of rolled shapes; they all conform to standards and good-

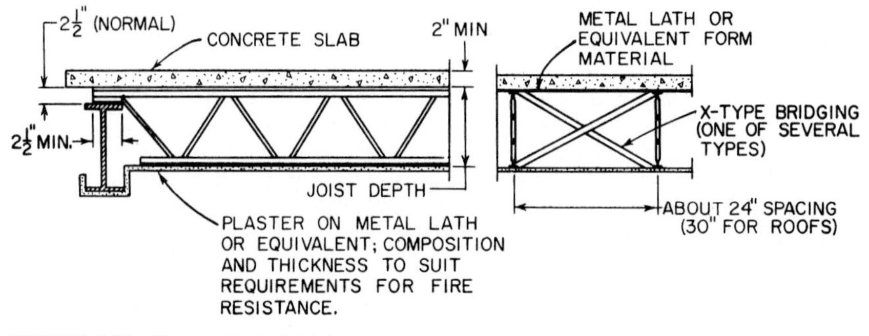

FIGURE 7.17 Open-web steel joist.

practice specifications promulgated by the Steel Joist Institute and the American Institute of Steel Construction (see Table 7.1). All joists conform to the standard loading tables and carry the same size designation so that designers need only indicate on project drawings the standard marking without reference to manufacturer, just as for a steel beam or column section.

Satisfactory joist construction is assured by adhering to SJI and AISC recommendations. Joists generally are spaced 2 ft c to c. They should be adequately braced (with bridging) during construction to prevent rotation or buckling, and to avoid "springy" floors, they should be carefully selected to provide sufficient depth.

This system has many advantages: Falsework is eliminated. Joists are easily handled, erected, and connected to supporting beams—usually by tack welding. Temporary coverage and working platforms are quickly placed. The open space between joists, and through the webs, may be utilized for ducts, cables, light fixtures, and piping. A thin floor slab may be cast on steel lath, corrugated-steel sheets, or wire-reinforced paper lath laid on top of the joists. A plaster ceiling may be suspended or attached directly to the bottom flange of the joists.

Lightweight beams, or so-called "junior" beams, are also used in the same manner as open-web joists, and with the same advantages and economy, except that the solid webs do not allow as much freedom in installation of utilities. Beams may be spaced according to their safe load capacity; 3- and 4-ft spacings are common. As a type, therefore, the lightweight-steel-beam floor is intermediate between concrete arches and open-web joists.

7.9.3 Cellular-Steel Floors

Cold-formed steel decking is frequently used in office buildings. One type is illustrated in Fig. 7.18. Other manufacturers make similar cellular metal decks, the primary difference being in the shape of the cells. Often, decking with half cells is used. These are open ended on the bottom, but flat sheets close those cells that incorporate services. Sometimes, cells are enlarged laterally to transmit air for air conditioning.

Two outstanding advantages of cellular floors are rapidity of erection and ease with which present and future connections can be made to telephone, computer, light, and power wiring, each cell serving as a conduit. Each deck unit becomes a working platform immediately on erection, thus enabling the several finishing trades to follow right behind the steel erector.

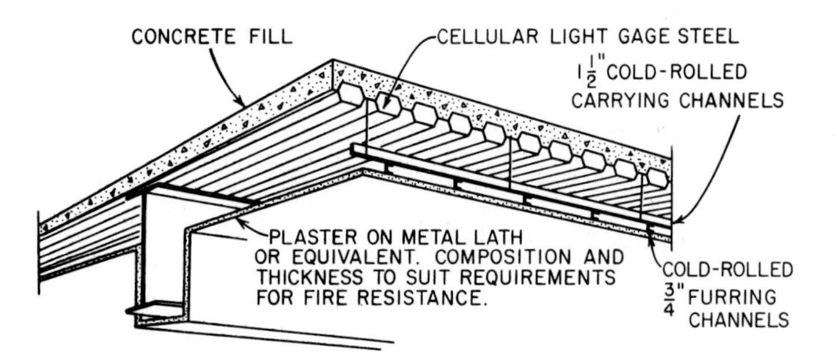

FIGURE 7.18 Cellular-steel floor.

Although the cost of the steel deck system may be larger than that of other floor systems, the cost differential can be narrowed to competitive position when equal consideration for electrical facility is imposed on the other systems; e.g., the addition of 4 in of concrete fill to cover embedded electrical conduit on top of a concrete flat-slab floor.

In earlier floors of this type, the steel decking was assumed to be structurally independent. In that case, the concrete fill served only to provide fire resistance and a level floor. Most modern deckings, however, are bonded or locked to the concrete, so that the two materials act as a unit in composite construction. Usually, only top-quality stone concrete (ASTM C33 aggregates) is used, although lightweight concrete made with ASTM C330 aggregates is an acceptable alternative.

Usage of cellular deck in composite construction is facilitated by economical attachment of shear connectors to both the decking and underlying beams. For example, when welded studs are used, a welding gun automatically fastens the studs through two layers of hot-dipped galvanized decking to the unpainted top flanges of the steel beams. This construction is similar to composite concrete-steel beams (Art. 7.9.4).

The total floor weight of cellular steel construction is low, comparable to open-web steel joists. Weight savings of about 50% are obtained in comparison with all-concrete floors; 30% savings in overall weight of the building. However, a big cost saving in a high-labor-rate area results from elimination of costly formwork needed for concrete slabs, since the steel decking serves as the form.

Fire resistance for any required rating is contributed by the fill on top of the cells and by the ceiling below (Fig. 7.18). Generally, removable panels for which no fire rating is claimed are preferred for suspended ceilings. In this case, fireproofing materials are applied directly to the underside of the metal deck and all exposed surfaces of steel floor beams, a technique often called spray-on fireproofing.

7.9.4 Composite Concrete-Steel Beams

In composite construction, the structural concrete slab is made to assist the steel beams in supporting loads. Hence, the concrete must be bonded to the steel to ensure shear transfer. When the steel beams are completely encased in the concrete, the natural bond is considered capable of resisting horizontal shear. But that bond generally is disregarded when only the top flange is in contact with the concrete. Consequently, shear connectors are used to resist the horizontal shear. Commonly used connectors are welded studs, hooked or headed, and short lengths of channels.

Usually, composite construction is most efficient for heavy loading, long spans, large beam spacing, and restricted depths. Because the concrete serves much like a cover plate, lighter steel beams may be used for given loads, and deflections are smaller than for noncomposite construction.

7.10 ROOF FRAMING SYSTEMS

These are similar in many respects to the floor types discussed in Arts. 7.9.1 to 7.9.4. In fact, for flat-top tier buildings, the roof may be just another floor. However, when roof loads are smaller than floor loads, as is usually the case, it may be economical to lighten the roof construction. For example, steel joists may be spaced farther apart. Where roof decking is used, the spacing of the joists is determined by the load-carrying ability of the applied decking and of the joists.

Most of the considerations discussed for floors in Art. 7.9 also are applicable to roof systems. In addition, however, due thought should be given to weather resistance, heat conductance and insulation, moisture absorption and vapor barriers, and especially to maintenance.

Many roof systems are distinctive as compared with the floor types; for example, the corrugated sheet-metal roofing commonly employed on many types of industrial or mill buildings. The sheets rest on small beams, channels, or joists, called **purlins,** which in turn are supported by trusses. Similar members on the sidewalls are called **girts.**

7.11 BRACING DESIGN CONSIDERATIONS

Bracing as it applies to steel structures includes secondary members incorporated into the system of main members to serve these principal functions:

1. Slender compression members, such as columns, beams, and truss elements are braced, or laterally supported, so as to restrain the tendency to buckle in a direction normal to the stress path. The rigidity, or resistance to buckling, of an individual member is determined from its length and certain physical properties of its cross section. Economy and size usually determine whether bracing is to be employed.

2. Since most structures are assemblies of vertical and horizontal members forming rectangular (or square) panels, they possess little inherent rigidity. Consequently, additional rigidity must be supplied by a secondary system of members or by rigid or semi-rigid joints between members. This is particularly necessary when the framework is subject to lateral loads, such as wind, earthquakes, and moving loads. Exempt from this second functional need for bracing are trusses, which are basically an arrangement of triangles possessing in their planes an inherent ideal rigidity both individually and collectively.

3. There frequently is a need for bracing to resist erection loads and to align or prevent overturning, in a direction normal to their planes, of trusses, bents, or frames during erection. Such bracing may be temporary; however, usually bracing needed for erection is also useful in supplying rigidity to the structure and therefore is permanently incorporated into the building. For example, braces that tie together adjoining trusses and prevent their overturning during erection are useful to prevent sway—even though the swaying forces may not be calculable.

7.11.1 Column Bracing

Interior columns of a multistory building are seldom braced between floor connections. Bracing of any kind generally interferes with occupancy requirements and architectural considerations. Since the slenderness ratio l/r in the weak direction usually controls column size, greatest economy is achieved by using only wide-flange column sections or similar built-up sections.

It is frequently possible to reduce the size of wall columns by introducing knee braces or struts in the plane of the wall, or by taking advantage of deep spandrels or girts that may be otherwise required. Thus the slenderness ratio for the weak and strong axis can be brought into approximate balance. The saving in column weight may not always be justified; one must take into account the weight of additional bracing and cost of extra details.

Column bracing is prevalent in industrial buildings because greater vertical clearances necessitate longer columns. Tall slender columns may be braced about both axes to obtain an efficient design.

Undoubtedly, heavy masonry walls afford substantial lateral support to steel columns embedded wholly or partly in the wall. The general practice, however, is to disregard this assistance.

An important factor in determining column bracing is the allowable stress or load for the column section (Art. 7.14). Column formulas for obtaining this stress are based on the ratio of two variables, effective length Kl and the physical property called radius of gyration r.

The question of when to brace (to reduce the unsupported length and thus slenderness ratio) is largely a matter of economics and architectural arrangements; thus no general answer can be given.

7.11.2 Beam Bracing

Economy in size of member dictates whether laterally unsupported beams should have additional lateral support between end supports. Lateral support at intermediate points should be considered whenever the allowable stress obtained from the reduction formulas for large l/r_t falls below some margin, say 25%, of the stress allowed for the fully braced condition. There are cases, however, where stresses as low as 4.0 ksi have been justified, because intermediate lateral support was impractical.

The question often arises: When is a steel beam laterally supported? There is no fixed rule in specifications (nor any intended in this discussion) because the answer requires application of sound judgment based on experience. Tests and studies that have been made indicate that it takes rather small forces to balance the lateral thrusts of initial buckling.

Figure 7.19 illustrates some of the common situations encountered in present-day practice. In general, **positive lateral support is provided by:**

(*a*) and (*b*) All types of cast-in-place concrete slabs (questionable for vibrating loads and loads hung on bottom flange).

(*c*) Metal and steel plate decks, with welded connections.

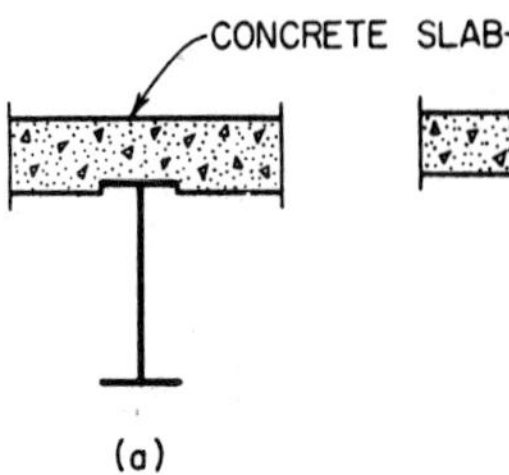

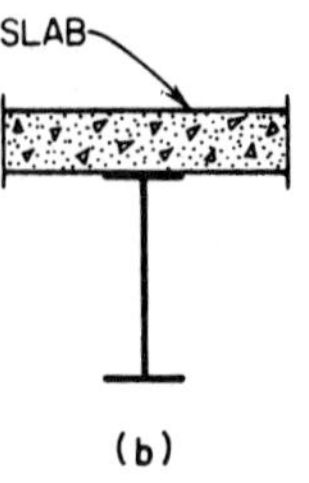

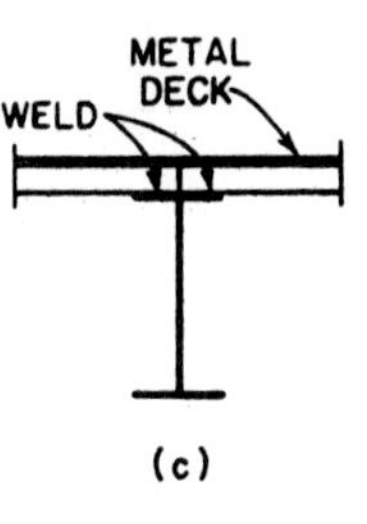

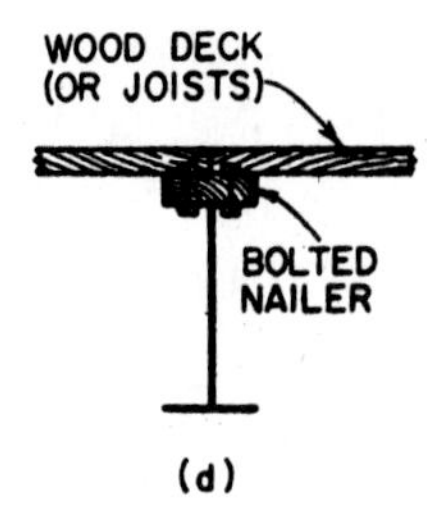

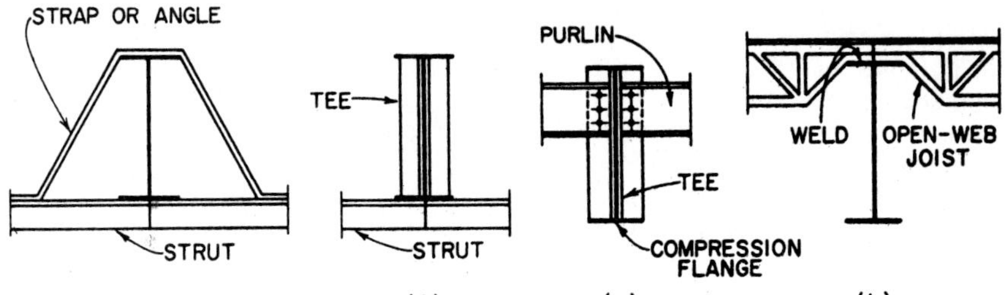

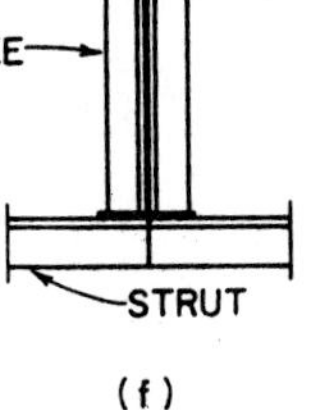

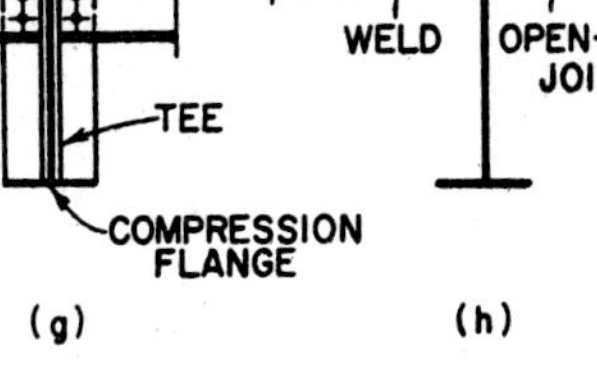

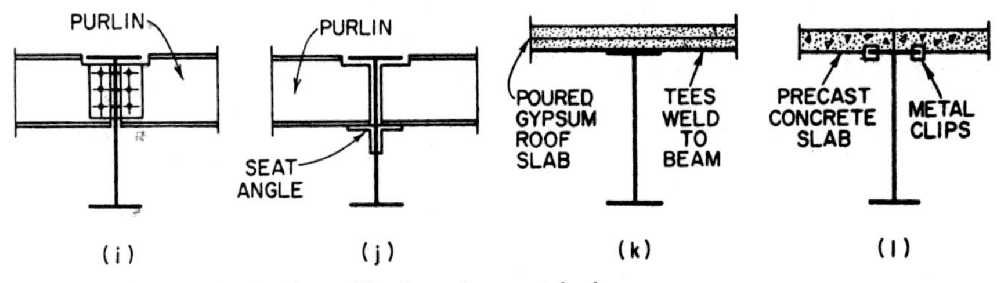

FIGURE 7.19 Methods of providing lateral support for beams.

(*d*) Wood decks nailed securely to nailers bolted to the beam.

(*e*) and (*f*) Beam flange tied or braced to strut system, either as shown in (*e*) or by means of cantilever tees, as shown in (*f*); however, struts should be adequate to resist rotation.

(*g*) Purlins used as struts, with tees acting as cantilevers (common in rigid frames and arches). If plate stiffeners are used, purlins should be connected to them with high-strength bolts to ensure rigidity.

(*h*) Open-web joists tack-welded (or the equivalent) to the beams, but the joists themselves must be braced together (bridging), and the flooring so engaged with the flanges that the joists, in turn, are adequately supported laterally.

(*i*) Purlins connected close to the compression flange.

(*k*) Tees (part of cast-in-place gypsum construction) welded to the beams.

Doubtful lateral support is provided by:

(*j*) Purlins seated on beam webs, where the seats are distant from the critical flange

(*l*) Precast slabs not adequately fastened to the compression flange.

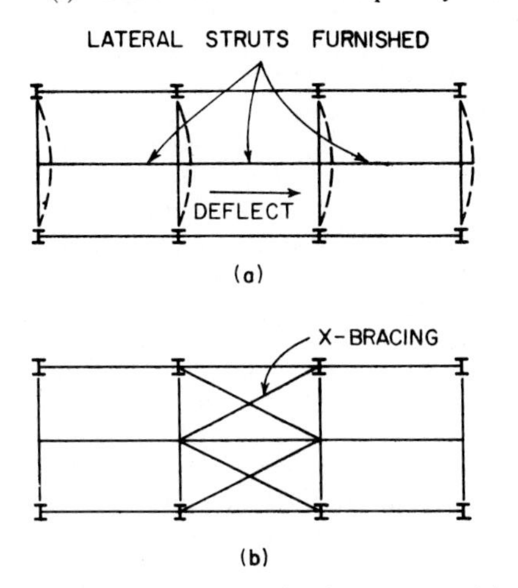

FIGURE 7.20 Lateral bracing systems: (*a*) without and (*b*) with X bracing.

The reduction formulas for large l/r, given in Fig. 7–31 do not apply to steel beams fully encased in concrete, even though no other lateral support is provided.

Introducing a secondary member to cut down the unsupported length does not necessarily result in adequate lateral support. The resistive capacity of the member and its supports must be traced through the system to ascertain effectiveness. For example the system in Fig. 7.20*a* may be free to deflect laterally as shown. This can be prevented by a rigid floor system that acts as a diaphragm, or in the absence of a floor, it may be necessary to X-brace the system as shown in Fig. 7.20*b*.

7.11.3 Load Capacity of Bracing

For an ideally straight, exactly concentrically loaded beam or column, only a small force may be needed from an intermediate brace to reduce the unbraced length of a column or the unsupported length of the compression flange of a beam. But there is no generally accepted method of calculating that force.

The principal function of a brace is to provide a node in the buckled configuration. Hence, rigidity is the main requirement for the brace. But actual members do contain nonuniform residual stresses and slight initial crookedness and may be slightly misaligned, and these eccentricities create deformations that must be resisted by the brace.

A rule used by some designers that has proved satisfactory is to design the brace for 2% of the axial load of columns, or 2% of the total compressive stress in beam flanges. Studies and experimental evidence indicate that this rule is conservative.

7.11.4 Lateral Forces on Building Frames

Design of bracing to resist forces induced by wind, seismic disturbances, and moving loads, such as those caused by cranes, is not unlike, in principle, design of members that support vertical dead and live loads. These lateral forces are readily calculable. They are collected at points of application and then distributed through the structural system and delivered to the ground. Wind loads, for example, are collected at each floor level and distributed to the columns that are selected to participate in the system. Such loads are cumulative; that is, columns resisting wind shears must support at any floor level all the wind loads on the floors above the one in consideration.

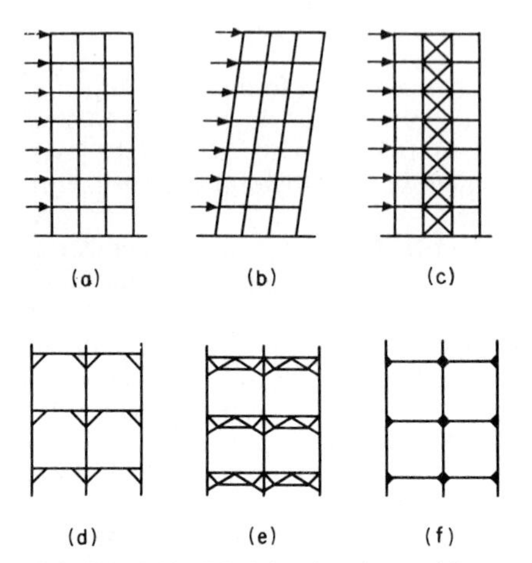

FIGURE 7.21 Wind bracing for multistory buildings.

7.11.5 Bracing Tall Buildings

If the steel frame of the multistory building in Fig. 7.21*a* is subjected to lateral wind load, it will distort as shown in Fig. 7.21*b*, if the connections of columns and beams are of the standard type, for which rigidity (resistance to rotation) is nil. One can visualize this readily by assuming each joint is connected with a single pin. Naturally, the simplest method to prevent this distortion is to insert diagonal members—triangles being inherently rigid, even if all the members forming the triangles are pin-connected.

Braced Bents. Bracing of the type in Fig. 7.21*c*, called X bracing, is both efficient and economical. Unfortunately, X bracing is usually impracticable because of interference with doors, windows, and clearance between floor and ceiling. Usually, for office buildings large column-free floor areas are required. This offers flexibility of space use, with movable partitions. But about the only place for X bracing in this type of building is in the elevator shaft, fire tower, or wherever a windowless wall is required. As a result, additional bracing must be supplied by other methods. On the other hand, X bracing is used extensively for bracing industrial buildings of the shed or mill type.

Moment-Resisting Frames. Designers have a choice of several alternatives to X bracing. Knee braces, shown in Fig. 7.21*d*, or portal frames, shown in Fig. 7.21*e*, may be used in outer walls, where they are likely to interfere only with windows. For buildings with window walls, the bracing often used is the bracket type (Fig. 7.21*f*). It simply develops the end connection for the calculated wind moment. Connections vary in type, depending on size of members, magnitude of wind moment, and compactness needed to comply with floor-to-ceiling clearances.

Figure 7.22 illustrates a number of bracket-type wind-braced connections. The minimum type, represented in Fig. 7.22*e*, consists of angles top and bottom: They are ample for moderate-height buildings. Usually the outstanding leg (against the column) is of a size that permits only one gage line. A second line of fasteners would not be effective because of the eccentricity. When greater moment resistance is needed, the type shown in Fig. 7.22*b* should be considered. This is the type that has become rather conventional in field-bolted construction. Figure 7.22*c* illustrates the maximum size with beam stubs having flange widths that permit additional gage lines, as shown. It is thus possible on larger wide-flange columns to obtain 16 fasteners in the stub-to-column connection.

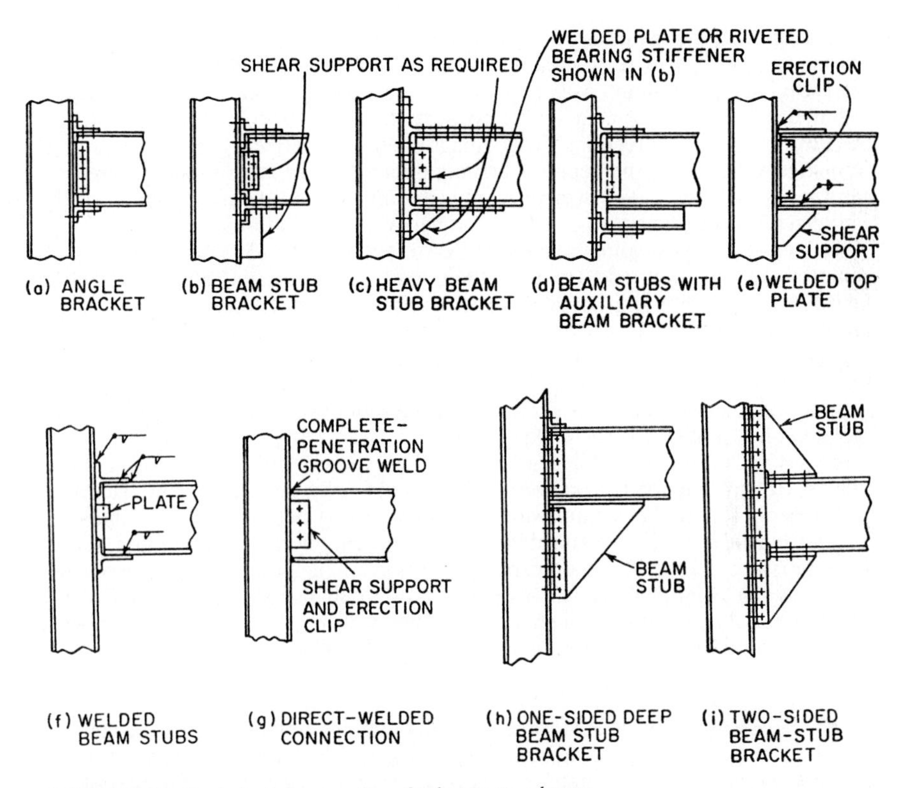

FIGURE 7.22 Typical wind connections for beams to columns.

The resisting moment of a given connection varies with the distance between centroids of the top and bottom connection piece. To increase this distance, thus increasing the moment, an auxiliary beam may be introduced as shown in Fig. 7.22*d*, if it does not create an interference.

All the foregoing types may be of welded construction, rather than bolted. In fact, it is not unusual to find mixtures of both because of the fabricator's decision to shop-bolt and field-weld, or vice versa. Welding, however, has much to offer in simplifying details and saving weight, as illustrated in Fig. 7.22*e*, *f*, and *g*. The last represents the ultimate efficiency with respect to weight saving, and furthermore, it eliminates interfering details.

Deep wing brackets (Fig. 7.22*h* and *i*) are sometimes used for wall beams and spandrels designed to take wind stresses. Such deep brackets are, of course, acceptable for interior beam bracing whenever the brackets do not interfere with required clearances.

Not all beams need be wind-braced in tall buildings. Usually the wind load is concentrated on certain column lines, called **bents,** and the forces are carried through the bents to the ground. For example, in a wing of a building, it is possible to concentrate the wind load on the outermost bent. To do so may require a stiff floor or diaphragm-like system capable of distributing the wind loads laterally. One-half these loads may be transmitted to the outer bent, and one-half to the main building to which the wing connects.

Braced bents are invariably necessary across the narrow dimension of a building. The question arises as to the amount of bracing required in the long dimension, since wind of equal unit intensity is assumed to act on all exposed faces of structures. In buildings of square or near square proportions, it is likely that braced bents will be provided in both directions. In buildings having a relatively long dimension, as compared with width, the need for bracing diminishes. In fact, in many instances, wind loads are distributed over so many columns that the inherent rigidity of the whole system is sufficient to preclude the necessity of additional bracing.

Column-to-column joints are treated differently for wind loads. Columns are compression members and transmit their loads, from section above to section below, by direct bearing between finished ends. It is not likely, in the average building, for the tensile stresses induced by wind loads ever to exceed the compressive pressure due to dead loads. Consequently, there is no theoretical need for bracing a column joint. Actually, however, column joints are connected together with nominal splice plates for practical considerations—to tie the columns during erection and to obtain vertical alignment.

This does not mean that designers may always ignore the adequacy of column splices. In lightly loaded structures, or in exceptionally tall but narrow buildings, it is possible for the horizontal wind forces to cause a net uplift in the windward column because of the overturning action. The commonly used column splices should then be checked for their capacity to resist the maximum net tensile stresses caused in the column flanges. This computation and possible heaving up of the splice material may not be thought of as bracing; yet, in principle, the column joint is being "wind-braced" in a manner similar to the wind-braced floor-beam connections.

Tubular Framing. Designs of some very tall buildings employ a structural concept that departs from the foregoing conventional method of wind-bracing steel frames. In these buildings, the perimeter columns and spandrel beams form a theoretical, vertical-cantilever hollow tube that resists the entire lateral load. The floors, through diaphragm action, transfer the lateral forces to the exterior columns. These members are closely spaced and either X-braced or rigidly connected to the spandrel beams to approximate an idealized solid-surface tube. Interior columns merely support their share of the gravity loads. The elimination of floorbeam-to-column moment connections and standardization of floorbeam sizes, freed from accumulative lateral forces, which are characteristic of conventional design, offer substantial savings. (See also Art. 3.2.4.)

("Manual of Steel Construction, Volume II Connections ASD/LRFD," American Institute of Steel Construction.)

7.11.6 Shear Walls

Masonary walls enveloping a steel frame, interior masonry walls, and perhaps some stiff partitions can resist a substantial amount of lateral load. Rigid floor systems participate in lateral-force distribution by distributing the shears induced at each floor level to the columns and walls. Yet, it is common design practice to carry wind loads on the steel frame, little or no credit being given to the substantial resistance rendered by the floors and walls. In the past, some engineers deviated from this conversatism by assigning a portion of the wind loads to the floors and walls; nevertheless, the steel frame carried the major share. When walls of glass or thin metallic curtain walls, lightweight floors, and removable partitions are used, this construction imposes on the steel frame almost complete responsibility for transmittal of wind loads to the ground. Consequently, windbracing is critical for tall steel structures.

In tall, slender buildings, such as hotels and apartments with partitions, the cracking of rigid-type partitions is related to the wracking action of the frame caused by excessive deflection. One remedy that may be used for exceptionally slender frames (those most likely to deflect excessively) is to supplement the normal bracing of the steel frame with shear walls. Acting as vertical cantilevers in resisting lateral forces, these walls, often constructed of reinforced concrete, may be arranged much like structural shapes, such as plates, channels, Ts, Is, or Hs. (See also Arts. 3.2.4 and 5.12.) Walls needed for fire towers, elevator shafts, divisional walls, etc., may be extended and reinforced to serve as shear walls, and may relieve the steel frame of cumbersome bracing or avoid uneconomical proportions.

7.11.7 Bracing Industrial-Type Buildings

Bracing of low industrial buildings for horizontal forces presents fewer difficulties than bracing of multistory buildings, because the designer usually is virtually free to select the most efficient bracing without regard to architectural considerations or interferences. For this reason, conventional X bracing is widely used—but not exclusively. Knee braces, struts, and sway frames are used where needed.

Wind forces acting on the frame shown in Fig. 7.23*a*, with hinged joints at the top and bottom of supporting columns, would cause collapse as indicated in Fig. 7.23*b*. In practice, the joints would not be hinged. However, a minimum-type connection at the truss connection and a conventional column base with anchor bolts located on the axis transverse to the frame would approximate this theoretical consideration of hinged joints. Therefore, the structure requires bracing capable of preventing collapse or unacceptable deflection.

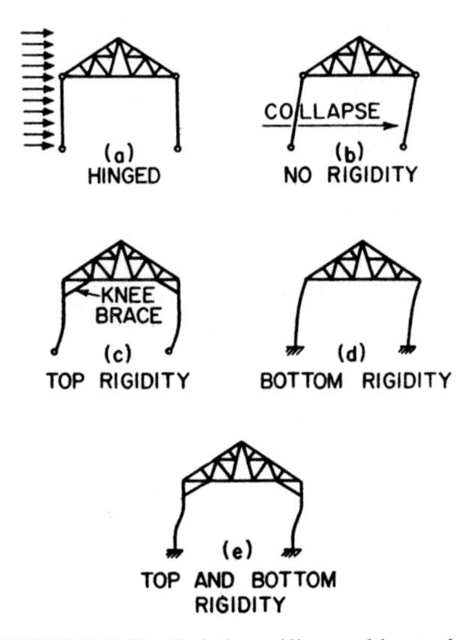

FIGURE 7.23 Relative stiffness of bents depends on restraints on columns.

In the usual case, the connection between truss and columns will be stiffened by means of knee braces (Fig. 7.23*c*). The rigidity so obtained may be supplemented by providing partial rigidity at the column base by simply locating the anchor bolts in the plane of the bent.

In buildings containing overhead cranes, the knee brace may interfere with crane operation. Then, the interference may be eliminated by fully anchoring the column base so that the column may function as a vertical cantilever (Fig. 7.23*d*).

The method often used for very heavy industrial buildings is to obtain substantial rigidity at both ends of the column so that the behavior under lateral load will resemble the condition illustrated in Fig. 7.23*e*. In both (*d*) and (*e*), the footings must be designed for such moments.

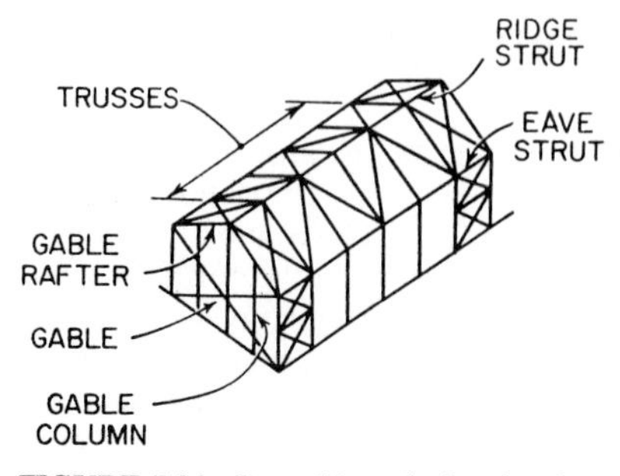

FIGURE 7.24 Braced bays in framing for an industrial building.

A common assumption in wind distribution for the type of light mill building shown in Fig. 7.24 is that the windward columns take a large share of the load acting on the side of the building and deliver the load directly to the ground. The remaining wind load on the side is delivered by the same columns to the roof systems, where the load joins with the wind forces imposed directly on the roof surface. Then, by means of diagonal X bracing, working in conjunction with the struts and top chords of the trusses, the load is carried to the eave struts, thence to the gables and, through diagonal bracing, to the foundations.

Because wind may blow from any direction, the building also must be braced for the wind load on the gables. This bracing becomes less important as the building increases in length and conceivably could be omitted in exceptionally long structures. The stress path is not unlike that assumed for the transverse wind forces. The load generated on the ends is picked up by the roof system and side framing, delivered to the eave struts, and then transmitted by the diagonals in the end sidewall bays to the foundation.

No distribution rule for bracing is intended in this discussion; bracing can be designed many different ways. Whereas the foregoing method would be sufficient for a small building, a more elaborate treatment may be required for larger structures.

Braced bays, or towers, are usually favored for structures such as that shown in Fig. 7.25. There, a pair of transverse bents are connected together with X bracing in the plane of the columns, plane of truss bottom chords, plane of truss top chords, and by means of struts and sway frames. It is assumed that each such tower can carry the wind load from adjacent bents, the number depending on assumed rigidities, size, span, and also on sound judgment. Usually every third or fourth bent should become a braced bay. Participation of bents adjoining the braced bay can be assured by insertion of bracing designated "intermediate" in Fig. 7.25*b*. This bracing is of greater importance when knee braces between trusses and columns

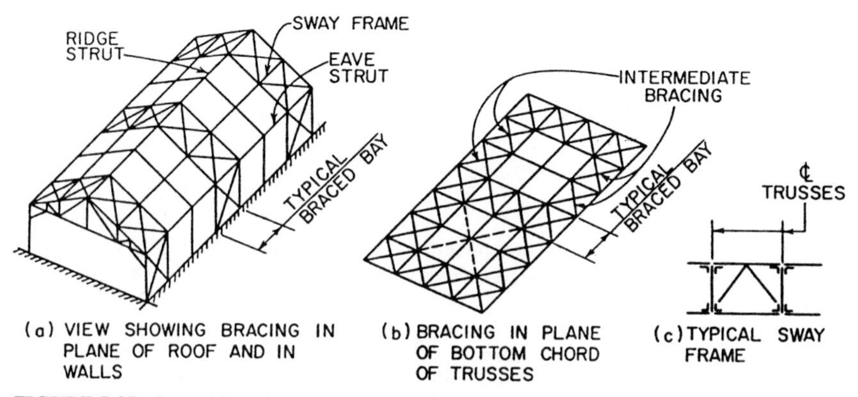

FIGURE 7.25 Braced bays in a one-story building transmit wind loads to the ground.

cannot be used. When maximum lateral stiffness of intermediate bents is desired, it can be obtained by extending the X bracing across the span; this is shown with broken lines in Fig. 7.25*b*.

Buildings with flat or low-pitched roofs, shown in Fig. 7.12*d* and *e,* require little bracing because the trusses are framed into the columns. These columns are designed for the heavy moments induced by wind pressure against the building side. The bracing that would be provided, at most, would consist of X bracing in the plane of the bottom chords for purpose of alignment during erection and a line or two of sway frames for longitudinal rigidity. Alignment bracing is left in the structure since it affords a secondary system for distributing wind loads.

7.11.8 Bracing Craneway Structures

All building framing affected by overhead cranes should be braced for the thrusts induced by sidesway and longitudinal motions of the cranes. Bracing used for wind or erection may be assumed to sustain the lateral crane loadings. These forces are usually concentrated on one bent. Therefore, normal good practice dictates that adjoining bents share in the distribution. Most effective is a system of X bracing located in the plane of the bottom chords of the roof trusses.

In addition, the bottom chords should be investigated for possible compression, although the chords normally are tension members. A heavily loaded crane is apt to draw the columns together, conceivably exerting a greater compression stress than the tension stress obtainable under dead load alone. This may indicate the need for intermediate bracing of the bottom chord.

7.11.9 Bracing Rigid Frames

Rigid frames of the type shown in Fig. 7.14 have enjoyed popular usage for gymnasiums, auditoriums, mess halls, and with increasing frequency, industrial buildings. The stiff knees at the junction of the column with the rafter imparts excellent transverse rigidity. Each bent is capable of delivering its share of wind load directly to the footings. Nevertheless, some bracing is advisable, particularly for resisting wind loads against the end of the building. Most designers emphasize the importance of an adequate eave strut; it usually is arranged so as to brace the inside flange (compression) of the frame knee, the connection being located at the midpoint of the transition between column and rafter segments of the frame. Intermediate X bracing in the plane of the rafters usually is omitted.

DESIGN OF MEMBERS

In proportioning of members, designers should investigate one or more or a combination of five basic stress or strength conditions: axial tension, axial compression, bending, shearing, and member element crippling. Other conditions that should be investigated under special conditions are local buckling, excessive deflection, and torsion. In the past, such analyses were based on allowable stress design (ASD). More recently, a method known as load and resistance factor design (LRFD) has come into use because it permits a more rational design. It takes into account the probability of loading conditions and statistical variations in the strength, or resistance capability, of members and connection materials.

7.12 BASES FOR ASD AND LRFD

ASD is based on elastic theory. Design limits the maximum unit stress a member is permitted to bear under service loads to a level determined by a judgmental, but experience-based, safety factor. Building codes establish allowable unit stresses, which are normally related to the minimum yield stress for each grade of steel.

Plastic design is based on the ultimate strength of members. A safety factor, comparable to that established for elastic design, is applied to the design load to determine the ultimate-load capacity required of a member.

LRFD is based on the concept that no applicable limit state should be exceeded when the structure, or any member or element, is subject to appropriate combinations of factored loads.

A **limit state** is defined as a condition in which a structure or structural component becomes unfit for further structural service. A structural member can have several limit states.

Strength limit states relate to maximum load-carrying capacity.

Serviceability limit states relate to performance under normal service conditions with respect to such factors as deflection and vibration.

Design specifications establish **load factors** to be applied to each type of service load, such as dead, live, and wind loads, the values of the factors depending on the specific combination of loads to be imposed on a structure (Art. 5.1.3).

The AISC "Load and Resistance Factor Design Specification for Structural Steel Buildings" requires that structures be designed so that, under the most critical combination of factored loads, the design strength of the structures or their individual elements is not exceeded. For each strength limit state, the design strength is the product of the nominal strength and a resistance factor ϕ, given in the specification. Derived with the use of probability theory, ϕ provides an extra margin of safety for the limit state being investigated. Nominal strength of a member depends on its geometric properties, yield or ultimate strength, and type of loading to be resisted, such as tension, compression, or flexure.

The AISC LRFD specification permits structural analysis based on either elastic or plastic behavior. Elastic theory is most commonly used. Where plastic theory is used for complex structures, all possible mechanisms that may form in the structure should be investigated. The collapse mechanism is the one that requires the lightest load for collapse to occur.

Numerous computer programs for analysis and design of members or structures are available. If data input describing the structure and loading are accurate, most of these programs yield a quick and accurate design. For complex structures, care should be taken in use of computer programs to check the results to ensure that they are logical, since a critical input error may not be easily found. If a program can produce a plot of the configuration of the loaded structure based on the data input, the plot should be used as a check, inasmuch as omission of a member or other errors in connectivity data can be readily discerned from the plot.

("Plastic Design in Steel—A Guide and Commentary," M & R No. 41, American Society of Civil Engineers.)

7.13 TENSION MEMBERS

These are proportioned so that their gross and net areas are large enough to resist imposed loads. The criteria for determining the net area of a tension member with bolt holes is the same for allowable stress design and load-and-resistance-factor design. In determination of net area, the width of a bolt hole should be taken 1/16 in larger than the nominal dimension of the hole normal to the direction of applied stress. Although the gross section for a tension member without holes should be taken normal to the direction of applied stress, the net section for a tension member with holes should be chosen as the one with the smallest area that passes through any chain of holes across the width of the member. Thus, the net section may pass through a chain of holes lying in a plane normal to the direction of applied stress or through holes along a diagonal or zigzag line.

Net section for a member with a chain of holes extending along a diagonal or zigzag line is the product of the net width and thickness. To determine net width, deduct from the gross width the sum of the diameters of all the holes in the chain, then add, for each gage space in the chain, the quantity

$$\frac{s^2}{4g}$$

where s = longitudinal spacing (**pitch,** in) of any two consecutive holes and g = transverse spacing (**gage,** in) of the same two holes.

The critical net section of the member is obtained from that chain with the least net width.

When a member axially stressed in tension is subjected to nonuniform transfer of load because of connections through bolts to only some of the elements of the cross section, as in the case of a W, M, or S shape connected solely by bolts through the flanges, the net area should be reduced as follows: 10% if the flange width is at least two-thirds the beam depth and at least three fasteners lie along the line of stress; 10% also for structural tees cut from such shapes; 15% for any of the preceding shapes that do not meet those criteria and for other shapes that have at least three fasteners in line of stress; and 25% for all members with only two fasteners in the line of stress.

7.13.1 ASD of Tension Members

Unit tensile stress F_t on the gross area should not exceed $0.60F_y$, where F_y is the minimum yield stress of the steel member (see Table 7.9). Nor should F_t exceed

TABLE 7.9 Tension on Gross Area

Allowable tensile stress (ASD)		Unit design tensile strength (LRFD)	
F_y, ksi	F_t, ksi	F_y, ksi	$\phi P_n/A_g$
36	21.6	36	32.4
42	25.2	42	37.8
45	27.0	45	40.5
50	30.0	50	45.0
55	33.0	55	49.5
60	36.0	60	54.0

$0.50F_u$, where F_u is the minimum tensile strength of the steel member, when the allowable stress is applied to the net area of a member connected with fasteners requiring holes. However, if the fastener is a large pin, as used to connect eyebars, pin plates, etc., F_t is limited to $0.45F_y$ on the net area. Therefore, for the popular A36 steel, the allowable tension stresses for gross and net areas are 22.0 and 29.0 ksi, respectively, and in the case of pin plates, 16.2 ksi.

7.13.2 LRFD of Tension Members

Design tensile strength ϕP_n, kips, of the gross area A_g, in^2, should not exceed $0.90F_y$, where F_y is the minimum yield stress of the steel (Table 7.9) and $P_n = A_gF_y$. Nor should the design tensile strength ϕP_n, kips, exceed $0.75F_u$ on the net area A_e, in^2, of the member. Other criteria control the design tensile strength of pin-connected members. (Refer to the AISC specification for LRFD.)

7.14 COLUMNS AND OTHER COMPRESSION MEMBERS

The principal factors governing the proportioning of members carrying compressive forces are overall column buckling, local buckling, and gross section area. The effect of overall column buckling depends on the slenderness ratio Kl/r, where Kl is the effective length, in, of the column, l is the unbraced length, and r is the least radius of gyration, in, of the cross section. The effect of local buckling depends on the width-thickness ratios of the individual elements of the member cross section.

W shapes with depths of 8, 10, 12, and 14 in are most commonly used for building columns and other compression members. For unbraced compression members, the most efficient shape is one where the value of r_y with respect to the minor axis approaches the value of r_x with respect to the major axis.

When built-up sections are used as compression members, the elements joining the principal load-carrying elements, such as lacing bars, should have a shear capacity of at least 2% of the axial load.

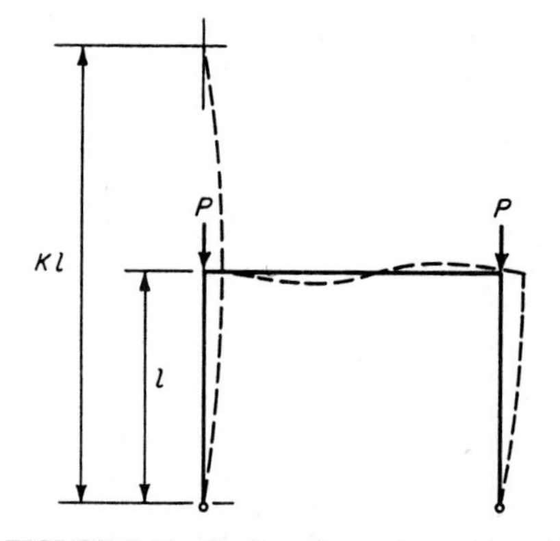

FIGURE 7.26 Configurations of members of a rigid frame caused by sidesway.

7.14.1 Effective Column Length

Proper application of the column capacity formulas for ASD or LRFD depends on judicious selection of K. This term is defined as the ratio of effective column length to actual unbraced length.

For a pin-ended column with translation of the ends prevented, $K = 1$. But in general, K may be greater or less than unity. For example, consider the columns in the frame in Fig. 7.26. They are dependent entirely on their own stiffness for stability against sidesway. If enough axial load is applied to them,

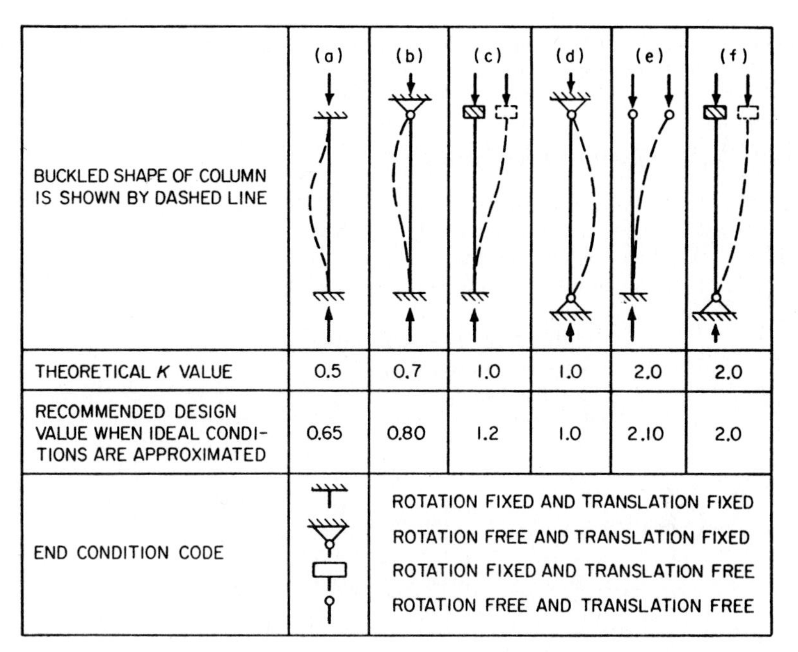

	(a)	(b)	(c)	(d)	(e)	(f)
BUCKLED SHAPE OF COLUMN IS SHOWN BY DASHED LINE						
THEORETICAL *K* VALUE	0.5	0.7	1.0	1.0	2.0	2.0
RECOMMENDED DESIGN VALUE WHEN IDEAL CONDITIONS ARE APPROXIMATED	0.65	0.80	1.2	1.0	2.10	2.0

END CONDITION CODE	
	ROTATION FIXED AND TRANSLATION FIXED
	ROTATION FREE AND TRANSLATION FIXED
	ROTATION FIXED AND TRANSLATION FREE
	ROTATION FREE AND TRANSLATION FREE

FIGURE 7.27 Values of effective column length *K* for idealized conditions.

their effective length will exceed their actual length. But if the frame were braced to prevent sidesway, the effective length would be less than the actual length because of the resistance to end rotation provided by the girder.

Theoretical values of *K* for six idealized conditions in which joint rotation and translation are either fully realized or nonexistent are given in Fig. 7.27. Also noted are values recommended by the Column Research Council for use in design when these conditions are approximated. Since joint fixity is seldom fully achieved, slightly higher design values than theoretical are given for fixed-end columns.

Specifications do not provide criteria for sidesway resistance under vertical loading, because it is impossible to evaluate accurately the contribution to stiffness of the various components of a building. Instead, specifications cite the general conditions that have proven to be adequate.

Constructions that inhibit sidesway in building frames include substantial masonry walls, interior shear walls; braced towers and shafts; floors and roofs providing diaphragm action—that is, stiff enough to brace the columns to shear walls or bracing systems; frames designed primarily to resist large side loadings or to limit horizontal deflection; and diagonal X bracing in the planes of the frames. Compression members in trusses are considered to be restrained against translation at connections. Generally, for all these constructions, *K* may be taken as unity, but a value less than one is permitted if proven by analysis.

When resistance to sidesway depends solely on the stiffness of the frames; for example, in tier buildings with light curtain walls or with wide column spacing, and with no diagonal bracing systems or shear walls, the designer may use any of several proposed rational methods for determining *K*. A quick estimate, however, can be made by using the alignment chart in an AISC "Manual of Steel Construction."

The effective length Kl of compression members, in such cases, should not be less than the actual unbraced length.

7.14.2 ASD of Compression Members

The allowable compressive stress on the gross section of axially loaded members is given by formulas determined by the effective slenderness ratios Kl/r of the members. A critical value, designated C_c, occurs at the slenderness ratio corresponding to the maximum stress for elastic buckling failure (Table 7.10). This is

TABLE 7.10 Slenderness Ratio at Maximum Stress for Elastic Buckling Failure

F_y, ksi	C_c	F_y, ksi	C_c
36.0	126.1	60.0	97.7
42.0	166.7	65.0	93.8
45.0	112.8	90.0	79.8
50.0	107.0	100.0	75.7
55.0	102.0		

illustrated in Fig. 7.28. An important fact to note: when Kl/r exceeds $C_c = 126.1$, the allowable compressive stress is the same for A36 and all higher-strength steels.

$$C_c = \sqrt{2\pi^2 E/F_y} \tag{7.1}$$

where E = modulus of elasticity of the steel = 29,000 ksi and F_y = specified minimum yield stress, ksi.

When Kl/r for any unbraced segment is less than C_c, the allowable compressive stress, ksi, is

$$F_a = \frac{[1 - (Kl/r)^2/2C_c^2]F_y}{\text{FS}} \tag{7.2}$$

where FS is the safety factor, which varies from 1.67 when $Kl/r = 0$ to 1.92 when $Kl/r = C_c$.

$$\text{FS} = \frac{5}{3} + \frac{3Kl/r}{8C_c} - \frac{(Kl/r)^3}{8C_c^3} \tag{7.3}$$

When Kl/r is greater than C_c:

$$F_a = \frac{12\pi^2 E}{23(Kl/r)^2} = \frac{149{,}000}{(Kl/r)^2} \tag{7.4}$$

This is the Euler column formula for elastic buckling with a constant safety factor of 1.92 applied.

Increased stresses are permitted for bracing and secondary members with l/r greater than 120. (K is taken as unity.) For such members, the allowable com-

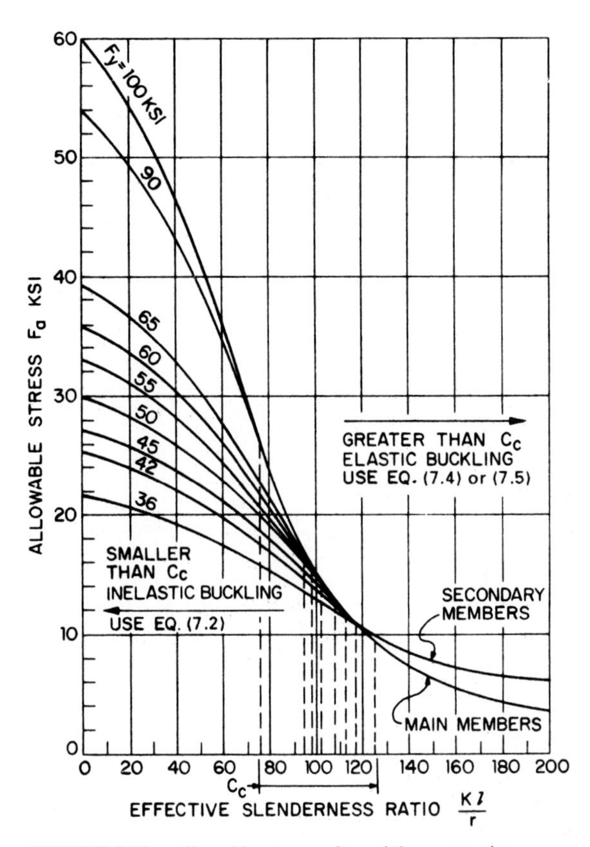

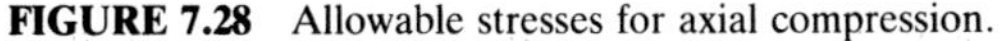

FIGURE 7.28 Allowable stresses for axial compression.

pressive stress is

$$F_{as} = \frac{F_a}{1.6 - l/200r} \tag{7.5}$$

where F_a is given by Eq. (7.2) or (7.4). The higher stress is justified by the relative unimportance of these members and the greater restraint likely at their end connections. The full unbraced length should always be used for l.

Tables giving allowable stresses for the entire range of Kl/r appear in the AISC ASD "Manual of Steel Construction." Approximate values may be obtained from Fig. 7.28. Allowable stresses are based on certain minimum sizes of structural members and their elements that make possible full development of strength before premature buckling occurs. The higher the allowable stresses the more stringent must be the dimensional restrictions to preclude buckling or excessive deflections.

The AISC ASD specification for structural steel buildings limits the effective slenderness ratio Kl/r to 200 for columns, struts, and truss members, where K is the ratio of effective length to actual unbraced length l, and r is the least radius of gyration.

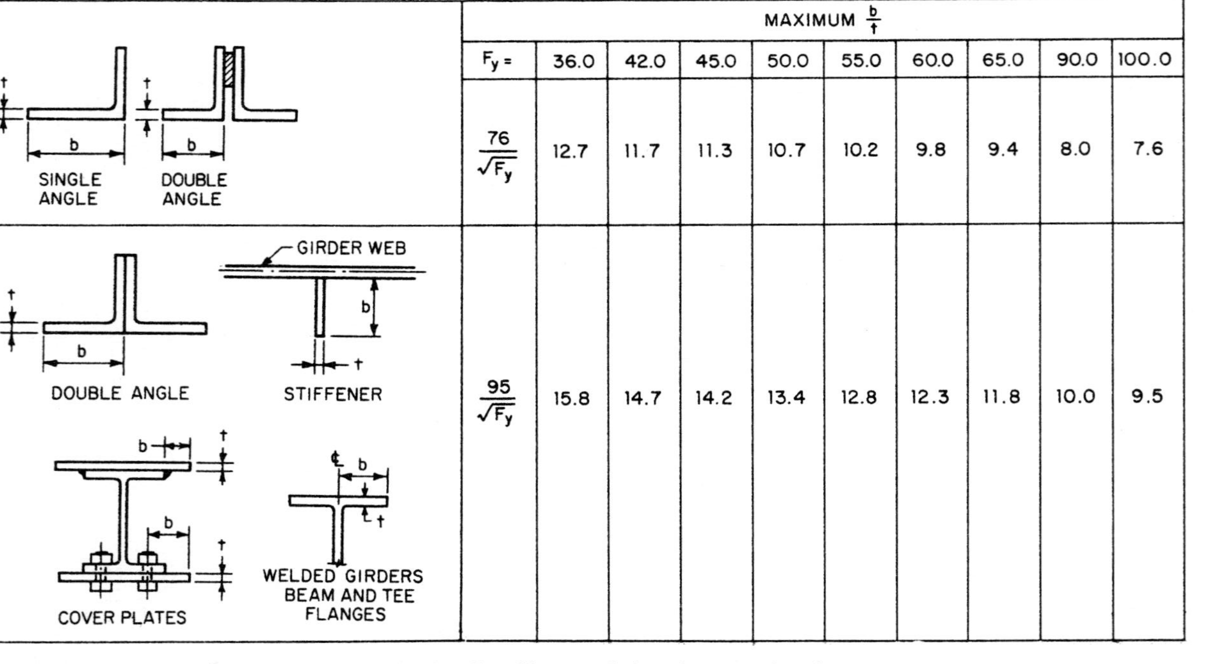

	MAXIMUM $\frac{b}{t}$								
$F_y =$	36.0	42.0	45.0	50.0	55.0	60.0	65.0	90.0	100.0
$\frac{76}{\sqrt{F_y}}$	12.7	11.7	11.3	10.7	10.2	9.8	9.4	8.0	7.6
$\frac{95}{\sqrt{F_y}}$	15.8	14.7	14.2	13.4	12.8	12.3	11.8	10.0	9.5

FIGURE 7.29 Maximum width-thickness ratios for allowable stress design of compression elements.

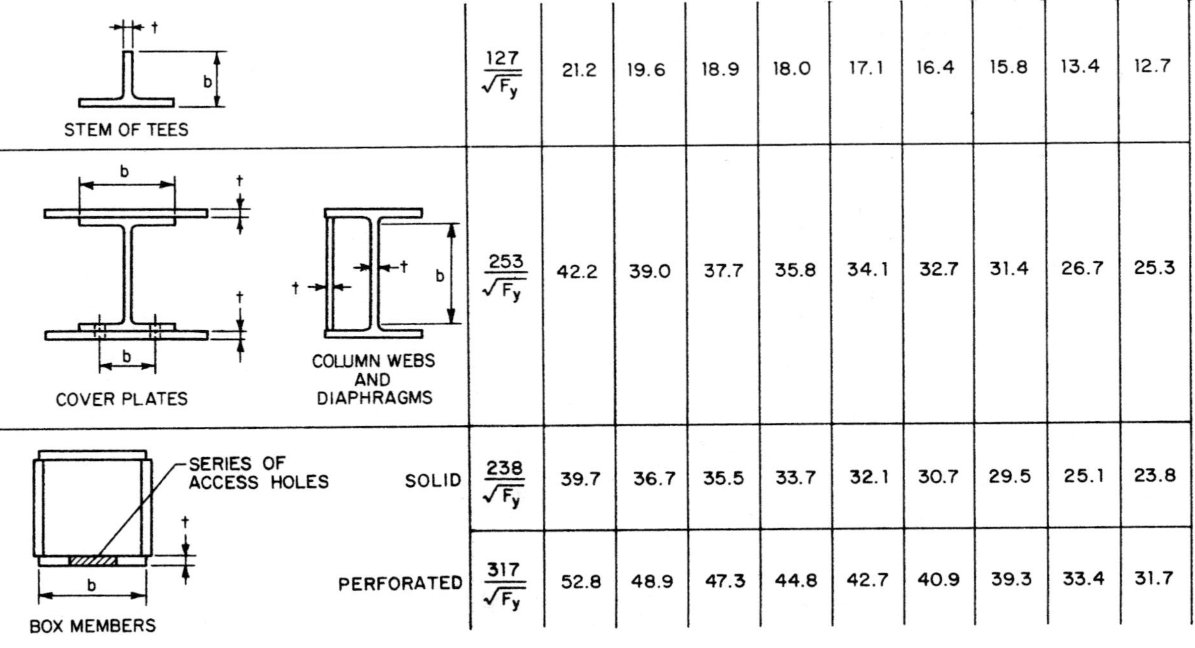

STEM OF TEES		$\frac{127}{\sqrt{F_y}}$	21.2	19.6	18.9	18.0	17.1	16.4	15.8	13.4	12.7
COVER PLATES; COLUMN WEBS AND DIAPHRAGMS		$\frac{253}{\sqrt{F_y}}$	42.2	39.0	37.7	35.8	34.1	32.7	31.4	26.7	25.3
BOX MEMBERS	SOLID	$\frac{238}{\sqrt{F_y}}$	39.7	36.7	35.5	33.7	32.1	30.7	29.5	25.1	23.8
	PERFORATED	$\frac{317}{\sqrt{F_y}}$	52.8	48.9	47.3	44.8	42.7	40.9	39.3	33.4	31.7

FIGURE 7.29 (*Continued*)

A practical rule also establishes limiting slenderness ratios l/r for tension members:

For main members	240
For bracing and secondary members	300

But this does not apply to rods or other tension members that are drawn up tight (prestressed) during erection. The purpose of the rule is to avoid objectionable slapping or vibration in long, slender members.

The AISC ASD specification also specifies several restricting ratios for compression members. One set applies to projecting elements subjected to axial compression or compression due to bending. Another set applies to compression elements supported along two edges.

Figure 7.29 lists maximum width-thickness ratios, b/t, for commonly used elements and grades of steel. Tests show that when b/t of elements normal to the direction of compressive stress does not exceed these limits, the member may be stressed close to the yield stress without failure by local buckling. Because the allowable stress increases with F_y, the specified yield stress of the steel, width-thickness ratios are less for higher-strength steels.

These b/t ratios should not be confused with the width-thickness ratios described in Art. 7.15. There, more restrictive conditions are set in defining compact sections qualified for higher allowable stresses.

7.14.3 LRFD of Compression Members

When the elements of the cross section of a compression member have width-thickness ratios that do not exceed the limits tabulated in Table 7.11, the design compressive strength is $\phi_c P_n$. The resistance factor ϕ_c should be taken as 0.85. The nominal strength is given by $P_n = A_g F_{cr}$, where A_g is the cross-sectional area, in^2, and F_{cr} is the critical compressive stress, ksi. Formulas for F_{cr} are based on a parameter λ_c.

$$\lambda_c = \frac{Kl}{r\pi}\sqrt{\frac{F_y}{E}} = \frac{Kl}{r}\sqrt{\frac{F_y}{286{,}220}} \tag{7.6}$$

where E = modulus of elasticity, ksi = 29,000 ksi. For $\lambda_c \leq 1.5$,

$$F_{cr} = 0.658^{\lambda_c^2} F_y \tag{7.7}$$

For $\lambda_c > 1.5$,

$$F_{cr} = (0.877/\lambda_c^2)F_y \tag{7.8}$$

Computations can be simplified by use of column load tables in the AISC LRFD "Steel Construction Manual."

For design of columns with elements having width-thickness ratios exceeding the limits in Table 7.11, refer to the AISC LRFD specification.

(T. V. Galambos, "Guide to Design Criteria for Metal Compression Members," 4th ed., John Wiley & Sons, Inc., New York.)

TABLE 7.11 **Limiting Width-Thickness Ratios for LRFD of Columns**

Compression elements	Width-thickness ratio	Limiting width-thickness ratio λ_r		
		General	A36 steel	A50 steel
Flanges of W and other I shapes and channels; outstanding legs of pairs of angles in continuous contact	b/t	$95/\sqrt{F_y}$	15.8	13.4
Flanges of square and rectangular box sections; flange cover plates and diaphragm plates between lines of fasteners or welds	b/t	$238/\sqrt{F_y - F_r^*}$	46.7(rolled) 53.9 (welded)	37.6 (rolled) 41.1 (welded)
Legs of single angle struts and double angle struts with separators; unstiffened elements (i.e., supported along one edge)	b/t	$76/\sqrt{F_y}$	12.7	10.7
Stems of tees	d/t	$127/\sqrt{F_y}$	21.2	18.0
All other stiffened elements (elements supported along two edges)	b/t h_c/t_w	$253/\sqrt{F_y}$	42.2	35.8

$^*F_r$ = compressive residual stress in flange: 10 ksi for rolled shapes, 16.5 ksi for welded sections.

7.15 BEAMS AND OTHER FLEXURAL MEMBERS

The capacity of members subject to bending depends on the cross-section geometry. AISC ASD and LRFD procedures incorporate the concept of compact and non-compact sections.

7.15.1 ASD of Flexural Members

Beams classified as compact are allowed a bending stress, ksi, $F_b = 0.66F_y$ for the extreme surfaces in both tension and compression, where F_y, is the specified yield stress, ksi. Such members have an axis of symmetry in the plane of loading, their compression flange is adequately braced to prevent lateral displacement, and they develop their full plastic moment (section modulus times yield stress) before buckling.

Compactness Requirements. To qualify as compact, members must meet the following conditions:

1. The flanges must be continuously connected to the web or webs.

2. The width-thickness ratio of unstiffened projecting elements of the compression flange must not exceed $65.0/\sqrt{F_y}$. For computation of this ratio, with b equals one-half the full flange width of I-shaped sections, or the distance from the free edge to the first row of fasteners (or welds) for projecting plates, or the full width of legs of angles, flanges of zees or channels, or tee stems.

3. The web depth-thickness ratio d/t_w must not exceed $640(1 - 3.74_a/F_y)/\sqrt{F}$ when f_a, the computed axial stress, is equal to or less than $0.16F_y$, or $257/\sqrt{F_y}$ when $f_a > 0.16F_y$.

4. The width-thickness ratio of stiffened compression flange plates in box sections and that part of the cover plates for beams and built-up members that is included between longitudinal lines of bolts or welds must not exceed $190/\sqrt{F_y}$.

5. For the compression flange of members not box shaped to be considered supported, unbraced length between lateral supports should not exceed $76.0b_f/\sqrt{F_y}$ or $20{,}000\ A_f/F_y d$, where b_f is the flange width, A_f the flange area, and d the web depth.

6. The unbraced length for rectangular box-shaped members with depth not more than 6 times the width and with flange thickness not more than 2 times the web thickness must not exceed $(1950 + 1200\ M_1/M_2)b/F_y$. The unbraced length in such cases, however, need not be less than $1200b/F_y$. M_1 is the smaller and M_2 the larger of bending moments at points of lateral support.

7. The diameter-thickness ratio of hollow circular steel sections must not exceed $300/F_y$.

Allowable Bending Stresses for Compact Beams. Most sections used in building framing, including practically all rolled W shapes of A36 steel and most of those with $F_y = 50$ ksi, comply with the preceding requirements for compactness, as illustrated in Fig. 7.30. Such sections, therefore, are designed with $F_b = 0.66F_y$. Excluded from qualifying are hybrid girders, tapered girders, and sections made from A514 steel.

Braced sections that meet the requirements for compactness, and are continuous over their supports or rigidly framed to columns, are also permitted a redistribution of the design moments. Negative gravity-load moments at supports may be reduced 10%. But then, the maximum positive moment must be increased by 10% of the average negative moments. This moment redistribution does not apply to cantilevers, hybrid girders, or members of the A514 steel.

Allowable Bending Stresses for Noncompact Beams. Many other beam-type members, including nearly compact sections that do not meet all seven requirements, are accorded allowable bending stresses, some higher and some considerably lower than $0.66F_y$, depending on such conditions as shape factor, direction of loading, inherent resistance to torsion or buckling, and external lateral support. The common conditions and applicable allowable bending stresses are summarized in Fig. 7.31. In the formulas,

l = distance, in, between cross sections braced against twist or lateral displacement of the compression flange

r_t = radius of gyration, in, of a section comprising the compression flange plus one-third of the compression web area, taken about an axis in the plane of the web

A_f = area of the compression flange, in^2

The allowable bending stresses F_b, ksi, for values often used for various grades of steel are listed in Table 7.12.

TABLE 7.12 Allowable Bending Stresses, ksi

F_y	$0.60F_y$	$0.66F_y$	$0.75F_y$
36.0	22.0	24.0	27.0
42.0	25.2	27.7	31.5
45.0	27.0	29.7	33.8
50.0	30.0	33.0	37.5
55.0	33.0	36.3	41.3
60.0	36.0	39.6	45.0
65.0	39.0	42.9	48.8

Lateral Support of Beams. In computation of allowable bending stresses in compression for beams with distance between lateral supports exceeding requirements, a range sometimes called laterally unsupported, the AISC ASD formulas contain a moment factor C_b in recognition of the beneficial effect of internal moments, both in magnitude and direction, at the points of support. For the purpose of this summary, however, the moment factor has been taken as unity and the formulas simplified in Fig. 7.31. The formulas are exact for the case in which the bending moment at any point within an unbraced length is larger than that at both ends of this length. They are conservative for all other cases. Where more refined values are desired, see Art. 7.15.2 or refer to the AISC ASD specification for structural steel buildings.

Limits on Beam Width-Thickness Ratios. For flexural members in which the width-thickness ratios of compression elements exceed the limits given in Fig. 7.31 and which are usually lightly stressed, appropriate allowable bending stresses are suggested in "Slender Compression Elements," Appendix C, AISC specification.

For additional discussion of lateral support, see Art. 7.11.2. Also, additional information on width-thickness ratios of compression elements is given in Fig. 7.29.

ASD for Shear in Flexural Members. The shear strength of a flexural member may be computed by dividing the total shear force at a section by the web area, the product of the web thickness and overall member depth. Whereas flexural strength normally controls selection of rolled shapes, shear strength can be critical when the web has cutouts or holes that reduce the net web area or when a short-span beam carries a large concentrated load. Also, in built-up members, such as plate girders or rigid frame elements, shear often controls web thickness.

The web depth-thickness ratio permitted without stiffeners, $h/t \leq 380/\sqrt{F_y}$ for ASD and $h/t \leq 418/\sqrt{F_y}$ for LRFD, is satisfied by all W shapes of A36 steel.

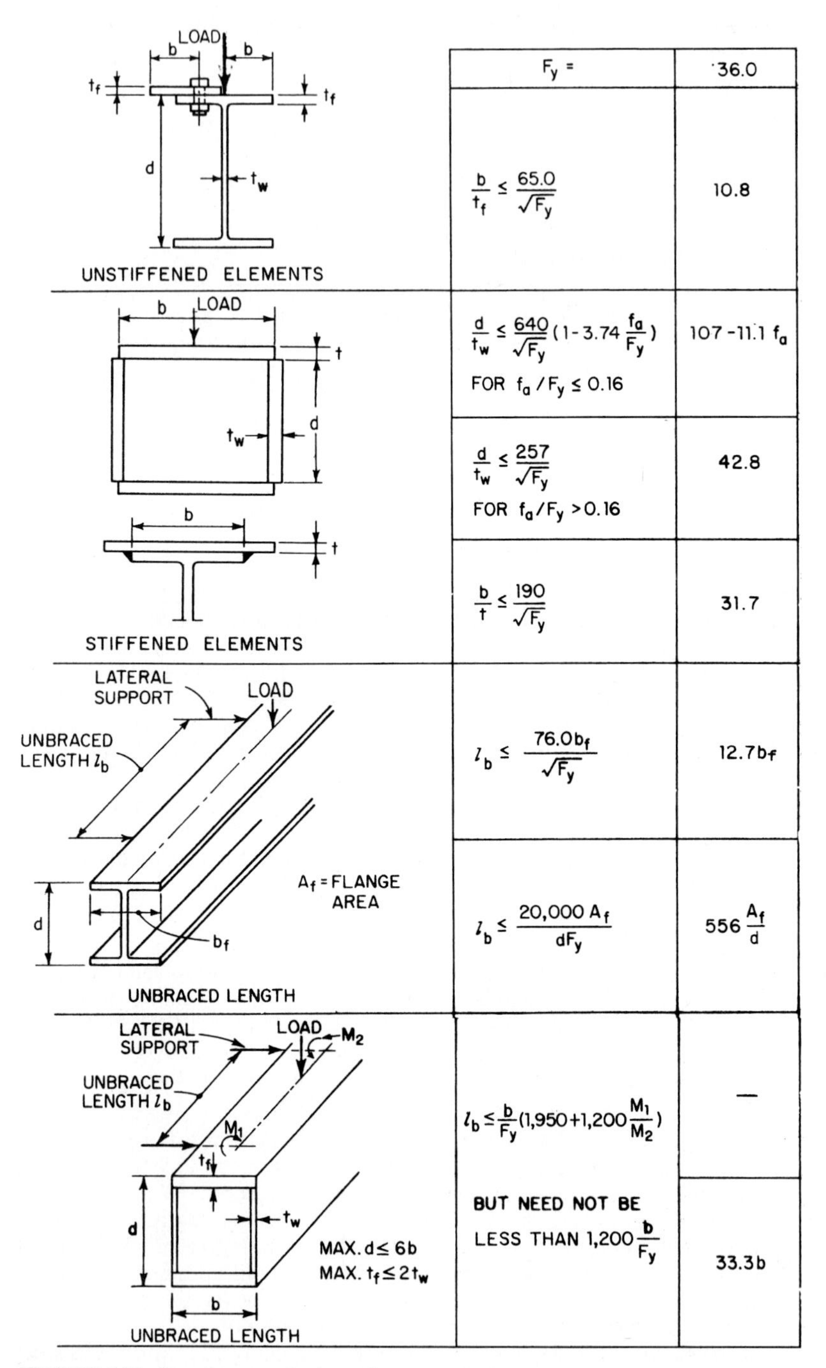

F_y =	36.0
$\frac{b}{t_f} \leq \frac{65.0}{\sqrt{F_y}}$	10.8
$\frac{d}{t_w} \leq \frac{640}{\sqrt{F_y}}\left(1 - 3.74\frac{f_a}{F_y}\right)$ FOR $f_a/F_y \leq 0.16$	$107 - 11.1 f_a$
$\frac{d}{t_w} \leq \frac{257}{\sqrt{F_y}}$ FOR $f_a/F_y > 0.16$	42.8
$\frac{b}{t} \leq \frac{190}{\sqrt{F_y}}$	31.7
$l_b \leq \frac{76.0 b_f}{\sqrt{F_y}}$	$12.7 b_f$
$l_b \leq \frac{20{,}000 A_f}{d F_y}$	$556 \frac{A_f}{d}$
$l_b \leq \frac{b}{F_y}\left(1{,}950 + 1{,}200\frac{M_1}{M_2}\right)$	—
BUT NEED NOT BE LESS THAN $1{,}200\frac{b}{F_y}$	33.3b

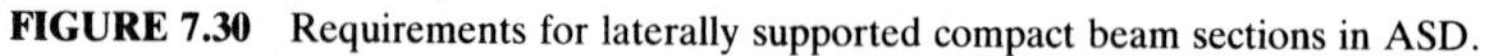
FIGURE 7.30 Requirements for laterally supported compact beam sections in ASD.

42.0	45.0	50.0	55.0	60.0	65.0
10.0	9.7	9.2	8.8	8.4	8.1
$98.8 - 8.8 f_a$	$95.4 - 7.9 f_a$	$90.5 - 6.8 f_a$	$86.3 - 5.9 f_a$	$82.6 - 5.2 f_a$	$79.4 - 4.6 f_a$
39.7	38.3	36.3	34.7	33.2	31.9
29.3	28.3	26.9	25.6	24.5	23.6
$11.7 b_f$	$11.3 b_f$	$10.7 b_f$	$10.2 b_f$	$9.8 b_f$	$9.4 b_f$
$476 \frac{A_f}{d}$	$444 \frac{A_f}{d}$	$400 \frac{A_f}{d}$	$364 \frac{A_f}{d}$	$333 \frac{A_f}{d}$	$308 \frac{A_f}{d}$
—	—	—	—	—	
28.6b	26.7b	24.0b	21.8b	20.0b	18.5b

FIGURE 7.30 *(Continued)*

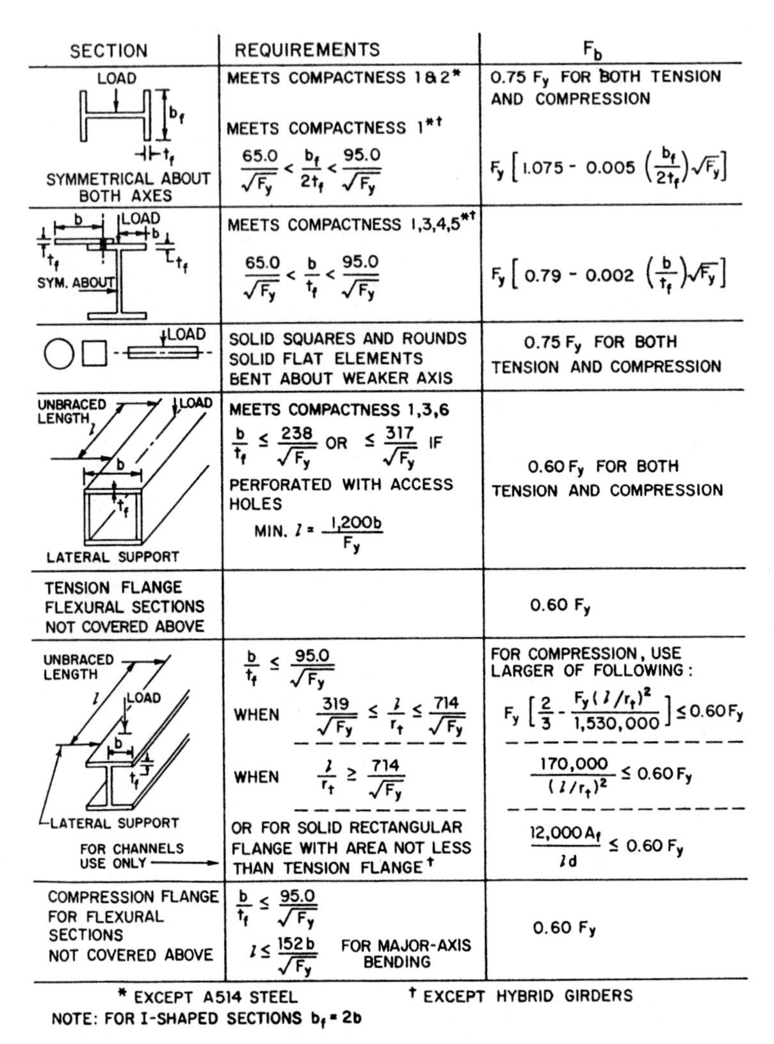

SECTION	REQUIREMENTS	F_b
LOAD; b_f; t_f; SYMMETRICAL ABOUT BOTH AXES	MEETS COMPACTNESS 1 & 2*	$0.75 F_y$ FOR BOTH TENSION AND COMPRESSION
	MEETS COMPACTNESS 1*† $\frac{65.0}{\sqrt{F_y}} < \frac{b_f}{2t_f} < \frac{95.0}{\sqrt{F_y}}$	$F_y \left[1.075 - 0.005 \left(\frac{b_f}{2t_f}\right)\sqrt{F_y}\right]$
b; LOAD; b; t_f; t_f; SYM. ABOUT	MEETS COMPACTNESS 1,3,4,5*† $\frac{65.0}{\sqrt{F_y}} < \frac{b}{t_f} < \frac{95.0}{\sqrt{F_y}}$	$F_y \left[0.79 - 0.002 \left(\frac{b}{t_f}\right)\sqrt{F_y}\right]$
LOAD	SOLID SQUARES AND ROUNDS SOLID FLAT ELEMENTS BENT ABOUT WEAKER AXIS	$0.75 F_y$ FOR BOTH TENSION AND COMPRESSION
UNBRACED LENGTH l; LOAD; b; t_f; LATERAL SUPPORT	MEETS COMPACTNESS 1,3,6 $\frac{b}{t_f} \le \frac{238}{\sqrt{F_y}}$ OR $\le \frac{317}{\sqrt{F_y}}$ IF PERFORATED WITH ACCESS HOLES MIN. $l = \frac{1,200b}{F_y}$	$0.60 F_y$ FOR BOTH TENSION AND COMPRESSION
TENSION FLANGE FLEXURAL SECTIONS NOT COVERED ABOVE		$0.60 F_y$
UNBRACED LENGTH l; LOAD; b; t_f; LATERAL SUPPORT	$\frac{b}{t_f} \le \frac{95.0}{\sqrt{F_y}}$ WHEN $\frac{319}{\sqrt{F_y}} \le \frac{l}{r_t} \le \frac{714}{\sqrt{F_y}}$	FOR COMPRESSION, USE LARGER OF FOLLOWING: $F_y \left[\frac{2}{3} - \frac{F_y (l/r_t)^2}{1,530,000}\right] \le 0.60 F_y$
	WHEN $\frac{l}{r_t} \ge \frac{714}{\sqrt{F_y}}$	$\frac{170,000}{(l/r_t)^2} \le 0.60 F_y$
FOR CHANNELS USE ONLY →	OR FOR SOLID RECTANGULAR FLANGE WITH AREA NOT LESS THAN TENSION FLANGE†	$\frac{12,000 A_f}{l d} \le 0.60 F_y$
COMPRESSION FLANGE FOR FLEXURAL SECTIONS NOT COVERED ABOVE	$\frac{b}{t_f} \le \frac{95.0}{\sqrt{F_y}}$ $l \le \frac{152b}{\sqrt{F_y}}$ FOR MAJOR-AXIS BENDING	$0.60 F_y$

* EXCEPT A514 STEEL † EXCEPT HYBRID GIRDERS

NOTE: FOR I-SHAPED SECTIONS $b_f = 2b$

FIGURE 7.31 Allowable bending stresses for sections not qualifying as compact.

Furthermore, only the lightest one or two W sections in each depth fail to satisfy these criteria for 50-ksi material.

For members with $h/t \le 380/\sqrt{F_y}$, the unit shear stress on the gross section should not be greater than $F_v = 0.40F_y$, where F_y is the minimum yield point of the web steel ksi (Table 7.13). Members with higher h/t ratios require stiffeners (see Art. 7.16.1).

Beams with web angle or shear-bar end connections and a coped top flange should be checked for shear on the critical plane through the holes in the web. In this case, the allowable unit shear stress is $F_v = 0.30F_u$, where F_u is the minimum tensile strength of the steel, ksi.

A special case occurs when a web lies in a plane common to intersecting members; for example, the knee of a rigid frame. Then, shear stresses generally are

TABLE 7.13 Allowable Shear on Gross Area, ksi

For ASD when $h/t \leq 380/\sqrt{F_y}$		For LRFD when $h/t \leq 418/\sqrt{F_y}$	
F_y	F_v	F_y	ϕV_n
36.0	14.5	36.0	19.4
42.0	17.0	42.0	22.7
45.0	18.0	45.0	24.3
50.0	20.0	50.0	27.0
55.0	22.0	55.0	29.7
60.0	24.0	60.0	32.4

high. Such webs, in elastic design, should be reinforced when the web thickness is less than $32M/A_{bc}F_y$, where M is the algebraic sum of clockwise and counter-clockwise moments (in ft-kips) applied on opposite sides of the connection boundary, and A_{bc} is the planar area of the connection web, in^2 (approximately the product of the depth of the member introducing the moment and the depth of the intersecting member). In plastic design, this thickness is determined from $23M_p/A_{bc}F_y$, where M_p is the plastic moment, or M times a load factor of 1.70. In this case, the total web shear produced by the factored loading should not exceed the web area (depth times thickness) capacity in shear. Otherwise, the web must be reinforced with diagonal stiffeners or a doubler plate.

For deep girder webs, allowable shear is reduced. The reduction depends on the ratio of clear web depth between flanges to web thickness and an aspect ratio of stiffener spacing to web depth. In practice, this reduction does not apply when the ratio of web depth to thickness is less than $380/\sqrt{F_y}$.

7.15.2 LRFD of Flexural Members

The AISC LRFD specification for structural steel buildings permits plastic analysis for steels with yield stress not exceeding 65 ksi. Negative moments induced by gravity loading may be reduced 10% for compact beams, if the positive moments are increased by 10% of the average of the negative moments.

Design strength in bending of flexural members is defined as $\phi_b M_n$, where the resistance factor $\phi_b = 0.90$ and M_n is the nominal flexural strength. M_n depends on several factors, including the geometry of the section, the unbraced length of the compression flange, and properties of the steel. Beams may be compact, non-compact, or slender-element sections. For compact beams, the AISC specification sets limits on the width-thickness ratios of section elements to restrict local buckling. These limits are listed in Table 7.14.

For a compact section bent about the major axis, the unbraced length L_b of the compression flange where plastic hinges may form at failure may not exceed L_{pd} given by Eqs. (7.9) and (7.10). For beams bent about the minor axis and square and circular beams, L_b is not restricted for plastic analysis.

For I-shaped beams that are loaded in the plane of the web and are symmetric about major and minor axes or symmetric about the minor axis but with the

TABLE 7.14 Limiting Width-Thickness Ratios for LRFD of Beams

Beam element	Width-thickness ratio	Limiting width-thickness ratio, λ_p		
		General	A36 steel	A50 steel
Flanges of W and other I shapes and channels	b/t	$65/\sqrt{F_y}$	10.8	9.2
Flanges of square and rectangular box sections; flange cover plates and diaphragm plates between lines of fasteners or welds	b/t	$190/\sqrt{F_y}$	31.7	26.9
Webs in flexural compression	h_c/t_w	$640/\sqrt{F_y}$	106.7	90.5

compression flange larger than the tension flange, including hybrid girders,

$$L_{pd} = \frac{3600 + 2200(M_1/M_p)}{F_{yc}} r_y \tag{7.9}$$

where F_{yc} = minimum yield stress, ksi, of compression flange
M_1 = smaller of the moments, in-kips, at the end of the unbraced length of the beam
M_p = plastic moment, in-kips
r_y = radius of gyration, in, about minor axis

For homogenous sections, $M_p = F_y Z$, where Z is the plastic section modulus, in^3. (For hybrid girders, Z may be computed from the fully plastic distribution.) M_1/M_p is positive for beams with reverse curvature, negative for single curvature.

For solid rectangular bars and symmetric box beams,

$$L_{pd} = \frac{5000 + 3000(M_1/M_p)}{F_y} r_y \tag{7.10}$$

The flexural design strength $0.90M_n$ is determined by the limit state of lateral torsional buckling and should be calculated for the region of the last hinge to form and for regions not adjacent to a plastic hinge. For compact sections bent about the major axis, M_n depends on the following unbraced lengths:

L_b = distance, in, between points braced against lateral displacement of the compression flange or between points braced to prevent twist
L_p = limiting laterally unbraced length, in, for full plastic bending capacity
= $300r_y/\sqrt{F_{yf}}$ for I shapes and channels
= $3750(r_y/M_p)/\sqrt{JA}$ for box beams and solid rectangular bars
F_{yf} = flange yield stress, ksi
J = torsional constant, in^4 (see AISC LRFD "Manual of Steel Construction")
A = cross-sectional area, in^2
L_r = limiting laterally unbraced length, in, for inelastic lateral buckling

For I-shaped beams symmetric about the major or minor axis or symmetric about the minor axis with the compression flange larger than the tension flange, and channels loaded in the plane of the web,

$$L_r = \frac{r_y X_1}{(F_{yw} - F_r)} \sqrt{1 + \sqrt{1 + X_2(F_{yw} - F_r)^2}} \tag{7.11}$$

where F_{yw} = specified minimum yield stress of web, ksi
F_r = compressive residual stress in flange = 10 ksi for rolled shapes, 16.5 ksi for welded sections
$X_1 = (\pi/S_x)\sqrt{EGJA/2}$
$X_2 = (4C_w/I_y)(S_x/GJ)^2$
E = elastic modulus of the steel = 29,000 ksi
G = shear modulus of elasticity = 11,200 ksi
S_x = section modulus about major axis, in^3 (with respect to the compression flange if that flange is larger than the tension flange)
C_w = warping constant, in^6 (see AISC Manual—LRFD)
I_y = moment of inertia about minor axis, in^4

7.15.3 Limit-State Moments

For the aforementioned shapes, the limiting buckling moment M_r, ksi, may be computed from

$$M_r = (F_{yw} - F_r)S_x \tag{7.12}$$

For compact beams with $L_b \leq L_r$, bent about the major axis,

$$M_n = C_b \left[M_p - (M_p - M_r)\frac{L_b - L_p}{L_r - L_p} \right] \leq M_p \tag{7.13}$$

where $C_b = 1.75 + 1.05(M_1/M_2) + 0.3(M_1/M_2)^2 \leq 2.3$, where M_1 is the smaller and M_2 the larger end moment in the unbraced segment of the beam; M_1/M_2 is positive for reverse curvature
= 1.0 for unbraced cantilevers and beams with moments over much of the unbraced segment equal to or greater than the larger of the segment end moments (see T. V. Galambos "Guide to Stability Design Criteria for Metal Structures," 4th ed., John Wiley & Sons, Inc., New York, for use of larger values of C_b)

For solid rectangular bars bent about the major axis,

$$L_r = 57{,}000(r_y/M_r)\sqrt{JA} \tag{7.14}$$

and the limiting buckling moment is given by

$$M_r = F_y S_x \tag{7.15}$$

For symmetric box sections loaded in the plane of symmetry and bent about the major axis, M_r should be determined from Eq. (7.12) and L_r from Eq. (7.14).

For compact beams with $L_b > L_r$, bent about the major axis,

$$M_n = M_{cr} \leq C_b M_r \tag{7.16}$$

where M_{cr} = critical elastic moments, kip-in. For shapes to which Eq. (7.11) applies,

$$M_{cr} = C_b\,(\pi/L_b)\,\sqrt{EI_yGJ + I_yC_w(\pi E/L_b)^2} \tag{7.17}$$

For solid rectangular bars and symmetric box sections,

$$M_{cr} = 57{,}000C_b\,\sqrt{JA}/(L_b/r_y) \tag{7.18}$$

Noncompact Beams. The nominal flexural strength M_n for noncompact beams is the least value determined from the limit states of

1. Lateral-torsional buckling (LTB)
2. Flange local buckling (FLB)
3. Web local buckling (WLB)

The AISC LRFD specification for structural steel buildings presents formulas for determining limit-state moments. In most cases, LRFD computations for flexural members can be simplified by use of tables in the AISC "Manual of Steel Construction—LRFD." See also Art. 7.16.

LRFD for Shear in Flexural Members. The design shear strength is $\phi_V V_n$, where ϕ_V = 0.90, and for rolled shapes and built-up members without stiffeners is governed by the web depth-thickness ratio. The design shear strength may be computed from

$$\phi V_n = 0.90 \times 0.6F_yA_w = 0.54F_yA_w \qquad \frac{h}{t} \leq \frac{418}{\sqrt{F_y}} \tag{7.19}$$

$$\phi V_n = 0.90 \times 0.6\,\frac{418/\sqrt{F_y}}{b/t} = 0.54F_yA_w\frac{418/\sqrt{F_y}}{h/t} \qquad \frac{418}{\sqrt{F_y}} < \frac{h}{t} \leq \frac{523}{\sqrt{F_y}} \tag{7.20}$$

$$\phi V_n = 0.90 \times A_w\,\frac{132{,}000}{(h/t)^2} = \frac{119{,}000}{(h/t)^2} \qquad \frac{h}{t} > \frac{523}{\sqrt{F_y}} \tag{7.21}$$

where V_n = nominal shear strength, kips
A_w = area of the web, in^2 = dt
d = overall depth, in
t = thickness of web, in
h = the following web dimensions, in: clear distance between fillets for rolled shapes; clear distance between flanges for welded sections
F_y = specified minimum yield stress, ksi, of web steel

See also Art. 7.16.2.

7.16 PLATE GIRDERS

Plate girders may have either a box or an I shape. Main components are plates or plates and angles, arranged so that the cross section is either singly or doubly symmetrical. Generally, the elements are connected by continuous fillet welds. In existing construction, the connection may have been made with rivets or bolts through plates and angles. Fig. 7.32 depicts typical I-shape girders.

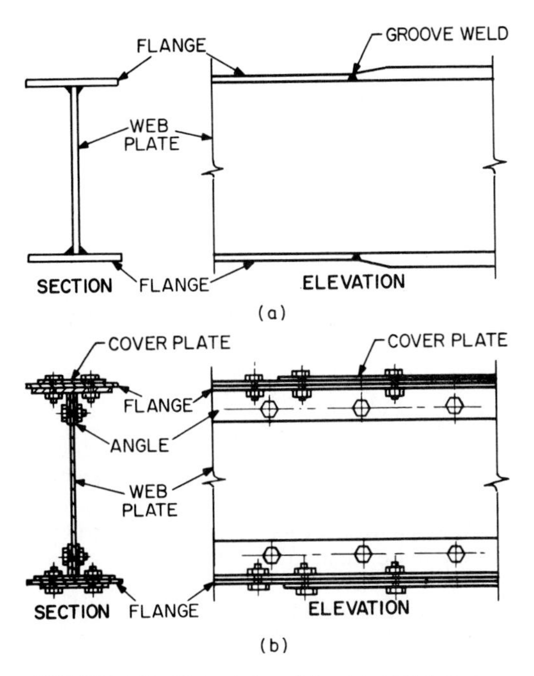

FIGURE 7.32 Plate girders: (*a*) welded (*b*) bolted.

Plate girders are commonly used for long spans where they cost less than rolled W shapes or where members are required with greater depths or thinner webs than those available with rolled W shapes. The AISC LRFD "Specification for Structural Steel for Buildings" distinguishes between a plate girder and a beam in that a plate girder has web stiffeners or a web with $h_c/t_w > 970/\sqrt{F_y}$, or both, where h_c is twice the distance from the neutral axis to (1) the inside face of the compression flange when it is welded to the web or (2) the nearest line of fasteners to the compression flange when the web-flange connection is bolted.

7.16.1 ASD Procedure for Plate Girders

Allowable stresses for tension, compression, bending, and shear are the same for plate girders as those given in Arts. 7.13 to 7.15, except where stiffeners are used. But reductions in allowable stress are required under some conditions, and there are limitations on the proportions of girder components.

Web Depth-Thickness Limits. The ratio of the clear distance h between flanges, in, to web thickness t, in, is limited by

$$\frac{h}{t} \le \frac{14{,}000}{\sqrt{F_y(F_y + 16.5)}} \tag{7.22}$$

TABLE 7.15 Limiting Depth-Thickness Ratios for ASD of Plate-Girder Webs

F_y, ksi	h/t Eq. (7.22)	h/t Eq. (7.23)	F_y, ksi	h/t Eq. (7.22)	h/t Eq. (7.23)
36.0	322	333	60.0	207	258
42.0	282	309	65.0	192	248
45.0	266	298	90.0	143	211
50.0	243	283	100.0	130	200
55.0	223	270			

where F_y is the specified yield stress of the compression flange steel, ksi (Table 7.15). When, however, transverse stiffeners are provided at spacings not exceeding 1.5 times the girder depth, the limit on h/t is increased to

$$\frac{h}{t} \leq \frac{2{,}000}{\sqrt{F_y}} \tag{7.23}$$

General Design Method. Plate girders may be proportioned to resist bending on the assumption that the moment of inertia of the gross cross section is effective. No deductions need be made for fastener holes, unless the holes reduce the gross area of either flange by more than 15%. When they do, the excess should be deducted.

Hybrid girders, which have higher-strength steel in the flanges than in the web, may also be proportioned by the moment of inertia of the gross section when they are not subjected to an axial force greater than 15% of the product of yield stress of the flange steel and the area of the gross section. At any given section, the flanges must have the same cross-sectional area and be made of the same grade of steel.

The allowable compressive bending stress F_b for plate girders must be reduced from that given in Art. 7.14, where h/t exceeds $760/\sqrt{F_b}$. For greater values of this ratio, the allowable compressive bending stress, except for hybrid girders, becomes

$$F_b' \leq F_b \left[1 - 0.0005 \frac{A_w}{A_f} \left(\frac{h}{t} - \frac{760}{\sqrt{F_b}}\right)\right] \tag{7.24}$$

where A_w = the web area, in^2 and A_f = the compression flange area, in^2.

For hybrid girders, not only is the allowable compressive bending stress limited to that given by Eq. (7.21), but also the maximum stress in either flange may not exceed

$$F_b' = F_b \left[\frac{12 + (A_w/A_f)(3\alpha - \alpha^3)}{12 + 2(A_w/A_f)}\right] \tag{7.25}$$

where a = ratio of web yield stress to flange yield stress.

Flange Limitations. The projecting elements of the compression flange must comply with the limitations for b/t given in Art. 7.15. The area of cover plates, where used, should not exceed 0.70 times the total flange area. Partial-length cover plates (Fig. 7.32*b*) should extend beyond the theoretical cutoff point a sufficient distance to develop their share of bending stresses at the cutoff point. Preferably for welded-plate girders, the flange should consist of a series of plates, which may differ in thickness and width, joined end to end with complete-penetration groove welds (Fig. 7.32*a*).

Bearing Stiffeners. These are required on girder webs at unframed ends. They may also be needed at concentrated loads, including supports. Set in pairs, bearing stiffeners may be angles or plates placed on opposite sides of the web, usually normal to the bending axis. Angles are attached with one leg against the web. Plates are welded perpendicular to the web. The stiffeners should have close bearing against the flanges through which they receive their loads, and should extend nearly to the edges of the flanges.

These stiffeners are designed as columns, with allowable stresses as given in Art. 7.14. The column section is assumed to consist of a pair of stiffeners and a strip of girder web with width 25 times web thickness for interior stiffeners and 12 times web thickness at ends. In computing the effective slenderness ratio Kl/r, use an effective length Kl of at least 0.75 the length of the stiffeners.

Intermediate Stiffeners. With properly spaced transverse stiffeners strong enough to act as compression members, a plate-girder web can carry loads far in excess of its buckling load. The girder acts, in effect, like a Pratt truss, with the stiffeners as struts and the web forming fields of diagonal tension. The following formulas for stiffeners are based on this behavior. Like bearing stiffeners, intermediate stiffeners are placed to project normal to the web and the bending axis, but they may consist of a single angle or plate. They may be stopped short of the tension flange a distance up to 4 times the web thickness. If the compression flange is a rectangular plate, single stiffeners must be attached to it to prevent the plate from twisting. When lateral bracing is attached to stiffeners, they must be connected to the compression flange to transmit at least 1% of the total flange stress, except when the flange consists only of angles.

The total shear force, kips, divided by the web area, in^2, for any panel between stiffeners should not exceed the allowable shear F_v given by Eqs. (7.26*a*) and (7.26*b*).

Except for hybrid girders, when C_v is less than unity:

$$F_v = \frac{F_y}{2.89}\left[C_v \frac{1 - C_v}{1.15\sqrt{1 + (a/h)^3}}\right] \leq 0.4F_y \tag{7.26a}$$

For hybrid girders or when C_v is more than unity or when intermediate stiffeners are omitted:

$$F_v = \frac{F_y C_v}{2.89} \leq 0.4F_y \tag{7.26b}$$

where a = clear distance between transverse stiffeners, in

h = clear distance between flanges within an unstiffened segment, in

$C_v = \dfrac{45{,}000k}{F_y(h/t)^2}$ when C_v is less than 0.8

$= \dfrac{190}{h/t}\sqrt{\dfrac{k}{F_y}}$ when C_v is more than 0.8

t = web thickness, in
$k = 5.34 + 4/(a/h)^2$ when $a/h > 1$
$= 4 + 5.34/(a/h)^2$ when $a/h < 1$

Stiffeners for an end panel or for any panel containing large holes and for adjacent panels should be so spaced that the largest average web shear f_v in the panel does not exceed the allowable shear given by Eq. (7.26*b*).

Intermediate stiffeners are not required when h/t is less than 260 and f_v is less than the allowable stress given by Eq. (7.26*b*). When these criteria are not satisfied, stiffeners should be spaced so that the applicable allowable shear, Eq. (7.26*a*) or (7.26*b*), is not exceeded, and in addition, so that a/h is not more than $[260/(h/t)]^2$ or 3.

Solution of the preceding formulas for stiffener spacing requires assumptions of dimensions and trials. The calculations can be facilitated by using tables in the AISC "Manual of Steel Construction." Also, Fig. 7.33 permits rapid selection of

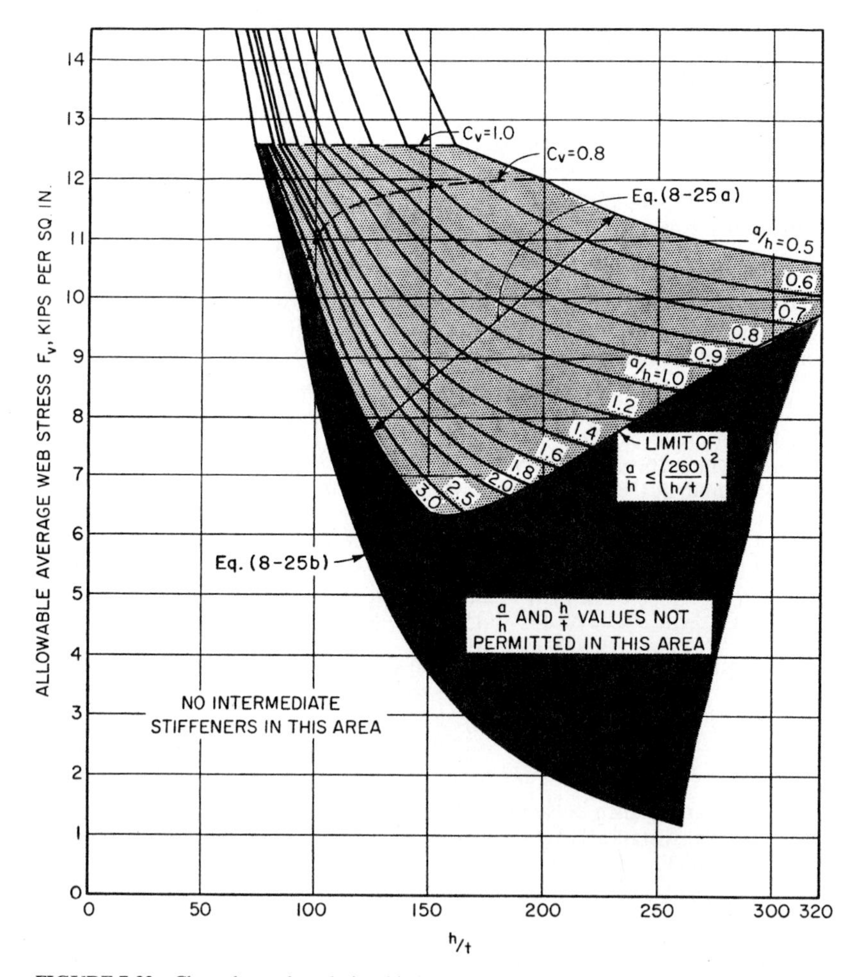

FIGURE 7.33 Chart shows the relationship between allowable shears in webs of plate girders, with yield stress $F_y = 36$ ksi, and web thickness, distance between flanges, and stiffener spacing.

the most efficient stiffener arrangement, for webs of A36 steel. Similar charts can be drawn for other steels.

If the tension field concept is to apply to plate girder design, care is necessary to ensure that the intermediate stiffeners function as struts. When these stiffeners are spaced to satisfy Eq. (7.26*a*), their gross area, in^2 (total area if in pairs) should be at least

$$A_{st} = \frac{1 - C_v}{2}\left[\frac{a}{h} - \frac{(a/h)^2}{\sqrt{1 + (a/h)^2}}\right] YDht \qquad (7.27)$$

where Y = ratio of yield stress of web steel to yield stress of stiffener steel
D = 1.0 for stiffeners in pairs
= 1.8 for single-angle stiffeners
= 2.4 for single-plate stiffeners

When the greatest shear stress f_v in a panel is less than F_v determined from Eq. (7.26*a*), the gross area of the stiffeners may be reduced in the ratio f_v/F_v.

The moment of inertia of a stiffener or pair of stiffeners, about the web axis, should be at least $(h/50)^4$. The connection of these stiffeners to the web should be capable of developing shear, in kips per lineal inch of single stiffener or pair, of at least

$$f_{vs} = h\sqrt{\left(\frac{F_{yw}}{340}\right)^3} \qquad (7.28)$$

where F_{yw} is the yield stress of the web steel (Table 7.16). This shear also may be reduced in the ratio f_v/F_v, as above.

TABLE 7.16 Required Shear Capacity of Intermediate-Stiffener Connections to Girder Web

F_{yw}, ksi	f_{vs}, kips per lin in	F_{yw}, ksi	f_{vs}, kips per lin in
36.0	0.034*h*	60.0	0.074*h*
42.0	0.043*h*	65.0	0.084*h*
45.0	0.048*h*	90.0	0.136*h*
50.0	0.056*h*	100.0	0.160*h*
55.0	0.065*h*		

Combined Stresses in Web. A check should be made for combined shear and bending in the web where the tensile bending stress is approximately equal to the maximum permissible. When f_v, the shear force at the section divided by the web area, is greater than that permitted by Eq. (7.26*a*), the tensile bending stress in the web should be limited to no more than $0.6F_{yw}$ or $F_{yw}(0.825 - 0.375f_v/F_v)$, where F_v is the allowable web shear given by Eq. (7.26*a*). For girders with steel flanges and webs with F_y exceeding 65 ksi, when the flange bending stress is more than 75% of the allowable, the allowable shear stress in the web should not exceed that given by Eq. (7.26*b*).

Also, the compressive stresses in the web should be checked (see Art. 7.17).

7.16.2 LRFD Procedure for Plate Girders

Plate girders are normally proportioned to resist bending on the assumption that the moment of inertia of the gross section is effective. The web must be proportioned such that the maximum web depth-thickness ratio h/t does not exceed h/t given by (7.29) or (7.30), whichever is applicable.

If $a/h \leq 1.5$,

$$\frac{h}{t} \leq \frac{2000}{\sqrt{F_{yf}}} \tag{7.29}$$

If $a/h > 1.5$,

$$\frac{h}{t} \leq \frac{14{,}000}{\sqrt{F_{yf}(F_{yf} + F_r)}} \tag{7.30}$$

where a = clear distance between transverse stiffeners, in
t = web thickness, in
F_{yf} = specified minimum yield stress of steel, ksi
F_r = compressive residual stress in flange = 16.5 ksi for plate girders

Web stiffeners are frequently required to achieve an economical design. However, web stiffeners are not required if $h/t < 260$ and adequate shear strength is provided by the web. The criteria for the design of plate girders are given in the AISC LRFD Specification.

Design Flexural Strength. The design flexural strength is $\phi_b M_n$, where $\phi_b = 0.90$. If $h_c/t \leq 970\sqrt{F_y}$, determine the nominal flexural strength as indicated in Art. 7.15, for either compact or noncompact shapes. If $h_c/t > 970\sqrt{F_y}$, M_n is governed by the limit states of tension-flange yielding or compression-flange buckling.

The design strength is the smaller of the values of $\phi_b M_n$ for yielding of the tension flange, which is

$$\phi_b M_n = 0.90 S_{xt} R_{PG} R_e F_{yt} \tag{7.31}$$

and for buckling of the compression flange, which is

$$\phi_b M_n = 0.90 S_{xc} R_{PG} R_e F_{cr} \tag{7.32}$$

where R_{PG} = plate-girder bending-strength reduction factor
= $1 - 0.0005a_r(h_c/t - 970/\sqrt{F_{cr}}) \leq 1.0$
R_e = hybrid girder factor
= $1 - 0.1(1.3 + a_r)(0.81 - m) \leq 1.0$
= 1 for nonhybrid girders
a_r = ratio of web area to compression-flange area
m = ratio of web yield stress to flange yield stress or to F_{cr}
F_{cr} = critical compression-flange stress, ksi
F_{yt} = yield stress of tension flange, ksi
S_{xt} = section modulus, in^3, with respect to the tension flange
S_{xc} = section modulus, in^3, with respect to the compression flange

The critical stress F_{cr} is different for different limit states. Its value is computed from the values of parameters that depend on the type of limit state: plate girder coefficient C_{PG}, slenderness parameter λ, limiting slenderness parameter λ_p for a compact element, and limiting slenderness parameter λ_r for a noncompact element. Thus, F_{cr} may be computed from one of Eqs. (7.31) to (7.33) for the limit states of lateral-torsional buckling and flange local buckling. The limit state of local buckling of web does not apply.

$$F_{cr} = F_{yf} \qquad \lambda \leq \lambda_p \tag{7.33}$$

$$F_{cr} = C_b F_{yf}\left[1 - \frac{1}{2}\left(\frac{\lambda - \lambda_p}{\lambda_r - \lambda_p}\right)\right] \leq F_{yf} \qquad \lambda_p < \lambda \leq \lambda_r \tag{7.34}$$

$$F_{cr} = C_{PG}/\lambda^2 \qquad \lambda > \lambda_r \tag{7.35}$$

where F_{yf} = specified minimum flange yield stress, ksi
C_b = bending coefficient dependent on moment gradient
= $1.75 + 1.05(M_1/M_2) + 0.3(M_1/M_2)^2$ for lateral-torsional buckling
= 1 for flange local buckling
C_{PG} = $286{,}000/C_b$ for lateral torsional buckling
= 11,200 for flange local buckling
λ = L_b/r_T for lateral-torsional buckling
= $b_f/2t_f$ for flange local buckling
L_b = laterally unbraced length of girder, in
r_T = radius of gyration, in, of compression flange plus one-sixth the web
b_f = flange width, in
t_f = flange thickness, in
λ_p = $300/\sqrt{F_{yf}}$ for lateral-torsional buckling
= $65/\sqrt{F_{yf}}$ for flange local buckling
λ_r = $756/\sqrt{F_{yf}}$ for lateral-torsional buckling
= $150/\sqrt{F_{yf}}$ for flange local buckling

Design Shear Strength. This is given by $\phi_v V_n$, where $\phi_v = 0.90$. With tension-field action, in which the web is permitted to buckle due to diagonal compression and the web carries stresses in diagonal tension in the panels between vertical stiffeners, the design shear strength is larger than when such action is not permitted.

Tension-field action is not allowed for end panels in nonhybrid plate girders, for all panels in hybrid girders and plate girders with tapered webs, and for panels in which the ratio of panel width to depth a/h exceeds 3.0 or $[260(h/t)]^2$, where t is the web thickness. For these conditions, the design shear strength is given by

$$\phi_n V_n = 0.90 \times 0.6 A_w F_{yw} C_v = 0.54 A_w F_{yw} C_v \tag{7.36}$$

where A_w = web area, in^2
F_{yw} = specified web yield stress, ksi
C_v = ratio of critical web stress, in the linear buckling theory, to the shear yield stress of the web steel

For tension-field action, the design shear strength depends on the ratio of panel width to depth a/h. For $h/t \leq 187\sqrt{k/F_{yw}}$,

$$\phi_v V_n = 0.54 A_w F_{yw} \tag{7.37}$$

For $h/t > 187\sqrt{k/F_{yw}}$,

$$\phi_v V_n = 0.54 A_w F_{yw} \left(C_v + \frac{1 - C_v}{1.15\sqrt{1 + (a/h)^2}} \right) \tag{7.38}$$

where k = web buckling coefficient
= 5 if $a/h > 3.0$ or $a/h > [260/(h/t)]^2$
= $5 + 5/(a/h)^2$ otherwise

$C_v = \dfrac{187\sqrt{k/F_{yw}}}{h/t}$ when $187\sqrt{k/F_{yw}} \leq h/t \leq 234\sqrt{k/F_{yw}}$

$= \dfrac{44{,}000}{(h/t)^2}\dfrac{k}{F_y}$ when $h/t > 234\sqrt{k/F_{yw}}$

Web Stiffeners. Transverse stiffeners are required if the web shear strength without stiffeners is inadequate, if $h/t > 418/\sqrt{F_{yw}}$, or if h/t does not meet the requirements of Eqs. (7.27) and (7.28). Where stiffeners are required, the spacing of stiffeners should be close enough to maintain the shear within allowable limits. Also, the moment of inertia I_{st}, in^4, of a transverse stiffener should be at least that computed from

$$I_{st} = at^3 j \tag{7.39}$$

where $j = 2.5/(a/h)^2 - 2$.

The moment of inertia for a pair of stiffeners should be taken about an axis through the center of the web. For a single stiffener, I_{st} should be taken about the web face in contact with the stiffener. In addition, for design for tension-field action, the stiffener area A_{st}, in^2, should be at least that computed from

$$A_{st} = \frac{F_{yw}}{F_{ys}} \left[0.15 Dht(1 - C_v) \frac{V_u}{\phi_v V_n} - 18t^2 \right] \geq 0 \tag{7.40}$$

where F_{ys} = specified yield stress of stiffener, ksi
V_u = required shear strength at stiffener, kips, calculated for the factored loads
D = 1.0 for a pair of stiffeners
= 1.8 for a single-angle stiffener
= 2.4 for a single-plate stiffener

Bending and Shear Interaction. Plate girders should also be proportioned to satisfy Eq. (7.40) if they are designed for tension-field action, stiffeners are required, and V_u/M_u lies between 60 and 133% of V_n/M_n.

$$\frac{M_u}{M_n} + 0.625 \frac{V_u}{V_n} \leq 1.24 \tag{7.41}$$

where M_n = design flexural strength
M_u = required flexural strength calculated for the factored loads but may not exceed $0.90M_n$
V_n = design shear strength
V_u = required shear strength calculated for the factored loads but may not exceed $0.90V_n$

7.17 WEB OR FLANGE LOAD-BEARING STIFFENERS

Members subject to large concentrated loads within their length or large end reactions should be proportioned so that the forces on the web or flange cannot cause local failure or the webs or flanges should be stiffened to carry the concentrated loads. Both ASD and LRFD procedures include design criteria.

7.17.1 ASD for Load-Bearing Stiffeners

Webs of rolled beams and plate girders should be so proportioned that the compressive stress, ksi, at the web toe of the fillets does not exceed

$$F_a = 0.66F_y \tag{7.42}$$

where F_y = specified minimum yield stress, ksi.

Web failure probably would be in the form of buckling caused by concentrated loading, either at an interior load or at the supports. The capacity of the web to transmit the forces safely should be checked.

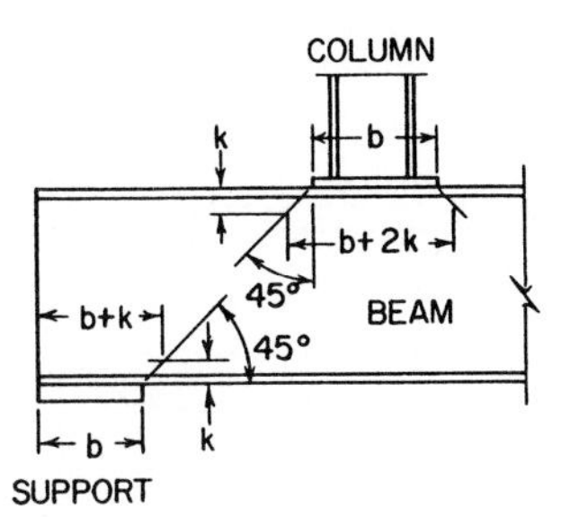

FIGURE 7.34 Web crippling in a simple beam. The critical web section is assumed to occur at the fillet.

Load Distribution. Loads are resisted not only by the part of the web directly under them but also by the parts immediately adjacent. A 45° distribution usually is assumed, as indicated in Fig. 7.34 for two common conditions. The distance k is determined by the point where the fillet of the flange joins the web; it is tabulated in the beam tables of the AISC "Manual of Steel Construction." F_a is applicable to the horizontal web strip of length $b + k$ at the end support or $b + 2k$ under an interior load. Bearing stiffeners are required when F_a is exceeded.

Bearing atop Webs. The sum of the compression stresses resulting from loads bearing directly on or through a flange on the compression edge of a plate-girder web should not exceed the following:

When the flange is restrained against rotation, the allowable compressive stress, ksi, is

$$F_a = \left[5.5 + \frac{4}{(a/h)^2}\right]\frac{10{,}000}{(h/t)^2} \tag{7.43}$$

When the flange is not restrained against rotation,

$$F_a = \left[2 + \frac{4}{(a/h)^2}\right]\frac{10{,}000}{(h/t)^2} \tag{7.44}$$

where a = clear distance between transverse stiffeners, in
h = clear distance between flanges, in
t = web thickness, in

The load may be considered distributed over a web length equal to the panel length (distance between vertical stiffeners) or girder depth, whichever is less.

Web Stiffeners on Columns. The web of a column may also be subject to crippling by the thrust from the compression flange of a rigidly connected beam, as shown at point a in Fig. 7.35. Likewise, to ensure full development of the beam plastic moment, the column flange opposite the tensile thrust at point b may require stiffening.

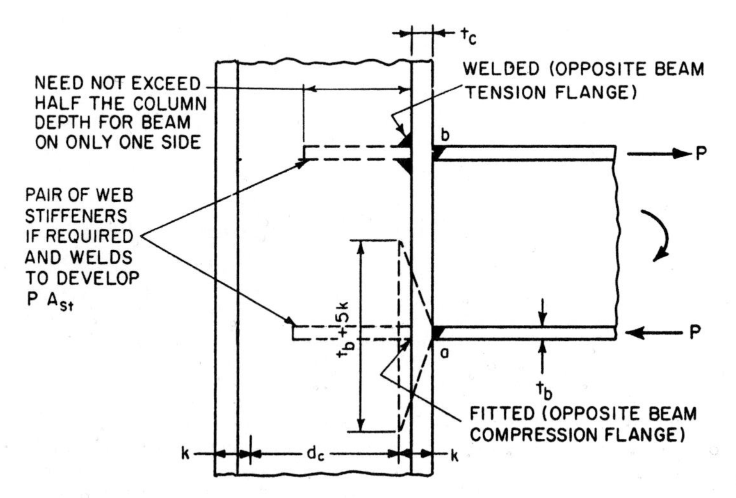

FIGURE 7.35 Web crippling in a column at a welded joint with a beam.

Web stiffeners having a combined cross-sectional area A_{st}, in^2, are required on the column whenever A_{st} computed from Eq. (7.45) is positive.

$$A_{st} = \frac{P - F_{yc}t(t_b + 5k)}{F_{ys}} \tag{7.45}$$

where t = thickness of column web, in
t_b = thickness, in, of beam flange delivering concentrated load
F_{yc} = column steel yield stress, ksi
F_{ys} = stiffener steel yield stress, ksi
P = computed force delivered by beam flange or connection plate multiplied by 5⁄3 when force is a result of dead and live loads, or by 4⁄3 when it is a result of wind or earthquake forces, kips
k = distance from face of column to edge of fillet on rolled sections (use equivalent for welded sections)

Regardless of the preceding requirement, a single or double stiffener is needed opposite the compression force delivered to the column at point a when

$$d_c > \frac{4100\, t^3 \sqrt{F_{yc}}}{P} \tag{7.46}$$

where d_c = clear distance, in, between column flanges (clear of fillets). Also, a pair of stiffeners is needed opposite the tension force at point b when

$$t_f < 0.4 \sqrt{\frac{P}{F_{yc}}} \tag{7.47}$$

where t_f = thickness of column flange, in.

The thickness of a stiffener should not be less than one-half the thickness of the beam flange or plate that delivers force P to the column. Stiffener width should not be less than one-third of the flange or plate width.

7.17.2 LRFD for Load-Bearing Stiffeners

Six limit states should be considered at locations where a large concentrated force acting on a member introduces high local stresses. These limit states are local flange bending, local web yielding, web crippling, sidesway web buckling, compression buckling of the web, and high shear in column web panels. Detailed requirements for determining the design strength for each of these limit states are contained in the AISC LRFD "Specification for Structural Steel for Buildings."

When web stiffeners are required to prevent web crippling or compression buckling of the web, they are designed as columns with an effective length of $Kl = 0.75h$, where h is the clear distance between flanges. The effective cross section is the area of the stiffeners plus $25t$ for interior stiffeners or $12t$ for stiffeners at the end of a member, where t is the web thickness.

7.18 BEARING

For bearing on finished surfaces, such as milled ends and ends of fitted bearing stiffeners, or on the projected area of pins in finished holes, the allowable stress in ASD is

$$F_p = 0.90F_y \tag{7.48}$$

where F_y is the specified minimum yield stress of the steel, ksi. When the parts in contact have different yield stresses, use the smaller F_y (Table 7.17).

The allowable bearing stress on expansion rollers and rockers, kip/in, is

$$F_p = \frac{F_y - 13}{20}\, 0.66d \tag{7.49}$$

where d is the diameter of roller or rocker, in (Table 7.18).

TABLE 7.17 Bearing on Finished Surfaces, ksi

Allowable stress (ASD)		Design strength (LRFD)	
F_y	F_p	F_y	$\phi R_n / A_{pb}$
36	32.4	36	54.0
42	37.8	42	63.0
45	40.5	45	67.5
50	45.0	50	75.0
55	49.5	55	82.5
60	54.0	60	90.0

TABLE 7.18 Allowable Bearing Loads on Expansion Rollers or Rockers, kips per in of Bearing

Allowable load (ASD)		Design strength (LRFD)	
F_y	F_p	F_y	ϕR_n*
36	0.76d	36	1.30d
42	0.96d	42	1.64d
45	1.06d	45	1.81d
50	1.22d	50	2.08d
55	1.39d	55	2.37d
60	1.55d	60	2.64d

*d is the diameter, in, of the roller or rocker.

Allowable bearing stresses on masonry usually can be obtained from a local or state building code, whichever governs. In the absence of such regulations, however, the values in Table 7.19 may be used.

TABLE 7.19 Allowable Bearing on Masonry, ksi

On sandstone and limestone	0.40
On brick in cement mortar	0.25
On the full area of concrete	$0.35f'_c$
On less than full concrete area	$0.35f'_c \sqrt{A_2/A_1} \leq 0.7f'_c$

where f'_c = specified compressive strength, ksi, of the concrete
A_1 = bearing area
A_2 = concrete area

LRFD Procedure for Bearing. The design strength in bearing on the projected bearing area for finished surfaces, such as milled ends and ends of bearing stiffeners, or on the projected area of pins in finished holes, is ϕR_n, where $\phi = 0.75$.

$$R_n = 2.0F_y A_{pb} \tag{7.50}$$

where F_y is the lesser minimum yield stress, ksi, of the steel (Table 7.17) and A_{pb} is the projected bearing area, in^2.

For expansion rollers and rockers, R_n, kips, is given by

$$R_n = 1.5(F_y - 13)Ld/20 \tag{7.51}$$

where L is the length, in, of bearing, and d is the diameter, in (see Table 7.18).

7.19 COMBINED AXIAL COMPRESSION AND BENDING

A member carrying both axial and bending forces is subjected to secondary bending moments resulting from the axial force and the displacement of the neutral axis. This effect is referred to as the P-Δ effect. Such secondary bending moments are more critical in members where the axial force is a compressive force, because the P-Δ secondary moment increases the deflection of the member. In ASD, the effects of these secondary moments may be neglected where the axial force is a tensile force or where the actual compressive stress is less than 15% of the allowable compressive stress. LRFD does not include this concept.

The following design criteria apply to singly and doubly symmetrical members.

7.19.1 ASD for Compression and Bending

When the computed axial stress, f_a, is less than 15% of F_a, the stress that would be permitted if axial force alone were present, a straight-line interaction formula may be used. Thus, when $f_a/F_a \le 0.15$:

$$\frac{f_a}{F_a} + \frac{f_{bx}}{F_{bx}} + \frac{f_{by}}{F_{by}} \le 1.0 \tag{7.52}$$

where subscripts x and y indicate, respectively, the major and minor axes of bending (if bending is about only one axis, then the term for the other axis is omitted), and

f_b = computed compressive bending stress, ksi, at point under consideration
F_b = compressive bending stress, ksi, that is allowed if bending alone existed

When $f_a/F_a > 0.15$, the effect of the secondary bending moment should be taken into account and the member proportioned to satisfy Eqs. (7.53a) and (7.53b) where, as before, subscripts x and y indicate axes of bending:

$$\frac{f_a}{F_a} + \frac{C_{mx} f_{bx}}{[1 - f_a/F'_{ex}]F_{bx}} + \frac{C_{my} f_{by}}{[1 - f_a/F'_{ey}]F_{by}} \le 1.0 \tag{7.53a}$$

$$\frac{f_a}{0.60F_y} + \frac{f_{bx}}{F_{bx}} + \frac{f_{by}}{F_{by}} \le 1.0 \tag{7.53b}$$

$$F'_e = \frac{12\pi^2 E}{23(Kl_b/r_b)^2} \tag{7.54}$$

where E = modulus of elasticity, 29,000 ksi
l_b = actual unbraced length, in, in the plane of bending
r_b = corresponding radius of gyration, in
K = effective-length factor in the plane of bending
C_m = reduction factor determined from the following conditions:

1. For compression members in frames subject to joint translation (sidesway), $C_m = 0.85$.

2. For restrained compression members in frames braced against joint translation and not subject to transverse loading between their supports in the plane of bending, $C_m = 0.6 - 0.4M_1/M_2$, but not less than 0.4. M_1/M_2 is the ratio of the smaller to larger moments at the ends of that portion of the member unbraced in the plane of bending under consideration. M_1/M_2 is positive when the member is bent in reverse curvature, and negative when it is bent in single curvature.

3. For compression members in frames braced against joint translation in the plane of loading and subjected to transverse loading between their supports, the value of C_m may be determined by rational analysis. Instead, however, C_m may be taken as 0.85 for members whose ends are restrained, and 1.0 for ends unrestrained.

In wind and seismic design F'_e may be increased one-third. The resultant section, however, should not be less than that required for dead and live loads alone without the increase in allowable stress.

Additional information, including illustrations of the foregoing three conditions for determining the value of C_m, is given in the AISC "Commentary" on the AISC ASD "Specification for Structural Steel for Buildings."

7.19.2 LRFD for Compression and Bending

Members subject to both axial compression and bending stresses should be proportioned to satisfy Eq. (7.55) or (7.56), whichever is applicable.

For $(P_u/\phi_c P_n) \geq 0.2$,

$$\frac{P_u}{\phi_c P_n} + \frac{8}{9}\left(\frac{M_{ux}}{\phi_b M_{nx}} + \frac{M_{uy}}{\phi_b M_{ny}}\right) \leq 1.0 \qquad (7.55)$$

For $(P_u/\phi_c P_n) < 0.2$,

$$\frac{P_u}{2\phi_c P_n} + \left(\frac{M_{ux}}{\phi_b M_{nx}} + \frac{M_{uy}}{\phi_b M_{ny}}\right) \leq 1.0 \qquad (7.56)$$

where P_u = required compressive strength, kips, calculated for the factored axial loads
M_u = required flexural strength, kip-in calculated for primary bending and P-Δ effects
$\phi_c P_n$ = design compressive strength (Art. 7.14.3)
$\phi_b M_n$ = design flexural strength (Art. 7.15.2)

M_u may be determined for the factored loads from a second-order elastic analysis. The AISC LRFD specification, however, permits M_u to be determined from Eq. (7.57) with the variables in this equation determined from a first-order analysis.

$$M_u = B_1 M_{nt} + B_2 M_{lt} \qquad (7.57)$$

where M_{nt} = required flexural strength, kip-in, with no relative displacement of the member ends; for example, for a column that is part of a rigid frame, drift is assumed prevented

M_{lt} = required flexural strength, kip-in, for the effects only of drift as determined from a first-order analysis

B_1 = magnification factor for M_{nt} to account for the P-Δ effects

$= \dfrac{C_m}{1 - P_u/P_e}$

C_m = reduction factor defined for Eq. (7.54)

B_2 = magnification factor for M_{lt} to account for the P-Δ effects

B_2 may be calculated from either Eq. (7.58) or (7.59), the former usually being the simpler to evaluate.

$$B_2 = \frac{1}{1 - (\Sigma P_u/\Sigma HL)\Delta_{oh}} \tag{7.58}$$

$$B_2 = \frac{1}{1 - \Sigma P_u/\Sigma P_e} \tag{7.59}$$

where ΣP_u = sum of the axial-load strengths, kips, of all the columns in a story

$P_e = A_g F_y/\lambda_c^2$

A_g = gross area of member, in^2

F_y = specified yield stress, ksi

$\lambda_c = \dfrac{Kl}{r\pi}\sqrt{\dfrac{F_y}{E}} = \dfrac{Kl}{r}\sqrt{\dfrac{F_y}{286{,}220}}$

K = effective column length factor in the plane of bending, to be determined by structural analysis, but not to exceed unity in calculation of B_1 and not to be less than unity in calculation of B_2

r = governing radius of gyration, in, about the plane of buckling

Δ_{oh} = drift, in, of the story in which the column is located

L = story height, in

ΣH = sum of all the horizontal forces on the story that cause Δ_{oh}

7.20 COMBINED AXIAL TENSION AND BENDING

For ASD, members subject to both axial tension and bending stresses should be proportioned to satisfy Eq. (7.52), with f_b and F_b, respectively, as the computed and allowable bending tensile stress. But the compressive bending stresses must not exceed the values given in Art. 7.15.1.

LRFD for Tension and Bending. Symmetric members subject to both axial tension and bending stresses should be proportioned to satisfy either Eq. (7.55) or Eq. (7.56), whichever is applicable.

7.21 COMPOSITE CONSTRUCTION

In composite construction, rolled or built-up steel shapes are combined with reinforced concrete to form a structural member. Examples of this type of construction

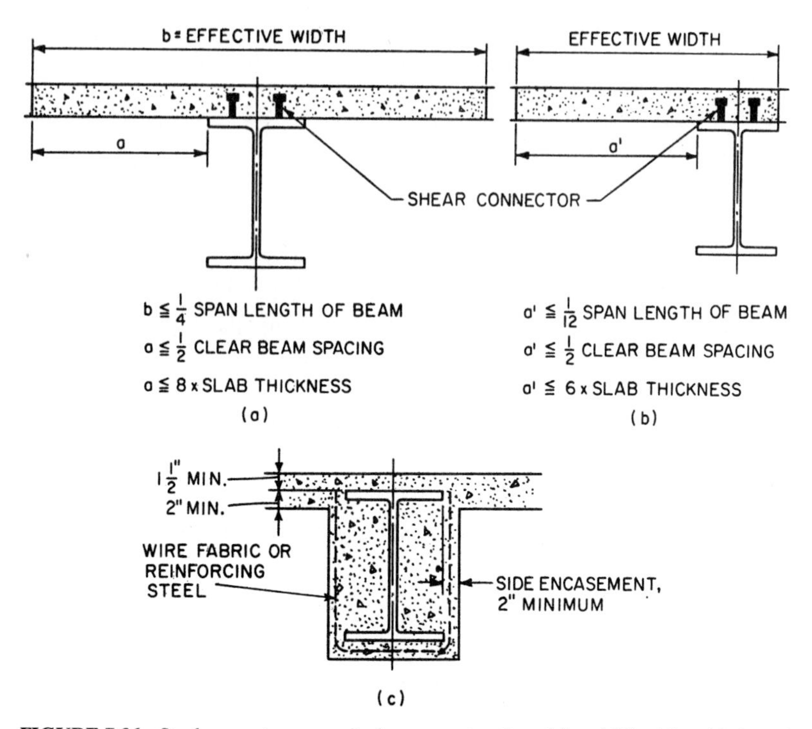

FIGURE 7.36 Steel-concrete composite-beam construction: (*a*) and (*b*) with welded-stud shear connectors; (*c*) with encasement in concrete.

include: (*a*) concrete-encased steel beams (Fig. 7.36*c*), (*b*) concrete decks interactive with steel beams (Fig. 7.36*a* and *b*), (*c*) concrete encased steel columns, and (*d*) concrete filled steel columns. The most common use of this type of construction is for composite beams, where the steel beam supports and works with the concrete slab to form an economical building element.

Design procedures require that a decision be made regarding the use of shoring for the deck pour. (Procedures for ASD and LRFD differ in this regard.) If shoring is not used, the steel beam must carry all dead loads applied until the concrete hardens, even if full plastic capacity is permitted for the composite section afterward.

The assumed composite cross section is the same for ASD and LRFD procedures. The effective width of the slab is governed by beam span and beam spacing or edge distance (Fig. 7.36*a* and *b*).

Slab compressive stresses are seldom critical for interior beams but should be investigated, especially for edge beams. Thickening the slab key and minimum requirements for strength of concrete can be economical.

Connector Details. In composite construction, shear connectors welded to the top flange of the steel beam are typically used to ensure composite action by transferring shear between the concrete deck and steel beam. Location, spacing, and size limitations for shear connectors are the same for ASD and LRFD procedures. Connectors, except those installed in ribs of formed steel decks, should

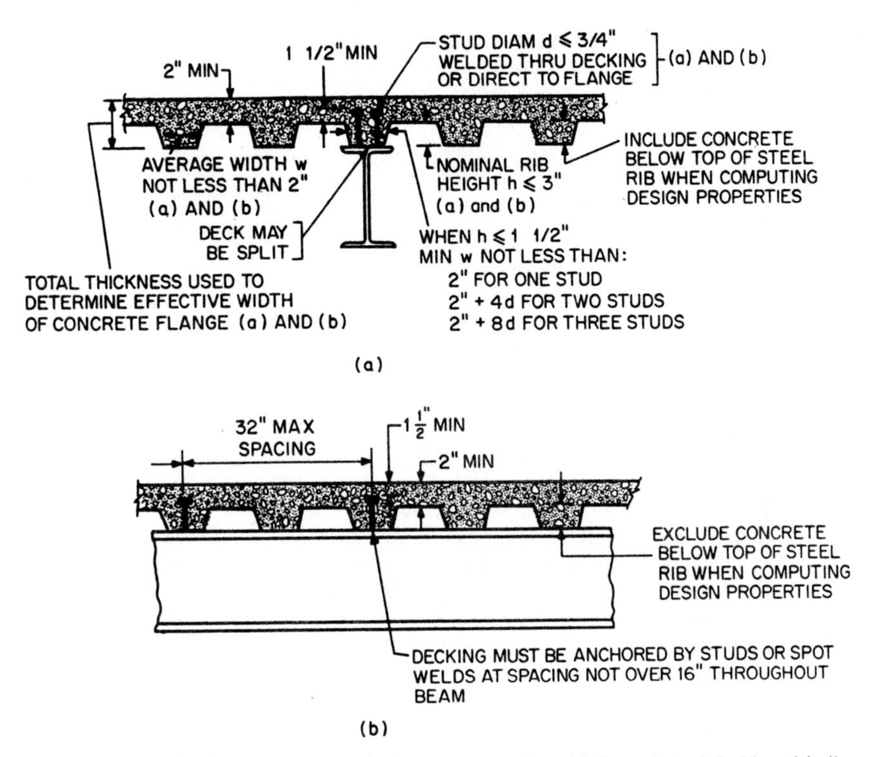

FIGURE 7.37 Steel-concrete composite-beam construction with formed steel decking: (*a*) ribs parallel to beam; (*b*) ribs transverse to beam [refer to (*a*) for applicable requirements].

have a minimum lateral concrete cover of 1 in. The diameter of a stud connector, unless located directly over the beam web, is limited to 2.5 times the thickness of the beam flange to which it is welded. Minimum center-to-center stud spacing is 6 diameters along the longitudinal axis, 4 diameters transversely. Studs may be spaced uniformly, rather than in proportion to horizontal shear, inasmuch as tests show a redistribution of shear under high loads similar to the stress redistribution in large bolted joints. Maximum spacing is 8 times the slab thickness.

Formed Steel Decking. Concrete slabs are frequently cast on permanent steel decking with a ribbed, corrugated, cellular, or blended cellular cross section (see Sec. 8). Two distinct composite-design configurations are inherent: ribs parallel or ribs perpendicular to the supporting beams or girders (Fig. 7.37). The design procedures, for both ASD and LRFD, prescribed for composite concrete-slab and steel-beam construction are also applicable for systems utilizing formed steel decking, subject to additional requirements of the AISC "Specification for Structural Steel for Buildings" and as illustrated in Fig. 7.37.

Shear and Deflection of Composite Beams In ASD and LRFD, shear forces are assumed to be resisted by the steel beam. Deflections are calculated based on composite section properties. It should be noted that, because of creep of the concrete, the actual deflections of composite beams under long-term loads, such as dead load, will be greater than those computed.

7.21.1 ASD of Encased Beams

Two design methods are allowed for encased beams. In one method, stresses are computed on the assumption that the steel beam alone supports all the dead load applied prior to concrete hardening (unless the beam is temporarily shored), and the composite beam supports the remaining dead and live loads. Then, for positive bending moments, the total stress, ksi, on the steel-beam bottom flange is

$$f_b = \frac{M_D}{S} + \frac{M_L}{S_t} \leq 0.66F_y \tag{7.60}$$

where F_y = specified yield stress of the steel, ksi
M_D = dead-load bending moment, kip-in
M_L = live-load bending moment, kip-in
S = section modulus of steel beam, in^3
S_t = section modulus of transformed section, in^3. To obtain the transformed equivalent steel area, divide the effective concrete area by the modular ratio n (modulus of elasticity of steel divided by modulus of elasticity of concrete). In computation of effective concrete area, use effective width of concrete slab (Fig. 7.36*a* and *b*)

The stress $0.66F_y$ is allowed because the steel beam is restrained against lateral buckling.

The second method stems from a "shortcut" provision contained in many building codes. This provision simply permits higher bending stresses in beams encased in concrete. For example,

$$f_b = \frac{M_D + M_L}{S} \leq 0.76F_y \tag{7.61}$$

This higher stress would not be realized, however, because of composite action.

7.21.2 ASD of Beams with Shear Connectors

For composite construction where shear connectors transfer shear between slab and beam, the design is based on behavior at ultimate load. It assumes that all loads are resisted by the composite section, even if shores are not used during construction to support the steel beam until the concrete gains strength. For this case, the computed stress in the bottom flange for positive bending moment is

$$f_b = \frac{M_D + M_L}{S_t} \leq 0.66F_y \tag{7.62}$$

where S_t = section modulus, in^3, of transformed section of composite beam. To prevent overstressing the bottom flange of the steel beam when temporary shoring is omitted, a limitation is placed on the value of S_t used in computation of f_b with Eq. (7.62):

$$S_t \leq \left(1.35 + 0.35 \frac{M_L}{M_D}\right) S_s \tag{7.63}$$

where M_D = moment, kip-in, due to loads applied prior to concrete hardening (75% cured)
M_L = moment, kip-in, due to remaining dead and live loads
S_s = section modulus, in^3, of steel beam alone relative to bottom flange

Shear on Connectors. Shear connectors usually are studs or channels. The total horizontal shear to be taken by the connectors between the point of maximum positive moment and each end of a simple beam, or the point of counterflexure in a continuous beam, is the smaller of the values obtained from Eqs. (7.64) and (7.65).

$$V_h = \frac{0.85 f'_c A_c}{2} \tag{7.64}$$

$$V_h = \frac{A_s F_y}{2} \tag{7.65}$$

where f'_c = specified strength of concrete, ksi
A_c = actual area of effective concrete flange, as indicated in Fig. 7.36*a* and *b*, in^2
A_s = area of steel beam, in^2

In continuous composite beams, where shear connectors are installed in negative-moment regions, the longitudinal reinforcing steel in the concrete slab may be considered to act compositely with the steel beam in those regions. In such cases, the total horizontal shear to be resisted by the shear connectors between an interior support and each adjacent inflection point is

$$V_h = \frac{A_{sr} F_{yr}}{2} \tag{7.66}$$

where A_{sr} = total area, in^2, of longitudinal reinforcing steel within the effective width of the concrete slab at the interior support
F_{yr} = specified yield stress of the reinforcing steel, ksi

These formulas represent the horizontal shear at ultimate load divided by 2 to approximate conditions at working load.

Number of Connectors. The minimum number of connectors N_1, spaced uniformly between the point of maximum moment and adjacent points of zero moment, is V_h/q, where q is the allowable shear load on a single connector, as given in Table 7.20. Values in this table, however, are applicable only to concrete made with aggregates conforming to ASTM C33. For concrete made with rotary-kiln-produced aggregates conforming to ASTM C330 and with concrete weight of 90 pcf or more, the allowable shear load for one connector is obtained by multiplying the values in Table 7.20 by the factors in Table 7.21.

TABLE 7.20 Allowable Horizontal-Shear Loads, q, for Connectors, kips
(Applicable only to concrete made with ASTM C33 aggregates)

Connector	$f'_c = 3.0$	$f'_c = 3.5$	$f'_c \geq 4.0$
½-in dia. × 2-in hooked or headed stud*	5.1	5.5	5.9
⅝-in dia. × 2½-in hooked or headed stud*	8.0	8.6	9.2
¾-in dia. × 3-in hooked or headed stud*	11.5	12.5	13.3
⅞-in dia. × 3½-in hooked or headed stud*	15.6	16.8	18.0
3-in channel, 4.1 lb	4.3*w*†	4.7*w*†	5.0*w*†
4-in channel, 5.4 lb	4.6*w*†	5.0*w*†	5.3*w*†
5-in channel, 6.7 lb	4.9*w*†	5.5*w*†	5.6*w*†

*Length given is minimum.
†w = length of channel, in.

If a concentrated load occurs between the points of maximum and zero moments, the minimum number of connectors required between the concentrated load and the point of zero moment is given by

$$N_2 = \frac{V_h}{q}\left(\frac{S_t M_c / M - S_s}{S_t - S_s}\right) \tag{7.67}$$

where M = maximum moment, in-kips
M_c = moment, in-kips, at concentrated load $< M$
S_s = section modulus, in^3, of steel beam relative to bottom flange
S_t = section modulus, in^3, of transformed section of composite beam relative to bottom flange but not to exceed S_t computed from Eq. (7.63).

The allowable shear loads for connectors incorporate a safety factor of about 2.5 applied to ultimate load for the commonly used concrete strengths. Not to be confused with shear values for fasteners given in Art. 7.26, the allowable shear loads for connectors are applicable only with Eqs. (7.64) to (7.66).

The allowable horizontal shear loads given in Tables 7.20 and 7.21 may have to be adjusted for use with formed steel decking. For decking with ribs parallel to supports (Fig. 7.37*a*), the allowable loads should be reduced when w/h is less than 1.5 by multiplying the tabulated values by

$$0.6\left(\frac{w}{h}\right)\left(\frac{H}{h} - 1\right) \leq 1 \tag{7.68}$$

TABLE 7.21 Shear-Load Factors for Connectors in Lightweight Concrete

Air dry weight, pcf, of concrete	90	95	100	105	110	115	120
Factors for $f'_c \leq 4.0$ ksi	0.73	0.76	0.78	0.81	0.83	0.86	0.88
Factors for $f'_c \geq 5.0$ ksi	0.82	0.85	0.87	0.91	0.93	0.96	0.99

where w = average width of concrete rib, in
h = nominal rib height, in
H = length of stud after welding, in, but not more than $(h + 3)$ for computations

For decking with ribs perpendicular to supports, the reduction factor is:

$$\left(\frac{0.85}{\sqrt{N}}\right)\left(\frac{w}{h}\right)\left(\frac{H}{h} - 1\right) \leq 1 \tag{7.69}$$

where N = number of studs on a beam and in one rib, but three studs are the maximum that may be considered effective.

7.21.3 LRFD of Encased Beams

Two methods of design are allowed, the difference being whether or not shoring is used. In both cases, the design strength is $\phi_b M_n$, where $\phi_b = 0.90$. M_n is calculated for the elastic stress distribution on the composite section if shoring is used or the plastic stress distribution on the steel section alone if shoring is not used.

7.21.4 LRFD of Composite Beams

As with ASD, the use of shoring to carry dead loads prior to the time the concrete has hardened determines which design procedures are used. For composite construction where the steel beams are exposed, the design flexural strength for positive moment (compression in the concrete) is $\phi_b M_n$. It is dependent on the depth-thickness ratio h_c/t_w of the steel beam, where t_w is the web thickness and, for webs of rolled or formed sections, h_c is twice the distance from the neutral axis to the toe of the fillet at the compression flange, and for webs of built-up sections, h_c is twice the distance from the neutral axis to the nearest line of fasteners at the compression flange or the inside face of a welded compression flange.

When $h_c/t_w \leq 640/\sqrt{F_y}$, $\phi_b = 0.85$ and M_n is calculated for plastic stress distribution on the composite section. If $h_c/t_w > 640/\sqrt{F_y}$, $\phi_b = 0.90$ and M_n is calculated for elastic stress distribution, with consideration of the effects of shoring. When the member is subject to negative moment, the practical design approach is to neglect the composite section and use the requirements for beams in flexure, as given in Art. 7.15.2.

LRFD of Beams with Shear Connectors. The concepts described in Art. 7.21.2 for ASD apply also to LRFD of beams with shear connectors. Inasmuch as factored loads are used for LRFD, however, the equations used in the two types of design differ.

In regions of positive moment, the total horizontal shear V_h, kips, to be carried by the shear connectors between the point of maximum moment and the point of zero moment is the smallest value computed from Eqs. (7.70) to (7.72).

$$V_h = 0.85f'_c A_c \tag{7.70}$$

$$V_h = A_s F_y \tag{7.71}$$

$$V_h = \Sigma Q_n \tag{7.72}$$

where f'_c = 28-day compressive strength of concrete, ksi
A_c = area of concrete slab within the effective width, in^2
A_s = area of the cross section, in^2, of steel beam
F_y = specified minimum yield stress of the steel
ΣQ_n = sum of the nominal strengths, kips, of the shear connectors between the point of maximum moment and the point of zero moment

The number of shear connectors n must equal or exceed V_h/Q_n, where Q_n is the nominal strength of one shear connector.

The nominal strength, kips, of one stud shear connector embedded in a solid concrete slab is

$$Q_n = 0.5A_{sc}\sqrt{f'_c E_c} \leq A_{sc}F_u \tag{7.73}$$

where A_{sc} = cross-sectional area of a stud shear connector, in^2
f'_c = 28-day compressive strength of concrete, ksi
F_u = minimum specified tensile strength of a stud shear connector, ksi
E_c = modulus of elasticity of the concrete, ksi

The nominal strength, kips, of one channel shear connector embedded in a solid concrete slab is

$$Q_n = 0.3(t_f + 0.5t_w)L_c\sqrt{f'_c E_c} \tag{7.74}$$

where t_f = flange thickness of channel shear connector, in
t_w = web thickness of channel, in
L_c = length of channel, in

As in ASD, the shear capacity of stud connectors may have to be reduced if they are used with formed metal decking. The reduction factors, Eqs. (7.68) and (7.69), also apply to LRFD.

7.22 MEMBERS SUBJECT TO TORSION

This is a special type of load application, since in normal practice eccentric loads on beams are counterbalanced to the point where slight eccentricities may be neglected. For example, spandrel beams supporting a heavy masonry wall may not be concentric with the load, thus inducing torsional stresses, but these will largely be canceled out by the equally eccentric loads of the floor, partitions, attached beams, and similar restraints. For this reason, one seldom finds any ill effects from torsional stresses.

It is during the construction phase that torsion may be in evidence, usually the result of faulty construction procedure. In Fig. 7.38 are illustrated some of the bad practices that have caused trouble in the field: when forms for concrete slabs are hung on one edge of a beam (usually the light secondary beam) the weight of the wet concrete may be sufficient to twist the beam. Figure 7.38*a* shows the correct method, which reduces torsion. Likewise for spandrels, the floor ties, if any, forms, or the slab itself should be placed prior to the construction of the eccentric wall (Fig. 7.38*b*). Connectors for heavy roofing sheets when located on one side of the purlin may distort the section; the condition should be corrected by staggering, as indicated in Fig. 7.38*c*.

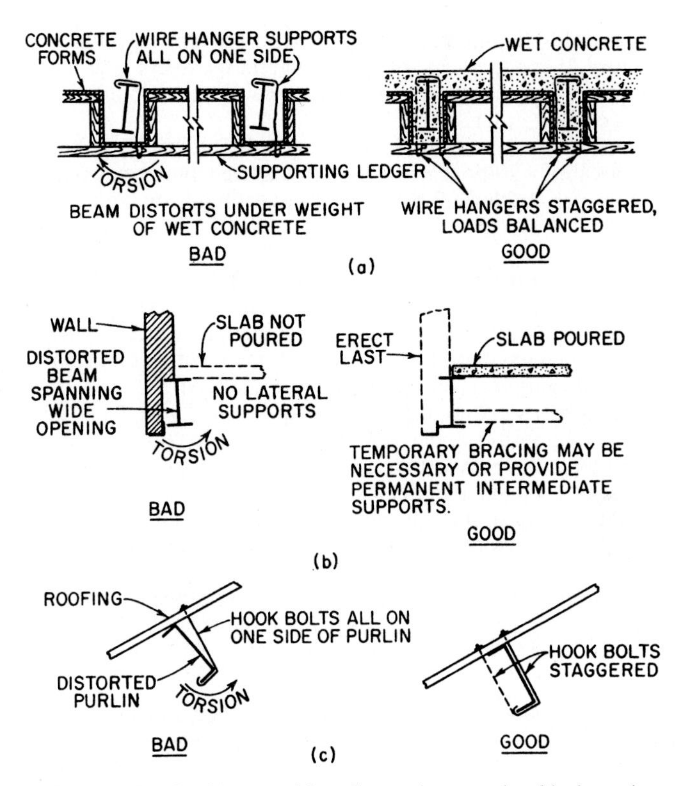

FIGURE 7.38 Steel beams subjected to torsion—good and bad practice.

Equations for computing torsion stresses are given in Art. 5.4.2. Also, see Bibliography, Art. 7.50.

7.23 MEMBERS SUBJECT TO CYCLIC LOADING

Relatively few structural members in a building are ever subjected to large, repeated variations of stress or stress reversals (tension to compression, and vice versa) that could cause fatigue damage to the steel. Members need not be investigated for this possibility unless the number of cycles of such stresses exceeds 20,000, which is nearly equivalent to two applications every day for 25 years.

7.24 SERVICEABILITY CRITERIA FOR STEEL DESIGN

The AISC ASD "Specification for Structural Steel for Buildings" restricts the maximum live-load deflection of beams and girders supporting plaster ceilings to

$\frac{1}{360}$ of the span. This requirement is intended to prevent cracking of monolithic plaster; therefore, it is not necessarily applicable to members with unfinished ceilings or ceilings of other materials. The LRFD specification contains no numerical limits on serviceability criteria.

Basically, deflection criteria are left to the designer. It is impracticable to specify limits for all possible variations of loads, occupancies, and types of construction. As a guide, however, Table 7.22 may be used to set limits on deflection of flexural members.

Minimum Depth-Span Ratios. Also, as a guide, Table 7.22 lists suggested minimum depth-span ratios for various loading conditions and yield strengths of steel up to $F_y = 50.0$ ksi. These may be useful for estimating or making an initial design selection. Since maximum deflection is a straight-line function of maximum bending stress f_b and therefore is nearly proportional to F_y, a beam of steel with $F_y = 100.0$ ksi would have to be twice the depth of a beam of steel with $F_y = 50.0$ ksi when each is stressed to allowable values and has the same maximum deflection.

Vibration of large floor areas that are usually free of physical dampeners, such as partitions, may occur in buildings such as shopping centers and department stores, where pedestrian traffic is heavy. The minimum depth-span ratios in Table 7.22 suggested for "heavy pedestrian traffic" are intended to provide an acceptable solution.

One rule of thumb that may be used to determine beam depth quickly is to choose a depth, in, not less than 1.5% of F_y times the span, ft. Thus, for A36 steel, depth, in, should be at least half the span, ft.

TABLE 7.22 Guide to Selection of Beam Depths and Deflection Limits

	Yield stress F_y, ksi				Maximum stress, ksi	
	36.0	42.0	45.0	50.0	$0.60F_y$	$0.66F_y$
Specific beam condition	Minimum depth-span ratio				Maximum ratio of deflection to span	
Heavy shock or vibration	$\frac{1}{18}$	$\frac{1}{15.5}$	$\frac{1}{14.5}$	$\frac{1}{13}$	$\frac{1}{357}$	$\frac{1}{324}$
Heavy pedestrian traffic	$\frac{1}{20}$	$\frac{1}{17}$	$\frac{1}{16}$	$\frac{1}{14.5}$	$\frac{1}{320}$	$\frac{1}{291}$
Normal loading	$\frac{1}{22}$	$\frac{1}{19}$	$\frac{1}{18}$	$\frac{1}{16}$	$\frac{1}{290}$	$\frac{1}{264}$
Beams for flat roofs*	$\frac{1}{25}$	$\frac{1}{21.5}$	$\frac{1}{20}$	$\frac{1}{18}$	$\frac{1}{258}$	$\frac{1}{232}$
Roof purlins, except for flat roofs*	$\frac{1}{28}$	$\frac{1}{24}$	$\frac{1}{22}$	$\frac{1}{20}$	$\frac{1}{232}$	$\frac{1}{210}$

*Investigate for stability against ponding.

Ponding. Beams for flat roofs may require a special investigation to assure stability against water accumulation, commonly called ponding, unless there is adequate provision for drainage during heavy rainfall. The AISC specification gives these criteria for stable roofs:

$$C_p + 0.9C_s \leq 0.25 \tag{7.75}$$

$$I_d \geq \frac{25S^4}{10^6} \tag{7.76}$$

where $C_p = 32L_sL_p^4/10^7\, I_p$
$C_s = 32SL_s^4/10^7\, I_s$
L_p = column spacing in direction of girder, ft (length of primary members)
L_s = column spacing perpendicular to direction of girder, ft (length of secondary member)
S = spacing of secondary members, ft
I_p = moment of inertia for primary members, in^4
I_s = moment of inertia for secondary member, in^4. Where a steel deck is supported on primary members, it is considered the secondary member. Use $0.85I_s$ for joists and trusses
I_d = moment of inertia of a steel deck supported on secondary members, in^4/ft

Uniform-Load Deflections. For the common case of a uniformly loaded simple beam loaded to the maximum allowable bending stress, the deflection in inches may be computed from

$$\delta = \frac{5}{24}\frac{F_b l}{Ed/l} \tag{7.77}$$

where F_b = the allowable bending stress, ksi
l = the span, in
E = 29,000 ksi
d/l = the depth-span ratio

Camber. Trusses of 80-ft or greater span should be cambered to offset dead-load deflections. Crane girders 75 ft or more in span should be cambered for deflection under dead load plus one-half live load.

DESIGN OF CONNECTIONS

Design of connections and splices is a critical aspect of the design process. Because each fabricator has unique equipment and methods, the detailed configuration of connections plays an important part in determining the cost of the fabricated product. Consequently, the detailed design of these elements is a part of the work performed by the fabricator. In the industry, this work is known as **detailing.**

Usually, the structural engineer indicates the type of connections and type and size of fasteners required; for example, "framed connections with ⅞-in A325 bolts in bearing-type joints," or the type of connection with reference to AWS D1.1 requirements. For beams, the design drawings should specify the reactions. If, however, the reactions are not noted, the detailer will determine the reactions from the uniform-load capacity (tabulated in the AISC Manual), giving due consideration to the effect of large concentrated loads near the connection. For connections resisting lateral loads, live, wind, or seismic, the design drawing should stipulate the forces and moments to be carried. Generally, the drawing should also include a sketch showing the type of moment connection desired.

Design Criteria for Connections. Either ASD or LRFD may be used to design the connections of a structure. Selection of the design procedure, however, must be consistent with the method used to proportion the members. When LRFD procedures are used, the loads and load factors discussed in Arts. 7.12 to 7.23 should be incorporated. The AISC Manual, Vol. II, Connections, provides many design aids for both design procedures.

7.25 COMBINATIONS OF FASTENERS

The AISC ASD and LRFD "Specification for Structural Steel for Buildings" distinguish between existing and new framing in setting conditions for use of fasteners in connection design.

In new work, A307 bolts or high-strength bolts in bearing-type connections should not be considered as sharing the load with welds. If welds are used, they should be designed to carry the load in the connection. However, when one leg of a connection angle is connected with one type of fastener and the other leg with a different type, this rule does not apply. The load is transferred across each joint by one type of fastener. Such connections are commonly used, since one type of fastener may be selected for shop work and a different type for field work.

High-strength bolts in slip-critical joints may share the load with welds on the same connection interface if the bolts are fully tightened before the welds are made.

For connections in existing frames, existing rivets and high-strength bolts may be used for carrying stresses from existing dead loads, and welds may be provided for additional dead loads and design live loads. This provision assumes that whatever slip that could occur in the existing joint has already occurred.

7.26 LOAD CAPACITY OF BOLTS

Under service conditions, bolts may be loaded in tension, shear, or a combination of tension and shear. The load capacities specified in AISC ASD and LRFD specifications are closely related and are based on the "Specification for Structural Joints Using ASTM A325 or A490 Bolts," Research Council on Structural Connections of the Engineering Foundation. Both bearing-type and slip-critical bolted connections are proportioned for the shear forces on the gross area of bolts.

7.26.1 ASD for Bolts

Allowable tension and shear stresses for bolts are listed in Table 7.23. The allowable bearing load at a bolt hole is $1.5F_u dt$, where F_u is the specified tensile strength, d is the nominal bolt diameter, and t = thickness of connected part.

Table 7.24 tabulates maximum sizes for standard, oversize, and slotted bolt holes. Oversize holes are permitted only in slip-critical connections. In slip-critical connections, slots may be formed without regard to the direction of loading; but in bearing-type connections, slot length should be placed normal to the direction of loading. Washers, hardened when used with high-strength bolts, should be placed over oversize and short-slot holes.

TABLE 7.23 Allowable Stresses, ksi, for Bolts and Threaded Parts[a]

	Shear in slip-critical connections F_v[b]				Bearing-type connections	
			Long-slot holes			
Fasteners	Standard-size holes	Oversize and short-slot holes	Transverse load[c]	Parallel load[c]	Shear F_v	Tension F_t, including reduction for shear stress f_v[d]
A307 bolts					10.0[e,f]	$26 - 1.8f_v \leq 20$
Threaded parts and A449 bolts, threads[g] not excluded from shear planes					$0.17F_u$[h]	$0.43F_u - 1.8f_v \leq 0.33F_u$[h,i]
Threaded parts and A449 bolts, threads excluded from shear planes[g]					$0.22F_u$[h]	$0.43F_u - 1.4f_v \leq 0.33F_u$[h]
A325 bolts, when threads are not excluded from shear planes	17.0	15.0	12.0	10.0	21.0[f]	$\sqrt{(44)^2 - 4.39f_v^2}$
A325 bolts, when threads are excluded from shear planes	17.0	15.0	12.0	10.0	30.0[f]	$\sqrt{(44)^2 - 2.15f_v^2}$
A490 bolts, when threads are not excluded from shear planes	21.0	18.0	15.0	13.0	28.0[f]	$\sqrt{(54)^2 - 3.75f_v^2}$
A490 bolts, when threads are excluded from shear planes	21.0	18.0	15.0	13.0	40.0[f]	$\sqrt{(54)^2 - 1.82f_v^2}$

[a]For wind or seismic loading, acting alone or in combination with design dead and live loads, allowable stresses in the table may be increased one-third, if the required section then is at least that required for design, dead, live, and impact loads without this increase. For tension combined with shear, the coefficients of f_v in the tabulated formulas should not be changed.

[b]Assumes clean mill scale and blast-cleaned surfaces with Class A coatings (slip coefficient 0.33). For special faying-surface conditions, see the Research Council on Structural Connections specification.

[c]Relative to the long axis of the slotted hole.

[d]Static loading only. For fatigue conditions, see the AISC ASD "Specification for Structural Steel for Buildings."

[e]Threads permitted in shear planes.

[f]Reduce 20% for bolts in bearing-type splices of tension members if the fastener pattern has a length, parallel to the line of force, exceeding 50 in.

[g]Applicable to threaded parts meeting the requirements of ASTM A36, A242, A441, A529, A572, A588, A709, or A852 and to A449 bolts in bearing-type connections requiring bolt diameters exceeding 1½ in.

[h]F_u = minimum tensile strength, ksi, of bolts.

[i]For the threaded portion of an upset rod, A_bF_t should be larger than $0.60A_sF_y$, where A_b is the area at the major thread diameter, A_s is the nominal body area before upsetting, and F_y is the specified yield stress, ksi.

TABLE 7.24 Maximum Bolt-Hole Sizes, in*

Bolt diameter, in	Diameter of standard hole	Diameter of oversize hole	Short-slot hole (width × length)	Long-slot hole (width × length)
½	9⁄16	⅝	9⁄16 × 11⁄16	9⁄16 × 1¼
⅝	11⁄16	13⁄16	11⁄16 × ⅞	11⁄16 × 19⁄16
¾	13⁄16	15⁄16	13⁄16 × 1	13⁄16 × 1⅞
⅞	15⁄16	11⁄16	15⁄16 × 1⅛	15⁄16 × 23⁄16
1	11⁄16	1¼	11⁄16 × 15⁄16	11⁄16 × 2½
1⅛	d + 1⁄16	d + 5⁄16	(d + 1⁄16) × (d + ⅜)	(d + 1⁄16) × (2.5 × d)

*Approval of the designer is required for use of oversize or slotted holes. Larger holes than those listed in the table, if required for tolerance in location of anchor bolts in concrete foundations, may be used in column base details.

Long-slot holes may be used in only one ply of the connected parts at an individual faying surface. When the slot is in an outer ply, plate washers or a continuous bar with standard holes should be installed to cover the entire slot. Washers or bars for A325 or A490 bolts should be 5⁄16 in or more thick but need not be hardened. If hardened washers are required, they should be placed over the outer surface of a plate washer or bar.

7.26.2 LRFD for Bolts

The design strength of bolts or threaded parts is ϕR_n (tabulated in Table 7.25) applied to the nominal body area of bolts and threaded parts except upset rods (see footnote *h* for Table 7.25). The applied load is the sum of the factored external loads plus the tension, if any, resulting from prying action caused by deformation of connected parts. If high-strength bolts are required to support the applied loads by direct tension, they should be proportioned so that the average required strength (not including initial bolt tightening force) applied to the nominal bolt area will not exceed the design strength.

The design strength in tension for a bolt or threaded part subject to combined tension and shear stresses is also listed in Table 7.25. The value of f_v, the shear caused by the factored loads producing tensile stress, should not exceed the values for shear alone given in Table 7.25.

Table 7.24 lists maximum dimensions for standard, oversize, and slotted bolt holes. The limitations on these are the same as those for ASD (Art. 7.26.1).

The design bearing strength at a bolt hole may be taken as $\phi R_n = \phi 3.0dtF_u$, or with $\phi = 0.75$, as $2.25dtF_u$, where d is the nominal bolt diameter, t is the thickness of the connected part, and F_u is the tensile strength of the connected part.

7.27 LOAD CAPACITY OF WELDS

For welds joining structural steel elements, the load capacity depends on type of weld, strength of electrode material, and strength of the base metal. Fillet or groove

TABLE 7.25 Design Strength, ksi, for Bolts and Threaded Parts

Fasteners	Shear in slip-critical connections F_v[a]				Bearing-type connections	
	Standard-size holes	Oversized and short-slot holes	Long-slot holes: Transverse loads[a]	Long-slot holes: Parallel load[b]	Design shear strength ϕP_n	Tension F_t, including reduction for shear stress f_v[c]
A307 bolts					16.2[d,e]	$39–1.8f_v \leq 30$
Threaded parts and A449 bolts, threads[f] not excluded from shear planes					$0.45F_u$[g]	$0.73F_u–1.8f_v \leq 0.56F_u$[g,h]
Threaded parts and A449 bolts, threads excluded from shear planes[f]					$0.60F_u$[g]	$0.73F_u–1.4f_v \leq 0.56F_u$[g]
A325 bolts, when threads are not excluded from shear planes	17.0	15.0	12.0	10.0	35.1[e]	$85–1.8f_v \leq 68$
A325 bolts, when threads are excluded from shear planes	17.0	15.0	12.0	10.0	46.8[e]	$85–1.4f_v \leq 68$
A490 bolts, when threads are not excluded from shear planes	21.0	18.0	15.0	13.0	43.9[e]	$106–1.8f_v \leq 84$
A490 bolts, when threads are excluded from shear planes	21.0	18.0	15.0	13.0	58.5[e]	$106–1.4f_v \leq 84$

[a]Assumes clean mill scale and blast-cleaned surfaces with Class A coatings (slip coefficient 0.33). For special faying-surface conditions, see the Research Council on Structural Connections LRFD specification for structural joints.

[b]Relative to the long axis of the slotted hole.

[c]Static loading only. For fatigue conditions, see the AISC ASD "Specification for Structural Steel for Buildings."

[d]$\phi = 0.60$. Threads permitted in shear planes.

[e]$\phi = 0.65$. Reduce design shear strength 20% for bolts in bearing-type splices of tension members if the fastener pattern has a length, parallel to the line of force, exceeding 50 in.

[f]Applicable to threaded parts meeting the requirements of ASTM A36, A242, A441, A529, A572, A588, A709, or A852 and to A449 bolts in bearing-type connections requiring bolt diameters exceeding 1½ in.

[g]F_u = minimum tensile strength, ksi, of bolts.

[h]For the threaded portion of an upset rod, A_bR_n should be larger than A_sF_y, where A_b is the area at the major thread diameter, A_s is the nominal body area before upsetting, F_y is the specified yield stress, ksi, and ϕR_n is the design tensile strength, where $\phi = 0.65$.

welds (Fig. 7.42) are commonly used for steel connections. Groove welds are classified as complete or partial penetration. (See Art. 7.3.5.)

A significant characteristic of fillet-welded joints is that all forces, regardless of the direction in which they act, are resolved as shear on the effective throat of the weld. For instance, when joining elements such as a girder flange to a web, fillet welds are designed to carry the horizontal shear without regard to the tensile or compressive stresses in the elements.

For computation of load capacity, the effective area of groove and fillet welds is the effective length times the effective throat thickness. The effective area for a plug or slot weld is the nominal cross-sectional area of the hole or slot in the plane of the faying surface.

Except for fillet welds in holes or slots, the effective length of a fillet weld is the overall length of weld, including the return. For a groove weld, the effective length should be taken as the width of the part joined.

The effective throat thickness of a fillet weld is the shortest distance from the root of the joint to the nominal face of the weld (Fig. 7.3). For fillet welds made by the submerged-arc process, however, the effective throat should be taken as the leg size for welds ⅜ in and smaller but as the theoretical throat plus 0.11 in for larger fillet welds.

For a complete-penetration groove weld, the effective throat is the thickness of the thinnest part joined. For partial-penetration groove welds, the effective throat thickness depends on the included angle at the root of the groove. For all J or U joints and for bevel or V joints with an included angle of 60° or more, the effective throat thickness may be taken as the depth of the chamfer. When the included angle for bevel or V joints is between 45° and 60°, the effective throat thickness should be the depth of chamfer minus ⅛ in. For flare bevel and flare V-groove welds when flush to the surface of a bar or a 90° bend in a formed section, the effective throat thickness is 5⁄16 and ½ the radius of the bar or bend, respectively. When the radius is 1 in or more, for gas metal arc welding, the effective thickness is ¼ the radius.

Welds subject to static loads should be proportioned by ASD for the allowable stresses and by LRFD for the design strengths in Table 7.26. If connections will be subject to fatigue from stress fluctuations, load capacity should be reduced as provided in the AISC "Specification for Structural Steel for Buildings."

7.28 BEARING-TYPE BOLTED CONNECTIONS

When some slip, although very small, may occur between connected parts, the fasteners are assumed to function in shear. The presence of paint on contact surfaces is therefore of no consequence. Fasteners may be A307 bolts or high-strength bolts or any other similar fastener not dependent on development of friction on the contact surfaces.

Single shear occurs when opposing forces act on a fastener as shown in Fig. 7.39*a*, the plates tending to slide on their contact surfaces. The body of the fastener resists this tendency; a state of shear then exists over the cross-sectional area of the fastener.

Double shear takes place whenever three or more plates act on a fastener as illustrated in Fig. 7.39*b*. There are two or more parallel shearing surfaces (one on each side of the middle plate in Fig. 7.39*b*). Accordingly, the shear strength of the fastener is measured by its ability to resist two or more single shears.

Bearing on Base Metal. This is a factor to consider; but calculation of bearing stresses in most joints is useful only as an index of efficiency of the net section of tension members.

Edge Distances. The AISC "Specification for Structural Steel for Buildings," ASD and LRFD, recommends minimum edge distances, center of hole to edge of connected part, as given in Table 7.27. In addition, the edge distance, in, when in the direction of force should not be less than $2P/F_u t$ for ASD or $P/\phi F_u t$ for LRFD,

TABLE 7.26 Design Shear Strength for Welds, ksi*

Types of weld and stress	Material	LRFD: Resistance factor ϕ	LRFD: Nominal strength† F_{BM} or F_w	ASD: Allowable stress
		Complete penetration groove weld		
Tension normal to effective area	Base	0.90	F_y	Same as base metal
Compression normal to effective area Tension or compression parallel to axis of weld	Base	0.90	F_y	Same as base metal
Shear on effective area	Base Weld electrode	0.90 0.80	$0.60F_y$ $0.60F_{EXX}$	0.30 × nominal tensile strength of weld metal
		Partial penetration groove welds		
Compression normal to effective area Tension or compression parallel to axis of weld†	Base	0.90	F_y	Same as base metal
Shear parallel to axis of weld	Base Weld electrode	0.75	$0.60F_{EXX}$	0.30 × nominal tensile strength of weld metal
Tension normal to effective area	Base Weld electrode	0.90 0.80	F_y $0.60F_{EXX}$	0.30 × nominal tensile strength of weld metal
		Fillet welds		
Shear on effective area	Base Weld electrode	0.75	$0.60F_{EXX}$	0.30 × nominal tensile strength of weld metal
Tension or compression parallel to axis of weld†	Base	0.90	F_y	Same as base metal
		Plug or slot welds		
Shear parallel to faying surfaces (on effective area)	Base Weld electrode	0.75	$0.60F_{EXX}$	0.30 × nominal tensile strength of weld metal

*Reprinted with permission from F. S. Merritt and R. L. Brockenbrough, "Structural Steel Designers Handbook," 2d ed., McGraw-Hill, Inc., New York.

†Design strength is the smaller of F_{BM} and F_u:

F_{BM} = nominal strength of base metal to be welded, ksi.
F_w = nominal strength of weld electrode material, ksi
F_y = specified minimum yield stress of base metal, ksi
F_{EXX} = classification strength of weld metal, as specified in appropriate AWS specifications, ksi

where P is the force, kips, transmitted by one fastener to the part for which the edge distance is applicable; $\phi = 0.75$; F_u is the specified minimum tensile strength of the part (not the fastener), ksi; and t is the thickness of the part, in.

A special rule applies to beams with framed connections that are usually designed for the shear due to beam reactions. The edge distance for the beam web, with

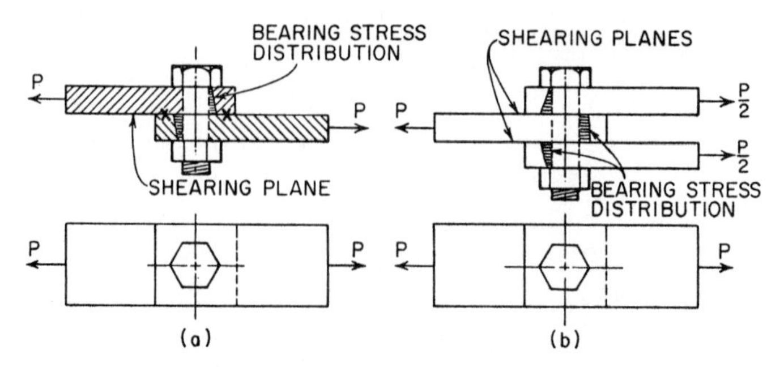

FIGURE 7.39 Bolted connection in shear and bearing; (*a*) with bolt in single shear; (*b*) with bolt in double shear (two shearing planes).

standard-size holes, should be not less than $2P_R/F_ut$ for ASD or $P_R/\phi F_ut$ for LRFD, where P_R is the beam reaction per bolt, kips. This rule, however, need not be applied when the bearing stress transmitted by the fastener does not exceed $0.90F_u$.

The maximum distance from the center of a fastener to the nearest edge of parts in contact should not exceed 6 in or 12 times the part thickness.

Minimum Spacing. The AISC specification also requires that the minimum distance between centers of bolt holes be at least 2⅔ times the bolt diameter. But at least three diameters is desirable. Additionally, the hole spacing, in, when along the line of force, should be at least $2P/F_ut + d/2$ for ASD or $P/\phi F_ut + d/2$ for LRFD, where P, F_u, and t are as previously defined for edge distance and d = nominal diameter of fastener, in. Since this rule is for standard-size holes, appropriate adjustments should be made for oversized and slotted holes. In no case should the clear distance between holes be less than the fastener diameter.

Eccentric Loading. Stress distribution is not always as simple as for the joint in Fig. 7.39*a* where the fastener is directly in the line of stress. Sometimes, the load

TABLE 7.27 Minimum Edge Distance for Punched, Reamed, or Drilled Holes, in

Fastener diameter, in.	At sheared edges	At rolled edges of plates, shapes or bars or gas-cut edges†
½	⅞	¾
⅝	1⅛	⅞
¾	1¼	1
⅞	1½*	1⅛
1	1¾*	1¼
1⅛	2	1½
1¼	2¼	1⅝
Over 1¼	1¾ × diameter	1¼ × diameter

*These may be 1¼ in at the ends of beam connection angles.

†All edge distances in this column may be reduced ⅛ in when the hole is at a point where stress does not exceed 25% of the maximum allowed stress in the element.

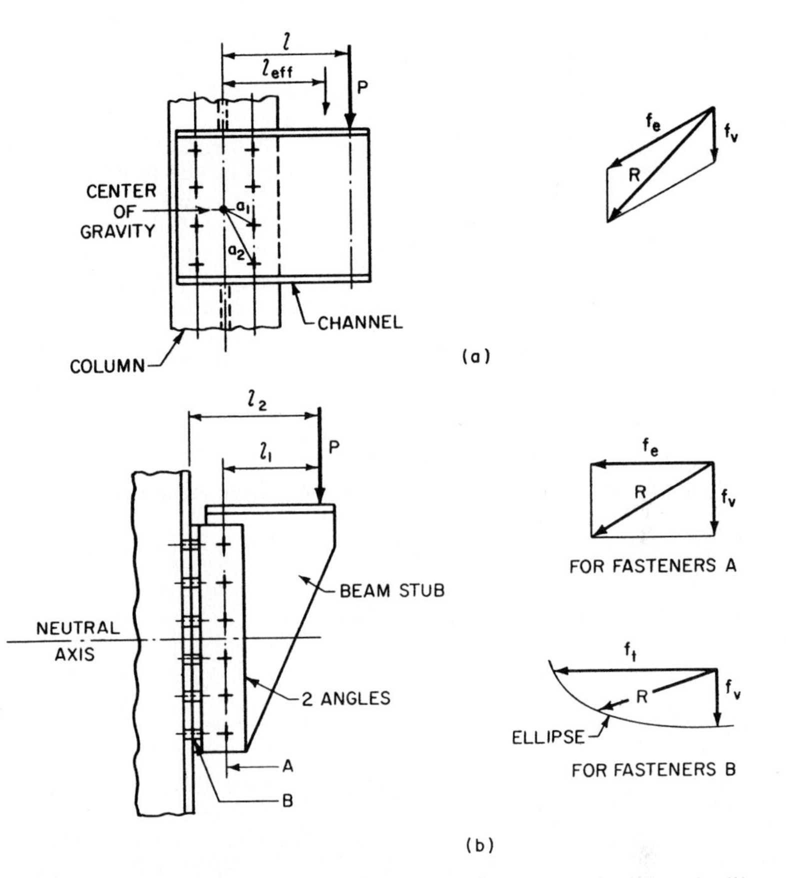

FIGURE 7.40 Eccentrically loaded fastener groups: (*a*) with bolts in shear only; (*b*) with bolts in combined tension and shear.

is applied eccentrically, as shown in Fig. 7.40. For such connections, tests show that use of actual eccentricity to compute the maximum force on the extreme fastener is unduly conservative because of plastic behavior and clamping force generated by the fastener. Hence, it is permissible to reduce the actual eccentricity to a more realistic "effective" eccentricity.

For fasteners equally spaced on a single gage line, the effective eccentricity in inches is given by

$$l_{\text{eff}} = l - \frac{1 + 2n}{4} \tag{7.78}$$

where l = the actual eccentricity and n = the number of fasteners. For the bracket in Fig. 7.40*b* the reduction applied to l_1 is $(1 + 2 \times 6)/4 = 3.25$ in.

For fasteners on two or more gage lines

$$l_{\text{eff}} = l - \frac{1 + n}{2} \tag{7.79}$$

when n is the number of fasteners per gage line. For the bracket in Fig. 7.40*a*, the reduction is $(1 + 4)/2 = 2.5$ in.

In Fig. 7.40*a*, the load P can be resolved into an axial force and a moment: Assume two equal and opposite forces acting through the center of gravity of the fasteners, both forces being equal to and parallel to P. Then, for equal distribution on the fasteners, the shear on each fastener caused by the force acting in the direction of P is $f_v = P/n$, where n is the number of fasteners.

The other force forms a couple with P. The shear stress f_e due to the couple is proportional to the distance from the center of gravity and acts perpendicular to the line from the fastener to the center. In determining f_e, it is convenient to first express it in terms of x, the force due to the moment Pl_{eff} on an imaginary fastener at unit distance from the center. For a fastener at a distance a from the center, $f_e = ax$, and the resisting moment is $f_e a = a^2 x$. The sum of the moments equals Pl_{eff}. This equation enables x to be evaluated and hence, the various values of f_e. The resultant R of f_e and f_v can then be found; a graphical solution usually is sufficiently accurate. The stress so obtained must not exceed the allowable value of the fastener in shear (Art. 7.26).

For example, in Fig. 7.40*a*, $f_v = P/8$. The sum of the moments is

$$4a_1^2 x + 4a_2^2 x = Pl_{eff}$$

$$x = \frac{Pl_{eff}}{4a_1^2 x + 4a_2^2 x}$$

Then, $f_e = a_2 x$ for the most distant fastener, and R can be found graphically as indicated in Fig. 7.40*a*.

Tension and Shear. For fastener group B in Fig. 7.40*b*, use actual eccentricity l_2 since these fasteners are subjected to combined tension and shear. Here too, the load P can be resolved into an axial shear force through the fasteners and a couple. Then, the stress on each fastener caused by the axial shear is P/n, where n is the number of fasteners. The tensile forces on the fasteners vary with distance from the center of rotation of the fastener group.

A simple method, erring on the safe side, for computing the resistance moment of group B fasteners assumes that the center of rotation coincides with the neutral axis of the group. It also assumes that the total bearing pressure below the neutral axis equals the sum of the tensile forces on the fasteners above the axis. Then, with these assumptions, the tensile force on the fastener farthest from the neutral axis is

$$f_t = \frac{d_{max} P l_2}{\Sigma A d^2} \tag{7.80}$$

where d = distance of each fastener from the neutral axis
d_{max} = distance from neutral axis of farthest fastener
A = nominal area of each fastener

The maximum resultant stresses f_t and $f_v = P/n$ are then plotted as an ellipse and R is determined graphically. The allowable stress is given as the tensile stress F_t as a function of the computed shear stress f_v. (In Tables 7.23 and 7.25, allowable stresses are given for the ellipse approximated by three straight lines.)

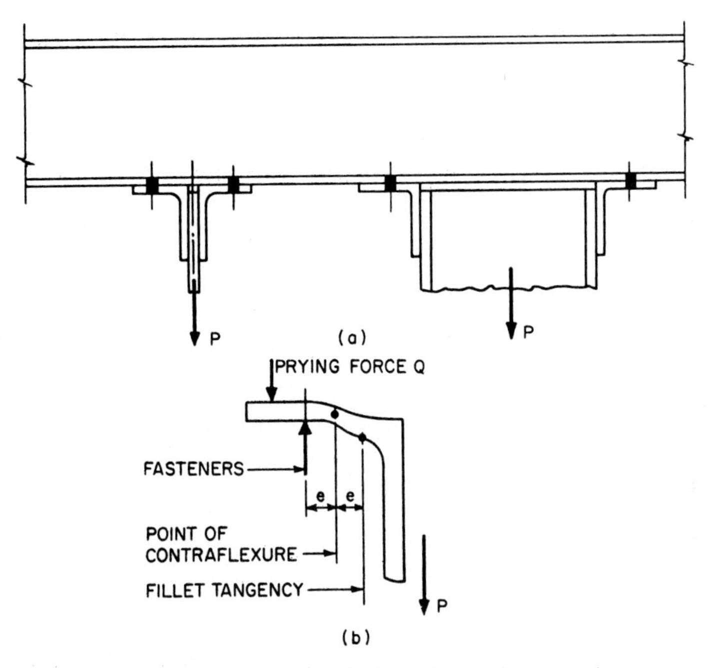

FIGURE 7.41 Fasteners in tension. Prying action on the connection causes a moment $M = Pe/n$ on either side, where P = applied load, e its eccentricity, as shown above, and n the number of fasteners resisting the moment.

Note that the tensile stress of the applied load is not additive to the internal tension (pretension) generated in the fastener on installation. On the other hand, the AISC Specification does require the addition to the applied load of tensile stresses resulting from prying action, depending on the relative stiffness of fasteners and connection material. Prying force Q (Fig. 7.42*b*) may vary from negligible to a substantial part of the total tension in the fastener. A method for computing this force is given in the AISC Manual.

The old method for checking the bending strength of connection material ignored the effect of prying action. It simply assumed bending moment equal to P/n times e (Fig. 7.41). This procedure may be used for noncritical applications.

7.29 SLIP-CRITICAL BOLTED CONNECTIONS

Design of this type of connection assumes that the fastener, under high initial tensioning, develops frictional resistance between the connected parts, preventing slippage despite external load. Properly installed A307 bolts provide some friction, but since it is not dependable it is ignored. High-strength steel bolts tightened nearly to their yield strengths, however, develop substantial, reliable friction. No slippage will occur at design loads if the contact surfaces are clean and free of paint or have only scored galvanized coatings, inorganic zinc-rich paint, or metallized zinc or aluminum coatings.

The AISC "Specification for Structural Steel for Buildings," ASD and LRFD, lists allowable shear for high-strength bolts in slip-critical connections. Though there actually is no shear on the bolt shank, the shear concept is convenient for measuring bolt capacity.

Since most joints in building construction can tolerate tiny slippage, bearing-type joints, which are allowed much higher shears for the same high-strength bolts when the threads are not in shear planes, may, for reasons of economy, lessen the use of slip-critical joints.

The capacity of a slip-critical connection does not depend on the bearing of the bolts against the sides of their holes. Hence, general specification requirements for protection against high bearing stresses or bending in the bolts may be ignored.

If the fasteners B in Fig. 7.40*b* are in a slip-critical connection, the bolts above the neutral axis will lose part of their clamping force; but this is offset by a compressive force below the neutral axis. Consequently, there is no overall loss in frictional resistance to slippage.

When it is apparent that there may be a loss of friction (which occurs in some type of brackets and hangers subject to tension and shear) and slip under load cannot be tolerated, the working value in shear should be reduced in proportion to the ratio of residual tension to initial tension.

Slip-critical connections subjected to eccentric loading, such as that illustrated in Fig. 7.40, are analyzed in the same manner as bearing-type connections (Art. 7.28).

7.30 ECCENTRICALLY LOADED WELDED CONNECTIONS

Welds are of two general types, fillet (Fig. 7.42*a*) and groove (Fig. 7.42*b*), with allowable stresses dependent on grade of weld and base steels. Since all forces on a fillet weld are resisted as shear on the effective throat (Art. 7.27), the strength of connections resisting direct tension, compression and shear are easily computed on the basis that a kip of fillet shear resists a kip of the applied forces. Many connections, some of which are shown in Fig. 7.43, are not that simple because of eccentricity of applied force with respect to the fillets. In designing such joints it is customary to take into account the actual eccentricity.

The underlying design principles for eccentric welded connections are similar to those for eccentric bolted connections (Art. 7.28). Consider the welded bracket

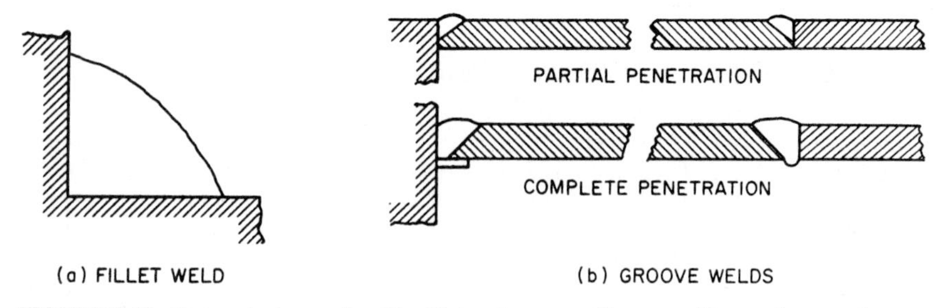

FIGURE 7.42 Two main types of weld—fillet and groove. Groove welds may be complete or partial penetration.

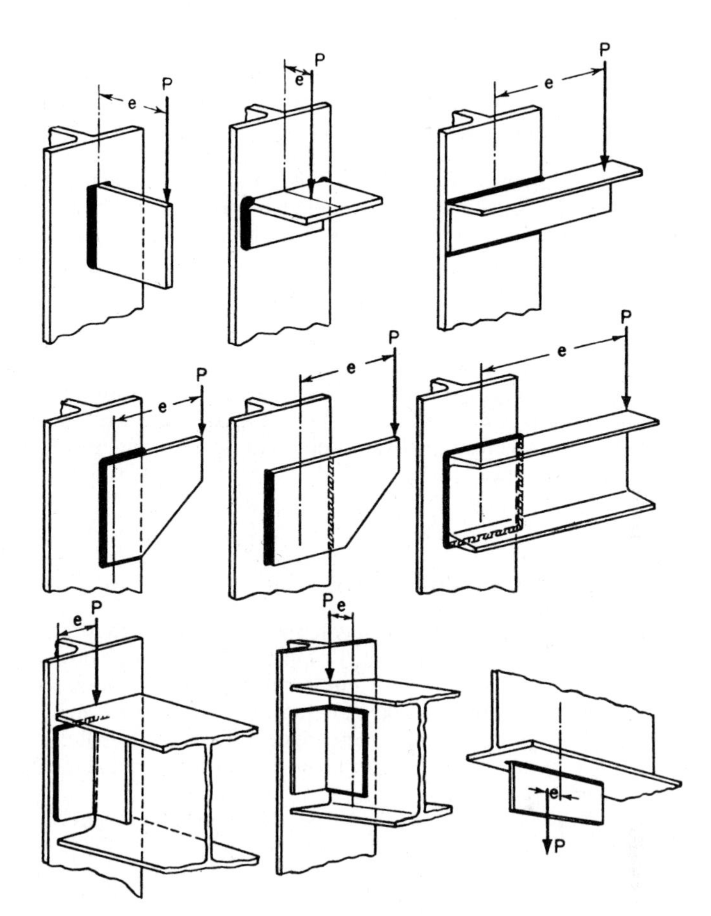

FIGURE 7.43 Typical eccentric welded connections.

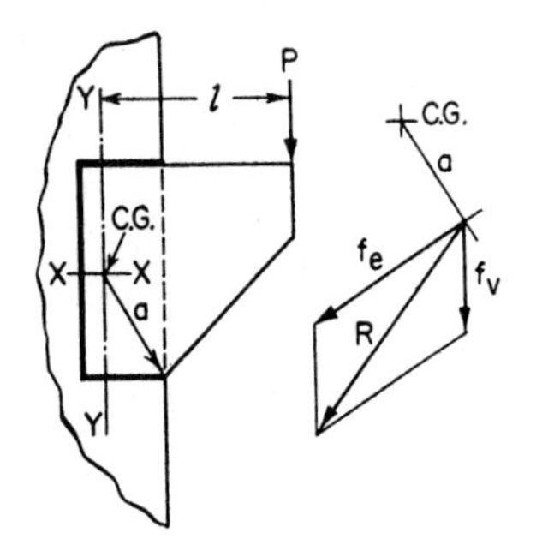

FIGURE 7.44 Stresses on welds caused by eccentricity.

in Fig. 7.44. The first step is to compute the center of gravity of the weld group. Then, the load P can be resolved into an equal and parallel load through the center of gravity and a couple. The load through the center of gravity is resisted by a uniform shear on the welds; for example, if the welds are all the same size, shear per linear inch is $f_v = P/n$ where n is the total linear inches of weld. The moment Pl of the couple is resisted by the moment of the weld group. The maximum stress, which occurs on the weld element farthest from the center of gravity, may be expressed as $f_e = Pl/S$, where S is the polar section modulus of the weld group.

To find S, first compute the moments of inertia I_x of the welds about the XX axis and I_Y about the perpendicular YY axis. (If the welds are all the same size, their lengths, rather than their relative shear capacities, can be conveniently used in all moment calculations.) The polar moment of inertia $J = I_X + I_Y$, and the polar section modulus $S = J/a$, where a is the distance from the center of gravity to the farthest weld element. The resultant R of f_v and f_e, which acts normal to the line from the center of gravity to the weld element for which the stress is being determined, should not exceed the capacity of the weld element (Art. 7.27).

7.31 TYPES OF BEAM CONNECTIONS

In general, all beam connections are classified as either **framed** or **seated.** In the framed type, the beam is connected to the supporting member with fittings (short angles are common) attached to the beam web. With seated connections, the ends of the beam rest on a ledge or seat, in much the same manner as if the beam rested on a wall.

7.31.1 Bolted Framed Connections

When a beam is connected to a support, a column or a girder, with web connection angles, the joint is termed "framed." Each connection should be designed for the end reaction of the beam, and type, size and strength of the fasteners, and bearing strength of base materials should be taken into account. To speed design, the AISC Manual lists a complete range of suitable connections with capacities depending on these variables. Typical connections for beam or channels ranging in depth from 3 to 30 in are shown in Fig. 7.45.

To provide sufficient stability and stiffness, the length of connection angles should be at least half the clear depth of beam web.

For economy, select the minimum connection adequate for the load. For example, assume an 18-in beam is to be connected. The AISC Manual (ASD) lists three- and four-row connections in addition to the five-row type shown in Fig. 7.45. Total shear capacity ranges from a low of 26.5 kips for ¾-in-diam A307 bolts in a three-row regular connection to a high of 263.0 kips for 1-in-diam A325 bolts in a

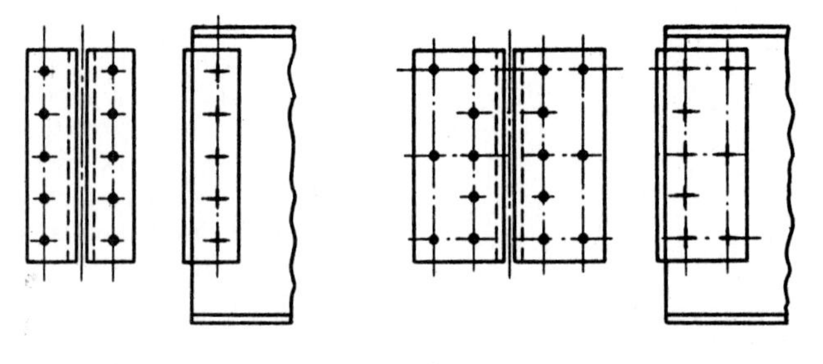

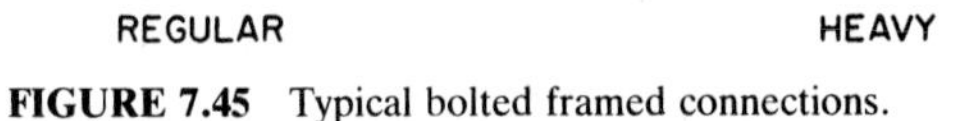

FIGURE 7.45 Typical bolted framed connections.

five-row heavy connection, bearing type. This wide choice does not mean that all types of fasteners should be used on a project, but simply that the tabulated data cover many possibilities, enabling an economical selection. Naturally, one type of fastener should be used throughout, if practical; but shop and field fasteners may be different.

Bearing stresses on beam webs should be checked against allowable stresses (Arts. 7.26.1 and 7.26.2), except for slip-critical connections, in which bearing is not a factor. Sometimes, the shear capacity of the field fasteners in bearing-type connections may be limited by bearing on thin webs, particularly where beams frame into opposite sides of a web. This could occur where beams frame into column or girder webs.

One side of a framed connection usually is shop connected, the other side field connected. The capacity of the connection is the smaller of the capacities of the shop or field group of fasteners.

In the absence of specific instructions in the bidding information, the fabricator should select the most economical connection. Deeper and stiffer connections, if desired by the designer, should be clearly specified.

7.31.2 Bolted Seated Connections

Sizes, capacities, and other data for seated connections for beams, shown in Fig. 7.46, are tabulated in the AISC Manual. Two types are available, stiffened seats (Fig. 7.46*a*) and unstiffened seats (Fig. 7.46*b*).

Unstiffened Seats. Capacity is limited by the bending strength of the outstanding, horizontal leg of the seat angle. A 4-in leg 1 in thick generally is the practical limit. In ASD, an angle of A36 steel with these dimensions has a top capacity of 60.5 kips for beams of A36 steel, and 78.4 kips when $F_y = 50$ ksi for the beam steel. Therefore, for larger end reactions, stiffened seats are recommended.

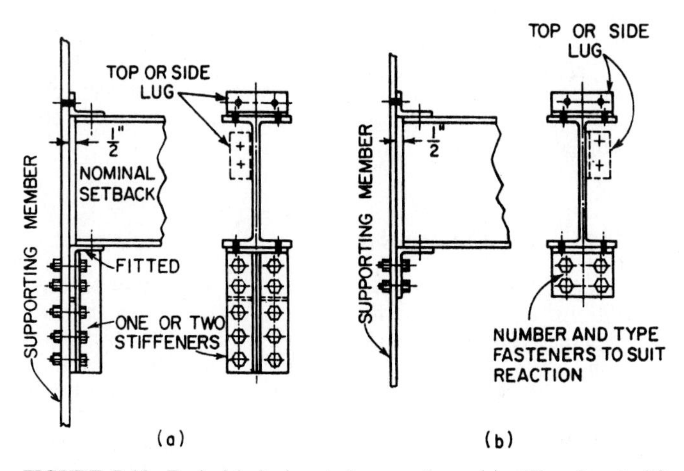

FIGURE 7.46 Typical bolted seated connections: (*a*) stiffened seat; (*b*) unstiffened seat.

The actual capacity of an unstiffened connection will be the lesser of the bending strength of the seat angle, the shear resistance of the fasteners in the vertical leg, or the bearing strength of the beam web. (See also Art. 7.17 for web crippling stresses.) Data in the AISC Manual make unnecessary the tedious computations of balancing the seat-angle bending strength and beam-web bearing.

The nominal setback from the support of the beam to be seated is ½ in. But tables for seated connections assume ¾ in to allow for mill underrun of beam length.

Stiffened Seats. These may be obtained with either one or two stiffener angles, depending on the load to be supported. As a rule, stiffeners with outstanding legs having a width less than 5 in are not connected together; in fact, they may be separated, to line up the angle gage line (recommended centerline of fasteners) with that of the column.

The capacity of a stiffened seat is the lesser of the bearing strength of the fitted angle stiffeners or the shear resistance of the fasteners in the vertical legs. Crippling strength of the beam web usually is not the deciding factor, because of ample seat area. When legs larger than 5 in wide are required, eccentricity should be considered, in accordance with the technique given in Art. 7.28. The center of the beam reaction may be taken at the midpoint of the outstanding leg.

Advantages of Seated Connections. For economical fabrication, the beams merely are punched and are free from shop-fastened details. They pass from the punching machine to the paint shed, after which they are ready for delivery. In erection, the seat provides an immediate support for the beam while the erector aligns the connecting hole. The top angle is used to prevent accidental rotation of the beam. For framing into column webs, seated connections allow more erection clearance for entering the trough formed by column flanges than do framed connections. A framed beam usually is detailed to within ¹⁄₁₆ in of the column web. This provides about ⅛ in total clearance, whereas a seated beam is cut about ½ in short of the column web, yielding a total clearance of about 1 in. Then, too, each seated connection is wholly independent, whereas for framed beams on opposite sides of a web, there is the problem of aligning the holes common to each connection.

Frequently, the angles for framed connections are shop attached to columns. Sometimes, one angle may be shipped loose to permit erection. This detail, however, cannot be used for connecting to column webs, because the column flanges may obstruct entering or tightening of bolts. In this case, a seated connection has a distinct advantage.

7.31.3 Welded Framed Connections

The AISC Manual tabulates sizes and capacities of angle connections for beams for three conditions: all welded, both legs (Fig. 7.47); web leg shop welded, outstanding leg for hole-type fastener; and web leg for hole-type fastener installed in shop, outstanding leg field welded. Tables are based on E70 electrodes. Thus, the connections made with A36 steel are suitable for beams of both carbon and high-strength structural steels.

Eccentricity of load with respect to the weld patterns causes stresses in the welds that must be considered in addition to direct shear. Assumed forces, eccentricities, and induced stresses are shown in Fig. 7.47*b*. Stresses are computed as in the

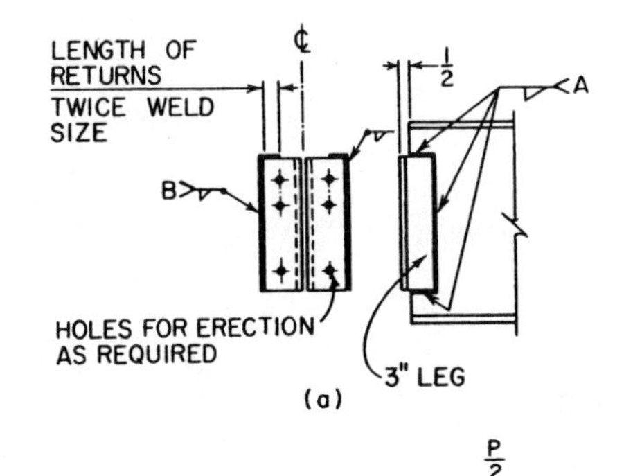

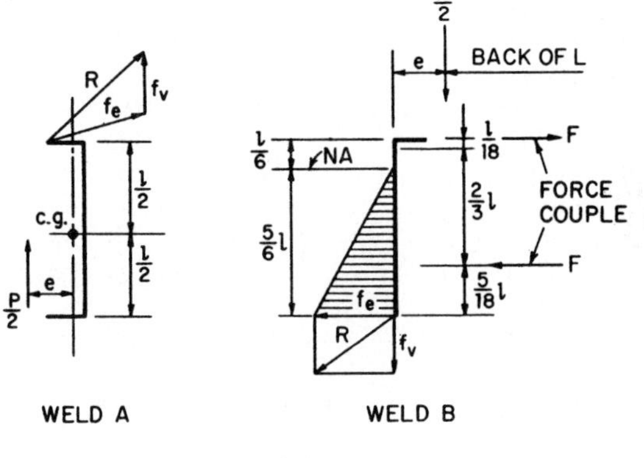

FIGURE 7.47 Welded framed connection on beam web: (*a*) weld locations along connection angles; (*b*) forces on welds.

example in Art. 7.30, based on vector analysis that characterizes elastic design. The capacity of welds A or B that is smaller will govern design.

If ultimate strength (plastic design) of such connections is considered, many of the tabulated "elastic" capacities are more conservative than necessary. Although AISC deemed it prudent to retain the "elastic" values for the weld patterns, recognition was given to research results on plastic behavior by reducing the minimum beam-web thickness required when welds A are on opposite sides of the web. As a result, welded framed connections are now applicable to a larger range of rolled beams than strict elastic design would permit.

Shear stresses in the supporting web for welds B should also be investigated, particularly when beams frame on opposite side of the web.

7.31.4 Welded-Seat Connections

Also tabulated in the AISC Manual, welded-seat connections (Fig. 7.48) are the welded counterparts of bolted-seat connections for beams (Art. 7.31.2). As for welded framed connections (Art. 7.31.3), the load capacities for seats, taking into account the eccentricity of loading on welds, are computed by "elastic" vector analysis. Assumptions and the stresses involved are shown in Fig. 7.48*c*.

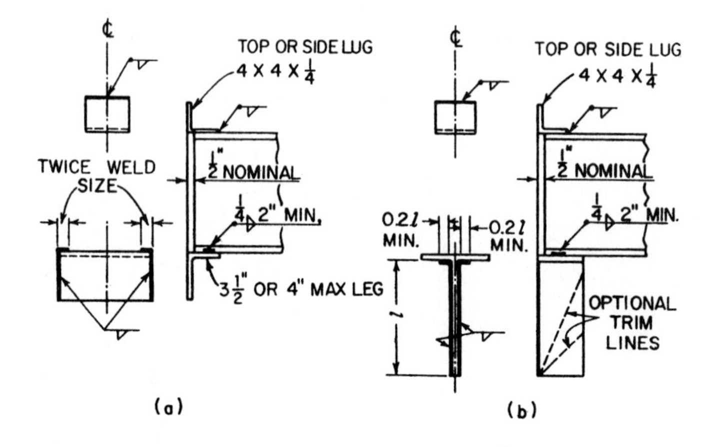

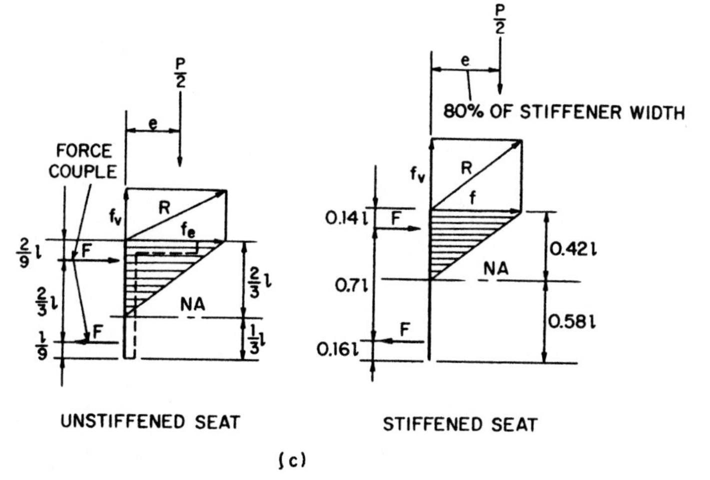

FIGURE 7.48 Welded-seat connections: (*a*) unstiffened seat; (*b*) stiffened seat; (*c*) stresses in the welds.

In ASD, an unstiffened seat angle of A36 steel has a maximum capacity of 60.5 kips for supporting beams of A36 steel, and 78.4 kips for steel with $F_y = 50$ ksi (Fig. 7.48*a*). For heavier loads, a stiffened seat (Fig. 7.48*b*) should be used.

Stiffened seats may be a beam stub, a tee section, or two plates welded together to form a tee. Thickness of the stiffener (vertical element) depends on the strengths of beam and seat materials. For a seat of A36 steel, stiffener thickness should be at least that of the supported beam web when the web is A36 steel, and 1.4 times thicker for web steel with $F_y = 50$ ksi.

When stiffened seats are on line on opposite sides of a supporting web of A36 steel, the weld size made with E70 electrodes should not exceed one-half the web thickness, and for web steel with $F_y = 50$ ksi, two-thirds the web thickness.

Although top or side lug angles will hold the beam in place in erection, it often is advisable to use temporary erection bolts to attach the bottom beam flange to the seat. Usually, such bolts may remain after the beam flange is welded to the seat.

7.31.5 End-Plate Connections

The art of welding makes feasible connections that were not possible with older-type fasteners, e.g., end-plate connections (Fig. 7.49).

Of the several variations, only the flexible type (Fig. 7.49*c*) has been "standardized" with tabulated data in the AISC Manual. Flexibility is assured by making the end plate ¼ in thick wherever possible (never more than ⅜ in). Such connections in tests exhibit rotations similar to those for framed connections.

The weld connecting the end plate to the beam web is designed for shear. There is no eccentricity. Weld size and capacity are limited by the shear strength of the beam web adjoining the weld. Effective length of weld is reduced by twice the weld size to allow for possible deficiencies at the ends.

As can be observed, this type of connection requires accurate cutting of the beam to length. Also the end plates must be squarely positioned so as to compensate for mill and shop tolerances.

The end plate connection is easily adapted for resisting beam moments (Fig. 7.49*b*, *c*, and *d*). One deterrent, however, to its use for tall buildings where column flanges are massive and end plates thick is that the rigidity of the parts may prevent drawing the surfaces into tight contact. Consequently, it may not be easy to make such connections accommodate normal mill and shop tolerances.

7.31.6 Special Connections

In some structural frameworks, there may be connections in which a standard type (Arts. 7.31.1 to 7.31.5) cannot be used. Beam centers may be offset from column centers, or intersection angles may differ from 90°, for example.

For some skewed connections the departure from the perpendicular may be taken care of by slightly bending the framing angles. When the practical limit for bent angles is exceeded, bent plates may be used (Fig. 7.50*a*).

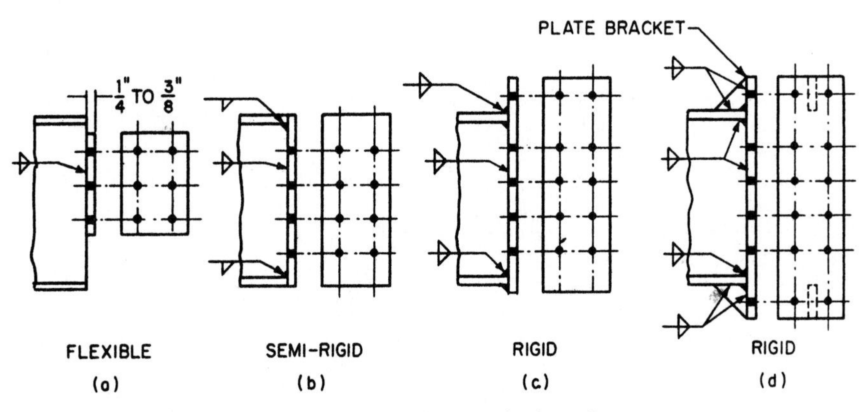

FIGURE 7.49 End-plate connection between beam and column flange.

Special one-sided angle connections, as shown in Fig. 7.50*b*, are generally acceptable for light beams. When such connections are used, the eccentricity of the fastener group in the outstanding leg should be taken into account. Length l may be reduced to the effective eccentricity (Art. 7.28).

Spandrel and similar beams lined up with a column flange may be conveniently connected to it with a plate (Fig. 7.50*c* and *d*). The fasteners joining the plate to the beam web should be capable of resisting the moment for the full lever arm l for the connection in Fig. 7.50*c*. For beams on both sides of the column with equal reactions, the moments balance out. But the case of live load on one beam only must be considered. And bear in mind the necessity of supporting the beam reaction as near as possible to the column center to relieve the column of bending stresses.

When spandrels and girts are offset from the column, a Z-type connection (Fig. 7.50*e*) may be used. The eccentricity for beam-web fasteners should be taken as l_1, for column-flange fasteners as l_2, and for fasteners joining the two connection angles as l_3 when l_3 exceeds 2½ in; smaller values of l_3 may be considered negligible.

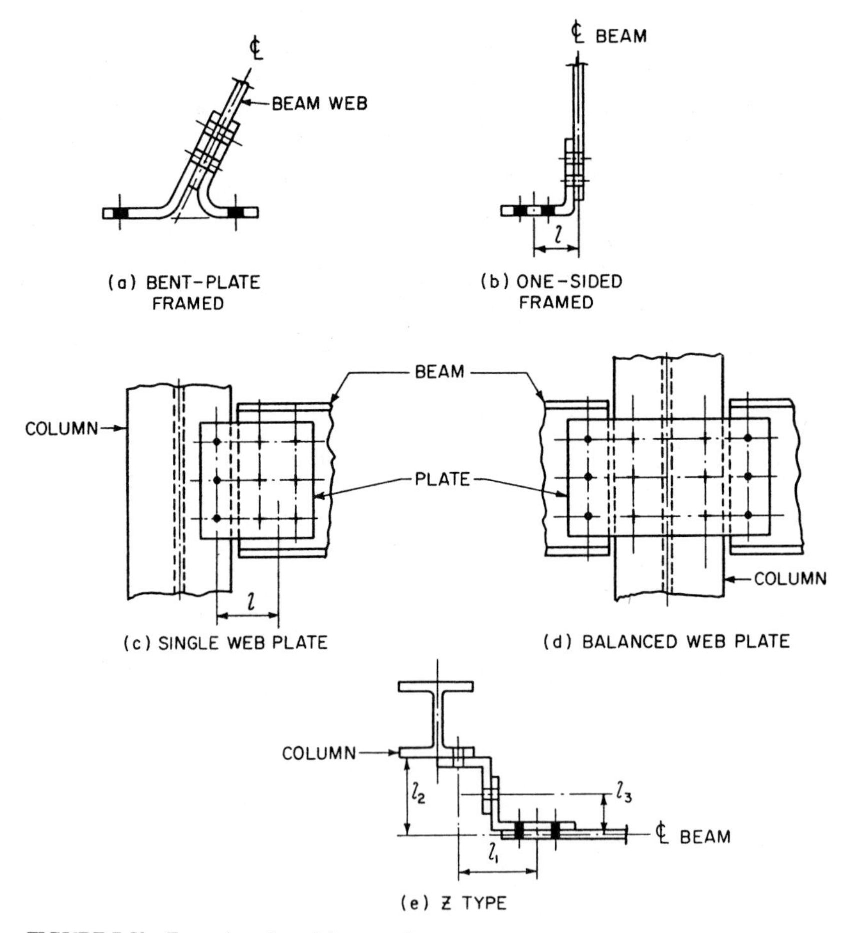

FIGURE 7.50 Examples of special connections.

7.31.7 Simple, Rigid, and Semirigid Connections

Moment connections are capable of transferring the forces in beam flanges to the column. This moment transfer, when specified, must be provided for in addition to and usually independent of the shear connection needed to support the beam reaction. Framed, seated, and end-plate connections (Arts. 7.31.1 to 7.31.5) are examples of shear connections. Those in Fig. 7.22, p. 7.39, are moment connections. In Fig. 7.22*a* to *g*, flange stresses are developed independently of the shear connections, whereas in *h* and *i*, the forces are combined and the entire connection resolved as a unit.

Moment connections may be classified according to their design function: those resisting moment due to lateral forces on the structure, and those needed to develop continuity, with or without resistance to lateral forces.

The connections generally are designed for the computed bending moment, which often is less than the beam's capacity to resist moment. A maximum connection is obtained, however, when the beam flange is developed for its maximum allowable stress.

The ability of a connection to resist moment depends on the elastic behavior of the parts. For example, the light lug angle shown connected to the top flange of the beam in Fig. 7.51*b* is not designed for moment and accordingly affords negligible resistance to rotation. In contrast, full rigidity is expected of the direct

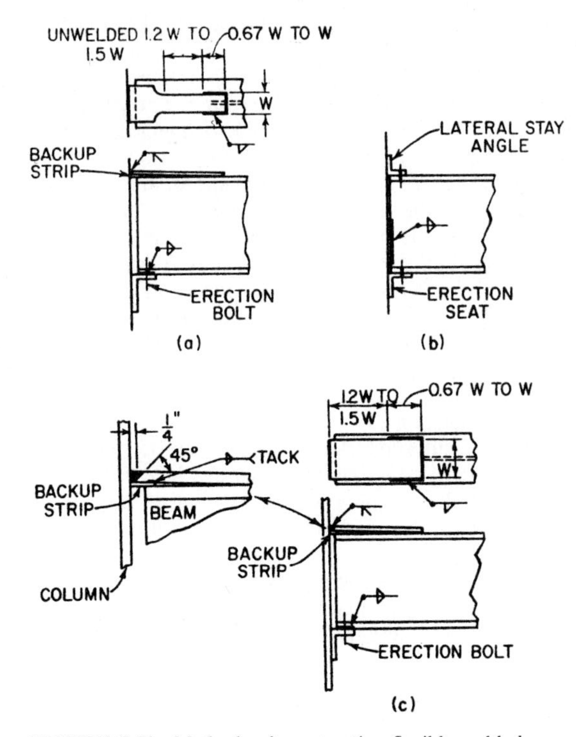

FIGURE 7.51 Methods of constructing flexible welded connections.

welded flange-to-column connection in Fig. 7.51*a*. The degree of fixity, therefore, is an important factor in design of moment connections.

Fixity of End Connections. Specifications recognize three types of end connections: simple, rigid, and semirigid. The type designated *simple* (unrestrained) is intended to support beams and girders for shear only and leave the ends free to rotate under load. The type designated *rigid* (known also as rigid-frame, continuous, restrained frame) aims at not only carrying the shear but also providing sufficient rigidity to hold virtually unchanged the original angles between members connected. *Semirigid*, as the name implies, assumes that the connections of beams and girders possess a dependable and known moment capacity intermediate in degree between the simple and rigid types. Figure 7.53 illustrates these three types together with the uniform-load moments obtained with each type.

Although no definite relative rigidities have been established, it is generally conceded that the simple or flexible type could vary from zero to 15% (some researchers recommend 20%) end restraint and that the rigid type could vary from 90 to 100%. The semirigid types lie between 15 and 90%, the precise value assumed in the design being largely dependent on experimental analysis. These percentages of rigidity represent the ratio of the moment developed by the connection, with no column rotation, to the moment developed by a fully rigid connection under the same conditions, multiplied by 100.

Framed and seated connections offer little or no restraint. In addition, several other arrangements come within the scope of simple-type connections, although they appear to offer greater resistance to end rotations. For example, in Fig. 7.51*a*, a top plate may be used instead of an angle for lateral support, the plate being so designed that plastic deformation may occur in the narrow unwelded portion. Naturally, the plate offers greater resistance to beam rotation than a light angle, but it can provide sufficient flexibility that the connection can be classified as a simple type. Plate and welds at both ends are proportioned for about 25% of the beam moment capacity. The plate is shaped so that the metal across the least width is at yield stress when the stresses in the wide portion, in the butt welds, and in the fillet welds are at allowable working values. The unwelded length is then made from 20 to 50% greater than the least width to assure ductile yielding. This detail can also be developed as an effective moment-type connection.

Another flexible type is the direct web connection in Fig. 7.51*b*. Figured for shear loads only, the welds are located on the lower part of the web, where the rotational effect of the beam under load is the least. This is a likely condition when the beam rests on erection seats and the axis of rotation centers about the seat rather than about the neutral axis.

Tests indicate that considerable flexibility also can be obtained with a properly proportioned welded top-plate detail as shown in Fig. 7.51*c* without narrowing it as in Fig. 7.51*a*. This detail is usually confined to wind-braced simple-beam designs. The top plate is designed for the wind moment on the joint, at the increased stresses permitted for wind loads.

The problem of superimposing wind bracing on what is otherwise a clear-cut simple beam with flexible connections is a complex one. Some compromise is usually effected between theory and actual design practice. Two alternatives usually are permitted by building codes:

1. Connections designed to resist assumed wind moments should be adequate to resist the moments induced by the gravity loading and the wind loading, at specified increased unit stresses.

2. Connections designed to resist assumed wind moments should be so designed that larger moments, induced by gravity loading under the actual condition of restraint, will be relieved by deformation of the connection material.

Obviously, these options envisage some nonelastic, but self-limiting, deformation of the structural-steel parts. Innumerable wind-braced buildings of riveted, bolted, or welded construction have been designed on this assumption of plastic behavior and have proved satisfactory in service.

Fully rigid, bolted beam end connections are not often used because of the awkward, bulky details, which, if not interfering with architectural clearances, are often so costly to design and fabricate as to negate the economy gained by using smaller beam sections. In appearance, they resemble the type shown in Fig. 7.22 for wind bracing; they are developed for the full moment-resisting capacity of the beam.

Much easier to accomplish and more efficient are welded rigid connections (Fig. 7.52). They may be connected simply by butt welding the beam flanges to the columns—the "direct" connection shown in Fig. 7.52*a* and *b*. Others may prefer the "indirect" method, with top plates, because this detail permits ordinary mill tolerance for beam length. Welding of plates to stiffen the column flanges, when necessary, is also relatively simple.

In lieu of the erection seat angle in Fig. 7.52*b*, a patented, forged hook-and-eye device, known as Saxe erection units, may be used. The eye, or seat, is shop

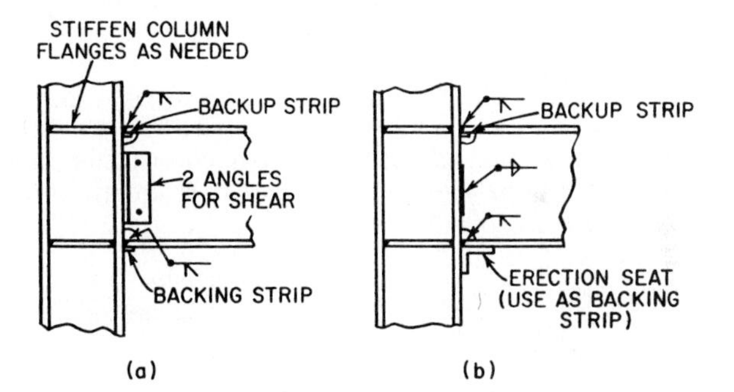

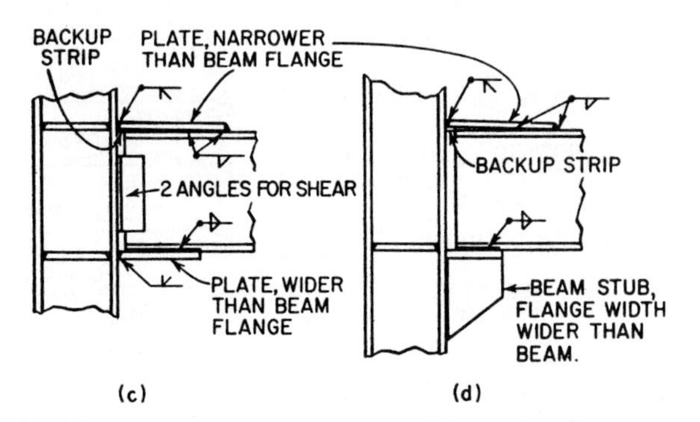

FIGURE 7.52 Methods of constructing welded rigid connections.

welded to the column, and the hook, or clip, is shop welded to the underside of the beam bottom flange. For deep beams, a similar unit may be located on the top flange to prevent accidental turning over of the beams. Saxe units are capable of supporting normal erection loads and deadweight of members; but their contribution to the strength of the connection is ignored in computing resistance to shear.

A comparison of fixities intermediate between full rigidity and zero restraint in Fig. 7.53 reveals an optimum condition attainable with 75% rigidity; end and center-span moments are equal, each being $WL/16$, or one-half the simple-beam moment. The saving in weight of beam is quite apparent.

Perhaps the deterrent to a broader usage of semirigid connections has been the proviso contained in specifications: "permitted only upon evidence that the connections to be used are capable of resisting definite moments without overstress of the fasteners." As a safeguard, the proportioning of the beam joined by such connections is predicated upon no greater degree of end restraint than the minimum known to be effected by the connection. Suggested practice, based on research with welded connections, is to design the end connections for 75% rigidity but to provide a beam sized for the moment that would result from 50% restraint; i.e., $WL/12$. ("Report of Tests of Welded Top Plate and Seat Building Connections," *The Welding Journal,* Research Supplement 146S–165S, 1944.) The type of welded connection in Fig. 7.51*c* when designed for the intended rigidity, is generally acceptable.

End-plate connections (Fig. 7.49) are another means of achieving negligible, partial, and full restraint.

7.32 BEAM SPLICES

These are required in rigid frames, suspended-span construction, and continuous beams. Such splices are usually located at points of counterflexure or at points

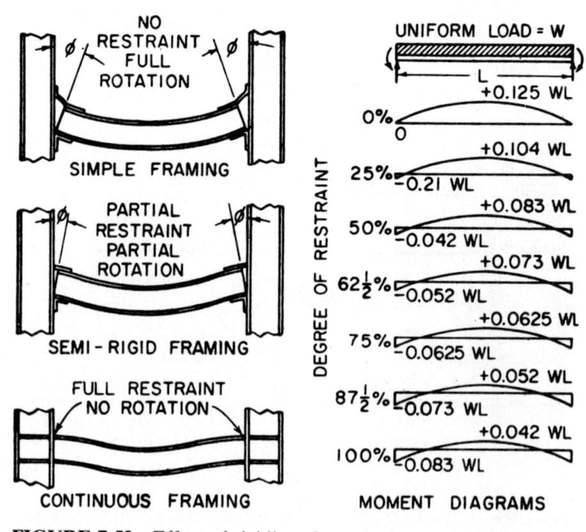

FIGURE 7.53 Effect of rigidity of connections on end moments.

where moments are relatively small. Therefore, splices are of moderate size. Flanges and web may be spliced with plates or butt welded.

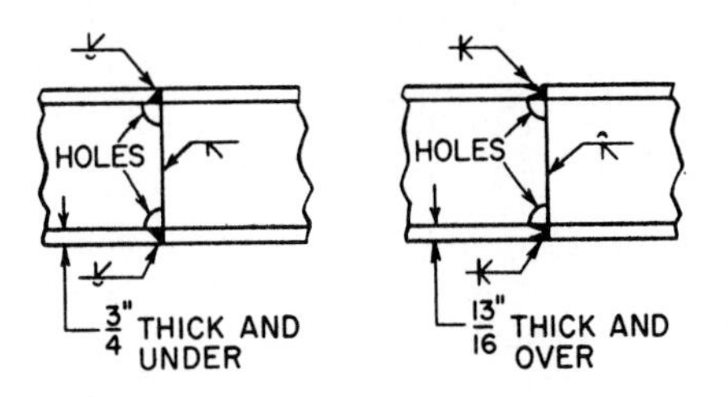

FIGURE 7.54 Welded beam splices.

For one reason or another it is sometimes expedient to make a long beam from two short lengths. A welded joint usually is selected, because the beams can be joined together without splice plates and without loss of section because of bolt holes. Also, from the viewpoint of appearance, the welded joint is hardly discernible.

Usually, the joint must be 100% efficient, to develop the full section. Figure 7.54 illustrates such a detail. The back side of the initial weld is gouged or chipped out; access holes in the beam webs facilitate proper edge preparation and depositing of the weld metal in the flange area in line with the web. Such holes are usually left open, because plugs would add undesirable residual stresses to the joint.

7.33 COLUMN SPLICES

Column-to-column connections are usually determined by the change in section. In general, a change is made at every second floor level, where a shop or field splice is located. From an erection viewpoint, as well as for fabrication and shipment, splices at every third floor may be more economical because of the reduced number of pieces to handle. This advantage is partly offset by extra weight of column material, because the column size is determined by loads on the lowest story of each tier, there being an excess of section for the story or two above.

Splices are located just above floor-beam connections, usually about 2 to 3 ft above the floor. Because column stresses are transferred from column to column by bearing, the splice plates are of nominal size, commensurate with the need for safe erection and bending moments the joint may be subjected to during erection. From the viewpoint of moment resistance, a conventional column splice develops perhaps 20% of the moment capacity of the column.

Figure 7.55 illustrates the common types of column splices made with high strength bolts. In Fig. 7.55*a* and *b*, the upper column bears directly on the lower column; filler plates are supplied in *(b)* when the differences in depth of the two columns are greater than can be absorbed by erection clearance.

As a rule, some erection clearance should be provided. When columns of the same nominal depth are spliced, it is customary to supply a ⅛-in fill under each splice plate on the lower column, or, as an alternate, to leave the bolt holes open on the top gage line below the finished joint until the upper shaft is erected. The latter procedure permits the erector to spring the plates apart to facilitate entry of the upper column.

When the upper column is of such dimension that its finished end does not wholly bear on the lower column, one of two methods must be followed: In Fig. 7.55*c*, stresses in a portion of the upper column not bearing on the lower column are transferred by means of flange plates that are finished to bear on the lower column. These bearing plates must be attached with sufficient single-shear bolts to develop the load transmitted through bearing on the finished surface.

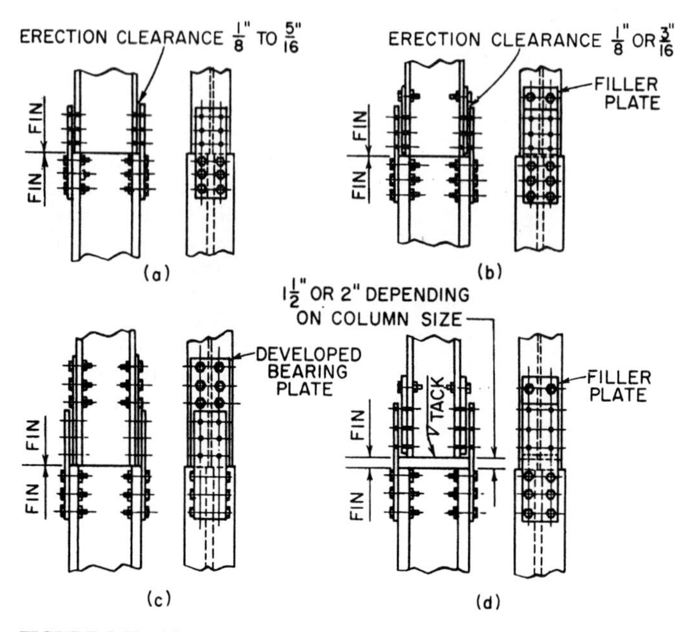

FIGURE 7.55 Slip-critical bolted column splices.

When the difference in column size is pronounced, the practice is to use a horizontal bearing plate as shown in Fig. 7.55*d*. These plates, known as **butt plates,** may be attached to either shaft with tack welds or clip angles. Usually it is attached to the upper shaft, because a plate on the lower shaft may interfere with erection of the beams that frame into the column web.

Somewhat similar are welded column splices. In Fig. 7.56*a*, a common case, holes for erection purposes are generally supplied in the splice plates and column flanges as shown. Some fabricators, however, prefer to avoid drilling and punching of thick pieces, and use instead clip angles welded on the inside flanges of the columns, one pair at diagonally opposite corners, or some similar arrangement. Figure 7.56*b* and *c* corresponds to the bolted splices in Fig.7.55*c* and *d*. The shop and field welds for the welded butt plate in Fig. 7.56*c* may be reversed, to provide erection clearance for beams seated just below the splice. The erection clip angles would then be shop welded to the underside of the butt plate, and the field holes would pierce the column web.

The butt-weld splice in Fig. 7.56*d* is the most efficient from the standpoint of material saving. The depth of the bevel as given in the illustration is for the usual column splice, in which moment is unimportant. However, should the joint be subjected to considerable moment, the bevel may be deepened; but a ⅛-in minimum shoulder should remain for the purpose of landing and plumbing the column. For full moment capacity, a complete-penetration welded joint would be required.

STEEL ERECTION

A clear understanding of what the fabricator furnishes or does not furnish to the erector, particularly on fabrication contracts that may call for delivery only, is all-

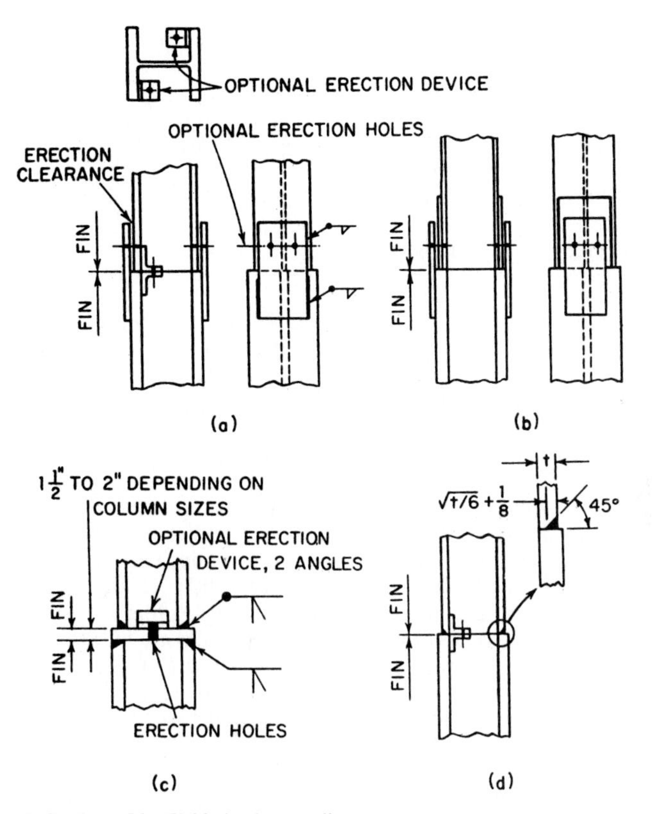

FIGURE 7.56 Welded column splices.

important—and in many instances fabricated steel is purchased on delivery basis only.

Purchasing structural steel is simplified by the "Code of Standard Practice for Buildings and Bridges," American Institute of Steel Construction. A provision in the construction contract making the code a part of the contract is often used, since it establishes a commonly accepted and well-defined line of demarcation between what is, and what is not, to be furnished under the contract. Lacking such a provision, the contract, to avoid later misunderstandings, must enumerate in considerable detail what is expected of both parties to the contract.

Under the code—and unless otherwise specifically called for in the contract documents—such items as steel sash, corrugated-iron roofing or siding, and open-web steel joists, and similar items, even if made of steel and shown on the contract design drawings, are not included in the category "structural steel." Also, such items as door frames are excluded, even when made of structural shapes, if they are not fastened to the structure in such way as to comply with "constituting part of the steel framing." On the other hand, loose lintels shown on design plans or in separate schedules are included.

According to the code, a fabricator furnishes with "structural steel," to be erected by someone else, the field bolts required for fastening the steel. The fabricator, however, does not furnish the following items unless specified in the invitation to bid: shims, fitting-up bolts, drift pins, temporary cables, welding electrodes, or thin leveling plates for column bases.

The code also defines the erection practices. For example, the erector does not paint field boltheads and nuts, field welds, or touch up abrasions in the shop coat, or perform any other field painting unless required in specifications accompanying the invitation to bid.

7.34 ERECTION EQUIPMENT

If there is a universal piece of erection equipment, it is the crane. Mounted on wheels or tractor treads, it is extremely mobile, both on the job and in moving from job to job. Practically all buildings are erected with this efficient raising device. The exception, of course, is the skyscraper whose height exceeds the reach of the crane. Operating on ground level, cranes have been used to erect buildings of about 20 stories, the maximum height being dependent on the length of the boom and width of building.

The guy derrick is a widely used raising device for erection of tall buildings. Its principal assest is the ease by which it may be "jumped" from tier to tier as erection proceeds upward. The boom and mast reverse position; each in turn serves to lift up the other. It requires about 2 h to make a two-story jump.

Stiff-leg derricks and gin poles are two other rigs sometimes used, usually in the role of auxiliaries to cranes or guy derricks. Gin poles are the most elementary—simply a guyed boom. The base must be secure because of the danger of kicking out. The device is useful for the raising of incidental materials, for dismantling and lowering of larger rigs, and for erection of steel on light construction where the services of a crane are unwarranted.

Stiff-leg derricks are most efficient where they may be set up to remain for long periods of time. They have been used to erect multistory buildings but are not in popular favor because of the long time required to jump from tier to tier. Among the principal uses for stiff legs are (1) unloading steel from railroad cars for transfer to trucks, (2) storage and sorting, and (3) when placed on a flat roof, raising steel to roof level, where it may be sorted and placed within reach of a guy derrick.

Less time for "jumping" the raising equipment is needed for cranes mounted on steel box-type towers, about three stories high, that are seated on interior elevator wells or similar shafts for erecting steel. These tower cranes are simply jacked upward hydraulically or raised by cables, with the previously erected steelwork serving as supports. In another method, a stiff-leg derrick is mounted on a trussed platform, spanning two or more columns, and so powered that it can creep up the erected exterior columns. In addition to the advantage of faster jumps, these methods permit steel erection to proceed as soon as the higher working level is reached.

7.35 CLEARANCE FOR ERECTING BEAMS

Clearances to permit tightening bolts and welding are discussed in Art. 7.3.7. In addition, designers also must provide sufficient field clearance for all members so as to permit erection without interference with members previously erected. The shop drafter should always arrange the details so that the members can be swung into their final position without shifting the members to which they connect from their final positions. The following examples illustrate the conditions most frequently encountered in building work:

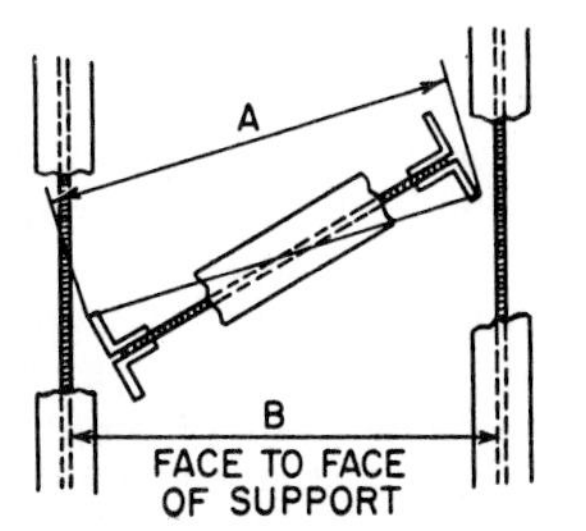

FIGURE 7.57 Erection clearance for beams.

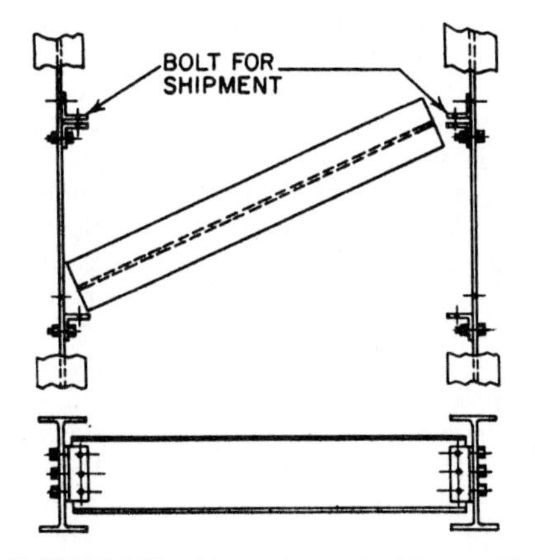

FIGURE 7.58 Alternative method for providing erection clearance.

In framed beam connections (Fig. 7.57), the slightly shorter distance out-to-out of connection angles (B—⅛ in), as compared with the face-to-face distance between supporting members, is usually sufficient to allow forcing the beam into position. Occasionally, however, because the beam is relatively short, or because heavy connection angles with wide outstanding legs are required, the diagonal distance A may exceed the clearance distance B. If so, the connection for one end must be shipped bolted to the framed beam to permit its removal during erection.

An alternative solution is to permanently fasten one connection angle of each pair to the web of the supporting beam, temporarily bolting the other angle to the same web for shipment, as shown in Fig. 7.58. The beam should be investigated for the clearance in swinging past permanently bolted connection angles. Attention must also be paid to possible interference of stiffeners in swinging the beam into place when the supporting member is a plate girder.

Another example is that of a beam seated on column-web connections (Fig. 7.59). The first step is to remove the top angles and shims temporarily. Then, while hanging from the derrick sling, the beam is tilted until its ends clear the edges of the column flanges, after which it is rotated back into a horizontal position and landed on the seats. The greatest diagonal length G of the beam should be about ⅛ in less than the face-to-face distance F between column webs. It must also be such as to clear any obstruction above; e.g., G must be equal to or less than C, or the obstructing detail must be shipped bolted for temporary removal. To allow for possible overrun, the ordered length L of the beam should be less than the detailing length E by at least the amount of the permitted cutting tolerance.

Frequently, the obstruction above the beam connection may be the details of a column splice. As stated in Art. 7.33, it may be necessary to attach the splice material on the lower end of the upper shaft, if erection of the beam precedes erection of the column in the tier above.

7.36 ERECTION SEQUENCE

The order in which steel is to be fabricated and delivered to the site should be planned in advance so as not to conflict with the erector's methods or construction

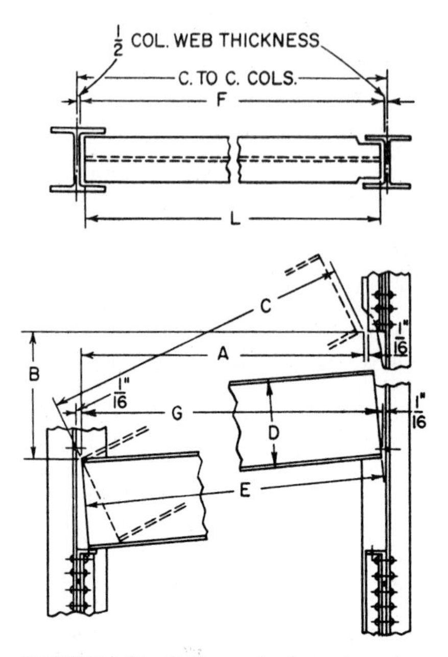

FIGURE 7.59 Clearance for beam seated on column-web connections.

schedule. For example, if steel is to be erected with derricks, the approximate locations at which the derricks will be placed will determine the shipping installments, or sections, into which the frame as a whole must be segregated for orderly shipment. When installments are delivered to the site at predetermined locations, proper planning will eliminate unnecessary rehandling. Information should be conveyed to the drafting room so that the shipping installments can be indicated on the erection plans and installments identified on the shipping lists.

In erection of multistory buildings with guy derricks, the practice is to hoist and place all columns in each story first, spandrel beams and wall bracing next, and interior beams with filler beams last. More specifically, erection commences with bays most distant from the derrick and progresses toward the derrick, until it is closed in. Then, the derrick is jumped to the top and the process is repeated for the next tier. Usually, the top of the tier is planked over to obtain a working platform for the erectors and also to afford protection for the trades working below. However, before the derrick is jumped, the corner panels are plumbed; similarly when panels are erected across the building, cables are stretched to plumb the structure.

There is an established sequence for completing the connections. The raising gang connects members together with temporary fitting-up bolts. The number of bolts is kept to a minimum, just enough to draw the joint up tight and take care of the stresses caused by deadweight, wind, and erection forces. Permanent connections are made as soon as alignment is within tolerance limits. Usually, permanent bolting or welding follows on the heels of the raising gang. Sometimes, the latter moves faster than the gang making the permanent connections, in which case it may be prudent to skip every other floor, thus obtaining permanent connections as close as possible to the derrick—a matter of safe practice.

Some erectors prefer to use permanent high-strength (A325 and A490) bolts for temporary fitting up. Because bolts used for fit-up are not tightened to specified minimum tension, they may be left in place and later tightened as required for permanent installation.

7.37 FIELD-WELDING PROCEDURES

The main function of a welding sequence is to control distortion due primarily to the effects of welding heat. In general, a large input of heat in a short time tends to produce the greatest distortion. Therefore, it is always advisable, for large joints, to weld in stages, with sufficient time between each stage to assure complete dispersal of heat, except for heat needed to satisfy interpass-temperature requirements (Art. 7.3.5). Equally important, and perhaps more efficient from the erector's viewpoint, are those methods that balance the heat input in such a manner that the distortional effects tend to cancel out.

Welding on one flange of a column tends to leave the column curled toward the welded side cooling, because of shrinkage stresses. A better practice for beams connecting to both sides of a column is to weld the opposite connections simultaneously. Thus the shrinkage of each flange is kept in balance and the column remain plumb.

If simultaneous welding is not feasible, then the procedure is to weld in stages. About 60% of the required weld might be applied on the first beam, then the joint on the opposite flange might be completely welded, and finally, welding on the first beam would be completed. Procedures such as this will go far to reduce distortion.

Experience has shown that it is good practice to commence welding at or near the center of a building and work outward. Columns should be checked frequently for vertical alignment, because shrinkage in the welds tends to shorten the distance between columns. Even though the dimensional change at each joint may be very small, it can accumulate to an objectionable amount in a long row of columns. One way to reduce the distortion is to allow for shrinkage at each joint, say, 1/16 in for a 20-ft bay, by tilting or spreading the columns. Thus, a spread of 1/8 in for the two ends of a beam with flanges butt welded to the columns may be built in at the fabricating shop; for example, by increasing the spacing of erection-bolt holes in the beam bottom flange. Control in the field, however, is maintained by guy wires until all joints are welded.

Shortening of bays can become acute in a column row in which beams connect to column flanges, because the shrinkage shortening could possibly combine with the mill underrun in column depths. Occasionally, in addition to spreading the columns, it may be necessary to correct the condition by adding filler plates or building out with weld metal.

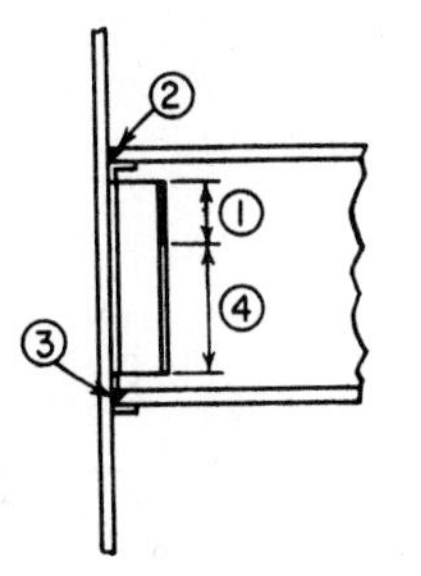

FIGURE 7.60 Indication of sequence in welding a connection.

Some designers of large welded structures prefer to detail the welding sequence for each joint. For example, on one project, the procedure for the joint shown in Fig. 7.60 called for four distinct operations, or stages: first, the top 6 in of the shear weld on the vertical connection was made; second, the weld on the top flange; third, the bottom-flange weld; and fourth, the remaining weld of the vertical connection. The metal was allowed to return to normal temperature before starting each stage. One advantage of this procedure is the prestressing benefits obtained in the connecting welds. Tensile stresses are developed in the bottom-flange weld on cooling; compressive stresses of equal magnitude consequently are produced in the top flange. Since these stresses are opposite to those caused by floor loads, welding stresses are useful in supporting the floor loads. Although this by-product assistance may be worthwhile, there are no accepted methods for resolving the alleged benefits into design economy.

Multistory structures erected with equipment supported on the steelwork as it rises will be subjected by erection loads to stresses and strains. The resulting deformations should be considered in formulating a field-welding sequence.

7.38 ERECTION TOLERANCES

Dimensional variations in the field often are a consequence of permissible variations in rolling of steel and in shop fabrication. Limits for mill variations are prescribed in ASTM A6, "General Requirements for Delivery of Rolled Steel Plates, Shapes, Sheet Piling, and Bars for Structural Use." For example, wide-flange beams are considered straight, vertically or laterally, if they are within ⅛ in for each 10 ft of length. Similarly, columns are straight if the deviation is within ⅛ in per 10 ft, with a maximum deviation of ⅜ in.

It is standard practice to compensate in shop details for certain mill variations. The adjustments are made in the field, usually with clearances and shims.

Shop-fabrication tolerance for straightness of columns and other compression members often is expressed as a ratio, 1:1000, between points of lateral support. (This should be recognized as approximately the equivalent of ⅛ in per 10 ft, and since such members rarely exceed 30 ft in length, between lateral supports, the ⅜-in maximum deviation prevails.) Length of fabricated beams have a tolerance of 1/16 in up to 30 ft and ⅛ in over 30 ft. Length of columns finished to bear on their ends have a tolerance of 1/32 in.

Erected beams are considered level and aligned if the deviation does not exceed 1:500. Similarly, columns are plumb and aligned if the deviation of individual pieces, between splices in the usual multistory building, does not exceed 1:500. The total or accumulative displacement for multistory columns cannot exceed the limits prescribed in the American Institute of Steel Construction "Code of Standard Practice." For convenience, these are indicated in Fig. 7.61. Control is placed only on the exterior columns and those in the elevator shaft.

Field measurements to determine whether columns are plumb should always be made at night or on cloudy days, never in sunshine. Solar radiation induces differential thermal strains, which cause the structure to curl away from the sun by an amount that renders plumbing measurements useless.

If beam flanges are to be field welded (Fig. 7.55*a*) and the shear connection is a high-strength-bolted, slip-critical joint, the holes should be made oversize or horizontal slotted (Art. 7.3.1), thus providing some built-in adjustment to accommodate mill and shop tolerances for beams and columns.

Similarly, for beams with framed connections (Fig. 7.45 and 7.46) that will be field bolted to columns, allowance should be made in the details for finger-type shims, to be used where needed for column alignment.

Because of several variables, bearing of column joints is seldom in perfect contact across the entire cross-sectional area. The AISC recommends acceptance if gaps between the bearing surfaces do not exceed 1/16 in. Should a gap exceed 1/16 in and an engineering investigation shows need for more contact area, the gap may be filled with mild steel shims.

Tolerance for placing of machinery directly on top of several beams is another problem occasionally encountered in the field. The elevation of beam flanges will vary because of permissible variations for mill rolling, fabrication, and erection. This should be anticipated and adequate shims provided for field adjustments.

7.39 ADJUSTING LINTELS

Lintels supported on the steel frame (sometimes called shelf angles) may be permanently fastened in the shop to the supporting spandrel beam, or they may be

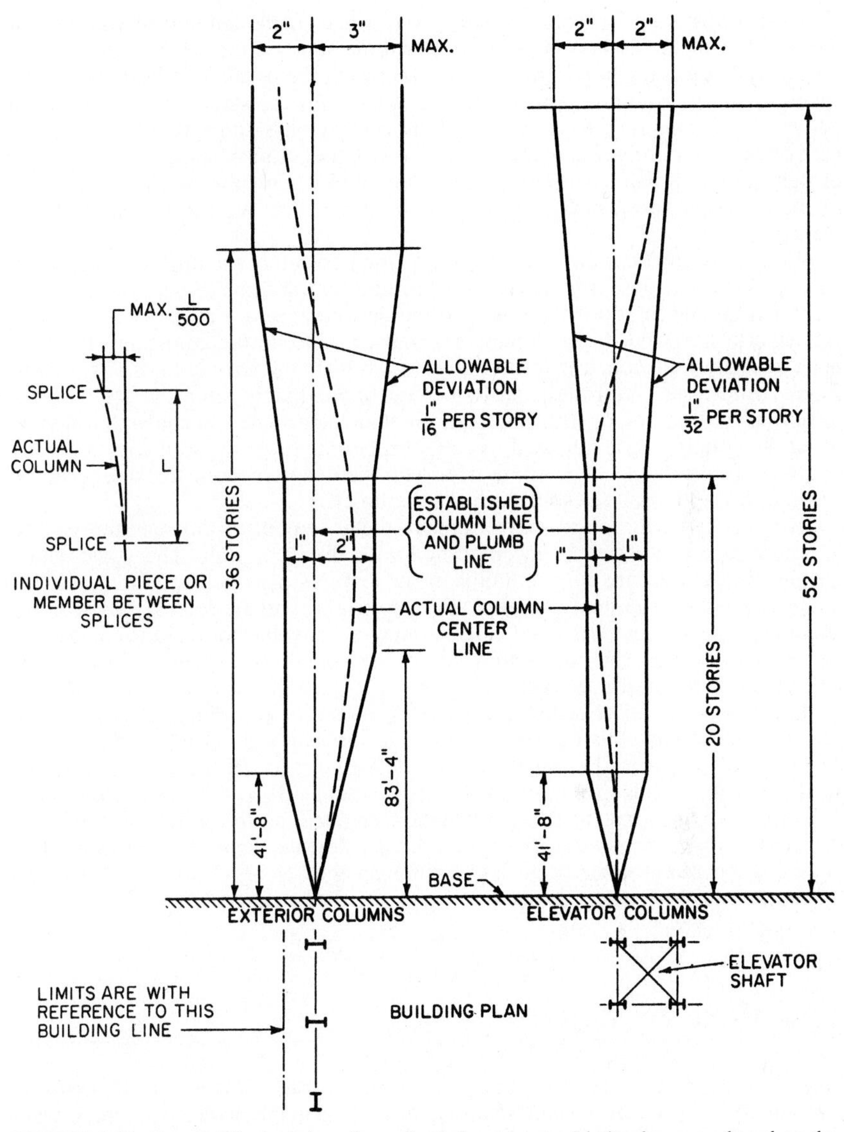

FIGURE 7.61 Permissible deviations from plumb for columns. Limits shown are based on the assumption that the center of the column base coincides with the established column line.

attached so as to allow adjustment in the field (see Fig. 7.9, p. 7.21). In the former case, the final position is solely dependent on the alignment obtained for the spandrel itself, whereas for the latter, lintels may be adjusted to line and grade independently of the spandrel. Field adjustment is the general rule for all multistory structures. Horizontal alignment is obtained by using slotted holes in the connection clip angles. Vertical elevation (grade) is obtained with shims.

When walls are of masonry construction, a reasonable amount of variation in the position of lintels may be absorbed without much effort by masons. So the erector can adjust the lintels immediately following the permanent fastening of the spandrels to the columns. This procedure is ideal for the steel erector, because it allows him to complete his contract without costly delays and without interference with other trades. Subsequent minor variations in the position of the lintels, because of deflection or torsional rotation of the spandrel when subjected to deadweight of the floor slab, are usually absorbed without necessitating further lintel adjustment.

With lightweight curtain walls, however, the position of the lintels is important, because large paneled areas afford less latitude for variation. As a rule, the steel erector is unable to adjust the lintels to the desired accuracy at the time the main framework is erected. If the erector has contracted to do the adjusting, this work must wait until the construction engineer establishes the correct lines and grades. In the usual case, floor slabs are concreted immediately after the steelwork is inspected and accepted. The floor grades then determined become the base to which the lintels can be adjusted. At about the same time, the wall contractor has scaffolds in place, and by keeping pace with wall construction, the steel erector, working from the wall scaffolds, adjusts the lintels.

In some cases, the plans call for concrete encasement of the spandrel beams, in which case concreting is accomplished with the floor slab. The construction engineer should ensure that the adjustment features provided for the lintels are not frozen in the concrete. One suggestion is to box around the details, thus avoiding chopping out concrete. In some cases, it may be possible to avoid the condition entirely by locating the connection below the concrete encasement, where the adjustment is always accessible.

The whole operation of lintel adjustment is one of coordination between the several trades. That this be carried out in an orderly fashion is the duty of the construction engineer. Furthermore, the desired procedure should be carefully spelled out in the job specifications so that erection costs can be estimated fairly.

Particularly irksome to the construction engineer is the lintel located some distance below the spandrel and supported on flexible, light steel hangers. This detail can be troublesome because it has no capacity to resist torsion. Avoid this by developing the lintel and spandrel to act together as a single member.

CORROSION PROTECTION

Protection of steel surfaces has been, since the day steel was first used, a vexing problem for the engineers, paint manufacturers, and maintenance personnel. Over the years, there have been many developments, the result of numerous studies and research activities. Results are published in the "Steel Structures Painting Manual." This work is in two volumes—Vol. 1, "Good Painting Practice," and Vol. II, "Systems and Specifications" (Steel Structures Painting Council, 4400 Fifth Avenue, Pittsburgh, PA 15213). Each of the paint systems covers the method of cleaning surfaces, types of paint to be used, number of coats to be applied, and techniques to be used in their applications. Each surface treatment and paint system is identified by uniform nomenclature, e.g., Paint System Specification SSPC-PS7.00-64T, which happens to be the identity of the minimum-type protection as furnished for most buildings.

7.40 CORROSION OF STEEL

Ordinarily, steel corrodes in the presence of both oxygen and water, but corrosion rarely takes place in the absence of either. For instance, steel does not corrode in dry air, and corrosion is negligible when the relative humidity is below 70%, the critical humidity at normal temperature. Likewise, steel does not corrode in water that has been effectively deaerated. Therefore, the corrosion of structural steel is not a serious problem, except where water and oxygen are in abundance and where these primary prerequisites are supplemented with corrosive chemicals such as soluble salts, acids, cleaning compounds, and welding fluxes.

In ideal dry atmospheres, a thin transparent film of iron oxide forms. This layer of ferric oxide is actually beneficial, since it protects the steel from further oxidation.

When exposed to water and oxygen in generous amounts, steel corrodes at an average rate of roughly 5 mils loss of surface metal per year. If the surface is comparatively dry, the rate drops to about ½ mil per year after the first year, the usual case in typical industrial atmospheres. Excessively high corrosion rates occur only in the presence of electrolytes or corrosive chemicals. Usually, this condition is found in localized areas of a building.

Mill scale, the thick layer of iron oxides that forms on steel during the rolling operations, is beneficial as a protective coating, if it is intact and adheres firmly to the steel. In the mild environments generally encountered in most buildings, mill scale that adheres tightly after weathering and handling offers no difficulty. In buildings exposed to high humidity and corrosive gases, broken mill scale may be detrimental to both the steel and the paint. Through electrochemical action, corrosion sets in along the edges of the cracks in the mill scale and in time loosens the scale, carrying away the paint.

Galvanic corrosion takes place when dissimilar metals are connected together. Noble metals such as copper and nickel should not be connected to structural steel with steel fasteners, since the galvanic action destroys the fasteners. On the other hand, these metals may be used for the fasteners, because the galvanic action is distributed over a large area and consequently little or no harm is done. When dissimilar metals are to be in contact, the contacting surfaces should be insulated; paint is usually satisfactory.

7.41 PAINTING STEEL STRUCTURES

Evidence obtained from dismantled old buildings and from frames exposed during renovation indicates that corrosion does not occur when steel surfaces are protected from the atmosphere. Where severe rusting was found and attributed to leakage of water, presence or absence of shop paint had no significant influence. Consequently, the AISC "Specifications for Structural Steel for Buildings" exempts from one-coat shop paint, at one time mandatory, all steel framing that is concealed by interior finishing materials—ceilings, fireproofing partitions, walls, and floors.

Structures may be grouped as follows: (1) those that need no paint, shop or field; (2) those in which interior steelwork will be exposed, probably field painted; (3) those fully exposed to the elements. Thus, shop paint is required only as a primer coat before a required coat of field paint.

Group (1) could include such structures as apartment buildings, hotels, dormitories, office buildings, stores, and schools, where the steelwork is enclosed by other materials. The practice of omitting the shop and field paint for these structures, however, may not be widely accepted because of tradition and the slowness of building-code modernization. Furthermore, despite the economic benefit of paint omission, clean, brightly painted steel during construction has some publicity value.

In group (2) are warehouses, industrial plants, parking decks, supermarkets, one-story schools, inside swimming pools, rinks, and arenas, all structures shielded from the elements but with steel exposed in the interior. Field paint may be required for corrosion protection or appearance or both. The severity of the corrosion environment depends on type of occupancy, exposure, and climatic conditions. The paint system should be carefully selected for optimum effectiveness.

In group (3) are those structures exposed at all times to the weather: crane runways, fire escapes, towers, exposed exterior columns, etc. When made of carbon steel, the members will be painted after erection and therefore should be primed with shop paint. The paint system selected should be the most durable one for the atmospheric conditions at the site. For corrosion-resistant steels, such as those meeting ASTM A242 and A588, field painting may be unnecessary. On exposure, these steels acquire a relatively hard coat of oxide, which shields the surface from progressive rusting. The color, russet brown, has architectural appeal.

7.42 PAINT SYSTEMS

The Steel Structures Painting Council has correlated surface preparations and primer, intermediate, and finish coats of paints into systems, each designed for a common service condition ("Steel Structure Painting Manual"). In addition, the Council publishes specifications for each system and individual specifications for surface preparations and paints. Methods for surface cleaning include solvent, hand-tool, power-tool, pickling, flame, and several blast techniques.

Surface preparation is directly related to the type of paints. In general, a slow-drying paint containing oil and rust-inhibitive pigments and one possessing good wetting ability may be applied on steel nominally cleaned. On the other hand, a fast-drying paint with poor wetting characteristics requires exceptionally good surface cleaning, usually entailing complete removal of mill scale. Therefore, in specifying a particular paint, the engineer should include the type of surface preparation, to prevent an improper surface condition from reducing the effectiveness of an expensive paint.

Paint selection and surface preparation are a matter of economics. For example, while blast-cleaned surfaces are conceded to be the best paint foundation for lasting results, the high cost is not always justified. Nevertheless, the Council specifies a minimum surface preparation by a blast cleaning process for such paints as alkyd, phenolic, vinyl, coal tar, epoxy, and zinc-rich.

As an aid for defining and evaluating the various surface preparations, taking into account the initial condition of the surface, an international visual standard is available and may be used. A booklet of realistic color photographs for this purpose can be obtained from the Council or ASTM. The applicable standard and acceptance criteria are given in "Quality Criteria and Inspection Standards," American Institute of Steel Construction.

The Council stresses the relationship between the prime coat (shop paint) and the finish coats. A primer that is proper for a particular type of field paint could be an unsatisfactory base for another type of field paint. Since there are numerous paint formulations, refer to Council publications when faced with a painting condition more demanding than ordinary.

In the absence of specific contract requirements for painting, the practice described in the AISC "Specification for Structural Steel for Buildings" may be followed. This method may be considered "nominal." The steel is brushed, by hand or power, to remove loose mill scale, loose rust, weld slag, flux deposit, dirt, and foreign matter. Oil and grease spots are solvent cleaned. The shop coat is a commercial-quality paint applied by brushing, dipping, roller coating, flow coating, or spraying to a 2-mil thickness. It affords only short-time protection. Therefore, finished steel that may be in ground storage for long periods or otherwise exposed to excessively corrosive conditions may exhibit some paint failure by the time it is erected, a condition beyond the control of the fabricator. Where such conditions can be anticipated, as for example, an overseas shipment, the engineer should select the most effective paint system.

7.43 FIELD-PAINTING STEEL

There is some question as to justification for protecting steelwork embedded in masonry or in contact with exterior masonry walls built according to good workmanship standards but not impervious to moisture. For example, in many instances, the masonry backing for a 4-in brick wall is omitted to make way for column flanges. Very definitely, a 4-in wall will not prevent penetration of water. In many cases, also, though a gap is provided between a wall and steelwork, mortar drippings fall into the space and form bridges over which water may pass, to attack the steel. The net effect is premature failure of both wall and steel. Walls have been shattered—sheared through the brick—by the powerful expansion of rust formations. The preventatives are: (1) coating the steel with suitable paint and (2) good wall construction.

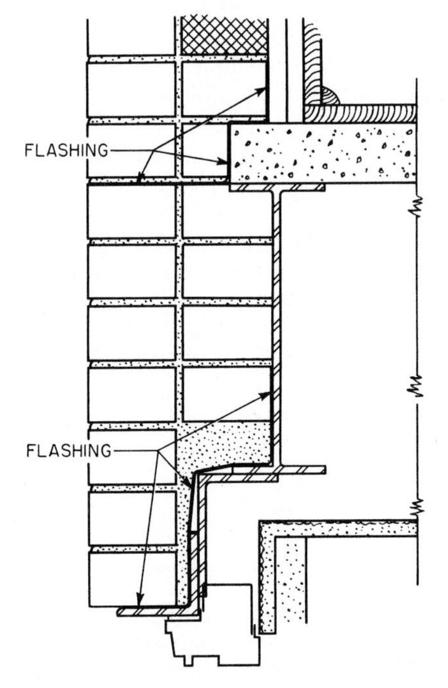

FIGURE 7.62 Flashing at spandrel and lintels.

A typical building code reads: "Special precautions shall be taken to protect the outer surfaces of steel columns located in exterior walls against corrosion, by painting such surfaces with waterproof paints, by the use of mastic, or by other methods or waterproofing approved by the building inspector."

In most structures an asphalt-type paint is used for column-flange protection. The proviso is sometimes extended to include lintels and spandrels, since the danger of corrosion is similar, depending on the closeness and contact with the wall. However, with the latter members, it is often judicious to supplement the paint with flashing, either metallic or fabric. A typical illustration, taken from an actual apartment-building design, is shown in Fig. 7.62.

In general, building codes differ on field paint; either paint is stipulated or the code is silent. From a practical viewpoint, the question of field painting cannot be properly resolved with a single broad rule. For an enclosed building in which the structural members are enveloped, for example, a field coat is sheer wastage, except for exterior steel members in contact with walls. On the other hand, exposed steel subject to high-humidity atmospheres and to exceptionally corrosive gases and contaminants may need two or three field coats.

Manufacturing buildings should always be closely scrutinized, bearing in mind that original conditions are not always permanent. As manufacturing processes change, so do the corrosive environments stimulated by new methods. It is well to prepare for the most adverse eventuality.

Special attention should be given to steel surfaces that become inaccessible, e.g., tops of purlins in contact with roof surfaces. A three-coat job of particularly suitable paint may pay off in the long run, even though it delays placement of the roof covering.

7.44 STEEL IN CONTACT WITH CONCRETE

According to the "Steel Structures Painting Manual," Vol. I, "Good Painting Practice" (Steel Structures Painting Council, Pittsburgh, PA):

1. Steel that is embedded in concrete for reinforcing should not be painted. Design considerations require strong bond between the reinforcing and the concrete so that the stress is distributed. Painting of such steel does not supply sufficient bond. If the concrete is properly made and of sufficient thickness over the metal, the steel will not corrode.

2. Steel that is encased in exposed lightweight concrete that is porous should be painted with at least one coat of good-quality rust-inhibitive primer. When conditions are severe, or humidity is high, two or more coats of paint should be applied, since the concrete may accelerate corrosion.

3. When steel is enclosed in concrete of high density or low porosity, and when the concrete is at least 2 to 3 in thick, painting is not necessary, since the concrete will protect the steel.

4. Steel in partial contact with concrete is generally not painted. This creates an undesirable condition, for water may seep into the crack between the steel and the concrete, causing corrosion. A sufficient volume of rust may be built up, spalling the concrete. The only remedy is to chip or leave a groove in the concrete at the edge next to the steel and seal the crack with an alkali-resistant calking compound (such as bituminous cement).

5. Steel should not be encased in concrete that contains cinders, since the acidic condition will cause corrosion of the steel.

FIRE PROTECTION FOR STRUCTURAL STEEL

Structural steel is a noncombustible material. It is therefore satisfactory for use without protective coverage in many types of buildings where noncombustibility is sufficient, from the viewpoint of either building ordinances or owner's preference. When structural steel is used in this fashion, it is described as "exposed" or "unprotected." Unprotected steel may be selected wherever building codes permit combustible construction.

Exposed or unprotected structural steel is commonly used for industrial-type buildings, hangars, auditoriums, stadiums, warehouses, parking garages, billboards, towers, and low stores, schools, and hospitals. In most cases, these structures contain little combustible material. In others, where the contents are highly combustible, sprinkler systems may be incorporated to protect the steelwork.

Steel building frames and floor systems should be covered with fire-resistant materials in certain buildings to reduce the chance of fire damage. These structures may be tall buildings, such as offices, apartments, and hotels, or low-height buildings, such as warehouses, where there is a large amount of combustible content. The buildings may be located in congested areas, where the spread of fire is a strong possibility. So for public safety, as well as to prevent property loss, building codes regulate the amount of fire resistance that must be provided.

The following are some of the factors that enter into the determination of minimum fire resistance for a specific structure: height, floor area, type of occupancy (a measure of combustible contents), fire-fighting apparatus, sprinkler systems, and location in a community (fire zone), which is a measure of hazard to adjoining properties.

7.45 EFFECT OF HEAT ON STEEL

A moderate rise in temperature of structural steel, say up to 500°F, is beneficial in that the strength is about 10% greater than the normal value. Above 500°F, strength falls off, until at 700°F it is nearly equal to the normal temperature strength. At a temperature of 1000°F, the compressive strength of steel is about the same as the maximum allowable working stress in columns.

Unprotected steel members have a rating of about 15 min, based on fire tests of columns with cross-sectional areas of about 10 in^2. Heavier columns, possessing greater mass for dissipation of heat, afford greater resistance—20 min perhaps. Columns with reentrant space between flanges filled with concrete, but otherwise exposed, have likewise been tested. Where the total area of the solid cross section approximates 36 in^2, the resistance is 30 min, and where the area is 64 in^2, the resistance is 1 hr.

The average coefficient of expansion for structural steel between the temperatures of 100 and 1200°F is given by the formula

$$C = 0.0000061 + 0.0000000019t \tag{7.81}$$

in which C = coefficient of expansion per °F and t = temperature, °F.

Below 100°F, the average coefficient of expansion is taken as 0.0000065.

The modulus of elasticity of structural steel, about 29,000 ksi at room temperature, decreases linearly to 25,000 ksi at 900°F. Then, it drops at an increasing rate at higher temperatures.

7.46 FIRE PROTECTION OF EXTERIOR MEMBERS

Steel members, such as spandrel beams and columns, on the exterior of a building may sometimes be left exposed or may be protected in an economical manner from fire damage, whereas interior steel members of the same building may be required to be protected with more expensive insulating materials, as discussed in Art. 7.47. Standard fire tests for determining fire-endurance ratings of exterior steel members are not available. But from many tests, data have been obtained that provide a basis for analytical, thermodynamic methods for fire-safe design. (See for example, "Fire-Safe Structural Steel—A Design Guide," American Iron and Steel Institute, 1000 16th St., N.W., Washington, DC 20036.)

The tests indicate that an exterior steel spandrel beam with its interior side protected by fire-resistant construction need only have its flanges fire protected. This may be simply done by application of fireproofing, such as sprayed-on mineral fibers, to the upper surface of the top flange and the under surface of the bottom flange. In addition, incombustible flame-impingement shields should enclose the flanges to deflect flames that may be emitted through windows. The shields, for example, may be made of ¼-in-thick weathering steel. This construction prevents the temperature of the spandrel beam from reaching a critical level.

Exposed-steel columns on the outside of a building may be made fire safe by placement at adequate distances from the windows. Such columns may also be located closer to the building when placed on the side of windows at such distances that the steel is protected by the building walls against flame impingement. Thermodynamic analysis can indicate whether or not the chosen locations are fire safe.

7.47 MATERIALS FOR IMPROVING FIRE RESISTANCE

Structural steel may be protected with any of many materials—brick, stone, concrete, gypsumboard, gypsum block, sprayed-on mineral fibers, and various fire-resistant plasters.

Concrete insulation serves well for column protection, in that it gives additional stability to the steel section. Also, it is useful where abrasion resistance is needed. Concrete, however, is not an efficient insulating medium compared with fire-resistant plasters. Normally, it is placed completely around the columns, beams, or girders, with all reentrant spaces filled solid (Fig. 7.63*a*). Although this procedure contributes to the stability of columns and effects composite action in beams and slabs, it has the disadvantage of imposing great weight on the steel frame and foundations. For instance, full protection of a W12 column with stone concrete weighs about 355 psf, whereas plaster protection weighs about 40 psf, and lightweight concretes made with such aggregates as perlite, vermiculite, expanded shale, expanded slag, pumice, pumicite and sintered flyash weigh less than 100 psf.

Considerable progress has been made in the use of lightweight plasters with aggregates possessing good insulating properties. Two aggregates used extensively are perlite and vermiculite. They replace sand in the sanded-gypsum plaster mix. A 1-in thickness weighs about 4 psf, whereas the same thickness of sanded-gypsum plaster weighs about 10 psf.

Typical details of lightweight plaster protection for columns are shown in Fig. 7.63*b* and *c*. Generally, vermiculite and perlite plaster thicknesses of 1 to 1¾ in afford protection of 3 and 4 h, depending on construction details. Good alternatives include gypsum board (Fig. 7.63*d* and *e*) or gypsum block (Fig. 7.63*f*).

For buildings where rough usage is expected, a hard, dense insulating material such as concrete, brick, or tile would be the logical selection for fire protection.

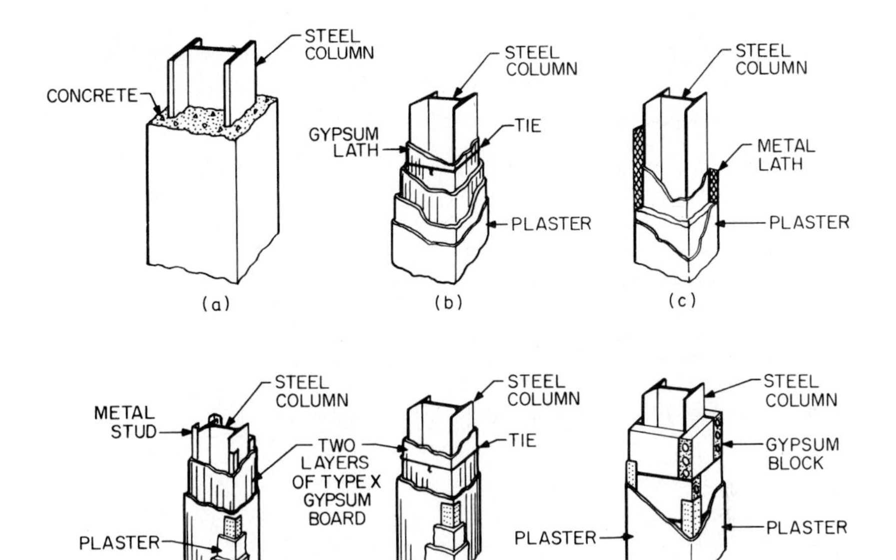

FIGURE 7.63 Fire protection of steel columns by encasement with (*a*) concrete, (*b*) plaster on gypsum lath, (*c*) plaster on metal lath, (*d*) furring and gypsumboard, (*e*) gypsumboard without furring, and (*f*) gypsum block and plaster.

For many buildings, finished ceilings are mandatory. It is therefore logical to employ the ceiling for protecting roof and floor framing. All types of gypsum plasters are used extensively for this dual purpose. Figure 7.64 illustrates typical installations. For 2-h floors, ordinary sand-gypsum plaster ¾ in thick is sufficient. Three- and four-hour floors may be obtained with perlite gypsum and vermiculite gypsum in the thickness range of ¾ to 1 in.

Instead of plastered ceilings, use may be made of fire-rated dry ceilings, acoustic tiles, or drop (lay-in) panels (Fig. 7.64*d* and *e*).

Another alternative is to spray the structural steel mechanically (where it is not protected with concrete) with plasters of gypsum, perlite, or vermiculite, proprietary cementitious mixtures, or mineral fibers not deemed a health hazard during spraying (Fig.7.65). In such cases, the fire-resistance rating of the structural system is independent of the ceiling. Therefore, the ceiling need not be of fire-rated construction. Drop panels, if used, need not be secured to their suspended supports.

Still another sprayed-on material is the intumescent fire-retardant coating, essentially a paint. Tested in conformance with ASTM Specification E119, a 3/16-in-thick coat applied to a steel column has been rated 1 h, a ½-in-thick coating 2 h. As applied, the coating has a hard, durable finish, but at high temperatures, it puffs to many times its original thickness, thus forming an effective insulating blanket. Thus, it serves the dual need for excellent appearance and fire protection.

Aside from dual functioning of ceiling materials, the partitions, walls, etc., being of incombustible material, also protect the structural steel, often with no additional assistance. Fireproofing costs, therefore, may be made a relatively minor expense in the overall costs of a building through dual use of materials.

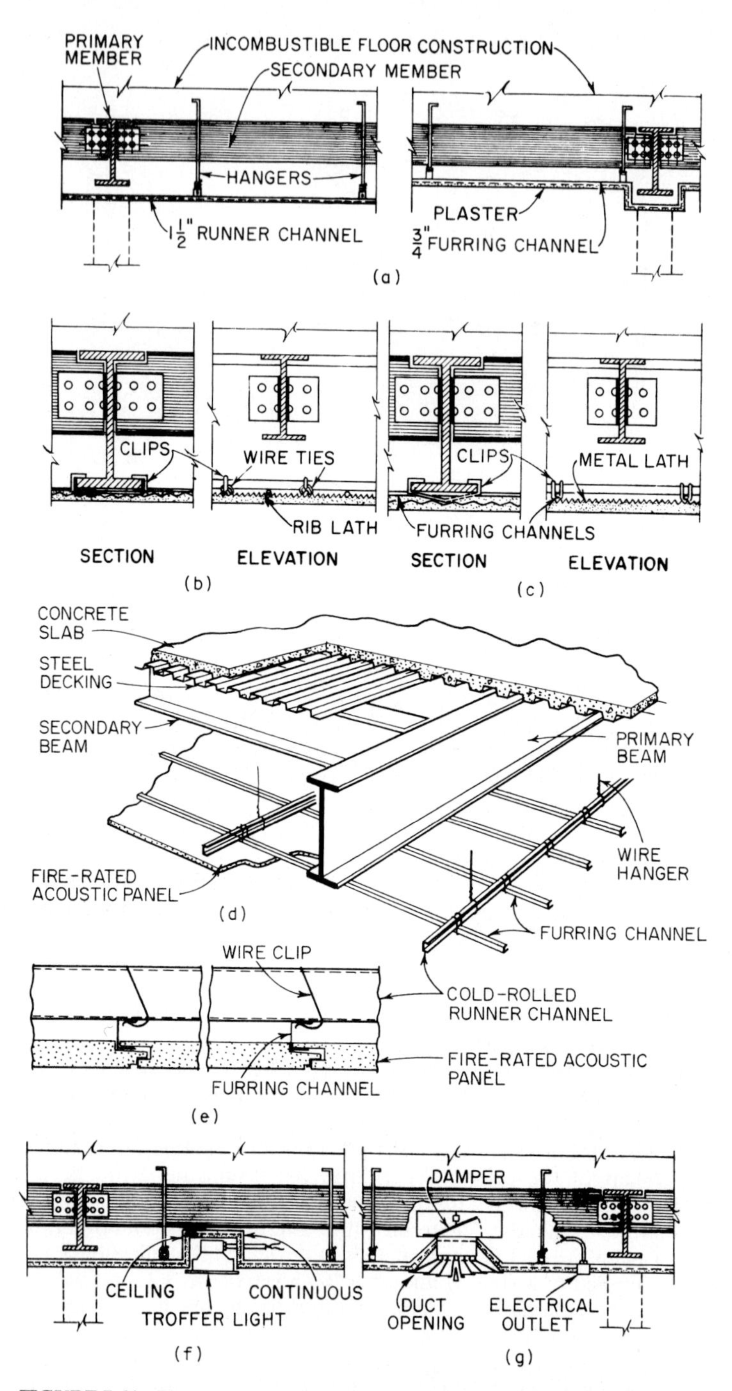

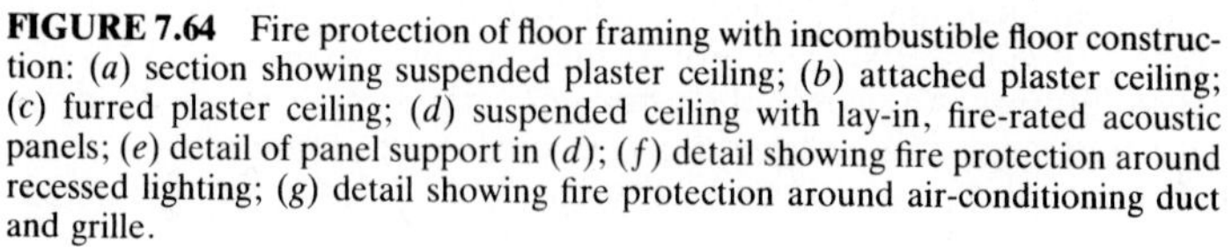

FIGURE 7.64 Fire protection of floor framing with incombustible floor construction: (*a*) section showing suspended plaster ceiling; (*b*) attached plaster ceiling; (*c*) furred plaster ceiling; (*d*) suspended ceiling with lay-in, fire-rated acoustic panels; (*e*) detail of panel support in (*d*); (*f*) detail showing fire protection around recessed lighting; (*g*) detail showing fire protection around air-conditioning duct and grille.

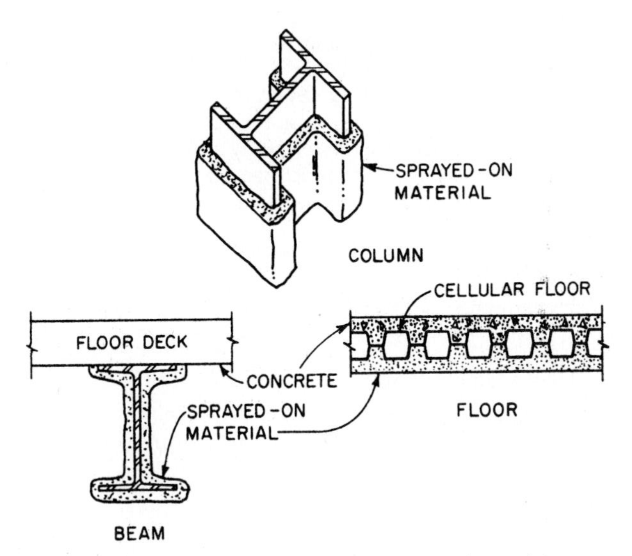

FIGURE 7.65 Typical fire protection with sprayed material.

7.48 PIERCED CEILINGS AND FLOORS

Some buildings require recessed light fixtures and air-conditioning ducts, thus interrupting the continuity of fire-resistive ceilings. A rule that evolved from early standard fire tests permitted 100 in^2 of openings for noncombustible pipes, ducts, and electrical fixtures in each 100 ft^2 of ceiling area.

It has since been demonstrated, with over 100 fire tests that included electrical fixtures and ducts, that the fire-resistance integrity of ceilings is not impaired when, in general:

Recessed light fixtures, 2 by 4 ft, set in protective boxes, occupy no more than 25% of the gross ceiling area.

Air-duct openings, 30 in maximum in any direction, are spaced so as not to occupy more than 576 in^2 of each 100 ft^2 of gross ceiling area. They must be protected with fusible-link dampers against spread of smoke and heat.

These conclusions are not always applicable. Reports of fire tests of specific floor systems should be consulted.

A serious infringement of the fire rating of a floor system could occur when pipes, conduit, or other items pierce the floor slab, a practice called "poke-through." Failure to calk the openings with insulating material results in a lowering of fire ratings from hours to a few minutes.

7.49 FIRE-RESISTANCE RATINGS

Most standard fire tests on structural-steel members and assemblies have been conducted at one of two places—the National Institute of Standards and Technology, Washington, D.C., or the Underwriters Laboratories, Northbrook, Ill. Fire-testing laboratories also are available at Ohio State University, Columbus, Ohio, and the University of California, Berkeley, Calif. Laboratory test reports form the basis for establishing ratings. Summaries of these tests, together with tabulation of recognized ratings, are published by a number of organizations listed below. The trade associations, for the most part, limit their ratings to those constructions employing the material they represent.

The American Insurance Association (formerly The National Board of Fire Underwriters), 85 John St., New York, NY 10038

The National Institute of Standards and Technology, Washington, DC 20234

Gypsum Association, 1603 Orrington Ave., Evanston, IL 60201

Metal Lath/Steel Framing Association, 600 Federal St., Chicago, IL 60605

Perlite Institute, 600 S. Federal St., Chicago, IL 60605

American Iron and Steel Institute, 1000 16th St., N.W., Washington, DC 20036

American Institute of Steel Construction, 1 E. Wacker Dr., Chicago, IL 60601

7.50 BIBLIOGRAPHY

Designing Fire Protection for Steel Columns; Designing Steel Protection for Steel Trusses; Fire-Safe Structural Steel, American Iron and Steel Institute, 1000 16th St., N.W., Washington, DC 20036.

Design Guide—Industrial Buildings, Roofs to Column Anchorage; Column Base Plates (ASD); Iron and Steel Buildings, 1873–1952; Guide to Shop Painting of Structural Steel; Low- and Medium-Rise Steel Buildings; Structural Steel Detailing, American Institute of Steel Construction, 1 E. Wacker Dr., Chicago, IL 60601.

Fundamentals of Welding; Structural Welding Code, D1.1; American Welding Society, 550 N.W. Le Jeune Rd., Miami, FL 33135.

E.H. Gaylord, Jr., et al., "Design of Steel Structures," 3d ed.; E. H. Gaylord, Jr., and C. N. Gaylord, "Structural Engineering Handbook," 3d ed.; F. S. Merritt and R. L. Brockenbrough, "Structural Steel Designers Handbook," 2d ed.; A. J. Rokach, "Structural Steel Design, LRFD," McGraw-Hill, Inc., New York.

T. V. Galambos, "Guide to Stability Design Criteria for Metal Structures," John Wiley & Sons, Inc., New York.

SECTION EIGHT
COLD-FORMED STEEL CONSTRUCTION

Don S. Wolford
Consulting Engineer
Middletown, Ohio

The term cold-formed steel construction, as used in this section, refers to structural components that are made of flat-rolled steel. This section deals with fabricated components made from basic forms of steel, such as bars, plates, sheet, and strip.

COLD-FORMED SHAPES

Cold-formed shapes usually imply relatively small, thin sections made by bending sheet or strip steel in roll-forming machines, press brakes, or bending brakes. Because of the relative ease and simplicity of the bending operation and the comparatively low cost of forming rolls and dies, the cold-forming process lends itself well to the manufacture of unique shapes for special purposes and makes it possible to use thin material shaped for maximum stiffness.

The use of cold-formed shapes for ornamental and other non-load-carrying purposes is commonplace. Door and window frames, metal-partition work, non-load-bearing studs, facing, and all kinds of ornamental sheet-metal work employ such shapes. The following deals with cold-formed shapes used for structural purposes in the framing of buildings.

There is no standard series of cold-formed structural sections, such as those for hot-rolled shapes, yet, although groups of such sections have been designed ("Cold-formed Steel Design Manual," American Iron and Steel Institute, 1101 17th St., NW, Washington, DC 20036). For the most part, however, cold-formed structural shapes are designed to serve a particular purpose. The general approach of the designer is therefore similar to that involved in the design of built-up structural sections.

Cold-formed shapes invariably cost more per pound than hot-rolled sections. They will be found to be more economical under the following circumstances:

1. Where their use permits a substantial reduction in weight compared to hot-rolled sections. This occurs where relatively light loads are to be supported over short spans, or where stiffness rather than strength is the controlling factor in the design.

2. In special cases where a suitable combination of standard hot-rolled shapes would be heavy and uneconomical.

3. Where quantities required are too small to justify the investment necessary to produce a suitable hot-rolled section.

4. In dual-purpose panel work, where both strength and coverage are desired.

8.1 MATERIAL FOR COLD-FORMED STEEL SHAPES

Cold-formed shapes are usually made from hot-rolled sheet or strip steel, which costs less per pound than cold-rolled steel. The latter, which has been cold-rolled to desired thickness, is used for thinner gages or where, for any reason, the surface finish, mechanical properties, or closer tolerances that result from cold-reducing is desired. Manufacture of cold-formed shapes from plates for use in building construction is possible but is done infrequently.

8.1.1 Plate, Sheet, or Strip?

The commercial distinction between steel plates, sheet, and strip is principally a matter of thickness and width of material. In some sizes, however, classification depends on whether the material is furnished in flat form or in coils, whether it is carbon or alloy steel, and, particularly for cold-rolled material, on surface finish, type of edge, temper or heat treatment, chemical composition, and method of production. Although the manufacturers' classification of flat-rolled steel products by size is subject to change from time to time, that given in Table 8.1 for carbon steel is representative.

Carbon steel is generally used. High-strength, low-alloy steel, however, may be used where strength or corrosion resistance justify it, and stainless steel may be used for exposed work.

8.1.2 Mechanical Properties

Material to be used for structural purposes generally conforms to one of the standard specifications of ASTM. Table 8.2 lists the ASTM specifications for structural-quality carbon and low-alloy sheet and strip, and their principal mechanical properties.

8.1.3 Stainless-Steel Applications

Stainless-steel cold-formed shapes, although not ordinarily used in floor and roof framing, are widely used in exposed components, such as stairs, railings, and balustrades; doors and windows; mullions, fascias; curtain walls and panel work; and other applications in which a maximum degree of corrosion resistance, retension of appearance and luster, and compatibility with other materials are primary

TABLE 8.1 Classification by Size of Flat-Rolled Carbon Steel

a. Hot-rolled

Width, in	Thicknesses, in			
	0.2300 and thicker	0.2299–0.2031	0.2030–0.1800	0.1799–0.0449
To 3½ incl.	Bar	Bar	Strip	Strip[a]
Over 3½ to 6 incl.	Bar	Bar	Strip	Strip[b]
Over 6 to 8 incl.	Bar	Strip	Strip	Strip
Over 8 to 12 incl.	Plate[c]	Strip	Strip	Strip
Over 12 to 48 incl.	Plate[d]	Sheet	Sheet	Sheet
Over 48	Plate[d]	Plate[d]	Plate[d]	Sheet

b. Cold-rolled

Width, in	Thicknesses, in		
	0.2500 and thicker	0.2499–0.0142	0.0141 and thinner
To 12, incl.	Bar	Strip[e,f]	Strip[e]
Over 12 to 23¹⁵⁄₁₆, incl.	Sheet[g]	Sheet[g]	Strip[h]
Over 23¹⁵⁄₁₆	Sheet	Sheet	Black plate[i]

[a]0.0255-in minimum thickness.
[b]0.0344-in minimum thickness.
[c]Strip, up to and including 0.5000-in thickness, when ordered in coils.
[d]Sheet, up to and including 0.5000-in thickness, when ordered in coils.
[e]Except that when the width is greater than the thickness, with a maximum width of ½ in and a cross-sectional area not exceeding 0.05 in^2, and the material has rolled or prepared edges, it is classified as flat wire.
[f]Sheet, when slit from wider coils and supplied with cut edge (only) in thicknesses 0.0142 to 0.0821 and widths 2 to 12 in, inclusive, and carbon content 0.25% maximum by ladle analysis.
[g]May be classified as strip when a special edge, a special finish, or single-strand rolling is specified or required.
[h]Also classifed as black plate[i], depending on detailed specifications for edge, finish, analysis, and other features.
[i]Black plate is a cold-rolled, uncoated tin-mill product that is supplied in relatively thin gages.

considerations. Stainless-steel sheet and strip are available in several types and grades, with different strength levels and different degrees of formability, and in a wide range of finishes.

Information useful in design of stainless-steel cold-formed members can be obtained from the "Specification for the Design of Cold-Formed Stainless Steel Structural Members," American Society of Civil Engineers (ASCE), 345 East 47th St., New York, NY 10017. The specification is applicable to material covered by ASTM A666, "Austenitic Stainless Steel, Sheet, Strip, Plate and Flat Bars for Structural Applications." It contains requirements for 201, 202, 301, 302, 304, and 316 types of stainless steels. Further information on these steels as well as steels

TABLE 8.2 Principal Mechanical Properties of Structural Quality Sheet, Strip, and Plate Steel

ASTM designation	Material	Grade	Minimum yield point, ksi	Minimum tensile strength, ksi		Minimum elongation, % in 2 in	Bend test, 180°, ratio of inside diameter to thickness
				Hot rolled	Cold rolled		
A446	Galvanized sheet steel, zinc-coated by the hot-dip process, structural quality	A	33	45		20	1½
		B	37	52		18	2
		C	40	55		16	2½
		D	50	65		12	†
		E	80	82			†
		F	50	70		12	†
A570	Hot-rolled sheet and strip, carbon steel	30	30	49		*	1
		36	36	53		*	1½
		40	40	55		*	2
		45	45	60		*	2½
		50	50	65		*	3
A606	Hot-rolled and cold-rolled sheet and strip, high-strength, low-alloy steel	Cut lengths	50	70		22	1
		Coils	45	65		22	1
		Annealed or normalized	45	65		22	1
		Cold rolled	45		65	22	1
						HR§ CR§	
A607	Hot-rolled and cold-rolled, high-strength, low-alloy columbium or vanadium steels, sheet and strip, cut lengths or coils	45	45	60	60	25 22	1
		50	50	65	65	22 20	1
		55	55	70	70	20 18	1½
		60	60	75	75	18 16	2
		65	65	80	80	16 15	2½
		70	70	85	85	14 14	3
A611	Cold-rolled sheet, structural carbon-steel sheet, cut lengths or coils	A	25		42	26	0
		B	30		45	24	1
		C	33		48	22	1½
		D	40		52	20	2
		E	[illegible]		[illegible]		[illegible]

A36	Structural steel (plates only)		36	58–80		23	‡
A242	High-strength, low-alloy structural steel (plates ¾ in and under)		50			†	‡
A441	High-strength, low-alloy structural manganese vanadium steel (plates ¾ in and under)		50	70		†	‡
A529	Structural steel with 42 ksi minimum yield point (½ in maximum thickness) (plates only)		42	60–85		†	‡
A572	High-strength, low-alloy columbium-vanadium steels of structural quality (plates only)	42	42	60		24	‡
		50	50	65		21	‡
		60	60	75		18	‡
		65	65	80		17	‡
A588	High-strength, low-alloy structural steel with 50 ksi minimum yield point to 4 in thick (plates only)	A	50	70		21	‡
		B	50	70		21	‡
		C	50	70		21	‡
		D	50	70		21	‡
		E	50	70		21	‡
		F	50	70		21	‡
		G	50	70		21	‡
		H	50	70		21	‡
		J	50	70		21	‡
A715	High-strength, low-alloy hot-rolled steel with improved formability	50	50	60		HR§ CR§	1
		60	60	70		22 20	1½
		70	70	80		22 18	
		80	80	90		18 16	
						18 16	
A792	Aluminum-zinc alloy coated steel sheet by the hot-dip process, general requirements	33	33	45		20	1½
		37	37	52		18	2
		40	40	55		16	2½
		50A	50	65		12	—
		50B	50	—		12	—
		80	80	82		12	

*Varies, see specification. †Not specified or required. ‡S14 bend test. §HR = hot rolled; CR = cold rolled.

covered by ASTM A176, A240, and A276 may be obtained from the American Iron and Steel Institute (AISI) and from the International Nickel Company, Inc.

8.1.4 Coatings

Material for cold-formed shapes may be either *black* (uncoated), galvanized, or aluminized. Because of their higher costs, metal-coated steels are used only where exposure conditions warrant paying more for the increased protection afforded against corrosion.

Low-carbon sheets suitable for coating with vitreous enamel are frequently used for facing purposes, but not as a rule to perform load-carrying functions in buildings.

8.1.5 Selection of Grade

The choice of a grade of material, within a given class or specification, usually depends on the severity of the forming operation required to make the required shape, strength desired, weldability requirements, and the economics involved. Grade C of ASTM A611, with a specified minimum yield point of 33 psi, has long been popular for structural use. Some manufacturers, however, use higher-strength grades to good advantage.

8.1.6 Gage Numbers

Thickness of cold-formed shapes was formerly expressed as the manufacturers' standard gage number of the material from which the shapes were formed. *Use of millimeters or decimal parts of an inch, instead of gage numbers,* is now the standard practice. However, for information, the relationships among gage number, weight, and thickness for uncoated and galvanized sheets are given in Table 8.3 for even gages.

8.2 UTILIZATION OF COLD WORK OF FORMING

When strength alone, particularly yield strength, is an all-important consideration in selecting a material or grade for cold-formed shapes (Table 8.2), it is sometimes possible to take advantage of the strength increase that results from cold working of material during the forming operation and thus use a lower-strength, more workable, and possibly more economical grade than would otherwise be required. The increase in cold-work strength is ordinarily most noticeable in relatively stocky, compact sections produced in thicker steels. Cold-formed chord sections for open-web steel joists are good examples (Fig. 8.22). Overall average yield strengths of more than 150% of the minimum specified yield strength of the plain material have been obtained in such sections.

The strengthening effect of the forming operation varies across the section but is most pronounced at the bends and corners of a cold-formed section. Accordingly, for shapes in which bends and corners constitute a high percentage of the whole

TABLE 8.3 **Gages, Weights, and Thicknesses of Sheets**

Steel manufacturer's standard gage No.	Weight, psf	Equivalent sheet thickness, in*	Galvanized sheet gage No.	Weight, psf	Thickness equivalent,† in
4	9.3750	0.2242			
6	8.1250	0.1943			
8	6.8750	0.1644	8	7.03125	0.1681
10	5.6250	0.1345	10	5.78125	0.1382
12	4.3750	0.1046	12	4.53125	0.1084
14	3.1250	0.0747	14	3.28125	0.0785
16	2.5000	0.0598	16	2.65625	0.0635
18	2.0000	0.0478	18	2.15625	0.0516
20	1.5000	0.0359	20	1.65625	0.0396
22	1.2500	0.0299	22	1.40625	0.0336
24	1.0000	0.0239	24	1.15625	0.0276
26	0.7500	0.0179	26	0.90625	0.0217
28	0.6250	0.0149	28	0.78125	0.0187
30	0.5000	0.0120	30	0.65625	0.0157
32	0.40625	0.0097	32	0.56250	0.0134
34	0.34375	0.0082			
36	0.28125	0.0067			
38	0.25000	0.0060			

*Thickness equivalents of steel are based on 0.023912 in/(lb-ft^2) (reciprocal of 41.820 psf per inch of thickness, although the density of steel is ordinarily taken as 489.6 lb/ft^3, 0.2833 lb/in^3, or 40.80 psf per inch of thickness). The density is adjusted because sheet weights are calculated for specified widths and lengths of sheets, with all shearing tolerances on the over side, and also because sheets are somewhat thicker at the center than at the edges. The adjustment yields a close approximation of the relationship between weight and thickness. ("Steel Products Manual, Carbon Steel Sheets," American Iron and Steel Institute.)

†Total thickness, in, including zinc coating. To obtain base metal thickness, deduct 0.0015 in per ounce coating class, or refer to ASTM A446.

section, cold working increases the overall strength more than for shapes having a high proportion of thin, wide, flat elements that are not heavily worked in forming. For the latter type of shapes, the strength of the plain, unformed sheet or strip may be the controlling factor in the selection of a grade of material.

Full-section tests constitute a relatively simple, straightforward method of determining as-formed strength. They are particularly applicable to sections that do not contain any elements that may be subject to local buckling. However, each case has to be considered individually in determining the extent to which cold forming will produce an increase in utilizable strength. For further information, refer to the AISI "Specification for the Design of Cold-Formed Steel Structural Members" and its "Commentary," 1986, American Iron and Steel Institute, 1101 17th St., Washington, DC 20036.

8.3 TYPES OF COLD-FORMED SHAPES

Many cold-formed shapes used for structural purposes are similar in their general configurations to hot-rolled structural sections. Channels, angles, and zees can be roll-formed in a single operation from one piece of material. I sections are usually made by welding two channels back to back or by welding two angles to a channel. All sections of this kind may be made with either plain flanges as in Fig. 8.1*a* to *d*, *j*, and *m* or with flanges stiffened by means of lips at outer edges, as in Fig. 8.1*e* to *h*, *k*, and *n*.

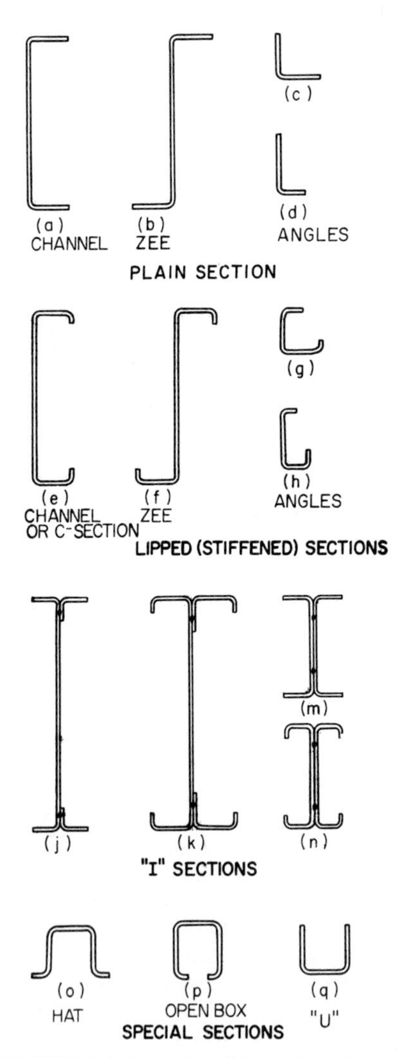

FIGURE 8.1 Typical cold-formed steel structural sections.

In addition to these sections, which follow somewhat conventional lines and have their counterparts in hot-rolled structural sections, the flexibility of the forming process makes it relatively easy to obtain inverted U, or hat-shaped, sections and open box sections (Fig. 8.1*o* to *q*). These sections are very stiff in a lateral direction and can be used without lateral support where other more conventional types of sections would fail because of lateral instability.

Other special shapes are illustrated in Fig. 8.2. Some of these are nonstructural in nature; others are used for special-purpose structural members. Figure 8.3 shows a few cold-formed stainless steel sections.

An important characteristic of cold-formed shapes is that the thickness of section is substantially uniform. (A slight reduction in thickness may occur at bends, but that may be ignored for computing weights and section properties.) This means that, for a specified thickness, the amount of flange material in a section, such as a channel, is almost entirely a function of the width of the section, except for shapes where additional flange area is obtained by doubling the material back on itself. Another distinguishing feature of cold-formed sections is that the corners are rounded on both the inside and the outside of the bend, since the shapes are formed by bending flat material.

Sharp corners, such as can be obtained with hot-rolled structural channels, angles, and zees, cannot be obtained in cold-formed shapes by simple bending, although they can be achieved in a coining or upsetting operation. This, however, is not customary in the man-

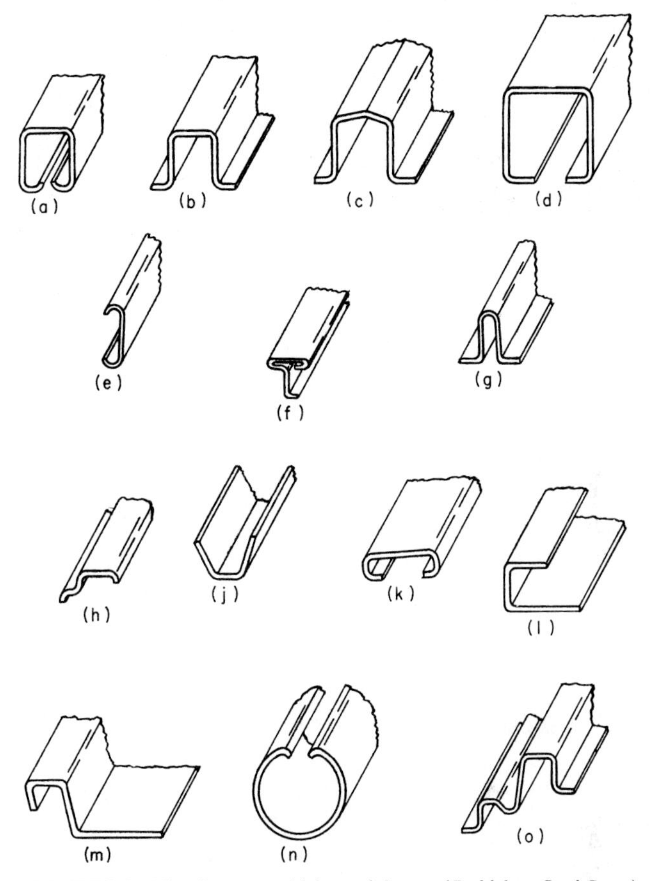

FIGURE 8.2 Miscellaneous cold-formed shapes. (*Bethlehem Steel Corp.*)

ufacture of structural cold-formed sections; and in proportioning such sections, the inside radius of bends should never be less, and should preferably be 33 to 100% greater, than specified for the relatively narrow ASTM bend-test specimens. Deck and panel sections, such as are used for floors, roofs, and walls, are as a rule considerably wider, relative to their depth, than are the structural framing members shown in Figs. 8.1 to 8.3.

DESIGN PRINCIPLES FOR COLD-FORMED STEEL SHAPES

The structural behavior of cold-formed shapes follows the same laws of structural mechanics as does that of conventional structural-steel shapes and plates. Thus, design procedures commonly used in the selection of hot-rolled shapes are generally applicable to cold-formed sections. Although only a portion of a section, in some

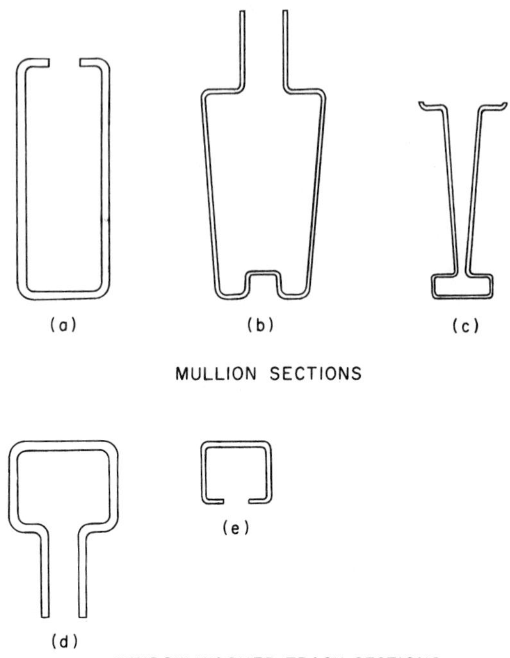

FIGURE 8.3 Cold-formed stainless steel sections. (*The International Nickel Co., Inc.*)

cases, may be considered structurally effective, computation of the structural properties of the effective portion follows conventional procedure.

8.4 SOME BASIC CONCEPTS OF COLD-FORMED STEEL DESIGN

The uniform thickness of most cold-formed sections, and the fact that the widths of the various elements composing such a section are usually large relative to the thickness, make it possible to consider, in computing structural properties (moment of inertia, section modulus, etc.) that such properties vary directly as the first power of the thickness. So, in most cases, section properties can be approximated by first assuming that the section is made up of a series of line elements, omitting the thickness dimension. Then, final values can be obtained by multiplying the line-element result by the thickness.

With this method, the final multiplier is always the first power of the thickness, and first-power quantities such as radius of gyration and those locating the centroid of the section do not involve the thickness dimension. The assumption that the area, moment of inertia, and section modulus vary directly as the first power of the thickness is particularly useful in determining the required thickness of a section after the widths of the various elements composing the section have been fixed. This method is sufficiently accurate for most practical purposes. It is advisable, however, particularly when a section is fairly thick compared to the widths of the elements, to check the final result through an exact method of computation.

Properties of thin elements are given in Table 8.4.

TABLE 8.4 Properties of Area and Line Elements

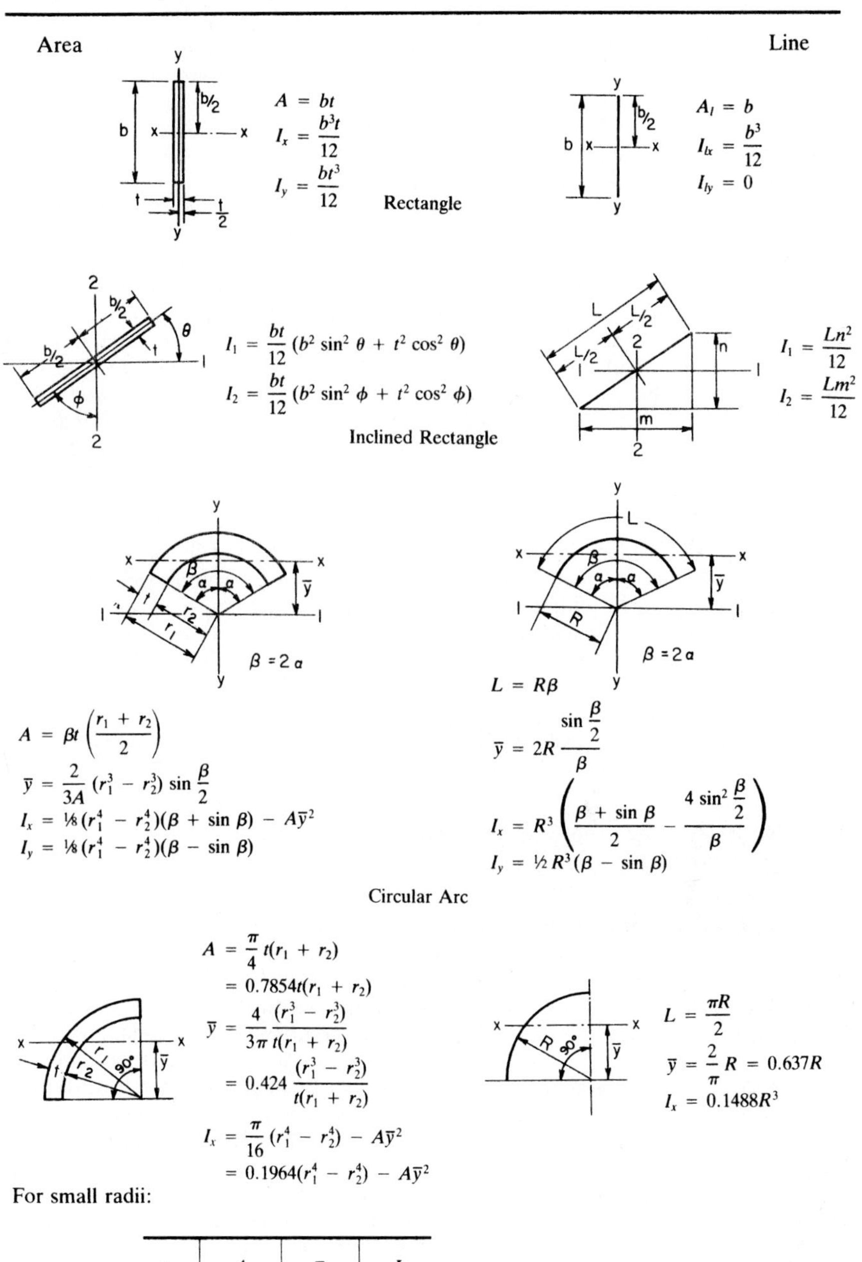

Rectangle

Area: $A = bt$, $I_x = \dfrac{b^3 t}{12}$, $I_y = \dfrac{bt^3}{12}$

Line: $A_l = b$, $I_{lx} = \dfrac{b^3}{12}$, $I_{ly} = 0$

Inclined Rectangle

Area:

$$I_1 = \frac{bt}{12}(b^2 \sin^2 \theta + t^2 \cos^2 \theta)$$

$$I_2 = \frac{bt}{12}(b^2 \sin^2 \phi + t^2 \cos^2 \phi)$$

Line: $I_1 = \dfrac{Ln^2}{12}$, $I_2 = \dfrac{Lm^2}{12}$

Circular Arc

Area ($\beta = 2\alpha$):

$$A = \beta t \left(\frac{r_1 + r_2}{2}\right)$$

$$\bar{y} = \frac{2}{3A}(r_1^3 - r_2^3) \sin \frac{\beta}{2}$$

$$I_x = \tfrac{1}{8}(r_1^4 - r_2^4)(\beta + \sin \beta) - A\bar{y}^2$$

$$I_y = \tfrac{1}{8}(r_1^4 - r_2^4)(\beta - \sin \beta)$$

Line ($\beta = 2\alpha$):

$$L = R\beta$$

$$\bar{y} = 2R \frac{\sin \frac{\beta}{2}}{\beta}$$

$$I_x = R^3 \left(\frac{\beta + \sin \beta}{2} - \frac{4 \sin^2 \frac{\beta}{2}}{\beta}\right)$$

$$I_y = \tfrac{1}{2} R^3 (\beta - \sin \beta)$$

90° Circular Corner

Area:

$$A = \frac{\pi}{4} t(r_1 + r_2) = 0.7854t(r_1 + r_2)$$

$$\bar{y} = \frac{4}{3\pi} \frac{(r_1^3 - r_2^3)}{t(r_1 + r_2)} = 0.424 \frac{(r_1^3 - r_2^3)}{t(r_1 + r_2)}$$

$$I_x = \frac{\pi}{16}(r_1^4 - r_2^4) - A\bar{y}^2 = 0.1964(r_1^4 - r_2^4) - A\bar{y}^2$$

Line:

$$L = \frac{\pi R}{2}$$

$$\bar{y} = \frac{2}{\pi} R = 0.637R$$

$$I_x = 0.1488R^3$$

For small radii:

r_2	A	$\bar{y}$	I_x
$2t$	$3.927t^2$	$1.613t$	$2.549t^4$
$1.5t$	$3.142t^2$	$1.300t$	$1.369t^4$
t	$2.356t^2$	$0.990t$	$0.635t^4$
$0.75t$	$1.963t^2$	$0.838t$	$0.400t^4$
$0.5t$	$1.571t^2$	$0.690t$	$0.235t^4$

Various Failure Modes. One of the distinguishing characteristics of lightweight cold-formed sections is that they are usually composed of elements that are relatively wide and thin. As a result, attention must be given to certain modes of structural behavior ordinarily neglected in dealing with heavier sections, such as hot-rolled structural shapes.

When thin, wide elements are in axial compression, as in the case of a beam flange or a part of a column, they tend to buckle elastically at stresses below the yield point of the steel. This local buckling is not to be confused with the general buckling that occurs in the failure of a long column or of a laterally unsupported beam. Rather, local buckling represents failure of a single element of a section, and conceivably may be relatively unrelated to buckling of the entire member. In addition, there are other factors, such as shear lag, which gives rise to nonuniform stress distribution; torsional instability, which may be more pronounced in thin sections than in thicker ones and requires more attention to bracing; and other related structural phenomena customarily ignored in conventional structural design that sometimes must be considered with thin material. Means of taking care of these factors in ordinary structural design are described in the "Specification for the Design of Cold-Formed Steel Structural Members."

[The Allowable Stress Design Method (ASD) is used currently in structural design of cold-formed steel members and described in the rest of this section. The Load and Resistance Design Method (LRFD), however, is gaining acceptance by designers as a means of obtaining more economical structures. See Sec. 7.]

The Committee on Specifications of the American Iron and Steel Institute has strived to put all formulas in the "Specification for the Design of Cold-Formed Steel Structural Members" on nondimensional bases so that their use with English or SI units is rigorous and convertible. (See Appendix.)

(AISI "Cold-Formed Steel Design Manual," American Iron and Steel Institute, 1101 17th St., Washington, DC 20036.)

8.5 STRUCTURAL BEHAVIOR OF FLAT COMPRESSION ELEMENTS

In buckling of flat, thin compression elements in beams and columns, the **flat-width ratio** w/t is an important factor. It is the ratio of width w of a single flat element, exclusive of any edge fillets, to the thickness t of the element (Fig. 8.4). Local buckling of elements with large w/t may be resisted with stiffeners or bracing.

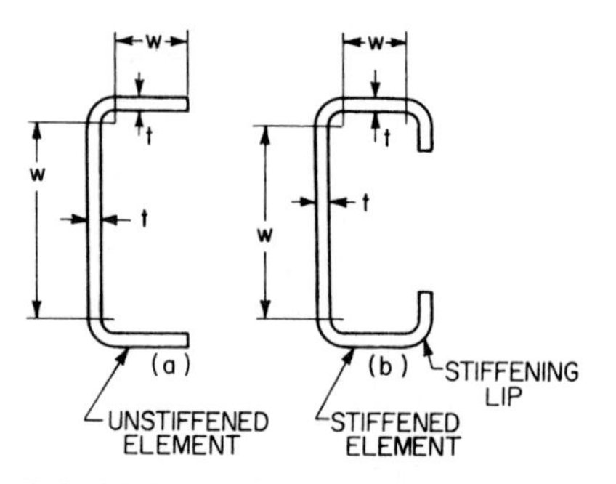

FIGURE 8.4 Compression elements.

Flat compression elements of cold-formed structural members are accordingly classified as stiffened or unstiffened. Stiffened compression elements have both edges of the element parallel to the direction of stress stiffened by a web, flange, or stiffening lip. If the sections in Fig. 8.1a to n are used as compression members, the webs are considered as stiffened compression elements. The wide, lipless flange elements and the lips that stiffen the outer edges, however, are unstiffened elements. Any section can be broken down into a combination of stiffened and unstiffened elements.

Only part of an element may be considered effective under compression in computation of net section properties. The portion that may be treated as effective depends on w/t for the element.

The cold-formed structural cross sections shown in Fig. 8.5 indicate that the effective portions b of the width of a stiffened compression element are considered to be divided equally into two parts, $b/2$, located next to the two edge stiffeners of that element. (A stiffener may be a web, another stiffened element, or a lip in beams. Lips in these examples are presumed to be fully effective.) In computation of net section properties, only the effective portions of stiffened compression elements are used and the ineffective portions are disregarded. For beams, because flange elements subjected to uniform compression may not be fully effective, reduced section properties, such as moments of inertia and section moduli, must be used. For computation of the effective widths of webs, see Art. 8.7. Effective areas of column cross sections are based on full cross-sectional areas less all ineffective

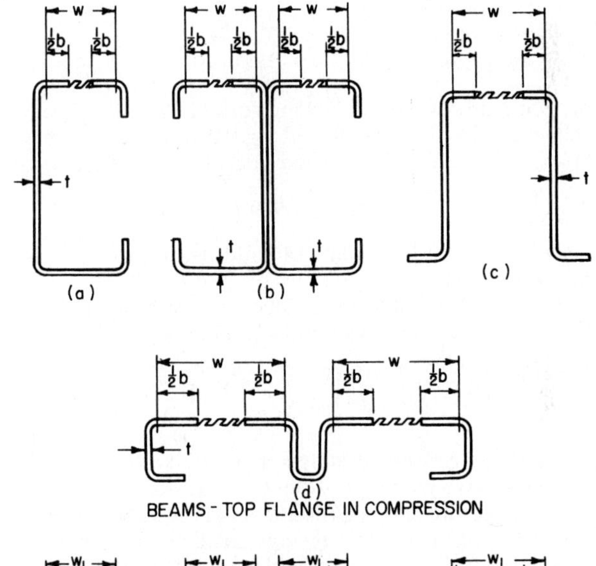

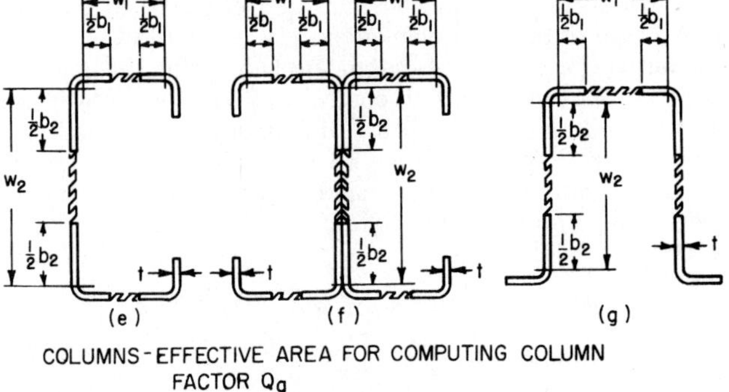

FIGURE 8.5 Effective width of stiffened compression elements with stiffening lips assumed to be fully effective.

portions for use in the formula for axially loaded columns, Eq. (8.22), in Art. 8.13.

The critical load, P_{cr}, kips, for elastic buckling of a bar of uniform cross section, concentrically end loaded as a column, is given by the Euler formula:

$$P_{cr} = \pi^2 EI/L^2 \tag{8.1}$$

where E = modulus of elasticity, 29,500 ksi for steel
I = moment of inertia of bar cross section, in^4
L = column length of bar, in

Bryan, in 1891, determined the critical buckling stress, f_{cr}, ksi, for a thin rectangular plate compressed between two opposite edges with the other two edges supported, to be given by

$$f_{cr} = k\pi^2 E(t/w)^2/12(1 - \nu^2) \tag{8.2}$$

where k = a coefficient depending on edge-support restraint
w = width of plate, in
t = thickness of plate, in
ν = Poisson's ratio

In 1932, von Karman gave the following formula for determining the effective width-to-thickness ratio b/t at yielding along the simply supported edges of a thin rectangular plate subjected to compression between the other two opposite edges:

$$b/t = 1.9t\sqrt{E/f_y} \tag{8.3}$$

where b = effective width for a plate of width w, in, and f_y = yield strength of plate material, ksi.

After extensive tests of cold-formed steel structural sections, Winter, in 1947, recommended that von Karman's formula be modified to

$$b/t = 1.9t\sqrt{E/f_{max}}\left(1 - \frac{0.475\sqrt{E/f_{max}}}{w/t}\right) \tag{8.4}$$

where f_{max} = maximum stress at simply supported edges, ksi. This formula for determining the effective widths of stiffened, thin, flat elements was first used in the AISI "Light-Gage Steel Design Manual," 1949. Subsequent studies showed that the factor 0.475 was unnecessarily conservative and that 0.415 was more appropriate. It was used in AISI specifications between 1968 and 1980 to evaluate postbuckling strength of thin, flat elements.

Until 1986, all AISI specifications based strength of thin, flat elements stiffened along one edge on *buckling stress*. In contrast, *effective width* was used for thin, flat elements stiffened along both edges. This treatment changed after Pekoz in 1986 presented a unified approach using effective width as the basis of design for both stiffened and unstiffened elements and even for web elements subjected to stress gradients. Pekoz proposed the following three equations to generalize Eq. (8.4) with a factor of 0.415:

$$\lambda = [1.052(w/t)\sqrt{f/E}]/\sqrt{k} \tag{8.5}$$

where k = 4.00 for stiffened elements
= 0.43 for unstiffened elements
f = stress in the compression element of the section computed on the basis of the design width, in

w = flat width of the element exclusive of radii, in
t = base thickness of element, in
λ = a slenderness factor

The effective width is computed from

$$b = w \qquad \lambda \leq 0.673 \tag{8.6a}$$

$$b = \rho w \qquad \lambda > 0.673 \tag{8.6b}$$

where ρ is a reduction factor to be computed from

$$\rho = \frac{1 - 0.22/\lambda}{\lambda} \tag{8.7}$$

These equations were adopted in the AISI "Specification for the Design of Cold-Formed Steel Structural Members," 1986. See also Arts. 8.6 to 8.8.

8.6 UNSTIFFENED COLD-FORMED ELEMENTS SUBJECT TO LOCAL BUCKLING

As indicated in Art. 8.5, the effective width of an unstiffened element in compression may be computed from Eqs. (8.5) to (8.7). By definition, unstiffened elements have only one edge in the direction of compression stress supported by a web or stiffened element while the other edge has no auxiliary support (Fig. 8.6*a*). The coefficient k in Eq. (8.5) is 0.43 for such an element. When the flat-width-to-thickness ratio does not exceed $72/\sqrt{f}$, where f = compressive stress, ksi, an unstiffened element is fully effective and $b = w$. Generally, however, Eq. (8.5) becomes

$$\lambda = \frac{1.052(w/t)\sqrt{f/E}}{\sqrt{0.43}} = 0.0093(w/t)\sqrt{f} \tag{8.8}$$

where E = 29,500 ksi for steel. Substitution of λ in Eq. (8.7) yields $b/w = \rho$. Fig. (8.7*a*) shows a nest of curves for the relationship of b/t to w/t for unstiffened elements for w/t between 0 and 60 with f between 15 and 90 ksi.

In *beam deflection* determinations requiring use of the moment of inertia of the cross section, the allowable stress f is used to calculate the effective width of an unstiffened element in a cold-formed steel member loaded as a beam. However, in *beam strength* determinations requiring use of the section modulus of the cross section, $1.67f$ is the stress to be used in Eq. (8.8) to calculate the effective width of the unstiffened element and provide an adequate margin of safety.

In determination of safe loads for a cold-formed steel section used as a column, the effective width for an unstiffened element should be determined for a stress of $1.92f$, to ensure an adequate margin of safety.

8.7 STIFFENED COLD-FORMED ELEMENTS SUBJECT TO LOCAL BUCKLING

As indicated in 8.5, the effective width of a stiffened element in compression may be computed from Eqs. (8.5) to (8.7). By definition, stiffened elements have one

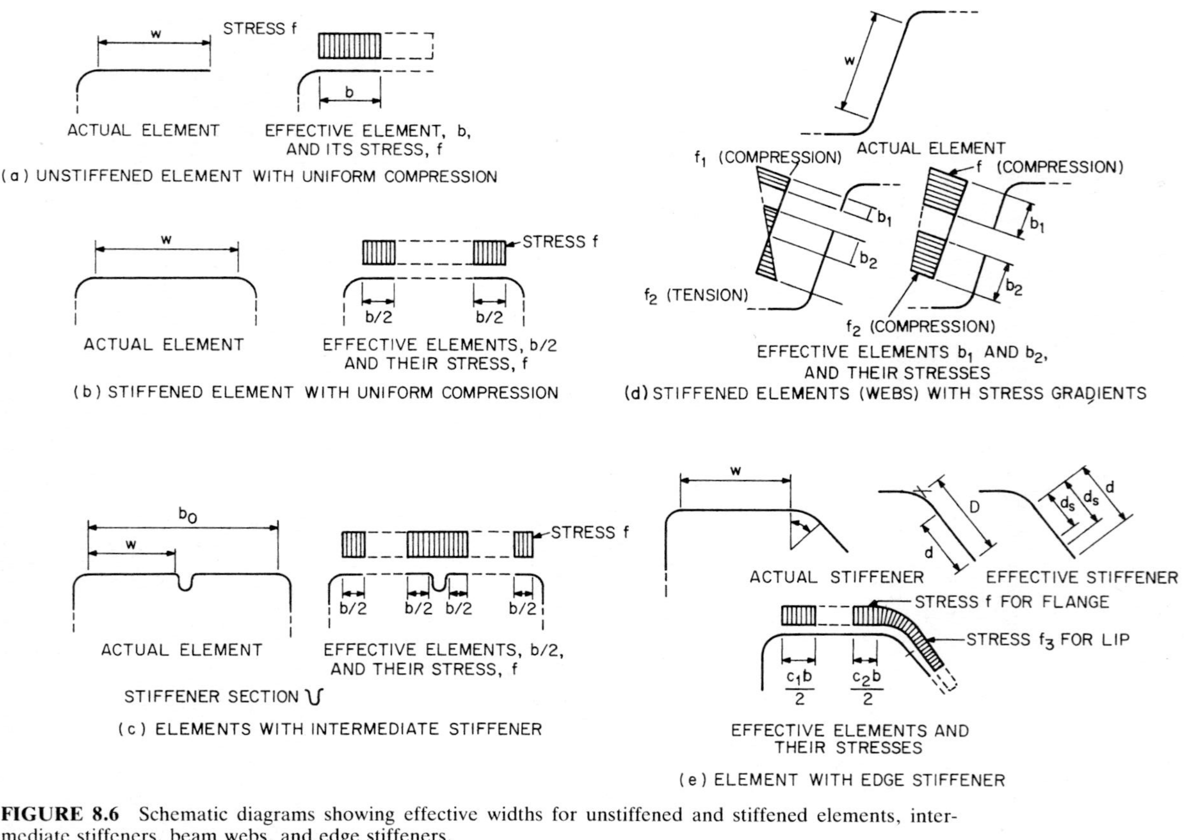

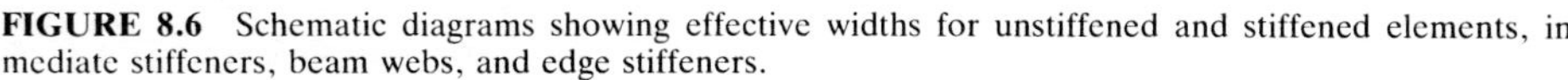

FIGURE 8.6 Schematic diagrams showing effective widths for unstiffened and stiffened elements, intermediate stiffeners, beam webs, and edge stiffeners.

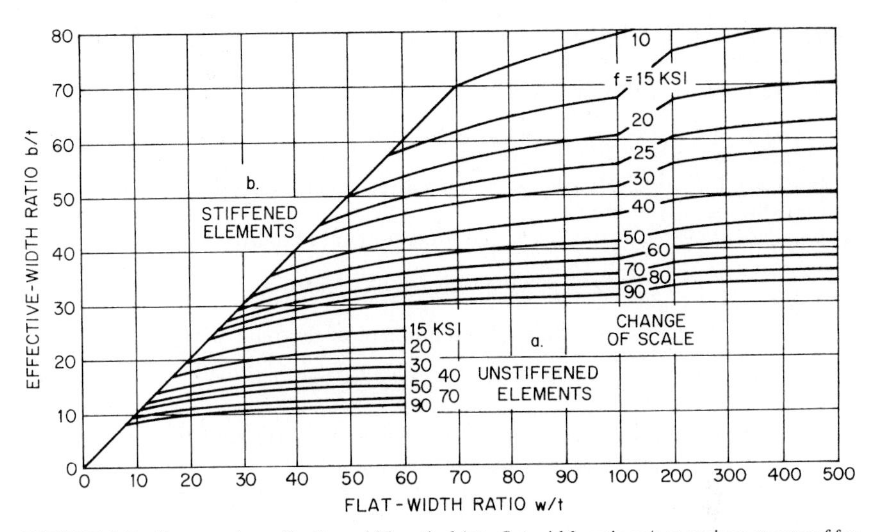

FIGURE 8.7 Curves relate effective-width ratio b/t to flat-width ratio w/t at various stresses f for (a) unstiffened elements and (b) stiffened elements.

edge in the direction of compression stress supported by a web or stiffened element and the other edge also supported by a qualified stiffener (Fig. 8.6b). The coefficient k in Eq. (8.5) is 4.00 for such an element. When the flat-width-to-thickness ratio does not exceed $220/\sqrt{f}$, where f = compressive stress, ksi, computed on the basis of the design width, a stiffened element is fully effective and $b = w$. Generally, however, Eq. (8.5) becomes

$$\lambda = \frac{1.052(w/t)\sqrt{f/E}}{\sqrt{4}} = 0.0031(w/t)\sqrt{f} \qquad (8.9)$$

where E = 29,500 ksi for steel. Substitution of λ in Eq. (8.7) yields $b/w = \rho$. Moreover, when $\lambda \leq 0.673$, $b = w$ and when $\lambda > 0.673$, $b = \rho w$. Figure 8.7b shows a nest of curves for the relationship of b/t to w/t for stiffened elements for w/t between 0 and 500 with f between 10 and 90 ksi.

In *beam deflection* determinations requiring use of the moment of inertia of the cross section, the allowable stress f is used to calculate the effective width of a stiffened element in a cold-formed steel member loaded as a beam. However, in *beam strength* determinations requiring use of the section modulus of the cross section, $1.67f$ is the stress to be used in Eq. (8.9) to calculate the effective width of the stiffened element and provide a margin of safety.

In determination of the safe loads for a cold-formed steel section used as a column, effective width for a stiffened element must be determined for a stress of $1.92f$, to ensure an adequate margin of safety.

Since effective widths are proportional to $\sqrt{k}$, the effective width of a stiffened element is $\sqrt{4.00/0.43} = 3.05$ times as large as that of an unstiffened element at applicable combinations of f and w/t. Thus, stiffened elements offer greater strength and economy.

Single Intermediate Stiffener. For uniformly compressed stiffened elements with a single intermediate stiffener, as shown in Fig. 8.6*c*, calculations for required moment of inertia I_a of the stiffener are based on a parameter S.

$$S = 1.28\sqrt{E/f} \tag{8.10}$$

For Case I, $S \geq b_o/t$, where b_o = flat width, in, including the stiffener. $I_a = 0$ and no stiffener is required.

For Case II, $S < b_o/t < 3S$. The required moment of inertia is determined from

$$I_a/t^4 = [50(b_o/t)/S] - 50 \tag{8.11a}$$

For Case III, $b_o/t \geq S$. The required moment of inertia is determined from

$$I_a/t^4 = [128(b_o/t)/S] - 285 \tag{8.11b}$$

Webs Subjected to Stress Gradients. Effective widths also are applicable to stiffened elements subject to stress gradients in compression, such as in the webs of beams. Figure 8.6*d* illustrates the application. The effective widths b_1 and b_2 are determined with the use of the following equations:

$$b_1 = b_e/(3 - \psi) \tag{8.12}$$

where $\psi = f_2/f_1$
f_1 = stress, ksi, in compression flange (Fig. 8.6*d*)
f_2 = stress, ksi, in opposite flange (Fig. 8.6*d*)
b_e = effective width b determined from Eqs. (8.5) to (8.7) with f_1 substituted for f and with k calculated from Eq. (8.14)

Stress f_2 may be tensile (negative) or compressive (positive). When both f_1 and f_2 are compressive, $f_1 \geq f_2$.

$$b_2 = \tfrac{1}{2}b_e \qquad \psi \leq -0.236 \tag{8.13a}$$

where $b_1 + b_2$ should not exceed the depth of the compression portion of the web calculated for the effective cross section.

$$b_2 = b_e - b_1 \qquad \psi > -0.236 \tag{8.13b}$$

$$k = 4 + 2(1 - \psi)^3 + 2(1 - \psi) \tag{8.14}$$

Uniformly Compressed Elements with Edge Stiffener. While a slanted lip, as depicted in Fig. 8.6*e*, may be used as an edge stiffener for a cold-formed steel section, calculation of stresses for such a section is complex. (See AISI "Specification for the Design of Cold-Formed Steel Structural Members.") Consequently, the following is primarily applicable to 90° lips.

Calculation of the required moment of inertia, I_a, falls into one of three cases:

For Case I, $w/t \leq S/3$. $b = w$, where b is the effective width, and no edge support is needed. S is defined by Eq. (8.10) and is the maximum w/t for full effectiveness of the flat width without auxiliary support.

For Case II, $S/3 < w/t < S$. The required moment of inertia of the lip is determined from

$$I_a/t^4 = 399\{[(w/t)/S] - 0.33\}^3 \tag{8.15}$$

When $S/3$ is substituted for w/t in Eq. (8.15), $I_a = 0$ and no support is needed at the edge for which a lip is being considered (see Case I). When $w/t = S$, a stiffening lip would be required to have a depth-thickness ratio d/t of 11.3. The maximum stress in a lip with this value of d/t, however, could be only 40.6 ksi, which corresponds to a maximum allowable stress of 24.3 ksi in bending and 21.1 ksi in compression, with safety factors of 1.67 and 1.92, respectively.

For Case III, $w/t \geq S$. The required moment of inertia of the edge stiffener is determined from

$$I_a/t^4 = 115[(w/t)/S] + 5 \tag{8.16}$$

Edge support required will be beyond the capability of a simple lip and have to be of the nature of a web, a stiffening element, or a multielement shape. For example, let $w/t = 500$, $E = 29{,}500$ ksi, and $f = 50$ ksi. Substitution in Eq. (8.16) yields

$$I_a/t^4 = 115 \times 500/1.28\sqrt{29{,}500/50} + 5 = 1854$$

For a slanted lip as shown in Fig. 8.6*e*, the moment of inertia provided is

$$I = (d^3t/12)\sin^2 \theta \tag{8.17}$$

where d = flat width of lip, in, and θ = angle between normals to stiffened element and its lip (90° for a right angle lip). (See Fig. 8.6*d*.)

For a 90° lip, $I = d^3/12$. Hence for $I_a = 1854t^4$, $d/t = 28.1$. Such a wide lip would itself be unstable at stresses exceeding $f = 6.6$ ksi and therefore would be completely impractical as a stiffener for a wide element. Only a web, a stiffened element, or a multielement shape could meet the need.

("Cold-Formed Steel Design Manual," American Iron and Steel Institute, 1101 17th St., NW, Washington, DC 20036.)

8.8 APPLICATION OF EFFECTIVE WIDTHS

The curves of Fig. 8.7 were plotted from values of Eqs. (8.8) and (8.9). They may be used to determine b/t for different values of w/t and unit stresses f. The effective width b is dependent on the actual stress f, which in turn is determined by reduced-section properties that are a function of effective width. Employment of successive approximations consequently may be necessary in using these equations and curves. A direct solution for the correct value of b/t can be obtained from the formulas, however, when f is known or is held to a specified maximum allowable value (20 ksi for $F_y = 33$ ksi, for example). This is true, though, only when compression controls; for example, for symmetrical channels and Z and I sections used as flexural members bending about their major axis (Fig. 8.1*e*, *f*, *k* and *n*) or for unsymmetrical channels and Z and I sections with neutral axis closer to the tension flange than to the compression flange. If w/t of the compression flange does not exceed about 60, little error will result in assuming that $f = 0.60 \times 33 = 20$ ksi for $F_y = 33$ ksi. This is so even though the neutral axis is above the geometric centerline. For wide, inverted, pan-shaped sections, such as deck and panel sections, a somewhat more accurate determination using successive approximations will prove necessary.

For computation of moment of inertia for deflection or stiffness calculations, properties of the full unreduced section can be used without significant error when w/t of the compression elements does not exceed 60. For greater accuracy, use Eqs. (8.8) and (8.9) to obtain appropriate effective widths.

Example. As an example of effective-width determination, consider the hat section of Fig. 8.8. The section is to be made of steel with a specified minimum yield strength $F_y = 33$ ksi. It is to be used as a simply supported beam with the top flange in compression, at a basic working stress of 20 ksi. Safe load-carrying capacity is to be computed; so $f = 20 \times 1.67 = 33$ ksi is used to obtain b/t.

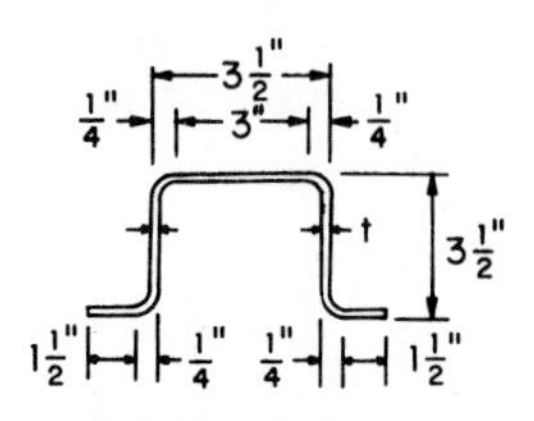

FIGURE 8.8 Hat section.

The top flange is a stiffened compression element with 3-in flat width. If the thickness is 1/16 in, then the flat-width-thickness ratio (w/t) is 48 (greater than $w/t = 220/\sqrt{33} = 38$), stiffening is required, and Eq. (8.9) applies. For $w/t = 48$ and $f = 33$ ksi, Eq. (8.9) gives $b/t = 41$. Thus, with $b/w = 41/48$, only 85% of the top-flange flat width can be considered effective. The neutral axis will lie below the horizontal center line, and compression will control. In this case, the assumption that $f = 20$ ksi, made at the start, controls maximum stress, and b/t can be determined directly from Eq. (8.9) without successive approximations. However, for a wide hat section in which the horizontal axis is nearer the compression than the tension flange, stress in the tension flange controls, and successive approximations are required for the determination of unit stress and effective width of the compression flange.

("Cold-Formed Steel Design Manual," American Iron and Steel Institute, 1101 17th St., NW, Washington, DC 20036.)

8.9 MAXIMUM FLAT-WIDTH RATIOS OF COLD-FORMED SHAPES

When the flat-width-thickness ratio (w/t) exceeds about 30 for an unstiffened element and about 250 for a stiffened element, noticeable buckling of the element may develop at relatively low stresses. Present practice is to permit buckles to develop in the sheet and to take advantage of what is known as post-buckling strength of the section. The effective-width formulas, Eqs. (8.5) to (8.7), are based on this practice. To avoid intolerable deformations, however, w/t, disregarding intermediate stiffeners and based on the actual thickness t of the element, should not exceed the following:

Stiffened compression element having one longitudinal edge connected to a web or flange, the other to a simple right-angle lip	60
Stiffened compression element having both edges stiffened by stiffeners other than a simple right-angle lip	250
Stiffened compression element with both longitudinal edges connected to a web or flange element, such as in a hat, U, or box-type section	500
Unstiffened compression element	60

8.10 UNIT STRESSES FOR COLD-FORMED STEEL

For sheet and strip of A611, Grade C steel with a specified minimum yield strength $F_y = 33$ ksi, use a basic allowable stress $f = 20$ ksi in tension and bending. For other strengths of steels, f is determined by taking 60% of the specified minimum yield strength F_y. (This procedure implies a safety factor of 1.67.) However, an increase of 33⅓% in allowable stress is customary for combined wind or earthquake forces with other loads.

8.11 LATERALLY UNSUPPORTED COLD-FORMED BEAMS

If cold-formed steel sections are not laterally supported at frequent intervals, the allowable unit stress must be reduced to avoid failure from lateral instability. The amount of reduction depends on the shape and proportions of the section and the spacing of lateral supports. (See AISI "Specification for the Design of Cold-Formed Steel Structural Members.")

Because of the torsional flexibility of lightweight channel and Z sections, their use as beams without close lateral support is not recommended. When a compression flange is fully connected to a deck or sheathing material, the flange is considered braced for its full length and bracing of the other flange may not be needed to prevent buckling of the beam. This depends on the collateral material and its connections, dimensions of the member, and the span.

When laterally unsupported beams must be used, or where lateral buckling of a flexural member is likely to occur, consideration should be given to the use of relatively bulky sections that have two webs, such as hat or box sections (Fig. 8.1*o*, *p*, and *u*).

8.12 ALLOWABLE SHEAR STRESS IN WEBS

The shear V, kips, at any section should not exceed the allowable shear V_a, kips, calculated as follows:

For $h/t \leq 1.38\sqrt{k_v E/F_y}$,

$$V_a = 0.38t^2\sqrt{k_v E F_y} \leq 0.4F_y ht \tag{8.18}$$

For $h/t > 1.38\sqrt{k_v E/F_y}$,

$$V_a = 0.53k_v Et^3/h \tag{8.19}$$

where t = web thickness, in

h = depth of the flat portion of the web measured along the plane of the web, in

E = modulus of elasticity of the steel = 29,500 ksi

k_v = shear buckling coefficient = 5.34 for unreinforced webs for which $(h/t)_{max}$ does not exceed 200

F_y = specified yield stress of the steel, ksi

For design of reinforced webs, especially when h/t exceeds 200, see AISI "Specification for the Design of Cold-Formed Steel Structural Members."

For a web consisting of two or more sheets, each sheet should be considered as a separate element carrying its share of the shear.

For beams with unreinforced webs, the moment M and shear V should satisfy the following interaction equation:

$$(M/M_{axo})^2 + (V/V_a)^2 \leq 1.0 \tag{8.20}$$

where M_{axo} = allowable moment about the centroidal axis, in-kips, when bending alone is present
V_a = allowable shear, kips, when shear alone exists
M = applied bending moment, in-kips
V = actual shear, kips

8.13 CONCENTRICALLY LOADED COMPRESSION MEMBERS

The following formulas apply to members in which the resultant of all loads acting on a member is an axial load passing through the centroid of the effective section (calculated at the nominal buckling stress F_n, ksi). The axial load should not exceed P_a, kips, calculated from

$$P_a = P_n/\Omega_c \tag{8.21}$$

Except for channels, single angles, and Z shapes when they have unstiffened flanges,

$$P_n = A_e F_n \tag{8.22}$$

where P_n = ultimate compression load, kips
Ω_c = factor of safety for axial compression, 1.92
A_e = effective area at stress F_n, in^2

The magnitude of F_n is a function of the elastic buckling stress F_e, ksi:

$$F_n = F_y[1 - (F_y/4F_e)] \qquad F_e > F_y/2 \tag{8.23a}$$

$$F_n = F_e \qquad F_e \leq F_y/2 \tag{8.23b}$$

where F_y is the yield stress of the steel, ksi.

F_e is the least of the elastic flexural, torsional, or torsional-flexural buckling stresses. For elastic flexural behavior,

$$F_e = \frac{\pi^2 E}{(KL/r)^2} \tag{8.24}$$

where K = effective length factor
L = unbraced length of member, in
r = radius of gyration of full, unreduced cross section, in
E = modulus of elasticity of the steel, ksi

When F_e is determined for fully effective sections having thickness equal to or greater than 0.09 in and $F_e > F_y/2$:

$$\Omega_e = [5/3 + (3R/8) - (R^3/8)] \tag{8.25}$$

where $R = \sqrt{F_y/2F_e}$.

When channels, Z shapes, and single-angle sections have unstiffened flanges, P_n should be taken as the smaller of P_n as calculated from Eq. (8.22) and P_n calculated from

$$P_n = \frac{A\pi^2 E}{25.7(w/t)^2} \tag{8.26}$$

where A = area of full, unreduced cross section of member, in^2
w = flat width of the unstiffened element, in
t = thickness of the unstiffened element, in

Moreover, angle sections should be designed for the applied axial load P acting simultaneously with a moment equal to $PL/1000$ applied about the minor principal axis and causing compression in the tips of the angle legs.

The slenderness ratio KL/r of all compression members preferably should not exceed 200, except that during construction only, KL/r preferably should not exceed 300.

Column design curves for flexural buckling of cold-formed steel shapes are shown in Fig. 8.9. For treatment of sections that may be subject to torsional or torsional-flexural buckling, refer to AISI "Specification for the Design of Cold-Formed Steel Structural Members," American Iron and Steel Institute, 1101 17th St., NW, Washington, DC 20036.

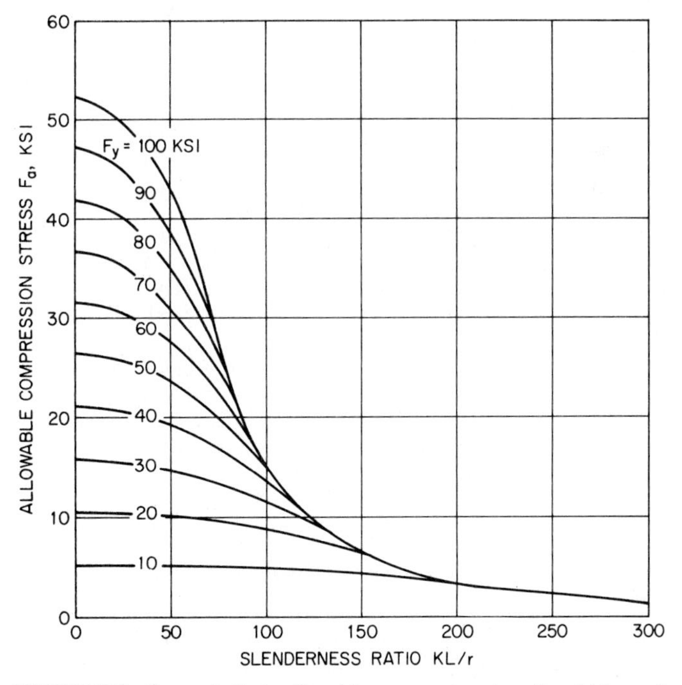

FIGURE 8.9 Curves indicate allowable compression stress in cold-formed steel column for various yield strengths F_y and slenderness ratios KL/r.

8.14 COMBINED AXIAL AND BENDING STRESSES

Combined axial and bending stresses in cold-formed sections can be handled exactly the same way as for structural steel. The interaction criterion to be used is given in the AISI "Specification for the Design of Cold-Formed Structural Members."

JOINING OF COLD-FORMED STEEL

Cold-formed members may be assembled into desired shapes or spliced or joined to other members with any of various types of fasteners. For the purpose, welds, bolts, and screws are most frequently used, but other types, such as rivets, studs, and metal stitching, can also be used.

8.15 WELDING OF COLD-FORMED STEEL

Electric currents are generally used in either of two ways to join cold-formed steel components, with electric-arc welding or resistance welding. The former method is described in Art. 8.16 and the latter in Art. 8.17.

Welding offers important advantages to fabricators and erectors in joining steel structural components. Welded joints make possible continuous structures, with economy and speed in fabrication; 100% joint efficiencies are possible.

Conversion to welding of joints initially designed for mechanical fasteners is poor practice. Joints should be specifically designed for welding, to take full advantage of possible savings. Important considerations include the following: The overall assembly should be weldable; welds should be located where notch effects are minimal; the final appearance should not suffer from unsightly welds; and welding should not be expected to correct poor fit-up.

Steels bearing protective coatings require special consideration. Surfaces precoated with paint or plastic are damaged by welding. Coatings may adversely affect weld quality. Metal-coated steels, such as galvanized (zinc-coated), aluminized, and terne-coated (lead-tin alloy), however, may be successfully welded using procedures tailored for the steel and its coating.

Generally, steel to be welded should be clean and free of contaminants such as oil, grease, paints, and scale. Paint should be applied only after the welding process. (See "Welding Handbook," American Welding Society, 550 NW LeJeune Rd., Miami, FL 33135 and O. W. Blodgett, "Design of Weldments," James F. Lincoln Welding Foundation, Cleveland, OH 44117.)

8.16 ARC WELDING OF COLD-FORMED STEEL

Arc welding may be done in the shop or in the field. The basic sheet-steel weld types are shown in Fig. 8.10. Factors favoring arc welding are portability and versatility of equipment as well as freedom in joint design. Only one side of a joint need be accessible, and overlap of parts is not required if joint fit-up is good.

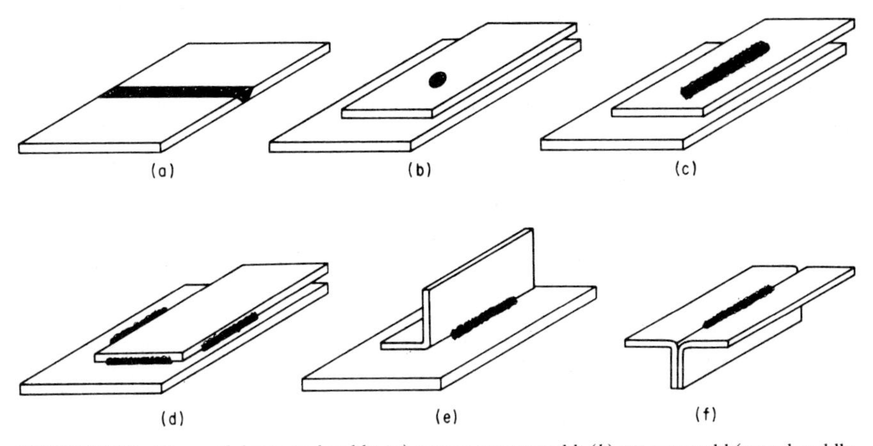

FIGURE 8.10 Types of sheet-steel welds: (*a*) square-groove weld; (*b*) arc spot weld (round puddle weld); (*c*) arc seam weld (oblong puddle weld); (*d*) fillet welds; (*e*) flare bevel-groove weld; (*f*) flare V-groove weld.

8.16.1 Helpful Hints for Welding

Distortion may occur with lightweight steel weldments, but it can be minimized by avoiding overwelding. Weld sizes should be matched with service requirements.

Always design welded joints to minimize shrinking, warping, and twisting. Jigs and fixtures for holding lightweight work during welding should be used to control distortion. Directions and amounts of distortion can be predicted and sometimes counteracted by preangling the parts. Discrete selection of weld sequence can also be used to control distortion.

Groove welds (made by butting sheet edges together, Fig. 8.10*a*) can be designed for 100% joint efficiency. Calculation of design stress is usually unnecessary if the weld penetrates 100% of the section.

Stresses in fillet welds should be considered as shear on the throat for any direction of applied stress. The dimension of the throat is calculated as 0.707 times the length of the shorter leg of the weld. For example, a 12-in-long, ¼-in-fillet weld has a leg dimension of ¼ in, a throat of 0.177 in, and an equivalent area of 2.12 in^2. For all grades of steel, fillet and plug welds should be proportioned so that the unit stresses do not exceed 13.2 ksi in shear on the throat.

8.16.2 Types of Arc Welding

Shielded metal arc welding, also called manual stick electrode, is the most common arc-welding process because of its versatility. The method, however, requires skilled operators. The welds can be made in any position, but vertical and overhead welding should be avoided when possible.

Gas metal arc welding uses special equipment to feed a continuous spool of bare or flux-cored wire into the arc. A shielding gas such as argon or carbon dioxide is used to protect the arc zone from the contaminating effects of the atmosphere. The process is relatively fast, and close control can be maintained over the deposit. The process is not applicable to materials ⅓₂ in thick but is extensively used for thicker steels.

Gas tungsten arc welding operates by maintaining an arc between a nonconsumable tungsten electrode and the work. Filler metal may or may not be added. Close control over the weld can be maintained. This process is not widely used for high-production fabrication, except in specialized applications, because of higher cost.

One form of spot welding is an adaptation of gas metal arc welding wherein a special welding torch and automatic timer are employed. The welding torch is positioned on the work and a weld is deposited by burning through the top layer of the lap joint. The filler wire provides sufficient metal to fill the hole, thereby fusing together the two parts. Access to only one side of the joint is necessary. Field welding by unskilled operators is feasible. This makes the process advantageous.

Another form of arc spot welding utilizes gas tungsten arc welding. The heat of the arc melts a spot through one of the sheets and partly through the second. When the arc is cut off, the pieces fuse. No filler metal is added.

Design of arc-welded joints of sheet steel is treated in the American Welding Society "Specification for Welding Sheet Steel in Structures," AWS D1.3.

8.16.3 Groove Welds in Butt Joints

The maximum load for a groove weld in a butt joint, welded from one or both sides, should be determined on the basis of the lower-strength base steel in the connection, provided that an effective throat equal to or greater than the thickness of the material is consistently obtained.

8.16.4 Arc Spot Welds

Arc spot welds (Fig. 8.10*b*), also known as puddle welds, are permitted for welding sheet steel to thicker supporting members in the flat position. Such welds, which result when coalescence proceeds from the surface of one sheet into one or more other sheets of a lapped joint without formation of a hole, should not be made on steel where the thinnest connected part is more than 0.15 in thick, or through a combination of steel sheets having a total thickness exceeding 0.15 in. Arc spot welds are specified by minimum effective diameter of fused area, d_e. Minimum effective allowable diameter is ⅜ in. The nominal shear load P_n, kips, on each arc spot weld between sheet or between sheets and a supporting member should not exceed the smaller of the values given by Eqs. (8.27) to (8.30).

$$P_n = 0.625 d_e^2 F_{xx} \tag{8.27}$$

For $d_a/t \leq 0.815\sqrt{E/F_u}$,

$$P_n = 2.20 t d_a F_u \tag{8.28}$$

For $0.815\sqrt{E/F_u} < d_a/t \leq 1.397\sqrt{E/F_u}$,

$$P_n = 0.280 \left[1 + \frac{5.59\sqrt{E/F_u}}{d_a\sqrt{F_u}} \right] t d_a F_u \tag{8.29}$$

For $d_a/t \geq 1.397\sqrt{E/F_u}$,

$$P_n = 1.40 t d_a F_u \tag{8.30}$$

where d_a = average diameter, in, of the arc spot weld at midthickness of sheet
= $d - t$ for a single sheet
= $d - 2t$ for multiple sheets (not more than four lapped sheets over a supporting member)
d = visible diameter of outer surface of arc spot weld, in
d_e = effective diameter of fused area, in
= $0.7d - 1.5t \leq 0.55d$
t = total combined base steel thickness, in (exclusive of coatings) of sheets involved in shear transfer
F_{xx} = stress-level designation in AWS electrode classification, ksi
F_u = tensile strength of the base steel as specified, ksi

The distance measured in the line of force from the centerline of a weld to the nearest edge of an adjacent weld or to the end of the connected part toward which the force is directed should be at least $e_{\min}$, in, as given by

$$e_{\min} = e\Omega_e \tag{8.31}$$

where $e = P/(F_u t)$
Ω_e = factor of safety for sheet tearing
= 2.0 when $F_u/F_{sy} \geq 1.15$
= 2.22 when $F_u/f_{sy} < 1.15$
P = force transmitted by weld, kips
F_{sy} = yield strength of sheet steel, ksi, as specified
t = thickness of thinnest connected sheet, in

In addition, the distance from the centerline of any weld to the end or boundary of the connected member should be at least $1.5d$. In no case should the clear distance between welds and the end of the member be less than d.

The nominal tension load P_n, kips, on an arc spot weld between a sheet and a supporting member should not exceed

$$P_n = 0.7td_aF_u \tag{8.32}$$

The following limitations also apply: $e_{\min} \geq d$, $F_{xx} \geq 60$ ksi, $F_u \leq 60$ ksi, and $t \geq$ 0.028 in.

As for arc spot welds (Art. 8.16.3), if measurements indicate that a given weld procedure will consistently give larger diameters d_e or d_a, as applicable, the larger diameter may be used to calculate the maximum allowable load, if that procedure will be used.

8.16.5 Arc Seam Welds

These are basically the same as arc spot welds but are made linearly without slots in the sheets (Fig. 8.10*c*). Arc seam welds apply to the following types of joints:

1. Sheet to a thicker supporting member in the flat position
2. Sheet to sheet in the horizontal or flat position

The shear load P_n, kips, on an arc seam weld should not exceed the values given by either Eq. (8.33) or (8.34).

$$P_n = 2.5F_{xx}(d_e^2/4 + d_eL/3) \tag{8.33}$$

$$P_n = 2.5tF_u(0.25L + 0.96d_a) \tag{8.34}$$

where d_a = average width, in, of arc seam weld
= $d - t$ for a single sheet
= $d - 2t$ for a double sheet
d = width, in, of arc seam weld
L = length, in, of weld not including the circular ends (in computations, L should not exceed $3d$)
d_e = effective width, in, of weld at fused surfaces
= $0.7d - 1.5t$

F_u and F_{xx} are defined as for arc spot welds (Art. 8.16.3). Minimum edge distances also are defined as for arc spot welds.

If measurements indicate that a given weld procedure will consistently give a larger effective width d_e or larger average diameter d_a, as applicable, these values may be used to calculate the maximum allowable load on an arc seam weld, if that welding procedure will actually be used.

8.16.6 Fillet Welds

These are made along the edges of sheets in lapped or T joints (Fig. 8.10*d*). The fillet welds may be made in any position and either sheet to sheet or sheet to thicker steel member.

The shear load P_n, kips, on a fillet weld in lapped or T joints should not exceed the value of P_n computed from Eqs. (8.35) to (8.38).

For longitudinal loading along the weld:

$$P_n = (1 - 0.01L/t)tLF_u \qquad L/T < 25 \tag{8.35}$$

$$P_n = 0.75tLF_u \qquad L/t \geq 25 \tag{8.36}$$

where t = smaller thickness of sheets being welded, in
L = length, in, of the fillet weld
F_u = specified tensile strength of base steel, ksi

For loading transverse to the weld:

$$P_n = tLF_u \tag{8.37}$$

For $t > 0.15$ in,

$$P_n = 0.75t_wLF_{xx} \tag{8.38}$$

where F_{xx} = stress-level designation in AWS electrode classification, ksi
t_w = effective throat of weld, in
= 0.707 times the smaller of the weld-leg lengths

8.16.7 Flare Groove Welds

These are made on the outsides of curved edges of bends in cold-formed shapes (Fig. 8.10*e* and *f*). The welds may be made in any position to join:

1. Sheet to sheet for flare V-groove welds
2. Sheet to sheet for flare bevel-groove welds
3. Sheet to thicker steel member for flare bevel-groove welds.

The shear load P_n, kips, on a weld is governed by the thickness t, in, of the sheet adjacent to the weld. The load should not exceed the values of P_n given by Eqs. (8.39) to (8.42).

For flare bevel-groove welds subject to transverse loading,

$$P_n = 0.833tLF_u \tag{8.39}$$

where L = length, in, of the weld and F_u = specified tensile strength, ksi, of the base steel.

For flare V-groove welds, subject to longitudinal loading,

$$P_n = 0.75tLF_u \qquad t \leq t_w < 2t \text{ or } h < L \tag{8.40}$$

where t_w = effective throat of the weld, in and h = lip height, in.

$$P_n = 1.50tLF_u \qquad t_w \geq 2t \text{ and } h \geq L \tag{8.41}$$

In addition, if $t > 0.15$ in,

$$P_n = 0.75t_wLF_u \tag{8.42}$$

where F_{xx} = stress-level designation in AWS electrode designation, ksi.

8.17 RESISTANCE WELDING OF COLD-FORMED STEEL

Resistance welding comprises a group of welding processes wherein coalescence is produced by the heat obtained from resistance of the work to flow of electric current in a circuit of which the work is part and by the application of pressure. Because of the size of the equipment required, resistance welding is essentially a shop process. Speed and low cost are factors favoring its selection.

Almost all resistance-welding processes require a lap-type joint. The amount of contacting overlap varies from ⅜ to 1 in, depending on sheet thickness. Access to both sides of the joint is normally required. Adequate clearance for electrodes and welder arms must be provided.

8.17.1 Spot Welding

Spot welding is the most common resistance-welding process. The weld is formed at the interface between the pieces being joined and consists of a cast-steel nugget. The nugget has a diameter about equal to that of the electrode face and should penetrate about 60 to 80% of each sheet thickness.

For structural design purposes, spot welds can be treated the same way as bolts, except that no reduction in net section due to holes need be made. Table 8.5 gives the essential information for design purposes for uncoated steel based on "Recommended Practices for Resistance Welding," American Welding Society, 1966. The maximum allowable loads per weld for design purposes are based on shear strengths of welds observed in tests after application of a safety factor of 2.5 to the lower bounds of data. Note that the thickest steel for plain spot welding is ⅛ in. Thicker material can be resistance welded by projection or by pulsation methods if high-capacity spot welders for material thicker than ⅛ in are not available.

TABLE 8.5 Design Data for Spot and Projection Welding of Low-Carbon Sheet Steel

Thickness t of thinnest outside piece, in	Min OD of electrode, D, in 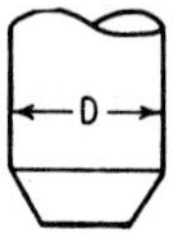	Min contacting overlap, in 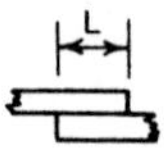	Min weld spacing c to c, in	Approx dia of fused zone, in 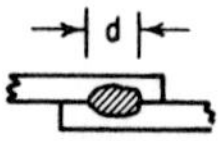	Min shear strength per weld, lb	Dia of projection, D, in 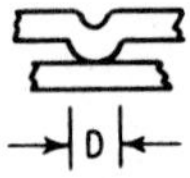
Spot welding						
0.021	3⁄8	7⁄16	3⁄8	0.13	320	
0.031	3⁄8	7⁄16	1⁄2	0.16	570	
0.040	1⁄2	1⁄2	3⁄4	0.19	920	
0.050	1⁄2	9⁄16	7⁄8	0.22	1,350	
0.062	1⁄2	5⁄8	1	0.25	1,850	
0.078	5⁄8	11⁄16	1 1⁄4	0.29	2,700	
0.094	5⁄8	3⁄4	1 1⁄2	0.31	3,450	
0.109	5⁄8	13⁄16	1 5⁄8	0.32	4,150	
0.125	7⁄8	7⁄8	1 3⁄4	0.33	5,000	
Projection welding						
0.125		11⁄16	9⁄16	0.338	4,800	0.281
0.140		3⁄4	5⁄8	7⁄16	6,000	0.312
0.156		13⁄16	11⁄16	1⁄2	7,500	0.343
0.171		7⁄8	3⁄4	9⁄16	8,500	0.375
0.187		15⁄16	13⁄16	9⁄16	10,000	0.406

8.17.2 Projection Welding

This is a form of spot welding in which the effects of current and pressure are intensified by concentrating them in small areas of projections embossed in the sheet to be welded. Thus, satisfactory resistance welds can be made on thicker steel using spot welders ordinarily limited to thinner stocks.

8.17.3 Pulsation Welding

Pulsation, or multiple-impulse, welding is the making of spot welds with more than one impulse of current, a technique that makes some spot welders useful for thicker materials. The tradeoffs influencing choice between projection welding and impulse welding involve the work being produced, volume of output, and equipment available.

8.17.4 Recommended Practices for Spot Welding

The spot welding of higher-strength steels than those contemplated under Table 8.5 may require special welding conditions to develop the higher shear strengths of which the higher-strength steels are capable.

All steels used for spot welding should be free of scale; therefore, either hot-rolled and pickled or cold-rolled steels are usually specified.

Steels containing more than 0.15% carbon are not as readily spot welded as lower-carbon steels, unless special techniques are used to ensure ductile welds. High-carbon steels such as ASTM A446, Grade D, which can have a carbon content as high as 0.40% by heat analysis, are not recommended for resistance welding. Designers should resort to other means of joining such steels.

Maintenance of sufficient overlaps in detailing spot-welded joints is important to ensure consistent weld strengths and minimum distortions at joints. Minimum weld spacings specified in Table 8.5 should be observed, or shunting to previously made adjacent welds may reduce the electric current to a level below that needed for welds being made. Also, the joint design should provide sufficient clearance between electrodes and work to prevent short-circuiting of current needed to make satisfactory spot welds. For further information on spot welding of coated steels, see "Recommended Practices for Resistance Welding of Coated Low-Carbon Steel," American Welding Society, 550 N.W. Lejeune Rd., Miami, FL 33135.

The nominal shear strength per spot, is a function of the thickness of the thinnest outside sheet, Table 8.6 lists spot shear strengths for sheets with thicknesses from 0.010 to 0.250 in, as recommended for design by the American Iron and Steel Institute.

TABLE 8.6 Nominal Shear Strength per Spot for Low-Carbon Sheet Steel

Thickness of thinnest outside sheet, in	Nominal shear strength per spot, kips	Thickness of thinnest outside sheet, in	Nominal shear strength per spot, kips
0.010	0.13	0.080	3.33
0.020	0.48	0.090	4.00
0.030	1.00	0.100	4.99
0.040	1.42	0.110	6.07
0.050	1.65	0.125	7.29
0.060	2.28	0.190	10.16
0.070	2.83	0.250	15.00

8.18 BOLTING OF COLD-FORMED STEEL MEMBERS

Bolting is convenient in cold-formed construction. Bolts, nuts, and washers should generally conform to the requirements of the ASTM specifications listed in Table 8.7. The maximum sizes of bolt holes are given in Table 8.8. Standard holes should be used in bolted connections when possible. If slotted holes are used, the length

TABLE 8.7 ASTM Bolt, Nut, and Washer Steels

A194	Carbon and alloy steel nuts for high-pressure and high-temperature service
A307	Type A carbon-steel, externally and internally threaded standard fasteners
A325	High-strength bolts for structural steel joints
A354	Grade BD quenched and tempered alloy-steel bolts, studs, and other externally threaded fasteners (for bolt diameter less than ½ in)
A449	Quenched and tempered steel bolts and studs (for bolt diameter less than ½ in)
A490	Quenched and tempered alloy-steel bolts for structural steel joints
A563	Carbon and alloy steel nuts
F436	Hardened steel washers
F844	Washers, steel, plain (flat), unhardened for general use
F959	Compressible washer-type, direct-tension indicators for use with structural fasteners

TABLE 8.8 Maximum Size of Bolt Holes, in

Nominal bolt diameter, in	Standard hole diameter d, in	Oversized hole diameter d, in	Short-slotted hole, in	Long-slotted hole, in
Less than ½	$d + 1/32$	$d + 1/16$	$(d + 1/32)$ by $(d + 1/4)$	$(d + 1/32)$ by $(2\frac{1}{2}d)$
½ or larger	$d + 1/16$	$d + 1/8$	$(d + 1/16)$ by $(d + 1/4)$	$(d + 1/16)$ by $(2\frac{1}{2}d)$

of the holes should be normal to the direction of the shear load. Washers should be installed atop oversized or slotted holes.

8.18.1 Spacing of Bolts

The distance e, in, measured in the direction of applied force, from the center of a standard hole to the nearest edge of an adjacent hole or to the end of the connected part toward which the force is directed should not be less than $e_{\min}$.

$$e_{\min} = e\Omega_e \tag{8.43}$$

$$e = P/F_u t \tag{8.44}$$

where Ω_e = safety factor for sheet tearing
= 2.00 when $F_u/F_{sy} \geq 1.15$
= 2.22 when $F_u/F_{sy} < 1.15$
P = force, kips, transmitted by a bolt
t = thickness, in, of thinnest connected part
F_u = tensile strength, ksi, of connected part
F_{sy} = yield strength, ksi, of connected part

In addition, the minimum distance between centers of bolt holes should provide sufficient clearance for bolt heads, nuts, washers, and wrench but be at least 3 times the nominal diameter d, in. The distance from the center of any standard hole to the end or boundary of the connecting member should be at least $1\frac{1}{2}d$.

8.18.2 Bolted Cold-Formed Members in Tension

Calculation of the allowable tension force on the net section of a bolted connection depends on the thickness t, in, of the thinnest connected part. When t exceeds $\frac{3}{16}$ in, design of the connection is governed by the AISC "Specification for the Design, Fabrication and Erection of Structural Steel for Buildings," American Institute of Steel Construction, One East Wacker Drive, Chicago, IL 60601. When t does not exceed $\frac{3}{16}$ in and washers are provided under the bolt head and nut, the following is applicable:

The tension force on the net section should not exceed P_a, kips, calculated from Eq. (8.45).

$$P_a = P_n/\Omega_t \tag{8.45}$$

$$P_n = A_n F_t \tag{8.46}$$

where Ω_t = safety factor for tension on net section
= 2.22 for single shear
= 2.00 for double shear
A_n = area of net section of thinnest sheet, in^2

The nominal limiting tension stress F_t, kips, is given by

$$F_t = (1 - 0.9r + 3rd/s)F_u \leq F_u \tag{8.47}$$

where s = bolt spacing, in, measured normal to line of stress
= width of sheet for a single bolt in the net section
F_u = tensile strength, ksi, of connected part
d = nominal diameter, in, of bolt
r = ratio of force transmitted by the bolts at the section to the tension force in the member at that section (if $r < 0.2$, it may be taken equal to zero)

8.18.3 Bearing Stresses and Bolt Tension

The bearing force should not exceed P_a, kips, calculated from Eq. (8.48).

$$P_a = P_n/\Omega_b \tag{8.48}$$

$$P_n = F_p dt \tag{8.49}$$

where Ω_b = safety factor for bearing = 2.22
F_p = nominal bearing stress, ksi, in connected part
d = nominal diameter of bolt, in
t = thickness, in, of thinnest connected part

Table 8.9 lists nominal bearing stresses for bolted connections.

TABLE 8.9 Nominal Bearing Stresses for Bolted Connections of Cold-Formed Steel Components[a]

Type of joint	Nominal bearing stress F_p, ksi	
	With washers under both bolt head and nut[b]	Without washers under bolt head and nut or with only one washer[c]
Inside sheet of double-shear connection	$3.33F_u$ $(F_u/F_{sy} \geq 1.15)^d$ $3.00F_u$ $(F_u/F_{sy} < 1.15)^d$	$3.00F_u^e$
Sheets in single shear and outside sheets of double-shear connection	$3.00F_u$	$2.22F_u^e$

[a]For joints with parts 3⁄16 in or more thick, see the "Specification for the Design, Fabrication and Erection of Structural Steel for Buildings," American Institute of Steel Construction.
[b]For joints with parts 0.024 in or more thick.
[c]For joints with parts 0.036 in or more thick.
[d]F_u/F_{sy} is the ratio of the tensile strength of a connected part to its yield strength.
[e]For $F_u/F_{sy} \geq 1.15$.

Table 8.10 lists allowable shear and tension stresses for various grades of bolts. The bolt force resulting in shear, tension, or combinations of shear and tension should not exceed the allowable force P_a, kips, calculated from Eq. (8.50).

$$P_a = A_b F \tag{8.50}$$

where A_b = gross cross-sectional area of bolt, in^2
F = allowable stress, ksi, F_v, F_t or F'_t in Tables 8.10 and 8.11

A safety factor of 2.22 may be used with Eq. (8.50) to compute ultimate loads on bolted joints.

Table 8.11 lists allowable tension stresses F'_t for bolts subjected to a combination of shear and tension. A safety factor of 2.22 may be used to compute ultimate loads for such joints.

8.18.4 Example—Tension Joints with Two Bolts

Assume that the bolted tension joints of Fig. 8.11 comprise two sheets of 3⁄16-in-thick, A611, Grade C steel. For this steel, $F_{sy} = 33$ ksi and $F_u = 48$ ksi. The sheets in each joint are 4 in wide and are connected by two 5⁄8-in-diameter, A325 bolts, with washers under both bolt head and nut.

Case 1 has the two bolts arranged in a single transverse row. A force $T/2$ is applied to each bolt and the total force T has to be carried by the net section of each sheet through the bolts. So, in Eq. (8.47), $r = 2(T/2)T = 1$. Spacing of the bolts $s = 2$ in and $d/s = 5/8/2 = 0.312$. The tension stress in the net section, computed from Eq. (8.47), is then

$$F_t = (1 - 0.9 \times 1 + 3 \times 1 \times 0.312)F_u = 1.04F_u \leq F_u$$

TABLE 8.10 Allowable Shear and Tension for Grades of Bolts

Description of bolts	Allowable shear stress F_v, ksi*		Allowable tension stress F_t, ksi†
	Thread not excluded from shear plane	Thread excluded from shear plane	
A325 bolts	21	30	44
A354, Grade B bolts $\frac{1}{4} \leq d < \frac{1}{2}$‡	24	40	49
A449 bolts $\frac{1}{4} \leq d < \frac{1}{2}$‡	18	30	40
A490 bolts	28	40	54
A307, Grade A $\frac{1}{4} \leq d < \frac{1}{2}$‡	9	9	18
A307, Grade A $d > \frac{1}{2}$‡	10	10	20

*Allowable shear stress × gross area of bolt = allowable bolt load
†Allowable tension stress × net area of bolt = allowable bolt load
‡d = nominal bolt diameter, in

TABLE 8.11 Allowable Tension Stresses, F_t' for Bolts Subjected to Combined Shear and Tension*

Description of bolts	Threads not excluded from shear planes	Threaded excluded from shear planes
A325 bolts	$55 - 1.8f_v \leq 44$	$55 - 1.4f_v \leq 44$
A354 bolts	$61 - 1.8f_v \leq 49$	$61 - 1.4f_v \leq 49$
A449 bolts	$50 - 1.8f_v \leq 40$	$50 - 1.4f_v \leq 40$
A490 bolts	$68 - 1.8f_v \leq 54$	$68 - 1.4f_v \leq 54$
A307 bolts		
When $\frac{1}{4}$ in $\leq d < \frac{1}{2}$ in	$23 - 1.8f_v \leq 18$	
When $d \geq \frac{1}{2}$ in	$26 - 1.8f_v \leq 20$	

*Allowable tension stress × net area of bolt = allowable bolt load

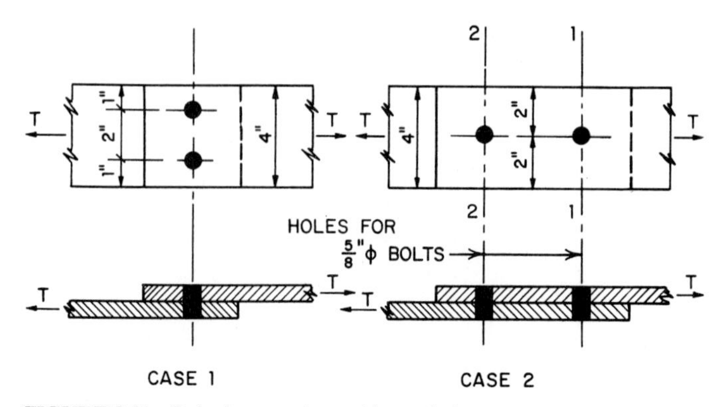

FIGURE 8.11 Bolted connections with two bolts.

Substitution in Eq. (8.46) with $F_u = 48$ yields the maximum allowable tension load on the net section:

$$P_n = [4 - (2 \times 5/8)] \times 3/16 \times 48 = 24.75 \text{ ksi}$$

This compares with the tensile strength of the full width of each sheet:

$$P_n = A_n F_{sy} = 4 \times 3/16 \times 33 = 24.75 \text{ ksi}$$

Case 2 has the two bolts, with 4-in spacing, arranged in a single line along the direction of applied force. For the top sheet at Section 1-1 then, $r = (T/2)/T = 1/2$, and for this sheet at Section 2-2, $r = (T/2)/(T/2) = 1$. For the top sheet at both sections, $d/s = 5/8/4 = 0.156$.

From Eq. (8.47), for the top sheet at Section 1-1,

$$F_t = (1 - 0.9 \times 1/2 + 3 \times 1/2 \times 0.156)F_u = 0.784F_u$$

The maximum load for that sheet would then be

$$P_n = [4 - 5/8] \times 3/16 \times 0.784 \times 48 = 23.81 \text{ kips}$$

For Section 2-2, top sheet,

$$F_t = (1 - 0.9 \times 1 + 3 \times 1 \times 0.156)F_u = 0.568F_u$$

Maximum load for Section 2-2, top sheet, would then be

$$P_n = (4 - 5/8) \times (3/16) \times 0.568 \times 48 = 17.25 \text{ kips}$$

The minimum distance between a bolt center and adjacent bolt edge or sheet edge is for Case 1 and Case 2

$$e = P/F_u t = (24.75/2)/(48 \times 3/16) = 1.37 \text{ in}$$

The combined load P_t of the two bolts in Case 2 is potentially 23.81 + 17.25 = 41.06 kips.

The bearing strength P_n of the 3/16-in-thick steel sheet is

$$P_n = F_p dt\Omega_b = 48 \times 5/8 \times 3/16 \times 2.22 = 12.49 \text{ kips}$$

This is adequate to carry the expected load on each bolt.

The A325 bolts with threads not excluded from the shear plane would be sufficiently strong in shear to carry the 24.75/2 = 12.37 kips needed per bolt inasmuch as the shearing strength of each bolt is

$$P_s = A_b F_s \Omega_s = (5/8)^2 \times 0.7854 \times 21 \times 2.22 = 14.30 \text{ kips}$$

It can be concluded that the Case 1 bolt arrangement is adequate and the Case 2 bolt arrangement has even more capacity.

8.19 SELF-TAPPING SCREWS FOR JOINING SHEET STEEL COMPONENTS

Self-tapping screws that are hardened so that their threads form or cut mating threads in one or both of the sheet steel parts being connected are frequently used for making field joints. Such screws provide a rapid and efficient means of making light-duty connections. The screws are especially useful for such purposes as fastening sheet-metal siding, roofing, and decking to structural steel; making attachments at joints, side laps, and closures in siding, roofing, and decking; fastening collateral materials to steel framing; and fastening steel studs to sill plates or channel tracks. The screws may also be used for fastening bridging to steel joists and studs, fastening corrugated decking to steel joists, and similar connections to secondary members.

There are no standard design rules for safe loads on such screws. They should not be used for load-carrying purposes unless justified by tests of mocked-up prototype details. The tests should show that allowable loads can be carried with a safety factor of 3.0 for a reasonable number of repetitions when repeated or reversed loads are expected. Otherwise, tapping-screw manufacturer's recommendations should be followed explicitly.

Several types of tapping screws are shown in Fig. 8.12. Other types are available. There are many different head styles—slotted, recessed, hexagonal, flat, round, etc. Some types, called *Sems*, are supplied with preassembled washers under the heads. Other types are supplied with neoprene washers for making watertight joints in roofing.

All the types of screws shown in Fig. 8.12 require prepunched or predrilled holes. Self-drilling screws, which have a twist drill point that drills the proper size of hole just ahead of threading, are especially suited for field work, because they eliminate separate punching or drilling operations. Another type of self-drilling screw, capable of being used in relatively thin sheets of material in situations where the parts being joined can be firmly clamped together, has a very sharp point that pierces the material until the threads engage.

Torsional-strength requirements for self-tapping screws have been standardized under American National Standards Institute B18.6.4, "Slotted and Recessed Head Tapping Screws and Metallic Drive Screws." Safe loads in shear and tension on such screws can vary considerably, depending on type of screw and head, tightening torque, and details of the assembly. When screws are used for structural load-carrying purposes, the user should rely on experience with the particular application, manufacturer's recommendations, or actual tests of the type of assembly involved.

Essential body dimensions of some types of self-tapping screws are given in Table 8.12. Complete details on these and other types, and recommended hole sizes, may be found in ANSI B18.6.4 and in manufacturers' publications.

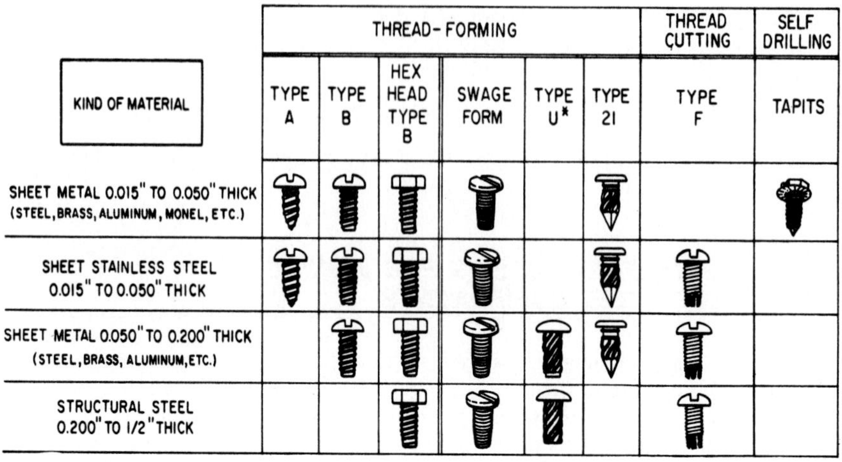

THREAD-FORMING
THREAD CUTTING
SELF DRILLING
KIND OF MATERIAL
TYPE A
TYPE B
HEX HEAD TYPE B
SWAGE FORM
TYPE U*
TYPE 21
TYPE F
TAPITS
SHEET METAL 0.015" TO 0.050" THICK (STEEL, BRASS, ALUMINUM, MONEL, ETC.)
SHEET STAINLESS STEEL 0.015" TO 0.050" THICK
SHEET METAL 0.050" TO 0.200" THICK (STEEL, BRASS, ALUMINUM, ETC.)
STRUCTURAL STEEL 0.200" TO 1/2" THICK

FIGURE 8.12 Tapping screws. NOTE: A blank space does not signify necessarily that the type of screw cannot be used for this purpose; it denotes that the type of self-tapping screw will not generally give the best results in this type of material. (*Parker-Kalon Corp., Emhart Corp., Campbellsville, Ky.*)

TABLE 8.12 Average Diameters of Self-Tapping Screws, in*

Number or size, in	Types AB and B		Type F†	Type U
	Outside	Root	Outside	Outside
No. 4	0.112	0.084	0.110	0.114
No. 6	0.137	0.102	0.136	0.138
No. 8	0.164	0.119	0.161	0.165
No. 10	0.186	0.138	0.187	0.180
No. 12	0.212	0.161	0.213	0.209
No. 14‡ or ¼	0.243	0.189	0.247	0.239‡
5/16	0.312	0.240	0.309	0.312
3/8§	0.376	0.304	0.371	0.375

*Averages of standard maximum and minimum dimensions adopted under ANSI B18.6.4-1966.

†Type F has threads of machine-screw type approximating the Unified Thread Form (ANSI B1.1-1960). The figures shown are averages of those for two different thread pitches for each size of screw.

‡Size No. 14 for Type U.

§Does not apply to Type AB.

8.20 SPECIAL FASTENERS FOR COLD-FORMED STEEL

Special fasteners, such as tubular rivets, blind rivets (capable of being driven from one side only), special bolts used for "blind insertion," special studs, lock nuts, and the like, and even metal stitching, which is an outgrowth of the common office stapling device for paper, are used for special applications. When such a fastener is required, refer to manufacturers' catalogs for design information, and base any structural strength attributed to the fastener on the results of carefully made tests or the manufacturer's recommendations.

COLD-FORMED STEEL FLOOR, ROOF, AND WALL CONSTRUCTION

Steel roof deck consists of ribbed sheets with nesting or upstanding-seam joints designed for the support of roof loads between purlins or frames. A typical roof-deck assembly is shown in Fig. 8.13. The Steel Deck Institute, P.O. Box 9506, Canton, OH 44711, has developed much useful information on steel roof deck.

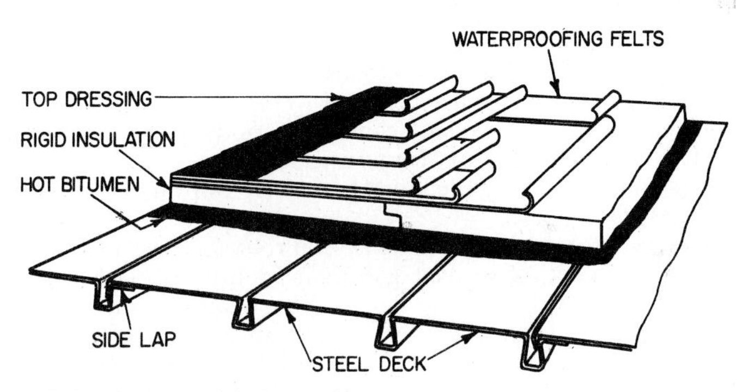

FIGURE 8.13 Roof-deck assembly.

8.21 Steel Roof Deck

Various types of steel roof deck are available and may be classified in accordance with recommendations of the Steel Deck Institute. All types consist of long, narrow sections with longitudinal ribs at least 1½ in deep and spaced about 6 in on centers (Fig. 8.14). Other rib dimensions are shown in Fig. 8.14*a* to *c* for some standard styles.

8.21.1 Types of Steel Roof Deck

Steel roof deck is commonly available in 24- and 30-in covering widths, but sometimes in 18- and 36-in widths, depending on the manufacturer. Thickness of steel

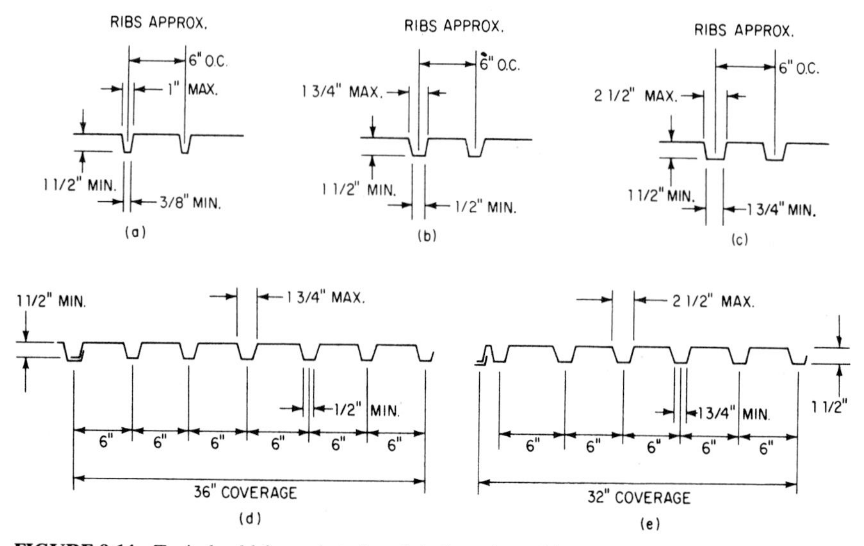

FIGURE 8.14 Typical cold-formed steel roof-deck sections. (*a*) Narrow rib; (*b*) intermediate rib; (*c*) wide rib; (*d*) intermediate rib in 36-in-wide sheets with nested side laps; (*e*) wide rib in 32-in-wide sheets with upstanding seams.

commonly used is 0.048 or 0.036 in, but most building codes permit 0.030-in-thick steel to be used. Figure 8.14*d* and *e* shows full-width decking in cross section. Usual spans, which may be simple, two-span continuous, or three-span continuous, range from 4 to 10 ft. The SDI "Design Manual for Floor Decks and Roof Decks" gives allowable total uniform loading (dead and live), lb/ft², for various steel thicknesses, spans, and rib widths.

Some manufacturers make special long-span roof-deck sections, such as the 3-in-deep, Type N roof deck shown in Fig. 8.15, in 24- to 16-ga black and galvanized.

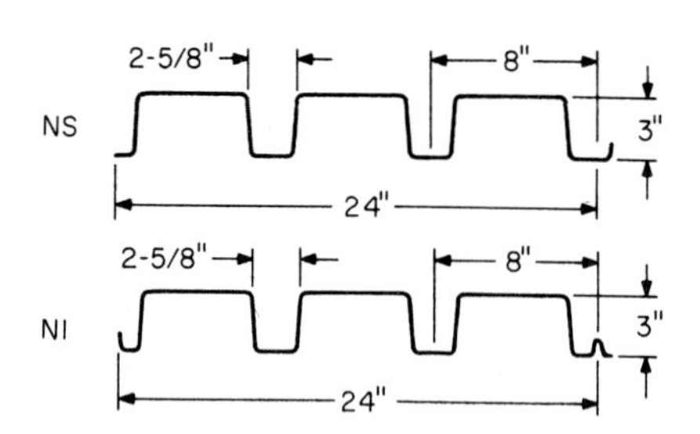

FIGURE 8.15 Cross sections of types NS and NI roof deck for 9- to 15-ft spans.

The weight of the steel roof deck shown in Fig. 8.14 depends on rib dimensions and edge details. For structural design purposes, weights of 2.8, 2.1, and 1.7 lb/ft² can be used for the usual design thicknesses of 0.048, 0.036, and 0.030 in, respectively, for black steel in all rib widths, as commonly supplied.

Steel roof deck is usually made of structural-quality sheet or strip, either black (ASTM A611, Grade C) or galvanized (A446, Grade A). Both steels specify minimum yield strengths of 33 ksi. Black steel is given a shop coat of priming paint by the roof deck manufacturer. Galvanized steel may or may not be painted; if painted, it should first be bonderized to ensure paint adherence. Aluminized steel is another metal-coated steel option.

SDI also publishes "Recommendations for Site Storage and Erection" and standard details for accessories.

8.21.2 Load-Carrying Capacity of Steel Roof Deck

The Steel Deck Institute has adopted a set of basic design specifications, with limits on rib dimensions, as shown in Fig. 8.14*a* to *c*, and publishes allowable uniform loading tables for narrow-, intermediate-, and wide-rib steel roof deck (Table 8.13, for example). These tables are based on section moduli and moments of inertia computed with effective-width procedures stipulated in the AISI "Specification for the Design of Cold-Formed Steel Structural Members" (Art. 8.8). SDI has banned compression flange widths otherwise assumed to be effective and also the use of testing to determine vertical load-carrying capacity of steel roof deck. Moreover, SDI "Basic Design Specifications" recommends the following:

Moment and Deflection Coefficients. Where steel roof decks are welded to supports, a moment coefficient of $1/10$ (applied to WL) should be used for three or more spans, and a deflection coefficient of $3/384$ (applied to WL^3/EI) should be used for all except simple spans. All other steel roof-deck installations should be designed as simple spans, with moment and deflection coefficients $1/8$ and $5/384$, respectively. (W = total uniform load, L = span, E = modulus of elasticity, I = moment of inertia.)

Maximum Deflections. The deflection under live load should not exceed $1/240$ of the clear span, center to center of supports. (Suspended ceiling, lighting fixtures, ducts or other utilities should not be supported by the roof deck.)

Anchorage. Steel roof deck should be anchored to the supporting framework to resist the following uplifts:

45 lb/ft^2 for eave overhang

30 lb/ft^2 for all other roof areas

The dead load of the roof-deck construction may be deducted from the above uplift forces.

8.21.3 Diaphragm Action of Decks

In addition to their normal function as roof panels under gravity loading, steel roof deck assemblies can be used as shear diaphragms under lateral loads, such as wind and seismic forces. When steel roof deck is used for these purposes, special attention should be paid to connections between panels and attachments of panels to building frames.

8.21.4 Details and Accessories of Steel Roof Deck

In addition to the use of nesting or upstanding seams, most roof-deck sections are designed so that ends can be lapped shingle fashion.

Special ridge, valley, eave, and cant strips are provided by roof-deck manufacturers (Fig. 8.16).

Roof decks are commonly arc welded to structural steel supports with puddle welds at least $1/4$ in in diameter or with elongated welds of equal perimeter. Elec-

TABLE 8.13 Allowable Total (Dead plus Live) Uniform Loads, psf, on Steel Roof Deck*

	Deck type	Span condition	Design thickness, in	Span—c to c joists or purlins, ft-in										
				4-0	4-6	5-0	5-6	6-0	6-6	7-0	7-6	8-0	8-6	9-0
NARROW RIB DECK TYPE NR; RIBS APPROX. 6" C-C; MAX 1"; MIN 3/8"; 1 1/2" MIN	NR 22	SIMPLE	0.0295	74	58	47								
	NR 20	SIMPLE	0.0358	90	72	58	48	40						
	NR 18	SIMPLE	0.0474	121	95	77	64	54	46					
	NR 22	2-SPAN	0.0295	80	64	51	42							
	NR 20	2-SPAN	0.0358	96	76	62	51	43						
	NR 18	2-SPAN	0.0474	124	98	79	66	55	47	40				
	NR 22	3 OR MORE	0.0295	100	79	64	53	45						
	NR 20	3 OR MORE	0.0358	120	95	77	64	53	46					
	NR 18	3 OR MORE	0.0474	155	122	99	82	69	59	51	44			
				4-0	4-6	5-0	5-6	6-0	6-6	7-0	7-6	8-0	8-6	9-0
INTERMEDIATE RIB DECK TYPE IR; RIBS APPROX. 6" C-C; MAX 1 3/4"; MIN 1"; 1 1/2" MIN	IR 22	SIMPLE	0.0295	86	68	55	45							
	IR 20	SIMPLE	0.0358	106	83	68	56	47	40					
	IR 18	SIMPLE	0.0474	141	112	90	75	63	54	46	40			
	IR 22	2-SPAN	0.0295	93	74	60	49	41						
	IR 20	2-SPAN	0.0358	112	88	72	59	50	42					
	IR 18	2-SPAN	0.0474	145	114	93	76	64	55	47	41			
	IR 22	3 OR MORE	0.0295	116	92	74	62	52	44					
	IR 20	3 OR MORE	0.0358	140	110	89	74	62	53	46	40			
	IR 18	3 OR MORE	0.0474	181	143	116	96	80	68	59	51	45	40	

WIDE RIB DECK
TYPE WR
RIBS APPROX. 6" C-C
MAX 2 1/2"
1 1/2" MIN
MIN 1 3/4"

			5-0	5-6	6-0	6-6	7-0	7-6	8-0	8-6	9-0	9-6	10-0
WR 22	SIMPLE	0.0295	89	70	56	46							
WR 20		0.0358	112	87	69	56	47	40					
WR 18		0.0474	154	118	94	76	63	53	45				
WR 22	2-SPAN	0.0295	98	81	68	58	50	43					
WR 20		0.0358	125	103	87	74	64	56	49	43			
WR 18		0.0474	165	136	115	98	84	73	64	57	51	45	40
WR 22	3 OR MORE	0.0295	122	101	85	70	58	49	42				
WR 20		0.0358	156	129	108	87	72	60	52	45			
WR 18		0.0474	207	171	143	120	98	81	69	59	51	45	40

*Load tables were calculated with sectional properties for minimum thicknesses of 0.028, 0.034, and 0.045 in, corresponding respectively to design thicknesses of 0.0295, 0.0358, and 0.0474 in, exclusive of coating on base metal.

Loads shown in tables are uniformly distributed total (dead plus live) loads, psf. Loads in shaded areas are governed by live-load deflection not in excess of $\frac{1}{240} \times$ span. The dead load included is 10 psf. All other loads are governed by the allowable flexural stress limit of 20 ksi for a 33-ksi minimum yield point.

Rib-width limitations shown are taken at the theoretical intersection points of flange.

Span length assumes c-to-c spacing of supports. Tabulated loads shall not be increased by assuming clear-span dimensions.

Bending moment formulas used for flexural stress limitation are: for simply supported and two-span decking, $M = wl^2/8$; for decking with three continuous spans or more, $M = wl^2/10$.

Deflection formulas for deflection limitation are: For simply supported decking, $\Delta = 5wl^4/384EI$; for two- and three-span decking, $\Delta = 3wl^4/384EI$.

Normal installations covered by these tables do not require midspan fasteners for spans of 5 ft or less.

From "Design Manual for Floor Decks and Roof Decks," Steel Deck Institute.

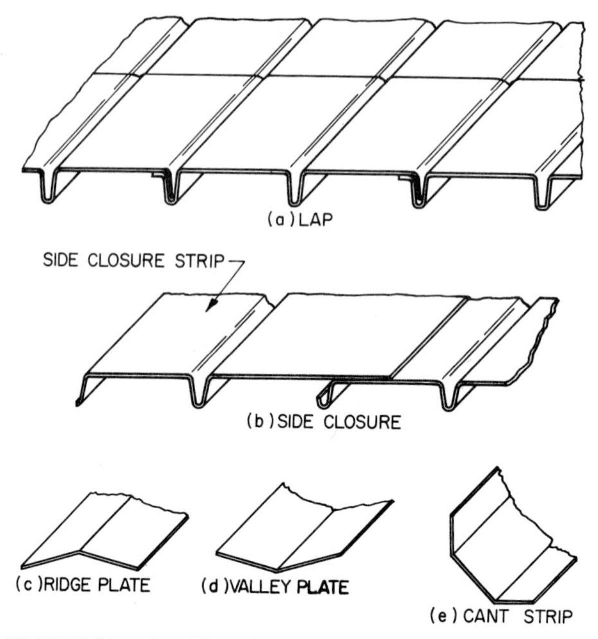

FIGURE 8.16 Roof-deck details.

trodes should be selected for amperage adjusted to fuse all layers of steel roof decking to supporting members without creating blowholes around the welds. Welding washers are recommended for thicknesses less than 0.030 in.

Fillet welds at least 1 in long should be used to connect lapped edges of roof deck.

Tapping screws are an alternative means of attaching steel roof deck to structural support members, which should be at least 1⁄16 in thick. All edge ribs and a sufficient number of interior ribs should be connected to supporting members at intervals not exceeding 18 in. When standard steel roof deck spans 5 ft or more, adjacent sheets should be fastened together at midspan with either welds or screws.

8.21.5 Roof Deck Insulation and Fire Resistance

Although insulation is not ordinarily supplied by the roof-deck manufacturer, it is standard practice to install 3⁄4- or 1-in-thick mineral fiberboard between roof deck and roofing. SDI further recommends that all steel decks be covered with a material of sufficient insulating value to prevent condensation under normal occupancy conditions. Insulation should be adequately attached to the steel deck by means of adhesives or mechanical fasteners. Insulation materials should be protected from the elements at all times during storage and installation.

The UL "Fire Resistance Directory," Underwriter's Laboratories, Inc., 333 Pfingsten Rd., Northbrook, IL 60062, lists fire-resistance ratings for steel roof-deck construction. Some systems with fire ratings up to 2 h are listed in Table 8.14.

TABLE 8.14 Fire Resistance Ratings for Steel Floor and Roof Assemblies*

Roof construction	Insulation	Underside protection	Authority
	2-h rating†		
Min. 1½-in-deep steel deck on steel joists or steel beams	Min. 1¾-in-thick listed mineral fiberboard	Min. 1¾-in-thick, direct-applied, sprayed vermiculite plaster, UL listed	UL design P711†
Min. 1½-in-deep steel deck on steel joists or steel beams	Min. 1 1/16-in-thick listed mineral fiberboard	Min. 1 9/16-in-thick, direct-applied, sprayed fiber protection, UL listed	UL design P818†
Floor construction	**Concrete**	**Underside protection**	**Authority**
	2-h rating‡		
1½-, 2-, or 3-in-deep steel floor units on steel beams	2½-in-thick normal-weight or lightweight concrete	Min. ⅜-in-thick, direct-applied, sprayed vermiculite plaster, UL listed	UL design D739‡
1½-, 2-, or 3-in-deep steel floor units on steel beams	2½-in-thick normal-weight or lightweight concrete	Min. ⅜-in-thick, direct-applied, sprayed fiber protection, UL listed	UL design D858†

*Based on "Fire Resistance Index," 1990, Underwriters Laboratories, Inc., 333 Pfingsten Rd., Northbrook, IL 60062.
†1½-h and 1-h ratings are also available.
‡1-h, 2½-h, 3-h, and 4-h ratings are also available.

8.22 CELLULAR STEEL FLOOR AND ROOF PANELS*

Several different designs of cellular steel panels and fluted steel panels for floor and roof construction are available. Sections of some of these panels are illustrated in Fig. 8.17.

*Courtesy of R. E. Albrecht, Engineer, H. H. Robertson Company, Ambridge, Pa.

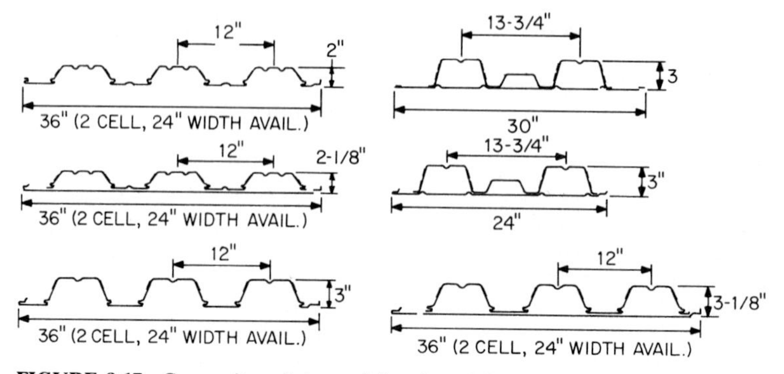

FIGURE 8.17 Composite cellular and fluted steel floor sections. (*Courtesy H. H. Robertson Co., Ambridge, Pa.*)

8.22.1 Cellular-Steel-Floor Raceway System

One form of cellular steel floor assembly with a distribution system for electrical wiring, telephone cables, and data cables is described below and is illustrated in Fig. 8.18. This system is used in many kinds of structures, including massive high-rise buildings for institutional, business, and mercantile occupancies.

The cellular-steel-floor raceway system is basically a profiled steel deck containing wiring raceways and having structural concrete on top. The cellular deck consists of closely spaced cellular raceways. These are connected to a main trench header duct with removable cover plate for lay-in wiring. Set on a repetitive module, the cellular raceways are assigned to electrical power, telephone, and data wiring.

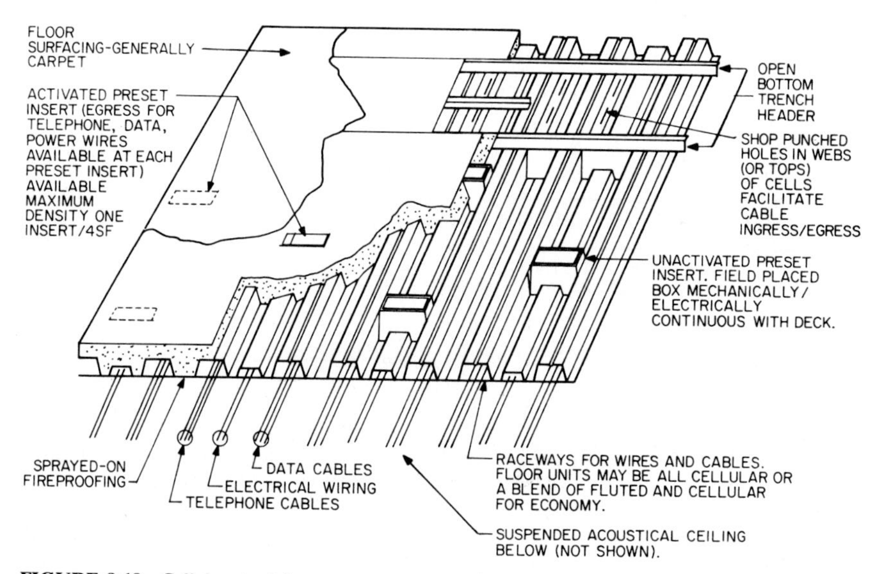

FIGURE 8.18 Cellular steel floor raceway system. (*Courtesy H. H. Robertson Co., Ambridge, Pa.*)

At prescribed intervals, as close as 2 ft longitudinally and 2 ft transversely over the floor, preset inserts may be provided for access to the wiring and activation workstations. When an insert is activated at a workstation, connections for electrical power, telephone, and data are provided at one outlet. Insert fittings may be flush with the top floor surface or project above it.

This system provides the required fire-resistive barrier between stories of a building. The cellular metal floor units also serve the structural purposes of acting as working platforms and concrete forms during construction and as tensile reinforcement for the concrete floor slab after the building is occupied.

Cellular steel floor raceways have many desirable features including moderately low cost, good flexibility, which contributes to lower life-cycle cost, and minimal limitations on placement of outlets. Little or no increase over floor depth required for strictly structural purposes is necessary to accommodate the system.

Wiring may penetrate the floor surface only at outlet fittings. Therefore, if carpet is used, it will have to be cut and a flap peeled back to provide access to the fittings. Use of carpet tiles rather than sheet carpet facilitates access to the preset inserts.

Where service outlets are not required to be as close as 2 ft on centers, a blend of fluted and cellular floor sections may be used. As an example, alternating 3-ft-wide fluted floor deck with 2-ft-wide cellular floor panels results in a module for service outlets of 5 ft in the transverse direction and as close as 2 ft in the longitudinal direction. Other modules and spacings are available.

8.22.2 Steels Used for Cellular and Fluted Decking

Cellular and fluted floor and roof sections (decking) usually are made of steel 0.030 in or more thick complying with the requirements of ASTM A611, Grade C, for uncoated steel or ASTM A446, Grade A, for galvanized steel, both having specified minimum yield points of 33 ksi. The steel may be either galvanized or painted.

8.22.3 Structural Design of Steel Floor and Roof Panels

Design is usually based on the "Specification for the Design of Cold-Formed Steel Structural Members," American Iron and Steel Institute, 1101 17th St., NW, Washington, DC 20036. Structural design of composite floor slabs incorporating sheet-steel floor and roof panels is usually based on "Specifications for the Design and Construction of Composite Slabs" and "Commentary on Specifications for the Design and Construction of Composite Slabs," ANSI-ASCE 3-84, American Society of Civil Engineers, 345 East 47th St., New York, NY 10017.

Details of design and installation vary with types of panels and manufacturers. In any particular instance, refer to the manufacturer's recommendations.

8.22.4 Fire Resistance of Cellular and Fluted Steel Decking

Any desired degree of fire protection for cellular and fluted steel floor and roof assemblies can be obtained with concrete toppings and plaster ceilings or direct-application compounds (sprayed-on fireproofing). Fire-resistance ratings for a considerable number of assemblies are available. (See "Fire-Resistant Steel-Frame Construction," American Institute of Steel Construction," and "Fire Resistance Directory," Underwriters Laboratories).

8.23 CORRUGATED SHEETS FOR ROOFING, SIDING, AND DECKING

Although the use of corrugated sheets of thin steel for roofing and siding leaves something to be desired for weathertightness and appearance, they are used for barns and similar buildings for some protection against weather elements. They are cheap, easy to install on a wood frame, and last for many years if galvanized. (Corrugated steel sheets are the oldest type of cold-formed steel structural members. They have been used since 1784, when Henry Cort introduced sheet rolling in England.)

The commonest form of corrugated sheet, the arc-and-tangent type, has the basic cross section shown in Fig. 8.19*a*. Its section properties are readily calculated with factors taken from Fig. 8.19*b* to *f* and substituted in the following formulas.

The area, in^2, of the corrugated sheet may be determined from

$$A = \lambda b t \tag{8.51}$$

where b = width of sheet, in

t = sheet thickness, in

$\lambda = \dfrac{(2/K + \alpha)\sin\alpha + (1 - 2\alpha/K)\cos\alpha - 1}{1 - \cos\alpha}$ (See Fig. 8.19*d*)

K = pitch-depth ratio of a corrugation = p/d

p = pitch, in, of corrugation

d = depth, in, of corrugation

α = tangent angle, radians, or angle of web with respect to the neutral axis of the sheet cross section

The moment of inertia, in^4, of the corrugated sheet may be obtained from

$$I = C_5 b t^3 + C_6 b d^2 t \tag{8.52}$$

where $C_5 = \dfrac{q(6\alpha + \sin 2\alpha - 8\sin\alpha) + 4\sin\alpha + K\cos\alpha}{12K}$ (See Fig. 8.19*b*)

$$C_6 = \frac{1}{K}\left[q^3\left(6\alpha + \sin 2\alpha - 8\sin\alpha - \frac{4}{3}\tan^3\alpha\sin^2\alpha\right)\right.$$
$$+ q^2(4\sin\alpha + K\tan^3\alpha\sin\alpha - 4\alpha)$$
$$\left. + q\left(\alpha - \frac{1}{4}K^2\tan^3\alpha\right) + \frac{K^3\tan^2\alpha}{48\cos\alpha}\right]$$ (See Fig. 8.19*c*)

$q = \dfrac{r}{d} = \dfrac{K\tan\alpha - 2}{4(\sec\alpha - 1)}$ (See Fig. 8.19*e*)

The section modulus of the corrugated sheet may be computed from

$$S = \frac{2I}{d + t} \tag{8.53}$$

The radius of gyration, in, is given by

$$\rho = \sqrt{\frac{I}{A}} \tag{8.54}$$

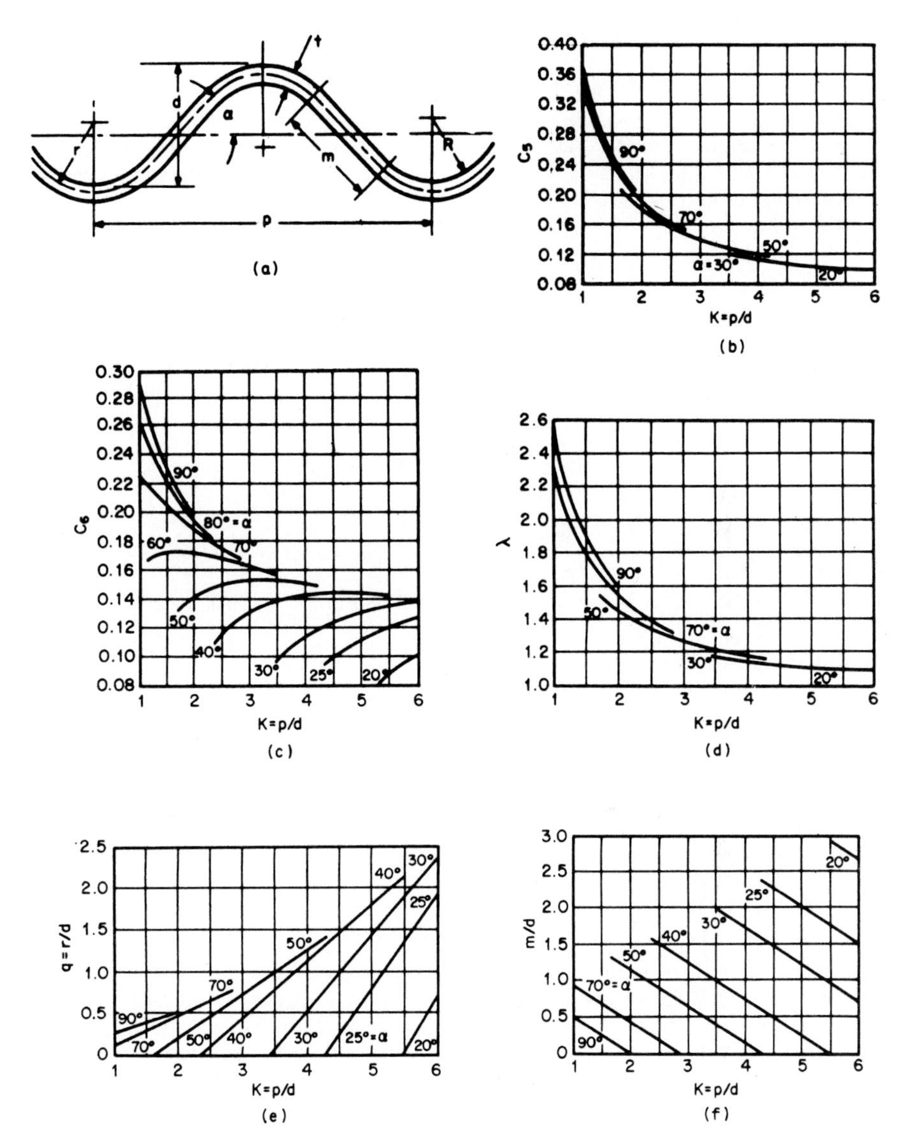

FIGURE 8.19 Factors for determining section properties of the arc-and-tangent type of corrugated steel sheet shown in (*a*).

and the tangent length-depth ratio is

$$\frac{m}{d} = \frac{\sin \alpha}{1 - \cos \alpha} - \frac{K}{2} \tag{8.55}$$

(See Fig. 8.19*f*.)

Example—Corrugated Sheet Properties. Consider a corrugated sheet with a 6-in pitch, 2-in depth, inside radius R of 1⅛ in, and thickness t of 0.135 in. The mean radius r is then $1.125 + 0.135/2 = 1.192$ in; $q = r/d = 1.192/2 = 0.596$ in, and $K = p/d = 6/2 = 3$. From Fig. 8.19e, angle α is found to be nearly 45°. For $p/d = 3$ and $\alpha = 45°$, Fig. 8.20b, c, d, and f indicate that $C_5 = 0.14$, $C_6 = 0.145$, $\lambda = 1.24$, and $m/d = 0.93$. Section properties per inch of corrugated width are then computed as follows:

From Eq. (8.51),

$$A = 1.24 \times 1 \times 0.135 = 0.167 \text{ in}^2$$

From Eq. (8.52),

$$I = 0.14 \times 1(0.135)^3 + 0.145 \times 1(2)^2 0.135 = 0.0786 \text{ in}^4$$

From Eq. (8.53),

$$S = \frac{2 \times 0.0786}{2 + 0.135} = 0.0736 \text{ in}^3$$

From Eq. (8.54),

$$\rho = \sqrt{\frac{0.0786}{0.167}} = 0.686 \text{ in}$$

and from Eq. (8.55),

$$m = 0.93 \times 2 = 1.86 \text{ in}$$

I, S, and A for corrugated sheets with width b are obtained by multiplying the per-inch values by b.

Unit Stresses. The allowable unit bending stress F_r, ksi, at extreme fibers of corrugated sections of carbon or low-alloy steel may be taken as $0.6F_y$, if r/t does not exceed $1650/F_y$. For $1650/F_y \leq r/t < 6500/F_y$,

$$F_r = 331t/r + 0.399F_y \tag{8.56}$$

where F_y = specified minimum yield point of the steel, ksi.

Section properties of corrugated sheets with cross sections composed of flat elements may be computed with the linear method given in Art. 8.4, by combining properties of the various elements as given in Table 8.4. (See also "Sectional Properties of Corrugated Sheets Determined by Formula," *Civil Engineering,* February 1954.)

8.24 LIGHTWEIGHT STEEL METRIC SHEETING

Metric sheeting, the cross section of which is shown in Fig. 8.20, has a corrugation-like conformation with locking side edges. It has a laying width of 500 mm or 0.5 m (19⅔ in), and is available in thicknesses of 5, 7, 8, 10, and 12 ga. Sheets are installed vertically in soil with edges of successive units interlocking. For additional corrosion protection, metric sheeting may be ordered galvanized after continuous cold forming in lengths of 4 to 40 ft.

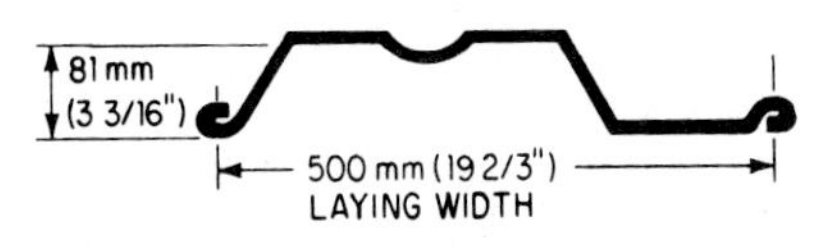

FIGURE 8.20 Steel metric sheeting.

Applications include checkdams, core walls, wingwalls, trench walls, excavations, low retaining walls, ditch checks, jetties and lagoon baffles. The sheeting often can be put into soft ground with the aid of a backhoe, although for harder subgrades, conventional drop, vibratory, or diesel hammers applied to a light driving head make emplacement easier. The tight metal-to-metal interlock at the edges of metric sheeting contains soil and controls water movement. Table 8.15 lists its structural properties.

Metric sheeting should not be confused with steel sheetpiling, which is a heavier hot-rolled steel product used for major construction projects, including breakwaters, bulkheads, cofferdams, and docks. Metric sheeting is nevertheless an economical product suitable for many less-demanding applications for both temporary and permanent uses. An advantage for contractors is that it can be withdrawn and reused on another job.

More information on lightweight steel construction is available from CONTECH Construction Products, P.O. Box 800, Middletown, OH 45042.

8.25 STAINLESS STEEL STRUCTURAL DESIGN

Cold-formed, stainless-steel structural members require different design approaches from those presented in Arts. 8.1 through 8.13 for cold-formed structural members of carbon and low-alloy steels. An exception is the stainless steels of the ferritic type that are largely alloyed with chromium and exhibit a sharp-yielding stress-strain curve. The austenitic types of stainless steel, incorporating substantial amounts of nickel as well as chromium, have stress-strain curves that are rounded, do not show sharp yield points, and exhibit proportional limits that are quite low. Because of excellent corrosion resistance, stainless steels are suitable for exterior wall panels and exterior members of buildings as well as for other applications subject to corrosive environments.

TABLE 8.15 Physical Properties of Metric Sheeting*

Thickness		Weight		Section properties			
				Section modulus, in^3		Moment of inertia, in^4	
Gage	in	lb/lin ft of pile	lb/ft² of wall	Per section	Per ft	Per section	Per ft
5	0.2092	19.1	11.6	5.50	3.36	9.40	5.73
7	0.1793	16.4	10.0	4.71	2.87	7.80	4.76
8	0.1644	15.2	9.3	4.35	2.65	7.36	4.49
10	0.1345	12.5	7.6	3.60	2.20	6.01	3.67
12	0.1046	9.9	6.0	2.80	1.71	4.68	2.85

*Based on "CONTECH Metric Sheeting," 1990, Contech Construction Products Inc., Middletown, Ohio.

The "Specification for the Design of Cold-Formed Stainless Steel Structural Members," ANSI-ASCE 8-90, American Society of Civil Engineers, 345 East 47th St., New York, NY 10017, presents treatments paralleling those of Arts. 8.1 through 8.13, except the primary emphasis is on the load resistance factor design (LRFD) method. The allowable strength design (ASD) method, however, is also mentioned. For detailed information on austenitic grades of stainless steel, see ASTM A666, "Austenitic Stainless Steel, Sheet, Strip, Plate and Flat Bar for Structural Applications."

(W. W. Yu, "Cold-Formed Steel Design," 2d ed., John Wiley and Sons, Inc., New York.)

PREENGINEERED STEEL BUILDINGS

Preengineered steel buildings may be selected from catalogs. They are fully designed by a manufacturer, who supplies them with all structural and covering material, and all fasteners.

8.26 CHARACTERISTICS OF PREENGINEERED STEEL BUILDINGS

These structures eliminate the need for engineers and architects to design and detail both the structures and the required accessories and openings, as would be done for conventional buildings with components from many individual suppliers. Available with floor areas of up to 1 million ft^2, preengineered buildings readily meet requirements for single-story structures, especially for industrial plants and commercial buildings (Fig. 8.21).

Preengineered buildings may be provided with custom architectural accents. Also, standard insulating techniques may be used with thermal accessories incorporated to provide energy efficiency. Exterior wall panels are available with durable factory-applied colors.

Many preengineered steel building suppliers are also able to modify their standard designs, within certain limits, while still retaining the efficiencies of predesign and automated volume fabrication. Examples of such modifications include the addition of cranes; mezzanines; heating, ventilating, and air-conditioning equipment; sprinklers; lighting; and ceiling loads with special building dimensions.

Preengineered buildings make extensive use of cold-formed steel structural members. These lend themselves to mass production, and their designs can be more accurately fitted to the specific structural requirements. For instance, a roof purlin can be designed with the depth, moment of inertia, section modulus, and thickness required to carry the load, as opposed to picking the next higher size of standard hot-rolled shape, with more weight than required. Also, because this purlin is used on many buildings, the quantity justifies investment in automated equipment for forming and punching. This equipment is nevertheless flexible enough to permit a change of thickness or depth of section to produce similar purlins for other buildings.

The engineers designing a line of preengineered buildings can, because of the repeated use of the design, justify spending additional design time refining and optimizing the design. Most preengineered buildings are designed with the aid of

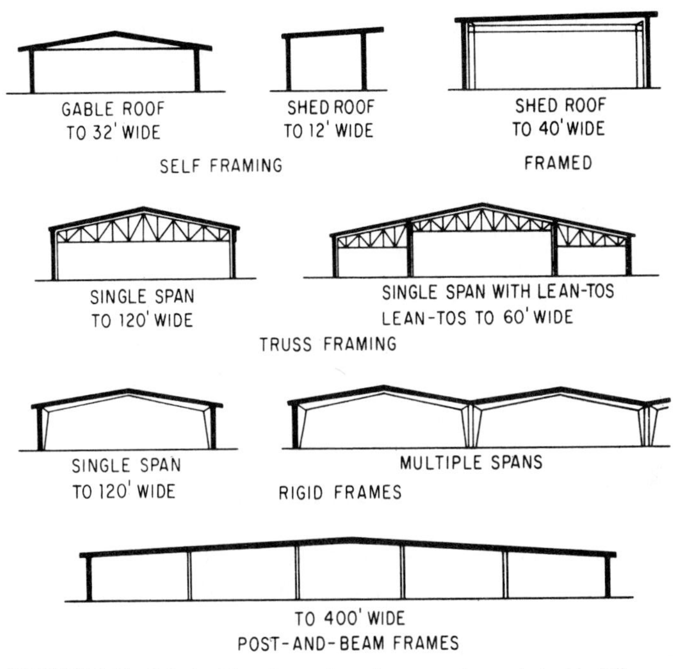

FIGURE 8.21 Principal framing systems for preengineered steel buildings.

computers. Their programs are specifically tailored to produce systems of such buildings. A rerun of a design to eliminate a few pounds of steel is justified, since the design will be used many times during the life of that building model.

8.27 STRUCTURAL DESIGN OF PREENGINEERED BUILDINGS

The buildings are designed for loading criteria in such a way that they may be specified to meet the geographical requirements of any location. Combinations of dead load, snow load, live load, and wind conform with requirements of several model building codes.

Standards in "Metal Building Systems," Metal Building Manufacturers Association, 1300 Sumner Ave., Cleveland, OH 44115 discuss methods of load application and maximum loading, for use where load requirements are not established by local building codes. Other appropriate design specifications include:

Structural Steel. "Specification for Design, Fabrication, and Erection of Structural Steel for Buildings," American Institute of Steel Construction, 400 North Michigan Ave., Chicago, IL 60607.

Cold-Formed Steel. "Specification for the Design of Cold-Formed Steel Structural Members," American Iron and Steel Institute, 1101 17th St., NW, Washington, DC 20036.

Welding. "Structural Welding Code," D1.3 and "Specification for Welding Sheet Steel in Structures," D1.3, American Welding Society, 550 NW LeJeune Rd., Miami, FL 33152.

The Systems Building Association promotes marketing of metal buildings and is located at 28 Lowery Dr., P.O. Box 117, West Milton, OH 45383.

OPEN-WEB STEEL JOISTS

The first steel joist was produced in 1923 and consisted of solid round bars for top and bottom chords and a web formed from a single continuous bent bar, thus simulating a Warren truss. The Steel Joist Institute (SJI) was organized to promote sales of such joists in 1925 and has sponsored further research and development since then.

8.28 DESIGN OF OPEN-WEB STEEL JOISTS

Currently, open-web steel joists are still relatively small, parallel-chord trusses, but *hot-rolled steel shapes* usually make up the components. (For a time, *cold-formed steel shapes* were preferred for chords to utilize higher working stresses available

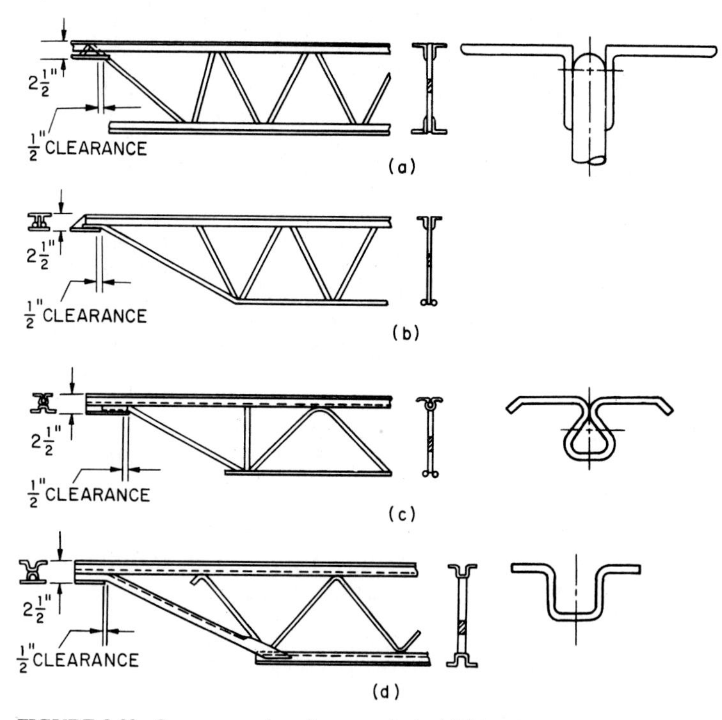

FIGURE 8.22 Some examples of open-web steel joists.

in cold-formed sections of ordinary carbon-steel grades. Unfavorable fabrication costs, however, led to a change to the hot-rolled steel chords.)

Joists are suitable for direct support of floors and roofs of buildings, when designed according to SJI "Standard Specifications, Load Tables and Weight Tables for Steel Joists and Joist Girders," Steel Joist Institute, Suite A, 205 48th Ave. North, Myrtle Beach, SC 29577. Moreover, since 1972, the American Institute of Steel Construction (AISC) has cooperated with SJI in producing an industry standard for steel joist design. However, exact forms of chords and webs, and their methods of manufacture, then as now, have continued to be in the provenance of SJI members. Figure 8.22 shows a number of proprietary steel joist designs.

Joists are designed primarily for use under uniform distributed loading with substantially uniform spacing of joists, as depicted in Fig. 8.23. They can carry concentrated loads, however, especially if loads are applied at joist panel points. Partitions running crosswise to joists usually can be considered as being distributed by the concrete floor slabs, thus avoiding local bending of joist top chords. Even so, joists must always be size-selected to resist the bending moments, shears, and reactions of all loads, uniform or otherwise. So joist loadings given in tables for uniform loading should be used with caution and modified when necessary.

(a)

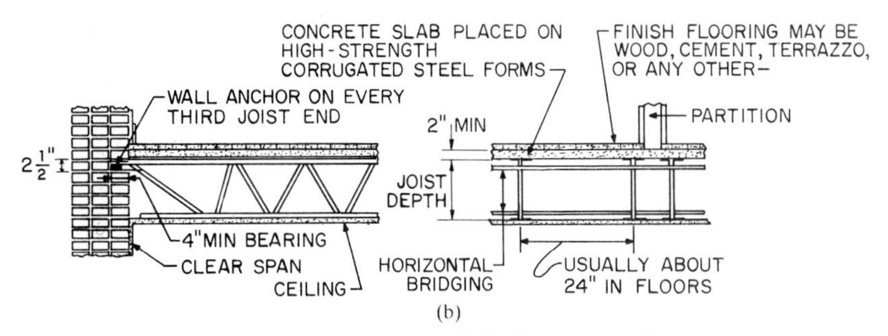

(b)

FIGURE 8.23 Some examples of open-web steel-joist floor construction.

TABLE 8.16 Typical Fire-Resistance Ratings of Floor-Ceiling Assemblies*

	Fire rating, h				
	1	1½	2	3	4
Ceiling construction					
Gypsum thickness, in†	½	½	½	⅝	¾
Metal lath and plaster‡					¾
Minimum thickness of reinforced concrete cover over top of joists, in	2	2	2½	2½	2½
Minimum chord size for steel joists	No. 4	No. 4	No. 3	No. 3	No. 5

*As recommended by the Steel Joist Institute, 1988.

†Underwriters' Laboratories or Factory Mutual approved. Also, see "Design of Fire-Resistive Assemblies with Steel Joists," *SJI Technical Digest*, No. 4, 1972. ⅛ in of plaster may be applied to the gypsumboard.

‡¾-in vermiculite plaster on fibered gypsum scratch and brown coats or unfibered brown coats, applied on metal lath on ¾-in cold-rolled-steel furring channels.

One cardinal rule is that the clear span of a joist should never exceed 24 times its depth. Another rule is that deflections should not exceed 1/360 of the joist span for floors and roofs to which plaster ceilings are attached or 1/240 of the span for all other cases.

SJI publishes loading tables for K-series (short span), LH-series (long span), and DLH-series (deep long span) joist girders. The K-series joists are available in depths of 8 to 30 in and spans of 8 to 60 ft in 11 different chord weights to sustain uniform loads along the span as high as 550 lb/ft. LH-series joists are available in depths from 18 to 48 in and spans of 25 to 80 ft in six different chord weights capable of supporting total loads of 12,000 to 57,600 lb. DLH-series joists are available in depths of 52 to 72 in and spans from 89 to 144 ft in 10 different chord weights with total-load capacities of 25,700 to 80,200 lb. Load capacities in the foregoing were based on a maximum allowable tensile strength of 30 ksi, which calls for high-strength, low-alloy steel having a specified minimum yield strength of 50 ksi or cold-formed steel having the same yield strength.

Table 8.16 lists fire-resistance ratings for several floor-ceiling assemblies.

8.29 CONSTRUCTION DETAILS FOR OPEN-WEB STEEL JOISTS

It is essential that bridging be installed between joists as soon as possible after the joists are placed and before application of any construction loads. The most commonly used type of bridging is continuous horizontal bracing composed of steel rods fastened perpendicular to the top and bottom chords of the joists. Diagonal bridging, however, is also permitted. The attachment of the floor or roof is expected to provide additional support of the joists against lateral buckling.

It is important that masonry anchors be used on wall-bearing joists. Where the joists rest on steel beams, they should be welded, bolted, or clipped to the beams.

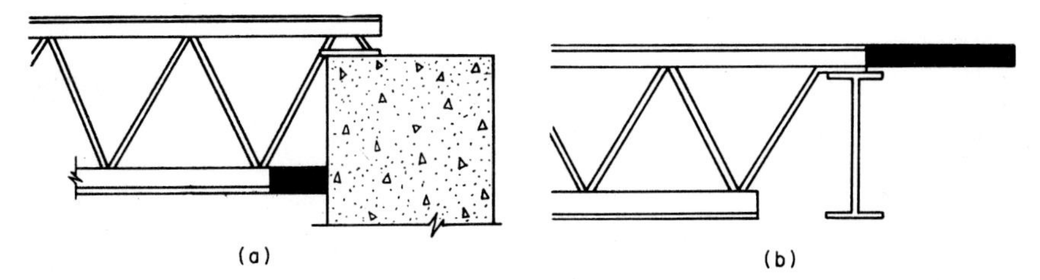

FIGURE 8.24 Open-web steel joist with (*a*) ceiling extension, (*b*) extended end. (*Courtesy of Steel Joist Institute.*)

Plastered ceilings attached directly to regular open-web steel joists are usually supported at underslung ends by means of *ceiling extensions,* as shown in Fig. 8.24*a*. *Extended ends,* as shown in Fig. 8.24*b*, allow floor and roof treatments beyond outer supporting stringers.

Relatively small openings between joists may usually be framed with angle, channel, and Z-shaped headers supported on adjacent joists. Larger openings should be framed in structural steel. Headers should preferably be located so that they are supported at trimmer-joist panel points.

SECTION NINE
CONCRETE CONSTRUCTION

Edward S. Hoffman
President, Edward S. Hoffman, Ltd.,
Structural Engineers, Chicago

David P. Gustafson
Technical Director
Concrete Reinforcing Steel Institute, Schaumburg, Illinois

Economical, durable construction with concrete requires a thorough knowledge of its properties and behavior in service, of approved design procedures, and of recommended field practices. Not only is such knowledge necessary to avoid disappointing results, especially when concrete is manufactured and formed on the building site, but also to obtain maximum benefits from its unique properties.

To provide the needed information, several organizations promulgate standards, specifications, recommended practices, and reports. Reference is made to these where appropriate throughout this section. Information provided herein is based on the latest available editions of the documents. Inasmuch as they are revised frequently, the latest editions should be used for current design and construction.

CONCRETE AND ITS INGREDIENTS

The American Concrete Institute "Building Code Requirements for Reinforced Concrete," ACI 318, contains the following basic definitions:

Concrete is a mixture of portland cement, cementitious materials, fine aggregate, coarse aggregate, and water.

Admixture is a material other than portland cement, cementitious material, aggregates, or water that is added to concrete to modify its properties.

In this section, unless indicated otherwise, these definitions apply to the terms concrete and admixture.

9.1 CEMENTITIOUS MATERIALS

The American Concrete Institute Building Code, ACI 318, defines cementitious materials as those that have cementitious value when used in concrete either by themselves, such as portland cement or blended hydraulic cements, or in combination with fly ash (ASTM specification C618), raw or calcined natural pozzolans (ASTM C618), ground granulated blast-furnace slag (ASTM C989), or silica fume (American Association of State Highway and Transportation Officials specification M307). Addition to a concrete mix of fly ash, silica fume, or slag decreases permeability, protects reinforcement, and increases strength. Concrete made with polymers, plastics with long-chain molecules, can have many qualities much superior to those of ordinary concrete. See also Sec. 4.

9.2 CEMENTS

Portland cements meeting the requirements of ASTM C150, "Standard Specification for Portland Cement," or air-entraining portland cements (ASTM C175) are available in types I to V and IA to IIIA, respectively, for use under different service conditions. Portland-type cements also include portland blast-furnace slag cement (ASTM C205) and portland-pozzolan cement (ASTM C340), which may be employed in concrete under the ACI Building Code, and may therefore be included under its definition of concrete. Proprietary "shrinkage-compensating" portland-type cements are also available for general use in concrete.

Although all the preceding cements can be employed for concrete, they are *not* interchangeable. Note that both tensile and compressive strengths vary considerably, at early ages in particular, even for the five types of basic portland cement. Consequently, although project specifications for concrete strength f'_c are usually based on a standard 28-day age for the concrete, the proportions of ingredients required differ for each type. For concrete strengths up to 19,000 psi for columns in high-rise buildings, specified compressive strengths are usually required at 56 days after initial set of the concrete. For the usual building project, where the load-strength relationship is likely to be critical at a point in strength gain equivalent to 7-day standard curing (Fig. 9.1), substitution of a different type (sometimes brand) of cement without reproportioning the mix may be dangerous.

The accepted specifications (ASTM) for cements do not regulate cement temperature nor color. Nevertheless, in hot-weather concreting, the temperature of the fresh concrete and therefore of its constituents must be controlled. Cement temperatures above 170°F are not recommended ("Hot Weather Concreting," ACI 305R).

For exposed architectural concrete, not intended to be painted, control of color is desirable. For uniform color, the water-cement ratio and cement content must be kept constant, because they have significant effects on concrete color. Bear in mind that because of variations in the proportions of natural materials used, cements from different sources differ markedly in color. A change in brand of cement therefore can cause a change in color. Color differences also provide a convenient check for substitution of types (or brands) of cement different from those used in trial batches made to establish proportions to be employed for a building.

9.3 AGGREGATES

Only material conforming to specifications for normal-weight aggregate (ASTM C330) or lightweight aggregate for structural concrete (ASTM C330) is accepted

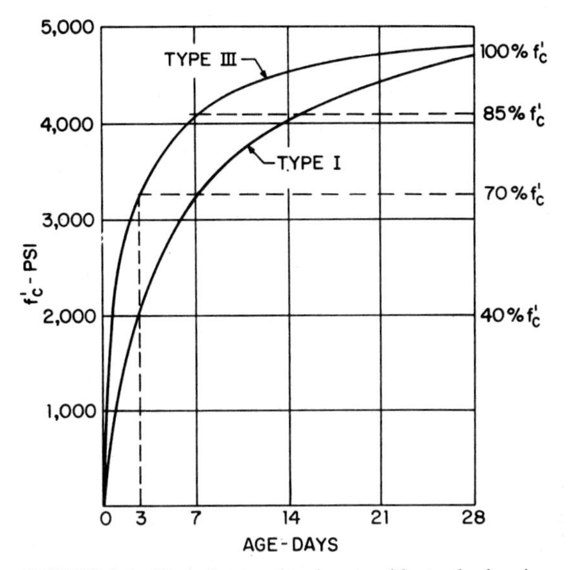

FIGURE 9.1 Typical strength-gain rate with standard curing of non-air-entrained concrete having a ratio of water to cementitious material of 0.50. (*Courtesy Materials Science Corp., Chicago.*)

under the American Concrete Institute Building Code, ACI 318, without special tests. When an aggregate for which no experience record is available is considered for use, the modulus of elasticity and shrinkage as well as the compressive strength should be determined from trial batches of concrete made with the aggregate. In some localities, aggregates acceptable under C33 or C330 may impart abnormally low ratios of modulus of elasticity to strength (E_c/f'_c) or high shrinkage to concrete. Such aggregates should not be used.

9.4 PROPORTIONING CONCRETE MIXES

Principles for proportioning concrete to achieve a prescribed compressive strength after a given age under standard curing are simple.

1. The strength of a hardened concrete mix depends on the water-cement ratio (ratio of water to cementitious material, by weight). The water and cementitious material form a paste. If the past is made with more water, it becomes weaker (Fig. 9.2).

2. The ideal minimum amount of paste is that which will coat all aggregate particles and fill all voids.

3. For practical purposes, fresh concrete must possess workability sufficient for the placement conditions. For a given strength and with given materials, the cost of the mix increases as the workability increases. Additional workability is provided by more fine aggregate and more water, but more cementitious material must also be added to keep the same water-cement ratio.

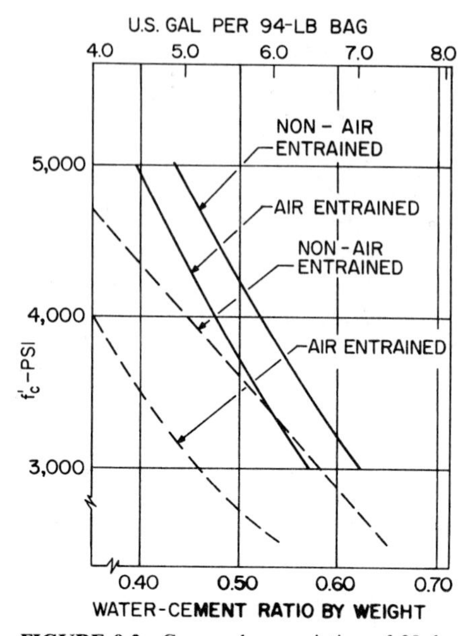

FIGURE 9.2 Curves show variation of 28-day compressive strength of normal-weight concrete with water-cement ratio. Solid lines indicate average results of tests by Materials Service Corp., Chicago. Dashed lines indicate relationship given in the American Concrete Institute Building Code, ACI 318-89, for maximum permissible water-cement ratio and specified 28-day strengths.

Because of the variations in material constituents, temperature, and workability required at job sites, theoretical approaches for determining ideal mix proportions usually do not give satisfactory results on the job. Most concrete therefore is proportioned empirically, in accordance with results from trial batches made with the materials to be used on the job. Small adjustments in the initial basic mix may be made as a job progresses; the frequency of such adjustments usually depends on the degree of quality control.

When new materials or exceptional quality control will be employed, the trial-batch method is the most reliable and efficient procedure for establishing proportions.

In determination of a concrete mix, a series of trial batches (or past field experience) is used to establish a curve relating the water-cement ratio to the strength and ingredient proportions of concrete, including admixtures if specified, for the range of desired strengths and workability (slump). Each point on the curve should represent the average of test results on at least three specimens, and the curve should be determined by at least three points. Depending on anticipated quality control, a demonstrated or expected coefficient of variation or standard deviation is assumed for determination of minimum average strength of test specimens (Art. 9.10). Mix proportions are selected from the curve to produce this average strength.

For any large project, significant savings can be made through use of quality control to reduce the overdesign otherwise required by a building code (law). When the owner's specifications include a minimum content of cementitious material, however, much of the economic incentive for the use of quality control is lost. See Fig. 9.3 for typical water-cement ratios.

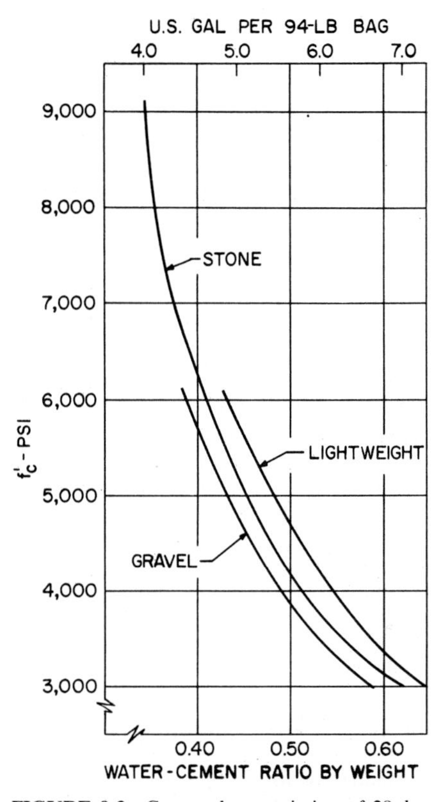

FIGURE 9.3 Curves show variation of 28-day compressive strength of non-air-entrained concrete with type of aggregate and water-cement ratio, except that strengths exceeding 7000 psi were determined at 56 days. All mixes contained a water-reducing agent and 100 lb/yd^3 of fly ash. Calculation of water-cement ratio included two-thirds of the fly-ash weight in the cement content.

Note that separate procedures are required for selecting proportions when lightweight aggregates are used, because their water-absorption properties differ from those of normal-weight aggregates.

("Building Code Requirements for Reinforced Concrete," ACI 318; "Standard Practice for Selecting Proportions for Normal, Heavyweight, and Mass Concrete," ACI 211.1; "Standard Practice for Selecting Proportions for Structural Lightweight Concrete," ACI 211.2; "Recommended Practice for Evaluation of Strength Test

Results of Concrete," ACI 214, American Concrete Institute, P.O. Box 19150, Redford Station, Detroit, MI 48219; "Design and Control of Concrete Mixtures," EB001T, Portland Cement Association, 5420 Old Orchard Road, Skokie, IL 60077.)

9.5 YIELD CALCULATION

Questions often arise between concrete suppliers and buyers regarding "yield," or volume of concrete supplied. A major reason for this is that often the actual yield may be less than the yield calculated from the volumes of ingredients. For example, if the mix temperature varies, less air may be entrained; or if the sand becomes drier and no corrections in batch weights are made, the yield will be under that calculated.

If the specific gravity (sp. gr.) and absorption (abs.) of the aggregates have been determined in advance, accurate yield calculations can be performed as often as necessary to adjust the yield for control of the concrete.

Example

Yield of Non-Air-Entrained Concrete. The following material properties were recorded for materials used in trial batches: fine aggregate (sand) sp. gr. = 2.65, abs. = 1%; coarse aggregate (gravel) sp. gr. = 2.70, abs. = 0.5%; and cement, sp. gr. = 3.15 (typical). These properties are not expected to change significantly as long as the aggregates used are from the same source. The basic mix proportions for 1 yd^3 of concrete, selected from the trial batches are

Cement: 564 lb (6 bags)

Surface-dry sand: 1170 lb

Surface-dry gravel: 2000 lb

Free water: 300 lb/yd^3 (36 gal/yd^3)

Check the yield.

$$\text{Cement volume} = \frac{564}{3.15 \times 62.4} = 2.87\ \text{ft}^3$$

$$\text{Water volume} = 300/62.4 = 4.81\ \text{ft}^3$$

$$\text{Sand volume} = \frac{1170}{2.65 \times 62.4} = 7.08\ \text{ft}^3$$

$$\text{Gravel volume} = \frac{2000}{2.70 \times 62.4} = \underline{11.87\ \text{ft}^3}$$

$$\text{Total volume of solid constituents} = 26.63\ \text{ft}^3$$

Volume of entrapped air = 27 − 26.63 = 0.37 ft^3 (1.4%)

Total weight, lb/yd^3 = 564 + 300 + 1170 + 2000 = 4034

Total weight, lb/ft^3 = 4034/27 = 149.4

Weight of standard 6 × 12 in cylinder (0.1963 ft^3) = 29.3 lb

These results indicate that some rapid field checks should be made. Total weight, lb, divided by the total volume, yd^3, reported on the trip tickets for truck mixers should be about 4000 on this job, unless a different slump was ordered and the proportions adjusted accordingly. If the specified slump for the basic mix was to be reduced, weight, lb/yd^3, should be increased, because less water and cement would be used and the cement paste (water plus cement) weighs $864/7.68 = 113$ $lb/ft^3 < 149.4$ lb/ft^3. If the same batch weights are used for all deliveries, and the slump varies erratically, the yield also will vary. For the same batch weights, a lower slump is associated with underyield, a higher slump with overyield. With a higher slump, overyield batches are likely to be understrength, because some of the aggregate has been replaced by water.

The basic mix proportions in terms of weights may be based on surface-dry aggregates or on oven-dry aggregates. The surface-dry proportions are somewhat more convenient, since absorption then need not be considered in calculation of free water. Damp sand and gravel carry about 5 and 1% free water, respectively. The total weight of this free water should be deducted from the basic mix weight of water (300 lb/yd^3 in the example) to obtain the weight of water to be added to the cement and aggregates. The weight of water in the damp aggregates also should be added to the weights of the sand and gravel to obtain actual batch weights, as reported on truck-mixer delivery tickets.

9.6 PROPERTIES AND TESTS OF FRESH (PLASTIC) CONCRETE

About 2½ gal of water can be chemically combined with each 94-lb sack of cement for full hydration and maximum strength. Water in excess of this amount will be required, however, to provide necessary workability.

Workability. Although concrete technologists define and measure workability and consistency separately and in various ways, the practical user specifies only one—slump (technically a measure of consistency). The practical user regards workability requirements simply as provision of sufficient water to permit concrete to be placed and consolidated without honeycomb or excessive water rise; to make concrete "pumpable" if it is to be placed by pumps; and for slabs, to provide a surface that can be finished properly. These workability requirements vary with the job and the placing, vibration, and finishing equipment used.

Slump is tested in the field very quickly. An open-ended, 12-in-high, truncated metal cone is filled in three equal-volume increments and each increment is consolidated separately, all according to a strict standard procedure (ASTM C143, "Slump of Portland Cement Concrete"). Slump is the sag of the concrete, in, after the cone is removed. The slump should be measured to the nearest ¼ in, which is about the limit of accuracy reproducible by expert inspectors.

Unless the test is performed exactly in accordance with the standard procedure, the results are not comparable and therefore are useless.

The slump test is invalidated if: the operator fails to anchor the cone down by standing on the base wings; the test is performed on a wobbly base, such as forms carrying traffic or a piece of metal on loose pebbles; the cone is not filled by inserting material in small amounts all around the perimeter, or filled and tamped in three equal increments; the top two layers are tamped deeper than their depth plus about 1 in; the top is pressed down to level it; the sample has been transported and permitted to segregate without remixing; unspecified operations, such as tapping the cone, occur; the cone is not lifted up smoothly in one movement; the cone tips over because of filling from one side or pulling the cone to one side; or if the measurement of slump is not made to the center vertical axis of the cone.

Various penetration tests are quicker and more suitable for untrained personnel than the standard slump test. In each case, the penetration of an object into a flat surface of fresh concrete is measured and related to slump. These tests include use of the patented "Kelley ball" (ASTM C360, "Ball Penetration in Fresh Portland Cement Concrete") and a simple, standard tamping rod with a bullet nose marked with equivalent inches of slump.

Air Content. A field test frequently required measures the air entrapped and entrained in fresh concrete. Various devices (air meters) that are available give quick, convenient results. In the basic methods, the volume of a sample is measured, then the air content is removed or reduced under pressure, and finally the remaining volume is measured. The difference between initial and final volume is the air content. (See ASTM C138, C173, and C231.)

Cement Content. Tests on fresh concrete sometimes are employed to determine the amount of cement present in a batch. Although performed more easily than tests on hardened concrete, tests on fresh concrete nevertheless are too difficult for routine use and usually require mobile laboratory equipment.

9.7 PROPERTIES AND TESTS OF HARDENED CONCRETE

The principal properties of concrete with which designers are concerned and symbols commonly used for some of these properties are:

f'_c = specified compressive strength, psi, determined in accordance with ASTM C39 from standard 6- × 12-in cylinders under standard laboratory curing; unless otherwise specified, f'_c is based on tests on cylinders 28 days old

E_c = modulus of elasticity, psi, determined in accordance with ASTM C469; usually assumed as $E_c = w^{1.5}(33)\sqrt{f'_c}$, or for normal-weight concrete (about 145 lb/ft³), $E_c = 57{,}000\sqrt{f'_c}$

w = weight, lb/ft³, determined in accordance with ASTM C138 or C567

f_t = direct tensile strength, psi

f_{ct} = average splitting tensile strength, psi, of lightweight-aggregate concretes determined by the split cylinder test (ASTM C496)

f_r = modulus of rupture, psi, the tensile strength at the extreme fiber in bending (commonly used for pavement design) determined in accordance with ASTM C78

Other properties, frequently important for particular conditions are: durability to resist freezing and thawing when wet and with deicers, color, surface hardness, impact hardness, abrasion resistance, shrinkage, behavior at high temperatures (about 500°F), insulation value at ordinary ambient temperatures, insulation at the high temperatures of a standard fire test, fatigue resistance, and for arctic construction, behavior at cold temperatures (−60 to −75°F). For most of the research on these properties, specially devised tests were employed, usually to duplicate or simulate the conditions of service anticipated. (See "Index to Proceedings of the American Concrete Institute.")

In addition to the formal testing procedures specified by ASTM and the special procedures described in the research references, some practical auxiliary tests, precautions in evaluating tests, and observations that may aid the user in practical applications follow.

Compressive Strength, f'_c. The standard test (ASTM C39) is used to establish the quality of concrete, as delivered, for conformance to specifications. Tests of companion field-cured cylinders measure the effectiveness of the curing (Art. 9.14).

Core tests (ASTM C42) of the hardened concrete in place, if they give strengths higher than the specified f'_c or an agreed-on percentage of f'_c (often 85%), can be used for acceptance of material, placing, consolidation, and curing. If the cores taken for these tests show unsatisfactory strength but companion cores given accelerated additional curing show strengths above the specified f'_c, these tests establish acceptance of the material, placing, and consolidation, and indicate the remedy, more curing, for the low in-place strengths.

For high-strength concretes, say above 5000 psi, care should be taken that the capping material is also high strength. Better still, the ends of the cylinders should be ground to plane.

Indirect testing for compressive strength includes surface-hardness tests (impact hammer). Properly calibrated, these tests can be employed to evaluate field curing. (See also Art. 9.14.)

Modulus of Elasticity E_c. This property is used in all design, but it is seldom determined by test, and almost never as a regular routine test. For important projects, it is best to secure this information at least once, during the tests on the trial batches at the various curing ages. An accurate value will be useful in prescribing camber or avoiding unusual deflections. An exact value of E_c is invaluable for long-span, thin-shell construction, where deflections can be large and must be predicted accurately for proper construction and timing removal of forms.

Tensile Strength. The standard splitting test is a measure of almost pure uniform tension f_{ct}. The beam test (Fig. 9.4*a*) measures bending tension f_r on extreme surfaces (Fig. 9.4*b*), calculated for an assumed perfectly elastic, triangular stress distribution.

The split-cylinder test (Fig. 9.4*c*) is used for structural design. It is not sensitive to minor flaws or the surface condition of the specimen. The most important application of the splitting test is in establishment of design values for reinforcing-steel development length, shear in concrete, and deflection of structural lightweight-aggregate concretes.

The values f_{ct} (Fig. 9.4*d*) and f_r bear some relationship to each other, but are not interchangeable. The beam test is very sensitive, especially to flaws on the surface of maximum tension and to the effect of drying-shrinkage differentials, even between the first and last of a group of specimens tested on the same day. The value f_r is widely used in pavement design, where all testing is performed in the same laboratory and results are then comparable.

Special Properties. Frequently, concrete may be employed for some special purpose for which special properties are more important than those commonly considered. Sometimes, it may be of great importance to enhance one of the ordinary

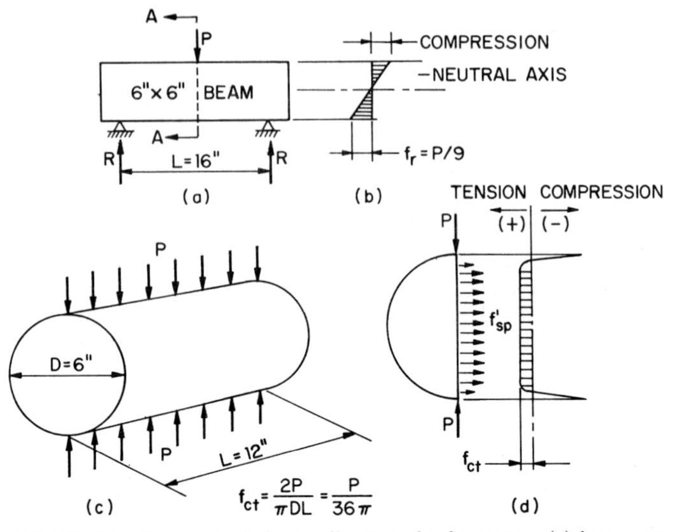

FIGURE 9.4 Test methods for tensile strength of concrete: (*a*) beam test determines modulus of rupture f_r; (*b*) stress distribution assumed for calculation of f_r; (*c*) split-cylinder test measures internal tension f_{ct}; (*d*) stress distribution assumed for f_{ct}.

properties. These special applications often become apparent as new developments using new materials or as improvements using the basic materials. The partial list of special properties is constantly expanding—abrasion and impact resistance (heavy-duty floor surfacings), heat resistance (chimney stacks and jet engine dynamometer cells), light weight (concrete canoes), super-high-compressive strength, over 10 ksi (high-rise columns), waterproof concrete, resistance to chemical attack (bridge decks, chemical industry floors, etc.), increased tensile strength (highway resurfacing, precast products, etc.), no-shrink or expansive concrete (grouting under base plates), etc. Some of these special properties are achieved with admixtures (see Art. 9.9). Some utilize special cements (high-alumina cement for heat resistance or shrinkage-compensating cement for no-shrink concrete). Some utilize special aggregates (lightweight aggregate, steel fiber, plastic fiber, glass fiber, and special heavy aggregate). (See "State-of-the-Art Report on Fiber Reinforced Concrete," ACI 544.1R.) Some special properties—increased compressive and tensile strength, waterproofing, and improved chemical resistance are achieved with polymers, either as admixtures or surface treatment of hardened concrete. (See "Guide for the Use of Polymers in Concrete," ACI 548.1R.)

9.8 MEASURING AND MIXING CONCRETE INGREDIENTS

Methods of measuring the quantities and mixing the ingredients for concrete, and the equipment available, vary greatly. For very small projects where mixing is performed on the site, the materials are usually batched by volume. Under these conditions, accurate proportioning is very difficult. To achieve a reasonable minimum quality of concrete, it is usually less expensive to prescribe an excess of cement than to employ quality control. The same conditions make use of air-entraining cement preferable to separate admixtures. This practical approach is preferable also for very small projects to be supplied with ready-mixed concrete. Economy with excess cement will be achieved whenever volume is so small that the cost of an additional sack of cement per cubic yard is less than the cost of a single compression test.

For engineered construction, some measure of quality control is always employed. In general, all measurements of materials including the cement and water should be by weight. The ACI Building Code (ACI 318) provides a sliding scale of overdesign for concrete mixes that is inversely proportional to the degree of quality control provided. In the sense used here, such overdesign is the difference between the specified f'_c and the actual average strength as measured by tests.

Mixing and delivery of structural concrete may be performed by a wide variety of equipment and procedures:

Site mixed, for delivery by chute, pump, truck, conveyor, or rail dump cars. (Mixing procedure for normal-aggregate concretes and lightweight-aggregate concretes to be pumped are usually different, because the greater absorption of some lightweight aggregates must be satisfied before pumping.)

Central-plant mixed, for delivery in either open dump trucks or mixer trucks.

Central-plant batching (weighing and measuring), for mixing and delivery by truck ("dry-batched" ready mix).

Complete portable mixing plants are available and are commonly used for large building or paving projects distant from established sources of supply.

Generally, drum mixers are used. For special purposes, various other types of mixers are required. These special types include countercurrent mixers, in which the blades revolve opposite to the turning of the drum, usually about a vertical axis, for mixing very dry, harsh, nonplastic mixes. Such mixes are required for concrete masonry or heavy-duty floor toppings. Dry-batch mixers are used for dry shotcrete (sprayed concrete), where water and the dry-mixed cement and aggregate are blended between the nozzle of the gun and impact at the point of placing.

("Guide for Measuring, Mixing, Transporting, and Placing Concrete," ACI 304R.)

9.9 ADMIXTURES

These are materials other than portland cement, cementitious material, fiber reinforcement, aggregate, or water that are added to concrete to modify its properties.

Air Entrainment. Air-entraining admixtures (ASTM C260) may be interground as additives with the cement at the mill or added separately at the concrete mixing plant, or both. Where quality control is provided, it is preferable to add such admixtures at the concrete plant so that the resulting air content can be controlled for changes in temperature, sand, or job requirements.

Use of entrained air is recommended for all concrete exposed to weathering or deterioration from aggressive chemicals. The ACI Building Code (ACI 318) requires air entrainment for all concrete subject to freezing temperatures while wet. Detailed recommendations for air content are available in "Standard Practice for

Selecting Proportions for Normal, Heavyweight, and Mass Concrete," ACI 211.1, and "Standard Practice for Selecting Proportions for Structural Lightweight Concrete," ACI 211.2.

One common misconception relative to air entrainment is the fear that it has a deleterious effect on concrete strength. Air entrainment, however, improves workability. This will usually permit some reduction in water content. For lean, low-strength mixes, the improved workability permits a relatively large reduction in water content, sand content, and water-cement ratio, which tends to increase concrete strength. The resulting strength gain offsets the strength-reducing effect of the air itself, and a net increase in concrete strength is achieved. For rich, high-strength mixes, the relative reduction in the ratio of water to cementitious material, water-cement ratio, is lower and a small net decrease in strength results, about on the same order of the air content (4 to 7%). The improved durability and reduction of segregation in handling, because of the entrained air, usually make air entrainment desirable, however, in all concrete except extremely high-strength mixtures, such as for lower-story interior columns or heavy-duty interior floor toppings for industrial wear.

Accelerators. Calcium chloride for accelerating the rate of strength gain in concrete (ASTM D98) is perhaps the oldest application of admixtures. Old specifications for winter concreting or masonry work commonly required use of a maximum of 1 to 3% $CaCl_2$ by weight of cement for all concrete. Proprietary admixtures now available may include accelerators, but not necessarily $CaCl_2$. The usual objective for use of an accelerator is to reduce curing time by developing 28-day strengths in about 7 days (ASTM C494).

In spite of users' familiarity with $CaCl_2$, a number of misconceptions about its effect persist. It has been sold (sometimes under proprietary names) as an accelerator, a cement replacement, an "antifreeze," a "waterproofer," and a "hardener." It is simply an accelerator; any improvement in other respects is pure serendipity. Recent discoveries, however, indicate possible corrosion damage from indiscriminate use of chloride-containing material in concrete exposed to stray currents, containing dissimilar metals, containing prestressing steel subject to stress corrosion, or exposed to severe wet freezing or salt water. For further information, see "Chemical Admixtures for Concrete," ACI 212.3R.

Retarders. Unless proper precautions are taken, hot-weather concreting may cause "flash set," plastic shrinkage, "cold joints," or strength loss. Admixtures that provide controlled delay in the set of a concrete mix without reducing the rate of strength gain during subsequent curing offer inexpensive prevention of many hot-weather concreting problems. These (proprietary) admixtures are usually combined with water-reducing admixtures that more than offset the loss in curing time due to delayed set (ASTM C494). See "Hot Weather Concreting," ACI 305R, for further details on retarders, methods of cooling concrete materials, and limiting temperatures for hot-weather concreting.

Superplasticizers. These admixtures, which are technically known as "high-range water reducers," produce a high-slump concrete without an increase in mixing water. Slumps of up to 10 in for a period of up to 90 min can be obtained. This greatly facilitates placing concrete around heavy, closely spaced reinforcing, or in complicated forms, or both, and reduces the need for vibrating the concrete. It is important that the slump of the concrete be verified at the job site prior to the addition of the superplasticizer. This ensures that the specified water-cement ratio required for watertight impermeable concrete is in fact being achieved. The superplasticizer is then added to increase the slump to the approved level.

Waterproofing. A number of substances, such as stearates and oils, have been employed as masonry-mortar and concrete admixtures for "waterproofing." Indiscriminate use of such materials in concrete without extremely good quality control usually results in disappointment. The various water-repellent admixtures are intended to prevent capillarity, but most severe leakage in concrete occurs at honeycombs, cold joints, cracks, and other noncapillary defects. Concrete containing water-repellent admixtures also requires extremely careful continuous curing, since it will be difficult to rewet after initial drying.

Waterproof concrete can be achieved by use of high-strength concrete with a low water-cement ratio to reduce segregation and an air-entraining agent to minimize crack width. Also, good quality control and inspection is essential during the mixing, placing, and curing operations. Surface coatings can be used to improve resistance to water penetration of vertical or horizontal surfaces. For detailed information on surface treatments, see "Guide to Durable Concrete," ACI 201.2R.

Cement Replacement. The term "cement replacement" is frequently misused in reference to chemical admixtures intended as accelerators or water reducers. Strictly, a cement replacement is a finely ground material, usually weakly cementitious (Art. 9.1), which combines into a cementlike paste replacing some of the cement paste to fill voids between the aggregates. The most common applications of these admixtures are for low-heat, low-strength mass concrete or for concrete masonry. In the former, they fill voids and reduce the heat of hydration; in the latter, they fill voids and help to develop the proper consistency to be self-standing as the machine head is lifted in the forming process. Materials commonly used are fly ash, silica fume, ground granulated blast-furnace slag, hydraulic lime, natural cement, and pozzolans.

Special-Purpose Admixtures. The list of materials employed from earliest times as admixtures for various purposes includes almost everything from human blood to synthetic coloring agents.

Admixtures for coloring concrete are available in all colors. The oldest and cheapest is perhaps carbon black.

Admixtures causing expansion for use in sealing cracks or under machine bases, etc., include powdered aluminum and finely ground iron.

Special admixtures are available for use where the natural aggregate is alkali reactive, to neutralize this reaction.

Proprietary admixtures are available that increase the tensile strength or bond strength of concrete. They are useful for making repairs to concrete surfaces.

For special problems requiring concrete with unusual properties, detailed recommendations of "Chemical Admixtures for Concrete," ACI 212.3R, and references it contains, may be helpful.

For all these special purposes, a thorough investigation of admixtures proposed is recommended. Tests should be made on samples containing various proportions for colored concrete. Strength and durability tests should be made on concrete to be exposed to sunlight, freezing, salt, or any other job condition expected, and special tests should be made for any special properties required, as a minimum precaution.

QUALITY CONTROL

9.10 MIX DESIGN

Concrete mixes are designed with the aid of information obtained from trial batches or field experience with the materials to be used. In either case, the proportions of ingredients must be selected to produce, for at least three test specimens, an average strength f'_{cr} greater than the specified strength f'_c.

The required excess $f'_{cr} - f'_c$, depends on the standard deviation σ expected. Strength data for determining the standard deviation can be considered suitable if they represent either a group of at least 30 consecutive tests representing materials and conditions of control similar to those expected or the statistical average for two groups totaling 30 or more tests. The tests used to establish standard deviation should represent concrete produced to meet a specified strength within 1000 psi of that specified for the work proposed.

$$\sigma = \sqrt{\frac{(x_1 - \bar{x})^2 + (x_2 - \bar{x})^2 + (x_3 - \bar{x})^2 + \cdots + (x_n - \bar{x})^2}{n - 1}} \tag{9.1}$$

where $x_1, x_2, \ldots, x_n$ = strength, psi, obtained in test of first, second, . . . , nth sample, respectively
n = number of tests
$\bar{x}$ = average strength, psi of n cylinders

Coefficient of variation is the standard deviation expressed as a percentage of the average strength. ("Recommended Practice for Evaluation of Strength Test Results of Concrete," ACI 214.)

The strength used as a basis for selecting proportions of a mix should exceed the required f'_c by at least the amount indicated in Table 9.1.

The values for f'_{cr} in Table 9.1 are the larger of the values calculated from Eqs. (9.2) and (9.3).

$$f'_{cr} = f'_c + 1.343\sigma \tag{9.2}$$

$$f'_{cr} = f'_c + 2.326\sigma - 500 \tag{9.3}$$

For an established supplier of concrete, it is very important to be able to document the value of σ. This value is based on a statistical analysis in which Eq. (9.1) is applied to at least 30 consecutive tests. These tests must represent similar materials and conditions of control not stricter than those to be applied to the proposed project. The lower the value of σ obtained from the tests, the closer the average strength is permitted to be to the specified strength. A supplier is thus furnished an economic incentive, lower cement content, to develop a record of good control (low σ). A supplier who does maintain such a record can, in addition, avoid the expenses of trial batches.

Where no such production record exists, trial batches must be used as a basis for selecting initial proportions. This condition is likely to occur when new sources of cement or aggregate are supplied to an established plant, to a new facility, such as a portable plant on the site, or for the first attempt at a specified strength f'_c more than 1000 psi above previous specified strengths.

TABLE 9.1 Recommended Average Strengths of Test Cylinders for Selecting Proportions for Concrete Mixes

Range of standard deviation σ, psi	Average strength f'_{cr}, psi
Under 300	$f'_c + 400$
300–400	$f'_c + 550$
400–500	$f'_c + 700$
500–600	$f'_c + 900$
Over 600	$f'_c + 1200$

If the specified $f'_c \leq 5000$ psi for non-air-entrained concrete or $f'_c \leq 4500$ psi for air-entrained normal-weight concrete, trial batches need not be employed. Table 9.2 lists maximum water-cement ratios that may be used to select *initial* proportions for normal-weight concrete. These ratios are expected to provide $f'_{cr} - f'_c \geq 1200$ psi and are economical only for small projects where the cost of trial batches is not justified.

Note that the term "initial proportions" includes cement. The initial proportions can be used during progress of a project only as long as the strength-test results justify them. The process of quality control of concrete for a project requires maintenance of a running average of strength-test results and changes in the proportions whenever the actual degree of control (standard deviation σ) varies from

TABLE 9.2 Maximum Water-Cement Ratios for Specified Concrete Strengths*

Specified compressive strength f'_c, psi†	Non-air-entrained concrete		Air-entrained concrete	
	Absolute ratio by weight	U.S. gal per 94-lb bag of cement	Absolute ratio by weight	U.S. gal per 94-lb bag of cement
2500	0.67	7.6	0.54	6.1
3000	0.58	6.6	0.46	5.2
3500	0.51	5.8	0.40	4.5
4000	0.44	5.0	0.35	4.0
4500	0.38	4.3	‡	‡
5000	‡	‡	‡	‡

*As specified by ACI 318-89, Table 5.4.

†28-day strengths for cements meeting strength limits of ASTM C150 Types I, IA, II, or IIA, and 7-day strengths for Types III or IIIA. With most materials, the water-cement ratios shown will provide average strengths greater than indicated.

‡For strengths above 4500 psi (non-air-entrained concrete) and 4000 psi (air-entrained concrete), proportions should be established by field experience or laboratory trial batches.

that assumed for the initial proportioning. Equations (9.2) and (9.3) are applied for this analysis. With contract specifications based on the ACI Building Code, no minimum cementitious-material content is required; so good control during a long-time project is rewarded by permission to use a lower cementitious-material content than would be permitted with inferior control.

Regardless of the method (field experience, trial batches, or maximum water-cement ratio) used, the basic initial proportions should be based on mixes with both air content and slump at the maximum permitted by the specifications.

Other ACI Building Code requirements for mix design are:

1. Concrete exposed to freezing and thawing while wet should have air entrained within the limits in Table 9.3, and the water-cement ratio by weight should not exceed 0.45. If lightweight aggregate is used, f'_c should be at least 4250 psi.

2. For watertight, normal-weight concrete, maximum water-cement ratios by weight are 0.50 for exposure to fresh water and 0.40 for seawater. With lightweight aggregate, minimum f'_c is 3750 psi for concrete exposed to fresh water and f'_c is 4750 psi for seawater.

TABLE 9.3 Required Air Entrainment in Concrete Exposed to Freezing and Thawing while Wet*

Nominal maximum size of coarse aggregate, in	Total air content, % by volume	
	Severe exposure	Moderate exposure
3/8	7½	6
½	7	5½
¾	6	5
1	6	4½
1½	5½	4½
2	5	4
3	4½	3½

*From ACI 318-89, Table 4.1.1. For $f'_c > 5000$ psi, air content may be reduced 1%.

Although the Code does not distinguish between a "concrete production facility" with in-house control and an independent concrete laboratory control service, the distinction is important. Very large suppliers have in-house professional quality control. Most smaller suppliers do not. Where the records of one of the latter might indicate a large standard deviation, but an independent quality-control service is utilized, the standard deviation used to select $f'_{cr} - f'_c$ should be based on the proven record of the control agency. Ideally, the overdesign should be based, in these cases, on the record of the control agency operating in the concrete plant used.

9.11 CHECK TESTS OF MATERIALS

Without follow-up field control, all the statistical theory involved in mix proportioning becomes an academic exercise.

The complete description of initial proportions should include: cement analysis and source; specific gravity, absorption, proportions of each standard sieve size; fineness modulus; and organic tests for fine and coarse aggregates used, as well as their weights and maximum nominal sizes.

If the source of any aggregate is changed, new trial batches should be made. A cement analysis should be obtained for each new shipment of cement.

The aggregate gradings and organic content should be checked at least daily, or for each 150 yd^3. The moisture content (or slump) should be checked continuously for all aggregates, and suitable adjustments should be made in batch weights. When the limits of ASTM C33 or C330 for grading or organic content are exceeded, proper materials should be secured and new mix proportions developed, or until these measures can be effected, concrete production may continue on an emergency basis but with a penalty of additional cement.

9.12 AT THE MIXING PLANT—YIELD ADJUSTMENTS

Well-equipped concrete producers have continuous measuring devices to record changes in moisture carried in the aggregates or changes in total free water in the contents of the mixer. The same measurements, however, may be easily made manually by quality-control personnel.

To illustrate: for the example in Art. 9.5, the surface-dry basic mix is cement, 564 lb; water, 300 lb; sand, 1170 lb; and gravel, 2000 lb. Absorption is 1% for the sand and 0.5% for the gravel. If the sand carries 5.5% and the gravel 1.0% total water by weight, the added free water becomes:

Sand: $1170\ (0.055 - 0.01) = 53$ lb

Gravel: $2000\ (0.010 - 0.005) = 10$ lb

Batch weights adjusted for yield become:

Cement: 564 lb

Water: $300 - 53 - 10 = 237$ lb

Sand: $1170 + 53 = 1223$ lb

Gravel: $2000 + 10 = 2010$ lb

Note that the corrective adjustment includes adding to aggregate weights as well as deducting water weight. Otherwise, the yield will be low, and slump (slightly) increased. The yield would be low by about

$$\frac{53 + 10}{2.65 \times 62.4} = 0.381\ \text{ft}^3/\text{yd}^3 = 1.4\%$$

9.13 AT THE PLACING POINT—SLUMP ADJUSTMENTS

With good quality control, no water is permitted on the mixing truck. If the slump is too low (or too high) on arrival at the site, additional cement must be added. If the slump is too low (the usual complaint), additional water and cement in the prescribed water-cement ratio can also be added. After such additions, the contents must be thoroughly mixed, 2 to 3 min at high speed. Because placing-point adjustments are inconvenient and costly, telephone or radio communication with the supply plant is desirable so that most such adjustments may be made conveniently at the plant.

Commonly, a lesser degree of control is accepted in which the truck carries water, the driver is on the honor system not to add water without written authorization from a responsible agent at the site, and the authorization as well as the amounts added are recorded on the record (trip ticket) of batch weights.

Note: If site adjustments are made, test samples for strength-test specimens should be taken only after all site adjustments. For concrete in critical areas, such as lower-floor columns in high-rise buildings, strictest quality control is recommended.

9.14 STRENGTH TESTS

Generally, concrete quality is measured by the specified compressive strength f'_c of 6- × 12-in cylinders after 28 days of laboratory curing.

Conventional Tests. The strength tests performed after various periods of field curing are typically specified to determine curing adequacy. For lightweight-aggregate concretes only, the same type of laboratory-cured test specimen is tested for tensile splitting strength f_{ct} to establish design values for deflection, development of reinforcing steel, and shear. Applicable ASTM specifications for these tests are

C21, "Making and Curing Concrete Compressive and Flexural Strength Test Specimens in the Field."

C39, "Test for Compressive Strength of Molded Concrete Cylinders."

C496, "Test for Splitting Tensile Strength of Molded Concrete Cylinders."

The specifications for standard methods and procedures of testing give general directions within which the field procedures can be adjusted to job conditions. One difficulty arises when the specimens are made in the field from samples taken at the job site. During the first 48 h after molding, the specimens are very sensitive to damage and variations from standard laboratory curing conditions, which can significantly reduce the strength-test results. Yet, job conditions may preclude sampling, molding, and field storage on the same spot.

If the fresh-concrete sample must be transported more than about 100 ft to the point of molding cylinders, some segregation occurs. Consequently, the concrete sample should be remixed to restore its original condition. After the molds for test cylinders have been filled, if the specimens are moved, high-slump specimens segregate in the molds; low-slump specimens in the usual paper or plastic mold are often squeezed out of shape or separated into starting cracks. Such accidental damage varies with slump, temperature, time of set and molding, and degree of carelessness.

If the specimen cylinders are left on the job site, they must be protected against drying and accidental impact from construction traffic. If a worker stumbles over a specimen less than 3 days old, it should be inspected for damage. The best practice is to provide a small, insulated, dampproofed, locked box on the site in which specimens can be cast, covered, and provided with 60 to 80°F temperature and 100% humidity for 24 to 72 h. Then, they can be transported and subjected to standard laboratory curing conditions at the testing laboratory. When transported, the cylinders should be packed and handled like fresh eggs, since loose rattling will have about an equivalent effect in starting incipient cracks.

Similarly, conditions for field-cured cylinders must be created as nearly like those of the concrete in place as possible. Also, absolute protection against impact or other damage must be provided. Because most concrete in place will be in much larger elements than a test cylinder, most of the in-place concrete will benefit more from retained heat of hydration (Fig. 9.5). This effect decreases rapidly, because the rate of heat development is greatest initially. To ensure similar curing conditions, field-cured test cylinders should be stored for the first 24 h in the field curing box with the companion cylinders for laboratory curing. After this initial curing, the field-cured cylinders should be stored near the concrete they represent and cured under the same conditions.

Exceptions to this initial curing practice arise when the elements cast are of dimensions comparable to those of the cylinders, or the elements cast are not

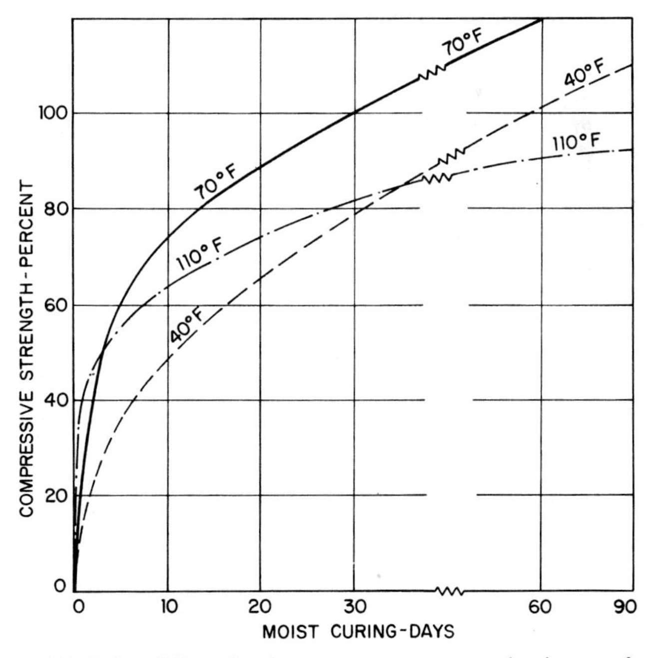

FIGURE 9.5 Effect of curing temperature on strength-gain rate of concrete, with 28-day strength as basis.

protected from drying or low temperatures, including freezing, or test cylinders are cured *inside* the elements they represent (patented system).

These simple, seemingly overmeticulous precautions will eliminate most of the unnecessary, expensive, job-delaying controversies over low tests. Both contractor and owner are justifiably annoyed when costly later tests on hardened concrete, after an even more costly job delay, indicate that the original fresh-concrete test specimens were defective and not the building concrete.

Special Tests. Many other strength tests or tests for special qualities are occasionally employed for special purposes. Those most often encountered in concrete building construction are strength tests on drilled cores and sawed beams (ASTM C42); impact tests (ASTM C805), e.g., Schmidt hammer; pullout tests (ASTM C900); penetration tests (ASTM C803); determination of modulus of elasticity during the standard compression test; and deflection measurements on a finished building element under load (Chap. 20, ACI 318-89). (See also "Commentary on ACI 318-89" and the "Manual of Concrete Inspection," (ACI SP-2.)

Newer methods for evaluating in-situ strength of concrete include the following: Methods, such as the one in which test cylinders are field-cured inside the in-situ concrete, measure compressive strength directly, refined even to measuring it in a desired direction. Others actually measure other properties, such as penetration, impact, or pullout, which are indirect measures of compressive strength, but may be employed because the property they measure is itself important. For example, in cantilevered form construction where forms for each new lift are bolted into the previous lift, pullout results may be more meaningful than standard compression tests. (See "Testing Hardened Concrete," ACI Monograph No. 9, 1976.) Most of the in-situ tests may also be classified as accelerated tests, although not all accelerated tests are performed in situ.

Because construction time is continually becoming a more important factor in overall construction economy, the standard 28-day strength becomes less significant. For example, the final strength at completion of a high-rise project requiring high-strength concrete in lower-story columns is often specified at 90 days. At the other extreme, a floor system may be loaded by the forms and concrete for the floor above in as little as 2 days. These conditions demand accelerated testing. (See "Use of Accelerated Strength Testing," ACI 214.1R.)

9.15 TEST EVALUATION

On small jobs, the results of tests on concrete after the conventional 28 days of curing may be valuable only as a record. In these cases, the evaluation is limited to three options: (1) accept results, (2) remove and replace faulty concrete, or (3) conduct further tests to confirm option (1) or (2) or for limited acceptance at a lower-quality rating. The same comment can be applied to a specific element of a large project. If the element supports 28 days' additional construction above, the consequences of these decisions are expensive.

Samples sufficient for at least five strength tests of each class of concrete should be taken at least once each day, or once for each 150 yd^3 of concrete or each 5000 ft^2 of surface area placed. Each strength test should be the average for two cylinders from the same sample. The strength level of the concrete can be considered satisfactory if the averages of all sets of three consecutive strength-test results equal or exceed the specified strength f'_c and no individual strength-test result falls below f'_c by more than 500 psi.

If individual tests of laboratory-cured specimens produce strengths more than 500 psi below f'_c, steps should be taken to assure that the load-carrying capacity of the structure is not jeopardized. Three cores should be taken for each case of a cylinder test more than 500 psi below f'_c. If the concrete in the structure will be dry under service conditions, the cores should be air-dried (temperature 60 to 80°F, relative humidity less than 60%) for 7 days before the tests and should be tested dry. If the concrete in the structure will be more than superficially wet under service conditions, the cores should be immersed in water for at least 48 h and tested wet.

Regardless of the age on which specified design strength f'_c is based, large projects of long duration offer the opportunity for adjustment of mix proportions during the project. If a running average of test results and deviations from the average is maintained, then, with good control, the standard deviation achieved may be reduced significantly below the usually conservative, initially assumed standard deviation. In that case, a saving in cement may be realized from an adjustment corresponding to the improved standard deviation. If control is poor, the owner must be protected by an increase in cement. Specifications that rule out either adjustment are likely to result in less attention to quality control.

FORMWORK

For a recommended overall basis for specifications and procedures, see "Guide to Formwork for Concrete," ACI 347R. For materials, details, etc., for builders, see "Formwork for Concrete," ACI SP-4.

9.16 RESPONSIBILITY FOR FORMWORK

The exact legal determination of responsibilities for formwork failures among owner, architect, engineer, general contractor, subcontractors, or suppliers can be determined only by a court decision based on the complete contractual arrangements undertaken for a specific project.

Generally accepted practice makes the following rough division of responsibilities:

Safety. The general contractor is responsible for the design, construction, and safety of formwork. Subcontractors or material suppliers may subsequently be held responsible to the general contractor. The term "safety" here includes prevention of any type of formwork failure. The damage caused by a failure always includes the expense of the form itself, and may also include personal injury or damage to the completed portions of a structure. Safety also includes protection of all personnel on the site from personal injury during construction. Only the supervisor of the work can control the workmanship in assembly and the rate of casting on which form safety ultimately depends.

Structural Adequacy of the Finished Concrete. The structural engineer is responsible for the design of the concrete structure. The reason for project specifications requiring that the architect or engineer approve the order and time of form removal, shoring, and reshoring is to ensure proper structural behavior during such removal and to prevent overloading of recently constructed concrete below or damage to the concrete from which forms are removed prematurely. The architect or engineer should require approval for locations of construction joints not shown on project drawings or project specifications to ensure proper transfer of shear and other forces through these joints. Project specifications should also require that debris be cleaned from form material and the bottom of vertical element forms, and that form-release agents used be compatible with appearance requirements and future finishes to be applied. None of these considerations, however, involves the safety of the formwork per se.

9.17 MATERIALS AND ACCESSORIES FOR FORMS

When a particular design or desired finish imposes special requirements, and only then, the engineers' project specifications should incorporate these requirements and preferably require sample panels for approval of finish and texture. Under competitive bidding, best bids are secured when the bidders are free to use ingenuity and their available materials ("Formwork for Concrete," ACI SP-4).

9.18 LOADS ON FORMWORK

Formwork should be capable of supporting safely all vertical and lateral loads that might be applied to it until such loads can be supported by the ground, the concrete structure, or other construction with adequate strength and stability. Dead loads on formwork consist of the weight of the forms and the weight of and pressures from freshly placed concrete. Live loads include weights of workers, equipment, material storage, and runways, and accelerating and braking forces from buggies and other placement equipment. Impact from concrete placement also should be considered in formwork design.

Horizontal or slightly inclined forms often are supported on vertical or inclined support members, called shores, which must be left in place until the concrete placed in the forms has gained sufficient strength to be self-supporting. The shores may be removed temporarily to permit the forms to be stripped for reuse elsewhere, if the concrete has sufficient strength to support dead loads, but the concrete should then be reshored immediately. Loads assumed for design of shoring and reshoring of multistory construction should include all loads transmitted from the stories above as construction proceeds.

9.18.1 Pressure of Fresh Concrete on Vertical Forms

This may be estimated from

$$p = 150 + 9000 \frac{R}{T} \tag{9.4}$$

where p = lateral pressure, psf
R = rate of filling, ft/h
T = temperature of concrete, °F

See Fig. 9.6*a*.

For columns, the maximum pressure p_{max} is 3000 psf or $150h$, whichever is less, where h = height, ft, of fresh concrete above the point of pressure. For walls where R does not exceed 7 ft/h, p_{max} = 2000 psf or $150h$, whichever is less.

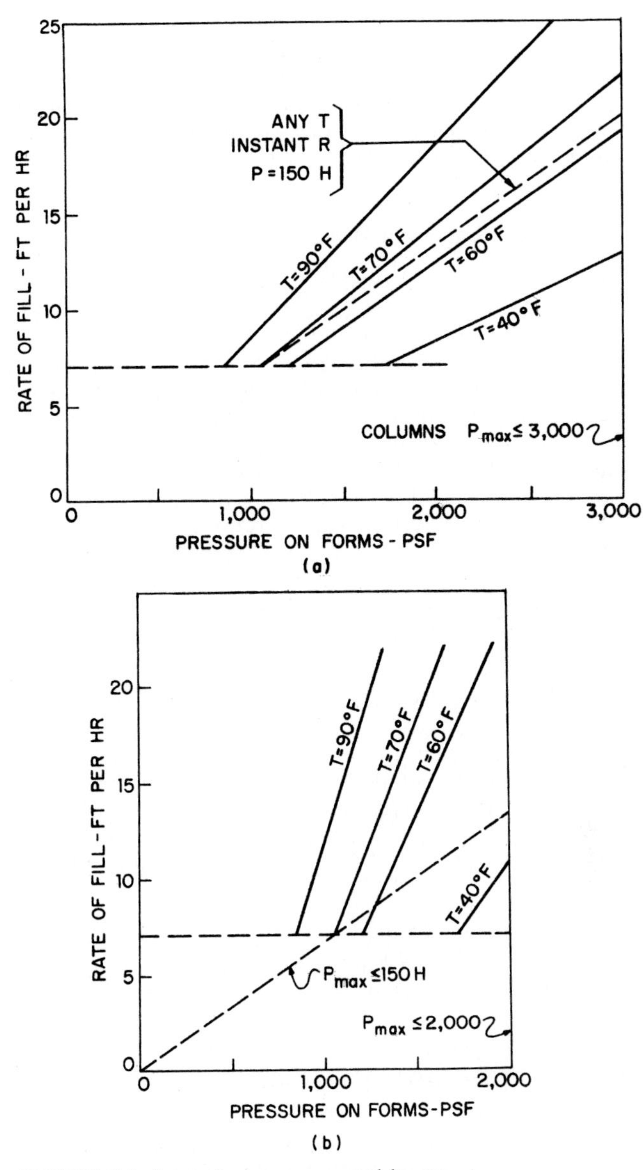

FIGURE 9.6 Internal pressures exerted by concrete on formwork: (*a*) column forms; (*b*) wall forms.

For walls with rate of placement $R > 7$,

$$p = 150 + \frac{43{,}400}{T} + 2800\,\frac{R}{T} \tag{9.5}$$

where $p_{\max} = 2000$ psf or $150h$, whichever is less. See Fig. 9.6*b*.

The calculated form pressures should be increased if concrete weight exceeds 150 psf, cements are used that are slower setting than standard portland cement, slump is more than 4 in with use of superplasticizers, retarders are used to slow set, the concrete is revibrated full depth, or forms are externally vibrated. Under these conditions, a safe design assumes that the concrete is a fluid with weight w and $p_{max} = wh$ for the full height of placement.

9.18.2 Design Vertical Loads for Horizontal Forms

Best practice is to consider all known vertical loads, including the formwork itself, plus concrete, and to add an allowance for live load. This allowance, including workers, runways, and equipment, should be at least 50 psf. When concrete will be distributed from overhead by a bucket or by powered buggies, an additional allowance of at least 25 psf for impact load should be added. Note that the weight of a loaded power buggy dropping off a runway, or an entire bucket full of concrete dropped at one spot, is not considered and might exceed designs based on 50- or 75-psf live load. Formwork should be designed alternatively, with continuity, to accept such spot overloads and distribute them to various unloaded areas, or with independently braced units to restrict a spot overload to a spot failure. The first alternative is preferable.

9.18.3 Lateral Loads for Shoring

Most failures of large formwork are "progressive," vertically through several floors, or horizontally, as each successive line of shoring collapses like a house of cards. To eliminate all possibility of a large costly failure, the overall formwork shoring system should be reviewed before construction to avoid the usual "house-of-cards" design for vertical loads only. Although it is not always possible to foresee exact sources or amounts of lateral forces, shoring for a floor system should be braced to resist at least 100 lb/lin ft acting horizontally upon any of the edges, or a total lateral force on any edge equal to 2% of the total dead loads on the floor, whichever is larger.

Wall forms should be braced to resist local building-code wind pressures, plus at least 100 lb/lin ft at the top in either direction. The recommendation applies to basement wall forms even though wind may be less, because of the high risk of personal injury in the usual restricted areas for form watchers and other workers.

9.19 FORM REMOVAL AND RESHORING

Much friction between contractors' and owners' representatives is created because of misunderstanding of the requirements for form removal and reshoring. The contractor is concerned with a fast turnover of form reuse for economy (with safety), whereas the owner wants quality, continued curing for maximum in-place strength, and an adequate strength and modulus of elasticity to minimize initial deflection and cracking. Both want a satisfactory surface.

Satisfactory solutions for all concerned consist of the use of high-early-strength concrete or accelerated curing, or substitution of a means of curing protection other than formwork. The use of field-cured cylinders (Arts. 9.7 and 9.14) in conjunction with appropriate nondestructive in-place strength tests (Art. 9.14) enables owner and contractor representatives to measure the rate of curing to determine the earliest time for safe form removal.

Reshoring or ingenious formwork design that keeps shores separate from surface forms, such as "flying forms" that are attached to the concrete columns, permits early stripping without premature stress on the concrete. Properly performed, reshoring is ideal from the contractors' viewpoint. But the design of reshores several stories in depth becomes very complex. The loads delivered to supporting floors are very difficult to predict and often require a higher order of structural analysis than that of the original design of the finished structure. To evaluate these loads, knowledge is required of the modulus of elasticity E_c of each floor (different), properties of the shores (complicated in some systems by splices), and the initial stress in the shores, which is dependent on how hard the wedges are driven or the number of turns of screw jacks, etc. ("Formwork for Concrete," ACI SP-4). When stay-in-place shores are used, reshoring is simpler (because variations in initial stress, which depend on workmanship, are eliminated), and a vertically progressive failure can be averted.

One indirect measure is to read deflections of successive floors at each stage. With accurate measurements of E_c, load per floor can then be estimated by structural theory. A more direct measure (seldom used) is strain measurement on the shores, usable with metal shores only. On large projects, where formwork cost and cost of failure justify such expense, both types of measurement can be employed.

9.20 SPECIAL FORMS

Special formwork may be required for uncommon structures, such as folded plates, shells, arches, and posttensioned-in-place designs, or for special methods of construction, such as slip forming with the form rising on the finished concrete or with the finished concrete descending as excavation progresses, permanent forms of any type, preplaced-grouted-aggregate concreting, underwater concreting, and combinations of precast and cast-in-place concreting.

9.21 INSPECTION OF FORMWORK

Inspection of formwork for a building is a service usually performed by the architect, engineer, or both, for the owner and, occasionally, directly by employees of the owner. Formwork should be inspected before the reinforcing steel is in place to ensure that the dimensions and location of the concrete conform to design drawings (Art. 9.16). This inspection would, however, be negligent if deficiencies in the areas of contractor responsibility were not noted also.

(See "Guide to Formwork for Concrete," ACI 347R, and "Formwork for Concrete," ACI SP-4, for construction check lists, and "Manual of Concrete Inspection," ACI SP-2.)

REINFORCEMENT

9.22 REINFORCING BARS

The term *deformed steel bars for concrete reinforcement* is commonly shortened to *rebars*. The short form will be used in this section.

TABLE 9.4 Standard United States Rebar Sizes

Bar size no.	Nominal weight, lb per ft	Nominal diameter, in	Nominal cross-sectional area, in^2
3	0.376	0.375	0.11
4	0.668	0.500	0.20
5	1.043	0.625	0.31
6	1.502	0.750	0.44
7	2.044	0.875	0.60
8	2.670	1.000	0.79
9	3.400	1.128	1.00
10	4.303	1.270	1.27
11	5.313	1.410	1.56
14	7.65	1.693	2.25
18	13.60	2.257	4.00

Standard United States rebars are produced in 11 sizes, designated on design drawings and project specifications by a size number (Table 9.4). In accordance with type of steel used and whether the bars are coated, rebars may be specified to conform with ASTM standard specifications as billet steel (ASTM A615), rail steel (A616), axle steel (A617), low-alloy steel (A706), epoxy-coated (A775), or galvanized (A767). The coated bars are used where corrosion protection is desired in reinforced concrete structures. Table 9.5 shows the bar sizes and strength grades

TABLE 9.5 Rebar Sizes and Grades Conforming to ASTM Standard Bar Specifications

Type of steel bars and ASTM designation	Bar numbers (sizes)	Grade*
Billet steel A615	3–6	40
	3–11, 14, 18	60
	11, 14, 18	75
Rail steel A616	3–11	50
	3–11	60
Axle steel A617	3–11	40
	3–11	60
Low-alloy steel A706	3–11, 14, 18	60

*Minimum-yield designation.

covered by each specification for bare rebars. The grade number indicates minimum yield strength, ksi, of the steel. Grade 60 billet-steel rebars, with a specified minimum yield strength of 60 ksi, are the most widely used type.

Rebars conforming to ASTM A706 are intended for applications where controlled tensile properties are essential, for example, in seismic design. The A706 specification also includes requirements to enhance ductility and bendability, and the rebars conforming to this specification may be welded in accordance with procedures specified in "Structural Welding Code—Reinforcing Steel," AWS D1.4, American Welding Society. Weldability is accomplished by setting limits or controls on chemical composition. Such bars may not be readily available. Before they are specified, their availability should be investigated.

Rebars conforming to ASTM A615, A616, and A617 are not produced to meet weldability requirements. They may be welded, however, if it can be shown that the steel has properties necessary to conform to procedures specified in AWS D1.4.

9.23 WELDED-WIRE FABRIC (WWF)

Welded-wire fabric is an orthogonal grid made with two kinds of cold-drawn wire: plain or deformed. The wires can be spaced in each direction of the grid as desired, but for buildings, usually at 12 in maximum. Sizes of wires available in each type, with standard and former designations, are shown in Table 9.6.

Welded-wire fabric usually is designated WWF on drawings. Sizes of WWF are designated by spacing followed by wire sizes; for example, WWF 6 × 12, W12/W8, which indicates plain wires, size W12, spaced at 6 in, and size W8, spaced at 12 in. WWF 6 × 12, D-12/D-8 indicated deformed wires of the same nominal size and spacing.

All WWF can be designed for Grade 60 material. Wire and welded-wire fabric are produced to conform with the following ASTM standard specifications:

ASTM A82, Plain Wire

ASTM A496, Deformed Wire

ASTM A185, Plain Wire, WWF

ASTM A497, Deformed Wire, WWF

Epoxy-coated wire and welded wire fabric are covered by the ASTM specification A884. Applications of epoxy-coated wire and WWF include use as corrosion-protection systems in reinforced concrete structures and reinforcement in reinforced-earth construction, such as mechanically stabilized embankments.

9.24 PRESTRESSING STEEL

Cold-drawn high-strength wires, singly or stranded, with ultimate tensile strengths up to 270 ksi, and high-strength, alloy-steel bars, with ultimate tensile strengths up to 160 ksi, are used in prestressing. The applicable specifications for wire and strands are:

ASTM A416, Uncoated Seven-Wire Stress-Relieved Strand

ASTM A421, Uncoated Stress-Relieved Wire

ASTM A722, Uncoated High-Strength Bar

TABLE 9.6 Standard Wire Sizes for Reinforcement

Size of deformed wire (A496)	Size of plain wire (A82)	Nominal dia, in	Nominal area, in^2	Size of deformed wire (A496)	Size of plain wire (A82)	Nominal dia, in	Nominal area, in^2
D-31	W31	0.628	0.310	D-12	W12	0.390	0.120
D-30	W30	0.618	0.300	D-11		0.374	0.110
D-29		0.608	0.290	D-10	W10	0.356	0.100
D-28	W28	0.597	0.280	D-9		0.338	0.090
D-27		0.586	0.270	D-8	W8	0.319	0.080
D-26	W26	0.575	0.260	D-7		0.298	0.070
D-25		0.564	0.250	D-6	W6	0.276	0.060
D-24	W24	0.553	0.240		W5.5	0.265	0.055
D-23		0.541	0.230	D-5	W5	0.252	0.050
D-22	W22	0.529	0.220		W4.5	0.239	0.045
D-21		0.517	0.210	D-4	W4	0.225	0.040
D-20	W20	0.504	0.200		W3.5	0.211	0.035
D-19		0.491	0.190	D-3		0.195	0.030
D-18	W18	0.478	0.180		W2.9	0.192	0.029
D-17		0.465	0.170		W2.5	0.178	0.025
D-16	W16	0.451	0.160	D2	W2	0.159	0.020
D-15		0.437	0.150		W1.4	0.134	0.014
D-14	W14	0.422	0.140		W1.2	0.124	0.012
D-13		0.406	0.130	D-1		0.113	0.010
					W0.5	0.080	0.005

The American Concrete Institute Building Code, ACI, 318, also permits use of other types of strand, wire, or bar, including galvanized wire or strand, that meet the minimum requirement of ASTM A416, A421, or A722.

Single strands are used for plant-made pretensioned, prestressed members. Posttensioned prestressing may be performed with the member in place, on a site fabricating area, or in a plant. Posttensioned tendons usually consist of strands or bars. Single wires, grouped into parallel-wire tendons, may also be used in posttensioned applications.

9.25 FABRICATION AND PLACING OF REBARS

Fabrication of rebars consists of cutting to length and required bending. The preparation of field placing drawings and bar lists is termed *detailing*. Ordinarily, the rebar supplier details, fabricates, and delivers to the site, as required. In the farwestern states, the rebar supplier also ordinarily places the bars. In the New York

City area, fabrication is performed on the site by the same (union) workers who place the reinforcement. (See "Details and Detailing of Concrete Reinforcement," ACI 315).

Standard Hooks. The geometry and dimensions of standard hooks that conform to the American Concrete Institute Building Code, ACI 318, and industry practice are shown in Table 9.7.

Fabrication Tolerances. These are covered in "Standard Specifications for Tolerances for Concrete Construction and Materials," ACI 117.

Shipping Limitations. Shipping widths or loading limits for a single bent bar and an L-shaped bar are shown in Fig. 9.7. Bundles of bars occupy greater space. The limit of 7 ft 4 in has been established as an industry practice to limit the bundle size to an 8-ft maximum load width. ("Manual of Standard Practice," Concrete Reinforcing Steel Institute.)

Erection. For construction on small sites, such as high-rise buildings in metropolitan areas, delivery of materials is a major problem. Reinforcement required for each area to be concreted at one time is usually delivered separately. Usually, the only available space for storage of this steel is the formwork in place. Under such conditions, unloading time becomes important.

The bars for each detail length, bar size, or mark number are wired into *bundles* for delivery. A *lift* may consist of one or more bundles grouped together for loading or unloading. The maximum weight of a single lift for unloading is set by the job crane capacity. The maximum weight of a *shop lift* for loading is usually far larger, and so shop lifts may consist of several separately bundled *field lifts.* Regional practices and site conditions establish the maximum weight of bundles and lifts. Where site storage is provided, the most economical unloading without an immediately available crane is by dumping or rolling bundles off the side. Unloading arrangements should be agreed on in advance, so that loading can be carried out in the proper order and bars bundled appropriately. Care must be exercised during the unloading and handling of epoxy-coated rebars to minimize damage to the coating. ("Placing Reinforcing Bars," CRSI.)

Placement Tolerances. The ACI Building Code prescribes rebar placement tolerances applicable simultaneously to effective depth d and to concrete cover in all flexural members, walls, and columns as follows:

Where d is 8 in or less, $\pm 3/8$ in; more than 8 in, $\pm 1/2$ in. The tolerance for the clear distance to formed soffits is $-1/4$ in. For additional information on tolerances, see "Standard Specifications for Tolerances for Concrete Construction and Materials," ACI 117. These tolerances may not reduce cover more than one-third of that specified. The tolerance on longitudinal position of bends or ends of bars is ± 2 in. At discontinuous ends, the specified cover cannot be reduced more than $1/2$ in.

Bundling. Rebars may be placed in concrete members singly or in bundles (up to four No. 11 or smaller bars per bundle). This practice reduces rebar congestion or the need for several layers of single, parallel bars in girders. For columns, it eliminates many interior ties and permits use of No. 11 or smaller bars where small quantities of No. 14 or 18 are not readily available.

TABLE 9.7 Standard Hooks*

Recommended end hooks—all grades, in or ft-in

Bar size no.	180° hooks D†	180° hooks A or G	180° hooks J	90° hooks A or G
3	2¼	5	3	6
4	3	6	4	8
5	3¾	7	5	10
6	4½	8	6	1–0
7	5¼	10	7	1–2
8	6	11	8	1–4
9	9½	1–3	11¾	1–7
10	10¾	1–5	1–1¼	1–10
11	12	1–7	1–2¾	2–0
14	18¼	2–3	1–9¾	2–7
18	24	3–0	2–4½	3–5

†D = finished beam diameter, in

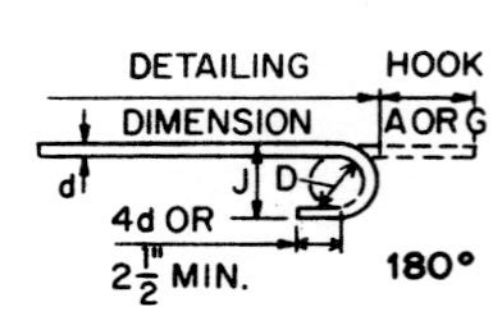

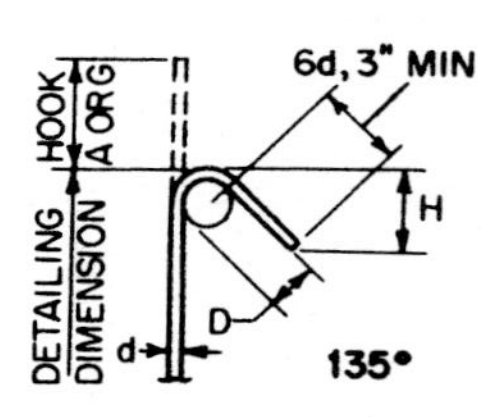

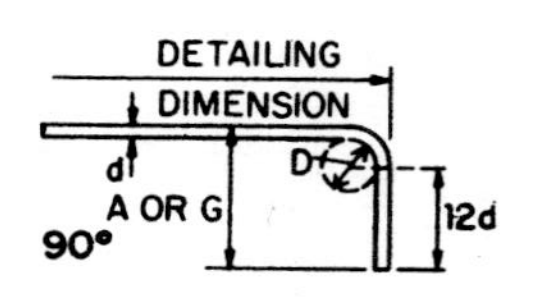

Stirrups (ties similar)

Stirrup and tie hook dimensions, in—Grades 40-50-60 ksi

Bar size no.	D, in	90° hook: Hook A or G	135° hook: Hook A or G	135° hook: H, approx.
3	1½	4	4	2½
4	2	4½	4½	3
5	2½	6	5½	3¾
6	4½	1–0	8	4½
7	5¼	1–2	9	5¼
8	6	1–4	10½	6

135° Seismic stirrup/tie hook dimensions (ties similar) in—Grades 40-50-60 ksi

Bar size no.	D, in	135° hook: Hook A or G	135° hook: H, approx.
3	1½	4¼	3
4	2	4½	3
5	2½	5½	3¾
6	4½	8	4½
7	5¼	9	5¼
8	6	10½	6

*Notes:

1. All specific sizes recommended by CRSI in this table meet minimum requirements of ACI 318.

2. 180° hook J dimension (sizes 10, 11, 14, and 18) and A or G dimension (Nos. 14 and 18) have been revised to reflect recent test research using ASTM/ACI bend-test criteria as a minimum.

3. Tables for stirrup and tie hook dimensions have been expanded to include sizes 6, 7, and 8, to reflect current design practices.

Courtesy of the Concrete Reinforcing Steel Institute.

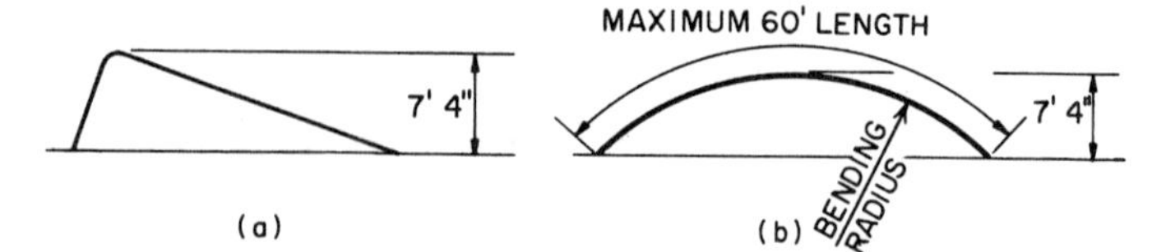

FIGURE 9.7 Shipping limitations: (*a*) height limit; (*b*) length and height limits.

Only straight bars should be bundled ordinarily. Exceptions are bars with end hooks, usually at staggered locations, so that the bars are not bent as a bundle ("Placing Reinforcing Bars," CRSI).

A bundle is assembled by wiring the separate bars tightly in contact. If they are preassembled, placement in forms of long bundles requires a crane. Because cutoffs or splices of bars within a bundle must be staggered, it will often be necessary to form the bundle in place.

Bending and Welding Limitations. The ACI Code contains the following restrictions:

All bars must be bent without heating, except as permitted by the engineer.

Bars partly embedded in concrete may not be bent without permission of the engineer.

No welding of crossing bars (tack welding) is permitted without the approval of the engineer.

For unusual bends, heating may be permitted because bars bend more easily when heated. If not embedded in thin sections of concrete, heating the bars to a maximum temperature of 1500°F facilitates bending, usually without damage to the bars or splitting of the concrete. If partly embedded bars are to be bent, heating controlled within these limits, plus the provision of a round fulcrum for the bend to avoid a sharp kink in the bar, are essential.

Tack welding creates a *metallurgical notch effect,* seriously weakening the bars. If different size bars are tacked together, the notch effect is aggravated in the larger bar. Tack welding therefore should never be permitted at a point where bars are to be fully stressed, and never for the assembly of ties or spirals to column verticals or stirrups to main beam bars.

When large, preassembled reinforcement units are desired, the engineer can plan the tack welding necessary as a supplement to wire ties at points of low stress or to added bars not required in the design.

9.26 BAR SUPPORTS

Bar supports are commercially available in four general types of material: wire, precast concrete, all-plastic, and fiber-reinforced cementitious material. Descriptions of the various types of bar supports, as well as recommended maximum spacings and details for use, are given in the CRSI "Manual for Standard Practice."

Wire bar supports are generally available in the United States in three classes of rust prevention: plastic-protected, stainless-steel-protected, and no protection (plain). Precast-concrete bar supports are normally supplied in three styles; plain block, block with embedded wires, and block with a hole for the leg of a vertical bar for top- and bottom-bar support.

Various types and sizes of all-plastic bar supports and sideform spacers are available. Consideration should be given to the effects of thermal changes, inasmuch as the coefficient of thermal expansion of the plastic can differ significantly from that of concrete. Investigation of this property is advisable before use of all-plastic supports in concrete that will be exposed to high variations in temperature.

Cementitious fiber-reinforced bar supports are commercially available in two types: plain or with wire. They are chemically inert and have a coefficient of thermal expansion about that of concrete.

Bar supports for use with epoxy-coated rebars should be made of dielectric material. Alternatively, wire bar supports should be coated with dielectric material, such as plastic or epoxy.

9.27 INSPECTION OF REINFORCEMENT

This involves approval of rebar material for conformance to the physical properties required, such as ASTM specifications for the strength grade specified; approval of the bar details and placing drawings; approval of fabrication to meet the approved details within the prescribed tolerances; and approval of rebar placing.

Approvals of rebar material may be made on the basis of mill tests performed by the manufacturer for each heat from which the bars used originated. If samples are to be taken for independent strength tests, measurements of deformations, bending tests, and minimum weight, the routine samples may be best secured at the mill or the fabrication shop before fabrication. Occasionally, samples for check tests are taken in the field; but in this case, provision should be made for extra lengths of bars to be shipped and for schedules for the completion of such tests before the material is required for placing. Sampling at the point of fabrication, before fabrication, is recommended.

Inspection of fabrication and placement is usually most conveniently performed in the field, where gross errors would require correction in any event.

Under the American Concrete Institute Building Code, ACI 318, the bars should be free of oil, paint, form coatings, and mud when placed. Rust or mill scale sufficiently loose to damage the bond is normally dislodged in handling.

If heavily rusted bars (which may result from improper storage for a long time exposed to rusting conditions) are discovered at the time of placing, a quick field test of suitability requires only scales, a wire brush, and calipers. In this test, a measured length of the bar is wire-brushed manually and weighed. If less than 94% of the nominal weight remains, or if the height of the deformations is deficient, the rust is deemed excessive. In either case, the material may then be rejected or penalized as structurally inadequate. Where space permits placing additional bars to make up the structural deficiency (in anchorage capacity or weight), as in walls and slabs, this solution is preferred, because construction delay then is avoided. Where job specifications impose requirements on rust more severe than the structural requirements of the ACI Code, for example, for decorative surfaces exposed to weather, the inspection should employ the special criteria required.

CONCRETE PLACEMENT

9.28 GOOD PRACTICE

The principles governing proper placement of concrete are:

Segregation must be avoided during all operations between the mixer and the point of placement, including final consolidation and finishing.

The concrete must be thoroughly consolidated, worked solidly around all embedded items, and should fill all angles and corners of the forms.

Where fresh concrete is placed against or on hardened concrete, a good bond must be developed.

Unconfined concrete must not be placed under water.

The temperature of fresh concrete must be controlled from the time of mixing through final placement, and protected after placement.

("Guide for Measuring, Mixing, Transporting, and Placing Concrete," ACI 304R; "Specifications for Structural Concrete for Buildings," ACI 301; "Guide for Concrete Floor and Slab Construction," ACI 302.IR.)

9.29 METHODS OF PLACING

Concrete may be conveyed from a mixer to point of placement by any of a variety of methods and equipment, if properly transported to avoid segregation. Selection of the most appropriate technique for economy depends on job conditions, especially job size, equipment, and the contractor's experience. In building construction, concrete usually is placed with hand- or power-operated buggies; drop-bottom buckets with a crane; inclined chutes; flexible and rigid pipe by pumping; *shotcrete,* in which either dry materials and water are sprayed separately or mixed concrete is shot against the forms; and for underwater placing, tremie chutes (closed flexible tubes). For mass-concrete construction, side-dump cars on narrow-gage track or belt conveyers may be used. For pavement, concrete may be placed by bucket from the swinging boom of a paving mixer, directly by dump truck or mixer truck, or indirectly by trucks into a spreader.

A special method of placing concrete suitable for a number of unusual conditions consists of grout-filling preplaced coarse aggregate. This method is particularly useful for underwater concreting, because grout, introduced into the aggregate through a vertical pipe gradually lifted, displaces the water, which is lighter than the grout. Because of bearing contact of the aggregate, less than usual overall shrinkage is also achieved.

9.30 EXCESS WATER

Even within the specified limits on slump and water-cement ratio, excess water must be avoided. In this context, excess water is present for the conditions of placing if evidence of water rise (vertical segregation) or water flow (horizontal segregation) occurs. Excess water also tends to aggravate surface defects by increased leakage through form openings. The result may be honeycomb, sand-streaks, variations in color, or soft spots at the surface.

In vertical formwork, water rise causes weak planes between each layer deposited. In addition to the deleterious structural effect, such planes, when hardened, contain voids through which water may pass.

In horizontal elements, such as floor slabs, excess water rises and forms a weak laitance layer at the top. This layer suffers from low strength, low abrasion resistance, high shrinkage, and generally poor quality.

9.31 CONSOLIDATION

The purpose of consolidation is to eliminate voids of entrapped air and to ensure intimate complete contact of the concrete with the surfaces of the forms and the reinforcement. Intense vibration, however, may also reduce the volume of desirable entrained air; but this reduction can be compensated by adjustment of the mix proportions.

Powered internal vibrators are usually used to achieve consolidation. For thin slabs, however, high-quality, low-slump concrete can be effectively consolidated, without excess water, by mechanical surface vibrators. For precast elements in rigid, watertight forms, external vibration (of the form itself) is highly effective. External vibration is also effective with in-place forms, but should not be used unless the formwork is specially designed for the temporary increase in internal pressures to full fluid head plus the impact of the vibrator ("Guide to Formwork for Concrete," ACI 347R).

Except in certain paving operations, vibration of the reinforcement should be avoided. Although it is effective, the necessary control to prevent overvibration is difficult. Also, when concrete is placed in several lifts or layers, vibration of vertical rebars passing into partly set concrete below may be harmful. Note, however, that revibration of concrete before the final set, under controlled conditions, can improve concrete strength markedly and reduce surface voids (bugholes). This technique is too difficult to control for general use on field-cast vertical elements, but it is very effective in finishing slabs with powered vibrating equipment.

Manual spading is most efficient for removal of entrapped air at form surfaces. This method is particularly effective where smooth impermeable form material is used and the surface is unward sloping.

On the usual building project, different conditions of placement are usually encountered that make it desirable to provide for various combinations of the techniques described. One precaution generally applicable is that the vibrators not be used to move the concrete laterally.

("Guide for Consolidation of Concrete," ACI 309R.)

9.32 CONCRETING VERTICAL ELEMENTS

The interior of columns is usually congested; it contains a large volume of reinforcing steel compared with the volume of concrete, and has a large height compared with its cross-sectional dimensions. Therefore, though columns should be continuously cast, the concrete should be placed in 2- to 4-ft-deep increments and consolidated with internal vibrators. These should be lifted after each increment has been vibrated. If delay occurs in concrete supply before a column has been completed, every effort should be made to avoid a cold joint. When the remainder of the column is cast, the first increment should be small, and should be vibrated to penetrate the previous portion slightly.

In all columns and reinforced narrow walls, concrete placing should begin with 2 to 4 in of grout. Otherwise, loose stone will collect at the bottom and form honeycomb. This grout should be proportioned for about the same slump as the concrete or slightly more, but at the same or lower water-cement ratio. (Some engineers prefer to start vertical placement with a mix having the same proportions of water, cement, and fine aggregate, but with one-half the quantity of coarse aggregate, as in the design mix, and to place a starting layer 6 to 12 in deep.)

When concrete is placed for walls, the only practicable means to avoid segregation is to place no more than a 24-in layer in one pass. Each layer should be vibrated separately and kept nearly level.

For walls deeper than 4 ft, concrete should be placed through vertical, flexible trunks or chutes located about 8 ft apart. The trunks may be flexible or rigid, and come in sections so that they can be lifted as the level of concrete in place rises. The concrete should not fall free, from the end of the trunk, more than 4 ft or segregation will occur, with the coarse aggregate ricocheting off the forms to lodge on one side. Successive layers after the initial layer should be penetrated by internal vibrators for a depth of about 4 to 6 in to ensure complete integration at the surface of each layer. Deeper penetration can be beneficial (revibration), but control under variable job conditions is too uncertain for recommendation of this practice for general use.

The results of poor placement in walls are frequently observed: sloping layer lines; honeycombs, leaking, if water is present; and, if cores are taken at successive heights, up to a 50% reduction in strength from bottom to top. Some precautions necessary to avoid these ill effects are:

Place concrete in level layers through closely spaced trunks or chutes.

Do not place concrete full depth at each placing point.

Do not move concrete laterally with vibrators.

For deep, long walls, reduce the slump for upper layers 2 to 3 in below the slump for the starting layer.

On any delay between placing of layers, vibrate the concrete thoroughly at the interface.

If concreting must be suspended between planned horizontal construction joints, level off the layer cast, remove any laitance and excess water, and form a straight, level construction joint, if possible, with a small cleat attached to the form on the exposed face (see also Art. 9.39).

9.33 CONCRETING HORIZONTAL ELEMENTS

Concrete placement in horizontal elements follows the same general principles outlined in Art 9.32. Where the surface will be covered and protected against abrasion and weather, few special precautions are needed.

For concrete slabs, careless placing methods result in horizontal segregation, with desired properties in the wrong location, the top consisting of excess water and fines with low abrasion and weather resistance, and high shrinkage. For a good surface in a one-course slab, low-slump concrete and a minimum of vibration and finishing are desirable. Immediate screeding with a power-vibrated screed is helpful in distributing low-slump, high-quality concrete. No further finishing should be undertaken until free water, if any, disappears. A powered, rotary tamping float can smooth very-low-slump concrete at this stage. Final troweling should be delayed, if necessary, until the surface can support the weight of the finisher.

When concrete is placed for deep beams that are monolithic with a slab, the beam should be filled first. Then, a short delay for settlement should ensue before slab concrete is cast. Vibration through the top slab should penetrate the beam concrete sufficiently to ensure thorough combination.

When a slab is cast, successive batches of concrete should be placed on the edge of previous batches, to maintain progressive filling without segregation. For slabs with sloping surfaces, concrete placing should usually begin at the lower edge.

For thin shells in steeply sloping areas, placing should proceed downslope. Slump should be adjusted and finishing coordinated to prevent restraint by horizontal bars from causing plastic cracking in the fresh concrete.

9.34 BONDING TO HARDENED CONCRETE

The surface of hardened concrete should be rough and clean where it is to be bonded with fresh concrete.

Vertical surfaces of planed joints may be prepared easily by wire brushing them, before complete curing, to expose the coarse aggregate. (The timing can be extended, if desired, by using a surface retarder on the bulkhead form.) For surfaces fully cured without earlier preparation, sandblasting, bush hammering, or acid washes (thoroughly rinsed off) are effective means of preparation for bonding new concrete. (See also Art. 9.33.)

Horizontal surfaces of previously cast concrete, for example, of walls, are similarly prepared. Care should be taken to remove all laitance and to expose sound concrete and coarse aggregate. (See also Art. 9.32. For two-course floors, see Art. 9.35.)

9.35 HEAVY-DUTY FLOOR FINISHES

Floor surfaces highly resistant to abrasion and impact are required for many industrial and commercial uses. Such surfaces are usually built as two-course construction, with a base or structural slab topped by a wearing surface. The two courses may be cast integrally or with the heavy-duty surface applied as a separate topping.

In the first process, which is less costly, ordinary structural concrete is placed and screeded to nearly the full depth of the floor. The wearing surface concrete, made with special abrasion-resistant aggregate, emery, iron filings, etc., then is mixed, spread to the desired depth, and troweled before final set of the concrete below.

The second method requires surface preparation of the base slab, by stiff brooming before final set to roughen the surface and thorough washing before the separate heavy-duty topping is cast. For the second method, the topping is a very dry (zero-slump) concrete, made with ⅜-in maximum-size special aggregate. This topping should be designed for a minimum strength, f'_c = 6000 psi. It must be tamped into place with powered tampers or rotary floats. (*Note:* If test cylinders are to be made from this topping, standard methods of consolidation will not produce a proper test; tamping similar in effect to that applied to the floor itself is necessary.) One precaution vital to the separate topping method is that the temperatures of topping and base slab must be kept compatible.

("Guide for Concrete Floor and Slab Construction," ACI 302.1R.)

9.36 CONCRETING IN COLD WEATHER

Frozen materials should never be used. Concrete should not be cast on a frozen subgrade, and ice must be removed from forms before concreting. Concrete allowed to freeze wet, before or during early curing, may be seriously damaged. Furthermore, temperatures should be kept above 40°F for any appreciable curing (strength gain).

Concrete suppliers are equipped to heat materials and to deliver concrete at controlled temperatures in cold weather. These services should be utilized.

In very cold weather, for thin sections used in building, the freshly cast concrete must be enclosed and provided with temporary heat. For more massive sections or in moderately cold weather, it is usually less expensive to provide insulated forms or insulated coverings to retain the initial heat and subsequent heat of hydration generated in the concrete during initial curing.

The curing time required depends on the temperature maintained and whether regular or high-early-strength concrete is employed. High-early-strength concrete may be achieved with accelerating admixtures (Art. 9.9) or with high-early-strength cement (Types III or IIIA) or by a lower water-cement ratio, to produce the required 28-day strength in about 7 days.

An important precaution in employing heated enclosures is to supply heat without drying the concrete or releasing carbon dioxide fumes. Exposure of fresh concrete to drying or fumes results in chalky surfaces. Another precaution is to avoid rapid temperature changes of the concrete surfaces when heating is discontinued. The heat supply should be reduced gradually, and the enclosure left in place to permit cooling to ambient temperatures gradually, usually over a period of at least 24 h.

("Cold Weather Concreting," ACI 306R.)

9.37 CONCRETING IN HOT WEATHER

Mixing and placing concrete at a high temperature may cause *flash set* in the mixer, during placing, or before finishing can be completed. Also, loss of strength can result from casting hot concrete.

In practice, most concrete is cast at about 70 ± 20°F. Research on the effects of casting temperature shows highest strengths for concrete cast at 40°F and significant but practically unimportant increasing loss of strength from 40°F to 90°F. For higher temperatures, the loss of strength becomes important. So does increased shrinkage. The increased shrinkage is attributable not only to the high temperature, but also to the increased water content required for a desired slump as temperature increases. See Fig. 9.5.

For ordinary building applications, concrete suppliers control temperatures of concrete by cooling the aggregates and, when necessary, by supplying part of the mixing water as crushed ice. In very hot weather, these precautions plus sectional casting, to permit escape of the heat of hydration, may be required for massive foundation mats. Retarding admixtures are also used with good effect to reduce slump loss during placing and finishing.

("Hot Weather Concreting," ACI 305R.)

9.38 CURING CONCRETE

Curing of concrete consists of the processes, natural and artificially created, that affect the extent and rate of hydration of the cement.

Many concrete structures are cured without artificial protection of any kind. They are allowed to harden while exposed to sun, wind, and rain. This type of curing is unreliable, because water may evaporate from the surface.

Various means are employed to cure concrete by controlling its moisture content or its temperature. In practice, curing consists of conserving the moisture within newly placed concrete by furnishing additional moisture to replenish water lost by evaporation. Usually, little attention is paid to temperature, except in winter curing and steam curing.

Most effective curing is beneficial in that it makes the concrete more watertight and increases the strength.

Methods for curing may be classified as:

1. Those that supply water throughout the early hydration process and tend to maintain a uniform temperature. These methods include ponding, sprinkling, and application of wet burlap or cotton mats, wet earth, sawdust, hay, or straw.
2. Those designed to prevent loss of water but having little influence on maintaining a uniform temperature. These methods include waterproof paper and impermeable membranes. The latter is usually a clear or bituminous compound sprayed on the concrete to fill the pores and thus prevent evaporation. A fugitive dye in the colorless compound aids the spraying and inspection.

A white pigment that gives infrared reflectance can be used in a curing compound to keep concrete surfaces cooler when exposed to the sun.

The criterion for judging the adequacy of field curing provided in the American Concrete Institute Building Code, ACI 318, is that the field-cured test cylinders produce 85% of the strengths developed by companion laboratory-cured cylinders at the age for which strength is specified.

9.39 JOINTS IN CONCRETE

Several types of joints may occur or be formed in concrete structures:

Construction joints are formed when fresh concrete is placed against hardened concrete.

Expansion joints are provided in long components to relieve compressive stresses that would otherwise result from a temperature rise.

Contraction joints (control joints) are provided to permit concrete to contract during a drop in temperature and to permit drying shrinkage without resulting uncontrolled random cracking.

Contraction joints should be located at places where concrete is likely to crack because of temperature changes or shrinkage. The joints should be inserted where there are thickness changes and offsets. Ordinarily, joints should be spaced 30 ft c to c or less in exposed structures, such as retaining walls.

To avoid unsightly cracks due to shrinkage, a dummy-type contraction joint is frequently used (Fig. 9.8). When contraction takes place, a crack occurs at this deliberately made plane of weakness. In this way, the crack is made to occur in a straight line easily sealed.

Control joints may also consist of a 2- or 3-ft gap left in a long wall or slab, with the reinforcement from both ends lapped in the gap. Several weeks after the wall or slab has been concreted, the gap is filled with concrete. By that time, most of the shrinkage has taken place. A checkerboard pattern of slab construction can be used for the same reason.

In expansion joints, a filler is usually provided to separate the two parts of the structure. This filler should be a compressible substance, such as corkboard or premolded mastic. The filler should have properties such that it will not be squeezed out of the joint, will not slump when heated by the sun, and will not stain the surface of the concrete.

To be waterproof, a joint must be sealed. For this purpose, copper flashing may be used. It is usually embedded in the concrete on both sides of the joint, and folded into the joint so that the joint may open without rupturing the metal. The flashing must be strong enough to hold its position when the concrete is cast.

Proprietary flexible water stops and polysulfide calking compounds may also be used as sealers.

Open expansion joints are sometimes employed for interior locations where the opening is not objectionable. When exposed to water from above, as in parking decks, open joints may be provided with a gutter below to drain away water.

The engineer should show all necessary vertical and horizontal joints on design drawings. All pertinent details affecting reinforcement, water stops, and sealers should also be shown.

Construction joints should be designed and located if possible at sections of minimum shear. These sections will usually be at the center of beams and slabs, where the bending moment is highest. They should be located where it is most convenient to stop work. The construction joint is often keyed for shearing strength.

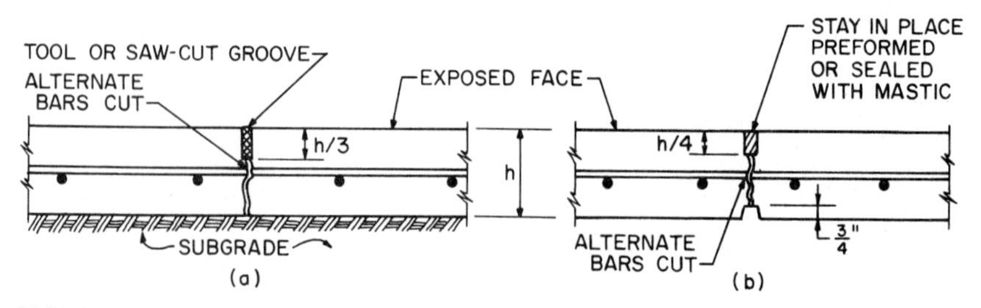

FIGURE 9.8 Control joints for restraining temperature and shrinkage cracks: (*a*) vertical section through a slab on grade; (*b*) horizontal section through a wall.

If it is not possible to concrete an entire floor in one operation, vertical joints preferably should be located in the center of a span. Horizontal joints are usually provided between columns and floor; columns are concreted first, then the entire floor system.

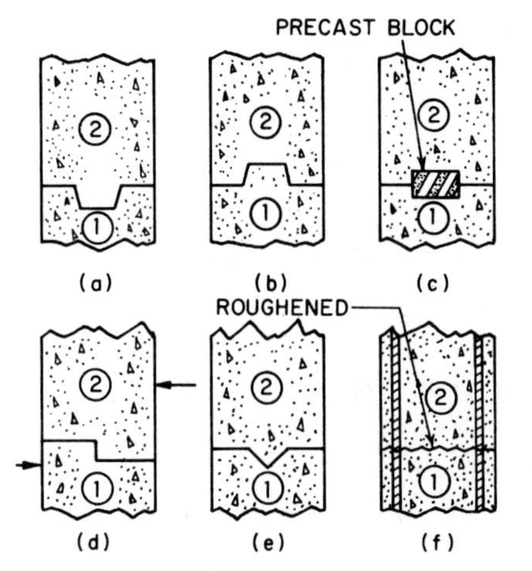

FIGURE 9.9 Types of construction joints. Circled numbers indicate order of casting.

Various types of construction joints are shown in Fig. 9.9. The numbers on each section refer to the sequence of placing concrete.

If the joint is horizontal, as in Fig. 9.9*a*, water may be trapped in the key of the joint. If the joint is vertical, the key is easily formed by nailing a wood strip to the inside of the forms. A raised key, as in Fig. 9.9*b*, makes formwork difficult for horizontal joints.

In the horizontal joint in Fig. 9.9*c*, the key is made by setting precast-concrete blocks into the concrete at intermittent intervals. The key in Fig. 9.9*d* is good if the shear acts in the directions shown.

The V-shaped key in Fig. 9.9*e* can be made manually in the wet concrete for horizontal joints.

The key is eliminated in Fig. 9.9*f*, reliance being placed on friction on the roughened surface. This method may be used if the shears are small, or if there are large compressive forces or sufficient reinforcement across the joint.

See also Arts. 9.32 to 9.34.

9.40 INSPECTION OF CONCRETE PLACEMENT

Concrete should be inspected for the owner before, during, and after casting. Before concrete is placed, the formwork must be free of ice and debris and properly coated with bond-breaker oil. The rebars must be in place, properly supported to bear any traffic they will receive during concrete placing. Conduit, inserts, and other items to be embedded must be in position, fixed against displacement. Construction personnel should be available, usually carpenters, bar setters, and other trades, if piping or electrical conduit is to be embedded, to act as form watchers and to reset any rebars, conduit, or piping displaced.

As concrete is cast, the slump of the concrete must be observed and regulated within prescribed limits, or the specified strengths based on the expected slump may be reduced. An inspector of placing who is also responsible for sampling and making cylinders, should test slump, entrained air, temperatures, and unit weights during concreting and should control any field adjustment of slump and added water and cement. The inspector should also ascertain that handling, placing, and finishing procedures that have been agreed on in advance are properly followed, to avoid segregated concrete. In addition, the inspector should ensure that any emergency construction joints made necessary by stoppage of concrete supply, rain, or other delays are properly located and made in accordance with procedures specified or approved by the engineer.

Inspection is complete only when concrete is cast, finished, protected for curing, and attains full strength.

("Manual of Concrete Inspection," ACI SP2.)

STRUCTURAL ANALYSIS OF CONCRETE STRUCTURES

Under the American Concrete Institute Building Code, ACI 318, concrete structures generally may be analyzed by elastic theory. When specific limiting conditions are met, certain approximate methods are permitted. For some cases, the Code recommends an empirical method.

9.41 ANALYSES OF ONE-WAY FLOOR AND ROOF SYSTEMS

The American Concrete Institute Building Code, ACI 318, permits an approximate analysis for continuous systems in ordinary building if:

Components are not prestressed.

Beams are continuous over two or more spans.

In successive spans, the ratio of the larger span to the smaller does not exceed 1.20.

The spans carry only uniform loads.

The ratio of live to dead service load(s) (not factored) does not exceed 3.

Members are prismatic.

This analysis determines the maximum moments and shears at faces of supports, and the midspan moments representing envelope values for the respective loading combinations. In this method, moments are computed from

$$M = Cw_u L_n^2 \tag{9.6}$$

where C = coefficient, given in Fig. 9.10
w_u = uniform factored load
L_n = clear span for positive moment or shear and the average of adjacent clear spans for negative moment

For an elastic ("exact") analysis, the spans L of members that are not built integrally with their supports should be taken as the clear span plus the depth of slab or beam but need not exceed the distance between centers of supports. For spans of continuous frames, spans should be taken as the distance between centers of supports. For solid or ribbed slabs with clear spans not exceeding 10 ft, if built integrally with their supports, spans may be taken as the clear distance between supports.

If an elastic analysis is performed for continuous flexural members for each loading combination expected, calculated moments may be redistributed if the ratio

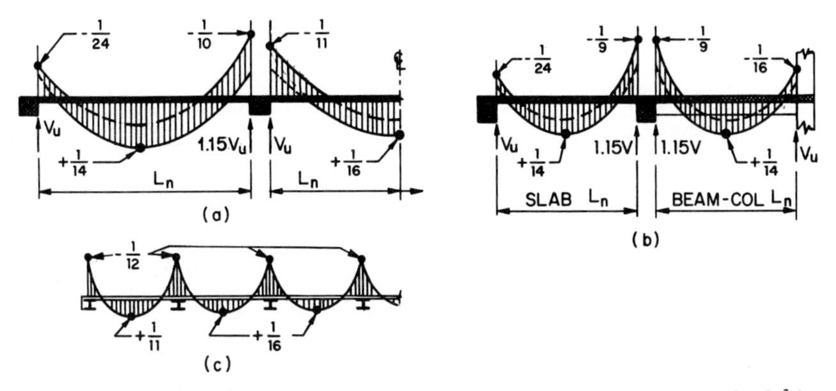

FIGURE 9.10 Coefficients C for calculation of bending moments from $M_u = Cw_uL_n^2$ in approximate analysis of beams and one-way slabs with uniform load w_u. For shears, $V_u = 0.5w_uL_n$. (*a*) More than two spans. (*b*) Two-span beam or slab. (*c*) Slabs—all spans, $L_n \leq 10$ ft.

ρ of tension-reinforcement area to effective concrete area or ratio of $\rho - \rho'$, where ρ' is the compression-reinforcement ratio, to the balanced-reinforcement ratio ρ_b, lie within the limits given in the ACI Building Code. Positive moments should be increased by a percentage γ and negative moments decreased by γ to reflect the moment redistribution property of underreinforced concrete in flexure. When ρ or $\rho - \rho'$ is not more than $0.5\rho_b$, the percentage is given by

$$\gamma = 20\left(1 - \frac{\rho - \rho'}{\rho_b}\right) \tag{9.7}$$

For example, suppose a 20-ft interior span of a continuous slab with equal spans is made of concrete with a strength f'_c of 4 ksi and reinforced with bars having a yield strength f_y of 60 ksi. Factored dead and live loads are both 0.100 ksf. The factored moments are determined as follows:

Maximum negative moments occur at the supports of the interior span when this span and adjacent spans carry both dead and live loads. Call this loading Case 1. For Case 1 then, maximum negative moment equals

$$M_u = -(0.100 + 0.100)(20)^2/11 = -7.27 \text{ ft-kips/ft}$$

The corresponding positive moment at midspan is 2.73 ft-kips/ft.

Maximum positive moment in the interior span occurs when it carries full load but adjacent spans support only dead loads. Call this loading Case 2. For Case 2, then, the negative moment is $-(10.00 - 5.00) = -5.00$ ft-kips/ft, and the maximum positive moment is 5.00 ft-kips/ft. Figure 9.11*a* shows the maximum moments.

For the concrete and reinforcement properties given, the balanced-reinforcement ratio computed from Eq. (9.26) is $\rho_b = 0.0285$. Assume now that reinforcement ratios for the top steel and bottom steel are 0.00267 and 0.002, respectively. If alternate bottom bars extend into the supports, $\rho' = 0.001$. Substitution in Eq. (9.7) gives for the redistribution percentage

$$\gamma = 20\left(1 - \frac{0.00267 - 0.001}{0.0285}\right) = 18.8\%$$

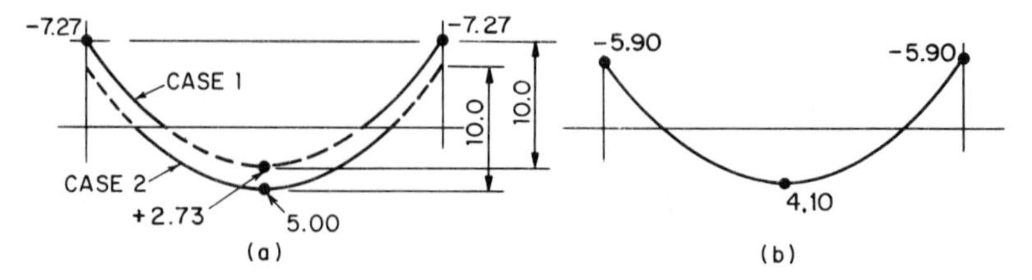

FIGURE 9.11 Bending moments in an interior 20-ft span of a continuous one-way slab: (*a*) moments for Case I (this and adjacent spans fully loaded) and Case II (this span fully loaded but adjacent spans with only dead load); (*b*) Case I moments after redistribution.

The negative moment (Case 1) therefore can be decreased to $M_u = -7.27(1 - 0.188) = -5.90$ ft-kips/ft. The corresponding positive moment at midspan is $10 - 5.90 = 4.10$ kips/ft (Fig. 9.11*b*).

For Case 2 loading, if the negative moment is increased 18.8%, it becomes $-5.94 \approx 5.90$ ft-kips/ft. Therefore, the slab should be designed for the moments shown in Fig. 9.11*b*.

9.42 TWO-WAY SLAB FRAMES

The American Concrete Institute Building Code, ACI 318, prescribes an *equivalent column* concept for use in analysis of two-way slab systems. This concept permits a three-dimensional (space-frame) analysis in which the "equivalent column" combines the flexibility (reciprocal of stiffness) of the real column and the torsional flexibility of the slabs or beams attached to the column at right angles to the direction of the bending moment under consideration. The method, applicable for all ratios of successive spans and of dead to live load, is an elastic ("exact") analysis called the "equivalent frame method."

An approximate procedure, the "direct design method," is also permitted (within limits of load and span). This method constitutes the direct solution of a one-cycle moment distribution. (See also Art. 9.58.)

(P. F. Rice et al., "Structural Design Guide to the ACI Building Code," 3d ed., Van Nostrand Reinhold Company, New York.)

9.43 SPECIAL ANALYSES

Space limitations preclude more than a brief listing of some of the special analyses required for various special types of reinforced concrete construction and selected basic references for detailed information. Further references to applicable research are available in each of the basic references.

Seismic-loading-resistant ductile frames: ACI 318; ACI Detailing Manual.

High-rise construction, frames, shear walls, frames plus shear walls, and tube concept: "Planning and Design of Tall Buildings," Vols. SC, CL, and CB, American Society of Civil Engineers.

Nuclear containment structures: ASME-ACI Code for Nuclear Containment Structures, ACI 359; also ACI 349 and 349R.

Environmental engineering structures: "Environmental Engineering Concrete Structures," ACI 350R.

Bridges: "Analysis and Design of Reinforced Concrete Bridge Structures," ACI 343R.

It should be noted that the ACI Building Code specifically provides for the acceptance of analyses by computer or model testing to supplement the manual calculations when required by building officials.

STRUCTURAL DESIGN OF FLEXURAL MEMBERS

9.44 STRENGTH DESIGN WITH LOAD FACTORS

Safe, economical strength design of reinforced concrete structures requires that their ultimate-load-carrying capacity be predictable or known. The safe, or service-load-carrying capacity can then be determined by dividing the ultimate-load-carrying capacity by a factor of safety.

The American Concrete Institute Building Code, ACI 318, provides for strength design of concrete members by use of **factored loads** (actual and specified loads multiplied by load factors). Factored axial forces, shears, and moments in members are determined as if the structure were elastic. Strength-design theory is then used to design critical sections for these axial forces, shears, and moments.

Strength design of reinforced concrete flexural members (Art. 9.62) may be based on the following assumptions and applicable conditions of equilibrium and compatibility of strains:

1. Strain in the reinforcing steel and the concrete is directly proportional to the distance from the neutral axis (Fig. 9.12).

2. The maximum usable strain at the extreme concrete compression surface equals 0.003 in/in.

3. When the strain, in/in in reinforcing steel is less than f_y/E_s, where f_y = yield strength of the steel and E_s = its modulus of elasticity (29,000,000 psi), the steel

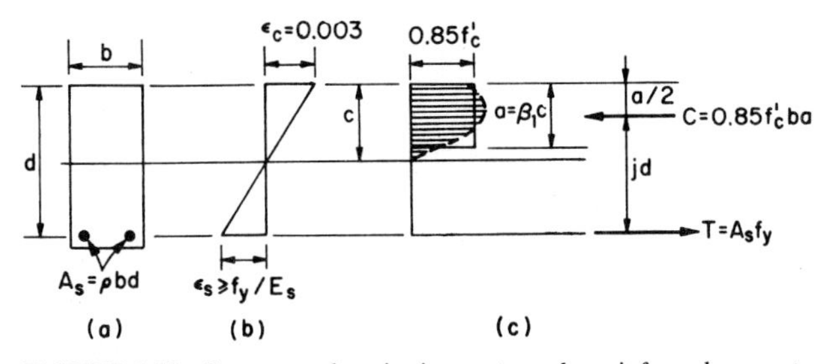

FIGURE 9.12 Stresses and strains in a rectangular reinforced-concrete beam, reinforced for tension only, at ultimate load: (*a*) cross section of beam; (*b*) strain distribution; (*c*) two types of stress distribution.

stress, psi, equals 29,000,000 times the steel strain. After the steel yield strength has been reached, the stress remains constant at f_y, though the strain increases.

4. Except for prestressed concrete (Art. 9.103) or plain concrete, the tensile strength of the concrete is negligible in flexure.

5. The shape of the concrete compressive distribution may be assumed to be a rectangle, trapezoid, parabola, or any other shape in substantial agreement with comprehensive strength tests.

6. For a rectangular stress block, the compressive stress in the concrete should be taken as $0.85f'_c$. This stress may be assumed constant from the surface of maximum compressive strain to a depth of $a = \beta_1 c$, where c is the distance to the neutral axis (Fig. 9.12). For $f'_c = 4000$ psi, $\beta_1 = 0.85$. For greater concrete strengths, β_1 should be reduced 0.05 for each 1000 psi in excess of 4000, but β_1 should not be taken less than 0.65.

(See also Art. 9.81 for columns).

9.44.1 Strength Reduction Factors

The ACI Code requires that the strength of a member based on strength design theory include **strength reduction factor** ϕ to provide for small adverse variations in materials, workmanship, and dimensions individually within acceptable tolerances. The degree of ductility, importance of the member, and the accuracy with which the member's strength can be predicted were considered in assigning values to ϕ:

ϕ should be taken as 0.90 for flexure and axial tension; 0.85 for shear and torsion; 0.70 for bearing on concrete; 0.65 for bending in plain concrete; and for axial compression combined with bending, 0.75 for members with spiral reinforcement, and 0.70 for other members.

9.44.2 Load Factors

For combinations of loads, a structure and its members should have the following strength U, computed by adding factored loads and multiplying by a factor based on probability of occurrence of the load combination:

Dead load D **and live load** L, plus their internal moments and forces:

$$U = 1.4D + 1.7L \tag{9.8}$$

Wind load W:

$$U = 0.75(1.4D + 1.7L + 1.7W) \tag{9.9}$$

When D and L reduce the effects of W:

$$U = 0.9D + 1.3W \tag{9.10}$$

Earthquake forces E:

$$U = 0.75(1.4D + 1.7L + 1.87E) \tag{9.11}$$

When D and L reduce the effects of E:

$$U = 0.9D + 1.43E \tag{9.12}$$

Lateral earth pressure H:

$$U = 1.4D + 1.7L + 1.7H \tag{9.13}$$

When D and L reduce the effects of H:

$$U = 0.9D + 1.7H \tag{9.14}$$

Lateral pressure F **from liquids** (for well-defined fluid pressures):

$$U = 1.4D + 1.7L + 1.4F \tag{9.15}$$

Impact effects, if any, should be included with the live load L.

Where the structural effects T of differential settlement, creep, shrinkage, or temperature change can be significant, they should be included with the dead load D, and the strength should not be less than $1.4D + 1.4T$, or

$$U = 0.75(1.4D + 1.4T + 1.7L) \tag{9.16}$$

9.45 ALLOWABLE-STRESS DESIGN AT SERVICE LOADS (ALTERNATIVE DESIGN METHOD)

Nonprestressed, reinforced-concrete flexural members (Art. 9.62) may be designed for flexure by the alternative design method of the American Concrete Institute Building Code, ACI 318 (working-stress design). In this method, members are designed to carry service loads (load factors and ϕ are taken as unity) under the straight-line (elastic) theory of stress and strain. (Because of creep in the concrete, only stresses due to short-time loading can be predicted with reasonable accuracy by this method.)

Working-stress design is based on the following assumptions:

1. A section plane before bending remains plane after bending. Strains therefore vary with distance from the neutral axis (Fig. 9.13c).

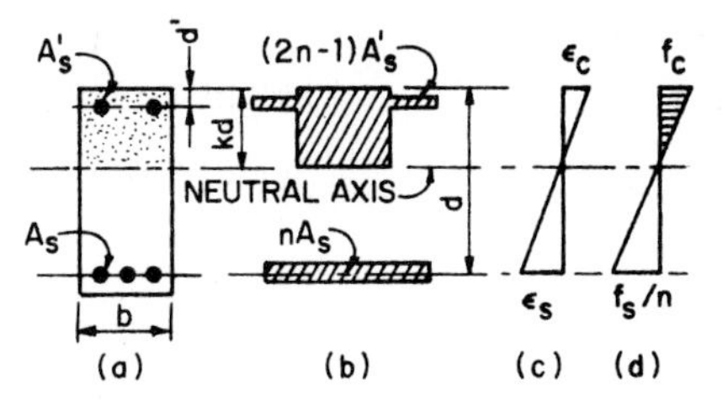

FIGURE 9.13 Stresses and strains in a beam with compression reinforcement, as assumed for working-stress design: (a) rectangular cross section of beam; (b) transformed section with twice the steel area, to allow for effects of creep of concrete; (c) assumed strains; (d) assumed distribution of stresses in the concrete.

2. The stress-strain relation for concrete plots as a straight line under service loads within the allowable working stresses (Fig. 9.13c and d), except for *deep beams*.

3. Reinforcing steel takes all the tension due to flexure (Fig. 9.13a and b).

4. The modular ratio, $n = E_s/E_c$, where E_s and E_c are the moduli of elasticity of reinforcing steel and conctete, respectively, may be taken as the nearest whole number, but not less than 6 (Fig. 9.13b).

5. Except in calculations for deflection, n for lightweight concrete should be assumed the same as for normal-weight concrete of the same strength.

6. The compressive stress in the extreme surface of the concrete must not exceed $0.45f'_c$, where f'_c is the 28-day compressive strength of the concrete.

7. The allowable tensile stress in the reinforcement must not be greater than the following:

Grades 40 and 50	20 ksi
Grade 60 or greater	24 ksi

For ⅜-in or smaller-diameter reinforcement in one-way slabs with spans not exceeding 12 ft, the allowable stress may be increased to 50% of the yield strength but not to more than 30 ksi.

8. For doubly reinforced flexural members, including slabs with compression reinforcement, an effective modular ratio of $2E_s/E_c$ should be used to transform the compression-reinforcement area for stress computations to an equivalent concrete area (Fig. 9.13*b*). (This recognizes the effects of creep.) The allowable stress in the compression steel may not exceed the allowable tension stress.

Because the strains in the reinforcing steel and the adjoining concrete are equal, the stress in the tension steel f_s is n times the stress in the concrete f_c. The total force acting on the tension steel then equals nA_sf_c. The steel area A_s, therefore can be replaced in stress calculations by a concrete area n times as large.

The transformed section of a concrete beam is a cross section normal to the neutral surface with the reinforcing replaced by an equivalent area of concrete (Fig. 9.13*b*). (In doubly reinforced beams and slabs, an effective modular ratio of $2n$ should be used to transform the compression reinforcement and account for creep and nonlinearity of the stress-strain diagram for concrete.) Stress and strain are assumed to vary with the distance from the neutral axis of the transformed section; that is, conventional elastic theory for homogeneous beams may be applied to the transformed section. Section properties, such as location of neutral axis, moment of inertia, and section modulus S, may be computed in the usual way for homogeneous beams, and stresses may be calculated from the flexure formula, $f = M/S$, where M is the bending moment at the section. This method is recommended particularly for T beams and doubly reinforced beams.

From the assumptions the following formulas can be derived for a rectangular section with tension steel only.

$$\frac{nf_c}{f_s} = \frac{k}{1 - k} \tag{9.17}$$

$$k = \sqrt{2n\rho + (n\rho)^2} - n\rho \tag{9.18}$$

$$j = 1 - \frac{k}{3} \tag{9.19}$$

where $\rho = A_s/bd$ and b is the width and d the effective depth of the section (Fig. 9.13).

Compression capacity:

$$M_c = \tfrac{1}{2}f_ckjbd^2 = K_cbd^2 \tag{9.20a}$$

where $K_c = \tfrac{1}{2}f_ckj$.

Tension capacity:

$$M_s = f_s A_s jd = f_s \rho j b d^2 = K_s b d^2 \tag{9.20b}$$

where $k_s = f_s \rho j$.

Design of flexural members for shear, torsion, and bearing, and of other types of members, follows the strength design provisions of ACI 318, because allowable capacity by the alternative design method is an arbitrarily specified percentage of the strength.

9.46 STRENGTH DESIGN FOR FLEXURE

Article 9.44 summarizes the basic assumptions for strength design of flexural members. The following formulas are derived from those assumptions.

The area A_s of tension reinforcement in a reinforced-concrete flexural member can be expressed as the ratio

$$\rho = \frac{A_s}{bd} \tag{9.21}$$

where b = beam width and d = effective beam depth = distance from the extreme compression surface to centroid of tension reinforcement. At nominal (ultimate) strength of a critical section, the stress in this steel will be equal to its yield strength f_y, psi, if the concrete does not first fail in compression. (See also Arts. 9.47 to 9.51 for additional reinforcement requirements.)

9.46.1 Singly Reinforced Rectangular Beams

For a rectangular beam, reinforced with only tension steel (Fig. 9.12), the total tension force in the steel at ultimate strength is

$$T = A_s f_y = \rho f_y bd \tag{9.22}$$

It is opposed by an equal compressive force

$$C = 0.85 f'_c b \beta_1 c \tag{9.23}$$

where f'_c = specified strength of the concrete, psi
c = distance from extreme compression surface to neutral axis
β_1 = a constant (given in Art. 9.44)

Equating the compression and tension at the critical section yields:

$$c = \frac{\rho f_y}{0.85 \beta_1 f'_c} d \tag{9.24}$$

The criterion for compression failure is that the maximum strain in the concrete equals 0.003 in/in. In that case:

$$c = \frac{0.003}{f_s/E_s + 0.003} d \tag{9.25}$$

where f_s is the steel stress, ksi, and E_s = 29,000,000 psi is the steel modulus of elasticity.

Tension-Steel Limitations. Under balanced conditions, the concrete will reach its maximum strain of 0.003 in/in when the steel reaches its yield strength f_y. Then, c as given by Eq. (9.25) will equal c as given by Eq. (9.24). Also, the steel ratio for balanced conditions in a rectangular beam with tension steel only becomes:

$$\rho_b = \frac{0.85\beta_1 f'_c}{f_y} \frac{87{,}000}{87{,}000 + f_y} \tag{9.26}$$

All structures should be designed to avoid sudden collapse. Therefore, reinforcement should yield before the concrete crushes. Gradual yielding will occur if the quantity of tensile reinforcement is less than the balanced percentage determined by strength design theory. To avoid compression failures, the American Concrete Institute Building Code, A318, therefore limits the steel ratio ρ to a maximum of $0.75\rho_b$.

The Code also requires that ρ for positive-moment reinforcement be at least $200/f_y$ to prevent sudden collapse when the design positive-moment strength is equal to or less than the cracking moment. This requirement does not apply, however, if the reinforcement area at every section of the member is at least one-third greater than that required by the factored moment.

For flexural members of any cross-sectional shape, without compression reinforcement, the tension steel is limited by the ACI Building Code so that $A_s f_y$ does not exceed 0.75 times the total compressive force at balanced conditions. The total compressive force may be taken as the area of a rectangular stress block of a rectangular member; the strength of overhanging flanges or compression steel, or both, may be included. For members with compression reinforcement, the portion of tensile reinforcement equalized by compression reinforcement need not be reduced by the 0.75 factor.

Design Strength: Tension Steel Only. For underreinforced rectangular beams with tension reinforcement only (Fig. 9.12) and a rectangular stress block with depth a ($\rho \leq 0.75\rho_b$), the flexural design strength may be determined from:

$$M_u = 0.90bd^2\rho f_y \left(1 - \frac{0.59\rho f_y}{f'_c}\right) \tag{9.27a}$$

$$= 0.90A_s f_y \left(d - \frac{a}{2}\right) \tag{9.27b}$$

$$= 0.90A_s f_y jd \tag{9.27c}$$

where $a = A_s f_y / 0.85 f'_c b$ and $jd = d - a/2$.

9.46.2 Doubly Reinforced Rectangular Beams

For a rectangular beam with compression-steel area A'_s and tension-steel area A_s, the compression-steel ratio is

$$p' = \frac{A'_s}{bd} \tag{9.28}$$

and the tension-steel ratio is

$$\rho = \frac{A_s}{bd} \tag{9.29}$$

where b = width of beam and d = effective depth of beam. For design, ρ should not exceed

$$0.75\left(\rho_b - \rho'\frac{f'_s}{f_y}\right) + \rho'\frac{f'_s}{f_y} \tag{9.30a}$$

for

$$\rho_b = \frac{0.85 f'_c \beta_1}{f_y}\frac{87{,}000}{87{,}000 + f_y} + \rho'\frac{f'_s}{f_y} \tag{9.30b}$$

where f'_s = stress in compression steel, psi, and other symbols are the same as those defined for singly reinforced beams (Art. 9.46.1). The compression force on the concrete alone in a cross section (Fig. 9.14) is

$$C_{1b} = 0.85 f'_c ba \tag{9.31}$$

where $a = \beta_1 c$ is the depth of the stress block and the compression steel carries $A'_s f'_s$. Forces equal in magnitude to these but opposite in direction stress the tension steel. The depth to the neutral axis c can be found from the maximum compressive strain of 0.003 in/in, or by equating the compression and tension forces on the section. (See also Art. 9.63.)

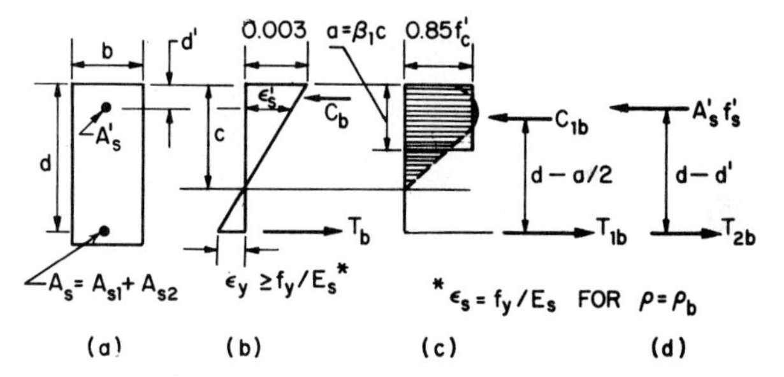

FIGURE 9.14 Stresses and strains, at ultimate load in a rectangular beam with compression reinforcement: (*a*) beam cross section; (*b*) strain distribution; (*c*) two types of stress distribution; (*d*) compression stress in reinforcement.

9.46.3 T Beams

When a T form is used to provide needed compression area for an isolated beam, flange thickness should be at least one-half the web width, and flange width should not exceed 4 times the web width.

When a T is formed by a beam cast integrally with a slab, only a portion of the slab is effective. For a symmetrical T beam, the effective flange width should not exceed one-fourth the beam span, nor should the width of the overhang exceed 8 times the slab thickness nor one-half the clear distance to the next beam. For a beam having a flange on one side only, the effective flange width should not exceed one-twelfth the span, 6 times the slab thickness, nor one-half the clear distance to the next beam.

The overhang of a T beam should be designed to act as a cantilever. Spacing of the cantilever reinforcing should not exceed 18 in or 5 times the flange thickness.

In computing the moment capacity of a T beam, it may be treated as a singly reinforced beam with overhanging concrete flanges (Fig. 9.15). The compression force on the web (rectangular beam) is

$$C_w = 0.85 f'_c b_w a \tag{9.32}$$

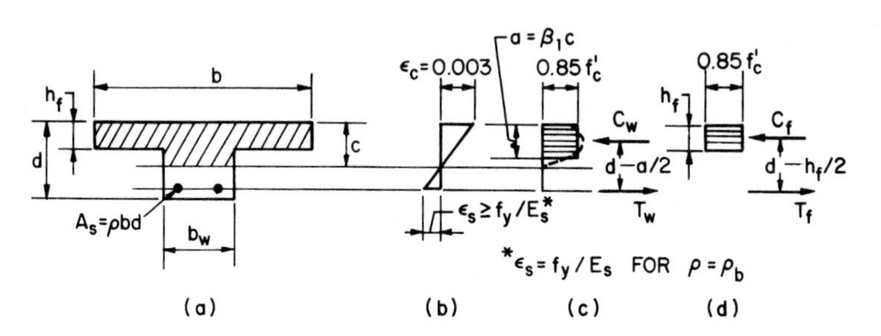

FIGURE 9.15 Stresses and strains in a T beam at ultimate load: (*a*) beam cross section; (*b*) strain distribution; (*c*) stress distributions in web; (*d*) block distribution of flange compression stresses.

where b_w = width of web. The compression force on the overhangs is

$$C_f = 0.85 f'_c (b - b_w) h_f \tag{9.33}$$

where h_f = flange thickness and b = effective flange width of T beam. Forces equal in magnitude to these but opposite in direction stress the tension steel:

$$T_w = A_{sw} f_y \tag{9.34}$$

$$T_f = A_{sf} f_y \tag{9.35}$$

where A_{sw} = area of reinforcing required to develop compression strength of web and A_{sf} = area of reinforcing required to develop compression strength of overhanging flanges. The steel ratio for balanced conditions is given by

$$\rho_b = \frac{b_w}{b} \left[\frac{0.85 f'_c \beta_1}{f_y} \frac{87{,}000}{87{,}000 + f_y} + \frac{A_{sf}}{b_w d} \right] \tag{9.36}$$

The depth to the neutral axis c can be found in the same way as for rectangular beams (Arts. 9.46.1 and 9.46.2).

9.47 SHEAR IN FLEXURAL MEMBERS

Design at a section of a reinforced-concrete flexural member with factored shear force V_u is based on

$$V_u \leq \phi V_n = \phi(V_c + V_s) \tag{9.37}$$

where ϕ = strength reduction factor (given in Art. 9.44)
V_u = factored shear force at a section
V_c = nominal shear strength of concrete
V_s = nominal shear strength provided by reinforcement

Except for brackets, deep beams, and other short cantilevers, the section for maximum shear may be taken at a distance d from the face of the support when the reaction in the direction of the shear introduces compression into the end region of the member.

For shear in two-way slabs, see Art. 9.58.

For nonprestressed flexural members of normal-weight concrete without torsion, the nominal shear strength V_c carried by the concrete is limited to a maximum of $2\sqrt{f'_c}b_u d$, where b_w is the width of the beam web, d = depth to centroid of reinforcement, and f'_c is the specified concrete strength, unless a more detailed analysis is made. In such an analysis V_c should be obtained from

$$V_c = \left(1.9\sqrt{f'_c} + \frac{2{,}500\rho_w V_u d}{M_u}\right) b_w d \leq 3.5\sqrt{f'_c}\, b_w d \tag{9.38}$$

where M_u = factored bending moment occurring simultaneously with V_u at the section considered, but $V_u d/M_u$ must not exceed 1.0
ρ_w = $A_s/b_w d$
A_s = area of nonprestressed tension reinforcement

For one-way joist construction, the American Concrete Institute Building Code, ACI 318, allows these values to be increased 10%.

For rectangular members with short cross-sectional dimension x and long cross-sectional dimension y for which the factored torsional moment T_u exceeds $0.5\phi\sqrt{f'_c}\,\Sigma x^2 y$, the nominal torsional strength that can be assigned to the concrete should be limited to the maximum value of V_c computed from

$$V_c = \frac{2\sqrt{f'_c}\, b_w d}{\sqrt{1 + (2.5C_t T_u/V_u)^2}} \tag{9.39}$$

where $C_t = b_w d/\Sigma x^2 y$.

For lightweight concrete, V_c should be modified by substituting $f_{ct}/6.7$ for $\sqrt{f'_c}$, where f_{ct} is the average splitting tensile strength of lightweight concrete, but not more than $6.7\sqrt{f'_c}$. When f_{ct} is not specified, values of $\sqrt{f'_c}$ affecting V_c should be multiplied by 0.85 for sand-lightweight concrete and 0.75 for all-lightweight concrete.

Shear Reinforcement. When V_u exceeds ϕV_c, shear reinforcement must be provided to resist the excess diagonal tension. The reinforcement may consist of stirrups making an angle of 45 to 90° with the longitudinal steel, longitudinal bars bent at an angle of 30° or more, or a combination of stirrups and bent bars. The nominal shear stress carried by the shear reinforcement V_s must not exceed $8\sqrt{f'_c}b_w d$.

Spacing of required shear reinforcement placed perpendicular should not exceed $0.5d$ for nonprestressed concrete, 75% of the overall depth for prestressed concrete, or 24 in. Inclined stirrups and bent bars should be spaced so that at least one intersects every 45° line extending toward the supports from middepth of the member to the tension reinforcement. When V_s is greater than $4\sqrt{f'_c}b_w d$, the maximum spacing of shear reinforcements should be reduced by one-half. (See Art. 9.108 for shear-strength design for prestressed concrete members.)

The area required in the legs of a vertical stirrup, in^2, is

$$A_v = \frac{V_s s}{f_y d} \tag{9.40a}$$

where s = spacing of stirrups, in, and f_y = yield strength of stirrup steel, psi. For inclined stirrups, the leg area should be at least

$$A_v = \frac{V_s s}{(\sin \alpha + \cos \alpha) f_y d} \tag{9.40b}$$

where α = angle of inclination with longitudinal axis of member.

For a single bent bar or a single group of parallel bars all bent at an angle α with the longitudinal axis at the same distance from the support, the required area is

$$A_v = \frac{V_s}{f_y \sin \alpha} \tag{9.41}$$

in which V_s should not exceed $3\sqrt{f'_c}\, b_w d$.

A minimum area of shear reinforcement is required in all members, except slabs, footings, and joists or where V_u is less than $0.5\phi V_c$. When the factored torsion does not exceed $0.5\phi\sqrt{f'_c}\,\Sigma x^2 y$, as indicated in Art. 9.48, the minimum area of shear reinforcement is given by $A_v = 50 b_w s/f_y$. For greater torsions, $A_v + 2A_t = 50 b_w s/f_y$, where A_t is the area of one leg of a closed stirrup resisting torsion within a distance s.

See also Art. 9.64.

9.48 TORSION IN CONCRETE MEMBERS

Under twisting or torsional moments, a member develops normal (warping) and shear stresses. Torsional design may be based on

$$T_u \leq \phi(T_c + T_s) \tag{9.42}$$

where T_u = factored torsional moment
ϕ = strength reduction factor = 0.85
T_c = nominal torsional moment strength provided by concrete
T_s = nominal torsional moment strength provided by torsion reinforcement

Torsional moment strength of the concrete may be computed from

$$T_c = \frac{0.8\sqrt{f'_c}\,\Sigma x^2 y}{\sqrt{1 + \left(\dfrac{0.4V_u}{C_t T_u}\right)^2}} \tag{9.43}$$

where C_t = $b_w d/\Sigma x^2 y$
x = shorter overall dimension of rectangular part of a cross section (Fig. 9.16)
y = longer overall dimension of the rectangular part (Fig. 9.16)

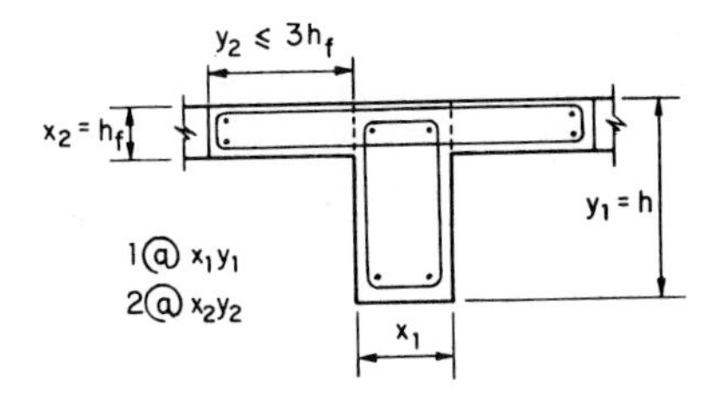

FIGURE 9.16 Resolution of T beam into component rectangles for torsional-shear calculations.

For computations, it is usually convenient to divide a cross section into n rectangles with sides x, and y_r, where r ranges from 1 to n. Calculation of $\Sigma x^2 y$ for flanged sections depends on how the rectangles are chosen. They can be selected to yield the greatest value for $\Sigma x^2 y$. But they should not overlap.

A rectangular box section may be taken as a solid section when wall thickness $h \geq x/4$. When $x/10 < h < x/4$, a solid section may be used, but $\Sigma x^2 y$ should be multiplied by $4h/x$. When $h < x/10$, divide the hollow section into separate rectangles. Fillets should be provided at all interior corners of box sections to reduce the effects of stress concentrations.

If T_u does not exceed $0.5\phi \sqrt{f'_c}\, \Sigma x^2 y$, where f'_c is the specified concrete strength, psi, torsional effects need not be included with those from shear and bending.

The maximum allowable torsional moment is

$$T_u \leq \phi 5 T_c \tag{9.44a}$$

In a statically indeterminate system, where a reduction of the distributed torsional moment can occur, torsional cracking reduces the initial stiffness of the element in torsion. A redistribution of internal forces can be assumed, provided that ductility is maintained by torsional reinforcement to develop the cracking moment. For these conditions, the maximum factored torsional moment may be taken as

$$T_u = \phi(4 \sqrt{f'_c}\, \Sigma x^2 y/3) \tag{9.44b}$$

Members should be designed for the torsion at a distance d from the face of support, where d is the distance from extreme compression surface to centroid of tension steel of the members.

Stirrups. Torsion reinforcement, when required, should be provided in addition to other reinforcement for flexure, axial forces, or shear. For this purpose, the required area A_t for one leg of a closed stirrup can be computed from

$$A_t = (T_u - \phi T_c) \frac{s}{\alpha_1 x_1 y_1 \phi f_y} \tag{9.45}$$

where α_t = $0.66 + 0.33 y_1/x_1 \leq 1.50$
s = spacing of torsion reinforcement measured parallel to longitudinal reinforcement
x_1 = shorter center-to-center dimension of a closed rectangular stirrup
y_1 = larger center-to-center dimension of a closed rectangular stirrup
f_y = yield strength of stirrup steel

To control crack widths, ensure development of the ultimate torsional strength of the member, and prevent excessive loss of torsional stiffness after cracking, the spacing of closed-stirrup torsion reinforcement should not exceed $(x_1 + y_1)/4$ or 12 in, whichever is smaller.

Longitudinal Reinforcement. Steel reinforcement parallel to the axis of the member should be provided in each corner of closed-stirrup torsion reinforcement to assist in development of the torsional design strength without excessive cracking. The amount of longitudinal reinforcement A_l, in², for resisting torsion should be the greater of

$$A_l = 2A_t(x_1 + y_1)/s \tag{9.46}$$

$$A_l = \left[\frac{400xs}{f_y}\left(\frac{T_u}{T_u + V_u/3C_t}\right) - 2A_t\right]\frac{(x_1 + y_1)}{s} \tag{9.47}$$

except that A_l need not be greater than the values given by Eqs. (9.46) and (9.47) when $50b_w s/f_y$ is substituted for $2A_t$.

The spacing of longitudinal torsional reinforcement around the perimeter of closed stirrups should not exceed 12 in.

See also Art. 9.65.

9.49 DEVELOPMENT, ANCHORAGE, AND SPLICES OF REINFORCEMENT

Steel reinforcement must be bonded to the concrete sufficiently so that the steel will yield before it is freed from the concrete. Despite assumptions made in the past to the contrary, bond stress between concrete and reinforcing bars is not uniform over a given length, not directly related to the perimeter of the bars, not equal in tension and compression, and may be affected by lateral confinement. The American Concrete Institute Building Code, ACI 318, requirements therefore reflect the significance of average bond resistance over a length of bar or wire sufficient to develop its strength (**development length**).

The calculated tension or compression force in each reinforcing bar at any section [Eqs. (9.48) to (9.54)] must be developed on each side of that section by a development length l_d, or by end anchorage, or both. Hooks can be used to assist in the development of tension bars only.

The critical sections for development of reinforcement in flexural members are located at the points of maximum stress and where the reinforcement terminates or is bent.

The following requirements of ACI 318 for the development of reinforcement were proposed to help provide for *shifts* in the location of maximum moment and for *peak stresses* that exist in regions of tension in the remaining bars wherever adjacent bars are cut off or bent. In addition, these requirements help minimize any loss of shear capacity or ductility resulting from flexural cracks that tend to open early whenever reinforcement is terminated in a tension zone.

9.49.1 Development for All Flexural Reinforcement

Reinforcement should extend a distance of d or $12d_b$, whichever is larger, beyond the point where the steel is no longer required to resist stress, where d is the effective depth of the member and d_b is the nominal diameter of the reinforcement. This requirement, however, does not apply at supports of simple spans and at the free end of cantilevers.

Continuing reinforcement should extend at least the development length l_d beyond the point where bent or terminated reinforcement is no longer required to resist tension.

Reinforcement should not be terminated in a tension zone unless *one* of the following conditions is satisfied:

1. Shear at the cutoff point does not exceed two-thirds of the shear permitted, including the shear strength of web reinforcement.

2. Stirrup area A_v not less than $60b_w s/f_y$ and exceeding that required for shear and torsion is provided along each terminated bar over a distance from the termination point equal to $0.75d$. (A_v = cross-sectional area of stirrup leg, b_w = width of member, and f_y = yield strength of stirrup steel, psi.) The spacing should not exceed $d/8\beta_b$, where β_b is the ratio of the area of the bars cut off to the total area of bars at the cutoff section.

3. For No. 11 bars and smaller, continuing bars provide double the area required for flexure at the cutoff point, and the shear does not exceed three-fourths of that permitted.

9.49.2 Development for Positive-Moment Reinforcement

A minimum of one-third the required positive-moment reinforcement for simple beams should extend along the same face of the member into the support, and in beams, for a distance of not less than 6 in.

A minimum of one-fourth the required positive-moment reinforcement for continuous members should extend along the same face of the member into the support, and in beams, for a distance of at least 6 in.

For lateral-load-resisting members, the positive-moment reinforcement to be extended into the support in accordance with the preceding two requirements should be able to develop between the face of the support and the end of the bars the yield strength f_y of the bars.

Positive-moment tension reinforcement at simple supports and at points of inflection should be limited to a diameter such that the development length, in, computed for f_y with Eqs. (9.49) to (9.51), (9.53), and (9.54) does not exceed

$$l_d \leq \frac{M_n}{V_u} + l_a \tag{9.48}$$

where M_n = nominal moment strength of a section, in lb, assuming all reinforcement at the section stressed to $f_y = A_s f_y(d - a/2)$

V_u = factored shear at the section, lb

l_a = embedment length, in, beyond center of support plus equivalent embedment length of any hook or mechanical anchorage, but not more than d or $12d_b$, whichever is greater

d = effective depth, in, of member

d_b = nominal bar diameter, in

A_s = area of tensile reinforcement, in^2

a = depth, in, of rectangular stress block (Art. 9.46.1)

9.49.3 Development for Negative-Moment Reinforcement

Negative-moment reinforcement in continuous, restrained, or cantilever members should be developed in or through the supporting member.

Negative-moment reinforcement should have sufficient distance between the face of the support and the end of each bar to develop its full yield strength.

A minimum of one-third of the required negative-moment reinforcement at the face of the support should extend beyond the point of inflection the greatest of d, $12d_b$, or one-sixteenth of the clear span.

9.49.4 Computation of Development Length

The basic development length l_{db} for deformed reinforcing bars and deformed wire **in tension** is

For No. 11 or smaller bars and deformed wire:

$$l_{db} = \frac{0.04A_b f_y}{\sqrt{f'_c}} \geq 0.0004 d_b f_y \tag{9.49}$$

For No. 14 bars:

$$l_{db} = \frac{0.085 f_y}{\sqrt{f'_c}} \tag{9.50}$$

For No. 18 bars:

$$l_{db} = \frac{0.125 f_y}{\sqrt{f'_c}} \tag{9.51}$$

where A_b = area of a bar or wire, in^2, and f'_c = specified concrete compressive strength, psi.

The required minimum development length in tension l_d is obtained by multiplication of l_{db} by modification factors, which may be mandatory under the ACI Building Code or optional. The factors generally are larger than unity. The mandatory factors account for bar spacing; thickness of concrete cover; transverse reinforcement, if present; top-bar effect; type of aggregate; and epoxy coating, if used. Optional factors may be used for the following conditions: for widely spaced bars, bars enclosed within spiral reinforcing of a prescribed size and pitch, or bars enclosed within ties or stirrups of size No. 4 or larger and spaced 4 in c to c or closer.

If excess flexural reinforcement is installed, the l_d computed may be reduced by the ratio of the area of the steel required to the area of steel used.

The development length l_d should be at least that computed from $0.03 d_b f_y / \sqrt{f'_c}$ and 12 in or more.

The modification factor for top reinforcement is 1.3. This applies to horizontal bars with more than 12 in of concrete below the bars, cast integral with the top cover.

For lightweight-aggregate concrete, when f_{ct} is specified, the modification factor is $6.7\sqrt{f'_c}/f_{ct} \geq 1$, where f_{ct} is the average concrete splitting tensile strength, psi. When f_{ct} is not specified, the factor is 1.3.

Two modification factors apply to epoxy-coated bars: When concrete cover is less than $3d_b$ or clear spacing between bars is less than $6d_b$, the factor is 1.5. For all other conditions, the factor is 1.2. The product of the factors for top reinforcement and the effects of epoxy coating need not exceed 1.7.

The development length l_d for uncoated bars embedded in normal-weight concrete are given in Table 9.8. Table 9.8*a* lists minimum development lengths for Grade 60 bars and concrete strengths of 4000 psi corresponding to the categories defined in Table 9.8*b*. Table 9.8*c* lists factors for adjusting the values of Table 9.8*a* for other concrete strengths.

9.49.5 Anchorage with Hooks

For rebars in tension, standard 90° and 180° end hooks can be used as part of the length required for development or anchorage of the bars. Table 9.9 gives the minimum tension embedment length l_{dh} required with standard end hooks (Fig. 9.17 and Table 9.7) and Grade 60 bars to develop the specified yield strength of the bars.

9.49.6 Development for Compression Reinforcement

Basic development length l_{db}, in, for deformed bars in compression may be computed from

$$l_{db} = \frac{0.02 d_b f_y}{\sqrt{f'_c}} \geq 0.0003 d_b f_y \geq 8 \text{ in} \tag{9.52}$$

Compression development length l_d is calculated by multiplying l_{db} by optional modification factors. When bars are enclosed by a spiral at least ¼ in in diameter and with not more than a 4-in pitch, or by ties at least size No. 4 with a spacing not more than 4 in, a modification factor of 0.75 may be used but the lap should be at least 8 in. If excess reinforcement is provided, l_{db} may be reduced by the ratio of the area of steel required to area of steel provided. For general practice, with concrete compressive strength $f'_c \geq 3000$ psi, use $22d_b$ for compression embedment of dowels (Table 9.10).

For bundled bars in tension or compression, the development length of each bar within the bundle should be increased by 20% for a three-bar bundle and 33% for a four-bar bundle.

9.49.7 Development for Welded-Wire Fabric

For deformed welded-wire fabric with at least one cross wire within the development length not less than 2 in from the point of critical section (Fig. 9.18), the basic tension development length, in, is

$$l_{db} = \frac{0.03 d_b (f_y - 20{,}000)}{\sqrt{f'_c}} \geq \frac{0.20 A_w f_y}{s_w \sqrt{f'_c}} \tag{9.53}$$

where d_b = nominal diameter of wire, in
A_w = area of a wire, in^2
s_w = spacing of the wires, in

The tension development length l_d for deformed-wire fabric is calculated as the product of l_{db} and the applicable modification factors, which were discussed in Art.

TABLE 9.8 Development Lengths for Reinforcement in Tension[a]

a. Lengths l_d, in, for Grade 60 rebars, f'_c = 4000 psi, normal-weight concrete[b]

Bar size no.	Category for top bars						Category for other bars					
	1	2	3	4	5	6	1	2	3	4	5	6
3	14	14	14	14	14	14	12	12	12	12	12	12
4	20	19	19	19	19	19	15	15	15	15	15	15
5	31	25	23	23	23	23	24	19	18	18	18	18
6	44	35	31	28	28	28	34	27	24	22	22	22
7	59	48	42	33	33	33	46	37	32	26	25	25
8	78	63	55	44	39	37	60	48	42	34	30	29
9	99	79	69	56	50	42	76	61	53	43	38	32
10	126	101	88	70	63	50	97	77	68	54	48	39
11	154	123	108	86	77	62	119	95	83	67	59	48
14	210	210	147	147	105	105	162	162	113	113	81	81
18	309	309	216	216	154	154	237	237	166	166	119	119

b. Definition of categories

Structural element	Concrete cover t_c	Category determined by bar spacing *s* center to center			
		$s \leq 3d_b$	$3d_b < s < 4d_b$	$4d_b \leq s < 6d_b$	$s \geq 6d_b$
Beams, columns, and inner layer of walls or slabs	$t_c \leq d_b$	1	1	1	2
	$t_c > d_b$	1	3	5	6
All others	$t_c \leq d_b$	1	1	1	2
	$d_b < t_c < 2d_b$	1	3	3	4
	$t_c \geq 2d_b$	1	3	5	6[c]

c. Correction factors for concrete strengths different from 4000 psi[d]

Concrete strength, psi	3000	5000	6000	7000	8000
Correction factor	1.155	0.894	0.817	0.756	0.707

[a]In beams and columns with transverse reinforcement meeting minimum requirements for stirrups given in Art. 9.47 or ties in Art. 9.82 and with the minimum concrete cover specified in Art. 9.65.

[b]For concrete strengths different from 4000 psi, multiply the values for l_d given in Table 9.8*a* by the factors in Table 9.8*c*.

[c]Category 5 applies instead of 6 when edge bars No. 11 or smaller have a side cover less than $2.5d_b$.

[d]To obtain development length for a Grade 60 bar with concrete strength f'_c listed, multiply the corresponding correction factor by l_d given in Table 9.8*a*.

TABLE 9.9 Minimum Embedment Lengths for Hooks on Steel Reinforcement in Tension

a. Embedment lengths l_{dh}, in, for standard end hooks on Grade 60 bars in normal-weight concrete*

Bar size no.	Concrete compressive strength f'_c, psi					
	3000	4000	5000	6000	7000	8000
3	6	6	6	6	6	6
4	8	7	6†	6†	6†	6†
5	10	9	8	7	7	6†
6	12	10	9	8	8	7†
7	14	12	11	10	9	9
8	16	14	12	11	10	10
9	18	15	14	13	12	11
10	20	17	15	14	13	12†
11	22	29	17	16	14	14†
14	37	32	29	27	25	23
18	50	43	39	35	33	31

b. Embedment lengths, in, to provide 2-in concrete cover over tail of standard 180° end hooks

No. 3	No. 4	No. 5	No. 6	No. 7	No. 8	No. 9	No. 10	No. 11	No. 14	No. 18
6	7	7	8	9	10	12	14	15	20	25

*Embedment length for 90° and 180° standard hooks is illustrated in Fig. 9.17. Details of standard hooks are given in Table 9.7. Side cover required is a minimum of 2½ in. End cover required for 90° hooks is a minimum of 2 in. To obtain embedment lengths for grades of steel different from Grade 60, multiply l_{dh} given in Table 9.9 by $f_y/60{,}000$. If reinforcement exceeds that required, multiply l_{dh} by the ratio of area required to that provided.

†For 180° hooks at right angles to exposed surfaces, obtain l_{dh} from Table 9.9*b* to provide 2-in minimum cover to tail (Fig. 9.16*a*).

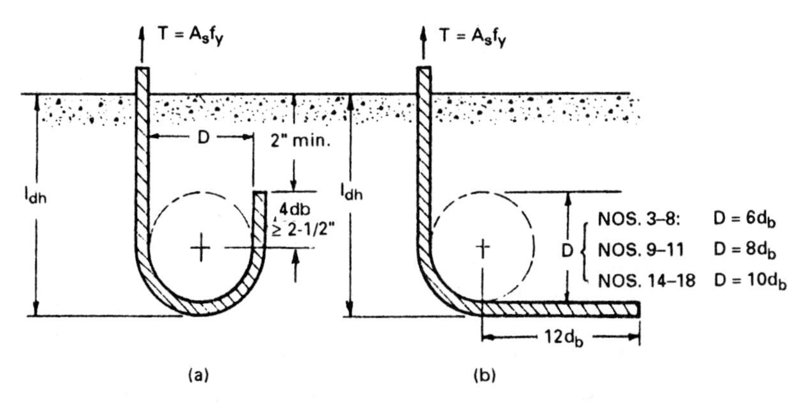

FIGURE 9.17 Embedment lengths for 90° and 180° standard hooks.

TABLE 9.10 Tension Lap-Splice Lengths, in, for Grade 60 Bars, f'_c = 4000 psi; Normal-Weight Concrete*

Bar size no.	Lap class†	Top bars						Other bars					
		Category‡						Category‡					
		1	2	3	4	5	6	1	2	3	4	5	6
3	A	14	14	14	14	14	14	12	12	12	12	12	12
	B	18	18	18	18	18	18	16	16	16	16	16	16
4	A	20	19	19	19	19	19	15	15	15	15	15	15
	B	26	24	24	24	24	24	20	19	19	19	19	19
5	A	31	25	23	23	23	23	24	19	18	18	18	18
	B	40	32	30	30	30	30	31	25	23	23	23	23
6	A	44	35	31	28	28	28	34	27	24	22	22	22
	B	57	45	40	36	36	36	44	35	31	28	28	28
7	A	59	48	42	33	33	33	46	37	32	26	25	25
	B	77	62	54	43	42	42	59	48	42	33	33	33
8	A	78	63	55	44	39	37	60	48	42	34	30	29
	B	102	81	71	57	51	48	78	63	55	44	39	37
9	A	99	79	69	56	50	42	76	61	53	43	38	32
	B	129	103	90	72	64	55	99	79	69	56	50	42
10	A	126	101	88	70	63	50	97	77	68	54	48	39
	B	163	131	114	92	82	65	126	101	88	70	63	50
11	A	154	123	108	86	77	62	119	95	83	67	59	48
	B	200	160	140	112	100	80	154	123	108	86	77	62

*Splice lengths l_s equal l_d for Class A splices and 1.3 l_d for Class B splices, where l_d is the development length listed in Table 9.8*a*.

†Classes A and B are defined in Art. 9.49.8.

‡Categories are defined in Table 9.8*b*.

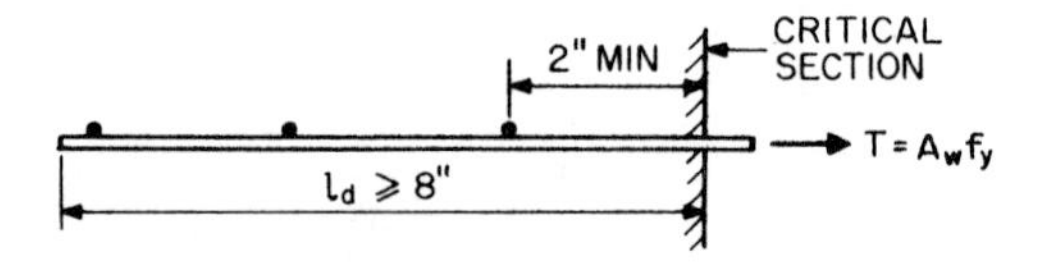

FIGURE 9.18 Minimum development length for deformed-wire fabric.

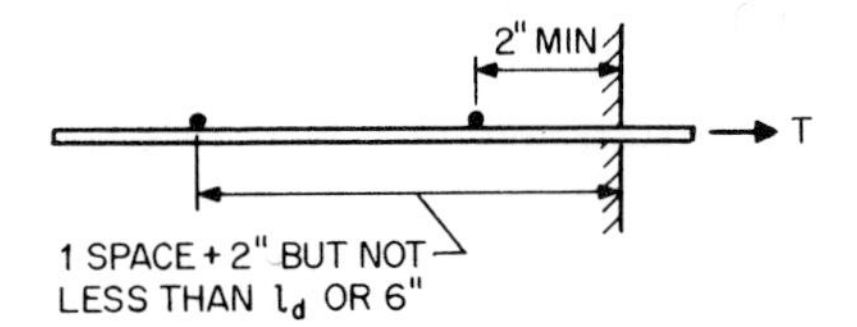

FIGURE 9.19 Minimum development length for plain-wire fabric.

9.49.4. The development length should be at least 8 in except in calculation of lap splice lengths and development of web reinforcement.

Plain welded-wire fabric is considered to be developed by embedment of two cross wires. The closer cross wire should be located not less than 2 in from the point of critical section (Fig. 9.19). The ACI Code also requires the basic development length l_{db} measured from the point of critical section to the outermost cross wire to be at least

$$l_{db} = \frac{0.27A_w f_y}{s_w \sqrt{f'_c}} \tag{9.54}$$

The basic length l_{db} has to be modified by the factor for lightweight-aggregate concrete, as indicated in Art. 9.49.4. If excess tension reinforcement is provided, l_{db} may be reduced by the ratio of area of steel required to the area of steel provided. The development length should be at least 6 in, except in calculation of lap splices.

9.49.8 Tension Lap Splices

Bar sizes No. 11 or less and deformed wire may be spliced by lapping. Tension lap splices are classified in two classes, A and B, depending on the stress in the bars to be spliced. The minimum lap length l_s is expressed as a multiple of the tension development length l_d of the bar or deformed wire (Art. 4.49.4).

Class A tension lap splices include splices at sections where the tensile stress due to factored loads does not exceed $0.5f_y$ and not more than one-half the bars at these sections are spliced within one Class A splice length of the section. For Class A splices,

$$l_s = l_d \geq 12 \text{ in} \tag{9.55}$$

Class B tension lap splices include splices at sections where the tensile stress exceeds $0.5f_y$ and where more than 50% of the bars at the section are spliced. For Class B splices,

$$l_s = 1.3l_d \geq 12 \text{ in} \tag{9.56}$$

Laps for tension splices for uncoated Grade 60 rebars in normal-weight concrete with $f'_c = 4000$ psi are given in Table 9.10.

The tension lap-splice lengths for welded-wire fabric are indicated in Figs. 9.20 and 9.21.

9.49.9 Compression Lap Splices

Minimum lap-splice lengths of rebars in compression l_s vary with nominal bar diameter d_b and yield strength f_y of the bars. For bar sizes No. 11 or less, the

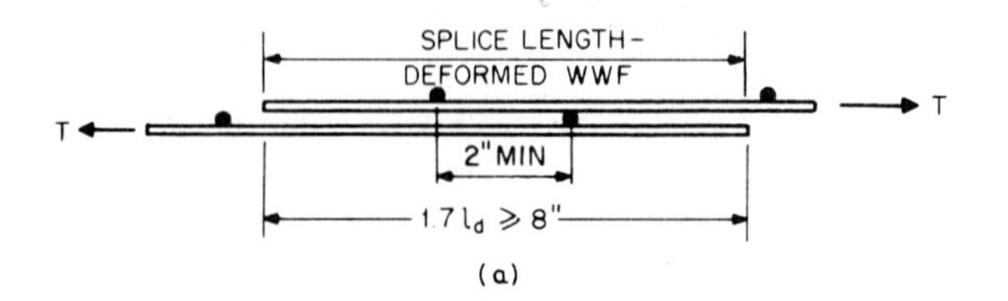

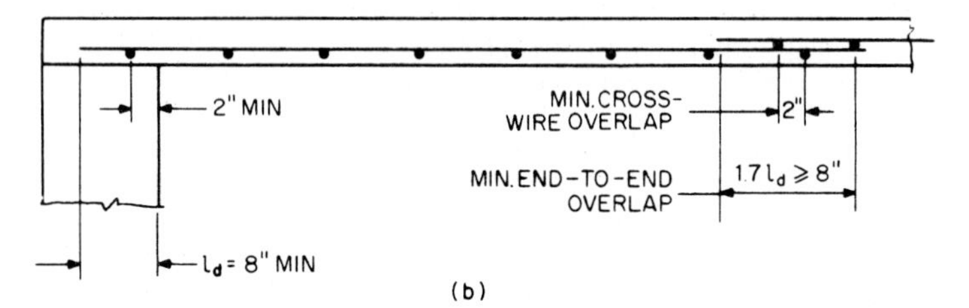

FIGURE 9.20 (*a*) Minimum splice length for deformed-wire fabric. (*b*) Slab reinforced with deformed-wire fabric.

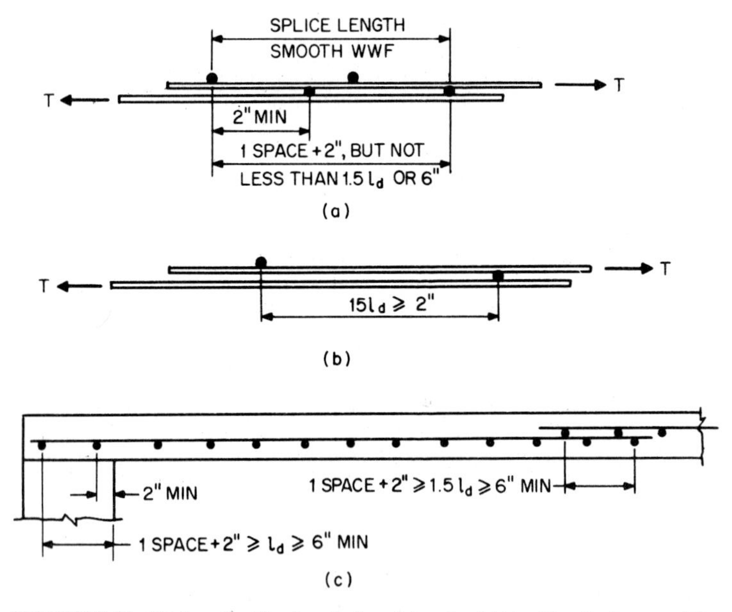

FIGURE 9.21 Minimum splice length for plain-wire fabric. Use the larger of the values shown in (*a*) and (*b*). In calculation of splice length, the computed value of development length l_d, not the minimum required value, should be used. (*a*) Splice length when steel area used is less than twice the required area. (*b*) Splice length when steel area used is two or more times the required area. (*c*) Slab reinforced with plain-wire fabric providing twice the required reinforcement area.

compression lap-splice length is the largest of 12 in or the values computed from Eqs. (9.57*a*) and (9.57*b*):

$$l_s = 0.0005 f_y d_b \qquad f_y \le 60{,}000 \text{ psi} \qquad (9.57a)$$
$$l_s = (0.0009 f_y - 24) d_b \qquad f_y > 60{,}000 \text{ psi} \qquad (9.57b)$$

When f'_c is less than 3000 psi, the length of lap should be one-third greater than the values computed from the preceding equations.

When the bars are enclosed by a spiral, the lap length may be reduced by 25%. For general practice, use 30 bar diameters for compression lap splices (Table 9.11). Spirals should conform to requirements of the American Concrete Institute Building Code, ACI 318: Spirals should extend from top of footing or slab in any story to the level of the lowest horizontal reinforcement in members supported above. The ratio of volume of spiral reinforcement to the total value of the concrete core (out to out of spirals) should be at least that given in Art. 9.82. Minimum spiral diameter in cast-in-place construction is ⅜ in. Clear spacing between spirals should be limited to 1 to 3 in. Lapped splices of spirals should have a lap of $48d_b$, but at least 12 in. Spirals should be anchored by 1½ extra turns of spiral bar or wire at each end.

TABLE 9.11 Compression Dowel Embedment and Compression Lap Splices, in, for Grade 60 Bars and All Concrete with $f'_c \ge 3000$ Psi

Bar size no.	Recommended dowel embedment $22d_b$	Minimum lap length	
		Standard lap $30d_b$	With column spirals* $22.5d_b$
3	9	12	12
4	11	15	12
5	14	19	14
6	17	23	17
7	20	26	20
8	22	30	23
9	25	34	25
10	28	38	29
11	31	42	32
14	38		
18	50		

*For use in spirally reinforced columns with spirals conforming to requirements in Art. 9.49.9. Regarding anchorage and splicing of spirals, the ACI Code requirements are:

Anchorage of spiral reinforcement shall be provided by 1½ extra turns of spiral bar or wire at each end of a spiral unit.

Splices in spiral reinforcement shall be lap splices of $48d_b$, but not less than 12 in, or welded.

The ACI Code contains provisions for lap splicing bars of different sizes in compression. Length of lap should be the larger of the compression development length required for the larger size bar or the compression lap-splice length required for the smaller bar. It is permissible to lap-splice the large bar sizes, Nos. 14 and 18, to No. 11 and smaller bars.

9.49.10 Mechanical Connections and Welded Splices

As an alternative to lap splicing, mechanical connections or welded splices may be used. When traditional lap splices satisfy all requirements, they are generally the most economical. There are conditions, however, where they are not suitable: The ACI Building Code does not permit lap splices of the large-size bars (Nos. 14 and 18) to No. 11 and smaller, except in compression. Lap splices cause congestion at the splice locations and their use then may be impracticable. Under certain conditions, the required length of tension lap splices for No. 11 and similar-size bars can be excessive and make the splices uneconomical. For these reasons, mechanical connections or welded splices may be suitable alternatives.

Mechanical connections are made with proprietary devices. The ACI Building Code requires a full mechanical connection to have a capacity, in tension or compression, equal to at least 125% of the specified f_y of the bar. End-bearing mechanical connections may be used where the bar stress due to all conditions of factored loads is compressive. For these types of compression-only splices, the ACI Code prescribes requirements for the squareness of the bar ends. Descriptions of the commercially available proprietary mechanical connection devices are given in "Mechanical Connections of Reinforcing Bars," ACI 439.3R, and "Reinforcement: Anchorages, Lap Splices and Connections," Concrete Reinforcing Steel Institute.

For a full welded splice, the ACI Building Code requires the butt-welded bars to have a tensile capacity of at least 125% of the specified f_y of the bar. Welding should conform to "Structural Welding Code—Reinforcing Steel" (AWS D1.4), American Welding Society.

9.49.11 Anchorage of Web Reinforcement

Stirrups are reinforcement used to resist shear and torsion. They are generally bars, wire or welded-wire fabric, either single leg or bent into L, U, or rectangular shapes.

Stirrups should be designed and detailed to be installed as close as possible to the compression and tension surfaces of a flexural member as cover requirements and the proximity of other steel will permit. They should be installed perpendicular or inclined with respect to flexural reinforcement and spaced closely enough to cross the line of every potential crack. Ends of single-leg, simple U stirrups, or transverse multiple U stirrups should be anchored by one of the following means:

1. A standard stirrup hook around a longitudinal bar for stirrups fabricated from No. 5 bars or D31 wire or smaller sizes. Stirrups fabricated from bar sizes Nos. 6, 7, and 8 in Grade 40 can be anchored similarly.

2. For stirrups fabricated from bar sizes Nos. 6, 7, and 8 in Grade 60, a standard stirrup hook around a longitudinal bar plus a minimum embedment of $0.014 d_b f_y / \sqrt{f'_c}$ between midheight of the member and the outside end of the hook.

Each leg of simple U stirrups made of plain welded-wire fabric should be anchored by one of the following means:

1. Two longitudinal wires located at the top of the U and spaced at 2 in.

2. One longitudinal wire located at a distance of $d/4$ or less from the compression face and a second wire closer to the compression face and spaced at least 2 in from the first wire. (d = distance, in, from compression surface to centroid of tension reinforcement.) The second wire can be located on the stirrup leg beyond a bend, or on a bend with an inside diameter of at least $8d_b$.

Each end of a single-leg stirrup, fabricated from plain or deformed welded-wire fabric, should be anchored by two longitudinal wires spaced at 2 in minimum. The inner wire of the two longitudinal wires should be located at least the larger of $d/4$ or 2 in from the middepth of the member $d/2$. The outer longitudinal wire at the tension face of the member should be located not farther from the face than the portion of primary flexural reinforcement closest to the face.

Between anchored ends, each bend in the continuous portion of a simple U or multiple U stirrup should enclose a longitudinal bar.

9.49.12 Stirrup Splices

Pairs of U stirrups or ties placed to form a closed unit may be considered properly spliced when the legs are lapped over a minimum distance of $1.3l_d$. In members at least 18 in deep, such splices may be considered adequate for No. 3 bars of Grade 60 and Nos. 3 and 4 bars of Grade 40 if the legs extend the full available depth of the member.

9.50 CRACK CONTROL

Because of the effectiveness of reinforcement in limiting crack widths, the American Concrete Institute Building Code, ACI 318, requires minimum areas of steel and limits reinforcement spacing, to control cracking.

Beams and One-Way Slabs. If, in a structural floor or roof slab, principal reinforcement extends in one direction only, shrinkage and temperature reinforcement should be provided normal to the principal reinforcement, to prevent excessive cracking. The additional reinforcement should provide at least the ratios of reinforcement area to gross concrete area of slab given in Table 9.12, but not less than 0.0014.

To control flexural cracking, tension reinforcement in beams and one-way slabs should be well distributed in zones of maximum concrete tension when the design yield strength of the steel f_y is greater than 40,000 psi. Spacing of principal reinforcement in slabs should not exceed 18 in or 3 times the slab thickness, except in concrete-joist construction.

TABLE 9.12 Minimum Shrinkage and Temperature Reinforcement

In slabs where Grade 40 or 50 deformed bars are used	0.0020
In slabs where Grade 60 deformed bars or welded-wire fabric, deformed or plain, are used (Table 9.17)	0.0018
In slabs reinforced with steel having a yield strength f_y exceeding 60,000 psi measured at a strain of 0.0035 in/in	$108/f_y$
This reinforcement should not be placed farther apart than 5 times the slab thickness or more than 18 in.	

Where slab flanges of beams are in tension, a part of the main reinforcement of the beam should be distributed over the effective flange width or a width equal to one-tenth the span, whichever is smaller. When the effective flange width exceeds one-tenth the span, some longitudinal reinforcement should be provided in the outer portions of the flange. Also, reinforcement for one-way joist construction should be uniformly distributed throughout the flange.

To control flexural cracking in beams, reinforcement should be so distributed that:

For interior exposures,

$$z \leq f_s \sqrt[3]{d_c A} \leq 175 \text{ kips/in} \tag{9.58}$$

For exterior exposures,

$$z \leq f_s \sqrt[3]{d_c A} \leq 145 \text{ kips/in} \tag{9.59}$$

where d_c = thickness, in, or concrete cover measured from extreme tension surface to center of bar

A = effective tension area, in^2, of concrete per rebar. It can be computed by dividing the concrete area surrounding the main tension reinforcing bars and having the same centroid as that reinforcement by the number of bars. If bar sizes differ, the number of bars should be computed as the total steel area divided by the area of the largest bar used

f_s = calculated stress in reinforcement at service loads, ksi, but may be taken as $0.60f_y$ in lieu of such calculations

The numerical limitations on z of 175 and 145 kips/in correspond to limiting crack widths of 0.016 and 0.013 in for interior and exterior exposures, respectively. For water-tight slabs or severe exposures, z should be smaller.

For one-way slabs, tests indicate that z should not exceed 155 kips/in for interior exposures, or 129 kips/in for exterior exposures.

The z values can be transformed into the following expressions for the maximum spacing of Grade 60 bars to control flexural cracking:

Interior exposure	Maximum spacing, in
Beams	$57.4/d_c^2$
One-way slabs	$39.9/d_c^2$
Exterior exposure	**Maximum spacing, in**
Beams	$32.7/d_c^2$
One-way slabs	$23.0/d_c^2$

See Tables 9.13 and 9.22.

Two-Way Slabs. Flexural cracking in two-way slabs is significantly different from that in one-way slabs. For control of flexural cracking in two-way slabs, such as solid flat plates and flat slabs with drop panels, the ACI Building Code restricts the maximum spacing of tension bars to twice the overall thickness h of the slab but not more than 18 in. In waffle slabs or over cellular spaces, however, reinforcement should be the same as that for shrinkage and temperature in one-way slabs (see Table 9.13).

TABLE 9.13 Maximum Spacing, in, of Grade 60 Bars for Control of Flexural Cracking

	Beams		One-way slabs				
	2-in cover*		Interior exposure			Exterior exposure	
			Cover, in			Cover, in	
Bar size no.	Interior exposure	Exterior exposure	¾	1	1½	1½	2
3			18.0	18.0		8.1	
4			18.0	18.0		7.5	
5	10.7	6.1	18.0	18.0		7.0	
6	10.2	5.8	18.0	18.0			4.1
7	9.7	5.5	18.0	18.0			3.9
8	9.2	5.2	18.0	17.7			3.7
9	8.7	5.0	18.0	16.3			3.5
10	8.3	4.7	18.0	14.9			3.3
11	7.8	4.5	18.0	13.7			3.1
14	7.1	4.0			7.2		
18	5.9				5.8		

*Cover to stirrups, if any, should be at least 1½ in.

9.51 DEFLECTION OF CONCRETE BEAMS AND SLABS

Reinforced-concrete flexural members must have adequate stiffness to limit deflection to an amount that will not adversely affect the serviceability of the structure under service loads.

Beams and One-Way Slabs. Unless computations show that deflections will be small (Table 9.14), the American Concrete Institute Building Code, ACI 318, requires that the depth h of nonprestressed, one-way solid slabs, one-way ribbed slabs, and beams of normal-weight concrete—with Grade 60 reinforcement—be at least the fraction of the span L given in Table 9.15.

When it is necessary to compute deflections, calculation of short-term deflection may be based on elastic theory, but with an effective moment of inertia I_e.

For normal-weight concrete,

$$I_e = \left(\frac{M_{cr}}{M_a}\right)^3 I_g + \left[1 - \left(\frac{M_{cr}}{M_a}\right)^3\right] I_{cr} \leq I_g \tag{9.60}$$

where M_{cr} = cracking moment = $f_r I_g / y_t$
M_a = service-load moments for which deflections are being computed
I_g = gross moment of inertia of concrete section
I_{cr} = moment of inertia of cracked section transformed to concrete (for solid slabs, see Fig. 9.22)
f_r = modulus of rupture of concrete, psi = $7.5\sqrt{f'_c}$

TABLE 9.14 Maximum Ratios of Computed Deflection to Span L for Beams and Slabs

Type of member	Deflection to be considered	Deflection limitation
Flat roofs not supporting or attached to nonstructural elements likely to be damaged by large deflections	Immediate deflection due to the live load	$L/180$*
Floors not supporting or attached to nonstructural elements likely to be damaged by large deflections	Immediate deflection due to the live load	$L/360$
Roof or floor construction supporting or attached to nonstructural elements likely to be damaged by large deflections	This part of the total deflection that occurs after attachment of the nonstructural elements (the sum of the long-term deflection due to all sustained loads and the immediate deflection due to any additional live load)†	$L/480$‡
Roof or floor construction supporting or attached to nonstructural elements not likely to be damaged by large deflections		$L/240$§

*This limit is not intended to safeguard against ponding. Ponding should be checked by suitable calculations of deflection, including the added deflections due to ponded water, and considering long-term effects of all sustained loads, camber, construction, tolerances and reliability of provisions for drainage.

†The long-term deflection may be reduced by the amount of deflection that occurs before attachment of the nonstructural elements.

‡This limit may be exceeded if adequate measures are taken to prevent damage to supported or attached elements.

§But not greater than the tolerance provided for the nonstructural elements. This limit may be exceeded if camber is provided so that the total deflection minus the camber does not exceed the limitation.

TABLE 9.15 Minimum Depths h of Reinforced-Concrete Beams and One-Way Slabs*

	One-way solid slabs	Beams and one-way ribbed slabs
Cantilever	$L/10 = 0.1000L$	$L/8 = 0.1250L$
Simple span	$L/20 = 0.0500L$	$L/16 = 0.0625L$
Continuous:		
End span	$L/24 = 0.0417L$	$L/18.5 = 0.0540L$
Interior span	$L/28 = 0.0357L$	$L/21 = 0.0476L$

*For members with span L (Art. 9.41) not supporting or attached to partitions or other construction likely to be damaged by large deflections. Thinner members may be used if justified by deflection computations. For structural lightweight concrete of unit weight w, lb/ft^3, multiply tabulated values by $1.65 - 0.005w \geq 1.09$, for $90 < w < 120$. For reinforcement with yield strength $f_y > 60{,}000$ psi, multiply tabulated values by $0.4 + f_y/100{,}000$.

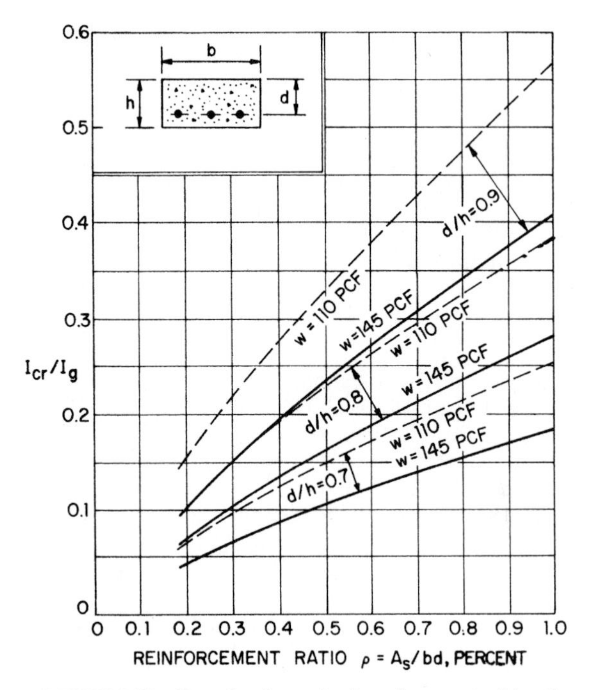

FIGURE 9.22 Chart for determination of moment of inertia I_{cr} of transformed (cracked) section of one-way solid slab, given the moment of inertia of the gross section, $I_g = bh^3/12$, reinforcement ratio $\rho = A_s/bd$, weight w of concrete, pcf, and ratio d/h of effective depth to thickness, for $f'_c = 4$ ksi.

f'_c = specified concrete strength, psi

y_t = distance from centroidal axis of gross section, neglecting the reinforcement, to the extreme surface in tension

When structural lightweight concrete is used, f_r in the computation of M_{cr} should be taken as $1.12 f_{ct} \leq 7.5\sqrt{f'_c}$, where f_{ct} = average splitting tensile strength, psi, of the concrete. When f_{ct} is not specified, f_r should be taken as $5.6\sqrt{f'_c}$ for all lightweight concrete and as $6.4\sqrt{f'_c}$ for sand-lightweight concrete.

For deflection calculations for continuous spans, I_c may be taken as the average of the values obtained from Eq. (9.60) for the critical positive and negative moments.

Additional long-term deflection for both normal-weight and lightweight concrete flexural members can be estimated by multiplying the immediate deflection due to the sustained load by $\zeta/(1 + 50\rho')$, where ζ = time-dependent factor (2.0 for 5 years or more, 1.4 for 12 months, 1.2 for 6 months, and 1.0 for 3 months, and ρ' = compression-steel ratio, the area of the compression reinforcement A'_s, in^2, divided by the concrete area bd, in^2.

The sum of the short-term and long-term deflections should not exceed the limits given in Table 9.14.

Two-Way Slabs. Unless computations show that deflections will not exceed the limits listed in Table 9.14, thickness h for nonprestressed two-way slabs with a ratio of long to short span not exceeding 2 may be computed from

$$h = \frac{L_n(0.8 + f_y/200{,}000)}{36 + 5\,[\alpha_m - 0.12(1 + 1/\beta)]} \tag{9.61}$$

The thickness, however, should not be less than

$$h = \frac{L_n(0.8 + f_y/200{,}000)}{36 + 9\beta} \tag{9.62}$$

and need not be more than

$$h = \frac{L_n(0.8 + f_y/200{,}000)}{36} \tag{9.63}$$

where L_n = clear span in long direction, in
α_m = average value of α for all beams along panel edges
α = ratio of flexural stiffness of beam section to flexural stiffness of a width of slab bounded laterally by the centerline of the adjacent panel, if any, on each side of the beam
β = ratio of clear span in long direction to clear span in short direction

The thickness, however, should not be less than the following:

For slabs with $\alpha_m < 2.0$	5 in
For slabs having beams on all four edges with α_m at least equal to 2.0	3½ in

The minimum thicknesses for flat slabs with standard drop panels may be reduced 10%.

Unless edge beams with $\alpha \geq 0.8$ are provided at discontinuous edges, the minimum thicknesses of panels at those edges should be increased at least 10%.

The computed deflections of prestressed-concrete construction should not exceed the values listed in Table 9.14.

ONE-WAY REINFORCED-CONCRETE SLABS

A one-way reinforced-concrete slab is a flexural member that spans in one direction between supports and is reinforced for flexure only in one direction (Art. 9.62). If a slab is supported by beams or walls on four sides, but the span in the long direction is more than twice that in the short direction, most of the load will be carried in the short direction; hence, the slab can be designed as a one-way slab.

One-way slabs may be solid, ribbed, or hollow. (For one-way ribbed slabs, see Arts. 9.54 to 9.57.) Hollow one-way slabs are usually precast (Arts. 9.97 to 9.104). Cast-in-place, hollow one-way slabs can be constructed with fiber or cardboard-cylinder forms, inflatable forms that can be reused, or precast hollow boxes or blocks. One-way slabs can be haunched at the supports for flexure or for shear strength.

9.52 ANALYSIS AND DESIGN OF ONE-WAY SLABS

Structural strength, fire resistance, crack control, and deflections of one-way slabs must be satisfactory under service loads.

Strength and Deflections. Approximate methods of frame analysis can be used with uniform loads and spans that conform to ACI Building Code requirements (see Art. 9.41). Deflections can be computed as indicated in Art. 9.51, or in lieu of calculations the minimum slab thicknesses listed in Table 9.15 may be used. In Fig. 9.22 is a plot of ratios of moments of inertia of cracked to gross concrete section for one-way slabs. These curves can be used to simplify deflection calculations.

Strength depends on slab thickness and reinforcement and properties of materials used. Slab thickness required for strength can be computed by treating a 1-ft width of slab as a beam (Arts. 9.45 and 9.46).

Fire Resistance. One-way concrete slabs, if not protected by a fire-resistant ceiling, must have a thickness and a concrete cover around reinforcement that conforms to the fire-resistance rating required by the building code. Table 9.16 gives typical slab thickness and cover around reinforcement for various fire-resistance ratings for normal-weight and structural-lightweight-concrete construction.

Reinforcement. Requirements for minimum reinforcement for crack control are summarized in Art. 9.50. Table 9.17 lists minimum reinforcement when Grade 60 bars are used. Reinforcement required for flexural strength can be computed by treating a 1-ft width of slab as a beam (Arts. 9.44 to 9.46).

Rebar weights, lb/ft^2 of slab area, can be estimated from Fig. 9.24*a* for one-way, continuous, interior spans of floor or roof slabs made of normal-weight concrete.

One-way reinforced concrete slabs with spans less than 10 ft long can be reinforced with a single layer of draped welded-wire fabric for both positive and negative moments. These moments can be taken equal to $wL^2/12$, where w is the uniform load and L is the span, defined in Art. 9.41, if the slab meets ACI Code requirements for approximate frame analysis with uniform loads.

For development (bond) of reinforcement, see Art. 9.49.

Shear. This is usually not critical in one-way slabs, but the ACI Building Code requires that it be investigated (see Art. 9.47).

TABLE 9.16 Typical Fire Ratings for Concrete Members*

Fire rating, h	Slabs						Cover, in, for beams and columns 12 in or larger in section
	Lightweight concrete		Normal-weight concrete				
			Carbonate aggregate		Siliceous aggregate		
	Depth h, in	Cover, in	Depth h, in	Cover, in	Depth h, in	Cover, in	
1	2½	¾	3	¾	3½	¾	1½
2	3½	¾	4½	¾	5	1	1½
3	4½	1	5¾	1	6	1	1½
4	5	1	6½	1	7	1¼	1½

*From "Uniform Building Code," International Conference of Building Officials.

TABLE 9.17 Minimum and Maximum Reinforcement for One-Way Concrete Slabs

Slab thickness, in.	Minimum reinforcement*			Maximum reinforcement†		
	Area A_s, in^2	Bar size and spacing, in	Weight, psf	Area A_s, in^2	Bar size and spacing, in	Weight, psf
4	0.086	No. 3 @ 12‡	0.30	0.555	No. 5 @ 9½	1.89
4½	0.097	No. 3 @ 13½	0.33	0.655	No. 6 @ 8	2.24
5	0.108	No. 3 @ 12½	0.36	0.750	No. 6 @ 7	2.56
5½	0.119	No. 3 @ 11	0.41	0.845	No. 6 @ 6	2.99
6	0.130	No. 4 @ 18	0.45	0.931	No. 7 @ 7½	3.26
6½	0.140	No. 4 @ 17	0.48	1.025	No. 7 @ 7	3.50
7	0.151	No. 4 @ 15½	0.53	1.110	No. 8 @ 8½	3.80
7½	0.162	No. 4 @ 14½	0.56	1.208	No. 8 @ 7½	4.30
8	0.173	No. 4 @ 13½	0.61	1.291	No. 9 @ 9	4.54
8½	0.184	No. 4 @ 13	0.63	1.382	No. 9 @ 8½	4.80
9	0.194	No. 4 @ 12	0.68	1.482	No. 9 @ 8	5.10

*For Grade 60 reinforcement. Minimum area $A_s \geq 0.0018bh$, where b = slab width and h = slab thickness.

†For f'_c = 3000 psi and no compression reinforcement.

‡This spacing for a 4-in slab is the maximum spacing for flexure but can be increased to 15 in for temperature and shrinkage reinforcement.

9.53 EMBEDDED PIPES IN ONE-WAY SLABS

Generally, embedded pipes or conduit, other than those merely passing through, should not be larger in outside dimension than one-third the slab thickness and should be spaced at least three diameters or widths on centers. Piping in solid one-way slabs is required to be placed between the top and bottom reinforcement unless it is for radiant heating or snow melting.

ONE-WAY CONCRETE-JOIST CONSTRUCTION

One-way concrete-joist construction consists of a monolithic combination of cast-in-place, uniformly spaced ribs (joists) and top slab (Fig. 9.23). (See also Art. 9.62.) The ribs are formed by placing rows of permanent or removable fillers in what would otherwise be a solid slab.

One-way joist construction was developed to reduce dead load. For long spans, the utility of solid-slab construction is offset by the increase in dead load of the slab. One-way concrete-joist construction provides adequate depth with less dead load than for solid slabs, and results in smaller concrete and reinforcement quantities per square foot of floor area.

Uniform-depth floor and roof construction can be obtained by casting the joists integral with wide, supporting band beams of the same total depth as the joists. This design eliminates the need for interior beam forms.

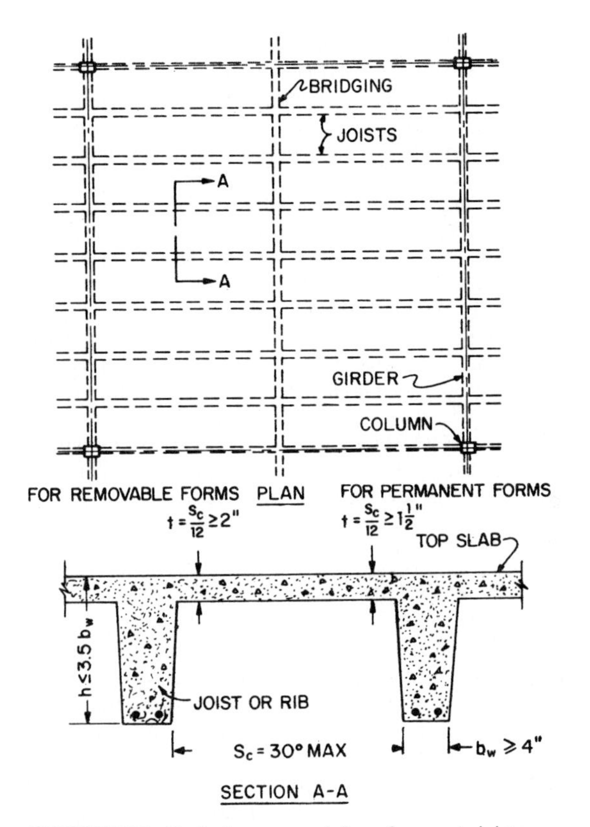

FIGURE 9.23 Typical one-way reinforced-concrete joist construction.

9.54 STANDARD SIZES OF JOISTS

One-way concrete-joist construction that exceeds the dimensional limitations of the American Concrete Institute Building Code, ACI 318, must be designed as slabs and beams. These dimensional limitations are:

Maximum clear spacing between ribs—30 in

Maximum rib depth—3.5 times rib width

Minimum rib width—4 in

Minimum top-slab thickness with removable forms—2 in, but not less than one-twelfth the clear spacing of ribs

Minimum top-slab thickness with permanent forms—1½ in, but not less than one-twelfth the clear spacing of ribs

Removable form fillers can be standard steel *pans* or hardboard, corrugated cardboard, fiberboard, or glass-reinforced plastic. Standard removable steel pans that conform to "Types and Sizes of Forms for One-Way Concrete-Joist Construction," (ANSI/CRSI A48.1-1986), American National Standards Institute, include

20- and 30-in widths and depths of 8, 10, 12, 14, 16, and 20 in. Standard steel square-end pans are available in 36-in lengths. Widths of 10, 15, and 20 in and tapered end fillers are available as special items. For forms 20 and 30 in wide, tapered end forms slope to 16 and 25 in, respectively, in a distance of 3 ft.

Permanent form fillers are usually constructed from structural-clay floor tiles (ASTM C57) or hollow load-bearing concrete blocks (ASTM C90).

Structural-clay tiles are usually 12 in square and can be obtained in thicknesses varying from 3 to 12 in. The tiles are laid flat, or end to end, in rows between and at right angles to the joists. The usual clear distance between rows is 4 in, making the center-to-center spacing of the rows 16 in.

Hollow load-bearing concrete blocks are usually 16 in long and 8 in high. They can be obtained in thicknesses of 4, 6, 8, 10, and 12 in. The blocks are laid flat, end-to-end, in rows between and at right angles to the joists. The usual clear distance between rows is 4 in making the center-to-center spacing of the rows 20 in.

If the clay-tile or concrete-block fillers have a compressive strength equal to that of the concrete in the joists, the vertical shells of the fillers in contact with the joist ribs can be included in the width of the joist rib.

9.55 DESIGN OF JOIST CONSTRUCTION

One-way concrete joists must have adequate structural strength, and crack control and deflection must be satisfactory under service loads. Approximate methods of frame analysis can be used with uniform loads and spans that conform to requirements of the American Concrete Institute Building Code, ACI 318 (see Art. 9.41). Table 9.15 lists minimum depths of joists to limit deflection, unless deflection computations justify shallower construction (Table 9.14). Load tables in the Concrete Reinforcing Steel Institute's CRSI "Handbook" indicate when deflections under service live loads exceed specified limits.

Economy can be obtained by designing joists and slabs so that the same-size forms can be used throughout a project. It will usually be advantageous to use square-end forms for interior spans and tapered ends for end spans, when required, with a uniform depth.

Fire Resistance. Table 9.18 gives minimum top-slab thickness and reinforcement cover for fire resistance when a fire-resistant ceiling is not used.

Temperature and Shrinkage Reinforcement. This reinforcement must be provided perpendicular to the ribs and spaced not farther apart than 5 times the slab thickness, or 18 in. The required area of Grade 60 reinforcement for temperature and shrinkage is 0.0018 times the concrete area (Table 9.19). For flexural reinforcement, see Art. 9.56. For shear reinforcement, see Art. 9.57.

Embedded Pipes. Top slabs containing horizontal conduit or pipes that are allowed by the ACI Code (Art. 9.53) must have a thickness of at least 1 in plus the depth of the conduit or pipe.

Bridging. Distribution ribs are constructed normal to the main ribs to distribute concentrated loads to more than one joist and to equalize deflections. These ribs are usually made 4 to 5 in wide and reinforced top and bottom with one No. 4 or 5 continuous rebar. One distribution rib is usually used at the center of spans of

TABLE 9.18 Required Top-Slab Thickness and Cover for Fire Resistance of One-Way Concrete Joist Construction*

Fire-resistance rating, h	Normal-weight concrete				Lightweight concrete	
	Carbonate aggregate		Siliceous aggregate			
	Top-slab thickness, in	Cover, in	Top-slab thickness, in	Cover, in	Top-slab thickness, in	Cover, in
1	3	¾	3½	¾	3	¾
2	4½	1	5	1	4½	1
3	5¾	1¼	6	1½	5¾	1¼

*From "Uniform Building Code," International Conference of Building Officials.

up to 30 ft, and two distribution ribs are usually placed at the third points of spans longer than 30 ft.

Openings. These can be provided in the top slab of one-way concrete joist construction between ribs without significant loss in flexural strength. Header joists must be provided along openings that interrupt one or more joists.

9.56 REINFORCEMENT OF JOISTS FOR FLEXURE

Reinforcement required for strength can be determined as indicated in Art. 9.46, by treating as a beam a section symmetrical about a rib and as wide as the spacing of ribs on centers.

TABLE 9.19 Temperature and Shrinkage Reinforcement for One-Way Joist Construction

Top-slab thickness, in	Required area of temperature and shrinkage reinforcement, in^2	Reinforcement	Reinforcement weight, psf
2	0.043	WWF 4 × 12, W1.5/W1	0.19
2½	0.054	WWF 4 × 12, W2/W1	0.24
3	0.065	WWF 4 × 12, W2.5/W1	0.29
3½	0.076	No. 3 @ 17½ in	0.26
4	0.086	No. 3 @ 15 in	0.30
4½	0.097	No. 3 @ 13½ in	0.33
5	0.108	No. 3 @ 12½ in	0.36
5½	0.119	No. 3 @ 11 in	0.41

Minimum Reinforcement. It is the opinion of the authors that all reinforcement (both positive and negative) with a yield strength f_y should have an area equal to or greater than $200/f_y$ times the concrete area of the rib $b_w d$, where b_w is the rib width and d = rib depth. Less steel can be used, however, if the areas of both the positive and negative reinforcement at every section are one-third greater than the amount required by analysis. (See also Art. 9.55.)

If bottom bars in continuous joists are not continuous through the support, top bars should meet the requirements for shrinkage and temperature reinforcement.

Maximum Reinforcement. Positive- and negative-moment steel ratios must not be greater than three-quarters of the ratio that produces balanced conditions (Art. 9.46). The positive-moment reinforcement ratio is based on the width of the top flange, and the negative-moment reinforcement ratio is based on the width of the rib b_w.

Reinforcement for one-way concrete-joist construction consists of straight top and bottom bars, cut off as required for moment.

For top-slab reinforcement, see Art. 9.55 and Table 9.13. Straight top- and bottom-bar arrangements are more flexible in attaining uniform distribution of top bars to control cracking in the slab than straight and bent bars.

Requirements for structural integrity included in the American Concrete Institute Building Code, ACI 318, affect detailing of the bottom bars in the ribs. Over supports, at least one bottom bar should be continuous or lap spliced to a bottom bar in the adjacent span with a Class A tension lap splice (Art. 9.49.8). At exterior supports, one bottom bar should be terminated with a standard hook.

For development (bond) of reinforcement, see Art. 9.49.

Figure 9.24*b* shows rebar quantities, lb/ft^2 of floor or roof area, for continuous interior spans of one-way concrete-joist construction made with normal-weight concrete for superimposed factored live load of 170 psf, for preliminary estimates.

9.57 SHEAR IN JOISTS

The factored shear force V_u at a section without shear reinforcement should not exceed

$$V_u = \phi V_c = \phi(2.2\sqrt{f'_c}\,b_w d) \tag{9.64}$$

where V_c = nominal shear strength of the concrete
ϕ = strength reduction factor (Art. 9.44) = 0.85
d = distance, in, from extreme compression surface to centroid of tension steel
b_w = rib width, in

(Based on satisfactory performance of joist construction, the ACI Building Code allows the nominal shear strength V_c for concrete in joists to be taken 10% greater than for beams or slabs.) The width b_w can be taken as the average of the width of joist at the compression face and the width at the tension steel. The slope of the vertical taper of ribs formed with removable steel pans can safely be assumed as 1 in 12. For permanent concrete block fillers, the shell of the block can be included as part of b_w, if the compressive strength of the masonry is equal to or greater than that of the concrete.

If shear controls the design of one-way concrete-joist construction, tapered ends can be used to increase the shear capacity. The Concrete Reinforcing Steel Insti-

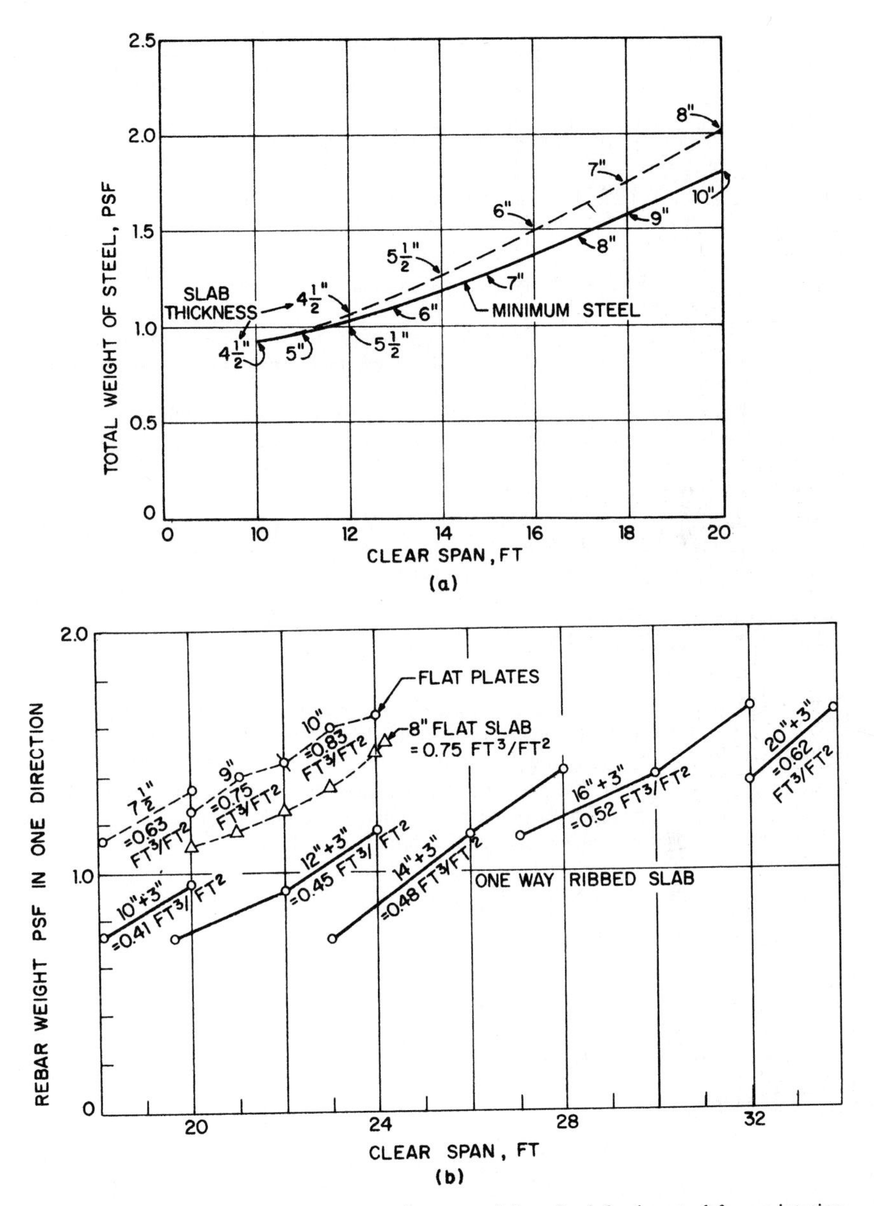

FIGURE 9.24 For use in preliminary estimates, weights of reinforcing steel for an interior span of a continuous slab: (*a*) for a one-way solid slab of 3000-psi concrete carrying 100-psf service live load (170-psf factored live load); (*b*) for flat-plate, flat-slab, and one-way joist construction. See also Fig. 9.31.

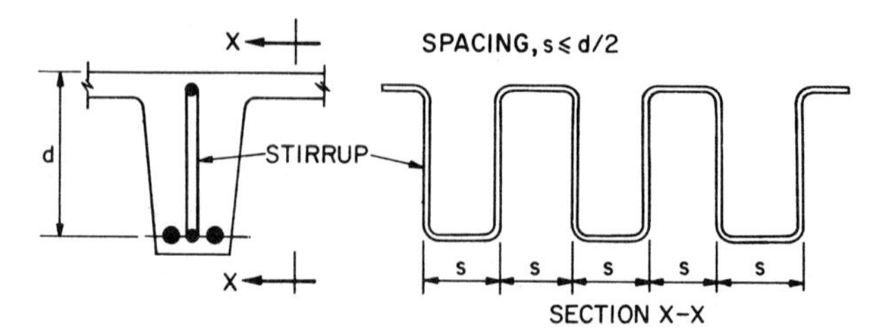

FIGURE 9.25 Stirrups for concrete joist construction.

tute's "CRSI Handbook" has comprehensive load tables for one-way concrete-joist construction that indicate where shear controls and when tapered ends are required for simple, end, and interior spans.

For joists supporting uniform loads, the critical section for shear strength at tapered ends is the narrow end of the tapered section. Shear need not be checked within the taper.

Reinforcement for shear must be provided when the factored shear force V_u exceeds the shear strength of the concrete ϕV_c. The use of single-prong No. 3 stirrups spaced at half depth, such as that shown in Fig. 9.25, is practical in narrow joists; they can be placed between two bottom bars.

TWO-WAY CONSTRUCTION

A two-way slab is a concrete panel reinforced for flexure in more than one direction. (See also Art. 9.62.) Many variations of this type of construction have been used for floors and roofs, including flat plates, solid flat slabs, and waffle flat slabs. Generally, the columns that support such construction are arranged so that their centerlines divide the slab into square or nearly square panels, but if desired, rectangular, triangular, or even irregular panels may be used.

9.58 ANALYSIS AND DESIGN OF FLAT PLATES

The flat plate is the simplest form of two-way slab—simplest for analysis, design, detailing, bar fabrication and placing, and formwork. A flat plate is defined as a two-way slab of uniform thickness supported by any combination of columns and walls, with or without edge beams, and without drop panels, column capitals, and brackets.

Shear and deflection limit economical flat-plate spans to under about 30 ft for light loading and about 20 to 25 ft for heavy loading. While use of reinforcing-steel or structural-steel shear heads for resisting shear at columns will extend these limits somewhat, their main application is to permit use of smaller columns. A number of other variations, however, can be used to extend economical load and span limits (Arts. 9.59 and 9.60).

The American Concrete Institute Building Code, A318, permits two methods of analysis for two-way construction: *direct design,* within limitations of span and load, and *equivalent frame* (Art. 9.42). Limitations on use of direct design are:

A minimum of three spans continuous in each direction

Rectangular panels with ratios of opposite sides not greater than 2

Successive clear span ratios not to exceed 2:3

Columns offset from centerlines of successive columns not more than 0.10 span in either direction

Specified ratio of live load to dead load not more than 3

All loads are due to gravity only and uniformly distributed over the entire panel

9.58.1 Design Procedures for Flat Plates

The procedure for either method of design begins with selection of preliminary dimensions for review, and continues with six basic steps.

Step 1. Select a plate thickness expected to be suitable for the given conditions of load and span. This thickness, unless deflection computations justify thinner plates, should not be less than h determined from Eq. (9.63). With Grade 60 reinforcement, minimum thickness is, from Eq. (9.63), for an interior panel

$$h = \frac{L_n}{32.7} \geq 5 \text{ in} \tag{9.65}$$

where L_n = clear span in direction moments are being determined. Also, as indicated in Art. 9.51, for discontinuous panels, h computed from Eq. (9.63) or (9.65) may have to be increased at least 10%.

Step 2. Determine for each panel the total static factored moment

$$M_o = 0.125 w_u L_2 L_n^2 \tag{9.66}$$

where L_2 = panel width (center-to-center spans transverse to direction in which moment is being determined)
w_u = total factored load, psf = $1.4D + 1.7L$, typically
D = dead load, psf
L = live load, psf

Step 3. Apportion M_o to positive and negative bending. In the direct-design method:

For interior spans, the negative factored bending moment is

$$M_u = -0.65 M_o \tag{9.67}$$

and the positive factored bending moment is

$$M_u = 0.35 M_o \tag{9.68}$$

For end spans (edge panels), the distribution of M_o may be based on α_{ec}, the ratio of the flexural stiffness of the edge equivalent column K_{ec} to the flexural stiffness of the slab K_s.

$$\alpha_{ec} = \frac{K_{ec}}{K_s} \tag{9.69}$$

Equivalent column consists of actual columns above and below the slab plus an attached torsional member transverse to the direction in which moments are being determined and extending to the centerlines on each side of the column of the bounding lateral panel. The torsional member should be taken as a portion of the slab as wide as the column in the direction in which moments are determined. (See also Art. 9.58.2.)

At the edge column, the negative factored moment is

$$M_u = -0.65M_o \frac{\alpha_{ec}}{\alpha_{ec} + 1} \tag{9.70}$$

Also, in an edge panel the factored positive moment is

$$M_u = M_o \left(0.63 - 0.28 \frac{\alpha_{ec}}{\alpha_{ec} + 1}\right) \tag{9.71}$$

and at the first interior column, the negative factored moment is

$$M_u = -M_o \left(0.75 - 0.10 \frac{\alpha_{ec}}{\alpha_{ec} + 1}\right) \tag{9.72}$$

As an alternative, M_o can be distributed in an end span as indicated in Table 9.20.

TABLE 9.20 Alternative Distribution of M_o for the End Span of a Flat Slab*

	Without edge beam	With edge beam
Negative factored moment at edge of column	0.26	0.30
Positive factored moment	0.52	0.50
Negative factored moment at first interior of column	0.70	0.70

*For moment distribution between the slab and an edge column, the nominal moment strength provided for the column strip should be used as the transfer moment for gravity load.

Step 4. Distribute panel moments M_u to column and middle strips.

Column strip is a design strip with a width of $0.25L_2 \leq 0.25L_1$ on each side of the column centerline, where L_1 is the center-to-center span in the direction in which moments are being determined (Fig. 9.26).

Middle strip is the design strip between two column strips (Fig. 9.26).

For flat plates without beams, the distribution of M_u become:

For positive moment, column strip 60%, middle strip 40%

For negative moment at the edge column, column strip 100%

For interior negative moments, column strip 75%, middle strip 25%

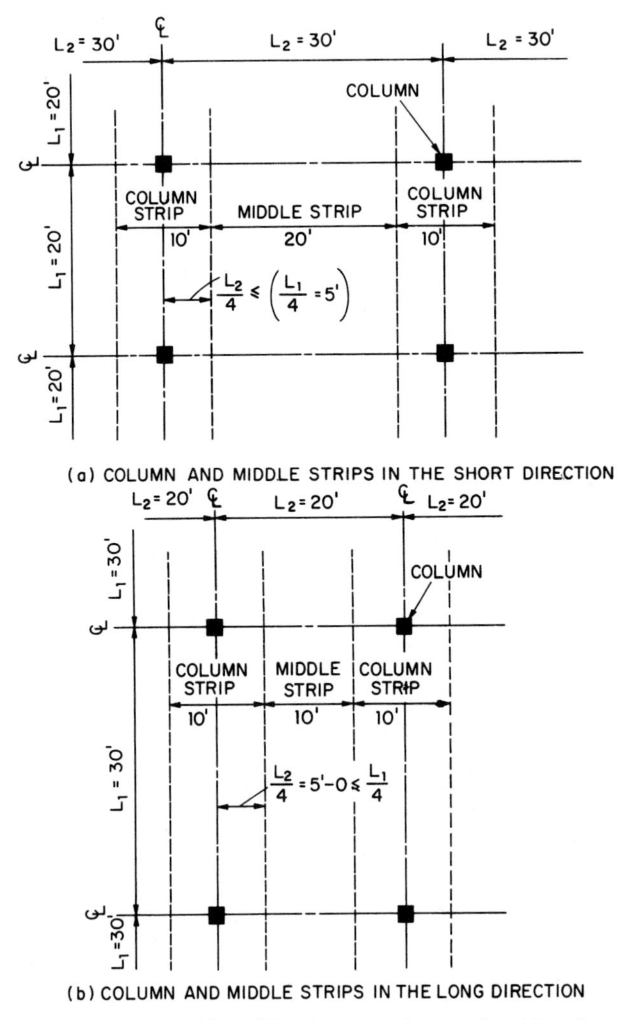

FIGURE 9.26 Division of flat plate into column and middle strips.

A factored moment may be modified up to 10% so long as the sum of the positive and negative moments in the panel in the direction being considered is at least that given by Eq. (9.66).

Step 5. Check for shear. Shear strength of slabs in the vicinity of columns or other concentrated loads has to be checked for two conditions: when the slab acts as a wide beam and when the load tends to punch through the slab. In the first case, a diagonal crack might extend in a plane across the entire width of the slab. Design for this condition is described in Art. 9.47. For the two-way action of the second condition, diagonal cracking might occur along the surface of a truncated cone or pyramid in the slab around the column.

The critical section for two-way action, therefore, should be taken perpendicular to the plane of the slab at a distance $d/2$ from the periphery of the column, where d is the effective depth of slab. Unless adequate shear reinforcement is provided, the factored shear force V_u for punching action must not exceed ϕV_c; i.e., $V_u \leq V_c$, where ϕ = strength reduction factor = 0.85 and V_c is the nominal shear strength of the concrete when it is not adequately reinforced for shear. V_c is the smallest of the values computed from Eqs. (9.73) to (9.75).

$$V_c = \left(2 + \frac{4}{\beta_c}\right) \sqrt{f'_c} b_o d \tag{9.73}$$

$$V_c = \left[\frac{\alpha_s d}{b_o} + 2\right] \sqrt{f'_c} b_o d \tag{9.74}$$

$$V_c = 4\sqrt{f'_c} b_o d \tag{9.75}$$

where β_c = ratio of long side to short side of the column
b_o = perimeter of critical section, in
d = distance from extreme compression surface to centroid of tension reinforcement, in
α_s = 40 for interior columns; 30 for edge columns; and 20 for corner columns
f'_c = specified compressive strength of the concrete, psi

When shear reinforcement is provided (Art. 9.47), $V_u \leq \phi V_n$, where V_n is the nominal shear strength of the reinforced section and equals the sum of V_c and the shear strength added by the reinforcement. V_n should not exceed $6\sqrt{f'_c} b_o d$. With shearhead reinforcement (steel shapes fabricated by welding with a full-penetration weld into identical perpendicular arms) at interior columns, V_n may be as large as $7\sqrt{f'_c} b_o d$.

Determine the maximum shear at each column for two cases: all panels loaded, and live load on alternate panels for maximum unbalanced moment to the columns. Combine shears due to transfer of vertical load to the column with shear resulting from the transfer of part of the unbalanced moment to the column by torsion. At this point, if the combined shear is excessive, steps 1 through 5 must be repeated with a larger column, thicker slab, or higher-strength concrete in the slab; or shear reinforcement must be provided where $V_u > \phi V_c$ (Art. 9.47).

Step 6. When steps 1 through 5 are satisfactory, select reinforcement.

9.58.2 Stiffnesses in Two-Way Construction

The "Commentary" to the ACI Building Code prescribes a sophisticated procedure for computation of stiffness to determine α_{ec} (Art. 9.58.1). Variations in cross sections of slab and columns, drop panels, capitals, and brackets should be taken into account. Columns should be treated as infinitely stiff within the joint with the slab. The slab should be considered to be stiffened somewhat within the depth of the column.

With the equivalent-frame method, the stiffnesses of the equivalent column K_{ec} (Art. 9.58) and the slab K_s are employed in a straightforward elastic analysis. With the direct-design method, use of the value of α_{ec} for determination of moments for edge spans comprises a one-step moment-distribution analysis.

In the direct-design method, certain simplifications are permissible in computation of stiffnesses (see "Commentary" on ACI 318-89).

Column stiffness may be taken as

$$K_c = \frac{4E_c I_c}{L_c} \tag{9.76}$$

where E_c = modulus of elasticity of column concrete
I_c = column moment of inertia based on uniform cross section
L_c = story height

Slab flexural stiffness may be taken as

$$K_s = \frac{4E_{cs} I_{cs}}{L_1} \tag{9.77}$$

where E_{cs} = modulus of elasticity of slab concrete
I_{cs} = moment of inertia of slab based on gross, uniform section
L_1 = center-to-center span in direction in which moments are being determined

Equivalent column stiffness is given by

$$K_{ec} = \frac{\Sigma K_c}{1 + (1/K_t)\Sigma K_c} \tag{9.78}$$

where K_t = torsional stiffness of portion of slab (Art. 9.58).

The summation ΣK_c applies to the column above and the column below the slab.

$$K_t = \sum \frac{9E_{cs}C}{L_2(1 - c_2/L_2)^3} \tag{9.79}$$

where $C = \Sigma(1 - 0.63x/y)x^3y/3$
c_2 = edge-column width transverse to direction in which moments are being determined
L_2 = center-to-center span transverse to L_1
x, y = dimensions of the section of flat plate in contact with side of column parallel to L_1, c_1 in width and h in depth ($x \leq y$)

In Eq. (9.79), the summation applies to the transverse spans of the slab on each side of the edge column.

Part of the unbalanced moment at the exterior-edge column is transferred to the column through torsional shear stresses on the slab periphery at a distance $d/2$ from the column, where d is the effective slab depth. The fraction of the unbalanced moment thus transmitted is

$$\gamma_v = 1 - \frac{1}{1 + \frac{2}{3}\sqrt{\frac{c_1 + d/2}{c_2 + d}}} \tag{9.80}$$

For preliminary design, with square columns flush at edges of the flat plate, a rapid estimate of the shear capacity to allow for effects of torsion can be made by using uniform vertical load w only, with nominal strength for factored load as follows:

For edge column, total shear $V_u = 0.5wL_2L_1$ and shear strength $V_c = 2\sqrt{f'_c}b_od$

For first interior column, $V_u = 1.15wL_2L_1$ and shear strength $V_c = 4\sqrt{f'_c}b_od$

where f'_c = specified concrete strength, psi.

Use of this calculation in establishing a preliminary design is a short cut, which will often avoid the need for repeating steps 1 through 5 in Art. 9.58.1, because it gives a close approximation for final design.

The minimum cantilever edge span of a flat plate so that all columns can be considered interior columns and the direct-design method can be employed without tedious stiffness calculations is 4/15 of the length of the interior span (Fig. 9.27). This result is obtained by equating the minimum cantilever moment at the exterior column to the minimum negative moment at the interior column.

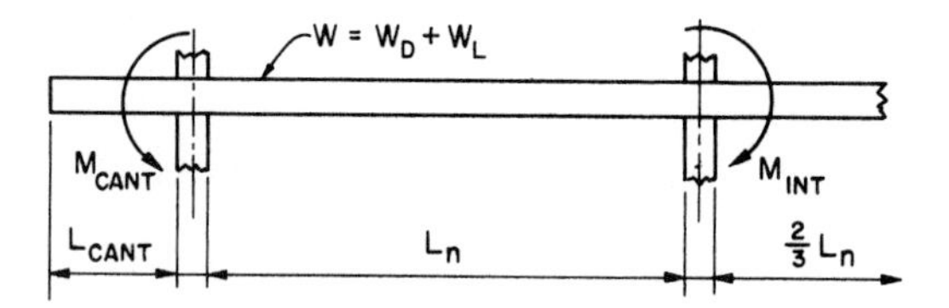

FIGURE 9.27 Length of cantilever (at left) determines whether the exterior column may be treated as an interior column.

9.58.3 Bar Lengths and Details for Flat Plates

The minimum bar lengths for reinforcing flat plates shown in Fig. 9.28, prescribed by the American Concrete Institute Building Code, ACI 318, save development (bond) computations. The size of all top bars must be selected so that the development length l_d required for the bar size, concrete strength, and grade of the bar is not greater than the length available for development (see Table 9.8).

The size of top bars at the exterior edge must be small enough that the hook plus straight extension to the face of the column is larger than that required for full embedment (Table 9.8).

Column-strip bottom bars in Fig. 9.28 are shown extended into interior columns so that they lap, and one line of bar supports may be used. This anchorage, which exceeds ACI minimum requirements, usually ensures ample development length and helps prevent temperature and shrinkage cracks at the centerline.

Figure 9.24*b* shows weights of steel and concrete for flat plates of normal-weight concrete carrying a superimposed factored weight of 170 psf, for preliminary estimates.

Minimum requirements for structural integrity for two-way slabs specified in the ACI Building Code require a minimum of two column-strip bottom bars in each direction to be made continuous through the column core. It is permissible to lap splice the bars at interior columns with a Class A tension lap splice (Art. 9.49.8). The bars must be anchored within exterior columns.

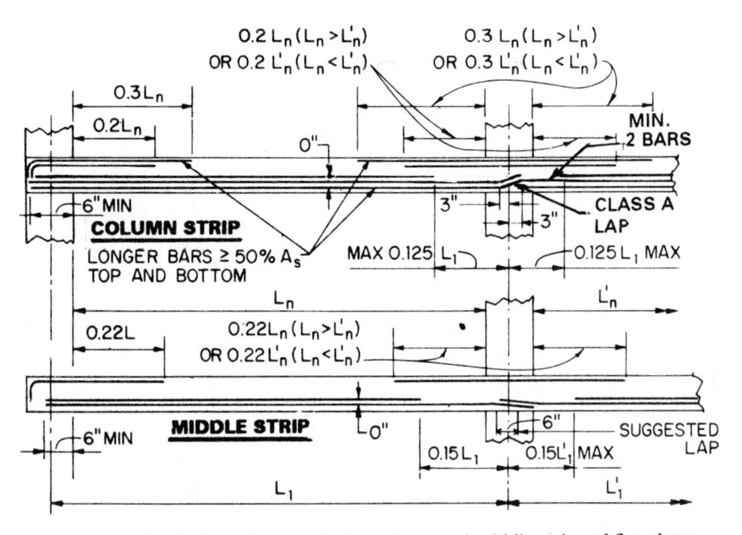

FIGURE 9.28 Reinforcing details for column and middle strips of flat plates.

Crack Control. The ACI Code prescribes formulas for determination of maximum bar spacings based on crack width considered suitable for interior and exterior exposure (Art. 9.50). These requirements apply only to one-way reinforced elements. For two-way slabs, bar spacing at critical sections should not exceed twice the slab thickness, except in the top slab of cellular or ribbed (waffle) construction, where requirements for temperature and shrinkage reinforcement govern.

9.59 FLAT SLABS

A flat slab is a two-way slab generally of uniform thickness, but it may be thickened or otherwise strengthened in the region of columns by a drop panel, while the column top below the slab may be enlarged by a capital (round) or bracket (prismatic). If a drop panel is used to increase depth for negative reinforcement, the minimum side dimensions of this panel are $L_1/3$ and $L_2/3$, where L_1 and L_2 are the center-to-center spans in perpendicular directions. Minimum depth of a drop panel is $1.25h$, where h is the slab thickness elsewhere.

A **waffle flat slab** or **waffle flat plate** consists of a thin, two-way top slab and a grid of joists in perpendicular directions, cast on square dome forms. For strengthening around columns, the domes are omitted in the drop panel areas, to form a solid head, which also may be made deeper than the joists. Other variations of waffle patterns include various arrangements with solid beams on column centerlines both ways. Standard sizes of two-way joist forms are given in Table 9.21.

The main complication introduced in the design of flat slabs and waffle flat plates is the change in cross section of the section considered effective for torsional stiffness K_t (Art. 9.58.2). The American Concrete Institute Building Code, ACI 318, permits this variation in depth to be neglected. This approximation results in an overestimation of flat-slab torsional stiffness, making it equal to that of a flat plate with thickness equal to that of the flat-slab drop panel or waffle-flat plate solid head.

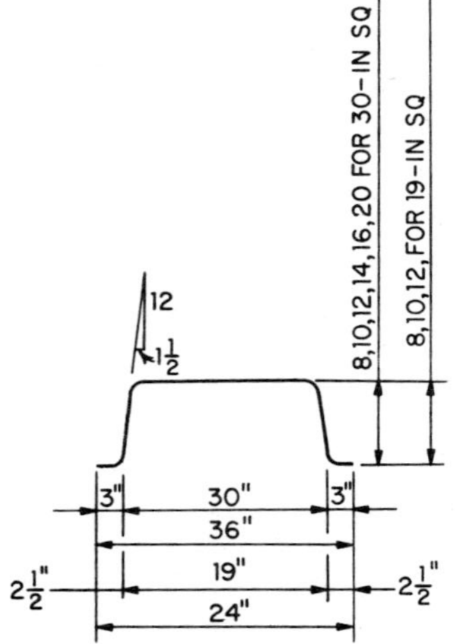

TABLE 9.21 Commonly Used Sizes of Two-Way Joist Forms

Depth, in	Volume, ft^3 per dome	Weight of displaced concrete, lb per dome	3-in top slab		4½-in top slab	
			Equiv. slab thickness, in	Weight,* psf	Equiv. slab thickness, in	Weight,* psf
30-in-wide domes						
8	3.85	578	5.8	73	7.3	92
10	4.78	717	6.7	83	8.2	102
12	5.53	830	7.4	95	9.1	114
14	6.54	980	8.3	106	9.9	120
16	7.44	1116	9.1	114	10.6	133
20	9.16	1375	10.8	135	12.3	154
19-in-wide domes						
			3-in top slab		4½-in top slab	
8	1.41	211	6.8	85	8.3	103
10	1.90	285	7.3	91	8.8	111
12	2.14	321	8.6	107	10.1	126

*Basis: $w = 150$.

Except for the increase in negative moments at exterior columns, however, this assumption changes moments only slightly. Thus, the procedures outlined in Art. 9.58.1 can also be used for flat slabs.

The drop panel increases shear capacity. Hence, a solid flat slab can ordinarily be designed for concrete of lower strength than for a flat plate. Also, deflection of a flat slab is reduced by the added stiffness that drop panels provide.

The depth of drop panels can be increased beyond 1.25h to reduce negative-moment reinforcement and to increase shear capacity when smaller columns are desired. If this adjustment is made, shear in the slab at the edge of the drop panel may become critical. In that case, shear capacity can be increased by making the drop panel larger, up to about 40% of the span. See Fig. 9.29 for bar details (column strip).

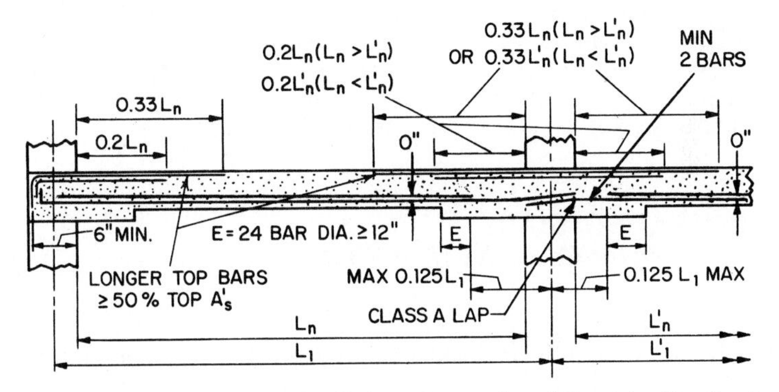

FIGURE 9.29 Reinforcing details for column strips of flat slabs. Details of middle strips are the same as for middle strips of flat plates (Fig. 9.28).

Waffle flat plates behave like solid flat slabs with drop panels. Somewhat higher-strength concrete, to avoid the need of stirrups in the joists immediately around the solid head, is usually desirable. If required, however, such stirrups can be made in one piece as a longitudinal assembly, to extend the width of one dome between the drop head and the first transverse joist. For exceptional cases, such stirrups can be used between the second row of domes also. See Fig. 9.30 for reinforcement details.

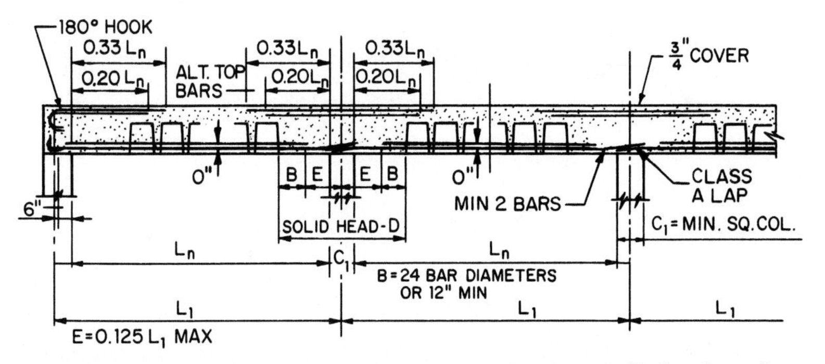

FIGURE 9.30 Reinforcing details for column strips of two-way waffle flat plates. B = 24 bar diameters or 12 in minimum. Details for middle strips are the same as for middle strips of flat plates (Fig. 9.28).

9.60 TWO-WAY SLABS ON BEAMS

The American Concrete Institute Building Code, ACI 318, provides for use of beams on the sides of panels, on column centerlines. (A system of slabs and beams supported by girders, however, usually forms rectangular panels. In that case, the slabs are designed as one-way slabs.)

Use of beams on all sides of a panel permits use of thinner two-way slabs, down to a minimum thickness $h = 3\frac{1}{2}$ in. A beam may be assumed to carry as much as 85% of the column-strip moment, depending on its stiffness relative to the slab (see the ACI Building Code). A secondary benefit, in addition to the direct advantages of longer spans, thinner slabs, and beam stirrups for shear, is that many local codes allow reduced live loads for design of the beams. These reductions are based on the area supported and the ratio of dead to live load. For live loads up to 100 psf, such reductions are usually permitted to a maximum of 60%. Where such reductions are allowed, the reduced total panel moment M_o (Art. 9.58.1) and the increased effective depth to steel in the beams offer savings in reinforcement to offset partly the added cost of formwork for the beams.

9.61 ESTIMATING GUIDE FOR TWO-WAY CONSTRUCTION

Figure 9.31 can be used to estimate quantities of steel, concrete, and formwork for flat slabs, as affected by load and span. It also affords a guide to preliminary

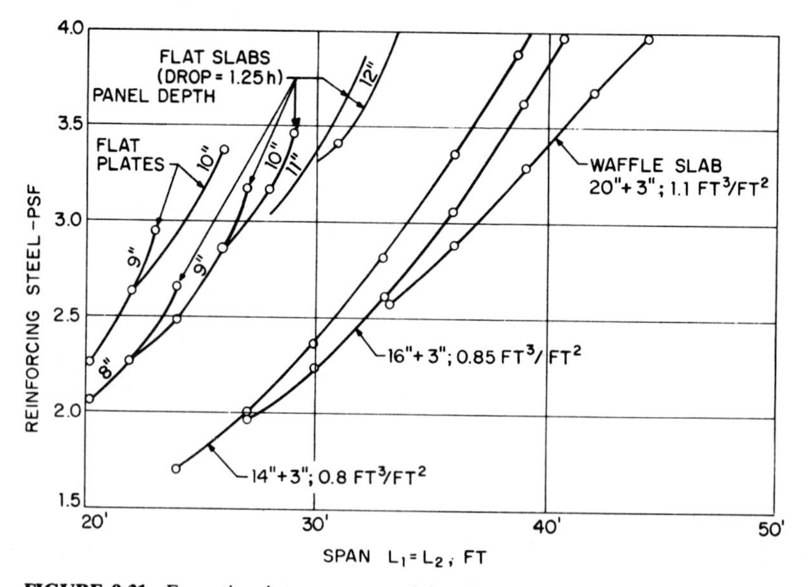

FIGURE 9.31 For estimating purposes, weight of reinforcing steel in square interior panels of flat plates, flat slabs, and waffle flat slabs of 4000-psi concrete, with rebars of 60-ksi yield strength, carrying a superimposed factored load of 200 lb/ft². See also Fig. 9.24, p. 9.79.

selection of dimensions for analysis, and can be used as an aid in selecting the structural system most appropriate for particular job requirements.

BEAMS

Most requirements of the American Concrete Institute Building Code, ACI 318, for design of beams and girders refer to flexural members. When slabs and joists are not intended, the Code refers specifically to beams and occasionally to beams and girders, and provisions apply equally to beams and girders. So the single term, beams, will be used in the following.

9.62 DEFINITIONS OF FLEXURAL MEMBERS

The following definitions apply for purposes of this section:

Slab. a flexural member of uniform depth supporting area loads over its surface. A slab may be reinforced for flexure in one or two directions.

Joist-slab. a ribbed slab with ribs in one or two directions. Dimensions of such a slab must be within the ACI Building Code limitations (see Art. 9.54).

Beam. a flexural member designed to carry uniform or concentrated line loads. A beam may act as a primary member in beam-column frames, or may be used to support slabs or joist-slabs.

Girder. a flexural member used to support beams and designed to span between columns, walls, or other girders. A girder is always a primary member.

9.63 FLEXURAL REINFORCEMENT

Nonprestressed beams should be designed for flexure as explained in Arts. 9.44 to 9.46. If beam capacity is inadequate with tension reinforcement only and the capacity must be increased without increasing beam size, additional capacity may be provided by addition of compression bars and more tension-bar area to match the compression forces developable in the compression bars (Fig. 9.14). (Shear, torsion, crack control, and deflection requirements must also be met to complete the design. See Arts. 9.49 to 9.64 and 9.66.) Deflection need not be calculated for ACI Code purposes if the total depth h of the beam, including top and bottom cover, is at least the fraction of the span L given in Table 9.15.

A number of interdependent complex requirements (Art. 9.49) regulate the permissible cutoff points or bend points of bars within a span, based on various formulas and rules for development (bond). An additional set of requirements applies if the bars are cut off or bent in a tensile area. These requirements can be satisfied for cases of uniform load and nearly equal spans for the top bars by extending at least 50% of the top steel to a point in the span $0.30L_n$ beyond the face of the support, and the remainder to a point $0.20L_n$, where L_n = clear span. For the bottom bars, all requirements are satisfied by extending at least 40% of the total steel into the supports 6 in past the face, and cutting off the remainder

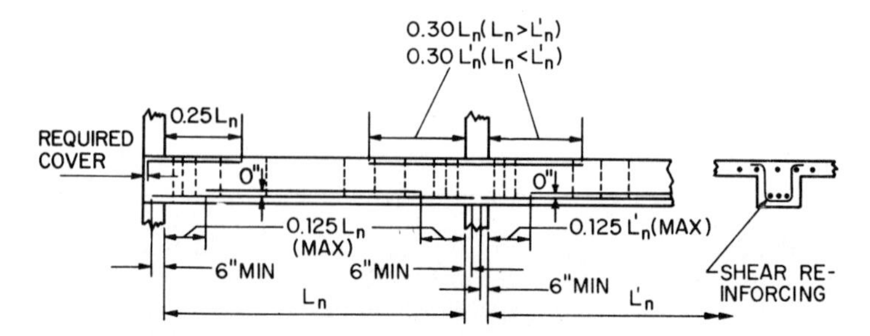

FIGURE 9.32 Reinforcing details for uniformly loaded continuous beams. At columns, embed alternate bottom bars (at least 50% of the tension-steel area) a minimum of 6 in, to avoid calculation of development length at $0.125L_n$.

at a distance $0.125L_n$ from the supports. Note that this arrangement does not cut off bottom bars in a tensile zone. Figure 9.32 shows a typical reinforcement layout for a continuous beam, singly reinforced.

The structural detailing of reinforcement in beams is also affected by ACI Building Code requirements for structural integrity. Beams are categorized as either *perimeter* beams or *nonperimeter* beams. (A spandrel beam would be a perimeter beam.) In perimeter beams, at least one-sixth of the tension-reinforcement area required for negative moment ($-A_s/6$) at the face of supports, and one-quarter of the tension-reinforcement area required for positive moment ($+A_s/4$) at midspan have to be made continuous around the perimeter of the structure. Closed stirrups are also required in perimeter beams. It is not necessary to place closed stirrups within the joints. It is permissible to provide continuity of the top and bottom bars by splicing the top bars at midspan and the bottom bars at or near the supports. Splicing the bars with Class A tension lap splices (Art. 9.49.8) is acceptable. (See Fig. 9.33*a*.)

For nonperimeter beams, the designer has two choices to satisfy the structural integrity requirements: (1) provide closed stirrups or (2) make at least one-quarter of the tension-reinforcement area required for positive moment ($+A_s/4$) at midspan continuous. Splicing the prescribed number of bottom bars over the supports with Class A tension lap splices is acceptable. At discontinuous ends, the bottom bars must be anchored with standard hooks. (see Fig. 9.33*b*.)

The limit in the ACI Building Code on tension-reinforcement ratio ρ that it not exceed 0.75 times the ratio for balanced conditions applies to beams (Art. 9.46). Balanced conditions in a beam reinforced only for tension exist when the tension steel reaches its yield strength f_y simultaneously with the maximum compressive strain in the concrete at the same section becoming 0.003 in/in. Balanced conditions occur similarly for rectangular beams, and for T beams with negative moment, that are provided with compression steel, or doubly reinforced. Such sections are under balanced conditions when the tension steel, with area A_s, yields just as the outer concrete surface crushes, and the total tensile-force capacity A_sf_y equals the total compressive-force capacity of the concrete plus compression steel, with area A'_s. Note that the capacity of the compression steel cannot always be taken as A'_sf_y, because the straight-line strain distribution from the fixed points of the outer concrete surface and centroid of the tension steel may limit the compression-steel stress to less than yield strength (Fig. 9.14).

For design of doubly reinforced beams, the force A_sf_y in the tension steel is limited to three-fourths the compression force in the concrete plus the compression

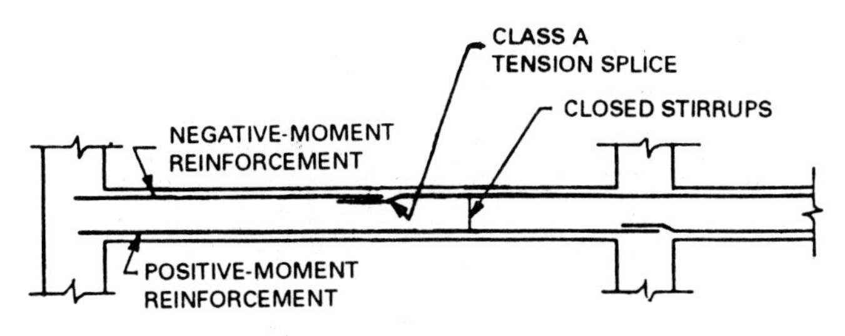

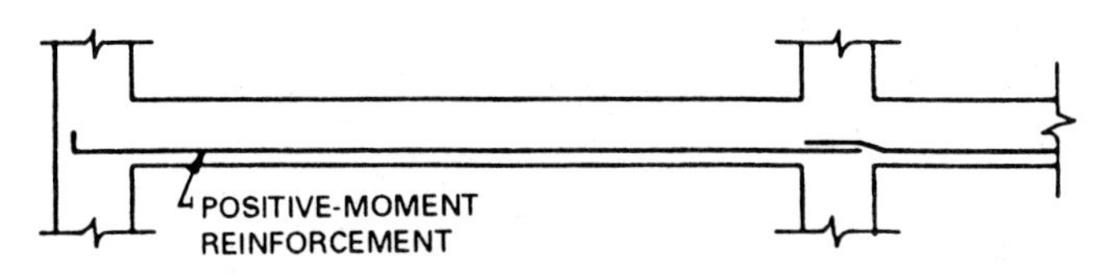

FIGURE 9.33 Steel reinforcement required to ensure structural integrity of beams. At least one-sixth of the negative-moment rebars and one-fourth of the positive-moment rebars should be continuous around the perimeter of the structure (*a*), with closed stirrups throughout, except at joints. Class A tension splices may be made at midspan. For nonperimeter beams (*b*), one-fourth the positive-moment rebars should be continuous. For clarity, other rebars are not shown in (*a*) or (*b*).

in the compression steel at balanced conditions. For a beam meeting these conditions in which the compression steel has not yielded, the design moment capacity is best determined by trial and error:

1. Assume the location of the neutral axis.
2. Determine the strain in the compression steel.
3. See if the total compressive force on the concrete and compression steel equals $A_s f_y$ (Fig. 9.34).

Example. Design a T beam to carry a factored negative bending moment of 225 ft-kips. The dimensions of the beam are shown in Fig. 9.34. Concrete strength $f'_c = 4$ ksi, the reinforcing steel has a yield strength $f_y = 60$ ksi, and strength reduction factor $\phi = 0.90$.

Need for Compression Steel. To determine whether compression reinforcement is required, first check the strength of the section when it is reinforced only with tension steel. For this purpose, compute the steel ratio ρ_b for balanced conditions from Eq. (9.26) with $\beta_1 = 0.85$:

$$\rho_b = \frac{0.85 \times 4000 \times 0.85}{60{,}000} \times \frac{87{,}000}{87{,}000 + 60{,}000} = 0.0285$$

The maximum steel ratio permitted by the ACI Code is

$$\rho_{\max} = 0.75\rho_b = 0.75 \times 0.0285 = 0.0214$$

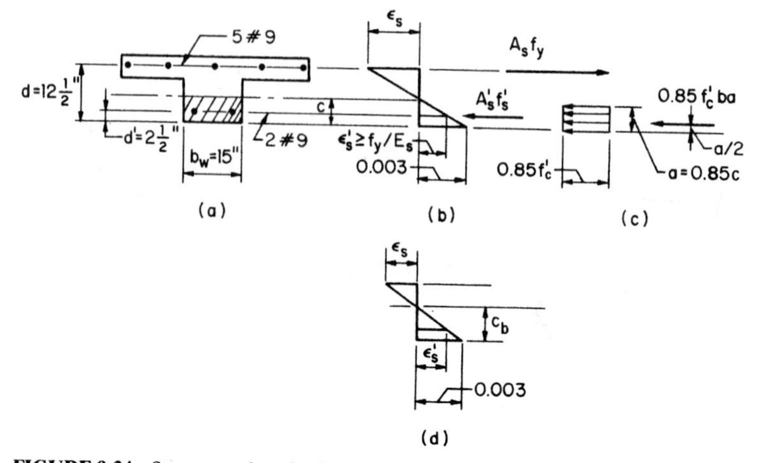

FIGURE 9.34 Stresses and strains in a T beam reinforced for compression: (*a*) beam cross section; (*b*) strain distribution; (*c*) block distribution of compression stresses; (*d*) balanced strains.

and the corresponding steel area is

$$A_s = 0.0214 \times 12.5 \times 15 = 4.01 \text{ in}^2$$

As noted in Art. 9.46.1, depth of the stress block is

$$a = \frac{A_s f_y}{0.85 f'_c b} = \frac{4.01 \times 60{,}000}{0.85 \times 4000 \times 12.5} = 5.66 \text{ in}$$

From Eq. (9.27*b*), the maximum moment capacity with tension reinforcement only is

$$M_{\max} = 0.90 \times 4.01 \times 60{,}000(12.5 - 5.66/2)/12 = 174{,}500 \text{ ft-lb}$$

The required strength, 225,000 ft-lb, is larger. Hence, compression reinforcement is needed.

Compression on Concrete. (Trial-and-error solution.) Assume that the distance c from the neutral axis to the extreme compression surface is 5.1 in. The depth a then may be taken as $0.85c = 4.33$ in (Art. 9.44). For a rectangular stress distribution over the concrete, the compression force on the concrete is

$$0.85 f'_c b_w a = 0.85 \times 4 \times 15 \times 4.33 = 221 \text{ kips}$$

Selection of Tension Steel. To estimate the tension steel required, assume a moment arm $jd = d - a/2 = 12.5 - 4.33/2 = 10.33$ in. By Eq. (9.27*c*), the tension-steel force therefore should be about

$$A_s f_y = 60A_s = \frac{M_u}{\phi jd} = \frac{225 \times 12}{0.9 \times 10.33} = 290 \text{ kips}$$

from which $A_s = 4.84\text{ in}^2$. Select five No. 9 bars, supplying $A_s = 5\text{ in}^2$ and providing a tensile-steel ratio

$$\rho = \frac{5}{15 \times 12.5} = 0.0267$$

The bars can exert a tension force $A_s f_y = 5 \times 60 = 300$ kips.

Stress in Compression Steel. For a linear strain distribution, the strain ϵ_s' in the steel 2½ in from the extreme compression surface can be found by proportion from the maximum strain of 0.003 in/in at that surface. Since the distance $c = a/\beta_1 = 4.33/0.85 = 5.1$,

$$\epsilon_s' = \frac{5.1 - 2.5}{5.1} \times 0.003 = 0.0015 \text{ in/in}$$

With modulus of elasticity E_s taken as 29,000 ksi, the stress in the compression steel is

$$f_s' = 0.0015 \times 29{,}000 = 43.5 \text{ ksi}$$

Selection of Compression Steel. The total compression force equals the 221-kip force on the concrete previously computed plus the force on the compression steel. If the total compression force is to equal to total tension force, the compression steel must resist a force

$$A_s' f_s' = A_s'(43.5 - 3.4) = 300 - 221 = 79 \text{ kips}$$

from which the compression-steel area $A_s' = 2\text{ in}^2$. (In the above calculation, the force on the steel is reduced by the force on the concrete, $\phi f_c' A_s' = 0.85 \times 4A_s' = 3.4A_s'$, replaced by the steel.)

Check the Balance of Forces ($\Sigma F_c = \Sigma F_t$). For an assumed position of the neutral axis at 5.10 in, with five No. 9 tension bars and two No. 9 compression bars, the total compression force C is

$$\begin{aligned} &\text{Concrete: } 0.85 \times 4 \times 4.33 \times 15 = 221 \text{ kips} \\ &\text{Steel: } 2(43.5 - 3.4) = \underline{80} \text{ kips} \\ &C = 301 \text{ kips} \end{aligned}$$

This compression force for practical purposes is equal to the total tension force: $5 \times 60 \times 1 = 300$ kips. The assumed position of the neutral axis results in a balance of forces within 1% accuracy.

Nominal Flexural Strength. For determination of the nominal flexural strength of the beam, moments about the centroid of the tension steel are added:

$$M_n = 0.85 f_c' ba \left(d - \frac{a}{2}\right) + A_s' f_s'(d - d') \tag{9.81}$$

Substitution of numerical values yields:

$$\begin{aligned} M_n &= 0.85 \times 4 \times 15 \times 4.33(12.5 - 4.33/2) + 2(43.5 - 3.4)(12.5 - 2.5) \\ &= 221 \times 10.33 + 80 \times 10 = 257 \text{ ft-kips} \end{aligned}$$

Check Design Moment Strength (ϕM_n)

$$\phi M_n = 0.90 \times 257 = 231 \text{ ft-kips} > M_u = 225 \text{ ft-kips}) \qquad \text{OK}$$

9.64 REINFORCEMENT FOR SHEAR AND FLEXURE

Determination of the shear capacity of a beam is discussed in Art. 9.47. Minimum shear reinforcement is required in all beams with total depth greater than 10 in, or 2½ times flange (slab) thickness, or half the web thickness, except where the factored shear force V_u is less than half the design shear strength ϕV_c of the concrete alone. Torsion should be combined with shear when the factored loads cause a torsional moment T_u larger than $\phi(0.5\sqrt{f'_c})\ \Sigma x^2 y$ (see Art. 9.48).

Shear strength should be computed at critical sections in a beam from Eq. (9.37). Open or closed stirrups may be used as reinforcement for shear in beams; but closed stirrups are required for torsion. The minimum area for open or closed stirrups for vertical shear only, to be used where $0.5\phi V_c \leq V_u \leq \phi V_c$ and the factored torsional moment $T_u \leq \phi(0.5\ \sqrt{f'_c})\ \Sigma x^2 y$, should be calculated from

$$A_v = \frac{50 b_w s}{f_y} \tag{9.82}$$

where A_v = area of all vertical legs in the spacing s, in parallel to flexural reinforcement, in²

b_w = thickness of beam web, in²

f_y = yield strength of reinforcing steel, psi

Note that this minimum area provides a capacity for 50-psi shear on the cross section $b_w s$.

Where V_u exceeds V_c, the cross-sectional area A_v of the legs of open or closed vertical stirrups at each spacing s should be calculated from Eq. (9.40*a*). A_v is the total area of vertical legs, two legs for a common open U stirrup or the total of all legs for a transverse multiple U. Note that there are three zones in which the required A_v may be supplied by various combinations of size and spacing of stirrups (Fig. 9.35):

1. Beginning 1 or 2 in from the face of supports and extending over a distance d from each support, where d is the depth from extreme compression surface to centroid of tension steel (A_v is based on V_u at d from support).

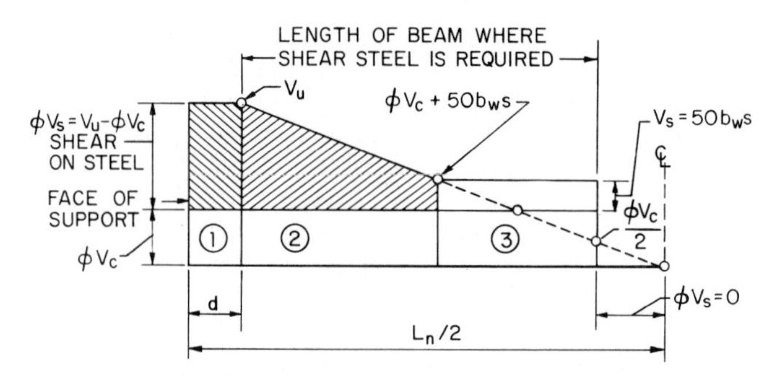

FIGURE 9.35 Required shear reinforcement in three zones of a beam between supports and midspan is determined by cross-hatched areas.

2. Between distance d from each support and the point where $\phi V_s = V_u - \phi V_c = 50b_w s$ (required A_v decreases from maximum to minimum).

3. Distance over which minimum reinforcement is required (minimum A_v extends from the point where $\phi V_s = 50b_w s$ to the point where $V_u = 0.5\phi V_c$).

9.65 REINFORCEMENT FOR TORSION AND SHEAR

Any beam that supports unbalanced loads that are transverse to the direction in which it is subjected to bending moments transmits an unbalanced moment to the supports and must be investigated for torsion. Generally, this requirement affects all spandrel and other edge beams, and interior beams supporting uneven spans or unbalanced live loads on opposite sides. The total unbalanced moment from a floor system with one-way slabs in one direction and beams in the perpendicular direction can often be considered to be transferred to the columns by beam flexure in one direction, neglecting torsion in the slab. The total unbalanced moment in the other direction, from the one-way slabs, can be considered to be transferred by torsional shear from the beams to the columns. (Determination of the torsional capacity of a beam is discussed in Art. 9.48.)

The factored torque T_u for rectangular or T beams should be less than ϕT_n, where T_n is the nominal torsional strength [Eq. (9.42)]. If $T_u \le \phi(0.5\sqrt{f'_c})\,\Sigma x^2 y$, T_u may be neglected. The portion of the torque carried by the concrete should not exceed T_c computed from Eq. (9.43).

The required area A_t of each leg of a closed stirrup for torsion should be computed from Eq. (9.45). Stirrup spacing should not exceed 12 in or $(x_1 + y_1)/4$, where x_1 and y_1 are the dimensions center to center of a closed rectangular stirrup.

Torsion reinforcement also includes the longitudinal bars shown in each corner of the closed stirrups in Fig. 9.36 and the longitudinal bars spaced elsewhere inside

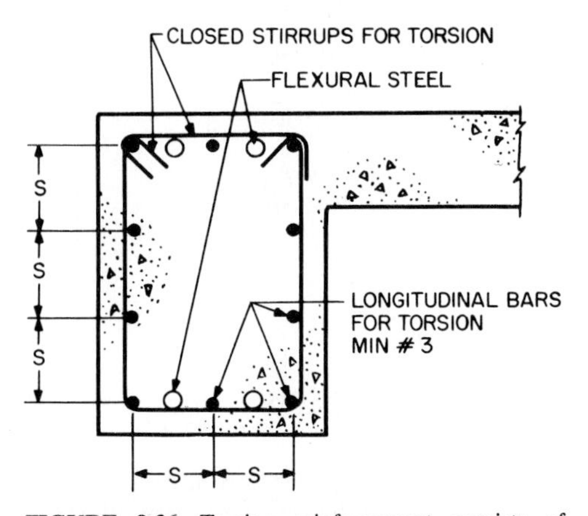

FIGURE 9.36 Torsion reinforcement consists of closed stirrups and longitudinal bars within the perimeter of the stirrups.

the perimeter of the closed stirrups at not more than 12 in. At least one longitudinal bar in each corner is required. [For required areas of these bars, see Eqs. (9.46) and (9.47).] If a beam is fully loaded for maximum flexure and torsion simultaneously, as in a spandrel beam, the area of torsion-resisting longitudinal bars A_t should be provided in addition to flexural bars.

For interior beams, maximum torsion usually occurs with live load only on a slab on one side of the beam. Maximum torsion and maximum flexure cannot occur simultaneously. Hence, the same bars can serve for both.

The closed stirrups required for torsion should be provided in addition to the stirrups required for shear, which may be the open type. Because the size of stirrups must be at least No. 3 and maximum spacings are established in the American Concrete Institute Building Code, ACI 318, for both shear and torsion stirrups, a closed-stirrup size-spacing combination can usually be selected for combined shear and torsion. Where maximum shear and torsion cannot occur under the same loading, the closed stirrups can be proportioned for the maximum combination of forces or the maximum single force; whichever is larger (Fig. 9.37).

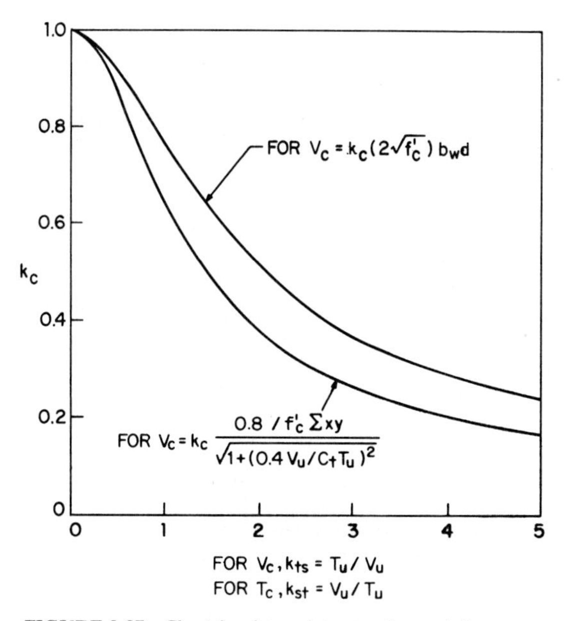

FIGURE 9.37 Chart for determining torsion and shear carried by concrete.

9.66 CRACK CONTROL IN BEAMS

The ACI Building Code contains requirements limiting reinforcement spacing to regulate crack widths when the yield strength f_y of the reinforcement exceeds 40,000 psi (Art. 9.50). Limits for interior and exterior exposures are prescribed. Since the computed crack width is made a function of the tensile area of concrete tributary

TABLE 9.22 Minimum Clear Concrete Cover, in, for Beams and Girders

	Exterior exposure			Interior exposure		
Bar size No.	Cast-in-place concrete	Precast concrete	Prestressed concrete	Cast-in-place concrete	Precast concrete	Prestressed concrete
3, 4, and 5	1½	1¼	1½	1½	⅝	1½
6 through 11	2	1½	1½	1½	d_b*	1½
14 and 18	2	2	1½	1½	1½	1½
Stirrups	Same as above for each size				⅜	1

*d_b = nominal bar diameter, in.

to and concentric with each bar, the required cover is important to crack control. (See Tables 9.13 and 9.22.)

WALLS

Generally, any vertical member whose length and height are both much larger than the thickness may be treated as a wall. Walls subjected to vertical loads are called **bearing walls.** Walls subjected to no loads other than their own weight, such as panel or enclosure walls, are called **nonbearing walls.** Walls with a primary function of resisting lateral loads are called **shear walls.** They also may serve as bearing walls. See Art. 9.88.

9.67 BEARING WALLS

Reinforced concrete bearing walls may be designed as eccentrically loaded columns or by an empirical method given in the American Concrete Institute Building Code, ACI 318. The empirical method may be used when the resultant of the applied load falls within the middle third of the wall thickness. This method gives the capacity of the wall as

$$P_u \leq \phi P_{nw} = 0.55\phi f'_c A_g \left[1 - \left(\frac{kL_c}{32h}\right)^2\right] \tag{9.83}$$

where f'_c = specified concrete strength
ϕ = strength reduction factor = 0.70
A_g = gross area of horizontal cross section of wall
h = wall thickness
L_c = vertical distance between supports
k = effective length factor

When the wall is braced against lateral translation at top and bottom:

$k = 0.8$ for restraint against rotation at one or both ends

$k = 1.0$ for both ends unrestrained against rotation

When the wall is not braced against lateral translation, $k = 2.0$ (cantilever walls).

The allowable average design stress f_c for a wall is obtained by dividing P_u in Eq. (9.83) by A_g.

Length. The effective length of wall for concentrated loads may be taken as the center-to-center distance between loads, but not more than the width of bearing plus 4 times the wall thickness.

Thickness. The minimum thickness of bearing walls for which Eq. (9.83) is applicable is one-twenty-fifth of the least distance between supports at the sides or top, but not less than 4 in. Exterior basement walls and foundation walls should be at least 7½ in thick. Minimum thickness and reinforcement requirements may be waived, however, if justified by structural analysis.

Reinforcement. The area of horizontal steel reinforcement should be at least

$$A_h = 0.0025A_{wv} \tag{9.84}$$

where A_{wv} = gross area of the vertical cross section of wall.

Area of vertical reinforcement should be at least

$$A_v = 0.0015A_{wh} \tag{9.85}$$

where A_{wh} = gross area of the horizontal cross section of wall. For Grade 60 bars, No. 5 or smaller, or for welded-wire fabric, these steel areas may be reduced to $0.0020A_{wv}$ and $0.0012A_{wh}$, respectively.

Walls 10 in or less thick may be reinforced with only one rectangular grid of rebars. Thicker walls require two grids. The grid nearest the exterior wall surface should contain between one-half and two-thirds the total steel area required for the wall. It should have a concrete cover of at least 2 in, but not more than one-third the wall thickness. A grid near the interior wall surface should have a concrete cover of at least ¾ in but not more than one-third the wall thickness. Minimum size of bars, if used, is No. 3. Maximum bar spacing is 18 in. (These requirements do not apply to basement walls, however. If such walls are cast against and permanently exposed to earth, minimum cover is 3 in. Otherwise, the cover should be at least 2 in for bar sizes No. 6 and larger, and 1½ in for No. 5 bars or ⅝-in wire and smaller.)

At least two No. 5 bars should be placed around all window and door openings. The bars should extend at least 24 in beyond the corners of openings.

Design for Eccentric Loads. Bearing walls with bending moments sufficient to cause tensile stress must be designed as columns for combined flexure and axial load. Minimum reinforcement areas and maximum bar spacings are the same as for walls designed by the empirical method. Lateral ties, as for columns, are required for compression reinforcement and where the vertical bar area exceeds 0.01 times the gross horizontal concrete area of the wall. (For column capacity, see Art. 9.81.)

Under the preceding provisions, a thin, wall-like (rectangular) column with a steel ratio less than 0.01 will have a greater carrying capacity if the bars are detailed as for walls. The reasons for this are: The effective depth is increased by omission of ties outside the vertical bars and by the smaller cover (as small as ¾ in) permitted for vertical bars in walls. Furthermore, if the moment is low (eccentricity less than one-sixth the wall thickness), so that the wall capacity is determined by Eq. (9.83), the capacity will be larger than that computed for a column, except where the column is part of a frame braced against sidesway.

9.68 NONBEARING WALLS

Nonbearing reinforced-concrete walls, frequently classified as panels, partitions, or cross walls, may be precast or cast in place. Panels serving merely as exterior cladding, when precast, are usually attached to the columns or floors of a frame, supported on grade beams, or supported by and spanning between footings, serving as both grade beams and walls. Cast-in-place cross walls are most common in substructures. Less often, cast-in-place panels may be supported on grade beams and attached to the frame.

In most of these applications for nonbearing walls, stresses are low and alternative materials, such as unreinforced masonry, when supported by beams above grade, or panels of other materials, can be used. Consequently, unless esthetic requirements dictate reinforced concrete, low-stressed panels of reinforced concrete must be designed for maximum economy. Minimum thickness, minimum reinforcement, full benefits of standardization for mass-production techniques, and design for double function as both wall and deep beam must be achieved.

Thickness of nonbearing walls of reinforced concrete should be at least one-thirtieth the distance between supports, but not less than 4 in.

The American Concrete Institute Building Code, ACI 318, however, permits waiving all minimum arbitrary requirements for thickness and reinforcement where structural analysis indicates adequate strength and stability.

Where support is provided, as for a panel above grade on a grade beam, connections to columns may be detailed to permit shrinkage. Friction between base of panel and the beam can be reduced by an asphalt coating and omission of dowels. These provisions will permit elimination or reduction of horizontal shrinkage reinforcement. Vertical reinforcement is seldom required, except as needed for spacing the horizontal bars.

If a nonbearing wall is cast in place, reinforcement can be nearly eliminated except at edges. If the wall is precast, handling stresses will often control. Multiple pickup points with rigid-beam pickups will reduce such stresses. Vacuum pad pickups can eliminate nearly all lifting stresses.

Where deep-beam behavior or wind loads cause stresses exceeding those permitted on plain concrete, the ACI Code permits reduction of minimum tension-reinforcement [$A_s = 200bd/f_y$ (Art. 9.46)] if reinforcement furnished is one-third greater than that required by analysis. (For deep-beam design, see Art. 9.87.)

9.69 CANTILEVER RETAINING WALLS

Under the American Concrete Institute Building Code, ACI 318, cantilever retaining walls are designed as slabs. Specific Code requirements are not given for cantilever walls, but when axial load becomes near zero, the Code requirements for flexure apply.

Minimum clear cover for bars in walls cast against and permanently exposed to earth is 3 in. Otherwise, minimum cover is 2 in for bar sizes No. 6 and larger, and 1½ in for No. 5 bars or ⅝-in wire and smaller.

Two points requiring special consideration are analysis for load factors of 1.7 times lateral earth pressure and 1.4 times dead loads and fluid pressures, and provision of splices at the base of the stem, which is a point of maximum moment. The footing and stem are usually cast separately, and dowels left projecting from the footing are spliced to the stem reinforcement.

A straightforward way of applying Code requirements for strength design is illustrated in Fig. 9.38. Soil reaction pressure p and stability against overturning

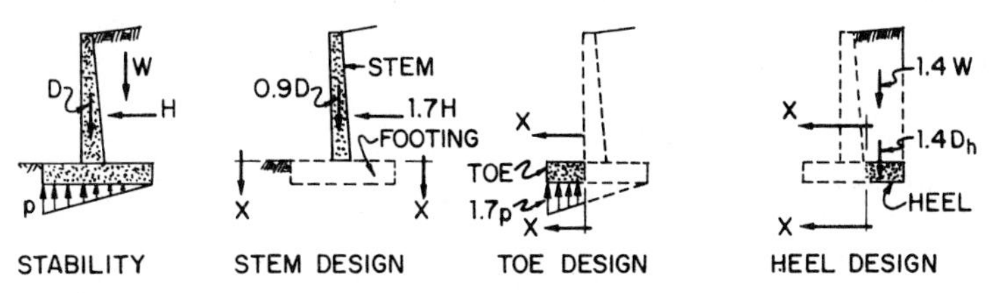

FIGURE 9.38 Factored loads and critical sections for design of cantilever retaining walls.

are determined for actual weights of concrete D and soil W and assumed lateral pressure of the soil H. The total cantilever bending moment for design of stem reinforcement is then based upon $1.7H$. The toe pressure used to determine the footing bottom bars is $1.7p$. And the top load for design of the top bars in the footing heel is $1.4(W + D_h)$, where D_h is the weight of the heel. The Code requires application of a factor of 0.9 to vertical loading that reduces the moment caused by H.

Where the horizontal component of backfill pressure includes groundwater above the top of the heel, use of two factors, 1.7 for the transverse soil pressure and 1.4 for the transverse liquid pressure, would not be appropriate. Because the probability of overload is about the same for soil pressure and water pressure, use of a single factor, 1.7, is logical, as recommended in the Commentary to the ACI Building Code. For environmental engineering structures where these conditions are common, ACI Committee 350 has recommended use of 1.7 for both soil and liquid pressure (see "Environmental Engineering Concrete Structures," ACI 350R). Committee 350 also favored a more conservative approach for design of the toe. It is more convenient and conservative to consider 1.7 times the entire vertical reaction uniformly distributed across the toe as well as more nearly representing the actual end-point condition (Fig. 9.39).

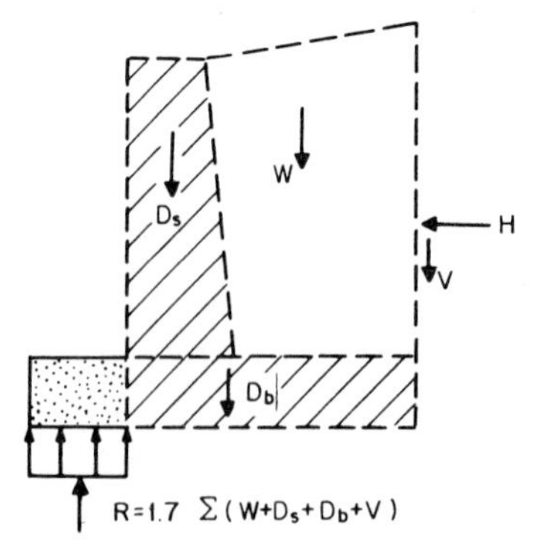

FIGURE 9.39 Loads for simplified strength design for toe of wall.

The top bars in the heel can be selected for the unbalanced moment between the factored forces on the toe and the stem, but need not be larger than for the moment of the top loads on the footing (earth and weight of heel). For a footing proportioned so that the actual soil pressure approaches zero at the end of the heel, the unbalanced moment and the maximum moment in the heel caused by the top loads will be nearly equal.

The possibility of an overall sliding failure, involving the soil and the structure together, must be considered, and may require a vertical lug extending beneath the footing, tie backs, or other provisions.

The base of the stem is a point of maximum bending moment and yet also the most convenient location for splicing the vertical bars and footing dowels. The ACI Code advises avoiding such points for the location of lap splices. But for cantilever walls, splices can be avoided entirely at the base of the stem only for low walls (8 to 10 ft high), in which L-shaped bars from the base of the toe can be extended full height of the stem. For high retaining walls (over 10 ft high), if all the bars are spliced at the base of the stem, a Class B tension lap splice is required (Art. 9.49.8). If alternate dowel bars are extended one Class A lap-splice length and the remaining dowel bars are extended at least twice this distance before cutoff, Class A lap splices may be used. This arrangement requires that dowel-bar sizes and vertical-bar sizes be selected so that the longer dowel bars provide at least 50% of the steel area required at the base of the stem and the vertical bars provide the total required steel at the cutoff point of the longer dowels (Fig. 9.40).

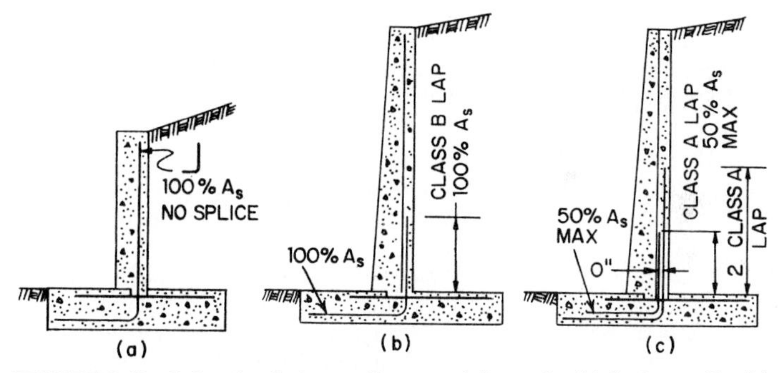

FIGURE 9.40 Splice details for cantilever retaining walls: (*a*) for low walls; (*b*) for high walls with Class B lap for dowels; (*c*) alternative details for high walls, with Class A lap for dowels.

9.70 COUNTERFORT RETAINING WALLS

In this type of retaining wall, counterforts (cantilevers) are provided on the earth side between wall and footing to support the wall, which essentially spans as a continuous one-way slab horizontally. Counterfort walls seldom find application in building construction. A temporary condition in which basement walls may be required to behave as counterfort retaining walls occurs though, if outside fill is placed before the floors are constructed. Under this condition of loading, each interior cross wall and end basement wall can be regarded as a counterfort. It is usually preferable, however, to delay the fill operation rather than to design and provide reinforcement for this temporary condition.

The advantages of counterfort walls are the large effective depth for the cantilever reinforcement and concrete efficiently concentrated in the counterfort. For very tall walls, where an alternative cantilever wall would require greater thickness and larger quantities of steel and concrete, the savings in material will exceed the additional cost of forming the counterforts. Accurate design is necessary for economy in important projects involving large quantities of material and requires refinement of the simple assumptions in the definition of counterfort walls. The analysis becomes complex for determination of the division of the load between one-way horizontal slab and vertical cantilever action.

(F. S. Merritt, "Standard Handbook for Civil Engineers," McGraw-Hill Publishing Company, New York.)

9.71 RETAINING WALLS SUPPORTED ON FOUR SIDES

For walls more than 10 in thick, the American Concrete Institute Building Code, ACI 318, requires two-way layers of bars in each face. Two-way slab design of this reinforcement is required for economy in basement walls or subsurface tank walls supported as vertical spans by the floor above and the footing below, and as horizontal spans by stiff pilasters, interior cross walls, or end walls.

This type of two-way slab is outside the scope of the specific provisions in the ACI Code. Without an "exact" analysis, which is seldom justified because of the uncertainties involved in the assumptions for stiffnesses and loads, a realistic design can be based on the simple two-way slab design method of Appendix A, Method 2, of the 1963 ACI Building Code.

FOUNDATIONS

Building foundations should distribute wall and column loads to the underlying soil and rock within acceptable limits on resulting soil pressure and total and differential settlement. Wall and column loads consist of live load, reduced in accordance with the applicable general building code, and dead load, combined, when required, with lateral loads of wind, earthquake, earth pressure, or liquid pressure. These loads can be distributed to the soil near grade by concrete spread footings, or to the soil at lower levels by concrete piles or drilled piers.

9.72 TYPES OF FOUNDATIONS

A wide variety of concrete foundations are used for buildings. Some of the most common types are illustrated in Fig. 9.41.

Spread wall footings consist of a plain or reinforced slab wider than the wall, extending the length of the wall (Fig. 9.41*a*). Plain- or reinforced-concrete **individual-column spread footings** consist of simple, stepped, or sloped two-way concrete slabs, square or rectangular in plan (Fig. 9.41*b* to *d*). For two columns close together, or an exterior column close to the property line so that individual spread or pile-cap footings cannot be placed concentrically, a reinforced-concrete, spread

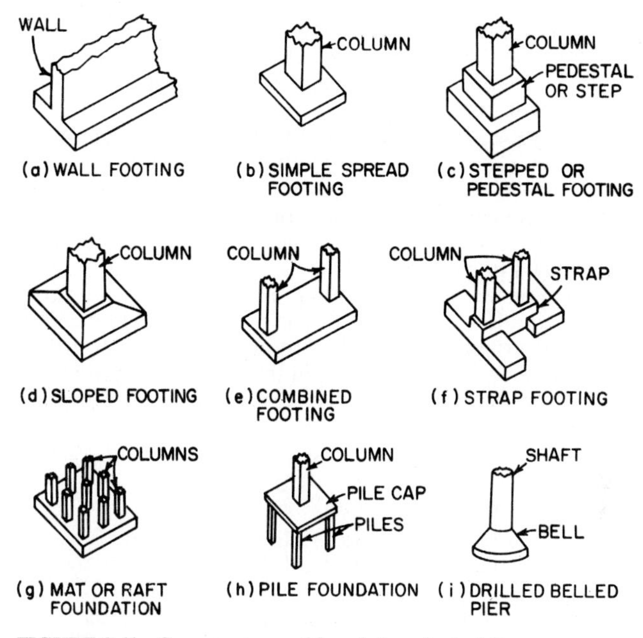

FIGURE 9.41 Common types of foundations for buildings.

combined footing (Fig. 9.41*e*) or a **strap footing** (Fig. 9.41*f*) can be used to obtain a nearly uniform distribution of soil pressure or pile loads. The strap footing becomes more economical than a combined footing when the spacing between the columns becomes larger, causing large bending moments in the combined footing.

For small soil pressures or where loads are heavy relative to the soil capacity, a reinforced-concrete **mat,** or **raft, foundation** (Fig. 9.41*g*) may prove economical. A mat consists of a two-way slab under the entire structure. Concrete cross walls or inverted beams can be utilized with a mat to obtain greater stiffness and economy.

Where sufficient soil strength is available only at lower levels, **pile foundations** (Fig. 9.41*h*) or **drilled-pier foundations** (Fig. 9.41*i*) can be used.

9.73 GENERAL DESIGN PRINCIPLES FOR FOUNDATIONS

The area of spread footings, the number of piles, or the number of drilled piers are selected by a designer to support actual unfactored building loads without exceeding *settlement limitations,* a safe soil pressure q_a, or a safe pile or drilled-pier load. A factor of safety from 2 to 3, based on the ultimate strength of the soil and its settlement characteristics, is usually used to determine the safe soil pressure or safe pile or drilled-pier load. See Arts. 6.13 to 6.22.

Soil Pressures. After the area of the spread footing or the number and spacing of piles or drilled piers has been determined, the spread footing, pile-cap footing, or drilled pier can be designed. The strength-design method of the ACI Building

Code (Art. 9.44) uses factored loads of gravity, wind, earthquake, earth pressure, and fluid pressure to determine *factored soil pressure* q_s, and factored pile or pier load. The factored loadings are used in strength design to determine factored moments and shears at critical sections.

For concentrically loaded footings, q_s is usually assumed as uniformly distributed over the footing area. This pressure is determined by dividing the concentric wall or column factored load P_u by the area of the footing. The weight of the footing can also be neglected in determining q_s because the weight does not induce factored moments and shears. The factored pile load for concentrically loaded pile-cap footings is determined in a similar manner.

When individual or wall spread footings are subjected to overturning moment about one axis, in addition to vertical load, as with a spread footing for a retaining wall, the pressure distribution under the footing is trapezoidal if the eccentricity e_x, of the resultant vertical load P_u is within the kern of the footing, or triangular if beyond the kern, as shown in Fig. 9.42. Thus, when $e_x < L/6$, where L is the

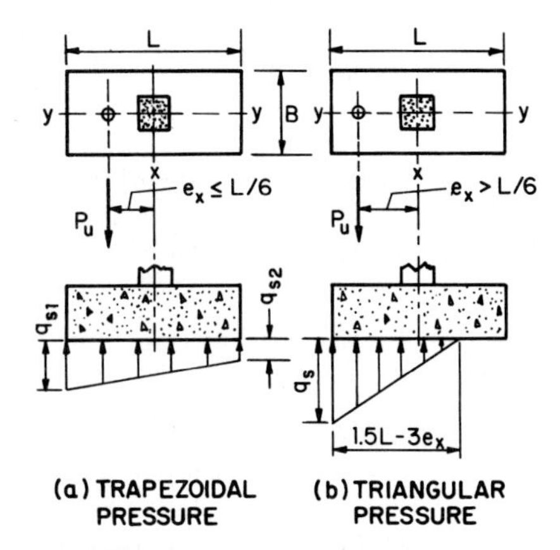

FIGURE 9.42 Spread footing subjected to moment pressures.

footing length in the direction of eccentricity e_x, the pressure distribution is trapezoidal (Fig. 9.42*a*) with a maximum

$$q_{s1} = \frac{P_u}{BL}\left(1 + \frac{6e_x}{L}\right) \tag{9.86}$$

and a minimum

$$q_{s2} = \frac{P_u}{BL}\left(1 - \frac{6e_x}{L}\right) \tag{9.87}$$

where B = footing width. When $e_x = L/6$, the pressure distribution becomes triangular over the length L, with a maximum

$$q_s = \frac{2P_u}{BL} \tag{9.88}$$

When $e_x > L/6$, the length of the triangular distribution decreases to $1.5L - 3e_x$ (Fig. 9.42*b*) and the maximum pressure rises to

$$q_s = \frac{2P_u}{1.5B(L - 2e_x)} \tag{9.89}$$

Reinforcement for Bearing. The bearing stress on the interface between a column and a spread footing, pile cap, or drilled pier should not exceed the allowable stress f_b given by Eq. (9.90), unless vertical reinforcement is provided for the excess.

$$f_b = 0.85\phi f'_c \sqrt{\frac{A_2}{A_1}} \qquad \frac{A_2}{A_1} \leq 4 \tag{9.90}$$

where ϕ = strength reduction factor = 0.70
f'_c = specified concrete strength
A_1 = loaded area of the column or base plate
A_2 = supporting area of footing, pile cap, or drilled pier that is the lower base of the largest frustrum of a right pyramid or cone contained wholly within the footing, with A_1 the upper face, and with side slopes not exceeding 2 horizontal to 1 vertical (Fig. 9.43).

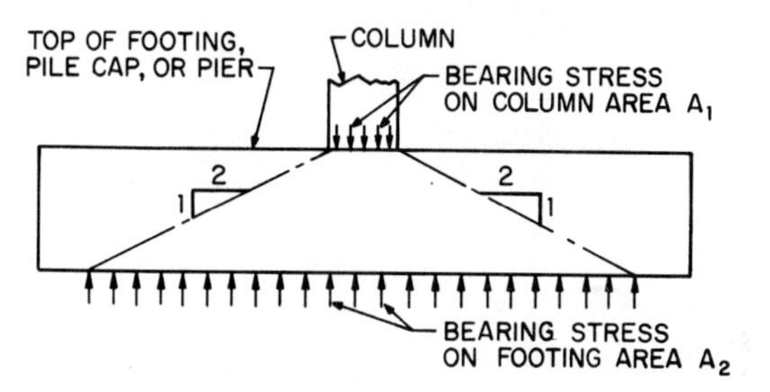

FIGURE 9.43 Bearing stresses on column and pressure against bottom of footing.

If the bearing stress on the loaded area exceeds f_b, reinforcement must be provided by extending the longitudinal column bars into the spread footings, pile cap, or drilled pier or by dowels. If so, the column bars or dowels required must have a minimum area of 0.005 times the loaded area of the column.

Provisions in the ACI Building Code assure that every column will have a minimum tensile capacity. Compression lap splices, which are permitted when the column bars are always in compression for all loading conditions, are considered to have sufficient tensile capacity so that no special requirements are needed. Similarly, the required dowel embedment in the footing for full compression development will provide a minimum tensile capacity.

Required compression-dowel embedment length cannot be reduced by hooks. Compression dowels can be smaller than column reinforcement. They cannot be larger than No. 11 bars.

If the bearing stress on the loaded area of a column does not exceed $0.85\phi f'_c$, column compression bars or dowels do not need to be extended into the footing, pile cap, or pier, if they can be developed within 3 times the column dimension (pedestal height) above the footing (Art. 9.49.6). It is desirable, however, that a minimum of one No. 5 dowel be provided in each corner of a column.

Footing Thickness. The minimum thickness allowed by the ACI Building Code for footing is 8 in for plain concrete footings on soil, 6 in above the bottom reinforcement for reinforced-concrete footings on soil, and 12 in above the bottom reinforcement for reinforced-concrete footings on piles. Plain-concrete pile-cap footings are not permitted.

Concrete Cover. The minimum concrete cover required by the ACI Code for reinforcement cast against and permanently exposed to earth is 3 in.

9.74 SPREAD FOOTINGS FOR WALLS

The critical sections for shear and moment for spread footings supporting concrete or masonry walls are shown in Fig. 9.44*a* and *b*. Under the soil pressure, the projection of footing on either side of a wall acts as a one-way cantilever slab.

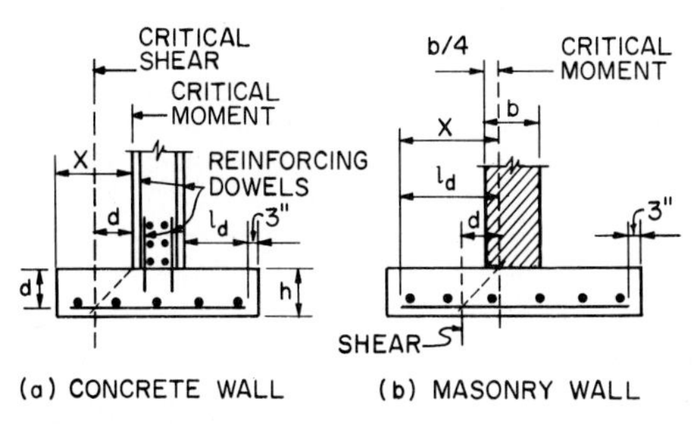

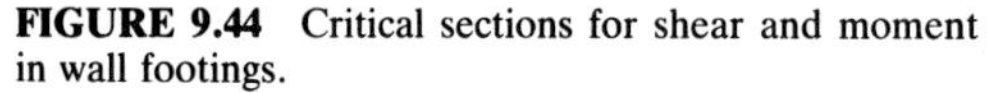

FIGURE 9.44 Critical sections for shear and moment in wall footings.

Unreinforced Footings. Recommendations for design of unreinforced concrete footings are presented in "Building Code Requirements for Structural Plain Concrete," ACI 318.1, and accompanying "Commentary," American Concrete Institute. For plain-concrete spread footings, the maximum permissible flexural tension stress, psi, in the concrete is limited by ACI 318.1 to $5\phi\sqrt{f'_c}$, where ϕ = strength reduction factor = 0.65 and f'_c = specified concrete strength, psi. For constant-depth, concentrically loaded footings with uniform design soil pressure, the thickness h, in, can be calculated from

$$h = 0.08X\sqrt{\frac{q_s}{\sqrt{f'_c}}} \tag{9.91}$$

where X = projection of footing, in, and q_s = net factored soil pressure, psf.

Shear is not critical for plain-concrete wall footings. The maximum tensile stress, $\phi 5\sqrt{f'_c}$, due to flexure controls the thickness.

Because it is usually not economical or practical to provide shear reinforcement in reinforced-concrete spread footings for walls, V_u is usually limited to the maximum value that can be carried by the concrete, $\phi V_c = \phi\, 2\sqrt{f'_c}b_w d$.

Flexural Reinforcement. Steel area can be determined from $A_s = M_u/\phi f_y jd$ as indicated in Art. 9.45. Sufficient length of reinforcement must be provided to develop the full yield strength of straight tension reinforcement. The critical development length is the shorter dimension l_d shown in Fig. 9.44*a* and *b*. The minimum lengths required to develop various bar sizes are tabulated in Table 9.8 (Art. 9.49.4). Reinforcement at right angles to the flexural reinforcement is usually provided as shrinkage and temperature reinforcement and to support and hold the flexural bars in position.

9.75 SPREAD FOOTINGS FOR INDIVIDUAL COLUMNS

Footings supporting columns are usually made considerably larger than the columns to keep pressure and settlement within reasonable limits. Generally, each column is also placed over the centroid of its footing to obtain uniform pressure distribution under concentric loading. In plan, the footings are usually square, but they can be made rectangular to satisfy space restrictions or to support rectangular columns or pedestals.

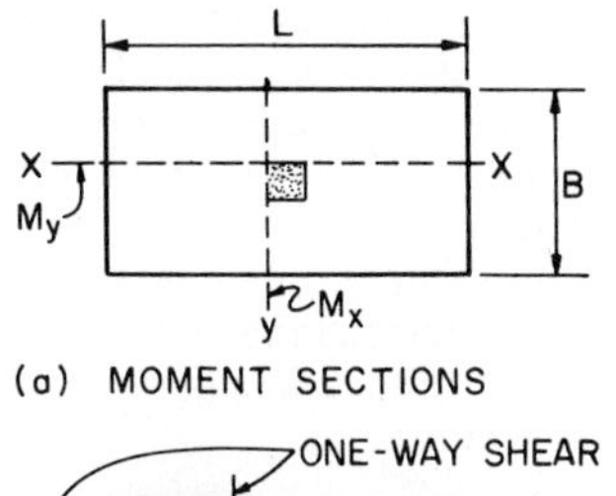

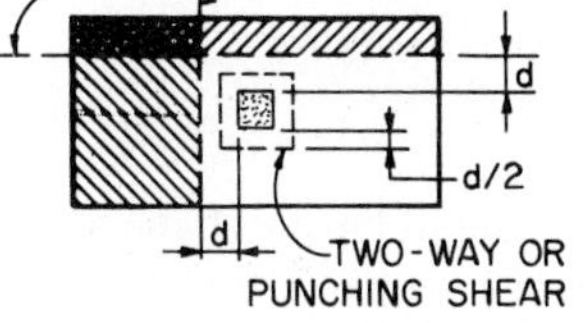

FIGURE 9.45 Critical sections for shear and moment in column footings.

Under soil pressure, the projection on each side of a column acts as a cantilever slab in two perpendicular directions. The effective depth of footing d is the distance from the extreme compression surface of the footing to the centroid of the tension reinforcement.

Bending Stresses. Critical sections for moment are at the faces of square and rectangular concrete columns or pedestals (Fig. 9.45*a*). For round and regular polygon columns or pedestals, the face may be taken as the side of a square having an area equal to the area enclosed within the perimeter of the column or pedestal. For steel columns with steel base plates, the critical section for moment may be taken halfway between the face of the column and the edges of the plate.

For plain-concrete spread footings, the flexural tensile stress must be limited to a maximum of $5\phi\sqrt{f'_c}$, where ϕ = strength reduction factor = 0.65 and $\sqrt{f'_c}$ = specified concrete strength psi. Thickness of such footings can be calculated with Eq. (9.91). Shear is not critical for plain-concrete footings.

Shear. For reinforced-concrete spread footings, shear is critical on two different sections.

Two-way or punching shear (Fig. 9.45*b*) is critical on the periphery of the surfaces at distance $d/2$ from the column, where d = effective footing depth.

One-way shear, as a measure of diagonal tension, is critical at distance d from the column (Fig. 9.45*b*), and must be checked in each direction.

It is usually not economical or practical to provide shear reinforcement in column footings. So the shear ϕV_c that can be carried by the concrete controls the thickness required.

For one-way shear for plain or reinforced concrete sections,

$$V_c = 2\sqrt{f'_c}b_w d \tag{9.92}$$

where b_w = width of section
d = effective depth of section

but Eq. (9.38) may be used as an alternative for reinforced concrete.

For two-way shear, V_c is the smallest of the values computed from Eqs. (9.73) to (9.75).

Flexural Reinforcement. In square spread footings, reinforcing steel should be uniformly spaced throughout, in perpendicular directions. In rectangular spread footings, the American Concrete Institute Building Code, ACI 318, requires the reinforcement in the long direction to be uniformly spaced over the footing width. Also, reinforcement with an area $2/(\beta + 1)$ times the area of total reinforcement in the short direction should be uniformly spaced in a width that is centered on the column and equal to the short footing dimension, where β is the ratio of the long to the short side of the footing. The remainder of the reinforcing in the short direction should be uniformly spaced in the outer portions of the footing. To maintain a uniform spacing for the bars in the short direction for simplified placing, the theoretical number of bars required for flexure must be increased by about 15%. The maximum increase occurs when β is about 2.5.

The required area of flexural reinforcement can be determined as indicated in Art. 9.46. Maximum-size bars are usually selected to develop the yield strength by straight tension embedment without end hooks. The critical length is the shorter dimension l_d shown for wall footings in Figs. 9.44*a* and *b*.

9.76 COMBINED SPREAD FOOTINGS

A combined spread footing under two columns should have a shape and location such that the center of gravity of the column loads coincides with the centroid of the footing area. It can be square, rectangular, or trapezoidal, as shown in Fig. 9.46.

Net design soil-pressure distribution can be assumed to vary linearly for most (rigid) combined footings. For the pressure distribution for flexible combined footings, see the report, "Suggested Analysis and Design Procedures for Combined Footings and Mats," ACI 336.2R, American Concrete Institute, and M. J. Haddadin, "Mats and Combined Footings—Analysis by the Finite Element Method," *ACI Journal Proceedings,* vol. 69, no. 12, December 1971, pp. 945–949. A computer program is available from the Portland Cement Association, Skokie, Ill.

If the center of gravity of the unfactored column loads coincides with the centroid of the footing area, the net soil pressure q_a will be uniform.

$$q_a = \frac{R}{A_f} \tag{9.93}$$

where R = applied vertical load
A_f = area of footing

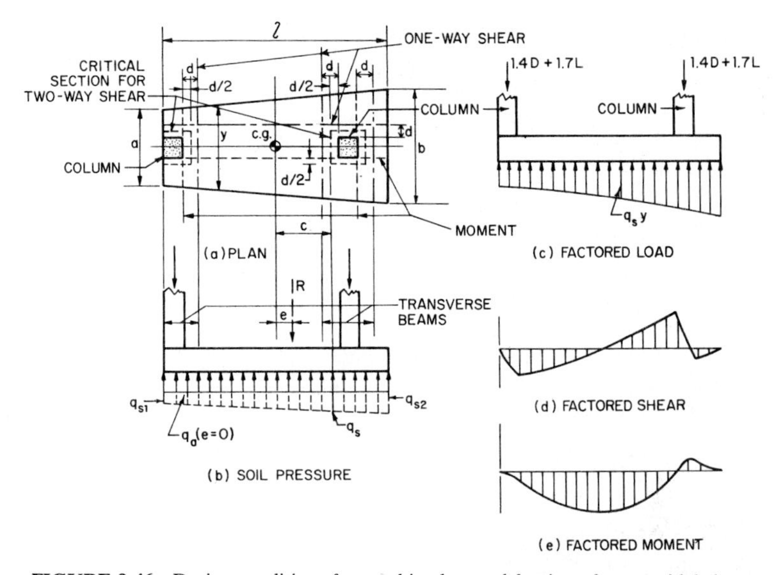

FIGURE 9.46 Design conditions for combined spread footing of trapezoidal shape.

The factored design load on the columns, however, may change the location of the center of gravity of the loads. If the resultant is within the kern for footings with moment about one axis, the design soil pressure q_s can be assumed to have a linear distribution (Fig. 9.46*b*) with a maximum at the more heavily loaded edge:

$$q_{s1} = \frac{P_u}{A_f} + \frac{P_u ec}{I_f} \tag{9.94}$$

and a minimum at the opposite edge:

$$q_{s2} = \frac{P_u}{A_f} - \frac{P_u ec}{I_f} \tag{9.95}$$

where P_u = factored vertical load
e = eccentricity of load
c = distance from center of gravity to section for which pressure is being computed
I_f = moment of inertia of footing area

If the resultant falls outside the kern of the footing, the net design pressure q_s can be assumed to have a linear distribution with a maximum value at the more heavily loaded edge and a lengthwise distribution of 3 times the distance between the resultant and the pressed edge. The balance of the footing will have no net design pressure.

With the net design soil-pressure distribution known, the design shears and moments can be determined (Fig. 9.46*d* and *e*). Critical sections for shear and moment are shown in Fig. 9.46*a*. The critical section for two-way shear is at a distance $d/2$ from the columns, where d is the effective depth of footing. The critical section for one-way shear in both the longitudinal and transverse direction is at a distance *d* from the column face.

Maximum design negative moment, causing tension in the top of the footing, will occur between the columns. The maximum design positive moment, with tension in the bottom of the footing, will occur at the face of the columns in both the longitudinal and transverse direction.

Flexural reinforcement can be selected as shown in Art. 9.46. For economy combined footings should be made deep enough to avoid the use of stirrups for shear reinforcement. In the transverse direction, the bottom reinforcement is placed uniformly in bands having an arbitrary width, which can be taken as the width of the column plus 2*d*. The amount at each column is proportional to the column load.

Size and length of reinforcing must be selected to develop the full yield strength of the steel between the critical section, or point of maximum tension, and the end of the bar (Art. 9.49.6).

9.77 STRAP FOOTINGS

When the distance between two columns to be supported on a combined footing becomes large, cost increases rapidly, and a strap footing, or cantilever-type footing, may be more economical. This type of footing, in effect, consists of two footings, one under each column, connected by a strap beam (Fig. 9.47). This beam distributes the column loads to each footing to make the net soil pressure with unfactored loads

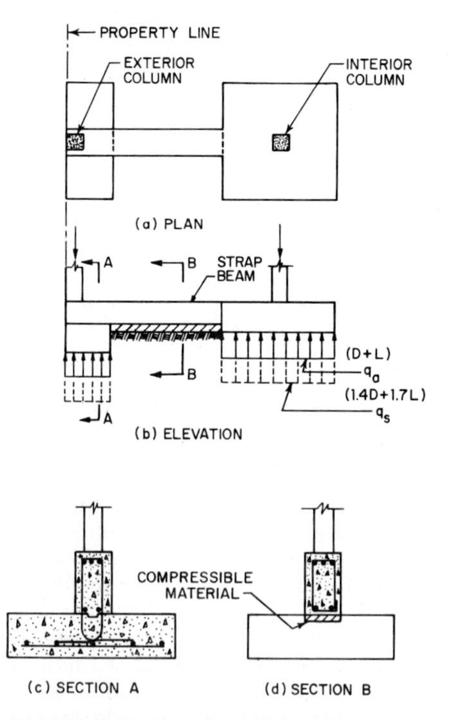

FIGURE 9.47 Design conditions for a strap footing.

uniform and equal at each footing, and with factored loads uniform but not necessarily equal. The center of gravity of the actual column loads should coincide with the centroid of the combined footing areas.

The strap beam is usually designed and constructed so that it does not bear on the soil (Fig. 9.47*b*). The concrete for the beam is cast on compressible material. If the concrete for the strap beam were placed on compacted soil, the resulting soil pressure would have to be considered in design of the footing.

The strap beam, in effect, cantilevers over the exterior column footing, and the bending will cause tension at the top. The beam therefore requires top flexural reinforcement throughout its entire length. Nominal flexural reinforcement should be provided in the bottom of the beam to provide for any tension that could result from differential settlement.

The top bars at the exterior column must have sufficient length to develop their full yield strength. Hence, the distance between the interior face of the exterior column and the property-line end of the horizontal portion of the top bar, plus the value of the hook, must provide tension development length (Art. 9.49.6).

Strength-design shear reinforcement [Eq. (9.40*a*)] will be required when $V_u > \phi V_c$, and requirements for minimum shear reinforcement [Eq. (9.82)] must be observed when $V_u > \phi V_c/2$, where V_c = shear carried by the concrete (Art. 9.47).

For the strap footing shown in Fig. 9.47, the exterior column footing can be designed as a wall spread footing, and the interior column footing as an individual-column spread footing (Arts. 9.74 and 9.75).

9.78 MAT FOUNDATIONS

A mat or raft foundation is a single combined footing for an entire building unit. It is economical when building loads are relatively heavy and the safe soil pressure is small. (See also Arts. 9.72 and 9.73.)

Weight of soil excavated for the foundation decreases the pressure on the soil under the mat. If excavated soil weighs more than the building, there is a net decrease in pressure at mat level from that prior to excavation.

When the mat is rigid, a uniform distribution of soil pressure can be assumed and the design can be based on a statically determinate structure, as shown in Fig. 9.48. (See "Suggested Analysis and Design Procedures for Combined Footings and Mats," ACI 336.2R, American Concrete Institute.)

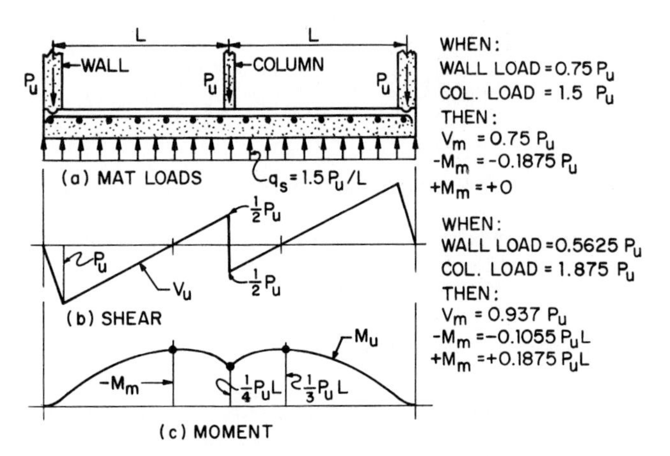

FIGURE 9.48 Design conditions for a rigid mat footing.

If the centroid of the factored loads does not coincide with the centroid of the mat area, the resulting nonuniform soil pressure should be used in the strength design of the mat.

Strength-design provisions for flexure, one-way and two-way shear, development length, and serviceability should conform to ACI Building Code requirements, ACI 318 (Art. 9.58).

9.79 PILE FOUNDATIONS

Building loads can be transferred to piles by a thick reinforced-concrete slab, called a pile-cap footing. The piles are usually embedded in the pile cap 4 to 6 in. They should be cut to required elevation after driving and prior to casting the footing. Reinforcement should be placed a minimum of 3 in clear above the top of the piles. The pile cap is required by the American Concrete Institute Building Code, ACI 318, to have a minimum thickness of 12 in above the reinforcement. (See also Art. 9.73.)

Piles should be located so that the centroid of the pile cluster coincides with the center of gravity of the column design load. As a practical matter, piles cannot be driven exactly to the theoretical design location. A construction survey should be made to determine if the actual locations require modification of the original pile-cap design.

Pile-cap footings are designed like spread footings (Art. 9.75), but for concentrated pile loads. Critical sections for shear and moment are the same. Reaction from any pile with center $d_p/2$ or more inside the critical section, where d_p is the pile diameter at footing base, should be assumed to produce no shear on the section. The ACI Code requires that the portion of the reaction of a pile with center within $d_p/2$ of the section be assumed as producing shear on the section based on a straightline interpolation between full value for center of piles located $d_p/2$ outside the section and zero at $d_p/2$ inside the section.

For pile clusters without moment, the pile load P_u for strength design of the footing is obtained by dividing the factored column load by the number of piles n. The factored load equals $1.4D + 1.7L$, where D is the dead load, including the weight of the pile cap, and L is the live load. For pile clusters with moment, the design load on the rth pile is

$$P_{ur} = (1.4D + 1.7L)\left(\frac{1}{n} + \frac{eC_r}{\sum_1^n C_r^2}\right) \qquad (9.96)$$

where e = eccentricity of resultant load with respect to neutral axis of pile group, in

C_r = distance between neutral axis of pile group and center of nth pile, in

n = number of piles in cluster

9.80 DRILLED-PIER FOUNDATIONS

A drilled-pier foundation is used to transmit loads to soil at lower levels through end bearing and, in some situations, side friction. (See also Art. 9.73.) It can be

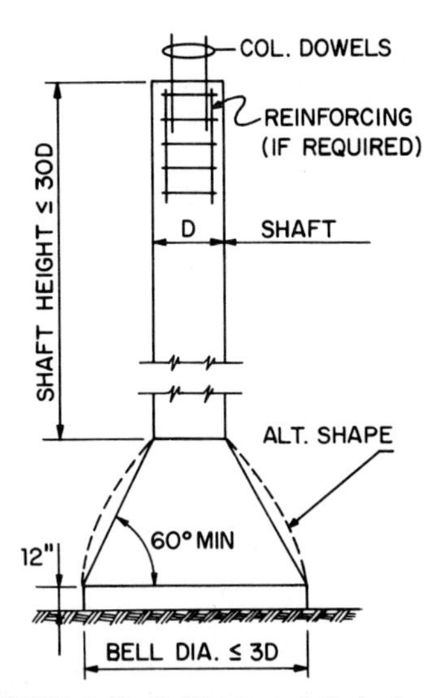

FIGURE 9.49 Bell-bottom drilled pier with dowels for a column.

constructed in firm, dry earth or clay soil by machine excavating an unlined hole with a rotating auger or bucket with cutting vanes and filling the hole with plain or reinforced concrete. Under favorable conditions, pier shafts 12 ft in diameter and larger can be constructed economically to depths of 100 ft and more. Buckets with sliding arms can be used to form bells at the bottom of the shaft with a diameter as great as 3 times that of the shaft (Fig. 9.49).

Some building codes limit the ratio of shaft height to shaft diameter to a maximum of 30. They also may require the bottom of the bell to have a constant diameter for the bottom foot of height, as shown in Fig. 9.49.

The compressive stress permitted on plain-concrete drilled piers with lateral support from surrounding earth varies with different codes. The ACI Code, ACI 318, limits the computed bearing stress based on service loads to $0.30f'_c$, where f'_c is the specified concrete strength, and the computed bearing stress based on factored loads to $0.55f'_c$.

Reinforced-concrete drilled piers can be designed as flexural members with axial load, as indicated in Art. 9.81.

Allowable unfactored loads on drilled piers with various shaft and bell diameters, supported by end bearing on soils of various allowable bearing pressures, are given in Table 9.23. For maximum-size bells (bell diameter 3 times shaft diameter) and a maximum concrete stress $f_{c1} = 0.30f'_c$ for unfactored loads, the required concrete strength, psi, is 20.83% of the allowable soil pressure, psf.

COLUMNS

Column-design procedures are based on a comprehensive investigation reported by American Concrete Institute Committee 105 ("Reinforced Concrete Column Investigation," *ACI Journal,* February 1933) and followed by many supplemental tests. The results indicated that basically the total capacity for axial load can be predicted, over a wide range of steel and concrete strength combinations and percentages of steel, as the sum of the separate concrete and steel capacities.

9.81 BASIC ASSUMPTIONS FOR STRENGTH DESIGN OF COLUMNS

At maximum capacity, the load on the longitudinal reinforcement of a concentrically loaded concrete column can be taken as the steel area A_{st} times steel yield strength f_y. The load on the concrete can be taken as the concrete area in compres-

TABLE 9.23 Allowable Service (Unfactored) Loads on Drilled Piers, kips*

Shaft dia, ft	Shaft area, in^2	$f'_c = 3000$	$f'_c = 4000$	$f'_c = 5000$	$f'_c = 6000$
1.5	254	229	305	382	458
2.0	452	407	543	679	814
2.5	707	636	848	1060	1272
3.0	1018	916	1221	1527	1832
3.5	1385	1247	1663	2078	2494
4.0	1810	1629	2171	2714	3257
4.5	2290	2061	2748	3435	4122
5.0	2827	2545	3393	4241	5099
5.5	3421	3079	4105	5132	6158
6.0	4072	3664	4886	6107	7329

		Safe allowable service-load bearing pressure on soil, psf					
Bell dia, ft	Bell area, ft^2	10,000	12,000	15,000	20,000	25,000	30,000
1.5	1.77	18	21	27	35	44	53
2.0	3.14	31	38	47	63	79	94
2.5	4.91	49	59	74	98	123	147
3.0	7.07	71	85	106	141	177	212
3.5	9.62	96	115	144	192	241	289
4.0	12.57	126	151	188	251	314	377
4.5	15.90	159	191	239	318	398	477
5.0	19.64	196	236	295	393	491	589
5.5	23.76	238	285	356	475	594	713
6.0	28.27	283	339	424	565	707	848
6.5	33.18	332	398	498	664	830	995
7.0	38.48	385	462	577	770	962	1155
7.5	44.18	442	530	663	884	1104	1325
8.0	50.27	503	603	754	1005	1257	1508
8.5	56.74	567	681	851	1135	1418	1702
9.0	63.62	636	763	954	1272	1590	1909
9.5	70.88	709	851	1063	1418	1772	2126
10.0	78.54	785	942	1178	1571	1963	2356
10.5	86.59	866	1039	1299	1732	2165	2598
11.0	95.03	950	1140	1425	1901	2376	2851
11.5	103.87	1039	1246	1558	2077	2597	3116
12.0	113.10	1131	1357	1696	2262	2827	3393

TABLE 9.23 (*Continued*)

Bell dia, ft	Bell area, ft²	Safe allowable service-load bearing pressure on soil, psf					
		10,000	12,000	15,000	20,000	25,000	30,000
12.5	122.72	1227	1473	1841	2454	3068	3682
13.0	132.73	1327	1593	1991	2655	3318	3982
13.5	143.14	1431	1718	2147	2863	3578	4294
14.0	153.94	1539	1847	2309	3079	3848	4618
14.0	165.13	1651	1982	2477	3303	4128	4954
15.0	176.15	1767	2121	2651	3534	4418	5301
15.5	188.69	1887	2264	2830	3774	4717	5661
16.0	201.06	2011	2413	3016	4021	5027	6032
16.5	213.82	2138	2566	3207	4276	5344	6415
17.0	226.98	2270	2724	3405	4540	5675	6809
17.5	240.53	2405	2886	3608	4811	6013	7216
18.0	254.47	2545	3054	3817	5089	6362	7634

*$f_{c1} = 0.30 f'_c$.
NOTE: Bell diameter preferably not to exceed 3 times the shaft diameter. Check shear stress if bell slope is less than 2:1. (Courtesy Concrete Reinforcing Steel Institute.)

sion times 85% of the compressive strength f'_c of the standard test cylinder. The 15% reduction from full strength accounts, in part, for the difference in size and, in part, for the time effect in loading of the column. Capacity of a concentrically loaded column then is the sum of the loads on the concrete and the steel.

The American Concrete Institute Building Code, ACI 318, applies a strength reduction factor $\phi = 0.75$ for members with spiral reinforcement and $\phi = 0.70$ for other members. For small axial loads ($P_u \leq 0.10 f'_c A_g$, where A_g = gross area of column), ϕ may be increased proportionately to as high as 0.90. Capacity of columns with eccentric load or moment may be similarly determined, but with modifications. These modifications introduce the assumptions made for strength design for flexure and axial loads.

The basic assumptions for strength design of columns can be summarized as follows:

1. Strain of steel and concrete is proportional to distance from neutral axis (Fig. 9.50*c*).
2. Maximum usable compression strain of concrete is 0.003 in/in (Fig. 9.50*c*).
3. Stress, psi, in longitudinal bars equals steel strain ϵ_s times 29,000,000 for strains below yielding, and equals the steel yield strength f_y, tension or compression, for larger strains (Fig. 9.50*f*).
4. Tensile strength of concrete is negligible.
5. Capacity of the concrete in compression, which is assumed at a maximum stress of $0.85 f'_c$, must be consistent with test results. A rectangular stress distribution (Fig. 9.50*d*) may be used. Depth of the rectangle may be taken as $a = \beta_1 c$,

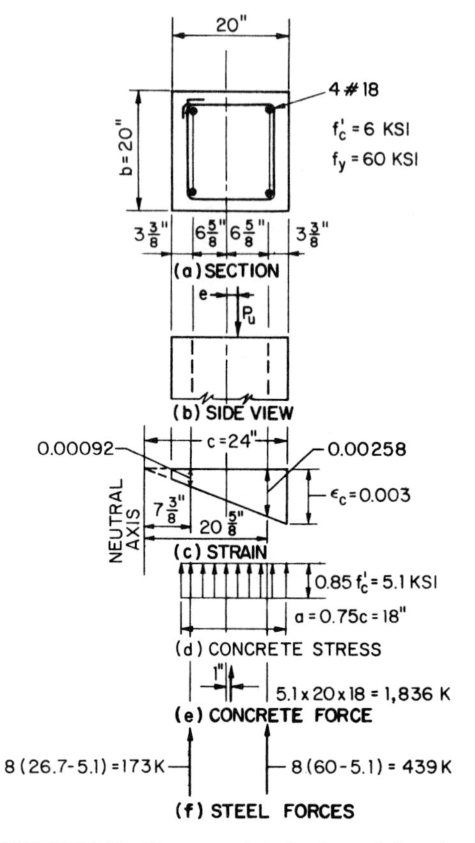

FIGURE 9.50 Stresses and strains in a reinforced-concrete column.

where c is the distance from the neutral axis to the extreme compression surface and $\beta_1 = 0.85$ for $f'_c \leq 4000$ psi and 0.05 less for each 1000 psi that f'_c exceeds 4000 psi, but β_1 should not be taken less than 0.65.

In addition to these general assumptions, design must be based on equilibrium and strain compatibility conditions. No essential difference develops in maximum capacity between tied and spiral columns, but spiral-reinforced columns show far more toughness before failure. Tied-column failures have been relatively brittle and sudden, whereas spiral-reinforced columns that have failed have deformed a great deal and carried a high percentage of maximum load to a more gradual yielding failure. The difference in behavior is reflected in the higher value of ϕ assigned to spiral-reinforced columns.

Additional design considerations are presented in Arts. 9.82 to 9.86. Following is an example of the application of the basic assumptions for strength design of columns.

Example. Determine the capacity of the 20-in-square reinforced-concrete column shown in Fig. 9.50*a*. The column is reinforced with four No. 18 bars, with $f_y =$

60 ksi, and lateral ties. Area of rebars total 16 in^2. Concrete strength is $f'_c = 6$ ksi. Assume the factored load P_u to have an eccentricity of 2 in and that slenderness can be ignored.

To begin, assume $c = 24$ in. Then, with $\beta_1 = 0.75$ for $f'_c = 6000$ psi, the depth of the compression rectangle is $a = 0.75 \times 24 = 18$ in. This assumption can be checked by computing the eccentricity $e = \phi M_n / \phi P_n$, where ϕM_n is the design moment capacity, ft-kips, and ϕP_n is the design axial load strength, kips.

Since the strain diagram is linear and maximum compression strain is 0.003 in/in, the strains in the reinforcing steel are found by proportion to be 0.00258 and 0.00092 in/in (Fig. 9.50*c*). The strain at yield is $60/29{,}000 = 0.00207 < 0.00258$ in/in. Hence, the stresses in the steel are 60 ksi and $0.00092 \times 29{,}000 = 26.7$ ksi.

The maximum concrete stress, which is assumed constant over the depth $a = 18$ in, is $0.85 f'_c = 0.85 \times 6 = 5.1$ ksi (Fig. 9.50*d*). Hence, the compression force on the concrete is $5.1 \times 20 \times 18 = 1836$ kips and acts at a distance $20/2 - 18/2 = 1$ in from the centroid of the column (Fig. 9.50*e*). The compression force on the more heavily loaded pair of reinforcing bars, which have a cross-sectional area of 8 in^2, is 8×60 less the force on concrete replaced by the steel 8×5.1, or 439 kips. The compression force on the other pair of bars is $8(26.7 - 5.1) = 173$ kips (Fig. 9.50*f*). Both pairs of bars act at a distance of $20/2 - 3.375 = 6.625$ in from the centroid of the column.

The design capacity of the column for vertical load is the sum of the nominal steel and concrete capacities multiplied by a strength reduction factor $\phi = 0.70$.

$$\phi P_n = 0.70(1836 + 173 + 439) = 1714 \text{ kips}$$

The capacity of the column for moment is found by taking moments of the steel and concrete capacities about the centerline of the column.

$$\phi M_n = 0.70 \left[1836 \times \frac{1}{12} + (439 - 173) \frac{6.625}{12} \right] = 209 \text{ ft-kips}$$

The eccentricity for the assumed value of $c = 24$ in is

$$e = \frac{209 \times 12}{1714} = 1.46 < 2 \text{ in}$$

If for a new trial, c is taken as 22.5 in, then $P_u = 1620$ kips, $M_u = 272$ ft-kips, and e checks out close to 2 in. If sufficient load-moment values for other assumed positions of the neutral axis are calculated, a complete load-moment interaction diagram can be constructed (Fig. 9.51).

The nominal maximum axial load capacity P_o of a column without moment equals the sum of the capacities of the steel and the concrete.

$$P_o = 0.85 f'_c (A_g - A_{st}) + f_y A_{st} \tag{9.97}$$

where A_g = gross area of column cross section and A_{st} = total area of longitudinal steel reinforcement. For the 20-in-square column in the example:

$$P_o = 0.85 \times 6(400 - 16) + 60 \times 16 = 2918 \text{ kips}$$

The maximum design axial-load strength permitted by the ACI Code is

$$\begin{aligned} \phi P_{n(max)} &= 0.80\, \phi\, [0.85 f'_c (A_g - A_{st}) + f_y A_{st}] \\ &= 0.80 \times 0.70\, [(0.85 \times 6(400 - 16) + 60 \times 16] = 1634 \text{ kips} \end{aligned} \tag{9.98}$$

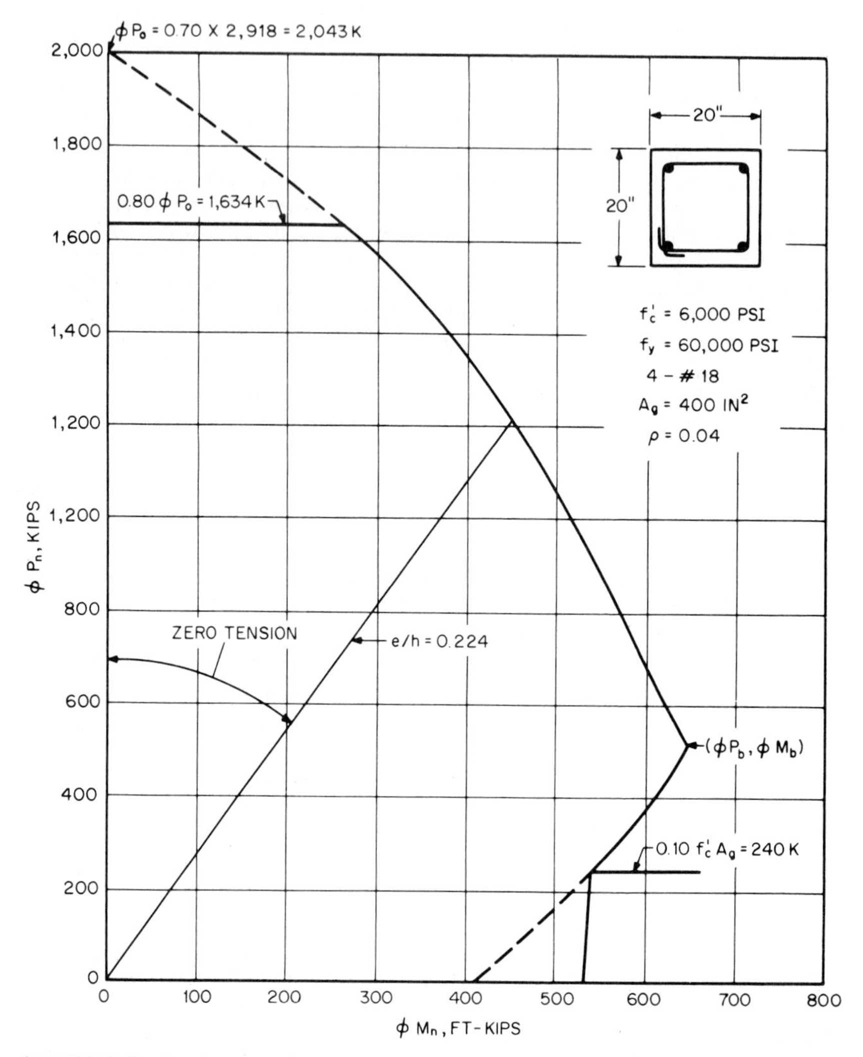

FIGURE 9.51 Load-moment interaction diagram for determination of strength of a rectangular reinforced-concrete column.

9.82 DESIGN REQUIREMENTS FOR COLUMNS

The American Concrete Institute Building Code, ACI 318, contains the following principal design requirements for columns, in addition to the basic assumptions (Art. 9.81):

1. Columns must be designed for all bending moments associated with a loading condition.

2. For corner columns and other columns loaded unequally on opposite sides in perpendicular directions, biaxial bending moments must be considered.

3. All columns are designed for an eccentricity of the factored load P_u because the maximum design axial load strength cannot be larger than $0.80P_o$ for tied columns, or $0.85P_o$ for spiral columns, where P_o is given by Eq. (9.97).

4. The minimum ratio of longitudinal-bar area to total cross-sectional area of column A_g is 0.01, and the maximum ratio is 0.08. For columns with a larger cross section than required by loads, however, a smaller A_g, but not less than half the gross area of the columns, may be used for calculating both load capacity and minimum longitudinal bar area. This exception allows reuse of forms for larger-than-necessary columns, and permits longitudinal bar areas as low as 0.005 times the actual column area. At least four longitudinal bars should be used in rectangular reinforcement arrangements, and six in circular arrangements.

5. The ratio of the volume of spiral reinforcement to volume of concrete within the spiral should be at least

$$\rho_s = 0.45 \left(\frac{A_g - A_c}{A_c}\right) \left(\frac{f'_c}{f_y}\right) \tag{9.99}$$

where A_g = gross cross-sectional area of concrete column, in^2
A_c = area of column within outside diameter of spiral, in^2
f'_c = specified concrete strength, psi
f_y = specified yield strength of spiral steel, psi (maximum 60,000 psi)

6. For tied columns, minimum size of ties is No. 3 for longitudinal bars that are No. 10 or smaller, and No. 4 for larger longitudinal bars. Minimum vertical spacing of sets of ties is 16 diameters of longitudinal bars, 48 tie-bar diameters, or the least thickness of the column. A set of ties should be composed of one round tie for bars in a circular pattern, or one tie enclosing four corner bars plus additional ties sufficient to provide a corner of a tie at alternate interior bars or at bars spaced more than 6 in from a bar supported by the corner of a tie.

7. Minimum concrete cover required for reinforcement is listed in Table 9.24. See also Art. 9.84.

TABLE 9.24 Minimum Cover, in, for Column Reinforcement*

Type of construction	Reinforcement	Interior exposure†	Exposed to weather*
Cast-in-place	Longitudinal	1½	2
	Ties, spirals	1½	1½
Precast	Longitudinal	$5/8 \leq d_b \leq 1\frac{1}{2}$‡	1½
	Ties, spirals	⅜	1¼
Prestressed	Longitudinal	1½	1½
	Ties, spirals	1	1½

*From ACI 318-89.
†See local code; fire protection may require greater thickness.
‡d_b = nominal bar diameter, in.

9.83 COLUMN TIES AND TIE PATTERNS

For full utilization, all ties in tied columns must be fully developed (for full tie yield strength) at each corner enclosing a vertical bar or, for circular ties, around the full periphery.

Splices. The American Concrete Institute Building Code, ACI 318, provides arbitrary minimum sizes and maximum spacings for column ties (Art. 9.82). No increase in size nor decrease in the spacings is required for Grade 40 materials. Hence, the minimum design requirements for splices of ties may logically be based on Grade 40 steel.

The ordinary closed, square or rectangular, tie is usually spliced by overlapping standard tie hooks around a longitudinal bar. Standard tie patterns require staggering of hook positions at alternate tie spacings, by rotating the ties 90 or 180°. ("Manual of Standard Practice," Concrete Reinforcing Steel Institute). Two-piece ties are formed by lap splicing or anchoring the ends of U-shaped open ties. Lapped bars should be securely wired together to prevent displacement during concreting.

Tie Arrangements. Commonly used tie patterns are shown in Figs. 9.52 to 9.54. In Fig. 9.53, note the reduction in required ties per set and the improvement in

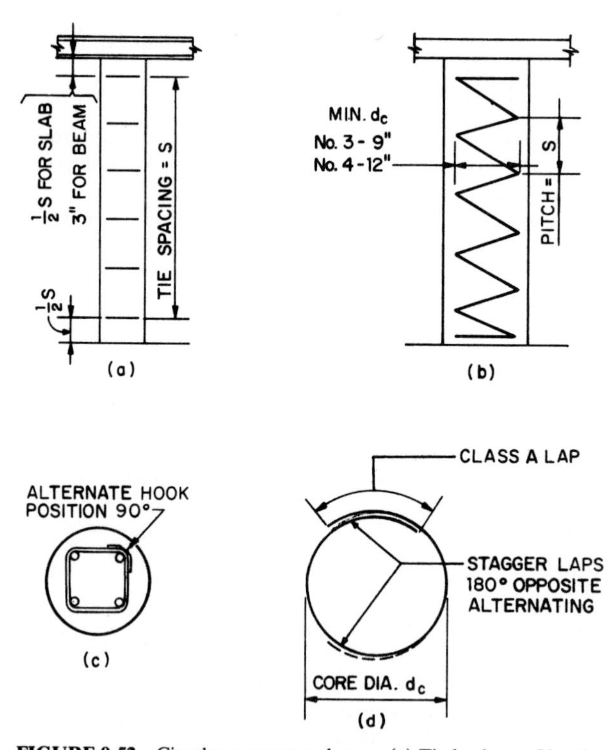

FIGURE 9.52 Circular concrete columns. (*a*) Tied column. Use ties when core diameter $d_c \leq s$. (*b*) Spiral-reinforced column, for use when $d_c > s$. (*c*) Rectangular tie for use in columns with four longitudinal bars. (*d*) Circular tie.

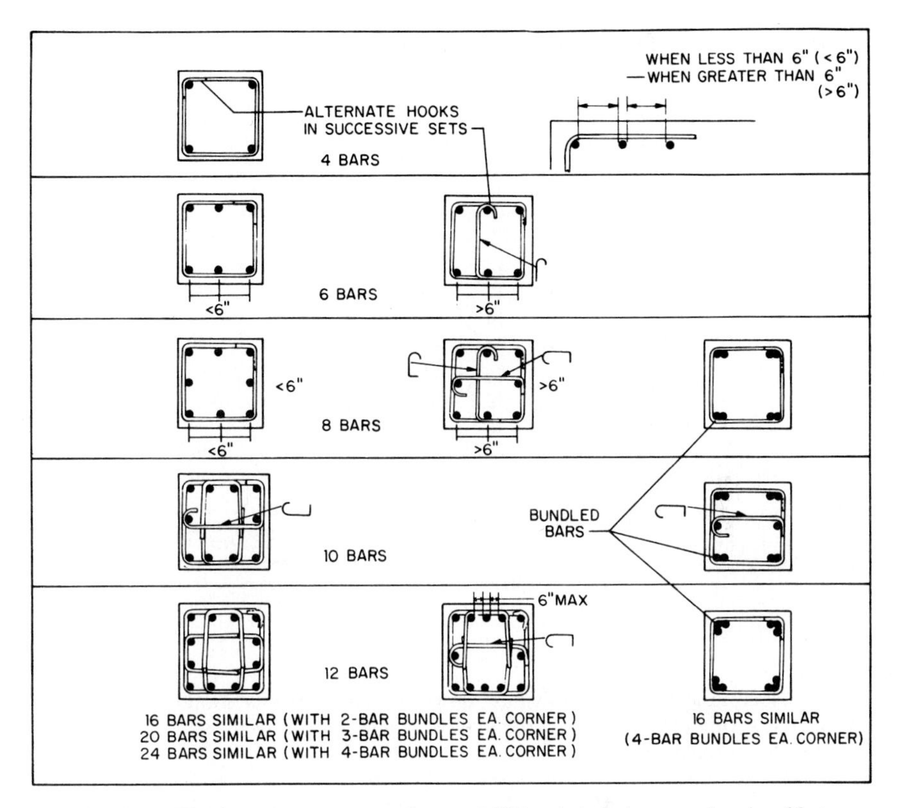

FIGURE 9.53 Ties for square concrete columns. Additional single bars may be placed between any of the tied groups, but clear spaces between bars should not exceed 6 in.

bending resistance about both axes achieved with the alternate bundled-bar arrangements. Bundles may not contain more than four bars, and bar size may not exceed No. 11.

Tie sizes and maximum spacings per set of ties are listed in Table 9.25.

Drawings. Design drawings should show all requirements for splicing longitudinal bars, that is, type of splice, lap length if lapped, location in elevation, and layout

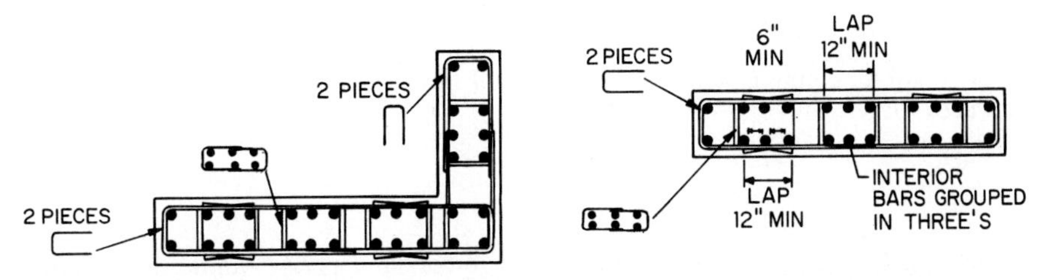

FIGURE 9.54 Ties for wall-like columns. Spaces between corner bars and interior groups of three bars may vary to accommodate average spacing not exceeding 6 in. A single additional bar may be placed in any of such spaces if the average spacing does not exceed 6 in.

TABLE 9.25 Maximum Spacing of Column Ties*

Vertical bar size, number	Size and spacing of ties, in		
	No. 3	No. 4	No. 5
5	10		
6	12		
7	14		
8	16	16	
9	18	18	
10	18	20	
11	†	22	22
14	†	24	27
18	†	24	30

*Maximum spacing not to exceed least column dimension.
†Not allowed.

in cross section. On detail drawings, dowel erection details should be shown if special large longitudinal bars, bundled bars, staggered splices, or specially grouped bars are to be used.

9.84 BIAXIAL BENDING OF COLUMNS

If column loads cause bending simultaneously about both principal axes of a column cross section, as for most corner columns, a biaxial bending analysis is required. For rapid preliminary design, Eq. (9.100) gives conservative results.

$$\frac{M_x}{M_{ox}} + \frac{M_y}{M_{oy}} \leq 1 \tag{9.100}$$

where M_x, M_y = factored moments about x and y axes, respectively
M_{ox}, M_{oy} = design capacities about x and y axes, respectively

For square columns with equal steel in all faces, $M_{ox} = M_{oy}$, and the relation reduces to:

$$\frac{M_x + M_y}{M_{ox}} \leq 1 \tag{9.101}$$

Because $M_x = e_x P_u$ and $M_y = e_y P_u$, the safe biaxial capacity can be taken from uniaxial load-capacity tables for the load P_u and the uniaxial bending moment $M_u = (e_x + e_y)P_u$. Similarly, for round columns, the moment capacity is essentially equal in all directions, and the two bending moments about the principal axes may be combined into a single uniaxial factored moment M_u which is then an exact solution.

$$M_u = \sqrt{M_x^2 + M_y^2} \quad (9.102)$$

The linear solution always gives a safe design, but becomes somewhat overconservative when the moments M_x and M_y are nearly equal. For these cases, a more exact solution will be more economical for the final design.

9.85 SLENDERNESS EFFECTS ON CONCRETE COLUMNS

The American Concrete Institute Building Code, ACI 318, requires that primary column moments be magnified to provide safety against buckling failure. Detailed procedures, formulas, and design aids are provided in the ACI Code and Commentary.

For most unbraced frames, an investigation will be required to determine the magnification factor to allow for the effects of sidesway and end rotation. The procedure for determination of the required increase in primary moments, after the determination that slenderness effects cannot be neglected, is complex. For direct solution, the requirements of Sec. 10.10, ACI 318-89 can be met by a P-Δ analysis. (See, for example, J. G. MacGregor and S. E. Hage, "Stability Analysis and Design of Concrete," *Journal of the Structural Division,* ASCE, Vol. 103, No. ST10, October 1977.)

The direct P-Δ method of MacGregor and Hage is based upon an equation for a geometric series that was derived for the final second-order deflection as a function of the first-order elastic deflection. This direct P-Δ analysis provides a very simple method for computing the moment magnifier δ when the stability index Q is greater than 0.04 but equal to or less than 0.22.

$$\delta = 1/(1 - Q) \qquad 0.04 < Q \leq 0.22 \quad (9.103)$$

where $Q = P_u\Delta_u/(H_u h_s)$
P_u = sum of the factored loads in a given story
Δ_u = elastically computed first-order lateral deflection due to H_u (neglecting P-Δ effects) at the top of the story, relative to the bottom
H_u = total factored lateral force (shear) within the story
h_s = height of story, center-to-center of floors or roof

The approximate method of ACI 318-89 may also be used to determine the moment magnifier. This approximate method is a column-by-column correction based upon the stiffnesses of the column and beams, applied primary design column end moments, and consideration of whether the entire structure is laterally braced against sidesway by definition. (See ACI 318-89, Secs. 10.11).

The ACI Code permits slenderness effects to be neglected only for very short, braced columns, with the following limitations for columns with square or rectangular cross-sections:

$L_u \leq 6.6h$ for bending in single curvature

$L_u \leq 10.2h$ for bending in double curvature with unequal end moments

$L_u \leq 13.8h$ for bending in double curvature with equal end moments and for round columns, five-sixths of the maximum lengths for square columns, where L_u is the unsupported length and h the depth or overall thickness of column in the direction being considered.

These limiting heights are based on the ratio of the total stiffnesses of the columns to the total stiffnesses of the flexural members, $\Sigma K_c/\Sigma K_B = 50$, at the joint at each end of a column. As these ratios become less, the limiting heights can be increased. When the total stiffnesses of the columns and the floor systems are equal at each end of the column (a common assumption in routine frame analysis), the two ratios = 1.00, and the limiting heights increase about 30%. With this increase, the slenderness effects can be neglected for most columns in frames braced against sidesway.

A frame is considered braced when other structural elements, such as walls, provide stiffness resisting sidesway at least 6 times the sum of the column stiffnesses resisting sidesway in the same direction in the story being considered.

9.86 ECONOMY IN COLUMN DESIGN

Actual costs of reinforced-concrete columns in place per linear foot per kip carrying capacity vary widely. The following recommendations based on relative costs are generally applicable:

Formwork. Use of the same size and shape of column cross section throughout a floor and, for multistory construction, from footing to roof will permit mass production and reuse for economy. Within usual practicable maximum building heights, about 60 stories or 600 ft, increased speed of construction and saving in formwork will save more than the cost of the excess concrete volume over that for smaller column sizes in upper stories.

Concrete Strength. Use of the maximum concrete strength required to support the factored loads with the minimum allowed steel area results in the lowest cost. The minimum size of a multistory column is established by the maximum concrete strength reliably available locally and the limit on maximum area of vertical bars. (Concrete with a compressive strength f'_c of 17,000 psi is commercially available in many areas of the United States.) If the acceptable column size is larger than the minimum possible at the base of the multistory stack, the steel ratio can begin with less than the maximum limit (Art. 9.82). At successive stories above, the steel ratio can be reduced to the minimum, and thereafter, for additional stories, the concrete strength can be reduced. Near the top, as loads reduce further, a further reduction in the steel area to 0.005 times the concrete area may be made (Art. 9.82).

Steel. Grade 60 vertical bars offer the greatest economy. Minimum tie requirements can be achieved with four-bar or four-bundle (up to four bars per bundle) arrangements, or by placing an intermediate bar between tied corners not more than 6 in (clear) from the corner bars. For these arrangements, no interior ties are required; only one tie per set is needed. (See Fig. 9.54 and Art. 9.82.) With no interior ties, low-slump concrete can be placed and consolidated more easily, and the cost and time for assembly of column reinforcement cages are greatly reduced. Note that, for small quantities, the local availability of Nos. 14 and 18 bars should be investigated before they are specified.

Details of Column Reinforcement. Where Nos. 14 and 18 bars are used in compression only, end-bearing splices usually save money. If the splices are staggered 50%, as with two-story lengths, the tensile capacity of the columns will also be adequate for the usual bending moments encountered. For unusually large bending moments, where tensile splices of No. 10 bars and larger are required, mechanical connections are usually least expensive in place. For smaller bar sizes, lap splices, tensile or compressive, are preferred for economy. Some provision for staggered lap splices for No. 8 bars and larger may be required to avoid Class B tension splices (Art. 9.49.8).

Where butt splices are used, it will usually be necessary to assemble the column reinforcement cage in place. Two-piece interior ties or single ties with end hooks for two bars (see Art. 9.84) will facilitate this operation.

Where the vertical bar spacing is restricted and lap splices are used, even with the column size unchanged, offset bending of the bars from below may be required. However, where space permits, as with low steel ratios, an additional saving in fabrication and erection time will be achieved by use of straight column verticals offset one bar diameter at alternate floors.

SPECIAL CONSTRUCTION

9.87 DEEP BEAMS

The American Concrete Institute Building Code, ACI 318, defines deep beams as flexural members with clear span-depth ratios less than 2.5 for continuous spans and 1.25 for simple spans. Some types of building components behave as deep beams and require analysis for nonlinear stress distribution in flexure. Some common examples are long, precast panels used as spandrel beams; below-grade walls, with or without openings, distributing column loads to a continuous slab footing or to end walls; and story-height walls used as beams to eliminate lower columns in the first floor area.

Shear. When the clear span-depth ratio is less than 5, beams are classified as deep for shear reinforcement purposes. Separate special requirements for shear apply when span-depth ratio is less than 2 or between 2 and 5. The critical section for shear should be taken at a distance from face of support of $0.15L_n \leq d$ for uniformly loaded deep beams, and of $0.50a \leq d$ for deep beams with concentrated loads, where a is the shear span, or distance from concentrated load to face of support, L_n the clear span, and d the distance from extreme compression surface to centroid of tension reinforcement. Shear reinforcement required at the critical section should be used throughout the span.

The nominal shear strength of the concrete can be taken as

$$V_c = 2\sqrt{f'_c}\, b_w d \tag{9.104}$$

where f'_c = specified concrete strength, psi
b_w = width of beam web
d = distance from extreme compression fiber to the centroid of the tension reinforcement

The ACI Building Code also presents a more complicated formula that permits the concrete to carry up to $6\sqrt{f'_c}\, b_w d$.

Maximum nominal shear strength when $L_n/d < 2$ should not exceed

$$V_n = V_c + V_s = 8\sqrt{f'_c}\, b_w d \tag{9.105}$$

where V_s = nominal shear strength provided by shear reinforcement. When L_n/d is between 2 and 5 maximum nominal shear strength should not exceed

$$V_n = \frac{2}{3}\left(10 + \frac{L_n}{d}\right)\sqrt{f'_c}\, b_w d \tag{9.106}$$

Required area of shear reinforcement should be determined from

$$f_y d\left[\frac{A_v}{s}\left(\frac{1 + L_n/d}{12}\right) + \frac{A_{vh}}{s^2}\left(\frac{11 - L_n/d}{12}\right)\right] = V_u/\phi - V_c \tag{9.107}$$

where A_v = area of shear reinforcement perpendicular to main reinforcement within a distance s

ϕ = strength reduction factor = 0.85

s = spacing of shear reinforcement measured parallel to main reinforcement

A_{vh} = area of shear reinforcement parallel to main reinforcement within a distance s_2

s_2 = spacing of shear reinforcement measured perpendicular to main reinforcement

f_y = yield strength of shear reinforcement

Spacing s should not exceed $d/5$ or 18 in. Spacing s_2 should not exceed $d/3$ or 18 in. The area of shear reinforcement perpendicular to the main reinforcement should be a minimum of

$$A_v = 0.0015 b_w s \tag{9.108}$$

where b_w = width of beam compression face. Area of shear reinforcement parallel to main reinforcement should be at least

$$A_{vh} = 0.0025 b_w s_2 \tag{9.109}$$

When $b_w > 10$ in, shear reinforcement should be placed in each face of the beam. If the beam has a face exposed to the weather, between one-half and two-thirds of the total shear reinforcement should be placed in the exterior face. Bars should not be smaller than No. 3.

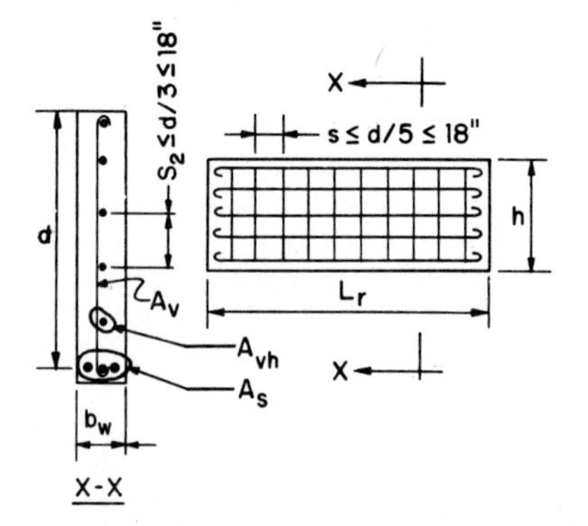

FIGURE 9.55 Reinforcement for deep beams. When the beam thickness exceeds 10 in, a layer of vertical rebars should be installed near each face of the beam.

Bending. The area of steel provided for positive bending moment in a deep beam should be at least

$$A_s = \frac{200 b_w d}{f_y} \tag{9.110}$$

where f_y = yield strength of flexural reinforcement, psi. This minimum amount can be reduced to one-third more than that required by analysis.

A safe assumption for preliminary design is that the extreme top surface in compression is 0.25 of the overall depth h below the top of very deep beams for computation of a reduced effective depth d for flexure (Fig. 9.55).

(J. G. MacGregor, "Reinforced Concrete Mechanics and Design," 2d ed., Prentice-Hall, Englewood Cliffs, NJ.)

9.88 SHEAR WALLS

Cantilevered shear walls used for bracing structures against lateral displacement (sidesway) are a special case of deep beams. They may be used as the only lateral bracing, or in conjunction with beam-column frames. In the latter case, the lateral displacement of the combination can be calculated with the assumption that lateral forces resisted by each element can be distributed to walls and frames in proportion to stiffness. For tall structures, the effect of axial shortening of the frames and the contribution of shear to lateral deformation of the shear wall should not be neglected. Figure 9.56 indicates the forces assumed to be acting on a horizontal cross section of a shear wall.

Reinforcement required for flexure of shear walls as a cantilever should be proportioned as for deep beams (Art. 9.89). Shear reinforcement is usually furnished as a combination of horizontal and vertical bars distributed evenly in each story (for increment of load). For low shear (where the factored shear force V_u at a section is less than $0.5\phi V_c$, where V_c is the nominal shear permitted on the concrete), the minimum shear reinforcement required and its location in a wall are the same as for bearing walls (Art. 9.67). Maximum spacing of horizontal shear reinforcement, however, should not exceed $L_w/5$, $3h$, or 18 in, where L_w is the horizontal length of wall and h the overall wall thickness (Fig. 9.56). Maximum spacing of the vertical reinforcement should not exceed $L_w/3$, $3h$, or 18 in.

A thickness of at least $L_w/25$ is advisable for walls with high shear.

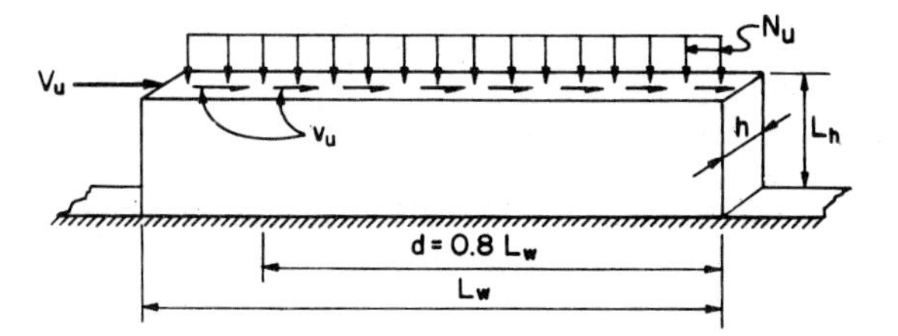

FIGURE 9.56 Shear and normal forces acting on a longitudinal section through a shear wall.

The factored horizontal shear force V_u acting on a section through the shear wall must not exceed the nominal shear strength V_n multiplied by $\phi = 0.85$.

$$V_u \leq (\phi V_n = \phi V_c + \phi V_s) \tag{9.111}$$

where V_c = nominal shear strength of the concrete and V_s = nominal shear strength provided by reinforcement. The horizontal shear strength at any section should not be taken larger than

$$V_n = 10\sqrt{F'_c}\,hd \tag{9.112}$$

where f'_c = specified concrete strength, psi
d = effective depth of wall, but not to be taken larger than 80% of the wall length
h = wall thickness

Shear carried by the concrete should not exceed the smaller of the values of V_c computed from Eq. (9.113) or (9.114).

$$V_c = 3.3\sqrt{f'_c}\,hd + \frac{N_u d}{4L_w} \tag{9.113}$$

where N_u = factored vertical axial load on wall acting with V_u, including tension due to shrinkage and creep (positive for compression, negative for tension).

$$V_c = \left[0.6\sqrt{f'_c} + \frac{L_w(1.25\sqrt{f'_c} + 0.2N_u/L_w h)}{M_u/V_u - L_w/2}\right]hd \tag{9.114}$$

where M_u = factored moment at section where V_u acts.

Alternatively, $V_c = 2\sqrt{f'_c}\,hd$ may be used if N_u causes compression. Shear strength V_c computed for a section at a height above the base equal to $L_w/2$ or one-half the wall height, whichever is smaller, may be used for all lower sections.

When $V_u > 0.5\phi V_c$, the area of horizontal shear reinforcement within a distance s_2 required for shear is given by

$$A_h = \frac{(V_u/\phi - V_s)s_2}{f_y d} \geq 0.0025hs_2 \tag{9.115}$$

where s_2 = spacing of horizontal reinforcement (max $\leq L_w/5 \leq 3h \leq 18$ in) and f_y = yield strength of the reinforcement.

Also, when $V_u > 0.5\phi V_c$, the area of vertical shear reinforcement with spacing s should be at least

$$A_{vh} = \left[0.0025 + 0.5\left(2.5 - \frac{L_h}{L_w}\right)\left(\frac{A_h}{nL_h} - 0.0025\right)\right] hs \geq 0.0025hs \tag{9.116}$$

where L_h is the wall height. But A_{vh} need not be larger than A_h computed from Eq. (9.115). Spacing s should not exceed $L_w/3$, $3h$, or 18 in.

9.89 REINFORCED-CONCRETE ARCHES

Arches are used in roofs for such buildings as hangars, auditoriums, gymnasiums, and rinks, where long spans are desired. An arch is essentially a curved beam with the loads, applied downward in its plane, tending to decrease the curvature. Arches are frequently used as the supports for thin shells that follow the curvature of the arches. Such arches are treated in analysis as two-dimensional, whereas the thin shells behave as three-dimensional elements.

The great advantage of an arch in reinforced concrete construction is that, if the arch is appropriately shaped, the whole cross section can be utilized in compression under the maximum (full) load. In an ordinary reinforced concrete beam, the portion below the neutral axis is assumed to be cracked and does not contribute to the bending strength. A beam can be curved, however, to make its axis follow the lines of thrust very closely for all loading conditions, thus virtually eliminating bending moments.

The component parts of a fixed arch are shown in Fig. 9.57. For a discussion of the different types of arches and the stress analyses required for each, see Art 5.14.

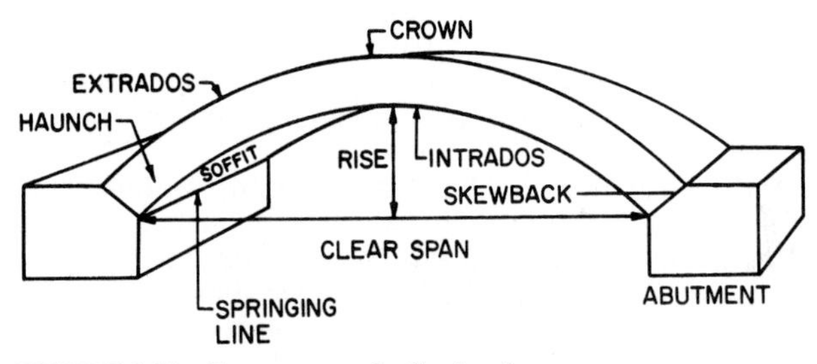

FIGURE 9.57 Components of a fixed arch.

Because the depth of an arch and loading for maximum moments generally vary along the length, several cross sections must be chosen for design, such as the crown, springing, haunches, and the quarter points. Concrete compressive stresses and shear should be checked at each section, and reinforcement requirements determined. The sections should be designed as rectangular beams or T beams subjected to bending and axial compression, as indicated in Arts. 9.81 to 9.83.

When an arch is loaded, large horizontal reactions, as well as vertical reactions, are developed at the supports. For roof arches, tie rods may be placed overhead, or in or under the ground floor, to take the horizontal reaction. The horizontal reaction may also be resisted externally by footings on sound rock or piles, by reinforced concrete buttresses, or by adjoining portions of the structure, for example a braced floor or roof at springing level.

Hinged arches are commonly made of structural steel or precast concrete. The hinges simplify the arch analysis and the connection to the abutment, and they reduce the indeterminate stresses due to shrinkage, temperature, and settlements of supports. For cast-in-place concrete, hingeless (fixed) arches are often used. They eliminate the cost of special steel hinges needed for hinged concrete arches and permit reduced crown thicknesses, to provide a more attractive shape.

Arches with spans less than 90 ft are usually constructed with ribs 2 to 4 ft wide. Each arch rib is concreted in a continuous operation, usually in 1 day. The concrete may be placed continuously from each abutment toward the crown, to obtain symmetrical loading on the falsework.

For spans of 90 ft or more, however, arch ribs are usually constructed by the alternate block, or voussoir, method. Each rib is constructed of blocks of such size that each can be completed in one casting operation. This method reduces the shrinkage stresses. The blocks are cast in such order that the formwork will settle uniformly. If blocks close to the crown section are not placed before blocks at the haunch and the springing sections, the formwork will rise at the crown, and placing of the crown blocks will then be likely to cause cracks in the haunch. The usual procedure is to cast two blocks at the crown, then two at the springing, and alternate until the complete arch is concreted.

In construction by the alternate block method, the block sections are kept separate by timber bulkheads. The bulkheads are kept in place by temporary struts between the voussoirs. Keyways left between the voussoirs are concreted later. Near piers and abutments where the top slopes exceed about 30° with the horizontal, top forms may be necessary, installed as the casting progresses.

If the arch reinforcement is laid in long lengths, settlement and deformation of the arch formwork can displace the steel. Therefore, depending on the curvature and total length, lengths of bars are usually limited to about 30 ft. Splicing should be located in the keyways. Splices of adjacent bars should be staggered (50% stagger), and located where tension is small.

Upper reinforcement in arch rings may be held in place with spacing boards nailed to props, or with wires attached to transverse timbers supported above the surface of the finished concrete.

Forms for arches may be supported on a timber falsework bent. This bent may consist of joists and beams supported by posts that are braced together and to solid ground. Wedges or other adjustment should be provided at the base of the posts so that the formwork may be adjusted if settlement occurs, and so that the entire formwork may be conveniently lowered after the concrete has hardened sufficiently to take its own load.

("Guide to Formwork for Concrete," ACI 347R, American Concrete Institute.)

9.90 REINFORCED-CONCRETE THIN SHELLS

Thin shells are curved slabs with thickness very small compared with the other dimensions. A thin shell possesses three-dimensional load-carrying characteristics. The best natural example of thin-shell behavior is that of an ordinary egg, which may have a ratio of radius of curvature to thickness of 50. Loads are transmitted through thin shells primarily by direct stresses—tension or compression—called membrane stresses, which are almost uniform throughout the thickness. Reinforced-concrete thin-shell structures commonly utilize ratios of radius of curvature to thickness about 5 times that of an eggshell. Because concrete shells are always reinforced, their thickness is usually determined by the minimum thickness required to cover the reinforcement, usually 1 to 4 in. Shells are thickened near the supports to withstand localized bending stresses in such areas. (See also Art 5.15.)

Shells are most often used as roofs for such buildings as hangars, garages, theaters, and arenas, where large spans are required and the loads are light. The advantages of reinforced-concrete thin shells may be summarized as follows:

Most efficient use of materials.

Great freedom of architectural shapes.

Convenient accommodation of openings for natural lighting and ventilation.

Ability to carry very large unbalance of forces.

High fireproofing value due to lack of corners, thin ribs, and the inherent fire resistance of reinforced concrete.

Reserve strength due to many alternative paths for carrying load to the supports. One outstanding example withstood artillery fire punctures with only local damage.

Common shapes of reinforced-concrete thin shells used include cylindrical (barrel shells), dome, grained vault, or groinior, elliptical paraboloid, and hyperbolic paraboloid (saddle shape).

Cylindrical shells may be classified as long if the radius of curvature is shorter than the span, or as short (Fig. 9.58). Long cylindrical shells, particularly the continuous, multiple-barrel version which repeats the identical design of each bay (and permits reuse of formwork) in both directions, are advantageous for roofing rectangular-plan structures. Short cylindrical shells are commonly used for hangar roofs with reinforced-concrete arches furnishing support at short intervals in the direction of the span.

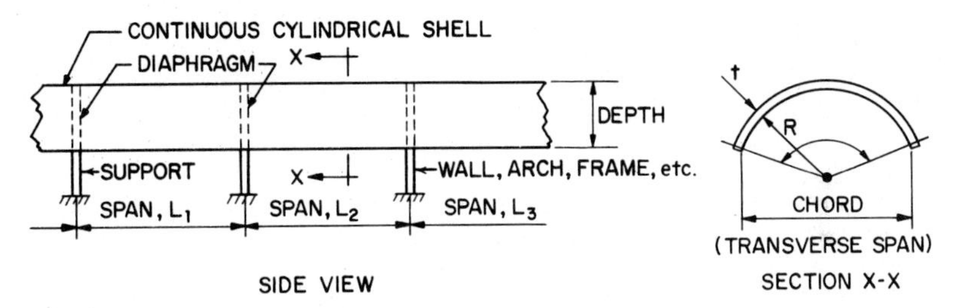

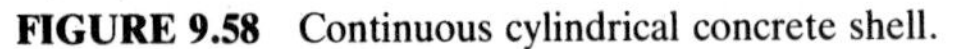

FIGURE 9.58 Continuous cylindrical concrete shell.

Structural analysis of these common styles may be simplified with design aids. ("Design of Cylindrical Concrete Shell Roofs," Manual No. 31, American Society of Civil Engineers; "Design Constants for Interior Cylindrical Concrete Shells," EB020D: "Design Constants for Ribless Concrete Cylindrical Shells," EB028D; "Coefficients for Design of Cylindrical Concrete Shell Roofs" (extension of ASCE Manual No. 31), EB035D; "Design of Barrel Shell Roofs," IS082D, Portland Cement Association; "Concrete Shell Structures—Practice and Commentary," ACI 334.1R, American Concrete Institute.

The ACI Building Code includes specific provisions for thin shells. It allows an elastic analysis as an accepted basis for design and suggests model studies for complex or unusual shapes, prescribes minimum steel, and prohibits use of the working-stress method for design, thus prescribing selection of all shear and flexural reinforcement by the strength-design method with the same load factors as for design of other elements. Figure 9.59 shows a typical reinforcement arrangement

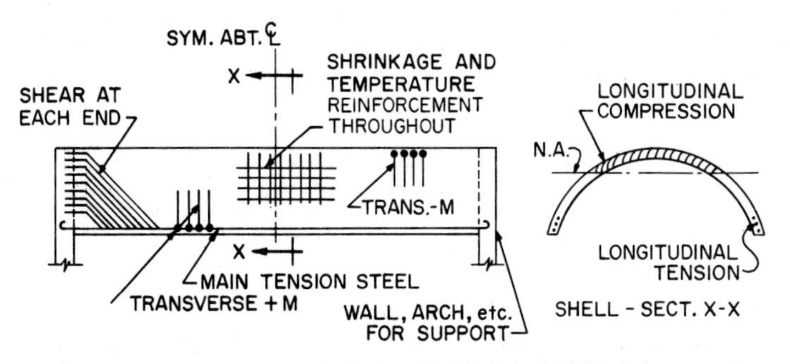

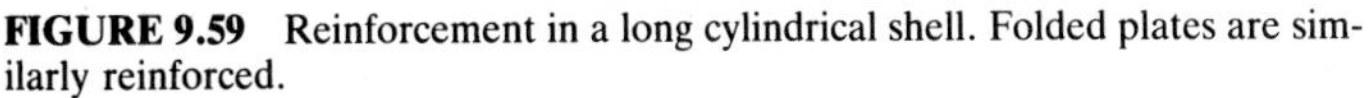

FIGURE 9.59 Reinforcement in a long cylindrical shell. Folded plates are similarly reinforced.

for a long cylindrical shell. (See also F. S. Merritt, "Standard Handbook for Civil Engineers," Sec. 8, "Concrete Design and Construction," and D. P. Billington, "Thin-Shell Concrete Structures," 2d ed., McGraw-Hill Publishing Company, New York.)

9.91 CONCRETE FOLDED PLATES

Reinforced-concrete, folded-plate construction is a versatile concept applicable to a variety of long-span roof construction. Applications using precast, simple V folded plates include segmental construction of domes and (vertically) walls (Fig. 9.60). Inverted folded plates have also been widely used for industrial storage bins. ("Standard Practice for Design and Construction of Concrete Silos and Stacking Tubes for Storing Granular Materials," ACI 313, and "Commentary," ACI 313R, American Concrete Institute.) Determination of stresses in folded-plate construction is described in Art. 5.15.5.

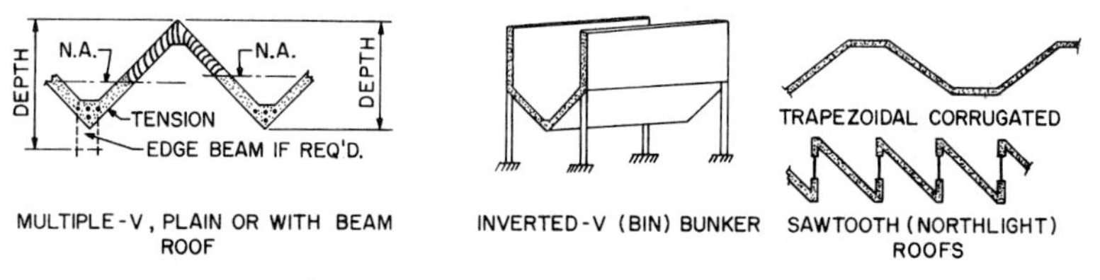

FIGURE 9.60 Typical shapes of concrete folded plates.

Formwork for folded plates is far simpler than that for curved thin shells. Precasting has also been a simpler process to save formwork, permit mass-production construction, and achieve sharp lines for exposed top corners (vees cast upside down) to satisfy aesthetic requirements. For very long spans, posttensioned, draped tendons have been employed to reduce the total depth, deflection, and reinforcing-steel requirements. The tendons may be placed in the inclined plates or, more conveniently, in small thickened edge beams. For cast-in-place, folded-plate construction, double forming can usually be avoided if the slopes are less than 35 to 40°.

Since larger transverse bending moments develop in folded plates than in cylindrical shells of about the same proportions, a minimum thickness less than 4 in creates practical problems of placing the steel. A number of areas in the plates will require three layers of steel and, near the intersections of plates, top and bottom bars for transverse bending will be required. Ratios of span to total depth are similar to those for cylindrical shells, commonly ranging from 8 to 15. (See also F. S. Merritt, "Standard Handbook for Civil Engineers," Sec. 8, "Concrete Design and Construction," McGraw-Hill Publishing Company, New York.)

9.92 SLABS ON GRADE

Slabs on ground are often used as floors in buildings. Special use requirements often include heavy-duty floor finish (Art 9.35) and live-load capacity for heavy concentrated (wheel) loads or uniform (storage) loads, or both.

Although slabs on grade seem to be simple structural elements, analysis is extremely complicated. Design is usually based on rules of thumb developed by experience. For design load requirements that are unusually heavy and outside common experience, design aids are available. Occasionally, the design will be controlled by wheel loads only, as for floors in hangars, but more frequently by uniform warehouse loadings. ("Guide for Design of Slabs on Grade," ACI 360R; American Concrete Institute; "Concrete Floors on Ground," EB075D, Portland Cement Association; "Design of Floors on Ground for Warehouse Loadings," Paul F. Rice, *ACI Journal*, August 1957, paper No. 54-7.)

A full uniform load over an entire area causes no bending moment if the boundaries of the area are simple construction joints. But actual loads in warehouse usage leave unloaded aisles and often alternate panels unloaded. As a result, a common failure of warehouse floors results from *uplift* of the slab off the subgrade, causing negative moment (top) cracking. In lieu of a precise analysis taking into account live-load magnitude, joint interval and detail, the concrete modulus of elasticity, the soil modulus, and load patterns, a quick solution to avoid uplift is to provide a slab sufficiently thick so that its weight is greater than one-fifth the live load. Such a slab may be unreinforced, if properly jointed, or reinforced for temperature and shrinkage stresses only. Alternatively, for very heavy loadings, an analysis and design may be performed for the use of reinforcement, top and bottom, to control uplift moments and cracking. ("Design of Floors on Ground for Warehouse Loadings," Paul F. Rice, *ACI Journal*, Aug., 1957, paper No. 53-7.)

Shrinkage and temperature change in slabs on ground can combine effects adversely to create warping, uplift, and top cracking failures with no load. Closely spaced joint intervals, alternate-panel casting sequence, and controlled curing will avoid these failures. Somewhat longer joint spacings can be employed if reinforcement steel with an area of about 0.002 times the gross section area of slab is provided in perpendicular directions.

With such reinforcement, warping will usually be negligible if the slab is cast in alternate lanes 12 to 14 ft wide, and provided with contraction joints at 20- to 30-ft spacings in the direction of casting. The joints may be tooled, formed by joint filler inserts, or sawed. One-half the bars or wires crossing the contraction joints should be cut accurately on the joint line. The warping effect will be aggravated if excess water is used in the concrete and it is forced to migrate in one direction to top or bottom of the slab, for example, when the slab has been cast on a vapor barrier or on a very dry subgrade. For very long slabs, continuous reinforcement, approximately 0.006 times the gross area, is used to eliminate transverse joints in highway and airport pavement. ("Design of Continuously Reinforced Concrete for Highways" and "Construction of Continuously Reinforced Concrete Pavements," Concrete Reinforcing Steel Institute; and "Suggested Specifications for Heavy-duty Concrete Floor Topping," IS021B; "Design of Concrete Floors on Ground," IS046B; "Suggested Specifications for Single-course Floors on Ground," IS070B, Portland Cement Association.)

9.93 SEISMIC-RESISTANT CONCRETE CONSTRUCTION

The American Concrete Institute Building Code contains special seismic requirements for design that apply only for areas where the probability of earthquakes capable of causing major damage to structures is high, and where ductility reduction factors for lateral seismic loads are utilized (ACI 318-89, Chap. 21). The general requirements of ACI 318-89 for reinforced concrete provide sufficient seismic resistance for seismic zones where only minor seismic damage is probable and no reduction factor for ductility is applied to seismic forces. Designation of seismic zones is prescribed in general building codes, as are lateral force loads for design. (See also Art. 5.18.7.)

Special ductile-frame design is prescribed to resist lateral movements sufficiently to create "plastic" hinges and permit reversal of direction several times. These hinges must form in the beams at the beam-column connections of the ductile frame.

Shear walls used alone or in combination with ductile beam-column frames must also be designed against brittle (shear) failures under the reversing loads ("Commentary on ACI 318-89").

Ductility is developed in reinforced concrete by:

Conservative limits on the net flexural tension-steel ratio $\rho \leq 0.025$, to ensure underreinforced behavior. At least two continuous bars must be provided at both top and bottom of flexural members.

Heavy confining reinforcement extending at joints through the region of maximum moment in both columns and beams, to include points where hinges may form. This confining reinforcement may consist of spiral reinforcing or heavy, closely spaced, well-anchored, closed ties with hooked ends engaging the vertical bars or the ties at the far face.

("ACI Detailing Manual," SP-66, American Concrete Institute.)

9.94 COMPOSITE FLEXURAL MEMBERS

Reinforced- and prestressed-concrete, composite flexural members are constructed from such components as precast members with cast-in-place flanges, box sections, and folded plates.

Composite structural-steel-concrete members are usually constructed of cast-in-place slabs and structural-steel beams. Interaction between the steel beam and concrete slab is obtained by natural bond if the steel beam is fully encased with a minimum of 2 in of concrete on the sides or soffit. If the beam is not encased, the interaction may be accomplished with mechanical anchors (shear connectors). Requirements for composite structural-steel-concrete members are given in the AISC "Specification for Structural Steel for Buildings—Allowable Stress Design and Plastic Design," American Institute of Steel Construction.

The design strength of composite flexural members is the same for both shored and unshored construction. Shoring should not be removed, however, until the supported elements have the design properties required to support all loads and limit deflections and cracking. Individual elements should be designed to support all loads prior to the full development of the design strength of the composite member. Premature loading of individual precast elements can cause excessive deflections as the result of creep and shrinkage.

The factored horizontal shear force for a composite member may be transferred between individual elements by contact stresses or anchored ties, or both. The factored shear force V_u at the section considered must be equal to or less than the nominal horizontal shear strength V_{nh} multiplied by $\phi = 0.85$.

$$V_u \leq \phi V_{nh} \tag{9.117}$$

When $V_u \leq 80b_v d$, where b_v is the section width and d the distance from the extreme compression surface to the centroid of tension steel, the factored shear force may be transferred by contact stresses without ties, if the contact surfaces are clean and intentionally roughened with a full amplitude of about ¼ in. Otherwise, fully anchored minimum ties [Eq. (9.82)], spaced not over 24 in or 4 times the least dimension of the supported element, may be used.

When V_u is between $80b_v d$ and $350b_v d$ the factored shear force may be transferred by shear-friction reinforcement placed perpendicular to assumed cracks. Shear force V_u should not exceed $800b_v d$ or $0.2f'_c b_v d$, where f'_c is the specified concrete strength. Required reinforcement area is

$$A_{vf} = \frac{V_u}{\phi f_y \mu} \tag{9.118}$$

where f_y = yield strength of shear reinforcement
μ = coefficient of friction
= 1.4λ for monolithic concrete
= 1.0λ for concrete cast against hardened concrete with surface intentionally roughened to a full amplitude of about 0.25 in
= 0.7λ for concrete anchored by headed studs or rebars to as-rolled structural steel (clean and without paint)
= 0.6λ for concrete cast against hardened concrete not intentionally roughened
λ = 1.0 for normal-weight concrete
= 0.85 for sand-lightweight concrete
= 0.75 for all-lightweight concrete

PRECAST-CONCRETE MEMBERS

Precast-concrete members are assembled and fastened together on the job. They may be unreinforced, reinforced, or prestressed. Precasting is especially advantageous when it permits mass production of concrete units. But precasting is also beneficial because it facilitates quality control and use of higher-strength concrete. Form costs may be greatly reduced, because reusable forms can be located on a casting-plant floor or on the ground at a construction site in protected locations and convenient positions, where workmen can move about freely. Many complex thin-shell structures are economical when precast, but would be uneconomical if cast in place.

9.95 DESIGN METHODS FOR PRECAST MEMBERS

Design of precast-concrete members under the American Concrete Institute Building Code, ACI 318, follows the same rules as for cast-in-place concrete. In some cases, however, design may not be governed by service loads, because transportation and erection loads on precast members may exceed the service loads.

Design of joints and connections must provide for transmission of any forces due to shrinkage, creep, temperature, elastic deformation, gravity loads, wind loads, and earthquake motion.

("Design and Typical Details of Connections for Precast and Prestressed Concrete," 2d ed., Precast/Prestressed Concrete Institute.)

9.96 REINFORCEMENT COVER IN PRECAST MEMBERS

Less concrete cover is required for reinforcement in precast-concrete members manufactured under plant control conditions than in cast-in-place members because the control for proportioning, placing, and curing is better. Minimum concrete cover for reinforcement required by ACI 318-89 is listed in Table 9.26.

For all sizes of reinforcement in precast-concrete wall panels, minimum cover of ¾ in is acceptable at nontreated surfaces exposed to weather and ⅜ in at interior surfaces.

TABLE 9.26 Minimum Reinforcement Cover for Precast Members, in

Concrete exposed to earth or weather:	
Wall panels:	
No. 14 and No. 18 bars	1½
No. 11 and smaller	¾
Other members:	
No. 14 and No. 18 bars	2
No. 6 through No. 11 bars	1½
No. 5 bars, ⅝-in wire and smaller	1¼
Concrete not exposed to weather or in contact with the ground:	
Slabs, walls, joists:	
No. 14 and No. 18 bars	1¼
No. 11 and smaller	⅝
Beams, girders, columns:	
Principal reinforcement:	
Diameter of bar d_b but not less than ⅝ in, and need not be more than 1½ in	
Ties, stirrups or spirals	⅜
Shells and folded-plate members:	
No. 6 bars and larger	⅝
No. 5 bars, ⅝-in wire and smaller	⅜

9.97 TOLERANCES FOR PRECAST CONSTRUCTION

Dimensional tolerances for precast members and tolerances on fitting of precast members vary for type of member, type of joint, and conditions of use. See "PCI Design Handbook," and "Design and Typical Details of Connections for Precast and Prestressed Concrete," Precast/Prestressed Concrete Institute, "ACI Detailing Manual," SP-66, American Concrete Institute.

9.98 ACCELERATED CURING

For strength and durability, precast concrete members require adequate curing. They usually are given some type of accelerated curing for economic reuse of forms and casting space. At atmospheric pressure, curing temperatures may be held between 125 and 185°F for 12 to 72 h. Under pressure, autoclave temperatures above 325°F for 5 to 36 h are applied for fast curing. Casting temperatures, however, should not exceed 90°F. See Fig. 9.5.

("Standard Practice for Curing Concrete," ACI 308; "Accelerated Curing of Concrete at Atmospheric Pressure—State of the Art," ACI 517.2R, American Concrete Institute.)

9.99 PRECAST FLOOR AND ROOF SYSTEMS

Long-span, precast-concrete floor and roof units are usually prestressed. Short members, 30 ft or less, are often made with ordinary reinforcement. Types of precast units for floor and roof systems include solid or ribbed slabs, hollow-core slabs, single and double tees, rectangular beams, L-shaped beams, inverted-T beams, and I beams.

Hollow-core slabs are usually available in normal-weight or structural lightweight concrete. Units range from 16 to 96 in in width, and from 4 to 12 in in depth. Hollow-core slabs may come with grouted shear keys to distribute loads to adjacent units over a slab width as great as one-half the span.

Manufacturers should be consulted for load and span data on hollow-core slabs, because camber and deflection often control the serviceability of such units, regardless of strength.

("PCI Design Handbook," Precast Prestressed Concrete Institute.)

9.100 PRECAST RIBBED SLABS, FOLDED PLATES, AND SHELLS

Curved shells and folded plates have a thickness that is small compared with their other dimensions. Such structures depend on their geometrical configuration and boundary conditions for strength.

Thickness. With closely spaced ribs or folds, a minimum thickness for plane sections of 1 in is acceptable.

Reinforcement. Welded-wire fabric with a maximum spacing of 2 in may be used for slab portions of thin-section members, and for wide, thin elements 3 in thick or less. Reinforcement should be preassembled into cages, using a template, and placed within a tolerance of +0 in or −⅛ in from the nearest face. The minimum clear distance between bars should not be less than 1½ times the nominal maximum size of the aggregate. For minimum cover of reinforcement, see Art. 9.98.

Compressive Strength. Concrete for thin-section, precast-concrete members protected from the weather and moisture and not in contact with the ground should have a strength of at least 4000 psi at 28 days. For elements in other locations, a minimum of 5000 psi is recommended.

Analysis. Determination of axial stresses, moments, and shears in thin sections is usually based on the assumption that the material is ideally elastic, homogeneous, and isotropic.

Forms. Commonly used methods for the manufacture of thin-section, precast-concrete members employ metal or plastic molds, which form the bottom of the slab and the sides of the boundary members. Forms are usually removed pneumatically or hydraulically by admitting air or water under pressure through the bottom form.

("Architectural Precast Concrete," Precast/Prestressed Concrete Institute.)

9.101 WALL PANELS

Precast-concrete wall panels include plain panels, decorative panels, natural stone-faced panels, sandwich panels, solid panels, ribbed panels, tilt-up panels, load-bearing and non-load-bearing panels, and thin-section panels. Prestressing, when used with such panels, makes it possible to handle and erect large units and thin sections without cracking.

Forms required to produce the desired size and shape of panel are usually made of steel, wood, concrete, vacuum-formed thermoplastics, fiber-reinforced plastics, or plastics formed into shape by heat and pressure, or any combination of these. For complicated form details, molds of plaster, gelatin, or sculptured sand can be used.

Glossy-smooth concrete finish can be obtained with forms made of plastic. But, for exterior exposure, this finish left untreated undergoes gradual and nonuniform loss of its high reflectivity. Textured surfaces or smooth but nonglossy surfaces obtained by early form removal are preferred for exterior exposure.

Exposed-aggregate monolithic finishes can be obtained with horizontal-cast panels by initially casting a thin layer containing the special surface aggregates in the forms and then casting regular concrete backup. With a thickness of exposed aggregate of less than 1 in, the panel can also be cast face up and the aggregate seeded over the fresh concrete or hand placed in a wet mortar. Variations of exposed surface can be achieved by use of set retardant, acid washes, or sandblasting.

Consolidation of the concrete in the forms to obtain good appearance and durability can be attained by one of the following methods:

External vibration with high-frequency form vibrators or a vibrating table.

Internal or surface vibration with a tamping-type or jitterbug vibrator.

Placing a rich, high-slump concrete in a first layer to obtain uniform distribution of the coarse aggregate and maximum consolidation, and then making the mix for the following layers progressively stiffer. This allows absorption of excessive water from the previous layer.

Tilt-up panels can be economical if the floor slab of the building can be designed for and used as the form for the panels. The floor slab must be level and smoothly troweled. Application of a good bond-breaking agent to the slab before concrete is cast for the panels is essential to obtain a clean lift of the precast panels from the floor slab.

If lifting cables are attached to a panel edge, large bending moments may develop at the center of the wall. For high panels, three-point pickup may be used. To spread pickup stresses, specially designed inserts are cast into the wall at pickup points.

Another method of lifting wall panels employs a vacuum mat—a large steel mat with a rubber gasket at its edges to contact the slab. When the air between mat and panel is pumped out, the mat adheres to the panel, because of the resulting vacuum, and can be used to raise the panel. The method has the advantage of spreading pickup forces over the mat area.

Panels, when erected, must be temporarily braced until other construction is in place to provide required permanent bracing.

Joints. Joint sealants for panel installations may be mastics or elastomeric materials. These are extensible and can accommodate the movement of panels.

Recommended maximum joint widths and minimum expansions for the common sealants are listed in Table 9.27.

TABLE 9.27 Maximum Joint Widths for Sealants

Type of sealant	Maximum joint width, in	Maximum movement, tension, and compression, %
Butyl	¾	±10
Acrylic	¾	±15–25
One-part polyurethane	¾	±20
Two-part polyurethane	¾	±25
One-part polysulfide	¾	±25
Two-part polysulfide	¾	±25

The joint sealant manufacturer should be asked to advise on backup material for use with a sealant and which shape factor should be considered. A good backup material is a rod of sponge material with a minimum compression of 30%, such as foamed polyethylene, polystyrene, polyurethane, polyvinyl chloride, or synthetic rubber.

("PCI Manual for Structural Design of Architectural Precast Concrete," Precast/Prestressed Concrete Institute.)

9.102 LIFT SLABS

Lift slabs are precast-concrete floor and roof panels that are cast on a base slab at ground level, one on top of the other, with a bond-breaking membrane between them. Steel collars are embedded in the slabs and fit loosely around the columns. After the slabs have cured, they are lifted to their final position by a patented jack system supported on the columns. The embedded steel collars then are welded to the steel columns to hold the lift slabs in place. This method of construction eliminates practically all formwork.

PRESTRESSED-CONCRETE CONSTRUCTION

Prestressed concrete is concrete in which internal stresses have been introduced during fabrication to counteract the stresses produced by service loads. The prestress compresses the tensile area of the concrete to eliminate or make small the tensile stresses caused by the loads.

9.103 BASIC PRINCIPLES OF PRESTRESSED CONCRETE

In the application of prestress, the usual procedure is to tension high-strength-steel elements, called tendons, and anchor them to the concrete, which resists the tendency of the stretched steel to shorten after anchorage and is thus compressed. If the tendons are tensioned before concrete has been placed, the prestressing is called **pretensioning.** If the tendons are tensioned after the concrete has been placed the prestressing is called **posttensioning.**

Prestress can prevent cracking by keeping tensile stresses small, or entirely avoiding tension under service loads. The entire concrete cross section behaves as an uncracked homogeneous material in bending. In contrast, in nonprestressed, reinforced-concrete construction, tensile stresses are resisted by reinforcing steel, and concrete in tension is considered ineffective. It is particularly advantageous with prestressed concrete to use high-strength concrete.

Loss of Prestress. The final compression force in the concrete is not equal to the initial tension force applied by the tendons. There are immediate losses due to elastic shortening of the concrete, friction losses from curvature of the tendons, and slip at anchorages. There are also long-time losses, such as those due to shrinkage and creep of the concrete, and possibly relaxation of the steel. These losses should be computed as accurately as possible or determined experimentally. They are deducted from the initial prestressing force to determine the effective prestressing force to be used in design. (The reason that high-strength steels must be used for prestresssing is to maintain the sum of these strain losses at a small percentage of the initially applied prestressing strain.) (See also Art. 9.106.)

Stresses. When stresses in prestressed members are determined, prestressing forces can be treated as other external loads. If the prestress is large enough to prevent cracking under design loads, elastic theory can be applied to the entire concrete cross section (Fig. 9.61).

Prestress may be applied to a beam by straight tendons or curved tendons. Stresses at midspan can be the same for both types of tendons, but the net stresses with the curved tendons can remain compressive away from midspan, whereas they become tensile at the top fiber near the ends with straight tendons. For a prestressing

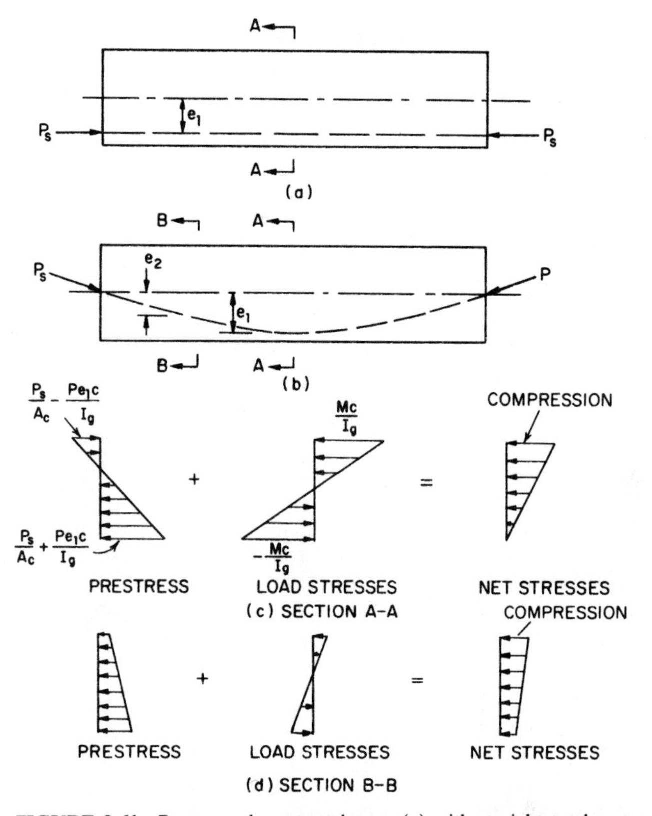

FIGURE 9.61 Prestressed-concrete beam: (*a*) with straight tendons; (*b*) with curved tendons; (*c*) midspan stresses with straight or curved tendons; (*d*) stresses between midspan and supports with curved tendons. Net stresses near the supports become tensile with straight tendons.

force P_s applied to a beam by a straight tendon at a distance e_1 below the neutral axis, the resulting prestress in the extreme surface throughout is

$$f = \frac{P_s}{A_c} \pm \frac{P_s e_1 c}{I_g} \tag{9.119}$$

where P_s/A_c is the compressive stress on a cross section of area A_c, and $P_s e_1 c/I_g$ is the bending stress induced by P_s (positive for compression and negative for tension), as indicated in Fig. 9.61. If stresses $\pm Mc/I_g$ due to moment M caused by external gravity loads are superimposed at midspan, the net stresses in the extreme fibers can become zero at the bottom and compressive at the top. Because the stresses due to gravity loads are zero at the beam ends, the prestress is the final stress there and the top surface of the beam at the ends is in tension.

If the tensile stresses at the ends of beams with straight tendons are excessive, the tendons may be draped, or harped, in a vertical curve. Stresses at midspan will be substantially the same as with straight tendons (if the horizontal component of prestress is nearly equal to P_s), and the stresses at the beam ends will be compressive, because the prestressing force passes through or above the centroid of the end sections (Fig. 9.61). Between midspan and the ends, the cross sections will also be in compression.

9.104 LOSSES IN PRESTRESS

Assumption in design of total losses in tendon stress of 35,000 psi for pretensioning and 25,000 psi for posttensioning to allow for elastic shortening, frictional losses, slip at anchorages, shrinkage, creep, and relaxation of steel usually gives satisfactory results. Losses greater or smaller than these values have little effect on the design strength but can affect service-load behavior, such as cracking load, deflection, and camber.

Elastic Shortening of Concrete. In pretensioned members, when the tendons are released from fixed abutments and the steel stress is transferred to the concrete by bond, the concrete shortens under the compressive stress. The decrease in unit stress in the tendons equals $P_sE_s/A_cE_c = nf_c$, where E_s is the modulus of elasticity of the steel, psi; E_c the modulus of elasticity of the concrete psi; n the modular ratio, E_s/E_c; f_c the unit stress in the concrete, psi; P_s the prestressing force applied by the tendons; and A_c the cross-sectional area of the member.

In posttensioned members, the loss due to elastic shortening can be eliminated by using the members as a reaction in tensioning the tendons.

Frictional Losses. In posttensioned members, there may be a loss of prestress where curved tendons rub against their enclosure. The loss may be computed in terms of a curvature-friction coefficient μ. Losses due to unintentional misalignment may be calculated from a wobble-friction coefficient K (per lin ft). Since the coefficients vary considerably, they should, if possible, be determined experimentally. A safe range of these coefficients for estimates is given in the "Commentary on ACI 318-89," American Concrete Institute.

Frictional losses can be reduced by tensioning the tendons at both ends, or by initial use of a larger jacking force which is then eased off to the required initial force for anchorage.

Slip at Anchorages. For posttensioned members, prestress loss may occur at the anchorages during the anchoring. For example, seating of wedges may permit some shortening of the tendons. If tests of a specific anchorage device indicate a shortening δl, the decrease in unit stress in the steel is equal to $E_s\delta l/l$, where l is the length of the tendon. This loss can be reduced or eliminated by overtensioning initially by an additional strain equal to the estimated shortening.

Shrinkage of Concrete. Change in length of a member caused by concrete shrinkage results in a prestress loss over a period of time. This change can be determined from tests or experience. Generally, the loss is greater for pretensioned members than for posttensioned members, which are prestressed after much of the shrinkage has occurred. Assuming a shrinkage of 0.0002 in/in of length for a pretensioned member, the loss in tension in the tendons is $0.0002E_s = 0.0002 \times 30 \times 10^6 =$ 6000 psi.

Creep of Concrete. Change in length of concrete under sustained load induces a prestress loss proportional to the load over a period of time depending greatly on the aggregate used. This loss may be several times the elastic shortening. An estimate of this loss may be made with an estimated creep coefficient C_{cr} equal to the ratio of additional long-time deformation to initial elastic deformation determined by test. The loss in tension for axial prestress in the steel is, therefore, equal to $C_{cr}nf_c$. Values ranging from 1.5 to 2.0 have been recommended for C_{cr}.

Relaxation of Steel. A decrease in stress under constant high strain occurs with some steels. Steel tensioned to 60% of its ultimate strength may relax and lose as much as 3% of the prestressing force. This type of loss may be reduced by temporary overtensioning, which artificially accelerates relaxation, reducing the loss that will occur later at lower stresses.

(P. Zia et al., "Estimating Prestress Loss," *Concrete International,* June 1979, p. 32, American Concrete Institute; "PCI Design Handbook," Precast/Prestressed Concrete Institute.)

9.105 ALLOWABLE STRESSES AT SERVICE LOADS

At service loads and up to cracking loads, straight-line theory may be used for computing stresses in prestressed beams with the following assumptions:

Strains vary linearly with depth through the entire load range.

At cracked sections, the concrete does not resist tension.

Areas of unbonded open ducts should not be considered in computing section properties.

The transformed area of bonded tendons and reinforcing steel may be included in pretensioned members and, after the tendons have been bonded by grouting, in posttensioned members.

Flexural stresses must be limited to ensure proper behavior at service loads. Limiting these stresses, however, does not ensure adequate design strength.

In establishing permissible flexural stresses, the American Concrete Institute Building Code, ACI 318, recognizes two service-load conditions, that before and that after prestress losses. Higher stresses are permitted for the initial state (temporary stresses) than for loadings applied after the losses have occurred.

Permissible stresses in the concrete for the initial load condition are specified as a percentage of f'_{ci}, the compressive strength of the concrete, psi, at time of initial prestress. This strength is used as a base instead of the usual f'_c, 28-day strength of concrete, because prestress is usually applied only a few days after concrete has been cast. The allowable stresses for prestressed concrete, as given in ACI 318-89, are tabulated in Table 9.28.

Bearing Stresses. Determination of bearing stresses at end regions around posttensioning anchorages is complicated, because of the elastic and inelastic behavior of the concrete and because the dimensions involved preclude simple analysis under the St. Venant theory of linear stress distribution of concentrated loads.

Lateral reinforcement may be required in anchorage zones to resist bursting, horizontal splitting, and spalling forces. Anchorages are usually supplied by a prestressing company and are usually designed by test and experience rather than theory.

TABLE 9.28 Allowable Stresses for Prestressed Concrete

Concrete:	
Temporary stresses after transfer of prestress but before prestress losses:	
Compression	$0.60f'_{ci}$
Tension in members without auxiliary reinforcement in tension zone	$3\sqrt{f'_{ci}}$*
Service-load stresses after prestress losses:	
Compression	$0.45f'_c$
Tension in precompressed tensile zone	$6\sqrt{f'_c}$†
Prestressing steel:	
Due to jacking force	$0.94f_{py}$‡
Pretensioning tendons immediately after transfer or posttensioning tendons immediately after anchoring	$0.70f_{pu}$‡

*Where the calculated tension stress exceeds this value, reinforcement should be provided to resist the total tension force on the concrete computed for assumption of an uncracked section.

†May be taken as $12\sqrt{f'_c}$ for members for which computations based on the transformed cracked section and on bilinear moment-deflection relationships show that immediate and long-time deflections do not exceed the limits given in Table 9.14.

‡f_{py} = specified yield strength of tendons but not greater than the lesser of 80% of the specified yield strength f_{pu} and the maximum value recommended by the manufacturer of the tendons or anchorages.

Concrete in the anchorage region should be designed to develop the guaranteed ultimate tensile strength of the tendons. The ACI Code formula for bearing stresses [Eq. (9.90)] does not apply to posttensioning anchorages, but it can be used as a guide by substitution of f'_{ci}, the compressive strength of the concrete at the time of initial prestress, for f'_c.

9.106 DESIGN PROCEDURE FOR PRESTRESSED-CONCRETE BEAMS

Beam design involves choice of shape and dimensions of the concrete member, positioning of the tendons, and selection of amount of prestress.

After a concrete shape and dimensions have been assumed, determine the geometrical properties—cross-sectional area, center of gravity, distances of kern and extreme surface from the centroid, moment of inertia, section moduli, and dead load of the member per unit of length.

Treat the prestressing forces as a system of external forces acting on the concrete.

Compute bending stresses due to service dead and live loads. From these, determine the magnitude and location of the prestressing force required at sections subject to maximum moment. The prestressing force must result in sufficient compressive stress in the concrete to offset the tensile stresses caused by the bending moments due to dead and live service loads (Fig. 9.61). But at the same time, the prestress must not create allowable stresses that exceed those listed in Table 9.28. Investigation of other sections will guide selection of tendons to be used and determine their position and profile in the beam.

After establishing the tendon profile, prestressing forces, and tendon areas, check stresses at critical points along the beam immediately after transfer, but before losses. Using strength-design methods (Art. 9.107), check the percentage of steel and the strength of the member in flexure and shear.

Design anchorages, if required, and shear reinforcement.

Finally, check the deflection and camber under service loads. The modulus of elasticity of high-strength prestressing steel should not be assumed equal to 29,000,000 psi, as for conventional reinforcement, but should be determined by test or obtained from the manufacturer.

9.107 FLEXURAL-STRENGTH DESIGN OF PRESTRESSED CONCRETE

Flexural design strength should be based on factored loads and the assumptions of the American Concrete Institute Building Code, ACI 318, as explained in Art. 9.44. The stress f_{ps} in the tendons at design load ($1.4D + 1.7L$, where D is the dead load and L the live load), however, should not be assumed equal to the specified yield strength. High-strength prestressing steels lack a sharp and distinct yield point, and f_{ps} varies with the ultimate strength of the prestressing steel f_{pu}, the prestressing steel percentage ρ_p, and the concrete strength f'_c at 28 days. A stress-strain curve for the steel being used is necessary for stress and strain compatibility computations of f_{ps}. For unbonded tendons, successive trial-and-error analysis of tendon strain for strength design is straightforward but tedious. Assume a deflection at failure by crushing of the concrete (strain = 0.003 in/in). Determine from the stress-strain curve for the tendon steel the tendon stress corresponding to the total tendon strain at the assumed deflection. Proceed through successive trials, varying the assumed deflection, until the algebraic sum of the internal tensile and compressive forces equals zero. The moment of the resulting couple comprising the tensile and compressive forces times $\phi = 0.90$ is the design moment strength.

Stress in Bonded Tendons. When such data are not available, and the effective prestress, after losses, f_{se} is at least half the specified ultimate strength f_{pu} of the tendons, the stress f_{ps} in bonded tendons at nominal strength may be obtained from

$$f_{ps} = f_{pu}\left(1 - \frac{\gamma_p}{\beta_1} R\right) \tag{9.120}$$

$$R = \rho_p \frac{f_{pu}}{f'_c} + \frac{df_y}{d_p f'_c}(\rho - \rho') \tag{9.121}$$

where γ_p = factor for type of tendon
= 0.55 for $f_{py}/f_{pu} \geq 0.80$
= 0.40 for $f_{py}/f_{pu} \geq 0.85$
= 0.28 for $f_{py}/f_{pu} \geq 0.90$

β_1 = 0.85 for $f'_c \leq 4000$ psi; for $f'_c > 4000$ psi, reduce β_1 by 0.05 for each 1000 psi that f'_c exceeds 4000 psi but not to less than 0.65

$\rho_p = A_{ps}/bd_p$

A_{ps} = area of tendons in tension zone

b = width of compression face of member

d_p = distance from extreme compression surface to centroid of tendons

$\rho = A_s/bd$

d = distance from extreme compression surface to centroid of nonprestressed tension reinforcement

A_s = area of nonprestressed tension reinforcement
ρ' = A_s'/bd
A_s' = area of compression reinforcement
f_y = specified yield strength of nonprestressed reinforcement

If the area of compression reinforcement is included in the calculation of f_{ps} from Eq. (9.120), R should not exceed 0.17 nor should the distance d' from the extreme compression surface to the centroid of the compression reinforcment exceed $0.15d_p$.

Stress in Unbonded Tendons. When the ratio of span to depth of a prestressed flexural member with unbonded tendons is 35 or less, the stress in the tendons at nominal strength is given by

$$f_{ps} = f_{se} + 10{,}000 + f_c'/100\rho_p \leq f_{se} + 60{,}000 \tag{9.122}$$

where f_{se} is the effective stress in the tendons after allowance for prestress losses, but f_{ps} should not exceed the specified yield strength f_{py} of the tendons.

When the ratio of span to depth is larger than 35,

$$f_{ps} = f_{se} + 10{,}000 + f_c'/300\ \rho_p \leq f_{se} + 30{,}000 \tag{9.123}$$

but f_{ps} should not exceed f_{py}.

Nonprestressed reinforcement conforming to ASTM A185, A615, A616, A617, A706, A496, or A497, when used, in combination with tendons, may be assumed equivalent, at factored moment, to its area times its yield strength, but only if

$$\omega_p \leq 0.36\beta_1 \tag{9.124}$$

$$\omega_p + (d/d_p)(\omega - \omega') \leq 0.36\beta_1 \tag{9.125}$$

$$\omega_{pw} + (d/d_p)(\omega_w - \omega_w') \leq 0.36\beta_1 \tag{9.126}$$

where ω_p = $\rho_p f_{ps}/f_c'$
ω = $\rho f_y/f_c'$
ω' = $\rho' f_y/f_c'$
ω_w, ω_{pw}, ω_w' = reinforcement indices for flanged sections, computed as the ω, ω_p, and ω', except that b is the web width, and the steel area is that required to develop the compressive strength of the web only

Nominal strength in flexure (without the strength reduction factor ϕ), as predicted by standard equations, does not correlate well with test results when ω exceeds 0.30.

Design and Cracking Loads. To prevent an abrupt flexural failure by rupture of the prestressing steel immediately after cracking without a warning deflection, the total amount of prestressed and nonprestressed reinforcement should be adequate to develop a factored load in flexure of at least 1.2 times the cracking load, calculated on the basis of a modulus of rupture f_r. For normal-weight concrete, this modulus may be taken as

$$f_r = 7.5\sqrt{f_c'} \tag{9.127}$$

and for lightweight concrete as

$$f_r = 1.12 f_{ct} \leq 7.5\sqrt{f_c'} \tag{9.128}$$

where f_{ct} = average splitting tensile strength of lightweight concrete. When the value for f_{ct} is not available, the modulus of rupture of lightweight concrete can be computed for sand-lightweight concrete from

$$f_r = 6.375 \sqrt{f'_c} \tag{9.129}$$

and for all-lightweight concrete from

$$f_r = 5.625 \sqrt{f'_c} \tag{9.130}$$

("PCI Design Handbook," Precast/Prestressed Concrete Institute.)

9.108 SHEAR-STRENGTH DESIGN OF PRESTRESSED CONCRETE

The American Concrete Institute Building Code, ACI 318, requires that prestressed concrete beams be designed to resist diagonal tension by strength theory. There are two types of diagonal-tension cracks that can occur in prestressed-concrete flexural members: flexural-shear cracks initiated by flexural-tension cracks, and web-shear cracks caused by principal tensile stresses that exceed the tensile strength of the concrete.

The factored shear force V_u computed from Eq. (9.37) can be used to calculate the diagonal-tension stress. The distance d from the extreme compression surface to the centroid of the tension reinforcement should not be taken less than 0.80 the overall depth h of the beam.

When the beam reaction in the direction of the applied shear introduces compression into the end region of the member, the shear does not need to be checked within a distance $h/2$ from the face of the support.

Minimum Steel. The ACI Code requires that a minimum area of shear reinforcement be provided in prestressed-concrete members, except where the factored shear force V_u is less than $0.5\phi V_c$, where ϕV_c is the assumed shear that can be carried by the concrete; or where the depth h of the member is less than 10 in, 2.5 times the thickness of the compression flange, or one-half the thickness of the web; or where tests show that the required ultimate flexural and shear capacity can be developed without shear reinforcement.

When shear reinforcement is required, the amount provided perpendicular to the beam axis within a distance s should be not less than A_v given by Eq. (9.82). If, however, the effective prestress force is equal to or greater than 40% of the tensile strength of the flexural reinforcement, a minimum area A_v computed from Eq. (9.131) may be used.

$$A_v = \frac{A_{ps}}{80} \frac{f_{pu}}{f_y} \frac{s}{d} \sqrt{\frac{d}{b_w}} \tag{9.131}$$

where A_{ps} = area of tendons in tension zone
f_{pu} = ultimate strength of tendons
f_y = yield strength of nonprestressed reinforcement
s = shear reinforcement spacing measured parallel to longitudinal axis of member

d = distance from extreme compression surface to centroid of tension reinforcement
b_w = web width

The ACI Code does not permit the yield strength f_y of shear reinforcement to be assumed greater than 60,000 psi. The Code also requires that stirrups be placed perpendicular to the beam axis and spaced not farther apart than 24 in or $0.75h$, where h is the overall depth of the member.

The area of shear reinforcement required to carry the shear in excess of the shear that can be carried by the concrete can be determined from Eq. (9.40a).

Maximum Shear. For prestressed concrete members subjected to an effective prestress force equal to at least 40% of the tensile strength of the flexural reinforcement, the shear strength provided by the concrete is limited to that which would cause significant inclined cracking and, unless Eqs. (9.133) and (9.134) are used, can be taken as equal to the larger of $2\sqrt{f'_c}b_w d$ or

$$V_c = \left(0.60\sqrt{f'_c} + 700\frac{V_u d}{M_u}\right) b_w d \leq 5\sqrt{f'_c}\, b_w d \tag{9.132}$$

where M_u = factored-load moment at section and V_u = factored shear force at section. The factored moment M_u occurs simultaneously with the shear V_u at the section. The ratio $V_u d/M_u$ should not be taken greater than 1.0.

If the effective prestress force is less than 40% of the tensile strength of the flexural reinforcement, or if a more accurate method is preferred, the value of V_c should be taken as the smaller of the shear forces causing inclined flexure-shear cracking V_{ci} or web-shear cracking V_{cw}, but need not be smaller than $1.7\sqrt{f'_c}\, b_w d$.

$$V_{ci} = 0.6\sqrt{f'_c}\, b_w d + V_d + V_i M_{cr} m M_{max} \tag{9.133}$$

$$V_{cw} = (3.5\sqrt{f'_c} + 0.3 f_{pc})\, b_w d + V_p \tag{9.134}$$

where V_d = shear force at section caused by dead load
V_i = shear force at section occurring simultaneously with M_{max}
M_{max} = maximum bending moment at the section caused by externally applied factored loads
M_{cr} = cracking moment based on the modulus of rupture (Art. 9.51)
b_w = width of web
f_{pc} = compressive stress in the concrete, after all prestress losses have occurred, at the centroid of the cross section resisting the applied loads, or at the junction of web and flange when the centroid lies within the flange
V_p = vertical component of effective prestress force at section considered

In a pretensioned beam in which the section $h/2$ from the face of the support is closer to the end of the beam than the transfer length of the tendon, the reduced prestress in the concrete at sections falling within the transfer length should be considered when calculating V_{cw}. The prestress may be assumed to vary linearly along the centroidal axis from zero at the beam end to the end of the transfer length. This distance can be assumed to be 50 diameters for strand and 100 diameters for single wire.

("PCI Design Handbook," Precast/Prestressed Concrete Institute.)

9.109 BOND, DEVELOPMENT, AND GROUTING OF TENDONS

Three- or seven-wire pretensioning strand should be bonded beyond the critical section for a development length, in, of at least

$$l_d = (f_{ps} - \tfrac{2}{3}f_{se})d_b \tag{9.135}$$

where d_b = nominal diameter of strand, in
f_{ps} = stress in tendons at nominal strength, ksi
f_{se} = effective stress in tendons after losses, ksi

(The expression in parentheses is used as a constant without units.) Investigations for bond integrity may be limited to those cross sections nearest each end of the member that are required to develop their full strength under factored load. When bonding does not extend to the end of the member, the bonded development length given by Eq. (9.135) should be doubled.

Minimum Bonded Steel. When prestressing steel is unbonded, the American Concrete Institute Code, ACI 318, requires that some bonded reinforcement be placed in the precompressed tensile zone of flexural members and distributed uniformly over the tension zone near the extreme tension surface. The amount of bonded reinforcement that should be furnished for beams, one-way slabs, and two-way slabs would be the larger of the values of A_s computed from Eqs. (9.136) and (9.137).

$$A_s = \frac{N_c}{0.5f_y} \tag{9.136}$$

$$A_s = 0.004A \tag{9.137}$$

where N_c = tensile force in the concrete under actual dead load plus 1.2 times live load
A = area of that part of the cross section between the flexural tension face and centroid of gross section
f_y = yield strength of bonded reinforcement, but not more than 60,000 psi

The minimum amount of bonded reinforcement for two-way slabs with unbonded tendons is the same as for one-way slabs, except that the minimum amount may be reduced when there is no tension in the precompressed tension zone under service loads.

Grouting of Tendons. When posttensioned tendons are to be bonded, a cement grout is usually injected under pressure (80 to 100 psi) into the space between the tendon and the sheathing material of the cableway. The grout can be inserted in holes in the anchorage heads and cones, or through buried pipes. To ensure filling of the space, the grout can be injected under pressure at one end of the member until it is forced out the other end. For long members, it can be injected at each end until it is forced out a vent between the ends.

Grout provides bond between the posttensioning tendons and the concrete member and protects the tendons against corrosion.

Members should be above 50°F in temperature at the time of grouting. This minimum temperature should be maintained for at least 48 hr.

Tendon Sheaths. Ducts for grouted or unbonded tendons should be mortar-tight and nonreactive with concrete, tendons, or filler material. To facilitate injection of the grout, the cableway or duct should be at least ¼ in larger than the diameter of the posttensioning tendon or large enough to have an internal area at least twice the gross area of the prestressing steel.

9.110 APPLICATION AND MEASUREMENT OF PRESTRESS

The actual amount of prestressing force applied to a concrete member should be determined by measuring the tendon elongation, also by checking jack pressure on a calibrated gage or load cell, or by use of a calibrated dynamometer. If the discrepancy in force determination exceeds 5%, it should be investigated and corrected. Elongation measurements should be correlated with average load-elongation curves for the particular prestressing steel being used.

9.111 CONCRETE COVER IN PRESTRESSED MEMBERS

The minimum thicknesses of cover required by the American Concrete Institute Building Code, ACI 318, for prestressed and nonprestressed reinforcement, ducts, and end fittings in prestressed concrete members are listed in Table 9.29.

TABLE 9.29 Minimum Concrete Cover in, in Prestressed Members

Concrete cast against and permanently exposed to earth	3
Concrete exposed to earth or weather:	
Wall panels, slabs, and joists	1
Other members	1½
Concrete not exposed to weather or in contact with the ground:	
Slabs, walls, joists	¾
Beams, girders, columns:	
Principal reinforcement	1½
Ties, stirrups, or spirals	1
Shells and folded-plate members:	⅜
Reinforcement ⅝ in and smaller	d_b but not less
Other reinforcement	than ¾ in

The cover for nonprestressed reinforcement in prestressed concrete members under plant control may be that required for precast members (Table 9.26). When the general code requires fire-protection covering greater than that required by the ACI Code, such cover should be used.

("PCI Design Handbook," Precast/Prestressed Concrete Institute.)

SECTION TEN
WOOD CONSTRUCTION

John "Buddy" Showalter
American Forest and Paper Association
Washington, D.C.

Thomas G. Williamson
American Plywood Association
Tacoma, Washington

Wood is the only renewable source for building materials. It comes from forests that are continually being replanted as they are harvested. This practice ensures a plentiful supply of wood for construction and for a myriad other uses.

Compared to other building materials, wood has a very large ratio of strength to weight. This makes it very economical for use in all types of construction. Wood also has an aesthetic quality and natural warmth unequalled by other building materials.

Wood has inherent characteristics with which construction users should be familiar. For example, as a consequence of its biological origin, it is nonhomogeneous. Also, properties of pieces of wood from different species of tree may be considerably different, and even properties of pieces of wood from the same tree may differ. In the past, determination of engineering properties depended heavily on visual inspection, keyed to averages, of wood pieces. Research, however, has made possible better estimates of these properties. It is no longer necessary to rely so heavily on visual inspection. Greater accuracy in determination of engineering properties has been made possible by mechanical grading procedures.

Improvements in adhesives for wood also have contributed to the betterment of wood construction. These advances in adhesion technology combined with a desire to utilize more efficiently available wood-fiber resources have led to increasing use of such products as oriented strand board, glued-laminated timber, wood I joists, and structural composite lumber.

10.1 BASIC CHARACTERISTICS OF WOOD

Wood differs in several significant ways from other building materials. Its cellular structure is responsible, to a considerable degree, for these differences. Because of this structure, structural properties depend on grain orientation. While most structural materials are essentially isotropic, with nearly equal propertites in all directions, wood has three principal grain directions—longitudinal, radial, and tangential. Loading in the longitudinal direction is referred to as parallel to the grain, whereas transverse loading is considered to be across the grain. Parallel to the grain, wood possesses high strength and stiffness characteristics. Across the grain, strength and stiffness are much lower. In tension, wood stressed parallel to the grain is 25 to 40 times stronger than when stressed across the grain. In compression, wood loaded parallel to the grain is 6 to 10 times stronger than when loaded perpendicular to the grain. Furthermore, a wood member has three moduli of elasticity, with a ratio of largest to smallest as large as 150:1.

Wood undergoes dimensional changes from causes different from those in most other structural materials. For instance, thermal expansion of wood is so small as to be unimportant in ordinary usage. Significant dimensional changes, however, occur because of gain or loss in moisture. Swelling and shrinkage caused by moisture changes vary in the three grain directions; these size changes are about 6 to 16% tangentially, 3 to 7% radially, but only 0.1 to 0.3% longitudinally. Radial swelling causes a decrease in the angle between the ends of a curved member; radial shrinkage causes an increase in this angle.

Such effects may be of great importance in three-hinged arches that become horizontal, or nearly so, at the crest of a roof. Shrinkage, increasing the relative end rotations, may cause a depression at the crest and create drainage problems. For such arches, therefore, consideration must be given to moisture content of the member at time of fabrication and in service, and to the change in end angles that results from change in moisture content and shrinkage across the grain. Table 10.1 gives shrinkage values for some commonly used species of wood.

Wood offers numerous advantages in construction applications—warmth and beauty, versatility, durability, workability, low cost per pound, high strength-to-weight ratio, good electrical insulation, low thermal conductance, and excellent strength at low temperatures. It has high shock-absorption capacity. It can withstand large overloads of short time duration with no overall strength degradation. It has good wearing qualities, particularly on end grain. It can be bent easily to relatively sharp curvature. A wide range of finishes can be applied for decorative or protective purposes. Wood can be used in both wet and dry applications. Preservative treatments are available for use when necessary, as are fire retardants (not appropriate for all wood products). Also, there is a choice of a wide range of species with a range of unique properties.

In addition, a wide variety of wood framing systems is available. The intended use of a structure, geographical location, configuration required, cost, and many other factors determine the best framing system to be used for a particular project.

Wood is naturally resistant to many chemicals that are highly corrosive to other materials. It is superior to many building materials in resistance to mild acids, particularly at ordinary temperatures. It has excellent resistance to most organic acids, notably acetic. However, wood is seldom used in contact with solutions that are more than weakly alkaline. Oxidizing chemicals and solutions of iron salts, in combination with damp conditions, should be avoided.

Wood is composed of roughly 50 to 70% cellulose, 25 to 30% lignin, and 5% extractives with less than 2% protein. Acids such as acetic, formic, lactic, and boric do not ionize sufficiently at room temperature to attack cellulose, and thus do not harm wood.

When the pH of aqueous solutions of weak acids is 2 or more, the rate of hydrolysis of cellulose is small and dependent on the temperature. A rough approximation of this temperature effect is that, for every 20°F increase, the rate of

TABLE 10.1 Shrinkage Values of Wood Based on Dimensions When Green

Species	Dried to 20% MC*			Dried to 6% MC†			Dried to 0% MC		
	Radial, %	Tangential, %	Volumetric, %	Radial, %	Tangential, %	Volumetric, %	Radial, %	Tangential, %	Volumetric, %
Softwoods:‡									
Cedar:									
Alaska	0.9	2.0	3.1	2.2	4.8	7.4	2.8	6.0	9.2
Incense	1.1	1.7	2.5	2.6	4.2	6.1	3.3	5.2	7.6
Port Orford	1.5	2.3	3.4	3.7	5.5	8.1	4.6	6.9	10.1
Western red	0.8	1.7	2.3	1.9	4.0	5.4	2.4	5.0	6.8
Cypress, southern	1.3	2.1	3.5	3.0	5.0	8.4	3.8	6.2	10.5
Douglas fir:									
Coast region	1.7	2.6	3.9	4.0	6.2	9.4	5.0	7.8	11.8
Inland region	1.4	2.5	3.6	3.3	6.1	8.7	4.1	7.6	10.9
Rocky Mountain	1.2	2.1	3.5	2.9	5.0	8.5	3.6	6.2	10.6
Fir, white	1.1	2.4	3.3	2.6	5.7	7.8	3.2	7.1	9.8
Hemlock:									
Eastern	1.0	2.3	3.2	2.4	5.4	7.8	3.0	6.8	9.7
Western	1.4	2.6	4.0	3.4	6.3	9.5	4.3	7.9	11.9
Larch, western	1.4	2.7	4.4	3.4	6.5	10.6	4.2	8.1	13.2
Pine:									
Eastern white	0.8	2.0	2.7	1.8	4.8	6.6	2.3	6.0	8.2
Lodgepole	1.5	2.2	3.8	3.6	5.4	9.2	4.5	6.7	11.5
Norway	1.5	2.4	3.8	3.7	5.8	9.2	4.6	7.2	11.5
Ponderosa	1.3	2.1	3.2	3.1	5.0	7.7	3.9	6.3	9.6
Southern (avg.)	1.6	2.6	4.1	4.0	6.1	9.8	5.0	7.6	12.2
Sugar	1.0	1.9	2.6	2.3	4.5	6.3	2.9	5.6	7.9
Western white	1.4	2.5	3.9	3.3	5.9	9.4	4.1	7.4	11.8
Redwood (old growth)	0.9	1.5	2.3	2.1	3.5	5.4	2.6	4.4	6.8
Spruce:									
Engelmann	1.1	2.2	3.5	2.7	5.3	8.3	3.4	6.6	10.4
Sitka	1.4	2.5	3.8	3.4	6.0	9.2	4.3	7.5	11.5
Hardwoods:‡									
Ash, white	1.6	2.6	4.5	3.8	6.2	10.7	4.8	7.8	13.4
Beech, American	1.7	3.7	5.4	4.1	8.8	13.0	5.1	11.0	16.3
Birch:									
Sweet	2.2	2.8	5.2	5.2	6.8	12.5	6.5	8.5	15.6
Yellow	2.4	3.1	5.6	5.8	7.4	13.4	7.2	9.2	16.7
Elm, rock	1.6	2.7	4.7	3.8	6.5	11.3	4.8	8.1	14.1
Gun, red	1.7	3.3	5.0	4.2	7.9	12.0	5.2	9.9	15.0
Hickory:									
Pecan§	1.6	3.0	4.5	3.9	7.1	10.9	4.9	8.9	13.6
True	2.5	3.8	6.0	6.0	9.0	14.3	7.5	11.3	17.9
Maple, hard	1.6	3.2	5.0	3.9	7.6	11.9	4.9	9.5	14.9
Oak:									
Red	1.3	2.7	4.5	3.2	6.6	10.8	4.0	8.2	13.5
White	1.8	3.0	5.3	4.2	7.2	12.6	5.3	9.0	15.8
Poplar, yellow	1.3	2.4	4.1	3.2	5.7	9.8	4.0	7.1	12.3

*MC = moisture content, as a percent of weight of oven-dry wood. These shrinkage values have been taken as one-third the shrinkage to the oven-dry condition as given in the last three columns of this table.

†These shrinkage values have been taken as four-fifths of the shrinkage to the oven-dry condition as given in the last three columns of this table.

‡The total longitudinal shrinkage of normal species from fiber saturation to oven-dry condition is minor. It usually ranges from 0.17 to 0.3% of the green dimension.

§Average of butternut hickory, nutmeg hickory, water hickory, and pecan.

hydrolysis doubles. Acids with pH values above 2, or bases with pH below 10, have little weakening effect on wood at room temperature, if the duration of exposure is moderate.

Design Recommendations. The following recommendations aim at achieving economical designs with wood framing:

Use standard sizes and grades of lumber. Consider using standardized structural components, whether lumber, stock glued-laminated beams, or other framing members designed for structural adequacy, efficiency, and economy.

Use standard details wherever possible. Avoid specially designed and manufactured connecting hardware.

Use as simple and as few joints as possible. Place splices, when required, in areas of lowest stress. Do not locate splices where bending moments are large, thus avoiding design, installation, and fabrication difficulties.

Avoid unnecessary variations in cross section of members along their length.

Use identical member designs repeatedly throughout a structure, whenever practicable. Keep the number of different arrangements to a minimum.

Consider using roof profiles that favorably influence the type and amount of load on the structure.

Specify allowable design stresses rather than the lumber grade or combination of grades to be used.

Select an adhesive suitable for the service conditions, but do not overspecify. For example, waterproof resin adhesives need not be used where less expensive water-resistant adhesives will do the job.

Use lumber treated with preservatives where service conditions dictate. Such treatment need not be used where decay hazards do not exist. Fire-retardant treatments may be used to meet a specific flame-spread rating for interior finish, but are not necessary for large-cross-sectional members that are widely spaced and have a natural resistance to fire because of their relatively large size.

Instead of long, simple spans, consider using continuous or suspended spans or simple spans with overhangs.

Select an appearance grade best suited to the project. Do not specify the highest quality appearance grade available for all members if it is not required.

Table 10.2 may be used as a general guide to typical ranges of spans for roof and floor framing members.

10.2 SECTIONAL PROPERTIES OF WOOD PRODUCTS

Dressed sizes of sawn lumber are given in the grading rules of agencies that formulate and maintain such rules and in Table 10.3. The nominal and dressed sizes are developed in accordance with the American Softwood Lumber Standard, Voluntary Product Standard PS20-70. These sizes are generally available, but it is good practice to consult suppliers before specifying sizes not commonly used to find out what sizes are on hand or can be readily secured.

TABLE 10.2 Typical Span Range for Wood Framing Members

Framing member	Typical span range, ft	Typical spacing, ft
Roof beams (generally used where a flat or low-pitched roof is desired):		
Simple span:		
Constant depth		
Solid-sawn	0–20	4–12
Glued-laminated	20–100	8–24
Tapered	25–100	8–24
Double-tapered pitched and curved beams	25–100	8–24
Curved beams	25–100	8–24
Simple beam with overhangs (usually more economical than simple span when span is over 40 ft):		
Solid-sawn	24	4–12
Glued-laminated	10–100	8–24
Continuous span:		
Solid-sawn	10–24	4–12
Glued-laminated	10–50	8–24
Arches:		
Three-hinged:		
Gothic	40–90	8–24
Tudor	30–120	8–24
A-frame	20–160	8–24
Three-centered	40–250	8–24
Parabolic	40–250	8–24
Radial	40–250	8–24
Two-hinged:		
Radial	50–200	8–24
Parabolic	50–200	8–24
Trusses:		
Flat or parallel chord	50–150	8–24
Triangular or pitched	50–90	8–24
Bowstring	50–200	8–24
Tied arches:		
Tied segment	50–100	8–24
Buttressed segment	50–200	14–24
Domes	50–350	*
Simple-span floor beams:		
Solid sawn	6–20	4–12
Glued laminated	6–40	4–16
Continuous floor beams	20–40	4–16
Roof sheathing and decking:		
1-in sheathing	1–4	
2-in sheathing	6–10	
3-in roof deck	8–15	
4-in roof deck	12–20	
Plywood sheathing	1–4	

*Not applicable.

TABLE 10.3 Nominal and Minimum Dressed Sizes of Sawn Lumber

Item	Thickness, in			Face width, in		
		Minimum dressed			Minimum dressed	
	Nominal	Dry*	Green†	Nominal	Dry*	Green†
Boards	1	¾	25⁄32	2	1½	1 9⁄16
	1¼	1	1 1⁄32	3	2½	2 9⁄16
	1½	1¼	1 9⁄32	4	3½	3 9⁄16
				5	4½	4⅝
				6	5½	5⅝
				7	6½	6⅝
				8	7¼	7½
				9	8¼	8½
				10	9¼	9½
				11	10¼	10½
				12	11¼	11½
				14	13¼	13½
				16	15¼	15½
Dimension	2	1½	1 9⁄16	2	1½	1 9⁄16
	2½	2	2 1⁄16	3	2½	2 9⁄16
	3	2½	2 9⁄16	4	3½	3 9⁄16
	3½	3	3 1⁄16	5	4½	4⅝
				6	5½	5⅝
				8	7¼	7½
				10	9¼	9½
				12	11¼	11½
				14	13¼	13½
				16	15¼	15½
	4	3½	3 9⁄16	2	1½	1 9⁄16
	4½	4	4 1⁄16	3	2½	2 9⁄16
				4	3½	3 9⁄16
				5	4½	4⅝
				6	5½	5⅝
				8	7¼	7½
				10	9¼	9½
				12	11¼	11½
				14		13½
				16		15½
Timbers	5 and thicker		½ in less	5 and wider		½ in less

*Dry lumber is defined as lumber seasoned to a moisture content of 19% or less.
†Green lumber is defined as lumber having a moisture content in excess of 19%.

The "National Design Specification for Wood Construction," American Forest and Paper Association (formerly the National Forest Products Association) presents tables of section properties of standard dressed sawn lumber and glued-laminated timber. Standard finished sizes of structural glued-laminated timber should be used to the extent that conditions permit. These standard finished sizes are based on lumber sizes given in Voluntary Product Standard PS20-70. Other finished sizes may be used to meet the size requirements of a design, or to meet other special requirements.

Nominal 2-in-thick lumber, surfaced to 1½ in before gluing, is used to laminate straight members and curved members having radii of curvature within the bending-radius limitations for the species. Nominal 1-in-thick lumber, surfaced to ¾ in before gluing, may be used for laminating curved members when the bending radius is too tight to permit use of nominal 2-in-thick laminations. Other lamination thicknesses may be used to meet special curving requirements.

Standard sizes and grades of structural panels are given in U.S. Product Standard PS 1-83 for Construction and Industrial Plywood and "Performance Standard for Wood-Based Structural-Use Panels," Voluntary Product Standard, PS 2-92. See also Art. 10.12.

Weight and Specific Gravity. Specific gravity is a reliable indicator of fiber content. Also, specific gravity and the strength and stiffness of solid wood or laminated products are interrelated. See Table 10.4 for weights and specific gravities of several commercial lumber species.

10.3 DESIGN VALUES FOR LUMBER AND TIMBER

Design values for an extensive range of sawn lumber and timber are tabulated in "National Design Specification for Wood Construction," (NDS), American Forest and Paper Association (AFPA), formerly National Forest Products Association (NFPA).

10.3.1 Lumber

Design values for lumber are contained in grading rules established by the National Lumber Grades Authority (Canadian), Northeastern Lumber Manufacturers Association, Northern Softwood Lumber Bureau, Redwood Inspection Service, Southern Pine Inspection Bureau, West Coast Lumber Inspection Bureau, and Western Wood Products Association. The rules and the design values in them have been approved by the Board of Review of the American Lumber Standards Committee. They also have been certified for conformance with U.S. Department of Commerce Voluntary Product Standard PS 20-70 (American Softwood Lumber Standard).

Design values for most species and grades of visually graded dimension lumber are based on provisions in "Establishing Allowable Properties for Visually Graded Dimension Lumber from In-Grade Tests of Full-Size Specimens," ASTM D1990. Design values for visually graded timbers, decking, and some species and grades of dimension lumber are based on provisions of "Establishing Structural Grades and Related Allowable Properties for Visually Graded Lumber," ASTM D245.

TABLE 10.4 Weights and Specific Gravities of Commercial Lumber Species

Species	Specific gravity based on oven-dry weight and volume at 12% moisture content	Weight, lb per ft³			Moisture content when green (avg), %	Specific gravity based on oven-dry weight and volume when green	Weight when green, lb per ft³
		At 12% moisture content	At 20% moisture content	Adjusting factor for each 1% change in moisture content			
Softwoods:							
Cedar:							
Alaska	0.44	31.1	32.4	0.170	38	0.42	35.5
Incense	0.37	25.0	26.4	0.183	108	0.35	42.5
Port Orford	0.42	29.6	31.0	0.175	43	0.40	35.0
Western red	0.33	23.0	24.1	0.137	37	0.31	26.4
Cypress, southern	0.46	32.1	33.4	0.167	91	0.42	45.3
Douglas fir:							
Coast region	0.48	33.8	35.2	0.170	38	0.45	38.2
Inland region	0.44	31.4	32.5	0.137	48	0.41	36.3
Rocky Mountain	0.43	30.0	31.4	0.179	38	0.40	34.6
Fir, white	0.37	26.3	27.3	0.129	115	0.35	39.6
Hemlock:							
Eastern	0.40	28.6	29.8	0.150	111	0.41	43.4
Western	0.42	29.2	30.2	0.129	74	0.38	37.2
Larch, western	0.55	38.9	40.2	0.170	58	0.51	46.7
Pine:							
Eastern white	0.35	24.9	26.2	0.167	73	0.36	35.1
Lodgepole	0.41	28.8	29.9	0.142	65	0.38	36.3
Norway	0.44	31.0	32.1	0.142	92	0.41	42.3
Ponderosa	0.40	28.1	29.4	0.162	91	0.38	40.9
Southern shortleaf	0.51	35.2	36.5	0.154	81	0.46	45.9
Southern longleaf	0.58	41.1	42.5	0.179	63	0.54	50.2
Sugar	0.36	25.5	26.8	0.162	137	0.35	45.8
Western white	0.38	27.6	28.6	0.129	54	9.36	33.0
Redwood	0.40	28.1	29.5	0.175	112	0.38	45.6
Spruce:							
Engelmann	0.34	23.7	24.7	0.129	80	0.32	32.5
Sitka	0.40	27.7	28.8	0.145	42	0.37	32.0
White	0.40	29.1	29.9	0.104	50	0.37	33.0
Hardwoods:							
Ash, white	0.60	42.2	43.6	0.175	42	0.55	47.4
Beech, American	0.64	43.8	45.1	0.162	54	0.56	50.6
Birch:							
Sweet	0.65	46.7	48.1	0.175	53	0.60	53.8
Yellow	0.62	43.0	44.1	0.142	67	0.55	50.8
Elm, rock	0.63	43.6	45.2	0.208	48	0.57	50.9
Gum	0.52	36.0	37.1	0.133	115	0.46	49.7
Hickory:							
Pecan	0.66	45.9	47.6	0.212	63	0.60	56.7
Shagbark	0.72	50.8	51.8	0.129	60	0.64	57.0
Maple, sugar	0.63	44.0	45.3	0.154	58	0.56	51.1
Oak:							
Red	0.63	43.2	44.7	0.187	80	0.56	56.0
White	0.68	46.3	47.6	0.167	68	0.60	55.6
Poplar, yellow	0.42	29.8	31.0	0.150	83	0.40	40.5

This standard specifies adjustments to be made in the strength properties of small clear specimens of wood, as determined in accordance with "Establishing Clear Wood Strength Values," ASTM D2555, to obtain design values applicable to normal conditions of service. The adjustments account for the effects of knots, slope of grain, splits, checks, size, duration of load, moisture content, and other influencing factors. Lumber structures designed with working stresses derived from D245 procedures and standard design criteria have a long history of satisfactory performance.

Design values for machine stress-rated (MSR) lumber and machine-evaluated lumber (MEL) are based on nondestructive tests of individual wood pieces. Certain visual-grade requirements also apply to such lumber. The stress rating system used for MSR lumber and MEL is checked regularly by the responsible grading agency for conformance with established certification and quality-control procedures.

10.3.2 Glued-Laminated Timber

Design values for glued-laminated timber, developed by the American Institute of Timber Construction (AITC) and American Wood Systems (AWS) in accordance with principles originally established by the U.S. Forest Products Laboratory, are included in the NDS. The principles are the basis for the "Standard Method for Establishing Stresses for Structural Glued-Laminated Timber (Glulam)," ASTM D3737. It requires determination of the strength properties of clear, straight-grained lumber in accordance with the methods of ASTM D2555 or as given in a table in D3737. The glulam test method also specifies procedures for obtaining design values by adjustments to those properties to account for the effects of knots, slope of grain, density, size of member, curvature, number of laminations, and other factors unique to glulam. The satisfactory performance of structures made with glulam members conforming to AITC specifications and manufactured in accordance with "Structural Glued-Laminated Timber," ANSI A190.1, demonstrates the validity of the methods used to establish glulam design values.

10.4 STRUCTURAL GRADING OF WOOD

Strength properties of wood are closely related to moisture content and specific gravity. Therefore, data on strength properties should be accompanied by corresponding data on these physical properties.

The strength of wood is actually affected by many other factors, including loading rate, load duration, temperature, grain direction, and position of growth rings. Strength is also influenced by inherent growth characteristics, including knots, slope of grain, shakes, and checks. Analysis and integration of available data have yielded a comprehensive set of principles for grading structural lumber (Art. 10.3.1).

The same characteristics that reduce the strength of solid timber also affect the strength of glued-laminated (glulam) members (Art. 10.3.2). There are, however, additional factors peculiar to glulam flexural members that should be considered. For example, knots located near the neutral axis, which is a region of low bending stress, have less effect on strength than knots closer to the outer surfaces, where bending stresses are higher. Thus, strength of a flexural member with low-grade laminations can be improved by substitution of a few high-grade laminations at the top and bottom of the member.

Dispersement of knots in laminated members has a beneficial effect on strength. With sufficient knowledge of the occurrence of knots within a grade, mathematical estimates of the effect may be established for members containing various numbers of laminations.

Allowable design stresses taking these factors into account are higher for glulam members than for solid timbers of comparable grade. But cross-grain limitations must be more restrictive than for solid timbers to justify these higher allowable stresses.

10.5 ADJUSTMENT FACTORS FOR STRUCTURAL MEMBERS

Design values obtained by the methods described in Art. 10.3 should be multiplied by adjustment factors based on conditions of use, geometry, and stability. The adjustments are cumulative, unless specifically indicated in the following.

The adjusted design value F'_b for extreme-fiber bending is given by

$$F'_b = F_b C_D C_M C_t C_L C_F C_V C_{fu} C_r C_c C_f \tag{10.1}$$

where F_b = design value for extreme-fiber bending
C_D = load duration factor (Art. 10.5.1)
C_M = wet service factor (Art. 10.5.2)
C_t = temperature factor (Art. 10.5.3)
C_L = beam stability factor (Arts. 10.5.5 and 10.7.2)
C_F = size factor—applicable only to visually graded, sawn lumber and round timber flexural members (Art. 10.5.4)
C_V = volume factor—applicable only to glulam flexural members (Art. 10.5.6)
C_{fu} = flat-use factor—applicable only to dimension-lumber beams 2 to 4 in thick and glulam beams (Art. 10.5.7)
C_r = repetitive member factor—applicable only to dimension-lumber beams 2 to 4 in thick (Art. 10.5.8)
C_c = curvature factor—applicable only to curved portions of glulam beams (Art. 10.5.9)
C_f = form factor (Art. 10.5.10)

For glulam beams, use either C_L or C_V, whichever is smaller, not both, in Eq. (10.1).

The adjusted design value for tension F'_t is given by

$$F'_t = F_t C_D C_M C_t C_F \tag{10.2}$$

where F_t = design value for tension.

For shear, the adjusted design value F'_V is computed from

$$F'_V = F_V C_D C_M C_t C_H \tag{10.3}$$

where F_V = design value for shear and C_H = shear stress factor ≥ 1—permitted for F_V parallel to the grain for sawn lumber members (Art. 10.5.13).

For compression perpendicular to the grain, the adjusted design value $F'_{c\perp}$ is obtained from

$$F'_{c\perp} = F_{c\perp}C_M C_t C_b \tag{10.4}$$

where $F_{c\perp}$ = design value for compression perpendicular to the grain and C_b = bearing area factor (Art. 10.5.11).

For compression parallel to the grain, the adjusted design value F'_c is given by

$$F'_c = F_c C_D C_M C_t C_F C_p \tag{10.5}$$

where F_c = design value for compression parallel to grain and C_P = column stability factor (Arts. 10.5.11 and 10.8.1.)

For end grain in bearing parallel to the grain, the adjusted design value F'_g is computed from

$$F'_g = F_g C_D C_t \tag{10.6}$$

where F_g = design value for end grain in bearing parallel to the grain. See also Art. 10.11.1.

The adjusted design value for modulus of elasticity E' is obtained from

$$E' = EC_M C_t C \cdots \tag{10.7}$$

where E = design value for modulus of elasticity
C_T = buckling stiffness factor—applicable only to sawn-lumber truss compression chords 2 × 4 in or smaller, when subject to combined bending and axial compression and plywood sheathing ⅜ in or more thick is nailed to the narrow face (Art. 5.10.12).

10.5.1 Load Duration Factor

Wood has the capacity to carry substantially greater loads for short periods of time than for long periods. Design values described in Art. 10.3 apply to normal load duration, which is equivalent to application of full design load for a cumulative duration of about 10 years. The full design load is one that stresses a member to its allowable design value. When the cumulative duration of the full design load differs from 10 years, design values, except $F_{c\perp}$ for compression perpendicular to grain and modulus of elasticity E, should be multiplied by the appropriate load duration factor C_D listed in Table 10.5.

TABLE 10.5 Frequently Used Load Duration Factors C_D

Load duration	C_D	Typical design loads
Permanent	0.9	Dead load
10 years	1.0	Occupancy live load
2 months	1.15	Snow load
7 days	1.25	Construction load
10 minutes	1.6	Wind or seismic load
Impact	2.0	Impact load

When loads of different duration are applied to a member, C_D for the load of shortest duration should be applied to the total load. In some cases, a larger-size member may be required when one or more of the shorter-duration loads are omitted. Design of the member should be based on the critical load combination. If the permanent load is equal to or less than 90% of the total combined load, the normal load duration will control the design. Both C_D and the modification permitted in design values for load combinations may be used in design.

The duration factor for impact does not apply to connections or structural members pressure-treated with fire retardants or with waterborne preservatives to the heavy retention required for marine exposure.

10.5.2 Wet Service Factor

Sawn-lumber design values apply to lumber that will be used under dry service conditions; that is, where moisture content (MC) of the wood will be a maximum of 19% of the oven-dry weight, regardless of MC at time of manufacture. When the MC of structural members in service will exceed 19% for an extended period of time, design values should be multiplied by the appropriate wet service factor listed in Table 10.6.

TABLE 10.6 Wet Service Factors C_M

Design value	C_M for sawn lumber*	C_M for glulam timber†
F_b	0.85‡	0.80
F_t	1.0	0.80
F_V	0.97	0.875
$F_{c\perp}$	0.67	0.53
F_c	0.80§	0.73
E	0.90	0.833

*For use where moisture content in service exceeds 19%.

†For use where moisture content in service exceeds 16%.

‡$C_M = 1.0$ when $F_bC_F \leq 1150$ psi.

§$C_M = 1.0$ when $F_cC_F \leq 750$ psi.

MC of 19% or less is generally maintained in covered structures or in members protected from the weather, including windborne moisture. Wall and floor framing and attached sheathing are usually considered to be such dry applications. These dry conditions are generally associated with an average relative humidity of 80% or less. Framing and sheathing in properly ventilated roof systems are assumed to meet MC criteria for dry conditions of use, even though they are exposed periodically to relative humidities exceeding 80%.

Glulam-timber design values apply when the MC in service is less than 16%, as in most covered structures. When MC of glulam timber under service conditions is 16% or more, design values should be multiplied by the appropriate wet service factor C_M in Table 10.6.

10.5.3 Temperature Factor

Design values apply to members used in ordinary temperature ranges. (Occasional heating to 150°F is permissible.) Strength properties of wood, however, increase when it is cooled below normal temperatures and decrease when it is heated. Members heated in use to temperatures up to 150°F return essentially to original strength when cooled. Prolonged exposure to temperatures above 150°F, however, may result in permanent loss of strength. Design values for structural members that will experience sustained exposure to elevated temperatures up to 150°F should be multiplied by the appropriate temperature factor C_t listed in Table 10.7.

10.5.4 Size Factor

For visually graded dimension lumber, design values F_b, F_t, and F_c for all species and species combinations, except southern pine, should be multiplied by the appropriate size factor C_F given in Table 10.8 to account for the effects of member size. This factor and the factors used to develop size-specific values for southern pine are based on the adjustment equation given in ASTM D1990. This equation, based on in-grade test data, accounts for differences in F_b, F_t, and F_c related to width and in F_b and F_t related to length (test span).

For visually graded timbers (5 × 5 in or larger), when the depth d of a stringer, beam, post, or timber exceeds 12 in, the design value for bending should be adjusted by the size factor

$$C_F = (12/d)^{1/9} \tag{10.8}$$

10.5.5 Beam Stability Factor

Design values F_b for bending should be adjusted by multiplying by the beam stability factor C_L specified in Art. 10.7.2. For glulam beams, the smaller value of C_L and the volume factor C_V should be used, not both. See also Art. 10.5.5.

10.5.6 Volume Factor

Design values for bending F_b for glulam beams should be adjusted for the effects of volume by multiplying by

$$C_V = K_L \left[\left(\frac{21}{L}\right) \left(\frac{12}{d}\right) \left(\frac{5.125}{b}\right) \right]^{1/x} \tag{10.9}$$

TABLE 10.7 Temperature Factors C_t

Design values and in-service moisture conditions	$T \leq 100°F$	$100°F < T \leq 125°F$	$125°F < T \leq 150°F$
F_t and E, wet or dry	1.0	0.09	0.9
F_b, F_V, F_c, and $F_{c\perp}$			
Dry	1.0	0.8	0.7
Wet	1.0	0.7	0.5

TABLE 10.8 Size Factors C_F

Grades	F_b			F_t	F_c
	Width, in	Thickness, in 2 and 3	4		
Select Structural, No. 1 and better, No. 1, No. 2, No. 3	2, 3 and 4	1.5	1.5	1.5	1.15
	5	1.4	1.4	1.4	1.1
	6	1.3	1.3	1.3	1.1
	8	1.2	1.3	1.2	1.05
	10	1.1	1.2	1.1	1.0
	12	1.0	1.1	1.0	1.0
	14 and wider	0.9	1.0	0.9	0.9
Stud	2, 3 and 4	1.1	1.1	1.1	1.05
	5 and 6	1.0	1.0	1.0	1.0
Construction and Standard	2, 3 and 4	1.0	1.0	1.0	1.0
Utility	4	1.0	1.0	1.0	1.0
	2 and 3	0.4		0.4	0.6

where L = length of beam between inflection points, ft
d = depth, in, of beam
b = width, in, of beam
= width, in, of widest piece in multiple piece layups with various widths (thus, $b \leq 10.75$ in)
x = 20 for southern pine
= 10 for other species
K_L = loading condition coefficient (Table 10.9)

For glulam beams, the smaller of C_V and the beam stability factor C_L should be used, not both.

10.5.7 Flat-Use Factor

Design values for beams adjusted by the size factor C_{fu} assume that load will be applied to the narrow face. When load is applied to the wide face (flatwise) of dimension lumber, design values should be multiplied by the appropriate flat-use factor given in Tables 10.10 and 10.11. These factors are based on the size-adjustment equation in ASTM D245. Available test results indicate that this equation yields conservative values of C_{fu}.

When glulam timber is loaded parallel to the wide face of the laminations and the member dimension parallel to that face is less than 12 in, the design value for bending for such loading should be multiplied by the appropriate flat-use factor in Tables 10.10 and 10.11.

TABLE 10.9 Loading Condition Coefficient K_L for Glulam Beams

Loading condition	K_L
Single-span beams	
Concentrated load at midspan	1.09
Uniformly distributed load	1.0
Two equal concentrated loads at third points of span	0.96
Continuous beams or cantilevers	
All loading conditions	1.0

TABLE 10.10 Flat-Use Factors C_{fu} for Dimension Lumber

	Thickness, in	
Width, in	2 and 3	4
2 and 3	1.0	
4	1.1	1.0
5	1.1	1.05
6	1.15	1.05
8	1.15	1.05
10 and wider	1.2	1.1

TABLE 10.11 Flat-Use Factors C_{fu} for Glulam Beams

Lamination width, in	C_{fu}
10¾ or 10½	1.01
8¾ or 8½	1.04
6¾	1.07
5⅛ or 5	1.10
3⅛ or 3	1.16
2½	1.19

10.5.8 Repetitive Member Factor

Design values for bending F_b may be increased when three or more members are connected so that they act as a unit. The members may be in contact or spaced up to 24 in c to c if joined by transverse load-distributing elements that ensure action of the assembly as a unit. The members may be any piece of dimension lumber subjected to bending, including studs, rafters, truss chords, joists, and decking.

When the criteria are satisfied, the design value for bending of dimension lumber 2 to 4 in thick may be multiplied by the repetitive member factor $C_r = 1.15$.

This factor applies to three or more essentially parallel members of equal size and with the same orientation that are in direct contact with each other. Transverse connecting elements may be mechanical fasteners, such as through nails, nail gluing, tongue-and-groove joints, or bearing plates, that ensure that the members act together to resist applied bending moments.

For spaced members, the transverse distributing elements should be acceptable to the applicable regulatory agency and should be capable, as demonstrated by test, analysis, or experience, of transmitting design loads without unacceptable deflections or indications of structural weakness. The load may be uniform or concentrated, or both, applied on the surface of the distributing element.

A transverse element attached to the underside of framing members and supporting no uniform load other than its own weight and other incidental light loads, such as insulation, qualifies as a load-distributing element only for bending moment associated with its own weight and that of the framing members to which it is attached. Qualifying construction includes subflooring, finish flooring, exterior and interior wall finish, and cold-formed metal siding with or without backing. Such elements should be fastened to the framing members by approved means, such as nails, glue, staples, or snap-lock joints.

Individual members in a qualifying assembly made of different species or grades are each eligible for the repetitive-member increase in F_b if they satisfy all the preceding criteria.

10.5.9 Curvature Factor and Radial Stresses

For the curved portions of glulam beams, the design value for bending should be multiplied by the curvature factor

$$C_c = 1 - 2000(t/R)^2 \tag{10.10}$$

where t = lamination thickness, in, and R = radius of curvature, in, of inside face of lamination. t/R should not exceed 0.01 for hardwoods and southern pine or 0.008 for other softwoods. The curvature factor does not apply to design values of F_b for the straight portions of a member, regardless of curvature elsewhere.

Radial Tension or Compression. The radial stress induced by a bending moment in a member of constant cross section may be computed from

$$f_r = \frac{3M}{2Rbd} \tag{10.11}$$

where M = bending moment, in-lb
R = radius of curvature at centerline of member, in
b = width of cross section, in
d = depth of cross section, in

When M is in the direction tending to decrease curvature (increase the radius), tensile stresses occur across the grain. For this condition, the allowable tensile stress across the grain is limited to one-third the allowable unit stress in horizontal shear for southern pine for all load conditions, and for Douglas fir and larch for wind or earthquake loadings. The limit is 15 psi for Douglas fir and larch for other types of loading. These values are subject to modification for duration of load. If these values are exceeded, mechanical reinforcement sufficient to resist all radial tensile stresses is required.

When M is in the direction tending to increase curvature (decrease the radius), the stress is compressive across the grain. For this condition, the allowable stress is limited to that for compression perpendicular to grain for all species.

(K. F. Faherty and T. G. Williamson, "Wood Engineering and Construction Handbook," and D. E. Breyer, "Design of Wood Structures," 2d ed., McGraw-Hill Publishing Company, New York.)

10.5.10 Form Factor

Design values for bending F_b for beams with a circular cross section may be multiplied by a form factor $C_f = 1.18$. For a flexural member with a square cross section loaded in the plane of the diagonal (diamond-shape cross section), C_f may be taken as 1.414.

These form factors ensure that a circular or diamond-shape flexural member has the same moment capacity as a square beam with the same cross-sectional area. If a circular member is tapered, it should be treated as a beam with variable cross section.

10.5.11 Column Stability and Bearing Area Factors

Design values for compression parallel to the grain F_c should be multiplied by the column stability factor C_P specified in Art. 10.8.1.

Design values for compression perpendicular to the grain $F_{c\perp}$ apply to bearing surfaces of any length at the ends of a member and to all bearings 6 in or more long at other locations. For bearings less than 6 in long and at least 3 in from the end of a member, $F_{c\perp}$ may be multiplied by the bearing area factor

$$C_b = \frac{L_b + 0.375}{L_b} \tag{10.12}$$

where L_b = bearing length, in, measured parallel to grain. Equation (10.12) yields the values of C_b for elements with small areas, such as plates and washers, listed in Table 10.12. For round bearing areas, such as washers, L_b should be taken as the diameter.

10.5.12 Buckling Stiffness Factor

The buckling stiffness of a truss compression chord of sawn lumber subjected to combined flexure and axial compression under dry service conditions may be increased if the chord is 2 × 4 in or smaller and has the narrow face braced by nailing to plywood sheathing at least ⅜ in thick in accordance with good nailing practice. The increased stiffness may be accounted for by multiplying the design value of the modulus of elasticity E by the buckling stiffness factor C_T in column stability calculations. When the effective column length L_e, in, is 96 in or less, C_T may be computed from

$$C_T = 1 + \frac{K_M L_e}{K_T E} \tag{10.13}$$

where K_M = 2300 for wood seasoned to a moisture content of 19% or less at time of sheathing attachment
= 1200 for unseasoned or partly seasoned wood at time of sheathing attachment

TABLE 10.12 Bearing Area Factors C_b

Bearing length, in	0.50	1.00	1.50	2.00	3.00	4.00	6 or more
Bearing area factor	1.75	1.38	1.25	1.19	1.13	1.10	1.00

K_T = 0.59 for visually graded lumber and machine-evaluated lumber
= 0.82 for products with a coefficient of variation of 0.11 or less

When L_e is more than 96 in, C_T should be calculated from Eq. (10.13) with L_e = 96 in. For additional information on wood trusses with metal-plate connections, see design standards of the Truss Plate Institute, Madison, Wis.

10.5.13 Shear Stress Factor

For dimension-lumber grades of most species or combinations of species, the design value for shear parallel to the grain F_V is based on the assumption that a split, check, or shake that will reduce shear strength 50% is present (Art. 4.34). Reductions exceeding 50% are not required inasmuch as a beam split lengthwise at the neutral axis will still resist half the bending moment of a comparable unsplit beam. Furthermore, each half of such a fully split beam will sustain half the shear load of the unsplit member. The design value F_V may be increased, however, when the length of split or size of check or shake is known and is less than the maximum length assumed in determination of F_V, if no increase in these dimensions is anticipated. In such cases, F_V may be multiplied by a shear stress factor C_H greater than unity.

In most design situations, C_H cannot be applied because information on length of split or size of check or shake is not available. The exceptions, when C_H can be used, include structural components and assemblies manufactured fully seasoned with control of splits, checks, and shakes when the products, in service, will not be exposed to the weather. C_H also may be used in evaluation of the strength of members in service. The "National Design Specification for Wood Construction," American Forest and Paper Association, lists values of C_H for lumber and timber of various species.

10.6 PRESSURE-PRESERVATIVE TREATMENTS FOR WOOD

Wood members are permanent without treatment if located in enclosed buildings where good roof coverage, proper roof maintenance, good joint details, adequate flashing, good ventilation, and a well-drained site assure moisture content of the wood continuously below 20%. Also, in arid or semiarid regions where climatic conditions are such that the equilibrium moisture content seldom exceeds 20%, and then only for short periods, wood members are permanent without treatment.

Where wood is in contact with the ground or with water, where there is air and the wood may be alternately wet and dry, a preservative treatment, applied by a pressure process, is necessary to obtain an adequate service life. In enclosed buildings where moisture given off by wet-process operations maintains equilibrium moisture contents in the wood above 20%, wood structural members must be preservatively treated. So must wood exposed outdoors without protective roof covering and where the wood moisture content can go above 18 to 20% for repeated or prolonged periods.

Where wood structural members are subject to condensation by being in contact with masonry, preservative treatment is necessary.

Design values for wood structural members apply to products pressure-treated by an approved process and with an approved preservative. (The "AWPA Book of Standards," American Wood Preservers Association, Stevensville, Md., describes these approved processes.) Design values for pressure-preservative treated lumber are modified with the usual adjustment factors described in Art. 10.5 with one exception. The load duration factor for impact (Table 10.5) does not apply to structural members pressure treated with waterborne preservatives to the heavy retentions required for "marine" exposure or to structural members treated with fire-retardant chemicals.

Each type of preservative and method of treatment has certain advantages. The preservative to be used depends on the service expected of the member for the specific conditions of exposure. The minimum retentions given in the applicable American Wood Preservers Association (AWPA) standards for specific products and end-use applications may be increased where severe climatic or exposure conditions are involved.

Creosote and creosote solutions have low volatility. They are practically insoluble in water, and thus are most suitable for severe exposure, contact with ground or water, and where painting is not a requirement or a creosote odor is not objectionable.

Oilborne chemicals are organic compounds dissolved in a suitable petroleum carrier oil, and are suitable for outdoor exposure or where leaching may be a factor, or where painting is not required. Depending on the type of oil used, they may result in relatively clean surfaces. While there is a slight odor from such treatment, it is usually not objectionable.

Waterborne inorganic salts are dissolved in water or aqua ammonia, which evaporates after treatment and leaves the chemicals in the wood. The strength of solutions varies to provide net retention of dry salt required. These salts are suitable where clean and odorless surfaces are required. The surfaces are paintable after proper seasoning. See also Art. 4.36.

("Design of Wood-Frame Structures for Permanence," WCD No. 6, American Forest and Paper Association, Washington, D.C.)

Fire-retardant treatment with approved chemicals can make wood highly resistant to the spread of fire. Although wood will char where exposed to fire or high temperatures, even if it is treated with a fire retardant, chemicals will retard transmission of heat and rate of destruction. Treated with adequate quantities of an approved chemical, wood will not support combustion nor contribute fuel to a fire and will cease to burn after the ignition source is removed. The fire retardant may be applied as a paint or by impregnation under pressure. The latter is more effective. It may be considered permanent if the wood is used where it will be protected from the weather.

The effects of fire-retardant impregnation treatments on strength should be considered in design. Design values, including those for connections, for lumber and structural glued-laminated timber pressure treated with fire-retardant chemicals should be obtained from the company providing the treatment and redrying service. The load duration factor for impact (Table 10.5) should not be applied to structural members pressure-treated with fire-retardant chemicals.

10.7 DESIGN PROVISIONS FOR FLEXURAL MEMBERS

Design of flexural members requires consideration primarily of bending and shear strength, stability, deflection, and end bearing.

10.7.1 Strength of Flexural Members

The stress induced in a beam (or other flexural member) when subjected to design loads should not exceed the strength of the member. The maximum bending stress f_b at any section of a beam is given by the flexural formula

$$f_b = M/S \tag{10.14}$$

where M is the bending moment and S the section modulus. For a rectangular beam, the section modulus is $bd^2/6$ and Eq. (10.14) transforms into

$$f_b = 6M/bd^2 \tag{10.15}$$

where b is the beam width and d the depth. At every section of the beam, f_b should be equal to or less than the design value for bending F_b adjusted for all end-use modification factors (Art. 10.5).

Shear stress induced by design loads in a member should not exceed the allowable design value for shear F_V. For wood beams, the shear parallel to the grain, that is, the horizontal shear, controls the design for shear. Checking the shear stress perpendicular to the grain is not necessary inasmuch as the vertical shear will never be a primary failure mode.

The maximum horizontal shear stress f_V in a rectangular wood beam is given by

$$f_V = 3V/2bd \tag{10.16}$$

where V is the vertical shear. In calculation of V for a beam, all loads occurring within a distance d from the supports may be ignored. This is based on the assumption that loads causing the shear will be transmitted at a 45° angle through the beam to the supports.

(K. F. Faherty and T. G. Williamson, "Wood Engineering and Construction Handbook," McGraw-Hill Publishing Company, New York.)

10.7.2 Beam Stability

Beams may require lateral support to prevent lateral buckling. Need for such bracing depends on the unsupported length and cross-sectional dimensions of the members. When buckling occurs, a member deflects in the direction of its least dimension b. In a beam, b usually is taken as the width. If bracing precludes buckling in that direction, deflection can still occur, but in the direction of the strong dimension. Thus, the unsupported length L, width b, and depth d are key variables in formulas for lateral support and for reduction of design values for buckling.

Design for lateral stability of flexural members is based on a function of Ld/b^2. For lumber beams of rectangular cross section, maximum depth-width ratios should satisfy the approximate rules, based on nominal dimensions, summarized in Table 10.13. When beams are adequately braced laterally, the depth of the members below the brace may be taken as the width.

No lateral support is required when the depth does not exceed the width of a beam. In that case also, the design value for bending does not have to be adjusted for lateral instability. Similarly, if continuous support prevents lateral movement of the compression flange, lateral buckling cannot occur and the design value need not be reduced.

TABLE 10.13 Approximate Lateral-Support Rules for Lumber Flexural Members*

Depth-width ratio (nominal dimensions)	Rule
2 or less	No lateral support required
3 to 4	Hold ends in position with full-depth blocking, bridging, hangers, or other structural members
5	Hold ends in position and compression edge in line, e.g., with direct connection of sheathing, decking, or joists
6	Hold ends in position and compression edge in line, as for 5 to 1, and provide adequate bridging or blocking at intervals not exceeding 8 ft
7	Hold ends in position and both edges firmly in line

If a beam is subject to both flexure and compression parallel to grain, the ratio may be as much as 5:1 if one edge is held firmly in line, e.g., by rafters (or roof joists) and diagonal sheathing. If the combined loads will induce tension on the unbraced face of the member, the ratio may be 6:1.

*From "National Design Specification for Wood Construction," American Forest and Paper Association.

When the beam depth exceeds the width, lateral support should be provided at end bearings. This support should be so placed as to prevent rotation of the beam about the longitudinal axis. Unless the compression flange is braced at sufficiently close intervals between supports, the design value should be adjusted for lateral buckling.

The slenderness ratio R_B for beams is defined by

$$R_B = \sqrt{\frac{L_e d}{b^2}} \tag{10.17}$$

The slenderness ratio should not exceed 50.

The beam stability factor C_L may be calculated from

$$C_L = \frac{1 + (F_{bE}/F_b^*)}{1.9} - \sqrt{\left[\frac{1 + (F_{bE}/F_b^*)}{1.9}\right]^2 - \frac{F_{bE}/F_b^*}{0.95}} \tag{10.18}$$

where F_b^* = design value for bending multiplied by all applicable adjustment factors except C_{fu}, C_V, and C_L (Art. 10.5)

F_{bE} = $K_{bE}E'/R_B^2$

K_{bE} = 0.438 for visually graded lumber and machine-evaluated lumber

= 0.609 for products with a coefficient of variation of 0.11 or less

E' = design modulus of elasticity multiplied by applicable adjustment factors (Art. 10.5)

The effective length L_e for Eq. (10.17) is given in Table 10.14 in terms of the unsupported length of beam. Unsupported length is the distance between supports or the length of a cantilever when the beam is laterally braced at the supports to prevent rotation and adequate bracing is not installed elsewhere in the span. When both rotational and lateral displacement are also prevented at intermediate points,

TABLE 10.14 Effective Length L_e for Lateral Stability of Beams*

Loading	For depth greater than width†	For loads from secondary framing‡
	Simple beam§	
Uniformly distributed load	$1.63L_u + 3d$	
Load concentrated at midspan	$1.37L_u + 3d$	$1.11L_u$
Equal end moments	$1.84L_u$	
Equal concentrated loads at third points		$1.68L_u$
Equal concentrated loads at quarter points		$1.54L_u$
Equal concentrated loads at fifth points		$1.68L_u$
	Cantilever§	
Uniformly distributed load	$0.90L_u + 3d$	
Concentrated load on the end	$1.44L_u + 3d$	

*As specified in the "National Design Specification for Wood Construction," American Forest and Paper Association.

†L_u = clear span when depth d exceeds width b and lateral support is provided to prevent rotational and lateral displacement at bearing points in a plane normal to the beam longitudinal axis and no lateral support is provided elsewhere.

‡L_u = maximum spacing of secondary framing, such as purlins, when lateral support is provided at bearing points and the framing members prevent lateral displacement of the compression edge of the beam at the connections.

§For a conservative value of L_e for any loading on simple beams or cantilevers, use $1.63L_u + 3d$ when $L_u/d < 14.3$ and $1.84L_u$ when $L_u/d > 14.3$.

the unsupported length may be taken as the distance between points of lateral support. If the compression edge is supported throughout the length of the beam and adequate bracing is installed at the supports, the unsupported length is zero.

10.7.3 Deflection of Wood Beams

The design of many structural systems, particularly those with long span, is governed by deflection. Strength calculations based on allowable stresses alone may result in excessive deflection. Limitations on deflection increase member stiffness.

Deflection of wood beams is calculated by conventional elastic theory. For example, for a uniformly loaded, simple-span beam, the maximum deflection is computed from

$$\Delta = 5wL^4/384EI \tag{10.19}$$

where w = the uniform load
L = span
E = elastic modulus
I = moment of inertia

Deflection should not exceed limitations specified in the local building code nor industry-recommended limitations. (See, for example, K. F. Faherty and T. G. Williamson, "Wood Engineering and Construction Handbook," McGraw-Hill Publishing Company, New York.) Deflections also should be evaluated with respect to other considerations, such as possibility of binding of doors or cracking of partitions or glass.

Table 10.15 gives recommended deflection limits, as a fraction of the beam span, for timber beams. The limitation applies to live load or total load, whichever governs.

TABLE 10.15 Recommended Beam Deflection Limitations, in* (In Terms of Span, *l*, in)

Use classification	Live load only	Dead load plus live load
Roof beams:		
Industrial	$l/180$	$l/120$
Commercial and industrial:		
Without plaster ceiling	$l/240$	$l/180$
With plaster ceiling	$l/360$	$l/240$
Floor beams:		
Ordinary usage†	$l/360$	$l/240$
Highway bridge stringers	$l/200$ to $l/300$	
Railway bridge stringers	$l/300$ to $l/400$	

*Camber and Deflection, AITC 102, Appendix B, American Institute of Timber Construction.

†Ordinary usage classification is intended for construction in which walking comfort, minimized plaster cracking, and elimination of objectionable springiness are of prime importance. For special uses, such as beams supporting vibrating machinery or carrying moving loads, more severe limitations may be required.

Beams may be cambered to offset the effects of deflections due to design loads. Glued-laminated beams are cambered during fabrication by creation of curvature opposite in direction to that of deflections under load. Camber, however, does not increase stiffness. Table 10.16 lists recommended minimum cambers for glulam beams.

Minimum Roof Slopes. Flat roofs have collapsed during rainstorms even though they were adequately designed for allowable stresses and definite deflection limitations. The failures were caused by ponding of water as increasing deflections permitted more and more water to collect.

Roof beams should have a continuous upward slope equivalent to ¼ in/ft between a drain and the high point of a roof, in addition to minimum recommended camber (Table 10.16), to avoid ponding. When flat roofs have insufficient slope for drainage (less than ¼ in/ft), the stiffness of supporting members should be such that a 5-lb/ft^2 load will cause no more than ½-in deflection.

TABLE 10.16 Recommended Minimum Camber for Glued-Laminated Timber Beams*

Roof beams†	1½ times dead-load deflection
Floor beams‡	1½ times dead-load deflection
Bridge beams:§	
Long span	2 times dead-load deflection
Short span	2 times dead-load plus ½ applied-load deflection

*Camber and Deflection, AITC 102, Appendix B, American Institute of Timber Construction.

†The minimum camber of 1½ times dead-load deflection will produce a nearly level member under dead load alone after plastic deformation has occurred. Additional camber is usually provided to improve appearance or provide necessary roof drainage.

‡The minimum camber of 1½ times dead-load deflection will produce a nearly level member under dead load alone after plastic deformation has occurred. On long spans, a level ceiling may not be desirable because of the optical illusion that the ceiling sags. For warehouse or similar floors where live load may remain for long periods, additional camber should be provided to give a level floor under the permanently applied load.

§Bridge members are normally cambered for dead load only on multiple spans to obtain acceptable riding qualities.

Because of ponding, snow loads or water trapped by gravel stops, parapet walls, or ice dams magnify stresses and deflections from existing roof loads by

$$C_p = \frac{1}{1 - W'L^3/\pi^4 EI} \tag{10.20}$$

where C_p = factor for multiplying stresses and deflections under existing loads to determine stresses and deflections under existing loads plus ponding
W' = weight of 1 in of water on roof area supported by beam, lb
L = span of beam, in
E = modulus of elasticity of beam material, psi
I = moment of inertia of beam, in^4

(Kuenzi and Bohannan, "Increases in Deflection and Stresses Caused by Ponding of Water on Roofs," Forest Products Laboratory, Madison, Wis.)

10.7.4 Bearing Stresses in Beams

Bearing stresses, or compression stresses perpendicular to the grain, in a beam occur at the supports or at places where other framing members are supported on the beam. The compressive stress in the beam $f_{c\perp}$ is given by

$$f_{c\perp} = P/A \tag{10.21}$$

where P = load transmitted to or from the beam and A = bearing area. This stress should be less than the design value for compression perpendicular to the grain $F_{c\perp}$ multiplied by applicable adjustment factors (Art. 10.5). (The duration-of-load factor does not apply to $F_{c\perp}$ for either solid sawn lumber or glulam timber.)

Limitations on compressive stress perpendicular to the grain are set to keep deformations within an acceptable range. An expected failure mode is excessive localized deformation rather than a catastrophic type of failure.

Design values for $F_{c\perp}$ are averages based on a maximum deformation of 0.04 in in tests conforming with ASTM D143. Design values $F_{c\perp}$ for glulam beams are generally lower than for solid sawn lumber with the same deformation limit. This is due partly to use of larger-size sections for glulam beams, length of bearing and partly to the method used to derive the design values.

Where deformations are critical, the deformation limit may be decreased, with resulting reduction in $F_{c\perp}$. For example, for a deformation maximum of 0.02 in, the "National Design Specification for Wood Construction," American Forest and Paper Association, recommends that $F_{c\perp}$, psi, be reduced to $0.73F_{c\perp} + 5.60$. For glulam beams, $F_{c\perp}$ may be taken as $0.73F_{c\perp}$.

10.7.5 Example of Design of a Glulam Beam

Standard beam formulas for bending, shear, and deflection may be used to determine beam sizes. Ordinarily, deflection governs design; but for short, heavily loaded beams, shear is likely to control.

Design values for bending are tabulated in the "National Design Specification for Wood Construction." These values should be adjusted for service conditions (Art. 10.5). Section properties for solid sawn lumber and timber and glulam members are listed in the American Institute of Timber Construction "Timber Construction Manual," 3d ed., John Wiley & Sons, Inc., New York.

With the following data, design a straight glued-laminated roof beam, simply supported and uniformly loaded: span, 28 ft; spacing, 9 ft c to c; live load, such as snow, 30 lb/ft^2; dead load, 5 lb/ft^2 for deck and 7.5 lb/ft^2 for roofing. Allowable design value for bending of glulam combination grade is 2400 psi, and for horizontal shear is 200 psi, for modulus of elasticity $E = 1{,}800{,}000$ psi. Deflection limitation for total load is $L/180$, where L is the span, ft. Assume the beam is laterally supported by the deck throughout its length.

With a 15% increase for duration loading, such as snow, the allowable bending stress F_b becomes 2760 psi, and the allowable horizontal shear F_V, 230 psi.

Assume the beam will weigh 22.5 lb/lin ft, averaging 2.5 lb/ft^2. Then, the total uniform load comes to 45 psf. So the beam carries $w = 45 \times 9 = 405$ lb/lin ft.

The end shear $V = wL/2$ and the maximum shearing stress $= 3V/2 = 3wL/4$. Hence, the required area, in^2, for horizontal shear is

$$A = \frac{3wL}{4F_v} = \frac{wL}{306.7} = \frac{405 \times 28}{306.7} = 37.0 \text{ in}^2$$

The required section modulus, in^3, is

$$S = \frac{1.5wL^2}{F_b} = \frac{1.5 \times 405 \times 28^2}{2760} = 172.6 \text{ in}^3$$

If $D = 180$, the reciprocal of the deflection limitation, then the deflection equals $5 \times 1728wL^4/384EI \leq 12L/D$, where I is the moment of inertia of the beam cross section, in^4. Hence, to control deflection, the moment of inertia must be at least

$$I = \frac{1.875DwL^3}{E} = \frac{1.875 \times 180 \times 405 \times 28^3}{1{,}800{,}000} = 1688 \text{ in}^4 \qquad (10.22)$$

Assume that the beam will be fabricated with 1½-in laminations. From a table in the AITC Manual, the most economical section satisfying all three criteria is 5⅛ × 16½, with $A = 84.6$, $S = 232.5$, and $I = 1919$. But it has a size factor of 0.97%. So the allowable bending stress must be reduced to 2760 × 0.97 = 2677 psi. And the required section modulus must be increased accordingly to 172.6/0.97 = 178. The selected section still is adequate.

10.7.6 Suspended-Span Construction

Cantilever systems may be composed of any of the various types and combinations of beam illustrated in Fig. 10.1. Cantilever systems permit longer spans or larger loads for a given size member than do simple-span systems, if member size is not controlled by compression perpendicular to grain at the supports or by horizontal shear. Substantial design economies can be effected by decreasing the depths of the members in the suspended portions of a cantilever system.

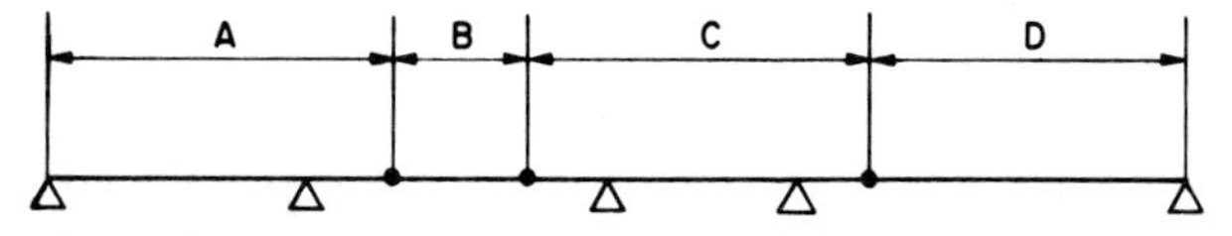

FIGURE 10.1 Cantilevered beam systems. *A* is a single cantilever. *B* is a suspended beam. *C* has a double cantilever, and *D* is a beam with the end suspended.

For economy, the negative bending moment at the supports of a cantilevered beam should be equal in magnitude to the positive moment.

Consideration must be given to deflection and camber in cantilevered multiple spans. When possible, roofs should be sloped the equivalent of ¼ in per foot of horizontal distance between the level of drains and the high point of the roof to eliminate water pockets, or provisions should be made to ensure that accumulation of water does not produce greater deflection and live loads than anticipated. Unbalanced loading conditions should be investigated for maximum bending moment, deflection, and stability.

(American Institute of Timber Construction, "Timber Construction Manual," John Wiley & Sons, Inc., New York; "Western Woods Use Book," Western Wood Products Association, Portland, Ore., "Wood Structural Design Data," American Forest and Paper Association, Washington, D.C.)

10.8 WOOD COMPRESSION MEMBERS

The design of wood columns or other types of compression members requires consideration of compressive strength parallel to the grain, end bearing, and stability, or resistance to buckling. Compressive strength considerations are the same regardless of the type of column, since the maximum compressive stress f_c induced by loads must not exceed the design value for compression parallel to the grain, F_c, multiplied by applicable adjustment factors for service conditions (Art. 10.5). (For design for end bearing, see Art. 10.11.1, and for stability, see Art. 10.8.1).

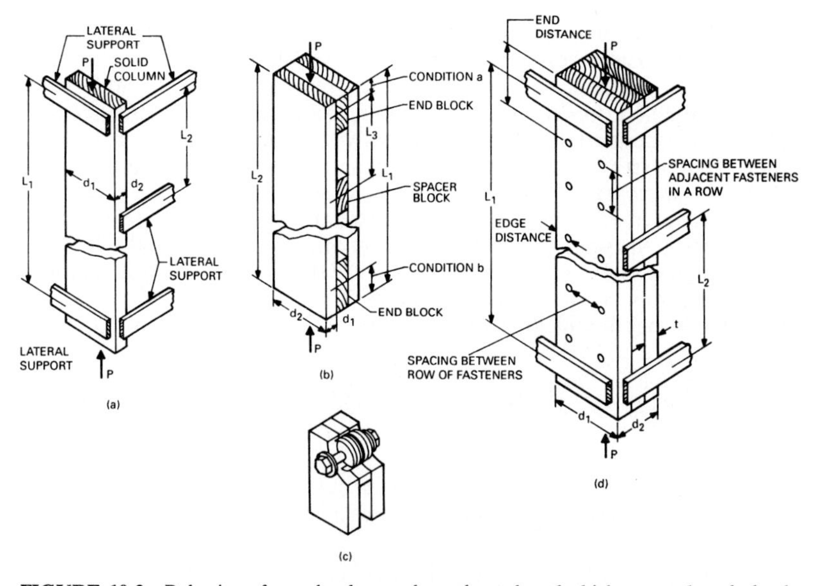

FIGURE 10.2 Behavior of wood columns depends on length-thickness or length-depth ratios: (*a*) solid wood column; (*b*) spaced column (the end distance for condition *a* should not exceed $L_1/20$ and for condition *b* should be between $L_1/20$ and $L_1/10$); (*c*) shear-plate connectors in the end block of the spaced column; (*d*) built-up column.

Wood compression members may be a solid piece of lumber or timber (Fig. 10.2*a*), or spaced columns, connector joined (Fig. 10.2*b* and *c*), or built-up (Fig. 10.2*d*).

10.8.1 Solid Columns

These consist of a single piece of lumber or timber or of pieces glued together to act as a single member. In general,

$$f_c = P/A_g \leq F_c' \tag{10.23}$$

where P = axial load on the column
A_g = gross area of column
F_c' = design value in compression parallel to grain multiplied by the applicable adjustment factors, including column stability factor C_P

There is an exception, however, applicable when holes or other reductions in area are present in the critical part of the column length most susceptible to buckling; for instance, in the portion between supports that is not laterally braced. In that case, f_c should be based on the net section and should not exceed F_c, the design value for compression parallel to grain, multiplied by applicable adjustment factors, except C_P; that is,

$$f_c = P/A_n \leq F_c \tag{10.24}$$

where A_n = net cross-sectional area.

The stability factor represents the tendency of a column to buckle and is a function of the slenderness ratio. For a rectangular wood column, a modified slenderness ratio, L_e/d, is used, where L_e is the effective unbraced length of column, and d is the smallest dimension of the column cross section. The effective column length for a solid column should be determined in accordance with good engineering practice. The effective length L_e may be taken as the actual column length multiplied by the appropriate buckling-length coefficient K_e indicated in Fig. 7.27. For the solid column in Fig. 10.2*a*, the slenderness ratio should be taken as the larger of the ratios L_{e1}/d_1 or L_{e2}/d_2, where each unbraced length is multiplied by the appropriate value of K_e. For solid columns, L_e/d should not exceed 50, except that during construction, L_e/d may be as large as 75.

The column stability factor C_P is given by

$$C_P = \frac{1 + (F_{cE}/F_c^*)}{2c} - \sqrt{\left[\frac{1 + (F_{cE}/F_c^*)}{2c}\right]^2 - \frac{F_{cE}/F_c^*}{c}} \qquad (10.25)$$

where F_c^* = design value for compression parallel to the grain multiplied by all applicable adjustment factors except C_P

$F_{cE} = K_{cE}E'/(L_e/d)^2$

E' = modulus of elasticity multiplied by adjustment factors

K_{cE} = 0.3 for visually graded lumber and machine-evaluated lumber

= 0.418 for products with a coefficient of variation less than 0.11

c = 0.80 for solid sawn lumber

= 0.85 for round timber piles

= 0.90 for glulam timber

For a compression member braced in all directions throughout its length to prevent lateral displacement, $C_P = 1.0$.

10.8.2 Built-up Columns

These often are fabricated by joining together individual pieces of lumber with mechanical fasteners, such as nails, spikes, or bolts, to act as a single member (Fig. 10.2*d*). Strength and stiffness properties of a built-up column are less than those of a solid column with the same dimensions, end conditions, and material (equivalent solid column). Strength and stiffness properties of a built-up column, however, are much greater than those of an unconnected assembly in which individual pieces act as independent columns. Built-up columns obtain their efficiency from the increase in the buckling resistance of the individual laminations provided by the fasteners. The more nearly the laminations of a built-up column deform together—that is, the smaller the slip between laminations, under compressive load—the greater is the relative capacity of the column compared with an equivalent solid column.

When built-up columns are nailed or bolted in accordance with provisions in the "National Design Specification for Wood Construction," American Forest and Paper Association, the capacity of nailed columns exceeds 60% and of bolted built-up columns, 75% of an equivalent solid column for all L/d ratios. The NDS contains criteria for design of built-up columns based on tests performed on built-up columns with various fastener schedules.

10.8.3 Spaced Columns

A wood spaced column consists of the following elements: (1) two or more individual, rectangular wood compression members with their wide faces parallel; (2) wood blocks that separate the members at their ends and one or more points between; and (3) steel bolts through the blocks to fasten the components, with split-ring or shear-plate connectors at the end blocks (Fig. 10.2*b*). The connectors should be capable of developing required shear resistance.

The advantage of a spaced column over an equivalent solid column is the increase permitted in the design value for buckling for the spaced-column members because of the partial end fixity of those members. The increased capacity may range from 2½ to 3 times the capacity of a solid column. This advantage applies only to the direction perpendicular to the wide faces. Design of the individual members in the direction parallel to the wide faces is the same for each as for a solid column. The NDS gives design criteria, including end fixity coefficients, for spaced columns.

10.9 TENSION MEMBERS

The tensile stress f_t parallel to the grain should be computed for the net section area. This stress should not exceed the design value for tension parallel to grain F_t.

Tensile stress perpendicular to the grain should be avoided when possible. The reason for this is that wood is weaker and more variable in tension perpendicular to the grain than in other properties. Furthermore, these tensile properties have not been extensively evaluated. When tension perpendicular to grain cannot be avoided, mechanical reinforcement sufficient to resist the stresses may be required. An example of a construction that induces critical tensile stress perpendicular to grain is a load hung from a beam from a point below the neutral axis. This practice should be avoided for medium to heavy loads.

10.10 COMBINED BENDING AND AXIAL LOADING

When a bending moment and an axial force act on a section of a structural member, the effects of the combined stresses must be provided for in design of the member.

10.10.1 Bending and Axial Tension

Members subjected to combined bending and axial tension should be proportioned to satisfy the interaction equations, Eqs. (10.26) and (10.27).

$$\frac{f_t}{F'_t} + \frac{f_b}{F^*_b} \leq 1 \tag{10.26}$$

$$(f_b - f_t)/F^{**}_b \leq 1 \tag{10.27}$$

where f_t = tensile stress due to axial tension acting alone
f_b = bending stress due to bending moment alone

F'_t = design value for tension multiplied by applicable adjustment factors
F^*_b = design value for bending multiplied by applicable adjustment factors except C_L
F^{**}_b = design value for bending multiplied by applicable adjustment factors except C_V

Adjustment factors are discussed in Art. 10.5.

The load duration factor C_D associated with the load of shortest duration in a combination of loads with differing duration may be used to calculate F'_t and F^*_b. All applicable load combinations should be evaluated to determine the critical load combination.

10.10.2 Bending and Axial Compression

Members subjected to a combination of bending and axial compression (beam-columns) should be proportioned to satisfy the interaction equation, Eq. 10.28.

$$\left(\frac{f_c}{F'_c}\right)^2 + \frac{f_{b1}}{[1 - (f_c/F_{cE1})]F'_{b1}} + \frac{f_{b2}}{[1 - (f_c/F_{cE2}) - (f_{b1}/F_{bE})^2]F'_{b2}} \leq 1 \quad (10.28)$$

where f_c = compressive stress due to axial compression acting alone
F'_c = design value for compression parallel to grain multiplied by applicable adjustment factors, including the column stability factor
f_{b1} = bending stress for load applied to the narrow face of the member
f_{b2} = bending stress for load applied to the wide face of the member
F'_{b1} = design value for bending for load applied to the narrow face of the member multiplied by applicable adjustment factors, including the column stability factor
F'_{b2} = design value for bending for load applied to the wide face of the member multiplied by applicable adjustment factors, including the column stability factor

For either uniaxial or biaxial bending, f_c should not exceed

$$F_{cE1} = K_{cE}E'/(L_{e1}/d_1)^2 \quad (10.29a)$$

Also, for biaxial bending, f_c should not exceed

$$F_{cE2} = K_{cE}E'/(L_{e2}/d_2)^2 \quad (10.29b)$$

and f_{b1} should not be more than

$$F_{bE} = K_{bE}E'/R_B^2 \quad (10.30)$$

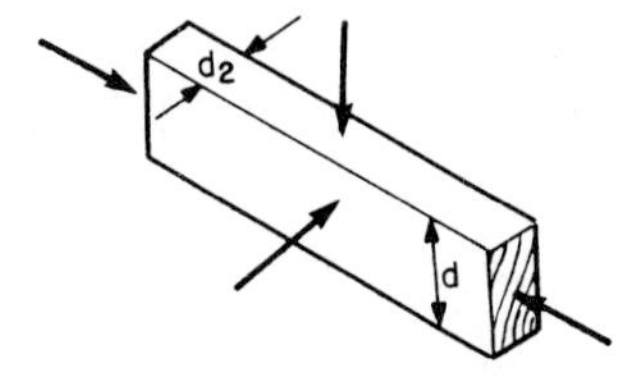

FIGURE 10.3 Beam-column.

where d_1 = width of the wide face (Fig. 10.3) and d_2 = width of the narrow face (Fig. 10.3). Slenderness ratio R_B for beams is given by Eq. (10.18). K_{bE} is defined for Eq. (10.19). The effective column lengths L_{e1} for buckling in the d_1 direction and L_{e2} for buckling in the d_2 direction, E', F_{cE1}, and F_{cE2} should be determined in accordance with Art. 10.8.1. Adjustment factors are discussed in Art. 10.5. The load duration factor C_D should be applied in calculation of F'_c, F'_{b1}, and F'_{b2} as indicated for combined bending and axial tension.

10.11 BEARING STRESSES

These may occur in a wood structural member parallel to the grain (end bearing), perpendicular to the grain, or at an angle to the grain.

10.11.1 Bearing Parallel to Grain

The bearing stress parallel to grain f_g should be computed for the net bearing area. This stress may not exceed the design value for bearing parallel to grain F_g multiplied by load duration factor C_D and temperature factor C_t (Art. 10.5). The adjusted design value applies to end-to-end bearing of compression members if they have adequate lateral support and their end cuts are accurately squared and parallel to each other.

When f_g exceeds 75% of the adjusted design value, the member should bear on a metal plate, strap, or other durable, rigid, homogeneous material with adequate strength. In such cases, when a rigid insert is required, it should be a steel plate with a thickness of 20 ga or more or the equivalent thereof, and it should be inserted with a snug fit between abutting ends.

10.11.2 Bearing Perpendicular to Grain

This is equivalent to compression perpendicular to grain. The compressive stress should not exceed the design value perpendicular to grain multiplied by applicable adjustment factors, including the bearing area factor (Art. 10.5.11). In the calculation of bearing area at the end of a beam, an allowance need not be made for the fact that, as the beam bends, it creates a pressure on the inner edge of the bearing that is greater than at the end of the beam.

10.11.3 Bearing at an Angle to Grain

The design value F_g for bearing parallel to grain and the design value for bearing perpendicular to grain $F_{c\perp}$ differ considerably. When load is applied at an angle θ with respect to the grain, where $0 \leq \theta \leq 90°$ (Fig. 10.4), the design value for bearing lies between F_g and $F_{c\perp}$. The "National Design Specification for Wood

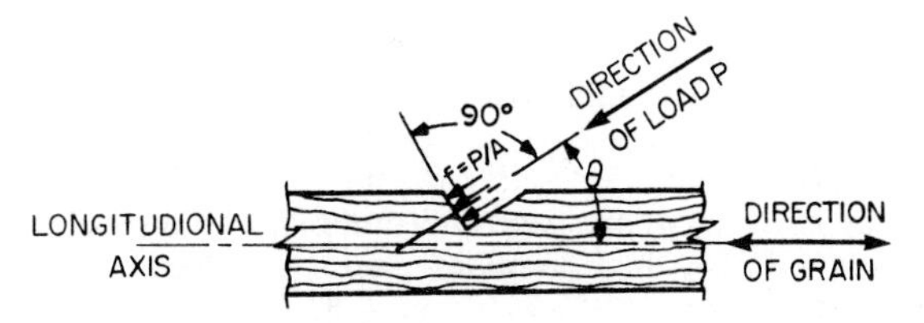

FIGURE 10.4 Load applied to a wood member at an angle to the grain.

Construction," American Forest and Paper Association, recommends that the design value for such loading be calculated from the Hankinson formula:

$$F'_n = \frac{F'_g F'_{c\perp}}{F'_g \sin^2\theta + F'_{c\perp} \sin^2\theta} \tag{10.31}$$

where F'_n = adjusted design value for bearing at angle θ to the grain (longitudinal axis)

F'_g = design value for end bearing multiplied by applicable adjustment factors

$F'_{c\perp}$ = design value for compression perpendicular to grain multiplied by applicable adjustment factors

10.12 STRUCTURAL PANELS

Wood-based structural panels are thin, flat, composite materials capable of resisting applied loads in specific applications. Structural panels fall into three basic categories based on the manufacturing process used: plywood, mat-formed panels (oriented strand board, or OSB), and composite panels.

Plywood—a flat panel built up of sheets of veneer, called plies. These are united under pressure by a bonding agent. The adhesive bond between plies is as strong as or stronger than solid wood. Plywood is constructed of an odd number of layers with the grain of adjacent layers perpendicular. Layers may consist of a single ply or two or more plies laminated with parallel grain direction. Outer layers and all odd-numbered layers generally have the grain direction oriented parallel to the long dimension of the panel. The odd number of layers with alternating grain direction equalizes strains, reduces splitting, and minimizes dimensional change and warping of the panel.

Mat-formed panel—any wood-based panel that does not contain veneer but is consistent with the definition of structural-use panels, including products such as waferboard and oriented strand board.

Oriented strand board—an engineered structural wood panel composed of compressed wood strands arranged in layers at right angles to one another and bonded with fully waterproof adhesive.

Composite panel—any panel containing a combination of veneer and other wood-based materials.

A structural panel may contain either softwoods or hardwoods. Panels approved for use in building-code-regulated construction carry the trademark of a code-approved agency, such as the American Plywood Association (APA). Most construction grades have either an Exterior or Exposure 1 durability classification and are made with fully waterproof adhesives. Exposure classifications are defined as follows:

Exterior—panels that are suitable for permanent exposure to weather or moisture.

Exposure 1—panels that are suitable for uses not permanently exposed to the weather but may be used where exposure durability to resist effects of moisture due to construction delays, high humidity, water leakage, or other conditions of similar severity is required.

Exposure 2—panels that are suitable for interior use where exposure durability to resist effects of high humidity and water leakage is required.

Interior—panels that are suitable for interior use where they will be subjected to only temporary, minor amounts of moisture.

10.12.1 Standards for Structural Panels

Structural panels approved for building-code-regulated construction are manufactured under one or more of three standards:

1. *U.S. Product Standard PS 1-83 for Construction & Industrial Plywood (PS 1).* It applies to plywood only. This voluntary product standard covers the wood species, veneer grading, glue bonds, panel construction and workmanship, dimensions and tolerances, marking, moisture content, and packing of plywood intended for construction and industrial uses. Also included are test methods to determine compliance and a glossary of trade terms and definitions. A quality certification program is provided, whereby qualified testing agencies inspect, sample, and test products identified as complying with the standard. Information regarding generally available sizes, methods of ordering, and reinspecting practices also is provided.

2. *Voluntary Product Standard PS 2-92, Performance Standard for Wood-Based Structural-Use Panels (PS 2).* It applies to all types of wood-based panels (typically plywood, OSB, and composite). It establishes requirements for assessing the acceptability of wood-based structural-use panels for construction sheathing and single-floor applications. It also provides a basis for common understanding among the producers, distributors, and users of these products. It covers performance requirements, adhesive bond durability, panel construction and workmanship, dimensions and tolerances, marking, and moisture content of structural-use panels. The standard also includes test methods to determine compliance and a glossary of trade terms and definitions. A quality certification program is provided, whereby qualified testing agencies inspect, sample, and test products for qualification under the standard.

3. *APA Performance Standards and Policies for Structural-Use Panels (PRP 108).* It is similar to PS 2 but also includes performance-based qualification procedures for siding panels.

10.12.2 Plywood Grades

Plywood grades are generally identified in terms of the veneer grade used on the face and back of the panel; for example, A-B, B-C, . . . , or by a name suggesting the panel's intended end use, such as APA Rated Sheathing or APA Rated Sturd-I-Floor.

Veneer grades define veneer appearance in terms of natural, unrepaired-growth characteristics and allowable number and size of repairs that may be made during manufacture (Table 10.17). The highest quality veneer grades are N and A. The minimum grade of veneer permitted in Exterior plywood is C grade. D-grade veneer is used in panels intended for interior use or applications protected from permanent exposure to weather.

Panels with B-grade or better veneer faces are always sanded smooth in the manufacturing process to fulfill the requirements of their intended end use—applications such as cabinets, shelving, furniture, and built-ins. Rated Sheathing panels are unsanded since a smooth surface is not a requirement of their intended end use. Still other panels, such as Underlayment, Rated Sturd-I-Floor, C-D

TABLE 10.17 **Veneer-Grade Designations**

Grade	Description
N	Smooth surface "natural finish" veneer. Select, all heartwood or all sapwood. Free of open defects. Allows not more than six repairs, wood only, per 4- × 8-ft panel, made parallel to grain and well matched for grain and color.
A	Smooth, paintable. Not more than 18 neatly made repairs, boat, sled, or router type, and parallel to grain, permitted. May be used for natural finish in less-demanding applications. Synthetic repairs permitted.
B	Solid surface. Shims, circular repair plugs, and tight knots up to 1 in across grain permitted. Some minor splits and synthetic repairs permitted.
C Plugged	Improved C veneer with splits limited to ⅛-in width and knotholes and borer holes limited to ¼ × ½ in. Admits some broken grain. Synthetic repairs permitted.
C	Tight knots to 1½ in. Knotholes up to 1 in across grain and some up to 1½ in if total width of knots and knotholes is within specified limits. Synthetic or wood repairs and discoloration and sanding defects that do not impair strength permitted. Limited splits allowed. Stitching permitted.
D	Knots and knotholes up to 2½-in wide across grain and ½ in larger within specified limits, limited splits, and stitching permitted. Limited to Exposure 1 or interior panels.

Plugged, and C-C Plugged, require only touch sanding for "sizing" to make the panel thickness more uniform.

Unsanded and touch-sanded panels, and panels with B-grade or better veneer on one side only, usually carry the trademark on the panel back. Panels with both sides of B-grade or better veneer, or with special overlaid surfaces, such as High-Density Overlay, usually carry the trademark on the panel edge.

10.12.3 Plywood Group Number

Plywood can be manufactured from over 70 species of wood. These species are divided on the basis of strength and stiffness into five groups under U.S. Product Standard PS 1-83. Strongest species are in Group 1; the next strongest in Group 2, etc. The group number that appears in the trademark on some APA trademarked panels, primarily sanded grades, is based on the species used for face and back veneers. Where face and back veneers are not from the same species group, the higher group number is used, except for sanded panels ⅜ in thick or less and decorative panels of any thickness. These are identified by face species if C or D grade backs are at least ⅛ in thick and are not more than one species group number larger. Some species are used widely in plywood manufacture, others rarely.

OSB panels, being composed of flakes or strands instead of veneers, are graded without reference to veneers or species, and composite panels are graded on an OSB performance basis by end use and exposure durability. Typical panel trademarks for all three panel types and an explanation of how to read them are shown in Fig. 10.5.

The"Design/Construction Guide—Residential & Commercial," American Plywood Association, Tacoma, Wash., contains a comprehensive summary of plywood grades and trademarks and their applications.

10.12.4 Span Ratings for Plywood

APA Rated Sheathing, APA Rated Sturd-I-Floor, and APA Rated Siding carry numbers in their trademarks called span ratings. These denote the maximum spacing, in, c to c of supports for plywood panels in construction applications. Except for Rated Siding panels, the span rating in the trademark applies when the long panel dimension is across supports, unless the strength axis is otherwise identified. The span rating in the trademark of Rated Siding panels applies when they are installed vertically.

The span rating in Rated Sheating trademarks appears as two numbers separated by a slash (Fig. 10.5*a*), such as 32/16 and 48/24. The left-hand number denotes the maximum recommended spacing of supports when the panel is used for roof sheathing with the long dimension or strength axis of the panel across three or more supports. The right-hand number indicates the maximum recommended spacing of supports when the panel is used for subflooring with the long dimension or strength axis of the panel across three or more supports. A panel marked 32/16, for example, may be used for roof decking over supports 32 in c to c or for subflooring over supports 16 in c to c. An exception is Rated Sheathing intended for use as wall sheathing only. The trademarks for such panels contain only a single number similar to the span rating for APA Rated Siding and Sturd-I-Floor.

The Span Ratings in the trademarks on APA Rated Sturd-I-Floor and APA Rated Siding panels appear as a single number. Rated Sturd-I-Floor panels are designed specifically for single-floor (combined subfloor-underlayment) applications under carpet and pad. They are manufactured with span ratings of 16, 20, 24, 32, and 48 in.

APA Rated Siding is available with span ratings of 16 and 24 in. Span-rated panels and lap siding may be connected directly to studs, or over nonstructural wall sheathing, or over nailable panel or lumber sheathing (double-wall construction). Panels and lap siding with a span rating of 16 in may be applied directly to studs spaced 16 in c to c. Those bearing a span rating of 24 in may be connected directly to studs 24 in c to c. All APA Rated Siding panels may be applied horizontally directly to studs 16 or 24 in c to c, if horizontal joints are blocked. The span rating of APA Rated Siding panels refers to the maximum recommended spacing of vertical rows of nails rather than to stud spacing when the panels are applied to nailable structural sheathing.

10.12.5 Availability of Plywood Grades

Some panel grades, thicknesses, span ratings, or species may be difficult to obtain in some areas. Check with your supplier for availability or include an alternative panel in specifications. Standard panel dimensions are 4 × 8 ft, although some

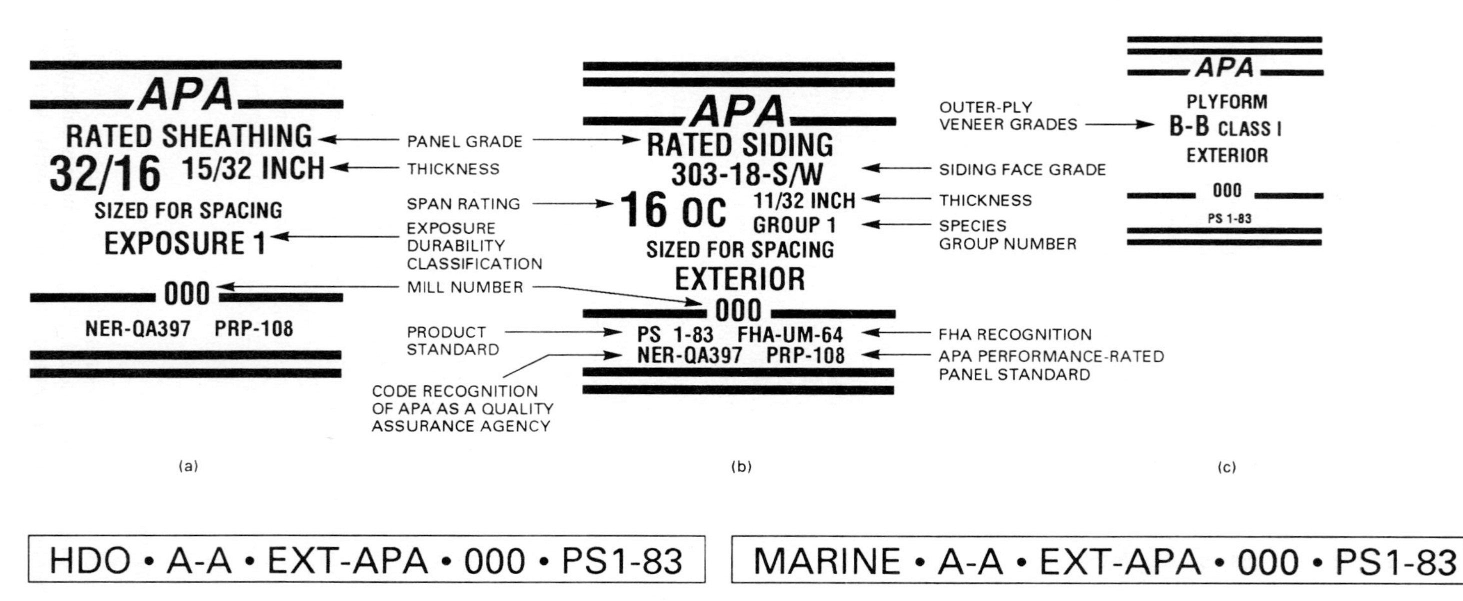

FIGURE 10.5 Typical trademarks for structural panels. (*a*) APA Rated Sheathing with a thickness of $^{15}/_{32}$ in and a span rating of 32 in for use as roof decking and 16 in for use as subflooring, suitable for Exposure 1 conditions (not permanently exposed to weather). (*b*) APA Rated Siding, grade 303-18-S/W, with a span rating of 16 in. (*c*) APA Plyform, intended for use in formwork for concrete. (*d*) APA high-density overlay (HDO), abrasion resistant and suitable for exterior applications (used for concrete forms, cabinets, countertops, and signs). (*e*) APA Marine, used for boat hulls.

mills also produce plywood panels 9 or 10 ft long or longer. OSB panels may be ordered in lengths up to 28 ft.

10.12.6 APA Rated Sturd-I-Floor

APA Rated Sturd-I-Floor (copyrighted name), or single-floor grade, is a span-rated product designed specifically for use in single-layer floor construction beneath carpet and pad. The maximum spacing of floor joists, or span rating, is stamped on each panel. Panels are manufactured with span ratings of 16, 20, 24, 32, and 48 in. These assume the panel continuous over two or more spans with the long dimension or strength axis across supports (Fig. 10.6). The span rating in the trademark applies when the long panel dimension is across supports unless the strength axis is otherwise identified.

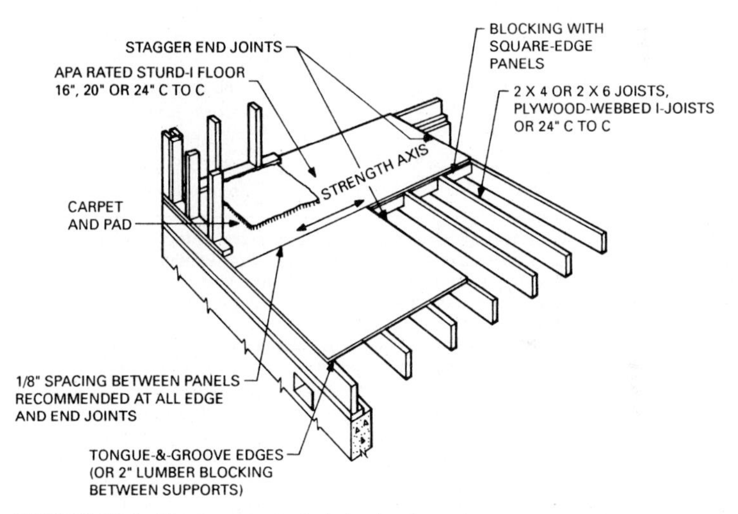

FIGURE 10.6 Floor constructed of structural panels.

Glue-nailing is preferred, though panels may be nailed only. Application provisions for both methods are given in Table 10.18. Uniform live loads are given in the APA "Design/Construction Guide—Residential & Commercial."

10.12.7 Panel Subflooring

The limiting factor in design of floors is deflection under concentrated loads at panel edges. Nailing provisions for APA panel subflooring (Fig. 10.7) are given in Table 10.19. Other code-approved fasteners, however, may be used. The span ratings in Table 10.19 applied to Rated Sheathing or sheathing grades only and are the minimum for the span indicated. The span ratings assume panels continuous over two or more spans with the long dimension or strength axis across supports.

TABLE 10.18 Fastener Size, Type, and Spacing for APA Rated Sturd-I-Floor[a]

Span rating (maximum joist spacing), in	Minimum panel thickness, in[b]	Glue Nailed[c]			Nailed Only		
		Nail size and type	Spacing, in		Nail size and type	Spacing, in	
			Supported panel edges	Intermediate supports		Supported panel edges	Intermediate supports[d]
16	19/32	6d ring- or screw-shank[e]	12	12	6d ring- or screw-shank	6	12
20	19/32	6d ring- or screw-shank[e]	12	12	6d ring- or screw-shank	6	12
24	23/32	6d ring- or screw-shank[e]	12	12	6d ring- or screw-shank	6	12
	7/8	8d ring- or screw-shank[e]	6	12	8d ring- or screw-shank	6	12
32	7/8	8d ring- or screw-shank[e]	6	12	8d ring- or screw-shank	6	12
48	13/32	8d ring- or screw-shank[f]	6	[g]	8d ring- or screw-shank	6	[g]

[a]Heavy traffic and concentrated loads may require construction in excess of the minimum values in the table.
[b]Panels of a specific thickness may have more than one span rating. Panels with a span rating larger than the joist spacing may be substituted for panels of the same thickness with a span rating equal to the joist spacing.
[c]Adhesives should conform to APA Specification AFG-01. Only solvent-based glues should be used for nonveneer panels with sealed surfaces and edges.
[d]Local building code may require nail spacing 10 in c to c at intermediate supports for floors.
[e]8d common nails may be used if these nails are not available.
[f]10d common nails may be used with 1⅛-in panels if supports are well seasoned.
[g]Nails should be spaced 6 in c to c for 48-in spans and 12 in c to c for 32-in spans.

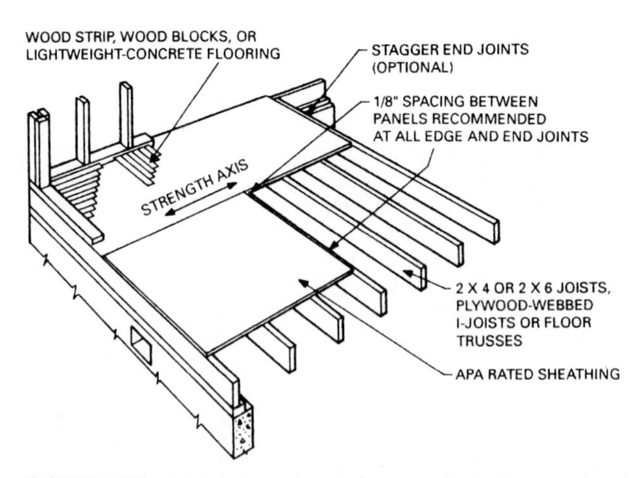

FIGURE 10.7 Subfloor constructed of structural panels.

TABLE 10.19 Spans and Nailing for APA Panel Subflooring[a]
(APA Rated Sheathing)

Span rating or group no.	Minimum panel thickness, in	Maximum span, in	Nail size and type[b]	Nail spacing, in	
				At supported edges	At intermediate supports[c]
24/16	7/16	16	6d common	6	12
32/16	15/32	16[d]	8d common[e]	6	12
40/20	19/32	20[d,f]	8d common	6	12
48/24	23/32	24	8d common	6	12

[a]For recommendations for subfloors under ceramic tile, see the APA "Design/Construction Guide—Residential and Construction." For subfloors under gypsum concrete, obtain data from topping producers.
[b]Other code-approved fasteners may be used.
[c]Local building codes may require nail spacing 10 c to c at intermediate supports for floors.
[d]A 24-in span may be used if ¾-in-thick wood strip flooring is installed perpendicular to joists.
[e]If the panel is ½ in or less thick, 6d common nails are permitted.
[f]A 24-in span may be used if at least 1½ in of lightweight concrete is applied over the panels.

Panel subflooring may also be glued for added stiffness and to reduce squeaks if it satisfied nailing provisions in Table 10.18. Long edges should be tongue-and-groove or supported with blocking unless:

1. A separate underlayment layer is installed with its joints offset from those in the subfloor. The minimum thickness of underlayment should be ¼ in for subfloors on spans up to 24 in and 11/32 in or more on spans longer than 24 in.

2. A minimum of 1½ in of lightweight concrete is applied over the panels.

3. A ¾-in wood strip flooring is installed over the subfloor.

In some nonresidential buildings, greater traffic and heavier concentrated loads may require construction in excess of the minimums given. Where joists are 16 in c to c, for example, panels with a span rating of 40/20 or 48/24 provide greater stiffness. For beams or joists 24 or 32 in c to c, 1⅛-in-thick panels provide additional stiffness.

10.12.8 Wall Systems

Rated siding (panel or lap) may be applied directly to studs or over nonstructural fiberboard, or gypsum or rigid-foam-insulation sheathing. Nonstructural sheathing is defined as sheathing not recognized by building codes as meeting both bending and racking-strength requirements.

A single layer of panel siding, since it is strong and rack resistant, eliminates the cost of installing separate structural sheathing or diagonal wall bracing. Panel sidings are normally installed vertically (Fig. 10.8*a*), but most may also be placed horizontally (long dimension across supports) if horizontal joints are blocked (Fig. 10.8*b*). Maximum stud spacings for both applications are given in Table 10.20.

Rated Sheathing meets building-code wall-sheathing requirements for bending and racking strength without let-in corner bracing. Installation provisions are given

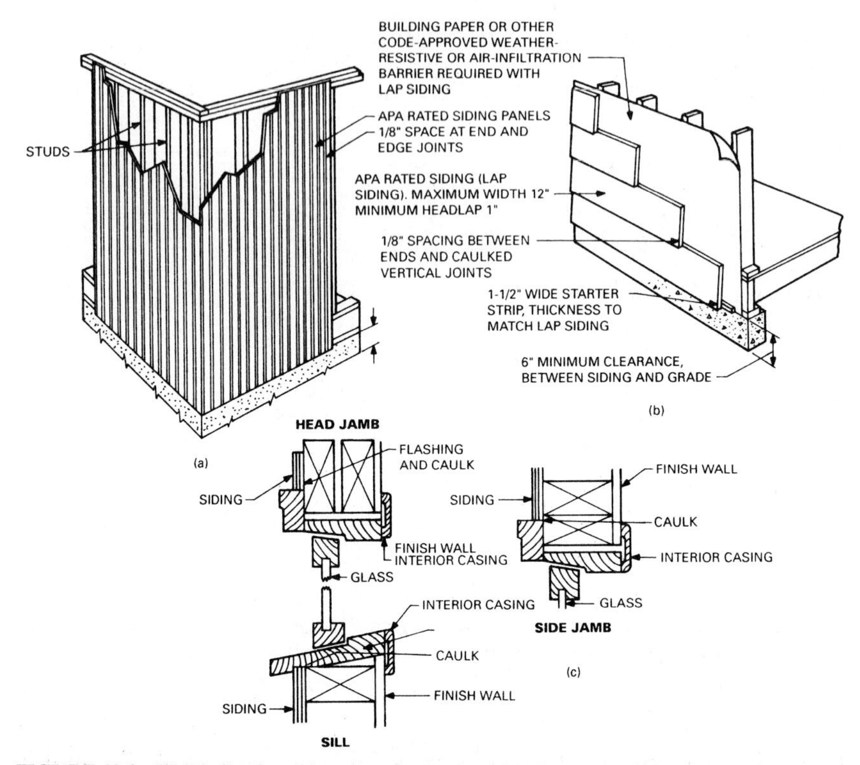

FIGURE 10.8 Wall built of wood studs and APA Rated Siding panels. (*a*) Vertical panel siding. If permitted by the local building and energy codes, no building paper is required when panel edges are shiplapped, battened, and caulked. If caulking is not used with unbattened square butt joints, apply a water repellent to panel edges. Caulk around windows and doors. (*b*) Horizontal lap siding. Diagonal bracing or other code-approved bracing methods for the wall should be provided. For engineered shear-wall segments, use APA Rated Sheathing under the lap siding. (*c*) Siding joint details at a window.

in Table 10.21 and Fig. 10.9. When ½-in gypsum or fiberboard sheathing is used, APA Rated Sheathing corner panels of the same thickness can also eliminate costly let-in bracing. The APA Rated Sheathing, $^{15}/_{32}$ or ½ in thick, should be nailed to studs spaced 16 or 24 in c to c with 6d common nails spaced 6 in c to c along panel edges and 12 in c to c at intermediate supports. When corner panels are PS 1 plywood, 1½-in roofing nails at 4 in along panel edges and 12 in at intermediate supports may be used.

Building paper is generally not required over wall sheathing, except under stucco or under brick veneer where required by the local building code. Recommended wall sheathing spans with brick veneer and masonry are the same as those for nailable panel sheathing.

10.12.9 Allowable Loads for APA Structural-Use Panels

Because it is sometimes necessary to have engineering design information for structural panel products for conditions not specifically covered in other literature, APA

TABLE 10.20 Recommended Installation of Sturd-I-Wall

	Siding Description[a]	Nominal thickness, in, or span rating	Maximum stud spacing, in: Long dimension vertical	Maximum stud spacing, in: Long dimension horizontal	Nail size (nonstaining box, siding or casing nails)[b,c]	Nail spacing,[d] in: At panel edges	Nail spacing,[d] in: At intermediate supports
Panel siding	APA MDO Exterior	11/32 and 3/8	16	24	6d for siding 1/2 in thick or less; 8d for thicker siding.	6[e]	12[f]
		1/2 and thicker	24	24			
	APA Rated Siding Exterior	16	16	16[g]			
		24	24	24			
Lap siding	APA Rating Siding—Lap Exterior	16		16	6d for siding 1/2 in thick or less; 8d for thicker siding.	16 along bottom edge	
		24		24		24 along bottom edge	

[a]Recommendations apply to all species groups for veneered APA Rated Siding, including APA 303 siding.

[b]Next regular size nailing should be used if panel siding is applied over foam-insulation sheathing. If lap siding is installed over such sheathing up to 1 in thick, 10d (3-in) nails should be used for 3/8- or 7/16-in siding, 12d (3 1/4-in) nails for 15/32- or 1/2-in siding, and 16d (3 1/2-in) nails for 19/32-in or thicker siding. Nonstaining box nails should be used for siding installed over foam-insulation sheathing.

[c]Hot-dipped or hot-tumbled galvanized steel nails are recommended for most siding applications, but electrically or mechanically galvanized nails are acceptable if the plating meets or exceeds thickness requirements of ASTM A641 Class 2 coatings and is protected by a yellow chromate coating. For best performance, stainless steel or aluminum nails are an alternative. Galvanized nails may react adversely under wet conditions with some wood species and cause staining if left unfinished. Such staining can be minimized if the siding is finished in accordance with APA recommendations or if the siding is protected by a roof overhang from direct exposure to moisture and weathering.

[d]Recommendations of siding manufacturers may vary.

[e]Nails should be spaced 3 in c to c along panel edges for a braced wall section with 11/32- or 3/8-in panel siding applied horizontally over studs spaced 24 in c to c.

[f]Where wind velocity exceeds 80 mph, nails attaching siding to intermediate studs within 10% of the width of the narrow side from wall corners should be spaced 6 in c to c.

[g]Stud spacing may be 24 in for veneer-faced siding panels.

publishes separate design-section capacities for the various span ratings for these products. These values are listed in APA Technical Note N375, "Design Capacities of APA Performance-Rated Structural-Use Panels." The APA "Plywood Design Specification" contains load-span tables that apply to APA trademarked structural-use panels qualified and manufactured in accordance with PS 2-92 or APA PRP-108, "Performance Standards and Policies for Structural-Use Panels." These panels include plywood, composite, and mat-formed products, such as oriented strand board. Loads are provided for applications where the panel strength axis is applied across or parallel to supports. For each combination of span L and Span Rating, loads are given for deflections of $L/360$, $L/240$ and $L/180$, and maximum loads controlled by bending and shear capacity. The values may be adjusted for panel type, load duration, span conditions, and moisture. Table 10.22 is provided to assist in selecting panel constructions for specific span ratings.

TABLE 10.21 Recommended Installation of Panel Wall Sheathing[a]

Panel span rating	Maximum stud spacing, in	Nail size[b]	Nail spacing, in	
			At supported panel edges	At intermediate supports
12/0, 16/0, 20/0 or Wall-16 in c to c	16	6d for panels ½ in thick or less; 8d for thicker panels	6	12
24/0, 24/16, 32/16 or Wall-24 in c to c	24			

[a]Applies to APA Rated Sheathing panels that are continuous over two or more spans. Different requirements may apply to nailable panel sheathing when the exterior covering is to be nailed to the sheathing. See the APA Design/Guide.

[b]For common, smooth, annular, spiral-thread, or galvanized box nails. Other code approved fasteners, however, may be used.

Some structural-panel applications are not controlled by uniform loads. Residential floors are a good example. They are commonly designed for 40-psf live load. The allowable uniform floor load on panels with maximum span according to APA recommendations is greatly in excess of typical design loads. This excess does not mean that floor spans for structural panels can be increased, but only that there is considerable reserve strength and stiffness for *uniform* loads. Actually, the recommendations for panel floors are based on performance under concentrated

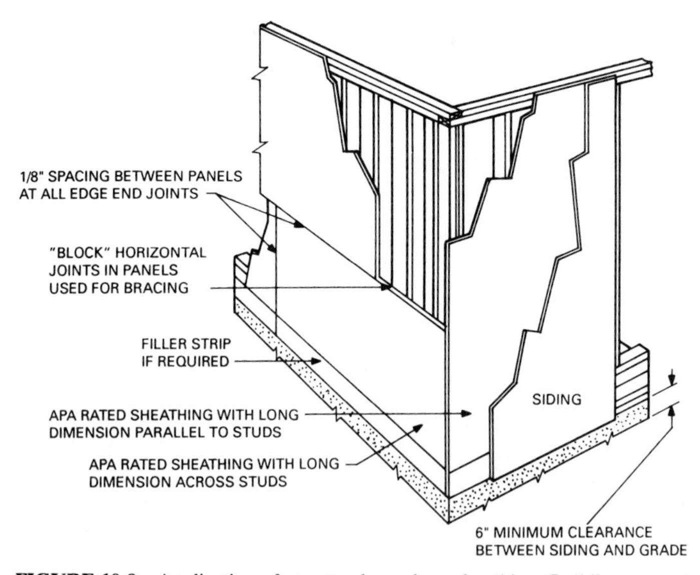

FIGURE 10.9 Application of structural panels as sheathing. Building paper is not required over the sheathing, except under stucco and brick veneer.

TABLE 10.22 Typical APA Panel Constructions*

	Plywood				
Span rating	3-Ply	4-Ply	5-Ply†	COM-PLY	OSB
APA Rated Sheathing					
24/0	X				X
24/16					X
32/16	X	X	X		X
40/20	X	X	X		X
48/24		X	X		X
APA Rated Sturd-I-Floor					
16 in c to c					
20 in c to c		X	X	X	X
24 in c to c		X	X	X	X
32 in c to c			X	X	X
48 in c to c			X	X	X

*Constructions may not be available in every area. Check with suppliers concerning availability.

†Applies to plywood with five or more layers.

loads, how the floor "feels" to passing foot traffic, and other subjective factors that relate to public acceptance. The maximum floor and roof spans for structural panels should always be checked before a final panel selection is made for these applications.

10.12.10 Panel Shear Walls

While the wall systems described in Art. 10.12.8 will provide sufficient strength under normal conditions in residential and light-frame construction, shear walls may be desirable or required in areas with frequent seismic activity or high wind loads. Shear walls are also advisable in commercial and industrial construction.

Either Rated Sheathing or all-veneer plywood Rated Siding can be used in shear walls, Table 10.23 gives maximum shears for walls with Rated Sheathing, with plywood Rated Siding installed directly to studs (Sturd-I-Wall), or with panels applied over gypsum sheathing for walls required to be fired rated from the outside.

To design a shear wall, follow these steps:

1. Determine the unit shear transferred by the roof diaphragm to the wall. This generally will be one-fourth the area of the adjacent wall, multiplied by the wind load, divided by the length of the shear wall being designed (subtract length of large openings).

TABLE 10.23 Maximum Shear, lb/ft, for APA Panel Shear Walls for Wind or Seismic Loading[a]

(For framing of Douglas fir, larch, or southern pine)[b]

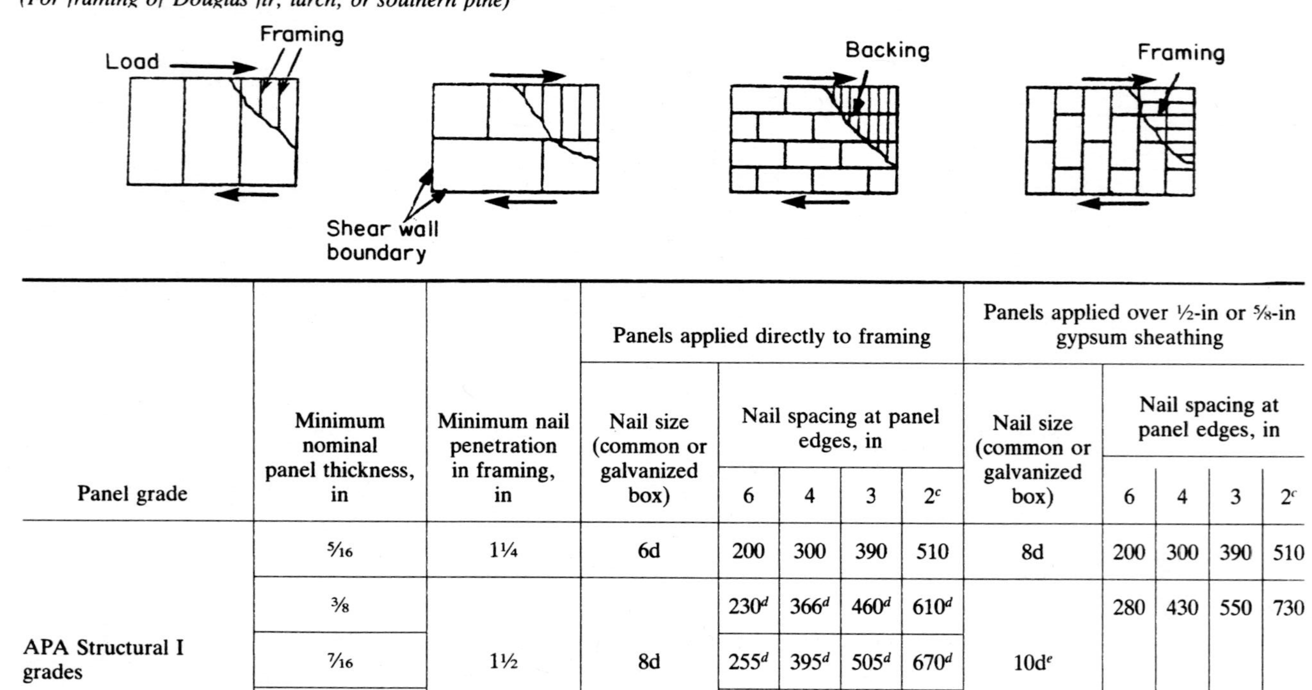

Panel grade	Minimum nominal panel thickness, in	Minimum nail penetration in framing, in	Panels applied directly to framing: Nail size (common or galvanized box)	Nail spacing at panel edges, in: 6	4	3	2[c]	Panels applied over ½-in or ⅝-in gypsum sheathing: Nail size (common or galvanized box)	Nail spacing at panel edges, in: 6	4	3	2[c]
APA Structural I grades	5/16	1¼	6d	200	300	390	510	8d	200	300	390	510
	3/8			230[d]	366[d]	460[d]	610[d]		280	430	550	730
	7/16	1½	8d	255[d]	395[d]	505[d]	670[d]	10d[e]				
	15/32			280	430	550	730					
	15/32	1⅝	10d[e]	340	510	665	870					

	5/16 or 1/4[f]	1¼	6d	180	270	350	450	8d	180	270	350	450
	3/8			200	300	390	510		200	300	390	510
APA Rated Sheathing; APA Rated Siding and other APA grades[g] except species Group 5	3/8	1½	8d	220[d]	320[d]	410[d]	530[d]	10d[e]	260	380	490	640
	7/16			240[d]	350[d]	450[d]	585[d]					
	15/32			260	380	490	640					
	15/32	1⅝	10d[e]	310	460	600	770					
	19/32			340	510	665	870					
APA Rated Siding and other APA grades[g] except species Group 5	5/16[f]	1¼	Nail size (galvanized casing) 6d	140	210	275	360	Nail size (galvanized casing) 8d	140	210	275	360
	3/8	1½	8d	160	240	310	410	10d	160	240	310	410

[a]All panel edges should be backed with framing, which should have a nominal width of 2 in or more. The plywood may be installed horizontally or vertically. Space nails 6 in c to c along intermediate framing members for 3/8-in and 7/16-in panels attached to studs that are spaced 24 in c to c. For other conditions and panel thicknesses, space nails 12 in c to c on intermediate supports.

[b]For framing of other species, determine the species group of the lumber from the AFPA National Design Specification. Then, proceed as follows: For common or galvanized box nails, find the shear value from the above table for the nail size for Structural I panels, regardless of the actual grade. For galvanized casing nails, use the shear value given in the table. Next, multiply this value by 0.82 for Lumber Group III or 0.65 for Lumber Group IV.

[c]Framing at adjoining edge panels should have a nominal width of 3 in or more and nails should be staggered.

[d]If studs are spaced at most 16 in c to c or if panels are applied with the long dimension across the studs, shears may be increased to the values shown for 15/32-in-thick sheathing with the same nail spacing.

[e]Framing at adjoining panels should have a nominal width of 3 in or more and nails with a penetration into the framing exceeding 1⅝-in and 3-in spacing c to c should be staggered.

[f]For 3/8-in-thick siding or APA Rated Siding when used as exterior siding applied directly to the framing, stud spacing of 16 in c to c is recommended.

[g]Values apply to all-veneer plywood APA Rated Siding panels only. For APA Rated Siding on framing spaced 16 in c to c, the plywood may be 11/32 in thick, 3/8 in thick, or thicker. Thickness at the point of nailing on panel edges governs shear values.

2. Determine the required panel grade and thickness and the nailing schedule from Table 10.23. Check the anchor bolts in the sill plate for shear.

3. Check wall framing on each end of the shear wall and design a foundation anchor if required (see Fig. 10.10).

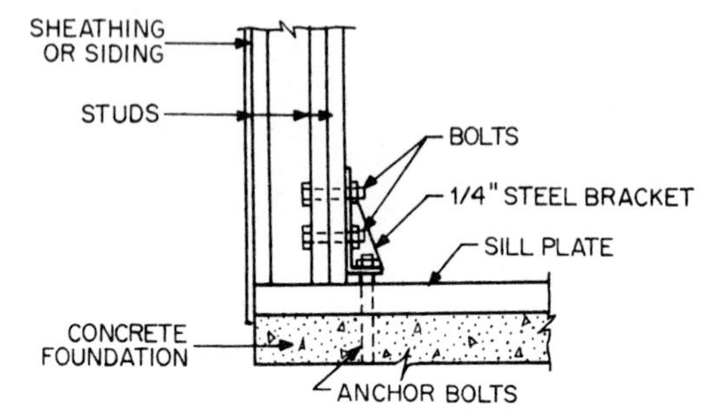

FIGURE 10.10 Foundation anchor for a wood shear wall.

10.12.11 Panel Roof Sheathing

Table 10.24 lists maximum uniform roof live loads for APA Rated Sheathing Exposure 1, and Structural I Rated Sheathing, Exposure 1 or Exterior. Uniform-load deflection limits are 1/180 of the span under live load plus dead load, and 1/240 under live load only. Panels are assumed continuous over two or more spans with the long dimension or strength axis across supports (Fig. 10.11). Special conditions, such as heavy concentrated loads, may require constructions in excess of these minimums, or allowable live loads may have to be decreased for dead loads greater than 10 psf, for example, for tile roofs.

Good performance of built-up, single-ply, or modified bitumen roofing applied on low-slope roofs requires a stiffer deck than does prepared roofing applied on pitched roofs. Although Span-Rated panels used as roof sheathing at maximum span are adequate structurally, an upgraded system is recommended for low-slope roofs. Table 10.25 lists maximum spans for low-slope roof decks. Live loads can be determined from Table 10.24, and minimum fastener requirements are given in Table 10.26.

Rated Sheathing is equally effective under built-up roofing, asphalt or glass-fiber shingles, tile roofing; or wood shingles or shakes. Roof trusses spaced 24 in c to c are widely recognized as the most economical construction for residential roofs, particularly when 3/8- or 7/16-in, 24/0 sheathing with panel clips is used. However, use of fewer supports with thicker panels, for example, 23/32- or 3/4-in, 48/24 panels over framing 48 in c to c, is also cost-effective for long-span flat or pitched roofs. Live loads are given in Table 10.24. Nailing provisions are given in Table 10.26.

When support spacing exceeds the maximum length of an unsupported edge, as given in Table 10.24, provide adequate block, tongue-and-groove edges, or other edge support such as panel clips. Some types of panel clips, in addition to edge support, automatically assure recommended panel spacing. When required, use one panel clip per span of less than 28 in and two for 48-in or longer spans.

TABLE 10.24 Maximum Uniform Roof Live Loads, psf,[a] for APA Rated Sheathing[b] and APA Rated Sturd-I-Floor

(Long dimension perpendicular to supports)[c]

(*a*) APA Rated Sheathing[b]

Panel span rating	Minimum panel thickness, in	Maximum span, in		Spacing of supports, in, c to c							
		With edge support[d]	Without edge support	12	16	20	24	32	40	48	60
12/0	5/16	12	12	30							
16/0	5/16	16	16	70	30						
20/0	5/16	20	20	120	50	30					
24/0	3/8	24	20[e]	170	100	60	30				
24/16	7/16	24	24	190	100	65	40				
32/16	15/32	32	28	220	155	120	70	30			
40/20	19/32	40	32		200	165	125	60	30		
48/24	23/32	48	36			210	165	95	45	35	

(*b*) APA Rated Sturd-I-Floor

Stud spacing, in	Minimum panel thickness, in	Maximum span, in		Spacing of supports, in, c to c							
		With edge support[d]	Without edge support	12	16	20	24	32	40	48	60
16 oc	19/32	24	24	185	100	65	40				
20 oc	19/32	32	32	270	150	100	60	30			
24 oc	23/32	40	36		240	160	100	50	30	25	
32 oc	7/8	48	40			250	185	100	60	40	
48 oc	1 3/32	60	48				290	160	100	65	40

[a]Maximum loads include an assumed 10 psf for dead load.
[b]Includes APA Rated Sheathing Ceiling Deck.
[c]Applies to panels 24 in or more wide.
[d]Edge support is provided by such means as tongue-and-groove edges, panel edge clips (generally one midway between each support but two equally spaced between supports that are 48 in c to c), or lumber or other blocking.
[e]Maximum span is 24 in for 15/32-in and 1/2-in panels.

10.12.12 Preframed Roof Panels

Spans of 8 to 12 ft are usually the most practical with preframed panel construction, although spans up to 30 ft are not uncommon. Unsanded 4 × 8-ft panels with stiffeners preframed at 16 or 24 in c to c are common. The long dimension of panels typically runs parallel to supports. Stiffeners and roof purlins provide support for all panel edges. Minimum nailing requirements for preframed panels are the same as for roof sheathing.

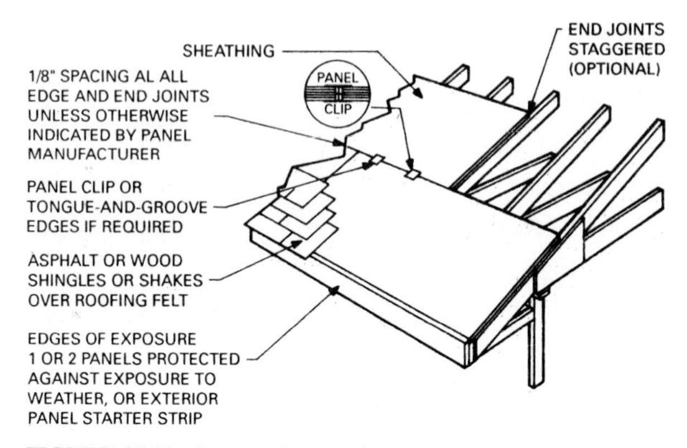

FIGURE 10.11 Structural panels installed as roof sheathing.

For preframed panels 8 × 8 ft or larger, the long panel dimension may run either parallel or perpendicular to stiffeners spaced 16 or 24 in c to c. Placing the long dimension across supports may require edge support such as panel clips or cleats between stiffeners at midspan in accordance with Table 10.24.

Deflection limits are 1/180 of the span for total load; 1/240 for live load only. Nailing requirements for preframed panels are the same as for roof sheathing. See "APA Design/Construction Guide—Residential and Commercial," American Plywood Association, for recommended maximum roof loads.

10.12.13 Panel Diaphragms

With only slight design modifications, any panel roof-deck system described in Arts. 10.12.11 and 10.12.12 will also function as an engineered roof or floor diaphragm to resist high wind and seismic loading.

TABLE 10.25 Maximum Spans for APA Panel Roof Decks for Low-Slope Roofs*

(Panels set with long dimension perpendicular to supports and continuous over two or more spans)

Grade	Minimum nominal panel thickness, in	Minimum span rating	Maximum span, in	Number of panel clips per span†
APA Rated Sheathing	15/32	32/16	24	1
	19/32	40/20	32	1
	23/32	48/24	48	2

*Built-up, single-ply, or modified bitumen roofing systems may be used for low-slope roofs. For guaranteed or warranted roofs, check with the membrane manufacturer for requirements for an acceptable deck.

†Edge support may also be provided by tongue-and-groove edges or solid blocking.

TABLE 10.26 Minimum Fastening for APA Panel Roof Sheathing

Panel thickness,† in	Nailing*		
	Size	Spacing, in	
		Panel edges	Intermediate
5/16 to 1/2	6d	6	12
19/32 to 1	8d	6	12‡
1 1/8	8d or 10d	6	12‡

*In general, use common smooth or deformed-shank nails for panels up to 1 in thick. For 1 1/8-in-thick panels, use 8d ring- or screw-shank or 10d common smooth-shank nails. Other approved fasteners, however, may be used.

†For stapling asphalt shingles to panels 5/16 in or more thick, use staples with a 15/16-in minimum crown width and a 1-in-long leg. Space the staples in accordance with the recommendations of the shingle manufacturer.

‡For spans of 48 in or more, space the nails 6 in c to c at all supports.

The ability of a diaphragm to function effectively as a deep beam, transferring lateral loads to shear walls, is related to the quality of the connections. Nailing is critical, since shears are transmitted through these fasteners. Common nails provide required strength. Other nail types may be used when their lateral bearing values are considered in the design. Load-carrying capacity is highest when the diaphragm is blocked.

Where 1 1/8-in roof panels are desired, such as for heavy timber construction, shear values for 19/32-in panels are used. Blocked shear values for 1 1/8-in panels may be obtained by specifying stapled tongue-and-groove (T&G) edges. Staples should be 16 ga, 1 in long, with a 3/8-in crown. They should be driven through the T&G edges 3/8-in from the joint so as to penetrate the tongue. Staples should be spaced at one-half of the boundary nail spacing for Cases 1 and 2, and at one-third the boundary nail spacing for Case 3 through 6, as illustrated in Table 10.27, which gives panel and fastening recommendations for roof diaphragms. Panels and framing are assumed already designed for perpendicular loads. To design a diaphragm, follow these steps:

1. Determine lateral loads and resulting shears.
2. Determine the nailing schedule (Table 10.27). Consider the load direction with respect to joints.
3. Compute the chord stress due to bending moment. Provide adequate splices. Check deflections. Check the anchorage of boundary framing, for example, to walls.

For situations where greater diaphragm capacities are necessary and framing with a nominal thickness of 3 or 4 in is available, diaphragms may be constructed using heavier nailing schedules, such as that given in the Uniform Building Code.

("Diaphragms," American Plywood Association, Tacoma, Wash.)

TABLE 10.27 Maximum Shear, lb/ft, for APA Panel Diaphragms for Wind or Seismic Loading

*(For Framing of Douglas fir, larch, or southern pine)**

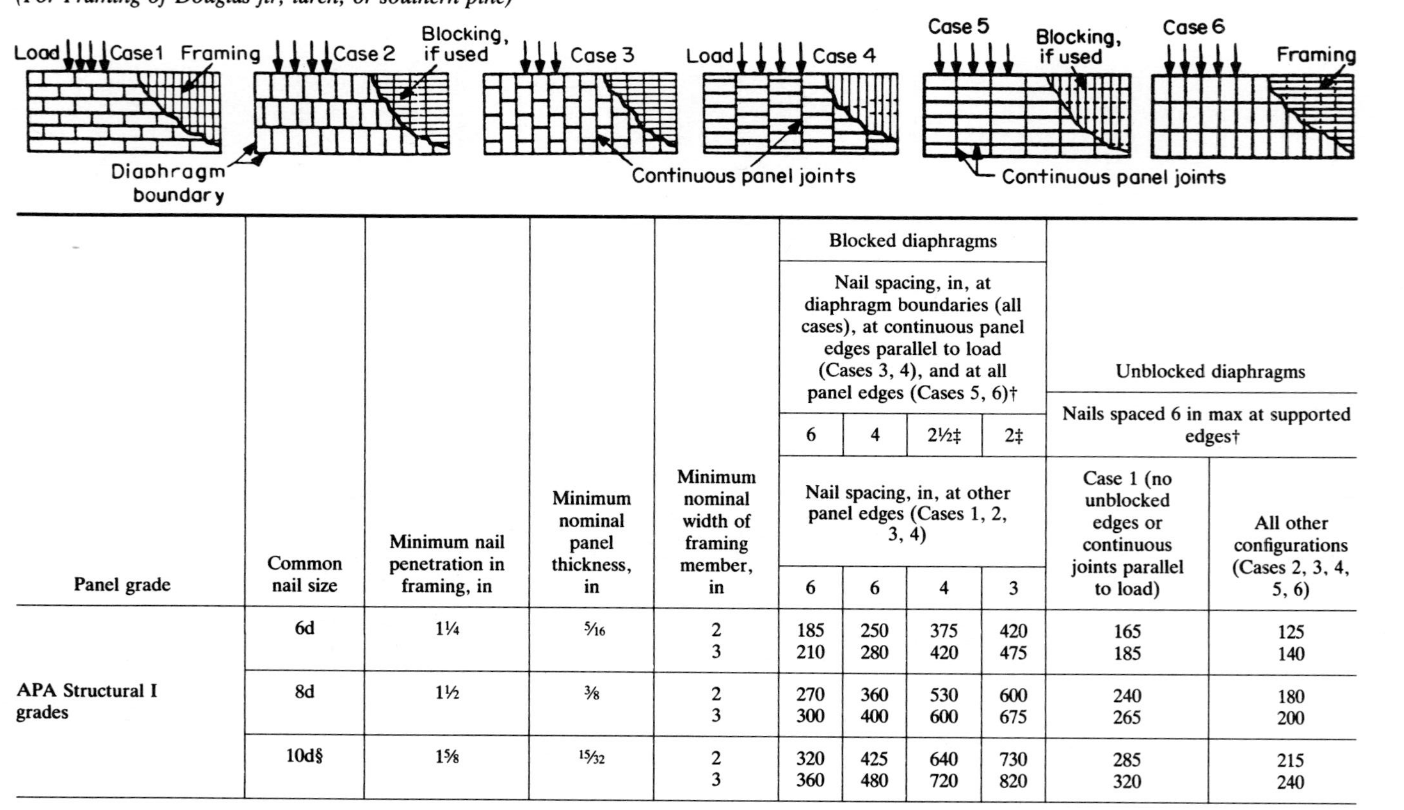

Panel grade	Common nail size	Minimum nail penetration in framing, in	Minimum nominal panel thickness, in	Minimum nominal width of framing member, in	Blocked diaphragms				Unblocked diaphragms	
					Nail spacing, in, at diaphragm boundaries (all cases), at continuous panel edges parallel to load (Cases 3, 4), and at all panel edges (Cases 5, 6)†				Nails spaced 6 in max at supported edges†	
					6	4	2½‡	2‡		
					Nail spacing, in, at other panel edges (Cases 1, 2, 3, 4)				Case 1 (no unblocked edges or continuous joints parallel to load)	All other configurations (Cases 2, 3, 4, 5, 6)
					6	6	4	3		
APA Structural I grades	6d	1¼	5/16	2	185	250	375	420	165	125
				3	210	280	420	475	185	140
	8d	1½	3/8	2	270	360	530	600	240	180
				3	300	400	600	675	265	200
	10d§	1⅝	15/32	2	320	425	640	730	285	215
				3	360	480	720	820	320	240

APA Rated Sheathing, APA Rated Sturd-I-Floor and other APA grades except Species Group 5	6d	1¼	5/16	2	170	225	335	380	150	110
				3	190	250	380	430	170	125
			3/8	2	185	250	375	420	165	125
				3	210	280	420	475	185	140
	8d	1½	3/8	2	240	320	480	545	215	160
				3	270	360	540	610	240	180
			7/16	2	255	340	505	575	230	170
				3	285	380	570	645	255	190
			15/32	2	270	360	530	600	240	180
				3	300	400	600	675	265	200
	10d§	1⅝	15/32	2	290	385	575	655	255	190
				3	325	430	650	735	290	215
			19/32	2	320	425	640	730	285	215
				3	360	480	720	820	320	240

*For framing of other species, determine the species group of the lumber from the AFPA National Design Specification. Then, proceed as follows: Find the shear value from the above table for the nail size for Structural I panels, regardless of the actual grade. Multiply this value by 0.82 for Lumber Group III or 0.65 for Lumber Group IV.

†In general, space nails 12 in c to c along intermediate framing members, but use 6 in when supports are 48 in c to c. Local building codes may require 10-in nail spacing at intermediate supports for floors.

‡Framing at adjoining panel edges should have a nominal width of 3 in or more and nails should be staggered.

§Framing at adjoining panel edges should have a nominal width of 3 in or more and nails with a penetration into the framing exceeding 1⅝ in and 3-in spacing c to c should be staggered.

Note: Designs based on diaphragm stresses depend on the direction of continuous panel joints with respect to the loads, rather than on the direction of the long dimension of the panels. For blocked diaphragms, continuous framing may be in either direction.

10.13 DESIGN VALUES FOR MECHANICAL CONNECTIONS

Nails, staples, spikes, wood screws, bolts, and timber connectors, such as shear plates and split rings, are used for connections in wood construction. Because determination of stress distribution in connections made with wood and metal is complicated, information for design of joints has been developed from tests and experience. The data indicate that design values and methods of design for mechanical connections are applicable to both solid sawn lumber and laminated members. The "National Design Specification for Wood Construction," (NDS) American Forest and Paper Association, lists design values for connections made with various types of fasteners. Design values for connections made with more than one type of fastener, however, should be based on tests or special analysis.

10.14 ADJUSTMENT OF DESIGN VALUES FOR CONNECTIONS

Nominal design values for laterally loaded fasteners Z, withdrawal of fasteners W, load parallel to grain P, and load perpendicular to grain Q should be multiplied by applicable adjustment factors to determine adjusted design values Z', W', P', and Q', respectively. Table 10.28 summarizes the adjustment factors that should be applied to the design values Z, W, P, and Q for connections made with commonly used types of fasteners. The load applied to a connection should not exceed the adjusted design value.

10.14.1 Load Duration Factor

Nominal design values should be multiplied by the load duration factor C_D specified in Art. 10.5.1, except that C_D may not exceed 1.6 or when the capacity of the connection is controlled by the strength of metal.

10.14.2 Wet-Service Factor

Nominal design values apply to wood that will be used under dry-service conditions; that is, where moisture content of the wood will be a maximum of 19% of the oven-dry weight, as would be the case in most covered structures. For connections in wood that is unseasoned or partly seasoned, or when connections will be exposed to wet-service conditions in use, nominal design values should be multiplied by the appropriate wet-service factor C_M in Table 10.29.

10.14.3 Temperature Factor

Nominal design values should be multiplied by the appropriate temperature factor C_t listed in Table 10.30 for connections that will experience sustained exposure to elevated temperatures up to 150°F.

TABLE 10.28 Adjusted Design Values for Connections*

Bolts:

$$Z' = ZC_DC_MC_tC_gD_\Delta$$

Split-ring and shear-plate connectors:

$$P' = PC_DC_MC_tC_gC_\Delta C_dC_{st}$$
$$Q' = QC_DC_MC_tC_gC_\Delta C_d$$

Lag screws:

$$W' = WC_DC_MC_tC_{eg}$$
$$Z' = ZC_DC_MC_tC_gC_\Delta C_dC_{eg}$$

Wood screws:

$$W' = WC_DC_MC_t$$
$$Z' = ZC_DC_MC_tC_dC_{eg}$$

Nails and spikes:

$$W' = WC_DC_MC_tC_{tn}$$
$$Z' = ZC_DC_MC_tC_dC_{eg}C_{di}C_{tn}$$

Metal plate connectors:

$$Z' = ZC_DC_MC_t$$

Drift bolts and drift pins:

$$W' = WC_DC_MC_tC_{eg}$$
$$Z' = ZC_DC_MC_tC_gC_\Delta C_dC_{eg}$$

Spike grids:

$$Z' = ZC_DC_MC_tC_\Delta$$

*The adjustment factors are as follows:

C_D = load duration factor, not to exceed 1.6 for connections (Art. 10.14.1)
C_M = wet-service factor, not applicable to toe-nails load in withdrawal (Art. 10.14.2)
C_t = temperature factor (Art. 10.14.3)
C_g = group-action factor (Art. 10.14.4)
C_Δ = geometry factor (Art. 10.14.5)
C_d = penetration-depth factor (Art. 10.14.6)
C_{eg} = end-grain factor (Art. 10.14.7)
C_{st} = metal-side-plate factor (Art. 10.14.8)
C_{di} = diaphragm factor (Art. 10.14.9)
C_{tn} = toenail factor (Art. 10.14.10)

10.14.4 Group-Action Factor

Nominal design values for split-ring connectors, shear-plate connectors, bolts with diameter D up to 1 in, and lag screws in a row should be multiplied by the group-action factor C_g given in Table 10.31. The NDS contains design criteria for determination of C_g for additional configurations.

For determination of C_g, a row of fasteners is defined as any of the following:

1. Two or more split-ring or shear-plate connectors aligned with the direction of the load.
2. Two or more bolts with the same diameter, loaded in shear, and aligned with the direction of the load.
3. Two or more lag screws of the same type and size loaded in single shear and aligned with the direction of the load.

TABLE 10.29 Wet-Service Factors C_M for Connections

Fastener type	Condition of wood*: At time of fabrication	Condition of wood*: In service	C_M
Split-ring or shear-plate connectors†	Dry	Dry	1.0
	Partially seasoned	Dry	‡
	Wet	Dry	0.8
	Dry or wet	Partially seasoned or wet	0.67
Bolts or lag screws	Dry	Dry	1.0
	Partially seasoned or wet	Dry	§
	Dry or wet	Exposed to weather	0.75
	Dry or wet	Wet	0.67
Wood screws	Dry or wet	Dry	1.0
	Dry or wet	Exposed to weather	0.75
	Dry or wet	Wet	0.67
Common wire nails, box nails:			
for withdrawal loads	Dry	Dry	1.0
	Partially seasoned or wet	Wet	1.0
	Partially seasoned or wet	Dry	0.25
	Dry	Subject to wetting and drying	0.25
for lateral loads	Dry	Dry	1.0
	Partially seasoned or wet	Dry or wet	0.75
	Dry	Partially seasoned or wet	0.75

*Conditions of wood for determining wet-service factors for connections:
Dry wood—moisture content up to 19%.
Wet wood—moisture content at or above 30% (approximate fiber saturation point).
Partly seasoned wood—moisture content between 19% and 30%.
Exposed to weather—wood will vary in moisture content from dry to partly seasoned but is not expected to reach the fiber saturation point when the connection is supporting full design load.
Subject to wetting and drying—wood will vary in moisture content from dry to partly seasoned or wet, or vice versa, with consequent effects on the tightness of the connection.

†For split-ring or shear-plate connectors, moisture-content limitations apply to a depth of ¾ in below the surface of the wood.

‡When split-ring or shear-plate connectors are installed in wood that is partly seasoned at time of fabrication but will be dry before full design load is applied, proportional intermediate wet-service factors may be used.

§When bolts or lag screws are installed in wood that is wet at the time of fabrication but will be dry before full design load is applied, the following wet service factors C_M apply:
For one fastener only or two or more fasteners placed in a single row parallel to grain, or fasteners placed in two or more rows parallel to grain with separate splice plates for each row, $C_M = 1.0$.
For all other arrangements, $C_M = 0.4$.

When bolts or lag screws are installed in wood that is partly seasoned at the time of fabrication but will be dry before full design load is applied, proportional intermediate wet-service factors may be used.

TABLE 10.30 Temperature Factor C_t for Connections

In-service moisture conditions*	$T \leq 100°F$	$100°F < T \leq 125°F$	$125°F < T \leq 150°F$
Dry	1.0	0.8	0.7
Wet	1.0	0.7	0.5

*Wet and dry service conditions are defined in a Table 10.29 footnote.

TABLE 10.31 Group-Action Factors

(*a*) For bolt or lag-screw connections with wood side members*

A_s/A_m†	A_s,‡ in²	Number of fasteners in a row: 2	3	4	5	6	7	8	9	10	11	12
0.5	5	0.98	0.92	0.84	0.75	0.68	0.61	0.55	0.50	0.45	0.41	0.38
	12	0.99	0.96	0.92	0.87	0.81	0.76	0.70	0.65	0.61	0.57	0.53
	20	0.99	0.98	0.95	0.91	0.87	0.83	0.78	0.74	0.70	0.66	0.62
	28	1.00	0.98	0.96	0.93	0.90	0.87	0.83	0.79	0.76	0.72	0.69
	40	1.00	0.99	0.97	0.95	0.93	0.90	0.87	0.84	0.81	0.78	0.75
	64	1.00	0.99	0.98	0.97	0.95	0.93	0.91	0.89	0.87	0.84	0.82
1	5	1.00	0.97	0.91	0.85	0.78	0.71	0.64	0.59	0.54	0.49	0.45
	12	1.00	0.99	0.96	0.93	0.88	0.84	0.79	0.74	0.70	0.65	0.61
	20	1.00	0.99	0.98	0.95	0.92	0.89	0.86	0.82	0.78	0.75	0.71
	28	1.00	0.99	0.98	0.97	0.94	0.92	0.89	0.86	0.83	0.80	0.77
	40	1.00	1.00	0.99	0.98	0.96	0.94	0.92	0.90	0.87	0.85	0.82
	64	1.00	1.00	0.99	0.98	0.97	0.96	0.95	0.93	0.91	0.90	0.88

(*b*) For 4-in split-ring or shear-plate connectors with wood side members§

A_s/A_m†	A_s‡ in²	Number of fasteners in a row: 2	3	4	5	6	7	8	9	10	11	12
0.5	5	0.90	0.73	0.59	0.48	0.41	0.35	0.31	0.27	0.25	0.22	0.20
	12	0.95	0.83	0.71	0.60	0.52	0.45	0.40	0.36	0.32	0.29	0.27
	20	0.97	0.88	0.78	0.69	0.60	0.53	0.47	0.43	0.39	0.35	0.32
	28	0.97	0.91	0.82	0.74	0.66	0.59	0.53	0.48	0.44	0.40	0.37
	40	0.98	0.93	0.86	0.79	0.72	0.65	0.59	0.54	0.49	0.45	0.42
	64	0.99	0.95	0.91	0.85	0.79	0.73	0.67	0.62	0.58	0.54	0.50
1	5	1.00	0.87	0.72	0.59	0.50	0.43	0.38	0.34	0.30	0.28	0.25
	12	1.00	0.93	0.83	0.72	0.63	0.55	0.48	0.43	0.39	0.36	0.33
	20	1.00	0.95	0.88	0.79	0.71	0.63	0.57	0.51	0.46	0.42	0.39
	28	1.00	0.97	0.91	0.83	0.76	0.69	0.62	0.57	0.52	0.47	0.44
	40	1.00	0.98	0.93	0.87	0.81	0.75	0.69	0.63	0.58	0.54	0.50
	64	1.00	0.98	0.95	0.91	0.87	0.82	0.77	0.72	0.67	0.62	0.58

*For fastener diameter $D = 1$ in and fastener spacing $s = 4$ in in bolt or lag-screw connections with modulus of elasticity for wood $E = 1{,}400{,}000$ psi. Tabulated values of C_g are conservative for $D < 1$ in, $s < 4$ in, or $E > 1{,}400{,}000$ psi.

†A_s = cross sectional area of the main members before boring or grooving and A_m = sum of the cross-sectional areas of the side members before boring or grooving. When $A_s/A_m > 1$, uses A_m/A_s.

‡When $A_s/A_m > 1$, use A_m instead of A_s.

§For spacing $s = 9$ in in connections made with 4-in split rings or shear plates with modulus of elasticity for wood $E = 1{,}400{,}000$ psi. Tabulated values of C_g are conservative for 2½-in split-ring connectors, 2⅝-in shear-plate connectors, $s < 9$ in, or $E > 1{,}400{,}000$ psi.

When fasteners in adjacent rows are staggered but close together, they may have to be treated as a single row in determination of C_g. This occurs when the distance between adjacent rows is less than one-fourth of the spacing between the closest fasteners in adjacent rows.

The group-action factor is necessary because of the following characteristics of a joint with more than two fasteners in a row: The two end fasteners carry a larger load than the interior fasteners. With six or more fasteners in a row, the two end fasteners carry more than 50% of the load. With bolts, however, a small redistribution of load from the end bolts to the interior bolts occurs due to crushing of the wood at the end bolts. If failure is in shear, though, a partial failure occurs before substantial redistribution of load takes place.

10.14.5 Geometry Factor

The NDS specifies minimum edge distance, end distance, and spacing required for full design value for bolts, lag screws, and split-ring and shear-plate connectors. The NDS also tabulates nominal design values for these fasteners based on the minimum distances. When the end distance or the spacing is less than the minimum required for full design value but larger than the minimum required for reduced design value, nominal design values should be multiplied by the smallest applicable geometry factor C_Δ determined from the end distance and spacing requirements for the type of connector specified. The smallest geometry factor for any connector in a group should be applied to all in the group. For multiple shear connections or for asymmetric three-member connections, the smallest geometry factor for any shear plane should be applied to all fasteners in the connection.

10.14.6 Penetration-Depth Factor

Nominal lateral design values for lag screws, wood screws, nails, and spikes are based on a specific penetration into the main member, as established by the shank diameter. The NDS also sets a minimum penetration into the main member for a reduced design value for each fastener. When the penetration is larger than the minimum but less than that assumed in establishment of the full lateral design value, linear interpolation should be used in determination of the penetration-depth factor C_d. In no case should C_d exceed unity. Table 10.32 lists minimum penetration and assumed penetration for full design value as well as C_d for each type of fastener.

TABLE 10.32 Penetration-Depth Factor*

Penetration, p	Lag screws	Wood screws	Nails or spikes
For full design value	$8D$	$7D$	$12D$
Minimun p	$4D$	$4D$	$6D$
C_d	$p/8D$	$p/7D$	$p/12D$

*D = fastener diameter

10.14.7 End-Grain Factor

Woods screws, lag screws, nails, and spikes are used in two types of connections, withdrawal and lateral load. In withdrawal connections, the load is applied parallel to the length of the fastener. In laterally loaded connections, the load is applied perpendicular to the length of the fastener. Either type of connection is weaker when fasteners are inserted in the ends of a member, parallel to the grain, rather than in the side grain.

Withdrawal Design Value. Wood screws, nails, and spikes should not be loaded in withdrawal from end grain. Tests show that splitting of the wood member causes erratic results relative to those for withdrawal from side grain.

When lag screws are loaded in withdrawal from end grain, the nominal withdrawal design value should be multiplied by the end-grain factor $C_{eg} = 0.75$.

Lateral Design Value. When lag screws, wood screws, nails, or spikes are inserted, with the axis in the direction of the wood grain, into the end grain of a main member, the nominal design value for lateral loading should be multiplied by the end-grain factor $C_{eg} = 0.67$.

10.14.8 Metal-Side-Plate Factor

Larger design values are permitted when metal side plates are used in lieu of wood plates. For joints made with nails, spikes, or wood screws, the design value for wood side plates may be multiplied by the metal-side-plate factor $C_{st} = 1.25$. For 4-in shear-plate connectors, the nominal design value for load parallel to grain P should be multiplied by the appropriate C_{st} given in Table 10.33. The values depend on the species of wood used in the connection, such as group A, B, C, or D listed in the NDS.

TABLE 10.33 Metal-Side-Plate Factors for Shear-Plate Connectors*

Species group†	C_{st}
A	1.18
B	1.11
C	1.05
D	1.00

*For 4-in shear plates loaded parallel to grain.
†For components of each species group, see the groupings in the NDS.

10.14.9 Diaphragm Factor

A diaphragm is a large, thin structural element that is loaded in its plane. When nails or spikes are used in a diaphragm connection, the nominal lateral design value should be multiplied by the diaphragm factor $C_{di} = 1.1$.

10.14.10 Toenail Factor

For such connections as stud to plate, beam to plate, and blocking to plate, toenailing is generally used. The NDS recommends that toenails be driven at an angle of about 30° with the face of the stud, beam, or blocking and started about one-third the length of the nail from the end of the member. For toenailed connections, the nominal lateral design values for connections with nails driven into side grain should be multiplied by the toenail factor $C_{tn} = 0.83$.

10.14.11 Adjustments for Fire-Retardant Treatment

For connections made with lumber or structural glulam timber pressure-treated with fire-retardant chemicals, design values should be obtained from the company providing the treatment and redrying service. The load-duration factor for impact does not apply to such connections.

10.15 BOLTS

The "National Design Specification for Wood Construction," American Forest and Paper Association, contains design provisions and design values for bolts with diameters up to 1 in conforming to the ANSI/ASME Standard B18.2.1. Bolt design values in the NDS apply to connections that have been snugly tightened and to connections that have loosened due to shrinkage of the wood components.

Following are some important NDS requirements for bolts: Bolt holes should have a diameter from 1/32 to 1/16 in larger than the bolt diameter. In establishment of design values, careful centering of holes in main members is assumed. Tight fit of bolts in the holes, requiring forced insertion, is not recommended. A metal plate, strap, or washer (not smaller than a standard cut washer) should be placed between the wood and bolt head and between the wood and the nut. The length of bolt threads subject to bearing on the wood should be kept to a practical minimum.

Two or more bolts placed in a line parallel to the direction of the load constitute a **row. End distance** is the minimum distance from the end of a member to the center of the bolt hole that is nearest to the end. **Edge distance** is the minimum distance from the edge of a member to the center of the nearest bolt hole. Figure 10.12 illustrates these distances, the spacing between rows, and the spacing of bolts

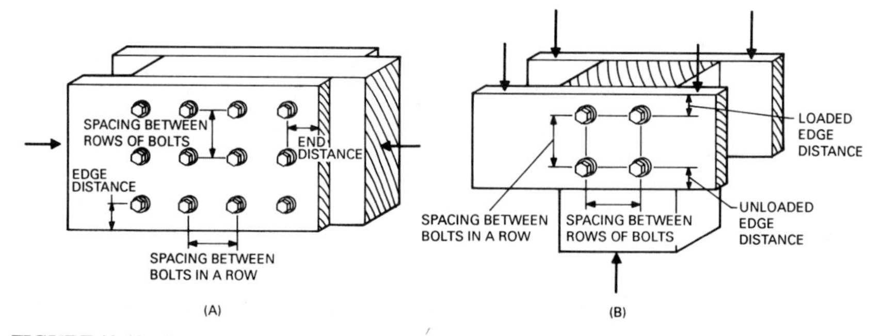

FIGURE 10.12 Bolt spacing and edge and end distances are defined with respect to load direction: (*a*) load parallel to the grain; (*b*) perpendicular to the grain.

in a row. NDS requirements are listed for minimum end distance in Table 10.34, for minimum edge distance in Table 10.35, and for minimum spacing between rows and between bolts in a row in Table 10.36. The geometry factor C_Δ discussed in Art. 10.14.5 is applied to the design value for a bolted connection when the end distance or spacing between bolts is less than that given in these tables for full design value.

TABLE 10.34 Minimum End Distances for Bolts*

Direction of loading	For reduced design value	For full design value
Perpendicular to grain	2*D*	4*D*
Compression parallel to grain (bolt bearing away from member end)	2*D*	4*D*
Tension parallel to grain (bolt bearing toward member end):		
For softwoods	3.5*D*	7*D*
For hardwoods	2.5*D*	5*D*

*D = bolt diameter

TABLE 10.35 Minimum Edge Distance for Bolts

Direction of loading*	Minimum edge distance
Parallel to grain:	
When $L/D \leq 6$	1.5*D*
When $L/D > 6$	1.5*D* or half the spacing between rows, whichever is greater
Perpendicular to grain:	
To loaded edge	4*D*
To unloaded edge	1.5*D*

*L = length of bolt in main member and D = bolt diameter.

10.16 LAG SCREWS

Also known as lag bolts, lag screws are large screws with a square or hexagonal bolt head. As is the case with bolts and timber connectors, lag screws are used where relatively heavy loads have to be transmitted in a connection. They are used in lieu of bolts where the components of a joint are so thick that an excessively long bolt would be needed, where one side of a connection is not accessible, or where heavy withdrawal loads have to be resisted. If desired, lag screws can be used in conjunction with split rings and shear plates.

Lag screws are turned with a wrench into prebored holes with total length equal to the nominal screw length. Soap or other lubricant may be used to facilitate insertion and prevent damage to screws. Two holes are drilled for each lag screw.

TABLE 10.36 Minimum Spacing for Bolts*

(*a*) For bolts in a row

Direction of loading	For reduced design value	For full design value
Parallel to grain	$3D$	$4D$
Perpendicular to grain	$3D$	Required spacing for attached member(s)

(*b*) Between bolts in a row

Direction of loading	Minimum spacing
Parallel to grain	$1.5D$
Perpendicular to grain:	
When $L/D \leq 2$	$2.5D$
When $2 < L/D < 6$	$(5L + 10D)/8$
When $L/D \geq 6$	$5D$

*L = length of bolt in main member and D = bolt diameter.

The first and deepest hole has a diameter, as specified in the NDS for various species, depending on the wood density, ranging from 40 to 85% of the shank diameter. The second hole should have the same diameter as the shank, or unthreaded portion of the lag screw, and the same depth as the unthreaded portion.

The NDS contains design values and design provisions for lag screws that conform to ANSI/ASME Standard B18.2.1. Lag screws loaded in withdrawal should be designed for allowable tensile strength in the net (root-of-thread) section as well as for resistance to withdrawal. For single-shear wood-to-wood connections, the lag screw should be inserted in the side grain of the main member with the screw axis perpendicular to the wood fibers. Minimum edge and end distances and spacing for a lag screw are the same as for a bolt with diameter equal to the shank diameter of the screw (Tables 10.34 to 10.36).

10.17 SPLIT-RING AND SHEAR-PLATE CONNECTORS

These are metal devices used with bolts or lag screws for producing joints with fewer fasteners without reduction in strength. Several types of connectors are available. Usually, they are either steel rings, called split rings, that are placed in grooves in adjoining members to prevent relative movement or metal plates, called shear plates, embedded in the faces of adjoining timbers. The bolts or lag screws are used with these connectors to prevent the timbers from separating. The load is transmitted across the joint through the connectors.

Split rings are used for joining wood to wood. They are placed in circular grooves cut by a hand tool in the contact surfaces. About half the depth of each ring is in each of the two members in contact. A bolt hole is drilled through the center of the core encircled by the groove. For economic reasons, split rings are seldom used, because of the accuracy required for properly fabricating the wood members and the relative difficulty of installation.

A single shear plate is used for wood-to-steel connections (Fig. 10.13*a* and *b*). When used in pairs, split rings may be used for wood-to-wood connections (Fig. 10.13*c*). Set with one plate in each member at the contact surface, they enable the members to slide easily into position during fabrication of the joint, thus reducing the labor needed to make the connnection. Shear plates are placed in precut daps and are completely embedded in the timber, flush with the surface. As with split rings, the role of the bolt or lag screw through each plate is to prevent the components of the joint from separating; loads are transmitted across the joint through the plates. They are manufactured in 2⅝- and 4-in diameters.

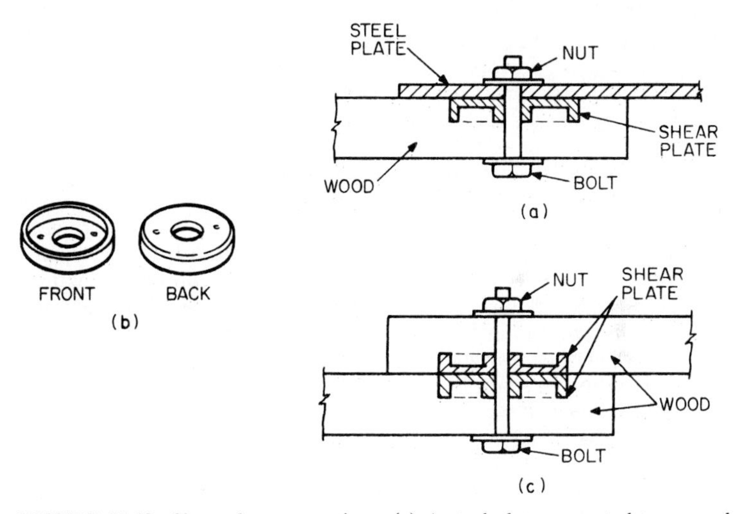

FIGURE 10.13 Shear-plate connections. (*a*) A steel plate connected to a wood member with a shear-plate connector. (*b*) Shear-plate connector. (*c*) Two wood members connected with a pair of shear plates and a bolt.

Shear plates are useful in demountable structures. They may be installed in the members immediately after fabrication and held in position by nails.

Toothed rings and spike grids sometimes are used for special applications; but split rings and shear plates are the prime connectors for joints in timber construction requiring transmission of very heavy loads.

The NDS contains tables that give design values for shear-plate connections. The tabulated values apply to seasoned timbers used where they will remain dry. If the timbers will be more or less continuously damp or wet in use, two-thirds of the tabulated values should be used.

Design values for split rings and shear plates for angles between 0° (parallel to grain) and 90° (perpendicular to grain) may be obtained from the Hankinson formula:

$$N = \frac{PQ}{P \sin^2 \theta + Q \cos^2 \theta} \tag{10.32}$$

where N, P, and Q are, respectively, the design value, lb, or stress, psi, at inclination θ with the direction of grain, parallel to grain, and perpendicular to grain.

Design values are based on the assumption that the wood at the joint is clear and relatively free from checks, shakes, and splits. If knots are present in the longitudinal projection of the net section within a distance from the critical section of half the diameter of the connector, the area of the knot should be subtracted from the area of the critical section. It is assumed that slope of the grain at the joint does not exceed 1 in 10.

The stress, whether tension or compression, in the net area, the area remaining at the critical section after subtracting the projected area of the connectors and the bolt from the full cross-sectional area of the member, should not exceed the design value of clear wood in compression parallel to the grain.

Tables in the NDS list the least thickness of member that should be used with the various sizes of connectors. The design values listed for the greatest thickness of member with each type and size of connector unit are the maximums to be used for all thicker material. Design values for members with thicknesses between those listed may be obtained by interpolation.

The NDS also lists minimum end and edge distances and spacing for timber connectors (Table 10.37). Edge distance is the distance from the edge of a member to the center of the connector closest to that edge and measured perpendicular to the edge. End distance is measured parallel to the grain from the center of the connector to the square-cut end of the member. If the end of the member is not cut normal to the longitudinal axis, the end distance, measured parallel to that axis from any point on the center half of the connector diameter that is perpendicular to the axis, should not be less than the minimum end distance required for a square-cut member. Spacing of connectors is measured between their centers along a line between centers.

10.18 WOOD SCREWS

The "National Design Specification for Wood Construction" contains design provisions and design values for wood screws that conform to ANSI/ASME Standard B18.6.2. When wood screws are loaded in withdrawal, the design value for tension in the screws should not be exceeded. For single-shear wood-to-wood construction, the screws should be inserted in the side grain of the main member with axis perpendicular to the wood fibers.

Edge and end distances and spacing of wood screws should be sufficient to prevent splitting of the wood. If building-code requirements for such distances are not available, Table 10.38 may be used to establish wood-screw patterns. Spacing, or pitch, between fasteners in a row is affected by species, moisture content, and grain orientation.

Screws may be inserted in such wood as spruce, pine, or fir, with a specific gravity less than 0.50, without preboring a hole for them. In denser wood, lead

TABLE 10.37 Minimum Edge and End Distances, Spacing, and Geometry Factors C_Δ for Split-Ring and Shear-Plate Connectors

		2½-in Split-ring connectors 2⅝-in Shear-plate connectors				4-in Split-ring connectors 4-in Shear-plate connectors			
		Loads parallel to grain		Loads perpendicular to grain		Loads parallel to grain		Loads perpendicular to grain	
		For reduced design value	For full design value	For reduced design value	For full design value	For reduced design value	For full design value	For reduced design value	For full design value
Edge	Unloaded edge, in C_Δ	1¾ 1.0	1¾ 1.0	1¾ 1.0	1¾ 1.0	2¾ 1.0	2¾ 1.0	2¾ 1.0	2¾ 1.0
distance	Loaded edge, in C_Δ	1¾ 1.0	1¾ 1.0	1¾ 0.83	2¾ 1.0	2¾ 1.0	2¾ 1.0	2¾ 0.83	3¾ 1.0
End	Tension member, in C_Δ	2¾ 0.625	5½ 1.0	2¾ 0.625	5½ 1.0	3½ 0.625	7 1.0	3½ 0.625	7 1.0
distance	Compression member, in C_Δ	2½ 0.625	4 1.0	2¾ 0.625	5½ 1.0	3¼ 0.625	5½ 1.0	3½ 0.625	7 1.0
Spacing	Spacing parallel to grain, in C_Δ	3½ 0.5	6¾ 1.0	3½ 1.0	3½ 1.0	5 0.5	9 1.0	5 1.0	5 1.0
	Spacing perpendicular to grain, in C_Δ	3½ 1.0	3½ 1.0	3½ 0.5	4¼ 1.0	5 1.0	5 1.0	5 0.5	6 1.0

TABLE 10.38 Minimum Edge and End Distances and Spacing for Wood Screws*

	Wood side members	
	Not prebored	Prebored
Edge distance	2.5*d*	2.5*d*
End distance:		
For tension load parallel to grain	15*d*	10*d*
For compression load parallel to grain	10*d*	5*d*
Spacing (pitch) between fasteners in a row:		
Parallel to grain	15*d*	10*d*
Perpendicular to grain	10*d*	5*d*
Spacing (gage) between rows of fasteners:		
In-line	5*d*	3*d*
Staggered	2.5*d*	2.5*d*
	Steel side members	
	Not prebored	Prebored
Edge distance	2.5*d*	2.5*d*
End distance:		
For tension load parallel to grain	10*d*	5*d*
For compression load parallel to grain	5*d*	3*d*
Spacing (pitch) between fasteners in a row:		
Parallel to grain	10*d*	5*d*
Perpendicular to grain	5*d*	2.5*d*
Spacing (gage) between rows of fasteners:		
In-line	3*d*	2.5*d*
Staggered	2.5*d*	2.5*d*

*d = shank diameter

holes should be drilled and screws inserted by turning, not by hammering. Holes for wood screws loaded in withdrawal should have a diameter of 90% of the screw root diameter in such wood as maple and oak, with a specific gravity of 0.6 or more, and of 70% of the root diameter in such wood as Douglas fir and larch, with a specific gravity between 0.5 and 0.6. For wood screws subjected to lateral loads, holes receiving the shanks in wood with a specific gravity of 0.6 or more should have the same diameter as the shank. Holes for the threaded portion should have a diameter about equal to the root diameter of the screws. For screws in less dense wood, the part of the hole for the shank should be about seven-eighths the diameter of the shank. Holes for the threaded portion should have a diameter about seven-eighths the root diameter. Soap or other lubricant may be used to facilitate insertion and to prevent damage to the screw.

Screws are designated by gage (diameter) of shank and overall length. For design purposes, it is adequate to assume two-thirds of the screw length is threaded.

10.19 NAILS AND SPIKES

The "National Design Specification for Wood Construction" (NDS) provides design values and design provisions for nails, spikes, box nails, and threaded, hardened-steel nails and spikes conforming to Federal Specification FF-N-105B. Sizes of nails are specified by the pennyweight, indicated by d. The pennyweight for different types of nails establishes lengths, shank diameter, and head size of the nail. Common wire nails and spikes are about the same, except that spikes have a larger diameter than nails with the same pennyweight designation. Box nails have a smaller diameter than common nails. For a specific pennyweight, the five types of nails have the same length.

Edge and end distances and spacing for nails and spikes should be sufficient to prevent splitting of the wood. If specific code requirements for these distances are not available, Table 10.38, although intended for use with wood screws, may be used to establish nailing patterns. Spacing, or pitch, between fasteners in a row is affected by species, moisture content, and grain orientation.

When a prebored hole is to be used to prevent splitting of the wood, the hole diameter should not exceed 90% of the nail or spike diameter for wood with a specific gravity exceeding 0.6, and should not exceed 75% of the diameter for less dense wood.

10.20 STRUCTURAL FRAMING CONNECTIONS

Standard and special preengineered metal hangers are used extensively in timber construction. Stock hangers are available from a number of manufacturers, but most manufacturers also provide hangers of special design. Where appearance is of prime importance, concealed hangers are frequently selected.

Figures 10.14 to 10.16 show structural framing details such as beam hangers and connectors and column anchors.

("Heavy Timber Connection Details," WCD No. 5, American Forest and Paper Association, Washington, D.C.)

10.21 GLUED FASTENINGS

Glued joints are generally made between two pieces of wood where the grain directions are parallel (as between the laminations of a beam or arch). Or such joints may be between solid-sawn or laminated timber and plywood, where the face grain of the plywood is either parallel or at right angles to the grain direction of the timber.

It is only in special cases that lumber may be glued with the grain direction of adjacent pieces at an angle. When the angle is large, dimensional changes caused by variations in wood moisture content set up large stresses in the glued joint. Consequently, the strength of the joint may be considerably reduced over a period of time. Exact data are not available, however, on the magnitude of this expected strength reduction.

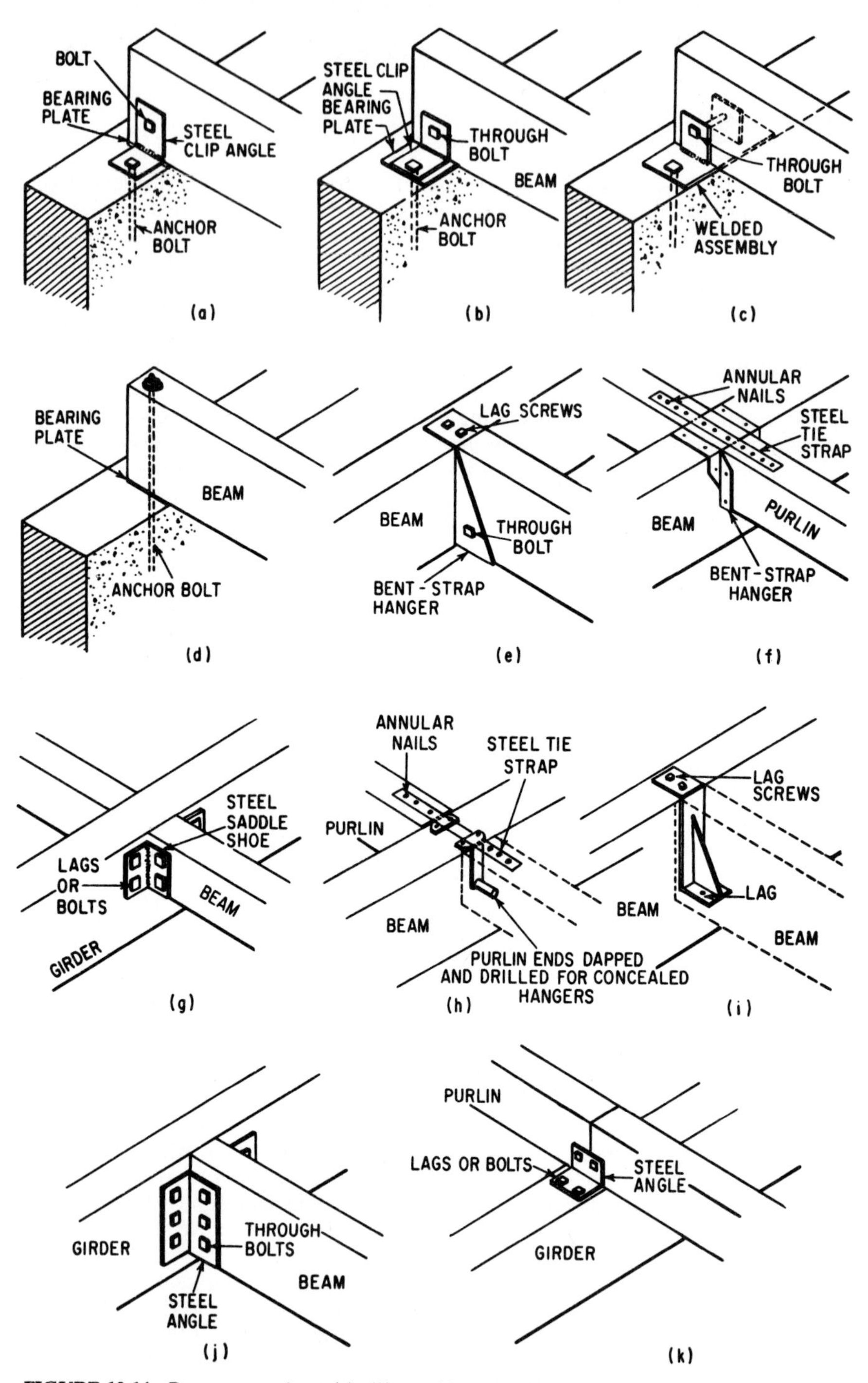

FIGURE 10.14 Beam connections: (*a*), (*b*) wood beam anchored on a wall with steel angles; (*c*) with welded assembly; (*d*) beam anchored directly with a bolt; (*e*) beam supported on a girder with bent-strap hanger; (*f*) similar support for purlins; (*g*) saddle connects beam to girder (suitable for one-sided connection); (*h*), (*i*) connections with concealed hangers; (*j*), (*k*) connections with steel angles.

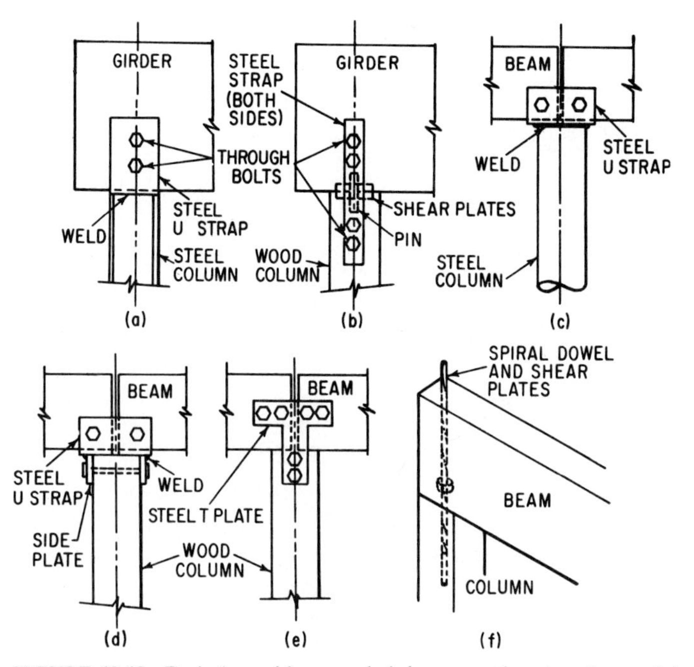

FIGURE 10.15 Typical wood beam and girder connections to columns: (*a*) wood girder to steel column; (*b*) girder to wood column; (*c*) beam to pipe column; (*d*) beam to wood column, with steel strap welded to steel side plates; (*e*) beam to wood column, with a T plate; (*f*) beam-column connection with spiral dowel and shear plates.

In joints connected with plywood gusset plates, this shrinkage differential is minimized, because plywood swells and shrinks much less than does solid wood.

Glued joints can be made between end-grain surfaces. They are seldom strong enough, however, to meet the requirements of even ordinary service. Seldom is it possible to develop more than 25% of the tensile strength of the wood in such butt joints. It is for this reason that plane sloping scarfs of relatively flat slope or finger joints with thin tips and flat slope on the sides of the individual fingers are recommended to develop a high proportion of the strength of the wood.

Joints of end grain to side grain are also difficult to glue properly. When subjected to severe stresses as a result of unequal dimensional changes in the members due to changes in moisture content, joints suffer from severely reduced strength.

For the preceding reasons, joints between end-grain surfaces and between end-grain and side-grain surfaces should not be used if the joints are expected to carry load.

For joints made with wood of different species, the allowable shear stress for parallel-grain bonding is equal to the allowable shear stress parallel to the grain for the weaker species in the joint. This assumes uniform stress distribution in the joint. When grain direction is not parallel, the allowable shear stress on the glued area between the two pieces may be estimated from Eq. (10.32).

(K. F. Faherty and T. G. Williamson, "Wood Engineering and Construction Handbook," McGraw-Hill Publishing Company, New York.)

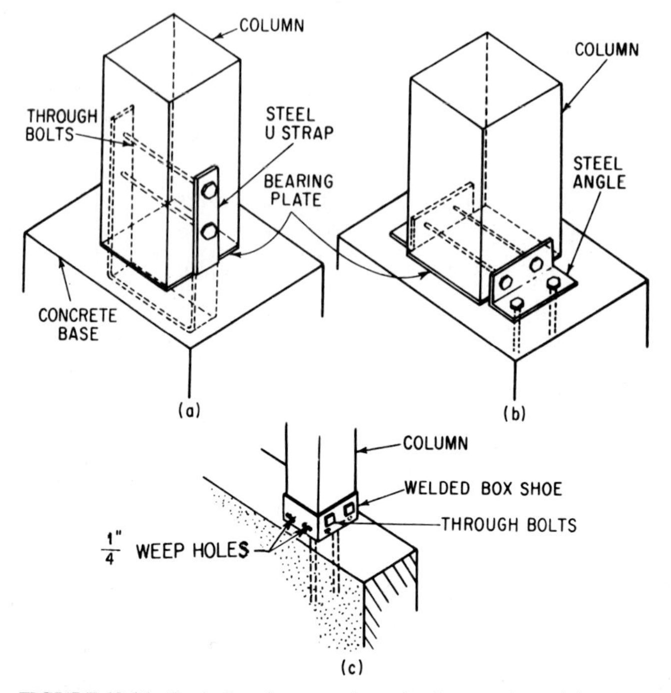

FIGURE 10.16 Typical anchorages of wood column to base: (*a*) column anchored to concrete base with U strap; (*b*) anchorage with steel angles; (*c*) anchorage with a welded box shoe.

10.22 WOOD TRUSSES

Used in all forms of building construction for centuries, wood trusses have appeared in a wide variety of forms. They utilized many different wood and nonwood structural components. Development of preengineered, prefabricated, lightweight wood and wood-steel composite trusses, however, significantly altered previous practices, which employed relatively complex and labor-intensive types of trusses, each truss requiring a detailed engineering analysis. As a result, repetitive designs are common and trusses are mass-produced in truss assembly plants.

10.22.1 Lightweight Truss Systems

Sheet-metal-gusset nail plates as a joint connection device for lumber trusses and the capability of mass producing wood trusses make practical lightweight, preengineered, preassembled wood trusses. These typically are installed at relatively close spacings (12 to 24 in c to c) and take advantage of repetitive-member design. Wider spacings (8 ft c to c or more) are also used, but these are designed without consideration of repetitive-member action.

The trusses generally use dimension lumber, either visually graded or machine stress rated, for the chords and web members. A connection at a truss joint utilizes two sheet-metal-gusset nail plates, which are physically pressed into the wood on opposite faces of the joint. The plates are sized to cover all the members at the joint with a sufficient plate area to transfer the load effectively from one member to the others. The load-transfer capacity is based on an allowable load per square inch of plate. Chord splices are accomplished in a similar manner by application of a plate on each side of the splice.

The ultimate and design load capacity of a plate depends on the steel gage and the number, size, and design of the steel projections (teeth) that are pressed into the wood. These characteristics are determined by the manufacturer. Consequently, metal gusset plates from different plate manufacturers are generally not interchangeable for a specific truss. Most plate manufacturers, in conjunction with their licensed truss assemblers, provide computerized truss design support for their products. Thus, designers need show only the required truss configuration and design loading conditions on the design drawings and the truss supplier will provide all applicable truss-member designs. Virtually any truss profile can be manufactured for metal-gusset-plate lumber trusses. More information on this type of truss can be obtained from the Truss Plate Institute (TPI) and the Wood Truss Council of America, both in Madison, Wis.

A variation of the lightweight metal-gusset-plate lumber truss employs much thicker metal-gusset nail plates in conjunction with heavier wood chord and web members, such as dimension lumber with a nominal width of 3 or 4 in or glued-laminated timbers. Use of the heavier components provides much greater load-carrying capacity. As a result, these trusses may economically be spaced much farther apart; for example, 8 ft c to c or more, and span much longer distances than the lighter trusses.

Another form of preengineered, preassembled, lightweight wood truss, commonly referred to as a composite wood-metal truss, consists of lumber chords and steel webs. One type uses metal-angle webs with nail plates formed on each end. At a joint, the teeth of the web end are pressed into a wood chord. Another configuration of composite wood-metal truss uses steel tubular web members, which are joined to the wood chords with patented steel-pin connections. The chords are typically machine stress-rated lumber or laminated veneer lumber (Art. 10.30.4). Because of the proprietary nature and uniqueness of the various types of composite wood-metal trusses, design data should be obtained from the manufacturers.

(K. F. Faherty and T. G. Williamson, "Wood Engineering and Construction Handbook," McGraw-Hill Publishing Company, New York.)

10.22.2 Heavy Timber Trusses

Type of heavy timber truss and arrangement of members may be chosen to suit the shape of structure, the loads, and stresses involved. The configurations most commonly used are bowstring, flat or parallel chord, pitched, triangular, and scissors (Fig. 10.17). For most commercial construction, trusses usually are spaced 8 to 24 ft apart.

Joints are critical in the design of a truss. Use of a specific truss type is often governed by joint considerations. Chords and webs may be single-leaf (or monochord), double-leaf, or multiple-leaf members. Monochord trusses and trusses with double-leaf chords and single-leaf web system are common. Web members may be attached to the sides of the chords, or web members may be in the same plane as the chords and attached with straps or gussets.

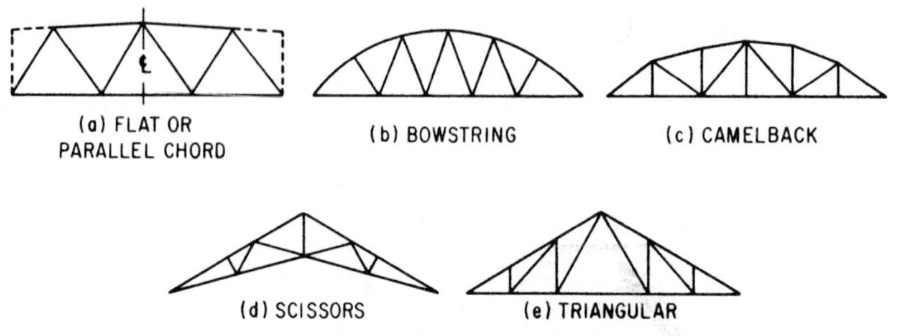

FIGURE 10.17 Types of wood trusses.

Individual truss members may be solid-sawn, glued-laminated, or mechanically laminated (rarely used in new construction). Steel rods or other steel shapes may be used as members of timber trusses if they meet design and service requirements.

Bowstring Trusses. Spans of 100 to 200 ft are common, with single top and bottom chords of glued-laminated timber, webs of solid-sawn timber, and steel heel plates, chord plates, and web-to-chord connections. This system is lightweight for the loads that it can carry. It can be shop or field assembled. Attention to the design of the top chord, bottom chord, and heel connections is of prime importance, since they are the major stress-carrying components. Since the top chord is nearly the shape of an ideal arch, stresses in chords are almost uniform throughout a bowstring truss and web stresses are low under uniformly distributed loads.

Parallel-Chord Trusses. Spans of 100 ft or more are possible with chords and webs of solid-sawn lumber or glulam timbers. Parallel-chord trusses are commonly used in floor systems and long-span commercial roofs. Advantages for this type of truss include ease of installation, relatively long spans, flexibility for installation of ductwork, plumbing, and electrical wiring, and overall economics.

Triangular and Scissor Trusses. Spans of 100 ft or more are possible with chords and webs of solid-sawn lumber or glulam timbers. These trusses are used extensively in residential roof construction and long-span applications in commercial roofs. Advantages are the same as those described above for parallel-chord trusses.

Truss Joints. Connectors used in truss joints include bolts, split rings, shear plates, lag screws, and metal connector plates, as described in Art. 10.22.1. The type of connector used depends on the type and size of wood truss components, loads to be transferred, fabrication capabilities, and aesthetics. Lag screws, bolts, and shear plates are suitable for field fabrication, whereas metal connector plates are almost always installed in a truss assembly plant.

Framing between Trusses. Longitudinal sway bracing perpendicular to the plane of the truss is usually provided by solid-sawn X bracing. Lateral wind bracing may be provided by end walls or intermediate walls, or both. The roof system and horizontal bracing should be capable of transferring the wind load to the walls. Knee braces between trusses and columns are often used to provide resistance to lateral loads.

Horizontal framing between trusses may consist of struts between trusses at bottom-chord level and diagonal tie rods, often of steel with turnbuckles for adjustment.

(K. F. Faherty and T. G. Williamson, "Wood Engineering and Construction Handbook," McGraw-Hill Publishing Company, New York.)

10.23 DESIGN OF TIMBER ARCHES

Arches typically are made of glued-laminated timber and may be two-hinged, with hinges at each base, or three-hinged with a hinge at the crown. Figure 10.18 illustrates typical forms of arches.

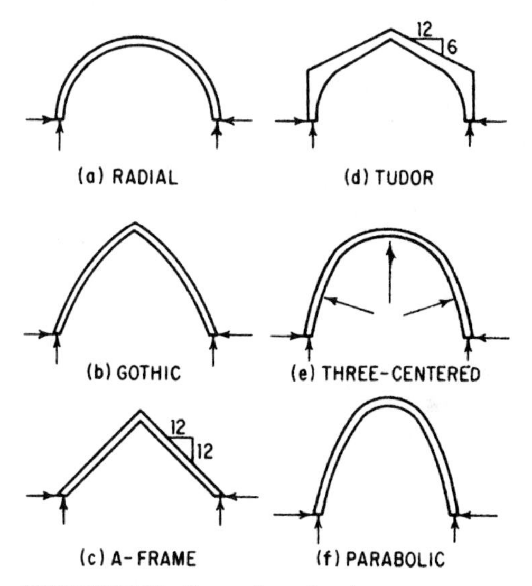

FIGURE 10.18 Types of wood arches.

Tudor arches are the most commonly used type of timber arch. They are gabled rigid frames with curved haunches. The half spans on each side of the crown usually are one piece of glued-laminated timber. This type of arch is frequently used in church and commercial construction.

A-frame arches are generally used where steep pitches are required. They may spring from grade, or concrete abutments, or other suitably designed supports.

Radial arches are often used where long clear spans are required. They have been employed for clear spans up to 300 ft.

Gothic, parabolic, and three-centered arches are often selected for architectural and aesthetic considerations.

Timber arches may be tied or buttressed. If an arch is tied, the tie rods, which resist the horizontal thrust, may be above the ceiling or below grade, and simple connections may be used where the arch is supported on masonry walls, concrete piers, or columns (Fig. 10.19).

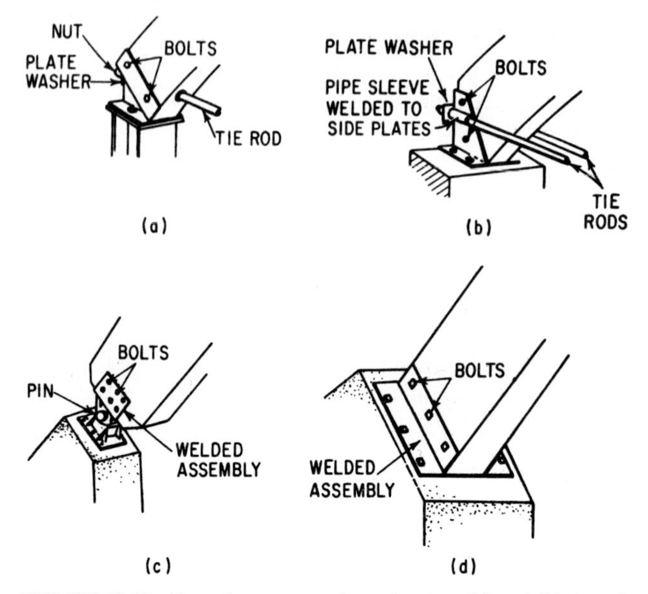

FIGURE 10.19 Bases for segmented wood arches: (*a*) and (*b*) tie rod anchored to arch shoe; (*c*) hinge anchorage for large arch; (*d*) welded arch shoe.

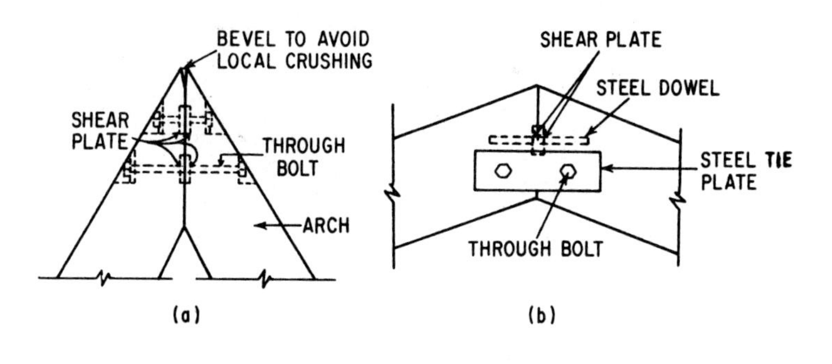

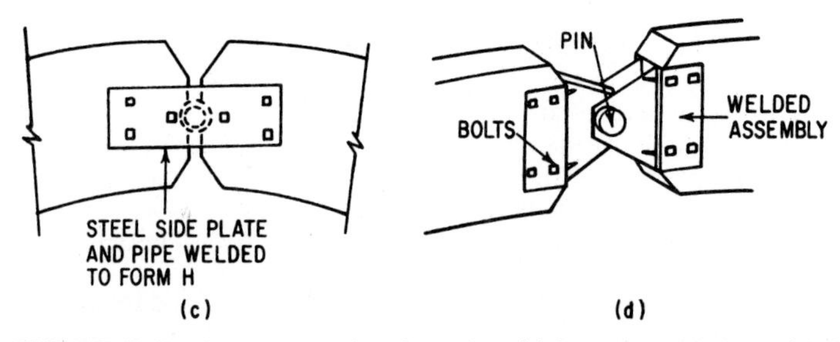

FIGURE 10.20 Crown connections for arches: (*a*) for arches with slope of 4:12 or greater, the connection consists of pairs of back-to-back shear plates with through bolts or threaded rods counterbored into the arch; (*b*) for arches with flatter slopes, shear plates centered on a dowel may be used in conjunction with tie plates and through bolts; (*c*) and (*d*) hinge at crown.

Arches are economical because of the ease of fabricating them and simplicity of field erection. Field splice joints are minimized; generally there is only one simple connection, at the crown (Fig. 10.20). Except for extremely long spans, they are shipped in only two pieces. Erected, they need not be concealed by false ceilings, as may be necessary with trusses. Inasmuch as arches have large cross sections, they are classified as heavy-timber construction.

A long-span arch may require a splice or moment connection to segment the arch to facilitate transportation to the job site. Figure 10.21 shows typical moment connections for wood arches.

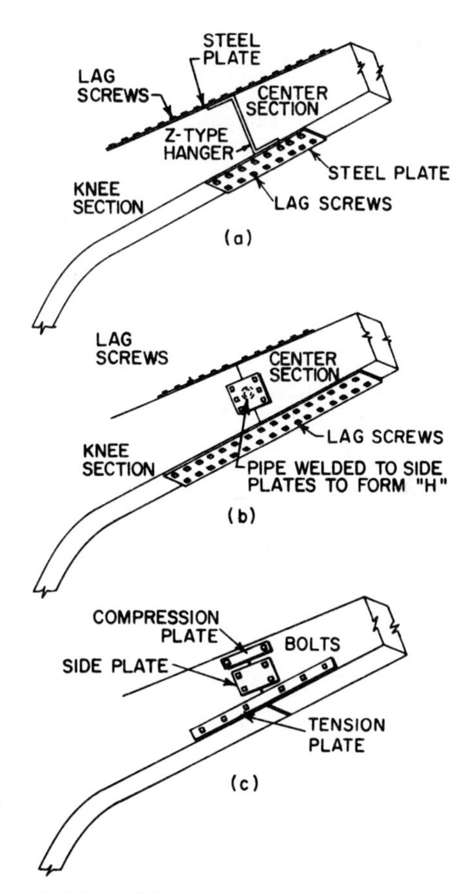

FIGURE 10.21 Schematics of some moment connections for timber arches: (*a*) and (*b*) connections with top and bottom steel plates; (*c*) connection with side plates.

(K. F. Faherty and T. G. Williamson, "Wood Engineering and Construction Handbook," McGraw-Hill Publishing Company, New York.)

10.24 TIMBER DECKING

Wood decking used for floor and roof construction may be sawn lumber with nominal thickness of 2, 3, or 4 in or glued-laminated. Panelized decking is made up of splined panels, usually about 2 ft wide.

For glued-laminated decking, two or more pieces of lumber are laminated into a single decking member, usually with nominal thickness of 2, 3, or 4 in. Other thicknesses may also be available. There are no consensus standards for glued-laminated decking. Deck manufacturers should be consulted for design information.

Solid-sawn decking usually is fabricated with edges tongued and grooved, ship-lap, or groove cut for splines, to provide transfer of vertical load between pieces. The decking may be end-matched, square end, or end-grooved for splines. As indicated in Fig. 10.22, the decking may be arranged in various patterns over supports.

In Type 1, the pieces are simply supported. Type 2 has a controlled random layup. Type 3 contains intermixed cantilevers. Type 4 consists of a combination of simple-span and two-span continuous pieces. Type 5 is two-span continuous.

In Types 1, 4, and 5, end joints bear on supports. For this reason, these types are recommended for thin decking, such as 2-in.

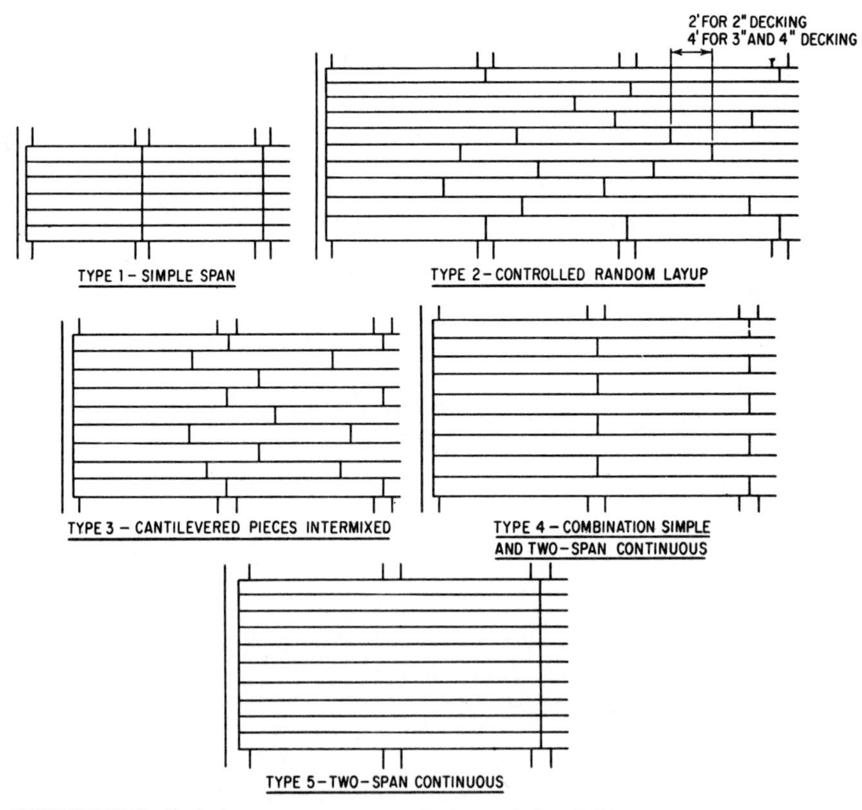

FIGURE 10.22 Typical arrangement patterns for heavy-timber decking.

Type 3, with intermixed cantilevers, and Type 2, with controlled random layup, are used for deck continuous over three or more spans. These types permit some of the end joints to be located between supports. Hence, provision must be made for load transfer at those joints. Tongue-and-groove edges, wood splines on each edge of the course, horizontal spikes between courses, and end matching or metal end splines may be used to transfer shear and bending stresses.

In Type 2, the distance between end joints in adjacent courses should be at least 2 ft for 2-in deck and 4 ft for 3- and 4-in deck. Joints approximately lined up (within 6 in of being in line) should be separated by at least two courses. All pieces should rest on at least one support, and not more than one end joint should fall between supports in each course.

In Type 3, every third course is simple span. Pieces in other courses cantilever over supports and end joints fall at alternate quarter or third points of the spans. Each piece rests on at least one support.

To restrain laterally supporting members of 2-in deck in Types 2 and 3, the pieces in the first and second courses and in every seventh course should bear on at least two supports. End joints in the first course should not occur on the same supports as end joints in the second course unless some construction, such as plywood overlayment, provides continuity. Nail end distance should be sufficient to develop the lateral nail strength required.

Heavy-timber decking is laid with wide faces bearing on the supports. Each piece must be nailed to each support. For 2-in decking a 3½-in (16d) toe and face nail should be used in each 6-in-wide piece at supports, with three nails for wider pieces. Tongue-and-groove decking generally is also toenailed through the tongue. For 3-in decking, each piece should be toenailed with one 4-in (20d) spike and face-nailed with one 5-in (40d) spike at each support. For 4-in decking, each piece should be toenailed at each support with one 5-in (40d) nail and face-nailed there with one 6-in (60d) spike.

Courses of 3- and 4-in double tongue-and-groove decking should be spiked to each other with 8½-in spikes not more than 30 in apart. One spike should not be more than 10 in from each end of each piece. The spikes should be driven through predrilled holes. Two-inch decking is not fastened together horizontally with spikes.

Deck design usually is governed by maximum permissible deflection in end spans. But each design should be checked for bending stress. For load-span tables and more information on sawn lumber decking, refer to AITC 112, Standard for Heavy Timber Roof Decking, and AITC 118, Standard for 2-in. Nominal Thickness Lumber Roof Decking for Structural Applications, American Institute of Timber Constuction, 11818 Mill Plain Blvd., Vancouver, WA 98684.

10.25 WOOD-FRAME CONSTRUCTION

This is the predominant method for building single- and multifamily dwellings. Wood-frame buildings can be erected speedily and are economical to build. For wood framing, the walls are conventionally built with studs spaced 16 or 24 in c to c. Similarly, joists and rafters, which are supported on the walls and partitions, are also usually spaced 16 or 24 in c to c. Facings, such as panel sheathing, wallboard, decking, floor underlayment, and roof sheathing, are generally available in appropriate sizes for attachment to studs, joists, and rafters with these spacings.

Wood studs are usually set in walls and partitions with wide faces perpendicular to the face of the wall or partition. The studs are nailed at the bottom to bear on a horizontal member, called the bottom or sole plate. At the top, they are nailed to a pair of horizontal members, called the top plate. These plates often are the same size as the studs. Joists or rafters may be supported on the top plate or on a header.

Studs are braced against racking by diagonal or horizontal blocking and facing materials, such as structural wood panels or in some instances by gypsum sheathing.

Three types of wood-frame construction are generally used: platform frame, balloon frame, and plank-and-beam frame.

10.25.1 Platform Frame

In this type of construction, first-floor joists are completely covered with subflooring to form a platform on which exterior walls and interior partitions are built (Fig. 10.23). This is the type of framing usually used for single-family dwellings.

10.25.2 Balloon Frame

Generally used for construction more than one story high, balloon frames employ wall studs that are continuous from story to story. First-floor joists and exterior wall studs bear on an anchored sill plate (Fig. 10.24). Joists for the second and higher floors bear on a 1- × 4-in ribband let-in to the inside edges and exterior wall studs. In two-story buildings, with brick or stone veneer exteriors, balloon framing minimizes variations in settlement of the framing and the masonry veneer.

10.25.3 Plank-and-Beam Frame

In contrast to conventional framing, which utilizes joists, rafters, and studs spaced 12 to 24 in c to c, plank-and-beam frames require fewer but larger-size piers, and wood components are spaced farther apart (Fig. 10.25). In plank-and-beam framing, subfloors or roofs, typically composed of members with a nominal thickness of 2 in, are supported on beams spaced 8 ft c to c. Ends of the beams are supported on posts or concrete piers. Supplemental framing is used between posts for attachment of exterior and interior wall framing and finishes. It also provides lateral support or bracing for the structure.

Plank-and-beam framing has several cost-saving and aesthetic advantages:

- Distinctive architectural effect provided by the exposed wood deck ceiling. In many cases, no further ceiling treatment is needed, except perhaps application of a sealer, stain, or paint.
- If well-planned, construction labor savings. Larger and fewer framing members require less handling and fewer mechanical fasteners.
- Elimination of cross bracing, which is often required in platform and balloon framing.

("Plank and Beam Framing for Residential Buildings," WCD No. 4, American Forest and Paper Association, Washington, D.C.)

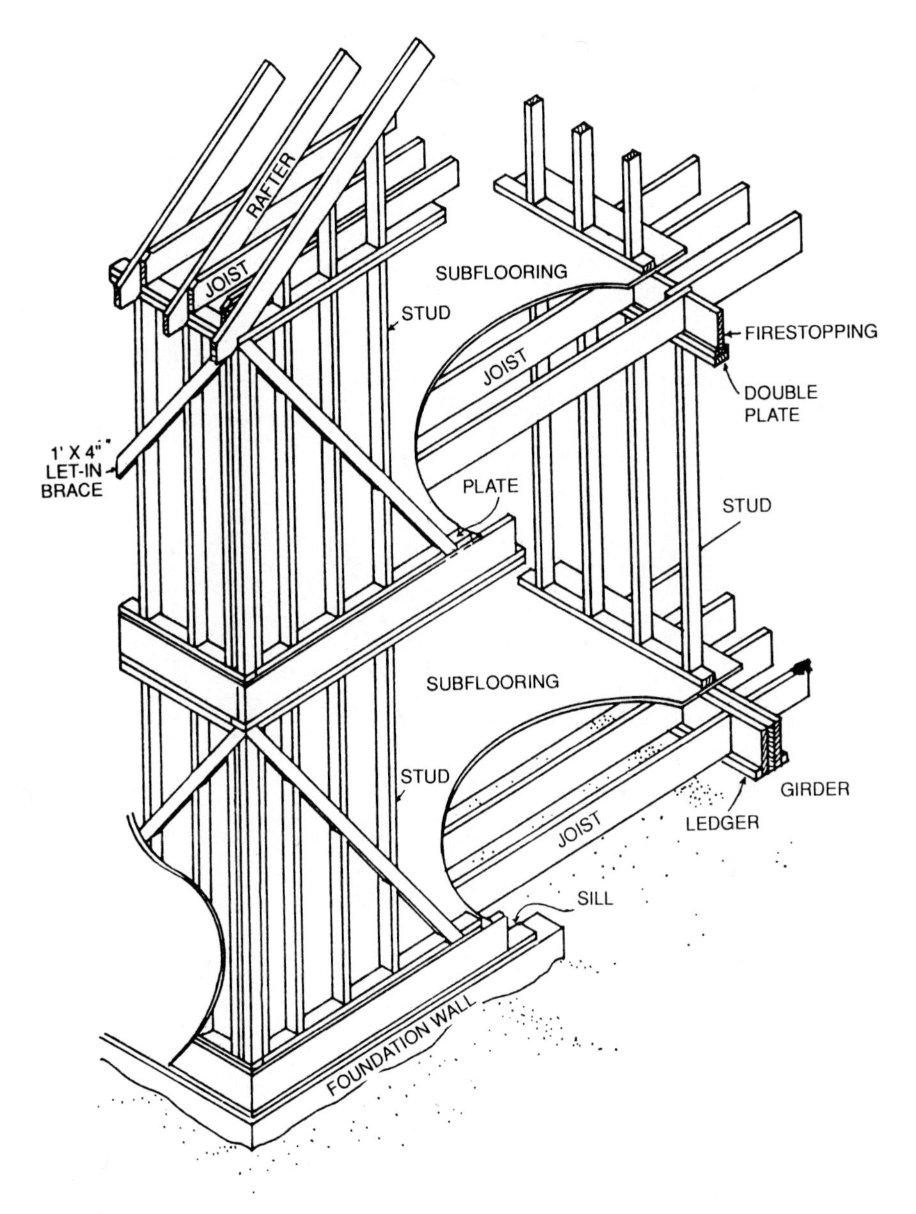

FIGURE 10.23 Platform framing for two-story building.

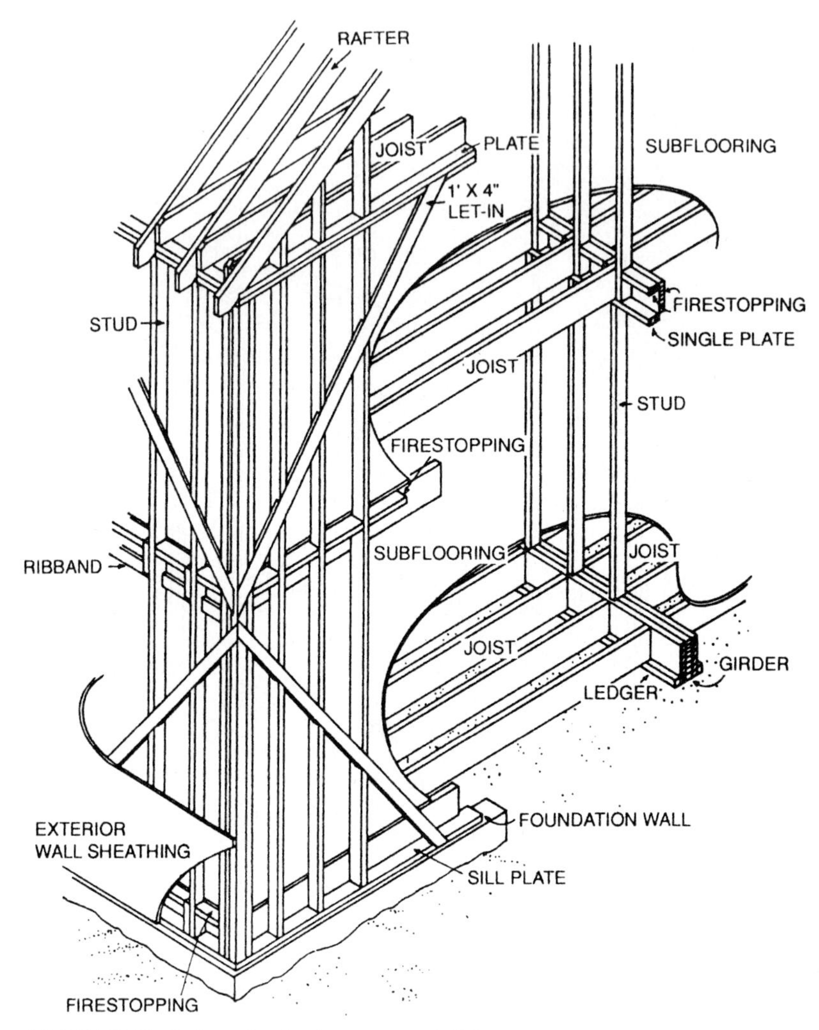

FIGURE 10.24 Balloon framing for two-story building.

10.26 PERMANENT WOOD FOUNDATIONS

Plywood and lumber walls are an alternative to concrete for foundation walls for one-story and multistory houses and other light-frame buildings. Main components of a wood foundation wall are plywood, ⅝ in or more thick, and wood studs, spaced 12 in or more on centers, both pressure-treated with preservative (Fig. 10.26). Some advantages of a wood foundation over concrete are faster construction, because there is no delay due to the wait for concrete or unit masonry to cure, easier interior finishing, because wood foundations provide nailable studs for the usual finishes, the ability to erect the system in virtually any weather, and generally

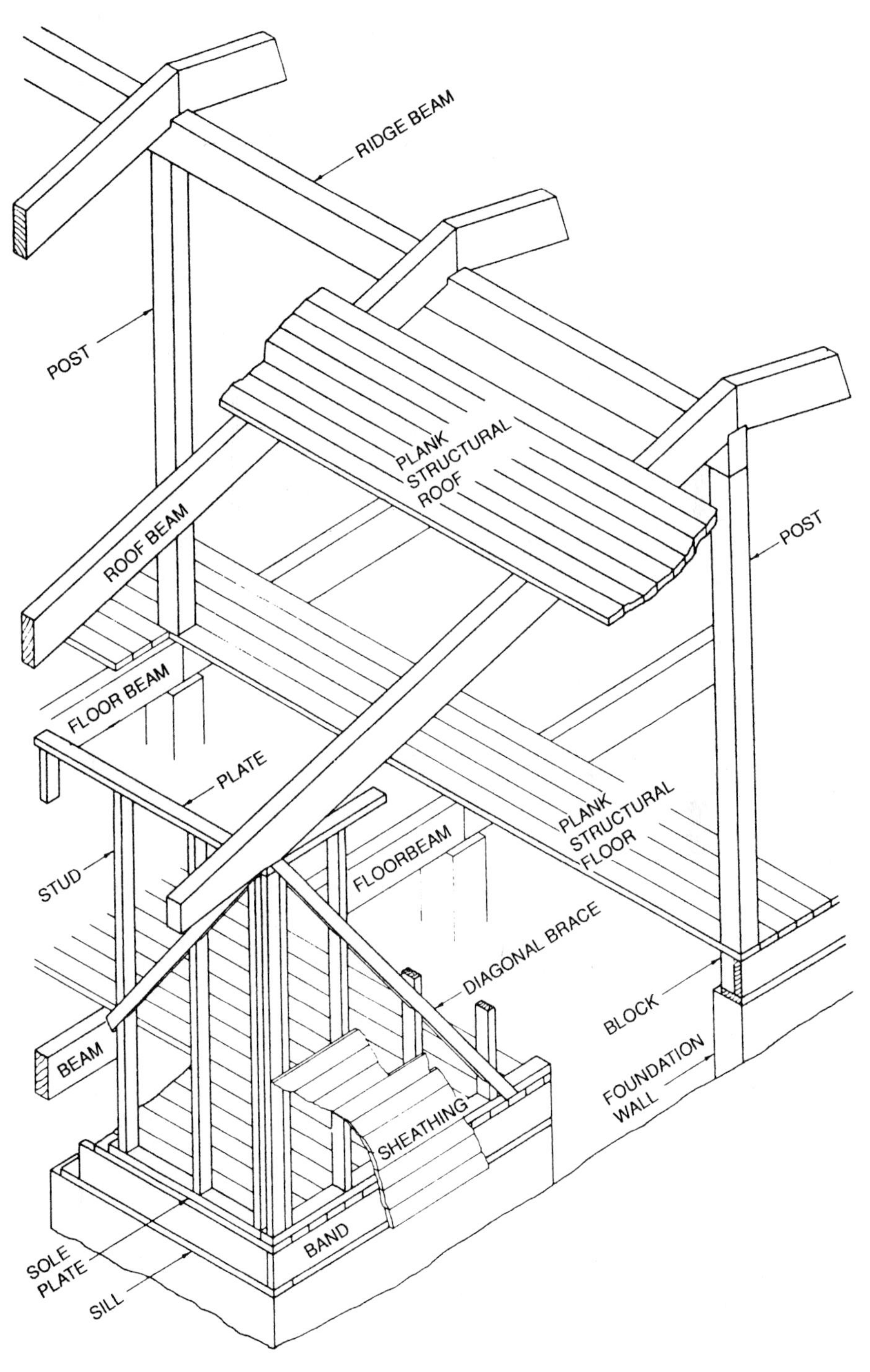

FIGURE 10.25 Plank-and-beam framing for one-story building.

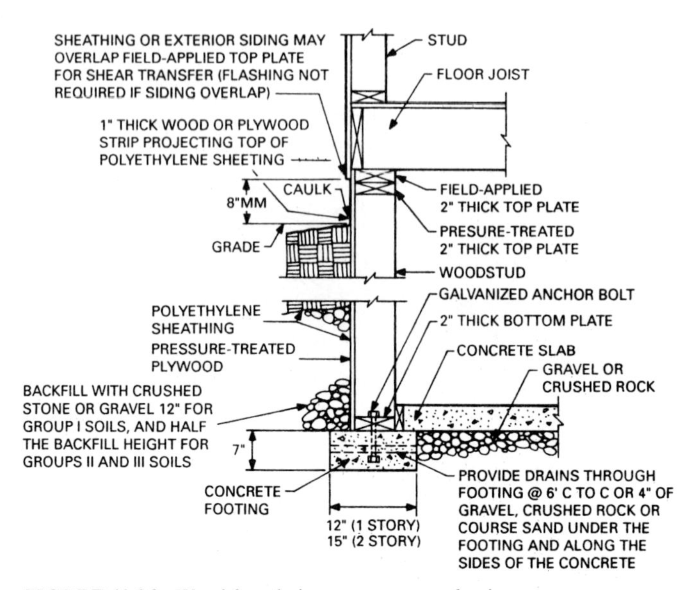

FIGURE 10.26 Wood foundation on a concrete footing.

drier basements due to the use of gravel backfill, which facilitates drainage of water away from the foundation. Wood basements also are much warmer and more comfortable for the occupant in cold weather.

Plywood should be an exterior type or an interior type that has been bonded with exterior glue, both types manufactured to meet the requirements of U.S. Product Standard PS 1 for Construction and Industrial Plywood or one of the American Plywood Association's performance-engineered proprietary specifications. Lumber should be grade marked by an approved inspection agency, should be capable of accepting pressure preservative treatment, and should be of a species for which allowable design values are given in Art. 10.3.

Treatment for the plywood and lumber involves impregnating into the wood under heat and pressure ammoniacal copper arsenate or chromated copper arsenate (A, B, or C). Salt retention should be at least 0.60 lb/ft^3 of wood, 50% more than building codes usually require for ground-contact applications. After pressure treatment, plywood should be dried to a moisture content of 18% or less, and lumber, to 19% or less. Portions of the wall more than 8 in above the ground, however, need not be pressure treated. If any of the materials have to be cut after treatment, the cut edges, unless they will be 8 in or more above grade, should be field treated with the preservative used in the original treatment but with a minimum concentration of 3% in solution. To minimize end cutting and field treatment of footing plates, these members may be extended past the corners of the foundation.

Only corrosion-resistant fasteners should be used in wood foundations. Plywood may be attached to the studs with Type 304 or 316 stainless-steel nails, hot-dipped or hot-tumbled galvanized nails, or silicon-bronze or copper fasteners. Lumber-to-lumber fasteners, whether above or below grade, should be hot-dipped or hot-tumbled galvanized nails, except that stainless-steel fasteners should be used for knee-wall assemblies.

To keep the interior of the foundation walls dry, it is advisable to enclose the exterior with a waterproofing membrane, such as polyethylene film, and to ensure good drainage. For the latter purpose, the ground surface should be sloped away from the building at ½ in/ft for a distance of at least 6 ft and gravel should be placed against the walls and under the cellar floor.

("The Permanent Wood Foundation System: Design, Fabrication and Installation Manual," American Forest and Paper Association, Washington, D.C.; "Under-Floor Plenum Manual," National Association of Home Builders, Washington, D.C.)

10.27 POST FRAME AND POLE CONSTRUCTION

Wood poles and posts are used for various types of construction, including flagpoles, utility poles, and framing for buildings. These employ preservatively treated round poles or posts with square or rectangular cross sections that are set into the ground as columns. The ground furnishes vertical and horizontal support and prevents rotation at the base.

Post frame construction is used extensively in agricultural buildings and in many commercial and industrial building applications.

In buildings with post frame and pole construction, a bracing system is often provided at the top of the poles or posts to reduce bending moments at the base and to distribute loads. Design of buildings supported by poles or posts without bracing requires good knowledge of soil conditions, to eliminate excessive deflection or sidesway. For allowable foundation and lateral pressures, see the local building code or Uniform Building Code, International Conference of Building Officials, Whittier, Calif.

Bearing values under the base of the vertical load-carrying elements should be checked. For backfilling the holes, well-tamped native soil, sand, or gravel may be satisfactory, but concrete or soil cement is more effective. Concrete fill reduces the required embedment depth and improves bearing capacity by increasing the skin-friction area of the pole.

Concrete footings used to increase bearing capacity under the base of posts should be designed to withstand the punching shear of the posts and bending moments. Thickness of concrete footings should be at least 12 in. Consideration should be given to use of concrete footings even in firm soils, such as hard dry clay, coarse firm sand, or gravel.

Methods for calculating required depth of embedment of poles and posts is provided in "Post and Pole Foundation Design," ASAE Engineering Practice, EP 486, American Society of Agricultural Engineers, St. Joseph, Mich.

Table 10.39 gives allowable stresses for treated poles. Table 10.40 lists standard dimensions for Douglas fir and southern pine poles, and Table 10.41 gives safe concentric column loads.

(See also "Design Properties of Round, Sawn and Laminated Preservatively Treated Construction Poles and Posts," ASAE Engineering Practice, EP 388.2.)

10.28 DESIGN FOR FIRE SAFETY

For buildings with structural lumber or timber framing, fire protection of the occupants and of the property itself can be enhanced by taking advantage of the fire-

TABLE 10.39 Stress Values for Treated Wood Poles*

Species	Modulus of rupture, extreme fiber in bending, psi†	Extreme fiber in bending F_b, psi‡	Modulus of elasticity E, psi‡	Compression parallel to grain F_c, psi‡
Cedar, northern white	4000	1540	600,000	740
Cedar, western red	6000	1850	900,000	1030
Douglas fir	8000	2700	1,500,000	1360
Hemlock, western	7400	2380	1,300,000	1250
Larch, western	8400	2940	1,500,000	1500
Pine, jack	6600	2100	1,100,000	1100
Pine, lodgepole	6600	1820	1,100,000	980
Pine, ponderosa	6000	1710	1,000,000	920
Pine, red or Norway	6600	2100	1,300,000	1020
Pine, southern	8000	2740	1,500,000	1360

*Air-dried prior to treatment.
†Based on *American National Standard Specifications and Dimensions for Wood Poles,* ANSI 05.1-1972.
‡Based on ASTM D 2899-86, "*Tentative Method for Establishing Design Stresses for Round Timber Piles.*"

endurance properties of wood in large cross sections and selecting details that make buildings fire-safe. Building materials or features alone or detection and fire-extinguishing equipment alone cannot provide maximum safety from fire in buildings. A proper combination of these measures, however, will provide the necessary fire protection (Art. 3.5).

When exposed to fire, wood forms a self-insulating surface layer of char, which provides fire protection. Even though the surface chars, the undamaged wood beneath retains most of its strength and will support loads in accordance with the capacity of the uncharred section. Heavy-timer members have often retained their structural integrity through long periods of fire exposure and remained serviceable after refinishing the charred surfaces. This fire endurance and excellent performance of heavy timber are attributable to the size of the wood members, and to the slow rate at which the charring penetrates.

The structural framing of a building, which is the criterion for classifying a building as combustible or noncombustible, has little to do with the hazard from fire to the building occupants. Most fires start in the building furnishings or contents and create conditions that render the inside of the structure uninhabitable long before the structural framing becomes involved in a fire. Thus, whether the building is of a combustible or noncombustible classification has little bearing on the potential hazard to the occupants. However, once the fire starts in the contents, the material of which the building is constructed can be of significant help in facilitating evacuation, fire fighting, and property protection. The most important protection factors for occupants, fire fighters, and the property, as well as adjacent exposed property, are prompt detection of the fire, immediate alarm, and rapid extinguishment of the fire.

TABLE 10.40 Dimensions of Douglas-Fir and Southern-Pine Poles*

Class		1	2	3	4	5	6	7	9	10
Min circumference at top, in		27	25	23	21	19	17	15	15	12
Length of pole, ft	Ground-line distance from butt,† ft	Min circumference at 6 ft from butt, in								
20	4	31.0	29.0	27.0	25.0	23.0	21.0	19.5	17.5	14.0
25	5	33.5	31.5	29.5	27.5	25.5	23.0	21.5	19.5	15.0
30	5½	36.5	34.0	32.0	29.5	27.5	25.0	23.5	20.5	
35	6	39.0	36.5	34.0	31.5	29.0	27.0	25.0		
40	6	41.0	38.5	36.0	33.5	31.0	28.5	26.5		
45	6½	43.0	40.5	37.5	36.5	32.5	30.0	28.0		
50	7	45.0	42.0	39.0	38.0	34.0	31.5	29.0		
55	7½	46.5	43.5	40.5	39.0	35.0	32.5			
60	8	48.0	45.0	42.0	40.5	36.0	33.5			
65	8½	49.5	46.5	43.5	41.5	37.5				
70	9	51.0	48.0	45.0	43.0	38.5				
75	9½	52.5	49.0	46.0	44.0					
80	10	54.0	50.5	47.0						
85	10½	55.0	51.5	48.0						
90	11	56.0	53.0	49.0						
95	11	57.0	54.0	50.0						
100	11	58.5	55.0	51.0						
105	12	59.5	56.0	52.0						
110	12	60.5	57.0	53.0						
115	12	61.5	58.0							
120	12	62.5	59.0							
125	12	63.5	59.5							

*See the latest edition of Standard Specifications and Dimensions for Wood Poles, ANSI 05.1.

†The figures in this column are intended for use only when a definition of ground line is necessary to apply requirements relating to scars, straightness, etc.

With member size of particular importance to fire endurance of wood, building codes classify buildings with wood framing as heavy-timber construction, ordinary construction, or wood-frame construction.

Heavy-timber construction is the type in which fire resistance is attained by placing requirements on the mimimum size, thickness, or composition of all load-carrying wood members; by avoidance of concealed spaces under floors and roofs; by use of approved connections, construction details, and adhesives; and by pro-

TABLE 10.41 Safe Concentric Column Loads on Douglas-Fir and Southern-Pine Poles, lb*

Top diam, in	8	7	6	5
Pole class	2	3	5	6
Unsupported pole length, ft (above ground line):				
0	68,500	52,500	38,500	26,000
10	68,500	51,000	28,500	14,000
12	61,500	36,500	20,500	10,000
14	46,000	27,500	15,500	8,000
16	36,000	22,000	12,500	6,500
18	29,500	17,500	10,000	5,000
20	24,500	15,000	8,500	4,500
25	16,500	10,000	6,000	
30	12,500	7,500		
35	10,000	6,000		

*See the latest edition of Standard Specifications and Dimensions for Wood Poles, ANSI 05.1.

viding the required degree of fire resistance in exterior and interior walls. (For design procedures and typical connector details, see "Design for Code Acceptance #2, Design of Fire-Resistive Exposed Wood Members," American Forest & Paper Association, Washington, D.C.)

Ordinary construction has exterior masonry walls and wood-framing membeis of sizes smaller than heavy-timber sizes.

Wood-frame construction has wood-framed walls and structural framing of sizes smaller than heavy-timber sizes.

Depending on the occupancy of a building or hazard of operations within it, a building of frame or ordinary construction may have its members covered with fire-resistive coverings. The fire endurance rating of protected wood-frame assemblies is measured in a standard fire endurance test, ASTM E119. Fire endurance ratings for various assembly designs can be found in listings that are published by recognized fire testing laboratories. These listings are updated on a regular basis. Alternatively, for many assemblies, a conservative estimate of the fire endurance rating can be obtained through a simplified calculation method ("Design for Code Acceptance #4, Component Additive Method (CAM) for Calculating and Demonstrating Assembly Fire Endurance," American Forest and Paper Association.) This method estimates the overall fire endurance rating as the sum of fire endurance times assigned to individual components of an assembly.

The interior finish on exposed surfaces of rooms, corridors, and stairways is important from the standpoint of its tendency to ignite and spread flame from one location to another. The fact that wood is combustible does not mean that it will spread flame at a hazardous rate. Code requirements to limit the flame-spread propensity of interior finishes are based on the flame-spread index (FSI) measured in the ASTM E84 tunnel test. Most wood products of sufficient thickness (¼ in or greater) have an FSI between 76 and 200 (Class III or C). Some wood products have an FSI between 26 and 75 (Class II or B). Fire-retardant treatment can improve performance to an FSI of 25 or less (Class I or A). For an extensive list of flame spread indices of wood products, see "Design for Code Acceptance #1, Flame Spread Performance of Wood Products," American Forest and Paper Association. In general, codes require Class I or A materials for enclosed vertical exits, Class II or B materials in exit access corridors, and Class III or C materials in other areas and rooms. Thus, wood products are accepted by the codes for a wide range of interior finish uses. Most codes exclude the exposed wood surfaces of heavy-timber structural members from flame-spread requirements, if the exposed area is sufficiently small.

Fire-retardant chemicals may be impregnated in wood with recommended retentions to lower the rate of surface flame spread. After proper surface preparation, the surface is paintable. Such treatments are accepted under several specifications, including federal and military. They are recommended only for interior or dry-use service conditions or locations protected against leaching. These treatments are sometimes used to meet a specific flame-spread rating for interior finish, or as an alternative to code requirements for noncombustible construction.

10.29 TIMBER FABRICATION AND ERECTION

During the fabrication and erection processes for wood construction, lumber should be handled and covered to prevent marring of the surfaces and moisture absorption. Overstressing of members and joints during handling and erection should be avoided. Competent inspectors should check materials and workmanship.

10.29.1 Fabrication of Structural Timber

Fabrication consists of boring, cutting, sawing, trimming, dapping, routing, planing, and otherwise shaping, framing, and finishing wood units, sawn or laminated, including plywood, to fit them for particular places in a final structure. Whether fabrication is performed in shop or field, the product must exhibit a high quality of workmanship.

Jigs, patterns, templates, stops, or other suitable means should be used for all complicated and multiple assemblies to ensure accuracy, uniformity, and control of all dimensions. All tolerances in cutting, drilling, and framing must comply with good practice and applicable specifications and controls. At the time of fabrication, tolerances must not exceed those listed below, unless they are not critical and not required for proper performance. Specific jobs, however, may require closer tolerances.

Location of Fastenings. Spacing and location of all fastenings within a joint should be in accordance with the shop drawings and specifications, with a maximum permissible tolerance of $\pm 1/16$ in. The fabrication of members assembled at any joint should be such that the fastenings are properly fitted.

Bolt-Hole Sizes. Bolt holes in all fabricated structural timber, when loaded as a structural joint, should be $1/16$ in larger in diameter than bolt diameter for $1/2$-in and larger-diameter bolts, and $1/32$ in larger for smaller-diameter bolts. Larger clearances may be required for other bolts, such as anchor bolts and tension rods.

Holes and Grooves. Holes for stress-carrying bolts, connector grooves, and connector daps must be smooth and true within $1/16$ in per 12 in of depth. The width of a split-ring connector groove should be within +0.02 in of and not less than the thickness of the corresponding cross section of the ring. The shape of ring grooves must conform generally to the cross-sectional shape of the ring. Departure from these requirements may be allowed when supported by test data. Drills and other cutting tools should be set to conform to the size, shape, and depth of holes, grooves, daps, etc., specified in the "National Design Specification for Wood Construction," American Forest and Paper Association.

Lengths. Members should be cut within ±1⁄16 in of the indicated dimension when they are up to 20 ft long, and ±1⁄16 in per 20 ft of specified length when they are over 20 ft long. Where length dimensions are not specified or critical, these tolerances may be waived.

End Cuts. Unless otherwise specified, all trimmed square ends should be square within 1⁄16 in per foot of depth and width. Square or sloped ends to be loaded in compression should be cut to provide contact over substantially the complete surface.

10.29.2 Timber Erection

Erection of timber framing requires experienced crews and adequate lifting equipment to protect life and property and to assure that the framing is properly assembled and not damaged during handling.

Each shipment of timber should be checked for tally and evidence of damage. Before erection starts, plan dimensions should be verified in the field. The accuracy and adequacy of abutments, foundations, piers, and anchor bolts should be determined. The erector must see that all supports and anchors are complete, accessible, and free from obstructions.

Job-Site Storage. If wood members must be stored at the site, they should be placed where they do not create a hazard to other trades or to the members themselves. All framing stored at the site should be set above the ground on appropriate blocking. Where practical, the members or bundles of material should be separated with strips, so that air may circulate around all sides. The top and all sides of each storage pile should be covered with a moisture-resistant covering that provides protection from the elements, dirt, and job-site debris. The use of clear polyethylene films is not recommended, since wood members may be bleached by sunlight. Individual wrappings should be slit or punctured on the lower side to permit drainage of water that accumulates inside the wrapping. Particular care should be taken with members such as glued-laminated timber that may be exposed to view in the completed structure.

Glued-laminated members of Premium and Architectural Appearance (and Industrial Appearance in some cases) are usually shipped with a protective wrapping of water-resistant paper. While this paper does not provide complete freedom from contact with water, experience has shown that protective wrapping is necessary to ensure proper appearance after erection. Though used specifically for protection in transit, the paper should remain intact during job-site storage. Removal of the paper is a designer or contractor option. For example, if the paper is removed from isolated areas to make connections from one member to another, either the paper should be replaced and should remain in position until all the wrapping is removed or all of the paper should be removed to minimize discoloration due to sun bleaching.

At the site, to prevent surface marring and damage to wood members, the following precautions should be taken:

Lift members or roll them on dollies or rollers out of railroad cars. Unload trucks by hand, forklift, or crane. Do not dump, drag, or drop members.

During unloading with lifting equipment, use fabric belts, or other slings that will not mar the wood. Provide additional protective blocking or padding at edges of members.

Guard against soiling, dirt, footprints, abrasions, or injury to shaped edges or sharp corners.

Equipment. Adequate equipment of proper load-handling capacity, with control for movement and placing of members, should be used for all operations. It should be of such nature as to ensure safe and expedient placement of the material. Cranes and other mechanical-devices must have sufficient controls that beams, columns, arches, or other elements can be eased into position with precision. Slings, ropes, and other securing devices must not damage the materials being placed.

The erector should determine the weights and balance points of the framing members before lifting begins, so that proper equipment and lifting methods may be employed. When long-span timber trusses are raised from a flat to a vertical position preparatory to lifting, stresses entirely different from design stresses may be introduced. The magnitude and distribution of these stresses depend on such factors as weight, dimensions, and type of truss. A competent rigger will consider these factors in determining how much suspension and stiffening, if any, is required and where it should be located.

Accessibility. Adequate space should be available at the site for temporary storage of materials from time of delivery to the site to time of erection. Material-handling equipment should have an unobstructed path from job-site storage to point of erection. Whether erection must proceed from inside the building area or can be done from outside will determine the location of the area required for operation of the equipment. Other trades should leave the erection area clear until all members are in place and are either properly braced by temporary bracing or are permanently braced in the building system.

Assembly and Subassembly. Whether done in a shop or on the ground or in the air in the field, assembly and subassembly are dependent on the structural system and the various connections involved.

Care should be taken with match-marking on custom materials. Assembly must be in accordance with the approved shop drawings. Any additional drilling or dapping, as well as the installation of all field connections, must be done in a workmanlike manner.

Heavy-timber trusses are usually shipped partly or completely disassembled. They are assembled on the ground at the site before erection. Arches, which are generally shipped in half sections, may be assembled on the ground or connections may be made after the half arches are in position. When trusses and arches are assembled on the ground at the site, assembly should be on level blocking to permit connections to be properly fitted and securely tightened without damage. End compression joints should be brought into full bearing and compression plates installed where intended.

Before erection, the assembly should be checked for prescribed overall dimensions, prescribed camber, and accuracy of anchorage connections. Erection should be planned and executed in such a way that the close fit and neat appearance of joints and the structure as a whole will not be impaired.

Field Welding. Where field welding is required, the work should be done by a qualified welder in accordance with job plans and specifications, approved shop drawings, and specifications of the American Institute of Steel Construction and the American Welding Society.

Cutting and Fitting. All connections should fit snugly in accordance with job plans and specifications and approved shop drawings. All cutting, framing, and boring should be done in accordance with good shop practices. Any field cutting, dapping, or drilling should be done in a workmanlike manner, with due consideration given to final use and appearance.

Bracing. Structural elements should be placed to provide lateral restraint and vertical support, to ensure that the complete assembly will form a stable structure. This bracing may extend longitudinally and transversely. It may comprise sway, cross, vertical, diagonal, and like members that resist wind, earthquake, erection, acceleration, braking, and other forces. And it may consist of knee braces, cables, rods, struts, ties, shores, diaphragms, rigid frames, and other similar components in combinations.

Bracing may be temporary or permanent. Permanent bracing, required as an integral part of the completed structure, is shown on the architectural or engineering plans and usually is also referred to in the job specifications. Temporary construction bracing is required to stabilize or hold in place permanent structural elements during erection until other permanent members that will serve the purpose are fastened in place. This bracing is the responsibility of the erector, who normally furnishes and erects it. Protective corners and other protective devices should be installed to prevent members from being damaged by the bracing.

In wood truss construction, temporary bracing can be used to plumb trusses during erection and hold them in place until other secondary framing and roof sheathing are installed. The major portion of temporary bracing for trusses is left in place, because it is designed to brace the completed structure against lateral forces.

Failures during erection occur occasionally and regardless of construction material used. The blame can usually be placed on insufficient or improperly located temporary erection guys or braces, overloading with construction materials, or an externally applied force sufficient to render temporary erection bracing ineffective. (See "Bracing Wood Trusses: Commentary and Recommendations," Truss Plate Institute, Madison, Wis., for guidance in erection of lightweight metal-gusset-plate, sawn-lumber trusses.)

Structural members of wood must be stiff, as well as strong. They must also be properly guyed or laterally braced, both during erection and permanently in the completed structure. Large rectangular cross sections of glued-laminated timber have relatively high lateral strength and resistance to torsional stresses during erection. However, the erector must never assume that a wood arch, beam, or column cannot buckle during handling or erection.

Specifications often require that:

1. Temporary bracing shall be provided to hold members in position until the structure is complete.
2. Temporary bracing shall be provided to maintain alignment and prevent displacement of all structural members until completion of all walls and decks.

3. The erector should provide adequate temporary bracing and take care not to overload any part of the structure during erection.

While the magnitude of the restraining force that should be provided by a cable guy or brace cannot be precisely determined, general experience indicates that a brace is adequate if it supplies a restraining force equal to 2% of the applied load on a column or of the force in the compression flange of a beam. It does not take much force to hold a member in line; but once it gets out of alignment, the force then necessary to hold it is substantial.

10.30 ENGINEERED GLUED WOOD PRODUCTS

These products, which include glued-laminated timber, prefabricated wood I joists, and structural composite lumber (SCL), are used extensively in residential and nonresidential building construction. They are used instead of nonwood products, such as steel framing elements, or as substitutes for conventional sawn-lumber products.

10.30.1 Characteristics of Glulam Timber

Glued-laminated timber, or glulam as it is often referred to, is the oldest type of the engineered glued wood products. With the employment of wet-use adhesives for laminating, glulam elements may be used in applications where they may be exposed to the elements. These applications include exterior building components, utility structures, marinas and wharfs, and bridge structures, such as pedestrian, highway, and railway bridges.

Preservative Treatment. For any exposed application, glulam members should be pressure impregnated with an improved preservative, such as creosote, pentachlorophenol (penta) in various carriers, and waterborne arsenicals. All treatments, however, are not compatible with all species; a specific treatment may not be available in the job-site region. Some treatments may be pressure impregnated into the laminations before the member is glued, whereas other treatments can be used only in conjunction with the finished product. (See Standard C-18, American Wood Preservers Association, Stevensville, MD, and technical notes on pressure treating of glulam of the American Institute of Timber Construction, Englewood, Colo.)

Versatility and Use. The most versatile of the engineered glued wood products, glulam can be fabricated in a wide variety of shapes, such as those shown in Fig. 10.27. Short-span glulam beams with constant rectangular cross sections, the most commonly used shape, are typically available as "stock beams" from distribution centers throughout the United States for use in residential and light commercial construction. Inventoried in a variety of sizes and lengths, stock beams are generally used for headers and floorbeams.

Straight or curved beams can be manufactured in lengths of over 100 ft and with large cross-sectional areas. Glulam arches have been erected to span 300 ft or more. For structures requiring very large spans, such as stadiums needing spans of 500 ft or more, glulam timber domes are often the most economical framing and are esthetically pleasing.

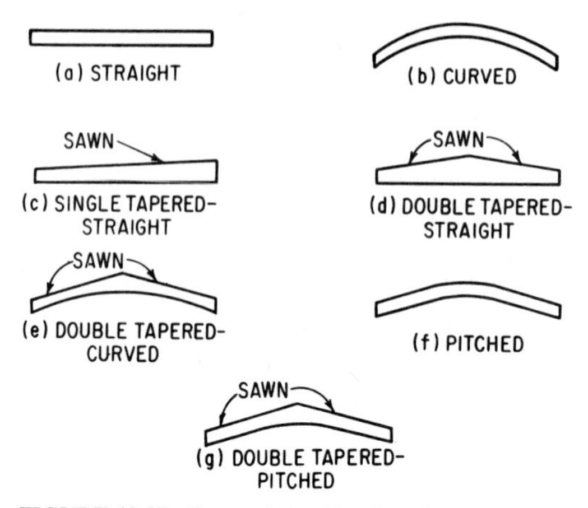

FIGURE 10.27 Types of glued-laminated beams.

The limitations on length for transporting glulam members from the manufacturing plant to the job site with available rail or truck facilities may control the size of glulam members that can be used for long spans without splicing. Designers and contractors should closely coordinate arrangements for transportation of long glulam components with the manufacturer.

("WOODCAD" computer program for design of beams and complete wood roof systems, American Plywood Association, Tacoma, Wash.)

10.30.2 Manufacture of Glulam Members

Structural glued-laminated timber is made by bonding layers of lumber together with adhesive so that the grain direction of all laminations is essentially parallel. Narrow boards may be edge-glued, and short boards may be end-joined to create greater lengths. The resultant wide and long laminations are then face-glued into large, shop-grown timbers.

Recommended practice calls for lumber of nominal 1- and 2-in thicknesses for laminating. The lumber is dressed to ¾- and 1½-in thicknesses, respectively, just before gluing. The thinner laminations are generally used in curved members.

Virtually any species of wood can be used in the laminating process if the design values have been determined. Different species can be intermixed within the depth of a section to achieve optimum resource utilization. Higher-strength species may be positioned in a beam in zones that will be subjected to high stresses under design loads. Lower-strength species can be placed in zones with lower in-service stresses. Similarly, manufacturer of glulam beams can be based on a graded layup concept. This requires that laminations with a higher lumber grading be set in zones subjected to high design stresses, and lower grades, in lower-stressed areas of the member. As a consequence, glulam members are a resource-efficient wood product, since varying grades and species can be used to achieve desired performance.

Constant-depth members normally are a multiple of the thickness of the lamination stock used. Variable-depth members, because of tapering or special assembly techniques, may not be exact multiples of these lamination thicknesses.

Nominal width of stock, in	4	6	8	10	12
Standard member finished width, in	3 or 3⅛	5 or 5⅛	6¾	8½ or 8¾	10½ or 10¾

Standard widths as listed above are most economical, since they represent the maximum width of board normally obtained from the stock used in laminating. Other widths, such as, 3½ or 5½ in, which fit well with conventional 2 × 4 and 2 × 6 framing, are also available in many market areas.

When members wider than the stock available are required, laminations may consist of two boards side by side. These edge joints must be staggered, vertically in horizontally laminated beams (load acting normal to wide faces of laminations), and horizontally in vertically laminated beams (load acting normal to the edge of laminations). In horizontally laminated beams, edge joints need not be edge-glued. Edge gluing is required in vertically laminated beams.

Edge and face gluings are the simplest to make, end gluings the most difficult. Ends are also the most difficult surfaces to machine into scarfs or finger joints.

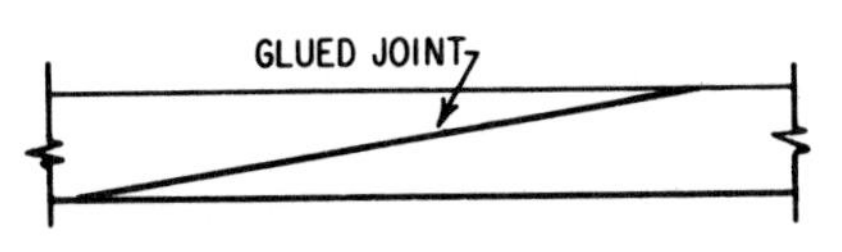

FIGURE 10.28 Plane sloping scarf.

A plane sloping scarf (Fig. 10.28), in which the tapered surfaces of laminations are glued together, can develop 85 to 90% of the strength of an unscarfed, clear, straight-grained control specimen. Finger joints (Fig. 10.29) are less wasteful of lumber. Quality can be adequately controlled in machine cutting and in high-frequency gluing. A combination of thin tip, flat slope on the side of the individual fingers, and a narrow pitch is desired. The length of fingers should be kept short for savings of lumber, but long for maximum strength.

The usefulness of structural glued-laminated timbers is determined by the lumber used and glue joint produced. Certain combinations of adhesive, treatment, and wood species do not produce the same quality of glue bond as other combinations, although the same gluing procedures are used. Thus, a combination must be supported by adequate experience with a laminator's gluing procedure.

The only adhesives currently recommended for wet-use and preservative-treated lumber, whether gluing is done before or after treatment, are resorcinol and phenol-resorcinol resins. The prime adhesive for dry-use structural laminating is casein. Melamine and melamine-urea blends are used in smaller amounts for high-frequency curing of end gluings.

Glued joints are cured with heat by several methods. Radio-frequency (RF) curing of glue lines is used for end joints and for limited-size members where there are repetitive gluings of the same cross section. Low-voltage resistance heating, where current is passed through a strip of metal to raise the temperature of a glue line, is used for attaching thin facing pieces. The metal may be left in the glue line as an integral part of the completed member. Printed electric circuits, in conjunction with adhesive films, and adhesive films impregnated on paper or on each side of a metal conductor placed in the glue line are other alternatives.

Preheating the wood to ensure reactivity of the applied adhesive has limited application in structural laminating. The method requires adhesive application as a wet or dry film simultaneously to all laminations, and then rapid handling of multiple laminations.

Curing the adhesive at room temperature has many advantages. Since wood is an excellent insulator, a long time is required for elevated ambient temperatures to reach inner glue lines of a large assembly. With room-temperature curing,

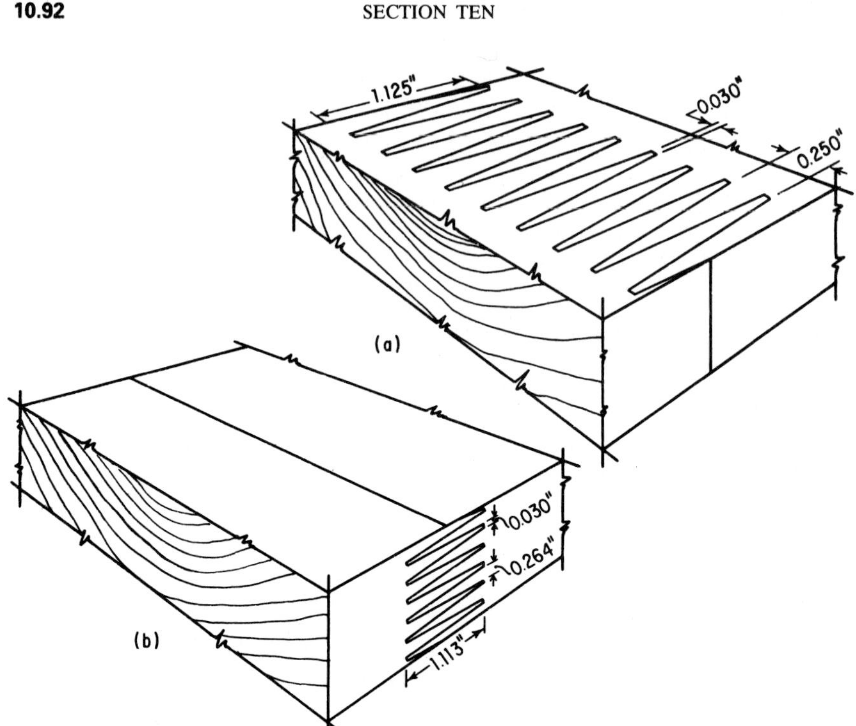

FIGURE 10.29 Finger joints: (*a*) fingers formed by cuts perpendicular to the wide faces of the boards; (*b*) fingers formed by cuts perpendicular to the edges.

equipment needed to heat the glue line is not required, and the possibility of injury to the wood from high temperature is avoided.

10.30.3 Prefabricated Wood I Joists

Flanges of prefabricated wood I-shaped joists are either sawn lumber, visually graded or machine stress rated, or some type of structural composite lumber product (Art. 10.30.4). The web members are either plywood or oriented strand board (OSB) (Art. 10.12).

Several manufacturers produce wood I joists, and the flange and web materials used depend on manufacturer preference. Inasmuch as these joists are proprietary products, the manufacturer provides design information, usually in the form of load/span tables, as well as installation and handling guidelines.

There are no nationally recognized manufacturing standards for wood I joists. Manufacturers establish design values for these products in accordance with the provisions of the "Specification for Establishing and Monitoring Structural Capacities of Prefabricated Wood I-Joists," ASTM D5055. Each manufacturer uses the design values thus determined to obtain National Evaluations Service (NER) building-code approval from model code sponsors and to generate proprietary load/span tables.

Wood I joists are available with a wide range of depths and load/span capabilities. Although joist sizes vary with manufacturer preference, most manufacturers produce joists with depths of 9¼ (or 9½) in or 11¼ (or 11⅞) in for direct substitution for 2 × 10 and 2 × 12 dimension lumber. Other depths—14 and 16 in and deeper—also are available. The deeper products are typically used in longer-span applications, such as for light commercial construction.

Prefabricated wood I joists have many advantages that make them an economical construction material. Some of these advantages are:

Manufactured product. They are shipped to the job site precut to length, thereby eliminating waste. Consistent product quality is assured by the manufacturing process so that all material arriving at the job site is usable. Wood I joists are manufactured from dry components, thus eliminating shrinkage, warping, and twisting.

Long lengths available. I joists can be manufactured and shipped to the job site in long lengths. This minimizes labor costs due to handling and allows the joists to be used in multispan applications.

Lightweight. Their low weight makes it very easy for construction workers to easily handle long lengths on the job site, whether for long clear spans or multiple spans.

Ease of fabrication. The structural panel webs are easily cut to permit passage of wiring, conduit, plumbing, and mechanical ductwork. Manufacturers provide charts that indicate the maximum permissible size of round or rectangular openings that can be cut in the web without adversely affecting structural performance of the joists.

10.30.4 Structural Composite Lumber

Structural composite lumber (SCL) comprises a family of secondary manufactured engineered glued wood products. The most widely available type of SCL, laminated veneer lumber (LVL), is similar to plywood in that dry sheets of veneer are structurally bonded together to create large panels, usually in widths of 2 or 4 ft. These panels are typically produced in thicknesses of 1½ or 1¾ in, and in long lengths. (Lengths vary, depending on the manufacturer). However, unlike plywood, which has the veneers cross-banded, all veneers in LVL products are oriented with grain approximately parallel, much like the positioning of sawn lumber in glulam timber.

Another type of SCL, parallel-strand lumber (PSL), is manufactured from long strands of veneer rather than veneer sheets as used in LVL manufacturing. PSL is manufactured in a variety of widths ranging from 1¾ in up to 7 in. Various depths and lengths are also produced. Designers should consult the manufacturers for information on size availability.

The large SCL sections can be resawn into a variety of smaller dimension lumber or timber products. An extensive use of this product is for beams and headers. Two pieces of 1¾-in-wide SCL can be nailed together to create a 3½-in-wide beam for use in conventional 2 × 4 framing. Wider beams of LVL can be created by nailing three or more pieces together or by cutting larger sections from PSL billets. SCL is used for scaffold plank, truss chords, flanges for wood I joists, ridge beams in mobile homes, and a myriad of other building and industrial uses.

Structural design properties for SCL are generally much higher than comparable values for sawn lumber. SCL is commonly available with allowable bending stresses up to 3000 psi with corresponding modulus of elasticity in the range of 2,000,000 psi. There are, however, no industry standards for establishing design stresses for SCL. Values vary among the various manufacturers.

In addition to exhibiting higher strength characteristics than other wood products, design properties for SCL products have less variability. This is largely due to the control in manufacturing of natural strength-reducing characteristics of wood, such as slope of grain, knots, and density. Also, the random dispersal of these strength-reducing characteristics throughout the finished member tends to offset the individual effects of these defects on the overall strength of the end product, much like the use of varying grades of sawn lumber in the manufacture of glulam timber. This combination of higher strengths and reduced variability makes SCL an economical wood structural material.

10.30.5 Environmental Considerations

With the increasing emphasis on more efficient uses of the available wood-fiber resource, engineered glued wood products are becoming more attractive. Each of the glued products described in the preceding articles makes optimum use of the base wood products in creating high-end, high-quality engineered products. Innovations in the engineered wood products industry are ongoing and it is these innovative engineered wood products that will allow wood to continue to be a viable construction material for building applications in the future.

SECTION ELEVEN
WALL, FLOOR, AND CEILING SYSTEMS

Frederick S. Merritt
Consulting Engineer,
West Palm Beach, Florida

This section discusses design and construction of systems generally used for enclosing buildings and the spaces within them. (Some such systems, such as roofs and foundations, however, are treated in other sections, because of their special functions in addition to enclosure of spaces.) The systems covered in this section, as described in Art. 1.7, include exterior walls; interior walls, or partitions; floors; and ceilings.

Each of these systems usually consists of one or more facing subsystems and a structural subsystem that supports them. The facing subsystems may be the surfaces of the structural subsystem or separate entities that enclose that subsystem. They serve esthetic purposes, provide privacy, and bar, or at least restrict, passage of people or other moving objects, water, air, sound, heat and also often light.

Wood structural subsystems are discussed in Sec. 10, and concrete is discussed in Sec. 9. Basic principles of waterproofing building exteriors are presented in Art. 3.4.2. This section describes techniques applicable to unit masonry and curtain walls.

Floors provide not only a horizontal separation of interior building spaces but also a surface on which human activities can take place and on which materials and equipment can be stored. The structural subsystem usually consists of a slab or deck and also often of beams that support it. These are described in Secs. 7 through 10. This section discusses constructions used for the upper facing, or floor coverings, which serve esthetic purposes and act as a wearing surface. The bottom facing, or ceiling, may be the bottom surface of the slab or deck or a separate entity, such as a gypsum-plaster membrane, which is also discussed in this section, or acoustical tile.

MASONRY WALLS

Masonry comprises assemblages of nonmetallic, incombustible materials, such as stone, brick, structural clay tile, concrete block, glass block, gypsum block, or adobe brick. Unit masonry consists of pieces of such materials, usually between 4 and 24 in in length and height and between 4 and 12 in in thickness. The units are bonded together with mortar or other cementitious materials.

Walls and partitions are classified as load-bearing and non-load-bearing. Different design criteria are applied to the two types.

Minimum requirements for both types of masonry walls are given in ANSI Standard Building Code Requirements for Masonry, A41.1 and ANSI Standard Building Code Requirements for Reinforced Masonry, A41.2, American National Standards Institute; Building Code Requirements for Engineered Brick Masonry, Brick Institute of America, and ACI Standard Building Code Requirements for Concrete Masonry Structures, ACI 531, American Concrete Institute.

Like other structural materials, masonry may be designed by application of engineering principles. In the absence of such design, the empirical rules given in this section and adopted by building codes may be used.

11.1 MASONRY DEFINITIONS

Following are some of the terms most commonly encountered in masonry construction:

Architectural Terra Cotta. (See Ceramic Veneer.)

Ashlar Masonry. Masonry composed of rectangular units usually larger in size than brick and properly bonded, having sawed, dressed, or squared beds. It is laid in mortar.

Bearing Walls. (See Load-Bearing Wall.)

Bonder. (See Header.)

Brick. A rectangular masonry building unit, not less than 75% solid, made from burned clay, shale, or a mixture of these materials.

Buttress. A bonded masonry column built as an integral part of a wall and decreasing in thickness from base to top, though never thinner than the wall. It is used to provide lateral stability to the wall.

Ceramic Veneer. Hard-burned, non-load-bearing, clay building units, glazed or unglazed, plain or ornamental.

Chase. A continuous recess in a wall to receive pipes, ducts, conduits.

Column. A compression member with width not exceeding 4 times the thickness, and with height more than 3 times the least lateral dimension.

Concrete Block. A machine-formed masonry building unit composed of portland cement, aggregates, and water.

Coping. A cap or finish on top of a wall, pier, chimney, or pilaster to prevent penetration of water to masonry below.

Corbel. Successive course of masonry projecting from the face of a wall to increase its thickness or to form a shelf or ledge (Fig. 11.3*f*).

Course. A continuous horizontal layer of masonry units bonded together (Fig. 11.3).

Cross-Sectional Area. Net cross-sectional area of a masonry unit is the gross cross-sectional area minus the area of cores or cellular spaces. Gross cross-sectional area of scored units is determined to the outside of the scoring, but the cross-sectional area of the grooves is not deducted to obtain the net area.

Eccentricity. The normal distance between the centroidal axis of a member and the component of resultant load parallel to that axis.

Effective Height. The height of a member to be assumed for calculating the slenderness ratio.

Effective Thickness. The thickness of a member to be assumed for calculating the slenderness ratio.

Grout. A mixture of cementitious material, fine aggregate, and sufficient water to produce pouring consistency without segregation of the constituents.

Grouted Masonry. Masonry in which the interior joints are filled by pouring grout into them as the work progresses.

Header (Bonder). A brick or other masonry unit laid flat across a wall with end surface exposed, to bond two wythes (Fig. 11.1*b*).

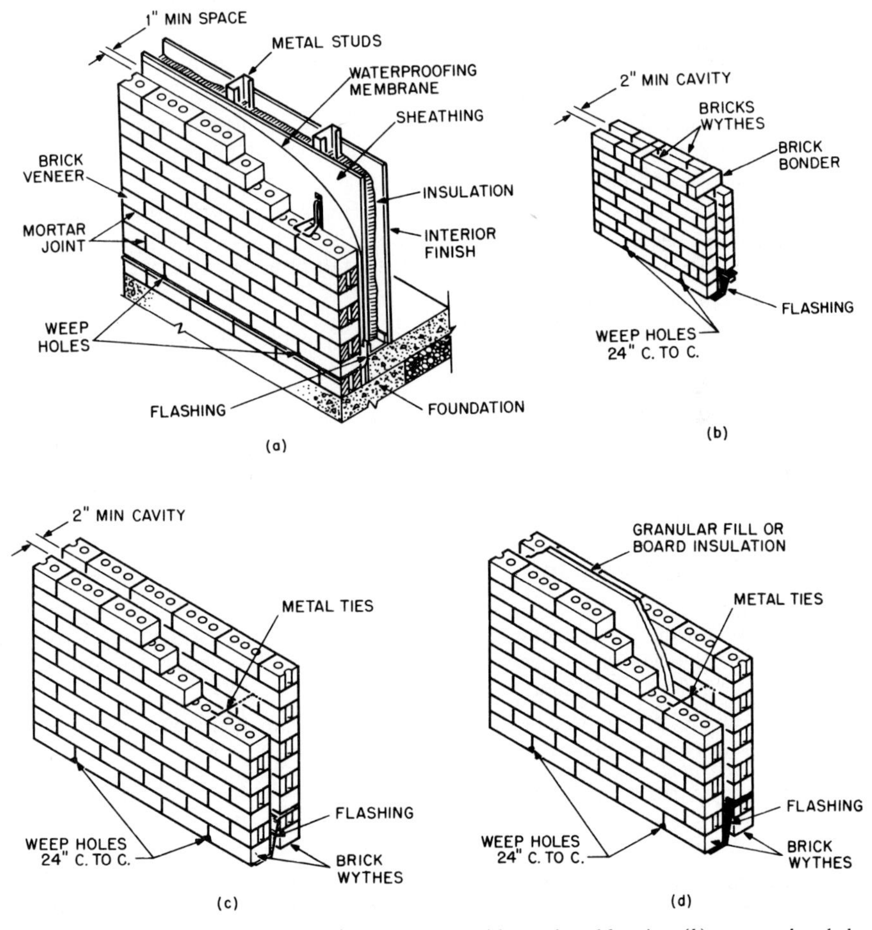

FIGURE 11.1 Brick exterior walls: (*a*) brick veneer with metal-stud framing; (*b*) masonry-bonded hollow walls; (*c*) hollow wall with metal ties between wythes; (*d*) insulated cavity wall.

Height of Wall. Vertical distance from top of wall to foundation wall or other intermediate support.

Hollow Masonry Unit. Masonry with net cross-sectional area in any plane parallel to the bearing surface less than 75% of its gross cross-sectional area measured in the same plane.

Lateral Support. Members such as cross walls, columns, pilasters, buttresses, floors, roofs, or spandrel beams that have sufficient strength and stability to resist horizontal forces transmitted to them may be considered lateral supports.

Load-Bearing Wall. A wall that supports any vertical load in addition to its own weight.

Masonry. A built-up construction or combination of masonry units bonded together with mortar or other cementitious material.

Mortar. A plastic mixture of cementitious materials, fine aggregates, and water.

Partition. An interior non-bearing wall one story or less in height.

Pier. An isolated column of masonry. A bearing wall not bonded at the sides into associated masonry is considered a pier when its horizontal dimension measured at right angles to the thickness does not exceed 4 times its thickness.

Pilaster. A bonded or keyed column of masonry built as part of a wall, but thicker than the wall, and of uniform thickness throughout its height. It serves as a vertical beam, column, or both.

Prism. An assemblage of brick and mortar for the purpose of laboratory testing for design strength, quality control of materials, and workmanship. Minimum height for prisms is 12 in, and the slenderness ratio should lie between 2 and 5.

Rubble:

Coursed Rubble. Masonry composed of roughly shaped stones fitting approximately on level beds, well bonded, and brought at vertical intervals to continuous level beds or courses.

Random Rubble. Masonry composed of roughly shaped stones, well bonded and brought at irregular vertical intervals to discontinuous but approximately level beds or courses.

Rough or Ordinary Rubble. Masonry composed of nonshaped field stones laid without regularity of coursing, but well bonded.

Slenderness Ratio. Ratio of the effective height of a member to its effective thickness.

Solid Masonry Unit. A masonry unit with net cross-sectional area in every plane parallel to the bearing surface 75% or more of its gross cross-sectional area measured in the same plane.

Solid Masonry Wall. A wall built of solid masonry units laid contiguously, with joints between units filled with mortar or grout.

Stretcher. A masonry unit laid with length horizontal and parallel with the wall face (Fig. 12.3).

Veneer. A wythe securely attached to a wall but not considered as sharing load or adding strength to it (Fig. 11.1*a*).

Virtual Eccentricity. The eccentricity of resultant axial loads required to produce axial and bending stresses equivalent to those produced by applied axial and transverse loads.

Wall. Vertical or near-vertical construction, with length exceeding three times the thickness, for enclosing space or retaining earth or stored materials.

Bearing Wall. A wall that supports any vertical load in addition to its own weight.

Cavity Wall. (See *Hollow Wall* below.)

Curtain Wall. A non-load-bearing exterior wall.

Faced Wall. A wall in which the masonry facing and the backing are of different materials and are so bonded as to exert a common reaction under load.

Hollow Wall. A wall of masonry so arranged as to provide an air space within the wall between the inner and outer wythes (Fig. 11.1*b, c,* and *d*). A cavity wall is built of masonry units or plain concrete, or of a combination of these materials, so arranged as to provide an airspace within the wall, which may be filled with insulation, and in which inner and outer wythes are tied together with metal ties (Fig. 11.1*d*).

Nonbearing Wall. A wall that supports no vertical load other than its own weight.

Party Wall. A wall on an interior lot line used or adapted for joint service between two buildings.

Shear Wall. A wall that resists horizontal forces applied in the plane of the wall.

Spandrel Wall. An exterior curtain wall at the level of the outside floor beams in multistory buildings. It may extend from the head of the window below the floor to the sill of the window above.

Veneered Wall. A wall having a facing of masonry or other material securely attached to a backing, but not so bonded as to exert a common reaction under load (Fig. 11.1*a*).

Wythe. Each continuous vertical section of a wall one masonry unit in thickness (Fig. 11.1).

11.2 QUALITY OF MATERIALS FOR MASONRY

Materials used in masonry construction should be capable of meeting the requirements of the applicable standard of ASTM.

Second-hand materials should be used only with extreme caution. Much salvaged brick, for example, comes from demolition of old buildings constructed of solid brick in which hard-burned units were used on the exterior and salmon units as backup. Because the color differences that guided the original masons in sorting and selecting bricks become obscured with exposure and contact with mortar, there is a definite danger that the salmon bricks may be used for exterior exposure and may disintegrate rapidly. Masonry units salvaged from chimneys are not recommended because they may be impregnated with oils or tarry material.

Design of load-bearing brick structures may be based on rational engineering analysis instead of the empirical requirements for minimum wall thickness and maximum wall height contained in building codes and given in Art. 11.8. Those requirements usually make bearing-wall construction for buildings higher than three to five stories uneconomical, and encourage use of other methods of support (steel or concrete skeleton frame). Since 1965, engineered brick buildings 10 or more stories high, with design based on rational structural analysis, have been built in the United States. This construction was stimulated by the many load-bearing brick buildings exceeding 10 stories in height constructed during the preceding two decades in Europe.

Design requirements for engineered brick structures given in this section are taken from "Building Code Requirements for Engineered Brick Masonry," promulgated by the Brick Institute of America, 11490 Commerce Park Drive, Reston, VA 22091.

11.2.1 General Requirements of Design Standard

"Building Code Requirements for Engineered Brick Masonry" provides minimum requirements predicated on a general analysis of the structure and based on generally accepted engineering analysis procedures. The standard contains a requirement for architectural or engineering inspection of the workmanship to ascertain, in general, if the construction and workmanship are in accordance with the contract drawings and specifications. Frequency of inspections should be such that an inspector can inspect the various stages of construction and see that the work is being properly performed. The standard requires reduced allowable stresses and capacities for loads when such architectural or engineering inspection is not provided.

Engineered brick bearing-wall structures do not require new techniques of analysis and design, but merely application of engineering principles used in the analysis and design of other structural systems. The method of analysis depends on the complexity of the building with respect to height, shape, wall location, and openings in the wall. A few conservative assumptions, however, accompanied by proper details to support them, can result in a simplified and satisfactory solution for most bearing-wall structures up to 12 stories high. More rigorous analysis for bearing-wall structures beyond this height may be required to maintain the economics of this type of construction.

11.2.2 Materials for Masonry Construction

Strength (compressive, shearing, and transverse) of brick structures is affected by the properties of the brick and the mortar in which they are laid. In compression, strength of brick has the greater effect. Although mortar is also a factor in compressive strength, its greater effect is on the transverse and shearing strengths of masonry. For these reasons, there are specific design requirements for and limitations on materials used in engineered brick structures.

Brick. These units must conform to the requirements for grade MW or SW, ASTM "Standard Specifications for Building Brick," C62. In addition, brick used in load-bearing or shear walls must comply with the dimension and distortion tolerances specified for type FBS of ASTM "Standard Specifications for Facing Brick," C216. Bricks that do not comply with these tolerance requirements may be used if the ultimate compressive strength of the masonry is determined by prism tests.

Mortar. Most of the test data on which allowable stresses for engineered brick masonry are based were obtained for specimens built with portland cement-hydrated lime mortars. Three mortar types are provided for: M, S, and N, as described in ASTM C270 (see Art. 4.16), except that the mortar must consist of mixtures of portland cement (type I, II, or III), hydrated lime (type S, non-air-entrained), and aggregate when the allowable stresses specified in "Building Code Requirements for Engineered Brick Masonry" are used. This standard provides, however, that "Other mortars . . . may be used when approved by the Building Official, provided strengths for such masonry construction are established by tests . . .".

For ordinary unit masonry, mortar should meet the requirements of ASTM C270 and C476. These define the types of mortar described in Art. 4.16. Each type is used for a specific purpose, as indicated in Table 11.1, based on compressive strength. However, it should not be assumed that higher-strength mortars are preferable to lower-strength mortars where lower strength is permitted for particular uses. The primary purpose of mortar is to bond masonry units together.

Mortars containing lime are generally preferred because of greater workability. Commonly used:

For concrete block, 1 part cement, 1 part lime putty, 5 to 6 parts sand

For rubble, 1 part cement, 1 to 2 parts lime hydrate or putty, 5 to 7 parts sand

For brick, 1 part cement, 1 part lime, 6 parts sand

For setting tile, 1 part cement, ½ part lime, 3 parts sand

11.3 CONSTRUCTION OF MASONRY

Compressive strength of masonry depends to a great extent on workmanship and the completeness with which units are bedded. Tensile strength is a function of the adhesion of mortar to a unit and of the area of bonding (degree of completeness with which joints are filled). Hence, in specifying masonry work, it is important

TABLE 11.1 Mortar Requirements of Masonry

Kind of masonry	Types of mortar
Masonry in contact with earth:	
Footings	M or S
Walls of solid units	M, S, or N
Walls of hollow units	M or S
Hollow walls	M or S
Masonry above grade or interior:	
Piers of solid masonry	M, S, or N
Piers of hollow units	M or S
Walls of solid masonry	M, S, N, or O
Grouted masonry	PL or PM
Walls of hollow units; load-bearing or exterior, and hollow walls 12 in or more in thickness	M, S, or N
Hollow walls less than 12 in in thickness where assumed design wind pressure:	
a. Exceeds 20 psf	M or S
b. Does not exceed 20 psf	M, S, or N
Glass-block masonry	M, S, or N
Nonbearing partitions of fireproofing composed of structural clay tile or concrete masonry units	M, S, N, O, or gypsum
Gypsum partition tile or block	Gypsum
Firebrick	Refractory air-setting mortar
Linings of existing masonry	M or S
Masonry other than above	M, S, or N

to call for a full bed of mortar, with each course well hammered down, and all joints completely filled with mortar. To minimize the entrance of water through a masonry wall, follow the practices recommended in Art. 3.4.2.

In particular, in filling head joints, a heavy buttering of mortar should be applied on one end of the masonry, and the unit should be pushed down into the bed joint into place so that the mortar squeezes out from the top and sides of the head joint. Mortar should correspondingly cover the entire side of a unit before it is placed as a header. An attempt to fill head joints by slushing or dashing will not succeed in producing watertight joints. Partial filling of joints by "buttering" or "spotting" the vertical edge of the unit with mortar cut from the extruded bed joint is likewise ineffective and should be prohibited. Where closures are required, the opening should be filled with mortar so that insertion of the closure will extrude mortar both laterally and vertically.

Mortar joints usually range from ¼ to ¾ in in thickness.

Tooling of joints, if done properly can help to resist penetration of water; but it is not a substitute for complete filling, or a remedy for incomplete filling of joints. A concave joint (Fig. 11.2) is recommended. Use of raked or other joints that provide horizontal water tables should be avoided. Mortar should not be too stiff at time of tooling, or compaction will not take place, nor should it be too fluid, or the units may move—and units should never be moved after initial contact with mortar. If a unit is out of line, it should be removed, mortar scraped off and fresh mortar applied before the unit is relaid.

The back face of exterior wythes should be back plastered, or parged, before backup units are laid. If the backup is laid first, the front of the backup should be parged. The mortar should be the same as that used for laying the masonry and should be applied from ¼ to ⅜ in thick.

The rate of absorption of water by unit masonry at the time of laying is important in determining the strength and resistance to penetration of water of a mortared joint. This rate can be reduced by wetting the unit before laying. Medium absorptive units may need to be thoroughly soaked with water. Highly absorptive units may require total immersion in water for some time before unit "suction" is reduced to the low limits needed.

The test for the rate of absorption of brick is described in ASTM Standard C67. For water-resistant masonry, the suction (rate of absorption) of the brick should not exceed 0.35, 0.5, and 0.7 oz, respectively, for properly constructed all-brick walls or facings of normal 4-, 8-, and 12-in thickness.

The amount of wetting that bricks will require to control rate of absorption properly when laid should be known or determined by measurements made before the bricks are used in the wall. Some medium absorptive bricks may require only frequent wetting in the pile; others may need to be totally immersed for an hour or more. While the immersion method is more costly than hosing in the pile, it ensures that all bricks are more or less saturated when removed from immersion. A short time interval may be needed before the bricks are laid; but the bricks are

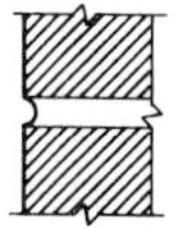
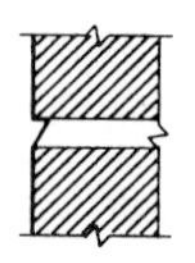

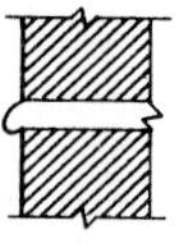
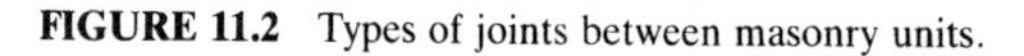

FIGURE 11.2 Types of joints between masonry units.

likely to remain on the scaffold, in a suitable condition to lay, for some time. Bricks on the scaffold should be inspected and moisture condition checked several times a day.

In general, method of manufacture and surface texture of masonry do not greatly affect the permeability of walls. However, water-resistant joints may be difficult to obtain if the units are deeply scored, particularly if the mortar is of a dry consistency. Loose sand should be brushed away or otherwise removed from units that are heavily sanded.

Mortar to be used in above-grade, water-resistant brick-faced and all-brick walls should be of as wet a consistency as can be handled by the mason and meet requirements of ASTM Standard C270, Type N. Water retention of the mortar should not be less than 75%, and preferably 80% or more. For laying absorptive brick that contain a considerable amount of absorbed water, the mortars having a water retention of 80% or more may be used without excessive "bleeding" at the joints and "floating" of the brick. The mortar may contain a masonry cement meeting the requirements of ASTM C91, except that water retention should not be less than 75 or 80%. Excellent mortar may also be made with portand cement and hydrated lime, mixed in the proportion of 1:1:6 parts by volume of cement, lime, and loose damp sand. The hydrated lime should be highly plastic. Type S lime conforming with the requirements of ASTM C207 is highly plastic, and mortar containing it, in equal parts by volume with cement, will probably have a water retention of 80% or more.

Since capillary penetration of moisture through concrete and mortar is of minor importance, particularly in above-grade walls, the mortar need not contain an integral water repellent. However, if desired, water-repellent mortar may be advantageously used in a few courses at the grade line to reduce capillary rise of moisture from the ground into the masonry. The mortar should be of a type that does not stiffen rapidly on the board, except through loss of moisture by evaporation.

Mortar should be retempered frequently if necessary to maintain as wet a consistency as is practically possible for the mason to use. At air temperatures below 80°F, mortar should be used or discarded within 3½ hr after mixing; for air temperatures of 80°F or higher, unused mortar should be discarded after 2½ hr.

11.3.1 Cold-Weather Construction of Masonry Walls

Masonry should be protected against damage by freezing. The following special precautions should be taken:

Materials to be used should be kept dry. Tops of all walls not enclosed or sheltered should be covered whenever work stops. The protection should extend downward at least 2 ft.

Frozen materials must be thawed before use. Masonry units should be heated to at least 40°F. Mortar temperature should be between 40 and 120°F, and mortar should not be placed on a frozen surface. If necessary, the wall should be protected with heat and windbreaks for at least 48 hr. Use of mortars made with high-early-strength cement may be advantageous for cold-weather masonry construction.

11.3.2 Bond between Wythes in Masonry Walls

When headers are used for bonding the facing and backing in solid masonry walls and faced walls, as shown in Fig. 11.3, not less than 4% of the wall surface of each

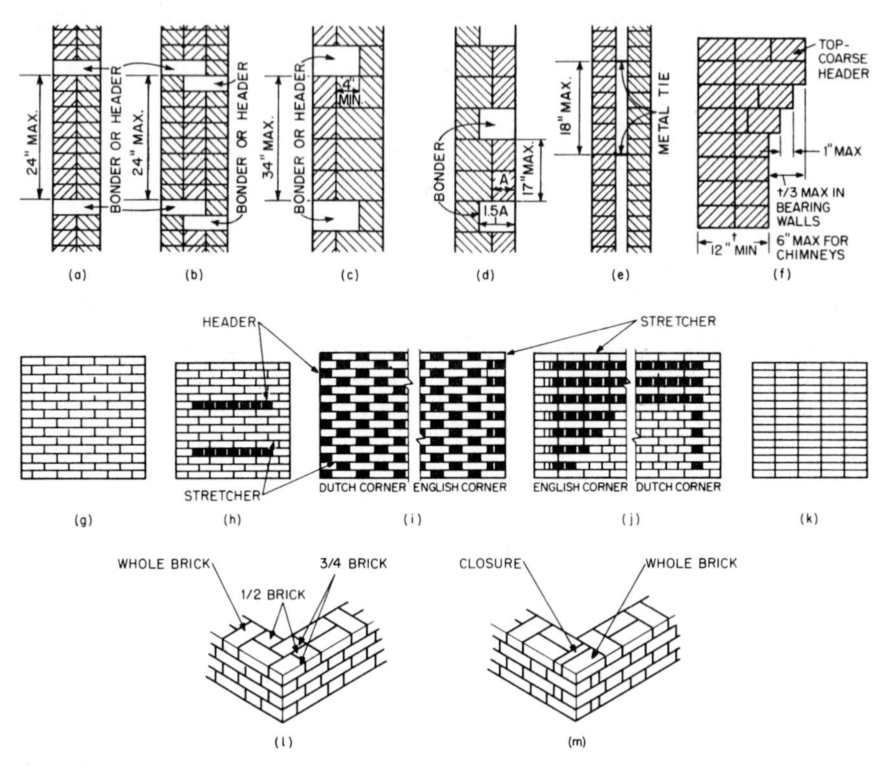

FIGURE 11.3 Types of unit-masonry construction. Cross sections through walls show: (*a*) two-wythe solid wall with bonders; (*b*) three-wythe solid wall with bonders; (*c*) and (*d*) hollow-unit walls; (*e*) hollow or cavity wall; (*f*) corbeled wall. Elevations of walls show types of masonry courses: (*g*) running bond; (*h*) common, or header, bond with bonders every sixth course; (*i*) Flemish bond with bonders in every course; (*j*) English bond; (*k*) stack bond. Types of corner bond: (*l*) Dutch; (*m*) English.

face should be composed of headers, which should extend at least 4 in into the backing. These headers should not be more than 24 in apart vertically or horizontally (Fig. 11.3*a* and *b*). In walls in which a single bonder does not extend through the wall, headers from opposite sides should overlap at 4 in or should be covered with another bonder course overlapping headers below at least 4 in.

If metal ties (Figs. 11.3 *e* and 11.4) are used for bonding, they should be corrosion-resistant. For bonding facing and backing of solid masonry walls and faced walls, there should be at least one metal tie for each 4½ ft^2 of wall area. Ties in alternate courses should be staggered, the maximum vertical distance between ties should not exceed 18 in, and the maximum horizontal distance should not be more than 36 in.

In walls composed of two or more thicknesses of hollow units, stretcher courses should be bonded by one of the following methods: At vertical intervals up to 34 in, there should be a course lapping units below at least 4 in (Fig. 11.3*c*). Or at vertical intervals up to 17 in, lapping should be accomplished with units at least 50% thicker than the units below (Fig. 11.3*d*). Or at least one metal tie should be incorporated for each 4½ ft^2 of wall area. Ties in alternate courses should be

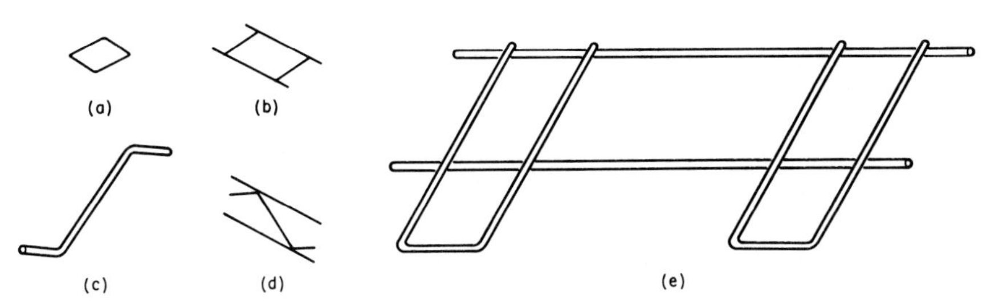

FIGURE 11.4 Types of metal ties for masonry walls: (*a*) rectangular; (*b*) ladder; (*c*) Z; (*d*) truss; (*e*) U.

staggered; the maximum vertical distance between ties should be 18 in and maximum horizontal distance, 36 in. Full mortar coverage should be provided in both horizontal and vertical joints at ends and edges of face shells of the hollow units.

In ashlar masonry, bond stones should be uniformly distributed throughout the wall and form at least 10% of the area of exposed faces.

In rubble stone masonry up to 24 in thick, bond stones should have a maximum spacing of 3 ft vertically and horizontally. In thicker walls, there should be at least one bond stone for each 6 ft ^{2} of wall surface on both sides.

For bonding ashlar facing, the percentage of bond stones should be computed from the exposed face area of the wall. At least 10% of this area should be composed of uniformly distributed bond stones extending 4 in or more into the backup. Every bond stone and, when alternate courses are not full bond courses, every stone should be securely anchored to the backup with corrosion-resistant metal anchors. These should have a minimum cross section of 3⁄16 × 1 in. There should be at least one anchor to a stone and at least two anchors for stones more than 2 ft long or with a face area of more than 3 ft^2. Larger facing stones should have at least one anchor per 4 ft^2 of face area of the stone, but not less than two anchors.

Cavity-wall wythes should be bonded with 3⁄16-in-diameter steel rods or metal ties of equivalent stiffness embedded in horizontal joints. There should be at least one metal tie for each 4½ ft^2 of wall area. Ties in alternate courses should be staggered, the maximum vertical distance between ties should not exceed 18 in (Fig 11.3*e*), and the maximum horizontal distance, 36 in. Rods bent to rectangular shape should be used with hollow masonry units laid with cells vertical. In other walls, the ends of ties should be bent to 90° angles to provide hooks at least 2 in long. Additional bonding ties should be provided at all openings. These ties should be spaced not more than 3 ft apart around the perimeter and within 12 in of the opening.

When two bearing walls intersect and the courses are built up together, the intersections should be bonded by laying in true bond at least 50% of the units at the intersection. When the courses are carried up separately, the intersecting walls should be regularly toothed or blocked with 8-in maximum offsets. The joints should be provided with metal anchors having a minimum section of ¼ × 1½ in with ends bent up at least 2 in or with cross pins to form an anchorage. Such anchors should be at least 2 ft long and spaced not more than 4 ft apart.

11.3.3 Grouted Masonry

Construction of walls requiring two or more wythes of brick or solid concrete block, similar to the wall shown in Fig. 11.3*a*, may be speeded by pouring grout between the two outer wythes, to fill the interior joints. Building codes usually require that, for the wythes, the mortar be type M or S, consisting of portland cement, lime, and aggregate (Art. 4.16). Also, they may require that, when laid, burned-clay brick and sand-lime units should have a rate of absorption of not more than 0.025 oz/in^2 over a 1-min period in the standard absorption test (ASTM C67). All units in the two outer wythes should be laid with full head and bed joints.

Low-Lift Grouting. The vertical spaces between wythes that are to be grouted should be at least ¾ in wide. Masonry headers should not project into the gap. One of the outer wythes may be carried up 18 in before grout is poured. The other outer wythe is restricted to a height up to 6 times the grouting space, but not more than 8 in, before grout is poured. Thus, in this type of construction, grout is poured in lifts not exceeding 8 in. The grout should be puddled with a grout stick immediately after it has been poured. If work has to be stopped for an hour or more, horizontal construction joints should be formed by raising all wythes to the same level and leaving the grout 1 in below the top. A suitable grout for this type of construction consists of 1 part portland cement, 0.1 part hydrated lime or lime putty, and 2¼ to 3 parts sand.

High-Lift Grouting. This type of construction is often used where steel reinforcement is to be inserted in the vertical spaces between wythes; for example, in the cavity of the wall shown in Fig. 11.3*e*. Grout is poured continuously in lifts up to 6 ft high and up to 30 ft long in the vertical spaces. (Vertical barriers, or dams, of solid masonry may be built in the grout space to control the horizontal flow of grout.) Building codes may require each lift to be completed within one day. The grout should be consolidated by puddling or mechanical vibrating as it is placed and reconsolidated after excess moisture has been absorbed but before plasticity has been lost. A suitable grout for gaps 2 or more inches wide consists of 1 part portland cement, 0.1 part hydrated lime or lime putty, 2 to 3 parts sand, and not more than 2 parts gravel, by volume.

In construction of the wall, the wythes should be kept at about the same level. No wythe should lay behind the others more than 16 in in height. The masonry should be allowed to cure for at least 3 days, to gain strength, before grout is poured. The grout space should be at least 2 in wide. If, however, horizontal reinforcement is to be placed in the gap, it should be wide enough to provide ¼ in clearance around the steel, but not less than 3 in wide.

Cleanouts should be provided for every pour. This may be done by omitting every other unit in the bottom course of the wall section being poured. Before grout is placed, excess mortar, mortar fins, and other foreign matter should be removed from the grout space. A high-pressure water jet may be used for the purpose. After inspection but before placement of grout, the cleanout holes should be plugged with masonry units, which should then be braced to resist the grout pressure.

Wire ties should be inserted in the mortar joints between masonry courses and span across each grout space, to bond the wythes (Fig. 11.5). The ties should be formed into rectangles, 4 in wide and with a length 2 in less than the distance between outer faces of the wythes being bonded. The wire size should not be less than No. 9. Spacing of ties should not exceed 24 in horizontally. For running-bond masonry (Fig. 11.3*f*), vertical tie spacing should not exceed 16 in, and for stack-bond masonry (Fig. 11.3*j*), 12 in.

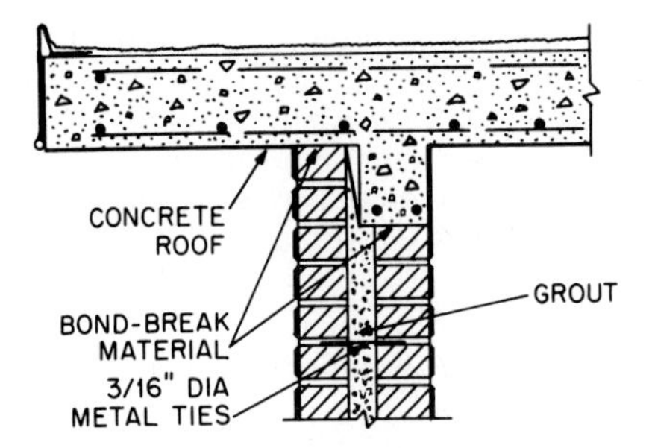

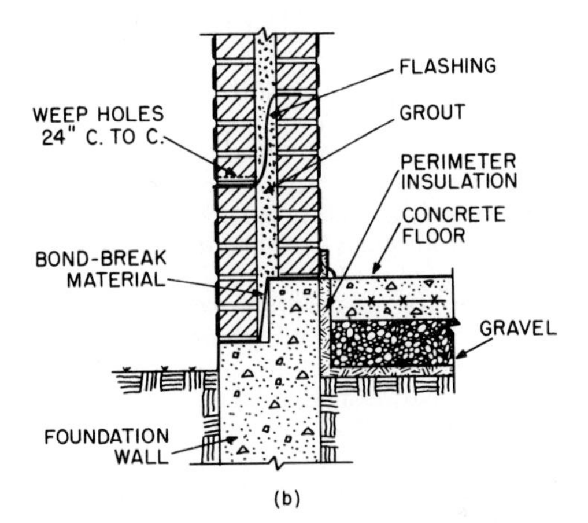

FIGURE 11.5 Grouted masonry wall: (*a*) cross section at the roof; (*b*) cross section at the base of the wall.

11.3.4 Support Conditions for Walls

Provision should be made to distribute concentrated loads safely on masonry walls and piers. Heavily loaded members should have steel bearing plates under the ends to distribute the load to the masonry within allowable bearing stresses. Length of bearing should be at least 3 in. Lightly loaded members may be supported directly on the masonry if the bearing stresses in the masonry are within permissible limits and if length of bearing is 3 in or more.

Masonry should not be supported on wood construction.

11.3.5 Corbeling

Where a solid masonry wall 12 in or more thick must be increased in thickness above a specific level, the increase should be achieved gradually by corbeling. In this method, successive courses are projected from the face of the wall, as indicated in Fig. 11.3*f*.

The maximum corbeled horizontal projection beyond the face of a wall should not exceed one-third the wall thickness for walls supporting structural members. In any case, projection of any course of masonry should not exceed 1 in.

Chimneys generally may not be coreled more than 6 in from the face of the wall. In the second story of two-story dwellings, however, corbeling of chimneys on the exterior of enclosing walls may equal the wall thickness.

11.3.6 Openings, Chases, and Recesses in Masonry Walls

Masonry above openings should be supported by arches or lintels of metal or reinforced masonry, which should bear on the wall at each end at least 4 in. Stone or other nonreinforced masonry lintels should not be used unless supplemented on the inside of the wall with structural steel lintels, suitable masonry arches, or reinforced-masonry lintels carrying the masonry backing. Lintels should be stiff enough to carry the superimposed load with a deflection of less than $1/720$ of the clear span.

In plain concrete walls, reinforcement arranged symmetrically in the thickness of the wall should be placed not less than 1 in above and 2 in below openings. It should extend at least 24 in on each side of the opening or be equivalently developed with hooks. Minimum reinforcement that should be used is one No. 5 bar for each 6 in of wall thickness.

In structures other than low residences, masonry walls should not have chases and recesses deeper than one-third the wall thickness, or longer than 4 ft horizontally or in horizontal projection. There should be at least 8 in of masonry in back of chases and recesses, and between adjacent chases or recesses and the jambs of openings.

Chases and recesses should not be cut in walls of hollow masonry units or in hollow walls but may be built in. They should not be allowed within the required area of a pier.

The aggregate area of recesses and chases in any wall should not exceed one-fourth of the whole area of the face of the wall in any story.

In dwellings not more than two stories high, vertical chases may be built in 8-in walls if the chases are not more than 4 in deep and occupy less than 4 ft^2 of wall area. However, recesses below windows may extend from floor to sill and may be the width of the opening above. Masonry above chases or recesses wider than 12 in should be supported on lintels.

Recesses may be left in walls for stairways and elevators, but the walls should not be reduced in thickness to less than 12 in unless reinforced in some approved manner. Recesses for alcoves and similar purposes should have at least 8 in of masonry at the back. They should be less than 8 ft wide and should be arched over or spanned with lintels.

If the strength of a wall will not be impaired, pipe or conduit may be passed horizontally or vertically through the masonry in a sleeve. Sleeves, however, should not be placed closer than three diameters center to center.

11.3.7 Flashing in Masonry Walls

Flashing should be used to divert to the exterior of a building water that may penetrate or condense on the interior face of masonry walls. Accordingly, flashing should be installed in exterior walls at horizontal surfaces, such as roofs, parapets,

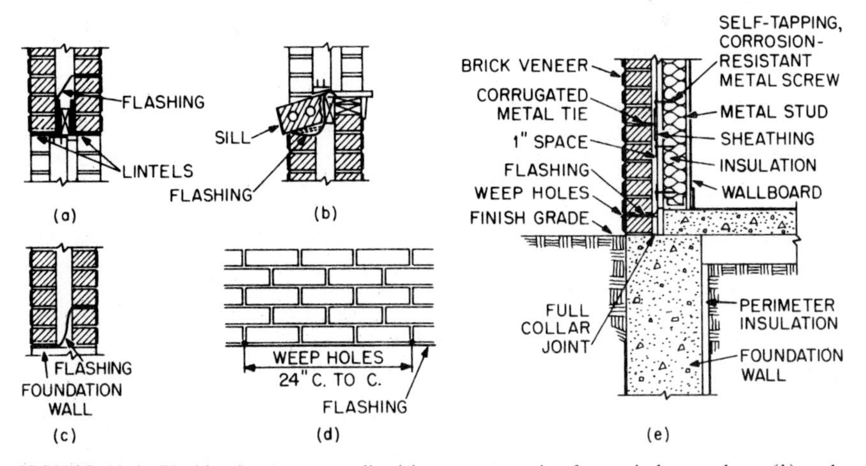

FIGURE 11.6 Flashing in masonry walls: (*a*) over an opening for a window or door; (*b*) under a window sill; (*c*) at the base of a wall; (*d*) and (*e*) below weep holes.

and floors, depending on type of construction; at shelf angles; at openings, such as doors and windows (Fig. 11.6*a* and *b*); and at the bases of walls just above grade (Fig. 11.6*c* and *e*). The flashing should extend through a mortar joint to the outside face of the wall, where it should turn down to form a drip.

Flashing in tooled mortar joints, however, would trap water unless some means is provided to drain it to the outside. Consequently, flashing should be used in conjunction with weep holes, which should be formed in head joints immediately above the flashing (Fig. 11.6*d*). When the weep holes are left open, spacing should not exceed 24 in c to c. If wicks of glass-fiber or nylon rope, cotton sash cord, or similar materials are left in the holes, spacing should not exceed 16 in c to c.

Materials used for flashing include sheet copper, bituminous fabrics, plastics, or a combination of these. Copper may be selected for its durability, but cost may be greater than for other materials. Combinations of materials, such as cold-formed steel and plastic or bituminous coating, may yield a durable flashing at lower cost.

11.4 LATERAL SUPPORT FOR MASONRY WALLS

For unreinforced solid or grouted masonry bearing walls, the ratio of unsupported height to nominal thickness, or the ratio of unsupported length to nominal thickness, should not exceed 20. For hollow walls or walls of hollow masonry units, the ratio should be 18 or less. For cavity or stone walls, the ratio should not exceed 14. See "ANSI Standard Building Code Requirements for Masonry," A41.1, American National Standards Institute.

In calculating the ratio of unsupported length to thickness for cavity walls, you can take the thickness as the sum of the nominal thickness of the inner and outer wythes. For walls composed of different kinds or classes of units or mortars, the ratio should not exceed that allowed for the weakest of the combinations. Veneers should not be considered part of the wall in computing thickness for strength or stability.

For nonbearing, unreinforced exterior walls, the thickness ratio should not exceed 20. For unreinforced partitions, the ratio should be 36 or less.

Cantilever walls and masonry walls in locations exposed to high winds should not be built higher than 10 times their thickness unless adequately braced or designed in accordance with engineering principles. Backfill should not be placed against foundation walls until they have been braced to withstand horizontal pressure.

In determining the unsupported length of walls, existing cross walls, piers, or buttresses may be considered as lateral supports, if these members are well bonded or anchored to the walls and capable of transmitting forces perpendicular to the plane of the wall to connected structural members or to the ground.

In determining the unsupported height of walls, the floors and roofs may be considered as lateral supports, if they can resist a lateral force of at least 200 lb/lin ft and provision is made to transmit the lateral forces to the ground. Ends of floor joists or beams bearing on masonry walls should be securely fastened to the walls (Fig. 11.7). (See also Arts. 11.6 and 11.11.) Interior ends of anchored joists should be lapped and spiked, or the equivalent, so as to form continuous ties across the building. When lateral support is to be provided by joists parallel to walls, anchors should be spaced no more than 6 ft apart and engage at least three joists, which should be bridged solidly at the anchors.

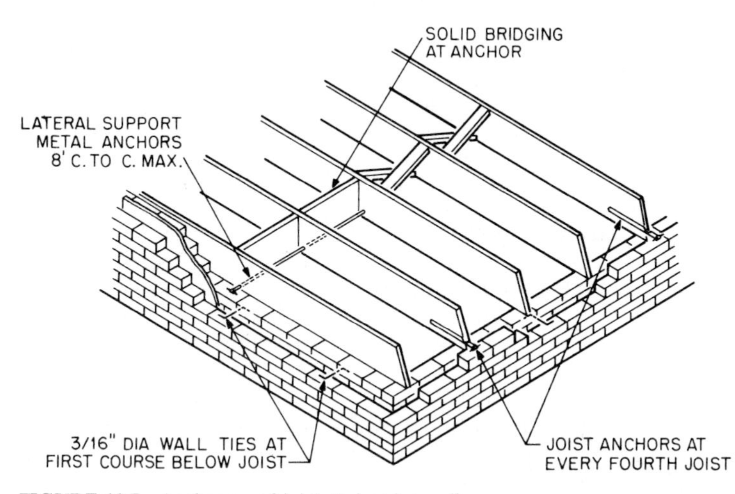

FIGURE 11.7 Anchorage of joists to bearing walls.

Unsupported height of piers should not exceed 10 times the least dimension. However, when structural clay tile or hollow concrete units are used for isolated piers to support beams or girders, unsupported height should not exceed 4 times the least dimension unless the cellular spaces are filled solidly with concrete or either Type M or S mortar (Art. 4.16).

Anchors for Masonry Facings. Support perpendicular to its plane may be provided an exterior masonry wythe, whether it is a veneer (non-load-bearing) or the outer wythe of a hollow wall, by anchoring it to construction capable of furnishing the required lateral support. Accordingly, a masonry veneer may be tied with

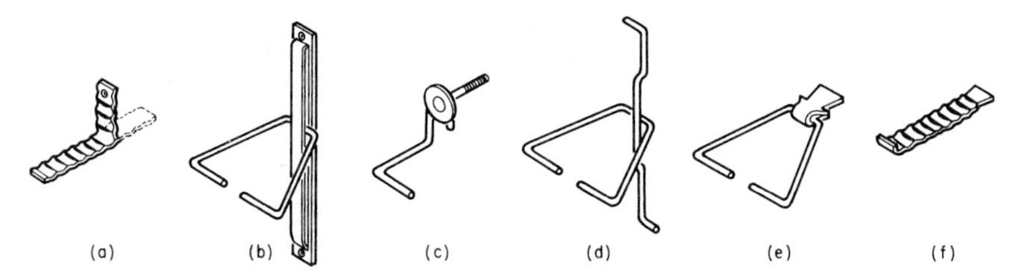

FIGURE 11.8 Anchors for use between masonry walls and structural framing: (*a*) corrugated metal for tie to wood studs; (*b*) and (*c*) wire ties for attachment to metal studs; (*d*) wire tie for anchorage to structural steel; (*e*) dovetail anchor for use with concrete; (*f*) corrugated metal for tie to cast-in-place concrete.

masonry bonders or metal ties to a backup masonry wall that is given lateral support or the veneer may be anchored directly to structural framing. Methods of bonding wythes together are described in Art. 11.3.2. The following applies to anchorage of masonry walls to structural framing.

Several types of anchors are illustrated in Fig. 11.8. They should be corrosion resistant. Also, they should be able to resist tension and compression applied by forces acting perpendicular to the wall. Yet, the anchors should be flexible enough to permit, between walls and framing, small differential horizontal and vertical movements parallel to the plane of the wall. The anchors should be embedded at one end in the mortar of bed joints and extend almost to the face of the wall. The other end should be securely attached to framing providing lateral support. The type of anchor to use depends on the construction to which the wall is to be anchored.

Figure 11.8*a* shows a corrugated metal tie for attachment of masonry walls to wood studs. Such ties should be fastened to studs with corrosion-resistant nails that are driven through sheathing to penetrate at least 1½ in into the studs. The ties should have a thickness of at least 22 ga, width of ⅞ in, and length of 6 in.

Anchors shown in Fig. 11.8*b* and *c* may be used to attach masonry walls to metal studs. The wires of these anchors should be at least 9 ga. The anchor shown in Fig. 11.8*d* is suitable for tying masonry walls to structural steel framing, as illustrated in Fig. 11.9*b*.

Dovetail anchors for anchorage of masonry walls into dovetail slots in concrete framing are illustrated in Fig. 11.8*e* and *f*. Applications of the type shown in Fig. 11.8*e* are shown in Fig. 11.9*c* and *d*. Wires in these anchors should be at least 6 ga and should be spread to a width of at least 4 in for embedment at least 2 in into bed joints in the wall. The flat-bar type (Fig. 11.8*f*) should have a minimum thickness of 16 ga and width of ⅞ in. The end to be embedded in a bed joint should be turned upward at least ¼ in.

11.5 CHIMNEYS AND FIREPLACES

Minimum requirements for chimneys may be obtained from local building codes or any model building code. In brief, chimneys should extend at least 3 ft above the highest point where they pass through the roof of a building and at least 2 ft higher than any ridge within 10 ft. (For chimneys for industrial-type appliances with discharge temperatures between 1400 and 2000°F, minimum height above the

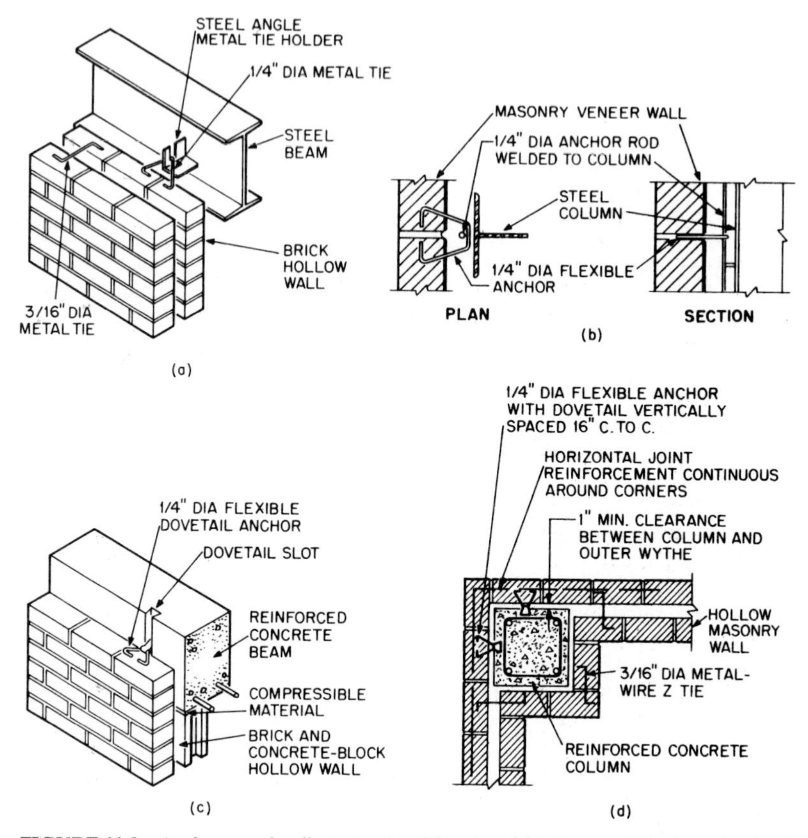

FIGURE 11.9 Anchorage of walls to structural framing: (*a*) hollow-wall ties to a structural steel beam; (*b*) masonry veneer wall anchored to a structural steel column; (*c*) masonry veneer wall anchored to a reinforced concrete beam; (*d*) hollow wall anchored to a concrete corner column.

roof opening or any part of the building within 25 ft should be 10 ft. For discharge temperatures over 2000°F, minimum height above any part of the building should be 20 ft.) Masonry chimneys should be constructed of solid masonry units or reinforced concrete and lined with firebrick or fire-clay tile. In dwellings, thickness of chimney walls may be 4 in. In other buildings, the thickness of chimneys for heating appliances should be at least 8 in for most masonry. Rubble stone thickness should be a minimum of 12 in. Cleanout openings equipped with steel doors should be provided at the base of every chimney.

When a chimney incorporates two or more flues, they should be separated by masonry at least 4 in thick.

In seismic zones where damage may occur, chimneys should be of reinforced masonry construction. They should be anchored to floors and ceilings more than 6 ft above grade and to roofs.

Fireplaces should have backs and sides of solid masonry or reinforced concrete, not less than 8 in thick. A lining of firebrick at least 2 in thick or other approved material should be provided unless the thickness is 12 in.

Fireplaces should have hearths of brick, stone, tile, or other noncombustible material supported on a fireproof slab or on brick trimmer arches. Such hearths should extend at least 20 in outside the chimney breast and not less than 12 in beyond each side of the fireplace opening along the chimney breast. Combined thickness of hearth and supporting construction should not be less than 6 in. Spaces between chimney and joists, beams, or girders and any combustible materials should be fire-stopped by filling with noncombustible material.

The throat of the fireplace should be not less than 4 in and preferably 8 in above the top of the fireplace opening. A metal damper (12 ga or thicker) extending the full width of the fireplace opening should be placed in the throat. The flue should have an effective area equal to one-twelfth to one-tenth the area of the fireplace opening.

11.6 PROVISIONS FOR DIMENSIONAL CHANGES

In design and construction of masonry walls, allowance should be made for relative movements of the masonry and contiguous construction. If this is not done, unsightly or troublesome cracking or even structural failure may result. In the past, such damage has occurred in masonry walls because of:

1. Restraint offered by contiguous construction to dimensional changes in the masonry. Such changes may be produced by temperature changes or by absorption of water by the masonry after construction.
2. Restraint offered by the masonry to movements of or dimensional changes in contiguous or bonded construction, such as concrete frames or backup walls. Such changes may be produced by drying shrinkage, elastic deformations under load, or creep of the concrete after construction.

To avoid such restraints, it is necessary to install in walls expansion joints with proper gaps, at appropriate intervals, and to break bond between the walls and construction that would restrain relative movements (Fig. 11.10).

Vertical expansion joints should be installed in masonry walls to permit horizontal movements of the masonry, and horizontal expansion joints, to permit vertical movements. In the absence of specific information on thermal and water-absorption properties, the unit strain may be assumed to be 0.0007 in/in in a brick wall when movement is restricted, for example, by bond to a concrete foundation. Thus, for 60-ft spacing of expansion joints in a straight brick wall, a joint width of $2 \times 60 \times 12 \times 0.0007$, or about 1 in, would be required. In general, spacing of vertical expansion joints should range between 50 and 100 ft, and a joint should be placed not more than 30 ft from wall intersections.

The width required for an expansion joint also depends on the maximum allowable strain of the sealant used to seal the gap. If the size of a joint is controlled by the elastic properties of the sealant, joint spacing should be adjusted to limit the joint size to accommodate the elasticity of the sealant. The sealant should be placed at the exterior wall face and inserted in the joint to a depth of at least ⅛ in but not deeper than one-half the joint width. This depth may be controlled by a backup material that is inserted in the joint before the sealant is applied and that will not adhere to the sealant.

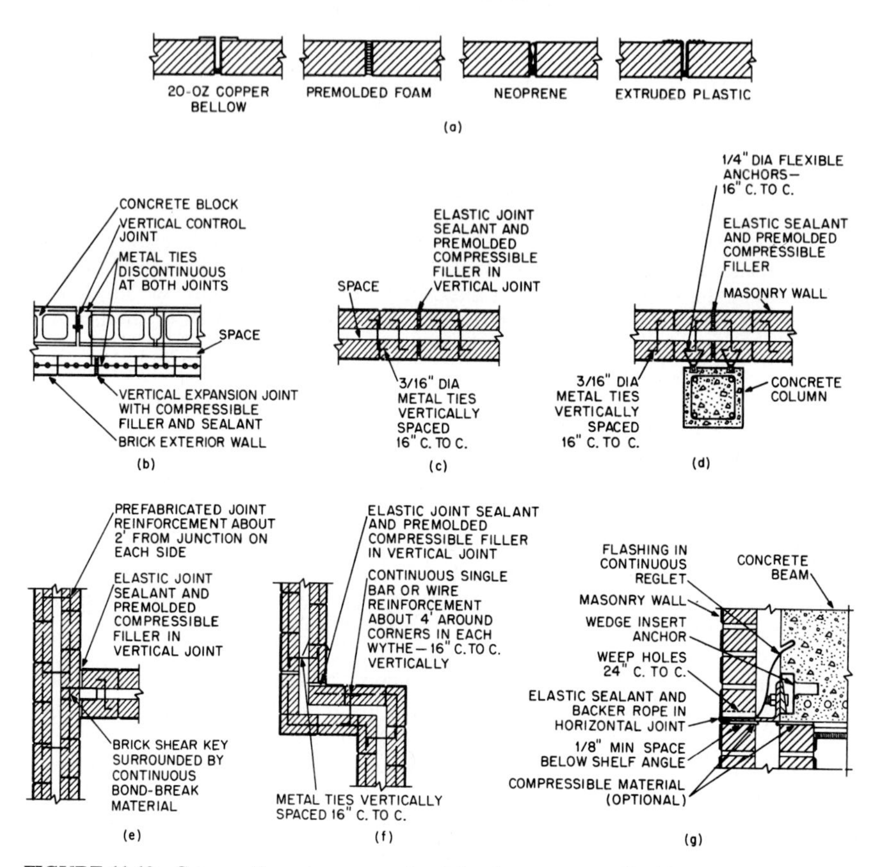

FIGURE 11.10 Cross sections show expansion joints in masonry walls: (*a*) types of fillers used; (*b*) expansion joint in brick veneer with a control joint in a concrete-block backup; (*c*) joint in a hollow wall; (*d*) joint at the anchorage of a wall to a column; (*e*) joint at a T intersection of a masonry wall; (*f*) joints at an offset in a hollow wall; (*g*) joint below a shelf angle.

In brick facades, horizontal expansion joints should be inserted directly under horizontal lintels that are supported on concrete frames (Fig. 11.10*g*). The joints should be sized for probable expansion of the masonry below the lintels plus probable shortening of the concrete frame produced by drying shrinkage, compressive loading, and creep. In the absence of specific information on the properties of the materials to be used, a relative vertical movement of 0.0014 in/in may be assumed. Thus, where a brick facade is supported on steel shelf angles, for example, spaced 15 ft apart vertically, a gap of $15 \times 12 \times 0.0014$, or about ¼ in, would be required. As for vertical joints, a sealant and backup material should seal the horizontal expansion joint.

Slip joints should be provided where abrupt changes in wall dimensions occur; for example, at panels bounding or included between openings in a masonry wall, such as those for windows and doors. Bond should be prevented by insertion of sheet metal, building paper, or other material that would permit sliding when thermal movements occur.

Similarly, to permit relative horizontal movements, slip planes should be provided between cast-in-place concrete floors or roofs and masonry bearing walls that support them (Fig. 11.5*a*). Flexible anchors that permit sliding may be installed between the slabs and walls to prevent uplift. Such anchors, however, should not be installed within a distance from a slab corner of one-tenth the slab length. The reason for this is that such corners tend to curl upward when shrinkage occurs, in which case the anchorages would apply tension to and crack the walls.

Particular care should be taken to provide for relative movements when dissimilar materials are combined in a wall. Preferably, they should be separated at least ½ in and joined with flexible ties (Fig. 11.10*b* and *d*). In particular, because of different thermal movements of the materials, bond should be prevented between brick walls and contiguous concrete foundation walls below that are exposed to the weather (Fig. 11.5*b*). Flexible anchors should be provided between the walls and foundations, to permit horizontal sliding yet prevent uplift. Also, foundations should be made sufficiently stiff to prevent deflections that would crack the walls above. In addition, for the same purpose, the walls may be designed as deep beams or Vierendeel masonry trusses. (A Vierendeel truss does not consist solely of triangular configurations of members as do conventional trusses, and thus the members are subjected to a combination of axial forces and bending moments. Such a truss would be formed by a wall with openings for doors and windows.)

When bearing-wall construction is used for a building, differential movements of adjacent supports of horizontal structural members should be kept very small. For this reason, bearing walls of dissimilar materials should not be used in the same structure, inasmuch as they are likely to have physical properties that cause unequal deformations. For example, either load-bearing brick walls or concrete walls, but not both, should be used in a structure.

11.7 REPAIR OF LEAKY JOINTS

Leakage of wind-driven rain through the joints in permeable brick masonry walls can be stopped by either repointing or grouting the joints. It is usually advisable to treat all the joints, both vertical and horizontal, in the wall face. Some "tuck-pointing" operations in which only a few, obviously defective joints are treated may be inadequate and do not necessarily ensure that the untreated joints will not leak.

Repointing consists of cutting away and replacing the mortar from all joints to a depth of about ⅝ in. After the old mortar has been removed, the dust and dirt should be washed from the wall and the brick thoroughly wetted with water to near saturation. While the masonry is still very damp but with no water showing, the joints should be repointed with a suitable mortar. This mortar may have a somewhat stiff consistency to enable it to be tightly packed into place, and it may be "prehydrated" by standing for 1 or 2 hr before retempering and using. Prehydration is said to stabilize the plasticity and workability of the mortar and to reduce the shrinkage of the mortar after its application to the joints.

After repointing, the masonry should be kept in a damp condition for 2 or 3 days. If the brick are highly absorptive, they may contain a sufficient amount of water to aid materially in curing.

Weathering and permeability tests described in C. C. Fishburn, "Effect of Outdoor Exposure on the Water Permeability of Masonry Walls," *National Bureau of Standards BMS Report 76* indicate that repointing of the face joints in permeable brick masonry walls was the most effective and durable of all the remedial treatments against leakage that did not change the appearance of the masonry.

Joints are **grouted** by scrubbing a thin coating of a grout over the joints in the masonry. The grout may consist of equal parts by volume of portland cement and fine sand, the sand passing a No. 30 sieve.

The masonry should be thoroughly wetted and in a damp condition when the grout is applied. The grout should be of the consistency of a heavy cream and should be scrubbed into the joints with a stiff bristle brush, particularly into the juncture between brick and mortar. The apparent width of the joint is slightly increased by some staining of the brick with grout at the joint line. Excess grout may be removed from smooth-textured brick with a damp sponge, before the grout hardens. Care should be taken not to remove grout from between the edges of the brick and the mortar joints. If the bricks are rough-textured, staining may be controlled by the use of a template or by masking the bricks with paper masking tape.

Bond of the grout to the joints is better for "cut" or flush joints than for tooled joints. If the joints have been tooled, they should preferably not be grouted until after sufficient weathering has occurred to remove the film of cementing materials from the joint surface, exposing the sand aggregate.

Grouting of the joints has been tried in the field and found to be effective on leaky brick walls. The treatment is not so durable and water-resistant as a repointing job but is much less expensive than repointing. Some tests of the water resistance of grouted joints in brick masonry test walls are described in *National Bureau of Standards BMS Report 76.*

The cost of either repointing or grouting the joints in brick masonry walls probably greatly exceeds the cost of the additional labor and supervision needed to make the walls water-resistant when built.

11.8 MASONRY-THICKNESS REQUIREMENTS

Walls should not vary in thickness between lateral supports. When it is necessary to change thickness between floor levels to meet minimum-thickness requirements, the greater thickness should be carried up to the next floor level.

Where walls of masonry hollow units or bonded hollow walls are decreased in thickness, a course of solid masonry should be interposed between the wall below and the thinner wall above, or else special units or construction should be used to transmit the loads between the walls of different thickness.

The following limits on dimensions of masonry walls should be observed unless the walls are designed for reinforcement, by application of engineering principles:

Bearing Walls. These should be at least 12 in thick for the uppermost 35 ft of their height. Thickness should be increased 4 in for each successive 35 ft or fraction of this distance measured downward from the top of the wall. Rough or random or coursed rubble stone walls should be 4 in thicker than this, but in no case less than 16 in thick. However, for other than rubble stone walls, the following exceptions apply to masonry bearing walls:

Stiffened Walls. Where solid masonry bearing walls are stiffened at distances not greater than 12 ft by masonry cross walls or by reinforced-concrete floors, they may be made 12 in thick for the uppermost 70 ft but should be increased 4 in in thickness for each successive 70 ft or fraction of that distance.

Top-Story Walls. The top-story bearing wall of a building not over 35 ft high may be made 8 in thick. But this wall should be no more than 12 ft high and should not be subjected to lateral thrust from the roof construction.

Residential Walls. In dwellings up to three stories high, walls may be 8 in thick (if not more than 35 ft high), if not subjected to lateral thrust from the roof construction. Such walls in one-story houses and one-story private garages may be 6 in thick, if the height is 9 ft or less or if the height to the peak of a gable does not exceed 15 ft.

Penthouses and Roof Structures. Masonry walls up to 12 ft high above roof level, enclosing stairways, machinery rooms, shafts, or penthouses, may be made 8 in thick. They need not be included in determining the height for meeting thickness requirements for the wall below.

Plain Concrete and Grouted Brick Walls. Such walls may be 2 in less in thickness than the minimum basic requirements, but in general not less than 8 in—and not less than 6 in in one-story dwellings and garages.

Hollow Walls. Cavity or masonry bonded hollow walls should not be more than 35 ft high. In particular, 10-in cavity walls should be limited to 25 ft in height, above supports. The facing and backing of cavity walls should be at least 4 in thick, and the cavity should not be less than 2 in or more than 3 in wide.

Faced Walls. Neither the height of faced (composite) walls nor the distance between lateral supports should exceed that prescribed for masonry of either of the types forming the facing and the backing. Actual (not nominal) thickness of material used for facings should not be less than 2 in and in no case less than one-eighth the height of the unit.

Nonbearing Walls. In general, parapet walls should be at least 8 in thick and the height should not exceed 3 times the thickness. The thickness may be less than 8 in, however, if the parapet is reinforced to withstand safely earthquake and wind forces to which it may be subjected.

Nonbearing exterior masonry walls may be 4 in less in thickness than the minimum for bearing walls. However, the thickness should not be less than 8 in except that where 6-in bearing walls are permitted, 6-in nonbearing walls can be used also.

Nonbearing masonry partitions should be supported laterally at distances of not more than 36 times the actual thickness of the partition, including plaster. If lateral support depends on a ceiling, floor, or roof, the top of the partition should have adequate anchorage to transmit the forces. This anchorage may be accomplished with metal anchors or by keying the top of the partition to overhead work. Suspended ceilings may be considered as lateral support if ceilings and anchorages are capable of resisting a horizontal force of 200 lb/lin ft of wall.

11.9 DETERMINATION OF MASONRY COMPRESSIVE STRENGTH

Allowable stresses are based, for the most part, on the ultimate compressive strength f'_m, psi, of masonry used. This strength may be determined by prism testing, or may be based on the compressive strength of the bricks and the type of mortar used. Compressive strength tests of brick should be conducted in accordance with ASTM "Standard Methods of Sampling and Testing Brick," C67.

TABLE 11.2 **Strength Correction Factors for Short Prisms**

Ratio of height to thickness h/t	2.0	2.5	3.0	3.5	4.0	4.5	5.0
Correction factor	0.82	0.85	0.88	0.91	0.94	0.97	1.00

Prism Tests. When the compressive strength of the masonry is to be established by preliminary tests, the tests should be made in advance of construction with prisms built of similar materials, assembled under the same conditions and bonding arrangements as for the structure. In building the prisms, the moisture content of the units at the time of laying, consistency of the mortar, thickness of mortar joints, and workmanship should be the same as will be used in the structure.

Prisms. Prisms should be stored in an air temperature of not less than 65°F and aged, before testing, for 28 days in accordance with the provisions of ASTM "Standard Method of Test for Compressive Strength of Masonry Assemblages," E447. Seven-day test results may be used if the relationship between 7- and 28-day strengths of the masonry has been established by tests. In the absence of such data, the 7-day compressive strength of the masonry may be assumed to be 90% of the 28-day compressive strength.

The value of f'_m used in the standard, "Building Code Requirements for Engineered Brick Masonry," is based on a height-thickness ratio h/t of 5. If the h/t of prisms tested is less than 5, which will cause higher test results, the compressive strength of the specimens obtained in the tests should be multiplied by the appropriate correction factor given in Table 11.2. Interpolation may be used to obtain intermediate values.

Brick Tests. When the compressive strength of masonry is not determined by prism tests, but the brick, mortar, and workmanship conform to all applicable requirements of the standard, allowable stresses may be based on an assumed value of the 28-day compressive strength f'_m computed from Eq. (11.1) or interpolated from the values in Table 11.3.

TABLE 11.3 **Assumed Compressive Strength of Brick Masonry, psi**

Compressive strength of brick, psi	Without inspection			With inspection		
	Type N mortar	Type S mortar	Type M mortar	Type N mortar	Type S mortar	Type M mortar
14,000 plus	2,140	2,600	3,070	3,200	3,900	4,600
12,000	1,870	2,270	2,670	2,800	3,400	4,000
10,000	1,600	1,930	2,270	2,400	2,900	3,400
8,000	1,340	1,600	1,870	2,000	2,400	2,800
6,000	1,070	1,270	1,470	1,600	1,900	2,200
4,000	800	930	1,070	1,200	1,400	1,600
2,000	530	600	670	800	900	1,000

$$f'_m = A(400 + Bf'_b) \tag{11.1}$$

where A = coefficient (2/3 without inspection and 1.0 with inspection)
B = coefficient (0.2 for type N mortar, 0.25 for type S mortar, and 0.3 for type M mortar)
f'_b = average compressive strength of brick, psi $\leq$ 14,000 psi

When there is no engineering or architectural inspection to ensure compliance with the workmanship requirements of the standard, the values in Table 11.3 under "Without inspection" should be used.

11.10 ALLOWABLE STRESSES IN MASONRY

In determining stresses in masonry, effects of loads should be computed on actual dimensions, not nominal. Except for engineered masonry, the stresses should not exceed the allowable stresses given in ANSI Standard Building Code Requirements for Masonry (A41.1), which are summarized for convenience in Table 11.4.

This standard recommends also that, in composite walls or other structural members composed of different kinds or grades of units or mortars the maximum stress should not exceed the allowable stress for the weakest of the combination of units and mortars of which the member is composed.

11.10.1 Allowable Stresses for Brick Construction

Allowable stresses—compressive f_m, tensile f_t, and shearing f_v—for brick construction should not exceed the values shown in Tables 11.5 and 11.6. For allowable loads on walls and columns, see Art. 11.10.2.

For wind, blast, or earthquake loads combined with dead and live loads, the allowable stresses in brick construction may be increased by one-third, if the resultant section will not be less than that required for dead loads plus reduced live loads alone. Where the actual stresses exceed the allowable, the designer should specify a larger section or reinforced brick masonry.

11.10.2 Allowable Loads on Brick Walls and Columns

Two stress-reduction factors are used in calculating allowable loads on walls and columns: slenderness coefficient C_s and eccentricity coefficient C_e. The eccentricity coefficient is used to reduce the allowable axial load in lieu of performing a separate bending analysis. The slenderness coefficient is used to reduce the allowable axial load to prevent buckling.

Allowable Axial Loads. Allowable loads, lb, on brick walls and columns can be computed from

$$P = C_e C_s f_m A_g \tag{11.2}$$

where C_e = eccentricity coefficient
C_s = slenderness coefficient

TABLE 11.4 Allowable Stresses in Unit Masonry*

Construction and grade of unit	Allowable compressive stress on gross cross section, psi†				
	Type M mortar	Type S mortar	Type N mortar	Type O mortar	Type PL or PM mortar
Solid masonry of brick or other solid units:					
8000+ psi	400	350	300	200	
4500 to 8000 psi	250	225	200	150	
2500 to 4500 psi	175	160	140	100	
1200 to 2500 psi	125	115	100	75	
Grouted solid masonry of brick or other solid units:					
8000+ psi					500
4500 to 8000 psi					350
2500 to 4500 psi					275
1200 to 2500 psi					225
Solid masonry of through-the-wall solid units:					
8000+ psi	500	400	300		
4500 to 8000 psi	350	275	200		
2500 to 4500 psi	275	200	150		
1200 to 2500 psi	225	175	125		
Masonry of hollow units:					
1000+ psi	100	90	85		
700 to 1000 psi	85	75	70		
Piers of hollow units, cellular spaces filled	105	95	90		
Hollow walls (cavity or masonry bonded):‡					
Solid units:					
4500+ psi	180	170	140		
2500 to 4500 psi	140	130	110		
1200 to 2500 psi	100	90	80		
Hollow units:					
1000+ psi	80	70	65		
700 to 1000 psi	70	60	55		
Stone ashlar masonry:					
Granite	800	720	640	500	
Limestone or marble	500	450	400	325	
Sandstone or cast stone	400	360	320	250	
Rubble, coursed, rough or random	140	120	100	80	

*See also Tables 11.5 and 11.6.

†Allowable bearing stress directly under concentrated loads may be taken 50% larger than the tabulated values.

‡On gross cross section of wall minus area of cavity between wythes. The allowable compressive stresses for cavity walls are based on the assumption that floor loads bear on only one of the two wythes. Increase stresses 25% for hollow walls loaded concentrically.

TABLE 11.5 Allowable Stresses for Nonreinforced Brick Masonry, psi

Description		Without inspection	With inspection
Axial compression			
Walls	f_m	$0.20\ f'_m$	$0.20\ f'_m$
Columns	f_m	$0.16\ f'_m$	$0.16\ f'_m$
Flexural compression			
Walls	f_m	$0.32\ f'_m$	$0.32\ f'_m$
Columns	f_m	$0.26\ f'_m$	$0.26\ f'_m$
Flexural tension			
Normal to bed joints			
M or S mortar	f_t	24	36
N mortar	f_t	19	28
Parallel to bed joints			
M or S mortar	f_t	48	72
N mortar	f_t	37	56
Shear			
M or S mortar	v_m	$0.3\sqrt{f'_m} \leq 40$	$0.5\sqrt{f'_m} \leq 80$
N mortar	v_m	$0.3\sqrt{f'_m} \leq 28$	$0.5\sqrt{f'_m} \leq 56$
Bearing			
On full area	f_m	$0.25\ f'_m$	$0.25\ f'_m$
On one-third area or less	f_m	$0.375\ f'_m$	$0.375\ f'_m$
Modulus of elasticity	E_m	$1000\ f'_m \leq 2{,}000{,}000$ psi	$1000\ f'_m \leq 3{,}000{,}000$ psi
Modulus of rigidity	E_v	$400\ f'_m \leq 800{,}000$ psi	$400\ f'_m \leq 1{,}200{,}000$ psi

f_m = allowable axial compressive stress, psi (Table 11.5 or 11.6)
A_g = gross cross-sectional area, in^2

To determine C_e and C_s, three constants are needed: end eccentricity ratio e_1/e_2, ratio of maximum virtual eccentricity to wall thickness e/t, and slenderness ratio h/t.

Eccentricity Ratio. At the top and bottom of any wall or column, a virtual eccentricity (Art. 11.1) of some magnitude (including zero) occurs. e_1/e_2 is the ratio of the smaller virtual eccentricity to the larger virtual eccentricity of the loads acting on a member. By this definition, the absolute value of the ratio is always less than or equal to 1.0. Where e_1 or e_2, or both, are equal to zero, e_1/e_2 is assumed to be zero. When the member is bent in single curvature (top and bottom virtual eccentricities occurring on the same side of the centroidal axis of a wall or column), e_1/e_2 is positive. When the member is bent in double curvature (top and bottom virtual eccentricities occurring on opposite sides of a wall or column centroidal axis), e_1/e_2 is negative (Fig. 11.11).

Eccentricity-Thickness Ratio. The ratio of maximum virtual eccentricity to wall thickness e/t is used in selecting the eccentricity coefficient C_e. Design of a nonreinforced member requires that e/t be less than or equal to ⅓. If e/t is greater

TABLE 11.6 Allowable Stresses for Reinforced Brick Construction, psi

Description		Without inspection	With inspection
Axial compression			
Walls	f_m	$0.25\ f'_m$	$0.25\ f'_m$
Columns	f_m	$0.20\ f'_m$	$0.20\ f'_m$
Flexural compression			
Walls and beams	f_m	$0.40\ f'_m$	$0.40\ f'_m$
Columns	f_m	$0.32\ f'_m$	$0.32\ f'_m$
Shear			
No shear reinforcement			
Flexural members	v_m	$0.43\sqrt{f'_m} \leq 25$	$0.7\sqrt{f'_m} \leq 50$
Shear walls	v_m	$0.3\sqrt{f'_m} \leq 50$	$0.5\sqrt{f'_m} \leq 100$
With shear reinforcement taking entire shear			
Flexural members	v	$1.2\sqrt{f'_m} \leq 60$	$2.0\sqrt{f'_m} \leq 120$
Shear walls	v	$0.9\sqrt{f'_m} \leq 75$	$1.5\sqrt{f'_m} \leq 150$
Bond			
Plain bars	u	53	80
Deformed bars	u	107	160
Bearing			
On full area	f_m	$0.25\ f'_m$	$0.25\ f'_m$
On one-third area or less	f_m	$0.375\ f'_m$	$0.375\ f'_m$
Modulus of elasticity	E_m	$1000\ f'_m \leq 2{,}000{,}000$ psi	$1000\ f'_m \leq 3{,}000{,}000$ psi
Modulus of rigidity	E_v	$400\ f'_m \leq 800{,}000$ psi	$400\ f'_m \leq 1{,}200{,}000$ psi

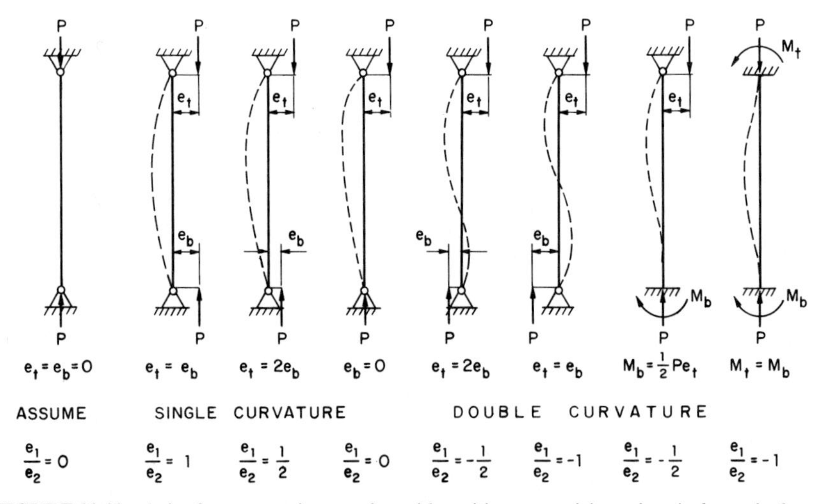

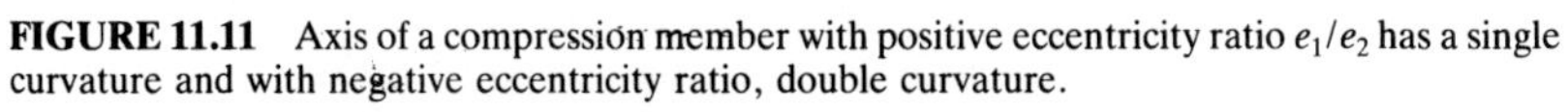

FIGURE 11.11 Axis of a compression member with positive eccentricity ratio e_1/e_2 has a single curvature and with negative eccentricity ratio, double curvature.

than ⅓, the designer should specify a larger section, different bearing details for load transfer to the masonry, or reinforced brick construction.

Slenderness Ratio. This is the ratio of the unsupported height h to the wall thickness t. It is used in selecting the slenderness coefficient C_s.

The unsupported height h is the actual distance between lateral supports. Not always the floor-to-floor height, h may be taken as the distance from the top of the lower floor to the bearing of the upper floor where these floors provide lateral support.

The effective thickness t for nonreinforced solid masonry is the actual wall thickness, except for metal-tied cavity walls. In cavity walls, each wythe is considered to act independently, thus producing two different walls. When the cavity is filled with grout, the effective thickness becomes the total wall thickness.

Eccentricity Coefficients. C_e may be selected from Table 11.7, or calculated from Eqs. (11.3) to (11.5). Linear interpolation is permitted within the table.

$$C_e = 1.0 \qquad 0 < \frac{e}{t} \leqq 0.05 \tag{11.3}$$

$$C_e = \frac{1.3}{1 + 6e/t} + \frac{1}{2}\left(\frac{e}{t} - \frac{1}{20}\right)\left(1 - \frac{e_1}{e_2}\right) \qquad 0.05 < \frac{e}{t} \leqq 0.167 \tag{11.4}$$

$$C_e = 1.95\left(\frac{1}{2} - \frac{e}{t}\right) + \frac{1}{2}\left(\frac{e}{t} - \frac{1}{20}\right)\left(1 - \frac{e_1}{e_2}\right) \qquad 0.167 < \frac{e}{t} \leqq 0.333 \tag{11.5}$$

Slenderness Coefficient. C_s may be selected from Table 11.8, or calculated from Eq. (11.6). Linear interpolation is permitted within the table.

$$C_s = 1.20 - \frac{h/t}{300}\left[5.75 + \left(1.5 + \frac{e_1}{e_2}\right)^2\right] \leq 1.0 \tag{11.6}$$

TABLE 11.7 Eccentricity Coefficients C_e

	e_1/e_2								
e/t	−1	−¾	−½	−¼	0	¼	½	¾	1
0–0.05	1.00	1.00	1.00	1.00	1.00	1.00	1.00	1.00	1.00
0.10	0.86	0.86	0.85	0.84	0.84	0.83	0.83	0.82	0.81
0.15	0.78	0.77	0.76	0.75	0.73	0.72	0.71	0.70	0.68
0.167	0.77	0.75	0.74	0.72	0.71	0.69	0.68	0.66	0.65
0.20	0.74	0.72	0.71	0.68	0.66	0.64	0.62	0.60	0.59
0.25	0.69	0.66	0.64	0.61	0.59	0.56	0.54	0.51	0.49
0.30	0.64	0.61	0.58	0.55	0.52	0.48	0.45	0.42	0.39
0.333	0.61	0.57	0.54	0.50	0.47	0.43	0.40	0.36	0.32

TABLE 11.8 Slenderness Coefficients C_s

	e_1/e_2								
h/t	−1	−¾	−½	−¼	0	¼	½	¾	1
5.0	1.00	1.00	1.00	1.00	1.00	1.00	1.00	1.00	1.00
7.5	1.00	1.00	1.00	1.00	1.00	0.98	0.96	0.93	0.90
10.0	1.00	0.99	0.98	0.96	0.93	0.91	0.88	0.84	0.80
12.5	0.95	0.94	0.92	0.90	0.87	0.83	0.79	0.75	0.70
15.0	0.90	0.88	0.86	0.83	0.80	0.76	0.71	0.66	0.60
17.5	0.85	0.83	0.81	0.77	0.73	0.69	0.63	0.57	0.50
20.0	0.80	0.78	0.75	0.71	0.67	0.61	0.55	0.48	0.40
22.5	0.75	0.73	0.69	0.65	0.60	0.54	0.47	0.39	
25.0	0.70	0.67	0.64	0.59	0.53	0.47	0.39		
27.5	0.65	0.62	0.58	0.53	0.47	0.39			
30.0	0.60	0.57	0.52	0.47	0.40				
32.5	0.55	0.52	0.47	0.41					
35.0	0.50	0.46	0.41						
37.5	0.45	0.41							
40.0	0.40								

11.10.3 Eccentrically Loaded Shear Walls

"Building Code Requirements for Engineered Brick Masonry" also provides a basis for design of eccentrically loaded shear walls. The standard requires that, in a nonreinforced shear wall, the virtual eccentricity e_L about the principal axis normal to the length L of the wall not exceed an amount that will produce tension. In a nonreinforced shear wall subject to bending about both principal axes, $e_t L + e_L t$ should not exceed $tL/3$, where e_t = virtual eccentricity about the principal axis normal to the thickness t of the shear wall. Where the virtual eccentricity exceeds the preceding limits, shear walls should be designed as reinforced or partly reinforced walls.

Consequently, for a planar wall the virtual eccentricity e_L, which is found by dividing the overturning moment about an axis normal to the plane of the wall by the axial load, should not exceed $L/6$. In theory, any virtual eccentricity exceeding $L/6$ for planar shear wall will result in development of tensile stresses. If, however, intersecting walls form resisting flanges, the classic approach may be used where the bending stress Mc/I is combined with the axial stress P/A. If the bending stress exceeds the axial stress, then tensile stresses are present. (Note that some building codes require a safety factor against overturning. It is usually 1.5 for nonreinforced walls. But the codes state that, if the walls are vertically reinforced to resist tension, the safety factor does not apply.)

For investigation of biaxial bending, the eccentricity limitation may, for convenience, be placed in the form:

$$\frac{e}{t} = \frac{e_i}{t} + \frac{e_L}{L} \leq \frac{1}{3} \tag{11.7}$$

Allowable vertical loads on shear walls may be computed from Eq. (11.2). The effective height h for computing C_s should be taken as the minimum vertical or horizontal distance between lateral supports.

Allowable shearing stresses for shear walls should be taken as the sum of the allowable shear stress given in Table 11.5 or 11.6 and one-fifth the average compressive stress produced by dead load at the level being analyzed, but not more than the maximum values listed in these tables.

In computation of shear resistance of a shear wall with intersecting walls treated as flanges, only the parts serving as webs should be considered effective in resisting the shear.

When non-load-bearing shear walls are required to resist overturning moment only by their own weight, the design can become critical. Consequently, positive ties should be provided between shear walls and bearing walls, to take advantage of the bearing-wall loads in resisting overturning. The shear stress at the connection of the shear wall to the bearing wall should be checked. The designer should exercise judgment in assumption of the distribution of the axial loads into the non-load-bearing shear walls.

11.10.4 Bearing Walls

Walls whose function is to carry only vertical loads should be proportioned primarily for compressive stress. Allowable vertical loads are given by Eq. (11.2). Effects of all loads and conditions of loading should be investigated. For application of Eq. (11.2), each loading should be converted to a vertical load and a virtual eccentricity.

In the lower stories of a building, the compressive stress is usually sufficient to suppress development of tensile stress. But in the upper stories of tall buildings, where the exterior walls are subject to high lateral wind loads and small axial loads, the allowable tensile stress may be exceeded occasionally and make reinforcement necessary.

When lateral forces act parallel to the plane of bearing walls that act as load-bearing shear walls, the walls must meet the requirements for both shear walls and biaxial bending (Art. 11.10.3). Such walls must be checked, to preclude development of tensile stress or excessive compressive or shearing stresses. If analysis indicates that tension requirements are not satisfied, the size, shape, or number of shear walls must be revised or the wall must be designed as reinforced masonry.

11.11 FLOOR-WALL CONNECTIONS

Bending moments caused in a wall by floor loading depend on such factors as type of floor system, detail of floor-wall connections, and sequence of construction. Because information available on their effects is limited, engineers must make certain design assumptions when providing for such moments. Conservative assumptions that may be used in design are discussed in the following.

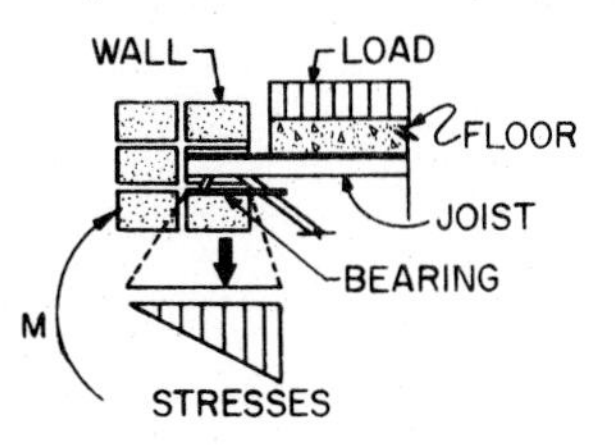

FIGURE 11.12 Stress distribution in a wall supporting a joist.

For a floor system that acts hinged at the floor-wall connection, such as steel joists and stems from precast-concrete joists, a triangular stress distribution can be assumed under the bearing (Fig. 11.12). The moment in the wall produced by dead and live loads is then equal to the reaction times the eccentricity resulting from this stress distribution.

For precast-concrete-plank floor systems, which deflect and rotate at the time of placing, a triangular stress distribution similarly can be assumed to result from the dead load of the plank, which also induces a moment in the wall. When the topping is placed as each level is constructed, a triangular stress distribution can still be assumed. The moment resulting from the dead load of the floor system, including the topping, then is that due to the eccentric loading. If, however, the topping is placed after the wall above has been built and the wall clamps the plank end in place, creating a restrained end condition, the moment in the wall will then be the sum of the moment due to the eccentric load of the plank itself and the fixed-end moment resulting from the superimposed loads of topping weight and live load (Fig. 11.13).

The degree of fixity and the resulting magnitude of the restrained end moments usually must be assumed. Full fixity of floors due to the clamping action of a wall under large axial loads in the lower stories of high- and medium-rise buildings appears a logical assumption. The same large axial loads that provide the clamping action in the lower stories also act to suppress development of tensile stresses in the wall at the floor-wall connection. Because axial loads are smaller in upper stories, however, the degree of fixity may be assumed reduced, with occurrence of slight rotation and elevation of the extreme end of the slab. Based on this assumption, slight, local stress-relieving in connections in upper stories could take place. Regardless of the assumption, the maximum moment transferred to the wall can never be greater than the negative-moment capacity of the floor system.

When full fixity is assumed, the magnitude of the moment in the wall will be approximately the distribution factor times the initial fixed-end moment of the slab at the face of the wall. As an approximation for precast-concrete plank with uniform load w and span L, $wL^2/36$ may be conservatively assumed as the wall moment. [Preliminary test results have indicated about 80% moment transfer from the slab into the wall sections (40% to the upper and 40% to the lower wall section) with flat, precast plank penetrating the full wall thickness.]

For a cast-in-place concrete slab, a fixed-end moment may be assumed for both dead and live loads, because usually the wall above the slab will be built before removal of shoring.

Because restrained end moments in a wall can become large, reduction of the eccentricity of the floor reaction is advantageous in limiting the moment in the wall. This may be accomplished by projecting only the stems of cast-in-place or precast-concrete systems into the wall (Fig. 11.14). In such cases, a bearing pad should be placed immediately under each stem.

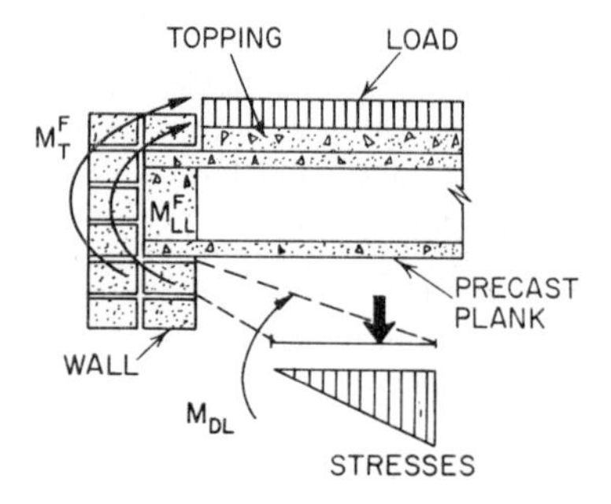

FIGURE 11.13 Stress distribution in a wall supporting a precast-concrete plank.

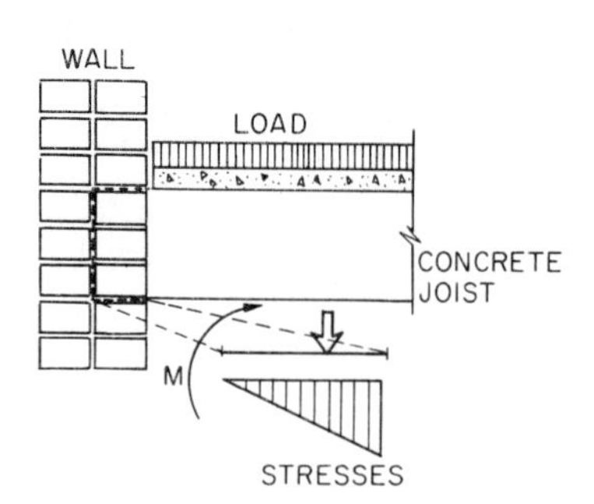

FIGURE 11.14 Stress distribution in a wall with only a joist stem embedded in it.

11.12 GLASS BLOCK

For control of light that enters a building and for better insulation than obtained with ordinary glass panes, masonry walls of glass block are frequently used (Fig. 11.15). These units are hollow, 3⅞ in thick by 6 in square, 8 in square, or 12 in square (actual length and height ¼ in less, for modular coordination, to allow for mortar joints). Faces of the units may be cut into prisms to throw light upward or the block may be treated to diffuse light.

Glass blocks may be used as nonbearing walls and to fill openings in walls. The glass block so used should have a minimum thickness of 3 in at the mortar joint. Also, surfaces of the block should be satisfactorily treated for mortar bonding.

For exterior walls, glass-block panels should not have an unsupported area of more than 144 ft^2. They should be no more than 15 ft long or high between supports.

For interior walls, glass-block panels should not have an unsupported area of more than 250 ft^2. Neither length nor height should exceed 25 ft.

Exterior panels should be held in place in the wall opening to resist both internal and external wind pressures. The panels should be set in recesses at the jambs so as to provide a bearing surface at least 1 in wide along the edges. Panels more

FIGURE 11.15 (*Top left*) First step in installation of a glass-block panel is to coat the sill with an asphalt emulsion to allow for movement due to temperature changes. Continuous expansion strips are installed at side and head jambs. (*Top right*) Blocks are set with full mortar joints. (*Bottom left*) Welded-wire ties are embedded in the mortar to reinforce the panel. (*Bottom right*) After all the blocks are placed, joints tooled to a smooth, concave finish, and the edges of the panel calked, the blocks are cleaned.

than 10 ft long should also be recessed at the head. Some building codes, however, permit anchoring small panels in low buildings with noncorrodible perforated metal strips.

Steel reinforcement should be placed in the horizontal mortar joints of glass-block panels at vertical intervals of 2 ft or less. It should extend the full length of the joints but not across expansion joints. When splices are necessary, the reinforcement should be lapped at least 6 in. In addition, reinforcement should be placed in the joint immediately below and above any openings in a panel.

The reinforcement should consist of two parallel longitudinal galvanized-steel wires, No. 9 ga or larger, spaced 2 in apart, and having welded to them No. 14 or heavier-gage cross wires at intervals up to 8 in.

Glass block should be laid in Type S mortar. Mortar joints should be from ¼ to ⅜ in thick. They should be completely filled with mortar.

Exterior glass-block panels should be provided with ½-in expansion joints at sides and top. These joints should be kept free of mortar and should be filled with resilient material (Fig. 11.15). An opening may be filled one block at a time, as in Fig. 11.15, or with a preassembled panel.

11.13 MASONRY BIBLIOGRAPHY

The following publications are available from the Brick Institute of America, McLean, Va.:

J. G. Gross, R. D. Dikkers, and J. C. Grogan, "Recommended Practice for Engineered Brick Masonry."

H. C. Plummer, "Brick and Tile Engineering."

"Building Code Requirements for Engineered Brick Masonry."

"Technical Notes on Brick and Tile Construction"—a series.

See also the following standards of the American Concrete Institute, P.O. Box 19150, Redford Station, Detroit, MI 48219:

"Building Code Requirements for Concrete Masonry Structures" and "Commentary. . . ," ACI 531.

"Specification for Concrete Masonry Construction," ACI 531.1.

STUD WALLS

Load-bearing walls in buildings up to three stories high and story-high partitions often are constructed of framing composed of thin pieces of wood or metal. When the main structural members of such walls are installed vertically at close spacing, the members are called **studs** and the walls are referred to as stud walls. Any of a wide variety of materials may be applied to the studs as facings for the walls.

Stud walls permit placement of insulation between studs, so that no increase in thickness is required to accommodate insulation. Also, pipe and conduit may be inexpensively hidden in the walls. Cost of stud construction is usually less than for all-masonry walls.

11.14 STUD-WALL CONSTRUCTION

Load-bearing and non-load-bearing stud walls may be built of wood, aluminum, or cold-formed steel. Basic framing consists of vertical structural members, or **studs,**

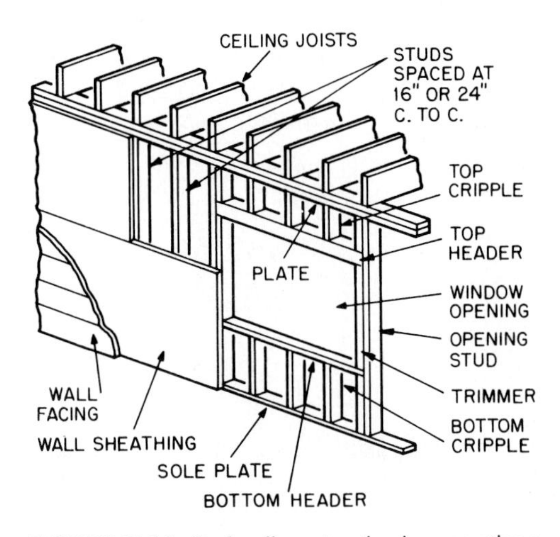

FIGURE 11.16 Stud-wall construction incorporating a window opening.

seated on a bottom, horizontal, bearing member, called a **sole plate,** and capped with a horizontal tie, called a **top plate** (Fig. 11.16). In addition, diagonal and horizontal bracing may be applied to the framing to prevent racking due to horizontal forces acting in the plane of the wall.

The studs usually are spaced 16 or 24 in on centers. Traditional surfacing materials are manufactured to accommodate these spacings; for example, panels to be attached to the framing usually come 48 in wide. (Inasmuch as the panels are fastened to each stud, panel thickness required, and hence cost, is determined by the stud spacing and generally is larger for 24-in spacing than for 16-in. Overall wall cost, however, may not be larger for the wider spacing, because it requires fewer studs.)

Wood stud walls are normally built of nominal 2 × 4-in lumber. This type of construction, usually used for residential buildings, is described in Art. 10.25. Advantages of wood construction include light weight and ease of fabrication and assembly, especially in the field.

Aluminum and cold-formed steel construction offer the advantages over wood of incombustibility and freedom from warping, shrinking, swelling, and attack by insects. Studs may be provided with punched openings, which not only reduce weight but also permit passage of pipe and conduit without the necessity of drilling holes in the field. Stud spacing usually is 24 in, rather than 16 in, to reduce the number of studs required.

Metal framing is not so easy to cut and fit in the field as wood. Hence, prefabrication of metal walls in convenient lengths is desirable.

Metal members are manufactured with a variety of widths, leg dimensions, lengths, and thicknesses. Steel studs, for example, are available as C shapes, channels and nailable sections; that is, attachments can be nailed to the flanges. Widths range from ½ to 6 in, and lengths, from 6 to 40 ft.

For partitions, a nonstructural interior finish, such as gypsum plaster, gypsumboard, fiberboard, or wood paneling, may be applied to both faces of stud-wall framing. For exterior walls, the interior face may be the same as for partitions,

FIGURE 11.17 Erection of a preassembled stud wall. (*U.S. Gypsum Company.*)

whereas the outer side must be enclosed with durable, weather-excluding materials, such as water-resistant sheating and siding or masonry veneer.

For quick assembly, stud walls may be prefabricated. Figure 11.17 illustrates erection of a cold-formed steel stud wall that has been preassembled with sheathing already attached.

11.15 SHEATHING

To deter passage of air and water through exterior stud walls, sheathing may be attached to the exterior faces of the studs. When sheathing in the form of rigid panels, such as plywood, is fastened to the studs so as to resist racking of the walls, it may be permissible to eliminate diagonal wall bracing, which contributes significantly to wall construction costs. Panels, however, may be constructed of weak materials, especially when the sheathing is also required to serve as thermal insulation, inasmuch as the sheathing is usually protected on the weather side by a facade of siding, masonry veneer, or stucco.

Materials commonly used for sheathing include plywood (see Art. 10.12), fiberboard, gypsum, urethanes, isocyanates, and polystyrene foams. Sheathing usually is available in 4-ft wide panels, with lengths of 8 ft or more. Available thicknesses range from ½ to 2½ in. Some panels require a protective facing of waterproofing paper or of aluminum foil, which also serves as reflective insulation.

CURTAIN WALLS

With skeleton-frame construction, exterior walls need carry no load other than their own weight, and therefore their principal function is to keep wind and weather out of the building—hence the name curtain wall. Nonbearing walls may be supported on the structural frame of a building, on supplementary framing (girts or studs, for example) in turn supported on the structural frame of a building, or on the floors.

11.16 FUNCTIONAL REQUIREMENTS OF CURTAIN WALLS

Curtain walls do not have to be any thicker than required to serve their principal function. Many industrial buildings are enclosed only with light-gage metal. However, for structures with certain types of occupancies and for buildings close to others, attractive appearance and fire resistance are important characteristics. Fire-resistance requirements in local building codes often govern in determining the thickness and type of material used for curtain walls.

In many types of buildings, it is desirable to have an exterior wall with good insulating properties. Sometimes a dead-air space is used for this purpose. Sometimes insulating material is incorporated in the wall or erected as a backup.

The exterior surface of a curtain wall should be made of a durable material, capable of lasting as long as the building. Maintenance should be a minimum; initial cost of the wall is not so important as the life-cycle cost (initial cost plus maintenance and repair costs).

To meet requirements of the owner and the local building code, curtain walls may vary in construction from a simple siding to a multilayer-sandwich wall. They may be job-assembled or be delivered to the job completely prefabricated.

Walls with masonry components should meet the requirements of Arts. 11.2 to 11.12.

11.17 WOOD FACADES

Wood is often applied on low buildings as an exterior finish in the form of siding, shingles, half timbers, or plywood sheets.

Siding may be drop, or novelty; lap, or clapboard; vertical boarding or horizontal flush boarding.

Drop siding can combine sheathing and siding in one piece. This type of siding consists of tongued-and-grooved individual pieces that are driven tightly up against each other when they are nailed horizontally in place to make the wall weathertight (Fig. 11.18*a* and *b*). It is not considered a good finish for permanent structures.

Lap siding or clapboard are beveled boards, thinner along one edge than the opposite edge, which are nailed horizontally over sheathing and building paper (Fig. 11.18*c*). Usually boards up to 6 in wide lap each other about 1 in; wider boards, more than 2 in. At the eaves, the top siding boards slip under the lower edge of a frieze board to make a weathertight joint.

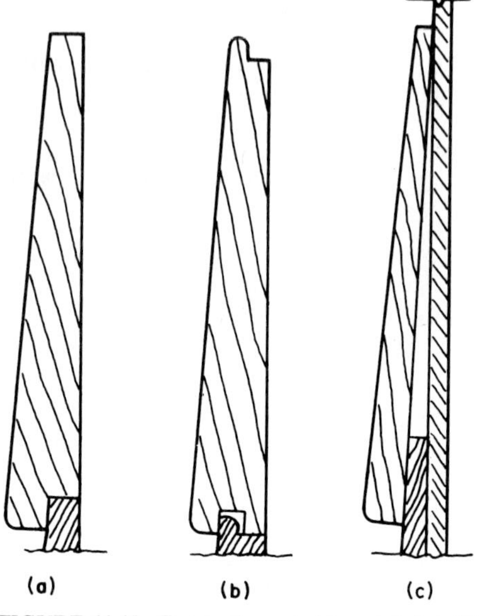

FIGURE 11.18 Types of wood siding: (*a*) and (*b*) drop siding; (*c*) lap siding.

When vertical or horizontal boards are used for the exterior finish, precautions should be taken to make the joints watertight. Joints should be coated with white lead in linseed oil just before the boards are nailed in place, and the boards should be driven tight against each other. Battens (narrow boards) should be applied over the joints if the boards are squared-edged.

In half-timber construction, timber may be used to form a structural frame of heavy horizontal, vertical, and diagonal members, the spaces between being filled with brick. This type of construction is sometimes imitated by nailing boards in a similar pattern to an ordinary sheathed frame and filling the space between boards with stucco.

Plywood for exterior use should be an exterior grade, with plies bonded with permanent waterproof glue (see Art. 10.12). The curtain wall may consist of a single sheet of plywood or of a sandwich of which plywood is a component. Also, plywood may be laminated to another material, such as a light-gage metal, to give it stiffness.

11.18 WALL SHINGLES AND SIDING

Wood, asphalt, and mineral fiber are frequently used for shingles over a sheathed frame. Shingles are made in a variety of forms and shapes and are applied in different ways. The various manufacturers make available instructions for application of their products.

Either in flat sheets or corrugated form, cold-formed metal, plastics or mineral-fiber panels may be used to form a lightweight enclosure. Corrugated sheets are stiffer than the flat. If the sheets are very thin, they should be fastened to sheathing or closely spaced supports.

When corrugated siding is used, details should be planned so that the siding will shed water. Horizontal splices should be placed at supporting members and the sheets should lap about 4 in. Vertical splices should lap at least 1½ corrugations. Sheets should be held firmly together at splices and intersections to prevent water from leaking through. Consideration should be given to sealing strips at openings where corrugated sheets terminate against plane surfaces. The bottommost girt supporting the siding should be placed at least 1 ft above the foundation because of the difficulty of attaching the corrugated materials to masonry. The siding should not be sealed in a slot in the foundation because the metal may corrode or a brittle siding may crack.

When flat sheets are used, precautions should be taken to prevent water from penetrating splices and intersections. The sheets may be installed in sash like window glass, or the splices may be covered with battens. Edges of metal sheets may be flanged to interlock and exclude wind and rain.

Pressed-metal panels, mostly with troughed or boxed cross sections, are also used to form lightweight walls.

Provision should be made in all cases for expansion and contraction with temperature changes. Allowance for movement should be made at connections. Methods of attachment vary with the type of sheet and generally should be carried out in accordance with the manufacturer's recommendations.

11.19 STUCCO

Applied like plaster, stucco is a mixture of sand, portland cement, lime, and water. Two coats are applied to masonry, three coats on metal lath. The finish coat may be tinted by adding coloring matter to the mix or the outside surface may be painted with a suitable material.

The metal lath should be heavily galvanized. It should weigh at least 2.5 lb/yd^2, even though furring strips are closely spaced. When supports are 16 in c to c, it should weigh 3.4 lb/yd^2. (See Table 11.9 in Art. 11.25.6). The lath sheets should be applied with long dimensions horizontal and should be tied with 16-ga wire. Edges should be lapped at least 1 in, ends 2 in.

The first, or scratch, coat should be forced through the interstices in the lath so as to embed the metal completely. In three-coat applications, the coat should be at least ½ in thick. Its surface should be scored to aid bond with the second, or brown, coat. That coat should be applied as soon as the scratch coat has gained sufficient strength to carry the weight of both coats, usually after about 4 or 5 hr from completion of the scratch coat. The second coat should be at least ⅜ in thick. It should be moist cured for at least 48 hr with fine sprays of water and then allowed to dry for at least 1 week. The finish coat should be at least ⅛ in thick. (When only two coats are used, for example, on a masonry base, the base coat should be a minimum of ⅜ in thick and the finish coat, ¼ in. Before application of the base coat, a bond coat, consisting of one part portland cement and one to two parts sand, should be dashed on the masonry with a stiff brush and allowed to set.)

For both the scratch and brown coats, the mix, by volume, may be 1 part portland cement to 3 to 5 parts sand, plus hydrated lime in amount equal to 25% of the volume of cement. Masonry cement may be used instead of portland cement, but without addition of lime, inasmuch as masonry cement contains lime. The finish coat may be a factory-prepared stucco-finish mix or a job mix of 1 part white portland cement, not more than 1/4 part of hydrated lime, 2 to 3 parts of a light-colored sand, and mineral oxide pigment, if desired.

Ingredients should be thoroughly mixed dry. Then, water should be added and the materials mixed for at least 5 min in a power mixer. The first two coats usually are applied with a trowel. The finish coat may be sprayed or manually applied.

("Plasterer's Manual," EBO49M, Portland Cement Association.)

11.20 PRECAST-CONCRETE OR METAL AND GLASS FACINGS

In contrast to siding in which a single material forms the complete wall, precast concrete or metal and glass are sometimes used as the facing, which is backed up with insulation, fire-resistant material, and an interior finish. The glass usually is tinted and is held in a light frame in the same manner as window glass. Metal panels may be fastened similarly in a light frame, attached to mullions or other secondary framing members, anchored to brackets at each floor level, or connected to the structural frame of the building. The panels may be small and light enough for one man to carry or one or two stories high, prefabricated with windows.

Provision for expansion and contraction should be made in the frames, when they are used, and at connections with building members. Metal panels should be shaped so that changes in surface appearance will not be noticeable as the metal expands and contracts. Frequently, light-gage metal panels are given decorative patterns, which also hide movements due to temperature variations ("canning") and stiffen the sheets. Flat sheets may be given a slight initial curvature and stiffened on the rear side with ribs, so that temperature variations will only change the curvature a little and not reverse it. Or flat sheets may be laminated to one or more flat stiffening sheets, like mineral-fiber panels or mineral-fiber panels and a second light-gage metal sheet, to prevent "canning."

It may be desirable in many cases to treat the metal to prevent passage of sound. Usual practice is to apply a sound-absorbing coating on the inside surface of the panel. Some of these coatings have the additional beneficial effect of preventing moisture from condensing on this face.

Metal panels generally are flanged and interlocked to prevent penetration of water. A good joint will be self-flashing and will not require calking. Care must be taken that water will not be blown through weep holes from the outside into the building. Flashing and other details should be arranged so that any water that may penetrate the facing will be drained to the outside. (See also Art. 11.21.)

11.21 SANDWICH PANELS

Walls may be built of prefabricated panels that are considerably larger in size than unit masonry and capable of meeting requirements of appearance, strength, durability, insulation, acoustics, and permeability. Such panels generally consist of an insulation core sandwiched between a thin lightweight facing and backing.

When the edges of the panels are sealed, small holes should be left in the seal. Otherwise, heat of the sun could set up sizable vapor pressure, which could cause trouble.

The panels could be fastened in place in a light frame, attached to secondary framing members (Fig. 11.19*b*), anchored to brackets at each floor level, or connected to the structural frame of the building. Because of the large size of the panels, special precautions should be taken to allow for expansion and contraction due to temperature changes. Usually such movements are provided for at points of support.

With metal curtain walls, special consideration also must be given to prevention of leakage, since metal and glass are totally nonabsorptive. It is difficult to make the outer face completely invulnerable to water penetration under all conditions; so a secondary defense in the form of an internal drainage system must be provided. Any water entering the wall must be drained to the exterior. Bear in mind that water can penetrate a joint through capillary action, reinforced by wind pressure. Running down a vertical surface, water can turn a corner to flow along a horizontal surface, defying gravity, into a joint.

When light frames are not used to support the units, adjoining panels generally interlock, and the joints are calked and sealed with rubber or rubberlike material to prevent rain from penetrating. Flashing and other details should be arranged so that any water that comes through will be drained to the outside. For typical details, see J. H. Callender, "Time-Saver Standards for Architectural Design Data," 6th ed., McGraw-Hill, Inc., New York; and "Tilt-up Concrete Walls," PA079.01B, Portland Cement Association.

Metal curtain walls may be custom, commercial, or industrial type. Custom-type walls are those designed for a specific project, generally multistory buildings. Commercial-type walls are those built up of parts standardized by manufacturers. Industrial-type walls are comprised of ribbed, fluted, or otherwise preformed metal sheets in stock sizes, standard metal sash, and insulation.

Metal curtain walls may be classified according to the methods used for field installation:

Stick Systems. Walls installed piece by piece. Each principal framing member, with windows and panels, is assembled in place separately (Fig. 11.19*a*). This type of system involves more parts and field joints than other types and is not so widely used.

Mullion-and-Panel Systems. Walls in which vertical supporting members (mullions) are erected first, and then wall units, usually incorporating windows (generally unglazed), are placed between them (Fig. 11.19*b*). Often, a cover strip is added to cap the vertical joint between units.

Panel Systems. Walls composed of factory-assembled units (generally unglazed) and installed by connecting to anchors on the building frame and to each other (Fig. 11.19*c*). Units may be one or two stories high. This system requires fewer pieces and fewer field joints than the other systems.

Ample provision for movement is one of the most important considerations in designing metal curtain walls. Movement continually occurs because of thermal expansion and contraction, wind loads, gravity, and other causes. Joints and connections must be designed to accommodate it.

When mullions are used, it is customary to provide for horizontal movement at each mullion location, and in multistory buildings, to accommodate vertical movement at each floor, or at alternate floors when two-story-high components are used. Common ways of providing for horizontal movements include use of split mullions, bellows mullions, batten mullions, and elastic structural gaskets. Split mullions

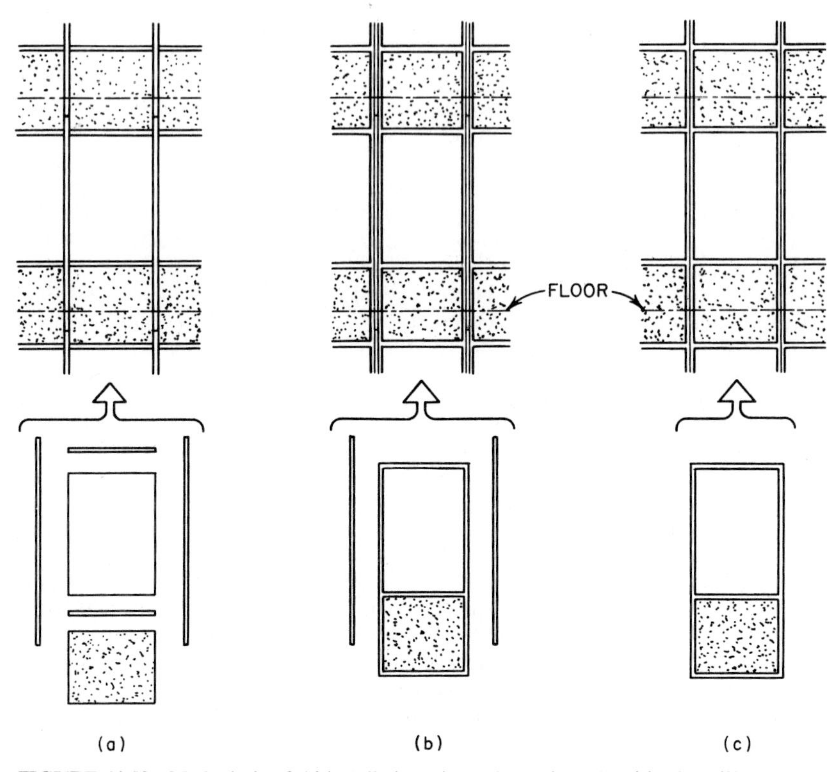

FIGURE 11.19 Methods for field installation of metal curtain walls: (*a*) stick; (*b*) mullion and panel; (*c*) panel system.

comprise two channel-shaped components permitted to move relative to each other in the plane of the wall. Bellow mullions have side walls flexible enough to absorb wall movements. Batten mullions consist of inner and outer cap sections that clamp the edges of adjacent panels, but not so tightly as to restrict movement in the plane of the wall. Structural gaskets provide a flexible link between mullions and panels. To accommodate vertical movement, mullions are spliced with a telescoping slip joint.

When mullions are not used and wall panels are connected to each other along their vertical edges, the connection is generally made through deep flanges. With the bolts several inches from the face of the wall, movement is permitted by the flexibility of the flanges.

Slotted holes are unreliable as a means of accommodating wall movement, though they are useful in providing dimensional tolerance in installing wall panels. Bolts drawn up too tightly or corrosion may prevent slotted holes from functioning as intended. If slotted holes are used, the connections should be made with shoulder bolts or sleeves and Bellville or nylon washers, to provide light but positive pressure and prevent rattling.

Since metals are good transmitters of heat, it is particularly important with metal curtain walls to avoid thermal short circuits and metallic contacts between inner and outer wall faces. When, for example, mullions project through the wall, the inner face should be insulated, or each mullion should comprise two sections separated by insulation.

For more details on curtain walls, see W. F. Koppes, "Metal Curtain Wall Specifications Manual," National Association of Architectural Metal Manufacturers, 600 S. Federal St., Chicago, IL 60605; "Curtain Wall Handbook," U.S. Gypsum Co., Chicago, IL 60606.

PARTITIONS

Partitions are dividing walls one story or less in height used to subdivide the interior space in buildings. They may be bearing or nonbearing walls. (See also Art. 1.7.)

11.22 TYPES OF PARTITIONS

Bearing partitions may be built of masonry or concrete or of wood or light-gage metal studs. These materials may be faced with plaster, wallboard, plywood, wood boards, plastic, or other materials that meet functional and architectural requirements. Masonry partitions should satisfy the requirements of Arts. 11.2 to 11.12. See also Art. 11.14.

Nonbearing partitions may be permanently fixed in place, temporary (or movable) so that the walls may be easily shifted when desired, or folding. Since the principal function of these walls is to separate space, the type of construction and materials used may vary widely. They may be opaque or transparent; they may be louvered or hollow or solid; they may extend from floor to ceiling or only partway; and they may serve additionally as cabinets or closets or as a concealment for piping and electrical conduit.

Fire resistance sometimes dictates the type of construction. If a high fire rating is desired or required by local building codes, the local building official should be consulted for information on approved types of construction or the fire ratings given in the following should be used: "Fire Resistance Design Manual," Gypsum Association, 1603 Orrington Ave., Evanston, IL 60201; "Approval Guide," Factory Mutual System, 1151 Boston-Providence Turnpike, Norwood, MA 02062; "Fire Resistance Directory," Underwriters Laboratories, 333 Pfingsten Road, Northbrook, IL 60062.

When movable partitions may be installed, the structural framing should be designed to support their weight wherever they may be placed.

Acoustics also sometimes affects the type of construction of partitions. Thin construction that can vibrate like a sounding board should be avoided. Depending on functional requirements, acoustic treatment may range from acoustic finishes on partition surfaces to use of double walls separated completely by an airspace or an insulating material.

Light-transmission requirements may also govern the selection of materials and type of construction. Where transparency or translucence is desired, the partition may be constructed of glass, or of glass block or plastic, or it may contain glass windows.

For installation of facings of ceramic wall tiles, see Arts. 11.28 and 11.29. For plaster or gypsumboard partitions, see Arts. 11.24 to 11.27.

Consideration should also be given to the necessity for concealing pipes, conduits, and ducts in partitions.

11.23 STRUCTURAL REQUIREMENTS OF PARTITIONS

Bearing partitions should be capable of supporting their own weight and superimposed loads in accordance with recommended engineering practice and should rest in turn on adequate supports that will not deflect excessively. Masonry partitions should meet the requirements of Arts. 11.2 to 11.12.

Nonbearing partitions should be stable laterally between lateral supports or additional lateral supports should be added. Since they are not designed for vertical loads other than their own weight, such partitions should not be allowed to take loads from overhead beams that may deflect and press down on them. Also, the beams under the partition should not deflect to the extent that there is a visible separation between bottom of partition and the floor or that the partition cracks.

Folding partitions, in a sense, are large doors. Depending on size and weight, they may be electrically or manually operated. They may be made of wood, lightgage metal, or synthetic fabric on a light collapsible frame. Provision should be made for framing and supporting them in a manner similar to that for large folding doors (Art. 11.57).

PLASTER AND GYPSUMBOARD

For walls or ceilings, an interior finish made of gypsum products may consist of materials partly or completely prepared in the field, or of prefabricated sheets (dry-type construction). Factors such as initial cost, cost of maintenance and repair, fire resistance, sound control, decorative effects, and speed of construction must be considered in choosing between them.

When field-prepared materials are used, the plaster finish generally consists of a base and one or more coats of plaster. When dry-type construction is used, one or more plies of prefabricated sheet may be combined to achieve desired results. For fire-resistance and sound-transmission ratings of plaster construction, see "Fire Resistance Design Manual," Gypsum Association, 1603 Orrington Ave., Evanston, IL 60201.

11.24 PLASTER AND GYPSUMBOARD CONSTRUCTION TERMS

Absorption. The rate of absorption of water into gypsumboard, as determined by the test described in ASTM C473.

Accelerator. Any material added to gypsum plaster that speeds the set.

Acoustical Plaster. A finishing plaster that corrects sound reverberations or reduces noise intensity.

Adhesive Spreader. A notched trowel or special tool that aids in application of laminating adhesives.

Adhesive Wall Clips. Special clips or nails with large, perforated bases for mastic application to a firm surface.

Adhesives:
Contact. An adhesive that forms a strong, instantaneous bond between two plies of gypsumboard when the two surfaces are brought together.
Laminating. An adhesive that forms a slowly developing bond between two plies of gypsumboard or between masonry or concrete and gypsumboard.
Stud. An adhesive suitable for attaching gypsumboard to framing.

Admixture. Any substance added to a plaster component or plaster mortar to alter its properties. (See also Dope.)

Arris. A sharp edge forming an external corner at the junction of two surfaces.

Back Blocking. Support provided for gypsumboard butt joints that fall between framing members.

Back Clip. A clip attached to the back of gypsumboard and designed to fit into slots in framing to hold the board in place. Used for demountable partitions.

Back Plastering. Application of plaster to one face of a lath system after application and hardening of plaster applied to the opposite face (used for solid plaster partitions and curtain walls).

Backing (Backer) Board. A type of gypsumboard intended to serve as a base layer in a multilayer gypsumboard system or for application with adhesive of acoustical tile or panels.

Band. A flat molding.

Banjo Taper. A mechanical device that dispenses tape and taping compound simultaneously.

Base (Baseboard). A plastic, wood, or metal trim or molding applied to a wall at the floor line to protect the wall from damage.

Base Coat. The plaster coat or combination of coats applied before the finish coat.

Batten. A predecorated strip or joint covering used to conceal the junction between two boards. It is often used with demountable systems.

Bead. A strip of sheet metal usually with a projecting nosing, to establish plaster grounds, and two perforated or expanded flanges, for attachment to the plaster base, for use at the perimeter of a plaster membrane as a stop or at projecting angles to define and reinforce the edge.

Beaded Molding. A cast plaster string of beads set in a molding or cornice.

Beading. (See Ridging.)

Bed, or Bedding, Coat. The first coat of joint compound over tape, bead, or fastener heads.

Bed Mold or Bed. A flat area in a cornice in which ornamentation is placed.

Bench, Hangers'. A low scaffold used by workers to reach the ceiling.

Binder. A chemical added during formulation of the gypsumboard core, often starch, to improve bond between the core and paper facings.

Bleeding. A discoloration, usually at a joint, on a finished wall or ceiling of gypsumboard.

Blister. Protuberance on the finish coat of plaster caused by application over too damp a base coat, or troweling too soon, or a loose, raised spot on the face of a gypsumboard, usually due to an airspace or void in the core. Also denotes a bulge under joint reinforcing tape, usually caused by insufficient compound under the tape.

Blow. Separation of a large area of paper facing from the core during the manufacturing process and usually appearing as a large, puffy blister or a full loose sheet of paper.

Board Knife. A hand tool that holds a replaceable blade for scoring or trimming gypsumboards.

Board Saw. A short handsaw with very coarse teeth used to cut gypsumboards for framed openings for windows and doors.

Bond Plaster. A plaster formulated to serve as a first coat applied to monolithic concrete.

Boss. A Gothic ornament set at the intersection of moldings.

Broad Knife. A wide, flexible finishing knife for applying joint-finishing compound.

Brown Coat. Coat of plaster directly beneath the finish coat. In two-coat work, brown coat refers to the base-coat plaster applied over the lath. In three-coat work, the brown coat refers to the second coat applied over a scratch coat.

Bubble. A large void in the core of gypsumboard caused by entrapment of air while the core is in a fluid state during manufacture.

Buckles. Raised or ruptured spots in plaster that eventually crack, exposing the lath. Most common cause for buckling is application of plaster over dry, broken, or incorrectly applied wood lath.

Bull Nose. This term describes an external angle that is rounded to eliminate a sharp corner and is used largely at window returns and door frames.

Butterflies. Color imperfections on a lime-putty finish wall, caused by lime lumps not put through a screen, or insufficient mixing of the gaging.

Caging. Framing, usually of metal, used to enclose pipes, columns, beams, or other components to be concealed by gypsumboard.

Capital or Cap. The ornamental head of a column or pilaster.

Case Mold. Plaster shell used to hold various parts of a plaster mold in correct position. Also used with gelatin and wax molds to prevent distortions during pouring operation.

Casing Bead. A bead set at the perimeter of a plaster membrane or around openings to provide a stop or separation from adjacent materials.

Casts. (See Staff.)

Catface. Flaw in the finish coat of plaster comparable to a pock mark.

Ceilings:

Coffered. Ornamental ceilings composed of recessed panels between ribs.

Contact. Ceilings attached in direct contact with the construction above, without use of runner channels or furring.

Cross Furred. Ceilings applied to furring members attached at right angles to the underside of main runners or other structural supports.

Furred. Ceilings applied to furring members attached directly to the structural members of the building.

Suspended. Ceilings applied to furring members suspended below the structural members of the building.

Chamfer. A beveled corner or edge.

Chase. A groove in a masonry wall to provide for pipes, ducts, or conduits.

Check Cracks. Cracks in plaster caused by shrinkage, but the plaster remains bonded to its base.

Chip Cracks. Similar to check cracks, except the bond is partly destroyed. Also referred to as fire cracks, map cracks, crazing, fire checks, and hair cracks.

Circle Cutter. An adjustable scribe for cutting circular openings in gypsumboard for lighting fixtures and other devices.

Cockle. A crease-like wrinkle or small depression in gypsumboard paper facing, usually extending in the long direction.

Compounds:

All-Purpose. A joint treatment material that can be used for bedding tape, finishing, laminating adhesive, and texturing.

Joint. A cementitious material for covering joints, corners, and fasteners in finishing of gypsumboard installations, to produce a smooth surface.

Setting-Type. A joint compound that hardens by chemical reactions before drying and that is used for shortening the time required for patching and completing joint finishing.

Taping. A joint compound specially formulated for embedment of joint tape.

Topping. A joint compound specially formulated to serve as the final finishing coat for a joint, but not intended for embedment of tape.

Core. The gypsum structure between face and back papers of gypsumboard.

Coreboard. A gypsumboard, usually 24 in wide and up to 1 in thick, with square, rounded or tongue-and-groove edges, and homogeneous or laminated.

Corner Bead. A strip of sheet metal with flanges and a nosing at the junction of the flanges; used to protect arrises.

Corner Cracks. Cracks in joint of intersecting walls or walls and ceilings.

Corner Floating. (See Floating Angles.)

Cornerite. Reinforcement for plaster at a reentrant corner.

Cornice. A molding, with or without reinforcement.

Cove. A curved concave, or vaulted, surface.

Crown. A buildup of joint compound over a joint to conceal the tape over the joint.

Cure. Treatment, usually of a portland-cement plaster, to ensure hydration after application.

Dado. The lower part of a wall usually separated from the upper by a molding or other device.

Darby. A flat wood tool with handles about 4 in wide and 42 in long; used to smooth or float the brown coat; also used on finish coat to give a preliminary true and even surface.

Dentils. Small rectangular blocks set in a row in the bed mold of a cornice.

Dimple. The depression in the surface of gypsumboard caused by a hammer in setting a nail head slightly below the surface, to permit concealment of the nail with joint compound.

Dope. Additives put in any type of mortar to accelerate or retard set.

Double-up. Applications of plaster in successive operations without a setting and drying interval between coats.

Double Nailing. A method of applying gypsumboard to framing with pairs of nails at intervals of about 12 in along the framing, to ensure firm contact. (The nails in each pair are usually set about 2 in apart.)

Dry out. Soft chalky plaster caused by water evaporating before setting.

Drywall Construction. Application of gypsumboard. (This is basically a dry process rather than a wet process, such as lath and plaster.)

Edges:

Beveled. The factory-formed edge of gypsumboard that has been sloped so that, where two boards abut, a V-groove joint is created.

Chisel. A slanted factory edge on gypsumboard.

Feathered. A thin edge formed by tapering joint compound at a joint to blend with adjoining gypsumboard surfaces; also denotes the skived edge of joint tape.

Featured. A configuration of the paper-bound edges of gypsumboard that provides special design or performance characteristics.

Floating. An edge that does not lie directly over framing and that will be unsupported after installation of gypsumboard or plaster.

Hard. A special core formulation used along the paper-bound edges of gypsumboard to improve resistance to damage during handling and application of the board.

Skive. The outside edges of joint tape that have been sanded or chamfered to improve adhesion and reduce waviness.

Tapered. A factory edge on gypsumboard that is progressively reduced in thickness to allow for concealment of joint tape below the plane of the gypsumboard surface.

Efflorescence. White fleecy deposit on the face of plastered walls, caused by salts in the sand or backing; also referred to as "whiskering" or "saltpetering."

Egg and Dart. Ornamentation used in cornices consisting of an oval and a dart alternately.

Eggshelling. Plaster chip-cracked concave to the surface, the bond being partly destroyed.

Enrichments. Any cast ornament that cannot be executed by a running mold.

Expanded Metals. Sheets of metal that are slit and drawn out to form diamond-shaped openings.

Fat. Material accumulated on a trowel during the finishing operation of plaster and used to fill in small imperfections. Also denotes a mortar that is not too stiff, too watery, or oversanded.

Feather Edge. A beveled-edge wood tool used to straighten reentrant angles in the finish plaster coat.

Fines. Aggregate capable of passing through a No. 200 sieve.

Finish Coat. Last and final coat of plaster; also denotes a thin coat of joint treatment to reduce variations in surface texture and suction.

Finisher. A tradesman with skill in finishing of gypsumboard joints.

Fire Taping. Taping of gypsumboard joints without subsequent finishing coats, usually used where esthetics is not important.

Fisheyes. Spots in plaster finish coat about ¼ in in diameter, caused by lumpy lime because of age or insufficient blending of material.

Float. A tool shaped like a trowel, with a handle braced at both ends and wood base for blade, used to straighten, level, and texture finish plaster coats.

Floating Angles (Corner Floating). Unrestrained surfaces intersecting at about 90°, usually with fasteners omitted near the intersection (Fig. 11.28).

Foil Back. A gypsumboard with a reflective aluminum-foil composite laminated to its back surface.

Furring. Strips that are nailed over studs, joists, rafters, or masonry to support lath or gypsumboard. This construction permits free circulation of air behind the plaster or gypsumboard.

Gaging. Mixing of gaging plaster with lime putty to acquire the proper setting time and initial strength. Also denotes type of plaster used for mixing with the putty.

Green Board. A gypsumboard with a tinted face paper, usually light green or blue, to distinguish special types of board; also denotes gypsumboard that is damp.

Green Plaster. Wet or damp plaster.

Grounds. A piece of wood, metal, or plaster attached to the framing to indicate the thickness of plaster to be applied.

Gypsum. Fully hydrated calcium sulfate (calcium sulfate dihydrate).

Gypsum Base. Gypsum lath used as a base for veneer plasters.

Gypsumboard. A noncombustible board with gypsum core enclosed in tough, smooth paper.
Type X. A gypsumboard specially formulated with high fire resistance for use in fire-rated assemblies.

Hardwall. Gypsum neat base-coat plaster.

Joint Treatment. Concealing of gypsumboard joints, usually with tape and joint compound.

Joints:
Butt. Joints in which gypsumboard ends with core exposed (usually in the direction of the board width) are placed together.
Crown (High or Hump). Protrusion of joint compound from gypsumboard surface at a joint.
Floating. (See Edges, Floating.)

Keene's Cement. A dead-burned gypsum product that yields a hard, high-strength plaster.

Lamination:
Sheet. A ply of gypsumboard attached to another ply with adhesive over the entire surface to be bonded.
Strip. A ply of gypsumboard attached to another ply by parallel strips of adhesive, usually 16 or 24 in apart.

Lath. A base to receive plaster.

Lime. Oxide of calcium produced by burning limestone. Heat drives out the carbon dioxide leaving calcium oxide, commonly termed "quicklime." Addition of water to quicklime yields hydrated or slaked lime.

Lime Plaster. Base-coat plaster consisting essentially of lime and aggregate.

Lime Putty. Thick paste of water and slaked quicklime or hydrated lime.

Marezzo. An imitation marble formed with Keene's cement to which colors have been added.

Mud. (See Compounds.)

Nail Popping. Protrusion above the face of gypsumboard of a nail used to attach the board to framing; usually caused by shrinkage due to drying of inadequately cured wood framing.

Nail Spotter. A small, box-type applicator used to cover with joint compound the heads of nails in gypsumboard.

Neat Plaster. A base-coat plaster to which sand is added at the job.

Niche. A small recess in a wall.

Ogee. A curved section of a molding, partly convex and partly concave.

Papers:
Calendered. Papers with a high glossy finish.
Cream (Ivory or Manila). Highly sized and calendered papers used as the face papers on gypsumboard.
Gray. Unsized, uncalendered papers used on the back side of regular gypsumboard and as the face and back papers of backing boards.
Sized. Paper treated with a sealant to equalize suction for paint and prevent rise of nap.

Perimeter Relief. A construction arrangement that permits building movements; also denotes gaskets that relieve stresses at intersections of walls and ceilings.

Pinhole. A small hole that appears in a plaster cast because of excess water in preparation of the plaster; also denotes a small perforation in gypsumboard paper or paper joint tape.

Plasterboard. (See Gypsumboard.)

Prefill. An application method used in preparation of joints of tapered- or beveled-edge gypsumboard to receive tape and joint treatment with the objective of reducing the possibility of ridging or beading.

Primer. A base coat of paint used to improve the bond and appearance of the finish coat of paint; usually referred to as an undercoat when tinted. (See also Sealer and Sizing.)

Punch out. A hole made in gypsumboard to fit closely around pipe that passes through the board.

Putty Coat. A smooth, troweled-finish coat containing lime putty and a gaging material.

Quicklime. (See Lime.)

Relief. Ornamental figures above a plane surface.

Retarder. Any material added to gypsum plaster that slows its set.

Return. The terminal of a cornice or molding that takes the form of an external miter and stops at the wall line.

Reveal. The vertical face of a door or window opening between the face of the interior wall and the window or door frame.

Ridging. A linear surface protrusion along treated joints.

Ripper. A narrow strip of gypsumboard used for soffits, window reveals, and finished openings.

Runner. A metal or wood track or strip placed at floor and ceiling to receive framing members for partitions.

Sanding. Smoothing a joint treatment for gypsumboard with sandpaper. In wet sanding, the joint is smoothed with a coarse wet sponge, so that less dust is produced than in dry sanding.

Scagliola. An imitation marble, usually precast, made with Keene's cement.

Score. A groove cut in the surface of a board with a sharp blade to expedite manual breaking of the board.

Scratch Coat. First coat of plaster in three-coat work.

Screeds. Long, narrow strips that serve as guides for plastering; also denotes tools, such as straightedges, used to shape an unhardened surface.

Sealer. A base coat of paint used to seal a surface and equalize differences in surface suction, to improve the bond and appearance of the finish coat. (See also Primer and Sizing.)

Seam. A treated gypsumboard joint.

Sheathing, Gypsum. A gypsumboard formulated for use as an enclosure for an exterior wall and a base for siding or other exterior facings, but not intended for long-time direct exposure to the weather.

Ship Lap. An offset lamination of two layers of gypsumboard.

Shoulder. The area between the tapered edge and the face of a gypsumboard.

Sizing. A surface sealant used to equalize suction of paint when applied to gypsumboard paper and prevent rising of the nap.

Skim Coat. (See Finish Coat.)

Skip Trowel. A method of texturing a surface that results in a rough *Spanish Stucco* effect.

Slaking. Adding water to hydrate quicklime into a putty.

Soffit. The underside of an arch, cornice, bead, or other construction.

Soffit Board. A gypsumboard formulated for use on the underside of exterior overhangs, carport ceilings, and other areas protected from the weather.

Splay Angle. An angle of more than 90°.

Spray Texture. A mechanically applied material, which may contain aggregates to produce various effects, used to form decorative finishes.

Staff (Casts). Plaster casts made in molds and reinforced with fiber; usually wired or nailed into place.

Stucco. Plaster applied to the exterior of a building.

Substrate. A surface capable of receiving additional finish or decoration; also denotes the base or concealed layer of gypsumboard in a composite assembly.

Suspension System. Construction, usually incorporating heavy-gage hanging wire, for supporting a ceiling set below structural floor framing, roof, subfloor, or floor deck.

Sweat out. A soft, damp wall area of plaster caused by poor drying conditions.

Swirl Texturing. A method of applying texturing material in a decorative circular pattern.

Tape:
Dry. A tape applied over gypsumboard joints with adhesive other than conventional joint compound.
Joint. A paper tape or fiber mesh for reinforcing joint compound to conceal and reinforce the joints of gypsumboard.

Tape Creaser. A hand-held tool for folding joint tape for use in reentrant corners.

Temper. Mixing of plaster to a workable consistency.

Template. A gage, pattern, or mold used as a guide to produce arches, curves, and various other shapes.

Veneer Plasters. Gypsum plasters meeting requirements of ASTM C587, and that may be applied in one or more coats to a maximum thickness of ¼ in.

Wadding. The act of hanging staff by fastening wads made of plaster of paris and excelsior or fiber to the casts and winding them around the framing.

Wainscot. The lower 3 or 4 ft of an interior wall when it is finished differently from the remainder of the wall.

Wallboard:
Gypsum. A gypsumboard used primarily as an interior surface.
Laminated. Two or more layers of gypsumboard held together with an adhesive.
Predecorated. A gypsumboard with a finished surface, such as paint, texturing material, vinyl film, or printed paper coverings, applied before the board is delivered to the building site.

White Coat. A gaged lime-putty troweled-finish coat.

11.25 PLASTER FINISHES

Prepared on the building site, plaster finishes are classified as wet-type construction. They take longer to complete than dry-type construction, because they are field mixed with water and require curing before decorative materials, such as paint or wallpaper, can be applied. Plaster finishes are selected nevertheless because they are hard, abrasion resistant, rigid, incombustible, and provide a monolithic (unseamed) surface, even at corners. They are relatively brittle, however, and must be properly applied to avoid cracking when movements due to drying shrinkage or thermal changes are restrained.

11.25.1 Components of Plaster

A plaster finish consists of a supporting base, such as masonry or lath, and one or more coats of plaster or mix of plaster and other ingredients with water that is troweled or machine sprayed over the base.

The principal ingredient of plaster usually is gypsum but may be portland cement. (Portland-cement plaster, or stucco, is discussed in Art. 11.19.) Gypsum plasters generally are formulated to meet the requirements of "Standard Specification for Gypsum Plasters," ASTM C28, or "Standard Specification for Gypsum Veneer Plaster," ASTM C587.

C28 gypsum plasters include ready-mixed, neat, wood-fiber and gaging plasters. They may be applied over a masonry or lath base, generally in two or more coats, with a total thickness exceeding ½ in. They are required to contain 66% or more by weight of $CaSO_4 \cdot \frac{1}{2}H_2O$.

Veneer plasters must be applied over a special gypsum base, meeting the requirements of ASTM C588, and are limited in thickness to a maximum of ¼ in. They are usually selected because of low cost, rapid installation (permitting application of decorative materials 24 hr after plastering), and high resistance to cracking, nail popping, and impact and abrasion failure. The finishes are, however, not so rigid as conventional lath and plaster. Also, veneer plasters are more susceptible than gypsum plasters to ridging and cracking at joints when dried too rapidly because of low humidity, high temperature, or exposure to drafts.

Application of gypsum and veneer plasters should meet the requirements of the following ASTM specifications:

C841. Installation of Interior Lathing and Furring

C842. Application of Interior Gypsum Plaster

C843. Application of Gypsum Veneer Plaster

C844. Application of Gypsum Base to Receive Gypsum Veneer Plaster

See also Art. 11.27.

11.25.2 Plaster Mixes

Plaster coats other than veneer plasters are generally composed of gypsum plaster, lime, an aggregate (sand, vermiculite, perlite), and water.

Sisal or synthetic fibers, such as nylon, may be added to some scratch-coat plasters for application to metal lath, to limit to what is needed for good bond the amount of plaster that passes through the lath meshes. Fibering, however, adds no strength.

Sand should comply with ASTM C35. It should be clean, free of organic material, more than about 5% clay, silt, or other impurities, and should not contain salt or alkali. The proportion of sand in the plaster has an important bearing on the characteristics of the product. Oversanding results in considerable reduction in strength and hardness. A mix as lean as 4:1 by weight should never be used.

The Gypsum Association suggests that a 1-ft^3 measuring box be used for preparing mixes. Some plasterers use a No. 2 shovel, which holds about 16 lb of moist sand, for maintaining proper proportions. Thus, with each 100-lb bag of plaster, a 1:2 mix requires 12 shovels of sand and a 1:3 mix 18 shovels ("Manual of Gypsum Lathing and Plastering").

Water should be clean and free of substances that might affect the rate of set of the plaster. It is not advisable to use water in which plasterers' tools have been washed because it might change the set. Excessive water is undesirable in the mix, because when the water evaporates, it leaves numereous large voids, which decrease the strength of the plaster. Hence, manufacturers' recommendations should be observed closely in determining water requirements.

Perlite and vermiculite are manufactured lightweight aggregates that are used to produce a lightweight plaster with relatively high fire resistance for a given thickness. Both aggregates should conform with ASTM C35, and the mix should be prepared strictly in accordance with manufacturers' recommendations.

11.25.3 Mixing Plaster

A mechanical mixer disperses the ingredients of a mix more evenly and therefore is to be preferred over box mixing. Recommended practice is as follows: (1) Place the anticipated water requirements in the mixer; (2) add about half the required sand (or all required perlite or vermiculite); (3) add all the plaster; (4) add the rest of the sand; (5) mix at least 2 min, but not more than 5 min, adding water, if necessary, to obtain proper workability; and (6) dump the entire batch at once.

The mixer should be thoroughly cleaned when it is not in use. If partly set material is left in it, the set of the plaster might be accelerated. For this reason also, tools should be kept clean.

For hand mixing, first sand and plaster should be mixed dry to a uniform color in a mixing box, water added, and the plaster hoed into the water immediately and thoroughly mixed. Undermixed plaster is difficult to apply and will produce soft and hard spots in the plastered surface.

Plaster should not be mixed more than 1 hr in advance. Nor should a new mix, or gaging, be mixed in with a previously prepared one. And once plaster has started to set, it should not be remixed or retempered.

11.25.4 Plaster Drying

A minimum temperature of 55°F should be maintained in the building where walls are to be plastered when outdoor temperatures are less than 55°F, and held for at least 1 week before plaster is applied and 1 week after the plaster is dry.

In hot, dry weather, precautions should be taken to prevent water from evaporating before the plaster has set. Plastered surfaces should not be exposed to drafts, and openings to the outside should be closed off temporarily. After the plaster sets, the excess moisture it contains evaporates. Hence, the room should be adequately ventilated to allow this moisture to escape.

("Architect Data Book—Construction Products and Systems," Gold Bond Building Products, a National Gypsum Division, 2001 Rexford Road, Charlotte, NC 28211; "Gypsum Products Design Data," Gypsum Association, 1603 Orrington Ave., Evanston, IL 60201; "Gypsum Construction Handbook," United States Gypsum, 101 South Wacker Drive, Chicago, IL 60606.)

11.25.5 Gypsum Bases for Plaster

One commonly used base for plaster is gypsum lath. This is a noncombustible sheet generally 16 × 48 in by ⅜ or ½ in thick, or 16 × 96 in by ⅜ in thick. It is composed principally of calcined gypsum that has been mixed with water, hardened, dried, and then sandwiched between two paper sheets (ASTM C37). Insulating gypsum lath is made by cementing shiny aluminum foil to the back of plain lath. It is used for vapor control, and for insulation against heat loss or gain. Also available is gypsum lath with a core specially formulated with minerals for high resistance to fire.

Gypsum lath has the advantage over metal lath that less plaster is required, because the first, or scratch, coat is applied only over the lath surface. Also, when used for suspended ceilings or hollow partitions, plaster on gypsum lath is less susceptible to cracking than on metal lath.

Installation of gypsum lath should meet the requirements of ASTM C841, "Installation of Interior Lathing and Furring." The lath should be applied to studs with long dimension horizontal, and vertical joints should be staggered (Fig. 11.20). In ceilings, the long sides should span supports. Ends should rest on or be nailed to framing, headers, or nailing blocks. Each lath should be in contact with adjoining

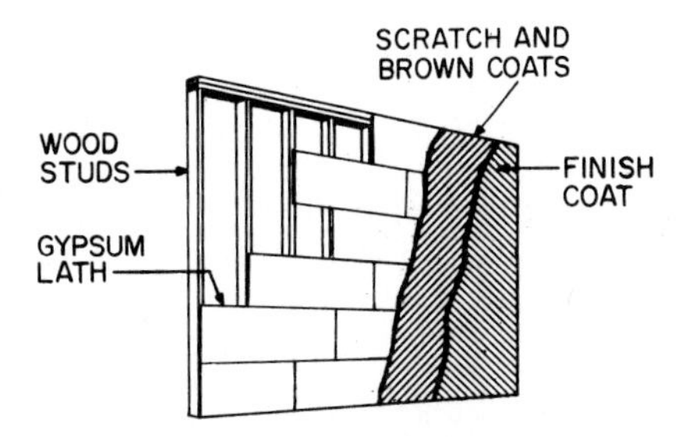

FIGURE 11.20 Two coats of plaster applied to perforated gypsum lath attached to wood studs.

sheets, but if spaces more than ½ in wide are necessary, the plaster should be reinforced with self-furring metal lath that is stapled or tied with wire to the gypsum lath.

When nailed to wood members, ⅜-in-thick gypsum lath should be attached with four nails, and ½-in lath with five nails, to each framing member covered. Nails should be blued gypsum lath nails, made of 13-ga wire, 1⅛ in long, with a 19/64-in-diam flat head. They should be driven until the head is just below the paper surface without breaking the paper. Lath also may be attached to wood framing with four or five 16-ga staples, 7/16 in wide, with ⅜-in divergent legs. With metal framing, screws should be used, as recommended by the lath manufacturer. Clips, however, are a suitable alternative for use with wood or metal framing. Fasteners should be driven at least ⅜ in away from ends and edges. Clips must secure the lath to framing at each intersection with the framing.

Studs or ceiling members supporting lath may be spaced up to 16 in c to c with ⅜-in gypsum lath, and up to 24 in c to c with ½-in lath.

Except at intersections that are to be unrestrained, reentrant corners should be reinforced with cornerite stapled or tied with wire to the gypsum lath. Exterior-angle corners should be finished with corner beads set to true grounds and nailed or tied with wire to the structural frame or to furring. Casing beads should be used around wall openings and at intersections of plaster with other finishes and of lath and lathless construction.

Gypsum Base for Veneer Plasters. Special gypsum lath meeting requirements of ASTM C588 is required as a base for veneer plasters. It is formulated to provide the strength and absorption necessary for proper application and performance of these thin coatings. The lath comes in thicknesses of ⅜ (for two-coat systems), ½, and ⅝ in, the last permitting 24-in spacing of wood framing. Installation should meet the requirements of ASTM C844. In general, gypsum base should be applied first to the ceiling, then to the walls. Maximum spacing of nails is 7 in on ceilings and 8 in on walls. Screw spacing should not exceed 12 in for wood framing 24 in c to c or for steel framing or for wood ceiling framing, or 16 in for wood studs spaced 16 in c to c.

11.25.6 Metal Lath

A metal base often is used in plaster construction because it imparts strength and resists cracking. Plaster holds to metal lath by mechanical bond between the initial coat of plaster and the metal. So it is important that the plaster completely surrounds and embeds the metal.

Basic types commonly used are expanded-metal, punched sheet-metal, and paper-backed welded-wire lath. Woven-wire lath may be used as a supplemental reinforcement over solid plaster bases, but not as the primary base for gypsum plaster. Wire lath should be made of galvanized, copper-bearing steel. The other types should be made of galvanized steel or copper-bearing steel with a protective coat of paint.

Expanded-metal lath is fabricated by slitting sheet steel and expanding it to form a mesh. Several types are available:

Diamond-mesh lath, with more than 11,000 meshes/yd^2, is an all-purpose lath, suitable as a base for flat or curved plaster surfaces (Fig. 11.21). The small meshes are helpful in reducing droppings of plaster during plastering. A self-furring type also is available. When it is attached to a backing, it is separated from the backing by at least ¼ in. Thus, self-furring lath is convenient for use as exterior stucco bases and bases for column fireproofing and replastering over old surfaces.

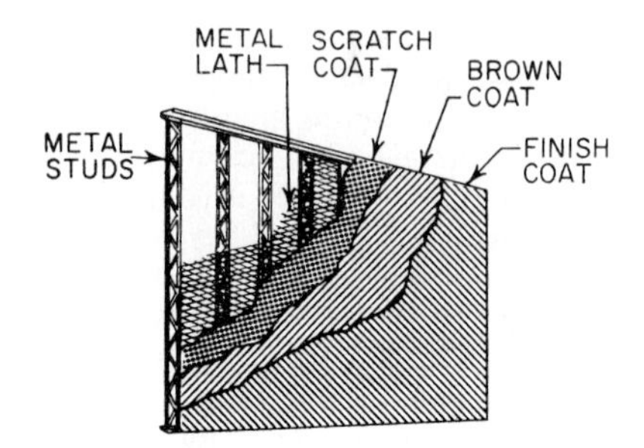

FIGURE 11.21 Three coats of plaster applied to metal lath attached to steel studs.

Flat-rib expanded-metal lath comes with smaller openings than diamond mesh, and it has ribs parallel to the length of the sheet that make it more rigid. Flat-rib lath is generally preferred for nailing to wood framing and tying to framing for flat ceilings, but it is not suitable for contour lathing.

High-rib expanded-metal lath is used when greater rigidity is desired, for example, for spacing supports up to 24 in c to c and for solid, studless plaster partitions. The lath has a herringbone mesh pattern and V-shaped ribs running the length of each sheet. For ⅜-in rib lath, ⅜-in-deep ribs are spaced 4½ in c to c, alternating with inverted $\frac{3}{16}$-in ribs. For ¾-in rib lath, ¾-in-deep ribs are spaced 6 in c to c. This type of high-rib lath may be used as a form and reinforcement for concrete slabs, but its thickness makes it generally unsuitable for plaster construction. The ⅜-in rib lath may also be used as a concrete form, but its rigidity makes it unsuitable for contour lathing.

Welded-wire lath should be made of wire 16 ga or thicker, forming 2 × 2-in or smaller meshes, stiffened continuously parallel to the long dimension of the sheet at intervals not exceeding 6 in. The paper backing should comply with Federal Specifications UU-B-790, "Building Paper, Vegetable Fiber (Kraft, Waterproofed, Water Repellant, and Fire Resistant)." Acting as a base to which plaster can adhere while hardening about the wire, the backing should permit full embedment in at least ⅛ in of plaster of more than half the total length and weight of the wires.

Table 11.9 lists limiting spans for various types and weights of metal lath for ceilings and walls.

Tying and Nailing. Installation of metal lath should meet the requirements of ASTM C841. Attachments of metal lath to supports should not be spaced farther apart than 6 in along the supports.

When metal framing or furring is used, metal lath should be tied to it with 18-ga, or heavier, galvanized soft-annealed wire. Rib lath, however, should be attached to open-web steel joists with single loops of 16-ga, or heavier, wire, or double loops of 18-ga wire, with the ends of each loop twisted together. Also, rib lath should be tied to concrete joists with loops of 14-ga, or heavier, wire or with wire hangers not less than 10 ga.

With wood supports, diamond-mesh, flat-rib, and welded-wire lath should be attached to horizontal framing with 1½-in, 11 ga, $\frac{7}{16}$-in-head, barbed, galvanized, or blued roofing nails, driven full length. For vertical wood supports the following may be used: 4d common nails; 1-in, 14-ga wire staples driven full length; and

TABLE 11.9 Limiting Spans for Metal Lath, in

		Vertical supports			Horizontal supports	
			Metal			
Type of lath	Min wt., lb/yd²	Wood	Solid partitions*	Other	Wood or concrete	Metal
Diamond mesh (flat expanded) metal lath	2.5	16	16	12	12	12
	3.4	16	16	16	16	13½
Flat- (⅛-in) rib expanded metal lath	2.75	16	16	16	16	16
	3.4	19	24	19	19	19
⅜-in rib expanded metal lath†	3.4	24	24†	24	24	24
	4.0	24	24†	24	24	24
¾-in rib expanded metal lath	5.4		†	24	36‡	36‡
Sheet-metal lath†	4.5	24	†	24	24	24
Welded-wire lath	1.16§	16	16	16	16	16
	1.95¶	24	24	24	24	24

*For paper-backed lath, only absorbent, perforated, or slotted paper separator should be used.
†Permitted for studless solid partitions.
‡Permitted only for contact or furred ceilings.
§Welded 16-ga wire, paper-backed lath.
¶Paper-backed lath with welded wire, face wires 16 ga, every third back wire parallel to line dimension of lath 11 ga.
Source: Based on "Uniform Building Code," International Conference of Building Officials, Inc.

1-in roofing nails driven at least ¾ in into the supports. Common nails should be bent over to engage a rib or at least three strands of lath. Alternatives of equal strength also may be used.

Metal lath should be applied with long sides of sheets spanning supports. Each sheet should underlap or overlap adjoining sheets on both sides and ends. Expanded-metal and sheet-metal lath should be lapped ½ in along the sides, or have edge ribs nested, and 1 in along the ends. Welded-wire lath should be lapped one mesh at sides and ends. All side laps of metal lath should be fastened to supports and tied between supports at intervals not exceeding 9 in.

Wherever possible, end laps should be staggered, and the ends should be placed at and fastened to framing. If end laps fall between supports, the adjoining ends should be laced or securely tied with 18-gage, galvanized, annealed steel wire.

Normally, metal lath should be applied first to ceilings. Flexible sheets may be carried down 6 in on walls and partitions. As an alternative, preferable for more rigid sheets, sides and ends of the lath may be butted into horizontal reentrant angles and the corner reinforced with cornerite. But for large ceilings (length exceeding 60 ft in any direction, or more than 2400 ft² in area) and other cases in which restraint should be avoided, and for portland-cement plaster ceilings, cornerite should not be used. Instead, the abutting sides and ends should terminate at a casing bead, control joint, or similar device that will isolate the ceiling lath and plaster from the walls and partitions.

Similar considerations govern installation of metal lath at vertical reentrant corners. Between partitions, flexible sheets may be bent around vertical corners and attached at least one support away from them; more rigid sheets may be butted and the corner reinforced with cornerite. But where restraint is undesirable at reentrant corners, for example, where partitions meet structural walls or columns, or where load-bearing walls intersect, cornerite should not be used. Instead, the walls, partitions, and structural members should be isolated from each other, as described for ceiling-wall corners.

11.25.7 Masonry Bases for Plaster

Gypsum partition tile has scored faces to provide a mechanical bond as well as the natural bond of gypsum to gypsum plaster. The 12 × 30-in faces of the tile present an unwarped plastering surface because the tile is dried without burning. This is done so that a mechanic can lay a straighter wall than with other types of units. The Gypsum Association's "Manual of Gypsum Lathing and Plastering" recommends that only gypsum plaster be applied to gypsum partition tile, since lime and portland cement do not bond adequately. Also, only gypsum mortar should be used for laying tile.

Brick and clay tile can be used as a plaster base if they are not smooth-surfaced or of a nonporous type. If the surface does not provide sufficient suction, it should offer a means for developing a mechanical bond, such as does scored tile.

Plaster should not be applied directly to exterior masonry walls because dampness may damage the plaster. It is advisable to fur the plaster at least 1 in in from the masonry.

Properly aged concrete block may serve in walls as a plaster base, but for block ceilings, a bonding agent or a special bonding plaster should be applied first.

For precast or cast-in-place concrete with smooth dense surfaces, a bonding agent or a special bonding plaster should be used first. But if a plaster thickness of more than ⅜ in is required for concrete ceilings, or ⅝ in for concrete walls, metal lath should be secured to the concrete before plastering, in which case sanded plaster can be used.

11.25.8 Plaster Base Coats

The base coat is the portion of the plaster finish that is applied to masonry or lath bases and supports the finish coat (Fig. 11.22).

Except for veneer plasters, plaster applications may be three-coat (Fig. 11.21) or two-coat (Fig. 11.20). The former consists of (1) a scratch coat, which is applied directly to the plaster base, cross-raked after it has "taken up," and allowed to set and partly dry; (2) a brown coat, which is surfaced out to the proper grounds, darbied (float-finished), and allowed to set and partly dry; and (3) the finish coat. Three-coat plaster is required over metal lath, ½-in gypsum lath spanning horizontal supports more than 16 in c to c, all gypsum lath attached by clips providing only edge support, and ⅜-in perforated gypsum lath on ceilings.

The two-coat application is similar, except that cross-raking of the scratch coat is omitted and the brown coat is applied within a few minutes to the unset scratch coat. Three coats are generally preferred, because the base coat thus produced is stronger and harder.

Veneer plaster applications, 1⁄16 in thick, may be one-coat or two-coat, both applied to special gypsum base (Arts. 11.25.1 and 11.25.5). The single coat is

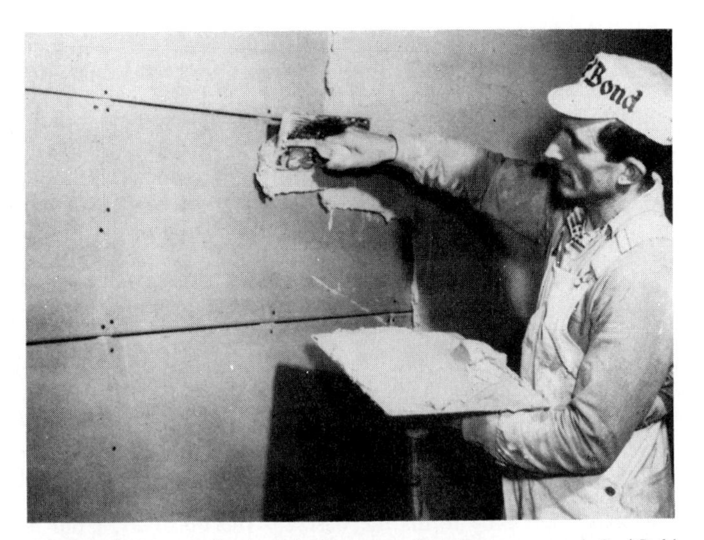

FIGURE 11.22 Application of a base coat of plaster to gypsum lath. (*Gold Bond Building Products, a National Gypsum Division.*)

composed of a scratch coat without cross-raking and a double-up coat immediately applied, then worked to a smooth or textured finish.

See also Art. 11.25.9 and "Manual of Gypsum Lathing and Plastering," Gypsum Association.

Plaster Grounds. Except for veneer plasters, thickness of base-coat plaster should be controlled with grounds—wood or metal strips applied at the perimeter of all openings and at baseboards or continuous strips of plaster applied at intervals along a wall or ceiling, to serve as screeds. Plaster screeds should be used on all plaster surfaces of large area.

Minimum Thicknesses. Grounds should be set to provide a minimum plaster thickness of ½ in over gypsum lath and gypsum partition tile; ⅝ in over brick, clay tile, or other masonry; and ⅝ in from the face of metal lath. A thickness of $^1/_{16}$ in is included for the finish coat.

Gypsum Base-Coat Plasters. Three types of gypsum base-coat plasters are in general use: gypsum heat plaster, gypsum ready-mixed plaster, and veneer plasters, which may be used in thin one-coat or two-coat systems. In two-coat veneer-plaster systems, the base-coat veneer plaster should be applied to a thickness of $^1/_{16}$ to $^3/_{32}$ in, and left with a rough surface to receive the finish coat. Veneer plasters should meet the requirements of ASTM C587 and application should be in accordance with ASTM C843.

Gypsum neat plaster, sometimes called hardwall or gypsum-cement plaster, is sold in powder form and mixed with an aggregate and water at the construction site. Mixed with no more than 3 parts sand by weight, it makes a strong base coat at low cost. Scratch coats generally consist of 1 part plaster powder to 2 parts sand by weight, fibered or unfibered; the base coat in two-coat work usually is a 1:2½ mix; brown coats are 1:3 mixes. With perlite or vermiculite instead of sand, a 1:2 mix may be used.

Gypsum ready-mixed plaster requires the addition only of water at the site, since it is sold in bags containing the proper proportions of aggregate and plaster. It is specified when good plastering sand is high cost or not available, or to avoid the possibility of oversanding. It costs a little more than neat plaster because of the extra cost of transporting the sand.

The water ratio for base coat neat and ready-mixed plasters should be such that slump does not exceed 4 in when tested with a 2 × 4 × 6-in cone at the mixer, for mixes with sand proportions not exceeding those given for gypsum neat plaster.

Application of gypsum plaster should meet the requirements of ASTM C842. The scratch coat (Figs. 11.20 and 11.21) applied to lath should be laid on with enough pressure to form a strong clinch or key. The coat should cover the lath to a thickness of ¼ in. For two-coat systems, the double-up brown coat is applied immediately. For three-coat systems, after the surface has been trued, the scratch coat should be scratched horizontally and vertically with a toothed tool to form a good bonding surface, then left to dry partly. When the surface is so hard that the edges of the scoring do not yield easily under the pressure of a thumbnail, the brown coat may be applied. Hardening may take at least 1 day, and sometimes as long as 1 week, depending on drying conditions.

The brown coat not only forms the base for the finish coat, but is also the straightening coat. The plaster should be laid on with a steel float, trued with rod or darby, and left rough in preparation for the finish coat.

11.25.9 Finish Plaster Coats

Several types of plasters are available for the finish coat (Figs. 11.20 and 11.21). Usually, lime is an important ingredient, because it gives plasticity and bulk to the coat.

Gaging plasters are coarsely ground gypsum plasters, which are available in quicksetting and slow-setting mixtures; so it is not necessary to add an accelerator or a retarder at the site. Gaging plasters also are supplied as white gaging plaster and a slightly darker local gaging plaster. Finish coats made with these plasters are amply hard for ordinary usages and are the lowest-cost plaster finishes. However, they are not intended for ornamental cornice work or run moldings, which should be made of a finer-ground plaster. Gypsum gaging plasters should conform with ASTM C28. Application should conform with ASTM C842.

Typical mixes consist of 3 parts lime putty to 1 part gaging plaster, by volume. If a harder surface is desired, the gaging content may be increased up to 1 part gaging to 2 parts lime putty.

The lime is prepared first, being slaked to a smooth putty, then formed on the plasterer's board into a ring with water in the center. Next, gaging plaster is gradually sifted into the water. Then, aggregates, if required, are added. Finally, all ingredients are thoroughly mixed and kneaded. Alternatively, materials, including Type S hydrated lime, may be blended in a mechanical mixer.

The lime-gaging plaster should be applied in at least two coats, when the brown coat is nearly dry. The first coat should be laid on very thin, with sufficient pressure to be forced into the roughened surface of the base coat. After the first coat has been allowed to draw a few minutes, a second or leveling coat, also thin, should be applied.

The base coat draws the water from the finish coats; so the finished surface should be moistened with a wet brush as it is being troweled. Pressure should be exerted on the trowel to densify the surface and produce a smooth hard finish. Finally, the surface should be dampened with the brush and clean water. It should be allowed to stand at least 30 days before oil paints are applied.

Prepared gypsum trowel finishes also are available that require only addition of water at the site. The resulting surface may be decorated as soon as dry. The plaster is applied in the same manner as lime-gaging plaster, but the base coat should be dry and, because the prepared plaster has a moderately fast set, it should be troweled before it sets. For best results, three very thin coats should be applied and water should be used sparingly.

Sand float finishes are similar to gypsum trowel finishes, except that these float finishes contain a fine aggregate to yield a fine-textured surface and the final surface is finished with a float. The base coat should be firm and uniformly damp when the finish coat is applied. These finishes have high resistance to cracking.

Molding plaster, intended for ornamental work, is made with a finer grind than other gaging plasters. It produces a smooth surface, free from streaks or indentations as might be obtained with coarser-ground materials. Equal parts of lime putty and molding plaster are recommended by the Gypsum Association for cornice moldings.

Veneer plasters, applied to a thickness of only 1/16 to 3/32 in, develop hard abrasion-resistant surfaces that can be decorated the day after application. Factory prepared, these plasters are easy to work, have high plasticity, and provide good coverage. The finish may be applied in one-coat systems over special gypsum base (Art. 11.25.5), or in two-coat systems over a veneer-plaster or sanded gypsum base coat. They require only addition of water on the job. Veneer plasters should meet the requirements of ASTM C587 and application should conform with ASTM C843.

11.26 GYPSUMBOARD FINISHES

To avoid construction delays due to the necessity of mixing plaster ingredients with water on the building site and, after application, waiting for the plaster to cure, dry-type construction with gypsum products may be used. (For discussion of other dry-type interior finishes, see Arts. 11.30 to 11.31.) For the purpose, gypsumboard is factory fabricated and delivered to the site ready for application.

Gypsumboard is the generic name for a variety of panels, each consisting of a non-combustible core, made primarily of gypsum, and a bonded tough paper surfacing over the face, back, and long edges. Different types of paper are used for specific purposes. Gypsumboard with a factory-applied, decorative face or with an aluminum-foil back for insulation purposes also is available.

The panels may be attached directly to framing, such as studs or joists. While gypsumboard also may be attached directly to unit masonry or concrete, use of furring strips between the panels and backing is desirable because of the possibilities of interference from surface irregularities in the backing or of moisture penetration through exterior walls. Successive panels are applied with edges or ends abutting each other. Depending on esthetic requirements, the joints may be left exposed, covered with battens, or treated to present an unseamed, or monolithic, appearance. Gypsumboard also can be used to construct self-supporting partitions spanning between floor and ceiling.

See also Art. 4.26.

11.26.1 General Application Procedures for Gypsumboard

Gypsumboards may be used in single-ply construction (Fig. 11.23) or combined in multiply systems (Fig. 11.27). The latter are preferred for greater sound control and fire resistance. Application and finishing should conform with ASTM C840.

Precautions. When outdoor temperatures are less than 55°F, the temperature of the building interior should be maintained at a minimum of 55°F for at least 24 hr before installation of gypsumboards. Heating should be continued until a permanent heating system is in operation, or outdoor temperatures stay continuously above 55°F. In warm or cold weather, gypsumboards should be protected from the weather.

Green lumber should not be used for framing or furring gypsumboard systems. Moisture content of the lumber should not exceed 19%, to avoid defects caused by shrinkage as the wood dries. For the same reason, dry lumber should be kept dry during storage and erection and afterward.

To prevent damage from structural movements or dimensional changes, control and expansion joints should be provided and floating-angle construction used, as described in Art. 11.27.

Many building components may be affected by the decision to use gypsumboard construction. For example, window and door frames should have the appropriate depth for the wall thickness resulting from use of wallboard. Therefore, walls and partitions and related installations, including mechanical and electrical equipment, should be carefully planned and coordinated.

Application. Wallboard preferably should be applied to studs with long dimension horizontal (Fig. 11.23), and vertical joints should be staggered. In ceilings, the long sides should span supports (often referred to as horizontal, or across, application). Board ends and edges parallel to framing or furring should be supported on those members, except for face layers of two-ply systems. Otherwise, back blocking should be used to reinforce the joints.

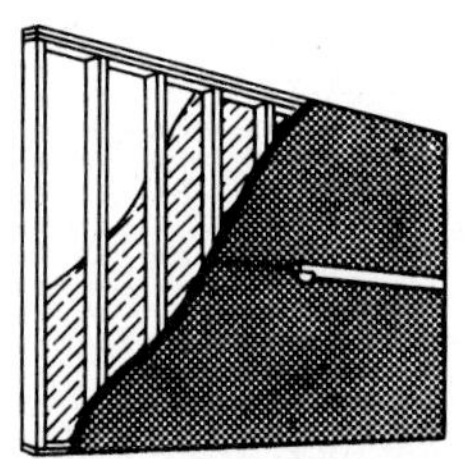

FIGURE 11.23 Single-ply application of gypsumboard to studs with long dimension horizontal.

Ceiling panels should be installed first, then the walls. Adjoining boards should be placed in contact, but not forced against each other. Tapered edges should be placed next to tapered edges, square ends in contact with square ends. (Joints formed by placing square ends next to tapered edges are difficult to conceal.)

Furring. Supplementary framing, or furring, should be used when framing spacing exceeds the maximum spacing recommended by the gypsumboard manufacturer for the thickness of board to be used (Table 11.10), or when the surface of framing or base layer is too far out of alignment.

At gypsumboard joints, wood furring should be at least 1½ in wide, and metal furring 1¼ in wide, to provide adequate bearing surface and space for attachment of the gypsumboard. The furring should be aligned to receive the board, and securely fastened to framing or masonry or concrete backing. For rigidity, lumber used for furring should be at least 2 × 2 in when nails are used and 1 × 3 in when screws are used for attachment of gypsumboard. Furring on masonry or concrete may be as small as ⅝ × 1½ in.

Fastening. Gypsumboards may be held in place with various types of fasteners or adhesives, or both. Special nails, staples, and screws are required for attachment of gypsumboards, because ordinary fasteners may not hold the boards tightly in place or countersink neatly, to be easily concealed. Clips and staples may be used only to attach the base layer in multi-ply construction.

To avoid damaging edges or ends, fasteners should be placed no closer to them than ⅜ in (Figs. 11.25 and 11.26). With a board held firmly in position, fastening should start at the middle of the board and proceed outward toward the edges or ends. Nails should be driven with a crown-headed hammer until a uniform depression, or dimple, not more than 1⁄32 in deep is formed around the nail head. Care should be taken in driving not to tear the paper or crush the gypsum core.

Nails. For use in attachment of gypsumboard, nails should conform with ASTM C514. (Some fire-rated systems, however, may require special nails recommended by the gypsumboard manufacturer.) For easy concealment, nail heads should be flat or slightly concave and thin at the rim. Diameter of nail heads should be ¼ or 5⁄16 in. Length of smooth-shank nails should be sufficient for penetration into wood framing of at least ⅞ in, and length of annular-grooved nails, for ¾-in penetration.

Screws. Three types of screws, all with cupped Phillips heads, are used for attachment of gypsumboard. Type W, used with wood framing, should penetrate at least ⅝ in into the wood. Type S, used for sheet metal, should extend at least ⅜ in beyond the inner gypsumboard surface. Type G, used for solid gypsum construction, should penetrate at least ⅜ in into supporting gypsumboard. These screws, however, should not be used to attach wallboard to ⅜-in backing board, because sufficient holding power cannot be developed. Nails or longer screws should be driven through both plies.

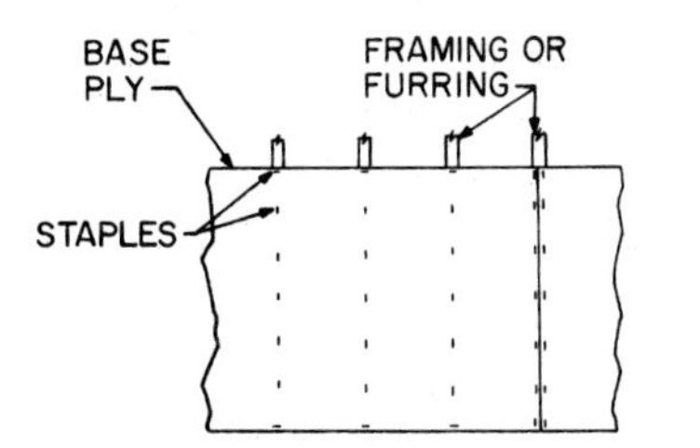

FIGURE 11.24 Gypsumboard base ply attached to wood supports with staples.

Staples. Used for attaching base ply to wood framing in multi-ply systems (Fig. 11.24) staples should be made of flattened, galvanized, 16-ga wire. The crown should be at least 7⁄16 in wide. Legs should be long enough to permit penetration into supports of at least ⅝ in, and should have spreading points.

When the face ply is to be laminated with adhesive to the base ply, the staples should be spaced 7 in c to c. When the face ply is to be nailed, the staples should be placed 16 in c to c.

Adhesives. These may be used to attach gypsumboard to framing or furring, or to existing flat surfaces. Nails or screws may also be used to provide supplemental support.

Adhesives used for bonding wallboard may be classified as stud, laminating, or contact or modified contact.

Stud adhesives are used to attach wallboard to wood or steel framing or furring. They should conform with ASTM C557. They should be applied to supporting members in continuous, or nearly so, beads.

Laminating adhesives are used to bond gypsumboards to each other, or to suitable masonry or concrete surfaces. They are generally supplied in powder form, and water is added on the site. Only as much adhesive should be mixed at one time as can be applied within the period specified by the manufacturer. The adhesive may be spread over the entire area to be bonded, or in parallel beads or a pattern of large spots, as recommended by the manufacturer. Supplemental fasteners or temporary support should be provided the face boards until sufficient bond has been developed.

Contact adhesives are used to laminate gypsumboards to each other, or to attach wallboard to metal framing or furring. A thin, uniform coat of adhesive should be applied to both surfaces to be joined. After a short drying time, the face board should be applied to the base layer and tapped with a rubber mallet, to ensure overall adhesion. Once contact has been made, it may not be feasible to move or adjust the boards being bonded.

Modified contact adhesives, however, permit adjustments, often for periods of up to ½ hr after contact. Also, they generally are formulated with greater bridging ability than contact adhesives. Modified adhesives may be used for bonding wallboard to all kinds of supporting construction.

See also Arts. 11.26.2 to 11.26.4 and 11.27.

11.26.2 Single-Ply Gypsumboard Construction

A single-ply system consists of one layer of wallboard attached to framing, furring, masonry, or concrete (Fig. 11.23). This type of system is usually used for residential construction and where fire-rating and sound-control requirements are not stringent. Maximum spacing of supports should not exceed the limits specified in Table 11.10.

TABLE 11.10 Maximum Spacing, in, of Framing for Single-Ply Gypsumboard

Board thickness, in	Orientation	Spacing c to c, in
	a. Applications in ceilings	
⅜*†	Across	16
½*	Across	24
½*	Parallel	16
⅝	Across	24
⅝	Parallel	16
	b. Applications in walls	
⅜	Across or parallel	16
½	Across or parallel	24
⅝	Across or parallel	24

*Gypsumboard for ceilings to receive a water-base-spray texture finish should be applied only across (perpendicular to) framing. For 16-in spacing of framing, board thickness should be increased from ⅜ to ½ in. For 24-in spacing of framing, board thickness should be increased from ½ to ⅝ in.

†Should not support thermal insulation.

Nail Attachment. Spacing of nails generally is determined by requirements for fire resistance and for firmness of contact between wallboard and framing necessary to avoid surface defects. Spacing normally used depends on whether single nailing or double nailing is selected. Double nailing provides tighter contact, but requires more nails. In either method, one row of nails is driven along each support crossed by a board or on which a board end or edge rests. The spacing applies to the center-to-center distance between nails in each row.

In the single-nailing method, nails should be spaced not more than 7 in apart for ceilings and 8 in apart for walls (Fig. 11.25).

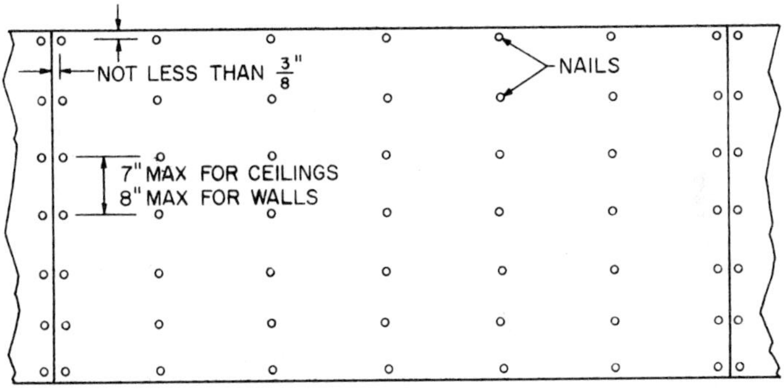

FIGURE 11.25 Wallboard attached to wood framing with single nailing.

In the double-nailing method, pairs of nails are driven 12 in c to c, except at edges or ends, in a special sequence (Fig. 11.26). First, one nail of each pair is placed, starting at the middle of the board and then proceeding toward edges and ends. Next, the second nail is driven 2 in from the first. Finally, the first set of nails placed should be given an extra hammer blow to reseat them firmly. Single-nailing spacing should be used for edges or ends at supports.

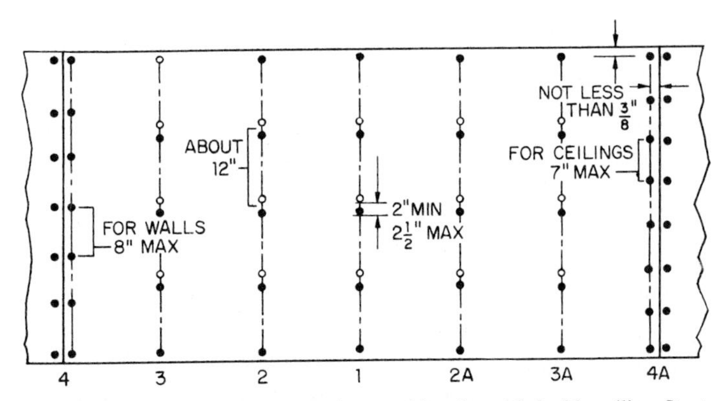

FIGURE 11.26 Wallboard attached to wood framing with double nailing. Starting at the middle of the board, one person nails rows 1 to 4, and a second person, rows 2*A* to 4*A* (solid dots). Then, again starting at the middle of the board, they drive the second set of nails (open dots). Finally, they reset the first set of nails with a blow on each nail.

Screw Attachment. Fewer screws than nails are needed for attachment of wallboard. With wood framing, Type W screws should be spaced 12 in c to c for ceilings. Spacing should not exceed 16 in for studs 16 in c to c, or 12 in for studs 24 in c to c. With metal framing, spacing of Type S screws should not exceed 12 in for walls or ceilings.

Adhesive Nail-on Attachment. With this method, at least 50% fewer nails are required in the middle of the wallboard than with conventional nailing. Also, a stiffer assembly results.

The first step is to apply stud adhesive to framing or furring in beads about 1/4 in in diameter. Each bead should contain enough adhesive so that it will spread to an average width of 1 in and thickness of 1/16 in when wallboard is pressed against it, but there should not be so much adhesive that it will squeeze out at joints. If joints are to be treated, an undulating or zigzag bead should be applied to supports under joints. If joints are to be untreated, as is likely to be the case for predecorated wallboards, two parallel beads should be applied near each edge of the framing. Along other supports a single, continuous, straight bead is adequate. Adhesive, however, should not be applied to members, such as diagonal bracing, blocking, and plates, not required for wallboard support.

The number and spacing of supplemental fasteners needed with adhesives depend on adhesive properties. Stud adhesives generally require only perimeter fasteners for walls. One fastener should be driven wherever each edge or end crosses a stud. For edges or ends bearing on studs, fasteners should be spaced 16 in c to c. The same perimeter fastening should be used for ceilings, but in addition, fasteners should be spaced 24 in c to c along all framing crossed by the wallboard.

Where perimeter fasteners cannot be used, for example, with predecorated wallboard for which joint treatment is unnecessary or undesirable, prebowing or temporary bracing should be used to keep the wallboard pressed against the framing until the adhesive develops full strength. Bracing should be left in place for at least 24 hr. Prebowing bends in wallboard so that the finish side faces the center of curvature. The arc is employed to keep the board in tight contact with the adhesive as the board is pressed into place starting at one end.

11.26.3 Multi-Ply Gypsumboard Construction

A multi-ply system consists of two or more layers of gypsumboard attached to framing, furring, masonry, or concrete. This type of system has better fire resistance and sound control than can be achieved with single-ply systems, principally because of greater thickness, but also because better insulation can be used for the base layer.

Face layers usually are laminated (glued), but may be nailed or screwed, to a base layer of wallboard, backing board, or sound-deadening board. Backing board, however, often is used for economy. When adhesives are used for attaching the face ply, some fasteners generally are also used to ensure bond.

Maximum support spacing depends principally on the thickness of the base ply and its orientation relative to the framing, as indicated in Table 11.11, and is the same for wood or metal framing or furring.

Nails, screws, or staples may be used to attach base ply to supports. For wood framing, when the face ply is to be laminated to the base ply, staples should be spaced 7 in c to c (Fig. 11.24), and nails and screws should be single-nailed as recommended for single-ply construction (Art. 11.26.2). If the face ply is to be

TABLE 11.11 Maximum Spacing of Framing for Two-Ply Gypsumboard

Base thickness, in	Face thickness, in	Orientation: Base	Orientation: Face	Spacing, c to c, in: Fasteners only	Spacing, c to c, in: Adhesive between plies
a. Application in ceilings					
⅜	⅜*	Across	Across	16	16
⅜	⅜*	Across	Parallel	†	16
½	⅜*	Parallel	Across	16	16
½	½	Parallel	Across	16	16
½	½*	Across	Across	24	16
⅝	½*	Across	Across	24	24
⅝	½*	Parallel	Across or parallel	†	24
⅝	⅝	Across	Across	24	24
⅝	⅝	Parallel	Across or parallel		24
b. Application in walls					
⅜, ½, ⅝	⅜	Across or parallel		16	24
⅜, ½, ⅝	½, ⅝	Across or parallel		24	24

*Gypsumboard for ceilings to receive a water-base-spray texture finish should be applied only across (perpendicular to) framing. For 16-in spacing of framing, board thickness should be increased from ⅜ to ½ in. For 24-in spacing of framing, board thickness should be increased from ½ to ⅝ in.
†Not recommended.

nailed, nails and staples for the base ply should be driven 16 in c to c, and screws 24 in c to c. For metal framing and furring, Type S screws for the base ply should be spaced 12 in c to c if the face ply is to laminated, and 16 in c to c if the face ply is to be attached with screws.

For attachment of the face ply with adhesive, sheet, strip, or spot lamination may be used. In **sheet lamination,** the entire back of the face ply is covered with adhesive, usually applied with a notched spreader. In **strip lamination,** adhesive is applied by a special spreader in grouped parallel beads, with groups spaced 16 to 24 in c to c. In **spot lamination,** adhesive is brushed or daubed on the back of the face ply at close intervals. Partitions with strip or spot lamination provide better sound control than those with sheet lamination.

Supplemental fasteners or temporary bracing is required to ensure complete bond between face and base plies. If a fire rating is not required, temporary fasteners may be placed at 24-in intervals. For fire-rated assemblies, permanent fasteners generally are required, and spacing depends on that used in the assembly tested and rated. Nails should penetrate at least 1⅛ in into supports. With sound-deadening base plies, fastener spacing should be as recommended by the base manufacturer.

In placing face ply on base ply, joints in the two layers should be offset at least 10 in. Face ply may be applied horizontally or vertically. Horizontal application (long sides of sheet perpendicular to supports) usually provides fewer joints, but vertical application may be preferred for predecorated wallboard that is to have joints trimmed with battens.

11.26.4 Finishing Procedures for Gypsumboard

The finish surface of wallboard may come predecorated or may require decoration. Predecorated wallboard may require no further treatment other than at corners, or may need treatment of joints and concealment of fasteners. Corner and edge trim are applied for appearance and protection, and battens often are used for decorative concealment of flush joints. Other types of wallboard require preparation before decoration can be applied.

While trim can be applied to undecorated wallboard, as for predecorated panels, usually, instead, joints are made inconspicuous.

Joint-Treatment Products. Materials used for treatment of wallboard joints to make them inconspicuous should meet the requirements of ASTM C475. Application of joint treatments should conform with ASTM C840. These materials include:

Joint tape, a strip of strong paper reinforcement with feathered edges, for embedment in joint compound. Sometimes supplied with small perforations, the tape usually is about 2 in wide and $\frac{1}{16}$ in thick.

Joint compound, and adhesive, with or without fillers, for bonding and embedding the tape. Two types may be used. One, usually referred to as joint, or taping, compound, is applied as an initial coat for filling depressions at joints and fasteners and for adhering joint tape. The second, called topping, or finishing, compound, is used to conceal the tape and for final smoothing and leveling at joints and fasteners. As an alternative, an all-purpose compound that combines the features of taping and topping compounds may be used for all coats. Compounds may be supplied premixed, or may require addition of water on the job.

Flush Joints and Fasteners. For concealing fastener heads and where wallboards in the same plane meet, at least three coats of joint compound should be used. The first coat, of taping compound, should be used for adhering tape at edges and ends of the boards and to fill all depressions over fastener heads and at tapered edges. Joint tape should be centered over each joint for the length or width of the wall or panel and pressed into the compound, without wrinkling, with a tape applicator, a broad knife. Excess compound should be redistributed as a skim coat over the tape. A skim coat of joint compound applied immediately after tape embedment reduces the possibility of edge wrinkling or curling, which may result in edge cracking.

A second, or bedding, coat should be applied with topping compound at fastener heads and joints after the taping compound has dried. The waiting period may be 24 hr if regular taping compound is used, and about 3 hr if a quick-setting compound is used. This bedding coat should be thin. At joints, it should completely cover the tape and should be feathered out 1 to 2 in beyond each edge of the tape at tapered edges, and over a width of 10 to 12 in at square edges.

When the coat has dried, a second, thin coat of the topping compound should be applied for finish at fasteners and joints. At joints, it should completely cover the preceding coat and should be feathered out 1 to 2 in beyond its edges at tapered edges, and over a width of 18 in at square edges. Irregularities may be removed by light sanding after each coat has hardened. But if joint treatment is done skillfully, such sanding should not be necessary.

Reentrant Corners. Treatment of interior corners differs only slightly from that of flush joints. Joint compound should first be applied to both sides of each joint. Joint tape should be creased along its center, then embedded in the compound to form a sharp angle. After the compound has dried, finishing compound should be applied, as for flush joints.

Trim. Exposed corners and edges should be protected with a hard material. Wood or metal casings may be used for protection and to conceal the joints at door and window frames. Baseboard may be attached to protect wallboard edges at floors. Corner beads may be applied at exterior corners when inconspicuous protection is desired. Casing beads may be similarly used in other places.

At least three-coat finishing with joint compound is required to conceal fasteners and obtain a smooth surface with trim shapes.

Decoration. Before wallboard is decorated, imperfections should be repaired. This may require filling with joint compound and sanding. Joint compound should be thoroughly dry before decoration starts.

The first step should be to seal or prime the wallboard surface. (Glue size, shellac, and varnish are not suitable sealers or primers.) An emulsion sealer blocks the pores and reduces suction and temperature differences between paper and joint compound. If wallpaper is used over such a sealer, the paper can be removed later without damaging the wallboard surface and will leave a base suitable for redecorating. A good primer provides a base for paint and conceals color and surface variations. Some paints, such as high-quality latex paints, often have good sealing and priming properties.

11.27 ISOLATION AND CONTROL JOINTS IN GYPSUMBOARD CONSTRUCTION

Plaster and gypsumboard construction have low resistance to stresses induced by structural movements. They also are subject to dimensional changes because of changes in temperature and humidity. If proper provision is not made in the application of plaster and gypsumboard systems for these conditions, cracking or chipping may occur or fasteners may pop out.

Generally, such defects may be prevented by applying these systems so that movement is not restrained. Thus, gypsumboard or lath and plaster surfaces should be isolated by expansion or control joints, floating angles, or other means where a partition or ceiling abuts any structural element, except the floor, or meets a dissimilar wall or partition assembly, or other vertical penetration. Isolation also is advisable where the construction changes in the plane of the partition or ceiling. In addition, expansion or control joints should be inserted where expansion or control joints, respectively, occur in structural components of the building and along the intersection of wings of L-, U-, and T-shaped ceilings.

In long walls or partitions, control joints should be inserted not more than 30 ft apart. In large ceiling areas, control joints should not be spaced more than 30 ft apart for plaster and 50 ft apart for gypsumboard. Control joints may be conveniently located along lighting fixtures, heating vents, or air-conditioning diffusers, where continuity would ordinarily be broken. Door frames extending from floor to ceiling may serve as control joints. With shorter frames, however, control joints should be inserted from the top corners of the frames to the ceiling.

Installation of Joints. Where a control joint is to be provided, the continuity of gypsumboard or lath in plaster installations should be interrupted (Fig. 11.27). The gap for gypsumboard should be at least ½ in and for lath at least ¼ in. A framing member should be inserted on each side of the gap to support the edges of the boards or lath. (Steel studs, however, should be fastened on only one side of the joint.)

The joint should be covered at the surface with a flexible material to prevent entrance of dirt into the wall. For the purpose, plastic or noncorrodible-metal, bellows-type stops, called expansion joints or control joints, may be used (Fig. 11.27). For plaster, the flanges of the control joint should be stapled or wire tied to the lath, and the surface of the gap should be protected with tape. Then, plaster should be applied over the lath, flush to grounds. After the finished surface has been completed, the protective tape should be removed. For gypsumboard, similarly, the control-joint flanges should be stapled to the board and the gap protected with tape. Next, the joint should be treated with joint compound or veneer plaster. Then, the protective tape should be removed. Where sound control or fire ratings are important, a sealant should be installed behind the control joints.

Floating Interior Angles. A floating angle is desirable at reentrant intersections of ceilings and walls or partitions (Fig. 11.28*a*). It is constructed by installing gypsum lath or gypsumboard first in the ceiling. The first line of ceiling fasteners should be spaced as shown in Fig. 11.28*b* and *c*. Normal fastener spacing should be used in the rest of the ceiling away from the wall. Gypsum lath or gypsumboard should then be applied to the wall with firm contact at the ceiling line to support the edges of the ceiling panels. The first line of fasteners for the wall panels also should be

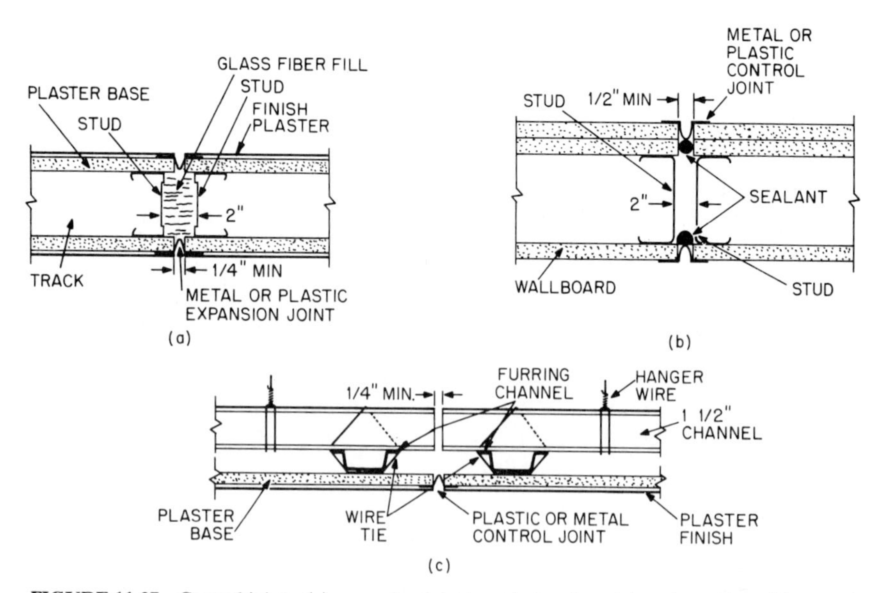

FIGURE 11.27 Control joints: (*a*) expansion joint in a plastered, metal-stud partition; (*b*) control joint in a wallboard, metal-stud wall; (*c*) control joint in a plastered, suspended ceiling.

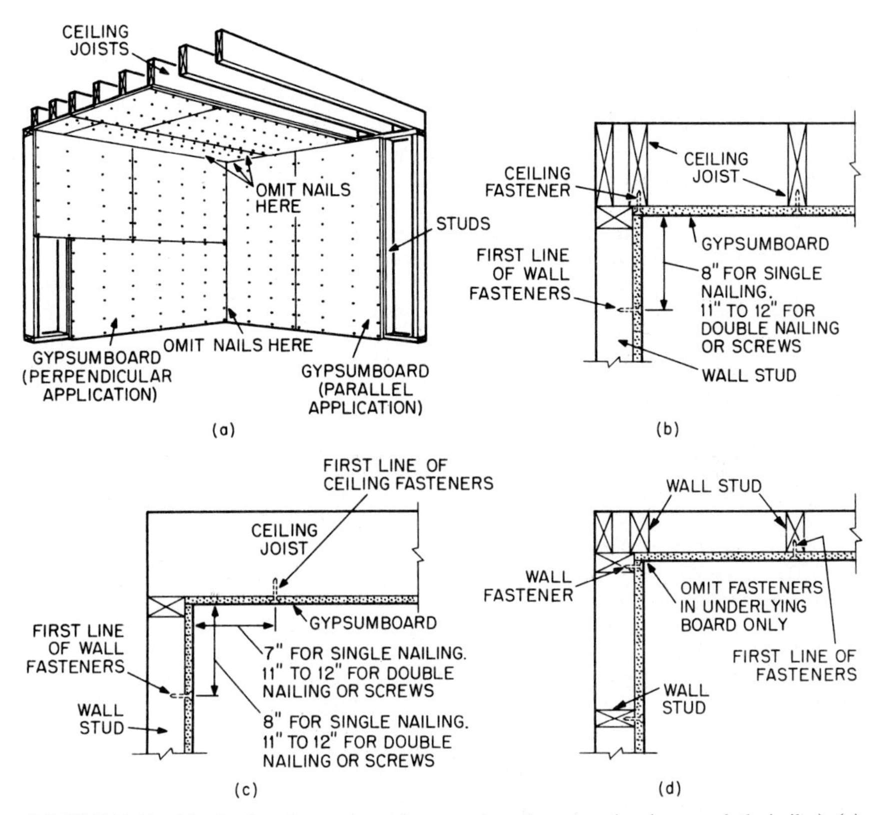

FIGURE 11.28 Floating interior angles with gypsumboard construction (gypsum lath similar): (*a*) omission of nails at intersections; (*b*) wall-ceiling intersection with ceiling boards applied parallel to joists; (*c*) wall-ceiling intersection with ceiling boards applied perpendicular to joists; (*d*) reentrant corner at walls.

spaced as shown in Fig. 11.28*b* and *c*. At wall intersections, fasteners should be omitted or inserted as indicated in Fig. 11.28*d*.

("Architect Data Book—Construction Products and Systems," Gold Bond Building Products, a National Gypsum Division, 2001 Rexford Road, Charlotte, NC 28211; "Gypsum Products Design Data," Gypsum Association, 1603 Orrington Ave., Evanston, IL 60201; "Gypsum Construction Handbook," United States Gypsum, 101 South Wacker Drive, Chicago, IL 60606.)

CERAMIC-TILE CONSTRUCTION

When a hard, incombustible, abrasion-resistant, easily cleaned finish is required for walls, ceilings, or floors, ceramic tile often is selected, although construction cost usually is higher than for other commonly used finishes. Ceramic tile is especially suitable where an attractive, water-resistant surface is desired, for instance, for shower enclosures or swimming pools.

11.28 TYPES OF CERAMIC TILE

Ceramic tile is a surfacing unit, relatively thin with respect to surface area, with a body of clay or a mixture of clay and other ceramic materials that has been fired above red heat. The tile should meet the requirements of the Recommended Standard Specification for Ceramic Tile, TCA 137.1, of the Tile Council of America, Inc., Princeton, NJ 08540.

Ceramic tile is referred to as **glazed** if it is given a surface finish composed of ceramic materials that have been fused into the body of the tile and that may be impervious, vitreous, semivitreous, or nonvitreous. [Glazed tiles tend to be slippery, especially when wet. If this characteristic is undesirable (for example, for floors), then unglazed tile, which has a homogenous composition, or tile incorporating a dispersed abrasive should be used.] **Impervious tile** is required to have water absorption of less than 0.5% when tested as described in ASTM C373. **Vitreous tile** is required to have water absorption between 0.5 and 3%.

Ceramic tile is classified as wall, ceramic mosaic, quarry, or paver tile. Mosaic, paver, and quarry tile are generally used as floor coverings.

Wall tile is glazed and usually nonvitreous (water absorption greater than 3% but less than 18%). It has a body that is suitable for interior applications, although it is not expected to withstand impact. Flat wall tile is nominally 5⁄16 in thick. Rectangular in shape, it generally comes in nominal sizes of 4¼ × 4¼, 6 × 4¼, and 6 × 6 in. Also, trim units are available with various shapes necessary for completing an installation. These units include bases, coves, caps, curbs, angles, and moldings.

Ceramic mosaic tile is made of porcelain or natural clay. It may be glazed or unglazed, and if glazed, may be either plain or contain a dispersed abrasive. Nominal thickness is ¼ in. Surface area is less than 6 in^2. Available in rectangular, hexagonal, and other shapes, flat mosaic tiles come in nominal sizes 1 × 1, 2 × 1, and 2 × 2 in. The tiles usually are mounted on sheets 1 ft by 2 ft, to facilitate tile setting.

Quarry tile is made by extrusion from natural clay or shale and is unglazed. It usually is ½ to ¾ in thick. Surface area generally exceeds 6 in^2.

Paver tile is similar to ceramic mosaic tile but is thicker, larger, and unglazed. Thickness usually is ⅜ or ½ in. Nominal sizes generally are 4 × 4, 6 × 6, and 8 × 4 in.

11.29 TILE INSTALLATION METHODS

Installation of ceramic tile should be in accordance with the appropriate specification in the A108 series of the American National Standards Institute, depending on the type of setting compound used. Specifications for materials commonly used for setting and grouting the tile are given in the series A118. These materials include portland-cement mortar, dry-set or latex portland-cement mortar, organic adhesive, and chemical-resistant, water-cleanable, tile-setting and grouting epoxy.

Tiles may be set into plastic mortar or adhesive or may be bonded to hardened mortar. (To ensure good alignment, the tile may be preassembled in groups, lightly adhered to sheets.) When the bond is sufficiently strong to prevent the tile from moving, the joints between tiles should be grouted. (Mounted tiles also are available pregrouted, usually with an elastomeric material.) The tiles may be face mounted, back mounted, or edge bonded. Face-mounted tiles are stuck face down on paper or other suitable material with a water-soluble adhesive so that the sheet can be removed easily after tile setting, before the joints are grouted. For back-mounted tiles, the back side, and for edge-bonded tiles, the edges, are adhered to perforated paper, fiber mesh, resin, or other suitable material, which remains in place after tile setting.

Quality and cost of tile installations are affected by the stability, permanence, and precision of installation of the backing or base material on which the tiles are set. The backing should be sound, clean, and dimensionally stable. Deflection of the supporting members under total load should not exceed $1/360$ of the span.

While the tile may not deteriorate when exposed to moisture, the backing may. For installations exposed to wetting, therefore, the backing should be concrete, masonry, or portland-cement mortar, or else a cleavage, or water-resistant, membrane should be inserted to protect the backing.

11.29.1 Backing for Tiles

When portland-cement mortar is to be applied, for tile setting, to a gypsum-product backing, a membrane and metal lath should be applied first to cover the backing. In contrast, sound concrete or masonry may have tile directly applied, after the surface has been prepared by sandblasting, chipping, or scarifying to expose an uncontaminated surface. (Without such preparation, use of metal lath is desirable.) For installations with organic adhesive or dry-set or latex portland-cement mortar, which are applied as a thin coating, it is especially desirable that the backing used be dimensionally stable and have negligible variation from a plane.

In all types of installation, control joints or other effective means of preventing cracking should be placed at appropriate intervals.

Exterior walls should be constructed to prevent moisture from collecting behind the tile. For the purpose, flashing, coping, and vapor barriers should be installed, and weep holes draining to the outside should be provided at the base of the exterior walls (see Art. 11.3.7).

Crack-free concrete floors with a screed finish may have portland-cement mortar directly bonded to them as a backing for tile. Concrete floors with a steel-troweled finish, including precast concrete, should be covered with a cleavage membrane before placement of the mortar, and steel reinforcement should be incorporated in the mortar bed. When mortar or adhesive is the type that is applied as a thin coating, the backing should have negligible variations from a plane, or else, a portland-cement mortar bed at least 1¼ in thick should be placed atop the concrete slab.

Metal Lath. Installation of metal lath for reinforcing a mortar bed for tile should be in accordance with the Standard Specification for Installation of Interior Lathing and Furring. ASTM C841. The lath should be galvanized for protection against corrosion. It should be the flat-expanded type and weigh at least 2.5 lb/yd^2.

For application of metal lath to wood or steel studs or to wood furring, a cleavage membrane—15-lb roofing felt or 4-mil polyethylene film—should be attached first, with joints lapped at least 2 in. Then, the lath should be fastened in place on the side to be mortared (Fig. 11.29*i*). (See also Art. 11.25.6.) Where sheets of lath are joined, the lath should be lapped at least 2 in at sides and ends. (If wire fabric is used, it should be lapped at least one full mesh.)

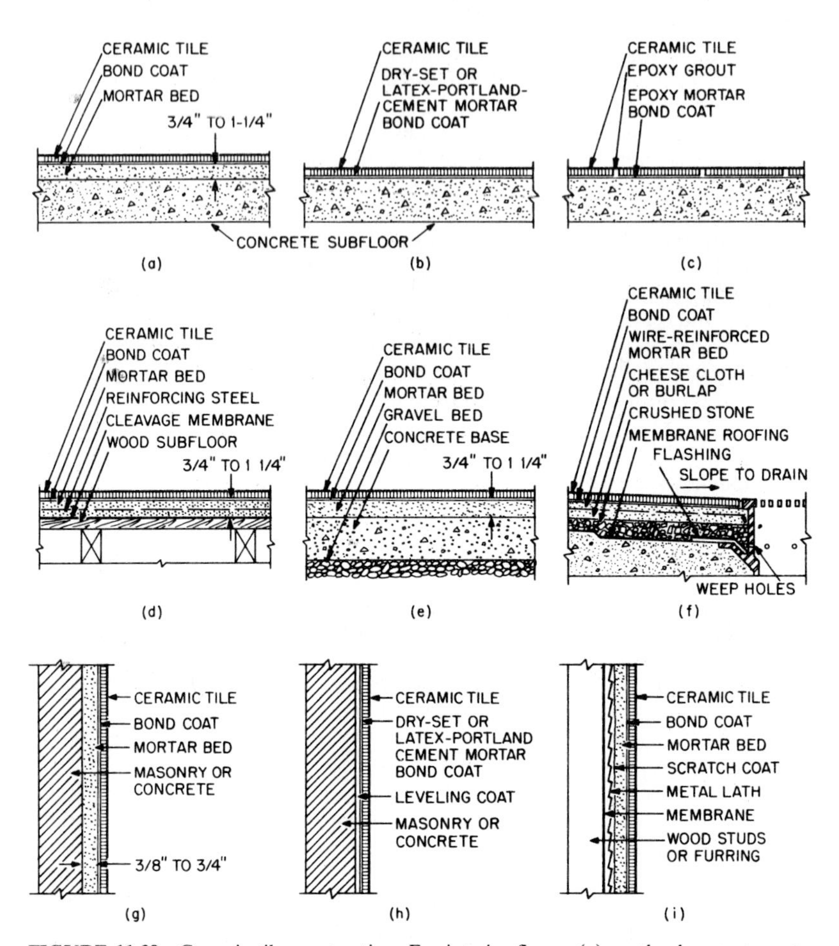

FIGURE 11.29 Ceramic tile construction. For interior floors: (*a*) portland-cement mortar bed on a concrete subfloor; (*b*) bond coat of dry set or latex portland-cement mortar on a concrete subfloor; (*c*) epoxy bond coat on a concrete subfloor; (*d*) reinforced mortar bed on a wood subfloor. For an outdoor walkway: (*e*) portland-cement mortar bed on a concrete base. For roofs: (*f*) reinforced mortar bed on crushed stone. For interior walls: (*g*) cement mortar on masonry or concrete; (*h*) dry set or latex portland-cement mortar on masonry or concrete; (*i*) cement mortar on metal lath attached to studs or furring.

11.29.2 Scratch and Leveling Coats

Over rough-surfaced backings, a level surface to receive tile may be built up with portland-cement mortar. For walls and ceilings, the mortar mix for the scratch and leveling coats for application to metal lath, concrete, or masonry should be 1 part portland cement and 3 parts dry sand or 4 parts damp sand, by volume. The ingredients should be mixed dry. Then, sufficient water should be added to make a stiff mix. The scratch coat should be cured for at least 24 hr.

A leveling coat is desirable when the surface of the scratch coat varies more than ¼ in in 8 ft from the required plane or when the required mortar-bed thickness exceeds ¾ in. As for the scratch coat, the leveling coat should be scratched and cured before the next bond coat is applied.

11.29.3 Mortars and Adhesives

For setting ceramic tile, portland-cement mortar is the only type recommended by the Tile Council of America for a thick bed. Other mortars and adhesives should be used only for thin beds. Setting materials such as epoxies and furans that do not contain portland cement offer special properties but are more expensive than cement-based mortars.

Portland-Cement Mortar. This is suitable for most surfaces and ordinary installations. For floors, the mix may be 1 part portland cement and about 6 parts sand. For walls, about ½ part of hydrated lime may be added to the floor mix. The bed may be ¾ to 1¼ in thick. Tiles may be set into the bed while the mortar is still plastic (ANSI A108.1). Or, after the mortar has cured, the tiles may be bonded to the mortar bed with a 1/16-in-thick coat of dry-set or latex portland-cement mortar or with neat cement. If a neat-cement bond coat is used and the tile is absorptive, it should be soaked before being set. (See Fig. 11.29*a, d, e, f, g,* and *i.*)

Dry-Set Mortar. This is a mix of portland cement and sand, with additives that impart water retentivity. The mortar should meet the requirements of ANSI A118.1 or A118.2. Applied as a coat as thin as 3/32 in, the mortar is suitable for use on stable backings with true surfaces (Fig. 11.29*b* and *h*). Tile installation should be in accordance with ANSI A108.5 or A108.7.

Latex Portland-Cement Mortar. This is a mix containing portand cement, sand, and a latex additive, prepared to meet requirements of ANSI A118.4. Tile applications and installation method with this mortar are similar to those for dry-set mortar and should be in accordance with ANSI A108.5. (See Fig. 11.29*b* and *h.*) Early exposure of latex mortar to water inhibits development of full strength and increases sensitivity to water. Hence, for applications such as swimming pools and showers, the mortar should be cured at least 14 days and allowed to dry thoroughly before exposure to water.

Organic Adhesive. This is an organic material that cures or sets by evaporation. It should meet the requirements of ANSI A136.1. Applied in a thin layer with a notched trowel, organic adhesive is suitable for use on stable backings with true surfaces, but not for swimming pools or exterior applications. Tile installation should be in accordance with ANSI A108.4.

Epoxy Mortar. Suitable for use where chemical resistance, high bond strength, or high impact resistance is needed, this mortar contains epoxy resin and hardener and meets requirements of ANSI A118.3. It should be applied as a thin coat in accordance with ANSI A108.6 (Fig. 11.29*c*).

Furan Mortar. Suitable for use where chemical resistance is important, this mortar consists of furan resin and hardener. It should be used in accordance with the manufacturer's recommendations.

Grouting Materials for Tile. After tile has been set, the joints should be grouted. Most grouts have a portland-cement base but are modified to provide specific qualities, such as whiteness, uniformity, hardness, flexibility, or water retentivity. Damp curing of such grouts for about 72 hr is required.

A grout mix may be prepared in the field with portland cement and a fine graded sand, with up to ⅕ part lime, if desired. The ratio of portland cement to sand, by volume, may be as follows: 1:1 for joints up to ⅛ in wide; 1:2 for joints up to ½ in wide, and 1:3 for joints wider than ½ in. Prepared cement-based mixes include commercial portland-cement grout and dry-set grout. A latex portland-cement grout may be prepared by adding latex to any of the preceding mixes and offers the advantages of lower permeability and no curing, but it is less rigid than regular cement grout.

Grouts also are available with mastic, furan, epoxy, or silicone rubber, to provide such properties as flexibility, stain resistance, high impact resistance, and capacity to withstand below-freezing temperatures and hot, humid conditions. Such grouts, however, are more expensive than the cement-based ones.

11.29.4 Expansion Joints

These generally are required where tiles abut restraints, such as walls, curbs, columns, or pipes, and where joints occur in structural backings, such as concrete floors and walls. For interior tile floors covering large areas, expansion joints generally should be inserted every 24 to 36 ft if the mortars are cement based. For roofs and outdoor floors, joint spacing may range from 12 to 16 ft. For interior walls, expansion joints are generally required directly over masonry control joints, where there are changes in materials, or every 24 to 36 ft.

Joint width should be at least 4 times any expected movement. The width should be at least ⅜ in for exterior joints that are 12 ft apart and ½ in for those 16 ft apart. These widths should be increased 1/16 in for each 15°F that the actual temperature range may exceed 100°F between summer high and winter low temperature. For interior joints, width should be at least ¼ in. In any case, if there is a structural joint in the backing, the tile expansion joint should be at least as wide as the structural joint.

The joint sealant should be a nonsagging type in vertical joints and a self-leveling type in horizontal joints. It should be inserted to a depth of not less than ⅛ in or more than one-half the joint width. The depth should be controlled by insertion of a backup strip, which should be flexible and compressible, with a rounded surface where it contacts the sealant. Closed-cell foamed polyethylene or butyl rubber is a suitable material for the backup.

("Handbook for Ceramic Tile," Tile Council of America, Inc., Princeton, NJ 08540.)

PANEL FINISHES

For speedy installation of interior finishes on walls or ceilings, rigid or semirigid boards, some of which require no additional decorative treatment, may be nailed directly to studs or masonry or to furring. For acoustic materials, see Arts. 11.79 to 11.82. Gypsumboard is discussed in Art. 11.26.

Joints between panels may be concealed or accentuated according to the architectural treatment desired. The boards may interlock with each other, or battens, moldings, or beads may be applied at joints.

When the boards are thinner than conventional lath and plaster, framing members may have to be furred out to conventional thicknesses unless stock doors and windows intended for use with panels can be used.

11.30 PLYWOOD FINISHES

Available with a large variety of surface veneers, including plastics, plywood is nailed directly to framing members or furring. Small finish nails, such as 4d, should be used, with a spacing not exceeding 6 in. All edges should be nailed down, intermediate blocking being inserted if necessary. The plywood should also be nailed to intermediate members. When joints are not to be covered, the nailheads should be driven below the surface (set). Under wood battens, ordinary flat-headed nails can be used. Water-resistant adhesive may be applied between plywood and framing members for additional rigidity. (See also, Art. 10.12.)

11.31 OTHER TYPES OF PANEL FINISHES

Like other types of panels, mineral-fiber panels can be nailed directly to framing members or furring. Thin sheets usually are backed with plywood or insulation boards to increase resistance to impact. Joints generally are covered with moldings or beads.

Fabricated with a wide variety of surface effects, fiber and pulp boards also are available with high acoustic and thermal insulation values. Application is similar to that for plywood. Generally, adjoining boards should be placed in moderate contact with each other, not forced.

FLOOR SYSTEMS

For proper selection of floor systems, designers should take into account many factors, including use of lightweight-concrete slabs, subfloors in direct contact with the ground, radiant heating, air conditioning, possible necessity for decontamination, dustlessness, traffic loads, and maintenance costs—all of which have an important bearing on floor selection. Consideration should be given to current standards of styling, comfort, color, and quietness.

The primary consideration of the designer of a flooring system is to select a floor covering that can meet the maximum standards at reasonable cost. To avoid the dissatisfaction that would arise from failure to select the proper flooring, designers must consider all the factors relevant to flooring selection.

This section contains information that can provide a guide toward this end. It summarizes the characteristics of the major types of floor coverings and describes briefly methods for the proper installation of these materials. (For ceramic-tile installations, see Arts. 11.28 and 11.29.)

11.32 ASPHALT TILES

These tiles are intended for use on rigid subfloors, such as smooth-finished or screeded concrete, structurally sound plywood, or hardboard floors not subject to excessive dimensional changes or flexing. The tiles can be satisfactorily installed on below-grade concrete subject to slight moisture from the ground.

Low cost and large selection of colors and designs make asphalt tile an economically desirable flooring.

Asphalt tile is composed of mineral fibers, mineral coloring pigments, and inert fillers bound together. For dark colors the binder is Gilsonite asphalt; for intermediate and light colors, the binder may consist of resins of the cumarone indene type or of those produced from petroleum. Tiles most commonly used are 9 × 9 in and ⅛ in thick.

Colors are classified into groups A, B, C, and D, graded from black and dark red (A) to cream, white, yellow, blue, and bright red (D). Cost is generally lower for the darker colors.

To avoid permanent indentations in asphalt tiles, contact surfaces of furniture or equipment should be smooth and flat to distribute the weight. This is particularly necessary for installations over radiant-heated floors and on areas near windows exposed to sun.

Never use on asphalt tiles waxes containing benzene, turpentine, or naphtha-type solvents and free fats or oils. Avoid strong detergents or cleaning compounds containing abrasives or preparations not readily soluble in water. These may soften the tiles and cause colors to bleed. Grease, oils, fats, vinegar, and fruit juices allowed to remain in contact with asphalt tiles will stain and soften them. Because of these restrictions, asphalt tiles are not recommended for use in kitchens or bathrooms.

See also Art. 11.36.

11.33 CORK TILES

Cork flooring is intended for use on rigid subfloors, such as smooth-finished or screeded concrete supported above grade and free of moisture, or on structurally sound plywood or hardboard. Cork tile is not recommended for application below grade. When it is installed at grade, moisture-free conditions must be ensured.

Cork tile is manufactured by baking cork granules with phenolic or other resin binders under pressure. Four types of finishes are produced: natural, factory-prefinished wax, resin-reinforced wax, and vinyl cork tile (Art. 14.8). The tiles are generally 6 × 6, 6 × 12, 9 × 9, 12 × 12, 12 × 24, or 36 × 36 in and ⅛, 3⁄16, 5⁄16, or ½ in thick.

Natural cork tile must be sanded (to level), sealed, and waxed immediately after installation.

Unless the exposed surface of cork floors is maintained with sealers and protective coatings, permanent stains from spillage and excessive soiling by heavy traffic will result.

Cork tiles are particularly suitable for areas where quiet and comfort are of paramount importance.

See also Art. 11.36.

11.34 VINYL FLOORING

Flooring of this type is unbacked. For a discussion of backed vinyl, see below. It is intended for use on rigid subfloors, such as smooth-finished or screeded concrete supported above grade, or structurally sound plywood or hardboard floors. Vinyl floors are not recommended for use below grade. They must be applied with an alkaline, moisture-resistant adhesive when used at grade.

Vinyl mats or runners may be laid without adhesive over relatively smooth surfaces. Large mats generally are installed in a recess in the concrete floor at building entrances. The mats are ribbed or perforated for drainage.

Vinyl flooring consists predominantly of polyvinyl chloride resin as a binder, plasticizers, stabilizers, extenders, inert fillers, and coloring pigments. Because of its unlimited color possibilities and opaqueness to transparent effects, it is widely used. Common thicknesses are 0.080, 3/32, and 1/8 in.

Since vinyl resins are tough synthetic polymers, vinyl flooring can withstand heavy loads without indentation, and yet is resilient and comfortable under foot. It is practically unaffected by grease, fat, oils, household cleaners, or solvents. But unless given a protective finish, it is easily scratched and scuffed.

See also Art. 11.36.

Backed Vinyl. The family of backed-vinyl flooring comprises vinyl wearing surfaces from 0.02 to 0.050 in thick, laminated to many different backing materials. In some products, the vinyl surfaces are unfilled transparent films placed over a design on paper, cork, or degraded vinyl. Filled vinyl surfaces with a 34% vinyl resin binder are placed over plastic composition backing or asphalt-saturated or resin-impregnated felt. The asphalt-felt type may be used in moist areas. Foamed rubber or plastic is incorporated in some of these materials to increase comfort and decrease impact noise.

Asphalt-felt-backed vinyl materials may be applied with a moisture-resistant adhesive on concrete at or below grade.

See also Art. 11.36.

11.35 RUBBER FLOORING

Rubber flooring is intended for use on rigid subfloors, such as smooth-finished or screeded concrete supported above grade, or on structurally sound plywood or hardwood subfloors. Rubber is not recommended for use below grade. When used at grade, it must be applied with an alkaline, moisture-resistant adhesive.

Rubber mats or runners may be laid without adhesive over relatively smooth surfaces. Large mats generally are installed in a recess in the concrete floor at building entrances. The mats are ribbed or perforated for drainage.

Most rubber flooring is produced from styrene-butadiene rubber. Reclaimed rubber is added to some floorings. The flooring also contains mineral pigments and mineral fillers, such as zinc oxide, magnesium oxide, and various clays. Another synthetic-rubber flooring, chlorosulfonated polyethylene (Hypalon), also is available.

Rubber floorings can be obtained in thicknesses of 3/32, 1/8, or 3/16 in. They have excellent resistance to permanent deformation under load. Yet they are resilient and quiet under foot.

See also Art. 11.36.

11.36 INSTALLATION OF THIN COVERINGS

Most manufacturers and trade associations make available instructions and specifications for installation and maintenance of their floorings.

The most important requirement for a satisfactory installation of a thin floor covering is a dry, even, rigid, and clean subfloor.

Protection from moisture is a prime consideration in applying a flooring over concrete. Moisture within a concrete slab must be brought to a low level before installation begins. Moisture barriers, such as 6-mil polyethylene, 55-lb asphalt-saturated and coated roofing felt, or 1/32-in butyl rubber, should be placed under concrete slabs at or below grade, and a minimum of 30 days' (90 in some cases) drying time should be allowed after placement of concrete before installation of the flooring.

Particular care should be taken with installations on lightweight concrete. It always has a higher gross water requirement than ordinary concrete and therefore takes longer to dry. So a longer drying period should be allowed before installation of flooring. Flooring manufacturers provide advice and sometimes also equipment to test for moisture.

As indication of dryness at any given time is no assurance that a concrete slab at or below grade will always remain dry. Therefore, protection from moisture from external sources must be given considerable attention. (See also Art. 3.4.)

Some concrete curing agents containing oils and waxes may cause trouble with adhesive-applied flooring. Commercial curing agents that have been shown to be satisfactory include styrene-butadiene copolymer or petroleum-hydrocarbon resin dissolved in a hydrocarbon solvent. These products dry to a hard film in 24 hr. Parting agents, used as slab separators in lift-slab and tiltup construction, may cause bond failure of adhesive-applied flooring if they contain nonvolatile oils or waxes. Concrete surfaces that have been treated with or have come in contact with oils, kerosene, or waxes must be cleaned as recommended by the adhesive manufacturer.

All concrete surfaces to receive adhesive-applied, thin flooring must be smooth. Also, they should be free from serious irregularities that would "telegraph" through the covering and be detrimental to appearance and serviceability. For rough or uneven concrete floors, a troweled-on underlayment of rubber latex composition or asphalt mastic is recommended. It can be applied from a thickness of 1/4 in to a featheredge. Small holes, cracks, and crevices may be filled with a reliable cement crack filler.

Wood subfloors, of sufficient strength to carry intended loads without deflection, should be covered with plywood or hardboard underlayments. These should be nailed 6 in c to c in perpendicular directions over the entire area with ringed or barbed nails. The nails should be driven flush with the underlayment surface without denting it around the nail head.

Adhesives commonly used to attach flooring to concrete and underlayment are listed in Table 11.12.

11.37 CARPETS

Extending from wall to wall, carpets are frequently used as floor coverings in residences, offices, and retail stores. They are often selected for the purpose because they offer foot comfort and, being available in many colors, patterns, and textures, attractive appearance. Rugs, often used as an alternative in residences, differ from carpets chiefly in being single pieces of definite shape and usually not covering an entire floor between walls.

A carpet is a thick, heavy fabric that is usually piled but could be woven or felted. Pile consists of closely placed loops of fiber, or tufts, that produce a raised

TABLE 11.12 Adhesives Used to Install Flooring*

Flooring	Concrete below grade	Concrete on grade	Concrete above grade	Plywood or hardboard
Asphalt and vinyl-asbestos tiles	Asphalt, cutback Asphalt, emulsion	Asphalt, cutback Asphalt, emulsion	Asphalt, cutback Asphalt, emulsion	Asphalt, cutback Asphalt, emulsion
Rubber and vinyl	Chemical set Latex	Latex	Latex	Latex
Linoleum, cork, and vinyl backed with felt or cork	Do not install	Do not install	Linoleum paste	Linoleum paste
Vinyl backed with asbestos felt	Latex	Latex	Latex Linoleum paste	Linoleum paste
Laminated wood block	Do not install	Asphalt, hot melt Asphalt, cutback Rubber base	Asphalt, hot melt Asphalt, cutback Rubber base	Asphalt, hot melt Asphalt, cutback Rubber base
Solid unit wood block	Do not install	Asphalt, hot melt Asphalt, cutback	Asphalt, hot melt Asphalt, cutback	Asphalt, hot melt Asphalt, cutback

*The adhesives listed are intended for each type of flooring; i.e., asphalt adhesives used for asphalt and vinyl tiles are not the same as asphalt adhesives used for wood blocks. There are a number of adhesives having special properties and characteristics, such as heat resistance, resistance to water spillage, and brush-on types not included in this table.

surface on a backing to which they are locked. The tufts may be sheared to produce a soft, velvety surface with a wide variety of patterns and textures.

Sheared or unsheared, the piled fabric is very resilient, thus contributing to foot comfort. Nevertheless, thin pads, generally of foam rubber or plastic, often are placed under carpets, to improve resilience. They offer the additional advantage of absorbing minor irregularities in the floor surface that otherwise would cause rapid wear in local areas.

Fibers used for tufting indoor carpets include wool, acrylic, polyester, continuous-filament or heat-set spun nylon and nylon with antistatic treatment for high resistance to soiling. Pile weight generally ranges between 15 and 40 oz/yd^2. The primary backing, to which the pile is tufted and which provides dimensional stability to the carpet, usually is polypropylene weighing 3.5 oz/yd^2. A secondary backing, which generally is attached to the primary backing with a latex adhesive, is used to protect the underside of the carpet and improve other characteristics, such as resilience and noise absorption. The secondary backing usually is woven jute weighing 7 oz/yd^2, nonwoven polypropylene weighing 3.5 oz/yd^2, or $^3/_{16}$-in-thick, high-density foam rubber weighing 38 oz/yd^2, which eliminates the need for an underlying pad.

In selection of carpeting, consideration should be given to the intensity of traffic to which the covering will be subjected; availability of desired colors, patterns, and textures; colorfastness; resistance to crushing and matting; soil resistance; cleanability; resistance to fuzzing, beading, and pilling, as measured by bundle wrap and latex penetration on the underside of the primary backing; subfloor conditions;

and installed cost of the carpet. Colorfastness may be judged by performance in an 80-hr xenon-arc fadeometer test and in wet-method cleaning in accordance with Federal Specification DDD-C-001559 and by negligible crocking (color transfer to a white cloth). Preference should be given to carpets that meet the following requirements:

Flammability. Flame spread rating (ASTM E84) of 75 or less; flame propagation index (floor chamber test) of 4.0 or less (see Underwriters Laboratories Subject 992); and corrected maximum specific optical density of 450 or less in the National Institute of Standards and Technology smoke-density-chamber test.

Static propensity. 3.0 kV or less.

Acoustical Properties. Impact noise rating of 30 or more; impact insulation class 81 or more; NC 30 masking level of 30 or less; and noise reduction coefficient, measuring airborne sound absorption, of 0.5 or more.

Tuft Bind. Average force of 10 lb or more required to pull out tufts from the face of the carpet (ASTM D1335).

Installation of Carpet. Carpet is supplied usually in widths of 12 or 15 ft in rolls in long lengths. It may be cut to desired sizes and shapes with a carpet knife. Strips of carpet are laid side by side to extend the covering from wall to wall. Joints may be stitched or taped.

Installation should conform to recommendations of the carpet manufacturer. In general, carpet may be laid on any firm, smooth floor.

Carpets with jute or synthetic secondary backing generally should be stretched over a good-quality pad, to eliminate bulges, and anchored at the walls with tacks or tackless strips. With this type of installation, carpets may be removed easily when replacement is necessary. However, they must still be cleaned in place, may require restretching, and can be difficult to repair. (Power stretchers should be used for carpets with synthetic secondary backing.) Alternatively, this type of carpet may be directly cemented to subfloors, eliminating an underlying pad and future restretching. But wearability may be lower and there may be a greater tendency to soil under heavy traffic.

Carpets with high-density foam-rubber backing also may be cemented directly to subfloors. Such carpets, however, are not suitable for carrying heavy traffic and may be difficult to remove when replacement is necessary.

In all cases, use of chair pads under castered chairs is desirable.

11.38 TERRAZZO

A Venetian marble mosaic, with portland cement matrix, terrazzo is composed of two parts marble chips to one part portland cement. Color pigments may be added. Three methods of casting in place portland-cement terrazzo atop structural concrete floor slabs are commonly used: sand cushion, bonded, and monolithic.

Sand-cushion (floating) terrazzo is used where structural movement that might injure the topping is anticipated from settlement, expansion, contraction, or vibration. This topping is at least 3 in thick. First, the underlying concrete slab is covered with a ¼- to ½-in bed of dry sand. Over this is laid a membrane, then wire-fabric reinforcing. The terrazzo underbed is installed to ⅝ in below the finished floor line. Next, divider strips are placed and finally, the terrazzo topping.

Bonded terrazzo has a minimum thickness of 1¾ in. After the underlying concrete slab has been thoroughly cleaned and soaked with water, the surface is slushed with neat portland cement to ensure a good bond with the terrazzo. Then, the underbed is laid, divider strips are installed, and terrazzo is placed.

Monolithic terrazzo is constructed by placing a ⅝-in topping as an integral part of a green-concrete slab. Adhesive-bonded monolithic terrazzo with an epoxy resin adhesive also has been used successfully, with a topping thickness of only ⅜ in.

Terrazzo may be precast. It generally is used in this form for treads, risers, platforms, and stringers on stairs.

Because of the large variety of color and surface textures that can be attained with terrazzo, it is used extensively as an exterior and interior decorative flooring. Portland-cement terrazzo, however, should not be used in areas subject to spillage, such as might be encountered in kitchens.

Details on selecting the proper type of terrazzo, marble-chip sizes, methods of applying, and finishing of the surface may be obtained from the National Terrazzo and Mosaic Association, 2-A West Loudon St., Leesburg, VA 22075. Also, specification sheets for terrazzo are published by the American Institute of Architects.

Other matrix materials used with marble chips include rubber latex, epoxy, and polyesters. Suppliers should be consulted for installation details.

11.39 CONCRETE FLOORS

A concrete topping may be applied to a concrete structural slab before or after the base slab has hardened. Integral toppings usually are ½ in thick; independent toppings about 1 in.

It is well to remember in specifying or installing a concrete floor that the difference between good and poor concrete lies in selection and grading of aggregates, proportioning of the mix, and the care with which the vital operations of placing, finishing, and curing are carried out. (See also Sec. 9.)

(ACI 302.1, "Guide for Concrete Floor and Slab Construction," American Concrete Institute.)

11.40 WOOD FLOORS

Both hardwoods and softwoods are used for floors. Hardwoods most commonly used are maple, beech, birch, oak, and pecan. Softwoods are yellow pine, Douglas fir, and western hemlock. The hardwoods are more resistant to wear and indentation than softwoods.

Hardwood strip floorings are available in thicknesses of $^{11}/_{32}$, $^{15}/_{32}$, and $^{25}/_{32}$ in and in widths of 1½, 2, 2¼, and 3¼ in.

Softwood strip flooring usually is $^{25}/_{32}$ in thick and can be obtained in widths of 2⅜, 3¼, and 5$^{3}/_{16}$ in.

Solid-unit wood blocks for floors are made from two or more units of stripwood flooring fastened together with metal splines or other suitable devices. A block usually is square. Tongued and grooved, either on opposite or adjacent sides, it is held in place with nails or an asphalt adhesive (Table 11.12). Blocks $^{25}/_{32}$ in thick are made up of multiples of 1½- and 2¼-in strips; blocks ½ in thick are composed of multiples of 1½- or 2-in strips.

A laminated block is formed with plywood comprising three or more plies of wood glued together. The core or cross bonds are laid perpendicular to the face and back of the block. Usually square, the block is tongued and grooved on either opposite or adjacent sides. The most common thickness is ½ in, but other thicknesses used are ⅜, 7⁄16, ⅝, and 13⁄16 in. Laminated blocks are installed with adhesives (Table 11.12).

Average moisture content of wood flooring at time of installation should be 6% in dry southwestern states, 10% in damp southern coastal states, and 7% in the rest of the United States ("Moisture Content of Wood in Use," U.S. Forest Products Laboratory Publication No. 1655, Madison, Wis.). The U.S. Forest Products Laboratory also recommends heating the building before installing any type of wood flooring, except when outdoor temperatures are high. Bundles of flooring should be opened and stored in the building before installation. Leave at least 1 in of expansion space at walls and columns.

Specifications for wood floors on concrete, gymnasium floors, or other special designs or conditions may be obtained from associations of wood flooring manufacturers and installers.

11.41 INDUSTRIAL FLOORS

The primary purpose of a floor covering in an industrial building is to protect the structural floor from foot and truck traffic and from corrosive affluents. The flooring also must provide a durable surface that will not be detrimental to plant operations.

A good-quality concrete floor is adequate if factory conditions are dry and the surface has to withstand only foot and truck traffic. A cast-iron grid filled with concrete on ¼-in-thick steel plates may be used in areas where heavy equipment or materials are dragged continually. End-grain wood blocks may be used in areas of heavy trucking of heavy castings, to protect both the structure and castings from impact damage. Asphalt mastic floors also are used in areas with heavy truck traffic.

If the factory process is wet and corrosive, the designer may have to choose between a less-resistant flooring that has to be replaced periodically and a flooring with low maintenance costs but high initial cost. Table 11.13 rates the relative resistance to liquids of some floor toppings and bedding and jointing materials for clay tiles and bricks. In addition to selecting the most suitable materials, the designer should provide adequate slope and drainage, a liquid-tight layer below the finished floor as a second line of defense, and a nonslip surface. Selection and laying of chemical-resistant materials are a specialized operation on which specialists should be consulted.

11.42 CONDUCTIVE FLOORING

Where explosive vapors are present, sparks resulting from accumulation of static electricity constitute a hazard. The most effective of several possible ways of mitigating this hazard is to keep electrical resistance so low that dangerous voltages never are attained.

Most objects normally rest or move on the floor and therefore can be electrically connected through the floor. Flooring of sufficiently low electrical resistance (con-

TABLE 11.13 Resistance of Floor Toppings and of Bedding and Jointing Material to Various Liquids*

Material	Water	Organic solvents	Oils	Acids	Alkalies
Floor topping:					
Portland-cement concrete	VG	G	F	VP	G
High-alumina cement concrete	G	G	F–G	VP	F
Pitch (coal tar) mastic	VG	F	F	F–G	G
Asphalt mastic	VG	VP	P	F–G	G
Epoxy	VG	G	G	G	G
Polyester	VG	G	G	G	F
Bedding and jointing materials for clay tiles and bricks:					
Portland-cement mortar			As above for concrete		
High-alumina cement mortar			As above for concrete		
Silicate cement	F	G	G	G	VP
Sulfur cement	VG	G	F	G	G
Epoxy			As above		
Furan	VG	VG	VG	VG	VG

VG = very good; G = good; F = fair; P = poor; VP = very poor.

*This table is intended for broad comparisons, as the items under various headings may need qualifications for particular conditions, such as temperature, and concentration of the liquids and acids that are oxidizing agents.

ductive flooring) thus is of paramount importance in elimination of electrostatic hazards. But the electrical resistance must be high enough to eliminate electric shock from faulty electrical wiring or equipment.

Electrical conductivity of flooring is improved by addition of acetylene black (carbon). In ceramic, rubber, and vinyl floors, the carbon is finely dispersed in the material during manufacture; in latex terrazzo, concrete terrazzo, and setting-bed cement for ceramic tile, the carbon is uniformly dispersed in the dry powder mixes, placed in containers, and shipped for on-the-job composition ("Conductive Flooring for Hospital Operating Rooms," Monograph 11, National Bureau of Standards, Washington, DC 20234).

For linoleum flooring, brass seam connectors with projecting points are used for electrical intercoupling between sheets. For vinyl tiles, copper foil is placed between the adhesive and tile for electrical intercoupling.

When flammable gases are in use, everyone and everything must be electrically intercoupled via a static conductive floor at all times. To ensure this, constant vigilance, inspection by testing, and a high standard of housekeeping are required. Wax or dirt accumulation on the floor and grounding devices can provide high resistance to electrical flow and cancel the effect of conductive flooring.

11.43 SPECIFICATIONS AND STANDARDS FOR FLOORING

Federal specifications are written to establish minimum standards for purposes of competitive bidding and to ensure that government agencies will receive quality material. Some flooring materials, however, cannot be classed under these specifications because only one manufacturer makes the products. Prices of the federal specifications may be obtained from the General Services Administration, Washington, DC 20405, or the nearest GSA regional office.

Commercial Standards are specifications that establish quality levels for manufacturerd products in accordance with the principal demands of the trade. The Office of Commodity Standards, National Bureau of Standards, Washington, DC 20234, will supply a price list.

WINDOWS

Selection of a proper window for a building entails many considerations. Functional requirements are usually most important. However, size, type, arrangement, and materials very often establish the character of the building, and this external show of fenestration is an integral part of design and expression.

11.44 WINDOW SELECTION

Distribution and control of daylight, desired vision of outdoors, privacy, ventilation control, heat loss, and weather resistance are all-important aspects of good design principles. Most building codes require that glass areas be equal to at least 10% of the floor area of each room. Nevertheless, it is good practice to provide glass areas in excess of 20% and to locate the windows as high in the wall as possible to lengthen the depth of light penetration. Continuous sash or one large opening in a room provides a more desirable distribution of light than separated narrow windows. Either arrangement eliminates the dark areas between openings.

Location, type, and size of window are most important for natural ventilation. The pattern of air movement within a building depends to a great extent on the angle at which the air enters and leaves. It is desirable, particularly in summer, to direct the flow of air downward and across a room at a low level. However, the type and location of windows best suited to ventilation may not provide adequately for admission of light and clear vision, or perhaps proper weather protection. To arrive at a satisfactory relationship, it may be necessary to compromise with these functional requirements.

While building codes may establish a minimum percentage of glass area, they likewise may limit the use of glass and also require a fire rating in particular locations. In hazardous industrial applications subject to explosions, scored glass is an added precaution for quick release of pressure.

Heat transmittance is of economic importance and can also affect comfort. Weather stripping or the use of windows with integral frame and trim can minimize air infiltration. Double glazing or heat-absorbing glazing makes large glass areas feasible by materially reducing heat transmittance. Solar heating may be achieved by placement of glass areas so that the rays of the sun can be admitted during the winter.

11.45 WINDOW DEFINITIONS

Bar. Either vertical or horizontal member that extends the full height or width of the glass opening.

Blind Stop. A thin strip of wood machined so as to fit the exterior vertical edge of the pulley stile or jamb and keep the sash in place.

Casing. Molding of various widths and thicknesses used to trim window openings.

Check Rails. Meeting rails sufficiently thicker than the window to fill the opening between the top and bottom sash made by the check strip or parting strip in the frame. They are usually beveled and rabbeted.

Dado. A rectangular groove cut across the grain of a frame member.

Drip Cap. A molding placed on the top of the head casing of a window frame.

Extension Blind Stop. A molded piece, usually, of the same thickness as the blind stop, and tongued on one edge to engage a plow in the back edge of the blind stop, thus increasing its width and improving the weathertightness of the frame.

Frame. A group of wood parts so machined and assembled as to form an enclosure and support for a window or sash.

Jamb. Part of frame that surrounds and contacts the window or sash the frame is intended to support.

Side Jamb. The upright member forming the vertical side of the frame (Fig. 11.32).

Head Jamb. The horizontal member forming the top of the frame (Fig. 11.32).

Rabbeted Jamb. A jamb with a rectangular groove along one or both edges to receive a window or sash.

Jamb Liner. A small strip of wood, either surfaced four sides or tongued on one edge, when applied to the inside edge of a jamb, increases its width for use in thicker walls.

Measurements:

Between Glass. Distance across the face of any wood part that separates two sheets of glass.

Face Measure. Distance across the face of any wood part, exclusive of any solid mold or rabbet.

Finished Size. Dimension of any wood part overall, including the solid mold or rabbet.

Wood Allowance. The difference between the outside opening and the total glass measurement of a given window or sash.

Muntin. Any short light bar, either vertical or horizontal.

Parting Stop. A thin strip of wood let into the jamb of a window frame to separate the sash.

Pulley Stile. A side jamb into which a pulley is fixed and along which the sash slides.

Rails. Cross, or horizontal, pieces of the framework of a sash or screen.

Sash. A single assembly of stiles and rails made into a frame for holding glass, with or without dividing bars. It may be supplied either open or glazed.

Sill. The horizontal member forming the bottom of the frame (Fig. 11.32).

Stiles. Upright, or vertical, outside pieces of a sash or screen.

Window. One or more single sash made to fill a given opening. It may be supplied either open or glazed.

Window Unit. A combination of window frame, window, weather strip, balancing device, and at the option of the manufacturer, screen and storm sash, assembled as a complete and properly operating unit.

11.46 MODULAR COORDINATION OF WINDOWS

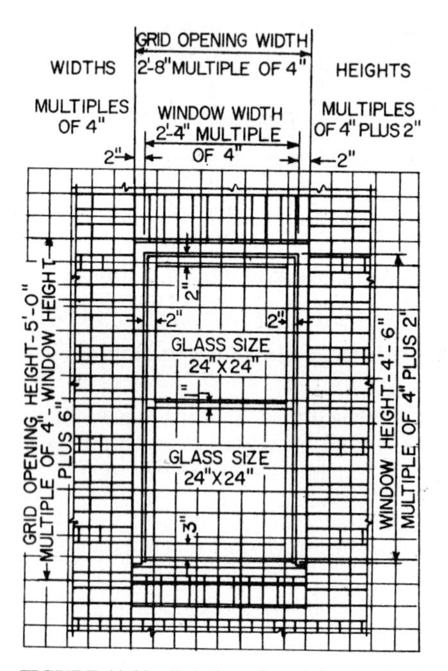

FIGURE 11.30 Relation of modular standard double-hung wood windows to grid opening in a brick wall.

Manufacturers have contributed substantially toward standardization of window sizes and greater precision in fabrication. The American National Standards Institute has established an increment of sizing generally accepted by the building industry ("Basis for the Coordination of Building Materials and Equipment," A62.1).

The basis for this standardization is a nominal increment of 4 in in dimension. However, it will be noted in Fig. 11.30, that for double-hung wood windows the 4-in module applies to the grid dimensions, whereas the standard window opening is 4 in less in width and 6 in less in height than the grid opening. Also, to accommodate 2- and 6-in mullions for multiple openings in brick walls, it is necessary to provide a 2-in masonry offset at one jamb (Fig. 11.31).

Use of modular planning effects maximum economies from the producer and simplification for the building industry. At the same time, a given layout does not confine acceptable products to such a limited range.

11.47 WINDOW SASH MATERIALS

Window frames are generally made of wood or metal. Plastics have been employed for special services.

Use of metal in window construction has been highly developed, with aluminum and steel most commonly used. Bronze, stainless steel, and galvanized steel are also extensively used for specific types of buildings and particular service. Use of metal windows is very often dictated by fire-resistance requirements of building codes.

11.47.1 Wood Windows

These can be used for most types of construction and are particularly adaptable to residential work. The most important factor affecting wood windows is weather

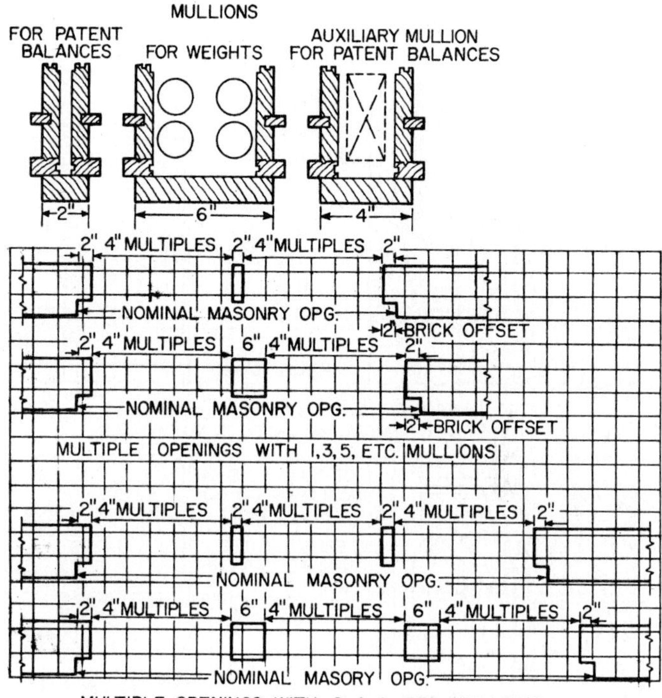

FIGURE 11.31 Dimensions for multiple modular openings for double-hung wood windows.

exposure. However, with proper maintenance, long life and satisfactory service can be expected. Sash and frames have lower thermal conductivity than metal.

The kinds of wood commonly used to resist shrinking and warping for the exposed parts of windows are white pine, sugar pine, ponderosa pine, fir, redwood, cedar, and cypress. The stiles against which a double-hung window slides should be of hard pine or some relatively hard wood. The parts of a window exposed to the inside are usually treated as trim and are made of the same material as the trim. (See also Fig. 11.32.)

11.47.2 Steel Windows

These are, in general, made from hot-rolled structural-grade new billet steel. However, for double-hung windows, the principal members are cold-formed from new billet strip steel. The principal manufacturers conform to the specifications of The Steel Window Institute, which has standardized types, sizes, thickness of material, depth of sections, construction, and accessories.

The life of steel windows is greatly dependent on proper shop finish, field painting, and maintenance. The more important aspects of proper protection against corrosion are as follows:

Shop Paint Finish. All steel surfaces should be thoroughly cleaned and free from rust and mill scale. A protective treatment of hot phosphate, cold phosphate-

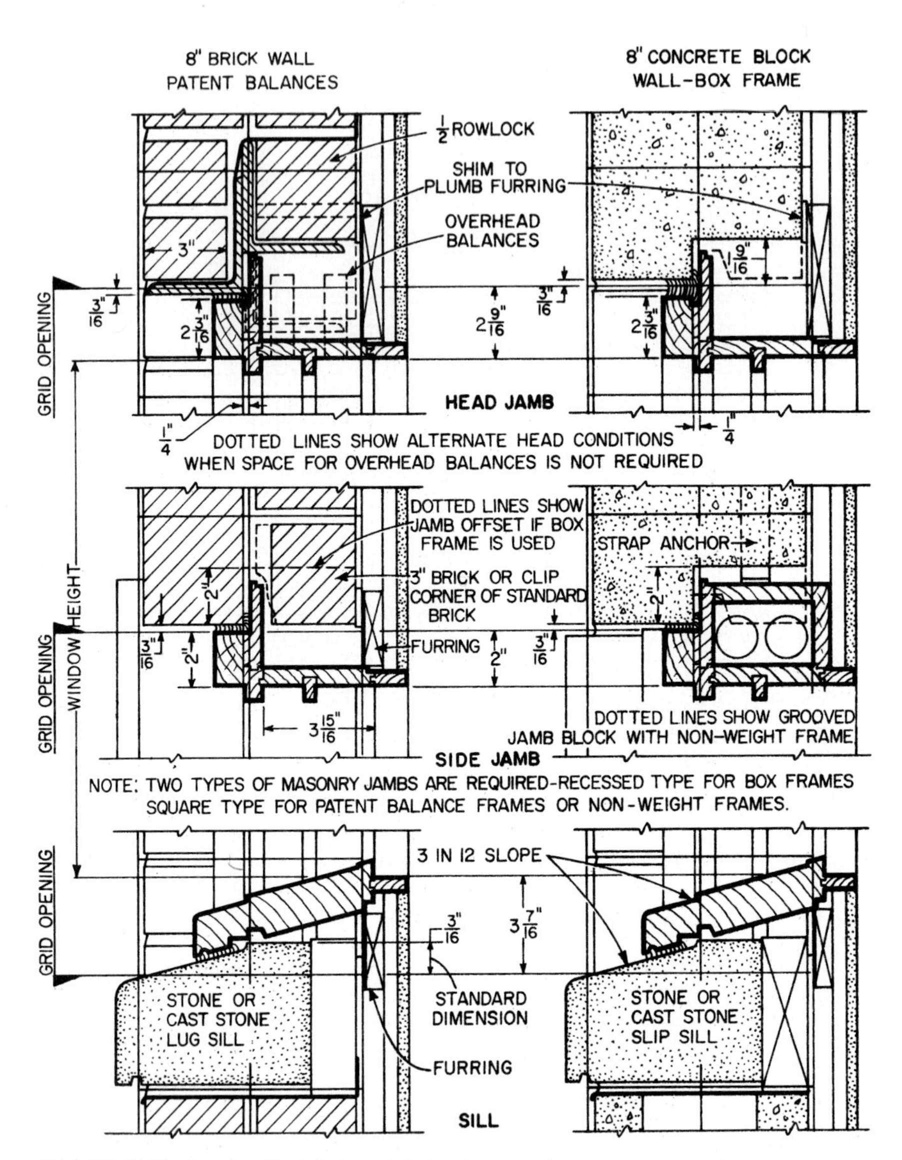

FIGURE 11.32 Details of installation of double-hung windows.

chromate, or similar method such as bonderizing or parkerizing is necessary to protect the metal and provide a suitable base for paint. All windows should then be given one shop coat of rust-inhibitive primer, baked on.

Field Painting. Steel windows should receive one field coat of paint either before or immediately after installation. A second coat should then be applied after glazing is completed and putty set.

Hot-Dipped Galvanized. Assembled frames and assembled ventilators should be galvanized separately. Ventilators should be installed into frames after galvanizing and bonderizing. Specifications must guard against distortion by the heat of hot dipping, abrasion in handling, and damage to the zinc galvanizing by muriatic acid used to wash brickwork. Painting is optional.

11.47.3 Aluminum Windows

Specifications set up by the Aluminum Window Manufacturers' Association are classified as residential, commercial, and monumental. These specifications provide for a minimum thickness of metal and regulate stiffness of the component parts, as well as limiting the amount of air infiltration.

The manufacturers offer about the same window types that are available in steel. However, in addition to these standard types, substantial progress has been made in the development of aluminum windows as an integral part of the walls of a building. With the exception of ventilator sections, which are shop-fabricated, some windows are completely assembled at the job. The wide range of extruded aluminum sections now manufactured has made it feasible to treat large expanses of glass as window walls. Picture windows with many combinations of ventilating sash are also furnished in aluminum. Many features are available, such as aluminum trim and casing, weather stripping of stainless steel or Monel metal, combination storm sash and screens, sliding wicket-type screens, and metal bead glazing.

Aluminum windows should be protected for shipment and installation with a coating of clear methacrylate lacquer or similar material able to withstand the action of lime mortar.

11.47.4 Stainless-Steel and Bronze Windows

These are high-quality materials used for durability, appearance, corrosion resistance, and minimum maintenance. Stainless steel is rustproof and extremely tough, with great strength in proportion to weight. Weather stripping of stainless steel is quite often used for wood and aluminum sash.

Bronze windows are durable and decorative and used for many fine commercial and monumental buildings.

11.47.5 Weather Stripping and Storm Sash

Weather stripping is a barrier against air leakage through cracks around sash. Made of metal or a compressible resilient material, it is very effective in reducing heat loss.

Storm sash might be considered an overall transparent blanket that shields the window unit and reduces heat loss. Weather stripping and storm sash also reduce condensation and soot and dirt infiltration, in addition to decreasing the amount of cold air near the floor.

Storm sash and screens are often incorporated in a single unit.

11.48 GLAZING

Article 4.31 describes various types of glass available for use in windows.

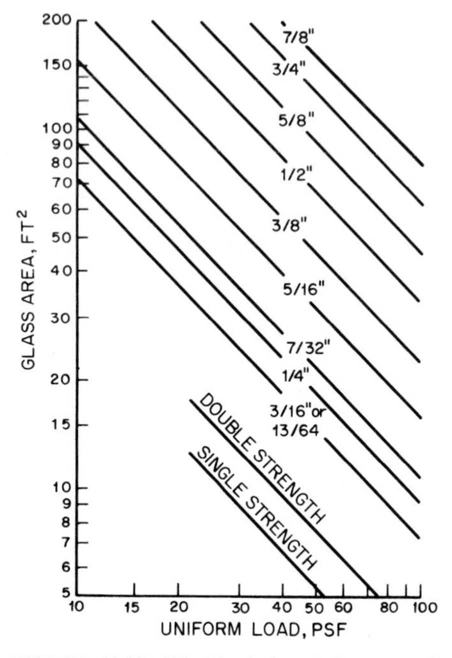

FIGURE 11.33 Wind-load chart indicates maximum glass area for various nominal thicknesses of plate, float, or sheet glass supported on four sides to withstand specified wind pressures, with design factor of 2.5.

Thickness. Figure 11.33 can be used as a guide to estimate thickness of plate, float, or sheet glass for a given glass area, or the maximum area for a given thickness to withstand a specified wind pressure. Based on minimum thickness permitted by Federal Specification DD-G-451c, the wind-load chart provides a safety factor of 2.5. It is intended for rectangular lights with four edges glazed in a stiff, weathertight rabbet.

For example, determine the thickness of a 108 × 120-in (90-ft^2) light of polished plate glass to withstand a 20-lb/ft^2 wind load. Since the 20-lb/ft^2 and 90-ft^2 ordinates intersect below the ⅜-in glass thickness line, the thickness to use is ⅜ in.

The correction factors in Table 11.14 also allow Fig. 11.33 to be used to determine the thickness for certain types of fabricated glass products. The table, however, makes no allowance for the weakening effect of such items as holes, notches, grooves, scratches, abrasion, and welding splatter.

The appropriate thickness for the fabricated glass product is obtained by multiplying the wind load, lb/ft^2, by the factor given in Table 11.14. The intersection of the vertical line drawn from the adjusted load and the horizontal line drawn from the glass area indicates the minimum recommended glass thickness. (See also Art. 11.48.3.) Glass producers should be consulted for more accurate thickness recommendations.

Table 11.15 gives the overall thermal conductance, or *U* factor, and weight of glass for several thicknesses. Table 11.16 presents typical sound-reduction factors for glass, and for comparison, for other materials. As a sound barrier, glass, inch

TABLE 11.14 Resistance of Glass to Wind Loads

Product	Factor*
Plate, float, or sheet glass	1.0
Rough rolled plate glass	1.0
Sand-blasted glass	2.5
Laminated glass†	2.0
Sealed double glazing:‡	
Metal edged, up to 30 ft^2	0.667
Metal edged, over 30 ft^2	0.556
Glass edged	0.5
Heat-strengthened glass	0.5
Fully tempered glass	0.25

*Enter Fig. 11.33 with the product of the wind load in psf multiplied by the factor.

†At 70°F or above, for two lights of equal thickness laminated to 0.015-in-thick vinyl. At 0°F, factor approaches 1.

‡For thickness, use thinner of the two lights, not total thickness.

for inch, is about equal to concrete and better than most brick, tile, or plaster. Double glazing (Fig. 11.34) is particularly effective where high resistance to sound transmission is required. Unequal glass thicknesses minimize resonance in this type of glazing, and a nonreflective mounting reduces sound transmission. For optimum performance, sound-isolation glazing should be carefully detailed and constructed, and should be mounted in airtight, heavy walls with acoustically treated surfaces.

11.48.1 Glazing Compounds

Putty, glazing compound, rubber, or plastic strips and metal or wood molds are used for holding glass in place in sash. Metal clips for metal sash (Fig. 11.35*e*) and glazing points for wood sash also are employed for this purpose.

TABLE 11.15 Overall Conductance and Weight of Glass

Thickness, in	*U* factor	Weight, psf
⅛	1.14	1.65
¼	1.13	3.29
⅜	1.12	4.94
½	1.11	6.58
1-in insulating glass	0.55	7.00

TABLE 11.16 Sound Reduction—Glass and Other Materials*

Material thickness, in	Sound transmission classification rating
Glass:	
0.087 (single strength)	26
0.118 (double strength)	29
1/8	29
3/16	29
1/4	27
5/16	29
3/8	30
1/2	33
1	32
1/4 laminated (up to 0.045 plastic layer)	33
1-in sealed double glazing, 1/2-in. air space	27
Other materials:	
1/4 plywood	22
1/2 fiberboard	21
1/4 lead	45
2 solid vermiculite plaster on metal lath	30
Hollow wall, 2 × 4 studs, 1/2-in gypsum plaster on metal lath on two sides	33

*Courtesy Libby-Owens-Ford Company, Toledo, Ohio.

Putty and glazing compounds are generally classified in relation to the sash material and should be used accordingly for best results.

Bedding of glass in glazing compound is desirable, because it furnishes a smooth bearing surface for the glass, prevents rattling, and eliminates voids where moisture can collect (Commercial Standards, U.S. Department of Commerce). Glazing terminology relative to putty follows:

Face Puttying (Fig. 11.35*a*). Glass is inserted in the glass rabbet and securely wedged where necessary to prevent shifting. Glazing points are also driven into the wood to keep the glass firmly seated. The rabbet then is filled with putty, the putty being beveled back against the sash and muntins with a putty knife.

Back Puttying (Fig. 11.35*b*). After the sash has been face-puttied, it is turned over, and putty is run around the glass opening with a putty knife, thus forcing putty into any voids that may exist between the glass and the wood parts.

Bedding (Fig. 11.35*c*). A thin layer of putty or bedding compound is placed in the rabbet of the sash, and the glass is pressed onto this bed. Glazing points are then driven into the wood, and the sash is face-puttied. The sash then is turned over, and the excess putty or glazing compound that emerged on the other side is removed.

Wood-Stop or Channel Glazing (Fig. 11.35*d*). A thin layer of putty or bedding compound is placed in the rabbet of the sash, and the glass is pressed onto this bed. Glazing points are not required. Wood stops are securely nailed in place. The

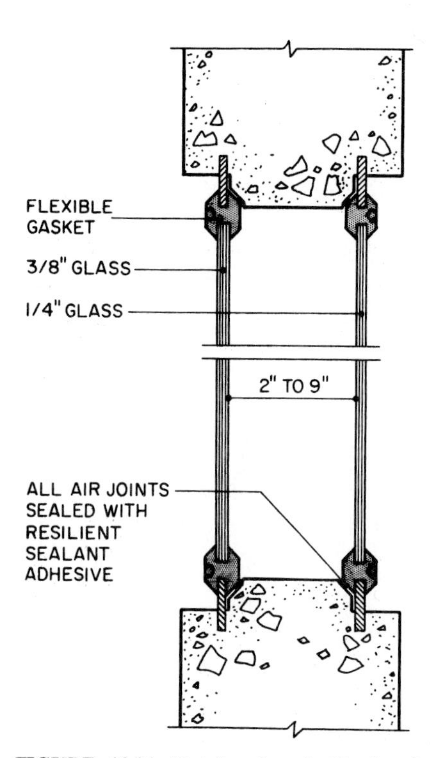

FIGURE 11.34 Details of a double-glazed window installed in a massive concrete wall to obtain high resistance to sound transmission.

sash then is turned over, and the excess putty or glazing compound that emerged on the other side is removed.

Glazing Beads (Fig. 11.35*f*). These are designed to cover exterior glazing compound and improve appearance.

Continuous Glazing Angles (Fig. 11.35*g*). These angles or similar supports for glass usually are required for "labeled windows" by the National Board of Fire Underwriters.

Table 11.17 lists some commonly used glazing compounds and indicates the form and method in which they usually are used. Other compounds used for special service include silicone rubber, butadiene-styrene (GR-S) rubber, polyethylene, nitrile rubbers, polyurethane rubbers, acrylics, and epoxy.

11.48.2 Factory-Sealed Double Glazing

This is a factory-fabricated, insulating-glass unit composed of two panes of glass separated by a dehydrated airspace. This type of sash is also manufactured with three panes of glass and two airspaces, providing additional insulation against heat flow or sound transmission. (See Fig. 11.35*e*.)

Heat loss and heat gain can be substantially reduced by this insulated glass, permitting larger window areas and added indoor comfort. Heat-absorbing glass often is used for the outside pane and a clear plate or float glass for the inside. However, there are many combinations of glass available, including several patterned styles. Thickness of glass and airspace between can be varied within prescribed limitations. In the selection of a window for double or triple glazing, accommodation of the overall glass thickness in the sash is an important consideration.

11.48.3 Gaskets for Glazing

Structural gaskets, made of preformed and cured elastomeric materials, may be used instead of sash for some applications to hold glass in place. Gaskets are extruded in a single strip, molded into the shape of the window perimeter, and installed against the glass and window frame. A continuous locking strip of harder elastomer is then forced into one side of the gasket as a final sealing component. Fit and compression of the gasket determine weathertightness. With proper installation, calking is not ordinarily required; but should it be necessary, sealants are available that are compatible with the material of which the gasket is made.

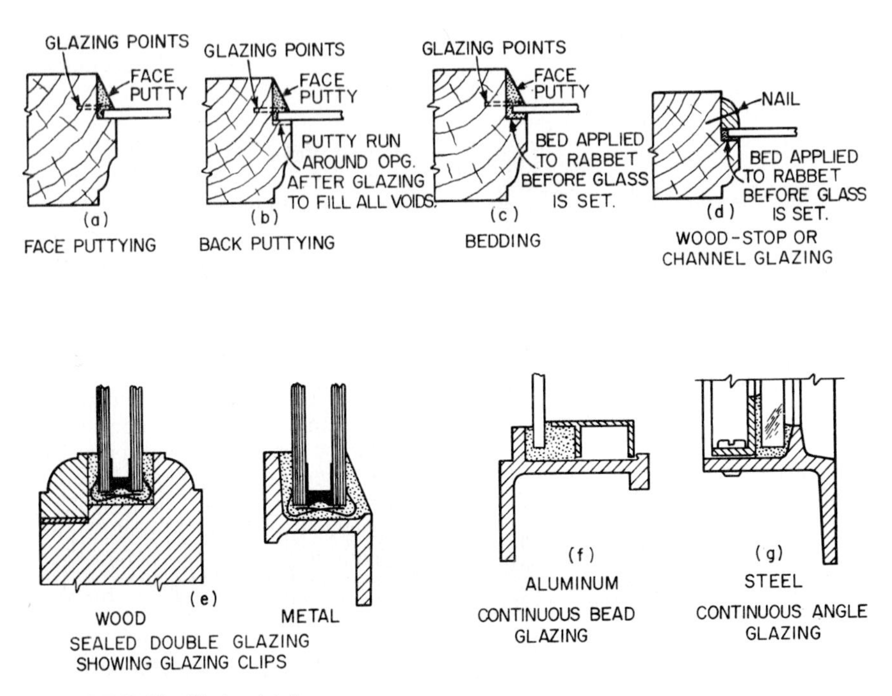

FIGURE 11.35 Glazing details.

There are many variations in the design of lockstrip gaskets, some of which are shown in Fig. 11.36. The two most commonly used types are the H and reglet types. The H gasket fits over a flange on the surrounding frame (Fig. 11.36*c* and *d*), whereas the reglet gasket fits into a recess in the frame (Fig. 11.36*a* and *b*). When gaskets are used in concrete wall panels, the entire opening should be cast within one panel. In all cases, the smoothness, tolerance, and alignment of the

TABLE 11.17 Commonly Used Glazing Compounds

Compound	Type of material	Final form	Type of glazing
Vegetable oil base (skin-forming type)	Gun, knife	Hard	Channel, bedding, face
Vegetable oil-rubber or nondrying oil blends (polybutene) etc.	Gun, knife	Hard, plastic	Face, channel, bedding
Nondrying oil types	Gun, knife, tacky tape	Plastic	Channel, bedding
Butyl rubber or polyisobutylene	Tacky tape	Plastic, rubber	Bedding, channel
Polysulfide rubber	Gun	Plastic, rubber	Bedding, channel
Neoprene (polychloroprene)	Gun, preformed gasket	Rubber	Channel
Vinyl chloride and copolymers	Preformed gasket	Hard, plastic	

contact surfaces are very important. The gaskets shown in Fig. 11.36*e* and *f* for use with double glazing require weepholes to allow water that may collect in the gaskets to drain.

The maximum glass area recommended for glazing with standard H-type lock-strip gaskets may be obtained from Fig. 11.37. Loads for the chart are based on minimum glass thicknesses allowed in Federal Specification DD-G-451c. Glass producers should be consulted to obtain the latest thickness recommendations.

11.49 WINDOW TYPES

A wide range of window types are available to satisfy the ever-increasing scope of architectural requirements. There are many materials, sizes, arrangements, details, and specific features from which to choose.

Types of windows (see symbols Fig. 11.38) include:

Pivoted windows, an economical, industrial window used where a good tight closure is not of major importance (Fig. 11.39). Principal members of metal window units are usually 1⅜ in deep, with frame and vent corners and muntin joints riveted. However, when frame and vent corners are welded, principal members may be 1¼ in deep. Vents are pivoted about 2 in above the center. These windows are adaptable to multiunits, both vertical and horizontal, with mechanical operation. Bottom of vent swings out and top swings in. No provision is made for screening. Putty glazing is placed inside.

Commercial projected windows, similar to pivoted windows (Fig. 11.40). These are an industrial-type window but also are used for commercial buildings where economy is essential. Vents are balanced on arms that afford a positive and easy control of ventilation and can be operated in groups by mechanical operators. Maximum opening is about 35°. Factory provision is made for screens.

Security windows, an industrial window for use where protection against burglary is important, such as for factories, garages, warehouses, and rear and side elevations of stores (Fig. 11.41). They eliminate the need for separate guard bars. They consist of a grille and ventilated window in one unit. Maximum grille openings are 7 × 16 in. The ventilating section, either inside or outside the grille, is bottom-hinged or projected. Muntins are continuous vertically and horizontally. Factory provision is made for screening. Glazing is placed from the inside.

Basement and utility windows, economy sash designed for use in basements, barracks, garages, service stations, areaways, etc. (Figs. 11.42 and 11.43). Ventilator opens inward and is easily removed for glazing or cleaning, or to provide a clear opening for passage of materials. Center muntin is optional. Screens are attached on the outside.

Architectural projected windows, a medium-quality sash used widely for commercial, institutional, and industrial buildings (Figs. 11.44 and 11.45). Ventilator arrangement permits cleaning from inside and provides a good range of ventilation control with easy operation. Steel ventilator sections are usually 1¼ in deep, with corners welded. Frames may be 1⅜ in deep when riveted or 1¼ in deep when welded. Screens are generally furnished at extra cost. Glazing may be either inside or outside.

Intermediate projected windows, high-quality ventilating sash used for schools, hospitals, commercial buildings, etc. (Figs. 11.46 and 11.47). When made of metal, corners of frames and ventilators are welded. Depth of frame and vent sections varies from "intermediate" to "heavy intermediate," depending on the manufac-

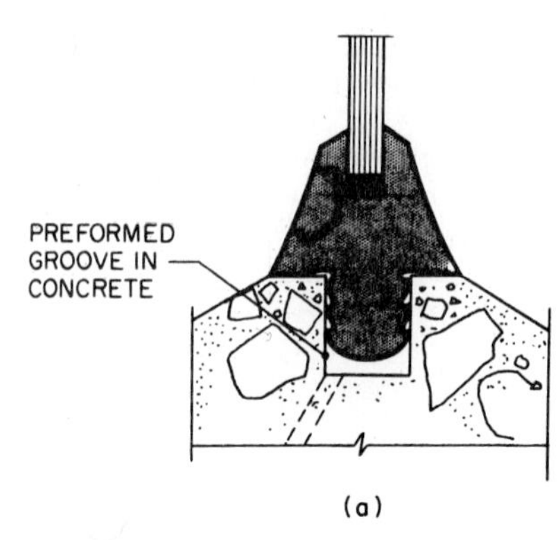

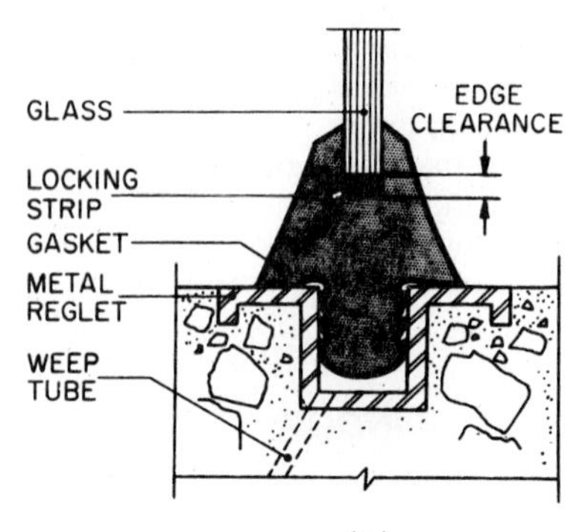

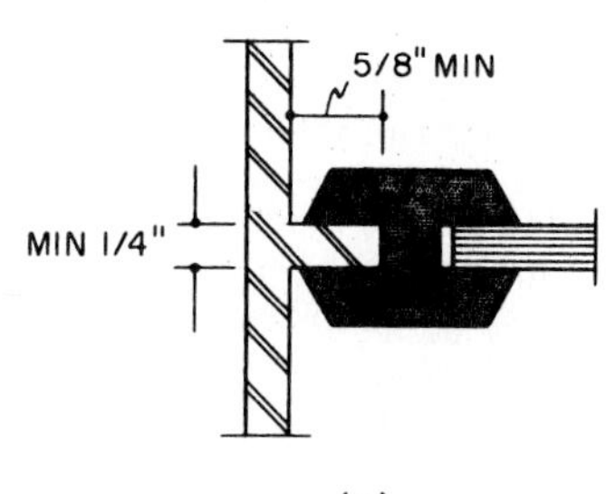

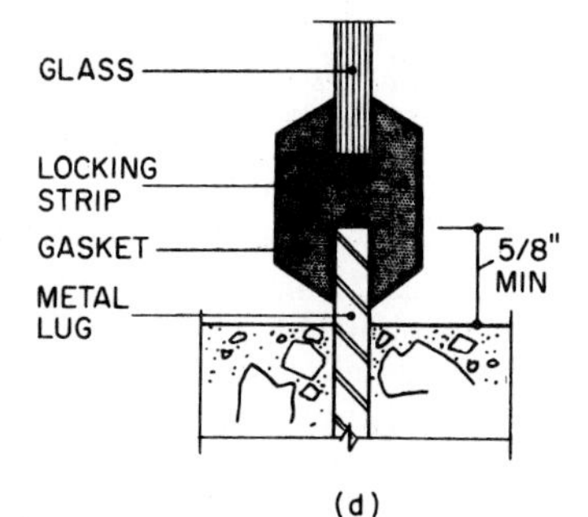

SETTING STRIPS
(NOT OVER
WEEPHOLES)

WEEPHOLE

WEEPHOLES

WEEPHOLE

PREFERRED
LOCATIONS

(e)

ALTERNATIVE
LOCATIONS

(f)

FIGURE 11.36 Structural gasket glazing: (*a*) and (*b*) gasket fitted into a groove; (*c*) and (*d*) H shape locked to continuous metal fin; (*e*) and (*f*) weep holes for factory sealed double-glazed window.

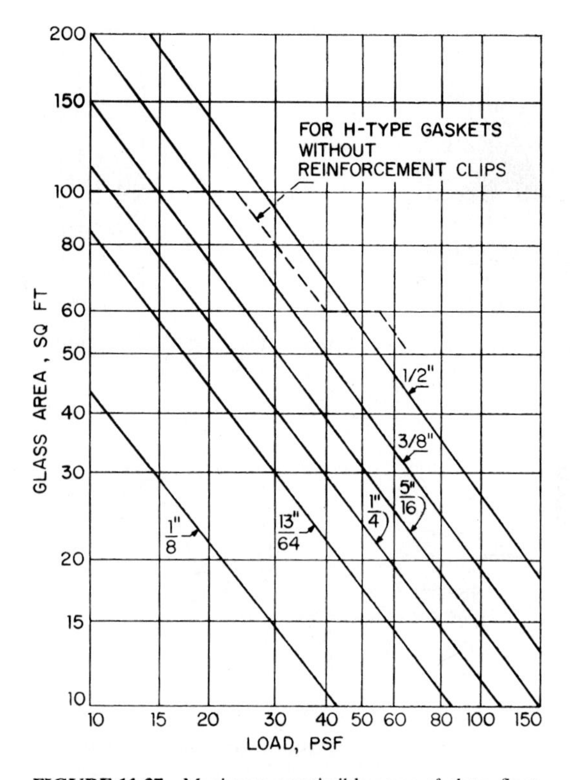

FIGURE 11.37 Maximum permissible areas of plate, float, or sheet glass, with design factor of 2.5, for various wind loads on vertical windows supported on four sides and having standard H-type lockstrip gaskets. Loads apply to glass lights with a length-to-width ratio of 3 or less.

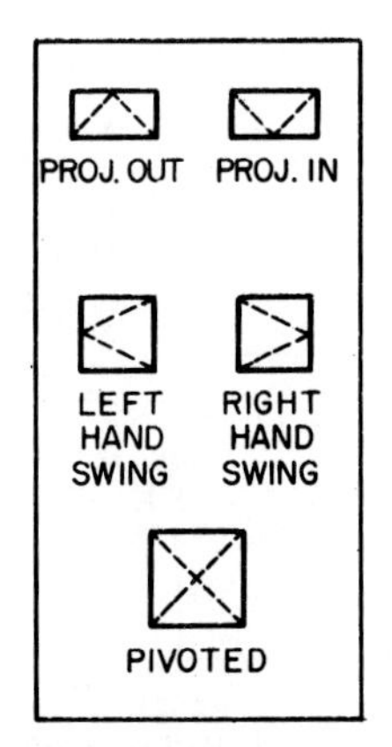

FIGURE 11.38 Symbols for common window types (viewed from outside).

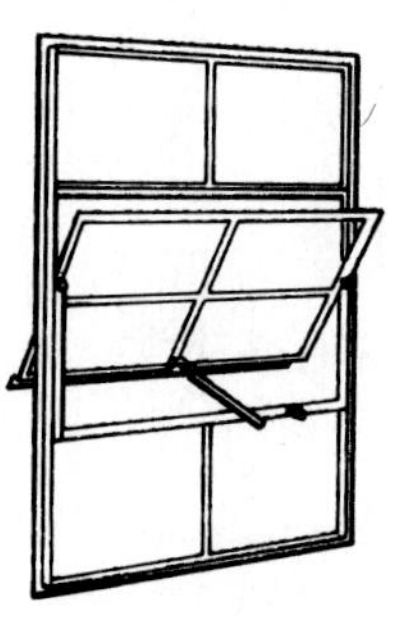

FIGURE 11.39 Pivoted window. Vent bottom swings out, top swings in.

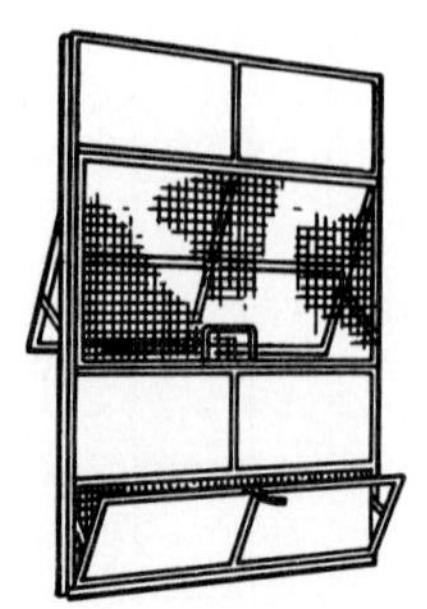

FIGURE 11.40 Commercial projected window.

FIGURE 11.41 Security window with continuous muntins.

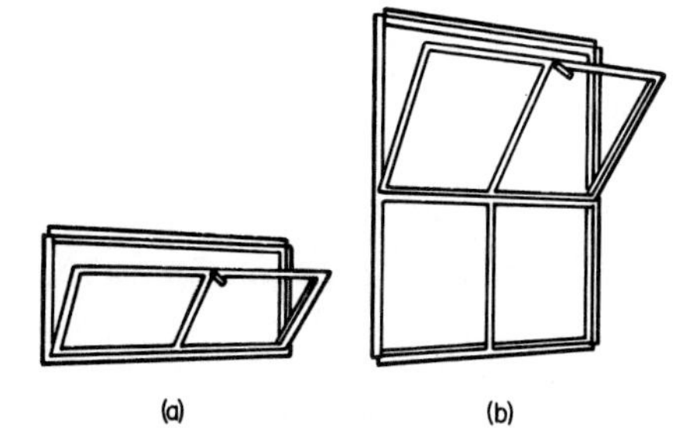

FIGURE 11.42 Basement and utility window.

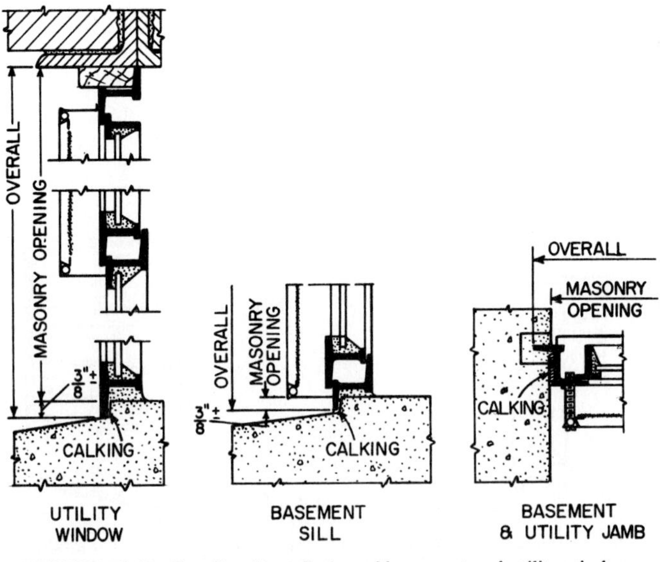

FIGURE 11.43 Details of installation of basement and utility windows.

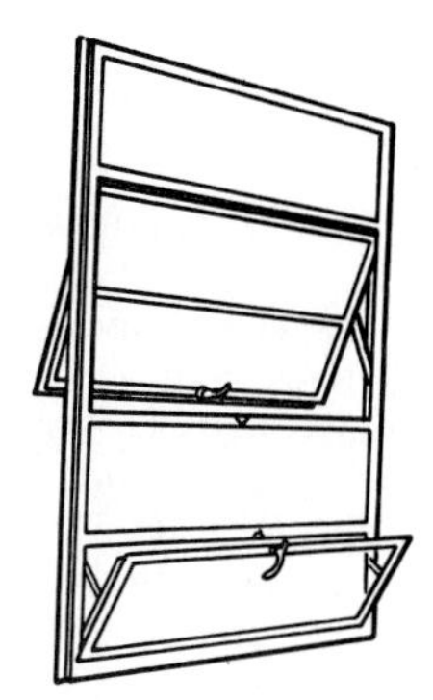

FIGURE 11.44 Architectural projected window.

turer. Each vent is balanced on two arms pivoted to the frame and vent. The arms are equipped with friction shoes arranged to slide in the jamb sections. Screens are easily attached from inside. All cleaning is done from the inside. Glazing is set from the outside. Double insulating glass is optional.

Psychiatric projected windows, for use in housing mental patients, to provide protection against exit but minimizing appearance of restraint (Fig. 11.48). Ventilators open in at the top, with a maximum clear opening of 5 to 6 in. Glazing is set outside. Screens also are placed on the outside but installed from inside. Metal casing can be had completely assembled and attached to window ready to install. Outside glass surfaces are easily washed from the inside.

Detention windows, designed for varying degrees of restraint and for different lighting and ventilating requirements. The guard type (Fig. 11.49) is particularly adaptable to jails, reform schools, etc. It provides security against escape through window openings. Ventilators are attached to the inner face of the grille and can be had with a removable key-locking device for positive control by attendants. Screens are installed from inside between vent and grille. Glazing is done from outside, glass being omitted from the grille at the vent section.

Residential double-hung windows (Figs. 11.30 to 11.32 and 11.50), available in different designs and weights to meet various service requirements for all types of buildings. When made of metal, the frame and ventilator corners are welded and weathertight. These windows are also used in combination with fixed picture windows for multiple window openings. They are usually equipped with weather strip-

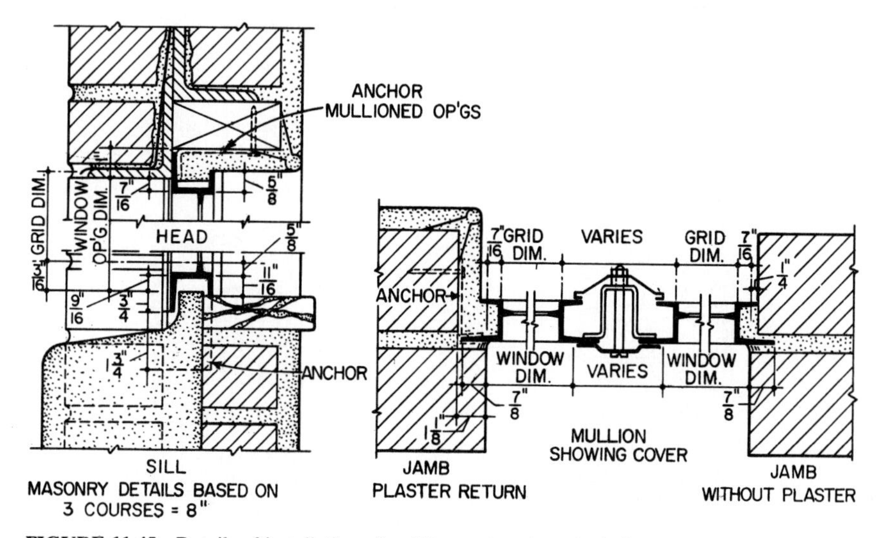

FIGURE 11.45 Details of installation of architectural projected windows.

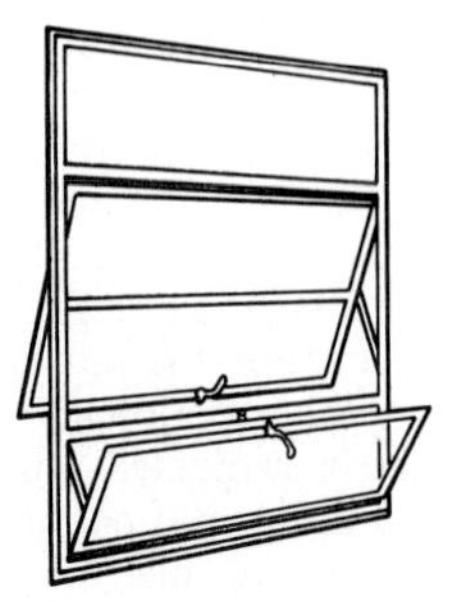

FIGURE 11.46 Intermediate projected window.

ping, which maintains good weathertightness. Screens and storm sash are furnished in either full or half sections. Glazing is done from outside.

Residence casements, available in various types, sizes, and weights to meet service requirements of homes, apartments, hotels, institutions, commercial buildings, etc. (Fig. 11.51). Rotary or lever operations hold the vent at the desired position, up to 100% opening. Screens and storm sash are attached to the inside of casement. Extended hinges on vents permit cleaning from inside. Glazing is done from outside.

Intermediate combination windows, with side-hinged casements and projected-in ventilators incorporated to furnish flexibility of ventilation control, used extensively for apartments, offices, hospitals, schools, etc., where quality is desired (Fig.

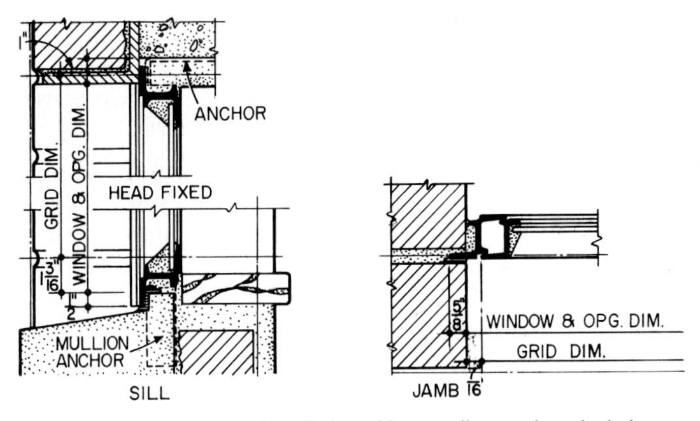

FIGURE 11.47 Details of installation of intermediate projected window.

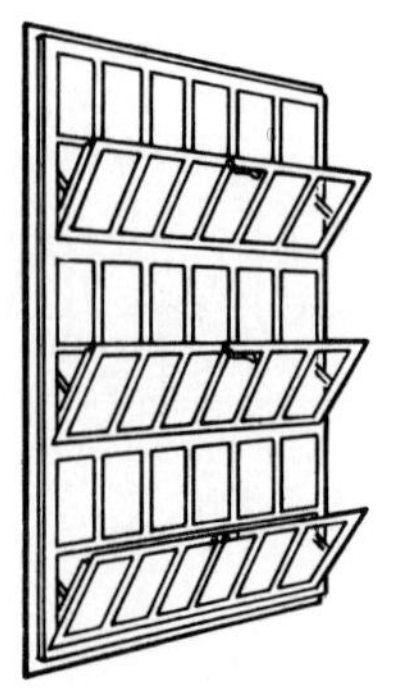

FIGURE 11.48 Psychiatric projected window.

FIGURE 11.49 Guard-type detention window.

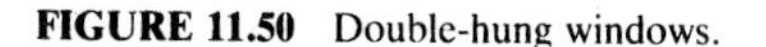

FIGURE 11.50 Double-hung windows.

11.52). They are available in several weights with rotary or handle operation. When made of metal, corners of vents and frames are welded. Factory provision is made for screens. All cleaning is done from inside. Glazing is set from outside. Special glazing clips permit use of double insulating glass.

Intermediate casements, a heavier and better quality than residence casements, used particularly for fine residences, apartments, offices, institutions, and similar buildings (Fig. 11.53). Frames and ventilators of metal units have welded corners. Easy control of ventilation is provided by rotary or handle operation. Extended hinges permit safe cleaning of all outside surfaces from inside. Screens are attached or removed from inside. Single or double glazing is set from outside.

Awning windows, suitable for residential, institutional, commercial, and industrial buildings (Fig. 11.54), furnishing approximately 100% opening for ventilation. Mechanical operation can provide for the bottom vent opening prior to the other vents, which open in unison. This is desirable for night ventilation. Manual operation can be had for individual or group venting. All glass surfaces are easily cleaned from inside through the open vents. Glazing is set from outside, storm sash and screens from inside.

Jalousie windows (Fig. 11.55) combine unobstructed vision with controlled ventilation and are used primarily for sunrooms, porches, and the like where protection from the weather is desired with maximum fresh air. The louvers can be secured in any position. Various kinds of glass, including obscure and colored, often are used for privacy or decoration. They do not afford maximum weathertightness but can be fitted with storm sash on the inside. Screens are furnished interchangeable with storm sash.

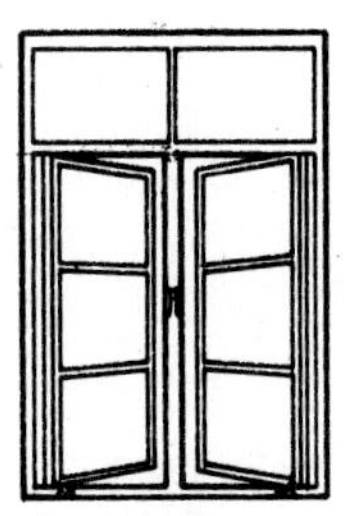

FIGURE 11.51 Residence casement window.

FIGURE 11.52 Intermediate combination window.

FIGURE 11.53 Intermediate casement window.

FIGURE 11.54 Awning window.

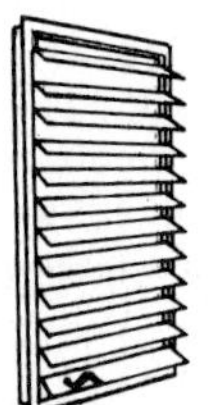

FIGURE 11.55 Jalousie window.

Ranch windows, particularly suited to modern home design and also used effectively in other types of buildings where more light or better view is desired (Fig. 11.56). When made of metal, corners of frames and vents are welded. Depth of sections vary with manufacturers. These windows are designed to accommodate insulating glass or single panes. Screens are attached from inside.

Continuous top-hung windows, used for top lighting and ventilation in monitor and sawtooth roof construction (Fig. 11.57). They are hinged at the top to the structural-steel framing members of the building and swing outward at the bottom. Two-foot lengths are connected end to end on the job. Mechanical operators may be either manual or motor-powered. Sections may be installed as fixed windows. Glazing is set from outside.

Additional types of windows (not illustrated) include:

Vertical pivoted, which sometimes are sealed with a rubber or plastic gasket.

Horizontal sliding sash usually in aluminum or wood for residential work.

Vertical folding windows, which feature a flue action for ventilation.

Double-hung windows with removable sash, which slide up and down, or another type which tilts for ventilation but does not slide.

Austral type, with sash units similar to double-hung but which operate in unison as projected sash. The sash units are counterbalanced on arms pivoted to the frame, the top unit projecting out at bottom and the bottom unit projected in at top.

Picture windows, often a combination of fixed sash with or without auxiliary ventilating units.

Store-front construction, usually semicustom-built of stock moldings of stainless steel, aluminum, bronze, or wood.

11.50 WINDOWS IN WALL-PANEL CONSTRUCTION

In the development of thin-wall buildings and wall-panel construction, windows have become more a part of the wall rather than units to fill an opening. Wall panels of metal and concrete incorporate windows as part of their general makeup; a well-integrated design will recognize this inherent composition.

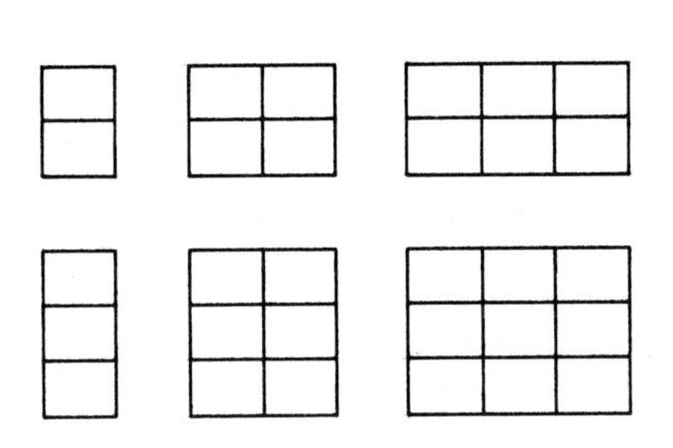

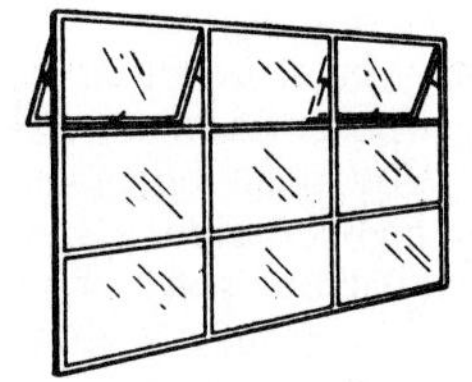

FIGURE 11.56 Ranch windows.

Manufacturers of metal wall panels can furnish formed metal window frames specifically adapted to sash type and wall construction.

In panel walls of precast concrete, window frames can be cast as an integral part of the unit or set in openings as provided (Fig. 11.58). The cast-in-place method minimizes air infiltration around the frame and reduces installation costs on the job. Individual units or continuous bands of sash, both horizontal and vertical, can be readily adapted by the proper forming for head, jamb, and sill sections.

11.51 MECHANICAL OPERATORS FOR WINDOWS

Mechanical operation for pivoted and projected ventilators is achieved with a horizontal torsion shaft actuated by an endless hand chain or by motor power. Arms attached to the torsion shaft open and close the vents.

Two common types of operating arms are the lever and the rack and pinion (Fig. 11.59). The lever is used for manual operation of small groups of pivoted vents where rapid opening is desirable. The rack and pinion is used for longer runs of vents and can be motor powered or manually operated. The opening and closing by rack and pinion are slower than by lever arm.

Usually one operating arm is furnished for each vent less than 4 ft wide and two arms for each vent 4 ft or more wide. Mechanical operators should be installed by the window manufacturer before glazing of the windows.

Continuous top-hung windows are mechanically controlled by rack-and-pinion or tension-type operators (Fig. 11.59).

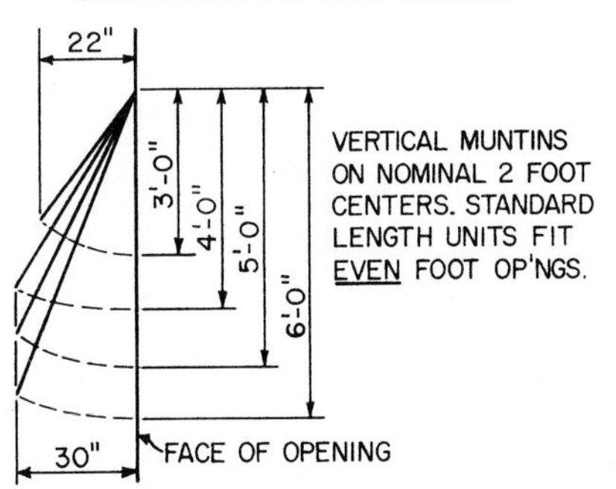

FIGURE 11.57 Continuous top-hung window.

DOORS

The purpose of doors is to close openings that are needed in walls and partitions for access to building interior spaces, when closing is required to prevent trespass by unauthorized persons, to provide privacy, to protect against weather, drafts, and noise, and to act as a barrier to fire and smoke. Many types of doors are available for this. They may be hinged on top or sides

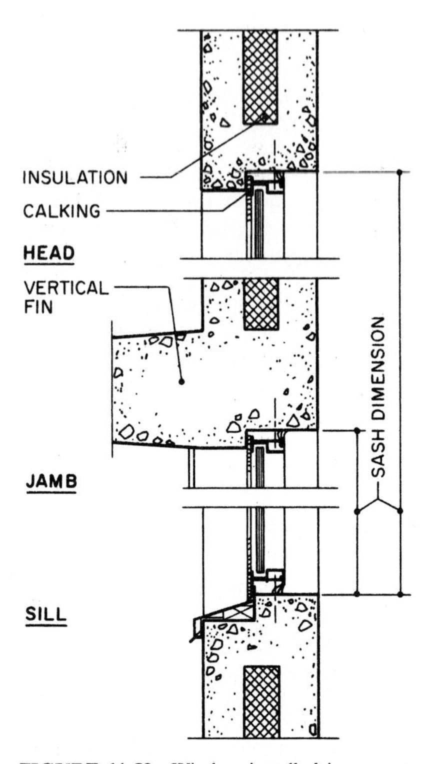

FIGURE 11.58 Window installed in precast-concrete sandwich wall panel.

to swing open and shut, they may slide horizontally or vertically, or they may revolve about a vertical axis in the center of the opening.

A large variety of materials also is available for door construction. Wood, metal, glass, plastics, and combinations of these materials with each other and with other materials, in the form of sandwich panels, are in common use.

Selection of a type of door and door material depends as much on other factors as on the primary function of serving as a barrier. Cost, psychological effect, fire resistance, architectural harmony, and ornamental considerations are but a few of the factors that must be taken into account.

Doors may be classified as ordinary or special purpose. Ordinary doors are those used to protect openings up to about 12 ft high or wide. Often, such doors are available in stock or standard sizes. Larger doors may be considered special purpose, because they are generally custom designed and built and require specially designed framing.

In either case, a door system consists of a door proper, hardware for control of door movement (Arts. 11.65 to 11.67), a door frame around the wall opening to support the door and trim the opening, and structural framing, such as a lintel, to support the wall and other building components directly over the opening.

11.52 TRAFFIC FLOW AND SAFETY

Openings in walls and partitions must be sized for their primary function of providing entry and exit to or from a building or its interior spaces, and doors must

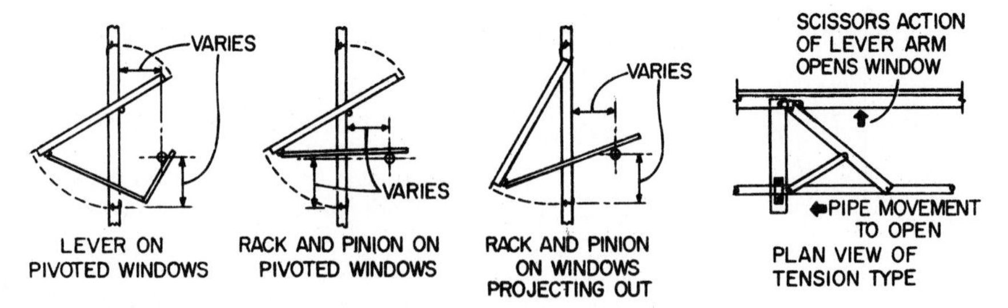

FIGURE 11.59 Levers, rack-and-pinion, and tension-type window operators.

be sized and capable of operating so as to prevent or permit such passage, as required by the occupants of the building. In addition, openings must be adequately sized to serve as an exit under emergency conditions. (See Art. 3.5.10.) In all cases, traffic must be able to flow smoothly through the openings.

To serve these needs, doors must be properly selected for the use to which they are to be put, and properly arranged for maximum efficiency. In addition, they must be equipped with suitable hardware for the application. (See also Arts. 11.65 to 11.67 and Art. 11.55.)

Safety. Exit doors and doors leading to exit passageways should be so designed and arranged as to be clearly recognizable as such and to be readily accessible at all times. A door from a room to an exit or to an exit passageway should be the swinging type, installed to swing in the direction of travel to the exit.

Code Limitations on Door Sizes. To ensure smooth, safe traffic flow, building codes generally place maximum and minimum limits on door sizes. Typical restrictions are as follows:

No single leaf in an exit door should be less than 28 in wide or more than 48 in wide. Minimum nominal width of opening should be at least:

36 in for single corridor or exit doors

32 in for each of a pair of corridor or exit doors with central mullion

48 in for a pair of doors with no central mullion

32 in for doors to all occupiable and habitable rooms

44 in for doors to rooms used by bedridden patients and single doors used by patients in such buildings as hospitals, sanitariums, and nursing homes

32 in for toilet-room doors

Jambs, stops, and door thickness when the door is open should not restrict the required width of opening by more than 3 in for each 22 in of width.

Nominal opening height for exit and corridor doors should be at least 6 ft 8 in. Jambs, stops, sills, and closures should not reduce the clear opening to less than 6 ft 6 in.

For a specific type of occupancy, the number of exit doors required in each story of a multistory building and the minimum width permitted for each door can be determined from the maximum capacity listed in Art. 3.5.10. The maximum sizes of openings permitted in fire barriers are given in Table 11.19, p. 11.117.

Safety at Entrances. Because of the heavy flow of traffic at building entrances, safety provisions at entrance doors are an important design consideration. Account must be taken of the location of such doors in the building faces, flow of outdoor traffic, and type and volume of traffic generated by the building. Following are some design recommendations:

Arc of a door swing should exceed 90°.

An entrance should always be set back from the building face.

Hinge jambs of swinging doors should be located at least 6 in from a wall perpendicular to the building face.

If hinge jambs for two doors have to be placed close together, there should be enough distance between them to permit the doors to swing through an arc of 110°.

If several doors swinging in the same direction are placed close together in the same plane, they should be separated by center lights and should also have sidelights, to enable the doors to swing through 110° arcs.

If doors hung on center pivots are arranged in pairs, they should be hinged at the side jambs and not at the central mullion.

The floor on both sides of an exit door should be substantially level for a distance on each side equal to at least the width of the widest single leaf of the door. If, however, the exit discharges to the outside, the level outside the door may be one step lower than inside, but not more than 7½ in lower.

11.53 STRUCTURAL REQUIREMENTS FOR OPENINGS AND DOORS

A wall or partition above a door or window opening must be adequately supported by a structural member. In design of such a member, stiffness as well as strength must be taken into account. Excessive deflection could interfere with door operation.

It is common practice to install a frame along the perimeter of a door opening, to support the door. Anchored to the wall, the frame usually also serves as trim around the opening. In design of door framing, wind loads are generally more critical than dead loads imposed by the wall or partition above. Wind load should be taken as at least 15 lb/ft^2. Under this loading, maximum deflection of framing members should not exceed ¾ in or $1/175$ the clear span. The design should take into account the fact that wind forces acting outward may be larger than those acting inward. Door framing should generally be made independent of other framing.

11.54 ORDINARY DOORS

Ordinary doors may be classified as exterior or interior doors. Exterior doors are those that are installed in an opening in a wall that separates an interior space from outdoors. Serving as an entrance or an exit door, or both, these doors must be capable of excluding weather but usually need not be fire rated. Interior doors are those installed in an opening in a wall or partition between two interior spaces. Such doors are not required to exclude weather but may be required to bar passage of fire and smoke.

11.54.1 Exterior Doors

Entrance and exit doors in exterior walls should be suitably selected and located for smooth traffic flow and safety. Choice of door movement, for example, should take into account in which direction persons are likely to turn immediately before or after they enter a doorway. Proper swing of door will not only smooth traffic flow but also reduce wear and tear on the doors by decreasing impact loading on the hardware from persons who lean on it to change directions.

Also, selection of entrance doors should take not only esthetics into account but also traffic volume. The heavier traffic volume will be, the more rugged the doors should be to withstand wear.

Water Exclusion at Exterior Doors. Exterior doors are subjected to all the effects of natural forces, as are walls and windows, including solar heat, rain, and wind. For ordinary installations, closed doors cannot be expected to completely exclude water or stop air movement under all conditions. One important reason for this is that clearances must be provided around each door. These are necessary to permit easy operation and thermal expansion and contraction of doors.

Exterior doors, however, can be made less vulnerable to water penetration by setting them back from the building face, or by providing overhead protection, such as a canopy, marquee, or balcony. These measures also will help reduce collection of snow and ice at door thresholds. In addition, provision of an entrance vestibule is desirable, because it can serve as a weather barrier. Also, weather-stripping around doors assists in preventing passage of water and air past closed doors.

At the bottom of a door opening for an exterior door is a sill (Fig. 11.60), which forms a division between the finished floor on the inside and the outside construction. The sill generally also serves as a step, for the door opening usually is raised above exterior grade to prevent rain from entering. The top of the sill is sloped to drain water away from the interior. Also, it may have a raised section in the plane of the door or slightly to the rear, so that water dripping from the door will fall on the slope. The raised section may be integral with the sill or a separate threshold. In either case, the rear portion covers the joint between sill and floor. In addition, all joints should be sealed.

Control of Air Movements at Entrances. In design of entrances to buildings, consideration should be given to the effects in those areas of outdoor winds. These may impose pressures or suctions on exterior doors. As a result, doors may be held partly open or may be violently opened beyond their design limits and severely damaged. To avoid such occurrences, doors should be set back from the building

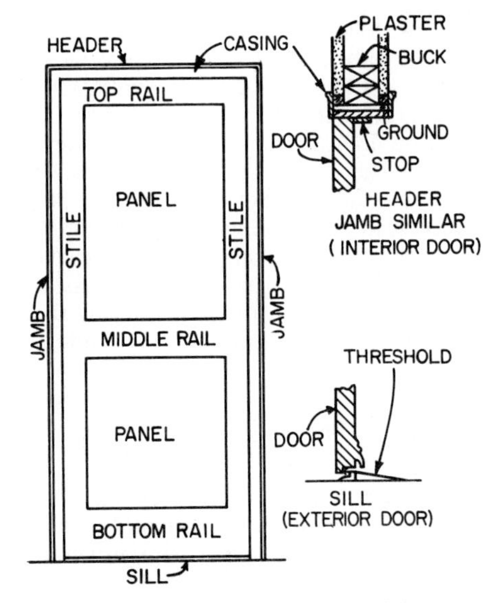

FIGURE 11.60 Typical paneled wood door.

face or provided with a wind shield. Also, design of the building facade should be checked, if necessary by wind-tunnel tests, to ensure that it will not create objectionable air movements.

Also, in tall buildings, there is likely to be a difference in air pressure between the inside and outside at entrances. This pressure results from air movements through stairways and shafts. When the building is heated in cold weather, warm air rises in the vertical passageways. When the building is cooled in warm weather, cool air flows downward through these passageways. The resulting pressure differential varies with building height and difference between indoor and outdoor temperatures. In many cases, this stack effect causes large airflows through entrance doors, significantly increases heating and cooling loads, and may make entrance-door operation difficult.

A commonly used means of reducing such loads is provision of an entrance vestibule with one set of doors leading outdoors and another set leading to the inside. In some cases, however, traffic flow is so great that at least one door of both sets of doors may be partly open simultaneously, permitting air to flow through the vestibule between the inside and outside of the building. Thus, the effectiveness of the vestibule is decreased. The loss of effectiveness may be reduced, however, by venting the vestibule to outdoors and providing compensating heating or cooling.

Another means of reducing air movements due to stack effect is installation of revolving doors (Art. 11.56). While in continuous contact with its enclosure, a revolving door permits entry and exit without extra force to offset the differential pressure between indoors and outdoors. (Building codes, however, require some swinging doors in conjunction with revolving doors.)

Where stack effects make operation of swinging doors difficult, operation may be made easier by hanging the doors on balanced pivots or equipping the doors with automatic operators. In some cases, it may be advantageous to replace the doors with automatic, horizontally sliding doors.

Automatic Entrances. Power-operated doors may be used to speed traffic flow at entrances to buildings with heavy traffic volume, such as retail stores, hospitals, and public buildings. Not only are such doors convenient for persons carrying packages or pushing carts, but they also are advantageous for physically handicapped persons. The doors are activated to open fully by persons approaching when their weight is applied to a floor mat or when they are detected by microwave or optical-electronic sensors. The doors are kept open until the persons are safely past, even if traffic stops or moves slowly. (ANSI A156.10, "Power-Operated Pedestrian Doors," American National Standards Institute.)

As an alternative to solid doors, an automatic entrance can be formed with an air curtain. At such an entrance, air is blown upward past a floor grate to the top of the door opening, to keep outdoor air from entering the building interior and prevent loss of interior air. Yet, human traffic can readily penetrate the curtain.

11.54.2 Door Materials

Wood is used in several forms for doors. When appearance is unimportant and a low-cost door is required, it may be made of boards nailed together. When the boards are vertical and held together with a few horizontal boards, the door is called a batten door. Better-grade doors are made with panels set in a frame or with flush construction.

Paneled doors consist of solid wood or plywood panels held in place by verticals called **stiles** and horizontals known as **rails** (Fig. 11.60). The joints between panels and supporting members permit expansion and contraction of the wood with atmospheric changes. If the rails and stiles are made of a single piece of wood, the paneled door is called solid. When hardwood or better-quality woods are used, the doors generally are veneered; rails and stiles are made with cores of softwood sandwiched between the desired veneer.

Tempered glass or plastic may be used for panels. In exterior doors, the lights should be installed to prevent penetration of water, especially in veneered doors. One way is to insert under the glass a piece of molding that extends through the door and is turned down over the outside face of the door to form a drip. Another way is to place a sheet-metal flashing under the removable outer molding that holds the glass. The flashing is turned up behind the inside face of the glass and down over the exterior of the door, with only a very narrow strip of the metal exposed.

Flush doors also may be solid or veneered. The veneered type has a core of softwood, while the flat faces may be hardwood veneers. When two piles are used for a face, they are set with grain perpendicular to each other.

Flush doors, in addition, may be of the hollow-core type. In that case, the surfaces are made of plywood and the core is a supporting grid. Edges of the core are solid wood boards.

Metal doors generally are constructed in one of three ways: cast as a single unit or separate frame and panel pieces; metal frame covered with sheet metal; and sheet metal over a wood or other type of insulating core. (See "ANSI Standard Nomenclature for Steel Doors and Steel Door Frames," A123.1, and "Recommended Standard Details, Steel Doors and Frames," Steel Door Institute, 712 Lakewood Center North, 14600 Detroit Ave., Cleveland, OH 44107.)

Cast-metal doors are relatively high-priced. They are used principally for monumental structures.

Hollow metal doors may be of flush or panel design, with steel faces having a thickness of at least 20 ga. Flush doors incorporate steel stiffeners; or an insulation core, such as hydrous calcium silicate, polyurethane foam, or polystyrene foam; or a honeycomb core as a lightweight support for the faces (Figs. 11.61 and 11.62). Voids between stiffeners may be filled with lightweight insulation. Panel doors may be of stile-and-rail or stile-and-panel construction with insulated panels. Also, a light-duty, 24-ga, steel-faced door with an insulated core is available.

Metal-clad (Kalamein) doors are of the swinging type only. They may be of flush or panel design, with metal-covered wood cores for stiles and rails and insulated panels covered with steel 24 ga or lighter.

Other Materials. Doors may be made wholly or partly transparent or translucent. Lights may be made of tempered glass or plastic. Doors made completely of glass are pivoted at top and bottom because the weight makes it difficult to support them with hinges or butts.

Sliding doors of the collapsible accordian type generally consist of wood slats or a light steel frame covered with textile. Plastic coverings frequently are used.

11.54.3 Swinging Doors

These are doors hinged near one edge to rotate about a vertical axis. Swinging doors are hung on butts or hinges (Art. 11.65). The part of a doorway to which a door is hinged and against which it closes is known as the door frame. It consists

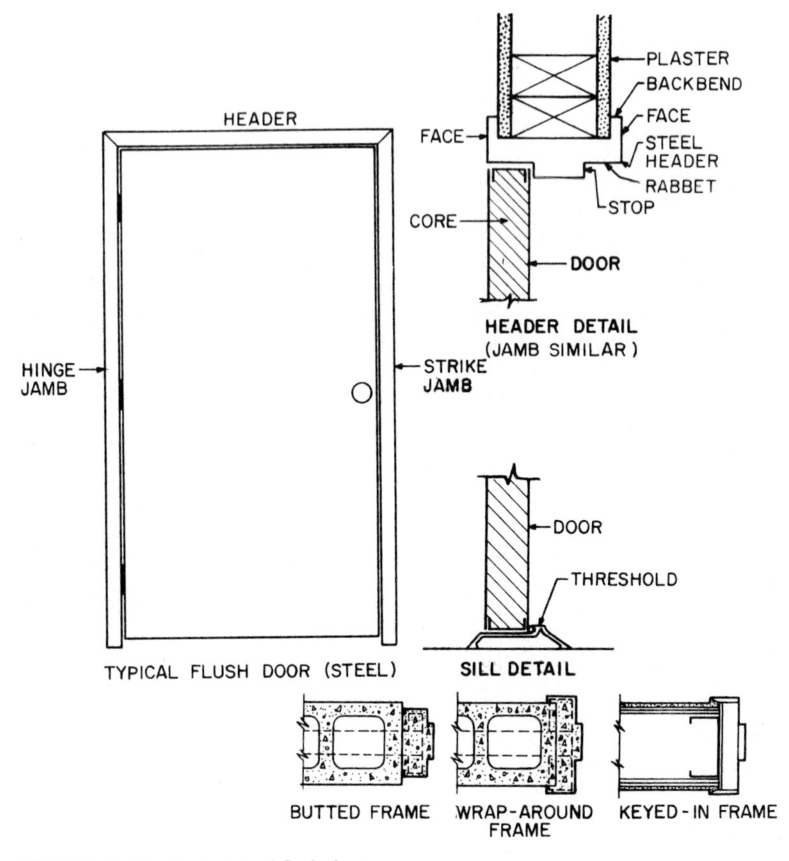

FIGURE 11.61 Typical steel flush door.

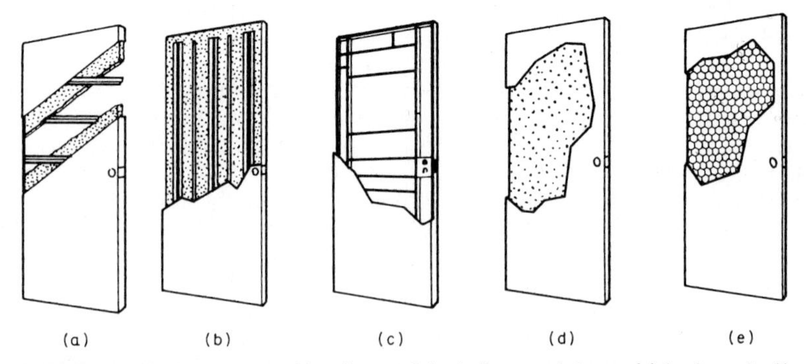

FIGURE 11.62 Some types of bracing used for hollow steel doors: (*a*) horizontal stiffeners; (*b*) vertical stiffeners; (*c*) grid of horizontal and vertical stiffeners; (*d*) rigid insulation core; (*e*) honeycomb core.

of two verticals, commonly called jambs, and a horizontal member, known as the header (Figs. 11.60 and 11.61). **Single-acting doors** can swing 90° or more in only one direction; **double-acting doors** can swing 90° or more in each of two directions.

To stop drafts and passage of light, the header and jambs have a stop, or projection, extending the full height and width, against which the door closes. The projection may be integral with the frame, or formed by attaching a stop on the surface of the frame, or inset slightly.

Door frames for swinging wood doors generally are fastened to rough construction members known as rough bucks, and the joints between the frame and the wall are covered with casings, or trim. For steel door frames, the trim is often integral with the frame, which is attached to the wall with anchors. ("Recommended Standard Details, Steel Doors and Frames," SDI 111, Steel Door Institute, 712 Lakewood Center North, 14600 Detroit Ave., Cleveland, OH 44107.)

Swinging doors are constructed in a variety of ways: usually flush (Fig. 11.61); stile and rail, with one or more recessed panels (Fig. 11.60); or stile and panel, with a wide center panel between hinge and lock stiles. Flush doors may be solid or hollow. Hollow doors usually are braced internally. Some of the types of bracing used for steel doors are shown in Fig. 11.62. Those with steel stiffeners (Fig. 11.62*a* to *c*) normally are thermally or acoustically insulated with a bat-type material or sprayed insulation. The core types (Fig. 11.62*d* and *e*) may have heavy-paper honeycomb, rigid foamed insulation, solid structural mineral blocking, or other bracing laminated to the door facings. (See also Art. 11.54.2.)

Wood or steel swinging doors and frames for them are available for ordinary applications in several stock sizes. Standard door thicknesses for these are 1⅜ in and 1¾ in. The doors are fabricated to fit in openings with heights of 6 ft 8 in, 7 ft, or 7 ft 2 in. Standard opening widths for single doors are 24, 28, 30, 32, and 36 in. Single 1¾-in stock doors are also available for opening widths of 40, 42, 44, and 48 in and heights of 8 ft. Standard opening widths for pairs of doors are twice as large as for single doors. (See SDI 100, "Recommended Specifications for Standard Steel Doors and Frames," Steel Door Institute.)

Nonstandard sizes are obtainable on special order, but certain precautions should be observed: Size and number of butts or hinges and offset pivots should be suitable for the door size. For rail-and-stile doors, check with the manufacturer to ensure that face areas will not be excessive for the stile width. Large doors should not be used where they will be exposed to strong winds. To prevent spreading of stiles of tall doors, push bars or other intermediate bar members should be attached to both stiles of doors over 8 ft high.

Package entrance doors are available from several manufacturers. The package includes a single door or pair of doors, door frame, and all hardware. Such doors offer the advantages of integrated design, assumption by the manufacturer of responsibility for satisfactory performance, usually quick delivery, and often cost savings.

Recommended mounting heights for hardware for swinging doors are shown in Fig. 11.63. For specific installations, refer to hardware templates supplied by door manufacturers.

Hand of Doors. The direction of swing, or hand, of each door must be known and specified in ordering such hardware as latches, locks, closers, and panic hardware, because the hand determines the type of operation required of them. The hand is determined with respect to the outside or key, or locking, side, following the conventions illustrated in Fig. 11.64.

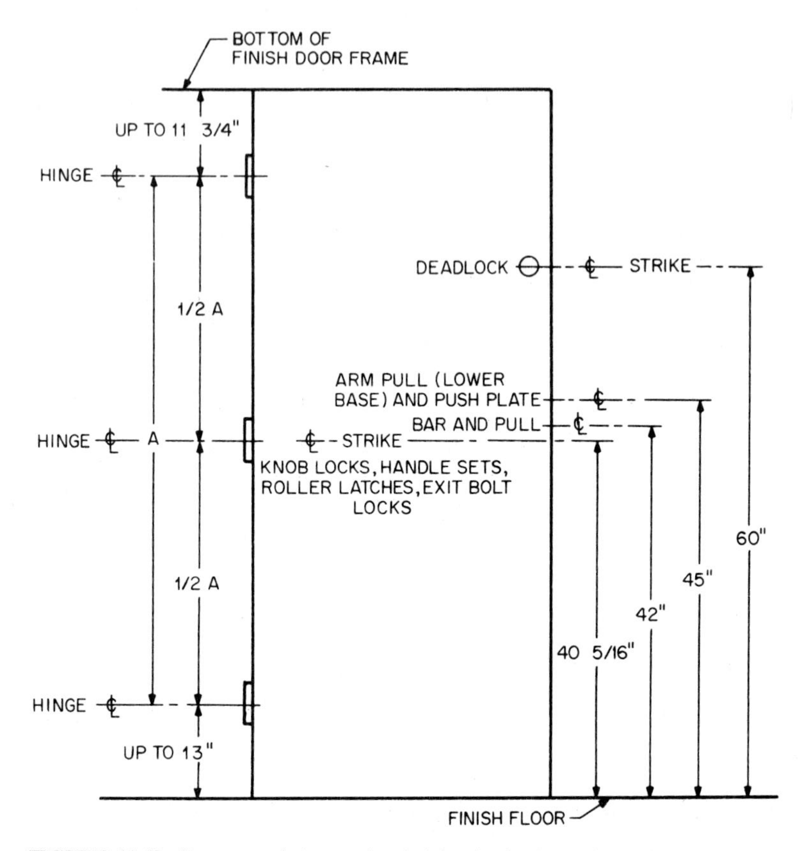

FIGURE 11.63 Recommended mounting heights for hardware for ordinary swinging doors.

Weatherstripping. To weatherproof the joint between the bottom of an exterior door and the threshold, a weatherstrip in the form of a hooked length of metal often is attached to the underside of the door. When the door is closed, the weatherstrip locks into the threshold to seal out water. Other types of weatherstripping, including plastic gaskets, generally are used for steel doors and may be

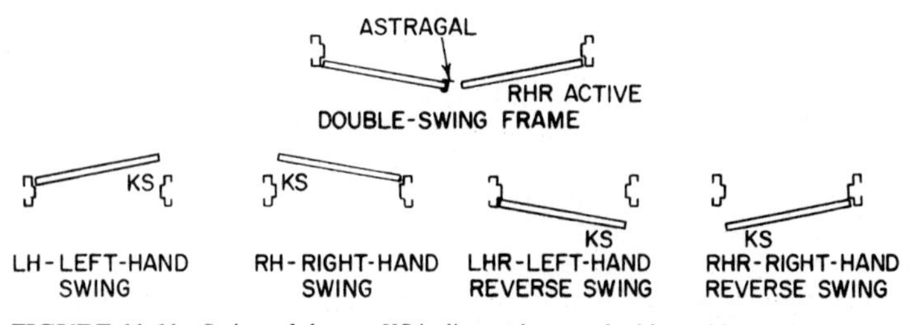

FIGURE 11.64 Swing of doors. *KS* indicates key, or locking, side.

installed on header and jambs ("Recommended Weatherstripping for Standard Steel Doors and Frames," SDI 111-E, Steel Door Institute).

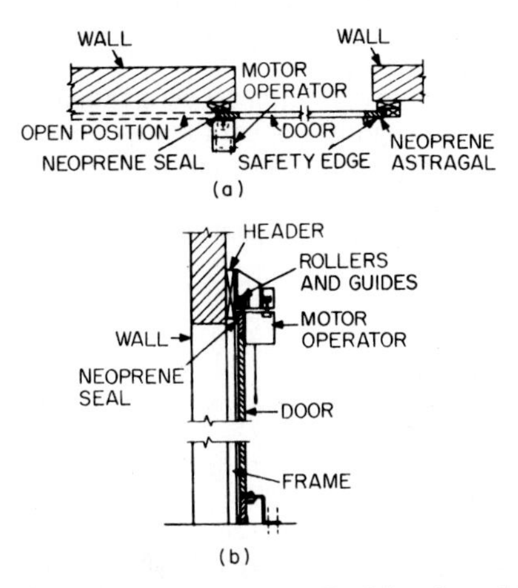

FIGURE 11.65 Horizontally sliding door: (*a*) horizontal section through the opening; (*b*) vertical section through the opening.

11.54.4 Ordinary Horizontally Sliding Doors

This type of door may roll on a track in the floor and have only guides at the top, or the track and rollers may be at the top and guides at the bottom (Fig. 11.65). Some doors fold or collapse like an accordion, to occupy less space when open. A pocket must always be provided in the walls on either or both sides, to receive rigid-type doors; with the folding or accordion types, a pocket is optional.

Horizontally sliding doors are advantageous for unusually wide openings, and where clearances do not permit use of swinging doors. Operation of sliding doors is not hindered by windy conditions or differences in air pressure between indoors and outdoors. Sliding doors, however, are not accepted as a means of egress.

11.54.5 Ordinary Vertically Sliding Doors

This type of door may rise straight up (Fig. 11.66), may rise up and swing in, or may pivot outward to form a canopy. Sometimes, the door may be in two sections, one rising up, the other dropping down. Generally, all types are counterweighted for ease of operation.

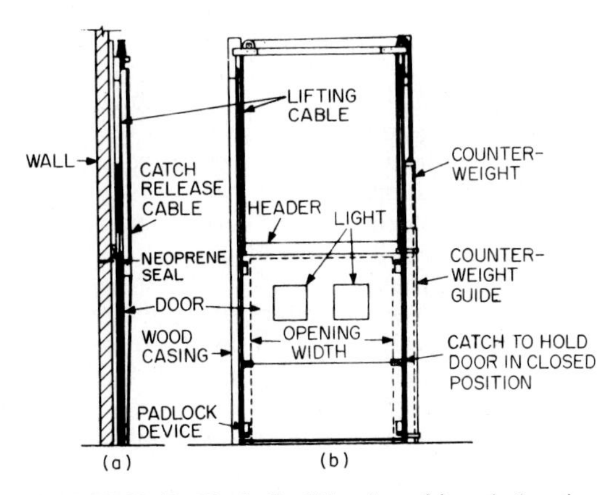

FIGURE 11.66 Vertically sliding door: (*a*) vertical section through the opening; (*b*) elevation of the door.

In design of structures to receive sliding doors, the clearance between the top of the doors and the ceilings or structural members above should be checked—especially the clearance required for vertically sliding doors in the open position. Also, the deflection of the construction above the door opening should be investigated to be certain that it will not cause the door to jam.

To keep out the weather, the upper part of a sliding door either is recessed into the wall above, or the top part of the door extends slightly beyond the inside face of the wall on the inside. Similarly, door sides are recessed into the walls or lap them and are held firmly against the inside. Also, the finished floor is raised a little above outside grade. Very large sliding doors require special study (Arts. 11.57 and 11.58).

11.55 FIRE AND SMOKESTOP DOORS

Building codes require fire-resistant doors in critical locations to prevent passage of fire. Such doors are required to have a specific minimum fire-resistance rating, and are usually referred to as fire doors. The codes also may specify that doors in other critical locations be capable of limiting passage of smoke. Such doors, known as smokestop doors, should have a fire-resistance rating of at least 20 min and should be tight fitting.

Fire protection of an opening in a wall or partition depends on the door frame and hardware, as well as on the door. All these components must be "labeled" or "listed" as suitable for the specific application. (See NFPA 80, "Fire Doors and Windows," National Fire Protection Association, 1 Batterymarch Park, Quincy, MA 02269.)

All fire doors must be self-closing or should close automatically when a fire occurs. In addition, they should be self-latching, so that they remain closed. Push-pull hardware should not be used. Exit doors for places of assembly for more than 100 persons usually must be equipped with fire-exit hardware capable of releasing the door latch when pressure of 15 lb, or less, is applied to the device in the direction of exit. Combustible materials, such as flammable carpeting, should not be permitted to pass under a fire door.

Fire-door assemblies are rated, in hours, according to ability to withstand a standard fire test, such as that specified in ASTM Standard E152. They may be identified as products qualified by tests by a UL label, provided by Underwriters Laboratories, Inc.; an FM symbol of approval, authorized by Factory Mutual Research Corp.; or by a self-certified label, provided by the manufacturer (not accepted by National Fire Protection Association and some code officials).

Openings in walls and partitions that are required to have a minimum fire-resistance rating must have protection with a corresponding fire-resistance rating. Typical requirements are listed in Table 11.18.

This table also gives typical requirements for fire resistance of exit doors, doors to stairs and exit passageways, corridor doors, and smokestop doors.

In addition, some building codes also limit the size of openings in fire barriers. Typical maximum areas, maximum dimensions, and maximum percent of wall length occupied by openings are given in Table 11.19.

Smokestop doors should be of the construction indicated in the footnote to Table 11.18.

They should close openings completely, with only the amount of clearance necessary for proper operation.

TABLE 11.18 Typical Fire Ratings Required for Doors

Door use	Rating, hr*
Doors in 3- or 4-hr fire barriers	3†
Doors in 2- or 1½-hr fire barriers	1½
Doors in 1-hr fire barriers	¾
Doors in exterior walls:	
Subject to severe fire exposure from outside the building	¾
Subject to moderate or light fire exposure from outside the building	⅓
Doors to stairs and exit passageways	1½
Doors in 1-hr corridors	⅓
Other corridor doors	0‡
Smokestop doors	⅓§

*Self-closing, swinging doors. Normally kept closed.

†A door should be installed on each side of the wall.

‡Should be noncombustible or 1¾-in solid wood-core doors. Some codes do not require self-closing for doors in hospitals, sanitariums, nursing homes, and similar occupancies.

§May be metal, metal covered, or 1¾-in solid wood-core doors (1⅜ in in buildings less than three stories high), with 1296-in² or larger, clear, wire-glass panels in each door.

["Standard for Fire Doors and Windows," NFPA No. 80; Life Safety Code, NFPA No. 101; "Fire Tests of Door Assemblies," NFPA No. 252, National Fire Protection Association, 1 Batterymarch Park, Quincy, MA 02269.

"Fire Tests of Door Assemblies," Standard UL 10(b); "Fire Door Frames," Standard UL 63; "Building Materials List" (annual, with bimonthly supplements), Underwriters Laboratories, Inc., 333 Pfingsten Road, Northbrook, IL 60062.

"Factory Mutual Approval Guide," Factory Mutual Research Corp., 1151 Boston-Providence Turnpike, Norwood, MA 02062.

"Hardware for Labeled Fire Doors," Door and Hardware Institute, 1815 N. Fort Myer Drive, Arlington, VA 22209.]

TABLE 11.19 Maximum Sizes of Openings in Fire Barriers

Protection of adjoining areas	Max area, ft²	Max dimension, ft
Unsprinklered	120*	12†
Sprinklers on both sides	150*	15*
Building fully sprinklered	Unlimited*	Unlimited*

*But not more than 25% of the wall length or 54 ft² per door if the fire barrier serves as a horizontal exit.

†But not more than 25% of the wall length.

SOURCE: Based on National Building Code, American Insurance Association, New York.

11.56 REVOLVING DOORS

This type of door is generally selected for entranceways carrying a continuous flow of traffic without very high peaks. They offer the advantage of keeping interchange of inside and outside air to a relatively small amount compared with other types of doors. They usually are used in combination with swinging doors because of the inability to handle large groups of people in a short period of time.

Revolving doors consist of four leaves that rotate about a vertical axis inside a cylindrical enclosure. Diameter of the enclosure generally is at least 6 ft 6 in, and the opening to the enclosure usually is between 4 and 5 ft.

Building codes prohibit use of revolving doors for some types of occupancies; for example, for theaters, churches, and stadiums, because of the limited traffic flow in emergencies. Where they are permitted as exits, revolving doors have limitations imposed on them by building codes. The National Fire Protection Association "Life Safety Code" allows only one-half unit of exit width per revolving door in computation of exit capacity, but some codes permit one unit per door. Also, revolving doors may not provide more than 50% of the required exit capacity at any location. The remaining capacity must be supplied by swinging doors within 20 ft of the revolving doors. Rotation speed must be controlled so as not to exceed 15 rpm. Each wing should be provided with at least one such bar and should be glazed with tempered glass at least 7⁄32 in thick. Some codes also require the doors to be collapsible.

11.57 LARGE HORIZONTALLY SLIDING DOORS

Door leaves in the horizontally sliding type are equipped with bearing-type bottom wheels and ride rails in the floor while top rollers operate in overhead guides. Two variations are in common use—telescoping and folding.

Telescoping doors (Fig. 11.67) are frequently used for airplane hangars. Normally composed of several leaves, they may be center parting or open to one side only, as illustrated in Fig. 11.67. Open doors are stacked in pockets at ends of the opening. They are built of wood, steel, or a combination of the two.

Telescoping doors are frequently operated by motors located in the end pockets. The motors drive an endless chain attached to tops of closing leaves. Remaining leaves are moved by a series of interconnecting cables attached to the powered leaf and arranged so that all leaves arrive at open or closed position simultaneously. Motor size ranges from 1 to 10 hp, travel speed of leaves from 45 to 160 ft/min.

The weight of the leaves must be taken by footings below the rails. Provision also must be made to take care of wind loads transmitted to the top guide channels by the doors and to carry the weight of these channels.

Folding doors (Fig. 11.68) are commonly used for subdividing gymnasiums, auditoriums, and cafeterias and for hangars with very wide openings. This type of door is made up of a series of leaves hinged together in pairs. Leaves fold outward, and when the door is shut, they are held by automatic folding stays. Motors that operate biparting doors usually are located in mullions adjacent to the center of the opening. The mullions are connected by cables to the ends of the opening, and when the door is to be opened, the mullions are drawn toward the ends, sweeping the leaves along. Travel speed may vary from 45 to 160 ft/min.

Chief advantage of folding over telescoping types is that only two guide channels are required, regardless of width. Thus, less metal is required for guide channels

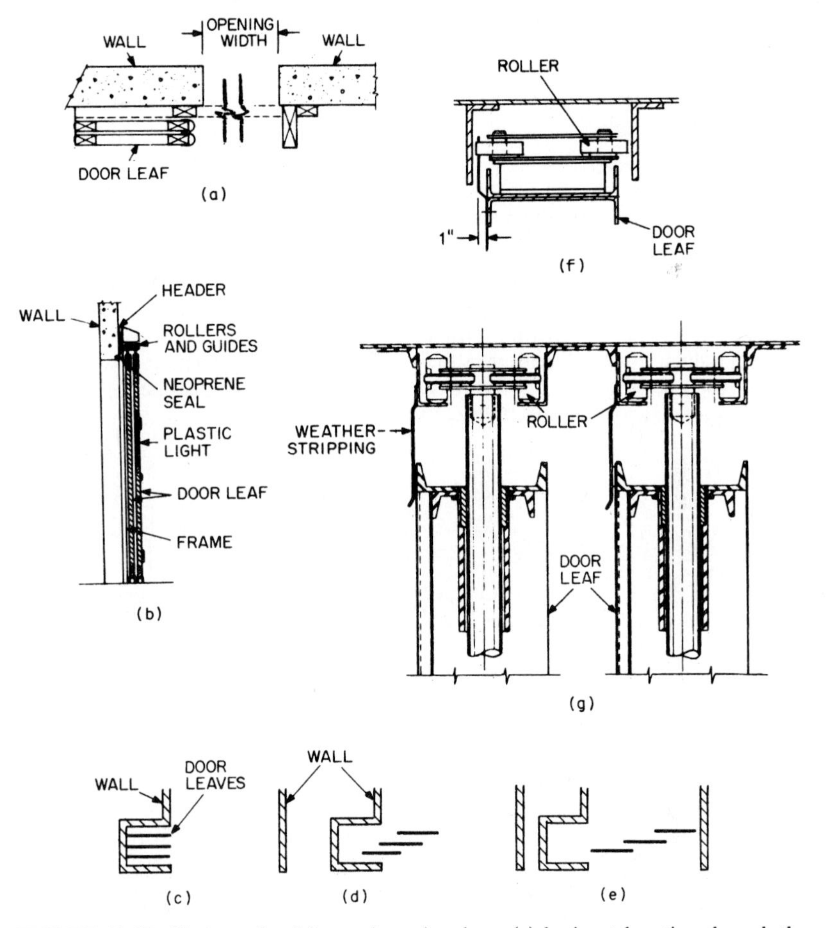

FIGURE 11.67 Horizontally sliding, telescoping door: (*a*) horizontal section through the opening; (*b*) vertical section through the opening; (*c*) door fully open; (*d*) door partly open (*e*) door closed; (*f*) and (*g*) alternative arrangements of top rollers.

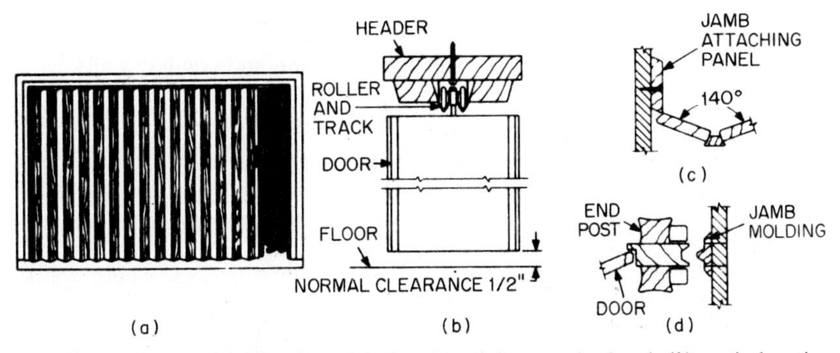

FIGURE 11.68 Wood folding door: (*a*) elevation of door nearly closed; (*b*) vertical section through the opening; (*c*) detail at the fixed end of the door; (*d*) detail at the strike jamb.

and rails, and less material for the supporting members above. Also, since wind loading is applied to leaves that are always partly in folded position, the triangular configuration gives the door considerable lateral stiffness. Hence, panel thickness may be less for folding than for telescoping doors, and a lighter load may be used for designing footings.

11.58 LARGE VERTICALLY SLIDING DOORS

When space is available above and below an opening into which door leaves can be moved, vertically sliding doors are advantageous. They may be operated manually or electrically. Leaves may travel at 45 to 60 ft/min. About 1½ ft in excess of leaf height must be provided in the pockets into which the leaves slide. So the greater the number of leaves the less space needed for the pockets. Figure 11.69 illustrates a vertically sliding, telescoping door.

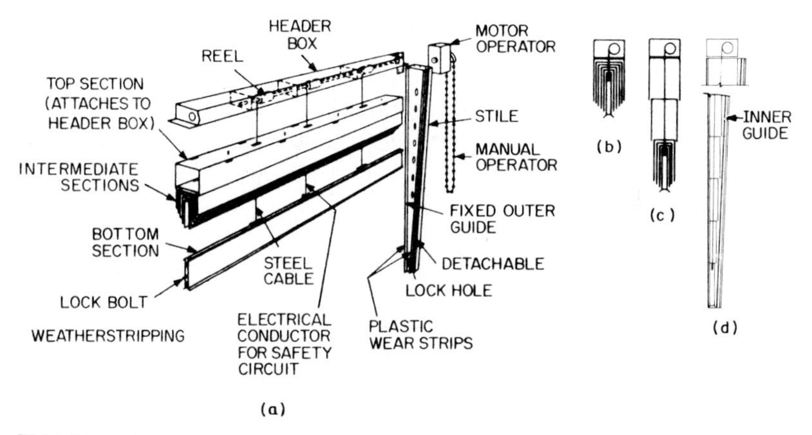

FIGURE 11.69 Vertically sliding, telescoping door: (*a*) door components; (*b*) door in open position; (*c*) door partly open; (*d*) door closed.

Vertically sliding doors normally are counterweighted. About 15 in of clear space back of the jamb line is needed for this purpose, but on the idler side only 4 in is required.

Vertical loads are transmitted to the door guides and then by column action to the footings. Lateral support is provided by the jamb and building walls.

11.59 LARGE SWINGING DOORS

When there is insufficient space around openings to accommodate sliding doors, swinging types may be used. Common applications have been for firehouses, where width-of-building clearance is essential, and railway entrances, where doors are interlocked with the signal system.

Common variations include single-swing (solid leaf with vertical hinge on one jamb), double-swing (hinges on both jambs), two-fold (hinge on one jamb and another between folds and leaves), and four-fold (hinges on both jambs and between each pair of folds).

The more folds, the less time required for opening and the smaller the radius needed for swing. Tighter swings make doors safer to open and allow material to be placed closer to supports.

11.60 HORIZONTALLY HINGED DOORS

Effective use of horizontally hinged swinging doors is made in such applications as craneway entrances to buildings. Widths exceeding 100 ft at or near the top of the building can be opened to depths of 4 to 18 ft. Frequently, horizontally sliding doors are employed below crane doors to increase the opening. If so, the top guides are contained in the bottom of the crane door; so the sliding door must be opened before the swinging door.

Top-hinged swinging doors are made of light materials, such as structural steel and exterior-grade plywood. The panel can be motor-operated to open in 1 min or less.

11.61 RADIATION-SHIELDING DOORS

These are used as a barrier against harmful radiation and atomic particles across openings for access to ''hot'' cells, and against similar radioactive-isotope handling arrangements and radiation chambers of high-energy x-ray machines or accelerators. Usually, they must protect not only personnel but also instruments even more sensitive to radiation than people.

Shielding doors usually are much thicker and heavier than ordinary doors, because density is an important factor in barring radiation. Generally, these special-purpose doors are made of steel plates, steel-sheathed lead, or concrete. To reduce thickness, concrete doors may be made of medium-heavy (240 lb/ft^2) or heavy (300 lb/ft^3) concrete, often made with iron-ore aggregate.

The heavy doors usually are operated hydraulically or by electric motor. Provision must be made, however, for manual operation if the mechanism should break down.

Common types of shielding doors include hinged, plug, and overlap. The hinged type is similar to a bank vault swinging door. The plug type, flush with the walls when closed, may roll on floor-mounted tracks or hang from rails. Overlap doors, surface-mounted, also may roll or hang from rails. In addition, vertical-lift doors sometimes are used.

BUILDERS' HARDWARE

Hardware is a general term covering a wide variety of fastenings and devices for operating or controlling the operation of movable building components, such as doors and windows. By common usage, the term builders' hardware generally covers only finishing hardware, but some rough hardware is discussed in this section.

In point of cost, the finishing hardware for a building represents a relatively small part of the finished structure. But the judicious selection of suitable items of hardware for all the many conditions encountered in construction and use of any building can mean a great deal over the years in lessened installation and maintenance costs and general satisfaction.

To make the best selection of hardware requires some knowledge of the various alternates available and the operating features afforded by each type. In this section, pertinent points relating to selecting, ordering, and installing some of the more commonly used builders' hardware items are discussed briefly.

11.62 SELECTION OF HARDWARE

Finishing hardware consists of items that are made in attractive shapes and finishes and that are usually visible as an integral part of the completed building. Included are locks, hinges, door pulls, cabinet hardware, door closers and checks, door holders, exit devices and lock-operating trim, such as knobs and handles, escutcheon plates, strike plates, and knob rosettes. In addition, there are push plates, push bars, kick plates, door stops, and flush bolts.

Rough hardware includes utility items not usually finished for attractive appearance, such as fastenings and hangers of many types, shapes, and sizes—nails, screws, bolts, studs secured by electric welding guns, studs secured by powder-actuated cartridge guns, inserts, anchor bolts, straps, expansion bolts and screws, toggle bolts, sash balances, casement and special window hardware, sliding-door and folding-door supports, and fastenings for screens, storm sash, shades, venetian blinds, and awnings.

Template hardware is used for hardware items that are to be fastened to metal parts, such as jambs or doors. Template items are made to a close tolerance to agree exactly with drawings furnished by the manufacturer. The sizes, shapes, location, and size of holes in this type of hardware are made to conform so accurately to the standard drawings that the ultimate fit of all associated parts is assured. In the case of hinges or butts, holes that are template-drilled usually form a crescent-shaped pattern (Fig. 11.73).

Hardware for stock may be nontemplate; however, certain lines are all template-made, whether for stock or for a specific order. Nontemplate items may vary somewhat and may not fit into a template cutout.

For use on wood or metal-covered doors, template drilling is not necessary; nontemplate hardware of the same type and finish can be used. Generally, there is no price difference between template and nontemplate items.

The operating characteristics of hardware items govern their selection, according to the particular requirements in each case. Then, the question of material, such as plastics, brass, bronze, aluminum, or steel, can be settled, as well as the finish desired. Selection of material and finish depends on the architectural treatment and decorative scheme.

Wrought, forged, and stamped parts are available for different items. Finishes include polished, satin, and oxidized. When solid metal is desired rather than a surface finish that is plated on metal, the order should definitely so specify.

National standards defining characteristics, sizes, dimensions, and spacings of holes, materials, and finishes for many items can assist greatly in identification and proper fit of hardware items to the parts on which they will be mounted.

An important point to bear in mind in connection with selection of hardware for a new building is that, in many instances, it is one of the earliest decisions that should be made, particularly when doors and windows are to be of metal. This is true despite the fact that the finishing hardware may not actually be applied until near the end of the construction period.

11.63 EFFECTS OF CODES AND REGULATIONS ON HARDWARE

Building codes, NFPA 80, "Standard for Fire Doors and Windows," and the "Life Safety Code" promulgated by the National Fire Protection Association contain many regulations directly affecting the use of builders' hardware. Some of these regulations permit tradeoffs in fire protection. For example, one regulation permits reduction or elimination of compartmentation of large floor areas in exchange for fully sprinklering a building.

Some of the major considerations of such regulations are life safety, security, and handicapped people (see Sec. 3). Life-safety requirements deal primarily with provision of safe egress in emergencies; requirements for disabled people deal with ease of circulation and use of facilities in buildings, and security requirements aim at making both unauthorized ingress and egress difficult. These requirements make conflicting demands on hardware. For example, local security ordinances often conflict with applicable building-code rules and may contain a clause exempting doors within means of egress from security requirements. Such an exemption, however, weakens security, because most doors are within a means of egress. The difficulty of designing or selecting hardware is further exacerbated when provisions for the disabled are superimposed on building-code and security ordinances.

Efforts are being made to find compromise solutions to the conflicting regulations. In the meantime, design professionals should be especially careful to keep abreast of the latest edition of all codes, regulations, and legislation governing the areas where their buildings are being erected.

11.64 STANDARDS FOR FINISHING HARDWARE

Several American National Standards present performance requirements for builders' hardware. These standards are available from the American National Standards Institute, Inc. (ANSI), 1430 Broadway, New York, NY 10018. The Builders Hardware Manufacturers Association, Inc., is the sponsor of the A156 series and submits additional standards or revisions of existing standards to ANSI for approval on a regular basis. The latest edition of the following should be consulted:

ANSI A156.1. Butts and Hinges

ANSI A156.2. Locks and Lock Trim

ANSI A156.3. Exit Devices

ANSI A156.4. Door Controls (Closers)

ANSI A156.5. Auxiliary Locks and Associated Products

ANSI A156.6. Architectural Door Trim

ANSI A156.7. Template Hinge Dimensions

ANSI A156.8. Door Controls (Overhead Holders)

ANSI A156.9. Cabinet Hardware

ANSI A156.10. Power-Operated Pedestrian Doors

ANSI A156.11. Cabinet Locks

ANSI/ASTM F 476. Standard Test Methods for Security of Swinging Door Assemblies

ANSI/ASTM F 571. Standard Practice for Installation of Exit Devices in Security Areas

In addition, there are a series of ANSI standards for the preparation of standard steel doors and frames to receive hardware—the A115 series, sponsored by the Door and Hardware Institute.

11.65 HINGES AND BUTTS

A hinge is a device permitting one part to turn on another. In builders' hardware, the two parts are metal plates known as leaves. They are joined together by a pin, which passes through the knuckle joints.

When the leaves are in the form of elongated straps, the device is usually called a hinge (Fig. 11.70 and 11.71). This type is suitable for mounting on the surface of a door.

When the device is to be mounted on the edge of a door, the length of the leaves must be shortened, because they cannot exceed the thickness of the door. The leaves thus retain only the portion near the pin, or butt end, of the hinge (Figs. 11.72 to 11.74). Thus, hinges applied to the edge of a door have come to be known as butts, or butt hinges.

Butts are mortised into the edge of the door. They are the type of hinge most commonly used in present-day buildings.

Sizes of butt hinges vary from about 2 to 6 in, and sometimes to 8 in. Length of hinge is usually made the same as the width; but they can be had in other widths.

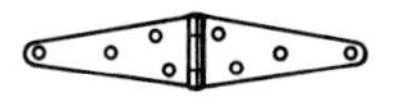

FIGURE 11.70 Heavy strap hinge.

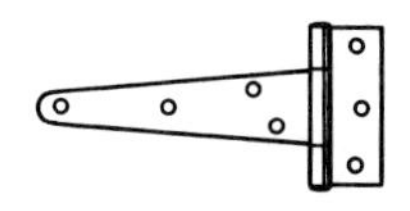

FIGURE 11.71 Heavy tee hinge.

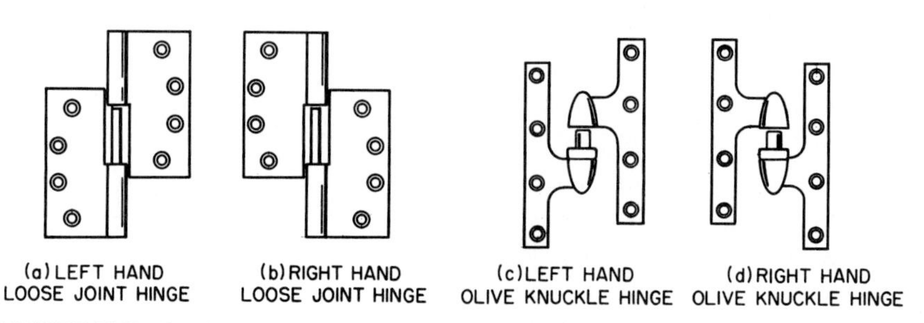

FIGURE 11.72 Special types of butt hinges that are handed. For hands of doors, see Fig. 11.79.

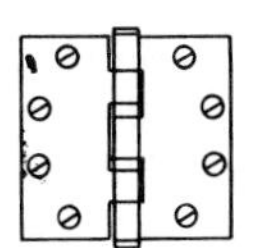

FIGURE 11.73 Bearing butt hinge.

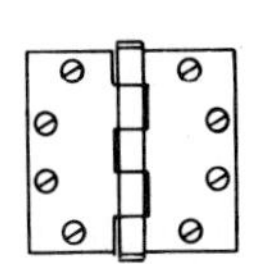

FIGURE 11.74 Plain bearing butt hinge.

Sometimes, on account of projecting trim, special sizes, such as 4½ × 6 in, are used.

For the larger, thicker, and heavier doors receiving high-frequency service, and for doors requiring silent operation, bearing butts (Fig. 11.73) or butts with Oilite bearings or other antifriction surfaces are generally used. It is also customary to use bearing butts wherever a door closer is specified.

Plain bearings (Fig. 11.74) are recommended for residential work. The lateral thrust of the pin should bear on hardened steel.

Unusual conditions may dictate the use of extra-heavy hinges or ball bearings where normal hinges would otherwise be used. One such case occurred in a group of college dormitories where many of the doors developed an out-of-plumb condition that prevented proper closing. It was discovered that the students had been using the doors as swings. Heavier hinges with stronger fastenings eliminated the trouble.

Two-bearing and four-bearing butt hinges should be selected, as dictated by weight of doors, frequency of use, and need for maintaining continued floating, silent operation. Because most types of butt hinges may be mounted on either right-hand or left-hand doors, it should be remembered that the number of bearing units actually supporting the thrust of the vertical load is only one-half the bearing units available. With a two-bearing butt, for example, only one of the bearings carries the vertical load, and with a four-bearing butt, only two carry the load. It should be noted, however, that some hinges (Fig. 11.72) are "handed" and must be specified for use on either a right-hand or left-hand door.

When butts are ordered for metal doors and jambs, "all machine screws" should be specified.

Location of Hinges. One rule for locating hinges for ordinary doors is to allow 5 in from rabbet of head jamb to top edge of top hinge and 10 in from finished floor line to bottom hinge. The third hinge should be spaced equidistant between top and bottom hinges. Another location for hinges is that of the standard steel frame and door (Fig. 11.63). The location varies a little among door manufacturers, but each is perfectly satisfactory from a functional standpoint.

Types of Hinge Pins. A very important element in the selection of hinges is the hinge pin. It may be either a loose pin or a fast pin.

Loose pins are generally used wherever practicable, because they simplify the hanging of doors. There are four basic pin types:

1. Ordinary loose pins
2. Nonrising loose pins
3. Nonremovable loose pins
4. Fast (or tight) pins

The **ordinary loose pin** can be pulled out of the hinge barrel so that the leaves may be separated. Thus, the leaves may be installed on the door and jamb independently, with ease and "mass-production" economy. However, these pins have a tendency—with resulting difficulties—of working upward and out of the barrel of the hinge. This "climbing" is caused by the constant twisting of the pin due to opening and closing the door. Present-day manufacture of hinges has tended to drift away from this type of pin, which is now found only in hinges in the lowest price scale.

The **nonrising loose pin** (or self-retaining pin) has all the advantages of the ordinary loose pin; but at the same time, the disadvantage of "climbing" is eliminated. The method of accomplishing the nonrising features varies with the type of hinge and its manufacture.

The **nonremovable loose pin** is generally used in hinges on entrance doors, or doors of locked spaces, which open out and on which the barrel of the hinge is therefore on the outside of the door. If such a door were equipped with ordinary loose-pin hinges or nonrising loose-pin hinges, it would be possible to remove the pin from the barrel and lift the door out of the frame and in so doing overcome the security of the locking device on the door. In a nonremovable loose-pin hinge, however, a setscrew in the barrel fits into a groove in the pin, thereby preventing its removal. The setscrew is so placed in the barrel of the hinge that it becomes inaccessible when the door is closed. This type of hinge offers the advantage of the ordinary loose-pin type plus the feature of security on doors opening out. Some manufacturers achieve the same results using other means, such as a safety stud.

The **fast (or tight) pin** is permanently set in the barrel of the hinge at the time of manufacture. Such pins cannot be removed without damaging the hinge. They are regularly furnished in hospital- or asylum-type hinges. The fact that the leaves of this type of hinge cannot be separated, however, makes the installation more difficult and costly. However, the difficulty is not too great, because with this type of hinge it is only necessary to hold the door in position while the screws for the jamb leaf are being inserted.

Ends of pins are finished in different ways. Shapes include flat-bottom, ball, oval-head, modern, cone, and steeple. They can be chosen to conform with type of architecture and decoration. Flat-button tips are generally standard and are supplied unless otherwise specified.

How to Select Hinges. One hinge means one pair of leaves connected with a pin. The number of hinges required per door depends on the size and weight of the door, and sometimes on conditions of use. A general rule recommends two butt hinges on doors up to 60 in high; three hinges on doors 60 to 90 in high; and four hinges on doors from 90 to 120 in high.

Table 11.20 gives general recommendations covering the selection of suitable hinges. Figure 11.75 indicates how hinge width is determined when it is governed by clearance.

The proper operating clearance between the hinged edge of a door and the jamb is taken care of in the manufacture of the hinges by "swaging" or slightly bending the leaves of the hinge near the pin. Since the amount of such bending required is determined by whether one or both leaves are to be mortised or surface-mounted, it is important in ordering hinges to specify the type of hinge needed to satisfy mounting conditions. If hinge leaves are to be mortised into both the edge of the door and the jamb, a full-mortise hinge is required (Fig. 11.76*a, c,* and *h*). If leaves are to be surface-applied to both the side of the door and the face of the jamb, a full-surface hinge is needed (Fig. 11.76*d*). If one leaf is to be surface-

TABLE 11.20 Hinges for Doors

Door thickness, in	Door width, in	Minimum hinge height, in
⅞ or 1	Any	2½
1⅛	To 36	3
1⅜	To 36	3½
1⅜	Over 36	4
1¾	To 41	4½
1¾	Over 41	4½ heavy
1¾ to 2¼	Any	5 heavy*

*To be used for heavy doors subject to high-frequency use or unusual stress.

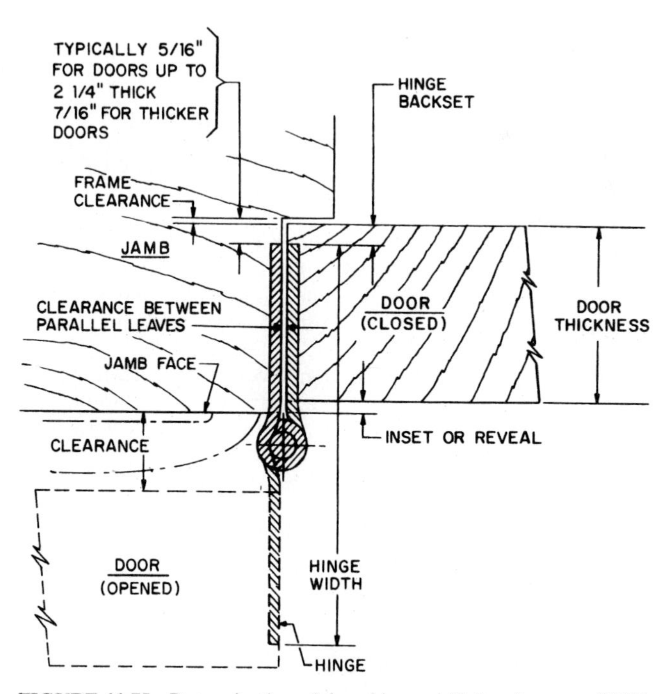

FIGURE 11.75 Determination of door hinge width for clearance. Width equals clearance plus inset (typically ½ in for doors up to 2¼ in thick and ⅜ in for thicker doors) plus twice the door thickness.

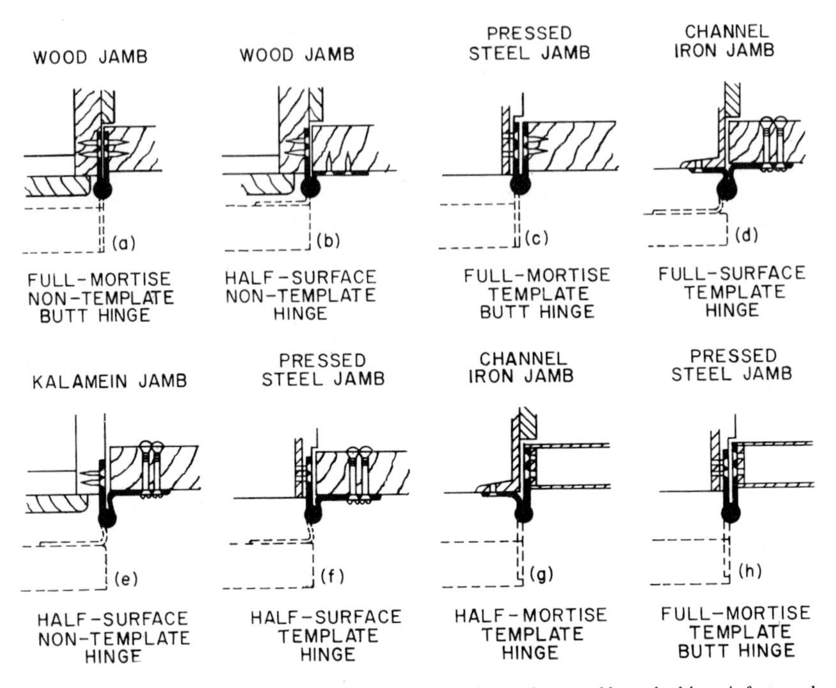

FIGURE 11.76 Hinge-mounting classification depends on where and how the hinge is fastened: (*a*), (*c*), and (*h*) full mortise; (*d*) full surface; (*b*), (*e*), and (*f*) half surface; and (*g*) half mortise.

applied and the other mortised, the hinge is a half-surface hinge or a half-mortise hinge, depending on how the leaf for the door is to be applied—half-surface if applied to door surface (Fig. 11.76*b*, *e*, and *f*) and half-mortise if mortised in the door end (Fig. 11.76*g*).

Exterior doors should have butts of nonferrous metal. Although chromium plating does not tarnish, it is not considered to be satisfactory on steel for exterior use. Interior doors in rooms where dampness and steam may occur should be of nonferrous metal and should also have butts of nonferrous metal. Butts for other interior work may be of ferrous metal.

Ferrous-metal butts should be of hardened cold-rolled steel. Where doors and door frames are to be painted at the job site, butts should be supplied with a prime-coat finish. For doors and trim that are to be stained and varnished, butts are usually plated.

Other types of hinges include some with a spring that closes the door. They may be either single- or double-acting. The spring may be incorporated in a hinge mounted on the door in the usual manner, or it may be associated with a pivot at the bottom of the door. In the latter case, the assembly may be of the type that is mortised into the bottom of the door, or it may be entirely below the floor.

11.66 DOOR-CLOSING DEVICES

These include overhead closers, either surface-mounted or concealed, and floor-type closers. These are some of the hardest-worked items in most buildings. To

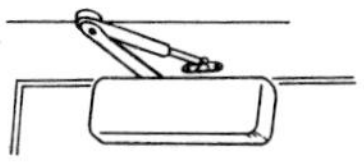

FIGURE 11.77 Door-closer spring closes the door, while a hydraulic mechanism (cylinder with piston) keeps the door from slamming.

get the most satisfactory operation at low first cost and low maintenance cost, each closer should be carefully selected and installed to suit the particular requirements and conditions at each door.

Most of these devices are a combination of a spring—the closing element—and an oil-cushioned piston, which dampens the closing action, inside a cylinder (Fig. 11.77). The piston operates with a crank or a rack-and-pinion action. It displaces the fluid through ports in the cylinder wall, which are closed or open according to the position of the piston in the cylinder. Opening of the door energizes the spring, thus storing up closing power. Adjustment screws are provided to change the size of the ports, controlling flow of fluid. This management makes the closer extremely responsive to the conditions of service at each individual door and permits a quiet closure, which at the same time ensures positive latching of the door.

While the fluid type of closer is preferred, pneumatic closers are also used, particularly for light doors, like screen doors.

Overhead door closers are installed in different ways, on the hinge side of the door or on the top jamb on the stop side of the head frame or on a bracket secured to the door frame on the stop side. Various types of brackets are available for different conditions. Also, when it is desired to install a closer between two doors hung from the same frame, or on the inside of a door that opens out, an arrangement with a parallel arm makes this possible. Other types of closers may be mortised into the door or housed in the head above the door.

Closers may be semiconcealed or fully concealed. Total concealment greatly enhances appearance but certain features of operation are limited.

An exposed-type closer should be mounted on the hinge side or stop side of the door unless there is real need for a bracket or parallel-arm mounting.

Whereas the use of brackets reduces headroom and may become a hazard, a parallel-arm closer mounted on the door rides out with the door, leaving the opening entirely clear.

When surface-applied door closers are used, careful consideration should be given to the space required in order that doors may be opened at least 90° before the closers strike an adjacent wall or partition.

Semiconcealed door closers are recommended for hollow metal doors. These closers are mortised into the upper door rail.

Various hold-open features also are available in different closer combinations to meet specific requirements. Another available feature is delayed action. This allows plenty of time to push a loaded vehicle through the opening before the door closes and also permits disabled people time to maneuver through a door opening.

When floor-type checking and closing devices are used, floor conditions should be carefully determined in order that there will be sufficient unobstructed depth available for their installation.

To get maximum performance from any door closer, it must be of ample size to meet the conditions imposed on it. If abnormal conditions exist, such as drafts or severe traffic, a closer of larger than the normal capacity should be employed. Installing too small a closer is an invitation to trouble. Manufacturers' charts should be used to determine the proper sizes and types of closers to suit door sizes and job conditions.

It is very important that door closers be installed precisely as recommended by the manufacturer. Experience has shown that a large percentage of troubles with closers results from disregard of mounting instructions.

In response to various code requirements for room-to-corridor protection, closer manufacturers have produced products that provide automatic door closing but still allow flexibility. These units permit the door to be held open in many locations of hold-open and yet upon a signal from a smoke detector, the hold-open mechanism disengages and the closer causes the door to close. A similar type of unit incorporates a device allowing the door to swing free as though it were not equipped with a door closer. As with the multiple-point hold-open device just described, the "swing-free' model, upon a signal from a smoke detector, cancels the closer nullifying device and the closer causes the door to close. These units are suitable in a variety of occupancies, particularly institutional.

Barrier-free provisions for disabled people dictate certain minimum opening resistance of doors. Depending upon the jurisdiction and location of doors, these forces range from a 5-lb force maximum to a 15-lb force maximum. There will be instances, because of the door size or air-pressure conditions, when the necessary closing force exerted by the closer to overcome these conditions will create opening resistance in excess of what is permitted. Power-operated doors or doors with a power assist specially made to solve this problem are recommended as a practical solution.

11.67 LOCKS, LATCHES, AND KEYS

The function of locks and latches is to hold doors in a closed position. Those known as rim locks or latches are fastened on the surface of the door. The ones that are mortised into the edge of the door are known as mortise locks (Fig. 11.78) or latches.

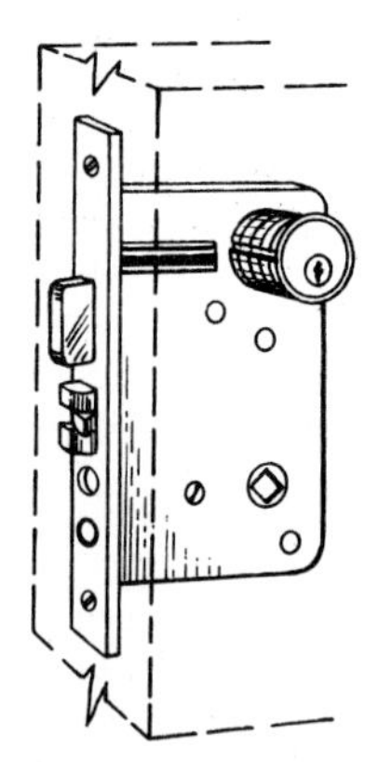

FIGURE 11.78 Mortise lock.

When the locking bolt is beveled, the device is usually referred to as a latch bolt; such a bolt automatically slides into position when the door is closed. A latch is usually operated by a knob or lever. Sometimes it may be opened with a key on the other side. Night latches came to be so called because they are generally used at night with other ordinary locks to give additional security.

Latches must take into account the **hand of the door** so the bevel will be right. A large percentage of latches are "reversible"—that is, they may be used on a right- or left-handed door (Art. 11.82). It is well, however, when ordering any lock to specify the hand of the door on which it is to be used.

If when you are standing outside the door the hinges are to your right and the door opens away from you, it is a right-hand, regular-bevel door; if the door opens toward you, it is a right-hand, reverse-bevel door. Similarly, if when you are outside the door, the hinges are to your left and the door swings away from you, it is a left-hand, regular-bevel door; if the door opens toward you, it is a left-hand, reverse-bevel door (Fig. 11.79).

When the locking bolt is rectangular in shape, it does not slide into position automatically when the door is closed; it must be projected or retracted by a thumb turn or key. This type of bolt is referred to as a dead bolt, and the lock as a dead lock. It may be worked with a key from one or both sides. Such a bolt is often used in conjunction with a latch. When the latch bolts and dead bolts are combined into one unit, it is known as a lock.

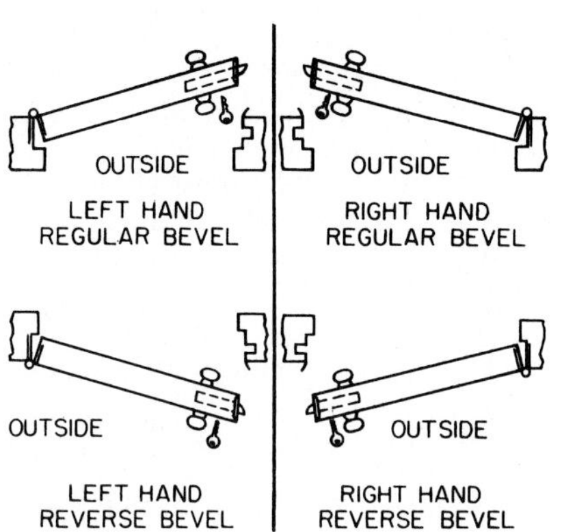

FIGURE 11.79 Conventional method of determining hand of doors and bevel.

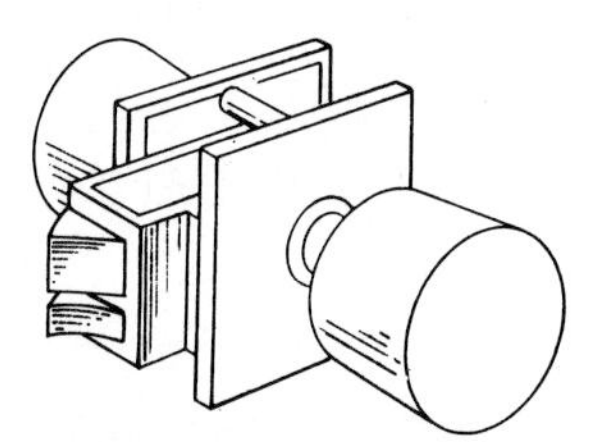

FIGURE 11.80 Unit lock.

For keyed locks that do not have dead bolts, it is desirable that the latch bolt be equipped with a dead latch (Fig. 11.82). This is a small plunger or an auxiliary dead latch (Fig. 11.80) that is held depressed when the door is closed and "deadlocks" the latch bolt so that it cannot be retracted by a shim, card, or similar device inserted between latch and door frame.

Various combinations of latches, dead bolts, knobs and keys, and locking buttons are applied to all types and kinds of doors. The exact combination most suitable for any given door is determined by a careful analysis of the use to which the door is to be put.

A point to bear in mind is that a uniform size should be selected, if practicable, for a project, no matter what the individual functions of the different locks may be. This makes possible the use of standard-size cutouts or sinkages on each installation. When this is done, not only is the cost of installation reduced, but any changes that may be made in the drawings as the job progresses, or any changes that may have to be made later are simplified, and special hardware is avoided.

Unit locks (Fig. 11.80) are complete assemblies that eliminate most of the adjustments during installation that would otherwise be necessary. These locks have merely to be slipped into a standard notch cut into a wood door or formed in a metal door.

Bored-in locks are another type that can be installed by boring standard-size holes in wood doors or by having uniform circular holes formed in metal doors. These bored-in locks are often referred to as tubular-lock sets or cylindrical-lock sets depending on how the holes have to be bored to accommodate them.

Tubular locks have a tubular case extending horizontally at right angles to the edge of the door. This type of case permits a horizontal hole of small diameter to be bored into the door at right angles to the vertical edge of the door; another small hole is required at right angles to the first hole to take care of the latching mechanism. Most tubular locks now produced fit the standard bores in steel doors conforming to ANSI A115.2 and A115.3, just as do cylindrical locks.

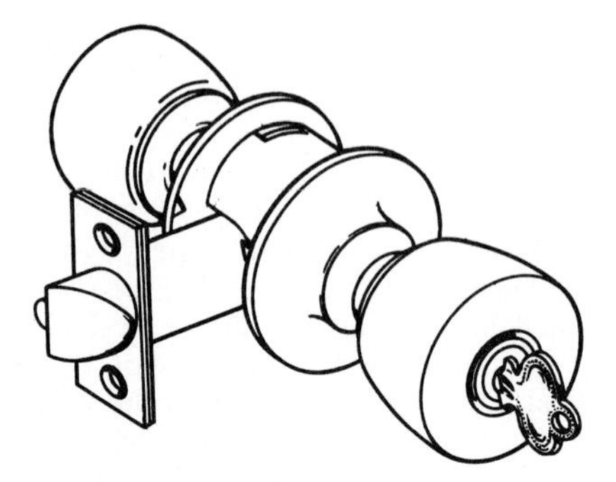

FIGURE 11.81 Cylindrical lock.

Cylindrical locks (Fig. 11.81), the other type of bored-in lock, have a cylindrical case requiring a relatively large-diameter hole in the door, bored perpendicular to the face of the door. This hole accommodates the main body of the lock. A hole of smaller diameter to take the latch bolt must be bored at right angles to the edge of the door.

When bored-in locks are used in hollow metal doors, a reinforcing unit is required in the door. This unit is generally supplied by the door manufacturer.

Exit devices are a special series of locks required by building codes and the National Fire Protection Association "Life Safety Code" on certain egress doors in public buildings. On the egress side, there is a horizontal bar running a minimum of one-half the width of the door. When pressed against, the bar releases the locking or latching mechanism, allowing the door to open. These devices are required to be labeled for safe egress by a nationally recognized independent testing laboratory. When used on fire doors, they must carry an additional label showing that the devices have also been investigated for fire. They then bear the name "fire exit hardware." These devices are available with various functions, including arrangements for having them locked from the outside and openable with a key. For all functions, however, they must be openable from the inside by merely depressing the cross bar.

Locking Bolts (Doors and Windows). Various types of bolts and rods are fastened to doors and windows for the purpose of securing them in closed position or to other doors and windows. Flush bolts and surface bolts, manual or automatic, are often used.

Top and bottom vertical bolts operated by a knob located at convenient height between them are known as **cremorne bolts.**

Keys. Locks are further classified according to the type of key required to retract the bolts. On all but the cheaper installations the principal type of key used is the cylinder key, which operates a pin-tumbler cylinder.

In the preceding locks, the locking cylinder, which is the assembly that supplies the security feature of the lock, is a cylindrical shell with rotatable barrel inside. The barrel has a longitudinal slot or keyway formed in it. The cross section of the keyway for every lock has a shape requiring a similarly designed key. Several holes (usually five or six) are bored through the shell and into the barrel at right angles to the axis of the barrel (Fig. 11.82). In each hole is placed a pair of pins—a driver pin in the shell end and a tumbler pin in the barrel end. Pins vary in length, and the combination of lengths differs from that of pins in other locks. A spring mounted in the shell end of each hole behind each driver pin forces these pins into the hole in the barrel, so that normally they are partly in the barrel portion of the hole and partly in the shell portion. Thus, the barrel is prevented from rotating in the shell to operate the bolt, and the door remains locked.

Keys have notches spaced along one side to correspond with the spacing and length of the pins. When a key is inserted in the slot in the barrel, each notch forces a tumbler pin back, against pressure from the spring. When the proper key

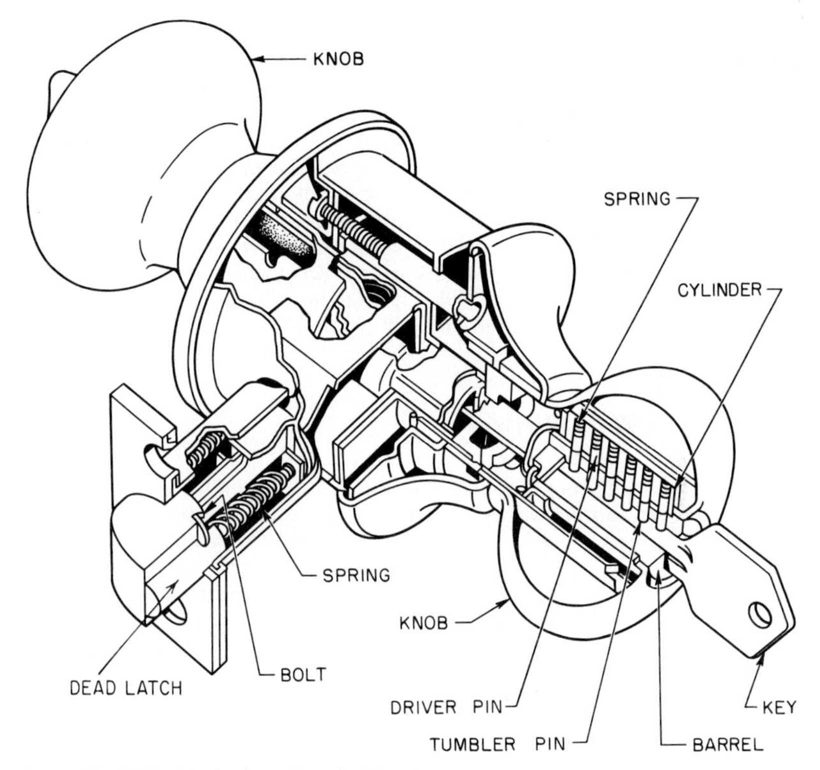

FIGURE 11.82 Mechanism of a cylindrical lock.

is inserted in the slot, each notch pushes its corresponding pair of pins just far enough back into the shell to bring the junction point of that pair exactly at the circumference of the barrel. The barrel then can be rotated within the shell by merely turning the key.

Turning the key operates a cam attached to the end of the barrel. The cam withdraws the bolt from its locked position.

The security feature of the cylinder lock results from the fact that only one series of notches in the key will correspond to the respective lengths of the several individual pins. With as many as five or six tumblers, it is apparent that many combinations are available.

By using split pins, different keyways, and various sizes and arrangements of the pins, master keying and grand master keying of cylinder locks are made possible. Thus one master key may be made to open all the separate locks on each floor of a building, while a grand master key will open any lock in any floor of that building. In case there are a number of such buildings in a group, a great-grand master key can be made that will open any door on any building in the group.

Security Precautions. The security of a building from the standpoint of unauthorized entry and life safety is affected by the proper selection of hardware. Although locks are important for security, the total system, including the building design, must be considered. A lock installed in a door that is loosely fitted in the frame may be ineffective. Hollow metal and aluminum frames that are not reinforced and anchored properly are easily "spread," and undetectable entry can result.

In high levels of master keying, more split pins must be used, and thus more "shear lines" are established. This makes the cylinder easier to pick. Hence, unnecessarily complicated master key systems should be avoided.

A good security lock is available with a dead bolt and a deadlocking latch bolt. Both bolts are retractable in one operation, merely by turning the inside knob. This satisfies requirements for security against unauthorized entry and life safety.

Hardware requirements for life safety and fire doors are covered by two National Fire Protection Association Standards, "Life Safety Code" and "Fire Doors and Windows." Both are incorporated in many building codes.

11.68 WINDOW HARDWARE

Sash balances are commonly used with double-hung windows as counterweights instead of weights and pulleys. Some balances have tape or cable with clock-type springs, which coil and uncoil as the sash is raised and lowered. Another type employs torsion springs with one end fixed to the side jamb of the window and the other arranged to turn as the sash goes up or down. The turning device in this type of balance may be a slide working in a rotatable spiral tube, or it may be a slotted bushing attached to the free end of the spring and fitted around a vertical rod attached to the sash. This rod (Fig. 11.83) is a flat piece of metal twisted into a spiral shape. The up-and-down movement of the sash causes the slotted bushing to revolve on its spiral sliding rod, thus winding and unwinding the spring. Still another type utilizes a vertical tension spring of the ordinary coil variety. One end of the spring is fixed and the other is fastened to the sash; the spring stretches or compresses in a vertical direction as the sash is moved up or down.

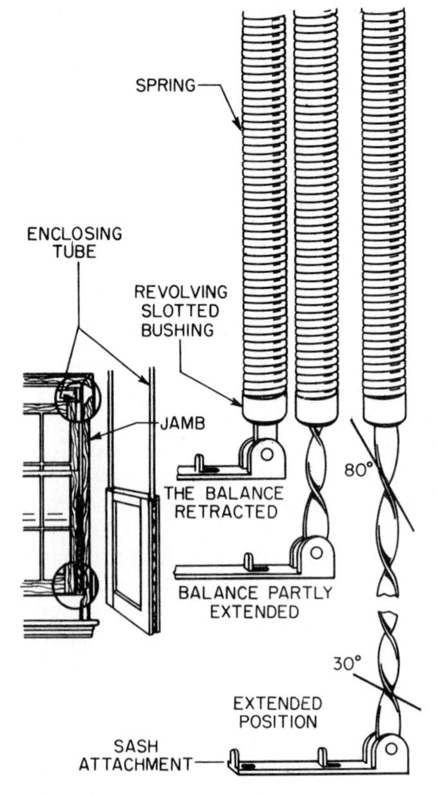

FIGURE 11.83 Constant-tension sash balance. (*Unique Balance Co.*)

Some sash balances combine weather-stripping with the balances. Others have friction devices to hold the sash in the desired position.

One patented type of sash balance known as the Unique sash balance incorporates a clever counterbalancing feature by making a change in the degree of pitch of the spiral rod from top to bottom, thereby controlling the increase and decrease of spring tension (Fig. 11.83). The pitch varies from 30 to 80°. As the spring turns, the changing spring tension is automatically compensated for by the variable pitch of the spiral rod sliding through the slotted bushing on the end of the spring. Thus the tension is equalized at every point of operation. This automatically prevents the sash from creeping or dropping of its own accord, without the necessity of introducing friction devices that interfere with easy operation.

Two sash balances are used per sash (one on each side) or four balances per double-hung window.

Other window hardware—locks, sash pulls, sash weights, and pulleys—are simple items, supplied as standard items with specific windows.

11.69 INSERTS, ANCHORS, AND HANGERS

Metal inserts of various types are cast into concrete floor slabs to serve as hangers or connectors for other parts of the structure that will be supported from the floor system. These inserts include electric conduit and junction boxes, supports for hung ceilings, slots for pipe hangers, fastenings for door closers to be set under the floor, and circular metal shapes for vertical pipe openings.

Metal anchors of various types are used in building construction. Each type is specifically shaped, according to the purpose served. These include anchors for securing stone facings to masonry walls, anchors for fastening marble slabs in place, column anchor bolts set in foundations, and anchor bolts for fastening sills to masonry.

Joist hangers are used for framing wood joists into girders and for framing headers into joists around stairwells and chimneys.

Various types of expansion bolts, screw anchors, and toggle bolts are available for securing fixtures, brackets, and equipment to solid material, such as masonry, brick, concrete, and stone. Anchors also can be had for fastening to hollow walls, such as plaster on metal lath in furred spaces and hollow tile. The best device to use depends on the requirements in each case.

For use with practically any materials, including soft and brittle ones, such as composition board, glass, and tile, fiber screw anchors with a hollow metal core find a universal application. These plugs with braided metal cores possess many advantages: They can be used with wood screws and with lag screws. The flexible construction permits the plug to conform to any irregularities. Because of this elastic compression, the fibers are compressed as the screw enters. Screws can be unscrewed and replaced. Shock and vibration have no effect on gripping power. The plugs come in about 40 sizes to fit anything from a No. 6 screw to a ⅝-in lag screw. In practice, a hole is drilled first, the plug is driven into it, and then the screw is inserted, expanding the plug against the sides of the hole.

For some fastenings, one-piece drive bolts are hammered like a nail into prepared holes in masonry or concrete. Other types of expansion bolts have expansion shields or are calked into place. The expansion shield is expanded in the hole by a tapered sleeve forced against a cone-shaped nut by the tightening of a bolt threaded into the cone. These types are not recommended for soft or brittle materials.

For thin hollow walls, toggle bolts equipped with spring-actuated wings are used (Fig. 11.89). The wings will pass into the hole in folded position. After entering the hollow space, the wings open out, thereby obtaining a secure hold.

For fastening lightweight materials to nailable supports, stapling machines are extensively used. In one patented system, for example, staples secure acoustic tile to wood furring strips. In this system, invisible fastenings are obtained by using a full-spline suspension for the kerfed pieces of tile and a special stapling machine adjusted to function at the proper distance below the furring strips. The machine staples the splines (at the joints of the tiles) to the supporting strips, giving a speedy, economical, and secure fastening.

11.70 NAILS

Wire nails, made of mild steel, are commonly used for most nailing purposes. Cylindrical in shape, they are stronger for driving than cut nails and are not so liable to bend when driven into hardwoods.

Cut nails, sheared from steel plate, are flat and tapered. They have holding power considerably greater than wire nails. They are usually preferred to wire nails for fastening wood battens to plaster; also in places where there is danger that the nails may be drawn out by direct pull. Cut nails are frequently used for driving into material other than wood. They are generally used for fastening flooring. When driven with width parallel to grain of wood, they have less tendency to split wood than wire nails.

The length of cut and wire nails is designated by the unit "pennies." Both cut and wire nails come in sizes from 2-penny (2d) which are 1 in long, to 60-penny (60d), which are 6 in long. Above 6 in, the fasteners are called **spikes.** They run in 7-, 8-, 9-, 10-, and 12-in lengths.

Various gages or thicknesses of nails, and different sizes and shapes of heads and points (Figs. 11.84 and 11.85), are available in both wire and cut nails. Certain types have distinguishing names. For example, the term **brads** is applied to nails with small heads, suitable for small finish work.

Common brads are the same thickness as common nails but have different heads and points. **Clout nails** have broad flat heads. They are used mostly for securing gutters and metalwork.

Casing nails are about half a gage thinner than common wire nails of the same length; **finishing nails,** in turn, are about half a gage thinner than casing nails of the same length. **Shingle nails** are half a gage to a full gage thicker than common nails of the same length.

Certain manufacturers have developed nails for special purposes that hold tighter and longer. Among these are threaded nails (Fig. 11.84), which combine the ease of driving of the ordinary nail with much greater holding power.

One type is a spiral-threaded flooring nail. These nails turn as they drive, minimizing splitting of the tongues of the floor boards. These nails are said to actually grip more firmly with the passage of time. A nailing machine is available for driving these nails. In one operation, it starts the nail, drives the joint between the flooring strips tight, and drives and sets the nail. With this machine, mashed tongues and marred edges of the wood are avoided.

Nails of aluminum and stainless steel are particularly useful in exposed locations where rust or corrosion of steel nails might cause unsightly stains to form on exterior finished surfaces. These nails are now made in most of the usual sizes, including special spiral-threaded nails.

Galvanized nails are used for fastening slate and shingles. These nails are sometimes used in exposed locations as protection against corrosion.

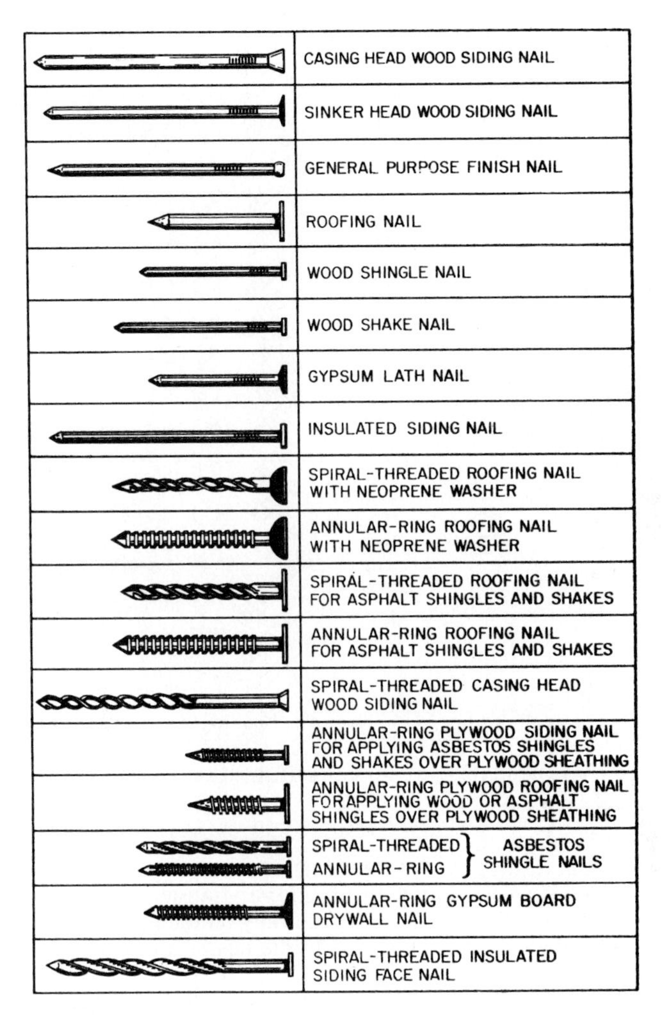

FIGURE 11.84 Nail heads and points.

11.71 SCREWS

These are used for applying hardware of all descriptions; also for panel work, cabinet work, all types of fine finish work, and support of electric and plumbing fixtures. Screws have greater holding power than nails and permit easy removal and replacement of parts without injury to the wood or finish. Screws avoid danger of splitting the wood or marring the finish, when the screw holes are bored with a bit.

Screws are made in a large variety of sizes and shapes to suit different uses (Fig. 11.86) and they are made of different metals to match various materials. A much-used type of head, other than the ordinary single-slot type, is the **Phillips head,**

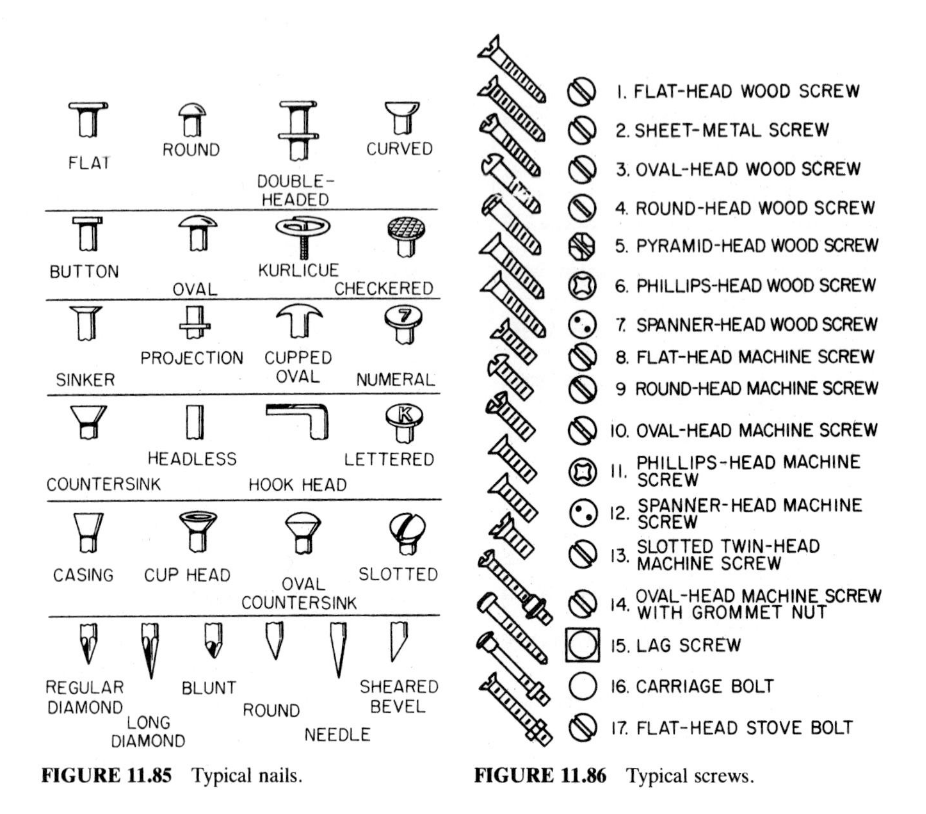

FIGURE 11.85 Typical nails.

FIGURE 11.86 Typical screws.

which has two countersunk slots at right angles to each other. The head keeps the screwdriver exactly centered during driving and also transmits greater driving power to the screw, while holding the screwdriver firmly on the head. Phillips heads are smoother at the edges, because the slots in the head do not extend to the outer circumference.

Steel screws for wood vary in length from ¼ to 6 in. Each length is made in a variety of thicknesses. Heads may be ordinary flat (for countersinking), round, or oval.

Parker screws for securing objects to thin metal are self-threading when screwed into holes of exactly the correct size.

Sizes of screws are given in inches of length and the gage of the diameter. Lengths vary by eighths of an inch up to 1 in, by quarters from there up to 3 in, and by halves from 3 to 5 in. Unlike wire gages, the smallest diameter of a screw gage is the lowest number; the larger the number, the greater is the diameter in a screw gage. Gage numbers range from 0 to 30.

Lag screws are large, heavy screws used for framing timber and ironwork (Figs. 11.86 and 11.89). Lengths vary from 1½ to 12 in and diameters from 5⁄16 to 1 in. Two holes should be bored for lag screws, one to take the unthreaded shank without binding and the other (a smaller hole) to take the threaded part. This smaller hole is usually somewhat shorter in length than the threaded portion. Lag screws usually have square heads and are tightened with a wrench.

11.72 WELDED STUDS

Studs electrically welded to the steel framework of a building are often used as the primary element for securing corrugated siding and roofing, insulation, metal window frames, ornamental outer skins, anchorages for concrete, and other items. The welded studs thus form an integral part of the basic structure.

Many types of studs or fasteners are available, each one being designed for a particular purpose. Most of the studs have threads formed on them, either externally or internally. Some of the studs are designed to have the material impaled over them and riveted to them.

The studs are designed so as to project the exact distance desired after they have been welded. Special sealing washers and nuts are usually placed on each stud over the flat sheet of material being fastened. Tightening the nut or expanding the head of the stud with a riveting hammer then makes the fastening complete, weathertight, and secure.

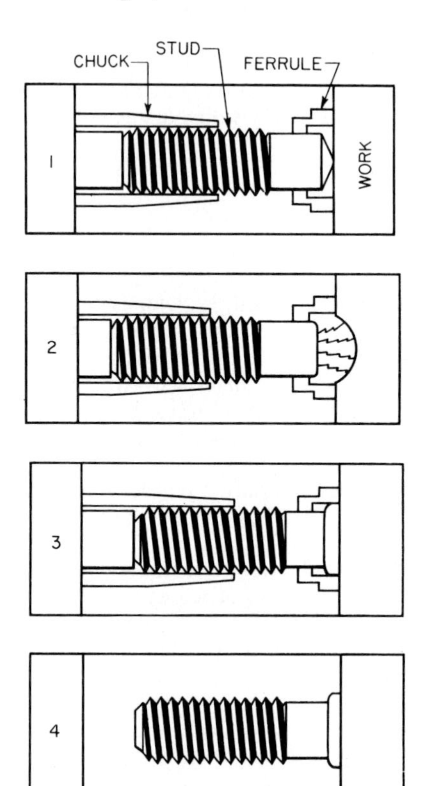

FIGURE 11.87 Steps in stud welding: (1) Press the welding end of the gun against the work plate. (2) Press the trigger to create an electric arc between the stud and the plate and melting portions of each. (3) The stud is automatically plunged into the molten pool. (4) Remove the gun and knock off the ferrule.

Studs are usually cadmium-plated mild steel or stainless steel. The latter is recommended for corrosive atmospheric conditions. Flux to assure a good weld is contained in the center of each stud at the welding end.

Equipment required for stud welding includes a stud-welding gun, a control unit for adjusting the amount of welding current fed to the gun, and a power source. The source of welding current may be a direct-current generator, a rectifier, or a battery unit. When a welding generator is used, the minimum National Electrical Manufacturers Association rating should be 400 A.

The welding gun usually has a chuck for holding the stud in position for welding (Fig. 11.87), and a leg assembly holding an adjustable-length extension sleeve into which the necessary arc-shielding ferrule is inserted. Expendable ceramic arc ferrules are generally used to confine the arc and control the weld fillet. After each weld, the arc ferrule is broken and removed by a light tap with any convenient metal object. In some cases where the required finished stud is short, an extra length is provided on the stud as furnished, for proper chucking in the gun. A groove is provided so the extra length can be easily broken off after the stud is welded.

A method of fastening corrugated-metal sheet to steel framing is shown in Fig. 11.88. This method permits application of siding and roofing entirely

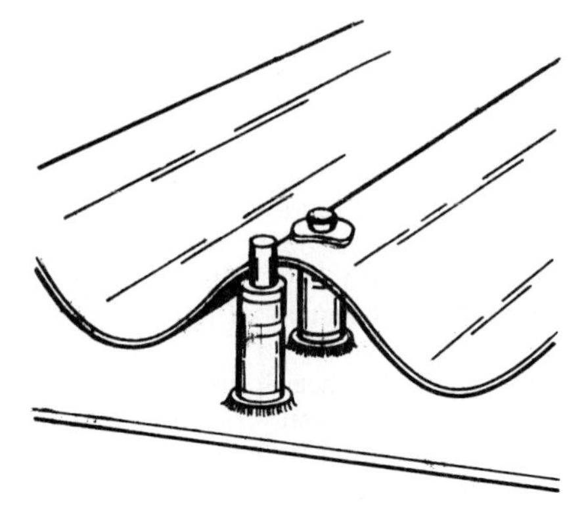

FIGURE 11.88 Corrugated metal impaled on and fastened in place with a welded stud.

from the outside. A worker is not required to be on the inside because there are no clips or fasteners on the interior. The expense of an interior scaffold is thus saved.

In securing corrugated metal, it is sometimes desirable to weld the studs to the steel frame in advance of placing predrilled sheets. In these cases, templates for quickly marking sheet and stud locations may be employed, as desired. Stud welding is applicable to any steel frame composed of standard structural steel. Steels of the high-carbon variety such as rerolled rail stock are not suitable for stud welding.

The manufacturer's recommendations should be followed in selecting the best type and size of stud for each specific installation. The leading manufacturers have direct field representatives in all areas who can supply valuable advice as to the best procedures to follow.

11.73 POWDER-DRIVEN STUDS

For many applications requiring the joining of steel or wood parts, or fastenings to concrete, steel, and brick surfaces, powder-actuated stud drivers are found to decrease costs because of their simplicity and speed. These drivers use a special powder charge to drive a pin or stud into relatively hard materials. The key to their efficiency is the proper selection of drive pins and firing charges. Because of the high velocity, the drive pin, in effect, fuses to the materials and develops the holding power.

Pullout tests of these driven studs prove remarkable holding power. Average pullout in 3500-psi concrete exceeds 1200 lb for 10-ga studs, and 2400 lb for ¼-in-diameter studs. In steel plate the pullout resistance is still greater.

All that is required is a stud driver, the correct stud, and the correct cartridge, as recommended by the manufacturer for each specific set of fastening conditions. There are about a half dozen different strengths of cartridge, each identified by color, and some two dozen varieties of studs. Some studs have external threads, others internal threads. A plastic coating protects the threads from damage while driving.

The drivers will force studs into steel up to about ½ in thick, into concrete through steel plates up to about ¼ in thick, into concrete through various thicknesses of woods, and into steel through steel up to ¼ in thick.

Powder-driven studs should never be driven into soft materials or into very hard or brittle materials, such as cast iron, glazed tile, or surface-hardened steel. Neither should they be used in face brick, hollow tile, live rock, or similar materials.

In driving studs a suitable guard must be used for each operation, and the driver must be held squarely to the work. If the driver is not held perpendicular, a safety device prevents firing of the charge. There also must be sufficient backup material to absorb the full driving power of the charge. Studs cannot safely be driven closer than ½ in from the edge of a steel surface, or closer than 3 in from the outside edge of concrete or brick surfaces.

11.74 BOLTS

A bolt is a cylindrical fastener that consists of a head shaped to facilitate turning with a wrench and a threaded shank of smaller diameter (Fig. 11.89). To fasten two or more building components together, the bolt is placed in a prepared hole that extends through the components and is slightly larger than the shank. A nut, usually of a type shown in Fig. 11.89, is threaded onto the shank and tightened to draw the parts together.

Bolts used in connections subjected to heavy loads, such as connections in structural framing, are described in Secs. 7 through 10.

A washer often is placed under the head of a bolt and under a nut. A washer can accomplish several purposes: (1) Distribute the compressive force exerted by

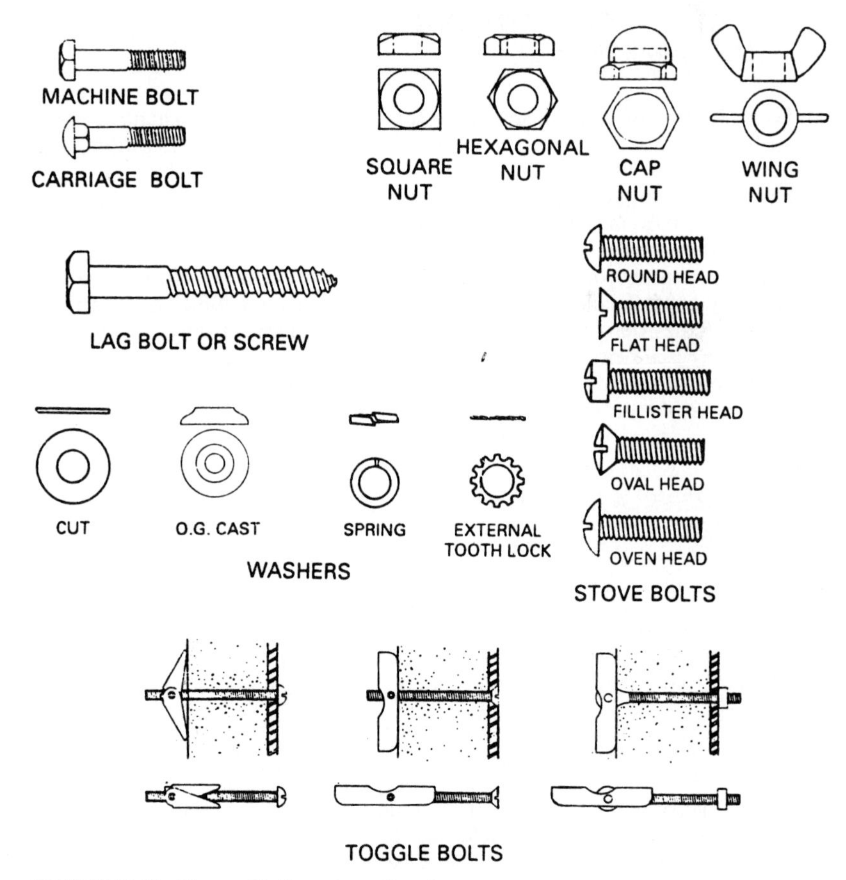

FIGURE 11.89 Types of bolts, nuts, and washers.

a tensioned bolt over a wider area than that of the contact surface of the bolt head or nut. (2) Prevent the head or nut from turning and thus prevent the components of a connection from separating. (3) Serve as a seal against moisture penetration. (4) Insulate incompatible materials from each other. (5) Prevent the head or nut from being pulled into the hole when the hole has about the same diameter as the head or nut.

ACOUSTICS

Acoustics, derived from a Greek word meaning to hear, refers to generation, detection, transmission, absorption, and control of sounds. Acoustics is part science and part art, but modern studies are stripping away much of its mystique. As a result, it is possible to plan and predict the acoustics of finished spaces with reasonable certainty.

11.75 SOUND PRODUCTION AND TRANSMISSION

Sound is a vibration in an elastic medium. It is a simple form of mechanical energy, and can be described by the mathematics associated with the generation, transmission, and control of energy.

Almost any moving, vibrating, oscillating, or pulsating object is a potential sound source. Usually, though, vibratory sources radiate enough energy to be audible to humans or felt by them.

Sound waves in air (or other gases or fluids) travel outward from the source, transmitted by air molecules, like a rapidly expanding soap bubble. Any particular group of molecules behaves like a pulsating balloon, moving only slightly, while the wave progresses swiftly to great distances. Transmission (flow), or prevention of transmission of sound, and conversion of sound energy to a nonaudible form are the function of so-called acoustical materials.

11.76 NOMENCLATURE FOR ANALYSIS OF SOUND

Cycle. A complete single excursion of a vibrating molecule.

Frequency. The number of cycles of vibration in a given unit of time, usually cycles per second (cps) or hertz (Hz).

Sound Wave. The portion of a sound between two successive compressions or rarefactions.

Wavelength. The distance between two successive rarefactions or compressions in a sound wave.

Amplitude. Maximum displacement, beyond its normal, or rest, position, of a vibrating element. In most audible sounds, these excursions are very small, although low-frequency sound may cause large excursions (as would be observed in the motion of a loudspeaker cone reproducing very low frequency sounds at audible level). Amplitude of motion is related to the increased pressure created in the medium, and to the intensity of the energy involved.

Velocity. Speed at which a sound impulse travels (not the speed of movement of any particular molecule). In a given medium, under fixed conditions, sound velocity is a constant. Therefore, the relationship between velocity, frequency, and wavelength can be expressed by the equation:

$$\text{Velocity} = \text{frequency} \times \text{wavelength} \tag{11.8}$$

Because velocity is constant in air, low-frequency sound has long wavelengths, and high-frequency sound has short wavelengths. This is important to remember in acoustical design.

In addition to the longitudinal (compression or "squeeze") waves by which sound travels, there is another type of vibrational motion to which most building materials and construction systems are subjected, a **transverse wave** (Fig. 11.90). This is the familiar motion of a vibrating string or reed. This type of wave, too, transmits energy that can be felt or heard, or both.

FIGURE 11.90 Transverse waves.

Sheets or panels, studs and joists, hangers and rods, and similar slender and somewhat flexible members are particularly apt to vibrate in a transverse mode and to transmit sound energy along their length, as well as radiating it from their surfaces to the surrounding air.

11.77 SOUND CHARACTERISTICS AND EFFECTS ON HEARING

Sound travels via elastic media. Hard, rigid, dense materials make excellent transmission paths, because they can accept much energy readily. Sound velocity is inversely proportional to the density of the medium, as can be seen from Eq. (11.9).

$$\text{Velocity} = k\sqrt{\frac{\text{modulus of elasticity}}{\text{density}}} \tag{11.9}$$

where k is a constant. Usually, however, dense materials often have a much larger modulus of elasticity than less dense materials, and the effect of the modulus more than offsets the effects of density. *Sound does not travel faster in dense media than in less dense media.* Table 11.21 lists the velocity of sound in some common materials.

Another characteristic of materials, very important to their acoustical performance, is acoustical **impedance.** This value is determined by multiplying the velocity of sound in the material by the density of the material. An examination of the units resulting from this operation will show that this is a way of measuring the rate at which the material will accept energy. Table 11.22 lists the acoustical impedance of several typical materials used in building construction.

In most cases of interest, a human is the "receiver" in the source-path-receiver chain. Hearing is the principal subjective response to sound. Within certain limits of frequency and energy levels, sound creates a sensation within the auditory equipment of humans and most animals.

TABLE 11.21 Velocity of Sound in Various Media

Material	Approximate sound velocity, ft per s
Air	1,100
Wood	11,000
Water	4,500
Aluminum	16,000
Steel	16,000
Lead	4,000

TABLE 11.22 Acoustical Impedance of Various Materials

Material	Acoustical impedance, psi per s
Rubber	100
Cork	165
Pine	1,900
Water	2,000
Concrete	14,000
Glass	20,000
Lead	20,500
Cast iron	39,000
Copper	45,000
Steel	58,500

At very low frequencies or at very high energy levels, additional sensations, ranging from pressure in the chest cavity to actual pain in the ears, are experienced. Human sense of feeling is also affected by vibrations in building structures, a subject closely related to ordinary building acoustics but outside the scope of this section.

From a strictly mechanical standpoint, the ear responds in a relatively predictable manner to physical changes in sound. Table 11.23 relates the objective characteristics of sound to our subjective responses to those characteristics.

TABLE 11.23 Subjective Responses to Characteristics of Sound

Objective (sound)	Subjective (hearing)
Amplitude Pressure Intensity	Loudness
Frequency	Pitch
Spectral distribution of acoustical energy	Quality

Loudness is the physical response to sound pressure and intensity. It is influenced somewhat by the frequency of the sound.

Pitch is the physical response to frequency. Low frequencies are identified as low in pitch, high frequencies as high in pitch. Middle C on the piano, for example, is 261 Hz; 1 octave below is 130 Hz; 1 octave above is 522 Hz.

An **octave** represents a 2:1 ratio of frequencies.

For practical purposes, humans can be considered to have a hearing range from about 16 Hz to somewhat less than 20,000 Hz. They usually are significantly deaf to low frequencies and very high frequencies. Human loudness response to sounds of identical pressure but varying frequency plots something like the curve in Fig. 11.91.

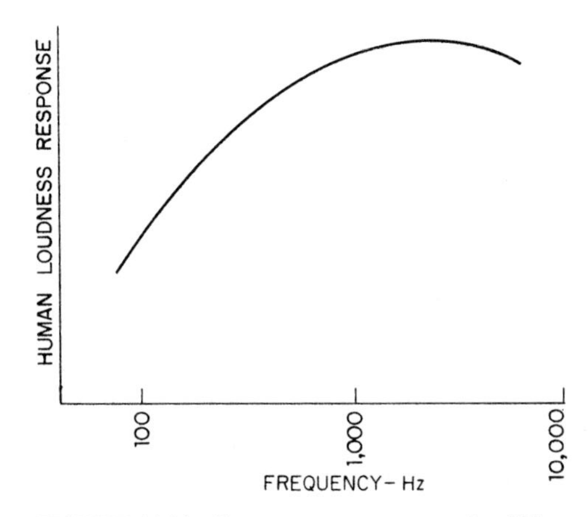

FIGURE 11.91 Human response to sounds of identical level but various frequencies.

Any measure of loudness must, in some way, specify frequency as well as pressure or intensity to have any real significance in human acoustics. As shown in Art. 11.78, measuring equipment and scales used are all modified to recognize this. Table 11.24 lists some of the significant frequency ranges.

Sounds generated by mechanical equipment may encompass the entire frequency range of human hearing. Jet aircraft, for example, have significant output throughout the entire range, and large diesel engines produce substantial energy from 30 to 10,000 Hz.

Wanted sound (signal), whatever its nature, communicates information to us. **Unwanted sound,** whatever its nature, is **noise.** When the signal becomes sufficiently louder than the noise, the signal can be detected and the information becomes available. Thus, the distinction between noise and communication is completely subjective, completely a matter of human desires and needs of the moment.

While very weak sounds can be heard in a very quiet background, the level of wanted sound may be increased to override the unwanted background (or noise); but there is a significant noise level above which even loud (shouting) speech is scarcely intelligible. Such a level, if long continued, can prove damaging to the hearing mechanism. At still higher levels, pain can be experienced, and immediate physical damage (often irreversible) may occur.

Complete silence (if such a thing could occur) would be most unpleasant. Humans would quickly become disoriented and distressed. Somewhere between silence and the uncomfortably loud, however, is a wide range that humans find acceptable.

TABLE 11.24 Significant Frequency Ranges

	Approximate frequency range, Hz
Range of human hearing	16–20,000
Speech intelligibility (containing the frequencies most necessary for understanding speech)	600–4,800
Speech privacy range (containing speech sounds that intrude most objectionably into adjacent areas)	250–2,500
Typical small table radio	200–5,000
Male voice	350*
Female voice	700*

*Frequency at about which energy output tends to peak.

11.78 MEASUREMENT OF SOUND

Generally, absolute numbers obtained from measurements have little significance in acoustics. Instead, measurements are almost always compared with some base or reference, and they are usually quoted as **levels** above or below that reference level. The levels are usually ratios of observed values to the reference level, such as 2:1, 1:2, 1:10, because rarely are simple, linear relationships found between stimuli and effects in humans.

Zero level of sound pressure is not a true physical zero; that is, the absence of any sound pressure at all. Rather, zero level is something of an average threshold of human hearing, of sound at about 1000 Hz. The physical pressure associated with this threshold level is very small, 0.00002 Pa (N/m^2).

Changes in human response tend to occur according to a ratio of the intensity of the stimuli producing the response. In acoustics, the ratio of 10:1 is called a **bel,** and one-tenth of a bel, a **decibel** (dB). Thus, power and intensity levels, in dB, are computed from

$$\text{Level} = 10 \log_{10} \frac{\text{quantity measured}}{\text{reference quantity}} \tag{11.10}$$

Intensity level (IL), dB, for example, represents the ratio of the intensity being measured to some reference level, and is given by

$$\text{IL} = 10 \log_{10} \frac{I}{I_o} \tag{11.11}$$

where I = intensity measured, W/cm^2
I_o = reference intensity = 10^{-16} W/cm^2

Since intensity varies as the square of the pressure, intensity level also is given by

$$IL = 10 \log_{10} \frac{p^2}{p_o^2} = 20 \log_{10} \frac{p}{p_o} \tag{11.12}$$

where p = pressure measured, Pa
p_o = reference pressure = 0.00002 Pa

Sound pressure level (SPL), dB, to correspond with intensity level, is defined by

$$SPL = 20 \log_{10} \frac{p}{p_o} \tag{11.13}$$

Sound power level refers to the power of a sound source relative to a reference power of 10^{-12} W. (*Note:* At one time, 10^{-13} W was used; thus, it is imperative that the reference level always be explicitly stated.)

The ear responds in a roughly logarithmic manner to changes in stimulus intensity, but approximately as shown in Table 11.25.

TABLE 11.25 Subjective Effect of Changes in Sound Characteristics

Change in sound level, dB	Change in apparent loudness
3	Just perceptible
5	Clearly noticeable
10	Twice as loud (or ½)
20	Much louder (or quieter)

Another comparison, which gives more meaning to various levels, is shown in Table 11.26.

Measurement Scales. Most measurements or evaluations of sound intensity or level are made with an electronic instrument that measures the sound pressure. The instrument is calibrated to read pressure levels in decibels (rather than volts). It can measure the overall sound pressure level throughout a frequency range of about 20 to 20,000 Hz, or within narrow frequency bands (such as an octave, third-octave, or even narrower ranges). Usually, the sound-level meter contains filters and circuitry to bias the readings so that the instrument responds more like the human ear—"deaf" to low frequencies and most sensitive to the midfrequencies (from about 500 to 5000 Hz). Such readings are called A-scale readings. Most noise level readings (and, unless otherwise specifically stated, most sound pressure levels with no stated qualifications) are A-scale readings (often expressed as dBA). This means that actual sound pressure readings have been modified electrically within the instrument to give a readout corresponding somewhat to the ear's response (Fig. 11.91).

TABLE 11.26 Comparison of Intensity, Sound Pressure Level, and Common Sounds

Relative intensity	SPL, dBA*	Loudness
100,000,000,000,000	140	Jet aircraft and artillery fire
10,000,000,000,000	130	Threshold of pain
1,000,000,000,000	120	Threshold of feeling
100,000,000,000	110	
10,000,000,000	100	Inside propeller plane
1,000,000,000	90	Full symphony or band
100,000,000	80	Inside automobile at high speed
10,000,000	70	Conversation, face-to-face
1,000,000	60	
100,000	50	Inside general office
10,000	40	Inside private office
1,000	30	Inside bedroom
100	20	Inside empty theater
10	10	
1	0	Threshold of hearing

*SPL as measured on A scale of standard sound level meter.

For various measurements and evaluations of performance for materials, constructions, systems, and spaces, see Art. 11.81.

11.79 SOUND AND VIBRATION CONTROL

This process consists of:

1. Acoustical analysis
  a. Determining the use of the structure—the subjective needs
  b. Establishing the desirable acoustical environment in each usable area
  c. Determining noise and vibration sources inside and outside the structure
  d. Studying the location and orientation of the structure and its interior spaces with regard to noise and noise sources

2. Acoustical design
  a. Designing shapes, areas, volumes, and surfaces to accomplish what the analysis indicates
  b. Choosing materials, systems, and constructions to achieve the desired result

Sound and vibration sources are usually speech and sounds of normal human activity—music, mechanical equipment sound and vibration, traffic, and the like. Characteristics of these sound sources are well known or easily determined. Therefore, the builder or designer is usually most interested in the transmission paths for sound and vibration. These are gases (usually air); denser fluids (water, steam, oil, etc.); and solids (building materials themselves). During sound transmission in a building, some of the sound energy is absorbed or dissipated, some is reflected from various surfaces, and some is transmitted through the building materials and furnishings.

Sound control is accomplished by means of barriers and enclosures, acoustically absorbent materials, and other materials and systems properly shaped and assembled. Vibration control is accomplished by means of various resilient materials and assemblies, and by damping materials (viscoelastic materials of various types).

Airborne and structure-borne energy are controlled by somewhat different techniques, described in the following.

11.79.1 Airborne Sound Transmission

A sound source in a room sets the air into vibration. The vibrating air causes any barrier it touches (partitions, floors, ceilings, etc.) to vibrate. The vibrating barrier, in turn, sets into vibration the air on the opposite side of the barrier (Fig. 11.92).

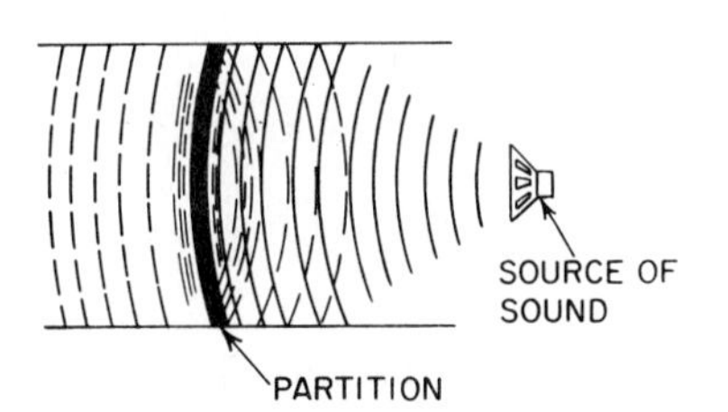

FIGURE 11.92 Sound transmission through a barrier.

The barrier, like any other body, resists motion because of its inherent inertia. Because it takes more force, and therefore more energy, to move a heavier (more massive) barrier, sound loss through the barrier depends directly on the mass of the barrier. The loss in energy level between the original signal striking the barrier and the level of the energy transmitted to the opposite side is called the **sound transmission loss** of the barrier.

Figure 11.93 shows graphically how a truly limp barrier of infinite width and height responds to sound energy of varying frequency (the *mass law,* straight-line vibration). In practice, however, no barrier is truly limp; a wall, floor, or ceiling always has a finite stiffness. Therefore, at low frequencies, barriers tend to have higher sound transmission losses than the mass law predicts. In some region of the spectrum, however, barriers tend to have lower sound transmission losses (pass sound more readily) than the mass law indicates (see curve for solid panel in Fig. 11.93).

This latter phenomenon results because the barrier, in addition to its back-and-forth motion, as shown in Fig. 11.92, moves in a simultaneous shear wave, like a rope being shaken (Fig. 11.90). At some frequency, the velocity of this shear wave in the barrier coincides with the velocity of the impinging sound wave in the air. Then, the partition is quite transparent to sound, and a deep coincidence dip in the sound transmission loss curve occurs (see curve for double wall in Fig. 11.93). At frequencies above the coincidence frequency, the curve tends to recover its preceding slope.

There are few practical limp materials useful for ordinary building panels. Soft sheet lead is occasionally used for specialized barriers, as described in Art. 11.80, or is applied to other panel material to increase the mass of the composite without increasing its stiffness. In addition, damping materials (Art. 11.79.4) can be applied to the panels to damp out their vibrations quickly and to increase the energy loss as the panels vibrate.

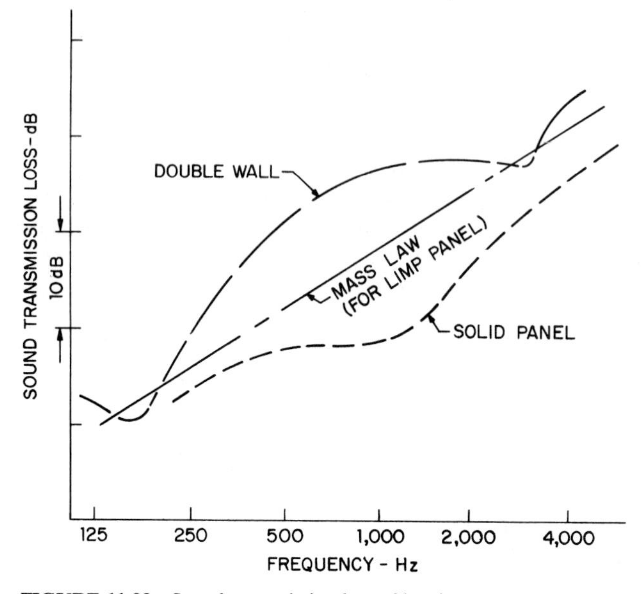

FIGURE 11.93 Sound transmission loss of barriers.

Figure 11.93 shows another characteristic of barriers that further complicates their performance. While single, solid panels have somewhat smaller transmission losses than the mass law predicts, a double wall, a barrier comprising separated wythes or leaves, of the same total weight produces greater transmission losses than the mass law predicts.

If the mass of a single, solid panel is divided into two separate, unconnected layers, the only energy transfer between them occurs via the air between the layers. Air is not stiff; it can sustain little of the shear wave, and it is not an efficient transmitter of the back-and-forth motion of one layer to the other (except at certain resonant frequencies). As a result, particularly in the midfrequencies, the sound transmission loss actually exceeds the mass-law values, often considerably. The greater the distance between layers, the better the performance. Theoretically, the sound transmission loss should increase about 6 dB for each doubling of the width of the airspace. In practice, however, the increase in decibels is somewhat less than this.

If a porous, sound-absorbent blanket is inserted in the void between layers, the standing waves in the air layers are minimized. Furthermore, such a blanket has the effect of reducing the stiffness of the air, as well as absorbing some of the energy of the air as it pumps back and forth through the blanket. The result is an additional increase in sound transmission loss.

The performance of various barriers, and rating systems for their performance, are shown in Table 11.27.

11.79.2 Bypassing of Sound Barriers

Rarely is a sound barrier the sole transmission path for the acoustic energy reaching it. Some energy invariably travels via the connecting structures (floors, ceilings, etc.), or through openings in or around the barrier.

TABLE 11.27 STC of Various Constructions

Construction	STC
¼-in plate glass	26
¾-in plywood	28
½-in gypsum board, both sides of 2 × 4 studs	33
¼-in steel plate	36
6-in concrete block wall	42
8-in reinforced-concrete wall	51
12-in concrete block wall	53
Cavity wall, 6-in concrete block, 2-in air space, 6-in concrete block	56

Structural flanking via edge attachments and junctions of walls, partitions, floors, and ceilings can seriously degrade the performance of a barrier. Figure 11.94 shows some typical flanking paths and how to avoid such flanking.

Even trivial openings through a sound barrier seriously degrade its performance. An opening with an area of only about 1 in^2 transmits as much acoustical energy as a 6-in-thick concrete block wall of 100-ft^2 area. Figure 11.95 shows the effect of openings of various sizes on walls of various effectiveness. (*Note:* Sound transmission class STC is a rating system described in Art. 11.80.) The better the wall, the more serious the effect of openings or leaks.

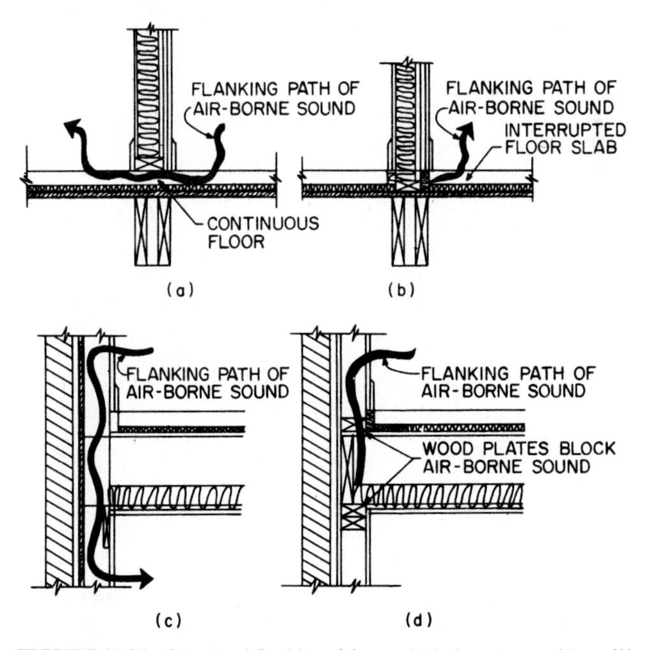

FIGURE 11.94 Structural flanking: (*a*) poor isolation at a partition; (*b*) adequate isolation at a partition; (*c*) poor isolation at an exterior wall (balloon framing); (*d*) adequate isolation (platform framing).

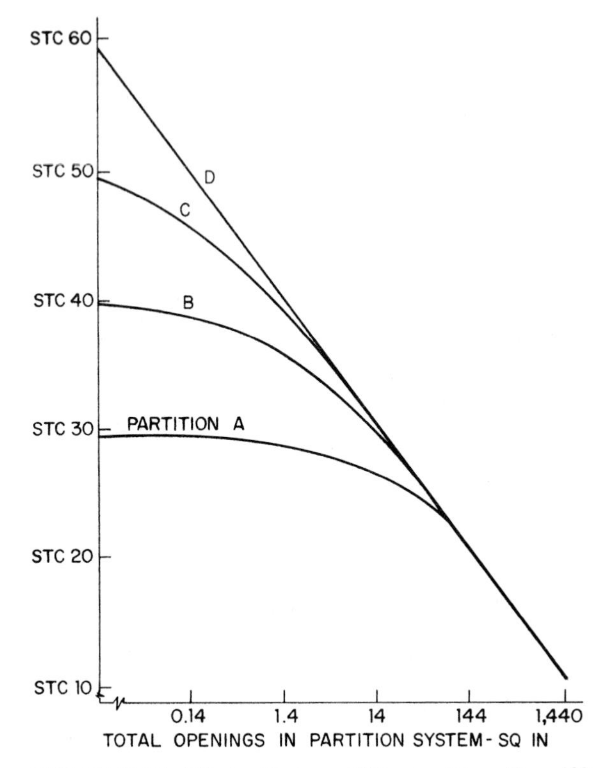

FIGURE 11.95 Effects of leaks on STC of partitions (for a 100 ft^2 test wall).

The more common points of leakage through or around barriers include perimeter of pipes, ducts, or conduits penetrating the barriers; relief grilles for return air; perimeter of doors or glazing; shrinkage or settlement cracks at partition heads or sills; joints between partitions and exterior curtain-wall mullions; joints around or openings through back-to-back electrical outlet boxes, medicine cabinets, etc.; common supply or return ducts with short, unlined runs between rooms; and operable windows opening to a common court. Such openings should be avoided, when possible, and all cracks, joints, and perimeters should be calked and sealed.

In some spaces (an open-plan school, "landscaped" office, etc.), screens, or partial-height barriers are used. Their effectiveness is much less than that of full-height partitions or walls.

See also Art. 11.79.7.

11.79.3 Structure-Borne Sound Transmission

Acoustical energy transmission from sound sources to distant parts of the structure via the structure itself is often a major annoyance in building acoustics. One reason, as discussed in Art. 11.79.2, is that structural flanking can seriously degrade the performance of sound barriers. Additional transmission paths include pipes, ducts, conduits, and almost any solid, continuous, rigid member in the building. Impact and vibrations can be transmitted through the floor-ceiling assemblies quite readily, if proper precautions are not taken.

Normally, in ordinary construction, a good carpet over a good pad will provide sufficient impact isolation against the sounds of footfalls, heel clicks, and dropped objects. Where carpet is not feasible, or in critical spaces, special floor-ceiling assemblies can be used. (See Art. 11.80 for various constructions and their impact isolation performance.)

Isolation against vibration or impact produced by machinery or other vibrating equipment is usually provided by use of special resilient or mounting systems. Springs, elastomeric pads, and other devices are used in such work. For a detailed discussion of such isolation, refer to the current ASHRAE "Guide," American Society of Heating, Refrigerating and Air-Conditioning Engineers, or other references on vibration and shock isolation.

Noise and vibration transmitted via plumbing and heating pipes, ducts, and conduits can be objectionable unless proper precautions are taken. Resilient connectors for rigid members, acoustically absorptive duct linings, and similar approaches are described in detail in the ASHRAE "Guide."

11.79.4 Damping of Vibrations

As a panel or object vibrates, it radiates acoustical energy to the air surrounding it and to solid surfaces touching it or attached to it. If the energy of vibration could be dissipated, the radiation would be reduced and the sound and vibration levels lowered. One way to do this is to attach firmly to a vibrating panel certain "lossy" substances (those with high internal friction or poor connections between particles) or viscoelastic materials (neither elastic nor completely viscous, such as certain asphaltic compounds). These damp the vibrations by absorbing the energy and converting it to heat.

In the assembly of barriers, damping can be accomplished with proper connections and attachments, use of viscoelastic adhesives, proper attachment of insulating materials, and similar means. Special viscoelastic materials for adhesive attachment or brush or spray application to panels are available.

11.79.5 Sound Absorption

The best-known acoustical materials are acoustical absorbents (although actually all materials are acoustical). Generally, these absorbents are lightweight, porous, "fuzzy" types of boards, blankets, and panels.

Acoustical absorbents act as energy transducers, converting the mechanical energy of sound into heat. The conversion mechanism involves either pumping of air contained within the porous structure of the material, or the flexing of thin panels or sheets. Most materials employ the first principle.

The internal construction of most absorbents consists of a random matrix of fibers or particles, with interconnected pores and capillaries (Fig. 11.96). It is necessary that the air contained within the matrix be able to move sufficiently to create friction against the fibers or capillaries. Nonconnected-cell or closed-cell porous materials are not effective absorbents.

Tuned chambers, with small openings and a restricted neck into the chambers, also are used for sound absorption; but they are somewhat specialized in design

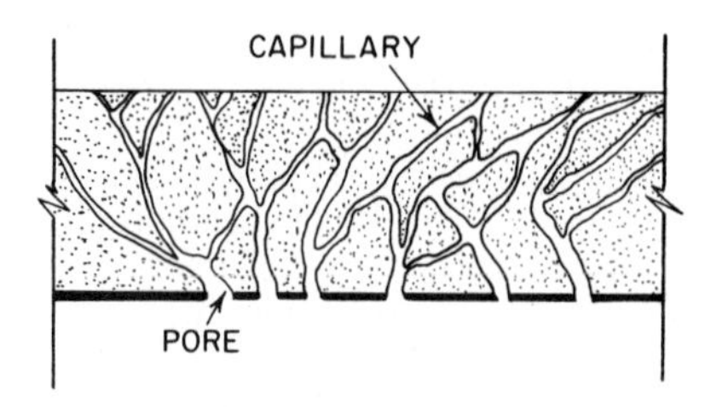

FIGURE 11.96 Cross section of sound absorbent (greatly enlarged).

and function, and their use requires expert design in most instances.

The surface of absorbents must be sufficiently porous to permit the pressures of impinging sound waves to be transferred to the air within the absorbent. Very thin, flexible facings (plastic or elastomeric sheets) stretched over panels and blankets do not interfere significantly with this pressure transfer, but thick, rigid, heavy coatings (even heavy coatings of paint) may seriously restrict the absorption process. Perforated facings, if sufficiently thin and with sufficient closely spaced openings, do not appreciably degrade the performance of most absorbents.

The visible surface of absorbent panels and tiles may be smooth or textured, fissured or perforated, or decorated or "etched" in many ways. Figure 11.97 shows a typical commercial tile. Figure 11.98 illustrates an example of the use of both acoustical tiles and panels in a ceiling.

Absorbents are normally produced from vegetable or mineral fibers, porous or granular aggregates, foamed elastomers, and other products, employing either added binders or their own structure to provide the structural integrity required. Because of their lightweight, porous structure, most such products are relatively

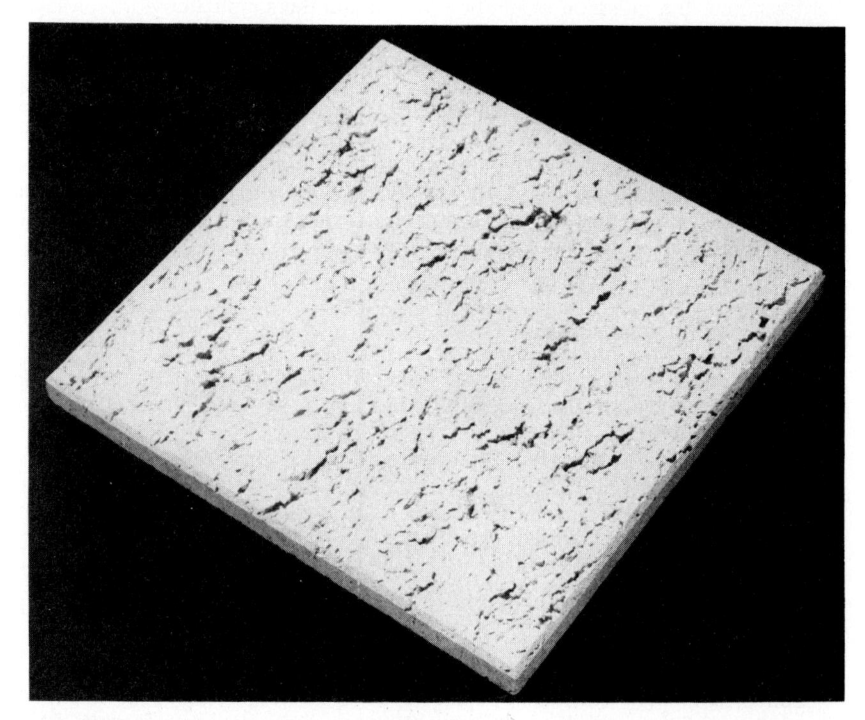

FIGURE 11.97 Fissured-surface acoustical tile with the appearance of travertine stone.

FIGURE 11.98 Library ceiling comprising both acoustical panels and acoustical tiles, integrated with lighting panels. (*U.S. Gypsum Company.*)

fragile and must be installed where they are not subject to abuse; or they may be covered with sturdy perforated or porous facings to protect them.

Absorbents are chosen for their appearance, fire resistance, moisture resistance, strength, maintainability, and similar characteristics. Their performance ratings in respect to these characteristics are usually published in advertising materials and in various bulletins and releases of trade associations.

The acoustical property of absorbents most important to designers and builders is absorptive efficiency. [Sound transmission through most absorbents takes place readily. They are very poor in this respect and should never be used to attempt to improve the airborne sound isolation of a barrier. Sound transmission over the top of a partition, through a lightweight, mechanically suspended acoustical panel ceiling, is frequently a serious annoyance in buildings (Art. 11.79.7).] The absorptivity of a tile or panel is usually expressed as the fraction or percentage of acoustical energy absorbed from an impinging plane wave. If a perfectly absorptive plane surface represents 1.00 or 100%, the ratio of absorption of a given product to this perfect absorber is called the **sound absorption coefficient** of the product.

The absorptivity of a material varies with its thickness, density, porosity, flow resistance, and other characteristics. Further, absorptivity varies significantly with the frequency of the impinging sound. Usually, very thick layers of absorptive material are required for good absorption of low-frequency sound, while relatively thin layers are effective at higher frequencies. Little or no absorption, however, can be obtained with thin layers or flocked surfaces, textured but nonporous surfaces, or other products often mistakenly thought to be absorptive or claimed to break up the sound.

Generally, there is an optimum density and flow resistance for any particular family of materials (particularly fibrous materials). Usually, absorptivity increases with thickness of the material.

Performance data for absorbent materials are usually readily available from manufacturers and trade associations. Performance of some typical products is shown in Table 11.30. See also Arts. 11.79.6, 11.79.7, and 11.80.

11.79.6 Control of Reflection and Reverberation

Normally, acoustical absorbents are used to prevent or minimize reflections of sound from the surfaces of rooms or enclosures. Distinct reflections—**echoes**—are usually objectionable in any occupied space. Rapid, repeated, but still partly distinguishable echoes, such as occur between parallel sidewalls of a corridor—**flutter**—are also objectionable.

Reverberation comprises very rapid, repeated, jumbled echoes, blending into an indistinct but continuing sound after the source that created them has ceased.

Usually, reverberation is one of the major causes of poor intelligibility of speech within a room; but, within limits, it may actually enhance the sound of music within a space. Reverberation control is a necessary and important aspect of good acoustic design, but it is often greatly overemphasized. Good room proportions and configuration, control of echoes, and absorption of noise usually assure an acceptable reverberation time within a space. Where careful determination and control of reverberation are required in a room, the services of a competent acoustical consultant are always advisable.

Reflections from strategically located and properly shaped room surfaces may be highly desirable, because such reflections may strongly enhance the source signal. But excessively delayed or highly persistent reflections are usually undesirable. (For most purposes, and within the normal frequency range of importance to human hearing, it may be assumed that the angle at which sound waves, like light waves, reflect from a surface equals the angle of incidence. Because of the enormously longer wavelength of sound compared with light, however, this assumption is inexact but nevertheless it is acceptable for most acoustic design.)

In most rooms, absorption of most of the acoustical energy impinging on many of the surfaces (the floor, distant walls, etc.) is desirable to prevent buildup or increase of unintelligible or useless sound. For this purpose, sound absorbents may be placed on some or all of the surfaces. The difference in sound pressure level (or noise level) caused within a space by the introduction of absorbents can be calculated, and from such a calculation, it is possible to determine how effective such treatment will be.

Noise reduction (NR), dB, provided by adding acoustical absorbents in a space can be determined from

$$NR = 10 \log_{10} \frac{A_o + A_a}{A_o} \tag{11.14}$$

where A_o = original acoustical absorption present
A_a = added acoustical absorption

Acoustical absorption equals the sum of the products of each area (in consistent units) in the space times the absorption coefficient of the material constituting the surface of the area; for example, the floor area, ft^2 × its absorption coefficient, plus ceiling area, ft^2 × its absorption coefficient, plus total wall area, ft^2 × its absorption coefficient.

[*Note:* An anomalous but useful term is often used in advertising data, the **noise reduction coefficient** (NRC). This is the arithmetic average of the sound absorption coefficients of a material as determined at 250, 500, 1000, and 2000 Hz (Art. 11.79.5). Since these frequencies include the most significant speech and intelligibility ranges, such a figure is a reasonably good means of comparing similar materials; that is, materials with absorption characteristics not differing widely from one another within this frequency range. Often, NRC is used, instead of the absorption coefficients at various frequencies, to determine an average noise reduction from Eq. (11.14).]

Equation (11.14) indicates that the more absorption present originally the less the improvement provided by added absorption. Thus, in a very "hard," bare room, addition of acoustical (sound-absorbent) tile to a full ceiling significantly reduces the noise level. But addition of the same ceiling tile in a room with a thick carpet, upholstered furniture, and heavy draperies would make little change.

Heavy carpet, upholstered furniture, heavy draperies, and similar materials are very effective absorbers. In residences, for example, rarely is additional acoustical absorption required in bedrooms, living rooms, and similar spaces. In kitchens, bathrooms, or recreation rooms, with normal hard floors and few additional furnishings or fabrics, however, an acoustical tile ceiling is helpful. This is equally true of offices and similar spaces. In theaters and auditoriums, the large expanse of upholstered seating and aisle carpets is normally adequate for most noise control, but added absorption on some wall surfaces may be required for control of echoes.

Reverberation Time. The reverberation within a space is usually expressed as the time required for a sound pulse to decay 60 dB (to one-millionth of its original level). For most purposes, reverberation time T, s, can be calculated from the simple Sabine formula:

$$T = \frac{0.049V}{A} \tag{11.15}$$

where V = volume of the space, ft^3
A = total acoustical absorption in the space

Equation (11.15) assumes a smooth, steady, logarithmic decay; random distribution of sound within the room, with the wave front striking every surface quickly and within the decay time; and no standing waves between surfaces that could support a persistent mode. These are idealized conditions and never exist, but the formula is sufficiently accurate for most purposes.

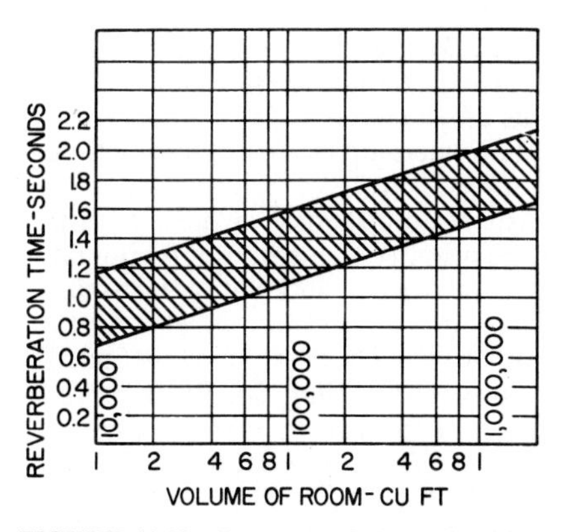

FIGURE 11.99 Recommended reverberation time, indicated by the shaded area, varies with the size of the room.

Because the absorption of an absorbent material varies with frequency of sound, it is necessary to calculate T for each significant frequency. For most reverberation calculations, determinations at 500 Hz are adequate. For concert halls, and critical spaces, calculations are usually made 2 octaves above and 2 octaves below 500 Hz as well.

Optimum reverberation time for a room is a subjective determination, governed by speech intelligibility and the fullness and richness of musical sound desired. Figure 11.99 shows, within the shaded area, the acceptable range of reverberation times for normal spaces of varying volume. For critical spaces (radio studios, concert halls, auditoriums, etc.), it is advisable to obtain the advice of competent acoustical consultants.

It is imperative that designers understand that a reverberation-time determination is not an acoustical analysis, and that, in many instances, reverberation time is a trivial part of an acoustical study.

11.79.7 Installation of Absorbents

The amount, location, and installation method for absorbents are all significant in a room. As discussed in Art. 11.79.6, the material is located on surfaces from which reflections are undesirable and where it is reasonably free from damage. The amount of material required can be determined from noise-reduction and reverberation-time calculations. (For most simple spaces, a full ceiling treatment is acceptable.)

Acoustical absorbents can be cemented directly to a smooth, solid surface, nailed or stapled to furring strips, or suspended by any of a number of mechanical systems, such as that shown in Fig. 11.100.

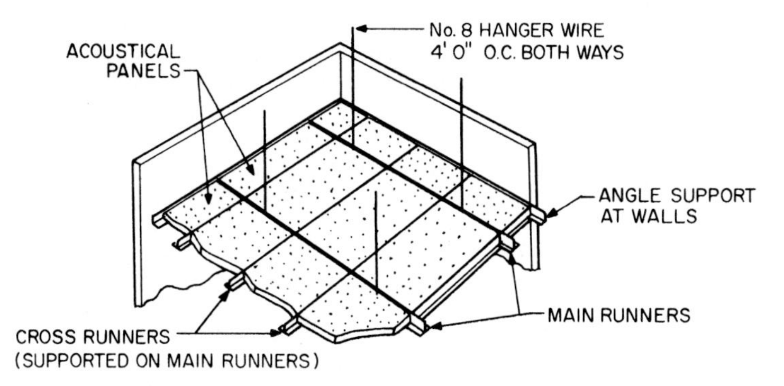

FIGURE 11.100 Typical suspended acoustical panels.

While acoustical absorbents are usually installed as, or on, flat, horizontal surfaces, various other configurations are used. Coffers, grids of hanging panels, and similar arrangements are often employed. The acoustical performance of each configuration must be tested to determine its effectiveness.

Sprayed-on and trowel-applied acoustical products are also employed, although their use is limited.

The structural, fire-resistance, and acoustical performance of most acoustical panels and tiles are significantly affected by the installation method. It is imperative that performance data be explicitly related to the specific installation method.

Partition Bypassing. When partitions or other sound barriers are constructed to (but not through) a suspended acoustical tile or panel ceiling, sound transmission through the tile from one space, over the top of the partition, into an adjacent space can be a serious annoyance. Figure 11.101 shows how sound can bypass a partition, and alternative methods of preventing this.

Similarly, structure-borne flanking transmission along the continuous metal runners of some suspension systems may seriously degrade the performance of an

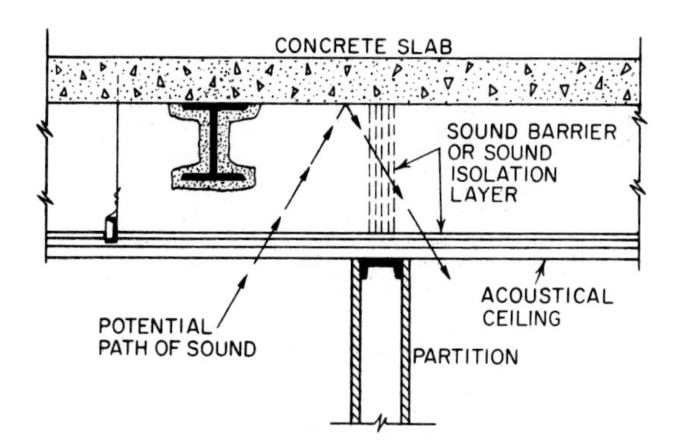

FIGURE 11.101 Prevention of sound transmission over partitions.

acoustical ceiling as a sound barrier. Performance data for most typical commercial ceilings are published in various bulletins and advertising matter.

11.80 ACOUSTICAL PERFORMANCE DATA

To simplify and standardize evaluation of the acoustical performance of materials and systems, various rating systems have been adopted. The best known and most widely used are those published by ASTM, 1916 Race St., Philadelphia, PA 19103.

Partitions, Floor-Ceiling Assemblies, and Barriers. Insulation (or isolation) of airborne sound provided by a barrier is usually expressed as its **sound transmission class** (STC). For a specific construction, STC is determined from a sound-transmission-loss curve obtained from a standardized test of a large-scale specimen. This curve is compared with a standard contour, and a numerical rating is assigned to the specimen (ASTM E90 test procedure and ASTM E413, Determination of Sound Transmission Class).

Table 11.27 lists typical STC ratings of several partition, wall, and floor-ceiling components or assemblies. Published data for almost any type of construction can be obtained from various sources.

A difference of one or two points between two similar constructions is rarely significant. Normally, constructions tend to fall into groups or classes with their median values about five points apart. In Table 11.28, the italic number represents the median of a performance group which includes the numbers on either side of the median.

The impact insulation (or isolation) provided by floor-ceiling assemblies is usually expressed as their impact noise rating (INR) or impact insulation (IIC). Like STC, INR and IIC values are obtained by comparing the curve of the sound spectrum obtained in a test with a standard contour (except that for INR and IIC the sound pressure level is measured in the room below the noise source). The entire procedure is controversial and far from widely accepted; but its use is so widespread that designers and builders should be aware of it. (See ASTM RM 14-4.)

Table 11.29 lists the impact isolation provided by several types of construction. Note particularly the enormous effect of certain floor coverings on the performance

TABLE 11.28 STC Performance Groups

30	31	*32*	33	34
35	36	*37*	38	39
40	41	*42*	43	44
45	46	*47*	48	49
50	51	52	53	54
55	56	57	58	59

TABLE 11.29 IIC of Various Floor Constructions

Construction	IIC
Oak flooring on ½-in plywood subfloor, 2 × 10 joists, ½-in gypsum board ceiling	23
With carpet and pad	48
8-in concrete slab	35
With carpet and pad	57
2½-in concrete on light metal forms, steel bar joists	27
With carpet and pad	50

of the construction. (While IIC values are shown in Table 11.29, they can be converted to INR values by subtracting 51 points; for example, IIC 60 = INR 9, and IIC 45 = INR −6, etc.)

Acoustical Absorbents. Sound absorption coefficients and noise reduction coefficients of acoustical absorbents, including carpets and other furnishings, are usually readily available from manufacturers. The coefficients are normally obtained from laboratory tests of panel specimens of about 72- to 80-ft^2 area, tested according to ASTM C423. It is imperative that the test specimens be as nearly identical as possible, in construction system and detail, with actual field installations, since construction details enormously affect acoustical performance.

Table 11.30 lists performance ranges of typical absorbent materials. For specific data on specific materials or systems, always refer to specific tests by accredited test agencies.

The sound transmission loss through an acoustical ceiling (up and over a barrier) when the ceiling is used as a continuous membrane is often an important rating. As might be expected, the effectiveness of acoustical absorbents as acoustical barriers is limited, and supplementary barriers or isolation are frequently required in normal construction.

Other Acoustical Materials. Performance ratings of damping materials, duct linings, vibration isolation materials and devices, etc., are available from various sources. Use of such materials is somewhat complex and specialized.

TABLE 11.30 Performance of Commonly Used Sound Absorbents

Absorbent	Thickness, in.	Density, lb per cu ft	Noise reduction coefficient
Mineral or glass fiber blankets	½–4	½–6	0.45–0.95
Molded or felted tiles, panels, and boards	½–1⅛	8–25	0.45–0.90
Plasters (porous)	⅜–¾	20–30	0.25–0.40
Sprayed-on fibers and binders	⅜–1⅛	15–30	0.25–0.75
Foamed, open-cell plastics, elastomers, etc.	½–2	1–3	0.35–0.90
Carpets	Varies with weave, texture, backing, pad, etc.		0.30–0.60
Draperies	Varies with weave, texture, weight, fullness		0.10–0.60

	Absorption coefficient per sq ft of floor area at frequencies, Hz:					
	125	250	500	1000	2000	4000
Seated audience	0.60	0.75	0.85	0.95	0.95	0.85
Unoccupied upholstered (fabric) seats	0.50	0.65	0.80	0.90	0.80	0.70

11.81 ACOUSTICAL CRITERIA

Acoustical performance criteria and environmental specifications are often part of building contract documents. Governmental agencies, lending institutions, owners, and tenants frequently require objective standards of performance.

To a large degree, acoustical criteria are subjective or are based on subjective response to acoustic parameters. This complicates attempts to provide objective specifications, but long experience has permitted acoustical experts to determine broad classes or ranges of criteria and standards that will produce satisfaction in most instances.

It is important to remember that one-point differences are normally insignificant in acoustical criteria. Usually, a tolerance of ±2½ points from a numerical value is acceptable in practice.

Tables 11.31 to 11.33 list some of the more common criteria for ordinary building spaces. They should, however, be used only as a guide. Obtain specific data or requirements whenever possible.

Acceptable Background Noise Levels. Steady, constant, unobtrusive sound levels that normally occur in typical rooms and that are acceptable as background noise are indicated in Table 11.31. For specific applications and in acoustically critical spaces, specific requirements should always be determined.

[*Note:* Levels are given in dBA, easily obtained numbers with simple measuring equipment (Art. 11.78). When noise criterion (NC) values are specified, they can

TABLE 11.31 Typical Acceptable Background Levels

Space	Background level, dBA
Recording studio	25
Suburban bedroom	30
Theater	30
Church	35
Classroom	35
Private office	40
General office	50
Dining room	55
Computer room	70

be determined by subtracting about 7 to 10 points from dBA values; that is, dBA 50 = NC 40 to NC 43.]

Use of electronically generated masking sound, usually by means of speakers located above the ceiling, has become commonplace in open-plan spaces as a means of establishing a uniform background level. However, it is an expensive and complex procedure and should be handled only by acoustics experts.

Sound-Transmission-Loss Requirements. Acoustical performance requirements of sound barriers separating various occupancies may vary widely, depending on the particular needs of the particular occupants. Typical requirements for common occupancies in normal buildings are shown in Table 11.32.

In highly critical spaces, or where a large number of spaces of identical use are involved (as in a large hotel or multiple-dwelling building), it is always advisable to obtain expert acoustical advice.

Impact Isolation Requirements. Because the only standard test method available is controversial, impact isolation performance specifications are only broad, general suggestions, at best. The values in Table 11.33, however, are reasonably safe and economically obtainable with available construction systems.

Acoustical Absorption Requirements. It is ironic that requirements for the most widely used acoustical materials are the least well defined. So-called optimum reverberation time requirements (Fig. 11.99) are a reasonably safe specification for most rooms, but many additional requirements are often involved.

In general, where noise control is the significant requirement, the equivalent absorption of a full ceiling of acoustical tile providing a noise-reduction coefficient of about 0.65 to 0.70 is adequate. This absorption may be provided by carpet, furnishings, or other materials, as well as by acoustical tile. In many instances, however, no absorption at all should be applied to the ceiling of a room, because the ceiling may be a necessary sound reflector.

For important projects, always obtain the advice of competent acoustical consultants.

TABLE 11.32 Sound Isolation Requirements between Rooms

Between: Room	and Adjacent area	Sound isolation requirement, STC
Hotel bedroom	Hotel bedroom	47
Hotel bedroom	Corridor	47
Hotel bedroom	Exterior	42
Normal office	Normal office	33
Executive office	Executive office	42
Bedroom	Mechanical room	52
Classroom	Classroom	37
Classroom	Corridor	33
Theater	Classroom	52
Theater	Music rehearsal	57

TABLE 11.33 Impact Isolation Requirements between Rooms

Between: Room	and Room below	Impact isolation requirement, IIC
Hotel bedroom	Hotel bedroom	55
Public spaces	Hotel bedroom	60
Classroom	Classroom	47
Music room	Classroom	55
Music room	Theater	62
Office	Office	47

11.82 HELPFUL HINTS FOR NOISE CONTROL

A building encloses a myriad of activities and the equipment used in those activities. It is imperative that designers and builders consider the control of noise and vibration associated with such activities. While it is not practical to give here detailed solutions to the many potential acoustical conditions that may arise in the design of even one building, the more common situations encountered and the most likely palliatives for such problems are indicated in Tables 11.34 and 11.35.

Sound originates at a source and travels via a path to a **receiver.** Sound control consists of modifying or treating any or all of these three elements in some manner.

TABLE 11.34 Available Options in Noise Control

Objectives of sound-control efforts	Sound-control procedures*					
	Quiet the source†	Barriers or enclosures	Vibration isolation or damping	Absorption	Masking	Personal protection
Reduce the general noise level to:						
Improve communication	X		X	X‡		
Increase comfort	X		X	X‡		
Reduce risk of hearing damage	X		X	X‡		
Reduce extraneous, intruding noise to:						
Increase privacy		X‡			X	
Increase comfort		X‡	X			
Improve communication		X‡	X			
Protect many persons against localized source producing damaging levels	X	X‡	X	X		X
Protect many persons against many distributed sources producing damaging levels	X	X	X	X		X‡
Protect one person against localized source producing damaging levels	X	X§	X			X‡
Protect a few persons against many distributed sources producing damaging levels		X§				X‡
Eliminate echoes and flutter				X¶		
Reduce reverberation				X‡		
Eliminate annoying vibration			X‡			

*No mention has been made of another option, reinforcement, because its purpose is to increase levels. It is the only available option when the signal level must be increased.

†Always the best and simplest means of eliminating noise, if practical and economical.

‡Indicates the most likely solution(s).

§A closed booth or small room for the person(s) is often feasible.

¶Assumes that the configuration of the reflecting surface(s) cannot be modified.

TABLE 11.35 Typical Acoustical Problems and Likely Solutions

Problem	Possible causes	Solution
"It's so noisy, I can't hear myself think"	High noise levels	Absorption
	Excessive reverberation	Absorption
	Excessive transmission	Sound isolation
	Excessive vibration	Vibration isolation
	Focusing effects	Eliminate cause of focusing effects
"Speech and music are fuzzy, indistinct"	Excessive reverberation	Absorption
"Little sounds are most distracting"	Background level too low	Masking
	Room too "dead"	Optimum reverberation
"There's an annoying echo"	Echo	Proper room shape
	Flutter	Proper room shape
	Focusing effects	Eliminate cause of focusing effects
	Excessive reverberation	Absorption
"I can hear everything the fellow across the office says"	Room too "dead"	Optimum reverberation
	Background level too low	Masking
	Focusing or reflection	Eliminate focusing or reflection
"It doesn't sound natural in here"	Flutter	Alter room shape Add absorption
	Distortion caused by improper sound system	Proper sound system
	Selective absorption	Proper type and amount of absorption
	Room too "dead," low reverberation time	
"It feels oppressive"	Reverberation time too low	Proper amount of absorption
	Background sound level too low	Use of background and masking sound
"It's not loud enough at the rear of the room"	Room too large	Electronic amplification
	Improper shape	Alter room shape
	Lack of reflecting surfaces	Add reflecting surfaces
	Poor distribution	Eliminate absorption on surfaces needed for reflection
	Too much absorption	
"There are dead spots in the room"	Poor distribution	Reflecting surfaces
	Improper shape	Eliminate cause of focusing effects
	Echoes	Alter room shape
"Sound comes right through the walls of these offices"	Sound leaks	Eliminate leaks
	Sound transmission	Sound isolation
	Vibration	Vibration isolation
	Receiving room too quiet	Masking
	Poor room location	Proper room layout
"I can hear machine noises and people walking around upstairs"	Vibration	Vibration isolation
	Sound transmission	Sound isolation
	Poor room location	Proper room layout
"Outside noises drive me crazy"	Poor room location	Proper acoustical environment
	Sound leaks	Eliminate leaks
	Sound transmission	Sound isolation

The most effective control measures often involve eliminating noise at the source. For example:

Balancing moving parts, lubricating bearings, improving aerodynamics of duct systems, etc.

Modifying parts or processes

Changing to a different, less noisy process

The most common sound control measures usually involve acoustical treatment to absorb sound; but equally important are:

Use of barriers to prevent airborne sound transmission

Interruption of the path with carefully designed discontinuities

Use of damping materials to minimize radiation from surfaces

Another important approach involves reinforcing the direct sound with controlled reflections from properly designed reflective surfaces.

Often, the most simple and effective approach involves protecting the receivers, enclosing them within adequate barriers, or equipping them with personal protection devices (ear plugs or muffs), rather than trying to enclose or modify huge sources or an entire room or building.

11.83 ACOUSTICS BIBLIOGRAPHY

L. L. Beranek, "Noise and Vibration Control," McGraw-Hill, Inc., New York.

P. D. Close, "Sound Control and Thermal Insulation of Buildings," Van Nostrand Reinhold Company, New York.

M. D. Egan, "Architectural Acoustics," McGraw-Hill, Inc., New York.

C. M. Harris, "Handbook of Acoustical Measurements and Noise Control," 3d ed., McGraw-Hill, Inc., New York.

P. M. Morse and K. U. Ingard, "Theoretical Acoustics," McGraw-Hill, Inc., New York.

L. F. Yerges, "Sound, Noise and Vibration Control," Van Nostrand Reinhold Company, New York.

SECTION TWELVE
ROOF SYSTEMS

Thomas Lee Smith
Director of Technology and Research
National Roofing Contractors Association (NRCA)
Rosemont, Illinois

Building owners and designers have many deck, insulation, and roof covering materials to choose from for low-slope and steep-slope roof systems. The various systems available can fulfill a wide range of functions, such as energy conservation, acoustical and thermal insulation, and water, fire, and wind resistance. Their ability to do this over their expected service life depends on good design, quality materials, good application, and a commitment by building owners to maintenance.

This section is intended to give an overview of the following: materials for low-slope and steep-slope roofing (including deck, insulation and roof coverings), key design considerations, application, warranties, maintenance, and reroofing. At the end of this section, a list of trade associations (Art. 12.21) and a list of publications (Art. 12.22) is given for those readers interested in further specific information.

ROOF MATERIALS

A **roof system** is an assembly of interacting roof components designed to weatherproof and, normally, to insulate a building's top surface. The roof assembly includes the roof deck, vapor retarder, and roof insulation (if they occur), and the roof covering.

12.1 ROOF DECKS

A good roof is dependent upon the structural integrity of the deck and compatibility of the deck with the roof covering and other materials attached to it. Following are descriptions of commonly used decks.

Cement-wood-fiber panels are composed of treated wood fibers that are bonded together with portland cement or other binder and compressed or molded in flat panels. These panels provide some acoustical attenuation and some thermal resistance.

Lightweight insulating concrete roof decks and fills are produced on the job site by combining insulating aggregates, such as perlite or vermiculite, with portland cement and water. Another variation of this type of deck is referred to as "cellular," lightweight insulating concrete. Rather than using aggregate, cellular concrete is produced with a foaming agent that creates small air cells within the matrix. The compressive strength and thermal resistance of lightweight insulating concrete decks depend on the mix design and composition.

Lightweight insulating concrete may be cast over metal or bulb-tee and formboard systems. Some types may also be cast atop concrete decks. For enhanced thermal resistance, molded expanded polystyrene (EPS) boards may be incorporated into lightweight insulating concrete.

Poured gypsum concrete decks, although widely used in the past, are now seldom used, except in a few locations in the United States. This type of deck is produced on the job site by combining gypsum with wood fibers or mineral aggregates and water. The mixture is then cast on formboards.

Structural concrete decks can either be cast-in-place, post-tensioned, or precast (tees, double tees, channel slabs, flat slabs, or hollow-core slabs).

Steel decks are fabricated by roll-forming cold-rolled sheets. They are available in a variety of depths, 1½ in being most common. The panels are available in narrow-rib (Type A), intermediate-rib (Type F), or wide-rib (Type B), the wide-rib being most common. Common thicknesses are 22, 20, 18, and 16 ga. The panels are available in a paint (prime coat or prime and finish coat) or galvanized finish. (See also Arts. 8.22 to 8.24.)

Steel decks can be fabricated with slots to allow downward-drying. Slotted decks are often used with certain types of wet-fill toppings. Acoustical decks, which have numerous small perforations, are also available. Batt insulation is usually installed in the flutes on the top side of the acoustical deck.

Thermosetting insulating fill is produced on the job site by mixing perlite aggregate with a hot asphalt binder. The mix is then placed over a structural deck. This fill provides some insulation, and it can be utilized to provide slope for drainage. Although more common years ago, this type of system is still available.

Wood planks or panels can be composed of solid wood planks (usually tongue-and-groove) or sheathing panels. Sheathing was originally composed of all-veneer plywood, but now, oriented strand board (OSB) also is used. OSB is composed of compressed strand-like particles arranged in layers oriented at right angles to one another. If sheathing is required for roof decking, sheathing intended for this purpose should be specified.

12.2 VAPOR RETARDERS

These comprise a wide range of materials used to control flow of water vapor from the building interior into wall or roof systems. Unless precautions are taken, water vapor in the interior of a building, especially if it has a high-moisture occupancy, may condense within the cold roof system, saturating the insulation and reducing its effectiveness, or will drip back into the building, staining the ceiling or wetting the floor.

A vapor retarder placed in an appropriate location however, can control such condensation. See also Art. 12.14.

The following general rules will assist in determining whether or not to install a vapor retarder under roofing:

Vapor retarders should generally be applied on all roof decks in cold climates and on all decks of heated buildings where the midwinter temperature may fall below 45°F.

Vapor retarders should be used on all roof decks in temperate climates wherever conditions of high inside humidity exist, such as in textile mills, laundries, canning factories, creameries, and breweries.

Vapor retarders should be used on all roof decks where the roof deck itself contains an appreciable amount of moisture. A vapor retarder should be used in all cases on top of concrete and poured gypsum decks.

Perm Ratings. The effectiveness of a vapor barrier is measured by its perm rating, which is a measure of porosity of material to passage of water vapor. Perm ratings are established by ASTM procedures. To be classified as a vapor barrier, the material should have a perm rating between 0.00 and 0.50 perms.

A perm rating for a material is the number of grains of water vapor (7000 grains equal 1 lb) that will pass through 1 ft^2 of the material in 1 hr when the vapor-pressure differential between the two sides of the material equals 1 in of mercury (0.49 psi).

Retarder Materials. Following are descriptions of some frequently used vapor retarder materials:

Bituminuous vapor retarders are constructed on the job site. They are composed of alternating layers of hot-applied asphalt and asphalt roofing felts (Art. 12.4.1). Generally, two plies of felt and two or three moppings of asphalt are applied.

Kraft paper retarders are typically factory fabricated by adhering two layers of kraft paper together with asphaltic adhesive and glass-fiber reinforcement. At the job site, the rolls of kraft paper are adhered to the substrate and to one another with a cold-applied asphalt adhesive.

Polyethylene sheets (typically 4, 6, or 8 mils thick) are employed in some types of roof systems. In some cases, they are loose-laid, or they may be attached with mechanical fasteners. The laps can be sealed with tape or sealant. In the past, polyethylene or similar types of plastic film materials were adhered with a cold-applied asphaltic adhesive. However, because of difficulties in obtaining secure attachment, plastic-sheet vapor retarders are no longer typically in this manner.

Aluminum foil used as a vapor retarder is typically applied to the face of an insulation product in the factory. Aluminum foil is also used as a reflective insulation system or a radiant barrier system (Art. 12.3). Aluminum-foil facers on rigid insulation boards are usually not considered a vapor retarder, because of the discontinuity at board joints.

12.3 ROOF INSULATION

Many of the insulation products described in the following are available in tapered configurations.

Cellular glass is a rigid insulation composed of heat-fused closed glass cells.

Cellulosic fibers are generally used as loose-fill insulation. They are made of recycled paper.

Glass-fiber batts or blankets (the only difference being the length of the product) are composed of glass fibers and a binding agent. The batts may be finished on one side with a kraft paper or aluminum-foil facer, or they may be left unfaced.

Glass-fiber board is a rigid insulation composed of glass fibers and a binding agent and is faced on the top surface with kraft paper.

Mineral-wool batts are similar to glass-fiber batts, except that they are composed of mineral fibers (produced from molten rock). Mineral-wool batts have high resistance to heat. Typically, they are used for fire-safing; for example, curtain walls or sealing at fire wall or floor penetrations, or steel fireproofing. Mineral batts are typically not used for roofing, except for insulating seismic joints or expansion joints, where enhanced fire resistance is desired.

Mineral-wool board is a rigid insulation similar to glass fiber boards except that it is composed of mineral fibers (produced from molten rock). These boards are available faced or unfaced with aluminum foil.

Perlite is used in rigid insulation composed of expanded perlite, cellulose, and a binding agent.

Phenolic resin has been formed into a rigid, plastic-foam insulation. It is no longer produced in the United States.

Polyisocyanurate resin is formed into rigid, plastic-foam insulation. It resembles, and has essentially replaced, polyurethane board insulation because of better fire resistance. Polyisocyanurate boards are produced with a variety of facers, including glass-fiber and asphalt-saturated organic or inorganic felts. The boards are also available as composites, which are factory produced by foaming the insulation to perlite, wood sheathing, or other types of substrates. The foam is produced by a CFC (chlorofluorocarbon) or HCFC (hydrochlorofluorocarbon) blowing agent, which initially is the gas that fills the cells. Over considerable time, oxygen and nitrogen diffuse into the cells, and the CFC or HCFC diffuses out, thereby decreasing the thermal resistance. This phenomenon is known as **thermal aging** or **thermal drift.** Polyisocyanurate insulation has the highest *R*-value (thermal resistance) per inch thickness of any insulation currently produced in the United States.

Polystyrene made into a rigid plastic foam has two distinctly different forms. Molded expanded polystyrene (EPS) has air-filled cells and hence is not subject to thermal aging. EPS is available with a variety of densities, and its *R*-value is a function of the density. Extruded polystyrene is blown with CFC or HCFC; thus it has a higher *R*-value than EPS. Extruded polystyrene insulation is very resistant to water and water vapor and is available in very high compressive strengths. Accordingly, it is the only type of insulation recommended for use in protected membrane roofs (PMR) or plaza decks (Art. 12.16).

Radiant barrier system (RBS) utilizes aluminum foil product with a low-emittance (high-reflectance) surface. An RBS is intended to reduce radiant heat transfer between a hot roof deck and cooler floor below (or vice versa).

Reflective insulation system (RIS) employs double-sided aluminum foil product, which is used in combination with bulk insulation, or in lieu of bulk insulation. The system incorporates an enclosed air space that contributes significantly to the thermal resistance.

Spray-applied polyurethane foam (PUF), in addition to providing thermal insulation, also functions as part of a roofing system (Art. 12.4.6).

Wood fiberboard is a rigid insulation manufactured from wood or cane fibers and binders.

12.4 LOW-SLOPE ROOF COVERINGS

Roof coverings may be classified into two main groups in accordance with the slope of the roof: Low-slope roof coverings are designed for roofs on which water can collect before proceeding slowly to drainage outlets. Steep-slope roof coverings are applied to roofs with swift drainage.

12.4.1 Built-Up Roofs (BUR)

This is the traditional low-slope membrane roof covering. It is composed of bitumen (either asphalt or coal tar), usually applied hot, felts (either organic, glass-fiber, or polyester), and a surfacing, such as aggregate, coating, or cap sheet (Fig. 12.1).

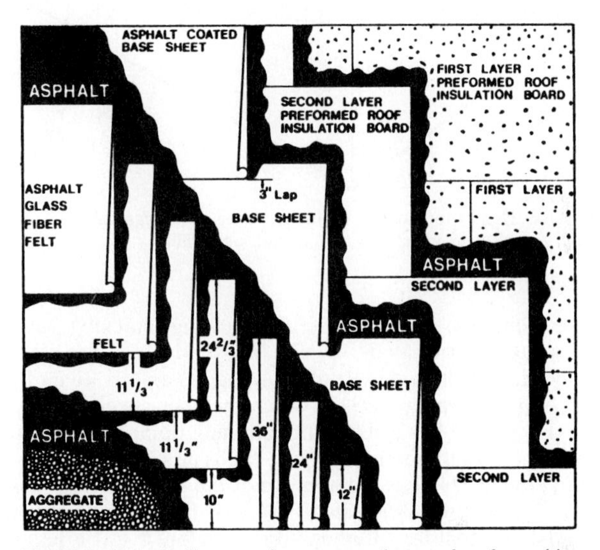

FIGURE 12.1 Built-up roofing over two layers of preformed insulation board. Three plies of asphalt-impregnated glass-fiber felt, embedded in a fluid, continuous application of asphalt, overlay a base sheet adhered with asphalt to the insulation. Aggregate surfacing, about ⅜ in in diameter, is spread over the top flood coat of asphalt.

The membrane is composed of three to five plies of felt (as few as two plies are sometimes specified when polyester felt is used). The first ply is typically either set in a continuous layer of hot bitumen or is nailed to the deck. Subsequent layers of felt are set in a continuous layer of hot bitumen.

An alternative to a traditional BUR is the protected-membrane roofing system. It consists of several layers installed in a sequence different from the usual one in which insulation is placed below the roof deck. First, standard built-up roofing is applied directly to the deck. Then, rigid insulation that is impervious to moisture, such as extruded polystyrene foam, is bonded to the top of the built-up roofing with a mopping of steep asphalt (Fig. 12.2). A layer of ¾-in crushed stone (1000 lb/square) or paving blocks or structural concrete on top of the insulation completes the assembly. Gravel or slag should not be used, because the sharp edges would damage the bare insulation underneath.

The theory is that the insulation, set above the roofing, both insulates the building and protects the built-up roofing from the harmful effects of thermal cycling, ultraviolet degradation, weathering, and roof traffic. Most common defects

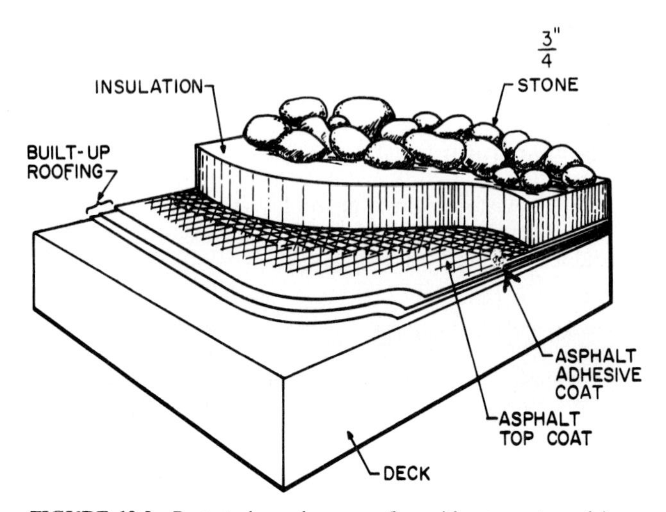

FIGURE 12.2 Protected-membrane roofing with aggregate and insulation placed over, instead of under, built-up roofing.

caused by these elements, such as blistering, ridging, cracking, alligatoring, and wrinkling, are virtually eliminated.

Bitumen may be asphalt or tar. Roofing asphalt is a derivative of petroleum. It is described in ASTM Standard D312, which includes specifications for Types I, II, III, and IV. Each type has a different softening-point range, which should be considered by the specifier when specifying the type of asphalt to be used. Coal tar, described in ASTM D450, is a derivative of the production of coke from coal. Type I is referred to as "old-style pitch." Type II is used for below-grade waterproofing. Type III, or coal-tar bitumen, was developed to be less of an irritant during application than other tars.

Felts are sheet materials used to reinforce membrane waterproofing or roof coverings. The predominant type of felt used is glass fiber although organic felts are still commonly used in the construction of coal-tar systems. Polyester is an alternative type of felt. Asbestos felts were used in the past but are no longer produced in the United States.

There are two primary categories of felt—base sheets and ply sheets. **Base sheets** are heavier felts that are often used for the first layer of felt to be installed. If the felt is to be nailed, a base sheet is recommended because of its greater strength. **Ventilating base sheets** are intended to allow for the venting of moisture-vapor pressure by lateral (horizontal) movement. However, if a ventilated base sheet is to be used, the design should take into account the small driving force for horizontal moisture transport and the small amount of moisture that can be moved horizontally.

Surfacings as applied to built-up membranes, are typically small pieces of aggregate or slag, liquid-applied coatings, or a cap sheet. Common coatings include cutbacks and emulsions, which are both cold-applied. **Cutbacks** are composed of asphalt and solvent and often include an aluminum pigment for reflectivity. **Emulsions** consist of clay and asphalt particles dispersed in water. Some emulsions include aluminum pigment or titanium dioxide for reflectivity. The cutback and emulsion coatings are available in fibrated or nonfibrated grades. Latex (acrylic) coatings are also available, but for built-up roofs. These coatings are not used as often as the other types of coatings. **Cap sheets** are heavily coated felts that are factory surfaced with mineral granules.

Cold-process roof coverings (also known as cold-applied) are similar to hot-applied BUR, except that instead of hot bitumen, asphalt-based cutbacks or emulsions are typically used. They are applied by sprayer, brush, broom, or squeegee.

12.4.2 Liquid-Applied Roof Coverings

Liquid-applied systems are supplied as either single or two-component elastomeric materials. They are applied by sprayer, brush, roller, or squeegee. Typically these systems are applied directly over concrete or plywood sheathing. Deck joints and cracks normally require special preparation. See also cold-process roof coverings (Art. 12.4.1) and coatings on polyurethane foam roofs (Art. 12.4.6).

12.4.3 Metal Roof Coverings

These are generally used for steep-slope roofs rather than for low-slope roofs. See Art. 12.5.3

12.4.4 Modified Bitumen Membranes

These are typically composed of prefabricated sheets of polymer-modified asphalt with polyester or glass-fiber reinforcement or a combination of these. The polymers most used for asphalt modification are atactic polypropylene (APP) or styrene-butadiene-styrene (SBS). These prefabricated sheets are commonly installed over a base sheet (Art. 12.4.1), which may or may not also be composed of modified bitumen. Sometimes the assembly also includes a ply sheet (Art. 12.4.1).

In the past, modified bitumen membranes were occasionally applied in a single layer. However, two or more layers are now the predominant system (Fig. 12.3).

SBS sheets are generally set in a continuous layer of hot asphalt, but some sheets may be torch-applied or set in cold adhesive. Self-adhering SBS sheets are also available. SBS sheets need protection from ultraviolet light (UV). Protection is typically provided by factory-applied mineral granules. They may also be surfaced with coatings (Art. 12.4.1).

APP sheets are generally torch-applied (Fig. 12.4). When APP sheets were introduced in the United States in the late 1970s, they were generally used without surfacing, since UV protection was reportedly provided by the APP modifier. While some such APP membranes weathered very well, others did not. Hence, coatings (cutbacks, emulsions, or latex) are now often used.

Instead of incorporating prefabricated modified bitumen sheets, membranes can also be constructed with modified mopping asphalt and felts (of the type used for BUR construction). For modification of asphalt for application by mopping or by mechanical spreaders, styrene-ethylene-butylene-styrene (SEBS) polymers are utilized.

12.4.5 Single-Ply Roof Coverings

The single-ply family of roofing materials includes some distinctly different products. (Modified bitumen products (Art. 12.4.4) are sometimes included in the single-

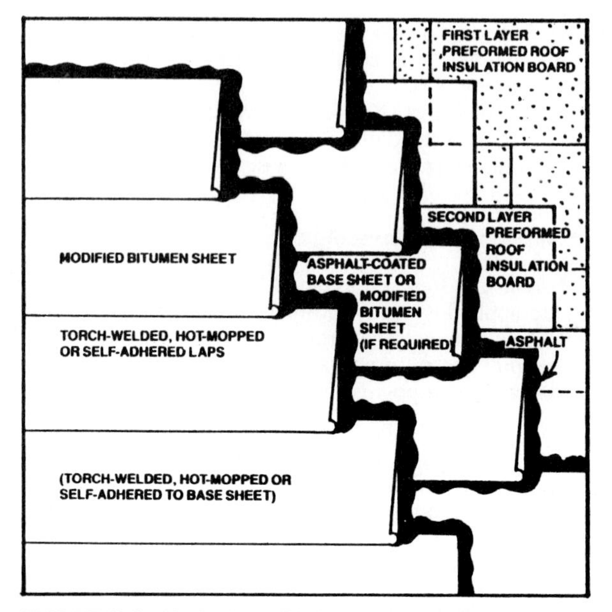

FIGURE 12.3 Single-ply modified-bitumen roof with a base sheet overlaying two layers of preformed insulation board.

FIGURE 12.4 Torch application of atactic polypropylene (APP) modified-bitumen membrane.

ply category.) The single-plies can be classified as either thermoset or thermoplastic materials. Thermoset materials normally cross-link (cure) during manufacturing. Once cured, these materials can only be bonded to themselves; for example, at a seam). Bonding is accomplished with an adhesive. Thermoplastic materials do not cross-link. Therefore, they should be capable of being welded together throughout their service life. During installation of a ply, welding is usually accomplished with hot air.

There are three primary methods for attachment of single-ply membranes to a roof deck or other substrate. In the ballasted system, the membrane is laid loose over the substrate and then covered with ballast to resist uplift from the wind (Fig. 12.5*a*). The ballast can either be large aggregate or concrete pavers. In the second method of attachment, the membrane is fully adhered in a continuous layer of adhesive (12.5*b*). In the third method, the membrane is mechanically attached to the substrate (Fig. 12.5*c*).

The mechanically attached system has three primary variations. In the first method, metal batten bars are placed at intervals on top of the membrane and then screwed to the deck. The bars are then covered with a stripping ply of the membrane material. Alternatively, the batten bars can be placed within the membrane lap (seam). In the second method, screws and plates (typically 2 in in diameter) are placed within the membrane lap. Another variation of this method utilizes larger disks that are placed on top of the membrane and screwed in place. They are then made watertight in any of a variety of ways. In the last method, a larger plate is screwed to the deck at prescribed spacing and the membrane is adhered to the plate.

Following are descriptions of materials used in single-ply membranes.

Chlorosulfonated polyethylene (CSPE) is commonly known by the trade name *Hypalon*. It is a thermoset plastic, but it cures after installation on a roof. This product is specified in ASTM D5019 (Type I). It is usually supplied in a white color.

Ethylene propylene diene terpolymer (EPDM) is a synthetic rubber membrane. It is a thermoset plastic. This product is specified in ASTM D4637 (Type I). It is available in a white color, but black is used most often.

Polyisobutylene (PIB) is a thermoplastic product, specified in ASTM standard D5019 (Type III). It is available in a black or white color.

Polyvinyl chloride (PVC) is a thermoplastic product, specified in ASTM D4434. It is available in a variety of colors.

PVC blends (also known as *copolymer alloys*) are based on PVC resin. They are similar to PVC membranes. The next revision of ASTM D4434 will probably also cover PVC blends.

("Single-Ply Roofing: A Professional's Guide to Specifications," Single-Ply Roofing Institute, Wellesley Hills, MA 02181.)

12.4.6 Spray-Applied Polyurethane Foam (PUF) Roof Coverings

These consist of polyurethane foam insulation, which is spray-applied to the substrate, and topped with a surfacing (Fig. 12.6). Traditionally, the foam is surfaced with a coating of latex (acrylic), polyurethane, or silicone. Mineral granules are sometimes applied to the wet coating for additional abrasion and impact resistance (at traffic walkway areas or throughout the entire roof).

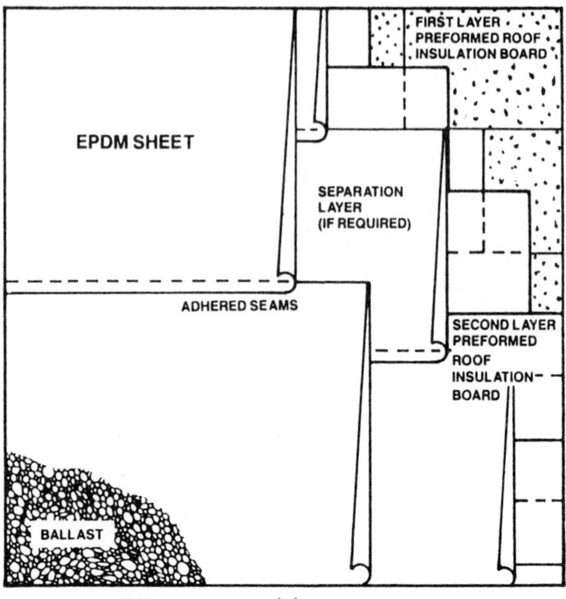

(a)

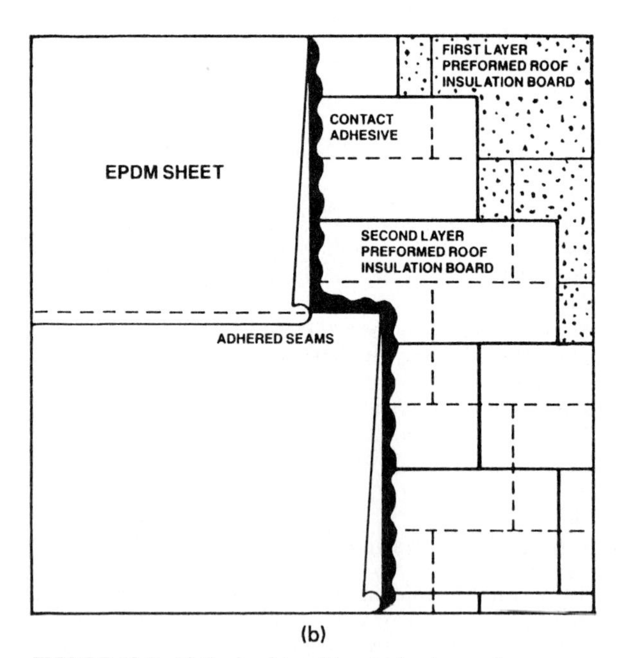

(b)

FIGURE 12.5 Methods of installing single-ply membrane over insulation board: (*a*) loose laid; (*b*) fully adhered; and (*c*) partly adhered.

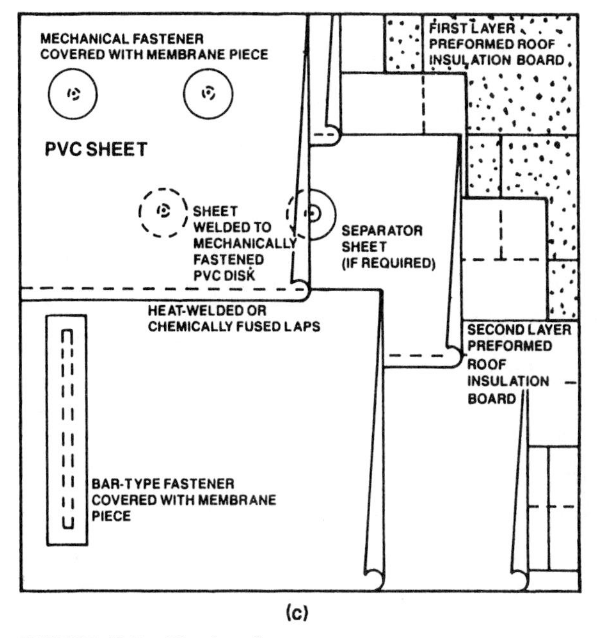

FIGURE 12.5 (*Continued*)

FIGURE 12.6 Robotic sprayer applying polyurethane foam.

An alternative surfacing is aggregate, similar to the type used for BUR placed directly over the foam. In this system, coatings are used only on vertical surfaces, such as parapets or equipment curbs.

See also Art. 12.22. Roof Systems Bibliography.

12.5 STEEP-SLOPE ROOF COVERINGS

These differ from low-slope roof coverings in that on steep-slope roofs water flows rapidly over exposed units to gutters and downspouts. Many of the low-slope roof coverings described in Art. 12.4 can be successfully used on steep slopes. Some of the low-slope materials, however, become slick when wet. This should be taken into account before they are specified for steep slopes.

12.5.1 Asphalt Shingles

These are composed of a reinforcing mat (organic or glass fiber), a specially formulated asphalt coating, and mineral granules. (Glass-fiber reinforced shingles are sometimes referred to as *fiberglass shingles*. Organic-reinforced shingles are sometimes referred to as *asphalt shingles*, which is confusing, for this term properly applies to both the organic and glass-fiber reinforced products.) Most asphalt shingles are manufactured with a self-seal adhesive for wind resistance.

Asphalt shingles are available in a variety of weights. (Weight, however, is not necessarily an indicator of product performance.) Also, they are available in a variety of styles and colors: three-tab shingles (Fig. 12.7); strip, or architectural, shingles which have no cutouts and which may be laminated and have a heavy texture, and shingles that also enhance the three-dimensional look by adding shadow lines or shading through use of colored granules. Glass-fiber reinforced

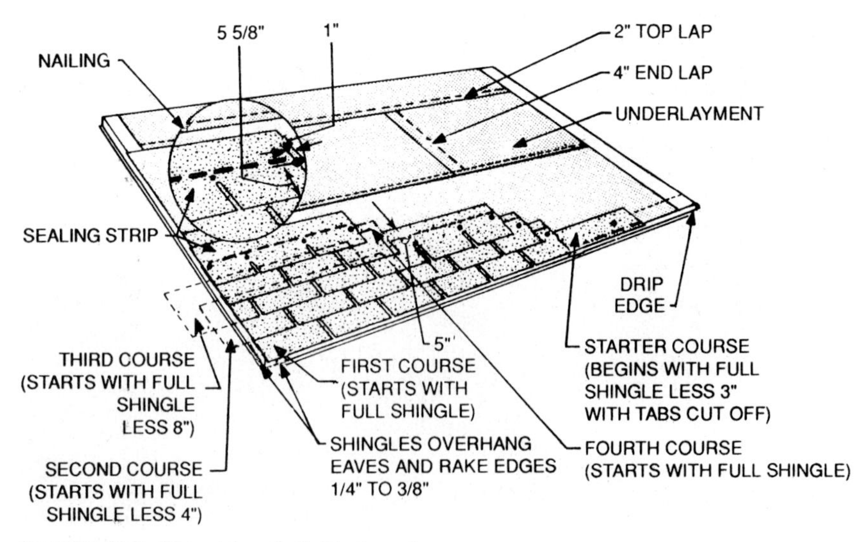

FIGURE 12.7 Three-tab asphalt-shingle roof.

shingles are specified in ASTM D3462. Organic-reinforced shingles are specified in ASTM D225.

12.5.2 Cement-Fiber Shingles

Formerly reinforced with asbestos fibers and hence known as *cement-asbestos shingles,* cement-fiber shingles are now reinforced with fibers other than asbestos. Some of these products are intended to visually simulate other products, such as slate.

Basically, there are four standard types of cement fiber roofing shingles, known as American-method, multiple-unit, Dutch-lap, and hexagonal.

American-method shingles are so called because of their shape and finish. Laid in a rectangular pattern and having a simulated wood grain, they present an appearance similar to conventional wood shingles.

Multiple-unit shingles are produced in large units, each of which covers an area equal to that of two to five standard-size shingles. Made in styles and sizes that vary with manufacturers, they retain the same general appearance, when installed, as the small units.

Dutch-lap shingles, sometimes called *Scotch-lap,* are larger than conventional shingles and are lapped at the top and one side, a method that effects savings in material and labor.

Hexagonal shingles, also known as shingles laid in the *French method,* are nearly square. They are laid in a diamond pattern with an overlap at the top and bottom. The effect is a hexagonal pattern.

Sizes of cement-fiber roof shingles vary from 16 × 16 in to 14 × 30 in.

12.5.3 Metal Roof Coverings

The metal category includes a large variety of products, such as metal shingles and panels. Metals used to form the products include aluminum, copper, and aluminum-zinc-alloy steel (*Galvalume*). Steel and aluminum panels are available with several different types of paint finishes and colors. Many of the products are formed in a factory, while others are formed on the job site by the roofing contractor.

Metal roof panels include four primary types: structural standing seam, architectural standing seam, exposed fastener, and traditional metal roofing. Standing-seam panels are available with or without battens. The standing-seam and exposed-fastener panels are roll-formed, while traditional panels are typically press-brake formed. The exposed-fastener panels have largely been replaced by standing-seam panels, except for very inexpensive construction.

Architectural and most traditional panels need to bear on a continuous structural substrate (deck) in order to carry live loads, such as snow. However, architectural panels should be designed to accommodate design wind loads.

Structural panels (Fig. 12.8) have the capability of spanning between supports. Accordingly, they can be placed over purlins (as is the case with preengineered metal buildings), as well as over continuous structural substrate. Some structural panels have the capability of being successfully used as low-slope coverings.

12.5.4 Roofing Slate

Slate is a dense durable rock that has a natural cleavage plane. The surface texture and color of slate after it is split depend on the characteristics of the rock from which it is quarried. Roofing slate is a long-lasting but very heavy material.

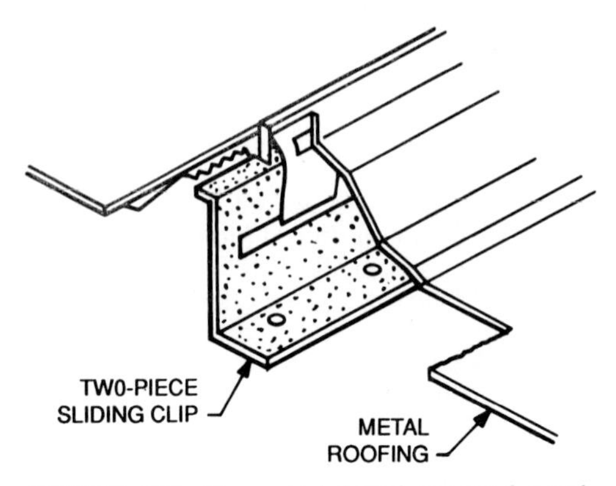

FIGURE 12.8 Trapezoidal-shaped structural metal-panel roofing system with concealed clips at seams.

Slate roofs may be divided into two classifications, standard commercial slate and textural slate, or random slating. The latter is the older form, in which slates are delivered to the job in a variety of sizes and thicknesses, to be sorted by slaters. The longer and heavier slates are placed at the eaves, medium sized at the center, and the smallest at the ridge.

Standard commercial slating results from grading of slates at the quarries and costs less in place. Slates are graded not only by length and width but also by thickness, the latter being about ¼ in. Figure 12.9 shows a typical slate installation.

12.5.5 Synthetics

This class of materials includes a variety of products that are intended to simulate other materials. Typically, synthetics simulate wood shingles or shakes, or slate. The imitation materials include cement-fiber and metal (Arts. 12.5.2 and 12.5.3), wood fiber, and polymer composites.

12.5.6 Roofing Tile

Produced from clay or concrete, tiles (Fig. 12.10) are available in a variety of shapes, colors, and textures. Clay tiles are specified in ASTM C1167.

12.5.7 Wood Shingles and Shakes

Shingles are sawn on both sides, while shakes are split on at least one surface. Cedar is typically used for wood roofs, but other species of wood are sometimes used. There are different grades of shingles and shakes, and they can be pressure-perservative treated or pressure treated for fire resistance. See also Art. 12.22, "Roof Systems Bibliography." Figure 12.11 shows a typical wood shake installation.

Key Design Considerations. A complete discussion of design considerations for the primary types of low-slope and steep-slope roofing systems is beyond the scope of this book. However, a few general key considerations are discussed in the following.

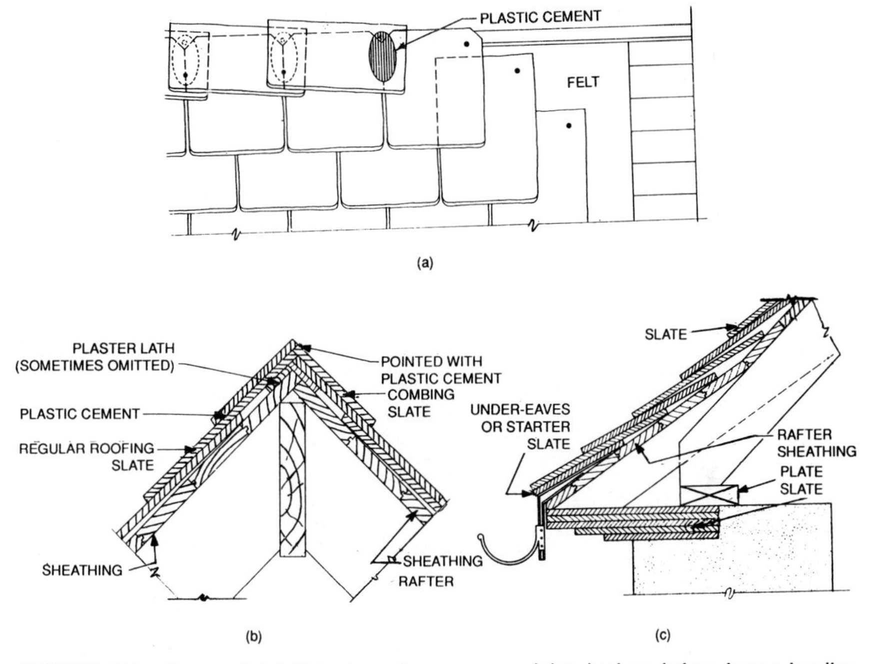

FIGURE 12.9 Slate roof. (*a*) Slates in each course cover joints in those below, have a headlap of 3 in over the lowest course below, and have an exposure (portion not covered by next course above) equal to $\frac{1}{2}(L - 3)$, where L is the slate length, in. (*b*) Section at roof ridge illustrates the saddle-ridge method, in which regular slates extend to the ridge so that slates on opposite sides of the ridge butt flush. Another course of slate, called combing slate, is set on top, butting flush at the ridge. (*c*) Section at the eaves.

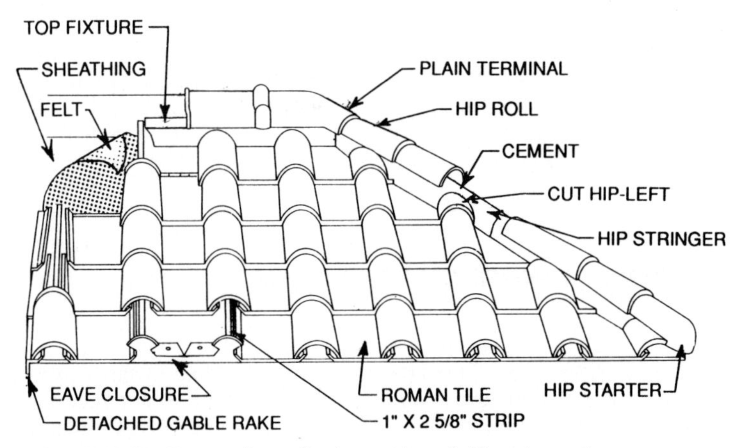

FIGURE 12.10 Roman tile application at hip and ridge of a roof.

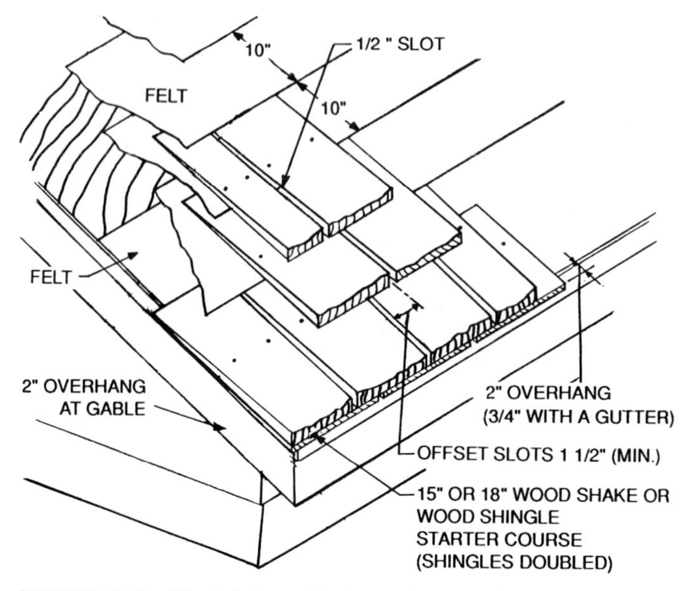

FIGURE 12.11 Wood shake application at the eaves of a roof.

12.6 NEED FOR FAMILIARITY WITH ROOF DESIGN

A common pitfall for designers is lack of familiarity with the principles of roof design. This is manifested in selection of inappropriate materials or systems or inadequate details, or use of poorly prepared specifications. To assist designers, several periodicals and books are available (Art. 12.22) and seminars are offered by the Roofing Industry Educational Institute and the National Roofing Contractors Association (Art. 12.21).

In addition to these resources, designers would find it helpful to establish a working relationship with a professional roofing contractor. Many contractors are willing to spend some time with designers. This can be particularly helpful if the designer is working on a project that has some unusual aspect, or if a material or system is being contemplated with which the designer is not familiar. For more extensive input, utilization of a professional roofing consultant will be advantageous.

12.7 BUILDING OWNERS' REQUIREMENTS

It is important for roof designers to determine if the building owner has any specific requirements, such as type of materials; what the building owner's expectations are regarding the roof's longevity; and to what extent the owner is committed to maintenance. If the owner is unwilling to allocate adequate funds for maintenance, a conservative durable system should be selected.

Also, designers should determine how detrimental leakage or a roof blow-off would be. For most buildings, these events are unpleasant but generally manageable. But, if the roof is over very expensive electronic equipment, or a critical facility such as a hospital, a conservative roof design may well be appropriate. It is also desirable to determine if the owner's insurer has specific requirements for the roof system.

Contract documents should be carefully designed and administered. A good roof design should be complemented with comprehensive, unambiguous specifications and drawings, so that the design intent is clearly communicated to the contractor.

12.8 BUILDING-CODE PROVISIONS FOR ROOFS

Designers should be aware of building-code requirements affecting design of the roof. Model building codes and many state and local codes have provisions related to roof systems (for new construction, as well as reroofing).

Energy efficiency may be one of the objectives of code requirements. Many local jurisdictions have such requirements for residential as well as for other types of construction.

12.9 EFFECTS OF CLIMATE

The climate in the region where the roof will be constructed often plays a key role in design of the roof system. Climate considerations should include the type of weather that will likely be encountered during application, as well as the climatic conditions that follow. For example, if the roof is to be constructed during cold weather or in a location that experiences frequent rains, selection of materials and a system that is more tolerant of these conditions can play a key role in obtaining a successful roof.

12.10 EFFECTS OF ROOF SIZE, SHAPE, AND SLOPE

Roof size and shape often dictate material and system selection. For example, if the roof is only 2 or 3 ft wide and several feet long, a system other than built-up roofing (BUR) would probably be easier to install than BUR.

Roof slope is a major factor in determining the rate of flow of rainwater over a roof to a drainage outlet, and good drainage is essential to good performance of a roof. Hence, regardless of the type of materials or system specified, adequate slope should be provided for drainage.

For low-slope roofs, a designed slope of ¼ in/ft is generally sufficient. When a roof is supported by structural members with long spans, however, deflection of the members can create excessive ponding for even a ¼-in slope. The "NRCA Roofing and Waterproofing Manual" (Art. 12.22) provides information for assessing slope requirements for these special conditions. See also Art 3.4.3.

For steep-slope roofs, the minimum recommended slope is a function of the type of roof-covering material used and the type of underlayment. For example, with asphalt shingles, the minimum recommended slope is 4:12, unless special underlayment provisions are made. With special provisions, the National Roofing Contractors Association (NRCA) recommends that the minimum slope be 2½:12. However, where possible, in most climates, and with most materials and systems, if a steeper slope can be designed, it is generally prudent to do so. Since most steep-slope materials are water-shedding, rather than waterproofing, a steeper slope typically decreases the likelihood of damaging or other undesirable conditions.

12.11 DECK SUITABILITY

In selection of a deck or the type of roof-covering material and system, it is important to develop a design that incorporates materials that are compatible. For example, if a mechanically attached single-ply membrane is desired, it would be appropriate to specify a deck that can readily accept the fasteners yet have sufficient capacity to hold the fasteners. In this case, a steel deck, wood plank, or thick plywood deck would normally be good options. But if the deck did not readily accept fasteners; for example, a concrete deck, or if the deck possessed minimal strength; for example, a lightweight, insulating-concrete fill, then a system other than a mechanically attached single-ply membrane should be selected.

12.12 EFFECTS OF ROOFTOP TRAFFIC

If there will be periodic traffic on a roof; for example, to maintain mechanical equipment, traffic walkway pads should be specified to protect the roof. Pads can also be beneficial around mechanical equipment, to protect against dropped tools. For aggregate-ballasted systems, use of concrete pavers for walkways can be helpful in protecting the roof and providing a comfortable surface on which to walk.

For heavier rooftop loading, greater protective measures; for example, installation of heavy concrete pavers may be needed. In some cases, window-washing equipment causes damage to roof coverings, because they do not have sufficient resistance to the high impact loads that can result as the equipment is moved around and dropped.

12.13 ESTHETIC CONSIDERATIONS

While esthetics typically is not of concern for low-slope roofs, it generally is with steep-slope roofs. For some low-slope roofs, an attractive roof is also desirable. Where esthetics is an issue, special effort is needed to achieve the desired goal.

For example, a clean white roof may look good for a year or two. But after it has desposits of wind-blown dirt, or from contaminants exhausted from mechanical equipment, or from sediment from ponded water, it may look awful. If a more uniform appearance is desired, an aggregate-surfaced built-up roofing or ballasted (with aggregate or pavers) single-ply membrane may be specified.

To be sure that a proposed design will produce the desired look, designers should visit an existing roof that has a similar roof covering to determine if expectations will likely be met.

12.14 EFFECTS OF CONDENSATION ON ROOFS

With some roofs, a vapor retarder is needed to avoid condensation damage within the roof system (Art. 12.2). Guidelines for determining when a vapor retarder is needed are given in the "NRCA Roofing and Waterproofing Manual." However, the U.S. Army's Cold Regions Research and Engineering Laboratory (CRREL) has reported that buildings with high interior humidity can experience unacceptably high levels of condensation, even when the buildings are located in warm climates, and has established criteria based on this conclusion. In such cases, the CRREL criteria may recommend a vapor retarder, whereas the NRCA criteria do not. The "NRCA Energy Manual" recognizes the validity of the CRREL refinement. The CRREL criteria are included in the "Proceedings of the Ninth Conference on Roofing Technology" (Art. 12.22).

12.15 EFFECTS OF WIND ON ROOFS

Particularly in those areas subjected to hurricanes or other high winds, provision for resistance to wind in design of roof systems is necessary for successful performance. Different roof systems are loaded differently, and they resist the loads in different ways. For example, loading and load resistance of asphalt shingles, a modified bitumen membrane, a ballasted single-ply membrane, and a mechanically attached single-ply membrane are all different.

An understanding of the loading and load response is needed to design wind-resistant roof coverings for use in high-wind environments. In particular, design of metal edge flashing (gravel stops) is a critical aspect for many types of roof systems. Further information regarding wind design is available from publications listed in Art. 12.22.

12.16 PROTECTED MEMBRANE ROOFS AND PLAZA DECKS

A special type of low-slope roof system is the protected membrane roof (PMR). In this system, extruded polystyrene is placed above the membrane. The insulating plastic is then covered with aggregate or concrete paver ballast. (If aggregate is used, a fabric mat is first placed over the insulation.) This construction protects the membrane from mechanical damage (dropped tools, wind-blown debris) and from direct exposure to the weather (UV, high and low temperatures).

Plaza decks are similar to protected membrane roofs, except that they are designed for high traffic loads (pedestrian, and in some cases, vehicular). Because of the difficulty in repairing these types of roof systems, they should be designed and constructed with care. If this is done, they can provide a long and successful service life.

Application of Roofing. If a roof has been adequately designed and quality materials have been specified and provided, the next step toward achieving a successful roof is to have it installed by a reputable roofing contractor. After a contractor has been retained to perform the work, the following items are of importance.

12.17 PREROOFING CONFERENCE

Well in advance of ordering materials and beginning work, the project designer or the general contractor should convene a preroofing conference. This meeting is normally attended by the roofing contractor (including the job-site person who will be in charge of the work), the building owner's representatives (including the project designer), the general contactor, and a representative of manufacturers who may be required to furnish a roof bond. If the project has a lot of rooftop mechanical or electrical equipment, these subcontractors would also normally attend the meeting.

The purpose of the meeting is to review the salient features of the drawings and specifications, to ensure there is understanding by all parties. If there are problems with the design or other aspects of the project, the intent is to identify and resolve them prior to commencement of the field work. As part of this meeting, the roof deck should be reviewed to verify that it is ready for roofing.

12.18 WARRANTIES

Long-term roofing warranties are quite common, but often their importance is overemphasized in selection of a roof. Many building owners and designers have focused on specifying warranties rather than on other aspects that are more likely to result in a successful roof. The "Consumer Advisory Bulletin on Roofing Warranties," 1992, National Roofing Contractors Association (NRCA), suggested the following:

> The length of a roofing warranty should not be the primary criterion in the selection of a roofing product or system because the warranty does not necessarily provide assurance of satisfactory roofing performance. The selection of a roofing system for a particular project application should be based upon the product's qualities and suitability for the prospective construction project. A long-term warranty may be of little value to a consumer if the roof does not perform satisfactorily and the owner is plagued by leaks. Conversely, if the roof system is well-designed, well-constructed and well-manufactured, the expense of purchasing a warranty may not be necessary.
>
> Manufacturers who use long-term warranties as a marketing tool have encountered a highly competitive roofing market and have found themselves compelled to meet or exceed warranties of competitive manufacturers. It is suspected that in some cases the length of the warranty was established without appropriate technical research or documentation of in-place field performance.
>
> NRCA believes that the roofing consumer, with the assistance of a roofing professional, should focus his purchase decision primarily on an objective and comparative analysis of proven roofing system options that best serve his specific roofing requirements and not on warranty time frames.
>
> NRCA further advises that the roofing consumer consult the membrane warranty section of the "Roofing Materials Guide" for a comparative analysis of the specific

provisions, remedies, limitations, and exclusions of the warranties of those roofing systems under consideration. All questions should be addressed to the respective roofing manufacturers for specific written clarification.

12.19 ROOF MAINTENANCE

After a good roof has been delivered to the building owner, it is important for the owner to commit to a program of periodic roof inspections and to follow-up maintenance or repairs as needed. While some systems require less maintenance than others, no roof should be forgotten about after it is installed!

Semiannual inspections are recommended for all roofs. The purpose of these inspections is to determine if debris removal is needed; for example, cleaning of roof drains, and if the roof is showing signs of distress. In many instances, undesirable conditions can be minimized if they are detected and corrected early. Without inspections, a small problem, such as a puncture, can go unnoticed until a large area of the roof is wet and in need of replacement.

To assist in the care of the roof, many professional roofing contractors offer maintenance contracts to building owners. By making a commitment to periodic inspections and appropriate maintenance and repair, the building owner can optimize the roof's service life and maximize the roofing investment.

12.20 REROOFING

Replacement of existing roofing or installation of a new roof covering over it is usually much more complicated than roofing a new building. Besides issues that are normally considered in design of a new roof, many additional issues arise when roof replacement is needed. For example, it is not uncommon to find existing undesirable building conditions including those that contributed to the demise of the existing roof. Unless these are adequately corrected in the new design (which may be difficult or expensive to do), they can also be expected to affect the new roof adversely.

Therefore, as part of the reroofing design process, it is important to determine the reason for the problems with the existing roof. Are they simply a consequence of old age, or are there fundamental conditions that could affect the new roof adversely; for example, migration of water from a curtain wall into the roof system? If there are inherent troubles, they need to be corrected as part of the reroofing work. Also, as part of the design process, the integrity of the roof deck should be assessed by an engineer, if deteriorated deck is suspected. And if the new roof system adds weight to the structure, the capacity of the structure should be evaluated by an engineer.

12.20.1 Reroofing Procedures

There are two primary approaches to reroofing: Either the existing system can be removed down to the deck, or it may remain in place and a new roof covering installed over it. The removal option is the most conservative approach, inasmuch as it allows a complete view of the top side of the deck. If there are deteriorated areas, they will be likely to be found. Also, this approach eliminates entrapment of water in the existing roof. However, this option is often more expensive, and it exposes the interior to water damage during the reroofing process.

The recover option has the advantage of retaining the thermal value of the existing insulation, and the existing roof provides protection against sudden rainfall during the application process. In some parts of the United States, this is not important, but in other areas, this advantage makes the recover option very desirable.

Prior to recovering, it is important to determine if there are areas of wet insulation. If so, it is recommended that it be removed. The utilization of nondestructive evaluation (NDE) can be very helpful in searching for wet insulation.

Unless the designer is very knowledgeable of techniques for assessing existing roof systems and of reroofing design, the designer would be well advised to work with a professional roofing contractor or other roofing professional.

12.20.2 Roof Moisture-Detection Surveys

Moisture within a roof membrane system tends to migrate from its point of entry. Thus, blisters, cracks, and splits caused by the moisture are often located at a distance from the source of a leak. Several methods are available for locating such sources. The equipment utilizes infrared photographic techniques, or nuclear or electronic devices, or variations of these systems. Each method has advantages and disadvantages but can be effective if properly used.

In the method employing an infrared roof scanner, an infrared photograph is taken of a roof at night from a helicopter. Because wet insulation emits vapors and loses heat, it can be detected by the scanner and photographed.

The electronic and nuclear methods use differing means for locating roof moisture, but the procedures are similar. First, a drawing to scale of the roof is made, showing all projections, roof equipment, etc. A grid is marked on the drawing, then physically reproduced on the roof with ropes or paint. Specially trained operators take a reading, with an instrument that can detect moisture, at every grid intersection and record the observation. At the same time, the roof is visually inspected and observations are carefully noted. The data collected then are evaluated by a specialist and adjusted for the effects of up to nine different factors, such as gravel depth, roof construction, etc. The information is entered on a final map of the roof, to provide a picture of water-damaged areas, under-membrane water-flow patterns, percentage of moisture in the roof, and locations of moisture. From this map, it is possible to determine where replacement is necessary, where repair will be sufficient, and what to budget for future work that may be needed to maintain the roof or whether to replace it section by section as required. Such moisture detection surveys are valuable tools for saving energy and for management planning.

12.21 ROOFING INDUSTRY ASSOCIATIONS AND RELATED ORGANIZATIONS

American Plywood Association (APA)
7011 South 19th St.
P. O. Box 11700
Tacoma, WA 98411
206-565-6600

ASTM
1916 Race Street
Philadelphia, PA 19103
215-299-5400

Asphalt Roofing Manufacturers Association (ARMA)
6000 Executive Boulevard, Suite 201
Rockville, MD 20852-3803
301-231-9050

Cedar Shake and Shingle Bureau
515 116th Avenue, NE, Suite 275
Bellevue, WA 98004
206-453-1323

Copper Development Association
P. O. Box 127
Boiling Springs, PA 17007
717-258-3904

Factory Mutual Research (FM)
1151 Boston-Providence Turnpike
P. O. Box 9102
Norwood, MA 02062
617-762-4300

Institute of Roofing and Waterproofing Consultants (IRWC)
4242 Kirchoff Road
Rolling Meadows, IL 60008
708-202-8500

Metal Building Manufacturers Association (MBMA)
1300 Sumner Ave.
Cleveland, OH 44115-2851
216-241-7333

Metal Construction Association (MCA)
1101 14th Street, N.W., Suite 1100
Washington, DC 20005
202-371-1243

Mineral Insulation Manufacturers Association (MIMA)
1420 King Street
Alexandria, VA 22314
703-684-0084

National Roofing Contractors Association (NRCA)
10255 W. Higgins Road, Suite 600
Rosemont, IL 60018-5607
708-299-9070

National Tile Roofing Manufacturers Association (NTRMA)
P. O. Box 947
Eugene, OR 97440
503-689-0366

Roof Consultants Institute (RCI)
7424 Chapel Hill Road
Raleigh, NC 27607
919-859-0742

Roofing Industry Educational Institute (RIEI)
14 Inverness Drive, E.
Building H, Suite 110
Englewood, CO 80112
303-790-7200

Sheet Metal and Air Conditioning Contractors National Association (SMACNA)
4201 Lafayette Center Dr.
Chantilly, VA 22021-1209
703-803-2980

Single-Ply Roofing Institute (SPRI)
20 Walnut Street, Suite 208
Wellesley Hills, MA 02181
617-237-7879

Steel Deck Institute (SDI)
P. O. Box 9506
Canton, OH 44711
216-493-7886

Underwriters Laboratories (UL)
333 Pfingsten Road
Northbrook, IL 60062
708-272-8800

12.22 ROOF SYSTEMS BIBLIOGRAPHY

"Architectural Sheet Metal Manual," 4th ed., Sheet Metal and Air Conditioning Contractors Association, Chantilly, VA 22021.

R. D. Herbert, "Roofing: Design Criteria, Options, Selection," R. S. Means Company, Inc., Duxbury, Mass.

H. O. Laaly, "Science and Technology of Traditional and Modern Roofing Systems," Laaly Scientific Publishing, Los Angeles, Calif.

"NRCA Roofing and Waterproofing Manual," "NRCA Energy Manual," and "Proceedings of the 9th Conference on Roofing Technology," 1989, National Roofing Contractors Association, Rosemont, IL 60018.

"Single-Ply Roofing: A Professional's Guide to Specifications," Single-Ply Roofing Institute, Wellesley Hills, MA 02181.

"Approval Guide," published annually, and "Loss Prevention Data Sheets," Factory Mutual Research Corporation, Norwood, MA 02062.

"Roofing Materials and Systems," and "Fire Resistance Directory," both published annually, Underwriters Laboratories Publications, Northbrook, IL 60062.

SECTION THIRTEEN
HEATING, VENTILATION, AND AIR CONDITIONING

Frank C. Yanocha
Vice President
STV Group
Pottstown, Pennsylvania

The necessity of heating, ventilation, and air-conditioning (HVAC) control of environmental conditions within buildings has been well established over the years as being highly desirable for various types of occupancy and comfort conditions as well as for many industrial manufacturing processes. In fact, without HVAC systems, many manufactured products produced by industry that are literally taken for granted would not be available today.

13.1 DEFINITIONS OF TERMS OF HEATING, VENTILATION, AND AIR CONDITIONING (HVAC)

Adiabatic Process. A thermodynamic process that takes place without any heat being added or subtracted and at constant total heat.

Air, Makeup. New, or fresh, air brought into a building to replace losses due to exfiltration and exhausts, such as those from ventilation and chemical hoods.

Air, Return (Recirculated). Air that leaves a conditioned spaced and is returned to the conditioning equipment for treatment.

Air, Saturated. Air that is fully saturated with water vapor (100% humidity), with the air and water vapor at the same temperature.

Air, Standard. Air at 70°F (21°C) and standard atmospheric pressure [29.92 in (101.3 kPa) of mercury] and weighing about 0.075 lb/ft^3 (1.20 kg/m^3).

Air Change. The complete replacement of room air volume with new supply air.

Air Conditioning. The process of altering air supply to control simultaneously its humidity, temperature, cleanliness, and distribution to meet specific criteria for a space. Air conditioning may either increase or decrease the space temperature.

Air Conditioning, Comfort. Use of air conditioning solely for human comfort, as compared with conditioning for industrial processes or manufacturing.

Air Conditioning, Industrial. Use of air conditioning in industrial plants where the prime objective is enhancement of a manufacturing process rather than human comfort.

Baseboard Radiation. A heat-surface device, such as a finned tube with a decorative cover, used in lieu of a room's baseboard trim.

Blow. Horizontal distance from a supply-air discharge register to a point at which the supply-air velocity reduces to 50 ft/min.

Boiler. A cast-iron or steel container fired with solid, liquid or gaseous fuels to generate hot water or steam for use in heating a building through an appropriate distribution system.

Boiler-Burner Unit. A boiler with a matching burner whose heat-release capacity equals the boiler heating capacity less certain losses.

Boiler Heating Surface. The interior heating surface of a boiler subject to heat on one side and transmitting heat to air or hot water on the other side.

Boiler Horsepower. The energy required to evaporate 34.5 lb/hr of water at 212°F, equivalent to 33,475 Btu/hr.

Booster Water Pump. In hot-water heating systems, the circulating pump used to move the heating medium through the piping system.

British Thermal Unit (Btu). Quantity of heat required to raise the temperature of 1 lb of water 1°F at or near 39.2°F, which is its temperature of maximum density.

Central Heating or Cooling Plant. One large heating or cooling unit used to heat or cool many rooms, spaces, or zones or several buildings, as compared to individual room, zone, or building units.

Coefficient of Performance. For machinery and heat pumps, the ratio of the effect produced to the total power of electrical input consumed.

Comfort Zone. An area plotted on a psychometric chart to indicate a combination of temperatures and humidities at which, in controlled tests, more than 50% of the persons were comfortable.

Condensate. Liquid formed by the condensation of steam or water vapor.

Condensers. Special equipment used in air conditioning to liquefy a gas.

Condensing Unit. A complete refrigerating system in one assembly, including the refrigerant compressor, motor, condenser, receiver, and other necessary accessories.

Conductance, Thermal *C*. Rate of heat flow across a unit area (usually 1 ft^2) from one surface to the opposite surface under steady-state conditions with a unit temperature difference between the two surfaces.

Conduction, Thermal. A process in which heat energy is transferred through matter by transmission of kinetic energy from particle to particle, the heat flowing from hot points to cooler ones.

Conductivity, Thermal. Quantity of heat energy, usually in Btu, that is transmitted through a substance per unit of time (usually 1 hr) from a unit area (usually 1 ft^2) of surface to an opposite unit surface per unit of thickness (usually 1 in) under a unit temperature difference (usually 1°F) between the surfaces.

Convection. A means of transferring heat in air by natural movement, usually a rotary or circulatory motion caused by warm air rising and cooler air falling.

Cooling. A heat-removal process usually accomplished with air-conditioning equipment.

Cooling, Evaporative. Cooling effect produced by evaporation of water, the required heat for the process being taken from the air. (This method is widely used in dry climates with low wet-bulb temperatures.)

Cooling, Sensible. Cooling of a unit volume of air by a reduction in temperature only.

Cooling Effect, Total. The difference in total heat in an airstream entering and leaving a refrigerant evaporator or cooling coil.

Cooling Tower. A mechanical device used to cool water by evaporation in the outside air. Towers may be atmospheric or induced- or powered-draft type.

Cooling Unit, Self-Contained. A complete air-conditioning assembly consisting of a compressor, evaporator, condenser, fan motor, and air filter ready for plug-in to an electric power supply.

Damper. A plate-type device used to regulate flow of air or gas in a pipe or duct.

Defrosting. A process used for removing ice from a refrigerant coil.

Degree Day. The product of 1 day (24 hr) and the number of degrees Fahrenheit the daily mean temperature is below 65°F. It is frequently used to determine heating-load efficiency and fuel consumption.

Dehumidification. In air conditioning, the removal of water vapor from supply air by condensation of water vapor on the cold surface of a cooling coil.

Diffuser (Register). Outlet for supply air into a space or zone. See also Grille below.

Direct Digital Control (DDC). An electronic control system that uses a computer to analyze HVAC parameters to operate control devices and to start or stop mechanical equipment.

Direct Expansion. A means of air conditioning that uses the concept of refrigerant expansion (through a thermostatic expansion valve) in a refrigerant coil to produce a cooling effect.

Ductwork. An arrangement of sheet-metal ducts to distribute supply air, return air, and exhaust air.

Efficiency. Ratio of power output to power input. It does not include considerations of load factor or coefficient of performance.

Emissivity. Ratio of radiant energy that is emitted by a body to that emitted by a perfect black body. An emissivity of 1 is assigned to a perfect black body. A perfect reflector is assigned an emissivity of 0.

Enthalphy. A measure of the total heat (sensible and latent) in a substance and which is equal to its internal energy and its capacity to do work.

Entropy. The ratio of the heat added to a material or substance to the absolute temperature at which the heat is added.

Evaporator. A cooling coil in a refrigeration system in which the refrigerant is evaporated and absorbs heat from the surrounding fluid (airstream).

Exfiltration. Unintentional loss of conditioned supply air by leakage from ductwork, rooms, spaces, etc., that is to be considered a load on the air-conditioning system.

Film Coefficient (Surface Coefficient). Heat transferred from a surface to air or other fluid by convection per unit area of surface per degree temperature difference between the surface and the fluid.

Furnace, Warm-Air. Heating system that uses a direct- or indirect-fired boiler to produce warm air for heating.

Grille. A metal covering, usually decorative, with openings through which supply or return air passes.

Head. Pressure expressed in inches or feet of water. A head of 12 in, or 1 ft, of water is the pressure equivalent to a column of water 12 in, or 1 ft, high. See also Inch of Water below.

Heat, Latent. Heat associated with the change of state (phase) of a substance, for example, from a solid to a liquid (ice to water) or from a liquid to a gas (water to steam vapor).

Heat, Sensible. Heat associated with a change in temperature of a substance.

Heat, Specific. Ratio of the thermal capacity of a substance to the thermal capacity of water.

Heat, Total. Sum of the sensible and latent heat in a substance above an arbitrary datum, usually 32°F or 0°C.

Heat Capacity. Heat energy required to change the temperature of a specific quantity of material 1°.

Heat Pump. A refrigerant system used for heating and cooling purposes.

Heat Transmission Coefficient. Quantity of heat (usually Btu in the United States) transmitted from one substance to another per unit of time (usually 1 hr) through one unit of surface (usually 1 ft^2) of building material per unit of temperature difference (usually 1°F).

Heater, Direct-Fired. A heater that utilizes a flame within a combustion chamber to heat the walls of the chamber and transfers the heat from the walls to air for space heating, as in a warm-air heater.

Heater, Unit. A steam or hot-water heating coil, with a blower or fan and motor, used for space heating.

Heating. The process of transferring heat from a heat source to a space in a building.

Heating, District. A large, central heating facility that provides heat from steam or hot water to a large number of buildings often under different ownership.

Heating, Radiant. Heating by ceiling or wall panels, or both, with surface temperatures higher than that of the human body in such a manner that the heat loss from occupants of the space by radiation is controlled.

Heating, Warm-Air. A heating system that uses warm air, rather than steam or hot water, as the heating medium.

Heating Surface. Actual surface used for transferring heat in a boiler, furnace, or heat exchanger.

Heating System, Automatic. A complete heating system with automatic controls to permit operation without manual controls or human attention.

Heating System, Hot-Water. A heating system that utilizes water at temperatures of about 200°F.

Humidity. Water vapor mixed with dry air.

Humidity, Absolute. Weight of water vapor per unit volume of a vapor-air mixture. It is usually expressed in grains/ft^3 or lb/ft^3.

Humidity, Percent. Ratio of humidity in a volume of air to the maximum amount of water vapor that the air can hold at a given temperature, expressed as a percentage.

Humidity, Relative (RH). Ratio of the vapor pressure in a mixture of air and water vapor to the vapor pressure of the air when saturated at the same temperature.

Humidity, Specific (Humidity Ratio). Ratio of the weight of water vapor, grains, or pounds, per pound of dry air, at a specific temperature.

Hygrometer. A mechanical device used to measure the moisture content of air.

Hygroscopic. Denoting any material that readily absorbs moisture and retains it.

Hygrostat. A mechanical device that is sensitive to changes in humidity and used to actuate other mechanical devices when predetermined limits of humidity are reached.

Inch of Water. A unit of pressure intensity applied to low-pressure systems, such as air-conditioning ducts. It is equivalent to 0.036136 psi.

Infiltration. Leakage into an air-conditioned area of outside air (usually unwanted), which becomes a load on the air-conditioning system.

Insulation, Thermal. Any material that slows down the rate of heat transfer (offers thermal resistance) and effects a reduction of heat loss.

Louvers. An arrangement of blades to provide air slots that will permit passage of air and exclude rain or snow.

MBH. 1000 Btu/hr (Btu/h).

Micron. 0.001 mm. It is frequently used to designate particle sizes of dust and the efficiency of filtration by air-conditioning filters.

Modulating. Process of making incremental adjustments, usually by an automatic device operating a valve or damper motor.

Pressure, Absolute. Pressure above an absolute vacuum. Absolute pressure equals the sum of gage and barometric pressures.

Pressure, Atmospheric. Air pressure indicated by a barometer. The standard atmospheric pressure is 29.92 in of mercury, or 14.696 psi (101.3 kPa).

Pressure, Head. Condensing pressure, often considered as the refrigerant compressor-discharge pressure.

Pressure, Saturation. The pressure that corresponds to a specific temperature that will permit simultaneous condensation and evaporation.

Pressure, Suction. The pressure in the suction line of a refrigeration system.

Pressure, Head. See Head above.

Psychrometer. A mechanical device utilizing a wet-bulb and dry-bulb thermometer and used to determine the humidity in an air-water vapor mixture, such as room air.

Psychrometric Chart. A chart used in air-conditioning design and analysis that indicates various properties of an air-water vapor mixture along with various relevant mathematical values.

Psychrometry. A branch of physics that concerns itself with the measurement and determination of atmospheric conditions, with particular emphasis on moisture mixed in the air.

Radiation. Transfer of energy in wave form, from a hot body to a colder body, independent of any matter between the two bodies.

Radiation, Equivalent Direct. Rate of steam condensation at 240 Btu/(hr)(ft^2) of radiator surface.

Refrigerant. A substance that will accept large quantities of heat, that will cause boiling and vaporization at certain temperatures, and that can be utilized in air-conditioning systems.

Register. See Diffuser.

Resistance, Thermal. The thermal quality of a material that resists passage of heat. Also, the opposite of conductance.

Resistivity, Thermal. The reciprocal of conductivity.

Split System. A separation of air-conditioning components, such as location of an air-blower-evaporator coil far from the compressor-condenser unit.

Steam. Water in gas or vapor form.

Steam Trap. A mechanical device that allows water and air to pass but prevents passage of steam.

Subcooling. Cooling at constant pressure of a refrigerant liquid to below its condensing temperature.

Suction Line. The low-temperature, low-pressure refrigerant pipe from an evaporator to a refrigerant compressor.

Sun Effect. Heat from the sun that tends to increase the internal temperature of a space or building.

Temperature, Absolute. Temperature measured on a scale for which zero is set at −273.16°C, or −459.69°F (presumably the temperature at which all molecular motion stops in a gas under constant pressure). The scale is called Kelvin, and 1°K = 1°C = 9/5°F.

Temperature, Design. An arbitrary design criterion used to determine equipment size to produce air conditioning, heating, or cooling capable of maintaining the designated temperature.

Temperature, Dew Point. Temperature of air at which its wet-bulb temperature and dry-bulb temperature are identical and the air is fully saturated with moisture. Condensation of water vapor begins at this temperature and will continue if the temperature is reduced further.

Temperature, Dry-Bulb. Temperature measured by a conventional thermometer. It is used to determine the sensible heat in air.

Temperature, Effective. A single or arbitrary index that combines into a single value the effects of temperature, humidity, and air motion on the sensation of comfort. This value is that of the temperature of still, saturated air that will induce an identical feeling of comfort.

Temperature, Wet-Bulb. Air temperature as indicated by a thermometer with a wet bulb. This temperature is less than the dry-bulb temperature, except when the air is fully saturated with water vapor, or at 100% relative humidity, when wet-bulb and dry-bulb temperatures will be equal.

Ton, Refrigeration. Refrigeration effect equivalent to 200 Btu/min, or 12,000 Btu/hr.

Vapor. The gaseous state of water and other liquid substances.

Vapor Barrier. An impervious material used to prevent the passage of water vapor through the walls or roofs of a building and to prevent the subsequent condensation within those spaces.

Velocity Pressure. The pressure caused by a moving airstream, composed of both velocity pressure and static pressure.

Ventilation. The process of supplying air to any space within a building without noticeable odors and without objectionable levels of contaminants, such as dusts and harmful gases, and of removing stale, polluted air from the space. Outside air is generally used as an acceptable source of ventilation air.

Ventilator, Unit. A type of unit heater with various modes of operation and degrees or percentages of outside air (frequently used for heating classrooms).

Volume, Specific. Volume, ft^3/lb, occupied by a unit weight of air.

Water, Makeup. Generally the water supplied to a cooling tower to replace the cooling water lost by evaporation or bleedoff.

Water Vapor. A psychrometric term used to denote the water in air (actually low-pressure, superheated steam) that has been evaporated into the air at a temperature corresponding to the boiling temperature of water at that very low pressure.

13.2 HEAT AND HUMIDITY

People have always struggled with the problem of being comfortable in their environment. First attempts were to use fire directly to provide heat through cold winters. It was only in recent times that interest and technology permitted development of greater understanding of heat and heating, and substantial improvements in comfort were made. Comfort heating now is a highly developed science and, in conjunction with air conditioning, provides comfort conditions in all seasons in all parts of the world.

As more was learned about humidity and the capacity of the air to contain various amounts of water vapor, greater achievements in environmental control were made. Control of humidity in buildings now is a very important part of heating, ventilation, and air conditioning, and in many cases is extremely important in meeting manufacturing requirements. Today, it is possible to alter the atmosphere or environment in buildings in any manner, to suit any particular need, with great precision and control.

13.2.1 Thermometers and Scales

Energy in the form of heat is transferred from one material or substance to another because of a temperature difference that exists between them. When heat is applied to a material or substance, there will be an increase in average velocity of its molecules or electrons, with an increase in their kinetic energy. Likewise, as heat is removed, there will be a decrease in the average molecular velocity and, therefore, also the electron or molecular kinetic energy.

A thermometer is used to measure the degree of heat in a substance or material. The thermometer includes an appropriate graduated scale to indicate the change in temperature of the substance. The change in temperature as read on a thermometer is a measure of heat transferred to or from the substance. A unit of temperature is called a degree and is equivalent to one graduation on the scale.

By convention, the scale is an interval scale. The Celsius thermometer is a metric system of measuring temperature; 0°C is assigned to the temperature at which water freezes and 100°C to the temperature at which water boils at normal atmospheric conditions. Hence, on a Celsius thermometer, there are 100 intervals or graduations, called degrees, between the freezing and boiling temperatures. Each interval or degree is called 1 Celsius degree.

In the Fahrenheit system, 32°F is used to designate the freezing temperature of water and 212°F the boiling temperature at normal atmospheric pressure. Hence, on the Fahrenheit scale, a degree is equal to $\frac{1}{180}$ of the distance on the scale between the freezing and boiling temperatures. Conversion formulas used for each scale are as follows:

$$°F = 1.8 \times °C + 32 \tag{13.1}$$

$$°C = \frac{5}{9}(°F - 32) \tag{13.2}$$

13.3.2 Thermal Capacity and Specific Heat

The thermal capacity of a substance is indicated by the quantity of heat required to raise the temperature of 1 lb of the substance 1°F. In HVAC calculations, thermal capacity is usually expressed by the British thermal unit (Btu).

One Btu is the amount of heat that is required to increase the temperature of 1 lb of water 1°F at or near 39.2°F, which is the temperature at which water has its maximum density. Conversely, if 1 Btu is removed from 1 lb of water, its temperature will be reduced by 1°F.

Various quantities of heat will produce changes of 1°F per pound of substances other than water. Thus, thermal capacity is entirely dependent on the specific heat of the substances.

The specific heat of a substance is the ratio of the heat content or thermal capacity of a substance to that of water. And by definition, the specific heat of water is unity.

It is customary in HVAC calculations to use specific heat in lieu of thermal capacity, because of the convenience of using the Btu as a unit of heat quantity without conversions. Specific heats of air and some common building materials are shown in Table 13.1. Data for other substances may be obtained from tables in the "ASHRAE Handbook—Fundamentals," American Society of Heating, Refrigeration and Air Conditioning Engineers. An examination of Table 13.1 indicates that the specific heat of these materials is less than unity and that, of all common substances, water possesses the largest specific heat and the largest thermal capacity.

13.2.3 Sensible Heat

When heat energy is added to or taken away from a substance, the resulting changes in temperature can be detected by the sense of touch, or sensibly. Therefore, this type of heat is called sensible heat. Since sensible heat is associated with a change in temperature, the quantity of sensible heat energy transferred in a heat exchange is usually calculated from

$$Q = Mc(t_2 - t_1) \tag{13.3}$$

TABLE 13.1 Specific Heats—Common Materials

Substance	Specific heat, Btu/(lb)(°F)
Air at 80°F	0.24
Water vapor	0.49
Water	1.00
Aluminum	0.23
Brick	0.20
Brass	0.09
Bronze	0.10
Gypsum	0.26
Ice	0.48
Limestone	0.22
Marble	0.21
Sand	0.19
Steel	0.12
Wood	0.45–0.65

where Q = sensible heat, Btu, absorbed or removed
M = mass, lb, of the substance undergoing the temperature change
c = specific heat of the substance
$(t_2 - t_1)$ = temperature difference of the substance, where t_2 is the final temperature after the heat exchange and t_1 is the temperature of the material before the heat exchange

13.2.4 Laws of Thermodynamics

The application of the laws of thermodynamics to HVAC calculations is usually limited to two well-known laws. These laws can be expressed differently, but in equivalent ways. A simplification of these laws as follows will permit an easier understanding.

The first law of thermodynamics states that when work performed produces heat, the quantity of the heat produced is proportional to the work performed. And conversely, when heat energy performs work, the quantity of the heat dissipated is proportional to the work performed. Work, ft-lb, is equal to the product of the force, lb, acting on the body for a distance, ft, that the body moves in the direction of the applied force.

Hence, this first law of thermodynamics can be expressed mathematically by the following equation:

$$W = JQ \tag{13.4}$$

where W = work, ft-lb
J = Joule's constant = mechanical equivalent of heat
Q = heat, Btu, generated by the work

Experiments have shown that the mechanical equivalent of heat, known as Joule's constant, is equivalent to 778 ft-lb/Btu. The first law is also known as the law of conservation of energy.

The second law of thermodynamics states that it is impossible for any machine to transfer heat from a substance to another substance at a higher temperature (if the machine is unaided by an external agency). This law can be interpreted to imply that the available supply of energy for doing work in our universe is constantly decreasing. It also implies that any effort to devise a machine to convert a specific quantity of heat into an equivalent amount of work is futile.

Entropy is the ratio of the heat added to a substance to the absolute temperature at which the heat is added.

$$S = \frac{dQ}{T_a} \tag{13.5}$$

where S = entropy
dQ = differential of heat (very small change)
T_a = absolute temperature

The second law of thermodynamics can be expressed mathematically with the use of the entropy concept.

Suppose an engine, which will convert heat into useful mechanical work, receives heat Q_1 from a heat source at temperature T_1 and delivers heat Q_2 at a temperature T_2 to a heat sink after performing work. By the first law of thermodynamics, the law of conservation of energy, Q_2 is less than Q_1 by the amount of work performed. And by the second law of thermodynamics, T_2 is less than T_1. The universe at the start of the process loses entropy $\Delta S_1 = Q_1/T_1$ and at the end of the process gains entropy $\Delta S_2 = Q_2/T_2$. Hence, the net change in the entropy of the universe because of this process will be $\Delta S_2 - \Delta S_1$.

Furthermore, this law requires that this net change must always be greater than zero and that the entropy increase is and must always be an irreversible thermodynamic process.

$$\Delta S_2 - \Delta S_1 > 0 \tag{13.6}$$

Because of the irreversibility of the process, the energy that has become available for performing work is

$$Q_u = T_2(\Delta S_2 - \Delta S_1) \tag{13.7}$$

13.2.5 Absolute Temperature

The definition of entropy given above involves the concept of absolute temperature measured on a ratio scale. The unit of absolute temperature is measured in degrees Kelvin (°K) in the Celsius system and in degrees Rankine (°R) in the Fahrenheit system. Absolute zero or zero degrees in either system is determined by considering the theoretical behavior of an ideal gas, and for such a gas,

$$P_aV = mRT_a \qquad \text{or} \qquad P_av = RT_a \tag{13.8}$$

where P_a = absolute pressure on the gas, psf
V = volume of the gas, ft^3
v = specific volume of the gas, ft^3/lb = reciprocal of the gas density

m = mass of the gas, lb
T_a = absolute temperature
R = universal gas constant

For a gas under constant pressure, the absolute temperature theoretically will be zero when the volume is zero and all molecular motion ceases. Under these conditions, the absolute zero temperature has been determined to be nearly −273°C and −460°F. Therefore,

$$\text{Kelvin temperature °K} = \text{Celsius temperature} + 273° \qquad (13.9)$$

$$\text{Rankine temperature °R} = \text{Fahrenheit temperature} + 460° \qquad (13.10)$$

In the Rankine system, the universal gas constant R equals 1545.3 divided by the molecular weight of the gas. For air, $R = 53.4$, and for water vapor, $R = 85.8$.

13.2.6 Latent Heat

The sensible heat of a substance is associated with a sensible change in temperature. In contrast, the latent heat of a substance is always involved with a change in state of a substance, such as from ice to water and from water to steam or water vapor. Latent heat is very important in HVAC calculations and design, because the total heat content of air almost always contains some water in the form of vapor. The concept of latent heat may be clarified by consideration of the changes of state of water.

When heat is added to ice, the temperature rises until the ice reaches its melting point. Then, the ice continues to absorb heat *without a change in temperature* until a required amount of heat is absorbed per pound of ice, at which point it begins melting to form liquid water. The reverse is also true: if the liquid is cooled to the freezing point, this same quantity of heat must be removed to cause the liquid water to change to the solid (ice) state. This heat is called the **latent heat of fusion** for water. It is equal to 144 Btu and will convert 1 lb of ice at 32°F to 1 lb of water at 32°F. Thus,

$$\text{Latent heat of fusion for water} = 144 \text{ Btu/lb} \qquad (13.11)$$

If the pound of water is heated further, say to 212°F, then an additional 180 Btu of heat must be added to effect the 180°F sensible change in temperature. At this temperature, any further addition of heat will not increase the temperature of the water beyond 212°F. With the continued application of heat, the water experiences violent agitation, called boiling. The boiling temperature of water is 212°F at atmospheric pressure.

With continued heating, the boiling water absorbs 970 Btu for each pound of water *without a change in temperature* and completely changes its state from liquid at 212°F to water vapor, or steam, at 212°F. Therefore, at 212°F,

$$\text{Latent heat of vaporization of water} = 970 \text{ Btu/lb} \qquad (13.12)$$

Conversely, when steam at 212°F is cooled or condensed to a liquid at 212°F, 970 Btu per pound of steam (water) must be removed. This heat removal and change of state is called **condensation.**

When a body of water is permitted to evaporate into the air at normal atmospheric pressure, 29.92 in of mercury, a small portion of the body of water evaporates from the water surface at temperatures below the boiling point. The latent heat of vaporization is supplied by the body of water and the air, and hence both become cooler. The amount of vapor formed and that absorbed by the air above the water surface depends on the capacity of the air to retain water at the existing temperature and the amount of water vapor already in the air.

Table 13.2 lists the latent heat of vaporization of water for various air temperatures and normal atmospheric pressure. More extensive tables of thermodynamic properties of air, water, and steam are given in the "ASHRAE Handbook—Fundamentals," American Society of Heating, Refrigerating and Air Conditioning Engineers.

TABLE 13.2 Thermal Properties—Dry and Saturated Air at Atmospheric Pressure

Air temperature, °F	Specific volume, ft³/lb		Pounds of water in saturated air per pound of dry air (humidity ratio)	Latent heat of vaporization of water, Btu/lb	Specific enthalpy of dry air h_a,* Btu/lb	Specific enthalpy of saturated air h_s,† Btu/lb	Specific enthalpy of saturation vapor h_g, Btu/lb
	Dry air	Saturated air					
0	11.58	11.59	0.0008		0	0.84	
32	12.39	12.46	0.0038	1075.2	7.69	11.76	1075.2
35	12.46	12.55	0.0043	1073.5	8.41	13.01	1076.5
40	12.59	12.70	0.0052	1070.6	9.61	15.23	1078.7
45	12.72	12.85	0.0063	1067.8	10.81	17.65	1080.9
50	12.84	13.00	0.0077	1065.0	12.01	20.30	1083.1
55	12.97	13.16	0.0092	1062.2	13.21	23.22	1085.2
60	13.10	13.33	0.0111	1059.3	14.41	26.46	1087.4
65	13.22	13.50	0.0133	1056.5	15.61	30.06	1089.6
70	13.35	13.69	0.0158	1053.7	16.82	34.09	1091.8
75	13.47	13.88	0.0188	1050.9	18.02	38.61	1094.0
80	13.60	14.09	0.0223	1048.1	19.22	43.69	1096.1
85	13.73	14.31	0.0264	1045.2	20.42	49.43	1098.3
90	13.85	14.55	0.0312	1042.4	21.62	55.93	1100.4
95	13.98	14.80	0.0367	1039.6	22.83	63.32	1102.6
100	14.11	15.08	0.0432	1036.7	24.03	71.73	1104.7
150	15.37	20.58	0.2125	1007.8	36.1	275.3	1125.8
200	16.63	77.14	2.295	977.7	48.1	2677	1145.8
212				970.2			1150.4

*Enthalpy of dry air is taken as zero for dry air at 0°F.
†Enthalpy of water vapor in saturated air = $h_s - h_a$, including sensible heat above 32°F.

13.2.7 Enthalpy

Enthalpy is a measure of the total heat (sensible and latent) in a substance and is equivalent to the sum of its internal energy plus its ability or capacity to perform work, or PV/J, where P is the pressure of the substance, V its volume, and J its mechanical equivalent of heat. Specific enthalpy is the heat per unit of weight, Btu/lb, and is the property used on psychrometric charts and in HVAC calculations.

The specific enthalpy of dry air h_a is taken as zero at 0°F. At higher temperatures, h_a is equal to the product of the specific heat, about 0.24, multiplied by the temperature, °F. (See Table 13.2.)

The specific enthalpy of saturated air h_s, which includes the latent heat of vaporization of the water vapor, is indicated in Table 13.2. The specific enthalpy of the water vapor or moisture at the air temperature may also be obtained from Table 13.2 by subtracting h_a from h_s.

Table 13.2 also lists the humidity ratio of the air at saturation for various temperatures (weight, lb, of water vapor in saturated air per pound of dry air). In addition, the specific enthalpy of saturated water vapor h_g, Btu/lb, is given in Table 13.2 and represents the sum of the latent heat of vaporization and the specific enthalpy of water at various temperatures.

The specific enthalpy of unsaturated air is equal to the sensible heat of dry air at the existing temperature, with the sensible heat at 0°F taken as zero, plus the product of the humidity ratio of the unsaturated air and h_g for the existing temperature.

13.2.8 Cooling by Evaporation

Evaporation of water requires a supply of heat. If there is no external source of heat, and evaporation occurs, then the water itself must provide the necessary heat of vaporization. In other words, a portion of the sensible heat in the liquid will be converted into the latent heat of vaporization. As a result, the temperature of the liquid remaining will drop. Since no external heat is added or removed by this process of evaporation, it is called **adiabatic cooling.**

Human beings are also cooled adiabatically by evaporation of perspiration from skin surfaces. Similarly, in hot climates with relatively dry air, air conditioning is provided by the evaporation of water into air. And refrigeration is also accomplished by the evaporation of a refrigerant.

13.2.9 Heating by Condensation

Many thermal processes occur without addition or subtraction of heat from the process. Under these conditions, the process is called **adiabatic.**

When a volume of moist air is cooled, a point will be reached at which further cooling cannot occur without reaching a fully saturated condition, that is, 100% saturation or 100% relative humidity. With continued cooling, some of the moisture condenses and appears as a liquid. The temperature at which condensation occurs is called the **dew point temperature.** If no heat is removed by the condensation, then the latent heat of vaporization of the water vapor will be converted to sensible heat in the air, with a resultant rise in temperature.

Thus, an increase in temperature is often accomplished by the formation of fog, and when rain or snow begins to fall, there will usually be an increase in temperature of the air.

13.2.10 Psychrometry

The measurement and determination of atmospheric conditions, particularly relating to the water vapor or moisture content in dry air, is an important branch of physics known as psychrometry. Some psychrometric terms and conditions have already been presented in this article. Many others remain to be considered.

An ideal gas follows certain established laws of physics. The mixture of water vapor and dry air does behave at normal atmospheric temperatures and pressures almost as an ideal gas. As an example, air temperatures, volumes, and pressures may be calculated by use of Eq. (13.8), $Pv = RT$.

Dalton's law also applies. It states:

When two or more gases occupy a common space or container, each gas will fill the volume just as if the other gas or gases were not present. Dalton's law also requires:

1. That each gas in a mixture occupy the same volume or space and also be at the same temperature as each other gas in the mixture.
2. That the total weight of the gases in the mixture equal the sum of the individual weights of the gases.
3. That the pressure of a mixture of several gases equal the sum of the pressures that each gas would exert if it existed alone in the volume enclosing the mixture.
4. That the total enthalpy of the mixture of gases equal the sum of the enthalpies of each gas.

An excellent example of the application of Dalton's law of partial pressures is the use of a liquid barometer to indicate atmospheric pressure. The barometer level indicates the sum of the partial pressure of water vapor and the partial pressure of the air.

Partial pressures of air and water vapor are of great importance in psychrometry and are used to calculate the degree of saturation of the air or relative humidity at a specific dry-bulb temperature.

13.2.11 Relative Humidity and Specific Humidity

Relative humidity is sometimes defined by the use of mole fractions, a difficult definition for psychrometric use. Hence, a more usable definition is desired. For the purpose, relative humidity may be closely determined by the ratio of the partial pressure of the water vapor in the air to the saturation pressure of water vapor at the same temperature, usually expressed as a percentage.

Thus, dry air is indicated as 0% relative humidity and fully saturated air is termed 100% relative humidity.

Computation of relative humidity by use of humidity ratios is also often done, but with somewhat less accuracy. Humidity ratio, or specific humidity W_a, at a specific temperature is the weight, lb, of water vapor in air per pound of dry air. If W_s represents the humidity ratio of saturated air at the same temperature (Table 13.2), then relative humidity can be calculated approximately from the equation

$$RH = \frac{W_a}{W_s} \times 100 \tag{13.13}$$

Dalton's law of partial pressures and Eq. (13.6) may also be used to calculate humidity ratios:

$$W_a = 0.622 \frac{p_w}{P - p_w} \tag{13.14}$$

where P = barometric pressure, atmospheric, psi
p_w = partial pressure of water vapor, psi

It is difficult to use this equation, however, because of the difficulty in measuring the partial pressures with special scientific equipment that is required and rarely available outside of research laboratories. Therefore, it is common practice to utilize simpler types of equipment in the field that will provide direct readings that can be used to convert into humidity ratios or relative humidity.

A simple and commonly used device is the wet- and dry-bulb thermometer. This device is a packaged assembly consisting of both thermometers and a sock with scales. It is called a **sling psychrometer.** Both thermometers are identical, except that the wet-bulb thermometer is fitted with a wick-type sock over the bulb. The sock is wet with water, and the device is rapidly spun or rotated in the air. As the water in the sock evaporates, a drop in temperature occurs in the remaining water in the sock, and also in the wet-bulb thermometer. When there is no further temperature reduction and the temperature remains constant, the reading is called the wet-bulb temperature. The other thermometer will simultaneously read the dry-bulb temperature.

A difference between the two thermometer readings always exists when the air is less than saturated, at or less than 100% relative humidity. Inspection of a psychrometric chart will indicate that the wet-bulb and dry-bulb temperatures are identical only at fully saturated conditions, that is, at 100% relative humidity. Commercial psychrometers usually include appropriate charts or tables that indicate the relative humidity for a wide range of specific wet- and dry-bulb temperature readings. These tables are also found in books on psychrometry and HVAC books and publications.

13.2.12 Dew-Point Temperatures

Dew is the condensation of water vapor. It is most easily recognized by the presence of droplets in warm weather on grass, trees, automobiles, and many other outdoor things in the early morning. Dew is formed during the night as the air temperature drops, and the air reaches a temperature at which it is saturated with moisture. This is the dew-point temperature. It is also equal to both the wet-bulb temperature and dry-bulb temperature. At the dew-point temperature, the air is fully saturated, that is, at 100% relative humidity. With any further cooling or drop in temperature, condensation begins and continues with any further reduction in temperature. The amount of moisture condensed is the excess moisture that the air cannot hold at saturation at the lowered temperature. The condensation forms drops of water, frequently referred to as *dew.*

Dew-point temperature, thus, is the temperature at which condensation of water vapor begins for any specific condition of humidity and pressure as the air temperature is reduced.

The dew-point temperature can be calculated, when the relative humidity is known, by use of Eq. (13.13) and Table 13.2. For the temperature of the unsaturated air, the humidity ratio at saturation is determined from Table 13.2. The product of the humidity ratio and the relative humidity equals the humidity ratio for the dew-point temperature, which also can be determined from Table 13.2. As an example, to determine the dew-point temperature of air at 90°F and 50% relative humidity, reference to Table 13.2 indicates a humidity ratio at saturation of 0.0312 at 90°F. Multiplication by 0.50 yields a humidity ratio of 0.0156. By interpolation in Table 13.2 between humidity ratios at saturation temperatures of 65 and 70°F, the dew-point temperature is found to be 69.6°F.

A simpler way to determine the dew-point temperature and many other properties of air-vapor mixtures is to use a psychrometric chart. This chart graphically relates dry-bulb, wet-bulb, and dew-point temperatures to relative humidity, humidity ratio, and specific volume of air. Psychrometric charts are often provided in books on psychrometrics and HVAC handbooks.

13.2.13 Refrigeration Ton

A ton of refrigeration is a common term used in air conditioning to designate the cooling rate of air-conditioning equipment. A ton of refrigeration indicates the ability of an evaporator to remove 200 Btu/min or 12,000 Btu/hr. The concept is a carry-over from the days of icemaking and was based on the concept that 200 Btu/min had to be removed from 32°F water to produce 1 ton of ice at 32°F in 24 hr. Hence,

$$\text{1 ton refrigeration} = \frac{2000\,\dfrac{\text{lb}}{\text{day}} \times 144\,\dfrac{\text{Btu}}{\text{lb}}}{24\,\dfrac{\text{hr}}{\text{day}}}$$

$$= 288{,}000 \text{ Btu/day}$$

$$= 12{,}000 \text{ Btu/hr}$$

$$= 200 \text{ Btu/min} \tag{13.15}$$

(ASHRAE Handbook—Fundamentals," American Society of Heating, Refrigerating and Air-Conditioning Engineers, 1791 Tully Circle, N. E., Atlanta, GA 30329.)

13.3 MAJOR FACTORS IN HVAC DESIGN

This article presents the necessary concepts for management of heat energy and aims at development of a better understanding of its effects on human comfort. The concepts must be well understood if they are to be applied successfully to modification of the environment in building interiors, computer facilities, and manufacturing processes.

13.3.1 Significance of Design Criteria

Achievement of the desired performance of any HVAC system, whether designed for human comfort or industrial production or industrial process requirements, is significantly related to the development of appropriate and accurate design criteria.

Some of the more common items that are generally considered are as follows:

1. Outside design temperatures:
   Winter and summer
   Dry bulb (DB), wet bulb (WB)
2. Inside design temperatures:
   Winter: heating °F DB and relative humidity
   Summer: cooling °F DB and relative humidity
3. Filtration efficiency of supply air
4. Ventilation requirements
5. Exhaust requirements
6. Humidification
7. Dehumidification
8. Air-change rates
9. Positive-pressure areas
10. Negative-pressure areas
11. Balanced-pressure areas
12. Contaminated exhausts
13. Chemical exhausts and fume hoods
14. Energy conservation devices
15. Economizer system
16. Enthalpy control system
17. Infiltration
18. Exfiltration

13.3.2 Design Criteria Accuracy

Some engineers apply much effort to determination of design conditions with great accuracy. This is usually not necessary, because of the great number of variables involved in the design process. Strict design criteria will increase the cost of the necessary machinery for such optimum conditions and may be unnecessary. It is generally recognized that it is impossible to provide a specific indoor condition that will satisfy every occupant at all times. Hence, HVAC engineers tend to be practical in their designs and accept the fact that the occupants will adapt to minor variations from ideal conditions. Engineers also know that human comfort depends on the type and quantity of clothing worn by the occupants, the types of activities performed, environmental conditions, duration of occupancy, ventilation air, and closeness of and number of people within the conditioned space and recognize that these conditions are usually unpredictable.

13.3.3 Outline of Design Procedure

Design of an HVAC system is not a simple task. The procedure varies considerably from one application or project to another, and important considerations for one project may have little impact on another. But for all projects, to some extent, the following major steps have to be taken:

1. Determine all applicable design conditions, such as inside and outside temperature and humidity conditions for winter and summer conditions, including prevailing winds and speeds.
2. Determine all particular and peculiar interior space conditions that will be maintained.
3. Estimate, for every space, heating or cooling loads from adjacent unheated or uncooled spaces.
4. Carefully check architectural drawings for all building materials used for walls, roofs, floors, ceilings, doors, etc., and determine the necessary thermal coefficients for each.
5. Establish values for air infiltration and exfiltration quantities, for use in determining heat losses and heat gains.
6. Determine ventilation quantities and corresponding loads for heat losses and heat gains.
7. Determine heat or cooling loads due to internal machinery, equipment, lights, motors, etc.
8. Include allowance for effects of solar load.
9. Total the heat losses requiring heating of spaces and heat gains requiring cooling of spaces, to determine equipment capacities.

13.3.4 Temperatures Determined by Heat Balances

In cold weather, comfortable indoor temperatures may have to be maintained by a heating device. It should provide heat to the space at the same rate as the space is losing heat. Similarly, when cooling is required, heat should be removed from the space at the same rate that it is gaining heat. In each case, there must be a heat-balance between heat in and heat out when heating and the reverse in cooling. Comfortable inside conditions can only be maintained if this heat balance can be controlled or maintained.

The rate at which heat is gained or lost is a function of the difference between the inside air temperature to be maintained and the outside air temperature. Such temperatures must be established for design purposes in order to properly size and select HVAC equipment that will maintain the desired design conditions. Many other conditions that also affect the flow of heat in and out of buildings, however, should also be considered in selection of equipment.

13.3.5 Methods of Heat Transfer

Heat always flows from a hot to a cold object, in strict compliance with the second law of thermodynamics (Art. 13.2). This direction of heat flow occurs by conduction, convection, or radiation and in any combination of these forms.

Thermal conduction is a process in which heat energy is transferred through matter by the transmission of kinetic energy from molecule to molecule or atom to atom.

Thermal convection is a means of transferring heat in air by natural or forced movements of air or a gas. Natural convection is usually a rotary or circular motion caused by warm air rising and cooler air falling. Convection can be mechanically produced (forced convection), usually by use of a fan or blower.

Thermal radiation transfers energy in wave form from a hot body to a relatively cold body. The transfer occurs independently of any material between the two bodies. Radiation energy is converted energy from one source to a very long wave form of electromagnetic energy. Interception of this long wave by solid matter will convert the radiant energy back to heat.

13.3.6 Thermal Conduction and Conductivity

Thermal conduction is the rate of heat flow across a unit area (usually 1 ft^2) from one surface to the opposite surface for a unit temperature difference between the two surfaces and under steady-state conditions. Thus, the heat-flow rate through a plate with unit thickness may be calculated from

$$Q = kA(t_2 - t_1) \tag{13.16}$$

where Q = heat flow rate, Btu/hr
k = coefficient of thermal conductivity for a unit thickness of material, usually 1 in
A = surface area, normal to heat flow, ft^2
t_2 = temperature, °F, on the warm side of the plate
t_1 = temperature, °F, on the cooler side of the plate

The coefficient k depends on the characteristics of the plate. The numerical value of k also depends on the units used for the other variables in Eq. (13.16). When values of k are taken from published tables, units given should be adjusted to agree with the units of the other variables.

In practice, the thickness of building materials often differs from unit thickness. Consequently, use of a coefficient of conductivity for the entire thickness is advantageous. This coefficient, called **thermal conductance,** is derived by dividing the conductivity k by the thickness L, the thickness being the length or path of heat flow.

***Thermal Conductance* C *and Resistance* R.** Thermal conductance C is the same as conductivity, except that it is based on a specific thickness, instead of 1 in as for conductivity. Conductance is usually used for assemblies of different materials, such as cast-in-place concrete and concrete block with an airspace between. The flow of heat through such an assembly is very complex and is determined under ideal test conditions. In such tests, conductance is taken as the average heat flow from a unit area of surface (usually 1 ft^2) for the total thickness of the assembly. In the case of 9-in-thick concrete, for example, the conductance, as taken from appropriate tables, would be 0.90 Btu/(hr)(ft^2)(°F). (It should be understood, however, that conversion of the conductance C to conductivity k by dividing C by the thickness will produce significant errors.)

Conductance C is calculated from

$$Q = CA(t_2 - t_1) \tag{13.17}$$

where $t_2 - t_1$ is the temperature difference causing the heat flow Q, and A is the cross-sectional area normal to the heat flow.

Values of k and C for many building materials are given in tables in "ASHRAE Handbook—Fundamentals" and other publications on air conditioning.

Thermal resistance, the resistance to flow of heat through a material or an assembly of materials, equals the reciprocal of the conductance:

$$R = \frac{1}{C} \tag{13.18}$$

Thermal resistance R is used in HVAC calculations for determining the rate of heat flow per unit area through a nonhomogeneous material or a group of materials.

Air Films. In addition to its dependence on the thermal conductivity or conductance of a given wall section, roof, or other enclosure, the flow of heat is also dependent on the surface air films on each side of the constructions. These air films are very thin and cling to the exposed surface on each side of the enclosures. Each of the air films possesses thermal conductance, which should always be considered in HVAC calculations.

The indoor air film is denoted by f_i and the outdoor film by f_o. Values are given in Table 13.3 for these air films and for interior or enclosed air spaces of assemblies. In this table, the effects of air films along both enclosure surfaces have been taken into account in developing the air-film coefficients. Additional data may be obtained from the "ASHRAE Handbook—Fundamentals."

TABLE 13.3 Thermal Conductance of Air, Btu/(hr)(ft²)(°F)

f_i	for indoor air film (still air)	
	Vertical surface, horizontal heat flow	1.5
	Horizontal surface	
	Upward heat flow	1.6
	Downward heat flow	1.1
f_o	for outdoor air film, 15-mi/hr wind (winter)	6.0
f_o	for outdoor air film, 7.5-mi/hr wind (summer)	4.0
C	for vertical air gap, ¾ in or more wide	1.1
C	for horizontal air gap, ¾ in or more wide	
	Upward heat flow	1.2
	Downward heat flow	1.0

Air-to-Air Heat Transfer. In the study of heat flow through an assembly of building materials, it is always assumed that the rate of heat flow is constant and continues without change. In other words, a steady-state condition exists. For such a condition, the rate of heat flow in Btu per hour per unit area can be calculated from

$$Q = UA(t_2 - t_1) \tag{13.19}$$

where U = coefficient of thermal transmittance.

***Coefficient of Thermal Transmittance* U.** The coefficient of thermal transmittance U, also known as the overall coefficient of heat transfer, is the rate of heat flow under steady-state conditions from a unit area from the air on one side to the air on the other side of a material or an assembly when a steady temperature difference exists between the air on both sides.

In calculation of the heat flow through a series of different materials, their individual resistances should be determined and totaled to obtain the total resistance R_t. The coefficient of thermal transmittance is then given by the reciprocal of the total resistance:

$$U = \frac{1}{R_t} \tag{13.20}$$

Tables of U values for various constructions are available in the "ASHRAE Handbook—Fundamentals," catalogs of insulation manufacturers, and other publications.

***Computation of* R *and* U.** An assembly that is constructed with several different building materials with different thermal resistances $R_1, R_2, R_3, \ldots, R_n$ provides a total thermal resistance

$$R_t = \frac{1}{f_i} + R_1 + R_2 + R_3 + \cdots + R_n + \frac{1}{f_o} \tag{13.21}$$

including the indoor and outdoor air film resistances f_i and f_o. The U value, coefficient of thermal transmittance, is then determined by use of Eq. (13.20). This coefficient may be substituted in Eq. (13.19) for calculation of the steady-state heat flow through the assembly.

As a typical example, consider an exterior wall section that is constructed of 4-in face brick, 4-in cinder block, ¾-in airspace, and lightweight ¾-in lath and plaster. The wall is 8 ft 6 in high and 12 ft long. The inside air temperature is to be maintained at 68°F, with an outdoor air temperature of +10°F and a 15-mi/hr prevailing wind. What will be the total heat loss through this wall?

From Table 13.3, the indoor air film conductance is 1.5. Its resistance is equal to 1/1.5 = 0.67. The outdoor air-film conductance for a 15-mi/hr wind is 6.0. Its resistance is equal to 1/6.0 = 0.17. Conductivity of the 4-in face brick is 5.0. Conductance of the 4-in cinder block is 0.90; of the ¾-in airspace, 1.1, and of the ¾-in lath and lightweight plaster, 7.70. The total resistance of the wall is then:

$$\begin{aligned} R_t &= \frac{1}{6.0} + 4 \times \frac{1}{5.0} + \frac{1}{0.90} + \frac{1}{1.1} + \frac{1}{7.70} + \frac{1}{1.5} \\ &= 0.17 + 0.8 + 1.11 + 0.91 + 0.13 + 0.67 \\ &= 3.79 \end{aligned}$$

and the coefficient of thermal transmittance is

$$U = \frac{1}{3.79} = 0.264$$

The heat flow rate through the entire wall will be, from Eq. (13.19),

$$Q = 0.264 \times (8.5 \times 12.0)(68 - 10) = 1562 \text{ Btu/hr}$$

13.3.7 Thermal Insulation

A substantial reduction in heating and cooling loads can be made by the judicious use of thermal insulation in wall and roof construction. Addition of insulation results in an increase in thermal resistance *R*, or a reduction in the coefficient of heat transfer *U* of the walls and roof.

Any material with high resistance to flow of heat is called insulation. Many kinds of insulation materials are used in building construction. See Art. 12.3.

Note that the maximum overall conductance *U* encountered in building construction is 1.5 Btu/(hr)(ft^2)(°F). This would occur with a sheet-metal wall. The metal has, for practical purposes, no resistance to heat flow. The *U* value of 1.5 is due entirely to the resistance of the inside and outside air films. Most types of construction have *U* factors considerably less than 1.5.

The minimum *U* factor generally found in standard construction with 2 in of insulation is about 0.10.

Since the *U* factor for single glass is 1.13, it can be seen that windows are a large source of heat gain, or heat loss, compared with the rest of the structure. For double glass, the *U* factor is 0.45. For further comparison, the conductivity *k* of most commercial insulations varies from about 0.24 to about 0.34.

13.3.8 Convection

Heating by natural convection is very common, because air very easily transfers heat in this manner. As air becomes warmer, it becomes less dense and rises. As it leaves the proximity of the heating surface, other cooler air moves in to replace the rising volume of heated air. As the warm air rises, it comes in contact with cooler materials, such as walls, glass, and ceilings. It becomes cooler and heavier and, under the influence of gravity, begins to fall. Hence, a circulatory motion of air is established, and heat transfer occurs.

When a heating device called a convector operates in a cool space, heat from the convector is transmitted to the cooler walls and ceiling by convection. The convection process will continue as long as the walls or ceiling are colder and the temperature difference is maintained.

Heating of building interiors is usually accomplished with convectors with hot water or steam as the heating medium. The heating element usually consists of a steel or copper pipe with closely spaced steel or aluminum fins. The convector is mounted at floor level against an exterior wall. The fins are used to increase greatly the area of the heating surface. As cool room air near the floor comes in contact with the hot surfaces of the convector, the air quickly becomes very warm and rises rapidly along the cold wall surface above the convector. Additional cold air at floor level then moves into the convector to replace the heated air. In this manner, the entire room will become heated. This process is called heating by natural convection.

13.3.9 Radiation

The most common form of heat transfer is by radiation. All materials and substances radiate energy and absorb radiation energy.

The sun is a huge radiator and the earth is heated by this immense source of radiated energy, which is often called solar energy (sunshine). Solar-collector devices are used to collect this energy and transfer it indoors to heat the interior of a building.

When radiation from the sun is intercepted by walls, roofs, or glass windows, this heat is transmitted through them and heats the interior of the building and its occupants. The reverse is also true; that is, when the walls are cold, the people in the space radiate their body heat to the cold wall and glass surfaces. If the rate of radiation is high, the occupants will be uncomfortable.

Not all materials radiate or absorb radiation equally. Black- or dark-body materials radiate and absorb energy better than light-colored or shiny materials. Materials with smooth surfaces and light colors are poor absorbers of radiant energy and also poor radiators.

Much of the radiation that strikes the surface of window glass is transmitted to the interior of the building as short-wave radiation. This radiation will strike other objects in the interior and radiate some of this energy back to the exterior, except through glass, as a longer wavelength of radiation energy. The glass does not efficiently transmit the longer wavelengths to the outside. Instead, it acts as a check valve, limiting solar radiation to one-way flow. This one-way flow is desirable in winter for heating. In summer, however, it is not desirable, because the longer-wavelength energy eventually becomes an additional load on the air-conditioning system.

The rate of radiation from an object may be determined by use of the **Stefan-Boltzmann law** of radiation. This law states that the amount of energy radiated from a perfect radiator, or a blackbody, is proportional to the fourth power of the absolute temperature of the body. Because most materials are not perfect radiators or absorbers, a proportionality constant called the hemispherical emittance factor is used with this law. Methods for calculating and estimating radiation transfer rates can be found in the "ASHRAE Handbook—Fundamentals."

The quantity of energy transferred by radiation depends on the individual temperatures of the radiating bodies. These temperatures are usually combined into a **mean radiant temperature** for use in heating and cooling calculations. The mean radiant temperature is the uniform temperature of a black enclosure with which a solid body (or occupant) would exchange the same amount of radiant heat as in the actual nonuniform environment.

13.3.10 Thermal Criteria for Building Interiors

There are three very important conditions to be controlled in buildings for human comfort. These important criteria are dry-bulb temperature, relative humidity, and velocity or rate of air movement in the space.

Measurements of these conditions should be made where average conditions exist in the building, room, or zone and at the breathing line, 3 to 5 ft above the floor. The measurements should be taken where they would not be affected by unusually high heat sources or heat losses. Minor variations or limits from the design conditions, however, are usually acceptable.

The occupied zone of a conditioned space does not encompass the total room volume. Rather, this occupied zone is generally taken as that volume bounded by levels 3 in above the floor and 6 ft above the floor and by vertical planes 2 ft from walls.

Indoor design temperatures are calculated from test data compiled for men and women with various amounts of clothing and for various degrees of physical exertion. For lightly clothed people doing light, active work in relatively still room air, the design dry-bulb temperature can be determined from

$$t = 180 - 1.4t_r \qquad (13.22)$$

where t = dry-bulb temperature, °F DB
t_r = mean radiant temperature of the space or room (between 70 and 80°F)

When temperatures of walls, materials, equipment, furniture, etc., in a room are all equal, $t = t_r = 75$°F. With low outside temperature, the building exterior becomes cold, in which case the room temperature should be maintained above 75°F to provide the necessary heat that is being lost to the cold exterior. In accordance with Eq. (13.22), the design dry-bulb temperature should be increased 1.4°F for each 1°F of mean radiant temperature below 75°F in the room. In very warm weather, the design temperature should be decreased correspondingly.

Humidity is often controlled for human comfort. Except in rare cases, relative humidity (RH) usually should not exceed 60%, because the moisture in the air may destroy wood finishes and support mildew. Below 20% RH, the air is so dry that human nostrils become dry and wood furniture often cracks from drying out.

In summer, a relative humidity of 45 to 55% is generally acceptable. In winter, a range of 30 to 35% RH is more desirable, to prevent condensation on windows and in walls and roofs. When design temperatures in the range of 75°F are maintained in a space, the comfort of occupants who are inactive is not noticeably affected by the relative humidity.

Variations from the design criteria are generally permitted for operational facilities. These variations are usually established as a number of degrees above or below the design point, such as 75°F DB ± 2°F. For relative humidity, the permitted variation is usually given as a percent, for example, 55% RH ± 5%.

Design conditions vary widely for many commercial and industrial uses. Indoor design criteria for various requirements are given in the "Applications" volume of the ASHRAE Handbook.

13.3.11 Outdoor Design Conditions

The outdoor design conditions at a proposed building site are very important in design of heating and cooling systems. Of major importance are the dry-bulb temperature, humidity conditions, and prevailing winds.

Outside conditions assumed for design purposes affect the heating and cooling plant's physical size, capacity, electrical requirements, and of considerable importance, the estimated cost of the HVAC installation. The reason for this is that in many cases, the differences between indoor and outdoor conditions have a great influence on calculated heating and cooling loads, which determine heating and cooling equipment capacities. Since in most cases the design outdoor air temperatures are assumed, the size of equipment will be greatly affected by assumed values.

Extreme outside air conditions are rarely used to determine the size of heating and cooling equipment, since these extreme conditions may occur, in summer or winter, only once in 10 to 50 years. If these extreme conditions were used for equipment selection, the results would be greatly oversized heating and cooling plants and a much greater installed cost than necessary. Furthermore, such oversized equipment will operate most of the year at part load and with frequent cycling of the machines. This results in inefficient operation and, generally, consumption of additional power, because most machines operate at maximum efficiency at full load.

On the other hand, when heating or cooling equipment is properly sized for more frequently occurring outdoor conditions, the plants will operate with less cycling and greater efficiency. During the few hours per year when outside conditions exceed those used for design, the equipment will run continuously in an attempt to maintain the intended interior design conditions. If such conditions persist for a long time, there will probably be a change in interior conditions from design conditions that may or may not be of a minor extent and that may produce uncomfortable conditions for the occupants.

Equipment should be selected with a total capacity that includes a safety factor to cover other types of operation than under steady-state conditions. In the midwest, for instance, the outdoor air temperature may fall as much as 45°F in 2 hr. The heating capacity of a boiler in this case would have to be substantially larger than that required for the calculated heat loss alone. As another example, many heating and cooling systems are controlled automatically by temperature control systems that, at a predetermined time, automatically reset the building temperature downward to maintain, say, 60°F at night for heating. At a predetermined time, for example, 7:30 a.m., before arrival of occupants, the control system instructs the boiler to bring the building up to its design temperature for occupancy. Under these conditions, the boiler must have the additional capacity to comply in a reasonable period of time before the arrival of the occupants.

Accordingly, design outdoor conditions should be selected in accordance with the manner in which the building will be used and, just as important, to obtain reasonable initial cost and low operational costs. Outdoor design conditions for a few cities are shown in Tables 13.4 and 13.5. Much more detailed data are presented in the "ASHRAE Handbook—Fundamentals."

The use of outdoor design conditions does not yield accurate estimates of fuel requirements or operating costs, because of the considerable variations of outdoor air temperature seasonally, monthly, daily, and even hourly. These wide variations must be taken into account in attempts to forecast the operating costs of a heating or cooling system.

Since most equipment capacities are selected for calculated loads based on steady-state conditions, usually these conditions will not provide acceptable estimates of annual operating costs. (Wide fluctuations in outside temperatures, however, may not always cause a rapid change in inside conditions as outdoor temperatures rise and fall. For example, in buildings with massive walls and roofs and small windows, indoor temperatures respond slowly to outdoor changes.) Hence, forecasts of fuel requirements or operating costs should be based on the average temperature difference between inside and outside air temperatures on an hourly basis for the entire year. Such calculations are extremely laborious and are almost always performed by a computer that utilizes an appropriate program and local weather tapes for the city involved. Many such programs are currently available from various sources.

13.4 VENTILATION

Ventilation is utilized for many different purposes, the most common being control of humidity and condensation. Other well-known uses include exhaust hoods in restaurants, heat removal in industrial plants, fresh air in buildings, odor removal,

TABLE 13.4 Recommended Design Outdoor Summer Temperatures

State	City	Dry-bulb temp, °F	Wet-bulb temp, °F	State	City	Dry-bulb temp, °F	Wet-bulb temp, °F
Ala.	Birmingham	95	78	Miss.	Vicksburg	95	78
Ariz.	Flagstaff	90	65	Mo.	St. Louis	95	78
Ariz.	Phoenix	105	75	Mont.	Helena	95	67
Ark.	Little Rock	95	78	Nebr.	Lincoln	95	78
Calif.	Los Angeles	90	70	Nev.	Reno	95	65
Calif.	San Francisco	85	65	N.H.	Concord	90	73
Colo.	Denver	95	65	N.J.	Trenton	95	78
Conn.	Hartford	95	75	N. Mex.	Albuquerque	95	70
D.C.	Washington	95	78	N.Y.	New York	95	75
Fla.	Jacksonville	95	78	N.C.	Greensboro	95	78
Fla.	Miami	95	79	N. Dak.	Bismarck	95	73
Ga.	Atlanta	95	76	Ohio	Cincinnati	95	75
Idaho	Boise	95	65	Okla.	Tulsa	100	77
Ill.	Chicago	95	75	Ore.	Portland	90	68
Ind.	Indianapolis	95	75	Pa.	Philadelphia	95	78
Iowa	Des Moines	95	78	R.I.	Providence	95	75
Kans.	Topeka	100	78	S.C.	Charleston	95	78
Ky.	Louisville	95	78	S. Dak.	Rapid City	95	70
La.	New Orleans	95	80	Tenn.	Nashville	95	78
Maine	Portland	90	73	Tex.	Dallas	100	78
Md.	Baltimore	95	78	Tex.	Houston	95	78
Mass.	Boston	95	75	Utah	Salt Lake City	95	65
Mich.	Detroit	95	75	Vt.	Burlington	90	73
Minn.	Minneapolis	95	75	Va.	Richmond	95	78

and chemical and fume hood exhausts. In commercial buildings, ventilation air is used for replacement of stale, vitiated air, odor control, and smoke removal. Ventilation air contributes greatly to the comfort of the building's occupants. It is considered to be of such importance that many building codes contain specific requirements for minimum quantities of fresh, or outside, air that must be supplied to occupied areas.

Ventilation is also the prime method for reducing employee exposure to excessive airborne contaminants that result from industrial operations. Ventilation is used to dilute contaminants to safe levels or to capture them at their point of origin before they pollute the employees' working environment. The Occupational Safety and Health Act (OSHA) standards set the legal limits for employee exposures to many types of toxic substances.

TABLE 13.5 Recommended Design Outdoor Winter Temperatures

State	City	Temp, °F	State	City	Temp, °F
Ala.	Birmingham	10	Miss.	Vicksburg	10
Ariz.	Flagstaff	−10	Mo.	St. Louis	0
Ariz.	Phoenix	25	Mont.	Helena	−20
Ark.	Little Rock	5	Nebr.	Lincoln	−10
Calif.	Los Angeles	35	Nev.	Reno	−5
Calif.	San Francisco	35	N.H.	Concord	−15
Colo.	Denver	−10	N.J.	Trenton	0
Conn.	Hartford	0	N. Mex.	Albuquerque	0
D.C.	Washington	0	N.Y.	New York	0
Fla.	Jacksonville	25	N.C.	Greensboro	10
Fla.	Miami	35	N. Dak.	Bismarck	−30
Ga.	Atlanta	10	Ohio	Cincinnati	0
Idaho	Boise	−10	Okla.	Tulsa	0
Ill.	Chicago	−10	Ore.	Portland	10
Ind.	Indianapolis	−10	Pa.	Philadelphia	0
Iowa	Des Moines	−15	R.I.	Providence	0
Kans.	Topeka	−10	S.C.	Charleston	15
Ky.	Louisville	0	S. Dak.	Rapid City	−20
La.	New Orleans	20	Tenn.	Nashville	0
Maine	Portland	−5	Tex.	Dallas	0
Md.	Baltimore	0	Tex.	Houston	20
Mass.	Boston	0	Utah	Salt Lake City	−10
Mich.	Detroit	−10	Vt.	Burlington	−10
Minn.	Minneapolis	−20	Va.	Richmond	15

13.4.1 Methods of Ventilation

Ventilation is generally accomplished by two methods: natural and mechanical. In either case, ventilation air must be air taken from the outdoors. It is brought into the building through screened and louvered or other types of openings, with or without ductwork. In many mechanical ventilation systems, the outside air is brought in through ductwork to an appropriate air-moving device, such as a centrifugal fan. With a network of ductwork, the supply air is distributed to areas where it is needed. Also, mechanical ventilation systems are usually designed to exhaust air from the building with exhaust fans or gravity-type ventilators in the roof or a combination-type system.

Many mechanical ventilation systems are installed for fire protection in buildings to remove smoke, heat, and fire. The design must be capable of satisfying the provisions of the National Fire Protection Association "Standard for Installation of Air-Conditioning Systems," NFPA 90-A. The standard also covers installation provisions of air intakes and outlets.

Natural ventilation in buildings is caused by the temperature difference between the air in the building and the outside air and by openings in the outside walls or by a combination of both. With natural ventilation, there should be some means for removing the ventilation air from the building, such as roof-mounted gravity vents or exhaust fans.

13.4.2 Minimum Ventilation Requirements

There are many codes and rules governing minimum standards of ventilation. All gravity or natural-ventilation requirements involving window areas in a room as a given percentage of the floor area or volume are at best approximations. The amount of air movement or replacement by gravity depends on prevailing winds, temperature difference between interior and exterior, height of structure, window-crack area, etc. For controlled ventilation, a mechanical method of air change is recommended.

Where people are working, the amount of ventilation air required will vary from one air change per hour where no heat or offensive odors are generated to about 60 air changes per hour.

At best, a ventilation system is a dilution process, by which the rate of odor or heat removal is equal to that generated in the premises. Occupied areas below grade or in windowless structures require mechanical ventilation to give occupants a feeling of outdoor freshness. Without outside air, a stale or musty odor may result. The amount of fresh air to be brought in depends on the number of persons occupying the premises, type of activity volume of the premises, and amount of heat, moisture, and odor generation. Table 13.6 gives the recommended minimum amount of ventilation air required for various activities.

TABLE 13.6 Minimum Ventilation Air for Various Activities

Type of occupancy	Ventilation air, cfm per person
Inactive, theaters	5
Light activity, offices	10
Light activity with some odor generation, restaurant	15
Light activity with moderate odor generation, bars	20
Active work, shipping rooms	30
Very active work, gymnasiums	50

The amount of air to be handled, obtained from the estimate of the per person method, should be checked against the volume of the premises and the number of air changes per hour given in Eq. (13.23).

$$\text{Number of air changes per hour} = \frac{60Q}{V} \tag{13.23}$$

where Q = air supplied, ft^3/min
V = volume of ventilated space, ft^3

When the number of changes per hour is too low (below one air change per hour), the ventilation system will take too long to create a noticeable effect when first put into operation. Five changes per hour are generally considered a practical minimum. Air changes above 60 per hour usually will create some discomfort because of air velocities that are too high.

Toilet ventilation and locker-room ventilation are usually covered by local codes—50 ft^3/min per water closet and urinal is the usual minimum for toilets and six changes per hour minimum for both toilets and locker rooms.

13.4.3 Heat, Odor, and Moisture Removal

Removal of concentrated heat, odor, or objectionable vapors by ventilation is best carried out by locating the exhaust outlets as close as possible to the heat source. Where concentrated sources of heat are present, canopy hoods will remove the heat more efficiently.

Figure 13.1 shows a canopy-hood installation over a kitchen range. Grease filters reduce the frequency of required cleaning. When no grease is vaporized, they may be eliminated.

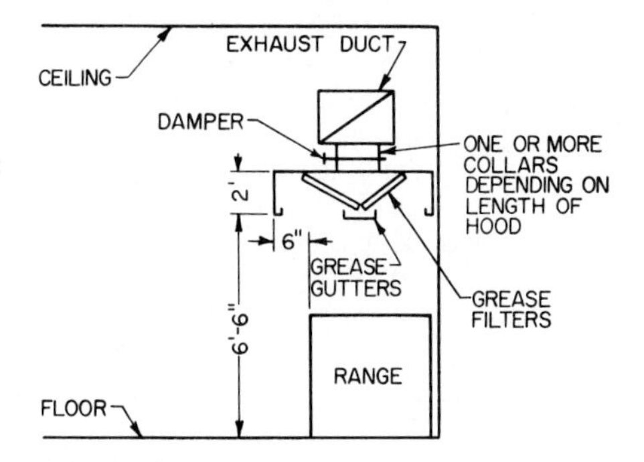

FIGURE 13.1 Canopy hood for exhausting heat from a kitchen range.

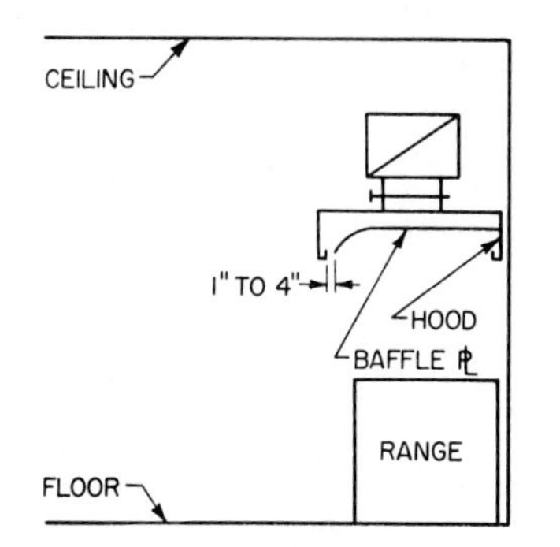

FIGURE 13.2 Double hood for exhausting heat.

Greasy ducts are serious fire hazards and should be cleaned periodically. There are on the market a number of automatic fire-control systems for greasy ducts. These systems usually consist of fusible-link fire dampers and a means of flame smothering—CO_2, steam, foam, etc.

Figure 13.2 shows a double hood. This type collects heat more efficiently; i.e., less exhaust air is required to collect a given amount of heat. The crack area is arranged to yield a velocity of about 1000 ft/min.

A curtain of high-velocity air around the periphery of the hood catches the hot air issuing from the range or heat source. Canopy hoods are designed to handle about 50 to 125 ft^3/min of exhaust air per square foot of hood. The total amount of ventilation air should not yield more than 60 changes per hour in the space.

Where hoods are not practical to install and heat will be discharged into the room, the amount of ventilation air may be determined by the following method:

Determine the total amount of sensible heat generated in the premises—lights, people, electrical equipment, etc. This heat will cause a temperature rise and an increase in heat loss through walls, windows, etc. To maintain desired temperature conditions, ventilation air will have to be used to remove heat not lost by transmission through enclosures.

$$q_v = 1.08Q(T_i - T_o) \qquad (13.24)$$

where q_v = heat, Btu/hr, carried away by ventilation air
Q = flow of ventilation air, ft^3/min
T_i = indoor temperature to be maintained
T_o = outdoor temperature

With Eq. (13.24), we can calculate the amount of ventilation air required by assuming a difference between room and outdoor temperatures or we can calculate this temperature gradient for a given amount of ventilation air.

The same method may be used to calculate the air quantity required to remove any objectionable chemical generated. For example, assume that after study of a process we determine that a chemical will be evolved in vapor or gas form at the rate of X lb/min. If Y is the allowable concentration in pounds per cubic foot, then $Q = X/Y$, where Q is the ventilation air needed in cubic feet per minute.

Where moisture is the objectionable vapor, the same equation holds, but with X as the pounds per minute of moisture vaporized, Y the allowable concentration of moisture in pounds per cubic foot above outdoor moisture concentration.

Once the amount of ventilation air is determined, a duct system may be designed to handle it, if necessary.

Ventilation air may be provided by installing either an exhaust system, a supply system, or both.

In occupied areas where no unusual amounts of heat or odors are generated, such as offices and shipping rooms, a supply-air system may be provided, with grilles or ceiling outlets located for good distribution. When the building is tight, a relief system of grilles or ducts to the outside should be provided. But when the relief system is too extensive, an exhaust fan should be installed for a combination supply and exhaust system.

All air exhausted from a space must be replaced by outside air either by infiltration through doors and windows or by a fresh-air makeup system. Makeup air systems that have to operate during the winter season are often equipped with heating coils to temper the cold outside air.

13.4.4 Natural Ventilation

Natural ventilation in buildings is accomplished by use of windows, louvers, skylights, roof ventilators, roof monitors, jalousies, intake hoods, etc. They should be located to admit fresh air only and not near sources of smoke, dust, odors, or polluted air from adjacent sources. Discharge vents should also be provided to eliminate vitiated air from the building. The outlet locations must not discharge toward other fresh-air intakes of the building or its neighbors. In multifloor buildings, vertical vent shafts, or risers, are used to supply ventilation air throughout the building.

13.4.5 Mechanical Ventilation

Mechanical ventilation is almost always preferred over natural ventilation because of reliability and the ability to maintain specific design requirements, such as air changes per hour and face velocities for exahust hoods. Natural ventilation permits wide variations in ventilation-air quantities and uncertain durations of ventilation. (In critical areas, such as in carcinogenic research laboratories, natural ventilation is never relied upon.) For this reason, mechanical ventilation systems are almost always used where ventilation requirements are critical and must be highly reliable.

Mechanical ventilation is often required by various building codes for various applications as follows:

1. Control of contaminants in the work area for health protection and compliance with OSHA standards for achieving the legal limits set on employee exposure to specific toxic and hazardous substances
2. Fire and explosion prevention for flammable vapors
3. Environmental protection
4. Reuse of valuable industrial materials
5. Human comfort—removal of heat, odors, and tobacco smoke
6. Humidity control
7. Corrosive fumes and noxious gases

Mechanical ventilation may be a single system without heating, cooling, filtration, humidification, dehumidification, etc., or it may include various combinations of these functions. In other words, the systems can be heating-ventilating units or heating–ventilating–air-conditioning (HVAC) units.

In many complex and specialized buildings, certain functional areas will be required to have various degrees of positive pressurization, negative pressurization, or balanced atmospheric conditions. The ventilation air as a part of the system supply air is used to provide the positive and balanced pressures. An exhaust system is utilized to maintain the negative-pressure areas. In many designs, air lost by pressurization is exfiltrated from the system and does not become part of the return-air stream.

Recirculation of ventilation air is prohibited from certain areas, such as toilets, bathrooms, biology labs, chemistry labs, hospital operating rooms, mortuary rooms, isolation rooms, and rooms with flammable vapors, odors, dust, and noxious gases.

In all ventilation systems, a quantity of air equal to the ventilation air should leave the building. If this is not accomplished, then the building will become pressurized, and the ventilation air will exfiltrate through available doors, windows, cracks, crevices, relief vents, etc. Since in many cases this is undesirable and unreliable, exhaust systems are usually employed. The exhaust, in many cases, may be part of a complete HVAC system.

13.5 MOVEMENT OF AIR WITH FANS

Inasmuch as most ventilation systems are designed as mechanical ventilation systems that utilize various kinds of fans, a knowledge of the types of fans in use will be of value in selection of ventilation fans. Fans are used to create a pressure differential that causes air to flow in a system. They generally incorporate one of several types of impellers mounted in an appropriate housing or enclosure. An electric motor usually drives the impeller to move the air.

Two types of fans are commonly used in air-handling and air-moving systems: axial and centrifugal. They differ in the direction of airflow through the impeller.

Centrifugal fans are enclosed in a scroll-shaped housing, which is designed for efficient airstream energy transfer. This type of fan has the most versatility and low first cost and is the workhorse of the industry. Impeller blades may be radial, forward-curved, backward-inclined, or airfoil. When large volumes of air are moved, airfoil or backward-inclined blades are preferable because of higher efficiencies. For smaller volumes of air, forward, curved blades are used with satisfactory results. Centrifugal fans are manufactured with capacities of up to 500,000 ft^3/min and can operate against pressures up to 30 in water gage.

Axial-flow fans are versatile and sometimes less costly than centrifugal fans. The use of axial fans is steadily increasing, because of the availability of controllable-pitch units, with increased emphasis on energy savings. Substantial energy savings can be realized by varying the blade pitch to meet specific duty loads. Axial fans develop static pressure by changing the velocity of the air through the impeller and converting it into static pressure. Axial fans are quite noisy and are generally used by industry where the noise level can be tolerated. When used for HVAC installations, sound attenuators are almost always used in series with the fan for noise abatement. Tubeaxial and vaneaxial are modifications of the axial-flow fan.

Propeller-type fans are also axial fans and are produced in many sizes and shapes. Small units are used for small jobs, such as kitchen exhausts, toilet exhausts, and air-cooled condensers. Larger units are used by industry for ventilation and heat removal in large industrial buildings. Such units have capacities of up to 200,000 ft^3/min of air. Propeller-type fans are limited to operating pressures of about ½ in water gage maximum, and are usually much noisier than centrifugal fans of equal capacities.

Vaneaxial fans are available with capacities up to 175,000 ft^3/min and can operate at pressures up to 12 in water gage. Tubeaxial fans can operate against pressures of only 1 in water gage with only slightly lower capacities.

In addition to the axial and centrifugal fan classifications, a third class for special designs exists. This classification covers tubular centrifugal fans and axial-centrifugal, power roof ventilators. The tubular centrifugal type is often used as a return-air fan in low-pressure HVAC systems. Air is discharged from the impeller in the same way as in standard centrifugal fans and then changed 90° in direction through straightening vanes. Tubular centrifugal fans are manufactured with capacities of more than 250,000 ft^3/min of air and may operate at pressures up to 12 in water gage.

Power roof ventilators are usually roof mounted and utilize either centrifugal or axial blade fans. Both types are generally used in low-pressure exhaust systems for factories, warehouses, etc. They are available in capacities up to about 30,000 ft^3/min. They are, however, limited to operation at a maximum pressure of about ½ in water gage. Powered roof ventilators are also low in first cost and low in operating costs. They can provide positive exhaust ventilation in a space, which is a definite advantage over gravity-type exhaust units. The centrifugal unit is somewhat quieter than the axial-flow type.

Fans vary widely in shapes and sizes, motor arrangements and space requirements. Fan performance characteristics (variation of static pressure and brake horsepower) with changes in the airflow rate (ft^3/min) are available from fan manufacturers and are presented in tabular form or as fan curves.

Dampers. Dampers are mechanical devices that are installed in a moving airstream in a duct to reduce the flow of the stream. They, in effect, purposely produce a pressure drop (when installed) in a duct by substantially reducing the free area of the duct.

Two types of dampers are commonly used by HVAC designers, parallel blade and opposed blade. In both types, the blades are linked together so that a rotation force applied to one shaft simultaneously rotates all blades. The rotation of the blades opens or closes the duct's free area from 0 to 100% and determines the flow rate.

Dampers are used often as opening and closing devices. For this purpose, parallel dampers are preferred.

When dampers are installed in ducts and are adjusted in a certain position to produce a desired flow rate downstream, opposed-blade dampers are preferred. When dampers are used for this purpose, the operation is called **balancing.**

Once the system is balanced and the airflows in all branch ducts are design airflows, the damper positions are not changed until some future change in the system occurs. However, in automatic temperature-control systems, both opening-and-closing and balancing dampers are commonly used. In complex systems, dampers may be modulated to compensate for increased pressure drop by filter loading and to maintain constant supply-air quantity in the system.

Filters. All air-handling units should be provided with filter boxes. Removal of dust from the conditioned air not only lowers building maintenance costs and creates a healthier atmosphere but prevents the cooling and heating coils from becoming blocked up.

Air filters come in a number of standard sizes and thicknesses. The filter area should be such that the air velocity across the filters does not exceed 350 ft/min for low-velocity filters or 550 ft/min for high-velocity filters. Thus, the minimum filter area in square feet to be provided equals the airflow, ft^3/min, divided by the maximum air velocity across the filters, ft/min.

Most air filters are of either the throwaway or cleanable type. Both these types will fit a standard filter rack.

Electrostatic filters are usually employed in industrial installations, where a higher percentage of dust removal must be obtained. Check with manufacturers' ratings for particle-size removal, capacity, and static-pressure loss; also check electric service required. These units generally are used in combination with regular throwaway or cleanable air filters, which take out the large particles, while the charged electrostatic plates remove the smaller ones. See also Art. 13.6.

13.6 DUCT DESIGN

After air discharge grilles and the air handler, which consists of a heat exchanger and blower, have been located, it is advisable to make a single-line drawing showing the duct layout and the air quantities each branch and line must be able to carry.

Of the methods of duct design in use, the equal-friction method is the most practical. It is considered good practice not to exceed a pressure loss of 0.15 in of water per 100 ft of ductwork by friction. Higher friction will result in large power consumption for air circulation. It is also considered good practice to stay below a starting velocity in main ducts of 900 ft/min in residences; 1300 ft/min in schools, theaters, and public buildings; and 1800 ft/min in industrial buildings. Velocity in branch ducts should be about two-thirds of these and in branch risers about one-half.

Too high a velocity will result in noisy and panting ductwork. Too low a velocity will require uneconomical, bulky ducts.

The shape of ducts usually installed as rectangular, because dimensions can easily be changed to maintain the required area. However, as ducts are flattened, the increase in perimeter offers additional resistance to airflow. Thus a flat duct requires an increase in cross section to be equivalent in air-carrying capacity to one more nearly square.

A 12 × 12-in duct, for example, will have an area 1 ft^2 and a perimeter of 4 ft, whereas a 24 × 6-in duct will have the same cross-sectional area but a 5-ft perimeter and thus greater friction. Therefore, a 24 × 7-in duct is more nearly equivalent to the 12 × 12. Equivalent sizes can be determined from tables, such as those in the "ASHRAE Handbook—Fundamentals," where rectangular ducts are rated in terms of equivalent round ducts (equal friction and capacity). Table 13.7 is a shortened version.

Charts also are available in the ASHRAE handbook giving the relationship between duct diameter in inches, air velocity in feet per minute, air quantity in cubic feet per minute, and friction in inches of water pressure drop per 100 ft of duct. Table 13.8 is based on data in the ASHRAE handbook.

TABLE 13.7 Diameters of Circular Ducts in Inches Equivalent to Rectangular Ducts

Side	4	8	12	18	24	30	36	42	48	60	72	84
3	3.8	5.2	6.2									
4	4.4	6.1	7.3									
5	4.9	6.9	8.3									
6	5.4	7.6	9.2									
7	5.7	8.2	9.9									
12		10.7	13.1									
18		12.9	16.0	19.7								
24		14.6	18.3	22.6	26.2							
30		16.1	20.2	25.2	29.3	32.8						
36		17.4	21.9	27.4	32.0	35.8	39.4					
42		18.5	23.4	29.4	34.4	38.6	42.4	45.9				
48		19.6	24.8	31.2	36.6	41.2	45.2	48.9	52.6			
60		21.4	27.3	34.5	40.4	45.8	50.4	54.6	58.5	65.7		
72		23.1	29.5	37.2	43.8	49.7	54.9	59.6	63.9	71.7	78.8	
84				39.9	46.9	53.2	58.9	64.1	68.8	77.2	84.8	91.9
96					49.5	56.3	62.4	68.2	73.2	82.6	90.5	97.9

TABLE 13.8 Sizes of Round Ducts for Airflow*

Friction, in H_2O per 100 ft	0.05		0.10		0.15		0.20		0.25		0.30	
Airflow, cfm	Diam, in	Velocity, ft/min	Diam, in	Velocity, ft/min	Diam, in	Velocity, ft/min	Diam, in	Velocity, ft/min	Diam, in	Velocity, ft/min	Diam, in	Velocity, ft/min
50	5.3	350	4.6	450	4.2	530	3.9	600	3.8	660	3.7	710
100	6.8	420	5.8	550	5.4	640	5.1	720	4.8	780	4.7	850
200	8.7	480	7.6	650	6.9	760	6.6	860	6.3	940	6.1	1020
300	10.2	540	8.8	730	8.2	850	7.7	960	7.3	1050	7.1	1120
400	11.5	580	9.8	770	9.0	920	8.5	1040	8.2	1130	7.8	1200
500	12.4	620	11.8	820	9.8	970	9.3	1080	8.8	1160	8.6	1270
1,000	15.8	730	13.7	970	12.8	1140	12.0	1280	11.5	1400	11.2	1500
2,000	20.8	870	18.0	1150	16.6	1370	15.7	1520	15.0	1660	14.5	1780
3,000	24.0	960	21.0	1280	19.7	1500	18.3	1680	17.5	1850		
4,000	26.8	1050	23.4	1360	21.6	1600	20.2	1800				
5,000	29.2	1100	25.5	1460	23.7	1700	22.2	1900				
10,000	37.8	1310	33.2	1770	30.3	2000						

*Based on data in "ASHRAE Handbook—Fundamentals," American Society of Heating, Refrigerating and Air-Conditioning Engineers.

In the equal-friction method, the equivalent round duct is determined for the required air flow at the predetermined friction factor. For an example illustrating the method of calculating duct sizes, see Art. 13.11.

(H. E. Bovay, Jr., "Handbook of Mechanical and Electrical Systems for Buildings," F. E. Beaty, Jr., "Sourcebook of HVAC Details," and "Sourcebook of HVAC Specifications," N. R. Grim and R. C. Rosaler, "Handbook of HVAC Design," D. L. Grumman, "Air-Handling Systems Ready Reference Manual," McGraw-Hill Publishing Company, New York; B. Stein et al., "Mechanical and Electrical Equipment for Buildings," 7th ed., John Wiley & Sons, Inc., New York.)

13.7 HEAT LOSSES

Methods and principles for calculation of heat losses are presented in Art 13.3. These methods provide a rational procedure for determination of the size and capacity of a heating plant.

Heat loads for buildings consist of heat losses and gains. Heat losses include those from air infiltration, ventilation air, and conduction through the building exterior caused by low temperatures of outside air. Heat gains include those due to people, hot outside air, solar radiation, electrical lighting and motor loads, and heat from miscellaneous interior equipment. These loads are used to determine the proper equipment size for the lowest initial cost and for operation with maximum efficiency.

Walls and Roofs. Heat loss through the walls and roofs of a building constitutes most of the total heat loss in cold weather. These losses are calculated with Eq. (13.19), $Q = UA(t_2 - t_1)$, with the appropriate temperature differential between inside and outside design temperatures.

Architectural drawings should be carefully examined to establish the materials of construction that will be used in the walls and roofs. With this information, the overall coefficient of heat transmittance, or U factor, can be determined as described in Art. 13.3. Also, from the drawings, the height and width of each wall section should be determined to establish the total area for each wall or roof section for use in Eq. (13.19).

Heat Loss through Basement Floors and Walls. Although heat-transmission coefficients through basement floors and walls are available, it is generally not practicable to use them because ground temperatures are difficult to determine because of the many variables involved. Instead, the rate of heat flow can be estimated, for all practical purposes, from Table 13.9. This table is based on groundwater

TABLE 13.9 Below-Grade Heat Losses

Ground water temp, °F	Basement floor loss,* Btu per hr per sq ft	Below-grade wall loss, Btu per hr per sq ft
40	3.0	6.0
50	2.0	4.0
60	1.0	2.0

*Based on basement temperature of 70°F.

temperatures, which range from about 40 to 60°F in the northern sections of the United States and 60 to 75°F in the southern sections. (For specific areas, see the "ASHRAE Handbook—Fundamentals.")

Heat Loss from Floors on Grade. Attempts have been made to simplify the variables that enter into determination of heat loss through floors set directly on the ground. The most practical method breaks it down to a heat flow in Btu per hour per linear foot of edge exposed to the outside. With 2 in of edge insulation, the rate of heat loss is about 50 in the cold northern sections of the United States, 45 in the temperature zones, 40 in the warm south. Corresponding rates for 1-in insulation are 60, 55, and 50. With no edge insulation the rates are 75, 65, and 60 Btu/(hr)(ft).

Heat Loss from Unheated Attics. Top stories with unheated attics above require special treatment. To determine the heat loss through the ceiling, we must calculate the equilibrium attic temperature under design inside and outside temperature conditions. This is done by equating the heat gain to the attic via the ceiling to the heat loss through the roof:

$$U_c A_c (T_i - T_a) = U_r A_r (T_a - T_o) \tag{13.25}$$

where U_c = heat-transmission coefficient for ceiling
U_r = heat-transmission coefficient for roof
A_c = ceiling area
A_r = roof area
T_i = design room temperature
T_o = design outdoor temperature
T_a = attic temperature

Thus

$$T_a = \frac{U_c A_c T_i + U_r A_r T_o}{U_c A_c + U_r A_r} \tag{13.26}$$

The same procedure should be used to obtain the temperature of other unheated spaces, such as cellars and attached garages.

Air Infiltration. When the heating load of a building is calculated, it is advisable to figure each room separately, to ascertain the amount of heat to be supplied to each room. Then, compute the load for a complete floor or building and check it against the sum of the loads for the individual rooms.

Once we compute the heat flow through all exposed surfaces of a room, we have the heat load if the room is perfectly airtight and the doors never opened. However, this generally is not the case. In fact, windows and doors, even if weather-stripped, will allow outside air to infiltrate and inside air to exfiltrate. The amount of cold air entering a room depends on crack area, wind velocity, and number of exposures, among other things.

Attempts at calculating window- and door-crack area to determine air leakage usually yield a poor estimate. Faster and more dependable is the air-change method, which is based on the assumption that cold outside air is heated and pumped into the premises to create a static pressure large enough to prevent cold air from infiltrating.

The amount of air required to create this static pressure will depend on the volume of the room.

If the number of air changes taking place per hour N are known, the infiltration Q in cubic feet per minute can be computed from

$$Q = \frac{VN}{60} \tag{13.27}$$

where V = volume of room, ft^3. The amount of heat q in Btu per hour required to warm up this cold air is given by

$$q = 1.08QT \tag{13.28}$$

where Q = ft^3/min of air to be warmed
T = temperature rise

13.8 HEAT GAINS

These differ from heat losses only by the direction of the heat flow. Thus, the methods discussed in (Art. 13.7) for heat losses can also be used for determining heat gains. In both cases, the proper inside and outside design conditions and wet-bulb temperatures should be established as described in Art. 13.3.

Heat gains may occur at any time throughout the year. Examples are heat from electric lighting, motor and equipment loads, solar radiation, people, and ventilation requirements. When heat gains occur in cold weather, they should be deducted from the heat loss for the space.

Ventilation and infiltration air in warm weather produce large heat gains and should be added to other calculated heat gains to arrive at the total heat gains for cooling-equipment sizing purposes.

To determine the size of cooling plant required in a building or part of a building, we determine the heat transmitted to the conditioned space through the walls, glass, ceiling, floor, etc., and add all the heat generated in the space. This is the cooling load. The unwanted heat must be removed by supplying cool air. The total cooling load is divided into two parts—sensible and latent.

Sensible and Latent Heat. The part of the cooling load that shows up in the form of a dry-bulb temperature rise is called sensible heat. It includes heat transmitted through walls, windows, roof, floor, etc.; radiation from the sun; and heat from lights, people, electrical and gas appliances, and outside air brought into the air-conditioned space.

Cooling required to remove unwanted moisture from the air-conditioned space is called latent load, and the heat extracted is called latent heat. Usually, the moisture is condensed out on the cooling coils in the cooling unit.

For every pound of moisture condensed from the air, the air-conditioning equipment must remove about 1050 Btu. Instead of rating items that give off moisture in pounds or grains per hour, common practice rates them in Btu per hour of latent load. These items include gas appliances, which give off moisture in products of combustion; steam baths, food, beverages, etc., which evaporate moisture; people; and humid outside air brought into the air-conditioned space.

Design Temperatures for Cooling. Before we can calculate the cooling load, we must first determine a design outside condition and the conditions we want to maintain inside.

For comfort cooling, indoor air at 80°F dry bulb and 50% relative humidity is usually acceptable.

Table 13.4, p. 13.26, gives recommended design outdoor summer temperatures for various cities. Note that these temperatures are not the highest ever attained; for example, in New York City, the highest dry-bulb temperature recorded exceeds 105°F, whereas the design outdoor dry-bulb temperature is 95°F. Similarly, the wet-bulb temperature is sometimes above the 75°F design wet-bulb for that area.

Heat Gain through Enclosures. To obtain the heat gain through walls, windows, ceilings, floors, etc., when it is warmer outside than in, the heat-transfer coefficient is multiplied by the surface area and the temperature gradient.

Radiation from the sun through glass is another source of heat. It can amount to about 200 Btu/(hr)(ft^2) for a single sheet of unshaded common window glass facing east and west, about three-fourths as much for windows facing northeast and northwest, and one-half as much for windows facing south. For most practical applications, however, the sun effect on walls can be neglected, since the time lag is considerable and the peak load is no longer present by the time the radiant heat starts to work through to the inside surface. Also, if the wall exposed to the sun contains windows, the peak radiation through the glass also will be gone by the time the radiant heat on the walls gets through.

Radiation from the sun through roofs may be considerable. For most roofs, total equivalent temperature differences for calculating solar heat gain through roofs is about 50°F.

Roof Sprays. Many buildings have been equipped with roof sprays to reduce the sun load on the roof. Usually the life of a roof is increased by the spray system, because it prevents swelling, blistering, and vaporization of the volatile components of the roofing material. It also prevents the thermal shock of thunderstorms during hot spells. Equivalent temperature differential for computing heat gain on sprayed roofs is about 18°F.

Water pools 2 to 6 in deep on roofs have been used, but they create structural difficulties. Furthermore, holdover heat into the late evening after the sun has set creates a breeding ground for mosquitoes and requires algae-growth control. Equivalent temperature differential to be used for computing heat gain for water-covered roofs is about 22°F.

Spray control is effected by the use of a water solenoid valve actuated by a temperature controller whose bulb is embedded in the roofing. Tests have been carried out with controller settings of 95, 100, and 105°F. The last was found to be the most practical setting.

The spray nozzles must not be too fine, or too much water is lost by drift. For ridge roofs, a pipe with holes or slots is satisfactory. When the ridge runs north and south, two pipes with two controllers would be practical, for the east pipe would be in operation in the morning and the west pipe in the afternoon.

Heat Gains from Interior Sources. Electric lights and most other electrical appliances convert their energy into heat.

$$q = 3.42W \tag{13.29}$$

where q = Btu/hr developed
W = watts of electricity used

For lighting, if W is taken as the total light wattage, it may be reduced by the ratio of the wattage expected to be consumed at any time to the total installed

TABLE 13.10 Rates of Heat Gain from Occupants of Conditioned Spaces*

Degree of activity	Typical application	Total heat adults, male, Btu/hr	Total heat adjusted, Btu/hr†	Sensible heat, Btu/hr	Latent heat, Btu/hr
Seated at theater	Theater—matinee	390	330	225	105
Seated at theater	Theater—evening	390	350	245	105
Seated, very light work	Offices, hotels, apartments	450	400	245	155
Moderately active office work	Offices, hotels, apartments	475	450	250	200
Standing, light work; walking	Department store, retail store	550	450	250	200
Walking; standing	Drugstore, bank	550	500	250	250
Sedentary work	Restaurant‡	490	550	275	275
Light bench work	Factory	800	750	275	475
Moderate dancing	Dance hall	900	850	305	545
Walking 3 mph; light machine work	Factory	1000	1000	375	625
Bowling§	Bowling alley	1500	1450	580	870
Heavy work	Factory	1500	1450	580	870
Heavy machine work; lifting	Factory	1600	1600	635	965
Athletics	Gymnasium	2000	1800	710	1090

*Tabulated values are based on 78°F room dry-bulb temperature. For 80°F room dry bulb, the total heat remains the same, but the sensible-heat value should be decreased by approximately 20% and the latent heat values increased accordingly. All values are rounded to the nearest 5 Btu/hr.

†Adjusted total heat gain is based on normal percentage of men, women, and children for the application listed, with the postulate that the gain from an adult female is 85% of that for an adult male, and that the gain from a child is 75% of that for an adult male.

‡Adjusted total heat value for eating in a restaurant includes 60 Btu/hr for food per individual (30 Btu sensible and 30 Btu latent).

§For bowling, figure one person per alley actually bowling, and all others as sitting (400 Btu/hr) or standing and walking slowly (550 Btu/hr).

Reprinted by permission of ASHRAE from "ASHRAE Handbook—Fundamentals," 1989, American Society of Heating, Refrigerating and Air-Conditioning Engineers.

wattage. This ratio may be unity for commercial applications, such as stores. Where fluorescent lighting is used, add 25% of the lamp rating for the heat generated in the ballast. Where electricity is used to heat coffee, etc., some of the energy is used to vaporize water. Tables in the "ASHRAE Handbook—Fundamentals" give an estimate of the Btu per hour given up as sensible heat and that given up as latent heat by appliances.

Heat gain from people can be obtained from Table 13.10.

Heat Gain from Outside Air. The sensible heat from outside air brought into a conditioned space can be obtained from

$$q_s = 1.08Q(T_o - T_i) \tag{13.30}$$

where q_s = sensible load due to outside air, Btu/hr
Q = ft³/min of outside air brought into conditioned space
T_o = design dry-bulb temperature of outside air
T_i = design dry-bulb temperature of conditioned space

The latent load due to outside air in Btu per hour is given by

$$q_l = 0.67Q(G_o - G_i) \tag{13.31}$$

where Q = ft^3/min of outside air brought into conditioned space
G_o = moisture content of outside air, grains per pound of air
G_i = moisture content of inside air, grains per pound of air

The moisture content of air at various conditions may be obtained from a psychrometric chart.

Miscellaneous Sources of Heat Gain. In an air-conditioning unit, the fan used to circulate the air requires a certain amount of brake horsepower depending on the air quantity and the total resistance in the ductwork, coils, filters, etc. This horsepower will dissipate itself in the conditioned air and will show up as a temperature rise. Therefore, we must include the fan brake horsepower in the air-conditioning load. For most low-pressure air-distribution duct systems, the heat from this source varies from 5% of the sensible load for smaller systems to 3½% of the sensible load in the larger systems.

Where the air-conditioning ducts pass through nonconditioned spaces, the ducts must be insulated. The amount of the heat transmitted to the conditioned air through the insulation may be calculated from the duct area and the insulation heat-transfer coefficient.

METHODS OF HEATING BUILDINGS

The preceding articles present the basic knowledge necessary for accurate determination of the heat losses of a building. Such procedures make possible sizing and selection of heating equipment that will provide reliable and satisfactory service.

13.9 GENERAL PROCEDURE FOR SIZING A HEATING PLANT

The basic procedure used in sizing a heating plant is as follows: We isolate the part of the structure to be heated. To estimate the amount of heat to be supplied to that part, we must first decide on the design indoor and outdoor temperatures. For if we maintain a temperature of, say, 70°F inside the structure and the outside temperature is, say, 0°F, then heat will be conducted and radiated to the outside at a rate that can be computed from this 70° temperature difference. If we are to maintain the design inside temperature, we must add heat to the interior by some means at the same rate that it is lost to the exterior.

Recommended design inside temperatures are given in Table 13.11. Recommended design outdoor temperatures for a few cities are given in Table 13.5, p. 13.27. (More extensive data are given in the "ASHRAE Handbook—Fundamentals," American Society of Heating, Refrigerating and Air-Conditioning Engineers.)

TABLE 13.11 Recommended Design Indoor Winter Temperatures

Type of building	Temp, °F
Schools:	
Classrooms	72
Assembly rooms, dining rooms	72
Playrooms, gymnasiums	65
Swimming pool	75
Locker rooms	70
Hospitals:	
Private rooms	72
Operating rooms	75
Wards	70
Toilets	70
Bathrooms	75
Kitchens and laundries	66
Theaters	72
Hotels:	
Bedrooms	70
Ballrooms	68
Residences	72
Stores	68
Offices	72
Factories	65

Note that the recommended design outdoor winter temperatures are not the lowest temperatures ever attained in each region. For example, the lowest temperature on record in New York City is −14°F, whereas the design temperature is 0°F. If the design indoor temperature is 70°F, we would be designing for (70 − 0)/(70 + 14), or 83.3%, of the capacity we would need for the short period that a record cold of −14°F would last.

Once we have established for design purposes a temperature gradient (indoor design temperature minus outdoor design temperature) across the building exterior, we obtain the heat-transmission coefficients of the various building materials in the exterior construction for computation of the heat flow per square foot. These coefficients may be obtained from the manufacturers of the materials or from tables, such as those in the "ASHRAE Handbook—Fundamentals." Next, we have to take off from the plans the areas of exposed walls, windows, roof, etc., to determine the total heat flow, which is obtained by adding the sum of the products of the area, temperature gradient, and heat-transmission coefficient for each item (see Art. 13.3).

Choosing Heating-Plant Capacity. Total heat load equals the heat loss through conduction, radiation, and infiltration.

If we provide a heating plant with a capacity equal to this calculated heat load, we shall be able to maintain design room temperature when the design outside temperature prevails, if the interior is already at design room temperature. However, in most buildings the temperature is allowed to drop to as low as 55°F during the night. Thus, theoretically, it will require an infinite time to approach design room temperature. It, therefore, is considered good practice to add 20% to the heating-plant capacity for morning pickup.

The final figure obtained is the minimum heating-plant size required. Consult manufacturers' ratings and pick a unit with a capacity no lower than that calculated by the above method.

On the other hand, it is not advisable to choose a unit too large, because then operating efficiency suffers, increasing fuel consumption.

With a plant of 20% greater capacity than required for the calculated heat load, theoretically after the morning pickup, it will run only 100/120, or 83⅓%, of the time. Furthermore, since the design outdoor temperature occurs only during a small percentage of the heating season, during the rest of the heating season the plant would operate intermittently, less than 83⅓% of the time. Thus it is considered good practice to choose a heating unit no smaller than required but not much larger.

If the heating plant will be used to produce hot water for the premises, determine the added capacity required.

13.10 HEATING-LOAD-CALCULATION EXAMPLE

As an example of the method described for sizing a heating plant, let us take the building shown in Fig. 13.3.

A design outdoor temperature of 0°F and an indoor temperature of 70°F are assumed. The wall is to be constructed of 4-in brick with 8-in cinder-block backup. Interior finish is metal lath and plaster (wall heat-transmission coefficient U = 0.25).

The method of determining the heat load is shown in Table 13.12.

Losses from the cellar include 4 Btu/(hr)(ft^2) through the walls [column (4)] and 2 Btu/(hr)(ft^2) through the floors. Multiplied by the corresponding areas, they yield the total heat loss in column (6). In addition, some heat is lost because of

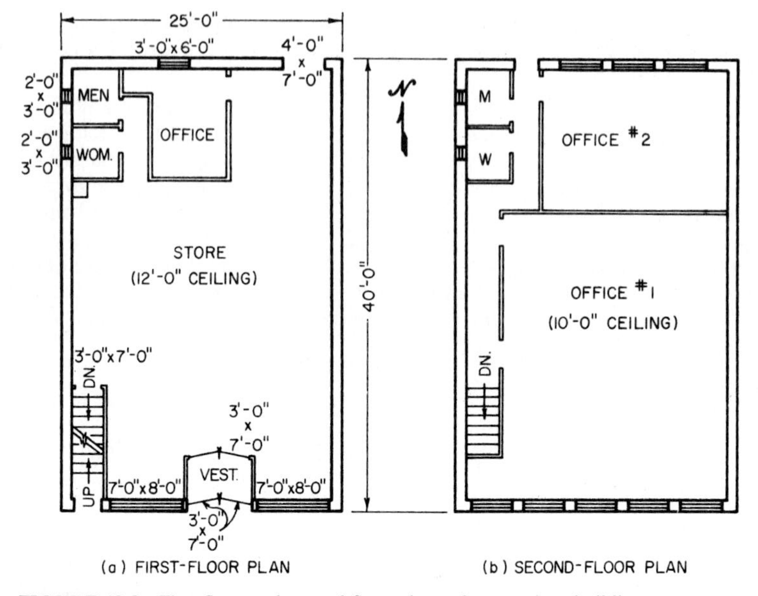

FIGURE 13.3 First-floor and second-floor plans of a two-story building.

infiltration of cold air. One-half an air change per hour is assumed, or 71.2 ft^3/min [column (3)]. This causes a heat loss, according to Eq. (13.28), of

$$1.08 \times 71.2 \times 70 = 5400 \text{ Btu/hr}$$

To the total for the cellar, 20% is added to obtain the heat load in column (7).

Similarly, heat losses are obtained for the various areas on the first and second floors. Heat-transmission coefficients [column (4)] were obtained from the "ASHRAE Handbook—Fundamentals," American Society of Heating, Refrigerating and Air-Conditioning Engineers. These were multiplied by the temperature gradient (70 − 0) to obtain the heat losses in column (6).

The total for the building, plus 20%, amounts to 144,475 Btu/hr. A heating plant with approximately this capacity should be selected.

13.11 WARM-AIR HEATING

A warm-air heating system supplies heat to a room by bringing in a quantity of air above room temperature, the amount of heat added by the air being at least equal to that required to counteract heat losses.

A gravity system (without a blower) is rarely installed because it depends on the difference in density of the warm-air supply and the colder room air for the working pressure. Airflow resistance must be kept at a minimum with large ducts and very few elbows. The result usually is an unsightly duct arrangement.

TABLE 13.12 Heat-Load Determination for Two-Story Building

Space (1)	Heat-loss source (2)	Net area or cfm infiltration (3)	*U* or coefficient (4)	Temp gradient, °F (5)	Heat loss, Btu per hr (6)	Total plus 20% (7)
Cellar	Walls	1,170	4		4,680	
	Floor	1,000	2		2,000	
	½ air change	71.2	1.08	70	5,400	14,500
First-floor store	Walls	851	0.25	70	14,900	
	Glass	135	1.13	70	10,680	
	Doors	128	0.69	70	1,350	
	1 air change	170	1.08	70	12,850	47,600
Vestibule	Glass	48	1.13	70	3,800	
	2 air change	12	1.08	70	905	5,650
Office	Walls	99	0.25	70	1,730	
	Glass	21	1.13	70	1,660	
	½ air change	9	1.08	70	680	4,900
Men's room	Walls	118	0.25	70	2,060	
	Glass	6	1.13	70	48	
	½ air change	3	1.08	70	226	2,800
Ladies' room	Walls	60	0.25	70	1,050	
	Glass	6	1.13	70	48	
	½ air change	3	1.08	70	226	1,590
Second-floor office No. 1	Walls	366	0.25	70	6,400	
	Glass	120	1.13	70	9,500	
	Roof	606	0.19	70	8,050	
	½ air change	50	1.08	70	3,780	33,300
Office No. 2	Walls	207	0.25	70	3,620	
	Glass	72	1.13	70	5,700	
	Roof	234	0.19	70	3,120	
	½ air change	19	1.08	70	1,435	16,650
Men's room	Walls	79	0.25	70	1,380	
	Glass	6	1.13	70	475	
	Roof	20	0.19	70	266	
	½ air change	3	1.08	70	226	2,820
Ladies' room	Walls	44	0.25	70	770	
	Glass	6	1.13	70	475	
	Roof	20	0.19	70	266	
	½ air change	3	1.08	70	226	2,080
Hall	Walls	270	0.25	70	4,720	
	Door	21	0.69	70	1,015	
	Roof	75	0.19	70	1,000	
	2 air change	40	1.08	70	3,020	12,585
						144,475

A forced-warm-air system can maintain higher velocities, thus requires smaller ducts, and provides much more sensitive control. For this type of system,

$$q = 1.08Q(T_h - T_i) \tag{13.32}$$

where T_h = temperature of air leaving grille
T_i = room temperature
q = heat added by air, Btu/hr
Q = ft^3/min of air supplied to room

Equation (13.32) indicates that the higher the temperature of the discharge air T_h, the less air need be handled. In cheaper installations, the discharge air may be as high as 170°F and ducts are small. In better systems, more air is handled with a discharge temperature as low as 135 to 140°F. With a room temperature of 70°F, we shall need (170 − 70)/(135 − 70) = 1.54 times as much air with the 135°F system as with a 170°F system.

It is not advisable to go much below 135°F with discharge air, because drafts will result. With body temperature at 98°F, air at about 100°F will hardly seem warm. If we stand a few feet away from the supply diffuser (register), 70°F room air will be entrained with the warm supply air and the mixture will be less than 98°F when it reaches us. We probably would complain about the draft.

Supply-air diffusers should be arranged so that they blow a curtain of warm air across the cold, or exposed, walls and windows. (See Fig. 13.4 for a suggested arrangement.) These grilles should be placed near the floor, since the lower-density warm air will rise and accumulate at the ceiling.

Return-air grilles should be arranged in the interior near unexposed walls—in foyers, closets, etc.—and preferably at the ceiling. This is done for two reasons:

1. The warm air in all heating systems tends to rise to the ceiling. This creates a large temperature gradient between floor and ceiling, sometimes as high as 10°F. Taking the return air from the ceiling reduces this gradient.
2. Returning the warmer air to the heating plant is more economical in operation than using cold air from the floor.

Duct sizes can be determined as described in Duct Design, Art. 13.6. To illustrate the procedure, duct sizes will be computed for the structure with floor plans shown in Figs. 13.3 and 13.4. The latter shows the locations chosen for the discharge grilles and indicates that the boiler room, which contains the heating unit, is in the basement.

Table 13.12 showed that a heating unit with 144,475 Btu/hr capacity is required. After checking manufacturers' ratings of forced-warm-air heaters, we choose a unit rated at 160,000 Btu/hr and 2010 ft^3/min.

If we utilized the full fan capacity and supply for the rated 160,000 Btu/hr and applied Eq. (13.32), the temperature rise through the heater would be

$$\Delta T = \frac{q}{1.08Q} = \frac{160{,}000}{1.08 \times 2010} = 73.6°\text{F}$$

If we adjust the flame (oil or gas) so that the output was the capacity theoretically required, the temperature rise would be

$$\Delta T = \frac{q}{1.08Q} = \frac{144{,}475}{1.08 \times 2010} = 66.5°\text{F}$$

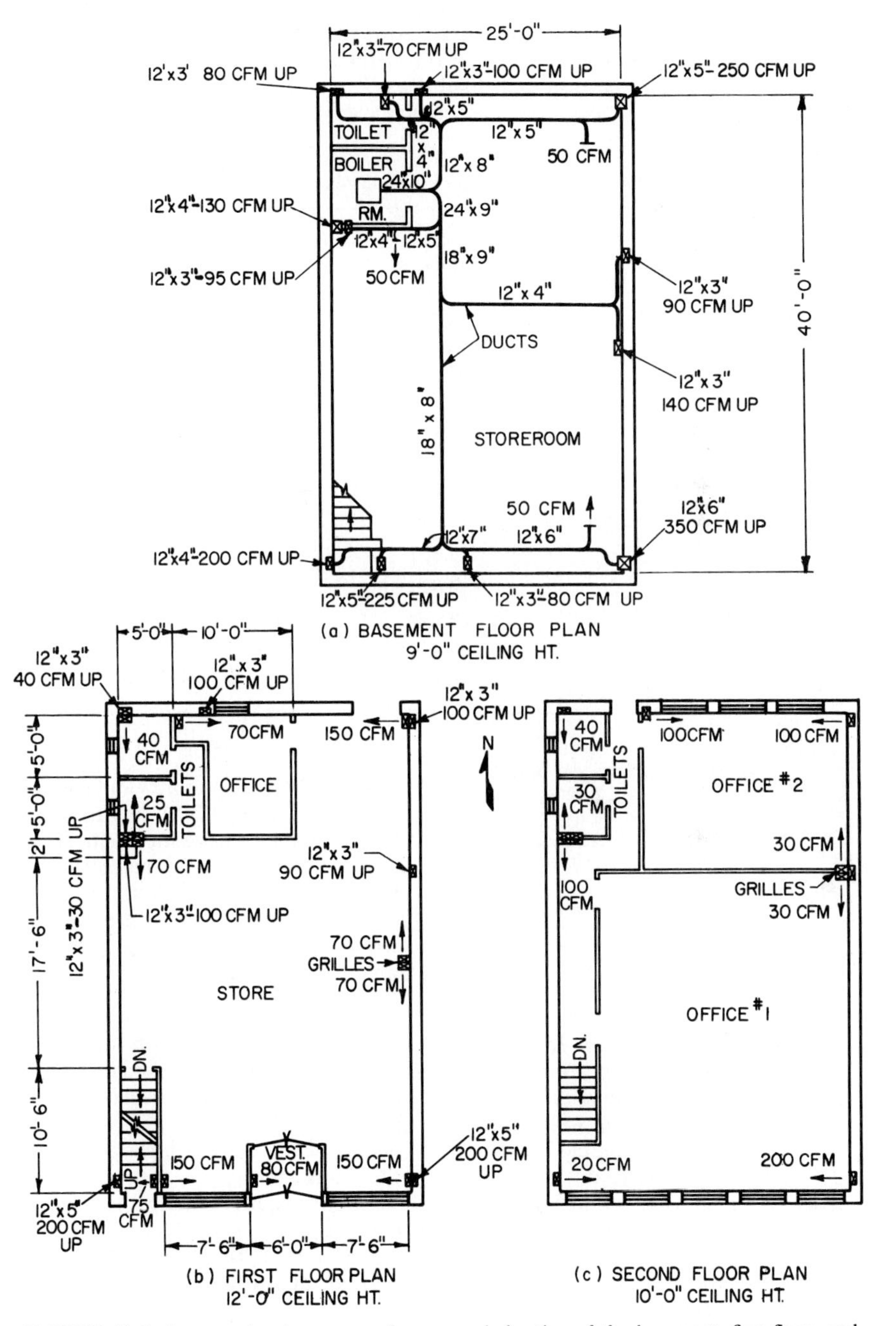

FIGURE 13.4 Layout of a duct system for warm-air heating of the basement, first floor, and second floor of the building shown in Fig. 13.3. The boiler room is in the basement.

In actual practice, we do not tamper with the flame adjustment in order to maintain the manufacturer's design balance. Instead, the amount of air supplied to each room is in proportion to its load. (See Table 13.13, which is a continuation of Table 13.12 in the design of a forced-air heating system.) Duct sizes can then

TABLE 13.13 Air-Supply Distribution in Accordance with Heat Load

Space	Load, Btu per hr	% of load	Cfm (% × 2,010)
Cellar	14,500	10.05	200
First-floor store	47,600	33.00	660
Vestibule	5,650	3.91	80
Office	4,900	3.39	70
Men's room	2,800	1.94	40
Ladies' room	1,590	1.10	25
Second-floor office No. 1	33,300	23.01	460
Office No. 2	16,650	11.51	230
Men's room	2,820	1.95	40
Ladies' room	2,080	1.44	30
Hall	12,585	8.70	175
	144,475	100.00	2,010

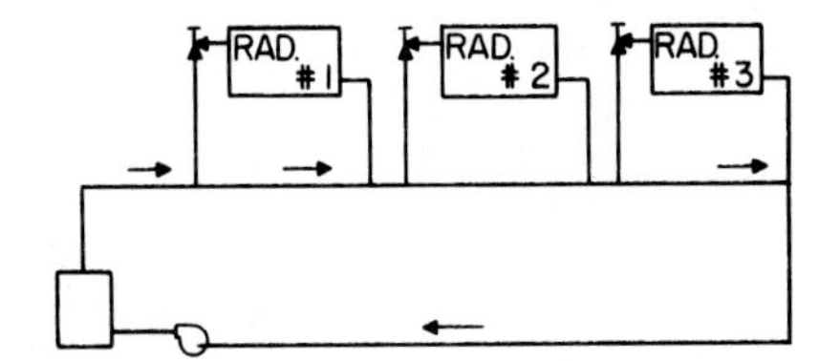

FIGURE 13.5 One-pipe hot-water heating system.

be determined for the flow indicated in the table (see Fig. 13.5). In this example, minimum duct size for practical purposes is 12 × 3 in. Other duct sizes were obtained from Table 13.8, for a friction loss per 100 ft of 0.15, and equivalent rectangular sizes were obtained from Table 13.7 and shown in Fig. 13.5.

Humidification in Warm-Air Heating. A warm-air heating system lends itself readily to humidification. Most warm-air furnace manufacturers provide a humidifier that can be placed in the discharge bonnet of the heater. This usually consists of a pan with a ball float to keep the pan filled.

Theoretically, a building needs more moisture when the outside temperature drops. During these colder periods, the heater runs more often, thus vaporizing more water. During warmer periods less moisture is required and less moisture is added because the heater runs less frequently.

Some manufacturers provide a woven asbestos-cloth frame placed in the pan to materially increase the contact surface between air and water. These humidifiers have no control.

Where some humidity control is desired, a spray nozzle connected to the hot-water system with an electric solenoid valve in the line may be actuated by a humidistat.

With humidification, well-fitted storm windows must be used, for with indoor conditions of 70°F and 30% relative humidity and an outdoor temperature of 0°F, condensation will occur on the windows. Storm windows will cut down the loss of room moisture.

Control of Warm-Air Heating. The sequence of operation of a warm-air heating system is usually as follows:

When the thermostat calls for heat, the heat source is started. When the air chamber in the warm-air heater reaches about 120°F, the fan is started by a sensitive element. This is done so as not to allow cold air to issue from the supply grilles and create drafts.

If the flame size and air quantity are theoretically balanced, the discharge air will climb to the design value of, say, 150°F and remain there during the operation of the heater. However, manual shutoff of grilles by residents, dirty filters, etc., will cause a reduction of airflow and a rise in air temperature above design. A sensitive safety element in the air chamber will shut off the heat source when the discharge temperature reaches a value higher than about 180°F. The heat source will again be turned on when the air temperature drops a given amount.

When the indoor temperature reaches the value for which the thermostat is set, the heat source only is shut off. The fan, controlled by the sensitive element in the air chamber, will be shut off after the air cools to below 120°F. Thus, most of the usable heat is transmitted into the living quarters instead of escaping up the chimney.

With commercial installations, the fan usually should be operated constantly, to maintain proper air circulation in windowless areas during periods when the thermostat is satisfied. If residential duct systems have been poorly designed, often some spaces may be too cool, while others may be too warm. Constant fan operation in such cases will tend to equalize temperatures when the thermostat is satisfied.

Duct systems should be sized for the design air quantity of the heater. Insufficient air will cause the heat source to cycle on and off too often. Too much air may cool the flue gases so low as to cause condensation of the water in the products of combustion. This may lead to corrosion, because of dissolved flue gases.

Warm-Air Perimeter Heating. This type of heating is often used in basementless structures, where the concrete slab is laid directly on the ground. The general arrangement is as follows: The heater discharges warm air to two or more underfloor radial ducts feeding a perimeter duct. Floor grilles or baseboard grilles are located as in a conventional warm-air heating system, with collars connected to the perimeter duct.

To prevent excessive heat loss to the outside, it is advisable to provide a rigid waterproof insulation between the perimeter duct and the outside wall.

Air Supply and Exhaust for Heaters. Special types of packaged, or preassembled, units are available for heating that include a direct oil- or gas-fired heat exchanger complete with operating controls. These units must have a flue to convey the products of combustion to the outdoors. The flue pipe must be incombustible and capable of withstanding high temperatures without losing strength from corrosion. Corrosion is usually caused by sulfuric and sulfurous acids, which are formed during combustion and caused by the presence of sulfur in the fuel.

Gas-fired heaters and boilers are usually provided with a draft hood approved by building officials. This should be installed in accordance with the manufacturer's recommendations. Oil-fired heaters and boilers should be provided with an approved draft stabilizer in the vent pipe. The hoods and stabilizers are used to prevent snuffing out of the flame in extreme cases and pulling of excessive air through the combustion chamber when the chimney draft is above normal, as in extremely cold weather.

Flues for the products of combustion are usually connected to a masonry type of chimney. A chimney may have more than one vertical flue. Where flue-gas temperatures do not exceed 600°F, the chimney should extend vertically 3 ft above the high point of the roof or roof ridge when within 10 ft of the chimney. When chimneys will be used for higher-temperature flue gases, many codes require that the chimney terminate not less than 10 ft higher than any portion of the building within 25 ft.

Many codes call for masonry construction of chimneys for both low- and high-temperature flue gases for low- and high-heat appliances. These codes also often call for fire-clay flue linings that will resist corrosion, softening, or cracking from flue gases at temperatures up to 1800°F.

Flue-pipe construction must be of heat-resistant materials. The cross-sectional area should be not less than that of the outlet on the heating unit. The flue or vent pipe should be as short as possible and have a slope upward of not less than ¼ in/ft. If the flue pipe extends a long distance to the chimney, it should be insulated to prevent heat loss and the formation of corrosive acids by condensation of the combustion products.

All combustion-type heating units require air for combustion, and it must be provided in adequate amounts. Combustion air is usually furnished directly from the outside. This air may be forced through ductwork by a fan or by gravity through an outdoor-air louver or special fresh-air intakes. If outside air is not provided for the heating unit, unsatisfactory results can be expected. The opening should have at least twice the cross-sectional area of the vent pipe leaving the boiler.

(H. E. Bovay, Jr., "Handbook of Mechanical and Electrical Systems for Buildings," and D. L. Grumman, "Air-Handling Systems Ready Reference Manual," McGraw-Hill Publishing Company, New York.)

13.12 HOT-WATER HEATING SYSTEMS

A hot-water heating system consists of a heater or furnace, radiators, piping systems, and circulator.

A gravity system without circulating pumps is rarely installed. It depends on a difference in density of the hot supply water and the colder return water for working head. Airflow resistance must be kept to a minimum, and the circulating piping system must be of large size. A forced circulation system can maintain higher water velocities, thus requires much smaller pipes and provides much more sensitive control.

Three types of piping systems are in general use for forced hot-water circulation systems:

One-pipe system (Fig. 13.5). This type has many disadvantages and is not usually recommended. It may be seen in Fig. 13.5 that radiator No. 1 takes hot water from the supply main and dumps the colder water back in the supply main. This causes the supply-water temperature for radiator No. 2 to be lower, requiring a corresponding increase in radiator size. (Special flow and return fittings are available to induce flow through the radiators.) The design of such a system is very difficult, and any future adjustment or balancing of the system throws the remainder of the temperatures out.

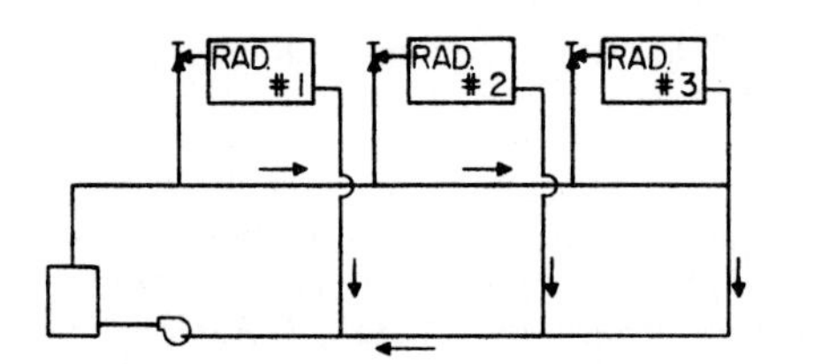

FIGURE 13.6 Two-pipe direct-return hot-water heating system.

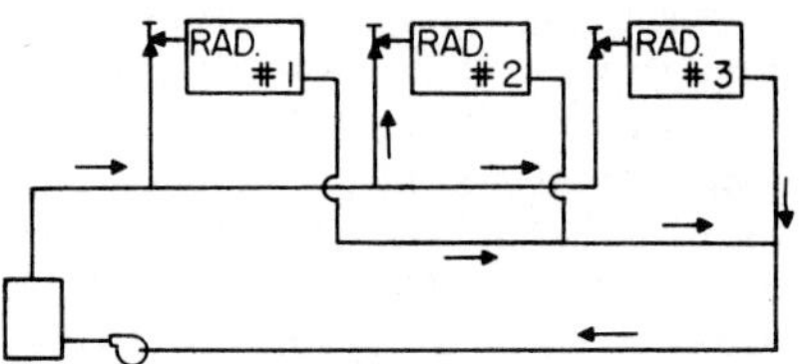

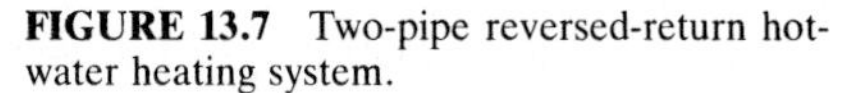

FIGURE 13.7 Two-pipe reversed-return hot-water heating system.

Two-pipe direct-return system (Fig. 13.6). Here all radiators get the same supply-water temperature, but the last radiator has more pipe resistance than the first. This can be balanced out by sizing the pump for the longest run and installing orifices in the other radiators to add an equivalent resistance for balancing.

Two-pipe reversed-return system (Fig. 13.7). The total pipe resistance is about the same for all radiators. Radiator No. 1 has the shortest supply pipe and the longest return pipe, while radiator No. 3 has the longest supply pipe and the shortest return pipe.

Supply design temperatures usually are 180°F, with a 20°F drop assumed through the radiators; thus the temperature of the return riser would be 160°F.

When a hot-water heating system is designed, it is best to locate the radiators, then calculate the water flow in gallons per minute required by each radiator. For a 20°F rise

$$q = 10{,}000Q \tag{13.33}$$

where q = amount of heat required, Btu/hr
Q = flow of water, gal/min

A one-line diagram showing the pipe runs should next be drawn, with gallons per minute to be carried by each pipe noted. The piping may be sized, using friction-flow charts and tables showing equivalent pipe lengths for fittings, with water velocity limited to a maximum of 4 ft/s. (See "ASHRAE Handbook—Fundamentals.") Too high a water velocity will cause noisy flow; too low a velocity will create a sluggish system and costlier piping.

The friction should be between 250 and 600 milinches/ft (1 milinch = 0.001 in). It should be checked against available pump head, or a pump should be picked for the design gallons per minute at the required head.

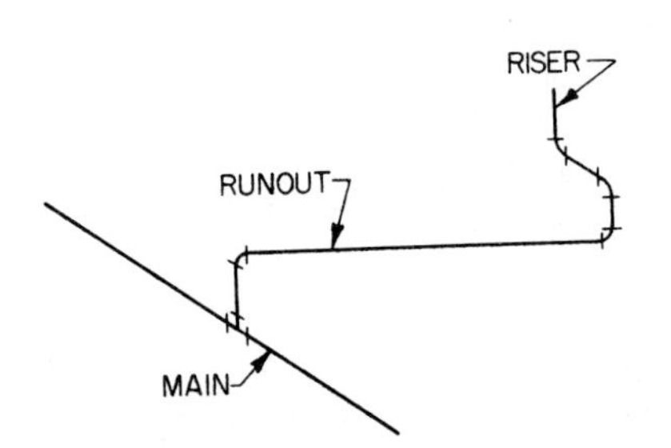

FIGURE 13.8 Swing joint permits expansion and contraction of pipes.

It is very important that piping systems be made flexible enough to allow for expansion and contraction. Expansion joints are very satisfactory but expensive. Swing joints as shown in Fig. 13.8 should be used where necessary. In this type of branch takeoff, the runout pipe, on expansion or contraction, will cause the threads in the elbows to screw in or out slightly, instead of creating strains in the piping.

(W. J. McGuinness et al., "Mechanical and Electrical Equipment for Buildings," John Wiley & Sons, Inc., New York.)

Hot-Water Radiators. Radiators, whether of the old cast-iron type, finned pipe, or other, should be picked for the size required, in accordance with the manufacturer's ratings. These ratings depend on the average water temperature. For 170°F average water temperature, 1 ft^2 of radiation surface is equal to 150 Btu/hr.

Expansion Tanks for Hot-Water Systems. All hot-water heating systems must be provided with an expansion tank of either the open or closed type.

Figure 13.9 shows an open-type expansion tank. This tank should be located at least 3 ft above the highest radiator and in a location where it cannot freeze.

The size of tank depends on the amount of expansion of the water. From a low near 32°F to a high near boiling, water expands 4% of its volume. Therefore, an expansion tank should be sized for 6% of the total volume of water in radiators, heater, and all piping. That is, the volume of the tank, up to the level of its overflow pipe, should not be less than 6% of the total volume of water in the system.

Figure 13.10 is a diagram of the hookup of a closed-type expansion tank. This tank is only partly filled with water, creating an air cushion to allow for expansion and contraction. The pressure-reducing valve and relief valve are often supplied as a single combination unit.

The downstream side of the reducing valve is set at a pressure below city watermain pressure but slightly higher than required to maintain a static head in the highest radiator. The minimum pressure setting in pounds per square inch is equal to the *height in feet* divided by 2.31.

The relief valve is set above maximum possible pressure. Thus, the system will automatically fill and relieve as required.

Precautions in Hot-Water Piping Layout. One of the most important precautions in a hot-water heating system is to avoid air pockets or loops. The pipe should be pitched so that vented air will collect at points that can be readily vented either automatically or manually. Vents should be located at all radiators.

Pipe traps should contain drains for complete drainage in case of shutdown.

Zones should be valved so that the complete system need not be shut down for repair of a zone. Multiple circulators may be used to supply the various zones at different times, for different temperature settings and different exposures.

Allow for expansion and contraction of pipe without causing undue stresses.

All supply and return piping should be insulated.

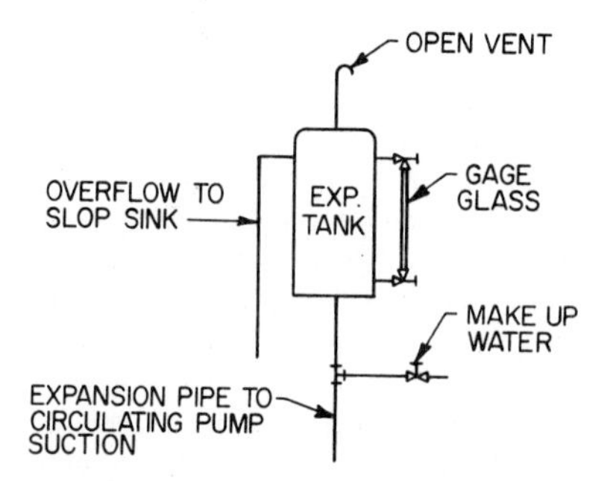

FIGURE 13.9 Open-type expansion tank.

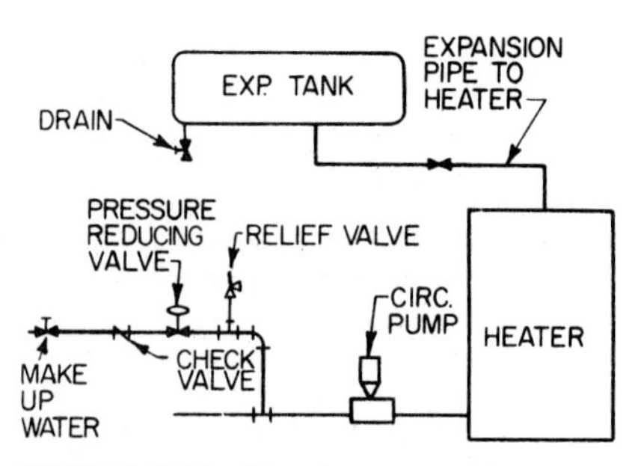

FIGURE 13.10 Closed-type expansion tank.

In very high buildings, the static pressure on the boiler may be too great. To prevent this, heat exchangers may be installed as indicated in Fig. 13.11. The boiler temperature and lowest zone would be designed for 200°F supply water and 180°F return. The second lowest zone and heat exchanger can be designed for 170°F supply water and 150°F return, etc.

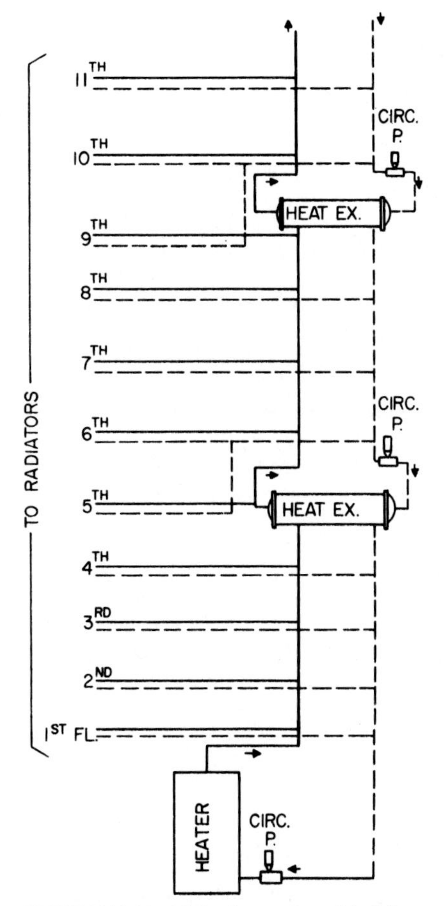

FIGURE 13.11 Piping layout for tall building, including heat exchangers.

Control of Hot-Water Systems. The control system is usually arranged as follows: An immersion thermostat in the heater controls the heat source, such as an oil burner or gas solenoid valve. The thermostat is set to maintain design heater water temperature (usually about 180°F). When the room thermostat calls for heat, it starts the circulator. Thus, an immediate supply of hot water is available for the radiators. A low-limit immersion stat, usually placed in the boiler and wired in series with the room stat and the pump, is arranged to shut off the circulator in the event that the water temperature drops below 70°F. This is an economy measure; if there is a flame failure, water will not be circulated unless it is warm enough to do some good. If the boiler is used to supply domestic hot water via an instantaneous coil or storage tank, hot water will always be available for that purpose. It should be kept in mind that the boiler must be sized for the heating load plus the probable domestic hot-water demand.

High-Temperature, High-Pressure Hot-Water Systems. Some commercial and industrial building complexes have installed hot-water heating systems in which the water temperature is maintained well above 212°F. This is made possible by subjecting the system to a pressure well above the saturation pressure of the water at the design temperature. Such high-temperature hot-water systems present some inherent hazards. Most important is the danger of a leak, because then the water will flash into steam. Another serious condition occurs when a pump's suction strainer becomes partly clogged, creating a pressure drop. This may cause steam to flash in the circulating pump casing and vapor bind the pump.

These systems are not generally used for heating with radiators. They are mostly used in conjunction with air-conditioning installations in which the air-handling units contain a heating coil for winter heating. The advantage of high-temperature hot-water systems is higher rate of heat transfer from the heating medium to the air. This permits smaller circulating piping, pumps, and other equipment.

(H. E. Bovay, Jr., "Handbook of Mechanical and Electrical Systems for Buildings," McGraw-Hill Publishing Company, New York; B. Stein et al., "Mechanical and Electrical Equipment for Buildings," 7th ed., John Wiley & Sons, Inc., New York.)

13.13 STEAM-HEATING SYSTEMS

A steam-heating system consists of a boiler or steam generator and a piping system connecting to individual radiators or convectors.

A **one-pipe heating system** (Fig. 13.12) is the simplest arrangement. The steam-supply pipe to the radiators is also used as a condensate return to the boiler. On startup, as the steam is generated, the air must be pushed out of the pipe and radiators by the steam. This is done with the aid of thermostatic air valves in the radiators. When the system is cold, a small vent hole in the valve is open. After the air is pushed out and steam comes in contact with the thermostatic element, the vent hole automatically closes to prevent escape of steam.

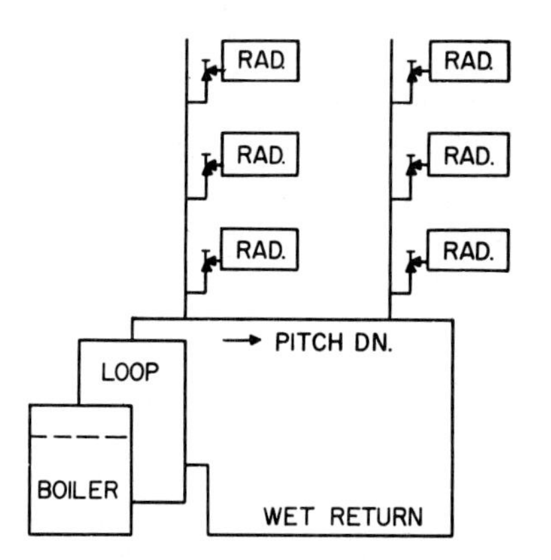

FIGURE 13.12 One-pipe steam-heating system. Condensate returns through supply pipe.

Where pipe runouts are extensive, it is necessary to install large-orifice air vents to eliminate the air quickly from the piping system; otherwise the radiators near the end of the runout may get steam much later than the radiators closest to the boiler.

Air vents are obtainable with adjustable orifice size for balancing a heating system. The orifices of radiators near the boiler are adjusted smaller, while radiators far from the boiler will have orifices adjusted for quick venting. This helps balance the system.

The pipe must be generously sized so as to prevent gravity flow of condensate from interfering with supply steam flow. Pipe capacities for supply risers, runouts, and radiator connections are given in the ASHRAE handbook (American Society of Heating, Refrigerating and Air-Conditioning Engineers). Capacities are expressed in square feet of **equivalent direct radiation (EDR):**

$$1 \text{ ft}^2 \text{ EDR} = 240 \text{ Btu/hr} \tag{13.34}$$

where capacities are in pounds per hour, 1 lb/hr = 970 Btu/hr.

Valves on radiators in a one-pipe system must be fully open or closed. If a valve is throttled, the condensate in the radiator will have to slug against a head of steam in the pipe to find its way back to the boiler by gravity. This will cause water hammer.

A **two-pipe system** is shown in Fig. 13.13. The steam supply is used to deliver steam to the supply end of all radiators. The condensate end of each radiator is connected to the return line via a thermostatic drip trap (Fig. 13.14). The trap is

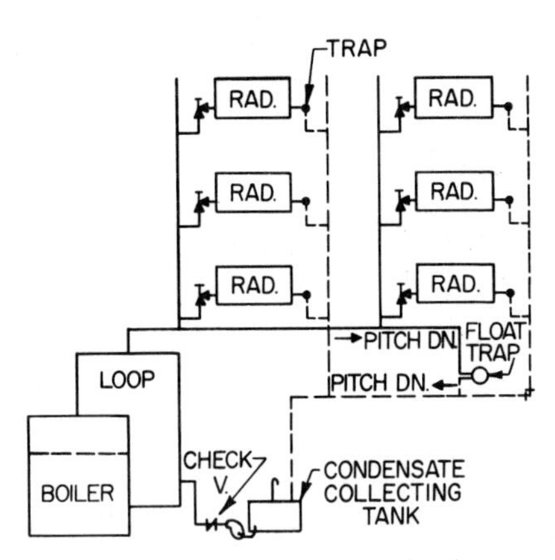

FIGURE 13.13 Two-pipe steam-heating system. Different pipes are used for supply and return processes.

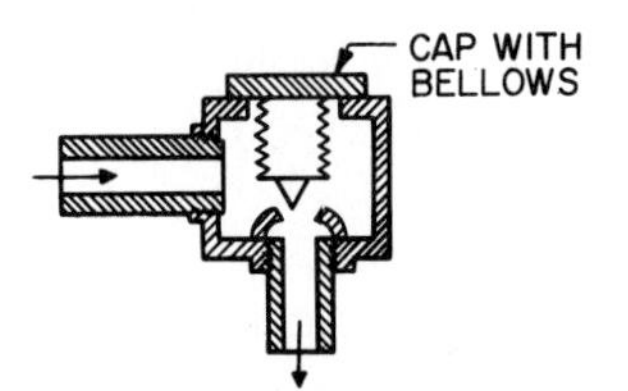

FIGURE 13.14 Drip trap for condensate return of a steam radiator.

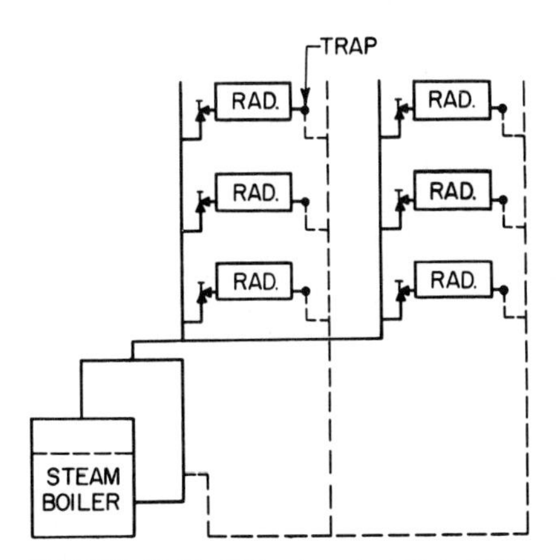

FIGURE 13.15 Wet-return two-pipe steam-heating system.

adjusted to open below 180°F and close above 180°F. Thus when there is enough condensate in the radiator to cool the element, it will open and allow the condensate to return to the collecting tank. A float switch in the tank starts the pump to return this condensate to the boiler against boiler steam pressure.

There are many variations in combining the two-pipe system and one-pipe system to create satisfactory systems.

In the **two-pipe condensate pump return system,** the pressure drop available for pipe and radiator loss is equal to boiler pressure minus atmospheric pressure, the difference being the steam gage pressure.

A **wet-return system,** shown in Fig. 13.15, will usually have a smaller head available for pipe loss. It is a self-adjusting system depending on the load. When steam is condensing at a given rate, the condensate will pile up in the return main above boiler level, creating a hydraulic head that forces the condensate into the boiler. The pressure above the water level on the return main will be less than the steam boiler pressure. This pressure difference—boiler pressure minus the pressure that exists above the return main water level—is available for pipe friction and radiator pressure drop.

On morning start-up when the air around the radiators is colder and the boiler is fired harder than during normal operation in an effort to pick up heat faster, the steam side of the radiators will be higher in temperature than normal. The air-side temperature of the radiators will be lower. This increase in temperature differential will create an increase in heat transfer causing a faster rate of condensation and a greater piling up of condensate to return the water to the boiler.

In laying out a system, the steam-supply runouts must be pitched to remove the condensate from the pipe. They may be pitched back so as to cause the condensate to flow against the steam or pitched front to cause the condensate to flow with the steam.

Since the condensate will pile up in the return pipe to a height above boiler-water level to create the required hydraulic head for condensate flow, a check of boiler-room ceiling height, steam-supply header, height of lowest radiator, height of dry return pipe, etc., is necessary to determine the height the water may rise in the return pipe without flooding these components.

The pipe may have to be oversized where condensate flows against the steam (see "ASHRAE Handbook—Fundamentals"). If the pitch is not steep enough, the steam may carry the condensate along in the wrong direction, causing noise and water hammer.

A **vacuum heating system** is similar to a steam pressure system with a condensate return pump. The main difference is that the steam pressure system can eliminate air from the piping and radiators by opening thermostatic vents to the atmosphere. The vacuum system is usually operated at a boiler pressure below atmospheric. The vacuum pump must, therefore, pull the noncondensables from the piping and radiators for discharge to atmosphere. Figure 13.16 shows diagrammatically the operation of a vacuum pump.

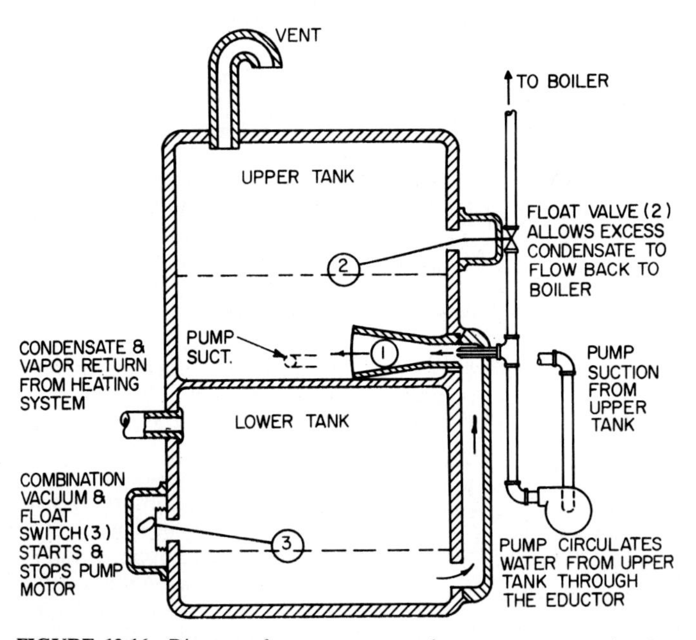

FIGURE 13.16 Diagram of a vacuum pump for a vacuum steam-heating system. The eductor (1) maintains the desired vacuum in the lower tank.

This unit collects the condensate in a tank. A pump circulates water through an eductor (1), pulling out the noncondensable gas to create the required vacuum. The discharge side of the eductor nozzle is above atmospheric pressure. Thus, an automatic control system allows the noncondensable gas to escape to atmosphere as it accumulates, and the excess condensate is returned to the boiler as the tank reaches a given level.

Vacuum systems are usually sized for a total pressure drop varying from ¼ to ½ psi. Long equivalent-run systems (about 200 ft) will use ½ psi total pressure drop to save pipe size.

Some systems operate without a vacuum pump by eliminating the air during morning pickup by hard firing and while the piping system is above atmospheric pressure. During the remainder of the day, when the rate of firing is reduced, a tight system will operate under vacuum.

(H. E. Bovay, Jr., "Handbook of Mechanical and Electrical Systems for Buildings," McGraw-Hill Publishing Company, New York.)

13.14 UNIT HEATERS

Large open areas, such as garages, showrooms, stores, and workshops, are usually best heated by unit heaters placed at the ceiling.

Figure 13.17 shows the usual connections to a steam unit-heater installation. The thermostat is arranged to start the fan when heat is required. The surface thermostat strapped on the return pipe prevents the running of the fan when insufficient steam is available. Where hot water is used for heating, the same arrangement is used, except that the float and the thermostatic trap are eliminated and an air-vent valve is installed. Check manufacturers' ratings for capacities in choosing equipment.

Where steam or hot water is not available, direct gas-fired unit heaters may be installed. However, an outside flue is required to dispose of the products of combustion properly (Fig. 13.18). Check with local ordinances for required flues from direct gas-fired equipment. (Draft diverters are usually included with all gas-fired heaters. Check with the manufacturer when a draft diverter is not included.) For automatic control, the thermostat is arranged to start the fan and open the gas solenoid valve when heat is required. The usual safety pilot light is included by the manufacturer.

Gas-fired unit heaters are often installed in kitchens and other premises where large quantities of air may be exhausted. When a makeup air system is not provided,

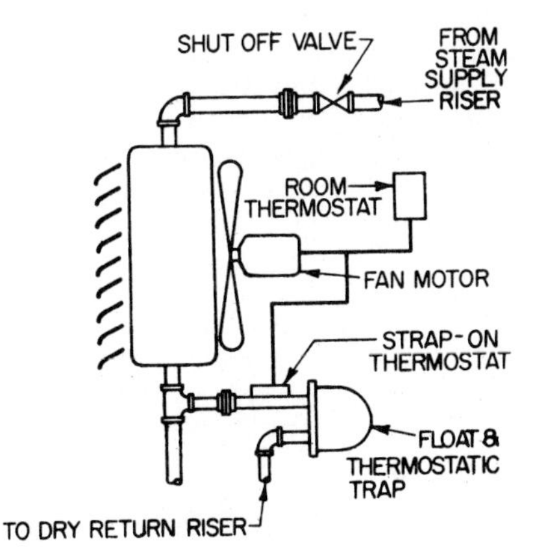

FIGURE 13.17 Steam unit heater. Hot-water heater has air vent, no float and trap.

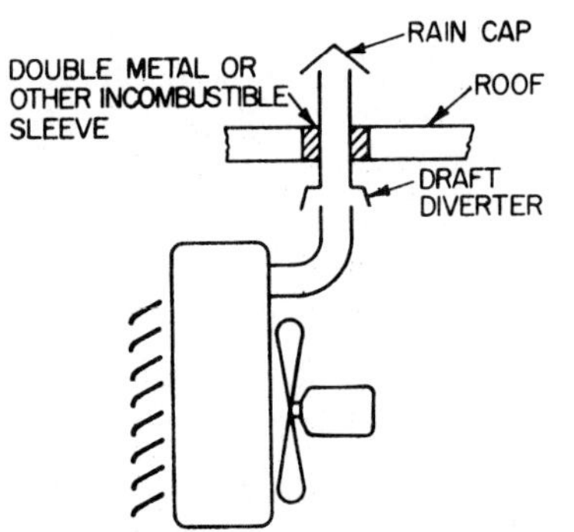

FIGURE 13.18 Connections to outside for gas-fired unit heater.

and the relief air must infiltrate through doors, windows, etc., a negative pressure must result in the premises. This negative pressure will cause a steady downdraft through the flue pipe from the outdoors into the space and prevent proper removal of the products of combustion. In such installations, it is advisable to place the unit heater in an adjoining space from which air is not exhausted in large quantities and deliver the warm air through ducts. Since propeller fans on most unit heaters cannot take much external duct resistance, centrifugal blower unit heaters may give better performance where ductwork is used. Sizes of gas piping and burning rates for gas can be obtained from charts and tables in the "ASHRAE Handbook—Fundamentals" for various capacities and efficiencies.

The efficiency of most gas-fired heating equipment is between 70 and 80%.

13.15 RADIANT HEATING

Radiant heating, or panel heating as it is sometimes referred to, consists of a warm pipe coil embedded in the floor, ceiling, or walls. The most common arrangement is to circulate warm water through the pipe under the floor. Some installations with warm-air ducts, steam pipes, or electric-heating elements have been installed.

Warm-air ducts for radiant heating are not very common. A modified system normally called the perimeter warm-air heating system circulates the warm air around the perimeter of the structure before discharging the air into the premises via grilles.

Pipe coils embedded in concrete floor slabs or plaster ceilings and walls should not be threaded. Ferrous pipe should be welded, while joints in nonferrous metal pipe should be soldered. Return bends should be made with a pipe bender instead of fittings to avoid joints. All piping should be subjected to a hydrostatic test of at least 3 times the working pressure, with a minimum of 150 psig. Inasmuch as repairs are costly after construction is completed, it is advisable to adhere to the above recommendations.

Construction details for ceiling-embedded coils are shown in Figs. 13.19 to 13.22. Floor-embedded coil construction is shown in Fig. 13.23. Wall-panel coils may be installed as in ceiling panels.

Electrically heated panels are usually prefabricated and should be installed in accordance with manufacturer's recommendations and local electrical codes.

The piping and circuiting of a hot-water radiant heating system are similar to hot-water heating systems with radiators or convectors, except that cooler water is used. However, a 20°F water-temperature drop is usually assumed. Therefore, charts used for the design of hot-water piping systems may be used for radiant heating, too. (See "ASHRAE Handbook—Fundamentals.") In radiant heating,

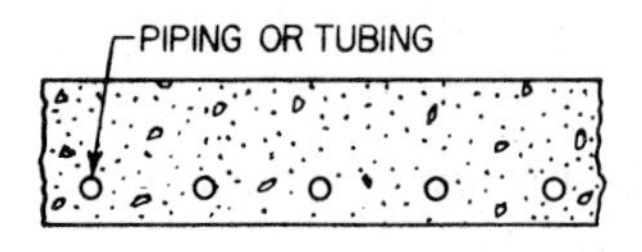

FIGURE 13.19 Pipe embedded in a concrete slab for radiant heating.

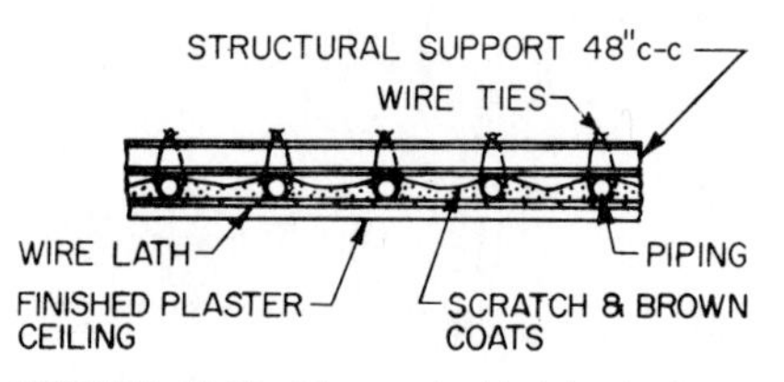

FIGURE 13.20 Pipe embedded in a plaster ceiling for radiant heating.

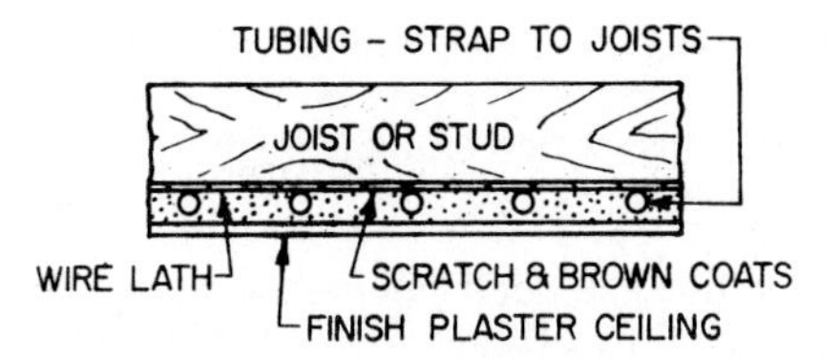

FIGURE 13.21 Pipe coil attached to joists or studs and embedded in plaster for radiant heating.

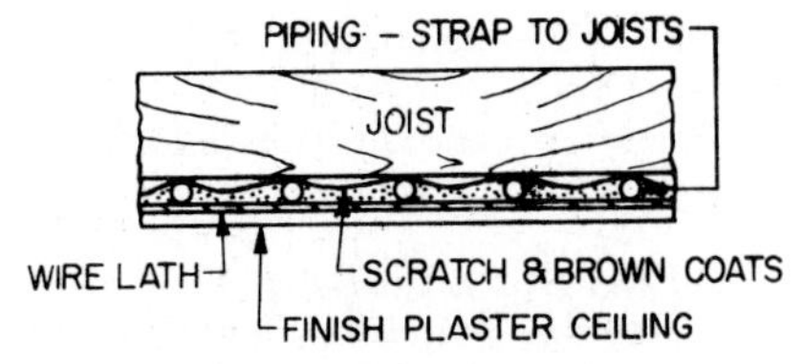

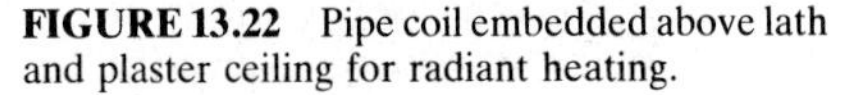

FIGURE 13.22 Pipe coil embedded above lath and plaster ceiling for radiant heating.

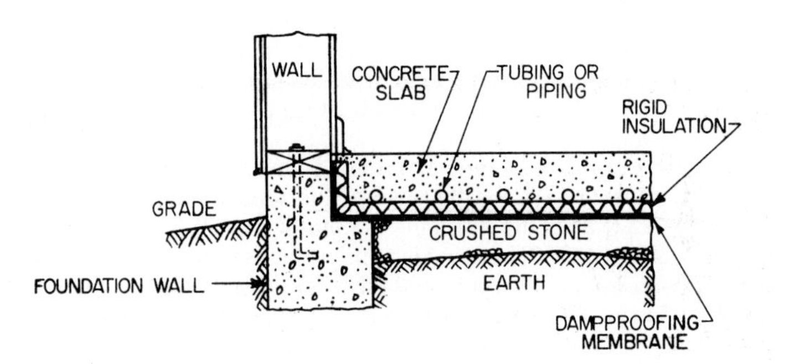

FIGURE 13.23 Pipe coil for radiant heating embedded in a floor slab on grade.

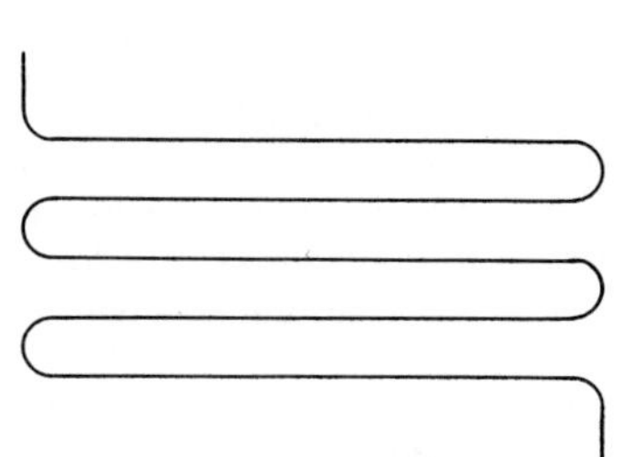

FIGURE 13.24 Continuous pipe coil for radiant heating.

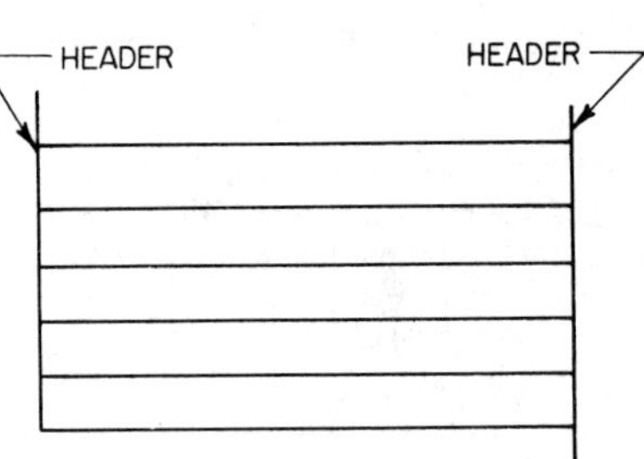

FIGURE 13.25 Piping arranged in a grid for radiant heating.

FIGURE 13.26 Combination of grid arrangement and continuous pipe coil.

the radiant coils replace radiators. Balancing valves should be installed in each coil, as in radiators. One wise precaution is to arrange coils in large and extensive areas so as not to have too much resistance in certain circuits. This can be done by using high-resistance continuous coils (Fig. 13.24) or low-resistance grid (Fig. 13.25) or a combination of both (Fig. 13.26).

One of the advantages of this system is its flexibility; coils can be concentrated along or on exposed walls. Also, the warmer supply water may be routed to the perimeter or exposed walls and the cooler return water brought to the interior zones.

Panel and Room Temperatures. Heat from the embedded pipes is transmitted to the panel, which in turn supplies heat to the room by two methods: (1) convection and (2) radiation. The amount of heat supplied by convection depends on the temperature difference between the panel and the air. The amount of heat supplied by radiation depends on the difference between the fourth powers of the absolute temperatures of panel and occupants. Thus, as panel temperature is increased, persons in the room receive a greater percentage of heat by radiation than by convection. Inasmuch as high panel temperatures are uncomfortable, it is advisable to keep floor panel temperatures about 85°F or lower and ceiling panel temperatures 100°F or lower. The percentage of radiant heat supplied by a panel at 85°F is about 56% and by one at 100°F about 70%.

Most advocates of panel heating claim that a lower than the usual design inside temperature may be maintained because of the large radiant surface comforting the individual; i.e., a dwelling normally maintained at 70°F may be kept at an air temperature of about 65°F. The low air temperature makes possible a reduction in heat losses through walls, glass, etc., and thus cuts down the heating load. However, during periods when the heating controller is satisfied and the water circulation stops, the radiant-heat source diminishes, creating an uncomfortable condition due to the below-normal room air temperature. It is thus considered good practice to design the system for standard room temperatures (Table 13.11) and the heating plant for the total capacity required.

Design of Panel Heating. Panel output, Btu per hour per square foot, should be estimated to determine panel-heating area required. Panel capacity is determined by pipe spacing, water temperature, area of exposed walls and windows, infiltration air, insulation value of structural and architectural material between coil and occupied space, and insulation value of structural material preventing heat loss from the reverse side and edge of the panel. It is best to leave design of panel heating to a specialist.

("ASHRAE Handbook—Fundamentals," American Society of Heating, Refrigerating, and Air-Conditioning Engineers, 1791 Tully Circle, N.E., Atlanta, GA 30329.)

13.16 SNOW MELTING

Design of a snow-melting system for sidewalks, roads, parking areas, etc., involves the determination of a design amount of snowfall, sizing and layout of piping, and selection of heat exchanger and circulating medium. The pipe is placed under the wearing surface, with enough cover to protect it against damage from traffic loads, and a heated fluid is circulated through it.

If friction is too high in extensive runs, use parallel loops (Art. 13.15). All precautions for drainage, fabrication, etc., hold for snow-melting panels as well as interior heating panels.

Table 13.14 gives a design rate of snowfall in inches of water equivalent per hour per square foot for various cities.

Table 13.15 gives the required slab output in Btu per hour per square foot at a given circulating-fluid temperature. This temperature may be obtained once we

TABLE 13.14 Water Equivalent of Snowfall

City	In. per hr per sq ft	City	In. per hr per sq ft
Albany, N.Y.	0.16	Evansville, Ind.	0.08
Asheville, N.C.	0.08	Hartford, Conn.	0.25
Billings, Mont.	0.08	Kansas City, Mo.	0.16
Bismarck, N. Dak.	0.08	Madison, Wis.	0.08
Boise, Idaho	0.08	Minneapolis, Minn.	0.08
Boston, Mass.	0.16	New York, N.Y.	0.16
Buffalo, N.Y.	0.16	Oklahoma City, Okla.	0.16
Burlington, Vt.	0.08	Omaha, Nebr.	0.16
Caribou, Maine	0.16	Philadelphia, Pa.	0.16
Chicago, Ill.	0.06	Pittsburgh, Pa.	0.08
Cincinnati, Ohio	0.08	Portland, Maine	0.16
Cleveland, Ohio	0.08	St. Louis, Mo.	0.08
Columbus, Ohio	0.08	Salt Lake City, Utah	0.08
Denver, Colo.	0.08	Spokane, Wash.	0.16
Detroit, Mich.	0.08	Washington, D.C.	0.16

TABLE 13.15 Heat Output and Circulating-Fluid Temperatures for Snow-Melting Systems

Rate of snowfall, in/(hr)(ft²) of water equivalent		Air temp, 0°F			Air temp, 10°F			Air temp, 20°F			Air temp, 30°F		
		Wind velocity, mi/hr											
		5	10	15	5	10	15	5	10	15	5	10	15
0.08	Slab output, Btu/(hr)(ft²)	151	205	260	127	168	209	102	128	154	75	84	94
	Fluid temp, °F	108	135	162	97	117	138	85	97	110	70	75	79
0.16	Slab output, Btu/(hr)(ft²)	218	273	327	193	233	274	165	191	217	135	144	154
	Fluid temp, °F	142	169	198	120	149	170	117	129	142	100	105	109
0.25	Slab output, Btu/(hr)(ft²)	292	347	401	265	305	346	235	261	287	203	212	221
	Fluid temp, °F	179	206	234	165	186	206	151	163	176	134	139	144

NOTE: This table is based on a relative humidity of 80% for all air temperatures and construction as shown in Fig. 13.27.

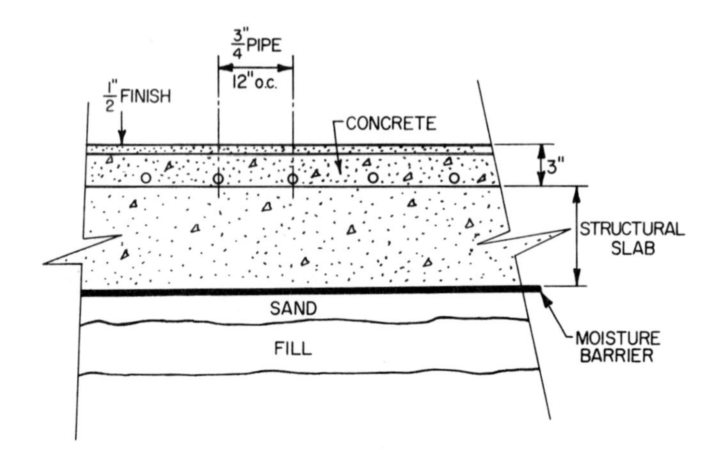

FIGURE 13.27 Pipe coil embedded in a concrete slab for outdoor snow melting.

determine the rate of snowfall and assume a design outside air temperature and wind velocity. The table assumes a snow-melting panel as shown in Fig. 13.27.

Once the Btu per hour per square foot required is obtained from Table 13.15 and we know the area over which snow is to be melted, the total Btu per hour needed for snow melting can be computed as the product of the two. It is usual practice to add 40% for loss from bottom of slab.

The circulating-fluid temperature given in Table 13.15 is an average. For a 20°F rise, the fluid temperature entering the panel will be 10°F above that found in the table, and the leaving fluid temperature will be 10°F below the average. The freezing point of the fluid should be a few degrees below the minimum temperature ever obtained in the locality.

Check manufacturers' ratings for antifreeze solution properties to obtain the gallons per minute required and the friction loss to find the pumping head.

When ordering a heat exchanger for a given job, specify to the manufacturer the steam pressure available, fluid temperature to and from the heat exchanger, gallons per minute circulated, and physical properties of the antifreeze solution.

13.17 RADIATORS AND CONVECTORS

In hot-water and steam-heating systems, heat is released to the spaces to be warmed by radiation and convection. The percentage transmitted by either method depends on the type of heat-dispersal unit used.

A common type of unit is the tubular radiator. It is composed of a series of interconnected sections, each of which consists of vertical tubes looped together. Steam-radiator sections are attached by nipples only at the base, whereas hot-water sections are connected at both top and bottom. Steam radiators should not be used for hot-water heating because of the difficulty of venting.

Pipe-coil radiators are sometimes used in industrial plants. The coils are usually placed on a wall under and between windows and are connected at the ends by branch trees or manifolds. Sometimes, finned-pipe coils are used instead of ordinary pipe. The fins increase the area of heat-transmitting surface.

Since radiators emit heat by convection as well as by radiation, any enclosure should permit air to enter at the bottom and leave at the top.

Convectors, as the name implies, transmit heat mostly by convection. They usually consist of finned heating elements placed close to the floor in an enclosure that has openings at bottom and top for air circulation.

Baseboard units consist of continuous heating pipe in a thin enclosure along the base of exposed walls. They may transmit heat mostly by radiation or by convection. Convector-type units generally have finned-pipe heating elements. Chief advantage is the small temperature difference between floor and ceiling.

13.18 HEAT PUMPS

A heat-pump cycle is a sequence of operations in which the heat of condensation of a refrigerant is used for heating. The heat required to vaporize the refrigerant is taken from ambient air at the stage where the normal refrigeration cycle (Art. 13.22) usually rejects the heat. In the summer cycle, for cooling, the liquid refrigerant is arranged to flow to the cooling coil through an expansion valve, and the hot gas from the compressor is condensed as in the standard refrigeration cycle (Fig. 13.29). During the heating season, the refrigerant gas is directed to the indoor (heating) coil by the use of multiport electric valves. The condensed liquid refrigerant is then directed to the "condenser" via an expansion valve and is evaporated. This method of heating is competitive with fuel-burning systems in warmer climates where the cooling plant can provide enough heat capacity during the winter season and where electric rates are low. In colder latitudes, the cooling plant, when used as a heat pump, is not large enough to maintain design indoor temperatures and is therefore not competitive with fuel-burning plants.

Other heat sources for heat pumps are well water or underground grid coils. These installations usually call for a valve system permitting warm condenser water to be piped to the air-handling unit for winter heating and the cold water from the chiller to be pumped to the air-handling unit for summer cooling. In winter, the well water or the water in the ground coil is pumped through the chiller to evaporate the liquid refrigerant, while in the summer this water is pumped through the condenser.

Packaged units complete with controls are available for both air-to-air and air-to-water use. Small residential air-to-air units are available in all-electric modes, with capacities up to 9 tons of cooling capacity. Units are made as roof-top or grade-mounted units. Special built-up systems with large capacity have also been constructed.

13.19 SOLAR HEATING

Solar radiation may be used to provide space heating, cooling, and domestic hot water, but the economics of the application should be carefully investigated. Returns on the initial investment from savings on fuel costs and from tax credits may permit a payback for solar domestic hot-water systems in about 6 to 10 years. Heating and cooling systems will take much longer. The advantage of solar heat is that it is free. Therefore, use of solar heat will overcome the continuing high cost of energy from other sources and conserve those fuels that are in limited supply.

There are many disadvantages in using solar radiation. For one thing, it is available only when the sun is shining and there are hourly variations in intensity, daily and with the weather. Also, the energy received per square foot of radiated surface is small, generally under 400 Btu/(hr)(ft^2). This is a very low energy flux. It necessitates large areas of solar collectors to obtain sufficient energy for practical applications and also provide a reasonable payback on the investment.

Solar heating or cooling is advantageous only when the cost of the solar energy produced is less than the cost of energy produced by the more conventional methods. In general, the cost of solar systems may be reduced by obtaining minimum initial costs, favorable amortization rates for the equipment required, and governmental investment tax credits. Also, continuous heat loads and an efficient heating-system design will keep costs low. Bear in mind that the efficiency of solar-system designs for heating or cooling depends on the efficiency of solar collectors, the efficiency of the conversion of the solar radiation to a more useful form of energy, and the efficiency of storage of that energy from the time of conversion until the time of use.

The simplest method of collecting solar energy is by use of a flat-plate-type collector (Fig. 13.28*b*). The collector is mounted in a manner that allows its flat surface to be held normal, or nearly so, to the sun's rays, thereby, in effect, trapping the solar radiation within the collector. Flat-plate-type collectors are used only in low-temperature systems (70 to 180°F). Evacuated-tube focusing/concentrating collectors generate much higher temperatures by minimizing heat losses and concentrating sunlight on a reduced absorber surface. Evacuated-tube type collectors operate in a range of 185 to 250°F, while the concentrating-type collectors operate in an even higher range of 250 to 500°F, or more.

Precautions should be taken in design and installation of solar collectors on low-sloped roofs to prevent excessive roof deflections or overloads. Also, care should be taken to avoid damaging the roofing or creating conditions that would cause premature roofing failures or lead to higher roofing repair, maintenance, or replacement costs. In particular:

1. Solar collectors should not be installed where ponding of rainwater may occur or in a manner that will obstruct drainage of the roof.
2. Solar collector supports and the roof should be designed for snow loads as well as for wind loads on the collectors, including uplift.
3. The installation should not decrease the fire rating of the roof.
4. The roofing should be protected by boards or other means during erection of the collector. If the roofing membrane must be penetrated for the supports or piping, a roofer should install pipe sleeves, flashing, and other materials necessary to keep out water. If bitumen is used, it should not be permitted to splatter on the collector cover plates.
5. At least 24-in clearance should be provided between the bottom of the collector frames and the roof, to permit inspection, maintenance, repair, and replacement of the roof. Similarly, clearance should be provided at the ends of the frames.
6. At least 14-in clearance should be provided between thermal pipes and the roof surface for the preceding reason. The pipes should be supported on the collector frames, to the extent possible, rather than on the roof.

A typical solar heating and cooling system consists of several major system components: collectors, heat storage, supplemental heat source, and auxiliary equipment.

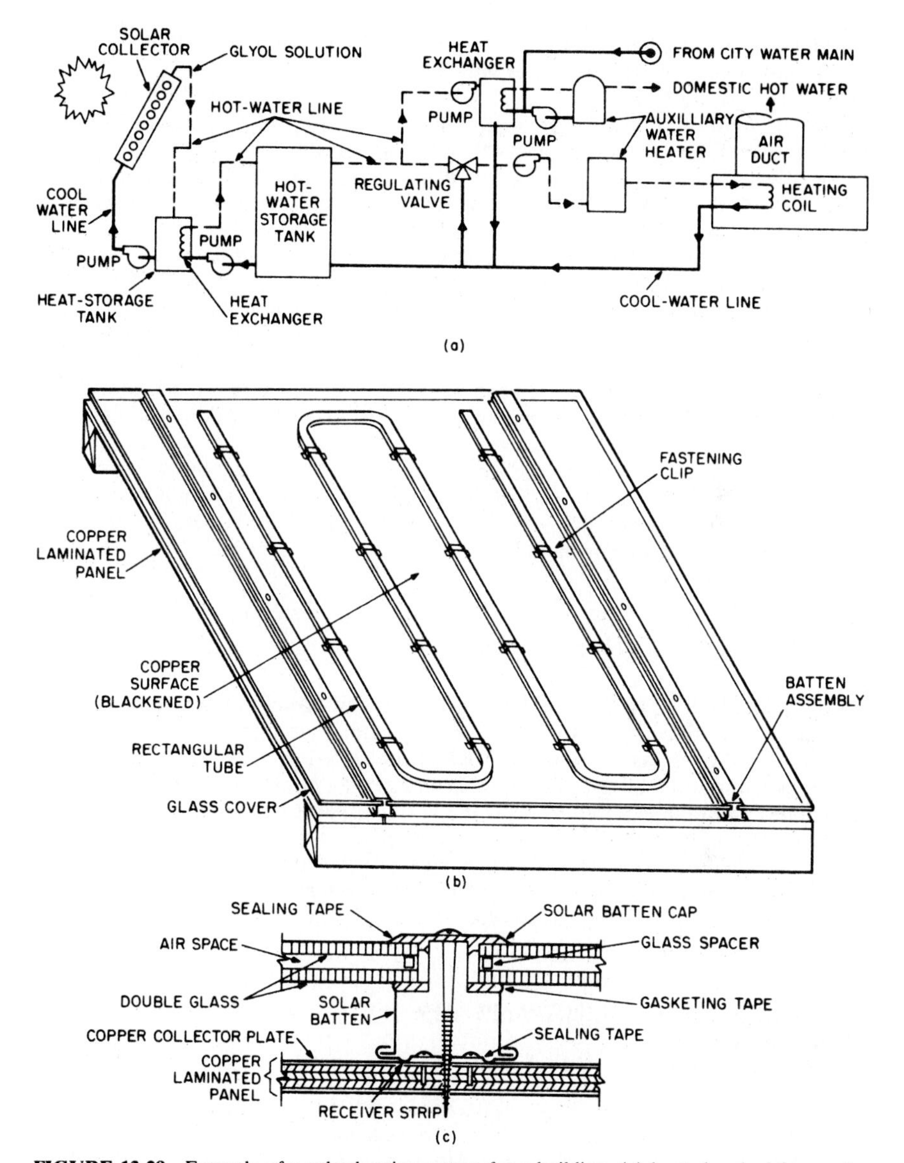

FIGURE 13.28 Example of a solar-heating system for a building: (*a*) heat absorbed from solar radiation by a glycol solution is transferred to water for heating a building and producing domestic hot water; (*b*) solar collector for absorbing solar radiation; (*c*) detail of the batten assembly of the solar collector. *(Reprinted with permission from F. S. Merritt and J. Ambrose, "Building Engineering and Systems Design," 2d ed., Van Nostrand Reinhold Company, New York.)*

Many liquid solutions are used for heat transmission in solar systems. A typical solution may be a mixture of water and 10% propylene-glycol antifreeze solution. It is pumped through the collector tubes in a closed-loop circuit to a heat exchanger, where the solution transfers its energy to a water-heating circuit (Fig. 13.28*a*). The cold glycol solution is then pumped back to the collectors.

The heated water travels in its own closed circuit to a hot-water storage tank and back to the exchanger. Hot water from the storage tank is withdrawn, as required, to satisfy building heating requirements.

The hot water may be pumped from the storage tank to another heat exchanger to produce hot water for domestic use. Hot water from the storage tank may also be pumped through another closed-loop circuit to a heating coil mounted in a central heating blower unit for space heating. In both cases, the cooled water is returned to the storage tank or to the heat exchanger served by the glycol solution.

For use when insufficient solar heat is available, an auxiliary water heater should be added to the system. Providing hot water for heating and domestic purposes, the auxiliary heater may be electrically operated or gas, oil, or coal fired.

When solar heat is to be used for cooling, the hot-water loop must be modified. A three-way valve is placed in the hot-water line to direct the hot water to an absorption-type water chiller. The chilled water discharge is then pumped to a cooling coil in the main air-handling unit.

13.20 INTEGRATED SYSTEMS

The proliferation of mechanical installations in new structures has resulted in the use of hitherto dead space for wireways, pipe runs, air supply or air return, etc. Raised floor space (as in computer rooms), hung-ceiling space, hollow structural columns (round, rectangular, or square tubing) may be used. The basic principles in the design of air requirements for the various spaces do not change.

Once the air quantities have been determined, it is then necessary to check the areas available to move these air quantities and the required horsepower for the job. Utilizing otherwise unused spaces will require reference to basic design manuals for relationship of velocity, friction factors, etc., and should be left to a specialist.

METHODS OF COOLING AND AIR CONDITIONING

The methods for establishing accurate heat losses for heating are also applicable for heat gains and air conditioning. It is mandatory that such procedures be used in order that the necessary cooling equipment can be sized and selected with the lowest first cost that will provide reliable and satisfactory service. (See Arts. 13.7 and 13.8.)

13.21 SIZING AN AIR-CONDITIONING PLANT

Consider the building shown in Fig. 13.3 (p. 13.44). The first and second floors only will be air conditioned. The design outdoor condition is assumed to be 95°F

DB (dry-bulb) and 75°F WB (wet-bulb). The design indoor condition is 80°F DB and 50% relative humidity.

The temperature gradient across an exposed wall will be 15°F (95°F − 80°F). The temperature gradient between a conditioned and an interior nonconditioned space, such as the cellar ceiling, is assumed to be 10°F.

Exterior walls are constructed of 4-in brick with 8-in cinder backup; interior finish is plaster on metal lath. Partitions consist of 2 × 4 studs, wire lath, and plaster. First floor has double flooring on top of joists; cellar ceiling is plaster on metal lath.

Lights may be assumed to average 4 W/ft^2. Assume 50 persons will be in the store, 2 in the first-floor office, 10 in the second-floor office No. 1, and 5 in second-floor office No. 2.

Cooling Measured in Tons. Once we obtain the cooling load in Btu per hour, we convert the load to tons of refrigeration by Eq. (13.15).

Air Requirements for First-Floor Store (Table 13.16*a*).

$$\text{Load in tons} = \frac{86{,}951}{12{,}000} = 7.25 \text{ tons}$$

TABLE 13.16*a* First-Floor Store Load

	Area, ft^2		$U(T_o - T_i)$, °F	Heat load, Btu/hr
Heat gain through enclosures:				
North wall	92	×	0.25 × 15	344
North door	28	×	0.46 × 15	193
South wall	412	×	0.25 × 15	1,540
South glass	154	×	1.04 × 15	2,405
East wall	480	×	0.25 × 15	1,800
West wall	228	×	0.25 × 15	854
Partition	263	×	0.39 × 10	1,025
Partition door	49	×	0.46 × 10	225
Floor	841	×	0.25 × 10	2,100
Ceiling	112	×	0.25 × 10	280
Load due to conduction				10,766
Sun load on south glass, 154 ft^2 × 98				15,100
Occupants (sensible heat) from Table 13.10, 50 × 315				15,750
Lights [Eq. (13.29)] 841 ft^2 × 4 W/ft^2 × 3.42				11,500
				53,116
Fan hp (4% of total)				2,125
Total internal sensible heat load				55,241
Fresh air (sensible heat) at 10 ft^3/min per person [Eq. (13.30), 500 ft^3/min × 1.08 × 15]				8,100
Total sensible load				63,341
Occupants (latent load) from Table 13.10, 50 × 325				16,250
Fresh air (latent heat) Eq. (13.31), 500 × 0.67 (99-77)				7,360
Total latent load				23,610
Total load				86,951

If supply air is provided at 18°F differential, with fresh air entering the unit through an outside duct, the flow required for sensible heat is [from Eq. (13.30)]:

$$Q = \frac{55{,}241}{1.08 \times 18} = 2842 \text{ ft}^3/\text{min}$$

If no fresh-air duct is provided and all the infiltration air enters directly into the premises, the flow required is computed as follows:

$$\text{Store volume} = 841 \times 12 = 10{,}100 \text{ ft}^3$$

$$\text{Infiltration} = \frac{10{,}100}{60} \times 1.5 = 252 \text{ ft}^3/\text{min}$$

$$\text{Sensible load from the infiltration air [Eq. (13.30)]} = 252 \times 1.08 \times 15 = 4090 \text{ Btu/hr}$$

$$\text{Internal sensible load plus infiltration} = 55{,}241 + 4090 = 59{,}331 \text{ Btu/hr}$$

$$\text{Flow with no fresh-air duct} = \frac{59{,}331}{1.08 \times 18} = 3052 \text{ ft}^3/\text{min}$$

Air Requirements for First-Floor Office (Table 13.16*b*).

$$\text{Load in tons} = \frac{4{,}172}{12{,}000} = 0.35 \text{ ton}$$

$$\text{Supply air required with fresh-air duct} = \frac{2671}{1.08 \times 18} = 137 \text{ ft}^3/\text{min}$$

TABLE 13.16*b* **First-Floor Office Load**

	Btu/hr	
Heat gain through enclosures:		
North wall, 102 × 0.25 × 15	382	
North glass, 18 × 1.04 × 15	281	
Partition, 30 × 0.39 × 10	117	
Floor, 79 × 0.25 × 10	198	
	978	
2 occupants (sensible load) at 255	510	
Lights, 79 × 4 × 3.42	1080	
	2568	
Fan hp 4%	103	
Total internal sensible load		2671
Fresh air (sensible load) at 2 changes per hour, 32 × 1.08 × 15		518
Total sensible load		3189
Occupants (latent load), 2 at 255	510	
Fresh air (latent load), 32 × 0.67(99 − 77)	473	
Total latent load		983
Total load		4172

$$\text{Supply air required with no fresh-air duct} = \frac{3189}{1.08 \times 18} = 164 \text{ ft}^3/\text{min}$$

Air Requirements for Second-Floor Office No. 1 (Table 13.16*c*).

$$\text{Load in tons} = \frac{41{,}888}{12{,}000} = 3.5 \text{ tons}$$

$$\text{Supply air required with fresh-air duct} = \frac{33{,}768}{1.08 \times 18} = 1737 \text{ ft}^3/\text{min}$$

$$\text{Supply air required with no fresh-air duct} = \frac{36{,}678}{1.08 \times 18} = 1887 \text{ ft}^3/\text{min}$$

TABLE 13.16*c* **Second-Floor Office No. 1**

	Btu/hr	
Heat gain through enclosures:		
South wall, 130 × 0.25 × 15	388	
South glass, 120 × 1.04 × 15	1,870	
East wall, 260 × 0.25 × 15	975	
West wall, 40 × 0.25 × 15	150	
Partition, 265 × 0.39 × 10	1,035	
Door in partition, 35 × 0.46 × 10	161	
Roof, 568 × 0.19 × 54	5,820	
	10,399	
Sun load on south glass, 120 × 98	11,750	
Occupants (sensible load), 10 × 255	2,550	
Lights, 568 × 4 × 3.42	7,770	
	32,469	
Fan hp 4%	1,299	
Total internal sensible load		33,768
Fresh air (sensible load), 180 × 1.08 × 15		2,910
Total sensible load		36,678
Occupants (latent load), 10 × 255	2,550	
Fresh air (latent load), 180 × 0.67(99 − 77)	2,660	
Total latent load		5,210
Total load		41,888

Air Requirements for Second-Floor Office No. 2 (Table 13.16*d*).

$$\text{Load in tons} = \frac{13{,}362}{12{,}000} = 1.13 \text{ tons}$$

$$\text{Supply air required with fresh-air duct} = \frac{9697}{1.08 \times 18} = 499 \text{ ft}^3/\text{min}$$

TABLE 13.16d Second-Floor Office No. 2

	Btu/hr	
Heat gain through enclosures:		
North wall, $103 \times 0.25 \times 15$	386	
North glass, $72 \times 1.04 \times 15$	1,110	
East wall, $132 \times 0.25 \times 15$	495	
Partition, $114 \times 0.39 \times 10$	445	
Door in partition, $18 \times 0.46 \times 10$	83	
Roof, $231 \times 0.19 \times 54$	2,370	
	4,889	
Occupants (sensible load), 5×255	1,275	
Lights, $231 \times 4 \times 3.42$	3,160	
	9,324	
Fan hp 4%	373	
Total internal sensible load		9,697
Fresh air (sensible load), $77 \times 1.08 \times 15$		1,250
Total sensible load		10,947
Occupants (latent load), 5×255	1,275	
Fresh air (latent load), $77 \times 0.67(99 - 77)$	1,140	
Total latent load		2,415
Total load		13,362

$$\text{Supply air required with no fresh-air duct} = \frac{10{,}947}{1.08 \times 18} = 563 \text{ ft}^3/\text{min}$$

Cooling and supply-air requirements for the building are summarized in Table 13.17.

TABLE 13.17 Cooling-Load Analysis for Building in Fig. 13.3 (p. 13.44)

Space	Tons	Flow with fresh-air duct, ft³/min	Flow with no fresh-air duct, ft³/min
First-floor store	7.25	2842	3052
First-floor office	0.35	137	164
Second-floor office No. 1	3.50	1737	1887
Second-floor office No. 2	1.13	499	563
	12.23	5215	5666

13.22 REFRIGERATION CYCLES

Figure 13.29 shows the basic air-conditioning cycle of the **direct-expansion type.** The compressor takes refrigerant gas at a relatively low pressure and compresses

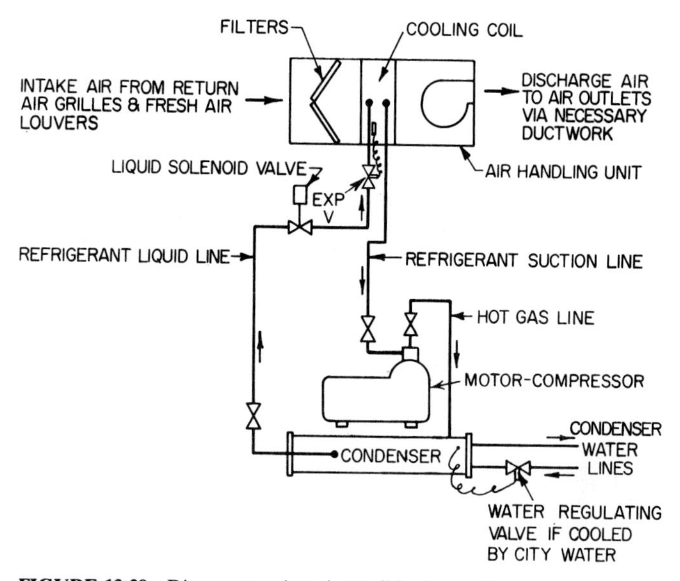

FIGURE 13.29 Direct-expansion air-conditioning cycle.

it to a higher pressure. The hot gas is passed to a condenser where heat is removed and the refrigerant liquefied. The liquid is then piped to the cooling coil of the air-handling unit and allowed to expand to a lower pressure (suction pressure). The liquid vaporizes or is boiled off by the relatively warm air passing over the coil. The compressor pulls away the vaporized refrigerant to maintain the required low coil pressure with its accompanying low temperature.

Chilled-Water Refrigeration Cycle. In some systems, water is chilled by the refrigerant and circulated to units in or near spaces to be cooled (Fig. 13.30), where air is cooled by the water.

In water-cooled and belt-driven air-conditioning compressors formerly used, motor winding heat was usually dissipated into the atmosphere outside the conditioned space (usually into the compressor rooms). For average air-conditioning service 1 hp could produce about 1 ton of cooling. When sealed compressor-motor units came into use, the motor windings were arranged to give off their heat to the refrigerant suction gas. This heat was therefore added to the cooling load of the compressor. The result was that 1 hp could produce only about 0.85 ton of cooling.

Because of water shortages, many communities restrict the direct use of city water for condensing purposes. As an alternative, water can be cooled with cooling towers and recirculated. But for smaller systems, cooling towers have some inherent disadvantages. As a result, air-cooled condensers have become common for small- and medium-sized air-conditioning systems. Because of the higher head pressures resulting from air-cooled condensers, each horsepower of compressor-motor will produce only about 0.65 ton of cooling. Because of these developments, air-conditioning equipment manufacturers rate their equipment in Btu per hour output and kilowatts required for the rated output, at conditions standardized by the industry, instead of in horsepower.

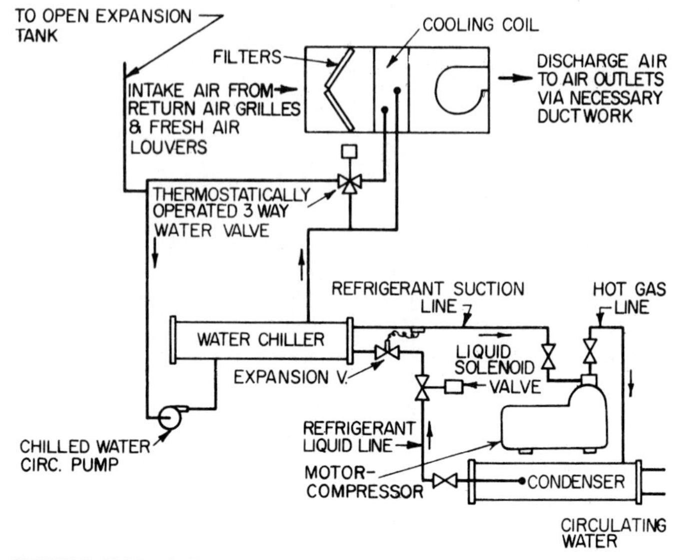

FIGURE 13.30 Chilled-water air-conditioning cycle.

13.23 AIR-DISTRIBUTION TEMPERATURE FOR COOLING

The air-distribution system is the most critical part of an air-conditioning system. If insufficient air is circulated, proper cooling cannot be done. On the other hand, handling large quantities of air is expensive in both initial cost and operation. The amount of cool air required increases rapidly the closer its temperature is brought to the desired room temperature.

If, for example, we wish to maintain 80°F DB (dry-bulb) in a room and we introduce air at 60°F, the colder air when warming up to 80°F will absorb an amount of sensible heat equal to q_s. According to Eq. (13.30), $q_s = 1.08Q_1(80 - 60) = 21.6Q_1$, where Q_1 is the required airflow in cubic feet per minute. From Eq. (13.30), it can be seen also that, if we introduce air at 70°F, with a temperature rise of 10°F instead of 20°F, $q_s = 10.8Q_2$, and we shall have to handle twice as much air to do the same amount of sensible cooling.

From a psychrometric chart, the dew point of a room at 80°F DB and 50% relative humidity is found to be 59°F. If the air leaving the air-conditioning unit is 59°F or less, the duct will sweat and will require insulation. Even if we spend the money to insulate the supply duct, the supply grilles may sweat and drip. Therefore, theoretically, to be safe, the air leaving the air-conditioning unit should be 60°F or higher.

Because there are many days when outside temperatures are less than design conditions, the temperature of the air supplied to the coil will fluctuate, and the temperature of the air leaving the coil may drop a few degrees. This will result in sweating ducts. It is good practice, therefore, to design the discharge-air temperature for about 3°F higher than the room dew point.

Thus, for 80°F DB, 50% RH, and dew point at 59°F, the minimum discharge air temperature would be 62°F as insurance against sweating. The amount of air to be handled may be obtained from Eq. (13.30), with a temperature difference of 18°F.

13.24 CONDENSERS

If a water-cooled condenser is employed to remove heat from the refrigerant, it may be serviced with city water, and the warm water discharged to a sewer. Or a water tower (Fig. 13.31) may be used to cool the condenser discharge water, which can then be recirculated back to the condenser.

Where practical, the water condenser and tower can be replaced by an evaporative condenser as in Fig. 13.32.

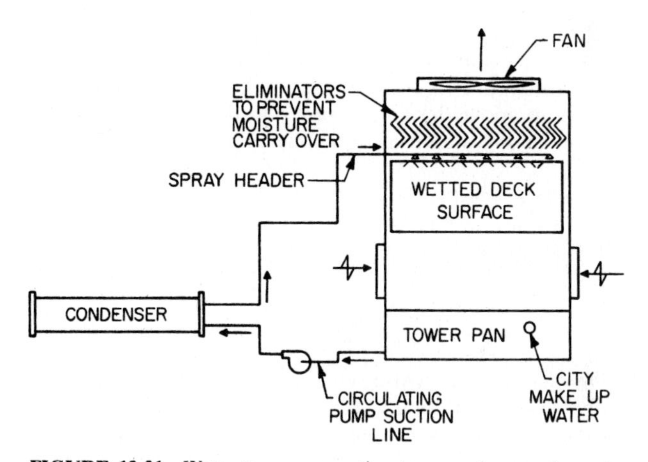

FIGURE 13.31 Water tower connection to a condenser of an air-conditioning system.

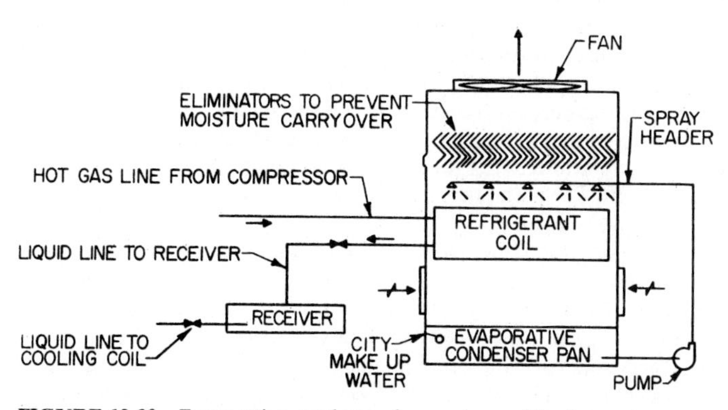

FIGURE 13.32 Evaporative condenser for an air-conditioning system.

The capacity of water savers, such as towers or evaporative condensers, depends on the wet-bulb temperature. The capacity of these units decreases as the wet-bulb temperature increases.

Such equipment should be sized for a wet-bulb temperature a few degrees above that used for sizing air-conditioning equipment.

As an example, consider an area where the design wet-bulb temperature is 75°F. If we size the air-conditioning equipment for this condition, we shall be able to maintain design inside conditions when the outside conditions happen to be 75°F WB. There will be a few days a year, however, when the outside air may register 79 or 80°F WB. During the higher wet-bulb days, with the air-conditioning equipment in operation, we shall balance out at a relative humidity above design. For example, if the design relative humidity is 50%, we may balance out at 55% or higher.

However, if the water towers and evaporative condensers are designed for 75°F WB, then at a wet-bulb temperature above 75, the water-saver capacity would be decreased and the compressor head pressure would build up too high, overload the motor, and kick out on the overload relays. So we use 78°F WB for the design of water savers. Now, on a 78°F WB day, the compressor head pressure would be at design pressure, the equipment will be in operation, although maintaining room conditions a little less comfortable than desired. At 80°F WB, the compressor head pressure will build up above design pressure and the motor will be drawing more than the design current but usually less than the overload rating of the safety contact heaters.

Water condensers are sized for 104°F refrigerant temperature and 95°F water leaving. When condensers are used with city water, these units should be sized for operation with the water at its warmest condition.

The amount of water in gallons per minute required for condensers is equal to

$$Q = \frac{\text{tons of cooling} \times 30}{\text{water-temperature rise}} \tag{13.35}$$

This equation includes the heat of condensation of the compressor when used for air-conditioning service. In many cities, the water supplied is never above 75°F. Thus 75°F is used as a condenser design condition.

When high buildings have roof tanks exposed to the sun, it will be necessary to determine the maximum possible water temperature and order a condenser sized for the correct water flow and inlet-water temperature.

When choosing a condenser for water-tower use, we must determine the water temperature available from the cooling tower with 95°F water to the tower. Check the tower manufacturer for the capacity required at the design WB, with 95°F water to the sprays, and the appropriate wet-bulb approach. (Wet-bulb approach is equal to the number of degrees the temperature of the water leaving the tower is above the wet-bulb temperature. This should be for an economic arrangement about 7°F.) Thus, for 78°F, the water leaving the tower would be 85°F, and the condenser would be designed for a 10°F water-temperature rise, i.e., for water at 95°F to the tower and 85°F leaving the tower.

Evaporative condensers should be picked for the required capacity at a design wet-bulb temperature for the area in which they are to be installed. Manufacturers' ratings should be checked before the equipment is ordered.

For small- and medium-size cooling systems, air-cooled condensers are available for outdoor installation with propeller fans or for indoor installation usually with centrifugal blowers for forcing outdoor air through ducts to and from the condensers.

13.25 COMPRESSOR-MOTOR UNITS

Compressor capacity decreases with increase in head pressure (or temperature) and fall in suction pressure (or temperature). Therefore, in choosing a compressor-motor unit, one must first determine design conditions for which this unit is to operate; for example:

Suction temperature—whether at 35°F, 40°F, 45°F, etc.

Head pressure—whether serviced by a water-cooled or air-cooled condenser.

Manufacturers of compressor-motor units rate their equipment for Btu per hour output and kilowatt and amperage draw at various conditions of suction and discharge pressure. These data are available from manufacturers, and should be checked before a compressor-motor unit is ordered.

Although capacity of a compressor decreases with the suction pressure (or back pressure), it is usual practice to rate compressors in tons or Btu per hour at various suction temperatures.

For installations in which the latent load is high, such as restaurants, bars, and dance halls, where a large number of persons congregate, the coil temperature will have to be brought low enough to condense out large amounts of moisture. Thus, we find that a suction temperature of about 40°F will be required. In offices, homes, etc., where the latent load is low, 45°F suction will be satisfactory. Therefore, choose a compressor with capacity not less than the total cooling load and rated at 104°F head temperature, and a suction temperature between 40 and 45°F, depending on the nature of the load.

Obtain the brake horsepower of the compressor at these conditions, and make sure that a motor is provided with horsepower not less than that required by the compressor.

A standard NEMA motor can be loaded about 15 to 20% above normal. Do not depend on this safety factor, for it will come in handy during initial pull-down periods, excess occupancy, periods of low-voltage conditions, etc.

Compressor manufacturers have all the above data available for the asking so that one need not guess.

Sealed compressors used in most air-conditioning systems consist of a compressor and a motor coupled together in a single housing. These units are assembled, dehydrated, and sealed at the factory. Purchasers cannot change the capacity of the compressor by changing its speed or connect to it motors with different horsepower. It therefore is necessary to specify required tonnage, suction temperature, and discharge temperature so that the manufacturer can supply the correct, balanced, sealed compressor-motor unit.

13.26 COOLING EQUIPMENT—CENTRAL PLANT PACKAGED UNITS

For an economical installation, the air-handling units should be chosen to handle the least amount of air without danger of sweating ducts and grilles. In most comfort cooling, an 18°F discharge temperature below room temperature will be satisfactory.

If a fresh-air duct is used, the required fresh air is mixed with the return air, and the mix is sent through the cooling coil; i.e., the fresh-air load is taken care of in the coil, and discharge air must take care of the internal load only. Then, from Eq. (13.30), the amount of air to be handled is

$$Q = \frac{q_s}{1.08\,\Delta T} \tag{13.36}$$

where q_s = internal sensible load
ΔT = temperature difference between room and air leaving coil (usually 18°F)

If a fresh-air duct is not installed, and the outside air is allowed to infiltrate into the premises, we use Eq. (13.30) with the sensible part of the load of the infiltration air added to q_s, because the outside air infiltrating becomes part of the internal load.

Once the amount of air to be handled is determined, choose a coil of face area such that the coil-face velocity V would be not much more than 500 ft^3/min. Some coil manufacturers recommend cooling-coil-face velocities as high as 700 ft^3/min. But there is danger of moisture from the cooling coil being carried along the air stream at such high velocities.

$$V = \frac{Q}{A_c} \tag{13.37}$$

where Q = air flow, ft^3/min
A_c = coil face area, ft^2
V = air velocity, ft/min

The number of rows of coils can be determined by getting the manufacturer's capacity ratings of the coils for three, four, five, six, etc., rows deep and choosing a coil that can handle not less than the sensible and latent load at the working suction temperature.

Multizone air-handling units (Fig. 13.33) are used to control the temperature of more than one space without use of a separate air-handling unit for each zone. When a zone thermostat calls for cooling, the damper motor for that zone opens

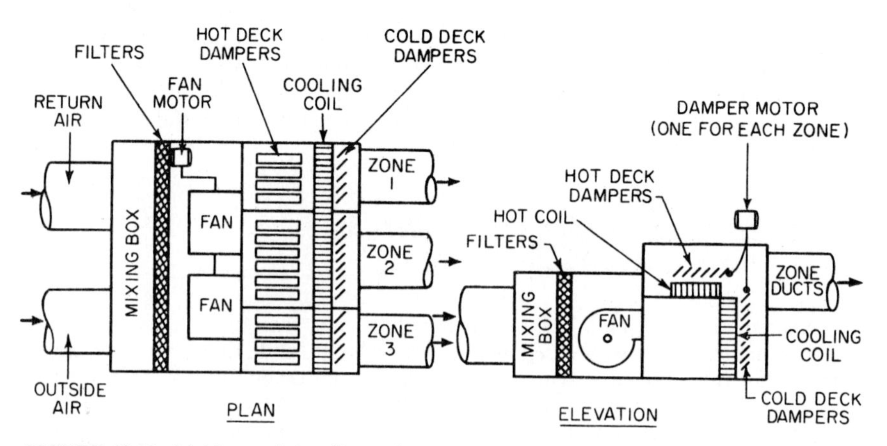

FIGURE 13.33 Multizone air-handling unit.

the cold deck dampers and closes or throttles the warm deck dampers. Thus, the same unit can provide cooling for one zone while it can provide heating for another zone at the same time.

13.27 ZONING

In large and complex buildings, there will be many spaces that have different inside design conditions. The reason for this is that it is very difficult, in many cases, to provide an air-handling unit that discharges supply air at a given set of conditions to satisfy the various spaces with different design conditions. Common practice is to combine areas with similar design requirements into a zone. Each zone is then served by a separate air-conditioning unit independently of the other zones.

In some cases, a zone may be satisfied by use of reheat coils to satisfy the zone requirements. However, reheating the low-temperature supply air consumes energy. Hence, this practice is in disfavor and in many cases prohibited by building codes and governmental policies. Other formerly well-established systems, such as multizone and dual-duct systems, are also considered energy consumers and are also in disfavor. Exceptions that permit these systems by codes and statutes are for specific types of manufacturing or processing systems, or for areas where the cooled space must be maintained at very specific temperatures and humidities, such as computer rooms, libraries, operating rooms, paper and printing operations, etc.

(H. E. Bovay, Jr., "Handbook of Mechanical and Electrical Systems for Buildings," McGraw-Hill Publishing Company, New York.)

13.28 PACKAGED AIR-CONDITIONING UNITS

To meet the great demand for cheaper air-conditioning installations, manufacturers produce packaged, or preassembled, units. These vary from 4000 to 30,000 Btu/hr for window units and 9000 to 1,200,000 Btu/hr for commercial units. Less field labor is required to install them than for custom-designed installations.

Packaged window units operate on the complete cycle shown in Fig. 13.30, but the equipment is very compactly arranged. The condenser is air cooled and projects outside the window. The cooling coil extends inside. Both the cooling-coil fan and condenser fan usually are run by the same motor.

These packaged units require no piping, just an electric receptacle of adequate capacity. Moisture that condenses on the cooling coil runs via gutters to a small sump near the condenser fan. Many manufacturers incorporate a disk slinger to spray this water on the hot condenser coil, which vaporizes it and exhausts it to the outside. This arrangement serves a double purpose:

1. Gets rid of humidity from the room without piping.
2. Helps keep the head pressure down with some evaporative cooling.

Floor-type and ceiling-type packaged units also contain the full air-conditioning cycle (Fig. 13.30). The air-cooled packaged units are usually placed near a window to reduce the duct runs required for the air-cooled condenser.

Roof-type packaged units are available for one-story structures or top-floor areas, such as industrial spaces and supermarkets. These units contain a complete cooling cycle (usually with an air-cooled condenser) and a furnace. All controls are factory prewired, and the refrigeration cycle is completely installed at the factory. Necessary ductwork, wiring, and gas piping are supplied in the field.

A combination packaged unit including an evaporative condenser is also available.

Most packaged units are standardized to handle about 400 ft^3/min of air per ton of refrigeration. This is a good average air quantity for most installations. For restaurants, bars, etc., where high latent loads are encountered, the unit handles more than enough air; also, the sensible capacity is greater than required. However, the latent capacity may be lower than desired. This will result in a somewhat higher relative humidity in the premises.

For high sensible-load jobs, such as homes, offices, etc., where occupancy is relatively low, air quantities are usually too low for the installed tonnage. Latent capacity, on the other hand, may be higher than desired. Thus, if we have a total load of 5 tons—4½ tons sensible and ½ ton latent—and we have available a 5-ton unit with a capacity of 4 tons sensible and 1 ton latent, it is obvious that during extreme weather the unit will not be able to hold the dry-bulb temperature down, but the relative humidity will be well below design value. Under such conditions, these units may be satisfactory. In the cases in which the latent capacity is too low and we have more than enough sensible capacity, we can set the thermostat below design indoor dry-bulb temperature and maintain a lower dry-bulb temperature and higher relative humidity to obtain satisfactory comfort conditions. Where we have insufficient sensible capacity, we have to leave the thermostat at the design setting. This will automatically yield a higher dry-bulb temperature and lower relative humidity—and the premises will be comfortable if the total installed capacity is not less than the total load.

As an example of the considerations involved in selecting packaged air-conditioning equipment, let us consider the structure in Fig. 13.4, p. 13.47, and the load analysis in Table 13.13, p. 13.48.

If this were an old building, it would be necessary to do the following:

First-floor store (load 7.25 tons). Inasmuch as a 5-ton unit is too small, we must choose the next larger size—a 7½-ton packaged unit. This unit has a greater capacity than needed to maintain design conditions. However, many people would like somewhat lower temperature than 80°F dry bulb, and this unit will be capable of maintaining such conditions. Also, if there are periods when more than 50 occupants will be in the store, extra capacity will be available.

First-floor office (load 0.35 tons). Ordinarily a 4000-Btu/hr window-unit would be required to do the job. But because the store unit has spare capacity, it would be advisable to arrange a duct from the store 7½-ton unit to cool the office.

Second-floor offices (No. 1, load 3.50 tons; No. 2, load 1.13 tons). A 5-ton unit is required for the two offices. But it will be necessary to provide a fresh-air connection to eliminate the fresh-air load from the internal load, to reduce air requirements to the rated flow of the unit. Then, the 2000 ft^3/min rating of the 5-ton packaged unit will be close enough to the 2236 ft^3/min required for the two offices.

Inasmuch as the total tonnage is slightly above that required, we will balance out at a slightly lower relative humidity than 50%, if the particular packaged unit selected is rated at a sensible capacity equal to the total sensible load.

A remote air-conditioning system would be more efficient, because it could be designed to meet the needs of the building more closely. A single air-handling unit for both floors can be arranged with the proper ductwork. However, local ordinances should be checked, since some cities have laws preventing direct-expansion systems servicing more than one floor. A chilled water system (Fig. 13.31) could be used instead.

Split Systems. Split systems differ from other packaged air-conditioning equipment in that the noisy components of the system, notably the refrigerant compressor and air-cooled condenser or cooling tower fans, are located outdoors, away from the air-handling unit. This arrangement is preferred for apartment buildings, residences, hospitals, libraries, churches, and other buildings for which quiet operation is required. In these cases, the air-handler is the only component located within the occupied area. It is usually selected for low revolutions per minute to minimize fan noise. For such installations, chilled-water piping or direct-expansion refrigerating piping must be extended from the cooling coil to the remote compressor and condensing unit, including operating controls.

Packaged Chillers. These units consist of a compressor, water chiller, condenser, and all automatic controls. The contractor has to provide the necessary power and condensing water, from either a cooling tower, city water, or some other source such as well or river. The controls are arranged to cycle the refrigeration compressor to maintain a given chilled-water temperature. The contractor need provide only insulated piping to various chilled-water air handling units.

Packaged chillers also are available with air-cooled condensers for complete outdoor installation of the unit, or with a condenserless unit for indoor installation connected to an outdoor air-colled condenser.

The larger packaged chillers, 50 to 1000 tons, are generally powered by centrifugal compressors. When the units are so large that they have to be shipped knocked down, they are usually assembled by the manufacturer's representative on the job site, sealed, pressure-tested, evacuated, dehydrated, and charged with the proper amount of refrigerant.

13.29 ABSORPTION UNITS FOR COOLING

Absorption systems for commercial use are usually arranged as packaged chillers. In place of electric power, these units use a source of heat to regenerate the refrigerant. Gas, oil, or low-pressure steam is used.

In the absorption system used for air conditioning, the refrigerant is water. The compressor of the basic refrigeration cycle is replaced by an absorber, pump, and generator. A weak solution of lithium bromide is heated to evaporate the refrigerant (water). The resulting strong solution is cooled and pumped to a chamber where it absorbs the cold refrigerant vapor (water vapor) and becomes dilute.

Condensing water required with absorption equipment is more than that required for electric-driven compressors. Consult the manufacturer in each case for the water quantities and temperature required for the proper operation of this equipment.

The automatic-control system with these packaged absorption units is usually provided by the manufacturer and generally consists of control of the amount of steam, oil, or gas used for regeneration of the refrigerant to maintain the chilled water temperature properly. In some cases, condenser-water flow is varied to control this temperature.

When low-cost steam is available, low-pressure, single-stage absorption systems may be more economical to operate than systems with compressors. In general, steam comsumption is about 20 lb/hr per ton of refrigeration. Tower cooling water required is about 3.7 gal/min per ton with 85°F water entering the absorption unit and 102°F water leaving it.

As an alternative, high-pressure steam, two-stage absorption units are available that operate much more efficiently than the low-pressure units. The high-pressure units use about 12 lb of steam per ton of refrigeration at full load. This is a 40% savings in energy.

13.30 DUCTS FOR AIR CONDITIONING

In designing a duct system for air conditioning, we must first determine air-outlet locations. If wall grilles are used, they should be spaced about 10 ft apart to avoid dead spots. Round ceiling outlets should be placed in the center of a zone. Rectangular ceiling outlets are available that blow in either one, two, three, or four directions.

Manufacturers' catalog ratings should be checked for sizing grilles and outlets. These catalogs give the recommended maximum amount of air to be handled by an outlet for the various ceiling heights. They also give grille sizes for various lengths of blows. It is obvious that the farther the blow, the higher must be the velocity of the air leaving the grille. Also the higher the velocity, the higher must be the pressure behind the grille.

When grilles are placed back to back in a duct as in Fig. 13.34, be sure that grille A and grille B have the same throw; for if the pressure in the duct is large enough for the longer blow, the short-blow grille will bounce the air off the opposite wall, causing serious drafts. But if the pressure in the duct is just enough for the short blow, the long-blow grille will never reach the opposite wall. Figure 13.35 is recommended for unequal blows because it allows adjustment of air and buildup of a higher static pessure for the longer blow.

In some modern buildings perforated ceiling panels are used to supply conditioned air to the premises. Supply ductwork is provided in the plenum above the suspended ceiling as with standard ceiling outlets. However, with perforated panels, less acoustical fill is used to match the remainder of the hung ceiling.

After all discharge grilles and the air-handling unit are located, it is advisable to make a single-line drawing of the duct run. The air quantities each line and branch must carry should be noted. Of the few methods of duct-system design in

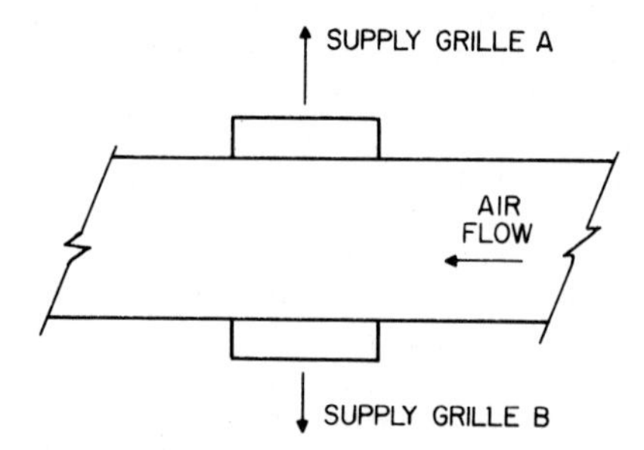

FIGURE 13.34 Grilles placed back to back on an air-conditioning duct with no provision for adjustment of discharge.

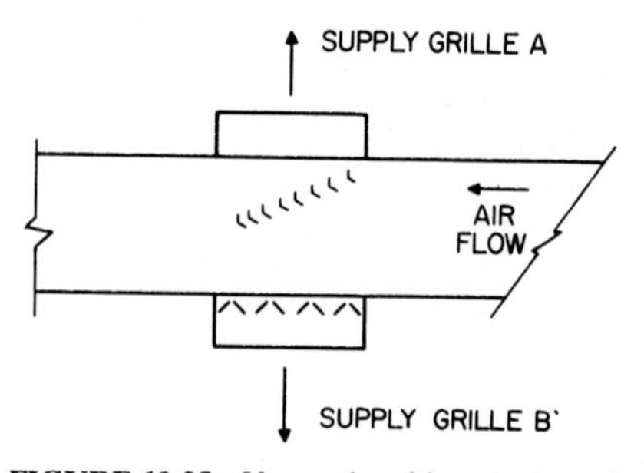

FIGURE 13.35 Vanes placed in a duct to adjust air flow to back-to-back grilles, useful when air throws are unequal.

use, the equal-friction method is most practical. For most comfort cooling work, it is considered good practice not to exceed 0.15-in friction per 100 ft of ductwork. It is also well to keep the air below 1500 ft/min starting velocity.

If a fresh-air duct is installed the return-air duct should be sized for a quantity of air equal to the supply air minus the fresh air.

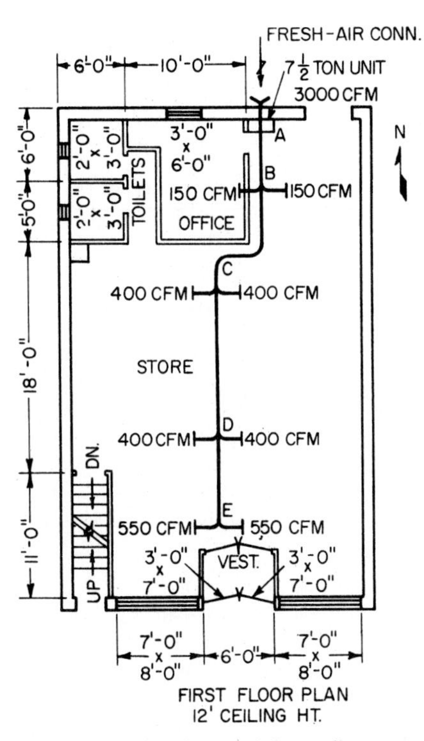

FIGURE 13.36 Ductwork for cooling a store.

It is advisable, where physically possible, to size the fresh-air duct for the full capacity of the air-handling unit. For example, a 10-ton system handling 4000 ft³/min of supply air—1000 ft³/min fresh air and 3000 ft³/min return air—should have the fresh-air duct sized for 4000 ft³/min of air. A damper in the fresh-air duct will throttle the air to 1000 ft³/min during the cooling season; however, during an intermediate season, when the outside air is mild enough, cooling may be obtained by operating only the supply-air fan and opening the damper, thus saving the operation of the 10-ton compressor-motor unit.

As an example of the method for sizing an air-conditioning duct system, let us determine the ductwork for the first floor of the building in Fig. 13.36. Although a load analysis shows that the air requirement is 2979 ft³/min, we must design the ducts to handle the full capacity of air of the packaged unit we supply. Handling less air will unbalance the unit, causing a drop in suction temperature, and may cause freezing up of the coil. If a 7½-ton packaged unit is used, for example, the ducts should have a capacity at least equal to the 3000 ft³/min at which this unit is rated.

Table 13.18 shows the steps in sizing the ducts. The 3000 ft³/min is apportioned to the various zones in the store in proportion to the load from each, and the flow for each segment of duct is indicated in the second column of the table. Next, the size of an equivalent round duct to handle each airflow is determined from Table 13.8 with friction equal to 0.15 in per 100 ft. The size of rectangular duct to be used is obtained from Table 13.7.

The preceding example of ductwork design falls into the category of low-pressure duct systems. This type of design is used for most air-distribution systems that are not too extensive, such as one- or two-floor systems, offices, and residences. In general, the starting air velocity is below 2000 ft/min, and the fan static pressure is below 3 in of water.

For large multistory buildings, high-velocity air-distribution duct systems often are used. These systems operate at duct velocities well above 3000 ft/min and above 3-in static pressure. Obvious advantages include smaller ducts and lower building cost, since smaller plenums are needed above hung ceilings. Disadvantages are high power consumption for fans and need for an air-pressure-reducing valve and

TABLE 13.18 Duct Calculation for 7½-ton Package Unit for Store (Fig. 13.37)

Duct	Cfm	Equivalent round duct (friction = 0.15 in per 100 ft, max velocity = 1500 fpm), diam, in	Rectangular duct, in
A-B	3000	19.7	28 × 12
B-C	2700	18.4	24 × 12
C-D	1900	16.1	18 × 12
D-E	1100	13.2	12 × 12

sound attenuation box for each air outlet, resulting in higher power consumption for operation of the system.

Some of the more elaborate heating and air-conditioning installations consist of a high-pressure warm-air duct system and a high-pressure cold-air duct system. Each air outlet is mounted in a sound attenuation box with pressure-reducing valves and branches from the warm- and cold-air systems (Fig. 13.37). Room temperature is controlled by a thermostat actuating two motorized volume dampers. When cooling is required, the thermostat activates the motor to shift the warm-air damper to the closed or throttled position and the cold-air damper to the open position.

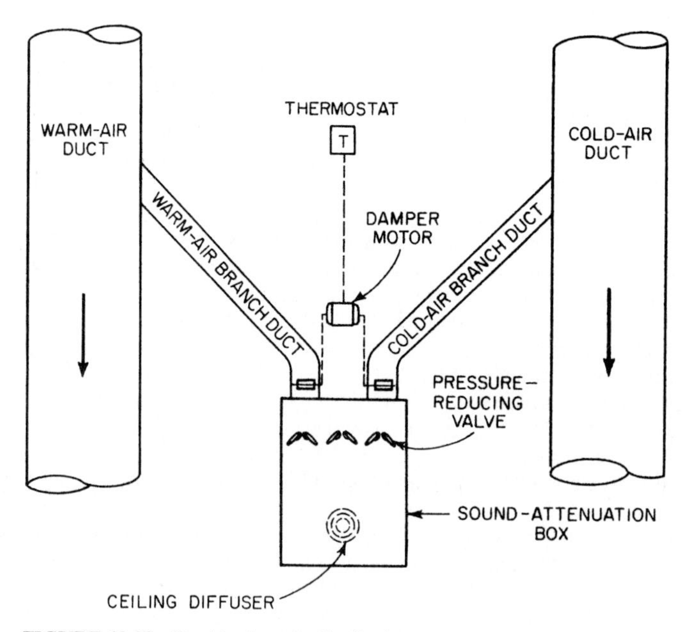

FIGURE 13.37 Double-duct air-distribution system.

13.31 BUILT-UP AIR-CONDITIONING UNITS

Built-up air-conditioning units differ from packaged units in that built-up units are assembled at the site, whereas components of packaged units are preassembled in a factory. Built-up units are usually limited to the larger-size units, with capacity of 50 tons and over. They provide cooling air in summer and heated air in winter. In intermediate seasons, the units may provide 0 to 100% outside air for ventilation, with economy control or enthalpy control. The units come complete with prefilters, final filters, return-air fan, dampers, and controls, as required. The units may be installed outdoors at grade or on rooftop, or indoors as a central-plant unit.

A built-up unit usually is enclosed with sandwich-type, insulated panels, which incorporate thermal insulation between steel or cement-asbestos sheets. Special details have been developed for connections of enclosure panels to provide rigidity and to make them weatherproof or airtight. Access doors should be provided between major in-line components.

Built-up units can be made as a complete system with return-air fan, refrigerant compressors and air-cooled condensers for outside installations. Other types without these components are available for split systems that utilize remote refrigerant compressors and condensing units. In chilled-water systems, chilled water is pumped to the cooling coil from the remote chilled-water unit. Heating water is also piped to the heating coil from a remote heat exchanger but with its own separate pumping system.

13.32 VARIABLE-AIR-VOLUME (VAV) SYSTEMS

The VAV concept as applied to air conditioning is based on the idea of varying the supply-air volume, with constant temperature, to meet changing load conditions of interior spaces or zones (Fig.13.38). Many installations utilize packaged roof units, although large or small built-up units may also be used.

Rooftop units may be used for the following VAV applications:

1. Interior VAV cooling and perimeter radiation
2. Interior VAV cooling and constant-volume perimeter heating

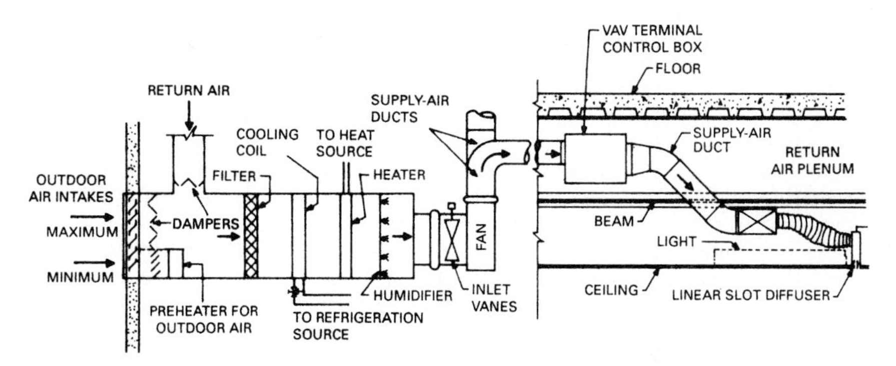

FIGURE 13.38 VAV system varies the supply-air volume to meet changing load conditions.

3. Interior VAV cooling with constant-volume, perimeter-heating, fan-coil units
4. Interior VAV cooling with fan-powered perimeter heating or cooling
5. Interior VAV cooling with perimeter VAV reheat
6. Interior VAV cooling, double-duct, with VAV perimeter cooling and heating

These systems vary widely in advantages and disadvantages and should be analyzed very carefully in the preliminary stages of design for each project.

13.32.1 Zoning of Building Interiors

Zoning is very important in the design stage and will contribute greatly to the success or failure of the HVAC system. An insufficient number of zones will lead to loss of temperature control in areas that do not have the controlling thermostat. The more zones utilized, the better will be the comfort conditions attained, but with some increase in initial costs.

13.32.2 Air Distribution

All types of diffusers (registers) may be used with VAV control units. Standard diffusers with reduced air-flow rates, however, often cause "dumping" of discharge air directly downward with wide variations in space temperature. Many authorities recommend linear-slot-type diffusers to negate dumping. Such diffusers discharge air into the room with a horizontal flow that hugs the ceiling. This is called the *coanda characteristic effect* and is very effective in distributing supply air throughout a room.

13.32.3 VAV Terminal Control Boxes

For producing variable volume other than by modulation of the supply fan, terminal control units of many types may be used:

Shut-off control diffusers are used for interior cooling. They are available with many features, such as shut-off operation, multiple slots, integral slot diffuser, electric-pneumatic or system-powered controls. They are available in 200- to 800-cfm capacities, with choice of aspirating, unit-mounted pneumatic, system powered or electric thermostats. Savings are attributable to modulations of supply-air fans, which also provide complete control flexibility with reduced control costs. These diffusers are easy to install and are compatible with most ceiling systems.

Fan-powered control units are generally used for perimeter and special-use areas. They feature VAV operation in the cooling mode. They are available with pressure-compensating controls with factory-installed hot-water or multistage electric coils that have pneumatic or electric controls. A built-in, side-mounted fan recirculates plenum air. Fan-powered units permit maximum modulation of the central fan as the control unit modulates to full shut-off VAV position. A terminal fan in the control unit is operated to maintain minimum flow in perimeter zones only. Heat generated by lighting fixtures is also used to assist in heating of the perimeter zones. Additional or supplemental heat is provided by unit-mounted hot-water or electric coils. No reheating of conditioned air is utilized.

Dual-duct VAV control units are specifically designed for dual-duct systems that require VAV for perimeter areas. These units feature pressure-compensated shut-off operation with pneumatic, electric, or system-powered controls, with a variable-volume cold deck and constant-volume or variable-volume hot deck. They are available with multiple outlets and factory-calibrated mixing points. Dual-duct control units maximize energy savings with complete control flexibility when fans modulate both heating and cooling air flows. Heating and cooling air can overlap to maintain minimum air flows with superior room-temperature control.

13.32.4 Dual-Duct VAV Control Diffusers

These assemblies consist of the dual-duct VAV control unit and linear-slot diffusers (Fig. 13.39). They are used for VAV dual-duct perimeters and special-use areas. They provide all the features and benefits described previously.

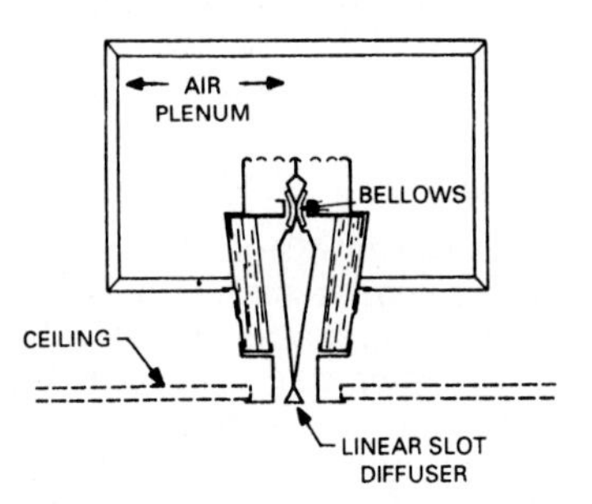

FIGURE 13.39 VAV terminal with linear-slot diffuser.

Linear-slot diffusers are available for all VAV systems. These diffusers maintain the desired coanda effect down to ±5% of air flow. They are available with one- or two-way throws, with up to four slots at 1000 cfm per diffuser. They may operate with aspirating electric, pneumatic or system-powered, unit-mounted thermostats. Factory-installed fire dampers are also available. Benefits include elimination of dumping at reduced flows, ease of relocation, and high capacity at low first cost, as well as reduced control costs with unit-mounted thermostats.

13.32.5 Noise Considerations

Rooftop packaged units used for VAV air conditioning may transmit high levels of noise to occupied spaces. Such noise is generally caused by compressor, condenser fans, or supply-duct exhaust fans. Computerized duct-design programs are available that indicate the sound generation that can be expected at any point in the duct distribution system. Examination of the predicted levels gives designers the opportunity to make design alterations to avoid adverse acoustical conditions.

Location of rooftop units is also important. They should not be installed over offices, conference rooms, or other critical areas. Units preferably should be located over storage rooms, corridors, or utility areas, where higher sound levels would be acceptable.

(H. E. Bovay, Jr., "Handbook of Mechanical and Electrical Systems for Buildings," N. R. Grimm and R. C. Rosaler, "Handbok of HVAC Design," McGraw-Hill Publishing Company, New York; B. Stein et al., "Mechanical and Electrical Equipment for Buildings," John Wiley & Sons, Inc., New York.)

13.33 AIR-WATER SYSTEMS

An alternative to the air-to-air systems described in Art. 13.32, air-water systems furnish chilled water from a remote chiller or central plant to the room terminal devices. These contain a cooling coil or a heating coil, or both. Room temperature is maintained by varying the flow of chilled water or heating fluid through the coils with valves that respond to the thermostat. Ventilation air is provided from a separate central plant directly to the room or the terminal device.

Two-pipe or four-pipe systems are used for distribution of chilled water and hot water to the room terminals from a central plant. In a two-pipe system, the supply pipe may carry either chilled water or hot water and the second pipe is used as a return. The four-pipe system provides two pipes for chilled-water supply and return and two pipes for hot-water supply and return. The installed cost of the two-pipe system is less than that of the four-pipe system, but the versatility is less. The major disadvantage of the two-pipe system is its inability to provide both heating and cooling with a common supply pipe on days for which both heating and cooling are desired. The four-pipe system has a major drawback in loss of temperature control whenever a changeover from cooling to heating is desired. To overcome this, thermostats are used that permit selection of either cooling or heating by a manual changeover at the thermostats.

Terminal devices for air-water systems are usually of the fan-coil or induction types.

Fan-Coil Terminal Units. A fan-coil terminal device consists of a fan or blower section, chilled-water coil, hot-water heating coil or electric-resistance heating elements, filter, return-air connection, and a housing for these components with an opening for ventilation air. The electric-resistance heating coil is often used with two-pipe systems to provide the performance of a four-pipe system without the cost of the two extra pipes for hot water, insulation, pumps, etc. Fan-coil units may be floor mounted, ceiling mounted-exposed or ceiling mounted-recessed, or ceiling mounted-recessed with supply- and return-air ductwork. When furnished with heating coils, the units are usually mounted on the outside wall or under a window, to neutralize the effects of perimeter heat losses.

Built-in centrifugal fans recirculate room air through the cooling coil. Chilled water circulating through the coil absorbs the room heat load. Ventilation air that is conditioned by another remote central plant is ducted throughout the building and supplied directly to the room or room terminal devices, such as a fan-coil unit. A room thermostat varies the amount of cooling water passing through the cooling coil, thus varying the discharge temperature from the terminal unit and satisfying the room thermostat.

Induction Terminal Units. These units are frequently used in large office buildings. The units are served by a remote air-handling unit that provides high-pressure conditioned air, which may be heated or cooled and is referred to as primary air. It is distributed to individual induction units that are located on the outside walls of each room or zone. At the terminal induction unit, a flow of high-pressure primary air through several nozzles induces a flow of room air through the heating or cooling coil in the unit. Room air mixes with the primary air to provide a mixed-air temperature that satisfies the thermal requirements of the space. In most systems, the ratio of induced air to primary air is about 4 to 1.

The induction system is a large energy consumer because of the extra power required to maintain the high pressure necessary to deliver the primary air to the room induction nozzles and induce room air to flow through the unit coils. Also, the induction terminal unit operates simultaneously with heating and cooling, wasting energy as in a terminal-reheat type of operation.

Air-water systems generally have substantially lower installed and operating costs than all-air systems. They do not, however, provide as good control over room temperature, humidity, air quality, air movement, and noise. The best control of an air-water system is achieved with a fan-coil unit with supplemental ventilation air from a central, primary-air system that provides ventilation air.

(H. E. Bovay, Jr., "Handbook of Mechanical and Electrical Systems for Buildings," and N. R Grimm and R. C. Rosaler, "Handbook of HVAC Design," McGraw-Hill Publishing Company, New York.)

13.34 CONTROL SYSTEMS FOR AIR CONDITIONING

One commonly used system has a thermostat wired in series with the compressor holding-coil circuit (Fig. 13.40). Thus the compressor will stop and start as called for by room conditions. The high-low pressure switch in series with the thermostat is a safety device that stops the compressor when the head pressure is too high and when the suction pressure approaches the freezing temperature of the coil. A liquid solenoid will shut off the flow of refrigerant when the compressor stops, to prevent flooding the coil back to the compressor during the off cycle. This valve may be eliminated when the air-handling unit and compressor are close together.

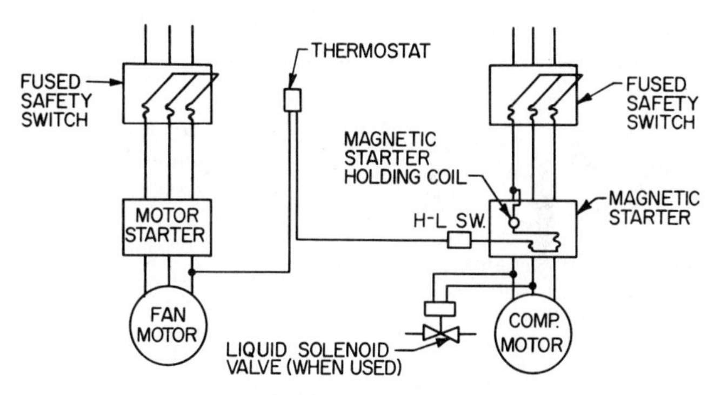

FIGURE 13.40 Control circuit for an air-conditioning system in which the thermostat controls compressor operation.

A second type of control is the pump-down system (Fig. 13.41). The thermostat shuts off the flow of the refrigerant, but the compressor will keep running. With the refrigerant supply cut off, the back pressure drops after all the liquid in the coil vaporizes. Then, the high-low pressure switch cuts off the compressor.

Either of these two systems is satisfactory. However, the remainder of this discussion will be restricted to the pump-down system.

Additional safety controls are provided on packaged units to reduce compressor burnouts and increase the average life of these units. The controls include crankcase heaters, motor-winding thermostats, and nonrecycling timers. These controls are usually prewired in the factory. The manufacturer supplies installation and wiring instructions for interconnecting the various components of the air-conditioning system.

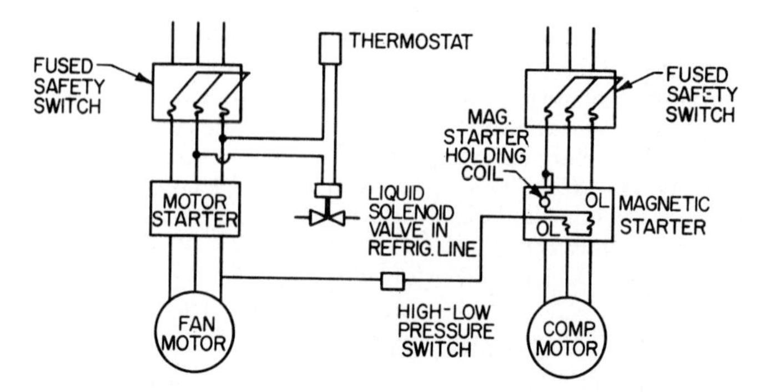

FIGURE 13.41 Pump-down system for control of air conditioning.

Where city water is used for condensing purposes, an automatic water-regulator valve is supplied (usually by the manufacturer). Figure 13.42 shows a cross section of such a valve. The power element is attached to the hot-gas discharge of the compressor. As the head pressure builds up, the valve is opened more, allowing a greater flow of water to the condenser, thus condensing the refrigerant at a greater rate. When the compressor is shut off, the head pressure drops, the flow of water being cut off by the action of a power spring working in opposition to the power element. As the water temperature varies, the valve responds to the resulting head pressure and adjusts the flow automatically to maintain design head pressure.

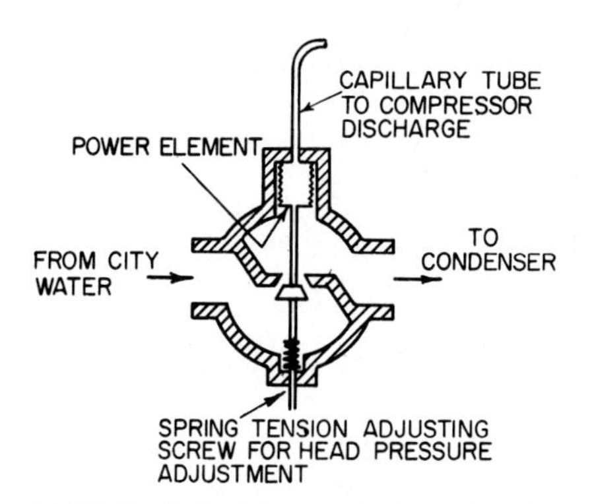

FIGURE 13.42 Water-regulating valve for condensing in an air-conditioning system.

When a water tower is used, the automatic water-regulating valve should be removed from the circuit, because it offers too much resistance to the flow of water. It is not usually necessary to regulate the flow of water in a cooling-tower system. For when the outside wet-bulb temperature is so low that the tower yields too low a water temperature, then air conditioning generally is not needed.

Figure 13.43 shows a simplified wiring diagram for a tower system. Because of the interconnecting wiring between the magnetic starters, the tower fan cannot run unless the air-conditioning fan is in operation. Also, the circulating pump cannot run unless the tower fan is in operation, and the compressor cannot run unless the circulating pump is in operation.

When the system is started, the air-conditioning fan should be operated first, then the tower via switch No. 1, the pump via switch No. 2, and the compressor via switch No. 3. When shutting down, switch No. 3 should be snapped off, then switch No. 2, then switch No. 1, and then the air-conditioning fan.

In buildings where one tower and pump are used to provide water for many units, pressure switches may be used in series with the holding coil of each compressor motor starter. No interconnecting wiring is necessary, for the compressor

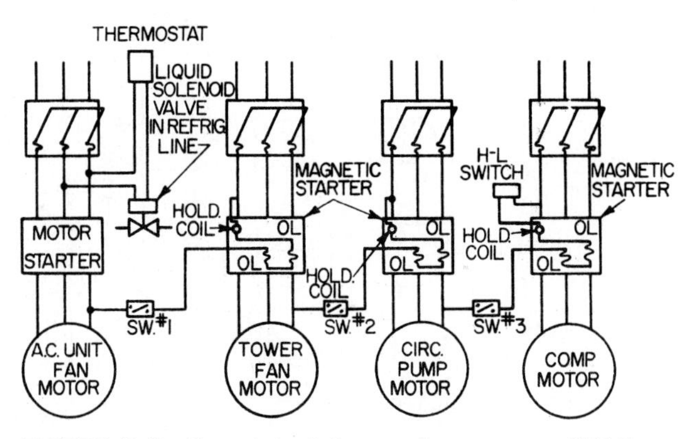

FIGURE 13.43 Control circuit for a cooling-tower-type HVAC system.

will not be able to run unless the pump is in operation and provides the necessary pressure to close the contacts on the pressure switch.

Some city ordinances require that indirect-expansion, or chilled-water, systems be installed in public buildings, such as theaters, night clubs, hotels, and depots. This type of system is illustrated in Fig. 13.30. The refrigerant is used to cool water and the water is circulated through the cooling coil to cool the air. The water temperature should be between 40 and 45°F, depending on whether the occupancy is a high latent load or a low latent load. Because the cost of equipment is increased and the capacity decreased as water temperature is lowered, most designers use 45°F water as a basis for design and use a much deeper cooling coil for high-latent-load installations.

The amount of chilled water in gallons per minute to be circulated may be obtained from

$$Q = \frac{24 \times \text{tons of refrigeration}}{\Delta T} \tag{13.38}$$

where ΔT is the temperature rise of water on passing through the cooling coil, usually 8 or 10°F. The smaller the temperature change, the more water to be circulated and the greater the pumping cost, but the better will be the heat transfer through the water chiller and cooling coil.

Coil manufacturers' catalogs usually give the procedure for picking the number of rows of coil necessary to do the required cooling with the water temperatures available.

The water chiller should be sized to cool the required flow of water from the temperature leaving the cooling coil to that required by the coil at a given suction temperature.

Assuming 45°F water supplied to the cooling coil and a temperature rise of 8°F, the water leaving the coil would be at 53°F. The chiller would then be picked to cool the required flow from 53 to 45°F at 37°F suction. The lower the suction temperature, the smaller will be the amount of heat-transfer surface required in the chiller. Also, the lower will be the compressor capacity. In no case should the suction temperature be less than 35°F, since the freezing point of water—32°F—is too close. A frozen and cracked chiller is expensive to replace.

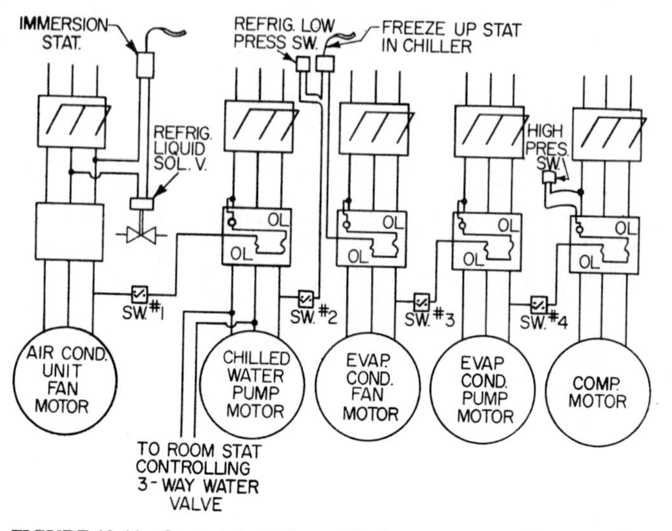

FIGURE 13.44 Control circuit for a chilled-water system with evaporative condenser.

A simplified control system for a chilled-water system with an evaporative condenser is shown in Fig. 13.44.

13.35 HEATING AND AIR CONDITIONING

Most manufacturers of air-conditioning equipment allow space in the air-handling compartment for the installation of a humidification unit and a heating coil for hot water or steam. These make it possible to humidify and heat the air in cold weather. Before a decision is reached to heat through the air-conditioning duct system, it is important to consider the many pitfalls present:

When a unit provides air conditioning in a single room, a heating coil may do the job readily, provided the Btu-per-hour rating of the coil is equal to or greater than the heating load. The fact that the supply grilles are high is usually no disadvantage; since winter heating is designed for the same amount of air as summer cooling, a small temperature rise results, and the large air-change capacity of the air-handling unit creates enough mixing to prevent serious stratification. Difficulties usually are encountered, however, in a structure with both exposed and interior zones. In the winter, the exposed zones need heating, while the interior zones are warm. If the heating thermostat is located in the exposed zone, the interior zones will become overheated. If the thermostat is located in the interior zone, the exposed zones will be too cold. Where some heat is generated in the interior zones, the system may require cooling of the interior and heating of the exterior zones at the same time. Thus, it is impossible to do a heating and cooling job with a single system in such structures.

Where heating and cooling are to be done with the same duct system, the air-handling equipment should be arranged to service individual zones—one or more units for the exposed zones, and one or more units for the interior zones.

When a heating system is already present and an air-conditioning system is added, a heating coil may be used to temper the outside air to room temperature. A duct-type thermostat may be placed in the discharge of the unit to control the steam or hot-water valve. When a room thermostat is used, the spare coil capacity may be used for quick morning pickup. Later in the day, the system may be used for cooling the premises with outside air if the building-heating system or other internal heat sources overheat the premises. In buildings with large window areas, it is advisable to place under the windows for use in cold weather some radiation to supply heat in addition to that from the heating coil, to counteract the down draft from the cold glass surfaces.

The heating-coil size should be such that its face area is about equal to the cooling-coil face area. The number of rows deep should be checked with manufacturers' ratings. When the unit is to be used for tempering only, the coil need be sized for the fresh-air load only [Eq. (13.30)].

When large quantities of outside air are used, it is usual practice to install a preheat coil in the fresh-air duct and a reheat coil in the air-conditioning unit. Install the necessary filters before the preheat coil to prevent clogging.

13.36 CONTROL OF COMPUTERIZED HVAC SYSTEMS

Control of HVAC systems ranges from simple thermostat "on-off" control to control by sophisticated electronic and computerized systems. Many variations of electronic and computerized systems are available for system control and energy management.

Programmable Thermostats. These are available for control of the total environment. They provide energy savings and improved comfort levels and are widely utilized to control many types of commercial unitary products. Different programmable types are available for heat pumps and for single- and two-stage heating and cooling equipment.

Programmable thermostats conserve energy by automatically raising or lowering the space temperature to preprogrammed settings several times each day. Building temperature adjustment is performed during occupied or unoccupied hours, again conserving energy.

Other types of thermostats are available that produce energy savings:

- Reduction of overshoot on recovery from the setback temperature, limiting the rate of temperature rise
- Selective recovery from the setback temperature by utilizing lower stages (single-stage and multistage only)
- Selective recovery from the setback temperature using a heat pump in lieu of an auxiliary heat source (heat-pump thermostat)
- Prevention of room temperature from deviating from a setpoint under varying load conditions

Microprocessor-Based Systems. Computerized energy-management systems are available to control complex HVAC systems. These microprocessor-based systems are capable of saving enough energy that they can pay for themselves in 2 years or less. Energy management systems are available in many variations and with many features. The most common include the following capabilities: equipment scheduling, duty cycling of equipment, temperature compensated control, demand limiting, optimum start/stop, and night setback.

Equipment Scheduling. Running HVAC equipment on a continuous basis, including periods when operation is not required, results in excessive and nonproductive energy consumption. These periods include lunch hours, evenings, weekends, and holidays. By using a scheduling program, the control panel promptly shuts down HVAC units and other units at preset times. A typical scheduling program may provide four start/stop times for each load on an 8-day cycle. It will also allow automatically for holidays, leap year, and daylight savings time.

Duty Cycling. Duty cycling applied to HVAC equipment utilizes an on-off control to reduce operating time and increase energy savings. The duty-cycling program will permit up to eight cycle patterns for each load. On-off protection timers are provided for each piece of equipment to assure equipment safety.

Temperature-Compensated Control. Maximum occupant comfort is ensured by use of a computerized control that provides equipment cycling with automatic temperature override. Cycling will be reduced or stopped, depending on whether room temperatures fall outside present comfort levels. Hence, the guesswork of selecting proper cycling strategies is eliminated.

Demand Limiting. As much as 50% of bills for use of electrical energy may be attributed to kW demand charges. A demand-limiting program may save a substantial amount of utility energy costs over a short period of time. A computerized control, in particular, can limit costly demand charges by use of a predictive demand program that anticipates electrical demand peaks. When the control panel senses that a peak is about to occur, it simply turns off selected loads on a priority basis until the peak subsides, thereby reducing the energy bill.

Optimum Start/Stop. Energy can be wasted by HVAC equipment that is scheduled to start and stop on worst-case conditions. This scheduling results in operations that begin too soon and run too long. Computerized control monitors indoor and outside temperatures and determines the most efficient start and stop times without sacrificing individual comfort. Optimum start/stop also allows independent control of individual comfort zones.

Night Setback. Utilization of night setback temperatures saves energy by automatically converting the specific comfort zone to a predetermined nighttime-temperature setpoint. The setpoint overrides the daytime setting for every load or zone. HVAC equipment will operate to maintain the nighttime setting until morning warm-up or cool-down is required. Night setback also allows independent control of each zone.

13.37 DIRECT DIGITAL CONTROL

Heating, ventilation, and air conditioning has usually been controlled by three basic systems: pneumatic, electric, electronic, but a newer technology, known as direct digital control (DDC), can lower costs and improve comfort in a vast range of complex applications. DDC is an electronic system that uses a digital microprocessor (computer) to analyze various parameters in HVAC, such as temperatures, pressures, air flows, time of day, and many other types of data. The information obtained is used to operate control devices, such as dampers and valves, and to start/stop mechanical equipment such as air-handling units and heating and cooling pumps, in accordance with instructions programmed into the microprocessors' memory.

DDC also offers features in addition to the basic functions, such as monitoring and alarm, data gathering, and energy monitoring and management. DDC systems also accommodate lighting-control and security-system functions and management. They can delegate all of the control logic, including determination of necessary output signals, to be transmitted to the control devices and to the microprocessor. Development of lower-cost chips (computers on a chip) may reduce limitations on the number of devices a single microprocessor can handle.

DDC technology tends toward distributed control, which permits multiple controllers to be used, each with its own microprocessor. This allows each controller to operate independently; after it has been programmed, it can operate independently. A twisted-wire pair is usually used as a data link for connecting the controllers together. This permits exchange of information with each other or with a host computer and, in many cases, with desktop personal computers (PCs).

The host computer is connected to the distributed controllers and is considered as the operator's link. It permits the user to exchange information on the status of any portion of the HVAC system and to revise control schedules, values, or setpoints. When the host computer is not being used for these tasks, it can be used with PCs for usual or conventional tasks, while the controllers continue to operate with their most current set of instructions.

Capacity of a controller is usually measured by its point capacity. Input or output points may be either analog or digital classifications. Analog inputs are those with values that change with time, such as temperatures and pressure. Digital outputs are those that describe a condition that may be of two types only such as on and off.

The advantages of DDC compared with pneumatic control systems are numerous. The greatest advantage is the ability to achieve impressive energy conservation without sacrificing comfort. DDC permits a system to be controlled in an optimum manner while maintaining space temperatures within desired limits. Also, DDC is an extremely flexible control system that permits an operator to control the parameters and control sequence to maintain maximum efficiency. On large installations with many buildings, air-handling units may be reprogrammed for readjustment of parameters for improved comfort conditions or optimal energy conservation, or both. These requirements are simpler to meet with a DDC system than with a pneumatic system.

An analog output signal is used for modulating dampers or varying flow quantities through a liquid-control valve. To start and stop electrical equipment, a digital output is used and is usually a dry type of contact.

As an alternative, a "mix" of DDC and pneumatic actuators may be used on the controlled devices, but electronically operated. In some cases, use of electronic actuation may be economical, particularly for larger valves and dampers. When a pneumatic output signal is required from the controller, a simple digital-to-pneumatic relay converts the electronic signal to a variable pressure signal. Pneumatic actuators provide the additional advantages of excellent reliability and low maintenance cost.

13.38 INDUSTRIAL AIR CONDITIONING

Certain manufacturing processes call for close control of temperature and relative humidity. A typical control system is shown in Fig. 13.45. The humidistat may be of the single-pole, double-throw type.

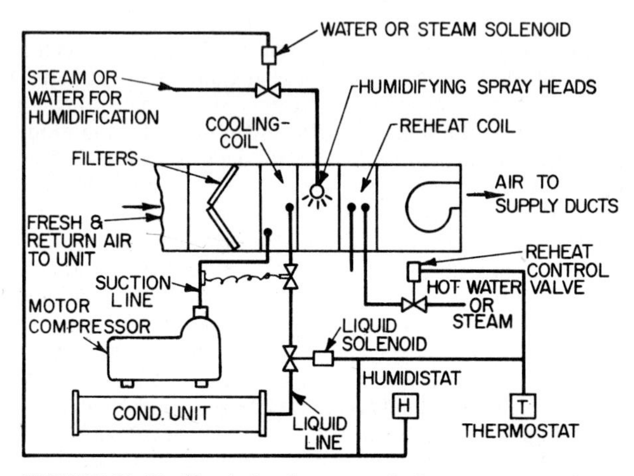

FIGURE 13.45 Circuit for close control of temperature and humidity.

On a rise in room humidity, the refrigeration compressor cuts in for the purpose of dehumidification. Since the sensible capacity of the cooling coil usually is much higher than the sensible load, whereas the latent capacity does not differ too much from the latent load, undercooling results. The room thermostat will then send enough steam or hot water to the reheat coil to maintain proper room temperature.

A small drop in room relative humidity will cut off the compressor. A further drop in room relative humidity will energize the water or steam solenoid valve to add moisture to the air. The thermostat is also arranged to cut off all steam from the reheat coil when the room temperature reaches the thermostat setting. However, there may be periods when the humidistat will not call for the operation of the refrigeration compressor and the room temperature will climb above the thermostat setting. An arrangement is provided for the thermostat to cut in the compressor to counteract the increase in room temperature.

When calculating the load for an industrial system, designers should allow for the thermal capacity and moisture content of the manufactured product entering and leaving the room. Often, this part of the load is a considerable percentage of the total.

13.39 CHEMICAL COOLING

When the dew point of the conditions to be maintained is 45°F or less, the method of controlling temperature and humidity described in Art. 13.38 cannot be used because the coil suction temperature must be below freezing and the coil will freeze up, preventing flow of air. For room conditions with dew point below 45°F, chemical methods of moisture removal, such as silica gel or lithium chloride, may be used. This equipment is arranged to have the air pass through part of the chemical and thus give up its moisture, while another part of the system regenerates the chemical by driving off the moisture previously absorbed. The psychrometric process involved in this type of moisture removal is adiabatic or, for practical purposes, may be considered as being at constant wet-bulb temperature. For example, if we take room air at 70°F and 30% relative humidity and pass the air through a chemical moisture absorber, the air leaving the absorber will be at a wet-bulb temperature of 53°F; i.e., the conditions leaving may be 75°F and 19% RH or 80°F and 11% RH, etc.

It may be noticed that the latent load here is converted to sensible load, and a refrigeration system will be necessary to do sensible cooling. See Fig. 13.46 for the control of such a system.

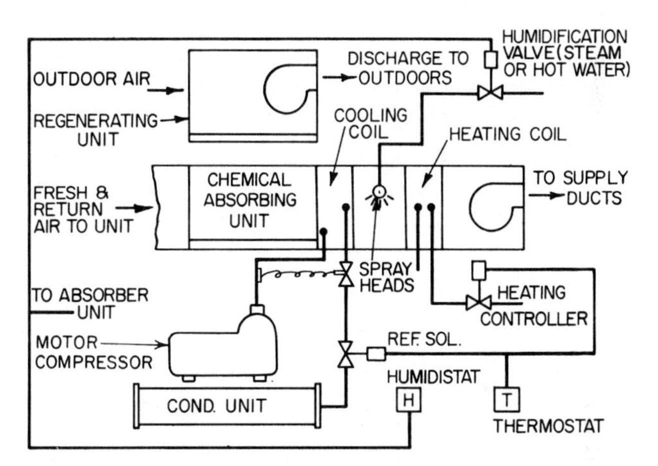

FIGURE 13.46 Controls for cooling with a chemical moisture absorber.

13.40 YEAR-ROUND AIR CONDITIONING

Electronic-computer rooms and certain manufacturing processes require temperature and humidity to be closely controlled throughout the year. Air-conditioning systems for this purpose often need additional controls for the condensing equipment.

When city water is used for condensing and the condensers are equipped with a water regulating valve, automatic throttling of the water by the valve automatically compensates for low water temperatures in winter. When cooling towers or evaporative condensers are used, low wet-bulb temperatures of outside air in winter will result in too low a head pressure for proper unit operation. A controller may be used to recirculate some of the humid discharge air back to the evaporative-condenser suction to maintain proper head pressure. With cooling towers, the water temperature is automatically controlled by varying the amount of air across the towers, or by recirculating some of the return condenser water to the supply. Figure 13.47 shows the piping arrangement.

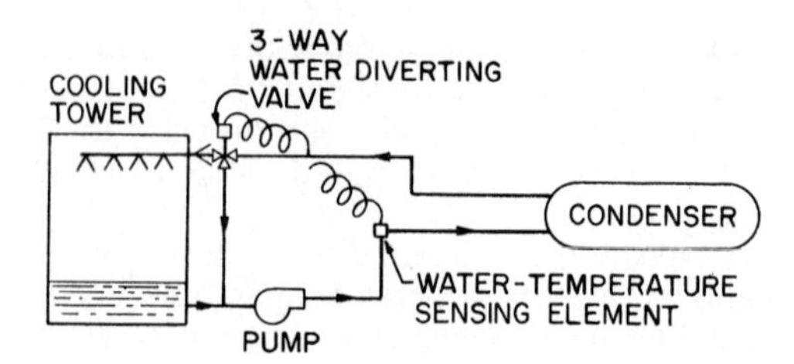

FIGURE 13.47 Piping layout for control of temperature of water from a cooling tower.

When air-cooled condensers are used in cold weather, the head pressure may be controlled in one of the following ways: for a single-fan condenser, by varying the fan speed; for a multifan condenser, by cycling the fans or by a combination of cycling and speed variation of fans for closer head-pressure control.

Some head-pressure control systems are arranged to flood refrigerant back to the condenser coil when the head pressure drops. Thus, the amount of heat-transfer surface is reduced, and the head pressure is prevented from dropping too low.

SECTION FOURTEEN
PLUMBING—WATER-SUPPLY, SPRINKLER, AND WASTEWATER SYSTEMS

Gregory P. Gladfelter
HNTB Architects/Engineers/Planners
Kansas City, Missouri

This section treats the major subsystems for conveyance of liquids and gases in pipes within a building. The pipes generally extend beyond the building walls to a supply source or a disposal means, such as a sewer.

14.1 PLUMBING AND FIRE PREVENTION CODES

Plumbing codes were created to prevent illness and death from unsanitary or unsafe conditions in supply of water and gases in buildings and removal of wastes in pipes. There are three commonly recognized model plumbing and fire prevention codes:

"Boca National Plumbing Code" and "Boca National Fire Prevention Code," Building Officials and Code Administrators International, Inc., Country Club Hills, Ill.

"Standard Plumbing Code" and "Standard Fire Prevention Code," Southern Building Code Congress International, Inc., Birmingham, Ala.

"Uniform Plumbing Code" and "Uniform Fire Code," International Association of Plumbing and Mechanical Officials, Walnut, Calif.

These codes are generally revised on 3-year cycles.

In addition to these model codes, several cities and states have adopted their own plumbing and fire prevention codes. The American National Standards Institute (ANSI) has also adopted the "National Plumbing Code," ANSI A.40.8, which is administered by the Mechanical Contractors Association of America, Rockville, Md. Also, numerous fire-safety codes and standards are contained in "National Fire Codes," National Fire Protection Association, Quincy, Mass.

Persons involved in the design and installation of plumbing systems should check with all local code authorities to determine which code is in effect prior to beginning a project. Also, local governmental authorities should be contacted about special regulations relating to sewer and water systems. Those involved in the design of plumbing systems should also be familiar with ANSI A117.1 and the Americans with Disabilities Act (ADA), which require that provision be made in buildings for accessibility and usability of facilities by the physically handicapped. Plumbing designers and architects should work together to assure strict compliance with these requirements.

14.2 HEALTH REQUIREMENTS FOR PLUMBING

Plumbing codes place strict constraints on plumbing installations in the interest of public health. Following are typical basic provisions:

All buildings must be provided with potable water in quantities adequate for the needs of their occupants. Plumbing fixtures, devices, and appurtenances should be supplied with water in sufficient volume and at pressures adequate to enable them to function properly. The pipes conveying the water should be of sufficient size to provide the required water without undue pressure reduction and without undue noise under all normal conditions of use.

The plumbing system should be designed and adjusted to use the minimum quantity of water consistent with proper performance and cleansing of fixtures and appurtenances.

Devices for heating and storing water should be designed, installed, and maintained to guard against rupture of the containing vessel because of overheating or overpressurization.

The wastewater system should be designed, constructed, and maintained to guard against fouling, deposit of solids, and clogging.

Provision should be made in every building for conveying storm water to a storm sewer if one is available.

Recommended tests should be made to discover any leaks or defects in the system. Pipes, joints, and connections in the plumbing system should be gastight and watertight for the pressure required by the tests.

Plumbing fixtures should be located in ventilated enclosures and should be readily accessible to users.

Plumbing fixtures should be made of smooth, nonabsorbent materials. They should not have concealed fouling surfaces. Plumbing fixtures, devices, and appliances should be protected to prevent contamination of food, water, sterile goods, and similar material by the backflow of wastewater. Indirect connections with the building wastewater system should be provided when necessary.

Every fixture directly connected to the wastewater system should be equipped with a **liquid-seal trap.** This is a fitting so constructed that passage of air or gas through a pipe is prevented while flow of liquid through the pipe is permitted.

Foul air in the wastewater system should be exhausted to the outside, through vent pipes. These should be located and installed to minimize the possibility of clogging and to prevent sewer gases from entering the building.

If a wastewater system is subject to the backflow of sewage from a sewer, suitable provision should be made to prevent sewage from entering the building.

The structural safety of a building should not be impaired in any way as a result of the installation, alteration, renovation, or replacement of a plumbing system.

Pipes should be installed and supported to prevent stresses and strains that would cause malfunction of or damage to the system. Provision should be made for expansion and contraction of the pipes due to temperature changes and for structural settlements that might affect the pipes.

Where pipes pass through a construction that is required to have a fire-resistance rating, the space between the pipe and the opening or a pipe sleeve should not exceed ½ in. The gap should be completely filled with code-approved, fire-stopping material and closed off with close-fitting metal escutcheons on both sides of the construction.

Pipes, especially those in exterior walls or underground outside the building, should be protected, with insulation or heat, to prevent freezing. Underground pipes should be placed below established frost lines to prevent damage from heaving and in high traffic areas should be encased in concrete or installed deep enough so as to not be damaged by heavy traffic. Pipes subject to external corrosion should be protected with coatings, wrappings, cathodic protection, or other means that will prevent corrosion. Dissimilar metals should not be connected to each other unless separated by a dielectric fitting. Otherwise, corrosion will result.

Each plumbing system component, such as domestic water, natural gas, and wastewaster pipes and fixtures, should be tested in accordance with the plumbing code. All defects found during the test should be properly corrected and the system retested until the system passes the requirements of the test.

WATER SUPPLY

Enough water to meet the needs of occupants must be available for all buildings. Further water needs for fire protection, heating, air conditioning, and possibly process use must also be met. This section provides specific data on all these water needs, except those for process use. Water needs for process use must be computed separately because the demand depends on the process served.

14.3 WATER QUALITY

Sources of water for buildings include public water supplies, groundwater, and surface water. Each source requires carerful study to determine if a sufficient quantity of safe water is available for the building being designed.

Water for human consumption, commonly called potable water, must be of suitable quality to meet local, state, and national requirements. Public water supplies generally furnish suitably treated water to a building, eliminating the need for treatment in the building. However, ground and surface waters may require treatment prior to distribution for human consumption. Useful data on water treatment are available from the American Water Works Association, Denver, Colo.

Useful data on water supplies for buildings are available in the following publications: American Society of Civil Engineers, "Glossary—Water and Sewer Control Engineering;" E. W. Steel, "Water Supply and Sewerage," McGraw-Hill Publishing Company, New York; G. Fair, J. C. Geyer, and D. A. Okun, "Water and Wastewater Engineering," John Wiley & Sons, Inc., New York; and E. Nordell, "Water Treatment for Industrial and Other Uses," Van Nostrand Reinhold, New York. The ASTM "Manual on Industrial Water" contains extensive data on process-water and steam requirements for a variety of industries. Data on water for fire protection are available from the American Insurance Association, New York, and the National Fire Protection Association (NFPA), Quincy, Mass.

Water for buildings is transmitted and distributed in pipes, which may be run underground or aboveground. Useful data on pipeline sizing and design are given in J. Church, "Practical Plumbing Design Guide," and C. E. Davis and K. E. Sorenson, "Handbook of Applied Hydraulics," McGraw-Hill Publishing Company, New York. The American Insurance Association promulgates a series of publications on water storage tanks for a variety of services.

Characteristics of Water. Physical factors of major importance for raw water are temperature, turbidity, color, taste, and odor. All but temperature are characteristics to be determined in the laboratory from carefully procured samples by qualified technicians utilizing current testing methods and regulations.

Turbidity, a condition due to fine, visible material in suspension, is usually due to presence of colloidal particles. It is expressed in parts per million (ppm or mg/L) of suspended solids. It may vary widely in discharges of relatively small streams of water. Larger streams or rivers tending to be muddy are generally muddy all the time. The objection to turbidity in potable supplies is its ready detection by the drinker. The U.S. Environmental Protection Agency (USEPA) limit is one nephelometric turbidity unit (NTU).

Color, also objectionable to the drinker, is preferably restricted to 15 color units or less. It is measured, after all suspended matter (turbidity) has been centrifuged out, by comparison with standard hues.

Tastes and odors due to organic material or volatile chemical compounds in the water should be removed completely from drinking water. But slight, or threshold, odors due to very low concentrations of these compounds are not harmful—just objectionable. Perhaps the most common source of taste and odor is decomposition of algae.

Chemical Content. Chemical constituents commonly found in raw waters intended for potable use and measured by laboratory technicians include hardness, pH, iron, and manganese, as well as total solids. Total solids should not exceed 500 ppm. Additionally, the USEPA is continually developing, proposing, and adopting new drinking water regulations as mandated by the Safe Drinking Water Act.

Hardness, measured as calcium carbonate, may be objectionable in laundries with as little as 150 ppm of $CaCO_3$ present. But use of synthetic detergents decreases its significance and makes even much harder waters acceptable for domestic uses. Hardness is of concern, however, in waters to be used for boiler feed, where boiler scale must be avoided. Here, 150 ppm would be too much hardness and the water would require softening (treatment for decrease in hardness).

Hydrogen-ion concentration of water, commonly called pH, can be a real factor in corrosion and encrustation of pipe and in destruction of cooling towers. A pH under 7 indicates acidity; over 7 indicates alkalinity; 7 is neutral. Tests using color can measure pH to the nearest tenth, which is of sufficient accuracy.

Iron and manganese when present in more than 0.3-ppm concentrations may discolor laundry and plumbing. Their presence and concentration should be determined. More than 0.2 ppm is objectionable for most industrial uses.

Organic Content. Bacteriological tests of water must be made on carefully taken and transported samples. A standard sample is five portions of 10 cm^3, each sample a different dilution of the water tested. A state-certified laboratory will use approved standard methods for analyses.

Organisms other than bacteria, such as plankton (free-floating) and algae, can in extreme cases be important factors in design of water treatment systems; therefore, biological analyses are significant. Microscopic life and animal and vegetable matter can be readily identified under a high-powered microscope.

Maintenance of Quality. It is not sufficient that potable water just be delivered to a building. The quality of the water must be maintained while the water is being conveyed within the building to the point of use. Hence, the potable-water distribution system must be properly designed to prevent contamination.

No cross connections may be made between this system and any portion of the wastewater-removal system. Furthermore, the potable-water distribution system should be completely isolated from parts of plumbing fixtures or other devices that might contaminate the water. Backflow preventers or air gaps may be used to prevent backflow or back siphonage.

Backflow is the flow of liquid into the distribution piping system from any source other than the intended water-supply source, such as a public water main.

Back siphonage is the suction of liquid back into the distribution piping system because of a siphonage action being applied to the distribution pipe system. The type of backflow preventer to use depends on the type of reverse flow expected (backflow or back siphonage) and the severity of the hazard. In general, double check-valve-type backflow preventers are normally approved for low-hazard backflow conditions and vacuum breakers are approved for low-hazard back-siphonage conditions. Where the hazard is great, reduced-pressure principal backflow preventers are normally required. The local code authorities should be consulted about local and state regulations pertaining to backflow prevention.

14.4 WATER TREATMENT

To maintain water quality within acceptable limits (Art. 14.3), water supplied to a building usually must undergo some form of treatment. Whether treatment should be at the source or after transmission to the point of consumption is usually a question of economics, involving hydraulic features, pumping energies and costs, and possible effects of raw water on transmission mains.

Treatment, in addition to disinfection, should be provided for all water used for domestic purposes that does not fall within prescribed limits. Treatment methods include screening, plain settling, coagulation and sedimentation, filtration, disinfection, softening, and aeration. When treatment of the water supply for a building is necessary, the method that will take the objectionable elements out of the raw water in the simplest, least expensive manner should be selected.

Softening of water is a process that must be justified by its need, depending on use of the water. With a hardness in excess of about 150 ppm, the cost of softening will be offset partly by the reduction of soap required for cleaning. When synthetic detergents are used instead of soap, this figure may be stretched considerably. But when some industrial use of water requires it, the allowable level for hardness must be diminished appreciably.

Since corrosion can be costly, corrosive water must often be treated in the interest of economics. In some cases, it may be enough to provide threshold treatment that will coat distribution lines with a light but protective film of scale. But in other cases—boiler-feed water for high-pressure boilers, for example—it is important to have no corrosion or scaling. Then, deaeration and pH control may be necessary. (The real danger here is the failure of boiler-tube surfaces because of overheating due to scale formation.)

(American Water Works Association, "Water Quality and Treatment," McGraw-Hill Publishing Company, New York; G. M. Fair, J. C. Geyer, and D. A. Okun, "Elements of Water Supply and Wastewater Disposal," John Wiley & Sons, Inc., New York.)

14.5 WATER QUANTITY AND PRESSURES

Quantity of water supplied must be adequate for the needs of occupants and processes to be carried out in the building. The total water demand may be calculated by adding the maximum flows at all points of use and applying a factor less than unity to account for the probability that only some of the fixtures will be operated simultaneously (Art. 14.8).

In addition, the pressure at which water is delivered to a building must lie within acceptable limits. Otherwise, low pressures may have to be increased by pumps and high pressures decreased with pressure-reducing valves. Table 14.1 lists minimum flow rates and pressures generally required at various water outlets. The pressure in Table 14.1 is the pressure in the supply pipe near the water outlet while the outlet is wide open and water is flowing.

In delivery of water to the outlets, there is a pressure drop in the distribution pipes because of friction. Therefore, water supplied at the entrance to the distribution system must exceed the minimum pressures required at the water outlets

TABLE 14.1 Required Minimum Flow Rates and Pressures during Flow for Fixtures

Fixture	Pressure, psi*	Flow, gpm
Basin faucet	8	3
Basin faucet, self-closing	12	2.5
Sink faucet, ⅜-in.	10	4.5
Sink Faucet, ½-in	5	4.5
Dishwasher	15–25	†
Bathtub faucet	5	6
Laundry tub cock, ¼-in	5	5
Shower	12	3–10
Water closet ball cock	15	3
Water closet flush valve	10–20	15–40
Urinal flush valve	15	15
Garden hose, 50 ft, and sill cock	30	5

*Residual pressure in pipe at entrance to fixture.
†As specified by fixture manufacturer.

by the amount of the pressure loss in the system. But the entrance pressure should not exceed 80 psi, to prevent excessive flow and damage to system components. Velocity of water in the distribution system should not exceed 10 ft/s.

A separate supply of water must be provided for fire-fighting purposes. This supply must be of the most reliable type obtainable. Usually, this requirement can be met with water from a municipal water supply. If the municipal water supply is not adequate or if a private water supply is utilized, pumps or storage in an elevated water tank should be provided to supply water at sufficient quantities and pressures. Generally, such water should be provided at a pressure of at least 15 psi residual pressure at the highest level of fire-sprinkler protection for light-hazard occupancies and 20 psi residual pressure for ordinary-hazard occupancies. Acceptable flow at the base of the supply riser is 500 to 700 gpm for 30 to 60 min for light-hazard occupancies and 850 to 1500 gpm for 60 to 90 min for ordinary-hazard occupancies.

If a building is so located that it cannot be reached by a fire department with about 250 ft of hose, a private underground water system, installed in accordance with NFPA 24, "Installation of Private Fire Service Mains and Their Appurtenances," may have to be provided. Many municipalities require that the water system for a building site be a type generally called a "loop-to-grid" system. It consists of pipes that loop around the property and has a minimum of two municipal-water-system connections, at opposite sides of the loop, usually at different water mains of the municipal system. Hydrants should be placed so that all sides of a building can be reached with fire hoses. The requirements for fire hydrants should be verified with the local code officials or fire marshal.

14.6 WATER DISTRIBUTION IN BUILDINGS

Cold and hot water may be conveyed to plumbing fixtures under the pressure of a water source, such as a public water main, by pumps, or by gravity flow from elevated storage tanks.

The water-distribution system should be so laid out that, at each plumbing fixture requiring both hot and cold water, the pressures at the outlets for both supplies should be nearly equal. This is especially desirable where mixing valves may be installed, to prevent the supply at a higher pressure from forcing its way into the lower-pressure supply when the valves are opened to mix hot and cold water. Pipe sizes and types should be selected to balance loss of pressure head due to friction in the hot- and cold-water pipes, despite differences in pipe lengths and sudden large demands for water from either supply.

Care should be taken to assure that domestic water piping is not installed in a location subject to freezing temperatures. When piping is installed in exterior walls in cold climate areas, the piping should be insulated and should be installed on the building side of the building wall insulation. Piping installed in exterior cavity walls or chases may require heat tracing, although the installation of high and low wall-mounted grilles, which allow heated air from the building to naturally flow through the cavity, will usually prevent the temperature in the cavity from falling below a temperature where water in the piping will freeze. Designers should thoroughly investigate local climatic conditions and building methods to assure proper installation. Designers should also specify freeze-proof-type hdyrants (hose bibs) for exterior applications.

14.6.1 Temperature Maintenance in Hot-Water Distribution

In large, central, hot-water distribution systems, many fixtures that require hot water are not located very close to the water-heating equipment. If some means of maintaining the temperature of the hot water in the piping is not provided, the water temperature will fall, particularly during periods of low demand. The supply to remote fixtures would have to run for a long period before hot water would be available at the outlet and wasting precious water. For this reason, designers should provide a temperature maintenance system whenever a fixture requiring hot water is over 25 ft away from the water heater.

One method of temperature maintenance is to use a hot-water recirculating system, which consists of a hot-water return piping system, a circulating pump, and a water-temperature controller to operate the pump. The return piping system starts at the end of each remote branch main and runs back to the water-heater cold-water-supply pipe connection. The circulating pump circulates hot water through the supply piping, return piping, and the water heater whenever the controller senses that the water temperature has fallen below a preselected set point. To reduce heat loss, all hot-water supply and return piping should be insulated.

Another method employs self-regulating, electric heat tracing that is applied directly to the hot-water supply piping prior to the installation of the piping insulation. The self-regulating heat tracing is made of polymers, which have variable electric resistances, depending on the surface temperature of the pipe. As the surface temperature of the pipe falls, the resistance increases and more heat is given off by the heat tracing. The opposite is true if the surface of the piping is hot. This type of system requires less maintenance once it is installed and less energy to maintain the hot-water temperature in the piping.

Horizontal pipe runs should not be truly horizontal. They should have a minimum slope of about ¼ in/ft toward the nearest drain valve when possible. An adequate number of drain valves should be provided to drain the domestic water system completely.

14.6.2 Up-Feed Water Distribution

To prevent rapid wear of valves, such as faucets, water should only be supplied to building distribution systems at pressures not more than about 80 psi. This pressure is large enough to raise water from 8 to 10 stories upward and still retain desired pressures at plumbing fixtures (Table 14.1). Hence, in low buildings, cold water can be distributed by the up-feed method (Fig. 14.1), in which at each story plumbing fixtures are served by branch pipes connected to risers that carry water upward under pressure from the water source.

In Fig. 14.1, cold water is distributed under pressure from a public water main. The hot-water distribution is by a discontinuous system. Hot water rises from the water heater in the basement to the upper levels under pressure from the cold-water supply to the water heater.

When an up-feed distribution system is desired, but the city water pressure is not sufficient to provide adequate water pressure, the water pressure may be boosted to desired levels by the installation of a packaged, domestic water-booster pump system. This equipment usually consists of a factory-built system with multiple pumps, a pressure tank, and all operating controls to maintain the required water pressure. This type of system may also be used in buildings in excess of 10 stories by proper zoning and the use of pressure-reducing valves at each zone.

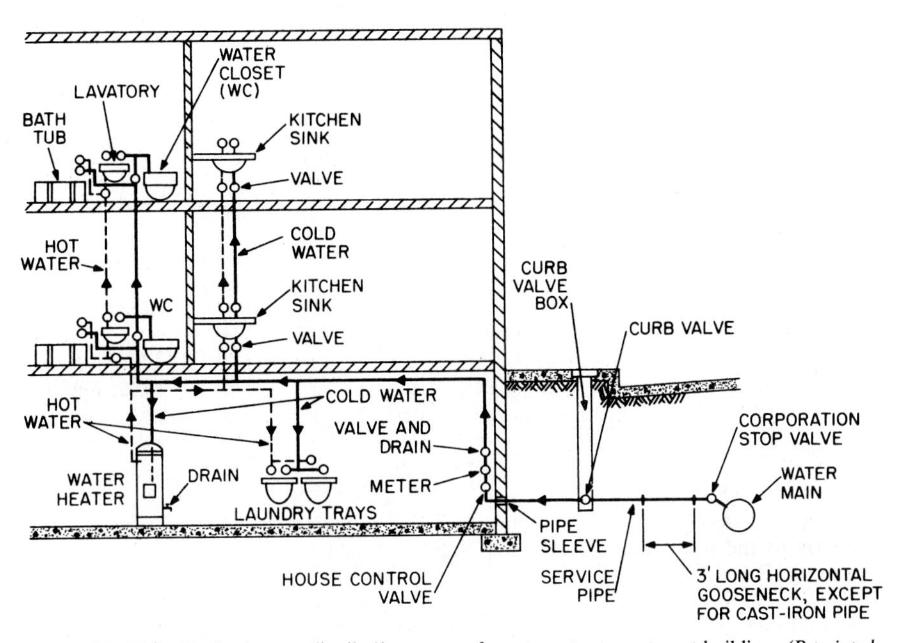

FIGURE 14.1 Up-feed water-distribution system for a two-story apartment building. *(Reprinted with permission from F. S. Merritt, "Building Engineering and Systems Design," Van Nostrand Reinhold Company, New York.)*

14.6.3 Down-Feed Water Distribution

For buildings more than 8 to 10 stories high, designers have the option to pump water to one or more elevated storage tanks, from which pipes convey the water downward to plumbing fixtures and water heaters. Water in the lower portion of an elevated tank often is reserved for fire-fighting purposes (Fig. 14.2). Generally, also, the tank is partitioned to provide independent, side-by-side chambers, each with identical piping and controls. During hours of low demand, a chamber can be emptied, cleaned, and repaired, if necessary, while the other chamber supplies water as needed. Float-operated electric switches in the chambers control the pumps supplying water to the tank. When the water level in the tank falls below a specific elevation, a switch starts a pump, and when the water level becomes sufficiently high, the switch stops the pump.

Usually, at least two pumps are installed to supply each tank. One pump is used for normal operation. The other is a standby, for use if the first pump is inoperative. For fire-fighting purposes, a pump must be of adequate size to fill the tank at the rate of the design fire flow.

When a pump operates to supply a tank, it may draw so much water from a public main that the pressure in the main is considerably reduced. To avoid such a condition, water often is stored in a suction tank at the bottom of the building for use by the pumps. The tank is refilled automatically from the public main. Because refilling can take place even when the pumps are not operating, water can be drawn from the public main without much pressure drop.

Figure 14.2 is a simplified schematic diagram of a down-feed distribution system of a type that might be used for buildings up to 20 stories high.

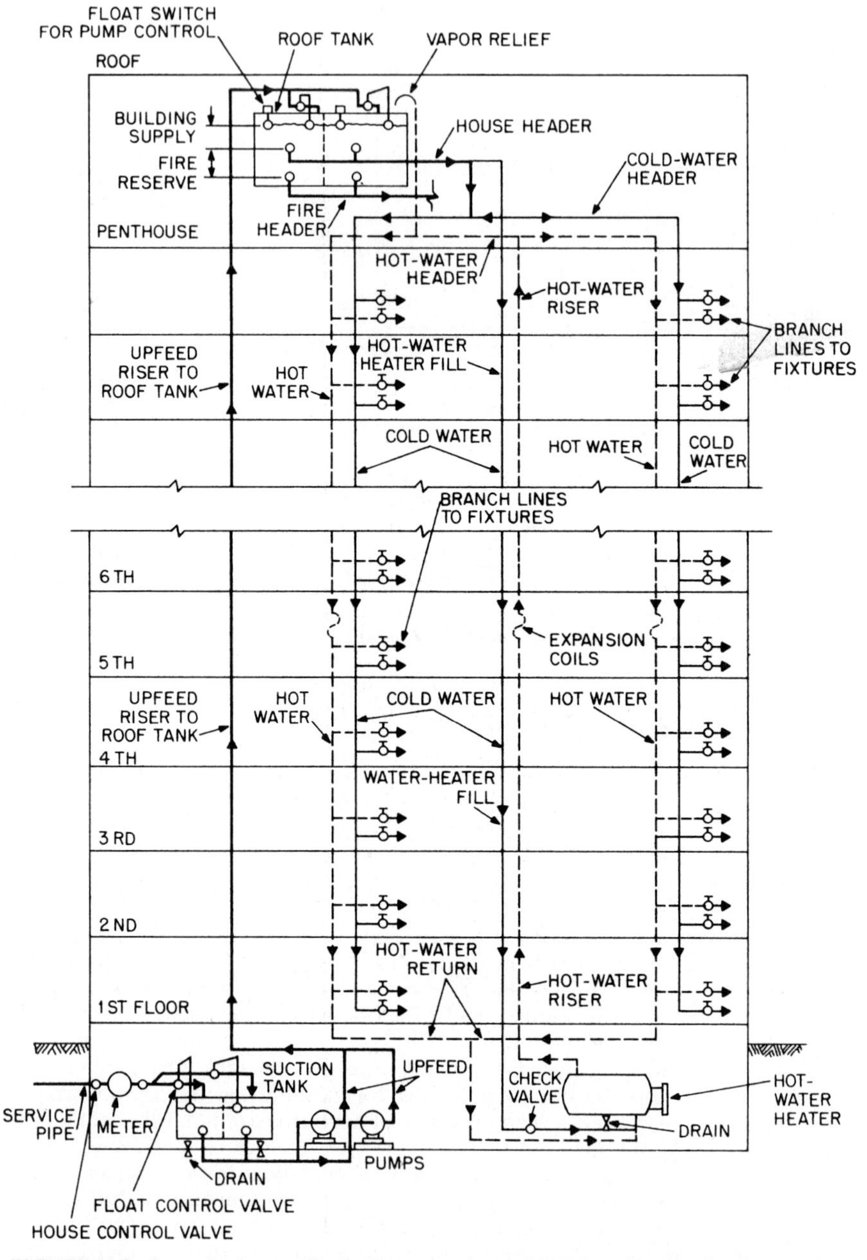

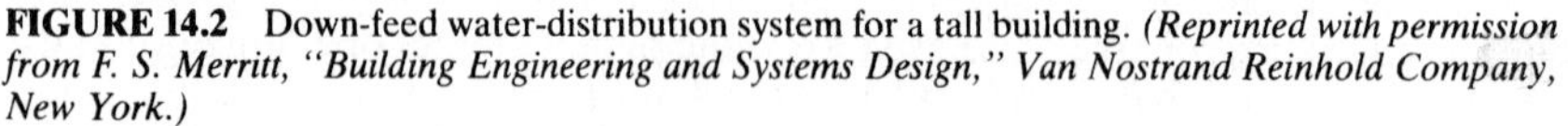

FIGURE 14.2 Down-feed water-distribution system for a tall building. *(Reprinted with permission from F. S. Merritt, "Building Engineering and Systems Design," Van Nostrand Reinhold Company, New York.)*

Tall buildings may be divided into zones, each of which is served by a separate down-feed system. (The first few stories may be supplied by an up-feed system under pressure from a public main.) Each zone has at its top its own storage tank, supplied from its own set of pumps in the basement. All the pumps draw on a common suction tank in the basement. Also, each zone has at its base its own water heater and a hot-water circulation system. In effect, the distribution in each zone is much like that shown in Fig. 14.2.

If space is not available to install storage tanks at the top of each zone, the main water supply from a roof-mounted storage tank may be supplied to the zones if pressure-reducing valves are utilized to reduce the supply-water pressure to an acceptable level at each zone.

14.6.4 Prevention of Backflow

All water-supply and distribution piping must be designed so there is no possibility of backflow at any time. The minimum code-required air gap (distance between the fixture outlet and the flood-level rim of the receptacle) should be maintained at all times. Domestic water systems that are subject to back siphonage or backflow should be provided with approved vacuum breakers or backflow preventers (Art. 14.3). Before any potable-water piping is put into use, it must be disinfected using a procedure approved by the local code authorities.

14.6.5 Pipe Materials

Pipes and tubing for water distribution may be made of copper, brass, polyvinyl chloride (PVC), polybutylene, ductile iron, or galvanized steel, if they are approved by the local code. When materials for potable-water piping are being selected, care should be taken to ensure that there is no possibility of chemical action or any other action that might cause a toxic condition.

14.6.6 Fittings

These are used to change the direction of water flow (because it usually is not practical to bend pipe in the field), to make connections between pipes, and to plug openings in pipes or close off the terminal of a pipe. In a water-supply system, fittings and joints must be capable of containing pressurized water flow. Fittings should be of quality equal to that of the pipes connected.

Standard fittings are available and generally may be specified by reference to an American National Standards Institute or a federal specification. Fitting sizes indicate the diameters of the pipes to which they connect. For threaded fittings, the location of the thread should be specified: A thread on the outside of a pipe is called a **male thread,** whereas an internal thread is known as a **female thread.**

Ductile-iron pipe is generally available with push-on mechanical joint or flanged fittings. Brass or bronze fittings for copper or brass pipe also may be flanged or threaded. Flanges are held together with bolts. In some cases, to make connections watertight, a gasket may be placed between flanges, whereas in other cases, the flanges may be machine-faced. Threaded fittings often are made watertight by coating the threads with an approved pipe compound or by wrapping the threads with teflon tape before the fittings are screwed onto the pipe.

14.6.7 Valves

These are devices incorporated in pipelines to control the flow into, through, and from them. Valves are also known as faucets, cocks, bibs, stops, and plugs. The term **cock** is generally used with an adjective indicating its use; for example, a sill cock (also called a hose bib) is a faucet used on the outside of a building for connection with a garden hose. A **faucet** is a valve installed on the end of a pipe to permit or stop withdrawal of water from the pipe.

Valves usually are made of cast or malleable iron, brass, or bronze. Faucets in bathrooms or kitchens are usually faced with nickel-plated brass.

The types of valves generally used in water-supply systems are gate, globe, angle, ball, and check valves.

Gate valves control flow by sliding a disk perpendicular to the water flow to fit tightly against seat rings when a handwheel is turned. This type of valve is usually used in locations where it can be left completely open or closed for long periods of time.

Globe valves control the flow by changing the size of the passage through which water can flow past the valves. Turning a handwheel moves a disk attached at the end of the valve stem to vary the passage area. When the valve is open, the water turns 90° to pass through an orifice enclosed by the seat and then turns 90° again past the disk, to continue in the original direction. Flow can be completely stopped by turning the handwheel to compress the disk or a gasket on it against the seat. This type of valve usually is used in faucets.

Angle valves are similar to globe valves but eliminate one 90° turn of the water flow. Water is discharged from the valves perpendicular to the inflow direction.

Check valves are used to prevent reversal of flow in a pipe. In the valves, water must flow through an opening with which is associated a movable plug. When water flows in the desired direction, the plug automatically moves out of the way; however, a reverse flow forces the plug into the opening, to seal it.

Ball valves are quick-closing (¼ turn to close) valves, which consist of a drilled ball that swivels on its vertical axis. This type of valve creates little water turbulence owing to its straight-through flow design.

14.6.8 Pipe Supports

When standard pipe is used for water supply in a building, stresses due to ordinary water pressure are well within the capacity of the pipe material. Unless the pipe is supported at short intervals, however, the weight of the pipe and its contents may overstress the pipe material. Generally, it is sufficient to support vertical pipes at their base and at every floor. Maximum support spacing for horizontal pipes depends on pipe diameter and material. The plumbing code should be consulted to determine maximum horizontal and vertical hanger spacings allowed.

While the supports should be firmly attached to the building, they should permit pipe movement caused by thermal dimensional changes or differences in settlement of building and pipe. Risers should pass through floors preferably through sleeves and transfer their load to the floors through tight-fitting collars. Horizontal pipe runs may be carried on rings or hooks on metal hangers attached to the underside of floors. The hangers and anchors used for plumbing piping should be metal and strong enough to prevent vibration.

Each hanger and anchor should be designed and installed to carry its share of the total weight of the pipe.

All piping installed should be restrained according to the requirements specific to the exact earthquake zone where the building is located. The local code authorities should be consulted about these requirements.

14.6.9 Expansion and Contraction

To provide for expansion and contraction, expansion joints should be incorporated in pipelines. Such joints should be spaced not more than 50 ft apart in hot-water pipe. While special fittings are available for the purpose, flexible connections are a common means of providing for expansion. Frequently, such connections consist of a simple U bend or a spiral coil, which permits springlike absorption of pipe movements.

14.6.10 Meters

These are generally installed on the service pipe to a building to record the amount of water delivered. The meters may be installed inside the building, for protection against freezing, or outside, in a vault below the frost line. Meters should be easily accessible to meter readers. Meter size should be determined by the maximum probable water flow, gal/min.

14.6.11 Water Hammer

This is caused by pressures developing during sudden changes in water velocity or sudden stoppage of flow. The result is a banging sound. It frequently results from rapid closing of valves, but it also may be produced by other means, such as displacing air from a closed tank or pipe from the top.

Water hammer can be prevented by filling a closed tank or pipe from the bottom while allowing the air to escape from the top. Water hammer also can be prevented by installing on pipelines air chambers or other types of water-hammer arresters. These generally act as a cushion to dissipate the pressures.

14.7 PLUMBING FIXTURES AND EQUIPMENT

The water-supply system of a building distributes water to plumbing fixtures at points of use. Fixtures include kitchen sinks, water closets, urinals, bathtubs, showers, lavatories, drinking fountains, laundry trays, and slop (service) sinks. To ensure maximum sanitation and health protection, most building codes have rigid requirements for fixtures. These requirements cover such items as construction materials, connections, overflows, installation, prevention of backflow, flushing methods, types of fixtures allowed, and inlet and outlet sizes. Either the building code or the plumbing code lists the minimum number of each type of fixture that must be installed in buildings of various occupancies (Table 14.2). Since these numbers are minimums, each project should be reviewed to determine if additional fixtures should be provided. This is especially true for assembly occupancies, where large numbers of people may utilize the restroom facilities in a short period of time; for example, at half-time at a football game.

TABLE 14.2 Minimum Plumbing Fixtures for Various Occupancies[a]

Type of building or occupancy	Water closets[b]	Urinals	Lavatories	Bathtubs or showers	Drinking fountains[c]
Dwelling or apartment house[d]	1 for each dwelling or apartment unit		1 for each dwelling or apartment unit	1 for each dwelling or apartment unit	
Office or public building	Male: 1 for 1–100; 2 for 101–200; 3 for 201–400 Female: 3 for 1–50; 4 for 51–100; 8 for 101–200; 11 for 201–400 Over 400, add one fixture for each additional 500 males and 2 for each 300 females	1 for 1–100 2 for 101–200 3 for 201–400 4 for 401–600 Over 600 add 1 fixture for each additional 300 males	Male: 1 for 1–200; 2 for 201–400; 3 for 401–750 Female: 1 for 1–200; 2 for 201–400; 3 for 401–750 Over 750, add one fixture for each additional 500 persons		1 per first 75[e]
Office or public buildings—for employee use	Male: 1 for 1–15; 2 for 16–35; 3 for 36–55 Female: 1 for 1–15; 3 for 16–35; 4 for 36–55 Over 55, add 1 fixture for each additional 40 persons	0 for 1–9 1 for 10–50 Add one fixture for each additional 50 males	Male: 1 per 40 Female: 1 per 40		
Schools—for staff use All schools	Male: 1 for 1–15; 2 for 16–35; 3 for 36–55 Female: 1 for 1–15; 2 for 16–35; 3 for 36–55 Over 55, add 1 fixture for each additional 40 persons	1 per 50	Male: 1 per 40 Female: 1 per 40		
Schools—for student use Nursery	Male: 1 for 1–20; 2 for 21–50 Female: 1 for 1–20; 2 for 21–50 Over 50, add 1 fixture for each additional 50 persons		Male: 1 for 1–25; 2 for 26–50 Female: 1 for 1–25; 2 for 26–50 Over 50, add 1 fixture for each additional 50 persons		1 per 75
Elementary	Male: 1 per 30 Female: 1 per 25	1 per 75	Male: 1 per 35 Female: 1 per 35		1 per 75
Secondary	Male: 1 per 40 Female: 1 per 30	1 per 35	Male: 1 per 40 Female: 1 per 40		1 per 75

Others (colleges, universities, adult centers)	Male 1 per 40 Female 1 per 30	1 per 35	Male 1 per 40 Female 1 per 40		1 per 75
Worship places—educational and activities unit	Male 1 per 125 2 for 126–250 Female 1 per 75 2 for 76–125 3 for 126–250	1 per 125	1 per 2 water closets		1 per 75
Dormitories—school or labor[f]	Male 1 per 10 Female 1 per 8 Add 1 fixture for each additional 25 males (over 10) and 1 for each additional 20 females (over 8)	1 per 25 Over 150, add 1 fixture for each additional 50 males	Male 1 per 12 Female 1 per 12 Over 12 add one fixture for each additional 20 males and 1 for each 15 additional females	1 per 8 For females, add 1 bathtub per 30. Over 150, add 1 per 20	1 per first 75[e]
Dormitories—for staff use	Male 1 for 1–15 2 for 16–35 3 for 36–55 Female 1 for 1–15 3 for 16–35 4 for 36–55 Over 55, add 1 fixture for each additional 40 persons	1 per 50	Male 1 per 40 Female 1 per 40	1 per 8	
Assembly places—theaters, auditoriums, convention halls—for permanent employee use	Male 1 for 1–15 2 for 16–35 3 for 36–55 Female 1 for 1–15 3 for 16–35 4 for 36–55 Over 55, add 1 fixture for each additional 40 persons	0 for 1–9 1 for 10–50 Add one fixture for each additional 50 males	Male 1 per 40 Female 1 per 40		
Assembly places—theaters, auditoriums, convention halls—for public use	Male 1 for 1–100 2 for 101–200 3 for 201–400 Female 3 for 1–50 4 for 51–100 8 for 101–200 11 for 201–400 Over 400, add one fixture for each additional 500 males and 2 for each 300 females	1 for 1–100 2 for 101–200 3 for 201–400 4 for 401–600 Over 600, add 1 fixture for each additional 500 males	Male 1 for 1–200 2 for 201–400 3 for 401–750 Female 1 for 1–200 2 for 201–400 3 for 401–750 Over 750, add one fixture for each additional 500 persons		1 per first 75[e]

TABLE 14.2 Minimum Plumbing Fixtures for Various Occupancies[a] (*Continued*)

Type of building or occupancy	Water closets[b]	Urinals	Lavatories	Bathtubs or showers	Drinking fountains[c]
Industrial, warehouses, workshops, foundries, and similar establishments—for employee use[g]	Male / Female 1 for 1–10 / 1 for 1–10 2 for 11–25 / 2 for 11–25 3 for 26–50 / 3 for 26–50 4 for 51–75 / 4 for 51–75 5 for 76–100 / 5 for 76–100 Over 100, add 1 fixture for each additional 30 persons		Up to 100, 1 per 10 persons Over 100, 1 per 15 persons[h]	1 shower for each 15 persons exposed to excessive heat or to skin contamination with poisonous, infectious, or irritating material	1 per 75
Institutional—other than hospitals or penal institution (on each occupied floor)	Male / Female 1 per 25 / 1 per 20	0 for 1–9 1 for 10–50 Add one fixture for each additional 50 males	Male / Female 1 per 10 / 1 per 10	1 per 8	1 per 75
Institutional—other than hospitals or penal institutional (on each occupied floor)—for employee use	Male / Female 1 for 1–15 / 1 for 1–15 2 for 16–35 / 3 for 16–35 3 for 36–55 / 4 for 36–55 Over 55, add 1 fixture for each additional 40 persons	0 for 1–9 1 for 10–50 Add one fixture for each additional 50 males	Male / Female 1 per 40 / 1 per 40	1 per 8	1 per 75
Hospitals Individual room Ward room	 1 per room 1 per 8 patients		 1 per room 1 per 10 patients	 1 per room 1 per 20 patients	 1 per 75
Hospital waiting rooms	1 per room		1 per room		1 per 75
Hospitals—for employee use	Male / Female 1 for 1–15 / 1 for 1–15 2 for 16–35 / 3 for 16–35 3 for 36–55 / 4 for 36–55 Over 55, add 1 fixture for each additional 40 persons	0 for 1–9 1 for 10–50 Add one fixture for each additional 50 males	Male / Female 1 per 40 / 1 per 40		
Penal Institutions—for prison use					1 per cell block floor

Cell Exercise room	1 per cell 1 per exercise room	 1 per exercise room	1 per cell 1 per exercise room	1 per exercise room
Penal Institutions—for employee use	Male / Female 1 for 1–15 / 1 for 1–15 2 for 16–35 / 3 for 16–35 3 for 36–55 / 4 for 36–55 Over 55, add 1 fixture for each additional 40 persons	0 for 1–9 1 for 10–50 Add one fixture for each additional 50 males	Male / Female 1 per 40 / 1 per 40	1 per 75
Worship places, principal assembly place	Male / Female 1 per 150 / 1 per 75 2 for 151–300 / 2 for 76–150 3 for 151–300 (Female)	1 per 150	1 per 2 water closets	1 per 75
Restaurants, pubs, and lounges[i]	Male / Female 1 for 1–50 / 1 for 1–50 2 for 51–150 / 2 for 51–150 3 for 151–300 / 4 for 151–300 Over 300, add 1 fixture for each additional 200 persons	1 for 1–150 Over 150, add 1 fixture for each additional 150 males	Male / Female 1 for 1–150 / 1 for 1–150 2 for 151–200 / 2 for 151–200 3 for 210–400 / 3 for 201–400 Over 400, add 1 fixture for each additional 400 persons	

[a]Based on "Uniform Plumbing Code," 1990, International Association of Plumbing and Mechanical Officials, Walnut, Calif.

The table lists the number of fixtures required for the number of persons indicated. Minimum exiting requirements determine the minimum number of occupants to be accommodated.

Every building should include provisions for the physically handicapped. (Refer to local authorities or "Specifications for Making Buildings and Facilities Accessible to, and Usable by, the Physically Handicapped," ANSI A117.1, American National Standards Institute.)

Building categories not listed in the table should be considered separately by the administrative authority.

Consideration should be given to the accessibility of the fixtures. Application of the table data strictly on a numerical basis may not produce an installation suited to the needs of building occupants. For example, schools should have toilet facilities on every floor on which there are classrooms.

Temporary facilities for workers: one water closet and one urinal for every 30 male workers and every 30 female workers, or fraction thereof. Trough urinals are prohibited. Walls and floors around every urinal should be lined with nonabsorbent materials. The lining should extend on the floor from the wall to 2 ft in front of the urinal lip, and on the wall, 4 ft above the floor and at least 2 ft on both sides of the urinal.

[b]The total number of water closets for females should be at least equal to the sum of the water closets and urinals required for men.

[c]There should be at least one drinking fountain per occupied floor in schools, theaters, auditoriums, dormitories, and office and public buildings. Drinking fountains should not be installed in toilet rooms. Where food is consumed indoors, water stations may be substituted for drinking fountains.

[d]One kitchen sink for each dwelling or apartment unit. One laundry tray or automatic washer standpipe for each dwelling and two laundry trays or two automatic washer standpipes or combination of these for every 10 apartments.

[e]One additional fountain for each additional 150 persons.

[f]One laundry tray for every 50 persons. One slop sink for every 100 persons.

[g]As required by local authorities or "Sanitation in Places of Employment," ANSI Z4.1.

[h]Where there is exposure to skin contamination from poisonous, infectious, or irritating materials, one lavatory should be provided for every 5 persons. A wash sink 24 in long or a circular basin 18 in in diameter, when equipped with water outlets for these dimensions, may be considered equivalent to one lavatory.

[i]Any business that sells food for consumption on the premises is considered a restaurant. Employee toilet facilities should not be counted toward meeting the restaurant requirements in the table. Hand washing must be available in the kitchen for employees. The number of occupants for a drive-in restaurant should be taken equal to the number of parking stalls.

The plumbing fixtures are at the terminals of the water-supply system and the start of the wastewater system. To a large extent, the flow from the fixtures determines the quantities of wastewater to be drained from the building.

Traps. Separate traps are required for most fixtures not fitted with an integral trap. The trap should be installed as close as possible to the unit served. More than one fixture may be connected to a trap if certain code regulations are observed. For specific requirements, refer to the governing code.

A water seal of at least 2 in, and not more than 4 in, is generally required in most traps. Traps exposed to freezing should be suitably protected to prevent ice formation in the trap body. Clean-outs of suitable size are required on all traps except those made integral with the fixture or those having a portion which is easily removed for cleaning of the interior body. Most codes prohibit use of traps in which a moving part is needed to form the seal. Double trapping is also usually prohibited. Table 14.4 lists minimum trap sizes for various fixtures.

Showers. Special care should be taken in selection of showers, especially shower valves. To ensure that a user is not scalded when pressure fluctuations occur in the water distribution system, pressure-balancing or temperature-limiting shower valves that prevent extreme variations in the outlet water temperature may be specified. In facilities with large numbers of showers, a central tempered water system may be used to serve the showers. As with the shower valves, the mixing valve serving a tempered water system should also be of a pressure-balancing or temperature-limiting type.

Water Closets. These consist of a bowl and integral trap, which always contains water, and a tank or a flushometer valve, which supplies water for flushing the bowl (Fig. 14.3). The passage through the trap to the discharge usually is large enough to pass a solid ball 2 to 3 in in diameter. Siphon-jet flushometer valves generally require a pressure of at least 15 psi for operation and blowout flushometer valves generally require 25 psi for operation. The water level in a tank of a tank-type water closet is raised above the water level in the bowl so that gravity provides sufficient pressure for flushing.

The cleansing action of water flow in the bowl may be achieved in any of several different ways. One method is illustrated by the **siphon jet** in Fig. 14.3*b*. The tank discharges water around the rim and also jets water into the up leg of the trap. As a result, the contents of the bowl are siphoned out of the down leg of the discharge pipe. Other types of action include the **reverse trap** (Fig. 14.3*c*), which is similar to the siphon-jet type but smaller; the **siphon vortex** (Fig. 14.3*a*), in which water from the rim washes the bowl, creates a vortex, becomes a jet, and discharges by siphonage; the **washdown** (Fig. 14.3*d*), in which pressure buildup causes the up leg to overflow and create a discharge siphon; and the **blowout** (Fig. 14.3*e*), used with a flushometer valve, which projects a strong jet into the up leg to produce

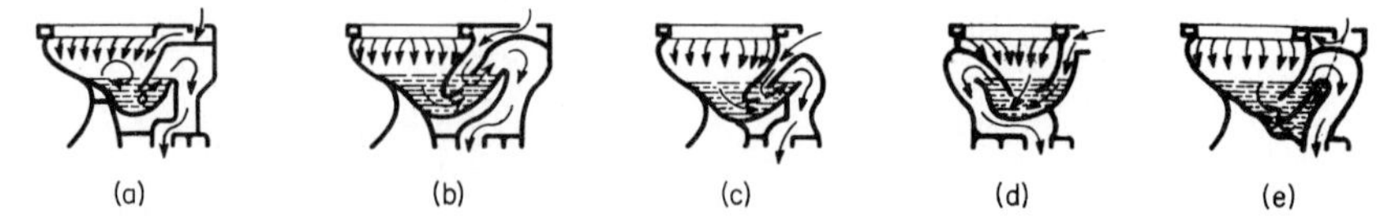

FIGURE 14.3 Typical water closets: (*a*) siphon-vortex; (*b*) siphon-jet; (*c*) reverse trap; (*d*) washdown; (*e*) blowout.

the discharge. Blowout-type water closets are generally reserved for use where clogs due to solids in the bowl are common, such as in penal institutions, stadiums, or arenas. Because of the large amount of water consumed during the flush of a blowout type of water closet, these types of fixtures are not used to the extent they once were. Siphon-jet type water closets are the most common type of water closets specified.

Because of insufficient supplies of water, some localities have mandated use of low-flush water closets, which require about 1.6 gpm per flush. Designers should carefully review the plumbing design in areas where low-flush water closets are required, to assure proper cleansing of the building waste systems.

Air Gaps. These should be provided to prevent backflow of wastewater into the water supply (Art 14.6.4). At plumbing fixtures, an air gap must be provided between the fixture water-supply outlet and the flood-level rim of the receptacle. Building codes usually require a minimum gap of 1 to 2 in for outlets not affected by a nearby wall and from 1½ to 3 in for outlets close to a wall. Table 14.3 lists minimum air gaps usually used.

In addition to the usual drain at the lowest point, receptacles generally are provided with a drain at the flood-level rim to prevent water from overflowing. The overflow should discharge into the wastewater system on the fixture side of the trap.

Floor and Equipment Drains. Floor drains should be installed at all areas where the possibility of water spillage occurs. Common areas that are provided with floor drains include restrooms, mechanical rooms, kitchens, and shower and locker rooms. Equipment that requires piped discharge from drains or relief devices, such as boilers, require recessed-type drains of adequate size, preferably with a funnel receptor. Large commercial kitchens often require deep, receptor floor sinks to receive indirect wastes from kitchen equipment.

14.8 WATER DEMAND AND FIXTURE UNITS

For each fixture in a building, a maximum requirement for water flow, gal/min, can be estimated. Table 14.1 indicates the minimum flow rate and pressure required by code. The maximum flow may be considerably larger. Branch pipes to each fixture should be sized to accommodate the maximum flow and minimum pressure the fixture will require. Mains serving these branches, however, need not be sized to handle the sum of the maximum flows for all branches served. It is generally unlikely that all fixtures would be supplying maximum flow simultaneously or even that all the fixtures would be operating at the same time. Consequently, the diameters of the mains need be sized only for the probable maximum water demand.

In practice, the probable flow is estimated by weighting the maximum flow in accordance with the probability of fixtures being in use. The estimate is based on the concept of fixture units.

Fixture unit is the average discharge, during use, of an arbitrarily selected fixture, such as a lavatory or water closet. Once this value is established, the discharge rates of other types of fixtures are stated in terms of the basic fixture. For example, when the basic fixture is a lavatory served by as 1¼-in trap, the average flow during discharge is 7.5 gal/min. So a bathtub that discharges 15 gal/min is rated as two fixture units (2×7.5). Thus, a tabulation of fixture units can be set up, based on an assumed basic unit.

TABLE 14.3 Minimum Air Gaps for Generally Used Plumbing Fixtures

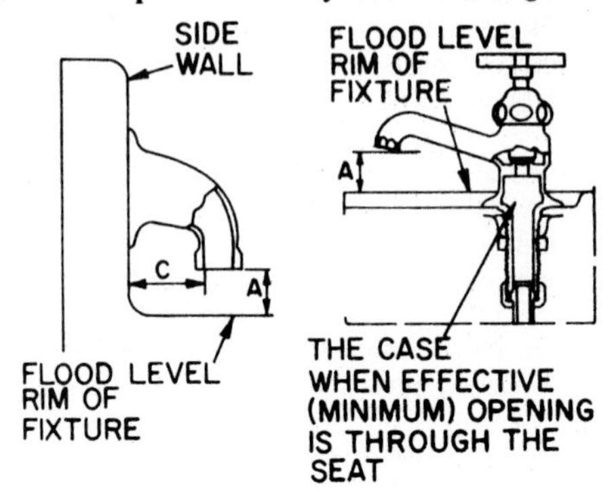

Fixture	Minimum air gap A, in	
	Away from a wall*	Close to a wall*
Lavatories with effective openings not greater than ½-in diameter	1.0	1.50
Sink, laundry trays, and goose neck bath faucets with effective openings not greater than ¾-in diameter	1.5	2.25
Overrim bath fillers with effective openings not greater than 1-in diameter	2.0	3.00
Drinking fountains with a single orifice not more than 7⁄16 in in diameter or multiple orifices with a total area of 0.150 in^2 (area of a 7⁄16-in-diameter circle)	1.0	1.5
Effective openings greater than 1-in	‡	§

*Side walls, ribs, or similar obstructions do not affect the air gaps when spaced from inside edge of spout opening a distance[c] greater than three times the diameter of the effective opening for a single wall, or a distance greater than four times the diameter of the effective opening for two intersecting walls (see figure).

†Vertical walls, ribs, or similar obstructions extending from the water surface to or above the horizontal plane of the spout opening require a greater air gap when spaced closer to the nearest inside edge of spout opening than specified in note* above.

‡2 × effective opening.

§3 × effective opening.

A specific number of fixture units, as listed in Table 14.4, is assigned to each type of plumbing fixture. These values take into account:

- Anticipated rate of water flow from the fixture outlet, gal/min
- Average duration of flow, min, when the fixture is used
- Frequency with which the fixture is likely to be used

The ratings in fixture units listed in Table 14.4 represent the relative loading of a water-distribution system by the different types of plumbing fixtures. The sum of the ratings for any part or all of a system is a measure of the load the combination of fixtures would impose if all were operating. The probable maximum water demand, gal/min, can be determined from the total number of fixture units served by any part of a system by use of graphs shown in Fig. 14.4.

The demand obtained from these curves applies to fixtures that are used intermittently. If the system serves fixtures, such as air-conditioning units, lawn sprinklers, or hose bibs, that are used continuously, the demand of these fixtures should be added to the intermittent demand. For a continuous or semicontinuous flow into a drainage system, such as from a pump, pump ejector, air-conditioning system, or similar device, two fixture units should be used for each gallon per minute of flow. When additional fixtures are to be installed in the future, pipe and drain sizes should be based on the ultimate load, not on the present load.

14.9 WATER-PIPE SIZING

The required domestic-water pipe sizes should be determined by application of the principles of hydraulics. While economy dictates use of the smallest sizes of pipe permitted by building-code requirements, other factors often make larger sizes advisable. These factors include:

1. Pressure at the water-supply source, usually the public main, psi
2. Pressure required at the outlets of each fixture, psi
3. Loss of pressure because of height of outlets above the source, pressure loss due to friction caused by the flow of water through water meters and backflow preventers, and friction from water flow in the piping
4. Limitations on velocity of water flow, ft/s, to prevent noise and erosion
5. Additional capacity for future expansion (normally 10% minimum)

14.9.1 Method for Determining Pipe Sizes

1. Sketch all the proposed risers, horizontal mains, and branch lines, indicating the number and the type of fixtures served, together with the required flow
2. Compute the demand weights of the fixtures, in fixture units, using Table 14.4
3. From Fig. 14.4 and the total number of fixture units, determine the water demand, gal/min
4. Compute the equivalent length of pipe for each stack in the system, starting from the street main
5. Obtain by test or from the water company the average minimum pressure in the street main. Determine the minimum pressure needed for the highest fixture in the system
6. Compute the pressure loss in the piping with the use of the equivalent length found in item 4
7. Choose the pipe sizes from a chart like that in Fig. 14.5 or 14.6, or from the charts given in the plumbing code being used

TABLE 14.4 Fixture Units and Trap and Connection Sizes for Plumbing Fixtures

Fixture type	Domestic water				Drainage	
	Fixture-unit value as load factors		Min size of connections, in			
	Private	Public	Cold water	Hot water	Fixture-unit value as load factors	Min size of trap, in
Bathtub† (with or without overhead shower)	2	4	½	½	2	1½
Bidet	2	4	½	½	2	Nominal 1½
Combination sink and tray	3		½	½	2	1½
Combination sink and tray with food-disposal unit	4				3	Separate traps 1½
Dental unit or cuspidor		1	⅜		1	1¼
Dental lavatory	1	2	½	½	2	1¼
Dishwater, domestic	2				2	1½
Drinking fountain	1	2	⅜		1	1¼
Floor drains‡	1				2	2
Kitchen sink	2	4	½	½	2 or 3	1½
Kitchen sink, domestic, with food-waste grinder	3				2	1½
Lavatory¶	1		⅜	⅜	1	Small P.O. 1¼

Lavatory¶		2	½	½	2	Large P.O.	1½
Lavatory, barber, beauty parlor		2			2		1½
Lavatory, surgeon's	2				2		1½
Laundry tray (1 or 2 compartments)	2	4	½	½	2		1½
Shower, per head	2	4	½	½	2		2
Sinks:							
Surgeon's	3		½	½	3		1½
Flushing rim (with valve)		2	¾	¾	6		3
Service (trap standard)	3		½	½	3		3
Service (P trap)	2	4	½	½	3		2
Pot, scullery, etc.		4			3		1½
Urinal, pedestal, siphon jet, blowout		10	1		6	Nominal	3
Urinal, wall lip		5	½		2		1½
Urinal stall		5	¾		2		2
Urinal with flush tank		3			2		1½
Wash sink (circular or multiple) each set of faucets		2	½	½	3	Nominal	1½
Water closet, tank-operated	3	5	⅜		4	Nominal	3
Water closet, valve-operated	6	10	1		6		3

*Fixture units listed in the table give the total water-supply demand of fixtures with both hot-water and cold-water supply. Fixture units for the maximum demand of either cold water or hot water alone may be taken as 75% of the fixture units in the table.

†A shower head over a bathtub does not increase the fixture value.

‡Size of floor drain should be determined by the area of surface water to be drained.

¶Lavatories with 1¼- or 1½-in trap have the same load value; larger P.O. (plumbing orifice) plugs have greater flow rate.

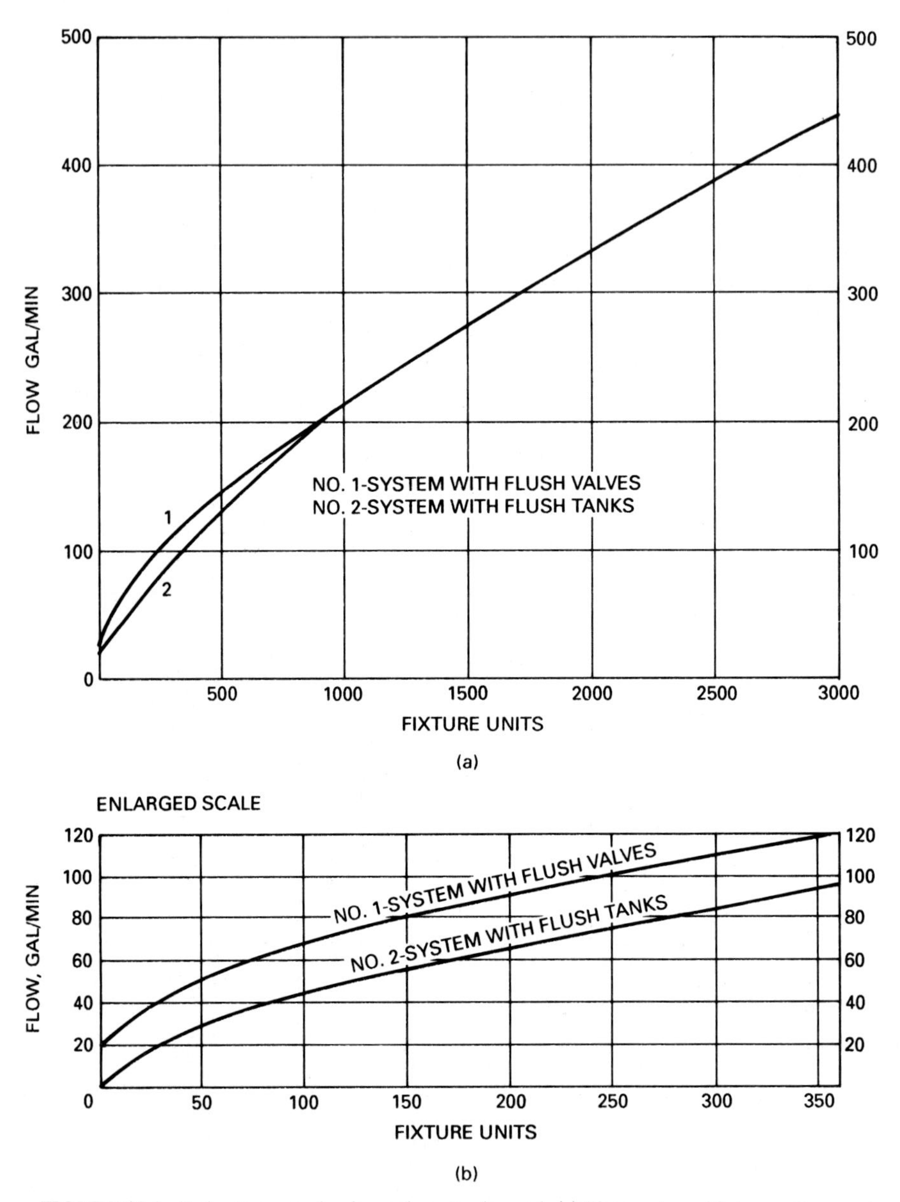

FIGURE 14.4 Estimate curves for domestic water demand. (*a*) The number of fixture units served determines the rate of flow. (*b*) Enlargement of the low-demand portion of (*a*).

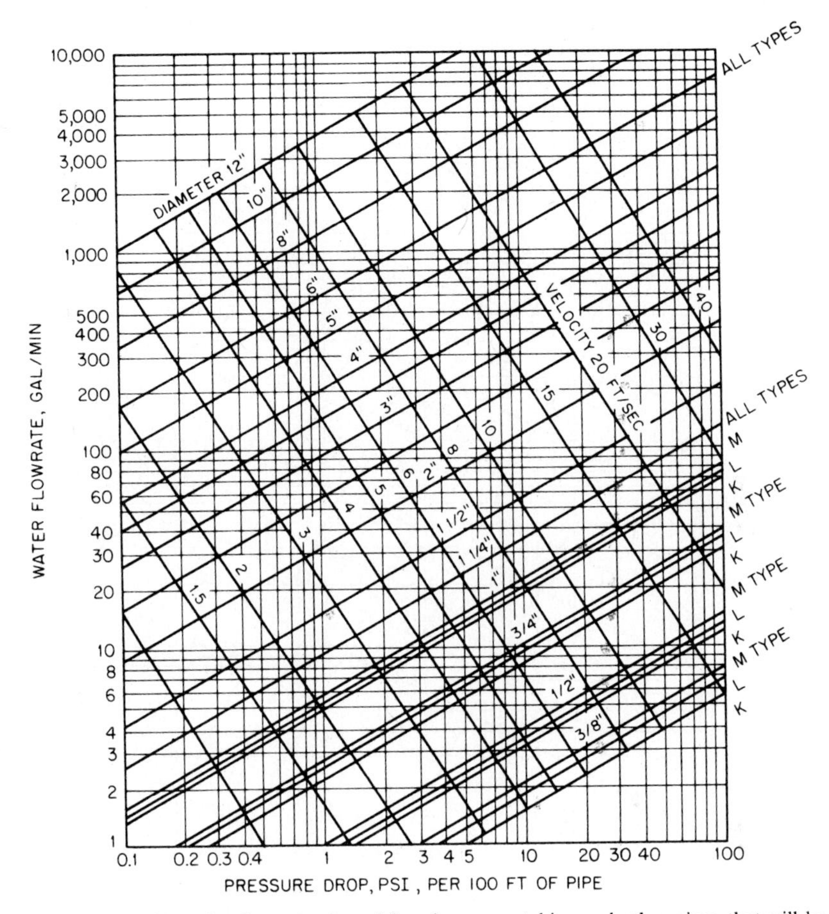

FIGURE 14.5 Chart for determination of flow in copper tubing and other pipes that will be smooth after 15 to 20 years of use.

14.9.2 Effects of Pressure

Rate of flow, ft^3/s, in a pipe is determined by

$$Q = AV \tag{14.1}$$

where A = pipe cross-sectional area, ft^2
V = water velocity, ft/s

In general, V should be kept to 8 ft/s or less to prevent noise and reduce erosion at valve seats. Hence, pipe area should be at least the flow rate Q divided by 8. Mains may be allowed to have a velocity of 10 ft/s, but lower velocities are preferred.

The minimum pressures at plumbing fixtures generally required by building codes are listed in Table 14.1. These pressures are those that remain when the

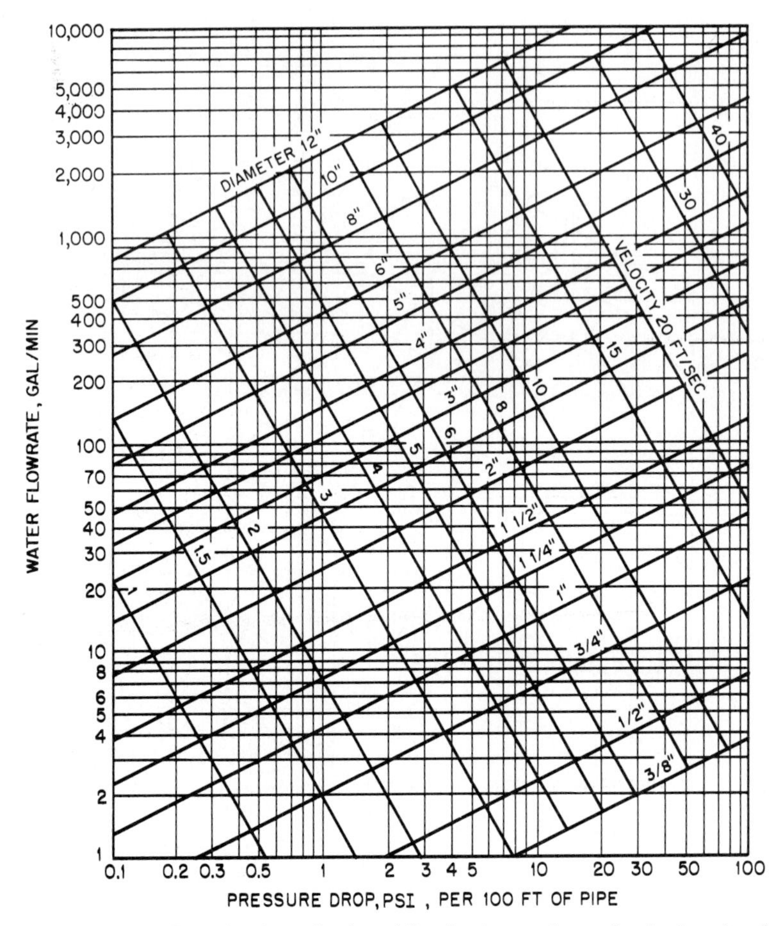

FIGURE 14.6 Chart for determination of flow in pipes such as galvanized steel and wrought iron that will be fairly rough after 15 to 20 years of use.

pressure drop due to height of outlet above the water source and the pressure lost by friction in pipes are deducted from the pressure at the water source. The pressure loss due to height can be computed from

$$p = 0.433h \tag{14.2}$$

where p = pressure, psi
h = height or pressure head, ft

The total head H, ft, on water at any point in a pipe is given by

$$H = Z + \frac{p}{w} + \frac{V^2}{2g} \tag{14.3}$$

where Z = elevation, ft, of the point above some arbitrary datum
p/w = pressure head, ft

w = specific weight of water = 62.4 lb/ft^3
$V^2/2g$ = velocity head, ft
g = acceleration due to gravity, 32.2 ft/s^2

When water flows in a pipe, the difference in total head between any two points in the pipe equals the friction loss h_f, ft, in the pipe between the points.

Any of several formulas may be used for estimating h_f. One often used for pipes flowing full is the Hazen-Williams formula:

$$h_f = \frac{4.727}{D^{4.87}} L \frac{Q}{C_1}^{1.85} \tag{14.4}$$

where Q = discharge, ft^3/s
D = pipe diameter, ft
L = length of pipe, ft
C_1 = coefficient

The value of C_1 depends on the roughness of the pipe, which, in turn, depends on pipe material and age. A new pipe has a larger C_1 than an older one of the same size and material. Hence, when pipe sizes are being determined for a new installation, a future value of C_1 should be assumed to ensure adequate flows in the future. Design aids, such as charts (Figs. 14.5 and 14.6) or nomograms, may be used to evaluate Eq. (14.4), but if such computations are made frequently, a computer solution is preferable.

In addition to friction loss in pipes, there are also friction losses in meters, valves, and fittings. These pressure drops can be expressed for convenience as equivalent lengths of pipe of a specific diameter. Table 14.5 indicates typical allowances for friction loss for several sizes and types of fittings and valves.

The pressure reduction caused by pipe friction depends, for a given length of pipe and rate of flow, on pipe diameter. Hence, a pipe size can be selected to create a pressure drop in the pipe to provide the required pressure at a plumbing fixture, when the pressure at the water source is known. If the pipe diameter is too large, the friction loss will be too small and the pressure at the fixture will be high. If the pipe size is too small, the friction loss will be too large and the pressure at the fixture will be too small.

14.9.3 Minimum Pipe Sizes

The minimum sizes for fixture-supply pipes are given for cold water and hot water in Table 14.4.

Sizes of pipes for small buildings, such as single-family houses, can usually be determined from the experience of the designer and applicable building-code requirements, without extensive calculations. For short branches to individual fixtures, for example, the minimum pipe diameters listed in Table 14.4 generally will be satisfactory. Usually also, the following diameters can be used for the mains supplying water to the fixture branches:

½ in for mains with up to three ¾-in branches

¾ in for mains with up to three ½-in or five ⅜-in branches

1 in for mains with up to three ¾-in or eight ½-in or fifteen ⅛-in branches

The adequacy of these sizes, however, depends on the pressure available at the water source and the probability of simultaneous use of the plumbing fixtures.

TABLE 14.5 Allowances for Friction Losses in Valves and Fittings, Expressed as Equivalent Length of Pipe, Ft*

Diameter of fitting, in	90° standard elbow	45° standard elbow	Standard 90° tee	Coupling or straight run of tee	Gate valve	Globe valve	Angle valve
3/8	1	0.6	1.5	0.3	0.2	8	4
1/2	2	1.2	3	0.6	0.4	15	8
3/4	2.5	1.5	4	0.8	0.5	20	12
1	3	1.8	5	0.9	0.6	25	15
1 1/4	4	2.4	6	1.2	0.8	35	18
1 1/2	5	3	7	1.5	1	45	22
2	7	4	10	2	1.3	55	28
2 1/2	8	5	12	2.5	1.6	65	34
3	10	6	15	3	2	80	40
4	14	8	21	4	2.7	125	55
5	17	10	25	5	3.3	140	70
6	20	12	30	6	4	165	80

*Allowances based on nonrecessed threaded fittings. Use one-half the allowances for recessed threaded fittings or streamlined older fittings.

14.10 DOMESTIC WATER HEATERS

The method of heat development for water heaters may be direct (heat from combustion of fuels or electrical energy directly applied to water) or indirect (heat from a remote heat source utilizing some other medium, such as steam, to heat water).

Direct-heat-type water heaters are classified as follows:

1. Automatic storage heaters, which incorporate burners or heating elements, storage tank, outer jacket, insulation, and controls as a packaged unit
2. Circulating tank heaters, which consist of what is essentially an instantaneous heater and an accessory storage tank. Hot water is circulated through the heating section by means of a circulating pump
3. Instantaneous heaters, which have little water storage capacity and generally have controls that modulate the heat output based on the demand
4. Hot-water supply boilers, which provide high-temperature hot water in a manner similar to hot-water heating boilers

Fuel for direct-fired water heaters is generally one of the fossil fuels, such as natural gas or oil, or electric power.

Indirect-type water heaters are classified as follows:

1. Storage type, which consists of a heat exchanger installed in a storage tank (Fig. 14.7) or in a separate storage tank and stand-alone heat exchanger provided with a circulating water system.
2. Indirect immersion type, a self-contained water heater, utilizing one of the fossil fuels as a heating medium for a horizontal fire tube containing a finned-tube bundle. Water, or some other heat-transfer fluid, is heated in the finned bundle in the burner section and is pumped to a water-heating bundle located in the shell or storage tank installed below the fire tube.

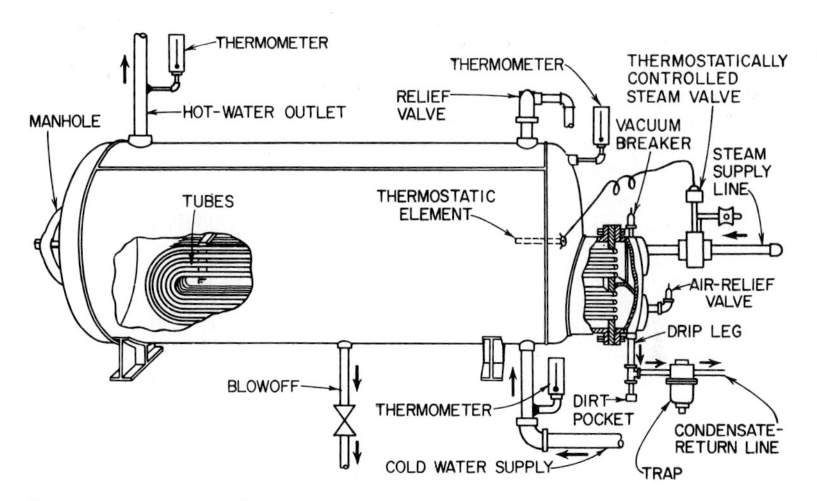

FIGURE 14.7 Storage heater for domestic hot water.

3. Instantaneous type, which is suited for facilities requiring steady, continuous supplies of hot water (Fig. 14.8). The rate of flow is indirectly proportional to the temperature of the water being supplied.

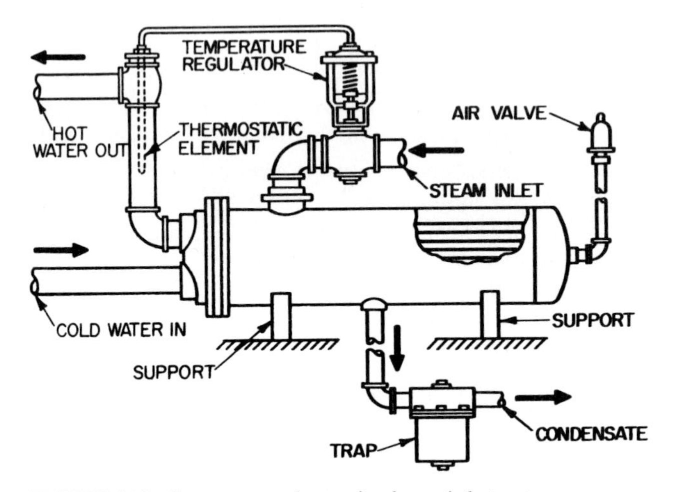

FIGURE 14.8 Instantaneous heater for domestic hot water.

4. Semi-instantaneous type, which have limited storage to meet momentary hot-water peak demands. These types of heaters consist of a heating element and a control system that closely controls leaving-water temperature. A hot-water storage tank provides additional hot water when required during periods of peak momentary hot-water demand.

The heat-transfer media normally utilized for indirect domestic hot-water heaters are steam and heating hot water. The heat-transfer media use heat provided by boilers and, in some instances, solar collectors, which collect heat from the sun.

(For detailed guidance in the sizing of domestic water heating systems, see "Service Hot-Water Systems," Chap. 4, ASPE Data Book, American Society of Plumbing Engineers, Westlake, CA 91362. Recovery versus storage curves that have been developed based on extensive research can be utilized to compare various combinations available.)

Plumbing designers should also assure that all required safety devices and controls have been provided to prevent an explosion of the storage vessel. There have been numerous instances of injury and death to occupants due to overfiring conditions caused by malfunctioning controls and safety-relief devices that did not operate properly. All storage vessels should be provided with AGA/ASME-rated pressure and temperature (P&T) relief valves, installed as directed by the vessel manufacturer. The rating of the P&T valves should meet or exceed the Btu input rating of the water-heating apparatus.

Most storage tanks are constructed of steel and therefore are subject to rusting when in direct contact with water. Various liners are available such as cement, glass, copper, and nickel. The designer should select a liner that best meets the needs of the building being designed. Storage tanks should be ASME certified.

The hot-water load for a given building is computed in a manner similar to that described in Art. 14.8 but with Table 14.6 and the tabulated demand factor for the particular building type. The heating-coil capacity of the heater must at least equal the maximum probable demand for hot water.

For storage-type heaters, the storage capacity is obtained by multiplying the maximum probable demand by a suitable factor, such as 1.25 for apartment buildings to 0.60 for hospitals. Table 14.7 lists representative hot-water utilization temperatures for various services. It should be noted that service-water temperatures in the 140°F range should be provided, to prevent the growth of *Legionella pneumophila* bacteria which causes Legionnaires' disease.

Example. Determine heater and storage tank size for an apartment building from a number of fixtures.

Solution. Calculation of the maximum possible demand with the use of Table 14.6 is shown in Table 14.8. Table 14.6 also gives a demand factor of 0.30 for apartment buildings.

$$\text{Probable maximum demand} = 2520 \times 0.30 = 756 \text{ gph}$$

This determines the minimum heater or coil capacity, 756 gph. From Table 14.6 also, the storage capacity factor is 1.25.

$$\text{Storage tank capacity} = 756 \times 1.25 = 945 \text{ gal}$$

WASTEWATER PIPING

Human, natural, and industrial wastes resulting from building occupancy and use must be disposed of in a safe, quick manner if occupant health and comfort are to be safeguarded. Design of an adequate plumbing system requires careful planning and adherence to the codes in effect and to state or municipal regulations governing these systems.

14.11 WASTEWATER DISPOSAL

There are three main types of wastewater: domestic, storm, and industrial. Separate plumbing systems are generally required for each type.

Domestic wastewater is primarily spent water from the building water supply, to which is added wastes from bathrooms, kitchens, and laundries. It generally can be disposed of by discharge into a municipal sanitary sewer, if one is available.

Storm water is primarily the water that runs off the roof or the site of the building. The water usually is directed to roof drains or gutters. These then feed the water to drainpipes, which convey it to a municipal or private storm-water sewer system. Special conditions at some building sites, such as large paved areas or steep slopes, may require the capture of storm water in retention areas or ponds to prevent the municipal storm sewer systems from being overloaded. From these areas or ponds, the storm water is generally conveyed to the storm sewers through outfall structures designed to delay and control the flow of storm water to the municipal storm sewer systems. Discharge into sanitary sewers is objectionable, because the large flows interfere with effective wastewater treatment and increase

TABLE 14.6 Hot-Water Demand per Fixture for Various Building Types*

[*Based on average conditions for the building type, gal/h of water per fixture at 140°F (60°C)*]

	Apartment buildings	Gymnasiums	Hospitals	Hotels	Industrial plants	Office buildings	Dwellings	Schools
Lavatory, private	2	2	2	2	2	2	2	2
Lavatory, public	4	8	6	8	12	6		15
Bathtubs	20	30	20	20			20	
Dishwashers†	15		50 190	50 190	20 76		15	20 76
Foot basins	3	12	3	3	12		3	3
Kitchen sink	10		20	30	20	20	10	20
Laundry, stationary tubs	20		28	28			20	
Pantry sink	5		10	10		10	5	10
Showers	30	225	75	75	225	30	30	225
Service sink	20		20	30	20	20	15	20

Hydrotherapeutic showers			400					
Hubbard baths			600					
Leg baths			100					
Arm baths			35					
Sitz baths			30					
Continuous-flow baths			165					
Circular wash sinks			20	20	30	20		30
Semicircular wash sinks			10	10	15	10		15
Factors								
Demand factor	0.30	0.40	0.25	0.25	0.40	0.30	0.30	0.40
Storage capacity factor‡	1.25	1.00	0.60	0.80	1.00	2.00	0.70	1.00

*Based on data in "ASPE Data Book," American Society of Plumbing Engineers, Westlake, Calif.

†Dishwasher requirements should be taken from this table or from manufacturers' data for the model to be used, if this is known.

‡Ratio of storage-tank capacity to probable maximum demand per hour. Storage capacity may be reduced where an unlimited supply of steam is available from street steam system or large boiler plant.

TABLE 14.7 Hot-Water Temperatures for Various Services

Use	Temperature °F	Temperature °C
Lavatory		
Hand washing	105	40
Shaving	115	45
Showers and tubs	110	43
Therapeutic baths	95	35
Commercial and institutional laundry	180	82
Residential dishwashing and laundry	140	60
Commercial spray-type dishwashing		
Single or multiple tank hood or rack type		
Wash minimum temperature	150	65
Final rinse	180–195	82–90
Single tank conveyor type		
Wash minimum temperature	160	71
Final rinse	180–195	82–90
Single tank rack or door type		
Single-temperature wash and rinse	165+	74+
Surgical scrubbing	110	43
Chemical sanitizing types (consult manufacturer for actual temperature required)	140	60
Multiple-tank conveyor type		
Wash minimum temperature	150	65
Pumped rinse minimum temperature	160	71
Final rinse	180–195	82–90
Chemical sanitizing glasswasher		
Wash	140	60
Rinse minimum temperature	75	24

TABLE 14.8 Hot-Water Demand for Apartment-Building Example

Number of fixtures	Flow per fixture, gal/h (from Table 14.6)	Hot-water demand, gal/hr
60 lavatories	2	120
30 bathtubs	20	600
30 showers	30	900
60 kitchen sinks	10	600
15 laundry tubs	20	300
Possible maximum demand		2520

treatment costs. If kept separate from other types of wastewater, storm water usually can be safely discharged into a large body of water. Raw domestic wastewater and industrial wastes, on the other hand, have objectionable characteristics that make some degree of treatment necessary before they can be discharged. Nevertheless, municipal combined sewers (sanitary and storm wastes) exist in some areas. Appropriate local authorities should be consulted to determine which type of system is available and specific regulations that relate to connection to these systems.

In areas where municipal sanitary sewers are not available, some form of wastewater treatment is required. Prefabricated treatment plants are available in various sizes and configurations. Most treatment systems are complex and require many steps. These include filtration and activated-sludge and aeration methods. The degree of treatment necessary generally depends on the assimilation potential of the body of water to receive the effluent, primarily the ability of the body to dilute the impurities and to supply oxygen for decomposition of organic matter present in the wastewater.

Industrial waste may present special problems because (1) the flow volume may be beyond the public sewer capacity, and (2) local regulations may prohibit the discharge of industrial waste into public sewers. Furthermore, many pollution regulations prohibit discharge of industrial waste into streams, lakes, rivers, and tidal waters without suitable prior treatment. Industrial wastes generally require treatments engineered to remove the specific elements injected by industrial processes that make the wastes objectionable. Often, these treatments cannot be carried out in public wastewater treatment plants. Special treatment plants may have to be built for the purpose. Treatment methods for a variety of industrial wastes are discussed in W. W. Eckenfelder, Jr., "Industrial Water Pollution Control," McGraw-Hill Publishing Company, New York. Specific design procedures for sewers, drains, and wastewater treatment, with accompanying numerical examples, are given in T. G. Hicks, "Standard Handbook of Engineering Calculations," McGraw-Hill Publishing Company, New York.

14.12 SEWERS

A sewer is a conduit for water carriage of wastes. For the purpose of this section, any piping for wastewater inside a building will be considered plumbing or process piping; outside the building, wastewater lines are called sewers.

Sewers carry wastewater. And a system of sewers and appurtenances is **sewerage.** Sanitary sewers carry domestic wastes or industrial wastes. Where buildings are located on large sites, or structures with large roof areas are involved, a storm sewer is used for fast disposal of rain and is laid out to drain inlets located for best collection of runoff.

14.12.1 Determination of Runoff

For figuring rates of runoff to determine storm-sewer requirements, the so-called rational method may be used. It employs the formula

$$Q = CIA \tag{14.5}$$

where Q = maximum rate of runoff, ft^3/s
C = runoff coefficient of the runoff area
I = rainfall intensity, in/hr
A = watershed area, acres

The runoff coefficient C indicates the degree of imperviousness of the land. It ranges from 0.6 to 0.9 for built-up areas and paved surfaces and from 0.30 to 0.50 for unpaved surfaces, depending on the surface slope. In storm-sewer design, however, it is necessary to know not only rate of runoff and total runoff, but also at what point in time after the start of a storm the rate of runoff reaches its peak. It is this peak runoff for which pipe must be sized and sloped. (The conduit designed to handle the peak runoff is for conveyance of runoff volume only and should not be considered for storage.)

14.12.2 Determination of Sewer Size

Sanitary sewers or lines carrying exclusively industrial wastes from a building to disposal must be sized and sloped according to best hydraulic design. The problem is generally one of flow in a circular pipe. (C. V. Davis and K. E. Sorenson, "Handbook of Applied Hydraulics," McGraw-Hill Publishing Company, New York.) Gravity flow is to be desired, but pumping is sometimes required.

Pipe should be straight and of constant slope between access holes, and access holes should be used at each necessary change in direction, slope, elevation, or size of pipe. Access holes should be no farther apart than 200 ft for pipes 24 in and smaller, and 500 ft for pipes 30 in and larger.

The sewer from a building must be sized to carry out all the water carried in by supply mains or other means. Exceptions to this are the obvious cases where losses might be appreciable, such as an industrial building where considerable water is consumed in a process or evaporated to the atmosphere. But, in general, water out about equals water in, plus all the liquid and water-borne solid wastes produced in the building.

Another factor to consider in sizing a sewer is infiltration. Sewers, unlike water mains, often flow at less pressure than that exerted by groundwater around them. Thus, they are more likely to take in groundwater than to leak out wastewater. An infiltration rate of 2000 to 200,000 gal/(day · mi) might be expected. It depends on diameter of pipe (which fixes length between joints), type of soil, groundwater pressure, and workmanship.

In an effort to keep infiltration down, sewer-construction contracts specify a maximum infiltration rate. Weir tests in a completed sewer can be used to check the contractor's success in meeting the specification; but unless the sewer is large enough for workers to traverse, prevention of excessive infiltration is easier than correction. In addition to groundwater infiltration through sewer-pipe joints, the entry of surface runoff through access hole covers and thus into sewers is often a factor. Observers have gaged as much as 150 gal/min leaking into a covered access hole.

Size and slope of a sanitary sewer also must satisfy a requirement that velocity under full flow be kept to at least 2.5 ft/s to keep solids moving and preventing clogging. In general, no drain pipe should be less than 6 in in diameter; an 8-in minimum is safer.

14.12.3 Sewer-Pipe Materials

Vitrified clay, concrete, ductile iron, polyvinyl chloride (PVC), acrylonitrile butylene styrene (ABS) composite pipe, or steel may be used for pipe to carry wastewater and industrial wastes. Vitrified clay up to about 24-in diameter and steel or reinforced concrete in larger sizes are used most often. PVC or ABS are used for the smaller diameters.

Choice of pipe material depends on required strength to resist load or internal pressure; corrosion resistance, which is especially vital for pipe carrying certain industrial wastes; erosion resistance in sewers carrying coarse solids; roughness factor where flat slopes are desirable; and cost in place. Sewer piping installed on the discharge side of pumps should have a pressure rating well in excess of the pressure that will be experienced.

Vitrified-clay pipe requires some care in handling to minimize breakage. Reinforced concrete pipe must be made well enough or protected to withstand effects of damaging sewer gas (hydrogen sulfide) or industrial wastes. Ductile-iron sewer pipe is good under heavy loads, exposed as on bridges, in inverted siphons, or in lines under pressure. Steel is used chiefly for its strength or flexibility. Corrugated steel pipe with protective coatings is made especially for sewer use; its long lengths and light weight and ease of handling and jointing. Plastic pipes are used because of corrosion resistance, light weight, and low installation cost.

14.13 WASTEWATER-SYSTEM ELEMENTS

The usual steps in planning a plumbing system are: (1) secure a sewer or waste-disposal plan of the site; (2) obtain architectural and structural plans and elevations of the building; (3) tabulate known and estimated occupancy data, including the number of persons, sex, work schedules, and pertinent details of any manufacturing process to be performed in the building; (4) obtain copies of the latest edition of the applicable codes, (5) design the system in accordance with code requirements, and (6) have the design approved by local authorities before construction is begun.

The typical plumbing layout in Fig. 14.9 shows the major elements necessary in most plumbing systems. **Fixtures** (lavatories, water closets, bathtubs, showers, etc.) are located as needed on each floor of the structure (Art. 14.7).

Each fixture is served by a **soil stack,** or **waste stack,** a **vent** or **vent stack,** and a **trap** (Fig. 14.9). Vertical soil or waste stacks conduct waste solids and liquids from one or more fixtures to a sloped **house drain,** or **building drain,** generally located below the lowest floor of the building. Each vent stack extends to a **stack vent** that projects above the building roof to a vent through roof (VTR). The vent stack may or may not have **branch vents** connected to it. Vents and vent stacks permit the entrance of fresh air to the plumbing system, diluting any gases present and balancing the air pressure in various branches. (See also Art. 14.20.)

Traps on each fixture provide a water seal, which prevents sewer gases from entering the working and living areas. In some areas, the plumbing regulations require installation of a building or house trap. The building drain delivers the discharge from the various stacks to the **house trap,** or **building trap** (Fig. 14.9), which is generally provided with a separate vent. Between the building trap and **public sewer,** or other main sewer pipe, is the **building sewer.** The building sewer is outside the building structure, while the building trap is just inside or outside of the building foundation wall.

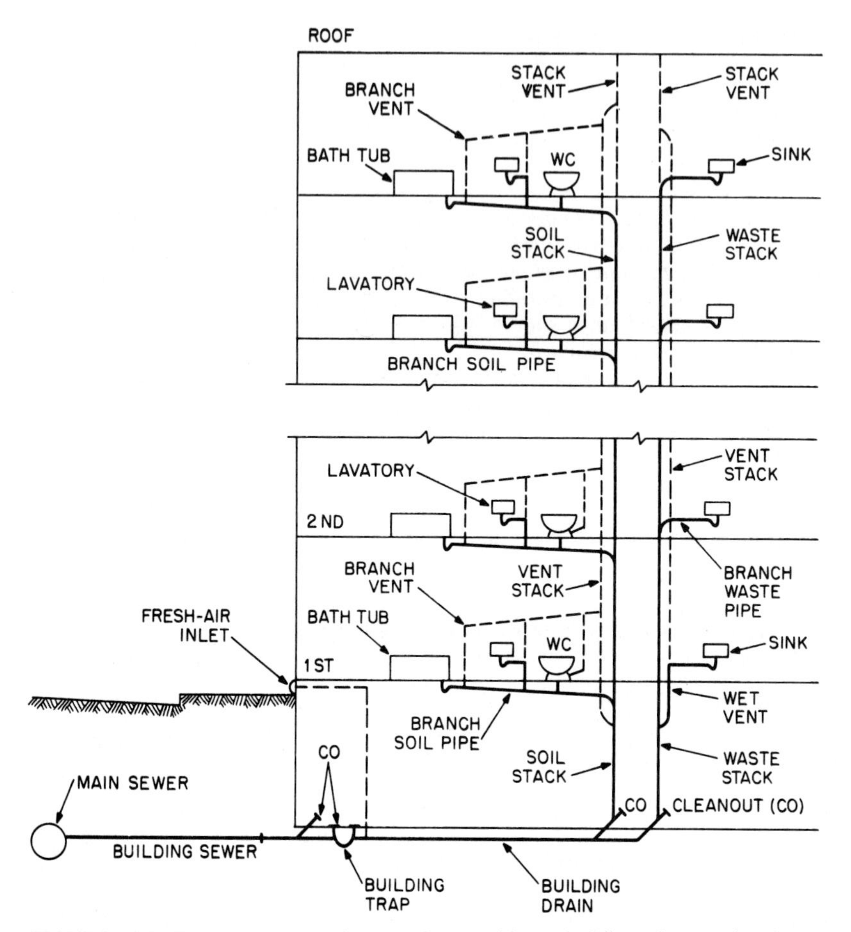

FIGURE 14.9 Wastewater-removal system for a multistory building. *(Reprinted with permission from F. S. Merritt, "Building Engineering and Systems Design," Van Nostrand Reinhold Company, New York.)*

Where the building drain is below the level of the public sewer line, some arrangement for lifting the wastewater to the proper level must be provided. This can be done by allowing the building drain to empty into a suitably sized **sump pit.** The wastewater is discharged from the sump pit to the public sewer by a pneumatic ejector or motor-driven sewage ejector pump.

Pipe Supports. Pipes of wastewater-removal systems should be supported and braced in the same way as pipes of water-supply systems (Art. 14.8). Vertical pipes generally should be supported at every floor. Horizontal pipes should be supported at intervals not exceeding the following: cast-iron soil pipe, 5 ft and behind every hub; threaded pipe, 12 ft; copper tubing, 10 ft. Supports also should be provided at the bases of stacks.

Consideration should be given to the possibility of building settlement and its effects on vertical pipes and to thermal expansion and contraction of pipes, especially when the pipes have a high coefficient of expansion or are made of copper.

Clean-outs. A clean-out is an opening that provides access to a pipe, either directly or through a short branch, to permit cleaning of the pipe. The opening is kept plugged, until the plug has to be removed for cleaning of the sewer. In horizontal drainage lines, at least one clean-out is required for each 100 ft of pipe. Clean-outs should be installed at the base of all stacks, at each change of direction in excess of 45°, and at the point where the building sewer begins. For underground drainage lines, the clean-out must be extended to the floor or ground level to allow easier cleaning. Clean-outs should open in a direction opposite to that of the flow in the pipe, or at right angles to it.

In pipes up to 4 in, the clean-out should be the same size as the pipe itself. For pipes larger than 4 in, the clean-out should be at least 4 in in diameter but may be larger, if desired. When underground piping over 10 in in diameter is used, an access hole is required at each 90° bend and at intervals not exceeding 150 ft.

14.14 WASTE-PIPE MATERIALS

Cast iron is the most common pipe material for systems in which extremely corrosive wastes are not expected. Polyvinyl chloride (PVC) is often used because it is inexpensive, lightweight, and easy to install. Vitrified-clay, galvanized steel, copper, and acrylonitrile butylene styrene (ABS) also are used.

Plumbing piping should conform to one or more of the accepted material standards approved by the plumbing code applicable in the area in which the building is located.

For cast-iron pipe, the fitting joints are calked (with oakum or hemp and filled with molten lead at least 1 in deep) or are no-hub (drawn stainless steel bands with neoprene gaskets). Clay pipe is generally joined by rubber or neoprene or push-on gaskets. Copper pipe is commonly soldered or brazed, while steel and wrought-iron pipe have screwed, flanged, or welded connections.

When planning a plumbing system, designers should check with the applicable code before specifying the type of joint to be used in the piping. Joints acceptable in some areas may not be allowed in others.

14.15 LAYOUT OF WASTE PIPING

Sanitary sewer systems should be sized and laid out to permit use of the smallest-diameter pipes capable of rapidly carrying away the wastewater from fixtures without clogging the pipes, without creating annoying noises, and without producing excessive pressure fluctuations at points where fixture drains connect to soil or waste stacks. Such pressure changes may siphon off the liquid seals in traps and force sewer gases back through the fixtures into the building. Positive or negative air pressure at the trap seal of a fixture should never be permitted to exceed 1 in of water.

Flow in Stacks. The drainage system is considered a nonpressure system. The pipes generally do not flow full. The discharge from a fixture drain is introduced to a stack through a stack fitting, which may be a long-turn tee-wye or a short-turn or sanitary tee. The fitting gives the flow a downward, vertical component. As the water accelerates downward under the action of gravity, it soon forms a sheet around the stack wall. If no flows enter the stack at lower levels to disrupt the sheet, it will remain unchanged in thickness and will flow at a terminal velocity, limited by friction, to the bottom of the stack. A core of air at the center of the stack is dragged downward with the wastewater by friction. This air should be supplied from outdoors through a vent through roof (Fig. 14.9), to prevent creation of a suction that would empty trap seals.

When the sheet of wastewater reaches the bottom of the stack, a bend turns the flow 90° into the buiding drain. Within a short distance, the wastewater drops from the upper part of the drain and flows along the lower part of the drain.

Slope of Horizontal Drainage Pipes. Plumbing codes generally require that horizontal pipes have a uniform slope sufficient to ensure a flow with a minimum velocity of 2 ft/s. The objective is to maintain a scouring action to prevent fouling of the pipes. Codes therefore often specify a minimum slope of ¼ in/ft for horizontal piping 3½ in in diameter or less and ⅛ or ¼ in/ft for larger pipes.

Because flow velocity increases with slope, greater slopes increase pipe-carrying capacity. In branch pipes, however, high velocities can cause siphonage of trap seals. Therefore, use of larger-size pipes is preferable to steeper slopes for attaining required capacity of branch pipes.

See also Art. 14.20.

14.16 INTERCEPTORS

These are devices installed to separate grease, oil, sand, and other undesirable matter from the wastewater and retain them, while permitting normal liquid wastes to discharge to the sewer.

Grease interceptors are used for kitchens, cafeterias, restaurants, and similar establishments where the amount of grease discharged might obstruct the pipe or interfere with disposal of the wastewater. Oil separators are used where flammable liquids or oils might be discharged to the sewer. Sand interceptors are used to remove sand or other solids from the wastewater before it enters the building sewer. They are provided with large clean-outs for easy removal of accumulated solids.

Other types of applications in which interceptors are usually required include laundries, beverage-bottling firms, slaughterhouses, and food-manufacturing establishments. The local authorities should be contacted to determine applicable local code or municipal regulations.

14.17 PIPING FOR INDIRECT WASTES

Certain wastes like those from food-handling, dishwashing (commercial), and sterile-materials machines should be discharged through an indirect waste pipe. This pipe is not directly connected with the building drainage pipes. Instead, it discharges waste liquids into a plumbing fixture or receptacle from where they flow directly to the building sanitary drainage system. Indirect-waste piping is generally required for the discharge from rinse sinks and such appliances as laundry washers, steam tables, refrigerators, egg boilers, iceboxes, coffee urns, dishwashers, stills, and sterilizers. It is also required for units that must be fitted with drip or drainage connections but are not ordinarily regarded as plumbing fixtures.

An air gap is generally required between the indirect-waste piping and the regular drainage system. The gap should be at least twice the effective diameter of the drain it services, but not less than 1 in. A common way of providing the required air gap is to lead the indirect-waste line to a floor drain, slop sink, or similar fixture that is open to the air and is vented or trapped in accordance with the governing code. To provide the necessary air gap, the indirect-waste pipe is terminated above the flood level of the fixture.

For a device that discharges only clear water, such as water from engine-cooling jackets, air-handling-unit coil condensate, sprinkler systems, or overflows, an indirect-waste system must be used. Clear water wastes from roof-mounted air-conditioning equipment can usually be discharged to roof drains or rainwater gutters. Although some jurisdictions require clear water wastes to be discharged to sanitary sewers, others allow or require clear water wastes to be discharged to the storm sewer system or dry wells.

Hot water above 140°F and steam pipes usually must be arranged for indirect connection into the building drainage system or into an approved interceptor.

To prevent corrosion of plumbing piping and fittings, any chemicals, acids, or corrosive liquids are generally required to be automatically diluted or neutralized before being discharged into the plumbing piping. Sufficient fresh water for satisfactory dilution, or a neutralizing agent, should be available at all times. A similar requirement is contained in most codes for liquids that might form or give off toxic or noxious fumes.

14.18 RAINWATER DRAINAGE

Exterior sheet-metal **gutters** and **leaders** for rainwater drainage are not normally included as part of the plumbing work. Interior leaders or storm-water drains, however, are considered part of the plumbing work. Depending on local codes or ordinances in the locality, rainwater from various roof areas may or may not be led into the sanitary sewer (Art. 14.11). Where separate rainwater leaders or storm drains are used, the building drains are then called **sanitary drains** because they convey only the wastes from the various plumbing fixtures in the building.

Interior storm-water drain pipes may be made of cast iron, steel, plastic, or wrought iron. All joints must be tight enough to prevent gas and water leakage. When a combined system is utilized, it is common practice to insert a cast-iron running trap between the storm drain and the building drain to maintain a trap seal on the storm drain at all times. Use of a combined system does not eliminate the need for separate drains and vents for wastewater. All codes prohibit use of storm drains for any type of wastewater.

Water falling on the roof may be led either to a gutter, from where it flows to a downspout (Fig. 14.10*a*), or it may be directed to a roof drainage device by means of a slope in the roof surface. Many different roof drainage devices, such as roof drains (Fig. 14.10*b*) and parapet drains, are available for different roof constructions and storm-water conditions.

Most plumbing codes include provisions to prevent the collapse of the building structure due to water ponding on the roof because of a clogged storm drainage

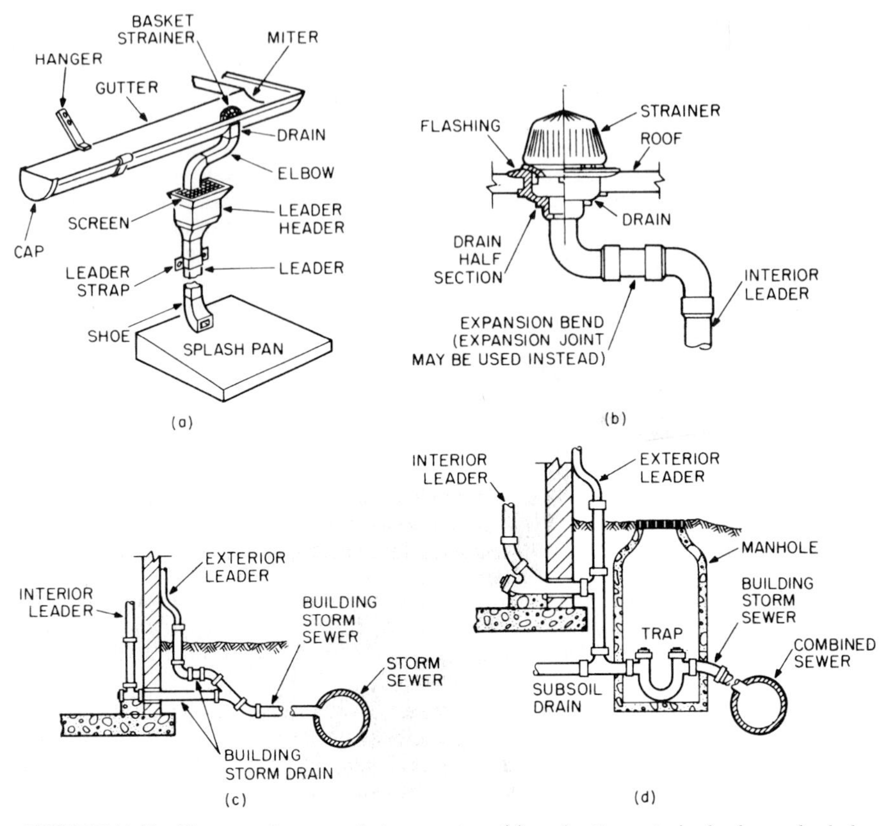

FIGURE 14.10 Elements of a storm-drainage system: (*a*) roof gutter, exterior leader, and splash pan; (*b*) roof drain and top portion of interior leader; (*c*) piping to a storm sewer; (*d*) piping to a combined sewer. *(Reprinted with permission from F. S. Merritt, "Building Engineering and Systems Design," Van Nostrand Reinhold Company, New York.)*

system. In most cases, these codes require installation of overflow roof drains or parapet overflow scuppers to relieve water from the roof in the event of such a condition. Local authorities should be contacted to determine what requirements apply in their jurisdiction.

When vertical leaders are extremely long, it is common practice to install an expansion joint between the leader inlet and the leader itself. Figure 14.10*c* shows an example of a connection of a building storm drain to a storm sewer. When the drain must be connected to a combined sanitary-storm sewer, a trap should be installed before the connection to the sewer (Fig. 14.10*d*).

Sizes of vertical leaders and horizontal storm drains depend on the roof area to be drained. Table 14.9 indicates the maximum horizontal projection of roof area permitted for various sizes of leaders and horizontal storm drains.

Semicircular gutters are sized on the basis of the maximum projected roof area served. Table 14.10 shows how gutter capacity varies with diameter and pitch. Where maximum rainfall is either more than, or less than, 1 in/hr, refer to the plumbing code for suitable correction factors.

TABLE 14.9 Sizes of Vertical Leaders and Horizontal Drains*

Vertical conductors and leaders

Size of leader or conductor, in†	Maximum projected area, ft²	Flow, gal/min
2	2,176	23
2½	3,948	41
3	6,440	67
4	13,840	144
5	25,120	261
6	40,800	424
8	88,000	913

Horizontal building storm drains and building storm sewers

Drain diameter, in	Maximum projected roof area, ft², and flow, gal/min, for various slopes					
	⅛ in per ft slope		¼ in per ft slope		½ in per ft slope	
	Area	Flow	Area	Flow	Area	Flow
3	3,288	34	4,640	48	6,576	68
4	7,520	78	10,600	110	15,040	156
5	13,360	139	18,880	196	26,720	278
6	21,400	222	30,200	314	42,800	445
8	46,000	478	65,200	677	92,000	956
10	82,800	860	116,800	1,214	165,600	1,721
12	133,200	1,384	188,000	1,953	266,400	2,768
15	238,000	2,473	336,000	3,491	476,000	4,946

*Roof areas and flows are based on a maximum rainfall intensity of 1 in/hr for a duration of 1 hr. For regions with different maximum rainfall intensity in storms with a 100-year recurrence interval, divide tabulated areas and flows by that intensity, in/hr.

†The area of rectangular leaders should equal or exceed that of the circular leader required. The ratio of width to depth of rectangular leaders should not exceed 3 to 1.

Drains for building yards, subsoil drainage systems, and exterior areaways may also be connected to the storm drainage system. Where this is not possible, these drains may be run to a dry well. When a dry well is used, only the discharge from these devices may be run to the dry well.

14.19 WASTE-PIPE SIZING

There are two ways of specifying the pipe size required for a particular class of plumbing service: (1) directly in terms of area served, as in roof-draining service (Table 14.9) and (2) in terms of fixture units (Table 14.11).

TABLE 14.10 Sizes of Semicircular Roof Gutters*

Gutter diameter, in	Maximum projected roof area, ft², and flow, gal/min, for gutters of various slopes							
	1/16 in per ft slope		1/8 in per ft slope		1/4 in per ft slope		1/2 in per ft slope	
	Area	Flow	Area	Flow	Area	Flow	Area	Flow
3	680	7	960	10	1,360	14	1,920	20
4	1,440	15	2,040	21	2,880	30	4,080	42
5	2,500	26	3,520	37	5,000	52	7,080	74
6	3,840	40	5,440	57	7,680	80	11,080	115
7	5,520	57	7,800	81	11,040	115	15,600	162
8	7,960	83	11,200	116	14,400	165	22,400	233
10	14,400	150	20,400	212	28,800	299	40,000	416

*See assumption of rainfall intensity and duration in note for Table 14.9.
†Gutters other than semicircular may be used if they have an equivalent cross-sectional area.

TABLE 14.11 Maximum Capacities of Building Drains and Building Sewers, Fixture Units*

Pipe diameter, in	Slope of pipe, in/ft			
	1/16	1/8	1/4	1/2
2			21	26
2½			24	31
3			42†	50†
4		180	216	250
5		390	480	575
6		700	840	1,000
8	1,400	1,600	1,920	2,300
10	2,500	2,900	3,500	4,200
12	2,900	4,600	5,600	6,700
15	7,000	8,300	10,000	12,000

*Maximum number of fixture units that may be connected to any portion of a building drain or building sewer. Consult the administrative authority for public sewers for sizing of on-site sewers that serve more than one building.

†A maximum of three water closets or three bathroom groups (water closet, lavatory, and bathtub or shower, or both) may be installed in single-family dwellings and two water closets or bathroom groups, in other types of construction.

As can be seen from these tables, the capacity of a leader or drain varies with the pitch of the installed pipe. The greater the pitch per running foot of pipe, the larger the capacity allowed, in terms of either the area served or the number of fixture units. The reason for this is that the steeper the pitch the larger is the static head available and hence the larger is the amount of liquid that the pipe can handle.

The steps in determination of pipe sizes by means of fixture units (Art. 19.8) are: (1) list all fixtures served by one stack or branch; (2) alongside each fixture list its fixture unit (Table 14.4, p. 14.22); (3) add the fixture units and enter the proper table (Tables 14.4, 14.11, or 14.13) to determine the pipe size required for the stack or the branch.

Fixture branches connecting one or more fixtures with a soil or waste stack are usually sized on the basis of the maximum number of fixture units for a given size of pipe or trap (Table 14.4). Where a large volume of water or other liquid may be contained in a fixture, such as in bathtubs or slop sinks, an oversize branch drain may be provided to secure more rapid emptying.

14.20 VENTING

Waste pipes are vented to the outside to balance the air pressure in various branches and to dilute any gases present. The availability of air prevents back pressure and protects traps against siphonage.

14.20.1 Types of Vents

The **main vent** is the principal artery of the venting system. It supplies air to **branch vents,** which, in turn, convey it to individual vents and wastewater pipes.

Every building should have at least one main **vent stack.** It should extend undiminished in size and as directly as possible from outdoor air at a level at least 6 in above the roof to the building drain. The main vent should be so located as to provide a complete loop for circulation of air through the wastewater-removal system. As an alternative to direct extension through the roof, a vent stack may be connected with a stack vent, if the connection is made at least 6 in above the flood-level rim of the highest fixture.

A **stack vent** is the extension of a soil or waste stack above the highest horizontal drain connected to the stack. This vent terminates above the roof.

A vent through roof (**VTR**) is any vent that extends through the roof to allow escape of sewer gases and to equalize pressures in the drainage system to prevent siphonage from trap seals. In colder climates, a VTR should be at least 4 in in diameter to prevent blockage from formation of frost and should terminate at least 12 in above the roof, but higher if the VTR is installed in regions with high snowfall rates.

An **individual vent,** or **back vent,** is a pipe installed to vent a fixture trap and is connected to the venting system above the fixture served or terminated outdoors. To ensure that the vent will adequately protect the trap, plumbing codes generally limit the distance downstream that the vent opening may be placed from the trap. This distance generally ranges from 2½ ft for a 1¼-in fixture drain to 10 ft for a 4-in fixture drain, but not less than two pipe diameters. The vent opening should be located above the bottom of the discharge end of the trap (Fig. 14.11). In general, all trapped fixtures are required to have an individual vent, although vents

TABLE 14.12 Size and Length of Vents

Size of soil or waste stack, in	Fixture units connected	Required vent diameter, in								
		1¼	1½	2	2½	3	4	5	6	8
		Maximum length of vent, ft								
1¼	2	30								
1½	8	50	150							
1½	10	30	100							
2	12	30	75	200						
2	20	26	50	150						
2½	42		30	100	300					
3	10		30	100	200	600				
3	30			60	200	500				
3	60			50	80	400				
4	100			35	100	260	1000			

4	200			30	90	250	900			
4	500			20	70	180	700			
5	200				35	80	350	1000		
5	500				30	70	300	900		
5	1100				20	50	200	700		
6	350				25	50	200	400	1300	
6	620				15	30	125	300	1100	
6	960					24	100	250	1000	
6	1900					20	70	200	700	
8	600						50	150	500	1300
8	1400						40	100	400	1200
8	2200						30	80	350	1100
8	3600						25	60	250	800
10	1000							75	125	1000
10	2500							50	100	500
10	3800							30	80	350
10	5600							25	60	250

TABLE 14.13 Horizontal Fixture Branches and Stacks

	Max number of fixture units that may be connected to			
			More than 3 branch intervals	
Pipe diameter, in	Any horizontal fixture branch*	One stack of 3 stories in height or 3 intervals, or less	Total for stack	Total at one story or branch interval
1¼	1	2	2	1
1½	3	4	8	2
2	6	10	24	6
2½	12	20	42	9
3	20†	30‡	72‡	20†
4	160	240	500	90
5	360	540	1100	200
6	620	960	1900	350
8	1400	2200	3600	600
10	2500	3800	5600	1000
12	3900	6000	8400	1500
15	7000			

*Does not include branches of the building drain.
†Not over two water closets or bathroom groups in each branch interval.
‡Not over six water closets or bathroom groups on the stack.

may be eliminated under some exceptional conditions. The plumbing code should be reviewed to determine where and how individual vents are to be installed.

To reduce the amount of piping required, two fixtures may be set back to back, on opposite sides of a wall, and vented by a single vent (**common vent**). In that case, however, the fixtures should discharge wastewater separately into a double fitting with inlets at the same level.

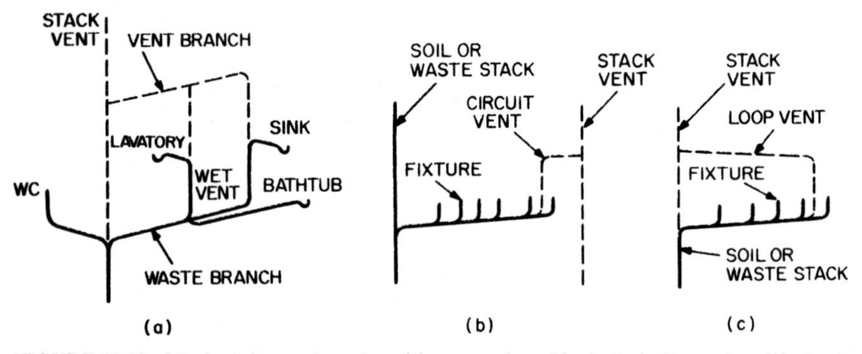

FIGURE 14.11 Venting of waste branches: (*a*) wet venting of bathtub drainage pipe; (*b*) circuit venting, and (*c*) loop venting of a battery of plumbing fixtures.

A **branch vent** is a pipe used to connect one or more individual vents to a vent stack or to a stack vent.

A **wet vent** is a pipe that serves both as a vent and as a drainage pipe for wastes other than those from water closets. This type of vent reduces the amount of piping from that required with individual vents. For example, a bathroom group of fixtures may be vented through the drain from a lavatory, kitchen sink, or combination fixture if such a fixture has an individual vent (Fig. 14.11*a*).

A battery of fixtures is any group of similar fixtures that discharges into a common horizontal waste or soil branch. A battery of fixtures should be vented by a circuit or loop vent. (Building codes usually set a limit on the number of fixtures that may be included in a battery.)

A **circuit vent** is a branch vent that serves two or more traps and extends from the vent stack to a connection to the horizontal soil or waste branch just downstream from the farthest upstream connection to the branch (Fig. 14.11*b*).

A **loop vent** is like a circuit vent but connects with a stack vent instead of a vent stack (Fig. 14.11*c*). Thus, air can circulate around a loop.

Soil and waste stacks with more than 10 branch intervals should be provided with a relief vent at each tenth interval installed, starting with the top floor. A **branch interval** is a section of stack at least 8 ft high between connections of horizontal branches. A **relief vent** provides circulation of air between drainage and venting systems. The lower end of a relief vent should connect to the soil or waste stack, through a wye, below the horizontal branch serving a floor where the vent is required. The upper end of the relief vent should connect to the vent stack, through a wye, at least 3 ft above that floor. Such vents help to balance the pressures that are continuously changing within a plumbing system.

14.20.2 Slopes and Connections for Vent Pipes

While the venting system is intended generally to convey only air to and from the drainage system, moisture may condense from the air onto the vent pipe walls. To remove the condensation from the venting system, all individual and branch vent pipes should be sloped and connected as to conduct the moisture back to soil or waste pipes by gravity.

14.20.3 Sizing of Vent Pipes

Fixture units (Art. 14.19) are also used for sizing vents and vent stacks (Table 14.12). In general, the diameter of a branch vent or vent stack should be one-half or more of that of the branch or stack it serves, but not less than 1¼ in. Smaller diameters are prohibited, because they might restrict the venting action.

14.20.4 Combined Draining and Vent Systems

These offer the possibility of considerable cost savings over the separate drainage and venting systems described in Art. 14.20.1.

One such system, introduced by the Western Plumbing Officials Association, employs horizontal wet venting of one or more lavatories, sinks, or floor drains by means of a common waste and vent pipe adequately sized to provide free movement of air above the flow line of the pipe. Relief vents are connected at the beginning

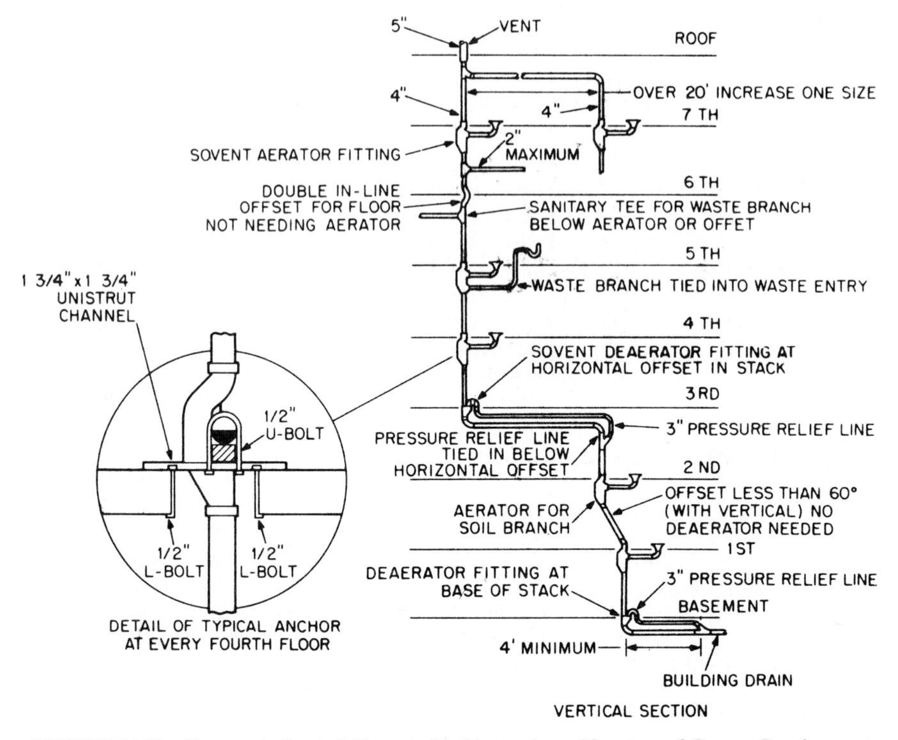

FIGURE 14.12 Copper single-stack Sovent plumbing system. *(Courtesy of Copper Development Association, Inc.)*

and end of the horizontal pipe. The traps of the fixtures are not individually vented. Some building codes permit such a system only where structural conditions preclude installation of a conventional system. Where this combined drainage and vent system may be used, it may require larger than normal waste pipes and traps. Each of the model codes has different requirements for this type of system and, therefore, the code in effect must be carefully reviewed during the design process.

The **Sovent system** is another type of combination system. It requires drainage branches and soil stacks, with special fittings, but no individual and branch vents and no vent stacks.

The system has four basic parts: a soil or waste stack with a stack vent extending through the roof, a Sovent aerator fitting on the stack at each floor, horizontal branches, and a Sovent deaerator fitting on the stack at its base and at horizontal offsets (Fig. 14.12). The aerator and deaerator provide means for self-venting the stack. In a conventional drainage system, a vent stack is installed to supply air to vent pipes connected to the drainage branches and to the soil or waste stack to prevent destruction of the trap seals. In the Sovent system, however, the vent stack is not needed because the aerator, deaerator, and stack vent avoid creation of a strong suction.

The aerator does the following: It reduces the velocity of both liquid and air in the stack. It prevents the cross section of the stack from filling with a plug of water.

And the fitting mixes the wastewater from the drainage branches with the air in the stack.

The deaerator separates the airflow in the stack from the wastewater. As a result, the wastewater flows smoothly into a horizontal offset or building drain. Also, air pressure preceding the flow at 90° turns is prevented from rising excessively by a pressure relief line between a deaerator and a stack offset or the building drain to allow air to escape from the deaerator.

An aerator is required on the stack at each level where a horizontal soil branch or a waste branch the same size as or one tube size smaller than the stack discharges to it. Smaller waste branches may drain directly into the stack. At any floor where an aerator fitting is not required, the stack should have a double in-line offset, to decelerate the flow (Fig. 14.12). No deaerators are required at stack offsets of less than 60°.

14.21 PLUMBING-SYSTEM INSPECTION AND TESTS

Plans for plumbing systems must usually be approved before construction is started. After installation of the piping and fixtures has been completed, both the new work and any existing work affected by the new work must be inspected and tested. The plumber or plumbing contractor is then informed of any violations. These must be corrected before the governing body will approve the system.

Most plumbing codes allow air or water to be used for preliminary testing of drainage, vent, and plumbing pipes. After the fixtures are in place, their traps should be filled with water and a final test made of the complete drainage system.

When a system is tested with water, all pipe openings are tightly sealed, except the highest one. The pipes are then filled with water until overflow occurs from the top opening. With this method, either the entire system or sections of it can be tested. In no case, however, should the head of water on a portion being tested be less than 10 ft, except for the top 10 ft of the system. Water should be kept in the system for at least 15 min before the inspection starts. During the inspection, piping and fixtures must be tight at all points; otherwise approval cannot be granted.

An air test is made by sealing all pipe outlets and subjecting the piping to an air pressure of 5 psi throughout the system. The system should be tight enough to permit maintaining this pressure for at least 15 min without the addition of any air.

The final test required of plumbing systems uses either smoke or peppermint. In the smoke test, all traps are sealed with water and a thick, strong-smelling smoke is injected into the pipes by means of a suitable number of smoke machines. As soon as smoke appears at the roof stack outlets, they should be closed and a pressure equivalent to 1 in of water should be maintained throughout the system for 15 min before inspection begins. For the peppermint test, 2 oz of oil of peppermint are introduced into each line or stack.

GAS PIPING

Natural and manufactured gases are widely used for heating in stoves, water heaters, and space heaters of many designs. Since gas can form explosive mixtures when mixed with air, gas piping must be absolutely tight and free of leaks at all times. Usual plumbing codes cover every phase of gas-piping size, installation, and testing. The local code governing a particular building should be carefully followed during design and installation.

("National Fuel Gas Code," ANSI Z223.1, American National Standards Institute and National Fire Protection Association; see also model plumbing codes and mechanical codes of the various building officials associations listed in Art. 14.1.)

14.22 GAS SUPPLY

The usual practice is for the public-service gas company to run its pipes to the exterior wall of a building, terminating with a brass shutoff valve and gas meter. The gas piping from the load side of the meter is generally extended to the inside of the building. From this point, the plumbing contractor or gas-pipe fitter runs lines through the building to the various fixture outlets. When the pressure of the gas supplied by the public-service company is too high for the devices in the building, a pressure-reducing valve can be installed near the point where the line enters the building. This valve is usually supplied by the gas company.

Besides municipal codes governing design and installation of gas piping and devices, the gas utility serving the area will usually have a number of regulations that must be followed. Typically, meters are required to be installed outside the building. The gas supply should not enter the building from below grade unless certain venting requirements are met, and gas pressure regulators installed inside the building must be vented to the outdoors. The local authorities and gas utility should be consulted as to special regulations relating to the installation of the gas piping system.

14.23 GAS-PIPE SIZES

Gas piping must be designed to provide enough gas to appliances without excessive pressure loss between the appliance and the meter. It is customary to size gas piping so the pressure loss between the meter and any appliance does not exceed 0.3 in of water during periods of maximum gas demand. Other factors influencing the pipe size include maximum gas consumption anticipated, length of pipe and number of fittings, specific gravity of the gas, and the diversity factor.

(C. M. Harris, "Handbook of Utilities and Services for Buildings," McGraw-Hill Publishing Company, New York.)

14.24 ESTIMATING GAS CONSUMPTION

Use the manufacturer's Btu rating of the appliances and the heating value of the gas to determine the flow required, ft^3/hr. When Btu ratings are not immediately available, the values in Table 14.14 may be used for preliminary estimates. The average heating value of gas in the area can be obtained from the local gas company, but when this is not immediately available, the values in Table 14.15 can be used for preliminary estimates.

TABLE 14.14 Minimum Demand of Gas Appliances, Btu/hr

Appliance	Demand
Barbecue (residential)	50,000
Bunsen burner	3,000
Domestic clothes dryer	35,000
Domestic gas range	65,000
Domestic recessed oven section	25,000
Domestic recessed top-burner section	40,000
Gas engines, per horsepower	10,000
Gas refrigerator	3,000
Steam boilers, per horsepower	50,000
Storage water heater:	
Up to 30-gal tank	30,000
30- to 40-gal tank	45,000
41- to 49-gal tank	50,000
50-gal tank	55,000
Water heater, automatic instantaneous:	
2 gal/min	142,800
4 gal/min	285,000
6 gal/min	428,400

TABLE 14.15 Typical Heating Values of Commercial Gases, Btu/ft³

Gas	Net heating value
Natural gas (Los Angeles)	971
Natural gas (Pittsburgh)	1021
Coke-oven gas	514
Carbureted water gas	508
Commercial propane	2371
Commercial butane	2977

Example. A building has two 50-gal storage hot-water heaters and 10 domestic ranges. What is the maximum gas consumption that must be provided for if gas with a net heating value of 500 Btu/ft³ is used?

Solution. From Table 14.14,

$$\text{Heat input} = 2(55{,}000) + 10(65{,}000) = 760{,}000 \text{ Btu/hr}$$

Maximum gas consumption is therefore 760,000/500 = 1520 ft^3/hr. The supply piping would be sized for this flow, even though all appliances would rarely operate at the same time.

(C. M. Harris, "Handbook of Utilities and Services for Buildings," and H. E. Bovay, Jr., "Handbook of Mechanical and Electrical Systems for Buildings," McGraw-Hill Publishing Company, New York.)

14.25 GAS-PIPE MATERIALS

The most common material used for gas piping is black steel pipe conforming to ASTM A53 or ATSM A106. Malleable-iron or steel fittings should be used, except for stopcocks and valves. Above 4-in nominal size, cast-iron fittings may be used.

Some local codes permit the use of brass or copper pipe of the same sizes as iron pipe if the gas handled is not corrosive to the brass or copper. Brazed or threaded fittings are generally used with these two materials.

Polyethylene (PE) and polybutylene (PB) with heat fusion joints and polyvinyl chloride (PVC) with solvent cement joints are used for outdoor underground installations, since these materials do not corrode and deteriorate as does black steel piping. If black steel is to be used underground, it must be provided with an exterior protective coating or tape or a cathodic-protection system, to prevent failure of the piping due to corrosion.

The usual gas supplied for heating and domestic cooking generally contains some moisture. Hence, all piping should be installed so it pitches back to the supply main, or drips should be installed at suitable intervals. Generally, unions or bushings are not permitted in gas piping systems owing to the danger of gas leakage and

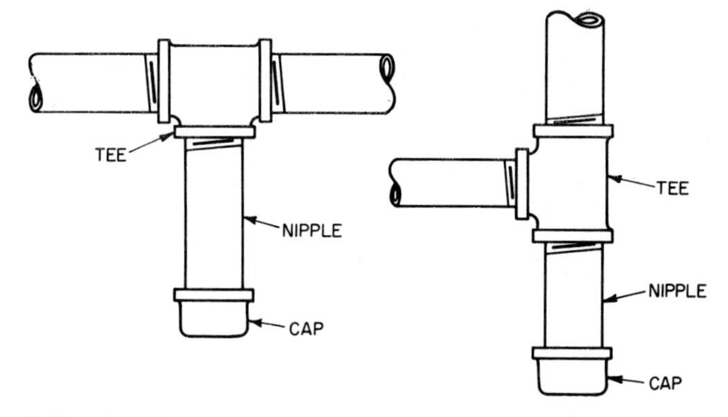

FIGURE 14.13 Drips for gas piping.

moisture trapping. To permit moisture removal, drips are installed at the lowest point in the piping, at the bottom of vertical risers at appliance connections, and at any other location where moisture might accumulate. Figure 14.13 shows typical drips for gas piping.

14.26 TESTING GAS PIPING

When installation of the gas piping in a building has been complete, it must be tested for leaks. Portions of the piping that will be enclosed in walls or behind other permanent structures should be tested before the enclosure is complete.

The usual code requirements for gas-pipe testing stipulates a test pressure of about 10 psig for 15 min, using air or one of several approved inert gases. Use of oxygen in pressure tests is strictly prohibited. Prior to connecting any appliances, the piping system should be purged, using the regular gas supply.

SPRINKLER SYSTEMS

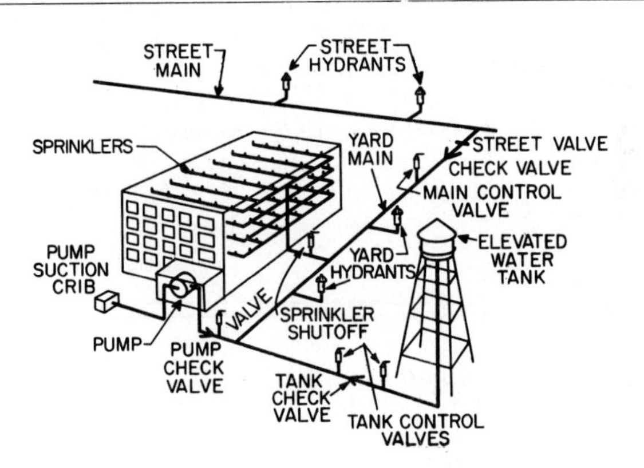

FIGURE 14.14 Water-supply piping for sprinklers.

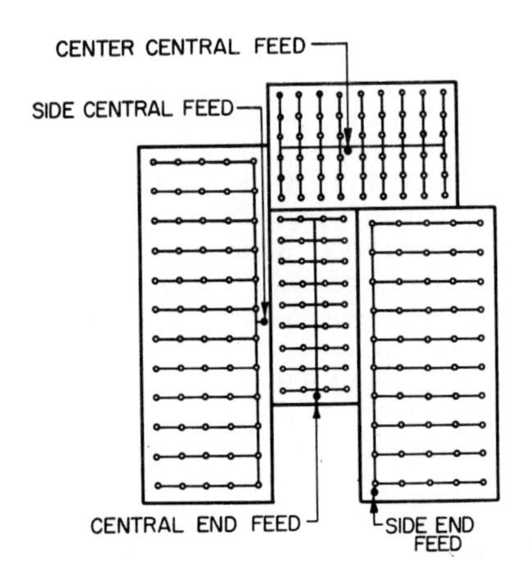

FIGURE 14.15 Typical layouts of sprinkler piping and risers.

Well-proved as effective in preventing the spread of fires in buildings of all types, sprinkler systems for fire protection are in wide use. Building codes generally dictate when and where sprinkler systems are required, and standards published by the National Fire Protection Association (NFPA), Batterymarch Park, Quincy, MA 02269, dictate how these systems are to be designed and installed. Sprinkler systems (Fig. 14.14) essentially consist of a series of horizontal parallel pipes, to which sprinkler heads are attached, placed near the building ceiling. Vertical risers (Fig. 14.15) supply the liquid—usually water—used for extinguishing fires. Sprinkler heads are designed (Fig. 14.16) to open automatically and discharge liquid when a fire is detected.

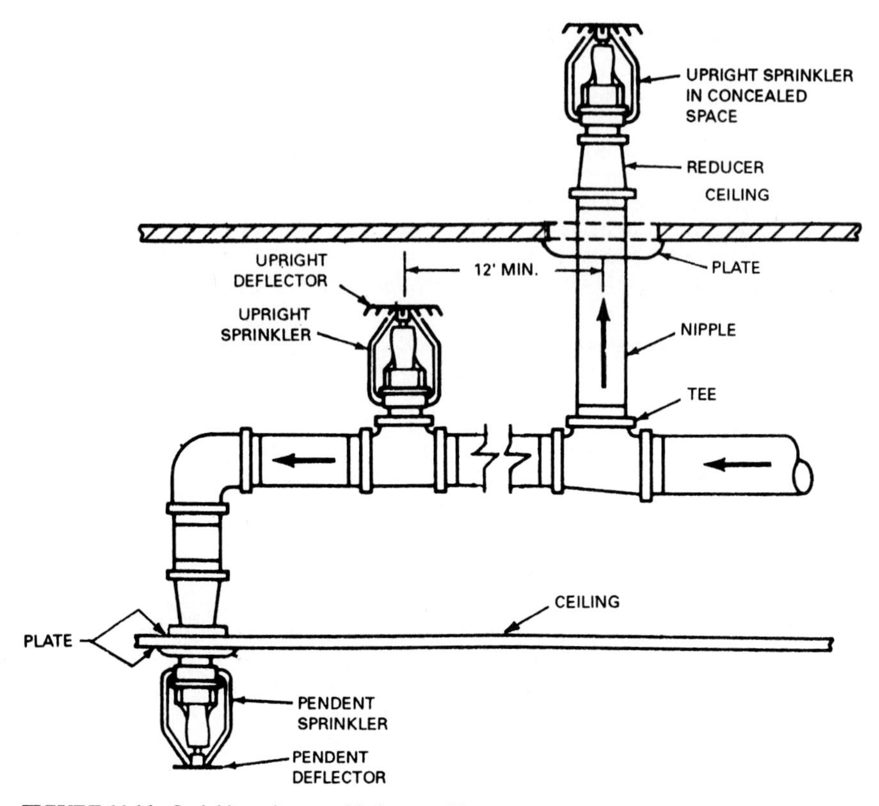

FIGURE 14.16 Sprinklers above and below a ceiling.

14.27 SPRINKLER SYSTEM STANDARDS AND APPROVALS

The most widely accepted standard for the design and installation of sprinkler systems is NFPA 13, "Installation of Sprinkler Systems," supplemented when necessary by NFPA 24, "Installation of Private Fire Services Mains and Their Appurtenances," NFPA 15, "Water Spray Fixed Systems," and NFPA 20, "Installation of Centrifugal Fire Pumps." Other NFPA standards are required for special hazards. These standards must be used with discretion when they are incorporated in specifications, because of the many references to the "authority having jurisdiction." Thus, before these standards are incorporated into a specification, rulings must be obtained from the "authority having jurisdiction" over the project at hand. The owner's insurance carrier can furnish the necessary information.

If the owner's insurance carrier is the Factory Mutual System, the FM standards will apply, and Factory Mutual will be the authority to be consulted.

In many cases the local building code will be the governing factor, and an insurance requirement a secondary matter. In such cases, conflicts must be reconciled to avoid a penalty for the owner and ambiguity for the construction bidders.

14.28 TYPES OF SPRINKLER SYSTEMS

The type of system that should be used depends chiefly on temperature maintained in the buiding, damageability of contents, expected propagation rate of a fire, and total fire load.

14.28.1 Wet-Pipe Systems

These always have water in the piping. Consequently, building temperature must be maintained above freezing. Wet-pipe systems are the simplest and most economical type that can be used. Other than a gate valve and an alarm valve, there are no other devices between the water supply and the sprinkler heads. Except for periodic tests of the alarm, no maintenance is required.

An alarm can be provided to indicate when a sprinkler head has started to operate and water is flowing in the piping. This is generally done by installing, in the main supply pipe, an alarm-check valve designed to distinguish between a momentary pressure surge and a steady flow. An alarm-check valve (Fig. 14.17) is essentially a swing check valve with an interior orifice that will admit water to a retard chamber that must fill completely to open a valve to the alarm supply pipe. The retarding chamber can also admit water to a pressure-actuated switch that can transmit a local or remote alarm electrically. In cases in which only a local alarm is considered adequate, water is admitted directly to a water-motor-driven gong, which requires no outside energy. Where no water-motor gong is needed or required, a vane-type water-flow indicator can be inserted in the supply pipe to give notification electrically of water flow.

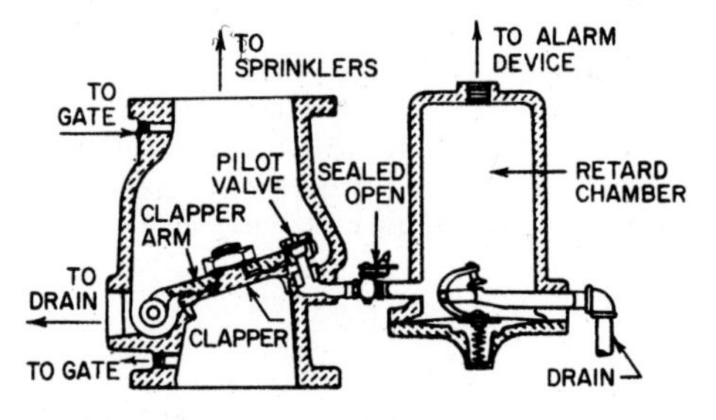

FIGURE 14.17 Alarm-check valve.

Pipe sizing for wet-pipe systems, as well as for all other types of systems, is based on the degree of the hazard covered, as defined in NFPA 13, "Installation of Sprinkler Systems," National Fire Protection Association. Pipe-sizing schedules are provided in NFPA 13 for light-hazard and ordinary-hazard occupancies (pipe sizing for extra-hazard occupancies must be calculated hydraulically). These schedules apply to all systems having sealed heads. A pipe-sizing schedule is also given for deluge systems, but this is only a guide. Determination of the schedule to be used must be made by the insurance or code authority. Frequently, a single building can require the use of all the schedules.

For buildings in which fire loads are unusually heavy, or stock is piled higher than 12 ft, or deluge systems are used, the authority can require that the schedule be disregarded and the sizing be calculated for a specific rate of discharge over a specific area. In this case, the authority will indicate the total area over which a specified discharge rate must be made. The area indicated will have no relation to the total area of the building or the total number of heads subject to the same fire. For example, in a 25,000-ft^2 building the authority might consider that the sprinkler system will bring a fire under control within an area of 2000 ft^2 with a water application rate (density) of 0.35 gal/min per square foot of floor area. This means that the maximum anticipated discharge of the sprinkler system would be 2000 × 0.35, or 700 gal/min. With heads spaced for a coverage of 100 ft^2 each, this means that each head would have to discharge 35 gal/min. With the discharge rate per head known, and the total area known, the pressure required to deliver this volume of water at the most distant point from the water supply can be calculated, in the manner described in NFPA 13, "Installation of Sprinkler Systems," and NFPA 15, "Water Spray Fixed Systems." Computer programs are also available to help designers perform required calculations.

In determining the area and density required, the authority considers the contents, type of building construction, heat venting, and type of system, among other factors. If high piling of stock is involved, the authority takes into account the pile size, pile height, aisle width, and combustibility of contents. Based on full-scale fire tests, a standard, NFPA 231C, describes the protection required for stockpiling from 12 to 20 ft, in racks. Storage heights less than 12 ft are covered by NFPA 13.

NFPA 231C contains design charts for sprinklers for 20-ft-high rack storage for different classifications of commodities defined in the standard. The curves enable determination of the area covered by sprinkler operation from the ceiling sprinkler density, gal/min per square foot.

14.28.2 Dry-Pipe Systems

These are used in locations and buildings where it is impractical to maintain sufficient heat to accommodate a wet-pipe system. The system of piping is the same, but it is normally empty, containing only air under pressure. A normally high water pressure is held by a normally low air pressure by use of a **dry-pipe valve.** This valve generally employs a combined air and water clapper (Fig. 14.18) where the

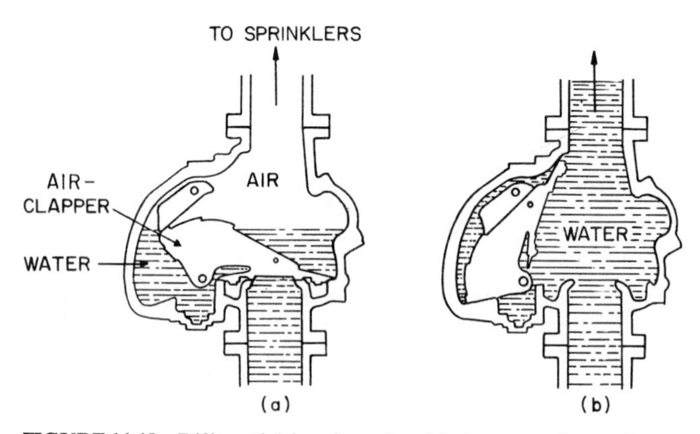

FIGURE 14.18 Differential dry-pipe valve: (*a*) air pressure keeps clapper closed; (*b*) venting of air permits clapper to open and water to flow.

area under air pressure is about 16 times the area subject to water pressure. When a sprinkler head is activated, air is released to the atmosphere, allowing the water to overcome the pressure differential and enter the piping. Air pressure must be checked periodically, but devices are available to maintain pressure automatically (typically an air compressor) and are equipped with an alarm that sounds in case of a malfunction which allows the air pressure to drop too low.

In dry-pipe systems of large capacity, the relatively slow drop in air pressure when a single head or a few heads are activated is overcome by use of an accelerator or exhauster. The former is a device, installed near the dry-pipe valve, to sense a small drop in pressure and transmit the system air pressure to a point under the air clapper. The water then can open the clapper quickly and push its way into the system without requiring a further pressure drop. An exhauster accomplishes the opening of the dry-pipe valve by sensing a small pressure drop and exhausting the pressurized air to the atmosphere through a large opening.

The dry-pipe valve must be installed in a heated area, or enclosure, because there is water in the piping up to the valve, and priming water in the valve itself.

14.28.3 Deluge Systems

In these systems, all heads open simultaneously to discharge water. The water is controlled by a **deluge valve,** which is operated by a temperature-sensitive detection system installed throughout the same area as the piping and heads. The detecting devices may be electrical, pneumatic, or mechanical, and may be connected to the deluge valve by wiring, tubing, or piping. Some valves are adaptable to all three means of operation.

A deluge system may be used for a few heads for a small, isolated industrial hazard, or in an entire, huge structure, such as an airplane hangar requiring several thousand heads. Deluge systems are also used to advantage in chemical plants where process vessels and tanks contain flammable materials. The purposes of such systems are to isolate a fire and to cool exposed equipment not originally involved in the fire.

In the calculation of pipe sizes for deluge systems consideration should be given to the probability of simultaneous operation of adjacent systems, if any may be involved. In addition, authorities frequently require that provision be made in the water supply system for operation of two or more fire hose streams while the deluge system is operating. In a multisystem installation, each system must be of manageable size, and good practice requires that no system be larger than one 6-in deluge valve can supply.

An electrically controlled system may employ fixed-temperature thermostats, rate-of-rise units, smoke detectors, combustion products detectors, or infrared or ultraviolet detectors. The nature and extent of the hazard determine the kind of detector required.

Operation of a hydraulically or pneumatically controlled deluge valve can be by means of a pilot line of small-diameter pipe on which are spaced automatic sprinkler heads at suitable intervals. These heads can be augmented, when necessary, by use of a mechanical air- or water-release device, which operates on the rate-of-rise principle as well as the fixed temperature of the sprinkler heads. Other pneumatic means include small copper air chambers, sensitive to rate-of-rise conditions, connected by small-diameter copper tubing to the release mechanism of the valve.

Most deluge valves are essentially check valves with a clapper that is normally latched in the closed position. The actuating system unlatches the valve, allowing water to enter the system and flow out of the heads.

For application of water from a deluge system, a variety of nozzles may be needed. If general area coverage is required, a standard sprinkler head, without a fusible element, may be used. A standard sprinkler orifice is ½ in, but larger or smaller orifices may be required. For specific equipment protection, orifices may be directional, or with a wide or narrow included angle, and with varying degrees of droplet size, including atomizing nozzles.

14.28.4 Preaction Systems

These employ the same components as deluge systems, but the sprinkler heads are sealed and the pipe sizing is in accordance with wet- and dry-pipe systems. A separate heat-detecting system with sealed heads is employed, because the detecting system is designed to operate before a head opens and at the same time to sound an alarm. If the fire can be extinguished before any heads open, the discharge of water from the preaction system can be aborted manually, and the total extent of the damage from the fire and water will be lessened. In addition, if any part of the piping is damaged, water will not flow unless heat is detected. The sealed-head piping system is usually supervised by air at low pressure, 1 or 2 psi, the loss of which will sound a trouble alarm.

This type of system lends itself well to use where highly damageable contents are involved. It is also a substitute for a dry-pipe system when the additional expense of the detecting system is justified.

14.28.5 Combined Dry-Pipe and Preaction Systems

These are suitable for large-area structures, such as piers, where more than one dry-pipe system would be required and where a water-supply pipe would be required in an unheated area.

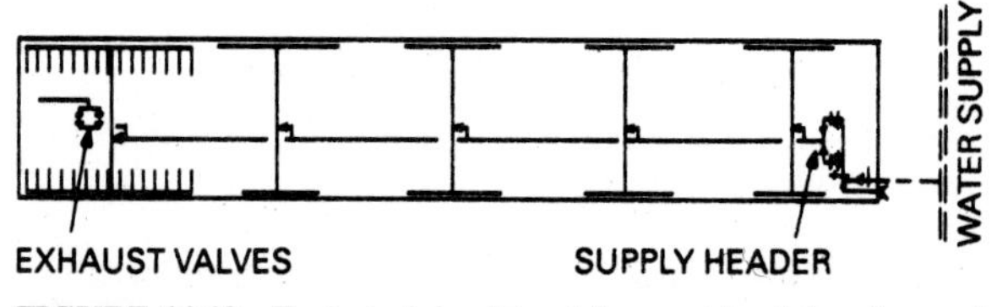

FIGURE 14.19 Typical piping layout for combined dry-pipe and preaction sprinkler systems. See Fig. 14.20 for supply header and Fig. 14.21 for exhaust header.

In this arrangement (Fig. 14.19), the heat-detecting system is the same as for a standard preaction system, but two dry-pipe valves in parallel are used instead of deluge valves. The system uses compressed air, as in a dry-pipe system. Valve arrangements at the supply end are shown in Fig. 14.20, and the exhauster system at the outboard end in Fig. 14.21.

When a fire is detected, the detecting system sounds an alarm and operates the exhausters at the valves and at the outboard end. The exhausters hasten the entry

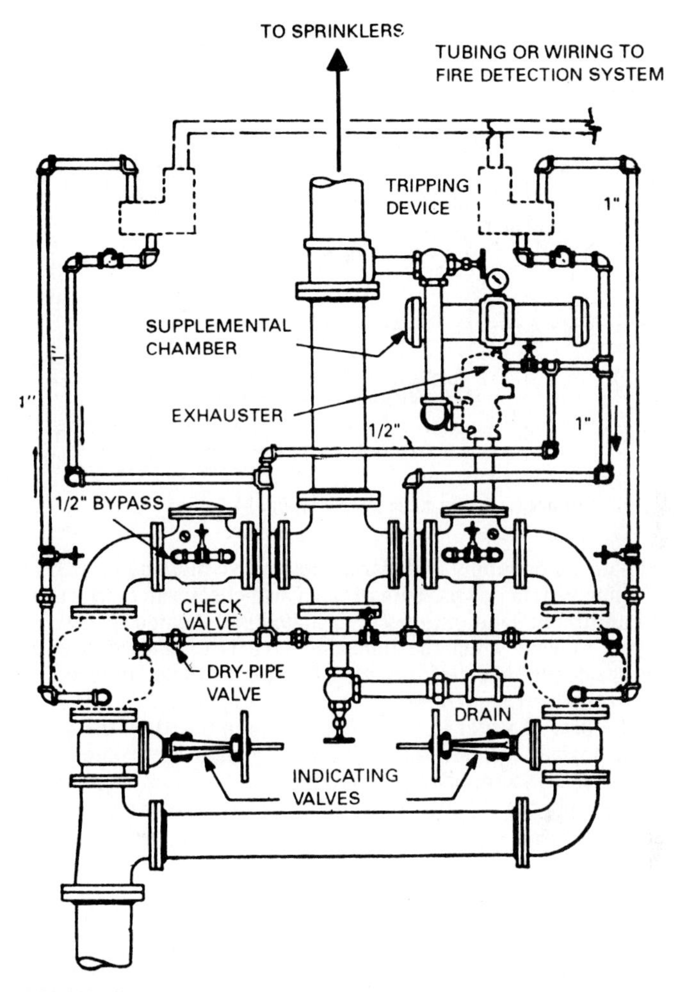

FIGURE 14.20 Supply header for sprinkler system in Fig. 14.19.

of water into the sprinkler pipes. Rubber-faced check valves on each subsystem avoid the need for exhausting air in the subsystems that are not involved in the fire. If there is a failure in the detecting system, it will still operate as a dry-pipe system, and if both dry-pipe valves fail, at least an alarm has been given.

In a large system, a number of dry-pipe valves, each with expensive heated enclosures, can be eliminated, as well as insulating and heat tracing of the supply line and the consequent maintenance of these items.

14.28.6 Antifreeze Systems

Like dry-pipe systems, these are used in locations and buildings where the possibility of temperatures sufficiently low enough to freeze water exist. With this type of

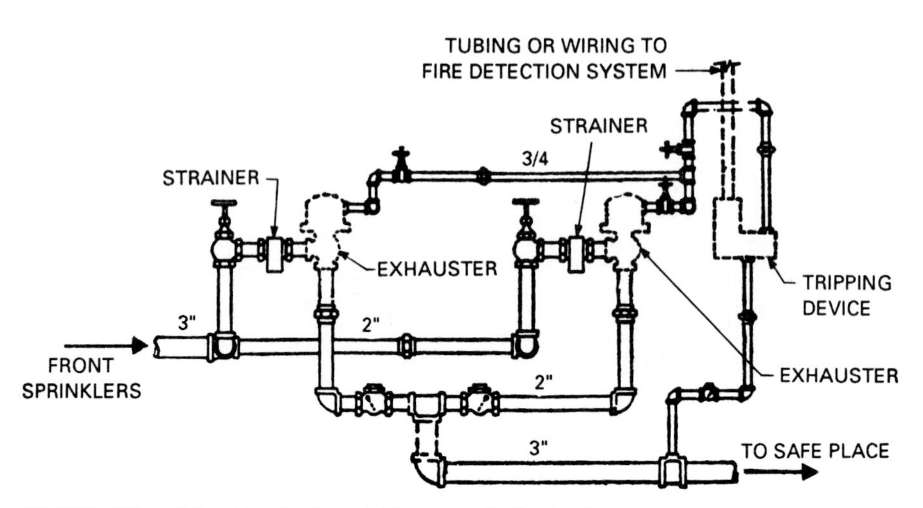

FIGURE 14.21 Exhaust header for sprinkler system in Fig. 14.19.

system, the main water supply and alarm-check valve must be located in a heated space. The piping system downstream of the alarm-check valve, which is filled with an approved antifreeze solution, may be installed in unheated spaces. This type of system is often utilized when the unheated area is of limited size, such as an unheated truck dock or an exterior cooling tower constructed of wood. In general, an antifreeze sprinkler system operates in a similar manner to a wet-pipe sprinkler system with the flow of water in the system being initiated by the activation of a sprinkler head. Installation of reduced-pressure backflow preventers on the water supplies to antifreeze systems is required to prevent backflow of antifreeze solution into the water supplies.

14.28.7 Fire-Cycle System

This is a recycling sprinkler system (patented by Viking Corporation, Hastings, Mich.) that will detect a fire, extinguish it with standard sprinklers, and turn off the water automatically when the fire is out (Fig. 14.22). If the fire is not completely out, or if it rekindles, the system will start again and repeat the performance as many times as necessary. This system can be valuable when there is highly damageable stock, or if it is likely to be a long time before there is a response to the alarm.

The basic design of the system is similar to the preaction system. It employs self-resetting, rate-of-rise, electrically operated detectors connected with conductors not subject to fire damage. The detector circuit operates a flow-control valve that is a modified form of the Viking deluge valve. As soon as the detecting system senses the fire, an alarm is actuated and the valve is opened, admitting water to the system. As ceiling temperature increases, sprinklers operate and discharge water over the fire area. This discharge will continue until the ceiling temperature is reduced to below the operating temperature of the detector. At that time, the detector circuit closes and a timing cycle starts. At the end of the timing cycle, the flow-control valve closes, and the system is ready for a repeat operation if the ceiling temperature increases again.

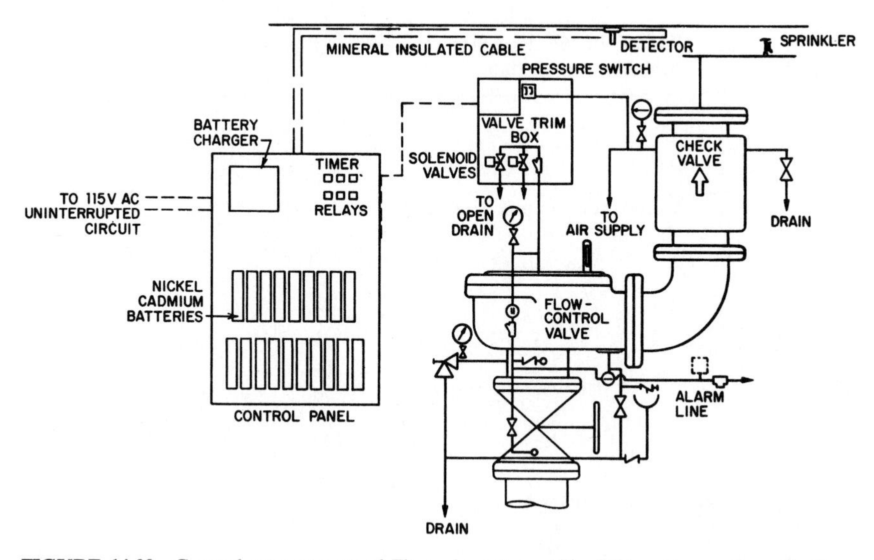

FIGURE 14.22 General arrangement of Firecycle system. *(The Viking Corporation, Hastings, Mich.)*

Central flow-control sprinkler heads (on/off-type heads) used in conjunction with a wet-pipe system function in a manner similar to a fire-cycle system, except that each sprinkler head works independently.

14.28.8 Outside Sprinklers

Sprinklers for exposure protection may be installed outside a building on any side exposed to an adjacent combustible structure or other potential fire hazard. Such systems have open nozzles directed onto the wall, windows, or cornices to be protected. The water supply may be taken from a point below the inside-sprinkler-system control valve if the building is sprinklered, otherwise from any other acceptable source, with the controlling valve accessible at all times. The system is usually operated manually by a gate valve but can be made automatic by use of a deluge valve actuated by suitable means on the exposed side of the building. The distribution piping is usually installed on the outside of the wall, with nozzles provided in sufficient numbers to wet the entire surface to be protected.

See also Art. 3.5.7.

14.29 SPRINKLER PIPING

The local fire-prevention authorities and the fire underwriters usually have specific requirements regarding the type, material, and size of pipe used in sprinkler systems. They should be consulted prior to the design of any system.

Sprinkler piping should be made of black or galvanized, welded and seamless steel pipe, wrought steel pipe, or seamless copper tube. "Installation of Sprinklers

TABLE 14.16 Pipe-Size Schedule for Typical Sprinkler Installations

Occupancy and pipe size, in.	Number of sprinklers	
	Steel	Copper
Light hazard		
1	2	2
1¼	3	3
1½	5	5
2	10	12
2½	30	40
3	60	65
3½	100	115
4	*	*
Ordinary hazard		
1	2	2
1¼	3	3
1½	5	5
2	10	12
2½	20	25
3	40	45
3½	65	75
4	100	115
5	160	180
6	275	300
8	*	*
Extra hazard		
1	1	1
1¼	2	2
1½	5	5
2	8	8
2½	15	20
3	27	30
3½	40	45
4	55	65
5	90	100
6	150	170
8	*	*

*Refer to NFPA 13.

Systems," NFPA 13, National Fire Protection Association, should be consulted to determine which ASTM or ANSI standards should be met. The pipe should be capable of withstanding a pressure of at least 175 psi. (Standard-wall pipe may be used for pressures up to 300 psi.) Table 14.16 gives the minimum diameters of pipe that may be used with specific numbers of sprinklers for various types of building-occupancy classifications. For exceptions and further details, see NFPA 13.

Fittings on sprinkler pipes should be capable of resisting the working pressures in the pipes, but at least 175 psi. Fittings may be made of cast or malleable iron or wrought copper.

Connections should be made with screwed or flanged mechanical joint or soldered fittings. Joints in risers and feed mains, however, may be welded.

Area Protected by Sprinklers. The maximum area protected by one sprinkler head should not exceed the areas listed in Table 14.17. Because of obstructions commonly found, the area ot coverage of each sprinkler head is usually significantly less than the maximum areas listed.

Sprinkler Position. In usual applications, sprinklers should be installed in the upright position (Fig. 14.16), where they are protected from damage by vehicles and other objects passing beneath them. When it is necessary to install sprinklers in a pendent position, they must be of a design approved for that position.

Sprinkler deflectors should be parallel to the building ceilings and roof, or the incline of stairs when installed in a stairwell. Deflectors for sprinklers in the peak of a pitched roof should be horizontal.

Sprinkler Alarms. These are often required, depending on what other types of alarms or fire-notification devices are installed in the structure. The usual sprinkler alarm is designed to sound automatically when the flow from the system is equal to or greater than that from a single automatic sprinkler. Dry-pipe systems use a water-actuated alarm; preaction and deluge systems use electric alarm attachments actuated independently of the water flow in the system (Art. 14.28).

TABLE 14.17 Maximum Protection Areas, Ft^2, for Sprinklers[a]

Occupancy or type of sprinkler	Noncombustible obstructed construction[b]	Combustible obstructed construction[b]
Light hazard	225[e]	168[f]
Ordinary hazard	130	130
Extra hazard[c]	100	100
High-piled storage[c]	100	100
Large-drop sprinklers[d]	130	130
Early-suppression, fast-response sprinklers[d]	100	N.A.

[a]For special protection areas, consult NFPA 13.

[b]Wood-truss construction is classified as obstructed construction for the purpose of determining sprinkler protection areas.

[c]The protection area per sprinkler for pipe schedule systems should not exceed 90 ft^2. For systems hydraulically designed (in accordance with NFPA 231 and 231C for high-piled storage), with discharges less than 0.25 gal/min, the area may exceed 100 ft^2 but not 130 ft^2.

[d]Maximum protection area is 80 ft^2.

[e]The protection area per sprinkler for pipe schedule systems should not exceed 200 ft^2.

[f]The maximum protection area for light framing members spaced less than 3 ft c to c is 130 ft^2 and for heavy combustible framing members spaced 3 ft or more c to c, 225 ft^2.

Sprinkler Layout. Horizontal branch pipes with sprinklers are usually laid out in parallel lines in each story (Figs. 14.14 and 14.15). In buildings with light- and ordinary-hazard occupancies, the maximum distance permitted between horizontal branch pipes and between sprinklers on those pipes is 15 ft. For high-piled storage, however, maximum spacing generally is limited to 12 ft. For extra-hazard occupancies, maximum spacing also is 12 ft. The distance from a wall to an end branch pipe or to an end sprinkler on a branch pipe should not exceed one-half the allowable spacing for the pipes or sprinklers, respectively.

Locations of sprinklers should be staggered on successive branch pipes in buildings with extra-hazard occupancies. Sprinklers also should be staggered in ordinary-hazard buildings when sprinkler spacing on branch pipes exceeds 12 ft or when the sprinklers are placed under solid beams spaced 3 to 7½ ft on centers. In areas where esthetics is important, however, the sprinklers need not be staggered.

Where the structure is exposed, sprinklers should be located to prevent structural members from interfering with the discharge pattern of the sprinklers. Where branch pipes are installed transverse to the lines of the beams, the sprinkler deflectors, which direct the discharge, should preferably be located above the bottom of the beam. "Installation of Sprinkler Systems," NFPA 13, National Fire Protection Association, should be consulted concerning adjustments to sprinkler head locations due to vertical and horizontal obstructions.

Drainage of Sprinkler Systems. Provision should be made for draining all or any part of a sprinkler system. For that purpose, air-inlet valves or plugged fittings should be installed at high points, and valve-controlled outlets at low points. A drain and valve connection should be provided for each group of up to 20 sprinkler heads. To ensure proper drainage, horizontal pipes installed in wet-pipe sprinkler systems should have a slope of at least ¼ in/ft downward toward a drain. In dry-pipe systems and portions of preaction systems subject to freezing, branch lines should be pitched at least ½ in per 10 ft, and mains should be pitched at least ¼ in per 10 ft.

For 4-in diameter and larger supply-pipe risers, drip or drain pipes should be at least 2 in in diameter. For risers between 2 and 4 in in diameter, drain pipes should be at least 1¼ in in diameter, and drainpipes should be at least ¾ in in diameter when the riser is 2 in in diameter or smaller.

Test Pipes. Sprinkler systems should be tested periodically to ensure that they will function when needed. A test connection for wet- and dry-pipe systems consists of a connection at least 1 in in diameter with a test valve terminating in a smooth-bore, corrosion-resistant orifice. This orifice connection should be sized to provide a test flow rate equivalent to the flow of one sprinkler head installed in the system.

Approvals of Sprinkler-System Design. Before a sprinkler system is installed or remodeled, the applicable drawings should be submitted to the local municipal agency and the underwriter of the building. Since beneficial reductions in insurance rates may be obtained by suitable installation of sprinkler systems, it is important that the underwriter have sufficient time for a full review of the plans before construction begins. Similarly, municipal approval of the sprinkler-system plans is necessary before the structure can be occupied. In actual construction, the piping contractor installing the sprinkler system generally secures the necessary municipal approval. However, existence of the approval should always be checked before construction is begun.

14.30 WATER SUPPLIES FOR SPRINKLERS

Water supply for sprinklers must be of the most reliable type obtainable (Art. 14.5). When a municipal water supply has been identified as not being sufficient to meet the needs of the building fire-suppression systems, fire pumps and water storage tanks or reservoirs may be required. "Fire Pumps," NFPA 20, and "Water Tanks for Private Fire Protection," NFPA 22, National Fire Protection Association, should be consulted as part of the design of such suppression system components.

The size and elevation of tanks should be determined by analysis of the conditions at the facility requiring a fire-suppression system. "Installation of Sprinkler Systems," NFPA 13, provides guidelines for determining flow and storage requirements. Where penstocks, flumes, rivers, or lakes supply the water, approved strainers or screens must be provided to prevent solids from entering the piping.

Water supply requirements for sprinklers are often given in municipal building codes. If, however, NFPA 13, is the governing standard, the rules are indefinite and the local authority should be consulted. (See Art. 14.5 for flow requirements.)

Where the occupancy requires a system with an unusual water demand, this demand must be met. In a calculated system, the maximum demand is known within limits, and at most must be supplemented for fire extinguishment with a reasonable flow from hose streams. Where there are deluge systems, attention must be given to the possible simultaneous operation of more than one system.

Each fire-suppression-system water supply should be provided with a fire department connection (siamese connection), which allows fire department personnel to provide an additional water supply through the use of a pumper truck connected to a fire hydrant.

14.31 STANDPIPES

Hoses supplied with water from standpipes are the usual means of manual application of water to interior fires in buildings. Standpipes are vertical pipes in which water is stored under pressure, or which can be rapidly supplied with water under pressure, primarily for use by a fire department. But they also may be used by building occupants or firefighters. As with sprinkler systems, building codes generally dictate when and where standpipes are required (Table 14.18).

Class of Service. Standpipe systems are classified by "Installation of Standpipe and Hose Systems," NFPA 14, National Fire Protection Association, into three groups. They are as follows:

Class I. Designed for use by the fire department or those trained in handling heavy hose streams. Hose stations consist of 2½-in hose valves, installed where required. Water supplies permit two hose streams to be fed simultaneously from a single riser. Each stream provides 250 gal/min at a minimum pressure of 65 psi.

Class II. Designed for use by the occupants or by the fire department during the initial response to a fire. Hose stations consist of 1½-in hose valves and hose racks with 100 ft of hose, installed where required. A water supply of 100 gal/min is provided at a station at a pressure of at least 65 psi.

TABLE 14.18 Requirements for Standpipes[a]

Type of occupancy	Buildings without sprinklers[b]		Buildings with sprinklers[c]	
	Standpipe class	Hoses required?	Standpipe class	Hoses required?
Buildings with two or more stories and over 150 ft high	III	Yes	I	No
Buildings with four or more stories and less than 150 ft high, except dwellings and lodging houses	I[d] and II[d] or III	Yes[e]	I	No
Assembly, for more than 1000 persons[f]	II	Yes	None	No
Assembly, with area more than 5000 ft² for exhibitions	II	Yes	II	Yes
Institutional, storage of hazardous materials, business, schools beyond the twelfth grade with less than 50 persons per room, open parking garages—less than four stories high but over 20,000 ft² per floor[g]	II[d]	Yes	None	No

[a]Based on requirements in the Uniform Building Code, International Conference of Building Officials.
[b]Class II standpipes need not be provided in basements having an automatic fire-extinguishing system throughout, except as specified for assembly occupancies with over 5000 ft² used for exhibitions.
[c]The standpipe system may be combined with an automatic sprinkler system. When buildings do not have automatic sprinklers throughout, the portions that do not have such protection should have Class II sprinklers installed.
[d]Where Class II sprinklers may be damaged by freezing in open structures, Class I standpipes located as required for Class II may be permitted.
[e]Hoses are required only for Class II standpipes.
[f]Applicable to theaters, auditoriums, stadiums, and reviewing stands. Class II standpipes need not be provided for assembly areas used exclusively for worship.
[g]Applicable to buildings in which no open flames, welding, or flammable liquids are used: office buildings, stores, restaurants and bars for less than 50 persons, police and fire stations, factories, storage and sales rooms for combustible goods, paint stores without bulk handling, printing plants, aircraft hangars, hospitals, sanitariums, nursing homes and nurseries accommodating more than five persons, and prisons.

Class III. A combination of Class I and Class II systems with both 2½-in hose valves and 1½-in hose valves and hose racks with 100 ft of hose, installed where required. Water supply at an outlet is the same as for a Class I system.

Riser Sizes. Risers for Class I and III services up to 100 ft high, measured to the highest hose outlet, should be at least 4 in in diameter. Risers in excess of 100 ft should be at least 6 in in diameter. NFPA 14 and the local building code should be consulted for exact requirements.

Number of Risers. Sufficient risers should be installed in a building so that every point of every floor can be reached by a 30-ft stream from a nozzle attached to 100 ft or less of hose connected to a riser outlet valve.

Maximum Pressure. The standpipe system should be zoned by the use of gravity tanks, automatic fire pumps, pressure tanks, and street pressure so that the maximum gage pressure at the inlet of any hose valve is not excessive. NFPA 14 limits the height of standpipes to 275 ft. Buildings more than 275 ft high require sufficient zones so that no riser exceeds 275 ft. This maximum height may be increased to 400 ft if pressure-limiting devices are installed at the hose outlets of all hose valves that are provided with water at pressures exceeding 100 psi. These pressure-limiting devices should limit the pressure on the downstream side of the hose valve to 100 psi.

Hose Stations. Standpipes generally are installed within stairway enclosures, or as near such an enclosure as possible, for safe access to hose outlets by firefighters. Hose stations should be located at the standpipes. Hoses to be stored should be placed on hose racks. The valves should be readily accessible from a floor or a stairway landing and should be placed 4 to 6 ft above the floor.

Water Supply. Standpipes can be supplied from municipal water mains when adequate pressure and volume are available. When necessary, this source can be supplemented with water from a tank or automatic fire pump.

NFPA requires that the minimum supply for a Class I or III service be 500 gal/min (for a period of 30 min) for one riser, and an additional 250 gal/min for each additional riser, up to a maximum of 2500 gal/min. A pressure of 65 psi is required at the highest outlet for all risers.

A **fire department connection** (siamese connection) is a wye to which two fire hoses can be connected. Standpipes with no water supply other than a fire-department pumper connection are permitted in open-air parking garages and other structures that experience temperatures below freezing.

All risers should be cross connected at, or below, the street entrance-floor level. In buildings that have zoned water-distribution systems, standpipes in each zone should be cross connected below, or in, the story with the lowest hose outlets from the water source in each zone. Cross connections should have a diameter at least as large as that of the largest riser supplied by the cross connection. The cross connection between standpipe risers at the street entrance-floor level and at the bottom of each zone should extend to the exterior walls of the building. The cross connection should terminate outside the building at a fire department connection located 18 to 36 in above grade. The hoses, when connected by firefighters to a street hydrant or to a mobile pump, can supply the standpipes with water in the event of a standpipe water-supply failure. One additional fire-department hose connection is required for each additional standpipe riser in excess of the first; that is, one additional fire-department connection is required for each additional 250 gal/min of system flow. Building codes may require a siamese connection for each 300 ft of building wall facing on a street or public place. This connection should be easily visible and accessible.

14.32 CENTRAL-STATION SUPERVISORY SYSTEMS

Any mechanical device or system is more reliable if it is tested and supervised regularly. Sprinkler systems are designed to be rugged and dependable, as shown by their satisfactory performance record over many years. Performance, however, is improved where systems have been connected to an approved central-station supervisory service that provides continuous monitoring of alarms, gate valves, pressures, and other pertinent functions.

A central station supplies the necessary attachments to a system or water supply and transmits appropriate signals over supervised circuits to a constantly monitored location. When any signal is received at the station, at whatever hour, suitable action should be taken by dispatch of investigators or notification of the fire department. Notification of a closed gate valve, for example, should be investigated immediately to avoid a possible loss. Any special extinguishing system, such as a carbon dioxide system, can also be supervised as to pressure, operation, or other vital condition.

Such services are available in most areas and are arranged by contract, usually with an installation charge and an annual maintenance charge. Requirements for such systems are in National Fire Protection Association standard NFPA 71, "Installation, Maintenance, and Use of Signaling Systems for Central Station Service." Where no such service is available, a local or proprietary substitute can be provided when such a facility is desirable. Standards for such installations are in "Installation, Maintenance, and Use of Protective Signaling Systems," NFPA 72.

SECTION FIFTEEN
ELECTRICAL SYSTEMS

James M. Bannon
Electrical Engineer
STV Group
Pottstown, Pennsylvania

Design of the electrical installations in a building used to be simple and straightforward. Such installations generally included electrical service from a utility company; power distribution within the building for receptacles, air conditioning, and other electrical loads; lighting; and a few specialty systems, such as fire alarm and telephone. There were, of course, some specialized installations for which this simple description did not apply, but such buildings were uncommon. Now, however, design of electrical systems has become more complex and sophisticated.

This development has been driven by rapid advances in technology, availability of computers and computerized equipment, more enlightened life-safety and security concerns, and changes in the philosophical outlook of workers toward their workplace and their need for a comfortable environment. To meet these needs, a new building will likely include in its electrical installation an access control system, intrusion detection system, computer data network, uninterruptible power supply, and numerous other systems not commonly installed in the past. Not only have these systems become common but the basic electrical systems have undergone drastic changes.

Advances in electrical-power-distribution materials and methods, which have occurred at a nearly uniform rate since the turn of the century, have accelerated rapidly under the influence of computers and microprocessor controls. New light sources give designers added opportunities to improve lighting and energy efficiency. Microprocessor-based fire-alarm systems with addressable devices offer greatly improved protection, flexibility, and economy. And establishment of more local telecommunication operating companies and competition between them, encouraging innovation, has brought designers new choices and challenges with respect to telecommunication systems for buildings.

Nevertheless, the basic principles of electrical design still apply, and they are described in this section. In addition, the section was developed to be helpful to those who must assume responsibility for applying, coordinating, integrating, and installing the many electrical systems now available for buildings.

15.1 ELECTRICAL POWER

In many ways, transmission of electricity in buildings is analogous to water-supply distribution. Water flows through pipes, electricity through wires or other conductors. Voltage is equivalent to pressure drop; wire resistance, to pipe friction; and electric current, or flow of electrons, to water droplets.

The hydraulic analogy is limited to only very elementary applications with electric flow like direct current, which always flows in the same direction. The analogy does not hold for alternating current, which reverses flow many times per second without apparent inertia drag. Direct-current systems are simple two-wire circuits, whereas alternating current uses two, three, or four wires and the formulas are more complex. Any attempt to apply the hydraulic analogy to alternating currents would be more confusing than helpful. The mathematical concepts are the only guides that remain true over the whole area of application.

Ampere (abbreviated A) is the basic unit for measuring flow of current. The unit flowing is an electric charge called a **coulomb.** An ampere is equivalent to a flow of one coulomb per second.

One source of direct current is the battery, which converts chemical energy into electric energy. By convention, direct current flows from the positive terminal to the negative terminal when a conductor is connected between the terminals. The voltage between battery terminals depends on the number of cells in the battery. For a lead-plate–sulfuric-acid battery, this voltage is about 1.5 to 2 V per cell.

For high voltages, a generator is required. A generator is a machine for converting mechanical energy into electrical energy. The basic principle involved is illustrated by the simple experiment of moving a copper wire across the magnetic field between a north pole and south pole of a magnet. In a generator, the rotor is wound with coils of wire and the magnets are placed around the stator in pairs, two, four, six, and eight. When the coil on the rotor passes through the magnetic field under a south pole, current flows in one direction. When the same coil passes through the north-pole field, the current reverses. For this reason, all generators produce alternating current. If direct current is required, the coils are connected to contacts on the rotor, which transfer the current to brushes arranged to pick up the current flowing in one direction only. The contacts and brushes comprise the commutator. If the commutator is omitted, the generator is an alternator, producing alternating current.

See also Conversion of AC to DC in Art. 15.3.

15.2 DIRECT-CURRENT SYSTEMS

Resistance of flow through a wire, measured in units called **ohms** (Ω), depends on the wire material. Metals like copper and aluminum have low resistance and are classified as conductors.

Resistance for a given material varies inversely as the area of the cross section and directly as the length of wire.

Ohm's law states that the voltage E required to cause a flow of current I, A, through a wire with resistance R, Ω, is given by

$$E = IR \tag{15.1}$$

Power P is measured in watts and is the product of volts and amperes:

$$P = EI = (IR)I = I^2R \tag{15.2}$$

Large amounts of power are measured in kilowatts (kW), a unit of 1000 W, or megawatts (MW), a unit of 1,000,000 W.

Electric Energy. The energy expended in a circuit equals the product of watts and time, expressed as watt-seconds or watt-hours (Wh). For large amounts of energy, a unit of 1000 watt-hours, or kilowatt-hours, kWh, is used.

Charges for electric use are usually based on two separate items. The first is total energy used per month, kWh, and the second is the peak demand, or maximum kW required over any short period during the month, usually 15 to 30 min.

Power Transmission. Power is usually transmitted at very high voltages to minimize the power loss over long distances. This power loss results from the energy consumed in heating the transmission cables and is equal to the square of the current flowing I, times a constant representing the resistance r of the wires, Ω/ft, times the length L, ft, of the wires. Measured in watts,

$$\text{Heat loss} = I^2rL \tag{15.3}$$

Series Circuits. A series circuit is by definition one in which the same current I flows through all parts of the circuit (Fig. 15.1*a*). In such a circuit, the resistance R of each part is the resistance per foot times the length, ft. Also, by Ohm's law, for each part of the circuit, the voltage drop is

$$E_1 = IR_1 \qquad E_2 = IR_2 \qquad \cdots \qquad E_n = IR_n \tag{15.4}$$

Kirchhoff's law for series circuits states that the total voltage drop in a circuit is the sum of the voltage drops:

$$\begin{aligned} E_T &= E_1 + E_2 + E_3 + \cdots + E_n \\ &= IR_1 + IR_2 + IR_3 + \cdots + IR_n \\ &= I(R_1 + R_2 + R_3 + \cdots + R_n) \end{aligned} \tag{15.5}$$

By Ohm's law the total resistance in a series circuit then is

$$R = R_1 + R_2 + R_3 + \cdots + R_n \tag{15.6}$$

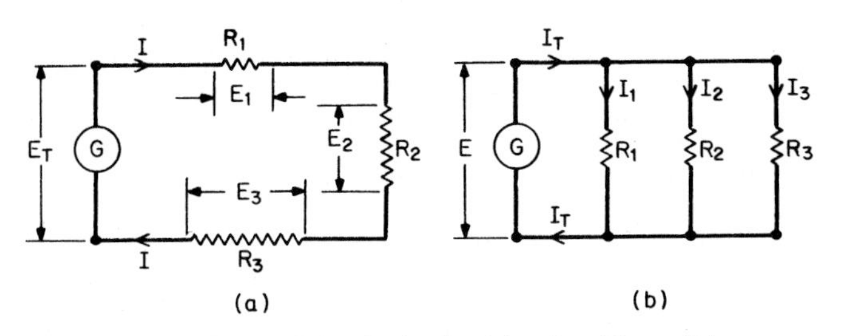

FIGURE 15.1 Types of electric circuits: (*a*) series; (*b*) parallel.

Parallel Circuits. These are, by definition, circuits in which the same voltage drop is applied to each circuit (Fig. 15.1*b*). The resistance of each circuit is obtained by multiplying the resistance per foot by the length, ft.

Kirchhoff's law for parallel circuits states that the total current in a circuit is equal to the sum of the currents in each part:

$$
\begin{aligned}
I_T &= I_1 + I_2 + I_3 + \cdots + I_n \\
&= \frac{E}{R_1} + \frac{E}{R_2} + \frac{E}{R_3} + \cdots + \frac{E}{R_n} \\
&= E\left(\frac{1}{R_1} + \frac{1}{R_2} + \frac{1}{R_3} + \cdots + \frac{1}{R_n}\right)
\end{aligned}
\tag{15.7}
$$

By Ohm's law, then, the total resistance R in a parallel circuit is given by

$$\frac{1}{R} = \frac{1}{R_1} + \frac{1}{R_2} + \frac{1}{R_3} + \cdots + \frac{1}{R_n} \tag{15.8}$$

It is sometimes convenient to use **conductance** G, siemens (formerly mhos), which is the reciprocal of resistance R:

$$G = \frac{1}{R} \tag{15.9}$$

By Ohm's law,

$$I_T = EG_1 + EG_2 + EG_3 + \cdots + EG_n \tag{15.10}$$

Series circuits are most commonly used in street and subway lighting. Most building systems use parallel circuits for both motor and lighting distribution.

Network systems consisting of a combination of series and parallel circuits are used for municipal distribution. With this type of system, a given building can be served from two different sides. This provides a second source of power, which is often accepted in lieu of an emergency generator.

15.3 ALTERNATING-CURRENT SYSTEMS

Any change in flow of current, such as that which occurs in alternating current, produces a magnetic field around the wire. With steady flow, as in direct current, there is no magnetic field.

One common application of magnetic fields is for solenoids. These are coils of wire, with many turns, around a hollow cylinder in which an iron pin moves in the direction of the magnetic field generated by the current in the coil. The movement of the pin is used to open or close electric switches, which start and stop motors. The pin returns to a normal position, either by gravity or spring action, when the current in the coil is stopped.

The motion of the pin can be predicted by the *right-hand rule*. If the fingers of the right hand are curled around the solenoid with the fingers pointing in the same direction as the current in the coil, the thumb will point in the direction of the magnetic field, or the direction in which the pin will move.

With direct current, a magnetic field exists only as the flow changes from zero to steady flow. Once steady flow is established in the wire, the magnetic field collapses. For this reason, all devices and machines that rely on the interaction of current and magnetic fields must use alternating current, which changes continuously. This equipment includes transformers, motors, and generators.

Transformers. These are devices used to change voltages. A transformer comprises two separate coils, primary and secondary, that wind concentrically around a common core of iron (Fig. 15.2). A common magnetic field consequently cuts both the primary and secondary windings. When alternating current (ac) flows in the primary coil, the changing magnetic field induces current in the secondary coil. The voltage resulting in each winding is proportional to the number of turns of wire in each coil. For example, a transformer with twice the number of turns in the secondary coil as in the primary will have twice the primary voltage across the secondary coil.

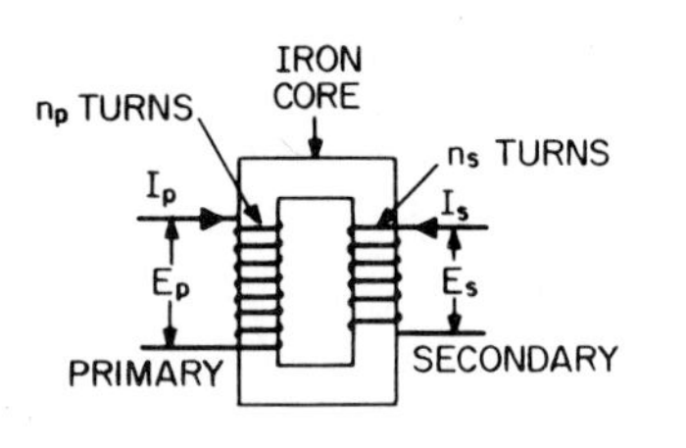

FIGURE 15.2 Transformer.

AC Generators and Motors. Just as changes in current flowing in a wire produce a magnetic field, movement of a wire through a magnetic field produces current in the wire. This is the principle on which electric motors and generators are built.

In these machines, a rotating shaft carries wire coils wound around an iron core, called an armature. A stationary frame, called the stator, encircling the armature, also carries iron cores around which are wound coils of wire. These cores are arranged in pairs opposite each other around the stator, to serve as poles of magnets. The windings are so arranged that if a north pole is produced in one core a south pole is produced in the opposite core. Current flowing in the stator, or field, coils create a magnetic field across the rotating armature.

When the machine is to be used as a motor, voltage is applied across the armature windings, and the reaction with the magnetic field produces rotary motion of the shaft. When the machine is to be used as a generator, mechanical energy is applied to rotate the shaft, and the rotation of the armature windings in the magnetic fields produces current in the armature windings. The current varies in magnitude and reverses direction as the shaft rotates.

Sine-Type Currents and Voltages. In generation of alternating current, rotation of the armature of a generator produces a current that starts from zero as a wire enters the magnetic field of a pole on the stator and increases as the wire moves through the field. When the wire is directly under the magnet, the wire is cutting across the field at right angles and the maximum flow of current results. The wire then moves out of the field and the current decreases to zero. The wire next moves into the magnetic field of the opposite pole, and the process repeats, except that the current now flows in the opposite direction in the wire. This variation from zero to a maximum in one direction (positive direction), down to zero, then continuing down to a maximum in the opposite direction (negative direction), and back to zero is called a **sine wave.**

The number of complete cycles per second of the wave is called the **frequency** of the current. This is usually 60 Hz (cycles per second) in the United States; 50 Hz in many European countries.

If P is the number of poles on the stator of a generator, the frequency of the alternating current equals $P \times \text{rpm}/120$, where rpm is the revolutions per minute of the armature. This relationship also holds for ac motors. Hence, for a frequency of 60 Hz, rpm = 60 × 120/P = 7200/P. This indicates that theoretically a standard four-pole motor would run at 1800 rpm, and a two-pole motor at 3600 rpm. Because of slippage, however, these speeds are usually 1760 and 3400 rpm, respectively.

Phases. Two currents or voltages in a circuit may have the same frequency but may pass through zero at different times. This time relationship is called phase. As explained in the preceding, the variation of the current (or voltage) from zero to maximum is a result of the rotation of a generator coil through 90° to a pole and back to zero in the subsequent 90°. The particular phase of a current is therefore given as angle of rotation from the zero start. If current (or voltage) 1 passes through its zero value just as another current (or voltage) 2 passes through its maximum, current 2 is said to *lead* current 1 by 90°. Conversely, current 1 is said to *lag* current 2 by 90°.

Effective Currents. The instantaneous value of an alternating current (or voltage) is continuously varying. This current has a heating effect on wire equal to the *effective current I* times the resistance R. Mathematically, the effective current is 0.707 times the maximum instantaneous current of the sine wave.

Ohm's law, $E = IR$, can be used in alternating circuits with E as the *effective voltage*, I, the effective current, A, and R the resistance, Ω.

Inductive Reactances and Susceptance. When alternating current flows through a coil, a magnetic field surrounds the coil. As the current decreases in instantaneous value from maximum to zero, the magnetic field increases in strength from zero to maximum. As the current increases in the opposite direction, from zero to maximum, the magnetic-field strength decreases to zero. When the current starts to decrease, a new magnetic field is produced that is continuously increasing in strength but has changed direction.

The magnetic field, in changing, induces a voltage and current in the wire, but the phase, or timing, of the zero and maximum values of this induced voltage and current are actually 90° behind the original voltage and current wave in the wire.

The induced voltage and current are proportional to a constant called the **inductive reactance** of the coil. This constant, unlike resistance, which depends on the material and cross-sectional area of the wire, depends on the number of turns in the coil and the material of the core on which the coil is wound. For example, a simple coil wound around an airspace has less inductive reactance than a coil wound around an iron core. Inductive voltage E_L and inductive current I_L, A, are related by

$$E_L = I_L X_L \tag{15.11}$$

where X_L is the constant for inductive reactance of the circuit, expressed in ohms, Ω. The reciprocal $1/X_L$ is called **inductive susceptance.**

When an inductive reactance is wired in a series circuit with a resistance, the inductive reactance does not draw any power (or heating effect) from the circuit. This occurs because the induced current I_L is 90° out of phase with the applied voltage E. In the variation of the instantaneous value of applied voltage, power is taken from the circuit in making the magnetic field. Then, as the magnetic field collapses, the power is returned to the circuit.

Capacitive Reactance and Susceptance. An electrostatic condenser, or capacitor, consists basically of two conductors, for example, flat metal plates, with an insulator between. Another familiar form in laboratory use is the arrangement of two large brass balls with an airspace between. Electrostatic charges accumulate on one plate when voltage is applied. When the voltage is high enough, a spark jumps across the air space. With direct current, the discharge is instantaneous and then stops until the charge builds up again. With alternating current, as one plate is being charged, the other plate is discharging, and the flow of current is continuous. In this case, the circuit is called capacitive.

$$E_C = I_C X_C \tag{15.12}$$

where X_C is the constant for **capacitive reactance, Ω.** The reciprocal $1/X_C$ is called **reactive susceptance.** The current I_C reaches its peak when the impressed voltage E is just passing through zero. Capacitive current is said to lead the voltage by 90°.

Impedance and Admittance. A circuit can have resistance and inductive reactance, or resistance and capacitive reactance, or resistance and both inductive and capacitive reactance. Resistance is present in all circuits. When there is any inductive or capacitive reactance, or both, in a circuit, the relation of the voltage E and current I, A, is given by

$$E = IZ \tag{15.13}$$

where Z is the **impedance, Ω,** the vector sum of the resistance, and the inductive and capacitive reactances. The reciprocal $Y = 1/Z$ is called the **admittance.** Electrical quantities such as E, I, Z, etc., can be represented graphically by **phasors.** These are the same as vectors used in other engineering disciplines and in mathematics but are called phasors because they are used to represent the phase relationship between different electrical quantities.

A **phasor** may be represented by a line and arrowhead. The length of the line is made proportional to the magnitude of E or I, and the arrowhead indicates plus or minus. In resistance circuits, the phasor E is indicated by a horizontal line with the arrowhead at the right:

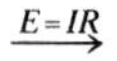

$E = IR$

In a circuit that contains inductive reactance, the current I_L lags behind the voltage E by 90°. This is indicated by phasors as follows:

$E = IR$ $E_L = IR_L$

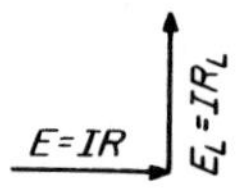

The phasor sum of these voltages is indicated by phasors as follows:

$E_Z = IZ$ θ_L $E = IR$ $E_L = IR_L$

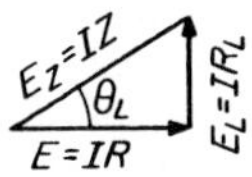

The diagram indicates that

$$E = IZ \cos \theta_L \tag{15.14}$$

where θ_L is the phase angle between voltage and current.

The relation between resistance R and inductance L is indicated by a similar phasor addition:

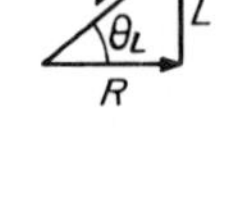

The diagram indicates that

$$R = Z \cos \theta_L \tag{15.15}$$

In a similar way, capacitive reactance in the circuit is indicated by

and the phasor sum is shown as follows:

The diagrams indicate that

$$E = IZ \cos \theta_C \tag{15.16}$$

$$R = Z \cos \theta_C \tag{15.17}$$

Note that the phasors for inductance and capacitance are in opposite directions. Thus, when a circuit contains both inductance and capacitance, they can be added algebraically. If they are equal, they cancel each other, and the Z value is the same as R. If the inductance is greater, Z will be in the upper quadrant. A greater value of capacitance will throw Z into the lower quadrant.

The diagram indicates that

$$\tan \theta = \frac{L - C}{R} \tag{15.18}$$

Kirchhoff's laws are applicable to alternating current circuits containing any combinations of resistance, inductance, and capacitance by means of phasor analysis:

In a **series circuit,** the current I is equal in all parts of the circuit, but the total voltage drop is the phasor sum of the voltage drops in the parts. If the circuit has resistance R, inductance L, and capacitance C, the voltage drops must be added phasorially as described in the preceding. Equations (15.14) to (15.18) hold for ac series circuits. To find the voltage drop in each part of the circuit, compute

$$E_R = IR \qquad E_L = IX_L \qquad E_C = IX_C$$

$$E_Z = E_R + E_L - E_C \qquad \text{(phasorially)} \tag{15.19}$$

In a **parallel circuit,** the voltage E across each part is the same and the total current I_Z is the vector sum of the currents in the branches,

$$\frac{E}{R} = I_R \qquad \frac{E}{X_L} = I_L \qquad \frac{E}{X_C} = I_C$$

For parallel circuits, it is convenient to use the reciprocals of the resistance and reactances, or susceptances, respectively S_R, S_L, and S_C. To find the current in each branch then, compute

$$ES_R = I_R \qquad ES_L = I_L \qquad ES_C = I_C$$

$$I_Z = I_R + I_L - I_C \qquad \text{(phasorially)} \tag{15.20}$$

Power in AC Circuits. Pure inductance or capacitance circuits store energy in either electric or magnetic fields and, when the field declines to zero, this energy. is restored to the electric circuit.

Power is consumed in an ac circuit only in the resistance part of the circuit and equals E_R, the effective voltage across the resistance, times I_R, the effective current. E_R and I_R are in phase. In a circuit with impedance, however, the total circuit voltage E_Z is out of phase with the current by the phase angle θ. In a series circuit, the current I is in phase with E_R; the voltage E_Z, on the other hand, is out of phase with E_R by the angle θ. In parallel circuits, the voltage E is in phase with E_R, but the current I_Z is out of phase with E_R. In both circuits, the power P is given by

$$P = E_R I_R \tag{15.21}$$

In series circuits, $E_r = E \cos \theta$ and $P = (E \cos \theta) I_R$. In parallel circuits, $I_R = I \cos \theta$ and $P = EI \cos \theta$. In any circuit with impedance angle θ, therefore, the power is given by

$$P = EI \cos \theta \tag{15.22}$$

Power Factor. The term $\cos \theta$ in Eq. (15.22) is called the power factor of the circuit. Because it is always less than 1, it is usually expressed as a percentage.

Low power factor results in high current, which requires high fuse, switch, and circuit-breaker ratings and larger wiring. Induction motors and electric-discharge-lamp ballasts are a common cause of low power factor. Since they are both inductive reactances (coils), the low power factor can be corrected by inserting capacitive reactances in the circuit to balance the inductive effects. This can be done with capacitors that are available commercially in standard kilovolt-ampere, kVA, capacities.

For example, a 120-V, 600-kVA circuit with a 50% power factor has a current of 5000 A. The actual power expended is only 300 kW, but the wire, switches, and circuit breakers must be sized for 5000 A. If a capacitor with a 300-kVA rating is wired into the circuit, the current is reduced to 2500 A, and the wiring, switches, and circuit breakers may be sized accordingly.

Conversion of AC to DC. Alternating current has the advantage of being convertible to high voltages by transformers. High voltages are desired for long-distance transmission. For these reasons, utilities produce and sell alternating current. However, many applications requiring accurate speed control need direct-current motors, for example, building elevators and railroad motors, including subways. In buildings, ac may be converted to dc by use of an ac motor to drive a dc generator, which, in turn, provides the power for a dc motor. The ac motor and dc generator are called a **motor-generator set.**

Another device used to convert ac to dc is a **rectifier.** This device allows current to flow in one direction but cuts off the sine wave in the opposite direction. The current obtained from the motor-generator set described previously is a similar unidirectional current of varying instantaneous value. The only nonvarying direct current is obtained from batteries.

Two-Phase and Three-Phase Systems. A single-phase ac circuit requires two wires, just like a dc circuit. One wire is the **live wire,** and the other is the **neutral,** so called because it is usually grounded (Fig. 15.3*a*).

A voltage commonly used in the United States is 240 V, which is obtained from the two terminals of the secondary coil of pole transformers on utility high-voltage transmission lines. If a third wire is connected to the midpoint of the secondary coil as a neutral, the voltage between either of the two terminal wires and the neutral will be 120 V, which also is a commonly used voltage (Fig. 15.3*b*).

The currents in the two terminal wires are 180° apart in phase. The neutral current from each is also 180° apart. These two currents, traveling in the same neutral wire, offset each other because of the phase difference. If the load currents in the two terminal wires are equal, the currents in the neutral will become zero.

In a similar way, three-phase electric service can be obtained directly from the utility company with three live wires and a grounded neutral (Fig. 15.3*c*). The currents in the three live wires, as well as their respective return flows in the neutral, are 120° apart in phase. If the currents are equal in the three live wires, the current in the neutral will be zero.

If the phase currents are not equal, the current in the neutral will be the phasor sum of the phase currents (Fig. 15.3*d*).

In many two-phase or three-phase systems, it is necessary therefore to balance the single-phase loads on each wire as much as possible. When the current in the neutral is zero, there is no voltage drop in the return circuit. Any voltage drop in the neutral subtracts from the voltage on the single-phase wires and affects the

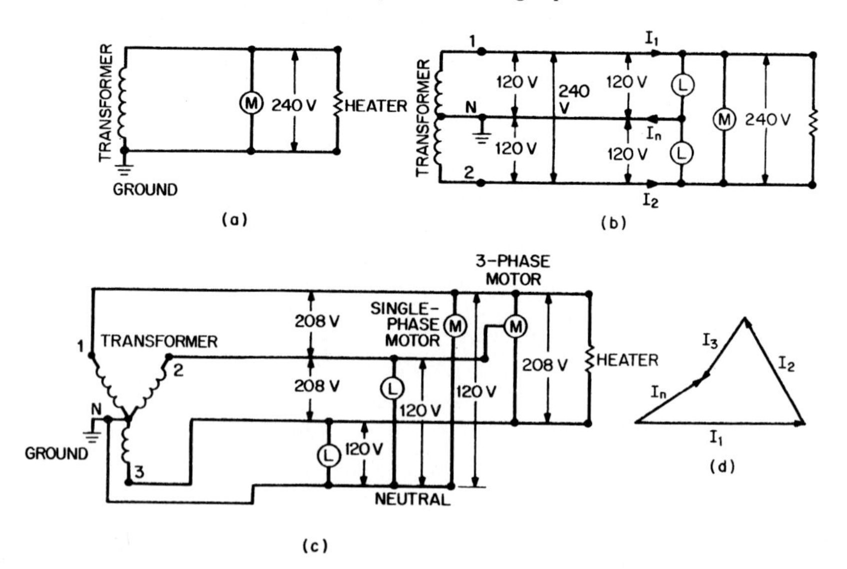

FIGURE 15.3 Examples of circuit wiring: (*a*) single-phase, two-wire circuit; (*b*) two-phase, three-wire circuit; (*c*) three-phase, four-wire circuit; (*d*) current I_n in the neutral wire of the three-phase circuit is the phasor sum of the currents in the phase legs.

loads on these circuits. The voltage drop times the current flowing in the neutral times the cosine of the phase angle is the power consumed in the neutral wire, and this adds to the total metered power on the utility bill.

15.4 ELECTRICAL LOADS

Electric services in a building may be provided for several different kinds of loads: lighting, motors, communication, equipment. These loads may vary in voltage and times of service, as for example, continuous lighting or intermittent elevator motors. Motors have high instantaneous starting currents, which can be four to six times the running current, but which lasts only a brief time.

It is highly improbable that all of the intermittent loads will occur at once. To determine the probable maximum load, **demand factors** and **coincidence factors** (diversity factors) must be applied to the total connected load (see Art. 15.8).

Lighting Loads. The minimum, and often the maximum, watts per square foot of floor area to be used in design are specified by building codes for various uses of the floor area. Maximum wattages are set to conserve energy and should be followed wherever possible. Electrical engineers, however, may exceed the minimum wattages if the proposed use requires more. For example, lighting may be designed to give a high intensity of illumination, which will require more watts per square foot than the code minimum. (Recommended lighting levels are given in the Illuminating Engineering Society "Lighting Handbook.")

Power Loads. In industrial buildings, the process equipment is normally the largest electrical power load. In residential, commercial, and institutional buildings, the power loads are mainly air-conditioning equipment and elevators. Some commercial and institutional buildings, though, contain significant computer and communication equipment loads, and special attention is required to properly serve these electronic equipment loads.

Electronic Equipment Loads. The electric power from the utility company is contaminated with electrical noise and spikes and is subject to sags, surges, and other power-line disturbances. The sensitivity of electronic equipment requires that the electrical system include equipment that will reduce the effect of these disturbances. Selection of this protection equipment should be based on the functions to be performed by the electronic equipment and a consideration of the consequences disturbance might cause, such as disruption of service, lost data, equipment damage, and

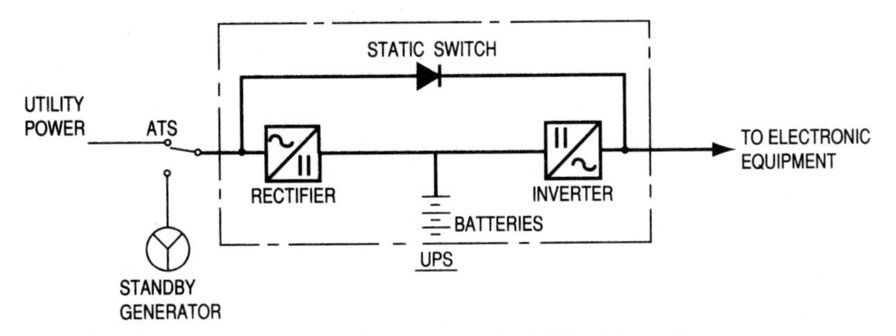

FIGURE 15.4 Typical uninterruptible power supply (UPS) with standby generator serving electronic equipment loads.

attendant costs. Most manufacturers specify the power quality needed for satisfactory operation of their electronic equipment. In fact, manufacturers of many computer systems furnish specific site-preparation instructions that address not only electrical power, but also lighting, air conditioning, grounding, and room finishes.

For protection purposes, for a personal computer or workstation, it may be only necessary to provide a good-quality plug-in strip with a transient-voltage surge suppressor (TVSS). A medical imaging system, such as a CAT-scan machine, may require a "power conditioner" that combines a voltage regulator, to eliminate sags and surges, with a shielded isolation transformer, to block spikes and noise. In critical installations, though, where equipment failure or an outage can have serious effects, more extensive steps must be taken. For example, loss of service to a satellite communication facility, a banking computer center, or an air-traffic control tower can have severe adverse consequences. To prevent this, such facilities should be provided with an uninterruptible power supply, backed up by either an alternate utility service or a stand-by generator.

A commonly used **uninterruptible power supply** (UPS) has a rectifier that is fed from the utility power line and delivers dc power to a large bank of batteries, keeping them fully charged, and to an inverter, which converts the dc back into high-quality ac power (Fig. 15.4). This arrangement isolates the electronic equipment from the utility power line, and, if the utility power fails, the batteries will instantly deliver power to the inverter, which will serve the load for 15 to 30 min. The stand-by generators will start and assume the load until the utility power is restored. If the generators also fail, the 15- to 30-min "protection time" of the batteries will allow an orderly shutdown of the equipment. The UPS includes a fast-acting, internal, static bypass switch, used in case the UPS itself fails.

Most electronic equipment will draw large amounts of harmonic current, principally odd-number harmonics (180 Hz, 300 Hz, . . .), which can overload and damage electrical equipment. Also, triple harmonics (3rd, 6th, . . .) add arithmetically and can overload a neutral conductor without tripping a breaker or blowing a fuse. To compensate for this, system neutral capacities must be increased. Empirical data indicate that, in systems with heavy electronic equipment loads, the neutral current will be about 1.8 times the phase current. So, use of a double-capacity neutral will usually suffice. Other system components, such as transformers and circuit breakers, also must be selected to operate satisfactorily for high harmonic loads.

Grounding. The National Electrical Code (NEC) requires an equipment ground system for every electrical installation to ensure personal safety. In sensitive electronic installations, a facility grounding system has the additional requirement of preventing damage to extremely sensitive computer equipment. Soil-resistivity measurements should be obtained at the site for use in design of a low-impedance ground system that will safely conduct lightning discharge currents to earth and allow sufficient ground current to flow to enable circuit-protective equipment to trip under fault conditions.

In addition, a low impedance (0.1 ohm or less) signal-reference ground system (often referred to as an equipotential plane) should be connected to the building ground. All electronic equipment and peripherals, as well as electrical equipment, HVAC equipment and ductwork, piping, raised floor system, and structural steel in proximity with the computer equipment should be connected to this system to ensure that all these items are at the same ground potential. This will ensure that ground current will not flow through the equipment. For this purpose, a manufactured copper grid with 2-ft by 2-ft spacing may be used. It should be located under the entire raised floor area.

15.5 EMERGENCY POWER

Local and national codes will dictate which electrical systems are required to be served by an emergency power system. NFPA 101, "Life Safety Code," and NFPA 99, "Health Care Facilities," published by the National Fire Protection Association, contain specific definitions of required emergency power loads but, in general, they include the following:

Emergency and egress (exit) lighting (90-min minimum is required by the National Electrical Code)

Elevator controls, lighting, communication, and signal systems

Alarm systems

Fire pumps

Communication systems, where used in emergencies

Generator area lighting and receptacles

Critical-patient-care systems in hospitals

Critical ventilation systems

In addition, users may require certain process equipment to be served by emergency power, to avoid loss of service to customers or equipment damage.

Small facilities may only need emergency power for emergency and egress lighting, in which case, fixtures with self-contained battery backup may be adequate. Larger facilities generally require an engine generator to provide emergency power. Emergency loads are connected to a dedicated panelboard or switchboard fed through an automatic transfer switch (ATS) that will detect loss of utility power, signal the generator to start, and transfer the emergency loads onto the generator, all in 10 s or less. After utility power returns, the ATS will retransfer the emergency loads back to utility power and then stop the generator. To prevent equipment damage and voltage surges, an ATS should be provided with an in-phase monitor that waits until the generator drifts into synchronism with the utility source before allowing retransfer.

Generator fuel source may be gasoline, LP gas, diesel fuel, or (if acceptable to the authority having jurisdiction) public utility gas. If a liquid fuel is chosen, any underground tanks and piping must be double-walled and equipped with electronic leak detection.

Though normally used only for emergency power, the generator, if large enough, may be used to reduce electric bills through demand peak shaving or as a cogenerator. Opportunities for these possibilities should be reviewed with the utility company.

15.6 ELECTRICAL CONDUCTORS AND RACEWAYS

All metals conduct electricity but have different resistances. Some metals, like gold or silver, have very low resistance, but they do not have the tensile strength required for electrical wire and are too costly. Consequently, only two metals are used extensively in buildings as conductors, copper and aluminum. The choice of copper or aluminum will be based on installed cost, and since aluminum conductors are less costly than copper conductors, one would expect that aluminum conductors would always be chosen. However, several factors should be considered: To carry a given amount of current, a larger aluminum wire is needed, and its raceway may also need to be larger. Because aluminum conductors expand more than copper as they warm up under load, they tend to move back and forth at terminals and, unless the proper termination methods and wiring devices (marked CO/ALR) are used, the conductors may work loose and create a fire hazard. Also, some local building codes and agency standards either do not permit aluminum conductors to be used, or restrict their use to larger wire sizes.

Conductors may be solid round wire, strands, or bus bars of rectangular cross section. Usually, conductors are wrapped in insulation of a type that prevents electric shock to persons in contact with it. The type of insulation also depends on the immediate environment surrounding the wire in its proposed use; dry or moist air, wetness, buried in earth, temperature, and exposure to mechanical or rodent damage.

Each size of commercial wire with a particular insulation is given by building codes a safe current-carrying capacity in amperes, called the **ampacity** of that wire. The code ampacity is based on the maximum heating effect that would be permitted before damage to the insulation.

The codes also require that wires installed in a building be protected from mechanical damage by encasement in pipes, called conduits, or other metal and nonmetallic enclosures, termed raceways.

15.6.1 Safety Regulations

The safety regulations for use of wire in buildings are given in local building codes, which are usually based on the National Fire Protection Association "National Electric Code" (see Art. 15.8). These codes are revised frequently; hence, the latest editions should be used.

Another agency, Underwriters Laboratories, Inc., tests electric materials, devices, and equipment. If approved, the item carries a UL label of approval.

15.6.2 Major Distribution Conductors

Most buildings receive their electrical power supply through service conductors from the street mains and transformers of a public utility.

The **service conductors** may be underground or above ground if taken from a utility-system pole. If a building is set back a great distance from the street poles, additional poles can be installed on the customer's property or the service conductors may be placed in an underground conduit extending to the building from the street pole. At the pole, the conduit should be extended 10 ft up to receive the service conductors.

At the building end, the service conductors come into a steel entrance box mounted on the building wall and then are brought to a service switch or circuit breaker. Service switches are commercially available up to 6000 A. Where the service load is greater, two or more service switches can be installed, up to a limit of six main service switches, called *drops,* on each service.

For large buildings, where six drops are not sufficient, the utility will install additional services with six drops available on each service.

Each service switch feeds a distribution center or groups of distribution centers, called panelboards. The connection between switch and panelboard is called a **feeder.** These main distribution panelboards consist of several circuit breakers or fused switches. Each of these breakers or switches feeds a load, either a motor or another remote panelboard or group of panelboards. The panelboards, in turn, serve branch circuits connected to lighting, wall receptacles, or other electrical devices.

Distribution systems in buildings are usually three-phase, four-wire. The final branch circuits are generally single-phase, two-wire. One wire in each circuit is grounded. The grounded-wire insulation in a feeder is colored white or natural gray, in accordance with the color code of the "National Electrical Code." This is commercially available up to No. 6 size wire. Larger feeders may be identified by white markings at the connections.

In a **branch circuit,** the equipment grounding conductor is colored green. When several grounded conductors are in one feeder raceway, one of the grounded conductors should be colored white or gray. The other grounded conductors should have a colored stripe (but not green) over the white or gray, and a different color should be used for the stripe on each wire. For four-wire systems, the colors for ungrounded conductors are usually blue, black, and red, with white used for the grounded conductor.

Conductor ampacity depends on the accumulative heating effect of the *IR* power loss in the wire. This loss is different for a given size wire with different insulations and depends on whether the wire is in open air and can dissipate heat or confined in a closed conduit with other heat-producing wires. Tables in the "National Electrical Code" give the safe ampacity for each type of insulation and the derated ampacity for more than three current-carrying wires in a raceway.

15.6.3 Types of Insulated Conductors

Following is a list of the various types of insulated conductors rated in the National Electrical Code:

Type MI. Mineral-insulated cable sheathed in a watertight and gastight metallic tube. Cable is completely incombustible and can be used in many hazardous locations and underground. MI cable can also be fire-rated, making it acceptable as a fire-pump feeder.

Type MC. No. 4 wire and larger, sheathed in an interlocking metal tape or a close-fitting, impervious tube. With lead sheath or other impervious jacket, Type MC may be used in wet locations.

Type AC. (Also known as BX cable.) This has an armor of flexible metal tape with an internal copper bonding strip in close contact with the outside tape for its entire length. This provides a grounding means at outlet boxes, fixtures, or other equipment. Type AC cable may be used only in dry concealed locations.

Type ACL. In addition to insulation and covering as for Type AC, Type ACL has lead-covered conductors. This makes this type suitable for wet or buried locations.

Type ACT. Only the individual conductors have a moisture-resistant fibrous covering.

Type NM or NMC. Nonmetallic-sheathed cables (also known as Romex). This type may be used in partly protected areas. The New York City Code permits BX (Type AC) but does not allow Romex because it is not rodentproof and is suject to nail damage in partitions.

Type SNM. The conductors are assembled in an extruded core of moisture-resistant, flame-resistant, nonmetallic material. The core is then covered with an overlapping metal tape and wire shield and sheathed in extruded nonmetallic material impervious to flames, moisture, oil, corrosion, fungus, and sunlight. This type may be used in hazardous locations.

Type SE or USE. Service-entrance cable has a moisture-resistant, fire-resistant insulation with a braid over the armor for protection against atmospheric corrosion. Type USE is the same as Type SE, except that USE has a lead covering for underground uses.

Type UF. This type is factory assembled in a sheath resistant to flames, moisture, fungus, and corrosion, suitable for direct burial in the earth. The assembly may include an uninsulated grounding conductor. Cables may be buried under 18 in of earth or 12 in of earth and a 2-in concrete slab.

15.6.4 Nonmetallic Extensions

Two insulated conductors within a nonmetallic jacket or extruded thermoplastic cover may be used for surface extensions on walls or ceilings or as overhead cable with a supporting steel cable made part of the assembly. Extensions may be used in dry locations within residences or offices.

Aerial cables may be used only for industrial purposes. At least 10 ft should be provided above the floor as clearance for pedestrians only, 14 ft for vehicular traffic.

15.6.5 Cable Bus and Busways

Busways are bare conductors of rectangular cross section, which are assembled in a sheet-metal trough. The conductors are insulated from the enclosure and each other. Busways must be exposed for heat dissipation. They are arranged with access openings for plug-in and trolley connections.

For heavy current loads, such as services, several insulated cables may be mounted in parallel, at least one diameter apart, within a ventilated metal enclosure with access facilities. Cable bus costs less than bus bars for the same load but generally takes up more space. Use is limited to dry locations.

15.6.6 Electrical Connections

A variety of devices are commercially available for connecting two or more wires. One type, a pressure connector, called a wire nut, may be screwed over two or three wires twisted together. Another type consists of end lugs attached to wires by squeezing them together under great pressure with a special tool. The lugs have a flat extension with a bolt hole for connection by bolts to a switch or busway. As an alternative, two wires may be joined together in a similar manner with a barrel-shaped splice.

All metal connectors should be insulated with either tape or manufactured insulated covers and should be enclosed in a metal box with cover. Several connections properly insulated can be enclosed in the same metal box if the box is adequate in size. The number of spliced conductors in a box is limited by building codes.

15.6.7 Raceways

A raceway is a general term used to describe the supports or enclosures of wires. For most power distribution systems in buildings, rigid conduit or tubing is used. The dimensions of such conduit or tubing and the number of wires of each size permitted is fixed by tables in the "National Electrical Code." Three or more conductors may not occupy more than 40% of the interior area, with some exceptions for lead-sheathed cable. All metallic raceways must be continuously grounded.

One wide use of **rigid steel conduit,** galvanized, is for branch circuits buried in the concrete slabs of multistory buildings.

Electrical metallic tubing is a thin-walled tube that is permitted by codes in locations where the raceway is not subject to physical damage.

For economy in industrial installations, a continuous, rigid structure may be designed to carry both power and signal wiring. This structure may be in the form of a trough, a ladder run, or a channel. It is limited in use to certain cables specifically approved by Underwriters Laboratories for such use.

Flexible metallic conduit, also known as Greenfield, is a continuous winding of interlocking metal stripping similar to that used for Type AC metal-clad cables (BX). These conduits are often used in short lengths at the terminal connection of a feeder to a motor. For wet locations, a watertight-type (Sealtite) is available.

Surface raceways are usually oval shaped and flat. When painted the same color as the wall or ceiling, they are less conspicuous than round pipe conduit. Surface raceways with a larger, rectangular cross section may be used to mount receptacles or telephone or data outlets, in addition to housing wiring.

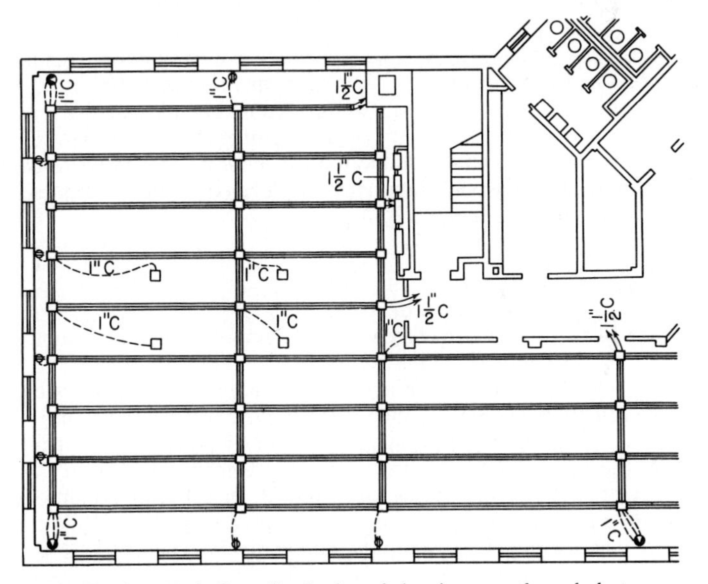

FIGURE 15.5 Underfloor distribution of electric power through ducts.

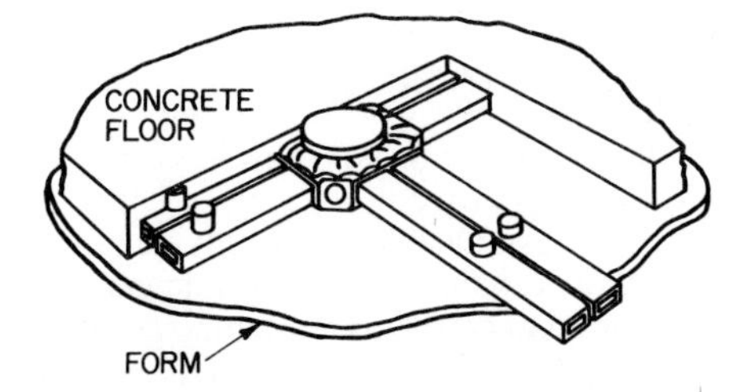

FIGURE 15.6 Raceways incorporated in a concrete floor, with outlet cover at the top of the floor.

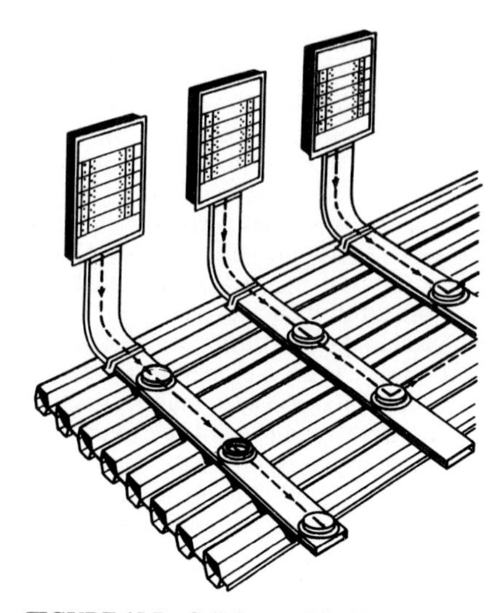

FIGURE 15.7 Cellular steel decking serves as underfloor electric ducts. Wires in headers distribute power to wires in the cells.

Underfloor raceways are ducts placed under a new floor in office spaces where desks and other equipment are frequently moved. Laid in parallel runs 6 to 8 ft apart, with separate ducts for power, signal, and telephone wires, these raceways may have flat-plate outlet covers spaced 4 to 6 ft along each run (Fig. 15.5). Large retail stores also find these installations a great convenience. The alternative is feeder runs above the hung ceiling of the story below, with fire-rated, *poke-through* construction to reach new outlets above the floor.

Underfloor raceways may be single-level (Fig. 15.6) or two-level (Fig. 15.7). In steel-frame buildings, with cellular steel decking, single-level raceways may be included in the structure of the floor itself. A concrete header across the cellular runs provides the means of entering from the finished floor. A similar arrangement can be used in cellular precast-concrete decks, with metal headers for connections. Wireways to carry large numbers of conductors carrying light-current signal or control circuits are commercially available in fixed lengths.

15.6.8 Access Floor Systems

In large computer rooms and in offices with heavy computer or communications usage, such as a brokerage or a customer service center, an access floor system may be used. This offers a false floor above the structural floor. The system consists of 2-ft by 2-ft removable panels, topped with a floor covering, which are supported from 6 to 36 in, or more, above the structural floor by pedestals and stringers. The space below the access floor is used for routing electrical, computer, and communication wiring. It is also used as a plenum for distributing conditioned air to the equipment and the occupied space. Since virtually the entire underfloor space is available and accessible, this system, though relatively expensive, offers flexibility for making changes in space use, such as adding equipment or rearranging room layouts.

15.6.9 Flat Conductor Cables (FCC)

These offer similar flexibility to that of an access floor system in that such cables permit outlets to be located anywhere in a room and allow easy relocation of an outlet. Flat conductor cables are available not only as power circuits but also in multiconductor, twisted pair, coaxial, and fiber-optic cables for use in communication and data systems. Manufacturers offer complete lines of power, data, and communication floor fittings for FCC system use. Use of FCC is limited to installation under carpet squares and is most commonly used in renovation work.

15.7 POWER SYSTEM APPARATUS

Most buildings, commercial, industrial, institutional, and residential, receive their power from a public utility. Usually, the customer is given a choice of voltages. For example, 240/120-V single-phase, three-wire service is very common in suburban and rural areas. This service comes from a single-phase, 240-V transformer on a utility pole, with one wire from each end of the secondary coil and with the neutral from the midpoint of its secondary coil. The voltage between the end terminal connections is 240 V and between each end wire and the neutral, 120 V (Fig. 15.3*b*).

In large cities, the service to large buildings is 208/120 V, three-phase, four-wire, with 208 V available between phase wires and 120 V between a phase wire and the neutral (Fig. 15.3*c*). Another choice is 480/277 V, three-phase, four-wire, with 277 V available between a phase leg and the neutral. It is more economical to use the higher voltage, 480/277 V, for motors and industrial lighting. The lower voltage 208/120 V is required for residential or commercial lighting and appliances.

In some areas, such as New Jersey, the utility will provide both voltage services on separate meters to a large building. But in other areas, such as New York City, the customer must choose one or the other voltage from only one meter and then use transformers to provide the second voltage service.

15.7.1 Transformers

Transformers may be dry or liquid-immersed type. The wet type is used for large installations. If the liquid is mineral oil, special fire-protection precautions are needed. Any liquid-filled transformer requires means for containing the liquid if the transformer tank should leak.

All transformers are rated in kVA, with primary and secondary voltages. Taps may be provided on the primary to compensate for variations in utility voltage as much as 10% below and 5% above nominal voltage, in 2½% increments. The manufacturer can also make available to the engineer the reactance and resistance of the coils and the noise rating. Noise can be minimized by use of vibration isolation mountings.

The power losses in a transformer create heat, which must be dissipated. Dry-type transformers are cooled by circulating air in the spaces enclosing the transformers. For liquid-filled transformers, which usually have very high capacity, the liquid may be circulated through coolers to transfer heat from the coils. Average losses in transformers used in buildings are about 2% of the rated capacity.

15.7.2 Meters

Consumption of electrical energy is measured by watt-hour meters. Utilities also include another charge, for demand, based on the maximum amount of power used in a specified time interval, usually about 15 to 30 min.

Three-wire meters are generally used for residences, either 208 V or, in some areas, 230/240 V. The 208-V service is usually taken from a three-phase, four-wire street or pole main. The voltage therefore differs 120° in phase from the current. There is a 120-V difference between the third, or neutral, wire and the phase leg. For the three 120-V, single-phase circuits, the total power, W, is computed from

$$P = 3EI \cos \theta \tag{15.23}$$

where E = voltage between phase legs and neutral
I = current, A
$\cos \theta$ = power factor

Industries and commercial installations with large motors require three-phase, four-wire meters. Distribution can be over one of three different types of circuits: 208 V, three-phase (motors); 208 V, single-phase (motors, appliances); or 120-V, single-phase (lighting, motors, appliances).

Meters for services supplied by a utility are provided and installed by the utility. The meter pans and current transformers must be provided by the customer in accordance with the utility's requirements.

All the service to one building may be measured by one meter, usually called a master meter. Buildings with rented spaces may have one meter for the owner's load and individual meters for each tenant.

15.7.3 Switches

These are disconnecting devices that interrupt electric current. Toggle switches (Fig. 15.8*a*) or snap switches are used for small currents like lighting circuits. They employ pressure contacts of copper to copper. Knife switches (Fig. 15.8*b*) are used for larger loads. A single-phase knife switch employs a movable copper blade hinged to one load terminal. To close a circuit, the blade is inserted between two fixed copper blades connected to the other terminal. The ground leg is usually continuous and unswitched, for safety reasons. For multiphase circuits, one hinged blade is used for each phase; thus, the switch may be double-pole (Fig. 15.9*a*) or three-pole (Fig. 15.9*b*), as the case may be.

A switch may be single-throw (Fig. 15.8*b*), as described, or double-throw (Fig. 15.9*c*). A double-throw switch permits the choice of connecting the load (always on the movable blade) to two different sources of power, each connected to opposite, fixed blades.

Once the blades of a switch are in solid contact, the heating effect at the contact surface is minimized. Opening and closing the switch, though, draws a hot arc, which burns the copper. This may cause an uneven surface of contact, with continuing small arcs across the separated points, and result in continual weakening of the contact switch and eventual breakdown.

Switches are carefully rated for load and classified for use by the National Electric Manufacturer's Association (NEMA) and the Underwriters Laboratories (UL).

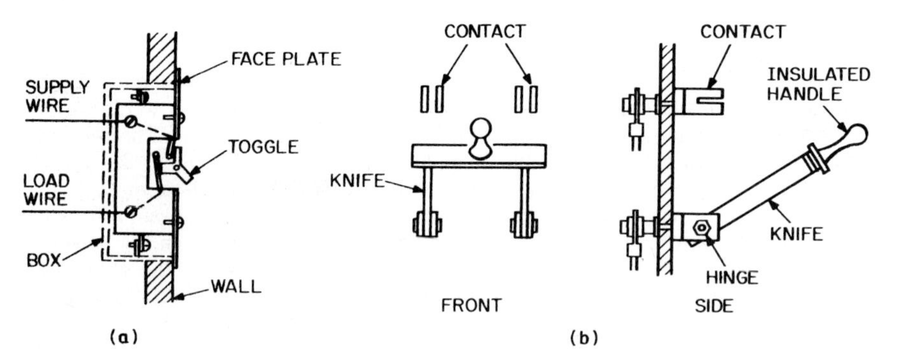

FIGURE 15.8 Switches: (*a*) snap; (*b*) blade. *(Reprinted with permission from F. S. Merritt and J. Ambrose, "Building Engineering and Systems Design," Van Nostrand Reinhold Company, New York.)*

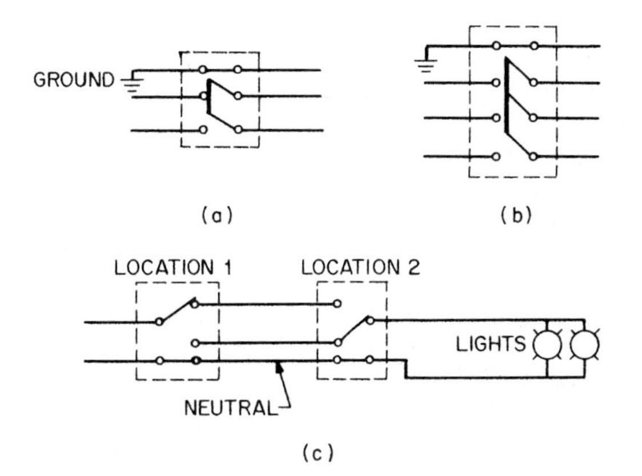

FIGURE 15.9 Single- and double-throw switches: (*a*) Double-pole, single-throw switch; (*b*) triple-pole, single-throw switch; (*c*) single-pole, double-throw switches used for remote control of lights from two locations.

For example, a motor-circuit switch, which carries a heavy starting current, is rated in maximum horsepower allowed for connection.

Many types of service-entrance switches are available to meet the requirements of utility companies. They may be classified as fuse pull switch, externally operated safety switch, bolted pressure contact-type switch, or circuit breaker.

An isolating switch may not be used to interrupt current. It should be opened only after the circuit has been interrupted by another general-use switch. Since isolating switches are very light, an arc will create high temperatures and can severely burn the operator.

For control of large, separate loads, the live copper blades of the various switches are concealed in steel enclosures, and the movable blades are operated by insulated levers on the front of the board. The equipment is called a dead-front switchboard.

15.7.4 Protective Devices for Circuits

In an electrical distribution system for a building, each electric service must have a means of disconnection, but it may not consist of more than six service switches. Each service switch may disconnect service to a panelboard from which lighter feeders extend to other distribution points, up to the final branch circuits with the minimum size wire, No. 12. This panelboard contains switches with lower disconnecting ratings than that of the service switch and that serve as disconnecting means for the light feeders. The rating and type of each switch must correspond to the size and kind of load and the wire size.

There must also be in every circuit some protective device to open the circuit if there is an unexpected overload, such as a short circuit or a jammed motor that prolongs a high inrush current. These protective devices may be fuses combined with knife switches, or circuit breakers, which provide both functions in one device. In addition, electrical systems should be protected against power surges caused by lightning strokes. See Art. 15.19.1.

Fuses in lighting and appliance circuits with loads up to 40 A may be the screwed plug type, with a metallic melting element behind a transparent top. For each rating, plug fuses are given different colors and the screw size is made intentionally different, to prevent errors.

Cartridge fuses are another type of fuse. They are cylindrical in form and are available in any size. They are classified for special purposes, and the rating is clearly marked on the cylinder. So-called HI-CAP fuses (high capacity) are used for fused service switches that have the capacity to interrupt very high short-circuit currents.

Short circuits in the heavy copper conductor immediately following a service switch can be very high, because of the high power potential of the street transformers. The currents may be 25,000 to 100,000 A. Upon inquiry, the utility will advise as to the maximum short-circuit current for an installation. The service switch fuse should be selected to suit this capacity.

15.7.5 Circuit Breakers

Circuit breakers operate on a different principle. A circuit breaker is essentially a switch that is provided with means to sense a short circuit or overload and then to open the circuit immediately. Circuit breakers have ratings that are equal to the current, A, that will cause the breaker to trip. When a circuit breaker is closed, a spring is compressed that provides the energy to "trip" or open the circuit breaker on overload. In its simplest form, a circuit breaker senses an overload by means of a bimetallic element that expands from the heat caused by excessive current. Also, it senses a short circuit by means of a solenoid or coil. In either case, an internal mechanism releases the spring to trip the breaker. After the cause of the trip has been removed, the circuit breaker is simply reset. More sophisticated circuit breakers use current-sensing coils and either microprocessor-based trip units or protective relays to initiate breaker trip.

Both fuses and circuit breakers have time-current characteristics; that is, both will operate in a predictable time for a given current, and both will operate more quickly for a higher value of current. The importance of this is that fuses and circuit breakers must be selected and their time-current characteristics coordinated so that, if a short circuit or overload occurs, only the fuse or circuit breaker directly upstream will operate. This selectivity will isolate the problem without causing an unnecessary outage elsewhere in the building.

Current limiters may also be used at the service connection. These are strips of metal that have a high rate of increase in electrical resistance when heated. If a short circuit occurs, this limits the flow of short-circuit current even before the short is cleared.

Current-Limiting Reactors. A coil with high inductive reactance may be placed in series with the service. If a heavy short-circuit current occurs, the impedance limits the current by temporarily storing energy in the magnetic field.

15.7.6 Protective Relays

There are several kinds of faults that can occur in an electrical system. Over- or undervoltage, reverse flow of power, and excessive currents are but a few. Protective relays are available for application at critical locations in the electrical system to protect against these faults. Originally, protective relays were, and most still are, intricate electromechanical devices. However, many solid-state devices have been introduced that match the performance and dependability of the electromechanical relays.

Protective relays do not directly trip a breaker; they close a contact to provide an electrical signal to the breaker trip circuit. When protective relays are used to trip a circuit breaker, it is always necessary to provide a source of electrical power, usually dc, from a battery bank at 48 V or 125 V, to provide tripping power.

15.7.7 Switchgear and Switchboards

The service switches and main distribution panelboards in large buildings are usually assembled in a specially designed steel frame housed in a separate electrical equipment room. The assembly is usually referred to as switchgear for large power units and switchboards for smaller assemblies.

A switchboard is defined in the National Electrical Code as a large single panel, frame, or assembly of panels, on which are mounted, on the face or back or both, switches, overcurrent and other protective devices, buses, and usually instruments. Switchboards are generally accessible from the rear as well as from the front and not intended to be installed in cabinets.

Switchboards are commonly divided into the following types:

1. Live-front.
2. Dead-front.
3. Safety enclosed switchboards.
   a. Unit or sectional.
   b. Draw-out.

Live-front switchboards have the current-carrying parts of the switch equipment mounted on the exposed front of the vertical panels. They are encountered only in existing installations and are now never used for new work.

Dead-front switchboards have no live parts mounted on the front of the board and are used in systems limited to a maximum of 600 V for dc and 2500 V for ac.

Unit safety-type switchboard is a metal-enclosed switchgear consisting of a completely enclosed self-supporting metal structure, containing one or more circuit breakers or switches.

Draw-out type switchboard is a metal-clad switchgear consisting of a stationary housing mounted on a steel framework and a horizontal draw-out circuit-breaker structure. The equipment for each circuit is assembled on a frame forming a self-contained and self-supporting mobile unit.

Metal-clad switchgear consists of a metal structure completely enclosing a circuit breaker and associated equipment such as current and potential transformers, interlocks, controlling devices, buses, and connections.

Because the switchboard room contains a heavy concentration of power, these rooms have special building-code requirements for ventilation and safety. The safety rules may require two exits remote from each other and minimum clear working spaces at front, top, back, and sides of the equipment. Also, the rules prohibit overhead piping and ducts above the equipment.

15.7.8 Substations

These are arrangements of transformers and switchgear used to step down voltages and connect to or disconnect from the mains. A master substation may be used to transform from utility-company high voltage down to 13,800 V or 4160 V for distribution. A load-center substation may be used to reduce to 600 V or less for customer use. The load-center substation may be located outside the building in an underground vault or on a surface pad. Where street space is limited, utilities sometimes permit inside substations in the cellar of the customer adjacent to the switchboard room. These inside vaults must comply with strict rules for ventilation and drainage set by the utility, and access should be available from the street through doors that are normally locked.

15.7.9 Panelboards

These are distribution centers that are fed from the service switches and switchgear. A panelboard is a single panel or a group of panel units designed for assembly in the form of a single panel in which are included buses and perhaps switches and automatic overcurrent protective devices for control of light, heat, or power circuits of small capacity. It is designed to be placed in a cabinet or cutout box placed in or against a wall or partition and accessible only from the front. In general, panelboards are similar to but smaller than switchboards.

A panelboard consists of a set of copper mains from which the individual circuits are tapped through overload protective devices or switching units.

Panelboards are designed for dead-front construction, with no live parts exposed when the door of the panelboard is opened. Panelboards also are designed for flush, semiflush, or surface mounting. They fall into two general classifications, those designed for medium loads, usually required for lighting systems, and those for heavy-duty industrial-power-distribution loads.

Distribution panelboards are designed to distribute current to lighting panelboards and power loads and panelboards. Lighting panelboards are generally used for distribution of branch lighting circuits. Power panelboards fall into the following types:

1. Dead-front, fusible switch in branches
2. Dead-front, circuit breaker in branches

Since motors fed from power panelboards vary in sizes, the switches and breakers in a power panelboard are available in several different sizes corresponding to the rating of the equipment.

Panelboards are designed with mains for distribution systems consisting of:

1. Three-wire, single-phase 240/120-V, solid-neutral, alternating current
2. Three-wire, 240/120-V, solid-neutral, direct current
3. Four-wire, three-phase, 208/120-V, solid-neutral, alternating current
4. Four-wire, three-phase, 480/277-V, solid-neutral, alternating current

The mains in the panelboard may be provided with lugs only, fuses, switch and fuses, or circuit breakers.

A single-phase, three-wire panelboard consists of two copper busbars set vertically in the center of the panel and horizontal strip connections on each side for branch circuit breakers or switches. The third wire, or neutral, is connected to a copper plate at the top of the panel with several bolted studs. Neutral-wire connectors from each circuit are connected to that plate.

A similar construction is used for three-phase, four-wire panelboards, which are used for lighting, receptacle, and motor circuits, but with three, instead of two, copper busbars set vertically in the center of the board. Single-phase circuits may be taken from both types of panelboards, but it is important to balance the loads as closely as possible on the two- or three-phase legs, to minimize the current in the neutral.

The following items should be taken into consideration in determining the number and location of panelboards:

1. No lighting panelboard should exceed 42 single-pole protective devices.
2. Panelboards should be located as near as possible to the center of the load it supplies.
3. Panelboards should always be accessible.
4. Voltage drop to the farthest outlet should not exceed 3%.
5. Panelboards should be located so that the feeder is as short as possible and have a minimum number of bends and offsets.
6. Spare circuit capacity should be provided at the approximate rate of one spare to every five circuits originally installed.
7. At least one lighting panelboard should be provided for each floor of a building.

Care should be taken when specifying panelboards to make sure that they are rated for the available short-circuit current. The panelboard bus bars must be physically braced to withstand the forces resulting from the flow of short-circuit current, and the fuses or circuit breakers must be capable of interrupting a downstream short circuit. Some panelboards are available with "integrated" or "series" short-circuit ratings, which indicate that even though the branch breakers cannot interrupt the available current, the panelboard main breaker can do so before any damage is done to the branch breakers. Such equipment can be less expensive than a fully rated panelboard, but the loss of selectivity for critical applications offsets the savings.

15.7.10 Motor Control Centers

These are an assembly, in one location, of motor controllers, devices that start and stop motors and protect them against overloads, and of disconnect switches for the motors. For safety reasons, the National Electrical Code requires that a disconnect switch be located within sight of the motor and its controller.

The controller basically is a contactor, operated by a solenoid and returned to its normal position by a spring. The initiating device contacts may either be normally closed or normally open, depending on the automatic function required. Usually, an on-off-automatic selector switch is installed to control the contactor and allow manual operation of the motor for testing purposes and then return to automatic for normal functioning. Overload protection is incorporated in the contactor. For the purpose, thermal, heat-operated relays are provided. A reset button is pushed to close the switch after the overload has been removed.

In addition to motor-starting contactors, motor control centers may also contain variable frequency controllers for motors requiring speed control. Containing electronics to convert the constant 60-Hz utility power to a variable frequency output ranging from 1.5 to 120 Hz, they can control motor speed over the same range since ac motor speed is proportional to the frequency.

For certain industrial process applications, motor control centers may be provided with microprocessor-based programmable logic controllers (PLC). These can control the operation of a single system or be integrated into a large, plantwide process control system.

15.8 ELECTRICAL DISTRIBUTION IN BUILDINGS

The National Fire Protection Association "National Electrical Code" is the basic safety standard for electrical design for buildings in the United States and has been adopted by reference in many building codes. In some cases, however, local codes may contain more restrictive requirements. The local ordinance should always be consulted.

The "National Electrical Code," or the "National Electrical Code Handbook," which explains provisions of the code, may be obtained from NFPA, 1 Battery March Park, Quincy, MA 02269-9101.

The American Insurance Association sponsors the Underwriters Laboratories, Inc., which passes on electrical material and equipment in accordance with standard test specifications. The UL also issues a semiannual List of Inspected Electrical Appliances, which can be obtained from the UL at 333 Pfingsten Road, Northbrook, IL 60062-2096.

Electrical codes and ordinances are written primarily to protect the public from fire and other hazards to life. They represent minimum safety standards. Strict application of these codes will not, however, guarantee satisfactory or even adequate performance. Correct design of an electrical system, over these minimum safety standards, to achieve a required level of performance, is the responsibility of the electrical designer.

15.8.1 Electrical Symbols

Table 15.1 illustrates the graphic symbols commonly used for electrical drawings for building installations. ANSI Y32.2, American National Standards Institute, contains an extensive compilation of such symbols.

15.8.2 Building Wiring Systems

The electrical load in a building is the sum of the loads, in kilowatts (kW), for lighting, motors, and appliances. It is highly unlikely, however, that all electrical loads in a building will be at full rated capacity at the same time. Hence, for economic selection of the electrical equipment in a building, demand and coincidence factors should be applied to the total connected load.

The **demand factor** is the ratio of the actual peak load of equipment or system to its maximum rating. An air-conditioning fan, for example, may require 8 hp at maximum load, but it will have a 10-hp motor (the standard available size). Therefore, its demand factor is 8/10. Lighting fixtures in a building, in contrast, can only operate at full load, or at a demand factor of 1.0.

The **coincidence factor** is the ratio of the maximum demand load of a system to the sum of the demand loads of its individual components and indicates the

TABLE 15.1 Electrical Symbols*

Wall	Ceiling	
-O	O	Outlet
-(B)	(B)	Blanked outlet
	(D)	Drop cord
-(E)	(E)	Electrical outlet–for use only when circle used alone might be confused with columns, plumbing symbols, etc.
-(F)	(F)	Fan outlet
-(J)	(J)	Junction box
-(L)	(L)	Lamp holder
-(L)$_{PS}$	(L)$_{PS}$	Lamp holder with pull switch
-(S)	(S)	Pull switch
(V)	-(V)	Outlet for vapor-discharge lamp
-(X)	(X)	Exit-light outlet
-(C)	-(C)	Clock outlet (specify voltage)
⊖		Duplex convenience outlet
⊖$_{1,3}$		Convenience outlet other than duplex. 1 = single, 3 = triplex, etc.
⊖$_{WP}$		Weatherproof convenience outlet
⊖$_{R}$		Range outlet
⊖S		Switch and convenience outlet
⊖[R]		Radio and convenience outlet
(▲)		Special-purpose outlet (designated in specifications)
⊙		Floor outlet
S		Single-pole switch
S_2		Double-pole switch
S_3		Three-way switch
S_4		Four-way switch
S_D		Automatic door switch
S_E		Electrolier switch
S_K		Key-operated switch
S_P		Switch and pilot lamp
S_{CB}		Circuit breaker
S_{WCB}		Weatherproof circuit breaker
S_{MC}		Momentary-contact switch
S_{RC}		Remote-control switch
S_{WP}		Weatherproof switch
S_F		Fused switch
S_{WF}		Weatherproof fused switch

TABLE 15.1 (*Continued*)

Wall	Ceiling	
$\ominus_{a,b,c}$ etc. $O_{a,b,c}$ etc. $S_{a,b,c}$ etc.		Any standard symbol as given above with the addition of a lower-case subscript letter may be used to designate some special variation of standard equipment of particular interest in a specific set of architectural plans. When used, they must be listed in the key of symbols on each drawing and if necessary further described in the specifications
		Lighting panel
		Power panel
		Branch circuit; concealed in ceiling or wall
		Branch circuit; concealed in floor
		Branch circuit; exposed
		Home run to panelboard. Indicate number of circuits by number of arrows. NOTE: Any circuit without further designation indicates a two-wire circuit. For a greater number of wires indicate as follows: -///- (three wires), -////- (four wires), etc.
		Feeders. NOTE: Use heavy lines and designate by number corresponding to listing in feeder schedule
		Underfloor duct and junction box. Triple system. NOTE: For double or single systems eliminate one of two lines. This symbol is equally adaptable to auxiliary-system layouts
G		Generator
M		Motor
I		Instrument
T		Power transformer (or draw to scale)
		Controller
		Isolating switch
		Push button
		Buzzer
		Bell
		Annunciator
		Outside telephone
		Interconnecting telephone
		Telephone switchboard
T		Bell-ringing transformer
D		Electric door opener
F		Fire-alarm bell
F		Fire-alarm station
		City fire-alarm station
FA		Fire-alarm central station

TABLE 15.1 (*Continued*)

Wall	Ceiling	
FS		Automatic fire-alarm device
W		Watchman's station
WI		Watchman's central station
H		Horn
N		Nurse's signal plug
TV		Television antenna outlet
R		Radio outlet
SC		Signal central station
		Interconnection box
		Battery
—·—·—		Auxiliary-system circuits. NOTE: Any line without further designation indicates a two-wire system. For a greater number of wires designate with numerals in manner similar to 12—No. 18W-¾″-C., or designate by number corresponding to listing in schedule.
□a,b,c		Special auxiliary outlets. Subscription letters refer to notes on plans or detailed description in specifications.

*Standard electrical symbols are compiled in ANSI Y32.2, American National Standards Institute.

largest portion of all the electrical loads likely to be operating at one time. **Diversity factor** is the multiplicative inverse of the coincidence factor. Demand factors and coincidence factors or diversity factors can be obtained from a number of sources, such as the NFPA "National Electrical Code."

Motor and appliance loads usually are taken at full value. Household and kitchen appliances, however, are exceptions. The National Electrical Code lists demand factors for household electric ranges, ovens, and clothes dryers. Some municipal codes allow the first 3000 W of apartment appliance load to be included with lighting load and therefore to be reduced by the factor applied to lighting.

For factories and commercial buildings, the electrical designer should obtain from the mechanical design the location and horsepower of all blowers, pumps, compressors, and other electrical equipment, as well as the load for elevators, boiler room, and other machinery. The load in amperes for running motors is given in Tables 15.8 and 15.9.

15.8.3 Plans

Electrical plans should be drawn to scale, traced or reproduced from the architectural plans. Architectural dimensions may be omitted except for such rooms as meter closets or service space, where the contractor may have to detail his equipment to close dimensions. Floor heights should be indicated if full elevations are not given. Locations of windows and doors should be reproduced accurately, and door swings shown, to facilitate location of wall switches. For estimating purposes, feeder or branch runs may be scaled from the plans with sufficient accuracy.

Electrical plans may be drawn manually or by using a computer-aided drafting and design (CADD) system. Although a significant initial investment is required, CADD can make the preparation of drawings fast and efficient and can make the interchange of information between electrical and the other engineering disciplines much easier.

Indicate on the plans by symbol the location of all electrical equipment (Table 15.1). Show all ceiling outlets, wall receptacles, switches, junction boxes, panelboards, telephone and interior communication equipment, fire alarms, television master-antenna connections, etc.

A complete set of electrical plans should include a diagram of feeders, panel lists, service entrance location, and equipment. Before these can be shown on the plans, however, wire sizes should be computed in accordance with procedures outlined in the following paragraphs.

Where there is only one panelboard in an area, and it is clear that all circuits in that area connect to that box, it is not necessary to number the panel other than to designate it as, for example, "apartment panel." In larger areas, where two or more panelboards may be needed, each should be labeled for identification and location; for example, L.P. 1-1, L.P. 1-2 . . . for all panelboards on the first floor; L.P. 2-1, L.P. 2-2 . . . for panels on the second floor.

15.8.4 Branch Circuits

It is good practice to limit branch runs to a maximum of 50 ft for 120-V circuits and 100 ft for 277-V circuits by installing sufficient panelboards in efficient locations.

Connect each outlet with a branch circuit and show the home runs to the panelboard, as indicated in Table 15.1. General lighting branch circuits with a 15-A fuse or circuit breaker in the panelboard usually are limited to 6 to 8 outlets, although most codes permit 12. No more than two outlets should be connected in a 20-A appliance circuit.

It is good practice to use wire no smaller than No. 12 in branch circuits, though some codes permit No. 14. Special-purpose individual branch circuits for motors or appliances should be sized to suit the connected load.

15.8.5 Electric Services

For economy, alternating current is transmitted long distances at high voltages and then changed to low voltages by step-down transformers at the point of service.

Small installations, such as one-family houses, usually are supplied with three-wire service. This consists of a neutral (transformer midpoint) and two power wires with voltage differing 180° in phase. From this service, the following types of interior branch circuits are available:

Single-phase two-wire 230-V—by tapping across the phase wires

Single-phase two-wire 115-V—by tapping across one phase wire and the neutral

Single-phase three-wire 115/230-V—by using both phase wires and the neutral

For larger installations, the service may be 480/277-V or 208/120-V, three-phase four-wire system. This has a neutral and three power wires carrying voltage differing

120° in phase. From this service, the following types of interior branch circuits are available:

Single-phase two-wire 480-V or 208-V—by tapping across two phase wires

Single-phase two-wire 277-V or 120-V—by tapping across one phase wire and the neutral

Two-phase three-wire 480/277-V or 208/120-V—by using two phase wires and the neutral

Three-phase three-wire 480-V or 208-V—by using three phase wires

Three-phase four-wire 480/277-V or 280/120-V–by using three phase wires and the neutral

15.9 CIRCUIT AND CONDUCTOR CALCULATIONS

The current in a conductor may be computed from the following formulas, in which

I = conductor current, A
W = power, W
f = power factor, as a decimal
E_p = voltage between any two phase legs
E_g = voltage between a phase leg and neutral, or ground

Single-phase two-wire circuits:

$$I = \frac{W}{E_p f} \quad \text{or} \quad I = \frac{W}{E_g f} \tag{15.24}$$

Single-phase three-wire (and balanced two-phase three-wire) circuits:

$$I = \frac{W}{2E_g f} \tag{15.25}$$

Three-phase three-wire (and balanced three-phase four-wire) circuits:

$$I = \frac{W}{3E_g f} \tag{15.26}$$

When circuits are balanced in a three-phase four-wire system, no current flows in the neutral. When a three-phase four-wire feeder is brought to a panelboard from which single-phase circuits will be taken, the system should be designed so that under full load the load on each phase leg will be nearly equal.

15.9.1 Voltage-Drop Calculations

Voltage drop in a circuit may be computed from the following formulas, in which

V_d = voltage drop between any two phase wires, or between phase wire and neutral when only one phase wire is used in the circuit
I = current, A

L = one-way run, ft
R = resistance, Ω/mil-ft
c.m. = circular mils

Single-phase two-wire (and balanced single-phase three-wire) circuits:

$$V_d = \frac{2RIL}{\text{c.m.}} \tag{15.27}$$

Balanced two-phase three-wire, three-phase three-wire, and balanced three-phase four-wire circuits:

$$V_d = \frac{\sqrt{3}\, RIL}{\text{c.m.}} \tag{15.28}$$

Equations (15.27) and (15.28) contain a factor R that represents the resistance in ohms to direct current of 1 mil-ft of wire. The value of R may be taken as 10.7 for copper and 17.7 for aluminum. Tables in the National Fire Protection Association "National Electrical Code Handbook" give the resistance, ohms per 1000 ft, for various sizes of conductors. For small wire sizes, up to No. 3, resistance is the same for alternating and direct current. But above No. 3, ac resistance is larger, and this value as given in the handbook should be applied.

Voltage drops used in design may range from 1 to 5% of the service voltage. Some codes set a maximum for voltage drop of 2.5% for combined light and power circuits from service entry to the building to point of final distribution at branch panels.

When this voltage drop is apportioned to the various parts of the circuit, it is economical to assign the greater part, say 1.5 to 2%, to the smaller, more numerous feeders, and only 0.5 to 1% to the heavy main feeders between the service and main distribution panels. Tables in the NFPA handbook give the maximum allowable current for each wire size for copper and aluminum wire and the area, in circular mils, to be used in the voltage-drop formulas.

First, select the minimum-size wire allowed by the building code, and test it for voltage drop. If this drop is excessive, test a larger size, until one is found for which the voltage drop is within the desired limit. This trial-and-error process can be shortened by first assuming the desired voltage drop, and then computing the required wire area with Eqs. (15.27) and (15.28). The wire size can be selected from the handbook tables.

For circuits designed for motor loads only, no lighting, the maximum voltage drop may be increased to a total of 5%. Of this, 1% can be assigned to branch circuits and 4% to feeders.

Tables in the handbook also give dimensions of trade sizes of conduit and tubing and permissible numbers of conductors that can be placed in each size.

15.9.2 Wiring for Motor Loads

Motors have a high starting current that lasts a very short time. But it may be 4 to 6 times as high as the rated current when running. Although motor windings will not be damaged by a high current of short duration, they cannot take currents much greater than the rated value for long periods without excessive overheating and consequent breakdown of the insulation.

Overcurrent protective devices, fuses and circuit breakers, should be selected to protect motors from overcurrents of long duration, and yet permit short-duration starting currents to pass without disconnecting the circuit. For this reason, the National Electrical Code permits the fuse or circuit breaker in a motor circuit to have a higher ampere rating than the allowable current-carrying capacity of the wire. Tables in the NFPA handbook give the overcurrent protection for motors allowed by the Code and data on time-delay fuses that permit smaller fuse holders for a given-size motor than with standard fuses.

The National Electrical Code requirements for motor circuit conductors and overcurrent protection are as follows:

Branch Circuits (One Motor). Conductors shall have an ampacity not less than 125% of the motor full-load current. Overcurrent protection, fuses or circuit breakers, must be capable of carrying the starting current of the motor. Maximum rating of such protection varies with the type, starting method, and locked-rotor current of the motor. For the great majority of motor applications in buildings, conductor and fuse protection may be selected from Tables 15.2 and 15.3

Feeder Circuits (More than One Motor on a Conductor). The conductor should have an allowable current-carrying capacity not less than 125% of the full-load current of the largest motor plus the sum of the full-load currents of the remaining motors on the same circuit. The rating of overcurrent protection, fuses or circuit breakers, shall not be greater than the maximum allowed by the code for protection of the largest motor plus the sum of the full-load currents of the remaining motors on the circuit.

If the allowable current-carrying capacity of the conductor or the size of the computed overcurrent device does not correspond to the rating of a standard-size fuse or circuit breaker, the next larger standard size should be used.

Amp-Traps and Hi-Caps are high-interrupting-capacity current-limiting fuses used in service switches and main distribution panels connected near service switches. This type of fuse is needed here because this part of the wiring system in large buildings consists of heavy cables or buses and large switches that have very little resistance. If a short circuit occurs, very high currents will flow, limited only by the interrupting capacity of the protective device installed by the utility company on its own transformers furnishing the service. Ordinary fuses cannot interrupt this current quickly enough to avert damage to the building wiring and connected electrical equipment. The interrupting-capacity value needed can be obtained from the utility company.

Fuses in service switches and connected main panels should have current-time characteristics that will isolate only the circuit in which a short occurs, without permitting the short-circuit current to pass to other feeders and interrupt those circuits too. The electrical designer should obtain data from manufacturers of approved fusing devices on the proper sequence of fusing.

15.9.3 Service-Entrance Switch and Metering Equipment

Fused switches or circuit breakers must be provided near the entrance point of electrical service in a building for shutting off the power. The National Electrical Code requires that each incoming service in a multiple-occupancy building be controlled near its entrance by not more than six switches or circuit breakers.

Metering equipment consists of a meter pan, meter cabinet, current transformer cabinet, or a combination of these cabinets, depending on the load requirements

TABLE 15.2 Protection of Single-Phase Motors and Circuits

Size of motor		Branch-circuit protection*		Motor-running protection†				Size of fused switch or fuse holder				
				Size of time-delay-cartridge or low-peak fuse								
						Maximum size						
Horsepower	Ampere rating	Maximum size fuse permitted by the code	Time-delay-cartridge or low-peak fuse that can be used	Ordinary service	Heavy service	40°C motor	All other motors	Maximum size switch that can be used	Size that can be used with time-delay-cartridge or low-peak fuses	Minimum size of starter: NEMA size	Minimum size and type of wire: AWG or MCM	Minimum size of conduit: diameter, in
115 V												
1/6	4.4	15	7	4½	5	5 6/10	5 6/10					
1/4	5.8	20	10	5 6/10	6¼	8	7	30	30	00	14 R	½
1/3	7.2	25	12	7	8	9	9					
½	9.8	30	17½	10	12	12	12	30	30	0	14 R	½
¾	13.8	45	25	15	17½	17½	17½	60	30			
1	16	50	30	17½	20	20	20	60	30	0	12 R	½
1½	20	60	30	20	25	25	25	60	30			
2	24.0	80	40	25	30	30	30	100	60	1	10 R	¾

230 V												
1/6	2.2	15	3½	2¼	2½	28/10	28/10					
¼	2.9	15	5	28/10	32/10	4	3½	30	30	00	**14 R**	½
⅓	3.6	15	6	3½	4	4½	4½					
½	4.9	15	8	5	56/10	6¼	6¼					
¾	6.9	25	12	7	8	9	8	30	30	00	**14 R**	½
1	8	25	15	8	9	10	10					
1½	10	30	17½	10	12	12	12	30	30			
2	12	40	20	12	15	15	15	60	30	0	**14 R**	½
3	17	60	25	17½	20	20	20	60	30	1	**10 R**	¾
5	28	90	45	30	35	35	35	100	60	2	**8 R**	¾
7½	40	125	60	40	45	50	50	200	60	2	**6 R**	1
10	50	150	80	50	60	70	60	200	100	3	**4 R**	1¼

*These do not give motor-running protection.
†On normal installations these also give branch-circuit protection.

TABLE 15.3 Protection of Three-Phase 208-V Motors and Circuits

Size and class of motor			Branch circuit protection*		Motor-running protection†				Size of fused switch or fuse holder				
					Size of time-delay-cartridge or low-peak fuse								
							Maximum size						
Horse-power	Ampere rating	Class	Maximum size fuse permitted by the code	Time-delay-cartridge or low-peak fuse that can be used	Ordinary service	Heavy service	40°C motor	All other motors	Maximum size switch that can be used	Size that can be used with time-delay-cartridge or low-peak fuses	Minimum size of starter: NEMA size	Minimum size and type of wire: AWG or MCM	Minimum size of conduit: diameter, in
½	2.1	Any	15	5	2¼	2½	2⁸⁄₁₀	2½					
¾	3	Any	15	8	3²⁄₁₀	3½	4	3½					
1	3.7	Any	15	8	4	4½	5	4½	30	30	00	14 R	½
1½	5.3	Any	15	10	5⁶⁄₁₀	6¼	7	6¼					
2	6.9	1	25	12	7	8	9	8					
		2	20	12	7	8	9	8	30	30	0	14 R	½
		3–4	15	12	7	8	9	8					
3	9.5	1	30	15	10	12	12	12					
		2	25	15	10	12	12	12					
		3	20	15	10	12	12	12	30	30	0	14 R	½
		4	15	15	10	12	12	12					
5	15.9	1	50	25	17½	20	20	20	60	30			
		2	40	25	17½	20	20	20	60	30			
		3	35	25	17½	20	20	20	60	30	1	12 R	½
		4	25	25	17½	20	20	20	30	30			

7½	23.3	1	80	35	25	30	30	30	100	60			
		2	60	35	25	30	30	30	60	60			
		3	50	35	25	30	30	30	60	60	1	10 R	¾
		4	40	53	25	30	30	30	60	60			
10	28.6	1	90	45	30	35	40	35	100	60			
		2	70	45	30	35	40	35	100	60			
		3	60	45	30	35	40	35	60	60	2	8 R	¾
		4	45	45	30	35	40	35	60	60			
15	42.3	1	125	60	45	50	50	50	200	60			
		2	110	60	45	50	50	50	200	60			
		3	90	60	45	50	50	50	100	60	2	6 R	1
		4	70	60	45	50	50	50	100	60			
20	55	1	175	90	60	70	70	70	200	100			
		2	150	90	60	70	70	70	200	100			
		3	110	90	60	70	70	70	200	100	3	4 R	1¼
		4	90	90	60	70	70	70	100	100			
25	68	1	225	100	70	80	90	80	400	100			
		2	175	100	70	80	90	80	200	100			
		3	150	100	70	80	90	80	200	100	3	2 R	1¼
		4	110	100	70	80	90	80	200	100			
30	83	1	250	125	90	100	110	100	400	200			
		2	225	125	90	100	110	100	400	200			
		3	175	125	90	100	110	100	200	200	3	1 R	1½
		4	125	125	90	100	110	100	200	200			
40	110	1	350	175	110	125	150	150	400	200			
		2	300	175	110	125	150	150	400	200			
		3	225	175	110	125	150	150	400	200	4	0 RH	2
		4	175	175	110	125	150	150	200	200			
50	132	1	400	200	150	175	175	175	400	200			
		2	350	200	150	175	175	175	400	200			
		3	300	200	150	175	175	175	400	200	4	00 RH	2
		4	200	200	150	175	175	175	200	200			

TABLE 15.3 (*Continued*)

Size and class of motor			Branch circuit protection*		Motor-running protection†				Size of fused switch or fuse holder				
					Size of time-delay-cartridge or low-peak fuse								
							Maximum size						
Horse-power	Ampere rating	Class	Maximum size fuse permitted by the code	Time-delay-cartridge or low-peak fuse that can be used	Ordinary service	Heavy service	40°C motor	All other motors	Maximum size switch that can be used	Size that can be used with time-delay-cartridge or low-peak fuses	Minimum size of starter: NEMA size	Minimum size and type of wire: AWG or MCM	Minimum size of conduit: diameter, in
60	159	1	500	250	175	200	200	200	600	400			
		2	400	250	175	200	200	200	400	400			
		3	350	250	175	200	200	200	400	400	5	000 RH	2
		4	250	250	175	200	200	200	400	400			
75	196	1	600	300	200	225	250	250	600	400			
		2	500	300	200	225	250	250	600	400			
		3	400	300	200	225	250	250	400	400	5	250 RH	2½
		4	300	300	200	225	250	250	400	400			
100	260	1		400	250	300	350	300		400			
		2		400	250	300	350	300		400			
		3	600	400	250	300	350	300	600	400	5	400 RH	3
		4	400	400	250	300	350	300	400	400			
125	328	1–3		500	350	400	450	400		600		2 sets‡	
		4	500	500	350	400	450	400	600	600	6	0000 RH	2½‡
150	381	1–3		600	400	450	500	450		600		2 sets‡	
		4	600	600	400	450	500	450	600	600	6	250 RH	2½‡

*These do not give motor-running protection.
†On normal installations these also give branch-circuit protection.
‡Indicates two sets of multiple conductors and two runs of conduit.

and other characteristics of the specific project. The meters and metering transformers for recording current consumed are furnished by the utility company. Unless otherwise permitted by the utility company, meters must be located near the point of service entrance. Sometimes, the utility company permits one or more tenant meter rooms at other locations in the cellar of an apartment house to suit economical building wiring design. Tenant meter closets on the upper floors, opening on public halls, also may be permitted. The most common form of tenant meters used is the three-wire type, consisting of two phase wires and the neutral, taken from a 208/120-V three-phase four-wire service.

The service switch and metering equipment may be combined in one unit, or the switch may be connected with conduit to a separate meter trough. For individual metering, the detachable-socket-type meter with prongs that fit into the jaws of the meter-mounting trough generally is used.

15.9.4 Switchboards and Panelboards

For low-capacity loads, wiring may be taken directly to a panelboard. For larger loads, wiring may be brought first to a switchboard and then to panelboards. This equipment is described in Art. 15.7. Branch circuits extend from panelboards to the various loads.

15.9.5 Sample Calculations for Apartment-Building Riser

A diagram of a light and power riser for a nine-story apartment building is shown in Fig. 15.10. Calculation of required wires and conduits may be carried out with the aid of tables in the NFPA "National Electrical Code Handbook."

1. Typical Meter Branch to Apartment Panel. Note that the meters are three-wire type, and three apartments are connected to the same neutral. Under balanced conditions, when each of the three identical apartments is under full load, no current will flow in the neutral. But at maximum unbalance, current in the neutral may be twice the current in the phase wire for any one apartment. The neutral wire must be sized for this maximum current, though the usual practice is to compute voltage drops for the balanced condition.

Assume that the apartment area is 900 ft^2. The one-way run from meter to apartment panel (apartment A) is 110 ft.

$$\begin{aligned} &\text{Apartment lighting load} = 900 \times 3\ \text{W/ft}^2 &&= 2700\ \text{W} \\ &\text{Apartment appliance load} &&= \underline{3000} \\ &\text{Total} &&= 5700\ \text{W} \end{aligned}$$

The electric service is three-phase four-wire 208/120 V. Thus, the voltage between phase wires is 208, and between one phase and neutral 120 V. Assume a 90% power factor. From Eq. (15.25):

$$\text{Current per phase} = \frac{5700}{2 \times 120 \times 0.9} = 26.4\ \text{A}$$

The local electrical code requires the minimum size of apartment feeder to be No. 8 wire. The allowable current in No. 8 RH wire is 45 A; so this wire would be adequate for the current, but it still must be checked for voltage drop. The

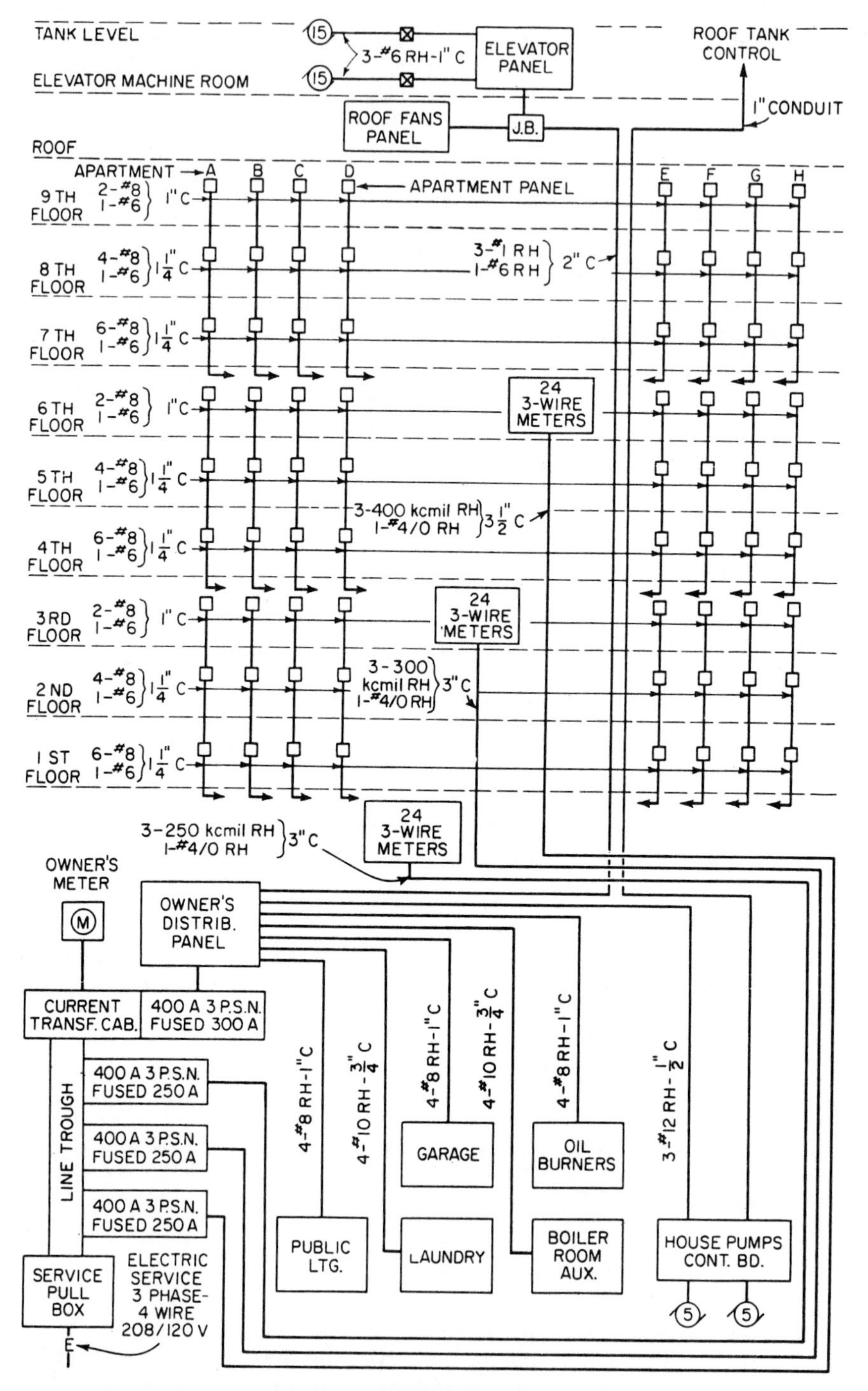

FIGURE 15.10 Diagram of a typical apartment-building electrical riser.

neutral must be sized for the maximum unbalance, under which condition the current in the neutral will be 2 × 26.4 = 52.8 A. This will require No. 6 wire.

The voltage drop between phase wires can be obtained from Eq. (15.28), with the area of No. 8 wire taken as 16,510 circular mils and length of wire as 110 ft:

$$V_d = \frac{\sqrt{3} \times 10.7 \times 26.4 \times 110}{16{,}510} = 3.24 \text{ V}$$

$$\% \text{ voltage drop} = \frac{3.24 \times 100}{208} = 1.56\%$$

2. *Feeder to Meter Bank on Sixth Floor.*

Total load (24 apartments) = 24 × 5700 W	=	136,800 W
Demand load: first 15,000 W at 100%	=	15,000
Balance, 121,800 W at 50%	=	60,900
Total demand load	=	75,900 W

Assume a power factor of 90% and apply Eq. (15.26):

$$\text{Current per phase} = \frac{75{,}900}{3 \times 120 \times 0.9} = 234 \text{ A}$$

Minimum-size type RH wire for this current is 250 kcmil. This has to be tested for voltage drop. For a one-way run of 150 ft from service switch to sixth-floor meter bank and an area of 250,000 circular mils, Eq. (15.28) gives:

$$V_d = \frac{\sqrt{3} \times 10.7 \times 234 \times 150}{250{,}000} = 2.60 \text{ V}$$

The correction factor for ac resistance of 250-kcmil wire is 1.06. Application of this factor yields a corrected voltage drop of 1.06 × 2.60 = 2.76 volts.

$$\% \text{ voltage drop} = \frac{2.76 \times 100}{208} = 1.33\%$$

Then, the total voltage drop from service switch to apartment panel A is

$$1.56 + 1.33\% = 2.89\%$$

which exceeds the 2.5% maximum voltage drop allowed by the local code. It is necessary, therefore, to increase the size of the meter bank feeder over the minimum size required for the current.

To bring the total drop down to 2.5% the meter bank feeder drop must be reduced to 0.94% or 1.96 volts. The required wire size may be found by proportion:

$$\text{Required area} = \frac{2.76}{1.96} \times 250{,}000 = 352{,}000 \text{ circular mils}$$

The nearest larger wire size is 400 kcmil.

For computing the size of the neutral for carrying 234 A, the local code allows a demand factor of 70% on lighting loads over 200 A. Hence, the net current in the neutral is 200 + 34 × 0.70 = 223.8 A. Minimum-size wire is No. 4/0 RH.

And the size of conduit required for the feeder is computed as follows with the aid of tables in the NFPA "National Electrical Code Handbook":

Three 400-kcmil RH wires	$= 3 \times 0.8365$	$= 2.5095\ \text{in}^2$
One 0000 RH wire		$= 0.4840$
Total area		$= 2.9935\ \text{in}^2$

Permissible raceway fill for four conductors is 40%. Hence the minimum area required for the conduit is $2.9935/0.40 = 7.484\ \text{in}^2$. And the required conduit size is 3½ in.

3. Feeder to Elevator and Roof Fans Panel. The total load on this feeder is

8 roof fans at ½ hp	= 16.8 A/phase
8 roof fans at ¼ hp	= 17.4
2 elevators at 15 hp	= 84.6
Total	= 118.8 A/phase

The minimum current-carrying capacity of the feeder must be 125% of the rated full-load current of the largest motor, an elevator motor, 42.3 A, plus the sum of the rated load currents of the other motors. Thus, the capacity must be

$$1.25 \times 42.3 + 42.3 + 16.8 + 17.4 = 129.4\ \text{A}$$

The minimum-size wire that may be used is No. 1 RH, with an area of 83,690 circular mils.

Check for voltage drop with a run of 150 ft:

$$V_d = \frac{\sqrt{3} \times 10.7 \times 118.8 \times 150}{83{,}690} = 3.95\ \text{V}$$

$$\%\ \text{voltage drop} = \frac{3.95 \times 100}{208} = 1.90\% \qquad \text{OK}$$

15.9.6 Computerized Analysis

Personal computers and available electrical design software with a wide range of sophistication and comprehensiveness offer a very efficient method for performing electrical design calculations. Computers can perform the calculations required for sizing wires, conduits, transformers, and panelboards speedily and with freedom from computational errors. For more intricate and involved calculations, such as short-circuit analysis on a complex, multisource system, or determination of the voltage drop occurring when a very large motor starts, computers can reduce several days of manual work to a few hours. Computer programs are available for calculating ground resistance, power load flow, wire-pulling tensions, and time-current coordination of circuit breakers, relays, and fuses. There also are programs that interface directly with CADD software to produce calculations automatically while drawing the single-line diagrams.

15.10 LIGHT AND SIGHT

Lighting is part of the environmental control system, which also includes sound control and heating, ventilation, and air conditioning (HVAC), within a building.

The prime purpose of a lighting system is to provide good visibility for execution of the tasks to be performed within the building. With good visibility, occupants can execute their tasks comfortably, efficiently, and safely.

Lighting also is desirable for other purposes. For example, it can be used to develop color effects for pleasure or accident prevention. It can be used to decorate select spaces or to accent objects. It can produce effects that influence human moods. And it can serve to illuminate an emergency egress system and as part of a security system.

Good lighting requires good quality of illumination (Art. 15.11), proper color rendering (Art. 15.12), and an adequate quantity of light (Art. 15.13). This result, however, cannot be achieved economically solely by selection and arrangement of suitable light sources. Lighting effects are also dependent on other systems and factors such as the characteristics of surrounding walls, floor, and ceiling; nature of tasks to be illuminated; properties of the backgrounds of the tasks; age and visual acuity of occupants; and characteristics of the electrical system. Design of a lighting system, therefore, must take into account its interfacing with other systems. Also, lighting design must take into account the influence of lighting requirements on other systems, including architectural systems; heating and cooling effects of windows provided for daylighting; energy supply required from the electrical system; and loads imposed by electric lighting on the HVAC system.

Sources of light within a building may be daylight or artificial illumination. The latter can be produced in many ways, but only the most commonly used types of electric lighting are discussed in this section.

Like other building systems, lighting design is significantly affected by building codes. These generally contain minimum requirements for illumination levels, for the safety and health of building occupants. In addition, electric lighting equipment and electrical distribution must conform to safety requirements in building codes and the National Electrical Code, which is promulgated by the National Fire Protection Association, and to standards of the Underwriters Laboratories, Inc. Also, the Illuminating Engineering Society has developed standards and recommended practices to promote good lighting design.

In the interests of energy conservation, federal and state government agencies have set limits on the amount of energy that may be expended (energy budget) for operation of buildings. These limits may establish maximum levels of illumination for specific purposes in buildings.

Because of the importance of good lighting, the need to control lighting costs and to conserve energy, and the multiplicity of legal requirements affecting lighting design, engagement of a specialist in lighting design is advisable for many types of buildings.

15.10.1 Visibility

A light source produces light by converting energy to electromagnetic waves. Light sources used in practical applications emit waves with a broad spectrum of frequencies or wavelengths. Light consists of those waves that the human eye normally perceives. A normal eye interprets the wavelengths as colors, the shortest wavelengths being recognized as blue, the longest wavelengths as red, and intermediate wavelengths as green, yellow, and orange. The eye also recognizes differences in intensity of light, or levels of illumination.

The eye sees an object because it receives light emitted by the object (if it is a light source) or reflected from it (if the object is not a light source). In the latter case, the eye can see the object if it reflects light received from a light source directly or from surfaces reflecting light. The total light reflected from the object equals the sum of the light from all sources that strike the object and is not absorbed. Thus, an object shielded from light sources will be revealed to the eye by light reflected from other surfaces, such as walls, floor, ceiling, or furniture. The amount of detail on the object that the eye can recognize, however, depends not only on the intensity of light that the eye receives from the object but also on the intensity of that light relative to the intensity of light the eye receives from the background (field of vision behind the object). Consequently, the eye can readily recognize details on a brightly lit object set against a darker background. But the eye perceives little detail or considers the object dark (in shadow) if the background is much brighter than the object.

15.10.2 Inverse Square Law

Consider now a point source radiating luminous energy equally in all directions and located at the common center of two transparent spheres of unequal diameter. Because each sphere receives the same amount of luminous energy from the light source, the quantity of light per unit of area is less on the larger sphere than that on the smaller one. In fact, the quantity per unit area varies inversely as the areas of the spheres, or inversely as the square of the radii. These considerations justify the inverse square law, which states:

> Illumination level at any point is inversely proportional to the square of distance from the point light source.

For large light sources, the law holds approximately at large distances (at least 5 times the largest dimension of the sources) from the sources.

15.10.3 Light Source Power

Analogous to a pump in a water system or a battery in an electrical system, a light source emits luminous power. The unit used to measure this power is **candlepower** (cp), or **candela** (cd) (metric unit). (At one time, 1 candlepower was assumed equivalent to the luminous intensity of a wax candle, but now a more precise definition based on radiation from a heated black body is used.) The unit used to measure luminous power at a distance from the light source is the lumen (lm).

A **lumen** is the luminous power on an area of 1 ft^2 at a distance of 1 ft from a 1-cp light source or since 1 cp = 1 cd, on an area of 1 m^2 at a distance of 1 m from a 1-cd light source.

Luminous efficacy is the unit used to measure the effectiveness of light sources. It is calculated by dividing the total lumen output of a light source by the total input, watts (W), and thus is measured in lm/W.

15.10.4 Level of Illumination

A major objective of lighting design is to provide a specified **illuminance,** or level of illumination, on a task. For design purposes, the task often is taken as a flat surface, called a **work plane.** If the task is uniformly illuminated, the level of illumination equals the lumens striking the surface divided by the area. The unit used to measure illuminance is the **footcandle** (fc) lm/ft^2. In accordance with the inverse square law, the illuminance on a work plane normal to the direction to a point light source is given by

$$fc = \frac{cp}{D^2} \tag{15.29}$$

where D = distance, ft, from work plane to light source
cp = candlepower of light source

For a work plane at an angle θ to the direction of the light source,

$$fc = cp \sin \frac{\theta}{D^2} \tag{15.30}$$

In the metric system, illuminance is measured in lux, or lm/m^2; 1 fc = 10.764 lux.

A **luminaire** is a lighting device that consists of one or more lamps, or light sources, a fixture that positions and shields them, components that distribute the light, and elements that connect the lamps to the power supply. In general, luminaires do not radiate light of equal intensity in all directions, because of the characteristics of the lamps or the geometry of the fixtures. The actual illuminance around a single luminaire is an important design consideration. This environment may be characterized by the candlepower distribution curve of the luminaire. A typical such curve for a light source symmetrical about two vertical, perpendicular planes is shown in Fig. 15.11.

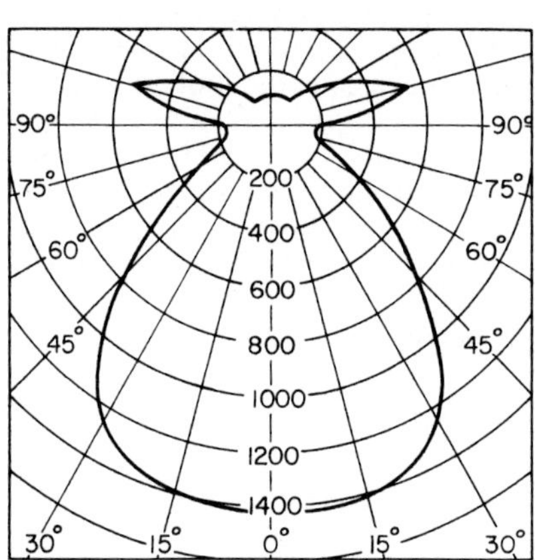

FIGURE 15.11 Candlepower distribution curve indicates variation in lighting intensity with direction from a light source.

To produce the curve, illuminance, fc, is measured, with a ganiophotometer, on a plane that is placed at various points at equal distances from the light source and that at each point is normal to the direction to the source. From Eq. (15.29), the candlepower corresponding to the illuminance at each point is computed. In practice, the candlepower is usually calculated for points in vertical planes along longitudinal and transverse axes through the luminaire. The results for each plane are then plotted with polar or rectangular coordinates relative to the light source. Such plots show the variation in illuminance with direction from the light source. For example, Fig. 15.11 shows in polar coordinates that the light source investigated directs light mainly downward. Because the source is symmetrical, only one curve is needed in this case to represent illuminance in longitudinal, transverse, and diagonal planes.

15.10.5 Equivalent Spherical Illumination

If proper care is not taken in positioning observer and task relative to primary light sources, one or more of these light sources may be reflected into the field of vision as from a mirror. The resulting glare reduces visibility of the task; that is, effective illuminance, fc, is less than the raw (actual) footcandles. For measurement of the effective illuminance of lighting installations, the concept of equivalent spherical illumination, ESI fc, was introduced. It is based on a standard sphere lighting that produces equal illumination from all directions (almost glarefree). In employment of the ESI concept, an actual lighting installation is assumed replaced by sphere illumination with task visibility equivalent to that of the actual installation.

15.10.6 Brightness

As mentioned previously, an observer sees an object because of light reflected from it. The observer interprets the intensity of the sensation experienced as brightness. The sensation of brightness usually is partly attributable to the general luminous environment, which affects the state of adaptation of the eye, and partly attributable to the intensity of light emanating from the object. The latter component is called luminance, or photometric brightness.

Luminance is the luminous power emitted, transmitted, or reflected by a surface in a given direction, per unit of area of the surface projected on a plane normal to that direction. The unit of measurement of luminance is the footlambert (cd/m^2 in the metric system).

A **footlambert** (fL) of luminance in a given direction is produced by one lumen per square foot emanating from a surface in that direction. Thus, a self-luminous surface emitting 10 lm/ft^2 has a luminance of 10 fL. For surfaces that reflect or transmit light, however, luminance depends both on the illuminance of light incident on the surface and characteristics of the surface.

For a reflecting surface, luminance is determined from

$$\text{fL} = \text{fc} \times \text{reflectance} \tag{15.31}$$

where fc = footcandles of incident light. A mirror (specular reflector) may give almost 100% reflection, whereas a black surface absorbs light and therefore has negligible reflectance. Most materials have an intermediate value of reflectance.

For a transmitting surface, luminance is determined from

$$\text{fL} = \text{fc} \times \text{transmittance} \tag{15.32}$$

Clear glass (transparent material) may have a transmittance of about 90%, whereas an opaque material has no transmittance. Transmittances of other transparent materials may be about the same as that for clear glass, while transmittances of translucent materials may be 50% or less. Light incident on a surface and not reflected or transmitted is absorbed by it.

In general, visibility improves with increase in brightness of a task. Because increase in brightness is usually accomplished at increase in operating cost caused by consumption of electric power, it is neither necessary nor desirable, however, to maintain levels of illumination higher than the minimum needed for satisfactory performance of the task. For example, tests show that speed of reading and comprehension are nearly independent of illuminance above a minimum level. This level depends on several factors, such as difficulty of the task, age of observers, duration of the task, and luminance relation between task and its surroundings.

The more difficult the task, the older the occupants, and the longer the task, the higher the minimum level of illumination should be.

High brightness also is useful in attracting visual attention and accenting texture. For this reason, bright lights are played on merchandise and works of art.

15.10.7 Contrast

This is created when the brightness of an object and its surroundings are different. The effects of contrast on visibility depend on several factors but especially on the ratio of brightness of object to that of its background. Ideally, the brightness of a task should be the same as that of its background. A 3:1 brightness ratio, however, is not objectionable; it will be noticed but usually will not attract attention. A 10:1 brightness ratio will draw attention, and a brightness ratio of 50:1 or more will accent the object and detract attention from everything else in the field of vision.

High background brightness, or low brightness ratios, may have adverse or beneficial effects on visibility. Such high contrast is undesirable when it causes glare or draws attention from the task or creates discordant light and dark patterns (visual noise). On the other hand, high contrast is advantageous when it helps the observer detect task details; for example, read fine print. High contrast makes the object viewed appear dark so that its size and silhouette can be readily discerned. But under such circumstances, if surface detail on the object must be detected, object brightness must be increased at least to the level of that of the background.

The reason for this is that the eye adapts to the brightness of the whole field of vision and visualizes objects in that field with respect to that adaptation level. If background brightness is disturbing, the observer squints to reduce the field of vision and its brightness and thus increase the brightness of the task. The need for squinting is eliminated, however, by increasing the illumination on the task.

15.10.8 Effects of Colored Lights

Color of light affects the color of an object (color rendering), because the surface of the object absorbs light of certain frequencies and reflects light of other frequencies. An object appears red because it reflects only red light and absorbs other hues. If a light source emits light that is only blue-green, the color complementary to red, the red object will reflect no light and will therefore appear to be gray. The eye, however, can, to some degree, recognize colors of objects despite the color of the illuminant; that is, the eye can adapt to colored light. It also becomes sensitive to the colors that would have to be added to convert the illuminant to white light.

Apparent color is also affected by high levels of illumination. For example, all colors appear less bright, or washed out, under high illumination. But brightness of color also depends on the hue. Under the same illumination, light colors appear brighter than dark colors. Thus, lower levels of illumination are desirable with white-, yellow-, and red-colored (warm-colored) objects than with black-, blue-, and green-colored (cool-colored) objects. (Warm-colored objects also have the psychological effect of appearing to be closer than they actually are, whereas cool-colored objects tend to recede. In addition, cool colors create a calm and restful atmosphere, conducive to mediation, but are not flattering to skin colors or food, whereas warm colors produce opposite effects.)

See also Arts. 15.12, Color Rendering, and 15.20, Bibliography.

15.11 QUALITY OF LIGHT

A good lighting system not only provides adequate quantities of light for safe, efficient visual performance but also good quality of light. Quality determines the visual comfort of building occupants and contributes to good visibility. In accordance with the relationships between light and sight described in Art. 15.10, quality and quantity therefore should be considered together in lighting design. For ease of presentation, however, the factors affecting these characteristics of light are discussed separately. Color rendering is treated in Art. 15.12 and quantity of light, in Art. 15.13.

The characteristics of a luminous environment that determine quality are contrast, diffusion, and color rendering. Contrast, created by shadows or by relatively bright areas in a field of vision, affects visibility, mood, comfort, and eyestrain (Art. 15.10). Diffusion, the dispersion of light in all directions, may be produced by transmission or reflection. Diffuse transmission occurs when light from a bright source passes through a material that disperses the incident light, with consequent reduction in brightness. Translucent materials or specially constructed lenses are often employed for this purpose in lighting fixtures. Diffuse reflection occurs when incident light on a surface is reflected almost uniformly in all directions by tiny projections or hollows. Such surfaces appear nearly equally bright from all viewing angles. Diffusion tends to reduce contrast and promote uniformity of lighting.

Brightness Ratios. For good quality of lighting, the degree of contrast of light and dark areas in the field of vision should be limited to provide for viewing angle changes. The reason for this is that the eye adapts to the luminance of a task after a period of time. When the eye leaves the task and encounters a field of different brightness, the eye requires an appreciable time to adapt to the new condition, during which eyestrain or visual discomfort may be experienced. To promote quick, comfortable adaptation, brightness ratios in the visual environment should be kept small.

For example, in offices, the ratio of brightness of task to that of darker immediate surroundings should not exceed 3:1, and to that of darker, more remote surroundings, 5:1. Similarly, the ratio of brightness of lighter, more remote surroundings to that of the task should be less than 5:1. (See "Office Lighting," RP 1, Illuminating Engineering Society of North America, 345 E. 47th St., New York, NY 10017-2377.)

Direct Glare. When background luminance is much greater than that of the task, glare results. It may take the form of direct glare or reflected glare. Direct glare is produced when bright light sources are included in the field of vision and cause discomfort or reduced visibility of the task. The intensity of glare depends on the brightness, size, and relative position of the light sources in the field of vision, and on the general luminance of the field of vision.

The brighter the light source, the greater will be the glare, other factors remaining substantially constant. Similarly, the larger the light source, the greater will be the glare. A small bright lamp may not be objectionable; in fact, in some cases, it may be desirable to provide sparkle and relieve the monotony of a uniformly lit space, whereas a large, bright luminaire in the field of vision would cause discomfort. In contrast, glare decreases as the distance of the light source from the line of sight increases. Also, glare decreases with increase in general luminance of the visual environment, or level of eye adaptation.

The Illuminating Engineering Society has established standard conditions for determining a criterion, called **visual comfort probability** (VCP), for rating discomfort glare. VCP indicates the percentage of observers with normal vision who will be visually comfortable in a specific environment. Tables of VCP values for various luminaires are available from their manufacturers. VCP values should be applied with caution, because they may not be applicable under conditions that depart significantly from the IES standard.

In general, direct glare should not be troublesome if all of the following conditions are satisfied for an overhead electric lighting system:

1. VCP is about 70 or more.
2. The ratio of maximum luminance of each luminaire to its average luminance is 5:1 (preferably 3:1) or less, at 45, 55, 65, 75, and 85° with respect to the vertical, when viewed lengthwise and crosswise.
3. The maximum luminance of each luminaire, when viewed lengthwise and crosswise, does not exceed the values given in Table 15.4 for the specified angles.

TABLE 15.4 Recommended Maximum Luminances of Light Sources

Angle with vertical, degrees	Luminance, fL
45	2250
55	1605
65	1125
75	750
85	495

Reflected Glare. Also called veiling reflection because of the effect on visibility, reflected glare results when incident light from a bright light source is reflected by the task into the eyes of the observer and causes discomfort or loss of contrast. Occurrence of glare depends on the brightness of the light source, overall luminance of the task, reflectance of the task surface, and relative positions of light source, task, and observer. When a bright light source is reflected into an observer's eyes, it casts an apparent veil over the image of the task. The result is a loss of contrast that would otherwise be useful in perception of task size and silhouette details; for example, print that would be readily legible without reflected glare would become difficult to read in the presence of a veiling reflection.

Several techniques have been found useful in maintaining an adequate quantity of light while limiting loss of contrast. These include the following:

Observers, tasks, and light sources should be positioned to reduce reflected glare. If there is only a single light source, positional change should remove it from the field of vision. When daylighting is used, occupants should be faced parallel to or away from windows, rather than toward them. When overhead luminaires are used, they should be positioned on either side of and behind the occupants, instead of in the general area above and forward of them. When continuous rows of linear luminaires are used, the occupants should be positioned between the rows with the line of sight parallel to the longitudinal axes of the luminaires.

Increase in luminance of the task will offset the loss of contrast, if the added illumination is provided at nonglare angles.

Decrease in the overall brightness of light sources or the brightness at angles that cause glare also is helpful. When windows are the sources of glare, it can be reduced or eliminated by tilting blinds. When luminaires are used, brightness of the source should be kept low to minimize glare. Low-brightness sources with large areas usually cause less glare than small sources with high luminance. As an alternative, the whole ceiling can be used as an indirect light source, which reflects light directed onto it by luminaires. These fixtures should be suspended at least 18 in below the ceiling to prevent high-brightness spots from being created on it. The ceiling should be white and clean and have a matte finish to diffuse the light.

Reflected glare also can be reduced by use of luminaires that distribute light mostly at nonoffending angles. Tasks are usually observed in a downward direction at angles with the vertical of 20 to 40°. Consequently, light incident on the task at those angles may cause reflected glare. Therefore, selection of a luminaire that produces little or no light at angles less than 40° cannot cause glare. Figure 15.12*a* illustrates the candlepower distribution curve for a luminaire that directs little light downward. (Note that it also directs little light near the horizontal, to prevent direct glare.) The curve is called a batwing, because of its shape. As indicated in Fig. 15.12*b*, a high percentage of the light from this luminaire cannot produce veiling reflections.

Specularity of the task can be a major cause of reflected glare. Consequently, high-gloss surfaces in the field of vision should be avoided. Often, tilting the task can reduce or eliminate veiling reflections.

Use of low-level illumination throughout a space, supplemented by local lighting on the task, offers several advantages, including adequate light with little or no glare and flexibility in positioning the local light source.

In many work environments, observers find themselves confronted with reflected glare both on horizontal surfaces, when reading material on a desktop, and on approximately vertical surfaces, when viewing a *visual display terminal (VDT)* screen. Designers should be aware of this and should be certain that the lighting system used to alleviate veiling reflections on horizontal materials does not aggravate the problem on the VDT screen. Luminaires selected to direct light outward beyond 40° from vertical to avoid veiling reflections may produce direct glare on

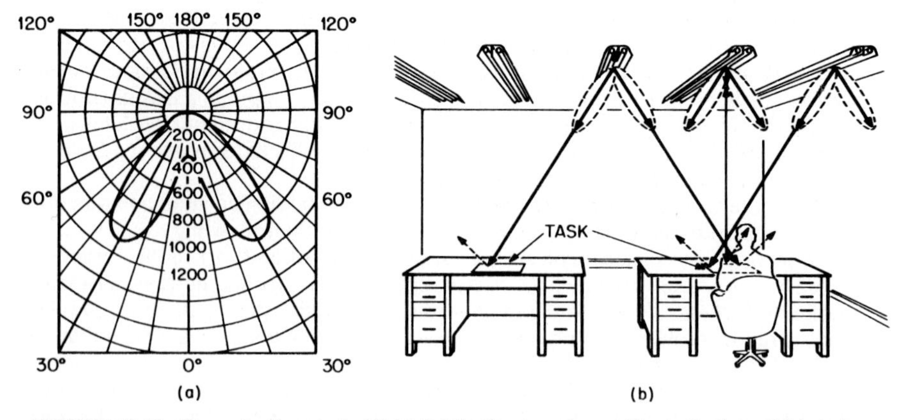

FIGURE 15.12 Example of control of light distribution to reduce veiling reflections: (*a*) batwing candlepower distribution curve for a luminaire; (*b*) arrangement of fluorescent lighting fixtures with distribution shown in (*a*), task and observer to limit veiling reflections.

the VDT. In general, in design of lighting systems for areas with VDTs, illuminance levels should be kept low, less than 75 fc. Illuminance ratios in the area also should be kept low, fixtures that have high VCPs should be selected and surface finishes that might reflect onto the VDT screen should have medium to low reflectances.

Luminaires with parabolic louvers, either small-cell, intermediate-cell, or deep-cell, are often used. They provide good illuminance on horizontal surfaces while contributing almost no direct glare onto the screen. They do, however, create a visual environment with relatively dark ceiling space that some may find objectionable.

Pendant-mounted, indirect-lighting luminaires or direct-indirect fixtures using fluorescent or HID sources can also be successfully applied to areas with VDTs. They produce a brighter feeling in the space for the occupants. Care must be taken to select luminaires with properly designed light control and to mount them sufficiently below the ceiling to create a uniform brightness across the ceiling surface. Failure to meet these requirements can result in "hot spots" on the ceiling that will be reflected onto the VDT screen. A relatively uniform, though somewhat bright, ceiling luminance reflected onto the screen is not objectionable to the viewer (VDT contrast and brightness controls can be adjusted to compensate) but the hot spots on the screen force the viewer's eyes to be constantly adjusting to the differences in reflected glare, causing eye fatigue and discomfort. (See "The IES Recommended Practice for Lighting Offices Containing Computer Visual Display Terminals (VDT's) RP-24," Illuminating Engineering Society of North America.)

See Art. 15.20, Bibliography.

15.12 COLOR RENDERING WITH LIGHTING

A black body is colorless. When increasing heat is applied to such a body, it eventually develops a deep red glow, then cherry red, next orange, and finally blue-white. The color of the radiated light is thus related to the temperature of the heated body. This phenomenon is the basis for a temperature scale used for the comparison of the color of light from different sources. For example, the light from an incandescent lamp, which tends to be yellowish, may be designated 2500° Kelvin (K), whereas a cool white fluorescent lamp may be designated 4500°K.

Light used for general illumination is mainly white, but white light is a combination of colors and some colors are more predominant than others in light emitted from light sources commonly used. When light other than white is desired, it may be obtained by selection of a light source rich in the desired hue or through use of a filter that produces that hue by absorbing other colors.

Color rendering is the degree to which a light source affects the apparent color of objects. **Color rendering index** is a measure of this degree relative to the perceived color of the same objects when illuminated by a reference source for specified conditions. The index actually is a measure of how closely light approximates daylight of the same color temperature. The higher the index, the better is the color rendering. The index for commonly used light sources ranges from about 20 to 99.

Generally, the color rendering of light should be selected to enhance color identification of an object or surface. This is especially important in cases where color coding is used for safety purposes or to facilitate execution of a task. Color enhancement is also important for stimulating human responses; for example, in a restaurant, warm-colored light would make food served appear more appetizing, whereas cool-colored light would have the opposite effect.

Sources producing white light are generally used. Because of the spectral energy distribution of the light, however, some colors predominate in the illumination. For example, for daylight, north light is bluish, whereas direct sunlight at midday is yellow-white; and light from an incandescent lamp is high in red, orange, and yellow. The color composition of the light may be correlated with a color temperature. For a specific purpose, a source with the appropriate color characteristics should be chosen. Lamp manufacturers provide information on the color temperature and color rendering index of their products.

Colored light, produced by colored light sources or by filtering of white light, is sometimes used for decorative purposes. Colored light also may be used to affect human moods or for other psychological purposes, as indicated in Art. 15.10.8. Care must be taken in such applications to avoid objectionable reactions to the colored light; for example, when it causes unpleasant changes in the appearance of human skin or other familiar objects.

Perceived color of objects also is affected by the level of illumination (Art. 15.10.8). When brilliant color rendition is desired and high-intensity lighting is to be used, the color saturation of the objects should be high; that is, colors should be vivid. Also, a source that would enhance the colors of the objects should be chosen.

See Art. 15.20, Bibliography.

15.13 QUANTITY OF LIGHT

As indicated in Art. 15.11, quantity and quality of light are actually inseparable in contributing to good lighting, although they are treated separately, for convenience of presentation, in this section. Illumination should meet the requirements of visual tasks for safe, efficient performance, esthetic reasons, and the purpose of attracting attention. Factors that affect visual performance of a task include:

Luminance, or brightness, of the task

Luminance relation between task and surroundings

Color rendering of the light

Size of details to be detected

Contrast of the details with their background

Duration and frequency of occurrence of the task

Speed and accuracy required in performance of the task

Age of workers

The influence of these factors on visibility is described in Art. 15.10.

Dim lighting is sometimes desirable for mood effects. For merchandising, however, a pattern of brightness is depended on to capture the attention of potential customers. For task lighting, sufficient lighting must be provided on the work area if the task is to be executed without eyestrain and fatigue. Higher than the minimum required level of illumination usually improves visibility but often with greater energy consumption and increased life-cycle costs for both lighting and cooling the building. Consequently, for task lighting, illumination should be kept to the minimum necessary for maintenance of adequate quantity and quality of lighting.

For many years, single values for minimum illuminance, or level of illumination, fc, developed by the Illuminating Engineering Society of North America (IES) and listed in tables in the "IES Lighting Handbook," have been widely used in the United States. In 1979, however, the IES revised its criteria to a recommended range of target illuminances. These recommended values take into account many of the factors listed above. Following is an example and abbreviated tables to illustrate the use of the tables in the IES handbook.

To determine the target illuminance, start by ascertaining the type of activity, or illuminance category, for the space to be illuminated. (The "IES Lighting Handbook" contains a detailed table correlating specific areas or activities with categories labeled A to I. For some of these categories, the effects of veiling reflections should be evaluated, for which purpose equivalent-sphere-illumination, ESI, calculations may be used, as indicated in Art. 15.10.5.) The first column of Table 15.7, which

TABLE 15.5 Weighting Factors for Determining Visual Conditions*

Task and worker characteristics	Weight W
Workers' ages	
Under 40	−1
40 to 55	0
Over 55	+1
Importance of speed or accuracy†	
Not important	−1
Important (errors costly)	0
Critical (errors unsafe)	+1
Reflectance of task background	
Greater than 70%	−1
30 to 70%	0
Less than 30%	+1

*Based on data in "Selection of Illuminance Values for Interior Lighting Design," RP-15A, Illuminating Engineering Society of North America. Calculate ΣW by adding the weighting factors W for the specific environment or required performance and determine the corresponding visual condition from Table 15.7.

†Use $W = 0$ for environment categories A to C.

TABLE 15.6 Visual Condition Number for Determining Recommended Illuminances

	ΣW*						
Category	3	2	1	0	−1	−2	−3
A to C		4	4	3	2	2	
D to I	4	4	3	3	3	2	2

*ΣW = sum of weighting factors given in Table 15.5 for a specific environment or required performance.

is based on descriptions in "Selection of Illuminance Values for Interior Lighting Design," IES RP-15A, gives a general description of these categories.

Next, determine the appropriate weighting factors W from Table 15.5 to adjust for loss of visual acuity with age, for importance of speed and accuracy in performing tasks, and for reflections of task background. For categories A to C, for which there are no task activities, use $W = 0$ as the weight for speed and accuracy. Then, add the weighting factors. From Table 15.6, determine the visual condition number corresponding to ΣW and the category. Decrease the condition number by one if

TABLE 15.7 Recommended Illuminances for Interior Lighting, fc[a]

Category of environment or required performance	Visual condition[b]			
	1. Short exposure	2. Moderate	3. Ordinary	4. Severe
A. Public areas with dark surroundings[d]		2	3	5
B. Simple orientation for short visits[d]		5	7.5	10
C. Working spaces for infrequent tasks[d]		10	15	20
D. Tasks with high contrast or large size[e,f]		20	30	50
E. Tasks with medium contrast or small size[e,g]	30	50	75	100
F. Tasks with low contrast or very small size[e,h]	75	100	150	200
G. Tasks with low contrast, very small size, and long duration[i,j]		200	300	500
H. Very prolonged and exacting tasks[i,k]		500	750	1000
I. Tasks with extremely low contrast and small size[i,l]		1000	1500	2000

[a]Based on data in "Selection of Illuminance Values for Interior Lighting Design," RP-15A, Illuminating Engineering Society of North America.

[b]Visual condition depends on workers' ages, importance of speed and accuracy in performance of task, and reflectance of task background. For tasks that occur infrequently or are of short duration and if conditions 2, 3, or 4 would be applicable, use conditions 1, 2, or 3, respectively. Where no value is given for condition 1, use condition 2. See text and Tables 15.5 and 15.6.

[c]Based on British IES "Code for Interior Lighting," 1977.

[d]General lighting throughout the building space.

[e]Illuminance on task.

[f]For example, reading printed material, typed originals, handwriting in ink or good xerography; rough bench or machine work; ordinary inspection; or rough assembly.

[g]For example, reading medium pencil handwriting, poorly printed or reproduced material; medium bench or machine work; difficult inspection, or medium assembly.

[h]For example, reading handwriting in hard pencil on poor-quality paper or very poor reproductions, or highly difficult inspections.

[i]Illuminance on task, obtained by combining general and local (supplementary) lighting.

[j]For example, fine assembly, unusually difficult inspection, or fine bench or machine work.

[k]For example, the most difficult inspection, extra-fine bench or machine work, or extra-fine assembly.

[l]For example, surgical procedures.

the task is of short duration or occurs infrequently. Finally, select the target illuminance, fc, from Table 15.7 corresponding to the category and visual condition.

The IES recommends that if information is available on the effect of changes in illuminance or equivalent sphere illumination, ESI, on task performance, the data may be used to determine if a variation from the design illuminance or ESI will be meaningful in terms of increased or decreased productivity, which may be used in a cost-benefit analysis.

See Art. 15.20, Bibliography.

15.14 LIGHTING METHODS

Interior lighting may be accomplished with natural or artificial illumination, or both. Natural illumination is provided by daylight. Artificial illumination usually is produced by consumption of electric power in various types of lamps and sometimes by burning candles or oil or gas in lamps. Usually, electric lighting for building spaces is produced by lighting devices called **luminaires,** which consist of one or more lamps, a fixture in which the lamps are held, lenses for distributing the light, and parts for supplying electricity. Fixtures may be portable or permanently set in or on ceilings or walls.

To meet specific lighting objectives, the following lighting methods may be used alone or in combination:

General Lighting. This provides uniform and, often, diffuse illumination throughout a space. This type of lighting is useful for performing ordinary activities and for reducing the relative luminance of surroundings when local lighting is applied to a work area.

Local or Functional Lighting. This provides a high level of illumination on the relatively small area in which a task is to be performed, such as reading, writing, or operation of tools.

Accent Lighting. This actually is a form of local lighting, but it has the objective of creating focal points for observers, to emphasize objects on display.

Decorative Lighting. This employs color or patterns of light and shadow to attract attention, hold interest, produce visual excitement or a restful atmosphere, or create esthetic effects.

Illumination may be classified as indirect, semiindirect, diffuse or direct-indirect, semidirect, or direct.

For indirect lighting, about 90 to 100% of the illumination provided in a space is directed at the ceiling and upper walls, and nearly all of the light reaches the task by reflection from them. The resulting illumination is, therefore, diffuse and uniform, with little or no glare.

For semiindirect lighting, about 60 to 90% of the illumination is directed at the ceiling and upper walls, the remaining percentage in generally downward directions. When overhead luminaires are used, the downward components should be dispersed by passage through a diffusing or diffracting lens to reduce direct glare. The resultant illumination on a task is diffuse and nearly glarefree.

General diffuse or direct-indirect lighting is designed to provide nearly equal distribution of light upward and downward. General-diffuse luminaires enclose the light source in a translucent material to diffuse the light and produce light in all directions. Direct-indirect luminaires give little light near the horizontal. Quality of the resulting illumination from either type depends on the type of task and the layout of the luminaires.

For semidirect lighting, about 60 to 90% of the illumination is directed downward, the remaining percentage upward. Depending on the eye adaptation level, as determined by overall room luminance, the upward component may reduce glare. Diffuseness of the lighting depends on reflectance of room enclosures and furnishings.

For direct lighting, almost all the illumination is directed downward. If such luminaires are spread out, reflections from room enclosures and furnishings may diffuse the light sufficiently that it can be used for general lighting, for example, in large offices. A concentrated layout of these luminaires is suitable for accent, decorative, or local lighting. Because direct lighting provides little illumination on vertical surfaces, provision of supplementary perimeter lighting often is desirable.

Lighting Distribution. Luminaires are designed for a specific type of lamp to distribute light in a way that will meet design objectives. For this purpose, luminaires incorporate various shapes of reflectors and various types of lenses (Fig. 15.13). Also, size and shape of openings through which light is emitted is controlled.

A luminaire may provide symmetrically or asymmetrically distributed light. With symmetrical distribution (Fig. 15.11), the level of illumination, fc, on a work plane is nearly the same at the same angle with the vertical and at equal distances from the light source. With asymmetrical distribution (Fig. 15.13*c*), the luminaire concentrates light in a specific direction. Symmetrical distribution is appropriate for general lighting. Asymmetrical distribution is advantageous for accent lighting.

See Art. 15.20, Bibliography.

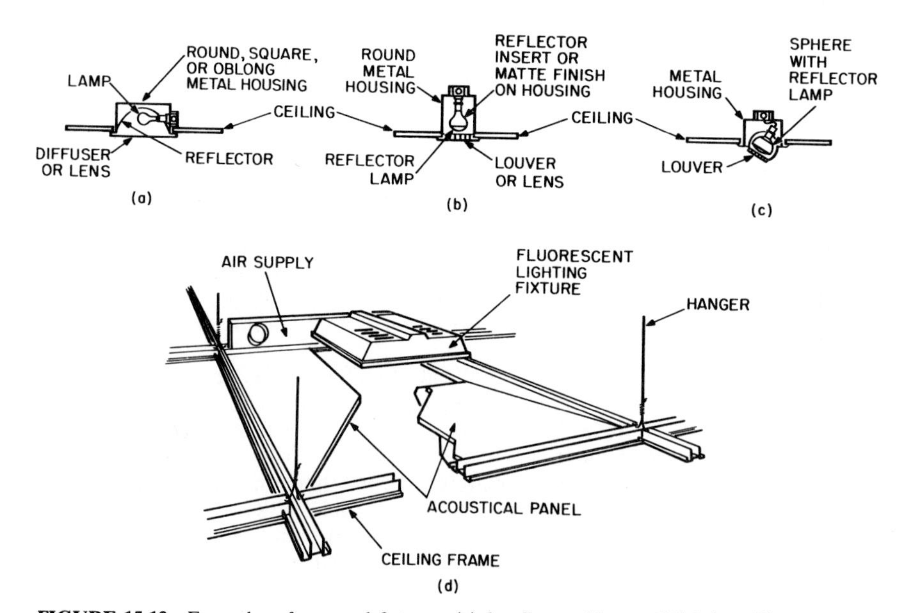

FIGURE 15.13 Examples of recessed fixtures: (*a*) for direct widespread lighting; (*b*) for direct, narrow-beam lighting; (*c*) for asymmetrical direct lighting; (*d*) for direct diffuse lighting.

15.15 DAYLIGHT

Use of natural light has the advantages, compared with artificial illumination, of not consuming fuel and not having associated operating costs. But daylight has the disadvantages of being dependent on the availability of windows and on the absence of light-blocking obstructions outside the windows, of not being available between sunset and sunrise, and of providing weak light on cloudy days and around twilight and dawn. When lighting is needed within a building at those times, it must be provided by artificial illumination. Also, for parts of rooms at large distances from windows, where adequate daylight does not reach, artificial illumination is needed to supplement the daylight. When supplementary lighting is required, initial, maintenance, and replacement costs of lamps and fixtures are not saved by use of daylight, although costs of power for lighting can be reduced by turning off lamps not needed when daylight is available. In addition, design for daylighting should be carefully executed so as not to introduce undesirable effects, for example, glare, intensive sunlight, or excessive heat gain.

Elements can be incorporated in building construction to control daylight to some extent to provide good lighting within short distances from windows. For example, to prevent glare, windows may be shielded, by blinds or by outside overhangs, against direct sunlight. To reduce heat gain, windows may be glazed with reflecting, insulating, or heat-absorbing panes. Ceiling, floor, and walls with high reflectance should be used to diffuse light and reflect it into all parts of rooms. To illuminate large rooms, daylight should be admitted through more than one wall and through skylights or other roof openings, if possible. In rooms that extend a long distance from windows, the tops of the windows should be placed as close to the ceiling as possible, to permit daylight to penetrate to the far end of the room.

Design procedures for daylighting are presented in "Recommended Practice for Daylighting," RP-5, Illuminating Engineering Society of North America.

15.16 CHARACTERISTICS OF LAMPS

Selection of the most suitable lamp consistent with design objectives is critical to performance and cost of a lighting system. This decision should be carefully made before a luminaire for the lamp is selected. Luminaires are designed for specific lamps.

Lamps are constructed to operate at a specific voltage and wattage, or power consumption. In general, the higher the wattage rating of a specific type of lamp, the greater will be its **efficacy,** or lumen output per watt.

15.16.1 Considerations in Lamp Selection

Greatest economy will be secured for a lighting installation through use of a lamp with the highest lumen output per watt with good quality of illumination. In addition to lumen output, however, color rendering and other characteristics, such as lighting distribution, should also be considered in lamp selection. Information on these characteristics can be obtained from lamp manufacturers. Latest data should be requested, because characteristics affecting lamp performance are changed periodically. The following information usually is useful:

Lamp life, given as the probable number of hours of operation before failure.

Lamp efficacy, measured by the lamp output in lumens per watt of power consumed.

Lamp lumen depreciation, as indicated by tests. Curves are plotted from data to show the gradual decrease in light output with length of time of lamp operation. The decrease occurs because of both aging and dirt accumulation (Art. 15.16.2). The latter can be corrected with good maintenance, but the possible effects nevertheless should be considered in lighting design.

Lamp warm-up time, which is significant for some fluorescent lamps and all high-intensity-discharge lamps, for which there is a delay before full light output develops.

Lamp restart time, or time that it takes some lamps to relight after they have been extinguished momentarily. The lamps may go out because of low voltage or power interruption. Use of lamps with long warm-up and restart times should be avoided for spaces where lights are to be turned on and off frequently.

Color rendering index and color acceptability, which are, respectively, measures of the degree to which illumination affects the perceived color of objects and the human reaction to perceived colors (see Art. 15.12).

Voltages and frequency of current at which lamps are designed to operate. In most buildings in the United States, electricity is distributed for light and power as an alternating current at 60 Hz. Lamp output generally increases at higher frequencies but capital investment for the purpose is higher. Voltage often is about 120, but sometimes, especially for industrial and large commercial buildings, voltages of 208, 240, 277, or 480 are used, because of lower transmission losses and more efficient operation of electrical equipment. Direct current from batteries often is used for emergency lighting, when the prime ac source fails. Low-voltage (generally 12-V) lamps may be used outdoors, for safety reasons, when conductors are placed underground or where the lamps are immersed in water.

Noise, which is significant for some types of lamp applications, such as fluorescent and high-intensity-discharge lamps. These depend for operation on a device, called a ballast, which may hum when the light operates. Whether the hum will be annoying under ordinary circumstances depends on the ambient noise level in the room. If the ambient level is high enough, it will mask the hum. Before a combination of lamp, ballast, and fixture is selected, ballast noise rating should be obtained from the luminaire manufacturer.

Ambient temperature, or temperature around a lamp when it is operating, which may affect lamp life, lumen output, and color rendering. If the ambient temperature exceeds the rated maximum temperature for the luminaire, the life of incandescent lamps may be considerably reduced. Consequently, lamps rated at a wattage greater than that recommended by the fixture manufacturer should not be used. Also, provision should be made for dissipation of the heat produced by lamps. Low ambient temperature slows starting of fluorescent and high-intensity-discharge lamps. Low temperature also reduces lumen output and changes the color of fluorescent lamps.

15.16.2 Maintenance of Lamp Output

The efficiency of a lighting system decreases with time because of dirt accumulation, decrease in lumen output as lamps age, lamp failures, and deteriorating lighting fixtures. Depending on type of fixtures, cleanliness of the environment, and time between cleanings of lamps and fixtures, lumen losses due to dirt may range from 8 to 10% in a clean environment to more than 50% under severe conditions. Also, the longer lamps operate, the dimmer they become; for example, a fluorescent lamp at the end of its life will yield only 80 to 85% of its initial lumen output. And when one or more lamps fail and are not replaced, the space being illuminated may suffer a substantial loss in light. Furthermore, in the case of lights operating with ballasts, the lamps, before burning out, overload the ballast and may cause it to fail. Consequently, a poorly maintained lighting installation does not provide the illumination for which it was designed and wastes money on the power consumed.

The design illumination level may be maintained by periodic cleaning and prompt replacement of aged or failed lamps. Relamping may be carried out by the spot or group methods. Spot relamping, or replacement of lamps one by one as they burn out, is more inefficient in use of labor and more costly. Group relamping calls for scheduled replacement of lamps at intervals determined from calculations based on expected lamp life and variation of lumen output with time. This method reduces labor costs, causes fewer work interruptions, maintains higher levels of illumination with no increase in cost of power, prevents the appearance of the lighting system from deteriorating, and reduces the possibility of damage to auxiliary equipment, such as ballasts, near the end of lamp life.

Because of the decrease in lamp output with time, in design calculations initial lamp output should be multiplied by a lamp lumen depreciation factor to correct for the effects of aging. This product equals the output after a period of time, usually the interval between group relampings. In addition, the output should be multiplied by a dirt depreciation factor. Both factors are less than unity. The two factors may be combined into a single maintenance factor M.

$$M = \text{LLD} \times \text{LDD} \tag{15.33}$$

where LLD = lamp lumen depreciation factor
LDD = dirt depreciation factor

After a period of operation, lamp output then is given by

$$L = L_i M \tag{15.34}$$

where L_i = initial output, lm

15.16.3 Lamp Control

Incoming power for a lamp normally is by a local switch in the power circuit. The switch turns the lamps on or off by closing or opening the circuit.

An alternative method may be used that is more economical when the lamps are distant from the switch and that reduces the possibility of personal injury or short-circuit damage at the switch. In this method, the main power circuit is opened and closed by a relay located near the lamps. The relay, in turn, may be activated by low-voltage power controlled by a remote switch. Control-system voltages usually range between 6 and 24 V, obtained by stepping down the normal distribution voltage with transformers.

To prevent waste of energy by lighting in unoccupied rooms, occupancy sensors may be installed. They sense entry of a person into a room and turn lights on. They also detect continued presence in the room and keep the lights on until after departure.

For control of light output from a luminaire, a control switch may be replaced with a dimmer. For incandescent lamps, this device can vary the voltage across the lamps from zero to the rated value and thus can be used to adjust the level of illumination. For fluorescent and high-intensity-discharge lamps, the dimmer is coordinated with the lamp ballast.

15.16.4 Types of Lamps

Lamps that are commonly used may be generally classified as incandescent, fluorescent, or high-intensity-discharge (HID). HID lamps include mercury-vapor, metal-halide, low-pressure sodium, and high-pressure sodium lamps. See Tables 15.8 and 15.9.

Incandescent Lamps. These lamps generate light by heating thin tungsten wires until they glow. The filaments are enclosed in a sealed glass bulb from which air is evacuated or that is filled with an inert gas, to prevent the heated tungsten from evaporating. In a tungsten-halogen incandescent lamp, for prolonged life, the filler gas contains halogens (iodine, chlorine, fluorine, and bromine), which restores to the filaments any metal that may evaporate.

Incandescent lamps are available in a variety of shapes. They also come with a variety of bases, making it necessary to ensure that selected luminaires provide sockets that can accommodate the desired lamps.

Incandescent lamps produce light mainly in the yellow to red portion of the spectrum. Color depends on the wattage at which the lamp is operated. Generally, the higher the wattage, the whiter is the color of light produced. A reduction in wattage or voltage results in a yellower light. See Tables 15.8 and 15.9 for other characteristics of incandescent lamps.

Often, the glass bulbs of these lamps are treated to obtain special effects. Usually, the effect desired is diffusion of the emitted light, to soften glare. For the purpose, the glass may be frosted, or etched, or silica coated (white bulb). Light diffusion and other effects can also be achieved with control devices incorporated in the fixtures, such as lenses or louvers (Figs. 15.13 to 15.16). Also, lamps may be treated with coatings or filters to produce any of a variety of colors.

Reflectorized incandescent lamps are made, with standard or special shapes, with a reflective aluminized or silver coating applied directly to part of the inside bulb surface. Such lamps are widely used for spot or flood lighting. Type PAR lamp has a parabolic shape to focus the light beam. Types EAR and ER lamps have elliptical shapes that cause the light beam to concentrate near the front of the lens, then to broaden into the desired pattern, yielding more usable light than other types of lamps, with little glare.

Compact fluorescent lamps often can be used, to reduce energy consumption significantly, in applications where incandescent lamps were used in the past.

Fluorescent Lamps. These lamps are sealed glass tubes coated on the inside surface with phosphors, chemicals that glow when bombarded by ultraviolet light. The tubes are filled with an inert gas, such as argon, and low-pressure mercury vapor. Passage of an electric arc through the mercury vapor causes it to emit the ultraviolet rays that activate the phosphors to radiate visible light. The electric arc is started and maintained by cathodes at the ends of the glass tubes.

The high voltage needed to form the arc is provided initially by a device called a **ballast.** After the arc has been formed, the ballast limits the current in the arc to that needed to maintain it. Ballasts also may be designed to decrease the stroboscopic effect of the lamp output caused by the ac power supply and to keep the

TABLE 15.8 Characteristics of Lamps Often Used for General Lighting

	Type of lamp						
Lamp characteristic	Incandescent	Tungsten-halogen incandescent	Fluorescent	Clear mercury	Clear metal halide	Clear high-pressure sodium	Clear low-pressure sodium
Efficacy, lm/W	15–25	6–23	25–84	30–63	68–125	77–140	137–183
Initial lumens	40–33,600	40–33,600	96–15,000	1,200–63,000	12,000–155,000	5,400–140,000	4,800–33,000
Lumen maintenance, %	80–90	75–97	75–91	70–86	73–83	90–92	75–90
Wattage range	5–1,500	6–1,500	4–215	40–1,000	175–1,500	70–1,000	35–180
Life, hr	750–1,000	750–8,000	9,000–20,000	16,000–24,000	1,500–15,000	20,000–24,000	18,000
Color temperature, K	2,400–3,400	2,800–3,400	2,700–6,500	3,300–5,900	3,200–4,700	2,100	1,780
Color rendering index	89–92	95–99	55–95	22–52	65–70	21	0
Color acceptability	Good	Good	Good	Fair	Good	Fair	Poor
Light control	Excellent	Excellent	Poor	Fair	Fair to good	Good	Poor
Initial cost per lamp	Low	Low	Moderate	Moderate	High	High	Moderate
Energy cost	High	High	Moderate	Moderate	Low	Low	Low

Courtesy of Sylvania Lighting.

TABLE 15.9 Comparison of Lamps Often Used for General Lighting*

Lamp characteristics / Relative rating of lamps

Type of lamp	Color rendering			Initial efficacy, lm/W			Lumen maintenance (mean lm)			Rated average life, hr			Degree of light control			Input power required for equal light				System operating cost for equal light			Initial equipment cost for equal light			Total owning and operating cost		
	Very important	Important	Unimportant	Highest (80 up)	Medium (50–80)	Lowest (15–50)	Highest (85% up)	Medium (75–85%)	Fair (65–75%)	Shortest (5,000 or less)	Intermediate (5,000–15,000)	Longest (15,000–25,000)	Highest	Intermediate	Lowest	Highest	High	Intermediate	Lowest	Highest	Intermediate	Lowest	Highest	Intermediate	Lowest	Highest	Intermediate	Lowest
Incandescent	●					●		●		●			●			●				●					●	●		

Tungsten-halogen	●					●	●			●			●			●				●					●	●		
Fluorescent	●				●			●				●			●			●			●			●			●	
Clear mercury			●		●			●				●	●				●				●			●			●	
Coated mercury		●			●			●				●		●			●				●			●			●	
Clear metal halide	●			●				●			●		●						●			●		●				●
Coated metal halide	●			●					●		●			●					●			●		●				●
Clear high-pressure sodium			●	●			●					●	●						●			●	●					●
Coated high-pressure sodium			●	●			●					●		●					●			●	●					●

*Dot indicates that the light source exhibits the listed characteristics.
Courtesy of Sylvania Lighting.

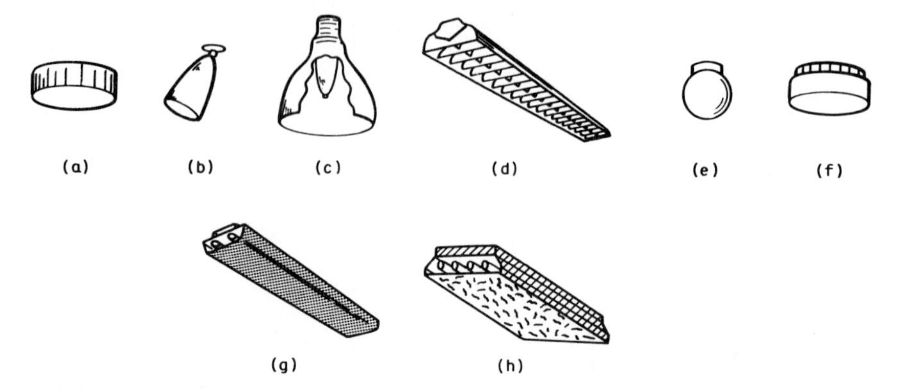

FIGURE 15.14 Examples of ceiling-mounted fixtures: (*a*) opaque-side drum for direct diffuse lighting; (*b*) spot lighting with incandescent lamp; (*c*) direct, widespread lighting with an HID lamp; (*d*) fluorescent fixture for direct diffuse lighting; (*e*) globe fixture with incandescent lamp for direct diffuse lighting; (*f*) small diffusing drum; (*g*) fluorescent fixture for semidirect diffuse lighting; (*h*) fluorescent fixture with diffusing lens for direct lighting.

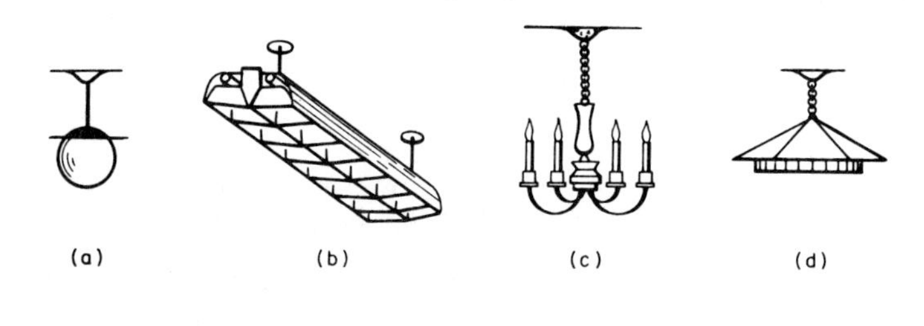

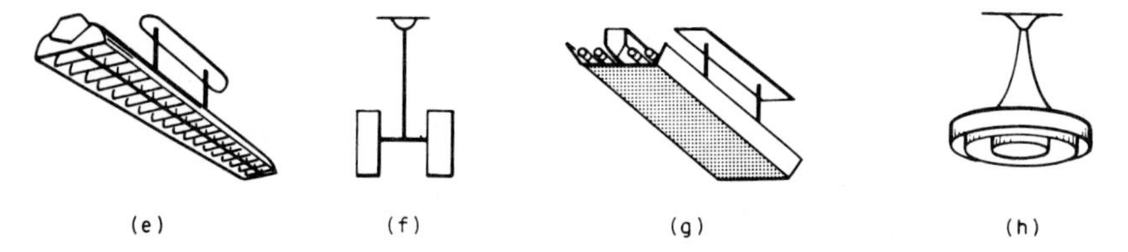

FIGURE 15.15 Examples of pendant fixtures: (*a*) globe fixture for direct diffuse light; (*b*) fluorescent fixture for general diffuse lighting; (*c*) exposed-lamp fixture for direct lighting; (*d*) direct downlight fixture; (*e*) fluorescent fixture for semidirect lighting; (*f*) fixture for direct-indirect lighting; (*g*) fluorescent fixture for semiindirect lighting; (*h*) fixture with high-intensity discharge (HID) lamp for indirect lighting.

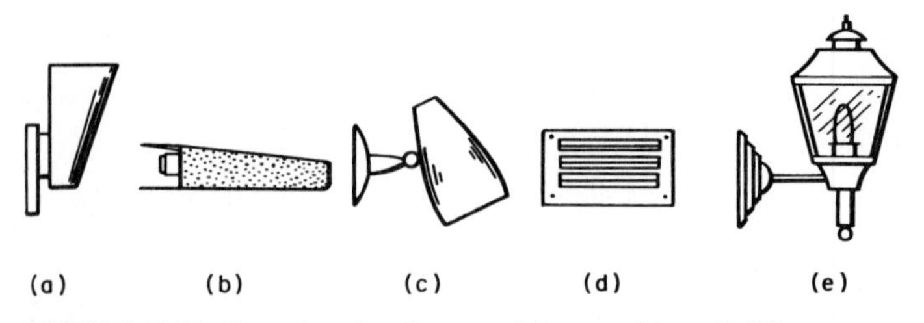

FIGURE 15.16 Examples of wall-mounted fixtures: (*a*) small diffuser type; (*b*) linear type, for example, four-lamp incandescent or a fluorescent lamp; (*c*) bullet-type, directional fixture, for accent lighting; (*d*) fixture for directional night lighting; (*e*) exposed-lamp fixture for direct lighting.

variation in current nearly in phase with the variation in voltage, thus maintaining a high power factor.

Fluorescent lamps generally are available as linear, bent U, or circular tubes, and luminaires are designed to be compatible with the selected shape.

Fluorescent lamps may be classified as preheat, rapid start, or instant start. They differ in the method used to decrease the delay in starting after a switch has been thrown to close the electrical circuit. The preheat type requires a separate starter, which allows current to flow for several seconds through the cathodes, to preheat them. For a rapid-start lamp, the cathodes are electrically preheated much more rapidly without a starter. For an instant-start lamp, high voltage from a transformer forms the arc, without the necessity of preheating the cathodes.

Instant-start lamps may be of the hot-cathode or cold-cathode type, depending on cathode shape and voltage used. Efficacy of cold-cathode lamps is lower than that of hot-cathode lamps. Both types are more expensive than rapid-start lamps and are less efficient in lumen output. Instant-start lamps, however, come in sizes that are not available for rapid-start lamps and can operate at currents that are not feasible for rapid-start lamps. Also, instant-start lamps can start at lower temperatures, for instance, below 50°F.

The life of most types of fluorescent lamps is adversely affected by the number of lamp starts. Cold-cathode lamps, however, have a long life, which is not greatly affected by the number of starts.

Fluorescent lamps last longer and have a higher efficacy than incandescent lamps. (See Tables 15.8 and 15.9.) Hence, fluorescent lamps cost less to operate. Initial cost of fixtures, however, may be higher. Fluorescent lamps require larger fixtures, because of tube length, and special equipment, such as ballasts and transformers (Figs. 15.13 to 15.16). Also associated with such lamps is ballast hum and possible interference with radio reception. Pattern control of light is better with incandescent lamps, but lamp brightness is low with fluorescent lamps, so that there is less likelihood of glare, even when the lamps are not shielded.

Color rendering of light emitted by a fluorescent lamp depends on the phosphors used in the tube. The best color rendering for general use may be obtained with the deluxe cool white (CWX) type. Check with manufacturers for color characteristics of lamps currently available, because new lamps designed with color rendering appropriate for specific purposes are periodically introduced.

High-Intensity-Discharge (HID) Lamps. These lamps generate light by passage of an electric arc through a metallic vapor, which is contained in a sealed glass or ceramic tube. The lamps operate at pressures and electrical current densities sufficient to produce desired quantities of light within the arc. Three types of HID lamps are generally available: mercury vapor, metal-halide, and high-pressure sodium. Major differences between them include the material and type of construction used for the tube and the type of metallic vapor. In performance, the lamps differ in efficacy, starting characteristics, color rendering, lumen depreciation, price, and life. (See Tables 15.8 and 15.9.) HID lamps are available with lumen outputs considerably greater than those of the highest-wattage fluorescent lamps available. HID lamps require ballasts that function like those for fluorescent lamps and that should be coordinated with the type and size of lamp for proper operation.

Each time an HID lamp is energized from a cold start, the lamp produces a dim glow initially and there is a time interval called *warm-up time* until the lamp attains its full lumen output. Warm-up time for metal-halide and sodium vapor lamps may range from 3 to 5 min and for mercury vapor lamps from 5 to 7 min. When the power to the lamp is interrupted, even momentarily, the lamp extinguishes immediately; and even if the power is restored within a short time, while the lamp is still hot, there is a delay, until the lamp cools to provide a condition that will permit the arc to restrike. Sodium vapor lamps have the fastest restart time, 1 min, compared with 3 to 6 min for mercury vapor lamps and as much as 10 min for metal-halide lamps. Because of this condition, it is desirable to employ supplemental lighting, usually incandescent, to provide minimal illumination during these intervals. (Lamp-ballast combinations that provide instant restart are available for metal-halide lamps. Though expensive, they have applications in security and sports lighting.)

The color of clear mercury lamps tends to be bluish. This type of lamp, however, also is available coated with phosphors that improve color rendering. The color of clear metal-halide lamps is stark white, with subtle tints ranging from pink to green. These lamps also may be coated for color correction. Light from clear sodium lamps, however, is yellowish. It strengthens yellow, green, and orange, but grays red and blue and turns white skin complexions yellow. Sodium lamps also may be coated to improve color rendering, but as color rendering is improved, efficacy decreases somewhat. Light from low-pressure clear sodium lamps is almost pure yellow. Use of such lamps, as a result, is limited to applications where color rendering is unimportant, such as freight yards and security lighting.

With respect to annual cost of light, high-pressure sodium lamps with ceramic (aluminum-oxide) arc tubes, with relatively small size, high efficacy, long life, and excellent lumen maintenance, appear to be the most economical HID type. Some variations of this type of lamp also offer improved color rendering.

HID lamps require special luminaires and auxiliaries. Some of these fixtures will accept replacement HID lamps of any of the three types in specific wattages. Others will accept only one type.

See Art. 15.20, Bibliography.

15.17 CHARACTERISTICS OF LIGHTING FIXTURES

A lighting fixture is that component of a luminaire that holds the lamps, serves as a protective enclosure, or housing, delivers electric power to the lamps, and incorporates devices for control of emitted light. The housing contains lampholders and usually also reflective inside surfaces shaped to direct light out of the fixture in controlled patterns. In addition, a fixture also incorporates means of venting heated air and houses additional light-control equipment, such as diffusers, refractors, shielding, and baffles. The power component consists of wiring and auxiliary equipment, as needed, such as starters, ballasts, transformers, and capacitors. The light-control devices include louvers, lenses, and diffusers. Fixture manufacturers provide information on construction, photometric performance, electrical and acoustical characteristics, installation, and maintenance of their products.

Some luminaires are sealed to keep out dust. Some are filtered and vented to dissipate heat and prevent accumulation of dust. Also, some are designed as part of the building air-conditioning system, which removes heat from the lamps before it enters occupied spaces. In some cases, this heat is used to warm spaces in the building that require heating.

Safety Requirements. Construction and wiring of fixtures should conform with local building codes and the National Electrical Code (NEC), recommendations of the National Electrical Manufacturers Association (NEMA) and "Standard for Lighting Fixtures," Underwriters Laboratories, Inc.

The NEC requires that fixtures to be installed in damp or wet locations or in hazardous areas containing explosive liquids, vapors, or dusts be approved by Underwriters Laboratories for the specific application. Auxiliary equipment for fluorescent and HID lamps should be enclosed in incombustible cases and treated as sources of heat.

The NEC specifies that fixtures that weigh more than 6 lb or are larger than 16 in in any dimension not be supported by the screw shell of a lampholder. The code permits fixtures weighing 50 lb or less to be supported by an outlet box or fitting capable of carrying the load. Fixtures also may be supported by a framing member of a suspended ceiling if that member is securely attached directly to structural members at appropriately safe intervals or indirectly via other adequately supported ceiling framing members. Pendent fixtures should be supported independently of conductors attached to the lampholders.

The NEC also requires that fixtures set flush with or recessed in ceilings or walls be so constructed and installed that adjacent combustible material will not be exposed to temperatures exceeding 90°C. (Thermal insulation should not be installed within 3 in of a recessed fixture.) Fire-resistant construction, however, may be exposed to temperatures as high as 150°C if the fixture is approved for such service. Screw-shell-type lampholders should be made of porcelain.

Lenses may be made of glass or plastic. In the latter case, the material should be incombustible and a low-smoke-density type. It should be stable in color and strength. The increase in yellowness after 1000 hr of testing in an Atlas FDA-R Fade-Ometer in accordance with ASTM G23 should not exceed 3 IES-NEMA-SPI units. Acrylics are widely used.

Considerations in Fixture Selection. Because fixtures are designed for specific types of lamps and for specific voltage and wattage ratings of the lamps, a prime consideration in choosing a fixture is its compatibility with lamps to be used. Other factors to consider include:

Conformance with the chosen lighting method (see Art. 15.14)

Degree to which a fixture assists in meeting objectives for quantity and quality of light through emission and distribution of light

Luminous efficiency of a fixture, the ratio of lumens output by the fixture to lumens produced by the lamps

Esthetics—in particular, coordination of size and shape of fixtures with room dimensions so that fixtures are not overly conspicuous

Durability

Ease of installation and maintenance

Light distribution from fixtures, to summarize, may be accomplished by means of transmission, reflection, refraction, absorption, and diffusion. Reflectors play an important role. Their reflectance, consequently, should be high—at least 85%. The shape of a reflector—spherical, parabolic, elliptical, hyperbolic—should be selected to meet design objectives; for example, to spot or spread light in a building space or to spread light over a fixture lens that controls light distribution. (The need for a curved reflector, which affects the size of the fixture, can be avoided by use of a Fresnel lens, which performs the same function as a reflector. With this type of lens, therefore, a smaller fixture is possible.) Light control also is affected by shielding, baffles, and louvers that are positioned on fixtures to prevent light from being emitted in undesirable directions.

A wide range of light control can be achieved with lenses. Flat or contoured lenses may be used to diffuse, diffract, polarize, or color light, as required. Lenses composed of prisms, cones, or spherical shapes may serve as refractors, producing uniform dispersion of light or concentration in specific directions.

Types of Installations. Luminaires may be classified in accordance with type and location of mountings, as well as with type of lighting distribution: flush or recessed (Fig. 15.13), ceiling mounted (Fig. 15.14), pendent (Fig. 15.15), wall mounted (Fig. 15.16) or structural.

Structural lighting is the term applied to lighting fixtures built into the structure of the building or built to use structural elements, such as the spaces between joists, as parts of fixtures. Structural lighting offers the advantage of a lighting system conforming closely to the architecture or interior decoration of a room. Some types of structural lighting are widely used in residences and executive offices. For the purposes of accent or decorative lighting, for example, cornices, valences, coves, or brackets are built on walls to conceal fluorescent lamps. For task lighting, fixtures may be built into soffits or canopies. For general lighting, large, low-brightness, luminous panels may be set flush with or recessed in the ceiling.

Lighting objectives can be partly or completely met with portable fixtures in some types of building occupancies. For the purpose, a wide variety of table and floor lamps are commercially available. Because the light sources in such fixtures are usually mounted at a relatively low height above the floor, care should be taken to prevent glare, by appropriate placement of fixtures and by selection of suitable lamp shades.

Number and Arrangement of Luminaires. With the type of lamp and fixture and the required level of illumination known, the number of luminaires needed to produce that lighting may be calculated and an appropriate arrangement selected. The **lumen method of calculation,** which yields the average illumination in a space, is generally used for this purpose.

The method is based on the definition of footcandle (Art. 15.10.4), in accordance with which the level of illumination on a horizontal work plane is given by

$$\text{fc} = \frac{\text{lumens output}}{\text{area of work plane, ft}^2} \tag{15.35}$$

Lamp manufacturers provide data on initial lumen output of lamps, but these values cannot be substituted directly in Eq. (15.35), because of light losses in fixtures and building spaces and the effects of reflection.

To adjust for the effects of fixture efficiency, distribution of light by fixtures, room proportions and surface reflectances, and mounting height and spacing of fixtures, the design lamp output, lm, is multiplied by a factor CU, called **coefficient of utilization,** to obtain lumens output for Eq. (15.35). (CU is the ratio of lumens striking the horizontal work plane to the total lumens emitted by the lamps. It may be obtained from tables available from fixture manufacturers.) Thus,

$$\text{fc} = \frac{\text{design lamp output, lm} \times \text{CU}}{\text{area of work plane, ft}^2} \tag{15.36}$$

To adjust for decreasing illumination with time, initial lamp output, lm, is multiplied by a light loss factor LLF to obtain the design lamp output. (LLF is the ratio of the lowest level of illumination on the work plane just before corrective action is taken, such as relamping and cleaning, to the initial level of illumination, produced by lamps operating at rated initial lumens. Maintenance of lighting installations thus is taken into account in LLF.) Substitution of this product in Eq. (15.36) yields

$$\text{fc} = \frac{\text{initial lamp output, lm} \times \text{CU} \times \text{LLF}}{\text{area of work plane, ft}^2} \tag{15.37}$$

Factors contributing to LLF include ballast performance, voltage to luminaires, luminaire reflectance and transmission changes, lamp outages, luminaire ambient temperature, provisions for removal of heat from fixtures, lamp lumen depreciation with use, and luminaire dirt depreciation.

The initial lamp output equals the product of the number of lamps by the rated initial lumens per lamp. Substitution in Eq. (15.37) and rearrangement of terms gives

$$\text{Number of lamps required} = \frac{\text{fc} \times \text{area}}{\text{initial lm per lamp} \times \text{CU} \times \text{LLF}} \tag{15.38}$$

$$\text{Number of luminaires required} = \frac{\text{number of lamps required}}{\text{lamps per luminaire}} \tag{15.39}$$

Layout of luminaires depends on architectural and decorative considerations, size of space, size and shape of fixtures, mounting height, and the effect of layout on quality of lighting. Different types of fixtures may be used in a space; for example, one type to provide general lighting, other types to provide supplementary local lighting, and still other types to produce accent or decorative lighting.

Details of the lumen method of calculation are given in books on lighting (see Art. 15.20, Bibliography).

Manufacturers list the maximum permissible spacing for each type of luminaire in photometric reports on their products. This spacing depends on the mounting height, relative to the work plane, for direct, semidirect, and general-diffuse luminaires, and relative to the ceiling height for indirect and semidirect luminaires. Spacings closer than the maximum improve uniformity of lighting and reduce shadows. Perimeter areas, however, require much closer spacing, depending on location of tasks and on the reflectance of the walls; generally, the distance between luminaires and the wall should not exceed half the distance between luminaires, and in some cases, supplementary lighting may have to be added. Computer programs are available for comparative analyses of different types and arrangements of luminaires.

15.18 SYSTEMS DESIGN OF LIGHTING

The objectives of and constraints on lighting systems and the interrelationship of lighting and other building systems are treated in this article. To design a lighting system for specific conditions, it is first necessary for the designer to determine the nature of and lighting requirements for the activities to be carried out in every space in the building. Also, the designer should cooperate with architects, interior designers, and structural, electrical, and HVAC engineers, as well as with the owner's representatives, to establish conditions for optimization of the overall building system. For example, where feasible, reflectances for ceiling, walls, and floor for each space may be selected for high lighting efficiency and visual comfort. Also, HVAC may be designed to remove and utilize heat from luminaires.

With tasks known, the designer should establish, for every space, criteria for illumination levels for task performance, safety, and visual comfort and also determine luminance ratios and light-loss factors. (For establishment of the light-loss factors, maintenance of the lighting system should be planned with the owner's representatives.) Based on the criteria and the lighting objectives, the designer can then decide how best to use daylighting and artificial lighting and select lamps and fixtures, luminaire mounting and layout, and lighting controls, such as switches and dimming. Because quality, color rendering, and quantity of light are interrelated, they should be properly balanced. This should be checked in an appraisal of the lighting system, which also should include comparisons of alternatives and studies of life-cycle costs and energy consumption. The analysis should compare alternatives not only for the lighting system but also for the other building components that affect or are affected by the lighting system.

Value analysis should examine illumination levels critically. A quantity of light that is sufficient for functional purposes is essential; but more light does not necessarily result in better lighting, higher productivity, or greater safety. Furthermore, higher illumination levels are undesirable, because they increase costs of lamps and fixtures, of lighting operation, of the electrical installation, and of HVAC installation and operation. Consequently, lighting should be provided, at the levels necessary, for visual tasks, with appropriate lower levels elsewhere, for example, for circulation areas, corridors, and storage spaces. Provision preferably should be made, however, for relocation or alteration of lighting equipment where changes in use of space can be expected.

The various types of lamps differ in characteristics important in design, such as color rendering, life, size, and efficacy. For each application, the most efficient type of lamp appropriate to it should be chosen. Consequently, prime consideration should be given to fluorescent and high-intensity-discharge lamps, which are highly efficient. Also, consideration should be given to high-wattage lamps of the type chosen, because the higher the wattage rating, the higher the lumen output per watt. Furthermore, in selection of a lamp, much more weight should be placed on life-cycle costs than on initial purchase price. Cost of power consumed by a lamp during its life may be 30 or more times the lamp cost. Consequently, use of a more efficient, though more expensive, lamp can save money because of the reduction in power consumption.

Similar consideration should also be given to luminaire selection. Efficient luminaires can produce more light on a task with less power consumption. Additional consideration, however, should be given to ease of cleaning and relamping, to prevention of direct glare and veiling reflections, and to removal of heat from the luminaires.

Control of lighting should be flexible. Conveniently located, separate switches or dimmers should be installed for areas with different types of activities. It should be easy to extinguish lights for areas that are not occupied and to maintain minimal emergency lighting, for safety.

Where feasible, daylighting should be used, supplemented, as needed, with artificial lighting. Light from windows can be reflected deeply into rooms, with venetian blinds, glass block, or other architectural elements. Glare and solar heat can be limited with blinds, shades, screens, or low-transmission glass. Provision should be made for decreasing or extinguishing supplementary artificial lighting when there is adequate daylight. Consideration should be given to use of photoelectric-cell sensors with dimmers for control of the artificial lighting.

Maximum use of daylighting and other energy conservation measures are often essential, not only to meet the owner's construction budget but also to satisfy the energy budget, or limitations on power consumption, set by building codes or state or federal agencies.

15.19 SPECIAL ELECTRICAL SYSTEMS

Enhancing the functions performed by the power and lighting systems, the special systems in a building serve the life safety, communication, and security needs of a facility and its occupants.

15.19.1 Lightning Protection

The design of lightning protection systems should conform with the standards of the American National Standards Institute, the National Fire Protection Association (Bulletin No. 78, "Lightning Code") and Underwriters Laboratories (Standard UL96A "Master-Labeled Lightning Protection Systems").

In deciding whether to provide a building lightning protection system, designers should first perform a risk assessment in accordance with NFPA 78, "Lightning Protection Code." This evaluates such factors as building height, terrain, building construction, proximity to other buildings, type of occupancy, and isoceraunic level (frequency of thunderstorms). If the risk assessment warrants, a lightning protection system is designed.

Protection for a building may be accomplished by several methods. An installation recognized by NFPA and UL consists of lightning rods or air terminals placed around the perimeter of the roof and on vertical projections, such as chimneys and the ridge of a peaked roof. These air terminals are bonded together and to a copper or aluminum conductor that extends down to a good electrical ground. There are also two less conventional types of installation that can be used. One involves the use of an air terminal containing a low-level radioactive source that produces a stream of ionized particles. This creates a low-resistance path that draws a lightning stroke to the air terminal, where it can be safely discharged to ground. The other type uses thousands of small air terminals spread along the high points of a structure to constantly dissipate any electrical charge in the air before it can build up high enough to induce a lightning stroke.

The electrical wiring in a building is especially susceptible to the effects of a lightning stroke. To minimize these effects, a multilevel protective approach is used. Lightning arresters, which are connected from the phase wires to ground, are provided on the utility company lines and at the various voltage levels down to utilization voltage. Usually of the metallic oxide varistor (MOV) type, the arresters present a very high resistance to ground at normal system voltage, but quickly collapse to zero resistance during a lightning discharge, dissipating the discharge to ground.

At sensitive electronic loads, it is necessary to provide a higher level of protection against the effects of lightning and other voltage disturbances by providing transient-voltage surge suppressors (TVSS). These devices utilize silicon-controlled rectifiers

(SCR) or combinations of SCRs and MOVs that closely limit the peak surge voltage and react within 5 nanoseconds to voltage surges.

15.19.2 Fire-Alarm Systems

These provide means of detecting a fire, initiating an alarm condition, either manually or via automatic detection, and responding to that alarm condition. A fire-alarm system consists of a central fire-alarm control panel; perhaps several remote subpanels; initiating devices, such as manual pull stations, smoke detectors, sprinkler-flow switches; and alarm devices, such as horns, gongs, and flashing lights. The control panel provides power to the system components and monitors the status of all of the initiating devices. It also monitors all fire-protection-system functions and supervises the condition of the wiring. In addition, the control panel provides outputs, under alarm conditions, to shut down air-conditioning fans, initiate smoke evacuation, close smoke doors, initiate elevator capture, release fire suppressants, activate alarm devices, and notify the fire department.

Larger systems are generally of the addressable type. The control panels are microprocessor based. Each device has a digital electronic identifier, or address. A control panel sequentially polls each device to check its status. This method allows as many as 30 devices to be connected to a single circuit and can greatly reduce the wiring costs of the fire-alarm system.

Design of the fire-alarm system must comply with the requirements of the National Fire Protection Association and local governing authorities. It is essential that fire-alarm systems be designed to interface with the HVAC system controls for unit smoke detection and shutdown and for smoke-exhaust-system control. Fire-alarm systems also should interface with the fire-protection system to monitor building sprinkler-system components and other fire-suppression systems. System design should also consider the building type and occupancy in selecting components and materials. Particular care should be taken in design of fire-alarm systems for high-rise buildings (over 75 ft high), which will require a firefighter's control panel, fire phone system, and voice-evacuation system as a minimum.

15.19.3 Communications Systems

These may include telephone, paging, and intercom systems. Telephone systems in large buildings generally have telephone service to a computerized business exchange (CBX), or switch, that controls the telephone system functions. It can offer numerous desirable features such as direct inward dialing, voice mail, speed dialing, system forward, conference, forward, message waiting, queuing, and transfer. The switch is located in a telephone service closet and requires power (preferably conditioned power) and air conditioning.

Telephone service is distributed via cables in conduit extending from telephone service room to telephone closets on each floor. Each closet should have a plywood backboard, for mounting equipment and punch-down boards, and two duplex receptacles. For distribution of telephone cables to their point of use, ladder-type cable trays can be routed above corridor ceilings from the telephone closets throughout each floor. Telephone cables must be suitable for running in cable trays and must be rated for use in air-handling spaces.

Telephone outlets consist of a single-gang outlet box with stainless steel cover plate, modular phone jack, and ¾ in electrical metallic tubing, which is run concealed to the cable tray. Telephones may be digital electronic type with programmable function buttons in addition to the 12-button tone pad, plus speaker phone and intercom features.

15.19.4 Intercom and Paging Systems

Intercom can be an integral function of the telephone system or a separate system.

Many buildings have a paging system of some sort. This can be a telephone system function, but it is most often a separate system consisting of receivers, playback equipment, amplifiers, speakers, telephone interface, and a microphone. The system may offer selective paging of certain areas, all-call paging of the entire facility, and background music.

15.19.5 Security Systems

These can range in sophistication from a combination lock (pad of numbered push buttons) or a simple card reader at the entry door to a comprehensive system integrating physical barriers, electronic access controls, surveillance, and intrusion detection systems (IDS). Although a card reader usually suffices for access control, high-value and high-security facilities often require biometric identification systems, such as retinal scan or hand geometry, for control of access.

Intrusion detection systems usually are of either of two types. One is a perimeter system, such as door switches, break-glass detectors, optical, or microwave beams, which creates an electronic envelope around a space. The other is a volumetric system, passive infrared or ultrasonic, which detects an intruder's presence in the protected space.

Closed-circuit television (CCTV) cameras may be provided to allow security persons to continuously watch for intruders at any of many locations from one point. Since several cameras may be required, video sequencers are provided that display the images from each of the cameras in turn on one or more monitors. Sequencers may be connected to the IDS to display and hold the image from a camera located where an alarm occurs.

A central security computer controls all of the security system functions. It monitors all of the devices and wiring, presents trouble and alarm messages to security staff, and keeps a historical record of all events such as authorized entries and alarms.

15.19.6 Television Distribution Systems

These are provided for apartment buildings, schools, and correctional facilities. There are two basic types, MATV (master antenna television), and CATV (community antenna television, usually referred to as cable TV). There are two principal differences between the systems. One is the source of signal; MATV systems require an antenna or a satellite dish to receive broadcast signals, whereas CATV systems receive signals from a cable system via a coaxial cable. The other difference is economic; MATV systems have a higher first cost to the builder but require only maintenance costs thereafter; CATV systems have lower first cost (the cable system may provide the wiring and equipment at no cost if the potential revenues are high enough) but require payment of a monthly fee by the users.

15.19.7 Data Network Systems

These are provided to allow free interchange of data between small groups of computers in a local area network (LAN) and between several LANs connected to a backbone network. Data transmission media may be wire, twisted-pair or coaxial cable, or fiber-optic cable. System topologies may be ring networks or radial (star) networks. Regardless of the type of network chosen, careful integration of the cabling system and hardware requirements into the building design is necessary.

Cabling closets are required for location of interface equipment and connector panels. A raceway system is required to distribute cables to the computer equipment. Often, the cabling conveyance used for the telephone system can be shared by the data network systems.

15.19.8 Intelligent Buildings

Design of an intelligent building integrates many or all of the systems listed above, plus HVAC systems, building operations, and even the building itself into a coordinated productivity tool for the occupants. The objective is to maximize efficiency today and adaptability for functional and technological changes tomorrow.

15.19.9 Special System Wiring

Circuits for the special systems previously discussed should meet the functional needs of the systems as well as the requirements of the National Electrical Code. The code divides these types of circuits into three classes and provides separate requirements for each.

Class 1 circuits are similar to power circuits in that they are permitted to carry up to 600 V and, in general, are subject to the same installation requirements. One exception is that use of No. 18 wire is permitted for Class 1 circuits. Special insulation and overcurrent protection devices are also permitted. Thermostats and other sensing devices controlling remote motor starts, available at 120 V, may be incorporated in Class 1 circuits and may use No. 18 wire, but with overcurrent protection not exceeding 20 A.

Class 2 and 3 circuits are limited to a maximum of 150 V. The conductors may not be placed in the same raceway as power circuits. Class 2 wire should provide protection against both fire and electric shock, whereas Class 3 wire should offer protection primarily against fire.

15.19.10 Power Distribution Management System

A power distribution management system (PDMS) is an integrated system of hardware and software that allows building operating personnel to monitor and control the building power-distribution system from a single location. PDMS systems are available from several manufacturers, either integral to a new electrical system or as an add-on in an existing building. Electronic interface devices, located in substations, switchgear, motor control centers, etc., provide real-time monitoring of system quantities like voltage, current, power factor, and frequency. They also enable remote control of circuit breakers, transfer switches, and starters. One or more remote terminal units (RTU) are provided in the field to tie-in the interface devices to the central processing unit (CPU) via a data highway. The data highway may use twisted-pair wiring, fiberoptic cable, or another transmission medium. It is also possible to interface the PDMS to other building systems, such as process-control or building automation systems.

The CPU is generally a personal computer (PC) which contains the PDMS software and, through its keyboard and screen, acts as the operator interface to the system. A single-line diagram representing the power distribution system is programmed into the CPU and is used by the operator to access system information, identify system faults, or remotely operate electrical equipment. The software allows continuous monitoring and archival storage of all electrical data. It also offers reporting functions that can be used to maximize operating efficiency and forecast impending faults. Should a fault occur, an alarm appears on the screen showing its location on the single-line diagram and all pertinent data. This information can be used to effect repairs and restore service quickly. Hand-held remote programmers can be connected to RTU's to adjust system settings in the field or to help repair efforts.

15.20 ELECTRICAL SYSTEMS BIBLIOGRAPHY

From Illuminating Engineering Society of North America, 345 East 47th St., New York, NY 10017:
"Lighting Handbook"
"Recommended Lighting Practices," RP-1 et al.
"Energy Management Series," EMS-1 et al.
From National Fire Protection Association, 1 Batterymarch Park, Quincy, MA 02269:
"National Electrical Code"
"National Electrical Code Handbook"
From John Wiley & Sons, Inc., New York:
B. Stein, et al., "Mechanical and Electrical Equipment for Buildings"
From McGraw-Hill, Inc., New York:
T. Croft and W. Summers, "American Electrician's Handbook"
M. D. Egan, "Concepts for Lighting in Architecture"
D. G. Fink and H. W. Beaty, "Standard Handbook for Electrical Engineers"
A. E. Fitzgerald and C. Kingsley, Jr., "Electric Machinery"
T. Gonen, "Electric Power Distribution System Engineering"
A. Kusko, "Emergency Standby Power Systems"
E. C. Lister, "Electrical Circuits and Machines"
J. F. McPartland and B. J. McPartland, "McGraw-Hill's National Electrical Code Handbook"
L. Watson, "Lighting Design Handbook"

SECTION SIXTEEN
VERTICAL CIRCULATION

Steven D. Edgett
Edgett-Williams Consulting Group, Mill Valley, California

Circulation, as usually applied in architecture, is the movement of people and goods between interior spaces in buildings and to entrances and exits. Safe, convenient, rapid circulation is essential for all buildings under both normal and emergency conditions. Such circulation may be channeled through any of several different types of passageways, such as lobbies, corridors, ramps, stairways, and elevator hoistways. General requirements for these have been discussed in previous sections. This section presents in more detail design and construction considerations in provision of means of vertical circulation, the movement of people and goods between floors of multistory buildings.

Vertical circulation of traffic in a multistory building is the key to successful functioning of the design, both in normal use and in emergencies. In fact, location of elevators or stairs strongly influences the floor plan. So in the design of a building, much thought should be given to the type of vertical circulation to be provided, number of units needed, and their location, arrangement, and design.

Traffic may pass from level to level in a multistory building by ramps, stairs, elevators, or escalators. The powered equipment is always supplemented by stairs for use when power is shut off, or there is a mechanical failure, or maintenance work is in progress, or in emergencies. In addition to conventional elevators, other types of human lifts are occasionally installed in residences, factories, and garages. For moving small packages or correspondence between floors, dumbwaiters, chutes, pneumatic tube systems, powered track conveyors, or vertical conveyors also may be installed. Ladders may be used for occasional access to attics or roofs.

16.1 CLASSIFICATION OF VERTICAL CIRCULATION SYSTEMS

Vertical circulation systems may be divided into two classes. Class I systems are intended for movement of both people and goods and include ramps, stairs, escalators, and elevators. Class II systems, including dumbwaiters and vertical conveyors, in contrast, may not be used for movement of people.

Class I systems may be divided into two subclasses, A and B. Class IA systems can be used by people both under normal and emergency conditions as a means of egress. This class includes ramps, stairs, and escalators (powered stairs) that meet requirements for means of egress specified in building codes or the National Fire Protection Association "Life Safety Code" (see Art. 3.5.10). Systems not acceptable as an emergency means of egress comprise Class IB. (Such systems nevertheless may be used for emergency evacuation of a building, but the capacity of Class IA systems alone must be sufficient for rapid, safe evacuation of the maximum probable building population.)

16.2 RAMPS

When space permits, a sloping surface, or ramp, can be used to connect different levels or floors (Fig. 16.1). As a means of saving space in some garages, every floor serves as a ramp. Each floor is split longitudinally, each section sloping gradually in opposite directions to meet the next level above and below.

Ramps are especially useful when large numbers of people or vehicles have to be moved from floor to floor. So they are frequently adopted for public buildings, such as railraod stations, stadiums, and exhibition halls. And they are either legally required or highly desirable for all buildings, especially to accommodate persons in wheelchairs. In all cases, ramps should be constructed with a nonslip surface.

Ramps have been built with slopes up to 15% (15 ft in 100 ft), but 8% is a preferred maximum. Some idea of the space required for a ramp may be obtained from the following: With the 8% maximum slope and a story height of, say, 8 ft, a ramp connecting two floors is 100 ft long (Fig. 16.1*a*). The ramp need not be straight for the whole distance, however. It can be curved, zigzagged (Fig. 16.1*b*), or spiraled. Level landings, with a length of at least 44 in in the direction of travel, should be provided at door openings and where ramps change slope or direction abruptly. Ramps and landings should be designed for a live load of at least 100 lb/ft^2. Railings should be designed for a load of 200 lb applied downward or horizontally at any point of the handrail or for a horizontal thrust of 50 lb/ft at top of

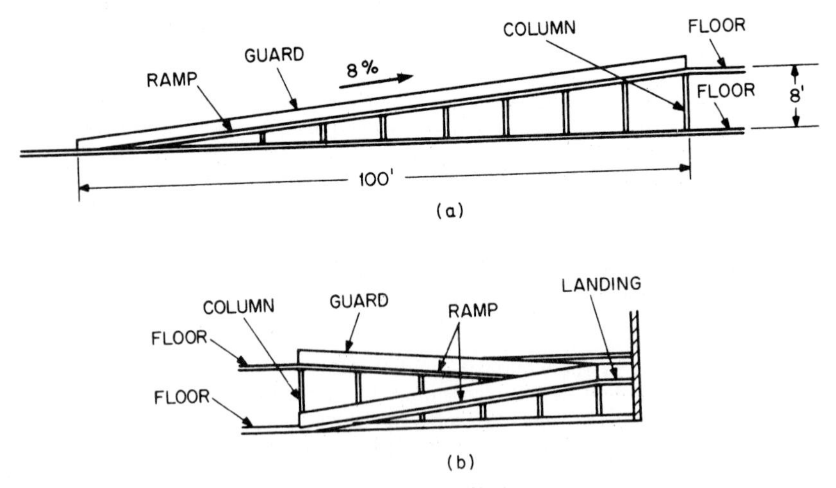

FIGURE 16.1 Types of ramps: (*a*) straight ramp; (*b*) zigzag ramp.

rail. Guards higher than the minimum required guard height of 42 in should be designed for 50 lb/ft applied 42 in above the floor.

Inside Ramps. Local building codes and the National Fire Protection Association "Life Safety Code" contain general requirements for acceptability of a ramp as a means of emergency egress (see Art. 5.10). Egress ramps are classified as Class A or Class B. The latter may be as narrow as 30 in, whereas Class A must be at least 44 in wide. (This width can accommodate two adults abreast.) Also, the "Life Safety Code" restricts Class A ramps to slopes of 10% or less and Class B ramps to slopes of not more than 1 in 8. In addition, for Class B, the vertical distance between landings may not exceed 12 ft, but no limit is placed on this distance for Class A. Building codes usually require Class A ramps only for places of assembly of more than 1000 persons. For other types of occupancy, codes may permit the choice of ramp class to be based on emergency exit capacity required.

The capacity, in persons per 22-in unit of ramp width, may be taken as 100 in the downward direction and 60 in the upward direction for Class B ramps. For Class A ramps, the capacity may be taken as 100 persons per unit of width in either direction.

To be acceptable as a means of egress, a ramp inside a building more than three stories high or a building of noncombustible or fire-resistant construction is required to be of noncombustible construction. The ramp also should be protected by separation from other parts of the building in the same way as other means of egress. There should be no enclosed usable space, such as closets, under the ramp, nor should the open space under the ramp be used for any purpose. (Other enclosed ramps, however, are permitted to be located under the ramp.)

For all inside ramps, guards—vertical protective barriers—should be provided along the edges of ramps and along the edges of floor openings over ramps, to prevent falls over the open edges. Requirements for type of construction and minimum height for such barriers are the same as those for stairs (see Art. 16.3). Handrails are required only for Class B ramps.

Outside Ramps. A ramp permanently installed on the outside of a building is acceptable as a means of egress if the life-safety requirements for inside egress ramps are met. For outside ramps more than three stories high, however, guards along ramp edges should be at least 4 ft high. Also, for such ramps, provision should be made to prevent accumulations of snow or ice.

Powered Ramps. In some buildings, such as air terminals, in which pedestrians have to be moved speedily over long distances, traffic may be transported on a moving walk, a type of passenger-carrying powered device on which passengers stand or walk. In the moving walk, the treadway, guards, and handrails are continuous and travel parallel to the direction of motion, which may be horizontal or on a slope up to 15°. (For greater slopes, an escalator should be used. See Art. 16.4.) Although moving walks can transport passengers at speeds up to 180 ft/min, speeds are generally between 90 and 120 ft/min.

Inclined moving walks are classified as powered ramps. Such ramps are acceptable as a means of egress if they meet the egress requirements of stationary ramps. (Moving walks are acceptable if they meet the requirements for exits. See Art. 3.10.5.) Powered ramps, however, must be incapable of operation in the direction opposite to normal exit travel.

The installation of moving walks and powered ramps should meet the requirements of the "American National Standard Safety Code for Elevators, Dumbwaiters, Escalators and Moving Walks," ANSI A17.1.

Basically, moving walks and powered ramps consist of a grooved treadway moved by a driving machine; a handrail on each side of the treadway that moves at the same speed as the treadway; balustrades, or guards, that enclose the treadway on each side and support the handrails; brakes; control devices; and threshold plates at the entrance to and the exit from the treadway. The purpose of the threshold plates is to facilitate smooth passage of passengers between treadway and landing. The plates are equipped with a comb, or teeth, that mesh with and are set into grooves in the treadway in the direction of travel. Their purpose is to provide firm footing and to prevent things from becoming trapped between the treadway and the landing.

The treadway may be constructed in one of the following ways:

1. Belt type—a power-driven continuous belt.
2. Pallet type—a series of connected, power-driven pallets. (A pallet is a short, rigid platform, which, when joined to other pallets, forms an articulated treadway.)
3. Belt pallet type—a series of connected, power-driven pallets to which a continuous belt is fastened.
4. Edge-supported belt type—a belt supported near its edges by rollers in sequence.
5. Roller-bed type—a treadway supported throughout its width by rollers in sequence.
6. Slider-bed type—a treadway that slides on a supporting surface.

Powered ramps resemble escalators in construction (see Art. 16.4). For example, both types of transporters are supported on steel trusses. The driving machine may be connected to the main drive shaft by toothed gearing, a coupling, or a chain. Movement of the treadway and handrails can be halted by an electrically released, mechanically applied brake, located either on the driving machine or on the main drive shaft and activated automatically when a power failure occurs, when the treadway or a handrail breaks, or when a safety device is actuated. For moving walks and ramps, safety devices required include switches for starting, emergency stopping, and maintenance stopping, and a speed governor that will prevent the treadway speed from exceeding 40% more than the maximum design speed.

Balustrades should be at least 30 in high, measured perpendicular to the treadway. Hand or finger guards are installed where the handrails enter the balustrades. The handrails should extend at normal height at least 12 in beyond each end of the exposed treadway, to facilitate entry and exit of passengers from or onto a level landing.

Information on passenger capacity of a moving walk or powered ramp should be obtained from the manufacturer. The capacity depends on treadway width and speed. Standard widths are 24, 32, and 40 in. With level entry and exit, ramp speeds generally are a maximum of 180 ft/min for slopes up to 8° and 140 ft/min for slopes between 8° and 15°.

("Life Safety Code," National Fire Protection Association, Batterymarch Park, Quincy, MA 02269.)

"American National Standard Safety Code for Elevators, Dumbwaiters, Escalators and Moving Walks," A17.1, and "Making Buildings and Facilities Accessible to and Usable by the Physically Handicapped," A117.1, American National Standards Institute, New York, NY 10018.)

16.3 STAIRS

Less space is required for stairs than for ramps, because steeper slopes can be used. Maximum slope of stairs for comfort is estimated to be about 1 on 2 (27°), but this angle frequently is exceeded for practical reasons. Exterior stairs generally range in slope from 20° to 30°, interior stairs from 30° to 35°.

16.3.1 Types of Stairs

Generally, stairs are of the following types: straight, circular, curved, or spiral, or a combination.

Straight stairs are stairs along which there is no change in direction on any flight between two successive floors. There are several possible arrangements of straight stairs. For example, they may be arranged in a **straight run** (Fig. 16.2*a*), with a single flight between floors, or a series of flights without change in direction (Fig. 16.2*b*). Also, straight stairs may permit a change in direction at an immediate landing. When the stairs require a complete reversal of direction (Fig. 16.2*c*), they are called **parallel stairs.** When successive flights are at an angle to each other, usually 90° (Fig. 16.2*d*), they are called **angle stairs.** In addition, straight stairs may be classified as **scissors stairs** when they comprise a pair of straight runs in opposite directions and are placed on opposite sides of a fire-resistive wall (Fig. 16.2*e*).

Circular stairs when viewed from above appear to follow a circle with a single center of curvature and large radius.

Curved stairs when viewed from above appear to follow a curve with two or more centers of curvature, such as an ellipse.

Spiral stairs are similar to circular stairs except that the radius of curvature is small and the stairs may be supported by a center post. Overall diameter of such stairs may range from 3 ft 6 in to 8 ft. There may be from 12 to 16 winder treads per complete rotation about the center.

16.3.2 Stairway Components

Among the principal components of a stairway are

Flight. A series of steps extending from floor to floor, or from a floor to an intermediate landing or platform.

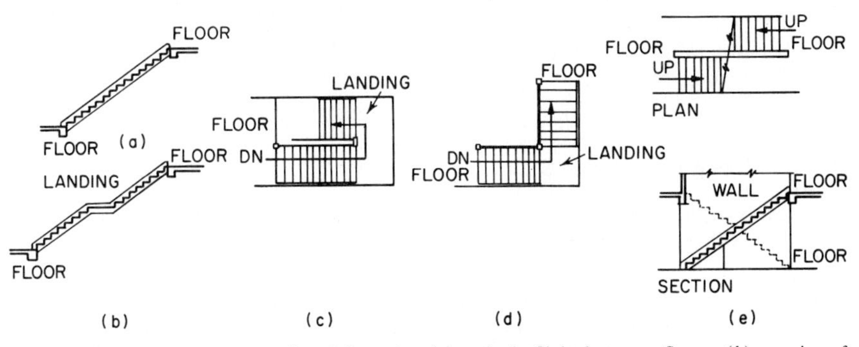

FIGURE 16.2 Arrangement of straight stairs: (*a*) a single flight between floors; (*b*) a series of flights without change in direction; (*c*) parallel stairs; (*d*) angle stairs; (*e*) scissors stairs.

Guard. Protective vertical barrier along edges of stairways, balconies, and floor openings.

Landings (platforms). Used where turns are necessary or to break up long climbs. Landings should be level, as wide as the stairs, and at least 44 in long in the direction of travel.

Step. Combination of a riser and the tread immediately above.

Rise. Distance from floor to floor.

Run. Total length of stairs in a horizontal plane, including landings.

Riser. Vertical face of a step. Its height is generally taken as the vertical distance between treads.

Tread. Horizontal face of a step. Its width is usually taken as the horizontal distance between risers.

Nosing. Projection of a tread beyond the riser below.

Soffit. Underside of a stair.

Header. Horizontal structural member supporting stair stringers or landings.

Carriage. Rough timber supporting the steps of wood stairs.

Stringers. Inclined members along the sides of a stairway. The stringer along a wall is called a wall stringer. Open stringers are those cut to follow the lines of risers and treads. Closed stringers have parallel top and bottom, and treads and risers are supported along their sides or mortised into them. In wood stairs, stringers are placed outside the carriage to provide a finish.

Railing. Framework or enclosure supporting a handrail and serving as a safety barrier.

Baluster. Vertical member supporting the handrail in a railing.

Balustrade. A railing composed of balusters capped by a handrail.

Handrail. Protective bar placed at a convenient distance above the stairs for a handhold.

Newel Post. Post at which the railing terminates at each floor level.

Angle Post. Railing support at landings or other breaks in the stairs. If the angle post projects beyond the bottom of the stringers, the ornamental detail formed at the bottom of the post is called the **drop.**

Winders. Steps with tapered treads in sharply curved stairs.

Headroom. Minimum clear height from a tread to overhead construction, such as the ceiling or next floor, ductwork, or piping.

16.3.3 Design Loads for Stairs

Stairs and landings should be designed for a live load of 100 lb/ft^2 or a concentrated load of 300 lb placed to produce maximum stresses.

Guards. To prevent people from falling over edges of stairs and landings, barriers, called guards, should be placed along all edges and should be at least 42 in high. These should support 2-in-diameter handrails, which should be set 30 to 34 in above the intersections of treads and risers at the front of the steps.

Interior stairs more than 88 in wide should have intermediate handrails that divide the stairway into widths of not more than 88 in, preferably into a nominal multiple of 22 in. Handrails along walls should have a clearance of at least 1½ in.

Guards should be designed for a horizontal force of 50 lb/ft, applied 42 in above the floor, or for the force transmitted by the handrail, whichever is greater. Handrails should be capable of supporting a load of at least 200 lb, downward or horizontally.

16.3.4 Dimensions for Stairs

Ample headroom should be provided not only to prevent tall people from injuring their heads, but to give a feeling of spaciousness. A person of average height should be able to extend his hand forward and upward without touching the ceiling above the stairs. Minimum vertical distance from the nosing of a tread to overhead construction should never be less than 6 ft 8 in and preferably not less than 7 ft.

Stairway Width. Width of a stairway depends on its purpose and the number of persons to be accommodated in peak hours or emergencies. Generally, the minimum width that can be used is specified in the local building code. For example, for interior stairs, clear width may be required to be at least 36 in in one- and two-family dwellings, and 44 in in hotels, motels, apartment buildings, industrial buildings, and other types of occupancy.

Step Sizes. Risers and treads generally are proportioned for comfort and to meet accessibility standards for the handicapped, although sometimes space considerations control or the desire to achieve a monumental effect, particularly for outside stairs of public buildings. Treads should be 11 to 14 in wide, exclusive of nosing. Treads less than 11 in wide should have a nosing of about 1 in. The most comfortable height of riser is 7 to 7½ in. Risers less than 4 in and more than 8 in high should not be used. The steeper the slope of the stairs, the greater the ratio of riser to tread. Among the more common simple formulas generally used with the preceding limits are:

1. Product of riser and tread must be between 70 and 75.
2. Riser plus tread must equal 17 to 17.5.
3. Sum of the tread and twice the riser must lie between 24 and 25.5.

In design of stairs, account should be taken of the fact that there is always one less tread than riser per flight of stairs. No flight of stairs should contain less than three risers.

16.3.5 Number of Stairways Required

This is usually controlled by local building codes. This control may be achieved by setting a minimum of two exits per floor, a restriction on the maximum horizontal distance from any point on a floor to a stairway, or a limitation on the maximum floor area contributory to a stairway. In addition, codes usually have special provisions for assembly buildings, such as theaters and exhibition halls. Restrictions usually also are placed on the maximum capacity of a stairway. For example, the National Fire Protection Association "Life Safety Code" sets a maximum capacity for stairways of 60 persons per 22-in unit of width, up or down.

16.3.6 Curved Stairways

Winders should be avoided when possible, because the narrow width of tread at the inside of the curve may cause accidents. Sometimes, instead, **balanced steps** can be used. Instead of radiating from the center of the curve, like winders, balanced steps, though tapered, have the same width of tread along the line of travel as the straight portion of the stairs. (Line of travel in this case is assumed to be about 20 in from the rail on the inside of the curve.) With balanced steps, the change in angle is spread over a large portion of the stairs.

16.3.7 Emergency Egress Stairway

In many types of buildings, interior exit stairways must be enclosed with walls having a fire-resistance rating, to prevent spread of smoke and flames. Wall construction and ratings must be in accordance with local code requirements. Openings in the walls should be protected by approved, self-closing fire doors. Stairs in buildings required by the code to be of fire-resistant construction should be completely made of noncombustible materials. Open space under stairs to be used as a means of egress should not be used for any purpose, including closets, except for another flight of stairs.

In buildings requiring such egress stairways, an alternative type of construction, called a **smokeproof tower,** may be used. A smokeproof tower is a continuous, vertical, fire-resistant enclosure protecting a stairway from fire or smoke that may develop elsewhere in a building. The intent is to limit the entrance into the stairway of products of combustion so that during a 2-h period the tower air will not contain smoke or gases with a volume exceeding 1% of the tower volume. All components of the tower should be made of noncombustible materials, and the enclosure should have a 2-h fire rating. Walls between the stairs and the building interior should not have any openings. If the exterior wall of the tower will not be subjected to a severe fire-exposure hazard, however, that wall may incorporate fixed or automatic fire windows.

Access to a smokeproof tower should be provided in each story through vestibules open to the outside on an exterior wall, or from balconies on an exterior wall, neither exposed to severe fire hazards. Doors should be at least 40 in wide, self-closing, and provided with a viewing window of clear, wired glass not exceeding 720 in^2 in area. It also is wise to incorporate some means of opening the top of the shaft, either with a thermally operated device or with a skylight, to let escape any heat that might enter the tower from a fire. Exits at the bottom of smokeproof towers should be directly to the outdoors, where people can remove themselves quickly to a safe distance from the building.

Stairs outside a building are acceptable as a required fire exit instead of inside stairs, if they satisfy all the requirements of inside stairs. Where enclosure of inside stairs is required, however, outside stairs should be separated from the building interior by fire-resistant walls with fire doors or fixed wire glass windows protecting openings. Some building codes limit the height of outside stairs to a maximum of six stories or 75 ft.

Fire escapes, outside metal-grating stairs, and landings attached to exterior walls with unprotected openings were acceptable at one time as required exits, but are generally unacceptable for new construction.

See also Art. 3.5.10.

("Life Safety Code," National Fire Protection Association, Batterymarch Park, Quincy, MA 02269.)

16.3.8 Wood Stairs

In wood-frame buildings, low nonfireproof buildings, and one- and two-family houses, stairs may be constructed of wood (Fig. 16.3*a*). They may be built in place or shop fabricated.

Construction of a built-in-place stairs starts with cutting of carriages to the right size and shape to receive the risers and treads (Fig. 16.3*b*). Next, the lower portion of the wall stringer should be cut out at least ½ in deep to house the steps (Fig. 16.3*c*). The stringer should be set in place against the wall with the housed-out profile fitted to the stepped profile of the top of the carriage. Then, treads and risers should be firmly nailed to the carriages, tongues at the bottom of the risers fitting into grooves at the rear of the treads. Nosings are generally finished on the underside with molding.

If the outer stringer is an open stringer (Fig. 16.3*c*), it should be carefully cut to the same profile as the steps, mitered to fit corresponding miters in the ends of risers, and nailed against the outside carriage. Ends of the treads project beyond the open stringer.

If the outer stringer is a curb or closed stringer, it should be plowed out in the same way as the wall stringer to house the steps. Ends of the treads and risers should be wedged and glued into the wall stringer (Fig. 16.3*d*).

("Manual for House Framing," National Forest and Paper Association, 1250 Connecticut Ave., NW, Washington, DC 20036.)

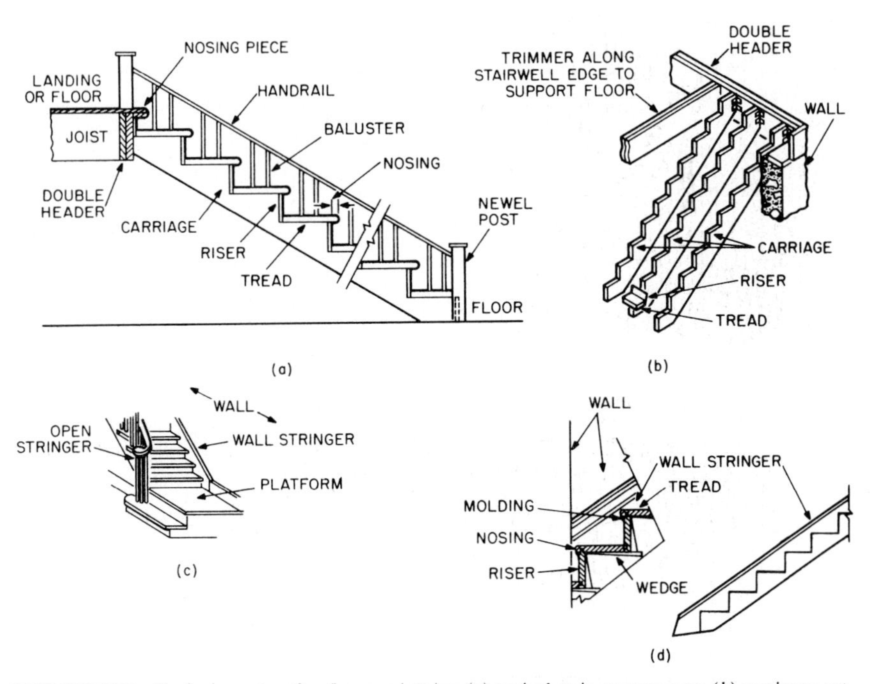

FIGURE 16.3 Typical construction for wood stairs: (*a*) typical stair components; (*b*) carriages cut to receive steps; (*c*) wall stringer cut to receive steps; (*d*) junction of steps with a closed stringer.

16.3.9 Steel Stairs

Cold-formed-steel or steel-plate stairs generally are used in fire-resistant buildings. They may be purchased from various manufacturers in stock patterns.

The steel sheets are formed into risers and subtreads or pans, into which one of several types of treads may be inserted (Fig. 16.4). Stringers usually are channel-shaped. Treads may be made of stone, concrete, composition, or metal. Most types are given a nonslip surface.

("Metal Stairs Manual," National Association of Architectural Metal Manufacturers, 600 S. Federal St., Chicago, IL 60605.)

16.3.10 Concrete Stairs

Depending on the method of support provided, concrete stairs may be designed as cantilevered or inclined beams and slabs (Fig. 16.5). The entire stairway may be cast in place as a single unit, or slab or T beams may be formed first and the steps built up later. Soffits formed with plywood or hardboard forms may have a smooth-enough finish to make plastering unnecessary. Concrete treads should have metal nosings to protect the edges. Stairs also may be made of precast concrete.

16.4 ESCALATORS

Escalators, or powered stairs, are used when it is necessary to move large numbers of people from floor to floor. They provide continuous movement of persons and

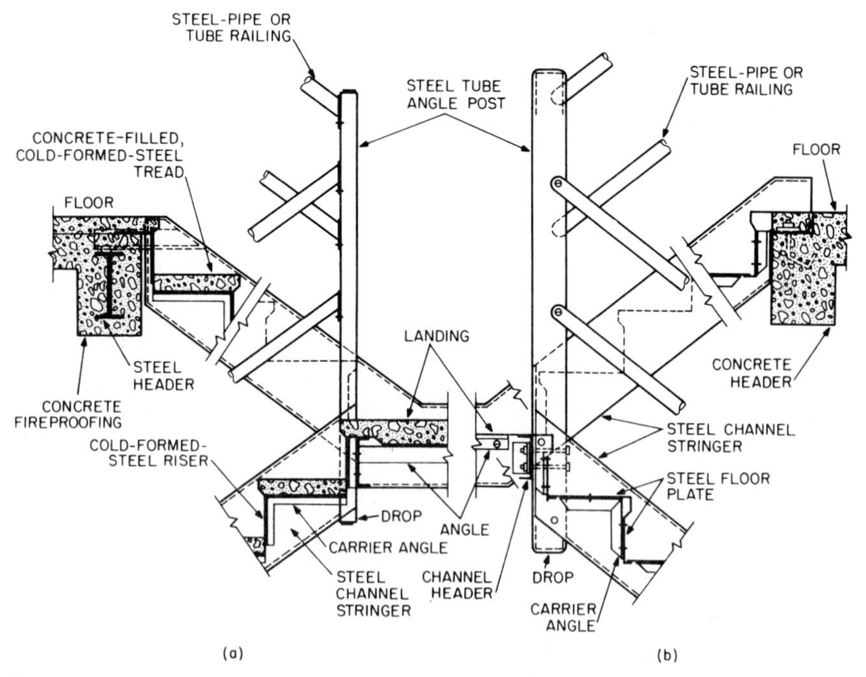

FIGURE 16.4 Types of metal stairs: (*a*) stairs made of cold-formed steel; (*b*) stairs made of steel plate.

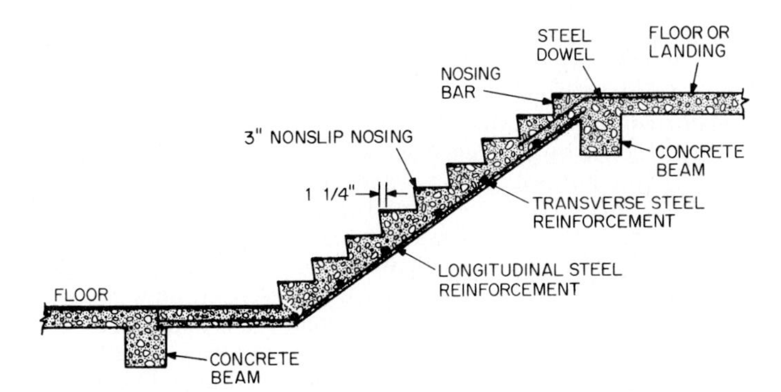

FIGURE 16.5 Reinforced concrete stairs.

can thus remedy traffic conditions that are not readily addressed by elevators. Escalators should be viewed as preferred transportation systems whenever heavy traffic volumes are expected between relatively few floors. Escalators are used to connect airport terminals, parking garages, sports facilities, shopping malls, and numerous mixed-use facilities.

Although escalators generally are used in straight sections (Fig. 16.6), spiral escalators (Fig. 16.7) also are available. Although expensive due to manufacturing complexities, they offer distinct advantages to both the designer and user because of their unique semicircular plan form.

16.4.1 Components of an Escalator

An escalator resembles a powered ramp in construction (Art. 16.2). The major difference is that a powered ramp has a continuous treadway for carrying passengers, whereas the treadway of an escalator consists of a series of moving steps. As

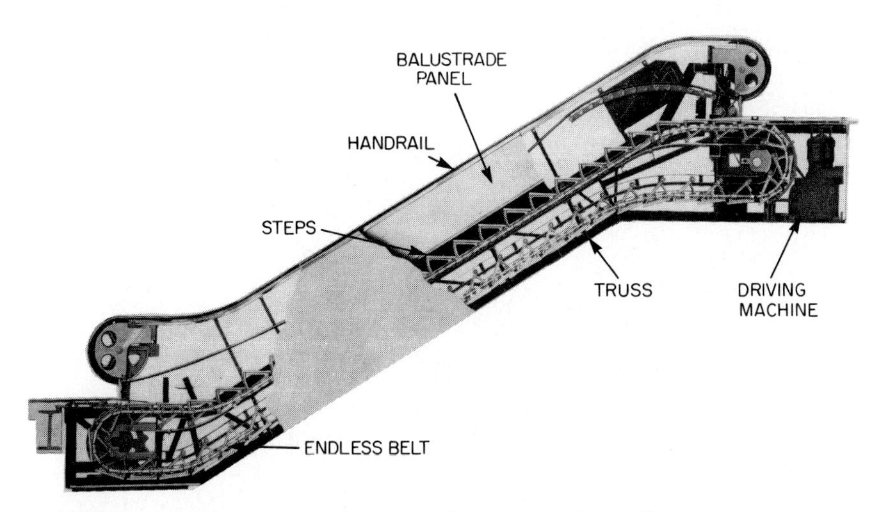

FIGURE 16.6 Details of a straight escalator. *(Courtesy Otis Elevator Co.)*

for a powered ramp, the installation of powered stairs should conform with the requirements of the "American National Standard Safety Code for Elevators, Dumbwaiters, Escalators and Moving Walks," ANSI A17.1.

An escalator consists of articulated, grooved treads and risers attached to a continuous chain moved by a driving machine and supported by a steel truss framework (Fig. 16.6). The installation also includes a handrail on each side of the steps that moves at the same speed as the steps; balustrades, or guards, that enclose the steps on each side and support the handrails; brakes; control devices; and threshold plates at the entrance to and the exit from the treadway. The purpose of the threshold plates is to facilitate smooth passage of passengers between the treadway and landing. The plates are equipped with a comb, or teeth, that mesh with and are set into grooves in the treadway in the direction of travel, so as to provide firm footing and to minimize the chance that items become trapped between treadway and the landing.

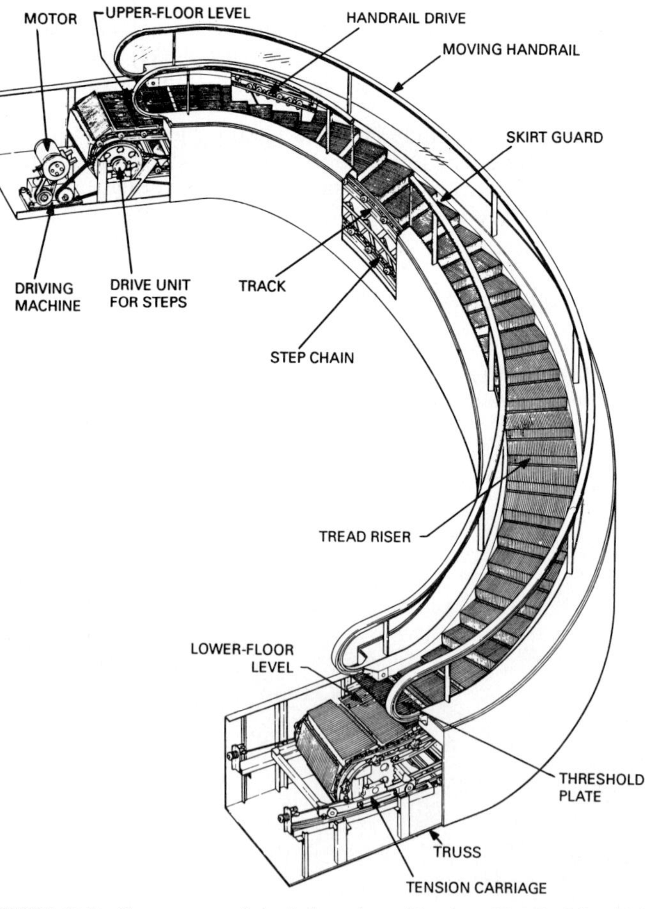

FIGURE 16.7 Components of a spiral escalator (developed by the Mitsubishi Electric Corporation).

Each step is formed by a grooved tread portion connected to a curved and grooved riser. The tread and riser assembly is either a single die-cast piece or is assembled to a frame. Both are suspended on resilient rollers whose axles are connected to the step chain that moves the steps. The step rollers ride on a set of tracks attached to the trussed framework. The tracks are shaped to allow the step tread to remain horizontal throughout its exposed travel.

16.4.2 Dimensions for Escalators

ANSI A17.1 sets the following limitations on escalator steps (Fig. 16.8):

Minimum depth of tread in direction of travel—15¾ in

Maximum rise between treads—8½ in

Minimum width of tread—16 in

Maximum width of tread—40 in

Maximum clearance between tread and adjacent skirt panel—⅜ in

Maximum distance between handrail centerlines—width between balustrades plus 6 in with not more than 3 in on either side of the escalator (see Fig. 16.8*b*)

The escalator width is measured on the incline between balustrades, as indicated in Fig. 16.8*b*.

It should be at least as wide as the step but not more than 13 in wider than the step.

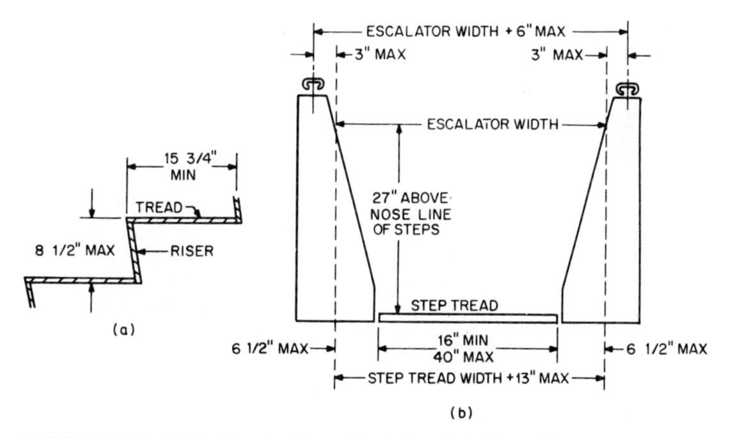

FIGURE 16.8 Limitations on dimensions of a straight escalator.

16.4.3 Safety Devices for Escalators

To provide a firm footing, treads are grooved in the direction of travel. The grooves mesh with the combs or teeth of the threshold plates at top and bottom of the escalator.

The handrails, which move in synchronization with the steps, should be between 30 and 34 in above the treads. The handrails should extend at normal height at least 12 in beyond the line of points of the combplate teeth. The balustrades carrying the handrails and acting as a guard to prevent passengers from falling off the sides of the moving steps should be designed to resist simultaneous application of a horizontal load of 40 lb/ft and a vertical load of 50 lb/ft, both applied to the top of the balustrades.

The driving machine may be connected to the main drive shaft by toothed gearing, a coupling, or a chain. Step movement is halted by an electrically released, mechanically applied brake, located either on the driving machine or on the main drive shaft. The brake is activated automatically when a power failure occurs, when the treadway or a handrail breaks, or when a safety device is activated.

Safety devices required for escalators include switches for starting, emergency stopping, and maintenance stopping and an electromechanical speed governor that will prevent the step speed from exceeding the maximum design speed. An emergency stop button, protected against accidental activation, is required to be set in the right-hand (when facing the escalator) newel at the top and bottom landings.

16.4.4 Escalator Speeds and Capacities

Escalators typically operate at 90 or 120 ft/min, as needed for peak traffic, and are reversible in direction. Slope of the stairs is standardized at 30° in the United States, although inclines of both 30° and 35° are used in other parts of the world.

Standard escalator widths are 32 and 48 in. Manufacturers rate their 90-ft/min units at corresponding capacities of 5000 and 8000 persons per hour, although observed capacities, even in heavy traffic, rarely exceed 2000 and 4000 persons per hour, respectively. Although 120-ft/min escalators will move about 30% more volume, they are rarely specified because of the potential for adverse litigation.

16.4.5 Planning for Escalators

The location of moving stairs should be selected only after a careful study of potential traffic flow within the planned project. They should be installed where most attractive to traffic and where convenient for passengers. The facility should be designed and signed in a manner that makes it apparent where the visitor will find the escalator. Since escalators are devices that will fail on occasion, the designer must provide alternative transportation (usually adjacent stairs) for times when the escalator is unavailable for passenger use. More importantly, where escalators will be operating at capacity as a result of specific programmatic considerations, the designer must plan alternative routing for times when one or more escalators is under repair. In retail applications, marketing needs generally motivate selection of escalator locations.

In design of a new building, adequate space should be allotted for escalators. Generous areas should be provided at both loading and unloading areas. Special consideration should be given to the possibility of a disaster resulting at a constricted exit from an escalator when pedestrian traffic is restricted below the escalator's capacity in the path of travel. Similarly, planning of landing areas should consider both queueing space and what happens when an escalator is stopped for some reason while pedestrian traffic continues. In addition, before stacked escalators are planned for an arena, stadium, or other facility having exit peaks, the potential for pedestrian traffic jams should be carefully weighed. If exiting traffic is very heavy in a stacked escalator system, upper levels can easily fill lower-level escalators, creating a jam at the escalator entries and leaving little space for lower-level pedestrians.

For an escalator installation in an existing building, careful study should be made to determine the necessary alterations to assure adequate space and supports.

16.4.6 Structural Considerations in Escalator Installation

Floor-to-floor height should be taken into account in determining loads on supporting members. Generally for floor-to-floor heights of less than 20 ft, the escalator truss need be supported only at top and bottom. Increased vertical rise can create the need for intermediate support points. A structural frame should be installed around the escalator well to carry the floor and wellway railing.

Inasmuch as an escalator is a mechanical device, careful consideration should be given to the potential for noise and vibration in design of the escalator structural supports. Where necessary, the escalator can be mounted on vibration-isolating devices to help reduce noise and vibration.

16.4.7 Escalator Installation

Design of escalators permits a vertical variation of ½ in in the level of the supporting beams from the specified floor-to-floor height. The escalator is shimmed to bring it level. If variations in elevation exceed ½ in, installation is difficult and much time will be lost. To allow for variations in overall escalator length, truss extensions can be provided.

Trusses generally are brought to the job in three sections. There, they are assembled and raised into position with chain hoists, either through an elevator shaft or on the outside of the building. Typically, the escalator manufacturer does not furnish either the exterior truss cladding or the wellway railings and accessories. Because of the need for economy, escalator manufacturers design for minimal weight in the truss cladding. Hence, care should be taken to coordinate carefully the desired design with the escalator manufacturer.

Escalators usually are installed in pairs—one for carrying traffic up and the other for moving traffic down. The units may be placed parallel to each other in each story (Fig. 16.9), or crisscrossed (Fig. 16.10). Crisscrossed stairs generally are preferred because they are more compact, reducing walking distance between stairs at various floors to a minimum. The curved characteristic of the spiral escalator allows for several alternative arrangements (Fig. 16.11).

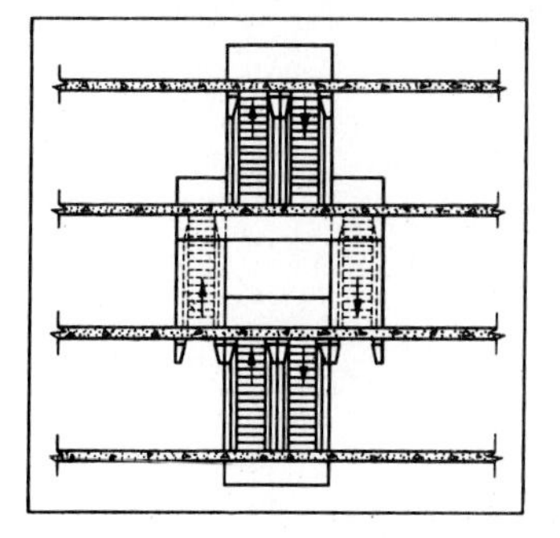

FIGURE 16.9 Parallel arrangement of up and down escalators.

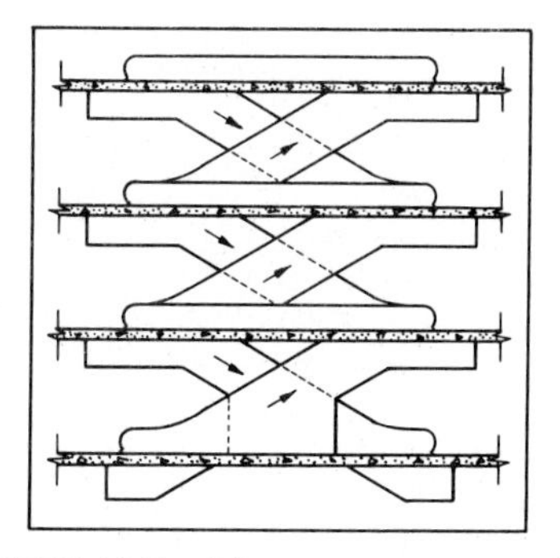

FIGURE 16.10 Crisscross arrangement of up and down escalators.

(a)

(b)

FIGURE 16.11 Arrangements of spiral escalators: (*a*) at main entrance or center of a building; (*b*) in a corner.

16.4.8 Fire Protection of Escalators

Escalators may be acceptable as required means of egress if they comply with the applicable requirements for exit stairs (Art. 16.3.7). Such escalators must be enclosed in the same manner as exit stairs. Escalators capable of reversing direction, however, may not qualify as required means of egress.

An escalator not serving as a required exit should have its floor openings enclosed or protected as required for other vertical openings. Acceptable protection, as an alternative, is afforded in buildings completely protected by a standard supervised sprinkler system by any of the following:

Sprinkler-vent method, a combination of automatic fire- or smoke-detection system, automatic air-exhaust system, and an automatic water curtain.

Spray-nozzle method, a combination of an automatic fire or smoke detection system and a system of high-velocity water-spray nozzles.

Rolling shutter method, in which an automatic, self-closing, rolling shutter is used to enclose completely the top of each escalator.

Partial enclosure method, in which kiosks, with self-closing fire doors, provide an effective barrier to spread of smoke between floors.

Escalator trusses and machine spaces should be enclosed with fire-resistant materials. Ventilation should be provided for machine and control spaces.

("Life Safety Code," National Fire Protection Association, Quincy, MA 02269; "American National Standard Safety Code for Elevators, Dumbwaiters, Escalators, and Moving Walks," A17.1, American National Standards Institute; G. R. Strakosch, "Vertical Transportation: Elevators and Escalators," John Wiley & Sons, Inc., New York.)

16.5 ELEVATOR INSTALLATIONS

An **elevator** is a hoisting and lowering mechanism equipped with a car or platform that moves along guides in a shaft, or hoistway, in a substantially vertical direction and that transports passengers or goods, or both, between two or more floors of a building. **Passenger elevators** are designed primarily to carry persons. **Hospital elevators** are also passenger elevators but employ special cars, suitable in size and shape for transportation of patients in stretchers or standard hospital beds and of attendants accompanying them. **Freight elevators** carry freight, which may be accompanied only by an operator and persons necessary for loading and unloading it.

Elevators are desirable in all multistory buildings for movement of passengers and freight. They may be required by local building codes for any buildings over two stories high or for transportation of disabled persons. Elevators, however, are not usually accepted as a means of egress, because no cohesive strategy has been established to assure proper operation of elevators in an emergency.

Most codes require automatic evacuation of all elevators if fire or smoke is detected on a served floor. These elevators can later be recaptured by emergency personnel. Nevertheless, elevators are vital for firefighting in a high-rise building. Also, they can be used for emergency evacuation of building occupants who cannot use the building stairs. The height of modern buildings makes it mandatory that elevators be included in emergency planning for fire or other disaster.

Most elevators are the roped electric or hydraulic type. For the roped electric elevator, the car is suspended from wire ropes and counterbalanced by a counterweight that mirrors the operation of the elevator. The electric elevator is moved via an electrically powered machine that drives a hardened steel traction **sheave** over which the wire ropes are suspended (Fig. 16.12*a* and *b*). Electric elevators are used exclusively in tall buildings and many low buildings (Art. 16.9). Hydraulic elevator cars (Fig. 16.12*c*) are raised and lowered by an oil pumping system, which actuates a plunger or piston (Art. 16.10). They are frequently used for passenger elevators serving up to five or six floors and for low-rise freight service. Their low performance when compared to electric-type elevators means that they cannot be substituted on a one-for-one basis and provide equivalent service. Where passenger-moving capability is paramount, the hydraulic elevator cannot compete with the more costly electric type.

Elevator installations should meet the requirements of the "American National Standard Safety Code for Elevators, Dumbwaiters, Escalators and Moving Walks," ANSI A17.1. Standardized elevator sizing has been developed by National Elevator Industries, Inc. (NEII). It is desirable that car sizes and shapes be in accord with NEII standards, such as "Elevator Engineering Standard Layouts" and "Suggested Minimum Passenger Elevator Requirements for the Handicapped," National Elevator Industry, Inc., 600 Third Avenue, New York, NY 10016.

Structural Considerations for Elevators. Elevators and related equipment, such as machinery, signal systems, ropes, and guide rails, are generally supplied and installed by the manufacturer. The general contractor has to guarantee the dimensions of the shaft and its freedom from encroachments. The owner's architect or engineer is responsible for the design and construction of components needed for supporting the plant, including buffer supports, machine-room floors, and guide-rail bracket supports. Magnitudes of loads generally are supplied by the manufacturer with a 100% allowance for impact.

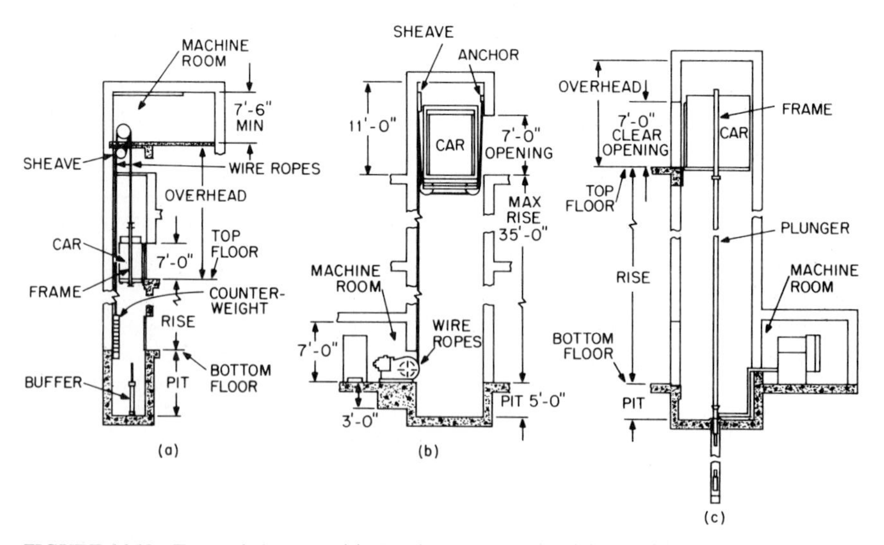

FIGURE 16.12 Types of elevators: (*a*) electric elevator with driving machine at top of hoistway; (*b*) electric elevator with driving machine in basement; (*c*) hydraulic elevator.

For design of machinery, sheave beams, and floor systems, unit stresses should not exceed 80% of those allowed for static loads in the design of usual building structural members. Importantly, deflections on machinery and sheave supporting structures may not exceed 1/1666 the span. This stiffness helps to minimize variations in leveling due to load-induced deflection. Where stresses due to loads other than elevator loads, supported on the beams or floor system exceed those due to elevator loads, 100% of the allowable unit stresses may be used.

Unit stresses, calculated without impact, in a steel guide rail or its reinforcement, caused by horizontal forces, should not exceed 15 ksi, and deflection should not exceed ¼ in. Guide-rail supports should be capable of resisting horizontal forces with a deflection of not more than ⅛ in.

(G. R. Strakosch, "Vertical Transportation: Elevators and Escalators," John Wiley & Sons, Inc., New York.)

16.6 DEFINITIONS OF ELEVATOR TERMS

(See also Figs. 16.12 to 16.16.)

Annunciator. An electrical device that indicates, usually by lights, the floors at which an elevator landing signal has been registered.

Buffer. A device for stopping a descending car or counterweight beyond its bottom terminal by absorbing and dissipating the kinetic energy of the car or counterweight. The absorbing medium may be oil, in which case the buffer may be called an **oil buffer,** or a spring, in which case the buffer may be referred to as a **spring buffer.**

Bumper. A device other than a buffer for stopping a descending car or counterweight beyond its bottom terminal by absorbing the impact.

Car. The load-carrying element of an elevator, including platform, car frame, enclosure, and car door or gate.

Car-Door Electric Contact. An electrical device for preventing normal operation of the driving machine unless the car door or gate is closed.

Car Frame. The supporting frame to which the car platform, guide shoes, car safety, and hoisting ropes or hoisting-rope sheaves, or the plunger of a hydraulic elevator are attached.

Car Platform. The structure on which the car and its floor are mounted.

Car Switch. A manual operating device in a car by which an operator actuates the control.

Control. The system governing the starting, stopping, direction of motion, acceleration, speed, and retardation of the car.

VVVF Control. A method of controlling the smooth starting and stopping of alternating-current motors, utilizing solid-state, *variable-voltage, variable-frequency* controls. This system is displacing dc motors for medium and high-speed elevators.

Generator-field control employs an individual generator for each elevator, with the voltage applied to a dc driving-machine motor adjusted by varying the strength and direction of the generator field.

Multivoltage control impresses successively on the armature of the driving-machine motor various fixed voltages, such as those that might be obtained from multicommutator generators common to a group of elevators.

Rheostatic control varies the resistance or reactance of the armature or the field circuit of the driving-machine motor.

Single-speed, alternating-current control governs a driving-machine induction motor that runs at a specified speed.

Two-speed alternating-current control governs a two-speed driving-machine induction motor, with motor windings connected to obtain various numbers of poles.

Dispatching Drive. A device that operates a signal in a car to indicate when the car should leave a designated floor or to actuate the car's starting mechanism when the car is at a designated floor.

Driving Machine. See *Machine.*

Emergency Stop Switch. A car-located device that, when operated manually, causes the car to be stopped by disconnecting electric power from the driving-machine motor.

Hoistway. A shaft for travel of one or more elevators. It extends from the bottom of the pit to the underside of the overhead machine room or the roof. A **blind hoistway** is the portion of the shaft that passes floors or other landings without providing a normal entrance.

Hoistway Access Switch. A switch placed at a landing to permit car operation with both the hoistway door at the landing and the car door open.

Hoistway-Door Electric Contact. An electrical device for preventing normal operation of the driving machine unless the hoistway door is closed.

Hoistway-Door Locking Device. A device for preventing the hoistway door or gate from being opened from the landing side unless the car has stopped within the landing zone.

Leveling Device. A mechanism for moving a car that is within a short distance of a landing toward the landing and stopping the car there. An **automatic maintaining, two-way, leveling device** will keep the car floor level with the landing during loading and unloading.

Machine (Driving Machine). The power unit for raising and lowering an elevator car.

Electric driving machines include an electric motor and brake, driving sheave or drum, and connecting gearing, belts, or chain, if any. A **traction machine** drives the car through friction between suspension ropes and a traction sheave. A **geared-drive machine** operates the driving sheave or drum through gears. A **gearless traction machine** has the traction sheave and the brake drum mounted directly on the motor shaft. A **winding-drum machine** has the motor geared to a drum on which the hoisting ropes wind. A **worm-geared machine** operates the driving sheave or drum through worm gears. A **helical-geared** machine operates the driving sheave through a helical-type gearbox.

Hydraulic driving machines raise or lower a car with a plunger or piston moved by a liquid under pressure in a cylinder.

Nonstop Switch. A device for preventing a car from making registered landing stops.

Operating Device. The car switch, push button, lever, or other manual device used to actuate the control.

Operation. The method of actuating the control.

Automatic operation starts the car in response to operating devices at landings, or located in the car and identified with landings, or located in an automatic starting mechanism, and stops the car automatically at landings. **Group automatic operation** starts and stops two or more cars under the coordination of a supervisory control system, including automatic dispatching means, with one button per floor in each car and up and down buttons at each landing. **Selective collective automatic operation** is a form of group automatic operation in which car stops are made in the order in which landings are reached in each direction of travel after buttons at those landings have been pressed. **Single automatic operation** has one button per floor in each car and only one button per landing, so arranged that after any button has been pressed, pushing any other button will have no effect on car operation until response to the first button has been completed.

Car-switch operation starts and stops a car in response to a manually operated car switch or continuous-pressure buttons in a car.

Parking Device. A device for opening from the landing side the hoistway door at any landing when the car is within the landing zone.

Pit. Portion of a hoistway below the lowest landing.

Position Indicator. Device displaying the location of a car in the hoistway.

Rise. See *Travel.*

Rope Equalizer. A device installed on a car or counterweight to equalize automatically the tensions in the hoisting ropes.

Runby. The distance a car can travel beyond a terminal landing without striking a stop.

Safety. A mechanical device attached to the counterweight or to the car frame or an auxiliary frame to stop or hold the counterweight or the car, whichever undergoes overspeed or free fall, or if the hoisting ropes should slacken.

Safety Bulkhead. In a cylinder of a hydraulic elevator, a closure, at the bottom of the cylinder but above the cylinder head, with an orifice for controlling fluid loss in case of cylinder-head failure.

Slack-Rope Switch. A device that automatically disconnects electric power from the driving machine when the hoisting ropes of a winding-drum machine become slack.

Terminal Speed-Limiting Device (Emergency). A device for reducing automatically the speed of a car approaching a terminal landing, independently of the car-operating device and the normal terminal stopping device if the latter should fail to slow the car as intended.

Terminal Stopping Device. Any device for slowing or stopping a car automatically at or near a terminal landing, independently of the car-operating device. A **final terminal stopping device,** after a car passes a terminal landing, disconnects power from the driving apparatus, independently of the operating device, normal terminal stopping device, or emergency terminal speed-limiting device. A **stop-motion switch,** or **machine final terminal stopping device,** is a final terminal stopping device operated directly by the driving machine.

Transom. One or more panels that close an opening above a hoistway entrance.

Travel (Rise). The vertical distance between top and bottom terminal landings.

Traveling Cable. A cable containing electrical conductors for providing electrical connections between a car and a fixed outlet in a hoistway.

Truck Zone. A limited distance above a landing within which the truck-zoning device permits movement of a freight-elevator car with its door or the hoistway door open.

Truck-Zoning Device. A device that permits a car operator to move, within a specified distance above a landing, a freight-elevator car with its door or the hoistway door open.

16.7 ELEVATOR HOISTWAYS

A hoistway is a shaft in which an elevator travels. To provide access to an elevator car, the shaft enclosure has openings, protected by doors with safety devices, at landings. In a pit at the bottom of the hoistway, buffers or bumpers must be installed to stop a descending car or counterweight beyond its normal limit of travel, by storing or by absorbing and dissipating its kinetic energy (Fig. 16.12). Construction of the hoistway and installation of the associated equipment should meet the requirements of the "American National Standard Safety Code for Elevators, Dumbwaiters, Escalators and Moving Walks," ANSI A17.1.

16.7.1 Hoistway Enclosure

The code requires that hoistways be enclosed throughout their height with fire-resistant construction, except for cases where no solid floors are penetrated. The enclosure should have a 2-h fire rating, and hoistway doors and other opening protective assemblies should have a 1½-h rating. Where fire-resistant construction is not required, laminated-glass curtain walls or unperforated metal, such as 18-ga sheet steel, should enclose the hoistway to a height of 6 ft above each floor and above the treads of adjacent stairways. Openwork enclosures may be used above that level, if openings are less than 2 in wide or high.

At the top of a hoistway, a metal or concrete floor should be provided (but is not necessary below secondary and deflection sheaves of traction-type driving machines located over the hoistway). If a driving machine is installed atop the hoistway, the floor should be level with or above the top of the beams supporting the machine. Otherwise, the floor should be set under the overhead sheaves. The floor should cover the entire top of the hoistway if its area would be 100 ft^2 or less. For larger hoistway cross-sectional areas, the floor should extend from the entrance to the machine space, at or above the level of the platform, for a distance at least 2 ft beyond the general contour of the driving machine, sheaves, or other equipment. In such cases, exposed floor edges should be protected with a toe board at least 4 in high and a railing at least 42 in high and conforming to the requirements of the "American National Standard Safety Code for Floor and Wall Openings, Railings and Toe Boards," ANSI A12.1.

16.7.2 Venting of Hoistways

In significant high-rise-building fires, the elevator hoistways have served as a flue for smoke and hot gases generated by fire. The prevailing thought has been that hoistway venting means could minimize the spread of smoke and hot gases throughout the building. As more has been learned about smoke movement in high-rise buildings, many alternatives have been developed to prevent migration of smoke from the fire floor to noninvolved floors of the structure. Among these alternatives are various systems for hoistway pressurization and mechanical-pressure sandwich systems, where building ventilating units are used to contain smoke during a fire. Although many codes continue to require specific means to address elevator hoistway venting, the overall design for smoke control in the building should be considered in design of elevator hoistways. Consideration for building occupants who may be threatened by fire requires designers to view the structure in a holistic fashion, where all systems can be integrated to maximize life-safety opportunities. The proposed design should be reviewed by the architect, mechanical engineer, and fire protection engineer to ensure that the finished result achieves goals set for the building's life-safety capabilities. The importance of proper elevator hoistway design in high-rise buildings cannot be overemphasized.

16.7.3 Machine Rooms

Construction of enclosures of spaces containing machines, control equipment, and sheaves should be equivalent to that used for the hoistway enclosure. To dissipate machinery heat and to preserve computerized elevator control equipment, the spaces should be air conditioned.

Due to the dangers involved in elevator machinery, nonelevator equipment is generally not permitted in elevator machine rooms. If the driving machine is located at the top of the hoistway, other machinery and equipment for building operation may also be installed in the machine room but must be separated from the elevator equipment by a substantial metal grille at least 6 ft high. The entrance to the elevator machine room should be guarded by a self-closing, self-locking door. If, however, the driving machine is not at the top of the hoistway, only elevator equipment is permitted in the machine room.

In machine rooms at the top of the hoistway, headroom of at least 7 ft above the floor should be provided. For spaces containing only overhead, secondary, or deflecting sheaves, headroom may be only 3½ ft, but 4½ ft is required if the spaces also contain overspeed governors, or other equipment.

16.7.4 Hoistway Doors

Each opening in a hoistway enclosure for access to elevator cars should be protected with a 1½-h fire-rated door for the full width and height of the opening. ANSI A17.1 lists types of doors that may be used and gives requirements for their openings. Generally, however, single-section swinging doors or horizontally sliding doors are used for freight and passenger elevators and vertically sliding doors are used exclusively for freight elevators.

Horizontally sliding or swinging doors for automatic elevators should be equipped with door closers. They should close open doors automatically if the car leaves the landing zone. (A **landing zone** is the space 18 in above and below a landing.) A horizontally sliding hoistway door may be kept open while a car is at a landing, but only while the car is being loaded and unloaded or when the door is under the control of an operator or an automatic elevator dispatching system.

For safety reasons, normal operation of the elevator car in a hoistway should not be possible unless all hoistway doors are closed and preferably also locked. For the purpose, the doors should be equipped with door-locking devices, hoistway access switches, and parking devices. A **locking device** holds a door closed, preventing it from being opened on the landing side, except for repair, maintenance, or emergency purposes. An **access switch** is placed at a landing and operated to permit movement of a car with the car door and the hoistway door at the landing open, for access to the top of the car or to the hoistway pit. A **parking device** is used to open or close a hoistway door from the landing side at any landing if a car is within the landing zone. Unless all hoistway doors automatically unlock as a car enters their landing zones, at least one landing of the hoistway should be provided with a parking device.

ANSI A17.1 also requires that hoistway doors be equipped with hoistway-unit-system interlocks. These consist of electric contacts or mechancial locks, or a combination of these devices, that prevent operation of the elevator driving machine by the normal operating device unless all hoistway doors are closed and locked.

Hoistway doors should be openable by hand from the hoistway side from a car within the interlock unlocking zone, except when the doors are locked out of service. (Doors at the main-entrance landing or at the top or bottom terminal landing should be incapable of being locked out of service, so that some means of access to the hoistway is always available.) Automatic fire devices controlled by heat should not lock any hoistway door so that it cannot be opened manually from inside the hoistway nor lock any exit leading from a hoistway door to the outside.

Vision panels of clear wired glass or laminated glass, with an area between 25 and 80 in^2, may be inserted in any type of hoistway door and car door, to enable passengers in a car to see if passengers at landings are waiting to enter. ANSI A17.1, however, specifically requires such a vision panel to be installed in all horizontally swinging hoistway doors and in manually operated, self-closing, sliding hoistway doors for elevators with automatic or continuous-pressure operation. But the code does not require a vision panel at landings for automatic elevators provided with a device that indicates the location of the car in the hoistway (hall position indicator).

16.7.5 Guide Rails

The paths of elevator cars and of counterweights, if used, are controlled by vertical guide rails installed in the hoistway. The rails usually are T shaped in cross section and have smooth guiding surfaces along which the car wheels roll. A rail is installed on each side of the shaft to guide the car (Fig. 16.13). When a counterweight is used with electric elevators, to reduce power requirements, a second pair of rails is placed along one wall of the shaft, to guide the counterweight.

The elevator manufacturer usually supplies and installs the guide rails. The owner is responsible for the building structure that supports them.

16.7.6 Buffers and Bumpers

Energy-absorbing devices are required at the bottom of a hoistway to absorb the impact from a car that descends below its normal limit of travel (Fig. 16.12*a*). ANSI A17.1 specifically requires buffers under cars and counterweights in hoistways over accessible spaces. The code also requires buffers or solid bumpers in the pits

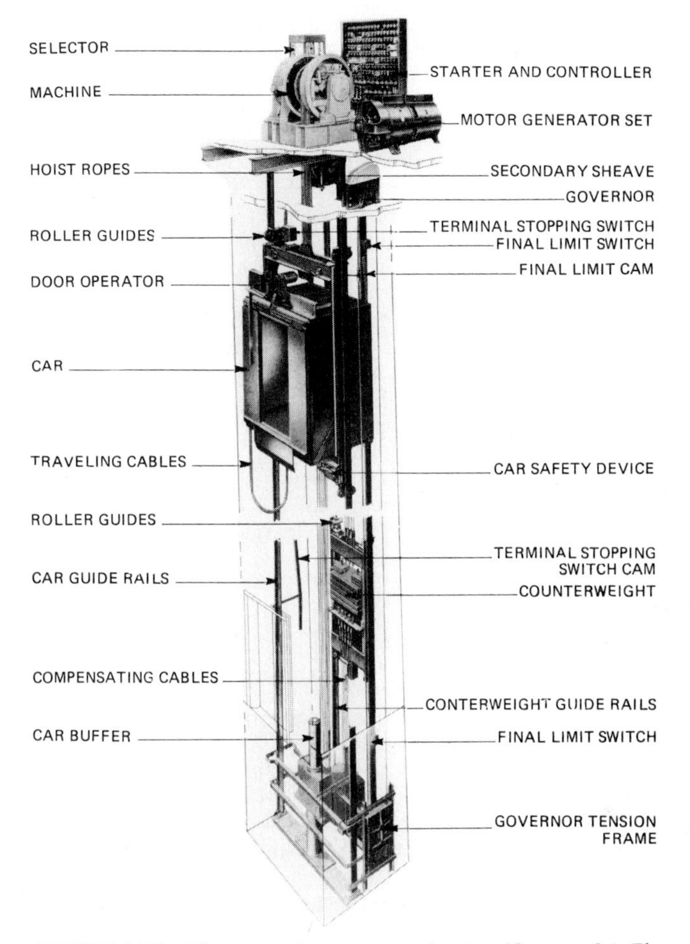

FIGURE 16.13 Electric traction passenger elevator. *(Courtesy Otis Elevator Co.)*

for passenger elevators with speeds up to 50 ft/min and for freight elevators with speeds up to 75 ft/min, but buffers are required for greater speeds. Oil buffers may be used under elevators at any rated speed, but spring buffers may be used only for speeds up to 200 ft/min.

Solid bumpers, which are permissible only for slow elevators, may be made of wood or other resilient material. As the name implies, spring buffers use springs, whereas oil buffers use the hydraulic pressure of oil against a plunger contacted by a descending car, to bring the car to a gradual stop.

16.7.7 Hoistway Dimensions

The National Elevator Industry standard, "Elevator Engineering Standard Layouts," lists clear inside dimensions required for hoistways for elevator cars covered by the standard. Standard sizes may be modified to meet specific building or structural requirements, so long as adequate clearances are maintained for guide rails and machinery installation. When actual sizing is outside of manufacturer's recommendations, it should be reviewed by the manufacturer or consultant to ensure sufficient room is permitted for the installation. The **Americans with Disabilities Act** lists specific elevator car sizes required for access to buildings.

For proper elevator operation and for safety, both maximum and minimum clearances between hoistways and cars and other moving equipment, as recommended by elevator manufacturers and ANSI A17.1, should be provided. The clearance between a car and the hoistway enclosure, for example, should be at least ¾ in, except on the sides used for loading and unloading. The clearance between the car-platform sill and vertically sliding hoistway doors or the hoistway edge of the landing sill should be at least ½ in where side door guides are used and ¾ in where corner guides are used, but not more than 1½ in. Maximum clearance between the loading side of the car platform and the hoistway enclosure generally is 5 in but may be as much as 7½ in when vertically sliding hoistway doors are used.

Clearance between a car and its counterweight should be at least 1 in. Between the counterweight and other components, the clearance should be at least ¾ in.

In multiple hoistways, a minimum of 2-in clearance should be provided between moving equipment.

16.8 ELEVATOR CARS

A car consists basically of a platform for transporting passengers and goods. The platform is raised or lowered by wire ropes or a hydraulic piston or plunger. The car is required to be completely enclosed. Car enclosures of sheet metal or plywood are common; some decorative elevators are enclosed in laminated glass. To provide access to the car, openings protected by doors are provided in one or two of the car walls. In addition, the platform and the car enclosures are supported on a structural steel frame (Fig. 16.12).

For electric elevators, the wire ropes that move the car are attached to the frame or threaded around sheaves connected to it. For hydraulic elevators, the frame is seated on the piston. The frame also supports the upper and lower wheels that roll along the vertical guide rails in the hoistway (Fig. 16.13). In addition, car frames of electric elevators carry safety devices that stop an overspeeding elevator mechanically.

Design, construction, and installation of elevator cars should meet the requirements of the "National Standard Safety Code for Elevators, Dumbwaiters, Escalators and Moving Walks," ANSI A17.1.

16.8.1 Door Controls

Car doors may be horizontally or vertically sliding. They usually are power operated. For safety, they should be equipped with devices that prevent them from opening while the car is moving or is outside the landing zones, the space 18 in above and below a landing. Also, ANSI A17.1 requires safety devices that will keep the car from moving while the doors are open. The Americans with Disabilities Act requires specific door-open dwell times in response to car and landing calls.

Additional devices are needed for power-operated doors to reopen the car and hoistway doors when they start to close on a passenger or other object. The National Elevator Industry standard, "Suggested Minimum Passenger Elevator Requirements for the Handicapped," recommends that the devices be capable of sensing a person or object in the path of a closing door, without requiring contact for activation, at a nominal 5 and 29 in above the floor. Also, the doors should be kept open for at least 20 s after reopening. Still other devices should be installed for other safety reasons, for example, to prevent car and hoistway doors from closing and the car from moving when it is overloaded.

16.8.2 Car Equipment

The interior of the car should be ventilated and illuminated with at least two electric lamps. Lighting provided at the landing edge of the car platform should be at least 5 fc for passenger elevators and 2.5 fc for freight elevators. In addition, an emergency electric-lighting power source should be installed, to operate immediately after failure of the normal power source. For a period of at least 4 h, this system should maintain at least 0.2 fc at a level 4 ft above the car floor and about 1 ft in front of a car station.

The car must also house an approved communication device consistent with rules outlined in the Americans with Disabilities Act. The communication device provides a means for two-way communication with persons outside the hoistway. To be available for use by persons in wheelchairs, an alarm button should be installed in the car. When pressed, this button should sound an alarm outside the hoistway, and an emergency stop switch should be installed about 35 in above the platform. The height of the highest push button or of a telephone should not exceed 54 in. A handrail should be provided about 32 in above the floor along the rear car wall.

It is also desirable that the car contain a car position indicator, located above the push buttons or the door. It should indicate the number of the floor that the car is passing or at which it has stopped. An audible signal should be given to advise passengers that the car is stopping or passing a floor that is served by the elevator.

Similarly, a visual and audible signal should be given at each hoistway door to indicate in the hall, or lobby, that the car is stopping at that floor in response to a call. The audible signal should sound once for the up direction and twice for the down direction of car travel. Call buttons for summoning cars should be located in the elevator lobbies about 42 in above the floor. A lamp should light when a call is registered and go out when the call has been answered by a car.

An emergency exit should be provided in the roof of each car. Also, means should be available for operating the car from its top during inspection, maintenance, and repair. In addition, an electric light and convenience outlet should be installed on the roof, with a switch near the fixture.

See also Art. 16.9.

16.8.3 Car Capacities and Sizes

Cars are rated in accordance with their load-carrying capacity. For passenger elevators, capacities generally range from 1500 lb for use in apartment buildings to 5000 lb for use in department stores and hospitals. (Approximate capacity in passengers can be estimated by dividing the rated capacity, in pounds, by 150.) Capacities of freight elevators usually range from 1500 lb for light duty up to 10,000 lb for general-purpose work or 20,000 lb for heavy duty.

The National Elevator Industry standard, "Elevator Engineering Standard Layouts," lists standard car platform sizes for various rated capacities for electric and hydraulic passenger, hospital, and freight elevators. The sizes give clear inside width and depth of the cars. To obtain the outside dimensions of a car, add 4 in to the clear width (parallel to car door) and the following to the clear depth:

10 in for passenger elevators with center-opening doors or a single sliding door

11½ in for passenger and hospital elevators with two-speed or two-speed, center-opening doors at one end only

19 in for hospital elevators with two-speed front and rear doors

7 in for freight elevators with front doors only

10 in for freight elevators with front and rear doors

Cars supplied by various manufacturers, however, may differ somewhat in sizes from those recommended in the standard. Consequently, it is advisable to obtain recommended car sizes from the car manufacturer or elevator installer for a specific installation.

16.9 ELECTRIC ELEVATORS

An electric-elevator installation requires, in addition to the car described in Art. 16.8 and the hoistway components described in Art. 16.7, wire ropes for raising and lowering the car and for other purposes, a driving machine, sheaves for controlling rope motion, control equipment for governing car movements, a counterweight, and safety devices (Fig. 16.13).

16.9.1 Driving Machines

Components of an electric driving machine include an electric motor, a brake, a drive shaft turned by the motor, a driving sheave or a winding drum, and gears, if used, between the drive shaft and the sheave or drum. The brake operates through friction on the drive shaft to slow or halt car movement. Hoisting-rope movement is controlled by the driving sheave or the winding drum around which the ropes are wound.

Traction machines are generally used for electric elevators. These machines have a motor directly connected mechanically to a driving sheave, with or without intermediate gears, and maintain and control motion of the car through friction between the hoisting ropes and the driving sheave. Also called a traction sheave, this wheel has grooves in its metal rim for gripping the ropes.

Geared-traction machines, used for slow- and medium-speed elevators, have gears interposed between the motor and the driving sheave. The gearing permits use of a high-speed ac or dc motor with low car speeds, for economical operation. Recently, helical gear machines have been employed effectively for variable-voltage, variable-frequency (VVVF) control ac elevators. Whereas conventional worm-geared machines limit car speeds to 450 or 500 ft/min, the dual efficiency of the helical gearbox coupled with an ac motor produces car speeds of up to 800 ft/min. Progress in solid-state design has virtually eliminated the classic single- and two-speed ac-drive systems.

Gearless traction machines, in contrast, are used with ac or dc motors for elevators that operate at speeds of 500 ft/min or more. This type of elevator machine is essentially a large motor with a traction sheave and brake mounted on a common shaft. Gearless dc-motored machines are effectively used for car speeds of 500 ft/min or more, whereas ac-motored (VVVF-control) gearless machines are typically employed for car speeds over 700 ft/min. Since the gearless traction machine consists of a custom-built motor, traction sheave, and brake on a custom motor frame, these machines are the most expensive elevator drive systems.

A **winding-drum machine** gear-drives a grooved drum to which the hoisting ropes are attached and on which they wind and unwind. For contemporary elevators, the winding-drum drive system is applied only to dumbwaiters and light-duty residential units.

16.9.2 Elevator Control

The system governing starting, stopping, direction of motion, speed, and acceleration and deceleration of the car is called control. Multivoltage control (also known as variable-voltage control) or rheostatic control has been commonly used for electric elevators, due largely to the relative simplicity of controlling the dc motor. The advent of larger power transistors has resulted in control systems known as VVVF control, that can be applied to ac motors to produce smooth starting and stopping equal to the classic dc elevator control system.

Multivoltage control usually is used with driving machines with dc motors. For elevator control, the voltage applied to the armature of the motor is varied. Because buildings usually are supplied with ac power, the variable voltage generally is obtained from a motor-generator set that converts ac to dc. This type of control commonly is used for passenger elevators because it combines smooth, accurate speed regulation with efficient motor operation. It also permits rapid acceleration and deceleration and accurate car stops, with low power consumption and little maintenance. But multivoltage control costs more initially than rheostatic control.

Variable-voltage, variable-frequency control is a means used to produce smooth acceleration, deceleration, and stopping of common ac motors at nonsynchronous speeds. VVVF control offers much higher efficiency than that realized through dc motors and is gradually replacing the various means used to control dc motors, along with the dc motor.

16.9.3 Car Leveling at Landings

Elevator installations should incorporate equipment capable of stopping elevator cars level with landings within a tolerance of ½ in under normal loading and unloading conditions. Because changing car loads vary the stretch of the hoisting ropes, provision should be made to compensate for this variation and keep the car platform level with the landing. Most elevators employ automatic leveling.

Automatic leveling controls the driving motor to level the car. Elevators typically employ a two-way, automatic leveling device to correct the car level on both overrun and underrun at a landing and hold the car level with the landing during car loading and unloading.

16.9.4 Terminal Stopping Devices

For safety, provisions should be made to control car movement as it approaches a terminal landing and to keep it from passing the terminal. For the purpose, special speed-limiting and stopping devices are needed.

An **emergency terminal speed-limiting device** is required to reduce car speed automatically as the car approaches the terminal landing. This should be done independently of the functioning of the operating device, which actuates the elevator control, and of the normal terminal stopping device if it should fail to slow the car down as intended.

The **normal terminal stopping device** slows down and stops the car at or near a terminal landing independently of the functioning of the operating device. It should continue to function until the final terminal stopping device operates.

The **final terminal stopping device** is required to interrupt automatically the electric power to the driving-machine motor and brake after the car has passed a terminal landing. But this device should not operate when the car has been stopped by the normal terminal stopping device. When the final terminal stopping device has been actuated, normal car operating devices should be rendered incapable of moving the car.

16.9.5 Car and Counterweight Safeties

A safety is a mechanically operated device that is capable of stopping and supporting the weight of an elevator car and its load when the device is actuated by a car-speed governor. The safety should be actuated when the car travels at more than 15% above its rated speed.

Car safeties are generally mounted on the *safety plank,* or bottom member of the car frame. When tripped, springs on the safeties push shoes against the guide rails hard enough to make the car slide to a stop. (When a hoistway is located above an accessible space, safeties, such as those used for cars, should also be provided on the counterweight frame.) The safeties are typically released by upward motion of the car.

The governor may be conveniently located in the machine space, where the device will not be struck by the car or the counterweight if either should overtravel. The governor may measure car speed from the rotation of a sheave around which is wound a wire rope connected to the car and held under tension. When the car goes too fast, the governor trips jaws that grip a wire rope connected through linkages to a safety and release a spring to actuate the safety. Also, electrical switches on the governor and the safety are opened to remove power from the driving machine and apply a friction brake to the drive shaft.

16.9.6 Counterweights

Power requirements of the driving machine for moving the car are reduced by hanging a counterweight on the hoisting ropes. Use of a counterweight also is advantageous for maintaining traction between the hoisting ropes and the driving sheave. The weight of the counterweight usually is made equal to the weight of the unloaded car and the ropes plus about 40% of the rated load capacity of the car (Fig. 16.14).

A counterweight usually is made up of cut steel plates set in a steel frame. Moving up as the car moves down and down when the car moves up, the coun-

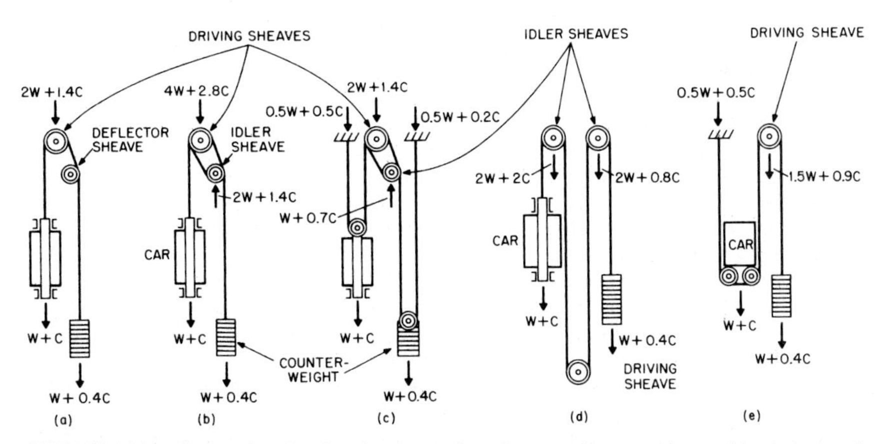

FIGURE 16.14 Types of roping for electric traction elevators. Rope tension and loads imposed on sheaves and supports depend on the type of roping, car weight *W*, and car capacity *C*.

terweight is kept in a fixed vertical path by upper and lower guide rollers that are attached to its frame and roll along a pair of guide rails.

16.9.7 Roping for Elevators

The "American National Standard Safety Code for Elevators, Dumbwaiters, Escalators and Moving Walks," ANSI A17.1, requires that a car be suspended from at least three hoisting ropes for traction-type machines and two ropes for winding-drum machines. At least two ropes are needed for a counterweight. All these ropes should be at least ½ in in diameter.

A wire rope for an elevator installation comprises a group of steel strands laid helically around a hemp core. Each strand, in turn, consists of steel wires placed helically around a central wire and has a symmetrical cross section.

For a given weight of a car and load, the method of roping an elevator has a considerable effect on car speed and loading on the hoisting ropes, machine bearings, and building structural members. The simple arrangement of hoisting ropes, cars, and counterweight shown in Fig. 16.14*a*, for example, is called **1:1 roping,** because car speed equals rope speed. The ropes are attached to the top of the car frame, wind around the driving sheave, bend around a deflector sheave, and then extend downward to the top of the counterweight. This rope arrangement is also known as **single-wrap roping** because the ropes pass over the driving sheave only once between the car and the counterweight. The 1:1 single-wrap roping often is used for high-speed passenger elevators.

For single-wrap roping, the rim of the driving sheave is given wedge-shaped or undercut grooves, to obtain sufficient traction. The sheave grips the ropes because of a wedging action between the sides of the grooves and the ropes. The pinching, however, tends to shorten rope life.

For good traction with less rope wear, **double-wrap roping** (Fig. 16.14*b*), is frequently used for high-speed passenger elevators instead of single-wrap. For double-wrap roping, the ropes are attached to the top of the car frame, wind twice around the driving sheave and secondary sheave, and are then deflected down to the counterweight by the secondary sheave. Because of the double wrap, less grip is needed at the driving sheave. As a result, its rim may be given U-shaped or round-seat grooves, which cause less rope wear. The 1:1 double-wrap roping shown in Fig. 16.14*b*, however, applies twice the load to the driving sheave for the same weight of car and counterweight as does the single-wrap roping in Fig. 16.14*a* and requires a heavier design for affected components.

For the double-wrap roping shown in Fig. 16.14*c*, rope speed is twice the car speed. The arrangement, called **2:1 roping,** is suitable for heavily loaded, slow freight elevators. For this arrangement, a higher-speed, less costly motor can be used for a given car speed than with 1:1 roping. The ropes in this case are not attached to the car and counterweight as for 1:1 roping. Instead, the ropes wind around idler sheaves on car and counterweight, and the ends of the ropes are anchored at the top of the hoistway at beams. As a result, the load on the driving and idler sheaves is only about one-half that for 1:1 roping.

In most buildings, driving machines are located in a penthouse. When a machine must be placed in a basement (Fig. 16.14*d*), the load on the overhead supports is increased, rope length is tripled and additional sheaves are needed, adding to the cost. Other disadvantages include higher friction losses and a larger number of rope bends, requiring greater traction between ropes and driving sheave for the same elevator loads and speeds; modestly higher power consumption; and potentially greater rope wear.

Figure 16.14*e* shows a type of 2:1 roping suitable for slow, low-rise elevators. In contrast to the roping in Fig. 16.14*c*, only one end of the hoisting ropes is dead-ended at the top of the hoistway. The other end is attached to the counterweight. Also, the ropes pass around idler sheaves at the car that are placed on the underside of the car frame.

16.9.8 Elevator Operating Systems

The method of actuating elevator control is called elevator operation. Many types of operation are available and some are complex and sophisticated. These may cost more than the simpler systems for installation and operation, but the sophisticated systems accomplish more automatically and handle traffic more efficiently. Following are descriptions of several types of operation:

Car-Switch Operation. With a manually operated car switch or continuous-pressure buttons in the car, an operator controls movement and direction of travel of the car. To ensure that the opertor controls car movement, the handles of lever-type operating devices should return to the stop position and latch there automatically when the operator's hand is removed. In automatic car-switch, floor-stop operation, the operator releases the lever or button to stop the car at a landing. Slowing and stopping are then accomplished automatically.

Signal Operation. The car can be started only by an operator pushing a start button in the car. The operator can register stops in advance by pressing and releasing a push button in the car corresponding to the predetermined floor number. Persons calling for the elevator can similarly register a stop by pushing an up or down hall button. The car automatically stops at landings for which signals were registered, regardless of the direction of car travel or of the sequence in which buttons at various floors were pressed. When a landing is served by two or more elevators, the first available car approaching the floor in the specified direction makes the stop automatically.

Automatic Operation. An operator is not needed for automatic operation. Starts and stops are signaled by passengers in the car or by hall buttons or by an automatic operating mechanism. The car starts either in response to this mechanism or when a passenger presses a car or hall button. Responding to signals from car or hall buttons, the car travels to and stops at the signaled landings, and car and hoistway doors open automatically. Following, in order of increasing sophistication, are descriptions of several types of automatic operation.

Single Automatic Operation. The car starts when a passenger presses and releases a car button corresponding to a landing. The car then travels to that floor and stops. The car also starts when a hall button is pressed, travels to that landing, and stops. After any button has been pressed, depression of any other button has no effect on car movement until the stop signaled by the first button has been made.

Selective Collective Automatic Operation. When a car button corresponding to a landing is pressed, the car travels to that floor but also, on the way, makes other stops signaled. Hall calls are answered in the order in which landings are reached in each direction of travel, regardless of the sequence in which signals were received. Up calls are served when the car is traveling upward, and down calls are answered when the car is traveling downward.

Group Automatic Operation. This is an extension of selective collective operation to a group of cars serving the same landings. A supervisory control system automatically dispatches cars to answer calls and coordinates the operation of the group. A call is answered by the first car to approach a landing in the proper direction. In response to a timer, the car leaves a terminal at predetermined intervals, which can be varied in accordance with traffic requirements. Group operation with automatic dispatching increases the number of passengers the elevators can carry in a given time.

16.9.9 Elevator Group Supervision

The supervisory control system for group automatic operation of elevators should be capable of making adjustments for varying traffic conditions. It should control car motion so that cars in the best location for answering calls do so. In the past, group control systems employed an elaborate relay network to "program" elevator motion. The advent of microprocessors, however, has dramatically changed elevator control systems.

The state-of-the-art group control system employs limited *artificial intelligence* features, which control the dispatching of high-speed elevator groups more efficiently. One such supervisory control system contains a database based on practical knowledge, traffic data, and experience of elevator group-control experts. The system is able to maximize every elevator operation by application of the database and through knowledge it has obtained from ongoing traffic monitoring functions, such as call quantity and car loading. Decisions also involve use of *fuzzy logic,* which allows the elevator group control to make decisions based on fragmentary and fuzzy intelligence concepts. For example, use of the incorporated "intelligence" and "common sense" in the decision-making process maximizes the effectiveness of the system and assists in determining whether or not potential car assignments will result in shorter waiting times or cause congestion in elevator lobbies.

In another approach, the system is designed to alleviate lobby congestion during heavy up-peak traffic periods. Elevators make fewer stops per round trip, with the result that cars return to the lobby faster. The floors about the lobby served by the group are divided into sectors of contiguous floors. The number of sectors is normally one less than the number of elevators in the group. As an elevator returns to the lobby during an up-peak period, it is assigned to service one of the sectors. Passengers can easily determine which floors each car is serving by checking the Information Display System screens located next to or above each elevator entrance. The same car will not necessarily serve the same sector on successive trips. Care is taken to ensure that each sector will receive equal service by the assignment of sectors in a round-robin manner.

In still another approach, unlike any other control system presently in use, the system is arranged to *know* the passenger quantity and destination before the call is answered by the elevator. This is achieved through the use of a keypad call-entry system installed at each landing. The prospective passenger enters the destination into the keypad, and the elevator control system immediately assigns an elevator to that call. The passenger is notified graphically which elevator will respond to the call. Since the efficiency of elevator movement is greatly improved, this system offers the opportunity to reduce the quantity of elevators used to meet specific traffic conditions.

16.9.10 Additional Elevator Safety Devices

Automatic elevators require several safety devices in addition to those usually installed in elevators with operators. These devices include:

An automatic load weigher to prevent doors from closing and the car from starting when it is overloaded

Car and hall buttons that passengers can push to stop the doors from closing and to hold them open

Means for preventing doors from closing when the entrance is obstructed

Emergency power system that is activated as soon as the primary system fails

Lights to indicate landings for which calls have been registered

Two-way communicaton with a supervisor outside the hoistway

(G. R. Strakosch, "Vertical Transportation: Elevators and Escalators," John Wiley & Sons, Inc., New York.)

16.10 HYDRAULIC ELEVATORS

For low-rise elevators, hydraulic equipment may be used to supply the lift. The car sits atop a plunger, or piston, which operates in a pressure cylinder (Fig. 16.15). Oil serves as the pressure fluid and is supplied through a motor-driven positive-displacement pump, actuated by an electric-hydraulic control system.

To raise the car, the pump is started, discharging oil into the pressure cylinder and forcing the plunger up. When the car reaches the desired level, the pump is stopped. To lower the car, oil is released from the pressure cylinder and is returned to a storage tank.

Single-bearing cylinders (Fig. 16.16*a*) are a simple type that operate like a hydraulic jack. They are suitable for elevator and sidewalk lifts where the car is guided at top and bottom, preventing eccentric loading from exerting side thrust on the cylinder bearing. A cylinder of heavy steel usually is sunk in the ground as

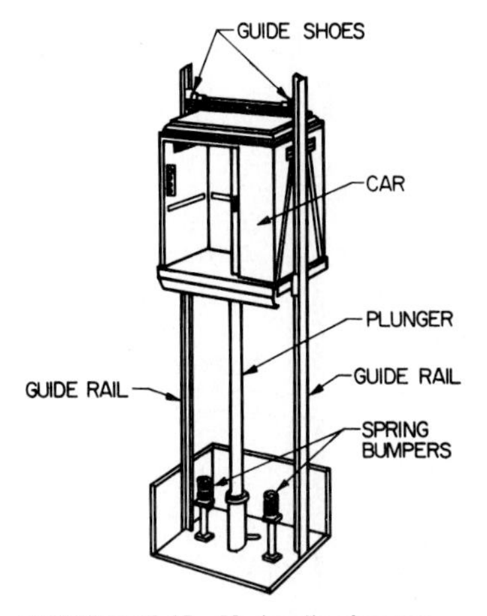

FIGURE 16.15 Hydraulic elevator.

far as the load rises. The plunger, of thick-walled steel tubing polished to a mirror finish, is sealed at the top of the cylinder with compression packing. Oil is admitted under pressure near the top of the cylinder, while air is removed through a bleeder.

A different cylinder design should be used where the car or platform does not operate in guides. One type capable of taking off-balance loads employs a two-bearing plunger (Fig. 16.16*b*). The bearings are kept immersed in oil.

Another type, suitable for general industrial applications, has a movable bearing at the lower end of the plunger to give support against heavy eccentric loads (Fig. 16.16*c*). At the top of the cylinder, the plunger is supported by another bearing.

For long-stroke service, a cage-bearing type can be used (Fig. 16.16*d*). The cage-bearing is supported by a secondary cylinder about 3 ft below the main cylinder head. Oil enters under pressure just below the main cylinder head, passes down through holes in the bearing, and lifts the plunger.

When the car or platform is not heavy enough to ensure gravity lowering, a double-acting cylinder may be used (Fig. 16.16*e*). To raise the plunger, oil is admitted under pressure below the piston; to lower it, oil is forced into the cylinder near the top, above the piston, and flows out below. Jack plunger sizes for the various types range from 2½ in in diameter for small low-capacity lifts to 18 in for large lifts, operating at 150 to 400 psi.

Hydraulic elevators have several advantages over electric elevators: They are cheaper and simpler. The car and its frame rest on the hydraulic plunger that raises and lowers them. There are no wire ropes, no overhead equipment, and no penthouse. Without heavy overhead loads, hoistway columns and footings can be smaller. Car safeties or speed governors are not needed, because the car and its load cannot fall faster than normal speed. Speed of the elevator is low; so the bumpers need be only heavy springs.

Capacity of hydraulic passenger elevators usually ranges from 1200 to 5000 lb at speeds from 75 to 150 ft/min. With gravity lowering, down speed may be 1.5 to 2 times up speed. So the average speed for a round trip can be considerably higher than the up speed. Standard hospital elevators have capacities of 3500 to 5000 lb at speeds of 75 to 150 ft/min.

Capacity of standard freight elevators ranges from 2500 to 8000 lb at 50 to 125 ft/min, but they can be designed for much greater loads.

(G. R. Strakosch, "Vertical Transportation: Elevators and Escalators," John Wiley & Sons, Inc., New York.)

16.11 PLANNING FOR PASSENGER ELEVATORS

Elevator service is judged by two primary criteria: quantitative, or the number of persons who can be moved by the system within a defined peak traffic period, and

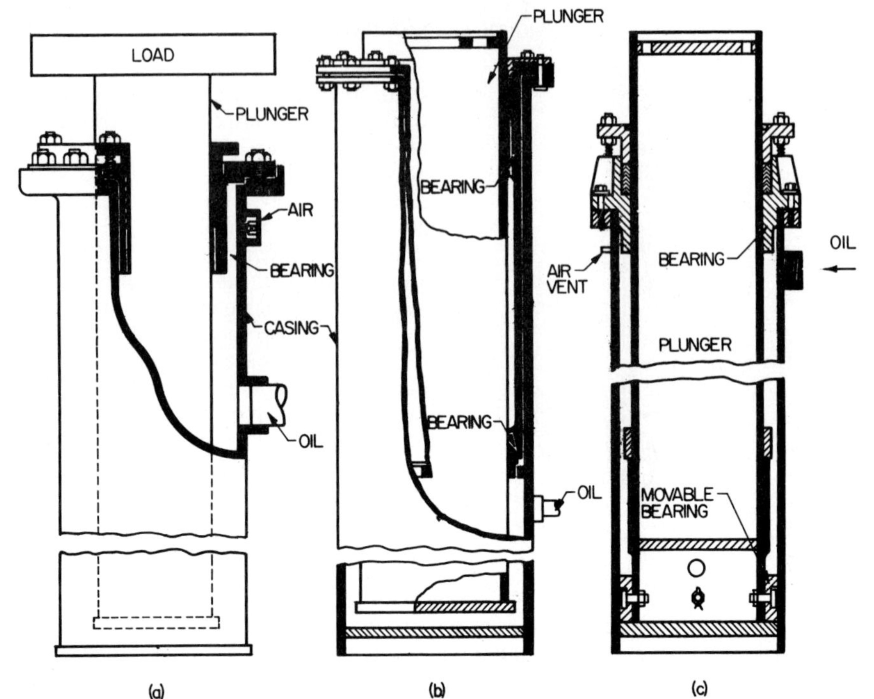

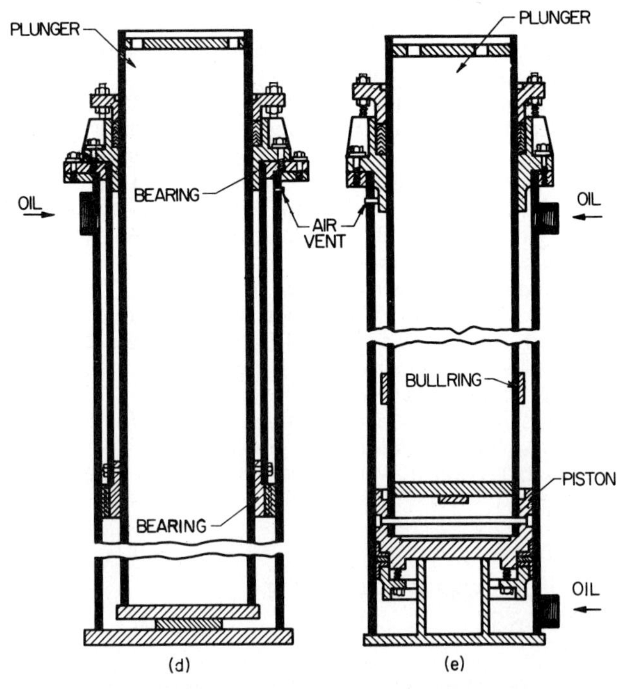

FIGURE 16.16 Jacks commonly used for hydraulic elevators: (*a*) single-bearing plunger for guided loads; (*b*) two-bearing plunger for off-balance loads; (*c*) movable-bearing plunger for heavy service; (*d*) cage bearing for long-stroke service; (*e*) double-acting plunger.

qualitative, which expresses the calculated time between departing elevators during the same heavy traffic period.

16.11.1 Number of Elevators Required

The number of passenger elevators required for a particular building depends on the number of persons expected to work or live in the building. Traffic is measured by the number of persons requiring service during a peak 5-min period. For proposed buildings, a population estimate is generated on the basis of occupancy trends for that specific building type. Peak-traffic projections are based on the type of tenancy expected for the building. From the population and peak-traffic projection, the demand is established as a peak 5-min traffic flow. While peak traffic in most buildings is a rather complex pattern of two-way and interfloor movement, most models assume a simplified traffic pattern in which traffic is primarily incoming or outgoing. The lack of a complex model is more a result of the poor understanding of the existing model than of the absence of sophisticated measuring devices.

After the peak 5-min traffic flow is established, an estimate may be made of the quantity of elevators required. The ability of a specific system to handle the traffic is tested against the projected traffic level. The 5-min handling capacity of an elevator is determined from the round-trip time.

$$HC = \frac{300}{T} P \tag{16.1}$$

where HC = handling capacity of car, persons in 5 min or 300 s
P = car capacity, persons
T = round-trip time of car, s

The minimum number of elevators n required can then be computed from

$$n = \frac{V}{HC} = \frac{VI}{300P} \tag{16.2}$$

where V = peak traffic, persons in 5 min. Equation (16.2) indicates that the minimum number of elevators required is directly proportional to the round-trip time for a car and inversely proportional to the car capacity.

Elevator-related space requirements may not be minimized through the use of the fewest elevators to serve a particular building, since large groups of high-capacity cabs must be employed to serve a large number of floors. Large groups of elevators increase cost of the overall system by increasing the average number of elevator entrances required for the building. For greatest efficiency and lowest cost, elevator group sizes should not exceed six elevators, with four elevators per group as a more practical approach. This method has the added advantage that passenger trip time—that is, the time it takes an individual to travel to a destination during peak traffic—is reduced due to use of smaller cabs and assignment of fewer floors to be served by a particular elevator group.

After an approximate quantity of elevators is found to meet quantitative traffic requirements, qualitative performance should be reviewed. The criteria for qualitative performance is generally based on the quality of service expected for a specific building, as well as the overall quality level of the project. Qualitative service is typically expressed as **interval,** or the calculated time between elevators departing the ground floor. Improving elevator service for a building, however, generally results in increased cost.

After the number of elevators has been computed on the basis of traffic flow, the average interval should be checked. It is obtained by dividing the round-trip time by the number of elevators.

Round-trip time is composed of all of the pieces of a projected elevator trip, including starting, running, and stopping of the elevator car, time for opening and closing doors, and time for passengers to move in and out. Often some factor is added to the round trip time to simulate normal use of the system.

Opening and closing of doors may contribute materially to lost time, unless the doors are properly designed. A 3-ft 6-in opening is excellent, because two passengers may conveniently enter and leave a car abreast. A slightly wider door would be of little advantage. Department stores, hospitals, and other structures served by larger passenger elevators (4000 lb and over) usually require 4-ft door openings.

Center-opening doors, preferred for power operation, are faster than either the single or two-speed type of the same width. The impact on closing is smaller with the center-opening door; hence, there is less chance of injuring a passenger. Also, transfer time is less, since passengers can move out as the door starts to open.

Another factor affecting passenger-transfer time is the shape of the car. The narrower and deeper a car, the greater is the time required for passenger entry and exit during peak-traffic conditions likely to be.

(See also Art. 16.9.8 and 16.9.9).

16.11.2 Elevators in General-Purpose Buildings

For a proposed diversified-tenancy, or general-purpose, office building, peak traffic may be estimated from the probable population computed from the net rentable area (usually 75 to 90% of the gross area). Net rentable area per person typically ranges from 170 to 200 ft^2. Some rare organizations occupy space at higher densities, but averages for specific floors rarely become as dense as one person to 100 ft^2.

Diversified-tenancy office buildings usually have important traffic peaks in the morning, at noon, and in the evening, and very little interfloor traffic. The 5-min morning peak generally is the controlling factor, because if the elevators can handle that peak satisfactorily, they can also deal with the others. In a well-diversified office building, the 5-min peak used is about 12.5% of the population.

For busy, high-class office buildings in large cities, time intervals between elevators may be classified as follows: 26 to 28 s, excellent; between 28 and 30 s, good; between 30 and 32 s, fair; between 32 and 35 s poor; and over 35 s may be unsatisfactory. In small cities, however, intervals of 30 s and longer may be satisfactory.

For express elevators, which make no intermediate stops, intervals of 30 to 35 s may be considered acceptable.

Car speeds used vary with height of building: 4 to 10 stories, 200 to 500 ft/min; 10 to 15 stories, up to 700 ft/min; 15 to 20 stories, up to 800 ft/min; 20 to 50 stories, up to 1200 ft/min; and over 50 stories, up to 2500 ft/min. Practically speaking, 200-ft/min elevators are generally not economically advantageous and have been replaced by 350-ft/min elevators for most passenger applications.

Elevators should be easily accessible from all entrances to a building. For maximum efficiency, they should be grouped near the center. Except in extremely large buildings, two banks of elevators located in different parts of the structure should not serve the same floors. Since one cannot guarantee equal use of the two groups, each group should be designed to handle 60 to 65% of the traffic.

Elevators cannot efficiently serve two lower terminal floors, inasmuch as cars stop twice to pick up passengers who are typically picked up once. The extra stop increases the round-trip time and decreases the handling capacity, resulting in the need for more elevators to satisfy the same traffic criteria. If there is sufficient traffic between the two lower floors, escalators or shuttle elevators should be installed, and one of the levels should be assigned as the sole terminal for the tower passenger elevators.

When laying out a local-travel elevator group, groupings should not exceed four elevators in line. This arrangement can be exceeded for groupings of express elevators where elevator arrivals can be preannunciated. Elevator core configurations must take into account the need for smoke control at elevator lobbies, as well as code limits on dead-end corridors. Lobby widths should be 9 to 12 ft, depending on the size of the elevators.

It is necessary to divide elevator groups into local and express banks in buildings of 15 floors and more, especially those with setbacks and towers, and in low buildings with large rental areas. In general, when more than six elevators are needed, consideration should be given to dividing the building into more elevator groups. In addition to improving service, the division into local and express banks has the advantage that corridor space on the floors where there are no doors can be used for toilets, closets, and stairs.

While the decision to include a dedicated service elevator is often market driven, office buildings of less than 250,000 gross square feet typically "swing" a passenger elevator for off-peak deliveries and moves. Buildings of up to 500,000 or 600,000 gross square feet frequently have only a single service elevator, whereas larger buildings are provided with two or more separate service cabs. Where dedicated service elevators are provided, at least one should be hospital shaped, with the capability of carrying end-loaded 9-ft-long gypsum wall board. A 12-ft ceiling allows easy movement of carpet rolls and long conduit.

16.11.3 Elevators in Single-Purpose Buildings

Elevator requirements and layouts are similar, in general, for both single-purpose and diversified-tenancy office buildings; but several different factors should be taken into consideration: Single-purpose buildings are occupied by one large organization. Generally, the floors that are occupied by the clerical staff are not subdivided into many offices; the net rentable area is about 80% of the gross area. Population densities are higher than for general-purpose buildings. Depending on the kind of business to be carried on, population density varies from 100 ft^2 per person for some life-insurance companies to about 300 ft^2 per person for some attorney's offices.

Although traffic peaks occur at the same periods as in the diversified-tenancy type, the morning peak may be very high, unless working hours are staggered. The maximum 5-min periods may be 13.5 to 16.0% of the population, depending on the type of occupancy. If traffic volumes are high, occupancy of the building should be carefully balanced against elevator requirements. Although many floors may be connected with an eight-car elevator bank, the time wasted on elevators becomes excessive as a result of the number of stops made during each elevator trip.

In the past, system designers specified more elevators to meet interfloor traffic demands of single-purpose buildings. The advent of microprocessor-based controls, however, has dramatically improved system response to complex traffic patterns.

With such controls, an elevator system designed to handle the incoming traffic rush will also provide satisfactory service in response to interfloor traffic.

Elevator service in single-purpose buildings is frequently hobbled by location of a cafeteria or similar high-density, facility at some level above the ground floor. If such facilities are served by the office passenger elevators, the total elevator requirement can increase by 15 to 20% as a result of the inefficiency introduced by the cafeteria.

16.11.4 Elevators in Government Buildings

Municipal buildings, city halls, state office buildings, and other government office buildings may be treated the same as single-purpose office buildings. Population density often may be assumed as one person per 140 to 180 ft^2 of net area. The 5-min maximum peak occurs in the morning and may be as large as 16% of the population.

16.11.5 Professional-Building Elevators

Population cannot be used as the sole basis for determining the number of elevators needed for buildings occupied by doctors, dentists, and other professional people, because of the volume of patient and visitor traffic. Peaks may occur in the forenoon and midafternoon. The maximum occurs when reception hours coincide. Traffic studies indicate that the maximum peak varies from two to six persons per doctor per hour up and down.

Since crowding of incapacitated patients is inadvisable, elevators should be of at least 3000-lb capacity. If the building has a private hospital, then one or two of the elevators should be hospital-type elevators.

16.11.6 Hotel Elevators

Hotels with transient guests average 1.3 to 1.5 persons per sleeping room and are typically populated based upon 90% occupancy. They have pronounced traffic peaks in morning and early evening. The 5-min maximum occurs during checkout hour and can be about 12.5 to 15% of the estimated population, with traffic moving in both directions.

Ballrooms and banquet rooms should be located on lower floors and served by separate elevators. Sometimes it is advisable to provide an express elevator to serve heavy roof-garden traffic. Passenger elevators should be of 3000-lb capacity or more to allow room for baggage carriers. Intervals for passenger elevators should not exceed 50 s.

Service elevators are very important in hotels. Hospital-shaped elevators are often preferred for handling linen and food service carts as well as baggage.

Typically, hotels are provided with one passenger elevator per 125 to 150 rooms. The service elevator quantity is 50 to 60% of the passenger elevator quantity. The ratio of rooms per elevator is lower for better-quality hotels and higher for more modest facilities.

Elimination of noise and vibration from medium- to high-speed elevators is virtually impossible, so hotels should be carefully planned to ensure that guest rooms are not adjacent to elevator hoistways. Rooms that adjoin elevator hoistways may generate complaints throughout the life of the building.

16.11.7 Apartment-Building Elevators

Multistory residential buildings do not have peaks so pronounced as other types of buildings. Generally, the evening peak is the largest. Traffic flow at that time may be 6 to 8% of the building population in a 5-min period. Building population should be estimated in consideration of the market for which the building is designed.

If only one elevator is selected to satisfy traffic conditions for a building of modest height, residents will be forced to use the stairs at times the elevator is out of service for repairs. Where the elevator is considered more than an amenity, two elevators should be provided. Market conditions may require that a separate service elevator also be provided in some urban settings. Typically, a 2500-lb elevator with a 9-ft clear ceiling height can be relied on to carry most furniture.

As is the case with hotels, the potential for noise and vibration should be considered in location of elevators in living units.

16.11.8 Department-Store Elevators

Department stores should be served by a coordinated system of escalators and elevators. The required capacity of the vertical-transportation system should be based on the transportation of merchandising area and the maximum density to which it is expected to be occupied by shoppers.

The **transportation area** is all the floor space above or below the first floor to which shoppers and employees must be moved. Totaling about 80 to 85% of the gross area of each upper floor, the transportation area includes the space taken up by counters, showcases, aisles, fitting rooms, public rooms, restaurants, credit offices, and cashiers' counters but does not include kitchens, general offices, accounting departments, stockrooms, stairways, elevator shafts, or other areas for utilities.

The **transportation capacity** is the number of persons per hour that the vertical-transportation system can distribute from the main floor to the other merchandising floors. The ratio of the peak transportation capacity to the transportation area is called the density ratio. This ratio is about 1 to 20 for a busy department store. So the required hourly handling capacity of a combined escalator and elevator system is numerically equal to one-twentieth, or 5%, of the transportation area. The elevator system generally is designed to handle about 10% of the total.

The maximum peak hour usually occurs from 12 to 1 pm on weekdays and between 2 and 3 pm on Saturdays.

The type of elevator preferred for use with moving stairs is one with 3500-lb capacity or more. It should have center-opening, solid-panel, power-operated car and hoistway doors, with at least a 4-ft 2-in opening and a platform 7 ft 8 in by 4 ft 7 in.

16.11.9 Hospital Elevators

Traffic in a hospital is of two types: (1) medical staff and equipment and (2) transient traffic, such as patients and visitors. Greatest peaks occur when visitor traffic is combined with regular hospital traffic. Waiting rooms should be provided at the main floor and only a limited number of visitors should be permitted to leave them at one time, so that the traffic peaks can be handled in a reasonable period and corridors can be kept from getting congested. In large hospitals, however, pedestrian and vehicular traffic should be separated.

For vehicular traffic or a combination of vehicular and pedestrian traffic, hospital elevators should be of stretcher size—5 ft 4 in to 6 ft wide and 8 or 9 ft deep, with a capacity of 4000 to 5000 lb. Speeds vary from 100 to 700 ft/min for electric elevators, depending on height of building and load. For staff, visitors, and other pedestrian traffic, passenger-type elevators, with wide, shallow platforms, such as those used for office buildings, should be selected (see Arts. 16.11.2 and 16.11.3.)

Elevators should be centrally located and readily accessible from the main entrance. Service elevators can be provided with front and rear doors, and, if desired, so located that they can assist the passenger elevators during traffic peaks.

16.11.10 Freight Elevators

In low-rise buildings, freight elevators may be of the hydraulic type (Art. 16.10), but in taller buildings (higher than about 50 ft) electric elevators (Art. 16.9) generally will be more practical. Figure 16.17 shows the components of an electric freight elevator.

In planning for freight elevators, the following should be considered:

1. Building characteristics, including the travel, number of floors, floor heights and openings required for a car. Also, structural conditions that may influence the size, shape, or location of the elevator should be studied.
2. Units to be carried on the elevator—weight, size, type, and method of loading.
3. Number of units to be handled per hour.
4. Probable cycle of operation and principal floors served during the peak of the cycle.
5. Freight elevators are not permitted to carry passengers.

For low-rise, slow-speed applications, especially where industrial trucks will be used, rugged hydraulic freight elevators generally will be more economical than electric freight elevators.

Classification of Freight Elevators. The "American National Standard Safety Code for Elevators, Dumbwaiters, Escalators and Moving Walks," ANSI A17.1, defines three classes of freight elevators. Class A applies to general freight loading. This is defined as a distributed load that is loaded or unloaded manually or by hand truck and no unit of which, including loading equipment, weighs more than one-quarter the rated load of the elevator. Class B elevators may handle only motor vehicles. Class C elevators may be subjected to heavy concentrated loads and fall into one of three subclasses: Class C1 applies to elevators that carry industrial trucks, Class C2 to elevators for which industrial trucks are used only for loading and unloading the cars and are not carried by them, and Class C3 to elevators carrying heavy concentrated loads other than trucks.

Car Capacity of Freight Elevators. The size of car to be used for a freight elevator is generally dependent on the dimensions of the freight package to be carried per trip and the weight of the package and loading equipment. Power trucks, for example, impose severe strains on the entire car structure and the guide rails than do hand trucks. As a power truck with palletized load enters an elevator, most of

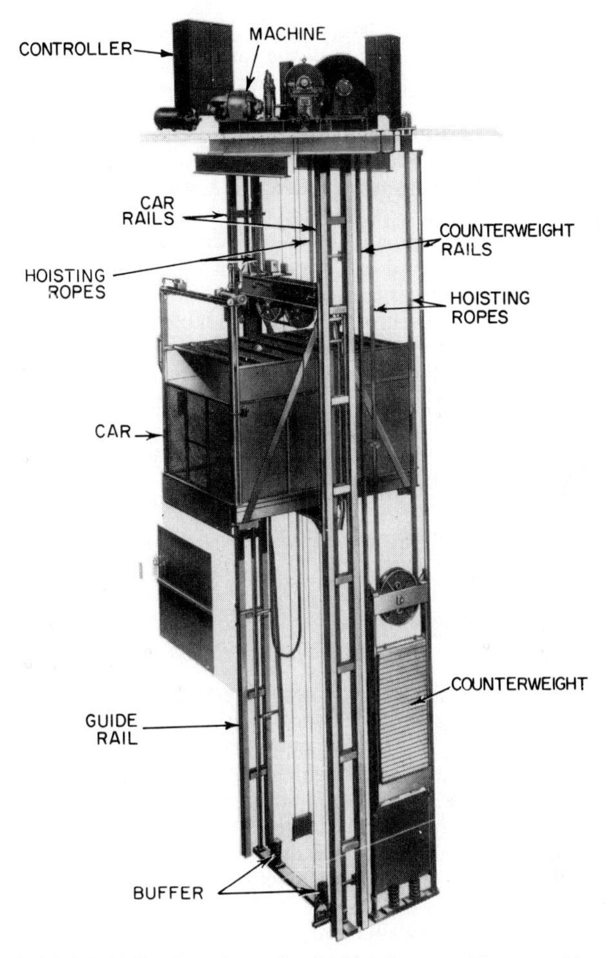

FIGURE 16.17 Electric traction freight elevator. *(Courtesy Otis Elevator Co.)*

the weight of the truck and its load are concentrated at the edge of the platform, producing heavy eccentric loading. Maximum load on an elevator should include most of the truck weight as well as the load to be lifted, since the truck wheels are on the elevator as the last unit of load is deposited.

The carrying capacity per hour of freight elevators is determined by the capacity or normal load of the elevator and the time required for a round trip. Round-trip time is composed of the following elements:

1. Running time, which may be readily calculated from the rated speed, with due allowance for accelerating and decelerating time (about 2¼ s for ac rheostatic control with inching, 1¾ s for multivoltage), and the distance traveled.
2. Time for operation of the car gate and hoistway doors (manual 16 s, power 8 s).

3. Loading and unloading time (hand truck 25 s, power truck 15 s). Wherever practical, a study should be made of the loading and unloading operations for a similar elevator in the same type of plant.

Operation of Freight Elevators. The most useful and flexible type of operation for freight elevators is selective-collective with fully automatic doors. Attendant operation requires an annunciator. When operated with an attendant, the car automatically answers the down calls as approached when moving down and similarly answers up calls when moving up. The elevator attendant, when present, has complete control of the car and can answer calls indicated by the annunciator by pressing the corresponding car button. The addition of fully automatic power-operated doors means the elevator is always available for use, unless taken out of service by the attendant.

The standard hoistway door is the vertical biparting, metal-clad wood type. For active elevators and openings wider than 8 ft, doors should be power-operated.

Automatic freight elevators can be integrated into material-handling systems for multistory warehouse or production facilities. On each floor, infeed and outfeed horizontal conveyers may be provided to deliver and remove loads, usually palletized, to and from the freight elevator. The elevator may be loaded, transported to another floor, and unloaded—all automatically.

16.12 DUMBWAITERS

These may be used in multistory buildings to transport small loads between levels. Generally too small to carry an operator or passenger, dumbwaiters are cars that are raised or lowered like elevators. They may be powered—controlled by push buttons—or manually operated by pulling on ropes. Powered dumbwaiters can automatically handle from 100 to 500 lb at speeds from 45 to 150 ft/min. They are available with special equipment for automatic loading and unloading. They also are designed for floor-level loading suitable with cart-type conveyances.

The "American National Standard Safety Code for Elevators, Dumbwaiters, Escalators and Moving Walks," ANSI A17.1, contains safety requirements for dumbwaiters. Powered dumbwaiters may be constructed like electric elevators with winding-drum or traction driving machines or like hydraulic elevators. Many of the safety requirements for elevators, however, are waived for dumbwaiters. Standard heights for a dumbwaiter are 3, 3½, and 4 ft. ANSI A17.1 restricts platform area to a maximum of 9 ft^2. Rated load capacity usually ranges from 20 to 500 lb. Like elevators, hoistways should have fire-resistant enclosures.

16.13 CONVEYERS AND PNEUMATIC TUBES

When there is a continuous flow of materials, such as mail or other documents to be distributed throughout a multistory building, conveyer systems may provide an economical supplement to elevators. In some installations, 200 lb or more of paper work and light supplies are circulated per minute.

Two types of conveyer systems are employed in commercial buildings. The **selective vertical conveyer** moves plastic tubs from one floor to another, automatically loading and unloading at preselected floors. The tubs typically are made to carry mail and small supplies and have payloads of up to 50 lb each. A typical selective vertical conveyer installation is similar to an escalator (Art. 16.4). A continuous roller chain is driven by an electric motor. Engaging sprockets at top and bottom, the chain extends the height of the building, or to the uppermost floor to be served. Carriers spaced at intervals along it transport trays from floor to floor at a speed of about 70 ft/min.

Another conveyer type, the **tracked conveyer system,** permits both vertical and horizontal document distribution. This system employs self-powered cars, which travel over a track system that allows "switching off" at selected station locations (Fig. 16.18). Where a specific floor may have a high volume of traffic, the track may be routed around the floor to one or more remote stations. The destination is programmed at the dispatching station, and the car is automatically switched onto the main track to begin its journey. Cars for this type of conveyer are generally limited to a maximum payload of 20 lb, although some are modified to carry up to 25 lb. The cars travel at about 100 ft/min.

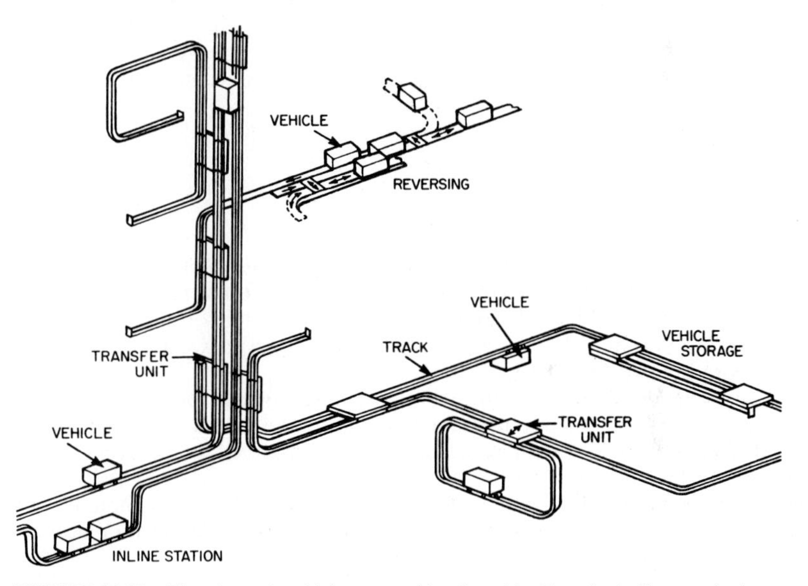

FIGURE 16.18 Electric track vehicle system (developed by Translogic Corporation).

For vertical track sections, a gear engages a continuous rack on the track for positive control in both the up and down directions. A friction-drive system is employed on horizontal track sections. Shaft-mounted machinery is minimized with the tracked conveyer system.

Like elevators, however, vertical conveyers must be enclosed in fire-resistant shafts. Generally, the only visual evidences of the existence of the installation are the wall cutouts for receiving and dispatching runoffs at each floor. In event of fire, vertical sliding doors, released by fusible links, should snap down over the openings, sealing off the conveyer shaft at each floor.

Vertical Conveyers. Operation of vertical conveyers is simple. When the tray or car is ready for dispatch, the attendant sets the floor-selector dial or presses a button alongside the dispatch cutout. For the selective vertical conveyer, trays are placed on the loading station, where they are automatically moved into the path of the traveling carriers. Each tray rides up and around the top sprocket and is automatically discharged on the downward trip at the preselected floor.

The best place to install a vertical conveyer is in a central location, next to other vertical shafts, to minimize horizontal runs in collecting and distributing correspondence at each level. The choice of conveyer types should be based on the needs of the user. The selective vertical conveyer is appropriate where the required movement is entirely vertical, while the tracked conveyer system lends itself to both horizontal and vertical layouts.

Pneumatic Tubes. These are also used to transport small loads within buildings. Units are moved through tubes under air pressure or suction, or both. Items to be transported are carried inside cylinders slightly smaller in diameter than the tubes.

In choosing between vertical conveyers and pneumatic tubes, the designer's first consideration should be the size of the load to be carried. The traveling cylinder is limited as to the size and weight of the material to be moved. Aside from that, an arterial system of pneumatic tubes may satisfy the requirements of a predominantly horizontal building, whereas a vertical conveyer is generally more advantageous in a tall building.

16.14 MAIL CHUTES

Used in multistory buildings for gravity delivery of mail from the various floors to a mailbox in the main lobby, a mail chute is simply a vertical, unpressurized, rectangular tube. With permission of the Post Office, one or more chutes may be installed in office buildings more than four stories high and in apartment buildings with more than 40 apartments.

Usually made of 20-ga cold-formed steel, with a glass front, and supported by vertical steel angles, a chute is about 3 × 8 in in cross section. In the front of the chute, available in a lobby in each story, a slot is provided for insertion of flat mail into the chute.

The mailbox usually is 20 in wide, 10 in deep, and 3 ft high. It should be placed with its bottom about 3 ft above the floor. The mailbox should be placed within 100 ft of the building entrance.

SECTION SEVENTEEN
CONSTRUCTION PROJECT MANAGEMENT

Robert F. Borg
Chairman, Kreisler Borg Florman Construction Company
Scarsdale, New York

and

Kiri J. Borg
Construction Professional

Construction project management encompasses organizing the field forces and backup personnel in administrative and engineering positions necessary for supervising labor, awarding subcontracts, purchasing materials, record keeping, and financial and other management functions to ensure profitable and timely performance of the job. The combination of managerial talents required presupposes training and experience, both in field and office operation of a construction job. Proper construction project management will spell the difference between a successful building or contracting organization and a failure.

This section outlines practical considerations in construction project management based on the operations of a functioning general contracting organization. Wherever possible, in illustrations given, the forms are from actual files for specific jobs. These forms, therefore, not only illustrate various management techniques, but also give specific details as they apply to particular situations.

17.1 TYPES OF CONSTRUCTION COMPANIES

The principles of construction project management, as outlined in this article, apply equally to those engaged in subcontracting and those engaged in general contracting.

Small Renovation Contractors. These companies generally work on jobs requiring small amounts of capital and the type of work that does not require much estimating or a large construction organization. They usually perform home alterations or small commercial and office work. Many small renovation contractors have their offices in their homes and perform the "paper work" at night or on weekends after working with the tools of their trade during the day. The ability to grow from this type of contractor to a general contractor depends mainly on the training and business ability of the individual. Generally, if one is intelligent enough to be a good small renovation contractor, that person may be expected to eventually move into the field of larger work.

General Contractors. These companies often are experts in either new buildings or alteration work. Many building contractors subcontract a major portion of their work, while alteration contractors generally perform many of the trades with their own forces.

Some general contractors specialize in public works. Others deal mainly with private and commercial work. Although a crossing of the lines by many general contractors is common, it is often in one or another of these fields that many general contractors find their niche.

Owner-Builder. The company that acts as an owner-builder is not a contractor in the strict sense of the word. Such a company builds buildings only for its own ownership, either to sell on completion, or to rent and operate. Examples of this type of company include giants in the industry, and many of them are listed on the various stock exchanges. Many owner-builders, on occasion, act in the capacity of general contractor or as construction manager (see below) as a sideline to their main business of building for their own account.

Real Estate Developer. This is a type of owner-builder who, in addition to building for personal ownership, may also build to sell before or after completion of the project. One- and two-family home builders are included in this category.

Professional Construction Manager. A professional construction manager may be defined as a company, an individual, or a group of individuals who perform the functions required in building a project as the agent of an owner, but do so as if the job was being performed with the owner's own employees. The construction management organization usually supplies all the personnel required. Such personnel include construction superintendents, expediters, project managers, and accounting personnel.

The manager sublets the various portions of the construction work in the name of the owner and does all the necessary office administration, field supervision, requisitioning, paying of subcontractors, payroll reports, and other work on the owner's behalf, for a fee. Generally, construction management is performed without any risk of capital to the construction manager. All the financial obligations are contracted in the name of the owner by the construction manager. (See Art. 17.9.)

Package (Turnkey) Builders. Such companies take on a contract for both design and construction of a building. Often these services, in addition, include acquisition of land and financing of the project. Firms that engage in package building usually are able to show prospective clients prototypes of similar buildings completed by them for previous owners. From an inspection of the prototype and discussion of possible variations or features to be included, an approximate idea is gained by the prospective owner of the cost and function of the proposed building.

Package builders often employ their own staff of architects and engineers, as well as construction personnel. Some package builders subcontract the design portion to independent architects or engineers. It is important to note that, when a package builder undertakes design as part of the order for a design-construction contract, the builder must possess the necessary professional license for engineering or architecture, which is required in most states for those performing that function.

Sponsor-Builder. In the field of government-aided or subsidized building, particularly in the field of housing, a sponsor-builder may be given the responsibility for planning design, construction, rental, management, and maintenance. A sponsor guides a project through the government processing and design stages. The sponsor employs attorneys to deal with the various government agencies, financial institutions, and real estate consultants, to provide the know-how in land acquisition and appraisal. On signing the contract for construction of the building, the sponsor assumes the builder's role, and in this sense functions very much as an owner-builder would in building for its own account.

17.2 CONSTRUCTION COMPANY ORGANIZATION

How a construction company organizes for its work depends on number and size of projects, project complexity, and geographical distribution of the work.

Sole Proprietor. This is a simple form of organization for construction contractors. It is often used by subcontractors, including those licensed in plumbing, electrical, or mechanical work. The advantage of operating as a sole proprietorship is that taxes on profits are much lower for individual owners. But there is the disadvantage of having the personal exposure to potential debts associated with a disastrous job.

Partnership. This is the joint ownership and operation of a company by two or more persons. Each partner, however, is personally liable for all the debts of the partnership. Profits and losses are shared in some manner predetermined by the partners. A partnership comes to an end with the death of one of the partners. (For typical provisions to be included in a partnership agreement, see Richard H. Clough, "Construction Contracting," John Wiley & Sons, Inc., New York.)

Corporation. This is the most common form of organization used by general contractors. A corporation is an entity that has the power to act as a separate body and enter into contracts. It has perpetual life and is owned by stockholders, each of whom has a share in the profits and losses of the corporation. An important advantage of the corporate form of ownership for general contractors is the absence of personal liability of the stockholders. This is desirable because of the risks of the contracting business, and is more than recompense for the additional burden of taxes that those taking part in corporate ownership must bear. (Small corporations can obtain some relief from Federal taxes, however.)

Corporations formed in one state must obtain, as a foreign corporation, a certificate of authority to do business in other states. This is important when bidding jobs in locations other than the home state of the contractor.

Some general contractor corporations are large enough to find it advantageous to raise capital by becoming public corporations, with shares sold over the counter or on the various stock exchanges. Such corporations publish financial reports yearly for the benefit of the stockholders, as required by law. A study of such reports is often helpful for those engaged in the contracting business.

Joint Venture. Often when an individual job is too large to be undertaken by one company, or the risks involved are too great for one company to want to assume (although it may be capable of doing so), a joint venture is formed. This is an association between two or more contracting firms for a particular project. It joins the resources of the venturers, who share the financing and management of the job and the profits or losses in some predetermined manner.

Generally, there are specific reasons for the formation of a joint venture between specific companies. For example, one may possess the equipment and the other the know-how for a particular job. Or one may possess the financing and the other the personnel required to perform the contract. Joint ventures do not bind the members to any debts of the coventurers other than for those obligations incurred for the particular jobs undertaken.

Staff Organization. An organization chart for a typical medium-size general building contracting company is shown in Fig. 17.1. The organization shown is for a company that subcontracts most of its work and is engaged mainly in new construction. For an organization with district offices, see the chart in Paul G. Gill, "Systems Management Techniques for Builders and Contractors," McGraw-Hill, Inc., New York.

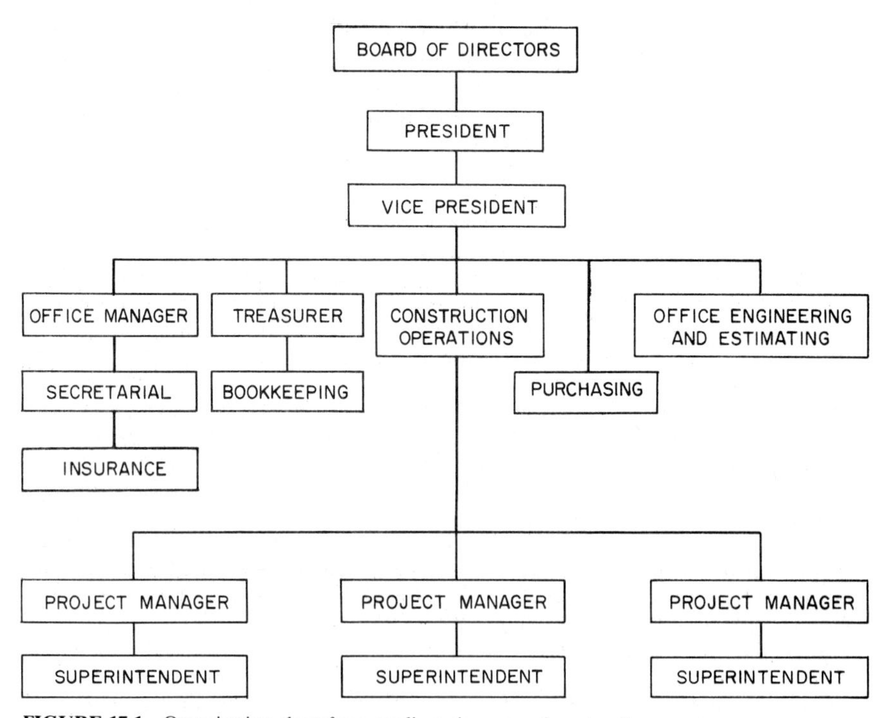

FIGURE 17.1 Organization chart for a medium-size general contracting company.

17.3 CONTRACTORS' BUSINESS CONSULTANTS

Construction contractors often have to engage experts from various disciplines to advise and assist them in conducting their business. In addition to architectural and engineering consultants, contractors usually consult the following:

Accountant. The accountant selected by a construction company should preferably be one who has had experience in contracting and construction accounting. This accountant will know the generally accepted principles of accounting applying to construction projects, such as costs, actual earnings, and estimated earnings on uncompleted construction contracts. In addition, the accountant should have an understanding of management's role to help formulate the financial picture of the firm. This role of management involves estimating the probable earnings on uncompleted jobs and the amounts of reserves that should be provided for by the accountant for contingencies on jobs that have already been completed but for which final settlements have not yet been made with all the subcontractors and suppliers.

Attorney. A construction company may find that it needs more than one attorney to handle all of its affairs. For example, it will in all likelihood have an attorney who will be retained for most of the routine matters of corporate business, such as formation of the corporation, registration by the corporation in other states, routine contract advice, and legal aid in the general affairs of the business. In addition to this, however, many construction companies require different attorneys for such phases of their activities as claims, personal affairs and estate work, real-estate and tax-shelter matters, and government programs and processing. Many of these functions are performed by attorneys who specialize in that type of work and can offer the most up-to-date advice.

Insurance and Bonding. An insurance broker who has a fairly sizable volume of business is the preferred source of insurance. Such a broker will have the greatest amount of leverage with insurance companies when questions arise about claims for losses or about requirements by insurance companies for premium deposits on policy renewals. Most general insurance brokers with a fairly large clientele should be able to handle construction company insurance.

When it comes to bonding, however, it may be advisable to deal with a firm that specializes in general contractors and their bonding problems. Often, general insurance brokers do not have much experience in this field. While surety companies, who issue bonds, are generally subsidiaries of insurance companies, bonding and insurance are guided by entirely different principles. A broker who has many clients requiring performance and payment bonds is to be preferred for meeting bonding requirements. Such a broker will be able to offer advice on the bonding companies best suited for the contractor's specific business. Also, the broker should be able to aid the contractor and the contractor's accountant in preparation of financial statements so as to show the contractor's financial position in the most favorable light for bonding purposes.

Banking. Bank accounts needed for the company's afairs should preferably be divided between two or more banking institutions. A banking relationship with more than one bank could be advantageous at times. For example, it could facilitate a request for banking references for credit; for a bank to be available for temporary loans or for writing letters of credit; for equipment loans; and for advice on investments or for custodian accounts for the handling of surplus funds of the contractor.

Trade Associations. Many contractors find it advantageous to become a member of a local trade association or local branches of national organizations, such as the Associated General Contractors of America and the Construction Management Association of America. Before doing so, however, the contractor should investigate and be thoroughly acquainted with the rules of such associations, particularly with regard to labor bargaining.

In most cases, membership in a local trade association binds the contractor to permit that association to do all of its labor-contract bargaining with local labor unions and the various construction trades. Furthermore, if there is a strike by any of the local labor organizations, the members must obey the dictates of the association and may be required to join in the association's stand against the strikers. On the other hand, a contractor who is not a member of the local association is free to bargain individually with the trade unions involved and can sign their contract as an *independent.* Often, labor unions, during a strike, will be willing to offer an interim contract to independent contractors under which terms the contractor will be able to resume work on agreeing to be bound by the terms of the labor agreement consummated to settle the strike, effective retroactively.

17.4 SOURCES OF BUSINESS

For continuity of operation, a construction organization needs a supply of new projects to build. After a company has been in existence for a long time and built up a reputation, new business may come to it with less effort. But most companies must work hard at obtaining new jobs. Furthermore, work that happens to come in may not be of a type that the organization prefers. To find that type requires serious, skillful efforts.

To be successful, a contracting company should have a person specifically assigned to attract new business. This person might be the proprietor of the construction company. In large firms, a complex organization with sales and public relations personnel, backed up by engineers and cost estimators, is used. The organization should be geared to follow up on all possible sources of new business.

Public Works. The following sources for leads to new jobs and submitting proposals can be used for public work bids:

Dodge Bulletin, or other construction industry newsletters.

ENR—"Pulse" and "Official Proposals" sections.

Bid invitations, as a result of requests to be placed on bid-invitation mailing lists of various government agencies.

Newspaper and trade-paper announcements and articles.

Official publications of government agencies that contain advertisements of contracts to be bid.

Private Contracts. All the sources for public works.

Contacts with and letters to architects and designers.

Contacts with and letters to owners and facilities personnel.

Personal recommendation.

Sponsorship. Applying for sponsorship of any of the following:

Government-encouraged housing programs.

Urban renewal.

Purchase of land, with financing of building construction to be provided by various government programs.

Owner-Builder/Developer

Construction and rental of apartment buildings.

Construction and rental of commercial and office facilities.

Construction and leasing of post offices and other government buildings.

Professional Construction Manager

Applications to city and state agencies or large corporations awarding this type of contract.

All the sources for obtaining private contracts.

Uses of Dodge Bulletin. From a typical *Dodge Bulletin* (McGraw-Hill Information Systems Company, New York), a subscriber to this daily information bulletin can gain the following information:

Contracts for which general contract and prime bids for mechanical and electrical work, etc., are being requested. A contractor interested in any of these types of work can obtain the plans and specifications from owners or designers, whose names and addresses are given, and submit a bid.

For contracts awarded, lists of names of contractors and amounts of contracts. Subcontractors or material suppliers who are interested in working for the contractors can communicate with those who have received the awards.

Lists of jobs being planned and estimates of job costs. Contractors and subcontractors who are interested in jobs in those locations and the sizes indicated can communicate directly with the owner.

Additional information that may be obtained from the *Dodge Bulletin* includes lists of subcontractors and suppliers being employed by general contractors on other jobs that are already under way, and tabulation of the low bidders on jobs bid and publicly opened.

17.5 WHAT CONSTITUTES THE CONTRACT DOCUMENTS?

Generally, plans and specifications for a project are completed before issuance to contractors. This, however, is not always the case. For example, during the course of design, the architect may distribute progress plans and specifications for review by government agencies or for pricing by contractors. Sometimes, the owner may be anxious to enter into a contract or start a job before the plans and specifications have been totally completed or approved. Therefore, to the general contractor, the most important thing to be alert to regarding plans and specifications is: What constitutes the agreed-upon plans and specifications and other contract documents that pertain to the project?

It is surprising how often this question is neglected by those entering into construction contracts and by attorneys and others concerned with signing of contracts. A clear understanding of what constitutes the contract documents and the revisions, if any, is one of the most urgent aspects of construction contracting. Furthermore, a precise list of what constitutes the contractor's obligation under the contract is essential to proper performance of the contract by the contractor.

In general, the contract documents should be identified and agreed to by both parties. A listing of the contract documents should be included as part of every contract. Contract documents generally include the following:

Plans (list each plan and revision date of each plan, together with title)

Specifications, properly identified, a copy initialed by each party and in the possession of each party

The agreement, or contract (Art. 17.8)

General conditions

Soil borings

Existing site plan

Special conditions

Original proposal (if it contains alternatives and unit prices, and these are not repeated in the contract)

Invitation (if it contains data on completion dates or other information that is not repeated in the contract)

Addenda (if any)

The contractor should repeat and list the contract documents in each subcontract and purchase order (Arts. 17.10.1 and 17.11).

It is essential to have a properly drawn and understood list of the contract documents agreed to by all parties if construction management is to proceed smoothly and if changes and disputes are to be handled in an orderly manner (Art. 17.14). To prevent misunderstandings and doubts as to which documents are in the possession of various subcontractors and suppliers for estimating, a properly drafted transmittal form should accompany each transmission of such documents (Fig. 17.2).

17.6 MAJOR CONCERNS WITH BUILDING CODES

Contractors should have a working knowledge of a variety of building codes. In most cities and municipalities, there is a local building code. It also may be the same as the state code.

TRANSMITTAL

Kreisler Borg Florman
CONSTRUCTION COMPANY
97 Montgomery Street, Scarsdale, New York 10583/Telephone (914) SC 5-4600

TO: Lipsky & Rosenthal, Inc.
155 Utica Avenue
Brooklyn, N Y 11213

ATT: Mr. D. G.
Mr. A. H.

DATE February 25,

RE: Combined School and Apartment, New York, N Y

WE ARE SENDING YOU THIS DATE THE FOLLOWING:

- [X] BLUEPRINTS
- [ ] SPECS
- [ ] SAMPLES
- [ ] BOOKLETS
- [ ] CATALOGUE CUTS
- [ ] LETTERS
- [ ] TEST REPORTS
- [ ]

SENT FOR THE FOLLOWING REASON:

- [ ] YOUR APPROVAL
- [ ] YOUR FINAL APPROVAL
- [ ] APPROVED
- [ ] APPROVED AS NOTED
- [ ] DISAPPROVED
- [ ] RESUBMIT
- [X] YOUR INFORMATION
- [ ] YOUR USE
- [ ]

NUMBER COPIES	DRAWING NUMBER	PREPARED BY	DESCRIPTION
1 Set	A-1 thru A-10 Dated 2/18/	Carl Puchall & Assoc.	School Preliminary Plans
1	7 of 12 Rev. 10/18/	" " "	Typical Apartment Floor Plan
1	9 of 12 Rev. 1/17/	" " "	Sub. Cellar Garage

REMARKS: We would appreciate your budget figure for referenced project, as soon as possible. Price must be broken down to School portion and Residential portion.

Thank you.

BY ________________

TITLE Estimator

FIGURE 17.2 Form for submittal of contract documents to subcontractors and suppliers.

Local and state codes usually govern most of the construction activities of a contractor, in addition to the requirements of the plans and specifications. The contractor's task of satisfying code regulations, however, is made more difficult because most building codes continually undergo significant changes. Consequently, contractors should be alert to such changes and have sources that provide them with information on the new regulations.

In addition, local and state building codes provide means for enforcing compliance of contractors with code requirements. Usually, such enforcement takes the form of inspections by building inspectors, affidavits by architects and engineers that construction has been performed consistent with the code, and submission of plans, specifications, and details to building officials for their approval before construction starts.

It is only after all code requirements have been complied with and all required approvals have been obtained that a contractor will receive a Certificate of Occupancy for a completed building. Before this, however, approval of the work may also have to be obtained from other local officials: Fire Department for sprinklers, fire alarms, and exit and elevator safety; Health Department for food-handling and kitchen equipment; Environmental Protection Agencies for boiler and incinerator emissions; Police Department for fire and sprinkler alarms.

See also Art. 1.10.

17.7 ESTIMATING, BIDDING, AND COSTS

Methods of preparing cost estimates for building construction are described in Sec. 19. It is advisable to have the routine to be followed in preparing cost estimates and submitting bids well established in a contractor's organization.

Particular attention should be given to the answers to the question: For whom is the estimate being prepared and for what purpose? The answers will influence the contractor as to the amount of time and effort that should be expended on preparation of the estimate, and also indicate how serious the organization should be about attempting to negotiate a contract at the figure submitted. Decision on the latter should be made at an early date, even before the estimate is prepared, so that the type of estimate can be decided.

Bid Documents. The documents should be examined for completeness of plans and specifications, and for the probable accuracy that an estimate will yield from the information being furnished. For example, sometimes contract documents are sent out for bid when they are only partly complete and the owner does not seriously intend to award a contract at that stage but merely wishes to ascertain whether construction cost will be acceptable.

Preparation of the Top Sheet. This is usually based on an examination of the specifications table of contents. If there are no specifications, then the contractor should use as a guide top sheets (summary sheets showing each trade) from previous estimates for jobs of a similar nature, or checklists.

Subcontractor Prices. Decide on which trades subbids will be obtained, and solicit prices from subcontractors and suppliers in those trades. These requests for prices should be made by postcard, telephone, or personal visit.

Decide on which trades work will be done by the contractor's own forces, and prepare a detailed estimate of labor and material for those trades.

Pricing. Use either unit prices arrived at from the contractor's own past records, estimates made by the members of the contractor's organization, or various reference books that list typical unit prices ("Building Construction Cost Data," Robert Snow Means Co., Inc., P.O. Box G, Duxbury, Mass. 02332). Computerized banks of unit prices for various types of work on different structures may be maintained by a contractor. These can be updated electronically with new wage and material costs, depending on the program used, so that prices can be applied nearly automatically.

Hidden Costs. Carefully examine the general conditions of the contract and visit the site, so as to have a full knowledge of all the possible hidden costs, such as special insurance requirements, portions of site not yet available, and complicated logistics.

Final Steps. Receive prices for materials and subcontracts.

Review the estimate and carefully note exclusions and exceptions in each subcontract bid and in material quotations. Fill in with allowances or budgets those items or trades for which no prices are available.

Decide on the markup. This is an evaluation that should be made by the contractor, weighing factors such as the amount of extras that may be expected, the reputation of the owner, the need for work on the part of the general contractor, and the contractor's overhead.

Finally, and most importantly, the estimate must be submitted in the form requested by the owner. The form must be filled in completely, without any qualifying language or exceptions, and must be submitted at the time and place specified in the invitation to bid. Figure 17.3 shows part of an estimate and bid summary produced for a multistory apartment building by a computer.

17.8 TYPES OF BIDS AND CONTRACTS

Contractors usually submit bids for a lump-sum contract or a unit-price-type contract, when based on complete plans and specifications.

When plans and specifications have not yet advanced to a stage where a detailed estimate can be made, the type of contract usually resorted to is a cost-plus-fee type. The bid may be based on either a percentage markup the contractor will receive over and above costs, or may include a lump-sum fee that the contractor charges over and above costs. Sometimes an incentive fee is also incorporated, allowing owner and contractor to share, in an agreed-on ratio, the cost savings achieved by the contractor, or to reward the contractor for completing the project ahead of schedule.

Evaluation of bids by an owner may take into consideration the experience and reputation of the contractor, as a result of which awards may not be made on a strictly low-bid basis.

Budget Estimate. Often, an owner in preparing plans and specifications will want to determine the expected construction cost while the plans are still in a preliminary stage. The owner, in that case, will ask contractors for a cost estimate for budget purposes. If the estimate from a specific contractor appears to be satisfactory to an owner, and if the owner is desirous of establishing a contractual relationship with the contractor early in the planning stage so as to benefit from the contractor's

KREISLER BORG FLORMAN
ST & BROADWAY, NY, NY — APRIL 28,19 — 24-Sep- REV = DJB
FILE :BWAYREV

APTS-	354	PLBG FIX	2074.00	RETAIL-	34685.00
ROOMS-	1364	FLUE	28.00	GARAGE-	33858.00
S.F.,GROSS=	496081	CHUTE	26.00	TOTAL	68543.00
APT-	427538	ELEV	113.00		
FDTN-	30665				
ROOF-	30665				

****

TRADE	QTY	ESTIMATE/UNIT	ESTIMATE	ADJUSTMENTS	ACTUAL BIDS		ACTUAL UNIT COST				
EXCAVATION	30665	0.00	0.00	0.00							
PILES			0.00	0.00							
CONCRETE FDTNS		W/EXCAV	0.00	0.00							
CONCRETE SUPER	496081	22.50	11,161,822.50	(867,083.50)	10,294,739.00	METRO	$20.75	FLUE EXTENSION $25,000			METRO
SOG	20956	0.00	0.00	0.00			$0.00				
SITE/W EXCAV	1	163,376.00	163,376.00	68,820.00	232,196.00	SITE CONSULTANTS	$232,196.00				SITE CONSULT
A.C. UNITS	0		0.00	0.00							
SPRINKLERS	33858	1.75	59,251.50	116,448.50	175,700.00	S&S	$5.19				S&S
CHUTE SPR.	1	10,000.00	10,000.00	0.00	INCL.		$0.00				
FLUE	28	2,000.00	56,000.00	105,100.00	161,100.00	J.K. ENVIRONMENTAL	$5,753.57	FLUE EXTENSIONS 96,600			J.K. ENVIRON
ELECTRIC	354	10,500.00	3,717,000.00	847,000.00	4,564,000.00	REMARK	$12,892.66				REMARK
GARAGE			0.00	0.00							
RETAIL(rental office)				500,000.00	500,000.00						
MISCELLANEOUS	1	50,000.00	50,000.00	134,000.00	184,000.00		$184,000.00				
SALES TAX	0.0825	2,100,670.00	173,305.28	11,133.46			$0.00				
SUBCONTRACTS			$37,476,162	$5,490,929	$42,967,091				121,375.96	86.61	
GENERAL CONDITIONS(@10%)			$3,747,616	$549,093	$4,296,709				12,137.60	8.66	
SUB-TOTAL			$41,223,778	$6,040,022	$47,263,800				133,513.56	95.27	
OVERHEAD & PROFIT(5%)			$2,061,189	$302,001	$2,363,190				6,675.68	4.76	
SUB-TOTAL			$43,284,967	$6,342,023	$49,626,990				140,189.24	100.04	
CONTINGENCY 6%				$2,977,619							
TOTAL COST(W/CONTING.)				$52,604,609						106.04	
COST/DU			$148,601								

FIGURE 17.3 Output from a computer spreadsheet program of an estimate and bid summary.

suggestions and guidance, a contract may be entered into after the submission of the estimate.

On the other hand, the owner may refrain from formally entering into the contract, but may treat the contractor as the "favored contractor." When requested, this contractor will assist the architect and engineers with advice and cost estimates and will expect to receive the contract for construction on completion of plans and specifications, if the cost of the project will lie within the budget estimate when plans and specifications have been completed.

Separate Prime Contracts. Sometimes an owner has the capability for managing construction projects and will take on some of the attributes of a general contractor. One method for an owner to do this is to negotiate and award separate prime contracts to the various trades required for a project. Administration of these trades will be done either by the owner's own organization or by a construction manager hired by the owner (Art. 17.9).

Sale Lease-Back. This is a method used by some owners and government agencies to obtain a constructed project. Prospective builders are asked to bid not only on cost of construction, but also on supplying a completed building and leasing it to the prospective user for a specified time. This type of bid requires a knowledge of real estate analysis and financing, as well as construction. Contractors who bid may have to associate with a real estate firm to prepare such a bid.

Developer/Sponsor-Builder. In this type of arrangement (Art. 17.1), the contractor may not only have to prepare a construction-cost estimate but may also need a knowledge of real estate and be prepared to act as owner of the completed project, in accordance with the terms of a sponsor-builder agreement with a government agency, or government-assisted neighborhood or nonprofit group.

The following types of contracts are used for general construction work:

Letter of Intent. This is used where a quick start is necessary and where there is not sufficient time for drafting a more detailed contract. A letter of intent also may be used where an owner wishes material ordered before the general contract is started, or where the commitment of subcontractors requiring extensive lead time must be secured immediately.

Lump-Sum Contract. (For example, Document A-101, American Institute of Architects, 1735 New York Ave., N.W. Washington, D.C. 20006.) Basis of payment is a stipulated sum. Progress payments, however, are made during the course of construction.

Cost-plus-Fixed-Fee Contract. (For example, A-111, American Institute of Architects.) This type of agreement is used generally with an "up-set" or "guaranteed maximum" price. The contractor guarantees that the total cost plus a fee will not exceed a certain sum. (See types of bids, preceding.) Generally, there are provisions for auditing of the construction costs by the owner.

Cost-plus-Percentage-of-Fee Contract. (For example, A201, American Institute of Architects.) Similar to cost-plus-fixed-fee contracts, but the fee paid, instead of being a lump sum in addition to the cost, is a percentage of the costs.

Unit-Price Contract. This type of agreement is used where the type of work involved is subject to variations in quantities and it is impossible to ascertain the total amount of work when the job is started. Bids will be submitted by the contractors on the basis of estimated quantities for each classification of work involved. On the basis of the unit prices submitted and the estimated quantities, a low bidder will be chosen for award of the contract. After the contract has been completed, the final amount paid to the contractor will be the sum of the actual quantities encountered for each class of work multiplied by the unit prices bid for that class.

Design-Construction Contracts. This type of agreement is used for turnkey projects and by package builders (Art. 17.1). Cost estimates must be prepared from preliminary plans or from similar past jobs. Preparation of these estimates requires high skill and knowledge of construction methods and costs, because the usual methods for preparing cost estimates do not apply.

Construction Management Contracts. Under this type of agreement, the construction entity serves as construction manager, acting as an agent of or consultant to the owner, for a fee plus reimbursement of General Conditions items. The construction manager may negotiate an additional fee for extras (Art. 17.9).

17.9 PROFESSIONAL CONSTRUCTION MANAGERS

A professional construction manager (CM) is an individual or organization specializing in construction management or practicing it on a particular project as part of a project management team that also includes an owner and a design organization. A prime construction contractor or a funding agency may also be a member of the team. (See Art. 1.13.)

As the primary construction professional on the project management team, the CM provides the following services or such portion thereof as may be appropriate to the specific project:

1. The CM works with the owner and the design organization from the beginning of the design concept through completion of construction and closeout, providing leadership to the construction team on all matters relating to construction, keeping the project management team informed, and making recommendations on design improvements, value engineering, construction technology, schedules, and construction economies.

2. The CM proposes construction and design alternatives to be studied by the project management team during the planning phase and analyzes the effects of these alternatives on the project cost, schedule, and life-cycle cost.

3. After the project budget, schedule, and quality requirements have been established, the CM monitors subsequent development of the project to ascertain that those targets are not exceeded without knowledge of the owner.

4. The CM advises on and coordinates procurement of equipment and materials and the work of all contractors on the project. Also, the CM may monitor payments to contractors, changes ordered in the work, contractors' claims for extra payments, and inspection for conformance with design requirements. In addition, the CM provides current cost and progress information as the work proceeds and performs other construction-related services required by the owner, such as furniture, fixtures, and equipment interfacing.

The CM normally engages other organizations to perform significant amounts of construction work but may provide some or all of the site facilities and services specified in the General Conditions of the construction contract and is usually reimbursed by the client for these costs.

The advantages of engaging a CM over conventional construction with a general contractor (see Art. 1.4) are as follows:

The CM treats project planning, design, and construction as integrated tasks, which are assigned to a project management team. The team works in the owner's best interests from the beginning of design to project completion. The contractual relationships between the members of the team are intended to minimize adversary relationships and to contribute to greater responsiveness within the team. In this way, the project can be completed faster and at lower cost. (See also Art. 2.19.)

(D. Barry and B. C. Paulson, "Professional Construction Management," Journal of the Construction Division, American Society of Civil Engineers, September 1976.)

17.10 CONTRACT ADMINISTRATION

Administration of construction contracts requires an intimate knowledge of the relationship of the various skills required for the construction, which involves labor, material suppliers, and subcontractors. Feeding into the job are all of the life-supplying services. Whether the contractor combines one or more of the jobs in more than one person is immaterial.

Construction project management that will result in a profitable, on-time job involves the organization and interplay of many talents. Activities of engineers, architects, field supervisors, construction labor, material and equipment suppliers, and subcontractors, aided by accountants, attorneys, insurance and bonding underwriters, design professionals, and the owner, must be organized and carefully coordinated.

Those who succeed in this complex and difficult business are the ones who familiarize themselves thoroughly with the daily operations of their jobs. They are constantly learning by reading the latest professional journals, keeping abreast of legislative developments and governmental regulations affecting the construction business, and attending seminars on industry functions. They are alert and open-minded about new ideas. They understand the needs of different clients and design professionals and are able to tailor services to them.

The task of the contractor, principal, or partner is to be familiar with and have responsibility for legal, bonding, insurance, and banking requirements of the firm. The contractor feeds into the job necessary organization and policy decisions. This contribution, when added to what is fed into the job by the project manager (progress), bookkeeping (money), superintendent (progress), clerical (correspondence and records), architect and engineers (plans and approvals), building department (approvals and inspection), and the owner (money), is essential for job progress.

Planning of the job is dealt with in Art. 17.10.2. Profit and loss of the job are controlled in the manner described in Art. 17.19.

17.10.1 Subcontracts

A contractor who engages others (subcontractors) to perform construction is called a general contractor. General contractors usually obtain subcontract bids as well as material-price solicitations during the general-contract bidding stage. Sometimes, however, general contractors continue shopping after award of the general contract to attain budget goals for the work that may have been exceeded during the initial bidding stages. In such cases, additional bids from subcontractors are solicited after the award of contract.

Purchasing Index. Contractors would find it advantageous to approach purchasing of subcontracts with a purchasing index (Fig. 17.4). This index should list everything necessary to be purchased for the job, together with a budget for each of the items. As subcontracts are awarded, the name of the subcontractor is entered, and the amount of the subcontract is noted in the appropriate column. Then, the profit or loss on the purchase is later entered in the last column; thus, a continuous tabulation is maintained of the status of the purchases.

Priority numbers are given to the various items, in order of preference in purchasing. The contractor should concentrate efforts on those subcontracts that must be awarded first. Those that follow in due course are given priority numbers that are appropriate. These priorities are indicated in Fig. 17.4 by the numbers in the left column.

Bid Solicitation. Bids are generally solicited through notices in trade publications, such as *Dodge Bulletin,* or from lists of subcontractors that the contractor maintains. Solicitation also can be by telephone call, letter, or postcard to those invited to bid. Where the owner or the law requires use of specific categories of subcontractors, bids have to be obtained from qualified members of such groups.

After subcontractor bids have been received, careful analysis and tabulation are needed for the contractor to compare bids fairly (Fig. 17.5).

KREISLER BORG FLORMAN CONSTRUCTION CO.
97 Montgomery Street
Scarsdale, New York 10583

PRIORITY NO.		ITEM	ESTIMATE	SUBCONTRACTOR	AMOUNT	PROFIT
1	2A	Excavation	77,000			
1	2B	Foundations	INCL.			
1	2H	Piling	INCL.			
1	3A	Concrete Superstructure	242,000			
2	4	Masonry	104,400			
3	4B	Cast Stone	INCL.			
3	4C	Ext. Cement Ash Sills	INCL.			
6	12A	Venetian Blinds	2,200			
2	14A	Elevators	29,600			

FIGURE 17.4 Purchasing index. Numbers in first column indicate priority in purchasing subcontracts, materials, and equipment.

Work	Bids Received			Comments	Low Bid	Difference from Budget
Roofing	Abbott Sommer	Kings County	Colonial			
Roof & terrace	146, 500		152, 896			
	Parapet		4-ply 133, 976			
	AL coping		AL coping 18, 920			
	Scupper		Sidewalk & tenant			
Pavers			garden			
Sidewalk-tenant	MRMRCO 23, 000	13, 950	waterproofing			
Garden waterprfg.	1/4 Prot. Bd.	3" Rig. Ins.	included.			
	Insulation?	1/8" Bd.	No insulation.		152, 896	7, 896
Moisture Protection	Abbott Sommer	Kings County				
Spandrel flashing	20 mil 21, 500	17, 050*		*Window sill flashing		
Dampproofing	8, 500	7, 800		If supply only		
Caulking		36, 600		-4, 200.		
Backer rod		13, 500				
		74, 950			74, 950	4, 950
Window	Mannix	Loxscreen	Traco			
	945, 000	1.014, 750	949, 700			
	(tax included)					
	Include louver					
	AL stools (inside)	AL stools?	AL stools?			
	Excludes AL copings	AL coping?	No			
	If windows are full					
	height (-100.000)				945, 000	20, 000
Doors & Bucks	Acme	Firedoor				
	Tower 46, 750	48, 750				
	YMCA 6, 250	6, 250				
	53, 000	55, 000 +				
	Seam on edge (-1, 900)	tax				
	Tower: 543 Fr.2x3 door					
	2 borrow lites					
	YMCA: 53 Fr. 19 doors				53, 000	3, 000
	3 panel				Tax 4, 372	

FIGURE 17.5 Analysis of subcontractor bids.

In a complicated trade, such as Moisture Protection, shown in Fig. 17.5, it is necessary to tabulate, from answers obtained by questioning each of the bidders, the exact items that are included and excluded. In this way, an evaluation can intelligently be made, not only of the prices submitted but also as to whether or not the subcontractors are offering a complete job for the section of work being solicited. Where an indication in the subcontractor's proposal as in Fig. 17.5 shows that a portion of the work is being omitted, it is necessary to cross-check the specifications and other trades to be purchased to ascertain that the missing items are covered by other subcontractors.

Subcontract Forms. Various subcontract forms are available for the written agreements. A commonly used form is the standard form of agreement between contractor and subcontractor (Contractor-Subcontractor Agreement, A401, American Institute of Architects). A short form of subcontract with all the information appearing on two sides of one sheet is shown in Fig. 17.6*a* and *b*. Changes may be made on the back of the printed form with the permission of both parties to the agreement.

Important, and not to be neglected, is a subcontract rider (Fig. 17.7), which is tailored for each job. Only one page of the rider is shown in Fig. 17.7. The rider takes into account modifications required to adapt the standard form to the specific project. The rider, dealing with such matters as options, alternatives, completion dates, insurance requirements, and special requirements of the owner or lending agency, should be attached to all copies of the subcontract and initialed by both parties.

Kreisler Borg Florman
CONSTRUCTION COMPANY (INC.)
97 Montgomery Street, Scarsdale, New York 10583
Telephone (914) SC 5-4600 DATE August 31,

TO: Brisk Waterproofing Co Inc
720 Grand Avenue
Ridgefield, New Jersey

PROJECT: East Midtown Plaza Stage II, Project #HRB 66-14B
24th & 25th Sts betw. 1st & 2nd Aves, New York, NY

SUBJECT TO CONDITIONS ON REVERSE SIDE HEREOF, YOU ARE AUTHORIZED TO PROCEED WITH THE WORK AS DESCRIBED BELOW IN CONNECTION WITH THE ABOVE PROJECT, THE COST OF WHICH IS TO BE $ 4,000.00

FOUR THOUSAND DOLLARS

DESCRIPTION

HYDROLITHIC IRON TYPE WATERPROOFING (Specification Division 5, Section 35) and all such work shown on the Plans, Specification, Specification Addenda 1, 2, 3 and 4, Construction Contract, General Conditions of the Contract, Invitation, Contractor's Loan Agreement, Bid Form, Performance and Payment Bond, and Subcontract Rider. If Performance and Payment Bond is required premium is to be paid by Contractor.
As a condition precedent to the duty of performance of the work hereunder the Contractor shall be awarded and execute a contract for the work of the project.
Subcontractor agrees to include at no additional cost any further details or corrections to the plans and specifications resulting from requirements of the Building Department or Housing and Development Administration and agrees to be bound by final plans and specifications when they are completed and approved.
This work includes the H.I.T. waterproofing of two elevator pits, 1 ejector pit and laundry troughs and curbs as indicated on plans. In connection with this work, light, water, heat, pumping and power to be furnished to us without charge. Floor slabs to be left raked by others.

PLEASE SIGN EXTRA COPY OF THIS ORDER AND RETURN TO KREISLER-BORG AT ONCE.
PLEASE SEND US CERTIFICATES OF INSURANCE FOR WORKMEN'S COMPENSATION, PUBLIC LIABILITY AND PROPERTY DAMAGE INSURANCE.

ALL OF THE ABOVE MATERIALS TO BE DELIVERED OR THE WORK COVERED BY THIS ORDER ARE TO BE COMPLETED IN ACCORDANCE WITH PLANS, SPECIFICATIONS AND ALL CONTRACT REQUIREMENTS BETWEEN US AND THE OWNER, BY ALL OF WHICH YOU AGREE TO BE BOUND UPON ACCEPTANCE OF THIS ORDER.

Kreisler Borg Florman (CONTRACTOR)
CONSTRUCTION COMPANY (INC.)

BY R Borg
Robert F. Borg, President

ACCEPTED:
BRISK WATERPROOFING CO INC
SUBCONTRACTOR

BY RW Ehrenberg
R. W. Ehrenberg, V.P.

DATE: 9/6/

(over)

FIGURE 17.6*a* Front side of short form for a subcontract.

CONDITIONS OF CONTRACT

Within five days after the date of this contract and before commencement of the work, the subcontractor agrees to furnish the contractor with a certificate showing that he is properly covered by Workmen's Compensation Insurance as required by the law of the State where the work is to be performed and with such other insurance that may be required by Contractor, the specifications and terms of contract between Contractor and the Owner.

Where the order covers the furnishing of labor and materials on a time and material basis, it is distinctly understood the subcontractor will furnish daily vouchers for verification and signature to an authorized representative of Contractor showing labor used and materials installed in the work. A copy of these signed vouchers to be presented with invoice, together with duplicate bills for materials furnished. Contractor shall have the right to examine all records of the subcontractor relating to said charges.

Where this order is issued to a subcontractor, and purports to cover labor and materials in addition to the original contract, it is given with the express understanding that, should it subsequently be proven that the work covered herein is in the subcontractor's original contract, this order becomes null and void.

Should the subcontractor or material dealer at any time refuse or neglect to supply an adequate number of properly skilled workmen or sufficient materials of the proper quality, or fail in any respect to prosecute the work with promptness and diligence, the Contractor shall in its exclusive opinion be at liberty after three days' notice to the subcontractor or material dealer to provide any such labor or materials in accordance with such notice and to deduct the cost thereof from any money then due or to become due to the subcontractor or material dealer under this contract, any excess cost to Contractor will be immediately paid by the subcontractor or material dealer.

Where this order covers materials only, it is agreed that the materials will be delivered F. O. B. job unless otherwise ordered, in such quantity and at such times as may be authorized by this company's representative. It is further understood that in all cases, quantities of materials mentioned herein are approximate only and the dealer agrees that deliveries will be based upon actual needs and requirements of the work.

It is understood that no claims for extra work performed or additional materials furnished shall be made unless ordered in writing by an officer or authorized representative of the Contractor.

TERMS OF PAYMENT: Payments to subcontractors will be made monthly to the amount of 85% of the value of the work and materials incorporated in the building during the previous calendar month. Final balance to become due and payable within sixty days after the subcontractor has completed his work to the entire satisfaction of the architects, engineers, other representatives of the Owner and Contractor.

All payments covering subcontractor's work and/or material shall be payable only after the Contractor has received corresponding payments from the Owner.

The subcontractor hereby accepts exclusive liability for the payment of all taxes now or which may hereafter be enacted covering the labor and material to be furnished hereunder, and any contributions under the New York State Unemployment Insurance Act, The Federal Social Security Act, and all legislation enacted either Federal, State or Municipal, upon the payroll of employees engaged by him or materials purchased for the performance of this contract, and agrees to meet all the requirements specified under the aforesaid acts or legislation, or any acts or legislation which may hereafter be enacted affecting said labor and/or materials. The subcontractor will furnish to the Contractor, any records the Contractor may deem necessary to carry out the intent of said acts or legislation and hereby authorizes the Contractor to deduct the amount of such taxes and contributions from any payments due the subcontractor and to pay same direct or take any such precaution as may be necessary to guarantee payment.

Samples and details are to be submitted to Contractor, if requested, and approval must be secured before proceeding with the work.

If inferior work or material is installed or furnished and allowed to remain, the Owner and/or Contractor at its option, may reject such work and/or material or are to be allowed the difference in value between cost of work and/or material installed or furnished and cost of work and/or material specified or ordered.

The subcontractor will furnish all labor, materials, tools, scaffolds, rigging, hoists, etc. required to carry on the work in the best and most expeditious manner and furnish protection for his and other work, and will do all necessary cutting and patching and also remove and replace any interfering work, for the proper installation of his work. The subcontractor will remove all rubbish from premises in connection with his work. The subcontractor agrees to perform work in a safe and proper manner and save the Owner and Contractor harmless against all penalties for violation of governing ordinances and all claims for damages resulting from negligence of the subcontractor, or his employees, or accidents in carrying on his work.

The subcontractor agrees to repair, replace or make good any damages, defects or faults resulting from defective work, that may appear within one year after acceptance of work or for such additional period as may be required by Owner or by the specifications relating to same.

Time is of the essence of this agreement.

All labor employed shall be Union labor of such type and character as to cause no Union or jurisdictional disputes at the site of the work.

The subcontractor shall furnish all labor, material and equipment and permits and pay all fees and furnish all shop drawings, templates, and field measurements incidental to the work hereby let to it that the Contractor is required to perform and furnish under the General Conrtact, and whatever the Contractor is required to do or is by the General Contractor bound in and about the work hereby let to the Subcontractor shall be done by the Subcontractor without any extra charge.

Any controversy or claim arising out of or relating to this contract, or breach thereof, shall be settled by arbitration in the City of New York in accordance with the Rules of the American Arbitration Association, and judgment upon the award rendered by the Arbitrator(s) may be entered in any Court having jurisdiction thereof.

The Subcontractor expressly covenants and agrees to file no lien of any nature or kind for any reason whatsoever arising out of this contract for matters and things related thereto and does hereby expressly and irrevocably constitute the Contractor as its agent to discharge as a public record any lien which may have been filed by it for any reason or cause whatsoever.

FIGURE 17.6*b* Back side of the short form for the subcontract.

One additional clause that should appear in all subcontracts is the following:

> This subcontract shall be subject to all the terms and conditions of the prime contract between the general contractor and the owner insofar as that contract shall be applicable to the work required to be performed under this subcontract, it being the intention that the subcontractor shall assume to the general contractor all obligations of the general contractor to the owner insofar as the work covered by this subcontract is concerned and that the subcontractor shall be bound by all rulings, determinations, and directions of the architect and/or owner to the same extent the general contractor is so bound.

Canaan IV Apartments
115th St. and Lenox Ave.

The printed part of this contract is hereby modified and supplemented as follows. Wherever there is any conflict between this rider and the printed part of the subcontract, the provisions of this rider are paramount and the subcontract shall be construed accordingly:

31. *ARTICLE 3 shall be amended by the addition of the following:*
The reference herein to payments to the contractor for the owner shall be deemed to be absolute conditions precedent to the contractor's obligation to pay the subcontractor herein, it being understood distinctly that said provision shall not be deemed to relate solely to the timing of the payments due from the contractor to the subcontractor, but, shall be deemed to relate to the contractor's liability therefor. The contractor shall not be liable to the subcontractor until and unless it receives the said payments from the owner.

32*a* *ARTICLE 10 OF STANDARD FORM OF SUBCONTRACT, eliminate and replace by:*
The subcontractor agrees to indemnify and save harmless the Owner and the Contractor against loss or expense by reason of the liability imposed by law upon the Owner, Contractor, their partners, representatives, agents, employees, officers, and directors, for damage because of bodily injuries, including death at any time resulting therefrom accidentally sustained by any person or persons or on account of damage to property arising out of this work, whether such injuries to person or damage to property are due or claimed to be due to any negligence of the subcontractor, the Owner, Contractor, their partners, representatives, agents, employees, officers and directors, or any other person. The foregoing, however, shall not be deemed a covenant, promise, agreement, or understanding that the subcontractor shall indemnify and save harmless the Owner, Contractor, their partners, representatives, agents, employees, officers, and directors, for damage resulting from their sole negligence.

32*b* The subcontractor agrees to obtain and maintain at his own expense and until the completion of this project the following insurance:
(I) Worker's Compensation with Employer's Liability coverage as required by State Statute.
(II) Comprehensive General Liability covering the Subcontractor's operations, sublet operations, completed operations and contractual [as described in 10*a* above]. This insurance shall be in the minimum amount of:
With respect to bodily injury $3,000,000 each occurrence
With respect to property damage $1,000,000 each occurrence
With respect to property damage, coverage for explosion, collapse and damage to underground utilities shall be included.
(III) Fire, extended coverage, theft insurance, etc. on subcontractor's equipment, forms, and temporary facilities.

FIGURE 17.7 Subcontract rider.

Also, the list of drawings (Fig. 17.8) should not be neglected. The content of the exact contract drawings should never be left in doubt. Without a dated list of the drawings that both parties have agreed to have embodied in the subcontract, disputes may later arise.

See also Arts. 17.11 and 17.14.

17.10.2 Project Management

In a small contracting organization, project management is generally the province of the proprietor. In larger organizations, an individual assigned to project management will be responsible for one large job or several small jobs.

East Midtown Plaza Stage II
24th & 25th Sts. between 1st & 2nd Aves.
New York, New York

DRAWING NO.	TITLE	DATE	REV.	REV.
T-1	Title & Drawing List			
T-2	Notes, Symbols & Abbreviations			
Z-1	Zoning			
Z-2	Zoning			
A-1	Cellar Plan - Parts G, GH & H			
A-2	First Floor Plan - Parts G, GH & H			
A-22	Elevators			
A-23	Doors			
	STRUCTURAL			
S-1	Foundation Plan - Bldgs. G & H			
S-11	Typical Floor Framing Plan-Bldg. H West half.	5/1/82		

R. B.
PLEASE INITIAL

FIGURE 17.8 List of drawings for construction of a project.

A project manager is responsible in whole or in part for the following:

Progress schedule (Fig. 17.14 or 17.15).

Purchasing (Arts. 17.10.1 and 17.11).

Arranging for surveys and layout.

Obtaining permits from government agencies from start through completion of the job.

Familiarity with contract documents and their terms and conditions.

Dealing with changes and extras.

Submission of and obtaining approvals of shop drawings and samples, and material certifications.

Conducting job meetings with the job superintendent and subcontractors and following up decisions of job meetings. Job meetings should result in assignments to various individuals for follow-up of the matters discussed at the meetings. Minutes kept of the plans of action discussed at each meeting should be distributed as soon as possible to all attending. At the start of the following job meeting, these minutes can be checked to verify that follow-up has been properly performed. Figure 17.9 illustrates typical job-meeting minutes. Figure 17.16 is a computer printout of a program "Expedition," registered by Primavera Systems, Inc., which shows the submittal status of drawings for a building project.

Construction Productivity. One of the principal tasks of the construction manager is coping with labor availability and productivity. Added to this is the increased complexity of modern construction, particularly in the structural and mechanical phases. These increasingly complex factors, however, can be kept within the contractor's control by efficient management tools, including scheduling and monitoring techniques.

KBF JOB MEETING

Kreisler Borg Florman — Date: August 19,

UNITED CEREBRAL PALSY

Present: Robert F. Borg, John V. Cricco, John A. Nowak, Robert W. Hewet, Barbara Brandwein

Spec. Section	Item of Work	Remarks	RWH	JVC	JAN	BB	GR
2A	Preliminary Work						
2B	Excavating, Bckfl., Grading/HOWNOR:	PH check subs & suppliers paid before check issued.					
2C	Site Improvement LA STRADA:	see page 3					
2D	Topsoil, Seed, Plant.	RWH push for decision w/Vella.	X				
3A	Concrete & Cement HOWNOR:	Make open items/omissions list.					X
3B	Precast Con. Plank KOEHLER:						
4A	Masonry KOEHLER:	see page 3					
5A	Misc. Metal UNITED:	One section stair rail still remains					X
5B	Struct. Steel UNITED:		---	---	---	---	---
6A	Carpentry BE-MI:	RWH call Bellona re: pulls. Install. bi-fold heads & shelves. Expedite 6th & 1st	X				X
6A	Millwork METALOC	fl. bi-folds. Measure handrail.					X
6B	Kitchen Cabinets BOISE CASCADE:	Follow up approval of pull.				X	
7A	Roofing & Sheet Metal/ COLONIAL:		---	---	---	---	---
7B	Dampproof KINGS COUNTY:	Waterprf. lobby flr. at main entrance					X
7C	Wall Wtpf. KINGS COUNTY:		---	---	---	---	---
7D	Caulking & Sealants KINGS COUNTY:	Make open items list.					X
8A	HM Door Frames WILLIAMSBURG:	Expedite 3 missing frames					X
8B	HM Doors WILLIAMSBURG:	Expedite 6 missing doors.					X
8C	Alum. Entrances LOXCREEN:	Expedite install. Q.T. & entrances.					X
8D	Alum. Windows LOXCREEN						
8E	Glass & Glazing LOXCREEN:	Make open items list.					X
8F	Finish Hardware SEECO:	Make open items list.					X

FIGURE 17.9 Minutes of a job meeting.

On a complex construction project for which thousands of items are to be procured, approved, manufactured, delivered, and installed, satisfactory use of computers can be made. An effective system for monitoring the time and cost of a project from design through procurement and installation, consisting of one data bank from which four reports are drawn, is as follows:

Purchasing/Cost Report. This report lists the various items to be procured and sets target dates for bidding and award of contracts. It keeps track of the budget and actual cost for each item. A summary prepared for top management provides totals in each category and indicates the status of the purchasing.

Expediting/Traffic Report. This report lists the items once they are purchased and gives a continual update of delivery dates, shop drawing and approval status, shipping information, and location of the material when stored either on or off the site.

Furniture, Fixture, and Equipment List. This report, which is normally used when the job involves a process or refinery, can also be used for lists of equipment in a complex building, such as a hospital or hotel. The report describes all the utility information for each piece of equipment, its size, functions, intent, characteristics, manufacturer, part number, location in the finished job, and guarantees, as well as information relating to its source, procurement, price, and location or drawing number of the plan it appears on.

Accounting System. The system consists of a comprehensive series of accounting reports, including a register for each supplier, and shows all disbursements. This information is used in preparing requisitions for progress payments. It also can be used to report costs of the job to date and to make predictions of probable costs to complete.

All of the above reports can be integrated from start to finish and synchronized with each other.

Project Time Management. After construction operations commence, there should be a continual comparison of field performance with the established schedule. When the schedule is not being met, some method should be used to make management aware of this and correct schedule lags. The corrective actions and rescheduling phases are known as project time management.

The monitoring phase of time management involves periodic measurement of actual job progress and comparison with the planned objectives. The only way that this can be done positively is to determine the work quantities put into place and to report this information for comparison with work quantities anticipated in the job schedule. Then, a determination can be made of the effect of the current status of the job on the completion date for the project. Any corrective actions necessary can then be planned and implemented, after which the schedule can be updated.

Computer-calculated network activities provide a convenient basis for measuring progress and for issuance of reports (Art. 17.12). The network diagram should be corrected as needed so that the current job schedule reflects actual job status.

(R. H. Clough and G. A. Sears, "Construction Project Management," John Wiley & Sons, Inc., New York.)

Computerized Project Management Control System. This system combines project scheduling with cost controls, resource allocation controls, and a contract-progress statistical reporting system so as to yield total control over time, cost, resources, and statistics.

Time. The time aspect of the system is designed to produce, through project scheduling, a set of time objectives, a visual means of presenting these objectives, and the devising and enforcing of a corrective method of adhering to the objectives in order that the desired results will be achieved, as described previously (Project Time Management).

Cost. These are summary costs monitored by budget reports, produced monthly and distributed to the owner. In addition, detailed reports for construction company management list costs under each class of construction activity and are used by project managers and field, purchasing, and top management personnel. A report on probable total cost to complete the project is intended for all levels of construction company personnel but is used primarily by those responsible for corrective action.

Resource Allocation. For the purpose of resource allocation, a graphical summary should be prepared of projected monthly manpower for individual activities and also of the estimated quantities of work to be in place for all trades on a cumulative basis. An update of these charts monthly will indicate which trades have low work quantities in place, so that these lagging trades may be augmented with the proper number of workers to permit them to catch up with and adhere to the schedule.

Statistics. From the information received from the preceding reports, an accurate forecast can be made of the probable construction completion date and total cost of the project.

Software and programs for computerized preparation of all the preceding reports can be developed by company personnel or by outside consultants.

17.10.3 Field Supervision

A field superintendent has the most varied duties of anyone in the construction organization. Responsibilities include the following: field office (establishment and maintenance); fencing and security; watchmen; familiarity with contract documents; ordering out, receiving, storing, and installing materials; ordering out and operation of equipment and hoists; daily reports; assisting in preparation of the schedule for the project; maintenance of the schedule; accident reports; monitoring extra work; drafting of backcharges; dealing with inspectors, subcontractors, and field labor; punch-list work; quality control; and safety.

Familiarity with contract documents and ability to interpret the plans and specifications are essential for performance of many of the superintendent's duties. (The importance of knowing the contract documents is discussed in Art. 17.5.) Should work being required by the architect, owner, or inspectors exceed the requirements of the contract documents, the superintendent should alert the contractor's office. A claim for pay for extra work, starting with a change-order proposal, may result (Art. 17.14.2).

The daily reports from the superintendent are a record that provides much essential information on the construction job. From these daily reports, the following information is derived: names of persons working and hours worked; cost code amounts; subcontractor operations and description of work being performed; materials received; equipment received or sent; visitors to the job site; other remarks; temperature and weather; accidents or other unusual occurrences. Figure 17.10 shows a typical daily report.

Back-Charges. Frequently, either at the request of a subcontractor or because of the failure of a subcontractor to perform, work must be done on behalf of a subcontractor and the subcontractor's account charged. If the work performed and the resultant back-charge is at the request of the subcontractor, then obtaining the information and the agreement of all parties to a back-charge order (Fig. 17.11) is easy.

DAILY REPORT

Kreisler Borg Florman
CONSTRUCTION COMPANY (INC.)
97 Montgomery Street, Scarsdale, New York 10583
Telephone (914) SC 5-4600

JOB No. 182 — SHEET 1 OF 1

JOB NAME Coney Island Site 17 — WEATHER Fair

SUPTS NAME H. L. — TEMP. A.M. 38° P.M. 49°

KREISLER BORG FLORMAN OPERATIONS

	CLASS	EMPLOYEE'S NAME	HRS.	COST CODE	EXPLANATION	TOTAL HOURS	RATE	AMOUNT
1	Proj. Mgr.	J. C.						
2	Secty	M. R.						
3	Secty	P. W.			UTSCO deliv. 500' of .018 S/S base flashing			
4	Supt.	H. L.			for Sepia			
5	Asst.	D. W.						
6	Lab.	J. F.	7		Sweeping & scraping 10th & 11th fl.			
7	Lab.	V. C.	7		Sweeping & cleaning floors for bathroom layout			
8	Lab.	B. M.	8		on the 17th & 18th fls.light salamanders and			
9	Lab.	C. T.	7		change propane cylinder on 2nd,3rd & 4th fls.			
10	Lab.	E. M. (1)	7		Cleaning "C" type window for installation of hopper frame on the 6th & 7th floors			

SUBCONTRACTOR OPERATIONS

SUB'S NAME & WORK FORCE	WORK PERFORMED
Island Security (4) 4 Watchmen	2 shifts, 3PM to 5AM
Alwinseal 2 Ironworkers	setting 19th floor windows;installing hopper frames on 6th fl. level Bldg "C"
Lieb 2 Ironworkers (1)	Setting roof level lintels;Stair platform frame at Mech.Rm.
Lipsky 1-15-3-17 Plb (1)(1)(1)	C.I.test 13th fl. thru roof Bldg B;installing gas risers 16th, 17th fls Blds.A,B,C; setting tubs 16th fl;maint.temp. water
Lefferts Waterproofer	Spandrel waterproofing south end "C" east side of B & A at roof level & mech.room slab level at bldg C.
Public Improvement 2-12-8 elec. (1) (3)	Installing conduits & corrective work 2nd & 3rd fl. working with sheetrock crew;misc.corrective work on 6th flr; pulling wire on 10th flr. bldgs B & C; nippling on 14th & 15th fl; layout for nippling on 15th & 16th flr; fabrication work;maint. temp. lights
Star-Circle 2-26 Carp (4)	Sheetrock 2nd flr; framing stairs 2 to 3 & 3 to 4 flr; framing kit.3 fl; sheetrock bathrms 3 fl; installing insul & sheetrock back of stairs 4 fl; fire cod'g duct risers 5 flr & framing 5 flr.apts; layout of bathrooms on 17 & 18 flrs.
Bafill 26-2-27 Brklyrs (7) (2)	Laying face brick and backup block to roof level at Building A West side of Building B
1-19 M.Tender 1 Hst.Egr. (7)	Brickwk. at bulkhead to mech.rm. slab level at building C; Block work at elev.shafts 7 & 8 flr. of buildings A,B & c
Dic 2 carp. 4 lab.	Framing & grouting pipe & duct openings on 10th flr; cleaning inserts at roof level.

REMARKS

Reid - 3 engr.	Took settlement readings with Raamot's Engr.
Active - 2 carp.	Installing storefront frames at 1st flr. of Bldgs B & C
Tropey	Continue excavating for Con Edison vault at 24th St near Bldgs B & C

MATERIAL RECEIVED	VISITORS
	F. B - UDC
153 men on jobsite	F. F - Bafill
40 minority	

DATE Jan. 16, 19__ DAY Tuesday

PROJ. MGR. INITIAL ____ REPORT No. 266 OVER ☐ SUPT. SIGNATURE Superintendent

FIGURE 17.10 Daily report, prepared by a construction superintendent.

BACK CHARGE ORDER

PAGE 1 OF 1

Kreisler Borg Florman
CONSTRUCTION COMPANY
97 Montgomery Street, Scarsdale, New York 10583
Telephone (914) SC 5-4600

NAME OF COMPANY TO BE BACKCHARGED Universal Ductwork Corporation

LOCATION Bronx Municipal Hospital Center Pelham Parkway So & Eastchester Rd, Bronx, NY (JOB NAME) JOB NO. 173

PERFORMED BY General Contractor (INSERT GENERAL CONTRACTOR OR NAME OF SUBCONTRACTOR) DATE January 22,

LABOR

NAME OF EMPLOYEE	OCCUPATION	HOURS WORKED	RATE OF PAY	AMOUNT
John M	Laborer	7		

MATERIALS, SUPPLIES AND/OR EQUIPMENT

QUANTITY	DESCRIPTION	UNIT PRICE	AMOUNT
2	Ballpoints	$7.50	

COMPLETE DESCRIPTION OF WORK DONE THIS DATE

Chopped holes in existing wall for new ducts, Room 301

THIS BACK CHARGE WORK HAS BEEN PERFORMED AND THE COST HAS BEEN DEDUCTED FROM THE MONEY DUE YOU ON THIS JOB

(SIGNED) ____ (SUBCONTRACTOR'S SUPERINTENDENT OR TRADE FOREMAN)

UNIVERSAL DUCTWORK CORP. (NAME OF SUBCONTRACTOR)

BY ____ (SIGNATURE OF K-B REPRESENTATIVE)

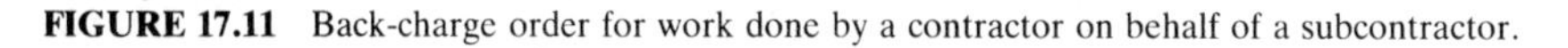

AT YOUR REQUEST ☒
BECAUSE OF FAULTY WORK BY YOU ☐
BECAUSE OF YOUR FAILURE TO DO THE WORK DESCRIBED ☐
OTHER, AS FOLLOWS: ☐

January 22, (DATE) Superintendent (TITLE)

NUMBER 173-6

UNLESS WE RECEIVE YOUR WRITTEN PROTEST WITHIN 3 DAYS IT WILL BE ASSUMED THAT YOU ACCEPT THE CHARGE.

FIGURE 17.11 Back-charge order for work done by a contractor on behalf of a subcontractor.

If, however, there is a dispute as to whether or not the work is part of the obligation of the subcontractor, then the task becomes more complicated. Back-charges in this situation should be used sparingly. Sending a back-charge to a subcontractor under circumstances that are controversial is at best only a self-serving declaration, and in all likelihood will be vigorously disputed by the subcontractor.

17.11 PURCHASE ORDERS

Issuance of a purchase order differs from issuance of a subcontract (Art. 17.10.1). A purchase order is issued for material on which no labor is expected to be performed in the field. A subcontract, on the other hand, is an order for a portion of the work for which the subcontractor is expected not only to furnish materials but also to perform labor in the field. An example of a purchase order form, front and back, is shown in Fig. 17.12*a* and *b*.

For the specific project, a purchase order rider (like Fig. 17.7) and list of contract drawings (Fig. 17.8) should be appended to the standard purchase order form. The rider describes special conditions pertaining to the job, options or alternates, information pertaining to shop drawings, or sample submissions and other particular requirements of the job.

Material-price solicitations are handled much in the same manner as subcontract-price solicitations (Art. 17.10.1). Material bids should be analyzed for complicated trades in the same manner as for subcontracts (Fig. 17.5).

To properly administer both the subcontract and the purchase orders, which may number between 40 and 60 on an average building job, it is necessary to have a purchasing log (Fig. 17.13) in which is entered every subcontract and purchased order after it has been sent to the subcontractor or vendor. The entry in the log is copied from the file copy of the typed document, and an initial in the upper right-hand corner of the file copy indicates that entry has been made. The log serves as a ready cross reference, not only to names of subcontractors and vendors but also to the amounts of their orders and the dates the orders were sent.

Computers in Purchasing. On a large-scale construction project, which costs many millions of dollars, thousands of items may have to be procured. Also, information, shop drawings, and part lists may have to be distributed and revised, redistributed, approved, and distributed again. Systems have been developed to use computers to keep track of all equipment and materials and related purchase information, such as specifications, quotations, final orders, shipment, and delivery dates.

The systems are based on the concept of critical-path items, and the various tasks that must be performed are assigned due dates. For example, a computer report could be by project and show all open purchase-order items for one project, or by buyer name, with all open purchase-order items for each buyer, including all projects.

In negotiating and awarding either a subcontract or a material purchase, the contractor should take into account the scope of the work, list inclusions properly, note exceptions or exclusions, and, where practicable, record unit prices for added or deleted work. Consideration should be given to the time of performance of units of work and availability of workers and materials, or equipment for performing the work. Purchase orders should contain a provision for field measurements by the vendor, if this is required, and should indicate whether delivery and transportation charges and sales taxes are included in the prices.

17.12 SCHEDULING AND EXPEDITING

Two common methods of scheduling construction projects are by means of the Gantt (bar) chart (Fig. 17.14) and by means of the critical path method (Fig. 17.15).

KBF Nº 3002 **PURCHASE ORDER**

97 Montgomery Street, Scarsdale, New York 10583 Telephone (914) SC 5-4600 **CONSTRUCTION COMPANY, INC.**
(herein referred to as "Purchaser")

TO Construction Products Co.
Route #7 Att: Mr. Alan Fishkin
Brookfield, Connecticut

DATE 180

(referred to herein as "Vendor") JOB NO.

Please enter our order for the following materials, subject to the terms and conditions and all of the provisions set forth herein and in accordance with all pertainent provisions of the contract dated 6/9/81 entered into between purchaser and Construction for Progress, Inc. (referred to herein as "The Owner") for the construction of: (Name of project and shipping address)

13-Story Apartment Building, 170th Street to 172nd Street and 93rd Avenue, Jamaica New York

(Which contract, together with all plans, specifications and all provisions and documents incorporated or referred to therein is referred to herein as the "Principal Contract").

QUANTITY	DESCRIPTION	PRICE (Unit or Lump Sum)

REFUSE CHUTE HOPPER DOORS, ACCESS DOOR AND SPARK ARRESTOR (Specification Division 23, Section 10), furnished, fabricated, and delivered, and all such work shown on the Plans, Specifications, AIA General Conditions, Modifications to AIA General Conditions, General Conditions, Contract and Rider.

This order is based on the following:

Twelve (12) 15"x18" hopper doors and frames
One (1) 12"x12", 1-1/2 hr. fireproof self-closing access door and sleeve
One (1) explosive vent

Five (5) hoppers to have sprinkler heads installed.

The price for this material FOB job site will be $1,350.00

Shop drawings and sample hopper door to be submitted immediately.

The price(s) set forth above shall constitute payment in full for the prompt and proper furnishing of the materials hereunder in accordance with the terms of this Purchase Order. If unit prices are set forth above, Vendor agrees to be bound by the quantities of such items which may be certified by the Owner as having been properly furnished under the terms of the Principal Contract, it being understood in such event that unit quantities set forth above are estimated and used only for convenience in determining the approximate amount of this Purchase Order and that the actual quantities required may substantially vary, upward or downward from the estimated quantity.

If no price is set forth above, it is agreed that Vendor's price will be the lowest prevailing market price and in no event is the order to be filled at higher prices than last previously quoted or charged without Purchaser's consent..

This order shall not become effective unless within ten (10) days from the date of execution hereof by Purchaser as set forth below, Purchaser receives the acknowledgment copy hereof, unconditionally executed by an officer of Vendor. If Purchaser does not so receive the executed acknowledgment hereof, the offer contained herein is withdrawn and shall thereafter be renewed only by written statement to such effect made by Purchaser.

TIME OF DELIVERY: approximately six to eight weeks.

THE ABOVE IS HEREBY ACCEPTED
VENDOR CONSTRUCTION PROD. CO. INC
BY J S Clarke VICE PRES
DATE 9/28/82
(Name and Title)

TERMS OF PAYMENT: Regular
Kreisler Borg Florman CONSTRUCTION COMPANY, INC.
BY R Borg
(Name and Title) President

1. ACKNOWLEDGMENT MUST BE SIGNED AND RETURNED BEFORE INVOICES WILL BE PAID. 2. PURCHASE ORDER NUMBER MUST APPEAR ON ALL INVOICES. 3. THERE MUST BE A SEPARATE INVOICE FOR EACH PURCHASE ORDER.

VENDOR

FIGURE 17.12*a* Front side of a purchase order.

Bar Chart. The bar chart is preferred by many contractors because of its simplicity, ease in reading, and ease of revision. The bar chart can show a great deal of information besides expected field progress. It can also show actual field progress; dates for required delivery; fabrications and approvals; percent of completion, both planned and actual; and time relationships of the various trades. Copies of bar

TERMS AND CONDITIONS

The purchase orders is subject to the following terms and conditions and by accepting this Purchase Order, Vendor agrees to be bound by each and every of said terms and conditions.

1. Acceptance of this Purchase Order by Vendor in time and manner hereinbefore specified shall constitute the only possible manner of acceptance hereof and Purchaser will in no way be responsible or indebted to Vendor for any materials or samples thereof furnished by Vendor or for any work or services performed by Vendor in the absence of such acceptances, it being understood that Vendor shall in such event make no claim against the Purchaser.

2. This Purchase Order may only be accepted by Vendor unconditionally and exactly as written. Any additional or different terms or conditions stated by Vendor are hereby objected to and rejected and the statement by Vendor of such additional terms and conditions shall be deemed to constitute a rejection of the offer contained therein, even if accompanied by the acknowledgment copy hereof executed, by Vendor.

3. The principal contract has been, is and will be available for review by Vendor at Purchaser's office during normal business hours. By accepting this order, Vendor acknowledges that it has read the Principal Contract and is fully aware of each and every of its provisions which in any way bear upon or relate to the materials to be furnished hereunder and Vendor understands and agrees that all of such provisions of the Principal Contract are applicable hereto and made a part hereof as if expressly set forth herein in full, including but not limited to the general provisions of the Principal Contract dealing with the subject project as a whole as well as the special provisions and parts thereof specially relating to the materials to be furnished hereunder and Vendor agrees to furnish said materials as if it were the contractor named in the principal contract, with Vendor hereby assuming the same obligations, undertakings, warranties, guarantees, risk and liabilities to the Purchaser as Purchaser has assumed to the Owner in addition to any other obligations, undertakings, warranties, guarantees, risks and liabilities provided for herein or otherwise provided as a matter of Law. In the event of any conflict between the Provisions of this Purchase Order and of the Principal Contract, this Purchase Order shall control. It is understood that if the Principal Contract affords to the Owner the right to approve a Vendor, then the Purchase Order is conditioned upon such approval and in the event the Owner does not so approve, then this Purchase Order shall be null and void and of no force or effect and Vendor shall have no rights whatsoever against Purchaser and of all the materials hereunder and/or any samples thereof, if any, which have previously been furnished by Vendor, may be returned to Vendor at Vendor's sole risk and expense and Vendor shall have no claim of any sort against Purchaser.

4. All materials hereunder shall be furnished in strict accordance with the Principal Contract and to the satisfaction of both Purchaser and the Owner. Both the Owner and Purchaser have the right to inspect the materials hereunder at Vendor's place of manufacture and/or at any time prior to delivery to the job and/or after delivery to the job and/or after installation in the project and in the event that the materials are rejected in whole or in part by the Owner and/or the Purchaser as not conforming to the requirements hereof, such materials may be returned to Vendor at Vendor's sole risk and expense. In such event all costs incurred by Purchaser in testing and/or inspecting the materials shall be for the account of Vendor. Purchaser reserves the right to elect to keep non-conforming materials or some portion thereof rather than returning them to Vendor. In such event, in addition to all other liabilities provided for hereunder and under the law otherwise, Vendor shall be charged for the difference between the value of the delivered non-confirming materials and the value of such materials would have had had they conformed to the requirements of this purchase order. Notwithstanding anything to the contrary hereunder or under the Law otherwise, Purchaser is not required to inspect and/or reject the materials hereunder within any particular period of time and delay in such inspection and/or rejection shall in no way limit Vendors obligation to furnish materials which strictly conform to the requirements hereof and shall in no way constitute an acceptance of non-conforming goods, nor shall payment to Vendor for or utilization by Purchaser of any or all of the materials hereunder be deemed an acceptance thereof. Vendor shall, at its own expense, arrange for an assist in all testing of materials hereunder as required under the Principal Contract or when requested by Purchaser and/or the Owner.

5. Unless hereinbefore expressly stated to the contrary, Purchaser shall, within ten (10) days after eceiving payment therefor from the Owner, pay to the Vendor eighty-five (85%) percent of the amount due in accordance with the prices set forth above, for materials furnished and delivered by Vendor and Vendor shall not be so entitled to said payment or any portion thereof unless and until Purchaser is paid by the Owner for such materials. The remaining fifteen (15%) percent of said payment shall be paid to Vendor within ten (10) days after Purchaser receives payment from the Owner of all retained percentages under the terms of the Principal Contract. Before any partial of final payment, Vendor shall furnish written proof satisfactory to Purchaser that all labor and all suppliers of material, equipment and/or services in connection with the materials hereunder have been paid in full and that there are no claims outstanding against it in connection with the materials hereunder.

6. Vendor shall be entitled to no price increases or escalators of any kind in connection with the furnishing of the materials hereunder nor shall Vendor be entitled to reimbursement for boxing or packing costs or cartage charges or taxes of any type (including any Federal, State or local taxes and including but not limited to Sales or Use Taxes), all of such costs and expenses and all other costs and/or expenses which might be incurred by Vendor in connection with the furnishing of the materials hereunder being included in the consideration set forth on the reverse side hereof, payment of which consideration shall constitute payment in full for the furnishing of the materials hereunder. Vendor represents that the price charged for the items or services covered by this Purchase Order, is the lowest price charged by Vendor to buyers of a class similar to Purchaser under conditions to those specified in this Purchase Order and that prices comply with applicable Government regulations in effect at time of quotation, sale or delivery. Vendor agrees that any price reduction made in merchandise covered by this order subsequent to the placement of this Purchase Order will be applicable to the Purchase Order.

7. The number of this Purchase Order must appear on all invoices and if more than one purchase order has been issued by Purchaser to Vendor, there must be separate invoices for each order. Written evidence (such as delivery receipts) of delivery of the materials hereunder to Purchaser must be attached to all invoices.

8. Unless a specific delivery date or period of delivery is expressly specified herein, Vendor shall within five (5) days after being notified so to do by Purchaser, deliver the subject material or any portion thereof in such amounts and at such times as Purchaser may direct. It is understood that such time of delivery is of the essence and shall be determined by the requirements of the Principal Contract and that Purchaser shall be the sole judge thereof. Delivery shall be made FOB job site. Vendor shall unload, uncrate and stock pile in the places and manner requested by Purcnaser.

9. Purchaser is under no obligation to accept any delivery exceeding the quantity specified in this Purchase Order and reserves the right to return, at Vendor's expense, part or all of any amount so delivered. Unless the entire order can be filled, Purchaser is under no obligation to accept delivery of any part thereof, unless previously consented to in writing and Purchaser reserves the right to return, at Vendor's expense, part or all of any such delivery. Material must be delivered as per instructions and if not so delivered, any extra handling costs incurred by Purchaser will be billed back to Vendor.

10. Vendor shall prepare and submit all samples and drawings and any other back-up data when and as required by the Principal Contract and/or when and as otherwise required by Purchaser or Owner and Vendor shall perform any and all other matters necessary in order to have the materials hereunder approved by the Owner. All of the preparation, submissions and other work, materials and/or services under or in connection with this paragraph shall be at the sole expense of Vendor. Unless otherwise agreed in writing, all special dies, molds, patterns, jigs, fixtures, and any other property furnished to the Vendor by the Purchaser, or specifically paid for by the Purchaser for use in the performance of this Purchase Order, shall be and remain the property of Purchaser, shall be subject to removal upon Purchaser's instructions; shall be used only in filling orders from Purchaser; shall be held at Vendor's risk, and shall be kept insured by Vendor at the Vendor's expense while in its custody or control in an amount equal to the replacement cost thereof, with loss payable to Purchaser. Copies of policies or certificates of such insurance will be furnished to Purchaser on demand.

11. Vendor agrees to employ workmen and to use materials and work practices and delivery methods which shall not in any manner cause or contribute to strikes or other labor disturbances by workmen employed by Purchaser and/or by subcontractors of Purchaser or by any other contractors involved in the subject project. Union labels shall be affixed to the materials hereunder if Purchaser so requires. Vendor will not discriminate against any employee or applicant for employment because of race, creed, color, or national origin. Vendor will take affirmative action to insure that applicants are employed, and that employees are treated during employment, without regard to their race, creed, color, or national origin.

12. If the Owner makes or causes to be made changes or issues or causes to be issued orders or directions, any of which effect the materials to be furnished hereunder, this Purchase Order shall be deemed to be modified accordingly. If such modification causes an increase or a decrease in the cost of furnishing the materials hereunder, then in lieu of all claims and demands which vendor might otherwise have or make, Vendor shall accept in full satisfaction and discharge of all such claims and demands, the amounts, if any, as may be allowed by the Owner to the Purchaser on account of Vendor's claims and demands. If such modification causes a decrease in the cost of furnishing the materials hereunder, Vendor shall accept and be bound by the Owner's determination as to the amount of money to be deducted on account thereof. Vendor agrees to accept the Owner's determination, as aforesaid, in full and complete discharge and release of Purchaser's liability to Vendor by reason of such modification and agrees to be bound and concluded thereby and agrees never to look to Purchaser on account thereof, except for such monies, if any, as the Owner may pay to Purchaser in satisfaction of Vendor's claims and demands. Vendor agrees to immediately comply with all orders and directions of Purchaser in any way bearing upon the material to be furnished hereunder, including but not limited to, orders or directions modifying the materials hereunder, irrespective of whether Vendor shall dispute same in any particular (including but not limited to Vendor claiming additional reimbursement therefor), it being understood that the furnishing of the materials hereunder is not to be held up pending the resolution of such dispute.

13. In addition and without prejudice to all other warranties expressed or implied by Law, Vendor warrants that all materials covered by this Purchase Order will strictly conform to the requirements set forth in the Principal Contract and will be of the best quality and workmanship and free from defects and fit for the intended purposes. All warranties, both expressed and implied, aslo constitute conditions and shall survive inspection, acceptance and payment and shall inure to Purchaser and the Owner. Without limitation of any rights by reason of any breach of warranty or otherwise, material or articles which are not as warranted may at any time be returned to Vendor at Vendor's expense for credit, correction or replacement, as Purchaser may direct. This warranty shall remain in effect for one (1) year from the date of acceptance of all work under the Principal Contract unless a longer period of warranty is contained in the Principal Contract, in which event such longer period shall govern.

14. Vendor shall observe and comply with all Federal, State, Municipal and Local Laws, orders, rules, ordinances and regulations and those of all pertient bureaus, agencies and departments, which in any way relate to the furnishing of the materials hereunder.

15. Vendor undertakes and agrees to defend at Vendor's own expense and to indemnify and hold Purchaser, its bonding company and the Owner harmless from and in connection with all claims, suits, actions, and proceedings in which Purchaser, and/or Purchaser's bonding company and/or the Owner are made defendants for actual or alleged infringement of any U.S. or foreign letters patent or copyright or trade secret or unfair competition resulting from the use or sale of the materials purchased hereunder and Vendor further agrees to pay and discharge any and all judgments or decrees which may be rendered in any such suit, action or proceedings against such defendants. The provisions of this paragraph shall apply whether or not Purchaser furnishes any specification to Vendor.

16. Vendor shall promptly pay for all labor, materials, equipment, services and all other charges it incurs in connection with the furnishing of the materials hereunder. All payments received by Vendor hereunder shall be held by Vendor as trust funds for the payment of such labor, materials, equipment, services and all other charges incurred by Vendor in connection with the furnishing of the materials hereunder and shall not be applied to any other purposes until all of Vendor's obligations in connection with the materials hereunder are satisfied. Vendor agrees that it shall file no lien in connection herewith and that it shall file no notice or claims with the Owner or with Purchaser's bonding company. Vendor further agrees that if a lien is filed by any third person or entity for whom Vendor obtains labor, materials, equipment or services in connection with the materials to be furnished hereunder, then Vendor shall at its own expense and immediately after being notified by Purchaser of the existence of such lien, take all steps necessary to discharge or vacate such lien including but not limited to, furnishing a lien bond if necessary to discharge such lien. If such lien is filed by a third party and is not discharged or vacated within five (5) days after Vendor is notified as to its existence, Purchaser may take the steps necessary to discharge or vacate such lien and charge to Vendor all costs in conncetion therewith, including attorney's fees. Further to the above, Vendor agrees to assume the entire responsibility and liability for and defense of and to indemnify and hold Purchaser and Purchaser's bonding company harmless from any and all demands, claims, liens and/or charges which may be filed, made or asserted against Purchaser and/or its bonding company and/or the Owner, for or in connection with the materials hereunder, which demands, claims, liens and/or charges may be filed and/or asserted by any third person or entity, including but not limited to demands, claims, liens and/or charges for payment for labor, equipment, materials or services as well as those for personal injury (including death) and/or property damage in any way related to the materials hereunder. If at any time there should be any such lien, demand, claim or charge, Purchaser shall have, in addition to all other rights, the right to retain out of any payment then due or which thereafter might become due to Vendor, an amount sufficient in Purchaser's opinion to fully indemnify it against any loss in connection therewith. If Purchaser so elects, it may make payment directly to such third persons or entities in order to resolve such third party claim and such payment shall be for the account of Vendor.

17. Vendor agrees that it shall not delegate or sublet or assign this Purchase Order, or any of its obligations or rights hereunder or any portion of such right or obligations, without prior written permission of Purchaser. Any such delegation, subletting or assignment without prior written permission of Purchaser shall vest no rights against Purchaser and any third party.

18. In the event that any provisions of the Bankruptcy Act of the United States of America shall be invoked by or against Vendor, or in the event Vendor shall become insolvent, or make an assignment for the benefit of creditors, or a Trustee or Receiver shall be appointed of his property, or any judgment is taken against Vendor and execution issued thereon, or any writ of attachment or the like is served upon Purchaser because of any pending litigation of Vendor, or if the Vendor shall in the opinion of Purchaser violate any of the terms of this Purchase Order, then upon the happening of any of the foregoing, Purchaser shall have the right to terminate the employment of Vendor under this Purchase Order, and in such event, may either employ its won forces or those of others to complete the work remaining to be performed under this Purchase Order either in whole or in part, and shall be entitled to have and use without cost to it, either through its own forces or through any other persons so employed by it, any plant, equipment or materials, whether on the premises or otherwise provided by Vendor in the performance of the work called for herein and the completion thereof (the right to possession of which plant, equipment and materials Vendor hereby assings and transfers to Purchaser), and Vendor shall receive no further payment until the work under the Purchase Order shall be finished, completed, accepted, and final payment made by the Owner. Then, if the unpaid balance that would be due under this Purchase Order shall exceed the cost to Purchaser of completing same, including Purchaser's overhead plus any damages or loss which the Purchaser may sustain by reason of Vendor's fault such excess shall be paid to Vendor, but if such cost plus all such damages shall exceed such unpaid balance, Vendor shall pay the difference to Purchaser. In addition to having the right to terminate this Purchase Order for cause, as aforesaid, Purchaser may at any time, terminate this Purchase Order in whole or in part, for Purchaser's own convenience. Upon such termination for convenience, the principals contained in Section 8-706 of the Armed Service Procurement Regulations relating to termination of subcontracts (as such is in effect as of the date of execution by Purchaser hereof) shall determine the rights of Vendor and Vendor shall have no other claims or rights against Purchaser.

19. None of the rights and remedies conferred upon or reserved by Purchaser under this Purchase Order or the liabilities or obligations hereby imposed or assumed by Vendor, shall be deemed exclusive, or impose any limitation upon or be in derogation of any right or remedy of Purchaser now existing or which hereafter may exist, at Law, in equity or by statute. Each and every right and remedy of Purchaser under this Purchase Order shall be cumulative and in addition to all of the rights and remedies of the Purchaser under this Purchase Order now existing, or hereafter to exist, at Law, in equity or statute. In the event that Vendor fails to perform any of its obligations hereunder, then in addition to all other obligations and/or liabilities imposed upon Vendor hereunder and/or under the Law otherwise, Vendor shall hold Purchaser harmless from and indemnify Purchaser for any and all losses, costs, damages and expenses (including, but not limited to, all incidental and consequential damages due to delay or otherwise) incurred by Purchaser as a result of any of such failures, it being understood by Vendor that any failure on its part may effect Purchaser to an extent in excess of the dollar amount of this Purchase Order.

20. Failure on the part of Vendor to perform any of the obligations called for hereunder shall be deemed to be a substantial breach of the within Purchase Order. It is understood that this Purchase Order constitutes an entire and indivisible agreement. No waiver by Purchaser of any breach of this Purchase Order shall be deemed to be a waiver of any other or any subsequent breach. Acquiescense by Purchaser in a course of performance by Vendor contrary to the terms of this Purchase Order shall not be construed as a waiver or modification of any term hereof. A waiver or modification of any term hereof may be accomplished only be a writing signed by Purchaser in which Purchaser so states its intent.

21. It is understood and agreed that each and every provision of this Purchase Order, including any alleged breach thereof, shall be interpreted in accordance with the Law of the State of New York.

22. This Purchase Order constitutes the entire agreement between Purchaser and Vendor and shall supersede and merge all prior understandings, whether oral or in form of correspondence heretofore had between said parties in connection with any and all matters pertaining to this purchase order and the performance thereof, and it is specifically agreed that the within Purchase Order shall be subject to modification only by a writing signed by an officer of Purchaser. Nothing contained in this Purchase Order is intended to inure to the benefit of any third party.

23. Purchase Order valid only if signed by an authorized officer of the Purchaser.

FIGURE 17.12*b* Back side of purchase order.

charts may be distributed to subcontractors, trade supervisors, and in many cases laypeople, with the expectation that it will be easily understood.

Steps in preparation of a bar chart should include the following:

1. On a rough freehand sketch, layout linearly and to scale horizontally the amount of time contemplated for the total construction of the job, based on either the contract or past experience.

2. List in the first column all the major trades and items of work to be performed by the contractor for the job.

PURCHASING LOG

KREISLER BORG FLORMAN CONSTRUCTION CO.
97 Montgomery Street, Scarsdale, N Y
FOR: 170 Street, Jamaica, N.Y. — TURNKEY PROJECT

SPEC	ITEM	NAME	DATE	AMOUNT
4.00	Excavation and Grading	Dedona Contracting	4/27/	$25,000.-
5.00	Concrete Foundations	AD&M General Cont.	4/26/	28,000.-
30.00	Heating & Ventilation	John A. Jones	6/11/	150,000.-
29.18, 19 &22	Asphalt Paving & Steel Curb	Strada Contracting	6/26/	4,500.-
22.0	Lath & Plaster	A. Palmese & Son	7/20/	1,600.-
12.00	Resilient Flooring	Staples Floorcraft	7/20/	20,000.-

FIGURE 17.13 Purchasing log lists every subcontract and purchase order after its submission to a subcontractor or vendor.

3. Based on past experience or on previous bar charts that give actual times of completion for portions of past jobs, block in the amount of time that will be needed for each of the major trades and items of work, and indicate their approximate starting and completion date in relationship to the other trades on the job.

4. On completion of Step 3, reexamine the chart in its entirety to ascertain whether the total amount of time being allocated for completion is realistic.

5. After adjusting various trades and times of completion and starting and completion dates on the chart, work backward on dates for those trades requiring fabrication of material off the site and for those trades that require approval and submission of shop drawings, samples, or schedules.

6. Block in the length of time necessary for fabrication and approval of items requiring those steps.

7. Using the contractor's trade-payment breakdown or schedule of payments for each trade and month-by-month analysis of which trades will be on the job, sketch in the percent-completion graph across the face of the chart.

8. After the chart has been completely checked and reviewed for errors, draw the chart in final form.

Critical Path Method (CPM). This is favored by some owners and government agencies because it provides information on the mutually dependent parts of a construction project. CPM also reveals the effect that each component has on the overall completion of the project and scheduling of other components. It permits a more realistic analysis of the daily problems that tend to delay work than does the customary bar chart. Strict adherence to the principles of CPM will materially aid builders to reduce costs substantially and to enhance their competitive position in the industry. Execution of the method can be expedited with the aid of a computer and CPM program.

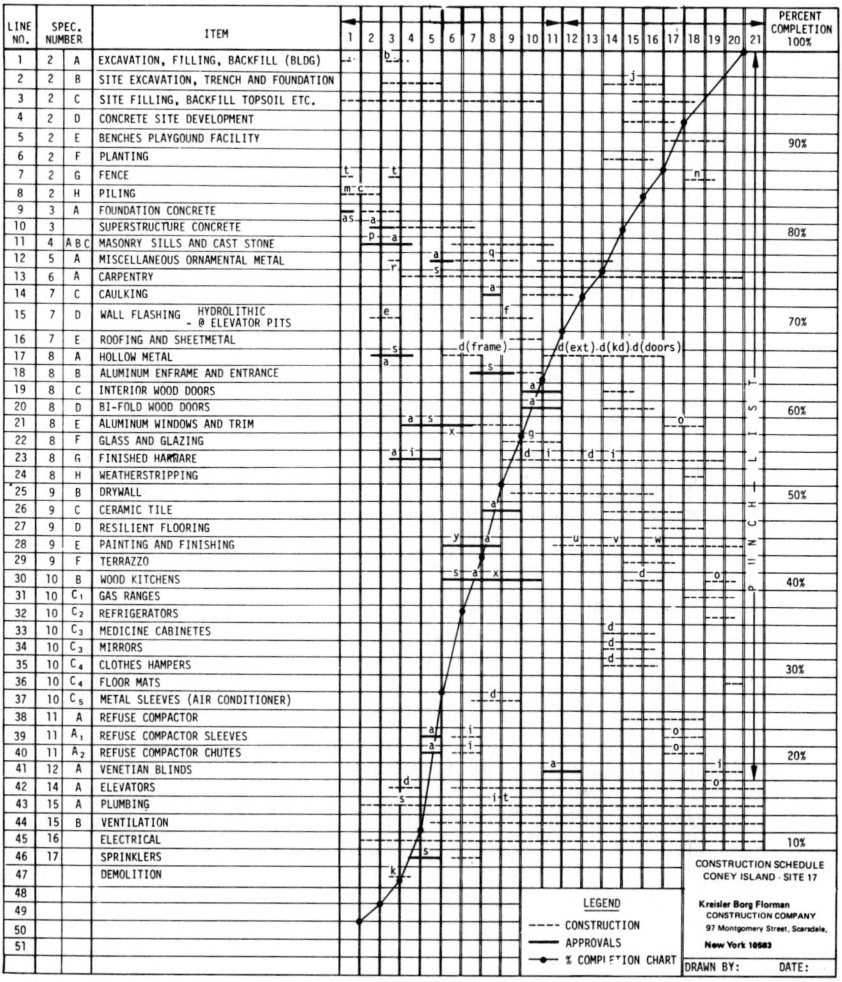

LINE NO.	SPEC. NUMBER		ITEM
1	2	A	EXCAVATION, FILLING, BACKFILL (BLDG)
2	2	B	SITE EXCAVATION, TRENCH AND FOUNDATION
3	2	C	SITE FILLING, BACKFILL TOPSOIL ETC.
4	2	D	CONCRETE SITE DEVELOPMENT
5	2	E	BENCHES PLAYGOUND FACILITY
6	2	F	PLANTING
7	2	G	FENCE
8	2	H	PILING
9	3	A	FOUNDATION CONCRETE
10	3		SUPERSTRUCTURE CONCRETE
11	4	A B C	MASONRY SILLS AND CAST STONE
12	5	A	MISCELLANEOUS ORNAMENTAL METAL
13	6	A	CARPENTRY
14	7	C	CAULKING
15	7	D	WALL FLASHING HYDROLITHIC - @ ELEVATOR PITS
16	7	E	ROOFING AND SHEETMETAL
17	8	A	HOLLOW METAL
18	8	B	ALUMINUM ENFRAME AND ENTRANCE
19	8	C	INTERIOR WOOD DOORS
20	8	D	BI-FOLD WOOD DOORS
21	8	E	ALUMINUM WINDOWS AND TRIM
22	8	F	GLASS AND GLAZING
23	8	G	FINISHED HARDWARE
24	8	H	WEATHERSTRIPPING
25	9	B	DRYWALL
26	9	C	CERAMIC TILE
27	9	D	RESILIENT FLOORING
28	9	E	PAINTING AND FINISHING
29	9	F	TERRAZZO
30	10	B	WOOD KITCHENS
31	10	C_1	GAS RANGES
32	10	C_2	REFRIGERATORS
33	10	C_3	MEDICINE CABINETES
34	10	C_3	MIRRORS
35	10	C_4	CLOTHES HAMPERS
36	10	C_4	FLOOR MATS
37	10	C_5	METAL SLEEVES (AIR CONDITIONER)
38	11	A	REFUSE COMPACTOR
39	11	A_1	REFUSE COMPACTOR SLEEVES
40	11	A_2	REFUSE COMPACTOR CHUTES
41	12	A	VENETIAN BLINDS
42	14	A	ELEVATORS
43	15	A	PLUMBING
44	15	B	VENTILATION
45	16		ELECTRICAL
46	17		SPRINKLERS
47			DEMOLITION
48			
49			
50			
51			

LEGEND
---- CONSTRUCTION
—— APPROVALS
—•— % COMPLETION CHART

CONSTRUCTION SCHEDULE
CONEY ISLAND - SITE 17

Kreisler Borg Florman
CONSTRUCTION COMPANY
97 Montgomery Street, Scarsdale,
New York 10583

DRAWN BY: DATE:

FIGURE 17.14 Bar chart for construction scheduling. Letters associated with bars indicate *a,* approvals; *b,* backfill; *c,* contract piles; *d,* delivery; *e,* elevator pits, hydrolithic; *f,* spandrel waterproofing; *g,* glazing; *h,* handrails, *i,* installation; *j,* site work; *k,* taxpayers; *l,* hardware schedule; *m,* load tests; *n,* permanent; *o,* adjustments; *p,* brick panels; *q,* lintel; *r,* saddles, wedge inserts; *s,* shop drawing; *t,* temporary; *u,* tape; *v,* prime, ceiling spray; *w,* finish; *x,* fabrication; *y,* finish schedule.

Briefly, CPM involves detailing, in normal sequence, the various steps to be taken by each trade, from commencement to conclusion. The procedure calls for the coordination of these steps with those of other trades with contiguous activities or allied or supplementary operations. The objective is to ensure completion of the tasks as scheduled, so as not to delay other work. And CPM searches out those trades that control the schedule. This knowledge enables the contractor to put pressure where it will do the most good to speed a project and to expedite the work at minimum cost.

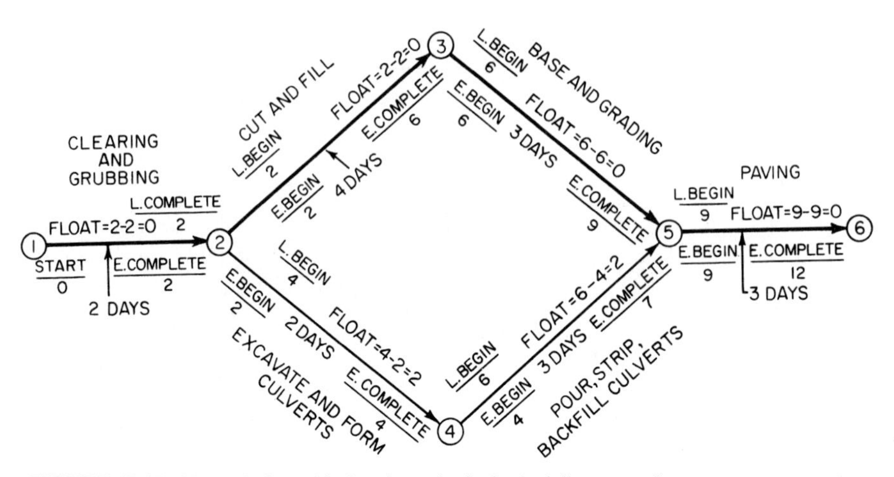

FIGURE 17.15 Network for critical-path method of scheduling comprises arrows representing steps, or tasks, and numbered nodes representing start or completion of those tasks. Heavy line marks critical path, the sequence of steps taking longest to complete.

An important, but not essential, element of CPM is a chart consisting of a network of arrows and nodes. Each arrow represents a step or task for a particular trade. Each node, assigned a unique number, represents the completion of the steps or tasks indicated by the arrows leading to it and signifies the status of the project at that point. The great value of this type of chart lies in its ability to indicate what tasks can be done concurrently and what tasks follow in sequence. This information facilitates expediting, and produces a warning signal for future activities.

The actual critical path in the network is determined by the sequence of operations requiring the most time or that would be considered the most important parts of the contract on which other trades would depend. This path determines the total length of time for construction of the project. Usually, it is emphasized by heavier or colored lines, to remind the operating staff that these tasks take precedence (see Fig. 17.15). Also, it is extremely important that the contractor be guided by the knowledge that a task leaving a node cannot be started until all tasks entering that node have been completed.

Another benefit of CPM scheduling is the immediate recognition of float time. Associated with noncritical activities, float time is the difference between time required and time available to execute a specific item of work. This information often enables a supervisor to revise planning and scheduling to advantage.

Float is determined in two steps, a forward and a backward pass over the network. The forward pass starts with the earliest begin time for the first activity. Addition of the duration of this task to the begin time yields the earliest complete time. This also is the earliest begin time for the next task or tasks. The forward pass continues with the computation of earliest complete times for all subsequent tasks. At nodes where several arrows meet, the earliest begin time is the largest of the earliest complete times of those tasks. The backward pass starts with the earliest complete time of the final task. Subtraction of the duration of that activity yields the latest allowable begin time for it. The backward pass continues with the computation of the latest allowable begin times for all preceding tasks. At nodes from which several arrows take off, the latest allowable complete time is the smallest of the latest allowable begin times, and the latest allowable begin times of preceding tasks are found by subtracting their duration from it. Float is the difference between earliest and latest begin (or complete) times for each task.

The computations can be done manually or with a computer. The latter is desirable for projects involving a large number of tasks, or for which frequent updating of the network is required.

Because of its complexity, the critical path method of scheduling construction has not been widely adopted by contractors. Two reasons for this are that computerized schedules often do not reflect the expectations or abilities of those who are actually doing the work, and the computer data lack simplicity and ease of interpretation and understanding.

As an alternative, the time-scale arrow diagram with the activities shown to scale may be used. Compared with bar charts, such arrow diagrams are almost as easy to read and understand. (S. S. Pinnell, "Critical Path Scheduling: An Overview and Practical Alternative," *Civil Engineering,* July 1980, pp. 66 to 69.) See also Art. 17.10.2.

Expediting. The task of keeping a job on schedule should be assigned to an expediter. The expediter must be alert to and keep on schedule the following major items: letting of subcontracts; securing of materials; and expediting of shop drawings, sample approvals, fabrication and delivery of materials, and building department and government agency submissions and approvals.

Records must be kept by the expediter of all these functions, so that the work can be properly administered. One method of record keeping for the expediting of shop drawings, change orders, samples, and other miscellaneous approvals is illustrated in Fig. 17.16. Information entered includes name of subcontractor, description of work being done, and dates of submissions and approvals received. Space is provided for shop-drawing and sample submissions, approvals and distributions, and other miscellaneous information, as well as contract and change orders. A page similar to the one shown in Fig. 17.16 should be used for every subcontractor and every supplier on the job.

Expediting requires detail work, follow-up work, and awareness of what is happening and what will be needed on the job. Its rewards are a completed job on or ahead of schedule.

17.13 FAST TRACKING

When time of completion is the most essential element in a construction job, superseding even the requirements for total coordination of the plans and specifications during design, and when competititve bids from general contractors are not essential, then a method of design and construction known as fast tracking may be utilized.

On a job that is fast tracked, design and construction begin almost simultaneously with groundbreaking and take place even while the designers are working on the foundations of the structure. Thereafter, work proceeds in the field directly after applicable portions of the design leave the drafting boards. This means that the normal time for design of the project and later bidding is telescoped by a procedure that, in effect, sets design and construction on separate tracks parallel to each other, rather than in consecutive sequence.

Hsg. Dev. Fund Co.Inc. — E X P E D I T I O N — Today's Date 18JUN9

KREISLER BORG FLORMAN

ST-24 — Submittal Status by Vendor — Page 1

PACKAGE	SUBMITTAL	TITLE	SUBMITTALS DRAWING	VENDOR	REQD STRT	REQD FIN	DAYS HELD	BALL-IN-COURT	DATE	STATUS
ACME STEEL DOOR										
	08111-001	Pg.1--15 Hollow Mtl	PG.1--15	ACME	08JAN9	18JAN9	125	At-ACME	14FEB9	ANR
ATLANTIC HARDWARE & SUPPLY CORP.										
	08700-001	Hardware Schedule	BKLET	ATLANT	25FEB9	12MAR9	1	At-DATNER	17JUN9	RES
AVANCD										
	07110-001	30 Mil.Spandrel Flash.	CUTS/SAMPLE	AVANCD	15JUN9	20JUN9	0	At-AVANCD	18JUN9	AAN
	07110-002	Tremco Sealants (Caulk)	CHART/DATA	AVANCD	15JUN9	22JUN9	3	At-DATNER	15JUN9	NEW
BRUNO APPLIANCE CORPORATION										
	11450-001	Roper Top Mount Refrig.	RT14DKXX	BRUNO	16APR9	27APR9	0	At-BRUNO	06MAY9	APP
	11450-002	Roper Vert.Freezer	RV15EFRXW	BRUNO	16APR9	27APR9	0	At-BRUNO	06MAY9	APP
	11450-003	Roper Strd.Clean Range	FGP215V/210V	BRUNO	16APR9	27APR9	0	At-BRUNO	06MAY9	APP
	11450-004	Roper Side by Side Refrig.	RS22ARXX	BRUNO	16APR9	27APR9	0	At-BRUNO	06MAY9	APP
	11450-005	Underctr.Dishwasher (Roper)	WU3000X	BRUNO	16APR9	27APR9	63	At-DATNER	16APR9	NEW
CARDOZA-WINER CORP.										
	15500-001	El Paso Bathtub	CUTS	CARDOZ	22JAN9	04FEB9	0	At-CARDOZ	02FEB9	APP
	15500-002	Pipe & Fitting Sched.	CUTS	CARDOZ	27JAN9	10FEB9	0	At-CARDOZ	05FEB9	APP
	15500-003	Roof Tank	CUTS	CARDOZ	27JAN9	10FEB9	0	At-CARDOZ	20MAR9	APP
	15500-004	Domestic&F.S.P.Wtr.Valve	CUTS	CARDOZ	27JAN9	10FEB9	0	At-CARDOZ	05FEB9	APP
	15500-005	Backflow Preventer	CUTS	CARDOZ	27JAN9	10FEB9	0	At-CARDOZ	05FEB9	APP
	15500-006	Domestic Water Heater	CUTS	CARDOZ	27JAN9	10FEB9	0	At-CARDOZ	05FEB9	APP
	15500-007	Drains	CUTS	CARDOZ	27JAN9	10FEB9	0	At-CARDOZ	05FEB9	APP
	15500-008	SL-1 Gr.Fl.Slv.	SL-1	CARDOZ	05FEB9	10FEB9	0	At-CARDOZ	28FEB9	APP
	15500-009	SL-2 2nd Fl.Slv.	SL-2	CARDOZ	05FEB9	10FEB9	0	At-CARDOZ	28FEB9	APP
	15500-010	SL-3 3rd-7thFl.Slv.	SL-3	CARDOZ	05FEB9	10FEB9	0	At-CARDOZ	28FEB9	APP
	15500-011	SL-4 8--10thFl.Slv.	SL-4	CARDOZ	05FEB9		0	At-CARDOZ	28FEB9	APP
	15500-012	Detail Bklet	BKLET	CARDOZ	05FEB9	10FEB9	0	At-CARDOZ	28FEB9	APP
	15500-013	Plumb.Fixtures	BKLET	CARDOZ	10FEB9	17FEB9	0	At-CARDOZ	06MAR9	AAN
	15500-014	P-G Part.Grnd/2Fl.Clg.Pip.	P-G	CARDOZ	10FEB9	17FEB9	0	At-CARDOZ	28FEB9	APP
	15500-015	SL-5 Roof/Bulkhd Slv./Piping	SL-5	CARDOZ	10FEB9	17FEB9	0	At-CARDOZ	28FEB9	APP
	15500-016	Croker Standpipe	BKLET	CARDOZ	24FEB9	10MAR9	0	At-CARDOZ	04MAR9	APP
	15500-017	Duplex Sump Pumps	BKLET	CARDOZ	24FEB9	10MAR9	0	At-CARDOZ	04MAR9	AAN
	15500-018	Duplex Booster Pump	BKLET	CARDOZ	24FEB9	10MAR9	0	At-CARDOZ	04MAR9	APP
	15500-019	Hot Water Circul.	BKLET	CARDOZ	24FEB9	10MAR9	0	At-CARDOZ	04MAR9	APP
	15500-020	Duplex Sewage Ejectors	BKLET	CARDOZ	24FEB9	10MAR9	0	At-CARDOZ	04MAR9	AAN
	15500-021	Duplex House Pumps	BKLET	CARDOZ	24FEB9	10MAR9	0	At-CARDOZ	04MAR9	AAN
	15500-022	Insulation	CUTS	CARDOZ	03MAR9	15MAR9	0	At-CARDOZ	06MAR9	APP
	15500-023	Roof Tk.Float Valve	CUTS	CARDOZ	03MAR9	15MAR9	0	At-CARDOZ	06MAR9	APP
	15500-024	P-U Undergrd Pipe.	P-U	CARDOZ	24MAR9	07APR9	66	At-CARDOZ	13APR9	ANR
	15500-025	P-C Cellar Clg.Pipe.	P-C	CARDOZ	24MAR9	07APR9	66	At-CARDOZ	13APR9	ANR
	15500-026	Hand Held Shower	BKLT	CARDOZ	29APR9	10MAY9	0	At-CARDOZ	05MAY9	APP
	15500-027	Holby Mixing Valve	CUTS	CARDOZ	20MAY9	30MAY9	0	At-CARDOZ	01JUN9	APP

FIGURE 17.16 Record for expediting approvals of shop drawings, change orders, samples, and miscellaneous items.

The advantage of fast tracking is that a considerable amount of time is saved if construction can indeed proceed simultaneously with, and in conjunction with, the design. Presumably, job completion will take place shortly after completion of the design or, certainly, much earlier than it otherwise would have.

The disadvantages of fast tracking are that in coordination of the work, the necessary input from various consultants will be largely lacking or after the fact.

This may necessitate redoing certain work, or at best, repairing deficiencies that are discovered only after portions of the construction are in place.

Another disadvantage of fast tracking is that much less control exists over costs than in a job where designs are complete when prices are solicited. This disadvantage can be partly overcome, however, if the work is given to reputable, carefully prequalified subcontractors and suppliers. Certain rules of thumb from past jobs certainly will give an indication of what the final cost will be. But because of the expected loss of efficiency and redoing of certain parts of the work, costs can be expected to be higher. Nevertheless, such cost increases could be offset by resultant savings in interest on construction loans, revenues accruing from earlier use of the building, and avoidance of the effects of monetary inflation on the cost of the job if it had been begun at a considerably later date.

Fast tracking lends itself particularly to professional construction management as a form of contract, although the cost-plus-fixed-fee or cost-plus-percentage-fee contract is applicable as well. (See also Art. 2.18.)

17.14 CHANGES, CLAIMS, AND DISPUTE RESOLUTION

After a construction contract has been awarded to a contractor, and usually after construction has begun, it may become necessary to make changes in the work that are not covered by the contract documents. To avoid writing a new contract whenever a work change is made, construction contracts usually include provisions for change orders. These are legal documents that provide a means by which an owner can order changes in the work or require extra work. Change orders may be issued for the following reasons:

1. *Change in scope.* Specifications for building construction include a "Scope of the Project," which includes a general verbal description of the project. The details are given in the technical sections of the specifications, each of which provides a "Scope of the Work." This is a statement of the work to be done under that section. If any change is to be made in the project, a scope becomes involved. Usually, the specifications give the owner the right to make changes in scope, with specified compensation to the contractor.

2. *Change in material or installed equipment.* For any of a variety of reasons, such as unavailability of a specified item or cost or time savings resulting from a substitution, the owner or the contractor may request a change in building materials or installed equipment.

3. *Change in expected conditions.* After the start of a project, a contractor may encounter conditions not anticipated by the building designers and not covered by the contract documents. For example, during excavation for the building foundations, subsurface conditions may be encountered that are different from those described in the plans and specifications. Or abnormal weather may interfere with progress of the work or may damage work already completed. Or labor strikes may occur. Change orders may be required to accommodate these unexpected conditions.

4. *Change to correct omissions.* During construction, the owner or the contractor may discover that certain necessary work or extra work desired by the owner is not covered by the contract documents. The owner will have to issue a change order for performance of that work.

17.14.1 Methods of Payment for Change Orders

Either an owner or a contractor may request a change order. Changes or extras may be priced in any of the following ways:

Unit Prices. At the time of either the bid or the signing of the contract, unit prices are listed by the contractor for various classes of work that may be subject to change. Usually, unit prices are easily administered for such trades as excavation, concrete, masonry, and plastering. The task of the purchaser is to obtain unit prices from subcontractors for various classes of work for the same trades that are in the contract. Although usually the same unit price is agreed for both added and deducted work, occasionally the unit prices for deducted work will be agreed to be 10% less than those for added work.

Cost of Labor and Materials, plus Markup. Another method of computing the value of changes or extra work is by use of actual certified costs, as derived from record keeping as the project proceeds. Rates for wages and fringe benefits must be verified, and the amount of percentage markup must be agreed to either in the contract or before the work is started. Usually, the general contractor is allowed a markup over and above subcontractors' costs and markup, but the general contractor's markup is less than the subcontractor's allowance in such cases.

When work is done on a cost-plus-markup basis, the contractor must maintain an accurate daily tabulation of all field costs. This document should be agreed to and signed by all parties responsible for the record keeping for the change. It will form an agreed-on certification that the work has been performed and of the quantities of labor and material used. A daily work-report certification is shown in Fig. 17.17.

Negotiation by Lump Sum. If the owner desires to have changes or extra work performed and does not want it done on a cost-plus or unit-price basis, owner and contractor may negotiate a lump-sum payment. In this situation, a cost estimate is prepared by the contractor or subcontractor involved, and a breakdown of costs is submitted, together with the estimate total. If the owner accepts the lump sum, the owner or the architect writes a change order, and the work is performed. Such a change order is shown in Fig. 17.18 as issued to a subcontractor.

17.14.2 Claims

Sometimes a dispute may arise as to whether or not the work is part of the contract documents and hence the obligation of the contractor. Such a dispute may result in a claim on the part of the contractor. In making a claim, the contractor may request that a change order be issued prior to proceeding with the work. Or the contractor may nevertheless proceed with the work so as not to delay the job, but request that a change order be issued. Such a request is made in the form of a change-order proposal (Fig. 17.19). Depending on the size of the claim and the attitude of the owner or architect, the contractor may decide whether to continue with the extra work or to press for a decision on the claim, through either arbitration (Art 17.14.4), or some other remedy available either under the contract or at law.

Kreisler Borg Florman
CONSTRUCTION COMPANY

97 Montgomery Street, Scarsdale, New York 10583
Telephone (914) SC 5-4600

DESCRIPTION OF WORK relocation site drainage line NO. 3 PAGE 1 OF 1
LOCATION Apartment House, 170th Street, Jamaica, New York JOB NO.
(JOB NAME)
PERFORMED BY General Contractor DATE January 23,
(INSERT GENERAL CONTRACTOR OR NAME OF SUBCONTRACTOR)

LABOR

NAME OF EMPLOYEE	OCCUPATION	HOURS WORKED	RATE OF PAY	AMOUNT
George M	Laborer	7	$7.50	

MATERIALS, SUPPLIES AND/OR EQUIPMENT

QUANTITY	DESCRIPTION
60 LF	4" Clay Drainage Pipe

COMPLETE DESCRIPTION OF WORK DONE THIS DATE

Relocated site drainage to catch basin B as shown on revised sketch A-301 dated January 10.

THE INFORMATION CONTAINED ABOVE IS CORRECT. ALL WORK SUBJECT TO TERMS OF SUBCONTRACT AND PRIME CONTRACT.

Kreisler Borg Florman CONSTRUCTION COMPANY	SUBCONTRACTOR (NAME)	OWNER'S OR REPRESENTATIVE CONSTRUCTION FOR PROGRESS, INC. (COMPANY)
BY	BY	BY
TITLE Superintendent	TITLE	TITLE Engineer

FIGURE 17.17 Daily work-report certification by a general contractor for changes made or extra work performed.

17.14.3 Who Pays for the Unexpected?

A well-drafted construction contract between contractor and owner should always contain a changed-conditions clause in its general conditions. (See "General Conditions of the Contract for Construction," AIA A201, American Institute of Architects, 1735 New York Ave., NW, Washington, DC 20006.) This clause should answer the question:

If a contractor encounters subsurface conditions different from what might normally be expected or from what is described in plans, specifications, or other job information provided by the owner or owner's agents, resulting in increased cost, should the contractor absorb the increased cost or should the owner pay it?

A properly drafted change-conditions clause should contain the following elements:

A requirement that the owner pay for the unexpected.

Arrangements so that the owner will not be the arbiter of whether the unexpected has occurred.

Recognition that the contract documents have been based on an assumed, described set of facts.

Indication that a changed condition can exist because of unanticipated difficulty of performance.

CHANGE ORDER

Kreisler Borg Florman
CONSTRUCTION COMPANY
97 Montgomery Street, Scarsdale, New York 10583
Telephone (914) SC 5-4600

DATE: March 2,

TO: National Tile & Marble Corporation
300 West 102nd Street
New York, New York 10025

RE: 13-Story Apartment Building
170th Street, Jamaica, New York

WE ARE INCREASING ~~(DECREASING)~~ YOUR SUBCONTRACT AMOUNT BY THE SUM OF $1100.00 FOR

Change tile in lobby and vestibule in accordance with letter from Clarence Lilien and Associates dated January 26,

Reference your proposal dated February 15,

ALL WORK MUST COMPLY WITH THE PLANS, SPECIFICATIONS AND ALL OTHER CONDITIONS AND REQUIREMENTS OF THE GENERAL CONTRACT. ALL TERMS AND CONDITIONS OF THE ORIGINAL SUBCONTRACT BETWEEN US SHALL APPLY AND REMAIN IN FULL FORCE.

KINDLY SIGN AND RETURN ONE COPY OF THIS ORDER.

ACCEPTED:
NATIONAL TILE & MARBLE CORPORATION

BY ____________ DATE ____________

OWNER'S CHANGE ORDER NO. 4

BASED ON CHANGE ORDER PROPOSAL NO. ____________

Kreisler Borg Florman
CONSTRUCTION COMPANY

BY ____________

TITLE Vice-President

FIGURE 17.18 Order issued by a general contractor to a subcontractor to proceed with a change in work not covered by the contract documents.

A requirement that the owner be made aware of a changed condition when it occurs.

Indication that changed conditions can include obstructions.

A requirement that the contractor stop work in the area of a changed condition until ordered to proceed.

Indication that either party can claim changed conditions.

Recognition that the method of procedure for handling changed conditions is provided for in the contract.

An arbitration clause or effective means of appeal other than to the courts.

Because of the comprehensiveness of the elements and because of the lack of uniformity in changed-conditions clauses, the American Society of Civil Engineers Committee on Contract Administration drafted the following recommended changed-conditions clause:

> The contract documents indicating the design of the portions of the work below the surface are based upon available data and the judgment of the Engineer. The quantities, dimensions, and classes of work shown in the contract documents are agreed

CHANGE ORDER PROPOSAL NO.

Kreisler Borg Florman
CONSTRUCTION COMPANY
97 Montgomery Street, Scarsdale, New York 10583
Telephone (914) SC 5-4600

TO Bond-Ryder Associates, Inc.
101 Central Park North
New York, New York 10026

DATE: January 15,

RE: Lionel Hampton Houses
UDC #29
New York, New York

WE SUBMIT OUR ESTIMATE IN THE AMOUNT OF $ 325.40 + Mark up FOR PERFORMING THE FOLLOWING extra WORK:

Furnishing and installing labor and material to remove and or eliminate receptacles that are directly over kitchen sinks and install blank cover plates in Buildings "A", "B", and "C".

Attached find back-up information from Meyerbank Electric Co., Inc., dated 12/13/ and letter of authorization to proceed from Mr. R. Germano, Project Manager, Urban Development Corporation dated December 11,

This proposal is our request for a Change Order to reimburse us for Change Order No. 56 issued to Meyerbank Electric Co., Inc., a copy of which we enclose.

This work is not part of the completed plans dated September 1, 19 , nor is it part of the Addenda, as agreed to by contractor. We are proceeding with this work so as not to delay the job.

~~BEFORE STARTING ANY WORK ON THIS CHANGE WE ARE AWAITING YOUR DECISION. IF YOU WISH US TO PROCEED WITH THIS WORK~~ PLEASE SIGN ONE COPY OF THIS PROPOSAL INDICATING YOUR APPROVAL. ALL OTHER CONDITIONS OF THE CONTRACT BETWEEN US SHALL REMAIN IN FULL FORCE AND EFFECT. ~~IN ACCORDANCE WITH THE APPLICABLE PROVISIONS OF THE CONTRACT WE REQUEST AN EXTENSION OF OUR CONTRACT COMPLETION DATE BECAUSE OF THIS CHANGE. A PROMPT DECISION IS REQUESTED.~~

APPROVED AS CHANGE ORDER NO.__________ TO THE CONTRACT

BY ______________________ DATE: ______________

DISAPPROVED; PROCEED IN ACCORDANCE WITH CONTRACT PLANS & SPECIFICATIONS.

BY ______________________ DATE: ______________

CC: Urban Development Corporation

Kreisler Borg Florman
CONSTRUCTION COMPANY

BY ______________________

TITLE Project Manager

FIGURE 17.19 Change-order proposal submitted by a contractor to an owner to request issuance of a change order.

upon by the parties as embodying the assumptions from which the contract price was determined.

As the various portions of the subsurface are penetrated during the work, the Contractor shall promptly, and before such conditions are disturbed, notify the Engineer and Owner, in writing, if the actual conditions differ substantially from those which were assumed. The Engineer shall promptly submit to Owner and Contractor a plan or description of the modifications which he or she proposes should be made in the contract documents. The resulting increase or decrease in the contract price, or the time allowed for the completion of the contract, shall be estimated by the Contractor and submitted to the Engineer in the form of a proposal. If approved by the Engineer, he or she shall certify the proposal and forward it to the Owner with recommendation for approval. If no agreement can be reached between the Contractor and the Engineer, the question shall be submitted to arbitration or alternate dispute resolution as provided elsewhere herein. Upon the Owner's approval of the Engineer's recommendation, or receipt of the ruling of the arbitration board, the contract price and time of completion shall be adjusted by the issuance of a change order in accordance with the provisions of the sections entitled, "Changes in the Work" and "Extensions of Time."

A contractor would be well advised not to enter into agreement for construction without a changed-conditions clause.

17.14.4 Alternative Dispute Resolution

Alternative dispute resolution is the term applied to methods of resolving disputes other than litigation. The American Arbitration Association, a nonprofit organization and pioneer in alternative dispute resolution, has established and is a financial supporter of the National Construction Industry Disputes Resolution Committee. This group comprises representatives of all the major organizations concerned with construction.

Assistance in Negotiation. When a dispute between owner and contractor arises on a construction project, the foremost method of resolving it is by immediate negotiation. One optional procedure for facilitating negotiation is to appoint before the start of a project a Dispute Resolution Board (DRB). It usually consists of three qualified persons whose role is to assist the parties in negotiating a settlement of a controversy or claim. If there cannot be a resolution of the dispute by such negotiation, then the DRB will issue nonbinding recommendations based on available information.

Besides negotiation, other methods used for dispute resolution include conciliation, facilitation, fact-finding, minitrials, and most importantly, arbitration and mediation.

Arbitration. Parties to a contract, either at the time of entering into the contract or when a dispute arises, agree to submit the facts of the dispute to impartial third parties who will hear a presentation of the claims by both parties and render a decision.

If the parties have agreed to submit a matter to arbitration, the decision of the arbitrators is binding on both parties. The decision can be enforced in any court of law having jurisdiction over the party against whom a claim is made.

Because of the large number of construction arbitration cases filed, the American Arbitration Association has, in conjunction with representatives from the construction industry, drafted special construction-industry arbitration rules ("Construction Contract Disputes—How They May Be Resolved under the Construction Industry Arbitration Rules," American Arbitration Association, 140 West 51st St., New York, NY 10020). These rules provide information for proceeding with construction arbitration.

Parties may agree beforehand to submit disputes to arbitration by inclusion of the standard arbitration clause in their agreement:

> Any controversy or claim arising out of or relating to this contract, or the breach thereof, shall be settled by arbitration in accordance with the Construction Industry Arbitration Rules of the American Arbitration Association, and judgment upon the award rendered by the Arbitrator(s) may be entered in any Court having jurisdiction thereof.

To the standard arbitration clause, it is sometimes best to add the words "in the city of _____ " immediately after "shall be settled by arbitration," to specify where the arbitration hearings should be held.

Mediation. In addition to arbitration, there is available through the services offered by the American Arbitration Association a system of mediation of disputes. Mediation differs from arbitration in that not only is it entered into voluntarily by the parties, but the results are nonbinding, as opposed to arbitration where the results are binding on both parties. When a dispute is submitted to mediation, a trained mediator, or several mediators, will assist the parties in reaching a settlement of the controversy.

The mediator participates impartially in the negotiations and advises and consults the various parties involved. The result of the mediation should be an agreement that the parties find acceptable. The mediator cannot impose a settlement and can only seek to guide the parties to a settlement.

The American Arbitration Association administers the mediation process by providing panels of trained and acceptable mediators. In addition, it will provide a place to hold the mediation hearings and will maintain records of cases and issues involved, the names and types of mediators appointed, and the results of the mediation efforts.

When drawing up a contract, the parties to the contract may, by mutual agreement, wish to insert into the contract the following mediation clause:

> If a dispute arises out of or relating to this contract, or the breach thereof, and if said dispute cannot be settled through direct discussions, the parties may agree to endeavor first to settle the dispute in an amicable manner by mediation under the Voluntary Construction Mediation Rules of the American Arbitration Association, before having recourse to arbitration or a judicial forum.

17.15 INSURANCE

Insurance policies are contracts under which an insurance company agrees to pay the insured, or a third party on behalf of the insured, should certain contingencies arise.

No business is immune to loss resulting from ever-present risks. It is imperative, therefore, that a sound insurance program be designed and that it be kept up to date.

Because few businesses can afford the services of a full-time insurance executive, it is important that a competent agent or broker be selected to: (1) prepare a program that will provide complete coverage against the hazards peculiar to the construction business, as well as against the more common perils; (2) secure insurance contracts from qualified insurance companies; (3) advise about limits of protection; and (4) maintain records necessary to make continuity of protection certain. While an executive of the business should oversee insurance coverage, much of the detail can be eliminated by utilizing the services of a competent agent or broker.

It is necessary, of course, that the responsibility for providing protection be placed on an insurance company whose financial strength is beyond doubt.

Another important point for the buyer of insurance is the service that the company selected may be in a position to render. Frequently, construction operations are conducted at a considerable distance from city facilities. It is necessary that the company charged with the responsibility of protecting the construction operations be in a position to render "on-the-job" service from both a claim and an engineering standpoint.

The interests of contractors, subcontractors, and building owners are very closely allied. Particular attention should be given to the definition of these respective interests in all insurance policies. Where the insurable interest lies may depend upon the terms of the contract. Competent advice is frequently needed in order that all policies protect all interests as required.

While the forms of protection purchased and the adequacy of limits are of great importance to a prime or general contractor, it is also of great importance that the insurance carried by subcontractors be written at adequate limits and be broad enough to protect against conditions that might arise as a result of their acts. Also, the policies should include the interests of the prime or general contractor insofar as much interests should be protected.

This section merely outlines those forms of insurance that may be considered fundamental (Table 17.1). It includes brief, but not complete, descriptions of coverages without which a contractor should not operate. It is not intended to take the place of the advice of experienced insurance personnel.

17.15.1 Fire Insurance

Fire insurance policies are well standardized. They insure buildings, contents, and materials on job sites against direct loss or damage by fire or lightning. They also include destruction that may be ordered by civil authorities to prevent advance of fire from neighboring property. Under such a policy, the fire insurance company agrees to pay for the direct loss or damage caused by fire or lightning and also to pay for removal of property from premises that may be damaged by fire.

Attention should be given to the computation of the amount of insurance to be applied to property exposed to loss. In addition, the cost of debris removal should be taken into consideration if the property could be subject to total loss. Under no circumstances will the amount paid ever exceed the amount stated in the policy. If the fire insurance policy has a coinsurance clause, the problem of valuation and adequate amount of insurance becomes even more important.

The form of fire insurance particularly applicable in the construction industry is known as **Builders Risk Insurance.** The purpose of this form is to insure an owner or contractor, as their interests may appear, against loss by fire while buildings are under construction. Such buildings may be insured under the Builders Risk form by the following methods:

1. The reporting form, under which values are reported monthly or as more buildings are started. Reports must be made regularly and accurately. If so, the form automatically covers increases in value.

TABLE 17.1 Typical Insurance Needs of Parties to Construction

Type	Purpose	Suggested limits for small-to-medium-size company	Basis for premium
Builders Risk Fire Insurance with extended coverage, vandalism, and malicious mischief	Protect building, fire and lightning, explosion, windstorm, etc. All-risk coverage is the broadest and most expensive	Amount of contract less demolition, construction below lowest basement slab and landscaping	Dollars per hundred
Commercial General Liability			
1*a*. Premises, operations of contractor	Third-party bodily injury and property damage	$3,000,000 B.I.* $1,000,000 P.D.†	Payroll classification
1*b*. Independent contractors contingent public liability and property damage	Protect contractor against subcontractor's negligence	Same as above	Amount of subcontract
1*c*. Completed operations insurance	Claims by third parties on projects already completed	Same as above	Amount of coverage
1*d*. (Hold-harmless) contractual liability	Contractually assumed liability	Same as above or as per contract	Amount of coverage
2. Owner's protective liability insurance	Third-party liability	As per contract	Amount of contract
3. Umbrella liability	Third-party liability	$3,000,000	Various
Owned and nonowned motor-vehicle insurance	Third-party liability and physical damage insurance	B.I., P.D., collision, fire and theft (no fault)*†	Number of cars and experience
Contractor's equipment insurance	Protects equipment, tools, etc.	$200,000	Coverage amount
Material floater	Losses to materials before installation in job (may be included in Builders Risk)	$200,000	Coverage amount
Workers' compensation and employers liability	Employee protection	Set by state	Payroll
Employers liability insurance	Employee suits outside of workers compensation	Unlimited in some states or $100,000	Payroll
Disability benefits insurance	(Where required by the state)	Set by the state	Payroll

TABLE 17.1 (*Continued*)

Type	Purpose	Suggested limits for small-to-medium-size company	Basis for premium
Fidelity insurance	Embezzlement of funds	$500,000	Number of employees
Valuable papers	Loss of valuable papers	$50,000	Amount of coverage
Key officer insurance	Life insurance policy to cover corporate officer in case of death or disability	$500,000 or amount to be determined	Amount of coverage
Group hospitalization	Employees' protection (Blue Cross or HMO‡)	120-day extended coverage	Number of employees and coverage
Major medical	Employees' protection	Limits beyond Blue Cross or HMO‡	Same as above
Group life insurance	All eligible employees	6 mo. to year salary	Age and number of employees
	Technically not insurance		
Pension and/or profit sharing plan	Employee retirement	25% salary when combined with Social Security	Salary of employees
Unemployment insurance	(Where required by the state)	Set by state and federal government	Payroll
Social Security	Employees' welfare	Set by federal government	Payroll
	Architects and engineers insurance		
Architects and engineers liability insurance	Damage to persons and property	$3,000,000 B.I.* $1,000,000 P.D.†	Amount of coverage
Errors and omissions insurance	Errors or omissions from the plans	Highest available	Amount of coverage
	Owner insurances		
Owner's protective liability insurance	Public liability and property damage as a result of construction operations	$1,000,000 $5,000,000	

TABLE 17.1 (*Continued*)

Type	Purpose	Suggested limits for small-to-medium-size company	Basis for premium
	Owner insurances		
Property fire insurance	When construction is being done on an existing building, fire, etc., general contractor and subcontractors	Full insurance value	
Loss of use insurance	Losses in property as result of fire or other loss	Approx. loss of income	
Steam boiler	Explosion, etc.	$500,000	
Nuclear incident	Nuclear reactor loss (if required)	High	

*B.I. = bodily injury
†P.D. = property damage
‡HMO = Health Maintenance Organization

2. The completed-value form under which insurance is written for the actual value of the building when it is completed. This is written at a reduced rate because it is recognized that the full amount of insurance is not at risk during the entire term of the policy. No reports are necessary in connection with this form.

3. Automatic Builders Risk Insurance, which insures the contractor's interest automatically in new construction, pending issuance of separate policies for a period not exceeding 30 days. This form generally is used for contractors who are engaged in construction at a number of different locations.

There are certain hazards which, though not quite so common as fire and lightning, are nevertheless real. The contractor should insist that these be included in the insurance, by endorsement. A few of the available endorsements are:

1. Extended coverage endorsement, which insures the property for the same amount as the basic fire policy against loss or damage caused by windstorm, hail, explosion, riot, civil commotion, aircraft, vehicles, and smoke.

2. Vandalism and malicious mischief endorsement, which extends the protection of the policy to include loss caused by vandalism or malicious mischief. There is a special extended coverage form that provides coverage on an all-risk basis and includes the peril of collapse.

Completed-value and reporting forms treat foundations of a building in the course of construction as a part of the Builders Risk Value for insurance purposes. But all work below the lowest basement slab, site work, and demolition are not covered by Builders Risk Insurance. The value of these items should be deducted from the amount of the policy. This will result in a saving of insurance premium.

Builders' machinery and equipment should be specifically insured as a separate item if coverage is desired under the policy.

Under the terms of the AIA General Conditions, the owner is to provide the Builders Risk Insurance and pay for it. Under those circumstances, the contractor should request that the following clause appear as an endorsement to the policy:

> Additional insureds under this policy: The Contractor and its Subcontractors, as their interests may lie.

This endorsement will assure the contractor and its subcontractors of a share in the insurance proceeds in case of claim. It further protects the contractor and subcontractors from suit by the insurance company under its subrogation clause, should either of these parties have caused or contributed to the loss.

17.15.2 Insurance for Contractor's Equipment

The so-called Inland Marine insurance market is the place to look for many coverages needed by contractors. From the insurance point of view, each contractor's problem is considered separately. The type of operation, the nature of equipment, the area in which the contractor works, and other pertinent factors are all points considered by an Inland Marine underwriter in arriving at a final form and rate.

Obvious contractors' equipment—mechanical shovels, hoists, bulldozers, ditchers, and all other mobile equipment not designed for highway use—is the primary subject of the **Contractors Equipment Floater.** Such protection is necessary because of the size of the investment in such equipment and of the multiplicity of perils to which the equipment is exposed.

Some companies will write the Contractors Equipment Floater Policy only on a named-perils basis, which ordinarily includes fire, collision, or overturning of a transporting conveyance, and sometimes theft of an entire piece of equipment. Other companies will write certain kinds of contractors' equipment on the so-called all-risk basis. Certain perils, such as collision during use, are subject to a deductible fixed amount. The rate for this broader insurance generally is higher than that for the named perils form. In the all-risk form, the customary exclusions, such as wear and tear, the electrical exemption clause, strikes, riots, and other similar exclusions, are present. Because of increased exposure to nuclear hazards, all Inland Marine policies that insure against fire must carry a nuclear exclusion clause that provides that the company shall not be liable for loss by nuclear reaction or radiation or radioactive contamination, whether controlled or uncontrolled and whether such loss is direct or indirect, approximate or remote.

In addition to the equipment, there is a need to provide protection for the materials and supplies en route to or from the site. If these materials and supplies are transported at the risk of a contractor and are in the custody of a common carrier, a **Transportation Floater** should be obtained. If, on the other hand, these materials and supplies are moved on the contractor's own trucks, a **Motor Truck Cargo-Owners Form** should be obtained. The premium for the transportation form is usually based on the value of shipments coming under the protection of the policy. The coverage is usually on an all-risk basis. The Motor Truck Cargo-Owners Form is generally on a named-perils basis at a flat rate applied against the limit of liability required by the insured's needs.

Occasionally, a contractor may be responsible for machinery, tanks, and other property of that nature until such times as they are completely installed, tested, and accepted. Exposures of this kind are usually covered under an **Installation Floater,** which would provide insurance to the site as well.

The Installation Floater Form is generally on a named-perils basis, including perils of loading and unloading, at a rate for the exposure deemed adequate by the underwriter. Large contractors should have such insurance written on a monthly reporting form to reflect increasing values as installation progresses. Small contractors generally can provide for the coverage under a stated amount on an annual basis subject to coinsurance.

There is also a form of policy known as a **Riggers Floater** that is designed for contractors doing that type of work. This policy is usually a named-perils form at rates based on the nature of the rigging operation.

Neither the forms nor rates in any of these classes are standard among the companies writing them, although in general they are all similar.

Some contractors building bridges might be required to take a **Bridge Builders Risk Form.** This is an exception to most of the insurance provided to contractors, in that it is required to be rated in accordance with forms and rates filed in most of the states and administered by a licensed rating bureau.

17.15.3 Motor-Vehicle Insurance

Loss and damage caused by or to motor vehicles should be separately insured under specific policies designed to cover hazards resulting from the existence and operation of such vehicles. Bodily injury or property damage sustained by the public as a result of the operation of contractor's motor vehicles or other self-propelled equipment is insured under a standard policy of insurance. To secure complete protection, a Comprehensive Automobile policy should be obtained to provide protection, in addition to the preceding, for hired cars, employers nonownership, and any newly acquired motor vehicles or self-propelled equipment during the term of the policy. Damage to owned vehicles can be added to the same policy on an automatic basis to provide Comprehensive Coverage and Collision Insurance.

All vehicles should be covered, for high limits.

Protection furnished by automobile liability and property damage insurance serves in two ways: (1) the insurance company agrees to pay any sum for bodily injury and property damage for which the insured is legally liable; (2) the policy agrees to defend the insured. It is important, therefore, that the limits be adequate to guarantee that the insured obtain full advantage of the service available. For example, suppose only $25,000 and $50,000 limits are carried, and action is brought against the contractor in the amount of $500,000. It may be necessary to employ an attorney to safeguard the insured's interests for the amount of the action that exceeds the limits of the policy.

A contractor should not operate any type of motor vehicle without insurance. Automatic coverage should be provided to include all motor vehicles owned or acquired. The cost for highest available limits is reasonable.

Damage to owned motor vehicles may be insured under a fire, theft, and collision policy. Comprehensive motor-vehicle protection covers physical damage sustained by motor vehicles because of fires, theft, and other perils, including glass breakage. Collision insurance insures against loss from collision or upset. While the latter is available on a full-coverage basis, it is generally written on a deductible basis (loss less a fixed sum).

To cover a contractor for liability arising from the use by employees of their own automobiles while on the contractor's business, nonownership or contingent liability coverage is necessary. This may be included in the policy by endorsement.

Frequently, contractors have occasion to hire trucks or other vehicles. Liability and property damage insurance to cover the contractor's liability when using hired vehicles should be included at the time automobile insurance is arranged.

17.15.4 Boiler and Machinery Insurance

Boilers and other pressure vessels and machinery require the protection provided by boiler and machinery insurance. These policies cover loss resulting from accidents to boilers or machinery, and in addition, cover contractor's liability for damage to the property of others. Policies may also include liability arising from bodily injuries sustained by persons other than employees. This is needed because of the exposure that many contractors have as a result of the interest of the public in construction work.

The service rendered by boiler and machinery companies is of great help. Nearly all insurance companies that write this form have staffs of competent and experienced inspectors whose job it is to see that boilers, pressure vessels, and other machinery are adequately maintained.

17.15.5 Liability Policies Covering Contractor's Operations

Anyone who suffers bodily injuries or whose property is damaged as a result of the negligence of another person can recover from that person, if the latter is legally liable. Every business should protect itself against claims and suits that may be brought against it because of bodily injuries or property damage suffered by third parties.

Maintenance of an office or yard, as well as the conduct of a construction job, presents exposures to the public. There may be no negligence, and consequently no legal liability on the part of the contractor, but should claim be brought or suit instituted for an injury, the contractor without insurance coverage will require trained personnel to investigate the claim and negotiate a settlement or defend a lawsuit if the claim goes to court.

17.15.6 Commercial General-Liability Insurance

This is expressly designed to serve contractors by providing insurance that will pay for bodily injuries and property damage suffered by third parties if the contractor is legally liable, but the policy will also serve by defending the interests of the contractor in court. Sometimes the litigation involves amount of damage; but frequently, the contractor being sued is not legally liable for the injuries or damage. It is fundamental, therefore, that a policy be obtained at substantial limits for both bodily injuries and property damage, that the policy cover all existing exposures and also provide for protection against exposures that may not exist or be contemplated on the inception date of the policy. The scope of operations conducted by most contractors is such that it is frequently difficult to visualize all the hazards that may exist or come about simply by being in the construction business.

The commercial general-liability policy that covers all liability of the insured, except that resulting from the use of automobiles, is a standard form available in all states at rates regulated by law. This policy protects the contractor under one insuring clause and with one limit against claims. Blanket coverage is provided at a premium based on actual exposures disclosed by an audit at the end of the policy term. As required by this policy, every contractor must maintain accurate records of payrolls, value of sublet work, dollar amount of sales, and other factors that will be important at the time an audit is made.

Under the bodily injury liability clause of this policy the insurance company agrees to pay on behalf of the insured "all sums that the insured shall become legally obligated to pay as damages because of bodily injury, sickness or disease, including death at any time resulting therefrom, sustained by any person and caused by an occurrence." This is a very broad insuring clause. It obviously includes the entire business operations of the insured.

Property-damage liability is also covered. Additionally, the policy provides Independent Contractors Contingent Public Liability and Property Damage protection, which insures the general contractor against subcontractors' negligence.

There are certain exclusions in the policy that should be noted. The policy does not include:

1. Ownership, maintenance, or use (including loading or unloading) of water craft away from premises owned, rented, or controlled by the insured; automobiles while away from the premises or the ways immediately adjoining; and aircraft under any condition. However, this exclusion does not apply to operations performed by independent contractors or to liability assumed by any contract covered by the policy.

2. Bodily injury sustained by employees while engaged in the employment of the insured.

3. Liability for damage to property occupied, owned, or rented to the insured or in the insured's care, custody, or control.

Contractors should be familiar with the provision of the commercial general-liability policy pertaining to contractual liability. The policy automatically provides coverage for the following written contracts: lease of premises; easement agreement, except in connection with construction or demolition operations on or adjacent to a railroad; undertaking to indemnify a municipality; side-track agreement; or elevator maintenance agreement. A premium is charged for such agreements as may be disclosed by audit.

There is no protection for the liability assumed in some very common types of agreements that include service, delivery, and work contracts. Many of these contracts are signed without full realization of the liability assumed. Each such agreement should be submitted to the insurance company at the time the policy is written in order that a premium charge may be computed and the agreement covered under the policy.

The most important contractual liability that may be assumed by the contractor is the so-called **hold-harmless clause.** A common type of hold-harmless clause as written by a general contractor to a subcontractor is illustrated in the subcontract rider in Fig. 17.7, Paragraph 32*a*. A similar clause written by the owner for inclusion in the general contract would be slightly modified and would substitute in appropriate places the word "Contractor" for "subcontractor" and the word "Owner" for "Contractor."

Hold-harmless clauses can be automatically included in a general contractor's commercial general-liability policy. However, the general contractor should include the previously mentioned subcontractor hold-harmless clause in each subcontract. Subcontractors will probably be required by their insurance companies to pay an additional premium for this coverage.

The commercial general-liability policy includes complete and automatic products-liability insurance, including completed-operations protection. The one exception to this complete coverage is that the policy does not include liability for damage to the work or to the goods themselves, such as the obligation of the contractor to repair or replace if there are defects. While the policy provides coverage, it is in fact an optional protection, which may be deleted. However, every contractor should take advantage of this coverage.

Most building contractors use elevators or hoists during construction. The policy automatically covers elevators, hoists, and other such hazards. Escalators may be covered for an additional premium.

The breadth of public-liability protection available, the numerous hazards to which a contractor may be exposed, both known and unknown, and the necessity for having complete coverage at all times indicate the need for the advice of trained insurance representatives.

17.15.7 Workers' Compensation Insurance

Every state requires an employer to secure a policy of workers' compensation to provide for an injured employee the benefits of the workers' compensation law of that state. An insurance company that has had extensive experience in the workers' compensation field is best suited to meet the requirements of most contractors. It is in the employer's, as well as the employee's, best interests to see that the company entrusted to provide workers' compensation insurance is equipped to provide loss-prevention service and prompt first aid and to settle compensation claims fairly and speedily.

Some states impose on contractors liability for injury to subcontractors and their employees unless insurance is specifically provided for subcontractors and their employees. Check the law of the state in which construction is to be performed.

Subcontractors' insurance should be carefully examined and deficiencies found should be corrected. Certificates of insurance should be required, including coverage for contractual liability (hold-harmless). Because many complex situations arise, it is important that the advice of a qualified agent or broker be obtained.

An injured employee may believe that it is advantageous to waive workers' benefits and sue the employer for negligence or failure to maintain a safe place to work. Since benefits under workers' compensation are limited by statute, the employee may believe that recovery in a lawsuit may be substantially higher. Insurance coverage for such suits is provided by Employers Liability Insurance.

17.15.8 Money and Securities Protection

Every contractor has cash, securities, a checking account, and payrolls that are vulnerable to attack by dishonest people, both on and off the payroll. The same hazards present in every business are present also in the construction business. And no contractor is immune to dishonesty, robbery of payroll, burglary of materials, or forgery of signature on check.

Employee dishonesty may be covered on a blanket basis, either under a Primary Commercial or Blanket Position Form of Bond. The fact that contractors generally entrust the maintence of payroll records and the payment of employees to subordinates demonstrates the necessity for blanket dishonesty protection.

While many contractors maintain an organization on a year-round basis, some may not. For those contractors who do have a permanent staff, the bond may be written on a 3-year basis at a saving. It is important that adequate limits be purchased. A blanket bond in an amount equal to 5% of the gross sales is desirable.

General funds, securities, and payroll funds should be covered on the broadest basis available that protects against burglary, robbery, mysterious disappearance, and destruction, on and away from any premises. The general funds may be covered in an amount sufficient to protect against the maximum single exposure. Payroll funds may be insured specifically and in a different amount.

Contractors who maintain inventories of materials should insure them against burglary and theft. It should be noted, however, that insurance companies are not willing to insure against loss by burglary or theft unless materials are under adequate protection. Insurance is not available to cover property on open sites or in yards, but only while within buildings that are completely secured when not open for business.

Every business that maintains a checking account, however small the balance may be, should insure against loss caused by the forgery of the maker's name or by the forgery of an endorsement of checks issued.

The policy to cover all these hazards is the Comprehensive Dishonesty, Disappearance, and Destruction Policy. Its several insuring agreements include employee-dishonesty coverage, broad-form money and securities protection, on and off the premises, and forgery. Other coverages to provide burglary and theft protection for merchandise and materials may be added by endorsement. Optional coverages available are numerous, and the contract may be designed specifically for all of a contractor's exposures.

17.15.9 Employee Group Benefits

There are many forms of group insurance for providing benefits for employees. Included in the following is a brief description of those forms of "group insurance benefits" that are of greatest current interest.

Group Life Insurance. A form of term life insurance written to cover in a specified amount a group of employees of a single employer. Frequently, the amount of insurance is related to an employee's earnings and increases as the earnings increase. However, the amount of insurance may be a flat sum. Usually, a specified number of employees must participate.

Group Disability Insurance. If an employee is away from work as a result of a nonoccupational disability, this insurance provides a continuing income during the period of absence. The amount of benefit is usually a precentage of earnings. Many states have adopted compulsory disability laws.

Group Hospitalization and Surgical Benefits. These forms of group protection are designed to protect an employee from the results of high hospital expense or the expense of a surgical operation. The protection may be written to cover an employee solely or it may be written to cover an employee and dependents.

Group Coverage for Major Medical Expenses. Rates vary greatly, as do forms of policy. The coverage is usually provided on a deductible basis (medical expense less a fixed sum). Frequently, a coinsurance feature is included so that the individual protected by such insurance bears part of the cost.

17.15.10 Amounts of Insurance

For a construction manager, the limits of insurance that should be carried by the contractor and the limits that should be carried by the subcontractors are of fundamental importance.

Usually, the contractor's limits are set forth in the agreement with the owner. When the owner's insurance limits are not high enough to afford the contractor full protection for the exposure the contractor anticipates, it is often worth the additional cost of increasing these limits. For example, insurance with limits of $500,000 to $1,000,000 for public liablity and $500,000 for property damage often may be less expensive in the long run than lower limits that may be the maximum required by the owner. Furthermore, if the general contractor requires subcontractors to carry insurance with the same limits, it may be provident for the general contractor to pay the additional cost to the subcontractor for the increase of limits above what the subcontractor normally carries. When insurance limits are raised from $100,000 to $500,000, or from $50,000 to $500,000, the increase in cost is not proportional to the increase in limits.

17.16 CONSTRUCTION CONTRACT BONDS

Agencies at all levels of government generally obtain competitive bids for construction. Awards are made to the lowest responsible bidder, who is required to furnish performance and payment bonds provided by a qualified corporate surety. In addition, all construction projects financed or insured by the Federal government, such as FHA-insured projects, require such bonds. Also, many private owners require bonds of contractors.

Generally, bidders are required to post a certified check or furnish a bid bond. A bid bond assures that, if a contract is awarded, the contractor will, within a specified time, sign the contract and furnish bond for its performance. If the contractor fails to furnish the performance bond, the measure of damage is the smaller of the following: the penalty of the bid bond or the amount by which the bid of the lowest bidder found to be responsible and to whom the contract is awarded exceeds the initial low bid.

Most surety companies follow the practice of authorizing a bid bond only after the performance bond on a particular contract has been underwritten and approved. For this reason, contractors are well advised against depositing a certified check with a bid unless there are assurances that the performance bond on that particular contract has been underwritten and approved.

There is no standard form of construction contract bond. The Federal government and each state, county, or municipal government has its own form. Private owners generally use the bond form recommended and copyrighted by the American Institute of Architects. Surety companies have developed a very broad form of bond, which is available to owners of private construction. Whatever form is used, the surety generally has a twofold obligation:

1. To indemnify the owner against loss resulting from the failure of the contractor to complete the work in accordance with the contract.
2. To guarantee payment of all bills incurred by the contractor for labor and materials.

Usually, two bonds are furnished, one for the protection of the owner, and another to protect exclusively those who perform labor or furnish materials. If one bond is furnished, the owner has prior rights.

Sureties underwrite construction contract bonds carefully. They are interested in determining whether a contractor has the capital to meet all financial obligations, the equipment to handle the physical aspects of the particular undertaking, and the construction experience to fulfill the terms of the contract.

A Performance and Payment Bond is not an insurance policy that will pool the premiums from those contractors who receive bonds and will pay the losses out of that pool, as is done with insurance. Rather, the bond essentially is a credit guarantee by the bonding company, and, as for any credit guarantee given in business, the bonding company expects to be reimbursed if this guarantee is enforced. Therefore, before providing a bond to a contractor, the surety requires the contractor to sign an Application for Surety Bond. This application, among other things, includes an agreement by the contractor to reimburse the surety for any losses that the surety may sustain as a result of having written the bond. The contractor, in order to receive the bond, therefore is indemnifying the surety. Cost of the bond is added by the contractor to the construction contract price.

Contractors should be aware of all the information that a surety will require and that is necessary to the underwriting of a contract bond. It is also important that a contractor take advantage of the services of a competent agent or broker who has close affiliations with a surety company that has the capacity to meet all the contractor's needs. Outlined below are several items of information that will be required by the surety:

1. A complete balanced financial statement with schedules of the principal items. This is a condition precedent to the approval of any contract bond. Sureties have forms on which a contractor may furnish financial information. They should be completed by the individual responsible for the financial operations of the company and the data taken directly from the company's books. It is preferable to have the financial statement prepared and certified by a public accountant.
2. A report on the contractor's organization. The surety is interested in knowing the length of time the contractor has been in business, whether the firm operates as an individual, a partnership, or a corporation, and certain specific details, depending upon the form of organization.
3. A report on the technical qualifications and experience of the individuals who will be in charge of work to be performed.
4. A report on the type of work undertaken in the past, together with information regarding jobs successfully completed.
5. An inventory of equipment, noting value and age of each piece and any existing encumbrance. An inventory of materials will also be helpful.

From the construction management point of view, the most important question involving bonding is: What avenues of business are open to the contractor who lacks sufficient bonding capacity to do bonded work?

In Art. 17.4 various sources of business are described, and in Art. 17.1 various types of construction companies are discussed. Many of these types of business and construction companies do not require bonds for their work. For example, it is very rare that a bond is required in a construction management contract. When a contractor lacks capacity for bonding, it is well to pursue the lines of work described in those articles for which a bond will not be required.

The second question confronting a construction manager is whether or not to require a bond of subcontractors. In general, if the financial capability or experience of a subcontractor is sufficiently doubtful as to require bonding, the job should not be awarded to that company. Exceptions to this can be made to assist young companies in starting and gaining experience.

There are alternatives to subcontract bonds. These alternatives include the following:

Personal guarantees by the principals of the subcontracting company

Personal guarantees of other individuals of substantial worth unconnected with the subcontracting company

Posting of a sum of money or of a security, such as a letter of credit, until performance of the subcontractor's work has been completed by the subcontractor.

17.17 TRADE PAYMENT BREAKDOWNS AND PAYMENTS

The Contractor's Trade Payment Breakdown (Fig. 17.20) is usually prepared by the contractor long in advance of purchase of materials or subcontracts. Therefore, the amounts shown may vary from the actual costs of the various items. In addition to this, the contractor usually has start-up costs for such expenses as mobilization, temporary structures, temporary utilities, and layout, which may not be reflected completely in the breakdown. To offset both of these conditions, the breakdowns may be unbalanced by giving greater value than true costs to the work of some of the trades done early than to the work of the trades done later. This unbalancing will enable the contractor to receive funds for these early start-up costs and thus to be compensated for having to pay various subcontractors and suppliers greater amounts than anticipated at the time the breakdown was prepared. The unbalancing will also make funds available for paying certain subcontractors and suppliers their retained percentages prior to payment of the general contractor's retained percentage. The reason why such payments are made earlier is that direct labor costs and many purchase commitments made by contractors, such as lumber, hardware, millwork, doors, and frames, are not subject to retained percentage. These materials, which are merely delivered and not contracted for to be both delivered and installed, must be paid for in full. Since there is a retained percentage being held on all of the work of the general contractor, funds must be available for the part of the retainage that is paid out by the contractor. The unbalancing of the Trade Payment Breakdown will serve this purpose.

Payments to subcontractors and suppliers are made on the basis of payments received by the general contractor on monthly requisitions. The general contractor should pay to these subcontractors and suppliers only the same percentage as approved on the contractor's requisition to the owner. In addition, the contractor should pay only extras for which corresponding payments have been received from the owner. In case of disputed extras, the contractor may have to pay a portion of the disputed amount to the subcontractor to keep the job running smoothly, despite the fact that the owner has not paid anything on these disputed claims.

Subcontractors should be paid only when the contractor gets paid for the subcontractors' portions of the work. Furthermore, subcontractors should be paid only after their subcontracts are signed and insurance certificates have been received. All subcontractors should be required to use the same requisition form. A typical Application for Payment on Account of Contract is illustrated in Fig. 17.21. Both the front and back must be filled out by the subcontractor and signed. Payments to a subcontractor should be shown cumulatively on the subcontractor's financial folder, which should be separately set up for each subcontractor.

U. S. DEPARTMENT OF HOUSING AND URBAN DEVELOPMENT
FEDERAL HOUSING ADMINISTRATION

CONTRACTOR'S and/or MORTGAGOR'S

COST BREAKDOWN
(SCHEDULES OF VALUES)

Date August 23	Project No. 012-38024-L8
Sponsor 80-86 Houses	Building Identification
Name of Project 80-86 Houses.	Location 80-86 Green Ave Brooklyn N.Y.

LINE	DIV	TRADE ITEM	COST	TRADE DESCRIPTION
1	3	Concrete	530,000	Foundations, superstructure plank, slab-on-grade
2	1	Masonry	410,000	Brick and block masonry and caulking
3	5	Metals	70,000	Lintels, stairs, railings
1	6	Rough carpentry	10,000	Block and rough framing
5	6	Finish carpentry	60,000	Millwork, material, and labor

LINE	DIV	TRADE ITEM	COST	TRADE DESCRIPTION
38	2	Site improvements	5,000	Benches, etc.
39	2	Lawns & planting	10,000	Topsoil, seeding, plants
10	2	Unusual site conditions		
11		Total land imprvts.	130,000	
12		Tot. struct. & land imprvts.	2,860,000	
13	1	General requirements	240,000	
14		Subtotal (lines 41 and 42)	3,100,000	
15		Builder's overhead	62,000	
16		Builder's profit		
17		Subtotal (lines 44 thru 46)	3,162,000	
18				
19		Other fees		
50		Bond premium		
51		Total for all improvements		
52		Builder's profit paid by means other than cash		
53		Total for all improvements less line 52		

NONRESIDENTIAL AND SPECIAL EXTERIOR LAND IMPROVEMENT (costs included in item breakdown)

DESCRIPTION	EST. COST
TOTAL $	

OTHER FEES

TOTAL $	

OFFICE COSTS (costs not included in item breakdown)

DESCRIPTION	EST. COST
TOTAL $	

DEMOLITION (costs not included in trade item breakdown)

DESCRIPTION	EST. COST
TOTAL $	

Mortgagor Kreigler Borg Florman By ______ Date ______

Contractor Construction Company (Inc.) By Samuel C. Florman, VP Date August 23,

FHA ______ Processing Analyst) ______ (Date) ______ (Chief, Cost Branch or Cost Analyst) ______ (Date)

FHA ______ (Chief Underwriter) ______ Date ______

FIGURE 17.20 Contractor's Trade Payment Breakdown.

After the first payment is made to a subcontractor, the contractor should verify that the subcontractor is paying suppliers. If suppliers are not being paid, the general contractor should insist on joint checks to suppliers and the subcontractor and require that the subcontractor endorse the check so that the supplier can receive the funds.

In making final payments to subcontractors and suppliers, the contractor may be able to get a reduction in their claims and extras. In addition to this, the records of backcharges and other charges for use of hoists, etc. should be shown to them. A close check should be kept on credits due. If a subcontractor omits portions of the work and if this is known, it is essential that a credit be requested of the subcontractor before final payment. All omitted work and shortcuts, especially in alteration work, should be recorded. Before final payment to a subcontractor, the general contractor should ensure that final waivers of liens have been signed and that all guarantees, warranties, operating manuals, as-builts, and application for final payment have been received. After final settlement between contractor and

Kreisler Borg Florman
CONSTRUCTION COMPANY (INC.)
97 Montgomery Street, Scarsdale, New York 10583
Telephone (914) SC 5-4600

APPLICATION FOR PAYMENT ON ACCOUNT OF CONTRACT

ALL REQUISITIONS MUST BE MAILED AND IN THIS OFFICE ON OR BEFORE THE LAST DAY OF THE MONTH.
BOTH SIDES OF THIS REQUISITION MUST BE COMPLETED AND SIGNED.

SUBCONTRACTOR Modern Pollution Control, Inc. PERIOD FROM Start TO September 30,

FOR 170th St @92rd Ave, Jamaica, New York WORK REQ. No. 1

PROJECT Apartment Building

ITEM	AMOUNT	DO NOT WRITE IN THIS SPACE
AMOUNT OF ORIGINAL CONTRACT	9,825	
APPROVED EXTRAS TO DATE (LIST ON REVERSE SIDE)	250*	
TOTAL CONTRACT AND EXTRAS	10,075	
CREDITS TO DATE (LIST ON REVERSE SIDE)	nil	
NET CONTRACT TO DATE	10,075	
VALUE OF WORK PERFORMED TO DATE	TOTAL	
LESS RESERVE OF ____% AS PER CONTRACT	1,475	
BALANCE	8,600	
LESS PREVIOUS PAYMENTS	nil	
AMOUNT OF THIS REQUISITION		

DO NOT WRITE IN THIS SPACE

JOB ____ DISTRIBUTION ____
CHECKED: QTY. ____ PRICE ____ EXT. ____
AMOUNT PAID ____ DATE ____
CHECK NO. ____ FOLIO NO. ____
ENTER ____ PAY ____

(OVER)

FIGURE 17.21*a* Front side of application for payment submitted by a subcontractor.

subcontractor of all claims, and while final payment is being awaited from the owner, the subcontractor may often be willing to discount the final payment for immediate receipt of the money, rather than wait.

17.18 COST RECORDS

Segregation of costs for each job is essential for proper construction project management. Only with such records can profits or losses for each job be calculated

THE FOLLOWING IS A FULL AND COMPLETE LIST OF ANY AND ALL PERSONS, FIRMS OR ENTITIES OF EVERY NATURE OR KIND AND DESCRIPTION WHO FURNISHED WORK, LABOR (OTHER THAN ON DIRECT PAYROLL OF APPLICANT), SERVICE AND/OR MATERIAL (INCLUDING BUT NOT LIMITED TO INSURANCE PREMIUMS, WATER, GAS, POWER, LIGHT, HEAT, OIL, GASOLINE, TELEPHONE SERVICE OR RENTAL OF EQUIPMENT AS WELL AS UNION WELFARE, PENSION AND OTHER FRINGE BENEFIT PAYMENTS AND FEDERAL, STATE & LOCAL TAXES) IN EXCESS OF $100.00 ARISING OUT OF OR IN CONNECTION WITH THE JOB FOR WHICH THIS PAYMENT IS REQUESTED TOGETHER WITH ANY AND ALL AMOUNTS NOW DUE AND OWING TO THEM AS WELL AS THE DATE THAT THEY LAST PERFORMED ANY WORK, LABOR, SERVICE OR DELIVERED ANY MATERIAL:

NAME	ADDRESS	DATE	AMOUNT DUE
NONE DUE AND OWING			

ALL CLAIMS FOR EXTRAS NOT HEREIN LISTED ARE WAIVED BY THE SUBCONTRACTOR

EXTRAS TO DATE				CREDITS TO DATE			
Order No.	Description	Date	Amount	Order No.	Description	Date	Amount
	Damages on electrical control box		$250.00				

THE FOREGOING REPRESENTATIONS ARE MADE IN ORDER TO INDUCE YOU TO MAKE AN INTERIM OR FINAL PAYMENT TO US, WELL KNOWING THAT YOU ARE RELYING ON THE TRUTH THEREOF

MODERN POLLUTION CONTROL INC	
APPLICANT'S NAME	SIGNATURE
36-50 38th St, L.I.C., New York	President
ADDRESS	TITLE

FIGURE 17.21*b* Back side of the application for payment.

and predictions made for future work costs. After award of subcontracts, in order for records to be up to date as to any extras or credits that are being claimed by or awarded to the subcontractors for changes (Art. 17.14), a monthly update of all anticipated costs for each subcontract is essential. This can be done by means of an Application for Payment Form (Fig. 17.21) from each subcontractor.

On this form, the approved extras are shown on the second line on the front of the sheet under Amount (Fig. 17.21*a*). Also, all extras claimed by the subcontractor must be listed on the back (Fig. 17.21*b*), as shown on the portion of the form Extras to Date.

Payrolls. From the daily reports received from the field (Art. 17.10.3), weekly payrolls can be prepared for each job. Labor for each job must be segregated and tabulated in a payroll report (Fig. 17.22). This report also provides statistics on

PAYROLL

Kreisler Borg Florman

CONSTRUCTION COMPANY (INC.)
97 MONTGOMERY STREET, SCARSDALE, N. Y., 10583, SC 5-4600

SHEET NO. 1 OF 1

PROJECT TWIN PARKS NORTHWEST

PERIOD - FROM 1/17/ TO 1/23/

	BADGE NUMBER	NAME OF EMPLOYEE	POSITION		HOURS WORKED DAILY WED.	THU.	FRI.	SAT.	SUN.	MON.	TUE.	TOTAL HOURS	RATE	AMOUNT EARNED STRAIGHT	PREMIUM	GROSS WAGES	W.T. EXEMP.	DEDUCTIONS O.A.B.	W.T.	N.Y.S. W.T.	N.Y.C. W.T.	OTHER	NET PAYMENT
1		J.R.		S.T. / O.T.												200.00		11.70	26.30	6.90	2.30		152.80
2		V.L.	Lab.	S.T. / O.T.												262.50		15.36	46.40	12.60	4.00		184.14
3		P.F.	Lab.	S.T. / O.T.												262.50		15.36	30.50	8.70	3.05		204.89
4				S.T. / O.T.																			
5				S.T. / O.T.																			
6				S.T. / O.T.																			
7				S.T. / O.T.																			
8				S.T. / O.T.																			
9				S.T. / O.T.																			
10				S.T. / O.T.																			
11				S.T. / O.T.																			
12				S.T. / O.T.																			
13				S.T. / O.T.																			
14				S.T. / O.T.																			
15				S.T. / O.T.																			
16				S.T. / O.T.																			
17				S.T. / O.T.																			
18				S.T. / O.T.																			
			TOTAL	S.T. / O.T.												725.00		42.42	103.20	28.20	9.35		541.83

CODE:
A—ABSENT: NOT PAID FOR
B—ABSENT: PAID FOR
H—HOLIDAY
V—VACATION WITH PAY
P—VACATION WITHOUT PAY
S—SICK

DEPT. HEAD APPROVAL RB

AMOUNT OF PREVIOUS WEEK'S PAYROLL $755.00

SIGNATURE OF SUPERINTENDENT ______

FIGURE 17.22 Payroll report for a construction project.

tax withholding to the government, as well as other payroll information. In addition, the report gives the gross wages week by week.

Monthly Cost Report. This report summarizes subcontracts and extras, material purchases, and labor costs encountered to date and expected to be encountered until completion. A monthly report prepared in a manner similar to Fig. 17.23 (computer spreadsheet) will yield information to the contractor, long before the job has been completed, for calculating anticipated profit for the job. It also will offer a method of monitoring job progress and costs to ascertain whether the anticipated profit is being maintained.

17.19 ACCOUNTING METHODS

The contractor together with an accountant should select a method of accounting best suited for the contractor's operations. The method selected will determine, to a great degree, the amount of taxes paid as well as the accuracy of the information that the contractor will be able to furnish to his bonding company.

One of two bases is generally used: the cash basis or accrual basis. The accrual basis is subdivided into (1) straight accrual basis; (2) completed-contract basis; (3) percentage-of-completion basis.

The **cash basis** is used mainly by small contracting companies. The gross income is reported on the basis of cash receipts. Contract costs are reported on the basis of actual expenses paid. This method has the advantages of simplicity, control of net income by the timing of requisitions and receipts, and payments of taxes *after* profits have been *earned* and *collected.*

The disadvantages of the cash basis are as follows: It is not an accurate reflection of the company's financial condition. It does not show monies earned and not collected, or debts incurred and not paid. It cannot be used for surety purposes.

The **straight-accrual method** can be used for both short-term and long-term contracts. In this method, the gross income is recorded when earned and expenses are recorded when incurred, regardless of when the cash is received or disbursed. It has the advantages that statements of income reflect actual operations during the period, all receivables and payables are recorded as incurred so that the balance sheet is more useful, and if there are more payables than receivables, less tax has to be paid.

The disadvantages are as follows: The gross income may not be accurate. It may not be the same as the gross profit because of advance billings (caused by unbalanced requisitioning, or front-loading) or job-site inventories. There is less flexibility in tax planning than in other methods. If there are more receivables than payables, the accrual method will result in a higher tax.

The **completed-contract basis** is used for long-term contracts. It is also very good for joint ventures. The income of each long-term contract is recognized only when that contract has been completed or substantially completed.

The advantages are as follows: Taxes are effectively deferred until completion of the project, thus augmenting the contractor's working capital. Contract profits are figured on the basis of actual results, rather than on estimates that could overlook unforeseen future costs or delays. All receivables and payables are recorded as they occur, making a more accurate balance sheet. Taxable income can be regulated between years by deliberately hastening or deferring contract completion. Profits as accrued can be invested and used to earn interest before taxes

Date – 08 JUL 199 Time – 06:32:57

Sample Company Number 99
J O B C O S T T O D A T E
REVISED ESTIMATE
Through Period 02 Year 9

JOB NUMBER – BAYS
Bayshore Medical Ctr
Bayshore Medical Center
North Beers Street, Holmdel, NJ
Remodeling, Renovation, Structural Changes

Job Supervisor: Bill Cook
Completion Date: 9/30/92
Trailer phone (908) 938-4533

PHASE	CODE	DESCRIPTION	ESTIMATE REVISED	ESTIMATE CHANGE ORDERS	ESTIMATE CURRENT	COST TO DATE MONTH	COST TO DATE YEAR	COST TO DATE JOB	VARIANCE
100-1000	-A	Job Overhead	12,450.00		12,450.00				12,450.00-
100-1300	-L	Job Supervis	13,500.00		13,500.00			2,284.59	11,215.41-
100-1400	-E	Bulldozer	2,500.00		2,500.00			1,791.30	708.70-
100-1810	-S	Grading – Ex	8,000.00	1,800.00	9,800.00			4,234.56	5,565.44-
100-1840	-L	Site Clearin	4,000.00		4,000.00			11,901.42	7,901.42
100-1843	-M	Footings	9,000.00		9,000.00			6,982.57	2,017.43-
100-1845	-M	Slabs	9,000.00		9,000.00			10,062.50	1,062.50
100-8230	-L	Seeding/Land	6,000.00		6,000.00			10,764.47	4,764.47
PHASE TOTALS			64,450.00	1,800.00	66,250.00			48,021.41	18,228.59-
200-2000	-M	Concrete Wal	10,000.00		10,000.00			16,365.36	6,365.36
200-2000	-S	Concrete Wal	10,000.00	7,500.00	17,500.00			7,500.00	10,000.00-
200-2200	-L	Framing	20,000.00	2,530.00	22,530.00			11,886.27	10,643.73-
200-2410	-S	Rough Electr	25,000.00	5,000.00-	20,000.00			22,000.00	2,000.00
200-2500	-S	Rough Plumbi	10,000.00	8,000.00	18,000.00			12,707.71	5,292.29-
200-2600	-M	Sheetrock	4,000.00		4,000.00			13,147.00	9,147.00
PHASE TOTALS			79,000.00	13,030.00	92,030.00			83,606.34	8,423.66-
300-3100	-L	Roofing	5,000.00		5,000.00			10,953.50	5,953.50
300-3100	-M	Roofing Slat	17,000.00		17,000.00			6,800.00	10,200.00-
300-3250	-M	Joists	6,985.00		6,985.00			4,000.00	2,985.00-
300-3300	-M	Flashing	3,275.00	4,000.00	7,275.00			859.83	6,415.17-
300-3350	-L	Install Flas	2,600.00		2,600.00			4,524.31	1,924.31
PHASE TOTALS			34,860.00	4,000.00	38,860.00			27,137.64	11,722.36-
400-4000	-L	Finishing Wo	16,850.00		16,850.00				16,850.00-
400-4100	-M	Carpeting	10,640.00	5,000.00	15,640.00				15,640.00-
400-4200	-M	Lighting Fix	6,950.00	7,000.00	13,950.00			6,030.00	7,920.00-
400-4300	-L	Finish Plumb	20,000.00		20,000.00			2,225.77	17,774.23-
400-4300	-M	Bathroom Fix	6,000.00		6,000.00			4,230.03	1,769.97-
PHASE TOTALS			60,440.00	12,000.00	72,440.00			12,485.80	59,954.20-
JOB TOTALS			238,750.00	30,830.00	269,580.00			171,251.19	98,328.81-

FIGURE 17.23 Monthly cost report as output by a computer spreadsheet program.

are paid. And payment of taxes has a better relationship to cash flow from accounts receivable and retained percentages.

The disadvantages of the completed-contract basis are that it does not reflect current performance when a contract spans more than one fiscal year and unincorporated contractors may be penalized, since net income may be irregular and taxed at higher rates one year and at lower rates in another year.

The **percentage-of-completion method** recognizes the gross income on each contract as work progresses. Annual income for each contract should be the same percentage of total income as contract costs to date are of the estimated total contract costs. In computation of contract costs, all or portions of costs may be temporarily excluded during the early stages—for example, start-up costs—if the exclusion would result in a more accurate allocation of income. The advantages of this method are: It is the most realistic annual determination of income. Receivables and payables are recorded when earned or incurred. Liability for taxes arises in the fiscal year when it is actually incurred. And it provides consistent data with very little distortion.

The advantages are: It is dependent on estimates from management of percentage of completion and cost to complete. Retained percentages and accounts receivable not yet billed are taken as income although they may not be collected until a much later date. It does not permit the deferral of income taxes, a procedure that may be available with the cash or completed-contract method.

17.20 SAFETY

Responsibility for job safety rests initially with the superintendent. Various safety manuals are available giving recommended practice for all conceivable types of construction situations; for example, see "Manual of Accident Prevention in Construction," Associated General Contractors of America, Inc., 1957 E St., NW, Washington, DC 20006.

Because of wide diversity in state safety laws, the Federal government in 1970 passed the Occupational Safety and Health Act (OSHA) (Title 29—Labor Code of Federal Regulations, Chapter XVII, Part 1926, U.S. Government Printing Office). Compared with state safety laws of the past, the Federal law had much stricter requirements. For example, in the past, a state agency had to take the contractor to court for illegal practices. In contrast, the Occupational Safety and Health Administration can impose fines on the spot for violations, despite the fact that inspectors ask the employers to correct their deficiencies. OSHA enforcement, however, may eventually be taken over by the states if they develop state regulations as strict as those of OSHA.

An essential aspect of job safety is fire prevention. In aiding in this endeavor, superintendents and project managers often have the advice of insurance companies who perform inspection of the jobs, free of charge. Such inspections by insurance companies often result in reports with advice on fire-prevention procedures. Contractors will benefit from adoption of these recommendations.

To deal most effectively with safety in the contractor's organization, the contractor should assign responsibility for safety to one person, who should be familiar with all Federal and state regulations in the contractor's area. This person should instruct superintendents and supervisors in safety requirements and, on visits to job sites, be constantly alert for violations of safety measures. The safety engineer or manager should ascertain that the construction superintendent holds weekly "toolbox" safety meetings with all supervisors and that the superintendent is writing accident reports and submitting them to the contractor's insurance administrator. In addition, the safety supervisor should maintain a file containing all the necessary records relative to government regulations and keep handy a copy of Record-Keeping Requirements under the Occupational Safety and Health Act, Occupational Safety and Health Administration (U.S. Department of Labor, Washington, D.C.). Management should hold frequent conferences with this individual and with the insurance company to review the safety record of the firm and to obtain advice for improving this safety record.

17.21 COMMUNITY RELATIONS

Community concerns with the results of new construction or disturbances from construction operations materially affect the construction industry. Some communities merely help shape projects that are being planned for construction in their environs. This aid consists of recommendations from community advisory boards and localization of planning. Other communities have assumed a more vigorous role in regulating construction, including the power of veto or costly delay over many projects.

Some of the areas of community relations that must be dealt with in construction management are discussed in the following:

Employment of Local Labor. Because many construction projects are built in inner city, or core areas, where there is much unemployment, communities may insist on utilization of unemployed local labor. This may be done in accordance with local plans or in the form of an Equal Opportunity Program of nondiscrimination, or by recruitment of local labor for employment on the job site. Additional equal opportunity must be given by contractors to women and the handicapped in both office and field positions.

Utilization of Local Subcontractors. Various government agencies may require or give preference to employment for construction work of local subcontractors, with special consideration for minority- and women-owned firms. In many cases, general contractors may be required to enter into written understandings with a government agency specifying goals to be set for local subcontractor employment on a job. As a result of such actions, poorly capitalized subcontractors have been able to make initial employment gains in fields requiring small capital investment. Payrolls for such subcontractors, however, in a number of these trades do present difficulties. In many instances, it may be necessary for the general contractor to make special arrangements for interim payments to these subcontractors, prior to the regular payment date, for work performed.

Among the routes used to bring about employment of subcontractors short of capital on construction jobs are the following:

Awarding a subcontract and orders to such firms.

Subdividing work into manageable-size subcontracts.

Encouragement of subcontractors to enter into joint ventures with better-financed subcontractors.

Awarding subsubcontracts to subcontractors by better-financed subcontractors who hold a large subcontract.

Awarding a pilot contract for a small job to a subcontractor, for example, tiling of one or two bathrooms, just to create an opportunity to begin to function.

Recruitment of local community labor and local subcontractors for a project requires maintenance by the contractor of an active program for the purpose. When the job is started, if there is a community group strongly organized and vocal in the area, the leaders of this group should be approached. If a request is made by the community group for employment of a member of the group as a community liaison or organizer, this request should be given earnest consideration. With a salaried liaison between the contractor's organization and the community, many pitfalls can be avoided.

Up-to-date lists of community subcontractors should be maintained by the contractor's office. These lists should be frequently updated, or they will rapidly become obsolete as these small subcontractors either expand or phase out. The contacts thus made with local firms are important, because an acquaintanceship with local conditions is essential in obtaining and executing contracts and dealing with communities.

Public Interest Groups. These also express their opinions and ask for a voice in planning and construction of proposed projects. They can promote projects they favor or seriously delay or cause to be canceled projects they oppose, by lobbying, court actions, presentation of arguments at public hearings, or influencing local officials in a position to regulate construction.

Environmental Impact Statements. An environmental impact statement is an analysis of the effect that proposed construction will have on the environment of the locality in which the project is to be built. The statement should take into consideration, among other things, the following factors: effect on traffic; potential noise, sound, and air pollution; effect on wild life and ecology; effect on population and community growth; racial characteristics; economic factors; and aesthetics and harmony with the appearance of the community.

For each project, contractors should ascertain whether an environmental impact statement is required. If such a statement is required, it should be begun early in the construction planning stage, if it is required to be drafted by the contractor. If the impact statement for a project is to be drafted by a government agency, the contractor should ascertain that the agency has drafted the statement and that it has been filed and approved.

17.22 RELATIONS WITH PUBLIC AGENCIES IN EXECUTING CONSTRUCTION OPERATIONS

A contractor must deal with numerous public agencies. In some localities, to obtain the necessary permits for start and approval of completed construction, a contractor, for example, may have business with the following agencies: building department, highway department, fire department, police department, city treasurer or controller, sewer and water department, and various government agencies providing the financing, such as Federal Housing Administration, State Division of Housing, or a public-interest corporation formed by the state or municipality for undertaking construction work.

The contractor must be familiar with the organization of the agency and the division of functions, that is, which portion of the agency does design, which does construction-cost approval, and which does inspection. The contractor should also determine whether financing is provided for completed construction, or merely by providing or guaranteeing a mortgage. In addition, the contractor should be knowledgeable on construction-code enforcement; source of payments, whether through a capital construction budget or merely a building mortgage; permits required, methods of obtaining them, and fees; record keeping; and methods of obtaining information from the agency's files.

The contractor should do the following to deal most effectively with public agencies. First, various members of the contractor's staff should concentrate their efforts and become experts on dealing with one or more agencies. For instance, the person in charge of field operations should be the one to deal with the building department and building inspectors. This person should become familiar with the organizational structure of the building department and know the inspectors. Others in the organization should be versed in dealing with city treasurers or comptrollers, and still others with state or Federal agencies.

Second, up-to-date files and information should be kept and segregated on agency regulations and procedures. For example, building department codes and regulations should be obtained, and revisions of these should be maintained in the contractor's offices.

Also, it is very important that a personal relationship be established between the contractor's personnel and the members of the agencies with which the personnel deal. Most agency personnel are willing to help when approached on a frank and open basis.

17.23 LABOR RELATIONS

Proper labor relations on a construction job involve many facets of a contractor's ability. Such relations often are affected by the type of labor organization involved.

Most craft labor on jobs in large cities and in the industrialized portions of the country is unionized. Most unionized employees on construction are members of American Federation of Labor building crafts unions. Open-shop contractors, however, often are able to perform work on a nonunion basis. In a few cases, local labor unions of an industrialized type perform construction work.

Strikes. Construction labor strikes after expiration of a labor contract can place a contractor in a difficult position. If the contractor is a member of a contractor's association, the association will handle the negotiations. If not an association member, the contractor has one or two alternatives: Sign up as an independent contractor on condition that the final terms of the agreement settled at the conclusion of the strike will apply to the contractor's agreement retroactively. Or the contractor can continue working around the affected trade and hope that the job will not be delayed or advanced too far from the normal job sequence.

Jurisdictional Disputes. Jurisdictional disputes between two crafts often make a contractor an innocent bystander. Appeals may be made to the national headquarters of the crafts involved, and machinery exists in such cases for resolution of such disputes in the National Joint Board for the Settlement of Jurisdictional Disputes and the National Appeals Board in Washington, D.C. A simpler way is for the contractor to file a charge with the nearest office of the National Labor Relations Board.

Standby Pay. In many union contracts, a requirement is imposed on contractors for standby pay for workers who may not actually be engaged in installing materials or performing work because of a requirement that members of the union be assigned to stand by for certain purposes. Examples of standby pay are pay for temporary water (plumbers), for temporary electric standby to maintain temporary power and lighting (electricians), and for steamfitters or electricians to maintain temporary heat. Standby pay is usually considered a normal part of the construction process, if in the union agreement, and must be allowed for by the contractor in cost estimates.

Prevailing Wages. Contracts for construction for governmental agencies often require that the labor on a job be paid prevailing wages. These are defined as the wages received by persons normally performing that trade in the locality where the job is being performed. Prevailing wages in localities where there is a strong union usually are the union wages paid to labor in that area. Also considered part of prevailing wages are all fringe benefits and allowances usually paid in addition to hourly wages. If a contractor enters into an agreement that provides for payment of prevailing wages, the contractor may also be responsible for a subcontractor who fails to pay prevailing wages.

Labor Recruitment. A contractor's labor recruitment takes many forms. Most importantly, the work force comes from a following of labor that has previously worked with the contractor. These workers are summoned by telephone or by word of mouth when they are needed on the job site. In instances in which the usual labor force must be vastly expanded, the contractor can resort to advertising in newspapers or recruitment by advising local labor-union officials that workers are needed.

Workers are assigned to tasks on the basis of their trade. In strong union jurisdictions, crossing-over by labor from one trade to another is prohibited when the union labor is organized along craft lines.

When there is strong competition for labor, the contractor may have to resort to overtime work to attract and hold capable people, with the expectation that overtime pay will prevent job hopping. In instances in which a great many jobs, or one tremendous job, will be under construction in a locality with a relatively small labor force, other problems may arise. Labor may have to be brought in from the outside, and to do this the contractor may find it necessary to pay travel time or living allowances for such labor.

Termination. When a project ends, generally all labor is discharged, except the contractor's key personnel, who may be moved to another job. Subcontractors, however, frequently, through judicious timing, are able to employ entire labor groups by moving them from one project to another.

Efficiency. Keeping production efficiency high is generally the chief task of the labor superintendent or supervisor. The means used may involve careful scheduling and planning of materials and equipment availability, weeding out of inefficient workers, training and instruction of those who may not be totally familiar with the work, and most importantly, skilled supervision.

17.24 SOCIAL AND ENVIRONMENTAL CONCERNS IN CONSTRUCTION

Construction project managers should seriously consider the social and environmental effects of their construction operations, job safety operations (Art 17.20), and for proper community relations, use of environmental impact statements and the opinions of public interest groups (Art. 17.22). Contractors should be familiar with operations that have resulted in criticism and restraints, so that they can avoid pitfalls and operate within desirable guidelines.

The Committee on Social and Environmental Concerns of the Construction Division of the American Society of Civil Engineers has defined the three main areas of concern for construction as follows:

Social. These areas cover:

Land usage, such as the visual aspect of the construction project, including housekeeping and security; avoidance of landscape defacement, such as needless removal of trees; prevention of earth cuts and borrow pits that would deface certain areas for a long time; protection of wildlife, vegetation, wetlands, and other ecological systems; and visual protection of surrounding residential areas through installation of proper fencing and plantings.

Historical and archaeological, including preservation of historical and archaeological items of an irreplaceable nature.

Crime. The construction process often creates temporary negative impacts on a community, resulting in a crime increase. This can include local crime as well as fraud and bribing of public officials.

Economics, including impact of a project on the economics of a region, such as a rapidly increased demand for labor far in excess of supply, with a negative effect on the wage structure in the area and economic harm to the area after construction has been completed.

Community involvement, including hiring practices and dealing with the leadership in the community, whether they be of different ethnic groups, income levels, or organizational affiliations (Art. 17.21); union hiring practices; and training programs and foreign language programs.

Safety (Art 17.20).

Physical Media. The effects of construction on land, air, water, and of release of pollutants and toxic substances are also a broad area of concern. Water is often altered in its purity and temperature, and wildlife often is destroyed on land and water by construction of such projects as dams, power plants, and river and harbor facilities.

Energy Conservation, Vibration, and Noise. Vibration has become of increasing concern through the increasingly frequent use in buildings of light construction materials. These are usually flexible and prone to vibrate. In addition, construction machinery has become larger and more powerful, with the result that vibration of this machinery requires strict control.

Identification of noise-producing construction operations and equipment, and control of building construction noise are a concern of contractors. Noise-abatement codes for construction exist in many municipalities. Unless the provisions of these codes are properly understood and enforced, they may result in prohibiting of two- or three-shift construction work and delaying work that requires overtime. Also, some Federal agencies have promulgated regulations requiring noise readings on projects (Table 17.2). In accordance with such readings, local officials may place a construction project in one of the following categories:

Unacceptable: Noise levels exceed 80 dB for 1 h or more per 24 h, or 75 dB for 8 h per 24 h.

Normally unacceptable (discretionary): Noise exceeds 65 dB for 8 h per 24 h, or loud repetitive noises on site.

Normally acceptable (discretionary): Noise does not exceed 65 dB for more than 30 min per 24 h.

Acceptable: Noise does not exceed 45 dB for more than 30 min per 24 h.

Energy conservation in construction projects is part of the overall problem of the conservation of energy resources of the nation as a whole. Contractors should

TABLE 17.2 Limits on Noise Levels of Construction Equipment

Equipment	Maximum noise level at 50 ft, dB(A)
Earthmoving	
Front loader	75
Backhoes	75
Dozers	75
Tractors	75
Scrapers	80
Graders	75
Trucks	75
Pavers	80
Materials handling	
Concrete mixers	75
Concrete pumps	75
Cranes	75
Derricks	75
Stationary	
Pumps	75
Generators	75
Compressors	75
Impact	
Pile drivers	95
Jackhammers	75
Rock drills	80
Pneumatic tools	80
Other	
Saws	75
Vibrators	75

be alert to and aware of any ways to bring this about. They can help by using recycled building materials, recycling materials used during demolition, and demonstrating sensitivity to depletion of endangered natural resources.

17.25 SYSTEMS BUILDING

The term systems building is used to define a method of construction in which use is made of integrated structural, mechanical, electrical, and architectural systems. The ultimate goal is integration of planning, designing, programming, manufacturing, site operation, scheduling, financing, and management into a disciplined method of mechanized production of buildings. Application of these systems should be controlled by an engineer-construction management firm rather than by use of prevailing contract building-management procedures.

Housing Production. The greatest concentration of effort has occurred in the realm of structural framing, leading to development of mass-production methods. These have, in general, been of the following types:

Panel type, consisting of floors and walls that are precast on site or at a factory and stacked in a house-of-cards fashion to form a building.

Volumetric type, consisting of boxes of precast concrete or preassembled steel, aluminum, plastic, or wood frames, or combinations of these, which are erected on the site after being produced in a factory.

Component type, consisting of individual members of precast-concrete beams and columns or prefabricated floor elements, which are brought to the site in volume, or mass produced.

These systems have displayed inherent disadvantages that came about through lack of opportunity to use the entire systems-building process, and because the systems often were not able to attain sufficient volume production to pay for the many fixed costs and start-up costs for the component factories that were built.

Experience has shown, however, that housing can be built efficiently and economically from standardized, modular designs. Some examples are as follows:

Mobile homes manufactured, complete with floor, roof, walls, electric wiring, plumbing, and cabinetry, by techniques similar to those used by automobile manufacturers. Built in factories, mobile homes are trucked to the sites and then installed and finished by on-site builders.

Packaged homes, also known as prefabricated or panelized. They are factory subassembled and, with an assortment of other building components, are delivered to the site for final assembly.

Modular homes, also known as sectional homes, manufactured in off-site factories, usually on an assembly line similar to that used by mobile-home manufacturers. Completely furnished, three-dimensional sections, including mechanical systems, of one or more rooms are factory assembled and then delivered to the sites by truck, railroad, or barge.

(A. D. Bernhardt, "Building Tomorrow: The Mobile/Manufactured Housing Industry," The MIT Press, Cambridge, Mass.)

The lessons learned from past experiences offer the following guidelines for construction managers:

Early commitment must be made to use of a specific system in the design stage.

The systems builder must control the design.

Shop drawings should be started early and refined during the design.

Preparations of schedules, cost estimate, and bids must involve all the members of the team, that is, the designers, owner, and contractor, because there will be many last-minute proposed changes and details that will have to be challenged and modified.

Construction must follow an industrialized-building sequence, rather than a conventional bar chart or CPM diagram. Unlike the procedure for conventional construction, the method of scheduling and monitoring for industrialized building requires that all trades closely follow the erection sequence and that every trade match the speed of erection. Thus, if the erection speed is eight apartments per day, every trade must automatically fall on the critical path. It is necessary that each of the subcontractors work at the rate of eight apartments per day, otherwise the scheduled date of occupancy will not be met.

Metal Building Systems. These systems, produced by several manufacturers, coordinate design and construction of buildings whose primary component is a structural steel frame of standard, modular size; whose secondary members are mass-produced, cold-formed steel shapes; whose roof is light-gage metal; and whose walls may be built of any material.

Metal building systems generally cost less than other construction systems because of in-plant automation, use of precut and prepunched components, and quick erection on the site. Furthermore, costs are predictable. Metal buildings usually are purchased under design-build contracts. In addition, these systems offer flexibility, in that expansion can be readily achieved by removal of end walls, erection of new modular framing, and addition of matching walls and roof.

(Metal Building Dealers Association and Metal Building Manufacturers Association, "Metal Building Systems," MBMA, 1300 Sumner Ave., Cleveland, OH 44115.)

17.26 BASICS OF SUCCESSFUL MANAGEMENT

As outlined in this section, construction project management that will result in a profitable on-time job involves the organization and interplay of many talents. Engineers, accountants, field supervisors, construction labor, suppliers, and subcontractors, all aided by attorneys, insurance and bonding underwriters, the design professional, and the owner, must be organized and carefully coordinated.

Those who succeed in this complex and difficult business are the ones who familiarize themselves thoroughly with the daily operations of their jobs. They are constantly learning by reading the latest literature and professional journals and by attending seminars and industry functions. They are alert and open-minded about new ideas. They understand the needs of the clients and the design professional and are able to tailor their services to them.

SECTION EIGHTEEN
SURVEYING FOR BUILDING CONSTRUCTION

John H. Baker
Adjunct Professor of Civil Engineering
Stevens Institute of Technology
Hoboken, New Jersey

Surveying for building construction lies within the domain of plane surveying. In that type of surveying, for measurement of horizontal distances, the curvature of the earth is neglected, a level line is considered straight, and the direction of a plumb line is assumed to be the same at all points within the limits of a building site. Elevations, however, are determined with respect to a spheroidal surface, such as mean sea level, called a **datum.**

This section deals with the practical application of plane surveying to building construction. For surveying theory, mathematical background required, field measurements and adjustments, applications of surveying instruments to many other types of surveys, and calculation of distances, angles, elevations, and control data, refer to books such as those listed in Art. 18.9, Bibliography.

18.1 NATURE OF BUILDING SURVEYS

Preliminary surveying for building construction consists of collecting and mapping all pertinent existing information about the building site. Construction surveying consists of laying out the corners of the foundation, setting the batter boards, from which control points are measured, and setting the finished-floor elevation. Multistory construction may require additional surveys to assure that the upper floors are at the correct elevation, are level, and are properly aligned vertically with the floor below.

18.2 PRECONSTRUCTION SITE INVESTIGATIONS

It is very important to investigate the history of the site with respect to prior uses as well as underlying existing conditions. For example, bentonite in the soil may cause differential settlement. A partially collapsed culvert can cause sink holes. Brick or concrete debris used to backfill the cellars of old buildings when they were demolished may present difficulties and additional unforeseen costs for new construction. Buried toxic wastes in a subdivision may cause serious health problems. All too frequently important underground utilities are located by the operator of a backhoe when they should have been located by the operator of a transit! The actual survey for a building is usually straightforward, but great care must be taken to be sure that it is done correctly and accurately.

18.2.1 Building-Code Requirements

Local zoning ordinances may contain numerous commercial and residential setback requirements that, in some cases, may be difficult to understand and apply. It is therefore important to obtain this information from the local building department and to study their requirements.

18.2.2 Property-Line Surveys

A licensed land surveyor with extensive knowledge of local land surveys and location of adjacent property markers should be engaged to do this work. All existing property corner monuments should be found and all missing corners should be located and monumented. Any property dedication lines should also be located. Only registered land surveyors are permitted to serve as expert witnesses in property-line proceedings.

One or two bench marks are also required to establish the datum, or reference point for elevation. It is a good policy when the excavation goes below groundwater level to call water level + 100, so as to eliminate minus elevations.

18.2.3 Location of Existing Utilities

Utility plans should be obtained from the local department of public works and local water, gas, electric, telephone, cable TV companies, as well as from other companies that may have installations in the area. The purpose is to learn the locations of those utilities, to avoid damaging them during construction, and to utilize them, if needed. Responsible officials of those companies should be kept informed, from planning stages through project completion, of project work that may affect their installations.

18.2.4 Preconstruction Inspection and Survey of Nearby Buildings

An inspection should be made of all nearby buildings that may possibly be damaged by construction of proposed buildings. All exterior walls should be inspected and all defects or irregularities in them that might later be attributed to recent construction activity should be recorded and photographed.

For example, the record of the inspection may be entered in the resident engineer's diary as follows:

"Lawrence Smith Building, River St. side—South elevation: granite water table requires pointing, no cracks in stone, a few spalls at corners (see Photo #1), limestone above 3rd Floor, no comment. The entablature at 3rd Floor has a large crack (see Photo #2), starting at the top of the 5th column from the west corner of the building and extending diagonally upward through the architecture and frieze to the cornice, open at the top about the thickness of a mortar joint. Above the 3rd floor . . ."

Bench mark elevations should be recorded at each corner of the buildings and intermediate readings should be taken at 25-ft intervals. Permission to enter on private property for the purpose of making a preconstruction inspection and survey should be requested in writing.

The examination must be supported by readings of elevations at the foundations of the various buildings to determine any settlement or lateral movement. The number and type of such reference marks will be controlled by the kind of new foundation. In general, reference points should be located and recorded in such a way that settlement or horizontal motion can be detected.

18.3 TOPOGRAPHIC MAPPING

As a prequisite for design of a building, the site should be mapped in detail (Fig. 18.1). Sufficient ground elevation readings should be taken to plot contours at 1- or 2-ft intervals. (A contour is a line connecting points at the same elevation on a site.) All ridges, knobs, depressions, and drainage courses should be shown on the map. In addition, the map should record the locations of important trees that might be saved and incorporated into the landscape design. All existing items and features also should be shown on the map. These include buildings, sidewalks, fences, power and telephone poles, fire hydrants, access holes, drop inlets, and anything else that may pertain to the planning and design of the project.

18.4 SITE PLANS

A site plan should show exactly where the proposed buildings are to be located on the property, including distances to adjoining streets and property lines. The plan should also show sidewalks, driveways, parking areas, and landscape features. In addition, it should indicate the location of any soil borings. The surveyor will use the site plan to lay out the building for construction. A building permit should be obtained before the buildings are laid out.

18.5 GENERAL LAYOUT FOR CONSTRUCTION

Before a survey is made to lay out a site for building construction, it is important to ascertain from the contract specifications what portion of the layout is the responsibility of the designers and what portion is the responsibility of the contractor.

A survey for building construction requires staking of the corners of the proposed buildings at the locations shown on the site plan. Buildings are usually located parallel to property or street lines and are set back specified distances from those lines. The elevation of the buildings is established by a grade stake. (This elevation

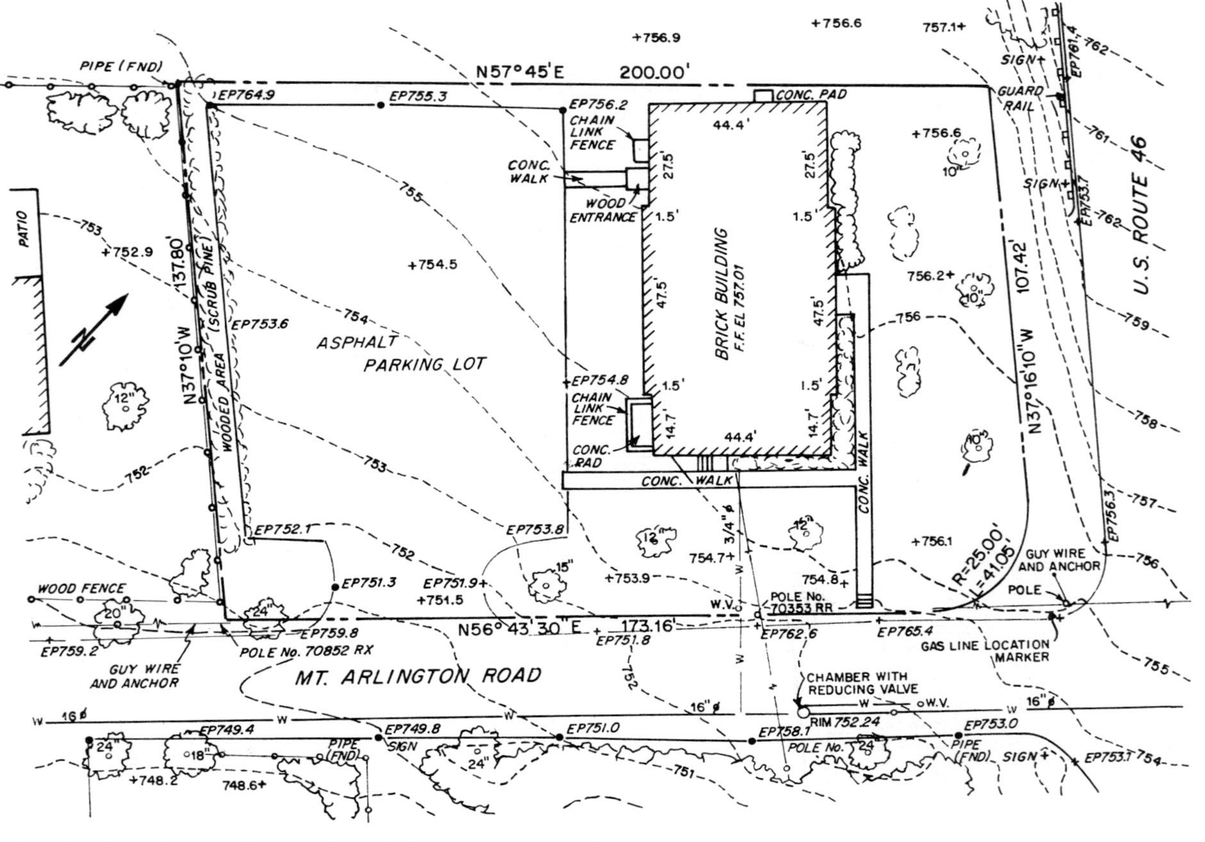

FIGURE 18.1 Topographic map of a building site. *(Courtesy of Langan Engineering and Environmental Services, Inc., Elmwood Park, N.J.)*

must be at the proper height for the building sewer to have an adequate height for gravity flow to the main sewer.)

18.5.1 Layout of Small Buildings

The simplest layout for building construction (Fig. 18.2) requires setting of offset stakes. Offset lines are laid out between the offset stakes, set back from the building corner stakes because otherwise these stakes would be lost when the foundation is excavated. Batter boards are nailed to the offset stakes and the tops are set with the aid of a level at the elevation of the finished floor. The batter boards serve as a nailer for string lines that are used to set the forms for the concrete foundation to line and grade.

Figure 18.3 illustrates use of the Pythagorean theorem to compute the diagonal distances between the corners of the building, for checking the layout.

18.5.2 Layout of Large Buildings

The principle of laying out large buildings is the same as that for small buildings, except that a high degree of precision is required to assure that prefabricated building components, such as floor beams and roof trusses, fit into place.

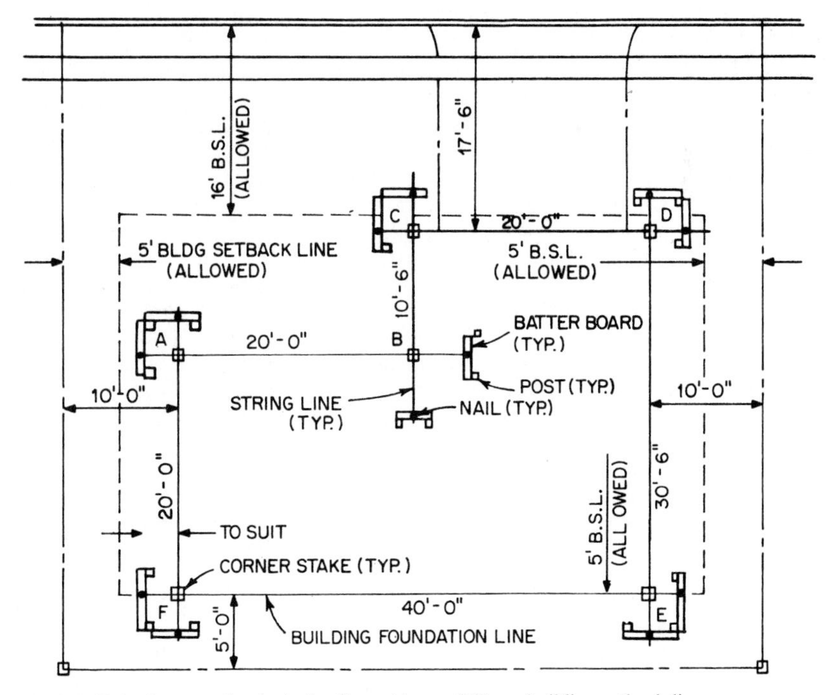

FIGURE 18.2 Layout of a single-family residence. BSL = building setback line.

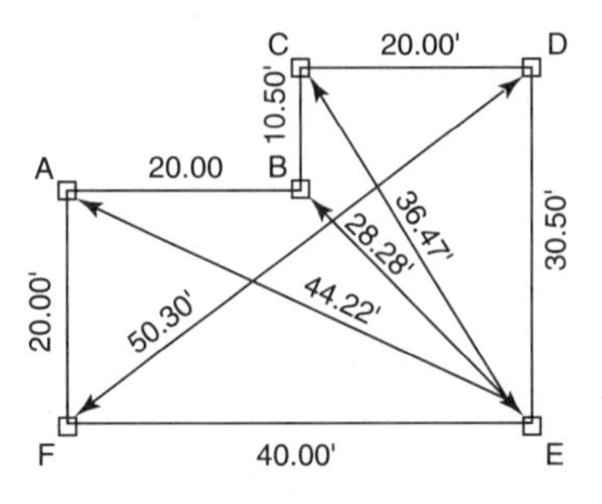

FIGURE 18.3 Diagonal distances between property corners may be used to check building layout:

$$AE = \sqrt{(AF)^2 + (EF)^2} = \sqrt{20^2 + 40^2} = 44.72$$

$$BE = \sqrt{(AF)^2 + (CD)^2} = \sqrt{20^2 + 20^2} = 28.28$$

$$CE = \sqrt{(DE)^2 + (CD)^2} = \sqrt{30.5^2 + 20^2} = 36.47$$

$$DF = \sqrt{(DE)^2 + (EF)^2} = \sqrt{30.5^2 + 40^2} = 50.30$$

In establishing building dimensions, temperature effects should be taken into account. Metal tapes used for measuring distances are calibrated at 68°F and expand and contract 0.01 ft per 100 ft for each 15°F change in temperature. Hence, it is difficult to make an accurate measurement with a metal tape on a hot day when part of the tape is in the sun and part is in the shade. It is advantageous to make precise measurements early in the morning, in the evening, or on a cloudy day.

Expansion and contraction of building materials used in the structure should also be taken into consideration.

Figure 18.4*a* shows a typical layout of main reference lines. The offset traverse or base line, which is wholly outside the building site, can vary in distance from the building line, but 6 ft is a good distance because the line will clear the conventional swing scaffold. The sidewalk bridge, which is necessary for protection, prevents sighting above it, unless the transit is set up outside the curb line at either end. The base line, therefore, must be produced well beyond the limits of the sidewalk bridge, so as to be of use above it.

After the base line is finished and checked, all points of use in layout should be projected to a permanent location on nearby existing buildings. Control points for base lines can in many cases be obtained by sighting on some architectural ornamentation a short distance away. Those for the inside building-line offset and column offset lines should be located as high as possible, so as to be seen from the bottom of the excavation. The roof cornice in Fig. 18.4*b* is ideal. Targets at such locations can be used to establish column lines by "bucking in" or double centering between control points on opposite sides.

The inside building line is useful in aligning foundation walls after the general excavation is completed. Placing removable windows in the sidewalk fence, which is part of the sidewalk bridge, permits transit lines to be dropped into the excavation.

18.5.3 Layout of Steel Framing

The centerlines of columns can be marked on batter boards set with the top a given number of feet above subgrade, footings, or pile caps. If piles are used, they can be staked out with a tape by measuring from the centerlines. Before columns are

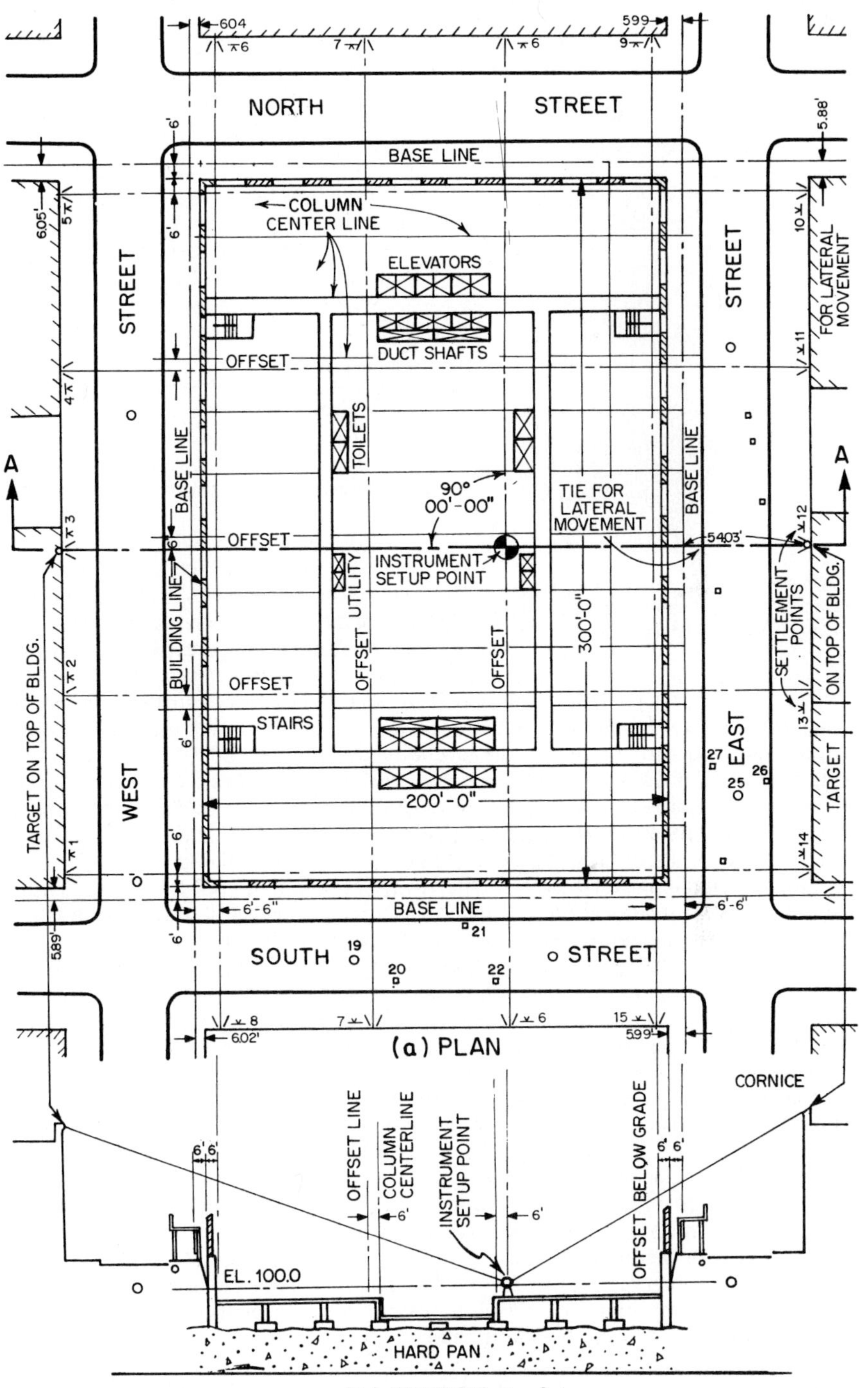

FIGURE 18.4 Base-line layout for a large commercial building.

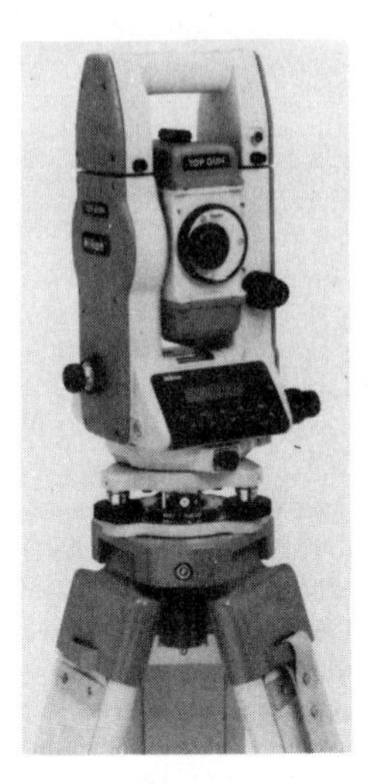

FIGURE 18.5 Electronic digital theodolite outputs horizontal and vertical angles with precision. *(Courtesy of Nikon, Inc., Melville, N.Y.)*

set, transit lines can be run on the centerlines, but after the columns are in place, the offset lines have to be used.

As soon as some concrete is placed, a permanent bench mark should be set in it. In a steel-frame building, an anchor bolt may be used; if not, set a special rod for this purpose.

When all the walls and columns are in place to ground level, at least two bench marks should be established some distance apart, preferably at each end of the building where they can be used to make vertical measurements as the building rises. Stair wells or elevator shafts are good locations for these control points.

Most mechanics work from a level mark 4 ft above each finished floor. Therefore, such marks should be placed on each column, or about 25 ft apart, in every story. They must be tied in accurately between floors, which is the reason for the vertical base lines. Two marks must be seen from the same level setup before the 4-ft marks are established on the columns. With an extension-leg tripod, set the horizontal hair of the level directly on the taped 4-ft marks. On the original run do not set above or below the mark and read a rule. Between setting the rule and reading the backsight, and reading the rule to set the foresight, errors can be made, especially when about ½ in from the 4-ft mark. If the rule is turned the wrong way, the error will be twice the amount of the reading.

Since the building starts to settle shortly after loading, do not try to check with an outside bench mark. Hold and watch the original vertical base lines, and from time to time take a quick check on the preestablished 4-ft marks on the lower floors to spot a differential settlement. If there is any, set the 4-ft marks on the upper floors so that, when a particular column that is slow or fast in settling catches up, all the marks will be level. In some factory buildings, settlement effects are unimportant, but when the interior finish is marble, trouble may be encountered if settlement is not taken care of in advance.

Extreme care must be used in setting columns around elevator shafts. Check each column independently with a heavy plumb bob. (It can consist merely of a 4-in pipe, 8 in long, filled with lead and suspended by piano wire.) Swinging can be reduced by suspending the bob in a pail of oil.

As the building rises above ground level, the exterior base line is the most important control line. The interior column offset lines, when used, must always be checked with the building line.

Accuracy of alignment of a steel-framed building stems from the anchor bolts and billets. It is important to have all horizontal measurements correct so that the lower section will go together without difficulty. Each tier of framing above depends on the accuracy of shop fabrication.

The elevation of the column base plates controls for the full height of the building; so it is obvious that extreme care must be used in setting them. If this is done, column splices, as a rule, will be as close to the true elevation as if established with a tape. This is very important so that the beams and girders will be in the correct vertical location. Clearances for mechanical work also could be affected by variation from the true location.

As columns are loaded, a shortening takes place. So vertical measurements must be made and tied in before this happens. Any total overall measuring done after loading will have to be corrected by calculated deformations based on the modulus of elasticity of the column material.

Fireproofing and the rough floor are located with respect to the steel. This is done because it is not practical to use an instrument on the frame until it is braced by some kind of deck, whether concrete or steel.

The frame is plumbed by the steel erector with a heavy bob and checked by the engineer using the base line produced beyond the sidewalk bridge. When the temperature in the field differs from that of the fabrication shop, the frame will be longer or shorter, an effect to be taken into account when checking. (Shop temperature is always a little above tape standard.)

After the structural floors are in place, the centerline is marked on each exterior column in each story for the mason and other trades to work from. This also is done from the base-line extension.

A straight piece of 1 × 1-in white pine, about 8 ft long, should be used for an offset rod, with a 6-in piece fastened to the end, forming a tee. Two nails are driven into the head of the tee to sight on. The transit operator will know the rod is at right angles to the line when the nails line up.

The column marks should be set accurately, but the mechanics building mason walls should be instructed to set masonry lines ¼ in inside the finish lines. This is advisable to compensate for the slight variations usually found in masonry work.

The column offset lines can be plumbed from the opposite side of the street to the floor edge at both ends or sides of the building. The lines can be carried through the building by setting up on the floor and "bucking in" between the end marks.

18.5.4 Layout of Concrete Framing

A concrete frame requires a layout for each floor. A level mark to work from is put on a column splice rod 1 ft above the finish floor. Exterior column centers are set from the base line. Intermediate columns may be located by taping.

After column forms are stripped, 4-ft marks are established as is done with a steel frame. Column centers for masonry also are established as for a steel frame. The mason uses the 4-ft marks to determine window head and sill heights.

A competent mason supervisor checks to the level mark on the floor above before getting there.

18.6 FOUNDATION SURVEYS

The local building department may make a foundation survey to make sure that setback requirements have been met. This is usually done when the foundation forms and reinforcing steel are inspected prior to placing concrete.

18.7 "AS BUILT" SURVEYS

This survey is made to record all changes that have been made during the course of construction. The contract drawings are then stamped "AS BUILT."

18.8 ELECTRONIC EQUIPMENT USED FOR BUILDING LAYOUT AND CONSTRUCTION

Electronic theodolites measure angles electronically with precision from 0.5 to 20 sec and are generally preferred to the engineer's transit (Fig. 18.5, p. 18.8). To determine the precision of any electronic instrument, check the manufacturer's specifications. Some instruments read out to a lesser number than the actual precision of the instrument. These instruments have zero-set buttons and digital angular readouts.

Electronic distance-measuring devices (EDMs) provide distance measurements that are accurate to within ± 5 mm (3⁄16 in). Accordingly, a steel tape is still the most precise measuring tool for short distances.

Self-leveling automatic-rotating levels with a rotating-laser-beam detector (300-ft radius range) are used for marking vertical control (4-ft marks) on steel columns. These instruments are self-leveling when set up within 4° of level and will automatically stop rotating if knocked off level. They are accurate to ± 0.01 ft in 100 ft. Figure 18.6 shows a rotating-laser-beam detector set up on its side so that the laser rotates in the vertical plane to keep the wall construction plumb and on line. One person can set up a rotating-laser-beam detector and mark or set the grades faster than two persons can using a level and a rod.

FIGURE 18.6 Rotating laser-beam detector positioned to align and plumb a wall. *(Courtesy of Laser Alignment, Inc., Grand Rapid, Mich.)*

18.9 BIBLIOGRAPHY

R. E. Davis et al. "Surveying Theory and Practice," 6th ed., McGraw-Hill Inc., New York.

B. F. Kavanagh, "Surveying with Construction Applications," Prentice-Hall, Englewood Cliffs, N.J.

F. S. Merrit, "Standard Handbook for Civil Engineers," 3d ed., McGraw-Hill Inc., New York.

J. A. Nathanson and P. C. Kissam, "Surveying Practice," 4th ed., McGraw-Hill Inc., New York.

P. W. Wolf and R. C. Brinker, "Elementary Surveying," Harper Collins Publishers, New York.

SECTION NINETEEN
CONSTRUCTION COST ESTIMATING

Colman J. Mullin
Senior Estimator
Bechtel Corporation
San Francisco, California

and

Robert H. Kantor
Editorial Manager, Bechtel Corporation,
San Francisco, California

Black magic. Bean counting. Guess work. Guestimates. Ask a group of engineers to describe construction cost estimating and that is what you are likely to hear. Not that they regard cost estimating as unimportant. They realize it is essential to all projects. But from their point of view, it is a mysterious process.

There is nothing esoteric, however, about cost estimating. It is an engineering discipline like any other, with its own rules and techniques, and a knowledge of these can be very helpful to those in other disciplines. Such knowledge can, for example, help designers, contractors, and building owners to determine whether an estimate adequately reflects their intentions and to understand how a change in design or construction can affect the schedule or total cost of a project. This can lead to consideration of appropriate alternatives and development of better quality, lower-cost projects.

19.1 COMPOSITION OF PROJECT PRICE

The total price of a construction project is the sum of direct costs, contingency costs, and margin.

Direct costs are the labor, material, and equipment costs of project construction. For example, the direct cost of a foundation of a building includes the following:

Costs of formwork, reinforcing steel, and concrete

Cost of labor to build and later strip the formwork, and place and finish the concrete

Cost of equipment associated with foundation activities, such as a concrete mixer

Contingency costs are those that should be added to the costs initially calculated to take into account events, such as rain or snow, that are likely to occur during the course of the project and affect overall project cost. Although the effects and probability of occurrence of each contingency event cannot be accurately predicted, the total effect of all contingencies on project cost can be estimated with acceptable accuracy.

Margin (sometimes called markup) has three components: indirect, or distributable, costs; company-wide, or general and administrative, costs; and profit.

Indirect costs are project-specific costs that are not associated with a specific physical item. They include such items as the cost of project management, payroll preparation, receiving, accounts payable, waste disposal, and building permits.

Company-wide costs include the following: (1) Costs that are incurred during the course of a project but are not project related; for example, costs of some portions of company salaries and rentals. (2) Costs that are incurred before or after a project; for example, cost of proposal preparation and cost of outside auditing.

Profit is the amount of money that remains from the funds collected from the client after all costs have been paid.

19.2 ESTIMATING DIRECT COSTS

Methods for preparing an estimate of direct costs may be based on either or both of two approaches: industry, or facility, approach, and discipline, or trade, approach. For any project, the approach that may be selected depends on user preference and client requirements. If used properly, the two approaches should yield the same result.

Industry, or Facility, Approach. Industry in this case refers to the specific commercial or industrial use for which a project is intended. For example, a client who wishes to build a factory usually is more concerned with the application to which the factory will be put than with the details of its construction, such as bricks, mortar, joists, and rafters. The client is interested in the specific activities that will be carried out and the space that will be needed. When information about these activities has been obtained, the designers convert this information into a total building design, including work spaces, corridors, stairways, restrooms, and air-conditioning equipment. After this has been done, the estimator uses the design to prepare an estimate.

Discipline, or Trade, Approach. This takes the point of view of the contractor rather than the client. The job is broken into disciplines, or trades, of the workers who will perform the construction. The estimate is arrived at by summing the projected cost of each discipline, such as structural steel; concrete; electrical; heating, ventilating, and air conditioning (HVAC); and plumbing.

19.2.1 Types of Estimates

Typical types of estimates are as follows: feasibility, order of magnitude, preliminary, baseline, definitive, fixed price, and claims and changes. These do not represent rigid categories. There is some overlap from one type to another. All the types can be prepared with an industry or discipline approach, or sometimes a combination of them.

Feasibility Estimates These give a rough approximation to the cost of the project and usually enable the building owner to determine whether to proceed with construction. The estimate is made before design starts and may not be based on a specific design for the project under consideration. For example, for a power plant, the estimate may involve only a determination of the energy density of the fuel; the altitude of the plant, which determines the amount of oxygen in the air and hence the efficiency of combustion; the number of megawatts to be produced; and the length of the transmission line to the grid. The feasibility estimate is inexpensive and can be made quickly. Not very accurate, it does not take into account creative solutions, new techniques, and unique costs. It can be prepared by the owner, the lender, or the designer.

Order-of-Magnitude Estimates. These are more detailed than feasibility estimates, because more information is available. For example, a site for the building may have been selected and a schematic design, including sketches of the proposed structure and a plot of its location on the site, may have been developed. Like the feasibility estimate, the order-of-magnitude estimate is inexpensive to prepare. Generally made by the designer, it is prepared after about 1% of the design has been completed.

Preliminary Estimates. These reflect the basic design parameters. For the purpose, a site plan and a schematic design are required. The schematic should show plans and elevations plus a few sections through the building. For buildings such as power plants and chemical refineries, it should also contain a process diagram, major equipment list, and an equipment arrangement diagram. Preliminary estimates can reflect solutions, identify unique construction conditions, and take into account construction alternatives. Usually, this type of estimate does not reveal design interferences.

Generally prepared by the designer, preliminary estimates are made after about 5 to 10% of the design has been completed. Several preliminary estimates may be made for a project as the design progresses.

Baseline Estimates. These are final preliminary estimates. For most buildings, requirements for preparation of an estimate include plans, elevations, and sections. For process plants, also necessary are complete flow diagrams, process and instrumentation diagrams (P&ID) in outline form, and a list of equipment selected and the location of the equipment. Subsequent changes in the estimate are measured with respect to the baseline estimate. Identifying all cost components, the estimate provides enough detail to permit price comparisons of material options and is sufficiently detailed to allow equipment quotations to be obtained.

The baseline estimate is generally prepared by the designer. It is made after about 10 to 50% of the design has been completed.

Definitive Estimates. From a definitive estimate, the client learns what the total project cost should be and the designer's overall intent. The estimate is based on plans, elevations, and sections; flow diagrams, P&IDs, and equipment and instrument lists (for process plants); design segments for each discipline; and outline specifications. It identifies all costs. It is sufficiently detailed to allow quotes to be obtained for materials, to order equipment, and to commit to material prices for approximate quantities.

This type of estimate is generally prepared by the designer and represents the end of the designer's responsibility for cost estimates. It is made after about 30 to 100% of the design has been completed.

Fixed-Price Estimates. Prepared by a general contractor, a fixed-price estimate, or bid, represents a firm commitment by the contractor to build the project. It is based on the contractor's interpretation of the design documents. It requires detailed drawings for each discipline, equipment lists, P&IDs, writing diagrams, and specifications. A fixed-price estimate is highly accurate. It should be in sufficient detail to enable the contractor to obtain quotes from suppliers and to identify possible substitutes for specific items. It is made after 70 to 100% of the design has been completed.

Claims and Changes Estimates. These are prepared when a difference arises between actual construction and the project as specified in the original contract. This type of estimate should identify the changes clearly and concisely. It should specify, whenever possible, the additional costs that will be incurred and provide strong and compelling support for the price adjustments requested. Generally, the estimate is reviewed by all parties involved (designer, contractor, and building owner) as soon as the need for change is identified. Claims and change estimates can be prepared by any or all of the parties to the contract.

19.2.2 Estimating Techniques

There are three estimating techniques: parametric, unit price, and crew development. In general, the parametric technique is the least expensive, least time-consuming, and least accurate. The crew development technique is the most expensive, most time-consuming, and most accurate. Of the three techniques, the parametric requires the most experience and the unit-price technique, the least. During the course of a typical project, all three of these techniques may be used.

Parametric Technique. For every type of project, there are certain key parameters that correlate strongly with cost; for example, for power plants, altitude, and hence the amount of oxygen in the air, is such a parameter. The parametric technique takes such a correlation into account. It is usually employed for preparing feasibility or order-of-magnitude estimates. Sometimes, it is used for preparing preliminary or baseline estimates or small portions of definitive or fixed-price estimates. It is often used for checking high-level estimates, such as definitive, fixed price, and claims and changes, that have been developed by the unit-price or crew development technique.

The parametric technique derives data from proprietary tables that incorporate historical data, or standard tables, or experience. Historical tables are compilations of data from numerous projects of various types. There are historical tables, for example, for the amount of pipe needed to process a barrel of oil in a refinery, the volume of fuel storage necessary for a given size of airport, and the optimum air-conditioning system for a given building size. Proprietary tables are updated as required. Standard tables may be either historical or calculated and tend to be updated more frequently than historical tables.

The industry approach to development of a feasibility cost estimate for a warehouse using the parametric technique typically proceeds as follows: The client supplies a list of the items to be stored in the warehouse, sizes of the items, and the number of types of items. The client also indicates the turnover, or shelf life, of each item. Given the preceding information, the estimator calculates the amount of storage volume and circulation area and obtains the total costs of materials, labor, and equipment from historical tables.

The discipline approach to development of a feasibility or preliminary cost estimate for a warehouse using the parametric technique generally proceeds as follows: Given the spacing between roof supports and the ceiling heights that are specified in the design, the estimator looks up the cost per square foot of ceiling in standard tables. (Disciplines involved are structural steel and concrete work.) From the design, the estimator determines the height and area of the exterior and interior walls. From the weather conditions for the location, the insulating properties required for the exterior walls and roof are determined using standard tables. From the preceding information, the estimator calculates the costs of the walls and roofs using standard tables. (Disciplines involved are carpentry, masonry, and roofing.) Also, from the design, the estimator computes the exterior area and volume of the building, the amount of sunlight falling on the building, and the internal lighting levels required and determines the cost of the mechanical and electrical work. (Disciplines involved are mechanical and electrical.) Finally, the preceding costs are added to arrive at the cost of materials, labor, and equipment for the warehouse.

Unit-Price Technique. This relates directly to specific physical entities in the design—square feet of office area, cubic yards of concrete, number of fixtures in rest rooms. Unlike the parametric technique, which often involves information that is not in the drawings (for example, barrels of oil to be processed) and may not pertain to a specific design, the unit-price technique is tied directly to the contract documents. The estimator employs the quantities given in these documents to determine costs.

The unit-price technique is frequently used for preparation of cost estimates. It can be used for any level of estimate but does require that some design be performed. Data for the technique are obtained from commercially available handbooks of unit prices, which are usually updated at least once a year.

The industry approach to development of preliminary or higher-level cost estimates for a warehouse using the unit-price technique usually proceeds as follows:

The warehouse is divided into categories, for example, loading dock, storage facilities, aisles, restrooms, and offices. The special equipment required, such as cranes, crane rails, and docks, is listed. Then, the estimator looks up in a unit-price book the cost of each of the items specified above. For each category, the unit-price book gives the total cost of materials, labor, and equipment to construct an item. For instance, for a loading dock, the unit price would be specified as either the cost per linear foot or the cost per truck accommodated by the dock; for rest rooms, the unit price would be specified as the total cost of all the fixtures needed or as the total cost per square foot. Finally, the estimator sums the preceding costs to arrive at the total cost of the warehouse.

The discipline approach to development of a preliminary or higher-level cost estimate for a warehouse using the unit-price technique typically proceeds as follows:

From the design documents, the estimator determines the ground area the building occupies (the footprint of the building). The costs of grading and the building floor slab are obtained from a unit-price handbook. With information from the contract documents, the estimator calculates the amount and cost of the structural materials and finishes needed. The unit cost of illumination and air-conditioning are also obtained from a unit-price handbook. Finally, the estimator adds the preceding costs to arrive at the total costs.

Crew Development Technique. This is used to prepare the estimate based on the costs for the specific personnel and equipment that would be needed to complete each item during each phase of construction. The crew development technique differs from the unit-price technique, where the activity is priced without assignment of specific workers and equipment.

For a specific project, the size and mix of crew selected depend on project needs. If early completion is the key consideration, a large crew working multiple shifts and much overtime might be advisable. If access to a site is difficult, a small crew might be necessary. Size and mix of crew can also vary during the course of construction. For example, for a typical high-rise structure, construction may start with personnel and equipment that provides the lowest cost per unit of production. As work progresses and access to work areas grows more difficult, a smaller crew using more equipment may be used. In the final construction stages, when the investment in the building is large and interest costs are high, the contractor may employ a large crew working shifts and overtime to finish as soon as possible, thereby minimizing total project costs.

Estimators tend to use the crew development technique for high-level estimates, the definitive and above. Unlike the unit-price technique, the crew development technique is based on the way the facility actually will be erected. Consequently, it is the most accurate of the estimating techniques. Hence, it is the principal technique for fixed-price estimates; where accuracy is critical.

The crew development technique is based on data from production handbooks. These may be organized in accordance with the use of a facility or by building trades.

The industry approach to development of a definitive cost estimate for a warehouse using the crew development technique generally proceeds as follows:

From the contract documents, the estimator determines the volume and footprint of the warehouse and the uses to which each area would be put, for example, offices, rest rooms, and loading docks. Assuming that one crew will be used to build the shell of the building and other crews to construct the interior areas, the estimator obtains the unit rates of production from standard production handbooks. (The production handbooks for facilities change only with the introduction of new equipment or materials.) Next, for each item taken off the contract documents, the estimator determines the unit costs of materials, labor, and equipment. Then, each unit cost is multiplied by the corresponding quantity of the item to be used. Finally, the estimator adds the products to obtain the total cost of materials, labor, and equipment for the warehouse.

Estimators tend to use the industry approach with the crew development technique where labor costs are low or differences between costs of different crafts are slight.

The discipline approach to preparation of a definitive cost estimate for a warehouse using the crew development technique usually proceeds as follows: From the contract documents, the estimator determines the exact quantities of materials—for example, for piping, linear feet of pipe, number of the various types of fittings, and amount of insulation; for electrical work, the number of fixtures and devices and linear feet of conduit and wire. Assuming the size and composition of the crew by trade (personnel plus equipment), the estimator obtains from production handbooks, for each discipline, the productivity of the crew and the length of time required for installation of the materials. Then, for each item taken off the contract documents, the estimator determines the unit costs of materials, labor, and equipment. Next, each unit cost is multiplied by the corresponding quantity of the item to be used. Finally, the estimator adds the results to arrive at the total cost of materials, labor, and equipment for the warehouse.

Estimators tend to use the crew development technique and discipline approach where labor costs and differences between the costs of the different crafts are high.

19.3 ESTIMATING CONTINGENCY COSTS

These are the costs that must be added to the initially calculated costs to take into account events that are likely to occur some time during the course of a project and that will affect project cost (Art. 19.1). Although the effect and the probability of occurrence of each contingency event cannot be predicted, the total effect of all the contingencies on the project cost can be estimated with an acceptable degree of accuracy. In this respect, contingency allowances are much like insurance. Contingency costs are usually expressed as a percentage of direct costs but they also may be expressed as a dollar amount.

These costs should not be considered a handy slush fund to compensate for inaccurate estimating. No matter how careful and expert the initial estimate, no matter how excellent the design, no matter how skilled the constructor, the unexpected is likely to occur and must be intelligently gaged in each estimate. If the contingency allowance is underestimated, all parties to the construction contract can suffer financial loss. If the contingency is overestimated, the contract may not be awarded or the client may not be able to finance the project.

Contingency should be evaluated for each estimate and will vary by project type, location, and level of estimate. A contingency cost as high as several hundred percent may be justified, for instance, for an experimental process plant, whereas the contingency cost for a prefabricated warehouse may be only 3 to 5%.

Owner's contingency covers the costs that the owner could incur during the course of a project. For example, if the project is delayed for any reason, there will be additional interest charges for the financing. If the city or state changes the building code during project execution, construction costs could increase. If an important commodity undergoes a sudden price increase, the overall cost of the project could be significantly impacted.

Designer's contingency covers the costs that the designer could incur during the course of a project, such as the cost of services that the designer renders and that were not originally anticipated and the additional construction costs due to changes in the design. Both types of contingency costs are illustrated in the following examples.

During the design phase, a designer finds that a portion of the structure being designed has an extremely congested area. This congestion requires changes in design of either the steel structure or the ventilation system. The steel might have to be reinforced or the ventilation might have to be redesigned. All additional design and construction costs will have to come out of the contingency funds.

As a second example, during the construction phase, the contractor learns that specified equipment has been discontinued by the manufacturer. A substitute will be needed, along with associated design modifications. Design and contingency costs needed for this modification will have to come out of contingency funds.

Contractor's contingency covers the costs that the contractor could incur during the course of a project. Suppose, for example, that rain occurs while excavation for the foundation is well under way. Water entrapped in the excavation must be pumped out and mud removed. Also, because of enlargement of the excavation, the amount of backfill required increases. The additional costs incurred must be covered by contingency funds.

Costs not normally covered by contingency allowances include: costs normally covered by insurance; substitution of better materials (should be covered by a change order); increases in project size or scope; and "acts of God," such as floods, tornadoes, and earthquakes.

19.4 ESTIMATING MARGIN (MARKUP)

Margin comprises three components: indirect costs, company-wide costs, and profit. These are defined in Art. 19.1.

19.4.1 Determining Indirect, or Distributable, Costs

The techniques used to calculate indirect costs (often called indirects) resemble those used to calculate direct costs (Art. 19.2).

Parametric Technique. The indirects calculated by this technique may be expressed in many ways, for example, as a percentage of the direct cost of a project, as a percentage of the labor cost, or as a function of the distance to the site and the volume of the construction materials that must be moved there. For a warehouse, for instance, the cost of indirects is often taken to be either one-third the labor cost or 15% of the total cost.

Unit-Price Technique. To determine indirects by the unit-price technique, the estimator proceeds as follows: The various project activities not associated with a specific physical item are determined. Examples of such activities are project management, payroll, cleanup, waste disposal, and provision of temporary structures. These activities are quantified in various ways: monthly rate, linear feet, cubic yards, and the like. For each of the activities, the estimator multiplies the unit price by the duration to obtain activity cost. The total cost of indirects is the sum of the products.

Crew Development Technique. To determine the cost of the indirects by this technique, the estimator proceeds as follows: The various project activities not associated with a specific physical item are determined. Next, the estimator identifies the specific personnel needed (project manager, project engineer, payroll clerks) to perform these activities and determines their starting and ending dates and salaries. Then, the estimator computes total personnel costs. After that, the estimator identifies the specific facilities and services needed, the length of time they are required, and the cost of each and calculates the total cost of these facilities and services. The total cost of indirects is the sum of all the preceding costs.

19.4.2 Determining Company-Wide Costs and Profit

Company-wide costs and profit, sometimes called gross margin, are usually lumped together for calculation purposes. Gross margin is generally a function of market conditions. Specifically, it depends on locale, state of the industry and economy, and type of discipline involved, such as mechanical, electrical, or structural.

To calculate gross margin, the estimator normally consults standard handbooks that give gross margin as a percent of project cost for various geographic areas and industries. The estimator also obtains from periodicals the market price for specific work. Then, the information obtained from the various sources is combined.

As an example, consider the case of a general contractor preparing a bid for a project in a geographic region where the company has not had recent experience. At the time that the estimate is prepared, the contractor knows the direct and indirect costs but not the gross margin. To estimate this item, the estimator selects from handbooks published annually the gross margin, percent of total cost, for projects of the type to be constructed and for the region in which the building site is located. Then, the estimator computes the dollar amount of the gross margin by multiplying the selected percentage by the previously calculated project cost and adds the product to that cost to obtain the total price for the project. To validate this result, the estimator examines reports of recent bids for similar projects and compares appropriate bids with the price obtained from the use of handbooks. Then, the estimator adjusts the gross margin accordingly.

19.5 SAMPLE ESTIMATE

As an example, the following illustrates preparation of an estimate for a trench excavation. The estimate can be regarded as a baseline type or higher type. The discipline approach and crew development technique is used.

The estimate begins with a study of information available for the project: From the design documents, the estimator takes off such information as trench depth, length, slopes, soil conditions, and type of terrain. Wages for the locality in which the trench is to be excavated are obtained from standard handbooks, local labor unions, and the U.S. Census Bureau. The wage figures influence determination of the level of mechanization to be used for the project.

Crew Operation Calculation Sheet. With the basic information on hand, the estimator can now prepare a crew operation calculation sheet (Fig. 19.1). This sheet indicates the work to be done, how it will be done, who will perform it, and duration of the tasks. (The crew operation calculation sheet is normally the first item developed in a cost estimate.)

Crew Worksheet. The items, quantities, and units for the first three columns of the crew worksheet (Table 19.1) are obtained from Fig. 19.1. Unit costs for materials and subcontractors for columns four and five are obtained by direct quotations from vendors and subcontractors or from standard price lists. The worker-hours listed in Table 19.1 are based on data in Fig. 19.1. The wages in Table 19.1 are part of the basic information.

To obtain equipment costs, the estimator either gets quotations from rental yards or performs equipment ownership and operating cost analyses (Table 19.2). These costs include labor and material costs for owning and maintaining the equipment. The equipment costs in Table 19.1 are assumed to be supplied by a subcontractor.

The total cost for one hour of production is calculated as the sum of the products of the quantities and the unit prices given in Table 19.1. The estimator obtains the

Calculation Sheet

Originator *C. Mallin*	Date *1-2-92*	Calc. No. *1*	Rev. No. *0*
Project *Demonstration*	Job No. *12345-678*	Checked *GRP*	Date *11/1/92*
Subject *Trench Excavation*			Sheet No. *1 of 1*

Scope

- *Excavate 3'-0" wide x 2'-0" deep trench in stable soil.*
- *Spoil excavated material.*
- *Backfill with clean fill.*

2'-0"
3'-0"

Production Description: *100 lf/hr (23 yd³/hr)*

- *Excavation and backfill proceed at same rate.*

Excavation

- *Excavator digs trench.*
- *Front-end loader loads trucks (2 each).*
- *One laborer trims bottom and sides.*
- *One laborer works with loader.*

Backfill

- *Purchased clean fill is delivered from quarry.*
- *Front-end loader dumps fill in trench.*
- *Two laborers spread fill in trench.*
- *Two laborers with plate tamper compact fill.*
- *One water truck sprinkles water on fill.*

Excavate and Backfill Crew

- *Labor*
  - *Equipment operators - 3 each*
  - *Teamsters - 3 each*
  - *Laborers - 6 each*
- *Equipment*
  - *Excavator - 1 each*
  - *Front-end loaders - 2 each*
  - *Water truck - 1 each*
  - *Dump trucks - 2 each*
  - *Plate tamper - 1 each*
- *Material*
  - *Clean fill - 26 yd³/hr (including 15% swell)*
- *Subcontract*
  - *Dump fee - 29 yd³/hr (including 30% swell)*

FIGURE 19.1 Crew operation calculation sheet for cost estimate.

unit cost for materials, labor, and equipment by dividing the cost for one hour of production by the length of the trench.

Estimate Worksheet. Table 19.3 gives the estimate for the total cost of the trench. The quantities and units for the trench are taken off the contract documents. The costs are derived from the crew worksheet (Table 19.1), to yield the total direct cost for the excavation. To the direct cost, the estimator adds contingency costs (for rain, striking concealed utility lines, excavating in unexpected subgrade conditions), indirects (30% of labor cost), and company-wide costs and profit (10%). The sum is the total price of the project.

TABLE 19.1 Crew Worksheet

JOB NO., TITLE & CLIENT *12345-678 Demonstration — McGraw-Hill*
DESCRIPTION *Crew Worksheet — Trenching*
JOB LOCATION *San Francisco, CA*

TAKEOFF *C. Mallin* APPROVED *GRP*
PRICED *C. Mallin* DATE *11/1/92*
CHECKED *RHK* SHEET *1* OF *1*

ITEM AND DESCRIPTION	QUANTITY	UNIT	UNIT COST MAT'L	UNIT COST S/C	WORKER HOURS BECHTEL / S/C UNIT	WORKER HOURS TOTAL	WORKER HOURS $/MH	TOTAL COST MATERIAL	TOTAL COST LABOR	TOTAL COST SUB-CONT	TOTAL COST TOTAL
1-hour production											
Material											
Select fill	26	yd^3	25.00					650			650
Subcontract fee											
Dumping fee	29	yd^3		10.00						290	290
Labor											
Equipment operators	2	hr			1	2	37.02		74		74
Teamsters	3	hr			1	3	32.71		98		98
Laborers	6	hr			1	6	29.12		175		175
Equipment											
Excavator	1	hr	14.76	7.06	0.1	0.1	S/C	15		7	22
Loaders	2	hr	9.72	4.34	0.1	0.2	S/C	19		9	28
Dump trucks	2	hr	19.22	11.17	0.3	0.6	S/C	38		22	60
Water truck	1	hr	12.16	12.40	0.2	0.2	S/C	12		12	24
Plate tamper	1	hr	1.02	0.62	0.2	0.0	S/C	1		1	2
Cost for 1 hour	100	ft	7.36	3.41	0.119	11.9	29.15	735	347	341	1423
CODE — Use for	1	ft	7.36	3.41	0.119		29.15				

TABLE 19.2 Equipment Ownership and Operating Cost Analysis Report

```
JOB NO : 12345-678                                                                                                                                   RUN NO  :
TITLE  : DEMONSTRATION PROJECT                                                                                                                       PAGE NO :   1
DATE   : 11/21/92
                                                                                                                                ELECT             MECH  MACHINE   TIRE
EQP SUB                                                                       WEIGHT-(TONS) ---VOLUME----                 FUEL  POWER  OIL  GREASE WAGE  ECON-LIFE LIFE
COD COD EQUIPMENT-DESCRIPTION          TIRE-DESCRIPTION    ACCESSORIES                U.S.  METRIC  CUFT   (M3)  CAPACITY UNIT $/GAL $/KWH $/GAL $/LB   $/HR  YRS HR/YR (HR)
=== === ============================== =================== ========================== ====== ====== ====== ====== ======== ==== ===== ===== ===== ====== ===== === ===== ====
102A    demonstration 1                                    UNIVERSAL BLADE            45.77  41.60  2,720     77   370.00 HP    1.10  0.08  0.80   0.60 62.00   7  1885    0

                                                            U.S.
OWNING COST                                               -- RATE --
***********                                               ==========
         LIST PRICE ........... $  390,000
         DISCOUNT ( 5%) ........ $  -19,500
         SALE TAX ( 7%) ........ $   25,935
         VALUE OF TIRES ........ $        0
         RESIDUE VALUE (20%) ... $  -78,000
                                 -----------
                                 $  318,435

         OWNERSHIP COST ($/HR)   $    24.13
         INTEREST COSTS ($/HR)   $    11.10             11.50 % / YR
         INSURANCE COSTS ($/HR)  $     0.96              1.00 % / YR
                                 -----------
    TOTAL HOURLY OWNING COST .... $   36.19

OPERATING COST
**************
    REPAIR LABOR COST ( $/OPERATING HR )
    ------------------------------------
         UNDERCARRIAGE ......... $    24.80              0.400 MH/HR
         REPAIR RESERVE ........ $    15.50              0.250 MH/HR
         SPECIAL ITEMS ......... $     2.48              0.040 MH/HR
                                 -----------
                                 $    42.78

    REPAIR SPARE PARTS COST ( $/OPERATING HR )
    ------------------------------------------
         UNDERCARRIAGE ......... $     0.00              0.00 $/HR
         REPAIR RESERVE ........ $    22.00             22.00 $/HR
         SPECIAL ITEMS ......... $     0.00              0.00 $/HR
                                 -----------
                                 $    22.00

         FUEL COST ............. $    19.80             18.00 GAL/HR
         LUBE OILS ............. $     2.40              3.00 QT/HR
         GREASE ................ $     1.20              2.00 LB/HR
         FILTERS ............... $     0.15              0.15 $/HR
         ELECT. POWER ...........$     0.00              0.00 KW/HR
                                 -----------
    SUBTOTAL FUEL,OIL,GREASE...  $    23.55

    DEPRECIATION OF TIRE .......  $    0.00

                                 -----------
    TOTAL HOURLY OPERATING COST  $    88.33

TOTAL OWNING & OPERATING COST.... $  124.52                              MONTHLY RATE FOR 1ST SHIFT ......  $   7,240
                                     ======                              MONTHLY RATE FOR TWO SHIFTS .....  $  12,308
```

TABLE 19.3 Estimate Worksheet

LUMP-SUM BIDDING GROUP

JOB NO. 12345-678	PROJECT Demonstration	TAKEOFF C. Mallin	APPROVED GRP
TYPE OF ESTIMATE Demonstration	CLIENT McGraw Hill	PRICED C. Mallin	DATE 11/1/92
	JOB LOCATION San Francisco, CA	CHECKED RHK	SHEET 1 OF 1

ITEM AND DESCRIPTION	QUANTITY	UNIT	UNIT COST		WORKER HOURS BECHTEL / S/C			TOTAL COST			
			MAT'L	S/C	UNIT	TOTAL	$/MH	MATERIAL	LABOR	SUB-CONT	TOTAL
Trenching 2'-0" x 3'-0" deep (direct costs)	1,750	ft	7.36	3.41	0.119	208	29.15	12877	6071	5969	24917
Contingency	10	%				21		1288	607	577	2492
Subtotal						229		14165	6678	6566	27409
30% of labor for indirects	1	lot				30%			2003		2003
Subtotal								14165	8681	6566	29412
Company-wide costs and profit	15	%									4412
Total	1,750	ft		19.33							33824
Parametric checks											
#1 Check by trench volume	389	yd³		73.65	OK						
#2 Clock time check @ 11.9 manhours/clock hour	19	hr	OK								
	2.4	days	OK								
CODE											

The estimator then makes two parametric checks to determine the reasonableness of the result:

1. *By trench volume.* The estimator compares the estimated price with that for similar projects with similar restrictions and requirements. A review of published bids for similar projects shows unit prices of $60 to $100 per cubic yard. This indicates that the estimated price of $73.65 is within that range.

2. *By time clock.* The estimator verifies that the specific equipment and personnel can be made available by the contractor. This is done by consulting the contractor's work schedule for the equipment and personnel.

19.6 REVIEWING ESTIMATES

All estimates should be reviewed by all responsible parties at every stage. An estimate review should begin with a survey of the verbal description of the work, including all or most of the following: scope statement, assumptions, clarifications, qualifications, and exclusions.

As an example, the estimate is to be reviewed for a warehouse to be built in an urban area as part of a redevelopment project. The scope statement should specify the location, refer to design drawings and specifications, and list applicable building codes. The assumptions might include such data as the number of persons who will work in the warehouse. This is an indication of the number of restrooms and fixtures needed, which can be listed as a clarification. If the price quotes are valid for 90 days, this should be stated as a qualification. Handling and disposing of any existing hazardous material found on the site might be listed as an exclusion.

This warehouse description may be reasonably complete from the viewpoint of designer and contractor and may be accurately priced. But because of the assumption regarding the number of occupants, it may not be suitable from the viewpoint of the intended users. The exclusion regarding hazardous materials may result in unacceptable financial exposure for the client. Issues such as these need to be addressed. The client may decide that the prospective tenants, or users, may employ more persons than the number assumed. Hence, either the estimate will have to allocate more money for rest rooms or the client will have to give the tenants an allowance to enable them to build the rest rooms they desire. The client may also decide that an analysis of soil samples may be necessary before any construction is done to determine the extent of contamination, if any, and cost of cleanup.

Bearing these issues in mind, the parties should now review the quantitative part of the estimate. This review should comprise the following:

A summary of the key quantities involved; for example, floor area, tons of steel, cubic yards of concrete.

As a cross check, a list of the key quantities—by discipline if the estimate has been prepared with the industry approach or by industry if the estimate has been prepared by the discipline approach.

A summary of the project, by industry or discipline.

At each step of the review, changes may be made, as required. After all parties agree to all parts of the estimate, it can be considered final. At this stage, the designer should be satisfied that enough money has been allocated to carry out the project. The client should have a clear idea as to what the project will entail and how much it will cost.

19.7 COMPUTER-AIDED ESTIMATING

There are essentially three types of commercial computer products useful in preparation of cost estimates:

Utilities. These are programs that arrange information or do arithmetic; for example, spreadsheets and database administrators. Most estimating programs fall into the utilities category.

Databases. These contain raw information, for example, prices of plumbing fixtures, that the estimator must analyze and choose from.

Expert Systems. These are programs that question the estimator, then use the answers to produce an estimate. (Expert systems are sometimes referred to as artificial intelligence systems.)

Some commercial packages may contain two or more of these.

19.7.1 Utilities

These enhance, but do not replace, the expertise and knowledge of estimators. They enable estimators to extract and summarize needed information rapidly and accurately.

Utilities are broad based; they can be applied to almost any estimating type, approach, and technique. Use of some programs is easy to learn; for example, spreadsheets. Others are difficult; for example, database administrators. In general, the more powerful a utility is, the more difficult it is to learn to use it.

19.7.2 Databases

Generally, databases are designed to be used with one specific utility and one specific approach, such as industry or discipline (Art. 19.2). Some are limited to a particular type of estimate and technique (Art. 19.2.2).

The estimator should always be aware that a database responds only to specific queries asked in a specific way and cannot interpolate. For instance, if a database has the prices of ½-in and 1-in bolts and the estimator requests the price of a ¾-in bolt, the computer will reply that the database contains no such price. At this stage, the estimator should devise the proper queries to solicit responses helpful in preparing the estimate.

19.7.3 Expert Systems

Generally, expert systems are even narrower in application than databases. Unlike databases, however, expert systems will respond to a query as long as the estimator operates within the limits of their area of expertise. They have several drawbacks: They do not take into account creative solutions or project-specific problems. They do not change with changing technology. And they tend to be very expensive. As a result, they are not widely used.

APPENDIX

Factors for Conversion to the Metric System (SI) of Units

Frederick S. Merritt
Consulting Engineer
West Palm Beach, Florida

Congress committed the United States to conversion to the metric system of units when it passed the Metric Conversion Act of 1975. This Act states that it shall be the policy of the United States to change to the metric system in a coordinated manner and that the purpose of this coordination shall be to reduce the total cost of the conversion. While conversion has already taken place in some industries and in some engineering disciplines, conversion is taking place in short steps at long time intervals in building design and construction. Consequently, conventional units are used throughout the preceding portion of this handbook. The metric system is explained and factors for conversion to it are presented in this Appendix, to guide and assist those who have need to apply metric units in design or construction.

The system of units that is being adopted in the United States is known as the International System of units, or SI, an abbreviation of the French *Le Système International d'Unités*. This system, intended as a basis for worldwide standardization of measurement units, was developed and is being maintained by the General Conference on Weights and Measures (CGPM).

For engineering, the SI has the advantages over conventional units of being completely decimal and of distinguishing between units of mass and units of force. With conventional units, there sometimes is confusion between use of the two types of units. For example, lb or ton may represent either mass or force.

SI units are classified as base, supplementary, or derived units. There are seven base units (Table A.1), which are dimensionally independent, and two supplementary units (Table A.2), which may be regarded as either base or derived units.

Derived units are formed by combining base units, supplementary units, and other derived units in accordance with algebraic relations linking the corresponding quantities. Symbols for derived units represent the mathematical relationships between the component units. For example, the SI unit for velocity, metre per second, is represented by m/s; that for acceleration, metres per second per second, by m/s^2, and that for bending moment, newton-metres, by N·m. Figure A.1 indicates how units may be combined to form derived units.

As indicated in Fig. A.1, some of the derived units have been given special names; for example, the unit of energy, N·m is called joule and the unit of pressure

TABLE A.1 SI Base Units

Quantity	Unit	Symbol
Length	metre	m
Mass	kilogram	kg
Time	second	s
Electric current	ampere	A
Thermodynamic temperature	kelvin	K
Amount of substance	mole	mol
Luminous intensity	candela	cd

TABLE A.2 Supplementary SI Units

Quantity	Unit	Symbol
Plane angle	radian	rad
Solid angle	steradian	sr

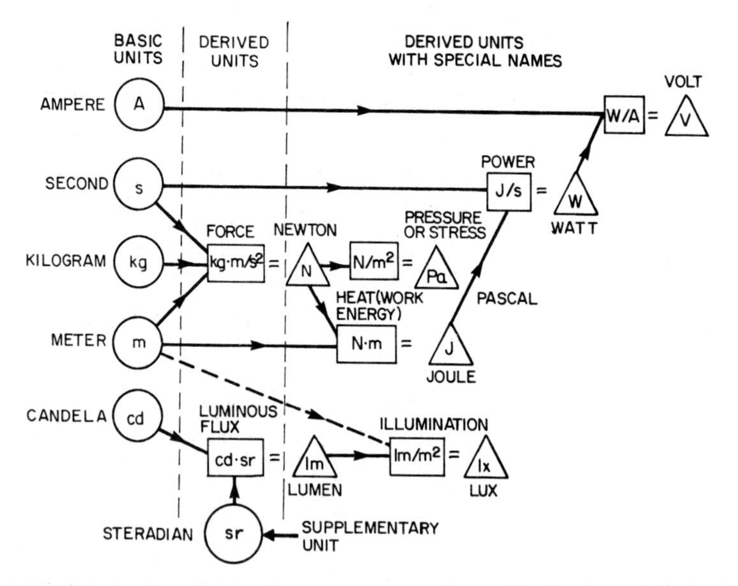

FIGURE A.1. How SI units of measurement may be combined to form derived units.

or stress, N/m^2, is called pascal. Table A.3 defines derived SI units that have special names and symbols approved by CGPM. Some such units used in building design and construction are given in Table A.4; others are listed with the conversion factors in Table A.6.

Prefixes. Except for the unit of mass, the kilogram (kg), names and symbols of multiples of SI units by powers of 10, positive or negative, are formed by adding

TABLE A.3 Derived SI Units with Special Names

Quantity	Unit	Symbol	Formula	Definition
Celsius temperature	degree Celsius	°C	K − 273.15	The *degree Celsius* is equal to the kelvin and is used in place of the kelvin for expressing Celsius temperature (symbol t) defined by the equation $t = T - T_0$ where T is the thermodynamic temperature and $T_0 = 273.15$ K by definition.
Electric capacitance	farad	F	C/V	The *farad* is the capacitance of a capacitor between the plates of which there appears a difference of potential of one volt when it is charged by a quantity of electricity equal to one coulomb.
Electric conductance	siemens	S	A/V	The *siemens* is the electric conductance of a conductor in which a current of one ampere is produced by an electric potential difference of one volt.
Electric inductance	henry	H	Wb/A	The *henry* is the inductance of a closed circuit in which an electromotive force of one volt is produced when the electric current in the circuit varies uniformly at a rate of one ampere per second.
Electric potential difference, electromotive force	volt	V	W/A	The *volt* (unit of electric potential difference and electromotive force) is the difference of electric potential between two points of a conductor carrying a constant current of one ampere, when the power dissipated between these points is equal to one watt.
Electric resistance	ohm	Ω	V/A	The *ohm* is the electric resistance between two points of a conductor when a constant difference of potential of one volt, applied between these two points, produces in this conductor a current of one ampere, this conductor not being the source of any electromotive force.
Energy	joule	J	N · m	The *joule* is the work done when the point of application of a force of one newton is displaced a distance of one metre in the direction of the force.

TABLE A.3 (*Continued*)

Quantity	Unit	Symbol	Formula	Definition
Force	newton	N	$kg \cdot m/s^2$	The *newton* is that force which, when applied to a body having a mass of one kilogram, gives it an acceleration of one metre per second squared.
Frequency	hertz	Hz	$1/s$	The *hertz* is the frequency of a periodic phenomenon of which the period is one second.
Illuminance	lux	lx	lm/m^2	The *lux* is the illuminance produced by a luminous flux of one lumen uniformly distributed over a surface of one metre.
Luminous flux	lumen	lm	$cd \cdot sr$	The *lumen* is the luminous flux emitted in a solid angle of one steradian by a point source having a uniform intensity of one candela.
Magnetic flux	weber	Wb	$V \cdot s$	The *weber* is the magnetic flux which, linking a circuit of one turn, produces in it an electromotive force of one volt as it is reduced to zero at a uniform rate in one second.
Magnetic flux density	tesla	T	Wb/m^2	The *tesla* is the magnetic flux density given by a magnetic flux of one weber per square metre.
Power	watt	W	J/s	The *watt* is the power which gives rise to the production of energy at the rate of one joule per second.
Pressure or stress	pascal	Pa	N/m^2	The *pascal* is the pressure or stress of one newton per square metre.
Quantity of electricity	coulomb	C	$A \cdot s$	The *coulomb* is the quantity of electricity transported in one second by a current of one ampere.

a prefix to base, supplementary, and derived units. Table A.5 lists prefixes approved by CGPM. For historical reasons, kilogram has been retained as a base unit. Nevertheless, for units of mass, prefixes are attached to gram, 10^{-3} kg. Thus, from Table A.5, 1 Mg = 10^3 kg = 10^6 g.

The prefixes should be used to indicate orders of magnitude without including insignificant digits in whole numbers or leading zeros in decimals. Preferably, a prefix should be chosen so that the numerical value associated with a unit lies between 0.1 and 1000. Preferably also, prefixes representing powers of 1000 should

TABLE A.4 Some Common Derived Units of SI

Quantity	Unit	Symbol
Acceleration	metre per second squared	m/s^2
Angular acceleration	radian per second squared	rad/s^2
Angular velocity	radian per second	rad/s
Area	square metre	m^2
Density, mass	kilogram per cubic metre	kg/m^3
Energy density	joule per cubic metre	J/m^3
Entropy	joule per kelvin	J/K
Heat capacity	joule per kelvin	J/K
Heat flux density	watt per square metre	W/m^2
Irradiance	watt per square metre	W/m^2
Luminance	candela per square metre	cd/m^2
Magnetic field strength	ampere per metre	A/m
Moment of force	newton-metre	$N \cdot m$
Power density	watt per square metre	W/m^2
Radiant intensity	watt per steradian	W/sr
Specific heat capacity	joule per kilogram kelvin	$J/(kg \cdot K)$
Specific energy	joule per kilogram	J/kg
Specific entropy	joule per kilogram kelvin	$J/(kg \cdot K)$
Specific volume	cubic metre per kilogram	m^3/kg
Surface tension	newton per metre	N/m
Thermal conductivity	watt per metre kelvin	$W/(m \cdot K)$
Velocity	metre per second	m/s
Viscosity, dynamic	pascal second	$Pa \cdot s$
Viscosity, kinematic	square metre per second	m^2/s
Volume	cubic metre	m^3

be used. Thus, for building construction, units of length should be millimetres, mm; metres, m; and kilometres, km. Units of mass should be milligrams, mg; gram, g; kilogram, kg; and megagram, Mg.

When values of a quantity are listed in a table or when such values are being compared, it is desirable that the same multiple of a unit be used throughout.

In the formation of a multiple of a compound unit, such as of velocity, m/s, only one prefix should be used, and, except when kilogram occurs in the denominator, the prefix should be attached to a unit in the numerator. Examples are kg/m and MJ/kg; do not use g/mm or kJ/g, respectively. Also, do not form a compound prefix by juxtaposing two or more prefixes; for example, instead of Mkm, use Gm. If values outside the range of approved prefixes should be required, use a base unit multiplied by a power of 10.

TABLE A.5 SI Prefixes

Multiplication factor	Prefix	Symbol
1 000 000 000 000 000 000 = 10^{18}	exa	E
1 000 000 000 000 000 = 10^{15}	peta	P
1 000 000 000 000 = 10^{12}	tera	T
1 000 000 000 = 10^{9}	giga	G
1 000 000 = 10^{6}	mega	M
1 000 = 10^{3}	kilo	k
100 = 10^{2}	hecto*	h
10 = 10^{1}	deka*	da
0.1 = 10^{-1}	deci*	d
0.01 = 10^{-2}	centi*	c
0.001 = 10^{-3}	milli	m
0.000 001 = 10^{-6}	micro	μ
0.000 000 001 = 10^{-9}	nano	n
0.000 000 000 001 = 10^{-12}	pico	p
0.000 000 000 000 001 = 10^{-15}	femto	f
0.000 000 000 000 000 001 = 10^{-18}	atto	a

*To be avoided where practical.

To indicate that a unit with its prefix is to be raised to a power indicated by a specific exponent, the exponent should be applied after the unit; for example, the unit of volume, $mm^3 = (10^{-3}\ m)^3 = 10^{-9}\ m^3$.

Units in Use with SI. Where it is customary to use units from different systems of measurement with SI units, it is permissible to continue the practice, but such uses should be minimized. For example, for time, while the SI unit is the second, it is customary to use minutes (min), hours (h), days (d), etc. Thus, velocities of vehicles may continue to be given as kilometres per hour (km/h). Similarly, for angles, while the SI unit for plane angle is the radian, it is permissible to use degrees and decimals of a degree. As another example, for volume, the cubic metre is the SI unit, but liter (L), mL, or μL may be used for measurements of liquids and gases. Also, for mass, while Mg is the appropriate SI unit for large quantities, short ton, long ton, or metric ton may be used for commercial applications.

For temperature, the SI unit is the kelvin, K, whereas degree Celsius, °C (formerly centigrade) is widely used. A temperature interval of 1°C is the same as 1 K, and °C = K − 273.15, by definition.

SI Units Preferred for Construction. Preferred units for measurement of length for relatively small structures, such as buildings and bridges, are millimetres and metres. Depending on the size of the structure, a drawing may conveniently note: "All dimensions shown are in millimetres" or "All dimensions shown are in metres." By convention, the unit for numbers with three digits after the decimal point, for example, 26.375 or 0.425 or 0.063, is metres, and the unit for whole numbers, for example, 2638 or 425 or 63, is millimetres. Hence, it may not be necessary to show unit symbols. For large-size construction, such as highways, metres and kilometres may be used for length measurements and millimetres for width and thickness.

For area measurements, square metres, m^2, are preferred, but mm^2 are acceptable for small areas. 1 m^2 = 10^6 mm^2. For very large areas, square kilometres, km^2, or hectares, ha, may be used. 1 ha = 10^4 m^2 = 10^{-2} km^2.

For volume measurements, the preferred unit is the cubic metre, m^3. The volume of liquids, however, may be measured in litres, L, or millilitres, mL. 1 L = 10^{-3} m^3. For flow rates, cubic metres per second, m^3/s, cubic metres per hour, m^3/h, and litres per second, L/s, are preferred.

For concentrated gravity loads, the force units newton, N, or kilonewton, kN, should be used. For uniformly distributed wind and gravity loads, kN/m^2 is preferred. (Materials weighed on spring scales register the effect of the force of gravity, but for commercial reasons, the scales may be calibrated in kilograms, the units of mass. In such cases, the readings should be multiplied by g, the acceleration of a mass due to gravity, to obtain the load in newtons.) For dynamic calculations, the force in newtons equals the product of the mass, kg, by the acceleration a, m/s^2, of the mass. The recommended value of g for design purposes in the United States is 9.8 m/s^2. The standard international value for g is 9.806650 m/s^2, whereas it actually ranges between 9.77 and 9.83 m/s^2 over the surface of the earth.

For both pressure and stress, the SI unit is the pascal, Pa (1 Pa = 1 N/m^2). Because section properties of structural shapes are given in millimetres, it is more convenient to give stress in newtons per square millimetre (1 N/mm^2 = 1 MPa). For energy, work, and quantity of heat, the SI unit is the joule, J (1 J = 1 N·m = 1 W·s). The kilowatthour, kWh (more accurately, kW·h) is acceptable for electrical measurements. The watt, W, is the SI unit for power.

Dimensional Coordination. The basic concept of dimensional coordination is selection of the dimensions of the components of a building and installed equipment so that sizes may be standardized and the items fitted into place with a minimum of cutting in the field. One way to achieve this is to make building components and equipment to fit exactly into a basic cubic module or multiples of the module, except for the necessary allowances for joints and manufacturing tolerances. For the purpose, a basic module of 4 in is widely used in the United States. Larger modules often used include 8 in, 12 in, 16 in, 2 ft, 4 ft, and 8 ft.

For modular coordination in the SI, Technical Committee 59 of the International Standards Organization has established 100 mm (3.937 in) as the basic module. In practice, where modules of a different size would be more convenient, preferred dimensions have been established by agreements between manufacturers of building products and building designers. For example, in Great Britain, the following set of preferences have been adopted:

1st preference	300 mm (about 12 in)
2d preference	100 mm (about 4 in)
3d preference	50 mm (about 2 in)
4th preference	25 mm (about 1 in)

Accordingly, for a dimension exceeding 100 mm, the first preference would be a multimodule of 300 mm. Second choice would be the basic module of 100 mm.

The preferred multimodules for horizontal dimensioning are 300, 600 (about 2 ft), and 1200 (about 4 ft) mm, although other multiples of 300 are acceptable.

Preferred modules for vertical dimensioning are 300 and 600 mm, but increments of 100 mm are acceptable up to 3000 mm. The submodules, 25 and 50 mm, are used only for thin sections.

Some commonly used dimensions, such as the 22 in used for unit of exit width, cannot be readily converted into an SI module. For example, 22 in = 558.8 mm. The nearest larger multimodule is 600 mm (23⅝ in), and the nearest smaller multimodule is 500 mm (19$^{11}/_{16}$ in). The use of either multimodule would affect the sizes of doors, windows, stairs, etc. For conversion of SI to occur, building designers and product manufacturers will have to agree on preferred dimensions.

Conversion Factors. Table A.6 lists factors with seven-digit accuracy for conversion of conventional units of measurement to SI units. To retain accuracy in a conversion, multiply the specified quantity by the conversion factor exactly as given in Table A.6, then round the product to the appropriate number of significant digits that will neither sacrifice nor exaggerate the accuracy of the result. For the purpose, a product or quotient should contain no more significant digits than the number with the smallest number of significant digits in the multiplication or division.

In Table A.6, the conversion factors are given as a number between 1 and 10 followed by E (for exponent), a plus or minus, and two digits that indicate the power of 10 by which the number should be multiplied. For example, to convert lbf/in^2 (psi) to pascals (Pa), Table A.6 specifies multiplication by $6.894\,757 \times 10^3$. For conversion to kPa, the conversion factor is $6.894\,757 \times 10^3 \times 10^{-3} = 6.894\,757$.

["Standard for Metric Practice," E 380, and "Practice for Use of Metric (SI) Units in Building Design and Construction," E 621, ASTM, 1916 Race St., Philadelphia, PA 19103; "NBS Guidelines for Use of the Metric System," NBS LC 1056, Nov. 1977, and "The International System of Units (SI)," NBS Specification Publication 330, 1977, Superintendent of Documents, Government Printing Office, Washington, DC 20402.]

TABLE A.6 Factors for Conversion to SI Units of Measurement

To convert from	to	multiply by
acre	square metre, m^2	4.046 873 E + 03
angstrom	metre, m	1.000 000*E − 10
atmosphere (standard)	pascal, Pa	1.013 250*E + 05
bar	pascal, Pa	1.000 000*E + 05
barrel (for petroleum, 42 gal)	cubic metre, m^3	1.589 873 E − 01
board-foot	cubic metre, m^3	2.359 737 E − 03
British thermal unit (mean)	joule, J	1.055 87 E + 03
Btu (International Table) · in/(h)(ft^2)(°F) (k, thermal conductivity)	watt per metre kelvin, W/(m · K)	1.442 279 E − 01
Btu (International Table)/h	watt, W	2.930 711 E − 01
Btu (International Table)/(h)(ft^2)(°F) (C, thermal conductance)	watt per square metre kelvin, W/(m^2 · K)	5.678 263 E + 00
Btu (International Table)/lb	joule per kilogram, J/kg	2.326 000*E + 03
Btu (International Table)/(lb)(°F) (c, heat capacity)	joule per kilogram kelvin, J/(kg · K)	4.186 800*E + 03
Btu (International Table)/ft^3	joule per cubic metre, J/m^3	3.725 895 E + 04
bushel (U.S.)	cubic metre, m^3	3.523 907 E − 02
calorie (mean)	joule, J	4.190 02 E + 00
cd/in^2	candela per square metre, cd/m^2	1.550 003 E + 03
chain	metre, m	2.011 684 E + 01
circular mil	square metre, m^2	5.067 075 E − 10
day	second, s	8.640 000*E + 04
day (sidereal)	second, s	8.616 409 E + 04
degree (angle)	radian, rad	1.745 329 E − 02
degree Celsius	kelvin, K	$T_K = t_c + 273.15$
degree Fahrenheit	degree Celsius	$t_C = (t_F - 32)/1.8$
degree Fahrenheit	kelvin, K	$T_K = (t_F + 459.67)/1.8$
degree Rankine	kelvin, K	$T_K = T_R/1.8$
(°F)(h)(ft^2) Btu (International Table) (R, thermal resistance)	kelvin square metre per watt, K · m^2/W	1.761 102 E − 01
(°F)(h)(ft^2)/(Btu (International Table) · in) (thermal resistivity)	kelvin metre per watt, K · m/W	6.933 471 E + 00
dyne	newton, N	1.000 000*E − 05
fluid ounce (U.S.)	cubic metre, m^3	2.957 353 E − 05
foot	metre, m	3.048 000*E − 01

TABLE A.6 (*Continued*)

To convert from	to	multiply by
foot (U.S. survey)	metre, m	3.048 006 E − 01
foot of water (39.2°F) (pressure)	pascal, Pa	2.988 98 E + 03
ft^2	square metre, m^2	9.290 304*E − 02
ft^2/h (thermal diffusivity)	square metre per second, m^2/s	2.580 640*E − 05
ft^2/s	square metre per second, m^2/s	9.290 304*E − 02
ft^3 (volume or section modulus)	cubic metre, m^3	2.831 685 E − 02
ft^3/min	cubic metre per second, m^3/s	4.719 474 E − 04
ft^3/s	cubic metre per second, m^3/s	2.831 685 E − 02
ft^4 (area moment of inertia)	metre to the fourth power, m^4	8.630 975 E − 03
ft/min	metre per second, m/s	5.080 000*E − 03
ft/s	metre per second, m/s	3.048 000*E − 01
ft/s^2	metre per second squared, m/s^2	3.048 000*E − 01
footcandle	lux, lx	1.076 391 E + 01
footlambert	candela per square metre, cd/m^2	3.426 259 E + 00
ft · lbf	joule, J	1.355 818 E + 000
ft · lbf/min	watt, W	2.259 697 E − 02
ft · lbf/s	watt, W	1.355 818 E + 00
ft-poundal	joule, J	4.214 011 E − 02
free fall, standard *g*	metre per second squared, m/s^2	9.806 650*E + 00
Gallon (Canadian liquid)	cubic metre, m^3	4.546 090 E − 03
gallon (U.K. liquid)	cubic metre, m^3	4.546 092 E − 03
gallon (U.S. dry)	cubic metre, m^3	4.404 884 E − 03
gallon (U.S. liquid)	cubic metre, m^3	3.785 412 E − 03
gallon (U.S. liquid) per day	cubic metre per second, m^3/s	4.381 264 E − 08
gallon (U.S. liquid) per minute	cubic metre per second, m^3/s	6.309 020 E − 05
grad	degree (angular)	9.000 000*E − 01
grad	radian, rad	1.570 796 E − 02
grain	kilogram, kg	6.479 891*E − 05
gram	kilogram, kg	1.000 000*E − 03

TABLE A.6 (*Continued*)

To convert from	to	multiply by
hectare	square metre, m^2	1.000 000*E + 04
horsepower (550 ft · lbf/s)	watt, W	7.456 999 E + 02
horsepower (boiler)	watt, W	9.809 50 E + 03
horsepower (electric)	watt, W	7.460 000*E + 02
horsepower (water)	watt, W	7.460 43 E + 02
horsepower (U.K.)	watt, W	7.457 0 E + 02
hour	second, s	3.600 000*E + 03
hour (sidereal)	second, s	3.590 170 E + 03
inch	metre, m	2.540 000*E − 02
inch of mercury (32°F) (pressure)	pascal, Pa	3.386 38 E + 03
inch of mercury (60°F) (pressure)	pascal, Pa	3.376 85 E + 03
inch of water (60°F) (pressure)	pascal, Pa	2.488 4 E + 02
in^2	square metre, m^2	6.451 600*E − 04
in^3 (volume or section modulus)	cubic metre, m^3	1.638 706 E − 05
in^4 (area moment of inertia)	metre to the fourth power, m^4	4.162 314 E − 07
in/s	metre per second, m/s	2.540 000*E − 02
kelvin	degree Celsius	$t_c = T_K - 273.15$
kilogram-force (kgf)	newton, N	9.806 650*E + 00
kgf · m	newton metre, N · m	9.806 650*E + 00
kgf · s^2/m (mass)	kilogram, kg	9.806 650*E + 00
km/h	metre per second, m/s	2.777 778 E − 01
kWh	joule, J	3.600 000*E + 06
kip (1000 lbf)	newton, N	4.448 222 E + 03
kip/in^2 (ksi)	pascal, Pa	6.894 757 E + 06
lambert	candela per square metre, cd/m	3.183 099 E + 03
liter	cubic metre, m^3	1.000 000*E − 03
maxwell	weber, Wb	1.000 000*E − 08
mho	siemens, S	1.000 000*E + 00
microinch	metre, m	2.540 000*E − 08
micron	metre, m	1.000 000*E − 06
mil	metre, m	2.540 000*E − 05
mile	metre, m	1.609 347 E + 03
mile (U.S. nautical)	metre, m	1.852 000*E + 03
mi^2 (U.S. statute)	square metre, m^2	2.589 998 E + 06

TABLE A.6 (*Continued*)

To convert from	to	multiply by
mi/h	metre per second, m/s	4.470 400*E − 01
mi/h	kilometre per hour, km/h	1.609 344*E + 00
millibar	pascal, Pa	1.000 000*E + 02
millimeter of mercury (0°C)	pascal, Pa	1.333 22 E + 02
minute (angle)	radian, rad	2.908 882 E − 04
minute	second, s	6.000 000*E + 01
minute (sidereal)	second, s	5.983 617 E + 01
ounce (avoirdupois)	kilogram, kg	2.834 952 E − 02
once (troy or apothecary)	kilogram, kg	3.110 348 E − 02
ounce (U.K. fluid)	cubic metre, m^3	2.841 307 E − 05
ounce (U.S. fluid)	cubic metre, m^3	2.957 353 E − 05
oz (avoirdupois)/ft^2	kilogram per square metre, kg/m^2	3.051 517 E − 01
oz (avoirdupois)/yd^2	kilogram per square metre, kg/m^2	3.390 575 E − 02
perm (0°C)	kilogram per pascal second metre, kg/(Pa · s · m)	5.721 35 E − 11
perm (23°C)	kilogram per pascal second metre, kg/(Pa · s · m)	5.745 25 E − 11
perm · in (0°C)	kilogram per pascal second metre, kg/(Pa · s · m)	1.453 22 E − 12
perm · in (23°C)	kilogram per pascal second metre, kg/(Pa · s · m)	1.459 29 E − 12
pint (U.S. dry)	cubic metre, m^3	5.506 105 E − 04
pint (U.S. liquid)	cubic metre, m^3	4.731 764 E − 04
poise (absolute viscosity)	pascal second, Pa · s	1.000 000*E − 01
pound (lb avoirdupois)	kilogram, kg	4.535 924 E − 01
pound (troy or apothecary)	kilogram, kg	3.732 417 E − 01
lb · in^2 (moment of inertia)	kilogram square metre, $kg \cdot m^2$	2.926 397 E − 04
lb/ft · s	pascal second, Pa · s	1.488 164 E + 00
lb/ft^2	kilogram per square metre, kg/m^2	4.882 428 E + 00
lb/ft^3	kilogram per cubic metre, kg/m^3	1.601 846 E + 01
lb/gal (U.K. liquid)	kilogram per cubic metre, kg/m^3	9.977 633 E + 01
lb/gal (U.S. liquid)	kilogram per cubic metre, kg/m^3	1.198 264 E + 02
lb/h	kilogram per second, kg/s	1.259 979 E − 04
lb/in^3	kilogram per cubic metre, kg/m^3	2.767 990 E + 04

TABLE A.6 (*Continued*)

To convert from	to	multiply by
lb/min	kilogram per second, kg/s	7.559 873 E − 03
lb/s	kilogram per second, kg/s	4.535 924 E − 01
lb/yd^3	kilogram per cubic metre, kg/m^3	5.932 764 E − 01
poundal	newton, N	1.382 550 E − 01
pound-force (lbf)	newton, N	4.448 222 E + 00
lbf · ft	newton-metre, N · m	1.355 818 E + 00
lbf/ft	newton per metre, N/m	1.459 390 E + 01
lbf/ft^2	pascal, Pa	4.788 026 E + 01
lbf/in	newton per metre, N/m	1.751 268 E + 02
lbf/in^2 (psi)	pascal, Pa	6.894 757 E + 03
quart (U.S. dry)	cubic metre, m^3	1.101 221 E − 03
quart (U.S. liquid)	cubic metre, m^3	9.463 529 E − 04
rod	metre, m	5.029 210 E + 00
second (angle)	radian, rad	4.848 137 E − 06
second (sidereal)	second, s	9.972 696 E − 01
square (100 ft^2)	square metre, m^2	9.290 304*E + 00
ton (long, 2240 lb)	kilogram, kg	1.016 047 E + 03
ton (metric)	kilogram, kg	1.000 000*E + 03
ton (refrigeration)	watt, W	3.516 800 E + 03
ton (register)	cubic metre, m^3	2.831 685 E + 00
ton (short 2000 lb)	kilogram, kg	9.071 847 E + 02
ton (long)/yd^3	kilogram per cubic metre, kg/m^3	1.328 939 E + 03
ton (short)/yd^3	kilogram per cubic metre, kg/m^3	1.186 553 E + 03
ton-force (2000 lbf)	newton, N	8.896 444 E + 03
tonne	kilogram, kg	1.000 000*E + 03
Wh	joule, J	3.600 000*E + 03
yard	metre, m	9.144 000*E − 01
yd^2	square metre, m^2	8.361 274 E − 01
yd^3	cubic metre, m^3	7.645 549 E − 01
year (365 days)	second, s	3.153 600*E + 07
year (sidereal)	second, s	3.155 815 E + 07

*Exact value.
From "Standard for Metric Practice," E380, ASTM.

INDEX